Accident
Prevention
Manual
for Industrial
Operations

Accident Prevention Manual for Industrial Operations

Seventh Edition

National Safety Council

Preface
to the Seventh Edition

This Seventh Edition has retained the larger type and page size of the Sixth Edition, but it is about 100 pages less in thickness. Much non-safety data that can readily be found elsewhere was eliminated and the order of chapters streamlined. The Editors hope that this will make this edition easier to use in preventing accidents.

The term *accident* is used in its broad scope and means "An unplanned, not necessarily injurious or damaging event, that interrupts the completion of an activity; it is invariably preceded by an unsafe act or an unsafe condition or both, or some combination or unsafe acts and/or unsafe conditions."

This definition broadens the scope of accident prevention activities beyond limited attention paid to injuries and prompts the safety professional to identify unsafe practices or defects in the environment that have accident-producing (and perhaps injury-producing) potential.

Although the term *safety professional* is used throughout this Manual, it is a general term. The Editors know that readers will have other titles—business managers, college professors and students, corporate safety officers, federal and state compliance officers, insurance engineers, personnel directors, and others who spend less than the majority of their time in safety work.

New material

Two new chapters, "The Occupational Safety and Health Act of 1970" and "Non-Employee Accident Prevention," have been added.

In addition to these, other chapters have been significantly rewritten—"Accident Investigation, Analysis, and Costs," "Accident Records and Injury Rates," "Human Behavior and Safety," "Human Factors Engineering," "Point-of-Operation and Transmission Guards," "Safety Program Organization" (including off-the-job safety), and "Workmen's Compensation."

"Exhaust and Ventilation" combines two Sixth Edition chapters. "Ionizing Radiation" has been made more practical and is now part of "Industrial Hygiene." "Fire Protection" combines material that was formerly in two chapters.

The arrangement of chapters has been improved so that those encompassing management techniques are in the first half of the Manual, and those involving the more technical duties of a safety professional comprise the second half.

v

Preface

The Manual continues to use the English (fps) system of measurement. The expanding use of the metric system is recognized, however. Detailed conversion tables have been included in the last chapter, "Safety Engineering Tables."

Because of many continuing changes in OSHA (U.S. federal) standards and because many users of this Manual will not be subject to them, references are minimized. The reader must check latest requirements of OSHA or other applicable standards or codes for specific procedures and equipment specifications.

Contributors

National Safety Council appreciates the help and guidance of many authorities —safety professionals and others, and the representatives of professional societies, who gave information, advice, and illustrations.

The following persons were on NSC Staff at the time of their contribution.

R. Belknap
J. Bryk
R. Coe
R. Currie
T. Curtin
D. Elwing
A. Finch
J. Flaherty
E. Koch
R. Lascoe
J. Mark
J. O'Donovan
J. Olishifski
F. Parker
B. Peszek
A. Phillips
C. Price
J. Recht

P. Schmidt
P. Sheppard
G. Shibley
L. Smith
R. Smith
L. Swift
J. VanSickle
T. Worhol

The following non-staff people made significant contributions.

R. Adams, Miller Davis Co.
J. Appel, Commonwealth Edison Co.
A. Baltzer, Consultant
C. Blankenship, A. O. Smith Corp.
J. Brooks, Goldkist Inc.
P. Ehrenfried, Allovio Service Corp.
P. Griffin, Portland Cement Association
R. Ketchmark, University of Illinois, Chicago Circle Campus
M. Krikorian, Brunswick Corp.
G. Lundie, Inland Steel Co.
J. O'Neill, Transcontinental Gas Pipe Line Corp.
J. Parmiter, The Boeing Co., Vertol Div.
R. Putney, Nationwide Insurance Co.
J. Radcliffe, Consultant
H. Rapp, Travelers Insurance Co.
A. Reed, Daniel Woodhead Co.
T. Reilly, Schirmer Engineering Co.
K. Robinson, General Motors Corp.
J. Romine, Phillips Petroleum Co.
C. Schirmer, Schirmer Engineering Co.
V. Sielert, GTE Service Corp.
D. Wenzel, Inland Steel Co.
W. Williford, The Chesapeake & Potomac Telephone Co.

G. Wineland, U.S. Industrial Chemicals Co.
F. Wischmeyer, Eastman Kodak Co.
T. Yoder, Eli Lilly & Co.

Members of the Technical Publications Committee of the Industrial Conference gave useful guidance. The following chaired this committee during the period of review.

R. Pryor, Phillips Petroleum Co., 1969–1972
W. Wood, Sun Oil Co., 1972–1973
F. Manuele, Marsh & McLennan, Inc., 1973–1974.

The editor of this edition was Frank McElroy, a registered professional engineer. James Weging and R. Pedroza assisted. Proofreading was handled by Friedrich Bishop, Harry Sharp, and Gloria Naurocki.

Contents

Chapter Page

Chapter	Page	
1	1	Occupational Safety: History and Growth
2	20	The Occupational Safety and Health Act of 1970
3	48	Safety Program Organization
4	70	Inspection and Control Procedures
5	104	Removing the Hazard from the Job
6	121	Accident Records and Injury Rates
7	150	Accident Investigation, Analysis, and Costs
8	172	Workmen's Compensation Insurance
9	196	Safety Training
10	217	Human Factors Engineering
11	239	Human Behavior and Safety
12	263	Maintaining Interest in Safety
13	300	Publicizing Safety
14	315	Audio-Visual Media
15	345	Office Safety
16	365	Industrial Buildings and Plant Layout
17	395	Building Construction and Maintenance
18	442	Planning for Emergencies
19	465	Personal Protective Equipment
20	528	Industrial Sanitation and Personnel Facilities
21	548	Occupational Health Services
22	570	Non-Employee Accident Prevention
23	591	Sources of Help

Contents

Chapter Page

24   631   Materials Handling and Storage

25   663   Hoisting Apparatus and Conveyors

26   711   Ropes, Chains, and Slings

27   738   Powered Industrial Trucks

28   762   Elevators and Plant Railways

29   794   Point-of-Operation and Transmission Guards

30   810   Woodworking Machinery

31   827   Machine Tools

32   856   Cold Forming of Metals

33   898   Hot Working of Metals

34   947   Welding and Cutting

35   976   Hand and Portable Power Tools

36   1006   Exhaust and Ventilation

37   1028   Industrial Hygiene

38   1108   Industrial Toxicology

39   1199   Chemical Hazards

40   1241   Noise and Hearing Conservation

41   1255   Electrical Hazards

42   1286   Flammable and Combustible Liquids

43   1324   Fire Protection

44   1387   Boilers and Unfired Pressure Vessels

45   1413   Motorized Equipment

46   1447   Safety Engineering Tables

  1493   Index

Dedication

To Roy G. Benson

(1907–1973)

Manager of the Industrial Department

(1959–1972)

and

To Paul E. Sheppard

(1921–1974)

Assistant Manager of the Industrial Department

(1973–1974)

NATIONAL SAFETY COUNCIL

FOUNDED SEPTEMBER 24, 1913
INCORPORATED IN ILLINOIS OCTOBER 1, 1930
INCORPORATED BY ACT OF CONGRESS AUGUST 13, 1953

Purposes and Powers. Certain essential provisions of the Act of Congress which incorporated the National Safety Council are as follows:

Objects and Purposes

"The objects and purposes of the corporation shall be --

to further, encourage, and promote methods and procedures leading to increased safety, protection, and health among employees and employers and among children in industries, on farms, in schools and colleges, in homes, on streets and highways, in recreation; and in other public and private places;

to collect, correlate, ... and disseminate educational and informative data, ... relative to safety methods and procedures;

to arouse and maintain the interest of the people of the United States, its Territories and possessions in safety and in accident prevention, and to encourage the adoption and institution of safety methods by all persons, ... and ... organizations;

to organize, establish, and conduct conduct programs, ... for the education of all persons, ... in safety methods and procedures;

to cooperate with, enlist, and develop the cooperation of and between all persons, ... and ... organizations ... both public and private, engaged or interested in, ... any or all of the foregoing purposes ..."

Powers

"The corporation shall have power --

to establish and maintain offices for the conduct of its business, and to charter local, State, and regional safety organizations, ... in appropriate places throughout the United States, its Territories and possessions;

to charge and collect membership dues, subscription fees, and receive contributions or grants of money or property to be devoted to the carrying out of its purposes;

to choose such officers, directors, trustees, managers, agents, and employees as the business of the corporation may require ...;

to adopt, amend, and alter a constitution and bylaws, ...

to organize, establish, and conduct conferences on safety and accident prevention;

to publish magazines and other publications and materials, ... consistent with its corporate purposes;

to adopt, alter, use, and display such emblems, seals, and badges as it may adopt."

Nonpolitical Nature

"The corporation, and its officers, directors, and duly appointed agents as such, shall not contribute to or otherwise support or assist any political party or candidate for office."

No Stock or Dividends

"The corporation shall have no power to issue any shares of stock nor to declare nor pay any dividends."

Audit and Congressional Report

The financial transactions shall be audited annually, ... by an independent certified public accountant ... A report of such audit shall be made by the corporation to the Congress not later than six months following the close of such fiscal year for which the audit is made."

Exclusive Right to Name and Emblem

"The corporation, and its subordinate divisions and regional, state, and local chapters, shall have the sole and exclusive right to use the name, National Safety Council. The corporation shall have the exclusive and sole right to use, or to allow or refuse the use of, such emblems, seals, and badges as it may legally adopt ..."

Transfer of Assets

"The corporation may acquire the assets of the National Safety Council, Incorporated, a corporation organized under the laws of the State of Illinois, upon discharging or satisfactorily providing for the payment and discharge of all of the liability of such corporation and upon complying with all laws of the State of Illinois applicable thereto."

Congressional and Presidential Approval. The Act which incorporated the National Safety Council was passed during the First Session of the 83rd Congress and was designated Public Law 259. Leadership in the Congress was as follows:

In the Senate	Bill introduced by *Arthur V. Watkins*, Senator from Utah	Approved by the Sub Committee on Federal Charters, of the Judiciary Committee, and Committee Chairmen, *John Marshall Butler*, Senator from Maryland	Approved by the Judiciary Committee, Chairman *William Langer*, Senator from North Dakota	Passed by the Senate, President of the Senate, *Richard M. Nixon*, Vice President of the United States
In the House of Representatives	Bill introduced by *Clifford Davis*, Representative from Tennessee	Approved by the Sub Committee on Federal Charters, of the Judiciary Committee, Acting Sub Committee Chairman, *John Robsion*, Representative from Kentucky	Approved by the Judiciary Committee, Chairman *Chauncey W. Reed*, Representative from Illinois	Passed by the House, Speaker of the House, *Joseph W. Martin, Jr.*, Representative from Massachusetts

The Act was signed by the President of the United States, Dwight D. Eisenhower, on August 13, 1953.

Transfer of Operations to the Federal Corporation. At the time the Act was passed granting the National Safety Council a federal charter, the Council was functioning as an Illinois corporation. This arrangement continued until on which date the Council's assets, operations and organizational structure were transferred from the Illinois corporation to the federal corporation.

THE NATIONAL SAFETY COUNCIL is the only national nongovernmental, privately supported, public service organization established solely for accident prevention. In 1953, the 83rd Congress of the United States issued a Federal Charter for the Council, thus recognizing its work as an integral part of the American way of life.

Occupational Safety

History and Growth

Chapter
1

Philosophy of Occupational Accident Prevention   **2**

History of the Safety Movement   **3**
Birth of NSC . . . Accident prevention discoveries . . . Acceleration of the drive for safety

Growth   **8**
Statistical evaluation . . . The dollar values . . . Industry and non-work accidents . . . The safety movement's resources . . . Summary of achievements

Safety Today   **13**
Large and small establishments . . . Labor-management cooperation . . . Research and standards . . . Safety and the law . . . Safety and occupational health . . . Psychology and "accident proneness" . . . Summary

Current Problems   **17**
Technology and public interest . . . Political problems . . . Organizational problems . . . A look to the future

References   **18**

1—Occupational Safety

Elimination of accidents is vital to the public interest. Accidents produce economic and social loss, impair individual and group productivity, cause inefficiency, and retard the advancement of standards of living.*

Philosophy of Occupational Accident Prevention

There is no question that accidents are costly to industry and society. Today, failure to try to prevent injuries to employees is indefensible.

The practical and moral aspects of accident prevention are interrelated, because accidents result both in a waste of manpower and resources, and in physical and mental anguish.

In Medieval days, the master craftsman tried to instruct his apprentices and journeymen to work skillfully and safely, because he could see the value of high quality and uninterrupted production, but it took the Industrial Revolution to create the conditions which led to the development of accident prevention as a specialized field.

The industrial safety philosophy developed as a result of the tremendous forces of production which were released. Without a deterrent to counter this waste of manpower and resources, the number of accidents and injuries would have challenged the imagination.

Once enlightened industrial management had accepted the responsibility for preventing accidents, the next step was workmen's compensation laws. This "new" line of thinking held the employer responsible for a share of the economic loss suffered by the employee because of an accident.

It was a rather short step from this to the realization that a large proportion of accidents could be prevented and that the same industrial brain power that could produce vast quantities of goods could also be used for accident prevention. Industry soon discovered that efficient production and safety were related. From this beginning grew the safety movement as it is known today.

The progress in reducing the number of accidents and injuries in this relatively short period of time has exceeded the highest expectations of the early safety pioneers. In less than one lifetime, safety has become a vital part of industry.

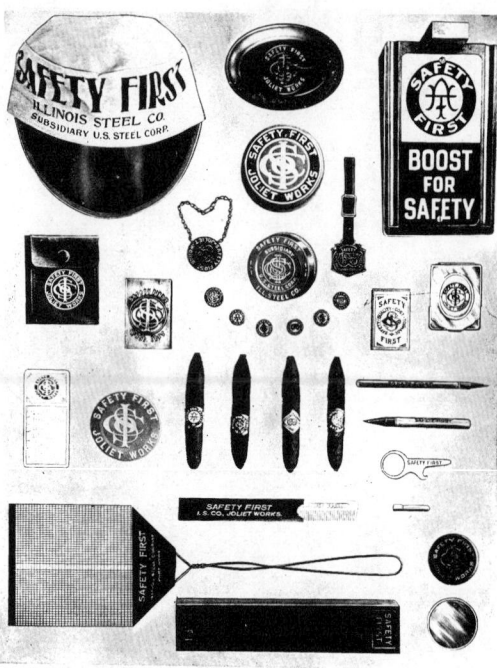

Fig. 1-1.—Give the man a cigar . . . or (in summer) a fly swatter! This 1913 display used by the Joliet Works of Illinois Steel Company, the first plant to organize a safety committee, offers a variety of safety incentives: watch fobs, paper weights, cigars, and a calendar.

Experience has shown that there is virtually no hazard or operation which cannot be overcome by practical safety measures. The future may introduce a type of unavoidable accident, but history indicates that practically all barriers can and will be surmounted.

In summary, here are the reasons for the continuing, concerted effort to prevent accidents:

1. Needless destruction of life and health is a moral evil

2. Failure to take necessary precautions against predictable accidents involves moral responsibility for those accidents

3. Accidents severely limit efficiency and productivity

---

*Excerpt from National Safety Council policy.

FIG. 1-2.—Old photo shows guarded arbor press gears, but drive belts from line shafts are left unguarded. First Law in U.S. compelling guarding of dangerous machinery was passed in Massachusetts in 1877. Inspection force with authority behind it was also created.

4. Accidents produce far-reaching social harm

5. The safety movement has already demonstrated that its techniques are effective in reducing accident rates and promoting efficiency

6. Nothing in the available data suggests that safety professionals are near a limit in their ability to extend the moral and practical values of accident prevention.

History of the Safety Movement

In the U.S. before the 19th Century, no industrial system existed. Families usually lived and worked on farms. No record was kept of injuries sustained by workers.

After 1800, when the effects of the Industrial Revolution were felt in the United States, factory work started.

During the last half of the 19th Century, American factories were expanding their product lines and producing at heretofore unimagined rates. While the factories were far superior in terms of production to the preceding small handicraft shops, they were often inferior in terms of human values, health, and safety.

These deficiencies were probably inevitable. The tools of mass production had to be invented and applied before anyone could begin to imagine the problems they would create,

3

FIG. 1-3.—The employee safety meeting was pioneered about the same time the National Safety Council was organized. This safety rally at the Falk Company, Milwaukee, was made more effective by use of flip chart (right) as visual aid.

and the problems had to be known before corrective measures could be considered, tested, and proved.

While this change in work environment was taking place, the thinking of the public, management, and the law was still reflecting the past, when the worker was an independent craftsman or a member of the family-owned shop.

In large industrial centers, the ugly results of industrial accidents and poor industrial health conditions became more and more obvious. Voices of protest were raised. Though there were employers who denied the existence of the problem, wiser management men began to try to meet specific aspects of it.

As early as 1867, Massachusetts had begun to use factory inspectors, and ten years later that state had a law requiring the safeguarding

of dangerous machinery.

From 1898 on, there were various efforts to make the employer financially liable for accidents. In 1911, the first effective workmen's compensation act was passed in Wisconsin (some authorities give this credit to New Jersey). This was followed by similar laws in many other states.

These laws were, at first, declared invalid because of conflict with the due process of law provisions of the 14th Amendment. After the U.S. Supreme Court in 1916 declared it to be constitutional, many states passed compulsory laws on workmen's compensation.

There had also been progress on the technical side of the problem. The railroads, which, perhaps, had suffered the most adverse publicity from accidents, adopted the air brake and the automatic coupler well before the turn

FIG. 1-4.—Clippings from October 17, 1913, issue of *The Chicago Tribune*.

of the century. Some progress was also made in such matters as guarding and fire prevention.

In the first decade of the 20th Century, two great industries, railroads and steel, began the first large-scale organized safety programs. From this period comes one of the great and historic documents of safety. In 1906, Judge Elbert Gary, president of the United States Steel Corporation, wrote:

"The United States Steel Corporation expects its subsidiary companies to make every effort practicable to prevent injury to its employees. Expenditures necessary for such purposes will be authorized. Nothing which will add to the protection of the workmen should be neglected."

The Association of Iron and Steel Electrical Engineers, organized soon after this announcement, devoted considerable attention to safety problems in its industry.

Birth of NSC

Then, in 1911, the year in which the Wisconsin Act was passed, a request came from the Association of Iron and Steel Electrical

ACCIDENT PREVENTION SCORE-BOARD
DODGE MANUFACTURING COMPANY.

Number Dept.	Name Department	Name Foreman	Deduction For Absence	Percentage Month	Year to Date	Rank For Month	Year to Date
FOUNDRIES							
30	CUPOLA & YARDS	L. KING.	19.	1000	1000	1	1
17	CORE ROOM	G. SCHNAU.	10.	1000	1000	1	1
12	SOUTH	W. MIDDLETON.	7.5	1000	1000	1	1
15	HANGER	G. F. LONG.	14.	976	943	5	17
14	PULLEY	J. BICKEL.	16.	1000	991	1	11
18	PATTERN	E. GARTNER.	38.	1000	1000	1	1
16	CHIPPING	J. STUFF.	6.	976	979	3	16
MACHINE SHOP							
5	BEARING	J. MILLER.	15.5	1000	996.1	1	6
3	SHEAVE	R. PRIEM.	20.	1000	995	1	
3A	MACHINE	J. ROUGH.	28.	1000	1000	1	1
3B	ERECTING	J. GONICK.	19.	1000	994	1	8
4	SHAFTING	W. MOURTS.	32.	1000	1000	1	1
6	CLUTCH	T. KENYON.	10.	1000	999	1	6
2	IRON PULLEY	G. FRIEDMAN.	10.	1000	991.5	1	10
20	TOOL	G. PETERSON.	28.	1000	981	1	14
WOOD SHOP							
1-A	SEGMENT	F. YOST.	115	1000	996.3	1	5
1-B	ARM	J. PHILION.	32.	1000	949	1	16
1-C	ASSEMBLING	C. MARSH.	7.7	1000	993	1	9
1-D	FINISHING	C. DeGROOTE.	28.	1000	1000	1	1
OTHER DEPARTMENTS							
7	STEEL SHOP	G. HUNT.	13.5	979	981	4	12
27	INSPECTION	G. McNEAL.	52.	1000	1000	1	1
25	SHIPPING	G. SHOBE.	8.7	1000	996.4	1	4
21	YARD & SAW MILL	E. DILS.	11.	1000	978	1	5
23	POWER	W. TUPPER.	38.	1000	1000	1	1
22	MILLWRIGHT	S. BRUBAKER.	26.	1000	987	1	13
19	METAL PATTERN	O. FORD.	104.	1000	1000	1	1

FOREMENS MONTHLY COMPETITION

All Departments Having a Score of 1000 or the Highest Three Scores, will Receive Special Prizes as per Monthly Prize List.

PRIZE LIST FOR MONTH OF ....
$25.00
Equally divided Among the Foremen having A Perfect Score

ANNUAL COMPETITION

GENERAL

All Departments Scoring 1000 For the Year or the One holding Rank one in Yearly Percentage will Receive Two Days Pay Extra
Second Highest will Receive One Days Pay Extra.

HEAD FOREMAN WILL PARTICIPATE IN FIRST PRIZE. IF WON BY A DIVISION IN HIS DEPARTMENT.

FIG. 1-5.—There were departmental safety contests as far back as 1913, as this score board illustrates. Compensation was made for sizes of the working groups. Degree of hazard was disregarded—"the differentiation being considered as equalized in the choice or selection of men with reference to their ability and fitness for their respective class of work."

Engineers to call a general industrial safety conference on a national scale. The result was the First Cooperative Safety Congress, which met in 1912 in Milwaukee. This gathering called for another meeting in New York the following year, and at that meeting the National Council for Industrial Safety was organized. Shortly afterward, the organization's name was changed to the National Safety Council, and its program was broadened to include all aspects of accident prevention. Yet it must be remembered that the Council was the creation of industry and that its activities have always been heavily concentrated on industrial safety.

The group that met in Milwaukee and New York was composed of a few "professional" safety men, some management leaders, public officials, and insurance men. Their one point in common was a desire to attack a problem which most people thought to be unimportant or could not be solved. Because these men were determined, the safety movement as we know it today was designed and built.

Accident prevention discoveries

As industry developed some experience in safety, it discovered that engineering could prevent accidents, that employees could be reached through education, and that safety rules could be established and enforced. Thus the "Three E's of Safety"—Engineering, Education, and Enforcement—were developed.

There were other discoveries, too. Safety departments had often argued that savings in compensation costs and medical expenses would many times repay safety expenditures. Thoughtful business leaders soon learned that these savings were only a fraction of the financial benefits to be derived from accident prevention work. Indirect financial savings are estimated to be several times larger than the direct savings in compensation and medical bills.

Acceleration of the drive for safety

Industrial safety received wide acceptance in the years between the two world wars. Conservation of manpower during World War II intensified the safety growth, and the federal government encouraged safety activities by its contractors. As industry expanded to meet the needs of the war effort, additional safety personnel were hastily trained to try to keep pace. The acceptance of safety as part of the industrial safety picture did not diminish with the end of the war. By then, the importance of safety to quality production was well established, and the small handful of dedicated people in 1912 had grown to millions.

In 1948, for example, Admiral Ben Moreell, then president of Jones and Laughlin Steel Corporation, wrote:

"Although safe and healthful working conditions can be justified on a cold dollars-and-cents basis, I prefer to justify them on the basic principle that it is the right thing to do. In discussing safety in industrial operations, I have often heard it stated that the cost of adequate health and safety measures would be prohibitive and that 'we can't afford it.'

"My answer to that is quite simple and quite direct. It is this: 'If we can't afford safety, we can't afford to be in business.'"

A discussion of current federal safety legislation follows later in this chapter under "Safety and the law."

A by-product of organized safety activity has been increased interest in safety engineering on the part of schools of higher learning. A number of schools now offer degrees and advanced courses in this subject and

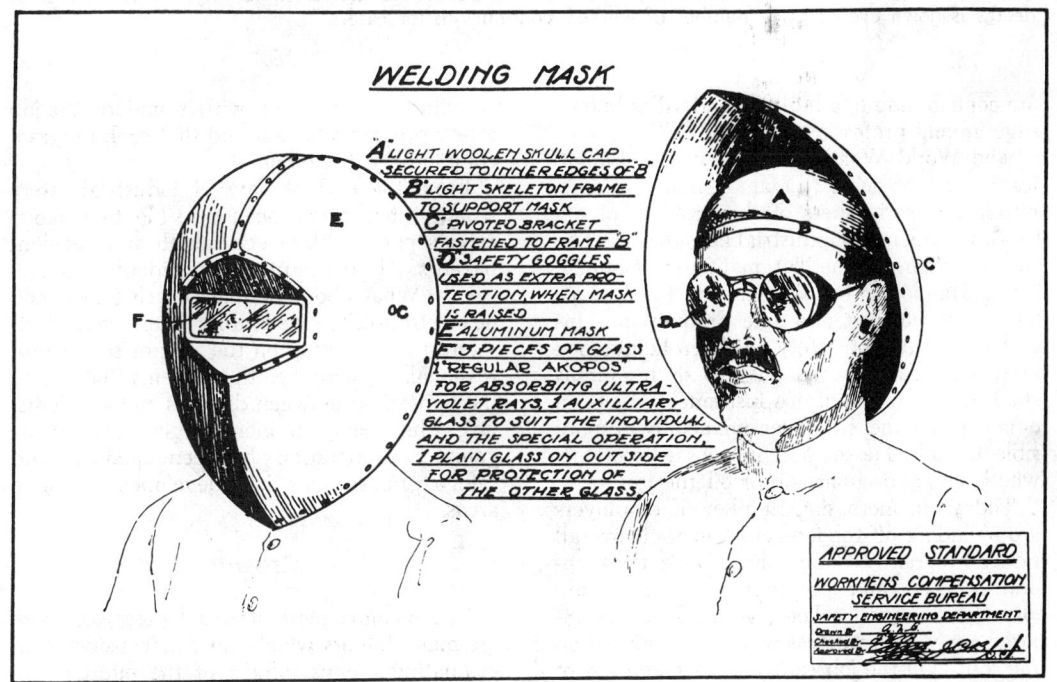

FIG. 1-6.—Workmen's Compensation Service Bureau gave "approval only of the principle" of protective equipment and clothing. Its "Universal Safety Standard" (dated 1913) did push for double protection even in those early days.

TREND OF ACCIDENTAL WORK DEATHS HAS BEEN DOWN

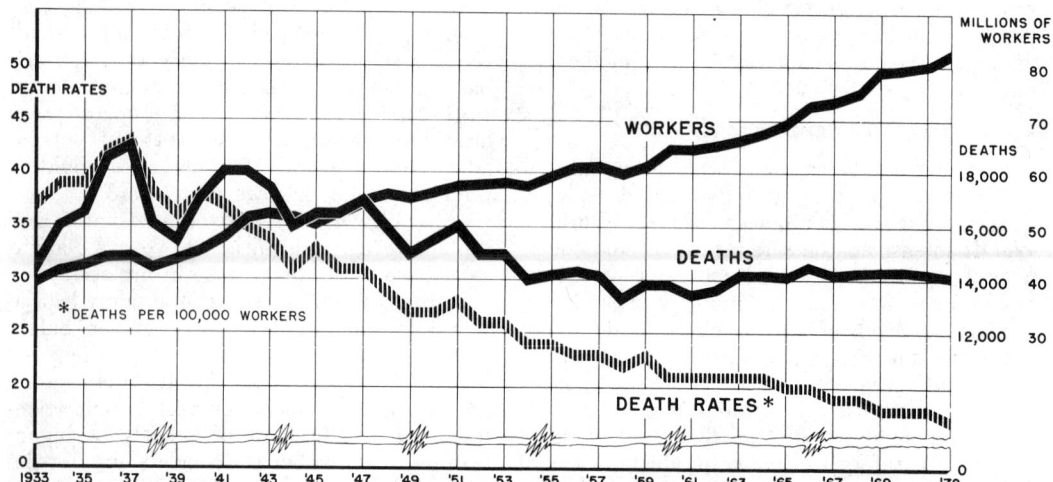

FIG. 1–7a.—The average American today has a better chance of avoiding accidental death than did the American of 30 or 40 years ago. This chart, based on figures for the entire country, shows number of deaths is down even though number of workers continues to increase.

are contributing to a higher standard of knowledge among professionals in the field.

The World War II labor shortage dramatically brought home to management the magnitude and seriousness of the problem of off-the-job accidents to industrial employees. The wartime theme of the National Safety Council, "Save Manpower for Warpower," focused attention on efficient and safe production. Interest in off-the-job safety has been heightened in recent years by the passage of numerous state laws on compulsory insurance which, in effect, make the employer financially responsible for all illnesses and injuries to workers, whether they originate on or off the job.

Today, an increasing number of employers are including off-the-job safety in their overall safety program. Companies realize their operating costs and production schedules are affected as much when an employee is injured away from work as when he is injured on the job. Off-the-job safety is an extension of a company's on-the-job safety program and is intended to educate the employee to follow the safe practices he uses on the job in his outside

activities. Companies with sound on-the-job safety programs have found that each program complements the other.

From the earliest days of industrial safety work, it has never been possible to make a clear separation between health and accident hazards. Is dermatitis an accident or a disease? What about hernias, hearing loss, and heart trouble? Inevitably, there has been interest and activity on the part of safety professionals in many health problems that are on the borderline between diseases and accidents. Over the years, an increasingly effective organized cooperation between medical and safety professionals has developed in these areas.

Growth

In any movement of social progress, there are many factors which must be considered in evaluating the usefulness of the effort.

In evaluating the work of the safety movement, particularly, a complex set of factors must always be kept in mind.

FIG. 1–7b.—Since 1933, death rates per 100,000 workers were at their highest for manufacturing in 1936 and for nonmanufacturing in 1937. Since those years, both rates decreased more than 50 percent and reached their lowest levels in the past five years. These lower rates resulted from deaths declining about one-fourth, and increased numbers of workers.

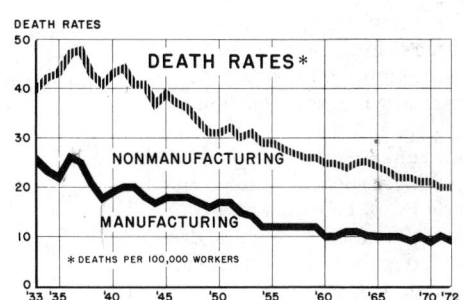

FIG. 1–7c.—The injury frequency rate of NSC reporters declined rapidly to 1932, increased in World War II, then reached a record low in 1962. The 67% increase to 10.17 in 1972 from the 6.10 average of 1961–63 would have been only 37% to 8.36 if 1972 industry representation had been the same as 1961–63.

The severity rate followed a similar pattern. The 1972 decline of 4% from 1961–63 was not affected by changes in industry representation.

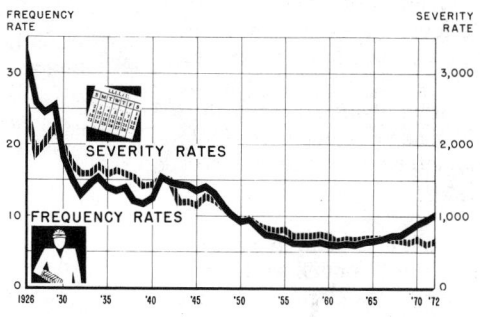

Safety professionals, unlike designers of a new piece of machinery, cannot look philosophically upon errors and breakdowns in the experimental stage. Since they deal with human lives, human health, and immediate efficiency and productivity, safety professionals must deal with today's problems today.

On the other hand, they cannot be content to run from emergency to emergency. They must consider the long-term effects of their work—the increase of knowledge, the improvement of techniques, the development of organizational forms which will serve them well next year and for many years to come.

Since the factors are complex, no simple rating scale can indicate an answer to the question, "What has the safety movement accomplished?" In the absence of such a rating scale, an attempt to answer the question must be made by assembling several kinds of data.

Statistical evaluation

First, the question must be asked, "Has the safety movement, in fact, done anything in the past to prevent accidents?" To that question can be answered a clear "Yes!"

If the annual accidental death rate per thousand of population which held in 1912 had continued, there would have been over 1.5 million more accidental deaths than actually occurred. Since 1912, the death rate for persons of normal working age—25 to 64 years—declined more than 67 percent while the rate of all ages of the entire population declined only half that much. Medical progress accounts for some of this gain, but the larger part is certainly the product of organized safety work.

The work accident figures probably understate the progress made in industrial safety, for they include a very large number of non-manufacturing work deaths. Farm work, trades, services, and government accounted for more than half the accidental work deaths in recent years. The remaining 6900 deaths occurred in manufacturing, public utilities, transportation, construction, and the extractive industries (mining, quarrying, and gas and oil wells).

Overall frequency and severity rates for all industries since 1926 are not available, but figures on the experience of concerns reporting to the National Safety Council provide a dramatic summary of the progress of organized safety work in the nation; a 66 percent cut in the frequency of disabling injuries.

Though these figures represent only companies involved in the work of the organized safety movement, they cannot be dismissed as representing a handful of firms with an exceptionally high level of safety enthusiasm. The latest figures are based on reports from over 13,000 units. These units were principally manufacturing concerns, which accounted for more than one-fourth of the total number of manufacturing man-hours worked in America.

It would be fair, perhaps, to call these figures representative of large-plant experience in this country, since the Council receives the majority of its reports from large concerns and only a minority from small ones.

Safety professionals cannot afford to underestimate the significance of the statistical data on their past performance. It is abundantly clear that the bitter problem confronting the safety pioneers of 1913 has been attacked with measurable success.

The dollar values

It has been estimated that the annual cost of occupational accidents in the United States exceeds $11.5 billion. If the 1912 accident rates had been left unchanged and if there had been no organized safety movement, this annual cost would have easily been two or three times as great.

Against such dollar savings, the relatively small expenditures for safety throughout America provide a striking contrast. Each dollar spent for safety by American industry is probably producing a clear profit of several hundred percent.

Industry and nonwork accidents

Directly and indirectly, industry is bearing a substantial part of the burden of the cost of nonwork accidents and their prevention. The National Safety Council is the creation of industry and largely supported by it. Statewide and local safety organizations play a major role in the fight against such accidents, and they, too, are largely industrial in origin and support. Thousands of safety professionals and their employers volunteer time and funds which are the mainstay of the National Safety Council's work.

Industry supports a large part of the job of informing the general public on these problems through the press, radio, and television.

The effectiveness of the nonwork accident-prevention campaign is shown by the fact that, from the time records were first kept in 1928, both home and public accident death rates (excluding motor vehicle accidents) have declined.

If industry has been a large contributor to this successful work, it has also been a heavy beneficiary of its fruits. Disruption of labor force, worry and hardship among employees, loss of purchasing power by consumers, and heavy tax burdens for the support of hospitals and relief agencies are all results of nonwork accidents which affect industry's pocketbook.

The safety movement's resources

Statistics measure what has been accomplished and also indicate the development of tools, methods, and knowledge which are the safety professional's capital and his resources for meeting future accident problems.

Know-how. A body of knowledge cannot be entirely translated into statistics, but it may be of use to note the outward signs of increased knowledge about industrial safety.

This Accident Prevention Manual is not a radical venture into new fields of knowledge. It is the cumulation of facts and opinions which have become a part of the safety movement's general heritage. Its purpose is to bring key points of specific, as well as general, value to safety workers.

Here are collected facts that took years to discover—years of searching and researching, of trial and error, of failure and success.

Today, an individual, whether an experienced safety professional, part-time safety administrator, or neophyte entering the field, can turn these pages and come up with better answers to a wider range of industrial safety problems than were available to the wisest and best-trained professional safetyman of a generation ago.

Yet this Manual contains only a fraction of the body of knowledge available to fight the never-ending war against accidents.

The volumes of National Safety Congress Transactions contain useful material and expert opinion on all phases of safety. In countless pamphlets and periodicals of safety organizations, government agencies, and insurance companies and in the studies and directives of individual industrial concerns is still more material. The literature of various trades and professions is likewise rich in safety information. A list of handbooks is given in Chapter 46, "Safety Engineering Tables."

The National Safety Council has a very useful series of training courses, at both the beginning and advanced levels, for safety professionals. Write the NSC Training Department for details.

Finally, tremendous stores of safety information in the heads of professional safety engineers, executives, supervisors, and rank-and-file employees are made available through exchange of information in safety conferences, technical seminars, safety newsletters, and other publications.

It may be argued that, of all the achievements which the safety movement has to its credit, the greatest is the accumulation and preservation of a body of know-how which the safety professional has at his fingertips in dealing with the problems that confront him now and may confront him in the future.

The heritage of cooperation. The safety movement would be a far less effective force than it is if, at the outset, its members had hoarded and concealed their discoveries from their colleagues in competitive companies.

It was teamwork that created the safety activities of the Association of Iron and Steel Electrical Engineers. It was broadened teamwork which was represented at the first Milwaukee Conference and which led to the formation of the National Safety Council and other safety organizations.

Effective accident prevention requires channels of cooperation. Through the Council and other safety organizations, safetymen found meeting places for the exchange of ideas, developed a safety press, and stimulated one another in friendly competition.

The tradition that there should be no secrets in safety, no denial of help to a competitor in life-saving, is one of the great elements of strength in the safety movement.

Good will. Like any new movement in industry, safety started with very little capital in the form of good will.

No small part of the safety professional's capital today is the prestige and good will built up for safety proposals and expenditures over the years. Where the pioneers had to battle every step of the budgetary way, safety professionals today have a far more receptive hearing from management.

Professionalism. Dedicated safety professionals continue to be accident prevention's most valuable asset. Their ranks have grown to the point where, in the mid-seventies, membership in the American Society of Safety Engineers is approaching 15,000. This organization, dedicated to both their interests and their professional development, has approximately one hundred chapters in the U.S. and Canada. Individual membership is worldwide.

There are many other qualified safety professionals, in addition to the ASSE members, who, together with thousands of specialists and technicians, carry out a limited scope of activities within the field, and numerous individuals who devote less than 50 percent of their time to safety functions.

In 1968, the ASSE was instrumental in forming a new organization, the Board of Certified Safety Professionals. Its purpose is to provide a means of giving professional status to qualified safety people by certification after meeting strict educational and experience requirements and passing an examination. Several hundred safety professionals have been certified, since the BCSP was formed.

Advancement of knowledge. There has always been an orderly development of knowl-

Fig. 1–8.—Atlantic Refining Company's fire brigade poses with what for the World War I era was up-to-date equipment. Compare this with the modern equipment shown in Chapter 18, "Planning for Emergencies."

edge, which, when applied with sufficient skill and judgment, has produced significant reductions in many types of accidents and accidental injuries. However, the tremendous increase in scientific knowledge and technological advancement since the close of World War II has added to the complexities of safety work.

The approach has oscillated between one that emphasizes environmental control or engineering, and one that emphasizes human factors. From this, several important trends in the pattern of the safety professional's development have emerged.

● First, increasing emphasis toward analyzing the loss potential of the activity with which the safety professional is concerned. Such analysis will require greater ability (a) to predict where and how loss- and injury-producing events will occur and (b) to find the means of preventing such events.

● Second, increased development of factual, unbiased, and objective information about loss-producing problems and accident causation, so that those who have ultimate decision-making responsibilities can make sound decisions.

● Third, increasing use of the safety professional's help in developing safe products. The application of the principles of accident causation and control to the product is becoming more important because of the increase in product liability cases, the sudden emphasis in law of the entire field of negligent design, and the obvious impact a safer product would have on the overall safety of the environment.

To identify and evaluate the magnitude of the safety problem, the safety professional must be concerned with all facets of the problem—personal and environmental, transient and permanent—to determine the causes of accidents or the existence of loss-producing conditions, practices, or materials. From the information he collects and analyzes, he proposes alternate solutions, together with recommendations based upon his specialized knowledge and experience, to those who have ultimate decision-making responsibilities.

Therefore, future application of this knowledge in all aspects of our civilization—whether to industry, to transportation, at home, or in recreation—makes it imperative that those in this field be trained to utilize scientific principles and methods to achieve adequate results. Of prime importance will be the knowledge, skill, and ability to integrate machines, equipment, and environments with man and his capabilities.

The safety professional in performing these functions will draw upon specialized knowledge in both the physical and social sciences. He will apply the principles of measurement and analysis to evaluate safety performance. He will be required to have fundamental knowledge of statistics, mathematics, physics, chemistry, and engineering.

He will utilize knowledge in the field of behavior, motivation, and communications. Knowledge of management principles as well as the theory of business and government organization will also be required. His specialized knowledge must include a thorough understanding of the causative factors contributing to accident occurrence as well as methods and procedures designed to control such events.

The safety professional will also need diversified education and training, if he is to meet the challenges of the future. The population explosion, the problems of urban areas and future transportation systems, as well as the increasing complexities of man's everyday life, will create many problems and extend the safety professional's creativity to its maximum, if he is to successfully provide the knowledge and leadership to conserve life, health, and property.

Training of the safety professional of the future can no longer be the "on-the-job," man-to-man type, only. It must contemplate specialized undergraduate level training, leading to a bachelor's degree or higher degree.

Training courses, such as those conducted by the National Safety Council, have and will continue to serve a very useful purpose for a large number of individuals who begin performing safety functions and must receive initial training or advanced training in certain specialized areas.

Approximately 200 four-year colleges and universities currently offer courses in safety, and several dozen offer a bachelor's degree or higher in safety. An increasing number of two-year, community colleges are offering associate degrees or certificates for courses designed for the safety technician or part-time administrator. Governmental agencies and ASSE are actively promoting and conducting such professional development programs.

This is the creative challenge of tomorrow—not only improved performance on the job, but also professionalizing and perpetuating this field of endeavor.

Summary of achievements

The safety movement has helped save many thousands of lives. It is saving industry and its employees billions of dollars a year.

It faces the future with great resources for eliminating accidents in the future—resources in know-how, teamwork, good will, and trained and dedicated safety workers.

It has, therefore, done much to meet the double challenge presented to it: to deal with accidents now and to build soundly for the long-range attack upon accidents in the future.

Safety Today

The answer to the question, "What has the safety movement accomplished?" is in purely positive terms—accomplishments, advances, achievements. The answer to the question, "How does the safety movement stand today?" demands a look at what is wrong, as well as what is right, with the present situation. The answer can be found in an appraisal of how the safety movement stands in relation to how it ought to stand. The first point to be considered is simple and grim:

- Accidents still bleed this country of more

Eye Accidents are Unexpected
But
They Happen Just the Same.

Who would have thought that this curved chip would fly ends first at his goggles and hit both lenses? But he was wearing his goggles.

Don't take a chance.

Wear Your Goggles

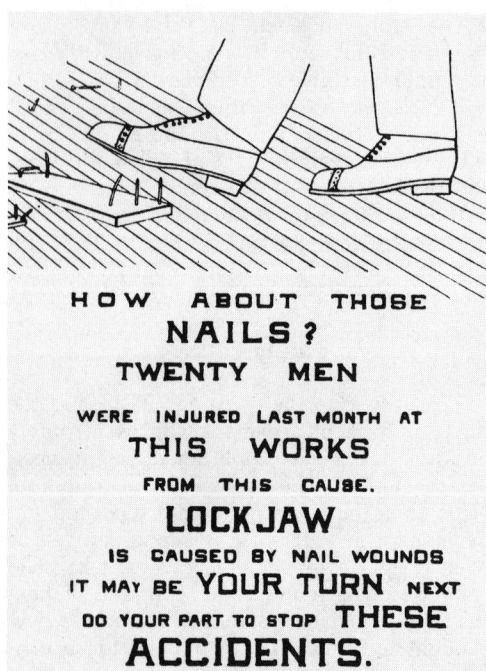

HOW ABOUT THOSE
NAILS?
TWENTY MEN
WERE INJURED LAST MONTH AT
THIS WORKS
FROM THIS CAUSE.
LOCKJAW
IS CAUSED BY NAIL WOUNDS
IT MAY BE **YOUR TURN** NEXT
DO YOUR PART TO STOP **THESE**
ACCIDENTS.

FIG. 1–9.—Two examples of early safety posters.

than 117,000 lives a year, cause more than 11.5 million disabling injuries, and account for a financial loss of more than $37 billion.

• Work accidents destroy more than 14,000 lives a year, about half these deaths occurring in what is normally considered industry. Work accidents injure 2.4 million persons annually and cost more than $11.5 billion.

In recent years, the ratio of off-the-job deaths to on-the-job deaths was about 3 to 1 and more than half of the injuries suffered by employees occurred off the job.

In terms of time loss, all injuries to workers, both on and off the job, caused a loss of about 120 million man-days of work directly and 200 million man-days indirectly.

Within the industrial community, there are very large variations in accident rates from industry to industry and from company to company.

Trade, service, and manufacturing are, on the average, low-accident businesses. Transportation, agriculture, construction, and the extractive industries are high-accident fields. These low and high rankings hold for both deaths and injuries. The communications industry has accidental death rates substantially below the national average and the lowest injury rates.

How much the variation from one industry to another reflects unavoidable differences in hazards and how much it reflects other factors, such as accident prevention activity or government regulation, has not been fully determined. Certainly, the mining industry, for example, has had a maximum amount of regulation and assistance from government, yet its rate remains high.

In construction and lumber operations, the high rates may be due in part to the fact that the typical operation is small, or a shifting operation.

Large and small establishments

It is generally assumed that small companies with, say, fewer than 100 employees have proportionately more work injuries than large corporations. However, since many small companies have not been accurately recording and reporting their experience, it is difficult to establish any valid ratio comparing work injury experience. However, it is safe to say that companies, large or small, that ignore systematic safety effort will, in the main, have more than their share of work accidents and injuries.

The seriousness of the small-enterprise problem has been widely recognized for many years, and the National Safety Council has devoted much effort to meeting it. Some promising steps have been taken to increase small-establishment participation in the organized safety movement, principally through the establishment of liaison between the National Safety Council and the trade associations representing many small companies.

Certain aspects of the small-company problem can be stated with assurance:

1. The small establishment may not need or cannot employ specialized safety personnel to deal with the accident problem.
2. The number of accidents in a significant amount of time or the financial position of many small concerns makes it difficult to convince them that spending the money necessary for proper equipment, layout, guarding, and other elements is important.
3. Managers of small operations are harried by a host of problems in all fields. Seldom do they have the expertise or can they find time for proper study of accidents and their causes.
4. In small units, statistical measures of performance are unreliable, so that it is difficult to produce clear-cut evidence as to the cost of accidents versus the effectiveness of accident prevention work. In other words, a small operation may have, by luck, a good or bad accident record over a few years, whether or not its safety program is sound.

These are obstacles to progress—real and serious ones. They are not, of course, excuses for failure to try to prevent accidents. The trade association approach offers some real hope for improvement on a group basis.

Labor-management cooperation

At top levels, and in some cases at the grass roots, progress has been made in enlisting the support of organized labor in management's campaign to reduce accident losses.

In 1949, the National Safety Council issued a policy statement declaring the common interest of labor and management in accident prevention. Even before this data, representatives of leading labor organizations served as members of the Council's governing boards. At a number of National Safety Congresses, reports were made by labor and management men on specific examples of local and company-wide cooperation between unions and corporations. In 1955, a Labor Department and a Labor Conference were formally established to function within the framework of the National Safety Council.

Labor has been represented at the various presidential and gubernatorial industrial safety conferences, and was instrumental in promoting legislation such as the Williams-Steiger Occupational Safety and Health Act of 1970 (OSHAct), described in the next chapter.

Some unions have independently done extensive safety work and have published printed matter and released films which are real contributions to safety.

It is the hope of the great majority of far-sighted safety professionals in both labor and management that every opportunity for cooperation will be pressed to the fullest extent.

Further, management must seek the multiple causes of accidents, not only the most obvious cause of disabling injuries. Workers must believe that safe performance benefits them, too.

Research and standards

Statistical data on industrial accidents have been compiled by the National Safety Council for more than 45 years. Analyses computed annually and published in industry rate pamphlets and *Accident Facts* have proven of utmost importance in evaluating the leading causes of accidents.

Some industries through their trade associations have recorded accident rates for almost 55 years. In most instances, even the divisions of an industry can establish their positions with regard to number and types of accidents and can determine their experience in comparison with national averages.

There are a large number of standards relating to safety. Continuing research has been necessary over the years to keep these standards in line with current industrial development and the development of new products and materials.

Special research projects, such as those making studies of walkway surfaces and safety belts, can be and have been financed by private sources and coordinated by the National Safety Council. Since the Council is not an approval agency, results of these projects are given as a summary of findings with no attempt to set minimum standards.

Safety and the law

The early legal action in industrial safety took the form of laws to regulate and investigate. The next phase was largely concerned with workmen's compensation payments.

The following years have seen a gradual growth in regulation of industry on safety matters by federal, state, and local governments. The Walsh-Healey Act, which deals with companies having supply contracts with the federal government, is an example of such regulation.

In certain industries—notably mining and transportation—federal government regulation and inspection have been extensive. The Construction Safety Act, which was passed in 1969, deals with the particular problems of that industry.

In 1970, the Williams-Steiger Occupational Safety and Health Act was passed and, for the first time, the United States had a *national* safety law. Every business, with one or more employees, which is affected by interstate commerce, is covered by the law. Safety in this country has taken on a new direction and meaning as a result of the Occupational Safety and Health Act. More details are in the chapter on OSHA-NIOSH, Chapter 2.

Today industry accepts almost without exception the idea of financial responsibility for work injuries. Not all of industry is convinced of the effectiveness of government regulation of safety procedures. Several states, with the cooperation of management and labor, have developed standards and regulations that have been effective in reducing the number of accidents.

A recent development in the laws of some states has been the establishment of health and accident insurance on a compulsory basis to cover employee disabilities from diseases or accidents which originate off the job.

This compulsory insurance might be considered either a drastic extension of the principle of workmen's compensation or an extension of social security legislation. It differs from workmen's compensation in that it puts a financial burden upon management for diseases and accidents which are products of conditions beyond its control.

Whatever the theory, the result of these laws is to give the employer a direct financial stake in dealing with the off-the-job accident problem.

Safety and occupational health

Even though medical and safety cooperation in accident prevention started back in the earliest days of the safety movement, interest in safety on the part of the medical profession is increasing. Part of this interest is the result of concern with occupational disease, noise, radiation, and other problems beyond the former concepts of occupational accident prevention.

Safety is, of course, the beneficiary of many medical advancements—notably prevention of industrial diseases and infections.

Industry's growing concern with problems of ionizing radiation has brought the physicist into partnership with medical and safety professionals.

Knowledge in the field of industrial hygiene has been greatly extended in a number of areas.

The utilization by industry of the physically handicapped has lead progressive companies to adjust their policies on pre-employment physical examinations, to make sure that they are screening out only the unfit and not those with defects that do not rule out useful work under proper conditions and limits.

Psychology and "accident proneness"

The safety professional who is thoughtfully looking for ways to improve his work encounters a great deal of challenging information in modern psychological writing—and also a great deal of careless and misleading generalizations.

Concern about the so-called "accident-prone" individual in industry is as old as the safety movement. Statistical information suggests existence of such individuals, though clear and sharp data proving this point are remarkably hard to come by. Too many alleged "proofs" turn out to be statistically deceptive, or based on inadequate samples, or the result of highly subjective diagnoses.

The elusiveness of statistical proof of the existence of accident-prone individuals suggests to some thoughtful safety professionals that accident proneness may be a passing phase in the individual rather than a permanent characteristic.

Realistically, objective analysis might disclose some supervisory deficiency or procedural weakness which may aggravate the hazard of certain operations or performance of individuals or groups of workers.

The same observation applies to psychological tests used as screening devices for new employees. Spectacular claims have been made from time to time for the effectiveness of such tests in predicting accident proneness, but none has established itself to the general satisfaction of the safety profession.

The work of psychologists like Dunbar and the Menningers has aroused great interest among safety professionals. However, great contributions to the practical day-to-day fight against accidents have not yet been made by psychologists—or have not been recognized if they have been made.

Summary

The present situation in the field of industrial safety is one of progress and improvement, largely through the continued application of techniques and knowledge slowly and painfully acquired through the years.

There appears to be no limit to the progress possible through the application of the universally accepted safety techniques of education, engineering, and enforcement.

Yet large and serious problems remain unsolved. A number of industries still have high accident rates. There are still far too many instances where management and labor are not working together or have different goals for the safety program.

The resources of the safety movement are great and strong—an impressive body of knowledge, a corps of able professional safety people, a high level of prestige, and strong organizations for cooperation and exchange of information.

Fig. 1–10.—Modern posters produced by the National Safety Council combine eye appeal and an easily understood message.

Current Problems

Some problems of the safety movement are less directly related to present strengths and weaknesses. Some of these problems are social and political. Still others are essentially organizational.

Technology and public interest

There is no reason for the safety professional to view the public's interest in product safety, a better environment, and general technological trends with alarm. Emphasis upon automation and more refined instrumentation will probably continue. New specific problems will arise, but they will be of a type which well established methods of safety engineering are competent to solve.

The use of new materials and techniques—particularly radioactive materials and lasers—is likely to present more serious difficulties to the safety professional. However, even here, there is considerable experience.

1—Occupational Safety

Political problems

On the political side remains the timeworn problem of industry-government relations. The difficult task of the serious safety professional is to objectively advise his management regarding legislation and what may be considered government interference in safety matters and still, at the same time, encourage good men in public service to work to improve safety standards.

If safety people simply fight government "on principle," they will lose their objectivity and professional stature, and this will probably only heighten the pressure from other quarters for stiff and inflexible regulation.

Organizational problems

On the national scale, a wide variety of organizations are attacking specific aspects of the safety problem. The National Safety Council is, of course, the giant in the field— a strong, constructive, and non-political, non-commercial giant. It has repeatedly sought and often achieved cooperative division of labor between itself and other organizations in the safety field.

One of the guiding principles of the Council has been that there was work enough and credit enough for all.

It remains to be seen whether the best organizational forms have been found for participation by all businesses in safety work. Safety professionals should be ready to consider new ideas and new forms.

A look to the future

The future is, as it always has been, most uncertain. Problems large and small, predictable and unpredictable, can be expected to crowd upon the safety professional.

Some of these problems will call for reapplication of established safety techniques. Others will call for radical departures and the creation of new methods and new organizational forms.

To be able to discriminate between the two situations will, perhaps, be the safety professional's greatest test.

References

American Engineering Council. *Safety and Production.* New York, Harper & Brothers Publishers, 1928.

American Insurance Association, Accident Prevention Department, 85 John St., New York, N.Y. *Handbook of Industrial Safety Standards.* 10th rev., 1962.

Andrews, E. W. "The Pioneers of 1912." *National Safety News,* 66:24–25, 64–65 (July 1952).

Beyer, David Stewart. *Industrial Accident Prevention,* 3rd ed. Boston and New York, Houghton Mifflin Co. 1928.

Blake, Roland P., editor. *Industrial Safety,* 3rd ed. Englewood Cliffs, N.J., Prentice-Hall, Inc., 1963.

Board of Certified Safety Professionals, 850 Busse Hy., Park Ridge, Ill. 60068.

Campbell, R. W. "The National Safety Movement." *Proceedings of the Second Safety Congress of the National Council for Industrial Safety,* pp. 188–192, 1913.

DeBlois, Lewis A. *Industrial Safety Organization for Executive and Engineer.* New York City, McGraw-Hill Book Company, 1926.

DeReamer, Russell. *Modern Safety Practices.* New York, John Wiley and Sons, Inc., 1958.

Eastman, Crystal. *Work Accidents and the Law.* New York, Charities Publication Committee, 1910.

Heinrich, H. W. *Industrial Accident Prevention,* 4th ed. New York, McGraw-Hill Book Company, 1959.

Holbrook, Stewart H. *Let Them Live.* New York, The Macmillan Co., 1939.

Menninger, K. A. *Man Against Himself.* New York, Harcourt, Brace and World, Inc., 1956.

National Safety Council, 425 North Michigan Ave., Chicago, Ill. 60611.
Accident Facts. Issued annually.
Family Safety. Issued quarterly.
National Safety News. Issued monthly.
Proceedings of the First Co-Operative Safety Congress. 1912.
Proceedings of the Second Safety Congress of the National Council for Industrial Safety. 1913.
Transactions (Proceedings) of the National Safety Congress. Issued annually from 1914.
"Golden Anniversary Issue." *National Safety News,* 87:6 (May 1963).

Schulzinger, Morris S. "Accident Syndrome—A Clinical Approach." *Archives of Industrial Health,* 11:66–71, 1955.

Schwedtman, Ferd., and James A. Emery. *Accident Prevention and Relief.* New York, National Association of Manufacturers in the United States of America, 1911.

Simonds, Rollin H., and John V. Grimaldi. *Safety Management.* Homewood, Ill., Richard D. Irwin, rev. 1963.

The Occupational Safety and Health Act of 1970

Chapter 2

Legislative History 21

Administration 21
Occupational Safety and Health Administration . . . Occupational Safety and Health Review Commission . . . National Institute for Occupational Safety and Health

Major Provisions of the OSHAct 24
Coverage . . . Employer and employee duties . . . The OSHA poster . . . Employee rights . . . Occupational safety and health standards . . . Recordkeeping and reporting requirements . . . Variances from standards . . . The enforcement process . . . Violations . . . Citations . . . Penalties

Contested Cases 40

Small Business Loans 42

Federal-State Relationships 42

What Does It All Mean? 43

Directory of Federal Agencies 44
National headquarters . . . Regional offices

References 46

A new national policy was established on December 29, 1970, when President Richard M. Nixon signed into law the Occupational Safety and Health Act of 1970 (Public Law No. 91–596).* The Congress of the United States declared that the purpose of this piece of legislation is "to assure so far as possible every working man and woman in the Nation safe and healthful working conditions and to preserve our human resources."

The OSHAct, which took effect April 28, 1971, covers approximately 55 million employees throughout the nation. The Act was coauthored by Senator Harrison A. Williams (Dem.-N.J.) and Congressman William Steiger (Rep.-Wisc.) and hence is sometimes designated as the Williams–Steiger Act. The Act is regarded by many as a landmark piece of legislation since it goes beyond the present workplace and considers the working environment of the future.

Legislative History

Historically, the enactment of safety and health laws has been left to the states. Prior to the 1960's only a few federal laws (such as the Walsh-Healey Public Contracts Act and the Longshoremen's and Harbor Workers' Compensation Act) directed any attention to occupational safety and health. The decade of the sixties, however, saw significant Congressional action in this arena. A number of pieces of legislation passed by the Congress during the sixties, including the Service Contract Act of 1965, the National Foundation on Arts and Humanities Act, the Federal Metal and Nonmetallic Mine Safety Act, the Federal Coal Mine Health and Safety Act and the Contract Workers and Safety Standards Act (Construction Safety Act), directed attention to occupational safety and health.

Each of these federal laws was applicable to a limited number of employers. These laws were directed at those who had obtained federal contracts or they zeroed-in on a specific industry. Even collectively, all the federal safety legislation passed prior to 1970 was not applicable to the majority of employers or employees. It becomes obvious, then, that until 1970, Congressional action related to occupational safety and health was, at best,

* 29 U.S.C. §§651–678.

sporadic, covering only specific sets of employers and employees with little attempt for an omnibus coverage that is a part of the OSHAct.

Proponents of more significant federal presence in occupational safety and health, mostly represented by organized labor, based their position primarily on the following:

• With few exceptions, the states failed to meet their obligation in regard to occupational safety and health. Only a few of the states had safety and health legislation that was considered reasonable or adequate. Many states legislated safety and health only in specific industries. In general, states had inadequate safety and health standards, inadequate enforcement procedures, inadequate staff with respect to quality and quantity, and inadequate budgets.

• Approximately 14,200 employees were killed annually on or in connection with their job and more than 2.2 million employees suffered a disabling injury each year as a result of work-related accidents. The injury/death toll was considered by most to be much too high and therefore not acceptable.

• The nation's work injury rates in most industries were increasing throughout the decade of the sixties. Since the trend was in the wrong direction, proponents of federal presence felt that federal legislation would assist in reversing this trend.

The act evolved amid a stormy atmosphere in both houses of Congress. Highly controversial issues were involved. Such issues were responsible for sharply drawn lines between political parties and between the business community and organized labor. After three years of political hassle, numerous compromises were made; this ultimately enabled the passage of the OSHAct of 1970 by both houses of Congress.

Administration

Administration and enforcement of the OSHAct are vested primarily with the Secretary of Labor and the Occupational Safety and Health Review Commission (discussed later). With respect to the enforcement function, the Secretary of Labor performs the investigation and prosecution aspects of the enforcement process and the Review Commission performs the adjudication portion of the enforcement process.

Research and related functions and certain educational functions are vested in the Secretary of Health, Education, and Welfare and are, for the most part, carried out by the National Institute for Occupational Safety and Health established within the Department of Health, Education, and Welfare (HEW). Compiling injury and illness statistical data is handled by the Bureau of Labor Statistics, U.S. Department of Labor.

To assist the Secretary of Labor, the Act authorizes the appointment of an Assistant Secretary of Labor for Occupational Safety and Health. This position is filled by Presidential appointment with the advice and consent of the Senate. The Assistant Secretary is the chief of the Occupational Safety and Health Administration (OSHA) established within the Department of Labor (DOL). The Assistant Secretary acts on behalf of the Secretary of Labor. For the purpose of this chapter, OSHA is also synonymous with the term Secretary or Assistant Secretary of Labor.

The primary functions of the four major governmental units assigned to carry out the provisions of the Act are described below.

Occupational Safety and Health Administration

The Occupational Safety and Health Administration (OSHA) is divided into three major program areas. Their functions are:

National programs.

• Develop and promulgate occupational safety and health standards and issue regulations.

• Provide programs for training OSHA personnel as well as employers and employees to improve unsafe and unhealthful working conditions.

• Assist in establishing federal agency plans and programs to assure effective compliance to standards within the federal agencies.

Regional programs.

• Provide for national support in development of state compliance programs and approve same.

• Assure the execution of an effective compliance program through inspection and technical assistance.

• Review and assess the effectiveness and efficiency of federal/state compliance activities.

Administrative programs.

• Provide financial, personnel, and administrative services for OSHA.

• Perform management reviews and analysis for improvement of services for OSHA.

• Provide for data collection, analysis, and design of information systems to facilitate the management of OSHA programs.

To assist in carrying out its responsibilities, OSHA has established ten regional offices. The primary mission of the regional office chief, known as the Assistant Regional Director, is to supervise, coordinate, evaluate and execute all programs of OSHA in the region. Assisting the Assistant Regional Director are Associate Assistant Regional Directors for (a) training, (b) technical support, and (c) state and federal programs.

Area offices have been established within each region, each headed by an Area Director. The mission of the Area Director is to carry out the compliance program of OSHA within designated geographic areas. The area office staff carries out its activities under the general supervision of the Area Director with guidance of the Assistant Regional Director, using policy instructions received from the national headquarters. The real action for implementing the enforcement portion of the OSHAct is carried out by the area offices.

The locations of the Washington headquarters and regional offices are listed in the Directory near the end of this chapter.

Occupational Safety and Health Review Commission

The Occupational Safety and Health Review Commission (OSHRC) is a quasi-judicial board of three members appointed by the President and confirmed by the Senate. The Commission is an independent agency of the Executive Branch of the U.S. Government and is not a part of the Department of Labor. The principal function of the Commission is to adjudicate cases resulting from an enforcement action initiated against an employer by OSHA when any such action is contested by the employer or by his employees or their representatives.

The Commission's actions are limited to contested cases. In such cases, OSHA notifies the Commission of the contested cases and the Commission hears all appeals on actions taken by OSHA concerning citations, proposed penalties, and abatement periods, and determines the appropriateness of such actions. When necessary, the Commission may conduct its own investigation and may affirm, modify, or vacate OSHA's findings.

There are two levels of adjudication within the Commission: (a) the administrative law judge, and (b) the three-member Commission. All cases not resolved on informal proceedings are heard and decided by one of the Commission's administrative law judges. The judge's decision can be changed by a majority vote of the Commission if one of the members, within 30 days of the judge's decision, directs that the judge's decision be reviewed by the Commission members. The Commission is the final administrative authority to rule on a particular case, but its findings and orders can be subject to further review by the courts. (For further information, see Contested Cases later in this chapter.)

The headquarters of the OSHRC is located at 1825 K Street, N.W., Washington, D.C. 20006.

National Institute for Occupational Safety and Health

The National Institute for Occupational Safety and Health (NIOSH) was established within the HEW under the provisions of the OSHAct. Administratively, NIOSH is located in HEW's Center for Disease Control. NIOSH is the principal federal agency engaged in research, education, and training related to occupational safety and health.

The primary functions of NIOSH are to (a) develop and establish recommended occupational safety and health standards, (b) conduct research experiments and demonstrations related to occupational safety and health, and (c) conduct education programs to provide an adequate supply of qualified personnel to carry out the purposes of the OSHAct.

Research and related functions. Under the OSHAct, NIOSH has the responsibility for conducting research for new occupational safety and health standards. NIOSH develops criteria for the establishment of such standards. Such criteria are transmitted to OSHA which has the responsibility for the final setting, promulgation, and enforcement of the standards.

The OSHAct also requires NIOSH to publish an annual listing of all known toxic substances and the concentrations at which such toxicity is known to occur. While the entry of a substance on the list does not mean that it is to be avoided, it does mean that the listed substance has a documented potential of being hazardous if misused and, therefore, care must be exercised to control the substance. Conversely, the absence of a substance from the list does not necessarily mean that a substance is nontoxic. Some hazardous substances may not qualify to be listed because the dose that causes the toxic effect is not known.

Education and training. NIOSH also has the responsibility to conduct (a) education and training programs which are aimed at providing an adequate supply of qualified personnel to carry out the purpose of the Act and (b) informational programs on the importance and proper use of adequate safety and health equipment. The long-term approach to an adequate supply of training personnel is occupational safety and health is found in the colleges and universities and other institutions in the private sector. NIOSH encourages such institutions, by contracts and grants, to expand their curricula in occupational medicine, occupational health nursing, industrial hygiene, and occupational safety engineering.

Employer and employee services. Of principal interest to individual employers and employees are the technical services offered by NIOSH. The five main services that are provided upon request to NIOSH's Division of Technical Services, Cincinnati, Ohio 45202, are:

1. Hazard evaluation—Provides on-site evaluations of potentially toxic substances used or found on the job.

2. Technical information—Provides technical information concerning health or safety conditions at workplaces, such as the possible hazards of working with specific solvents, and when to use protective equipment.

3. Accident prevention—Provides technical assistance for controlling on-the-job injuries including the evaluation of special problems and recommendations for corrective action.

4. Industrial hygiene—Provides technical assistance in the areas of engineering and industrial hygiene, including the evaluation of special health-related problems in the workplace and recommendations for control measures.

5. Medical service—Provides assistance in solving occupational medical and nursing problems in the workplace including the assessment of existing medically related needs and the development of recommended means for meeting such needs.

The locations of the headquarters and regional offices of NIOSH are listed in the Directory toward the end of this chapter.

Bureau of Labor Statistics

The responsibility for conducting statistical surveys and establishing methods used to acquire injury and illness data is placed in the Bureau of Labor Statistics (BLS). Questions regarding recordkeeping requirements and reporting procedures can be directed to any of the OSHA regional or area offices or the BLS regional offices. The locations of the headquarters and regional offices of BLS are listed in the Directory.

Major Provisions of the OSHAct

Coverage

Except for specific exclusions, the Act is applicable to every employer who has one or more employees and who is engaged in a business affecting commerce. The law applies to all 50 states, the District of Columbia, Puerto Rico, and all U.S. possessions.

Specifically *excluded* from coverage are all federal, state, and local government employ-

ees. There are, however, special provisions in the Act for federal employees and potential coverage for state and local government employees.

The OSHAct is also not applicable to those operations where a federal agency (and state agencies acting under the Atomic Energy Act of 1954) other than the Department of Labor has statutory authority to prescribe or enforce standards or regulations affecting occupational safety or health. An example of this exclusion are in the operations covered by the Federal Metal and Nonmetallic Mine Safety Act or the Federal Coal Mine Health and Safety Act, where statutory authority is vested in the Department of the Interior.

OSHA clarified its interpretations of coverage with respect to certain employees by issuing a regulation.* It has been declared that churches and religious organizations, with respect to their religious activities, are not regarded as employers. Likewise, persons who in their own residences employ others to perform domestic household tasks are not regarded as employers. Further, any person engaged in agriculture who is a member of the immediate family of the farmer is not regarded as an employee and hence is not covered by the Act.

Employer and employee duties

Each employer covered by the Act:

1. Has the general duty to furnish each of his employees employment and places of employment which are free from recognized hazards that are causing or likely to cause death or serious physical harm (this is commonly known as the "general duty clause"); and

2. Has the specific duty of complying with safety and health standards promulgated under the Act.

Each employee, in turn, has the duty to comply with the safety and health standards and all rules, regulations, and orders which

---

* The regulation clarifying policy regarding "Coverage of Employees Under the Williams-Steiger Occupational Safety and Health Act of 1970" is contained in the *Code of Federal Regulations* (C.F.R.), Title 29, Chapter XVII, Part 1975.

FIG. 2–1—OSHA poster, "Safety and Health Protection on the Job," must be posted conspicuously at every plant, job site, or other establishment. At left is the annual Summary of Occupational Injuries and Illnesses, OSHA Form 102, that must be posted by February 1 of the following year and remain in place for the next 30 consecutive days.

are applicable to his own actions and conduct on the job.

For employers, the general duty provision is used only where there are no specific standards applicable to a particular hazard involved. A hazard is "recognized" if it is a condition that is generally recognized as a hazard in the particular industry in which it occurs and is detectable (a) by means of the human senses, or (b) there are accepted tests known in the industry to determine its existence which should make its presence known to the employer. An example of a "recognized hazard" in the latter category is excessive concentrations of a toxic substance in the work area atmosphere, even though such concentrations could only be detected through use of measuring devices.

During the course of an inspection a compliance officer is concerned primarily with determining whether the employer is complying with the promulgated safety and health standards. However, he will also direct some attention to determining whether the employer is complying with the general duty clause. By far the majority of alleged violations cited by compliance personnel are concerned with the promulgated standards. Only a limited number of citations are issued alleging a violation of the general duty clause.

The law provides for sanctions against the employer in the form of citations and civil and criminal penalties if the employer fails to comply with his two duties. However, there is no provision for government sanctions against an employee for failure to comply with the employee's duty. While some may view the latter as unjust, significantly, it was not one of the controversial issues in the formative stages of the Act.

Both management and organized labor have long agreed that safety and health on the job is a management responsibility. The business community generally did not want the law structured to provide for government sanctions against an erring employee because there are measures which management can invoke against an employee who obstructs the employer's efforts to provide a safe workplace.

While the law expressly places upon each employee the obligation to comply with the standards, final responsibility for compliance with the requirements of the Act remains with the employer. Employers thus should take all necessary action to assure employee compliance with the promulgated standards and establish within their safety system a means whereby they become aware of situations where employees are not complying with applicable standards.

The OSHA poster

The OSHA poster (see Fig. 2–1) must be prominently displayed in a conspicuous place in the workplace where notices to employees are customarily posted. The poster informs employees of their rights and responsibilities under the Act.

Employee rights

While the employee has the legal duty to comply with all the standards and regulations issued under the OSHAct, there are many employee rights that are also incorporated in the Act. Since these rights may affect labor relations as well as labor negotiations, employers as well as employees should be aware of the employee rights contained in the Act. Employee rights fall into three main areas and are related to (*a*) standards, (*b*) access to information, and (*c*) enforcement.

With respect to standards:

1. Employees may request OSHA to begin proceedings for adoption of a new standard or to amend or revoke an existing one.

2. Employees may submit written data or comments on proposed standards and may appear as an interested party at any hearing held by OSHA.

3. Employees may file written objections to a proposed federal standard and/or appeal the final decision of OSHA.

4. Employees must be informed when an employer applies for a variance of a promulgated standard.

5. Employees must be afforded the opportunity to participate in a variance hearing as an interested party and have the right to appeal OSHA's final decision.

With respect to access to information:

1. Employees have the right to information from the employer regarding employee protections and obligations under the Act.

2. Affected employees have a right to information from the employer regarding the toxic effects, conditions of exposure and precautions for safe use of all hazardous materials in the establishment by means of labeling or other forms of warning where such information is prescribed by a standard.

3. If employees are exposed to harmful materials in excess of levels set by the standards, the affected employees must be so informed by the employer and the employer must also inform the employees thus exposed what corrective action is being taken.

4. If a compliance officer determines that an alleged imminent danger exists, he must inform the affected employees of the danger and that he is recommending that relief be sought by court action if the imminence of such danger is not eliminated.

5. Upon request, employees must be given access to records of their history of exposure to toxic materials or harmful physical agents which are required to be monitored or measured and recorded.

6. If a standard requires monitoring or measuring hazardous materials or harmful physical agents, employees must be given the

opportunity to observe such monitoring or measuring.

7. Employees have the right of access to (*a*) the list of toxic materials published by NIOSH, (*b*) criteria developed by NIOSH describing the effects of toxic materials or harmful physical agents, and (*c*) industry-wide studies conducted by NIOSH regarding the effects of chronic, low-level exposure to hazardous materials.

8. On written request to NIOSH, employees have the right to obtain the determination of whether or not a substance found or used in the establishment is harmful.

With respect to enforcement:

1. Employees have the right to confer with the compliance officer in connection with an inspection of an establishment.

2. An authorized employee representative must be given an opportunity to accompany the compliance officer during an inspection for the purpose of aiding such inspection. (This is commonly known as the "walk-around" provision.)

3. An employee has the right to make a written request to OSHA for a special inspection if the employee believes a violation of a standard threatens physical harm, and the employee has the right to request OSHA to keep his identity confidential.

4. If an employee believes any violation of the Act exists, he has the right to notify OSHA or a compliance officer in writing of the alleged violation, either before or during an inspection of the establishment.

5. If a request is made for a special inspection and it is denied by OSHA, the employee must be notified in writing by OSHA, together with the reasons, that the complaint was not valid. The employee has the right to object to such a decision and may request a hearing by OSHA.

6. If a written complaint concerning an alleged violation is submitted to OSHA and the compliance officer responding to the complaint fails to cite the employer for the alleged violation, OSHA must furnish the employee or his authorized representative a written statement setting forth the reasons for its final disposition.

7. If OSHA cites an employer for a violation, employees have the right to review a copy of the citation which must be posted by the employer at or near the place where the violation occurred.

8. Employees have the right to appear as an interested party or to be called as a witness in a contested enforcement matter before the Occupational Safety and Health Review Commission.

9. If OSHA arbitrarily or capriciously fails to seek relief to counteract an imminent danger and an employee is injured as a result, that employee has the right to bring action against OSHA for relief as may be appropriate.

10. An employee has the right to file a complaint to OSHA within 30 days if he believes he has been discriminated against because he asserted his rights under the Act.

11. An employee has the right to contest the abatement period fixed in the citation issued to his employer by notifying the OSHA Area Director that issued the citation.

Occupational safety and health standards

The Act authorizes OSHA to promulgate, modify, or revoke occupational safety and health standards.*

In order to get the initial set of standards in place without undue delay, the Act authorized OSHA to promulgate any existing federal standard or any national consensus standard without regard to the usual rulemaking procedures prior to April 28, 1973. The initial set of standards, Part 1910, appeared in the *Federal Register* of May 29, 1971. Subscriptions to the *Federal Register* are obtained through the Government Printing Office,

---

* The rules of procedure for promulgating, modifying or revoking standards are codified in the *Code of Federal Regulations* (C.F.R.), Title 29, Chapter XVII, Part 1911.

2—The Occupational Safety and Health Act of 1970

Washington, D.C. 20402.

Standards* contained in Part 1910 are applicable to general industry. Those contained in Part 1926 are applicable to the construction industry. Standards applicable to ship repairing, shipbuilding, shipbreaking and long-shoring are contained in Parts 1915 through 1918 respectively. Since standards cannot remain static due to use of new equipment, methods, and materials, all are subject to updating via modification.

OSHA also has the authority to promulgate emergency temporary standards where it is found that employees are exposed to grave danger. Emergency temporary standards can take effect immediately upon publication in the *Federal Register*. Such standards will remain in effect until superseded by a standard promulgated under the procedures prescribed by the Act.

Any person adversely affected by any standard issued by OSHA has the right to challenge its validity by petitioning the U.S. Court of Appeals within 60 days after its promulgation.

Input from the private sector. Occupational safety and health standards promulgated by OSHA will never cover every conceivable hazardous condition that could exist in any workplace. Nevertheless, new standards and modification of existing standards are of significant interest to employers and employees alike. Industry organizations as well as individuals and employee organizations should participate in the development of new and revised standards since it is within the private sector that most of the expertise and technical competence lies.

Much of the standards development work is carried on by the nationally recognized concensus standard producing organizations. Those so recognized by OSHA include the American National Standards Institute (ANSI), American Society for Testing and Materials (ASTM), and National Fire Protection Association (NFPA). By far the majority of the promulgated standards now in place were initially adopted by ANSI and NFPA. Revision of standards by such organizations can lead to revision of OSHA standards. Prudent employers and employees should urge their respective representative organizations not only to participate in the activities of the aforementioned standards-producing organizations, but also to provide the leadership and initiate actions for new standards. To do less means that industry and employees are willing to let the standards-development process rest in the hands of government agencies such as OSHA and NIOSH.

Other sources utilized by OSHA for the revision of existing occupational safety and health standards or development of new standards are (*a*) trade associations that produce proprietary standards, (*b*) standards advisory committees appointed by OSHA to develop a recommended standard on a specific subject, and (*c*) NIOSH criteria documents.

In order to promulgate, revise, or modify a standard, OSHA must first publish in the *Federal Register* a notice of any proposed rule that will adopt, modify, or revoke any standard and invite interested persons to submit their views on the proposed rule within 30 days after publication. Interested persons may file objections to the rule and are entitled to a bearing on their objections if they request a hearing be held. However, objections must specify the parts of the proposed rule to which they object and the grounds for such objection. If a hearing is requested, OSHA must hold one. Based on (*a*) the need for control of an exposure to an occupational injury or illness, and (*b*) the reasonableness, effectiveness and feasibility of the control measures required, OSHA may issue a rule promulgating an additional standard or modify or revoke an existing standard.

Recordkeeping and reporting requirements

Most employers covered by the Act are required to maintain in each establishment records of recordable occupational injuries and illnesses.** Such records consist of:

• A log of occupational injuries and illnesses, OSHA Form 100.

---

* The Occupational Safety and Health Standards, Title 29, C.F.R., Chapter XVII, Parts 1910, 1926, and 1915–1918 are available at all OSHA regional and area offices.

** Regulations pertaining to recording and reporting injuries and illnesses are codified in Title 29, C.F.R., Chapter XVII, Part 1904.

- A supplementary record of each occupational injury or illness, OSHA Form 101.

- An annual summary of occupational injuries and illnesses, OSHA Form 102.

- Records of employee exposure to toxic substances or harmful physical agents as are required by the standards and regulations.

- Other records such as logs of inspections and tests as required by the standards.

OSHA Forms 100, 101, and 102 are available at all BLS regional offices. If your state has an OSHA-approved plan, be sure to check for any special recordkeeping requirements.

In an effort to relieve small businesses from the recordkeeping requirements, OSHA has ruled that employers who had no more than seven employees during the calendar year immediately preceding the current calendar year need not maintain the log, the supplementary record, nor prepare or post the summary. However, if an employer, regardless of size, has been notified in writing by the Bureau of Labor Statistics that he has been selected to participate in the statistical survey of occupational injuries and illnesses, then he will be required to maintain the log and make reports for the period of time specified in the notice.

Log of occupational injuries and illnesses, OSHA Form 100. Each employer is required to maintain in each establishment a log of all recordable occupational injuries and illnesses for that establishment. However, the log may be maintained at a place other than the establishment or by means of data processing equipment provided that:

1. There is available at the place where the log is maintained sufficient information to complete the log to a date within six working days after receiving information that a recordable case has occurred; and

2. At each of the employer's establishments there is available a copy of the log of that establishment complete and current within 45 calendar days.

Every employer is required to enter each recordable occupational injury and illness on the log as early as practicable, but no later than six working days after receiving information that the recordable case has occurred.

OSHA Form 100 or any private equivalent may be used. The log is to be established on a calendar year basis.

Supplementary record, OSHA Form 101. Each employer is required to have available for inspection at each establishment a supplementary record of each occupational injury or illness for that establishment. This form is to be completed within six working days after receiving information that the recordable case has occurred. Other report forms are acceptable if they contain at least the information required by OSHA Form 101. (NSC's Supervisor's Accident Report form, shown in Chapter 6, "Accident Records and Injury Rates," is one.)

Annual summary, OSHA Form 102. Each employer is required to compile an annual summary of occupational injuries and illnesses for each establishment. The annual summary is to be based on information contained in the log for the given establishment. OSHA Form 102 must be used for this purpose (no substitutes or alternatives are allowed), and it must be completed by February 1 of each year. The summary must be certified—the employer or officer or employee who supervises the preparation of the summary must put his signature in the lower right-hand corner of the summary. The employer is required to post a copy of the establishment's summary in each establishment no later than February 1 and it is to remain in place for 30 consecutive calendar days after posting. (See Fig. 2–1.)

Retention of and access to records. All three record forms must be maintained in each establishment for five years following the end of the year to which they relate.

Records for personnel who do not primarily report or work at a single establishment, such as traveling salesmen, technicians, engineers, and the like, are to be maintained at the location from which they are paid or the base from which they operate to carry out their activities.

Employers of employees engaged in physically dispersed operations such as occur in construction, installation, repair or service activities who do not report to any fixed establishment on a regular basis, but are subject to common supervision, may satisfy the record-keeping requirements by (a) maintaining records of each operation subject to common supervision (field superintendent or supervisor) in an established central place, and (b) having the address and telephone number of the central place available at each worksite, and (c) having personnel available at the central place during normal business hours to provide information from the records maintained there by telephone or by mail.

The records are to be available for inspection and copying by compliance officers, by any representative from the Bureau of Labor Statistics, by any representative of NIOSH or by any representative of a state agency accorded jurisdiction.

Reporting fatal or catastrophic events. One of the reporting requirements is that within 48 hours of the occurrence of an event which results in a fatality to one or more employees or results in the hospitalization of five or more employees, the employer must report the event, either orally or in writing, to the nearest OSHA office. The report should contain an explanation of the circumstances pertinent to the accident, the number of fatalities, and the extent of any injuries.

The purpose of the notice is to enable a compliance officer to be dispatched to the scene if an accident investigation is deemed necessary. The purpose of the accident investigation is to determine if there is a violation of any standard which may have contributed to the accident.

Occupational injury and illness surveys, OSHA Form 103. The Bureau of Labor Statistics is charged with the responsibility for the collection, compilation, and analysis of occupational injury and illness data. To discharge this responsibility, BLS conducts statistical surveys, collecting information from employers on a selective basis. Those employers selected to participate receive a written notice. The survey, which can change from year to year, utilizes OSHA Form 103, "Occupational Injuries and Illnesses Survey."

OSHA Form 103, supplied by BLS, is distributed to a designated state agency and the state agency forwards the form to employers selected for inclusion in the survey. These employers must complete the form within three weeks of receipt of the form and return it to the designated state agency. Failure to comply with the reporting requirements may result in the issuance of citations and proposed penalties. The state agency will assemble the data and report designated information to the BLS headquarters in Washington, D.C. Questions or problems concerning the survey should be directed to the state agency indicated on the survey form.

Variances from standards

There will be some occasions when, for various reasons, standards cannot be met by some employers. In other cases, the protection already afforded by an employer to employees is equal to or superior to the protection that would be granted if the standard were followed strictly to the letter. The Act provides an avenue of relief from these situations by empowering OSHA to grant variances* from the standards, providing the granting of such variances would not degrade the purpose of the Act.

There are two types of variances—temporary and permanent. An employer may apply for an order granting a temporary variance provided he establishes that (a) he cannot comply with the applicable standard because of unavailability of personnel or equipment or time to construct or alter facilities; (b) he is taking all available steps to protect his employees against exposure covered by the standard; and (c) his program will effect compliance with the standard as soon as possible.

Employer applications for an order for a temporary variance must contain at least the:

1. Name and address of the applicant.

2. Address(es) of the place(s) of employment involved.

3. Identification of the standard from which the applicant seeks a variance.

4. A representation by the applicant that he is unable to comply with the standard and a detailed statement of reasons therefor.

---

* The detailed "Rules of Practice for Variances, Limitations, Variations, Tolerances, and Exemptions" are codified in Title 29, C.F.R., Chapter XVII, Part 1905.

5. A statement of the steps the applicant has taken and will take, with dates, to protect employees against the hazard covered by the standard.

6. A statement of when the applicant expects to be able to comply with the standard and what steps he has taken, with dates, to come into compliance with the standard.

7. A certification that he has informed his employees of the application. A description of how employees have been informed is to be included in the certification. Information to employees must also inform them of their right to petition for a hearing.

An employer may also apply for a permanent variance from a standard. A variance order can be granted if OSHA determines that an employer has demonstrated by a preponderance of evidence that he will provide a place of employment as safe and healthful as that which would prevail if he complied with the standard.

Employer applications for an order for a permanent variance must contain at least the following:

1. Name and address of the applicant.

2. Address(es) of the place(s) of employment involved.

3. A description of the countermeasures used or proposed to be used by the applicant.

4. A statement showing how such countermeasures would provide a place of employment which is as safe and healthful as that required by the standard for which the variance is sought.

5. Certification that he has informed his employees of the application.

6. Any request for a hearing.

7. A description of how employees were informed of the application and of their right to petition for a hearing.

The model shown in Fig. 2–2 is intended to serve as a guide for a permanent variance application.

An employer may request an interim order permitting either kind of variance until his formal application can be acted upon. Again, the request for an interim order must contain statements of fact or arguments why such interim order should be granted. If a request for an interim order is denied, the applicant will be notified promptly and informed of the reasons for the decision. If the order is granted, all concerned parties will be informed and the terms of the order will be published in the *Federal Register*. In such cases, the employer must inform the affected employees regarding the interim order in the same manner used to inform them of the variance application.

Upon filing an application for a variance, OSHA will publish a notice of such filing in the *Federal Register* and invite written data, views, and arguments regarding the application. Those affected by the petition may request a hearing. After review of all the facts, including those presented during the hearing, OSHA publishes its decision regarding the application in the *Federal Register*.

The enforcement process

The Department of Labor compliance officers (CO) may enter, at any reasonable time and without delay, any establishment covered by the OSHAct to inspect the premises and all its facilities.* CO's also have the right to question privately any employer, owner, operator, agent or employee.

The OSHAct authorizes an employer representative as well as an authorized employee representative to accompany the CO during the official inspection of the premises and all its facilities. Usually the authorized employee representative will be the union steward or the chairman of the employee safety committee. Occasionally there may be no authorized employee representative, especially in those establishments which are nonunion shops. In the absence of an employee representative, the CO will confer with employees picked at random.

Representatives of the Department of Health, Education and Welfare, although not authorized to enforce the OSHAct, are authorized to make inspections and to question

---

* The regulations governing enforcement procedures, including inspections, citations, and proposed penalties, are codified in Title 29, C.F.R., Chapter XVII, Part 1903.

MODEL OF AN APPLICATION FOR VARIANCE FROM STANDARDS

Assistant Secretary for Occupational Safety and Health
U.S. Department of Labor
Washington, D.C. 20210

Dear Sir:

Pursuant to Section 6(d) of the Williams-Steiger Occupational Safety and Health Act of 1970 (84 Stat. 1596: 29 U.S.C. 655), XYZ Company respectfully requests a permanent variance from the requirements of 29 C.F.R. 1910.179 (b) (4), concerning wind indicator on the outdoor bridge of a gantry crane.

(1) Applicant—XYZ Company (address).

(2) Place of employment—Same as (1).

(3) XYZ Company has a gantry crane which operates beside a slip for unloading shallow draft barges. It currently is used only for unloading salt barges, usually about twice a year. Because of the location of our operation and the experience we have had with high wind conditions, XYZ Company recognizes the importance for monitoring wind conditions. As a result, our plant has two wind-indicating devices with visual and automatic recording devices. We have a well established plan for monitoring Federal Weather Bureau reports and emergency plans of action for pending high winds. The key wind indicating device is located in the shift superintendent's office. A shift superintendent is on duty around the clock, seven days a week and is responsible for the plant. He keeps a close watch on the wind conditions and also monitors weather forecasts at least three times during each eight-hour period. A written record is kept of the weather conditions. If there are any changes in wind conditions or anticipated changes due to approaching bad weather, the wind indicators and weather forecast are monitored continuously. Because of these well established policies and the limited use of the gantry, we feel that our plan would offer a safer operation. In addition, we feel that the installation of a wind indicating device on the gantry crane would not be feasible because of conditions which tend to cause malfunction. Our past experience has verified this consistently. The gantry crane is very close to the molten sulfur pits and salt water. The close proximity to these agents creates a corrosive condition which would make the wind indicator inoperative soon after installation. We feel our monitoring and early warning system is more effective and safe than is the use of an infrequently used and difficult to maintain wind velocity instrument on the crane.

(4) Due to the corrosive conditions present around the gantry crane, XYZ Company feels that its established procedures would give more reliable information than an indicator on the crane. Therefore, we feel that this would ensure the safety of the crane and operator better than standard procedures.

Respectfully submitted,

_____
Manager

(5) This is to certify that a copy of the variance from Standard 1910.179(b) (4) requested by XYZ Company has been posted in an appropriate place for employees to examine. Also, an explanation of their right to send comments or petition for a hearing

are explained on an attached sheet. A copy of the variance application was also given to the authorized employee representative.

Posted on _____ , 19—

_____
Manager

(6)   **TO ALL EMPLOYEES**

The attached form is a copy of the variance application submitted by XYZ Company for a variance from Standard 1910.179(b) (4). This is to inform you that you may request a hearing with the Assistant Secretary for Occupational Safety and Health, U.S. Department of Labor, according to Standard 1905.15, Request for Hearing, if you object to the variance requested. This request must be filed in writing.

_____
Manager

(7) XYZ Company also respectfully requests that an interim order be granted to allow operation of the gantry crane without the indicator until action is taken on the variance application. We feel that the safety of the employees will be guaranteed because we do have wind indicators being monitored at our plant and the gantry operator would be promptly informed before wind velocities rose or approached a dangerous level.

_____
Manager

Fig. 2–2.—Copy to appear on establishment letterhead.

Courtesy The Bureau of National Affairs, Inc.

employers and employees in order to carry out those duties assigned to HEW under the Act.

Inspection priorities. OSHA has established priorities for assignment of manpower and resources. Such priorities are as follows:

1. Investigation of a fatality or catastrophic event (an event that results in the hospitalization of five or more employees).

2. Response to valid complaints from employees.

3. Target Industry Program. The industries that were initially included in the Target Industry Program are: longshoring, lumber and wood products, roofing and sheet metal, meat and meat products, and miscellaneous transportation (manufacture of mobile homes and other transportation equipment).

4. General inspections and related activities to provide broad representative coverage. Random selection factors include geographical distribution, size distribution, and cross-section of industry.

Overall, the compliance officer spends his time approximately as follows:

	Percent
Catastrophe and/or fatality investigation	10
General inspection and related activity	25
Complaint investigation	30
Target industry program	35

General inspection procedures. The primary responsibility of the CO, who is under the supervision of the OSHA area director, is to conduct an effective inspection to determine if employers and employees are in compliance with the requirements of the standards, rules, and regulations promulgated under the OSHAct.

To enter an establishment, the CO will present his credentials to a guard, receptionist, or other person acting in such a capacity. There have been instances where some credentials have been fraudulent. Employers, therefore, should check the CO's credentials carefully before letting the individual enter their establishment for the purpose of an inspection. (See Fig. 2–3.)

The CO will usually ask for the person with the highest authority that is on the premises. It is recommended that employers furnish written instructions to the security, receptionist, or other affected personnel regarding the right of entry, treatment, who should be notified, and to whom and where the CO should be directed so that undue delay can be avoided.

Opening conference. The employer or his designated representative will be required to participate in the opening conference conducted by the CO. Since the CO will want to talk with safety personnel, such personnel should also be invited by the employer to the opening conference. The employer representative who accompanies the CO through the establishment should also participate in the opening conference.

At the opening conference, the CO will:

1. Inform the employer that the purpose of his visit is to make an investigation to ascertain if the establishment, procedures, operations, and equipment are in compliance with the requirements of the OSHAct.

2. Give the employer copies of the Act, standards, regulations, and promotional material, as necessary.

3. Outline in general terms:
   a) The scope of the inspection
   b) The records he wants to review
   c) His obligation to confer with employees
   d) The physical inspection of the workplace
   e) The closing conference

4. If applicable, furnish a copy of the complaint(s).

5. Answer questions those in attendance might have.

During the course of the opening conference the CO may want to review OSHA Form 100, Log of Occupational Injuries and Illnesses, and OSHA Form 101, Supplementary Record of Occupational Injuries and Illnesses. Such records should be made readily available to him. In addition, the CO will want to obtain information regarding the safety and health program that the employer now has in place so that he can evaluate such a program. Naturally, a comprehensive safety and health program that shows evidence of effective performance in accident prevention will be impressive to all concerned.

The CO will also ascertain from the employer whether employees of another employer (for example, a contracting employer for maintenance or remodeling) are working in or on the establishment. If so, the CO will afford the authorized representative of those employees a reasonable opportunity to accompany him during the inspection of the workplaces where they are working.

During the conference the CO will explain the employee representative's rights and ask for the authorized employee representative. Generally the employee representative will be an employee of the establishment inspected. However, if, in the judgment of the CO, good cause has been shown that accompaniment of a third party (such as an industrial hygienist or safety consultant) who is not an employee of the employer (but is, indeed, an authorized employee representative) is reasonably necessary to conduct an effective and thorough inspection, such a third party may accompany the CO during the inspection. The final decision will rest with the CO.

The employer is not permitted to designate the employee representative. Employee representatives may change as the inspection process moves from department to department. The CO may deny the right of accompaniment to any person whose conduct interferes with a full and orderly inspection. If there is no authorized employee representative, the CO will consult with a reasonable number of employees concerning matters of safety and health in the workplace during the course of the inspection.

FIG. 2–3.—Bona-fide OSHA compliance officers are equipped with official identification as shown here. The credentials are signed by George C. Guenther, former Assistant Secretary of Labor, or John H. Stender, Assistant Secretary of Labor, or his successor. If in doubt about the validity of the credentials, it is recommended that the employer contact the nearest OSHA area office and determine whether or not the area office has scheduled an inspection at the establishment in question.

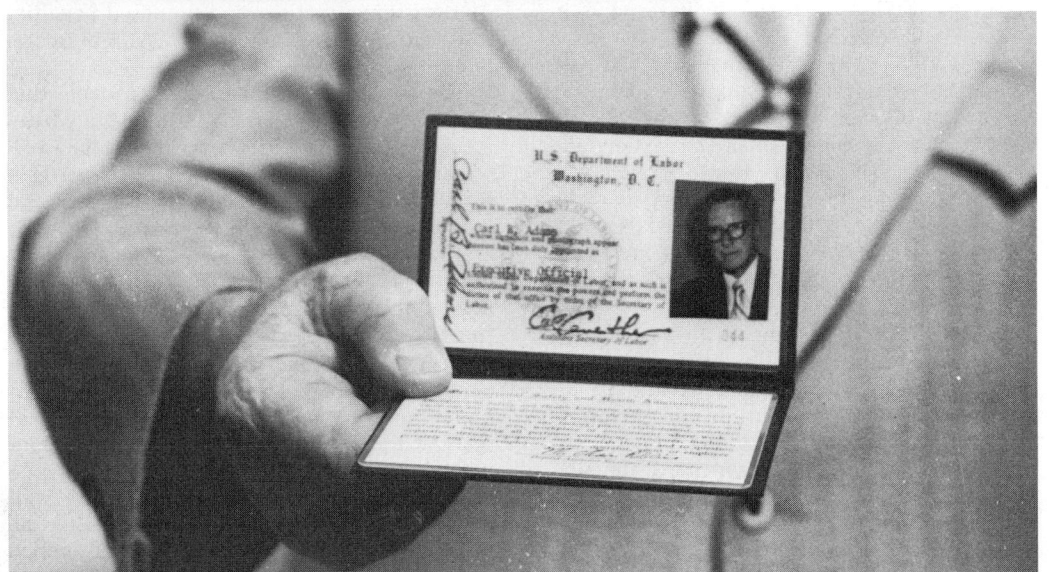

Inspection of facilities. The CO will take the time necessary to inspect all of the operations in the establishment. The inspections have as their primary objective the enforcement of the occupational safety and health standards as well as the enforcement of other promulgated regulations such as the posting of the OSHA poster (Fig. 2–1).

The CO will have the necessary instruments for checking certain items, such as noise levels, certain air contaminants and toxic substances, grounding, and the like. During the course of inspection, the CO will note any apparent violation of the standards and will normally record any apparent violation, including its location, and any comments that he has regarding the violation. He will do the same for any apparent violation of the general duty clause. His notes will serve as the basis of information for the area director for issuing citations or proposed penalties. For these reasons, the employer representative should ascertain any apparent violations from the CO during the actual inspection of the facilities. The employer representative should make notes identical to the CO's during the actual inspection so that he will have precisely the same information that the CO has.

It should be noted that the CO is only required to record apparent violations and is not required to present a solution or method of correcting, minimizing, or eliminating the violation. OSHA, however, will respond to requests for technical information concerned with complying with given standards. In such cases, the employer is urged to contact the regional or area office.

If, during the course of an inspection, the CO receives a complaint from an employee regarding a condition which is alleged to be in violation of an applicable standard, the CO, even though the complaint is brought to him via an informal process, will normally inspect for the alleged violation.

In the course of his normal inspection, the CO may make some preliminary judgments with respect to environmental conditions affecting occupational health. In such cases, he will generally use direct-reading instruments. Should this occur, and if proper instrumentation is available, it would be prudent for the employer to have qualified personnel at the establishment make duplicate tests in the same area at the same time under the same conditions. In addition, the employer representative should again take careful notes on the CO's methods as well as the results. If the inspection indicates a need for further investigation by an industrial hygienist, the CO will notify the Area Director who may assign a qualified industrial hygienist to investigate further. If a laboratory analysis is required, samples are sent to OSHA's laboratory in Salt Lake City and the results will be reported back to the Area Director.

At the completion of the inspection, but prior to the beginning of the closing conference, the employee representative is usually excused.

Closing conference. Upon completion of the inspection the CO will confer with the employer or his representative. Again, the employer's safety personnel should be present at the closing conference. It is at this time that the CO will advise the employer and his representatives of all conditions and practices which may constitute a safety or health violation. He should also indicate the applicable section or sections of the standards which may have been violated.

The CO will normally advise the employer that citations may be issued for alleged violations and that penalties may be proposed for each violation. Administratively, the authority for issuing citations and proposed penalties rests with the Area Director or his representative.

The employer will also be informed that citations will fix a reasonable time for abatement of the violations alleged. The CO will attempt to obtain from the employer an estimate of the time he feels would be required to abate the alleged violation and take such estimate into consideration when recommending a time for abatement. The CO should also explain the appeal procedures with respect to any citation or any notice of a proposed penalty.

Followup inspections. Followup inspections will always be made for those situations involving imminent danger or where citations have been issued for serious, repeated or willful violations. Followup inspections for all other cases will be conducted at the discretion of the Area Director.

The followup inspection is intended to be limited to verifying compliance of those conditions which were alleged to be in violation. The followup inspection is conducted with all of the usual formality of the original inspection, including the opening and closing conferences, and the walk-around rights of the employer and employee representative.

Violations

In addition to the general duty clause, the occupational safety and health standards promulgated under the OSHAct are used as a basis for determining alleged violations. There are four types of violations: imminent danger, serious, non-serious, and de minimis (very minor).

Imminent danger. The OSHAct defines imminent danger as "Any conditions or practices in any place of employment which are such that a danger exists which could reasonably be expected to cause death or serious physical harm immediately or before the imminence of such danger can be eliminated through the enforcement procedures otherwise provided by this Act." Therefore, for conditions or practices to constitute an imminent danger situation, it must be determined that there is a reasonable certainty that immediately or within a short period of time such conditions or practices could result in death or serious physical harm. Normally a health hazard would not constitute an imminent danger except in *extreme* situations, such as the presence of potentially lethal concentrations of airborne toxic substances that are an immediate threat to the life or health of employees.

If, during the course of inspection, the CO deems that the existing set of conditions appears to constitute an imminent danger situation, he will immediately advise the employer or his representative that such a danger exists and will attempt to have the danger corrected immediately through voluntary compliance. Further, if any employees appear to be in imminent danger, they will be informed of the danger and the employer will be requested to remove them from the area of imminent danger.

An employer will be deemed to have abated the imminent danger if he eliminates the im-minence of the danger by (*a*) removing employees from the danger area or (*b*) eliminating the conditions or practices which constitute the imminent danger. Normally abatement is achieved in either of these two ways. If the employer refuses to voluntarily abate the alleged imminent danger, the CO will inform the affected employees of the danger involved and will inform the employer as well as the affected employees that he will recommend a civil action (in the form of a court order) for appropriate relief (i.e., to shut down the operation). In such cases, the CO will personally post the imminent danger citation at or near the area in which the exposed employees are working. The federal CO has no authority to order the closing down of an operation or to direct employees to leave the area of the imminent danger or the workplace.

Serious violation. To determine if a violation is serious, the CO must decide:

1. Is there a substantial probability that death or serious physical harm could result? And if so,

2. Did the employer know, or with the exercise of reasonable diligence, should he have known of the hazard?

If the answer to both questions is "yes," then a serious violation exists.

The term "serious physical harm" is similar to the "permanent total disability" or "permanent partial disability" utilized in the ANSI Z16.1, *Method of Recording and Measuring Work Injury Experience.* (See Chapter 6 for more details.)

Serious physical harm includes impairment of the body where:

1. A part of the body would be permanently removed or rendered functionally useless (such as an amputation or loss of an eye) or substantially reduce efficiency on or off the job; or

2. A part of an internal bodily system would be inhibited in its normal performance to such a degree as to shorten life or cause reduction in physical or mental efficiency (such as severe loss of hearing).

It is obvious that the CO must make an

evaluation as to the *likelihood* that death or serious physical harm could result from a condition which is an alleged violation. For example, a violation involving an inadequate guard rail for workers at a substantially high level from the ground would normally be a serious violation since it is more probable than not that the result of such a condition would be death or serious physical harm.

Note that the emphasis in deciding whether or not a condition represents a serious violation is based on the *seriousness* or *severity* of the most likely injury that would arise out of the potential accident, rather than on the *probability* that the accident will occur as a result of the violation. In all cases, the decision in determining whether a violation is serious or not will require professional judgment.

Non-serious violation. If the more likely consequence of a violation is something less than death or serious physical harm, or the employer did not know of the hazard, then the violation will be considered a non-serious violation. For example, a violation of housekeeping standards that might result in a tripping accident would be classified as a non-serious violation since the more probable consequence of such a condition would be strains or contusions which are not classified as serious physical harm.

De minimis violations. De minimis* violations are those that have no immediate or direct relationship to safety or health.

Special types. There are two other special types of violations: a willful violation, and a repeated violation.

A willful violation exists where evidence shows that:

1. The employer committed an intentional and knowing violation of the Act and knows that such action constitutes a violation, or

2. Even though the employer was not consciously violating the Act, he was aware that a hazardous condition existed and made no reasonable effort to eliminate the condition.

A repeated violation is where a second citation is issued for a violation of a given standard or the same condition which violates the

general duty clause. A repeated violation differs from a failure to abate in that repeated violations exist where the employer has abated an earlier violation, and upon later inspection, is found to have violated the same standard.

The vast majority of citations that have been issued by OSHA so far have been for alleged violations of the occupational safety and health standards and have been classified as non-serious. Relatively few citations are issued for alleged violations of the general duty clause.

Citations

When an investigation or inspection reveals a condition which is alleged to be in violation of the standards or general duty clause, the employer may be issued a written citation which will describe the specific nature of the alleged violation, the standard allegedly violated and will fix a time for abatement. *Each citation,* or copy thereof, *must be prominently posted by the employer at or near the place where the alleged violation occurred.* All citations will be issued by the Area Director or his designee and will be sent to the employer by certified mail.

A "Citation for Serious Violation" will be prepared to cover those violations which fall into the "serious category." This type of violation *must* be assessed a monetary penalty.

A "Citation" is used for non-serious violations which *may* or *may not* carry a monetary penalty. A citation may be issued to the employer for employee actions which violate the safety and health standards.

A notice, in lieu of a citation, is issued for de minimis violations which have no direct relationship to safety and health. Unlike the citation, the employer is not required to post this notice.

Penalties

In proposing civil penalties for citations, a distinction is made between serious violations and all other violations. There is no requirement that a penalty be proposed when a violation is not a serious one, but a penalty *must*

---

* "De minimis" is short for the legal maxim, *De minimis non curat lex,* "The law does not concern itself with trifles."

be proposed for a serious violation. In either case the maximum penalty that may be proposed is $1000. In case of willful or repeated violations, a civil penalty of up to $10,000 may be proposed. Criminal penalties may be imposed on any employer who, among other things, willfully violates a standard and that violation causes death to any employee. There are no penalties for de minimis violations.

Penalties may be proposed for an alleged violation even though the employer immediately abates or initiates steps to abate the alleged violation. However, actions to abate should be favorably considered when determining the amount of adjustment applied for "good faith."

The information that follows describes the system that OSHA uses to arrive at its proposed penalties. Since it is unlikely that most employers comply with all of the promulgated standards, the employer can utilize a similar strategy to establish his own priorities for voluntary compliance with the standards.

Non-serious violations. For non-serious violations, the penalty may range from 0 to $1000 for each violation. An "unadjusted penalty" is based on the gravity of the alleged violation. Three factors are used to determine gravity and all factors require professional judgment.

1. The *severity* of injury or illness most likely to result. The severity factors are rated as follows:

   "A"—For conditions in which the injury would require first aid treatment or less, such as minor cuts, bruises, or splinters

   "B"—For conditions in which the injury would require treatment by a doctor, such as sutures or setting broken bones in a finger

   "C"—For conditions in which the injury would require hospitalization for 24 hours or more.

2. The *probability* or likelihood that an injury or illness would result from the alleged violation. Consideration is given to the extent that such a condition has already resulted in injury or illness and the number of employees exposed to the substandard condition. The probability is rated as follows:

   "A"—If the likelihood is low

   "B"—If the likelihood is moderate

   "C"—If the likelihood is high

3. The *extent* to which the standard is violated. Here there are two factors involved — (a) standards pertaining to the workplace and (b) standards pertaining to employee procedures. The rating scheme for standards pertaining to the workplace is as follows:

   "A"—If any isolated violations are observed; that is, no more than 15 percent of the units covered by the standards are in violation.

   "B"—If from 15 to 50 percent of the affected units are in violation.

   "C"—If over 50 percent of such units are in violation.

With respect to standards pertaining to employee procedures, the rating system is as follows:

   "A"—If the violation occurs occasionally

   "B"—If the violation occurs frequently

   "C"—If the violation occurs regularly

It is obvious from these measurement schemes that an "A" rating is the least severe and that the "C" rating is the most severe. These ratings are averaged to determine the final rating. (An "X" rating is used when the employer clearly demonstrates a blatant disregard for the violation in question.) Each final letter rating has been assigned the following dollar ranges for the purpose of establishing an "unadjusted penalty."

   "A" = none

   "B" = $100 to $200

   "C" = $201 to $500

   "X" = $501 to $1000

Penalty reductions. The "unadjusted penalty" may then be adjusted downward up to 50 percent depending on the employer's "good faith," size of business, and history of previous violations. A reduction of up to 20 percent may be given for "good faith." Evidence of good faith includes awareness of the OSHAct and any overt indications of the employer's desire to comply with the Act. A reduction of up to 10 percent may be given for business size measured in terms of the number of employees employed by the employer. A reduction of up to 20 percent may be given for a favorable history regarding previous violations. Normally, such history is based on the employer's past experience under the OSHAct. However, in certain cases, the employer's past history under other federal or applicable state safety and health statutes may be considered. The penalty adjustment factors are applied to the "unadjusted penalty" to determine the "adjusted penalty."

The "adjusted penalty" is further reduced by 50 percent (the abatement credit) if the employer corrects the violation within the abatement period specified in the citation. This reduction is made at the time the proposed penalty is calculated to determine the proposed penalty to be assessed for each violation.

Serious violations. The law requires that any employer who has received a citation for a serious violation must be assessed a proposed civil penalty of up to $1000 for each such violation. Due to the severity of a serious violation, the amount of the proposed penalty for each cited serious violation is usually calculated from a base of $1000 (the unadjusted penalty), which is the maximum penalty allowed. The unadjusted penalty may then be adjusted downward by up to 50 percent, depending on the employer's good faith, size of business, and history of violations, just as in the case of non-serious violations. However, the additional 50 percent "abatement credit" applicable to non-serious violations is *not* applicable to serious violations.

Imminent danger. Penalties may be proposed in cases of imminent danger even though the employer immediately eliminates the imminence of such danger or initiates steps to abate such danger. If the danger is abated, the situation can be reduced in gravity to the "serious" category, and some to the "non-

serious" category—all dependent on what was done to remove the imminence and how much of the hazard is removed.

Notice of proposed penalty. Once the proposed penalties for the serious and/or non-serious violations have been calculated, the "Notification of Proposed Penalty" is prepared and sent to the employer by certified mail either with the citation or soon thereafter.

Notice of failure to correct. The Act provides that any employer who fails to correct an uncontested violation within the abatement period may be assessed a proposed penalty of up to $1000 for each day that the violation continues after the expiration of the abatement period. This penalty provision can be applied when a followup inspection discloses that the employer has not abated a violation for which a citation has been issued and the citation and proposed penalty have become final. In such cases, a "Notification of Failure to Correct Violation and of Proposed Additional Penalty" is used to notify the employer; this again is sent by certified mail.

Time for payment of penalties. When a citation and/or proposed penalty is uncontested, the payment is due after the lapse of 15 working days following receipt of the penalty notice. When a citation and/or penalty are contested, the payment (if any) is not due until the final order of the Occupational Safety and Health Review Commission or the appropriate Circuit Court of Appeals is issued.

Contested Cases

An employer has the right to contest an OSHA action if he feels that such action is not justified. The employer may contest a citation, a proposed penalty, a notice of failure to correct a violation, the time allotted for abatement of an alleged violation, or any combination of these.* An employee or

---

* The regulations concerning the rules of procedure for contested cases adopted by the Occupational Safety and Health Review Commission are codified in Title 29, C.F.R., Chapter XX, Part 2200.

authorized employee representative may contest only the time allotted for an abatement of an alleged violation.

Prior to going through the formality of initiating a contest, employers should request an informal hearing with the Area Director or the Assistant Regional Director. Many times such informal sessions will resolve the questions and the issues; therefore, the formal contested case proceedings can be avoided.

If the informal conference fails to resolve the dispute between OSHA and the employer and the latter elects to contest the case, it must be remembered that affected employees or the authorized employee representative are automatically deemed to be parties to the proceeding. In contesting an OSHA action, the employer must comply with the following which are applicable to the specific case:

1. Notify the Area Office which initiated action that he is contesting. *This must be done within 15 working days from receipt of OSHA's notice of proposed penalty;* it must be sent by certified mail. If the employer does not contest within 15 working days after receipt of the notice of proposed penalty, the citation and proposed assessment of penalties are deemed to be a final order of the Occupational Safety and Health Review Commission and are not subject to review by any court or agency and the alleged violation must be corrected within the abatement period specified in the citation.

2. If any of the employees working on the site of the alleged violation are union members, a copy of the notice of contest must be served upon their union.

3. If any employees who work on the site are not represented by a union, a copy of the notice of contest must either be posted at a place where the employees will see it or be served upon them personally.

4. The notice of contest must also contain a listing of the names and addresses of those parties who have been personally served a notice and, if such notice is posted, the addresses of the place the notice was posted.

5. If the employees at the site of the alleged violation are not represented by a union and have not been personally served with a copy of the notice to contest, posted copies must specifically advise the unrepresented employees that they may be prohibited from asserting their status as parties to the case if they fail to properly identify themselves to the Commission or the Hearing Examiner prior to the commencement of the hearing or at the beginning of the hearing.

6. There is no specific form for the notice of contest. However, such notice should clearly identify what is being contested— the citation, the proposed penalty, the notice of failure to correct a violation, or the time allowed for abatement of the violation or any combination of these.

If the employer contests an alleged violation in good faith, and not solely for delay or variance of penalties, the abatement period does not begin until the entry of the final order by the Review Commission.

When a notice of contest is received by an Area Director from an employer or from an employee or an authorized employee representative, he will file with the Review Commission the notice of contest and all contested citations, notice of proposed penalties, or notice of failure to abate.

Upon receipt of the notice of contest from the Area Director, the Commission will assign the case a docket number. Ultimately, a judge (hearing examiner) will be assigned to the case and will conduct a hearing at a location reasonably convenient to those concerned. OSHA presents its case and is subject to cross examination by other parties. The party contesting then presents his case and is also subject to a cross examination by other parties. Affected employees or an authorized employee representative may participate in the hearings. The decision by the judge will be based *only* on what is in the record. Therefore, if statements are unchallenged, the statements will be assumed to be fact.

Upon completion of the hearings, the judge will submit the record and his report to the Review Commission. If no Commissioner orders a review of a judge's recommendation, such recommendation will stand as the Review Commission's decision. If any Commissioner orders a review of the case, the Commission itself must render a decision to affirm, modify, or vacate the judge's recommendation. The Commission's orders become final 15 days after issuance, unless stayed by a court order.

Any person adversely affected or aggrieved by an order of the Commission may obtain a review of such order in the U.S. Court of Appeals if sought within 60 days of the order's issuance.

— Small Business Loans

The Act enables economic assistance for small businesses. It amends the Small Business Act to provide for financial assistance to small firms for changes that will be necessary to comply with the standards promulgated under the OSHAct or standards promulgated by a state under a state plan. Before approving any such financial assistance, the Small Business Administration (SBA) must first determine that the small firm is likely to suffer substantial economic injury without such assistance.

Federal-State Relationships

The OSHAct encourages the states to assume the fullest responsibility for the administration and enforcement of their own occupational safety and health laws. Any state may assume responsibility for the development and enforcement of occupational safety and health standards relating to any occupational safety and health issue covered by a standard promulgated under the OSHAct. However, in order to assume this responsibility, such state must submit a state plan* to OSHA for approval. If such a plan satisfies designated conditions and criteria, OSHA must approve the plan.

The basic criteria for approval of state plans is that the plan must be "at least as effective as" the federal program. There was no Congressional intent to require the state programs to be a "mirror image" of the federal program. Congress believed rules for developing state plans should be flexible to allow consideration of local problems, conditions, and resources.

The Act provides for the funding of the development of a state program, up to 90 percent of such costs. It also provides for funding the implementation of the state program, up to 50 percent of such costs.

A state plan must include any occupational safety and health "issue" (industrial, occupational, or hazard group) for which a corresponding federal standard has been promulgated. A state plan cannot be less stringent, and may include subjects not covered in the federal standards. However, state plans which do not include those issues covered by the federal program, in effect, surrender such issues to OSHA. For example, a state plan may cover all industry except construction. If such is the case, the state surrenders its jurisdiction for safety and health programs over the construction industry to OSHA and it is then OSHA's obligation to enforce the federal standards for that industry not covered by the state plan.

Following approval of a state plan, OSHA will continue to exercise its enforcement authority until it determines on the basis of actual operations that the state plan is indeed being satisfactorily carried out. If the implementation of the state plan is satisfactory during the first three years after the plan's approval, then the federal standards and federal enforcement of such standards under the OSHAct can become inapplicable with respect to issues covered under the plan. This means that for the interim period of dual jurisdiction, employers must comply with the state standards as well as the federal standards.

While the state agencies administering the state plan are vitally concerned with its success, this is not always the case with the members of the state legislature. The state legislature must appropriate not only an adequate budget, but in many cases must pass legislation that will ultimately enable the state agency to carry out all the functions incorporated in the state plan. Should the state agency responsible fail to fully implement the state plan and the state's performance falls short of the mark of being "at least as effective as" the federal program, OSHA has the right and the obligation to withdraw its

---

* The regulations pertaining to state plans for the development and enforcement of state standards are codified in C.F.R., Title 29, Chapter XVII, Part 1902.

approval of the state plan and once again reassume full jurisdiction in that state.

What Does It All Mean?

Congressional action in the form of the OSHAct is only a limited step in achieving the full purpose of the Act. Getting it to work with reasonable efficiency is the second and more difficult task. Achieving the *purpose* will depend on the willingness and cooperation of all concerned—employees and organized labor as well as business and industry.

No one yet knows the effect the OSHAct has had on the nation's work injury experience. There is no doubt, however, that the Act has given new visibility to the whole realm of occupational safety and health. Since there are many employee rights incorporated in the OSHAct, it has given the employees a significant part of the action related to occupational safety and health matters. It has moved the laggards from "little or no safety" to "some safety," but not to "optimum safety." It has raised occupational safety and health to a higher priority in business management. It has given new status and responsibility to the occupational safety and health professional. Management is now relying more heavily on the safety profession for advice. And, it has given a new status to nationally recognized consensus standards-producing organizations.

New impetus is given to the field of occupational health, a much more difficult discipline with which to work when compared to occupational safety. Much more needs to be done to determine what kind of exposures are indeed hazardous to humans and under what conditions. Further, much more needs to be done to determine what countermeasures are not only adequate, but also reasonable and feasible to eliminate or minimize exposures to occupational health hazards. There exists a great need for much more research and data in occupational health to achieve optimum occupational safety and health programming.

The OSHAct has encouraged greater training for professionals in occupational safety and health. New curricula and university programs leading to various degrees in safety and health have been inaugurated by several universities and more are yet to come.

The OSHAct added new impetus to the product safety discipline. Until the passage of the Consumer Product Safety Act, the OSHAct was the most significant piece of legislation affecting product safety ever passed by the Congress. Designers and manufacturers of equipment now used by industry have a moral (but not legal) obligation to design, deliver, and install such equipment in accordance with the applicable standards.

The OSHAct, as well as the Occupational Safety and Health Administration, is not without limitations. Mere compliance with the requirements of the Act will not achieve optimum safety and health in terms of cost, benefits and human values. All concerned must recognize that occupational safety and health cannot be handed to the employer or to the employee by legislative enactment or administrative decree. At best, state or federal occupational safety and health standards can cover only those things that are enforceable—namely control over physical conditions and environment.

As a matter of hard reality, enforcement standards simply do not adequately relate to the man in the man-machine-environment system. Important elements of a complete safety program, such as (*a*) establishment of work procedures to limit risk, (*b*) supervisory training, (*c*) job instruction training for employees, (*d*) job safety analysis, and (*e*) human factors engineering, to name a few, have not, for the most part, been included in the standards promulgated under the OSHAct—nor do the standards relate to employee attitudes, morale, or teamwork.

For the most part, the occupational safety and health standards promulgated under the OSHAct are minimal criteria and represent a floor rather than a goal to achieve. Thus, to rely on mere compliance with the occupational safety and health standards is to invite disaster since the residual risk after compliance remains unacceptable. Effective accident prevention and control of occupational health hazards must go beyond the OSHAct.

A violation of a standard is only symptomatic of something wrong with the management safety system. Only complete occupational safety and health programming as described elsewhere in this Manual will achieve a level of risk that is acceptable to employers as well as employees. The *real objective* and the purpose of the OSHAct is better occupational safety and health performance and not simply complying with the promulgated set of standards.

Directory of Federal Agencies Concerned With the Occupational Safety and Health Act

National headquarters

OSHA
Occupational Safety and Health Administration U.S. Department of Labor, Department of Labor Building, 14th Street and Constitution Avenue, N.W., Washington, D.C. 20210

NIOSH
National Institute for Occupational Safety and Health, U.S. Department of Health, Education, and Welfare, Parklawn Building, 5600 Fishers Lane, Rockville, Md. 20852

BLS
Bureau of Labor Statistics, U.S. Department of Labor, 441 G Street, N.W., Washington, D.C. 20212

OSHRC
Occupational Safety and Health Review Commission, 1825 K Street, N.W., Washington, D.C. 20006

Regional offices

Region I

Connecticut
Maine
Massachusetts
New Hampshire
Rhode Island
Vermont

OSHA
18 Oliver Street
Boston, Mass. 02110

NIOSH
John F. Kennedy Federal Building
Room 1401-B-3, Government Center
Boston, Mass. 02203

BLS
John F. Kennedy Federal Building
Room 1603-A, Government Center
Boston, Mass. 02203

Region II

New York
New Jersey
Puerto Rico
Virgin Islands

OSHA
1515 Broadway (1 Astor Plaza)
New York, N.Y. 10036

NIOSH
26 Federal Plaza
New York, N.Y. 10007

BLS
1515 Broadway
New York, N.Y. 10036

Region III

Delaware
District of Columbia
Maryland
Pennsylvania
Virginia
West Virginia

OSHA
3535 Market Street (15220 Gateway Center)
Philadelphia, Pa. 19104

NIOSH
3535 Market Street
Philadelphia, Pennsylvania 19104
(P.O. Box 13716, Philadelphia, Pa. 19101)

BLS
3535 Market Street, Room 14100
Philadelphia, Pa. 19104
(P.O. Box 13309, Philadelphia, Pa. 19101)

Region IV

Alabama
Florida
Georgia
Kentucky
Mississippi
North Carolina
South Carolina
Tennessee

OSHA
1375 Peachtree Street, N.E., Suite 587
Atlanta, Ga. 30309

NIOSH
50 Seventh Street, N.E.
Atlanta, Ga. 30323

BLS
1371 Peachtree Street, N.E., Suite 540
Atlanta, Ga. 30309

Region V

Illinois
Indiana
Michigan
Minnesota
Ohio
Wisconsin

OSHA
300 South Wacker Drive, Room 1201
Chicago, Ill. 60606

NIOSH
300 South Wacker Drive
Chicago, Ill. 60606

BLS
300 South Wacker Drive, 8th Floor
Chicago, Ill. 60606

Region VI

Arkansas
Louisiana
New Mexico
Oklahoma
Texas

OSHA
1512 Commerce Street (Texaco Building)
7th Floor
Dallas, Texas 75201

NIOSH
1100 Commerce Street, Room 8-C-53
Dallas, Texas 75202

BLS
1100 Commerce Street, Room 6-B-7
Dallas, Texas 75202

Region VII

Iowa
Kansas
Missouri
Nebraska

OSHA
823 Walnut Street
(Waltower Building), Room 300
Kansas City, Mo. 64106

NIOSH
601 East 12th Street
Kansas City, Mo. 64106

BLS
911 Walnut Street
(Federal Office Building), 10th Floor
Kansas City, Mo. 64106

Region VIII

Colorado
Montana
North Dakota
South Dakota
Utah
Wyoming

OSHA
1961 Stout Street
(Federal Building), Room 15010
Denver, Colo. 80202

NIOSH
19th and Stout Streets
(Federal Building), Room 9017
Denver, Colo. 80202

BLS
911 Walnut Street
(Federal Office Building), 10th Floor
Kansas City, Mo. 64106

Region IX

Arizona
California
Hawaii
Nevada

OSHA
450 Golden Gate Avenue
(Federal Building), Room 9470
San Francisco, Calif. 94102
(P.O. Box 36017, San Francisco, Calif. 94102)

NIOSH
50 Fulton Street
(Federal Office Building), Room 254
San Francisco, Calif. 94102

BLS
450 Golden Gate Avenue (Federal Building)
San Francisco, Calif. 94102
(P.O. Box 36017, San Francisco, Calif. 94102)

Region X

Alaska
Idaho
Oregon
Washington

OSHA
506 Second Avenue
(Smith Tower Building), Room 1808
Seattle, Wash. 98104

NIOSH
1321 Second Avenue (Arcade Building)
Seattle, Wash. 98101

BLS
450 Golden Gate Avenue (Federal Building)
San Francisco, Calif. 94102
(P.O. Box 36017, San Francisco, Calif. 94102)

References

Bureau of National Affairs, Inc., 1231 25th Street, N.W., Washington, D.C. 20037. *Occupational Safety and Health Reporter.*

Commerce Clearing House, Inc., 4025 West Peterson Avenue, Chicago, Ill. 60646. *Employment Safety and Health Guide.*

National Institute for Occupational Safety and Health, 5600 Fishers Lane, Rockville, Md. 20852. "The Advisor" (newsletter).

National Safety Council, 425 North Michigan Avenue, Chicago, Ill. 60611.
National Safety News (magazine).
"OSHA Up-to-Date" (newsletter).
Self-Evaluation Checklists for General Industry (Part 1910).
Self-Evaluation Checklists for Construction (Part 1926).

Superintendent of Documents, U.S. Government Printing Office, Washington, D.C. 20402.
Annual List of Toxic Substances.
"The Employee and OSHA" (OSHA 2099)
Federal Register.
Field Operations Manual.
Job Safety and Health (magazine).
"Occupational Safety and Health Act of 1970 (P.L. 91–596)."
Occupational Safety and Health Act Regulations, Title 29, C.F.R., Chapter XVII.
Part 1901—Procedures for State Agreements.

Part 1902—State Plans for the Development and Enforcement of State Standards.
Part 1903—Inspections, Citations and Proposed Penalties.
Part 1904—Recording and Reporting Occupational Injuries and Illnesses.
Part 1905—Rules of Practice for Variances, Limitations, Variations, Tolerances, and Exemptions.
Part 1906—Administration Witnesses and Documents in Private Litigation.
Part 1907—Accreditation of Testing Laboratories.
Part 1910—Occupational Safety and Health Standards.
Part 1911—Rules of Procedure for Promulgating, Modifying, or Revoking Occupational Safety or Health Standards.
Part 1912—Advisory Committees on Standards.
Part 1913—Disclosure of Information.
Part 1915—Safety and Health Regulations for Ship Repairing.
Part 1916—Safety and Health Regulations for Shipbuilding.
Part 1917—Safety and Health Regulations for Shipbreaking.
Part 1918—Safety and Health Regulations for Longshoring.
Part 1926—Safety and Health Regulations for Construction.
Part 1950—Development and Planning Grants for Occupational Safety and Health.
Part 1951—Grants for Implementing Approved State Plans.
Part 1952—Approved State Plans for Enforcement of State Standards.
Part 1953—Changes to State Plans for the Development and Enforcement of State Standards.
Part 1954—Procedures for the Evaluation and Monitoring of Approved State Plans.
Part 1975—Coverage of Employees Under the Williams-Steiger Occupational Safety and Health Act of 1970.
Part 1977—Discrimination Against Employees Exercising Rights Under the Williams-Steiger Occupational Safety and Health Act of 1970.
Part 2200—Rules of Procedure (Chapter XX—Occupational Safety and Health Review Commission).

Safety
Program
Organization

Chapter
3

Organizing the Safety Program   49
Safety beyond OSHA . . . Basic elements of a safety organization . . .
Assumption of responsibility . . . Declaration of company policy . . .
Implementing the safety policy

Assignment of Responsibilities   53
Supervisors and foremen . . . Staff safety direction

Organizational Setups   55
Small company organization . . . Scattered operations . . . Public
employee organizations . . . Staff *vs.* line status

Training Employees   58
Safety attitudes . . . Safety rule enforcement

Safety Committees   59
Types of safety committees . . . Function of safety committees . . .
Selection and term of office . . . Policies and procedures . . .
Effective committee meetings . . . Motivation of committees . . .
Suggestion systems . . . Special committees

Off-the-Job Safety Programs   66

References   69

Organizing the Safety Program

Safety program organization may be defined as a method employed by management to share and to assign responsibility for accident prevention and to ensure performance under that responsibility.

A safety program is not something that is imposed on company organization. Safety must be built into every process or product design and into every operation. It must be an integral part of company operations.

The prevention of accidents and injuries is basically achieved through control of the working environment and control of people's actions. Only management can implement such control.

A company that has an effective safety program will have a working environment in which operations can be conducted economically, efficiently, and safely.

Safety beyond OSHA

The OSHAct, discussed in the previous chapter, has stimulated many organizations to get into activities intended to bring the organization into compliance with the Act. This new awareness of safety is beneficial if it results in a sincere effort by top management to take effective action to eliminate work injuries. Such organized activity goes beyond mere compliance with OSHA standards which largely involve equipment and physical conditions.

The stated objective of the OSHAct is to reduce occupational injuries and illnesses. Therefore, evidence of organized safety activities and proof of injury reduction should be brought to the attention of the OSHA Compliance Officer at the time of the inspection. This will demonstrate management's parallel objective to eliminate accidents. Such evidence may also be the basis for a reduction in the number of OSHA citations or in the amount of the proposed penalties, if any.

Conversely, superficial activities and minimal compliance cannot be considered as a viable safety program. Not only will such activities be ineffective in controlling accidents and occupational health hazards, but they will give a false sense of security and expose management's lack of control over more basic accident causes.

Basic elements of a safety organization

Analysis of safety programs in plants or concerns with outstanding safety records shows that invariably the programs are built around the seven basic elements shown in the illustration (Fig. 3–1). These elements or principles of accident prevention are the same in any industry and in any organization, large or small. Thus, although there is wide variation in the methods used in individual organizations, each incorporates most if not all of the seven basic elements shown in Fig. 3–1.

Assumption of responsibility

Top management's attitude and approach toward accident prevention is almost invariably reflected in the attitude of the supervisory force. Similarly, the worker's attitude is usually the same as his supervisor's. Thus, if top management is not genuinely interested in preventing accidents, no one else is likely to be. Since this basic fact applies to every level of management and supervision, an accident control program must result from top management's announced and demonstrated interest if employee co-operation and participation are to be obtained.

The details for carrying out an accident prevention program may be assigned, but the responsibility for the basic policy cannot be delegated. Delegation of responsibility for safety cannot be made merely by appointing a safety director or some safety committees and expecting them to function in an effective manner. The top executive officer of a company is responsible for overall safe performance. He must constantly review such performance; this is best done through regular lines of supervision.

In the beginning, the safety program requires close executive attention because it is generally a new activity. Consequently, the overall policy should be prepared so it will state clearly the objectives to be achieved.

Declaration of company policy

A company—large or small—that attempts to stop accidents without a definite guiding policy—one which is planned, publicized, and promoted—will find itself continuously "fighting fires."

Management, if it wants acceptable safety performance, must first write a safety policy.

BASIC ELEMENTS OF SAFETY ORGANIZATION

  I. MANAGEMENT LEADERSHIP
     (Assumption of responsibility—declaration of policy)

 II. ASSIGNMENT OF RESPONSIBILITY
     (To operating officials—safety directors—supervisors—committees)

III. MAINTENANCE OF SAFE WORKING CONDITIONS
     (Inspectors—engineering revisions—purchasing—supervisors)

 IV. ESTABLISHMENT OF SAFETY TRAINING
     (For supervisors—for workers)

  V. AN ACCIDENT RECORD SYSTEM
     (Accident analysis—reports on injuries—measurement of results)

 VI. MEDICAL AND FIRST AID SYSTEMS
     (Placement examinations—treatment of injuries—first aid services—periodic
     health examinations)

VII. ACCEPTANCE OF PERSONAL RESPONSIBILITY BY EMPLOYEES
     (Training—maintenance of interest)

Fig. 3–1.

It should be brief, to the point, and define management attitude.

Reasons for having a good policy are:

1. A good policy makes it easier to enforce safe practices and conditions.

2. It makes it easier for supervisors to implement company policy.

3. It also makes it easier for employees to follow safety rules and instructions.

4. It makes it easier to obtain good preventive maintenance of equipment or selection of proper equipment when purchased.

Writing a safety policy is a lot like writing a speech—top management writes its own policy, but can use material from many sources.

Basic to a policy declaration are these statements:

1. That safety of employees, the public, and company operations are paramount.

2. That safety will take precedence over expediency or short cuts.

3. That every attempt will be made to reduce the possibility of accident occurrence.

4. That the company intends to comply with OSHAct and all other safety laws.

The following statements on safety policies have been expressed by and for top managements:

• When a man enters the employ of this company, he has a right to expect that he will be provided with a proper place in which to work, and proper machines and tools with which to do his job, so that he will be able to devote his energies to his work without fear of possible harm to his life and health.

Only under such circumstances can the relationship between employer and employee be mutually profitable and harmonious. It is our desire (*a*) to provide a safe work place, safe equipment, proper materials, and (*b*) to establish and insist upon safe methods and practices at all times.

It is a basic responsibility of all executives to make the safety of human beings a part of their daily, hourly concern. This responsibility must be accepted by everyone who has a part in the affairs of the corporation, no matter in what capacity he may function.

• Management considers no phase of operation or administration as being of greater importance than accident prevention. It is the policy of the company, therefore, to provide and maintain safe and heathful working conditions, and to follow operating practices that will safeguard all employees and result in safe working conditions and efficient operation.

• We believe in the dignity and importance of the individual employee and in his right to derive personal satisfaction from his employment. Also spelled out in this creed is our belief that the safety of employees continues to be the first consideration in the operation of the business.

• Safety is *our* responsibility in management. Without question it is our *number one* responsibility, taking precedence over everything else.

• Accident prevention and efficient production go hand-in-hand. All levels of management have a primary responsibility for the safety and well-being of all employees. This responsibility can be met only by working continuously to promote safe work practices among all employees and to maintain property and equipment in safe operating condition.

• The supervisor is the key man in the safety program because he is in constant contact with employees. No foreman, supervisor, or operating head may ever be relieved of any part of his responsibility for safety. Safety is an operating function and cannot be transferred to a staff organization.

• Safe practices on the part of employees must be part of all operations. No job shall be considered efficiently completed unless the worker has followed every precaution and safety rule to protect himself and his fellow workers. The ideals of production and safety must be inseparable.

• "Total safety" extends into three important areas—company personnel, products and customers, and the public. This policy is implemented in these vital areas by:

1. Development and application of safety standards both for production facilities (equipment, tools, work methods, and safe guarding) and for products, based on applicable legal and voluntary codes, rules, and standards as a minimum.

2. Safety inspection to identify potential hazards, both in production and in products. Packaging, labeling, and instruction sheets are designed to minimize hazards, or alert users to hazards inherent in the product.

3. Accident investigation to determine cause of accidents and prevent recurrence.

4. Accident records and accident-cause analysis to determine accident trends and provide targets for corrective action.

5. Education and training in general safety principles and techniques. On-the-job safety instruction by the supervisor and periodic supervisory contacts for new instructions, followup, and general safety motivation.

6. Protective equipment to provide personal protection in hazardous areas.

7. Industrial hygiene studies to identify potential health hazards and develop necessary protective measures.

8. Safety publicity and promotion to step up program interest and participation.

9. Off-the-job accident prevention in cooperation with public and private agencies to promote the application of accident prevention principles to nonwork activities.

Implementing the safety policy

Once a safety policy statement is established, it should be publicized so every employee becomes familiar with it, particularly with those aspects which apply directly to him. Meetings, letters, pamphlets, and bulletin boards can be used to accomplish this. It should also be posted in all management offices to remind them of their obligation in this important aspect of company operations. The effectiveness of any safety policy and program

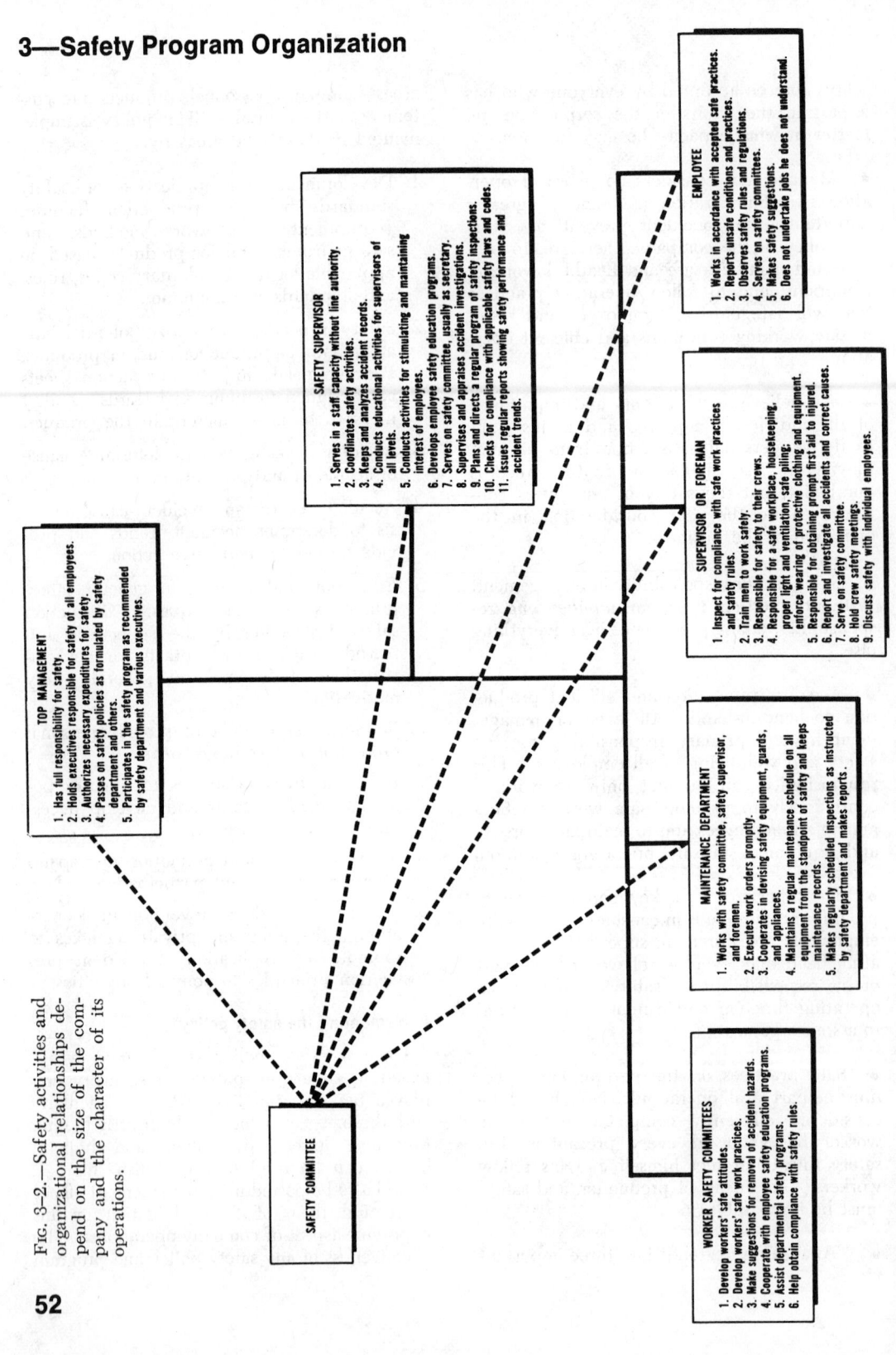

Fig. 3-2.—Safety activities and organizational relationships depend on the size of the company and the character of its operations.

TOP MANAGEMENT
1. Has full responsibility for safety.
2. Holds executives responsible for safety of all employees.
3. Authorizes necessary expenditures for safety.
4. Passes on safety policies as formulated by safety department and others.
5. Participates in the safety program as recommended by safety department and various executives.

SAFETY SUPERVISOR
1. Serves in a staff capacity without line authority.
2. Coordinates safety activities.
3. Keeps and analyzes accident records.
4. Conducts educational activities for supervisors of all levels.
5. Conducts activities for stimulating and maintaining interest of employees.
6. Develops employee safety education programs.
7. Serves on safety committee, usually as secretary.
8. Supervises and appraises accident investigation.
9. Plans and directs a regular program of safety inspections.
10. Checks for compliance with applicable safety laws and codes.
11. Issues regular reports showing safety performance and accident trends.

EMPLOYEE
1. Works in accordance with accepted safe practices.
2. Reports unsafe conditions and practices.
3. Observes safety rules and regulations.
4. Serves on safety committees.
5. Makes safety suggestions.
6. Does not undertake jobs he does not understand.

SUPERVISOR OR FOREMAN
1. Inspect for compliance with safe work practices and safety rules.
2. Train men to work safely.
3. Responsible for safety to their crews.
4. Responsible for a safe workplace, housekeeping, proper light and ventilation, safe piling; enforce wearing of protective clothing and equipment.
5. Responsible for obtaining prompt first aid to injured.
6. Report and investigate all accidents and correct causes.
7. Serve on safety committee.
8. Hold crew safety meetings.
9. Discuss safety with individual employees.

MAINTENANCE DEPARTMENT
1. Works with safety committee, safety supervisor, and foremen.
2. Executes work orders promptly.
3. Cooperates in devising safety equipment, guards, and appliances.
4. Maintains a regular maintenance schedule on all equipment from the standpoint of safety and keeps maintenance records.
5. Makes regularly scheduled inspections as instructed by safety department and makes reports.

WORKER SAFETY COMMITTEES
1. Develop workers' safe attitudes.
2. Develop workers' safe work practices.
3. Make suggestions for removal of accident hazards.
4. Cooperate with employee safety education programs.
5. Assist departmental safety programs.
6. Help obtain compliance with safety rules.

SAFETY COMMITTEE

varies directly with the active support given it by management.

Expressions of management interest include enforcement of policy and program, recognition of good safety records, review of safety reports, and participation in meetings, safety banquets, and other events to show their support for company safety efforts.

Management must take the lead in keeping interest alive by:

1. Emphasizing that production and safety go together for an efficient operation.

2. Setting a good example. For instance, if plant rules require employees to wear goggles or other personal protective equipment in certain areas, management should observe these regulations when visiting these areas.

3. Attending safety meetings.

4. Reviewing and acting upon accident reports.

5. Reviewing departmental safety records through group conferences with department heads.

6. Motivating interest in safety through general letters, bulletin board announcements, and discussing the accident record in the plant.

Assignment of Responsibilities

While top management has the ultimate responsibility for safety, it delegates authority for safe operation all the way down through all management levels. The supervisor is the key man in a safety program because he is in constant contact with employees. The safety professional acts in a staff capacity to help administer policy, to provide technical information, to help train, and to supply program material.

Supervisors and foremen

Heads of operating units can provide personal leadership. In achieving satisfactory protection against accidents, they can interpret the company's policy and actively support it. This sets an example to those responsible to them, and gives safety equal emphasis and weight with matters of production, cost, and quality. Top management should give prompt and fair consideration to recommendations made for ways to reduce hazards.

The first-line supervisor should be responsible to see that:

● Each of his employees understands the properties and hazards of the material stored, handled, or used by them.

● The necessary precautions are observed when using equipment, including the use of proper safeguards and proper personal protective equipment.

● That they understand and properly follow the established work procedures for their safety.

Some supervisors have a strong feeling for safety. Others must be trained by management to carry out the safety program.

Here is where the safety professional can help. The *Supervisors Safety Manual,* and its accompanying "Home Study Course," available through the National Safety Council, have been widely accepted by industry as practical training aids for foremen and supervisors.

Staff safety direction

For advice and help in administering a safety policy and ensuring the continuity of the safety program, top management usually places the administration of the program in the hands of a safety department or safety professional, with the title of safety director or manager of safety.

It is important that management definitely assign full staff responsibility for the safety activities to one responsible individual. The decision concerning proper placement of responsibility should be based upon the size of the company and the character of its operations (Fig. 3–2).

The administration of the safety program is dependent on these factors:

1. Size of organization

2. Nature of operations (potential hazards, accident costs, etc.)

3. Management interest in the kind of the safety program desired.

3—Safety Program Organization

Generally, the following guidelines are suggested. In either case the person responsible for safety should report to an officer of the company.

1. Organizations employing 500 or more people and/or with moderate or high hazards or costs should place the safety program in the hands of a full-time safety professional.

2. Organizations employing up to 500 people and/or with less severe hazards can place the program in the hands of an industrial relations manager, personnel manager, superintendent, or other responsible person in the organization who has some knowledge of safety procedures and standards.

The number of people employed should not be the main factor in determining whether the safety program will be in the hands of a full-time safety professional or not. In some cases, the nature of the operation may indicate a full-time man is needed regardless of the number of people employed.

The trend is to employment of full-time safety professionals for the following reasons:

1. The passage of the Occupational Safety and Health Act of 1970 requires that certain safety standards be met and maintained.

2. A better understanding of the safety professional's services and functions is developing. To effectively administer a safety program, the person responsible must be highly trained and/or have many years of experience in the safety field.

3. The trend towards the expansion of safety includes involvement in occupational health, product safety, machine design and plant layout, security, and fire prevention —in another term, "the total safety concept."

A careful study and analysis should be made to determine the type of program best suited for the organization's needs. If there is a person in the organization who shows an interest in this field, he can be of great help to management in establishing the best course. Otherwise, it may be necessary to seek outside help to develop a program.

An important item that should not be overlooked or underplayed is that an effective safety program requires money to operate— the safety professional's salary, the salary of staff to help him, and cost of safety equipment, safety training materials, awards, and meetings are just a few of the expense items connected with a good safety program. A budget should be set based on best estimates of safety needs. This budget, once approved, should not be tampered with.

The safety program under the direction of the assigned person should enjoy the same position as other established activities of the organization such as sales, production, engineering, or research.

The duties of the safety professional ordinarily include:

1. Formulating, administering, and making necessary changes in the accident prevention program.

2. Submitting, directly to the officer in charge, regular monthly, weekly, or daily reports on the status of safety.

3. Acting in an advisory capacity on all matters pertaining to safety as required for the guidance of management, the general manager, superintendents, foremen, and such departments as purchasing, engineering, and personnel.

4. Maintaining the accident record system, making necessary reports, personal investigation of fatal or serious accidents, investigating accidents through his staff, securing supervisors' accident reports, and checking corrective action taken by supervisors to eliminate accident causes.

5. Supervising or closely cooperating with the training supervisor in the safety training of employees.

6. Correlating safety work with the work of the medical department. Include proper selection and placement of employees.

7. Making personal inspections and supervising inspections by his staff and by special employee committees, for the purpose of discovering and correcting unsafe conditions or unsafe work practices before they cause accidents.

8. Maintaining outside professional contacts to exchange information with others and to keep the program up-to-date.

9. Making certain that OSHA requirements and state or local laws, ordinances, or orders bearing on industrial safety are complied with.

10. Securing necessary help or advice from organizations, such as the National Safety Council, governmental agencies, or insurance carriers on matters pertaining to safety and health.

11. Starting activities that will stimulate and maintain employee interest.

12. Directing the activities of his staff so that the accident prevention program will be efficiently operated. It is expected that the safety professional may delegate certain responsibilities to his staff engineers, such as acting as secretary for certain safety committees.

13. Controlling or supervising fire prevention and fire fighting activities where they are not responsibilities of other departments.

14. Setting standards for safety equipment to be used by plant personnel.

15. Approving designs of new equipment to be used in plant work areas.

16. Recommending provisions for safety in plans and specifications of new building construction and repair or remodeling of existing structures.

The amount of help needed by the safety director to discharge his responsibilities properly depends upon the degree of responsibility, the size and operating policies of the company, and the type of operations.

Organizational Setups

Much of what has been said so far applies to large organizations. One organizational setup suitable for a large company is shown in Fig. 3–3, on the next page.

Other set-ups are now discussed.

Small company organization

Active management and control of the small company safety program may be vested in the chief executive himself, the general manager, or in an experienced and qualified supervisor who has both authority and standing. It is neither fair to the man nor productive of results to assign such responsibility, however, without giving specified time and facilities for carrying it out. With a planned program, for which ample time is allotted, a small establishment need not lack safety leadership.

There are several advantages to safety inherent in small-scale operations, such as closer contact with the working force, more general acquaintance with the problems of the whole plant by foremen and supervisors, and, frequently, less labor turnover.

Aided by these inherent advantages, the small company can carry on the same kind of on-the-job safety training that is carried on by the large one. The accident record and analysis system can be complete. The program for winning acceptance of personal responsibility for safety on the part of employees can be complete and effective. Also, the small company may find it easier than the large one to develop personal loyalty.

The owner of a small company, or the manager of a small plant, may have special problems with engineering and medical services. He is not likely to be in a position to hire full-time safety engineers nor a full-time physician, and, possibly, not a full-time nurse.

For the small operator, technical safety knowledge is available through his insurance carrier, the National Safety Council, his state industrial commission, the United States Department of Labor and other agencies, safety consultants, and organizations. For certain problems, he can refer to engineering consultants in his locality. Through use of outside agencies, his total cost for technical service may be less in proportion to his total safety expenditure than that of the plant with a full-time safety staff.

The medical setup for the small establishment commonly provides for periodic visits by a physician under contract or on call, with emergency cases referred to the physician or nearby clinic or hospital for treatment. The small plant has special need for enough persons thoroughly trained in first aid to take care of the working force during each working shift.

In the industrial organization with few em-

3—Safety Program Organization

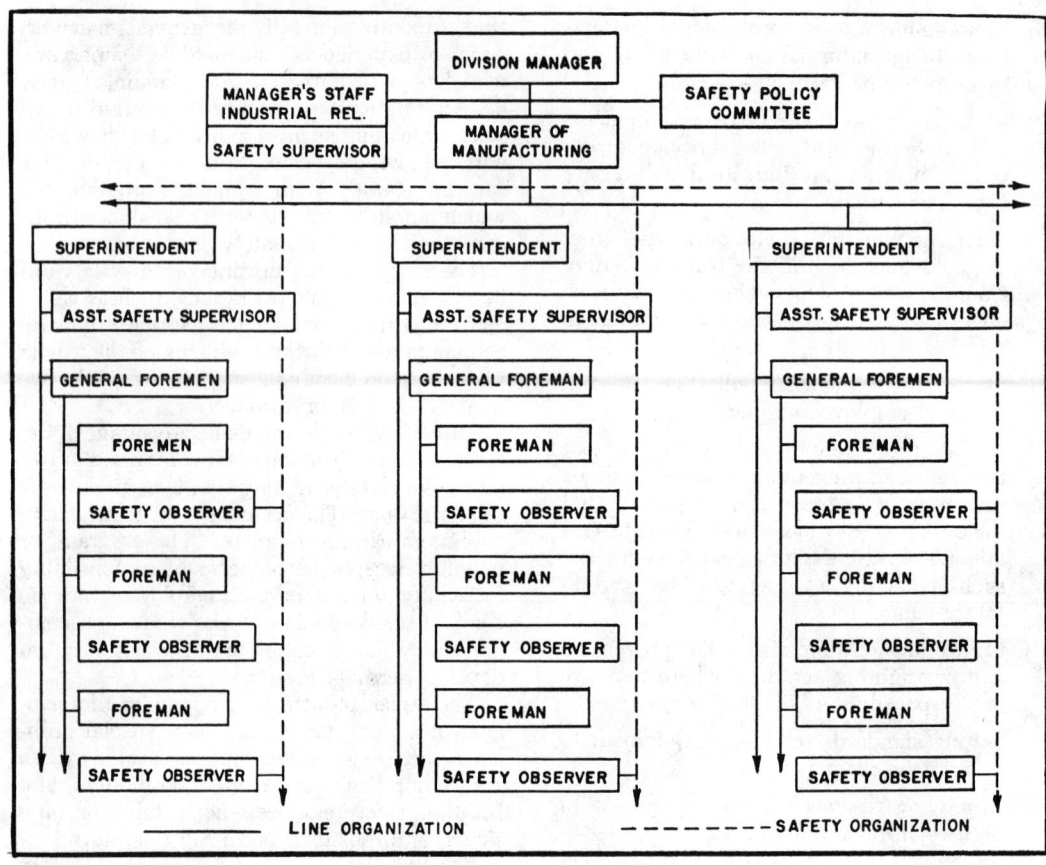

FIG. 3–3.—Organization chart showing the direct supervisory type of safety setup for a large company. *Courtesy Westinghouse Electric Corp.*

ployees, the full-time or part-time nurse becomes even more significant to the total safety program than she does in a large plant with a medical department. She carries more responsibility, is in a more strategic spot to be of help in the public relations aspect of safety, and becomes indispensable, almost in inverse proportion to the lack of medical service provided by a licensed physician. A full-time registered nurse should be considered for operations employing 200 or more persons, where the work is not unusually hazardous. The number of employees should not be the determining factor, however. Sometimes a nurse is needed where fewer people are employed, but the work is hazardous. See Chap-

ter 21, "Occupational Health Services."

For improved medical service, small companies and plants in the same locality can pool the services of a qualified physician.

Scattered operations

Organizations with scattered operations requiring relatively few employees, as well as scattered construction operations, face special problems of organization. Their operations may be seasonal or intermittent and there may not be a sufficiently stable working force to maintain an effective safety organization. A manager may need to adapt his program to local conditions, which may be highly variable.

Reports from numerous companies which

work under these conditions indicate that their method of safety operation is generally as outlined below:

1. Top management adopts a policy which provides a general plan to be followed by all locations.
2. A staff safety department at the main location acts in an advisory capacity.
3. Local managers are independent in that they determine the methods of fulfilling the general policy.
4. A local safety professional may be employed by the manager, or safety responsibility may be delegated to someone on a part-time basis.
5. The chief safety professional checks application of the general policy at all locations and gives assistance as required.

Public employee organizations

Many city and state organizations do much to promote public safety and organized safety programs, but often there is little uniformity in the programs affecting their own employees. City governments, for example, operate water works and power distribution systems, and construct and repair streets. A large metropolis may be involved in a variety of industrial operations, from heavy construction to horticulture.

It is not uncommon to find that some departments have effective safety programs whereas others have none at all. The same lack of uniformity appears inevitably in their accident records. Many municipalities are trying to develop effective safety programs.

Agencies of government, many of them with official responsibility for the safety of the public, can reasonably do no less than to protect their own employees, both on and off the job. Safety for public employees has all of the justifications which apply to private employment.

Safety programs in municipal government are often carried out through a general or central safety committee. Such a committee is usually composed of department heads and supervisors.

To carry out the work of a general committee of this type, small special subcommittees are formed, such as an eye protection committee, an attendance committee, and a rules committee. The function of these subcommittees is to create interest in and carry out the various phases of the safety program. For example, the eye protection committee may specify when eye protection is to be worn and put on drives when there is a letdown in the wearing of eye protection. They can make recommendations for those who need prescription glasses.

Public employee organizations can improve safety performance, increase operating efficiency, and reduce costs by utilizing the type of safety organization found effective in multiplant industrial organizations. Basically, the pattern should contain these elements:

1. A statement of policy by the top executive (mayor or city manager), including a general plan to be followed by all departments.
2. A staff safety department, responsible to the top executive, to serve all departments in an advisory capacity and to check application of the general plan. Such a safety department could gather accident facts for all other departments and effect economies by such activities as centralized planning of training programs and purchase or development of safety materials.
3. A joint safety committee to determine safety policies.
4. Independent operation in each department or division, subject to conformance with the general policy.
5. Employment by each department or division head of a safety professional qualified in the specific field, or employment of a competent technical advisor for the entire municipality.
6. Assistance from the central safety department as required.
7. Establishment of a safety training program for supervisors and employees.

It is not possible to recommend a standard pattern of safety organization for a city or for state departments. However, the organization is secondary in importance to the principles,

57

and the principles can be applied by any responsible public official who wants to apply them and who has the power to do so.

Staff *vs.* line status

In general the safety program is administered by supervisors or other persons holding line positions in a small company, and staff positions in a large company. In a large corporation, the safety professional and his organization should have staff status and authority. The exact organizational status of the safety staff must be determined by each firm in terms of its own operating policies.

A survey conducted by the National Safety Council tried to determine the organizational position of the safety professional in American and Canadian industry. Of the respondents, 44.8 percent reported directly to a member of top management, such as the president, vice-president, general manager, finance officer, or works manager. Another 19.8 percent reported to the plant manager or superintendent, and 30.4 percent reported to the industrial relations director or personnel manager. The remaining 5.0 percent reported to the managers of various departments (labor, insurance, security, medical, etc.) or to the chief engineer. Importantly, another survey, conducted by the American Society of Safety Engineers, showed frequency rates and severity rates of disabling injuries are considerably lower in those plants where the safety professional reported directly to a member of top management.

These surveys show the safety function to be most effective when the safety professional reports to someone in management.

The safety program as a staff function should have the following objectives:

1. To keep direct responsibility for safety at the supervisory level.
2. To maintain the advisory status of the safety professional.
3. To enable the safety professional to spread his efforts over a greater area.
4. To keep the authority of supervisors from being undermined.
5. To assure more ready compliance with safety orders, since workers are accustomed to direction by supervisors.

Sometimes the safety professional is given authority usually limited to line officials. On fast-moving and rapidly changing operations or those on which delayed action would endanger the lives of workers or others, as in construction and demolition work, fumigation, some phases of manufacture of explosives, chemicals, or other dangerous substances, or emergency work, it is not uncommon to find that the safety professional has authority to order immediate changes. This authority, where granted, must be used with discretion, because the safety professional will be accountable to management for errors in judgment.

Training Employees

No matter how well safety is engineered into a plant or a job, much of the safety of employees depends upon their own conduct. Some people work safely in dangerous surroundings whereas others have accidents on jobs that seem quite safe. Motivating people is, therefore, a necessary part of an accident prevention program.

The ordering of everything people do is neither possible nor desirable. In addition to providing direct supervision, it is necessary to influence the voluntary acts of workers by education and motivation. Much of the effort put into an industrial safety program is, therefore, directed toward educating and influencing people.

The training of the employee begins the day he goes to work. Whether or not the firm has a formal induction program, the employee starts to learn about his job and to form attitudes about many things—including safety—the first day.

The all-important problem of training employees is covered in Chapter 9, "Safety Training."

Safety attitudes

Training in safe practices has, as a side result, the forming of attitudes favorable to safety. Many organizations with good safety records believe that such training—supported by management's sincere, positive attitude—is the best means of influencing employee and supervisor attitudes. In addition, however, most employers with good safety records carry on a vigorous program of activities with no other purpose than to influence attitudes toward safety or to maintain employee interest in it.

A favorable safety attitude is somewhat intangible, but may be expressed to some extent as follows:

ACCIDENTS ARE CAUSED AND CAN BE PREVENTED.

SAFETY TRAINING IS A MARK OF SKILL AND OF GOOD SENSE.

THE COMPANY IS SINCERELY INTERESTED IN SAFETY AND IS WILLING TO PAY, IN TIME AND MONEY, WHATEVER IT COSTS TO MAINTAIN AN EFFECTIVE SAFETY PROGRAM.

THE FOREMAN INSISTS ON SAFE WORK PRACTICES. HE DOES NOT TOLERATE UNSAFE WORK METHODS OR VIOLATIONS OF SAFETY RULES.

Activities to foster safe attitudes are of three general kinds: (*a*) training or educational activities in which formal teaching is done, (*b*) cooperative work in which the employees actively participate in the safety program, and (*c*) general safety advertising or propaganda.

In addition, company publications and meetings can be used to continually train employees. Details on how to work with company publications, and the press in general, are discussed in Chapter 13, "Publicizing Safety."

Safety rule enforcement

Enforcement of safety rules is actually a matter of education. Employees must understand the rules and the importance of following them; language barriers should be considered. To set a good example for employee education, top management and supervisors must know and believe in the rules, and must religiously follow them.

Education often succeeds where discipline fails, and it is usually used when supervisors find violations.

Where labor bargaining groups represent employees, they should be consulted on proposed methods of enforcing rules. Such an agreement should prevent later misunderstanding.

Administering reprimands for the violation of rules demands tact and good judgment. This duty should be performed by someone in authority who commands respect and who is sympathetic without being lax.

When management feels the worker is deliberately disobeying the rules, and where he continues through unsafe acts to endanger his own life and lives of others, prompt and firm action is justified. It is far better to use extreme measures than to allow accidents to happen because of laxness in the enforcement of safety rules.

Many companies feel that a spirit of cooperation, or mutual understanding and agreement, make drastic measures of enforcement unnecessary. Indeed, this philosophy should be the goal of enforcing safety rules. See Chapter 9, "Safety Training."

Safety Committees

Safety committees of various types are found in many successful safety organizations. They in themselves should not be considered as a "safety program," but they do have two basic functions:

1. Creating and maintaining an active interest in safety, and

2. Serving as a means of safety communication.

Types of safety committees

Among the types of safety committees (Fig. 3–4) commonly used are:

1. Company or interplant committee
2. Plant central committee
3. Departmental safety committee
4. Foremen's committee
5. Workmen's committee
6. Joint labor-management committee
7. Inspection committee
8. "Get it done" committee

A large establishment having scattered operations may use all the above committees, but a small company may use only one.

Committees can be either harmful or helpful. The harmful aspect of safety committees

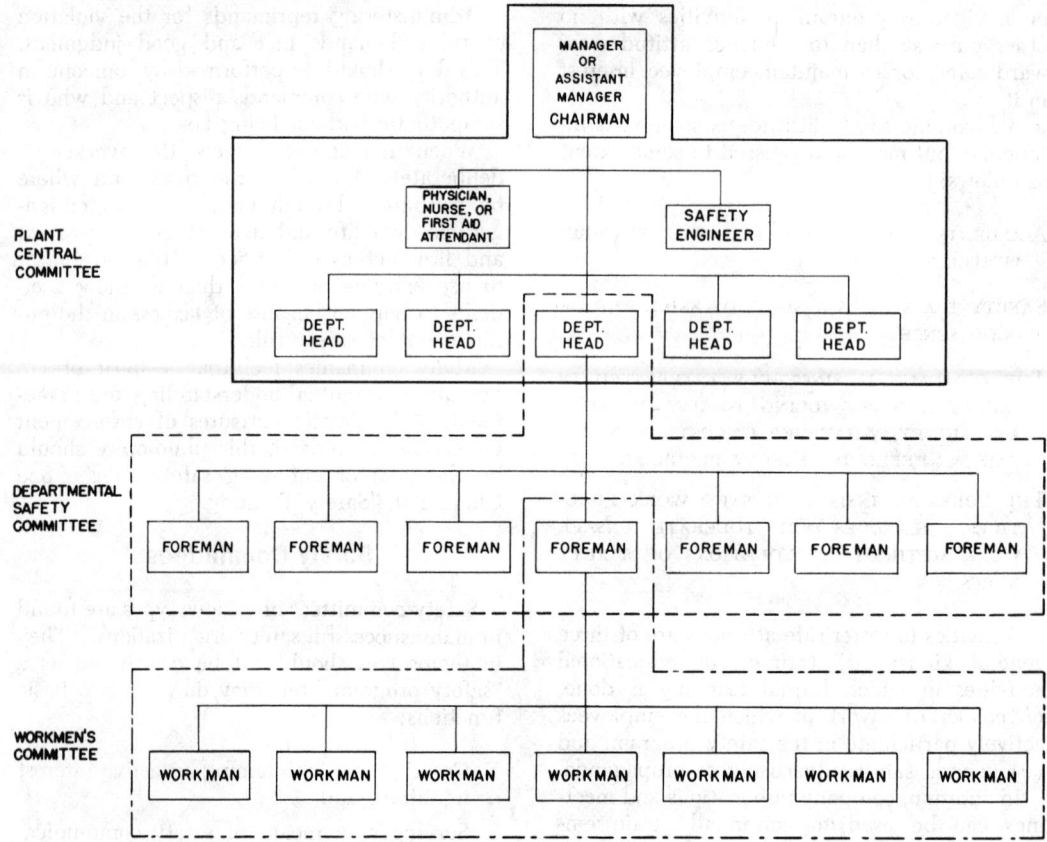

Fig. 3–4.—Typical organization of safety committees in a company, showing the interrelationship between three types.

results from management's frequent inclination to delegate its responsibility for safety to a committee. To avoid this, management should serve actively on a safety committee. Moreover, many committees are ineffective because of lack of businesslike direction or disregard of their constructive recommendations.

The helpful aspect of safety committees is that it permits employees to share in the work of accident prevention.

Management, supervisors and safety engineers give their best efforts to detecting and correcting unsafe conditions and unsafe practices, but employees are in an especially good position to observe such hazards.

Although excellent results are obtained by some organizations that do not use safety committees, such committees have been notably successful in furthering the cause of accident prevention.

Function of safety committees

A safety committee can be set up to carry out the following basic functions:

1. Discuss safety policies and recommend their adoption by management.
2. Discover unsafe conditions and practices and determine their remedies.
3. Work to obtain results by having its management-approved recommendations put into practice.
4. Teach safety to the committee members, who in turn will teach safety to the entire personnel of the company.

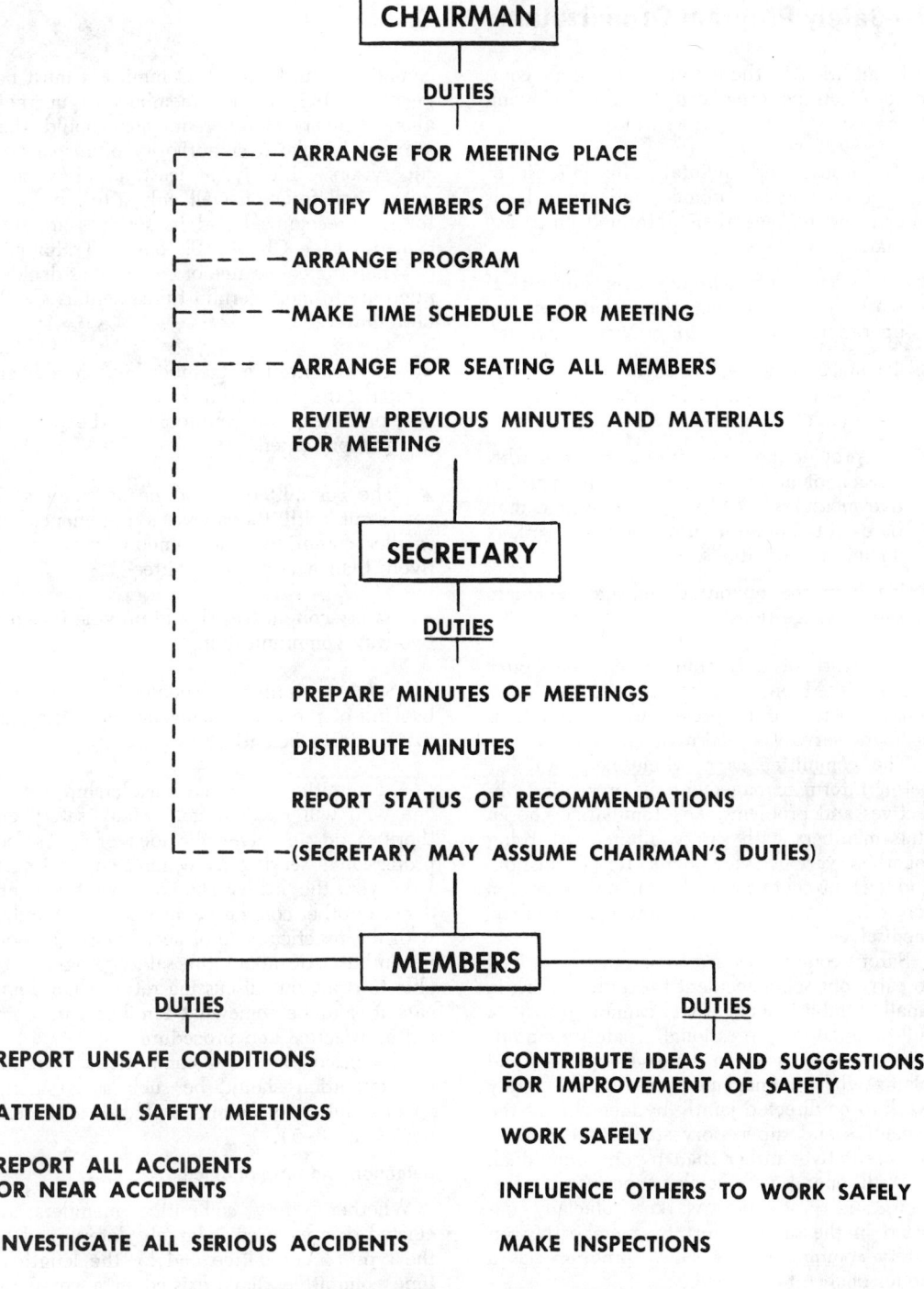

FIG. 3–5.—Duties of chairman, secretary, and members of a typical safety committee. All should know their duties and be trained to perform them efficiently. (Dashed lines show overlapping functions.)

3—Safety Program Organization

In addition to these functions, safety committees can be organized for the following purposes:

1. To arouse and maintain the interest of superintendents, foremen, and group leaders, and to keep them informed on safety matters.

2. To arouse and maintain the interest of workers and convince them that their co-operation is needed to prevent accidents.

3. To make safety activities an integral part of operating policies and methods and a function of operation.

4. To provide an opportunity for free discussion of accident problems and preventive measures. (A suggestion system may be used to obtain information from workers on unsafe conditions.)

5. To help the operating manager evaluate safety suggestions.

The work of a committee may be better accomplished in a prompt, effective, and orderly manner if a representative of top management serves as chairman.

The committee as a whole helps to (a) better inform management by presenting objectives and problems, (b) familiarize chosen staff members with safety affairs, (c) bring together various viewpoints for discussion, and (d) promote interest in and cooperation for safety among the committee members themselves.

Safety committees are sometimes organized to carry out some accident prevention work in small establishments that cannot justify a full-time safety professional. Safety committees are also used in large companies and plants where management prefers the safety work to be directed jointly by members of the executive and supervisory staff and employee representatives rather than by one individual.

With operating officials, supervisors, foremen, and production workers officially enlisted in the safety effort by membership on safety committees, the safety program has a better chance to succeed.

Without careful development of the committee setup and proper control of committee activities, much of the potential value of the committee can be lost. Committees must not become unwieldy nor membership in them allowed to be indifferent, nor should they attempt to take over authority of responsible supervisors. Ideally, at least, a safety committee will function well only after the need for it is recognized and its services are welcomed. (See Chapter 9, "Safety Training.")

When the committee or committees decided upon are formed, certain fundamentals should be followed.

• The committee membership should encompass the maximum knowledge of methods, practices, and conditions in the plant or group represented.

• The committee should be as small as is consistent with the above requirements. A smaller committee may function more effectively than a larger committee.

• Safety committees should provide for good two-way communication.

• Safety committees should have the full backing of the entire management if they are to function efficiently.

A committee must have leadership—someone who will make it go. Many safety authorities do not favor the idea of the safety professional serving as committee chairman. However, the leader should have the confidence of other committee members, be familiar with the practical side of accident prevention, be enthusiastic about the safety program, be able to draw out discussion rather than dominate it, and be somewhat familiar with committee practice and procedure.

The basic plan of a committee's structure and operation should be such as to assign specific duties to committee officials and members (Fig. 3–5).

Selection and term of office

Whether safety committee members are elected or appointed is largely determined by the type of committee and by the length of time committees have existed in a company.

Members of interplant and governing committees are more likely to be appointed by management. However, in some firms, agreements with unions provide for elections or appointments by the union. Selection of qualified persons is assured by the appointive method.

Employees should never be selected for committee work solely on the basis of popularity, since the man most liked might not be entirely sold on the safety program. For that reason, it is best for management to appoint committee representatives rather than to have them elected, at least in the early stages of the program. To gain maximum participation and interest of supervisors and workers it is desirable to rotate their representation on the committee.

In some companies, a foreman or first-line supervisor is selected by the safety committee to serve on the workmen's safety committee, and he automatically becomes its chairman. He usually serves for six months or for some other set term, after which another foreman is appointed for this work. In this way, all the plant foremen, as successive chairmen of the workmen's committee, are rotated and given this experience. In many companies, the foreman who serves as chairman of the workers' safety committee is also a member of a supervisory or management safety committee.

Policies and procedures

When a committee is formed, certain policies and procedures should be set forth in writing, and should cover at least:

1. Scope of committee activity.

2. Extent of committee authority.

3. Procedure:
   a) Time and place of meetings.
   b) Frequency of meetings.
   c) Order of business.
   d) Records to be kept.
   e) Attendance requirements.

Although records of activities should be kept and the procedure to be followed should be put in writing, the paperwork should be kept to a minimum. Too much system is a waste of valuable time and effort. Likewise, when little thought is given to the system, it often leads to confusion and duplication of effort.

The scope of committee activity and the authority of the committee will vary from one company to another and with the type of committee. Much depends also upon the size of the organization, the type of problems faced, and the character of employee relations. Plant central committees, for example, are often the executive or policy-making group.

Effective committee meetings

Good safety meetings require thorough planning and effort. Notices of meetings, preferably accompanied by an agenda, should be sent to each member of the committee.

The frequency of meetings varies, depending upon the type of committee and the program. There should be sufficient items of business for at least one meeting a month. When the interval between meetings is longer than one month, the members tend to lose interest.

Where possible, the meeting place should be comfortable and cheerful. Each person attending the meeting should be provided with a seat and be in a position to see and hear the speakers.

Meetings should be conducted according to the generally accepted rules of order. Formality should not be allowed to overwhelm the meeting and inhibit free and frank discussions.

The following is presented as a suggested order of business that may be adopted for safety committee meetings in general:

1. CALL TO ORDER. The meeting should be called to order promptly at the appointed time.

2. ROLL CALL BY THE SECRETARY. Names of members and others present should be recorded. Members who cannot attend should notify the secretary in advance, and the reasons for absence should be noted in the minutes.

3. INTRODUCTION OF VISITORS.

4. MINUTES of the previous meeting should be read and corrections made. (This item can sometimes be waived.)

5. UNFINISHED BUSINESS. All matters on which definite decisions have not been made are brought up for reconsideration.

6. REVIEW OF ACCIDENTS AND STATISTICS. Classification by cause should be deter-

mined and approved. Responsibility should be determined for every accident, and preventive measures discussed.

7. SAFETY EDUCATION. When it is desired and time permits, the chairman should request a member to speak at the next meeting. The subject to be discussed should be recorded in the minutes. Other programs can be scheduled.

8. INSPECTION AND RECOMMENDATIONS. An inspection of the facility should be made at regular intervals, sometimes by a subcommittee. A record of the inspection time, territory covered, unsafe conditions found, and recommendations made should be included in the minutes. Definite action, not necessarily favorable, should be taken on recommendations and reported to the committee.

9. POSTERS. The chairman should question each member as to the condition of bulletin boards in the jurisdiction of the committee. Posters are useful in obtaining subject matter for meetings.

10. NEW BUSINESS. The chairman should appoint subcommittees to arrange for:
    a) Competition between departments or plants.
    b) Special no-accident weeks or months.
    c) Safety rally programs.
    d) Speakers from outside the plant.
    e) Accident statistics.
    f) Revision of safety rules and shop practices.

11. ADJOURNMENT.

Minutes should be taken, prepared, and circulated by the secretary, after approval by the chairman. The minutes are of great importance since they are often sent to others besides committee members, especially top management. The minutes must record accurately all decisions made and actions taken, since they serve as a means of keeping management informed of the group's work and as a followup.

Motivation of committees

There are many ways of maintaining interest among safety committeemen. To begin with, the men selected to serve on the committee must have the interest of the safety work at heart. Then, to maintain that interest requires constant effort on the part of the chairman, the safety professional, and the central organization. Members of safety committees should be given something to do other than mere reporting of unsafe conditions and practices. They should be given a more active part in the work and made to feel that the success of the committee depends upon the individual support of each member. The primary aim is to arouse interest in the work among all employees to the extent that they will realize that safety is just as important to them as the work they perform.

Among the ways of arousing and maintaining interest among committeemen, none is more effective than supplying members with informative material, such as data from outside sources, frequency of certain classes of accidents, and comparative statements with other months of the year, other years, and other organizations. Pamphlets published by the National Safety Council and information gained through the columns of the *National Safety News* provide helpful information for committee members and may be used to advantage in promoting a safety spirit.

Monthly newsletters, consisting of contributions from members of safety committees, information concerning certain classes of accidents noticeable by their frequency, their causes and methods for overcoming them, and interesting articles taken from other magazines and plant publications, are often used to advantage.

Knowledge in every occupation is the surest basis of success. The value of educating employees in accident prevention lies in their automatically watching out for ways in which men are injured, and coming to realize that the personal causes of accidents are just as serious as the mechanical causes.

There are many other ways of maintaining interest among committeemen, such as well-prepared safety bulletins, safety meetings, proper and prompt handling of suggestions by chairmen, rotation of members, membership cards, safety buttons and emblems, and awards for special accomplishments in connection with the safety program.

More details are in Chapter 12, "Maintaining Interest in Safety."

It is important that management indicate in some way or other its appreciation of the service rendered by safety committee members. The method employed to do this need not involve undue expense or difficulty. The following suggestions are worthy of careful consideration.

1. Letter of appreciation signed by the general superintendent, manager, or an officer of the company.
2. Certificate of service that can be framed and displayed in the worker's home.
3. Card or badge of membership.
4. Expression of appreciation at a general employee group meeting.
5. Special dinner or small gifts for safety committee members.

Carrying out one or more of the above suggestions is one definite way to encourage the members of a safety committee. Other methods should be employed, too, to convince these men—and through them the whole body of workers—that safety is a serious matter, that the management is sincerely interested, that the executives are more than eager to do their part, and that worker cooperation is essential for success.

Suggestion systems

One of the many good methods used to encourage interest, thought, and cooperation for safety among employees may well be the adoption of a suggestion system. The men who operate the machines know the operation and condition of those machines and are often better able to suggest practical improvements in guards and safe practices than are inspectors, committeemen, and others. What is true about machine operators is also true of most of the other workers in a plant. It requires only a little encouragement to secure from them valuable suggestions, the adoption of which will prevent many accidents.

Accident prevention is closely allied to efficiency. Suggestions are valuable not only in the prevention of accidents, but also in lowering the cost of production, improving conditions and methods, and bettering the health and increasing the well-being of employees.

Employees should be encouraged to make suggestions—that will:

1. Decrease the danger of accidents to themselves and their fellow workmen, and reduce risks of damage to equipment and materials.
2. Eliminate fire hazards and increase the effectiveness of fire-extinguishing methods and equipment.
3. Improve the sanitary and health conditions in the work area.

Special committees

Special committees are sometimes created to make inspections. Such a committee aims at stopping accidents before they occur by discovering where they are likely to occur. A persistent, alert, and aggressive inspection committee is a potent weapon against accident losses. It is also an effective device to bring rank-and-file employees actively into the safety work, to train them in accident prevention methods, and to persuade them of the value of such methods.

Depending upon company size, the inspection committee may have from one to five members, with three being optimum. There may be an inspection committee for the whole company, or one for each department, with perhaps a central inspection committee to make general or special inspections.

Since a main contribution of the inspection committee is that of bringing a fresh viewpoint to an old familiar scene, and, therefore, of catching details overlooked by people too close to the job, some companies have departmental inspections by committees from other departments (Fig. 3–6).

Committee membership should rotate at intervals long enough so that each member has time to make a positive contribution and to gain personal experience and interest, and yet short enough so that the opportunity to serve can be passed around. Three months' service is usually the minimum for inspection committee members.

When inspection committees are started on their work, they should understand that their job is essentially a helpful and constructive one. It is important that the foreman accompany the committee on its inspection of his

Fig. 3–6.—Technical personnel can assist in committee activities. Here equipment is checked against blueprint for positive operation of safety devices.

area. The foreman can provide information and help required by the inspection committee and should be kept advised of its actions.

In making inspections, the committee should keep a watchful eye for unsafe practices. When they find these practices, they should report them immediately to the supervisor and thus give him an opportunity to correct them.

Those making inspections should wear the protective equipment required in the areas they enter. On some jobs and in some locations special equipment is needed. Safety hats, acid goggles, rubber gloves and aprons and respirators are examples. If the inspecting team members do not have and cannot get the necessary special protective equipment,

they should stay out of the area until this equipment is provided.

Findings and observations are cleared through the committee report when the inspection has been completed. *Under no circumstances does the committee or its members interfere with the work of employees or with the condition of the department, nor usurp any of the supervisor's authority.*

Off-the-Job Safety Programs

There is a certain amount of confusion as to what off-the-job (OTJ) safety really includes. Essentially, off-the-job safety involves employees, and is a term used by employers to designate that part of their safety program directed to the employee when he is not at work.

The principal aim of off-the-job safety is to get an employee to follow the same safe practices in his outside activities as he uses on the job. Experience indicates, however, that many individuals tend to leave their safety training at the work place when they go home. Therefore, off-the-job safety should not be a separate program, but rather an extension of a company's on-the-job safety program.

One of the basic reasons for a company to become involved in off-the-job safety is manpower. While companies now have a legal responsibility to prevent injuries on the job, they have a moral responsibility to try to prevent injuries away from the job. All injuries are a waste of a valuable resource— people. Injuries and fatalities happen to people who call on customers, make a product, service equipment, keep the books, and do many other jobs involved in running a business.

The other reason for an off-the-job safety program is cost. Operating costs and production schedules are affected as much when employees are injured away from work as when they are injured on the job. (These costs are discussed in detail in Chapter 7, "Accident Investigation, Analysis, and Costs.")

Although accidents occur off the job, a large part of the cost is borne by employers. Some cost is paid directly in the form of wages to absent workers and the cost of hiring and training replacement workers. Some of the cost is hidden. For example, a skilled tool maker, or a salesman injured off the job may not be replaced immediately. Their absences may result in lost sales, late deliveries, and loss of customers.

Some of the cost is hidden even deeper, although it is still very real. As accidents in a community increase, so do insurance costs, taxes, and welfare contributions. Probably no company is fully aware of all the costs that result from off-the-job accidents, and the impact they have on operations and profits. Enough experience has been accumulated, however, to develop a simplified plan for estimating such costs (see Fig. 3–7).

There is nothing special about techniques for promoting off-the-job safety. The same principles and techniques used to put across safety on the job are employed. From a safety standpoint, operating power tools at home is the same as operating the same equipment at work; driving the family car is the same as driving a company vehicle.

A company *does* have to depend more on education and persuasion to get its message across, because once an employee leaves his office, plant, or job site he is on his own and the supervision factor is no longer available. An employee must realize that accidents do not always happen to other people.

As with any other program—whether it be attendance, quality control, waste reduction— management support and guidance is essential. Once management has been shown the seriousness of the problem (through experience and cost records), there should be little problem in obtaining support.

The approach to off-the-job safety should not be negative—"Don't do this" or "Don't participate in that activity"—but positive. That is done by pointing out that any activity can be performed safely if the employee only thinks through the activity first, finds out what hazards are involved, takes instruction if needed, and uses the correct procedure and equipment.

An important element of off-the-job safety programming is to keep the activities seasonal. Generally, employees are more receptive to subjects when they coincide with their normal routine. Examples are: water safety in summer, and home fires in the winter. Of course some subjects, safety belts, falls, poison prevention, to name a few, can be used any time.

Take advantage of national programs, if possible. Activities such as Fire Prevention Week, Poison Prevention Week, and Safe Boating Week get a great deal of national publicity and can be used as a springboard for an activity.

Another important element in off-the-job safety programming is the help available in the community. That is particularly true for recreational activities. Local gun clubs, powerboat squadrons, the Red Cross, YMCA, health, police and fire departments, to name a few, have proven to be of valuable assistance in promoting off-the-job safety. Most of those organizations have materials that can be distributed; some have speakers and films avail-

OTJ Cost Categories

1. Direct

Wages paid to injured workers while off the job.

2. Indirect (disabling)

a. Wage cost due to decreased output of injured worker after he returns to work.

b. Personnel cost of hiring replacement workers.

c. Wage cost of supervisors for time spent in training replacement workers.

d. Wage cost due to lower output of replacement workers during break-in period.

e. Products, materials, tools, etc., spoiled by replacement workers during break-in period.

f. Wage cost of time lost by other workers who were delayed getting started because injured worker was a member of a team, his output was needed, or other workers discussed the accident.

g. Wages paid during time spent by non-injured workers for visiting the injured, attending funerals.

3. Indirect (nondisabling)

This cost arises in connection with the following (many workers will lose some hours from work even though they do not lose a full day at any one time).

a. Wages paid during time lost by workers for doctor or dispensary visits.

b. Wage cost due to decreased output of worker because of his injury.

c. Wage cost of other workers who may be slowed down, either because the injured worker was slow, temporarily absent, or needed help of other workers.

d. Spoilage of product or materials due to less efficient work because of the injury.

4. Insurance

Each company can determine its own insurance cost covering off-the-job accidents. Generally this will be included in some form of health and accident coverage, and the insurance company will be able to state that portion of the premium which is for the accident portion of the policy. While this cost is not as flexible as others listed above, most policies do provide for some form of credit for improved experience.

Note. Every one of these costs will not arise in connection with every off-the-job accident, but each is a potential cost, and during a period of time, many or all of them will arise whether actually identified or not.

Fig. 3–7.

able for company meetings. All can serve as sources of ideas for activities.

A company does not necessarily need actually to conduct activities itself to have an effective off-the-job safety effort. One important way to promote off-the-job safety is to encourage employees to take advantage of swimming and life saving classes, safe boating courses, safe hunter classes, Defensive Driving Courses, first aid courses, and others available in the community.

The same methods used to sell safety on the job can also be used to sell safety off the job—meetings, a company magazine, bulletin board notices, films, displays, and posters. (See Chapter 12, "Maintaining Interest in Safety.")

While some companies conduct elaborate open houses, safety fairs, and other large programs that have good results, others obtain equally good results with rather simple activities. Many companies work off-the-job safety activities right into their regular plant safety programs. One company, for example, selected the month of May to cover lifting and material handling. In-plant problems were covered in departmental meetings and notices. It was also pointed out that plant safety principles and practices for safe lifting of heavy objects could be used at home, while moving furniture and doing other spring housecleaning chores.

There are three benefits a company can realize from expanding its safety program to include off-the-job safety. The first is a reduction in lost production time and operating costs from both on- and off-the-job injuries.

Second, companies have found that efforts in off-the-job safety have produced an increased interest by employees in their on-the-job safety program. The third benefit, often overlooked, is that of better public relations.

The aim of safety education, changing the employee's attitude, is especially true for off-the-job safety. No asset is more important to a company than the employees. They should be protected not only during working hours, but also be given every incentive to be safe off the job.

References

National Association of Suggestion Systems, 435 N. Michigan Ave., Chicago, Ill. 60611.
    Journal (quarterly).

National Safety Council, 425 N. Michigan Ave., Chicago, Ill. 60611.
    Industrial Data Sheets
        Management Policies on Occupational Safety, No. 585.
        Off-the-Job Safety, No. 601.
        Safety Committees, No. 631.
    "Public Employee Safety Guides" (series).

Tainter, Sarah A., and Monro, Kate M. *The Secretary's Handbook.* New York, The Macmillan Co.

Inspection and Control Procedures

Inspection Techniques   71
Inspection checklists

Inspection of Work Areas   71
Periodic inspections . . . Intermittent inspections . . . Continuous inspections or observations . . . Special inspections

Inspection of Work Practices   83
Safety observation plan . . . "Safety sampling" technique . . . Reports

Inspection Personnel   86
Safety professionals and inspectors . . . Company or plant management . . . First-line supervision . . . Engineering and maintenance . . . Employees . . . Safety committees

Inspection Procedures   89
Starting the inspection . . . Taking notes . . . Making the report . . . Handling recommendations . . . Condemning equipment . . . Making night inspections . . . Making photo inspections . . . Checking plans and specifications

Total Loss Control and Damage Control   94
Total loss control . . . Damage control

System Safety   97
What is a system? . . . Accidents and errors degrade performance . . . Analyzing systems . . . Methods of analysis . . . Deviation caused by change . . . Value of systems approach

Critical Incident   100
What the technique can do

References   102

Inspection and control procedures have two basic objectives:

1. Maintaining a safe work environment and controlling the unsafe actions of people, and

2. Maintaining operational profitability.

The ultimate responsibility for this rests with top management. The authority for carrying it out extends down through all levels. Not only are all production management and workers included, but so are all the staff functions of a company—engineering, purchasing, industrial engineering, quality control, personnel, and medical departments.

With this background, let us discuss, in turn, inspection techniques and loss-control procedures. (Product quality inspection is not included in this discussion.)

No attempt is made here to specify the frequency of inspections, the equipment needed for certain kinds, or the exact form and manner for securing the best results. Such problems are properly the business of the individual company or plant and of those who supervise the inspections. A company that develops a program immediately applicable to its own situations will have a better program and will be more interested in it than one which attempts to use a ready-made setup.

Inspection Techniques

Safety inspections, one of the principal means of locating accident causes, help determine what safeguarding is necessary to protect against hazards before accidents and personal injuries occur.

As inspections of the manufacturing process are important functions in quality control, safety inspections are similarly of vital importance in accident control.

Inspections should not be limited to search for unsafe physical conditions, but should also try to detect unsafe practices.

Finding unsafe conditions and work practices by means of inspection and promptly correcting them is one of the best methods for management to prevent accidents and safeguard employees. Management also demonstrates to employees its interest and sincerity in accident prevention. Likewise, failure to find and correct promptly unsafe conditions weakens worker confidence in the employer's sincerity.

Inspections help to sell the safety program to employees. Each time an inspector or an inspection committee passes through the work area, management's interest in safety is advertised. Regular plant inspections encourage individual employees to inspect their immediate work areas.

In addition, inspections enable the personnel of the safety department to come in contact with individual workmen and to enlist their help in eliminating accidents. Frequently, the workmen are able to point out unsafe conditions that may otherwise go unnoticed and uncorrected. When an employee's suggestions are acted upon, he realizes that he has contributed to the safety of his company and that his cooperation is appreciated.

Safety inspections should not be conducted primarily to find how many things are wrong, but rather to determine if everything is satisfactory. Their purpose should be to discover conditions which, if corrected, will bring the plant up to accepted and approved standards, and result in making it a safer and more healthful place in which to work.

Inspectors should tactfully mention any unsafe act observed to the employees involved, pointing out the hazards. The supervisor should be notified promptly.

Inspection checklists

Systematic inspection is the basic tool for maintaining safe conditions and checking unsafe practices. Each company, plant, or department should develop its own checklist. Sample checklists, stressing work areas or work practices or both, are shown in Figs. 4–1 through 4–4.

Inspection of Work Areas

In preparing for an inspection, it is advisable to analyze all accidents (including no-injury accidents and near misses, if possible) for the last several years so that special attention can be given those conditions and those locations known to be accident producers. Tables and charts can be prepared as necessary to show the number of accidents,

SAFETY CHECK LIST

KOPPERS COMPANY INC.—TAR PRODUCTS DIVISION

MONTHLY SAFETY INSPECTION REPORT

PLANT _____

INSPECTED BY _____

DATE _____

NUMERIC RATING VALUES	0=POOR OR DEFICIENT (EXPLAIN)
	3=FAIR OR AVERAGE
	4=GOOD
	5=EXCELLENT

KP 204 2M 10-51

RATING SUMMARY

DEPARTMENT	GOOD HOUSEKEEPING	MATERIAL HANDLING	MATERIAL PILING AND STORAGE	AISLES AND WALKWAYS	MACHINERY AND EQUIPMENT	ELECTRICAL AND WELDING EQUIPMENT	TOOLS	LADDERS AND STAIRS	FLOORS, PLATFORMS AND RAILINGS	EXITS	LIGHTING	VENTILATION	OVERHEAD VALVES	PROTECTIVE CLOTHING AND EQUIPMENT	DUST, FUMES, GASES AND VAPORS	EXPLOSION HAZARDS	UNSAFE PRACTICES	FIRST AID FACILITIES	WASH ROOM AND LOCKER ROOM	DRINKING FOUNTAINS	HAND AND POWER DRIVEN TRUCKS	FIRE FIGHTING EQUIPMENT	VEHICLES	GUARDS AND SAFETY DEVICES	HORSE PLAY	MAINTENANCE	ACT. TOTAL	MAX. POSS.	% OF MAX.
PLANT TOTAL																													

DISTRIBUTION

1 _____ COMPANY SAFETY DIRECTOR
2 _____ DIVISION SAFETY DIRECTOR
3 _____ DIVISION SAFETY ENGINEER
4 _____ INSURANCE SECTION
5 _____ PLANT SAFETY DIRECTOR

SEND FIRST FOUR COPIES ABOVE TO DIV. SAFETY ENGINEER FOR DISTRIBUTION

REMARKS:

NOTE: REPORT HERE ALL UNSAFE ACTS OR CONDITIONS AND REMEDIAL ACTION PLANNED OR TAKEN SINCE LAST REPORT. CONTINUE ON OTHER SIDE IF NECESSARY.

Courtesy Tar Products Division, Koppers Company, Inc.

the departments in which they occurred, the types of accidents, the agencies involved, the nature of the injuries, and, most important, the accident causes.

A checklist of fifteen basic hazardous work areas can be used as a guide:

1. Pinch points
2. Catch points
3. Shear points
4. Squeeze points
5. Flying objects
6. Falling objects
7. "Run-in" points
8. Electricity
9. Gases
10. Heavy objects
11. Chemicals and flammables
12. Hot and cold objects and radiation
13. Sharp and pointed objects
14. Slippery surfaces
15. Trip-fall hazards

The safety professional can find research and reference material in subsequent chapters of this Manual, and in such publications as:

1. *Federal Register*—OSHAct (29 C.F.R. 1900–1950)
2. Building codes
3. National Fire Protection Association
4. Modern building inspection books
5. Guides to building and plant maintenance.
6. Walsh-Healey Public Contracts Act (Safety and Health Standards for Federal Supply contracts, 41 C.F.R. 50)
7. Insurance company's survey—A checklist to determine condition of buildings

Reference should also be made to the accident statistics published by state and federal governments and to *Accident Facts* published

◄ Fig. 4–1.—One company's monthly inspection report form shows plant areas subject to periodic inspection, and provides for rating.

annually by the National Safety Council. These statistics frequently disclose accident causes that may exist in the plant but have not as yet resulted in injuries.

A well planned inspection depends upon knowing where to look and what to look for.

Once the check points have been determined and a systematic procedure worked out, there are four ways that the work area can be inspected. Safety inspections can be classified as follows:

- Periodic inspections
- Intermittent inspections
- Continuous inspections
- Special inspections

Inspection ties in closely with plant maintenance; see Chapter 17 for additional details.

Periodic inspections

Periodic inspections are those scheduled to be made at regular intervals. It is advisable to schedule inspections for the entire plant, for certain operations, or for certain types of equipment. Such inspections may be made monthly, semi-annually, annually, or at other suitable intervals. Some types of equipment, such as elevators, boilers, unfired pressure vessels, cranes, power presses, and fire extinguishing equipment are required by law to be inspected at regular intervals.

All periodic inspections should be well planned so they can be made systematically and efficiently.

General inspections. In many companies or plants, it is an established policy to make a general inspection of the entire premises each year, except for those departments or equipment that are scheduled for more frequent inspections.

A general inspection should cover those places that "no one ever visits" and "where no one ever gets hurt." Many of these out-of-the-way places (where lives could be lost) are located overhead where it is difficult to see a hazard from the shop floor.

Inspections should not be confined to those places where serious injuries have occurred. This is why no-injury accidents and near-accidents should be checked—they will often

SAFETY INSPECTION CHECKLIST

Plant or Department_____*SHIPPING*_____ Date____ *4/1*

This list is intended only as a reminder. Look for other unsafe acts and conditions, and then report them so that corrective action can be taken. Note particularly whether unsafe acts or conditions that have caused accidents have been corrected. Note also whether potential accident causes, marked "X" on previous inspection, have been corrected.

(∨) indicates *Satisfactory*   (X) indicates *Unsatisfactory*

1. FIRE PROTECTION
- Extinguishing equipment................. ☑
- Standpipes, hoses, sprinkler heads and valves .... ☑
- Exits, stairs and signs.................... ☑
- Storage of flammable material .... ☒

2. HOUSEKEEPING
- Aisles, stairs and floors.................. ☑
- Storage and piling of material........... ☑
- Wash and locker rooms.................. ☒
- Light and ventilation.................... ☑
- Disposal of waste .... ☒
- Yards and parking lots.................. ☑

3. TOOLS
- Power tools, wiring .... ☑
- Hand tools............................. ☒
- Use and storage of tools .... ☑

4. PERSONAL PROTECTIVE EQUIPMENT
- Goggles or face shields .... ☒
- Safety shoes........................... ☒
- Gloves .... ☒
- Respirators or gas masks .... ☐
- Protective clothing .... ☐

5. MATERIAL HANDLING EQUIPMENT
- Power trucks, hand trucks .... ☑
- Elevators .... ☐
- Cranes and hoists .... ☐
- Conveyors .... ☐
- Cables, ropes, chains, slings .... ☐

6. BULLETIN BOARDS
- Neat and attractive .... ☑
- Display changed regularly .... ☑
- Well illuminated........................ ☐

7. MACHINERY
- Point of operation guards.............. ☐
- Belts, pulleys, gears, shafts, etc......... ☐
- Oiling, cleaning and adjusting .... ☐
- Maintenance and oil leakage............ ☑

8. PRESSURE EQUIPMENT
- Steam equipment .... ☐
- Air receivers and compressors.......... ☐
- Gas cylinders and hose.... .... ☐

9. UNSAFE PRACTICES
- Excessive speed of vehicles .... ☑
- Improper lifting .... ☒
- Smoking in danger areas .... ☑
- Horseplay .... ☑
- Running in aisles or on stairs........... ☑
- Improper use of air hoses.............. ☐
- Removing machine or other guards...... ☐
- Work on unguarded moving machinery..... ☐

10. FIRST AID
- First aid kits and rooms................ ☑
- Stretchers and fire blankets............. ☐
- Emergency showers .... ☐
- All injuries reported .... ☑

11. MISCELLANEOUS
- Acids and caustics...................... ☐
- New processes, chemicals and solvents.... ☐
- Dusts, vapors, or fumes................ ☑
- Ladders and scaffolds.................. ☑

Signed___*A. M. Baldwin*___

USE REVERSE SIDE FOR DETAILED COMMENTS OR RECOMMENDATIONS

Fig. 4–2.—Detailed inspection checklist helps inspector find unsafe acts and hazards before they can trigger an accident.

SUMMARY OF UNSAFE PRACTICES						

Area or Division _____ Force _____

District _____ Period Covered _____

	GROUP (FORCE, DISTRICT, DIVISION OR AREA)					TOTAL
1	SUPERVISORS REPORTING UNSAFE PRACTICES					
2	SUPERVISORS NOT REPORTING UNSAFE PRACTICES					
3	TOTAL NO. OF SUPERVISORS					
4	TOTAL NO. OF EMPLOYEES IN GROUP					
5	NUMBER OF UNSAFE PRACTICES REPORTED					

CAUSE OF UNSAFE PRACTICES

6	SUPERVISION	LACK OF ANALYZING, OR PLANNING THE WORK					
7		INADEQUATE BASIC TRAINING					
8		LACK OF DEFINITE OR SPECIFIC INSTRUCTIONS					
9		IMPROPER ASSIGNMENT OF EMPLOYEE					
10		FAILURE TO SEE THAT INSTRUCTIONS WERE FOLLOWED					
11		OTHER					
12	EMPLOYEE	LACK OF ANALYZING, OR PLANNING THE WORK					
13		DISREGARD FOR KNOWN SAFE PRACTICES					
14		LACK OF EXPERIENCE					
15		ABSTRACTION OR FORGETFULNESS					
16		HASTE					
17		OTHER					
18	TOTAL						

UNSAFE PRACTICES

19	MOTOR VEHICLES — OPERATION AND MAINTENANCE					
20	POLES — WORKING ALOFT					
21	ACTION. BOTH IN AND OUT OF BUILDINGS THAT MIGHT RESULT IN SLIPS OR FALLS					
22	LADDERS — EXTENSION C.O. AND STEPLADDERS					
23	BODY BELTS AND SAFETY STRAPS					
24	CLIMBERS, PADS, AND STRAPS					
25	GUARDING EMPLOYEES AND PUBLIC					
26	TOOLS AND MATERIALS					
27	GOGGLES					
28	USE OF RUBBER GLOVES AND OTHER PROTECTIVE DEVICES AND PRECAUTIONS TAKEN AROUND LIVE WIRES					
29	MANHOLES, CONDUIT AND EXCAVATIONS					
30	FIRST AID FOR AND CARE OF INJURIES					
31	MISCELLANEOUS					
32	TOTAL					

(Note: Each organization should substitute in this section the types of unsafe practices which are most likely to be encountered in its own operations.)

RECOMMENDATIONS _____

Observed by _____

Title _____

FIG. 4–3a.—Form for summarizing unsafe practices. Reverse side shown on next page.

point to causes of future injuries.

General inspections are particularly valuable before reopening a plant after a long shutdown or previous to a safety campaign or contest. However, they should not take the place of other types of inspections. A general inspection should not signal that all responsibility has ended for the continual

UNSAFE PRACTICES OBSERVED

(See Note under "UNSAFE PRACTICES" on other side of page.)

19. MOTOR VEHICLES - OPERATION AND MAINTENANCE	REFERENCE NUMBER	
IMPROPER PARKING		
FAILURE TO CONFORM TO TRAFFIC LAWS		
STEPPED OFF OR ON VEHICLE IN MOTION		
BACKING WITHOUT TAKING PROPER PRECAUTIONS		
POOR HOUSEKEEPING ON TRUCK OR CAR		
LIGHTS, BRAKES, HORN, ETC. NOT TESTED		
UNSAFE HANDLING OF DERRICK		
UNDER DERRICK IN OPERATION		
HANDS ON OR NEAR WINCH LINE SHEAVE OR DRUM		
UNSAFE HANDLING AND USE OF TRAILER		
UNSAFE PERFORMANCE OF MAINTENANCE OPERATIONS		
20. POLES - WORKING ALOFT		
FAILURE TO TEST OR MAKE SAFE BEFORE CLIMBING		
FAILURE TO COMPENSATE FOR UNBALANCED LOAD		
WORKING IN DANGEROUS POSITION		
LACK OF CARE IN CLIMBING		
STANDING UNDER WORKMAN ALOFT		
LACK OF CARE-APPROACHING OR LEAVING POLE		
21. ACTION BOTH IN AND OUT OF BUILDINGS, THAT MIGHT RESULT IN SLIPS OR FALLS		
UNNECESSARY RUNNING, STAIRS AND ACROSS FLOORS		
LACK OF ORDINARY CARE OUTSIDE BUILDINGS		
SITTING ON TILTED CHAIRS		
STANDING ON CHAIRS, BOXES, CANS, ETC.		
22. LADDERS - EXTENSION, C. O. AND STEP LADDERS		
FAILURE TO MAKE SECURE (FOOTING LASHING, HOLDING)		
FAILURE TO PLACE LEG THRU LADDER OR USE SAFETY		
PULL ROPE NOT TIED		
DEFECTIVE LADDER, SPURS NOT TURNED, ETC.		
OVERREACHING TOO HIGH UP ON LADDER, ETC.		
IMPROPER ANGLE FOR CLIMBING		
FAILURE TO INSPECT		
UNGUARDED AT HAZARDOUS LOCATION		
FAILURE TO USE WHERE REQUIRED OR USED WRONG KIND		
NOT LASHED TO MOTOR VEHICLE OR CARRIED PROPERLY		
FAILURE TO EXTEND SPREADERS OF STEPLADDER		
TOOLS OR MATERIAL LEFT ON STEPS OR TOP		
SHOVED C.O. LADDER ENDANGERING PERSON ON LADDER		
23. BODY BELTS AND SAFETY STRAPS		
FAILURE TO USE WHILE ALOFT (ON POLE, PLATFORM, ETC.)		
FAILURE TO LOOK, FEEL AND KNOW SNAPHOOK IS SECURE		
NOT INSPECTED OR PROPERLY MAINTAINED		
PUSHING KEEPER AGAINST OBJECTS		
24. CLIMBERS, PADS AND STRAPS		
WEARING IN TREES, VEHICLES, ON LADDER, GROUND, ETC.		
FAILURE TO INSPECT AND MAINTAIN PROPERLY		
UNSAFE CLIMBING HABITS		
25. GUARDING EMPLOYEES AND PUBLIC		
FAILURE TO USE ADEQUATE WARNING SIGNS OR FLAGS		
KEEPING CHILDREN AND OTHERS AWAY FROM OPERATION		
GUARDING PUBLIC AT DANGEROUS LOCATION		

26. TOOLS AND MATERIALS	REFERENCE NUMBER	
IMPROPER USE OF TOOLS AND MATERIALS		
USE OF DEFECTIVE OR UNAUTHORIZED TOOLS		
TOSSING TOOLS AND MATERIALS		
DROPPING TOOLS AND MATERIALS FROM ALOFT		
UNSAFE LIFTING OR HANDLING		
UNSAFE CARRYING OF TOOLS, NAILS, TACKS, ETC.		
IMPROPERLY STORED OR PLACED		
UNSAFE USE OF CABLE CAR		
HANDLINE DANGLING		
TOOLS AND MATERIALS LEFT LYING AROUND		
FAILURE TO USE TREE SLING		
UNSAFE CUTTING OF WIRE - FLYING ENDS OR PIECES		
NOT USING FLASHLIGHT WHERE REQUIRED		
27. GOGGLES - FAILURE TO USE WHEN:		
DRILLING CONCRETE BRICK OR OTHER MASONRY		
GRINDING, CHIPPING, HANDLING BRUSH, ETC.		
28. USE OF RUBBER GLOVES ETC. AND PRECAUTIONS TAKEN AROUND LIVE WIRES		
FAILURE TO USE RUBBER GLOVES WHEN REQUIRED		
FAILURE TO TEST AND MAINTAIN PROPERLY		
WORKING TOO CLOSE, LIVE WIRES WHEN ALOFT		
LACK OF PRECAUTIONS TO PREVENT WIRE FROM	X X X	X X X
FLIPPING UP INTO LIVE WIRES		
SAGGING OR FALLING ONTO LIVE WIRES		
REEL TENDER NOT PROTECTED		
UNSAFE USE OF CHAIN HOIST, TENT, STEEL TAPE, ETC.		
INCOMPLETE SURVEY AFTER SUSPECTED E. L. CONTACT		
CARELESSNESS AROUND ELECTRIC CIRCUITS AND EQUIP.		
FAILURE TO USE ADEQUATE PROTECTIVE DEVICES		
29. MANHOLES, CONDUIT AND EXCAVATIONS		
FAILURE, PROPER TESTS FOR GAS OR OXYGEN DEFICIENCY		
INSUFFICIENT VENTILATION (WITH SAIL OR BLOWER)		
ENTERED MANHOLE WITHOUT MANHOLE GUARD		
FAILURE TO USE LADDER IN MANHOLE OR EXCAVATION		
INSUFFICIENT GUARDING, MANHOLES OR EXCAVATIONS		
UNSAFE REMOVAL AND HANDLING OF COVER		
SMOKING OR OPEN FLAMES NEAR OR IN MANHOLES		
30. FIRST AID FOR AND CARE OF INJURIES		
FAILURE TO GIVE PROPER FIRST AID		
DID NOT CONTINUE PROPER CARE OF INJURY		
FAILURE TO REPORT INJURY		
FIRST AID KIT NOT PROPERLY MAINTAINED		
31. MISCELLANEOUS		
HORSEPLAY ON-THE-JOB		
DEBRIS LEFT LYING AROUND		
UNSAFE POSITION ON ST., HIGHWAY, RD., R. R. TRACK, ETC.		
HANDLING OF HOT SOLDER, PARAFFIN, ETC.		
POISON IVY OR OAK — LACK OF PROTECTION OR CARE		
CLOTHING AND SHOES — POOR, INSUFFICIENT		
TREES OR BRUSH — CUTTING AND HANDLING		

Fig. 4–3b.—Reverse side of previous form. Unsafe practices peculiar to the specific industry in which safety professional is working should be substituted, as required.

inspection throughout the year, nor should known hazards be neglected until such general inspections are made.

Inspecting buildings and physical plant should be conducted yearly.

1. Do the physical conditions noted contain potential accident hazards?

2. What practical measures can and should be taken to remove such hazards or failure points?

3. Is there any imminent danger hazard present that warrants immediate corrective measures?

An adequate checklist will call attention to the following items and conditions:*

● GROUNDS—Parking lots, roadways, and sidewalks need frequent inspection for cracks, holes, breaks and tripping and falling hazards.

● LOADING AND SHIPPING PLATFORMS—Loading and shipping platforms and docks get severe use from trucks and heavy traffic. Bumpers are needed to prevent damage from tail ends of trucks.

● OUTSIDE STRUCTURES—Small isolated buildings should be inspected the same way as the larger buildings. Fencing should be inspected for damage by tanks, cars, and other causes as well as corrosion and need for painting.

● RAILROAD TRACKS—Railroad sidings that are company-owned require regular inspection for washouts, condition of rails, clearance, switches, car stops, missing joint plates and spikes, rotted ties and the need for additional ballast.

● FLOORS—Floors, regardless of construction, should be carefully inspected especially in areas subject to heavy traffic. Slipperiness of floors should receive special study and floor treatment.

1. Are the surfaces wearing too rapidly?
2. Is shrinkage present?
3. Is the surface damaged or worn?
4. Are there slippery surfaces?
5. Are there hazards to trucking or walking, such as holes or unguarded openings?
6. Are there indications of cracks, sagging, or warping?
7. Are replacements necessary because of deterioration, etc.?

● STAIRWAYS—Stairways should be checked to determine:

1. Are treads and risers in good condition and of uniform width and height?

HAZARD SPOT CARD

WORK AREA

Footing: uneven ☐ obstructed ☐ slippery ☐
Cramped quarters ☐
Exposure to traffic ☐
Insecure piles or overhead material ☐
Inadequate illumination ☐ glare ☐
Temperature: too hot ☐ too cold ☐
Exposure to gases ☐ dust ☐ fumes ☐
Hazards from nearby operations ☐

MACHINERY

Point of operation: cutting ☐ shearing ☐
punching ☐ abrading ☐ flying material ☐
Power transmission: shafts ☐ belts ☐
gears ☐ pulleys ☐ electrical conductors ☐
Unsafe starting and stopping mechanisms ☐

TOOLS

Wrong tool for the job ☐
Tool in unsafe condition ☐
Tool placed in unsafe position ☐

HANDLING MATERIAL

Material or objects: heavy ☐ unwieldy ☐
rough ☐ sharp ☐ hot ☐ corrosive ☐
Unsafe handling equipment: trucks ☐
conveyors ☐ hoists ☐ containers ☐

FIG. 4-4.—Spot cards remind inspectors that a safety committee or inspection team has spotted a hazard that needs correction.

2. Are standard handrails provided and are they in good condition and secure?

3. Is lighting on stairs satisfactory?

4. Is there any material stored on stairs?

● HOUSEKEEPING—The general housekeeping throughout the plant should be checked to see if it is satisfactory. Aisles should be marked off with painted lines and material kept out of aisle width.

● ELECTRICAL INSTALLATIONS—Every year an average of 1000 persons are fatally shocked by accidents involving less than 600 volts.

● WIRING—The physical plant survey should include a periodic inspection of all plant wiring.

---

*Robert L. Moore, 1963 Congress Transactions. (See References.)

4—Inspection and Control Procedures

Fig. 4–5.—Detection of unsafe conditions and unsafe acts should not depend only on formalized inspections. Any observed hazard should be corrected.

● Transformers-switchboards — The survey should include an examination of transformer and switchboard installations.

● Elevators — The efficient and safe operation of plant elevators is an important physical survey determination.

● Roofs — A specialized item in the survey of the physical plant is roof condition.

● Factory chimney stacks — A thorough periodic annual inspection of chimneys and stacks by a competent person will uncover minor damages before they become critical.

● Imminent danger hazards should receive special consideration — The following five causes can develop into a catastrophe.

Foundation failure
Structural deterioration

Overloading
Alterations
Fire and explosion

● Overloading of floors — Experience has shown that the most serious floor loading problem encountered arises from the tendency of plant management to look upon warehouse or storage space as an unproductive evil.

See additional details in Chapters 16 and 17 in this Manual.

Fire inspection. One of the greatest hazards is fire. Consequently, a rigid system should be set up for periodic inspection of all types of fire protective equipment. Such inspections should include water tanks, sprinkler systems, standpipes, hose, fire plugs, extinguishers, and all other equipment used for fire protection. The schedule of inspections should be closely followed and an accurate record kept of each piece of equipment inspected and tested (Fig. 4–6).

Along with this scheduled inspection, a careful survey should be made of new equipment needed. Recommendations should be made for replacement of defective and obsolete equipment, as well as the purchase of any additional equipment. As new processes and products are added to the manufacturing system, new fire hazards may be introduced that require individual treatment and possibly special extinguishing devices. Be sure to follow through.

Surveys should also include all means of egress from the building. All exits, stairs, fire towers, fire escapes, halls, fire alarm systems, emergency lighting systems, and places seldom used should be thoroughly inspected to determine their adequacy and readiness for emergency use. More details in Chapter 43, "Fire Protection."

Elevators and pressure vessels. Another type of periodic inspections is that required by state and local laws include elevators, boilers, and unfired pressure vessels. Such equipment, however, is not usually inspected by company employees, but rather by the insurance carrier. A prearranged schedule of inspections as required by law should be followed. (See Chapters 28 and 44.)

BUTLER MANUFACTURING COMPANY
WEEKLY INSPECTION OF FIRE PROTECTIVE EQUIPMENT
Kansas City Plant

Instructions: Fill out this blank while making inspection. Do not report a valve open unless you personally have inspected and tested it. Every valve controlling sprinklers or
water supplies to sprinklers should be listed. When the blank is filled out, it should be
sent to the Safety Department.

SPRINKLER VALVES

Valve No.	AREA CONTROLLED	Location	Open	Shut	Sealed	Pressure
1	Entire West System	Bldg. 57				
2	Bldg. 43-43B	Bldg. 57				
3	Valves 4-5-6-7-8-9-10-11	Bldg. 43				PIV
4	East End Bldg. 2-2B	Bldg. 2				
5	Center 2-2B, Paint Line	Bldg. 2				
6	West End Bldg. 2	Bldg. 2				
7	Valves 8-9-10-11	Bldg. 2				PIV
8	Valves 9-10-11	Bldg. 62				
9	Bldg. 3	Bldg. 62				
10	Bldg. 62-62B-6A-5C-5	Bldg. 62				
11	Bldg. 4-4A-5-5A-5C	Bldg. 62				
12	Paint Booth P34KC	Bldg. 1				
13**	Oven	Bldg. 1				
14	Paint Booth P57KC	Bldg. 1				
15	Paint Booth P58KC	Bldg. 1				
16	Paint Booth P59KC	Bldg. 1				
17	Paint Booth P56KC	Bldg. 1				
18	Locker Room Offices	Bldg. 67				
19	Laboratory Paint Room	Bldg. 58A				
20	Paint Shop Bldg. 23	Bldg. 23				
21	Bldg. 65	Bldg. 65				
22	Paint Booth P49KC	Bldg. 63				
23	Paint Booth P710G	Bldg. 63				
24	Bldg. 17	Bldg. 17				
25	Bldgs. 53-63	Bldg. 17				
26*	Bldg. 57	Bldg. 57				
27	Bldg. 17A Offices	Bldg. 15				
28	Bldg. 17A Balcony	Bldg. 15				
29	Entire East System	12th St.				PIV
30*	Bldg. 52	Driveway				PIV
31*	Bldg. 46	Driveway				PIV
32	Valves 21-22-23-24-25	Driveway				PIV

* Controls Dry System
**Manually Operated, always shut

GENERAL CONDITIONS

HYDRANTS: In good condition?_____

Clear?_____ Remarks_____

_____

AUTOMATIC SPRINKLERS: Any heads miss-

ing?_____Disconnected?_____

Obstructed by high-piled stock?_____

Any rooms not sufficiently heated

to prevent freezing? _____ How

many extra heads available? _____

SPRINKLER ALARMS: Tested?_____ In

good condition?_____Do not test

hydraulic alarms when temperatures

are below freezing.

EXTINGUISHERS, SMALL HOSE: In good

condition? _____

FIRE DOORS: All inspected?_____

In good order?_____

HOUSEKEEPING: Good throughout?_____

Combustible waste removed before

night? _____

REMARKS on other matters relating to

fire hazard: _____

_____

_____

Date _____ Signed _____

FIG. 4-6.—Report form covers condition and accessibility of fire protection equipment.

Courtesy Butler Manufacturing Company.

4—Inspection and Control Procedures

The proper persons should be notified well in advance of such inspections so arrangements can be made to take the equipment out of service until the inspection is completed. It is wise to have supplementary inspections made more frequently by a qualified company employee.

Chains and ropes. Chains, wire and fiber ropes, and other equipment subject to severe strain in handling heavy materials should be inspected at regular intervals. A record should be kept of each inspection. OSHA regulations require such inspection and record-keeping. This type of equipment should be stenciled or marked to show the date and result of the latest inspection. (See Chapter 23, "Ropes, Chains, and Slings.")

Other periodic inspections. Floors and flooring should be regularly checked. At the same time, attention should be paid to the conditions causing falls, the second largest cause of accidental death. Fall-causing hazards include: slippery, wet, oily, and worn floors; ice and snow on walks and platforms; tripping hazards; loose material underfoot; worn or broken treads on stairs; insecure scaffolds and platforms; stairs, scaffolds, and platforms with no handrails; defective ladders or ladders not suited to the job; open elevator shaftways; and unguarded floor openings and manholes.

Many types of equipment and processes require periodic inspections if they are to be operated safely and efficiently. Some companies require that all tools and appliances be returned to a central storeroom after use each day, where they are carefully checked and repaired before being reused.

To maintain a schedule of inspections, some companies give all portable electric hand tools a visual or external inspection each time a tool is returned to the storeroom, and a thorough "knockdown" inspection every few weeks.

One company requires that all portable electric tools and extension cords be sent to the electrical department between the first and tenth of each month. The electrical department inspects the tools and makes necessary repairs, and attaches a colored tag to the tool or cord showing the month the equip-ment was last inspected. A different colored tag is used for each month. Any tool or cord found without the proper tag is sent to the electrical department at once.

Other types of equipment, such as cranes, hoists, presses, ladders, and power trucks require periodic inspection (Fig. 4–7). Equipment used in the field, such as in the construction and public utility industries, also requires frequent and periodic inspections. Such inspections should be determined by the proper plant executives. The safety director should prepare a working schedule so that the correct intervals of inspection can be maintained. Operators of all powered materials-handling equipment should be required to make a daily inspection of their equipment.

It is desirable to have periodic inspections by some definitely organized group or an individual of every department. These should be made preferable at not less than one-month intervals. The report of these inspections should be made up by departments for the superintendent of the division, or in small plants, for the plant manager. In some plants, foremen are required to make such inspections and reports.

Intermittent inspections

The inspection made at irregular intervals is very common and is the type of inspection found in most plants. It may include an unannounced inspection of a particular department, piece of equipment, or small work area. Such inspections made by the safety department tend to keep the supervisory staff alert to find and correct unsafe conditions before being found by the safety inspector.

The need for intermittent inspections is frequently indicated by accident tabulations and analysis. If the analysis shows an unusual number of accidents for a particular department or location, or an increase in certain types of injuries, inspections should be made to determine the reasons for the increase and what is necessary to make corrections.

Intermittent inspections may be made not only by the safety department, but also by supervisors, safety committees, and workmen.

Continuous inspections or observations

Many companies have set up a system of continuous inspection so that selected em-

PITTSBURGH ROLLS CORPORATION
CRANE INSPECTION REPORT

Crane No.............................Type............................... Capacity....................................

RUNWAY AND CONDUCTORS

Track Alignment.....................Spread...................... Fastenings ....

Line Conductors..................................Conductor Supports................................

TRUCKS AND MAIN COLLECTORS

Truck Wheels, Flat Spots?.....................Flanges.........................End Play...............

Axle Bearings...............................Lubrication ....

Truck Drive Bearings.................Lubrication.................... Gears ....

Gear Screws................Pinion.......................Key................ Collectors ....

GIRDERS AND DRIVE

Drive Shaft................Couplings..............Bearings.......... Lubrication ....

Foot Brake Shaft.........Couplings..............Bearings.......... Lubrication ....

Bridge Brake Case...................Adjustment..................... Lubrication ....

Walkway.......................... Railing .... Ladder ....

Bridge Motor Support..................Shaft Extension................... Couplings ....

Bridge Drive Gear Case...........................Gears........... Lubrication ....

MOTORS

Location	Armature	Commutator	Brushes	Brush Holders	Bearings	Lubrication
Bridge						
Hoist						
Aux. Hoist						
Trolley						

CONTROLLERS

	Brushes	Brush Holders	Contacts	Wiring	Springs	Resistance
Bridge						
Hoist						
Aux. Hoist						
Trolley						

Trolley Wheels...................Trolley Wheel Bearings..................... Lubrication ....

Trolley Gear Case...............Case Support..............Gears........... Lubrication ....

Hoist Gear Case............Main Gear Train............Comp. Train..........Lubrication............

Mech. Brake............Drift............Elec. Brake............ Adjustment ....Lubrication............

Drum.............Cable or Chain................Cable Pin................Limit Switch.................

Limit Switch Adjustment...................Hook............ Sheaves ....Lubrication............

Trolley Conductors...................Trolley Collectors..................Cage Roof.................Door..............

Windows.............Foot Brake Treadle.................Cont. Levers............Load Test.............

Bell or Signal...

Inspected by.. Date...........................

KEY WORDS—G=Good; F=Fair; W=Worn; A=Need Attention;
C=Need Cleaning; T=Too Tight

Fig. 4–7.—An example of a detailed rating form for cranes. Forms can also be devised for other types of equipment, such as presses, ladders, and power trucks.

Courtesy Pittsburgh Rolls Corporation.

ployees spend all their time observing certain equipment and operations. Maintenance men, electricians and others, whose job it is to keep equipment in good mechanical condition, are assigned to specific groups of machines. They have no other duties than to roam about the department, continuously observing operations and making adjustments and minor repairs.

These men are responsible also for the safe operation of assigned machines and help train employees in safe practices. This method of inspection makes these men familiar with the exact condition of each part of the machines covered. It enables them to point out weaknesses and defects long before they become serious or of a hazardous nature.

Some companies have roving troubleshooters selected from employees at regular intervals. This job is rotated so that in the course of time each employee has had an opportunity to inspect the entire operation or his particular department. The job is rotated on the theory that each employee will make a definite contribution to the safety of the entire plant and to the safety education of fellow employees, as well as himself.

In one case, the roving inspectors are given a free hand to investigate all operations and to report their findings to the foremen. Sometimes the continuous inspection is limited to key men in the department who have had long experience and a background of accident prevention work.

A continuous system of inspection of personal protective equipment is especially desirable. A constant check on goggles, respirators, safety shoes, gloves, and other protective clothing, as well as any other personal protective devices, will ensure safe maintenance of this equipment.

Systematic inspection by supervisors. Supervisors should continuously make sure that tools, machines, and other department equipment are maintained properly and are safe to use. To do this effectively, they use systematic inspection procedures, and may delegate authority to others in a department.

Toolroom employees should inspect all hand tools to see that they are kept in safe condition. Some companies require portable electric tools to be turned in to the electrical department for monthly checkups.

Inspection programs should be set up for new equipment, material, and processes. Nothing should be put into regular operation until it has been checked for hazards, its operation studied, additional safeguards installed (if necessary), and safety instructions or procedures developed. Serious injury has occurred because this routine was not followed. This is also a good time to train employees in safe operation. It takes less time and effort now than if done later.

Crew leaders are often given the responsibility for inspecting equipment and for seeing that their men observe safe practices. The supervisor should make certain that all inspections by persons other than himself are up to his standards.

By systematic planning and organization, the supervisor can carry out his primary function—supervision. He spot checks periodically to make sure assignments are being carried out, that safety precautions are being observed, and that equipment is running efficiently and safely.

Special inspections

Special inspections are sometimes necessary because of the installation of new equipment or new processes, the construction of new buildings or remodeling of old ones, or because of new hazards. Also, special inspections may be made during special campaigns, such as fire prevention week or waste elimination campaigns.

Accident investigation requires special inspections by the investigation committee and the safety professional. These should be made with the same thoroughness and determination to control accident causes, as are periodic inspections.

Health surveys. Wherever there is a suspected health hazard, a special inspection should be made to determine the extent of the hazard and the precautions or mechanical safeguarding needed to provide and maintain safe conditions. These inspections usually require air sampling for the presence of toxic fumes, gases and dusts, testing of materials for toxic properties, or the testing of ventilation and exhaust systems for proper operation.

Also, physical examinations should be made

of employees exposed to occupational disease hazards. This is a form of special inspection conducted by the medical staff as a disease-control measure.

Testing for toxic substances and physical examinations require special equipment not always available to the plant inspector and medical staff. In such cases, assistance can frequently be secured from the industrial hygiene division of the state department of labor or health, and from industrial hygienists employed by insurance companies.

In connection with the health survey, industrial sanitation and food service facilities should also be inspected with special attention given to personal service facilities. This is particularly important where there is exposure to certain toxic substances, such as lead, and where personal cleanliness is a large factor in controlling occupational disease. (See Chapter 20.)

Overhead inspections. Special inspections for overhead hazards are very important in the control of accident causes. Hazardous conditions often exist because of loose objects that may fall from building structures, cranes, roofs, and other overhead locations.

During such inspections, it is not uncommon to find loose tools, bolts, pipe lines, shafting, pieces of lumber, windows, electrical fixtures, and other objects that are potential accident-makers. Overhead inspections frequently disclose the need for repairs to skylights, windows, cranes, roofs, and other installations that affect the safety not only of employees but of the physical plant as well.

Also, the existence of overhead devices, which require frequent attention, adjustment, cleaning, oiling, and repairing, demands that employees occasionally check such locations. Inspections are necessary to determine that all reasonable safeguards are provided and safe practices observed.

Every overhead job, including inspections at elevated locations, should be carefully considered and the safe means of doing the job provided. Suitable staging, safety belts, and lifelines should be provided.

Accident investigations. It is obvious that every accident and near-accident that occurs should be thoroughly investigated as soon as possible to find its actual and contributing causes in order to prevent a recurrence. During the investigation, a special inspection of the scene of the accident is essential. The inspection should be made by persons who are painstaking and safety-minded. A photograph of the scene may be helpful for future reference. A full discussion of accident investigation is in Chapter 7.

Other special inspections. Numerous other types of special inspections are necessary from time to time, including inspections of hand tools, scaffolds, personal protective equipment, point-of-operation guards, lighting facilities, general ventilation equipment, excavations, and construction work. Such inspections may be made on the request of supervisors, groups of workmen, or because of special need indicated by increasing accident trends.

Inspection of Work Practices

The safety department should assist supervisors in instructing employees in the safe way of doing each job. But accidents can be prevented only when these recommended procedures are based upon a thorough analysis of the job and when these procedures are followed without exception. Hence, to prevent accidents, continuous observation is necessary to ensure that the task is done in the safest way, and that workers comply with all procedures.

The safety department can provide most effective assistance to the supervisors by observing all workers in the performance of their tasks to eliminate unsafe procedures. In this way, foremen and supervisors are apprised of all weaknesses in the safety program within their domain and can have a common reference point for monitoring these occurrences. (See preceding section Continuous Inspections or Observations.)

These are some of the unsafe work practices to look for when inspecting.

1. Do employees operate machinery, or use tools, appliances, or other equipment without authority?

2. Are they working or operating at unsafe speeds?

3. Have guards been removed, or have guards or other safety devices been rendered ineffective?

4. Do people use defective tools or equipment; or use tools or equipment in unsafe ways; or use hands or body instead of tools?

5. Do they overload, or crowd, or arrange, or handle objects or materials unsafely?

6. Do people stand or work under suspended loads, open hatches, or shafts, or scaffolds; or ride loads; or get on or off equipment or vehicles in motion; or walk on railroad tracks, or cross car tracks or vehicular thoroughfares except at crossings?

7. Do they repair or adjust equipment in motion, under pressure, electrically charged, or containing dangerous substances?

8. Does anyone or any thing distract the attention of, or startle, workers?

9. Failure to use, or use of inadequate, badly adapted, or wrongly chosen, personal protective equipment or safety devices.

10. Poor housekeeping and failure to remedy unsanitary or unhealthful conditions.

11. Horseplay.

The most common method in use today is the "safety inspection." However, there is another method considered by many safety experts to be superior to this. It is called "safety observation." This new method is the key to a modern technique known as "safety sampling." Both are discussed in the following paragraphs.

Safety observation plan

A safety inspection of work practices requires the active participation of all supervisory personnel. Basically, this inspection is a safety observation plan by which supervisors at the various organizational levels, and particularly the first-line supervisors directly responsible for employee safety, observe work while it is in progress. These observations cover the use of tools, material, and equipment, as well as any unsafe method or order of procedure in performing an act which indicates a lack of planning or failure to take into account all the circumstances surrounding the particular job.

Safety observations serve to: [*]

1. Check adequacy of training programs

2. Promote on-the-spot correction

3. Provide opportunities to compliment

4. Develop proper safety attitudes

5. Help supervisors learn about workers, and to identify problem workers

6. Sometimes suggest better job methods

The kinds of employees who warrant special observations are:

1. Inexperienced men

2. Accident repeaters

3. Chronic unsafe workers

In fact, all employees, even experienced workers, should be checked periodically.

One reason why safety observation has not been more fully utilized is that planned safety observation involves a little more effort than the incidental safety observation.

There are several factors necessary for effective safety observation. The inspector must:

1. BE SELECTIVE. An inspector might look over a department first for safety, second for improvement of operations, third for training needs, and so on.

2. KNOW WHAT TO LOOK FOR. The more a supervisor or safety professional knows about a job and a worker's responsibilities, the better an observer he will be.

3. PRACTICE OBSERVING. The more often a person looks with the conscious intention to observe, the more he will see at each fresh trial. Like all skills, observation improves with practice.

4. KEEP AN OPEN MIND. One way to increase open-mindedness is not to judge facts in advance. The inspector must not deny the

------

[*]A. T. Waad, *1963 Congress Transactions*.

fact, no matter what conclusion it may seem to lead to. The inspector must keep his mind open, at least until he has all the facts.

5. Do not be satisfied with general impressions. A clean shop, or a careful routine, may still contain hidden hazards.

6. Guard against habit and familiarity. Asking the questions What, Where, When, How, and (especially) Why will often help uncover the real meaning of the situation.

7. Record observations systematically. All notes should be dated, with space for comment on action taken and on results of the action. The notebook can serve both as a reminder and as a record of progress.

8. Prepare a checklist. A systematic check for litter, obstructions, handling of flammables, condition of fire fighting equipment, and so on, will uncover tangible problems that can be corrected. (Making a checklist was discussed earlier in this chapter.)

Here are a few pointers that might help.

1. Be firm, but friendly

2. Explain what and why and how

3. Review the safe alternative

4. Make your contact private

5. Get his reasons for acting unsafely

6. Get agreement on future practice

It is sometimes necessary to watch men at work, not casually in passing, but closely. This can be unnerving to the men under observation—and is one of the reasons for having a small inspection team. When it is necessary to observe a man closely, a member of the team should talk to him and explain what is going on, then ask his permission to watch him as he works. Usually, the man will continue his work naturally after such an introduction. If, however, he is nervous or uncooperative, it is better to postpone the observation.

When a person is observed performing an unsafe practice or act, it should be called to his attention and the correct method fully explained.

Scheduling inspections. In large plants, where constant observation by supervisors or members of the safety team is impossible, supervisors should set up an observation schedule that will permit observation of the maximum number of employees during a month. In addition, no opportunity should be overlooked to make routine or unscheduled observations whenever possible. Unsafe practices may occur at any time, and supervisors should record them whenever they appear.

"Safety sampling" technique

One comprehensive way of determining unsafe practices is a "safety sampling" technique, outlined by Vincent Pollina (see "References").

This technique is based on the random sampling application of probability theory, which is used in production control. As a result of applying the "safety sampling" technique to accident prevention, management is provided with a statement of how safe a plant is operating; where trouble spots are (by areas and supervisors); and the types and frequency of infractions.

As indicated by the title, "safety sampling" requires that a statistical sample be determined. The number of observations required to implement the "safety sampling" technique is based on a preliminary survey and the degree of accuracy desired. In the preliminary survey, a member of the safety department should walk rapidly through the entire plant being studied and observe all employees *who are working.*

It is important in random sampling that only instantaneous impressions be recorded and that no attempt be made to correct an unsafe act during compiling time. During this preliminary tour, two types of impressions are recorded: (*a*) count each worker who is performing his task safely and (*b*) count each worker who is performing his task unsafely. It is important that each unsafe operation be recorded; therefore, one worker's operations may be recorded several times. Each unsafe task is categorized by type of infraction; thus the analysis also indicates the types and locations of unsafe operations.

From these preliminary data, the percentage of unsafe operations is determined by dividing the number of unsafe operations by the total number of observations (both safe and unsafe). The percentage of unsafe operations (P) is then used in the formula

$$N = \frac{4(1 - P)}{Y^2(P)}$$

to calculate the number of observations (N) required to obtain a desired accuracy (Y).

For example, assume a total of 200 observations of which 50 are unsafe observations, then $P = (50/200)(100) = 25$ percent.

The quantity Y is assigned a value of accuracy considered satisfactory by the safety department. For most cases $Y = \pm 10$ percent is sufficient. Then by substitution into the formula

$$N = \frac{4(1 - 0.25)}{0.10^2(0.25)} = 1200 \text{ observations.}$$

What does 1200 observations mean? It means that a total of 1200 observations must be made if the safety sampling technique is to yield valid percentages of the types of unsafe practices and of how safe a plant is operating. Hence, in this example, six tours of the plant are needed since each tour supplies about 200 observations (the nearness to this number depends on the number of employees working and how many unsafe acts each is performing at the instant the inspection is conducted).

Reports

In both small plants where overall observation is possible by those responsible for safety, and in large plants where a form of safety sampling is more practical, formal reports of the observations are necessary. These reports should be made on a form which is designed specifically for the organizational structure and work operations of each industry, plant, or operation. Recording and reporting can be simplified by the use of forms similar to those shown in Fig. 4–3.

Written reports are prepared to point the ways to necessary action, such as revision of company rules, training or retraining, changes in processes or operations, and improvement in supervisory practices. Once designed, the report form can be used by safety group members, by supervisors, and by foremen for recording their observations. Then the same form can be used as a summary sheet for a monthly, quarterly, or annual report to management.

One major corporation uses the report form shown in Fig. 4–8. Supervisors find this report is easy to prepare, especially if entries are made daily, at the end of the shift. Only one number is entered in each column daily, and the information is quickly available.

The foreman knows how many injuries there were, how many Job Safety Analysis Conferences he held (see Chapter 5), and how many housekeeping, tool and equipment inspections he conducted during the day. All of the other items—individual and general contacts, safety observations, violations, unsafe acts, and disciplinary actions—are easily tabulated from his employee record cards.

At the end of the week, the report is submitted to the general foreman, who uses it in his weekly meeting with the foreman, to review his past safety activities, to see if he is meeting standards, and to help him plan his safety activities for the following week.

Inspection Personnel

Many people may be involved in safety inspections. Their duties are outlined in this section.

Inspectors should know how to locate hazards and have the authority to act and make recommendations. Four qualifications of a good inspector are knowledge of the company's accident experience, familiarity with accident potentials, ability to make intelligent recommendations for corrective action, and diplomacy in handling situations and personnel.

Safety inspectors set an example. It is of utmost importance that the safety inspector be properly equipped with personal protective equipment, protective clothing, and necessary apparatus to carry on his varied duties. It is difficult for an inspector to persuade the employee to wear goggles when he himself is not wearing them, or to "sell" safety shoes to workmen exposed to foot injuries if he does not wear them, or to have workmen use respirators or gas masks unless he wears one. It is essential that the inspector "practice what he preaches."

FOREMAN'S SAFETY ACTIVITY REPORT

DEPARTMENT **Pipe Shop**   FOREMAN **J.R. Kelly**   NO. MEN SUPERVISED **20**   WEEK OF **Aug. 9, 1964**

DATES	NO. OF INJUR.	CONTACTS		OBSERVATIONS			DISCIPLINE		JSA CONFERENCES		UNSAFE CONDIT.	INSPECTIONS	
		INDIVIDUAL	GENERAL	PLANNED	VIOLATION Rules, Proc. Instruction	UNSAFE ACTS	WRITTEN	TIME OFF	HELD	NO. EMPL. ATTENDING		HOUSE KEEPING	EQUIP.
8-10	0	5		1		1							1
8-11	0	3	1	4	2		1					1	
8-12	0	7		2							1		
8-13	1	4		2	1			1				1	
8-14	0	6	3	1					1	3			1
TOTALS	1	25	4	10	3	1	1	1	1	3	1	2	2

Fig. 4-8.—Supervisor's form should be easy to use. Entries can be made daily at the end of a shift.

Courtesy U.S. Steel Corp.

Safety professionals and inspectors

The safety professional must be in the group to spearhead the activity. He is in his most productive role during a safety inspection project because it is here that he coordinates the safety program and teaches by first-hand contact with supervisors and by using on-the-spot examples. The fire protection representative and the industrial hygienist are others who can be involved in this activity.

The number of safety professionals and inspectors needed depends a great deal on the size of the company or plant and the type of industry. Large companies with well organized accident prevention programs usually employ a staff of full-time safety professionals and inspectors who work directly under the safety director or safety supervisor. Large companies may also have a number of specially designated employees who spend part of their time on inspections. Also, there may be employee inspection committees who assist in this type of work. (This is discussed in the next section.)

Companies or plants too small to employ a full-time safety professional depend a great deal on inspections made by maintenance men and supervisors. Frequently, an employee carries out the duties on a part-time basis and makes periodic inspections. Many plants depend entirely on inspection service supplied by casualty insurance inspectors and state factory inspectors. More frequent inspections are usually necessary, however, than can be provided by outside agencies.

In employing inspectors, it should be recognized that a person qualified for general types of inspection work may not be qualified to make inspections of a special type. For instance, pressure equipment, such as boilers, autoclaves, digesters, air receivers, etc., and materials handling equipment, such as cranes, hoists, elevators, chains, and slings, are often among the most hazardous pieces of equipment used in industry. The inspection and testing of such equipment requires engineering knowledge and training. It should be determined that the inspector understands properly the hidden physical hazards involved and is competent to pass judgment on such equipment.

When chemicals are used, it is important that they measure up to exact standards, as otherwise they may cause fires or other accidents. In such cases, the chief chemist and safety inspector should work in close cooperation.

Also, in industries where toxic and corrosive substances are present, such as dusts, gases, vapors, and liquids, the safety professional or inspector should take special training to make sure he is familiar with the hazardous properties of these substances and with the methods of control. An industrial hygienist should be consulted, if needed for involved problems.

The safety professional should be in full charge of inspection activities and all inspectors, except special departmental inspectors who report directly to their foremen. The safety professional should either make safety inspections personally or supervise the inspectors in their work. Although the safety professional may have considerable desk work, he should get out into the production and maintenance areas as often as possible and make general as well as specific periodic inspections. If there is more than one plant under his supervision, he should plan to make at least an annual inspection survey of each plant.

The safety professional should also assist in conducting special surveys and in making accident investigations. He is often the most experienced and best qualified person for such work.

Company or plant management

Safety inspections should be part of the duties of all company or plant supervisors, including top management or the superintendent. Interest in the safety program is stimulated if a member of top management stops to make a casual inspection as he passes through various departments. Should an unguarded hazard or an unsafe practice be discovered, top management or the superintendent should confer as soon as possible with the supervisor to see that the conditions are corrected.

First-line supervision

The first-line supervisor or foreman is one of the most important inspectors in the entire industrial organization. He is as important to safety work as the safety inspector. He is "the key man." The foreman spends practically all of his time in the shop or plant. He is in constant contact with his employees and thoroughly familiar with all the hazards that may develop in his department. He should be on the alert at all times to discover and correct unsafe conditions and practices.

Because the foreman's duties require him to give close supervision over all his department, most of his inspections should be the continuous type. However, he should make general inspections at regular intervals to assure himself that all hazards have been properly safeguarded and that the safeguards are being used. Also, he should keep a close check on the periodic inspection of elevators, pressure vessels, and other special equipment in his department.

The foreman's responsibility requires that he continuously observe surrounding conditions and the tools used by the men in his crew. He must realize that conditions change continually and that unsafe conditions can arise at any time.

Engineering and maintenance

The mechanical engineer and the maintenance superintendent should make frequent trips through the plant. Necessary work orders for guards or for correcting faulty equipment can be written up on the spot. These men should be encouraged to be on the watch for unsafe conditions and unsafe practices and to report such conditions to the foreman for correction. If, unfortunately, these two officials are not safety-minded, their attitude could have a detrimental effect on all safety activities.

Employees

Employees who are constantly on the alert can be of great value in preventing accidents. In some companies, the rank-and-file employee is encouraged to inspect his workplace each day and to report any hazardous conditions to his supervisor. Employees who are safety conscious will always look for conditions that may cause an injury to themselves or to others.

Each operator may be required to inspect his workplace and the equipment or machinery he uses. He does this each day and immediately reports defects that he is not authorized to correct. In departments having a number of similar machines, a specialist should make regular inspections and perform the necessary adjustments and repairs.

When maintenance employees are working in various departments and observe hazards which should be corrected, they can be of great help if they report such hazards to the foreman of the department. Management should convince foremen that maintenance men can be of great help in locating and correcting hazardous conditions. Similarly, foremen should encourage mechanics to come to them with their suggestions.

Inspections of a more or less continuous type are necessary in plant storehouses, warehouses, and toolrooms. Here new machinery, tools, appliances, and materials enter the plant and should be checked before being requisitioned. The storekeeper and toolroom men can, therefore, contribute a great deal to the safety of a plant by inspecting and checking for simple defects.

Safety committees

Special inspections by the safety professional can be supplemented by regular inspections made by the committee (see Chapter 3).

The group which makes inspections should give equal consideration to accident, fire, and health exposures. If the committee is familiar with the history of accidents in the company or plant, the approach to the inspection can be planned all the more intelligently. The group should, therefore, review all accidents which have occurred in the area involved before it makes an inspection.

Safety committee members, or subcommittees appointed by the general committee, should make frequent inspection trips, either for periodic or intermittent inspections. This not only uncovers hazards and advertises safety, but members can discuss accident problems more intelligently in committee meetings if they make such trips from time to time. These inspections should be repeated at frequent intervals, preferably prior to each meeting because of the common practice of rotating committee members.

It is common practice to have accidents investigated by the safety committee or special accident investigation committee. Where departments are sufficiently large to justify such procedure, a standing committee for the investigation of accidents in these departments will not only get more details concerning the accidents, but will also become more safety-minded. Such a committee should make a complete inspection of the department involved as well as the scene of the accident.

Paper work for the inspection committee should be cut to a minimum. Checklists are useful for helping the inspection committee to cover details, but since checklists are not all inclusive, they should not be relied upon entirely in general inspections.

The inspection committee can also develop special report forms to simplify the problem of summarizing findings at the end of a tour. When these findings have been thoroughly discussed with the safety professional, recommendations can then be made with the understanding that the safety professional will follow through to a decision.

Technical problems. The inspection committee should not attempt to deal with technical problems best analyzed by trained observers or special testing methods. It should, however, be able to recognize the need for expert assistance where the possibility of danger exists, as in exposure to dusts, gases, vapors, or fumes which may create either health or fire and explosion hazards, and which call for scientific testing methods to measure the hazards and determine remedies. (For examples, see Chapter 37).

Inspection Procedures

Inspection procedures vary considerably in different companies and plants and with the types of inspections made.

Inspectors, of course, should know thoroughly all company safety and health rules and policies. They should also be familiar with OSHAct standards, state laws and regulations, and municipal ordinances affecting the safety and health of workers. The fire protection requirements that are applicable to the particular plant should also be known. Usually codes, federal and state laws, and regulations set up minimum requirements only. It is frequently necessary to exceed these requirements to comply with company policy and secure maximum safety.

Before making periodic or special inspections, it is highly important to make a complete analysis of company and department accident statistics and accident causes, as outlined early in this chapter under "Inspection of Work Areas." This material should be made available to the inspector by the safety professional.

Other aids for the inspector may include:

- Inspection checklists

- Inspection report forms

- Job safety analysis cards (see the discussion in the next chapter).

Starting the inspection

Three basic steps of how to inspect are: (a) contact the department head and solicit his help, (b) observe all conditions for compliance with established standards (use checklists), and (c) observe all operations for any unsafe acts or violations of safety rules.

Step 1. Before an inspection is started in any department, the inspector should first contact the foreman or the person in charge. Because the department head is responsible for the safety of his workers and whatever unsafe conditions that may exist, his participation and cooperation in making an inspection can expedite any needed corrections.

It is good practice, therefore, for the supervisor to accompany the inspector while covering his department.

Steps 2 and 3. Inspections should be systematic and thorough. No location that may contain a hazard should be overlooked. This can best be accomplished by following the manufacturing operations from the receiving of raw material until the finished product is shipped or stored. Occasionally, it is advisable to vary the route followed. In one plant, where inspections follow the flow of material, the route is reversed on reinspections.

Taking notes

Notes should be made at the time unsafe conditions and practices are discovered, to form the basis of a complete report to be prepared later. The inspector should not depend on his memory and write notes after he has left the particular department or returned to his office. A checklist, such as those shown earlier in this chapter, helps here.

It is important to secure all the information needed to describe the hazard found. The exact location in the department, and other data necessary to make a complete report, such as suggestions for correcting the condition, should also be given.

It should not be the aim of the inspector to pick up numerous trivial items merely to make his report look good. But it is not the inspector's prerogative to pass up any condition that may result in an injury. A long list should not ordinarily result in a plant that is regularly inspected. If safety work has been neglected, however, a long list can be expected.

If the inspector is accompanied by the supervisor, each recommendation should be discussed so that the supervisor will be fully informed concerning what recommendations will be included. Otherwise, the inspector should contact the supervisor before leaving the department and go over his list of recommendations. In this manner, an agreement can usually be reached as to the relative importance of recommendations.

When the supervisor fully understands what is required, he may be able to make corrections in a short time. Sometimes, the supervisor's promise to correct the hazard is not carried out; so it is advisable to include all items in the written report.

Making the report

Every inspection must be followed by a clearly written report. Inspection reports are usually of three types:

1. THE EMERGENCY REPORT—made without delay whenever immediate corrective action is necessary.

2. THE ROUTINE REPORT—covers those observations of unsatisfactory (nonemergency) conditions that should be corrected. This

report should be made no later than one day after the inspection.

3. THE PERIODIC REPORT—lists all items of previous routine reports for a given period. This report is actually a summary of safety activity and accomplishments.

The following descriptive details should identify and distinguish each report: name of department or area inspected (boundaries or location if needed), date and time of the inspection, name and title of the inspector (also names of any others who participated in the inspection), date of the report, and the names of those to whom the report was made. This information provides an excellent record and reference.

Particularly, locations must be accurately named or numbered. Machines and operations must be identified by their correct names. An unsafe condition or action must be described in detail. For example, if the checklist shows an "X" (for unsatisfactory) opposite "Hand Tools," the report might read like this: "Some of the chisels being used by chippers had mushroomed heads with sharp sections that could easily fly off. This is an eye hazard. We found some hammers in this department with cracked and splintered handles. These hammers were not in use when we were there, but would be dangerous to use."

"Bad housekeeping" is often listed in a report, but it means nothing unless details are given. Instead of merely reporting "bad housekeeping," the inspector should give full details—empty pallets left in the aisles, scrap from machines in piles on the floor around the machines, a ladder lying across empty stock boxes, slippery spots on the floor from oil leaks, for example.

Generally, the inspection reports are directed to the head of the department or area where the inspection was made. Copies of the reports are also directed to the executive management and/or to the plant manager to whom the department head reports.

Handling recommendations

Recommendations should be listed in the order in which they were discovered, or grouped by the department according to the individual responsible for their compliance.

After the recommendations have been approved by the proper officials, copies of the recommendations should be sent to each foreman or superintendent affected. Where possible, a definite time limit for compliance should be set for each recommendation, or they should be grouped according to their importance and marked "urgent," "important," or "desirable."

It may be required to have inspection reports reviewed by the local plant safety committee to get their reaction, particularly as it applies to educational features and those which directly affect the employees. All engineering problems, of course, should be referred to the engineering department. Make sure that conditions needing correction are listed in order of their importance. This is one of the surest ways of getting things done promptly.

Recommendations approved by the management should become a part of the plant improvement or maintenance program, be assigned job numbers, and pass through the regular channels. Supervisors should report progress in complying with the recommendations to management (or the general safety committee) at regular intervals. Also, the inspector should make periodic checks of all recommendations until they have been completed. In this manner, he may be able to make suggestions for the best method of eliminating the hazard and, at the same time, prepare a progress chart for his own information.

Checkup work may be simplified by entering individual safety recommendations on special forms of convenient size, which can be in a follow-up file. Such files can be arranged as a tickler file (Fig. 4–9).

Condemning equipment

A practical system can be established for taking certain materials and equipment from service because of unsafe conditions. Special danger tags (Fig. 4–10) can be employed effectively in preventing the use of equipment or materials that have become unsafe through wear, abuse, or defects. (A tag for use in condemning an unsafe ladder is shown in Chapter 17.) However, such systems should be employed only under the strictest supervision. Before working on equipment, be sure to lock out the main switch as well as tag it.

```
┌─────────────────────────────────────────────────────────────────┐
│                                                      No. 1053      │
│           SAFETY RECOMMENDATION                                    │
│                                                                    │
│                      Date Issued_____19__             │
│                                                                    │
│                      Date Ret'd. _____19__             │
│  To_____              │
│  PLEASE HAVE THE FOLLOWING UNSAFE CONDITION                        │
│  CORRECTED:                                                        │
│  _____ │
│                                                                    │
│  _____ │
│                                                                    │
│  _____ │
│                                                                    │
│  _____ │
│                                                                    │
│  _____ │
│                                                                    │
│     Please sign and return to Safety Department within ten days,   │
│     indicating below what disposition was made of this             │
│     recommendation.                                                │
│                         _____           │
│                              SAFETY DEPARTMENT                      │
│  RECOMMENDATION FOLLOWED ( )   WORK COMPLETED_____       │
│                                                  DATE              │
│  RECOMMENDATION REJECTED ( )   FOR FOLLOWING REASON:_____      │
│                                                                    │
│  _____ │
│                                                                    │
│  Copy of this recommendation is on                                 │
│  file in the Safety Department. The                                │
│  Safety Department is instructed to                                │
│  send a detailed list of all un-                                   │
│  answered recommendations more                                     │
│  than ten days old to the General    _____     │
│  Superintendent the first of each    (SIGNED) DEPARTMENT           │
│  period.                                      SUPERINTENDENT        │
└─────────────────────────────────────────────────────────────────┘
```

FIG. 4–9.—Safety department "field men" can carry a pad of these forms, which have a provision for making a copy of each sheet. Each pair is numbered—this not only saves office work later, but also permits follow-up until recommended work is satisfactorily completed.

Courtesy Tennessee Eastman Corporation.

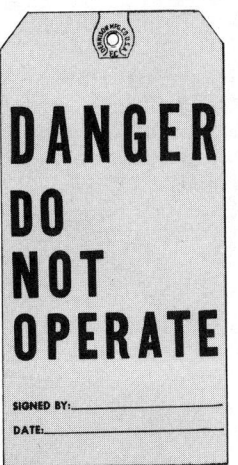

FIG. 4–10.—Front and back views of typical tag used when equipment is taken out of service because it has become unsafe.

When danger tags are used, provisions should always be made for the signatures of the inspectors who are authorized to condemn equipment. Only the inspector who places the tag should be permitted to remove it and only when he is satisfied that the hazardous condition has been removed.

No equipment or materials should be placed out of service without notifying the person in authority in the department affected. A shutdown to avoid what might appear to be a possible hazard might interrupt work at great expense without actually affording any protection. Consequently, authority to condemn equipment should be exercised with a great deal of care by the inspector.

Making night inspections

When a firm or plant is working on a two- or three-shift schedule, it is desirable occasionally to have inspectors working with each shift. Safety conditions vary considerably after dark, because of artificial illumination. The inspectors should make sure that adequate illumination is provided and that the lighting system is maintained in a satisfactory condition (Fig. 4–11). Use a sufficiently sensitive light meter.

When regular night shifts are not employed, night working conditions frequently go un-observed. An inspector should be assigned to make occasional plant visits to check on the safety of maintenance men, firemen, and others who may be required to work after dark. Also, make an occasional trip with the night watchman to observe the conditions under which he works.

Making photo inspections

A combined safety inspection and supervisor training session can be the result of a quick tour through a plant with an instant-developing camera, a camera equipped to take 2×2-inch color slides, or even a lightweight video tape recorder.

To add variety to supervisor safety meetings, arrive a little early for the meeting after making a quick inspection. An instant-developing camera easily records unsafe conditions or practices, and the pictures then become a quiz for the supervisors' meeting.

The pictures are numbered and passed around the table (or the slides are projected). The supervisors are told to write down the unsafe acts or conditions that they can see in the pictures. Occasionally put one in that doesn't have any noticeable hazard. This keeps the audience alert.

Many details are given in NSC Data Sheet 619, *Photography for the Safety Professional*.

Checking plans and specifications

An important duty of the safety professional should be that of checking plans and specifications both for new plant buildings and equipment and for products to be made. This important function affords an opportunity to discover and correct conditions that may otherwise be built into the building and its equipment and that would later result in injuries. Many companies will not permit a drawing or specification to be used until it has been approved by the safety department. Hazards involved in making products should be minimized insofar as possible. Instructions and warnings should be reviewed for safe manufacturing procedures.

The safety department should also make sure that company policies and applicable standards are followed in purchase specifications for new materials and equipment or for modification to existing equipment. Some companies have arranged for the purchasing department to notify the safety department when new materials or equipment are to be purchased. For instance, when a new chemical is requested, the safety department should see that information on flammability, toxicity, and similar properties is obtained from the manufacturer. This Manual outlines requirements and performance standards that should be used to evaluate the safety specifications for purchases of industrial equipment and raw materials.

Making safety an integral part of any modification or addition to the industrial plant means that, to list a few examples, each new machine will be received with its moving parts guarded, that all raw materials will be shipped in a manner to minimize preprocess handling hazards, and that selection of a flammable solvent will be based on control by using safe handling procedures. Such planning should impress employees with the fact that the company considers safety as a basic part of its operation, and not as an afterthought.

A safety program for outside contractors should be tied into any contract documents. See Chapter 17.

The remainder of this chapter is devoted to discussing some techniques for making safety part of every company's basic operations.

Total Loss Control and Damage Control

An important development in industrial safety in recent years has been that of total loss control and of damage control.

Many progressive companies have investigated and applied a concept of total accident control based on studies of "near misses" (non-injury accidents), and on detailed analyses of hidden as well as direct accident costs.

The emphasis now is on the total safety of the employee, not only on the job but also at home and off the job.

Total loss control

Cost-conscious companies find they must control accident costs if they are to do business in a highly competitive market. They do not consider safety a nuisance, but a necessity—not a fringe benefit, but a prerequisite.

Activities related to safeguarding corporate investment and continuity of operation need

coordination with the safety effort. Such related functions as, for example, personnel safety and fire protection must be closely integrated. Those activities which are related to loss control must be functionally integrated.

One large company uses an approach consisting of five closely related steps—each following the other in logical order to give a coordinated program.

1. HAZARD IDENTIFICATION—To prevent accidents and control losses, it is first necessary to identify all hazards—to determine those areas or activities in an operation where losses can occur. This requires studying processes at the research stage, reviewing design during engineering, checking pilot plant operations and startup, and regularly monitoring normal production.

2. HAZARD ELIMINATION—Toxic, flammable, or corrosive chemicals can sometimes be replaced by safer materials. Machines can be redesigned to remove danger points. Plant layouts can be improved—by eliminating blind corners or limited visibility crossings, for example.

3. HAZARD PROTECTION—Hazards that cannot be removed must be protected. There are many familiar examples—mechanical guards keep fingers from pinch-points, safety shoes safeguard toes against dropped objects, and automatic sprinkler systems can stop fires fast. Industry is concerned with all losses, injury to personnel, damage to products, and destruction of property.

4. MAXIMUM POSSIBLE LOSS—This step involves the determination of the maximum loss that could occur if *everything* went wrong. It is known from experience that entire buildings or areas can be lost by fires or explosions. It can be estimated how much a company could lose under the most adverse conditions.

5. LOSS RETENTION—Having some idea of the amount that could be lost under a combination of unfavorable circumstances, one can then determine what portion of such a loss a company is willing to bear itself. Industrial companies can afford to retain a portion of each loss themselves. The re-

PLANT ILLUMINATION SURVEY FORM

Production, safety, and personal comfort demand good plant lighting. Use this survey form to check on the general efficiency of the present lighting system. A "No" answer to any of the questions listed below will indicate a need for a more detailed inspection and, possibly, the services of a competent illumination engineer.

QUESTION	LOCATION	YES OR NO	REMARKS
1. Are proper levels of light maintained in different areas and for different operations? (Direct light meter readings should be taken and checked against established level for a particular area or operation.)			
2. Are "task zones" adequately illuminated? (harsh shadows and reflections eliminated)			
3. Are lights mounted properly so that employee will not have to perform work in his own shadow?			
4. Are all lights properly shielded, louvered, or mounted to prevent direct glare?			
5. Is there efficient control of indirect glare from work surfaces, machine control dials and knobs, and from the piece on which work is being performed?			
6. Is general lighting uniform outside "task zones"?			
7. Have "high contrast" areas been eliminated or "toned down" through use of color and balanced lighting?			
8. Are special lamps and fixtures provided where unusual conditions prevail? (vibration resisting, weather resistant, explosion resistant, etc.)			
9. Are lighting fixtures receiving proper maintenance? (periodic cleaning, replacement, etc.)			
10. Are windows and skylights cleaned regularly?			
11. Is an adequate emergency lighting system maintained for use in the event of power failure?			

Additional remarks_____

_____

_____

Date_____ Signed_____ Department_____

FIG. 4–11.—Illumination survey form. Because lighting fixtures are often overlooked, special instruction in this area may be required to maintain safe operations and to guard employees' eyesight. (See Table 46-G, "Levels of Illumination," in Chapter 46.)

maining loss potential is then insured against through the company's insurance carrier. When this is done, it proves a rather good incentive to strong loss prevention programs.

This major firm has found it desirable to consolidate these activities in one department. Called the Loss Prevention Department, this group brings together the safety professional, the fire protection manager, the security and plant protection manager, and the insurance manager. The administrator of a total loss control program does not himself need to know all the details of each function in his group, but he should be able to develop an atmosphere in which there is harmonious cooperation and mutual understanding. He should strive, through constant communication with management and supervisors, to coordinate loss prevention efforts in personnel safety, fire protection, insurance, and plant security.

Damage control

A program of damage control has been developed by a metal industry firm, in cooperation with other companies.

A strong case is made for investigating all accidents, not just those that produce injuries. It is possible that here the safety movement has another approach of potential value. This company believes that studying accidents instead of injuries does not downgrade human values. Such study recognizes that a so-called "no-injury accident" might have resulted in some personal injury, property damage, or both. Ferreting out the causes of accidents— whether or not they result in injury—determines what unsafe conditions, what unsafe acts (or combination of the two) were responsible for the accident.

H. W. Heinrich (see "References") observed there were about 300 "no-injury" accidents to one that resulted in a disabling injury. These also involved considerable property damage.

A company that requires reporting of all accidents (that result in injury or property damage or both) found that of 90,000 incidents reported over a 7-year period, more than 15,000 injuries occurred (of which 145 were classified as disabling). As shown in Fig. 4–12, property damage accidents are about

five times the number of injury-type accidents for the same period; their cost is great, and, they imply a huge injury potential.

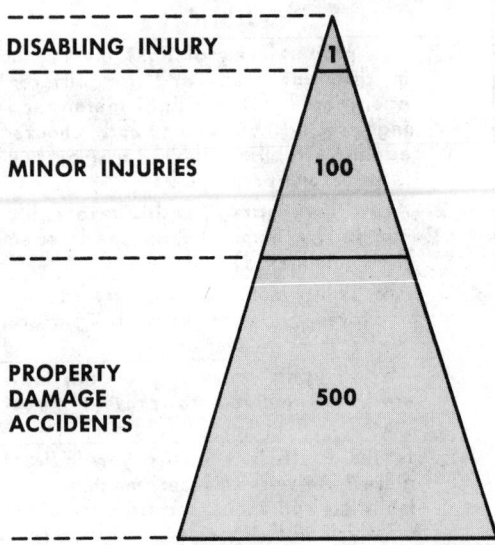

Fig. 4–12.—Survey of 90,000 incidents resulting in injury or property damage or both showed these ratios between disabling and minor injuries and property damage accidents.

Courtesy F. E. Bird and G. L. Germain, Damage Control.

Three basic steps are used successfully to reduce property damage (and injuries). They are (*a*) spot checking, (*b*) reporting by repair control centers, and (*c*) auditing.

Step one: Spot checking. Spot checking repair centers, making observations, and taking notes for reference permit damage estimates to be made for certain items by comparing total costs for a repair period with those found during sample observations.

Step two: Reporting by repair control centers. This step involves the development of a system in which the repair or cost control center records property damage. It is very important that the system be designed with the least amount of paperwork. No one system will work in all companies, because repair cost accounting methods vary greatly from company to company, and even from plant to plant.

Step three: Auditing. The final step to a completely effective reporting program is more complete auditing.

Safety personnel receive a copy of every original work order processed through the maintenance, planning, and cost control center. Safety professionals make on-the-spot checks to see if accidental damage was involved.

Measuring effectiveness. While a variety of methods could be used to measure the effectiveness of property damage control, the cost severity rate technique has proved to be of significant value. This method expresses total costs in relationship to production and maintenance man-hours.

A simple formula then is:

$$\frac{\text{Total costs of accidents} \times 1,000,000}{\text{Total production and maintenance man-hours}} = \text{Damage costs per million production and maintenance man-hours}$$

For the metal industry firm mentioned earlier, the cost of property damage accidents was near $138,000 per million hourly rated man-hours worked—this was a reduction of $188,000 per million hourly rated man-hours, or 58 percent, for the year as compared with the company's base year.

On the basis of the experience gained thus far, the following conclusions can be drawn as to some of the values of incorporating the missing elements—property damage—into the industrial accident prevention program.

1. Management will recognize that the investigation of the larger group on no-injury-producing accidents will remedy many unsafe practices and conditions that also have been common causes of injury-producing accidents.

2. Management is interested in this program, as it is in any program that points the way to improve quality and decrease production delays.

3. A broadening of the accident control effort, with the potential of accident cost reduction, can be the vehicle for economic justification of any valid need for additional safety personnel.

4. The stature of the safety organization, as a member of the corporate team is enhanced when its full worth is recognized by management. The addition of the property damage element can assist appreciably in accomplishing this recognition.

5. The realistic possibility of measurable and appreciable cost reduction is a dynamic tool to motivate management interest.

System Safety

Recent years have seen increased "system talk"—that is, discussion and exploration of system approaches to industrial accident prevention.

Safety professionals will not only hear about these techniques, they will have to understand them, for many will be called on to find ways of implementing them. And although complete "system safety analysis" requires specially trained engineers and rather sophisticated mathematics, safety professionals will find that some knowledge of these techniques can be of direct benefit in helping codify and direct their accident prevention programs.

Systems methods developed in the exacting aerospace field hold promise for all industrial safety professionals.

Today, scientists can successfully orbit an astronaut and get him back home safely. But yet we cannot reliably get the employee back home from the plant safely. What's different in these two situations? The first difference is the amount of money spent on the orbit task. The second difference is the systems approach to the orbit task, with the extremely high reliability and safety specifications required.

A system analysis can clarify a *complex* process by devising a chart or model that provides a comprehensive, overall view of the process by showing its principal elements and the ways in which they are inter-related.

Until recently, when a new aircraft was developed, it was first designed, then an experimental model was built, and finally it was test-flown to determine its capabilities and

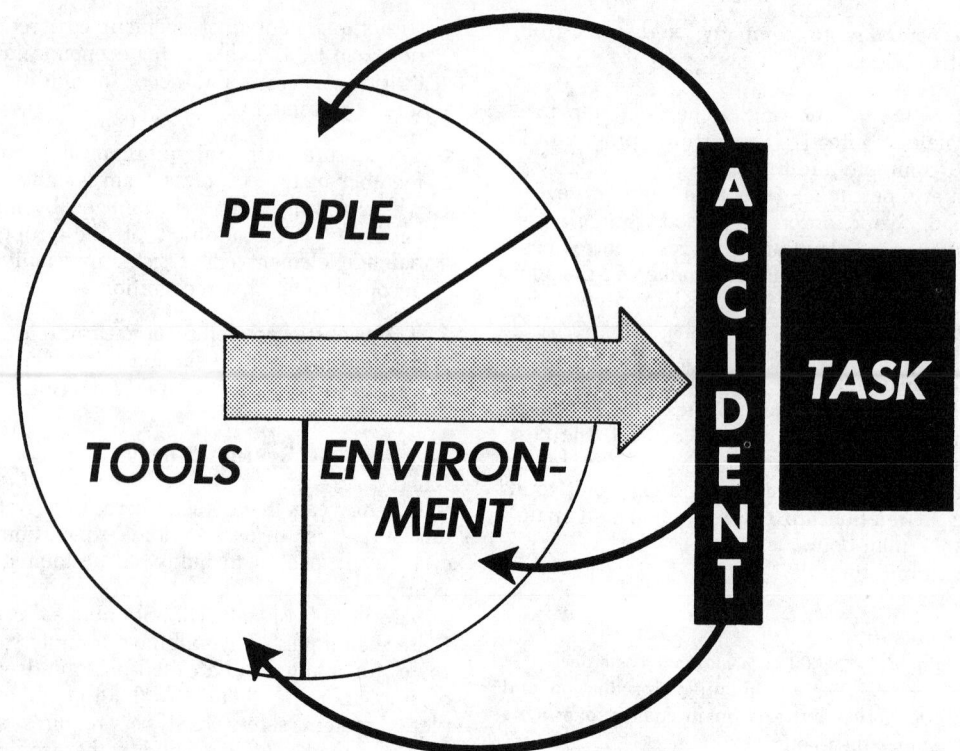

Fig. 4–13.—A "system" is an orderly arrangement of Components that are interrelated and act and Interact with one another to perform a Task or function in a particular Environment. An Accident results from deficiencies in People, Tools, and/or Environment (most accidents have deficiencies in all three), and sometimes deficiencies in the Task.

flaws; the information obtained indicated the necessary design changes and the cycle was repeated until the performance specifications were met. Today's aircraft and missiles are so complex and costly and the specifications are set so high that this procedure had to be changed. Moreover, missile flight tests involve loss of the model with only limited telemetry data obtained. Today the "bugs" must be found and corrected as far as possible in the design stage using analytical techniques.

The result is the development of the systems approach to safety.

What is a system?

A simple illustration of the basic elements of a system is shown in Fig. 4–13. People, Tools, and Environment are joined to per-

form a Task. Systems are task-oriented; they *do* something.

In any system, there are interactions between parts, for example:

- A fisherman uses tackle and boat on a lake to have fun,

- A man drives a truck along a road to deliver goods, or

- A workman uses metal and a press in a shop to make costume jewelry.

Keep in mind that a system is defined in terms of a task or function (it is task-oriented), and that the components of a system are interrelated, that is, each part affects the others.

The components of a system can cover a wide range including machines, tools, material (hardware, chemicals, etc.), environmental factors, people, documents (such as operating instructions, training manuals, or computer programs), and so on. As parts of a system, the components usually complement each other, but it is essential to recognize that a failure or malfunction of any component can affect the other components and thus degrade the performance of the task.

The environment is an important consideration in a system since most systems will perform their task properly only under a given set of conditions.

Accidents and errors degrade performance

As shown in Fig. 4–13, accidents degrade or stop task performance when they injure people, damage tools and equipment, or damage the environment. A serious accident could destroy the system, such as by burning down a building.

Accidents result from deficiencies in people, in tools, and in environment (most accidents have deficiencies in all three). And, sometimes, deficiencies in the task or objective. But more importantly, accidents result from interactions like an overtired young machine operator trying to operate a fast-closing press while his mind is focused on his sick son at home. His hand contacts a moving part of his machine and he is injured.

Possible errors include:

Mechanical failure

Defective materials

Electrical failure

Environmental conditions

Human failure: curiosity, distraction, fatigue, indolence, worry, anger, illness, chance-taking, improper attitude, lack of skill, poor physical condition, intoxication.

Analyzing systems

Having established the concept of a system and explained how it can be degraded, the next step is the analysis of systems—especially complex systems such as aircraft, communications networks, or production lines. It is in this area—the analysis of complex systems—that great progress has been made in the aerospace industry that holds promise for application throughout industry.

It might be helpful to indicate briefly the main techniques in order to clarify the nature of the systems approach to safety.

No matter which method of analysis is used, it is important to have a model of the system. Most models take the form of a diagram showing *all* the components. This makes it easier to grasp the interrelationships and simplifies tracing the effects of malfunctions.

Systems analysis, as applied to human factors engineering, is discussed in Chapter 10, "Human Factors Engineering"; and as applied to accident investigations, is discussed in Chapter 7, "Accident Investigation, Analysis, and Costs."

Methods of analysis

There are four principal methods of analysis: failure mode and effect, fault tree, THERP, and cost-effectiveness. Each has a number of variations and more than one may be combined in a single analysis.

Failure mode and effect. In the failure mode and effect method, failure or malfunction of each component is considered, including the mode of failure (such as, switch jammed "on"), the effects of the failure are traced through the system, and the ultimate effect on the task performance is evaluated. Failure mode and effect analysis is straightforward, assuming that the analyst is thoroughly informed about the system. One drawback of this method, however, is that it considers only one failure at a time and thus some possibilities may be overlooked.

Fault tree. In the fault tree method an undesired event is selected and all the possible happenings that can contribute to the event are diagrammed in the form of a tree. The branches of the tree are continued until independent events are reached. Probabilities are determined for the independent events and after simplifying the tree, both the probability of the undesired event and the most likely chain of events leading up to it can be computed.

This is a very powerful analysis technique but has the drawback of requiring a fairly heavy mathematical background and a good computer to obtain the maximum benefits of the method. Boeing Company has refined the fault tree method to a high degree and has found it practical for analyzing aerospace products.

THERP, a technique for human error prediction, developed by Sandia Corporation, provides a means for quantitatively evaluating the contribution of human error to the degradation of product quality. It can be used for human components in systems and thus can be combined either with the failure mode and effect or the fault tree methods.

Cost-effectiveness. In the cost-effectiveness method, the cost of system changes made to increase safety are compared with either the decreased costs of fewer serious failures, or with the increased effectiveness of the system to perform its task, to determine the relative value of these changes. Ultimately all system changes have to be costed, but this method makes such cost comparisons explicit. Moreover, cost-effectiveness is frequently used to help make decisions concerning the choice of one of several systems which can perform the same task.

In all of these analytical methods, the main point is to measure quantitatively the effects of various failures within a system. In each case probability theory is an important element.

Deviation caused by change

In any process or system, there is some risk. When something *changes,* accompanied by an *error,* there is an *accident.* The error can be either active or latent. For example, in a highway crash, a dog might run across a highway (the change), the first driver slams on his brakes, but there was too little space between him and the car behind (a latent error). The deviation is a two-car accident. The accident, a deviation, interrupted the task performance.

When a safety change is planned, unwanted changes must be anticipated. (*a*) An offsetting safety change can be made to channel unwanted change aside or get rid of it. (*b*) Or the change can be kept from affecting the error situation. (*c*) Or the error can be kept from resulting in an accident. (*d*) If there is an accident, its adverse results can be minimized by providing personal protective equipment. (*e*) If there is an injury, then it can be treated. In some cases, a new and better way

can be found to do the task, thus avoiding the trouble-spot entirely.

The objective of a plan for safety changes must be to help task performance. The objective of accident investigation (discussed in Chapter 7) is to determine the facts that produced the "change," and to take countermeasures to offset this change so that the desired task may be completed.

Value of systems approach

The industrial safety professional might well ask what all this "systems analysis" has to do with him. The answer to that question, and the major point of this discussion, is that anyone can use and profit from the systems approach to safety. The systems approach helps to enlarge one's viewpoint. Becoming oriented in terms of task performance and being forced to visualize the interrelationships of all the components of a system helps to bring most accident possibilities into consideration automatically and in an orderly manner.

The systems approach to safety can help to change the safety profession from an art to a science by codifying much of our knowledge. It helps change the application of safety from piece-meal problem solving (such as putting a pan under the leak) to a safely designed operation (avoiding the leak itself).

We can apply the question "what can happen if this component fails" to the various elements of the systems and come up with adequate safety answers *before* the accident occurs instead of after the damage has been done.

The Critical Incident

The critical incident technique is a method for identifying human errors and unsafe conditions which contribute to both injury and no-injury accidents within a given plant group. The technique can identify mechanical and human causes of potential accidents before a loss occurs.

The technique involves selecting a random sample of participant-observers from various portions of a plant population. These persons are selected from the major plant departments, in order to obtain a representative sample of workers exposed to various hazards.

TYPICAL LIST OF INCIDENTS

An "incident" is any observable human activity sufficiently complete in itself to permit references and predictions to be made about the person performing the act.

1. Adjusting and gauging (calipering) work while the machine is in operation.
2. Cleaning a machine or removing a part while the machine is in motion.
3. Using air hose to remove metal chips from table or work (a brush or other tool should be used for this purpose, except on recessed jigs).
4. Using compressed air to blow dust or dirt off of clothing or out of hair.
5. Using excessive pressure on air hose.
6. Operating machine tools (turning machines, knurling and grinding machines, drill presses, milling machines, boring machines, etc.) without proper eye protection (including side shields).
7. Not wearing safety glasses in a designated eye-hazard area.
8. Failing to use protective clothing or equipment (face shield, face mask, ear plugs, safety hat, cup goggles, etc.).
9. Failure to wear proper gloves or other hand protection when handling rough or sharp-edged material.
10. Wearing gloves, ties, rings, long sleeves, or loose clothing around machine tools.
11. Wearing gloves while grinding, polishing, or buffing.
12. Handling hot objects with unprotected hands.
13. No work rest or poorly adjusted work rest on grinder (1/8 in. maximum clearance).
14. Grinding without the glass eye-shield in place.
15. Making safety devices inoperative (removing guards, tampering with adjustment of guard, "beating" or "cheating" the guard, failing to report defects).
16. Using an ungrounded (or uninsulated) portable electric hand tool.
17. Improperly designed safety guard (for example, a wide opening on a barrier guard which will allow the fingers to reach the cutting edge).

FIG. 4–14.

Next, an interviewer questions a number of persons who have performed particular jobs within certain environments. He asks them to describe unsafe errors they have made or observed in the past, or unsafe conditions that have come to their attention in connection with plant operations. The incidents described by a number of participant-observers are classified into hazard categories, from which accident problem areas can be seen. A partial list of typical incidents is shown in Fig. 4–14. An accident prevention program can then be organized and directed toward these problem areas.

The technique is repeated using a new sample periodically, in order to:

1. Detect new problem areas (or people), or

2. Measure the effectiveness of an accident prevention program.

4—Inspection and Control Procedures

By carefully selecting a representative sample of persons, the safety professional can judge the level of safety performance of the entire system at that particular time. The object is to discover factors which have contributed to an actual or potential loss-producing accident.

What the technique can do

A two-year study evaluating the usefulness of the critical incident technique was recently conducted at a large manufacturing plant in New Jersey. The 187 different unsafe acts and conditions that were identified revealed that the causes of no-injury accidents were the same as the causes of injury type accidents and that future injury type accidents can be predicted from an analysis of "near-misses." It was also found that 52 percent more unsafe acts and conditions were revealed by the critical incident technique than were identified by the accident records compiled during the two-year period studied.

The results of this study show that:

1. The critical incident technique dependably reveals causal factors in terms of errors and/or unsafe conditions that lead to industrial accidents.

2. The technique is able to identify factors associated with both injury and no-injury accidents.

3. The technique reveals a greater amount of information about accident causes than do presently available methods of accident study. The technique provides a more sensitive measure of total accident performance.

4. The causes of no-injury accidents as identified by the critical incident technique can be used to identify sources of potentially injurious accidents.

5. Use of the critical incident technique to identify accident causes is feasible.

That the critical incident technique is sensitive to accident problems which have a potential for accident loss, but have not yet produced a loss, is an important advancement in our ability to measure industrial safety performance. The existence of an injury or property damage loss is no longer a necessary condition for appraising safety performance. Accident situations can now be identified and examined "before-the-fact" instead of "after-the-fact," in terms of their injury-producing or property-damaging consequences.

"Behavior sampling," similar to the critical incident technique, is described in Chapter 11, "Human Behavior and Safety."

References

Bethlehem Steel Co., Bethlehem, Pa. *Our Next Step to Zero,* "Instructor's Guide to Supervisors Safety Program," 1964. pp. 122–133.

Bird, Frank E. and George L. Germain. *Damage Control.* New York, N.Y., American Management Assn., 1966.

Firenze, Robert J. *Guide to Occupational Safety & Health Management.* Dubuque, Ia., Kendall/Hunt Publishing Co., 1973.

Fletcher, John A. *The Industrial Environment.* Willowdale, Ontario, Canada, National Profile Ltd., 1972.

Fletcher, John A., and Douglas, Hugh M. *Total Environmental Control.* Willowdale, Ontario, Canada, National Profile Ltd., 1970.

Heinrich, H. W. *Industrial Accident Prevention,* 4th ed. New York, N.Y., McGraw-Hill Book Co., 1959.

Hopkins, Stuart K. "Voluntary Standards at the Crossroads," *National Safety News* (April 1967).

Johnson, W. G. "The Role of Change in Accidents," *National Safety News* (November 1973).

————. *MORT—The Management Oversight and Risk Tree.* Washington, D.C., U.S. Atomic Energy Commission, Division of Operational Safety, 1973.

Kysor, Harley D. "Safety Management System," *National Safety News* (November and December 1973).

Moore, Robert L. "How to Inspect for Accident Prevention—Physical Condition of Buildings," *National Safety Congress Transactions,* vol. 12, 1963.

National Safety Council, 425 N. Michigan Ave., Chicago 60611
  Accident Facts (published annually).
  Industrial Data Sheets
      Management Policies on Occupational Safety, 585.
      Photography for the Safety Professional, 619.
  Self-Evaluation Checklists (OSHA)
      Construction Industry—Part 1926.
      General Industry—Part 1910.

Pollina, Vincent. "Safety Sampling," *Journal of the American Society of Safety Engineers* (August 1962).

————. *Safety Sampling Procedure.* Milwaukee, Wis., A. O. Smith Corp.

Recht, J. L. *Systems Safety Analysis, National Safety News* Reprint No. 12. Chicago, National Safety Council, 1967.

Stone, J. R. "Safety Inspections—A Safety Catalyst," *National Safety Congress Transactions,* vol. 12, 1964.

Surry, J. *Industrial Accident Research—A Human Engineering Appraisal.* Toronto, Toronto University, 1968.

Tarrants, William E. "Research" (Critical Incident Technique), *National Safety Congress Transactions,* vol. 12, 1966.

Waad, A. Timothy. "Observation—Perception," *National Safety Congress Transactions,* vol. 11, 1963.

Removing the Hazard from the Job

Chapter
5

Reducing Exposure to Injury 105
Design for safety . . . production hinderances . . . Role of the safety professional . . . unsafe acts, unsafe conditions

Machine Design 107
Basic considerations . . . Machine safety checklist . . . Built-in safety . . . Human factors

Job Safety Analysis 110
Select the job . . . Break the job down . . . Identify hazards and potential accidents . . . Develop solutions . . . Use JSA effectively

Purchasing 116
Specifications . . . Codes and standards . . . Purchasing-safety liaison . . . Safety considerations . . . Price considerations

References 120

Fig. 5–1.—The drafting board is where safety begins as a part of the designer's concept. Safety "designed" into equipment, processes, and plants reduces need for training and supervision of personnel.

Courtesy R. G. Burkhardt & Assoc., Consulting Engineers, Chicago.

The basic measures for preventing accidental injury, in order of effectiveness and preference, are:

1. Eliminate the hazard from the machine, method, material, or plant structure.

2. Control the hazard by enclosing or guarding it at its source.

3. Train personnel to be aware of the hazard and to follow safe job procedures to avoid it.

4. Prescribe personal protective equipment for personnel to shield them against the hazard.

This chapter will be primarily concerned with engineering the hazards out of operations before work is performed and, therefore, before accidents can occur. If all possibilities have been exhausted and the hazard is still not removed, then every effort should be made to enclose, or guard the hazard at its source so that exposure to injury is controlled. In some cases, this measure can be just as effective as elimination of the hazard.

If either of the first two measures can be successfully employed, the need for on-the-job training to protect personnel against hazards is greatly reduced, and use of personal protective equipment may not be necessary.

Reducing Exposure to Injury

Modern industry knows how to engineer the hazards out of jobs. However, on too many jobs, accidents occur because insufficient effort has been put forth to determine the hazards involved. Obviously, until a job has been studied to make these determinations,

105

the hazards cannot be engineered out of it.

Among the jobs or operations that have not been analyzed for hazards are many of a non-repetitive nature, the type of job on which accidents are likely to occur.

Design for safety

The ultimate goal is to design environments and equipment and to set up job procedures so that employee exposure to injury will be either eliminated or controlled as completely as possible during manufacture. When a high degree of safety is incorporated into the design of the equipment or the planning of the process, the need for training and supervision to control unsafe acts is reduced.

Company policy should specify that safety must be designed and built into the job before the job is executed. To add safety features after work on a job has begun is usually more costly, less efficient, and less effective.

The most efficient time to engineer hazards out of the plant, product, process, or job is prior to building or remodeling, while a product is being designed, before a change in a process is put into effect, or before a job is started. Every effort, therefore, should be made to find and remove potential hazards at the blueprint or planning stage (Fig. 5–1).

Each level of engineering should be given the responsibility for building safety into the job, right through to the production phase. This responsibility should extend to product design, machine design, plant layout and condition of premises, selection and specification of materials, production planning, time study, methods, duties of the production foreman, and the work of employees assigned to the job.

Production hindrances

In industrial operations, usually the same causes that result in employee injuries are also responsible for damage to materials and equipment and for other hindrances to efficient production, such as:

1. Reduced output

2. Excessive scrap and need for rework

3. Unnecessary materials handling

4. Excessive man-hours per unit of production

5. Excessive machine-hours per unit of production

6. Poor employee morale

7. High labor turnover

These hindrances to production result from accidents—if the term "accident" is used to mean "any unexpected event that interrupts or interferes with the orderly progress of the production activity or process." Thus an accident may cause damage to equipment or material or a production delay, without resulting in an injury. Although an injury may or may not result from a given mishap, interference with the smooth flow of production can be expected.

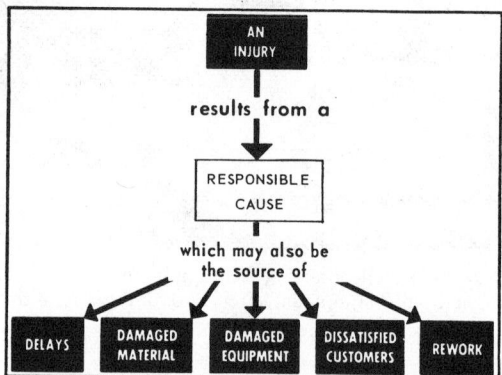

FIG. 5–2.—An injury is significant because it points to its basic cause, which may also be responsible for other operating hindrances (shown at bottom of chart).

Courtesy Hartford Accident and Indemnity Co.

The significance of an injury (Fig. 5–2), aside from humanitarian considerations, is that it points to a cause that may also be responsible for other operating hindrances. When this cause is identified and corrected, the basic source of both the injury and the related production hindrances is eliminated.

Role of the safety professional

In the removal of hazards from the job, cooperation between the safety professional and engineers in the company is essential. More often than not, the safety professional is not a registered professional engineer. He may have had no academic training in engineering. In such cases, it is especially important that the safety professional seek the advice and help of personnel within the company who are responsible for establishing engineering design criteria and for setting up job procedures.

The ultimate objective of all the company's engineering personnel is to design equipment and processes and to plan job procedures so that exposure to injury is eliminated or controlled. It is the safety professional's job to see that engineering personnel are acquainted with the particular hazards involved and the methods of eliminating them.

The safety professional should be able to discuss with management and the various supervisors involved the conditions responsible for accidents or potential accidents. He might well remember that, in making his corrective suggestions, he will be more successful if he can demonstrate that an injury is only one of *many* hindrances to efficient production.

The safety professional who has the necessary knowledge and skills for preventing accidents should be able to make recommendations for effective use of the facilities at hand for more efficient as well as safer production.

Unsafe acts, unsafe conditions

In most industrial accidents, *both* an unsafe condition and an unsafe act are contributing factors. In almost 80,000 work injuries reported in Pennsylvania in 1960, an unsafe condition was identified as a contributing factor in 98.4 percent of the nonfatal manufacturing cases.* In the same study, an unsafe act was identified as a contributing factor in 98.2 percent of the nonfatal manufacturing cases.

It must be remembered that an unsafe condition, in addition to being a direct cause of accidents in itself, often can lead people to perform unsafe acts. Many times, an unsafe act is the result of poor machine design, inadequately planned methods, and other engineering deficiencies. Thus, elimination of a hazard caused by an unsafe condition may also reduce the likelihood of injury from an unsafe act.

When an injury occurs, the unsafe condition is often not as glaringly evident as the unsafe act. Unless a careful study is made of the accident, the correctible physical hazard may escape notice.

Engineering for safety, therefore, should have as objectives both the elimination of hazardous conditions and the elimination of unsafe acts.

Machine Design

Machinery ranks fourth as a source of disabling work injuries, accounting for about 10 percent of all such injuries. Further pointing to the need for safe design of machines is the fact that they rank second as a source of permanent partial injuries, according to NSC's *Accident Facts*.

The design of machinery and equipment is an evolutionary process. It is always changing and dynamic, because design engineers constantly acquire wider experiences in the course of their everyday work. These experiences give them a broader scope and more initiative at the drawing board. This initiative, however, will be disciplined by the practical consideration of designing the most effective means for controlling hazards in the operation of machines or equipment.

Basic considerations

Evidently, the design of machinery must be further improved if the number of injuries caused by machines is to be reduced. How hazards can be eliminated from machines in the planning stage is illustrated by much of the evolution in machine design that has thus far occurred (Fig. 5–3). Behind this evolution has been the search for ever-increased efficiency and safety in machines of all types. In repeated instances, innovations in design which improved efficiency also eliminated or lessened a hazard. Conversely, measures taken to prevent accidents also improved production efficiency.

Policy in designing or purchasing should make sure that a machine is so designed that it will meet the requirements promulgated under authority of the OSHAct. Adding guards to control exposure to injury after the machine has been installed is usually expensive and second best.

---

*Pennsylvania Department of Labor and Industry, Harrisburg, Pa. *Industrial Injuries in Pennsylvania.* 1960.

5—Removing the Hazard from the Job

Fig. 5–3.—As a tool is being built, toolmakers and designers discuss production and safety features.

• A good example of elimination of hazards by design is the single-point lubricating system installed on many large machines. Use of this system not only eliminates the need to reach remote, and perhaps hazardous, oil points, but also ensures better lubrication.

• The evolution of safe and efficient design of a machine is well exemplified by the engine lathe. This machine was formerly driven by an overhead line shaft with a flat belt, with the spindle speed controlled by changing the belt on the cone pulleys. The feeds were controlled by changing the gears in the gear box. Since none of these danger points was guarded, the operator or other person standing nearby was exposed to all the hazards.

The modern engine lathe has an individual motor designed to be installed in the frame or structure of the machine. The drive mechanism, fully enclosed in the machine, is operated by V-belts. The speed of the spindle is controlled either by regulating a variable speed motor or by shifting a lever which acts as a clutch (as well as a gear shift). The feed is controlled by a shifting lever also. All the gears, drive belts, and sheaves are fully en-closed. Lubrication is done from a single point rather than at various points on the machine. The new machine is much safer than its predecessor and much more efficient.

Machine safety checklist

A safety checklist for the use of machine design engineers might well include the following points, to be incorporated in machine design where applicable and possible:

1. Design the machine so that it is impossible for the operator to get at the point of operation or any other hazard point while the machine is operating.

2. Design the machine so that corners and edges are rounded.

3. Locate machine controls so that the operator will not be in the vicinity of the point of operation while actuating the controls.

4. Place the controls so that the operator will not have to reach too far or move his body off balance in order to operate the machine.

5. Build power transmission and drive mechanisms as integral parts of the machine.

6. Build overload devices into the machine.

7. Design the machine for single-point lubrication.

8. Design mechanical, instead of manual, holding devices.

9. Design a mechanical device for feeding and ejecting parts so as to eliminate the use of hands for such operations.

10. Minimize motor drift time.

11. Provide fail-safe interlocks so that the machine cannot be started when it is being loaded or unloaded or being worked on.

12. Provide a grounding system for all electrical equipment.

13. Provide standard access platforms and ladders for inspection and maintenance of equipment.

14. Design component parts of equipment for easy and safe removal and replacement to facilitate maintenance.

15. Reduce sources of excessive noise.

In the past, many investigators of machine accident problems have thought that the practical, perhaps the only, solution was to add a guard here or there. However, installation of an external guard is not always the final, or the best, answer.

When the addition of a guard to a machine is being considered, this question should be asked: "Will this guard afford protection from a hazard but, in so doing, interfere with or defeat the machine's function?" In answer, it can be said that the best solution lies not in a guard that impairs production but rather in a change to a basic design that eliminates the hazard and increases efficiency. There can be little prospect for safe operation of a machine unless the idea of building safety into the machine's function is applied right on the drawing board.

A good illustration of the principle of designing safety into a mechanism is provided by the automatic coupler used on railroad cars. The automatic coupler is a great improvement over the old link and pin coupler, which required an employee to go between the cars to drop the pin and thus be exposed to a crushing hazard when the cars came together. Use of the automatic coupler eliminates exposure to a serious hazard. The device contributes to efficiency, too, by speeding train operation.

Built-in safety

The machine manufacturer, like any other businessman, wants to have satisfied customers. If his machines cause accidents and thus lead to economic losses, his customers will be unsatisfied. If the purchaser's order for a machine specifies that the machine must meet OSHA regulations and have safety built into it so that the operator will not be exposed to any working hazards, the manufacturer's designers will regard such a specification as a design requirement which they must meet.

In many instances, guards added to a machine after it has been installed in the plant are removed and not replaced. Accidents frequently result from this practice.

If a guard were an aid to production and efficiency rather than a hindrance, it is unlikely that the operator would run the machine without having the guard in place. Machine safety must be improved without either hindering the worker or reducing the efficiency of the machine.

Human factors

When safety is being designed into machines and equipment, special consideration must be given to human factors. Almost every machine requires an operator, and the mechanisms of many machines have become too complex to be understood by the average operator. Therefore, machines and equipment should be designed in terms of human limitations and capabilities, both mental and physical. (Also see Chapter 10, "Human Factors Engineering.")

Reflex actions. The employee who reaches into a danger zone, such as a point of operation, to adjust or remove material in process often does so by reflex action. Equipment must be designed so that an operator cannot endanger himself in this way.

Sensory limitations. The design engineer

must be aware of the limitations of the sensory organs. He must have some idea of the probable ability of the prospective operator to correlate and use the information to be supplied him by the sounds, lights, and other signal devices to be used in the operation of the equipment.

The design engineer must also remember that an operator may react slowly to sensory stimuli simply because he has normal human limitations or is disturbed by special problems, such as trouble at home, poor health, or poor attitude.

The equipment designer must not expect too much of the operator in regard to the sensory responses necessary in operating the machine. For example, each control lever should have a readily identifiable shape and location so that the operator can easily find the one he wants. Occasionally, an operator will inadvertently actuate a wrong control lever because he cannot differentiate between it and the correct one simply by touch. The same point applies to STOP, START, and INCH buttons on electrical controls. (See the discussion in Chapter 10, "Human Factors Engineering.")

Physical limitations. In addition to considering sensory limitations, the operator of the equipment may be suffering from temporary ailments, impairment of functions causing inefficiency, or fatigue. The operator's age may also affect his efficiency.

The physical environment in which the operator works must be taken into consideration, too. Environmental factors may include temperature, humidity, toxic gases, altitude, clothing, and many others. The designer should consider all factors and design the machine or equipment so it will not place unusual demands upon the physical and mental capabilities of the operator.

Advance analysis. If there are any defects in the design of a piece of machinery, it is usually only a matter of time before it fails and an accident ensues. Therefore, all possible faults in the design of equipment and all possible hazards in the work area of a machine, as well as the physical and mental capabilities of the operator, should be studied in advance for the purpose of establishing a safe work environment.

An advance analysis can, according to McFarland (see "References"), be based on the following considerations:

1. Operational job analysis should include a survey of the nature of the task, the work surroundings, the location of controls and instruments, and the way the operator performs his duties.

2. A functional concept of accidents is implied; that is, the errors that may occur while the operator is working on the machine are anticipated. The repetition or recurrence of near or real accidents clearly indicates a need for redesign.

3. From the "human limitations" point of view, it should be assumed that no worker is perfect. In fact, he may be far below the ability adjudged by the machine designer. If the worker's duties are too complex, the cumulative burden is great when he reaches or exceeds his limits of attention and ability.

4. A wide margin of safety should be provided to eliminate any possible situation that places the operator near his maximum ability with regard to aptitude or effort, especially when adverse factors enter the picture.

Job Safety Analysis

Job safety analysis (JSA) is a procedure used to review job methods and uncover hazards (a) that may have been overlooked in the layout of the plant or building and in the design of the machinery, equipment, and processes, or (b) that may have developed after production started. It is one of the first steps in safety training (see Chapter 9, "Safety Training").

Once the hazards are known, the proper solutions can be developed. Some solutions may be physical changes that control the hazard, such as placing a safeguard over exposed moving machine parts. Others may be job procedures that eliminate or minimize the hazard, for example, safe piling of materials. These will require training and supervision.

A job safety analysis can be written up in the manner shown in Fig. 5–4. In the left-

JOB SAFETY ANALYSIS WORK SHEET
JOB: Using a Pressurized Water Fire Extinguisher

WHAT TO DO (Steps in sequence)	HOW TO DO IT (Instructions) (Reverse hands for left-handed operator.)	KEY POINTS (Items to be emphasized. Safety is always a key point)
1. Remove extinguisher from wall bracket.	1. Left hand on bottom lip, fingers curled around lip, palm up. Right hand on carrying handle palm down, fingers around carrying handle only.	1. Check air pressure to make certain extinguisher is charged. Stand close to extinguisher, pull straight out. *Have firm grip, to prevent dropping on feet.* Lower, and as you do remove left hand from lip.
2. Carry to fire.	2. Carry in right hand, upright position.	2. Extinguisher should hang down alongside leg. (This makes it easy to carry and reduces possibility of strain.)
3. Remove pin.	3. Set extinguisher down in upright position. Place left hand on top of extinguisher, pull out pin with right hand.	3. Hold extinguisher steady with left hand. Do not exert pressure on discharge lever as you remove pin.
4. Squeeze discharge lever.	4. Place right hand over carrying handle with fingers curled around operating lever handle while grasping discharge hose near nozzle with left hand.	4. Have firm grip on handle to steady extinguisher.
5. Apply water stream to fire.	5. Direct water stream at base of fire.	5. Work from side to side or around fire. After extinguishing flames, play water on smoldering or glowing surfaces.
6. Return Extinguisher. Report Use.		

Fig. 5–4.—A job safety analysis is only as good as the details recorded.

hand column, the basic steps of the job are listed in the order in which they occur. The middle column describes how each step in handling is to be carried out. The right-hand column gives the safe procedures that should be followed to guard against the hazards and to prevent potential accidents.

For convenience, both the job safety analysis procedure and the written description are commonly referred to as JSA.

The four basic steps in making a job safety analysis are:

1. Select the job to be analyzed

2. Break the job down into successive steps

3. Identify the hazards and potential accidents

4. Develop ways to eliminate the hazards and prevent the potential accidents

111

JOB SAFETY ANALYSIS TRAINING GUIDE	JOB:		DATE:
DEPARTMENT:	TITLE OF MAN WHO DOES JOB:	FOREMAN/SUPR:	ANALYSIS BY:
	SECTION:		REVIEWED BY:
REQUIRED AND/OR RECOMMENDED PERSONAL PROTECTIVE EQUIPMENT:			APPROVED BY:

SEQUENCE OF BASIC JOB STEPS	POTENTIAL ACCIDENTS OR HAZARDS	RECOMMENDED SAFE JOB PROCEDURE
Break the job down into its basic steps, e.g., what is done first, what is done next, and so on. You can do this by 1) observing the job, 2) discussing it with the operator, 3) drawing on your knowledge of the job, or 4) a combination of the three. Record the job steps in their normal order of occurrence. Describe what is done, not the details of how it is done. Usually three or four words are sufficient to describe each basic job step. For example, the first basic job step in using a pressurized water fire extinguisher would be: 1) Remove the extinguisher from the wall bracket.	For each job step, ask yourself what accidents could happen to the man doing the job step. You can get the answers by 1) observing the job, 2) discussing it with the operator, 3) recalling past accidents, or 4) a combination of the three. Ask yourself: can he be struck by or contacted by anything; can he strike against or come in contact with anything; can he be caught in, on, or between anything; can he fall; can he overexert; is he exposed to anything injurious such as gas, radiation, welding rays, etc.? for example, acid burns, fumes.	For each potential accident or hazard, ask yourself how should the man do the job step to avoid the potential accident, or what should he do or not do to avoid the accident. You can get your answers by 1) observing the job for leads, 2) discussing precautions with experienced job operators, 3) drawing on your experience, or 4) a combination of the three. Be sure to describe specifically the precautions a man must take. Don't leave out important details. Number each separate recommended precaution with the same number you gave the potential accident (see center column) that the precaution seeks to avoid. Use simple do or don't statements to explain recommended precautions as if you were talking to the man. For example: "Lift with your legs, not your back." Avoid such generalities as "Be careful," "Be alert," "Take caution," etc.

Fig. 5-5.—Job safety analysis training guide.

112

Select the job

A job is a sequence of separate steps or activities that together accomplish a work goal. Some jobs can be broadly defined in general terms of what is accomplished. Making paper, building a plant, mining iron ore are examples. Such broadly defined jobs are not suitable for JSA. Similarly, a job can be narrowly defined in terms of a single action. Pulling a switch, tightening a screw, pushing a button are examples. Such narrowly defined jobs also are not suitable for JSA.

Jobs suitable for JSA are those job assignments that a line supervisor may make. Operating a machine, tapping a furnace, piling lumber are good subjects for job safety analyses. They are neither too broad nor too narrow.

Jobs should not be selected at random—those with the worst accident experience should be analyzed first if JSA is to yield the quickest possible results. In fact, some companies make this the focal point of their accident prevention program.

In selecting jobs to be analyzed and in establishing the order of analysis, top supervision of a department should be guided by the following factors:

1. FREQUENCY OF ACCIDENTS. A job that has repeatedly produced accidents is a candidate for a JSA. The greater the number of accidents associated with the job, the greater its priority claim for a JSA.

2. PRODUCTION OF DISABLING INJURIES. Every job that has produced disabling injuries should be given a JSA. The injuries prove that preventive action taken prior to their occurrence was not successful.

3. SEVERITY POTENTIAL. Some jobs may not have a history of accidents but may have the potential for severe injury.

4. NEW JOBS created by changes in equipment or in processes obviously have no history of accidents, but their accident potential may not be fully appreciated. A JSA of every new job should be made as soon as the job has been created. Analysis should not be delayed until accidents or near misses occur.

Break the job down

Before the search for hazards begins, a job should be broken down into a sequence of steps, each describing what is being done. Avoid the two common errors: (*a*) making the breakdown so detailed that an unnecessarily large number of steps results, or (*b*) making the job breakdown so general that basic steps are not recorded.

The technique of making a job safety analysis involves these steps:

1. Selecting the right person to observe

2. Briefing him on the purpose

3. Observing him perform the job, and trying to break it into basic steps

4. Recording each step in the breakdown

5. Checking the breakdown with the person observed

Select an experienced, capable, and cooperative person who is willing to share ideas. If the employee has never helped on a job safety analysis, explain the purpose—to make a job safe by identifying hazards and eliminating or controlling them—and show him a completed JSA. Tell him that he is not being watched to see if he works safely or not, but that the job is being studied, not him. Reassure him that he was selected because of his experience and capability.

To determine the basic job steps, ask "What step starts the job?" Then, "What is the next basic step?" and so on.

To record the breakdown, number the job steps consecutively as illustrated in the first column of the JSA training guide, illustrated in Fig. 5–5. Each step tells what is done, not how.

The wording for each step should begin with an "action" word, like "remove," "open," or "weld." The action is completed by naming the item to which the action (expressed by the verb) applied, for example, "remove extinguisher," "carry to fire."

In checking the breakdown with the person observed, obtain his agreement of what is done and the order of the steps. Thank the employee for his cooperation.

113

5—Removing the Hazard from the Job

Identify hazards and potential accidents

Before filling in the next two columns of the JSA—Potential Accidents or Hazards and Recommended Safe Job Procedure—begin the search for hazards. The purpose is to identify all hazards—both those produced by the environment and those connected with the job procedure. Each step, and thus the entire job, must be made safer and more efficient. To do this, ask yourself these questions about each step:

1. Is there a danger of striking against, being struck by, or otherwise making injurious contact with an object?

2. Can the employee be caught in, on, or between objects?

3. Can he slip or trip? Can he fall on the same level or to another?

4. Can he strain himself by pushing, pulling, or lifting?

5. Is the environment hazardous (toxic gas, vapor, mist, fume, or dust, heat or radiation)? (See discussion in Chapter 37, "Industrial Hygiene.")

Close observation and job knowledge are required. The job observation can be repeated as often as necessary until all hazards and potential accidents have been identified.

Include hazards that might result. Record the type of accident and the agent involved. To note that a man might injure a foot by dropping a fire extinguisher, for example, write down "struck by extinguisher."

Again check with the observed employee after the hazards and potential accidents have been recorded. The experienced employee will probably suggest additional ideas. You should also check with others experienced with the job. Through observation and discussion, you will develop a reliable list of hazards and potential accidents.

Develop solutions

The final step in a JSA is to develop a recommended safe job procedure to prevent occurrence of potential accidents. The principal solutions are:

1. Find a new way to do the job.

2. Change the physical conditions that create the hazards.

3. To eliminate hazards still present, change the job procedure.

4. Try to reduce the necessity of doing a job, or at least the frequency that it must be performed. This is particularly helpful in maintenance.

• To find an entirely new way to do a job, determine the work goal of the job, and then analyze the various ways of reaching this goal to see which way is safest. Consider work-saving tools and equipment.

• If a new way cannot be found, then ask this question about each hazard and potential accident listed: "What change in physical condition (such as change in tools, materials, equipment, or location) will eliminate the hazard or prevent the accident?"

When a change is found, study it carefully to find what other benefits (such as greater production or time saving) will accrue. These benefits should be pointed out when proposing the change to higher management. They make good selling points.

• The third solution in solving the job-hazard problem is to investigate changes in the job procedure. Ask of each hazard and potential accident listed: "What should the employee do—or not do—to eliminate this particular hazard or prevent this potential accident?" Where appropriate, ask an additional question, "How should he do it?" In most cases, the supervisor can answer these questions from his own experience.

Answers must be specific and concrete if new procedures are to be any good. General precautions—"be alert," "use caution," or "be careful"—are useless. Answers should precisely state what to do and how to do it. This recommendation—"Make certain the wrench does not slip or cause loss of balance"—is only "half good." It does not tell how to prevent the wrench from slipping.

Here, in contrast, is an example of a good recommended safe procedure that tells both "what" and "how": "Set wrench securely. Test its grip by exerting a slight pressure on it. Brace yourself against something immovable, or take a solid stance with feet wide apart, before exerting full pressure. This prevents loss of balance if the wrench slips."

● Often a repair or service job has to be repeated frequently because a condition needs correction again and again. To reduce the necessity of such a repetitive job, ask "What can be done to eliminate the cause of the condition that makes excessive repairs or service necessary?" If the cause cannot be eliminated, then ask "Can anything be done to minimize the effects of the condition?"

Machine parts, for example, may wear out quickly and require frequent replacement. Study of the problem may reveal excessive vibration is the culprit. After reducing or eliminating the vibration, the machine parts last longer and require less maintenance.

Reducing frequency of a job contributes to safety only in that it limits the exposure. Every effort still should be made to eliminate hazards and to prevent potential accidents through changing physical conditions or revising job procedures or both.

● Finally, check or test the proposed changes by reobserving the job and discussing the changes with the men who do the job. Their ideas about the hazards and proposed solutions may be of considerable value. They can judge the practicality of proposed changes and perhaps suggest improvements. Actually these discussions are more than just a way to check a JSA. They are safety contacts that promote awareness of job hazards and safe procedures.

Use JSA effectively

The major benefits of a job safety analysis come after its completion. However, benefits are also to be gained from the development work itself.

While making job safety analyses, supervisors learn more about the jobs they supervise. When employees are encouraged to participate in job safety analyses, their safety attitudes are improved and their safety knowledge is increased. As a JSA is worked out, safer and better job procedures and safer working conditions are developed.

But these important benefits are only a portion of the total benefits to be derived from the JSA program. The principal benefits were listed at the beginning of this discussion.

When a JSA is distributed, the supervisor's first responsibility is to explain its contents to employees and, if necessary, to give them further individual training. The entire JSA must be reviewed with the employees concerned so that they will know how the job is to be done—without accidents.

The JSA can furnish material for planned safety contacts. All steps of the JSA should be used for this purpose. The steps that present major hazards should be emphasized and reviewed again and again in safety contacts.

New employees on the job must be trained in the basic job steps. They must be taught to recognize the hazards associated with each job step and must learn the necessary precautions. There is no better guide for this training than a well-prepared JSA.

Occasionally, the supervisor should observe his employees as they perform jobs for which job analyses have been developed. The purpose of these observations is to determine whether or not the employees are doing the jobs in accordance with the safe job procedures. Before making such observations, the supervisor should prepare himself by reviewing the JSA in question so that he will have firmly in mind the key points that should be part of his observations.

Many jobs, such as certain repair or service jobs, are done infrequently or on an irregular basis. The employees who do them will benefit from pre-job instruction that reminds them of the important hazards and the necessary precautions. Using the JSA for the particular job, the supervisor should give this instruction at the time he makes the job assignment.

Whenever an accident occurs on a job covered by a job safety analysis, the JSA should be reviewed to determine whether or not it needs revision. If the JSA is revised, all employees concerned with the job should be informed of the changes and instructed in any new procedures.

When an accident results from failure to follow JSA procedures, the facts should be discussed with all the men who do the job. It should be made clear that the accident would not have occurred had the JSA procedures been followed.

All supervisors are concerned with improv-

Fig. 5–6.—Each year, thousands of persons glean product and service information at the National Safety Congress and Exposition where hundreds of companies exhibit their wares and tell about their services.

ing job methods to increase safety, reduce costs, and step up production. The job safety analysis is an excellent starting point for questioning the established way of doing a job. And study of the JSA may well suggest definite ideas for improvement of job methods.

Purchasing

The safety department should have excellent liaison not only with the engineering department but also with the purchasing department.

It should be the duty of the safety department to devise and put in writing the safety standards that will guide the purchasing department. These standards should be set up so that the hazards involved in a particular kind of equipment or material being purchased are eliminated (as by substitution of

a safe material for a dangerous one) or safeguarded for the protection of the worker, the machine, and the product.

The purchasing agent is not concerned closely with educational and enforcement activities, but he is vitally concerned with many phases of the engineering activities. It is his duty to select and purchase the various items of machinery, tools, equipment, and materials used in the organization; and it is his responsibility—at least in part, and often to a considerable degree—to see that in design, manufacture, and particulars of shipment of all these items, safety has received adequate attention.

In one example, a lead hazard occurred in the unloading of litharge (PbO), which was shipped in 10-gallon paint pails with covers. These pails arrived, either in trucks or in boxcars, with a film of litharge on the outside.

When they were moved, a lead concentration in the air 30 to 40 times the permissible limit was produced.

Several possible solutions were considered and tried, but the fundamental answer was to eliminate the hazard by having the purchasing department specify a rubber gasket under the pail lid as a part of the purchasing requirements. Thus the leakage, which created a serious health hazard, was easily controlled.

Specifications

The engineering department, with the help of the safety department, should specify to the purchasing department all the necessary safeguarding to be built into a machine before it is purchased.

Persons responsible for purchasing in an industrial plant are necessarily cost conscious. Consequently, the safety professional must have as complete a grasp as possible of the accident losses to the company in terms of specific machines, materials, and processes. If he is to recommend the expenditure of several thousand dollars for a superior grade of tool to be used throughout the plant, for instance, he should have evidence that the investment is justified.

Because of highly competitive marketing, manufacturers of machine tools and processing equipment often list safety devices designed for the protection of operators as separate auxiliary equipment. It is important that the safety professional be familiar with such auxiliary equipment and satisfy the purchasing agent of the need for its inclusion in the original order.

In some large organizations, the safety professional is charged with checking all plans and specifications for machinery and other equipment. In many organizations, particularly where certain items, such as goggles or safety shoes, are to be reordered from time to time, standard lists have been prepared through the cooperation of various operating officials, and purchases are selected only from among the types and from the companies shown on these approved lists. In still other establishments, the responsibility for design, quality, safety, and other features rests entirely on the man or men who are to use the articles. In such cases, the purchasing agent is responsible only for price, date of delivery, and similar items.

In many companies where purchases are made in enormous quantities and at a great investment of money, important duties are placed in three coordinate departments: (a) engineering department, where plans and specifications are prepared for all machinery and equipment to be purchased; (b) safety department, where these plans and specifications are carefully checked for safety and final inspections of articles purchased are carried out; and (c) purchasing department, which still has much latitude in making selections as well as in determining standards of quality, efficiency, and price.

Still another variable must be mentioned. Many companies have both a full-time purchasing agent and a full-time safety professional. However, in many other companies, especially smaller ones, these important duties are assumed by executives who devote part of their time to other activities. Nevertheless, the measures that should be taken to prevent accidents in the small plant are substantially the same as those in the large plant. The part that the purchasing agent can take in the safety program is similar; his interest will be the same and his activities will vary only in degree. His opportunity will lie in accepting as fully as possible all the suggestions that are presented here and in cooperating as closely as possible with others for the safety of all the workers in the establishment.

Specification of shipping methods. When materials are ordered, it may be desirable to specify that they be shipped in a particular manner. If safe and efficient shipping methods are worked out and then specified in the orders, the suppliers will be better able to deliver materials on time, in good condition, and in a shape or form that can be easily and safely handled by employees. Labeling of hazardous materials should be specified. Use DOT-authorized shipping labels.

Codes and standards

In purchasing, as in no other aspect of his job, the safety professional will find a need for thorough knowledge of the accident history of his plant, the costs involved in accidents, and the probable benefits of the changes he advocates.

5—Removing the Hazard from the Job

To fulfill his function in cooperation with the purchasing department, it is important that the safety professional be familiar with codes and standards. When he recommends a specific item of equipment, he should be able to state that it is a type approved by authoritative bodies and that it meets OSHA requirements.

It is usually necessary for the safety professional to consult with everyone concerned before setting up company standards for the guidance of the purchasing department.

There are many guides and standards that can be used as models. Accordingly, the safety director (and all others concerned with setting company standards) should be familiar with the following:

1. Codes and standards approved by the American National Standards Institute and other standards and specifications groups—see Chapter 23, "Sources of Help."

2. Codes and standards adopted or set by federal, state, and local governmental agencies, such as the Occupational Safety and Health Administration, the Bureau of Mines, and the National Bureau of Standards.

3. Codes, standards, and lists of approved or tested devices published by such recognized agencies as Underwriters Laboratories Inc., and fire protection organizations. For fire protection, the standards of the National Fire Protection Association should be followed. (See Chapter 23.)

4. Safe practice recommendations of such agencies as the National Safety Council, insurance carriers or their associations, and trade and industrial organizations.

Purchasing-safety liaison

With a background of knowledge gathered from such materials as those listed above, the safety professional should be well prepared to advise the purchasing department when required to do so.

What purchasing can expect. The purchasing agent can reasonably expect that the safety professional will:

1. Give specific information about machine and process hazards that can be eliminated by change in design or by guarding by the manufacturer.

2. Supply similar information about other equipment, tools, and materials, with the facts about injuries suffered and their causes.

3. Give specific information about health and fire hazards in the workplaces.

4. Provide information on federal and state safety requirements.

5. Supply on request additional special information on accident experience with machines, equipment, or materials when such articles are about to be reordered.

6. Request his assistance in the investigation of accidents that may have been caused by faulty equipment or material.

What the safety department can expect. Where there is effective liaison between the safety and purchasing departments, the safety professional can expect that the purchasing agent will:

1. Familiarize himself with the departmental and plant process hazards, especially in relation to machinery, other equipment, and materials.

2. Ask the safety department for information on hazards and accident costs, for federal and state safety requirements, and for lists of approved devices and appliances that may be helpful to him in considering purchases.

3. Acquaint himself with the specific location and departmental use of machinery or equipment that is about to be ordered.

4. Participate in accident investigations where injuries may have been caused through the failure of machinery, equipment, or materials.

Safety considerations

In the purchase of many articles, there is no need to consider safety. Some items, however, have a more important bearing upon safety than was at first suspected.

Utmost caution will be observed, of course, in the purchase of personal protective equipment, such as eye protection, respirators, masks, and the like; of equipment for the movement of suspended loads, such as ropes, chains, and cables; of equipment for the movement and storage of materials; of miscellaneous substances and fluids for cleaning and other purposes that might constitute or aggravate a fire or health hazard. Adequate labeling, identifying contents and calling attention to hazards, should be specified.

Investigation, however, may show that unsuspected hazards also lie in the purchase of very ordinary items, such as common hand tools, reflectors, tool racks, cleaning rags, paint for shop walls and machinery, and even filing cabinets. Among the factors to be considered by the purchasing agent are, for example, maximum load strength; long life without deterioration; sharp, rough, or pointed characteristics of articles; need for frequent adjustment; ease of maintenance; production of excessive fatigue; and hazard to the worker's health.

Here are a few examples of hazards created by purchased items that were thought to be safe. Goggles supplied to one group of workers were found to have imperfections in the lenses that caused eyestrain and headache, which led to fatigue and accidents. The toes of a laborer were crushed because the safety shoe he was wearing had an inferior metal cap and collapsed under a weight that should have been supported easily by a well made shoe. Men were supplied with wooden carrying boxes in another plant, when a proper type of metal box could have eliminated the hazard of splinters and perhaps an infected hand.

It is in the purchase of larger items, however, and especially in the buying of machines, that the more spectacular examples of purchasing for safety are found. Today, machines of many types are manufactured and may be purchased with adequate safeguards in place as integral parts of the machines. The enclosed motor drive is an outstanding example of such engineering for safety in machine construction.

When an order for equipment is about to be placed, the purchasing agent, if possible, will not consider any machine which has been only partly guarded by the manufacturer and which, therefore, will need to be fitted with makeshift safeguards after it has been installed. He will be in frequent consultation with the safety department in making all purchases where safety is a factor. He will also be particularly careful to see that every purchased machine complies fully with the safety regulations of the state in which it is to be operated.

Price considerations

When considering plant purchases from the standpoint of safety, the item of cost cannot be minimized. There is a constant struggle in the mind of the purchasing agent to reconcile quality, work efficiency, and safety with the price of an item.

Sometimes it may seem that the cost of an adequately safeguarded machine is out of all proportion to the cost of an unguarded machine plus the estimated cost of adding homemade safeguards. But the experience of many industrial plants has proved again and again that the best time to safeguard a machine or process is in the design stage. Safeguards planned and built as integral parts of a machine are the most efficient and durable.

The purchasing agent, through accident information supplied by the safety department, will be familiar with the costs of specific accidents in the plant. He will know that every accident has both direct and indirect costs. He will remember specific examples of accidents and their costs.

Of course, the purchasing agent must be able to defend his judgment. If he has made a study of the accident experience of his plant over a period of years, and if he has familiarized himself with accident details and accident costs, he will be able to justify his decisions. He will be firmly convinced that the safety program of his plant is necessary for production efficiency and profit, and that the safety department, with which he is cooperating, is really a production division, producing wealth by the elimination of accident wastes.

These are arguments that must appeal to all executives who are responsible for the success of the industrial organization. The executive who is already sold on safety will be favorable to expenditures reasonably justified in the interest of accident prevention. If an executive does not have this attitude, means of increasing his interest and of broadening his understanding of the accident situation must be sought. In this undertaking, the purchasing agent can undoubtedly count upon the active cooperation of the safety professional.

In some instances, the purchase of machinery or equipment involves engineering details of great importance. For such purchases, the industrial establishment will undoubtedly have a system whereby definite specifications, perhaps including drawings, will first be prepared by engineers. These plans and specifications will then be carefully checked by the safety professional before bids and estimates of cost are solicited.

The purchasing agent will have the plans and specifications at hand when he asks for prices. However, he will want to keep in close touch with the safety professional throughout the negotiations in order to use the latter's knowledge and experience in accident prevention.

After the purchase order has been made out and before it is signed, one other most important detail should not be overlooked. This is a statement, in language that cannot possibly be misinterpreted, that the articles ordered must comply fully with the federal and state safety laws and regulations of the locality in which they are to be used.

This statement must be made a part of the purchase order.

References

Bethlehem Steel Company, Bethlehem, Pa. "Job Safety Analysis." *Bethlehem Steel's Supervisory Safety Manual.* Chapter 5.

McFarland, Ross A. "Human Engineering Aspects of Safety." *Mechanical Engineering,* 76:407–10 (May 1954).

National Safety Council, 425 North Michigan Ave., Chicago, Ill. 60611
Accident Facts (published annually).
Self-Evaluation Checklists (OSHA).

Pennsylvania Department of Labor and Industry, Harrisburg, Pa. *Industrial Injuries in Pennsylvania.* 1960.

U.S. Department of Transportation, Office of Hazardous Materials, Washington, D.C. 20590. "Newly Authorized Hazardous Materials Warning Labels." Rev. January 1974. (Based on Title 49, C.F.R., sections 173.402, –403, and –404; import or export shipments are covered in Title 14, C.F.R., section 103.13.)

Accident Records and Injury Rates

Chapter 6

Accident Records   **122**
Uses of records . . . Recordkeeping systems . . . Accident reports
and injury records . . . Periodic reports . . . Use of reports . . .
The concept of bilevel reporting

ANSI Z16.1 and OSHA Recordkeeping   **134**

ANSI Z16.1 Recordkeeping and Rates   **134**
Use of Z16.1 . . . Standard formulas for rates . . . Calculation of
employee-hours . . . Disabling injuries . . . Classification of special
cases . . . Days charged . . . Definition of employment . . .
Interpretation of injury rates

Significance of Changes in Work-Injury Experience   **145**

Occupational Safety and Health Act (OSHAct) Recordkeeping   **146**
Definitions

Off-the-Job and Patron Injuries   **148**
Accidental injury experiences of employees . . . Patron and
nonemployee injury statistics

References   **149**

6—Accident Records and Injury Rates

In this chapter, the terms *accident, incident,* and *injury* are restricted to those that include occupational injuries and illnesses. In other chapters, accident and incident are used in their broad meanings—"unplanned events that interrupt the completion of an activity, and that may (or may not) include property damage or injury."

With the Williams-Steiger Occupational Safety & Health Act of 1970, a majority of employers are required by law to maintain certain records of work-related employee injuries and illnesses. In addition to these records, many employers are also required to make reports to state compensation authorities. Insuring agencies may also require reports. For contest and award programs, further reports based upon the American National Standards Institute Z16.1 Standard may be filed. Occupational injury and illness reports and records are now required of nearly every establishment by management or government.

Safety personnel are faced with two tasks—maintaining those records required by law and by their management, and maintaining records that are useful to an effective safety program. In general, unfortunately, the two are not always synonymous. A good recordkeeping system necessitates more data than that contained in almost all required forms.

This chapter deals with both aspects of recordkeeping. Specific legal requirements are beyond the scope of this volume because they differ from industry to industry and state to state and are subject to change with time. The appropriate federal and state authorities need to be contacted in order to obtain the most current requirements. An outline of the general recordkeeping requirements under the OSHAct, as it exists at this time, is presented in the last section of this chapter. The basic definitions and the method of keeping records under the Z16.1 Standard are also presented.

Although this chapter covers injuries and illnesses occurring to employees while on the job, standards exist both for off-the-job injuries to employees (ANSI Z16.3) and for injuries occurring to non-employees in establishments (ANSI Z108). Both standards are briefly summarized in the off-the-job and patron injuries section. The standards themselves should be consulted by the safety personnel concerned with those aspects of the overall safety program.

The first section deals with recordkeeping systems in general and contains sample forms and recommendations for establishing a good recordkeeping system.

Accident Records

Records of accidents and injuries are essential to efficient and successful safety programs, just as records of production, costs, sales, and profits and losses are esssential to efficient and successful operation of a business. Records supply the information necessary to transform haphazard, costly, ineffective safety work into a planned safety program that controls both conditions and acts that contribute to accidents. Good recordkeeping is the foundation of a scientific approach to occupational safety.

Uses of records

A good recordkeeping system can help the safety professional in the following ways:

1. Provide the safety personnel with the means for an objective evaluation of the magnitude of his accident problems and with a measurement of the overall progress and effectiveness of his safety program.

2. Identify high-rate units, plants, or departments and problem areas so that extra effort can be made in those areas.

3. Provide data for an analysis of accidents and illnesses that can point to specific circumstances of occurrence which can then be attacked by specific countermeasures.

4. Create interest in safety among supervisors by furnishing them with information about the accident experience of their own departments.

5. Provide supervisors and safety committees with hard facts about their safety problems so that their efforts can be concentrated.

6. Measure the effectiveness of individual countermeasures and determine if specific programs are doing the job that they were designed to do.

Recordkeeping systems

The system presented in this section is a model that can be used to provide the basic

```
Case No. 164                                    Date  2-12-
                    First Aid Report

Name  S. D. Smith                    Department  Shipping
Male ☒  Female ☐  Occupation  Packer      Foreman  Miller
Date of                    a.m.  Date of                        a.m.
Occurrence 2-12  Time 10   p.m.  First Treatment 2-12  Time 10  p.m.
Nature of
Occurrence  Splinter in index finger of left hand


Sent:  Back to Work ☒      Doctor ☐      Home ☐      Hospital ☐
Estimated Disability  0  days
Employee's Description of Occurrence  Handling wooden crates
  without gloves, ran splinter into finger

                            Signed  Mr. Miller
                                    First Aid

Issued by National Safety Council, Inc., 425 N. Michigan, Chicago, Ill. 60611
Form IS-6              Printed in U.S.A.              STOCK No. 129.26
```

FIG. 6–1.—A "First Aid Report" (4 × 6 in.) is prepared by the first aid attendant at the time an injured or ill person comes for treatment. A report should be prepared for each case, whether minor or serious. The report serves as a record and permits quick tabulation of such data as department, occupation, and the key facts of the occurrence.

items necessary for good recordkeeping. It is designed to dovetail with the present record-keeping requirements of the OSHAct and attempts to avoid a duplication of effort on the part of the personnel responsible for keeping records and filing reports. Provision is also made for easily entering the data necessary for using the Z16.1 Standard and computing those frequency and severity rates without the necessity of a separate set of records. Some of the forms presented in this section are also constructed with modern data processing methods in mind. In general, a self-coding, check-off form can save time for both the person who fills out the report and the person who is responsible for tabulating and processing the data on the forms.

A well designed form takes into account the person who will fill it out and the way in which the forms will be processed. It more likely will be filled out accurately and will present fewer problems for those who process and analyze the data. Care in the choice and design of forms will pay dividends in better, more reliable data.

The recordkeeping system in this section is not presented as the only way to keep records but rather as an example. The accident problems of individual establishments are unique and no one form or set of forms can possibly provide every establishment with all the data for solving all of its individual problems. A system that does a good job of collecting the basic facts, however, makes it easier to zero-in later on the specific data about a specific problem.

The following sections deal with occupational injuries and illnesses; property-damage accidents are covered in Chapter 7. Non-employee accidents are covered in Chapter 22.

6—Accident Records and Injury Rates

Accident reports and injury records

To be effective, preventive measures must be based on complete and unbiased knowledge of the causes of accidents. The primary purpose of an accident report is to obtain such information and not to fix blame. Since the completeness and accuracy of the entire accident record system depend upon information in the individual accident reports, be sure that the forms and their purpose are understood by those who must fill them out. Necessary training or instruction should be made available to these personnel.

The first aid report. The collection of injury data generally begins in the first aid department. The first aid attendant or nurse fills out a first aid report for each new case. Copies are sent to the safety department or safety committee, the worker's foreman, and other departments as management may wish. See Fig. 6–1.

The first aid attendant or the nurse should know enough about accident analysis and investigation to be able to record the principal facts about each case. The physician engaged or authorized by the employer to treat injured employees also should be informed of the basic rules for classifying cases since, at times, his opinion of the seriousness of an injury may be necessary to record the case accurately.

Supervisor's accident report. It is recommended that the supervisor make a detailed report about each accident, even when only a minor injury or no injury is the result. For purposes of ANSI Z16.1 or OSHA summaries, only those reports that meet the minimum severity level can be separated and tallied. Minor injuries occur in greater numbers than serious injuries and records of these injuries can be helpful in pinpointing problem areas. By working to alleviate these problems, serious injuries can sometimes be prevented. Furthermore, complications may arise out of the less serious injuries and their end result may be quite serious.

The supervisor's accident report form should be completed as soon as possible after an accident occurs. Copies of these reports should be sent to the safety department and to other designated persons. Information concerning unsafe acts and unsafe conditions is important in the prevention of future accidents, but often of even greater importance is information which shows *why* the unsafe conditions existed and *why* the injured persons acted unsafely. This type of information is particularly difficult to get unless it is obtained promptly after the accident occurs.

Generally, analyses of accidents are made only periodically, and often long after the accidents have occurred. Because it is often impossible to recall with accuracy the details of an accident, if details are not recorded accurately and completely at once, they may be lost forever.

Although all information may not be available at the time that the accident report is being filled out, items such as total time lost and amount of dollars of damage can be added later. This should not, however, prevent the other items from being answered as soon as possible after the accident occurs.

Two different supervisor's report forms are presented. Both of them fulfill all of the information requirements of the present OSHA 101 form. Provision is also made on both forms to easily enter ANSI Z16.1 information, which allows both types of recordkeeping to be accomplished in one form, without any duplication of effort. The two forms also include questions in addition to those contained in the present OSHA 101 form. These questions ask for additional basic data that should be known about each accident.

The first supervisor's report form (Fig. 6–2) is an open end, narrative type of form. The second form (Fig. 6–3) is a self-coding form that allows key-punching of data items directly from the form without the extra step of recoding this information for modern data processing equipment. By using self-coding forms, data processing equipment can be easily used to process the information and produce a variety of summary reports (such as summaries by department, by type of accident, etc.). Detailed cross tabulations can also be produced with little effort.

For definitions of the terms regarding severity of injury, what cases are recordable, etc., please consult the Z16.1 and OSHA sections of this chapter. For further clarification of OSHA definitions, it may be necessary to consult federal or state authorities (if your state has an approved plan in effect).

Injury and illness record of employee.
After cases are closed, the first aid report and the supervisor's report are filed by agency of injury (type of machine, tool, material, etc.), type of accident, or other factor that will facilitate use of the reports for accident prevention. Another form, therefore, must be used to record the injury experience of individual employees. (See Fig. 6–4.)

This form helps supervisors remember the experience of individual employees. Particularly in large plants where supervisors may have many people working for them, they probably cannot recall from memory the total number of injuries—especially if the injuries are minor—which are suffered by individual employees.

The employee injury card, therefore, fills a real need. It has space for recording such factors about the injury as date, classification, days charged, costs, and OSHA lost workdays.

Because of the importance of the personal factor in accidents, much may be learned about accident causes from studying employee injury records. If certain employees or job classifications have frequent injuries, a study of employee working habits, physical and mental abilities, training, job assignments, working environment, and the instructions and supervision given them may reveal more than a study of accident locations, agencies, or other factors.

Filing reports. After injury reports have been used to compile monthly summaries, the incomplete reports may be kept in a temporary file for convenient reference as later information about the injuries becomes available.

After the injury reports are complete, they should be filed in a way that will permit ready use of them for making special studies of accident conditions. To facilitate this work, the reports may be filed by agency of the injury, by occupation of the injured person, by department, or by some similar item.

The employee injury card should be cross referenced to the file location of the detailed accident report.

Periodic reports

The forms discussed in the preceding para-

graphs are prepared when the accidents occur; they are used to record the accidents and preserve information about contributing circumstances. Periodically, this information should be summarized and related to department or plant exposure so that the safety work can be evaluated and the principal accident causes brought into proper focus.

Monthly summary of injuries and illnesses.
The monthly summary of injury and illness cases (Fig. 6–5) allows for tabulating monthly and cumulative totals and the computation of ANSI Z16.1 frequency and severity rates as well as OSHA incidence rates. Space is also provided for yearly totals and rates. This form would be filled out on the basis of the individual report forms that were processed during the month.

The monthly summary should be prepared as soon after the end of each month as the information becomes available, but not later than the 20th of the following month. Because this report is prepared primarily to reveal the current status of accident experience, it is essential that this information be determined as soon as possible.

If an accident report is still incomplete on the 20th of the following month because the employee has not returned to work, or if the classification of an injury is still in doubt at that time, an estimate of the outcome should be made by the company physician and the report included with the completed cases in the monthly summary.

When definite information becomes available for estimated cases, any change in extent of disability or difference in estimated and actual time charge (or OSHA lost workdays) should be entered in the appropriate columns in the month of closing of the case, and the adjustment included in the cumulative figures for the year through that month. This procedure provides reliable monthly data and an easy method of adjusting cumulative data.

For example, for purposes of ANSI Z16.1 recordkeeping, an employee injured his arm on March 15, and on April 20 was still unable to work. The physician estimated that the employee would be able to work on April 27 and that there would be no permanent dis-

(*Text continues on page 130.*)

6—Accident Records and Injury Rates

SUPERVISOR'S ACCIDENT REPORT   Incident No. _____

(To be completed immediately after accident, even when there is no injury)

Company name and address _____

Plant or location address _____
(if different from above)

1. Name and address of injured_____ SSN_____ 2. Age_____
   (or ill) person

   _____ 3. Sex_____

4. Years of service_____ 5. Time on present job_____ 6. Title/occupation_____

7. Department_____ 8. Date of accident_____ 9. Time_____

10. Accident category (check) ☐ Motor Vehicle; ☐ Property Damage; ☐ Fire; ☐ Other_____

11. Severity of injury or illness ☐ Non-disabling; ☐ Disabling; ☐ Medical Treatment; ☐ Fatality

12. Amount of damage $_____ 13. Location_____

14. Estimated number of days away from job _____

15. Nature of injury or illness?_____

16. Part of body affected?_____

17. Degree of disability?_____
   (Temporary total; permanent partial; permanent total)

18. Causative agent most directly related to accident? (Object, substance, material, machinery, equipment, conditions)

   _____

   _____

   _____

   Was weather a factor?_____

19. Unsafe mechanical/physical/environmental condition at time of accident? (Be specific)

   _____

   _____

   _____

20. Unsafe act by injured and/or others contributing to the accident. (Be specific, must be answered)

   _____

   _____

   _____

21. Personal factors (improper attitude, lack of knowledge or skill, slow reaction, fatigue)

   _____

   _____

   _____

(over)

Fig. 6–2a.—"Supervisor's Accident Report" (8½ × 11 in.) provides record of contributing circumstances, as basis for specific remedial action. Foremen should be trained to fill it out properly.

22. Personal protective equipment required? (Protective glasses, safety shoes, safety hat, safety belt)__________

Was injured using required equipment?__________

23. What can be done to prevent a recurrence of this type of accident?
(Modification of machine; mechanical guards; correct environment; training)

____

____

24. Detailed narrative description (How did accident occur; why; objects, equipment, tools used, circumstance, assigned duties.

Be specific)__________

____

____

____

(Use additional sheets, as required)

25. Witnesses to accident__________

____

Date prepared__________ Signature of Foreman / Supervisor__________

Department__________

SUPERINTENDENT'S APPRAISAL AND RECOMMENDATION

a. In your opinion what action on the part of injured (or ill) person or others contributed to this accident?

____

____

b. Your recommendation__________

____

Date__________ Signature of Superintendent__________

FOR SAFETY OFFICE USE ONLY

Temporary Total ☐	Permanent Partial ☐	Death or Permanent Total ☐
Started losing time______	Part of Body______	
Returned to work______	Per cent loss or loss of use______	
Time charge______	Time charge______	Time charge: 6,000 days

Compensation $______ Medical $______ Other $______ Total $______

Name and address of hospital__________ Name and address of physician__________

Issued by NATIONAL SAFETY COUNCIL • 425 North Michigan Avenue • Chicago, Illinois 60611.

A NONGOVERNMENTAL, PRIVATELY SUPPORTED, PUBLIC SERVICE ORGANIZATION

This form meets information requirements of OSHA form 101.

Form IS-IA-25M-87301 Printed in U.S.A. Stock No. 129.21

FIG. 6–2b.—Central portion is filled in by higher level of management. Bottom portion of form is filled in by the safety department (as the facts become available) and contains data for computing injury rates and costs.

6—Accident Records and Injury Rates

SERVICE NO. (NSC)
► 1-9 _____

CASE OR FILE NO.
► 10-15 _____

OSHA No. 101   NSC revision
(Meets OSHA requirements
when Instruction 1. has
been followed.)

**SUPPLEMENTARY RECORD OF
OCCUPATIONAL INJURIES AND ILLNESSES**

THIS REPORT IS

► 16. 1 ☐ First report   2 ☐ Revised report

EMPLOYER

1. NAME _____

2. MAIL ADDRESS _____
_____

3. LOCATION, if different
from mail address _____

INJURED OR ILL EMPLOYEE

4. NAME _____

SOCIAL SECURITY NO. _____

► EMPLOYEE NO. 17-26 _____

5. HOME ADDRESS _____
_____

► 6. AGE 27-28 _____

► 7. SEX 29, 1 ☐ Male   2 ☐ Female

► 8. OCCUPATION (specify) _____

30-31, 01 ☐ Manager, official, proprietor
02 ☐ Professional, technical
03 ☐ Foreman, supervisor
04 ☐ Sales worker
05 ☐ Clerical worker
06 ☐ Craftsman—construction
07 ☐ Craftsman—other
08 ☐ Machinist
09 ☐ Mechanic
10 ☐ Operative (production worker)
11 ☐ Motor vehicle driver
12 ☐ Laborer
13 ☐ Service worker
14 ☐ Agricultural worker
15 ☐ Other
16 ☐ Unknown

9. DEPARTMENT _____
(Enter the name of department or division in which
the injured person is regularly employed.)

CLASSIFICATION OF CASE

A. **INJURY OR ILLNESS** (see code on Log, OSHA No. 100)

► 32, 1 ☐ Injury (10)
2 ☐ Occupational skin disease or disorder (21)
3 ☐ Dust disease of the lungs (pneumoconioses) (22)
4 ☐ Respiratory conditions due to toxic agents (23)
5 ☐ Poisoning (systemic effects of toxic materials) (24)
6 ☐ Disorder due to physical agents
(other than toxic materials) (25)
7 ☐ Disorder due to repeated trauma (26)
8 ☐ All other occupational illnesses (29)

B. **EXTENT OF INJURY OR ILLNESS**

► 33, 1 ☐ Fatality
2 ☐ Lost workday case
3 ☐ Nonfatal case without lost workdays

► C. Number of workdays lost 34-36 _____

D. Permanently transferred or terminated

► 37, 1 ☐ Yes   2 ☐ No

128

INSTRUCTIONS

1. Type or print the narrative where requested.
2. Check the one box which most clearly describes
each narrative statement.
3. See also original OSHA No. 101 for more details.
4. Complete form in duplicate. Retain original.
Mail duplicate to: National Safety Council,
425 N. Michigan Ave., Chicago IL 60611.

THE ACCIDENT OR EXPOSURE TO OCCUPATIONAL ILLNESS

10. **PLACE OF ACCIDENT
OR EXPOSURE**
(mail address) _____

11. **WHERE DID ACCIDENT OR EXPOSURE OCCUR?**

a. On employer premises

► 38, 1 ☐ Yes   2 ☐ No   3 ☐ Unknown

b. Place (specify) _____

► 39-40, 01 ☐ Office
02 ☐ Plant, mill
03 ☐ Shipping, receiving, warehouse
04 ☐ Maintenance shop
05 ☐ General or public area of employer premises
(corridor, washroom, lunchroom, parking lot, etc.)
06 ☐ Retail establishment
(store, restaurant, gasoline station, etc.)
07 ☐ Farm
08 ☐ Motor vehicle accident
09 ☐ Other
10 ☐ Unknown

12. **WHAT WAS THE EMPLOYEE DOING WHEN INJURED?** (Be specific)

_____
_____
_____
_____

a. Task performed at time of accident

► 41-42, 01 ☐ Operating machine
02 ☐ Operating hand tool (power or nonpower)
03 ☐ Materials handling
04 ☐ Maintenance & repair—machinery
05 ☐ Maintenance & repair—building & equipment
06 ☐ Motor vehicle driver, operator or passenger
07 ☐ Office and sales tasks, except above
08 ☐ Service tasks, except above
09 ☐ Other
10 ☐ Not performing task
11 ☐ Unknown

b. Activity at time of accident

► 43-44, 01 ☐ Climbing
02 ☐ Driving
03 ☐ Jumping
04 ☐ Kneeling
05 ☐ Lying down
06 ☐ Lifting
07 ☐ Reaching, stretching
08 ☐ Riding
09 ☐ Running
10 ☐ Sitting
11 ☐ Standing
12 ☐ Walking
13 ☐ Other
14 ☐ Unknown

13. HOW DID THE ACCIDENT OCCUR? (Describe fully the events)

_____

_____

_____

_____

a. AGENCY. (Object or substance involved)

  ACCIDENT AGENCY (1st column). The first object or substance involved in accident sequence.

  INJURY AGENCY (2nd column). The agency inflicting the injury. See also section 15.

  (Example: Worker fell from ladder and struck head on machine. Check "Ladder" under accident and check "Machine" under injury.)

ACCIDENT	INJURY	(Check one box in each column)
45-46, 01 ☐	47-48, 01 ☐	Machine
02 ☐	02 ☐	Conveyor, elevator, hoist
03 ☐	03 ☐	Vehicle
04 ☐	04 ☐	Electrical apparatus
05 ☐	05 ☐	Hand tool
06 ☐	06 ☐	Chemical
07 ☐	07 ☐	Working surface, bench, table, etc.
08 ☐	08 ☐	Floor, walking surface
09 ☐	09 ☐	Bricks, rocks, stones
10 ☐	10 ☐	Box, barrel, container (empty or full)
11 ☐	11 ☐	Door, window, etc.
12 ☐	12 ☐	Ladder
13 ☐	13 ☐	Lumber, woodworking materials
14 ☐	14 ☐	Metal
15 ☐	15 ☐	Stairway, steps
16 ☐	16 ☐	Other
17 ☐	17 ☐	Unknown
18 ☐	18 ☐	None

b. ACCIDENT TYPE. (First event in the accident sequence)

49-50, 01 ☐ Fall from elevation
02 ☐ Fall on same level
03 ☐ Struck against
04 ☐ Struck by
05 ☐ Caught in, under or between
06 ☐ Rubbed or abraded
07 ☐ Bodily reaction
08 ☐ Overexertion
09 ☐ Contact with electrical current
10 ☐ Contact with temperature extremes
11 ☐ Contact with radiations, caustics, toxic and noxious substances
12 ☐ Public transportation accident
13 ☐ Motor vehicle accident
14 ☐ Other
15 ☐ Unknown

ANSI Z16.1 INFORMATION

A. DEGREE OF DISABILITY UNDER Z16.1

60, 1 ☐ Not a recordable case under Z16.1
2 ☐ Temporary total disability
3 ☐ Permanent partial disability
4 ☐ Permanent total disability
5 ☐ Fatality

B. DAYS CHARGED UNDER Z16.1   61-64 _____

OCCUPATIONAL INJURY OR ILLNESS

14. DESCRIBE THE INJURY OR ILLNESS in detail and indicate the part of the body affected.

_____

_____

_____

a. NATURE OF INJURY OR ILLNESS. (Check most serious one)

51-52, 01 ☐ Amputation
02 ☐ Burn and scald (heat)
03 ☐ Burn (chemical)
04 ☐ Concussion
05 ☐ Crushing injury
06 ☐ Cut, laceration, puncture, abrasion
07 ☐ Fracture
08 ☐ Hernia
09 ☐ Bruise, contusion
10 ☐ Occupational illness
11 ☐ Sprain, strain
12 ☐ Other

b. PART OF BODY. (Check most serious one)

53-54, 01 ☐ Eyes
02 ☐ Head, face, neck
03 ☐ Back
04 ☐ Trunk (except back, internal)
05 ☐ Arm
06 ☐ Hand and wrist
07 ☐ Fingers
08 ☐ Leg
09 ☐ Feet and ankles
10 ☐ Toes
11 ☐ Internal and other

15. NAME THE OBJECT OR SUBSTANCE WHICH DIRECTLY INJURED THE EMPLOYEE. Also check one box in injury column under 13a.

_____

_____

16. DATE OF INJURY OR INITIAL DIAGNOSIS OF OCCUPATIONAL ILLNESS.

a. MONTH

55-56, 01 ☐ Jan.   07 ☐ July
02 ☐ Feb.   08 ☐ Aug.
03 ☐ March   09 ☐ Sept.
04 ☐ April   10 ☐ Oct.
05 ☐ May   11 ☐ Nov.
06 ☐ June   12 ☐ Dec.

b. DATE OF MONTH   57-58 _____

17. DID EMPLOYEE DIE?

59, 1 ☐ Yes   Date of Death _____
2 ☐ No

OTHER

18. NAME AND ADDRESS OF PHYSICIAN _____

_____

19. IF HOSPITALIZED, NAME AND ADDRESS OF HOSPITAL _____

_____

DATE OF REPORT _____

PREPARED BY _____

OFFICIAL POSITION _____

NSC will use for statistics only and hold report confidential.

Fɪɢ. 6–3

6—Accident Records and Injury Rates

INJURY AND ILLNESS RECORD OF EMPLOYEE							

S. D. Smith   845
_____   _____
(Name)   (Employee Number)

Occupation __Packer__ Department __Shipping__ Date Employed __1-23-69__

Case Number	Injury or Illness	Date of Occurrence	Z16 Type (Fatal, Permanent, Temporary, Non-disabling)	Z16 Days Charged	Comp. and Other Costs	OSHA Type (Fatal, Lost Workday, Non-Lost Workday)	OSHA Lost Workdays
164	Inj.	2-12-70	Non-disab.	0	0		
192	Ill.	8-3-71	Temporary	8	0	Lost Workday	6
349	Inj.	3-16-73	Temporary	3	0	Lost Workday	1

(Reverse side may be used for remarks)

Issued by NATIONAL SAFETY COUNCIL, 425 N. Michigan Ave., Chicago, Ill. 60611

IS3 Rev.   Printed in U.S.A.   Stock No. 129.23

Fig. 6–4.—The "Injury and Illness Record of Employee," a 4 × 6 in. card for recording injuries.

ability. Therefore, one temporary total disability and 42 days of lost time would be entered in the March summary.

The employee returned to work on April 27 as estimated, and treatment for stiffness of the arm was begun. On September 10, the physician decided that the injury had caused 20 percent permanent disability of the arm. The entry for March would be changed, and the cumulative through September would be computed to reflect the change in the March estimate. That is, the injury was changed in the ANSI Z16.1 records to a permanent partial disability, and the 42 days charge was changed to 900 days (20 percent of 4500 days).

Annual report. Whereas monthly summaries of injuries are prepared primarily to show the trend of safety performance during the year, annual reports are prepared so that comparisons for the longer period may be made with the experience of previous years, and with the experience of similar companies and of the industry as a whole.

Especially in smaller companies, monthly injury rates often show wide variations which make it difficult to evaluate the safety performance correctly. If a small company has only two or three injuries in a year, in the months in which these injuries occur the rates will jump to extreme highs, but in the other months, the rates will be zero. These variations will be smoothed out in annual totals, however, and the rate for the longer period will have increased significance.

Annual reports should be prepared as soon after the close of the year as information becomes available, but again not later than the twentieth of the month following. If any injury report is still incomplete twenty days after the close of the year because the employee has not returned to work, or if the classification of any injury is in doubt at that time, an estimate of the outcome should be made by the company physician and the report included with the completed cases.

When annual records are closed, the estimated charges or OSHA lost workdays on

130

cases still pending become final for that year, and any injuries or time charges reported later are considered only in rates for two or three years of which that year is a part. They need not be added to the record of that year, and they should *not* be added to the record of the succeeding year.

If time charges or OSHA lost workdays which occurred in one year were included in the record for the next year, the latter year's experience would be biased and the record would fail to give a true picture of the severity of injuries or OSHA lost workdays which actually happened during that time.

On the other hand, absolute accuracy in rates does not justify delaying the annual report more than twenty days after the end of the year in order to get exact time charges or OSHA lost workdays, or to make sure that there are no delayed cases. It is better to omit such delayed cases from the rates entirely, unless a two- or three-year rate is calculated, or unless the company wishes to make revisions for its own information.

Use of reports

Reports to management. Management is increasingly interested in the accident experience of its companies. Therefore, monthly and other periodic summary reports, which show the results of the safety program, should be furnished to the responsible executive. Such reports need contain no details nor technical language and may be supplemented by simple charts or graphs to show the recent accident experience in relation to that of the preceding period and that of companies engaged in similar work.

In a large company, departmental data help the executive visualize accident experience in various plant operations and provide a yardstick for better evaluation of progress made in the elimination of accidents. If cost figures are obtained, comparisons of such figures for different periods are of particular interest.

Bulletins to supervisors. A supervisor is primarily interested in his own department and workmen. One of the most effective ways to create and maintain the interest of supervisors and foremen in accident prevention is to keep them informed about the accident records of their departments. Department injury rates based on sufficient amounts of exposure reflect the effectiveness of the supervisors' safety activities.

Since interest increases with knowledge, bulletins containing analyses of the principal causes of accidents in each department not only will maintain the supervisors' interest at a high level, but will provide the supervisors with the type of information which will help them to effect further reductions in injuries.

The agenda for employee safety meetings should particularly include information about the outstanding injury and illness problems, frequent unsafe practices, hazardous types of equipment, and similar data disclosed by analysis of the accidents that have occurred in the department and plant.

Bulletin board publicity. Posting a variety of materials on bulletin boards is one of the best ways to maintain the interest of employees in safety. Accident records furnish many items, such as the following:

No-injury records

Unusual Accidents

Frequent causes of accidents

Charts showing reductions in accidents

Simple tables comparing departmental records

Standings in contests

Reports to National Safety Council. Each year the National Safety Council requests an annual summary report of occupational injury and illness experience from each member. These reports are tabulated to determine accident rates by industry.

An annual "Work Injury Rates" pamphlet containing these frequency and severity rates by industry is published and distributed to members, so that each company may compare its experience with the average experience of other companies in the same industry. These reports (along with data from the Bureau of Labor Statistics, state and national vital statistics authorities, state compensation authorities, and other federal, state, and private agencies) are also tabulated and shown in *Accident Facts*, the Council's annual statistical publication.

(*Text continues on page 134.*)

6—Accident Records and Injury Rates

Company __ABC Mfg. Co.__ Plant _____

Period	Average Number of Employees	Number of Man-Hours Worked	Fatal, Permanent Total	Permanent Partial	Temporary Total	Total Z16 Cases	Frequency Rate*	Fatal, Permanent Total	Permanent Partial	Temporary Total
			Disabling Injuries and Illnesses					Time Charges		
Jan.	2,060	345,000	0	1	1	2	5.80	0	150	18
Feb.	2,010	298,000	0	0	3	3	10.07	0	0	22
Cum.		643,000	0	1	4	5	7.78	0	150	40
Mar.	2,080	353,000	0	0	1	1	2.83	0	0	42
Cum.		996,000	0	1	5	6	6.02	0	150	82
Apr.	2,000	332,000	0	0	4	4	12.05	0	0	47
Cum.		1,328,000	0	1	9	10	7.53	0	150	125
May	2,150	375,000	0	0	5	5	13.33	0	0	63
Cum.		1,703,000	0	1	14	15	8.81	0	150	188
June	1,900	303,000	0	1	0	1	3.30	0	1,000	0
Cum.		2,006,000	0	2	14	16	7.98	0	1,150	203
July	1,825	295,000	0	1	3	4	13.56	0	250	23
Cum.		2,301,000	0	3	17	20	8.69	0	1,400	226
Aug.	1,800	285,000	0	0	4	4	14.04	0	0	31
Cum.		2,586,000	0	3	21	24	9.28	0	1,400	257
Sept.	1,875	301,000	0	0	0	0	0.00	0	0	0
Cum.		2,887,000	0	3	21	24	8.31	0	1,400	257
Oct.	1,795	302,000	0	0	1	1	3.31	0	0	14
Cum.		3,189,000	0	4	21	25	7.84	0	2,300	229
Nov.	1,665	280,000	0	0	2	2	7.14	0	0	17
Cum.		3,469,000	0	4	23	27	7.78	0	2,300	246
Dec.	1,620	275,000	1	0	0	1	3.64	6,000	0	0
YEAR		3,744,000	1	4	23	28	7.48	6,000	2,300	246

*Z16 Rates: Frequency rate is the total number of Z16 cases per 1,000,000 man-hours worked. Severity rate is the total Z16
†OSHA Rate: Incidence rate is the total number of OSHA cases per 200,000 man-hours worked.
Issued by NATIONAL SAFETY COUNCIL, 425 N. Michigan Ave., Chicago, Illinois 60611

Printed in U.S.A.

FIG. 6–5.—Results of the safety program may be gaged from data on this "Monthly Summary of Injuries and Illnesses" form (8½ × 14 in.). Rates computed for the month, year to date (cumulative), and year permit com-

INJURIES AND ILLNESSES, 19___

~~Dayton, Ohio~~ _____ Department ___ All _____

Total Z16 Charges	Severity Rate*	COSTS (Compensation, Other)	OSHA CASES						FIRST AID CASES ONLY
			Fatals	Lost Workday Cases	Non Lost Workday Cases	Total OSHA Cases	Incidence Rate†	Total Lost Workdays	
168	487	#284.50	0	3	18	21	12.2	42	20
22	74	42.65	0	4	27	31	20.8	34	36
190	295	327.15	0	7	45	52	16.2	76	56
42	119	77.82	0	1	9	10	5.7	12	10
~~228~~ 232	229	404.97	0	8	54	62	12.4	88	66
47	142	92.64	0	5	35	40	24.1	64	42
275	207	497.61	0	13	89	102	15.4	152	108
63	168	123.24	0	7	45	52	27.7	88	55
~~353~~ 338	207	620.85	0	20	134	154	18.1	240	163
1,000	3,300	985.56	0	1	13	14	9.2	25	15
1,353	674	1,606.41	0	21	147	168	16.7	265	178
273	925	368.18	0	5	40	45	30.5	91	51
1,626	707	1,974.59	0	26	187	213	18.5	356	229
31	109	63.24	0	5	43	48	33.7	123	53
1,657	641	2,037.83	0	31	230	261	20.2	479	282
0	0	843.66	0	0	9	9	6.0	0	12
~~2,515~~ 1,657	871	2,881.49	0	31	239	270	18.7	479	294
14	46	45.60	0	1	5	6	4.0	10	6
2,529	793	2,927.09	0	32	244	276	17.3	489	300
17	61	36.76	0	2	23	25	17.9	13	26
2,546	734	2,963.85	0	34	267	301	17.4	502	326
6,000	21,818	6,785.25	1	1	7	9	6.5	2	8
8,546	2,283	9,749.10	1	35	274	310	16.6	504	334

charges per 1,000,000 man-hours worked.

Stock No. 129.25

parisons between time periods, and departments, plants, and companies during the same period. Changes in the classification of injuries and other adjustments can be made easily. Details on this form are on page 125.

These annual summary reports are also used by the National Safety Council to evaluate members' experience for the presentation of awards under the Council's "Award Plan for Recognizing Good Industrial Safety Records."

National Safety Council members may also, according to a vote by their section membership, establish and compete in safety contests that are administered by the Council. Monthly bulletins are sent to contestants so that they can compare their experience with other competitors. Awards are made to the leaders of the individual contests at the end of each contest year.

The concept of bilevel reporting

As mentioned earlier in this chapter, each establishment has accident problems that are different from those of establishments in other industries and, in many cases, different from those of other establishments in the same industry. No individual form or set of forms can possibly include all of the information necessary to fully investigate the causes of all accidents. With this in mind and because very long forms are rarely completed accurately and are very often received with much resistance by those persons who must fill them out, the concept of bilevel reporting has arisen.

The basic idea of bilevel reporting is that in addition to the general information contained in the standard report form (such as the Supervisor's Report Form), further facts are necessary concerning specific types of accidents. To obtain this additional data, a supplementary form, containing a few specific questions about the accident type under investigation, is prepared and made available. This supplementary form is then filled out, and attached to the regular report—only for those accidents about which the investigator is trying to obtain in-depth data.

When a sufficient number of the bilevel forms have been collected, the supplementary form is discontinued and the results can be analyzed. In this way, a minimum of time needs to be spent by the persons who fill out the forms in order to obtain useful information. Several bilevel forms, each a different color for easy handling, can be used at any one time. They can be discontinued when their job is done and replaced by other supplementary forms, while the basic report form

remains unchanged and in use.

ANSI Z16.1 and OSHA Recordkeeping

During the period that industry is becoming accustomed to the OSHA recordkeeping system, no link to historical records and therefore measures of progress or lack of it will be available. The National Safety Council, at this time, supports the maintenance of American National Standard Z16.1 records by establishments, at least until the OSHA recordkeeping stabilizes. Also many people feel that ANSI Z16.1 should be retained because OSHA recordkeeping does not provide an adequate measure of the severity of accidents.

Work on the ANSI Z16.1 standard dates back to Bulletin No. 276, "Standardization of Industrial Accident Statistics," published by the U.S. Bureau of Labor Statistics in 1920. Although developed by a body representing governmental statistical agencies, the rate provisions of this Bulletin were widely followed in whole or part by private agencies. At a national conference on Industrial Accident Prevention, called by the U.S. Secretary of Labor in Washington in 1926, a resolution was adopted in favor of a revision of Bulletin 276 by a committee set up under the procedures of the American Standards Association (presently the American National Standards Institute). As a result of the work of this committee, the first edition of this standard was completed and approved in 1937 as Z16.1-1937. Since then this standard has been reviewed and revised.

Using the forms presented in this chapter, ANSI Z16.1 records can be easily kept along with OSHA records. By using a simple device, such as a red check mark or asterisk alongside OSHA Log entries to indicate the ANSI Z16.1 cases, little effort will be necessary to go to the individual reports and obtain the data for making annual summaries of ANSI Z16.1 cases.

ANSI Z16.1 Recordkeeping and Rates

Safety performance is *relative*. Only when a company compares its injury experience with that of similar companies, or with that of the entire industry of which it is a part, or with its own previous experience can it obtain a meaningful evaluation of its safety accomplishments.

Use of Z16.1

To make such comparisons, a method of measurement is needed which will adjust for the effects of certain variables which contribute to differences in injury experience. For two reasons, injury totals alone cannot be used.

First, a company with many employees may be expected to have more injuries than a company with few employees. Second, if the records of one company include all the injuries treated in the first aid room, whereas the records of a similar company include only injuries serious enough to cause lost time, obviously the first total will be larger than the second.

A standard procedure for keeping records, which provides for these variables, is included in ANSI Z16.1. First, this procedure uses frequency and severity rates which relate disabling injuries, and days charged as a result of them, to the number of man-hours worked; thus these rates automatically adjust for differences in the hours of exposure to injury. Second, this procedure specifies the kinds of injuries which should be included in the rates.

These standardized rates, which are easy to compute and to understand, have been accepted generally as uniform procedure in industry and permit the necessary and desired comparisons.

A chronological arrangement of these rates for a company will show whether its level of safety performance is improving or getting worse. Within a company, the same sort of arrangement by departments will not only show the trend of safety performance for each department, but may reveal to management information which will make safety work more efficient.

If it is found, for example, that the trend of injury rates in a company is up, a review of the rate trends by departments may reveal that this adverse change is accounted for by the rates of just a few departments. With the source of the highest company rates thus isolated, safety efforts may be concentrated at the points of worst experience.

A comparison of current injury rates with those of similar companies and with those of the industry as a whole will provide the safety professional with a more reliable evaluation of the safety performance of his company than he could obtain merely by reviewing numbers of injuries.

Standard formulas for rates

The injury frequency rate and the injury severity rate are based on standard formulas as set forth in ANSI Z16.1.

Frequency rate. The disabling injury frequency rate relates the injuries to the hours worked during the period and expresses them in terms of a million-hour unit by use of the following formula:

$$\frac{\text{Number of disabling injuries} \times 1{,}000{,}000}{\text{Employee-hours of exposure}}$$

Severity rate. The disabling injury severity rate relates the days charged to the hours worked during the period and expresses them in terms of a million-hour unit by use of the following formula:

$$\frac{\text{Total days charged} \times 1{,}000{,}000}{\text{Employee-hours of exposure}}$$

Average days charged. The frequency and severity rates show, respectively, the rate at which disabling injuries occur and the rate at which time is charged. A third measure included in the standard procedure shows the average severity of the disabling injuries. It is called the average days charged per disabling injury and may be calculated by either of the following formulas:

$$\text{I.} \quad \frac{\text{Total days charged}}{\text{Total disabling injuries}}$$

or

$$\text{II.} \quad \frac{\text{Severity rate}}{\text{Frequency rate}}$$

An example will show how injury rates are calculated. A certain plant worked 365,000 man-hours during a year in which there were five disabling injuries with a total of 175 days charged. To determine its rates, the appropriate figures are substituted in the formulas, as follows:

$$\text{Frequency rate} = \frac{5 \times 1{,}000{,}000}{365{,}000} = 13.70$$

$$\text{Severity rate} = \frac{175 \times 1,000,000}{365,000} = 479$$

Average days charged per disabling injury

$$= \frac{175}{5} = 35, \text{ or } \frac{479}{13.70} = 35$$

Dates for compiling rates. Injury rates should be determined as soon after each period (month or year, for example) as the information becomes available. A reasonable time may be allowed for completion of reports. However, absolute accuracy in rates does not justify long delays.

The ANSI Z16.1 Standard suggests the following schedule for compiling injury rates:

1. Annual frequency rates should be based on all disabling injuries occurring within the year and reported within twenty days after the close of the year. Monthly frequency rates should be based on all disabling injuries occurring within the month and reported within twenty days after the close of the month.

2. Days charged for reported cases in which disability continues beyond the closing dates in (1) should be estimated on the basis of medical opinion of probable ultimate disability.

3. Cases first reported after closing dates stated in (1) need not be included in the rates for that period, or for any similar subsequent period. However, they should be included, and should replace estimates, in rates for longer periods of which that period is a part.

A disabling injury, and all days lost or charged because of it, should be charged to the date on which the injury occurred, except that for injuries, such as bursitis, tenosynovitis, pneumoconiosis ("black lung"), or silicosis, which do not arise out of specific accidents, the date of the injury should be the date when the injury is first reported. (Remember, OSHA calls these "occupational illnesses.")

Disabling injury index. As an aid to those companies expressing a desire to combine frequency and severity into a single measure, the following index is shown:

$$\text{Disabling injury index} = \frac{F \times S}{1,000}$$

where F is the disabling injury frequency rate and S is the disabling injury severity rate. This measure combines both frequency and severity to yield an index of total disabling injury experience.

In the form shown it can be correctly applied only for ranking total experience from "best" to "worst." If the measure is to be used to determine percentage of improvement, or for comparing the degree of difference between two units, the square root of the index must be taken before such comparisons are made.

Calculation of employee-hours

Employee-hours used in calculating injury rates is the total number of hours worked by all employees, including those of operating, production, maintenance, transportation, clerical, administrative, sales, and other departments.

Employee-hours should be calculated from the payroll or time clock records. If this method cannot be used, they may be estimated by multiplying the total employee-days worked for the period covered by the number of hours worked each day.

The experience of a central administrative office or central sales office of a multi-establishment should not be included in the experience of any one establishment, nor should it be prorated among the establishments, but it should be included in the overall experience of the concern.

In calculating hours for employees who live on company property, only those hours are counted during which the employee is on duty.

For traveling personnel, such as salesmen, executives, and others whose working hours are not defined, an average of 8 hours per day should be used in computing the hours worked.

For standby employees who are restricted to the confines of the employer's premises, including seamen aboard vessels, all stand-by hours should be counted, as well as all work injuries occurring during such hours.

Disabling injuries

The standard specifies that a work injury is any injury, including occupational disease and other work-connected disability, which arises out of and in the course of employment. The following descriptions paraphrase the standard. For full details, refer to the standard.

Occupational disease is a disease caused by exposure to environmental factors associated with employment. Work-connected disability includes such ailments as silicosis, pneumoconiosis, tenosynovitis, bursitis, and loss of hearing. Even though there is no traumatic injury in such disabilities, if they are work connected, they are considered work injuries.

Definitions. To ensure uniformity in the computation of injury rates and thereby provide for comparison among rates, the standard specifies that only *disabling injuries* shall be counted in the computation of standard injury rates. In general terms, a disabling injury is one which results in death or permanent impairment or which renders the injured person unable to work for a full day on any day after the day of injury. Disabling injuries are of four classes, as follows:

1. *Death* is any fatality resulting from a work injury, regardless of the time intervening between injury and death.

2. *Permanent total disability* is an injury other than death which permanently and totally incapacitates an employee from following any gainful occupation, or which results in the loss (or the complete loss of use) of any of the following in one accident: (*a*) both eyes; (*b*) one eye and one hand, or arm, or foot, or leg; (*c*) any two of the following not on the same limb: hand, arm, foot, leg.

3. *Permanent partial disability* is any injury other than death or permanent total disability which results in the complete loss or loss of use of any member or part of a member of the body, or any permanent impairment of functions of the body or part thereof, regardless of any pre-existing disability of the injured member or impaired body function.

4. *Temporary total disability* is any injury which does not result in death or permanent impairment, but which results in one or more days of disability.

Temporary total disability—key points. Of the four classes of disabling injuries used in calculating standard injury rates, temporary total disability is the most difficult to interpret uniformly. Key points in the definition are:

- *Day of disability.* A day of disability is any day on which an employee is unable, because of injury, to perform effectively throughout a full shift the essential functions of a regularly established job which is open and available to him. Days include Sundays, days off, plant shutdowns, and other nonwork days subsequent to the day of injury. (Note: Disability days for an injury are not counted when scheduled charge(s) apply.)

Here is how the standard describes it.

1.5.1.1. The day of injury and the day on which the employee was able to return to full-time employment shall not be counted as days of disability; but all intervening calendar days, or calendar days subsequent to the day of injury (including weekends, holidays, other days off, and other days on which the plant may be shut down), shall be counted as days of disability provided they meet the criteria of the preceding paragraph.

1.5.1.2. Time lost on a workday, or on a non-workday, subsequent to the day of injury, ascribed solely to the unavailability of medical attention or of necessary diagnostic aids, shall be considered disability time, unless in the opinion of the physician authorized by the employer to treat the case the person was able to work on all days subsequent to the day of injury.

1.5.1.3. If the physician authorized by the employer to treat the case is of the opinion that the injured employee is actually capable of working a full normal shift at a regularly established job, but has prescribed certain therapeutic treatments, the employee may be excused from work for those treatments without counting the excused time as disability time, provided: (*a*) the time required to obtain the treatments does not, on any workday, prevent him from performing effectively the essential functions of his job assignment on that day, and (*b*) the treatments are professionally administered and constitute more than simple rest.

1.5.1.4. If the physician authorized by the employer to treat the case is of the opinion that the injured employee was actually capable of working a full normal shift at a regularly established job, but because of transportation problems, associated with his injury, the employee is forced to arrive at his place of work late, or to leave the workplace

137

before the established quitting time, such lost time may be excused and not counted as disability time, provided: (a) that the excused time does not materially reduce his working time, and (b) that it is clearly evident that this failure to work the full shift hours is the result of a bona fide transportation problem and not a deviation from the "regularly established job."

1.5.1.5. If the injured employee receives medical treatment for his injury, the determination of his ability to work shall rest with the physician authorized by the employer to treat the case. If the employee rejects medical attention offered by the employer, the determination may be made by the employer based upon the best information available to him. If the employer fails to provide medical attention, the employee's decision shall be controlling.

• *Specific disabilities.* Following are examples of specific disabilities and the standard procedures for handling them. In all these cases, the regular shift is assumed to be 8:00 A.M. to 5:00 P.M.

1. Injury occurred at 4:30 P.M. on Monday. The employee worked on Tuesday and Wednesday, but delayed infection prevented him from working from Thursday through the following Monday. He returned to work on Tuesday at 11:00 A.M. after seeing his doctor. This is a disabling injury with five chargeable days of disability (Thursday through Monday).

2. Injury occurred at 1:00 P.M. on Friday. Saturday was not a workday, but the doctor certified that the employee could not have worked on that day if it had been, although he could have worked on Sunday and thereafter. The employee returned to work on Monday. This is a disabling injury with one day of disability (Saturday), even though the employee lost no actual work time except Friday afternoon.

3. Injury occurred at 4:00 P.M. on Tuesday. The employee was hospitalized for treatment of his injury until noon on Wednesday when the doctor released him with permission to resume work on Thursday morning. This is a disabling injury with one day of disability.

4. Injury occurred at 11:00 A.M. on Tuesday. The employee was treated by the doctor and sent home with instructions not to work Wednesday, but to report to the doctor's office Wednesday afternoon. At 5:30 P.M. Wednesday the employee was seen by the doctor, who discharged him as fit to work immediately. This is a disabling injury with one day of disability.

5. The doctor determined that an injury would not allow an employee to return to his own job for at least a week. On checking with the personnel department, the doctor learned that another job was open which the employee could handle, and the employee was cleared for this work. This is not a disabling injury because the employee could perform another regularly established job which was open and available to him.

• *Additional key points* with regard to all the classifications of disabling injuries are:

1. An injury need not be traumatic—there does not have to be a visible wound. More precisely, the standard covers work-connected *disabilities*, which may include such conditions as bursitis, tenosynovitis, loss of hearing, and similar ailments.

2. An injury need not arise out of an accident, although it must be work connected. Examples of such injuries would be dermatitis, silicosis, and bursitis. A strain resulting from overlifting would be a work injury even though there was no accident.

3. The classification of an injury is entirely independent of workmen's compensation laws, and rulings of workmen's compensation agencies. This provision is necessary to promote uniformity in injury classification. Otherwise, similar injuries, might be classified differently in different states.

As may readily be seen, minor injuries are excluded in the calculation of the standard injury rates. The reason is that it is impossible to obtain standardization of minor cases. The standard does provide classification for minor injuries, though, as *medical treatment injuries,* and identifies them as injuries that do not result in death, permanent impairment, or temporary total disability, but which require medical treatment, including first aid. Although these nondisabling injuries are not included in the standard injury rates, they should be given consideration and attention by the safety professional. He may well keep totals of them, and note their frequency, the departments in which they occur, the types of injuries, and the relationship between them and disabling injuries.

Classification of special cases

In addition to the usual injuries and occupational diseases, the following are specifically identified as work injuries for statistical purposes:

1. **Inguinal hernia,** if it is precipitated by an impact, sudden effort, or severe strain and meets *all* of the following conditions:

   a). There is a clear record of an accident or an incident, such as a slip, trip, fall, sudden effort, or overexertion;

   b). There is actual pain in the hernial region at the time of the accident or incident;

   c). The immediate pain was so severe that the injured employee was forced to stop work long enough to draw the attention of his foreman or fellow employee to his condition, or the attention of a physician was secured within twelve hours.

2. **Back injury,** if:

   a). There is a clear record of an accident or an incident, such as a slip, trip, or fall, sudden effort or blow on the back; or

   b). The employee was engaged in a work activity which, in the opinion of the physician authorized by the employer to treat the case, produced a physical condition resulting from overexertion.

   A back condition which is revealed in the course of employment, but which does not satisfy *either* of these conditions, should not be considered a work injury.

3. Aggravation of a pre-existing physical deficiency, if the aggravation arises out of and in the course of employment.

4. Aggravation of a minor work injury,

whether due to improper diagnosis or treatment or to infection either on the job or off the job.

5. Animal and insect bites incurred in the performance of duties of employment.

6. Skin irritations and infections, such as poison ivy or dermatitis, when the employee is exposed to irritants in connection with his work.

7. Exposure to temperature extremes in the course of employment.

8. Muscular disability, such as bursitis or tenosynovitis, if it arises out of duties of employment.

The following cases resulting in lost time are not classified as disabling:

1. **Hospitalization for observation,** for a period not to exceed 48 hours from the time of injury, or of suspected injury known to have a delayed effect, from such accidents as:

   a) A blow on the head
   b) A blow to the abdomen
   c) The inhalation of harmful gases

   provided that in every such case, the physician determines that the injury was in reality slight and that the injured person could have returned to work without permanent impairment or temporary total disability. In cases of exposure to ionizing radiation, the period of observation may be extended to ten days. NOTE: if *any* treatment or medication is given for a confining injury (or the suspected confining injury) after the first 24 hours of observation, the injury shall be classified as a work injury.

2. Illness from antitoxin, provided that the illness results *solely* from the antitoxin, vaccines, or drugs used in the treatment of a nondisabling injury.

3. Disability arising solely out of a physical deficiency, provided that a worker without such physical deficiency would not have experienced the incident which resulted in the injury. (An injury which results from the work activity or the environment of the employment shall be considered a work injury, even though the employee had an

existing physical deficiency.)

4. Injury from an external event of such proportions and character as to be beyond the control of the employer, such as a tornado, twister, hurricane, earthquake, flood, conflagration, or explosion originating outside of employment, or from an immediate secondary event, such as a fire, boiler explosion, or falling electric wire. (An injury would be reportable, though, if the victim were a policeman, fireman, member of a disaster or emergency squad, utility lineman, or other employee who is assigned duties in connection with such events. An injury also would be reportable if it arose out of activities necessitated by an external event, such as fighting a fire, cleaning up debris, or repairing equipment.)

Days charged

Losses from work injuries are evaluated in terms of days of disability, or inability to produce, either actual or potential. These losses are referred to simply as "days charged." For the first three classes of injuries—death, permanent total disability, and permanent partial disability—the number of days charged is a predetermined total. For permanent partial disability, the predetermined total usually exceeds the actual time lost to reflect potential future losses of productive capacity. The predetermined totals are referred to as "scheduled charges."

This procedure is based on the philosophy of economic loss which reasons, for example, that a man who has a hand amputated will produce less during his remaining working years than a man who completely recovers from a hand injury, even though both injuries resulted in the same number of actual days lost at the time of the injury. If both injuries resulted in, say, 60 days lost at the time of the injury, the injury from which the victim completely recovers would be charged only the 60 days, whereas the amputation would be charged 3000 days, the scheduled charge for this kind of injury.

Death and permanent total disabilities. For death and permanent total disabilities, a scheduled charge of 6000 days is made in each case. There are no variations in this amount. If the injury is fatal, or if it results in any of the

TABLE 6–A.* SCHEDULED CHARGES
(See Fig. 6–6 for bone location)

A. FOR LOSS OF MEMBER – TRAUMATIC
OR SURGICAL

(For loss of use of member, see footnote°°)

Fingers, Thumb, and Hand

Amputation Involving All or Part of Bone°°

	Thumb	Fingers			
		Index	Middle	Ring	Little
Distal phalange	300	100	75	60	50
Middle phalange	..	200	150	120	100
Proximal phalange..	600	400	300	240	200
Metacarpal	900	600	500	450	400

Hand at wrist..3000

Toe, Foot, and Ankle

Amputation Involving All or Part of Bone°°	Great Toe	Each of Other Toes
Distal phalange	150	35
Middle phalange	. .	75
Proximal phalange	300	150
Metatarsal	600	350

Foot at ankle 2400

Arm

Any point above°°° elbow, including
  shoulder joint 4500
Any point above wrist and at or
  below elbow..3600

Leg

Any point above°°° knee...........................4500
Any point above ankle and at or
  below knee 3000

B. IMPAIRMENT OF FUNCTION

One eye (loss of sight), whether or not
  there is sight in the other eye.................1800
Both eyes (loss of sight), in one accident 6000
One ear (complete industrial loss of
  hearing), whether or not there is
  hearing in the other ear 600
Both ears (complete industrial loss of
  hearing), in one accident........................3000
Unrepaired hernia (for repaired
  hernia, use actual days lost)...................... 50

C. FATAL OR PERMANENT TOTAL DISABILITY
............6000

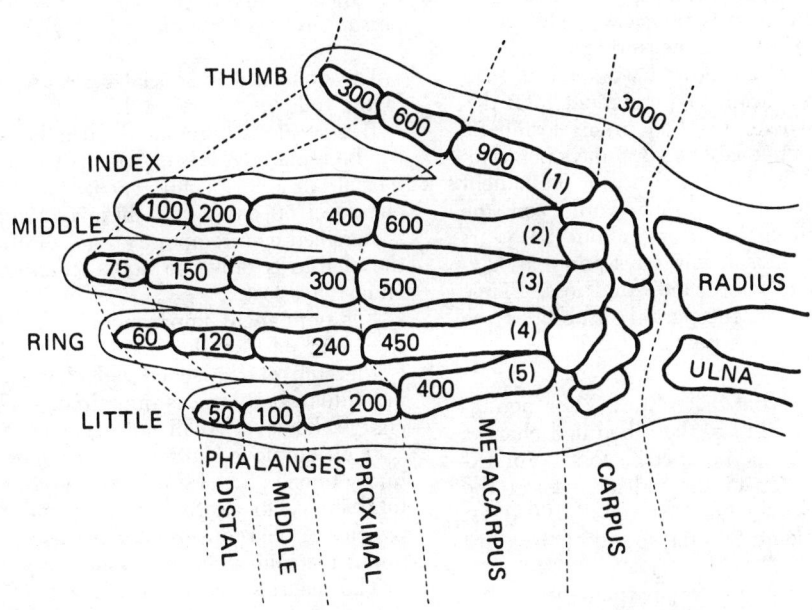

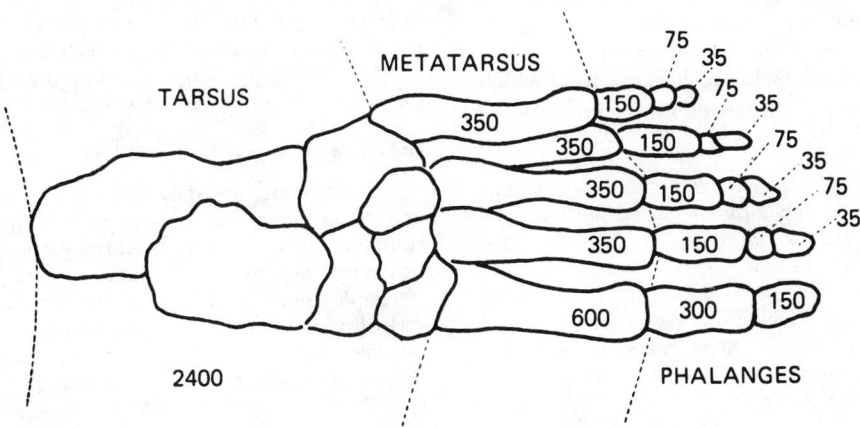

FIG. 6–6.—Chart of scheduled charges for hand (*above*) and foot (*below*). Numbers on the bones are the charges (in days) for loss involving part or all of the bone. (See also Table 6–A at left.)

Courtesy American National Standards Institute.

Footnotes to Table 6–A

° Source: ANSI Z16.1, *Method of Recording and Measuring Work Injury Experience.*

°° For loss of use, without amputation, use a percentage of the scheduled charge corresponding to the loss of use as determined by the physician authorized to treat the case. If the bone is not involved, use actual days lost and classify as temporary total disability.

°°° The term "above" when applied to the arm means toward the shoulder, and when applied to the leg means toward the hip.

losses specified as constituting permanent total disability, the charge is the same—6000 days.

The original basis for measuring the permanent disability of an injured worker was that, on the average, death or permanent total disability of a worker resulted in his losing 20 years of productive labor at 300 days per year, or 6000 days. Today, the death or permanent total disability of a worker would, on the average, result in his losing about 24 years of productive labor at 250 days per year, giving approximately the same total loss. Time charges for permanent partial disabilities are partly related to this possible total.

Permanent partial disabilities. For permanent partial disabilities, the scheduled charges vary depending on the specific loss. For example, amputation of the index finger at the first joint has a scheduled charge of 100 days; at the second joint, 200 days; and at the third joint, 400 days.

Scheduled charges for permanent partial injuries are shown in Table 6-A. Fig. 6–6 (previous page) illustrates pictorially the charges for various hand and foot losses.

For permanent partial injuries which result in loss of use of an injured member, a percentage of the scheduled charge is used which corresponds to the percentage loss of use, as determined by the physician authorized to treat the injury.

For fatal, permanent total, and permanent partial disabilities, only the scheduled charges are used—the actual days of disability are disregarded. In some permanent partial injuries, there may be no losses at all, or there may be losses which exceed the scheduled charge. In either of these cases, though, disregard the number of days lost and *use only the scheduled charge.*

Temporary total disabilities. For temporary total disabilities, the fourth class of injuries, the number of days charged is the total number of full calendar days on which the injured person was unable to work as a result of the injury. The total does not include the day the injury occurred nor the day the injured person returned to work, but it does include all intervening calendar days (including Sundays, days off, or plant shutdowns). It also includes any other full days of inability to work because of the specific injury, subsequent to the injured person's return to work.

Days charged in special cases are indicated in the following paragraphs.

If a hernia is unrepaired (whether or not it can be repaired), it is classified as a permanent partial disability and carries a scheduled charge of 50 days. If a hernia is repaired, it is classified as a temporary total disability and the charge is only the actual number of calendar days lost.

For permanent impairments affecting more than one part of the body, the total charge is the sum of the scheduled charges for the individual body parts impaired. The total charge, however, shall not exceed 6000 days.

If an employee suffers a permanent partial injury to one part of the body and a temporary total injury to another part in one accident, whichever charge is greater is used and determines the injury classification.

The charge for a permanent injury not identified in the schedule of charges (such as damage to internal organs, lungs, or back, or loss of speech) is a percentage of 6000 days corresponding to the percentage of permanent total disability which results from the injury. This percentage is determined by the physician authorized to treat the case.

Definition of employment

The Standard specifies that a work injury is one which arises out of and in the course of employment, i.e., results from the work activity or environment of employment.

Employment is defined in the standard as: (a) all work or activity performed in carrying out an assignment or request of the employer, including incidental and related activities not specifically covered by the assignment or request; (b) any voluntary work or activity undertaken while on duty with the intent of benefiting the employer; or (c) any other activities undertaken while on duty with employer's consent or approval.

For statistical purposes, an employee is considered to be in the course of employment while he is:

1. Riding in special, company-furnished transportation from a designated meeting place to a workplace that is inaccessible to ordinary transportation. The employee

would be considered in the course of employment from the time he was picked up at such a meeting place until he was returned to it.

2. A member of a crew that does not have a regular place of employment, such as a public utility line crew, from the time he reaches a designated meeting place for the crew until he is dismissed from duty at the point where the crew disbands.

3. Traveling in connection with his work, from the time his travel starts (either at his place of work or at his home), except:
   a) During normal living activities, such as eating, sleeping, and the like;
   b) During deviations from a reasonably direct route of travel, such as a side trip for personal reasons;
   c) During other activities neither necessitated by the travel nor in the interest of the employer.

4. Being entertained by, or as, a customer or client for the purpose of transacting, discussing, or promoting business.

5. Going from the entrance of the employer's premises to his place of work, or from his place of work to the exit of the employer's premises before or after working hours, or going from one part of the employer's premises to another for any purpose associated with his employment.

6. Absent from company premises if such absence is authorized by the employer or his agent and is in the interest of the employer or his agent.

7. Taking a coffee or other rest break.

8. Going to or from washroom, toilet, or shower facilities before, during, or after working hours; using toilet facilities at any time; taking a shower or otherwise using washroom facilities on company premises before, during, or after working hours, if use of the facilities is occasioned by the employee's work.

9. Engaged in company-sponsored athletic events for which he is paid directly or indirectly.

10. Participating in or the victim of horseplay during working hours.

11. Engaged in a fight if the dispute involves performance of duties or is otherwise connected with employment or the protection of company property.

12. Performing voluntary work with the intention of benefiting the employer, either in emergencies such as fire or flood or in routine duties.

An employee is *not* considered to be in the course of employment while he is:

1. Going to or from his regular place of employment during normal, routine travel. Normal or routine travel includes travel at irregular hours due to late shifts, overtime, special or emergency work.

2. Outside company property during working hours for personal reasons, not in the interest of his employer or the agent of his employer, nor in the performance of duties of employment.

3. Going to or from his home to a designated place where his crew meets, or where he will be met by special company transportation if his workplace is inaccessible to ordinary transportation.

4. On a company parking lot provided for his convenience to park his car, and not performing duties of employment.

5. Engaged in company-sponsored athletic events for which he receives no pay directly or indirectly.

6. Engaged in activities not connected with his employment while living on company property.

7. Engaged in a fight or other dispute over matters not pertaining to his or his antagonist's duties of employment.

8. Eating his lunch during a specifically defined lunch period or off-duty period. (However, if an injury arises out of hazards of the work area at such a time, it would be considered a work injury.)

Interpretation of injury rates

Frequency rate. This rate shows the rate of occurrence of disabling injuries. Since a relationship is needed to give the rate meaning, disabling injuries are related to 1,000,000

143

employee-hours of work.

A rate of 20.0 means that disabling injuries were incurred at the rate of 20 for each 1,000,000 employee-hours worked. Occasionally, this interpretation may be difficult to understand, particularly when a company does not work 1,000,000 employee-hours during the period for which the rate is determined. However, as with a vehicle which can travel 60 miles per hour without being operated for an entire hour, a plant can have a rate of 20 injuries per 1,000,000 employee-hours without actually working this number of hours.

On the average an employee works about 2000 hours per year (in the United States). One million employee-hours, then, represents a year's work for about 500 employees. In general terms, therefore, a rate of 20.0 may be interpreted as 20 disabling injuries per year for each 500 employees, or 1 disabling injury for each 25 employees. Expressed in another way, if a plant has an injury rate of 20.0 during the year, 1 of every 25 employees was injured to the extent that he lost at least one full day of work or suffered some permanent impairment.

Severity rate. This rate shows the rate at which days are lost or charged in relation to 1,000,000 employee-hours of work.

A severity rate of 500 means that 500 days were lost or charged for every 1,000,000 employee-hours worked. As with the frequency rate, the base may be interpreted more generally in terms of employees, and since 1,000,000 employee-hours represents the yearly experience of about 500 employees, a severity rate of 500 may be interpreted as 500 days lost or charged for each 500 employees, or about one day lost or charged for each employee.

Included in the severity rate are both actual days lost and scheduled charges, with the total heavily weighted by the latter. In most cases in which scheduled charges are assessed, these charges exceed the actual days lost.

For example, amputation of the index finger carries a scheduled charge of 400 days, which is used in place of the actual days lost. It is improbable, that a disability of this type will cause the worker to lose 400 days.

The severity rate, then, must not be interpreted as showing the number of days *lost* because of injuries. Actually, it shows the number of days lost or charged, and the charges which represent potential losses of production outweigh the actual days lost at the time of the injury.

Average days charged per disabling injury. This measure shows how serious the injuries were, on the average, and thus may reveal conditions not readily apparent from a review of frequency or severity rates alone. It thus makes possible a more complete evaluation of injury experience.

To illustrate how this measure may be used, in one year, forty-four large meat packing plants which reported their experience to the National Safety Council were classified in the "large plants" group, and the ranks of their frequency and severity rates were determined from 1 to 44.

In the plant with the second lowest frequency and severity rates, the injuries, *on the average*, were more serious than those which occurred in 15 of the other 43 plants. This fact clearly showed the need for directing additional attention toward the more serious cases.

On the other hand, the plant which ranked 40th in frequency and 29th in severity had an average charge per disabling injury that ranked only 11th. In other words, in this plant whose frequency and severity rates were among the highest, the injuries, on the average, were less serious than those which occurred in 33 of the other plants. The low average charge per disabling injury in this plant indicated that the less serious injuries accounted principally for the high plant rates, and that improvement in the injury rates depended primarily on more effective control over minor cases.

Since the average charge per disabling injury can be computed quickly for any plant, group, or industry for which frequency and severity rates are known (merely by dividing the frequency rate into the severity rate), the average charge per disabling injury for a plant may be compared directly with a similar figure computed for the entire industry of which the plant is a part. For example, the plant with the high frequency and severity rates discussed above had an average charge per disabling injury of only 21 days, whereas the average for the entire group of 44 plants was 49 days.

This simple comparison between the average severity of the disabling injuries in this plant and the average severity of those occurring in similar plants indicated that the high frequency rate was the result of a high frequency of minor disabling injuries.

The measure may be used further to determine the average severity of temporary total and permanent partial injuries separately. Average charges are computed for each type of disability and compared with similar averages for other plants or for the entire industry.

Computation of these measures might reveal that a high average charge for all disabling injuries resulted principally from a high average charge for permanent partial injuries, and that the temporary total injuries were relatively minor. All these refinements help to isolate the principal sources of high injury rates and thus permit concentration of effort and the most effective use of time.

Significance of Changes in Work-Injury Experience

The disabling injury frequency is the best measure for comparing the work-injury experience between companies of various sizes although it is often not sufficient for determining the significance of month-to-month changes in the actual number of injuries within a single company. For comparing month-to-month changes within a company it is usually necessary to use all injuries, not only the disabling cases. This provides a more objective measure for determining the significance of month-to-month fluctuations in the number of injuries, especially when there is an apparently large increase or decrease from the average number of injuries per month.

Since the average number of injuries per month is calculated from numbers of injuries which are larger and smaller than the average itself, variation from the average is expected. The variation can be either random or caused; caused variation is significant and random variation is nonsignificant. Therefore, the significance of variations can easily be determined by distinguishing between those which are random and those which are caused.

Manufacturing organizations are faced with the task of determining the significance of variations in such things as dimensions, weight, or performance of their products. To ease this task they frequently employ a tool known as a "quality control chart." The quality control chart identifies and distinguishes between random variation which is said to be "in control" and caused variation which is said to be "out of control." Being able to distinguish between the two types of variation permits management to concentrate its efforts on those variations which are "out of control."

A similar control chart can be developed to evaluate the significance of changes in work-injury experience. The first step in developing the chart is to calculate the average number of injuries per month. Because this monthly average will fluctuate from year to years, several years' experience should be used to develop a stable average. Preferably, the average should be calculated using 60 months' experience. After the average number of injuries per month has been determined, the upper and lower control limits are calculated using the following equation:

$$n \pm 2 \sqrt{n}$$

where n is the average number of injuries per month.

Fig. 6–7 shows a control chart developed for one company. Using 60 months' experience, it was determined that the company had an average of 25 injuries per month. Substitution of 25 for n in the equation yielded the upper and lower control limits shown below:

$$25 \pm 2 \sqrt{25} = 25 \pm 10 =$$
$$\begin{cases} 25 + 10 = 35 \text{ Upper limit} \\ 25 - 10 = 15 \text{ Lower limit} \end{cases}$$

The control chart was then constructed and the company recorded the actual number of injuries each month. Note that the actual monthly number of injuries, with the exception of February, fall within the upper and lower control limits. These variations are random and do not represent significant changes from the monthly average of 25 injuries. The variation for February, probably, is a caused variation indicating that the injury experience for February is "out of control" and that corrective measures should be taken. Suppose, for example, an investigation re-

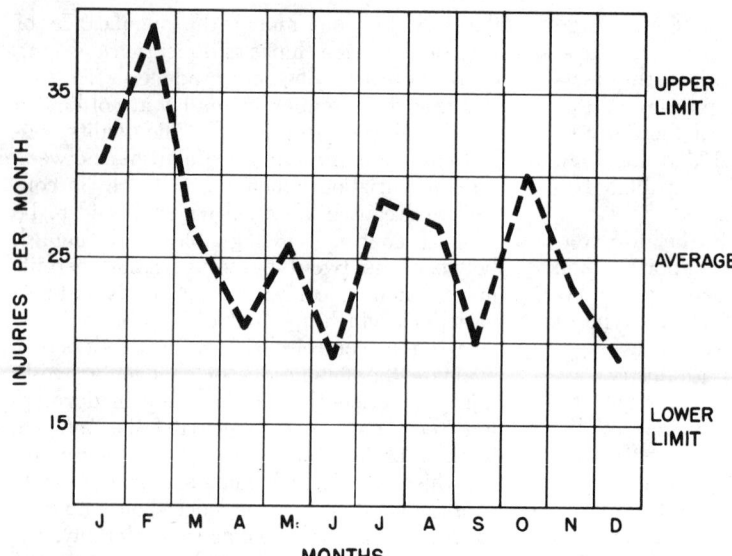

FIG. 6–7.—Upper and lower control limits of this control chart were based on 60 months' experience, and show points at which the number of injuries can statistically be shown to have a "contributing cause" rather than be "by chance." Thus action can be taken where it will be most effective.

vealed that an unguarded machine was the cause of the increase in injuries for the month of February. It can be assumed that the installation of a mechanical guard corrected this cause and brought the work-injury experience for March back "in control."

The control chart can be a useful tool when it is used properly. As with any tool, it will depend on the skill of the user to yield good results. Follow these rules when constructing a control chart:

1. Always use several years' experience when calculating the average number of injuries per month to assure that a stable average will be developed. Sixty months' experience is preferable, except as noted in (3), (4), and (5) following.

2. Count *all* injuries when constructing a control chart. The use of disabling injuries alone may not provide enough data. A control chart cannot be constructed if the average number of injuries per month is less than four because the lower limit will be negative. Since a plant cannot experience a negative number of injuries per month, a chart based on an average of less than four would not yield meaningful results.

3. Calculate a new average each year—add-

ing the latest year's injuries and subtracting the oldest year's injuries—to reflect any change in the monthly average number of injuries.

4. Construct more than one chart if there are seasonal changes in employment. For example, if an operation employs one hundred people for the first 6 months of each year and four hundred people for the last 6 months, a separate chart must be constructed for each 6-month period.

5. Construct a new control chart if external factors, such as a change in the level of employment, a change in the work environment, or a change in work hazards, bring about a permanent change in work-injury experience. When calculating a new average in such a situation, *do not* use experience prior to the change.

Occupational Safety and Health Act (OSHAct) Recordkeeping

The following material has been abstracted from an Occupational Safety and Health Administration pamphlet. It provides the basic definitions and recordkeeping requirements of the OSHAct, as they existed at the time this volume was prepared. The requirements and definitions are subject to change; the individual states, which may implement the Act in their own jurisdiction, may make modifications. To determine the most current definitions and requirements, safety personnel should contact the appropriate state or federal authorities.

Regulations issued under the Occupational Safety and Health Act of 1970 require all establishments subject to the Act to maintain records of recordable occupational injuries and illnesses occurring on or after July 1, 1971. Such records must consist of: (*a*) a log of occupational injuries and illnesses; (*b*) a supplementary record of each occupational injury and illness; and (*c*) an annual summary of occupational injuries and illnesses.

Definitions

Recordable occupational injuries and illnesses. These cases are recordable that result in:

Fatalities. Regardless of the time between the injury and death, or the length of illness.

Lost workday cases. Cases, other than fatalities, that result in lost workdays.

Nonfatal cases without lost workdays are cases of occupational injury or illness which did not involve fatalities or lost workdays but did result in: (*a*) transfer to another job or termination of employment, or (*b*) medical treatment, other than first aid, or (*c*) diagnosis of occupational illness, or (*d*) loss of consciousness, or (*e*) restriction of work or motion.

Occupational injury is any injury such as a cut, fracture, sprain, amputation, etc., which results from a work accident or from exposure in the work environment.

Occupational illness of an employee is any abnormal condition or disorder, other than one resulting from an occupational injury, caused by exposure to environmental factors associated with his employment. It includes acute and chronic illnesses or diseases which may be caused by inhalation, absorption, ingestion, or direct contact, and which can be included in the categories listed below.

For purposes of information, examples of each category are given. These are typical examples, however, and are not to be considered to be the complete listing of the types

of illnesses and disorders that are to be counted under each category.

• Occupational skin diseases or disorders. Examples: Contact dermatitis, eczema, or rash caused by primary irritants and sensitizers or poisonous plants; oil acne; chrome ulcers; chemical burns or inflammations; etc.

• Dust diseases of the lungs (pneumoconioses). Examples: Silicosis, asbestosis, coal worker's pneumoconiosis, byssinosis, and other pneumoconioses.

• Respiratory conditions due to toxic agents. Examples: Pneumonitis, pharyngitis, rhinitis or acute congestion due to chemicals, dusts, gases, or fumes; farmer's lung; etc.

• Poisoning (systemic effects of toxic materials). Examples: Poisoning by lead, mercury, cadmium, arsenic, or other metals, poisoning by carbon monoxide, hydrogen sulfide or other gases; poisoning by benzol, carbon tetrachloride, or other organic solvents; poisoning by insecticide sprays such as parathion, lead arsenate; poisoning by other chemicals such as formaldehyde, plastics and resins, etc.

• Disorders due to physical agents (other than toxic materials). Examples: Heatstroke, sunstroke, heat exhaustion and other effects of environmental heat; freezing, frostbite and effects of exposure to low temperatures; caisson disease; effects of ionizing radiation (isotopes, X rays, radium); effects of nonionizing radiation (welding flash, ultraviolet rays, microwaves, sunburn), etc.

• Disorders due to repeated trauma. Examples: Noise-induced hearing loss; synovitis, tenosynovitis, and bursitis; Raynaud's phenomena; and other conditions due to repeated motion, vibration, or pressure.

• All other occupational illnesses. Examples: Anthrax, brucellosis, infectious hepatitis, malignant and benign tumors, food poisoning, histoplasmosis, coccidioidomycosis, etc.

Lost workdays. Those days which the employee would have worked but could not because of occupational injury or illness. The number of lost workdays should not include the day of injury. The number of days includes all days (consecutive or not) on which, because of injury or illness: (*a*) the employee would have worked but could not, or (*b*) the employee was assigned to a temporary job, or (*c*) the employee worked at a permanent job less than full time, or (*d*) the employee worked at a permanently assigned job but could not perform all duties normally assigned to it.

Medical treatment includes treatment administered by a physician or by registered professional personnel under the standing orders of a physician. Medical treatment does NOT include first aid treatment.

First aid. One-time treatment and subsequent observation of minor scratches, cuts, burns, splinters, and so forth, which do not ordinarily require medical care even though provided by a physician or registered professional personnel.

Establishment. A single physical location where business is conducted or where services or industrial operations are performed. (For example: a factory, mill, store, hotel, restaurant, movie theater, farm, ranch, bank, sales office, warehouse, or central administrative office.) Where distinctly separate activities are performed at a single physical location (such as contract construction activities operated from the same physical location as a lumber yard), each activity shall be treated as a separate establishment.

For firms engaged in activities such as agriculture, construction, transportation, communications, and electric, gas and sanitary services, which may be physically dispersed, records may be maintained at a place to which employees report each day.

Records for personnel who do not primarily report or work at a single establishment, such as traveling salesmen, technicians, engineers, etc., shall be maintained at the location from which they are paid or the base from which personnel operate to carry out their activities.

Work environment. Comprised of the physical location, equipment, materials processed or used, and the kinds of operations performed by an employee in the performance of his work, whether on or off the employer's premises.

Incidence rate. Number of recordable injuries and illnesses × 200,000/total hours

worked by all employees during period covered. The 200,000 is equivalent to 100 full-time workers at 40 hours per week for 50 weeks.

Off-the-Job and Patron Injuries

Accidental injury experience of employees

In recent years, off-the-job disabling injuries of employees have exceeded on-the-job disabling injuries by 30 to 40 percent. Since the unscheduled absence of employees for any reason can cause production slowdowns and delays, costly retraining and replacement, or costly overtime by remaining employees, many safety specialists are concerned with the off-the-job injuries that occur to their employees. Moreover, activity to reduce off-the-job accidents should help promote interest in on-the-job safety.

The ANSI Z16.3 Standard provides a means for recording and measuring these off-the-job injuries—those injuries suffered by an employee which do not arise out of and in the course of employment. Definitions and rates used under the ANSI Z16.3 Standard are very similar to those used under ANSI Z16.1. Because the data on off-the-job injuries is not as easy to obtain, however, certain simplifications are introduced in ANSI Z16.3. Exposure (for use in rates per million employee-hours) is standardized at 312 employee-hours per employee per month (equal to 4 and one-third man-weeks less forty hours per week at work and 56 hours per week for sleeping). For the calculation of severity rates, each permanent partial disability is recorded at 390 days of disability (based upon Bureau of Labor Statistics averages for all permanent partial disabilities). Provision is also made in ANSI Z16.3 for recording home, public, and transportation injuries separately to allow for concentrated effort in problem areas.

Patron and nonemployee injury statistics

Many safety professionals concerned with employee injuries and workmen's compensation have a similar or even greater concern about injuries to customers, patients, hotel guests, diners, tenants or others not employed by the reporting facility. For this reason the Trades and Services Section of the National Safety Council initiated and sponsored the

American National Standard Z108.1. See Chapter 22, "Non-Employee Accident Prevention," for a discussion of applicable control activities.

A reportable patron injury is defined as one which occurs to a patron in the service environment (such as a store, parking lot, hotel, office, or restaurant) or from a "service"-connected accident or illness resulting from the use of a service or product, and which requires professional medical or dental treatment or examination. Property damage or injuries not serious enough to warrant professional attention (or which upon examination reveal no actual injury) need not be recorded. Accidents involving street and highway vehicles and passenger carrying operations subject to federal reporting systems would also be excluded.

Since the exposure base would vary widely between services such as theaters, with fairly consistent exposure time, and bus terminals, with inconsistent exposure, the ANSI Z108.1 Standard provides that the trade association or other national representative group determine its exposure base. In other words, the patron injury frequency rate may use as an exposure base 10,000 sales transactions, 100,-000 patron-days, $1 million gross sales, or other common denominator.

The measure of severity (patron loss injury rate) is also chosen by the industry or business group. It might be the dollar value of incurred losses (claims) per $1 million sales or some other basis, generally over a twelve-month time span.

References

American National Standards Institute, 1430 Broadway, New York, N.Y. 10018.
   Method of Recording and Measuring Work Injury Experience, Z16.1–1973.
   Method of Recording Basic Facts Relating to the Nature and Occurrence of Work Injuries, Z16.2–1969.
   Method of Recording and Measuring Off-the-Job Disabling Accidental Injury Experience of Employees, Z16.3–1972.
   Method of Measuring and Recording Patron and Non-Employee Injury Statistics, Z.108.1–1971.

National Safety Council, 425 N. Michigan Ave., Chicago, Ill. 60611. "Injury-Illness Reporting Requirements, A Report on the OSHA Briefing Session Sponsored by the National Safety Council in cooperation with U.S. Bureau of Labor Statistics." Chicago, Ill. July 23, 1971.

Recht, J. L., "Bilevel Reporting," *Journal of Safety Research,* Vol. 2, No. 2 (June, 1970), pp. 51–54.

U.S. Dept. of Labor, Occupational Safety and Health Administration, Washington, D.C. 20212. "Recordkeeping Requirements under the Williams-Steiger Occupational Safety and Health Act of 1970."

Walker, Helen M., and Lev, Joseph. *Elementary Statistical Methods,* rev. ed. New York, N.Y., Holt, Rinehart and Winston, 1958.

Accident Investigation, Analysis, and Costs

Chapter 7

Accident Investigation and Analysis 151
Types of investigation and analysis . . . Persons making the
investigation . . . Cases to be investigated . . . The key facts in
accidents . . . Identifying the key facts . . . Examples of identifying
key facts . . . Classifying the key facts . . . Making the analysis
. . . Using the analysis

Estimating Accident Costs 162
Definition of work accidents for cost analysis . . . Method for
estimating . . . Example of a cost estimate . . . Items of uninsured
(indirect) cost . . . Making a pilot study . . . Development of final
cost estimate

References 171

This chapter deals with costs of non-injury accidents, as well as injury accidents. Therefore the term "accident" is used in its broadest sense to cover occurrences, and their causes, which may lead to property damage and work injuries.

Accident Investigation and Analysis

Successful accident prevention requires a minimum of four fundamental activities:

1. A study of all working areas to detect and eliminate or control physical or environmental hazards which contribute to accidents.

2. A study of all operating methods and practices.

3. Education, instruction, training, and discipline to minimize human factors which contribute to accidents.

4. For cause analyses, a thorough investigation of at least every accident which results in a disabling injury (under ANSI Z16.1) or OSHA lost workdays to determine contributing circumstances. Accidents that do not result in personal injury (so-called "near-accidents" or "near-misses") are warnings. They should not be ignored.

This fourth activity, accident investigation and analysis, is a defense against hazards that are overlooked in the first three activities, those that are not obvious, or hazards that are the result of combinations of circumstances that are difficult to foresee.

Accident investigation and analysis is one of the means used to prevent accidents. As such, the investigation or analysis must produce information that leads to countermeasures which prevent or reduce the number of accidents. The more complete the information, the easier it will be for the safety professional to design effective countermeasures. For example, knowing that 40 percent of a plant's accidents involve ladders is not as useful as knowing that 80 percent of the plant's ladder accidents involve broken rungs.

A good recordkeeping system, as discussed in Chapter 6, is essential to accident investigation in that it allows the basic facts about an accident to be recorded quickly, efficiently, and uniformly. An investigation of at least every ANSI Z16.1 disabling injury or every OSHA lost workday case should be made. Accidents resulting in nondisabling injuries, or no injuries, and also "near-accidents," should be investigated if time and facilities permit, especially if there is frequent recurrence of certain types of nondisabling injuries, or if the frequency of accidents is high in certain areas or operations.

For purposes of accident prevention, investigations must be for fact-finding, not fault-finding; otherwise, they may do more harm than good. This is not to say that responsibility may not be fixed where personal failure has caused injury, or that such persons should be excused from the consequences of their actions. What this does mean is that the investigation itself should be concerned only with facts. The investigating individual, board, or committee is best kept free from involvement with the punitive aspects of their investigation.

Types of investigation and analysis

There are a variety of accident investigation and analysis techniques available to the investigator. Some of these techniques are more complicated than others. The choice of a particular method will depend upon the purpose and orientation of the investigation. The Failure Mode and Effect approach discussed in Chapter 4 could be very useful for investigating situations where large complex and interrelated machinery and procedures are involved, but may be of limited value in the investigation of accidents involving hand tools. If management procedures and communications and their relationship to accidents are of great interest, the Management Oversight and Risk Tree analysis (MORT, see references at the end of Chapter 4) could prove to be very helpful.

The accident investigation and analysis procedure explored in this chapter is based upon the ANSI Z16.2 Standard, and focuses primarily on unsafe acts and unsafe conditions, and is the most often used technique. Other similar techniques involve investigation within the framework of defects in man, machine, media, and management (the "4 M's"), or education, enforcement, and engineering (the "3 E's"). For analysis purposes, these techniques involve classifying the data about a group of accidents into various categories. This has been referred to as the statistical method of analysis. Countermeasures are designed on the basis of most frequent patterns of occurrence.

Other techniques discussed in Chapter 4 come under the systems approach to safety. Systems safety stresses an enlarged viewpoint that takes into account the interrelationships between the various events that could lead to an accident. As accidents will rarely have one cause, the systems approach to safety can point to more than one place in a system where effective countermeasures can be introduced. This allows the safety professional to choose the countermeasure that best meets his criteria for effectiveness, speed of installment, and the like. Systems safety techniques also have the advantage of application before accidents have occurred and can be applied to new procedures and operations.

Persons making the investigation

Depending on the nature of the accident and other conditions, the investigation may be made by the supervisor or foreman, the safety engineer or inspector, the workmen's safety committee, the general safety committee, or an engineer from the insurance company. If the accident involves special features, consultation with an engineer from the state labor department or U.S. Bureau of Mines, or with a union representative, may be warranted.

The supervisor or foreman. The supervisor or foreman should make an immediate report of every disabling injury (under ANSI Z16.1) and other accidents he may be directed to investigate. He is on the scene, he probably knows more about the accident than anyone else, and it is up to him, in most cases, to put into effect whatever measures may be adopted to prevent similar accidents. Fig. 6–2 in the previous chapter illustrates one form that can be used to record the findings of an accident investigation.

The safety professional. A representative of the safety department should verify the findings of the foreman and make an investigation of every important accident for his own information, and in most cases he should make a written report to the proper official or to the general safety committee.

Nowhere are the safety professional's value and ability better shown than in the investigation of an accident. His specialized training and analytical experience enable him to search for all the facts, apparent and hidden, and to submit a report free from bias or prejudice. He has no interest in the investigation other than to get information which can be used to prevent a similar accident.

Special investigative committee. In some companies, a special committee is set up to investigate and report on all serious accidents. This function is particularly important where a contributing factor was an unsafe act on the part of the worker.

If this committee, composed of the injured employee's fellow workmen, investigates the accident immediately, finds that the injured man did engage in an unsafe act which contributed to the accident, and so reports to the general safety committee or to other authority, and if this report is then publicized, it will be more generally accepted among the workers than a similar report made by the supervisor or the safety professional.

On the other hand, many companies disagree strongly with this concept of the duties of the workmen's safety committee, believing that the task of assigning responsibility for an accident is sometimes an unpleasant one and should not be foisted upon workers.

The general safety committee. In many companies, especially those of small or moderate size, a number of safety activities are handled by a general safety committee, one of whose activities is accident investigation. Ordinarily, such investigation would be handled in a routine manner, but in important cases the chairman might call an extra meeting of the committee to conduct a special investigation.

Safety committees are discussed in detail in Chapter 3, "Safety Program Organization."

Cases to be investigated

An accident that causes death or serious injury obviously should be thoroughly investigated. The "near-accident" that might have caused death or serious injury is equally important from the safety standpoint and should be investigated; e.g., the breaking of a crane hook or a scaffold rope, or an explosion associated with a pressure vessel.

Each investigation should be made as soon after the accident as possible. A delay of only a few hours may permit important evidence to be destroyed or removed, intentionally or unintentionally. Also, the results of the inquiry should be made known quickly, inasmuch as their publicity value in the safety education of employees and supervisors is greatly increased by promptness.

Any epidemic of minor injuries demands study. A particle of emery in the eye or a scratch from handling sheet metal may be a very simple case; the immediate cause may be obvious, and the loss of time may not exceed a few minutes. However, if cases of this or any other type occur frequently in the plant, or in any one department, an investigation should be made to determine the underlying causes.

The chief value of such an investigation lies in uncovering contributing causes. The energetic safety professional or manager is constantly alive to the advantage of this kind of accident investigation, which may prove more valuable, though less spectacular, than the "inquest" following a fatal injury.

Fairness and impartiality are absolutely essential. The value of the investigation is largely destroyed if there is any suspicion that its purpose is to place the blame or pass the buck. No one should be assigned to investigation work unless he has earned a reputation for fairness and is experienced in gathering evidence. It should be made quite clear that accident investigations are conducted entirely for the purpose of obtaining information which will help to prevent recurrence of accidents.

In the early years of the safety movement, accident prevention usually was a hit-or-miss activity. This approach has been replaced by a more scientific technique—see System Safety in Chapter 4, "Inspection and Control Procedures."

In the earlier years, a reduction in accident rates was prompted primarily by humanitarian appeal to management and workers. Although this appeal is still important, methods today are aimed at isolating and identifying accident causes in order to permit direct, positive action (countermeasures) to prevent their recurrence.

Like other phases of modern business management, accident prevention must be based on facts which clearly identify the problem. An approach to the accident prevention problem on this basis not only will result in more effective control over accidents, but will permit this objective to be accomplished with savings in time, effort, and money.

Accident analysis of individual cases will identify the plants, locations, or departments in which injuries occur most frequently, and will suggest countermeasures necessary to reduce accidents in those areas.

Sometimes an overall high rate is not identified with one or a few departments, but instead represents a high frequency of accidents throughout the plant. Under such circumstances, it is even more important that an analysis of the accidents be made.

Similar accidents may occur frequently but at widely separated locations, so that their high incidence is not always apparent. Accidents may be more numerous in the operation of some machines than in the operation of others, or in the performance of certain repetitive tasks. Some unsafe practices which cause accidents may be committed repeatedly but at different times and in different places, so that their importance as accident causes is not immediately recognized.

Analysis of the circumstances of accidents according to ANSI Z16.2 can produce these results:

1. Identify and locate the principal sources of accidents by determining, from actual experience, the materials, machines, and tools most frequently involved in accidents, and the jobs most likely to produce injuries.

2. Disclose the nature and size of the accident problem in departments and among occupations.

3. Indicate the need for engineering revision by identifying the principal unsafe conditions of various types of equipment and materials.

4. Disclose inefficiencies in operating processes and procedures where poor layout, for example, contributes to accidents, or

153

where outdated methods or procedures which overtax the physical capacities of the workers can be avoided, for example by using mechanical handling methods.

5. Disclose the unsafe practices which necessitate training of employees.

6. Disclose improper placement of personnel in instances in which inabilities or physical handicaps contribute to accidents.

7. Enable supervisors to use the time available for safety work to the greatest advantage by providing them with information about the principal hazards and unsafe practices in their departments.

8. Permit an objective evaluation of the progress of a safety program by noting in continuing analyses the effect of counter measures, educational techniques, and other methods adopted to prevent injuries.

The key facts in accidents

As explained in ANSI Z16.2, *Method of Recording Basic Facts Relating to Nature and Occurrence of Work Injuries,* the purpose of the Standard is to identify certain key facts about each injury and the accident that produced it and to record those facts in a form which will permit summarization to show general patterns of injury and accident occurrence in as great analytical detail as possible. These patterns are intended to serve as guides to the areas, conditions, and circumstances to which accident prevention efforts most profitably may be directed.

For a complete recording of an injury case, one item for each key fact should be selected from the injury report. Whether or not all the key facts are present in a case will be determined by the circumstances of the case. The items should be selected according to the following definitions:

1. **Nature of injury**—the type of physical injury incurred.

2. **Part of body**—the part of the injured person's body directly affected by the injury.

3. **Source of injury**—the object, substance, exposure, or bodily motion which directly produced or inflicted the injury.

4. **Accident type**—the event which directly

resulted in the injury.

5. **Hazardous condition**—the physical condition or circumstance which permitted or occasioned the occurrence of the accident type.

6. **Agency of accident**—the object, substance, or part of the premises in which the hazardous condition existed.

7. **Agency of accident part**—the specific part of the agency of accident that was hazardous.

8. **Unsafe act**—the violation of a commonly accepted safe procedure which directly permitted or occasioned the occurrence of the accident event.

Supplementary items of information closely related to the key facts, such as age, sex, occupation, and type of work being performed at the time the injury was incurred, are recorded and included in an analysis so that all the facts will be available for taking the proper preventive steps. Contributory factors should be indicated.

The principal source of information for an analysis is the supervisor's accident report (Fig. 6–2 in the previous Chapter). Complete data regarding all the key facts should be fully and accurately recorded on this form at the time of the accident.

Injury and accident reports, however, commonly consist of a few specific statements relating to the injury plus a narrative account of how and why the accident occurred. The reports vary widely in the amount of detail given and in the clarity and coherence with which the facts are presented. Therefore, the analyst rarely will find the key facts—the items needed for statistical recording—precisely stated. Usually, he must review all the data given in the report, select pertinent items, and fit them into a predetermined recording pattern.

Identifying the key facts

Since the reliability of an analysis depends greatly on selection of the correct key facts, the analyst must have a clear understanding of the key facts and of the method for identifying them. The bases on which identifications should be made are described and illustrated in the following paragraphs adapted from ANSI Z16.2.

Nature of injury. The type of physical injury incurred should be designated. If two or more injuries were incurred and one injury obviously was more severe than any of the others, that injury should be selected. For example, an injury involving permanent impairment should be selected in preference to a temporary injury. If there were several injuries of different natures, such as cuts and sprains, and no one of them was more serious than the others, the term "multiple injuries" should be used.

Part of body. If the injury was localized in one part of the body, that part should be named. If the injury extended to several sections of a major body part, that major body part should be named. For example, if a burn affected the fingers, the hand, the wrist, and the forearm, *upper extremities multiple* should be given as the body part affected. If the injury was internal, the body system affected should be named. For example, drowning or asphyxia would be considered injuries to the respiratory system.

Source of injury. The object, substance, exposure, or bodily motion which directly produced the injury should be identified as the source of injury. Sometimes an injury results from forcible contacts with two or more objects, occurring either simultaneously or in rapid sequence, and it is impossible to determine which object directly produced the injury. In such cases, the source of injury should be determined as follows. When the choice is between a moving object and a stationary object, the moving object should be selected. When the choice is between two moving objects or between two stationary objects, the one contacted last should be selected. For example, if a person fell from an elevation, struck one or more objects in the course of the fall and finally struck the floor, the floor should be named as the source of injury.

If the injury resulted solely from the stress or strain induced by a free movement of the body or its parts, for example, in reaching, twisting, or bending, *bodily motion* should be indicated as the source of injury.

Accident type. The accident type classification is directly related to the source of injury classification and explains how that source produced the injury. If the injury resulted from contact with an object or substance, the action that best describes that contact should be named as the accident type. If exposure, for instance, to extreme heat or cold, produced the injury, contact with temperature extremes would be the accident type. If the source of injury was bodily motion, the personal action or movement during which the bodily motion occurred should be chosen as the accident type.

Hazardous condition. The hazardous physical condition or circumstance which directly caused or permitted the occurrence of the accident should be named. The hazardous condition is related directly to both the accident type and the agency of accident. Generally, therefore, the hazardous condition selected will determine the agency of accident to be named. Since the hazardous condition classification represents the physical or environmental causes of accidents, tabulations of the data in this category properly may be labeled "accident causes."

Agency of accident. The object, substance, or part of the premises in which the hazardous condition existed should be named. The agency of accident may or may not be identical with the source of injury. These two classifications are entirely unrelated to each other.

The distinguishing characteristic of the *source of injury* is that it directly inflicted the injury. Its selection is based strictly upon this fact without consideration of whether or not it was hazardous in any way. The distinguishing characteristic of the *agency of accident*, on the other hand, is that it was significantly hazardous and for that reason contributed to the occurrence of the accident. Its selection is based strictly upon this fact without consideration of whether or not it inflicted the injury.

Agency of accident part. If the agency of accident had a specific hazardous part that contributed to the occurrence of the accident,

155

that part should be named. If, for example, a person attempted to climb on a ladder that had a defective rung and fell because the rung broke when he put his weight on it, the agency of accident part would be the defective rung.

Unsafe act. The unsafe personal action which directly caused or permitted the occurrence of the accident event should be designated. The selected unsafe act may be something a person did which he should not have done, something he should have done differently, or his failure to do something which he should have done. The person who committed the unsafe act may or may not have been the person who was injured. The person who acted unsafely may have done so deliberately, or he may not have known that he was acting unsafely. Since the unsafe act classification represents the personal causes of accidents, tabulations of data in this category properly may be labeled "accident causes."

Selection of *no* unsafe act, *no* hazardous condition, and *no* contributing factor for any injury report is impossible, because an accident could not have occurred unless one or more of these conditions or events were involved.

Examples of identifying key facts

The first step in analyzing an injury or accident report for statistical purposes consists of selecting from the report the answers to the following questions (see Table 7–A):

1. **Nature of injury.** What was the injury?

2. **Part of body.** What part of the body was affected by the injury named in (1)?

3. **Source of injury.** What object, substance, exposure, or bodily motion inflicted the injury named in (1)?

4. **Accident type.** How did the injured person come in contact with the object, substance, or exposure named in (3), or during what personal movement did the bodily motion named in (3) occur?

5. **Hazardous condition.** What hazardous physical or environmental condition or circumstance caused or permitted the occurrence of the event named in (4)?

6. **Agency of accident.** In what object, sub-

stance, or part of the premises did the hazardous physical or environmental condition named in (5) exist?

7. **Agency of accident part.** In what specific part of the agency of accident named in (6) did the hazardous condition named in (5) exist?

8. **Unsafe act.** What unsafe act of a person caused or permitted the occurrence of the event named in (4)? **Contributing factors** should be noted, when they can be determined.

Table 7–A gives typical checklists that can be used when making the report. For easier statistical handling of the data, the factors can be individually numbered, and the numbers inserted on the form. (See "Classifying the key facts," next.)

The following examples show how to identify the key facts in an injury case for purposes of classification.

Accident No. 1

The operator of a circular saw reached over the running saw to pick up a piece of scrap. His hand touched the blade, which was not covered, and his thumb was severely lacerated.

1. *Nature of injury*—laceration.

2. *Part of body*—thumb.

3. *Source of injury*—circular saw.

4. *Accident type*—struck against.

5. *Hazardous condition*—unguarded.

6. *Agency of accident*—circular saw.

7. *Agency of accident part*—blade.

8. *Unsafe act*—cleaning moving machine.

Accident No. 2

A forklift truck went out of control when one wheel hit a piece of stock lumber which projected into the aisle. The truck ran out of the aisle and struck a machine operator, breaking his leg between the ankle and the knee.

1. *Nature of injury*—fracture.

2. *Part of body*—lower leg.

TABLE 7–A. CHECKLIST FOR IDENTIFYING KEY FACTS

1. NATURE OF INJURY

Foreign body	Strain and sprain	Amputation	Dermatitis
Cut	Fracture	Puncture wound	Ganglion
Bruises and contusions	Burns	Hernia	Abrasions
			Others......................

2. PART OF BODY

Head and Neck	**Upper Extremities**	**Body**	**Lower Extremities**
Scalp	Shoulder	Back	Hips
Eyes	Arms (Upper)	Chest	Thigh
Ears	Elbow	Abdomen	Legs
Mouth, teeth	Forearm	Groin	Knee
Neck	Wrist	Others......................	Ankle
Face	Hand		Feet
Skull	Fingers and thumb		Toes
Others......................	Others......................		Others......................

4. ACCIDENT TYPE

Struck against (rough or sharp objects, surfaces etc. exclusive of falls)	Struck by sliding, falling or other moving objects	Overexertion (resulting in strain, hernia, etc.)	Inhalation, absorption, ingestion, poisoning, etc.
Struck by flying objects	Caught in (on or between)	Slip (not a fall)	Contact with electric current
	Fall on same level	Contact with temperature extremes, burns	Others......................
	Fall to different level		

5. HAZARDOUS CONDITION

Improperly or inadequately guarded	Defective tools, equipment, substances	Hazardous arrangement	Poor housekeeping
Unguarded	Unsafe design or construction	Improper illumination	Congested area
		Improper ventilation	Others......................
		Improper dress	No unsafe condition

6. AGENCY OF ACCIDENT

Machine	Can and end conveyors (belt, cable, can dividers, chain, twisters, drops, can elevators, etc.)	Hoists and Cranes	Chemicals
Vehicles		Elevators (passenger and freight)	Ladders or scaffolds
Hand tools		Building (door, pillar, wall, window, etc.)	Electrical apparatus
Tin and black plate (sheet, stock, or scrap)	Conveyors (chutes, belt, gravity)	Floors or level surfaces	Boilers, pressure vessels
Material work handled (other than tin and black plate)		Stairs, steps, or platforms	Others......................

8. UNSAFE ACT

Operating without authority	Using equipment, tools, materials or vehicles unsafely	Unsafe loading, placing and mixing	Adjusting, clearing jams, cleaning machinery in motion
Failure to warn or secure	Failure to use personal protective equipment	Unsafe lifting and carrying (including insecure grip)	Distracting, teasing
Operating at unsafe speed	Failure to use equipment provided, (except personal protective equipment)	Taking an unsafe position	Poor housekeeping
Making safety devices inoperative			Others......................
Using defective equipment, materials, tools or vehicles			No unsafe act

CONTRIBUTING FACTORS

Disregard of instructions	Lack of knowledge or skill	Failure to report to medical department	Others......................
Bodily defects	Act of other than injured		No contributing factor

3. *Source of injury*—lift truck.

4. *Accident type*—struck by.

5. *Hazardous condition*—
improperly placed lumber.

6. *Agency of accident*—lumber.

7. *Agency of accident part*—none.

8. *Unsafe act*—unsafe placement
of material.

Accident No. 3

A warehouse employee jumped from the loading platform to the ground instead of using the steps. As he landed, he sprained his ankle.

1. *Nature of injury*—sprain.

2. *Part of body*—ankle.

3. *Source of injury*—ground.

4. *Accident type*—fall from elevation.

5. *Hazardous condition*—none indicated.

6. *Agency of accident*—none indicated.

7. *Agency of accident part*—none indicated.

8. *Unsafe act*—jumping from elevation.

Accident No. 4

A laborer working in a trench was suffocated under a mass of earth when the unshored wall of the trench caved in.

1. *Nature of injury*—asphyxia.

2. *Part of body*—respiratory system.

3. *Source of injury*—earth.

4. *Accident type*—caught under.

5. *Hazardous condition*—lack of shoring.

6. *Agency of accident*—trench.

7. *Agency of accident part*—none.

8. *Unsafe act*—none indicated.

Accident No. 5

A salesman stopped his car at an intersection and was waiting for the traffic light to change from red to green. Another car struck his car in the rear. The whiplash effect fractured a vertebra in the salesman's neck.

1. *Nature of injury*—fracture.

2. *Part of body*—neck.

3. *Source of injury*—auto.

4. *Accident type*—collision.

5. *Hazardous condition*—traffic hazard.

6. *Agency of accident*—environment.

7. *Agency of accident part*—none.

8. *Unsafe act*—none.

Classifying the key facts

Even in large company operations in which hundreds of accidents may occur annually, only rarely do two accidents occur in exactly the same way. Accidents do follow general patterns, however, and grouping them according to pattern is necessary for purposes of analysis.

Setting up classifications. Before the actual analysis work is begun, classifications must be set up for grouping the various data. For each key fact, general classifications should be established in which similar data may be grouped. Then, more specific classifications should be set up within each general classification to preserve as many of the details as possible.

For example, in ANSI Z16.2, among the general classifications recommended for the key fact "Hazardous condition" are the following:

1. Defects of agencies.

2. Dress or apparel hazards.

3. Environmental hazards.

FIG. 7–1.—Original tabulations of hazardous conditions and unsafe acts which contributed to ladder accidents. The safety professional (or his clerk) compiled the information from Supervisor's Accident Reports (shown in Fig. 6–2, in the previous chapter) and set up the classifications on the basis of actual experience, such as recorded in these reports.

Analysis of Ladder Accidents

Code no.	Unsafe Conditions		Total no. of cases
1	Slippery rungs	ЖЖ I	6
2	Broken rungs	III	3
3	Lack of hooks for fastening ladder at top	I	1
4	Worn ladder shoes	I	1
5	Weak, worn, cracked rails	II	2

Unsafe Acts

Code	Act	Tally	Total
10	Failure to secure ladder at top	ЖЖ ЖЖ ЖЖ ЖЖ ЖЖ IIII	29
20	Failure to secure ladder at bottom	ЖЖ ЖЖ II	12
30	Placing ladder unsafely		22
31	On boxes or other equipment	ЖЖ ЖЖ IIII	14
32	Near moving equipment	II	2
33	On planking over opening	I	1
34	On inclined or irregular surface	II	2
35	At too great an angle	III	3
40	Working in unsafe position		10
41	Overreaching	ЖЖ I	6
42	Improper stance while using leverage tools	II	2
43	Straddling space between ladder and nearby object	II	2
50	Ascending or descending improperly		35
51	Improper or insecure grip	ЖЖ IIII	9
52	With back to ladder	I	1
53	Running up or down ladders	ЖЖ I	6
54	Jumping from lower steps to ground	ЖЖ	5
55	Oily or wet shoes	III	3
56	Carrying too heavy loads	ЖЖ ЖЖ I	11
60	Using ladders known to be defective	ЖЖ III	8

4. Placement hazards.

5. Inadequate guarding.

6. Public hazards.

Within each one of these general classifications, more specific classifications are set up. Under "Defects of agencies," for example, are listed:

1. Composed of unsuitable materials.

2. Dull.

3. Improperly constructed, assembled, etc.

4. Improperly designed.

5. Rough.

6. Sharp.

7. Slippery.

8. Worn, cracked, broken, etc.

9. Other.

It is not always possible to set up classifications before the analysis is begun. In this case, classifications can be developed as reports are reviewed and situations are revealed.

For example, if an analysis is being made of ladder accidents and it is found that in a number of cases broken rungs caused the accidents, a specific group "Broken rungs" should be set up under the classification "Defects of agencies."

ANSI Z16.2 recommends general and specific classifications for all the key facts. These classifications are presented principally as suggestions to guide the analyst in setting up classifications to fit his own problems. The point must be emphasized that for an analysis to be of maximum usefulness, classifications must be set up to encompass the situations which are pertinent to the particular company.

Use of a numerical code. Regardless of the method which eventually will be used to sort and tabulate the various key facts, the work will be facilitated if code numbers are assigned to the different classifications. With this method, each case need be read only once, at which time code numbers are assigned to the different facts, and subsequent sorting of the various facts can be quickly completed merely by reference to the code numbers.

A numerical code is simply the assigning of numbers in sequence to a list of similar facts. For each key fact (agency of accident, accident type, hazardous condition, etc.), there should be no duplication of numbers; but for the different facts, the numbering series may be repeated. Fig. 7–1 shows how code numbers were assigned for two key facts in an analysis of ladder accidents.

After the cases have been reviewed and code numbers have been assigned to the different key facts, the reports can be easily and quickly sorted or arranged by any of the facts to reveal the principal data concerning the accidents.

Numerical codes are already assigned to all the classifications which are included in ANSI Z16.2, and if an analyst uses this standard, he can use the code numbers, too. If he uses the standard as a starting point and adds other classifications to cover the specific accident experience of his own company, he can code the additional items to fit into the code of the standard.

Making the analysis

Experience has proved that the most effective way to reduce accidents is to concentrate on one phase of the accident problem at a time rather than attempting to stop all accidents at once. There are different ways in which the problem can be approached on this basis, any one of which should prove effective.

The reports may be grouped by occupation of the injured person. Each group of reports, then, may be reviewed to determine what accident types, sources of injury, and agencies of accident are most prevalent among different occupations. Such information is particularly helpful in planning employee training and in developing educational materials and programs.

Injury frequency rates computed by departments may reveal that injuries occur at sharply higher rates in some departments than in others. If this is the case, an analysis should be made of the accidents in the high-rate departments to find out the sources of the accidents and their causes. This method will permit concentration of effort in the locations in which accidents occur most frequently.

If injury frequency rates reveal that a high

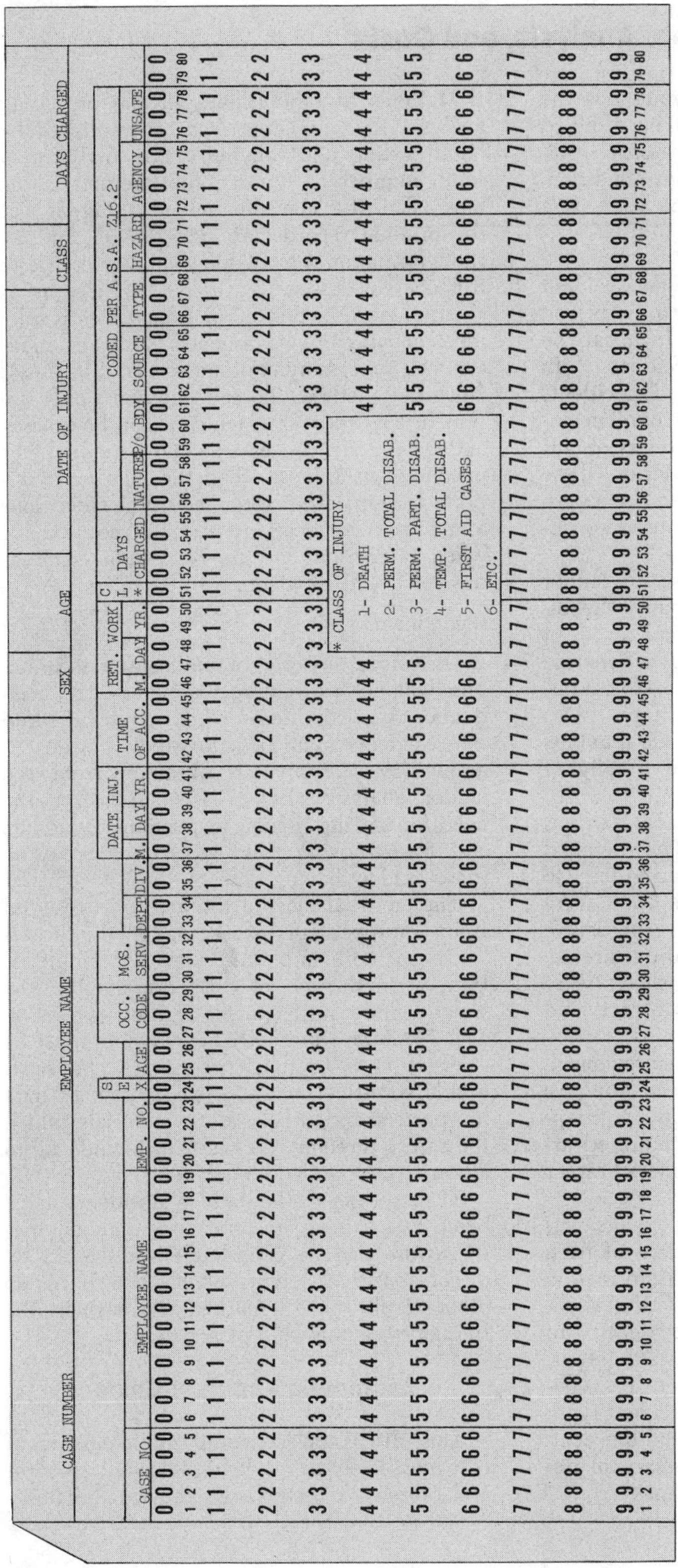

Fig. 7-2.—A punched card, developed for use in making an analysis of the circumstances of injuries, permits utilization of mechanical tabulating equipment.

rate of occurrence is general throughout the plant, the accident reports might be grouped by agency of accident, source of injury, accident type, unsafe act, or hazardous condition. Any one of the key facts may be used as a starting point for an analysis.

Steps to follow in an agency-of-accident analysis. An analysis by agency of accident will always reveal information which can be used effectively in reducing accidents. In one large company, a grouping of accidents in this manner revealed a large number involving ladders, and because accidents of this type often resulted in serious injuries, these cases were analyzed as described below to obtain information that could be used to decrease their occurrence.

Since this was the first analysis of ladder accidents undertaken in this company, classifications for the various key facts had to be set up as the cases were reviewed. Separate sheets were used to list the classifications which were developed for each fact. Also shown on these sheets were the code numbers assigned these classifications, and a tally of occurrences.

Fig. 7–1 shows a collation of hazardous conditions and unsafe acts. The detailed classifications are not in order according to frequency of occurrence because they were set up on the sheets as they were observed. Each classification is specific, relating directly to the ladder accidents. Similar sheets were developed for accident type, unsafe act, and nature of injury.

After each accident report was reviewed, the proper code number was placed opposite each fact on the report. Subsequent groupings of the reports, then, merely required arranging them by the code numbers, rather than re-reading each case.

Since the analysis of ladder accidents was relatively simple, single classifications of each fact revealed practically all the information necessary to permit correction of the situations which contributed to the accidents. In a more involved agency of accident analysis, cross classifications might be made, e.g., between hazardous conditions and accident type; but in this instance the majority of the accidents were falls, and it was necessary to determine only their principal causes.

Methods of tabulating. In the preceding analysis, the tabulation was accomplished by hand sorting and tallying. For analyzing a small number of reports, this method is the most efficient. The principal advantage is that the original records are being used, and all the information is available should reference to it become necessary.

Another method of tabulation, which is fully mechanical, employs punched cards and modern data processing equipment (see Fig. 7–2). With this method, the code numbers only are punched into cards, which can then be quickly and accurately processed. This method has maximum usefulness when the number of reports is very large, when many classifications and cross classifications will be required, or when tabulation of numerical data, such as days lost, will be made.

Using the analysis

Of course, merely obtaining the information will not prevent recurrence of the accidents. The conditions which contributed to the accidents must be corrected. As an example, the information obtained from the ladder analysis (Fig. 7–1) emphasizes the need for making regular inspections of ladders and for removing from service or repairing those ladders having defects.

The fact that most of the accidents resulted from improper use of the ladders indicates the even more important need for employee training in the use of this equipment. The very definite and specific causes shown in Fig. 7–1 indicate exactly what points must be stressed in a training program. The irrefutable statistical evidence also furnishes data for posters and other safety materials which the safety director will want to include in an all-out assault on ladder accidents.

The causes of the ladder accidents have been ascertained in terms so definite that guesswork can be completely supplanted by direct, positive action, which can have no other result than a marked decrease in the frequency of this kind of accident.

Estimating Accident Costs

This discussion concerns the elements of cost most likely to result from a work accident and presents a method whereby an organization can obtain an accurate estimate of the total costs of its work accidents.*

Reliable cost information is a basis for decisions upon which efficiency and profit depend. Even in so obviously desirable an activity as accident prevention, some proposed measures must be accepted or rejected on the basis of their probable effect on profits.

Although most executives want to make their company a safe place in which to work, they also feel a responsibility for running their business profitably. Consequently, they may be reluctant to spend money for accident prevention unless they can see a prospect for saving at least as much as they spend. *Without information on the cost of accidents, it is practically impossible to estimate the savings which are effected through expenditures for accident prevention.*

Annual reports expressed in terms of dollar savings are as meaningful to higher management as those which use frequency and severity figures. Facts about the costs of accidents may be used effectively in securing the active cooperation of foremen. Foremen are usually cost conscious because they are expected to run their departments profitably. Monthly reports showing the cost of accidents or the savings resulting from good accident records are an important motivation to achieve safe operating procedures.

Definition of work accidents for cost analysis

Work accidents, for the purpose of cost analysis, are unintended occurrences arising out of employment. These accidents fall into two general categories: (a) accidents resulting in work injuries and (b) accidents that cause property damage or interfere with production in such a manner that personal injury might result.

The inclusion of the no-injury accidents makes "work accident" roughly synonymous with the type of occurrences a safety department strives to prevent.

Method for estimating

To be of maximum usefulness, cost figures should represent as accurately as possible the specific experience of the company itself. A fixed ratio of indirect to direct costs developed from experience representing many different

companies in many different industries does not serve such a purpose. Estimated costs of accidents in general do not take into account differences in hazards from one industry to another or the more important differences in safety performance from one company to another.

Since the distinctions between "direct" and "indirect" costs are difficult to maintain they have been abandoned in favor of the more precise terms "insured" and "uninsured" costs. Using these data, a company can estimate its accident cost with reasonable accuracy.

Insured costs. Every organization paying compensation insurance premiums recognizes such expense as part of the costs of accidents. In some cases, medical expenses, too, may be covered by insurance. These costs are definite, and they are known. They comprise the insured element of the total accident cost.

In addition to these costs, many other costs arise in connection with accidents. The cost of damaged equipment is easily identified. Others, such as wages paid to the injured employee for hours during which he is not producing, are hidden. These items comprise the uninsured element of the total accident cost.

Uninsured—Indirect costs. Insured costs can be determined easily from accounting records. The difficult part is determining uninsured (frequently called "indirect") costs, and the method described here will serve that purpose.

The first step is to make a pilot study to ascertain approximate averages of uninsured costs for each of the following four classes of accidents:

Class 1—Permanent partial disabilities and temporary total disabilities.

Class 2—Medical treatment cases requiring the attention of a physician outside the plant.

Class 3—Medical treatment cases requiring

---

*This procedure for estimating costs was developed by Rollin H. Simonds, Ph.D., Professor, Michigan State College, under the direction of the Statistics Division, National Safety Council.

only first aid or local dispensary treatment and resulting in property damage of less than $20.00 or loss of less than 8 hours' working time.

CLASS 4—Accidents which either cause no injury or cause minor injury not requiring the attention of a physician, and which result in property damage of $20.00 or more, or loss of 8 or more man-hours.

Once average costs have been established for each accident class, they may be used as multipliers to obtain total uninsured costs in subsequent periods. These costs then may be added to known insurance premium costs to determine the total cost of accidents.

Example of a cost estimate

An estimate of costs made by one company is given in the following example. First, a pilot study was made to get the average cost of each class of accident. Included in the study were 20 Class 1 accidents, 30 Class 2 accidents, 50 Class 3 accidents, and 20 Class 4 accidents. Costs were determined and averages developed as in Table 7-B.

TABLE 7–B

PREDETERMINED AVERAGE COSTS

Class of Accident	Number of Accidents Reported	Average Uninsured Cost
Class 1	20	$ 90.00
Class 2	30	28.95
Class 3	50	5.60
Class 4	20	181.75

During the entire year, the company had 34 Class 1 accidents, 148 Class 2 accidents, and 4000 Class 3 accidents. No record was kept of the Class 4 accidents after the pilot study was completed. Instead, the ratio of the number of Class 4 to Class 1 accidents found in the pilot study was used. This ratio was shown to be about 1 to 1, and since there were 34 Class 1 accidents during the year, it was assumed there were about 34 Class 4 accidents. (A separate record could be kept of

the number of Class 4 accidents.)

The average cost for each accident class was applied to these totals to secure the results shown in Table 7-C.

TABLE 7–C

ESTIMATE OF YEARLY ACCIDENT COSTS

Class of Accident	Number of Accidents	Average Cost Per Accident (from pilot study)	Total Uninsured Cost
Class 1	34	$ 90.00	$ 3,060.00
Class 2	148	28.95	4,284.60
Class 3	4,000	5.60	22,400.00
Class 4	34	181.75	6,179.50
Total Uninsured Cost			35,924.10
Insurance Premiums			19,500.00
Total Accident Cost for the Period			54,424.10

Since the final total is the sum of many estimates, it should not be implied that the total figure suggests absolute accuracy. Whether $54 or $55 thousand is chosen as the final figure depends largely on the analyst's judgment of whether the various elements may have been overestimated or underestimated. In this case, the analyst judged that the pilot study represented conservative estimates of the average costs. So he reported to the plant manager, "During the past year, accidents cost this company about $55 thousand in compensation, medical expense, lost time, and property damage."

The average costs determined in this pilot study represent the actual experience of this particular company. Until important changes take place in this company's safety program, in the kind of machinery used or persons employed, or in other aspects which affect costs, the same average costs may continue to be used.

Items of uninsured (indirect) cost

Important to a pilot study is a careful investigation of each accident to determine all the costs arising from it. The following items

(*Text continues on page 168.*)

DEPARTMENT SUPERVISOR'S ACCIDENT COST REPORT

Injury Accident_____X_____

No-Injury Accident_____

Date _____6-23_____ Name of injured worker___Thomas Black_____

1. How many other workers (not injured) lost time because they were talking, watching, helping at accident? _____1_____

   About how much time did most of them lose? _____0_____ hours _____20_____ minutes

2. How many other workers (not injured) lost time because they lacked equipment damaged in the accident or because they needed the output or aid of the injured worker? ___30___

   About how much time did most of them lose? _____0_____ hours _____20_____ minutes

3. Describe the damage to material or equipment ___Hydraulic cylinder crushed by fork on lift truck. Cylinder scrapped and replaced with new part._____
   Estimate the cost of repair or replacement of above material or equipment $___280.00___

4. How much time did injured worker lose on day of injury for which he was paid? ___2_____
   hours _____45_____ minutes

5. If operations or machines were made idle: Will overtime work probably be necessary to make up lost production? Yes ☒, No ☐. Will it be impossible to make up loss of use of machines or equipment? Yes ☐, No ☒.

   Demurrage or other special non-wage costs due to stopping an operation $___118.00 (est.)___

6. How much of supervisor's time was used assisting, investigating, reporting, assigning work, training or instructing a substitute, or making other adjustments _____1_____ hours ___30___ minutes.

Name of supervisor_____Alan Hoskin_____

Fill in and send to the safety department not later than day after accident.

Published by National Safety Council
425 North Michigan Avenue
Chicago, Illinois 60611

Stock No. 129.27   Printed in U.S.A.   Form IS7-Rep. 2M86699

FIG. 7-3.—This cost form (8½ × 11 in.) should be prepared by the department supervisor as soon after the accident as information becomes available on the amount of time lost by all persons and the extent of damage to product and equipment.

INVESTIGATOR'S COST DATA SHEET

Class 1 _____
(Permanent partial or temporary total disability)

Class 2 _____
(Temporary partial disability or medical treatment case requiring outside physician's care)

Class 3 __X_____
(Medical treatment case requiring local dispensary care)

Class 4 _____
(No injury)

Name _____Thomas Black_____

Date of injury ___6-23_____   Its nature ___Abrasion on right leg_____

Department ___31_____   Operation ___Trucking____   Hourly wage ___$4.29___

Hourly wage of supervisor $___$6.75___

Average hourly wage of workers in department where injury occurred $___$4.10___

1. Wage cost of time lost by workers who were not injured, if paid by employer $__42.70_.

   a. Number of workers who lost time because they were talking, watching, helping __1_.
   Average amount of time lost per worker ___0_____ hours ___20_____ minutes.

   b. Number of workers who lost time because they lacked equipment damaged in accident or because they needed output or aid of injured worker _30_. Average amount of time lost per worker ____0____ hours ___20____ minutes.

2. Nature of damage to material or equipment ___Hydraulic cylinder crushed by___ fork on lift truck. Cylinder replaced with new unit.

   Net cost to repair, replace, or put in order the above material or equipment $_280.00_.

3. Wage cost of time lost by injured worker while being paid by employer $__11.80_.
   (other than workmen's compensation payments)

   a. Time lost on day of injury for which worker was paid ___2___ hrs. __45__ mins.

   b. Number of subsequent days' absence for which worker was paid ___----___ days
   (other than workmen's compensation payments) ___----___ hours per day.

   c. Number of additional trips for medical attention on employer's time on succeeding days after worker's return to work ___----___.
   Average time per trip ___----___ hrs. ___----___ min. Total trip time ___----___ hrs. _____ mins.

   d. Additional lost time by employee, for which he was paid by company ___----___ hrs. ___----___ mins.

(over)

FIG. 7–4.—This 8½ × 11 in. form can be used to convert time losses into money losses. Initial time losses are obtained from the "Department Supervisor's Accident Cost Report" (Fig. 7–4), and subsequent time losses are obtained from first aid and other de-

4. If lost production was made up by overtime work, how much more did the work cost than if it had been done in regular hours? (Cost items: wage rate difference, extra supervision, light, heat, cleaning for overtime.) $ 40.50

5. Cost of supervisor's time required in connection with the accident $ 15.20

    a. Supervisor's time shown on Dept. Supervisor's Report ___1___ hrs. ___30.___ mins.

    b. Additional supervisor's time required later _____ hrs. ___45___ mins.

6. Wage cost due to decreased output of worker after injury if paid old rate $ 6.86

    a. Total time on light work or at reduced output ___1___ days ___8___ hours per day.

    b. Worker's average percentage of normal output during this period ___80___%.

7. If injured worker was replaced by new worker, wage cost of learning period $ 0

    a. Time new worker's output was below normal for his own wage _----_ days _----_ hours per day. His average percentage of normal output during time _-----_%. His hourly wage $_----_.

    b. Time of supervisor or others for training _----_ hrs. Cost per hour $_----_.

8. Medical cost to company (not covered by workmen's compensation insurance) $ 15.00

9. Cost of time spent by higher supervision on investigation, including local processing of workmen's compensation application forms. (No safety or prevention activities should be included.) $ 22.50

10. Other costs not covered above (e.g., public liability claims; cost of renting replacement equipment; loss of profit on contracts cancelled or orders lost if accident causes net reduction in total sales; loss of bonuses by company; cost of hiring new employee if the additional hiring expense is significant; cost of *excessive* spoilage by new employee; demurrage). $ 20.00

Explain fully.

    Total uninsured cost... $454.56

Name of company___Midwest Mfg.___

Published by National Safety Council

425 North Michigan Avenue

Chicago, Illinois 60611

Printed in U.S.A.   Stock No. 129.28

partments as necessary. Wage rate information is obtained from the accounting department. The reverse side of the form, shown at the right, contains space for additional costs pertinent to the accident under study. See text discussion, under section Making a Pilot Study, page 169.

of uninsured or indirect cost may clearly be shown to result from work accidents and are subject to reasonably reliable measurement. Less tangible losses, such as the effect of accidents on public relations, employee morale or on the wage rates necessary to secure and retain employees, are not included in this method of estimating costs but may be an important factor in some cases.

Information on some of the items is derived from the Department Supervisor's Accident Cost Report, Form IS-7 (Fig. 7–3).

The items are discussed in the order in which they appear on the Investigator's Cost Data Sheet, Form IS-8 (Fig. 7–4, shown on the previous pages).

1. **Cost of wages paid for time lost by workers who were not injured.** These are employees who stopped work to watch or assist after the accident or to talk about it, or who lost time because they needed equipment damaged in the accident or because they needed the output or the aid of the injured worker.

2. **Cost of damage to material or equipment.** The validity of property damage as a cost can scarcely be questioned. Occasionally, there is no property damage, but a substantial cost is incurred in putting back in order material or equipment which has been thrown into a state of disorder. The charge should, however, be confined to the net cost of repairing or putting in order material or equipment that has been damaged or displaced, or to the current worth of the equipment less salvage value if it is damaged beyond repair.

An estimate of property damage should have the approval of the cost accountant, particularly if the current worth of the damaged property used in the cost estimate differs from the depreciated value established by the accounting department.

3. **Cost of wages paid for time lost by the injured worker,** other than workmen's compensation payments. Payments made under workmen's compensation laws for time lost after the waiting period are not included in this element of cost.

4. **Extra cost of overtime work necessitated by the accident.** The charge against an accident for overtime work necessitated by the accident is the difference between normal wages and overtime wages for the time needed to make up lost production, and the cost of extra supervision, heat, light, cleaning, and other extra services.

5. **Cost of wages paid supervisors for time required for activities necessitated by the accident.** The most satisfactory way of estimating this cost is to charge the wages paid to the foreman for the time spent away from normal activities as a result of the accident.

6. **Wage cost caused by decreased output of injured worker after return to work.** If the injured worker's previous wage payments are continued despite a 40 percent reduction in his output, the accident should be charged with 40 percent of his wages during the period of such low output.

7. **Cost of learning period of new worker.** If a replacement worker produces only half as much in his first two weeks as the injured worker would have produced for the same pay, then half of the new worker's wages for the two weeks' period should be considered part of the cost of the accident that made it necessary to hire him.

A wage cost for time spent by supervisors or others in training the new worker also should be attributed to the accident.

8. **Uninsured medical cost borne by the company.** This cost is usually that of medical services provided at the plant dispensary. There is no great difficulty in estimating an average cost per visit for this medical attention.

The question may be raised, however, whether this expense may properly be considered a variable cost. That is, would a reduction in accidents result in lower expenses for operating the dispensary?

9. **Cost of time spent by higher supervision and clerical workers** on investigations or in the processing of compensation application forms. Time spent by supervision (other than the foreman or supervisor covered in Item 5) and by clerical em-

ployees in investigating an accident, or settling claims arising from it, is chargeable to the accident.

10. **Miscellaneous usual costs.** This category includes the less typical costs, the validity of which must be clearly shown by the investigator on individual accident reports. Among such possible costs are public liability claims, cost of renting equipment, loss of profit on contracts canceled or orders lost if the accident causes a net long-run reduction in total sales, loss of bonuses by the company, cost of hiring new employees if the additional hiring expense is significant, cost of *excess* spoilage (above normal) by new employees, and demurrage. These cost factors and any others not suggested above would need to be well substantiated.

Miscellaneous costs were found in less than 2 percent of the cases in a group of several hundred disabling injury and medical treatment cases which were reviewed in connection with this study.

Making a pilot study

The purpose of the pilot study is to develop for different classes of accidents average uninsured costs which can be applied to future accident totals. Therefore, it is desirable not to include the costs of deaths and permanent total disabilities. Such accidents occur so seldom that the costs should be calculated individually and not estimated on the basis of averages.

Some flexibility in the grouping of classes of cases is desirable. If no distinction is made in the records between medical-treatment cases requiring a physician's attention and those not requiring a physician's attention, the pilot study may combine Classes 2 and 3. (See "Method for estimating" earlier in this chapter.)

The following discussion assumes that the study of costs will be made with the injuries grouped in the recommended classes. The discussion covers Classes 1, 2, and 4. A different method must be applied to Class 3 injuries, and it will be discussed later.

Classes 1, 2, and 4. To analyze uninsured costs for accidents in Classes 1, 2, and 4, the supervisor in charge of the department where an accident occurs should secure for each accident the information indicated on the Department Supervisor's Accident Cost Report form (Fig. 7–3). These data can be obtained during the supervisor's regular investigation of the accident. As soon as each report form is completed, it should be sent to the safety department.

In the safety department, the information from the department supervisor's report will be transferred to the Investigator's Cost Data Sheet (Fig. 7–4). The safety department then assumes responsibility for securing the supplementary information from the accounting department, industrial relations department, and other departments where records on lost time and other necessary information are kept.

As an alternative, a member of the safety department could secure all information needed on the data sheet. In this case, the supervisor's report form is not used, and the supervisor is required only to report each accident in Classes 1, 2, or 4 to the safety department as soon as it occurs.

Before he computes averages, the investigator should be certain that the pilot study has covered a sufficient number of cases of Classes 1, 2, and 4 to be representative. This number will rarely be less than 20 cases. However, more cases should be studied if the costs of the cases in a particular class vary widely. Information should be secured on enough cases of each class so that the average cost per case in each class is fully representative of past experience and therefore, by inference, will be applicable to future experience.

Once a sufficient number of cases has been accumulated, the investigation of individual cases can be discontinued. For the data thus collected, separate averages should be calculated for the cases of each class. It is recognized that these costs are averages of the uninsured costs only.

Class 3 injuries. These injuries are the common first aid cases in which no significant property damage results from the accident. They are the most difficult to analyze from the standpoint of cost, because such loss of time is likely to occur repeatedly and for only short periods, and the injuries may occur so frequently as to place an undue burden on the supervisor and safety director if a complete report form and data sheet are filled out for each case.

The points of essential information needed are the average amount of working time lost per trip to the dispensary, the average dispensary cost per treatment, the average number of visits to the dispensary per case, and the average amount of supervisor's time required per case.

The following method of developing averages for each of these items is recommended:

1. Secure an estimate of average working time lost per trip to the dispensary for first aid. Departmental time records should be consulted as they may show the amount of time each worker is absent from his job for first aid. If so, a random sample of 50 to 100 records of persons known to have received first aid should be selected from different departments. The average time lost per dispensary visit is calculated by adding the absence time for all visits in the sample and dividing by the total number of visits.

   If departmental records do not contain this information, it will be necessary to assign an investigator to observe a random sample of 50 or more persons visiting the dispensary.

   As before, to secure the average time, all the estimated time intervals of absence are added and then divided by the total number of persons observed.

2. Make an estimate of the average cost of providing medical attention for each visit by dividing the total cost of operating the dispensary for a year by the total number of treatments given during the year.

3. Calculate the average number of visits to the dispensary per case by dividing the number of treatments of Class 3 injuries in a representative period, perhaps a month or six weeks, by the number of Class 3 injuries reported during the same period of time.

4. Calculate the average amount of supervisor's time required per case, where pos-

sible, by observing the activities of representative supervisors in connection with first aid cases.

When a sufficient number of cases has been studied to be representative both of the activities of supervisors in different departments and of different types of first aid cases, the average time spent by a supervisor is computed by adding all the time intervals recorded and dividing by the number of cases.

If it is impossible to make a time study of the supervisor's activities in connection with first aid cases, the only alternative is to secure from each supervisor an estimate of the time he spends on the usual first aid case, and to average these estimates by adding them and dividing by the number of supervisors.

The average total uninsured cost of a case in Class 3 is estimated from the data accumulated above as follows: the average amount of time lost for a trip to the dispensary (1, above) is multiplied by the plant's average wage rate, secured from the payroll department, to get the average cost per trip for the worker's time lost. To this figure is added the estimated cost of providing medical attention for a single visit (2). This figure is then multiplied by the average number of dispensary visits per medical treatment case (3), and to this result is added the average amount of supervisor's time required (4).

This method of recording costs is designed to provide estimates of the average uninsured cost per case for accidents causing localized property damage or, at most, a few injuries.

The method of cost investigation for accidents resulting in deaths, permanent total disabilities, or extraordinarily extensive property damage is the same as for others, but the difference is that every one is investigated separately and should be included in the final cost estimate as a separate item. In estimating the cost of a fire, the investigator should bear in mind that the company's fire insurance may cover property damage that would appear as an uninsured cost in other work accidents.

Development of final cost estimate

Once the average for each class of case has been established, costs for any period in which a sufficiently large number of accidents has occurred to be representative can be estimated with considerable accuracy by multiplying the average uninsured cost per case for each of the four classes by the number of cases occurring in that class during the period.

If any deaths, permanent total disabilities, or extraordinarily extensive property damage accidents have occurred, the specific uninsured costs of these should be added to the estimated costs of the four classes of accidents.

To these uninsured cost totals should be added the cost of workmen's compensation and insured medical expense. For companies which are self-insured, this will be the total amount paid out in settlement of claims plus all expenses of administering the insurance. For companies not carrying their own insurance, it will be the amount of their insurance premiums.

The method will have to be modified in accordance with the recordkeeping systems of different companies. For example, most self-insurers will find it impossible to separate compensated medical expense from dispensary care. In that case, these items should be combined into one, and the dispensary cost omitted from the analysis of noncompensated costs on the data sheets.

For an illustration of the development of a final cost estimate, see the example in Tables 7-B and -C.

References

American National Standards Institute, 1430 Broadway, New York, N.Y. 10018. *Standard Method of Recording Basic Facts Relating to the Nature and Occurrence of Work Injuries,* Z16.2–1969.

Firenze, Robert J. *Guide to Occupational Safety & Health Management.* Dubuque, Ia., Kendall/Hunt Publishing Co., 1973.

Simonds, Rollin H., and Grimaldi, John V. *Safety Management—Accident Cost and Control.* Homewood, Ill., Richard D. Irwin, Inc., 1963.

Workmen's Compensation Insurance

Chapter
8

Economic Losses   173

Workmen's Compensation in the United States   173
Early laws . . . Current acts

Objectives of Workmen's Compensation   175
Income replacement . . . Restoration of disabled workers . . .
Accident prevention and reduction . . . Proper cost allocation

Major Characteristics   176
Covered employment . . . Covered injuries and diseases . . . Benefits
. . . Income replacement . . . Medical benefits . . . Rehabilitation

Administration   181
Objectives . . . Handling cases

Security requirements   182
Types of insurers . . . Compliance checks . . . Regulation of
insurers

Financing   183
Financing insured benefits . . . Financing self-insured benefits
. . . Insurer administrative costs . . . Other administrative costs

Occupational Safety   185
Safety activities of insurers . . . Insurance price incentives

Covered Employment   186
Limitations on coverage . . . Other criteria . . . Conclusion

Rehabilitation   187
Medical rehabilitation . . . Vocational rehabilitation

Degree of disability   189
Temporary disability . . . Permanent partial disability . . . Permanent total disability . . . The exclusive remedy doctrine and third party liability

Private Insurer Programs   192
Classifications of insurers . . . Self-insurers

Cost Levels and Allocation   193
Cost levels . . . Workmen's compensation incentives for safety

References   195

In the early 1970's, about 14,000 workers died each year from work-connected injuries or diseases, another 90,000 were permanently disabled, and more than 2,400,000 were temporarily disabled—and these deaths and disabilities were only one-fifth of the total accidental injuries from all causes. Injured workers and their families suffered substantial economic losses as well as bodily injuries. Their employers and society also suffered sizable economic losses.

When a worker dies, is disabled, or merely requires medical attention because of a work-connected injury or disease, the economic consequences affect the worker, his family, his employer, and society.

Economic Losses

The worker and his family may suffer two types of economic losses: (*a*) a loss of earnings and (*b*) extra expenses.

If a worker dies because of work-related injury or sickness, his survivors lose the income he would otherwise have earned, less the amount that he would have spent to maintain himself during the remainder of his working career and his retirement years. This loss can be substantial.

Total and permanent disability cause an even greater earnings loss than death because the worker must be maintained.

Permanent partial disability causes some fraction of the permanent total disability loss, depending upon the proportion of the annual earnings lost. A worker who is totally disabled for a temporary period loses his income for a specific number of weeks or months. Loss of even a month's earnings is a serious loss for the typical worker. In addition to these earnings losses, the deceased or disabled worker may no longer provide valuable household services that must now be forgone or replaced at some additional expense.

Not all injured workers are disabled but almost all require some form of medical attention. For all injuries combined, medical expenses are less than the total earnings loss; but for many workers, their medical expenses exceed their earning loss.

In addition to these losses, society loses the taxes that would have been paid by the injured employees and the products or services they would have produced. Some injured employees and families become public assistance beneficiaries and must be supported by other members of society.

Workmen's Compensation in the United States

In the United States, efforts to implement a system of compensation for industrial injuries lagged far behind the countries of Europe. As work-related injuries and diseases and their consequences grew less and less tolerable towards the end of the 19th century, the situation became ripe for a radical change. The first evidence of interest in workmen's compensation was seen in 1893 when legislators seized upon John Graham Brooks' account of the German system as a clue to the direction of efforts at reform. This interest was further stimulated by the passage of the British Compensation Act of 1897.

Early laws

In 1902 Maryland passed an act providing for a cooperative accident insurance fund; this represented the first legislation embodying to any degree the compensation principle. The scope of the act was restricted. Benefits, which were quite meager, were provided only for fatal accidents. Within three years, the courts declared the act unconstitutional. In 1908, a Massachusetts act authorized establishment of private plans of compensation upon approval of the state board of conciliation and arbitration. This law had no practical significance; it was a dead letter from the start.

By 1908, there was still no workmen's compensation act in the United States. President Theodore Roosevelt, realizing the injustice, urged the passage of an act for federal employees in a message to Congress in January. He pointed out that the burden of an accident fell upon the helpless man, his wife, and children. The President declared that this was "an outrage." Later in 1908, Congress passed a compensation act covering certain federal employees. Though utterly inadequate, it was the first real compensation act passed in the United States.

During the next few years, agitation continued for state laws. A law passed in Montana in 1909, applying to miners and laborers in coal mines, was declared unconstitutional. Nevertheless, many states appointed commissions to investigate the feasibility of compensation acts and to propose specific legislation. The greater number of compensation acts were the result of these commissions' reports, all of which favored some form of compensation legislation, combined with recommendations from various private organizations. Widespread agreement on the need for compensation legislation unfortunately did not end all conflict over reform. Interest groups clashed over specific bills and over questions of coverage, waiting periods, and state versus commercial insurance.

In 1910, New York became the first state to adopt a workmen's compensation act of general application which was compulsory for certain especially hazardous jobs, and optional for others. None of the early state compensation acts expressly covered occupational diseases. Statutes which provided compensation for "injury" were frequently interpreted to include disability from disease, but those acts which limited compensability to "injury by accident" excluded occupational disease. All except Oregon's act required uncompensated waiting periods of one to two weeks; several states provided retroactive payments after a prescribed period.

The 1911 Wisconsin workmen's compensation act was the first law to become and remain effective. The laws of four other states (Nevada, New Jersey, California, and Washington) also became effective that year. Although 24 jurisdictions had enacted such legislation by 1925, workmen's compensation was not provided in every state until Mississippi enacted its law in 1948.

Current acts

Today there are compensation acts in the 50 states, the District of Columbia, and Puerto Rico. In addition, the Federal Employees' Compensation Act covers the employees of the U.S. Government, and the Longshoremen's and Harbor Workers' Compensation Act covers maritime workers, other than seamen, and workers in certain other groups. This latter act provided compensation for workers in the "twilight zone" between ship and shore, since they were not covered under existing state compensation laws.

While economic changes and public policy have prompted increases in benefits and scope of the laws, the basic concepts have not undergone any radical changes. Employers and labor are both dissatisfied with certain aspects of workmen's compensation. Labor attacks the system for inadequate benefits, coverage limitations, and exclusion of many injuries, illnesses, and disabilities that they consider job-related. Employers are critical because the system covers some injuries and diseases they do not consider job-related and is costly relative to its apparent benefits. Thus, while the early advocates of workmen's compensation conceived it as a simple, speedy, efficient, equitable remedy that would reduce litigation over industrial injuries, many doubt their hopes have been realized.

Objectives of Workmen's Compensation

Workmen's compensation programs can be evaluated by the extent to which they satisfy the following commonly accepted objectives:

1. Income replacement

2. Restoration of earning capacity and return to productive employment

3. Industrial accident prevention and reduction

4. Proper allocation of costs

5. Achievement of the other four objectives in the most efficient manner possible

Not all of these objectives are equally important or accepted. The first two generally are considered most important. These objectives sometimes conflict with one another but in most ways they are linked by the design of the program.

Income replacement

The first objective listed for workmen's compensation is to replace the wages lost by workers disabled by a job-related injury or illness. According to this objective, the replacement should be adequate, equitable, prompt, and certain.

To be adequate, the program should pay the entire wage loss (present and projected, including fringe benefits), less those expenses such as taxes and job-related transportation costs that do not continue. The worker, however, may be asked to share a small proportion of the loss in order to provide some incentives for rehabilitation and accident prevention. The two-thirds replacement ratio that is found in about one-half the state statutes is generally considered acceptable where it is not undercut by maximums.

To be equitable the program must treat all workers fairly. According to one concept of fairness, most workers should have the same proportion of their wages replaced. However, a worker with a low wage may need a high proportion of his lost wage in order to sustain himself and his family. If a guaranteed minimum income plan existed, there would be less need to favor low-income workers. A high-income worker who can afford to purchase private individual protection may have his weekly benefit limited to some reasonable maximum. If workmen's compensation insurance is regarded primarily as a wage replacement program however, relatively few persons should be affected by this maximum. An alternative philosophy would argue in favor of a more substantial welfare component with a higher minimum benefit, low maximum benefit, and extra benefits when there are dependents.

Ideally, workers would be treated the same regardless of the jurisdiction in which they are injured. This criterion, therefore, implies a minimization of interstate differences in statutory provisions and their administration.

The program should pay all disabled persons an income starting as soon after their disability commences as possible. Finally, workers should know in advance what benefits they will receive if they are injured on the job and that these benefits will be paid regardless of the continued solvency of the employers.

Under the whole-man theory, the system would be required to indemnify the worker or his family for the effect on all his personal activities, not his earning capacity alone.

Restoration of disabled workers

The second listed objective is medical and vocational rehabilitation and return to productive employment. To achieve this objective, the worker should receive quality medical care at no cost to himself, care which will restore him as well as possible to his former physical condition. If complete restoration is impossible, he should receive vocational rehabilitation that will enable him to maximize his earning capacity. Finally the system should include incentives to disabled workers and prospective employers so the workers will return to productive employment as quickly as possible.

Accident prevention and reduction

Occupational accident prevention and reduction is a third commonly accepted objective of workmen's compensation. Those who consider this objective to be important believe that the system should and can provide significant financial and other incentives for employers to introduce measures that will decrease the frequency and severity of accidents. More specifically, the pricing of workmen's compensation should reward good safety practices and penalize dangerous operations. Employees should also have some incentive to follow safe work practices by sharing some of the losses. Injured workers should have the opportunity and be encouraged to return to work as soon as they are physically able.

Those who minimize this objective of workmen's compensation recognize the importance of safe work places and procedures but believe that workmen's compensation rates and other incentives have little effect upon how employers behave. Consequently they favor other approaches, such as public safety inspectors and criminal penalties.

Proper cost allocation

The fourth objective of workmen's compensation, which has a narrower support than the first three, is to allocate the costs of the program among employers and industries according to the extent to which they are responsible for the losses to employees and other expenses. Such an allocation is considered equitable by supporters of this objective because each employer and industry pays its fair share of the cost. The economic effects are considered desirable because this allocation tends in the long run in a competitive economy to shift resources from hazardous industries to safe industries and from unsafe employers within an industry to safe employers. Higher workmen's compensation costs will force employers with hazardous operations to consider raising their prices. To the extent that consumers will not accept the price increase, employer profits and their willingness to commit resources to this use will decline.

Critics of this objective argue that workmen's compensation costs are such a small part of the cost of production that they have little, if any, effect on resource allocations. Consequently, they would avoid the complicated pricing practices necessary to achieve this objective.

Major Characteristics

Covered employment

While most of the state workmen's compensation laws apply to both private and public employment, none of the laws covers all forms of employment. For various historical, political, economic, or administrative reasons, each of the laws has certain gaps. Laws that are elective rather than compulsory permit the employer to reject coverage; but in the event he does, he loses the customary common law defenses: assumed risk of the employment, negligence of a fellow servant, and contributory negligence.

A few states still restrict compulsory coverage to so-called hazardous occupations. Many laws exempt employers having fewer than a specified number of employees. The most common exception is for employers having fewer than three employees; the range goes from fewer than two employees in two states to fewer than 15 in one state. Most of the laws exclude farmwork, domestic service, and casual employment. Many laws also contain other exemptions, such as employment in charitable or religious institutions.

Two other major groups outside the coverage of the compensation laws are interstate railroad workers and maritime employees. Railroad workers, any part of whose duties involve the furtherance of interstate commerce, are covered by the Federal Employers' Liability Act (FELA). Maritime workers are subject to the Jones Act, which applies provisions of the FELA to seamen. The Federal Employers' Liability Act is not a workmen's compensation law. It gives an employee an action in negligence against his employer and provides that the employer may not plead the common law defenses of fellow servant or assumption of risk; moreover, the principle of comparative negligence is substituted for the common law concept of contributory negligence.

As to the state and local employees, the actual number of these employees subject to workmen's compensation or provided with such protection voluntarily is not available. All states (as well as Puerto Rico and the District of Columbia) have some coverage of public employees but with marked variations. Some laws specify no exclusions or exclude only such groups as elected or appointed officials. Others limit coverage to employees of specified political subdivisions or to employees engaged in hazardous occupations.

In still others, coverage is entirely optional with the state or with the city or political subdivision.

Certain other groups, such as the self-employed, unpaid family members, volunteers, and trainees, generally are not protected by workmen's compensation.

Gradual extension of coverage over the years has been achieved by piecemeal actions: replacement of elective laws by compulsory provisions, elimination or reduction of numerical exemptions, and adoption of amendments granting protection to farm workers and other previously excluded groups. States still must strive for complete coverage.

Covered injuries and diseases

Workmen's compensation is presently intended to provide coverage only for certain work-related conditions, not all of the worker's health problems. Statutory definitions and tests have been adopted to provide the line of demarcation between those conditions which are compensable and those which are not. Because, in drafting workmen's compensation laws, all jurisdictions relied to some extent on the English system (or other statutes that in turn relied upon the English model), their statutory language is remarkably similar. Nevertheless, as there are variations in language as well as differences in interpretation, a condition considered compensable in one state may be held non-compensable in others.

The statutes usually limit compensation benefits to personal injury caused by accident arising out of and in the course of the employment. Although this presents four distinct tests which must be met, in practice they are often considered in pairs: The "personal injury" and "by accident" requirements in one set, and the "arising out of" and "in the course of" requirements in the other.

Personal injury and by accident. If interpreted narrowly, personal injury would deal solely, with bodily harm, such as a broken leg or a cut, while the by accident test would refer to the cause, such as a blow to the body or an episode of excessive or improper lifting. In practice, however, the distinctions are blurred.

The by accident concept is a carryover from the English law. Early judicial interpretations of the English law made it quite clear that for their purposes the by accident requirement was intended to do little more than deny compensation to those who injured themselves intentionally. A number of U.S. jurisdictions, however, have applied the test so as to narrow the range of unintentional injuries which can be compensated.

One of the early victims of the by accident requirement was occupational disease coverage. As the typical judicial holding was that occupational disease and accidental injury were mutually exclusive, special legislation was required in order to provide disease coverage. At present occupational diseases are almost always treated separately in compensation law.

Although most jurisdictions cover all occupational diseases, about nine states limit the coverage to scheduled diseases. Even those jurisdictions that employ a general definition of occupational disease often attach other limitations. For example, they may require that the disease not be an ordinary disease of life. Thanks largely to relaxation of the by accident concept and dissatisfaction with the restrictive provisions of most occupational disease statutes, awards can be and have been made for diseases as injuries by accident when there was something particularly unusual about the cause of the disease and the mode of conveyance was specific, such as the entry of bacteria through a slight cut.

The by accident concept was also used in many jurisdictions to deny compensation unless the injury was caused by some sort of unusual, traumatic occurrence, generally requiring the application of outside agency. Obviously this would and did drastically limit the kinds of cases which could be compensated. At present, this use of the by accident test is limited to a few narrow areas.

Impairment involving psychological difficulties has been the source of much controversy based on application of the personal injury requirement. In some cases, a mental stimulus such as fear can produce a physical lesion, such as a cerebral hemorrhage. In event of a physical lesion, the courts have not encountered much difficulty in conceding personal injuries. Compensation is usually approved also if, as a result of a clear physical injury, the patient suffers psychological disorder.

8—Workmen's Compensation Insurance

As might be expected, disagreement is most likely when it is alleged that mental stimulation has resulted in a mental illness without obvious physical change. Although many jurisdictions award compensation in such cases, others still are reluctant to perceive that psychological disorder is a physical injury.

Work-related impairment. The term "arising out of and in the course of the employment," applied by almost every jurisdiction, is meant to define a certain level of relationship between the employment and injury or disease as a condition of eligibility for workmen's compensation. The phrase obviously lacks certainty. Often it is quite difficult to determine whether a given set of facts will support an award of compensation.

The "course of the employment" aspect of this test refers primarily to the time frame of the injury. Virtually every jurisdiction holds that an employee is within the course of his employment, barring certain types of unusual circumstances or unreasonable conduct, from the moment he steps onto the employer's premises at the beginning of the work day to the moment he leaves the premises at the end of the day.

Although this test appears to be relatively simple to apply, it has not been so. One uncertain issue is, what are the premises? Injuries which clearly occur off premises but appear to deserve compensation lead to a search for exceptions and encourage courts to modify the basic rules. Many workers are not attached to particular premises. Even though an injury occurs off premises, as in travel to and from work, the employee may be compensated if a sufficient employment relationship can be found, such as payment for time or expense of travel or the provision of a company vehicle for transportation. In these circumstances, the period of travel time to and from home may be incidental in the course of employment.

The "arising out of" segment of the test is intended to provide a causal relationship between the employment and the injury. For example, it is not enough that an employee suffer a heart attack while at work. He must show that the heart attack arose out of the employment or, in other words, that it was causally related to the employment.

This means that at a minimum (some states have more stringent rules) it must be shown that it was the stress and strain or exertion of the employment that caused the heart attack, not merely a spontaneous breakdown of the cardiovascular system.

The degree of employment relationship necessary varies from state to state and has been modified as workmen's compensation law has evolved. In earlier years, it was generally felt that the hazard-causing injury must be peculiar to the particular employment or be increased by the employment before the injury could be said to "arise out of the employment." This rather narrow view of compensability has been modified and to some extent abandoned in recent years.

Although it is difficult to place each jurisdiction in a particular category as to what it will hold sufficient to meet the "arising out of" test, two additional theories have been developed and followed. The first and more widespread is the "actual risk doctrine," which requires that the hazard resulting in injury be a risk of the particular employment, without regard to whether it was also a risk to which the general public is exposed. The second or "positional risk doctrine" could also be called the "but for" test. Here, if the employment places the worker in a position where he is injured ("but for" the employment the injury would not have occurred), the injury "arises out of the employment."

Benefits

Almost $3 billion in cash and medical benefits were received by workers in 1970 through the workmen's compensation system. Benefits include medical services, cash benefit payments to the worker while totally disabled, payments for residual partial disability, burial allowances (in all but one state) for work-related deaths and benefits to the worker's dependent survivors.

Some states provide special benefits also to cover attendants or prostheses; about three-fourths of the states provide maintenance and other services for rehabilitation. The largest proportion of benefits are in cash, either as periodic payments or as lump sums in settlement of claims. About $1.9 billion, almost two-thirds of the $3 billion 1970 benefit total, were paid to workers or their survivors in cash.

Benefits are paid through three channels: commercial insurance policies; publicly operated state insurance funds; and self-insured employers. In 1970, more than $1.8 billion in workmen's compensation was paid by private insurers, $0.7 billion by state funds, and $0.4 billion by self-insurers.

Income replacement

Of the $1.9 billion benefits paid in 1970 as cash income, almost 90 percent went to disabled workers and the other 10 percent to survivors of workers killed on the job. Although 70 percent or more of recent workmen's compensation cases are for temporary total disablement, such cases have accounted roughly for only one-fourth of cash benefits. At the same time, income benefits in the last few years to workers for permanent partial disabilities accounted for two-thirds of the total dollar amount.

Basic features. In general, the cash benefits provided for temporary total disability, permanent total disability, permanent partial disability, and death are payable as a wage-related benefit—the weekly amount is computed as a percentage of the worker's wage. The benefit varies by state and by type of disability but most commonly is set at two-thirds of wages. In some states, the statutory percentage varies with the worker's marital status and the number of dependent children, especially for survivors' benefits, which in a majority of states pay 50 percent or less of the deceased worker's wage to surviving widows without dependent children.

The benefit rate is limited to less than two-thirds of wages for many beneficiaries by another statutory provision—the maximum ceiling on the weekly benefit payable. Disabled workers whose wages are at or above the statewide average receive benefits below the statutory benefit rate in almost all states because of this ceiling, despite an occasional increase in the weekly ceiling by amendments to the law.

As inflation boosts wage levels, some states have attempted to prevent the deterioration in effective benefit-wage rates by providing for future increases in the maximum without need for further legislation—14 automatically adjust the maximum (for new beneficiaries only) in relation to changes in the state's average weekly wage. Much less common but gaining more interest in the last few years are provisions that, as wage levels of workers rise, will raise benefits for beneficiaries already on the rolls—five states plus the Federal Employees' Compensation Act provide such automatic increases.

Another type of limitation on benefits sets maximum time periods or aggregate dollar amounts. Such limitations in permanent total disability and death cases may cut off benefits even though the income need continues. Nevertheless, a majority of states limit 'the duration or total dollar benefits to widows and orphans.

In order to reduce administrative costs and to discourage malingering, benefits in all states are payable only after a waiting period following the report of disability. This delay in payment applies to the cash indemnity payments, not to medical and hospital care. The waiting period ranges from 2 days to 7. In all states, workers who are disabled beyond a specified minimum period of time receive payment retroactively for the waiting period. For more than three-fourths of covered workers, the minimum period for retroactive payment is a disability exceeding two weeks.

Benefits by type of disability. Most compensation cases concern workers who incur temporary disability but recover completely. The maximum weekly benefit for temporary total disability is at least $65 in more than half the states. In only eight states is the maximum benefit as high as two-thirds the state's average weekly wage. Although about two-thirds of the states have provisions limiting the duration of temporary benefits, these limits do not seriously affect adequacy of benefits because few temporary injuries persist beyond such limits, typically set between 6 and 8 years. For workers with dependents, about one-third of the states augment the weekly benefit for temporary disability, usually by some dollar amount for each dependent up to a specified total.

Benefits for permanent total disability are for disabilities that preclude any work or regular work in any well known branch of the labor market and that can be of indefinite duration. These are similar to benefits for temporary total disability benefits. In a few states, the weekly payment for permanent disability benefits is less than for temporary.

About one-third of the states restrict the duration of benefits for permanent disabilities, typically to from 6 to 10 years.

Residual limitations on earning capacity after recovery (that is after a permanent partial disability) are awarded compensatory benefits on a relatively complex basis. Partial disabilities are divided into two categories—"schedule" injuries, those listed in the law such as loss of specific bodily members; and "non-schedule" injuries, those which are of a more general nature, such as back and head injuries.

Weekly benefits for schedule injuries are a percentage of average weekly wages, often the same as the benefit rate for permanent total disability. The maximum weekly benefit is for the most part the same as or lower than that for total disability.

Non-schedule injuries are paid at the same or similar rate but as a percentage of wage loss, the difference between wages before injury and the wages the worker is able to earn after injury.

The schedule benefits are paid for fixed periods varying according to the type and severity of the injury. For example, most state laws call for payments ranging from 200 to 300 weeks for loss of an arm and 20 to 40 weeks for loss of a great toe.

The maximum period for non-schedule injuries for each state is the same as or, more generally, less than the duration limits established for permanent total disability.

In the majority of states, compensation payable for permanent partial disability is in addition to that payable during the healing period or while the worker is temporarily and totally disabled. In some states, lower benefits (or no benefits) are payable for permanent partial disability due to occupational disease than for disability due to accidental injury.

Death benefits are intended to furnish income replacement for families dependent upon the earnings of an employee whose death is work-related. As is true for the other types of benefits, the amount of survivor benefits and the length of time they are paid vary considerably from state to state. Benefits computed as a percentage of the deceased

worker's wage often are less than that for permanent total disability benefits if the survivor is a widow without dependents. If there are dependent children, the benefit in many states will be augmented. In most states, the duration of benefits is limited usually in a range from 250 to 600 weeks. In 28 states, payments to widows continue usually as long as they do not remarry and to children until they are no longer dependent, usually to age 18. Benefits may be terminated earlier in the 10 of these 28 states which also limit total dollar benefits.

In addition to benefits for widows and children, some states pay survivor benefits to dependent invalid widowers, parents, or siblings of the dead worker. The program for federal civilian employees pays survivor benefits to widowers under the same circumstances as to widows (i.e., invalidity is not required). Burial expenses are payable in all states except Oklahoma.

Medical benefits

For many years, disbursements for medical services provided under workmen's compensation have comprised about one-third of total outlay for benefits. Care includes first aid treatment, services of a physician, surgical and hospital services, nursing and drugs, supplies, and prosthetic devices. Some large employers, in addition to first aid facilities, employ staff physicians for workers. Most employers insure their medical care responsibility as they do the income benefits under workmen's compensation.

Every state law requires the employer to provide for medical care to the injured worker. In most jurisdictions, such treatment is provided without limit either through explicit statutory language or administrative interpretation. Only nine states limit the total medical care available for work-related injuries by specified maximum dollar amounts or maximum periods. In some of these nine, the initial ceiling may be exceeded by administrative decision. Even among states which provide unlimited medical care following on-the-job accidents, almost a third limit medical services for specified occupational diseases. Also, if specified types of injuries or disease are denied cash benefits, medical care also is denied.

A major issue concerning medical benefits for workmen's compensation is the procedure for choosing the physician who is to furnish the care. About half the states give the employer the right to designate the physician. In practice the insurance company of the employer will ordinarily select the physician since it is the insurer that handles the claim for benefits. Where the doctor is chosen in this way, the medical care furnished may be more highly skilled and effective because of the selected physician's specialized experience. On the other hand, workers feel that a more important consideration is the emphasis that their personal family physician is likely to put on their health and well-being. They feel other considerations may influence a physician they do not select.

Rehabilitation

Along with industrial safety, medical care, and cash compensation, rehabilitation of workers is recognized at least theoretically as one of the primary goals of the workmen's compensation system. At present the most widespread benefits offered through workmen's compensation laws to restore a worker to his fullest economic capacity are the special maintenance benefits authorized in about three-fifths of the states. These benefits usually are paid (sometimes in addition to the regular disability compensation) for various training, education, testing, and other services designed to aid the injured person to return to work. In addition, some state programs provide for travel expenses and for books and equipment needed for the training.

Ohio, Oregon, Rhode Island, Washington, and Puerto Rico directly operate rehabilitation facilities under the workmen's compensation program. Some insurance companies also have in-house facilities for rehabilitating workers.

Probably the main source of retraining and rehabilitation is the federal-state vocational program. The facilities operated by this program accept individuals with work-related disabilities as well as others. In all states, these institutions are directed by state vocational rehabilitation agencies. They provide medical care, counseling, training, and job placement. Unfortunately, not all workmen's compensation cases referred for vocational rehabilitation can be accepted promptly. Many others are never brought to their attention.

One notable drawback preventing full utilization of available rehabilitation facilities is the often protracted, adversary proceedings for determining a worker's right to benefits. Because the determination of whether there should be an award for permanent partial or total disability (and how large an award should be for partial disability) is conditioned on the worker's lack of ability to work, the claim may be a strong disincentive for rehabilitation. Further, in the many compromise settlements, the employer's (or insurer's) motivation is to pay an agreed amount of money and foreclose future responsibility for medical, vocational, or other needs arising from the injury. Such settlements also work against a full-fledged effort to restore the worker to full health and productivity.

Administration

The goal of workmen's compensation is to provide for quick, simple, and inexpensive determination of all claims for benefits and to provide such medical care and rehabilitation services as are necessary to restore the injured worker to employment. Nearly all of the states have agencies to carry out these administrative responsibilities. In about 20 jurisdictions, the agency is in the labor department; in 27, it is a separate workmen's compensation board or commission; in five, administration is left to the courts. Several states have separate, independent appeals boards to review claims when agency decisions are appealed.

Objectives

An agency's many correlated responsibilities include close supervision over the processing of cases. The primary objective is to assure compliance with the law and to guarantee an injured worker's rights under the statute. Administration by a division within the labor department or by a board or commission has been found to be more effective in achieving the full purpose of the law than administration by the courts. The courts are not organized and equipped to render the services needed.

One criticism of state agencies concerns the

delays in the first payment of compensation to the disabled worker. Although in most states insurers mail the checks, the state administrative agency has the responsibility to see that payments commence promptly. Full and prompt payment is essential because few workers can afford to wait long for benefits due. In one state, perhaps the best, the claims are paid within 15 days; in most states, however, it appears that the first payment comes as much as 30 days late.

Another responsibility of the agency is to see that the injured worker gets the full benefit due. To do this, it is important to follow an injury from the first report to the final closing of the case. Some states not only check the accuracy of total payments, but also require signed receipts for every compensation payment. Some require the filing of a final receipt which itemizes the purpose of each element in the total outlay, to permit a complete audit of individual payments.

Frequently, however, the legislation itself requires a workmen's compensation agency to operate on the assumption that each injured worker is responsible for securing his rights and that its primary function is to adjudicate contested claims. Even where the law does not favor this policy, lack of staff may force the agency to this restricted role.

Although it is known that many workers are not familiar with the provisions of their workmen's compensation act, in only a few states does the administrator (as soon as possible after the injury is reported) advise the worker of his rights to benefits, medical and rehabilitation services, and assistance available at the commission's office. Too many states fail to insist on prompt reporting of accidents by employers, on prompt payment of benefits, or on final reports which spell out the amounts paid and how these amounts were computed. Although prompt reporting is usually required, sometimes no penalty is imposed for violation.

Handling cases

Workmen's compensation claims may be either uncontested or contested. In uncontested cases, the two main methods followed are the direct payment system and the agreement system.

Under the direct payment system, the employer or his insurer takes the initiative and begins the payment of compensation to the worker or his dependents. The injured worker does not need to enter into an agreement and is not required to sign any papers before compensation starts. The laws prescribe the amount of benefit. If the worker fails to receive this, the administrative agency can investigate and correct any error. Jurisdictions whose laws provide the direct payment system include Arkansas, Michigan, Mississippi, New Hampshire, Wisconsin, and the District of Columbia; it is also provided for in the Longshoremen's and Harbor Workers Compensation Act.

Under the agreement system, in effect in a majority of the states, the parties (that is, the employer or his insurer, and the worker) agree upon a settlement before payment is made. In some cases, the agreement must be approved by the administrative agency before payments start.

In contested cases, most workmen's compensation laws provide for a hearing by a referee or hearing officer, with provisions for an appeal from the decision of the referees or hearing officer to the commission or appeals board and from there to the courts. As the administrative agency usually has exclusive jurisdiction over the determination of facts, appeals to the courts usually are limited to questions of law. In some states, however, the court is permitted to consider issues both of fact and law anew.

Security Requirements

All states except Louisiana require their employers in private industry to demonstrate that they are able to pay the benefits required under the workmen's compensation law. About two-thirds of the states have a similar provision for public employers. These security provisions in effect require that active steps be taken by employers to guarantee that workers, when they are disabled, will receive the benefits called for by the law.

Types of insurers

Most laws allow employers to satisfy the security requirement by insuring with private companies or to self-insure. As of January 1,

1972, only six states (Nevada, North Dakota, Ohio, Washington, West Virginia, Wyoming), Puerto Rico, and the Virgin Islands required employers to purchase protection from exclusive state operated funds. Three of these states (plus one that has no state fund) and Puerto Rico and the Virgin Islands also prohibit self-insurance. Besides the jurisdictions with exclusive state funds, 12 others have publicly operated programs in competition with private insurers. Regardless of the method of protection purchased, the same statutory benefits must be provided to the injured worker in that state.

Compliance checks

In order to make sure that workmen will receive benefits as intended, states need a method of checking that employers do in fact meet the security requirements. The workmen's compensation agency ordinarily requires not only notice of insurance secured by employers but of cancellations of such insurance. In more than a fifth of the states, however, no formal procedures are in effect to make sure that all employers have given proper notice.

Generally, it is believed that most employers comply with security requirements. In part, a high degree of compliance may be expected because of sanctions available to the State for noncompliance. In almost four-fifths of the states, noncomplying employers become liable to worker suits with the employer's traditional common law defenses abrogated; in some states, the business may be stopped from operating. In addition some statutes call for fines against the employer or imprisonment or both.

Regulation of insurers

Where employers are allowed to self-insure, they must generally demonstrate sound financial condition. An employer may have to make a deposit of a specified amount with the workmen's compensation agency or post a surety bond. In at least a third of the states, all applicants for self-insurance must meet this requirement; in a similar proportion of states, at the discretion of the agency, this deposit may not be required. Other types of requirements imposed on self-insurers in various states are minimum payroll size, minimum number of employees, type of business, safety record, and proof of proper facilities for administering claims.

Besides restrictions imposed upon employers directly, activities of workmen's compensation insurers also are regulated. Such regulation serves in part to assure that workers receive benefits when disabled. In order to write workmen's compensation insurance, insurers must conform to rules and regulations of both the state insurance department and the state agency administering the workmen's compensation act (usually the industrial commission).

The insurance department primarily regulates the conditions for establishment of insurance companies in the state, their continuing solvency, and their business practices. Like self-insurers, insurance companies (in many states) must post bond or make a deposit with the state insurance department.

Generally the role of the industrial commissions in regulating insurers is limited. Few have either the authority concerning companies' rights to underwrite workmen's compensation in their state or the information about financial status and operations of insurers. Further, although industrial commissions would seem to have a direct interest and concern in the claims-handling performance of insurers, few state agencies collect data on promptness of payment, amount of benefits paid, number of beneficiaries currently receiving benefits, and similar aspects of benefit operations. Generally, industrial commissions (to the extent they supervise claims operations) do so through review of individual cases, often only in the event of a dispute.

Financing

The total cost of workmen's compensation to employers has increased slightly over the years and now is slightly above 1 percent of covered payroll (1.13 percent in 1970). Since insurance is the main vehicle for meeting the statutory requirements of the workmen's compensation acts, the programs are financed mostly through insurance premiums.

Financing insured benefits

For both public funds and commercial insurance, class premiums are established by an elaborate system of rates that take into account the general occupational classifications or industrial activities of the insured. About 15 percent of the employers, paying about 85 percent of the premiums, are experience rated. That is, their premiums are modified to reflect their loss experience in the past relative to others in the same class. Also, the statistical reliability of that experience is taken into account; the larger the business, the more credible its experience. Since employers with a small number of workers are likely to experience volatile changes in injury rates from year to year, only employers of large numbers are experience rated.

Another factor in the premium setting procedure is that discounts are given according to the size of the risk; this is an advantage to large companies. Their rates thus reflect the economies of scale which result from spreading certain fixed costs over a larger amount of premium. Finally, large companies by retrospective rating may have their premiums adjusted at the end of a policy year to match their actual experience.

Most insurers use rates developed by a rating bureau. In some states, the rates developed by the bureau are mandatory; in others, advisory only. Almost half the premiums are written on a participating basis. Participating policyholders receive periodic dividends that reflect insurer experience and sometimes their own.

Financing self-insured benefits

Firms that cover workmen's compensation risks through private insurance companies or state funds pay a premium in advance. In contrast, self-insurers have several options for financing. They may simply pay for liabilities as they are experienced, directly from operating funds, or they may provide some advance funding in one or more ways. In those states requiring deposits of funds by self-insurers, part or all of the funding for outstanding liabilities is provided for in advance mandatorily. Even if not required, a self-insurer may set aside reserves, or even formally insure its risk through a wholly owned subsidiary insurance company created for this purpose.

Such advance funding prevents severe disruptions in cash flow from unforeseen loss experience or accumulated liabilities.

Insurer administrative costs

One of the recurring issues in evaluating workmen's compensation is the financial efficiency of the insurance mechanism for providing benefits. A major part of the issue is the comparison between private and state fund insurance.

The premiums collected by private insurers are used not only to pay benefits but also for expenses associated with claims such as investigation and legal fees; for sales, supervision, and collection; for administration; for safety programs; and for taxes, licenses, and other mandatory fees as well as for earnings. In 1970, stock insurers that do not pay dividends to policyholders had expenses totaling 31 percent of premiums earned; their underwriting gain was 5 percent of premiums paid. For stock insurers that pay dividends to policyholders, the expense ratio was 25 percent; the underwriting gain, 14 percent. Mutual insurers had an expense ratio of 24 percent and an underwriting gain of 13 percent. The dividend-paying stocks and mutuals returned part of their underwriting gain to their policyholders. In addition to their underwriting gains, these insurers had investment profits.

State funds have much the same costs as private insurers with these exceptions: lower (or no) taxes and fees to the state government, no margin for private profit, and lower selling costs. Consequently, although the variation among individual funds is great, expenses have averaged less than 10 percent of total premiums paid, well below the ratio for private insurers. Some state funds incur smaller expenses for administrative and legal services, which may be financed from other government funds. On the other hand, state funds in some instances may insure greater proportions of high-risk companies than private carriers and incur proportionately heavier charges for benefits.

Other administrative costs

Another aspect of financing workmen's compensation relates to the cost of supporting the public agency that administers the program. The cost of operating the industrial commission (or other administering agency) is borne either by assessments upon insurers and self-insurers or through appropriations from public funds. In the former event, the cost of administering the program is simply one more expense item in the premium charge to the employer. Where funds for the industrial commission are obtained from legislative appropriations, this part of the program is paid out of general taxes. More than one-third of state agency administrative costs nationally are funded by legislative appropriations.

In addition to that part of administrative costs financed by state general revenues, other elements in the workmen's compensation system may not be financed through insurance premiums paid for by employers. For example, many employers provide medical services at their establishment or by direct payments to medical facilities. Second- or subsequent-injury funds, which bear part of the cost of injuries to handicapped workers, are financed sometimes through assessments on insurers, reflected in premiums; in some states, as direct charges upon employers; as appropriations from state funds; in a few states, as joint employee and employer contributions to the fund; or by other means. Other special funds, paid for by general revenues, have been established for such purposes as supplementing benefits depreciated by inflation or paying benefits for specified occupational diseases.

Occupational Safety

From the beginning, the workmen's compensation movements in the United States have been associated with the movement to prevent occupational injury or disease. Although some interest in this work was manifested by various employers before the enactment of workmen's compensation laws, the organized safety movement, as we know it, began shortly after the first compensation laws went into effect. This movement was due in large part to an assumption on the part of industrial leaders that one of the best ways to reduce compensation costs would be to reduce the number of accidents.

The first move toward an organized effort came at the convention of the Association of Iron and Steel Electrical Engineers at Milwaukee in 1912, as discussed in Chapter 1. A session devoted to safety set up a committee on organization, which called a meeting of all interested groups and individuals in New York City the following year. This meeting resulted in the formation of the National Council for Industrial Safety, which since 1915 has been called the National Safety Council.

Safety activities of insurers

Although the insurance business had been extended into the industrial accident field before the first workmen's compensation act was passed in this country, industrial accident insurance received great impetus from this legislation. As specific schedules of payments for all work-connected injuries made the risk more definitely calculable, the business became more attractive to the underwriter.

From the first, insurance companies writing workmen's compensation policies have had a large part in the movement to prevent accidents. They have developed or aided in the development of safety standards and safe practices and have contributed to the development of methods and techniques of accident prevention. Much of the basic data of safety engineering has been supplied by insurance engineers. An important motive for their accomplishment is, of course, the fact that their business thrives on a declining injury rate. Unduly large losses jeopardize the financial solvency of an insurance company. Progressively lower losses make it easier for the stock company to show a profit to its stockholders and for the mutual to pay dividends to its policyholders.

The effective work of insurers, however, has been confined mainly to large establishments. The cost of providing technical assistance in accident prevention makes it difficult for an insurer to provide service adequately to plants whose premiums are small. Owners of small plants, moreover, cannot expect to receive much reduction in premium rates either through dividends or experience rating, no matter how effective their safety program.

Precise statistics are not available showing the total amount of money invested by the insurance industry in safety. Data compiled by the National Council on Compensation Insurance show that private insurance companies reported about $37.8 million or 1.1 percent of net premiums earned were applied to safety in 1970. The data also show that selected state insurance funds spent about 1.4 percent of net premiums earned, or $3.9 million.

Insurance price incentives

In addition to offering technical assistance, insurers have also tried, by various means, to make safety pay the policyholder in the form of immediately reduced premium rates. This monetary benefit for accident prevention is to some extent inherent in mutual insurance, since the surplus remaining after losses and expenses is distributed among the policyholders. To offer a similar inducement to policyholders in stock companies and to give further incentive to mutual policyholders, a system of merit rating was adopted to obtain reductions in premium rates for policyholders.

The first type of merit rating used was schedule rating, under which the reduction in premium rate was computed on the basis of the policyholder's performance in providing physical safeguards. This tactic proved unsatisfactory because safeguards, while vital, are only one part of prevention. The system now in general use is experience rating, described earlier in this chapter.

Historically, it has been assumed by many authorities that merit rating provided a powerful stimulus to the safety movement. However, safety professionals do not rely solely on the merit rating system to stimulate accident prevention efforts by industry. In fact, some recent studies have questioned the value of merit rating as a strong impetus to safety.

Covered Employment

Although many controversies have divided students of workmen's compensation, one point on which there is broad agreement is that coverage under the acts should be virtually if not completely universal. With few exceptions, employers and workers alike agree on the desirability of this basic protection. For the employer, it represents a relatively inexpensive way to protect himself against the possibility of lawsuits for injuries to his employees. For the worker it represents an important segment of his protection against income loss.

The principle of virtually universal coverage of all gainfully employed workers is basic. Yet, even today, none of the state laws meets this goal, though a few come close. Although it is believed generally that coverage has progressed and although the public has been educated on the justification for including all employment in the workmen's compensation system, for the past 20 years the proportion of covered civilian wage and salary workers included has hovered around four-fifths of the potential. Recently, it has edged up to about 83 percent, largely as a result of a shift of workers to covered employment. In 1970, according to the Social Security Administration, average weekly coverage under state and federal workmen's compensation totaled 58.8 to 59.0 million, out of the 70.6 million civilian wage and salary workers in the country.

Limitations on coverage

Most of the arguments originally brought against extension of coverage have lost their force. Nevertheless, in view of the persistence of some of the exemptions or exclusions in many state laws, a review of some of the reasons behind the original limitations may be helpful. Nearly all state acts were prepared and enacted in the face of constitutional challenges and outright opposition of certain interests. Thus, each act was the result of compromises rather than the outcome of a consistent, ideal program, even if, in some instances, much weight was given to a carefully studied plan.

Other criteria

Workmen's compensation was hailed as an innovation that would introduce a great deal of certainty in the calculation and payment of benefits, in contrast to the common law system. A worker could sue under common law. If he won, he might be assured of an adequate payment; those who lost would be left with nothing but debts. To reduce uncertainty, the workmen's compensation law specified the benefits that would be paid to all regardless of fault. Although the outcome of workmen's compensation cases is far more certain than the ordinary suit where negligence must be shown, the law is not "automatically" applied.

In part, this uncertainty stems from the variety of the permanent partial disability cases which the schedules do not cover satisfactorily. Two factors give rise to compensation litigation. One is the uncertainty as to whether an accident did or did not arise out of and in the course of employment; the other is the extent of disability. As workmen's compensation comes to encompass more and more of the ailments to which the general population might be susceptible, it becomes difficult to separate impairments that are work-connected from those that are not. In addition, it requires an exercise of legal skills and medical judgment to assess the extent of disability in occupational diseases, injuries to the soft tissue of the back, heart conditions, or cases where the only evidence before the commission may be a subjective complaint.

Conclusion

Workmen's compensation permanent partial disability benefits are the least duplicated benefit of any paid to injured workmen. There is a variety of benefit levels for temporary disability, permanent total disability, and death. Nowhere is the variety more apparent than in permanent partial disability. States differ in benefit levels, in relationships among benefits paid to the various types of disabilities, in minimums, in maximums, in weeks scheduled for particular losses, and above all in benefit payment philosophies. Not only are twice as many weeks allotted for loss of the use of a member in one jurisdiction than in another, but also what is normally and usually considered, say, a 50 percent loss of use of a member in one jurisdiction may be rated at 25 percent of loss of use in another. A particular residual impairment may rate 50 percent total disability in one state and go uncompensated in another because the employee has lost no wages. In a third jurisdiction, the award may be at a different percentage because of an administrative judgment about the estimated loss in wage-earning capacity. Such variety is in addition to the differences in statutory replacement ratios and minimums and maximums on a weekly or aggregate basis.

Because of the difficulty of predicting an award for a standard impairment and because of a lack of knowledge about the relationship between a given impairment and wage loss, we have no consistent and reliable estimates of the adequacy of permanent partial awards. Some estimates were presented on the basis of a 50 percent disability, but these estimates are possible only in states where aggregate limits or rating philosophies permit one to specify the number of weeks which will be paid.

Death benefits are payable to widows and other eligible survivors at a rate which usually is less than that paid in the permanent disability cases. Although a number of jurisdictions provide for payments during widowhood and until children reach a specified age, some limit the duration of payments.

Adequacy comparisons were based on a hypothetical worker making the average wage in his jurisdiction. The percentage of wage loss replaced in each jurisdiction was shown in the analysis. Certainty of payment, a prime objective of early compensation laws, has not been attained because of the litigation which persists over issues of liability and extent of disability. Promptness of payment, also an early goal, has been attained in some jurisdictions but, for most states, data on this aspect of administration are not available.

In the compromise and release cases, certainty is attained but often at the expense of closing out all possibility of future recovery should a worker's condition worsen. As most states do not maintain records on postsettlement developments, it is difficult to know how much compromise and release settlements interfere with the basic objectives of the law. In the usual case, the workmen's compensation benefit is paid in periodic installments as wages are paid so that, if the employee's condition changes, the case can be reopened within the period designated in the statute of limitations. The compromise and release settlements, of course, obviate this possibility.

Rehabilitation

Most employees injured in work accidents return to their jobs after minor medical attention with little if any worktime lost. As the effects of the injury are transient, the incident usually fades from memory. Even those who suffer days or weeks of disability and possibly endure substantial medical treatment may find the injury is not permanent. Although the loss of income and the medical expenses are distressing, eventually, when workers resume their jobs, they recover economically, too.

A minority, unfortunately as much as 10 percent of the total injured, according to a California report (see References), suffer injuries that disrupt their lives. Even when

187

these workers receive effective medical care so that eventually they return to productive jobs, their lives are physically and emotionally scarred. Injuries for some are so severe that prolonged medical treatment and convalescence fail to restore them completely. Residual handicaps prevent their acceptable performances in their former jobs. Only retraining and education combined with special treatment offer a prospect for future employment.

Some never return to work. If they do not die from their injuries, they live with such severe disabilities that they barely can manage for themselves. Often, the most that health services can do is to lighten the burden on those who take care of these persons.

The treatment for workers whose livelihood is threatened by work-related impairments consists of medical rehabilitation and vocational rehabilitation.

Medical rehabilitation

It is easier to discuss individual programs than to review medical rehabilitation in the United States as a whole. Each program, whether set up by an insurance company or workmen's compensation agency, contains its own requirements for treatment, qualifications for eligibility and definitions of service.

The delivery system. The worker who requires medical rehabilitation often receives it much as he receives other medical care. Some workmen's compensation laws obligate employers and insurers to pay costs of medical care for injured workers. The worker receives whatever medical care is needed to treat the impairment and restore lost function. He may report first to the plant nurse or physician for immediate attention. If the injury is serious, he may go to a hospital. Costs are covered by having health service workers on salary, by contractual arrangements with health personnel, or by payment of hospital and doctor bills. The insurer may or may not have much influence in the selection or course of treatment.

For injuries associated with chronic disabilities, the insurer usually attempts to control the selection of the rehabilitation services, frequently by directing the worker to a particular specialist or facility with a particular expertise. Often the insurer pays for transportation to the specialist or facility as well as for rooms during treatment.

Some insurance companies operate rehabilitation facilities, under individual or joint ownership, with medical personnel on salary, at least parttime.

When insurers contract to share rehabilitation programs or facilities, they may pay expenses case by case or through a rental agreement.

Some workmen's compensation agencies may be totally isolated from and unaware of rehabilitation procedures. Others keep relatively close tabs on the services rendered.

When informed of the potential need for rehabilitation, some agencies do little more than notify the worker and insurer that medical rehabilitation is worth considering. Other agencies conduct formal evaluations of the need for further medical care and recommend action. They seek to convince disabled workers of the wisdom of rehabilitation. When the workers agree, the insurers can be required to finance the care.

Rehabilitation in insurance. In 1972, a questionnaire concerning rehabilitation programs in workmen's compensation insurance was sent to 25 major insurance companies in the United States. The answers from 22 revealed a variety of policies and practices in rehabilitation under workmen's compensation so that generalizations are difficult. The following comments on various rehabilitation programs, therefore, are not to be regarded as typical of the entire industry.

In the insurance industry, the concern some carriers have for both medical care and medical rehabilitation is termed "medical management." It is the attempt to minimize the total costs of compensation through emphasis on well timed, high quality medical treatment. The concept tends to focus attention on the physical condition of the worker rather than on the monetary compensation due. The ultimate goal is to reduce the degree of disability. The insurer would like to see the worker's earning abilities fully restored rather than pay compensation indefinitely.

Vocational rehabilitation

Vocational rehabilitation prepares the injured worker for a new occupation or for ways of continuing in his old one. Usually, vocational rehabilitation is assigned when medical treatment fails to restore the worker to the job he held when injured. The worker's injury may be so severe or his work requirements such that residual impairment prohibits effective performance. Workers with such impairment must be trained to surmount or by-pass the residual limitations. Many will enter new occupations. In practice, the more effective the medical rehabilitation, the less the need for vocational rehabilitation.

This definition of vocational rehabilitation distinguishes it from medical rehabilitation more than it should. While the difference in kind of treatment seems clear enough—retraining as opposed to medical care—the categories overlap. In the public vocational rehabilitation programs in each state, services include medical diagnosis and evaluation, surgery, psychological support, the fitting of prostheses, and other health services along with education, vocational training, on-the-job training, and job placement.

The two programs blend also on the record. Recordkeeping by workmen's compensation insurers does not separate claimants who receive medical rehabilitation from those who receive vocational rehabilitation, although some distinguish between medical rehabilitation and acute medical care. In contrast, records kept by workmen's compensation agencies usually separate vocational rehabilitation from other benefits.

The delivery system. The relatively few injured workers who need vocational rehabilitation are served by several means. An employer or insurer may channel the worker to whatever sources he thinks will provide satisfactory service. Some workers are referred to the public vocational rehabilitation program where services may be financed by taxes, although insurers may reimburse the public agency. Other insurers direct workers into private facilities where vocational training is conducted by technical schools or on the job. For such services, insurers always pay the costs.

As with medical rehabilitation, some workmen's compensation agencies support vocational rehabilitation so that, if the insurer does not direct the worker into a program, the agency often will. Several jurisdictions select candidates either in conjunction with screening for medical rehabilitation or separately. Workers with serious injuries, permanent disabilities, or those who receive extended compensation payments are reviewed by the agency for referral to the state's public vocational rehabilitation agency or to the insurer.

Some workers obtain vocational rehabilitation through their own efforts. If no one refers them, they may go directly to the public vocational rehabilitation office. Since 1920, the federal government and the states have cooperated financially in supporting a vocational rehabilitation program, 80 percent federal and 20 percent state, which can be utilized by anyone with a vocational handicap. Rehabilitation counselors, who usually determine a referral's acceptability, simply look for a vocational handicap without regard to the source and consider the possibilities of overcoming the handicap. If the candidate shows relatively good prospects, a plan is designed for his restoration. For those who cannot return to a paying job, the objective of vocational restoration may be to enable clients to care for themselves and to free other members of the family to earn wages.

The worker may be referred also by his physician, a friend, or a member of his family.

Once a worker is established in a vocational rehabilitation program, he is aided by whatever sources the counselors think best fit his needs. Generally, the sources are not owned and operated by the vocational rehabilitation agency but are private vendors or other public agencies. A worker may be sent to a private rehabilitation center or school or a sheltered workshop such as those run by Goodwill Industries of America, or he may be enrolled in a public institution.

Degree of Disability

Determination of the extent of disability is perhaps responsible for more litigation than any other single issue in workmen's compensation. It requires not only correct application of legal principles but also evaluation of facts, subjective complaints and opinions, and attempts to predict the future.

As a general proposition (some jurisdictions

189

use different terminology and slightly different classifications), disability can be categorized in one of four classes: temporary total disability, temporary partial disability, permanent partial disability, and permanent total disability.

Temporary disability

Temporary total disability occurs when an injured worker, incapable of gainful employment, has a possibility or probability of improvement to the degree he will be able to return to work with either no disability or merely partial disability. Temporary partial disability is similar to temporary total in that it assumes a physical condition which has not stabilized and is expected to improve. The difference lies in the worker's current abilities. When temporarily partially disabled, the worker is capable of some employment, such as light duties or part-time work, but is expected to improve to the degree that he will attain much of his former capability.

Permanent partial disability is reached when the injured worker has attained maximum improvement without full recovery. That is, the worker has benefited from medical and rehabilitative services as much as possible and still suffers a partial disability.

Permanent total disability represents the same physical situation except that the disability is total.

The determination of temporary disability, either total or partial, is least difficult as it requires merely a determination of the employee's present physical condition in comparison with the work opportunities available. In practice, evaluation of temporary disability is concerned only with the ability of the employee to return to work for his last employer. Assuming that an employee will be able at some point to return to work for his employer, and given the difficulties involved in obtaining employment for workers still under medical care, compounded by the probability that the new employment will be temporary, most adjudicators have either expressly or in practice adopted the proposition that unless the worker can return to his last job, or can be supplied with temporary light or part-time duties with his last employer, he remains temporarily totally disabled, even though he might be able to perform another job in-

volving duties within his temporary physical limitations.

Permanent partial disability

The determination of the extent of permanent partial disability depends on what the jurisdiction chooses to label "permanent partial disability." Three basic theories have been discussed in conjunction with the payment of workmen's compensation benefits for such disability. Their underlying philosophies differ somewhat, as do the factors to be considered in applying each to a specific situation.

The "whole-man" theory is concerned solely with functional limitations. Here, the only considerations are whether the worker has in fact sustained a permanent physical impairment and, if he has, to what extent does it interfere with his usual functions and abilities. Age, occupation, educational background, and other factors are not considered.

In the "wage loss" theory, the aim is to determine what wages the worker would have been able to earn had he not suffered a permanent impairment. When, owing to impairment, his earnings dip below the estimated wage figure, he is paid compensation equal to some percentage of the difference between the wages that he should have earned and those he is actually earning. Here the actual degree of physical impairment is of little or no importance. The only concern is the actual wage loss which has been sustained and whether it is due to the impairment.

The "loss of wage-earning capacity" theory requires a peek into the future. After the worker has reached his maximum physical improvement, many factors, such as impairment, occupational history, age, sex, educational background, and other elements which affect one's ability to obtain and retain employment are all considered in an effort to estimate, as a percentage, how much of the worker's eventual capacity to earn has been destroyed by his work-related impairment. The worker is awarded benefits on the basis of this computation. Benefits may be paid at the maximum weekly rate for a limited number of weeks or they may be based upon a percentage of the difference between wage-earning capacity before and after disability, to be paid until a dollar or time maximum has been reached.

Combinations. These three basic theories are capable of being used also in combination. For example, some states expressly or in practice provide for the use of either the "wage loss" theory or the "loss of wage-earning capacity" theory but also provide a benefit floor determined by the actual medical impairment. Thus, a worker who sustains impairment but no loss of wage or of wage-earning capacity would still receive some permanent disability benefits.

The tedious or controversial aspects of rating for a significant proportion of permanent partial disability cases have been relieved by the use of schedules. The typical schedule covers injuries to the eyes, ears, hands, arms, feet, and legs. It states that for 100 percent loss (or loss of use) of that member, compensation at the claimant's weekly rate will be paid for a specified number of weeks. If loss or loss of use is less than total, the maximum number of weeks is reduced in proportion to the percentage of loss or loss of use. Only physical impairment is considered and the effect of the injury on wages or wage-earning capacity is ignored. If an injury is confined to a scheduled member, the benefits provided by the schedule are exclusive, even though disability rating on a wage loss or loss of wage-earning capacity basis might result in greater benefits. While this statement is true generally, some states provide additional benefits if use of one of the other theories does result in higher benefits being paid, if diminished wage-earning capacity continues after the scheduled amount is paid, if the scheduled injury results in permanent total disability, or if several scheduled injuries are sustained in the same accident. Use of the schedule may be avoided also by showing that the effect of the scheduled injury, such as radiating pain, extends into other parts of the body. A few jurisdictions limit the use of schedules to amputation or 100 percent loss of use of a member, as opposed to partial loss of use. Another group not only makes the schedule exclusive for permanent disability awards but requires that the weeks for which benefits are paid during the healing period be deducted from the number of weeks authorized by the schedule before an award is made for permanent partial disability.

Despite commentary and statutory language to the contrary, the workmen's compensation systems of the United States, with a few exceptions, operate primarily on the "loss of earning capacity" theory. Even where statutory language seems to indicate clearly that only functional impairment is to be considered, the courts have managed to hold that loss of earning capacity is the real consideration. Even the use of schedules has been justified on an earning capacity basis as merely a legislative determination of presumed wage 'loss resulting from the impairment listed in the schedule.

It is questionable that any legislative history would back up this rationale. A consideration of its practical day-to-day application shows that in individual cases the presumption is without basis in fact. When one considers that a concert pianist and a laborer who have both lost two fingers would receive exactly the same compensation under a schedule award, the justification for the use of schedules, that of administrative efficiency, is questionable by all who hold equity is a basic aim of workmen's compensation.

Permanent total disability

Permanent total disability evaluation is, in most respects, merely an extension of the determination of permanent partial disability. In fact, it is a part of the same process, since the factfinder's only additional task is to determine whether the worker's wage-earning capacity is so destroyed that he is unable to compete in the job market.

Two aspects of the permanent total disability question warrant special attention.

First, most states employ presumptions that make the factfinder's job much easier. For example, it may be presumed that the loss of sight of both eyes or the loss of any two limbs will constitute permanent total disability and thereby relieve the factfinder of the difficult task of evaluating all the factors previously mentioned. These presumptions in some cases may be rebutted by evidence of an established wage-earning capacity or may only apply for a limited period of time.

Second is the concept of what permanent total disability actually means. The injured

8—Workmen's Compensation Insurance

employee need not be completely helpless nor unable to earn a single dollar at a job. His limitations need only prevent him from competing in a practical way in the open job market and are such that no stable job market exists for him.

The exclusive remedy doctrine and third party liability

Before workmen's compensation laws were enacted in the states, an employee, in order to recover damages for a work-connected injury, always was required to show some degree of fault on the part of his employer. Under what is now known as the "quid pro quo of workmen's compensation law," employers accepted, or were required to accept, responsibility for injuries arising out of and in the course of employment without regard to fault. In exchange, employees gave up the right to sue employers for unlimited damages. These agreements are usually referred to in the state acts as "exclusive remedy" provisions, a term that is quite misleading. In no state are workmen's compensation benefits necessarily the only remedy available to an injured worker. Depending upon the working of the applicable statute, the worker may bring a negligence action against a fellow worker, another contractor on the same job, or some other entity or individual who caused the compensable injury. From the employer's viewpoint, it is best to refer to the doctrine as the "exclusive liability rule." As the employee sees the rule, it remains an "exclusive remedy" for obtaining compensation from the employer. But neither liability nor remedy are perfectly exclusive.

Private Insurer Programs

Private insurers dominate the coverage. The ten leading groups wrote almost half the business; the top twenty about 68 percent, up from 62 percent in 1950, although the share of the top ten has changed little. The ten leaders wrote as much as 88 percent of the coverage in Hawaii to as little as 51 percent in Kansas and Nebraska. In four states, one insurer wrote one-fourth of the business.

Only 36 of the 383 insurers earned workmen's compensation premiums of $20 million or more, but these 36 cornered more than 78 percent of the total. Companies with nationwide operations accounted for 83 percent. Insurers licensed in only one state, more than 20 percent of all insurance companies, wrote less than 4 percent of the premiums.

Workmen's compensation is the second largest property-liability insurance line; it is topped only by automobile insurance. Workmen's compensation premiums are about 11 percent of the total premium income. Among the ten leading private insurance groups, four derive at least one-third of their business from workmen's compensation.

Classifications of insurers

Private insurers can be classified according to their legal form of organization, their marketing methods, and their pricing policies.

Legally, insurers may be classified as proprietary or cooperative insurers. Proprietary insurers have owners who bear the risks of the insurer and whose representatives manage the operations. The leading example by far is the stock insurer owned by stockholders who elect the board of directors. In 1970 of the 383 private groups in workmen's compensation insurance, about 68 percent were stock companies, with about 70 percent of the total premium volume.

Cooperative insurers have no owners other than their policyholders. The leading example is the advance premium mutual whose board of directors is elected by those policyholders who exercise their right to vote.

Unlike stock insurers, these insurers have no capital stock. Instead retained earnings serve as a cushion against adverse experience. In 1970, mutual insurers, about 32 percent of the private workmen's compensation insurers, wrote 30 percent of the premiums earned.

Almost all of the 1970 workmen's compensation insurance premiums not written by stock or mutual insurers were written by reciprocal exchanges which, in their modern form, closely resemble advance premium mutuals.

The relative importance of stocks, mutuals, and reciprocal exchanges varies among states. In several states, stock insurers write over 75 percent of the business. In a few states, mutual insurers dominate the private insurance field.

Since 1951, when their share was 59 percent, stock insurers have steadily increased their proportion of the business.

Self-insurers

In all jurisdictions (except Nevada, North Dakota, Puerto Rico, Texas, the Virgin Islands, and Wyoming), employers are permitted under certain conditions to self-insure workmen's compensation. Although most jurisdictions reported 200 or fewer self-insurers, self-insured employers in 1970 paid 14.1 percent of the workmen's compensation benefits.

Self-insurance is becoming popular among those who qualify. In 1960 self-insurers paid only 12.4 percent of the benefits. Most states report an increase also in the number of self-insurers. Possible explanations are increasing cost consciousness, sales activities of agencies who seek to manage self-insurance programs, and the business merger movement which increases the size of firms and their ability to self-insure.

Self-insurance is attractive primarily because it may be less costly than insurance. An insurance premium is designed to pay the losses and expenses of the insurer and provide a margin for profit or contingencies. The self-insurer hopes to save money on the loss or the expense and profit components of the premium.

The loss component equals the average loss the insurer expects the employer and others like him to experience. If the employer is so much better than the average employer in his class that his expected loss may be less, he would save money by self-insuring. In the short run, however, the loss experience may differ substantially from the expected loss. Indeed the loss in a single year might be catastrophic. The larger the number of employees, the less the risk of fluctuation in annual losses. For most employers, the risk is such that self-insurance is out of the question. For others, the comparison between actual and expected loss is an important consideration.

By self-insuring, the employer can save that part of the premium charged to cover selling expenses, some general administration expenses, and profits. There may also be savings on loss prevention and loss adjustment services even though these services must still be performed.

Other considerations include the relative quality of the safety and claims services provided by insurers, by management service organizations, and by employers themselves; by tax factors; and by the opportunity cost of paying an insurer a premium instead of paying losses and expenses as they occur.

Cost Levels and Allocation

Workmen's compensation costs and other costs of industrial accidents impose a burden on industry, workers, and society generally. The magnitude of these costs and their distribution have important economic implications and pose several critical issues of policy.

Cost levels

Workmen's compensation in 1970 cost employers almost $5 billion or more than $1.13 per $100 of payroll. These costs include the premiums paid to private insurers on state funds and the benefits and administrative costs paid by self-insurers. Other employer costs of industrial accidents or the losses to employees not covered by workmen's compensation are not available.

Variation over time. Costs per $100 of payroll were less in 1970 than in 1940, but have increased since the late fifties. The 1970 dollar costs were 11.6 times the 1940 costs, 4.8 times the 1950 costs, and 2.4 times the 1960 costs.

Variation among industries. In its 1969 sample survey of employee benefits, the Chamber of Commerce of the United States found that workmen's compensation costs were 0.9 percent of gross payroll. In manufacturing industries the costs were 1.2 percent; in nonmanufacturing industries 0.5 percent. These rates ranged from 0.1 percent for insurance companies to 1.8 percent for primary metal industries.

A 1970 sample survey by the Bureau of Labor Statistics found workmen's compensation costs equal to 0.9 percent of gross payroll plus employer payments for legally required insurance programs and employee benefit plans. In manufacturing, compensation costs were 0.8 percent, in nonmanufacturing, 0.9 percent. The percentage for office workers

193

was 0.3; for nonoffice or production workers, 1.3 percent.

Workmen's compensation incentives for safety

The asserted safety incentive of workmen's compensation is based on the merit-rated pricing policy. "Merit rating" includes both the experience rating and retrospective rating systems. All state funds use merit rating of some sort. In most states, although private insurers are not required to rate employers on merit, they are permitted to. Under this procedure, the firm is charged a premium that is related to the dollar amount of claims for which it is liable. Consequently a merit-rated firm has an incentive to reduce the amount of its claims through loss preventive measures.

The strength of this incentive has been challenged. Only about one-fourth of insured firms, usually large ones, are eligible for merit rating.

The yearly accident record of firms with only a few employees is not a sufficiently reliable indication of their characteristic experience to be considered in establishing premium rates. On the other hand, merit-rated firms account for 85 percent of the dollar volume of premiums paid. In addition, self-insured firms which pay approximately 14 percent of all benefits are implicitly merit rated. Nevertheless, if incentive effects are inherent in experience rating, they are not available to a large number of small firms and their employees.

Firms not eligible for merit rating are class rated. Under this procedure, all employers engaged in similar business operations within a state pay the same rate per $100 of payroll. These employers have strong incentives to reduce the rates paid by their industry and may therefore exert efforts to reduce accidents within the industry. The only accident prevention incentive generated for individual employers within an industry is that as poor risks, they may land in an assigned risk pool.

Other problems, even under merit rating, moderate the incentive for safety. As benefit levels do not reflect the full costs of accidents, the premiums paid are less than adequate; consequently, any savings from safety programs are proportionally minimized. A saving in premium costs could be a significant reward for success in accident prevention if benefits were higher because premiums would then more truly measure the cost of accidents at work. The higher the benefit levels, the larger the premium costs to be avoided and the larger the incentive for prevention.

Even for the merit-rated firm, the functional relationship between injury rates and premium levels is not as direct as might be desirable. The sensitivity of premium to accident experience is dependent on the firm's payroll. As firms increase in size, the premium rate more nearly reflects the individual firm's experience. It has been suggested that more credibility be assigned to the experience of smaller merit-rated companies to increase their safety incentives.

The premium rate for firms of all sizes is more dependent on the frequency rate than the severity rate on the assumption that loss frequency is more within the control of the employer. Thus, firms are encouraged to be more concerned with the number of accidents than with the consequences of accidents. Some critics believe too much emphasis has been placed on loss frequency.

Merit rating suffers the further criticism that, since premiums are related to the level of claims paid, some firms may try to reduce costs by fighting claims rather than by preventing accidents.

Finally, workmen's compensation safety incentives have been questioned because the costs usually amount to little more than 1 percent of payroll and many feel that a firm is insensitive to any cost that small. Other costs of worker accidents that are not insurable provide even stronger incentives. In evaluating this criticism, it should be remembered that the costs of workmen's compensation for employers in hazardous industries or with unsafe operations are much greater than 1 percent of payroll.

Unfortunately, there has been little research on this question, but there is little evidence to indicate that any substantial connection exists between merit rating and accident prevention in most states. Conceptually, however, workmen's compensation should impel firms to operate at an optimal level of accident prevention activities in a least-cost fashion.

An improvement in functioning is what is needed.

References

California Department of Education, San Francisco, Calif. *The Vocational Rehabilitation of Industrially Injured Workers.* A report to the California Legislature, 1961.

Chamber of Commerce of the United States, Washington, D.C. *Analysis of Workmen's Compensation Laws,* 1966.

National Commission on State Workmen's Compensation Laws, Washington, D.C. Available through Government Printing Office, Washington, D.C. 20402.
Compendium on Workmen's Compensation, 1973.
Supplemental Studies, 1973.

National Council on Compensation Insurance, 200 E. 42nd St., New York, N.Y. 10017. Rate and rating plan manuals.

National Safety Council, 425 N. Michigan Ave., Chicago, Ill. 60611. *Accident Facts* (annually).

Safety
Training

Chapter
9

Developing the Training Program   197
Training needs . . . Program objectives . . . Course outlines and
materials . . . Training methods . . . The lesson plan

Safety Training for Supervisors   200
Responsibilities of the supervisor . . . Supervisors must accept their
duties . . . Objectives of supervisor safety training

A Basic Course for Supervisors   202
Subject outline of course . . . Tips on running a course . . .
Providing instructors . . . Conducting the "in-house" course

Special Training for Supervisors   204

On-the-Job Training   205
Over-the-shoulder coaching . . . Job safety analysis . . . Job
instruction training

Conference Method of Teaching   207
Problem-solving conferences . . . Conference leading . . . Misuse
of the conference method

Other Methods of Supervisory Training   209
Personal instruction and coaching . . . Programmed instruction . . .
Independent study . . . Videotape recordings . . . Seminars and
short courses . . . Reading material

Policies and Attitudes in Supervisory Training   210
Teaching at the adult level . . . Integrating safety into all training . . .
Continuing programs get best results

Training New Employees   **211**
Indoctrination . . . Preliminary instruction . . . Make rule books
logical, enforceable . . . Departmental induction and training . . .
Good supervision—consistent instruction and "discipline" . . . First
aid courses . . . The so-called "accident-prone" individual

Conclusion   **214**

References   **215**

An effective accident prevention and occupational health hazard control program is based on proper job performance. When people are trained to do their jobs properly, they will do them safely. This, in turn, means that supervisors must (a) know how to train an employee in the safe, proper way of doing a job, as well as (b) know how to supervise. It also means the safety professional should be familiar with sound training techniques. He may, or may not, become directly involved in the training effort, but he should be able to recognize the elements of a sound training program.

This chapter is devoted to safety training. Although training and education cannot be separated completely, safety education is broader in scope and usually covers a number of subjects not normally included in a training program. The "human relations" value of education was covered in the section on "Learning" in the previous chapter.

Training is only one way to influence human behavior. Safe performance is encouraged by the example of an employer who spares no effort to create safe working conditions. Safe performance is encouraged by developing safe work procedures, by teaching the procedures effectively, and by insisting that they be followed. Safe performance is also encouraged by teaching people the facts about accident causes and preventive measures.

A well planned training program will not only train employees, but will also help change these other influences so they will complement the effect of the training.

Developing the Training Program

When developing a training program, consider the training needs, program objectives, course outlines and materials, and training methods.

Training needs

A training program is needed (a) for new employees, (b) when new equipment or processes are introduced, (c) procedures have been revised or updated, (d) when new information must be made available, and (e) when employee performance needs to be improved. Unless a training program is needed however, there is little justification for spending time and money just to have one.

Here are some indications of a need for a good training program:

1. Proportionately more accidents and injuries, or insurance rates higher than other companies in the same type of work, or a rate that is on the upswing.

2. High labor turnover.

3. Excessive waste and scrap.

4. Company expansion of plant and equipment.

Program objectives

Training programs should be based on clearly defined objectives that determine the scope of the training and guide the selection and preparation of the training materials. Objectives should be planned carefully and written down. They should indicate what the trainee is to know or do by the end of the training period.

To make sure the objectives really cover the needs of those to be trained, the duties and responsibilities of the trainees should be determined. Job descriptions and job analyses should be reviewed. These, along with personal observations and performance tests

197

FIG. 9–1.—Typical training session for supervisors. Prepared course materials are being used.

Courtesy Black Hawk College, Moline, Ill.

will reveal where training is needed. For example, if a job description (or list of duties) for supervisors includes training workers, then supervisors should be trained to train. This now becomes one of the objectives of a supervisory training program.

Course outlines and materials

After defining the training program objectives, the next step is developing the outline of what is to be covered. Often, an outline can meet the program objectives by using existing texts and course materials, or by combining parts of several texts or courses already in use. (A later section in this chapter, A Basic Course for Supervisors, describes the National Safety Council's 12-session safety course outline for supervisors.)

Sometimes, a completely new program must be designed. Either way, the outline must be developed in detail. Major topics and subtopics should be arranged logically and the

time allotted to each should be in proportion to the importance of each. So arranged, the course outline is the framework that determines the method of training.

Training methods

At this stage in planning a training program, a decision must be made as to the training method that will best reach the stated objective. For job (or skill) training, the best method is the four-step method called "job instruction training," or JIT.

Full details of this system—ideal for training new workers or upgrading more experienced ones—are given under the section On-the-job Training, later in this chapter.

The lesson plan

The safety professional and others who must often teach safety subjects should be familiar with lesson plans. They are the blueprint

198

SELF-CHECK TEST FOR INSTRUCTORS

A good instructor should be able to answer "Yes" to at least 18 of these questions. Under 15 would be below average.

	Yes	No
1. Do you check your classroom for ventilation, lighting, seating arrangement?	☐	☐
2. Do you prepare a lesson plan?	☐	☐
3. Do you use training aids when possible?	☐	☐
4. Do you preview all films before showing them?	☐	☐
5. Do you use test results to find weak points in your teaching?	☐	☐
6. Do you vary your type of presentation according to the material?	☐	☐
7. Do you stay on the subject?	☐	☐
8. Do you cover the material in each lesson?	☐	☐
9. Do you make any attempt to know your students and to learn their names?	☐	☐
10. Do you introduce yourself to each class?	☐	☐
11. Do you make clear the objectives of the lesson?	☐	☐
12. Do you summarize each lesson?	☐	☐
13. Do you introduce each new subject and explain its importance?	☐	☐
14. Do you refrain from using sarcasm in your class?	☐	☐
15. Do you start and dismiss your class on time?	☐	☐
16. Do you have all your equipment ready and tested at class time?	☐	☐
17. Do you talk directly to the class and avoid such practices as staring at the floor, pacing the floor, juggling a piece of chalk?	☐	☐
18. Do you keep your meeting place orderly?	☐	☐
19. Do you use words that are easily understood by the class?	☐	☐
20. Do you make assignments, and are they clear?	☐	☐

FIG. 9–2.—A self-check test can help a supervisor rate his own training methods.

for presenting material contained in a course outline. In addition to standardizing training, lesson plans help the instructor:

1. Present material in proper order

2. Emphasize material in relation to its importance

3. Avoid omission of essential material

4. Run the classes on schedule

5. Provide for trainee (student) participation

6. Increase his confidence, especially if he is new

Names for the parts of a lesson plan may vary, even the order may not always be the same. The following is a good example of arrangement for a lesson plan:

1. **Title:** Must indicate clearly and concisely subject matter to be taught.

2. **Objectives:**

   a) Should state what the trainee should know or be able to do at the end of the training period.

   b) Should limit the subject matter.

   c) Should be specific.

   d) May be divided into a major and several minor objectives for each session.

3. **Training aids:** Should include such items as actual equipment or tools to be used, and charts, slides, films, etc.

4. **Introduction:**

   a) Should give the scope of the subject.

   b) Should tell the value of the subject.

   c) Should stimulate thinking on the subject.

5. **Presentation:**

   a) Should give the plan of action.

   b) Should indicate the method of teaching to be used (lecture, demonstration, class discussion, or a combination of these).

   c) Should contain suggested directions for instructor activity (show chart, write key words on chalkboard).

6. **Application:** Should indicate, by example, how trainees will apply this material immediately (problems may be worked; a job may be performed; trainees may be questioned on understanding and procedures).

7. **Summary:**

   a) Should restate main points.

   b) Should tie up loose ends.

   c) Should strengthen weak spots in instruction.

8. **Test:** Tests help determine if objectives have been reached. They should be announced to the class at the beginning of the session.

9. **Assignment:** Should give references to be checked or indicate materials to be prepared for future lessons.

The self-checklist shown in Fig. 9–2 (on the previous page) will aid the instructor in improving his teaching techniques.

Safety Training for Supervisors

The immediate job of preventing accidents and controlling work health hazards falls upon the supervisor, not because it has been arbitrarily assigned to him, but because safety and production control are closely associated supervisory functions.

Responsibilities of the supervisor

Whether or not a company has a safety program, the supervisor has these principal responsibilities:

1. Establish work methods
2. Give job instruction
3. Assign people to jobs
4. Supervise people at work
5. Maintain equipment and the workplace

These principal responsibilities of the supervisor are the very activities through which the work of preventing accidents is carried out. A brief examination of these jobs and their relation to safety will make this fact apparent.

Establishing work methods that are well understood and consistently followed is essential to orderly and safe operation. Many injuries and health hazards have been reported to result from "unsafe method or procedure," when later investigation disclosed that no standard method or procedure had been set up for those jobs. The method was declared hazardous only after it resulted in an accident. Making sure that safe procedures are established is a supervisory responsibility.

Giving job instruction, with necessary emphasis on safety aspects of the job, will help eliminate one of the most frequent causes of accidents—lack of knowledge or skill. If employees are expected to do their work safely, supervisors must show them exactly how to do the work and must make sure that the employees have the knowledge and skill to do it in exactly that manner.

Assigning people to jobs is closely related to job instruction. Whenever a supervisor makes a work assignment, safety, as well as good job performance, requires that he be sure the worker is qualified to do the job and thoroughly understands the work method.

Even an experienced worker needs some direction.

Supervising people at work is necessary even after a safe work method has been established and workers have been instructed according to that method. People deviate from established safe practices, and injuries result. Usually, it is then found that the injured employees have been neglecting safe practices. In order to prevent injuries from this cause, supervisors must watch for unsafe work methods and correct them as soon as they are observed.

Maintaining equipment and workplace in safe condition is no different from maintaining them in efficient condition. Accidents result from tools and equipment in poor condition, from a disorderly workplace, or from makeshift tools used because the right tools are not available. The supervisor who keeps his department and equipment in top condition helps to prevent accidents as well as to improve efficiency.

Supervisors must accept their duties

Not only are these five functions a normal part of the supervisor's job but, unless lines of organization and authority are seriously disregarded, nobody else but the supervisor can perform them. Sometimes this fact is overlooked. Since all these functions are closely allied with safety, it is only natural that safety should be included as one of the duties of the supervisor. If however, management wants to be sure that supervisors accept this role, it should be included in a policy statement and lines of authority should be established.

A clear and positive executive order defining the safety duties of all persons and departments should be issued. A careful program of education should be instituted to help supervisors understand and accept their role and to give them specific help in their work of preventing accidents.

Many supervisors have acquired their present positions in organizations where some sort of safety program was already in existence, and their understanding of the program as it has existed is firmly established. However, a safety professional undertaking the safety training of supervisors will almost invariably find that his first major job is to get supervisors of all levels to understand and accept their role in accident prevention. This job cannot be done in a single meeting or by a single communication.

Simply getting supervisors to agree in theory that safety is one of their duties is not enough. They must come to understand the many ways in which they can prevent accidents, and they must become interested in improving their safety performance.

Objectives of supervisor safety training

Before the safety training of supervisors is undertaken, the objectives of that training must be understood and stated specifically. Determining objectives for supervisory training is so important and involves so many kinds of activity that it should be made the subject of careful study and consultation. Members of the safety department should not attempt to set up objectives alone, but should confer with representative supervisors and members of higher management.

The following statement of objectives should be used only as a guide to be studied and modified to fit specific situations and should be accepted only to the extent that management persons concerned agree to it. Stated in general terms, the objectives of safety training for supervisors may be any or all of the following:

1. To involve supervisors in the company's accident prevention program

2. To establish the supervisor as being the key person in preventing accidents

3. To get supervisors to understand the nature of their safety duties

4. To provide supervisors with information on causes of accidents and occupational health hazards and methods of prevention

5. To give supervisors an opportunity to consider current problems of accident prevention and to develop solutions based on their own and others' experience.

6. To help supervisors gain skill in safety activities

7. To help supervisors do the safety job in their own departments

201

9—Safety Training

A Basic Course for Supervisors

The knowledge and philosophy of accident prevention is not "just common sense," as some slogans proclaim. It is a specialized body of information accumulated over a period of many years. It is the job of the safety professional to help supervisors gain whatever information is available that will make their safety efforts more productive.

The most direct way to develop the desired attitudes and to impart the necessary information about safety to supervisors is to provide a course of instruction. Courses in safety and other management subjects, conducted for supervisors by many companies, follow a fairly well established pattern.

Subject outline of course

The following outline based on the National Safety Council's 12-hour "Key Man Development" course shows the subjects that should usually be included in a safety course for supervisors. Visual aids are available on all the subjects and should be used at every meeting. Titles of films and other aids are not included here because new ones are being produced constantly. The person conducting a course should secure and use those best suited to his needs. (See catalog of materials published by the National Safety Council.)

Session 1

SAFETY AND THE SUPERVISOR

Safety and efficient production go together. Accidents affect morale and public relations. The duties of a supervisor under OSHA.

Session 2

KNOW YOUR ACCIDENT PROBLEMS

Elements of an accident. Unsafe acts, unsafe conditions, accident investigations, measurements of safety performance. Accident costs.

Session 3

HUMAN RELATIONS

Basic needs of workers. The supervisor as a leader.

Session 4

MAINTAINING INTEREST IN SAFETY

Committee functions, maintaining good employee relations. The supervisor's role in off-the-job safety.

Session 5

INSTRUCTING FOR SAFETY

Importance of job instruction. Making a job safety analysis (JSA). Job instruction training (JIT).

Session 6

INDUSTRIAL HYGIENE

Environmental health hazards. Skin diseases. Lighting, noise, ventilation, temperature effects.

Session 7

PERSONAL PROTECTIVE EQUIPMENT

Eye protection, face protection, foot and leg protection, hand protection. Respiratory protective equipment. Protection against ionizing radiation. OSHA requirements.

Session 8

INDUSTRIAL HOUSEKEEPING

Results of good housekeeping. The responsibility of the supervisor. OSHA requirements.

Session 9

MATERIALS HANDLING AND STORAGE

Lifting and carrying, handling specific shapes. Hand tools for materials handling. Motorized equipment. Hazardous liquids and compressed gases.

Session 10

GUARDING MACHINES AND MECHANISMS

Principles of guarding. Benefits of good guarding. Types of guards. Standards and codes. OSHA regulations.

Session 11

HAND AND PORTABLE POWER TOOLS

Selection and storage. Training in the safe use of tools.

Session 12

FIRE PROTECTION

Determining fire hazards. Understanding fire chemistry. Fire brigades. Reviewing the supervisor's job.

Tips on running a course

All supervisors should be given a basic course of this type. It should be repeated from time to time for new supervisors and prospective supervisors.

It should be given with considerable formality. Attendance records should be kept, and certificates or diplomas issued upon completion. Some companies give a diploma that may be framed and displayed; others favor a pocket-size card. Both may well be given.

It adds to the dignity of the training program if, at the beginning of a course, a company executive meets with the group and explains the importance of the supervisor's role in the safety program and the value of the training. At the end of the course, a formal graduation ceremony is often held.

Some large companies with wide-spread operations have made up their own courses to be given throughout the organization. The preparation of such a course is, however, a major and costly project. It is not recommended except for very large organizations with proportionately large safety and training staffs.

Prepared courses, such as the National Safety Council's "Key Man" course, are available, complete with text book, the "Supervisors Safety Manual," "Instructors' Guide," student kit, certificate of completion, and visual aids. Courses can be adapted to local needs. They have been thoroughly tested by use in hundreds of companies and many kinds of operations. Use of a prepared course saves a great deal of time and effort. Let the instructor give it the company slant.

Providing instructors

One problem in giving a basic in-house safety course to supervisors is providing qualified instructors. A considerable burden is placed on the safety personnel if they must conduct every class. Many companies have found that the best instructors are general supervisors or division managers. In addition to serving usefully as instructors, these men also learn a great deal about safety in the process of teaching.

It is well for those who are to serve as instructors to take a short course on how to instruct or, at least, hold two or three meetings on the subject of how best to give the instruction. After preparing his own lesson plan for each meeting by following the format outlined in this chapter and by using company situations and problems as examples, each instructor then takes charge of one class.

Safety professionals may feel that there is some risk of not getting good instruction by this method, but those who have tried it have found that the experience and understanding of men who are themselves supervisors more than offset a lack of formal teaching experience.

In addition, acceptance by the group is improved because the members know that the instructor is a man who has done the things he is talking about. In the case of a general supervisor who is teaching a group of his own supervisors, the meetings can become planning sessions in which agreement is reached as to how the group is to handle situations in the future. Training linked with actual operation can be most effective.

Courses conducted locally should be planned and instructed by safety professionals experienced in training and in the subject to be covered. Criteria for such instructors might be:

- Professional member or member of the American Society of Safety Engineers, or

- Certified Safety Professional (CSP), or

- Holder of NSC's "Advanced Safety Certificate," or

- Two years' recent experience in safety training, or equivalent experience.

Regardless of the experience of the person selected to do the teaching, each instructor should make a self-evaluation of his own teaching methods. A self-check test for instructors is an ideal guide for such evaluation.

Fig. 9–3.—Over-the-shoulder coaching. The trainee is expected to develop and apply skills in the work situation under the guidance of a qualified person.

Conducting the "in-house" course

Best results have been obtained in one-hour training sessions. Longer sessions are likely to become tiresome, and supervisors will become apprehensive about their work during a prolonged absence.

It is desirable to hold sessions during working hours, but this arrangement is possible only if there are two or more groups so that members can relieve one another on the job during sessions. Regular attendance and freedom from interruptions are of great importance and should be required by management.

If courses are held after working hours or between shifts, sessions are usually two hours; occasionally dinner or refreshments are provided. In any event, courses should be conducted in a professional, businesslike manner.

A supervisors safety course can produce many benefits—an increased understanding of safety, acceptance of responsibility for the prevention of accidents, and greater interest in all supervisory duties.

Special Training for Supervisors

Success in supervision requires skill fully as much as it requires knowledge and understanding. Basic training in the elements of a safety program is aimed toward imparting knowledge about safety and creating an attitude toward safety.

Many accidents, however, result from unskillful handling by supervisors of some of the important tasks of supervision—giving orders, assigning jobs, correcting workers for wrong performance, and the like.

If training in accident investigation has been given, remedial measures will have been assigned for specific accidents, but no detailed consideration may have been given as to how remedies should be applied.

To be able to do a good safety job, supervisors also need training in supervisory skills.

Among the most important of these skills are giving job instruction, supervising workers at work, determining accident causes, and building safety attitudes among workers. Special courses may be given in these subjects.

On-the-Job Training

On-the-job training (OJT) is widely used because the trainee can be producing while he is being trained. Whether the supervisor does the instructing himself or has someone else do the teaching, the training should be carefully planned and organized. Again the safety professional should be familiar with his organization's training program so that appropriate and correct safety training is integrated into it.

In too many cases, on-the-job training is a hit-or-miss procedure where the trainee is told to follow another worker around and learn his job. In situations of this type, the worker may be too busy to do any training, or may not know good training procedures. He may even be reluctant to train another to do his work.

Remember, it is the supervisor's responsibility to make sure the person doing the training can, in fact, train. On-the-job training includes many techniques and approaches and there is no one method that will fit all situations. Job safety analysis (JSA, discussed in the next section and in Chapter 5, "Removing the Hazard from the Job") is one method, job instruction training (JIT) is another, and over-the-shoulder coaching is still another widely used method. These methods may be used separately or in combination depending upon the complexity of the job and the time element.

Over-the-shoulder coaching

Over-the-shoulder coaching is perhaps the most flexible and direct of the three training methods. In coaching, the trainee is expected to develop and apply his skills in typical work situations under the guidance of a qualified person. The person to whom the trainee has been assigned should be one who knows the job thoroughly, is a safe operator, and has the patience, time, and desire to help others. (See Fig. 9–3.)

The advantages of training of this type are:

1. The worker is more likely to be highly motivated because the guidance is personal.

2. The instructor can identify specific performance deficiencies and take immediate and proper corrective action.

3. Results of the training are readily apparent since real equipment is being used and finished work can be judged by existing standards.

4. The training is practical and realistic and can be applied at the proper time.

Timing is important; not only do the trainees like to get help when needed, but also the instructor can judge the trainee's progress continually so he can present the next unit or phase of instruction when the trainee is ready.

To help keep track of the progress of each individual, a training chart is valuable. On the top of the chart are listed the various tasks required of a person in a particular job classification. By observation and direct discussions, the instructor can determine whether the employee is qualified to perform the tasks necessary to fill the job. The instructor determines the degree of skill and knowledge of the trainee and estimates his training needs. He makes notes and comments on the chart. Such a chart prevents neglecting training that is important; it also prevents unnecessary training. (See Fig. 9–4, next page.)

Job safety analysis

A job safety analysis (JSA) is a procedure to make a job safe by:

1. Identifying the hazards or potential accidents associated with each step of a job, and

2. Developing a solution for each hazard that will either eliminate or control the exposure.

Benefits of JSA. The principal benefits that arise from job safety analysis are these phases of a supervisor's work:

- Giving individual training in safe, efficient procedures

- Making employee safety contacts

- Instructing the new man on the job

- Preparing for planned safety observations

- Giving pre-job instructions on irregular jobs

Date March 1
Supervision L. O.
Department Office

Employee	KNOW CUSTOMERS	CUSTOMER RELATIONS	ACCURACY IN CASH	MAKE UP DEPOSIT	BANK DEPOSIT	TAKE DICTATION STENO	STENCILS	COMP. OPERATOR	FILE	MAIL	TELEGRAMS	PLANT ORDERS TELE PHONE BOARD	Notes
1. Mrs A.M.	X	X	X	X	X		⊗			?			FREQUENTLY ABSENT. MISS B.C. HAS TO TAKE OVER WORK
2. Miss B.C.	X		X	X	X	NQ	⊗						✓
3. Miss W.E.			X	X	X	X	X	X				X	TRAIN HER IN CASH AND DISBURSEMENT JOURNALS
4. John D.				X	X			X	X	X	X		CAUTION ON MORE CARE IN CHECK WRITING ON PLANT PAYROLL
5. Mrs W.C.	⊗	⊗	⊗	⊗	⊗	⊗							TRAIN HER TO HELP ON FREIGHT BILLS
6. Miss E.M.						⊗⊗		⊗	⊗	⊗		⊗	START HER ON CHECKING BANK STATEMENT - ASK W.E. TO HELP
7. Miss V.O.				?		X		X	X			3/5	GET HER STARTED TO HELP ON TELEPHONE BOARD
8. Miss M.O.							⊗	X	X				TRAIN ON STENCILS
9. Miss K.B.								⊗	⊗	X			GOOD PERSON?! - HOW ABOUT WORKING TOWARD CASHIER AIDE
10. Miss L.B.								X	X	X			
11. Bill K.					⊗			⊗ B/W	B/W	X			WORK WITH HIM ON MAIL - RE-ARRANGE - FIND BETTER WAY

CODE
NQ—Not qualified
X—can do
B/W—Better way?
?—Check if can do
⊗—Regular Job OK
%—Target date

FIG. 9-4.—Training chart keeps track of who needs (and who has) what training.

HOW TO GET READY TO INSTRUCT

Have a Timetable—
how much skill you expect him to have, by what date

Break Down the Job—*
list important steps, pick out the key points. (Safety is always a key point.)

Have Everything Ready—
the right equipment, materials and supplies.

Have the Workplace Properly Arranged—
just as the trainee will be expected to keep it.

*Use JSA, Job Safety Analysis breakdown to locate hazards.

JOB INSTRUCTION TRAINING (JIT)

HOW TO INSTRUCT

1. Prepare

Put trainee at ease.
Define the job and find out what he already knows about it.
Get him interested in learning job.
Place in correct position.

2. Present

Tell, show, and illustrate one IMPORTANT STEP at a time.
Stress each KEY POINT.*

3. Try Out Performance

Have him do the job—coach him.
Have him explain each key point to you as he does the job again.
Make sure he understands.
Continue until YOU know HE knows.

4. Follow-Up

Put him on his own.
Designate to whom he goes for help.
Check frequently. Encourage questions.
Taper off extra coaching and close follow-up.

*Safety is always a key point.

**SAFETY TRAINING INSTITUTE
NATIONAL SAFETY COUNCIL**

FIG. 9–5.—Every supervisor should follow the JIT format when teaching job skills.

- Reviewing job procedures after accidents occur
- Studying jobs for possible improvement in job methods

For further discussion and application of job safety analysis, see Chapter 5, "Removing the Hazard from the Job."

Job instruction training

The job instruction training (JIT) method described here is often called the Four-Point Method, because the instructing job is broken into four parts, each of which is shown in detail in Fig. 9–5:

1. Preparation
2. Presentation
3. Application
4. Testing

The four-point method is intended to help an instructor teach a learner to do a specific job. It aims at faster learning and better learning. When combined with a JSA, it becomes an excellent method for teaching safety along with job skills. (The reader will be relieved to know that it takes longer to describe some of these steps than it does to do them.)

Conference Method of Teaching

To aid understanding and stimulate participation in instruction courses, industrial educators many years ago developed the conference method of teaching. This method is widely used in teaching management subjects to people in business and industry.

The conference leader should be skilled in this method of teaching. He should know how to draw out information and opinions from the conferees, and sum up their conclusions. As is desirable in a participation class, the number of people in a conference group should be small enough that free discussion can occur.

The leader's part is mainly that of asking questions that will provoke thinking and discussion. Learning can take place in a conference if the conferees have a background of experience that enables them to discuss the subject intelligently.

Problem-solving conferences

Since a conference is a device for getting a job done, it is particularly useful in an area like safety where many people are involved. A conference yields big returns in education for all who participate, even though its immediate purpose is to solve a current problem, not merely a hypothetical one.

Frequently, the safety professional has problems relating to the operation of the safety program which he needs to discuss with production supervisors. The problem may have to do with the occurrence of a number of similar accidents, with some new feature of the safety program, with a new type of work to be undertaken, with a method of operation or plan to be worked out, or with any other subject in which both the foremen and the safety department are interested. To find a solution to the problem, the safety professional may call a conference with a group of supervisors.

The conference leader should be familiar with the relationships of the members of the conference and with the general area of the subject matter.

It is of the utmost importance that the leader and members of the conference know the scope and limitations of what they are expected to do. If, for example, a conference is called to decide how to put into practice a policy or a directive that has been issued, the members should understand that it is not within their authority to have the policy or directive changed. If they meet to discuss improvement of procedures for the elimination of a hazard, they should know whether or not they may consider extensive alternations of buildings and installation of new machinery. If they do not know the limitations within which they must work, then the conference

becomes a source of dissatisfaction and frustration or is regarded as simply play-acting.

If the job of a conference group is to recommend action and they do recommend action, they should, of course, know what becomes of their recommendation.

Often, however, a conference group will discuss a matter that affects only the members, in which case their conclusions are drawn for their own guidance. This is probably the most usual and most satisfactory kind of conference.

Conference leading

Every safety professional and supervisor should attempt to become a skilled conference leader. Since such skill comes mainly from practice, he should accept opportunities for holding conferences even if he has not had a great deal of experience. The application of good sense and an understanding of what a conference is supposed to accomplish will go far toward making the conference successful.

The sequence of conference leading is as follows:

1. The leader states the problem.

2. He attempts to break the problem into segments to keep the discussion orderly.

3. He encourages free discussion.

4. He makes sure that members have given adequate consideration to all the significant points raised.

5. He notes any conclusions that are reached.

6. He states the final conclusion in such a way that it truly represents the findings of the group.

One who attends many conferences has opportunities to observe examples of successful and unsuccessful conference leading. This should help him when he assumes the role of conference leader. The new leader should also seek help from training people and others who may give him sound advice on making his conferences productive.

Misuse of the conference method

If a conference leader, in his eagerness to get across his ideas or in order to cover ground quickly, departs from true conference procedures and steers the discussion in the way he wants it to go, then a meeting becomes a conference in name only.

A closely guided or controlled "conference" is not an effective teaching method. Worse still, because it is called a conference, it may establish in the minds of conferees and leader a pattern of conduct that makes it difficult for them to serve usefully as members of a true conference. Conferences are almost indispensable in business, and any process that tends to make them unproductive should be avoided.

Other Methods of Supervisory Training

Many methods of instruction and many teaching aids are available for training supervisors in accident prevention. Using a variety of methods in a single course or even in a single session makes for interest and understanding, so long as the methods chosen are appropriate for both the subject and the learners.

Teaching methods include individual coaching, lecture, discussion, case study, incident process, role playing, drill, demonstration, panels, and simulation. All have value as educational methods if they encourage participation by the learners.

Lecture. A discourse on a subject to impart knowledge or a talk used to present new information or new materials to a group. A lecture works well with large groups and where time is limited.

Discussion. A procedure involving an exchanging of ideas and the standardizing of procedures and techniques. A discussion allows students to pool their knowledge and become active participants in a controlled program.

Case study. A report of a real situation that has occurred, or a structured situation—both are designed to develop basic concepts and principles. The study describes what has happened in a particular case and the events leading up to the situation, but it leaves to the group the task of deciding the nature of the problem or problems, their significance, and the probable solution.

Incident process. A mini case study. This is a method of learning where an incident is presented to a group in written form. The individuals ask questions concerning relevant facts, clues, and details. The instructor supplies the answers to these questions, and the group assembles the facts, learns what happened and arrives at a decision.

Role playing. An instructional method in which incidents based on real-life situations are re-enacted by selected members of the class playing roles and making their own decisions. The decisions are discussed by the entire class and the instructor to bring out and highlight behavior patterns.

Drill. Repetition and guided practice to develop skill. Drills are used primarily for the most important and fundamental skills of a trade, job, or task.

Demonstration. A method widely used to teach skills. The operation is demonstrated by the instructor and then performed by the student as in job instruction training (JIT), discussed earlier in this chapter.

Panel. A planned session consisting of two or more qualified persons, each discussing an assigned topic or subject.

Simulation. Simulation often involves a simulator, such as those used for driver training and pilot training. Railroads have also developed simulators for training engineers. Simulation can also be achieved by the use of various types of management games, such as the "in basket technique," "war games" developed by the military, and, more recently, simulation techniques for the training of astronauts in the proper operation of space capsules. Simulation is a good method for paralleling real-life or actual conditions, thus affording opportunities for decision making without risk.

Personal instruction and coaching

Often overlooked as a method of instruction is personal discussion with individual supervisors. When the safety professional and the supervisor meet and talk about the job, there can be a complete and free interchange of ideas that is not possible in a group. In private conversation, the supervisor will often express reservations or doubts that he would not voice in a meeting, and the safety professional will have the opportunity to clear up misunderstandings.

The safety professional who meets with the individual supervisor not only instructs and coaches, but also learns many things he needs to know. Supervisors appreciate his interest in their work and are usually eager to teach him about the processes of their departments. The safety professional can learn much from them about the mechanical processes and also about problems of supervision. Possibly most important is that he becomes personally acquainted with the supervisors. Every safety professional ought to allot a part of his time to these personal contacts.

Programmed instruction

Programmed instruction may be used as a substitute or supplement for classroom and textbook methods of supervisory training. Using self-contained teaching materials (including so-called "teaching machines"), programmed instruction permits the trainee to set an individual pace and to absorb knowledge in easy-to-take bits. The learning process is reinforced by requiring the trainee to answer questions and correct his own errors before progressing with the course.

The National Safety Council offers one type of programmed instruction course for the safety training of supervisors. Courses available from other sources use self-education techniques for training in such subjects as management skills and communications.

Independent study

Courses offered through correspondence are often called home study courses or independent study courses. These have some advantages over other methods described—the supervisor can set his own pace, and he can study on his own time (which makes home study an ideal training method where a company's operations are scattered). See listing of correspondence courses in Chapter 23, "Sources of Help."

The National Safety Council offers two courses of this type for training supervisors— "Supervising for Safety," and "Supervisors Guide to Human Relations."

Videotape recordings

Videotape training uses television's "instant replay" techniques. The basic technique is to record the visual procedure and the directions on a videotape and then play it back instantly by means of a monitor. Processes or manual skills, once recorded, can be replayed many times. This technique is good for training employees as well as supervisors.

Seminars and short courses

A number of seminars and short courses are offered by colleges and universities on all phases of supervisory know-how. Insurance companies and private organizations also offer supervisory training courses. Check locally for what is available.

Reading material

Companies should provide supervisors with reading material—safety newsletters, safety magazines, magazines on supervision, booklets, and reprints. Many companies subscribe to the National Safety Council's 16-page monthly magazine *Industrial Supervisor,* and some also issue a monthly bulletin devoted to safety which is distributed to all supervisors.

In addition to periodical material made available to supervisors, it is desirable to have a management library from which supervisors may borrow books on safety and related subjects and which can furnish information about books that are worth reading.

Policies and Attitudes in Supervisory Training

The person conducting a training program for supervisors should remember a number of facts which experienced educators know.

Teaching at the adult level

Supervisors are adults who may well resent being treated like school children. Then too, the supervisor is intensely aware of and proud of the fact that his job is to get out production. He is often impatient at being asked to do work that takes his attention from that main purpose.

At the same time, the supervisor usually reflects the attitude of his immediate superior as to what is important and what can be ignored. If a superintendent lets it be known that he is strong for safety and that he judges his supervisors on the safety of their departments as well as on their production, the supervisors are likely to give attention to any instruction that will help them operate more safely. This fact is a strong argument for starting safety education at the upper levels of supervision.

Integrating safety into all training

In a company that has a formal training program, safety training should fit into that program; the director of training and the safety professional should work together in the planning of various kinds of safety training activity.

In the training of supervisors—usually the first concern of a training department—practically everything that is taught about good supervision helps to promote safety. Likewise, anything taught specifically for the purpose of promoting safety generally improves supervision in other ways.

In the arrangement of formal courses, safety subjects may be integrated with other instruction on general problems of supervision, or one or two safety meetings may be included in the course.

The number of participants in a class of supervisors should be small enough to allow free discussion; fifteen is about the upper limit. The leader uses carefully planned questions when he wants to get people to talk. There is no limit to the visual aids and demonstrations and other devices he can use to make his material interesting and educational, and the discussions easy to follow.

Continuing programs get best results

Safety training for supervisors needs to be a continuing program if it is to accomplish the best results. The program should cover many subjects, presented interestingly and in different ways. A limited effort, such as holding a few meetings, may cause supervisors to do a better job for a short time, but interest will lag if the initial effort is not followed up.

Industrial management must be dynamic because change is continually taking place. New methods and new ideas get attention. To hold its own in this atmosphere of change and progress, safety, too, must not be allowed to become static.

The safety professional who strives constantly to help line supervisors take over direct responsibility for safety in their operations finds the results of his efforts multiplied many fold. He becomes a true leader who gets the job done better, not by extending the scope of his own activities, but by helping supervisors assume their natural safety responsibilities.

Training New Employees

Safety training begins at the time of employment, before the employee starts work. An effective safety training program will include a carefully prepared and presented introduction to the company.

When a new employee comes to work, he immediately begins to learn things and to form attitudes about his company, his job, his boss, and his fellow employees. He does so whether or not the employer makes an effort to train him. So that the new employee may learn the things he needs to know and may form good attitudes, it is desirable for the employer to give him the right kind of start.

Indoctrination

At the beginning of his employment, each employee should know the company's safety policy, but the amount he can learn during the induction procedure is limited. Unfamiliarity with his surroundings, interest in many matters of seemingly more immediate concern, the detailed procedure of getting onto the payroll—all make it difficult for the employee to absorb and retain much safety instruction. It is necessary, therefore, to consider what safety information must be given first, and the best way to present it.

Each employee needs to learn the following things if he is to have a good start in safety training:

1. Management is sincerely interested in preventing accidents.

2. Accidents may occur, but it is possible to prevent them.

3. Safeguarding equipment and the workplace has been thoroughly done, and management is willing to go further as needs and methods are discovered.

4. Each employee is expected to report to his supervisor unsafe conditions which he encounters in his work.

5. The supervisor will give job instructions. No employee is expected to undertake a job until he has learned how to do it and is authorized to do it by his supervisor.

6. No employee should undertake a job that appears to him to be unsafe.

7. If an employee suffers an injury, even a slight one, he is required to report it at once.

In addition to these points, any safety rules which are a condition of employment, such as wearing of eye protection or safety hats, should be understood and enforced at once.

Preliminary instruction

This preliminary safety instruction is most often given to individuals or small groups by the employment department. Sometimes the safety professional or a management executive gives the safety instruction. This method may add force and interest, but it has the practical disadvantage that safety professionals and executives are busy people and may not always be available or may be so pressed with other duties that they skimp on the induction job or delegate it to subordinates less able to handle it.

More important than who gives the talk is how it is given. The talk should be prepared and presented with the utmost regard for the effect it will have on the new employees. Verbal instruction should be given earnestly and with an attitude of good will and friendly cooperation.

A safety film may do a good job of interesting and instructing new employees. It changes the pace and relieves the monotony of much talking. The film should be brief and should be limited to such essential points as those previously mentioned. Some companies produce their own films. Although the production of a film may seem costly, if it is shown to all new employees, the cost per employee showing may be only a few cents. Films for the purpose may also be purchased.

One important advantage of an induction film is that it can present a carefully planned message in a consistent and effective manner.

Management can be sure that its safety message is going to be told in exactly the same way to every new employee. A few employers have attempted to gain this advantage by presenting a recorded talk by a company executive, but a record alone generally has low interest value.

Charts illustrating points made in the induction talk add interest and aid both understanding and conviction. Charts should be large enough and simple enough to be seen and understood easily by every member of the group.

If a company has preplacement physical examinations, the doctor or nurse should establish good relations with each new employee when he is examined. The doctor or nurse should tell about the work of the medical department as it relates to the employees and should encourage them to make use of its services. Medical or nursing personnel, as well as the person who gives general safety information, should emphasize the importance of reporting all injuries.

The final step in the employment office safety procedure should be to emphasize the importance of the position of the supervisor in the safety program. The employee must understand that the supervisor is responsible for job training, and that such training will include safe work procedures.

In order that there will be no gap and no contradiction between the information given in the employment office and that given later, the supervisor should know what has been said in the induction talks.

Make rule books logical, enforceable

The use of rule books or manuals* has long been a part of the early training given to new employees. Such rule books should be prepared so the rules are presented in terms that are easily understood. Only logical and enforceable rules should be included. Employees cannot be expected to respect and

---

*The development, approval, and distribution of printed safety rules by industrial concerns are discussed in NSC Data Sheet 535, *Industrial Safety Rules*. The National Safety Council also publishes a number of rule booklets and leaflets, on both general and specific subjects, aimed at employees.

follow rules that are illogical, unfair, or unrealistic.

Rules that supervisors have had a chance to review are likely to be more effective than those they have had no part in formulating.

Rule books or manuals should contain general instructions as to the employee's responsibility in safety. The rules should cover such items as first aid, personal protective equipment, work clothing, fire fighting, electrical equipment, and housekeeping.

Rule books alone cannot be counted upon to accomplish much in the way of influencing attitudes. Well prepared, illustrated booklets or cards, however, can add much to the start of a good training program if the contents are briefly reviewed and discussed when the booklets are presented and if the employee is not loaded with a great mass of printed matter.

Departmental induction and training

When a new employee reaches his own department, his supervisor should give him additional safety instruction. This may cover some of the points made in the employment office interview, but now applied specifically to the kind of work he is going to do.

The supervisor may repeat the earlier instruction about reporting unsafe conditions, not undertaking a job without instruction and authorization, and other matters of policy. He may tell the new man about the safety record and the safety program of the department and if there are safety committeemen, he will introduce the new man to them.

The supervisor will explain general safety regulations of the department and will see that the new employee is provided with the personal protective equipment furnished by the company for his job. The supervisor will also make provisions for training in safe work procedures. He will make certain to follow up to make sure that safe procedures are followed.

Some safety departments have followup interviews with employees from one week to one month after employment. At this time, the safety professional reviews the points discussed at the time of employment and encourages the employee to talk about his experiences on the job.

Companies following this plan report that discussion with an employee after a few days on the job is more profitable than information given before the employee starts work. The new man has overcome his feeling of strangeness and uneasiness and can relate the discussion to what he has already experienced. If the employment department handles the initial safety instruction, it is especially desirable for a representative of the safety department to talk with the employee a short time after he goes to work.

Of course, any direct contact by the safety department with an employee should be with the knowledge and approval of the employee's supervisor.

This induction procedure is the least a new employee should receive if he is to get a real feeling that safety is important to his job, and if he is to understand the company's attitude toward safety.

Some companies have more elaborate induction programs. These may include a plant tour, discussion of the company's products, viewing a film, and listening to talks by representatives of various departments. The program may take one-half to a full day.

Companies having formal induction training programs are convinced that they pay off in lower labor turnover, in good employee relations, and in prevention of accidents.

In these programs, it is the job of the safety department to make sure that safety is presented as interestingly and effectively as any other part of the program. Safety must not be submerged in a mass of information so that it will be remembered only vaguely, if at all. *The employee must carry away a deep conviction that safety is important to himself and his company.*

Good supervision – consistent instruction and "discipline"

A consistent training program should include the supervisory function as a part of job instruction.

If the supervisor observes men taking short cuts or otherwise departing from safe methods, he should correct them at once. If he does not correct them, the unsafe method soon becomes standard practice.

Much has been said about "discipline" for violations of safe practices. "Discipline" in this connection means penalties, usually following a verbal or written warning and in the form of time off without pay. Probably every employer recognizes the theoretical necessity for penalties to punish willful misconduct, whether the offense has to do with safety rules or other company regulations. Actually, such penalties are rarely assessed and many companies never use them at all.

Often, violations of safety rules are overlooked until an accident occurs. If employees are corrected for every infraction of a safety rule or safe practice as soon as it is observed, there will be few occasions which require "discipline." In any event, if a penalty is to be assessed, it should be for the act and not for the accident.

First aid courses

First aid courses for employees have been conducted in some industries for many years. Such courses should be standard Red Cross or Bureau of Mines. The need for personnel trained in first aid is spelled out in OSHA requirements if there is no infirmary, clinic, hospital, or physician in proximity or reasonably accessible.

The direct value of first aid training is greatest in companies which have night shifts, or skeleton crews working, when medical facilities are closed or where field crews are working at points far removed from professional medical help. Public utilities, mines, oil-well drilling, and logging are a few operations where employees have become proficient in caring for seriously injured persons and in transporting them to the hospital. Providing instruction in first aid is a necessity in these industries.

Even in companies with complete medical departments, trained first-aiders are a potential asset because they can stop dangerous bleeding, administer artificial respiration, and transport injured workers safely. Not only can they render these services, which may be needed at any time, but their value in a disaster would be great.

In addition to its possible direct benefit on the job, first aid training has inherent interest for employees. On the theory that first aid students learn something about the causes of accidents and acquire an accident awareness that makes them cautious, some companies which sponsor courses believe that instruction in first aid makes an employee less likely to have accidents.

Companies sponsoring first aid courses usually arrange for a qualified instructor, provide a meeting place, pay for textbooks and supplies, and often reward the graduates with a dinner or other celebration. Classes are usually held on trainees' own time.

The so-called "accident-prone" individual

Probably no phrase in safety causes so much disagreement as to what it exactly means than does "accident proneness." Most definitions hinge upon the idea that a person with certain personality traits is very likely to have accidents. When a person is said to be "accident-prone," it is generally meant that some psychological characteristics he has predisposes him toward having accidents.

Too often the term is loosely applied to anyone who has more accidents than others who do the same type work. A person could, however, have more than his share of accidents because he never was trained properly, or because he needs new glasses, or simply because he is working in cramped quarters or where he can be jostled. Behavior of this type may also be due to poor supervision or the attitude of management toward accident prevention. A language barrier may be a factor in other cases.

It is true that a small group of people often account for more than their expected share of accidents during a given period, but over a long time period, the composition of the group changes—the accident repeaters of one time period do not usually show up during the next time period. Much more research has to be done in this field before any person can be positively identified as being accident prone. Above all, such a label should not be placed in the personnel file of any individual.

Conclusion

The emphasis on job instruction in this chapter may seem to indicate disregard of many mental and emotional states that lead to unsafe acts. Lack of knowledge or skill is, of course, but one cause. Some of the other causes of unsafe acts are:

Physical or mental handicap or inability	Overconfidence Absent-mindedness

Disregard of danger	Undue haste
Resentment of	Distraction
authority	Anger
Inattention to	Impatience
instruction	Playfulness
Indifference	Fatigue
	Boredom

General estimates as to which of these factors most frequently cause accidents are unreliable. The original report in any given case has to be made by an investigator who makes a judgment about why another person acted as he did. Often the investigator's judgment is influenced by his own feelings or his own experience. Often, too, he tries to select one cause for the unsafe act when there may have been two or more causes, all interrelated.

For example, a worker who lacks skill at a job of loading heavy parts into a car may become fatigued from his clumsy efforts to do what a more skilled worker would do easily. He may fall behind and then try to hurry in order to catch up. Encountering a minor difficulty, he may lose patience and throw his weight heedlessly into the work, with the result that he falls or suffers a back sprain.

It is easy to see that one cause of the unsafe act was anger or impatience. Another was undue haste. Still another was fatigue and, back of that, lack of skill. It is unlikely that any two investigators would report the same causes. Even the injured person himself often does not know why he acted unsafely.

There seems to be no specific training aimed at such human failings as impatience, boredom, distraction, and so on. Good job instruction, however, will prevent many of the harmful acts that could arise from these mental or emotional states. A worker sufficiently skilled does not break the pattern of good work performance even when he is seriously disturbed, at least not as readily as a person less skilled.

Good job instruction not only produces more skilled workers; it also impresses the man receiving the instruction with the high value the employer places on safety. Frequent follow-up and attention on the part of the supervisor to correct work practices also help to create understanding and to eliminate resentment, which is a source of some of the undesirable attitudes.

References

Programmed instruction

"A Bibliography of Programs and Presentation Devices." Carl Hendershot, 4114 Ridgewood Dr., Bay City, Mich. 48706.
A listing of programmed instructional materials and devices with quarterly supplements.

"Library of Programmed Instruction Courses." E. I. du Pont de Nemours and Co., Inc., Education and Applied Technology Division, Wilmington, Del. 19898.
A listing of vocational training courses. Also available on request is a list of safety training courses.

"STEP—Self Teaching Education Program." National Safety Council, 425 N. Michigan Ave., Chicago, Ill. 60611.
Six units of linear programmed instruction for supervisory training. The six units cover Safety and Production, Accident Causes, Influencing People, Materials Handling, Housekeeping, and How to Instruct.

Management games

"Simulation Series for Business and Industry." Science Research Associates, Inc., Department of Management Services, 259 East Erie St., Chicago, Ill. 60611.
This series includes Decision Making, Collective Bargaining, Equipment Evaluation, Supervisory Skills, Purchasing, Production, Control Inventory and Interviewing.

9—Safety Training

"The In-Basket Method." Bureau of Industrial Relations, Department of Training Materials for Industry, The University of Michigan, Graduate School of Business Administration, Ann Arbor, Mich. 48104.

A series of packaged courses, each set consisting of letters, notes, memos, and reports.

"The In-Basket Kit." Allan A. Zoll, Addison-Wesley Publishing Co., Inc., Reading, Mass. 01867.

A kit of materials covering management practices that involve the learner.

"Accident Case Studies Kit." National Safety Council, 425 N. Michigan Ave., Chicago, Ill. 60611

A kit containing a series of six incidents following the format of the Paul Pigors' incident method.

"A Catalog of Ideas for Action Oriented Training." Didactic Systems, Inc., P.O. Box 4, Cranford, N.J. 07016.

A listing of simulation games of all types, programmed instruction materials, and a listing of books on effective training.

Books

Analyzing Performance Problems. Mager and Pipe, Fearon Publishers, #6 Davis Dr., Belmont, Calif. 94002.

Preparing Instructional Objectives. Robert F. Mager, Fearon Publishers, #6 Davis Dr., Belmont, Calif. 94002.

Training Development Handbook. American Society for Training and Development, Inc., P.O. Box 5307, Madison, Wis. 53705.

Supervisors Safety Manual, 4th ed. National Safety Council, 425 N. Michigan, Chicago, Ill. 60611. Course materials are available.

"Training Requirements of the Occupational Safety and Health Standards." U.S. Department of Labor. Available from U.S. Government Printing Office, Washington, D.C. 20402, or local OSHA regional office (see Chapter 2).

Human
Factors
Engineering

Chapter
10

Man-Machine Systems 220
A scientific approach . . . HFE and training

Applications of HFE 222
Allocation of tasks . . . Task analysis . . . Anthropometric
considerations

Function 1 — Man as Sensor 225
Information displays . . . Visual displays . . . Auditory displays
. . . HFE display evaluation

Function 2 — Man as Information Processor 231

Function 3 — Man as Controller 231
Design principles . . . Arrangement . . . Control evaluations

Conclusions 236

References 237

217

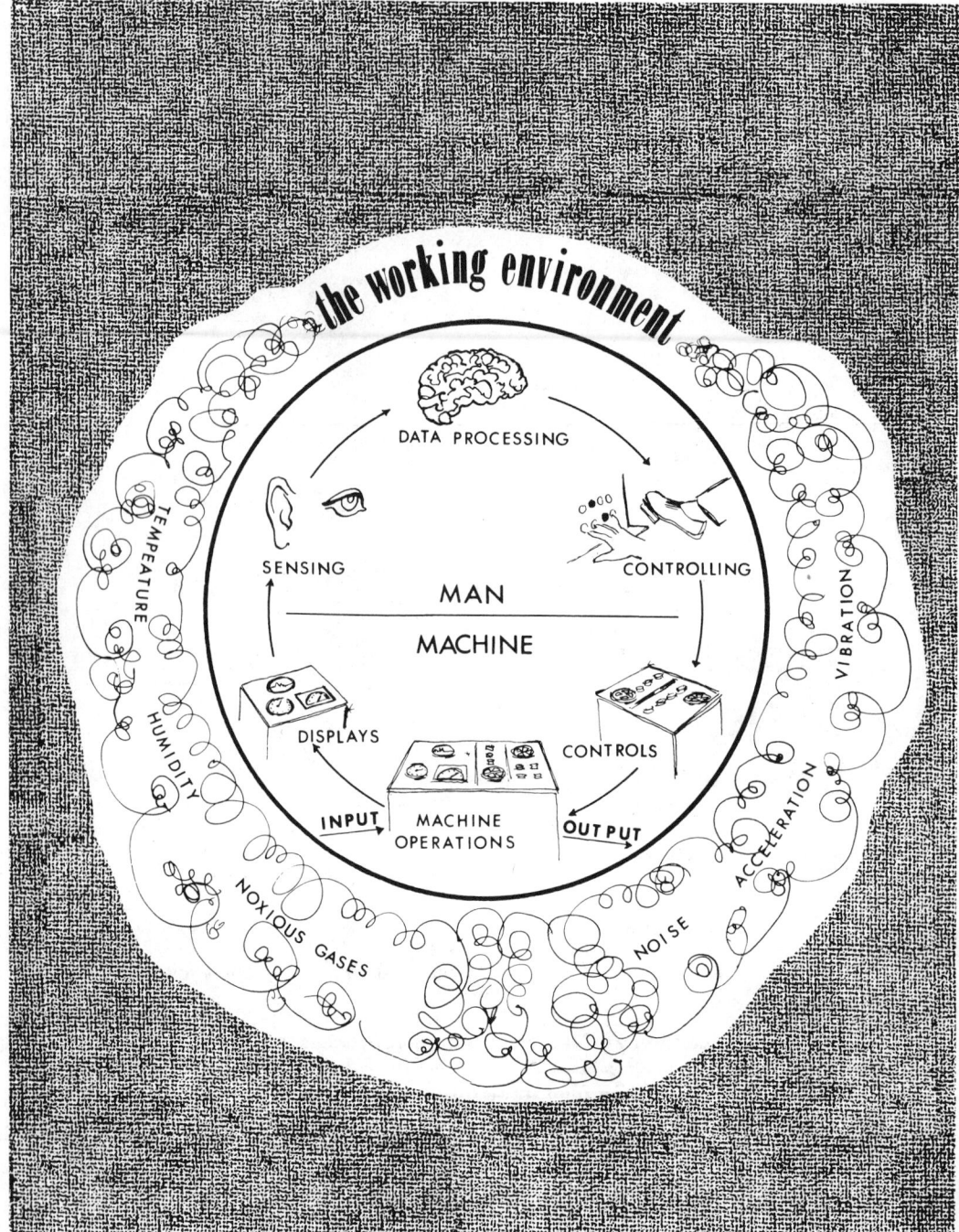

FIG. 10–1.—In any man-machine system, man serves three functions—sensor, information processor, or controller. Man interacts with the machine at two points—displays and controls. The purpose of human factors engineering is to minimize errors made at these two points.

MAN VS. MACHINE

Man Excels in	Machines Excel in
Detecting certain stimuli of low-energy levels	Monitoring (both men and machines)
Sensing an extremely wide variety of stimuli	Performing routine, repetitive, or very precise operations
Perceiving patterns and making generalizations about them	Responding very quickly to control signals
Detecting signals in high-noise levels	Exerting great force, smoothly and with precision
Storing large amounts of information for long periods — and recalling relevant facts at appropriate moments	Storing and recalling large amounts of information in short time-periods
Exercising judgment when events cannot be completely defined	Performing complex and rapid computation with high accuracy
Selecting own inputs	Sensitivity to stimuli beyond the range of human sensitivity (such as infrared, radio waves)
Improvising and adopting flexible procedures	Doing many different things at one time
Reacting to unexpected low-probability events	Reasoning deductively — going from general to specifics
Applying originality in solving problems: i.e., coming up with alternate solutions	Being insensitive to extraneous factors
Profiting from experience and alter course of action	Operating very rapidly, continuously, and precisely the same way over a long period
Performing fine manipulation, especially where misalignment appears unexpectedly	Operating in environments which are hostile to man or beyond human tolerance.
Continuing to perform even when overloaded	
Reasoning inductively — specifics to general	

Fig. 10–2.

Source: W. E. *Woodson,* Human Engineering Guide for Equipment Designers.

Designing-in safety—*in* the job, *in* the machine, and *in* the environment—and not trying to make man perform other than by "what comes naturally" is a major goal of human factors engineering. This chapter tells how controls and machines can be made more convenient and more comfortable and less confusing, less exasperating, and less fatiguing to the user.

Human factors engineering is vital to system analysis—a technology aimed at optimizing system performance (see Chapter 4, "Inspection and Control Procedures.").

By definition, a system is an orderly arrangement of components that act to perform some task in a given environment. Not mutually exclusive, components interact with each other. And this interaction is always carried out in some environment.

It is easy to see that the vast majority of systems are composed of men and machines (although there are wholly machine systems, such as the carburation systems in cars, and wholly human systems, such as a supervisor giving a 5-minute safety talk to his group). Human factors engineering, since its inception during World War II, has been concerned with the interaction (or interface) between man and machine—and man and his environment.

Although human factors engineering was originally limited to applications in the aviation and aerospace fields, it has more recently been applied to the problem of safety in industrial environments. Recent sources (see the Surry and Jones citations in References) have provided numerous examples of the application of human factors engineering to occupational safety questions. Some general principles of human factors engineering, applicable to a wide variety of industrial tasks, will be presented in the remainder of the chapter.

Before this can be done, however, it is necessary to examine a system in detail.

Man-Machine Systems

Fig. 10–1 is a schematic of a man-machine system. A number of things should be noted about it.

First, input can enter the system at any point (as depicted by the arrow entering the circle at "machine operation").

The subsystem "machine" has displays and controls. Reading the displays (which may be of any variety—visual, audible, tactual), the "man" component decides how he should use the controls. When an adjustment is required, it is done by the human muscle (effector) system—with such adjustments serving as new input.

The entire man-machine system operates in an environment of heat, stress, humidity, noise, and the like. The environment, to one degree or another, affects the performance of the system's components.

(As an aside, it should be pointed out now, the system illustrated is called a "closed-loop" system, one that allows the operator to correct the system's performance. An "open-loop"

system does not allow for corrective action; once activated, no further control is possible. A rifle shot is an example of an "open-loop" system.)

Fig. 10–1 also shows that in any man-machine system, man serves three functions—sensor, information processor, and controller—and interacts with the machine at two points—displays and controls. Each of these three functions will be discussed separately. Fig. 10–2 compares the relative abilities of both man and machine.

Here is the crux of the matter: *The purpose of human factors engineering is to minimize errors in using displays and controls by designing systems that will be compatible with both the "man" and "machine" components, while considering man as fulfilling one or all three of his functions—sensing, processing information, controlling. Human factors engineering, then, is concerned with the interaction (or interface) between man and machine— and man and his environment.*

Additionally, much of human factors engineering is concerned with the environment within which the man works. Thus, concern must be directed to such elements as:

• The atmospheric environment (including the effects of altitude, temperature, humidity, and toxicants).

• The mechanical environment (including the effects of acceleration, vibration, and noise).

In this Manual, specific consideration to a number of common environmental conditions, which affect human performance, is given in Chapters 37 through 40.

A scientific approach

Human factors engineering is not an exact science. It is, however, a scientific approach to problems of designing and constructing things, which people are expected to use—so the user will be more efficient and less likely to make errors resulting in accidents.

The discipline has various names. In Great Britain, it is called "ergonomics." Others call it bio-mechanics, bio-technology, biophysics, human engineering, human factors, and engineering psychology. Despite the variety of names, there is general agreement that human

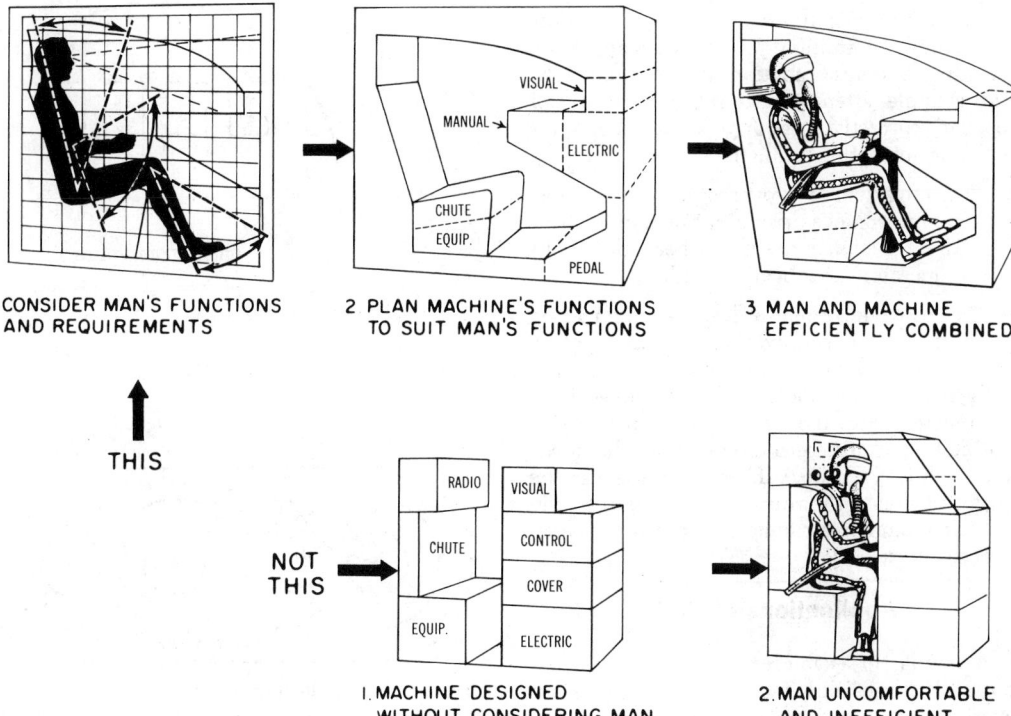

1. CONSIDER MAN'S FUNCTIONS AND REQUIREMENTS

2 PLAN MACHINE'S FUNCTIONS TO SUIT MAN'S FUNCTIONS

3 MAN AND MACHINE EFFICIENTLY COMBINED

THIS

NOT THIS

1. MACHINE DESIGNED WITHOUT CONSIDERING MAN

2. MAN UNCOMFORTABLE AND INEFFICIENT

FIG. 10–3.—Example of human factors engineering in equipment design.

From Human Engineering Guide to Equipment Design *by C. T. Morgan.* Copyright 1963 by McGraw-Hill Book Co. *Used with permission.*

factors engineering is concerned with the interaction of a number of disciplines including psychology, physiology, and anthropology.

There are several scholarly definitions of human factors engineering, but the three following simplified ones will serve the purpose here:

1. Engineering something for the population that will use it.

2. Designing a system so that machines, human tasks, and the environment are compatible with the capabilities and limitations of people—to minimize error.

3. Designing the system to fit the characteristics of people rather than retrofitting people into the system.

Figure 10–3 demonstrates the difference between an engineering solution which includes

and omits human factors in the case of an equipment design problem.

HFE and training

Much of the effort in occupational safety has been directed toward altering the man by training. The approach to the machine has been to guard the obvious hazards—many of which are simply products of deficient engineering design.

There are at least three limitations to altering the man by means of training:

● The long-term, high cost of training in terms of dollars and time. Obviously, it is not possible to dispense with some type of training for virtually all work situations. But in certain man-machine systems, human factors engineering principles applied in the design of the machine will significantly reduce training requirements.

● Training sometimes fails as a solution. No amount of training will make some man-machine systems function more effectively. For example, attempting to retrain an operator to read correctly a poorly designed display will not solve the problem.

● Training will not overcome poor or disrupted performance arising out of undue stress caused by machine design—where the limits of the operator have been exceeded.

The problem, from a human engineering point of view, is simply that training is often not the most efficient technique for dealing with the man-machine interface. Indeed both the machine and the environment in industry can be largely structured to suit both the needs and abilities of man. Then, training can be used to increase the probability of successfully reducing human error and increasing system effectiveness.

Applications of HFE

A couple of examples help to clarify the possibilities of applying human factors engineering to displays and controls.

Fig. 10–4 shows two electric meters. To read the meter on the bottom (an old, but still-used meter), a man relies heavily on previous knowledge of such meters. But the meter at the top has been redesigned to print-out directly the reading, which requires no special knowledge or interpretation to record.

Two heat-regulation controls are shown in Fig. 10–5. Control A is not designed with the human operator in mind; Control B, on the other hand, is designed to eliminate errors in the information-processing and machine-controlling functions.

Although these two illustrations are elementary, they present the possibilities that human factors engineering holds.

Allocation of tasks

In any man-machine system, there are tasks that are better performed by man than by machine—and, conversely, tasks that are better handled by machines. See Fig. 10–2.

In general, machines can usually perform more efficiently on those tasks that must be performed routinely and rapidly with a high degree of accuracy. Men perform better the

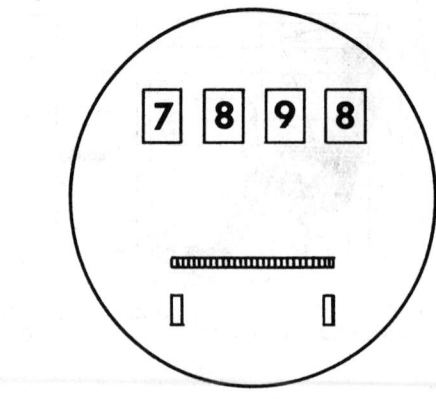

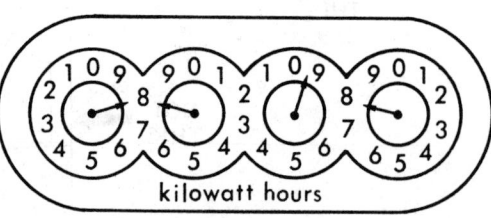

Fig. 10–4.—Application of HFE to meter reading. Reading on top meter requires no special knowledge for interpretation.

tasks calling for responsibility and flexibility (adaptability), in addition to tasks that cannot be anticipated.

Man is generally *excluded* from tasks that are likely to result in a high probability of error. Such tasks are:

● Perceptual requirements near or beyond the physiological limits or that conflict with established perceptual patterns.

● Response requirements that are physically difficult, conflict with established patterns, or cannot be readily checked or monitored for adequacy.

● Decisions that require undue reliance on short-term memory or must be accomplished within too short a time interval in view of other necessary tasks.

● Tasks that overload the human, resulting in an imbalanced workload/time distribution, or do not permit adequate or timely monitoring of the system.

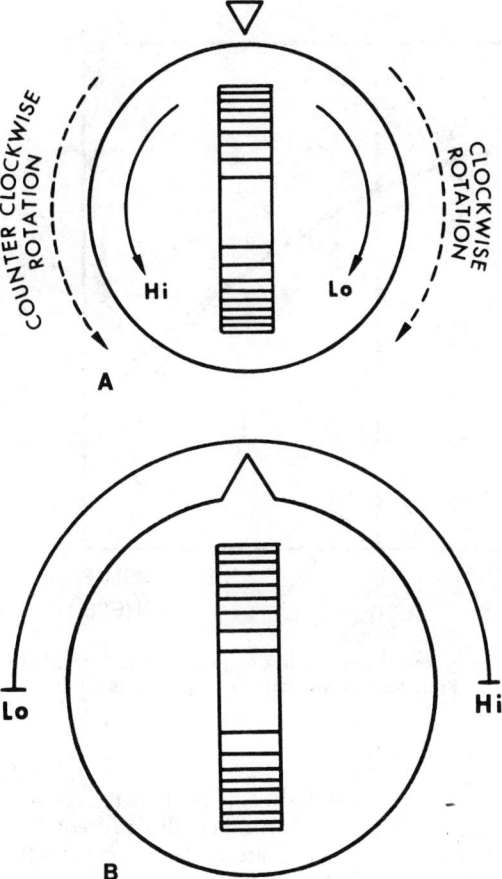

A

B

FIG. 10–5.—Control B is designed to eliminate errors in the information-processing and machine controlling functions. (See also Fig. 10–16.)

- Communication requirements that conflict with other activities.

Often, the man's contribution to a system is to provide a counter-measure in the event of system malfunction or component failure. To do this he must know that a failure has occurred and what to do about it.

Generally, displays provide the failure information and, therefore, must be designed to communicate such information to the operator. The operator must then perform the proper response with the controls provided with a minimum of error—yet it is at this point that breakdowns in the system frequently occur.

(Displays are discussed later, under "Function 1—Man as Sensor.")

Task analysis

Just as equipment can be designed to fit human limitations and capabilities, so, too, can jobs (or tasks) be designed for humans.

Human factors engineering research has shown that man needs to be challenged but not overburdened. If a job is too easy and too routine it is possible that monotony, boredom, and eventually errors (and accidents) will occur.

Machines have a built-in upper tolerance limit. If an electrical circuit, for example, is overloaded, the fuse will blow and no harm to the system will result.

Man, on the other hand, does not have a "safety fuse box." He can work for short periods under overloaded conditions, e.g., high production demands; however, when such an overload reaches some undefinable point, the human may completely break down. Such stress overloading may account for the co-worker who suddenly "flies off the handle."

The task of the job designer, then, is to find the happy blend between "easy" and "difficult" jobs. With very low levels of psychological stress (boring jobs) performance is also low; as stress increases, however, performance also increases—to a point.

The task is to design jobs that will be centered around optimum performance. (See Fig. 10–6.)

Current research in human factors engineering is considering the problem of the distribution of work with respect to the length of the work period. For example, the question of distributing the work load over five (short) days or four (long) days requires investigation especially with respect to the type of work involved. (See the Yoder and Botzum citation in References for an example of such research.)

How to predict task requirements. Human tasks are predicted from the design of the equipment and from the tentative organizational and procedural setup. A breakdown of the task requirements can be used for determining training requirements, modification of hardware, and for qualitative and quantitative personnel estimates.

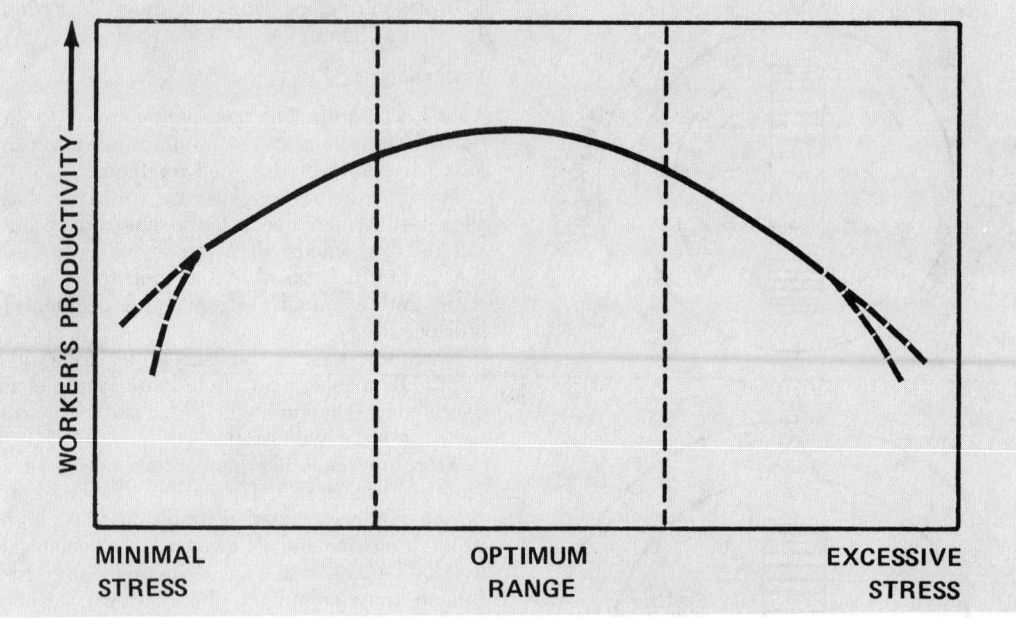

Fɪɢ. 10–6.—In general, long continued work under extremely low stress conditions produces lack of alertness; work under high stress conditions produces fatigue. Both factors are known to play a leading role in causing accidents.

The general steps of task analysis are:

• Identification of the broad functions that the human will perform in the system (e.g., detection, processing data, decision making, and maintenance).

• Selection of the types of information and control that the human will require in order to perform the function (e.g., information to make a decision and the response requirements).

• Detailed specification of the controls, displays, and auxiliary equipment (e.g., layout, size, lighting, display brightness, and control movements).

Some of this, however, cannot be done in the design phase, but must wait until mockup development.

However, in any system or product development, management's constraints are cost, schedule, and performance of the system. Design engineers will also be concerned with reliability, quality, maintainability, and like factors. It must be recognized that management constraints and other design requirements may, unfortunately, take precedence over human factors engineering.

Anthropometric considerations

Including anthropometric measurements in system design is another approach to occupational and product safety.

This means considering and applying dynamic and static body measurements as design criteria to improve the ease, efficiency, and safety of the human in the system. Such data is abundant in the literature.

Tables 10–A and 10–B present representative anthropometric information on height and weight for male civilian populations.

Anthropometric data is useful for deriving optimum and limiting dimensions for a wide variety of operator positions. (See Fig. 10–7.)

A human factors engineering checklist for system or product design considerations will be found in Fig. 10–8.

Function 1—Man as Sensor

Previously, it was pointed out that one of the functions man serves in a man-machine system is that of sensor, or information seeker. Contrary to the popular notion, man has something like 12 to 13 senses—not just five. (Some of these are given in Fig. 10–9, along with the sensing organ and physical energy of various stimuli.)

As the senses are used as communications channels, they can all be used as signaling or communicating inputs, even though one generally thinks of communication only by means of sight or sound. Of importance to the safety practitioner is protecting all of the operator's senses by holding energy levels within a safe range.

Information displays

An information display is a device used to gather needed information and to translate such information into inputs that the human brain can perceive.

Two general classes of information displays —pictorial and symbolic—are utilized.

- In PICTORIAL DISPLAYS, the geometrical and spatial relationships are shown as they exist. Maps, pictures, and TV are examples of pictorial displays.

- SYMBOLIC DISPLAYS present the information in a form that has no resemblance to what it is measuring. Some examples are a speedometer, a thermometer, a pressure gauge, and an altimeter.

The two most common types of symbolic displays are the visual and auditory (see Fig. 10–11). Much study has been given to the design characteristics of these types of displays and some general principles have emerged.

Visual displays

Principles of parsimony. Visual displays are used for one of three purposes:

- QUANTITATIVE READINGS—to determine the exact quantity involved, such as a thermometer.

- QUALITATIVE READING—to determine the state or condition at which the machine or system is functioning—usually three conditions, such as "hot," "safe," or "cold."

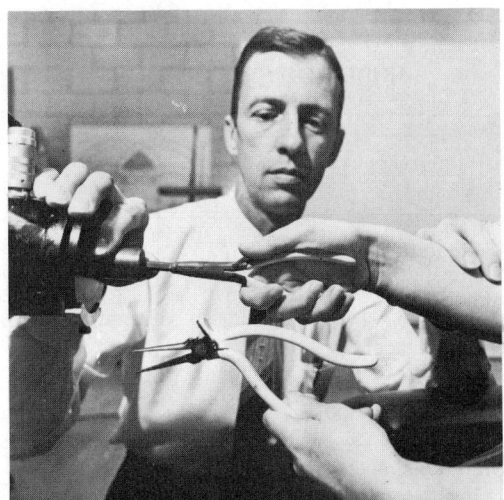

Fig. 10–7a.—Conventional pliers (top) require the wrist to be bent. Redesigned tool (bottom) has contour handles, spring, and thumb stop to reduce worker fatigue.

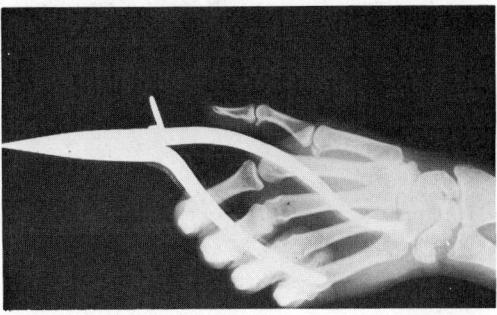

Fig. 10–7b.—X ray shows how contour handles fit easily into a natural hand position.

Courtesy Western Electric Company, Inc.

- DICHOTOMOUS (check) READINGS—to check operations or to identify one or two levels, such as "off" or "on."

The purpose for which the display is to be read will dictate its design. But as a general principle, the simplest design is the best.

Fig. 10–10 shows three types of dials appropriate for the three purposes just stated. The dial on the left is suitable for check readings; the one in the center, for qualitative readings; and the one on the right, for quantitative readings.

TABLE 10–A

NUDE-BODY HEIGHT (STANDING) OF MALE CIVILIAN POPULATIONS

Population	Percentiles (in.)					Standard Deviation
	1st	5th	50th	95th	99th	
Railroad travelers[1]	62.5*	64.5*	69.0*	73.8*	75.6*	
Truck and bus drivers[2]	63.0	64.6	68.4	72.5	74.1	
Airline pilots[3]	64.4	66.0	70.0	73.9	75.6	2.40
Industrial workers[4]	64.4*	66.1*	70.3*	74.4*	76.2*	2.46
College students[5]	62.5	64.4	68.7	73.1	74.9	2.68
Eastern, 18 yr old[6]	64.5	66.1	69.9	73.8	75.4	2.38
Eastern, 19 yr old[7]	65.0	66.5	70.2	74.0	75.5	2.30
Midwest, 18 yr old[8]	63.2	65.0	69.1	73.3	75.0	2.60
Midwest, 18–22 yr old[9]	64.2	65.9	70.0	74.1	75.8	2.49
Draft registrants[10]						
18–19 yr old	62.0	63.8	68.0	72.3	74.1	2.61
20–24 yr old	62.1	63.9	68.2	72.4	74.2	2.60
25–29 yr old	61.9	63.7	68.1	72.4	74.2	2.63
30–34 yr old	61.7	63.5	67.8	72.1	73.9	2.66
35–37 yr old	61.3	63.2	67.6	72.0	73.8	2.64
Spanish-American-War veterans	61.1	62.6	66.1	69.7	71.2	2.15
Canadians[11]						
18–19 yr old	62.4	64.1	68.2	72.1	73.7	
20–24 yr old	62.0	63.8	68.3	72.5	74.3	
25–29 yr old	60.6	62.9	68.3	74.0	76.2	
30–34 yr old	61.5	63.4	68.1	72.8	74.8	
35–44 yr old	60.5	62.7	67.6	72.6	74.7	
45–54 yr old	59.7	61.8	66.8	72.0	74.1	
55–64 yr old	58.4	60.6	66.0	71.3	73.6	
More than 64 yr old	58.6	60.6	65.1	69.8	71.8	

[1] Hooton, 1945.
[2] McFarland, et al., 1958.
[3] McCormick, 1947.
[4] Tyroler, 1958.
[5] Diehl, 1933a.
[6] Bowles, 1932.

[7] Heath, 1945.
[8] Damon, 1955.
[9] Elbel, 1954.
[10] Karpinos, 1958.
[11] Pett and Ogilvie, 1957.

* Including shoes (subtract 1 in. for nude height).

From C. T. Morgan, et al. Human Engineering Guide to Equipment Design. New York, McGraw-Hill Book Co., 1963. Used with permission of the publisher.

Principle of compatibility. The principle of compatibility holds that the motion of the display should be compatible with (or in the same direction as) the motion of the machine and its control mechanism.

For example, a display increasing in numerical value should indicate that the mechanism being measured is also increasing.

Furthermore, a pointer that moves to the right to show an increase should have its corresponding control mechanism designed so that a rightward movement of the control will increase the machine value and the corresponding display output value.

Principle of arrangement. As the design of the display is important, so too is its location or arrangement with other displays. A poor arrangement of displays can be the source of error.

Sometimes dials must be arranged in groups on a large control panel. If all the dials must

TABLE 10–B

NUDE-BODY WEIGHT OF MALE CIVILIAN POPULATIONS

Population	Percentiles [1] (lb)					Standard Deviation
	1st	5th	50th	95th	99th	
Railroad travelers[2]		132 *	167*	218 *		
Truck and bus drivers[3]		129	164	213	247	
Airline pilots[4]		(134)	168	(201)		20.3
Industrial workers[5]		(130)*	170*	(210)*		24.5
College students[6]		(112)	142	(172)		18.1
Eastern, 18 yr old[7]		(122)	150	(178)		17.2
Eastern, 19 yr old[8]		(132)	159	(187)		16.2
Midwest, 18 yr old[9]		(115)	148	(180)		19.7
Midwest, 18–22 yr old[10]		(118)	156	(195)		23.5
Draft registrants[11]						
18–19 yr old		(106)	141	(176)		21.1
20–24 yr old		(109)	146	(183)		22.4
25–29 yr old		(110)	151	(192)		24.8
30–34 yr old		(110)	153	(195)		25.8
35–37 yr old		(111)	154	(197)		26.1
Spanish-American-War veterans	110	118	153	190	200	22.1
Canadians[12]						
18–19 yr old			140			
20–24 yr old			151			
25–29 yr old			157			
30–34 yr old			168			
35–44 yr old			165			
45–54 yr old			161			
55–64 yr old			159			
More than 64 yr old			156			

[1] Percentiles in parentheses were computed from the 50th percentile using the S.D. Because of the skewed distribution of weight, these values might differ somewhat from the true values and should be used with caution.

[2] Hooton, 1945.
[3] McFarland, et al., 1958.
[4] McCormick, 1947.
[5] Tyroler, 1958.
[6] Diehl, 1933a.
[7] Bowles, 1932.
[8] Heath, 1945.
[9] Damon, 1955.
[10] Elbel, 1954.
[11] Karpinos, 1958.
[12] Pett and Ogilvie, 1957.

* Including shoes and indoor clothing (subtract 5 or 6 lb for nude weight).

From C. T. Morgan, et al. Human Engineering Guide to Equipment Design. New York, McGraw-Hill Book Co., 1963. Used with permission of the publisher.

be read at the same time, they should be pointing the same direction when in the desired range. This will reduce check-reading time and increase accuracy.

Principle of coding. All displays should be coded (or labeled) so that the operator can tell immediately just what mechanism the display refers to, what units are measured, and what the critical range is.

Labeling is especially important if operators are unfamiliar with the equipment.

The effectiveness of labels is greatly affected by the environment. If the equipment is being used in a dimly lighted area, illumination must be provided. Glare, of course, may be a problem in a brightly lighted room.

Other problems may be caused by vibration,

**CHECKLIST
HUMAN FACTORS ENGINEERING EVALUATION**

Task:

CONTROL DESIGN
  Compatibility of movement with display
  Movement required:
    (push, pull, turn, move left, move right, move up, move down, combination)
  Critical controls coded
  Critical controls labeled
  Controls coded
  Controls labeled
  Coding used (size, color, shape, movement)
  Location of controls (accessibility to operator, frequency of use, critical to the system)
  Control resistance
  Anthropometric requirements

DISPLAY DESIGN
  Type of displays (visual, auditory, other)
  Control/display ratio
  Control/display movement compatible

TASK DESIGN
  Monitoring (vigilance)   Receive communications
  Information processing   Memory
  Decision making   Recording
  Relay information   Overload
  Transmit communications   Underload
                                        Anthropometric requirements

---

SIZE DESIGN
  Operator (seated, standing, both)
  Control size
  Display size
  Accessibility for movement
  Accessibility for maintenance
  Work space allocation

ENVIRONMENTAL FACTORS
  Atmospheric pressure   Space limitation
  Heat   Noise
  Cold   Vibration
  Acceleration   Light
  Deceleration   Glare

NOTE: This is not intended to be a comprehensive checklist for all systems. Other items, equally
  important, should be added depending on the system.

FIG. 10–8.

Man's Senses and the Physical Energies that Stimulate Them			
Sensation	**Sense Organ**	**Stimulated by**	**Originating**
Sight	Eye	Some electromagnetic waves	Externally
		Mechanical pressure	Externally or internally
Hearing	Ear	Some amplitude and frequency variations of the pressure of surrounding media	Externally
Rotation	Semi-circular canals	Change of fluid pressures in inner ear	Internally
	Muscle receptors	Muscle stretching	Internally
Falling and rectilinear movement	Semi-circular canals	Position changes of small, bony bodies in the inner ear	Internally
Taste	Specialized cells in tongue and mouth	Chemical substances dissolvable in saliva	Externally on contact
Smell	Specialized cells in mucous membrane at top of nasal cavity	Vaporized chemical substances	Externally
Touch	Skin mainly	Surface deformation	On contact
Vibration	None specific	Amplitude and frequency variations of mechanical pressure	On contact
Pressure	Skin and underlying tissue	Deformation	On contact
Temperature	Skin and underlying tissue	Temperature changes of surrounding media or of objects contacted	Externally and on contact
		Mechanical movement Some chemicals	
Cutaneous pain	Unknown but thought to be free nerve endings	Intense pressure, heat, cold, shock, chemicals	Externally on contact
Subcutaneous pain	Thought to be free nerve endings	Extreme pressure and heat	Externally and on contact

FIG. 10–9.

Source: H. W. Sinaiko, Selected Papers on Human Factors in the Design and Use of Control Systems.

acceleration, and, in the case of auditory displays, (described next), noise.

Auditory displays

Auditory displays should follow the principles outlined for visual displays.

In addition, special problems are posed by auditory displays.

The most immediate problem faced by the system designer is that of deciding whether

229

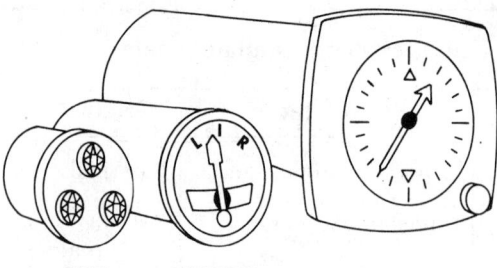

CHECK   DIRECTIONAL   QUANTITATIVE

FIG. 10–10.—Three dials illustrate the principles of parsimony: Check readings (for one- or two-level operations), directional readings (qualitative reading, usually of three conditions), and quantitative readings (show exact quantity involved).

he should use an auditory or visual display. Fig. 10–11 compares the relative advantages of auditory and visual displays.

Other considerations, though, are equally important. The following principles can act as a guide:

● SITUATIONALITY. The design of auditory displays should consider other relevant characteristics of the environment in which the system is to function (e.g., noise levels, types of responses controlled by the auditory signal).

● COMPATIBILITY. Where feasible, signals should "explain" and exploit learned or natural relationships on the part of the users, such as high frequencies being associated with "up" or "high" and wailing signals indicating emergency.

● APPROXIMATION. Two-stage signals should be considered when complex information is to be displayed and a verbal signal is not feasible. The two stages should consist of (*a*) attention-demanding signals to attract attention and identify a general category of information, and (*b*) designation signals to follow the attention-demanding signals to designate the precise information within the general category.

● DISSOCIABILITY. Auditory signals should be easily discernible from other sounds (be they meaningful or noise).

● PARSIMONY. Input signals to an operator should not provide more information than is necessary to carry out the proper response.

● FORCED ENTRY. When more than one kind of information is to be presented, the signal must prevent the receiver from listening to just one aspect of the total signal.

● INVARIANCE. The same signal should designate the same information at all times.

RELATIVE MERITS OF AUDITORY AND VISUAL PRESENTATIONS

Use Auditory Presentation if:	Use Visual Presentation if:
Message is simple	Message is complex
Message is short	Message is long
Message will not be referred to later	Message will be referred to later
Message deals with events in time	Message deals with location in space
Message calls for immediate action	Message does not call for immediate action
Receiving location is too bright	Receiving location is too noisy
Person's job requires him to move continually	Person's job allows him to remain in one position
Visual system of person is overburdened	Auditory system of person is overburdened

FIG. 10–11.

HFE display evaluation

When designing a display, a number of human factors should be considered. But a few questions about any existing displays quickly evaluates them:

• Has the threshold level for that sense been reached by the display? (Because each sense has its own threshold level, energy intensities below it cannot be perceived.)

• Is the sense overloaded? What other demands are made on this sense at the time the display in question is to be read?

• Is the display compatible with similar displays, controls, and machine movements?

• What environment factors, if any, will mask the display?

Function 2— Man as Information Processor

Much research is presently being performed to learn more about man as an information processor. (See the McCormick citation in References.)

Human judgements may be classified as either relative or absolute. A relative judgement is one that is made when an opportunity to compare two or more objects presents itself.

An absolute judgement is made in the absence of any standard or comparison. It has been estimated that most people can differentiate as many as 10,000 to 300,000 different colors on a relative basis—but only 11 to 15 on an absolute basis. As a general rule, therefore, a system should have more relative than absolute judgements.

Function 3—Man as Controller

The third function man serves in the man-machine system is that of controller. Just as principles exist for designing displays for man to use more readily, so, too, can controls be designed to eliminate error.

The control function in the man-machine system can be considered as the response to a given stimulus.

For many situations, a generalized response is given. Most Americans, for example, expect a light switch to be turned on by flipping the switch "up" and off by a "down" movement. A clockwise motion generally refers to an increase.

Such responses are called "population stereotypes," a behavioral response common to nearly everyone in the population. Some examples are given in Fig. 10–12.

In occupational safety, population stereotypes are particularly important from the point of view of hazard identification and recognition through various warning systems. Ideally, a visual or auditory warning system should make use of known associations for words (Danger, Caution, Warning) and colors (red, yellow, green, blue) in specifying the degree of hazard associated with a specific industrial condition. For example, research on accident prevention signs as specified in American National Standard *Specifications for Accident Prevention Signs, Z35.1* (see the Bryk and Bresnahan citation in References), has demonstrated that workers do associate different degrees of hazard with various visual hazard alert cues. Thus, DANGER signs (color-coded red) elicit higher amounts of hazard association than CAUTION signs (color-coded yellow). Likewise, THINK signs (color-coded green) elicit higher amounts of hazard association than NOTICE signs (color-coded blue).

Any display-response that calls for a movement contrary to the established stereotype is likely to produce errors. The designer is calling for errors by asking the operator to change, in this or that unique situation, a behavior pattern that can be described as habit.

If an operator misreads a poorly designed display and operates the wrong control or the right control in the wrong direction, safety may be jeopardized, and the system effectiveness degraded, if not lost entirely.

Despite the fact that many accident reports would classify this as an "unsafe act" or human error, it is, in fact, a design error. Retraining of the operator would not prevent reoccurrence of the series of events that lead to the accident.

Several guidelines for design of displays and controls have been published. (See the McCormick and Morgan citations in References.)

Design principles

Through research, these principles of con-

POPULATION STEREOTYPES—BEHAVIORAL RESPONS[E]

CONTROL MOVEMENT	SYSTEM (OR EQUIPMENT COMPONENT) RESPONSE				
	DIRECTIONAL				NONDIRECTIONAL
	UP	RIGHT	FORWARD	CLOCKWISE	INCREASE*
UP	RECOMMENDED	NOT RECOMMENDED	RECOMMENDED	NOT RECOMMENDED	RECOMMENDED
RIGHT	NOT RECOMMENDED	RECOMMENDED	NOT RECOMMENDED	RECOMMENDED	RECOMMENDED
FORWARD	RECOMMENDED	NOT RECOMMENDED	RECOMMENDED	NOT RECOMMENDED	RECOMMENDED
CLOCKWISE	NOT RECOMMENDED	RECOMMENDED	NOT RECOMMENDED	RECOMMENDED	RECOMMENDED

*Increase refers to increase in power output, brightness, rpm, etc., and to "on" or "start" as opposed to "off" or "stop".

Fig. 10–12.—General population stereotype control expectancy. When the control is moved as shown at left, most people expect a response as shown at right. Additional stereotypes are shown in table on facing page.

Adapted from C. T. Morgan, et al. Human Engineering Guide to Equipment Design. *New York, N.Y., McGraw-Hill Book Co. 1963.*

trol design have emerged:

Compatibility. Just as in the design of displays, control movement should be designed to be compatible with the display and machine movement. A lift truck, for example, that has the lift controls move right to left to raise or lower the lift is bound to have a number of errors associated with its operation.

Coding. Whenever possible, all controls should be coded in some way. A good coding system can reduce many errors by shape and texture, location, color, and operation. A summary and comparison of various visual coding methods is presented in Table 10–C.

● SHAPE AND TEXTURE. Controls can be coded by their shape or their texture (see Fig.

- Handles used for controlling liquids are expected to turn clockwise for off and counter-clockwise for on.

- Knobs on electrical equipment are expected to turn clockwise for on, to increase current, and counter-clockwise for off or to decrease current. (Note this is opposite to the stereotype for liquid.)

- Toggle switches are expected to turn "on" when flipped up, "off" when flipped down.

- Certain colors are associated with traffic, operation of vehicles, and safety.

- For control of vehicles in which the operator is riding, the operator expects a control motion to the right or clockwise to result in a similar motion of his vehicle, and vice versa.

- Sky-earth impressions carry over into colors and shadings. Light shades and bluish colors are related to the sky or up, whereas dark shades and greenish or brownish colors are related to the ground or down.

- Things which are further away are expected to look smaller.

- Coolness is associated with blue and blue-green colors, warmth with yellows and reds.

- Very loud sounds or sounds repeated in rapid succession, and visual displays that move rapidly or are very bright, imply urgency and excitement.

- Very large objects or dark objects imply heaviness. Small objects or light-colored objects appear light in weight. Large, heavy objects are expected to be at the bottom. Small, light objects are expected at the top.

- People expect normal speech sounds to be in front of them and approximately head height.

- Seat heights are expected to be at a certain level when a person sits down.

Source: Woodson and Conover, Human Engineering Guide for Equipment Design.

10–13). The desirable features of shape-coded controls are:

1. Useful where illumination is low or where device may be identified and operated by feel only

2. Supplement to visual identification

3. Useful in standardizing controls for identi-fication purposes

Some undesirable features are:

1. Limited number of controls that can be identified

2. Use of glove reduces sensitivity of hand

- LOCATION. Controls can be identified by

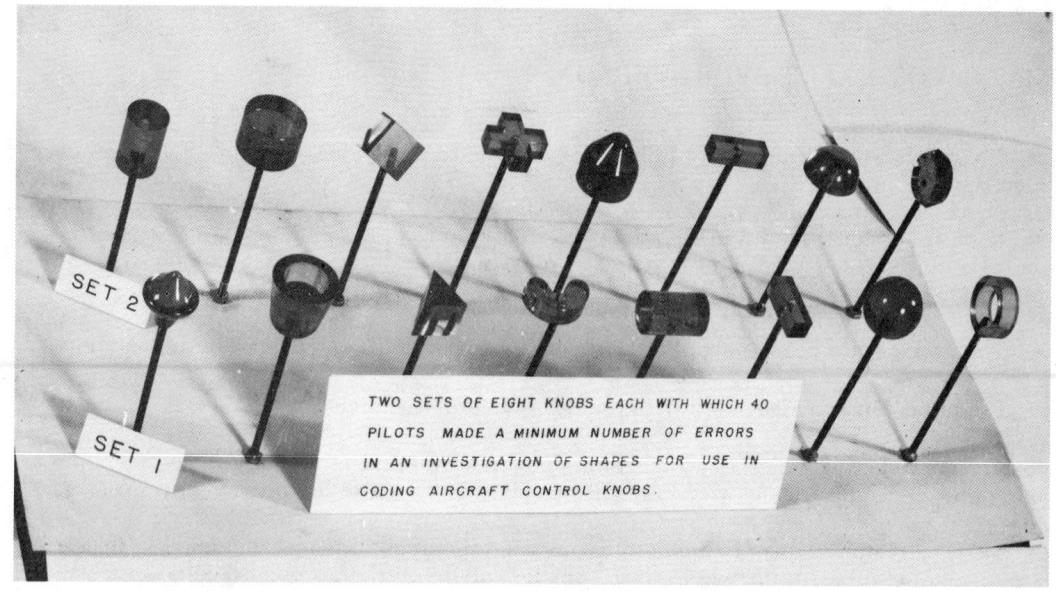

TWO SETS OF EIGHT KNOBS EACH WITH WHICH 40
PILOTS MADE A MINIMUM NUMBER OF ERRORS
IN AN INVESTIGATION OF SHAPES FOR USE IN
CODING AIRCRAFT CONTROL KNOBS.

SET 2

SET 1

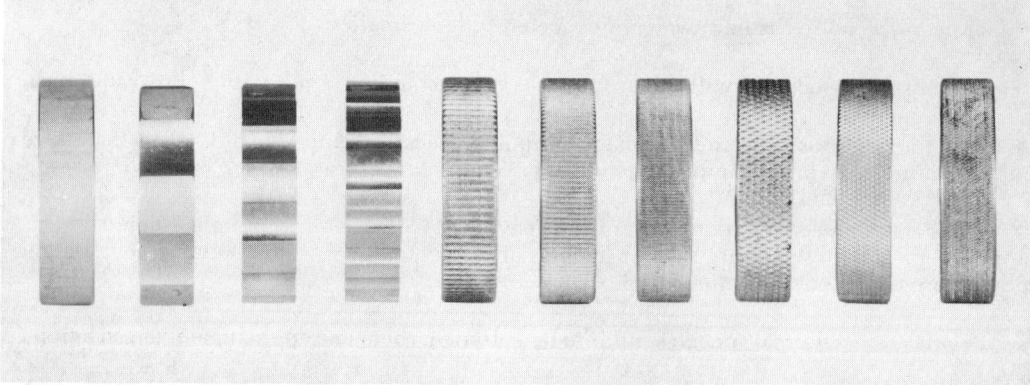

FIG. 10–13.—*Top:* Shape coding. Two sets of knobs for levers that are distinguishable by touch alone. *Bottom:* Texture coding. Ten texture codings that can be distinguished.

Courtesy Department of the Air Force, Wright-Patterson AFB. Top: W. O. Jenkins, "Psychological Research on Equipment Design," Research Report 19, 1947. Bottom: J. V. Bradley, "Tactual Coding of Cylindrical Knobs," Tech. Report 59–182, 1959.

their location. For example, all brakes on forklift trucks can be placed on the left side regardless of model. Location coding can also be achieved by providing a minimum distance between controls.

The advantages of coding by location are the same as those for shape and texture. Disadvantages include:

1. Limited number of controls that can be identified

2. Increased space requirements

3. Identification not as certain as with other types of coding

• COLOR. Color may also be used as a cod-

TABLE 10–C

COMPARISON OF CODING METHODS

Code	Maximum number of items *	Evaluation	Comment
Color	11	Good	Little space required Location time short
Numerals and letters	Unlimited for combinations of symbols	Good	Little space required if contrast and resolution is good Location time longer than for color
Geometric shapes	~15	Good	Little space required if resolution is good
Size	5	Fair	Considerable space required Location time longer than for color or shapes
Number of dots	6	Fair	Considerable space required Easily confused with other coded items
Orientation of line	12	Fair	For special purposes
Length of line	4	Fair	Will clutter display with many signals
Brightness	4	Poor	Poor contrast effects will reduce visibility of weaker signals
Flash rate	4	Poor	Interacts poorly with other codes
Stereoscopic depth	Unknown	Fair	Requires complex electronic displays and special viewing equipment

* That generally will give overall accuracies of 95 percent or better.

From C. T. Morgan, et al. Human Engineering Guide to Equipment Design. *New York, McGraw-Hill Book Co., 1963. Used with permission of the publisher.*

ing technique for various controls. Color codes can:

1. Be useful for visual identification

2. Be useful for standardizing controls for identification purposes

3. Offer a moderate number of coding categories

On the other hand the undesirable features associated with the use of color as a code are:

1. Controls must be viewed directly

2. Illumination cannot be poor or restricted

3. People must have adequate color vision

• OPERATION. Some controls make use of an operational method of coding; that is, the mode of operation will be different for different controls—for example, automobile windshield wiper controls turn clockwise to activate, while headlights must be pulled. The desirable features associated with such a system are:

1. Usually controls cannot be operated incorrectly

2. System designers can usually capitalize on compatible relationships

With such a system the following undesirable features are associated:

1. The control must be activated before operator knows if correct control has been selected

2. Specific design might have to incorporate incompatible relationships

Regardless of the type of coding used, all controls and displays should be labeled. Labeling is crucial where the operators change often or equipment is shared. The use of labels may also reduce operator training time.

Arrangement

Remember that a system is task-oriented and that its components act and interact with each other to perform this task. Consequently, the various elements and components of the system need to be arranged with these considerations in mind:

● FUNCTIONAL PRINCIPLE. This principle provides for the grouping of elements or components according to their function—those having related functions are grouped together.

● IMPORTANCE PRINCIPLE. Components can be arranged by their importance. Items of some type (displays, controls, components) should be grouped in terms of how critical they are in carrying out a set of operations. The important controls should be positioned in the best locations for rapid and easy use.

Relative importance, of course, is largely a matter of judgement. So, to apply this principle one must be in a position to obtain judgements of persons who are knowledgeable about the equipment. This can be done by either interview or questionnaire.

● OPTIMUM-LOCATION PRINCIPLE. This principle provides for the arrangement of items so that each one is in its "optimum" location in terms of some criterion of usage (convenience, accuracy, speed, strength to be applied, etc.).

● SEQUENCE-OF-USE PRINCIPLE. In using controls, sequences or patterns of relationship typically or frequently occur. In applying this principle, then, items can be so arranged as to take advantage of such patterns; thus, items used in sequence typically would be in close physical relationship with each other.

● FREQUENCY-OF-USE PRINCIPLE. To arrange items in terms of frequency of use, first obtain information about how often different items might be expected to be used. Then place the less frequently used items in more distant locations.

In the event there is conflict among principles some trading-off must be done. Although no one principle should be held rigorously, frequency of use and sequence of use should be given major consideration.

Seek to avoid arrangements on which frequent transfers (of the entire body, or of the eye, hand, or other body member) from place to place would be required.

Control evaluations

The following questions should be considered in assessing the human element in the design of controls:

● What bodily limbs are involved? Is any one muscle overloaded?

● Where are the controls placed? Can they be reached? Are they spaced far enough apart? Are they labeled and coded?

● What type of control is used? Is it compatible?

● Do the controls themselves present a hazard?

● Are similar control operations similar in design and function? How standardized are the controls?

Conclusions

Every organization is obliged to improve its safety performance where it can. The safety professional and his management are evading the issue if either ignores the smaller accident problems and continually harps upon the total accident problem that cannot be solved by a single action.

Most occupational safety counter-measures deal with a bit of the entire occupational safety problem. Slowly, but surely, the improvement that results from solving bits of the whole problem will be significant. If solutions to small bits of the total are ignored, then nothing will be accomplished.

Human factors engineering will help solve a bit of the whole problem. The lack of response to the human factors engineering approach is one of the outstanding failures in occupational and product safety efforts. Human factors engineering considerations have not been explored as at least a partial answer to safety problems. This neglect is partially due to lack of understanding the role of human factors engineering in occupational and product safety.

The examination of human factors engineering, combined with the traditional approach, clearly establishes its role in the safety movement. Among benefits that may be expected are:

1. Greater system effectiveness.

2. Fewer performance errors.

3. Fewer accidents resulting in injury or damage to property.

4. Minimizing redesign and retrofit after the system is operational—if applied at the design phase.

5. Reduced training time and cost.

6. More effective use of personnel with less restrictive selection requirements.

The role of human factors engineering will become more significant as systems become more and more complex and automated. The application of human factors engineering is indispensable as a basic consideration in the design of future systems if optimum safety and system effectiveness are to be achieved.

References

Bennett, E., et al. (eds.) *Human Factors in Technology.* New York, N.Y., McGraw-Hill Book Co., 1963.

Bryk, J. A., and Bresnahan, T. "The Hazard Association Values of Accident Prevention Signs." Paper presented at the meeting of the Human Factors Society, New York, 1971.

Chapanis, A., *Man-machine Engineering.* Belmont, Calif., Wadsworth, 1965.

Christensen, J. M. "An Overview of Human Factors Engineering," *National Safety Congress Transactions,* 1967.

Davis, H. L. (ed.) "Human Factors in Industry," *Human Factors* (Special Issue), Vol. 15, 1973.

Haddon, W. "Energy Damage and the Ten Counter-measure Strategies," *Human Factors,* Vol. 15, 1973.

Jones, D. F. *Human Factors—Occupational Safety.* Toronto, Ontario Department of Labour, 1969.

McCormick, E. J. *Human Factors Engineering.* New York, N.Y., McGraw-Hill Book Co., 1964.

McFarland, R. A., "Application of Human Factors Engineering to Safety Engineering Problems," *National Safety Congress Transactions,* 1967.

Meister, D., and Rabideau, G. F. *Human Factors Evaluation in System Development,* New York, N.Y., John Wiley, 1965.

Morgan, C. T., et al. *Human Engineering Guide to Equipment Design.* New York, N.Y., McGraw-Hill Book Co., 1963.

National Safety Council. "Safety Performance Measurement in Industry," *Journal of Safety Research* (Special Issue), Vol. 2, 1970.

Powell, P. I., et al. *2000 Accidents: A Shop Floor Study of Their Causes.* London, National Institute of Industrial Psychology, 1971.

10—Human Factors Engineering

Sinaiko, H. W. (ed.) *Selected Papers on Human Factors in the Design and Use of Control Systems.* New York, N.Y., Dover, 1961.

Surry, J. *Industrial Accident Research: A Human Engineering Appraisal.* Toronto, University of Toronto, Department of Industrial Engineering, 1969.

Tarrants, W. E. "The Role of Human Factors Engineering in the Control of Industrial Accidents," *Journal of the American Society of Safety Engineers,* Vol. 8, 1963.

Woodson, W. E. *Human Engineering Guide for Equipment Designers.* Berkeley, Calif., University of California Press, 1964.

Yoder, T. A., and Botzum, G. D. The Long-Day Short-Week in Shift Work: A Human Factors Study. Indianapolis, Ind., Eli Lilly and Company, 1971.

Human
Behavior
and Safety

Chapter
11

Psychological Factors in Safety   240

Individual Differences   241
Factors contributing to personality . . . The average-man fallacy . . .
Fitting the man to the job . . . Influences of differences on safety
. . . Methods of measuring characteristics

Motivation   244
Complexity of motivation . . . Job satisfaction . . . Management
theories of motivation . . . Theory X and Theory Y . . . Job
enrichment theory

Frustration and Conflict   250

Approach-approach . . . Avoidance-avoidance . . . Approach-
avoidance . . . Reaction to frustration . . . Emotion and frustration
. . . Non-directive counseling

Attitudes   254
Determination of attitudes. . . Changing attitudes . . . Structural
change

Learning   257
Motivational requirements . . . Principles of learning . . . Forgetting

Summary   261

References   262

11—Human Behavior and Safety

The function of a safety professional in industry is to assist line management in achieving maximum production by limiting the number of disabling accidents which occur as a result of the failure of some unit in the industrial process. This unit could be the equipment, materials, or methods used, the safety devices, or the individual worker. To do his job properly, the safety professional must make the worker conscious of the importance of safety, either as a goal in itself or as a means to some other goal. Consequently, a major portion of safety work lies in the realm of human relations.

This does not reduce the importance of the other facets of a sound safety program—the safety professional is also concerned with safe plant layouts, safety devices on machines and use of such devices by employees, the wearing of safe clothing and use of protective equipment—all of which contribute to the reduction of disabling accidents.

When, however, in spite of every precaution on the part of the manufacturer of equipment, the supplier of materials, the supervisor, and the safety professional, accidents occur, the human element emerges as an important factor. It takes working with supervisors, foremen, and other line management as well as individual employees if accidents are to be reduced.

This chapter of the Manual is designed to promote understanding of human behavior in the work environment. It complements the previous chapter, which emphasized designing equipment, controls, and jobs to fit the limitations of the human being.

Psychological Factors in Safety

To work effectively with people, one must first understand them. To gain such understanding, one must examine those individual personality factors that set one individual apart from all others. This exploration must find factors which are common to all and upon which the safety director and the supervisor can capitalize to promote safe conduct on the part of all workers.

The basic question is, "What forces in people can be utilized to make a safety program effective?" In seeking an answer, one must consider the individual as he is.

The person most likely to know and understand the employee should be the supervisor. The safety director should, therefore, make sure that every supervisor has a working knowledge of human relations and can use it effectively so that good patterns of concerted action can be developed.

Much of the success of a safety program depends upon its acceptance by those to whom it is directed. Program acceptance in turn is dependent upon an understanding of the psychological factors which influence program success.

- **Individual differences.** First to be considered are individual differences, one of the ever-existent problems within industry. These differences are seen constantly. Yet, within the framework of differences in people are factors that are common to all, and therefore useful in dealing with work groups.

- **Motivation.** Second, to understand the motivations of people is a must. To want something is motivation, but not to want something also requires motivation. To use a safety device to protect one's fingers from a saw is, perhaps, indicative of motivation for safe practices, but the desire to ignore a safety device because it might decrease production is also motivated. Conflicting motivations should also be considered in any attempt to understand human relations.

- **Emotion.** Humans frequently act at the emotional level. While emotions can be constructive at times, they can also be destructive—working to the detriment of the individual and the safety program. Emotion can so disrupt or confuse the thought process that an individual will do *not* what he knows he should do, but rather the opposite merely because of the way he feels.

- **Attitudes.** Industry has recognized the effect that attitudes can have on production, plant morale, turnover, absenteeism, plant safety, and the like. As a result, management has spent much time and money to determine the attitudes of workers. Measuring, developing, and changing attitudes constitute a major problem for personnel men and psychologists—one of extreme importance to the safety professional.

● **Learning processes.** Finally, there should be concern with learning processes. Learning has gone on from the first day of birth, and in all of the topics previously mentioned, it has played a major role. One cannot understand motivation, attitudes, emotions, or even individual differences without some consideration of the learning process involved in bringing them about.

Many topics with which the industrial psychologist is concerned cannot be covered in this chapter. Some of them play a direct part in the success or failure of sound personnel procedures in industry, but they are not all directly related to safety, or they do not fall within the assigned duties of the safety professional. They may, as part of the regular personnel procedure, contribute indirectly to safety in the shop.

Individual differences, motivation, emotions, attitudes, and learning are important in understanding accidents. Because the safety professional must work with them directly, they are important enough to discuss separately.

Individual Differences

When a chemist analyzes a chemical compound, he can accurately specify its exact nature and composition. When he uses this compound, he knows exactly what behavior he can expect. The action and reaction of this sample is usually the same as other samples of the compound he has analyzed.

When the psychologist studies human behavior, he is not dealing with the same degree of certitude as is the chemist. The psychologist does not know the composition of the agent he is dealing with—in many instances he has very little knowledge of his subject's past history.

In addition, the behavior of one person is not the same as the behavior of another person. Person A in the sample is not equivalent to person B, in the way 1 cubic centimeter of distilled water equals all other cubic centimeters of distilled water.

The known fact that people differ has been referred to as the "personal equation" or, more commonly, "individual differences." The personal equation presents many problems for the safety professional and supervisor.

The case is not all hopeless, for within the framework of individual differences, certain general patterns common to all men do exist.

For example, all behavior is motivated. Regardless of what the individual does, he does it for some purpose. Often, the purpose is a reduction of some basic underlying tension which must be resolved. The degree and nature of this tension depends in part upon his individual values and his internal perceptions. More about motivation later.

Here is another example. Even though each child develops at his own rate and to different levels of efficiency, each crawls before he walks, forms words before sentences, etc.

The general model used to describe such modes of development is the following:

$$B = f(S, E).$$

Behavior (B) is a function (f) of the present situation (S) and all previous experiences (E). Whether or not an employee works safely depends (a) upon his present situation—is he rushed? fatigued? in poor health?—and (b) his past experiences—did he avoid accidents in the past? What amount of training does he have? In our general model, the E variable could be termed the individual's personality.

Factors contributing to personality

The personality of the individual is his total make-up, the sum of all his traits and experiences. To a large extent, personality can be expressed as some function of heredity and environment. What an individual inherits from his parents will set the upper limits of the individual's personality. For example, a person can have no more native intelligence than that which he inherited from his parents. However, whether or not he ever reaches this upper limit depends upon his environment and the interaction of other traits. The individual who inherits a great potential intelligence, but is unstimulated, may never reach his potential. On the other hand, the child who does not inherit a great intellect, but is extremely studious, may surpass the former in terms of IQ.

The average-man fallacy

In the previous paragraphs, the point was stressed that each person differs from each

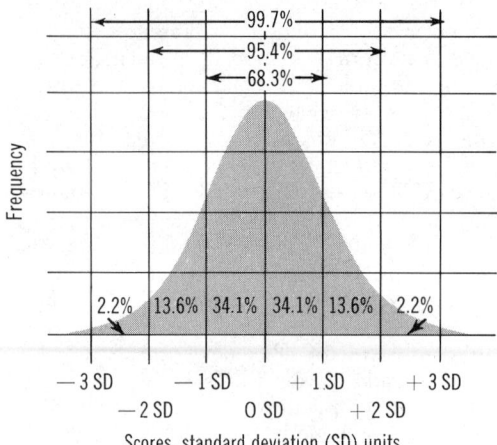

FIG. 11–1.—The distribution of scores in a normal curve. Since the normal curve has a known shape, it is possible to state the percentage of scores that lie between +1 and −1 standard deviation, or between any other two points expressed in SD units. This normal curve might represent the distribution of IQ scores in the United States; in this case, the mean would be 100 and the standard deviation would be approximately 15. Given the mean and standard deviation, therefore, the complete distribution of scores in a normal distribution can be known.

other person (although there are certain common characteristics). The fact that individual differences exist should not be news. What is important is how one deals with individual differences.

Fig. 11–1 presents the distribution of scores in a normal curve. Many human traits (for example IQ) are assumed to be distributed according to the normal curve distribution. Note that in such a distribution, half of the persons are below the mean (the arithmetical average) and half of the persons are above the mean. While the trait is relatively systematic and the extremes are relatively rare, nevertheless only a portion of the population has an average amount of the trait.

Often the manager of an enterprise or director of a safety program realizes that individual differences exist and that it is physically impossible to handle every one individually. Therefore, he will do what he feels is the next best thing—appeal to the average-man. There is, unfortunately, no such thing as an average-man. By definition, average means that a measured trait is possessed by 50 percent and absent in the remaining 50 percent of those measured. Therefore, when an appeal is aimed at the average, by definition it misses much of the population.

A better approach is the use of percentiles. When designing the system, instead of setting the standards to fit the average, design it to fit all but the upper 5 percent and the lower 5 percent. Then the system will fit 90 percent of the population.

The use of percentiles, rather than averages, has long been the technique for relating individual test scores to a group as a whole, and also a guiding rule in the design of man-machine relations.

Fitting the man to the job

Very often when managers are replacing or relocating personnel, they strive to find the individual possessing most of the characteristics deemed necessary. Job specifications and employee requirements are given in terms of a minimum, but very seldom are psychological factors expressed within a minimum *and* maximum.

When looking for someone, often the best qualified is considered to be the one who is most intelligent, most loyal, biggest, or whatever. In other words, they look for the perfect rather than the right person. However, a person with a below-average IQ might be more desirable for a monotonous job than will a person with an average or above-average IQ. The amount of boredom, monotony, etc., may be less for the less bright, thereby eliminating carelessness.

The same is true for loyalty and devotion. The person who comes to work under any condition, whether ill or not, may be as undesirable as the one who uses every excuse to take off. The worker with a high fever, bad cold, or sore back is a potential hazard, especially if he is under heavy medication.

Influences of differences on safety

The problem of individual differences has been related directly to the problem of safety in industry. For some years, safety professionals and psychologists alike explored the concept of accident proneness. It was believed for a time that herein was the final solution to the whole problem. Such has not been the case, and this concept has lately received less attention. This fact does not, however, minimize the importance of the relationship between individual differences—both physical and psychological—and industrial safety.

See Chapter 9, "Safety Training," for more on accident proneness.

Physical characteristics of the individual. Reaction time, manual dexterity, and visual abilities seem to have at least some bearing on safe performance. While the extent to which they are directly responsible for accidents is neither clear nor constant, it appears that a certain minimum degree of physical competence is required for successful, accident-free performance. Some types of jobs demand superior physical abilities while others do not. It is also possible to redesign a job to allow a person with certain physical handicaps to perform efficiently and safely.

Most of the research to date has considered these factors individually. Recently a few studies have examined the connection between combinations of physical shortcomings and accidents. Apparently such studies were initiated because investigators realized that physical and psychological abilities operate not as single, discrete items, but rather in interacting combination. This research is concerned, for example, with general perceptive capabilities and accidents, and not with just the separate abilities that contribute to perception.

Individual personality. (Accident proneness) The other explanation for accident proneness was that it stemmed from the basic personality structure of the individual, from two characteristics in particular: impulsiveness and a desire to escape from authority. These two characteristics are determined not by heredity, but rather by interaction of the individual with society.

If this latter explanation for accident proneness were complete, one would need then only to measure personality differences through tests or inventories, disregarding physical characteristics entirely. However, research has given little evidence to support

this concept, so that only one conclusion is acceptable at this time—that a minor share of accidents can be attributed to personality.

It is the function of personnel people in industry to screen candidates on the basis of all the characteristics required by the specific job for which application is made. In many cases, both physical characteristics, such as size, visual acuity, and steadiness, and personality characteristics are important, so that many individuals will not be considered because they do not have all the necessary physical qualifications. As a result, many who might be accident prone do not even get a chance to work on the job.

In the design of equipment, human engineering experts take into consideration physical limitations as well as other human characteristics in an effort to make machines as nearly perfect as possible. See Chapter 10 for a discussion of human factors engineering in relation to occupational safety.

Where hazards cannot be eliminated, guarding provides protection against potential failure of people to utilize equipment correctly. Both safe design and guarding minimize the effect of individual differences on accident frequency and severity.

Methods of measuring characteristics

Regardless of what technique is used to screen, place, and motivate employees, a method of checking program effectiveness is necessary. Techniques used to obtain feedback range from the across company accident rates, to the within-company approach of safety sampling, or critical incident technique, described in Chapter 4, "Inspection and Control Procedures."

All measuring techniques must possess the qualities of reliability and validity.

Reliability of measurement refers to the degree to which the test or instrument produces the same results over repeated uses. Reliability is simply the degree of correlation the instrument has with itself.

Reliability is expressed as a coefficient with a range of −1.00 to +1.00. The sign of the coefficient tells the nature of the relationship. A negative sign means that as one dimension is increased, the other dimension measured decreases; a positive value means that as one dimension is increased, the other dimension also increases. Because the number deals with the magnitude of relationship, a reliability of $-.60$ is greater in magnitude than a coefficient of $+.45$, although the variables are inversely related in the former.

Although it is theoretically possible to have a reliability coefficient as low as -1.00, in practice the value rarely goes below $+.20$. How high can a coefficient be? This is difficult to answer, since the answer depends upon the use of the testing instrument, the nature and manner in which reliability was obtained, and other factors.

Validity refers to the test's or instrument's power or ability to measure what it is intended to measure—the degree of correlation the instrument has with the external criterion it is supposed to measure. Once again validity is expressed as a correlation coefficient. A negative sign, however, is as useful as a positive coefficient. A negative validity indicates that those possessing more of the measured trait are less desirable. An illustration might be testing for a job that requires very little mental ability—it may be negatively correlated with an IQ test so that the desirable workers for the job are those with the lower test scores.

A measurement may be reliable without being valid, but a valid measurement must also be reliable. Because reliability is self correlation and validity is a correlation with some external criteria, a test cannot be correlated with something else and not correlate with itself. To illustrate that a reliable measure may not also be valid, consider a yardstick. A yardstick is very reliable—it gives a consistent measurement every time it is used. It is also valid for measuring the length of a table. But if a yardstick is used for weighing the table, it is no longer valid for measuring.

In addition to reliability and validity, a measuring technique must be practical. A technique may possess high validity but be so cumbersome and intricate that it can only be used in special situations, and then only by highly skilled technicians. In spite of its statistical value, such a technique is almost worthless.

Two sampling techniques used for evaluating potential accident-producing behavior are (*a*) the critical incident technique, described in Chapter 4, and (*b*) behavior sampling.

The behavior sampling or activity sampling technique involves the observation of worker behaviors at random intervals and the instantaneous classification of these behaviors according to whether they are safe or unsafe. Calculations are then made to determine either (*a*) the percent of time the workers are involved in unsafe acts or (*b*) the percent of workers involved in unsafe acts during the observation period. Various components of a safety program (i.e., safety lectures, posters, 5-minute safety talks, safety inspections, motion picture films, supervisory training) can be applied and an immediate indication of their influence on unsafe behavior can be obtained.

It is important that the safety professional approach any psychological evaluation with caution. Texts and references in industrial, social, and personnel psychology will provide guidelines for the development and use of test batteries for measuring individual differences. These tests or evaluative instruments are best developed and administered by competent professionals who specialize in this work.

With respect to the problem of screening employees or potential employees in regard to their accident potential, caution must be exercised. To date, no systematic screening procedures have been developed which meet both reliability and validity criteria and which are adequate for use in all industries or even in a specific industry. While in theory such a screening procedure is possible, the state of scientific knowledge in occupational safety research is too limited to support the development of such screening measures.

Motivation

Through the interaction of hereditary and environmental factors, each worker is an individual personality. The safety professional must be continually aware of individualities when dealing with human beings.

There are instruments by which various aspects of human behavior can be evaluated, many of which are already in use in industry.

Through psychological tests, interviews, rating scales, and allied aids and techniques, personnel departments have for a long time been evaluating individual differences. Insofar as they are doing adequate jobs, personnel departments are working with safety programs to eliminate job candidates who obviously would be unsatisfactory.

Each day millions of men and women work in the manufacture and distribution of industrial products. It hardly seems logical that if all these people were individual personalities they could work together in harmony. Individual differences alone are not all there is to human behavior. There are some factors operating in all people which allow supervisors, safety professionals, and plant managers to obtain work and cooperation for a common cause. The psychologists, when they try to predict and control human behavior, are also concerned with these factors that all individuals have in common.

To have all personnel in a company from the president down to the lowest-paid employees working together productively and safely is one of the goals of a safety program. Such cooperative effort must be motivated as an appeal to achieve a common goal, or as a means to another goal which is of greater importance to the individual. In either case, the result is of value to the safety program.

It is understandable, therefore, that the question most asked by safety professionals and supervisors alike is, "How do we best motivate our workers?"

To answer this question, one must first be familiar with motivational theory.

Complexity of motivation

The motivational problem is perhaps the most complex one in the field of human behavior. It is not possible at the present stage of development to give clear-cut, concise answers to all the questions that might be asked about other people's motivations. Rather, attempts are made to set forth some basic factors and point out where the complexities exist. Hierarchical motivation, multiple motivation, continuing psychosocial need, and conflicting motives must be considered.

Hierarchical motivation means simply that some needs take a higher priority—there is a hierarchy of motivational factors. It has been pointed out that the psychosocial needs (for example, recognition, affection, social approval) take precedence over the biological ones (for example, hunger, thirst, sex) when the latter are relatively well satisfied. This is but one aspect of the hierarchy of needs concept.

The safety professional who plans carefully and takes each detail of his program to his boss for approval may be exhibiting an overwhelming desire for achievement and recognition, a desire much stronger than his need for affection. This is true particularly if he forgets his human relations skills in dealing with workers. He may be so impressed with his own personal achievements, and perhaps the recognition given for them, that he forces plans upon workers which they would not accept from free choice.

Another example is the worker who considers recognition as the major need that he must satisfy. This individual perhaps would go all out to become a strong militant leader of the union if he were to lose a promotion in the company. The safety director may himself be in conflict with this individual who may not care about the safety director's disapproval and instead values the affection and recognition of his fellow workers. This hierarchy changes over the years.

What is desired more early in life may assume less importance later. The need for achievement and recognition perhaps is greatest in youth, thus giving a drive toward accomplishment. Later in life, perhaps, the warm affection of friends or the security of belonging to groups may assume major importance if one has already had recognition for past work.

Multiple motivation is the second facet that complicates the analysis of behavior. People are seldom motivated by just one need—many forces operate at any single moment. For example at lunch time, one may eat because he is in need of food, or just because it is the usual time. However, an individual might delay his lunch thirty minutes so that he can join several friends and have lunch with them. Thus, he combines the satisfaction of his need to belong and to eat.

In another case an employee could desire

recognition from his fellow workers and so might engage in practical jokes and harmful horseplay. At the same time, he may be seeking recognition from his foreman by working hard on the safety committee. The need is the same but the means of satisfaction are in direct conflict.

This is multiple motivation. It would be to the safety professional's advantage, in the second case, to recognize what the need is that the worker is trying to satisfy, and find ways to channel the behavior so he can achieve recognition from fellow employees and the foreman.

Continuing psychosocial need is akin to biological need. However, people often assume that in the psychosocial needs, satisfaction at one time should suffice for the future and thus they need not be concerned any longer with giving recognition, or affection, or social approval.

For all people, these needs continue throughout life. Satisfaction is always sought for these needs, but not always attained. It should be apparent also that the satisfactions sought by individuals will not necessarily be for the same needs. At one time, the motivation may be for social approval while at another time the need most requiring satisfaction might be affection. Thus, in dealing with people, one must recognize that what was effective yesterday may not work today, although the satisfaction-seeking behavior is to some degree similar.

Being aware that these needs continually want satisfaction makes it somewhat easier for the safety professional or anyone else to deal effectively with people in the work situation. There may be variance from day to day so the safety professional must therefore cultivate his own ability, and the ability of supervisors, to work with people so that he can sense which needs require satisfaction at a given time.

Since they spend so much time together and with workers, the safety professional and foreman (or supervisor) come to know each other and the workers very well. This should provide the key to determining what a worker's behavior at the moment means in terms of need satisfaction. This key is based on little cues in the individual's behavior which they have learned over the years to recognize, if they have taken the time and made the effort to know their people.

Careful observation is required. Safety professionals should make certain that any foreman training program includes methods of effective observation of workers.

Conflicting motives constitute another major problem in motivation—needs themselves can be in conflict with one another. Seeking affection could lead to behavior which might be different with fellow workers than with supervisors. It may be necessary to determine which source of affection is the more important before doing anything.

People can internalize these problems to such a degree that the resultant physical stress, worry, anxious periods, or even behavior is completely inconsistent with that expected of them normally. For example, a young man who is assistant safety director in a given company may strongly desire to become the safety director and yet at the same time be fearful of the duties, the responsibilities, and obligations of the job. The promotion would mean more prestige, more money —in all, a better way of life. Here then is a serious conflict. The young man's answer will depend greatly on his background experiences. One of the things he might do would be to quit his job and seek employment elsewhere. He might, on the other hand, seek training to better qualify for the position. These are but two of the many choices available.

The same applies to the worker on the line. He may strongly desire the approval of his supervisor and at the same time desire to remain an accepted member of his work group. If the work group makes fun of or minimizes the importance of the safety program, this worker now is in conflict. He may follow the group or he may seek the approval of the management. He may do one or the other, or he may do something entirely inconsistent with either.

Some conflicts can be solved rather simply, for the alternatives lead to positive need satisfactions regardless of which way the individual goes. These are really no problem. Others may have at one pole a positive satisfaction and at the other a negative or un-

TABLE 11-A

SUMMARY OF DIFFERENT SURVEYS ON JOB SATISFACTION
In Order of Importance of Different Factors

	Women Factory Workers	Union Workers	Nonunion Workers	Men	Women	Employees of Five Factories
Steady work	1	1	1	1	3	1
Type of work				3	1	3
Opportunity for advancement	5	4	4	2	2	4
Good working companions	4			4	5	
High pay	6	2½	2	5½	8	2
Good boss	3	5½	5	5½	4	6
Comfortable working conditions	2	2½	3	8½	6	7
Benefits		5½	6	8½	9	5
Opportunity to learn a job	8					
Good hours	9	7½	7	7	7	
Opportunity to use one's ideas	7	7½	8			
Easy work	10					

wanted result. This is no problem either for the obvious choice is the one which is satisfying to the individual. Those choices that *cause* the problems are those that put an individual in a dilemma.

The solution depends upon how the individual has learned to work out such situations. Whether he runs away or faces up to the problem will give the safety professional or supervisor a clue as to what will occur in the future in a similar situation.

Job satisfaction

In the interest of further understanding of motivation in the work environment, many studies have been conducted to determine what constitutes job satisfaction in the workforce. Generally, these studies have sought to assess what workers claim to be the elements of their job that contribute to their satisfaction (or dissatisfaction) with their jobs. The results of these investigations suggest that the satisfaction of psychosocial needs rather than physiological needs may be the major motivational aspect of job satisfaction.

Table 11-A presents the results of a number of different surveys of job satisfaction. The numbers represent the rankings of the factors which were considered in each study. While different language and alternatives were used in each survey, the factors have been paraphrased to represent the elements covered in the surveys.

The results of these surveys suggest that high pay is not in itself a primary job motivator. Although workers expect a just and equitable income, they appear to expect only what others would be paid for comparable work. While the worker might feel dissatisfied if he were underpaid, higher pay alone does not guarantee job satisfaction.

On the other hand, steady work or job security does appear to be a primary job motivator. Workers want the security of knowing that on the basis of their performing their job well, they will have a job in the future. Job security, as a component of job satisfaction, may explain the willingness of a worker to maintain a low-paying, stable job instead of accepting a higher-paying, less-stable job.

Other factors which appear important as job satisfiers include—type of work, opportunities for advancement, and good working companions. Note that all of these factors seem related to the psychosocial needs of feeling important and belonging to a peer group which is acceptable to the worker. Likewise, comfortable working conditions (rated high by a number of employees) are probably associated with a desire of the worker to be treated humanely by the employer.

The results of these numerous surveys on job satisfaction are important when considering the safety program within the context of personnel policy. Inasmuch as the safety program is designed to ensure the well-being of the employee, it helps to maintain the employee's continued ability to do the work (which, in turn, gives job security).

Likewise, the safety program represents management's interest in the working companions and working conditions of the employee. All of these aspects of safety programming should be anticipated and incorporated into the approach which is taken with both supervisory staff and the employees. Honest and sincere positioning of the safety program within the context of the employee's welfare makes practical sense in light of our current knowledge regarding job satisfaction.

Management theories of motivation

The literature on the psychology of human motivation abounds with theories which seek to provide a unifying conceptualization of how people are motivated. Within this literature, specific theories have evolved with special reference to management as it exists in industrial organizations. Two such theories are presented here, although other equally cogent points of view could be discussed.

Because theories of human motivation lack sufficient data to support all their tenets, they might best be viewed as philosophies of management. They are important to the safety professional because, if accepted, they can influence the direction in which management seeks to develop and implement a safety program.

Theory X and Theory Y

In an attempt to analyze how management personnel view human motivation, McGregor (see References) has evolved the notion that there are two basic ways in which management can view the worker. According to which view management accepts, essentially different management practices will be observed.

Theory X, according to McGregor, assumes that the worker is essentially uninterested and unmotivated to work. In order to resolve this condition, the motivation must be instilled into the worker by the adoption of a variety of external motivation agents. In effect, the worker becomes motivated to work by virtue of the external rewards and punishments which are offered to him.

For example, in order to create motivation, a Theory X manager might use any and all of the following—introduction of rules to constrict the worker's behavior, pay incentives based on production, and threats to job security associated with performance failure.

Thus, under Theory X policy, management uses control and direction as the means of worker motivation.

Theory Y, according to McGregor, assumes that the worker is basically interested and motivated to work. In fact, work is assumed to be as natural and desirable as other forms of human activity, such as sleep and recreation. Under such circumstances, management is confronted with the role of organizing work so that the worker's job coincides with the goals and objectives of the organization. Thus, a Theory Y manager views his task as constructively using the worker's self-control and self-direction as the instrumentality for accomplishing the work to be done.

By emphasizing responsibility and goal orientation, management capitalizes upon the inherent motivation already present within the worker. If conflicts occur between the worker's goals and management's goals, they are resolved through mutual exploration and discussion. Always, under Theory Y policy,

Hygiene Approach	Job-Enrichment Approach
Company policies and administration	Achievement
Supervision	Recognition
Working conditions	Work itself
Interpersonal relations	Responsibility
Money, status, security	Professional growth

FIG. 11–2.—Contrast of the hygiene approach to motivation and the job-enrichment approach.

it is assumed that the worker's inherent motivation is essential to the completion of the organization's goals.

Both Theory X and Theory Y proponents exist, and apparently management systems operating on the basis of each of these theories can be found throughout American industry. What seems important is that whichever system is operating within an organization, it is necessary to recognize that safety programming can be initiated and implemented. While the technique of implementation may differ, Theory X and Theory Y approaches to human motivation can both be amenable to enhancing a worker's motivation to safe behavior.

Job-enrichment theory

Another analysis of human motivation in occupational environments has been developed by Herzberg (see References). While being quite comparable to the Theory X and Theory Y distinction, Herzberg has been explicit in both the detail and philosophy which he has evolved. His concept of job enrichment, in many ways an extension of Theory Y, is a major current force in management theory.

The classic approach to motivation concerns itself with changing the environment in which a person works—the circumstances that surround him while he works (good or poor lighting, an agreeable or offensive supervisor), the incentives he is given in exchange for his work (money, a pat on the back, etc.).

Herzberg believes that the concern for environment is important—but not all important. He says that it is not sufficient in itself for effective motivation. That, he contends, requires experiences that are inherent in the work itself.

Herzberg holds that there is no conflict between the classic (environmental) approach to motivation and his approach to motivation through work itself. He regards both as important. The classic approach is called hygiene whereas Herzberg's approach is called job enrichment.

The hygiene approach may be understood by the following analogy—a person is provided with pure drinking water and waste disposal; both are necessary to keep a man healthy, but neither makes him any healthier. By extension, environmental factors always need replenishment. Good human relations, for an example, are important to keep a man happy, *but* no one can love a man into increased creativity or safer working habits.

Fig. 11–2 presents a contrast of the classic hygiene approach to motivation and the job-enrichment approach.

Further, treating a man better does not enrich his job, although he would be unhappy if not treated well. Again, a salary increase may keep a man from becoming dissatisfied for a time, but sooner or later another increase will be required.

Although there may be inherent hazards in a work environment, such as coal mining or bridge building, the worker has a right to expect controls to prevent the environment from becoming unreasonably unsafe, for example testing and removal of explosive gases, or safety nets and life lines. Such protection might not motivate him because it makes the job safer, but he might be very unhappy if he knew no effort was made to protect him.

Herzberg's idea that work itself can be a motivator represents an important behavioral science breakthrough. Traditionally, work has been regarded as an unpleasant necessity but it has not been thought of as a potential motivator.

Accordingly, the potential motivating power of work is obscured because most jobs are not at all stimulating, and, therefore, some kind of external pressure (either positive or negative) has to be applied if people are to do the jobs.

Although automation is helping to phase out the unstimulating aspect of many jobs, a job should provide an opportunity for personal satisfaction or growth. When it does, it becomes a powerful motivating force.

People, Herzberg further theorizes, must be given the opportunity to do work that they think is meaningful. Merely telling a man who is doing a routine job that he is happy and that he is doing something meaningful accomplishes nothing. But job rotation is not the answer, either; it does not enrich a job—it only makes it bigger.

Another point. Herzberg observes, "Resurrection is more difficult than giving birth." Obsolescence must, therefore, be eliminated by continued retraining—not just a once-in-a-while effort. Jobs should be kept up-to-date, and people doing the jobs must be kept up-to-date.

Even though a company may provide the hygiene factors, they must also provide a task that has challenge, meaning, and significance. If an unchallenged man does not quit, he stays on—but with poor morale. That, says Herzberg, is the price a company pays for not motivating people.

In summary, seven principles of job enrichment can be itemized.

• Organize the job to give each worker a complete natural unit of work

• Provide new and more difficult tasks to each worker

• Allow the worker to perform specialized tasks in order to provide a unique contribution

• Increase the authority of the worker in his job

• Eliminate unneeded controls on the worker while maintaining accountability

• Require increased accountability of the worker for his own work

• Provide direct feedback through periodic reports to the worker himself

Frustration and Conflict

A motivated individual is one who is attempting to reach a goal. Often, however, a barrier is placed between the goal and the one who is seeking it. When this occurs, the

individual becomes frustrated. This differs from mere *lack of satisfaction* of a need, which is called deprivation. The *thwarting of behavior* directed toward a goal results in frustration.

The barrier may come from within the in-

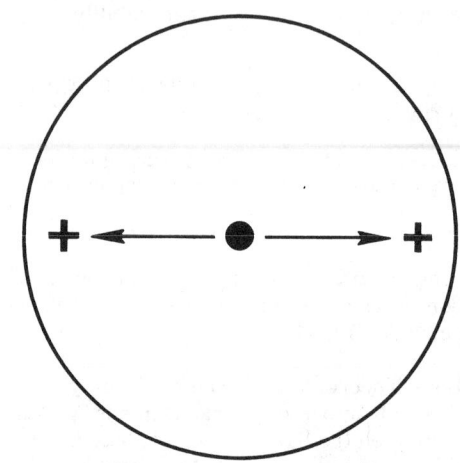

FIG. 11–3.—Approach-approach conflict. The individual is attracted to two equally appealing goals.

dividual himself. The person who sets impossible goals for himself may become frustrated when he is unable to reach the goals because he cannot work that fast. The "way out" may be an accident.

The barrier may also arise from the environment. An example—if the person mentioned above (who set a very high production output goal for himself) is unable to reach such limits because of faulty equipment.

A third type of frustration is caused by conflict. If two motives somehow conflict, the satisfaction of one means the frustration of the other. For example, the worker who sets high production and safety standards for himself, but is unable to meet both under the present system, must satisfy one and sacrifice the other.

It is to the conflict-caused frustration that we will now address ourselves. Basically, there are three types of conflicts—called "approach-approach," "avoidance-avoidance," and "approach-avoidance."

Approach-approach

As the label implies the "approach-approach" conflict arises when an individual is faced with two goals which are equally attractive, but only one of which is obtainable at the time (Fig. 11–3). An example is the young girl who has been asked to go both to a dance and to a swimming party on the same evening.

The approach-approach conflict is the easiest to resolve. No matter what goal the individual selects, a need will be satisfied, without much loss to another need. Usually the individual will resolve such a conflict by satisfying one need first and then satisfying the other. If the person is both hungry and sleepy, for example, he might eat first and then retire.

The approach-approach conflict does not present a serious problem to the safety professional.

Avoidance-avoidance

A second type of conflict is the "avoidance-avoidance" or double-negative conflict (Fig. 11–4). This conflict may arise when an em-

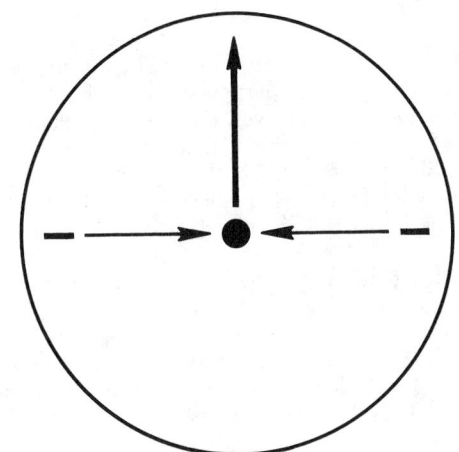

Fig. 11–4.—Avoidance-avoidance conflict. The individual is repelled by two equally negative goals.

ployee is told to wear heavy fire-retardant outer clothing in the summer heat or else risk the possibility of catching his clothes on fire. This employee is, as the saying goes, caught

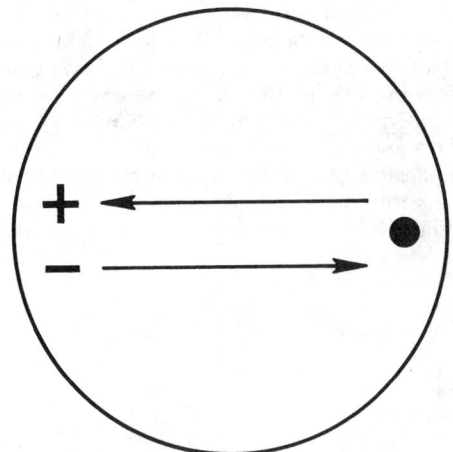

Fig. 11–5.—Approach-avoidance conflict. A single goal has both positive and negative aspects.

"between the devil and the deep blue sea."

Two kinds of behavior generally result from such a conflict—vacillation and flight.

In vacillation, the individual approaches one goal; retreats, approaches the other, retreats, and so on. As he approaches a goal, the unpleasant portions of the goal increase, so he withdraws. In the example, the worker may stall off doing the job requiring wearing the heavy clothing until the quitting bell sounds. And then, of course, it is too late for that day, and he can agonize over the decision the next morning.

Another result from the avoidance-avoidance conflict is flight. The athlete may leave the field or the worker may quit his job. This of course has serious consequences in lost prestige or lost income.

A more common type of fleeing, therefore, is in a figurative sense—like daydreaming, fantasy, etc. The worker faced with an unpleasant task may repress the reality by daydreaming.

The avoidance-avoidance conflict is not as easy to resolve as the approach-approach, nor is it as difficult as this next conflict type.

Approach-avoidance

The "approach-avoidance" conflict is the most difficult to resolve. The individual is both attracted and repelled by the same goal (Fig. 11–5). The worker who is striving for top production may be forced to take unnecessary risks to achieve that goal. The only way out of such a conflict is the erasing or weakening one of the signs, thereby altering the individual's internal motivational system.

The approach-avoidance conflict is further complicated by the notion of a goal gradient for positive and negative goals, as pictured in Fig. 11-6. Other things being equal, the

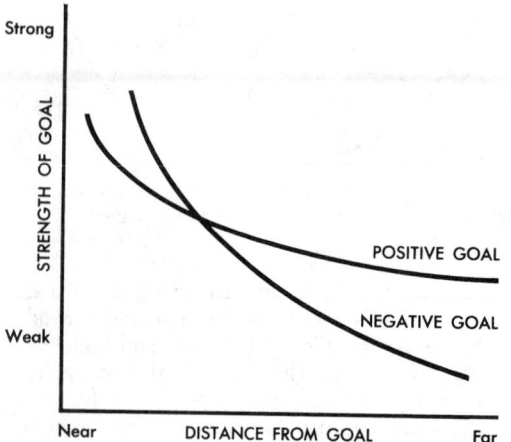

FIG. 11-6.—The approach-avoidance conflict is complicated by the different goal gradients for positive and negative goals.

closer one is to a goal, the stronger it is. Negative goals, however, are stronger than positive ones, the closer one is to the goal. Conversely, positive goals appear stronger than negative ones the farther away one is from the goal. So as the individual approaches an approach-avoidance goal, the negative or repelling aspects of the goal appear stronger. This may lead to retreat which will result in vacillation and indecision. The approach-avoidance situation is common.

Reaction to frustration

In most cases, the conflicts described will lead to frustration. This frustration may often lead to positive, constructive resolution of the situation. For example, the individual who is facing the conflict of top production, but at increased risk, may develop a new system for processing the product at a faster *and* safer rate.

On the other hand the frustration often leads to some negative form of behavior, a few examples of which are listed. This is by no means an exhaustive list, nor are they mutually exclusive; one may appear in combination with another.

Aggression. This form of behavior is characterized by some type of attack toward another person or object. The four-year-old boy who beats his playmate over the head with a baseball bat is obviously exhibiting aggressive behavior. The forms of aggression of adults can be overt, physical, or verbal. They can also be disguised; the tidbit of shop gossip, or the subtle comment by an aspirant for a promotion about one of his possible competitors, could be just as devastating to the victim as the baseball bat.

Regression. Regression is the tendency for an individual who finds himself in a frustrating circumstance to revert to an earlier form of behavior, such as putting on a temper tantrum or pouting. Such behavior most likely would occur in the case of an individual who, during his formative years, found that such behavior worked; in other words, he got his way by a display of temper.

Fixation. In fixated behavior, the individual persists in a particular kind of nonconstructive behavior even though it is clear that the behavior is inadequate to resolve the problem. Thus, a mechanic may persist in trying to fit a bolt in place, even though he realizes that it is the wrong size.

Resignation. Resignation is the tendency for an individual to give up—to withhold any sense of emotional or personal involvement in the situation. Failing to achieve some goal, this person loses any positive concern about his job and adopts an apathetic attitude toward the situation.

Negativism. In this form of behavior, the individual adopts a negative, resistive position with respect to the situation. A person whose suggestions have not been accepted may take a dim view of any other ideas.

Repression. This behavior is characterized

by blocking out from consciousness those cognitive associations that are disturbing. It is an unrealistic form of behavior because it implies that the problem will simply go away if one does not think about it.

Withdrawal. In this type of behavior, the frustrated individual simply removes himself from the situation in question—either physically or psychologically. A person who is not able to cope with a business adversary may avoid situations which would put him in contact with the individual. An individual who is the butt of jibes and jokes may become a loner.

Emotion and frustration

Emotions can have a disturbing influence upon a person's behavior. Anxiety is one emotional reaction which, because of its distracting influence, its stimulus to heightened reaction, and its generally upsetting effect, can make an individual more susceptible to accidents. This general upsetting feeling can be spread to others working in the same situation and create an atmosphere not conducive to safe procedures.

Anxiety, however, is but one kind of emotional pattern shown by individuals in the face of frustration. The anxious individual is worried, circular in his thinking, and fearful in such a way as to make his behavior inadequate to reach his goal. Even though he himself may realize that his behavior is getting him nowhere, he is unable to find a method that will solve his problem.

In frustration, some people become angry, some fearful, and some accept frustration as a challenge and attempt to solve the problem. Most people who react emotionally to frustration and threat find it extremely difficult to cope with life situations. It is not necessary to discuss the physiological pattern which develops in emotion for everyone is familiar with the feelings experienced during fear, anxiety, anger, shame, and other emotions. Rather, one needs to recognize that over the years people have developed accepted social expressions for emotions. To some degree, one individual can be aware of how another is feeling. This is not to say that one cannot be fooled or that the outward expression always truly indicates the feelings of the individual. Many people learn to mask their real feelings.

However, after day-to-day interaction over a long period of time, one may learn to tell the difference between true feelings and those shown. This is useful to the foreman and the safety professional in his dealings with individuals in the shop. As they go about the plant carrying out their normal duties, they may discover those individuals having difficulties and take some extra precautions to avoid accidents.

It seems important to make note of the problem of behavior disorders within the context of employee problems. Individuals whose normal behavior is seriously disrupted or whose ability to react to frustration is seriously impaired may require professional services beyond the scope of the normal supervisory-employee relationship. Although a supervisor should assist in directing the employee to in-plant services or outside agencies, he should avoid attempting to deal directly with psychological problems requiring professional care. While the current movement toward company-sponsored psychological services may grow, it is unlikely to be the standard in all industries in the near future. For a detailed discussion of industry programs for drug abuse and alcoholism, see Chapter 21, "Occupational Health Services."

Non-directive counseling

Although no attempt can be made here to train safety professionals or supervisors to become personnel counselors, they can do much to help reduce the immediate result of strong emotion. The supervisor is in the best position to evaluate the emotional level of an individual worker and to do something about it.

Everyone is familiar with the emotional outbursts of people who have had a trying and upsetting experience. Such expression of emotion may take many forms from an outburst of invective, a torrent of tears, to physical assault on inanimate objects which cannot strike back. The end result is the same—relief from the intolerable tension, and emotional relaxation.

In order to deal with a strong emotional response, the supervisor can engage in a type of client-centered counseling in which the employee is given an opportunity to talk about whatever is bothering him. The freedom to say what the individual pleases, without the evaluation of the supervisor, can reduce tension and open communication channels.

The stresses which cause emotional upset can occur within the work environment or away from it, but each kind has its effect on the other. A simple statement, such as "How are you today?" or "What's bothering you?," may well be sufficient to set off the verbalization of pent-up emotion. After the discussion, the employee can often evaluate more clearly and realistically the things that upset him in the first place.

To attack the problem at the roots calls for permissive listening on the part of the employee's immediate supervisor. In his normal routine every supervisor has an opportunity to talk with individual employees. Through the use of open-ended questions (those which do not allow for a simple yes-or-no answer) an employee is encouraged to discuss his problems. The supervisor needs to listen not only to words but the feelings they carry. He must give his attention to the unspoken language of facial expressions, gestures, bodily postures, and the like. This will bring him closer to understanding the attitudes and strength of feeling attached to them.

In such face-to-face discussions or interviews, it is inappropriate for the supervisor to argue, caution, evaluate, or cajole the employee, but rather he should try to reverbalize and reflect the expressions of feelings and attitudes. Having ventilated the emotion, this reflection will give the employee a chance to reevaluate without emotion his own position. He will thereby come to make a new appraisal of the situation more in line with the requirements of objectivity.

Frequently it is impossible to discuss such matters at length in the work area, so the supervisor should set a time and a place away from other employees to allow the individual freedom of expression. Even though the supervisor may not have a completely private office in which to engage in such kinds of interviewing, some place can be found for a quiet face-to-face discussion.

This relationship between the supervisor and the worker must be one of the supervisor's normal everyday functions. Diligent observation and effective and sensitive listening should be tools of every supervisor.

Attitudes

Attitudes and attitudinal responses are closely related to emotions. Strictly defined, an attitude is a predisposition on the part of a person to respond in a given way. In other words, an attitude is manifest when an individual is likely to respond to an initiating cause in a predicted manner.

What the reaction will be depends upon the previous experiences of the individual. Seeing an individual on the street who resembles a friend, one immediately begins to smile, make friendly gestures, and have a warm quality in his voice. When a person is perceived to be a stranger, there is an immediate change to another facial expression, gesture, voice quality, and so on. Anticipated reaction can be set off by a certain look from another person, a manner of speaking, a mustache or the lack of one, the color of hair, or kind of handclasp. All individuals have certain feelings about these and will react in accordance to their attitudes, as these have been formed over the years.

Some of these attitudes are latent. The predisposition to respond will lie dormant within the individual until called forth. Given the appropriate stimulus, the attitude shows itself and the behavior exhibited is in accordance with the feelings of the individual. A certain word, for example, evokes different feelings in different people. Some individuals have given little or no thought to what the real meanings are, yet, they may react negatively or positively. The words "union," "management," "labor," and even "safety" carry with them certain connotations that touch off different attitudinal reactions in individuals, depending upon the kinds of experiences they have had with the subject.

Because attitudes play such an important role in everyday relationships, consideration should be given to their development, their effects on individuals, and what can be done to change them.

Determination of attitudes

In attitudinal responses, "feelings" are involved. They are determined by the experiences that individuals have had in the past, experiences of a quite emotional character. The attitude of a worker toward management or his supervisor may be fearful and hostile if the worker has lost jobs for no apparent reason, as he sees it. There may have been sound reasons for his dismissals or layoffs, but to admit this to himself would be a threat to his own pride, and so he puts the blame elsewhere. Because of the effects that these dismissals have had on him and his family, such as causing financial stress, he may become very angry. Management to him now is a menacing thing, and thus the cause of his present hostile and fearful attitude.

Many people have had experiences that caused fear, sorrow, pain, or happiness. All of these will tend to make them react the same way to anything that is similar to the original emotion-provoking situation. Although there is only one positive emotion listed among those just mentioned, it is not to be concluded that most attitudes are negative, for there are many positive ones.

Changing attitudes

The first step is to recognize that an individual's attitude reflects a deep-seated conviction that what he feels is right and true. It does not matter whether or not objective data and facts support a contrary position. What the individual sees as truth is, to him, true. Much can be done in an effort to effect changes in attitude starting with the individual as he is now.

Some attitudes are hard to deal with when they are shared with others to the degree that there seems to be near unanimity in a group. Substantial research has been done which leads to the belief that through participatory group discussions in which there is freedom of expression and easy communication, attitudes can be changed. In such situations, the supervisor or safety professional serves as a resource person and discussion leader to make sure that objective data is explored and contrasted with all the various opinions and ideas held by the group. In this situation, it is necessary that each individual has an opportunity to express his feeling about the matter. It is up to the leader to protect those who find themselves expressing a minority opinion from derisive comments by others.

The final solution in such group discussions should be by general consensus rather than majority vote to best utilize social approval as a means to involve and gain acceptance by all participating members.

The group discussion is not only a method to reduce hostilities and emotional feelings, but also an opportunity to capitalize on an individual's need for belonging, recognition, participation, and knowledge. Finally, it helps develop the willingness to accept responsibility for the final decision and the success of the action required to implement it.

When films, explanations, group · discussions, and the like are used in an atmosphere of participation and acceptance, attitudes typically influenced by the group within which individuals work and live can be changed.

The safety professional and the supervisor must be sure that they themselves do not create attitudes which will be negative to the safety program. If the safety professional and foreman insist upon workers wearing protective equipment, they themselves must also do so. A worker who gets the impression that safety is a "don't-do-as-I-do-but-do-as-I-say" routine, will, despite all other reeducative processes, develop a negative attitude toward the whole program.

Organizational development. Recent concern for the amount and rate of change within our technological society has focused on the industrial environment and its ability to withstand such change. Behavioral scientists have evolved an approach, called organizational development, which attempts to assess a corporation's ability to adjust to such conditions as rapid and unexpected change, growth in size, increasing diversity, and management system problems. Organization development, while variously defined, generally has as its goal the implementation of an educational strategy designed to alter the attitude and structure of organizations so that they can better adapt to the changing technology.

Generally, organizational development uses initial feedback of the employees and management to determine the climate and capacity of the organization to adapt its objectives to the technological environment. Based upon such feedback, an attempt is made to develop organic systems to replace mechanical systems within the organization. Organic systems are characterized by a preoccupation with people as they operate together whereas mechanical systems are characterized by a preoccupation with the structure that operates within a system. Fig. 11–7 presents a summary of the

Mechanical Systems	Organic Systems
Emphasis upon individual performance	Emphasis on relationships in group
Chain of command concepts	Confidence and trust among everyone
Adherence to delegated responsibility	Adherence to shared responsibility
Division of labor and management	Participation in multi-member teams
Centralized management control	De-centralized sharing of control
Resolution of conflict through grievance procedures	Resolution of conflict through problem solving

FIG. 11–7.—Organization development seeks to implement organic systems in place of mechanical systems within organizations.

differences between mechanical and organic systems. In the final analysis, organizational development rejects bureaucracy as an organizational model and substitutes a model based on interpersonal competence.

In order to implement the ideas represented by organizational development, a number of procedures are used to effect changes in the organization. Since the changes are typically people-oriented, due consideration is given to the need to motivate acceptance of the changes within both management and work groups. Typically, such techniques as sensitivity training, confrontation groups, and transactional analysis are used to effect reorientations within the organization's staff. These techniques have merit in implementing change only so long as top management concurs with the changes and provides incentives within the organization for their adoption.

Ultimately, organizational development seems to provide the necessary means for preparing an organization for orderly planning for the future. Included in such an effort should be the recognition of how changes in an organization will affect the current safety program. By adopting widespread involvement of all elements of the organization in safety programming, it is likely that safety programs will be able to meet the future needs of modern organizations.

Structural change

The discussion of attitudes and attitude change presented above assumes that there is a direct relationship between attitude change and behavior change. Safety personnel are interested, for example, in changing an employee's attitude regarding the wearing of protective equipment only so long as such an attitude change ultimately results in the employee's actually wearing the equipment. While the research literature indicates that a variety of influence efforts (training and counseling for example) can change attitudes, there is much less support for a direct relationship between attitude changes and subsequent behavior changes.

In effect, it is incumbent upon the safety professional to consider techniques other than attitude change when considering changes in the employees' behavior. Recent evidence suggests that it is possible to change behavior by the introduction of structural changes within the work environment. Structural changes are procedures designed to change the organizational constraints which operate within work groups. Examples include: changing job contents, modifying the physical arrangements of work, changing worker interaction patterns, and rearranging work procedures.

In each case, other than an appropriate introduction of the change, it is not necessary to expend the time and effort to change attitudes prior to changing behavior directly. Rather, the introduction of the structural change can modify behavior directly and possibly, in turn, attitudes may change.

While there has not been an extensive attempt to apply the structural change model in occupational safety, it warrants consideration in the future. Human behavior can be changed by eliciting the change through the very circumstances under which the individual works. Unsafe acts, for example, cannot occur where the conditions of the work and work groups preclude their occurrence.

Learning

Learning underlies much of what makes for differences and similarities among people. Through learning, people have developed certain kinds of psychosocial needs, habitual patterns of behavior, ways of reacting to emotion, and the attitudes which they bring with them to industry. It is important to consider learning and the laws that affect it. This is especially vital in any discussion of safety, because training is a major consideration in safety programs.

To change attitudes, one must substitute new learning for old concepts and ideas. To change behavior in need satisfaction sequences, one must teach better means by which the goal can be achieved. In each case, some new learning must be substituted for the old.

Motivational requirements

Repeatedly in everyday living, people recall many things which they did not set out to learn. While unintentional learning does occur, it is the intentional kind of learning in which the safety professional is primarily interested. This requires motivation.

In educational systems, great emphasis is placed upon making the individual want the knowledge which is available to him. Materials are designed to relate to practical situations. Teachers attempt to make the individual interested in the material as such. To teach a boy something about angles or distances, the teacher may use a baseball diamond as an example. To get a young lady interested in yardgoods, in terms of both its makeup and durability, the teacher uses an example involving clothes. These teachers are tapping motivation to make sure that learning will take place.

The safety professional must ever keep in mind that if his workers are going to learn safe procedures they must be so motivated. To merely point out that accidents cost the company money will not motivate them. Rather, point out the hazards which are risked when using unsafe work procedures: the probability of serious and painful injury and the possible loss of earning power. The cost, not only in dollars but also psychologically to both the worker and his family, is of paramount importance to him and will motivate him to learn safe work methods.

It is not wise to assume that, because management sees the value of safe procedures, the worker will also. Management may be motivated to start a safety program because it will reduce insurance costs, reduce the amount of waste, and increase the number of units produced. Workers cannot be expected to desire it for the same reasons. In selling a safety program to the employees, one must capitalize on the things which will motivate them. Here the safety professional can capitalize on the needs discussed previously, and he probably can find many more which are consistent with the aims of the safety program. Everything that will motivate the worker to learn the right procedures should be used. (The next chapter discusses this in more detail.)

Principles of learning

Some consideration of basic principles is valid whether the learning is to be done in a college classroom or in a work area. When training procedures utilize these principles, learning is more efficient and thorough.

Reinforcement. Through experimentation, psychologists have found that learning reinforced by reward is more efficient than learning which is not reinforced by reward. In the practical situation, there is some reinforcement when learning makes work more efficient and more productive. When a worker's pay increases because of more units of production, he is receiving reward for his learning. This can have negative aspects as well—the worker who figures out a hazardous short cut, which produces more units, may be rewarded for an unsafe act, as was mentioned earlier in this chapter.

It is apparent that the employee must be rewarded only for safe work methods. Higher productivity because of safe work methods satisfies the need for achievement and recognition. This in itself reinforces the learning of the employee, but also spreads its effect to other employees who see this take place. A supervisor by recognizing the same needs, however, can reinforce this learning through praise for greater productivity and telling the worker how much better he is doing, for example.

The publicity given to a safety record, a bonus, a promotion, or anything which satisfies individual needs would serve as a reward to reinforce whatever learning has brought about the desired behavior.

A reward must closely follow the behavior to be rewarded. The praise, for example, should be given at the time the desired behavior occurs, not some long time later. This does not mean it must be instantaneous, but it should be within a reasonable length of time. If delay is necessary, certainly recall to the individual the reason for the reward.

Reward reinforces desired learning. Short-cut hazardous methods or any deliberate unsafe acts must not be rewarded. In fact, in such situations reprimands or punishment should be given. Action must concentrate on the error and not on the personality involved, for to do otherwise would lead to trial-and-error by the individual which is no more satisfactory than the wrong method. In short, wrong methods should be corrected by full and complete explanation and demonstration of the correct work procedures. Such being the case, it also follows that there should be reinforcement of the new correct work pattern as soon as feasible.

This principle also applies to participation by employees in a safety program whether it be through suggestion systems, safety committees, discussion groups or training sessions. In all these areas a man gains personal worth if his opinions are asked for and graciously received. He feels, too, that he is more than just a cog, and that he is giving of his creative self as an important member of the team.

A safety program that gathers the ideas of all, either individually or by representation, satisfies the need people have for being "in the know." In this way, the safety professional creates that positive atmosphere of ego-involvement in the program and a sense of obligation and responsibility for its success. Research has demonstrated that when employees do not feel that programs are top-level-planned, but rather come from all, there

is far more chance for success.

Much research on the effectiveness of punishment suggests that it possesses limited utility in the learning process. While punishment may temporarily inhibit a specific response, it does not appear to have enduring effects upon learning. When used to control behavior, punishment (and the threat of punishment) must be present continuously rather than intermittently. Thus, it does not appear to be an efficient technique for controlling behavior.

The rewarding of correct procedures will lead to a more positive attitudinal response on the part of the worker than any punishment. A positive attitude toward training procedures is much to be desired. Both the positive and the negative aspects of reinforcement may generalize over the whole work situation. The foreman's praise for a particular thing well done may spread over other aspects of the total situation, including training procedures, safety devices, and the safety program. The proper use of rewards thus can lead to efficient methods.

Much of the practice underlying programmed instruction (see Chapter 9, "Safety Training") is based upon the reinforcement concept as well as the other principles of learning which follow.

Knowledge of results. Closely allied to reward—in fact, one aspect of reward—is knowledge of results. Everyone likes to know how he is doing on a particular job. Letting the individual employee know how well he is getting along in his training program will likely motivate him to continue training and do a better job. To train a worker and not inform him of any improvement is defeating one's own purpose. It cannot be contended that the only one who needs to know about the effects of training is the foreman or the safety professional. When the worker knows that this new procedure is helping him get greater production, he is getting reward reinforcement for his learning.

One of the factors which appears in some kinds of learning is a plateau at which learning levels off for some time before again showing an increase. Often some individuals become discouraged and learning can be retarded. If the trainer understands this phenomenon and indicates to the individual that a leveling off had been expected and that an increase will come later, the discouraging aspect of such a plateau may be avoided, with learning then proceeding more easily and efficiently. Demonstrating to the worker the achievements he is making through production curves, which are in effect learning curves, gives the worker knowledge of results, and this is a motivating factor for future learning.

Practice. The safety professional is interested in developing in an individual safe habit patterns which will become almost automatic in his work methods. Merely putting a worker through sufficient training sessions is not enough. Despite his apparent mastery, the next time through the work he may make one or more mistakes. To ensure firmly entrenched habit patterns, the worker must practice. The Job Instruction Training programs include one of the important aspects of training, namely, follow-up by the foreman. This is for no other purpose than to ensure mastery. Within a reasonable time, depending on the job complexity, this follow-up may become unnecessary.

Take, for example, a simple task such as bicycle riding. A youngster, in learning to ride a bicycle, needs to know how to balance, how to get on, how to get off, and how to stop, and he must learn all of these things as part of the total process. Perhaps he begins with balancing, then the start and stop, and then how to get on and off. Once he has accomplished all of these, he is not immediately left to himself. He needs several more sessions to ensure that he is doing it correctly. This, in effect, is reinforcement by practice. Furthermore, the youngster himself keeps practicing each time he goes out to ride. So it is also with the individual on the job.

Whole vs. part learning. Whole or part learning has been a knotty problem for industrial trainers for many years. Whether trainers should teach the procedure as a whole or break it down and teach it part-by-part is the question. There is no best answer. Both methods have advantages and disadvantages depending upon: job complexity, the trainee, and the kind of job breakdown used. Perhaps a combination of the two is best, using the whole method, but with sufficient flexi-

bility to emphasize meaningful parts of the task wherever necessary.

Meaningfulness. Studies in verbal learning have demonstrated the importance of the meaningfulness of the material to be learned. Meaningfulness is important in safety education because the worker needs to understand why a certain procedure is better than another. Adequate explanation of a given movement or change in position, in terms of the hazards eliminated, with no decrease in production, gives meaningfulness to the procedure. With this understanding, the worker will be motivated to learn the safe procedure. Without it, he will be inclined to utilize his own method until he learns, perhaps by an accident, the inadequacy of it.

The safety professional and line supervisor in safety programs should not forget the advantage of workers' understanding the reason for protective clothing, safety devices, safety meetings, and discussions, as well as the need for full and complete accident reports.

Meaningfulness to management and workers is understanding the value of a reduction of accidents, fewer disabilities, and retention of earning power.

Selective learning. Out of each day's many experiences, people select those which they desire to retain. This probably is related to motivation more than to anything else, and for that reason motivational aspects of a training program need to be considered. Safety trainers must be sure that the workers retain the most important facts. Relating subject matter to individual needs will ensure the proper selection.

Frequency. Everyone is aware of the fact that he does best those things which he practices the most.

This principle is certainly important to the safety professional for it emphasizes the necessity of frequent applications of safety rules and regulations in the training program. Frequent reference to the various kinds of problems, hazards, and procedures to eliminate accidents ensures greater effect than just a one-exposure routine. Giving the worker a copy of the safety rules and regulations in hopes that he will learn them and use them is not enough. Means should be adopted which will bring them to his attention frequently and regularly.

A major railroad practiced this when its supervisors discussed with the men a safety regulation each day before work began. The men were responsible to know, understand, and be able to apply it to situations if the supervisor were to ask about it during the day. Here was learning, reinforcement, and followup all operating at one time.

A trainer must insist not only on frequent practice, but on the trainee's following the correct method. Day-after-day use of safe methods will create safe habit patterns that will later be followed almost automatically. The supervisor and safety director must make sure that the work method practiced is the safe one.

Recency. Closely allied to the principle of frequency is that of recency. That which is learned last can usually be most easily recalled. As has been indicated, handing a worker a set of safety rules and regulations does not ensure learning. Those who received printed instructions or a few training demonstrations a long time ago may not be able to recall now what the rules and regulations are. Safety professionals must devise means by which workers have constant contacts with these regulations through such continuing activities as contests, reviews of safety regulations, and committee work.

Primacy. The law of primacy must be taken into consideration in two aspects of the safety program. The worker's initial contact with procedures should be one of major importance. If this initial contact is of a negative nature—such as being tossed a book of rules, accompanied by a shouted, "Make sure you learn 'em"—the worker is left with the impression that safety is unimportant. From the very beginning of employment, the worker must get the impression that not only the rules but the whole program is of great importance. This will help assure the positive response desired.

Second, in the training program, habit patterns utilizing safe methods should assume primary importance. The supervisor must be certain that the worker does not have an op-

portunity to work by any other than the safe method. It becomes harder to establish good patterns after having first learned the poor ones. This principle applies to the golfer who has picked up bad techniques and then has to unlearn them as well as to the worker on the job. His training must be *right* from the very beginning. Old habits are hard to break and, to the safety professional, expensive in training costs and accident costs.

Intensity. Those things which are made most vivid to the worker will be retained the longest. Safety programs already utilize this principle in safety publicity with eye-catching posters and the like.

It is part of the safety professional's job to dramatize the program to workers. In this way, it will be a long remembered experience. Under other conditions, it might well be forgotten minutes after it is over. To some degree, this is in effect positioning the safety program in the same way that advertisers position their commercials to catch the public's attention.

Transfer of training. All new learning occurs within the context of previous learning experiences. The fact that current learning (or performance) can be influenced by previous learning is known as transfer of training. Positive transfer of training occurs when the previous learning facilitates the current learning or enhances current performance. Negative transfer of training occurs when the previous learning makes the current learning experience more difficult or in some way inhibits current performance.

Inasmuch as the safety professional desires to facilitate the learning of correct responses in new situations, he should attempt to maximize the positive transfer which occurs within the industrial environment.

Generally, learning to make identical responses to new stimuli results in positive transfer. For example, learning to drive in a new (or different) automobile is facilitated by the fact that identical responses (such as accelerating or steering) are required to the new stimuli (accelerator, steering wheel). If the new stimuli are very similar to previous cars (for example, the location of the controls, their texture, and their direction of movement), positive transfer should be high. In such situations, positive transfer increases as a function of increased similarity of the stimuli in the two situations.

Learning to make new responses to identical stimuli results in negative transfer. For example, if the controls of the cars appear identical but each car requires a different response in order to be operated correctly, negative transfer can be expected. After driving a car in which the "park" position appears on the extreme right of the indicator panel, much difficulty (negative transfer) can be anticipated when driving a car in which the "park" position appears on the extreme left of the indicator panel. Or, if the "off" position of one toggle switch is the same as the "on" position of another, the potential for accidents is serious.

Such negative transfer can take the form of errors, delayed reactions, and generally inefficient performance on the new task. In such situations the amount of negative transfer increases as a function of increases in the similarity of the stimuli in the two situations.

With an understanding of transfer phenomena, it is possible to maximize the opportunities for positive transfer and minimize those for negative transfer within the industrial environment. This requires careful planning of machine purchases and work procedures to make sure that the new tasks required of a worker make use of (and do not conflict with) his previous learning experiences.

Forgetting

This discussion of learning would not be complete without some consideration of the process of forgetting, which goes on as learning takes place. Never assume that learning and forgetting are mutually exclusive, for such is not the case. As one learns, one also forgets what has been previously learned. Curves of forgetting indicate that most is lost immediately after learning has taken place. Depending on the complexity of the job, the amount of learning lost will vary after each day's training session. There should be less forgetfulness or initial mistakes in successive days of training, and more need for patient reteaching.

Consideration of the various principles of learning and the motivational aspects of the problem are essential if one is to retard this process of forgetting. The safety professional needs to consider all of them from the initial stages of employment right on through to the everyday work situations in the company.

Summary

The human factor operates at all levels in industry, and is perhaps the most potent factor for success or failure of a safety program. It makes a difference whether the president of a company approves or drags his feet, whether the safety professional works hard or coasts along, whether the supervisor emphasizes safety or subordinates it to production, whether the janitor cleans well or does only the minimum—attitudes are important to safety in the company. Safety can be achieved only by working through all these people. In every area of industry the human factor must be dealt with.

To achieve some mark of success in dealing with people, individuals can best be considered within the framework established in this discussion. Each person is an individual, to some degree different from every other one. The differences are for the most part obvious, but there are subtle ones too, which must be recognized, if only to the extent that their existence is acknowledged.

Despite great differences in people, reasons for their activities are common to all. Many needs are the same, particularly at the biological level and to a large degree at the psychosocial level. It is upon these needs that the safety professional and others in positions of leadership in industry can capitalize to most effectively promote safety.

People become frustrated when their goals cannot be achieved. There are many ways of reacting to such frustrations, but the emotional reactions are of great concern to the safety professional. These reactions, as well as the attitudes formed during the reactions, can be highly disruptive to safety precautions and procedures.

Training programs are established to teach safe work methods. These learning situations, if they are to operate efficiently, call for knowledge of the basic principles of learning that cause people to learn and act as they do. These principles should be used to make new learning more efficient.

In no way is this chapter intended to minimize or ignore the work already being carried on to reduce accidents. Safety devices, safer machines, safe work layout, and many other aids and measures all are a part of the total program. The human factor is one more aspect of the whole system.

Men, machines, and materials are still the three components of industry that can contribute to safety.

Machines and materials can be controlled, but the human factor must be guided in the interests of accident prevention.

References

Argyris, C. *Management and Organizational Development.* New York, McGraw-Hill, 1971.

Bennis, W. G. *Organization Development: Its Nature, Origins, and Prospects.* Reading, Mass., Addison-Wesley Publishing Company, 1969.

Deese, J. & Hulse, S. H. *The Psychology of Learning.* 3rd Ed. New York, McGraw-Hill, 1967.

Drucker, P. F. "New Templates for Today's Organizations." *Harvard Business Review,* Jan.–Feb. 1974.

Fishbein, M. & Ajzen, I. Attitudes and Opinions. *Annual Review of Psychology,* Vol. 23, 1972.

Gausch, J. P. *Balanced Involvement: Safety, Production, Motivation,* Monograph No. 3. Park Ridge, Ill., American Society of Safety Engineers, 1973.

Hale, A. R. & Hale, M. *A Review of the Industrial Accident Research Literature.* London, Her Majesty's Stationery Office, 1972.

Hannaford, E. S. *Supervisors Guide to Human Relations.* Chicago, National Safety Council, 1967.

Herzberg, F. *Work and the Nature of Man.* Cleveland, World Publishing Co., 1966.

————. "One More Time: How Do You Motivate Employees?" *Harvard Business Review,* Jan.–Feb. 1968.

Hinricks, J. R. "Psychology of Men at Work." *Annual Review of Psychology.* Vol. 21, 1970.

McGregor, D. *The Human Side of Enterprise.* New York, McGraw-Hill, 1960.

Morgan, C. T. & King, R. A. *Introduction to Psychology.* 3rd Ed. New York, McGraw-Hill, 1966.

National Institute for Occupational Safety and Health. *The Present Status and Requirements for Occupational Safety Research.* Rockville, Md., U.S. Department of Health, Education, and Welfare, 1972.

Ruch, F. L. *Psychology and Life.* 7th Ed. Glenview, Ill., Scott, Foresman and Company, 1967.

Shafai-Sahrai, Y. *Determinants of Occupational Injury Experience: A Study of Matched Pairs of Companies.* East Lansing, Mich., Michigan State University (Business Studies), 1973.

Tyler, L. E. *The Psychology of Human Differences.* 3rd Ed. New York, Appleton-Century-Crofts, 1965.

Zaleznick, A. *The Human Dilemmas of Leadership.* New York, Harper & Row, 1966.

Maintaining Interest in Safety

Chapter
12

Reasons for Maintaining Interest   264
Indications of need for a program . . . Program objectives and
benefits

Selection of Program Activities   265
Basis for the program . . . Factors to be considered

Staff Functions   268
Role of the safety professional . . . Role of the supervisor

Safety Committees and Observers   271

Meetings   272
Types of meetings . . . Planning programs

Contests Stimulate Interest   275
Purpose and principles . . . Injury rate contests . . . The "XYZ
Company" contest explained . . . Interdepartmental contests . . .
Intergroup competitions . . . Intraplant or intradepartmental contests
. . . Personalized contests . . . Overcoming difficulties . . . Non-
injury rate contests . . . Contest publicity . . . Meaningful awards
. . . Award presentations

Posters and Displays   289
Purposes of posters . . . Effectiveness of posters . . . Types of
posters . . . Changing and mounting posters . . . Bulletin boards
. . . Displays and exhibits

Other Promotional Methods   294
Campaigns . . . Off-beat safety ideas . . . Courses and demonstrations . . . Publications . . . Public address systems . . . Suggestion
systems

References   299

FIG. 12–1.—A sticker like this one (actual size 4¼ by 5½ in.) can be placed on bulletin boards, entrance doors, and in employment offices to remind everyone of a company's continuous efforts to remove hazards.

This chapter deals primarily with promoting and maintaining interest in safety on the part of supervisors and employees—but that does not mean that the interest of *top* management is always assured. This apparent lack of interest should not be construed as indifference or even opposition to safety; usually it can be traced to lack of awareness of the basic benefits of an organized safety program. The safety professional should be certain that his management is sincerely interested in the program by providing necessary information to key executives and managers.

When top management has demonstrated its interest and actively supports a solid safety program, then, and only then, can activities to promote employee's interest be undertaken.

Reasons for Maintaining Interest

Maintaining interest in safety is necessary even if the work place has been "engineered" for safety, even when work procedures have been made as safe as possible, and even after supervisors train their men thoroughly and continue to enforce safe work procedures. Why is it still necessary to maintain interest? Because even with these optimum work conditions, accident prevention basically depends upon the *desire* of people to work safely.

Because all possible hazardous conditions, unsafe acts, and loss-control problems cannot

be anticipated, each employee must frequently use his own imagination, common sense, and self-discipline to protect himself. Each employee must be stimulated to think *beyond* his immediate work procedures in order to act safely in questionable situations when he is "on his own."

The techniques used in modern advertising and merchandising have much in common with those used to "sell" safety. Just as most products and services require steady and imaginative sales promotion, safety likewise requires constant and skillful promotion.

Workers are accustomed to the modern techniques of advertising and sales promotion, so the basic elements of accident prevention can be made more understandable and acceptable if they are presented in a similarly interesting fashion.

Indications of need for a program

Various yardsticks can indicate the attitude of supervisors and employees toward accident prevention. These can be used to point out areas of greatest need for a program to create and maintain interest in safety.

● Increased frequency of injuries, accidents, and near accidents may be one indication that a program is needed. If no explanation for such an increase can be found in engineering methods, training, or supervision, the reason is likely to be that employees are forgetting or ignoring work rules, failing to stay alert, or taking chances. A program to develop and maintain their interest will help reverse this trend.

● If housekeeping is deteriorating, protective equipment is not being used, and guards are not being replaced, it is time to tighten up on supervision and to promote more interest in safety on the part of supervisors.

● Incomplete or missing accident reports also indicate a slackening of interest on the part of supervisors and, perhaps, even the failure of employees to report minor accidents and injuries. Motivation to assure better reporting is then in order.

Program objectives and benefits

A well planned program can create and maintain interest in safety, although it cannot be expected to do *everything*. For example, it can:

1. Help develop safe work habits and safe attitudes, but it cannot compensate for unsafe conditions and unsafe procedures.

2. Focus attention on specific causes of accidents, although by itself it cannot eliminate them.

3. Supplement safety training, yet it cannot be considered a substitute for a good training program.

4. Give employees a chance to participate in accident prevention activities, such as suggesting safety improvements in job procedures.

5. Provide a channel of communication between workers and management, because accident prevention is certainly a common meeting ground.

6. Improve employee, customer, and public relations, because it is evidence of management's sincerity with regard to accident prevention. (See Fig. 12–1).

The ultimate objective of a program to maintain interest in safety is to prevent accidents. Usually, though, it is as difficult to determine the degree of success achieved by an interest-maintaining program as it is to isolate the effectiveness of an advertising campaign separate from the entire marketing program. The reason is that generally companies with such programs also have sound basic safety programs: working conditions are safe, employees are well trained and safety minded, and supervision is of a high caliber.

However, one prominent company that already had a good basic program attributed a reduction in its work injury rate to a stepped-up program to maintain interest. The program was based on an idea submitted by an employee: Each month candy bars were distributed to injury-free employees. Wrapped with some of the candy bars were slips that could be traded for free pairs of safety shoes.

Another company gave each employee who had an injury during the month a package of gum with the slogan "Something to chew on" and a friendly safety message and wishes for an injury-free future.

A meat-packing firm also was able to assess the value of its program to maintain interest. Several safety bulletin boards were installed, and posters and safety contest reports were displayed on them. These displays were credited with an impressive cut in the number of injuries and with a workmen's compensation insurance refund of more than $1200.

Selection of Program Activities

Safety directors frequently undertake promotional activities with no preliminary planning or determining of objectives. Some men are prone to use, for example, films or posters for no other reason than that they happen to be available at low cost.

All too often, safety professionals spend a disproportionate amount of time on committee work, contests, or "homemade" visual aids because of their bosses' or their own personal interests. Sometimes these activities are substituted for a sound, well rounded program based on supervisory responsibility.

Basis for the program

To be effective, a program for maintaining interest in safety must be based on needs. Select activities so they yield the desired results, not just because they will be popular. To develop suitable activities and promotional material, the needs of supervisors and employees must be known.

To determine what the employees really thought of the safety program and just how interested they were, one company inaugurated a safety inventory plan. Each year after the regular *stock inventory* had been taken, *safety inventory* cards were distributed to all employees—salaried workers as well as hourly.

The cards were distributed by foremen or supervisors, who asked each employee to take stock of his job and environment with regard to safety. The following year's safety program was planned on the basis of the questionnaire returns, which ran better than 90 percent. Many suggestions for improvement of the safety program were received and subsequently put into practice.

Factors to be considered

Company policy and experience. If a company ordinarily uses activities, such as com-

12—Maintaining Interest in Safety

MISSABE DIV.	Jan.	Feb.	Mar.	Apr.	May	June	July	Aug.	Sept.	Oct.	Nov.	Dec.	Extra Balls	TOTAL
1 NORTH END	9✓ / 19	9✓ / 38	9✓ / 53	5- / 58	7- / 65	4- / 69	0- / 69	8- / 77	4- / 81	8✓ / 101	X / 121	X / 131		131
2 ROAD	9✓ / 20	X / 50	X / 78	X / 96	8- / 104	X / 124	9✓ / 143	9✓ / 163	X / 180	7- / 187	9✓ / 203	6 / 209		209
3 PROCTOR	X / 18	8- / 26	9✓ / 46	7✓ / 66	7✓ / 83	7✓ / 102	9✓ / 121	9✓ / 135	4- / 139	6- / 145	9✓ / 164	9✓ / 173		173
4 STEELTON–M.J.	7- / 7	9✓ / 26	9✓ / 45	9✓ / 62	7- / 69	6- / 75	X / 95	9✓ / 110	5- / 115	9✓ / 135	X / 153	8 / 161		161
5 DOCKS	9✓ / 19	9✓ / 39	X / 59	8✓ / 79	9✓ / 93	4- / 97	5✓ / 115	8✓ / 133	8✓ / 151	8✓ / 171	X / 191	X / 201		201
IRON RANGE DIV.	4 / 5	8 / 12	14 / 17	19 / 22	21 / 24	23 / 27	27 / 31	31 / 34	33 / 39	36 / 43	41 / 48	44 / 52	945	875
1 NORTH END	8✓ / 20	X / 37	7- / 44	X / 64	8✓ / 78	8✓ / 82	8- / 102	8✓ / 122	8✓ / 142	X / 172	X / 192	X / 202		202
2 ROAD	X / 20	8✓ / 40	X / 60	8✓ / 75	5- / 80	X / 100	6- / 112	2- / 114	8✓ / 126	2- / 128	7- / 135	X / 145		145
3 TWO HARBORS	X / 28	8✓ / 48	8✓ / 66	8✓ / 84	8✓ / 102	8✓ / 120	7- / 137	8✓ / 144	8✓ / 162	8✓ / 180	8✓ / 196	X / 202		202
4 ELY–ENDION	8✓ / 20	X / 40	8✓ / 60	X / 76	6- / 82	8✓ / 102	X / 122	8✓ / 142	X / 162	8✓ / 182	X / 202	X / 212		212
5 DOCKS	X / 30	X / 58	X / 78	8✓ / 88	0- / 88	2- / 90	4- / 94	X / 114	X / 134	X / 154	6✓ / 174	X / 184		184

FIG. 12–2.—This Safety Bowling Sweepstakes is based on the safety performance of a railroad's transportation department. Two divisions of the department form the teams; there are five areas within each division. "Strikes" are scored for each frame (month) that an area has no personal injury. "Spares" are marked when an area goes through a month with some personal, but no disabling injuries. Disabling injuries reduce the monthly (frame) points acording to a system of handicapping that offsets the varying manhour exposures of the different divisional areas.

Courtesy Duluth, Missabe & Iron Range Railway Company

mittees, mass meetings, and contests in areas other than safety, then the safety director can consider them for his own program. He would be unwise to spend much time on activities that are foreign to company policy and experience, unless he was convinced the new "sales pitch" was justified.

On the other hand, if supervisors and employees are over-involved in committee work and such activities as sales promotion, quality control, and tool damage programs so that similar activities for safety would be lost or burdensome, then other approaches often prove more effective.

Once a promotional program is under way, it should not be permitted to unbalance other aspects of the accident prevention program or other company activities. For example, safety meetings should not take a great deal more time than meetings for quality control, sales, or industrial relations.

Budget and facilities. Plans for a safety promotion program will be affected by budget considerations. At first the program will require extra effort, time, and money, but this expenditure is justified because it is an investment that will produce direct as well as indirect benefits. If the program is to be successful, the budget will have to be sufficient to carry it out.

In the selection of program activities, consider the facilities available. Sound films, for instance, require not only a sound projector and screen, but also a darkened room free of background noise.

In many companies, facilities, publications, or services may be available from the industrial or public relations department, which also may be a valuable source of help in the planning of promotional activities.

Types of operations. The nature and organization of company operations affect the choice of activities and materials for maintaining interest in safety. When operations are widely scattered and diversified, as in the construction, railroad, marine, motor transport, and air transport industries, the job of selecting and disseminating safety information becomes considerably complicated.

When operations are decentralized, the safety director must rely upon materials that can be used readily in the field, such as publications and films. He should depend on local supervisors to conduct meetings, present material, and handle posters.

The needs of employees doing widely different kinds of work at far-flung locations also must be considered. Where a poster program is used as one means for maintaining interest in safety at each outlying location, a trustworthy employee may be designated by the local supervisor to receive posters and take care of their distribution and posting. He should see to it that poster boards and display cases are kept clean, attractive, and free of extraneous paper.

Another method is to use a trailer equipped with permanent displays to carry the company safety story to far-flung locations. Self-contained rear-screen projectors are also well suited for use in scattered locations and in the field.

In the same organization, the types of educational materials used for different groups of employees may vary considerably. For instance, a movie scheduled for a day shift because a large number of employees could be taken off the job would not be suitable for a small night shift because no one could leave his job.

However, when activities are planned, night crews and maintenance employees should not be overlooked. Their work is as vital as that of other employees to the overall accident prevention effort. Programs for them will have to tie in to their specific requirements.

Types of employees. The types and backgrounds of employees have a bearing on the choice of safety promotion activities. For example, migrant workers frequently do not receive sufficient job training, particularly those who do not understand English. Material for these employees should present basic safe practices for their jobs in brief and easy-to-understand form. A similar approach should be used with temporary workers or those assigned from union halls. Employees who have difficulty understanding English

need visual material. Material in Spanish is available from the Inter-American Safety Council (see Chapter 23, "Sources of Help").

Basic human interest. If employees seem bored or uninterested in safety activities, an extra push must be given to pep up pallid programs. One approach is to base promotional activities on interests, such as bowling or fishing, that are shared by a large number of the workers. (See Fig. 12–2.)

Wise choice of promotional activities depends upon an understanding of basic human needs and emotions. The basic interest factors listed in Fig. 12–3 are forms of motivation common to all employees. The activities suggested, therefore, should be of general appeal.

Other considerations. Tasteful use of sex appeal, material featuring children and animals (human interest), and activities and contests based on a moderate amount of competition play a big part in the safety promotion programs of many companies. These elements are as effective in safety promotion as they are in sales and advertising promotion. Many successful safety directors capitalize on them without compromising company policy or offending anyone.

An Ohio company capitalized on two basic interest catchers—wagering and pretty girls. Employees in carpools were urged to participate in a little wager in which the rider who forgot to buckle his safety belt had to pay for lunch or dinner. The idea proved so popular the company decided to spread the word by means of an ad campaign featuring a pretty girl (Fig. 12–4).

Ideas for maintaining interest often use humor with telling effect. The "light touch" is essential, and should be good-natured. Ridicule should not be used; it is likely to arouse only resentment.

A positive, constructive approach is generally better than a negative approach. However, the latter sometimes is preferable if it is more dramatic. A picture showing the consequences of an accident, such as a fall on a slippery floor, will have a greater impact than one depicting the safe act that could have prevented the accident.

Variety is essential. Often a simple change,

BASIC HUMAN INTERESTS AND CORRESPONDING ACTIVITIES

Basic Interest Factors	Ways To Use These Factors
Fear of painful injury, death, loss of income, family hardship, group disapproval or ridicule, supervisory criticism.	**Visual material:** emotional or shocker posters, dramatic films, pictures and reports of serious injuries on bulletin boards, in company papers.
Pride in safe workmanship, in good records, both individual and group.	**Recognition** for individual and group achievement; trophies, personal awards, letters of appreciation.
Recognition: desire for approval of others in group and family, for praise from supervisors.	**Publicity:** photos and stories in company and community papers, on bulletin boards.
Participation: desire to be "one of the gang," "to get in the act."	**Group and individual activities:** safety committees, suggestion plans, safety stunts, campaigns.
Competition: desire to win over others, such as shown in sports.	**Contests** with attractive awards.
Financial gain through increased departmental or company profits.	**Monetary awards** through suggestion systems, profit-sharing plans, promotions, increased responsibility.

FIG. 12–3.—Ways to put six basic human interest factors to effective use in promoting safety: Basic needs and desires that motivate people are left; right column lists direct appeals safety promotional programs can make.

such as a different type contest, redesign of a bulletin board, or revising the format of safety meetings, can result in renewed interest. The activity itself may not be more effective, but its new form stimulates thought, discussion, and interest. Although safe practices should become routine, their presentation should not.

Activities that require participation generate more interest than do those that involve only seeing and hearing. Companies that have allowed the National Safety Council to make movies in their plants report an upsurge in interest in safety because some of their employees got a chance to act in a film which emphasized the importance of what to them might have seemed routine.

Employees who are asked to submit suggestions for equipment guards or to help in the selection of personal protective equipment are more inclined to use the guards and the personal equipment than they would be if they had no opportunity to make their opinions known.

For the same reasons, helping to draft the safety rules encourages compliance on the part of those workers who participate in the project, and serving on a safety committee leads to increased awareness of safety responsibilities.

Staff Functions

The job of creating and maintaining employee interest places certain demands upon the safety director, who has the basic responsibility for planning the safety program, and upon the supervisors, who have the responsibility for carrying it out.

Role of the safety professional

The safety professional should be a well informed specialist if he is to maintain interest in safety. He coordinates the program and supplies the ideas and inspiration, while enlisting the wholehearted support of manage-

Fig. 12–4.—"Wearing a safety belt may earn you a free lunch." Company used this photograph and slogan to promote the idea that when two employees drive in the same vehicle together, if one fails to buckle up, he must pay for the lunch or dinner of both. The idea even caught on with some of the company's customers.

Courtesy Orr Felt Company, Piqua, Ohio.

ment, supervision, and employees. He may work with local safety councils, chapters of the American Society of Safety Engineers, and other civic or technical groups interested in accident prevention.

The safety professional can gain much by attending the National Safety Congress and regional safety conferences. He can learn of the ideas and programs of thousands of other companies. He can translate many of those ideas into practical activities that will be useful to his own organization. After participating in round-table discussions and in open and sectional meetings, he should return to his own job with renewed enthusiasm (Fig. 12–5).

Because the safety professional often may

be called upon to address groups, he should be able to present his ideas clearly, effectively, and convincingly. He should cultivate his ability to speak because this will help his ability to deal with people.

The safety professional should be well versed in the use of visual aids and familiar with the techniques of advertising, sales, and publicity, particularly those used in his own organization. Ideas picked up from these sources often can be applied in the safety program.

Showmanship tactics, however, should be used with discretion. They should not consist of questionable activities which could reflect unfavorably on the safety professional, the company, or the safety program. Show-

Fig. 12–5.—National Safety Congress workshop sessions give delegates an opportunity to "brainstorm" for answers to their various safety problems.

manship tactics can do more harm than good if they are insincere or if the basic safety program is weak or unsound.

Feel free to submit interesting data, difficult problems, or "gimmicks" of any nature to the National Safety Council. Through this clearinghouse, problems can be solved or given to others for solution. The Council has information on every phase of safety, gleaned from the experience of members in various industries. One man's solution to a problem may be of real help to many others.

A vast amount of program material of use to the safety professional is available in National Safety Council publications. For example, many of the illustrations in this chapter were selected from winning entries of the monthly feature "Ideas That Worked" in the Council's *National Safety News*. Each month a committee of staff members and volunteer officers judges the entries and every six months selects the best of the past six win-

ners. Every individual submitting a winning entry receives a gift certificate. For information, contact the Industrial Department, National Safety Council.

Role of the supervisor

The supervisor is the key man in any program to create and maintain interest in safety because he is responsible for translating management's policies into action and for promoting safety activities directly among the employees. How well he meets this responsibility will determine to a large extent how favorably the employees receive the safety activities.

The supervisor's attitude toward safety is a significant factor in the success, not only of specific promotional activities, but also of the entire safety program, because his views will be reflected by the employees in his department.

The supervisor who is sincere and enthusiastic about accident prevention can do more than the safety director to maintain interest. Conversely, if the supervisor pays only lip service to the program or ridicules any part of it, his attitude offsets any good that might be done by the safety professional.

Many supervisors are reluctant to change their mode of operation or to accept new safety engineering ideas, much less to regard with enthusiasm contests, safety stunts, committee projects, and other activities used to promote and maintain interest in safety. It is the safety director's task to sell these supervisors on the benefits of accident prevention, to convince them that promotional activities are not "frills," but rather projects that can help prevent injuries, and to persuade them that their wholehearted cooperation is essential to the success of the entire program. A supervisors' safety meeting can stimulate cooperation.

Setting a good example, for instance, wearing safety glasses and other personal protective equipment whenever it is required, is one of the most effective ways in which the supervisor can promote safety.

Teaching safety is an important function of the supervisor. He cannot depend upon safety posters, a few warning signs, or even general rules to do his job of training and supervision. A good balance of basic training and supervision and judicious use of promotional material proves effective. However, the supervisor himself must first be trained if he is to be competent. (See Chapter 9, "Safety Training" for details on training courses and techniques.)

The safety professional should help educate the supervisor so that he sees that working conditions are kept as safe as possible and insists that his workers follow safe procedures consistently, simply as a part of good job performance. The supervisor should not have to suddenly adopt a "get tough" approach to enforce safety rules. He should be consistently firm and fair. If workers have the impression that the supervisor either cannot recognize unsafe conditions and unsafe acts or does not care whether or not they exist, they too will become lax.

The supervisor is entitled to all the help the safety department can give through correspondence, supplies of educational material for distribution, and through as frequent visits as circumstances permit. He should also receive adequate recognition for independent and original activity.

Supervisors can be most effective in giving facts and personal reminders on safety to employees. This procedure is particularly necessary in the transportation and utility industries, where crews are on their own from terminal to terminal.

In any case, supervisors should be encouraged to take every opportunity to exchange ideas on accident prevention with workers, to commend them for their efforts to do the job safely, and to invite them to submit safety suggestions.

Safety Committees and Observers

There are many different types of safety committees having many different functions. (For further details, see Chapter 3, "Safety Program Organization.") However, *the basic function of every safety committee is to create and maintain interest in safety and thereby help reduce accidents.*

In some cases, other types of employee participation are preferred over formal safety committees. Some companies report that safety committees require a disproportionate amount of administrative time, that they generally tend to pass the buck, that they frequently stir up more trouble than they are worth, and that some supervisors try to unload their responsibilities onto the safety committee.

The answer to these objections is not to abolish the committees but rather to reexamine their duties, responsibilities, and methods of operation. Such analysis often can lead to constructive changes that will enable a committee to fulfill its original objective—that of stimulating and maintaining interest in safety.

The safety observation plan (discussed in Chapter 4) is used by some companies instead of, or in addition to, formal safety committees. The basic objective is the same: to get more employees actively involved and interested in the safety program. Planning, publicizing, and following definite procedures will streamline the work of both committees and observers and help ensure effective results.

Fig. 12–6.—The conclusion of another injury-free year calls for celebration! Here, employees, families, and community VIP's enjoy a family safety picnic—complete with balloons, refreshments, and special souvenirs to commemorate the event.

Courtesy Carling Brewing Company.

Meetings

Safety meetings may be conducted for supervisors, employees, or other groups, but in every case the purpose is to stimulate and maintain interest. If meetings fail to achieve this, they should be dropped, or their format or content should be changed sufficiently to make them effective.

Types of meetings

Among the various types of safety meetings commonly held to arouse and maintain interest in accident prevention are the following. (Also see Chapter 3.)

1. Meetings of operating executives and supervisors to formulate policies, initiate a safety program, or plan special activities.

2. Mass meetings of all employees, sometimes including families, or even the entire community to serve special purposes.

3. Departmental meetings to discuss special problems, plan campaigns, or analyze accidents.

4. Small group meetings to plan the day's work so that it can be done safely, to discuss specific accidents, or to review safety instructions.

Meetings of executives. When a safety program is to be inaugurated, it is especially important that the top executive officer of the company or plant should call a meeting to announce the general accident prevention plans and policies to all his superintendents, foremen, and other operating executives.

If these men meet at regular intervals to discuss operating problems, this announcement can be made at one of these regular meetings. Otherwise, the manager should call a special meeting for this purpose.

After this first meeting, the group may hold sessions periodically to evaluate the safety program, to check on the progress being made in accident prevention, and to appraise proposed activities.

Mass meetings. Large mass meetings are held for special purposes, such as the launching of a contest, the presentation of awards, the introduction of interesting new equipment, the explanation of a change in company policy, or the announcement of an exceptionally fine safety record. Fig. 12–6 illustrates how a company, prominent in a community, strengthens public relations with a family safety picnic.

In companies with plants in different cities, a top executive may call a meeting of employees when he visits a plant. His talk may cover safety as well as other subjects. One company president makes an annual round of plants with the safety director and speaks at a safety rally of all employees at each plant.

Under certain conditions, particularly in smaller communities, large meetings can be held in a local theater or public hall. It then is necessary to make fairly elaborate arrangements and to give the meetings considerable publicity to assure good attendance.

A mass meeting in a public hall naturally has one advantage over a plant meeting. It makes possible the attendance not only of employees, but also of their wives or husbands, families, and friends.

"Family safety nights" of this sort are very popular in the railroad industry. A considerable number may be held at several different points on the railroad during the year.

In addition to the address or two around which the program is centered, there should be some worthwhile entertainment. Often good talent can be found right in the plant or shop.

This type of meeting affords an excellent opportunity for using an outside speaker who can talk authoritatively and convincingly on general accident prevention work. Such a speaker can be obtained from a nearby plant, an insurance company, the city administration, an automobile club, or a community safety council.

If movies relating to accident prevention

are desired, a suitable selection can be made from films available through the National Safety Council, or through regional film service organizations, whose locations may be obtained from the Council.

Departmental meetings have many safety uses. Their purpose may be to discuss the company safety program so that employees will better understand what is going on. They may be held to provide information about accident causes and accident types. They may be purely inspirational, to create an awareness of hazards and a desire to prevent accidents.

In many company programs, departmental safety meetings are held monthly. The trend is toward meetings conducted by the foreman, who may receive assistance in planning as well as materials, such as visual aids, from the safety department.

The program for a departmental meeting may include the following:

1. Report of injuries in the department since the past meeting; report of a safety inspection in the department; and report of the department's standing in a contest. (The total time spent on reports must not be so great that this part of the meeting becomes tiresome.)

2. Discussion by the foreman of where observance of safe practices needs to be improved.

3. Talk, demonstration, or visual-aid presentation on an appropriate accident prevention subject. The speaker may be the foreman, a member of the department, the company safety director, an outside expert, or an executive of the company.

Departmental meetings give the foreman an opportunity to point out in a forceful manner the dangers of certain unsafe practices. By condemning those practices, he practically binds himself to set a good example for his men. In addition, most workers welcome an opportunity to "get off their chests" whatever safety ideas they have.

At the conclusion of departmental meetings, the foremen should be required to prepare written reports for presentation to the plant (or company) safety committees and

273

Fɪɢ. 12–7.—An on-site group meeting of persons involved in a construction project. A flip chart (left) is used to illustrate solutions to the problems at hand.

review by the managers.

Small group meetings with men doing similar kinds of work can be held at or near the work place. (See Fig. 12–7.) The supervisor or foreman may discuss the causes of an accident of which the men have personal knowledge or in which they have personal interest. The men should be encouraged to join in the discussion, and a conclusion should be reached as to show the accident might have been prevented.

The supervisor may present a problem that has developed because of new work or new equipment. Again the men should participate and offer their views.

At times the supervisor may present a film or chart talk on a subject related to the work of the group members. Other visual aids such as models or exhibits may be used. Safety devices or pieces of equipment or material may be shown and discussed.

"Production huddles" are instruction sessions about a specific job being done that includes safety. Such meetings are particularly useful with maintenance crews when an unusual job is about to start. The plans for doing the job safely and efficiently are gone over and a procedure is agreed upon. Public utility line crews use this type of meeting and call it a tail board conference. Before starting a job, the crew gathers around the truck and discusses the job, laying out the tools and materials they will need and agreeing upon the part each man is to do.

A particular advantage of small group meetings is that they provide excellent opportunities for presenting all types of information, including safety information, directly to employees and stimulate exchange of ideas that can benefit the accident prevention program. To be successful, the same safety meeting must include a tangible message, originality of presentation, opportunity for audience participation, and a conclusion that spurs action toward an attainable goal.

Planning programs

Making the safety meeting interesting is of the utmost importance. There should be no complaining or scolding. Talks should be definitely limited in time and they should start and end on time. The subject matter of a talk should be considered in advance to make sure that it is pertinent and does not repeat other talks recently presented.

274

Large occasional meetings need even more careful planning and timing than do small meetings. People who are to speak, including company executives, should review what they intend to say with the person planning the meeting to assure that their remarks will serve the desired purpose. Films and other visual aids should be checked in advance.

Persons responsible for employee meetings should observe them critically to see whether or not they are accomplishing the purpose for which they are run. When meetings are held periodically, there is always danger that they will become dull routine. Only continual effort and planning will prevent this.

A plan of action to develop a successful safety meeting includes these points:

PREPARE IN ADVANCE. The preliminary arrangement determines the results. Do not conduct a meeting without preparation.

SELECT A MAJOR TOPIC. Make it timely and practical—one that the group can discuss.

OBTAIN FACTS AND FIGURES. Be sure they are correct and complete. Make up a visual aid, such as a simple chart or table, whenever possible.

MAP THE PRESENTATION. Decide on the best way to present the subject of the meeting. Try to anticipate the group's reaction and questions. Outline results you hope to accomplish.

SET A TIMETABLE. Allow adequate time, but set a reasonable limit.

BE SINCERE. Your sincerity and your interest in the employees' welfare must be unmistakable.

INTRODUCE THE TOPIC. Tell in simple terms what the meeting is all about. Use a punch line or other good lead-in.

PRESENT FACTS, AROUSE INTEREST. State highly pertinent facts in an interesting manner.

PROMOTE GROUP DISCUSSION. Ask questions that cannot be answered "Yes" or "No." Prompt members of the group to think individually and collectively. Let *them* talk.

AGREE ON DOING SOMETHING. Try for group agreement on methods of correction

and improvement. Write these down.

SUMMARIZE THE MEETING. Review briefly what has been discussed and decided . . . follow up in the various departments.

Contests Stimulate Interest

Many safety professionals say that contests (such as housekeeping contests, interdepartment contests, and many other types) are inadequate substitutes for management interest, safe procedures, and "built in" safety. They will also agree that while a good accident prevention program is one and the same with good management, good training, and efficient operation, some *special* effort may be needed to maintain interest in good housekeeping, reporting hazards, and the like. Moreover, the interest value of contests (such as maintaining a high level of good housekeeping and efficiency) has direct bonus values in good publicity and improved employee morale.

A competition usually is held, therefore, only after the basic steps in a safety program have been taken—a policy statement made, a record system adopted, equipment safeguarded, a first aid department installed. Such substantial demonstration of management's interest, sincerity, and responsibility greatly helps to obtain the active participation of foremen and workers in a contest.

Purpose and principles

Safety contests are operated purely for their interest-creating value. A contest that creates favorable interest is valuable; one that does not create interest is worthless. In the usual type of contest, the competing groups are departments of the same plant or divisions of the same company. Generally, contests are based on accident experience and are operated over a stated period, with a prize for the group with the best record according to the contest rules.

Contests have been important almost from the time of the first safety programs, and a fairly well established group of operating principles has been developed:

1. A contest should be planned and conducted by a committee representative of the competing groups.

2. Competing groups should be natural units,

not arbitrary divisions.

3. Methods of grading must be simple and easily understood.

4. The grading system must be fair to all groups.

5. Awards must be worth winning and of the sort that create interest.

6. Good publicity and enthusiasm are important.

Contests may run for various periods—from a few months to a year. Those who recommend longer periods believe that if workers are kept on their toes for a longer time, safe working is more likely to become a habit. Some safety professionals, however, believe that greater interest can be aroused and maintained during a short period and therefore prefer short and frequent competitions. Contests of different duration can be tried to see which is most effective.

A safety contest stock certificate idea was developed by the Maxwell House Division of General Foods Corp. and ran for one year.

For each week a department works without a disabling work injury, the department receives a stock certificate worth 50 cents. Dividends are paid on this stock at the rate of: Ten cents for the first 1000 consecutive safe hours worked. Twenty-five cents for the first 10,000 consecutive safe hours worked. Fifty cents for the first 50,000 consecutive safe hours worked. One dollar for the first 100,000 consecutive safe hours worked. This means that if a department works 100,000 consecutive man-hours without a disabling injury it receives a total of $1.84 in dividends for each share of stock held.

There are penalties, however. If a disabling injury occurs in a department, that department will be penalized one month's stock earnings. This means that during that particular month the department could not be awarded any stock or dividends. It would also mean that if an accident should occur in the latter few days of the month the department would lose all dividends and stock certificates previously issued for that month.

Injury rate contests

In a contest based on injury rates, the measure of safety performance should be the frequency of disabling injuries per million man-hours. The rate should be easy to figure and explain.

Contests should not be based upon severity, because severity data cannot be determined promptly and because severity frequently is a matter of luck and contributes little to knowing how to prevent the accident in the future. Using a combination of frequency and severity is not good for the same reason. Contests should not be based upon reduction of reported first aid cases because people therefore may fail to report such injuries.

If in-plant contests are to create interest, they must have variety and originality. The most effective contests generally do not run continuously, with a new one beginning as the old one ends. They are more in the nature of special campaigns to run for a specified time and are launched with advance publicity and fanfare. Often the president or other high official makes the original announcement, presents awards, and otherwise lends his prestige to the contest.

Competition, if properly organized, can do much to develop teamwork. Some workers who apparently give no thought to their own work habits can be influenced to cooperate with their fellow workers if they know that their unsafe acts and resulting accidents will discredit their department or "team."

The NSC contests and awards. National Safety Council members firmly believe in the value of contests for maintaining interest. The Council has about two dozen Industrial Section Contests. All are open to Council members, and several include nonmember participants through specially arranged contests cosponsored with trade associations.

In the Council contests, companies are grouped according to size and operation so that competing units will be comparable with one another. The definitions and rules set forth in American National Standard Z16.1, *Method of Recording and Measuring Work Injury Experience,* are followed. (See Chapter 6 for details.)

Many Council members consider the Sectional Contest to be one of the most important features of Council service. Awards are presented at Sectional meetings, local safety council meetings, and occasionally at trade association conventions.

Most of the Council's Sectional Contests are integrated with the Council's Award Plan. Under this plan, four levels of awards are set up to provide some recognition for every good safety record of a member company or unit. In order of importance, the awards are:

AWARD OF HONOR

AWARD OF MERIT

CERTIFICATE OF COMMENDATION

PRESIDENT'S LETTER

The Award Plan recognizes perfect records (no disabling injuries) covering an entire calendar year.

Association contests. A number of associations conduct their own contests, which also adhere to the definitions and rules in American National Standard Z16.1.

Statistics from association contests are submitted to the National Safety Council, which uses them to give the corresponding industries' injury rates in its *Accident Facts* booklet. In addition to stimulating the interest of the associations' members, such contests give the Council a broad and reliable accident reporting base.*

The 'XYZ Company' contest explained

The "XYZ Company" is engaged in steel fabrication and erection. Rules for one of its safety contests are given in Fig. 12–8, but some explanation is needed.

RULE 1. Although contests can run for any period of time, the company chose a one year period to allow development of safe working habits—the useful objective of the contest—and, since some of the departments were relatively small, to eliminate random (chance) factors from influencing an individual department's experience record.

RULE 2a. Hazards in steel fabrication and erection vary greatly; therefore, fabricating units compete in one division and erection units in another. Operations are similar enough to have a common basis for determining standings—see Rule 4.

Other companies may find that hazards differ sharply from one plant or operation to

another, and the similar units may be too few to group. Such plants or operations may compete on an equitable basis in several ways.

● The participant achieving the largest percentage reduction in its frequency rate in comparison with a base period, such as the previous year, may be the winner. Since each unit competes against its past record as par, this method provides a fair basis for comparison of rates.

● The compensation insurance rates for different types of units in the same state have been used to establish a handicap factor to compensate for differences in hazards. If the rates per $100 payroll are $3.00 for Plant A, $2.00 for Plant B, and $1.50 for Plant C, factors for the units are 3, 2, and 1.5, respectively. The frequency rate of each plant is adjusted by dividing the rate by its factor. Plants are ranked from the lowest to the highest on the basis of the adjusted rates.

● The national average rates for different types of units may be used similarly for determining standings. If Plant A achieved a frequency rate of 10.0 and the national average for units in this industry was 12.0, Plant A's rate is 0.83 of the national average. Plant B's rate in comparison with its national average rate is 0.75. Therefore, Plant B would rank nearer the top than Plant A. Average rates for most industries are published in the National Safety Council's booklet *Accident Facts*.

RULE 2b. Separation of large and small units is essential because a small unit finds it easier to go through an entire contest period without a disabling injury than does a larger unit. If the number of units is sufficient, three size classifications may be set up. Unequal size groups can be competitive if they include units that can attain no-injury records. For this reason, the erection departments of the company in this example were divided into Group A—the five largest departments—and Group B—the remaining units. Five contestants in a group are usually considered minimum.

* For the time being and for all practical purposes, the National Safety Council will continue to gather information under ANSI Z16.1.

RULES FOR THE 'XYZ COMPANY' SAFETY CONTEST

Rule 1. The contest shall begin January 1, and end December 31, 19 . . . .

Rule 2. The contest shall consist of two divisions.
   a. Fabricating units shall participate in Division I.
   b. Field erection departments under the direction of a superintendent shall participate as separate units in Division II, which shall be divided on the basis of size into two groups: Group A shall consist of the five erection units working the largest number of man-hours, and Group B shall consist of all other units. Units shall be tentatively grouped by size during the first three months, and the final classification shall be made on the basis of total man-hours worked at the end of four months. No further changes in size groups will be made after April 30, 19 . . . .

Rule 3. Recognition awards shall be:
   a. Trophies to the winners in Division I and Groups A and B of Division II.
   b. Engraved certificates to plants and erection units ranking second and third in Division I and in Groups A and B of Division II.

Rule 4. a. The winners shall be the contestants having the lowest weighted frequency rates.
   b. In the event that two or more contestants in any classification have had no chargeable injuries during the contest period, the winner shall be the contestant who has worked the largest number of man-hours since the last chargeable injury.

Rule 5. All injuries resulting in death, permanent total, permanent partial, or temporary total disabilities shall be counted, as defined by ASI Standard Z16.1, *Method of Recording and Measuring Work Injury Experience.*

Rule 6. A sum of $50.00 shall be presented to the units that have had no disabling injuries during the first six months of the contest or have reduced their average frequency rates for the first six months 50 per cent in comparison with the average rate for the preceding six months' period. The award shall be divided into prizes of $12, 10, 8, 6, 4, 2, and eight $1 prizes and raffled to employees. No employee may win more than one prize.

Rule 7. Standings shall be compiled monthly and published in a bulletin that will be distributed to all managers, superintendents, and foremen.

Rule 8. All questions pertaining to the definitions of injuries and rules shall be referred to the Contest Committee, whose decisions shall be final.

Rule 9. Awards shall be presented at an appropriate ceremony to be announced at the end of the contest.

FIG. 12–8.—A typical set of rules for a safety contest. (See text for point-by-point discussion.)

STANDINGS IN THE 'XYZ COMPANY' SAFETY CONTEST
JANUARY–JUNE

Oakland leads at the halfway mark!

Tulsa moved into second place.

Chicago slipped from second to third place in June.

Corbin, in last place, had no disabling injuries during June. Good work!

Tulsa still leads for the President's Award for largest improvement over the previous year's record.

Plant	Rank	January–June Frequency rates°	% Increase + or decrease − over last year
Oakland	1	14.1	−30%
Tulsa	2	14.6	−40%
Chicago	3	17.5	+ 5%
Cincinnati	4	21.9	−20%
Corbin	5	24.8	+22%

*Frequency rate is number of disabling injuries per 1,000,000 hours worked.

Tips on How to Win, No. 6

One-sixth of our disabling injuries occur in the use of cranes and hoists. Have foremen hold a safety meeting on safe practices in hitching loads and other crane operations for shop men. We'll furnish a film. Use posters on the subject. Require safe methods. See the enclosed bulletin for suggestions on how to solve this major accident problem.

Fig. 12–9.—Simple monthly contest bulletin gives essential information about current standings and also includes a suggestion for improvement in "Tips on How To Win."

RULE 3. Awards should be specified. "XYZ Company" follows the general practice of giving first, second, and third place awards.

RULE 4a. The frequency rate is most often used as the basis for determining standings as it can be computed promptly and easily.

RULE 4b. There should be a satisfactory method of determining the winner between two or more units having perfect records in order to give the smallest contestant a fair chance. Selecting the winner on the basis of the largest number of man-hours worked since the last chargeable injury is fair to all units, regardless of size.

RULE 5. A standard method of counting injuries is essential in order to avoid controversies and maintain confidence in standards. American National Standard Z16.1 is described in Chapter 6, "Accident Records and Injury Rates."

First aid and other minor injuries generally

are excluded because, if they are counted, workers may fail to obtain treatment so that cases will not be put on the record. The result could be an increase in infections.

RULE 6. Various "special rules" may be used in a contest that runs six or more months' duration in order to stimulate and maintain employee interest.

RULE 7. Frequent contest bulletins keep everyone informed about standings. Bulletins can be posted and standings discussed at safety meetings. Contest results can be announced in plant publications and in other ways to build and maintain interest.

RULE 8. Questions about rule interpretation will arise and must be settled fairly. This is an important function of the contest committee, which, in turn, may refer decisions about disputable injuries to outside judges. A committee of the American National Standards Institute has been set up to interpret standard definitions.

RULE 9. (See section Awards Should Be Meaningful, later in this chapter, for award ideas.)

Interdepartmental contests

Interdepartmental competitions get "close to home" and place responsibility for a good showing on supervisors. Since workers have a greater personal interest in the standing of their department than in the record of the entire plant or other unit, this type of competition has proved popular for creating interest among both supervisors and workers.

An interplant contest plan often may be adapted to an interdepartmental competition. A company operating a number of similar plants may take advantage of workers' interest in their departmental records by conducting a competition among the same departments in various plants. Public utilities may conduct a contest among districts, and other nonmanufacturing companies may follow a similar plan. Since hazards in similar operating units are about the same, the basis of standings may be the frequency rate. Departments or other units may be grouped according to size. (A contest between divisional areas of a railroad is illustrated in Fig. 12–2.)

Most departmental contests, however, are conducted among dissimilar departments in one plant, and the departments often differ greatly in size. The difficulties due to variations in hazards and number of employees from one department to another may be overcome in the same manner as with similar differences between plants, just discussed.

One plan that produced excellent results was aimed particularly at foremen and supervisors in charge of departments having five to sixty employees. The principal rules were:

1. This contest will cover the period from July 1 to December 31.

2. Each department will compete against its own previous six months' accident record. Only disabling injuries are counted.

3. The department's accident experience will be judged by the frequency rate developed during the contest as compared with its frequency rate of the previous six months.

4. A suitable prize will be given each foreman who reduces his department's frequency rate in the contest months by 50 percent or more over his frequency rate in the previous six months.

5. If a foreman or supervisor had no disabling injuries in his department during the previous six months, the prize will be given if he meets his previous six months' record.

A bronze trophy was given to each foreman and supervisor who met the requirements for an award, and a dinner party was held. When the rates of the contest period were compared with the rates of the previous six months, the results were reductions of 30 percent in frequency and 62 percent in severity.

It is important that competing plants or groups be kept posted on the lasted standings —Fig. 12–9 shows a typical contest bulletin.

Intergroup competitions

Intergroup contests are particularly suitable for units that employ fewer than 400 people and have small departments in which hazards vary sharply. Employees are divided into teams of from 20 to 50 workers. To

Fig. 12–10.—A popular type of contest pits a plant or department against its own injury-free record—in this case, a 25-year record and almost 5 million man-hours without a disabling injury.

Courtesy Pennwalt Corporation.

equalize factors of size and difference in hazards, each team has a proportionate number of employees from the most and least hazardous occupations. Each group is led by a captain, whose principal duty is to contact members of the team and create interest in winning. Team members may be members of contest committees.

Interest is promoted by naming teams after prominent baseball, football, or other outstanding sports teams, and the entire competition may be named after a league or other sports organization. Team names can be drawn from a hat and membership can be shown by colored pin-on buttons.

Identification of workers by colored buttons helps to overcome difficulties in scoring. Since the members of different teams work together, they make sure that every case involving a competitor is charged against his group record.

Intraplant or intradepartmental contests

By setting standards of performance, an intraplant or intradepartmental safety contest lays particular emphasis on the responsibilities of supervision and employees for avoiding accidents. These standards are often no-injury records for varying periods, achievement of lower injury rates in comparison with a previous period, and improvement over the average injury rates of similar units or of the industry. In this type of contest, units of an organization do not compete with one another; rather, each unit attempts to match or surpass established standards. See Fig. 12–10.

Personalized contests

Some plans single out for acknowledgement employees who have safe records. Certificates are given to those who have worked one, two, five, and ten years without an injury.

281

Holders of certificates have found them useful in obtaining promotion and even in seeking employment with other organizations.

Periodic raffles of merchandise or cash, or the use of a new car for three or four months, regardless of the injury record of a department or plant, also have been used successfully to encourage and acknowledge the efforts of safety-conscious workers. Employees who are involved in disabling accidents (or drivers who have had a "preventable accident") during the period become ineligible for drawings.

Various sweepstakes plans have proved popular and effective in maintaining interest in a good record from month to month. One such plan was operated successfully by a branch plant of a well known paper company.

On the payday prior to the beginning of each month all hourly employees received cards with serial numbers at the pay office window. The workers wrote their names and departments on the cards, tore off the stubs and put them in a box, and retained their portions of the cards. Names of employees in departments having no disabling injuries during the month then were drawn for prizes ranging from $5 to $25. The company contributed a total of $75 per month.

Since eligibility for the drawing depended on a perfect departmental record, each worker had to be careful about his actions. Workers frequently corrected others for unsafe practices that might spoil chances for prizes.

If no accidents occurred during a three-month period, foremen participated in a drawing for prizes ranging from $5 to $15. This feature proved helpful in enlisting their cooperation.

Some companies give awards like first aid kits or saving stamps to those in every department that have worked a given period without accident. This approach is effective only if going a month, for instance, without accident is unusual.

Overcoming difficulties

A few difficutlies in operating contests are rather easily overcome. One is that some departments are inherently more hazardous than others.

In some contests handicaps have been established, based on manual rates of insurance companies or on average accident frequency for the different kinds of work. This has been discussed previously.

Another method is to base standings on improvement over past records. Thus, a department with a past average frequency of 20 and a current frequency of 15 would be rated as having made 25 percent improvement and would win over a department with a past frequency of 15 and a current frequency of 12— a 20 percent improvement.

Usually, in both methods an average of rates over three to five years is used as the base.

Another problem is that a department may have so many accidents at the beginning of the contest period that it is out of the running and loses interest. Having shorter contest periods helps to overcome this difficulty. Another remedy is to have different awards for different achievements. An award for the department having the longest run of injury-free man-hours is an example.

Division of responsibility for the cause of an accident may become a problem; in other words, an unsafe act of one supervisor's worker (or an unsafe condition in this supervisor's department) may affect another supervisor's record. These situations must be anticipated and the contest planned to deal with them fairly, thus creating a favorable interest.

Noninjury rate contests

Noninjury rate contests, such as safety slogan, poster, housekeeping, and community contests, can be just as effective in maintaining interest as contests based on injury rates. The objective in any case is to get the maximum number of people talking, thinking, and participating in safety. They are especially effective in promoting off-the-job safety.

Descriptions of various kinds of contests frequently can be found in the Council's *National Safety News* and in the *National Safety Congress Transactions*. On occasion, staff members or officers of the Council help judge such contests or assist in other ways.

Slogan, limerick, and poster contests. Safety slogan contests may be of various kinds. One can be for the best safety slogan sub-

Fig. 12–11.—In-plant presentation of the "bronzed safety shoe" to the plant foreman for maintaining an excellent safety record.

Courtesy Jorgensen Sheet & Strip Division.

mitted by an employee. Another may be run in which employees or their wives are rewarded if they can repeat the "slogan of the week" or the message on a certain safety poster.

The National Safety Council's big safety limerick contest, which is tied in with the use of the Council's Safety Calendars, attracts thousands of entries each month, an indication of its popularity.

Company magazines may conduct contests to "finish the limerick" or "write a rhyme" or write "twenty-five words on the best way to be safe." Often these are open to both employees and family members.

The value of homemade posters is in their special application to a particular industry or company. If the posters are the result of an employee contest, their interest-creating value will be increased, possibly exceeding that of the tailor-made variety of posters. An important ingredient of such a contest is to get employees and their families participating in the planning and judging stages too.

Frequently, employee poster ideas—aside from the quality of the art work—are so good, that companies even submit the winning contest entries to the National Safety Council for possible conversion into printed safety posters.

A housekeeing contest often is conducted among departments. This type of competition is fundamental because it is aimed at accident *causes* and usually tries to eliminate unsafe practices and conditions.

Housekeeping contest plans differ from one company to another. The following plan is used successfully by a metals firm.

Once a week a committee of three representatives of management inspects each department and reports unsafe conditions to the superintendent. A copy is furnished to the works manager, and another is kept for the use of later inspection committees. A demerit for each unsatisfactory condition is charged to the department. If the condition is not rectified within one week, an additional demerit is chalked up.

At the end of the month the demerits for each department are totaled, and departments are rated on the basis of the proportion of demerits to the total number of employees in the department. If Department A employed 175 people and had 25 demerits, its rating would be 85.7. This figure is obtained by dividing 25 (number of demerits) by 175 (number of employees), multiplying by 100, and subtracting the product from 100. Standings are posted monthly on the bulletin boards in each department.

Awards are made at a mass meeting held after the lunch period. Names of employees in winning departments are placed in a box from which is drawn the name of the winner for the month. The winner's picture is posted on a special bulletin board, and a short talk on safety by a representative of management is broadcast throughout the plant. (A general rule prohibits an employee from winning more than one award during the contest.)

The name of the winning department is inscribed on a plaque each month. The head of the winning department receives the plaque from the previous month's winner at the mass meeting, and at the end of the year the department that has won the plaque the greatest number of times receives it permanently.

Community and family contests. Many companies have stimulated interest by sponsoring safety essay or poster contests for children of employees, local school children, or young art students. The publicity before and after such contests plus the interest generated by the posters themselves and the judging not only stimulates the interest of employees but also promotes the company's community and public relations.

More than one company has launched a safety poster or essay contest for the children of employees with the full knowledge that the employees would give their children considerable help and that there would be much favorable discussion about the contest in locker rooms, lunchrooms, and car pools.

Miscellaneous contests. There is an endless number of different types of contest possibilities. Often they can be combined with injury reduction contests. Contests can be held for attending safety meetings, for wearing safety shoes, for reporting unsafe conditions or unsafe acts, or for off-the-job or public safety activities of individuals, departments, or branch plants.

Although contests are popular with both management and employees, the safety director always should attempt to determine before starting them whether or not they will require time and effort that should be spent on providing safer equipment or better training for supervisors and employees (Fig. 12–11).

Contest publicity

All stages of a contest should be played up as dramatically as possible. Placards and news stories should prepare for it. Standings should be announced at frequent intervals. Special posters can be used for this purpose. Handout bulletins can be sent to individual employees urging care in keeping the record perfect. The company magazine and even the local newspaper and radio station can make good copy of a contest. Trade journals and National Safety Council newsletters are other outlets for contest publicity. In a smaller community, outstanding safety performance by a well known company deserves —and usually gets—excellent publicity.

The publicity value of a successful contest is considerable, although difficult to estimate. It is a fact that large companies invest literally thousands of dollars and hundreds of hours of planning time to set up awards, judge, and make elaborate presentations.

Publicity should be commensurate with the real significance of the occasion. The presentation of a watch or other personal award to an employee who has gone 25 years without a disabling injury has human interest value to a company paper and perhaps to a local newspaper or trade journal.

Recognition of an exceptionally fine "no injury" record made by a corporation (one million injury-free manhours would probably be a minimum), or presentation of a National Safety Council award to a company deserves a different and more impressive type of publicity.

Publicity and the photograph shown in Fig. 12–12 appeared in a Sacramento, California, newspaper. The information was forwarded

Fig. 12–12.—A. W. Koppes, plant manager, and A. Andrews, president of the Independent Oil and Chemical Workers' Union receive NSC's "Award of Honor" from Mrs. Sandra Smoley, member of the Sacramento County (California) Board of Supervisors.

Courtesy The Procter and Gamble Manufacturing Company.

to the media in the form of a press release, which indicated that the company's local plant had received the National Safety Council's highest industrial award, the "Award of Honor," for operating five-million manhours without a lost-time injury. In addition to the information pertaining to the local operation, the news release also noted that the company's nationwide manufacturing operation had maintained an impressive industrial safety record throughout the year and had received eight "Awards of Honor" for "outstanding safety performance."

To further promote the company's good safety record, the news item also further noted that the companies plant in the Phillippines (Manila) also holds the world industrial safety record for the soap and glycerine industry—more than five-million manhours without a disabling injury. See Chapter 13, "Publicizing Safety," for details of preparing a news release.

Some companies pay for radio or television time to announce the results of a contest. Others arrange for photographs and stories to go in their company papers, on their bulletin boards, and perhaps in local or trade association papers. One company had large campaign-type buttons made, and photos of children wearing them appeared in the local press.

Fig. 12–13.—This eye-catching display of thousands of dollars of U.S. Savings Bonds was used in a campaign promoting safe driving.

Courtesy Aero Mayflower Transit Company.

Meaningful awards

An award serves several purposes. It is an inducement, a builder of good will, a continuing reminder, and a basis for publicity. To serve these purposes, however, an award must be meaningful.

The winning and display of a multiplicity of awards may detract from the true value of the program, particularly if the awards are given too freely by sales-minded donors. Employees sense when awards are given only for sales or publicity purposes and are based on little or no safety effort.

The value of awards lies in their appeal to basic interest factors, such as pride, need for recognition, urge to compete, and desire for financial gain. Monetary awards, however, should not appear to be bribes. In general, select awards which are worthy of good publicity, which will photograph well, and which provoke conversation. The distribution of U.S. Savings Bonds for safe driving records gets away from the appearance of "bribery" and, at the same time, promotes a sense of patriotism and thrift. (See Fig. 12–13.)

The originality or cleverness of an award or of its method of presentation is an important factor. A cup of coffee awarded to all employees in a department after the completion of one million injury-free hours probably would create more favorable comment than the presentation of a fancy plaque to the department supervisor. The drawing of a small cash prize or a grab bag prize would attract more interest than a routine presentation of the same award. An award to an employee's wife for completing a home safety checklist, for identifying a safety slogan, or for contributing to the company paper would create more interest than the same award given on the job.

● One of the Council's members reported an award idea that received an unusual amount of publicity, both within the plant and locally. A local automobile agency loaned the company a luxury limousine that was driven for a week by an injury-free employee whose name was drawn from the hat. The employee and the company received excellent publicity in the local papers; the employee had a special "reserved for John Doe" parking space in the company lot. The only cost to the employer was a few dollars to cover special insurance. Even a special parking space alone can be effective. (See Fig. 12–14.)

● Another way to gain interest is to let the employees participate in selecting the award, planning its presentation, and helping with publicity. Frequently, the employees will suggest a humorous or novel award or publicity approach that may attract more interest than one planned by management. In any case, the employees' participation is a wise investment in safety promotion.

● Payment of bonuses as a type of award for good safety records evokes considerable difference of opinion. Some managements and safety people feel that this approach is unwarranted; and some plans using it have proved unsuccessful or have been abandoned.

● Sums of money are divided into various amounts, or government savings bonds are purchased and raffled, particularly for outstanding achievements by departments and small plants. Some companies raffle household or sports merchandise. Many employees value attractive pins or engraved cards commending them for years of employment without an accident. One company places a safety record sticker on the employee's hard hat. Others provide special badges, pins or shoulder patches to recognize safety achievement or service on a safety committee (Fig. 12–15).

● Interest in safety among foremen and workers often is developed by personal awards like wallets, knives, or key cases, often suitably inscribed.

In addition to contest awards, recognition should be given to those who have saved lives, served on safety committees, submitted valuable suggestions to the management, or made other significant contributions to accident prevention.

Fig. 12–14.—A "reserved parking" privilege at a choice location is an inexpensive but very effective award for contest winners and for employees who achieve outstanding safety records.

Courtesy Kaiser Aluminum and Chemical Corp., Chalmette, La.

Fig. 12–15.—Shoulder patch can be used to recognize safety achievement or service on a safety committee.

Fig. 12–16.—Displays and contests are effective ways to maintain employee enthusiasm and to obtain participation in various phases of a safety program. For maximum publicity, photographs can be posed against a meaningful background.

Courtesy Underseas Division, Westinghouse Corporation.

Award presentations

Make an award presentation something special, and recognize it by well planned publicity. One company rents an auditorium and invites civic and labor leaders, "main office" executives, other dignitaries, and employees' families to a large celebration party. The presentation of even a modest award to an individual worker or supervisor would call for the participation of company officials and perhaps family members or other workers.

The president of the company himself, or some other high official, can present the awards at a general meeting, a picnic, or a dinner (or breakfast) that may even include entertainment. The reason for inviting VIP's is not only to add prestige to the presentation, but also to promote their interest and commitment to safety.

The presentation requires planning, and must be in keeping with the importance of the occasion. The place of the event should be appropriate and comfortable, not noisy or crowded. An award to an individual might be made in an executive office; a group award might be made in a conference room, private dining room, or in a company lounge or cafeteria (during a non-rush period).

Brief the participants on the agenda. Familiarize those who make the presentation with the significance of the award, the achievement it recognizes and the background of the individual(s) who earned it. Arrange for reporting and photographing the event in order to get maximum publicity. Photographs can be specially posed after the actual presentation to take advantage of desirable backgrounds, such as the plant or company name, some prominent trademark, or other interest-catching effect (Fig. 12–16). (See NSC Data Sheet 619, *Photography for the Safety Professional,* for other ideas.)

For a group award—such as a company, a plant, or a department's completing an injury-free year—free refreshments, such as coffee and cold drinks, can be offered for a specified time—ranging from one coffee break (per shift) to a full 24-hour period.

Another, more elaborate award was given by the president of a company employing about 300. At the end of a year in which there had been no disabling injuries, he took the group to a major league baseball game.

The following year, after his company maintained its no-injury record, he invited the group, plus their families, to an all-day picnic and cruise.

Such activities help build better employee and public relations, as well as promote more interest in safety.

fluence, making employees feel that the company is not sincere. (See NSC Data Sheet 616, *Posters, Bulletin Boards, and Safety Displays.*)

Purposes of posters

Posters properly used have great value in a

Fig. 12–17.—Poster messages should be brief and simple.

Posters and Displays

Posters and displays are meant to reach large numbers of people on the move with brief, simple messages, designed to accomplish one or more missions—to convey information, to change attitudes, and/or to change behavior. They are designed to communicate with people going about their normal activities and consequently the audience must be seized, the message conveyed, and the contact finished in a very brief span of time. (See Fig. 12–17.)

Safety posters are one of the most visible evidences of accident prevention work. Because of this, perhaps, some companies have mistakenly assumed that posters alone would do the safety job and have negelcted such essentials as real management support, guarding, and job instruction. In fact, hit-and-miss use of posters in a plant where no other safety work is done is likely to have a negative in-

safety program through their influence on attitudes and behavior. One has only to see the efforts that commercial advertisers make to acquire space in business areas or near factory gates in order to appreciate the value of posters *inside* the work place.

When posters are selected, it is well to have in mind their specific purposes:

1. To remind employees of common human traits that cause accidents

2. To impress people with the good sense of working safely

3. To suggest behavior patterns that help prevent accidents

4. To inspire a friendly interest in the company's safety efforts

5. To foster the attitude that accidents are mistakes and safety is a mark of skill

6. To remind employees of specific hazards

289

Posters are useful also in supporting special campaigns, for instance, using guards, wearing eye protection, maintaining good housekeeping, or offering safety suggestions. Posters promote traffic, home, and even pedestrian safety by encouraging safe work habits.

Effectiveness of posters

A number of studies have been made on the effectiveness of posters for training and motivating.

• One study was conducted by the British Iron and Steel Research Association. Three posters that reminded workers to hook cable slings were displayed in six plants over a period of six weeks. A seventh plant was used as a control. Tallies made in the six plants before and after display of the posters showed about an 8 percent increase in compliance with the rule. The seventh plant, in which the posters were not used, showed a very slight decrease in compliance.

Although use of the posters merely supplemented previous training, plants that originally had the lowest rates of compliance showed some of the best gains. Use of the test posters separately, on a biweekly basis, proved slightly more effective than simultaneous use of all three posters during the entire six-week test.

• In a survey conducted by a prominent casualty insurance company, over 200 employees were interviewed in depth on the effectiveness of safety posters, films, and leaflets. Results indicated that all the media were instrumental in bringing workers to a high level of safety awareness and that all were effective in sustaining that awareness. Employees were found to prefer posters to leaflets, although they acknowledged the value of leaflets for more detailed coverage of, for example, "off-the-job safety."

• A survey of Council members indicated that about three-fourths of the nearly 800 respondents use a variety of poster subjects with one-third preferring cartoons and all-industry posters. Horror or shocker posters were least preferred.

Fifty-four percent of the respondents use posters to influence general attitudes; 27 percent to cover special operations; 18 percent, to meet special or seasonal problems; and 14 percent, to promote off-the-job safety.

Sixty-nine percent of the respondents in the Council's poster survey preferred posters of 8½ by 11½ in. and 17 by 23 in. size. Of these two sizes, the smaller was preferred six-to-one over the larger.

While there can be no question about the interest stimulated by a "sexy" photo, there may be some question as to the effectiveness of the safety message, particularly if the photo is unrelated to the message. Frequently, an unusually striking photo or overly elaborate (and expensive) artwork actually detracts from the safety message. This is not to say that "eye-popping" illustrations should never be used; occasional use may serve to attract attention to the more conservative, serious messages on other posters.

Types of posters

Posters available from the National Safety Council, insurance companies, associations, and other sources fall into two broad categories: general and special industry (or special hazard).

The general posters are concerned with such subjects as chance-taking, disregarding safety rules, forgetting to replace guards, and other human failures.

Special industry posters, as the term indicates, have application only in specific industries, such as mining or logging.

Special hazard posters, for example on lifting, ladders, the storage or handling of flammable liquids, are useful in every industry where the particular hazards are encountered. In some cases, a hazard is so serious that a special poster is developed because of the severity and not the frequency of exposure.

The National Safety Council carries more than 1000 different safety posters in stock at any one time. These range from pocket-size pressure-sensitive stickers to billboard-size jumbo posters. About 15 new posters are added each month.

Subject matter is roughly in proportion to the occurrence of certain types of accidents. For example, more posters are concerned with materials handling than with chemicals and gases. (For illustrations of the many posters available from the National Safety Council, see the *Catalog/Poster Directory*.)

Other locations. Posters and stickers can be mounted on delivery trucks, industrial trucks, mail trucks and carts, in elevators, and even on doors. Pocket cards or plastic pocket protectors, such as those available from the National Safety Council, might be called "walking safety posters." (See Fig. 12–18.)

FIG. 12–18.—"Walking safety poster." Construction superintendent lets workers select own safety message, place it on their own safety hat.

Courtesy Stone & Webster Engineering Corp.

Other materials. Safety messages need not be limited to printing and artwork. They can be very effective when used in illuminated or changable signs. One company paints a safety message on a plywood welding screen.

Homemade posters. A company can develop its own posters to deal with special hazards not covered by posters available from outside sources. Even the smallest company can make an occasional special poster inexpensively, using colored paper, crayons, or felt marking pens, to call attention to a special hazard, to commemorate the winning of a safety award, or to point up a problem not likely to be covered by a commercial poster. See Chapter 14, "Audio-Visual Media," for details.

Effective posters can be made using photo-graphs of local conditions or accidents, even if the situations must be posed. A common type of homemade poster is the "testimonial" showing a photo of an employee and a close-up of his damaged safety glasses or safety shoes, with a brief statement explaining how this equipment protected him.

Homemade posters on new processes, new guards, or new rules personalize the safety program and augment even the best selection of commercial posters.

Changing and mounting posters

No specific rule can be given for the frequency of changing posters because of varying definitions of the term "poster."

Some types of posters may well be mounted permanently. For example, a poster on artificial respiration can be kept in the first aid room, or one on the use of a certain kind of fire extinguisher can be posted near it.

Most companies change general interest posters at definite intervals, usually weekly, perhaps rotating them from one area to another or filing them and then reusing them after a year or so.

The type of posters displayed should be varied. Consecutive posting of several infection posters, for instance, or of machinery posters is not desirable unless a special campaign is being conducted. To secure proper balance, it is better to use an eye poster, then a machinery poster, next an infection poster, and so on.

For maximum effectiveness, posters not only must be selected carefully and changed on a definite schedule, but also displayed attractively in well lighted locations where they will be seen by the greatest number of people. They should be placed on safety bulletin boards, near time clocks, in cafeterias, and at points of special hazard, such as paint storage rooms, rubbish cans, hazardous machinery, or dangerous intersections.

Bulletin boards

Bulletin boards should permit convenient change of posters and should be placed where employees can see them when they are momentarily at leisure, near drinking fountains (Fig. 20–2). They should be centered at eye level, about 63 in. from the floor. They should be in a well lighted place, especially

12—Maintaining Interest in Safety

Fig. 12–19.—In addition to inspirational and educational safety bulletins, this "information station" carries pertinent chemical safety data as well as NSC booklets for employee use.

Courtesy Borg-Warner Corporation, Washington, W. Va.

lighted if possible. A good size for a bulletin board is about 22 in. wide by 30 in. long.

Boards should be attractively painted and glass-covered. One board at a location in the work place is usually desirable, but in lunchrooms or locker rooms several panels may be used effectively. Flashing lights, sometimes desirable in nonproduction areas, are likely to be objectionable in workplaces.

A bulletin board should be used for only one display at a time, but need not carry safety posters exclusively. Any program of mutual interest to company and employees may legitimately use the bulletin boards. In fact, safety posters may have a stronger appeal if they appear on a board on which employees occasionally see displays on other subjects.

Bulletin boards in the same company may range from large, enclosed, illuminated boards with special sections for posters, safety bulletins, and other messages to a number of small frames or other inexpensive poster mounts installed at strategic points.

The National Safety Council has available black enameled poster frames to which clip-on literature racks can be added. This permits convenient distribution of leaflets and other pickup literature which support the safety message. (See Fig. 12–19.)

Fig. 12–20.—An eye-catching seasonal display for shoes.

Fig. 12–21.—Safety professional points out some of the items in his company's off-the-job safety display, which emphasizes the need for automobile maintenance scheduling.

Courtesy Allen-Bradley Company.

Displays and exhibits

Personal protective devices, tools, and pieces of fire fighting equipment can be used to make up displays or exhibits on bulletin boards, with or without corresponding posters, if design of the bulletin boards permits.

Another good interest catcher is a combination of an NSC poster, a seasonal topic, and a safety display. (See Fig. 12–20.)

Displays can also be used to promote off-the-job safety as well, for example, the need for a continuing automobile maintenance schedule. (See Fig. 12–21.)

Signs with changeable letters, electric tape messages, or eye-catching lighting can be used for safety displays. (See Fig. 12–22.)

Many simple and attractive displays have been devised for presenting statistical data to workers. One is a safety clock, the face of which is marked off to indicate the frequency of disabling injuries. Twin clocks or dials often are used, one recording the present rate and the other the rate for the corresponding period of the past month or year.

One company used large thermometer-like boards, placed at every gate and clock-house. Arrows indicated the present and previous month's records. The comparative standings of departments were shown below.

An auto race was the theme of another display; each car represented a department. The cars moved daily to denote progress being made. Airplanes can be used similarly. Another exhibit featured race horses participating in a "Safety Derby" and named after items of personal protective equipment.

FIG. 12–22.—This sign helped employees count down each day of the final 100, before plant reached 4000 injury-free days.

Courtesy Ideal Cement Co., Trident, Mont.

Other Promotional Methods

Other methods that can be used effectively to arouse and maintain interest in safety are campaigns, safety stunts, courses and demonstrations, publications, public address systems, and suggestion systems.

Campaigns

Campaigns serve to focus the attention of the entire plant personnel on one specific accident problem. They are, of course, additions to and not substitutes for persistent accident prevention effort the year round.

Campaigns may be undertaken to promote the use of safety shoes, home safety, vacation safety, or fire safety. A "Clean-Up Week" may be held, or a "Stop Accidents" campaign

may be run to promote safe attitudes both on and off the job.

The National Safety Council's nationwide campaigns "Zero In on Safety," "Take Time To Be Safe," and others included posters, films, booklets, and specialty items to promote and maintain the interest of employees. In addition, special campaign materials have been developed in collaboration with trade associations and for special problem areas such as "Stop Shock," and "Fight Falls." Information on current campaign material is given in NSC catalogs, and other publications, and is available from NSC headquarters.

Moreover, large corporations have conducted extensive campaigns to promote safety on and off the job. Much of this safety-awareness material is aimed at families of employees, local citizens, and even groups outside the U.S.

The Dow Chemical Company's campaign, for example, was built upon the theme "Life is Fragile—Handle With Care." It included a series of 26 full-color safety posters, a sound-slide presentation, and a movie using local citizens acting without a written script. Bumper stickers, shoulder patches, buttons, pocket protectors, and anti-litter bags all supported the awareness program's theme. (See Fig. 12–23.)

Suitable publicity should be planned for the campaign from kickoff to conclusion, similar to that discussed for safety contests earlier in this chapter. Signs, flags, desktop symbols, and other items can be used to dramatize the campaign. To wind it up, a special event can be scheduled, such as giving each employee an inexpensive novelty item, free coffee, or a free breakfast or dinner.

Many of the same promotion stunts or ideas that help maintain interest in contests also can be effective in special campaigns. For example, a first aid drill or a demonstration of artificial respiration may be given. Some companies use safety parades, exhibits of unsafe and safe tools and equipment, pledge cards, and other such features.

Timeliness may be an important factor in the way employees respond to a campaign. Successful safety campaigns have been linked to elections, World Series, the football season, Thanksgiving, and other special events.

See NSC Data Sheet 616, *Posters, Bulletin*

Boards, and Safety Displays, for more ideas.

Off-beat safety ideas

Off-beat safety ideas or "stunts" capitalize on all the effective aspects of showmanship and thrive on an endless variety of ideas. They can be developed as separate devices for maintaining interest or can be used to supplement contests and campaigns. *National Safety News* and other publications regularly give details on various stunts.

Stunts that criticize or ridicule seem to belong to the past. However, if handled with a light touch and in a group that accepts them good-naturedly, stunts with a negative approach may have their uses.

Typical of such stunts is the giving of a goat or an old broom to the department with the most demerits for poor housekeeping. It would be better, however, to give a clean broom to the cleanest department.

Most companies agree that constructive stunts help inspire employees to high standards while stunts that ridicule may do more harm than good, particularly if the employees resent that they have been treated unfairly. More important, employees and supervisors who are the objects of ridicule may have just cause to blame management for not setting up safe procedures or providing safe facilities and equipment.

Safety stunts can involve an entire company, a department, a small group, or just the individual. A stunt may be humorous, novel, or dramatic, and occasionally even shocking.

A simple stunt is often most effective. A pivoted hammer, mounted over a pair of safety glasses in a display case, can be operated by a string to demonstrate the impact resistance of the glasses. To dramatize the importance of eye protection, the "let's pretend" test can be used. Several volunteers are blindfolded and then asked to light a cigarette, eat, write, and move around.

Stunts developed for the company safety program often can be used at company open houses or safety picnics and in community safety projects as well. Such stunts, when supported by visual aids, signs, and printed material, demonstrate the company's interest in accident prevention and give the employees a chance to participate in programs that help create safer attitudes on the job, too.

Off-beat "posters" are interest-getting. Fig. 12–24 shows a hand that reaches out of a pocket.

Courses and demonstrations

Most safety professionals agree that courses in first aid, lifesaving, water safety, civil defense, and disaster control have bonus values that help prevent work injuries, too.

The worker who has gone through a course in first aid and has learned to give artificial respiration will be more mindful of the hazards of electric shock and more likely to help maintain electrical equipment in safe condition. Likewise, the employee who learns how to stop arterial bleeding better appreciates the consequences of using a saw or a power press without the guard.

Fig. 12–23.—The "Life is fragile" theme (on the plaque held by one of company's employees) received widespread publicity throughout Texas when one of the company executives was issued this special license plate.

Courtesy Dow Chemical U.S.A.

295

Home study and extension courses, although designed primarily for training purposes, also serve to stimulate and maintain interest. They give the employee a better understanding of the job and do a great deal to dispel unsafe attitudes. Most safety training courses, in fact, are designed specifically to improve the attitudes of both supervisors and employees. The use of good visual aids will enhance the effectiveness of the courses; the use of video tape for a technical discussion demonstrates the progressive nature of the company as well. See Chapter 14, "Audio-Visual Media."

The participation of the public in courses taught or attended by employees promotes community good will. Many industrial safety men are doing an excellent job of promoting safety and fire prevention through arranging courses on these subjects for the Boy Scouts and Explorer Scouts, the Girl Scouts, Junior Achievement groups, and other youth or school groups.

The National Safety Council's "Defensive Driving" Course provides an excellent way to promote good employee and public relations. They also stimulate safer attitudes both on the job as well as off.

Demonstrations of fire equipment have a practical value beyond that of teaching employees how to react in an emergency. The mere fact that the equipment is provided for their use reminds them of management's concern about their welfare. Moreover, the demonstrations make employees more aware of the dangers of fire and point up the need for obeying fire prevention rules.

Demonstrations of fire equipment by local fire departments or distributors of fire equipment are easily arranged. Many companies conduct their own demonstrations, using extinguishers that require recharging, or "not-in-service" extinguishers kept specifically for this purpose.

Publications

Reports. The safety professional should make reports on safety program progress interesting to his superiors and to supervisors. Visual aids can be effective. (See Chapter 14.)

Once a procedure for such reporting is set up, it may be administered routinely by an insurance department or an accounting de-

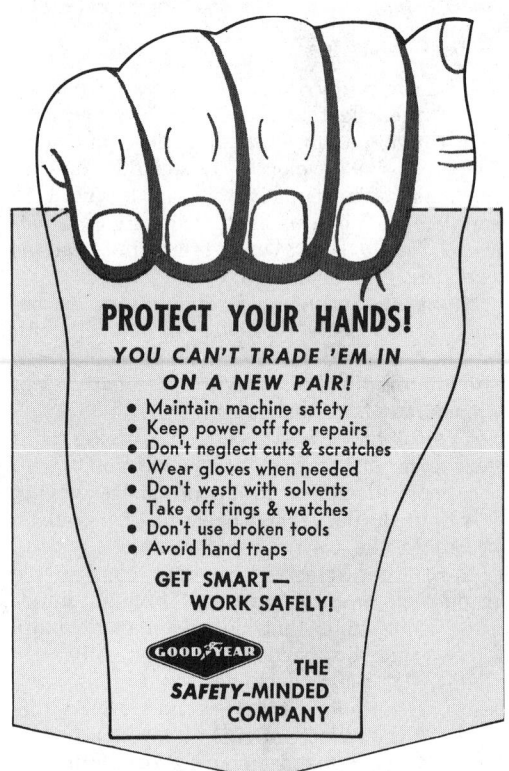

PROTECT YOUR HANDS!

YOU CAN'T TRADE 'EM IN ON A NEW PAIR!
- Maintain machine safety
- Keep power off for repairs
- Don't neglect cuts & scratches
- Wear gloves when needed
- Don't wash with solvents
- Take off rings & watches
- Don't use broken tools
- Avoid hand traps

GET SMART— WORK SAFELY!

GOOD/**YEAR**   THE **SAFETY**-MINDED COMPANY

Fig. 12–24.—This die-cut pocket card has "fingers" that seemingly reach out of the pocket, tempt viewer to inquire about the message (which is revealed as card is pulled out).

Courtesy Goodyear Tire & Rubber Co., Inc.

partment or even may be made a part of production cost figures.

The cost of accidents and, perhaps, the cost of prevention should be given in terms that are significant to management, such as medical and compensation costs, production losses, sales losses, increased maintenance costs, and the less tangible but perhaps more important hidden costs involved in administrative problems and in impaired public, customer, or employee relations. Reports need not be dull. Photographs, for example, can pin-point a company's four major sources of disabling work accidents.

In one company a statement of accident losses and safety achievements may be included in the annual report, whereas in another a special annual or monthly safety report may be issued to top executives and supervisors. If departmental accident losses, like injury frequency rates, can be charged on an equitable basis, such as "per hundred thousand dollars of sales" or "per one thousand man-days of production," the comparative standings of departments and improvement in departments or units are easy to evaluate.

The fact that such information is recorded and publicized is in itself an incentive to supervisors. It reminds everyone concerned that accident costs are just as much an integral part of profit and loss as production, sales, maintenance, distribution, and advertising.

Special charts, graphs, and statistical reports can be used to give the facts about accidents. One chart can show the number of disabling injuries, others the number of days lost, injury causes, accident causes, or body location of injuries. It cannot be too strongly emphasized that unless such charts are kept up to date they can do more harm than good.

Annual reports. In recent years a great many companies have gone to considerable lengths to make their annual reports to stockholders interesting and clearly understandable. In many cases, annual reports are distributed to employees so that they can become better acquainted with the company's purposes and problems. A section on aims and accomplishments in accident prevention attracts employee interest and further serves to emphasize the interest of management in the safety of its employees.

Publicity regarding a good safety record may be arranged in local newspapers or in trade journals. Such publicity is particularly valuable in a smaller community where the company is well known or where the quality of a product, so far as the public is concerned, is reflected somewhat in its safety record. (This discussion is continued in depth in the next chapter, "Publicizing Safety.")

Newsletters. Monthly or weekly newsletters are especially important as a means of maintaining interest. They keep employees and supervisors informed, particularly in centralized or field operations where bulletin boards are not feasible. Such newsletters can give detailed information on standings in a safety contest and publicize unusual accidents or serious hazards. They can help explain safety rules, remind employees of safe work practices, and support the safety program in general. If the employees can serve as "reporters" or help produce such a newsletter, so much the better. (More details in the next chapter.)

Booklets, leaflets, and personalized messages take many forms: safety rule booklets, special "one-shot" leaflets, monthly publications such as the National Safety Council's *Industrial Supervisor* magazine and its *Safe Worker* and *Safe Driver* for employees, and letters from management.

The content of an employee rule booklet, except for material involving company policy, may be developed with the help of safety committees or selected workers as a means of stimulating interest and helping ensure compliance with the rules.

Larger companies may have their own editors and artists, even their own printing facilities, and produce publications of professional caliber. Smaller companies, however, also can issue attractive booklets, leaflets, and personalized messages, and at negligible expense.

The National Safety Council, trade associations, and professional organizations publish a wide variety of booklets and leaflets that are authoritative, attractive, and inexpensive. They cover a large range of subjects—handling materials, first aid, housekeeping, fire prevention, vacation safety, safe driving, and the like. Such materials, carefully selected and regularly distributed, effectively supplement company prepared publications.

Letters commending meritorious service, signed by the manager and addressed to individuals, make an excellent impression upon workers.

Safety calendars, published by the National Safety Council, together with Christmas letters from the manager have a direct appeal that reaches the workers' homes. Such mailing lists should include each employee. *Family Safety*, a quarterly publication of the National Safety Council, is sent to over one million homes by managements who are interested in the welfare of their employees and their employees' families.

Buttons, blotters, book matches, pencils, and other small novelties, all conveying a safety message, also may be used. Booklets of silicone tissues for cleaning glasses, imprinted with brief messages or safety rules, serve to remind employees of the rules and to encourage proper use of safety equipment.

Public address systems

Public address systems often are used to broadcast announcements and page employees. Many companies have taken advantage of these installations to broadcast safety information.

Such messages should be planned carefully. Employees might readily lose interest in long speeches or too-frequent safety reminders. When the public address system is used for broadcasting music, safety announcements can be made between numbers.

Suggestion systems

Because accident prevention is closely associated with efficient operation, many suggestions help not only to prevent accidents but also to lower production cost; not only to improve manufacturing conditions and methods, but also to better the health and change the outlook of workers.

Effective suggestion systems, like many other things in life, cost money. Many companies pay considerable sums for employee suggestions—one firm awards more than $10 million annually, but feels the money is well spent because of the value of the suggestions themselves.

Improved employee interest and personal involvement are additional benefits. Many companies prefer the positive, more enduring motivation of recognition and awards, to the "negative admonitions" of some safety publicity. Safety suggestions are also regarded as a highly desirable way to avoid safety grievances.

If a company does not have a general suggestion system, it is probably better not to establish one for safety suggestions alone. Setting safety apart from ordinary operating procedures may de-emphasize its importance.

Getting good suggestions is important and must be encouraged by all ranks of management. Posters, contests, campaigns, merchandise incentives, direct mail, printed handouts, personal appeal, supervisory training, safety clubs, and press releases are employed to motivate employees to submit safety suggestions.

To merit an award, a safety suggestion, like a production suggestion, should be substantial, be practical, and be a real solution. Changing a method or material, guarding a hazard, inventing a safety tool or device are examples of suggestions worthy of awards. Erecting a sign, cautioning workers, or publishing slogans are examples of ideas usually not considered eligible for an award.

Suggestion awards. It is easy to measure the monetary value of suggestions that result in greater efficiency, less material cost, decreased labor cost, or reduced waste. Usually awards for suggestions in these categories are in proportion to the savings derived by the company. Although some safety suggestions also have a monetary value, they are hard to evaluate; hence, payments for suggestions that contribute to the welfare of the employees, but result in no direct savings to the company, are most often estimates or composite judgments. Some firms have developed guidelines which consider such factors as the degree of hazard, originality, extent of application, etc. One company has also developed an award guide based upon disability cost experience.

Distinguishing between "safety awards" and others is a mistake that can result only in a feeling that safety is regarded by the company as a sideline of no great importance. Payment for a safety suggestion must be on the same basis as that for other suggestions—it should be based upon its real worth if it can be determined. If a suggested safety device enables an operation to be run at a speed that would be dangerous without the device, a saving may be measured. If a number of accidents have occurred on an operation and a suggested device will eliminate them, the cost of those accidents can be projected and a saving calculated.

Awards should reflect the merit of the suggestions. Most companies award cash and/or bonds. Some award merchandise, all-expenses-paid trips, company stock, certificates of merit, medals, gifts for the suggester's wife and family, or recognition luncheons.

Some companies exclude superintendents, foremen, designers, methods and systems men, and other supervisory or technical personnel from receiving awards, so that the other workers will have someone to whom they can go for assistance. Some companies feel supervisory personnel should not be excluded and several firms have separate plans and award schedules for salaried, supervisory, technical, and management personnel.

Suggestion committee. If necessary, a special subcommittee can be set up to determine the monetary value of safety suggestions so that employees will be rewarded for them exactly as they would be for other money-saving suggestions. However, some firms believe such special treatment sets safety ideas apart from other ideas.

Many old, established suggestion systems now in operation are producing excellent results in savings. No company should start a suggestion plan or decide not to start one without first studying carefully the plans now in existence. Further information can be obtained by contacting the Executive Director, National Association of Suggestion Systems. See References.

Boxes and forms. Suggestion boxes should be attractive and well placed, and stocked with special blank submission forms. It is essential that management acknowledge and resolve all suggestions promptly, to increase the interest of the employees and establish a spirit of cooperation and importance.

Commercial suggestion forms also are available.

References

National Association of Suggestion Systems, 435 N. Michigan Ave., Chicago, Ill. 60611.

National Safety Council, 425 N. Michigan Ave., Chicago, Ill. 60611.
 Accident Facts (annual).
 Catalog/Poster Directory.
 Family Safety Magazine.
 Industrial Supervisor Magazine.
 Industrial Data Sheets
  Motion Pictures for Safety, 556.
  Nonprojected Visual Aids, 564.
  Photography for the Safety Professional, 619.
  Posters, Bulletin Boards, and Safety Displays, 616.
  Projected Still Pictures, 574.
 National Safety News Magazine.
 101 Ideas That Worked
 Safe Driver Magazine.
 Safe Worker Magazine.

Publicizing Safety

Chapter
13

Public Relations and Publicity   301
Basis for success

The Voice of Safety   302
Working within the company . . . Sixteen ways to make safety news

Some Basics of Publicity   304
Select the publicity audience . . . Use humor and human interest . . .
Names, not statistics, are news . . . Friendly rivalry . . . Publicity
techniques . . . Publicity by the safety office . . . Safety on the air
. . . Hints for TV interviews . . . Handling an accident story

Working with Company Publications   311
Producing a publication

References   314

The preceding chapter covered "internal" publicity. This chapter discusses how to influence the way a company looks to people on the outside, and how to keep people on the "inside" informed of what is going on. Favorable publicity is an unmistakable bonus to a good safety program. Why it is so often left uncashed is difficult to understand.

Any company likes to have someone—especially a prospective customer—say, "I like what I hear about this company. I understand that it really takes care of its employees. So I figure it must treat its customers right; therefore, I'll be treated right."

One good way for a company to get a reputation for taking care of its employees is to be known as a really safe place to work.

Yet an amazing number of companies do little or nothing to let their public—customers, stockholders, the community—know that the safety and welfare of their employees are important to them.

That is what this chapter is all about. It is an effort to present a simple and sensible formula for letting people know that "at my company, the welfare and safety of the workers are important."

Most companies have a professional public relations department which handles the communication program. In smaller companies, the safety director may have to generate his own publicity. In both cases, the information in the following pages should prove useful. The safety professional should be aware of overall company policies and programs and know when to turn over routine portions of his safety publicity to specialists, and when to ask them for creative help.

Public Relations and Publicity

There is a difference between public relations and publicity—a big difference.

Public relations is the "management function which evaluates public attitudes, identifies the policies and procedures of an individual or an organization with the public interest, and plans and executes a program of action to earn public understanding and acceptance," according to the magazine *Public Relations News*. Every employee, every activity, every facility of a company contributes in many ways to the overall feeling that persons outside the company have about that company. This is true public relations.

Anyone concerned with accident prevention in any way—safety professional, foreman, member of the plant safety committee, or officer of the school or community safety council—should realize that any time he communicates with someone outside his committee, department, or company, he is involved in public relations.

Publicity is a specialized tool of public relations. It is the technique used to acquaint the public with something an organization or an individual is doing via editorial time or space in the mass or trade media. As such, it merely brings to light what is already happening. Although important to good public relations, publicity cannot do it alone.

Any public relations program has to be backed up by a sound organization. Public relations reflects the quality of an organization—but it cannot create that quality. Successful safety achievement merits and can result in good publicity, but canned publicity or publicity based upon inflated facts, or specious statistics will be recognized for what it is—and can do more harm than good.

Publicity, to click, need not always be red-hot news, but it must have an element of spot news, or human interest, or self-help. Then it will have feature value.

Activity—real, honest, legitimate activity—makes news. Of course, urgent need or dramatic circumstances help make news, too. Until they turn up, however, genuine effort will go a long way toward giving a program news and publicity value.

Basis for success

The basis of a successful public relations program is a successful management—management that makes sure that staff and employees produce good products safely and efficiently, that they cooperate with each other and with the customers, and that all give the best and friendliest service humanly possible—and give it at all times.

The plain fact is that poor public relations is costing individuals and organizations in this country millions of dollars each year.

The remedy is simple; a better understanding and use of fundamental public relations on the part of everyone—and a sincere effort to

13—Publicizing Safety

Fig. 13–1.—A plant tour—often an important part of a public relations program—can demonstrate a company's concern for the safety and well-being of visitors as well as employees. Providing in-plant transportation for visitors in one way of showing such concern.

Courtesy Pontiac Division, General Motors Corporation

put it into practical use.

For lack of good public relations, many a worthy cause has failed to get the support it deserves, and many an organization has failed.

A good public relations program need not cost a great deal of money. But it is worth time, effort, and a reasonable budget.

The Voice of Safety

In any genuine, effective public relations (PR) program, emphasizing safety can be a real help. In fact, it is hard to imagine a PR program where sincere and effective concern for the protection of employees from accidents is not a top priority.

If a company does not have a safety program, it misses vital opportunities for good public relations and dramatically increases the chance for adverse public attitudes.

The safety professional should not only welcome publicity for his safety efforts, but should energetically seek it.

Working within the company

First, it is essential to talk a little about the

basic facts of public relations and publicity. There are two questions to be answered:

1. Does the company have a public relations department?

2. Is there an employee publication in the company?

If both answers are "yes," the safety professional should get in touch with both these units before he does anything about publicizing the safety program.

This step is important. It not only assures professional skill and consistency of efforts to publicize safety activities, but it will save confusion, avoid duplication, and possibly prevent misunderstanding.

Why does such an obvious procedure have to be mentioned? The reason too often is that there is little, if any, communication between a safety professional and the publicity man and publications editor.

Communication between the safety engineer and the publicity man and editor is indispensable, for these three must work together, or safety is not going to get the attention it deserves and needs.

The safety professional should tell the publicity man and editor, if he has not done so before, that he is more determined than ever to cut accidents in the company and that he realizes their help and advice are needed to reach this goal.

He should point out to them that he is fully aware of the necessity of employee and public acceptance of the safety program and that they are the people who can help get it.

There is a wealth of real news in safety, and there is a strong possibility that everyone in the safety business has been too quick in assuming that safety must by its very nature be on the dull side. In recent times, there has been more and more recognition by more and more writers and others that safety can be made interesting. It just takes the combined efforts of safety professionals, publicity men, and editors to turn the trick.

Sixteen ways to make safety news

It might be useful to list some of the things that can make safety news in an organization and that editors and publicity men ought to know.

1. No-accident records for the entire company or for any one unit—in terms of either days or man-days.

2. Improved safety records for the company or any one unit, even if no prolonged no-accident period is involved.

3. An interplant safety contest, or an inter-company contest—especially if anyone has dreamed up an unusual angle (like an out-of-the ordinary prize).

4. Any unusual safety record for safety performance by an officer or employee of the company—either in length of time or character of the job done.

5. Innovations in safety programs of the company that will prevent accidents. An invention, too, as special news value if the company has been plagued with accidents the new gadget may prevent.

6. An unusual or highly valuable safety suggestion by an employee.

7. Safety conventions or meetings, either those held by the company or those held elsewhere to which company representatives will go. A digest of such a report should be publicized.

8. Other special safety events besides conventions—a safety banquet, a safety training course, fire and first aid demonstration, a special meeting, or an award ceremony.

9. Some unusual stunt intended to get the employees to take their safety training home to their families, or something the company is doing directly with the families of workers in an effort to promote around-the-clock safety. Open-house tours, local water safety shows or public showings of safety films are examples.

10. Some pronouncement or statement by the president or other high official of the company on some unusual or new safety device or company safety service, such as free inspection of employees' cars.

11. A speech by the head of the company or the safety professional at a local, regional, state, or national safety convention or conference. The editors or publicity men should have advance copies of it. The person making the speech should be sure to say something worthy of public attention.

12. The company's annual report is the foundation for corporate communications. Stockholders *do* read these reports. A good paragraph or two on the safety record for the past year will go a long way in achieving sound publicity within the corporate family, as well as inform the analysts, who recommend stocks, what the company is doing beyond its financial performance.

13. Any act of heroism by someone in the company. This is a sure-fire story for local papers as well as company publications. Maybe this type of news is not pure safety, but newspapers regard it as part of safety, and it can always be tied in with an indirect safety message.

14. A survey or study of some phase of accident prevention in the company. If the investigators discover that married men who own their own homes are safer than their bachelor brethren, they have pro-

13—Publicizing Safety

FIG. 13–2.—Pretty girls in an industrial situation can help publicize a safety activity. Here, safety award plaques are made even prettier.

Courtesy Ford Motor Co.

vided a ready-made story.

15. Election or appointment of a company official or safety professional to an important post as a volunteer officer of the National Safety Council, American Society of Safety Engineers, Board of Certified Safety Professionals, local safety organization or governmental agency.

16. A company or employee winning an award in a National Safety Council contest. Winners, not losers are publicized.

A good rule of thumb is to stress the positive, rather than the negative side of an event. For example, instead of a story that says "bachelors are less safe," it could say that "a company study shows a need for special safety efforts by bachelors."

Some Basics of Publicity

If a company does not have a publicity department, this fact need not prevent its chances of getting publicity into local papers or on the air.

It hurts, of course, because publicity men are more experienced in getting publicity and naturally know their way around in media circles better than the safety professional does. However even in the absence of a company publicity department, the safety professional can get publicity for safety activities by going directly to the newspapers, magazines, and radio and TV. He might solicit and secure advice and help from local safety organizations, or even business or trade associations with such service. He should certainly, however, keep his manager or vice president informed of what is going on.

The safety professional should not pretend to the editors and the program managers that he is a publicity expert. On the contrary, he should cash in on his innocence of any of the wiles professional publicity men employ to get space or time.

Editors and program people are usually not difficult to approach—provided that the safety professional is courteous and friendly and admits that he lacks specialized knowledge of publicity techniques. Naturally, no editor or anyone else likes to have someone come charging in and pretend that he is doing a big favor by delivering the story the world has been waiting for.

Although the common sense and salesmanship needed for success in the safety field certainly are enough to enable the safety professional to present his case clearly and effectively to the paper or radio or TV station, he should, nonetheless, be willing to accept advice from publicity professionals on how to best tell his story.

Select the publicity audience

"Who must be reached with safety publicity?" This seems like a fundamental question, but many companies never try to answer it.

Industrial or manufacturing companies, of course, would scarcely mind if the whole populace insisted on reading or hearing or looking at the company's publicity and taking it deeply to heart.

Because publicity cannot reach everyone, the audience must always be chosen carefully, especially if the budget is tight or time is limited. In that case, it is logical to assume that—in addition to the in-plant (company) audience—the company would prefer to reach people who might be in a position to buy the product, or help the company in some other direct and profitable manner.

Use humor and human interest

It is worthwhile to try to brighten safety, to make it positive, rather than ponderous and dreary. It is even possible to evoke a chuckle now and then.

Editors are familiar with the solemn pronouncement that "Safety is a serious subject, and must be taken seriously—safety is no laughing matter." No one can argue with that position. Of course, safety is a serious subject. Of course, an accident is no laughing matter. But does it follow, therefore, that no one can put into safety—the enemy of accidents— some of the same techniques, the same sales appeal, the same sparkle that are used so successfully to sell all the things people need to keep them shipshape?

If those techniques can sell shampoo or a personal care product or an automobile, is it unreasonable to expect they can also sell safety? There is nothing wrong with using a little "sex appeal" to publicize safety, as long as it is done with good taste. Overdoing it will have a negative effect. Simply remember that you are promoting safety, not the pretty girls (see Fig. 13–2).

Or how about a cartoon treatment? This just may brighten what might otherwise be a slightly dull and drab presentation.

The safety professional should not be too disturbed if someone points out that a cartoon has treated safety negatively. It may well have done just that. This is the very thing that gives a cartoon its punch. A "prat-fall" cartoon will draw attention to the slippery, icy sidewalk in a way that cannot be shown by a person walking and not falling. There is no need to dread being negative now and then. However, care should be taken to avoid ridiculing or negatively portraying ethnic or minority groups and victims of accidents.

It is even possible to get a cute child or baby into the act, or even a faithful, shaggy dog, in order to get that spark, that punch, that human touch that lifts safety activities out of an impersonal rut.

Names, not statistics, are news

Remember that facts and figures about injuries and their frequency and severity are not really interesting in themselves. There must be a good-sized injection of human interest in safety news, and human interest means people.

The safety professional who wants publicity must talk more about people and safety, and less about things and safety. The quickest way in the world to drum up interest in a stuffy safety meeting, for example, is to develop a discussion or even an argument as to whether men or women can drive better or work better—"better" in this instance means safer.

Everyone knows this question can never be settled, but it can be used to keep interest in safety alive.

Friendly rivalry

Safety awards, safety records, safety contests, safety inventions, and "gimmicks"— these are only a few of the many things that make good safety news.

If the company is trying for a new injury-free record in its industry, it is headed for headlines. The editor and the publicity department must be kept informed all along

Fig. 13–3.—Make the most of getting a well earned award. Photographs of the key people involved can be published in company and local publications.

Courtesy Metropolitan Sanitary District of Greater Chicago.

the way. They will help arouse public interest in the performance, and also stimulate greater interest and greater effort among the employees themselves.

An award is worthless if kept a secret. It is worth only what is made out of it. Photos of award presentations are commonly used for publicity but they should be interesting and even unusual to attract special attention. (See Fig. 13–3.)

In some instances, top safety awards have been accepted by some companies as if they were only a dime a dozen. On the other hand, other companies have made similar awards the occasion for some of the biggest, bell-ringing celebrations ever seen—and "safety stock" took a big rise as a result.

A public utility company in Michigan, for example, made so much of its intercity rivalry over the safety record of the various units throughout the state that an outsider might think the winning city had won the World Series.

At one banquet marking the celebration of such a victory, more than 1500 employees, from the top brass on down, jammed a big hall to "whoop it up."

During the closing months of this intercity contest, the excitement among employees was akin to pennant fever in the baseball leagues, and no one dared to violate a safety rule.

This victory got tremendous coverage in the papers and on the air throughout the entire state of Michigan. Here was publicity—and public relations—that any organization would welcome. Safety had made news; the publicity had made safety.

Publicity techniques

These pointers are offered to the safety professional who wants to make the most of his public relations and publicity opportunities:

1. Be honest in what you say. Never exaggerate. Underplay instead of overplay, if you have to make a choice.

2. Deliver what you promise. If you say to the press that something is going to happen, make certain it happens—and as you said it would. This often calls for a "run-through" in advance.

3. If for any reason there is a change in plans from what you have announced, notify the papers and radio and TV stations at once.

4. Be scrupulously accurate in your names, places, and other facts. There is no such thing as being too careful in this respect. If an editor misspells names, the only thing to do is to give him the names correctly spelled again and hope for the best.

5. Be reasonable in your requests for space and time. Complaining about your company PR department, or complaining that the local paper or station has treated your company shabbily will not only accomplish exactly nothing, but will make for a bad relationship.

6. Do not alert your PR department or publications editor (or put out anything yourself) unless you have real news or features to offer. You must not issue material just to be issuing it. Be reasonable with the amount of material you send out. You can wear out your welcome.

7. Tip off your local or industry association, safety councils, and your publicity department (or if you do not have one, the papers or radio and TV stations) to anything worthwhile you run across that might make an item or program for them, even though it has no relation to you or your company. They will appreciate it.

8. Above all, do things that make news. Almost every routine safety item can, with a little extra effort by the safety professional and the editor, become a more readable, more constructive piece. News will be published only if something is being done for safety that makes news. News can always be heightened by intelligent, imaginative treatment, but it must be there in the first place to be worth telling. Advertising space can be purchased for special items.

9. Use good sense and an honest approach if the news is bad. Prove to press representatives and the public that you can roll with the punches. (Check with legal counsel and public relations officials on how far you need go, however.)

Publicity by the safety office

If a safety professional must handle his own publicity with the local press, he should also

OFFICE OF THE

COMMISSIONER OF STREETS AND SANITATION

ROOM 700, CITY HALL

RICHARD J. DALEY
MAYOR

CHICAGO 60602

JAMES J. McDONOUGH
COMMISSIONER

July 1, 1974

NEW EQUIPMENT AND TRAINING
FOR STREETS AND SANITATION

(For immediate release.)

Refuse collection personnel on a national level sustain 109.95 disabling injuries per 1 million man hours. The rate for all other industries is 10.17. Chicago's refuse collectors do their job with about 65 per cent fewer injuries than the national average. But even with this lower rate, the City of Chicago lost 4,833 days because of accidents in 1973 with direct costs of more than $650.00.

To cut these costs and to improve collection services, the Bureau of Sanitation and the Department's Safety and Training Division are attacking the problem from two directions: equipment and training.

A personal protective equipment program is being launched to provide the City's 2000 refuse collection employees with reflective safety vests, aprons, gloves, palm guards, shoes, and glasses.

FIG. 13–4.—News release should be double-spaced typewritten. Ordinary letterhead can be used.

know the following. These hints might seem unnecessary, but many stories have died because someone failed to observe them.

1. Timing is vital when calling on the editor or program director. It is considerate to call up first, suggest that the item might interest him, and ask him when it would be most convenient to drop in. One should never suggest he send someone out.

2. Generally, the man to contact is the city

Fig. 13–5.—Cooperative planning with editors and program directors is the basis for effective news coverage. Here a camera crew covers the announcement of Vincent L. Tofany as president of the National Safety Council on October 29, 1973. At right is Howard Pyle, former NSC president.

editor of the paper or the program director of the radio or TV station. Of course, if an item is specifically written for a certain columnist or commentator, it is better to contact him directly. If it applies only to a specialized area (finance or sports, for example), it should be brought to the attention of that editor.

3. Write not for your boss, but for your reader or listener. Answer objectively the questions: who, what, when, where, how and why. Do not load your releases with propaganda for the company. There is no

surer way to kill your positive relations with the media.

4. Make your releases just as professional in style, appearance, and general quality as you possibly can. (Fig. 13–4 shows an example.) Note the lead time that is required. Often one must ask a speaker, in advance, what he will say.

5. In writing a release, be brief and to the point. Newspaper space is limited, and costly. Try to "hold down" the piece to a page or a page and a half at most. Papers receive thousands of releases each month.

These are skimmed, and only the best get into print.

6. If you are sending a picture with the release, the caption should be typed on a piece of plain white paper and pasted to the back or bottom of the photo. *Never* use paper clips, and *never* write on the back of the picture. Either of these will likely damage the photo and make it difficult to reproduce clearly.

Safety on the air

Newspaper publicity is only half the battle. The safety professional who would mold opinion must get air-minded and see what can be done to get safety on radio and TV.

In the first place, it is amazing to find how much of a safety program lends itself to radio and TV. The publicity department can, of course, provide guidance, but the publicity people must know what they have to work with before they can offer it to the radio and television stations. A few possibilities are:

1. TV goes in for femininity. There is no choice but to play along with this basic, age-old approach. If there is an attractive damsel, it should be easy to unearth some way for her to model a new safety cap or hairnet. TV will like that.

2. Why not a safety fashion show, modeled by working girls? If the station prefers to use its own models, the pros can model the safety clothing. TV viewers will then be able to see what the well-dressed working woman wears on the job.

3. The publicity department can stage a wrong vs. right program on what the men and women working for the company should wear. This subject is a natural for company publications, and chances are, TV will nibble at it, too—especially if they can use a lot of girls and manage to "mug it up" enough so that the "wrong" examples are a little exaggerated.

4. If an employee has come up with an idea or a device for preventing accidents, and can demonstrate it visually, he is a possibility for a TV spot.

5. When one of the company's officials has been chosen for a state or national safety post, such as a director or officer of the National Safety Council, radio and TV may be happy to salute him as a local personality who has been tapped for a top job.

6. Any time the company can hang up a fine safety record, the local radio stations should get a chance to interview some of the men responsible for it.

7. If an important person (in the safety field) comes to visit, the safety professional should see that advance word of his visit reaches the media. (See Fig. 13–5.)

8. Leave script writing to the script writers but check facts. If a radio or TV station requests material, send them the facts, figures, and whatever narrative is necessary. The people at the station will put it into the proper form.

Hints for TV interviews

The safety professional must often be the spokesman for his company, not only for newspaper coverage, but also for radio and television. Although getting the facts correct and watching legal implications may be adequate for a newspaper interview, a television or radio interview reflects more of the company than merely what the facts show. It projects a company image through the company spokesman. If you do not feel that you project a good image over the radio or television, pick someone who will, in your department, or in the public relations department.

Here are some tips that will help you give a better radio-TV appearance.

1. Remember that you are being interviewed for your knowledge, not for your personality, entertainment value, or good looks. Be yourself. Don't put on a special voice or worry how the lavaliere microphone looks with your suit.

2. Don't worry about being nervous. Just don't panic. Don't back out of the interview after the station crew has set up its equipment.

3. Go over with the interviewer in advance just what areas he wants to discuss. You can steer him away from areas you cannot discuss, and you can get help on questions that you might not be able to answer.

4. Using notes is OK. If you must read, read normally, but well. Do not rush or drag. Be sure to maintain eye contact with the interviewer or camera. Do not memorize a statement and rattle it off. You will waste everyone's time.

5. If you have a bad cold or sore throat, turn over the interview to a colleague.

Handling an accident story

In any public relations or publicity program, it is just as important to know what not to do as to know what to do. In fact, it can be even more important.

The foremost warning is this: do not cover up bad news. Good press relations are of utmost importance. It is at such a time that a sound public relations program "pays off."

Every safety professional hopes the day will never come that an accident—a bad one—occurs and knocks the props out from under him and his safety record, but it has happened. In some instances, the repercussions of the way the accident was handled have been even more tragic than the accident itself—at least, to the company as a whole.

Here is an example of how not to handle a press representative: In a midwestern city some time ago, two workers were killed by a crane in a big plant.

This company enjoyed a first-rate relationship with the newspapers and radio and TV people in that city. It worked hard at safety and at public relations. It was good to its employees and had a fine reputation for playing its cards fairly and on the table.

On this particular occasion, however, someone in the company's higher echelons got "buck fever." So, when a young reporter came out to the plant to get what was to his paper a routine story of the accident, he ran into censorship at the plant.

The safety professional shoved him off to the personnel manager. The personnel man switched him to the general manager, who gave him some "double talk" and tossed him to the company doctor. The doctor said the safety director was the man to talk to.

By this time the reporter's righteous wrath was rising. He knew he was getting the treatment, and what had started out to be just a routine assignment now had become a challenge to dig up something that, for some reason, appeared to be covered up by the company.

The reporter could not lose in a contest like this. Since the workers had been killed, the coroner would have all the facts. If they had been badly hurt, one of the hospitals would have the information. If the men had not been killed or hurt badly, it was no story in the first place.

So the reporter got the facts from the coroner's office, and he wrote a story that was just as nasty toward the company as he could make it without committing libel.

The story was edited, headed, set in type, and lay in the composing room, awaiting its turn to get into the paper.

Now in this paper, as in every other paper, there is usually more news set in type than the paper can print. Each day dozens of items get left out—the "overset," as it is called in newspaper parlance.

The story of the accident might well have ended up as "overset" and, if printed, might not have been played up. These no longer were normal circumstances, unfortunately. The cover-up and run-around the reporter had received at the plant had changed all that.

This little story had been marked "must" when it was sent to the composing room. It thereupon became something very special—a story that was now given front-page prominence.

When there is bad news to report, the safety director will just have to swallow hard, grit his teeth, and back up his publicity department 100 percent in giving out the news as straight and fast and completely as if the tidings were all in the company's favor.

Along with the grief, a mention of the good things—that this is the first accident in months or years, that the company has a safety record far better than the national average for its type of operation, and that it has won a number of safety awards—will help take the curse off the story. The newsmen are usually happy to include these facts, too.

It is not only fair and honorable, but downright smart to "lay it on the line" for press representatives whenever there is news, regardless of whether it is pleasant or unpleasant news. This principle is vital to a good public relations program.

It would be wise for a safety professional

to anticipate that some day he may have to serve as a company spokesman at an accident or disaster scene. He should, therefore, seek legal counsel to make certain he knows how much he can say in a press interview.

News media can actually help during a big emergency. Families, friends, and neighbors will be clamoring for news and the media can get it to them fast. Details of any casualties must first be given to next-of-kin.

Working with Company Publications

If the safety professional thinks that safety has been neglected in his company publication, it is time to correct the situation. He should ask the editor frankly how to get more news value and human interest into safety stories. He should tell the editor he wants the program to be just as newsworthy as it can possibly be and that he realizes he needs something besides cold facts and colder figures to make safety articles and pictures attractive to his readers.

The editor is just as eager as anyone to publish interesting news and features, and he will go more than halfway to think up ways to put news value and reader interest into safety doings.

Here is an example of how the safety professional and the editor can team up to make a routine safety happening more newsworthy.

Suppose one of the employees, Oscar B—, reaches his twenty-fifth anniversary of steady work without a day's lost time due to an injury. This achievement probably entitles Oscar to a button or a badge or a plaque or something.

The public relations-conscious safety professional asks, "Well, instead of just pinning this button on Oscar with a hearty handclasp and a few words of commendation, why not make a real thing out of it? Take the occasion to tell Oscar—and all the world—that at this plant there is nothing more important than recognizing the contribution to a safer, better way of work that Oscar has made through his personal example of safe practices over the years."

Spurred by the talk the safety professional and the editor have had recently about perking up safety news—or maybe just because he is an energetic sort, anyway—the editor does not merely publish a picture of Oscar and his award along with one flabby little item.

The editor finds Oscar, sits down with him over a cup of coffee, and asks him a few questions about his career, about his opinions on safety "way back yonder" and now, and about any ideas he may have for making things even safer at the plant.

Oscar, either on his own or inspired by a leading question, might observe that "Women workers are, for my money, a darned sight safer workers than men," or vice versa.

Now the editor is on the trail of a story. Oscar has put a little human interest into what could have been just another item. The editor, his nose for news twitching a little, not only writes a sprightly piece on Oscar's provocative comment, with a picture, for the company magazine, but tells the company publicity department what the star employee has said.

So not only do the workers at the plant get the quotable quote on safety from Oscar, but the publicity department may also give the quote to the papers and to the radio and TV.

In the next issue of the employee publication, the editor permits some other employee—masculine, of course—to take issue with Oscar on the women vs. men thing. This gentleman says that although he reveres womanhood, nevertheless, he is constrained in the interest of accuracy to say that Oscar must be out of his mind.

The uninitiated will ask, "But what's this sort of folderol got to do with preventing accidents?"

The answer is—a lot. In the first place, this little exchange between Oscar and his male pals—and, of course, the women libs will get into it, too—puts the safety program into the minds of the employees.

In the second place, the women are going to try harder to live up to Oscar's faith in them. At least some of the men are going to say, "Oscar's flipped his lid," and that men are much, much safer on the job than women are.

By this time, word has leaked out to the community—perhaps with the help of the public relations man—that Oscar's allegation about women vs. men is stirring up some good natured rivalry at the plant, which is already noted for its safety program. The publicity department arranges to have a female and male worker appear on a local TV show, or on a radio program, and debate the issue—mentioning where they work, of course, and getting in a good word for safety in general and for their company in particular.

Now the foregoing is only one little example of what can happen when the safety man and the editor of the company publication get together to do a more imaginative and energetic job of publicizing the safety program.

The same system can be used when one department of the company wins an interdepartmental safety contest. Instead of merely recording the results of the contest, the editor can dig into the program of the winning department, interview the people responsible for its success, and, perhaps, come up with a piece for his magazine that will give every department some hints on how to improve its safety activities.

Producing a publication

Materials, such as safety newsletters, instruction cards, bulletins, broadsides, booklets and manuals for communicating safety rules, information, and ideas in print, require careful planning and preparation. Among steps to be taken in planning both internal (to a company audience) and external (to the general public or other out-of-company group) publications are:

1. Clearly define the objectives of the publication. Consider the type of audience to be reached by those objectives.

2. Determine how general or how restricted the message is to be.

3. Decide what form of publication will best convey the message.

4. Estimate cost of preparing and printing the publication in whatever forms, sizes, and quantities needed. An expenditure for a new publication must, of course, be provided for in the budget, whether or not the item is produced "in house."

If the objective is to place in the hands of the worker the specific rules he is to follow in doing his job safely and efficiently, an instruction card may be suitable. To stimulate general safety-consciousness, a broadside (single sheet printed on one side) may be effective.

If a series of short reminders, for example, on fire prevention, is needed, posters may be the answer. To treat a topic of general interest, such as methods of materials handling, a leaflet may be used. Here, posters or leaflets from the National Safety Council, insurance company, or other organizations may be more effective, and more economical than "in-house" produced material. For highly technical jobs or for more thorough coverage of a plant's safety policies and rules, manuals may be required. Even a company-wide (or plant-wide) public-address system would be appropriate.

When the form of the publication is being decided, it should be remembered that there is a direct relationship between the appearance of a printed piece and the degree of interest which it arouses. Most readers will react unfavorably to a bulletin, newsletter, or booklet with text in very small type, few or no illustrations, narrow margins, and long paragraphs.

Reasonably large type (10 point or 12 point), selected to fit the size of the page and, of course, to accommodate the volume of material, will help readability. For comparison, this column is set in 9-point type. Elite typewriter type is 10-point size. The Safety Newsletters, published by the various divisions of the National Safety Council, are set in 9-point type in 2⅛-inch wide columns. In addition, judicious use of white space and variety in size and placement of illustrations help make a publication both pleasing to the eye and easy to read. In safety, as in other fields, ideas conveyed in print are best received and best absorbed if they are well organized and attractively presented.

Illustrations serve to break up the text and help to get points across to the reader. Photographs which show action described in the copy add realism in instructional materials such as manuals. Human interest photos are desirable in newsletters. Line drawings and sketches are valuable to clarify technical points on instruction cards, in manuals, and in other training materials.

If the printing process permits reproduction of photos and other illustrations, pictures of award winners, safety devices, and safe and unsafe practices can be used. To avoid em-

```
                                            RELEASE NO. _____


                                    Date _____

                                   Place _____

       For the consideration of _____ , the undersigned grants permission to
       _____ and its assigns, to publish and reproduce the attached
       photographs of persons or objects shown therein.  It is understood that my name
       will not be used in connection with the aforementioned pictures.

       It is further declared that the undersigned has legal authority to sign this
       document.



                                    _____


                                    _____
       Description of photographs                            Witness
```

FIG. 13–6.—One form of model release. Some companies have each employee sign a release when he starts to work. It is suggested that legal counsel be sought before setting up any company procedure.

barrassing or ridiculing employees who have been injured or caught in an unsafe act, their features can be blocked out, or pictures specially posed (and so identified), by other employees can be taken of similar situations.

Since some states have laws that forbid publication of a person's photograph without his written permission, a signed release should be obtained from every person who appears in recognizable form in any picture. Often having a new employee sign a photo release is part of the employment routine. Asking for a photo release is just good manners. See Fig. 13–6 for sample.

Details on illustrations are given in Chapter 14, "Audio-Visual Media."

Preparation of material. Once the objectives, scope, and form of a publication are determined, the person preparing it should make an outline of the subject or subjects to be covered. For most types of material, the outline need not be elaborate, but it should be logical and complete, showing how each topic is a part of the overall plan.

Before gathering material, the writer might well spend some time studying the people for whom the message is intended so that he will know something of the knowledge and comprehension of the readers-to-be. In the interests of accuracy, completeness, and balance, material should be gathered from several sources—including articles, books, and especially foremen, workers, and others in the company who have had experience in the matters to be treated. To ensure technical accuracy it may be necessary to solicit help from specialists in specific areas.

No matter what the form of the publication, the writer should keep in mind certain basic rules of good writing. To get ideas across quickly and easily, short sentences, simple words, and brief paragraphs are recommended. Try to avoid being so simple that the copy reads like a second grade primer, however.

In a piece of some length, such as a booklet, a system of headings, kept as informal as

313

possible, will both arouse the reader's interest and guide his thinking as he reads. In a piece designed to instruct, numbered lists of job steps, for instance, will prove helpful. In any case, the writer should follow closely the line of logical thought developed in his outline.

Copy should be written in a positive, constructive style. When the nature of the material and the form of publication permit, a friendly—but never condescending—tone can be used effectively. Personal references and names, as in a newsletter, will increase readership. Humor tied to the message and pitched to the employees' sense of what is funny can add a great deal to some types of publications. For instance, cartoon illustrations and a light touch in copy may be particularly effective in a rule booklet.

Readability of the proposed publication can be gaged by having a few of the people to whom it will be addressed test-read it for understanding.

Production of publications. For the technical details of printing, the advertising department or experts in the publishing field can be consulted. In layout and typography, readability should be the first consideration.

How the piece is to be used will determine its size, paper, binding, cover and similar details. For materials that are to be filed or for insertion of revised pages, loose-leaf binders may be used.

The in-company or outside editor or printer who will handle the job should be asked for technical advice.

Getting ideas. Everyone in the public relations and publicity business runs dry of ideas now and then. Anyone who is suffering from this affliction should not hesitate to call on others for help.

Employee publications do not compete with one another; so ideas can be borrowed freely from them.

Some national agencies produce and supply safety material. See listings in Chapter 23, "Sources of Help."

The National Safety Council issues a "Safety Features Portfolio," which contains stories and illustrations for editors of employee publications. The NSC Sectional Newsletters, *National Safety News, National Safety Congress Transactions,* and other NSC publications contain a wealth of interesting and informative material. The NSC reprint, "How to Run a Newsletter," single copies available on request, has additional ideas. NSC poster miniatures and other Council materials usually are released for general use if the customary credit is given.

References

Adams, Alexander B. *Handbook of Practical Public Relations.* New York, Thomas Y. Crowell Co., 1965.

Ashley, Paul P. *Say It Safely.* Seattle, Wash. 98105, University of Washington Press, 1969.

Flesch, Rudolph, and Lass, A. H. *A New Guide to Better Writing.* New York, Popular Library, 1963.

Lee, Irving J. *How To Make the Safety Speech.* Chicago, National Safety Council, 1939.

Lesly, Phillip, editor. *Public Relations Handbook.* Englewood Cliffs, N.J., Prentice-Hall, Inc., 1962.

National Safety Council, 425 N. Michigan Ave., Chicago, Ill. 60611. *Photography for the Safety Professional,* Data Sheet 619.

Schutte, William M., and Steinberg, Erwin R. *Communication in Business and Industry.* New York, Holt, Rinehart and Winston, Inc., 1960.

Turabian, Kate L. *A Manual for Writers,* 4th ed. Chicago, The University of Chicago Press, 1973.

Audio-Visual Media

Chapter
14

Effectiveness of Audio-Visual Media   316
Selection of media . . . Commercial *vs.* Homemade aids . . .
Making motion pictures

Preparation of Audio-Visual Media   323
Preparing a script . . . Lettering . . . Use of color . . . Drawings
and graphs . . . Photographic illustrations . . . Adding sound

Presentation of Audio-Visual Media   328
Room lighting . . . Pointers . . . Amplifying and recording systems
. . . Teleprompters . . . Screens . . . Seating . . . Mobile
presentations . . . Rehearsal

Nonprojected Visuals   334
Chalk boards and paper pads . . . Flip charts and show cards
Flannel boards . . . Hook and loop boards . . . Magnetic boards
. . . Photographs . . . Exhibits and working models . . .
Demonstrations

Projected Visuals   338
Slides . . . Filmstrips . . . Overhead projectors . . . Opaque projectors . . . Motion pictures . . . Television

Audio Aids   342
Tape recorders . . . Radio . . . Commercial recordings . . . Public
address systems

References   343

14—Audio-Visual Media

"Audio-visuals" encompass several dozen communications tools, ranging from purely audio aids (such as recordings and amplifying systems) to purely visual aids (such as silent films and posters). The bulk, of course, depends on both sound and sight.

Audio-visuals serve as a communications media for transmitting messages used to inform, persuade, inspire, and entertain people, and to solve problems. They are the bridge through which a message travels between a sender and a receiver—between safety professionals and management, between safety professionals and workers, between management and workers and safety professionals. When a communications gap exists, it is usually not because there is nothing to say, it is usually because the right message did not get to the right person in the right way.

Direct experience is the best teacher. Looking, listening, smelling, tasting, touching, and manipulating all stamp realism and meaning into a communication, and make a training program a learning program. These direct experiences frequently cannot be used because of practical considerations—lack of time, money, or skill. The communicator must, therefore, use more abstract means—sometimes contrived, reconstructed, or simulated.

Audio-visuals are more than mere aids to communication, they are the medium of communication itself. Hence the name, "audio-visual media."

Effectiveness of Audio-Visual Media

One should be aware of the range of audio-visual resources and their appropriateness both to transmitting the message content and to appealing to and reaching the other person or the audience.

Figure 14–1 rates various communications media on a relative scale of concrete-to-abstract experience. The more concrete the communication, the more effective it is. Mere words and sentences, whether spoken or written, not only provide relatively little meaning, but can often be misconstrued by the other person or persons. Designing a communication, or a training program, to take advantage of more concrete resources, as they are needed, takes more ability than merely know-

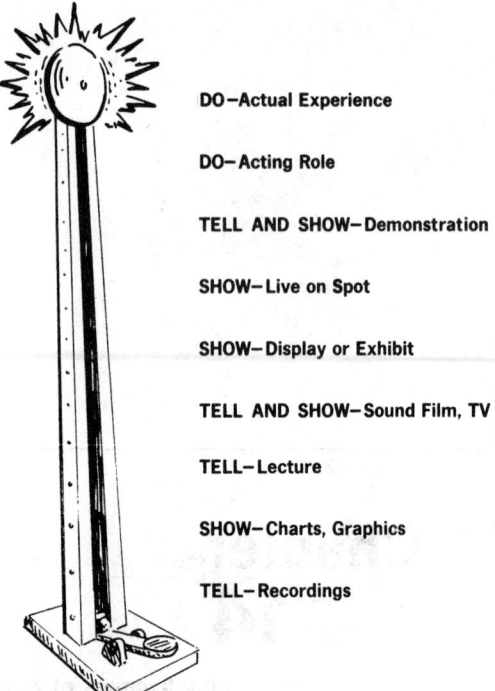

DO—Actual Experience

DO—Acting Role

TELL AND SHOW—Demonstration

SHOW—Live on Spot

SHOW—Display or Exhibit

TELL AND SHOW—Sound Film, TV

TELL—Lecture

SHOW—Charts, Graphics

TELL—Recordings

FIG. 14–1.—Direct, actual experience "rings the bell." In measuring the impact of a presentation, the more concrete the medium of communication is, the more effective it is.

ing the subject matter. It demands an ability to use audio-visual media effectively, and to know when each can be most effective.

Another advantage for using audio-visual media is that it is estimated that about 90 percent of learning takes place through simultaneous use of seeing and hearing. In addition to this reason, both management and workers are exposed to movies and television constantly, and they actually expect the use of audio-visual media.

Frequently, too much is expected of audio-visual materials. Just having an audio-visual available is not sufficient justification for using it. Whether an audio-visual should be used at all depends on the objective, the effectiveness of the audio-visual in carrying or supplementing the message, the type and size of the audience, and the budget and facilities available.

One audio-visual cannot do the entire com-

TABLE 14-A
SURVEY OF A/V EQUIPMENT
THAT SAFETY DEPARTMENTS OWN

Type of Equipment	Number	Per Cent
Motion picture projector	733	66.2
Slide projector	544	49.1
Filmstrip projector	341	30.8
Overhead projector	190	17.1
Opaque projector	207	18.7
Flannel board	246	22.2
Magnetic board	144	13.0
Chalk board	646	58.3
Hook and loop board	53	4.8
Tape recorder	391	35.3
Motion picture camera	189	17.1
Camera (still picture)	713	64.4

munications job. Each is part of the total communications program and must fit into the overall outline and presentation plan.

This is one reason why safety departments use many types of audio-visual products. A survey by *National Safety News* of 1108 safety departments, summarized in Table 14–A, showed that a variety of equipment was owned. To the question, "Does the safety department in your organization have a regular, continuing training program for supervisors," 61.7 percent answered "yes"; 56.9 percent had special training rooms for conducting this program.

In another *National Safety News* survey, of 146 member companies which responded, 13 percent planned to use filmstrips to a greater extent than previously, 48 percent indicated increased use of slides, and 63 percent intended to use more movies.

Specific conclusions should not be drawn from these survey results unless one knows the exact purpose for which each type is used. With many audio-visuals, it is difficult to say whether the primary objective is motivation or training. Often, safety professionals use a number of visuals to meet both objectives. Different people frequently have different opinions as to just what constitutes, "training" or "motivation." Also, the immediate availability of projection equipment, and the wide variety of films and slides may help make these aids more popular.

Use in training and reporting. Audio-visual materials are important conveyors of information. Photos of hazards or of unusually good conditions provide visual evidence useful in reports to supervisors and to top management. These, plus graphical presentation, can make routine reports and statistical analyses interesting and easy to understand. More than one safety professional has found a simple bar chart or colored graph to be more meaningful to his boss or to other supervisors than pages of detailed statistics.

A visual aid can emphasize the points of information in a safety talk; it can even provide a convenient "outline" for the speaker. Visual materials can be used to organize group thinking and to summarize safety committee action.

Use as a motivational tool. Aids that appeal to the emotions can help change attitudes, encourage safe work habits and compliance with safety rules, and remind workers of special rules or hazards.

Audio-visual materials of all types are used widely to promote interest and obtain cooperation in special campaigns, safety contests, and similar activities.

Selection of media

To be most effective, audio-visuals must be selected with care, after considering many factors. (See Table 14–B.)

• What is the purpose of the communication—motivating, training, reporting, fact-finding, entertaining? What result is wanted? Which medium (or combination) will serve this purpose best, within budget limits.

317

• Which medium will best convey the content of the message? For example, detailed technical figures can be communicated by a chart that can be held up, or projected in front of the audience, and held for some time while it is explained. A tape recorder or a movie projector would be of little help.

• What is the size and type of audience? What is their attitude toward you and toward your subject? How knowledgeable are they? How good are their communications abilities?

• How capable are the communicators? Do they need special training in either the subject matter or in the effective use of the audio-visual? Do they need other help?

• Where is the audio-visual to be used—in a training room, at a meeting, in the office or plant, in the field, or at home? Use of audio-visuals requires scheduling, preparation or purchase, distribution, and storage. Suitable facilities must be made available.

• How flexible (or how formal) must the audio-visual be? In some cases, a flexible type (which each speaker can adapt to his particular uses) may be desirable. In other cases, a formalized aid which offers careful planning, conformity of message with company policy, and uniformity of presentation may be preferable.

When a formalized aid is being considered, a number of questions, such as the following, should be asked: Will the entire message apply to many different audiences, even though they are located in different geographical areas or are confronted with different hazards? Will the material become dated, or can it be used almost indefinitely? Is it likely, for instance, that changes will be made in machines, processes, job layouts, or even personal protective equipment illustrated?

In some cases, a combination of flexible and formalized aids is desirable. For example, some speakers, including National Safety Council staff engineers, use a carrying case containing three-dimensional exhibits, charts, flannel boards, and other nonprojected material, as well as slides in a portable projector, to give road show presentations which can be changed to meet specific needs.

• How do the costs of the various audio-visual media compare? In the selection of

TABLE 14-B—MAJOR FEATURES AND LIMITATIONS OF VARIOUS VISUAL AIDS

Type and Popular Size	Audience Size	Shipping and Handling	Limitations	Strong Points	Comments
MOTION PICTURES 16mm sound	Medium to large	Film easy and cheap to ship; projectors heavy, expensive to ship.	Camera and projector expensive; require trained operator, except for self-threading models. Film not easily changed or updated.	Effective for training and motivating. Uniform professional message. Optical sound nonerasable. Sharper image than 8 mm for given projection size. Single-frame, stop motion projectors are available.	Silent version less costly, but less effective.
MOTION PICTURES 8mm sound and "Super-8"	Small to large	Low shipping cost; projector is lighter.	Safety subjects not as widely available as 16 mm. Not too suitable for large audiences. Sound is available on Super-8.	Lower cost, lighter weight equipment than 16mm. Homemade movies more feasible. Instant cartridge-loading types are easy to use.	Useful in rear screen projection and automatic continuous showings.
FILMSTRIPS 35mm sound	Small to large	Strips and recordings easy to handle. Projectors light to medium.	Strips and records not easily changed or updated. "Canned" message may not be effective or paced suitably for user.	Effective for training and motivating. Message uniform. Sound easily added on tape or disc.	Silent strips with script less expensive, but still effective.
SLIDES 2 × 2 in. (35 mm, 126, or 127) 2¼ × 2¼ in. (120 film)	Small to large	Slides easy to store, handle, and ship. Projectors light to heavy.	Slides may get out of sequence, reversed, etc. Cardboard mounts not durable. Slides bulkier to store and ship than strips.	Effective for training and motivating. Less of a "canned" show since slides may be rearranged. Slides can be made and processed quickly. Color inexpensive.	Taped message or reading script easily added or changed. Remote control and multiple projection possible.

Medium	Audience size	Portability / Storage	Limitations	Advantages / Uses	Remarks
OVERHEAD PROJECTORS 10 × 10 in. 7 × 7 in.	Small to large	Shipping costs of transparencies higher. Projectors light to heavy.	Transparencies positioned by hand. Projector close to screen; it or user may block view unless screen is raised or set at an angle. Ready-made material not widely available.	Effective for training. User can write on transparency while facing audience. No need to darken room. Transparencies easily made and filed. Presentation informal and flexible.	Color transparencies or overlays easily made.
OPAQUE PROJECTORS 10 × 10 in. max.	Small to medium	Projectors heavy and bulky.	Projectors require manual operation. Material in books may be difficult to store or ship. Room must be darkened. Copy may be too small.	Effective for training. No transparencies required; small objects, printed material, drawings, and photographs used "as is."	Copies or originals can be hinged or put on rolls to maintain sequence.
CLOSED-CIRCUIT TELEVISION	Small to medium	Camera, recorder, and monitor require dolly or handtruck if they are to be moved about, except for light-weight models.	Expensive. Requires adequate lighting. Presently black and white. Film copies must be made one at a time.	Instant replay. Excellent for training situation where trainee must "see himself in action"—such as driver training, or where he must observe action from a remote location.	Small number of people can view screen. Projector, or additional screens, required for larger audience.
FLANNEL, HOOK AND LOOP, MAGNETIC 12 × 36 in. to 48 × 72 in.	Small to medium	Larger boards are bulky	Presentation requires advance preparation. Few ready-made presentations available. Flannel board material may fall off if not applied correctly or if board too nearly vertical.	Effective for training. Message easily changed, yet can be filed and reused. Permits informal presentation with desirable audience contact. Dramatic, "slap-on" effect builds interest.	Boards suitable for heavier displays; cost slightly higher than cards or pads.
FLIP CHARTS AND CARDS 36 × 48 in. 18 × 24 in.	Small to medium	Easels or charts may be bulky and heavy, but usually portable.	Limited to small groups. Limited as to amount of copy. Good lighting necessary.	Effective for training and informing. Prepared material can be arranged in sequence. Good audience contact. Material easily prepared; can be added during talk and can be saved.	Ready-made letters, color, sketches, cut-outs easily added. Colored paper effective.
PAPER SHEETS AND PADS 28 × 36 in.	Small	Pads usually disposable.	Speaker must print legibly. Good lighting necessary. Ink from felt markers may bleed onto adjacent sheets.	Effective for training or discussion; informal. Permits reference to other sheets both during the discussion and for later writing of minutes. Low-cost pads easily obtainable.	Used in place of chalk boards, no erasing.
CHALK BOARDS (*portable and wall mounted*) 36 × 48 in., *larger for wall mounted*	Small	Portable boards bulky and heavy.	Board must be erased before reuse and recall not possible. Good lighting necessary. Ordinary chalk marks hard to see. Chalk may not be on hand. Dust from chalk and erasers annoying.	Effective for training or for discussing a limited number of points. Presentation informal. Portable chalk boards also useful for holding charts or displays.	Colored or fluorescent chalk adds life to talk. Magnetic boards available.
POSTERS AND BANNERS 8½ × 11½ in., 17 × 23 in., and larger	Small or large	Easily filed, handled, and mailed.	Only one or two ideas can be presented at a time; considerable time needed for changing.	Effective for motivating; support training. Specific messages can be posted at points of hazard or to meet timely situations. Ample posters available.	Homemade posters supplement general posters.
WORKING MODELS, EXHIBITS, AND DEMONSTRATIONS	Small or large	May be hard to handle, store, and ship.	May require special training to use. Live action is subject to errors.	Action can closely simulate actual conditions. Permit group participation.	

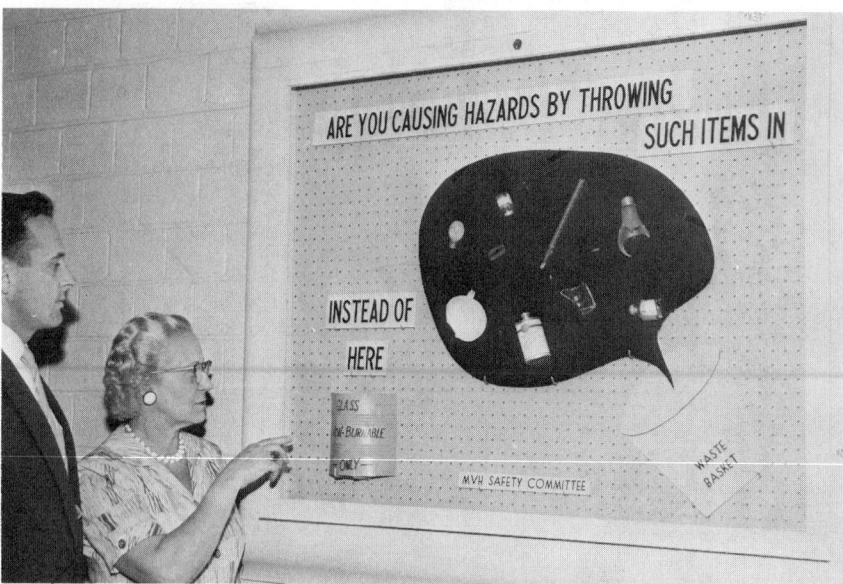

Fig. 14–2.—This home-made display makes its "point" out of junk that was picked up from the floor.

audio-visual equipment, this point is especially important. The speaker must answer this question to *his* satisfaction—and to his employer's.

Whether or not the cost of an audio-visual is justified must be considered in the light of what it will buy. An investment of $50,000 in a well planned sound movie might be justified for a long-range training program or public relations campaign. One of the advantages of such an aid is that it can be used over a relatively long period of time as a means of communicating the same message to many people. The repeated showing of a $50,000 film to large audiences over several years might well bring the cost per viewer down to a few pennies.

In contrast, the apparently modest expenditure of $500 on a homemade movie developed without sufficient planning and applicable to only a handful of workers could be excessively high and perhaps ineffective. Moreover, the same amount of time, money, and effort devoted to a training program, individual job instruction, or perhaps production of an inexpensive safety rule booklet might get better results.

Cost is only one of a number of factors to be weighed in the selection of an aid. An expensive aid is not necessarily the best one. For example, a simple paper pad or chalk board may be more effective than an elaborate printed brochure for presenting a safety report to a group of executives or for training employees in safe practices.

● How is the message to be supported? Not only must there usually be follow up, but other people must often know what was communicated to whom, in order that they can reinforce the message, or at least not contradict it "accidentally."

Commercial *vs.* homemade aids

The survey of 1108 subscribers to *National Safety News,* discussed earlier in Table 14–A, also indicated that safety departments often make their own training materials. The survey indicated that:

14.4% made motion pictures,

9.5% made film strips, and

34.9% made slide presentations.

To determine whether it is more economical and practical to make audio-visual aids than it is to buy them, the same factors that affect selection of aids must be considered: the purpose for which the aid is to be used, the type and size of audience, and the degree of flexibility desired.

For an informal supervisors' meeting, for a report to a safety committee, or for a presentation to company officials, a homemade aid, such as a chalk board or hand-lettered flip chart, would be appropriate. However, for more formal talks or for a number of meetings at decentralized locations, a commercially prepared aid—a set of slides with a script to be read, a filmstrip with a record or a tape, or a set of commercially painted charts— might be a wise investment.

Other points to be examined when a choice is being made between a commercial aid and a homemade aid are the costs involved and the availability of facilities, talent, and time. See Fig. 14–2.

Still another factor is that participation of individuals or of committees in the planning and development of an aid may be highly desirable as a means of arousing and maintaining interest in accident prevention. The net effect of employee participation, in fact, may compensate for lack of the professional touch, provided that the quality of the finished product is not seriously affected.

When a final decision is being made, the various factors must be considered in terms of one another. For example, slides may be selected as the type of visual aid to be produced. The speaker may have a suitable camera and lights, plus the ability and the time to do the job himself. However, the number of showings, the size of the audience, or the importance of the message may warrant the expense of professional treatment.

The same comparison can be made between professional and homemade charts, signs, and other visuals.

Even if a safety department alone could not justify a professional visual aid staff, the needs of the department, when added to those in sales, training, and other communications areas, might justify such in-company, or in-plant, facilities and staff. The combined benefits and savings would make this worthwhile, not to mention the more effective visual aids that would be available.

Making motion pictures

When a film is to be made, the advantages of having a commercial producer do the job should be carefully assessed. Having the necessary technical knowledge and equipment and knowing the tricks of the trade, the commercial producer can make a film that probably will be far more effective than a homemade film and, when the indirect costs of a homemade job are taken into account, no more expensive.

When a homemade movie is to be produced, it is wise to obtain professional advice before and during production. Professional laboratories can add sound or special effects if desired.

When considering a professional production, first define the job that the film is to do and the audience it must reach, and then prepare a scenario (the story or script, with instructions to the actors and photographic crew) or have one prepared. Send the scenario to studios for bidding, remembering that studio reputation and quality are as important as price. Some sponsors, however, dispense with competitive bids if they are satisfied with the capabilities, performance, and good reputation of one producer.

Prod the studio to make suggestions for improving the script, recommending talent and music, and suggesting money-saving ideas. Be sure to hire an "art source," not merely a "film factory."

Making a motion picture is a team effort. Be sure to enlist the cooperation of not only all the people who can be of direct assistance, but also those who must approve the expenditure of funds.

Cost. How much does a good film cost? No one can tell how much a good film will cost until the purpose is specified and the necessary ingredients for achieving that purpose are decided upon in the light of the budget. Some films must be shot in the studio, others in remote outdoor locations, and still others in difficult-to-light factories. There are cost advantages and disadvantages in each case.

Once there is a working blueprint—a script or a detailed treatment—an established film studio can tell you the exact cost within a few dollars. Cost can vary from $750 to

Fig. 14–3.—Members of the Monsanto maintenance training department prepare the camera and video-tape recording equipment for on-the-scene coverage of a safety training presentation on the operation and maintenance of instruments and machinery. Shooting of tapes or motion pictures may disrupt production, but it can also stimulate employee interest by their participation.

Courtesy Ampex Corporation.

$1450 *per minute* and upward from there in special cases. This cost includes the first print; the cost of additional prints must be added.

State universities often cooperate in making a film. They may require joint approval of the script and final shooting plans. They may permit no advertising other than the co-operating company's name at film "heads" and "tails." The university may hold copyright to the film, and can sell it to companies. The advantage, however, is that cost can be kept down to about $600 per minute. Universities and local drama groups can also suggest near-professional actors, narrators, and artists.

In either case, the safety professional can expect to devote several weeks (half of which might be on location) to helping make the film. A competent film crew requires 6 to 12 weeks to complete a job.

If a good camera and lighting is readily available, the cost of a self-made film might be only ten percent of a commercial film, particularly if professional actors, models, travel and expensive sets are not required. However, a company production would still require considerable time for planning, script writing, editing, and photography. The actual shooting and arrangements for lighting might disrupt production schedules or tie up key people and expensive equipment. On the other hand, production on the job, with or without direct employee participation, will usually stimulate interest and serve to help employee relations beyond the primary objective of accident prevention. See Fig. 14–3.

The film showing and discussion will require a certain amount of time from the job. To be truly effective, the show should have advance publicity and post-showing followup, preferably with the distribution of recall literature if it is a training film. A leader's guide and other lesson material can also supplement the film itself. All this requires additional editorial and printing expense.

The cost of production might be shared with other departments or even outside agencies such as trade associations, insurance companies, manufacturers, distributors, or safety councils. The cooperation or even co-sponsorship of outside groups can range from editorial and technical assistance (recognized sufficiently by a credit line) to actual division or sharing of cost of writing, photography, duplicate prints, distribution, and promotion. As far as a company is concerned, some of the production costs might be allocated to public relations, employee relations, sales, or advertising with the safety department supplying the technical and safety know-how.

Professional actors. Professional performers and models are usually preferable for a sponsored film because they have the experience, poise and skills required to give the whole production a polished appearance. Their services (if required) are included as part of the estimated price submitted by a reputable producer. Being skilled in the medium, they quite often shorten production time, as compared to an amateur cast. Actors charge $500 per week to $1500 per week.

Animation. In the process of developing the best filmic content, there may well be decisions about whether to include or exclude animation. Based on the content of the majority of films in use today, the chances are that animation will not be included, but if it is, it probably will not be an important factor. What is loosely called "animation" can cost $500 a minute, $2000 a minute, or much more —but the isolated figures for a brief animation portion of the film, as with any other special scene or effect, are meaningful only if related to the total film budget in their proper proportion.

Sets. A reputable, competent film production organization ordinarily includes all the essentials in its estimated price for a film. If it turns out that an effective film requires sets, lavish or modest, the studio will supply them as part of the contracted package price. A large proportion of films today are shot on location, therefore no sets are required.

Music. Music is included among the essential film ingredients supplied by the producer. Good professional producers have all sorts of appropriate atmospheric theme music in their extensive tape libraries—and they are associated with competent union musicians who can create and perform excellent film music to order for your purposes. Be sure contracts include all possible uses and have all clearances documented; i.e., ordinary commercial records cannot be used.

Announcer or narrator. Most radio and TV announcers and narrators work for a standard film producer's fee, but there are some big-name national personalities who ask a price that is many times the usual scale. If a company wanted one of the latter types, the extra cost would be reflected in the estimated price submitted for the production of the film.

Preparation of Audio-Visual Media

Once an aid that is suitable for the job and within the limits of budget and time has been selected, the details of production must be worked out. The general procedure is to make an outline, develop script and picture descriptions, have pictures taken or art work made, and check content for length.

Many types of aids include written or spoken words, and coordination between the illustrations and the message is essential. Both should complement each other for maximum effectiveness. Also, art work used for aids might also be effective in printed brochures or magazine articles.

Preparing a script

A film, a set of slides, or a filmstrip requires good organization and a script suitable to the audience and the action desired from it. Here are some guidelines:

1. Start by jotting down the objective of the

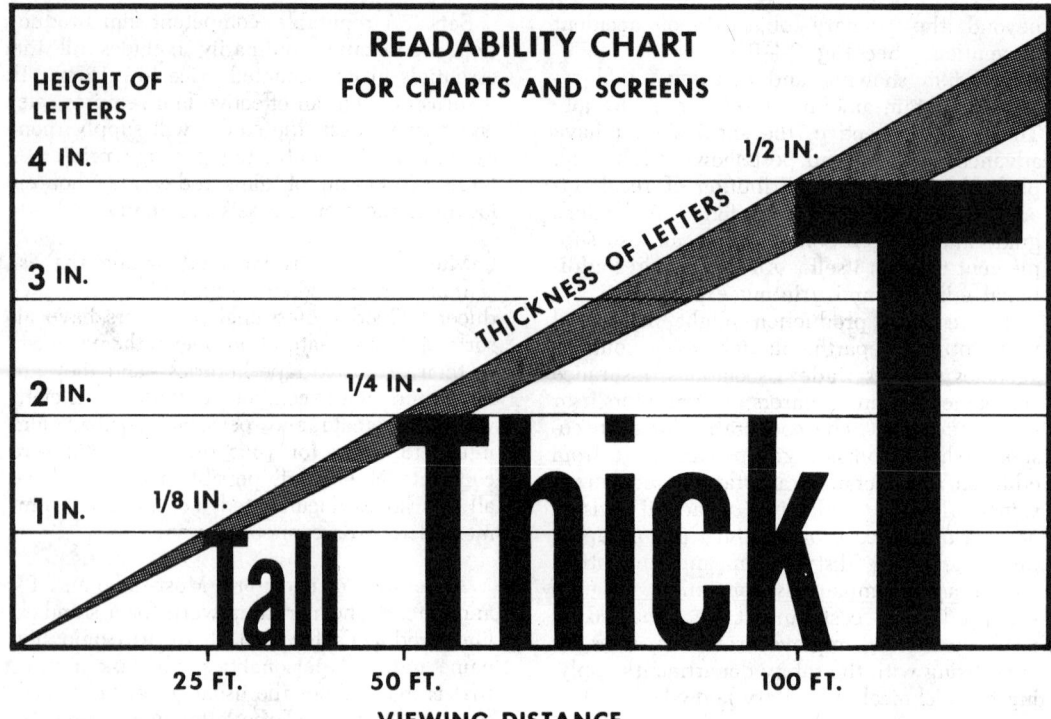

FIG. 14–4.—Readability chart for determining size and thickness of letters for nonprojected visuals. Size of lettering on projected visuals should be such that lettering is of adequate size when projected on screen. Make sure that thickness is proportional to height, that spacing between lines is about letter height, and that color aids readability, not fights it.

visual; perhaps discuss it with colleagues or with a committee assigned to help with the project.

2. Develop a simple outline of the subjects to be covered, indicating the approach (humorous or serious, for example) to be used, and the props or types of illustrations needed.

3. Avoid the temptation to crowd too many ideas into the outline. Concentrate on the chief objective, and do not dilute it with trivia, or lose it with a subtle, "cute," or over-complicated plot.

4. Determine how you will open and close the story. In a training script, main points should be repeated and summarized.

5. Follow the outline, write a rough draft of the script and of the picture descriptions, and then a final draft.

For a script to have maximum effectiveness, short words and simple sentences are recommended. The script should be kept brief and to the point. As an example, a script for a set of slides might have only a few words or as many as 30 to 80 words per slide. A change of pace is sometimes desirable, providing none of the frames has too long a script. If a lengthy description for one slide is necessary, variations of the scene or subject could be developed, thus obviating the need to hold one slide too long.

Story board technique. In preparing a script for a set of slides or for a filmstrip, the story board technique is recommended.

Usually, the copy is typed double space,

down the right side of the page or illustration board, frame by frame, with the corresponding illustrations or picture descriptions placed opposite.

Or each frame can be represented by a small card with the illustration on one side and the script and production instructions on the other side. The cards can be mounted on a large piece of cardboard, laid out on a desk top, or placed in an illuminated planning board rack. This technique makes it easy to rearrange the art work and to visualize the entire finished product.

When a set of charts or even a chalk-talk is being prepared, rough sketches or notes can be made on a paper pad and scaled to size. Even in miniature, a rough sketch will give a good idea of the amount and size of lettering that can be used, the effect of color, and other aspects.

If the script is to be reviewed by safety committee members, company officials, or other persons, it can be typed and duplicated. Deadlines for reviews must be set and followed. Of course, important points should be approved by key executives or other authorities.

Lettering

The most common complaint regarding both nonprojected and projected visuals is that lettering is difficult to read—too small, too thin, too crowded—or even illegible. The best rule-of-thumb is: Design lettering so that it can be read from the back row of the audience. Simplicity is the keynote.

Block letters show up better than handwritten copy. To be easily read at a distance of 50 ft, letters should be 2 in. high and ¼ in. thick, as they are projected on the screen or show on a visual. See Fig. 14–4.

For use with overhead or opaque projectors, material typed with characters at least ¼ in. high, with spacing of ¼ in. between lines, will give a letter height of 2 in. on a screen 6 ft wide and will be clearly visible at a distance of six to eight times the screen width.

The space between rows of letters should be at least one-half the height of the letters, preferably the same as the full height. For example, there should be at least ½ in. (and preferably 1 in.) spacing between letters 1 in. high.

Printed or typed material on 8½ by 11 in. sheets, such as record forms, will require larger lettering or typing on a machine with oversize (¼ in.) characters. Material typed all-caps in an area 3 in. high by 4½ in. wide will be legible when converted to a 2-in. square slide. If possible, all illustrations and titles (art and lettering) should fit the horizontal format of the screen. That is, the width should be 1½ times the height. See Fig. 14–5.

IF YOU CAN READ 2X2" SLIDES WITHOUT A MAGNIFIER, PEOPLE IN REAR SEATS CAN PROBABLY READ THEM ON THE SCREEN

Fig. 14–5.—Rough rule-of-thumb for lettering to be shown on a 2 by 2 in. slide.

Courtesy Eastman Kodak Co.

Material on a visual should not be crowded, should be well organized, and kept simple. A simple rule for the amount of copy is: no more than 10 lines with five to seven words of typed or lettered material per line. Typewriters which permit half-spacing (technically, "one-and-a-half spacing") are of special value here.

A growing variety of ready-made lettering material is available in camera and art supply stores. Examples are plastic stick-on letters, rub-on transfer letters, gummed letters, and ceramic, cork, cardboard, or other letters which give a three-dimensional effect when lighting is from one side.

Most companies have stenciling equipment —either templates that are painted through or letters that are painted around—that can be used with either an ordinary stenciling brush

325

Fig. 14–6.—It is easy to get variety in layout color and lettering. Hand lettering, colored tapes, stick-on letters, and stencils can be used. To keep chalk or charcoal dust from smudging, spray the finished artwork with a "workable fixative."

or an inexpensive spray paint. Also, lettering guides are available for use with special lettering pens and felt-tipped marking pens. Often many styles of lettering can be made with one lettering set.

For a visual to be shown to a small group, a large black or colored crayon may be used. For a visual intended for larger groups, instructor's chalk, broad tip felt marking pens, stencils, cutout letters, or brush-painted letters are preferable. Background colors can be varied also. Too much of any one color can be tiring. Try white on black or black on yellow.

Use of color

Color enhances both nonprojected and projected visuals. Contrasting colors always should be used. For example, black lettering shows up well on a white, yellow, or light-orange background, and worst on a dark blue.

With projected visuals, color adds realism, provides contrast values that can bring out important points, and gives a professional look to the completed visual.

With color film, for instance, title frames can be made attractive and closeups can be shot against black or colored backgrounds for good contrast. Textured backgrounds and colored lighting effects lend a professional touch. Full-color motion pictures are not always necessary. Because emphasis is focused on the action, black and white may even be more effective.

For posters, dramatic effects can be obtained with the high visibility fluorescent paints, chalks, and papers, particularly if "black" light is used.

Color can be added with large blocks of instructor's colored chalk or with colored felt-tipped pens that make a broad, heavy line. Colored designs easily can be made by spraying through stencils or simple cutouts. Spray cans can be used also to give overall color, and powdered colored chalk can be daubed lightly over lettered material to give a tint.

Colored tape and ready-made arrows, circles, and other stock designs can be used to make charts and graphs, and also to mark important parts in equipment photographs.

Drawings and graphs

Because of the size of type and the amount of copy, a graph, chart, or other line drawing may become illegible when reproduced on film or viewed from a distance. Therefore reduce the details to those fundamentals required to illustrate the point.

The same rules that apply to lettering apply to drawings: the line work should be broad and opaque. Frequently a complex item can be constructed from areas cut out of colored paper. Colored paper is also effective for bars of a bar chart or areas under a curve. Graphs may be simply constructed from colored tapes (⅛ to 1 in. wide). Remember in presentations, graphs are frequently used to show a trend rather than specific points; hence the grid background should be omitted or should consist of only a few fine lines. Where accuracy is important, the actual numbers should be used.

Another inexpensive do-it-yourself technique is to draw a cartoon or chart. Varicolored chalk on a contrasting background lends an interesting fillip that is often lacking in conventional black-and-white material.

Try to show only one point at a time in building your overall story.

Depending upon the size of type and the amount of copy, a page of printed material may become illegible when reproduced on film. If such material must be used, it should be converted to a form more suitable for the purpose. The information given on a page of statistics, for instance, might be expressed in a few simple charts or graphs. Reduce the details to those fundamentals required to illustrate the point. Then do it effectively. See Fig. 14–7.

Photographic illustrations

The value of a visual aid depends to a considerable extent upon selecting the right illustrations and the quality of the illustrations. So far as photographs are concerned, the general principles of good photography are the same regardless of the type of camera or film used. Here are a few suggstions for

taking good pictures.*

● The important part of the picture should be highlighted by means of a closeup, a supplementary sketch, a contrasting background, or by an arrow or sign placed by the item. For example, if a guard or a piece of safety equipment is being photographed, it can be painted (spray cans are handy) or shot against a colored background that provides effective contrast.

● Both with motion and with still pictures, use of long shots, then medium shots, followed by close-ups help establish the scene or situation.

● If material far in the background must be shown in detail, extra lighting must be used. A single flashbulb will not suffice. If background detail is not important, it can be kept out of the original picture, cropped out of the negative or the finished photo, or touched out of the print. Off-to-one-side lighting will give a pleasant three-dimensional effect and will keep light off the background, thus playing it down in the photo.

● If there is doubt as to the possible result, it is a good idea to take two or three different exposures at the time of the original shooting. Probably, there then will be no need to come back later to get a better picture. Use of a Polaroid® photo to determine exposure for black-and-white or color film in difficult situations is often a help.

● Use of an exposure meter will help ensure good results with both natural or existing light and with floodlights or spotlights. An exposure meter, of course, cannot be used for flashbulbs or electronic flash. In this case, exposures should be calculated on the basis of guide numbers provided by the manufacturer of the bulbs, camera flash unit, or in the instructions for the film. Numbers are for average rooms. Large industrial areas require lower numbers (more light). Where possible, photos should be made outdoors, to take advantage of natural lighting, or in a

---

*For a detailed discussion, see National Safety Council, *Photography for the Safety Professional,* Data Sheet 619.

ANNUAL SAFETY REPORT			
Accident Cost Factors by Departments (Incurred losses per 1,000 hours)			
Group	Base Period 3 Years	Last Year	This Year
Production Division	$5.40	$7.10	$2.50
Sales Division	7.20	3.25	6.70
Company Average	8.50	5.10	3.60

FIG. 14–7.—For use as a visual aid, complicated statistics should be converted into simple graph for greater clarity. Typed slide (left) also shows poor spacing:

studio under controlled light. If actual job situations are wanted, then shooting on location is called for.

• Because some states have laws that forbid publication of a person's photograph without his written permission, a signed release should be obtained from everyone who appears in recognizable form in any picture. Standard model release forms are available from photo supply houses and commercial studios. This is particularly important when photos are to be used for advertising purposes, but does not apply to newsphotos used for strictly editorial purposes.

Adding sound

With either a set of slides or a filmstrip, sound can be added by means of a record player or a tape recorder on which a prepared sound track, with or without commentary, is played. The slide or filmstrip operator can follow a marked reading script, or a bell sound, "beep," or other audible signal can be used to tell him when to move the next picture into position. Using special equipment, in-

audible signals can be inserted on the record or the tape to advance the slides or the strip automatically. Frequently, records are made with the audible signal on one side and the inaudible signal on the other.

If the slides are updated or rearranged or the filmstrip is revised, the tape easily can be remade.

A motion picture film can be made with either optical sound or magnetic sound. The magnetic sound requires a projector that plays it, but it is more easily applied and is finding increased use, especially on standard-8mm where optical sound is not available. Unless original standard-8mm (or super-8mm) film is to be shot and projected, however, it is best to film on 16mm and reduce to 8mm optically for the print.

Presentation of Audio-Visual Media

Even the best planned audio-visual aid will miss its mark unless the necessary facilities are at hand and unless the speaker checks them and rehearses with them.

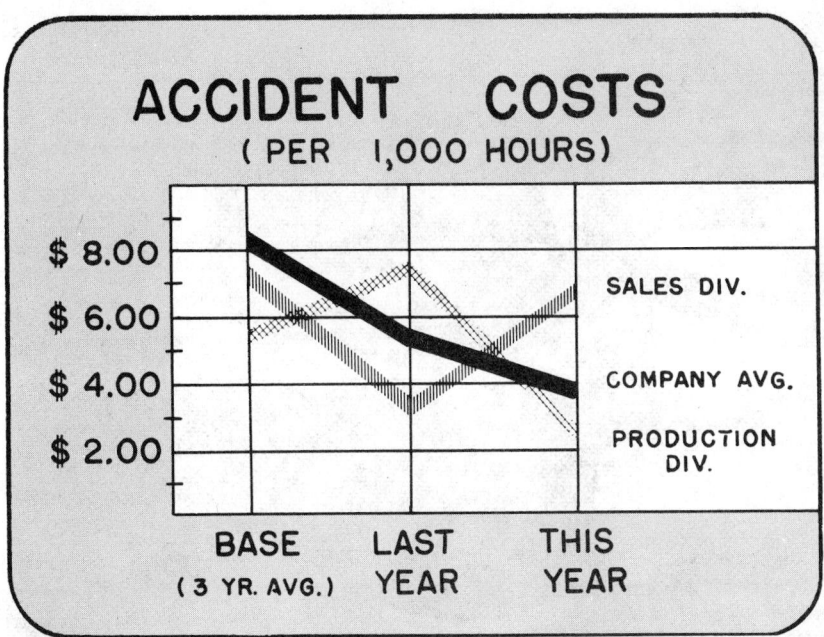

ACCIDENT COSTS
(PER 1,000 HOURS)

$ 8.00

$ 6.00

$ 4.00

$ 2.00

SALES DIV.

COMPANY AVG.

PRODUCTION DIV.

BASE (3 YR. AVG.)   LAST YEAR   THIS YEAR

top part is crammed while bottom portion is left blank. Slide at right is better balanced, facts are easier to grasp.

If a company is planning on building a training room, it is best that the safety professional work with the designers at the earliest stages to make sure that the room incorporates all the features deemed necessary for effective use of audio-visual aids.

Room lighting

Nonprojected visuals require good general lighting. If a room has only indirect illumination, a portable floodlight or spotlight can be used on charts, exhibits, chalk boards, and other visuals. Such a light can be clamped to a chair or to a portable stand immediately in front of the visual, but placed so it will not interfere with the view of the audience. See Fig. 14–8.

A spotlight can be used also to put some light on the speaker so that he can maintain "eye contact" with the audience of the visual —slides or a film, for instance—requires that the room be semidarkened or darkened.

Colored spotlights can heighten the dramatic effect of a visual. Revolving colored lights, like those used for Christmas displays, are suitable for more permanent exhibits, signs, or displays. "Black" or ultra-violet lighting used with fluorescent paint, paper, chalk, or ink gives a vivid effect.

If lighting must be dimmed for the showing of a projected visual, preparations must be made for darkening the room at the proper time, but "killing house lights" should not cut off power to the projector and reading lights. The location of the light switches must be noted, and someone should be asked to darken the room at a given signal. A shielded, reading light will be needed when the house lights are turned off for the presentation. A small flashlight may come in handy if the reading light is not operative.

Window blinds or drapes should permit shutting out daylight.

Well in advance of the meeting, electric outlets should be located and electric equipment checked for safe working condition. Having spare bulbs and extra extension cords on hand may save embarrassment or prevent delay. Extension cords should be marked with high-visibility tape or placed so that

Fig. 14–8.—A portable spot light helps illuminate visuals when used in a poorly lit area. Spot can be clamped to back of chair or mounted on a light-weight tripod stand. A 500-watt bulb is sufficient for an 8-ft wide area. Also note now unused chart (right) is kept covered in order not to distract the attention of the audience.

they do not create tripping hazards.

Pointers

A pointer is a necessary item when a speaker is using a chalk board or charts and may be helpful with other types of aids as well. A pencil or a finger is a poor substitute for a pointer.

Use of a pointer enables the speaker to face the audience and at the same time easily relate his words to the visual material. Do not play with the pointer—this distracts from the presentation. Do not touch the visual or the projection screen with the pointer—it may move the visual or screen, or even mar it.

If visibility is a problem, a pointer with a fluorescent painted tip will be helpful. In a darkened or semidarkened room, a battery-operated or 110-volt pointer can be used to project a spot of light or a bright arrow onto a screen from a considerable distance.

A high visibility electric pointer is useful even in a fully lighted room when the speaker must stand at some distance from his chart or screen.

A telescoping, pocket-size pointer is useful for speakers who must carry a pointer with them.

Amplifying and recording systems

If the acoustics in the room are bad or if outside noise makes hearing difficult, an amplifying system will be needed for successful presentation of an audio-visual aid. This is particularly important if there are a number of speakers and some may not be heard in far corners of the room.

If the speaker must move around, a lavaliere (chest-type), wireless, or lapel microphone is necessary. (Be sure the lavaliere microphone does not rub against a tie clip as this makes a lot of noise.) If the speaker can remain in one place, a pedestal or lectern microphone is satisfactory.

If audience participation is desired from a large group, floor microphones placed in the aisles or roving microphones carried by assistants will enable members of the audience to be heard throughout the room. Otherwise, the speaker must repeat questions and comments from the floor, using his microphone, so that the entire audience will know what has been said.

Before each use, an amplifying system should be checked for good operating condition and to be sure that reception is satisfactory throughout the room. At all times while the system is being used, it should be supervised by an individual familiar with electronic equipment to assure control of volume and of annoying acoustic feedback.

Also see the discussion of public address systems at the end of this chapter.

Teleprompters®

Teleprompters may be rented for dramatic or formal presentations in which actors, technicians, or executives are required to follow a prepared script.

Of course, technical help is needed to set up the Teleprompter, and those using it must be familiar with the technique so that their presentation will have the desired natural effect.

Screens

There are several basic types of screens used for projection:*

1. *Glass-beaded screens.* Usually portable, but often of the large pull-down variety, these screens have a high reflectance value but within a narrow angle of projection. (45 deg).

2. *Matte finish screens* do not give quite as bright an image as beaded screens. How-

ever, since they have a wider viewing angle (60 deg), they are more suitable for larger audiences, and for any room where some of the audience must sit at a considerable angle to the screen.

3. *Lenticular-surfaced screens* have embossed surfaces that reflect a high percentage of projected light on a viewing angle wider than that of beaded screens (70 deg), and reject stray incident light.

4. *Permanently mounted aluminum foil screens* are the brightest obtainable. Solid backed and slightly curved, they cannot be rolled up. The viewing angle is only 30 deg, maximum. The screen is designed for use with room light on; it is too bright for use in a darkened room, unless projector illumination is reduced.

5. *Rear projection screens* can be used in partially lighted rooms and have a wide viewing angle. However, any stray light behind the rear screen must be held to an absolute minimum.

The image on the screen should be neither dazzlingly bright nor dim. Generally, a 500-watt projector is satisfactory for small and medium groups in either a darkened or partially darkened room, provided that the film, slides, or transparencies have good color and well defined material. A 1000-watt bulb in a 16mm movie projector is better for medium and larger groups. Special longer-burning bulbs are available.

Where possible, screens should be set slightly above the heads of the audience for maximum visibility. With some projectors, principally the overhead type that is set close to the screen, a keystone effect may be created, whereby the top of the image is noticeably wider than the bottom. This distortion can be reduced by slanting the screen slightly so that its plane is more nearly perpendicular to the projected light. Some screens now are equipped with a clip or bar which permits this adjustment. Such bars can also be purchased separately and attached to portable or wall screens.

---

*Detailed information can be supplied by screen or projector manufacturers.

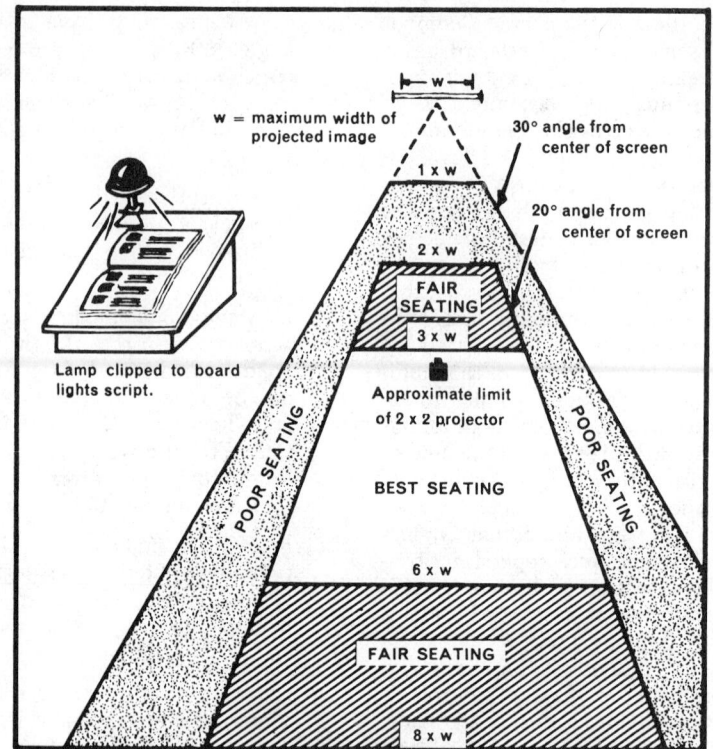

w = maximum width of projected image

30° angle from center of screen

20° angle from center of screen

1 x w

2 x w

FAIR SEATING

3 x w

Lamp clipped to board lights script.

Approximate limit of 2 x 2 projector

POOR SEATING

POOR SEATING

BEST SEATING

6 x w

FAIR SEATING

8 x w

FIG. 14–9.—Shading indicates both good and poor seating areas when ordinary (beaded) projection screens are used. A lenticular screen would widen the angle, which is measured from centerline. Shaded reading light permits reading script without distracting audience.

Be sure that the area of the projected image fills the screen, if possible, but does not extend over on the background.

Although modern projectors equipped with powerful bulbs do not require complete room darkening, room lights should be dimmed (or some lights switched off) and windows shaded as needed. This leaves sufficient light for note-taking or script reading, if necessary. There is no need to completely darken a room for showing non-continuous-tone images (such as graphs, diagrams, or lettering). A reverse-projection unit or a shadow box permits showing filmstrips, slides, or movies to a small group without dimming lights.

Slide projectors that must be located in or behind the audience should have a remote-control device, or be operated by someone other than the speaker (if the presentation does not have a recorded narration). There should be a subtle signal agreed upon for changing slides. When the speaker must say "Next slide please," or snap a cricket, he distracts the audience. Some offhand, but clear, gesture with the pointer or flashlight should suffice.

Seating

A room in which an audio-visual is to be presented should be checked for safety features and for seating arrangement. Aisle space should be adequate, exits ready for emergency use, and ash trays provided if smoking is permitted.

The seating arrangement should be planned so that every member of the audience will have an unobstructed view of the visual. For meeting room or auditorium seating, allow

21 to 24 in. width for each chair and 36 in. for each row of chairs, or about 5 to 6 sq ft per viewer. In classrooms or conference rooms, allow twice as much space.

For viewing a projected visual, the most desirable seating area is within a 30-degree angle from the projection axis with a matte screen and within a 20-degree angle from the axis with a beaded screen. See Fig. 14–9.

The recommended minimum viewing distance is at least twice the screen width, and the maximum no more than eight times the screen width, although a distance no more than six times the screen width is desirable. A 6-foot-wide image, therefore, could be viewed at a *maximum* distance of about 50 ft by approximately 100 persons.

For extremely wide rooms, and if the seats cannot be arranged so the audience sits within the recommended viewing angle, it is possible to project images from slide projectors or overhead projectors simultaneously on two or three screens, separated by at least 25 ft.

Mobile presentations

Where good projection facilities are lacking, such as in the field or at branch terminals or plants, mobile presentations should be considered. Flip charts, flannel boards, and demonstrations can be used outdoors or in quarters not well suited to projected visuals.

Trucks, trailers, converted busses, and other large vehicles may be used as mobile classrooms. Rear screen projectors, mounted in station wagons or van wagons, facilitate the use of projected visuals in the field.

Rehearsal

With even the simplest aid, practice before use is imperative. It will help prevent the speaker from running overtime and will help assure smoothness of presentation.

Unrehearsed use of an aid may reduce its effectiveness considerably. A set of charts, for instance, may be well prepared, easy to understand, and attractive, but if they are shown in random fashion or must be fumbled with by the speaker, much of their impact will be lost.

Moreover, if the speaker wanders from the subject, the set of charts or transparencies, instead of serving as an "aid," may even

prove distracting. If other material than that illustrated must be discussed, the speaker would be wise to cover the charts (or turn off the projector) until he is ready to return to them.

When items such as chalk, an eraser, and a pointer are needed, the speaker should make sure in advance that they are at hand. A person who needs a marking pen should carry two pens in case one runs dry.

Training and rehearsal is particularly important with more complicated equipment such as a movie projector. Well in advance of the showing, a trained operator should make sure that the equipment is "ready to roll." The projector should be threaded or slides inserted in correct projection position, the motion picture header run off, the sound adjusted, picture focused, and the screen set up at the proper distance. When a tape or a record is to be used with a film, a set of slides, or a filmstrip, the sound and picture should be synchronized. Arrangements for turning lights on and off should always be checked.

A film, a set of slides, or a filmstrip should be previewed and checked for good condition, and, in the case of slides, for proper sequence and right side up. When slides are in proper sequence, draw a diagonal line across the top of the pack—an out-of-place or mis-turned slide shows up at once.

The need for introductory remarks, discussion questions, and recall or follow-up materials should be considered.

If a script is to be read, the speaker should not indulge in lengthy ad-libs. He should stick to his plan of presentation, giving each chart, slide, or frame the time and attention it deserves, but no more.

The speaker should face the audience as much of the time as possible, particularly if he is using charts or a chalk board. He should not talk when he is moving about or not facing his audience. Not only would the distraction be bad, but the audience would have difficulty hearing him.

If the speaker wishes to face the audience the entire time, he can arrange for an assistant to write on the chalk board or turn the flip charts. Be sure that this doesn't become more distracting than useful. If the speaker is using a projected visual, he can operate the

FIG. 14–10.—"Silent sermon" on safety goggles uses photographs of actual case history.
Courtesy Lear Siegler, Inc.

projector with a remote control cord or have another person run it for him.

It is more effective if the speaker has the full information of the slide in front of him so that he does not have to turn continually toward the screen and away from the audience.

Nonprojected Visuals

Nonprojected visuals* include graphics, three-dimensional exhibits and models, and live presentations. Among the various types of graphics are chalk boards (formerly called blackboards), paper pads, flip charts, display cards, flannel boards, magnetic boards, and hook and loop boards.

Chalk boards and paper pads

Chalk boards are a well known, basic visual. They come in several colors and charcoal, but light green is considered standard. A dustless chalk should be used, preferably a yellow or other bright color for maximum visibility since white chalk does not show up well on a dusty chalk board.

Large blocks of instructors' chalk or the side of stick chalk gives a heavier and wider line

---

*Posters, the most commonly used type of visual, are discussed fully in Chapter 12, "Maintaining Interest in Safety," and in NSC Data Sheet No. 616 "Posters, Bulletin Boards, and Safety Displays."

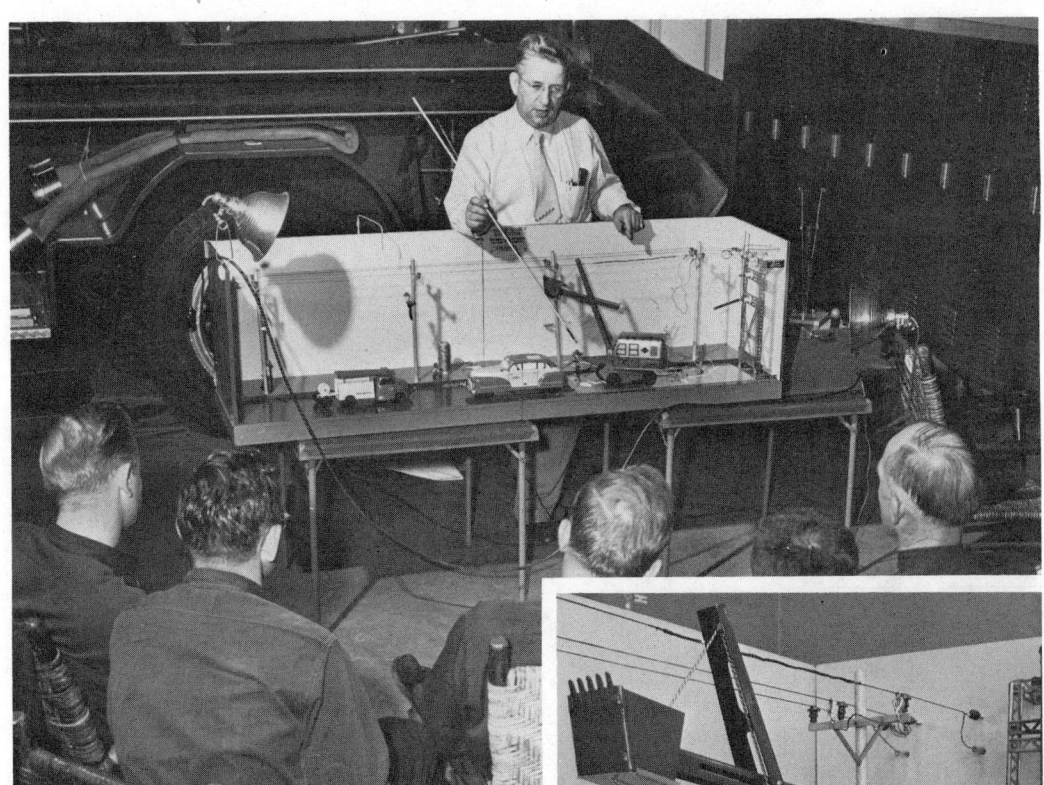

for greater visibility. Different colors of chalk, particularly the fluorescent chalks with black light, are very effective.

Paper pads (usually 2 by 3 ft) provide inexpensive visuals for small groups. Pads, or a number of sheets of paper, clamped on lightweight wooden or aluminum folding easels (as flip charts are) are easily portable and always ready to go. The speaker can keep training material on the pad so that he can refer to it, or throw it away, as he wishes. Some commercially available pads have faint cross-hatching to facilitate lettering and layout of artwork on the paper.

Pads are of real value to those who must lead discussions or run "brainstorming" sessions. As each chart is filled, it can be tacked to a corkboard or clipped to a wire running

Fig. 14–11.—High-line hazard model uses 15 kv, 3 ma circuitry (which will produce a 3-in. arc) to demonstrate dangers inherent in high-line work. Smaller photograph shows a closeup of the model power shovel and operator. A spark is drawn by operator's left hand to illustrate how someone touching cab may be electrocuted if the boom is in contact with the high line.

Courtesy Indiana & Michigan Electric Company.

335

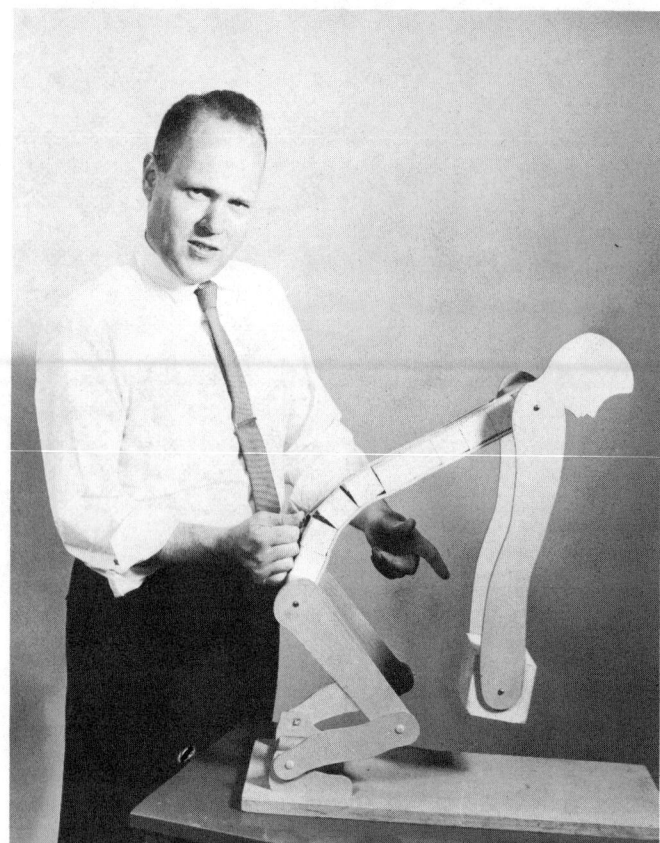

There is a proper and an improper way to pick up a heavy object. The proper way is to keep the back straight, the knees bent and spread, and the load close to the body. The improper way is to reach way over and lift. (Twisting the back complicates the bad effect.)

To demonstrate this effectively, a special model can be used. If used with its block "spine" "locked," the model (dubbed "Junior") simulates lifting with strong leg muscles; the ribbon on the "spine" remains limp, indicating very little tension of the back muscles. It demonstrates that the back cannot be kept straight without bending the knees.

To demonstrate improper lifting, "Junior" is used as is shown in the photograph. The legs are bent only slightly (or held straight). One hand lifts the handle just ahead of the fulcrum at the "hips" in order to lift the weight in "Junior's" "hands." The back arches under the strain and visibly pulls each block apart.

Model can be made to show proper foot placement (see drawing at right).

along one side of the conference room. This (a) provides the group with a continuous record of what has been discussed, (b) lets latecomers catch up to the discussion, without having to stop the discussion and have a review, and (c) helps in writing a good report of the meeting—the notes are right on the pad sheets.

Inexpensive rolls of white paper can be used for homemade charts. Even brown wrapping paper will do if material such as lettering is added in a contrasting color which will stand out. Material can be written so that it can be read as the chart is unrolled like a scroll.

Flip charts and show cards

Flip charts or show cards may be produced ahead of time or may include blank sheets or blank spaces for on-the-spot additions.

A hint for using flip charts is given by David Ogilvy, a highly successful advertising man. "The best tool ever devised for explaining complicated plans to committees is the flip-over easel, which the presenter reads aloud. It has the effect of riveting the attention of everyone in the room on what you are saying. Here I have some advice to offer. It may sound trivial, but it can be crucial to the success of the presentation: *as you read aloud, never depart from your printed text by a single word.* The trick lies in assaulting your audience simultaneously through their eyes and their ears. If they see one set of words, and hear a different set, they become confused and inattentive."[*]

---

[*]David Ogilvy, *Confessions of an Advertising Man.* New York, N.Y., Atheneum Publishers, 1963.

Construction Specifications

The spine is made with 9 blocks, each 2-in. square and 1⅝-in. high. Each is drilled at the center to accomodate a standard screen door spring. The T-shaped head and shoulder piece is about 8-in. long, 2-in. thick, and supports the arms on shoulders about 5-in. apart. The hip block is 2-in. square, with sloping sides so legs will spread open in front. Add a ⅜-in. spacer between the hip block and each leg.

Assemble the body by using wood screws to attach ends of spring to the head and hip pieces. Tack a piece of 2-in.-wide belting to the *front* of the spine blocks to hold them in alignment.

The arms and legs can be shaped from ¼-in. plywood.

A block, approximately 5-in. square and 3½-in. high, represents the lifting weight. Elastic tape or multiple rubber bands are stretched from the shoulder to the hip, along the *back* of the spine blocks to represent the spine muscles.

A metal handle can be secured to the lower end of the hip block, as shown in the drawing.

"Junior" can be mounted on a 1x12-in. board, 18 to 24-in. long.

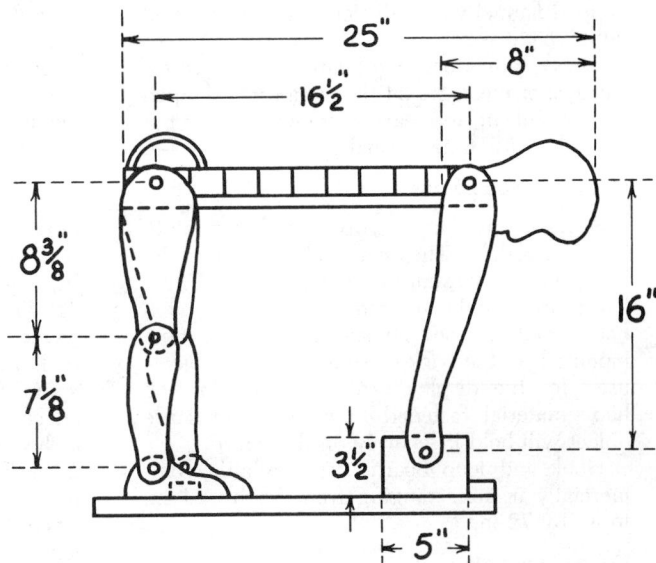

Fig. 14–12.—Model for demonstrating correct lifting.

This does not preclude the speaker's use of additional material or interpolating the printed text.

As with other nonprojected visuals, the speaker adds personality and interest to the presentation. The use of assistants may be justified for more dramatic effect. Similarly, the "gradual reveal" technique can be used to uncover material just as it is presented.

Flip charts are made on heavy paper that has little show-through, but has sufficient flexibility to flip back over the easel. Show cards are on a stiffer stock and must be hand held, placed individually on an easel or chalkboard tray, or spiral bound (or notebook-ring bound) so that they can be flipped over. For use on an easel, 2 ft wide by 3 ft high is an optimum size; for use on a desk top, 2 ft wide by 1½ ft high is optimum.

Each sheet or card can have one to four illustrations, depending on how far the audience is away. To help in making the presentation, explanatory material can be placed on the back of each show card or chart for the instructor to read, or brief notes can be written in lightly in blue pencil on the face of the card or flip chart. Notes are written small enough so that the instructor can read them, but yet they are "invisible" to anyone in the audience, if he is more than 6 or 7 ft away.

Flannel boards

A flannel board is a plywood board, commonly 3 by 4 ft, covered with dark flannel. A number of frames can be fastened together or placed close together to form multiples of the original size.

Roll flannel can be tacked, tied, or otherwise stretched over any large, flat, slightly inclined surface, including a chalk board. Art work or lettering can be made directly on a special velour-backed paper or on light-weight cardboard or heavy-weight paper to which flocking paper with felt adhesive or strips is affixed. In an emergency, sand paper can be used instead of the special flocking paper. The flocking or sandpaper grips the long-napped flannel with sufficient strength to hold light-weight cards.

These individual cardboards, with words, designs, or messages on them, can be "slapped on" or built up to create a dramatic effect not possible with other visuals. See Fig. 14–8.

Hook and loop boards

A hook and loop board is a heavy duty "slap board." The material covering the board contains countless nylon loops. These loops are caught by almost invisible nylon hooks that are on the small pieces of tape mounted on the back of signs or other objects used for the display. A tiny patch of the hook material fastened to a heavy or large object will hold it securely on the board.

Hook and loop boards are available commercially in sizes ranging from 18 by 24 in. to 48 by 72 in.

Magnetic boards

A magnetic board can be made of either a spray-painted sheet metal plate or a steel-backed chalk board. Small objects or cutouts mounted on small magnets or on magnetic tape can be placed on the board and then moved at will.

This type of visual often is used for training operators of vehicles such as forklift trucks. The mobility of the objects—toy vehicles or cutouts of trucks, together with cutouts of aisles and loads—enables the instructor to give a realistic demonstration of safe practices.

Photographs

Photographs need not be projected to be useful. Candid shots of safe and unsafe practices or conditions, photos of award presentations, meetings, new equipment, and the like, have excellent news value on bulletin boards, in company (or plant) papers (house organs or safety magazines), and are even welcomed by national trade and professional magazines. (See Fig. 14–10.)

The "instant" photograph can be another excellent tool. It provides a record of hazards and is useful in accident investigation. See NSC Data Sheet No. 619.

Exhibits and working models

Exhibits and models can make effective three-dimensional presentations. Exhibits can feature safety equipment, first aid equipment, or fire protection equipment. Some displays include mechanical animation; others permit operation by the viewer.

Miniature models, no matter how crude, can be used to demonstrate various hazards, such as those presented by moving machine parts, or to show safe procedures, such as those required for safe piling of materials. Fig. 14–11 shows a model for demonstrating high-line electrical hazards.

One of the most widely used models is a small wooden dummy that has an articulated "spine" for demonstrating correct lifting. This device is available commercially but also can be homemade. The model illustrates action of the human back and legs more realistically than is possible with other types of visuals. See Fig. 14–12 for construction specifications.

Demonstrations

Safety skits, plays, role playing, and the modeling of safety equipment have been used with good results. Frequently, the safety professional will combine a live presentation with the use of a film or other visual aid.

One safety director reported good results from asking people in his audience to demonstrate safe lifting and to give part of a flip chart talk. Some safety professionals have had "actors" carry safety placards, signs, statistical charts, or portray injured workers. Others have used pretty girls to enliven their presentations.

Projected Visuals

Projected visuals include slides, filmstrips, transparencies for overhead projectors, objects used in opaque projectors, motion pictures, and television.

This chapter will not go into great detail regarding slides, filmstrips, and movies, because excellent information is available from the manufacturers of film and projectors, camera stores, libraries, schools, and publications. The *National Safety News*, for instance, frequently carries detailed articles on producing slide shows, using slides, and related topics. The National Safety Council also publishes a series of Data Sheets on the subject (see References).

Slides

The most versatile visual aid probably is the 2 by 2 in. slide, made from 35 mm (or 126 or 127) film in color or in black and white. Among the advantages of this aid are the low cost of the original film and the ease with which the pictures can be made and shown and that sound can be added. Programmers and other accessories permit special effects, sound, and automatic features which lend greater versatility.

The larger 3¼ by 4 in. slides, which require a powerful auditorium-type projector, are used more often than the 2 in. slides by instructors and speakers appearing before large audiences.

"Instant" slides 2¼ in. square or 3¼ in. by 4 in. can be made directly from Polaroid® Land camera transparencies. The slides can be mounted quickly in plastic frames for use in either an overhead projector fitted with a slide attachment or in a special lantern slide projector.

Slide projectors range from inexpensive single slide projectors that can be held in the hand and pointed at the screen, to remote control or automatic change projectors that have trays or magazines holding up to 140 slides. A small rear projection portable machine in a self-contained case, designed for table top use, is suitable for very small groups. With a taped or recorded message, this type is well suited to training one or more employees.

Filmstrips

The term "filmstrip" is preferred to the term "slide film" because the latter may be mistakenly understood to mean the separate slides. A filmstrip is merely a strip of standard 35mm film on which has been photographed a series of single frame pictures whose area is about one-half that of a standard 35 mm double frame 2 in. slide.

Because the pictures in a filmstrip never can get reversed, lost, or out of sequence and because a strip can be rolled and placed in a small can, filmstrips are widely used by decentralized or multi-unit organizations in which a single copy must be shipped back and forth. Also, although the first cost is higher than that of the 2-in. square slides, it is not expensive to have a large number of copies made. Fig. 14–13 shows dimensions

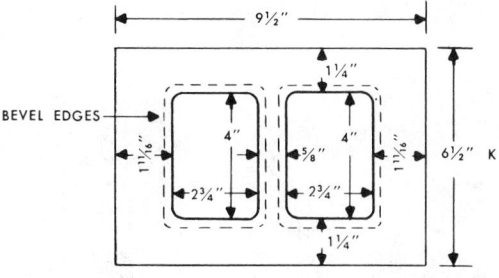

FILMSTRIP AND SLIDE MASKS — ⅛" HARD BOARD (TEMPERED MASONITE)

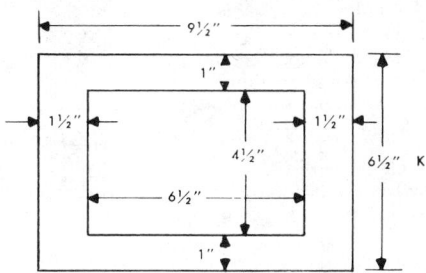

Fig. 14–13.—Filmstrip and slide masks can be used in conjunction with a copy stand to make filmstrips (above diagram) at two exposures per exposure on standard 35mm film, or 2 by 2-in. slides (lower diagram). Masks can be made of ⅛-in. hard board with beveled edges. Inexpensive, commercially avaliable copying stands make it easy for anyone to transfer printed material or artwork to 35mm film.

Courtesy Eastman Kodak Co.

of masks that can be used for making both filmstrips and 2 by 2 slides by copying with a 35mm camera.

Filmstrips require even more careful planning than slides because it is impossible to rearrange the frames or to add new material to suit specific situations or changing times. Also, unless a single-frame camera is used,

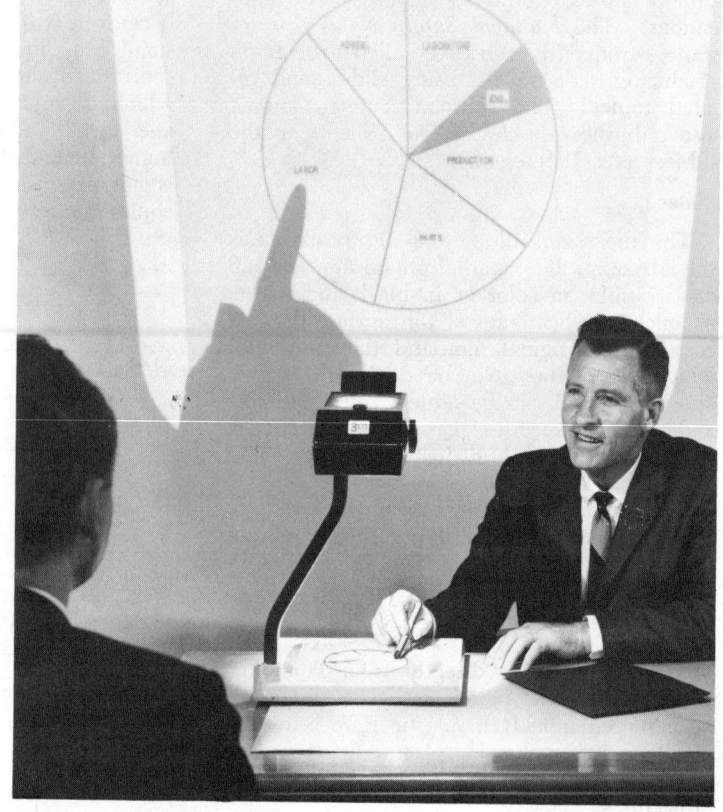

FIG. 14–14.—An overhead projector permits the speaker to face his audience, as he can see his visual without having to turn his back on them. Pencil or pointer in speaker's hand shows as shadow "pointer" on projected image behind him. He can cover portion of transparency with ordinary paper if he wants to use "1-2-3" revelation technique.

Courtesy 3M Business Products Co.

filmstrip material, unlike the material for slides, must be reduced to standard size reflection copy. For a single copy or a few copies, an ordinary 35mm camera and a copying stand can be used to photograph art work or color prints. Quantity production of filmstrips is best done by a commercial laboratory.

If no record script is to be provided with a filmstrip, captions can be inserted over the pictures or in separate frames. Generally, however, the speaker gives a commentary that follows the sequence of the strip. His timing can vary to suit the detail or importance of the individual picture, since each picture can be held to the length of the spoken message.

Filmstrip projectors for continuous showing or for showing on rear projection units are also available.

Overhead projectors

In an overhead projector a powerful light shines through a transparency, which is a transparent plastic sheet placed on a glass projection table, to a mirror over the table and then past the speaker to a screen behind him.

Hand lettering and art work can be done directly on the transparencies either before or during the presentation. However, unless special *transparent* color is applied, the marking will show up as black. Also, typing, hand lettering, or drawings can be done directly on special black "carbon paper" through which the projected image appears in white on black background.

Printed or typewritten material or art work can be transferred to clear or colored background transparencies in a matter of seconds and at negligible cost by the dry-process copy method, diazo method, or photographic method.

Full-color transparencies can be made directly from clay-coated stock originals such as magazines and brochures by means of a special color-lift-off film. For repeated showings of an important subject, the additional expense of full-color transparencies of original subjects might be justified.

Color can also be added by the use of tinted plastic overlay sheets, either loose sheets or self-adhering. The latter can be cut to cover and add color to portions of the art work or can be cut to form interesting art or background designs.

The transparencies can be as large as 10 in. square and can be mounted in cardboard frames. "Commentator cues" or a script can be added to the frame for convenient reference. Unmounted plastic sheets or a large roll of plastic film, which can be rolled to the other side of the projector and either discarded or rolled back for reference, also can be used. Both mounted and unmounted sheets can be filed for reuse.

This aid has a number of important advantages. The speaker needs no assistant, he can face the audience, and the room can be either partially or fully lighted. He can use any opaque material as a cover sheet, which can be withdrawn to reveal data point by point, and he can point to items on the transparency—and therefore on the screen—with a pencil or other object. See Fig. 14–14.

Opaque projectors

Opaque projectors are useful for showing printed material and even three-dimensional objects. Printed material up to 10 in. square or an object up to 2½ in. thick is placed in the machine, and the image is projected onto the screen by means of a powerful light and a mirror. The room must be darkened for effective viewing.

With an opaque projector, material that cannot conveniently be transferred for overhead projectors or photographed for 2 in. slide projectors can be quickly shown. An example of such material would be a safety catalog or a safety poster in full color. Of course, to be suitable for use in an opaque projector,

printed material must have type large enough to be legible to the entire audience when the material is projected.

Motion pictures

Movies are rated excellent for training and motivating. Because of their higher cost compared with that of other visuals, they should be planned with special care. Usually, movies are important enough to justify commercial production or at least professional advice before and during production. Some homemade movies, however, have proved to be effective. See previous section on Making Motion Pictures.

The three principal sizes of motion picture film commonly in use are 35mm, 16mm, and 8mm, both standard-8 and super-8 sizes. The 35mm film is found almost exclusively in theaters. The 16mm size is widely used by industrial organizations, and the 8mm size is gaining attention because of its lower cost.

The development of magnetic and optical sound, improved color stock, rear-projection screen, self-threading cartridge, single-frame viewing, and super-8 equipment has resulted in greater use of motion pictures for in-plant training.

The standard-8 and super-8 projectors and cameras are smaller and lighter than the 16mm equipment—an advantage where portability or size is a factor. These projectors can be used for continuous showing of safety films in such locations as cafeterias and lounges, with either a regular screen or small rear-screen self-contained unit.

A wide selection and variety of 16mm safety films are available from insurance companies, local safety organizations, commercial film libraries, industrial producers, and the National Safety Council.

Television

Industrial safety professionals can promote community safety by participating in television programs that feature the prevention of accidents when such programs are scheduled on local stations or national networks. Such participation by an individual can lend emphasis and prestige to his company's own off-the-job safety program.

Moreover, public relations benefits accrue to the company, local safety council, or local

FIG. 14–15.—Both sight and sound can be recorded on magnetic tape and played back immediately. Technique is especially effective for improving a trainer's effectiveness.

Courtesy General Electric Corp.

chapter of the American Society of Safety Engineers that sponsors televised programs of such a nature.

Companies currently using closed-circuit TV for training or other kinds of communication might consider this medium as a means of furthering accident prevention, too. Closed-circuit TV has been employed to advantage in special situations involving extreme hazards or through relay or by the use of several monitor sets, to multiply its effectiveness.

Recent developments in video-tape recording, kinescope production, and conversion to 8mm for playback through cartridge-loading rear-screen projectors make closed-circuit TV a medium worth investigating and using. Once the equipment is available, the convenience and low cost, plus instant replay,

give certain advantages in training and accident investigation. (See Fig. 14–15.)

Videotape recording and playback is playing an important role in safety and fire protection training. One company produces a number of video tapes on various procedures which includes replay discussions with small groups of all of the plant's employees. See Fig. 14–3.

Audio Aids

Purely audio aids include tape recorders, radio, commercial recordings, and public address systems.

Tape recorders

A tape recorder can be used many ways:

• One safety professional, using a battery-operated recorder, dictates a running commentary while photographing safe or unsafe operating conditions, new processes or equipment that require subsequent study. This method proves easier than writing notes. After being edited, the tape is used with the slides to make up a training tool.

• A tape recorder can record minutes in safety meetings, valuable discussion in training sessions, and on-the-job interviews which may prove useful later for bulletins, newsletters, and future meetings.

• Tape recorders and dictating machines are frequently used to record speeches and conferences. Pedestal or table microphones generally pick up extraneous sounds, as well as the desired one; also it is sometimes difficult to identify a speaker. Those planning such discussion recordings should arrange for one-at-a-time discussion or for placement of recorders at different parts of the discussion table. It is a good idea to mention names frequently as one addresses a conferee, and to rephrase discussion points that may be unclear, either to meaning or to clarity (someone else talking at the same time, for example).

• A safe, efficient job procedure can be taped. Sufficient time is allowed for performing each task. The trainee, with the tape player nearby, follows the instructions as he performs the work.

A tape recording can be used with slides or other visual training material and used individually to train new employees at a remote location—for example, one restaurant of a chain of restaurants.

Foreign languages can be recorded on a tape for instruction of non-English-speaking employees. If visual aids are to accompany presentations that are given in two or more languages, it is best that they be strictly pictorial, having no languages written or printed on them.

• A well known speaker on audio-visual aids even uses a tape recorder to "argue with himself" at meetings. This requires close timing.

Radio

Radio can serve as a medium to promote accident prevention through scheduled programs on various aspects of home, traffic, and community safety. Spot announcements concerned with traffic safety, for instance, commonly are broadcast at frequent intervals by local stations during long holiday periods such as the Labor Day weekend.

Commercial recordings

Music, special sound effects, and even dramatic episodes on commercial recording can be added to safety talks, slide scripts, and the like, to give a professional touch.

Public address systems

A public address system is a useful tool for making safety announcements, directing emergency evacuations, and perhaps even publicizing unusual safety achievements.

There is some question as to whether public address systems should be used to broadcast safety messages on the job. Some administrators feel that music or messages might prove dangerously distracting to workers at moving machinery or on other work which requires full attention. This technique may not be effective and may even draw complaints if used excessively.

Portable public address systems have been used by supervisors and safety professionals as an aid in on-the-job meetings, either to help overcome extraneous noise or to create a dramatic effect.

References

Association of National Advertisers, 155 East 44th St., New York, N.Y. 10017. *Advertiser Practice in the Production and Distribution of Business Films.*

Auger, B. Y. *How To Run Better Business Meetings.* St. Paul, Minn., Business Service Press, 1966.

14—Audio-Visual Media

Bowman, William J. *Graphic Communication.* New York, N.Y., John Wiley & Sons, Inc., 1967.

Herman, Lewis. *Educational Films: Writing, Directing, and Producing.* New York, N.Y., Crown Publishers, Inc., 1965.

Hodnett, Edward. *Effective Presentations.* West Nyack, N.Y., Parker Publishing Co., Inc., 1967.

Horn, George F. *Visual Communication: Bulletin Boards, Exhibits, Visual Aids.* Worcester, Mass. 01608, Davis Publications, Inc., 1973.

Mambert, W. A. *Presenting Technical Ideas.* New York, N.Y. John Wiley & Sons, 1968.

National Audio-Visual Association, Inc. 3150 Spring St., Fairfax, Va. 22030. *Audio-Visual Equipment Directory.*

National Education Association, Dept. of Audio-Visual Instruction, 1201 16th St. NW., Washington, D.C. 20036.

National Safety Council, 425 N. Michigan Ave., Chicago, Ill. 60611.
Catalog/Poster Directory.
Industrial Data Sheets
   Motion Pictures for Safety, 556.
   Nonprojected Visual Aids, 564.
   Photography for the Safety Professional, 619.
   Posters, Bulletin Boards, and Safety Displays, 616.
   Projected Still Pictures, 574.

Shefter, Harry. *How to Prepare Talks and Oral Reports,* Pocket Books, Inc., 1 West 39th St. New York, N.Y. 10018.

U.S. Office of Education, Bureau of Adult and Vocational Education, 400 Maryland Ave. SW., Washington, D.C. 20202.

Wittich, Walter A., and Schuller, F. C. *Instructional Technology: In Nature and Use.* New York, N.Y., Harper and Row, 1967.

(Other data books and pamphlets are available from distributors and manufacturers of cameras, film, and other visual media equipment. Trade journals in the fields of photography, education, sales management, advertising, and training also contain excellent information.)

Office
Safety

Chapter
15

Seriousness of Office Injuries   347
Complacency—prime cause of injury . . . Who gets injured? . . .
Types of disabling-injury accidents

Controls for Office Hazards   352
Office layout . . . Safe office equipment . . . Printing services . . .
Enforced safety procedures . . . Fire protection

Safety Organization in the Office   362
Safety committee

References   364

345

LEADING ACCIDENT TYPES
OFFICE EMPLOYEES COMPARED WITH OTHER INJURED WORKERS

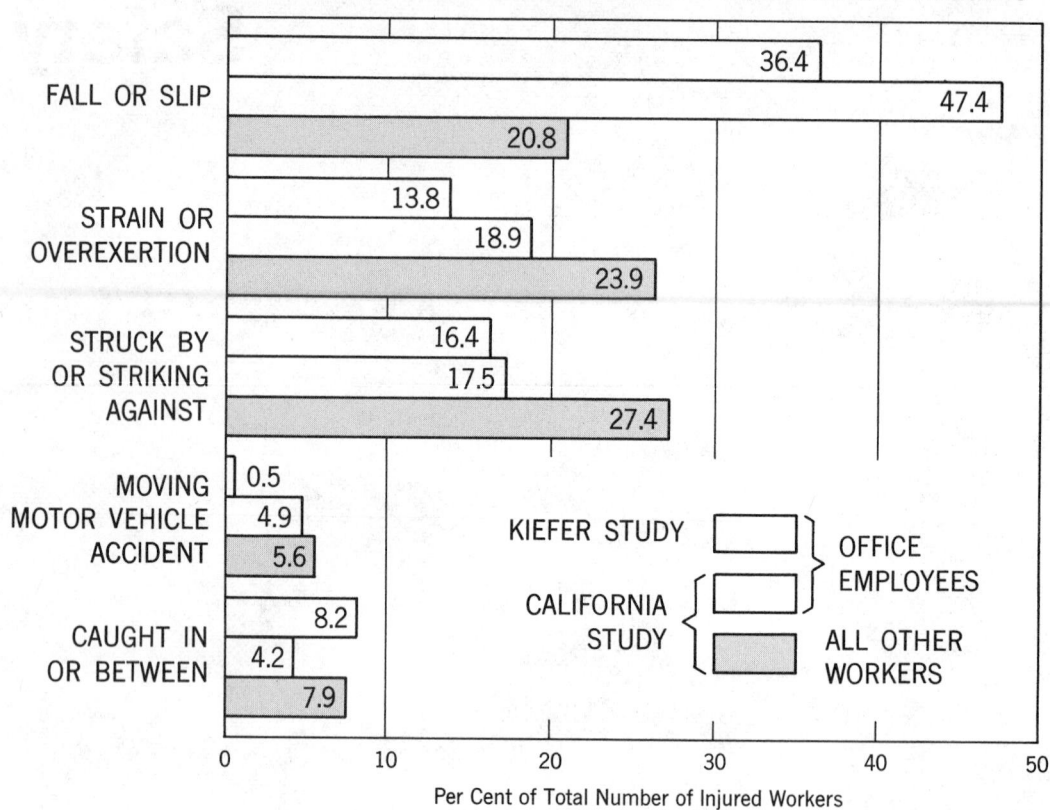

Fig. 15–1.—A comparison of the California study with the Equitable study, and of office employees with all other workers.

Office safety programs are necessary for several reasons:

• Many large organizations today consist almost entirely of office workers (insurance companies and banks are good examples). Accidental injuries are just as painful, severe, and expensive to their workers as to production workers. Unless these organizations have effective office safety programs, however, accidental injuries are far more likely to occur.

• An injury to an office worker can break the safety record of a manufacturing plant just as effectively as can an injury to a production worker. In the aerospace industry,

for example, the injury frequency rates for office workers have sometimes been greater than those for production workers.

• A company safety program cannot be fully effective if it only covers a portion of a company's employees. (a) A safety program not vigorously pursued in company offices will probably not be vigorously pursued in the factory, shop, or plant. If office workers are "exempt," then production workers often feel that following rules to avoid hazards is an unnecessary burden, and, perhaps, an unnecessary exercise of authority by management. (b) "Exempt" office workers seldom

TABLE 15-A

DISABLING ACCIDENTS FROM FALLS IN OFFICE WORK

Kiefer Study—8,000 employees (a five-year study)		Disabling Accidents	Days Lost
In hallways and work areas, caused by running, slipping, tripping over wires, desk drawers, file cabinet drawers, etc.		34	377
From chairs		19	116
Stairs		11	95
Escalators or elevators		7	54
	Total	71	642

California Survey—1,000,000 employees (a one-year study)	Disabling Accidents By Category	Totals
FALLS OR SLIPS ON WALKING SURFACES		941
Slippery surface	245	
Large object in path	164	
Uneven surface	105	
Small object (litter) in path	55	
Fainting, seizure	30	
Loss of balance in pushing/pulling object	14	
Surface condition not specified	328	
FALLS OR SLIPS ON STAIRS OR STEPS		384
Uneven surface	133	
Slippery surface	15	
Littered surface	5	
Condition unspecified	231	
FALLS OR SLIPS FROM CHAIRS		192
FALLS OR SLIPS FROM ELEVATION		76
Standing on office furniture (including chairs)	31	
While climbing ladders	24	
Other	21	
FALLS WHEN GETTING INTO OR ALIGHTING FROM ELEVATOR		10

understand the importance of safety and may scoff at or criticize production-oriented safety activities, especially in front of production employees.

• A safety professional who expects to "sell safety" to management must get management involved in a total safety program—get them involved in preventing office accidents as well as production accidents. Any management that preaches safety must also practice it.

Seriousness of Office Injuries

One reason that office safety programs are not more widespread is that many people believe office injuries are inconsequential.

This is not true. One aerospace firm, for example, paid out $102,000 over eight years at one facility just for injuries incurred by people falling out of chairs. Approximately 25,000 people worked at the plant, on the average, and more than half were office work-

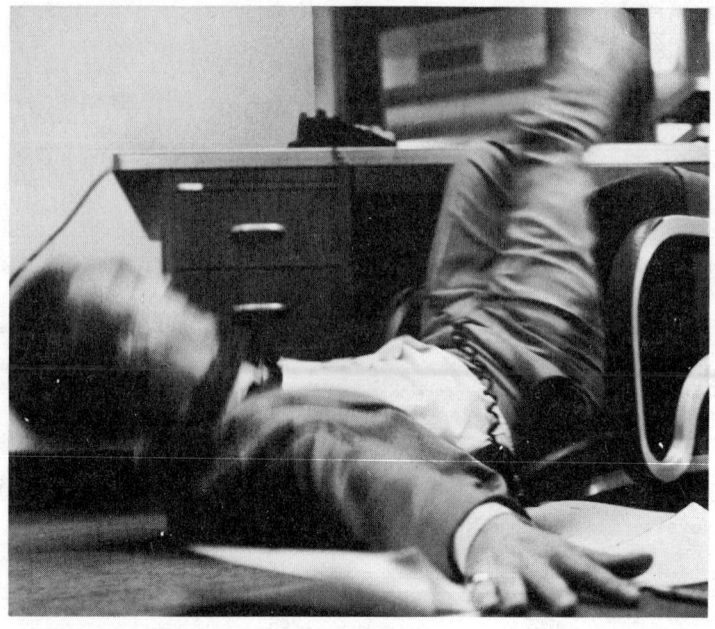

Fig. 15–2.—Case history. Vice president was talking to a customer at his desk. The slight movement of the chair as he sat down caused a caster to come out of its socket; man fell to the floor. Result: midback fracture, collapsed vertebrae, and 9 months lost time.

ers. Fourteen chair accidents occurred, and the two worst totaled $97,000.

Studies made by the State of California Department of Industrial Relations, and by Norvin C. Kiefer, M.D., Chief Medical Director, Equitable Life Assurance Society of the U.S. (see "References"), show that this is not an isolated incident (Fig. 15–1).

The California State Department of Industrial Relations analyzed reports filed by more than 3000 California employers (employing more than one million office workers) on disabling injuries to employees. When extrapolated to nationwide scale, on-the-job office accidents would annually amount to about 40,000 disabling injuries at a *direct cost* (indemnity benefits and medical expenses) of about $100 million. (This figure does not include any indirect costs for employers, workers, or the nation.)

In addition, of the estimated 25,000,000 office workers in the U.S., 300 to 400 deaths might be caused each year by office accidents. Of these, 150 to 200 deaths per year may occur from work-connected automobile accidents alone, a category that can be controlled by a good "defensive driving" safety program, such as that developed by NSC.

"Office worker" is defined as (in the California study) "a person primarily engaged in performing clerical, administrative, or professional tasks indoors in an office at his employer's place of business." This definition would *not* cover salespersons, claims adjusters, social workers, medical and teaching personnel (other than clerical or administrative), and certain stock, order, and inventory clerks. The California study did not cover employees of the Federal government, maritime workers, and railroad workers in interstate commerce. Dr. Kiefer's study, on the other hand, *did* include salespeople, claims adjusters, medical personnel, and supply and warehouse personnel.

The insurance company study covered approximately 8000 employees working in one building, about 5 percent of whom were maintenance personnel. During the five year study period, the injury frequency rate was 2.6, and severity rate was 15 for the office workers, using ANSI Z16.1 method of calculation. The average days charged per disabling injury was 5.9.

These figures compare with a frequency of 1.6 found in the California study, and a frequency of 0.5 for office workers of paper manufacturers, in a survey made by the American Mutual Insurance Alliance.

Complacency – prime cause of injury

Office injuries may seem to be inconsequential because they lack dramatic impact.

The person who is injured when he falls off his chair (Fig. 15–2) seems to merit little sympathy or attention. The picture of an office worker slipping and crashing to the floor on his buttocks seems amusing.

Industrial accidents are commonly considered to occur more frequently and with greater severity—amputations, lost eyes, and broken bones.

But who is likely to be absent from work the longer? The office worker who sustains a severe compression fracture of one or more lumbar vertebrae from a bad fall, or the man whose hand is amputated?

Complacency—the attitude that office accidents do not amount to much—is one of the prime causes of office accidents. The average office worker gives little thought to safety because he just does not believe that accidents happen in offices.

On the other hand, the well instructed worker in a plant manufacturing flammable, toxic chemicals knows that he is in a risky business—he understands why he must use safe procedures and wear safety equipment. His safety attitude and training is his best defense. As a result, his production department may have less accidents than the plant office.

The office worker, therefore, has to be informed of the hazards he must guard against, and of safe work procedures that he must use. He must also be shown that management provides a safe environment and safe equipment if he is to develop a proper safety attitude. He must be willing to adopt safe procedures, and be encouraged to do so. Even more important, office supervisors must understand the nature of office hazards and unsafe practices and take necessary preventive measures.

Who gets injured?

The two studies pointed to whom most injuries occurred, and how they occurred.

New surroundings. These studies showed the importance of teaching office workers to look for new hazards and to correct them. It was also found that there was a substantial increase in the number of injuries in the first year a company moved into a new office building. The change upset established routines and presented unknown hazards. Even going to and from work became more hazardous as employees had to explore new routes.

New workers. New workers require time to "get acquainted" with the concept of safety. The injury rate among employees with less than one year's service was double that of employees with one to four years' service, and this latter group's rate was three to four times as high as that among employees with five or more years' service.

Younger employees. The studies indicated that 10 to 12 percent of the office employees were youngsters, just graduated from high school and in their first year of employment. Most of them were girls, and although many of them were excellent, conscientious workers, they were not as interested in their job and career as they were in "finding the right man." This is the group that, each year, most needs prompt and special safety orientation.

Sex of employee. In the California study, 80 percent of the disabling injuries occurred to women, although they comprised only 66 percent of the office labor force. Compare this percentage with that of California industry generally—women accounted for 14 percent of the job injuries, even though they represented about 33 percent of the work force.

This study shows that, for office occupations, the estimated disabling work injury rate for women was about twice as high as that for men. The rate was approximately 4 disabling injuries per 1000 women employed in office work compared with 2 for the same number of men.

The injury statistics were substantially different in the Equitable study, which was compiled from carefully made reports. The rate of injury accidents per thousand male employees was about the same as it was for female employees. However, the rate of total days lost from disabling injuries was 2½ to 3 times higher for the men than for the women.

Types of disabling-injury accidents

Falls are the most common office accident, and account for the most disabling injuries, as shown in both surveys, Fig. 15–1. They cause from 2 to 2¼ times as high a disabling injury rate among office as among nonoffice employees. Falls were the most severe office accident and were responsible for 55 percent of the total days lost because of injuries.

Women were the most commonly injured by slips and falls, Fig. 15–3. The California survey showed that 53 percent of all disabling injuries to women were caused by falls; men sustained only 25 percent of their injuries from this source.

The California survey showed that three out of four falls occurred on walking surfaces, such as floors, stairs, or sidewalks. About one out of four falls was from a chair. Refer to Table 15–A.

● About 8 percent of the injured people were hurt by falls on slippery floors; almost all of these falls occurred to women.

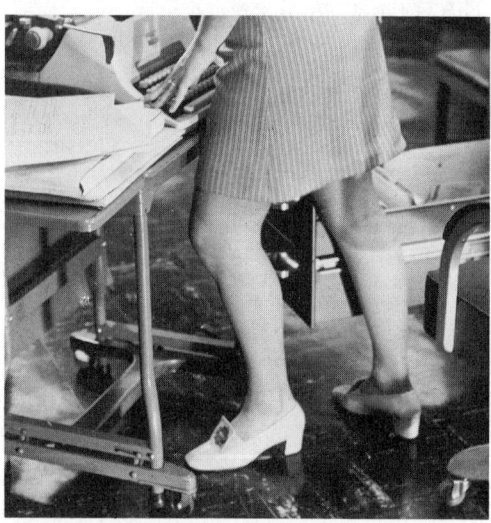

FIG. 15–3.—Case history. Being in a hurry to deliver a report, secretary did not notice the open left drawer of her desk until she tripped over it. Unable to regain her balance, she fell, striking against the typewriter table. Results: a chipped bone in the left elbow, bruised left hip, and 14 days lost time.

FIG. 15–4.—Case history. Statistician used a stool with casters to replace books on a shelf—as she had often done before without mishap. This time, the stool rolled backward; she lost her balance and fell to the floor, landing on her hands and knees. Results: fractured right knee and 2 months lost time.

● About 7 percent of the falls were caused by uneven surfaces. Tripping over, or catching a heel on stairs or a door sill accounted for about one-fourth of these cases; the other three-quarters were due to defective surfaces —women would catch their heels in small holes or on loose floor coverings.

Office people were involved in 46.5 percent of all job-related accidents, compared to 36.4 percent for blue-collar workers, according to a report issued by Prentice-Hall Publishing Co. The major damage was done by falls due to objects in a worker's path and littered floors. The usual object was a loose telephone wire or electric cord on floor-mounted outlets, open drawers, and waxed floors.

• Most chair falls came when a person was sitting down, rising, or moving about on a chair. A few were caused by people leaning back and tilting their chairs in the office or cafeteria, or putting their feet up on the desk. Although stairs seem more hazardous than chairs, people seem to recognize the hazard and are more cautious. Furthermore, people are not as often exposed to the stair hazard as they are to a chair.

• A final category was falls from elevations (Fig. 15–4), caused by standing on chairs or other office furniture and by falls from ladders, loading docks, or other miscellaneous elevations. These falls accounted for approximately two percent of the disabling injuries in the California study, whereas falls on stairs accounted for almost four percent.

Strains from over-exertion. In California,

Fig. 15–6.—Case history. Girl, returning tray to company cafeteria, was walking on left side of aisle (note guide line on floor). She turned a corner and struck man walking in the right lane. Result: No lost-time injury; man had to go home to change clothes so he could attend a meeting that afternoon; he later learned he had bruised his left side.

Fig. 15–5.—Case history. Assistant librarian was hurrying to finish her day's work, in anticipation of an evening of bowling. Unable to see where she was going with her load of books, she struck and broke a glass partition. Results: severely lacerated right foot and 2 days lost time.

strains and over-exertion occurred almost entirely to men, possibly because of the lesser weights normally lifted by women.

Almost three-fourths of the strain or exertion mishaps occurred while employees were trying to move objects—carrying or otherwise moving office machines, supplies, file drawers and trays, office furniture, heavy books (Fig. 15–5), or other loads. Often, the employees were making unauthorized moves without the knowledge or consent of their supervisors.

A significant number resulted from the sudden or awkward movement of the employee himself, and did not involve any specific agency. Reaching, stretching, twisting, bending down and straightening-up were often associated with these injuries.

Striking against objects caused approximately nine percent of the office injuries in both studies discussed here. Two out of three of these injuries were the result of bumping into doors, desks, file cabinets, open drawers, and even other people (Fig. 15–6) while walking. Hitting open desk drawers or the desk itself, while seated at a desk, or striking open file drawers while bending down or straightening-up caused most of the rest of these injuries.

Other incidents of striking against objects included bumping against sharp objects such as office machines, spindle files, staples, and pins. Also, cuts a worker would receive in handling paper, file drawers, and supplies often became infected.

Objects striking workers accounted for about eight percent of injuries to office workers in both studies. Most of these injuries were sustained when the employee was struck by a falling object—file cabinets that became over-

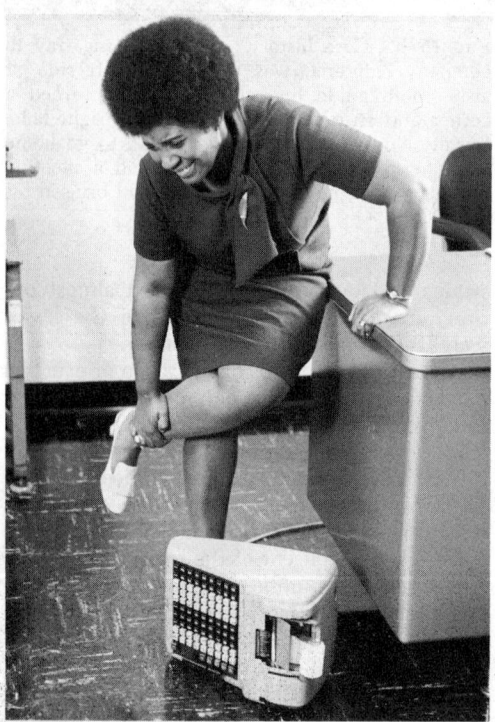

FIG. 15–7.—Objects striking workers accounted for about 8 percent of injuries, according to both studies.

balanced when two or more drawers were open at the same time, file drawers that fell when pulled out too far, office machines and other objects that employees dropped on their feet when attempting a move (Fig. 15–7), or typewriters that fell from a folding pedestal or rolling stand.

In addition, a number of employees were struck by doors being opened from the other side. Office supplies or other material and equipment sliding from shelves or cabinet tops caused a few injuries in this classification.

Caught in or between. The final major classification was accidents where the worker was caught in or between machinery or equipment (Fig. 15–8). Mostly, this was getting caught in a drawer, door, or window. However, a number of employees got caught in duplicating machines, copying machines, addressing machines, and fans. Several got their fingers under the knife edge of a cutter.

Miscellaneous office accidents included foreign substances in the eye, spilled hot coffee or other hot liquid, burns from fire, insect bites, and electric shocks.

Controls for Office Hazards

To control office accidents, the first necessary steps are to eliminate hazards or reduce exposure to them. The best times to take these steps is when the office is laid out, when equipment is purchased, or when office procedures are set up.

Office layout

Offices should be laid out for efficiency, convenience, and safety. The principles of work flow apply to offices as well as to factories.

Employees having frequent callers should be near entrances. Those doing paper work should be so located that work can flow through the office with a minimum of back tracking. Particularly, necessity to travel up and down stairs should be eliminated, as far as possible. Transportation distance of the work should be at a minimum.

Desks should be arranged so that each worker will receive his work from the person behind or beside him, if possible. Desks should be facing in the same direction for most operations. Where two employees are working together, they may face each other.

For desks facing in the same direction, distance between the back of one desk and the front of another should not be less than three feet. More space per employee is highly desirable. Not less than 50 sq ft per employee should be allowed in large open areas.

Employees using the same machine should be grouped, and employees should be placed in front of or around the persons having supervision over them.

Office machines should not be placed near the edges of tables or desks on which they are used. Machines that creep during operation should be secured. The maintenance department should either put rubber feet on them or fasten them down, because office machines have fallen off tables and onto people's feet. And, particularly, typewriters on folding pedestals should be fastened to the pedestal.

Heavy equipment and files should be placed against walls or columns; files can also be placed against railings. File cabinets should be bolted together or fastened to the floor or wall so that they cannot be tipped over. If they are again moved, they can be temporarily fastened together by stout adhesive tape and then rebolted as soon as possible.

Floors in offices are the major agency connected with office accidents. They should be as durable and maintenance free as possible. Floor finishes should be selected for anti-slip qualities. Well maintained carpet should provide good protection against slips and falls. Defective tiles or boards or carpet should be repaired immediately. Worn or warped mats under office chairs and rubber or plastic floor mats with curled edges or tears should be replaced or repaired as these conditions can create tripping hazards.

Wax as a cause of slippery floors is vastly exaggerated. Slipperiness is characteristic of highly polished and extremely hard but unwaxed surfaces such as marble, terrazzo, and steel plates. Slip-resistant floor wax can give these materials a higher coefficient of friction than they already have and can reduce their slipping hazard. However, wax must not be applied so thickly that a smeary coating results. Also, an oil mop should not be used

Fig. 15–8.—Case history. Secretary's typewriter was removed for routine servicing. Maintenance man lowered the spring-loaded shelf into its usual place and closed the door. Secretary, uninformed about the spring (which normally accommodates a 31 lb typewriter), decided to dust the drawer space and snapped the shelf out. Result: fractured metacarpal, 3 days lost time, and inability to type for 2 months.

on a wax floor because a soft and smeary coating will result. See page 579–582.

Special anti-slip protection should be used on stairways and at elevator entrances, and these specially hazardous areas should always be maintained in the best possible condition. Floor mats and runners often provide a better, more slip-resistant walking surface. Their use is discussed in NSC Industrial Data Sheet 595, *Floor Mats and Runners.*

Aisles and stairs. A suggested minimum width for aisles is four feet. Passages through the work area should be unobstructed. Waste baskets should be kept where people do not trip over them. Telephone and electrical outlets should not protrude into passages that people use. These and other obstructions, such as low tables and office equipment should be protected by being placed against walls or partitions, under desks or in corners. Stepoffs from one level to another in an office should be avoided. If one exists it should be well marked and guarded with a railing.

353

File drawers should not open into aisles unless extra space is provided for this. Particularly, file drawers should not open into narrow aisles. Pencil sharpeners and typewriter carriages must not jut out into aisles.

Stairways and exits (including access and discharge) should comply with NFPA Std. 101, *Life Safety Code;* floor and wall openings should comply with ANSI *Safety Requirements for Floor and Wall Openings, Railings, and Toe Boards,* A.12.1. Hand rails are specified for one side of stairs up to 44 in. wide, and both sides for stairs wider than 44 in. For stairs wider than 88 in., add an intermediate (center) rail. Exits should be frequently checked to be sure that stairways are unobstructed and well illuminated. Exit doors, if locked, shall not require the use of a key for operation from inside the building.

Doors are another frequent source of accidents in offices. Glass doors should have some conspicuous design, either painted or decal, about four and a half feet above the floor and centered on the door so that people will not walk into it.

Safety glass complying with American National Standard Z97.1 should be installed rather than plate glass. Sometimes local codes specify the type that must be installed.

Solid doors also present a hazard, because people often approach both sides and one of them can be struck when the door opens toward him. Frosted glass in doors gives a view through for accident prevention, but still preserves privacy. If it is necessary that a door be solid (such as those of a fire tower), the hazardous area that the door swings over can be marked by yellow-and-black tape, or painted a bright color (protected by clear varnish), or the path of the swinging door can be outlined by colored plastic circles. On carpeted floors, a quarter- or half-circle of different-colored carpet can be used.

Another hazard is the door that opens directly onto a passageway. If the door opens directly into the path of on-coming traffic, somebody might bump into the edge of the door. If doors in hallways cannot be recessed, they should be protected with short angled, deflector rails or U-shaped guardrails which protrude about 18 in. into the passage way (Fig. 15–9), or the area they swing over can

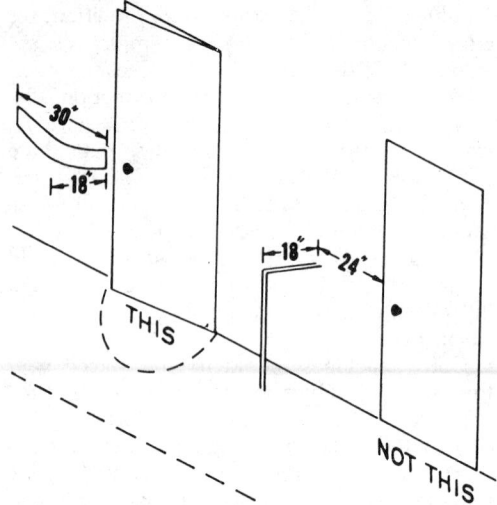

FIG. 15–9.—Short U-shaped guard rail, protruding about 18 in. into aisle, protects pedestrians from edge of door should it open.

be marked as suggested in the paragraph preceding. Another procedure is to place storage lockers or benches along the wall near the door, which (in effect) recesses the door.

Some offices paint white or yellow stripes or apply commercially available tape to floors to separate traffic or to guide people away from a rapidly opening door. Floors in front of swinging doors can also be marked or painted to warn against swinging doors. (See Figs. 15–6 and 15–9.) As a final precaution, it is good practice to have the door hinges on the upstream side of the traffic; that is, on the right hand side as one faces the door from the hallway.

Adequate light, ventilation, washrooms, and other employee services have an important influence on employee morale. Growth of a business sometimes results in installing more desks and other equipment than original plans called for. Over-crowding is bad from the standpoints of both appearance and psychological effect on employees, especially if it overtaxes ventilation facilities. Smaller offices can be made to appear larger and less crowded, if walls, woodwork and furniture placed against the walls are the same color.

Illumination levels recommended by the Illuminating Engineering Society for an office are listed in Table 15-B. Also see p. 1449.

Employees should not face windows, unshielded lamps, or other sources of glare. Indirect, shielded fluorescent lamps are particularly desirable to produce high levels of illumination without glare.

Walls and other surfaces should conserve

contributing causes of office accidents. Some of these are: direct glare, reflected glare from the work and harsh shadows, all of which hamper seeing.

Excessive visual fatigue itself may be an element leading toward accidents. Accidents may also be prompted by the delayed eye

TABLE 15–B

LEVELS OF ILLUMINATION

	Currently Recommended Illumination* Footcandles
OFFICES	
Cartography, designing, detailed drafting	200
Accounting, auditing, tabulating, bookkeeping, business machine operation, reading poor reproductions, rough layout drafting	150
Regular office work, reading good reproductions, reading or transcribing handwriting in hard pencil or on poor paper, active filing, index references, mail sorting	100
Reading or transcribing handwriting in ink or medium pencil on good quality paper, intermittent filing	70
Reading high contrast or well-printed material, tasks and areas not involving critical or prolonged seeing such as conferring, interviewing, inactive files, and washrooms	30
Corridors, elevators, escalators, stairways	20 (or not less than ⅕ level in adjacent areas)

* Minimum on task at any time.

From Illuminating Engineering Society.

light, while avoiding annoying reflections.

If offices depend largely on daylight, employees engaged in the visual tasks should be located near windows. North light is preferred by draftsmen and artists. Ceiling walls and floor act as secondary large area light sources and if finished with the recommended reflectances, will increase the utilization of light and reduce shadows.

Some accidents may be attributed to poor illumination. However, many less-tangible factors associated with poor illumination are

adaption a person experiences when moving from bright surroundings into dark ones and vice versa. Some accidents which are attributed to the individual's "carelessness" can be traced to difficulty in seeing, from one or more of these mentioned causes.

Illumination evaluation. As a uniform means of evaluation, a standard procedure, entitled "How to Make a Lighting Survey" has been developed in cooperation with the U.S. Public Health Service. (This Appendix is not

355

part of the American National Standard *Practice for Office Lighting,* A132.1, but is presented as background material for the use of the standard.) Table 15-B shows the illumination levels currently recommended by the Illuminating Engineering Society and approved by the American National Standard *Practice for Office Lighting.*

Ventilation. Window ventilation is often unsatisfactory compared with comfort (or "air") conditioning. Persons near windows may feel cold; those farther away may be warm. Where there are large interior spaces, forced ventilation is needed, usually, if the space is to be used for office purposes.

All mechanical ventilation and air conditioning systems require careful planning and installation by experts. Private offices installed around the outer walls of a large office space should not cut off light and ventilation from the other employees.

Fans in offices should be placed where they cannot fall on anybody and they probably should be secured in place.

Adequate ventilation should be supplied for duplicating machines, particularly spirit duplicating machines, or others that use ammonia, methanol, or other toxic liquids. Concentrations of methanol around poorly ventilated duplicating machines can reach 200–300 ppm. Duplicating processes should not be confined to a separate small room, unless it is vented to the outside.

Electrical. A number of hazards from electrical equipment should be provided against in the layout of the office. Electric key switches should be avoided because people try to pick the lock with hair pins or paper clips. Dictating machines, electric typewriters, desk lamps, and other equipment require outlets and extension cords arranged to avoid tripping hazards.

Where possible, outlets (receptacles) should be installed to eliminate extension cords. Cords that are necessary should be clipped to backs of desks or taped down. If cords *must* cross the floor, cover them with rubber channels designed for this purpose.

Outlets should accommodate 3-wire grounding plugs to prevent electric shock to operators. Outlets should not only be located under desks to eliminate tripping hazards, but they should also be placed where they will not be accidentally kicked or used as a foot rest. As they loosen or wear, outlets can become sources of electric shock.

Electric equipment in photographic laboratories, particularly, should be grounded because the photographer is close to both water and electricity in the dark room.

Caution should be exercised in the use of poorly maintained or unsafe, poor quality non-U.L.-listed coffee makers, radios, lamps, etc., provided by or used by employees, particularly in out-of-the-way locations. Such appliances can create fire and shock hazards.

Cords for electrically operated office machines, fans, lamps, and other equipment should be properly installed and inspected frequently to prevent the occurrence of defects which may cause shocks or burns. Switches should be provided, either in the equipment or in the cords, so that it is not necessary to pull the plugs to shut off the power. Extension cords should be free from splices and they should be disconnected by grasping the plug, not by pulling on the cord. The fact that office equipment is operated at relatively low voltage is no assurance that serious injury will not occur. There are on record instances of fatalities caused by circuits of 110 volts and less.

Keyed metal light sockets may be especially dangerous when located near plumbing or other grounded equipment. Such sockets should be grounded or replaced with a nonconducting type, such as porcelain, plastic, or rubber. Wall receptacles should be so designed and installed that no current-carrying parts will be exposed, and outlet plates should be kept tight to eliminate possibility of shock or collision injury. Extension cords should not rest on steam pipes or other metallic surfaces.

In the installation or repair of any electrical equipment, the work should be done by qualified workmen using only approved materials. Since defective wiring may constitute both shock and fire hazards, all recommendations of the NFPA *National Electrical Code* should be observed.

Materials stored in offices sometimes cause problems also. In general, materials should be stored where heavy traffic does not have to be crossed to reach them, and they should be stored where they are not likely to fall on anybody. Also, nothing should be allowed on the floor in a passageway where it could be a tripping hazard.

Pile materials neatly in stable piles that will not fall over. The heaviest and largest pieces should be on the bottom of the pile. Where materials are stored on shelves, the heavy objects should be on the lower shelves.

No smoking should be allowed in mailing, shipping, or receiving rooms, or other areas where there may be large quantities of loose paper and other combustible packing material, or in areas where flammable fluids are used, such as duplicating rooms or artists' supply areas.

Flammable fluids and similar materials should be stored in safety cans, preferably in locked cabinets. Only minor quantities should be left in the office and bulk storage should be in properly constructed fireproof vaults. (See Chapter 42, "Flammable and Combustible Liquids.")

Safe office equipment

A good quality of office furniture not only contributes to the safety of the office but also to its appearance. This, in turn, improves the attitudes of both employees and visitors.

Chairs, especially, should be comfortable and sturdily built with a wide enough base to prevent easy tipping. The casters on swivel chairs should be on at least a 20-in. diameter base, and a 22-in. base is preferred. The casters should be securely fixed to the base of the chair and well constructed because loose or broken casters are a frequent cause of chair falls (Fig. 15–2). About 20 percent of the chair falls in the California study were due to chair defects.

Adjustment features on chairs should be well designed and well made so that they will work properly. Chairs with poor adjustments sometimes must be welded or bolted into a fixed position so that they can be used at all. Also, springs on chairs should be guarded so that if they do break, they will not shoot out and injure the occupant's leg.

Desks and files. Spring-loaded typing desks should be investigated carefully. If some models are opened without due care, the typewriter table will snap out and cause a bruise or a cut.

Also, even if good quality desks and file cabinets are purchased, it is still possible that occasionally one will have a sharp burr or corner on it. Office furniture should be inspected when received and such burrs or corners should be removed immediately.

Drawers on desks and file cabinets should have safety stops.

Purchase well guarded office machines such as rotary files and copying machines. It is worthwhile to put a guard on the paper cutter.

Glass tops on desks and tables crack and cause safety hazards. Durable synthetic surfaces are free from this trouble.

Sufficient non-combustible waste baskets should be furnished to catch litter. Also enough safety-type ash trays should be available and they should be large enough and stable enough to safely contain smoking materials.

Office fans should have substantial bases and convenient attachments for moving and carrying. They should also be well guarded, front and back, with not over ½-in. mesh to prevent fingers getting inside guard.

Many cut fingers have resulted when people try to move fans by grasping the guard, or tried to catch falling fans. Fans should not be handled until the power is turned off and the blades stop turning.

Computers. If a computer is to be installed in a sprinklered building, keep sprinkler protection in service throughout the area, but get advice on necessary protection against both fire and water damage. Keep combustible materials such as paper, tapes, and cards at an absolute minimum in the room with the computer. When safeguarding such an investment, this is a good time to call in your fire protection adviser.

Rolling ladders and stands used for reaching high storage should have brakes that operate automatically when weight is applied to them.

Chemical products. If possible, substitute

non-toxic and non-flammable solvents for those used in printing and duplicating or other operations. If chlorinated bleaches are purchased for cleaning purposes, make sure that they will not be mixed with strongly acidic or easily oxidized materials. Purchase a good grade of slip-resistant floor wax.

Purchasing equipment. The company safety professional should work with the purchasing agent in buying office furniture and equipment. Both should be aware that sometimes office equipment is advertised in such a way that safety features are stressed, but the machines may be delivered without them. Mechanical hazards of heavy office equipment can be ascertained by careful, expert inspection before purchase. These hazards can almost always be eliminated or minimized, although sometimes at substantial expense.

The purchasing division must also be made aware of precautions to be taken in connection with chemicals, dyes, inks, and other supply items. Particular attention should be paid to toxic, irritant or flammable properties. Where hazards are unavoidable, specific instructions for careful handling should be issued when the material is received.

The purchasing division should gather all pertinent information from the manufacturer on equipment design, and should try to determine the composition of proprietary compounds. This they can forward to the safety professional (or safety department) for an opinion concerning inherent safety hazards before purchasing new equipment or supplies.

Single or multiple-copy office duplicating equipment can injure as severely as large presses. They should only be operated by employees thoroughly trained in their operation, and maintained by competent service personnel. Nobody should be permitted to interfere with interlocks on guards, or to clean or adjust machines while they are running.

All office duplicators should be grounded, well guarded, and interlocked so that removal of guards for cleaning will stop the machine. These machines should have all drive mechanisms guarded. The operators should be dressed so that loose clothing and hair present no problem. Paper stocks should be handled in such manner that heavy lifting is not nec-

essary. Ream packages can be used.

With the high number of female employees running office duplicators, special attention should be paid to clothing, instruction, and layout of the area. All machines of this sort should have enforced safety procedures; constant training and retraining is necessary.

Employees should be told that if any office machine gives a shock, appears defective, or if it sparks or smokes—turn it off, pull the plug, and tell the supervisor.

Printing services

Larger offset presses. Check operation of offset presses. Is the operator putting his thumb on the blade or blanket while the press is in motion? One offset press department had seven finger injury accidents in the first two weeks of operation, all caused by pressmen who put their fingers in the running press to remove hickies from the plate.

Presses should meet all guarding regulations imposed by local, state and Federal agencies.

The area around the presses should be free from clutter and be well lighted, the flooring should be resilient or rubber mats should be provided to minimize operator fatigue and to prevent slipping.

Only qualified operators should run presses. Loose clothing and long hair are dangerous around these machines. Low flashpoint flammable liquids or toxic solvents should not be used to clean the presses; office supervisors and press operators should understand the fire and health hazards involved and follow all instructions for safe use, storage and disposal of such flammables.

Gathering and stitching machines. Guards should be installed on open sprockets and collector chain drives of gathering and stitching machines to protect employees from hand and body injury. The operating arm on the end of the gathering machine should be guarded.

Hinged drop-guards should be installed to cover any exposed operating mechanism which creates nipping hazards under the machine and along the working area where operators fill the pockets. The floors and work platform at this area should be covered with non-skid material.

Operators should be trained to open signatures in the middle and place them on the saddle or rod between the hooks on the moving chain. If the hook is not put on the rod or chain correctly, *no attempt should be made to straighten it out until the machine has been shut off*. The machine should also be shut down when threading stitcher heads, making any adjustments or removing jams.

Folding machines. Following are specific points to be stressed for safe operation of folding machines:

1. When jammed-up paper is pulled from the machine, shut the motor off to avoid getting hands caught in the feed rollers.

2. Finger clearance at folding knife should be checked before pulling out paper, putting tape on rollers, or adjusting plates and roller pressure.

3. Workers should walk down the steps of folder feeder platforms facing forward, never backing down.

4. On large-size folders, all steps and platforms should be protected by railings.

Defective stitches protruding out of books should be removed to avoid cuts from them while books are being jogged, trimmed or wrapped. Workers should be trained to cup their hands over the work when removing defective wire stitches.

Employees engaged in this operation should wear safety glasses or face protection, and passers-by should be protected against flying stitches by screens or by isolation of this work.

Enforced safety procedures

Because the major category of office accidents is slips and falls, running in offices, particularly for elevators, should be prohibited.

A number of office accidents could be prevented if everybody walking in passageways would keep to the right. A white stripe down the middle of the passageways encourages the following of this rule, and can prevent collisions at blind corners and at doors, as mentioned earlier in this chapter. Collisions at doors also can be prevented if people do not stand directly in front of the door, but out of the path of its swing when they go to open it. (Details were discussed earlier in this chapter. See Fig. 15–9.)

People carrying a stack of light material must be sure they can see over and around it when walking through the office. They should not carry stacks of materials on stairs; they should use the elevator. People should not have both arms loaded when using stairs; one hand should be free to use the hand rail.

Workers leaving the office at night should be instructed to go single file on stairs, to keep to the right, and to always hold the hand rail. They should not crowd or push, and they should pay attention. Commonly, falls on stairs occur when the person is talking, laughing, and turning to friends while going downstairs.

Other safety rules for stairs are: do not congregate on stairs or landings, and do not stand outside doors at the head or foot of stairways.

Good housekeeping is essential to prevent falls. Littering of the floors should not be allowed. Wipe up spilled liquids immediately, and pick up pieces of paper, paper clips, rubber bands, pencils, and other loose objects as soon as they are spotted.

Broken glass should be swept up immediately. It should not be placed loose in a waste basket but it should be wrapped in heavy paper and marked "Broken Glass." Glass which shatters into fine pieces should be blotted up with damp paper towels.

Other tripping hazards such as defective floors, rugs, floor mats, should be reported to the maintenance department and fixed immediately.

Chair falls. Habits that lead to chair falls must be discouraged. Scooting across the floor while sitting on a chair should be forbidden. Leaning out from the chair to pick up objects on the floor is also dangerous and should be discouraged. Leaning back in the chair and placing the feet on the desk is not to be allowed either.

People should seat themselves properly in their chairs. They should form the habit of placing their hand behind them to make sure the chair is in place. Sitting down on the edge of the seat rather than in the center, or backing too far without looking, or kicking the chair out from under one can result in a sudden fall to the floor. Particularly, standing on a chair to reach an overhead object must be forbidden.

359

Filing cabinets are a major cause of injuries including bumped heads from getting up too quickly under open drawers, mashed fingers from closing drawers improperly, and hand injuries and strains from moving the cabinets around.

Some precautions are necessary against these accidents. First, people should never bump file drawers closed with their chest or any other part of their body. They should place their hand on the drawer handle to close the cabinet, making sure their fingers are not curled over the edge when the drawer closes. All file drawers should be closed immediately after use.

Second, only one file drawer at a time should be opened in order to prevent the cabinet from toppling over. As previously indicated, where possible, file cabinets should be bolted together or otherwise secured to safeguard against this chance of human failure.

Third, when one person has a file drawer open, he should warn other persons working in the area so that they do not turn around or straighten up quickly and bump an open drawer.

Fourth, climbing on open file drawers must be forbidden.

Fifth, small stools used in filing areas are tripping hazards when left in passageways. Any person who sees one out of place should put it where it cannot cause a fall.

Sixth, filing personnel should wear rubber finger guards to eliminate cut fingers from metal fasteners or paper edges.

Desks or files should never be moved by office personnel; they should be moved by maintenance men preferably using special dollies or trucks made for such moving. In general, furniture should not be rearranged without authorization or checking with the office management. When desks or cabinets are moved, thought should be given to floor obstructions and necessary aisle space before making the move. When a floor-mounted telephone or electrical outlet box is exposed after moving furniture, mark the box with a tripping hazard sign until it is removed. The outlet must be removed and, if it is needed, relocated; it is far cheaper to do this than to pay for a fall.

Electric cords laid under rugs sometimes come out because of traffic movement and form tripping hazards. Extension cords can be the cause of fire, also. New outlets should be installed to eliminate the necessity for extension cords.

Materials storage. There are a number of precautions to be taken when storing materials. Neat storage makes it easy to find and recover materials without dropping or knocking over other materials. Supervisors must keep employees from stacking boxes, papers, and other heavy objects on file cabinets, desks, and window ledges, and from storing heavy objects on file cabinets, desks, and window ledges, or from placing these materials carelessly on shelves so that they could spill off like in an avalanche. If heavy objects are spilled toward a window, the glass might even break and cause a serious accident to pedestrians below.

Card index files, dictionaries, or other heavy objects should be kept off the top of file cabinets and other high furniture. Movable objects such as flower pots, vases, and bottles should not be allowed on window sills or ledges.

Razor blades, thumb tacks, and other sharp objects should not be thrown loosely into drawers. They should be carefully boxed, or blades and points kept stuck in foamed-polystyrene blocks.

Other hazards. It is important that safety instructions be given to employees in order that other hazards can be ascertained by avoiding the following practices (*a*) never allow a spindle (spike) file in the office; (*b*) Never store pencils in a glass on the desk with points outward; (*c*) Never leave knives or scissors on a desk with a point toward the user or hand them to somebody else with the point toward them; (*d*) Trim boards (paper cutters) should be equipped with a guard that affords maximum protection (bar guards or single-rod barriers found in some trim boards are not considered full protection); (*e*) Also, do not leave glass objects on the edge of desks or tables where they can easily be pushed off.

Office machinery should be operated only by authorized persons. This was discussed in an earlier section.

Office supervisors should make sure that all materials are kept in their proper places. All litter must be placed in waste baskets, and all drawers not being used must be kept closed. Supervisors should check that drawers are closed promptly and properly.

When it is noticed that people are seating themselves carelessly, the supervisor should take some opportunity to point out that people should hold chairs in place with their hand before seating themselves.

Supervisors are just as responsible for training their people in the safe procedure as they are in training them for efficiency.

Supervisors should encourage employees to report all broken chairs, or missing casters, stuck drawers, cracked glass, and other hazards to the maintenance department for correction. A policy of immediate correction of these defects should be followed.

Fire protection

Fire hazards. Solvent-soaked or oily rags

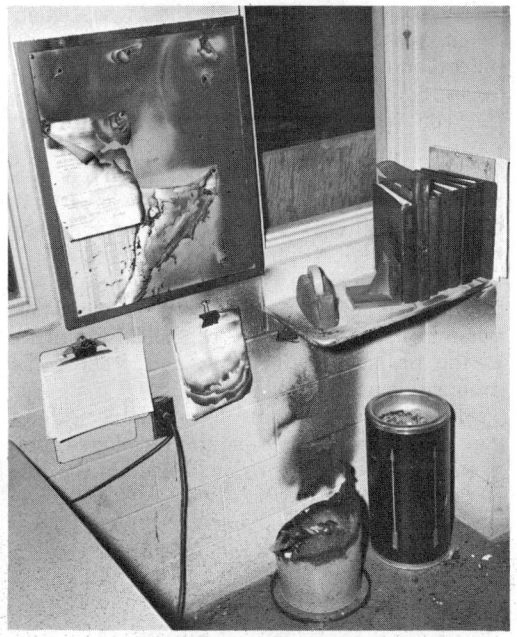

Fig. 15–10.—Still-warm ashes from a cigar or cigarette butt were dumped into an office wastebasket containing paper—with these results.

Courtesy Printing and Publishing Newsletter.

used for cleaning duplicating equipment should be kept in a metal safety container. No smoking should be allowed within 10 ft of where flammable solvents are used in duplicating or any other office operation. Solvents should not be handled carelessly because there is danger of splashing them in the eyes.

Never allow smoking on elevators.

Do not throw matches or cigarettes into waste baskets. The contents of these are usually very combustible (Fig. 15–10).

Procedures should be established in order that cleaning and maintenance personnel avoid collecting possible smoldering combustible material from ash trays in other combustible containers, such as cardboard boxes or cloth bags.

Some waste containers of plastic or other flame-resistant material may actually be combustible if subjected to fire or intense heat and thereby produce toxic gases or dense smoke which could endanger personnel. In order to avert this hazard, metal or fire-safe tested materials, designed to contain the fire should be used.

Fire extinguishers. Portable fire extinguishers in a fully charged operable condition should be kept in their designated places at all times when they are not used. (See Chapter 43, "Fire Protection," for correct type of extinguishers for specific office hazard areas.

It is important that certain employees be trained to operate extinguishers and know how to react in case of fire or other emergency. Employees, in general, should know what to do in case of fire.

When a fire is discovered, an employee should do three things: (*a*) turn in the alarm (no matter how small the fire is), (*b*) alert fellow workers, and (*c*) if trained to do so, use the proper fire fighting equipment.

Most small fires can be controlled easily. But there is always a danger that they may spread through the office. Panic and confusion can be as dangerous as flame and smoke.

Emergency plan. Every office should have an emergency plan. Monitors should be appointed in every area to guide people safely out of the building. Every department should

361

be assigned a specific route—and also an alternate in case the exit is blocked. A surprise fire drill may save many lives.

When the alarm sounds, fire monitors should direct the show, but every employee must play his part. The group should move calmly along in a businesslike manner, without hurrying or pushing, and wait outside the building for the signal to return. In a real emergency, the officials in charge would authorize return to the building.

Many large offices cooperate with local Civil Defense authorities by arranging emergency evacuation procedures as well as fire drills. Certain sections of the office building are designated as shelter areas—usually the basement or central rooms away from outside walls. (See Chapter 18, "Planning for Emergencies.")

Safety Organization in the Office

The supervisor is, of course, the key person in the office safety program. However, even the hardest-working supervisor will have difficulty maintaining full-time interest in safety all by himself. The office safety committee can serve as a work horse in maintaining interest in the accident prevention program, but it cannot substitute for good management.

Safety committee

In planning an office safety program, make office representation on the safety committee on the same basis as that of any other company department or division. The office should be on the inspection itinerary of the company's safety engineer. Arrange for the office supervisor or his assistant to accompany the safety engineer on every inspection, along with the office safety committee member.

The committee assists all supervisors in maintaining safety, reporting directly to the safety director or whoever is in charge of the program. Along with the department head, it can make periodic inspections of the office to look for accident or fire hazards. It makes recommendations, many of them based on suggestions from supervisors and other employees. It also can help prepare and revise company safety rules.

Often the committee is in charge of office-wide communication and incentive programs designed to maintain peak interest in safety, using such means as posters, bulletins, and contests.

The organization of the office safety committee can be the same as that of the safety committees discussed in Chapter 3, "Safety Program Organization."

Accident records are absolutely necessary if a safety program is to succeed. Not only do accident investigations and analysis of records spotlight problems that must be corrected, but the records show whether or not progress is being made in accident prevention.

Office employees should report every accident, no matter how minor the injury. The reports should be detailed, and made as quickly as possible following the accident or near accident. Unsafe conditions or procedures that are indicated in the reports should be corrected as quickly as possible, because near-miss accidents are warnings of worse accidents to come.

Records are the concrete foundation of the safety structure. They tell the "who, what, when, why, and how" of accidents in your office—and help you prevent repeat performances. Accurate records also provide guidelines on which company insurance rates are based.

The average office will not have enough major injuries to warrant extensive investigation and analysis. However, it's urgent that records be kept to pinpoint problems.

If, for instance, a large number of falls are injuring workers, supervisors can double-check possible hazards and devote special attention to the problem in meetings and other communications.

Standard report forms are available. They can be used, or others may be developed, but even the simplest form should contain the following information:

What was the employee doing?

What was the nature of his injury?

What caused the accident?

What was done to prevent repetition?

How much time was lost (if any)?

What were the compensation and medical costs (if any)?

Such reports, although brief, should be specific. A lot of concrete information can be packed into a few words. For instance, instead of saying "She cut herself," report that "Employee cut left thumb on paper cutter by holding hand too close to cutting edge." That pins down the unsafe act, may point to the need for a guard, and gives you the chance to discuss safe use of the paper cutter. Remember, though, not to make the injured employee a goat in the process.

Instruction. In order to develop proper safety attitudes, safety instructions should be properly given to new office employees, regardless of age. The personnel or industrial relations department can provide an accident prevention brochure or a set of printed rules. They should also arrange for all explanations of procedures as quickly as possible during the employee's early work days. The employee should have "settled down" enough, however, so that he can absorb the information.

Unfamiliar surroundings, new equipment, or altered work tasks increase the likelihood of accidents, even among veteran employees. Therefore, these people too should be instructed upon entering a new job. There should be specific instructions for each piece of equipment. No one should ever be permitted to use a machine unless he has been fully instructed in its operation, and also he should know the location and use of fire equipment as well as how to summon medical aid.

Instruction for supervisors in safe office operation is necessary because they are likely to have the same lack of recognition of accident potentials as the employees. However, prevention of accidents requires the dedicated vigilance of the supervisor throughout every working day. If he fails to carry out this function, accidents due to unsafe acts by his employees will continue, undiminished.

In developing a good attitude toward safe behavior, it is important to promote off-the-job safety also. Safe attitudes and behavior are not merely "put on" when an employee enters the office, and "taken off" when he walks out the front door.

In cooperation with the personnel or industrial relations department, a continual educational program in on-the-job and off-the-job safety should be established for all employees, beginning with the supervisors. Many firms have formal safety programs for office employees—movies, talks, first aid classes. Whether formally or informally presented, safety can always be present: it can be discussed in merit-rating reviews; the NSC "Travel Safety Guide" can be given to an employee a few weeks before he goes on vacation; safety can be covered during training and in procedure manuals; it can be mentioned when discussing on-the-job and off-the-job activities. It can even be reflected in the personal habits and example of the boss, no matter what line level.

Good example. Another aspect of company safety is the requirement that all office employees who must enter production areas where safety hats, eye protection, and ear protection are necessary should be provided with these items and should be required to wear them. Every employee who visits the plant should have a card of the general safety rules that apply to the plant, and he should be familiar with them. The same requirement should be enforced for all visitors.

Rigorous enforcement, coupled with a good overall safety program, builds respect for the company in both employee's and visitor's minds. To build management cooperation with the program of wearing required protection when entering general areas, the safety departments of some firms supply executives with prescription ground safety glasses and safety hats personalized with conspicuous decals showing their name and position in the company. Of course, some members of the top management will still prefer a plain safety hat, and they should have them if they wish. The point is that safety rules should apply to everybody if the program is to work at all. Involving the executive staff in the office safety program will strengthen the safety program throughout the company.

Full management backing. Finally, a company safety program cannot succeed unless it has the wholehearted backing of its top management. The supervisor must know that his accident prevention performance is watched and that good performance is appreciated.

15—Office Safety

Supervisors should personally investigate each accidental injury and they should report what caused the accident and what steps have been taken to prevent a recurrence. Also, when safety inspections are made and accident hazards are found, the report of these hazards should go to the vice president or other top officer in charge of the department. This top executive should then request a report of corrective measures directly from the supervisor responsible for the hazard. This procedure will go a long way to making a safety expert of every company supervisor, and to reducing the office accident problem to negligible size.

References

American National Standards Institute, 1430 Broadway, New York, N.Y. 10018.
> *Performance Specifications and Methods of Test for Safety Glazing Material Used in Buildings*, Z97.1.
> *Safety Requirements for Floor and Wall Openings, Railings, and Toe Boards*, A12.1.

Kiefer, Norvin C. "The Nature and Prevention of On-the-job Accidents to Office Workers," *1967 National Safety Congress Transactions*, v. 12, pp. 22–28.

———. "Office Safety," *Journal of Occupational Medicine*, 9:11 (Nov. 1967).

National Fire Protection Assn., 60 Batterymarch St., Boston, Mass. 02110. "Life Safety Code," Standard 101 (ANSI A9.1).

National Safety Council, 425 N. Michigan Ave., Chicago, Ill. 60611.
> Industrial Data Sheets
>> *Falls on Floors*, 495.
>> *Floor Mats and Runners*, 595.
>> *Power Guillotine Cutters*, 298.
>> *Power Lawn Mowers*, 464.
>> *Snow and Ice Removal in Industry*, 402.
> "A Safety Handbook for Office Supervisors,"
> *Safety Manual for the Graphic Arts Industry.*
> "Travel Safety Guide."

"Safe at the Office???" *National Safety News*, 89:5 (May 1964).

State of California, Dept. of Industrial Relations, Div. of Labor
> Statistics and Research, 455 Golden Gate Drive, San Francisco.
> "Disabling Work Injuries to Office Employees," August 1963.

Industrial
Buildings and
Plant Layout

Chapter
16

Design for Safety   366
Buildings, equipment, and processes . . . Codes and standards . . .
Planning checklist

Site Planning   367
Location, climate, and terrain . . . Space requirements

Outside Facilities   368
Enclosures . . . Shipping and receiving . . . Railroad sidings . . .
Roadways and walkways . . . Parking lots . . . Landscaping . . .
Waste disposal . . . Air pollution . . . Use of waterways . . .
Trestles . . . Bins . . . Pits and openings . . . Outside lighting . . .
Docks and wharves

Plant Layout   373
Location of buildings and structures . . . Flow sheet . . . Layout of
equipment . . . Electrical equipment . . . Ventilating, heating, and
air conditioning . . . Inside storage

Lighting   378
Electric lighting . . . Quantity of illumination . . . Quality of illumination . . . Glare . . . Safety . . . Luminaires . . . Hazardous location . . . Protective lighting . . . Maintenance . . . Electrical
distribution systems . . . Light sources

Use of Color   385
Interiors and equipment . . . Color-coding . . . Accident prevention
signs

Building Structures   386
Runways and ramps . . . Stairs and walkways . . . Exits . . .
Windows . . . Flooring materials . . . Special flooring . . .
Floor loads

References   393

Design for Safety

Efficiency and safety in industrial operations can be greatly increased by careful planning of the location, design, and layout of a new plant or of an existing one in which major alterations are to be made. Numerous accidents, occupational diseases, explosions, and fires are preventable if suitable measures are taken right from the earliest planning stages.

Buildings, equipment, and processes

Major factors determining the size, shape, and type of buildings and structures are the nature of the processes and materials, maintenance, mechanical handling equipment, and working conditions. These can vary greatly.

Catastrophes resulting in large loss of life and heavy property damage often are due to inadequate planning-stage consideration of the physical and chemical properties of dangerous substances and their processing methods.

High-hazard processes should be located in small isolated buildings of limited occupancy, or in areas cut off by fire-resistant materials appropriate to the hazard involved. Buildings can be designed so that internal explosions will produce minimum damage and minimum broken glass. Lower-hazard operations can justify larger units.

Even less serious injuries resulting from poor housekeeping facilities and congestion of workplaces can be reduced substantially by preplanning for safety. Layout of equipment and buildings so that production lines do not cross and so that operators have enough working space saves time, increases production, and helps prevent accidents.

Personnel facilities, such as lunch rooms and medical, safety, and disaster services, are essential for reasons of health and treatment of injuries. They can be planned and located for maximum effectiveness.

Codes and standards

The policy of modern companies is to have their safety, health, and fire departments review and check plans and specifications for new or remodeled plants. This saves costly alterations and installations that must be made after a plant is in operation in the event of failure to meet local and state requirements covering fire, accident, and health hazards.

Most ordinances and state laws require that authorized persons approve the means of both normal and emergency egress from buildings. In some states, plans for the installation of emergency lighting, fire alarm, and automatic sprinkler systems must be satisfactory to the authorities. Also, in certain states, exhaust and ventilating installations must be approved.

Many local codes require means of controlling air-polluting industrial contaminants, smoke and vapors. Disposal of raw wastes into public streams and waterways is also being controlled by federal and local statute. The Environmental Protection Agency, set up under the Clean Air Act of 1970, has set certain emission standards and these should be checked to determine that one isn't in violation.

The voluntary safety codes developed by various organizations establish standards for certain structures and equipment. Specifications for the construction of floor and wall openings and railings, for example, are given in American National Standard A12.1, *Safety Code for Floor and Wall Openings, Railings, and Toe Boards.* Proper electric wiring and electrical installations are covered in the *National Electrical Code,* issued by the National Fire Protection Association.

Requirements for fire extinguishing equipment and fire protection standards and codes for flammable liquids and gases, combustible solids, dusts, chemicals, and explosives are provided in the *National Fire Codes*, developed by the National Fire Protection Association. See Chapter 43, "Fire Protection."

Be sure to check the latest editions of the standards and codes.

It must be remembered that "design by the code" is no substitute for intelligent engineering, and that codes only establish a minimum requirement which in many situations must be exceeded.

Planning checklist

The following important features should be investigated, checked, and approved in planning a new plant or building:

Site planning

Transportation facilities: docks and wharves, railroad, and roadways

Exits and other wall openings

Floors, walkways, stairs, ramps, platforms

Storage facilities, including those for explosives and flammable materials, for harmful substances, for finished products, and yard storage

Electric wiring and installation

Illumination

Mechanical handling equipment: cranes, conveyors, industrial trucks

Elevators

Boilers and other pressure equipment

Ventilation, heating, and air conditioning

Fire control

Health and safety: water, waste disposal, first aid, personal protective equipment, distribution and repair facilities, isolation of excessively noisy operations

Water supply adequate for fire fighting

Personal service facilities: parking, restaurants, rest rooms, employment, and training

Site Planning

Selection of a site involves consideration of a number of factors: possible hazards to the community and their relationship to climate and terrain, space requirements, type and size of buildings, necessary facilities, transportation, market, and labor supply. Of the factors listed, those of direct concern to the safety professional are discussed next. While the safety professional is not directly responsible for specialized plant and process engineering and design, still it is his duty to keep the engineers and architects as safety conscious as possible, and to anticipate potential safety hazards and try to have them "designed out" of the system before the construction stage begins.

Location, climate, and terrain

Plants producing or using highly flammable or toxic substances in bulk should be located so as not to cause damage to a city or town, either generally or in case of a disaster or some unexpected happening.

Study of the climate of the area may be required. It may reveal that, if a certain site is selected, the prevailing winds will cause noxious fumes or dusts to settle on a neighboring community or plant, thereby doing severe harm to people or property or creating a nuisance. Plants, adjacent to roads and highways, and which emit large volumes of steam must be certain that when certain winds prevail the steam is not blown across the highway causing poor visibility. The prevailing winds may also determine the best location for processing equipment in relation to administration offices—especially their air conditioning air intakes. Drainage and waste disposal problems must also be anticipated.

In a hurricane, tornado, or flood area, plans and specifications should include suitable protective measures, and safety factors must be designed into the operation. (See Chapter 18, "Planning for Emergencies.")

If site plans call for bridges over streams, ditches, and other hazards, be sure they are adequately protected by a fence or by handrails 42 in. high and intermediate rails (Fig. 16–1).

Space requirements

Size of a site is determined by both present space requirements and possible future expansion. In the past the growth of business and population has resulted in serious crowding, and the resulting congestion of buildings and other facilities has increased fire, accident,

FIG. 16-1.—A bridge provides safe crossover of a freight yard. Be sure construction and personnel protection are adequate.

Courtesy Guardian Engineering & Development Co.

and health hazards.

New developments may figure in the size of the plant site. Some companies, anticipating increased use of air transportation, are allowing space for landing fields or heliports as they acquire new property. Plans for such installations should include all necessary safety precautions.

Fire prevention codes specify minimum distances between buildings according to their size, type, and occupancy. Laws governing storage of explosives and of highly flammable materials specify minimum distances between manufacturing and storage facilities for various quantities of such materials, as well as minimum distances of the materials from adjoining property.

Ample space for outdoor storage areas is essential. When the area for storing such materials as steel, pipe, and lumber adjacent to shops proves insufficient, space must be provided elsewhere. This necessitates added

handling and transportation and increases both costs and accident risk.

Parking lots are best located inside the plant fence for convenience, protection, and safety. Since a considerable area may be necessary to accommodate employee and visitor cars, space requirements are important in site planning.

In some cases, well located disposal areas for solid wastes also must be provided when a site is being laid out.

Scale relief models of the site in addition to maps can be of much help to the planner in predesign, and will help in spotting potential safety problems.

Outside Facilities

In the planning of outside facilities, safety considerations must be kept in mind in order to reduce the future accident potential. Here are a number of them.

Enclosures

Fenced yards and grounds have many advantages: Fencing keeps out trespassers who may interfere with work or be injured on the property. Fencing protects transformer stations, pits, sumps, stream banks under certain circumstances, and similar dangerous places. A galvanized woven wire fence makes a good enclosure. Enough entrances should be provided to accommodate the volume of traffic, with clearance for loaded trucks and for switchmen riding on the sides of railroad cars.

Since it is unsafe for pedestrians to use the entrances for railroads and motor vehicles, separate gates for pedestrians, convenient to their transportation and to their work places, should be provided. If a pedestrian entrance must be located near railroad tracks, part of the right of way should be fenced to keep employees from shortcutting along the tracks. Good visibility in all directions is essential at entrances.

If pedestrian entrances must be located on busy thoroughfares or if workers cross railroad tracks on which trains are operated frequently, traffic signals should be installed and subways or pedestrian bridges built. Such precautions are especially important where parking lots must be located opposite entrances.

Shipping and receiving

Shipping and receiving facilities should fit in with overall material flow within the company or plant, and should aid efficient flow of materials into and out of production areas. Shipping and receiving areas should be designed to keep building heat and cooling losses at a minimum.

Self-leveling dock boards, truck levelers, and cranes can facilitate loading and unloading.

Railroad sidings

This commonly used shipping and receiving facility requires planning, especially if bulk receiving of raw, process and maintenance materials is an economic advantage. Tank car lots of hazardous materials such as chlorine, caustics, acids, and other volatile, flammable, explosive and toxic materials, require proper consideration for pressure piping, breakaway piping, valves, pumps, derails, etc.

Each sidetrack should be guarded away from main line, public thoroughfares and proper clearance between main plants and cars observed.

More details are given in Chapter 28, "Elevators and Plant Railways."

Roadways and walkways

The safety engineer should always be alert to opportunities where he can assist the civil engineer in designing for maximum safety. Roadways in plant yards and grounds are sources of frequent accidents unless they are carefully laid out, substantially constructed, and well surfaced and drained, and kept in good condition. Recommendations of the *Asphalt Handbook,* published by the Asphalt Institute, or of *Concrete Pavement Design,* published by the Portland Cement Association, should be followed.

Heavy duty motor truck hauling requires roadways up to 50 ft wide for two-way traffic with ample radii at curves. Grades, in general, are limited to a maximum of 8 percent. A slight crown is necessary for drainage, with ditches to carry off water.

Roadways should be located at least 35 ft from buildings, especially at entrances. At loading docks an allowance of 1½ truck lengths is desirable to facilitate backing.

The regulation and control of traffic, signs, road layout, and markings should conform to federal and state practices. Guidance is provided in the *Manual on Uniform Traffic Control Devices for Streets and Highways,* ANSI D6.1, published by U.S. Dept. of Transportation, Bureau of Public Roads, Washington, D.C.

Traffic signs and signals regulating speed and movements at hazardous locations are essential. Stop signs are specified for railroad crossings and entrances to main thoroughfares, and SOUND YOUR HORN signs are necessary at sharp curves (blind corners) where view is obstructed and at entrances to buildings. Mirrors mounted to afford views around sharp turns or corners of buildings are of great help in preventing accidents if roadways must be built this close to buildings. Barricades and MEN WORKING signs are needed for construction and repair work. If roadways are used at night, traffic signs should be made of reflective or luminous materials.

PARKING ANGLE	DIMENSIONS A	B	C	D
Parallel	12′	9′-0″ 9′-6″	23′	(Equals B)
45°	13′	19′-10″ 20′-2″	12′-9″ 13′-6″	9′-0″ 9′-6″
60°	18′	21′-0″ 21′-3″	10′-5″ 11′-0″	9′-0″ 9′-6″
90°	24′	20′		9′-0″ 9′-6″ (Equals C)

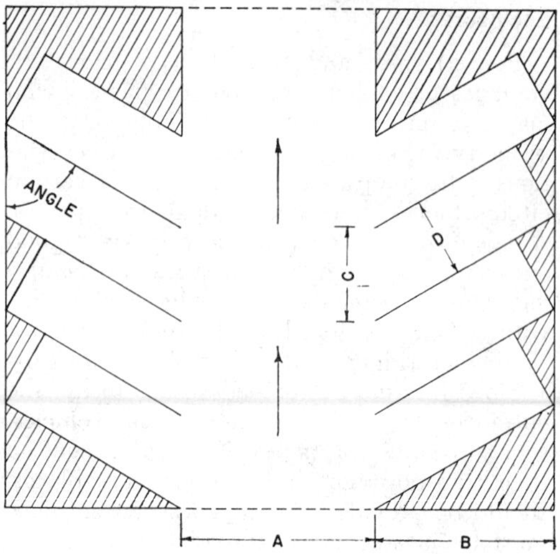

Fig. 16-2. – Parking lots, aisles, and stalls should be designed for maximum safety and efficiency both within the lot and on approaches to adjacent streets. Detailed diagram shows how dimensions vary with the parking angle. Parking angles may be varied within one lot in order to best use the space, if parking stalls are clearly marked and aisle space (A) is adequate for parking angle of largest degree. Double-line markings would increase stall widths (D) by 18 to 24 in.

Good walkways between outside facilities prevent injury from employees' stepping on round stones or into holes and ruts in rough ground. As far as possible, walkways should be the shortest distance from one building to another to discourage short-cutting. A walkway that must be next to railroad tracks should be separated from them by a fence or railing. Warning signs should be installed at railroad crossings and other hazardous places. If walkways are located clear of the eaves of buildings, danger from falling icicles is reduced.

Concrete is preferred for sidewalks, especially in principal areas like entrances and between main buildings. Crushed rock surfaced with gravel or asphalt makes a good walk for less used locations.

Walkways should be kept in good condition, especially where they cross railroad tracks, and should be cleared of ice and snow.

Parking lots

To reduce travel in the plant grounds, a desirable location for a parking lot is between an entrance and the locker room. The entire parking area should be fenced. The surface of the parking lot should be smooth and hard to eliminate the frequent injuries that would occur from falls on stony or rough ground. Good drainage is essential.

The use of the white lines 4 to 6 in. wide to designate stalls reduces confusion and the number of backing accidents. Standard stalls are 9 ft wide and 20 ft long. The center-to-center distance between parked cars varies according to the method of parking. (See Fig. 16–2.) Be sure parking does not encroach on fire hydrant zones, approaches to corners, bus stops, loading zones, and clearance spaces for islands. Driveways should be a minimum of 25 ft for two-way traffic and there should be no obstructions to viewing.

Angle parking has both advantages and disadvantages. The smaller the angle, the fewer the number of cars that can be parked in the same area. Aisle widths can be narrower but traffic is usually restricted to one way. On the other hand, angle parking is easier for customers and it does not require a lot of room for sharp turns.

The area allowed per car in parking lots varies from 200 to more than 300 sq ft, if aisles are included. Large, economically laid out lots may approach the 200-sq-ft figure; small or poorly configured lots may have a higher percentage of aisle space and may approach 300 sq ft per car. A large, commercial, attended lot is considered efficient if the layout keeps the space requirements to 240 sq ft per car.

Details for signs and marking pavement are given in U.S. Department of Transportation's *Manual on Uniform Traffic Control Devices for Streets and Highways.* See also National Safety Council Industrial Data Sheet 597, *Parking Lots—Self-Service.*

Separate entrances for incoming and outgoing cars facilitate orderly traffic movements. Entrances should be designated with suitable signs. Speed signs and signs limiting areas to employees or visitors should be installed as circumstances require. These signs should conform to recommended standards, and should be similar to other street and highway signs (see *Manual for Uniform Traffic Control Devices*).

Traffic at exits to heavily traveled streets should either be controlled by a traffic light, or there should be an acceleration or merging lane.

If the lot is used at night, adequate lighting should be provided for safety and for the prevention of theft; about 1 to 5 footcandles per square foot at a height of 36 in. should be an adequate level. See page 1449.

Plans for ice and snow removal should include parking lots.

Landscaping

An increasing number of companies are landscaping the grounds of both old and new plants. Landscaping should be designed so that trees and shrubbery do not create blind spots at roadway or walkway intersections.

Proper maintenance to keep bushes at proper height is important if their growth will otherwise create blind spots.

Waste disposal

Unsafe methods of waste disposal may cause injury to workers and the public and damage to property. Knowledge of the nature of the wastes is essential in planning suitable disposal methods, which must also conform to applicable municipal and state regulations—which are being enforced to an increasing extent. Since investigation, treatment, and disposal of wastes require specialized knowledge and training, qualified engineers should be consulted.

If it is planned to use a city or district sewage system, the officials in charge of the facility should be informed of the kind and amount of the wastes. If the wastes have properties that will interfere with operation of the sewage disposal plant, the officials may refuse to accept them.

Under no circumstances should toxic, corrosive, flammable, or volatile materials, or radiation wastes be drained into a public sewage system. It is especially important that plants handling or processing radioactive materials—even in small quantities—conform exactly to local and state regulations for disposal of radioactive wastes.

Many wastes can be burned or disposed of on dumps, if state and local laws permit. However, precautions are necessary. Pick a day on which a slight wind is blowing and there is no inversion. Be sure wind blows smoke away from populated areas.

Combustible materials, such as wood, scraps, and paper, can be burned in an incinerator. An incinerator should conform to local laws, be safely located, and properly attended.

Chemical wastes should be rendered harmless before being dumped; strong acids, for example, should be neutralized. Poisonous materials, magnesium chips, explosives, and similar substances require special procedures for safe disposal.

In some cases it might be necessary to contract a private scavenger service to dispose of waste materials.

Air pollution

Smoke from power plants and other sources and inert dusts from various types of plants may be nuisances or even sources of danger to the public. Provision for complying with smoke control ordinances should be made before construction is begun. The Environmental Protection Agency emission standards should also be checked to be certain that they are not being violated.

Toxic smoke, fumes, or dusts are a serious problem in some industries. Tall stacks often are used to diffuse gases in the atmosphere. The effectiveness of this method depends on the nature and volume of the gases, the location of the plant, prevailing wind direction, and atmospheric conditions. Rain may absorb harmful gases and cause heavy damage to crops. A combination of prolonged fog or other unfavorable atmospheric conditions and a contaminant in a valley has resulted in serious injuries to animals and people. Such conditions causing poor diffusion occasionally have temporarily shut down plants with high stacks.

The possibility of recovering usable or marketable materials from wastes should always be investigated. Dusts and fumes often are recovered by filters, cyclones, electronic precipitators, or similar equipment. Useful gases and vapors may often be economically recovered by means of spray towers or other recovery equipment.

Smoke from heating units may be eliminated entirely by electrically driven heat pumps, or reduced by gas-engine driven units. When buildings have a relatively high heat load, are in not too severe climates, and when favorable electrical or gas rates can be obtained, the heat pump may be considered. A mechanical engineer should be consulted.

Use of waterways

Profitable salvage may be obtained by treating liquid and solid wastes which frequently cause problems when dumped into streams and lakes.

Public opposition to contamination by industrial wastes is growing and more stringent regulations are being adopted. Some wastes are objectionable because of offensive odors, resultant water pollution, and deleterious effect on animal and vegetable life.

When considering disposal of wastes in a waterway, a study of the kind and amount of contaminant and of the use of the waterway for drinking water, recreation, and other purposes may be necessary. Even disposal in the ground must first be studied so as not to pollute nearby wells.

In some cases, liquid wastes must be diluted, neutralized, filtered, settled, or otherwise chemically treated before they are disposed of.

Trestles

If employees are required to perform duties on trestles, a footwalk 5 ft, 1 in. wide, measured from the nearest rail, should be provided on at least one side. The footwalk

should have a railing 42 in. high and, on the exposed side, toe boards 6 in. high.

If employees travel on both sides of the track, crosswalks should be provided at convenient locations.

Metal gratings or screens installed over walkways or passages under trestles furnish protection from falling materials.

Openings for conveyors or hoppers require gratings or a grizzly with bars spaced not more than 12 in. apart, to prevent employees from falling into the openings.

Bins

Many employees who have entered bins to loosen materials have been carried down by slides and suffocated. This danger often can be entirely eliminated if the bottom of the bin is designed so that the slope exceeds the angle of repose of the material. Also, a heavy grating or screen, on which an employee can work without danger, can be installed above the bottom. The openings in the grating should be big enough to permit the material to pass through readily and to push a pole through. Vibrating screens, or surface-mounted vibrators, are recommended to keep materials moving if they "hang up" at the angle of repose.

When necessary that employees enter bins, they should use a safety harness and lifeline, and have another employee (similarly equipped) remaining topside.

Stairs provide the safest means of access to bins, tanks, and similar structures. If fixed ladders are installed, they should conform to the requirements of American National Standard A10.2, *Safety Code for Building Construction*. The platform at the top of a tank should have standard railings and toeboards. Standard railings also should be installed along the edges adjoining the platform.

Pits and openings

Pits for storage, for access to pipelines, and for other purposes are sources of frequent and serious injuries unless precautions are taken to make them safe. If plans call for a pit in which employees may be exposed to dangerous gases or vapors while working on pipelines or equipment, means of ventilating the pit should be provided. Preferably the plans should be changed to eliminate the pit.

When openings require covers, the design should specify metal covers with hand rings folding flush with the top. Covers should fit snugly and be provided with lugs. Open pits may be safeguarded with fences or rails.

Stairs or fixed ladders provide safe access to pits. Drains reduce the hazard of falls on muddy, wet, or icy pit floors.

Outside lighting

Outside lighting must serve not only as a production tool and as a safety factor but should also function as part of the plant security system. Luminaires using many different types of light sources are available for all types of specialized applications. Considerations should be given to lamp life and ease of maintenance of luminaires selected. Lighting should be provided for:

Wharves, docks, and loading platforms

Roadways, walkways, and railroad crossings

Entrances and exits

Production areas

Stairways and landings

Ramps

Active storage areas

Lighting equipment used in outdoor locations must withstand exposure to the elements without deterioration. American National Standard A11.1, *Practice for Industrial Lighting*, and a later section in this chapter, "Lighting."

Docks and wharves

The characteristic of the bottom of the sea, lake, or river is an important factor in design and construction of docks and wharves. A soft, deep bottom limits the use of concrete and of heavy fire-resistive materials. Wood piles must be protected if marine borers are present. Flexibility and elasticity are essential where waves force vessels against piers and in tidal waters.

Safety requirements include good illumination for night work, a floor that will withstand heavy trucking, and traffic control equipment. Fatigue and numerous injuries to stevedores can be greatly reduced by the use of mechanical handling equipment, such as forklift trucks.

Design of piers should take into account

the speed and size of vehicles to be operated on them. Provision for removing snow and treating icy surfaces should be provided.

Plant Layout

Size, shape, location, construction, and layout of buildings and other facilities should permit the most efficient utilization of materials, processes, and methods.

Materials and processes may favor gravity flow and the construction of a multistory building, for example, in the case of mills for ore treatment.

Other uses, such as automobile assembly plants, may favor a one-story building because floor loads will be heavy and because excessive handling of materials by vertical and inclined conveyors would be required in a multistory building.

Location of buildings and structures

The segregation of raw materials storage, processing buildings, and storage for finished products warrants thorough study in laying out a plant to minimize fire and explosion hazards. Storage of volatile flammable liquids in an area apart from processing buildings reduces the fire hazard. In the event of fire, control is more easily achieved. Moreover, the cost of separate storage eventually may be less than the investment for storage in a processing building.

Ample space should be provided between segregated units, from such flame sources as boilers, and from shops, streets, and adjoining property. Fig. 16–3 shows a well planned layout of the structures in a lacquer manufacturing plant.

The codes of local and state authorities and of the National Fire Protection Association should be followed in planning the location of the units of a plant.

Where no legal restrictions govern the storage of explosives, magazines should be located and constructed in accordance with the recommendations of the Institute of Makers of Explosives. Table 16-A specifies the quantities of explosives that may be stored safely at various distances from inhabited buildings, passenger railways, and public highways.

The type of retardant required in relation

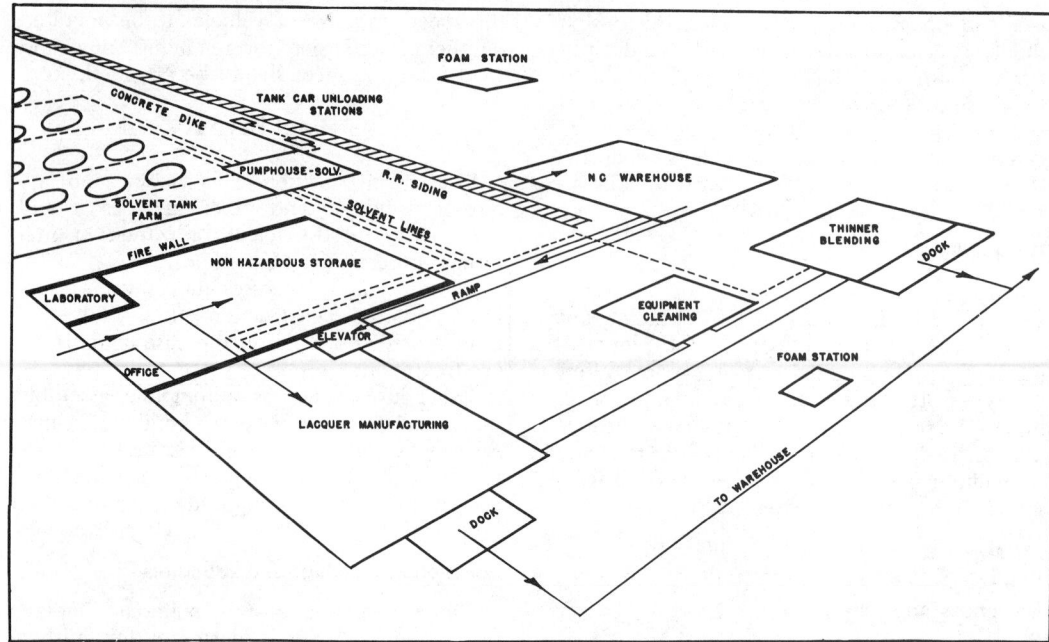

Fig. 16-3.—A combination flow sheet and layout showing a plant having hazardous operations.
Courtesy The Arco Company

to the horizontal distance between buildings of frame, brick, and fire-resistant construction, and minimum separation distances between buildings are given in NFPA Standard 80A, *Fire Resistive Walls, Protection of Openings Against Fire Exposure.*

Many plants use and store flammable liquids having flash points below 200 F. Plans and layouts in such plants should conform strictly to specifications for handling and storing flammable liquids developed by the NFPA *Flammable and Combustible Liquids Code, Standard No. 30,* and to the requirements of local fire prevention authorities. Also see Chapter 42, "Flammable and Combustible Liquids," in this Manual.

NFPA Standard No. 30 specifies the conditions under which flammable liquids of various classes can be stored in and around buildings.

Exacting specifications must be met in storing flammable liquids outside of buildings in underground or aboveground tanks. The amount of liquid that can be stored in an underground tank depends upon the proximity of the tank to any building, depth in the ground in relation to the building, and flash point of the liquid. Standards for tanks include type and thickness of material, provisions for relieving excessive internal pressure, grounding, insulation, piping, and other appurtenances, and similar details.

The required distance between the line of adjoining property which may be built upon and the location of an aboveground tank depends upon the content, construction, fire extinguishing equipment, and greatest dimension (diameter or height) of the tank. NFPA Standard No. 30 specifies four different groups of tanks and the minimum distances apart for each.

Tanks must be built of steel unless the properties of the flammable or combustible liquid are such that some other material must be used for its corrosion resistance.

Flow sheet

A detailed flow sheet is a useful guide in laying out plants, particularly those using dangerous and harmful materials and complicated processes. The nature of the ma-

TABLE 16-A

AMERICAN TABLE OF DISTANCES FOR STORAGE OF EXPLOSIVES*
(Condensed)

Explosives		Distances in feet When Storage Is Barricaded°°			
Pounds Over	Pounds Not Over	Inhabited Buildings	Passenger Railways	Public Highways	Separation of Magazines
10	20	110	45	45	10
40	50	150	60	60	14
100	125	200	80	80	18
150	200	235	95	95	21
300	400	295	120	120	27
500	600	340	135	135	31
900	1,000	400	160	160	36
1,400	1,600	470	190	175	43
2,000	2,500	545	220	190	49
5,000	6,000	730	295	235	65
8,000	9,000	835	335	255	75
10,000	12,000	875	370	270	82
16,000	18,000	940	420	285	94
20,000	25,000	1,055	470	315	105
35,000	40,000	1,275	550	380	124
50,000	55,000	1,460	610	440	140
70,000	75,000	1,665	675	500	160
95,000	100,000	1,815	745	545	185
140,000	150,000	1,900	850	570	235
190,000	200,000	2,030	935	610	285
275,000	300,000***	2,275	1,075	690	385

Adapted from table of same name approved by the Institute of Makers of Explosives, June 5, 1964.

NOTES:

*This table applies only to the manufacture and permanent storage of commercial explosives. It is not applicable to the transportation of explosives or to any handling or temporary storage necessary or incident thereto. It is not intended to apply to bombs, projectiles, or other heavily encased explosives.

**Barricaded, as here used, signifies that the building containing explosives is screened from other buildings, railways, and highways, by either natural or artificial barriers. Where such barriers do not exist, the distances shown in the table should be doubled.

***The Institute of Makers of Explosives does not approve the permanent storage of more than 300,000 pounds of commercial explosives in one magazine.

terials and processes in each manufacturing stage can be studied and provision made to eliminate or control hazards.

A simple flow sheet is shown in Fig. 16-3. In this operation, a solvent and nitrocellulose are used in a lacquer manufacturing department, thus making special safety and fire prevention measures essential. Plans include wall ventilating fans capable of producing an air change every 3 to 4 minutes. Since solvent vapors are heavier than air, inlet and pickup ducts are located at floor level.

An automatic sprinkler system is considered primary protection for hazardous installations. However, fire prevention measures in this flow sheet consist of a CO_2 total flooding automatic system, a steel and concrete structure, a ¾-in. coating of sprayed asbestos, and first-aid fire

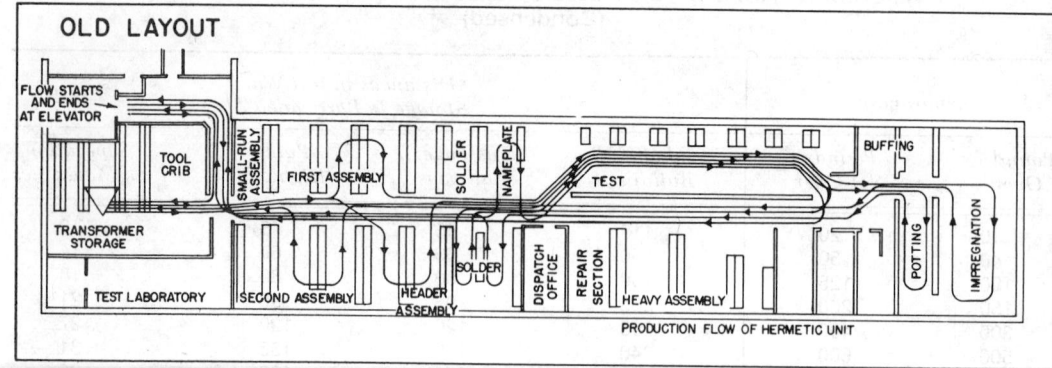

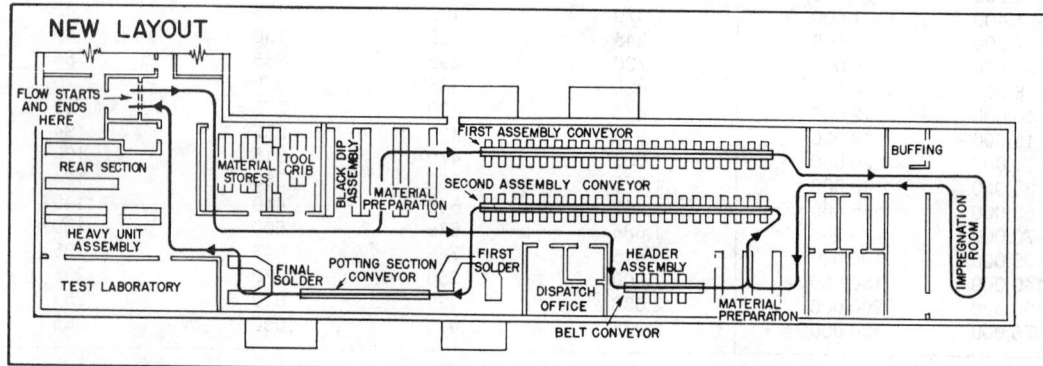

Fig. 16-4. — The new layout greatly reduced the hazards incidental to materials handling in this transformer manufacturing operation.

equipment.

Safety features include spark-resistant conductive floor surfaces and grounding of equipment and structures. Electrical motors, lighting, and apparatus are Class 1, Group D explosion-proof. Emergency exit doors are provided, and all tools, trucks, and similar equipment are made from spark-resistant metal.

Layout of equipment

Various methods are used to determine the safest and most efficient layout of production machines and equipment. The two-dimensional method consists of templates made to scale and fitted into a plan of the site or floor area.

The most effective method is to use three-dimensional models, made to scale and set up on a scaled floor plan. The models can be rearranged until the safest and most efficient

layout has been devised.

Results to be obtained through layout studies are shown in Fig. 16–4. Here the new layout reduced unit travel from 2532 ft to 807 ft, of which 208 ft were conveyorized. The saving in distance amounted to about 76 percent. This reduction is particularly significant to the safety professional because the largest number of injuries in most industries occurs in materials handling.

Layout studies show the most suitable locations for spray painting, welding, and similar work generally requiring segregation as they relate to other operations.

By using the three-dimensional method, congested areas can be anticipated and avoided. These pile-up areas mean frequent handling of materials, many unnecessary movements, and result in bad housekeeping —in itself a prolific source of accidents.

Insufficient headroom at aisles, platforms,

pipelines, overhead conveyors, other parts of the structure, and other installations is disclosed in studying the models. A vertical distance of at least 7 ft generally is specified between passageways and stairways and overhead structures to provide ample clearance. Overhead cranes and conveyors require at least 24 in. of vertical and horizontal clearance.

Trucks and other vehicles require enough room for movement without endangering men and equipment and need space to back and turn near machines. Aisles should be wide enough to permit trucks to pass without colliding. For one-way traffic, aisles should be not less than 3 ft wider than the widest vehicles. Aisles for two-way traffic should be not less than 3 ft wider than twice the width of the widest vehicle. These minimum widths are exceeded considerably in some new buildings where, because of anticipated heavy traffic, aisles from 12 to 20 ft wide have been specified.

Safe layout of aisles requires that there be no blind corners and that adequate radius be provided to allow for vehicle turns. A 6-ft radius is large enough for small industrial trucks.

Aisles should be clearly defined with approved markings of either traffic paint or striping material. Follow American National Standard Z53.1, *Safety Color Code for Marking Physical Hazards*. See the recommendations on pages 385–386. Plastic buttons that are glued or fastened with metal fasteners to the floor are used by many because of their durability.

Parking areas for both hand and power trucks should be designated. Large plants often provide garages with room for both storage and maintenance. See Chapter 27, "Powered Industrial Trucks," in this Manual.

If building plans include ramps for use by both pedestrians and trucks, a 3-ft wide section should be reserved as a walkway. Sharp turns into aisles at the top and bottom of ramps are hazardous and should be avoided. Provide an abrasive coating where slippery floor conditions may exist.

Electrical equipment

Complete metal-enclosed unit substations have been developed for industrial plants and are grounded. Such rooms, however, are not proof against the hazards of oil-filled transformers. This equipment may give off flammable gases under certain conditions, which may form an explosive mixture with the atmosphere in the room. Nonflammable transformers should be installed in confined areas and near flammable materials.

Short circuit protective devices should be large enough to carry the load and should be rated to open without any danger from the maximum short circuit current. Circuit breakers, fuses, and safety switches that fail under short circuit current may explode and cause serious damage and injuries.

The potential of every grounding system should be measured with a meter to determine whether or not the system is capable of conducting the necessary amount of return.

When direct current voltage is supplied from batteries, isolate the battery room from the work area. Battery room should be well ventilated; no smoking should be permitted.

In an up-to-date industrial electrical system, sections may be de-energized for maintenance and other work without shutting down the entire system. In addition, all electrical installations should conform to the NFPA *National Electrical Code* as well as local ordinances. Also see Chapter 41, "Electrical Hazards," in this Manual.

Major switching apparatus should have metal enclosures connected to the grounding system, and cable circuits likewise should be enclosed in rigid conduit or interlocked armor cable. Metal-enclosed plug-in bus-ways should provide an easy method of connecting and disconnecting machine tools.

Ungrounded portable electric tools and equipment constitute a common source of fatalities in industry. The hazard can be largely eliminated if plans include a three-wire system, of which one is a ground wire, and if the equipment is provided with three-conductor cables.

(For additional information, see Chapter 41, "Electrical Hazards.")

Ventilating, heating, and air conditioning

Ventilating, heating, and air conditioning are needed both for personal comfort and very often to meet process conditions. Personal comfort is very important because it affects employee efficiency. All effort should be made not only to make general office and plant conditions comfortable, but to eliminate —or at least reduce—poor conditions which can contribute to excessive employee fatigue or discomfort. If machinery or processes radiate heat or generate bad fumes, extra ventilation should be supplied to the work area.

Boilers, fans, and air conditioning equipment are usually kept separate from the general work areas because of their noise and vibration and the fact that their adjustment is very fine. It is important to make certain that boilers receive adequate air and that combustion by-products are exhausted safely. In locating incinerators, too, be sure that a negative pressure differential in a building doesn't cause an incinerator stack to serve as an air source.

Maintenance should be considered. Not only must authorized employees have easy access to the machinery, but there must be sufficient space around the equipment to replace parts as needed. There must be room behind a boiler to pull the tubes if necessary.

Chapter 36, "Exhaust and Ventilation," provides more specific suggestions, as does NFPA Standard No. 90A, *Air Conditioning and Ventilating Systems (Non-Residential)*.

Inside storage

Industrial buildings seldom have "enough" storage room. Sufficient space for raw materials and finished products may be estimated on the basis of maximum production requirements—with allowances for shortages, seasonal shipping, and quantity purchases.

Modern mechanical handling and stacking equipment permits extensive use of vertical space through double and triple decking. If this method is anticipated in the future, basic live floor requirements in the initial building design must allow for the possible increase in square-foot load.

Plans for storage of supplies, the finished product, and empty or full pallets should permit ease of access, stability of piling or stacking, and proper functioning of the emergency sprinkler system. Design density (gpm/sq ft) and area of application are dependent on type of material and the height

that it is piled. When piling height exceeds 12 ft, double check the sprinkler system's ability to put out any fire.

Space for storing supplies, tools, and infrequently used equipment near working areas seldom is included in plans for departmental layouts. The result is that such items often are left in unsafe positions and locations. To discourage employees from leaving long materials and parts against walls or other places where they can fall, portable ladders, tools such as pry bars, and machine parts such as shafts should have definite storage places.

Space should be provided for racks, bins, and shelves. Materials that project from such places should be stored above eye level. Metal baskets or special racks provided with drip pans can be used for storing machine parts covered with cutting oils.

Closets should be built for the storage of janitorial supplies such as waxes and soaps.

Waste storage may take considerable room, especially if the waste is bulky or is produced in large quantities. In buildings with basements, chutes to basement storage bins help prevent accumulations in working areas. To dispose of small quantities of sharp-edged waste, boxes with handles that are protected from contact with sharp materials, or similar safe containers may be specified. Chapter 24, "Materials Handling and Storage," gives additional suggestions.

Lighting

Daylight is an uncertain and unpredictable type of illumination. To use it to advantage the following design factors should be taken into account:

Variations in the amount and direction of incident daylight.

Brightness distribution of clear, cloudy, partly cloudy, and overcast skies.

Variations in sunlight intensity and direction.

Effect of local terrain, landscaping, and nearby buildings on the available light.

The daylight that enters and is made available for use inside the building depends upon the architectural design of the windows (fenestration) and the decoration and furnishings of the interiors.

Window areas serve at least three useful purposes in industrial buildings: (a) they can provide for the admission, control, and distribution of daylight for seeing; (b) provide a distant focus and thus relax eye muscles; and (c) eliminate the dissatisfaction some people experience in completely closed-in structures. However, an adequate electric lighting system should always be provided because of the wide variation, with time and weather conditions, in the amount of daylight available. This variation can range from several thousand foot candles down to zero.

Brightness-control devices adapted to the problems encountered should be provided for all window areas. These must be designed for the specific direct sunlight and sky brightness conditions. These are largely dependent upon the latitude and orientation of the building.

For a more comprehensive treatment of this subject see Illuminating Engineering Society's *Recommended Practice for Daylighting*.

Electric lighting

The prime requirements for industrial lighting are sufficient quantity and high-quality illumination on all work planes. Under these conditions, personnel will be able to observe and control effectively the operation and maintenance of various types of machines and processes.

For most industrial areas, a sufficient quantity of daylight is often not available, even under optimum daylight conditions. To maintain good seeing conditions therefore, electric lighting is required.

It is essential that the electric lighting system be so designed and installed as to continue the general level of illumination in areas adjacent to the windows or walls, thus ensuring good lighting over the entire working area.

Distribution of light from a luminaire is important. Highly concentrated distributions make high mounting heights economically feasible. Low mounting heights on the other hand allow a wide-spread type of distribution.

There are three forms of electric lighting used in industrial areas: general, localized general, and supplementary.

General lighting should produce uniform illumination throughout the area involved. Uniform illumination is defined as a distribution of light where the maximum and minimum illumination at any point is not more than one-sixth above or below the average level in the area. Care must be taken not to exceed the spacing-to-mounting-height ratios established for the lighting equipment used.

Localized general lighting reinforces the general lighting in specific areas through the use of additional general lighting equipment in the areas involved.

Supplementary lighting is used to provide higher illumination levels for small or restricted areas where such levels cannot readily be obtained by general lighting methods. Supplementary lighting is also used to furnish a specific brightness, or color, or to permit special aiming or positioning of light sources.

Quantity of illumination

The desirable quantity of light for any particular installation depends primarily upon the work that is being done. Investigations show that as the illumination of the task is increased, the ease, speed, and accuracy of accomplishing it are also increased. The current minimum levels of illumination for industrial areas as recommended by the Illuminating Engineering Society (IES) are given in the American National Standard A11.1, *Practice for Industrial Lighting*. See Table 46-G.

Quantity of illumination is stated in foot candles and is measured with a light meter to give a direct reading of the number of foot candles of light reaching the working plane.

Quantity of light required is measured by an instrument called the "visual task evaluator." The information gathered with this instrument is then interpreted by means of experimental laboratory data to secure recommended foot candles.

Quality of illumination

Quality of illumination pertains to the distribution of brightness in the visual environment. Glare, diffusion, direction, uniformity, color, brightness, and brightness ratios all have a significant effect on visibility and the ability to see easily, accurately, and quickly.

Very poor quality installations are uncomfortable and possibly hazardous. Moderate deficiencies are not readily detected, although the cumulative effect of even slightly glaring conditions can result in material loss of seeing efficiency and in fatigue.

Glare

Glare may be defined as any brightness within the field of vision of such character as to cause discomfort, annoyance, interference with vision, or eye fatigue. Glare may be direct or reflected.

When glare is caused by the source of lighting within the field of view (whether it be daylight or electric), it is described as direct glare.

When glare is caused by high-brightness images or brightness differences reflected from shiny ceilings, walls, desk tops, materials, and machines or surfaces within the visual field, it is described as reflected glare.

Reflected glare is frequently more annoying than direct glare because it is so close to the line of vision that the eye cannot avoid it. Furthermore, reductions in contrast often occur which can drastically reduce the task contrast and hence the ability to discern detail. Reduced brightness of light sources will reduce both direct and reflected glare.

Glare reduces the efficiency of the eye and may cause discomfort and fatigue. It may reduce the detail of the visual task to such an extent as to seriously impair vision, thus increasing accident hazards. See page 1456.

Direct glare may be reduced by: (*a*) decreasing the brightness of the light source; (*b*) positioning the light source so that it no longer falls within the normal field of vision; (*c*) increasing the brightness of the area surrounding the glare source and against which it is seen.

Reflected glare may be reduced by: (*a*) decreasing the brightness of the light source; (*b*) positioning the light source or the visual task so that the reflected image will be directed away from the eye of the observer; (*c*) increasing the level of illumination by increasing the number of sources in order to reduce the relative brightness of the glare; or (*d*) in special cases, changing the character of the offending surface to eliminate the specular reflection and the resultant reflected glare.

Soft shadows from general illumination can accent the depth and form of various objects. Avoid harsh shadows, however, since they may obscure hazardous conditions or interfere with visibility at the work area because of undesirable brightness contrasts.

Safety

Because safe working conditions are essential in any industrial plant, the effect of light on safety must be considered. The environment of a plant should be designed to compensate for the limitations of human capability. Any factor that aids seeing increases the probability that a workman will detect the potential cause of an accident and act to avert it.

In most cases where accidents are attributed to poor illumination, the cause is marked down as "very noticeable poor quality of illumination" or "practically no illumination at all." Many less tangible factors associated with poor illumination are, however, important contributing causes of industrial accidents. Some of these are: direct glare, reflected glare from the work, dark shadows—all of which hamper seeing and together cause after-images —and excessive visual fatigue, which alone may lead to accidents.

Accidents may also be caused by delayed eye adaption when coming from bright surroundings into dark ones. Some accidents which are attributed to "individual carelessness" can be traced to difficulty in seeing, from one or more of these causes.

Seeing tasks tend to become more difficult and vital to profitable operation. Close observations of equipment and instruments, and quick physical response to the minute changes indicated will call for more and better lighting to protect major investments in machines and highly trained personnel.

For extensive lighting installations, a qualified illuminating engineer should always be consulted. The safety director, however, should be familiar generally with the lamps and reflectors, and lighting requirements for industrial environments.

For a more extensive treatment of industrial lighting see pages 1449 and 1456 in Chapter 46, and the following references:

1. ANSI A11.1, *Standard Practice for Indus-*

trial Lighting, and

2. IES *Lighting Handbook.*

Luminaires

A wide range of types of luminaires permits a choice of designs for industrial applications. When selecting specific types for a proposed installation, consider:

1. Candle-power distribution;

2. Design of luminaire (*a*) to avoid objectionable glare under normal seeing conditions and (*b*) to produce highest initial and sustained light outputs;

3. Mechanical construction permitting convenient installation and servicing; and

4. Environmental suitability for use in normal, classified (hazardous), or special areas, indoor or out.

All interior lighting systems are included in one of five classifications: direct, semi-direct, general diffuse, semi-indirect, and indirect.

No *one* system can be recommended to the exclusion of all others, for each has characteristics that may or may not match the requirements of a given application. The performance of each should be evaluated to make sure it will efficiently provide the area with lighting both in quantity and with comfort. Luminaire maintenance and the character of the task being performed must be carefully considered.

As the percentage of upward light increases, the system generally becomes more comfortable. On the other hand, the higher percentages of uplight tend to reduce the overall utilization of the system; this, coupled with more difficult maintenance, generally causes these systems to be uneconomical. Although all systems may find use to some degree, in most average production areas direct or semi-direct equipment is used.

Hazardous location

Many areas in industrial plants are classified as hazardous locations by the NFPA/American National Standard *National Electrical Code.* Such areas require the use of specialized lighting equipment (vapor-proof, explosion-proof, and dust-tight luminaires), which provide required illumination without introducing hazards to life and property. Because each type of equipment is designed to meet certain requirements, the types are not interchangeable. Chapter 5 of the *National Electrical Code* should be studied in detail to determine requirements for hazardous-location lighting equipment. In case of question or doubt, the local electrical inspector should be consulted. See pages 1269–1274.

Protective lighting

Protective lighting is necessary for nighttime policing of outdoor areas to discourage would-be intruders, or to render them visible to plant guards should they attempt entry. It may also reduce fire risk. Illumination for policing, however, is not usually adequate for efficient plant operation; therefore, protective lighting is generally treated as an auxiliary to productive lighting.

Protective lighting is achieved by adequate light upon bordering areas of buildings and by producing glaring light in the eyes of the trespasser with no light on the guard. Equipment should be so located that concealing shadows are eliminated.

In general there are four types of lighting units used in protective lighting systems. These are: floodlights, street lights, Fresnel lens units, and searchlights. Evacuation of personnel also influences choice. Battery capacity of units and the number of lamps and their wattages should be correlated to provide lighting at least for the brief length of time required for complete evacuation. Where longer durations of emergency lighting are required, generator sets are used as the power source. These are driven by a prime mover, started automatically upon failure of the normal power supply. Power transfer from the normal to the emergency power supply is made by an automatic transfer switch which reverses the procedure upon re-establishment of the normal power supply.

Maintenance

A regular, scheduled system of maintenance should be established to make sure that luminaires and room surfaces are kept clean and in proper condition. Even if this is done well and at appropriate periods, an average lighting level from 25 to 35 percent less than the initial level frequently occurs under normal operating conditions. The amount of depreciation depends upon: (*a*) the decrease of light output of the lamps with age, (*b*) the design of the luminaires with reference to the rate at which dirt accumulates, and (*c*) the severity of dirt conditions in the area. If maintenance is not well performed or is performed infrequently, the light at the working surfaces is likely to be below 50 percent of the initial value. Because it is essential to the efficient operation of an industrial lighting system, frequent maintenance is more economical than allowing the lighting system to operate at low efficiency.

Electrical distribution systems

Efficient utilization of lighting equipment depends on the distribution system that carries the necessary electrical energy. The term "distribution system" includes the methods and materials used for getting the energy from the main supply source in any given building or area to the light source that converts the electrical energy to light. In many cases this includes distribution transformers, feeders, panelboards, overcurrent protection and branch circuits. In "total energy" systems, it includes generation of the power itself.

In most communities, installations of electrical distribution systems within a building or plant are governed by electrical installation codes that are given the force of law by local or state ordinances. While not uniform everywhere, these codes have much in common since they are usually based on the *National Electrical Code* (NEC), a standard compiled by the National Fire Protection Association and approved by the American National Standards Institute as Standard C1. (Construction and alteration of power supply transmission and distribution lines and equipment is covered in subpart V of OSHA Safety and Health Regulations for Construction, Part 1926.)

The essential function of the code is to provide reasonable safety to persons and property. The NEC covers the general lighting loads by broad occupancies; it does not give a breakdown for specific uses of space. The NEC requirements should be considered as minimum safety requirements, not as design criteria.

The system should be designed to provide economical distribution of electrical energy. Every conductor of electricity resists the free flow of energy, thus causing energy losses. Large conductor sizes are the best protection against losses, therefore a careful balance between energy loss and installation cost must be considered. Satisfactory voltage conditions must be maintained, since all electrical devices are designed for specific operating voltages. Failure to supply the specified voltage may seriously affect the efficient performance of the lighting system. Both resistance and reactance cause a reduction of the voltage delivered to any utilization point and must be considered in computing voltage drop. NEC recommends a maximum of one percent voltage drop in feeders and two percent voltage drop in branch circuits.

Distribution systems should be of sufficient capacity to take care of reasonable and anticipated increases in future loads. Local codes and ordinances, in general, consider wiring only from a safety standpoint, with no thought for adequacy of lighting.

The electrical engineer is best qualified to handle details of electrical distribution system design and should be consulted to make sure the system meets both capacity and safety requirements.

Light sources

Light sources having widespread application in industry may be classified under the general headings of incandescent, fluorescent, and mercury lamps. These groups of lamps differ considerably in physical dimensions, electrical characteristics, and operating performance. Some have certain applications for which they are better suited than any other type; however, in many cases two or more sources are qualified to fulfill a particular lighting requirement.

Incandescent lamps. An incandescent lamp produces light when a wire or filament is heated to incandescence by the flow of electric current. Initial efficacy of typical incandescent lamps (for sizes 25 to 1000 watts) ranges from approximately 10 to 23 lumens per watt, depending on size and design. Incandescent lamps for industry are commonly designed for approximately 1000 hours of life. Longer-life lamps are available in some cases, but only with relatively lower efficiency since light output and life have an inverse relationship. It is important that incandescent lamps conform to the supply voltage, because the change of only a few volts will seriously affect both life and light output.

• Inside-frosted and "white-coated" lamps provide diffusion and thereby reduce high filament brightnesses. Light outputs are approximately equal to those of clear lamps; however, the effective source sizes are increased, which may affect luminaire distribution. Many incandescent lamps are available in colors, or in tints of colors. For instance, the daylight incandescent lamp has a bluish glass bulb that reduces the percentage of red light and gives a color closer to that of daylight. Daylight incandescent lamps are roughly 65 percent as efficient as other comparable types.

• Reflectorized (R and PAR) lamps having self-contained reflectors are manufactured in a number of bulb types in sizes from 30 to 1500 watts, furnishing various distributions of light. PAR lamps are a special form of reflectorized lamps with pressed glass construction, more accurate beam control, and greater resistance to breakage. While efficiencies of reflectorized lamps are lower than for other incandescent lamps, it must be remembered that the bulb also acts to direct the light and that the overall utilizations approach those of conventional lamps in luminaires. Because of economic factors, reflectorized lamps are commonly designed to provide approximately 2000 hours service life. This is equivalent to twice the life of most other incandescent lamps. Reflectorized lamps provide better maintained illumination than other types not only because the sealed-in reflecting surfaces are protected from airborne dust and dirt, but also because dirt and dust do not tend to adhere to the bulb face (in base-up burning position). When used for general light applications, suitable housings should be furnished to protect the lamps against mechanical damage and to provide adequate shielding of the bright light source.

• "Rough service" and "vibration service" lamps are special types that find considerable application in industry. Rough service lamps (in sizes from 25 to 500 watts) are made with extra filament supports to withstand mechanical shocks. One of the principal applications is for use on extension cords. Vibration service lamps (in sizes from 25 to 150 watts) are made with a flexible filament support for use where vibration would cause early failure of general-service lamps. Vibration and shock frequently accompany each other; sometimes only experiment will determine the better lamp for the purpose. Lower-voltage lamps, generally operated from transformers, are much more resistant to both shock and vibration than standard-voltage types. Shock- and vibration-absorbing socket mounts, utilizing a coiled spring or other flexible material to dampen vibration, have been employed where general-service lamps are used under conditions of severe vibration.

• "Extended-service" lamps are designed to produce approximately two to three times the normal rated life. Available for both regular and high voltage circuits in almost all wattage ratings, they are recommended for use where the cost of lamp replacement is high and cost of power is low. Otherwise their reduced efficacy makes it more economical to use a lamp of standard design.

• "Thermal shock resistant" or "special-service" lamps are available in various wattages and bulb shapes. They are recommended for use in applications where moisture may fall on the hot bulb.

• Quartz-iodine lamps employing iodine to keep the tubular envelope free from blackening have extremely good lumen maintenance in the order of 97 percent over a life of 2000 hours or more. These lamps are made in a variety of sizes from less than 50 watts to 1500 watts. The small linear shape of the lamps enables the luminaire to provide excellent beam control in the plane perpendicular to the lamp axis. Quartz incandescent lamps resist thermal shock.

• "Standard-voltage general-lighting" lamps are designed for use on nominal 120-volt circuits. Lamps designed for 115, 125, and 130 volts are available.

• "High-voltage general-lighting" lamps for

230- and 250-volt circuits are available in wattages from 100 to 1500. Some types are available in 260, 277, and 300-volt design. High-voltage lamps necessarily have filaments that are less rugged, require more supports and are considerably less efficient than those of equal wattage 120-volt lamps.

• "Infrared" lamps are available in various types and sizes up to 1000 watts. These lamps are used in baking, drying, and heat processing, as well as for therapeutic purposes. They also have fire protection value: a great amount of heat can be exchanged without flame. Such lamps cannot be used in hazardous location applications.

Fluorescent lamps. The fluorescent lamp is an electric discharge source in which light is produced predominantly by the fluorescence of phosphors activated by ultraviolet energy from a low-pressure mercury arc. The lamp requires a ballast to limit the current and in many instances to transform the supply voltage to a suitable level. Lamp performance is influenced by the character of the ballast and luminaire, line voltage, ambient temperature, burning hours per start, and air movement.

Fluorescent lamps vary in efficiency from approximately 30 to 80 lumens per watt, for lamps of standard cool-white color, depending on bulb size and shape. (Lamps range in size from 6 to 96 in. long). This does not include power losses in the ballast, typically in the order of 20 percent. While most fluorescent lamps are of conventional tubular design, there are some special types such as circular reflectorized, jacketed and panel-shaped.

Fluorescent lamps are available in many different shades of white and in a number of colors. "Standard cool-white" is most popular for industrial lighting because of its cool appearance, even though it is not quite as efficient as "standard warm-white." "Daylight" and "white" are also used considerably for industrial lighting.

Plastic safety sleeves for fluorescent lamp tubes offer excellent protection against lamp breakage in critical areas where such breakage would endanger workers or cause broken glass and phosphor contamination of proc-

esses, foods, and the like.

Since voltage changes affect the performance of fluorescent lamps, voltages at the luminaire should be kept at the specified voltage rating of the ballast. It should be noted that low voltage is as undesirable as high voltage. Low voltage, as well as high voltage, may reduce lamp and ballast life somewhat. Low voltage may also cause instability in the arc and starting difficulty.

Ambient temperature and air movement around the lamp are important factors in the performance of fluorescent lamps. The bulb wall temperature has a substantial effect on the amount of ultraviolet energy generated by the arc. Where fluorescent lamps are to be used in cold temperatures, as in unheated buildings or in refrigerated rooms, special luminaires or special lamps are recommended, otherwise light output may be greatly reduced. The light output of fluorescent lamps also falls off with high-temperature operation, the amount depending on lamp type.

Light output is also a function of the length of time a lamp is burned. When a fluorescent lamp is first lit, its output is abnormally high. During the first 100 hours of burning, this level drops about five percent. For this reason, the published "initial lumens" is the value obtained after 100 hours burning. Thereafter the drop in output is more gradual. The published "mean lumen" value is the approximate average of the output throughout lamp life. Lamp life is affected by the number of hours per start. A minimum number of starts favors lamp life. In general, fluorescent lamps are rated at three hours per start, and have 7 to 18 times the life of incandescent lamps.

Mercury lamps. Mercury lamps are electric discharge sources. The basic difference between mercury and fluorescent lamps is that the operating pressure of the mercury arc is much higher in mercury lamps. Spectral characteristics are different because the higher-pressure arc emits a large portion of its energy as visible light. Mercury lamps produce full light output only when full operating pressure has been reached, which generally is several minutes after starting. Most mercury lamps contain both an inner and outer bulb. The inner commonly is made of quartz while the outer bulb is generally made of thermal shock-resistant glass. Light output is practically unaffected by surrounding temperatures, except for a few types made with only a single envelope. Like fluorescent lamps, mercury lamps also require devices that limit the current.

Mercury lamps for general lighting are available with either "clear" or phosphor-coated bulbs, of various sizes and shapes, from 75 to 1000 watts. There is also a 3000 watt clear mercury lamp. Typical efficiencies range from 40 to 60 lumens per watt, not including 10 to 20 percent power loss in the ballast. "Clear" mercury lamps produce light that is rich in yellow and green tones and almost entirely lacking in red. Phosphor-coated (mercury-fluorescent) lamps afford several different steps of color improvement and are, therefore, more popular than clear lamps for industrial lighting.

A number of special mercury lamps are available, including semi-reflector, reflectorized, and self-ballasted.

The performance of mercury lamps depends to large extent on the characteristics of the ballasts. In particular, ballast design will govern the ability of the lamp to start at low temperatures and will control the time required for the lamp to reach full output. It will determine to a large extent the tolerance of a lamp to voltage dips. Serious voltage dips or any power interruption will cause the lamp to become extinguished. If for any reason the lamp is extinguished, it requires several minutes to cool before it will restart.

Initial output ratings commonly apply after 100 hours operation to allow for some early drop in light output. Thereafter the decline is very gradual. The life and maintained output of a mercury lamp are relatively little affected by the number of operating hours per start. Overwattage and underwattage generally have an adverse effect on both life and lumen maintenance.

Because of their high brightness, mercury lamps (including reflectorized and semi-reflector types) require suitable luminaires to provide adequate shielding of the light source. Luminaires also protect the lamps from physical damage.

When fluorescent and mercury lamps are operated on alternating current circuits, the carry-over of light on each half cycle depends on the phosphorescent qualities of the coating. Since the light, in effect, goes on and off 120 times a second, it creates a stroboscopic effect on moving objects such as flywheels and punch press parts, which may even appear to be at a standstill if the speed is in synchronism with current frequency.

Stroboscopic effect can be overcome by operating adjacent fixtures on different phases of a multi-phase system.

Use of Color

Interiors and equipment

Perception and visibility are improved by use of suitable colors on walls, ceiling, floor, and equipment. Light-reflecting qualities of surfaces contribute to fuller utilization of available light, and properly chosen color helps eliminate sharp contrasts in brightness in the worker's field of vision, thus contributing to good seeing.

White ceilings give maximum brightness. If floors and equipment are rather dark, reflecting 25 to 40 percent of the light, then upper walls should have a reflectance of 50 to 60 percent.

By judicious use of colors, an interior can be made attractive and thus have a good psychological effect on employees. Light shades are appropriate for most parts of a plant. Green and blue tints give a cool effect, and are psychologically valuable where temperatures are relatively high. A soft blue-green color is commonly used on walls. Ivory and cream are warm colors. Rose shades are suitable for female rest rooms, while blue is preferred by men. Light gray is effective for machinery—parts at the point of operation being painted orange to highlight any dangerous parts.

Color-coding

Color is used extensively for safety purposes. While never intended as a substitute for good safety measures and use of mechanical safeguards, standard colors are used to identify specific hazards. Standards have been developed and are given in American National Standard Z53.1, *Safety Color Code for Marking Physical Hazards and the Identification of Certain Equipment.* Be sure to check latest regulations for in-plant use, shipping, or consumer protection. In summary they are as follows:

RED identifies fire protection equipment, danger, and emergency stops on machines.

YELLOW is the standard color for (a) marking hazards that may result in accidents from slipping, falling, striking against, etc.; (b) flammable liquid storage cabinets; (c) a band on red safety cans; (d) materials handling equipment, such as lift trucks and gantry cranes; and (e) radiation hazard areas or containers (AEC still requires purple). Black stripes or "checker board" patterns can be used.

GREEN designates the location of first aid and safety equipment (other than fire fighting equipment). (Also see blue, below.)

BLACK AND WHITE and combinations of them in stripes or checks are used for housekeeping and traffic markings. They are also permitted as contrast colors.

ORANGE is the standard color to highlight dangerous parts of machines or energized equipment, such as exposed edges of cutting devices and the inside of (a) movable guards and enclosure doors, and (b) transmission guards.

BLUE is used on informational signs and bulletin boards not of a safety nature. (If of a safety nature, use green.) Has railroad uses.

REDDISH-PURPLE identifies radiation hazards; check A.E.C. regulations.

The piping in a plant may carry harmless, valuable, or dangerous contents, and therefore it is highly desirable to identify different piping systems. American National Standard A13.1, *Scheme for Identification of Piping Systems,* specifies standard colors for identifying pipelines and describes methods of applying these colors to the lines. The contents of pipelines are classified:

Classification	Color
Fire protection	Red
Dangerous	Yellow
Safe	Green
Protective materials (e.g., inerting gases)	Bright blue

The proper color may be applied to the entire length of the pipe or in bands 8 to 10 in. wide near valves, pumps, and at repeated intervals along the line. The name of the specific material is stenciled in black at readily visible locations such as valves and pumps.

Piping less than ¾ in. diameter is identified by enamel-on-metal tags.

The code also recommends highly resistant colored substances for use where acids and other chemicals may affect paints.

Accident prevention signs

Accident prevention signs (Fig. 16–5) are among the most widely used safety measures in industry so that uniformity in the color and design of signs is essential. Employees may be unable to speak English or may be color-blind and yet react correctly to standard signs. ANSI Z35.1, *Specifications for Industrial Accident Prevention Signs,* should be referred to.

The following is a digest of requirements:

DANGER—Immediate and grave danger or peril. Red oval in top panel; black or red lettering in lower panel.

CAUTION—Against lesser hazards. Yellow background color; black lettering.

GENERAL SAFETY—Green background on upper panel; black or green lettering on white background on lower panel.

FIRE AND EMERGENCY—White letters on red background. Optional for lower panel: red on white background.

INFORMATION—See BLUE at left.

IN-PLANT VEHICLE TRAFFIC—Standard highway signs (ANSI D6.1).

EXIT MARKING—See NFPA *Life Safety Code,* section 5–11.

Building Structures

Stairs, runways, ramps and other access structures are principal sources of injuries. One-fifth of all industrial injuries result from falls; and of those that take place from one level to another, the largest number occurs from stairs and ladders. (See Chapter 17, "Building Construction and Maintenance.") Many serious injuries can be prevented by

Fig. 16–5.—Accident prevention signs should conform to ANSI Z35.1 specifications. Posters and safety bulletins can be of any size or design.

Courtesy West Point Manufacturing Co., West Point, Ga.

careful design and construction. Standards of state and municipal governments and the American National Standards Institute should be followed. A convenient means of access should be provided to most locations more than 4 ft above the floor. Pointers in their selection, construction and use follow.

Runways and ramps

Construction is discussed in Chapter 17, "Building Construction and Maintenance." Width should be adequate for traffic to be handled; open sides should be protected with standard railings. Platforms, four or more feet above the floor or ground level, should be guarded by a standard railing, 36 to 42 in. high with intermediate rail and toeboard. (See Fig. 16–7.)

Ramps should be of the least slope practicable—some states specify a slope of 1 in 10 (5° 43′). Fifteen degrees (a slope of 2.68 in 10) is a recommended maximum, and a slope should never exceed 20 degrees (3.64 in 10). (See Fig. 16–6.)

Except where dislodged abrasives would be detrimental to equipment or process materials, ramps should have abrasive coatings or pressure-sensitive adhesive strips to help provide safe footing.

Toeboards should be installed where a ramp extends over a workplace or a passageway. Cleats 16 in. apart are needed on steep inclines.

Planks should not overlap and should run the long way of the ramp. Ramps to be used for wheelbarrows should have an odd number of planks and no cleats on the center plank.

Wire screen enclosures are recommended where materials must be stored on platforms or where persons below will be endangered by falling objects or fragments of materials. Screening can be made of wire netting of No. 16 U.S. gage wire with 1½-in. mesh. Adequate plywood or other fully closed enclosures also may be used.

Stairs and walkways

Stairs are taking the place of ladders in boiler rooms and other places both inside and outside of buildings. Circular stairways should be avoided, but if absolutely necessary, should be designed with a minimum variation in tread width. Treads should be covered with a durable antislip material.

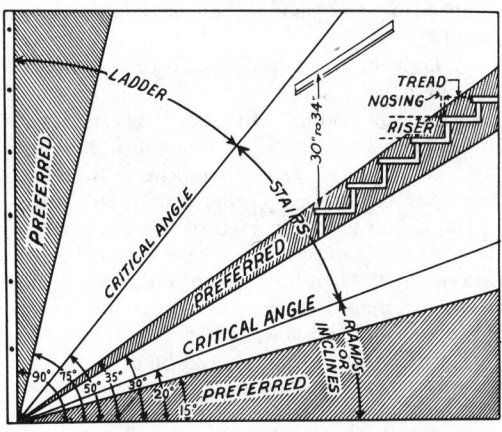

Fig. 16–6.—Preferred angles for fixed ladders, stairs, and ramps. Details are on the next page.

387

The preferred slope for a stairway is between 30 and 35 degrees from the horizontal (Fig. 16–6 and Table 16–B). The most suitable slope for a fixed ladder is from 75 to 90 degrees.

A tread width of not less than 9½ in. plus

TABLE 16–B

SLOPE AND DIMENSIONS OF TREADS AND RISERS

Angle of Stairway with Horizontal	Riser (inches)	Tread and Nosing (inches)
30° 35′	6½	11
32° 08′	6¾	10¾
33° 41′	7	10½
35° 16′	7¼	10¼
36° 52′	7½	10
38° 29′	7¾	9¾
40° 08′	8	9½
41° 44′	8¼	9¼
43° 22′	8½	9
45° 00′	8¾	8¾
46° 38′	9	8½
48° 16′	9¼	8¼
49° 54′	9½	8

From OSHA Standards § 1910.24.

a nonslip nosing of 1 in. are recommended. Riser height should not be more than 8 or less than 5 in., and should be constant for each flight.

A flight of stairs having two or more risers should have a standard handrail as specified in American National Standard A12.1, *Safety Code for Floor and Wall Openings, Railings, and Toe Boards.* Rails should be 30 to 34 in. from the top surface of the stair tread, measured in line with the face of the riser. (Up to 42 in. is permitted on steep-angles.) An intermediate handrail is recommended for stairways more than 88 in. wide. Consult applicable local and state codes.

Hardwood handrails should be at least 2 in. diameter. If standard black iron pipe is used, the diameter should be at least 1½ in. outside diameter. Clearance between the handrail and the wall should be at least 1½ in.

Rails are mounted directly on a wall or partition by means of brackets attached to the lower part of the handrail, to provide a smooth surface along the top and both sides of the handrail. Brackets should be spaced not more than 8 ft apart, and the mounting should be able to withstand a 25 pound per linear foot load applied on the handrail.

Because of space limitations, a permanent stairway sometimes has to be installed at an angle above 50 degrees. Such an installation (commonly called an inclined ladder or "ship's ladder") should have handrails on both sides and open risers.

Design should not include long flights. Landings every tenth or twelfth tread are recommended by some authorities. At least one state limits the vertical distance between landings in factories to 12½ ft.

Adequate illumination by lights located so they do not cause glare is important. Outside stairways should be covered to keep off rain, snow and ice, if possible.

It is good practice to enclose all inside stairs with partitions of fireproof or fire-resistive material and to install approved fire doors to prevent the spread of smoke or flames from one floor to another. Check codes and insurance men about this.

Stairs and landings should be able to sustain a live load of not less than 100 pounds per square foot with a factor of safety of four.

The original plans should specify elevated walkways and platforms on tanks, bins, big machinery, and other places where men must go during normal operations or for maintenance purposes. Conveyors should be provided with crossovers having nonslip surfaces. Railings and toeboards should also be provided.

Walkways and ramps should be equipped with a standard railing. A standard railing shall consist of a top rail, intermediate rail, and posts, and shall have a vertical height of 36 to 42 in. nominal from upper surface of top rail to floor, platform, runway, or ramp level. A standard toeboard shall be 4 in. nominal in vertical height from its top edge to the level of the floor, platform, runway, or ramp. (See Fig. 16–7.)

Many injuries are due to standing on improper devices to reach valves and regulators or to read instruments. The danger can be

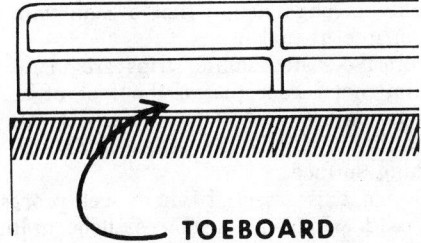

TOEBOARD

Fig. 16–7.—Open-sided floors or platforms more than 4 ft above floor or ground level, and scaffolds more than 10 ft above floor or ground level, should be guarded by a 36- to 42-in.-high railing (with midrail). If persons can pass beneath or if there is moving machinery or other equipment with which falling materials could create a hazard, the guardrail should also have a 4-in.-high toeboard. Screening can also be added.

eliminated by relocating the regulator or instrument at an accessible location, or equipping it so it can be adjusted or read from the floor. Otherwise, a permanent or adequate portable platform or ladder should be provided.

Exits

Exits must be sufficient both in number and size and so located that in case of fire or emergency the building can be quickly evacuated without loss of life. Plans should be adequate and conform to NFPA, federal, state, and local requirements, because changing or adding exits after a building is constructed is very costly.

The NFPA *Life Safety Code*, Standard No. 101 (also issued as American National Standard A9.1) generally requires two exits be provided on each floor, including basements. One exit from an upper floor may be an inside stairway or smokeproof tower and the other may be a moving stairway or horizontal exit. Some governing authorities require two or more exits that are remote from each other to be provided for all floors of industrial buildings two or more stories high. One must be a stair tower made of fire-resistive material, leading directly to the outside at grade (ground) level.

To a considerable extent, number and width of exits are determined by the building occupancy. In high hazard occupancy, no part of a building should be farther than 75 ft from an exit. For medium and low hazard occupancy, 100 to 150 ft is permissible. NFPA Std. No. 101 also specifies that access to exits provided by aisles, passageways, or corridors shall be convenient to every occupant and that the aggregate width of passageways and aisles shall be at least the required width of the exit. Exit doors should be clearly visible, illuminated, provided with signs, and must open in the direction of exit travel.

Exterior door openings used for hoisting equipment or material on the outside of buildings should be protected by guardrails or gates. Because such doors are a serious hazard when open and not in use, they should be posted with NOT AN EXIT signs.

Windows

Windows involve various hazards, of which the most common is that men may fall while cleaning them. Window sashes that swing back into the room eliminate the necessity for washing windows from outside the building. Some types of these windows provide good control over ventilation and can be opened and closed readily.

It is recommended that awning type sash not be installed adjacent to walkways unless it is more than 7 ft above them.

A monorail with cage may be used for window cleaning and other outside work. The American National Standard window cleaning code (*Safety Code for Window Cleaning*, A39.1) specifies the design of belts, belt terminals, and anchors when this work is to be done by standing on the outside sill. The code requires that the belt be so designed and constructed that, if in case one terminal is loosened from the window anchor, the terminal cannot pass through its fastening on the belt. The entire belt assembly must be capable of holding a suspended load of 1000 pounds.

The standards for design and use of fastening anchors should be strictly followed. Heavy forged metal anchors of brass, Monel, or copper-nickel alloy equivalent to Monel metal are recommended for their strength, toughness, and resistance to corrosion. Bolts for securing the anchors to the frames must be of the same material. In hollow metal frame construction, the bolts on combination anchor fittings are at least ⅜ in. in diameter

389

and must pass through a wrought iron or steel plate not less than $\frac{5}{16}$ in. thick by 6 in. long.

The plate is in the form of a "Z," or equivalent. One portion reaches behind the exposed face of the frame and is bolted or riveted by at least two $\frac{1}{4}$-in. bolts, screws, or rivets of the same material as the anchor to the part of the frame protected by masonry or concrete. The code also contains standards for swinging scaffolds, boatswain's chairs, and ladders.

Flooring materials

Comfort, health, and safety are closely related to the design and specifications for floors. Requirements often vary sharply from department to department, so that a careful study of many factors is essential to determine the best type of floor for a particular location. These factors include:

Load	Illumination
Durability	Maintenance
Noise	Dustiness
Drainage	Heat conductivity
Resilience	Electrical conductivity
Appearance	Chemical composition

Since a principal cause of floor accidents is slipperiness, the inherent slipping hazard of various types of floor surfaces should be ascertained. See also pages 578–582.

Ordinary smooth iron or steel plates are unsafe. Marble, slate, asphalt tile, linoleum, cement, and wood are reasonably safe if properly maintained and kept clean. The safest floor materials are cork, dry rubber, creosoted wood blocks, asphalt, roughened concrete, and checkered plate. The anticipated traffic and the nature of the process will influence floor choice.

Inserts of various materials can be used to reduce slipperiness in specific areas or to combat conditions that cause rapid deterioration of flooring. For example, cast metal inserts are used around woodworking machines. Sheets of soft lead may cover the floor surface where acid is spilled occasionally. Drainage is provided so that the acid can be washed away with water.

Inserts should be installed flush with the surface of the floor. An insert placed on top of the floor requires a bevel on every side from which it can be approached. The bevel

should be long enough that a man will not trip or lose his balance.

Abrasive-coated fabric strips are made for use on metal floor plates, the foot of stairs, stair nosings, and other high-hazard locations. An adhesive binds the strips firmly to the walking surface.

Drainage is essential where wet processes are used. In some instances, floor gratings are installed to reduce slipperiness due to water and other liquids, especially along passageways.

The importance of selecting floor materials with suitable characteristics for the specific location is illustrated by plans for two new textile plants. In one plant, the floor in the spinning department was concrete overlaid with maple parquet for resilience and durability. The dye house floor was acid-proof brick. For the office, asphalt tile was laid over concrete; the laboratory floor was rubber tile, also laid over concrete. Creosoted wood block floor, which is easy on the feet and withstands trucking, was used in a large machine shop.

Plans for the other plant specified a floor that would be resilient, moisture-resistant, and free of vibration. To get these qualities, 3-in. creosoted square timbers were imbedded in a base made of 3 in. of tar concrete under $1\frac{1}{2}$ in. of tarred sand. The subfloor was $\frac{3}{4}$-in. salt-treated pine with a layer of waterproof paper. The floor surface was 25/32-in. hard maple.

Asphalt is used in various flooring materials. It is dustless, plastic, odorless, and warm to the feet. If the aggregate is siliceous, the mixture is resistant to acid. Ordinary grades soften in hot weather and are not recommended for heavy industrial trucking. Asphalt tile, often used in offices, may become slippery when washed or not properly waxed.

Paving brick, if laid on a solid foundation like concrete, is satisfactory for heavy traffic. Cement mortar joints make a smooth surface. Foundry floors have been made from hard-burned bricks laid face up on a concrete base with sand-filled joints.

Concrete floors are used widely in warehouses and factories. A wood float finish on a mixture of pea gravel, sand, and cement gives a roughened surface that will not crack or "dust." A smooth finish is slippery when wet.

Concrete does not withstand acids, and in some types of work employees consider concrete too hard and too cold. Resilient nonslip mats with low heat conductivity can be placed over the concrete where workmen must stand in one position for considerable periods of time.

Cork tile is desirable for its insulation, resiliency, quietness, high nonslip rating, and ability to withstand light traffic for long periods. It is not suitable for wet locations or where there are heavy loadings, and it is expensive.

Cast iron plates with checkered or otherwise roughened surfaces laid in cement or asphalt are suitable for some types of rough wear such as that in warehouses. The slipping hazard at the door sills of elevators may be reduced by installing steel grating filled with concrete.

Lead is used for floors exposed to acids and as a mat for secure footing near wood-working or other machines involving serious danger to the operator if he slips and falls. Lead is heat conductive, spark resistant, and quiet. It should not be used for heavy trucking.

Zinc also is used for its spark-resistant qualities; but, compared to lead, it is attacked more readily by alkalies. Floors in compounding rooms where fire and explosion hazards exist are zinc-covered in some plants.

Asphalt-base, vinyl-base or linoleum-type tile often is used in offices, laboratories, and workrooms where cleanliness and good looks are important. The material is easily cleaned, noiseless, and a poor conductor of heat.

Magnesite is suitable for light traffic or where light oils are used. It must be laid on a rigid base and should not be used where there is excessive moisture or hydrostatic pressure, as in basements. A coating of bituminous paint is necessary to protect metals such as pipe from contact, since magnesite corrodes some metals.

Checkered steel plate wears well, but it is noisy and highly conductive of heat and electricity. It is relatively nonslippery unless worn smooth or wet with oil.

Metal grille floors and gratings will not collect dust, dirt, or liquids but, because of noise, are not often used where hand trucking is done regularly. This type of floor material is particularly useful in boiler rooms and over openings.

Parquet is laid on an underfloor and is suitable for offices. If the material is sealed properly, little maintenance is necessary. Parquet wears well but may be noisy.

Rubber flooring is resilient. It also has high dielectric strength, which is undesirable where static electricity is a problem. However, conductive types of rubber flooring are available. Abrasive rubber flooring can be used to overcome slipperiness.

Since terrazzo flooring has no joints in the surface, its use eliminates some of the difficulties that may be encountered with other types of flooring. It can be made electrically conductive by grounded grilles. It is a conductor of heat. Suitable sealers are necessary to make the floor impervious to most acids. The terrazzo mixture is slippery unless it includes abrasive aggregates.

Plants requiring extreme cleanliness and sanitary conditions, such as dairies, use ceramic glazed tile. It should be laid in Trinidad asphalt or in cement with a low lime content. Two or three layers of asphalt roofing felt may be laid beneath the tile.

Wood floors of the proper material and construction are suitable for various services except for the obvious objection of a fire hazard. Plank or board floors of the softer woods are generally unsatisfactory and a source of numerous injuries from slipping and falling as well as from splinters and truck accidents.

A good floor for light manufacturing can be made from matched or jointed hardwood flooring nailed to a subfloor structure or to sleepers in concrete. The thickness of the hardwood flooring will depend on the service to which it will be subjected. It is laid with the grain parallel to the line of truck travel.

Wood blocks meet many of the requirements for a good floor. Properly made, a floor of this type is relatively noiseless and does not become slippery or cause fatigue. If the blocks are laid on a smooth, rigid base, the floor is not likely to crack and will withstand heavy service.

Blocks impregnated with creosote are necessary for floors in contact with liquids or moisture. Expansion joints are required along walls, columns, and similar places. If blocks are laid with a high melting point pitch, hot weather does not create problems. Oils and

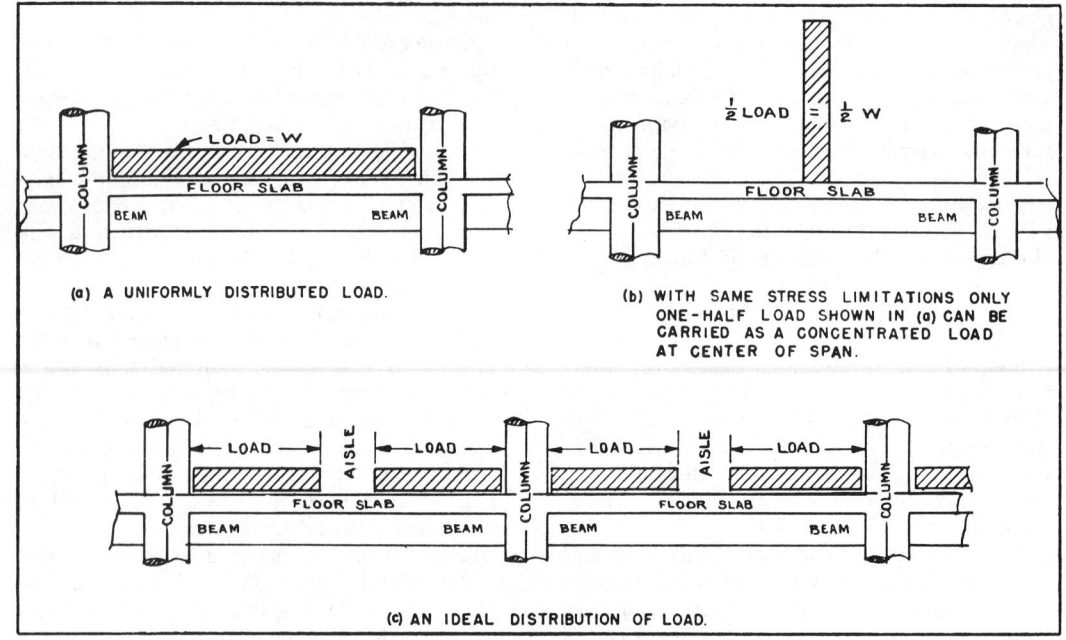

Fig. 16–8.—Various types of loadings. Changing the location of a load changes the total load a floor can carry.

organic solvents, however, cause trouble because of their solvent effect on bituminous fillers and coatings.

To reduce tracking of mud and dirt into a building, moisture-absorbing mats or runners can be installed at entrances.

Special flooring

Special types of flooring are necessary where fires or explosions may occur. Spark-resistant and conductive flooring should be specified where flammable vapors, dusts, or gases are present.

Conductive floor resistance (as measured between two electrodes placed 3 ft apart at any points on the floor) may vary from not less than 25,000 ohms to not more than 1,000,000 ohms, according to NFPA's *Inhalation Anesthetics Code*, Standard No. 56A, to assure protection against electric shock and still provide for removal of static. These floors can be grounded at two or more points by means of copper strips stapled to the material or copper wires imbedded in it.

In areas where welding is done regularly and in oven, furnace, boiler and similar areas,

noncombustible flooring should be installed.

Openings in floors should be protected by railings or barriers at exposed edges, or otherwise guarded. There should be a top rail, intermediate rail, and toeboard, according to American National Standard A12.1, *Safety Requirements for Floor and Wall Openings, Railings, and Toe Boards*. Sheeting or woven wire can be installed around the opening to prevent material from falling in.

Floor loads

Floors should be designed for carrying anticipated loads safely. A registered structural engineer should be consulted. Also check American National Standard A58.1, *Building Code Requirements for Minimum Design Loads in Buildings and Other Structures*.

Fig. 16–8 presents some floor loading fundamentals. The ideal load is uniformly distributed over the floor area (as shown in A). The same load concentrated at the center of the span (shown in B) will require twice the structural strength. Conversely stated—if a floor is designed for a given uniform load, only one-half this amount can be concentrated at the center of the span. Illustration C shows the ideal location of aisles and loads.

In estimating, weights of men are figured at 160 pounds and women at 138 pounds. Equipment weights can be obtained from manufacturers; weights of bulk materials from handbooks. (In piling bulk materials, often air spaces result which reduce the overall density of the material.)

Foundations which distribute loads and vibration over larger areas, or cushion shocks with springs or vibration mounting, help reduce structural reinforcement. Live loads should be figured and future loads anticipated. Industrial trucks when loaded may weigh as much as 60,000 pounds.

References

American Insurance Assn., 85 John St., New York, N.Y. 10038. *Recommended Good Practice Requirements of the National Board of Fire Underwriters for the Construction and Protection of Piers and Wharves.*

American Society of Heating, Refrigerating, and Air-Conditioning Engineers, 345 E. 47th St., New York, N.Y. 10017. *Guide and Data Book.*

American National Standards Institute, 1430 Broadway, New York, N.Y. 10018.
Building Code Requirements for Minimum Design Loads in Buildings and Other Structures, A58.1.
Code for Safety to Life from Fire in Buildings and Structures, A9.1 (NFPA Std. 101).
Minimum Requirements for Sanitation in Places of Employment, Z4.1.
National Electrical Code, C1.
Practice for Industrial Lighting, A11.1.
Practice for Protective Lighting, A85.1.
Requirements for Fixed Industrial Stairs, A64.
"Safety Requirements for Construction," A10 Series.
Safety Requirements for Floor and Wall Openings, Railings, and Toe Boards, A12.1.
Safety Code for Portable Wood Ladders, A14.1.
Safety Requirements for Window Cleaning, A39.1.
Safety Color Code for Marking Physical Hazards, Z53.1.
Scheme for the Identification of Piping Systems, A13.1.
Specifications for Accident Prevention Signs, Z35.1.

Asphalt Institute, University of Maryland, College Park, Md. 20740 *Asphalt Handbook.*

Associated Factory Mutual Fire Insurance Companies, Engineering Division. *Handbook of Industrial Loss Prevention.* New York, McGraw-Hill Book Co.

Faber Birren and Company, 500 Fifth Ave., New York. *Specifications of Illumination and Color in Industry.* Reprinted from *Transactions,* American Academy of Ophthalmology and Otolaryngology.

Illuminating Engineering Society, 345 E. 47th St., New York, N.Y. 10017.
Glare and Lighting Design.
Lighting Handbook.
Recommended Practice for Daylighting.

Institute of Traffic Engineers, 2029 K Street, N.W., Washington, D.C. 20006 *Handbook.*

Jones, Charles L. *Safety in Lacquer Plants.* Hercules Powder Company, 917 Market St., Wilmington, Del.

National Fire Protection Association, 470 Atlantic Ave., Boston, Mass. 02210.
Air Conditioning and Ventilating Systems (Non-Residential), Standard No. 90A.
Construction and Protection of Piers and Wharves, Standard No. 87.

Fire Protection Handbook.
Flammable and Combustible Liquids Code, Standard No. 30.
Flammable Anesthetics Code, Standard No. 56.
Life Safety Code, Standard No. 101 (ANSI A9.1).
National Electrical Code, Standard No. 70.
National Fire Codes, Ten volumes.
Operation of Marine Terminals, Standard No. 307.
Protection from Exposure Fires, Standard No. 80A.

National Safety Council, 425 North Michigan Ave., Chicago, Ill. 60611
Industrial Data Sheets
Emergency Lighting, Data Sheet 248.
Firearms for Plant Protection, Data Sheet 413.
Parking Lots—Self-Service, Data Sheet 597.
Poison Ivy, Poison Oak, and Poison Sumac, Data Sheet 304.
Tree Trimming, Data Sheet 244.

Portland Cement Association, 5420 Old Orchard Rd., Skokie, Ill. 60076. (General.)

U.S. Department of Transportation, Federal Highway Administration Bureau of Public Roads. *Manual on Uniform Traffic Control Devices for Streets and Highways.* (ANSI Standard D6.1). Washington, D.C., U.S. Government Printing Office.

U.S. Environmental Protection Agency, 401 M St., SW., Washington, D.C. 20460.

Building Construction and Maintenance

Chapter
17

Construction on Company Premises   396
Analyzing the construction job . . . Contract documents . . .
Contractor safety program . . . Protection of employees and
equipment . . . Trucking . . . Miscellaneous machinery and
equipment . . . Steel erection . . . Lateral bracing . . . Temporary
flooring . . . Torches and salamanders . . . Temporary heating
equipment . . . Demolition of structures

Excavation   411
Machine excavation . . . Open excavation . . . Trench excavation

Ladders   413
Nonslip bases and safety tops . . . Ladder maintenance . . . Use
of ladders

Scaffolds   418
Railings and toeboards. . . Overhead protection . . . Means of
access . . . Types of scaffolds . . . Wooden scaffolds . . . Working
platforms . . . Swinging scaffolds . . . Boatswain's chairs

Hoists   424
Inside material hoistways . . . Outside material hoistways . . .
Material hoist platforms . . . Signal systems . . . Personnel hoists
. . . Wire ropes and sheaves . . . Hoisting engines

Plant maintenance   427
Foundations . . . Structural members . . . Walls . . . Floors . . .
Roofs . . . Tanks and towers . . . Stacks . . . Platforms and loading

docs . . . Canopies . . . Sidewalks and drives . . . Underground utilities . . . Lighting systems . . . Stairs and exits

Maintenance Crews   **435**
Training . . . Scheduling preventive maintenance . . . Equipment inspections . . . Personal protective equipment . . . Lock-out of mechanical and electrical equipment . . . Piping . . . Crane runways . . . Final check—tools and guards . . . Lubrication . . . Shop equipment maintenance . . . Special tools . . . Keeping up-to-date

References   **440**

Construction on Company Premises

Construction is a specialized industry with many specialized divisions. The material in this chapter is intended primarily for industrial concerns that undertake incidental construction or demolition operations and those that need standards for the control of construction operations done on their property.

Industrial concerns doing construction work themselves have complete control over their employees and therefore can require adherence to safe practices. When a company hires an outside contractor or subcontractor, the company and the contractor must have a complete understanding of the legal relationship between them. The company can insist upon safe practices that affect its own employees and plant property, but can only suggest safe practices that pertain to the construction work itself. This is the contractor's responsibility.

Blind reliance on building codes and construction safety standards is not the answer— codes and standards cannot guarantee that a company's own men, or a contractor will work safely. The main purpose of building codes is to provide safety to future building occupants, and at best, they represent a minimum requirement for materials and construction. The federal government, many states, and many localities have construction safety standards as well, but these—like building codes— are only as effective as the enforcement behind them.

One of the best times to "think safety" is during the designing and specifying steps— designing safety into the final product, as well as into its construction. With this in mind, the safety professional should be brought in at the earliest planning stages (as discussed in Chapters 4 and 16), and also should discuss the proposed job with the construction department or with outside contractors. (Many safety-conscious independent contractors try to get a running start on a good safety program and will even insist on a pre-job safety conference.)

Analyzing the construction job

Advance analysis of construction jobs is not new; in fact, some sort of an analysis is made on every construction job. A contractor preparing a bid must make a reasonable, accurate analysis if he hopes to compete with other bidders and if he is to avoid a monetary loss. Practical experience in running construction jobs proves one cannot get very far without a set of plans and specifications which clearly defines the work to be done by the interested trades. Accident prevention, while not exactly a job trade, is a necessary function, yet, many contractors start a job without determining the rightful place accident prevention holds financially and do little advance planning to eliminate or reduce job exposures.

Most contractors make up their bids under five major divisions: material, labor, plant, overhead, and profit. In the "overhead" is buried an item designated as insurance. The amount placed in the bid is usually a per-

396

centage of the cost of material, labor, and plant. The amount will vary with the type of job and is affected to some extent by the requirements of owners, architects, and the state in which the job is located. This insurance item includes the expense for the performance and payment bonds, workmen's compensation, Social Security, unemployment insurance, employer's liability, property damage, automobile insurance, fire insurance, builders' risk insurance, and other types of coverage that the contractor may select. Further, this insurance item is considered one of the fixed charges against the job. That is, like death and taxes, little can be done about it.

This reasoning that all insurance is a fixed charge is very misleading and indicates lack of understanding of one very important item—the compensation insurance cost. Smart contractors claim that there is no one item in the entire bid which can more easily produce a substantial saving.

If such profits can be made on insurance, the next questions are why has it been overlooked and why has there not been more enthusiastic interest by contractors generally. The answer lies with the safety professional and may be due to inadequate analysis of the construction job and lack of proper presentation to management.

Safety professionals think and talk mostly in terms of accidents, usually involving personal injury. They use terms like "accident frequencies," usually meaning disabling injury frequency rates. Much of this information is based on hindsight, and, while it has a certain value during the progress of the job, it is unobtainable in practical form at the time the job is bid. Frequency figures cannot easily be translated into the terms of the trade, such as the "cubic feet," the "square yards," or "the dollar," and, therefore, lose much of their value because of lack of understanding. Chapters 5, 6, and 7 give a detailed discussion of how to understand accident statistics and use them to prevent accidents.

Accident prevention data has its greatest value when it is of the advance type and includes:

1. A translation of exposures into terms of cost to establish its position of importance with other items in the bid.

2. Defines the job exposure in specific terms.

3. Sets up practical and effective safeguards to control these exposures.

Accident prevention is a real factor in the economic success of any construction job. Fortunately, the means by which it can be controlled are known. Further, there is ample evidence that it is controllable and can be very profitable to do so. To society, advance analysis of a job means lives saved, misery and suffering spared, natural human resources more fully utilized. To the construction industry, it means lower cost and greater efficiency, which, in turn, can be passed on to the owner. To the safety professional, it means a job well done.

Contract documents

For work within a plant operating area, contract documents or specifications should include a provision that the outside contractor conform to certain safety requirements. The contract documents should spell out that the contractor meet certain minimum safety, health, and equipment requirements, including provisions for protecting company employees (and the public) from construction hazards.

In order to enforce subsequent recommendations, it is important to spell them out in the contract documents. This makes them binding, otherwise they are merely loosely enforced agreements.

Under no circumstances should a company undertake supervision of the contractor's employees. The usual procedure is to point out matters of safety to the contractor's superintendent or top supervisor so that he can exercise the proper authority.

It is important to note that in all construction operations, applicable codes, laws, and ordinances must be complied with.

Contractor safety program

The documents should also spell out that the contractor set up an effective safety program (if he does not already have one). By doing this, a company emphasizes that the human element must be reckoned with—maintaining a good safety record is basically a cooperatively directed accomplishment of the employees.

Take, for example, the way Ebasco Services Inc. makes an active safety program part of everything a workman does on a construction job. Ebasco, a firm that designs, specifies, and constructs electric power-generating facilities throughout the world, and which is a member of the National Safety Council, spells this out in their comprehensive construction safety program.

Even though Ebasco directly employs the majority of its work force, it does contract for some specialty trades. Ebasco's safety program applies to all Ebasco employees, and through the cooperation of contractors on Ebasco projects, also to all contractors' employees.

The safety program for Ebasco employees on a project is set up by Ebasco's project manager who maps out a program of safety measures commensurate with the size of the project and the conditions and hazards peculiar to it. A typical Ebasco construction safety program involves:

1. A full- or part-time safety professional. Adequate first aid facilities and trained personnel.

2. A project safety committee comprised of Ebasco's key personnel (in which all key contractor supervisors are also invited to participate). On some jobs workmen are also on this committee, but membership is rotated so that the maximum number of persons will have some opportunity to participate actively in committee work.

3. The committee meets weekly, and submits a review copy of the minutes (which includes accident statistics) to the safety department at Ebasco's New York office. The safety committee discusses accidents and near accidents of the previous week. Other hazards are pointed out, housekeeping is evaluated, and plans are made for correcting unsafe conditions and unsafe practices before they lead to an accident. Periodically a portion of some meetings is dedicated to first aid instruction, resuscitation techniques, fire-fighting, and other emergency procedures.

4. A safety representative of the week, appointed on a rotating basis at each committee meeting, acts as the project safety

inspector and submits a report of the committee.

5. Crew foremen who hold brief meetings with workers under their supervision at least once a week to discuss the safe operation of their crew, the safety of other workers, and specific problems. This meeting is usually held after the project's safety committee meeting.

6. Safety instructions given all new workers as part of their first day indoctrination.

7. Special mass safety meetings called by the project engineer. All workers attend these meetings.

8. Safety devices designed to prevent injury, such as safety hats, eye protection, safety belts, and similar protective equipment are provided as necessary.

9. Facilities that will help prevent accidents (such as suitable roads, lights, barricades, signs, warning devices, guardrails) are provided.

10. Safety releases are forwarded periodically from Ebasco headquarters.

11. Periodic visits to projects by Ebasco's construction managers and the safety supervisor. While there, these men inspect the job and participate in the safety program.

The enthusiasm generated by this program gives a real awareness of job safety as a whole and to making each workman individually safety conscious.

The project manager and his staff talk with each contractor's supervisory personnel before work is started in order to explain how the contractor's work will proceed in relation to the work of others. The project safety program and the part of the contractor are discussed. The contractor's key supervisor is invited to serve as a member of the project safety committee.

Prejob safety conference. A preconstruction conference should be held between company (or plant) management (including the safety department) and the contractor's superintendent or other equally authoritative representative. Ways of access (for the contractor's employees, construction materials, and equipment delivery), storage space, and parking areas should be established. Also, the contractor's representative should be acquainted with the plant safety program, first aid facilities, special safety equipment that will be required because of hazards due to plant operations, and how this safety equipment can be obtained.

The U.S. Army Corps of Engineers uses the following outline when planning a preconstruction safety conference. Prior to commencement of work, written proposals should be submitted for effectuating the provision of accident prevention. All interested top supervision should then meet in a preconstruction safety conference to discuss the proposed program and to develop mutual understandings relative to the over-all aspects of the plan. Provisions should be made for holding periodic staff meetings of top supervision for evaluating and revising the program as required by changing conditions and new problems that arise. Following is a suggested outline for the preconstruction safety conference.

1. Purpose
    a) Evaluation of proposed program
    b) Discussion of job organization and operating procedures
    c) Preplanning the work and agreement of a means for practical application of standard procedures
2. Notification to all parties
3. Evaluation of proposed program
4. Conference facilities
5. Meeting attendance
6. Conference record
7. Agenda for conference
    a) Orientation
        (1) Explain why we have a program
        (2) Advantages in terms of economy and efficiency
        (3) Prescribed safety standards
        (4) Review of:
            (a) Accident prevention agreements
            (b) General conditions of specifications on safety
            (c) Special conditions of specifications on safety
        (5) Other requirements—local, state, federal
        (6) Supervision
            (a) Organization at project
            (b) Functions of personnel
            (c) Responsibilities
            (d) Delegated authorities
            (e) Relations regarding enforcement
    b) Discussion of proposed program
        (1) Plans as to layout of temporary construction, site, buildings, etc.
        (2) Action taken toward planning and coordinating activities between different operations and crafts
        (3) Access to work areas
        (4) Safety indoctrination and safety education
        (5) Delegation of safety responsibilities to supervisors
        (6) Integration of safety into operating methods and procedures
        (7) Housekeeping program
        (8) Safety factors in job built appurtenances
        (9) Traffic control
        (10) Fire protection
        (11) Lighting, ventilation, protective apparel, and medical care
        (12) Safe operating condition of equipment and maintenance
8. General
    a) Methods for meeting objectives
    b) Understandings for periodic readjustment of safety objectives
    c) Handling of safety deficiencies
    d) Arrangements for additional meetings and periodic staff meetings
    e) Follow-up of agreements in preconstruction meeting
    f) Three cardinal rules to observe for a workable safety program
        (1) All agreements must be fair
        (2) Paper work should be kept at a minimum
        (3) The program should be simple— it should deal with facts

Protection of employees and equipment

When construction work is being done in or around an industrial plant, the company employees and equipment should be pro-

READY MIX TRUCK
SIGNALS
STANDARD SIGNALS FOR MIXER DRIVERS & CONTRACTORS GUIDES

DRIVE IN	BACK IN	BACK UP
PULL FORWARD	STOP	RAISE CHUTE / LOWER CHUTE
START POURING	MORE WATER	STOP POURING

Fig. 17–1.—Suggested signals for mixer drivers and contractor's guides.

Courtesy Northern California Ready Mixed Concrete & Materials Assn.

tected from all construction hazards, including open excavations, falling objects, welding operations, dust, and dirt. The construction work should be isolated from company operations if at all possible. Barricades, fences, and guard rails should be set up, and appropriate warning signs should be posted. Warning signs are detailed in ANSI Standard Z35.1, *Specifications for Accident Prevention Signs.*

When a construction job is being done in an area that must be kept in operation, a sheeted bulkhead can be erected to keep out dust and dirt and to isolate the operation as much as possible. Studs are put up from floor to ceiling or to the underside of the roof, then sheeted with plain lumber, wallboard panels, plastic or canvas tarpaulins, or other material

to make the bulkhead dust-tight. Flame-retardant materials may be required in some cases.

When construction operations are to be done where flammable vapors and liquids may be present or in other hazardous areas, the contractor's men should be subjected to the same rigid requirements that apply to the plant employees. Before cutting, burning, or welding is done by men employed by the contractor, a clearance should be obtained from the plant engineer or safety department to assure that the necessary fire and safety precautions have been taken. The usual requirements for ventilation and health protection also should be observed. (See NSC Data Sheet 522, *Hot Work Permits (Flame or*

400

Spark), and Chapters 34 and 36 in this Manual).

In an area where no flame is permitted, either the manufacturing process will have to be shut down while welding and cutting are done, or screwed or bolted fittings will have to be used instead of welded connections.

Night lighting should be provided where necessary, especially in areas where open trenches or ditches create hazards in aisles and roadways. A minimum of 5 footcandles is recommended by the Illuminating Engineering Society for vital exterior locations or structures. See page 1449.

The contractor and plant management should cooperate closely at all times in order to determine what precautions are necessary to prevent accidents. Particularly close liaison should be maintained during tie-in of new piping or equipment to existing lines and equipment to ensure the safety of subsequent operation. The contractor should maintain all his work in an orderly, well-kept manner.

Trucking

Contractors working either inside or outside the industrial building or area should take great care to prevent trucks and other mobile equipment from colliding with pipelines and power lines, and with equipment in order to avoid interrupting manufacturing or processing operations.

One method of handling construction deliveries is to have a signalman who serves as the eyes for the truck driver. Suggested signals for ready-mix concrete trucks are shown in Fig. 17–1. Be sure barricades, guard rails, and warning signs are placed to assure maximum safety.

When trucks, bulldozers, powered wheelbarrows, and other mechanized construction equipment are to be operated within a plant, it is a good idea for the contractor and the plant management to agree upon the lines of traffic flow so that the areas in which the construction equipment will be operated will be known. When such agreement is reached, drawings of the areas can be made available to both the contractor's key personnel and the plant personnel affected.

It may be necessary to haul men in trucks from one location to another within the work area, and, unless controlled, this operation can become a major source of serious injuries. Men should not be allowed to stand on the running board or bed of a truck. They should not be permitted to sit at the side or end of a flat-bed truck. Instead, temporary (removable) plank seats should be installed in trucks that will be used for moving employees. Safety belts should be supplied for all passengers. Men must not be permitted to ride on a loaded truck. (See NSC Data Sheet 330, *Motor Trucks for Mines, Quarries, and Construction*.)

When men are boarding or alighting from a truck, the truck should be standing still. A boarding ladder should be provided. If necessary, a bus should be provided to transport the employees to the work site.

Miscellaneous machinery and equipment

Modern construction requires a variety of machines: tractors and bulldozers for site preparation, power shovels and draglines for excavating, cranes and derricks for placing structural members, concrete mixers, compressors, generators, and many others.

No machine or piece of equipment should be placed in operation until it has been inspected by a qualified person and found to be in safe operating condition. Regular inspections should be made thereafter, and the inspection reports should be kept on file and necessary action taken to remedy unsafe conditions.

Causes of accidents. In a study (by the U.S. Army Corps of Engineers) of accidents resulting from the operation of general construction and related mechanical equipment, it was found that a great majority of accidents was caused by inadequate maintenance of equipment, insufficient instruction in safe practices, or lack of insistence on their observance. Unsafe practices contributed to 80 percent of all injuries.

Guarding, safety devices, platforms, and means of access. Belts, pulleys, sheaves, gears, chains, shafts, clutches, drums, flywheels, and other reciprocating or rotating parts of equipment should be guarded.

No guard or safety appliance or device should be removed or made ineffective unless immediate repairs or adjustments are required, and then only after the power has been shut off. Guards and devices should be replaced as soon as repairs and adjustments have been completed.

Current-carrying parts of electrically operated equipment should be properly insulated or guarded. All non-current-carrying metal parts should be properly grounded.

High-temperature lines and equipment, located where they endanger employees or create a fire hazard, should be covered with suitable insulating material.

Exhausts from all equipment powered by steam or internal combustion engines should be properly released and so located that they do not endanger workmen or obstruct the view of the operator.

Platforms, footwalks, steps, ladders, hand holds, guard rails, and toeboards should be installed on all equipment where they are needed to provide safe access. Suitable operating floors or platforms, surfaced with non-slip material, should be provided for all equipment operators. (See details in the previous chapter.)

Operators of equipment should have protection against the elements, falling objects, swinging loads, and similar hazards.

Windows in cabs or enclosures on equipment should be made of safety glass, and should be kept in good repair at all times.

Reverse alarms should be installed on all heavy mobile equipment and trucks. Warning devices should be provided on all equipment where there is danger to workmen in moving the equipment or from swinging loads, buckets, booms, or crane counterweights.

Positive means should be provided to prevent the starting of equipment by unauthorized persons. This could take the form of key ignition systems or simply blocking the starting apparatus and locking it.

At the end of a work shift, equipment should be set and locked so that it cannot be released, dropped, or activated in any way. Front-end loader buckets should be lowered to the ground as should bulldozer blades; scraper bowls should be dropped; and crane booms should be secured against movement.

Flammable liquids. Refueling gasoline-operated equipment while the motor is run-ning should be prohibited. Continuously operating equipment should be fueled from properly protected tanks located outside the operating room. This should be adequately grounded to prevent static electric buildup.

Smoking or the use of open flames on or in the immediate vicinity of gasoline-operated equipment while it is being refueled should be prohibited.

No solvent with a flash point below 100 F should be used for cleaning equipment or parts.

Gasoline and other highly flammable fluids used on equipment should be handled by pumps or in approved safety cans. Gasoline, fuel oil, and other flammable liquids should not be stored on equipment except in fuel tanks or approved safety cans with a capacity for only one day's requirements.

Fuel tank filler openings should not be located in such a position that spills or overflows can run down on a hot motor, exhaust pipes, or battery.

A suitable fire extinguisher should be located on or close to each piece of equipment. See Chapter 42 for more details.

Repairs. All "out of order" equipment should be shut down for repairs. Suitable signs should be posted and not removed until repairs have been completed. Mobile equipment should, if possible, be removed to a safe location where operations will not interfere with the repair work. Equipment suspended in slings or supported by hoists or jacks for repairs should be blocked or cribbed before men are permitted to work underneath it.

When repairs are made remote from the source of power on such equipment as conveyors and cable ways, use should be made of chains, blocking, or similar devices to prevent injury in case of accidental starting.

Before repairs on electrically powered equipment are begun, the main switch should be locked in the open (or OFF) position. The key to the switch lock should be retained by the repairman. If there is more than one repairman on a circuit, each should lock the main switch with his lock to which only he has the key. Switch boxes should have this provision.

Miscellaneous. Accumulations of debris,

oil, grease, oily rags, and waste on equipment should not be permitted.

Safe load capacity and operating speeds should be posted on all equipment and should not be exceeded.

Equipment should be placed on an adequate foundation and properly secured.

Before mobile equipment is moved, a survey of the area in which it is located should be made to check for overhead wires, pipelines, excavations, and similar hazards. Equipment with high clearances (such as cranes) should not be moved into or out of, or operated in, any area containing electric power lines until the approval of the superintendent has been obtained. No part, including the load, may reach within 10 ft of electric lines, unless power in the lines is shut off. (Local laws may specify greater distances, in which case these should prevail.)

A proximity warning device that sounds an alarm to the operator and workmen when the crane boom is brought near electrically charged equipment can be installed on the crane. The device operates at varying distances, depending on the line voltage and the setting of the sensitivity control of the warning system. It uses its own special circuits and is activated by ac or dc current in power lines, active telephone lines, street car lines, other charged equipment, and some shallow-buried cables.

Another device that reduces the hazards involved in crane contacts with electric lines is a cage-like insulating guard that can be attached to the top side of the boom. Also available is an insulated safety link that can be installed between the load hook and load attachment cables, or the line hook and sling, to provide protection for the hookup men. This equipment is no substitute for an alert, well trained operator. Sole reliance should not be placed in this equipment as in the event of failure the results could be deadly.

Equipment should not be located or operated so that slides, blasts, or the collapse of trenches or excavations can endanger people.

Steel erection

Steel erection involves extensive use of cranes, derricks, hoists, ropes, and slings. (See Chapters 25, 26, and 34, "Hoisting Apparatus and Conveyors," "Ropes, Chains, and Slings," and "Welding and Cutting" in this Manual.)

For lifting heavy loads, wire rope slings are preferable to chains. With either chain or wire rope, the manufacturer's capacity rating should not be exceeded. At points where rope slings pass around sharp corners, padding should be provided.

Eye protection should be provided workmen who are reaming or drilling, or driving wedges, shims, or pins. Containers should be provided for storing or carrying rivets, bolts, and driftpins.

Air tools. Pneumatic hand tools should be disconnected from the power source and pressure in hose lines should be released before any adjustments or repairs to the tools are made. Air hose sections should be tied together except when automatic cutoff couplers are used. If air hose must extend across a roadway, it should be protected so vehicles will not damage it.

Bolting. When bolts or driftpins are being knocked out, they should be kept from falling. Bolts, nuts, washers, and pins should never be thrown; rather, they should be placed in a bolt basket or other good container and raised or lowered by a line.

The use of high-tensile machine bolts or structural rib bolts is becoming increasingly popular for field assembly of structural steel, and in many cases replaces hot riveting, eliminating many of the old hazards. However, contractors must follow the manufacturer's instructions carefully for proper installation, adequate torque application, and prevention of nut back-off.

Impact wrenches should be provided with a locking device to retain the socket.

Riveting. Rivets should not be driven above flammable or combustible materials unless these materials are protected from hot rivets that might fall. When rivet heads are being backed out or knocked off, a shield should also be provided.

Air hammers should have a safety wire installed on the snap and handle. The wire size should be not less than No. 9 (B&S Wire Gage) leaving the handle, and annealed No. 14 on the snap.

Welding. Precautions to be taken when securing steel by welding are covered in Chapter 34, "Welding and Cutting," later in this Manual.

Drilling and reaming machines should be operated by two employees unless the handle is firmly secured to resist the torque created by the machine if the reaming or drilling bit should foul.

Plumbing-up. Hooks or lashings used for plumbing-up should be attached securely before stressing the turnbuckle. Once the turnbuckle is under stress, a device should be used to keep the turnbuckle from unwinding.

Plumbing-up guys should be placed so that the bolters, riveters, or welders can get at the connection points. Guys should not be removed without first getting permission from the job superintendent.

A definite set of directional signals must be established before starting to plumb a structure.

Connecting. When connectors are working together, one should give the signals. This person should make sure that his partner and others working on the job are in the clear. They should select positions that avoid their being struck by a swinging beam.

When connectors are working in pairs, one end of the piece shall be bolted before going out to connect the other end; only one connector should go out to fasten the other end. Whenever possible, an employee should straddle the beam instead of walking along the top.

Connect a beam with a minimum of two bolts at each end. If only one bolt is used, it should be in the top hole and pulled up tight with a hand wrench so that the beam will not start to roll when walked on. If connecting lungs are bent, send the piece back to the ground to be dressed properly.

When setting columns before lifting falls are unhitched, either draw down the nuts tightly on the anchor bolts or affix temporary guys. Never set a piece loose until the required maximum number of bolts have been installed; do not rely on a wrench or driftpin.

Work should be discontinued during rain, high wind, or weather of any sort that might increase the hazard to workmen or to others.

Erection under plant operation. Steel erection work under mill or plant operation is especially hazardous as there is usually much congestion from plant personnel and from employees and materials of other contractors. The operating plant supervisor should be responsible for the various phases of the work; he and others who act as coordinators should be identified to the steel erector and other contractors.

The work area should be clearly defined. The identification and location of gas lines, oxygen lines, electric utilities and electrified rails should be fully established. Responsibility should be established for preliminary work—closing passageways and cleaning grease or other material from crane runways, for example.

All existing mill or plant safety regulations should be observed. Areas should be restricted to operating personnel; others should enter only on permission from the superintendent or foreman.

The foreman should be responsible for obtaining clearance from the mill or plant supervisor for all phases of his work; he should not delegate this responsibility to a member of his crew.

Where electric wires are near the work, supervision should determine their voltage and set the necessary clearances, or deenergize them. All wires should be assumed to be "hot" until proven otherwise. Electric wires should not be cut or removed except by the maintenance department of the mill or plant. When persons must work near hot rails (and current cannot be turned off), they must be provided with adequate insulation and protection. It is best if the current can be turned off and the switches labeled and locked out.

No one should work in an operating crane runway until the supervisor has been notified and the foreman has obtained permission. When work is being done on or near crane runways that are operating, rail stops should be placed between the worker and the operating crane. If operating conditions do not permit such stops, a safety observer should be in the cab of the crane to protect the worker. Flasher lights are recommended to define clearly the work area and warn the crane op-

TABLE 17-A

FORCE OF WIND FOR GIVEN VELOCITIES

Miles per hour (V)	Feet per minute	Feet per second	Force in pounds per square foot $(0.004V^2)$	Description
1	88	1.47	0.004	Hardly perceptible
2	176	2.93	0.014 ⎤	Just perceptible
3	264	4.40	0.036 ⎦	
4	352	5.87	0.064 ⎤	Gentle breeze
5	440	7.33	0.1 ⎦	
10	880	14.67	0.4 ⎤	Pleasant breeze
15	1,320	22.0	0.9 ⎦	
20	1,760	26.6	1.6 ⎤	Brisk gale
25	2,200	29.3	2.5 ⎦	
30	2,640	44.0	3.6 ⎤	High wind
35	3,080	51.3	4.9 ⎦	
40	3,520	58.6	6.4 ⎤	Very high wind
45	3,960	66.0	8.1 ⎦	
50	4,400	73.3	10.0	Storm
60	5,280	88.0	14.4 ⎤	Great storm
70	6,160	102.7	19.6 ⎦	
80	7,040	117.3	25.6 ⎤	Hurricane
100	8,800	146.6	40.0 ⎦	

From Kidder-Parker. Architects and Builders Handbook.

erator. Do not permit loose items to remain on cranes or crane runway girders without fastening them.

For additional information, refer to American National Standard *Safety Requirements for Steel Erection,* A10.13. See References.

General safe practices. Suggested precautions that should be followed wherever practicable in steel erection work are:

1. Require proper protective equipment to be used, including safety hats.

2. Do not permit employees to ride loads, hooks, or "headache" balls.

3. Do not permit men to work underneath electric wires unless the wires are fully insulated or first disconnected.

4. Take precautions to keep workmen who are under the influence of liquor or in poor physical health from working on the job.

5. Do not allow men to work on freshly painted steel construction that is still wet, or is slippery for any other reason.

6. Have workmen wear goggles while cutting out rivets, chipping, and doing similar work. Keep adjacent areas clear of personnel or screen such operations.

7. When men must throw rivets across shaftways or toward the outside of a building, install screens of wire wesh or flameproof tarpaulins, to protect the public and other workmen.

8. Where it is impractical to provide temporary floors, suspend safety nets below points where men are working, or have the men wear safety belts. (See NSC Data Sheet 608, *Safety Nets for Construction Projects.*)

9. Where guy cables or braces are used to hold steel during erection, be sure they are guarded to prevent trucks or other equipment from hooking into them and pulling the steel down.

Fig. 17–2.—Wall bracing must be adequate for anticipated wind loads. This shows the result of not supplying bracing for a partially constructed wall.

Lateral bracing

Incomplete buildings for which designed lateral support is not yet in place, and all free-standing walls should be adequately braced against the maximum anticipated wind pressures.

Exterior masonry walls, especially of either load-bearing or nonload-bearing type whatever their height, are subject to wind loads beyond their designed capacity prior to the final set of the mortar or final tie into the structure. These wind loads have caused walls or sections of walls to break off and fall, causing both personal injury (sometimes fatal) and property damage. See Figure 17–2.

Masonry walls should follow the erection of permanently installed structural members so that adequate lateral stability is provided. If this is impossible or impractical, temporary bracing should be placed until structural members can be installed.

Usually the architect will include in his design pilasters as well as the requirements for anchors and ties. Also local building codes require that during erection of walls the proper bracing and supporting shall be provided. The question is, what is proper.

During construction of exterior masonry, two external forces should be considered—the weight (vertical) and the wind load (horizontal). Because the vertical load is supported by spandrels and relieving angles, the critical consideration is the horizontal load. Protection or bracing for this load must be provided either by screening or by simple shoring.

Codes and other engineering data indicate the thickness and height of walls which will withstand specific wind loads while unsupported. These specifications should be checked against the local wind conditions and bracing provided when required. (See R. W. Armstrong citation in "References.")

The National Building Code (1967 Edition) requires the following:

During erection, masonry walls shall not be built higher than 10 times their thickness unless adequately braced or until provision is made for the prompt installation of permanent bracing at

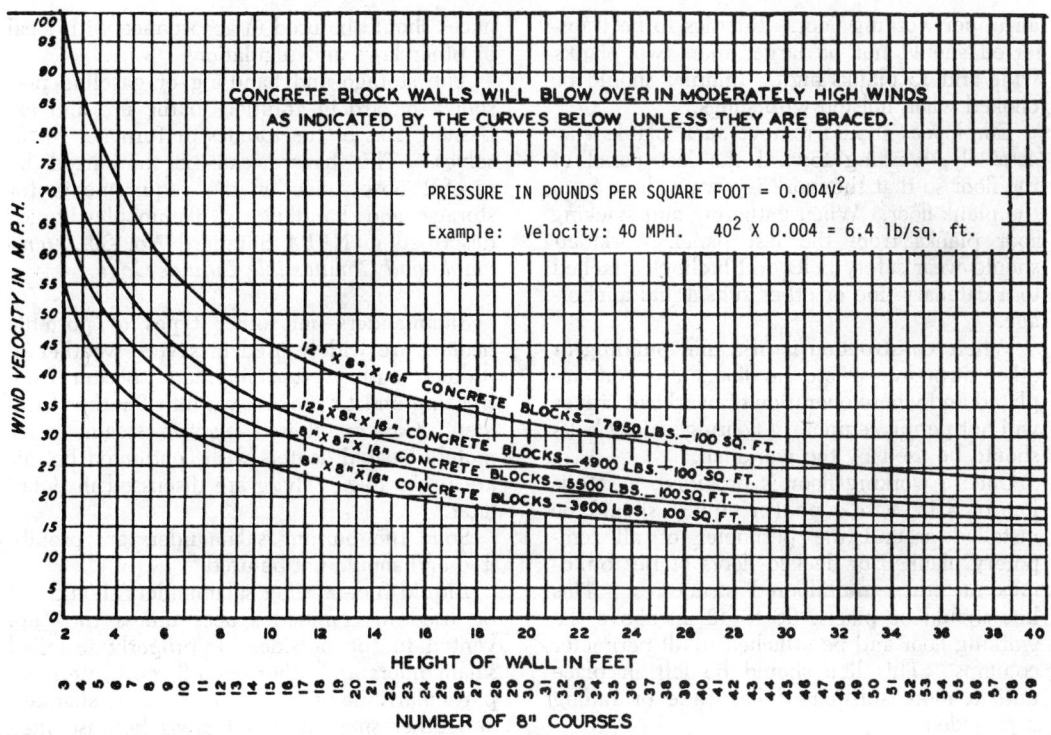

FIG. 17–3.—Wind velocities that can be withstood by concrete block walls of varying height.

Courtesy The Travelers Insurance Companies.

the floor or roof level immediately above the story under construction.

OSHA regulations state, "Masonry walls shall be temporarily shored and braced until the design level strength is reached to prevent collapse due to wind or other forces."

Table 17-A can be used to check bracing that will resist pressures developed by wind at different velocities. The graph shown in Fig. 17–3 shows wind velocities that can be withstood by concrete block walls of varying heights.

Temporary flooring

Where skeleton steel construction in tiered buildings is used, permanent floors should be installed as the erection of the steel progresses. There should not be more than eight stories beyond the erection floor and the uppermost permanent floor, except where the structural integrity is maintained as a result of the de-

sign. At no time, however, should there be more than four floors (or 48 ft) of unfinished bolting or welding above the foundation or uppermost permanently secured floor.

The derrick (or erection) floor should be solidly planked or decked over its entire surface, except for access openings. Planking, or decking of equal strength, should be of proper thickness to carry the working load, a minimum of 50 pounds per square foot (psf). Planking should be not less than 2 in. thick, full size undressed, and should be laid tight and secured to prevent movement, especially displacement by wind.

There should be a tight and substantial floor within two floors (or 25 ft, whichever is less) directly under the portion of each tier of beams on which bolting, riveting, or welding is being done, except when gathering and stacking temporary floor planks on a lower floor in preparation for taking these planks to an upper working floor. Bundles to be transferred should not be larger than two planks wide and 15 planks high. Bundles should be choked when hoisted with slings.

Employees should remove such planks successively, working toward the last panel of the floor so that the work is always done from the plank floor. When gathering and stacking floor planks from the last panel, workmen should wear safety belts and lifelines attached to a catenary line or other substantial anchorage.

Where construction is in a mill building or other structure where no floors are contemplated and where operation of overhead cranes will not permit temporary flooring, safety belts should be used by the workmen.

Once a working floor is provided, a safety line of ⅜ in. wire rope (or equal) should be installed around the perimeter of all temporary planked or decked floors of tier buildings or other multifloored structures. This line should be placed 36 to 42 in. above the working floor and be attached to all perimeter columns. This line should be left in place until a more substantial barricade or railing is provided.

Metal decking used in place of planks should be of sufficient strength, should be laid tight, and should be secured to prevent movement.

Planks should overlap the bearing ends by a minimum of 12 in. Wire mesh or exterior plywood should be used around columns where planks leave an unprotected gap and do not fit tightly. All unused floor openings should be planked over or barricaded until they are needed. Floor planks removed to perform work should be replaced as soon as possible or the open area should be guarded.

Torches and salamanders

Gasoline blowtorches and plumbers' furnaces involve the hazards of fire and explosion. Safer equipment should be used, if possible, such as electric soldering irons, electrically heated paint-remover irons, electric glue pots, and other electric heating devices. Proper gas or oil space heaters listed by Underwriters' Laboratories are recommended for use in areas containing combustibles.

Where flammable or explosive dust or vapors may be present, such torches or fur-

naces should be used in accordance with local or other laws and regulations.

The storage and handling of gasoline present a hazard in addition to the fire and explosion risks of the torches or furnaces themselves. This hazard can be minimized by careful observance of the requirements for storage and handling of flammable liquids described in NFPA Standard No. 30, *Flammable and Combustible Liquids Code.*

Salamanders and other types of portable heaters are widely used in severe weather to protect masonry, concrete, and plaster from freezing and to provide warmth for the workmen. Gas- or oil-fired, electric, steam, or remotely located heaters with conducted hot air are preferable. These are discussed in detail later.

Solid fuel-burning salamanders are prohibited and should not be used.

Liquid fuel-burning salamanders should not be used in confined spaces unless they are vented to the outside. Improperly installed salamanders and other open-flame heaters are particularly dangerous in tool sheds, shanties, and other small enclosed areas because they can give off large amounts of carbon monoxide, as well as consume much oxygen.

If the concentration of carbon monoxide (CO) is greater than 50 ppm at worker breathing levels, the heater should be turned off or additional ventilation provided. Tests for CO should be made about one hour after starting each shift and at least four hours later.

The horizontal clearance between salamanders and combustibles should be at least 2½ ft; overhead clearance should be at least 6 ft. Keep tarpaulins and canvas or plastic coverings at least 10 ft away. Make sure these are securely fastened to keep them from blowing toward the salamander. Use flameproofed materials.

To keep people away from their hot surfaces, salamanders should be surrounded by a noncombustible railing at least 18 in. away. When in use, heaters should be horizontally level, unless otherwise permitted by manufacturer's instructions.

Temporary heating equipment

When using temporary heating equipment,

assign a qualified employee to its operation and maintenance. Be sure to follow all instructions of the manufacturer.

Each time the heater is placed in operation, check it to make sure it is functioning properly; operation should also be checked periodically when the heater is in use. Heaters should be equipped with an automatic flame-loss device to stop fuel flow if the flame is extinguished.

Thermostatically controlled heaters should be identified by a warning label advising that the unit may start up at any time.

Fueling. Check and follow manufacturer's instructions as well as applicable regulations. Turn off all flames, including the pilot (if any). Use only the type of fuel specified for the unit. The unit should be cool to the touch.

During fueling, check all fuel lines, hoses, and connections for leaks.

Only one day's supply of fuel should be stored inside a building in the vicinity of the heater; this should be at least 25 ft away from a source of ignition. General fuel storage should be outside the structure.

Fan-assisted units. Use only heaters that are designed so that a power failure (or the failure of any electrical components) will not create a fire or electrical shock. Only power supply circuits with three- or four-wire grounding should be used. Grounding continuity should be provided to all parts of the heater unit, including connection to a grounded power supply.

Natural gas-fueled heaters. All piping, tubing, or hose should be leak-tested after assembly and proven free of leaks at normal operating pressure by use of soapsuds or other noncombustible means. Never use a flame.

When placing a unit in operation, make sure it is working properly. Before disconnecting a heater, shut off fuel supply at the source in order to purge the line.

A flexible gas supply line should be no longer than 25 ft, and shorter if possible. Check all hoses and fittings to make sure they are designed for the pressure and capacity and type of fuel being used. Hoses should have a minimum working pressure of 250 psig

and a minimum burst pressure of 1250 psig. All hose connectors should be capable of withstanding a test pressure of 125 psig without leaking, and a pull test of 400 lb. Hoses should be securely connected to the heater by mechanical means; never use "slip-end" connectors.

Protect hose and fittings from damage and check for deterioration. Hoses should not be allowed to contact surfaces above 125 F (50 C). Hoses should be placed to minimize any physical damage.

Normal maintenance includes inspection of the hose supply system for cracks, checks, abrasions, and rupture; and leak testing the hose, pipe, or tubing connections. Disconnect electric power supply before repairing heaters.

Liquefied petroleum gas (LP-gas) heaters. Follow the precautions for carbon monoxide concentration, testing operating capability, and leak testing as outlined earlier.

Use only hose labeled "LP-Gas" or "LPG." Minimum working pressure should be 250 psig and minimum burst pressure, 1250 psig. The hose should be at least 10 ft long, but no longer than 25 ft. All hose should be protected from damage, deterioration, and hot surfaces.

Hose connectors should be capable of withstanding a test pressure of 500 psig without leaking, and a pull test of 400 lb. Hoses should be securely connected to the heater by mechanical means; never use "slip-end" connectors and ring keepers tightened over the hose to give increased force to the metal fitting.

Heaters should be equipped with an approved regulator in the supply line between the fuel cylinder and the heater unit. Cylinder connectors should be provided with an excess flow check valve to minimize the flow of gas in the event of a fuel line rupture.

The maximum water capacity of individual containers is 245 lb (nominal 100 lb LP-gas capacity). For temporary heating, such as in concrete curing, heaters should be located at least 6 ft away from LP-gas containers. This does not, however, prohibit the use of heaters specifically designed for attachment to the container or to a supporting standard with connecting hose less than 6 ft, provided the design and installation prevents the direct application of radiant heat on to the container. Blower type or radiant heaters should not be directed toward any LP-gas container within 20 ft. If two or more heater-container units (of either the integral or nonintegrated type) are located in an unpartitioned area on the same floor, the container(s) of each unit should be separated by at least 20 ft. The total water capacity of containers manifolded together in an unpartitioned area should not be greater than 735 lb (nominal 300 lb LP-gas capacity). These containers should also be separated by at least 20 ft.

On floors on which heaters are not connected together for use, containers may be manifolded together for connection to heaters on another floor, provided that (a) the total water capacity of the containers connected to any one manifold is not greater than 2450 lb (nominal 1000 lb LP-gas capacity), and (b) where more than one manifold having a total water capacity greater than 735 lb (nominal 300 lb LP-gas capacity) is located in the same unpartitioned area, the manifolds should be separated by at least 50 ft.

LP-gas cylinders should not be refilled inside buildings or structures. Cylinders should be stored outside of buildings, should stand on a firm and substantially level surface, and should be secured in an upright position.

Demolition of structures

Only minor demolition should be done by plant personnel. Specialists in the field should be employed if structures to be removed look as though they will present a problem, such as being over 25 ft high, being of steel or reinforced concrete construction, or adjoining other structures that are not to be damaged.

Wrecking specialists are familiar with the procedures and precautions necessary to do the work safely, to protect the public and adjacent property, and to comply with applicable federal, state, and municipal codes and regulations.

Following are some fundamental safety procedures and suggestions for minor demolition operations to guide the inexperienced in safe removal of unwanted structures:

1. Make provision to keep the public and unauthorized plant employees at least 15 ft away from the structure.

2. Make an engineering survey, by a competent person, of the structure. This is to determine the condition of the framing, floors and walls and check for any unanticipated conditions. Check for hazardous chemicals, gases, explosives or flammable materials.

3. Disconnect utility services (gas, steam, electricity) outside the building. Maintain water lines as long as possible. Install a temporary water source for fire protection and for wetting down the site to reduce dust.

4. Remove all glass doors and windows throughout the structure.

5. Strip off lath and plaster to eliminate excessive dust during succeeding operations.

6. Remove chimneys and extensions of walls above the roof down to roof level while working from the roof.

7. Remove the roof.

8. Remove walls by picking them apart, either by machine or by hand tools. Work from scaffolds supported independently outside of the walls.

9. Remove all debris promptly, using chutes or thru internal holes.

10. Avoid subjecting walls to lateral pressure from stored material or to lateral impact from falling material.

11. Barricade any area where material is being dumped, and place barricades where necessary to protect workmen from flying pieces.

12. Permit no employee to work below others.

13. Require safety hats, goggles, safety shoes, respirators, and gloves if needed for all workers. (See Chapter 19, "Personal Protective Equipment.")

When the conventional methods for altering or removing concrete installations are infeasible or undesirable, sometimes "powder cutting," a process that substitutes penetration by intense heat for concussion breakage, can be employed, or demolition by explosives may also be utilized. For a description of powder cutting, see Chapter 34, "Welding and Cutting," in this Manual.

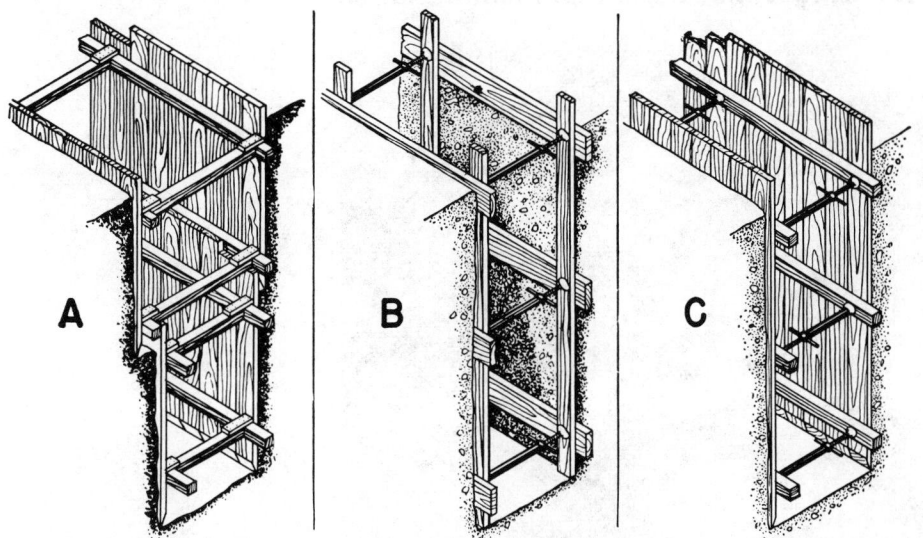

FIG. 17–4.—Trench bracing: **A**—bracing used with two lengths of sheet piling: **B**—bracing with screw jacks in hard soil; and **C**—screw jacks used with complete sheet piling.

Excavation

To prevent injury and property damage, during excavation work, make adequate protective measures part of the job. Study pre-excavation conditions (superimposed loads, soil structure, hydrostatic pressure, and the like) in order to evaluate changes that might occur, or situations that might develop, and in order to plan the job ahead, based on these findings.

A major hazard in urban or built-up areas is the presence of underground facilities, such as utility lines (water, electricity, gas, or telephone), tanks, process piping, and sewers. If these are dug into, undercut, or damaged in any way, there may be injury or death to workmen, and interruption of service, contamination of water, disruption of processes, and expensive delays.

Before starting operations, it is important that the company or plant engineer, and utility and city or town engineers be consulted, and that the location of various facilities and their approximate depth below ground be determined and marked by stakes in the ground or by markings on the floor.

Electronic locators can be especially helpful where an excavation would cross numerous buried obstacles. If the facilities are to be left in place, they must be protected against damage, and sometimes also against freezing.

Contents of buried tanks and piping should be indicated on the location markings. If the contents are flammable or toxic, proper protective equipment should be readily available in case of rupture. The bottom depth of the tank should also be indicated.

Machine excavation

Before any attempt is made to excavate, check public utilities for locations of underground pipe or electric lines.

No shovel, dragline, or other digging machine should be allowed to excavate close to underground facilities that must be left in place. Establish a proximity limit for machine operations and complete the excavation by hand digging. According to OSHA, if the excavation is deeper than 5 ft, adequate bracing and shoring must be provided or sloping of the trench must be done.

When hand excavation is being done, workmen must be warned about driving picks, paving breakers, or other powered tools through the envelope of buried facilities. Attaching the tool air hose to a driven ground

411

Fig. 17–5.—Sliding trench shield is moved forward by back-hoe. Bottom and top sections are joined by large hinge pins.

Courtesy Safer Oregon, Oregon State Industrial Accident Commission.

will give the workman protection in the event of sudden contact with an underground electric line. The air hose should be of the conductive type.

Open excavation

Whenever an open excavation must be made within or adjacent to a building and lower than wall or column footings and machinery or equipment foundations, the job should be given to a specialty contractor who has qualified personnel to make a thorough study to determine the amount and strength of shoring required before work on an open excavation is begun. Such a study will include nature of the soil, hydrostatic pressure, superimposed loads (both static and live), and other factors. The depth and location of the excavation and the other characteristics will determine the need for sheet piling, shoring, and bracing, which should be designed by an engineer or other person with experience in this type of design.

If underpinning (deeper support under an existing column, wall, or machine) is necessary, it should be done before the open excavation is carried down to final grade.

Material excavated by machine should be thrown at least 24 in. from the edge of the excavation but not into aisles or work areas in the plant. Tarpaulins, sheeted barricades, or low built-up board barricades can be used to confine the excavated material to the immediate area under construction. Excavated material should not be permitted to accumulate in a busy work area, but should be trucked or otherwise removed from the building.

Pick-and-shovel men working in excavations should be kept far enough apart to prevent injury to one another. Excavated material should be placed at least 24 in from the side of the excavation, unless toeboards or other effective barricades have been installed to prevent fallback.

Excavations should be barricaded to prevent employees and others from falling into them. When an excavation must remain open for the duration of the construction work, barricades, fences, horses, and warning signs are necessary. In some cases, watchmen and flagmen may be needed. The work area should be guarded by flares, lanterns, or flashing lights at night. (For more information, see NSC Data Sheet 482, *General Excavation*.)

Trench excavation

A trench 4 ft or more deep should be provided with ladders to facilitate safe entrance and exit. The ladders should be so spaced that no worker in the trench will ever be more than 25 ft from one of them. The ladders should extend from the bottom of the trench to at least 3 ft above the surface of the ground.

It is recommended that the side of trenches more than 4 ft deep be shored unless they are sloped to the angle of repose or unless the trench is in solid rock. Shoring should be adequate to prevent trench wall collapse in whatever soil condition encountered. See Fig. 17–4 for three trench-bracing methods.

In hand-excavated trenches, wooden cleats should be spiked or bolted to join the ends of braces to stringers to prevent the braces from being knocked out of place. (For more detail, see NSC Data Sheet 254, *Trench Excavation*.)

In a long machine-excavated trench, a sliding trench shield may be used instead of shoring. Sliding trench shields generally are custom made to size for a specific job. They must be designed and fabricated strong enough to withstand the pressures that will be encountered. Fig. 17–5 illustrates a well made sliding trench shield.

Also available are metal, portable hydraulic shoring systems that are light, quick and inexpensive.

Ladders

Construction of all ladders should conform to the provisions of the applicable ladder or safety code of the locality or the state, whichever is more restrictive. Special-use climbing equipment, such as a combination stepladder-work platform, should comply with the applicable codes.

For complete ladder information consult the American National Standards Institute publications: *Safety Requirements for Portable Wood Ladders*, A14.1; *Safety Requirements for Portable Metal Ladders*, A14.2; *Safety Requirements for Fixed Ladders*, A14.3; and *Safety Requirements for Job-Made Ladders*, A14.4. Also see NSC Data Sheets 568, *Job-Made Ladders*, and 606, *Fixed Ladders and Climbing Devices.*

The fixed ladder safety code permits use of safety devices, such as those illustrated in Fig. 17–6, in lieu of cage guards on tower, water tank, and chimney ladders over 20 ft in unbroken length. Platforms are not required when such devices are used.

Devices of this type allow a climber to attach his safety belt to a sliding fixture that travels along a carrier rail or cable anchored to the ladder. The traveling fixture will lock and suspend a man if he should lose his grasp. Many safety professionals consider such devices preferable to cage guards, where allowed by the code. (See NSC Data Sheet 606, *Fixed Ladders and Climbing Devices*.) A variation is a ladder belt with one or two snap hooks attached. This is not primarily a climbing safety device but provides a means of securing a climber to a ladder and freeing his hands if he is to work from a ladder.

Nonslip bases and safety tops

It is recommended that all portable ladders be equipped with nonslip bases or that the bottoms of the ladders be held, tied, or securely anchored to prevent slipping.

Some companies replace the top rung on a portable ladder with a chain for work on cylindrical objects like poles and round col-

FIG. 17-6.—Fixed ladder safety device.

Drawing courtesy Safety Tower Ladder Co.
Photograph courtesy Meyer Industries.

CARRIER
RAIL

umns. Such an arrangement will help keep the ladder from slipping sideways. Other companies use a rope lashing to tie the top of the ladder to the pipeline or other object being worked on.

Ladder maintenance

Inspection. After receipt, ladders should be inspected promptly for conformity to purchase order specifications and applicable codes (already mentioned).

An inspection program should assure that all ladders are inspected once every three months. An accurate record of each inspection should be kept. A routine inspection form is shown in Fig. 17-7.

Coating. The *Safety Code for Portable Wood Ladders*, ANSI A14.1, states that "ladders should be kept coated with a suitable protective material. The painting of ladders is satisfactory providing the ladders are carefully inspected prior to painting by competent and experienced inspectors acting for and responsible to the purchaser, and the ladders are not for resale."

One large company has this policy:

All ladders upon receipt from vendor should be delivered to the paint shop where, after approval by the purchasing department inspector, they should be given a treatment of water-repellant preservative.

At the discretion of the department using the ladders, they may or may not then be coated with paint, varnish or enamel.

If initial inspection shows that ladders are free from such defects as knots, cross grain, compression wood, and pitch pockets, these defects will not develop after the ladders are coated or painted. Checks, cracks, splits, and compression failures that may occur subsequently can ordinarily be detected through a transparent coating such as clear varnish, shellac, or other clear preservative.

On ladders coated with ordinary paint, varnish, and other commonly used coatings, moisture may enter the wood by capillary attraction at the uncoated joints of rung holes and tenons and cannot leave the wood because of the more or less impervious coating. Where such joints are used, the protective coating may favor rather than retard decay.

After a ladder has been given a thorough visual inspection, a suggested method of treatment is to heat the ladder in an oven or small room and then immerse it in a cool solution of NSP preservative, a nonswelling and paintable solvent. The nonaqueous volatile solvent carries a toxic chemical and may be used with or without other water-repelling ingredients. Use suction hoods and ventilation.

Markings. Each ladder should be marked with the name of the department to which it belongs. Some companies number their ladders consecutively so that none will be overlooked during inspection, while others stencil the date on each ladder as it is put into service. Proper identification assists in inspection procedures and also in storage. (Also

LADDER INSPECTION CHECKLIST

General	Item To Be Checked	Needs Repair	Condition O.K.
	Loose steps or rungs (considered loose if they can be moved at all with the hand)	☐	☐
	Loose nails, screws, bolts, or other metal parts	☐	☐
	Cracked, split, or broken uprights, braces, steps, or rungs	☐	☐
	Slivers on uprights, rungs, or steps	☐	☐
	Damaged or worn nonslip bases	☐	☐

Stepladders

		Needs Repair	Condition O.K.
	Wobbly (from side strain)	☐	☐
	Loose or bent hinge spreaders	☐	☐
	Stop on hinge spreaders broken	☐	☐
	Broken, split, or worn steps	☐	☐
	Loose hinges	☐	☐

Extension Ladders

		Needs Repair	Condition O.K.
	Loose, broken, or missing extension locks	☐	☐
	Defective locks that do not seat properly when the ladder is extended	☐	☐
	Deterioration of rope, from exposure to acid or other destructive agents	☐	☐

Trolley Ladders

		Needs Repair	Condition O.K.
	Worn or missing tires	☐	☐
	Wheels that bind	☐	☐
	Floor wheel brackets broken or loose	☐	☐
	Floor wheels and brackets missing	☐	☐
	Ladders binding in guides	☐	☐
	Ladder and rail stops broken, loose, or missing	☐	☐
	Rail supports broken or section of rail missing	☐	☐
	Trolley wheels out of adjustment	☐	☐

Trestle Ladders

		Needs Repair	Condition O.K.
	Loose hinges	☐	☐
	Wobbly	☐	☐
	Loose or bent hinge spreaders	☐	☐
	Stop on hinge spreader broken	☐	☐
	Center section guide for extension out of alignment	☐	☐
	Defective locks for extension	☐	☐

Sectional Ladders

		Needs Repair	Condition O.K.
	Worn or loose metal parts	☐	☐
	Wobbly	☐	☐

Fixed Ladders

		Needs Repair	Condition O.K.
	Loose, worn, or damaged rungs or side rails	☐	☐
	Damaged or corroded parts of cage	☐	☐
	Corroded bolts and rivet heads on inside of metal stacks	☐	☐
	Damaged or corroded handrails or brackets on platforms	☐	☐
	Weakened or damaged rungs on brick or concrete slabs	☐	☐
	Base of ladder obstructed	☐	☐

Fire Ladders

		Needs Repair	Condition O.K.
	Markings illegible	☐	☐
	Improperly stored	☐	☐
	Storage obstructed	☐	☐

Fig. 17–7.

see the paragraph on Electrical hazards, under Use of Ladders.)

Storage. Ladders should be stored where they will not be exposed to the weather and where there is good ventilation. They should not be stored near radiators, stoves, or steam pipes or in other places subjected to excessive heat or dampness.

Ladders can be hung on brackets against a wall, with more than two supports for long ladders to prevent warping, or placed on edge on racks or on rollers, rather than stored flat. These methods will facilitate removal of ladders.

Ladder storage space should be kept free of obstructions and accessible at all times, so that ladders can be obtained quickly in case of emergency.

Use of ladders

Placement. Workmen should observe the following practices when placing ladders:

1. Place a ladder so that the horizontal distance from the base to the vertical plane of the support is approximately one-fourth the ladder length between supports. (For example, place a 12 ft ladder so that the bottom is 3 ft away from the object against which the top is leaning.) See Fig. 17–8.

2. Do not use ladders in a horizontal position as runways or as scaffolds. Single and extension ladders are designed for use in a nearly vertical position and cannot be safely used in a horizontal position or with the base at a greater distance from the support than that indicated in the preceding paragraph.

3. Never place a ladder in front of a door that opens toward the ladder unless the door is locked, blocked, or guarded.

4. Do not place a ladder against a window pane or sash. Securely fasten a board (not with nails) across the top of the ladder to give a bearing at each side of the window. On wide windows with metal sash, the bearing may be across the mullions or between window jambs.

5. Place a portable ladder so that both side rails have secure footing. Provide solid

footing on soft ground to prevent the ladder from sinking.

6. Place the ladder feet on a substantial and level base, not on movable objects.

7. Never lean a ladder against unsafe backing, such as loose boxes or barrels.

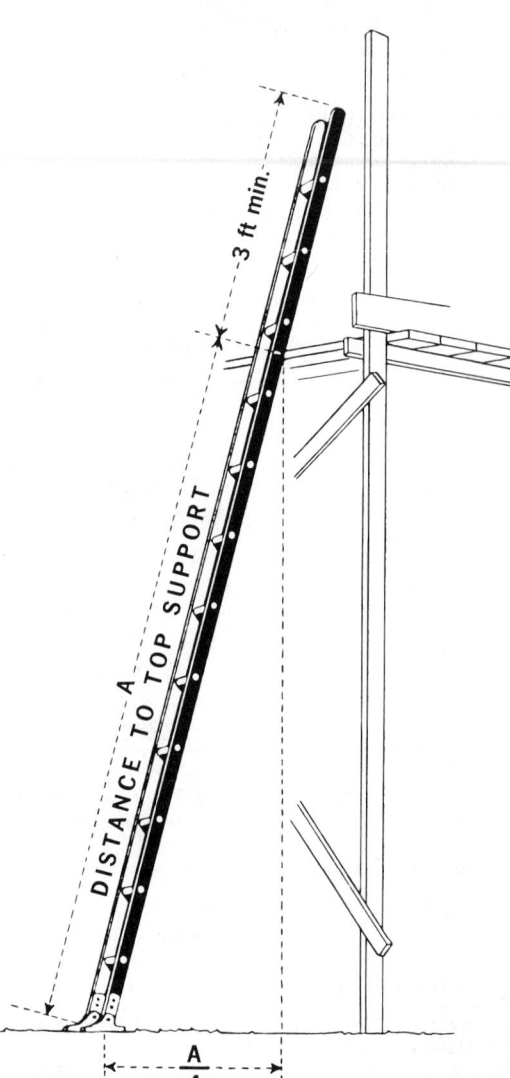

FIG. 17–8.—Safe procedure in setting up a ladder. The base should be one-fourth the ladder length from the vertical plane of the top support. Where the rails extend above the top landing, ladder length to the top support only is considered.

8. When you use a ladder for access to high places, securely lash or otherwise fasten the ladder to prevent its slipping.

9. Secure both bottom and top to prevent displacement when using a ladder for access to a scaffold.

10. Extend the ladder side rails at least 3 ft above the top landing.

11. Do not place a ladder close to live electric wiring or against any operational piping (acid, chemical, sprinkler system, etc.) where damage may be done.

Ascending or descending ladders. Workers should follow these safe practices when ascending or descending ladders:

1. Hold on with both hands when going up or down. If material must be handled, raise or lower it with a rope either before going down or after climbing to the desired level.

2. Always face the ladder when ascending or descending.

3. Never slide down a ladder.

4. Be sure that your shoes are not greasy, muddy, or slippery before you climb.

5. Do not climb higher than the third rung from the top on straight or extension ladders or the second tread from the top on stepladders.

Other safe practices. All should observe these additional safe practices when using ladders:

1. Do not use makeshift ladders, such as cleats fastened across a single rail.

2. Be sure that a stepladder is fully open and the divider locked before you start to climb it.

3. Before using a ladder, inspect it for defects.

4. Never use a defective ladder. Tag or mark it so that it will be repaired or destroyed (Fig. 17–9).

5. If a ladder is to be discarded, cut it in half immediately to prevent reuse.

6. Do not splice short ladders together. They are designed for use in their original

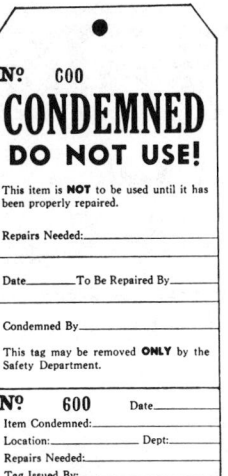

Fig. 17–9.—"Condemned—Do Not Use" tag shows that a ladder or other piece of equipment should not be used until repaired.

lengths and are not strong enough for use in greater lengths. Also, most splicing methods, particularly "on-the-job methods," are not safe.

7. Keep ladders clean and free from dirt and grease, which might conceal defects.

8. Do not use ladders during a strong wind except in emergency, and then only when they are securely tied.

9. Do not leave placed ladders unattended, especially outdoors, unless they are anchored at top and bottom.

Electrical hazards and metal ladders. Since metal ladders are electrical conductors, they should not be used around electrical circuits or in places where they may come in contact with such circuits. The importance of these electrical hazards cannot be overemphasized, and those using metal ladders should be warned of the danger.

In addition to this warning, metal ladders should be marked with signs or decals reading CAUTION—DO NOT USE NEAR ELECTRICAL EQUIPMENT. These decals may be placed on the inside of the side rails at about eye-level from the bottom of the ladder. (See Fig. 17–10.) Glass fiber ladders, as well as wood,

FIG. 17–10.—Place decals (like this one) on the side rails of metal ladders.

should be considered for use near electrical hazards.

Scaffolds

A scaffold is an elevated working platform for supporting both men and materials. It is usually temporary, its main use being in construction work. Scaffolds should be designed to support at least four times the anticipated weight of men and materials which will use them.

Scaffolding is the structure (made of wood or metal) that supports the working platform.

Railings and toeboards

All temporary elevated working platforms (especially those more than 10 ft from the ground or floor) should be guarded on all exposed sides. This should consist of top rail, midrail, and toeboard.

Where scaffolds are erected above walks or work areas, the space between toeboard and railing should be screened.

Railings and toeboards should conform to the provisions of ANSI A10.8.

Overhead protection

Whenever work is being done over men who are working on a scaffold, overhead protection should be provided on the scaffold. This protection should be not more than 9 ft above the working platform and should be of planking or other suitable material.

Means of access

A safe and convenient means of access should be provided to the platform level.

This requirement does not apply to swinging scaffolds or those with convenient access from adjacent floors. Means of access may be a portable ladder, fixed ladder, ramp or runway, or stairway.

Ladders used for access to scaffolds should conform to the requirements of the applicable ladder code. (See the previous section. Also see Chapter 16 for information concerning runways and ramps.)

Types of scaffolds

Most scaffolds fall into one of three primary categories: tubular, suspended, and rolling. They are also classified according to their intended use: light duty (working load must not exceed 25 lb/sq ft for platform surface), medium duty (50 lb/sq ft), and heavy duty (75 lb/sq ft).

- **Tubular scaffolds** are an assembly consist-

FIG. 17–11.—Folding sections with fixed platforms, guard rails, and built-in stairs that swing down and snap into place are features of this built up type rolling scaffold. Casters are adjustable and have locks.

Courtesy Up-right Scaffolds, Inc.

Fig. 17–12.—Powered suspension scaffolds are the backbone of the trades in high-rise construction and maintenance. Note guardrails and wire rope locking device.

Courtesy Sky Climber, Inc., Gardenia, Ga.

ing of tubing which serves as posts, bearers, braces, ties, and runners, a base supporting the posts and special couplers which serve to connect the uprights and to join the various members. (Fig. 17–11).

- **Suspended scaffolds** carry the working platform on beams and ropes secured to structural members or to thrust-outs from the structure (Fig. 17–12).

- **Mobile (rolling) scaffolds** are castor-mounted sections of tubular metal scaffolding, or are made of components specifically made for the purpose (Fig. 17–13).

Because tubular metal scaffolding is readily available, versatile, adaptable to all scaffolding problems, and economical to use, it is generally used. Most tubular metal scaffold-

ing manufacturers and suppliers provide engineering service to help in the design of adequate scaffolding for any situation. Many suppliers also furnish erection and dismantling service.

Following are some common-sense rules for erecting, disassembling, and using steel scaffolding, recommended by the Steel Scaffolding and Shoring Institute.

General

1. Post scaffolding safety rules in a conspicuous place and make sure people follow them.

2. Abide by all state, local, and federal codes, ordinances, and regulations.

3. Inspect all equipment before use. Never use equipment that is damaged or de-

419

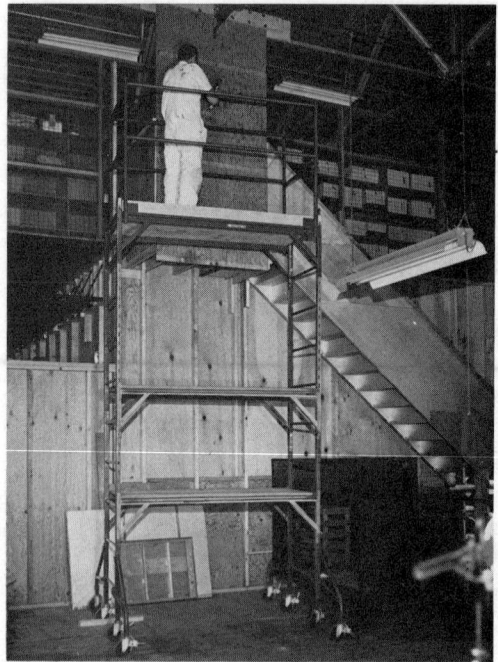

Fig. 17–13.—Mobile (rolling) scaffold has guard-rails and locking wheels.

Courtesy Baker-Roos Div., Hugh J. Baker & Co., Indianapolis, Ind.

teriorated in any way.

4. Keep equipment in good repair. Avoid using rusted equipment; its strength is not known.

5. Inspect erected scaffolds regularly to be sure they are maintained in safe condition.

6. Consult the scaffolding supplier when in doubt. Never take chances.

SPECIFIC

1. Provide adequate sills for scaffold posts and use base plates.

2. Use adjusting screws instead of blocking to adjust to uneven grades.

3. Plumb and level scaffolds as erection proceeds, so that braces will fit without forcing.

4. Fasten all braces securely.

5. Do not climb cross braces.

6. Anchor wall scaffolds securely between structure and scaffold, at least every 30 ft of length and 26 ft of height.

7. Restrain free-standing scaffold towers from tipping by guying or other means.

8. Equip all planked or staged areas with proper guard rails and toeboards.

9. Use caution when working near power lines. Consult the power service company.

10. Do not use ladders or makeshift devices on top of scaffolds to increase the height.

11. Do not overload scaffolds.

12. Follow these suggestions for using planking:
    a) Use only properly graded and inspected lumber.
    b) Have at least 12 in. overlap and 6 in. extension beyond centerline of support, or cleat both ends to prevent sliding.
    c) Do not allow unsupported ends of planking to extend an unsafe distance beyond supports.
    d) Secure plank to scaffold when necessary.

13. Seat all brackets correctly—side brackets should be parallel to the frames and end brackets at 90 degrees to the frames. Brackets must not be bent or twisted from normal position.

14. Use and install scaffolding accessories in accordance with manufacturer's recommended procedure. Do not alter accessories in the field.

MOBILE SCAFFOLDS

1. Do not ride rolling scaffolds.

2. Remove all material and equipment from platform before moving scaffold.

3. Apply castor brakes at all times when a scaffold is not being moved.

4. Attach castors with plain stems to the panel or adjustment screw by pins or other suitable means.

5. Do not try to move a rolling scaffold without sufficient help. Watch out for holes in the floor and for overhead obstructions.

6. Do not extend adjusting screws more than 12 in.

7. Use horizontal bracing near the bottom,

TABLE 17-B

SAFE CENTER LOADS FOR PLANKS (POUNDS)

Douglas fir (Rocky Mountain region), Sitka spruce, white spruce, red pine and Port Orford or white cedar. Based on extreme fiber stress of 1,300 psi

Size of Plank in Inches

Span in feet	2x8 Dressed to 1⅝x7½	2x10 Dressed to 1⅝x9½	2x12 Dressed to 1⅝x11½	3x8 Dressed to 2⅝x7½	3x10 Dressed to 2⅝x9½	3x12 Dressed to 2⅝x11½
6	230	290	355	610	775	935
8	170	210	260	450	570	690
10	130	165	200	355	450	540
12	105	130	160	285	365	440
14	80	105	130	235	305	365
16	70	90	105	200	255	310

SPECIFICATIONS AND NOTES:

1. For other species of woods, determine safe center loads by multiplying tabular load above by constant indicated for each group.

Species	Extreme Fiber Stress (psi)	Constant
a. Balsam fir, ponderosa pine, sugar pine and Western white fir	1,050	0.80
b. Tamarack	1,400	1.07
c. White ash, red oak, white oak and bald cypress	1,500	1.15
d. Douglas fir (Coast region) and Southern yellow pine	1,750	1.34
e. Dense Douglas fir and dense Southern yellow pine	2,050	1.57

2. Above table applies to planks surfaced 4 sides, seasoned to moisture content between 15 and 8 per cent, free from shakes, decay, or other defects. Slope of grain not steeper than 1 in 15. Knots in edge not larger than ⅛ of plant thickness. Knots in center of wide face not larger than ¼ of plank width, with permissible size decreasing as they approach edge to ⅛ of plank width at edge. Sum of sizes of all knots in wide face in length equal to plank width not to exceed ¼ of plank width. A better grade planking could hold up to 25 per cent more load.
3. Values in table apply to material unlikely to be wet except for brief periods. If wetness for prolonged periods is anticipated, loads should be reduced.
4. Table shows total weight of workmen, materials, and equipment that may be placed on the span. Allowance for weight of planks, based on average weight of 30 pounds per cubic foot, has been made.

Table based on data supplied by Forest Products Laboratory

---

top, and at intermediate levels of 20 ft.

8. Do not use brackets on rolling scaffolds without first considering the overturning effect.
9. Do not let the working platform height exceed four times the smallest base dimension, unless guyed or otherwise stabilized.

PUTLOGS AND TRUSSES

1. Do not cantilever or extend putlogs/ trusses as side brackets, without thoroughly considering the loads to be applied.
2. Make sure putlogs/trusses extend at least 6 in. beyond point of support.
3. Place proper bracing between putlogs/ trusses where the span of putlogs/trusses is more than 12 in.

It is also recommended that the manufacturer or supplier be consulted whenever use of medium- or heavy-duty suspended scaffolding is contemplated.

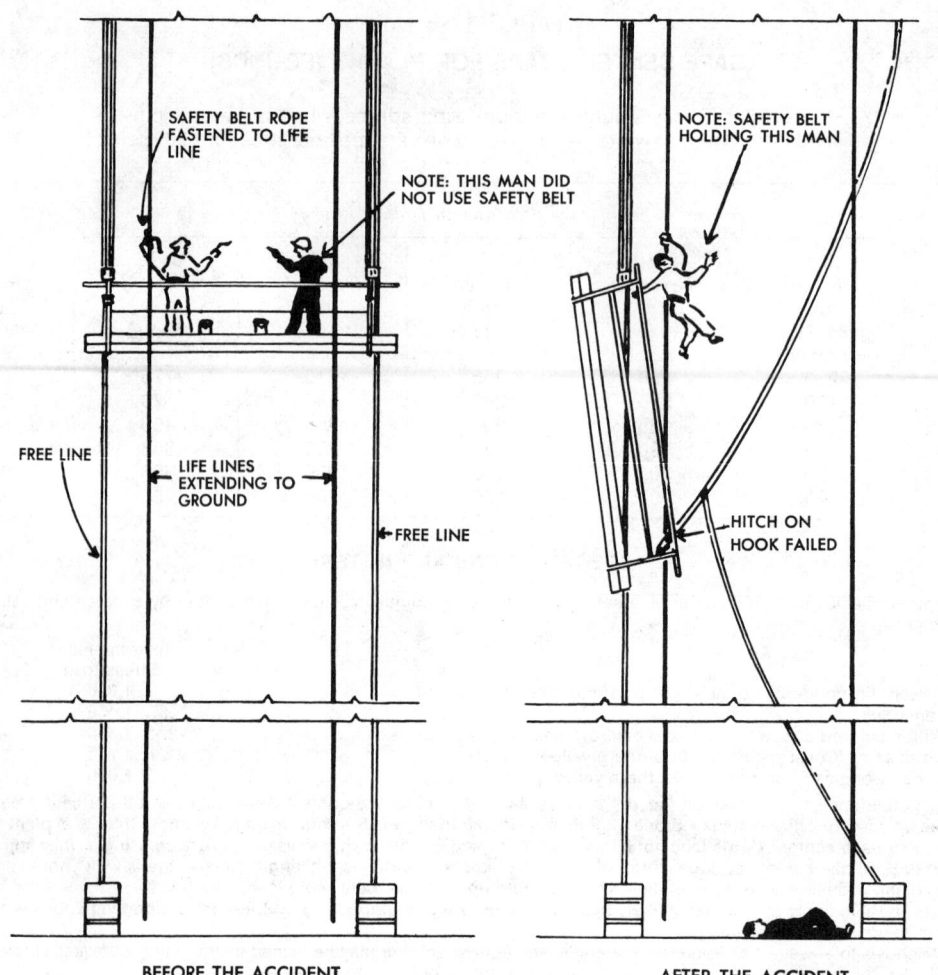

FIG. 17–14.—Swing stage scaffold failure. Each man on a swinging scaffold should wear a safety belt tied off to a separate lifeline. Drawing at right shows accident to man who was not wearing safety belt.

Wooden scaffolds

Built up and suspended scaffolds may be made of wooden structural members, but, as previously suggested, steel tubing should be the first choice.

For details on construction and design of wooden scaffolds, refer to American National Standard *Safety Requirements for Scaffolding,* A10.8.

Working platforms

The working platform of all built up, sus-

pended, and rolling scaffolds (except those manufactured with a platform provided) should be solidly planked.

The platforms should be provided with guard rails and toeboards, and overhead protection, when needed. (See the discussions earlier in this chapter.)

Working platform planking should be 2 x 10's with one-foot overlaps on bearer. Loads that planks can carry depend on the size of the planks, the type of wood from which they are made, and the span between supports.

Table 17-B gives safe center loads for planks of various recommended species and sizes of wood.

Testing scaffold planks. Scaffold planks should not be proof-tested with loads several times as great as they may be expected to carry in service because this method may result in concealed or unrecognized damage that may subsequently cause failure.

To proof-test scaffold planks, the following procedure is recommended:

1. Examine the plank for large knots, excessive grain slope, shakes, decay, and other defects that may render it unfit.

2. Determine the safe load for a plank of its size and species. (See Table 17-B.)

3. Place the plank on supports spaced the same as the anticipated span and about 1 ft. off the ground; then place twice the tabular load on it (as near the center as possible) and leave this load for not more than five minutes. Do not jump on the plank.

4. Discard the plank upon visible or audible evidence of failure or if obvious deflection remains after the load is removed.

5. Clearly and permanently mark the top side of a plank which has passed this test and use it with that side upward.

6. Do not load the plank in excess of the tabular safe load after it has been tested.

Swinging scaffolds

Swinging scaffolds (stages) usually are factory built. The three types of platforms in common use are ladder, plank, and beam. They are designed for light duty, primarily for men using hand tools.

Ropes supporting swinging scaffolds should be comparable to first grade manila fiber, not less than ¾ in. diameter. They should be properly rigged into a set of standard 6-in. blocks, consisting of at least one double and one single block.

If the scaffold is to be used for acid cleaning or similar purpose, wire rope not less than ⁵⁄₁₆ in. diameter should be substituted for manila rope. Where wire rope is used, a hoisting mechanism must be provided on the end of the scaffold platform.

Swinging scaffolds must be hung securely from eaves, cornices, or other reliable supports with properly placed hooks of sufficient strength to provide a safety factor of 4.

Anchorages should be carefully inspected before the hooks and hangers are placed.

Swinging scaffold hooks should be tied off with at least 1-in. manila rope secured to a stack or other stable roof-mounted structure, with safety lines tied off to a lifeline reaching the ground. Not more than two men should be allowed to work on a swinging scaffold at one time.

Other safe practices to be followed in the use of swinging scaffolds are:

1. Test the installation before using it by raising platform about 12 in. and having two men stand in the middle.

2. Inspect the raising and lowering mechanism frequently.

3. Remember that this is a light duty scaffold, designed primarily for operations involving men and hand tools, with a minimum of materials; so do not overload it.

4. Be sure there are at least three turns of wire rope on drums at all times. Do not lower the scaffold below this point.

5. Equip the stage with toeboards, a guard rail, properly supported at both ends and in the center, and wire mesh between mid and toe rail.

6. Supply each man working on the swinging scaffold with a safety belt and a lanyard securely attached to a separate line extending from roof to ground. Each lanyard should be tied to the lifeline with a triple sliding hitch or with a mechanical rope grab. (See Figs. 17–14 and –15, and NSC Data Sheet 606, *Fixed Ladders and Climbing Devices.*)

Boatswain's chairs

Boatswain's chairs should be erected with care by the men who will use them. The seat should be not less than 2 ft long by 1 ft wide. One-inch cleats should be nailed to the underside of each end of the chair and should project at least 9 in. in front of the seat. The chair should be supported by a sling attached to a suspension rope. Ordinarily,

423

Fig. 17–15.—How to tie a triple sliding hitch is shown in left photograph; hitch has been loosely tied to show clockwise lower turns and counterclockwise upper turns. Dead end should be at least 12 in. long. Photograph at right shows appearance of hitch when pulled tight. Dead end and live end of safety strap are on same side of vertical portion connecting upper and lower turns. Hitch must be tightened after every movement on the life line to which it is attached.

Courtesy Workmen's Compensation Board of British Columbia.

¾-in, rope, spliced into 6-in. blocks, is used for support.

Where blow torches, cutting torches, or open flames are used, slings should be made of wire rope. A safety belt secured to the tackle to prevent the man from falling out of the seat should be provided. The slings should be at least ⅝-in. manila rope (or its equivalent in strength if a substitute must be used), doubled and passed through holes bored in the seat.

For information on other types of scaffolds, see ANSI A10.8, *Safety Requirements for Scaffolding.*

Hoists

Material hoists and personnel hoists are made of tubular steel structural members. Be sure to consult with tubular steel manufacturers or suppliers for current technical data.

They may be erected in hoistways inside the building or in outside towers. Personnel hoists are generally used just for the transport of people; however, if they conform to the material hoist design standards, they may be used to transport material. Personnel should never be permitted to ride at the same time.

Work in or on the hoistway while the hoist is in operation should not be permitted. For additional information, be sure to check the latest ANSI Standards (A10.4, *Safety Requirements for Personnel Hoists;* and A10.5, *Safety Requirements for Material Hoists.*

Inside material hoistways

If the material hoist is installed inside the building, the hoistway should be enclosed. Solid enclosure is preferable, heavy wire screening (½-inch mesh, No. 18 U.S. Gage wire) is often substituted. Adjacent hoistways should be partitioned.

Entrances should be protected by solid or slatted wood gates at least 5½ ft high and within 4 in. of the hoistway. Gates should be counterweighted and have latching or locking mechanisms. A single wood bar, such as a 2 x 4, sometimes is used at entrances.

Such a bar should be attached permanently by a hinge at one end with a substantial socket at the other. The bar should be between 36 and 42 in. above the floor and at least 2 ft from the hoistway line.

Protective covering of heavy planking should be provided below the cathead of all hoists to prevent objects from falling down the hoistway.

Outside material hoistways

Hoisting towers are usually of tubular steel construction. Design should be based on a safety factor of at least 5. The tower should be on a level and solid foundation and should be well guyed or fastened to the building (Fig. 17–16).

The tower should be enclosed with heavy wire screening and equipped with a fixed ladder extending the full height of the tower. Runways connecting the tower to the building should have standard railings and toeboards.

Material hoist platforms

Material hoist platforms should be built with a safety factor of at least 5 (Fig. 17–17). Flooring should be of not less than 2-in. timber, and sides not used for loading should be enclosed with heavy wire screening and should have 6-in. toeboards. There should be an overhead plank covering at the crosshead. The covering can be built in hinged sections to permit handling of long material. Where wheelbarrows are handled, stop cleats should be attached to the floor. The car should be equipped with a broken-rope type of safety device. If car cable breaks, the car safety clamps or dogs are thrown into position on the guide rails to stop the car.

Signal systems

A good signal system is necessary for safe operation. Electrically operated lights or bells, bells operated by pull cords, a combination of bells and lights, or a telephone system can be used. Standard signals should be adopted and posted at each entrance and at the engine.

Personnel hoists

Any hoist used for carrying passengers should be made to conform to the safety requirements of American National Standard A10.4, *Safety Requirements for Personnel Hoists.*

Temporary use of permanent elevators. Permanent passenger or freight elevators in buildings under construction, modification, or demolition may be used for carrying workers or materials or both, provided they are approved for such use and a temporary permit is issued for the class of service. Elevators should conform to ANSI *Safety Code for Elevators, Dumbwaiters, Escalators, and Moving Walks,* A17.1 and A17.1a.

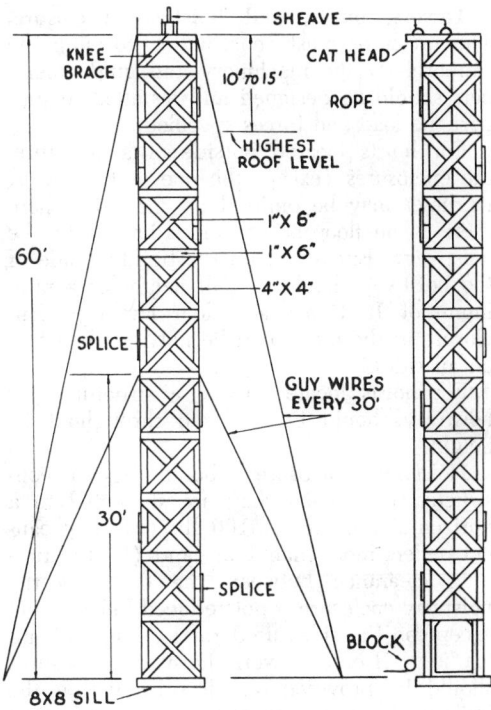

Fig. 17–16.—Wooden tower for bucket or cage capacities not exceeding 1000 pounds. Additional protection against falling material is obtained by enclosing the tower with No. 16 U.S. Gage wire screen with 1½-in. mesh.

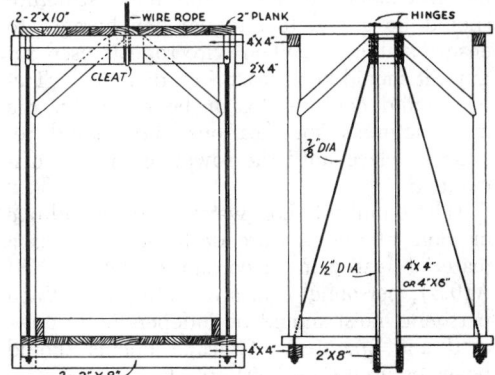

Fig. 17–17.—Typical design of material hoist platform. For additional safety, the sides not used for loading can be enclosed with wire screening.

425

Towers, masts, and hoistway enclosure. The tower or mast construction forming the supports for the machinery and guide members should be designed and installed to support the load and forces specified.

For hoists located outside of a structure, the enclosures (except the one at the lowest landing) may be omitted on the side where there is no floor or scaffold adjacent to the hoistway. Enclosures on the building side of the hoistway should be full height or a minimum of 10 ft at each floor landing. Enclosure at the pit should be not less than 8 ft on all sides.

For hoists located inside a structure, the hoistway should be enclosed throughout its height.

Hoistway enclosures should be so constructed that when they are subjected to a horizontal pressure of 100 lb, (a) they cannot deflect more than 1 in., and (b) the running clearance between the car and the hoistway enclosure is not reduced below ¾ in., except on the sides used for loading and unloading. If on openwork, hoistway enclosures should be provided on all sides within the building or structure with an unperforated kickplate extending not less than 12 in. above the level of each floor above the lowest.

Foundations of hoists should distribute the transmitted load so as not to exceed the safe load-bearing capacity of the ground upon which it is set. Hoist structures should be anchored to the building (or other structure) at vertical intervals not exceeding 25 ft. If tie-ins cannot be made, the hoist structure should be guyed to adequate anchorages to ensure stability. When wire rope is used for guys, it should be at least ½-in. diameter. Tie-ins should conform to, or be equal to, the manufacturer's specifications and should remain in place until the tower or mast is dismantled.

Where multiple hoistways are used and one or more of the cars are designed solely as a material car (in accordance with ANSI A10.5), personnel cars are prohibited. Each personnel hoist should be independently powered and operated. Chicago booms should never be used on a hoist structure.

Doors for hoistways should not be less than 6 ft–6 in. high. If a solid door is used, it should have a vision panel, not wider than 6 in. and not larger than 80 sq in., covered with expanded metal.

Landing doors should lock mechanically so they cannot be opened from the landing side. At the landings other than the lowest one, these locks should be of a type that can be released only by a person in the car. If the door at the lowest terminal landing is locked automatically when closed with the car at the landing, it should be provided with a means to unlock it from the landing side to permit access to the car. A hook-and-eye should never be used as a door-locking device.

Car platforms. Each car should have a platform extending over the entire area of the car enclosure. It should be nonperforated and fire retardant, and supported by the car frame. Both the frame and the floor should be designed to handle the anticipated loads.

Car enclosures. Car enclosures and linings should be made of metal or fire-retardant wood. Personnel hoist cars should be permanently enclosed on the top and all sides, except the entrance and exit. This enclosure should be securely fastened to the car platform and so supported that it cannot become loosened or displaced when the car safety or buffer is engaged. The enclosure walls should be strong enough so that their running clearance is reduced by no more than ¾ in. when a force of 100 lb is applied horizontally to the walls of the enclosure.

An emergency exit with an outward-opening cover should be provided in the top of all cars. The opening should be not less than 400 sq in. area, with a minimum dimension of 16 in. It should provide a clear passageway unobstructed by fixed hoist equipment on or in the car.

Do not locate a working platform, or place equipment that is not required for the operation of the hoist or its appliances, on the top of the hoist car, unless specifically provided in ANSI A10.4.

Wire glass (or the equivalent) should be used for vision panels. Plain glass should be used only for car operating appliances.

Wire ropes and sheaves

In no case should hoisting ropes less than

½ in. diameter be used except on such equipment as small winches, like those on gin poles. In any case, they should provide a factor of safety conforming to the requirements of applicable elevator codes. Ropes should be inspected frequently and kept lubricated. They should be replaced when inspection discloses that wear, breakage, or corrosion has reduced their strength below the permissible limit of safety.

Where clip fastenings are used, there should be at least three clips, with the U-side on the dead end of the rope. Ropes should be guarded at points where persons may come in contact with them and where they may strike or rub against objects. Sheaves should be well aligned, and bearings should be kept lubricated. In general, sheave diameter should be at least twenty times the rope diameter.

For additional details, see Chapter 25, "Hoisting Apparatus and Conveyors," and Chapter 26, "Ropes, Chains, and Slings."

Hoisting engines

Hoisting engines should not be located in public streets. Where they must be so located, they should be enclosed with barricades for protection of the public. In any case, a roof to protect equipment and operator from the elements is advisable.

Engines should have brakes capable of stopping and holding 150 percent of the rated safe load; in addition, there should be a pawl for holding suspended loads.

Exposed gears, shafting, and couplings should be enclosed. Exposed steam pipes should be covered, and exhaust pipes should be placed where steam cannot strike nearby persons.

Where electric hoists are used, switches should be of the enclosed safety type, and all current-carrying parts should be enclosed or guarded to prevent personal contact. Installation should be grounded.

Plant Maintenance

A sound, efficient maintenance program is essential in any industrial plant. Such a program will keep the physical plant in good condition and will reflect in the safety record.

Maintenance should include proper, long-term care of the building, as well as of the equipment. It should include routine care to maintain service and appearance, as well as repair work required to restore or improve service and appearance.

Too often, maintenance is thought to mean only repair. In this discussion, considerable emphasis will be placed on *preventive* maintenance and on the type of inspection that will discover conditions pertaining to the building or equipment which, if uncorrected, might result in accidents.

Since the maintenance program usually is under the supervision of the plant engineer or maintenance superintendent, this discussion will be of particular interest to such people. However, because the safety of employees is closely tied in with the condition of buildings and equipment and because there is a close relationship between maintenance and safety, the safety professional will find that the items covered have an important bearing on the safety program as well. He should not hesitate to point out—to the proper authorities—items that need correcting or replacement.

Foundations

Footings and columns. It is rather difficult to detect structural irregularities of footings, but it is possible to check unusual settlement of building columns and footings. Level marks of known elevation, placed about 5 ft above the basement floor, can be checked periodically for settlement.

Excessive settlement may threaten the stability of a building, as well as the effective working of the machines and equipment in it. Excessive settlement should be reported at once to a structural engineer for immediate action. Footings or columns may need to be temporarily shored, underpinned, or repaired, or piles may need to be driven to correct the difficulty.

Dry rot around the bottom of wood columns at the basement floor level results if the basement floor is damp, subject to water seepage, or for other reasons alternately dry and wet. A 1-ft concrete curb around a column will keep water away. Rust at the base of steel columns should be scraped away, and the columns given a coating of rust-resistive paint.

Foundation walls. The inside of foundation walls should be inspected for cracks, which may result from settlement of the building and shrinkage of concrete. Since these cracks are below grade, they may admit water to the basement area. If enough water comes in through large cracks, it may cause settlement of the backfilled earth around the outside of the foundation walls. Sidewalks and adjacent roadways may be damaged.

Minor or small cracks can be repaired from the interior by application of a water-proofing material. If the cracks are relatively large, it may be necessary to dig down outside the building to the bottom of the wall, clean off all earth and other foreign material, and apply a waterproofing compound. A membrane covering should then be applied directly on the compound and covered with another coating of compound. When unusual settlement is noted, a check should be made and settlement readings taken like those for footings and columns.

Pits. Inspection should also include pits, with cracks noted and repaired. No debris or rubbish should be allowed to collect in pits. Guard rails or covering should be supplied where needed.

Structural members

Joists, beams, and girders. In many instances, joists, beams, and girders are covered by suspended or sealed ceilings and are relatively inaccessible. In such cases, excessive deflection may be indicated only by a badly sagged floor. At least once a year the entire floor system on each floor level should be examined.

Joists, beams, and girders should be checked, and deflection, twisting, tipping, or other unusual conditions corrected. If major repairs are necessary, the floor should be cleared immediately of stored materials. Before repairs are made, a qualified engineer should be consulted.

Columns. Building columns should be examined for unusual distortion or buckling. Excessive eccentric column loadings should be avoided, and a check should be made for holes cut in or through columns. If holes are found, they should be called to the attention of a qualified engineer.

Steel. Steel I-beams, channels, columns, angles, girders, and other steel structural members should be checked for rust once a year, more often in corrosive atmospheres. Where rust is noted, the steel member should be scraped and a rust-resistive paint applied.

Concrete. Floor slabs, beams, girders, and columns should be checked regularly for cracks, spalling, and sections of concrete chipped away from the reinforcing steel. Since rust may form on exposed reinforcing, repairs should be made at once. Guniting will provide a protective coating for the exposed reinforcing. This spalling may be evidence of more serious problems and a competent engineer should investigate.

Wood. A wood floor system should be inspected for shakes, checks, and splits in joists, planks, beams, stringers, posts, and columns. Decay or dry rot in wood columns, joists, and other members should be looked for. It is important that beams, joists, and girders have full bearing, and every evidence of movement or slippage should be fully investigated.

Walls

Exterior walls of brick, concrete, terra cotta, stone, cement or cinder block, and stucco need inspection for cracks or joint separation. Cracks may be caused by expansion, contraction, vibration, or settlement. When cracks are found, they should be filled immediately because water may freeze in them and cause additional damage.

Windows. Masonry buildings require periodic minor repairs of walls and windows. Mortar joints will loosen and disintegrate from settling of the building and from weathering. Such joints should be raked and pointed. If such joints are not repaired, moisture eventually will get into interior wall surfaces and cause progressive damage.

A similar precaution applies to window openings. Because of settlement of the building, drying out of the wood, or improper setting of metal frames, caulked joints will crack open, allowing moisture to enter.

Before recaulking is done, all loose material should be removed and all cracks cut out so that the new compound will be well

FIG. 17–18.—Heavy pipe and angle railing extend about 18 in. from the floor to prevent damage to door jambs, walls, and building columns. Standard 3- or 4-inch black iron pipe is welded to angle iron and fastened to the floor with heavy strap iron brackets.

Courtesy Kraft Foods Company

bonded. A high-grade caulking compound with a gun that will force the compound well into the openings (and not just cover the surface joints) should be used.

Parapet repairs include the maintenance and repair of masonry, metal, and wood parapets. Serious repairs can be avoided if masonry walls are checked carefully once a year for cracks or spalling.

Brick parapets and walls above grade should never be coated on either side with any material such as pitch, roof paper, or asphalt roof coating that will not allow the wall to "breathe." Such coating, which is commonly misapplied to brick parapets, causes spalling of the brick and disintegration of the mortar

joints, particularly in areas where freezing temperatures occur.

Sometimes brick walls are painted on the outside, either for decoration or for a sign. If there are high humidities inside a building, excessive moisture will penetrate the brick and condense under the impervious paint film, again causing spalling and joint disintegration during the winter.

Stone caps and other stone work on brick walls should be checked for cracks at all mortar joints. Cracks might allow moisture to enter and eventually loosen the stone. They should be filled with a cement grout or mastic filler. If these are on high buildings, their falling off could cause injuries and damage.

Interior walls and ceilings. The same rigid inspection should be given to partitions, cross walls, interior sides of main walls, and ceilings. Inspectors should look for such defects as cracks in interior walls, holes, loose mortar in joints, broken or missing brick, and spalled or worn areas where power trucks may have frequently scraped tile or brick walls. To prevent damage from trucks, standard 3-in. or 4-in. black iron pipe railings can be put near the floor level as barriers (Fig. 17–18).

Ceilings may need painting, cleaning, or repair. Unusual sag should be investigated immediately and corrected. If sag is found in a suspended ceiling, hangers and fastenings should be checked. However, the sag may be the result of excessive loading of the floor above, which should be corrected.

Floors

Accidents due to inadequate maintenance of floors are a major source of injuries in many plants. Slippery conditions, particularly, account for numerous falls by workers. Holes and other irregularities in wood and concrete floors, both inside and outside plant buildings, result in frequent injuries from stumbling and falling and, in addition, cause many truck accidents.

Repair procedures. Wood flooring should be inspected for rot, wear, and unusual stress as indicated by sag. In many cases, a section of wood flooring may need to be replaced. New flooring should be installed flush with the existing flooring. Badly worn or loose

429

wood block flooring should be replaced by anchored wood blocks.

To anchor loose wood finish flooring, holes are drilled at an angle through the finished floor into the subfloor, and flooring nails larger than the holes then are driven into the subfloor.

For concrete floors, the damaged area should be chipped out, cleaned thoroughly and wet down. Cement mortar (1 part Portland cement and 3 sand) is troweled in. Minimum patch thickness is 1 in. Patches 2 in. thick or more may require wire mesh or reinforcing steel. For finishing, a wood float gives a less slippery surface than does a steel trowel.

For proper curing, traffic should be kept off the patch for at least three days, unless a quick-setting cement has been used.

If mixed and applied properly, epoxy resin repair materials give excellent results. A thickness of as little as $\frac{1}{8}$ to $\frac{3}{16}$ in. will give an extremely tough wearing surface. Be sure to take adequate precautions when using these mixes. (See National Safety Council's *Epoxy Resin Systems,* Data Sheet 533.)

Maintenance procedures. Use of the wrong cleaning materials, methods, and surfacing often causes even the most suitable types of flooring to deteriorate and become slippery. Alkaline cleaners should not be used on terrazzo, but mild alkaline cleaners may be used on asphalt tile. Oils are unsuitable for rubber tile and, when applied to wood floors, increase the fire hazard. To keep floors clean, safe, and sanitary, the recommendations of the flooring manufacturer should be followed.

In general, the routine maintenance procedure for linoleum, marble, terrazzo, asphalt tile, and other types of flooring used in offices, institutions, and similar buildings is to clean the floors with a soft floor brush or vacuum cleaner and, when necessary, mop them with a mop dampened with clean, cold water.

One section of floor is cleaned at a time, and, if traffic in the area is heavy, that section is roped off. When soap is used, any soapy film must be removed by thorough rinsing, to avoid a slippery condition.

Ordinary wax for polishing wood, tile, and similar floor surfaces is unsuitable because of its inherently slippery nature. However, according to the manufacturers, floor oils and waxes can be used to surface various types of floors without adding unduly to their slipperiness, provided that instructions regarding their application are followed.

Oils and greases, water, paper, sawdust, and numerous other foreign materials on floors create slipping hazards. Leaks of oil from machines and of water and other liquids from pipelines and spillage from processing equipment often can be eliminated by good maintenance of the equipment, such as promptly tightening loose connections.

Pans and absorbent materials can be used when leakage cannot be readily eliminated at the source. However, a study of such sources often reveals ways of keeping slippery materials from getting on the floors, such as the installation of splash guards on machines using cutting oils.

Slippery materials spilled on floors should be cleaned up promptly. To remove grease and oils, the area can be covered with airslaked lime to a depth of about ¼ in. After two or three hours, the lime then is removed with a scraper or stiff brush. Sand and various commercial cleaners also can be used.

Even as innocuous a substance as coffee can cause an accident if spilled in a high traffic area. In National Safety Council offices, cups containing beverages must either be covered or be on a tray if carried to an employee's desk.

Aisles should be kept clear of machinery, equipment, raw and manufactured materials. In many cases, the allowable floor loading has been figured on clear aisle space with no allowance made for power trucks using aisles. It is advisable for efficient, safe operation to determine whether floors and aisles are capable of sustaining loads of power trucks. Lines to indicate aisle width should be well maintained.

Overloading of floors, which may be caused by the installation of heavy equipment, excessive weight and unequal distribution of stored raw and finished materials, and heavy truck transportation, is particularly hazardous. Signs stating allowable floor loads and horizontal lines showing the maximum height to which materials may be piled should be painted on the walls (Fig. 17–19).

430

Fig. 17–19.—Where the same type of material is stored regularly, a line can be marked on the wall to indicate the height to which material can be piled without exceeding the allowable floor load. In addition to safe floor load limit signs, a warning sign reading Do Not Pile Above Line can also be placed on the wall.

Strength and load determinations. Accurate data on floor load capacity are a prime requisite for a physical plant survey. If these data are not already known or readily obtainable from building plans, a structural analysis by a qualified engineer may be necessary. Rough estimates based upon experience or conclusions arrived at by a casual glance at a handbook are dangerous. Use accurate weight data.

Load distribution is often a complex problem. Most buildings are designed to carry uniform loads. A concentrated load places twice as much stress on supporting members as a uniform load of equal weight. For this reason, most heavily concentrated loads, such as machines, are placed directly over beams or girders rather than on slabs or joists.

As is the case with floor load capacity, the safety professional should consult plant engineering specialists for an accurate determination of concentrated loads.

Overloading determinations. The relationship between design load capacity and actual floor loading will indicate accurately whether a floor is safely loaded or is overloaded.

Although visual evidence of overloading is not always to be found, the inspector should look for it. Deflection is the most common evidence of overloading in wood and steel beams. A sag or deflection greater than 1/360th of the span length is a warning that the floor may be overloaded.

To measure the deflection, a cord is stretched between two columns of the span 5 ft down from the underside of the beams or girders at ceiling level. The distance from the center of the taut cord to the bottom of the beam or girder is measured, and the difference between this measurement and 5 ft is the deflection.

Overloaded wood beams will check and crack. In reinforced concrete beams and girders, concrete will spall and fall away from the tensile side. Wood floors will be punctured, and flat concrete slabs will crack and spall. Timber columns will split and crack. Concrete columns will spall, the concrete falling away from the reinforcing. On steel columns, flanges will be twisted.

Bearing walls of masonry will show extensive cracking, disintegration of the bricks, bulges, and pulling away of floor joists.

Whenever there is any doubt as to the amount of deterioration of floor supporting members and therefore the load carrying capacity of the floor, inspection and determination should be done by a qualified structural engineer.

Roofs

Types of roofs. There are several common types of roofs. The flat roof usually is covered with roofing paper and hot pitch, then coated with pea-size gravel. Except for purposes of inspection, this type of roof should not be walked on; never, if possible, during hot weather, when holes in the paper may result. On flat decks coated with roofing paper and hot asphalt, cracks in the joints of the paper can be repaired with muslin strips and cold mastic roofing cement.

Pitched roofs, too, should be periodically checked to make certain that they are in good condition. Gutters should be kept clean, and gutter areas checked for ice damage in the early spring months.

Abnormal loading. Roofs are usually de-

431

signed to carry the maximum snow load expected for the locale. Roof configuration and wind frequently combine to deposit heavy drifts over portions of a roof. Where practical, these accumulations should be removed as quickly as possible to prevent roof collapse or lesser damage. Ice is a serious problem and can also cause overloading.

Inspection. All roofs should be inspected periodically, perhaps once every six months. Roof flashings should be checked for cracks at the parapet wall. Roof gutters and drain connections should be checked for cracks at the roof line. Areas around dormers, chimneys, and valleys, where metal is used, must be checked to see that the metal is tight with the roof as well as with the drain. Roof drains and overflow gutters through parapet walls should be kept open; leaks will occur if water should rise above the flashings.

It is often difficult to find the source of a leak, since the fault may be distant from the place where the leak shows inside the building. On sloping roofs, a check should be made above and to the sides of the place where the leak shows. Flat decks, laid with concrete, are difficult to check because water frequently follows a crack and shows up at a distant point.

Repairs. A leaky roof should be patched as soon as possible after the leak is first apparent. Complete recoating should be done as soon as the roof paper has lost resilience, as shown by cracks or a dried-out appearance.

A new roof is indicated when maintenance becomes excessive and leaks occur after each rain. Cold mastic applications usually are good for about 5 years, whereas an application of hot pitch covered with pea gravel may last from 5 to 10 years. Whoever does the recoating must be sure he knows how to apply the material; otherwise, much will be wasted. Reputable roofing manufacturers will provide a guaranteed-performance bond for a number of years commensurate with the quality of the job specified.

On a repair job, a roof can be punctured by tools, boards with nails, stones, or other sharp objects. Runways and protective runboard coverings should be used. If hot pitch or roofing compound is used, workmen should wear gloves, goggles, and knee-length asbestos, leather, or fire-resistant duck leggings.

Roof anchorage is a very important point to check during inspection. The lifting of unanchored roofs accounts for a large percentage of the total wind damage loss to American industry.

The roofs of all buildings should be securely anchored. Anchors can be readily installed, and the cost is reasonably low. In most cases, bolting one end of a strap, plate, or rod to the roof and the other end to the building provides adequate anchorage.

Roof-mounted structures. Roof inspection should include penthouses, stacks, vents, and supports for water tanks where these structures are flashed at the main roof level. The roofs of penthouses should be checked, since most penthouses contain elevator machinery which may be damaged by water seepage.

Penthouse windows, skylights, and monitor sash should have necessary reglazing, reputtying, frame caulking, or painting. When operating mechanisms are being repaired or repainted, workmen should work on safe scaffolds and should have safety lines tied off to lifelines. They should never work on or over unprotected skylights.

Tanks and towers

If a tank is more than 20 ft above the ground or building roof, a wood or steel balcony should be placed around the base of the tank, designed to support a weight of at least 100 pounds per square foot. For a tank not more than 15½ ft in diameter, the width of the balcony should be at least 18 in.; for tank diameters greater than 15½ ft, the balcony width should be 24 in., with railings. See ANSI A12.1, *Safety Requirements for Floor and Wall Openings, Railings, and Toe Boards.*

Maintenance of tanks and towers is important, not only for fire protection, but also because structural failures may cause serious accidents.

When tanks are to be cleaned or painted on the inside, all precautions relating to tank entry should be strictly observed. (For a detailed discussion of the required precautions, see Cleaning Tanks in Chapter 42, "Flammable and Combustible Liquids.")

Stacks

Stacks should be inspected at least once every six months. They are subject to deterioration, both inside and out, from weathering, high winds, lightning, foundation settlement, and the action of corrosive flue gases.

A brick or concrete chimney can be protected by lightning rod conductors of low resistance and ample carrying capacity if they are installed from the top of the chimney to a good electrical ground. Lightning rods can be readily installed on chimneys when repairs necessitate the erection of scaffolding. Underground water pipes or buried copper plates afford good ground connections.

Stack ground wires should be checked for effective grounding.

Platforms and loading docks

Mechanized traffic on platforms and docks often causes damage to platform surfaces. With no edge protection, concrete may become spalled or chipped, and ruts in the concrete finish may cause power or hand trucks to swerve and run off the dock or into other employees or material. For these reasons, angle iron or channel iron protection should be provided at the edge of the platform, and kept well maintained. Badly rutted platforms may be resurfaced with concrete or epoxy cement.

Wood platforms should be checked for decay or dry rot, loose or uneven planking, weakened or broken supporting members, and repairs made immediately.

Canopies

A canopy roof should receive the same careful inspection as that given to the roof of the main building. Evidence of pulling away from the building should be noted and corrected. Drainage is important, and downspouts and gutters should be kept open and in good repair.

Supporting members, if wood or steel, should be scraped and painted periodically.

Sidewalks and drives

Concrete sidewalks and drives should be repaired as soon as spalling or cracking of the concrete creates a hazardous condition. Inspections should be made in the spring after the ground has thoroughly thawed. Sections of sidewalks can be relaid, but bituminous driveways will need patching with hot tar or similar material.

Drainage of a driveway is important, whether it is made of concrete, asphalt, or gravel. Repairs should be made as soon as possible so as to keep damage to a minimum. To keep asphalt drives in good condition, they should be recoated periodically by someone who understands paving techniques.

Colored plastic disks can be used to mark parking areas, to eliminate the need for frequent repainting of stripes.

Underground utilities

Sewers. At least two men should work on a sewer maintenance job, whether or not dangerous gases are suspected. Before men go into a manhole structure or sewer, tests should be made for oxygen deficiency, methane, hydrogen sulfide, carbon monoxide, carbon dioxide, and nitrogen. If any of these gases are found, the necessary precautions must be strictly observed.

Proper respiratory equipment (see Chapter 19, "Personal Protective Equipment") should be used, or complete ventilation should be provided with either blowers or suction fans. Blowers are preferable since their source of supply is known. Suction fans may draw poisonous gases from unseen pockets or crevices into the area.

Utility trenches and tunnels. Where men must work in trenches over 5 ft deep, shoring is required. If the trench is adjacent to machinery foundations, or other superimposed loading, a thorough design of the bracing and shoring to be installed should be made before work is started.

It is important to check atmospheric conditions in utility trenches, tunnel trenches, tunnels, and manholes before work is started. Proper ventilation is necessary for the safety of the workmen, who should work in pairs. Manholes, tunnels, and trenches known to be contaminated should be tagged or otherwise identified for the information of work crews.

Smoking should not be permitted in or near manholes, tunnels, or trenches known to be contaminated. In fact, it is good practice to prohibit smoking around and in all underground structures.

The use of open-flame devices, such as solder pot furnaces and welding equipment, should not be permitted in or near manholes, tunnels, or trenches, in which tests indicate the presence of flammable gas.

Waste disposal facilities. Waste disposal trenches of reinforced concrete have been known to settle, the side walls separating from the bottom slab of the trench and allowing water to seep out. Such seepage carries earth with it so that the trench and building footings may be undermined and the support may be removed from the underside of the floor slab at ground level.

Structural failures may result also, if waste disposal trenches are not kept in good repair. If trenches are lined with lead, acid-proof brick, or other lining, they should be checked for leakage.

Other underground pipelines. When men are to work on an underground pipeline, it is important that its approximate location be known before digging is begun. The sides of all trenches more than 5 ft in depth should be shored unless they are sloped to the angle of repose or unless the trench is in solid rock. No excavation should be made immediately adjacent to or below the level of a pipeline because of the danger of cave-in and of failure of the line, which may be carrying liquids or gases vital to plant operation. Workmen cutting into or working on a pipeline should know what the line is carrying and the dangers involved.

Before repairs are begun, pipelines should be completely drained, the connecting systems blanked off, and the valves closed and locked. Whenever a workman opens a line or valve, he should watch for back pressure in the line.

Workmen should be cautioned to open valves slowly and to equalize pressures slowly. Sudden changes in pressure can wreck equipment and endanger lives.

After a steam valve has been opened completely, it should be backed off at least one-half turn so that thermal expansion will not lock the valve in the open position.

When pipe or trenches must remain open overnight, they must be protected by barricades, signs, and lanterns.

Lighting systems

All lighting systems—fluorescent, incandescent, or mercury vapor—require regular maintenance to maintain maximum light output. (For design factors, see section on Lighting in the preceding chapter.)

To reduce the number of persons who might be exposed to flying glass and dust if lamps are broken, fixtures can be relamped on weekends or at other times when personnel are not in the area. Gloves should be worn by those who handle the lamps.

To dispose of a few lamps, put them in special containers for regular rubbish removal or wrap in several thicknesses of newspaper or wrapping paper, place outdoors in a disposal container, and crush with a heavy stick or shovel. Large numbers of tubes should be broken in commercially available machines and in a ventilated enclosure equipped with an exhaust and a dust collector. Captured dust should be wet down and removed.

Lamp burn-outs and depreciation, dirt accumulation, voltage drops, and light absorption by dirty walls and ceilings will cause a loss of light. Because dark or dirty surfaces absorb as much as 80 percent of the light that strkes them, a regular cleaning schedule is important. For dirty areas, this may be every two weeks.

A light meter can be used to check conditions, and the readings then can be checked with the minimum standards of illumination for industrial interiors, ANSI A11.1, *Practice for Industrial Lighting.* These standards, recommended by the Illuminating Engineering Society, are summarized in the table entitled Levels of Illumination in Chapter 46, "Safety Engineering Tables." These values can be used as a guide to formulate a regular maintenance program. If cleaning lamps and replacing burnouts fail to increase the illumination to standard levels, then a qualified illuminating engineer should be called upon to make a complete survey of the present system.

Fluorescent fixtures should be of the type that locks the tubes in place. Older fixtures should be equipped with shields or grids beneath the tubes to prevent the tubes from falling if vibration should loosen them.

Where reflectors or glassware cannot be taken down, the current should be shut off and a cleaner that requires no rinsing should be used, followed by wiping with a cloth.

Maintenance problems can be simplified by laborsaving devices. When possible, disconnecting-type reflectors should be used. They permit cleaning and relamping from the floor rather than at the outlet. For relatively low mounting heights, stepladders should be used because of their convenience and portability. Clips and hooks which hold spare lamps and cleaning rags enable a man to do an entire cleaning and relamping job with one trip up the ladder.

In some cases, ordinary ladders can be slightly modified for lamp maintenance, by adding built-in lamp compartments and trays to support cleaning buckets. Where the entire installation is cleaned at frequent intervals, special cleaning trucks can be used which have separate compartments for cleaning solutions, for warm rinse water, and for clean rags.

Portable maintenance platforms can be used where there are a great many luminaires at a given mounting height. These platforms are made of lightweight material and are equipped with castors so that one man can handle them easily. This type of equipment permits a man to reach several units safely and not reposition the platform.

Manufacturers have thought of various means to reach all types and styles of lamps. One of the simplest devices is the clamp grip mounted on the end of a pole and formed to fit fluorescent tubes. In many industrial plants with installations having open bottoms and exposed lamps, these pole-changing devices are used between periods of regular maintenance for emergency lamp replacement. They are well suited for recessed reflector lamps.

Stairs and exits

The following items should be noted when the condition of stairways and exits is checked:

Lack of handrails—handrails too low—handrails rough.

Improper illumination.

Poor housekeeping.

Improper or inadequate design, construction, or location.

Wet, slippery, or damaged surfaces.

Faulty treads or mats on stairs.

Lack of curbing on ramps.

Whenever any one of these items is detected in the maintenance inspection, it should be repaired or corrected immediately. Various types of nonslip surface materials that can be applied directly to stairways are on the market.

Passageways should not serve as storage areas. The floor surface should be smooth and the area well lighted. Hazardous conditions should be corrected as soon as possible after inspection.

Operation of exit doors should be checked. There should be no obstruction to their free movement, nor should the paths to them be blocked in any way.

Exit signs and lights should be kept in good repair. Exit signs designed to show in the dark in case of failure of the lighting systems should be tested in darkness.

For details, see National Fire Protection Association Standard No. 101, *Life Safety Code* (ANSI A9.1).

Maintenance Crews

Maintenance men should be selected for their experience, alertness, and mechanical ability, and for their capacity to learn the essential safety principles of machines or operations for which they may be made responsible. Repairmen must be willing to accept orders, to work well without direct supervision, and to carry out a job safely.

Training

A maintenance man should have more thorough training in accident prevention than the regular worker. Safety for him involves not a set pattern of activity, but a complex and constantly changing set of problems. Furthermore, he must know how to use not only ordinary tools but also ladders, various kinds of protective equipment, chains, slings, and ropes, and many others.

Maintenance crews must be made aware of job hazards and be receptive to proper training in order to protect themselves and others working close by. Their training program should include first aid and life-saving techniques. In industries where irritating, toxic, or corrosive dusts, gases, vapors, or fluids are present, the maintenance man should be given special training to make sure that he is familiar with the properties of these substances and with the methods of controlling the hazards.

Some companies have arrangements for the purchasing department to notify crews when new chemicals are purchased so that the necessary precautions can be planned.

The men must be trained to inspect chains, blocks, falls, and ropes and to discard those that show signs of excessive wear. They must know methods of securing loads and must understand stresses.

The crew should be called together at the start of a nonroutine job to discuss the hazards involved and the method of doing it safely. A two-man safety committee may be appointed for such a job to check the equipment and to call the supervisor before work that looks unsafe is started. For especially complicated or hazardous jobs, the safety professional may be called upon to help in the job planning. Scale models can be constructed to determine clearances, the best methods of moving, and sequences of action. After several trials have been made and the men have agreed on a safe procedure, the various steps should be recorded as a guide for each worker.

The tools and tackle required to do such special jobs should be inspected for wear and defects before they are used. When special tools can be devised to make a job safer, the engineering department can provide design and construction specifications.

In the course of their daily work, the maintenance men travel throughout the plant, becoming familiar with every machine and process. If properly selected and trained, they can do much to locate and correct unsafe conditions in both plant and equipment.

In smaller companies, the maintenance man may also be responsible for inspection and care of portable power tools, extension cords, and the like. If so, special procedures and training are suggested.

Scheduling preventive maintenance

Because the function of plant maintenance is to keep physical equipment in top operating condition, a good maintenance system must catch breakdowns before they happen. That is, it must be a program of *preventive* maintenance.

Even if funds are limited, a preventive maintenance plan can be set up at least for critical equipment and for machinery that might seriously affect the safety of workers if it should break down. A good preventive maintenance program starts with a list of buildings, machinery, and equipment, that require periodic inspection, adjustment, cleaning, and lubrication, as well as adjustment of guards or changes in the types of guards.

Detailed engineering specifications should be acquired for each machine or structure. For machinery, these give dimensions, weights, sizes of plant service connections, lubrication requirements, details on bearings, and power transmission or motor drive data. For buildings, they include data on general layout, services available, floor load capacities, ceiling clearances, column spacing, and many other features.

After a supervisor has made an analysis of his maintenance responsibilities, an organization chart should be formulated for the selection and safe assignment of workers. The safety professional helps in this. Such a chart gives the supervisor a comprehensive picture of his force, pointing out the status of trained keymen and reserves. Outside specialists who can be called in for unusual or very hazardous jobs should be listed. In emergencies when men are needed in a hurry, this information is quite valuable.

Equipment inspections

An inspection schedule based on the potential trouble points listed on maintenance records is essential to preventive maintenance. The schedule can be determined by the number of inspection reports turned in by regular full-time inspectors or by maintenance crew personnel.

An inspector must know what to look for. He must be something of an expert on, or at least thoroughly familiar with, the equipment under his care. This knowledge is particularly important for electrical equipment inspectors. Electrical equipment usually gives few obvious indications of impending trouble, and is likely to stop altogether with little advance warning. Mechanical equipment often warns of deterioration by such easily recognized signs as noise, unusual appearances, or substandard output. But there are some warnings that are more subtle.

It is important that inspectors be equipped with suitable instruments for making their observations. While there is no difficulty in recognizing burned contacts on a motor starter or in hearing the pounding of a worn gear, checking insulation resistance of a motor or measuring the wear on a shaft requires the proper instruments.

Personal protective equipment

Maintenance men should dress properly for their specific job. Garments should be snug-fitting, with a few small pockets. Breast pockets are often sewn closed or removed to prevent items from dropping into machinery or inaccessible places when the wearer leans over. Rings, watch chains, wristwatches, and other jewelry should not be worn.

Neckties and loose clothing should not be worn by maintenance men. Loose rags must be kept clear of moving machinery.

If a man carries so few tools that he does not need a tool bag, he should wear a special belt fitted with tool carriers. To prevent back and spine injuries in case the worker should fall, tools should be carried at the side instead of in the back portion of the belt.

Gloves or hand leathers should be worn by a man handling rough or sharp objects. Welding gloves, rubber gloves for electrical insulation, and rubber gloves for handling acids should be used as needed, but never worn around moving machinery.

Goggles should be in every maintenance tool kit. Moreover, unless the repairman knows the exact conditions he is to work under, his equipment should include necessary types of goggles to provide protection against flying objects and molten metal, injurious heat and light rays, dust and wind, and acid splashes.

When a man needs to work in high places or to enter a manhole, bin, or tank, he should wear a life belt, with the lifeline either attached to some permanent support or held by another man. In any case, on jobs involving such risks an extra man should be stationed close by.

Every repairman's kit should include an explosion-proof flashlight, since he is often called on for work in dark or gaseous places. See Chapter 19 for details.

Lock-out of mechanical and electrical equipment

Maintenance of power presses, various other types of machinery, electrical equipment, and boilers is dealt with in detail in the respective chapters of this Manual.

Because maintenance men are frequently working on machines or electric equipment which might be inadvertently energized, special lock-out procedures should be established, and locks and other "shut-down" equipment provided. See Chapter 41, "Electrical Hazards."

Piping

Proper identification. Accidents have occurred because of improper identification of lines. In factories and power plants where high and low pressure steam lines run adjacent to compressed air, sprinkler system, and sanitary lines, maintenance men must be especially protected against opening wrong valves or disconnecting wrong pipes.

Piping must be identified. Color schemes, tags, and stencils can be worked out so line contents can be identified at a glance. Identification is particularly important when outsiders are called in for service or when emergencies occur. Standardized identification does not present a new learning problem to newcomers. To prevent confusion, use a consistent piping color code, such as ANSI A13.1, *Scheme for the Identification of Piping Systems.* Also see the section Use of Color in the preceding chapter.

Proper isolation and handling. Before work is done on a pipeline, the line must be shut off, valves locked and tagged, the section of the line relieved of pressure and drained. When hazardous materials are encountered, the foreman should ascertain the need for special protective equipment, such as chemical-type goggles, rubber suits, rubber gloves.

To prevent hands from slipping, maintenance men should carry a piece of waste or a rag in their pockets to wipe off excessive oil on pipes and fittings. Gloves should be worn when handling pipes and fittings, especially when ends are threaded. Pipes should be checked for burrs and these should be filed off immediately.

Shutoff valves are usually of the gate type, but if globe valves are used, the pressure-side should be under the valve seat so that the packing will not have to hold the pressure when the valve is closed.

Plants using chemicals should have pipelines which carry them installed in trenches or tunnels. If they must run overhead, they should be isolated or covered so they will not drip on men or materials underneath. Emergency showers should be provided and plainly marked. Full instructions should be given to maintenance men (as well as operating personnel) in their location and use.

There are many special safety precautions which should be taken in maintenance of pipelines, valves, and bolted flanges—especially when hazardous materials are involved. The section on hazardous materials in Chapter 24, "Materials Handling and Storage," contains pertinent recommendations for both operating and maintenance personnel.

Industrial gas lines. When industrial gases such as propane and butane are encountered, mechanics should understand their behavior, storage characteristics, pipeline arrangement, and have a copy of the piping layout which shows equipment location and safety features, such as soft heads and backfire preventers. Sectional and main shutoff valves throughout the entire plant should be located so in case of mishap, the maintenance man can find these valves quickly.

Handling pipe. When maintenance involves a considerable amount of pipe work, lengths of pipe, valves, and fittings to be used should be placed so that the floor is not overloaded. If possible, the material should be moved to strategic points as the job progresses, to eliminate spot accumulation and reduce the amount of handling.

When lengths of pipe are transported by mechanical means, red warning flags should be attached to material that extends beyond the conveyance. If the load is allowed to stand near passageways, the flags should remain and additional warning signs or barricades should be used to prevent other workers from bumping into the load.

Long lengths of pipe being run overhead and continuously joined should be pulled up from the floor with overhead rigging to the proper location and secured with tie ropes,

wires, or fixture straps. Use of overhead rigging will prevent back strains and possible falls from ladders and will keep the material itself from dropping.

Crane runways

Before a repairman works on or near an overhead crane runway, he should notify the crane operator and make sure that the operator understands what is being done and what part he is to take in the work. The repairman also should provide a temporary rail stop between the crane and the point where the work is to be done. See Chapter 25, "Hoisting Apparatus and Conveyors."

When work must be done on the crane itself, the man in charge should be held responsible for locking out safety switches, placing warning signals to indicate that men are working above, and seeing that a careful inspection is made after the job is completed. This inspection should include a check for parts or tools left on the crane, which later might drop into the mechanism or fall on persons.

A handline should be used to eliminate the need for carrying material up a ladder. The line should be tied to a crane part, not attached to the worker's body. In the latter case, he might become entangled or pulled off.

Final check—tools and guards

After a machine has been repaired, it should, if possible, be turned over first by hand. Mislaid tools or materials thus may be discovered in time to prevent their wrecking the machinery and perhaps causing injury. Guards should be replaced and securely fastened, tools picked up, and the work area left as nearly in its original condition as possible. The repairman can aid greatly in preserving good housekeeping.

Too much emphasis cannot be given to the fact that guards are a part of a machine and must be replaced before ending a job.

To prevent tools from being left on or near machinery after a repair job has been completed, a tool box with a special rack for each personal tool should be used so that the repairman can quickly see when a piece is missing. Also, a tool check system will help him keep track of special tools used for the job.

Lubrication

As a rule, lubrication is handled by a separate schedule. Nevertheless, it is an important part of a preventive maintenance program. Improper lubrication sets up special problems; lack of lubrication can cause overheating of bearings, shutdowns, and fires.

Over-oiling of motor bearings causes oil to drop or to be thrown onto the insulation of the electrical windings. The oil deteriorates the insulation, exposing live conductors that will arc and cause fires or electrically charged ungrounded surfaces. If the maintenance man works on this equipment, he may accidentally complete the circuit to ground with his body and be fatally shocked.

If the windings become sticky from oil and dirt accumulates on them, defects may be covered up and go unnoticed until serious trouble or a complete breakdown occurs.

A complete survey should be made of the plant to determine lubrication requirements, and the information should be entered on the machinery records. The machinery then should be checked for missing fittings, missing oil cups, and plugged oil holes.

If possible, improvements should be made at this time. Automatic feed oil cups, mechanical force feed lubrication systems, and special fixtures that will enable the lubrication man to oil parts in remote locations without danger from moving parts should be installed.

The oiler should be provided with a simple diagram of each machine with all the parts that require lubrication plainly marked. The diagram should include a table that tells the oiler the kind of lubricant to use and how often it is to be applied. Some companies devise a color code that indicates the various lubricants and the frequency of application. The point of lubrication, the oil cups, oil feeds, and oil holes are painted according to the color code, and the oiler is supplied with a color code chart.

Each oiler should have a supply of necessary oils, greases, guns, fittings, and wiping cloths before he starts out for the day's work. In large plants, a special cart should be provided to hold all the equipment and supplies needed for the day.

Special precautions or instructions to the oiler should be transferred from the master sheet and entered on his personal schedule. These notations would concern certain types of machines that must be stopped to assure the oiler's safety or those that involve exposure to electrical elements and must have power lines shut off.

Since it is sometimes necessary for the oiler to remove a guard from the lubrication point, he should be trained to replace the guard in its fittings and holding devices properly so that it will clear all moving parts. If the guard or fittings are worn badly or damaged, he should make a report immediately, requesting repairs or replacement.

If possible, shafting should be oiled while the machinery is at rest. Special pump oil cans with long spouts can be used to reach overhead hangers or out-of-the-way bearings. Some of these oiling devices can be used without a ladder.

Ladders should not straddle machines in operation, nor should the oiler attempt to reach several countershafts from one position. Makeshift arrangements, such as chairs and boxes, should never be used instead of ladders.

Shop equipment maintenance

Repair facilities should be provided for maintenance men. The stock of spare parts and units should be large enough to meet probable demand. Machine tools should be modern and arranged to assist the maintenance man in easy flow of work. Hoists should be provided to handle heavy machinery, and power or hand trucks to transport material from one department to another.

The maintenance shop should be well ventilated, especially if welding, paint spraying, or cleaning of metal parts of machinery is done in the shop. Fire extinguishers should be located in this area, and the men should be trained in the types to use on grease, oil, electrical, and other types of fires.

Special tools

Spark-resistant tools of nonferrous materials sometimes are advised for use in areas where flammable gases, highly volatile liquids, and other explosive substances are stored or handled. There is some question as to the danger of friction sparks igniting gasoline or other petroleum vapors but intensified sparks from steel tools are capable of igniting substances such as explosives, lint, carbon disulfide, and ethyl ether.

Nonferrous tools need inspection before each use to be certain that they have not picked up steel particles, which could produce friction sparks.

Keeping up-to-date

The maintenance department should not overlook the potentialities of new products, cleaners, lubricants, paints, wood preservatives, insulation, floor repair materials, protective coatings, alloys, and other developments. It should also keep up with new applications of existing products. New products or applications usually lead to better methods and safer practices.

Increased mechanization requires careful assessment of new potential hazards, especially in high-speed equipment and processes. Use of color throughout the plant, as specified

in the previous chapter will contribute to accident prevention and help develop high standards of operation.

Sometimes mechanical means help to offset potential hazards. Centralized lubrication, floor cleaning machines, centralized spray painting equipment, steam cleaners, and other devices will help make a maintenance program safer. For frequently performed maintenance jobs, permanent accessories, such as hoists, fixed ladders, and catwalks, should be provided.

Maintenance procedures should be reexamined periodically for safer ways to do the job. A special suggestion system should be set up for maintenance men so that they will be able to present new ideas or corrective measures.

Engineering books and service manuals should be a part of the maintenance program. Workers should be familiar with them and use them when necessary.

References

American Institute of Architects, 1735 New York Ave. NW., Washington, D.C. 20006. *Standard Form of Agreement Between Owner and Contractor.*

American National Standards Institute, 1430 Broadway, New York, N.Y. 10018.
"Cranes, Derricks, Hoists, Hooks, and Jacks," B30 Series.
Practice for the Inspection of Elevators, A17.2.
Safety Code for Elevators, Dumbwaiters, Escalators, and Moving Walks, A17.1.
Safety Code for Portable Wood Ladders, A14.1.
Safety Color Code for Marking Physical Hazards, Z53.1.
Safety in Welding and Cutting, Z49.1.
Safety Requirements for Demolition, A10.6.
Safety Requirements for Fixed Ladders, A14.3.
Safety Requirements for Floor and Wall Openings, Railings, and Toeboards, A12.1.
Safety Requirements for Job-Made Ladders, A14.4.
Safety Requirements for Personnel Hoists, A10.4.
Safety Requirements for Portable Metal Ladders, A14.2.
Safety Requirements for Scaffolding, A10.8.
Safety Requirements for Steel Erection, A10.13.
Safety Requirements for Temporary and Portable Space Heating Devices and Equipment Used in the Construction Industry, A10.10.
Scheme for the Identification of Piping Systems, A13.1.
Specifications for Industrial Accident Prevention Signs, Z35.1.

Armstrong, Ralph W. "Masonry Walls Against Wind," *Construction Section Safety Newsletter,* September 1967.

Associated General Contractors of America, Inc., 1957 E St. NW., Washington, D.C. *Manual of Accident Prevention in Construction.*

Huntington, Whitney Clark. *Building Construction: Materials and Types of Construction,* 3rd ed. New York, N.Y., John Wiley & Sons, Inc. 1963.

Laborers' International Union, 905 16th St. NW., Washington, D.C. 20006. (General.)

National Fire Protection Association, 470 Atlantic Ave., Boston, Mass. 02210.
Flammable and Combustible Liquids Code, Standard No. 30.
Life Safety Code, Standard No. 101.

National Safety Council, 425 North Michigan Ave., Chicago, Ill. 60611.
"Contractor's Outline for Accident Prevention."
Industrial Data Sheets.
Acetylene, 494
Aerial Baskets, 572
Asphalt, 215
Atmospheres in Subsurface Structures and Sewers, 550
Barricades & Warning Devices for Highway Construction Work, 239
Blowtorches and Plumbers' Furnaces, 470
Chains (Alloy Steel) for Overhead Lifting, 478
Conveyors, Belt (Equipment), 569
Conveyors, Belt (Operation), 570
Conveyors, Roller, 528
Conveyors, Underground Belt, 447
Cutting and Clearing Vegetation, 575
Diving in Construction Operations, 555
Electromagnets Used with Crane Hoists, 359
Excavation, General, 482
Flexible Insulated Protective Equipment for Electrical Workers, 598
Fixed Ladders and Climbing Devices, 606
Hauler-Loaders, 576
Highways and Municipal and Industrial Roadways, Center Striping of, 221
Hoists, Construction Material, 511
Hot Work Permits (Flame or Spark), 522
Job-Made Ladders, 568
Lift-Slab Concrete Construction, 514
Motor Graders, Bulldozers and Scrapers, 256
Motor Trucks for Mines, Quarries and Construction, 330
Paving with Portland Cement Concrete, 541
Ready-Mix Concrete Trucks, 617
Safety Hats, 561
Safety Nets for Construction Projects, 608
Saws, Masonry (Stationary, Single-Blade Type), 506
Sidewalk Sheds, 368
Silicon Diodes, 581
Steel Plates, Handling for Fabrication, 565
Temporary Electric Wiring for Construction Sites, 515
Tilt-Up Concrete Construction, 513
Tools, Live Line, 498
Tractor Operation and Roll-over Protective Structures, 622
Trench Excavation, 254
Wire Rope Slings, Recommended Loads for, 380
"Self-Evaluation Checklists (OSHA)."

Reigeluth, R. J. *Safety and Economy in Heavy Construction.* New York, N.Y., McGraw-Hill Book Company.

Rossnagel, W. E. *Handbook of Rigging.* New York, N.Y., McGraw-Hill Book Company.

Sack, Thomas F. *A Complete Guide to Building, and Plant Maintenance.* New York, N.Y., McGraw-Hill Book Company, 1963.

Steel Scaffolding and Shoring Institute, 2130 Keith Bldg., Cleveland, Ohio 44115, "Steel Scaffolding Safety Rules."

U.S. Army, Corps of Engineers, Washington, D.C. *General Safety Requirements,* Manual EM385-1-1.

Planning for Emergencies

Chapter

18

Types of Emergencies   444
Fire and explosion . . . Floods . . . Hurricanes and tornados . . .
Earthquakes . . . Civil strife and sabotage . . . Work accidents and
rumors . . . Shutdowns . . . Industrial civil defense . . . Hazardous
materials . . . Weather extremes

Plan-of-Action Considerations   448
Program considerations . . . Chain of command . . . Training . . .
Command headquarters . . . Emergency equipment . . . Personnel
shelter areas . . . Alarm systems . . . Fire and emergency brigades
. . . Plant protection and security . . . First aid and medical
. . . Warden service and evacuation . . . Transportation

Outside Help   462
Mutual aid plans . . . Contracting for disaster services . . .
Municipal fire and police departments . . . Industry and medical
agencies . . . Governmental and community agencies

References   463

No industrial plant or office, and no commercial or mercantile organization is immune from a disaster. Emergencies can arise at any time and from many causes, but the potential harm is the same—people and property. Planning for emergencies, like other management functions, must be done in advance. Only in this way can the potential harm to people and property be minimized.

Planning in advance is necessary—it is not a luxury, rather it is good insurance. Where professionally trained emergency help and assistance may not be available, the need for emergency planning is intensified. A comprehensive management plan is intended to take care of all expected emergency situations. This includes both the spectacular (such as a tornado) and the common accident situation. Quite often emergency planning is assigned to the safety professional. This is fine, but there is a real need for the corporate management to be fully involved in the many decisions that must be made.

The safety of employees, visitors, and customers must be the first concern in planning for an emergency. Care for the injured must be available immediately. In some disasters, evacuation may be necessary.

Next, consideration should be given to protecting the property and the operation. In a new plant, consideration should be given to arranging and locating certain facilities and operations to provide greater inherent safety to the entire operation. In general, all emergency plans will include cleanup details necessary for the situation.

Finally, planning may be concerned with restoring business to normal. In emergencies likely to damage or wipe out a unit or plant, the question of resuming operations under conditions of temporary wiring, lack of heating, or repair and construction work should be considered.

Regardless of the size or type of organization, management is responsible for developing and operating a program, which is designed to meet these eventualities. An effective plan requires the same good organization and administration as any business undertaking. There is no one emergency plan that will do all things for all organizations. Each company must therefore decide on a plan that fits its needs and can be afforded.

Emergency plans involve organizing and training of small groups of people to perform specialized services, such as fire fighting or first aid. Small, well trained groups can serve as a nucleus to be expanded to any size needed to meet any kind of emergency. Even with outside help available, a self-help plan is the best assurance that losses will be kept to a minimum.

An organization will need to develop several plans to control different types of emergencies. While certain basic elements would be common to all plans, the same complete plan could not, for example, be used for both a tornado and a nuclear attack.

Before an organization initiates an emergency plan, it is necessary to evaluate the potential disasters that might occur. The next section, Types of Emergencies, discusses these in detail.

There are many sources of information to help determine the possibility that such an event might occur in your locality—weather records, accident and fire statistics, and industry or local authorities. After listing the potential disasters, some reasonable assessment must be made of likelihood of occurrence. Some adjustments might be made to allow for the seasonableness of certain events.

The next step is to assess the potential harm to people and property. Again some adjustments, perhaps a range in the extremes of most likely and most unlikely, could be completed. The time of day and the event itself are other factors that should be considered in assessing the potential damage. Planning should include all shifts and even for catastrophes which might occur during weekends or holidays when no one or only a skeleton staff may be on hand.

In trying to estimate potential damage to property one would follow the same general procedure. A building may be strong enough to resist a tornado, but a sudden 7-inch rain storm might cause dangerous or damaging flooding. On the other hand a boiler located in an adjacent building, in the event of an explosion, would not harm the main plant.

Next, probable warning time should be considered. For example, a flood may build up over a period of several days while a "bomb scare" affords only a few seconds warning from a telephone call. This warning time

Fig. 18–1.—Refinery installations subject to high water are protected by a barrier made of piled sandbags.

should permit some chance to alert personnel and mobilize the plan. It may be desirable to have two different plans, depending upon the actual time available.

The amount of change that must be made in the operations is another factor. For example, in anticipation of a heavy snow storm, it may or may not be necessary to send employees home early. And some equipment may be left turned on or idling, instead of shutting down completely.

There needs to be consideration to power supplies and utilities that may be involved, particularly those controlling fire protection, lighting, ventilation, and communications.

A basic emergency preparedness plan will usually include—a chain of command, an alarm system, medical treatment plans, communications systems, shutdown and evacuation procedures.

This chapter points out the various elements involved in developing emergency plans. Not every element discussed will apply to every organization. Also several of the functions may well be combined and handled by one person, particularly in a smaller company. Generally the text is directed to the

more elaborate and expanded type of organization and planning.

Types of Emergencies

Before a company begins extensive planning, organizing, and training, it is necessary to determine just what disasters are most likely to occur. In some areas, floods are no problem; hurricanes or earthquakes are not of concern in other regions. However, work accidents, fire or explosion, sudden shutdowns, or acts of aggression might occur anywhere.

Fire and explosion

Except where fires result from large-scale explosions, warfare, or civil strife, the fire emergency usually allows a short time for marshaling of fire fighters and organizing an evacuation if necessary. Many conflagrations originate as small fires; therefore, prompt action by a small, trained group can usually handle the situation. However, plans should include the marshaling of extensive fire fighting forces upon the first indication of any fire growing beyond the "small fire" stage (that is, fires that could positively be controlled by in-house personnel).

The main point is this: *small fires must be checked as soon as they start.* The first five minutes are considered the most important. Good housekeeping, prompt action by trained people, proper equipment and common-sense precautions will prevent a small fire from becoming a disaster.

Specific information on fire extinguishment and control is in Chapter 43, "Fire Protection."

Floods

When a company or plant is located in an area that can flood, it should have the permanent protection of dikes of earth, concrete, or brick-wall construction. Management can obtain the probable high-water mark from the U.S. Weather Bureau or the U.S. Army Corps of Engineers. The latter group also provides valuable assistance in planning flood water control.

Floods—except "flash" floods caused by torrential cloudbursts, bursting of a storage tank, dam, or water main—do not strike suddenly. Ordinarily, there is enough time to take protective measures when a flood seems imminent.

Hurricanes and tornados

Areas most frequently exposed to winds of destructive hurricane force are the Atlantic and Gulf coasts. However, inland locations are not immune to this type of disaster.

The U.S. Weather Bureau and other agencies have developed improved methods of detecting and tracking hurricanes; thus ample warning can be given for maximum protection of property and evacuation from threatened areas.

Companies regularly exposed to this hazard have developed a system of tracking the hurricanes on a map. At predetermined locations, a specified alert condition becomes effective and each supervisor completes a checklist for that alert. As the hurricane progresses through the 100-mile circle, 50-mile circle, etc., the plant is shut down in an orderly manner.

Buildings constructed in areas where hurricanes occur should be built strong enough to withstand these destructive winds and storm tides.

Basic preventive measures include equipping with storm shutters or battens which can be promptly applied, at least on the side from which the storm is expected to approach. If this is not done, failure of windows may lead to lifting of the roof and destruction of the building. Where the roof is lost or damaged, building contents are drenched by the heavy rain that accompanies the storm and by water from broken sprinkler pipes. In addition, roofs should be securely anchored and tall structures (such as chimneys, water towers, and flag poles) designed to withstand high wind velocities.

Although the central Mississippi Valley is considered the tornado area of the country, almost every state has experienced them. The damage is inflicted quickly and is usually restricted to a small area, but the destruction can be massive.

Although the U.S. Weather Bureau has effectively increased its forecasting of tornado conditions and determining possible areas of danger, it is not possible for them to give as much advance warning or to pinpoint the strike area as accurately as they can with hurricanes. Therefore, a company must be prepared to protect its personnel on short notice and to take corrective action to protect and restore undamaged equipment and materials.

U.S. National Weather Service radio bands can be monitored during likely days. In one mid-west city, several large companies have set up a cooperative warning network. A lookout is stationed atop the city's tallest building. Through a central network, not only the member companies, but also the city's radio stations and civil defense, are alerted in the event of an approaching tornado.

Tornado and hurricane experience indicates that emergency plans should include:

1. Procedure for getting personnel to a safe place. If the building is not constructed to withstand the forces, emergency shelters should be located close to the work area. All personnel should be instructed in the procedure to follow, with and without advance warning.

2. Assignment of trained men to take care of power lines—dangling wires are a serious hazard.

3. Assignment of trained men to remove wreckage to prevent injury to salvage and repair workers.

4. Schedule regular meals and rest for the repair crews.

Earthquakes

Most seismic areas in the United States are around the Pacific Coast. Earthquakes generally occur without warning and affect the entire community or large areas thereby making community services unavailable for assistance.

"Earthquake-resistant" construction consists of building a structure so that it "floats" above the bedrock, ballasting it as a ship is ballasted, by making lower stories heavy and upper stories light. Utility lines and water mains should be flexible and laid in trenches that are free of the building, rising in open shafts and connected to fixtures by flexible joints.

The principal dangers from earthquakes are the collapse of buildings, fire originating from broken gas mains, and lack of water to fight any fire. Water reservoirs or emergency water sources should be provided for fighting fire if the municipal supply mains are broken or water pressure is otherwise likely to be disrupted.

Civil strife and sabotage

Riot or civil strife is a recent addition to the list of reasons why a company should plan in advance for an emergency.

Civil strife. While the same considerations for organizing against and combating natural forces are involved, particularly fire, there is one additional element which makes this type of emergency different. The difference is that people are involved. This raises the questions of the right to protect property and the individual's legal right to assemble. A company should obtain from an elected legal authority in the community (district attorney) a statement explaining the company's rights in protecting its property and the company's legal responsibility for the safety of employees and other people—such as customers, supplier salesmen, and visitors—who may be on the company property. A company's legal department can be helpful in determining such a position, but its opinion does not have the force of law.

Some of the problems involved are: business disrupted when an office or plant area is invaded by outsiders; protection against a mob intent on destroying company property; neighboring companies request for assistance of your personnel during a riot; rights and responsibilities of armed company guards.

This type of emergency can be just as disastrous as any other type and should receive advance planning by manufacturing, mercantile, or commercial establishments.

Sabotage. Protection against sabotage is also an important consideration. The saboteur may be a highly trained professional or an amateur. He may be anyone—usually one of the least suspected members of the organization. Because physical sabotage is frequently an inside job or requires the assistance—knowingly or unknowingly—of someone inside the plant, the principal measures of defense must be against entry of persons bent on sabotage. Evidence of sabotage should be reported to the FBI and if defense work is involved, to the Department of Defense.

Work accidents and rumors

The "chain reaction" from a so-called "routine" work accident can result in an emergency situation. (Examples would be a break in a chemical line or toxic vapors from outside the plant entering the ventilating system.) Panic caused by a rumor or lack of knowledge can also create an emergency.

Some of the points to be considered are: auxiliary areas in the building to be used for medical treatment, method of notifying employees of the actual situation, method of quickly taking a head count, and sources of oxygen supplies available on short notice.

Shutdowns

Although a shutdown is not an emergency per se, it can result from an unscheduled action, such as a disaster or strike; hence a fast shutdown procedure should be covered under an emergency plan. This plan should be based on a priority checklist. That is, all of the tasks to be assigned and functions to be performed should be arranged in order of importance so that if time is short, at least the most vital precautions are completed. This "crash" procedure is usually an adaptation of the routine procedure used for scheduled shutdowns, such as for vacation or renovation. Naturally the amount of warning time controls the speed of shutdown. Whenever a plant or other unit or building must be shut down, safeguards against fire take on added importance. Few employees, if any, will be present to discover and deal with a fire which might start. The extent of these measures will vary with the size and purpose of the plant. It is important to organize a formal program for instructing personnel. Examples of items which need attention include removal of lint, dirt, and rubbish; draining and cleaning of dip and mixing tanks and other equipment where flammables have been used; cleaning spray booths, ducts, and flammable liquid storage; closing of gas and fuel line valves; opening switches on power circuits which will be out of service; checking for serviceable condition of sprinkler systems, fire extinguishers, hydrants, alarms, and other protective apparatus; and anchoring cranes.

Prior to the closing, employees are alerted by special instructions to keep their work stations clean and fire-safe.

During the shutdown, continuous inspection of any maintenance or special operations, such as remodeling, must be maintained. Gas cutting and welding should be carefully supervised. The men who remain on duty—the plant protection force, watchmen, maintenance workers, supervisors, or executives—should be briefed in effective countermeasures in case a fire breaks out.

If there has not been sufficient notice to effect a normal shutdown, it may become necessary to allow personnel into the area to perform necessary functions.

Company management should designate someone to authorize the admittance of personnel necessary to handle emergencies arising within the area. The chief of protection or the fire chief should arrange with local police and fire department officials for assistance if an emergency gets beyond local control. It is especially important that arrangements be completed for expediting the admittance of fire fighters and their equipment.

Some companies use plant protection service agencies to prevent loss from theft, fire, and accident hazards during shutdowns. Similar plans should be worked out with these men so that police and fire assistance is expedited when it is needed.

Industrial civil defense

One of the main differences between planning for peace-time emergencies and planning for emergencies resulting from warfare conditions is that war may cripple an entire community. This difference makes it more important that emergency plans take into account self-sufficiency because the outside sources of help—fire and police departments, hospitals and doctors, regular sources of supply for materials and equipment—would not be so readily available.

Industrial civil defense consists of the plans and preparations engaged in by managements of business and industry to achieve a state of readiness. This would enable their plants and facilities, and their employees, to cope with the effects of nuclear attack.

Even if a particular area is not attacked,

plants in the area that has been spared may be requested to furnish transportation to evacuate the injured from damaged areas and to house and feed the evacuees. Plant emergency squads may also be required to go to the assistance of stricken plants. In a major catastrophe there would probably not be enough hospital space available, making it necessary to keep the injured in temporary shelter for a considerable time. In such cases, employees with the proper training might be required to administer sedatives and plasma and to treat minor and major injuries.

Under general plans of the Office of Preparedness, Government Service Administration, authorities in that group would handle the investigation of areas dangerously contaminated with radioactivity, if this were the cause of the trouble. However, it is advisable that plant personnel with responsibility for health and decontamination have some knowledge of this work. Company personnel can be trained in this subject, as well as many others, by the National Staff College, Defense Civil Preparedness Agency, Battle Creek, Mich. 49016.

This chapter cannot give detailed survival plans for a nuclear attack or for protection against chemical and biological warfare. However, it is suggested that the disaster program director consult with the Office of Civil Defense (state, or city offices), which has material and trained personnel to assist a company in formulating its own program. The main point, as far as this Manual is concerned, is that top management must initiate and actively support the program. (Additional program details are given in the next section. Specific hazards of ionizing radiation are discussed in Chapter 37, "Industrial Hygiene.")

It is incumbent upon industry to engage in preparations to protect itself and its employees in event of nuclear attack so as to ensure continued economic production or early resumption of that production.

Hazardous materials

Because there are many chemical substances being used today, there must be concern with the potential usage and handling problems. There are many rules and procedures to be observed, but again ask the question, what if a safeguard fails? What if the container cracks and substances leak out?

In addition to normal hazards, are there potential chemical reactions with other substances that cause still further dangers to people and property?

Chemical hazards are discussed in Chapters, 37, 38, and 39.

Weather extremes

Throughout a year there may be some unusually severe and unexpected weather events that may require some changes in normal operations. Some examples follow.

In North Dakota, the temperature may occasionally drop to 35 degrees below zero, yet most activities and travel are not normally affected. But, if the wind increases in strength or the temperature drops suddenly, there may be a need to assist people in travel or other outside activity. A special alert might be made to employees prior to leaving from work. And often when or how they are to be notified about the company opening in the morning.

On the other hand, in the event of extremely heavy snowfall, what changes might be made in operations? What should employees be told prior to leaving for home?

Or suppose an unusually heavy rain strands hundreds of customers in a store just a few minutes before closing. Are supervisors and clerks prepared to handle the situation? May they allow telephone calls in and out? How do they control the crowd?

Hail or wind may start breaking glass windows while customers are shopping. What is the immediate action?

Suppose that adverse weather caused a power failure or someone suddenly shut off all power and lights while crowds were shopping? The emergency lighting system may operate as intended but employees, particularly key supervisors, must understand emergency plans and be prepared to act responsibly.

Plan-of-Action Considerations

Following the assessment of potential emergencies, the next step is to translate these needs into a plan of action. Management should be in charge of drafting a policy and

getting the plan underway. It will usually be necessary that the union leaders (if any) be involved in the planning process. Generally, someone should be appointed emergency planning director or coordinator, perhaps with help from an advisory committee. Usually because of their experience and training, the safety, medical, fire, and security departments will be involved. Of course, because production and maintenance will be affected, they must be consulted. Also, the legal staff needs to be aware of the plan.

And finally, contacts with local law-enforcement agencies, fire, and civil defense are necessary.

Program considerations

The cost and effort involved in giving immediate attention to emergency planning can be justified by weighing the cost of preparedness against the possibility of contributing to the yearly losses from accidents, fires, floods and other catastrophes.

The preliminary aspects of emergency planning have now been discussed—the need for advance planning and an evaluation of the type of emergencies and their potential harm to people and property. The next step, then, is to translate this need into a working plan within the organizational structure. In some cases, this requires working with other local agencies to most fully protect your operations.

Advance planning is the key. It is necessary to develop a written set of plans for action. The plans should be developed locally within the company (and corporate structure) and be in cooperation with other neighboring or similar organizations and with governmental agencies. It may not always be possible for them to fully cooperate or participate, but through planned action each organization should be aware of certain available assistances. And as a result, the company may need to plan to be largely dependent upon its own resources to provide the internal safety.

Often an emergency manual or handbook will be developed for the plant or organization. The following outline covers many of the items that might be included, but other items may be needed as dictated by the expected emergencies and the available resources.

1. Company policy, purposes, authority, prin-

cipal control measures, and emergency organization chart showing positions and functions.

2. Some description of the expected disasters with a risk statement.

3. A map of the plant, office, or store showing equipment, medical and first aid, fire control apparatus, shelters, command center, and evacuation routes.

4. A list (which may also be posted) of co-operating agencies and how to reach them.

5. A plant warning system, type of signals.

6. A central communications center, including home contacts of employees.

7. A shutdown procedure, including security guard.

8. How to handle visitors and customers.

9. Other related and necessary items needed locally.

Some of these items will be discussed in more detail on the following pages.

The plans should be rehearsed. Realistic conditions should be used so to further learn the effectiveness of the plan. For example, maybe the emergency lights failed when needed, or the telephone service failed; but such are also the conditions that might occur in a real disaster. Therefore planning should include all possible, as well as probable contingencies.

Chain of command

Once the decision has been made to establish a disaster plan, a director or coordinator should be appointed and an advisory committee, representing various departments established.

The director should be a member of top management, whether it be a one-building or one-plant company or a national corporation, because he will have to be able to delegate authority, and speak for the company. The head of the disaster-control organization must be a cool, quick-thinking person and should be sufficiently robust to withstand the arduous duties that will fall upon him if an emergency arises. The emer-

gency director's regular duties should be such that the greater part of his time will normally be spent at the unit he is responsible for. However, an alternate is always named in the plan—and the alternate should be a person who has the authority and qualifications similar to the director and he should be trained with the director.

The director (and his alternate) should be the first to be trained in his duties. Continuous liaison should be maintained with local Civil Defense authorities, if possible, to make sure that the company plans are coordinated with those of the community and to keep the company informed on new developments.

The director may be responsible for:

Communications.
Fire fighting.
Rescue service.
Guard service.
First aid and medical service.
Warden service.
Demolition and repair.
Transportation.
Investigation.
Public relations.

All of these functions are likely to be essential although some may be combined. The person (and alternates) responsible for each function should be selected with great care and trained by the director. These chiefs should be familiar with all parts of the plan and should have experience in the fields in which they are to serve.

Assigned personnel must be trained to carry out their duties in accordance with the overall emergency plan. In small operations, where there may be no regular guards or fire fighters, the operating personnel will be the people trained to take care of these duties. Of course, the number of members on each of the teams depends on the circumstances of each plant. Each team captain should select his own personnel from the available volunteers, supervise their training, and procure their equipment.

Women can work with first aid squads, as wardens, and in other emergency posts. Although they would not normally be assigned to service in rescue squads because the work usually demands strenuous physical effort,

---

IN CASE OF FIRE OR OTHER EMERGENCY

✔ **KEEP YOUR HEAD** — avoid panic and confusion.

✔ **KNOW THE LOCATION OF EXITS** — be sure you know the safest way out of the building no matter where you are.

✔ **KNOW THE LOCATION OF NEARBY FIRE EXTINGUISHERS** — learn the proper way to use all types of extinguishers.

✔ **KNOW HOW TO REPORT A FIRE OR OTHER EMERGENCY** — send in the alarm without delay; **notify the CHIEF OF EXIT DRILLS.**

✔ **FOLLOW EXIT INSTRUCTIONS** — stay at your work place until signaled or instructed to leave; complete all emergency duties assigned to you and be ready to march out rapidly according to plan.

✔ **WALK TO YOUR ASSIGNED EXIT** — maintain order and quiet; take each drill seriously — It may be "the real thing."

REMEMBER — IT IS PART OF YOUR
JOB TO PREVENT FIRES

Fig. 18–2.—Sample emergency exit notice for general posting.

they could be used in light salvage operations

Because wholehearted cooperation of personnel is necessary to the successful operation of an emergency plan, shop stewards or other representatives of labor should take part in the planning. They must be made to see that whatever measures are taken are for the protection of the lives and jobs of the workers as well as for protecting property.

Training of key people will be of little value unless it reaches down to all employees. The better informed and prepared the work force, the less chance of panic and confusion during the emergency.

Provisions should be made for emergency reporting centers so that employees will know where to report should the disaster occur while the employees are away from the plant. Reporting centers give employees a feeling of security and continuity, and aid the company in taking a "roll call." To facilitate these arrangements, each employee should carry an identification card which gives specific instructions on where to report, list of other reporting centers, basic employment record, and designation of the employee's next of kin in case they must be contacted or receive money due him. The reporting center will keep a duplicate record for each employee assigned to report there.

A one-plant company or a small company can consider using the home of a member of management, a supervisor, or an employee.

Training

One of the most important functions of the director and his staff, on both the corporate and plant levels, is training. Training for each type of disaster is essential in developing a disaster-control plan and keeping it functioning. Employees must be taught to realize that an emergency plan is vital and real — it cannot exist usefully if it remains a remote idea. Training and rehearsals are time consuming, but they keep the program in good working order.

EMERGENCY EXIT INSTRUCTIONS
MACHINE SHOP—DAY SHIFT

Read Carefully

The following persons will be in command in any emergency, and their instructions must be followed:

CHIEF OF EXIT DRILL—H. C. Gordon, General Sup't.
MACHINE SHOP EXIT DRILL CAPTAIN—R. L. Jones, Foreman
MACHINE SHOP MONITORS—Dave Thomas and A. L. Smith

In event of FIRE in machine shop

✔ **NOTIFY THE GENERAL SUPERINTENDENT'S OFFICE**

✔ **PUT OUT THE FIRE, IF POSSIBLE**—If the fire cannot quickly be controlled, follow instructions given by Exit Drill Captain R. L. Jones or by the shop monitors. Leave by the exit door at the south end of the shop; if it is blocked by fire, use the door through the toolroom to the outside stairway.

In event of FIRE or EMERGENCY in other sections of building

The general alarm gong will ring for two 10-second periods as an "alert" signal. Continue work, but be on the alert for the "evacuation" signal, which will be a series of three short rings. At the evacuation signal:

✔ SHUT OFF ALL POWER TO MACHINES AND FANS

✔ TURN OFF GAS UNDER HEAT TREATING OVENS

✔ CLOSE WINDOWS AND CLEAR THE AISLES

✔ FORM A DOUBLE LINE IN THE CENTER AISLE AND FOLLOW MONITORS AND EXIT DRILL CAPTAIN TO EXIT—Walk rapidly, but do not run or crowd; do not talk, push, or cause confusion!

After leaving the building, do not interfere with the work of the plant fire brigade or the city fire department. Await instructions from the General Superintendent or your foreman.

Returning to the building

Return-to-work instructions will be given over the loudspeaker system or by telephone from the Superintendent's office.

FIG. 18–3.—Sample individual instruction notice for general posting.

FIG. 18–4.—Underground emergency command headquarters has direct lines to company and municipal emergency service centers.

Courtesy Western Electric Company

Practice alerts should be conducted to make sure that the employees know where to report and what their duties are. Even the most carefully prepared plans can develop flaws when put into practice, and only rehearsals can show them up. The first one or two practices should be announced—lest there be a panic—but there should be no warning of succeeding ones. Officials should determine to their complete satisfaction that the disaster plan will work under emergency conditions. Once this has been determined, the plan should be maintained with periodic tests, staff discussions, and an occasional disaster problem. If this is not done, all the planning effort will have been wasted (see Figs. 18–2 and 18–3).

Management should assure employees that the company is doing everything possible to prevent injury to them, that every employee is an essential and necessary part of the team, and that the disaster-control organization is ready for any emergency. Such assurance will go a long way toward developing a state of mind that will not panic. Then when disaster strikes, emergency forces snap into action, workers file quietly into their shelters or other designated areas, firemen are ready with hoses and equipment, and first aid squads stand by ready to aid the wounded.

Such planning is further evidence of management's concern for employees.

Command headquarters

The average command headquarters will not withstand a direct nuclear attack, but it should still be planned for any of the other emergencies which may occur.

Coordination of the disaster control organization should come from a well equipped and well protected control room. The headquarters should be equipped with telephones, sound-powered phones, public address system, maps of the plant, emergency lighting and electric power, sanitary facilities, a second exit, and two-way radios for communication both locally and with Civil Defense authorities. (See Fig. 18-4.)

Good communications are necessary for effective control and flexibility in a disaster situation. Communications include the telephone, radio, messengers, and the plant's alarm system (discussed separately later in this chapter). The disaster plan should provide for adequate telephones in emergency headquarters to handle both incoming and outgoing calls. Panic and disintegration of the organization will develop quickly if these calls are not handled with dispatch. An accurate log is kept of all incoming and outgoing messages (Fig. 18-5).

Some means of communication independent of normal telephone service must be available during an emergency, such as battery-operated radio. That is, the disaster plan must anticipate the possibility of losing normal telephone communications and electric power.

Emergency equipment

This checklist will include equipment and material to be ordered as well as shutdown actions to be performed.

● For example, where it is not feasible to keep on hand the necessary emergency equipment and materials, a list should be maintained of sources from which these items can be obtained on short notice. These sources must be outside of the immediate area because of the rush for such material which would occur after receiving a flood alert. This equipment and material would consist of sandbags, battens for windows and doorways, boats, tarpaulins, fuel-driven generating equipment (such as gasoline-powered arc welding machines or motor-generator sets), standby pumping equipment, a supply of gasoline in safety containers to fuel this equipment, lubricating oil and grease, rope, life belts, portable battery-operated radio equipment, and audio speakers.

● Some of the items on the shutdown part

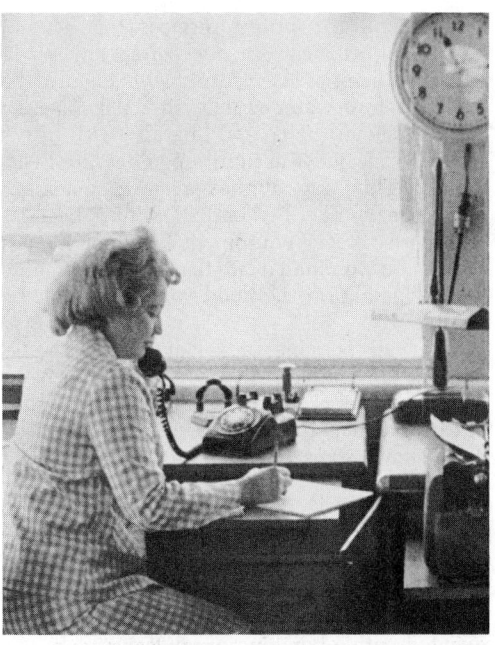

Fig. 18-5.—An important function of the communications setup is keeping an accurate log of in-coming and out-going messages, both radio and telephone, during the emergency.

Courtesy Uniroyal Incorporated.

of the checklist would be: valves to be closed; protection of equipment which cannot be moved; doors, windows, ventilators to be closed and battened to keep out looters as well as water; vents and breather pipes to be plugged. Included with the checklist would be a list of telephone numbers of supervisors and key employees to be notified.

Provision should be made for moving tank cars to higher ground and anchoring them if there is any possibility that they still might not be out of the flood area. Portable containers should also be moved above the high-water mark, as should buoyant materials and chemicals that are soluble in water.

Storage tanks under the probable high-water mark (including underground tanks) should be specially anchored to prevent floating. Auxiliary dikes of sandbags or dirt should be built around key areas (see Fig. 18-1).

Other procedures that must be included:

electric and gas utility services that should be shut off at the *main line* before any water reaches them. Hot equipment should be cooled before water reaches it. All machine surfaces should be coated liberally with heavy grease, especially around openings to bearings. (This step applies even to machines which may not be under water, because dampness affects equipment.) Open flames should be eliminated so that any flammable liquid floating on the flood waters will not be ignited.

If at all possible, a salvage crew should remain at the site to continue preventive operations after the operation has been shut down and take further necessary steps if the flood shows signs of exceeding the estimated high-water level.

Personnel shelter areas

While there could be times during natural disasters when employees would be moved to shelter areas, shelters are more often associated with a bombing or nuclear attack.

The average company would find it impractical to build a shelter to provide protection against a direct hit or "near miss," but consideration should be given to shelters for protection against fallout or other disasters. A shelter designed and equipped to provide protection against radioactive fallout will also serve as a personnel refuge for other types of emergencies a company might face. Shelters are designated by the sign shown in Fig. 18–6.

Basements, tunnels (if they cannot be flooded), and inside areas of multistory buildings of concrete construction are examples of possible shelter areas in existing buildings. The denser the shield, the better it protects.

Many companies have not seriously considered constructing employee shelters because they do not think that the expense could be justified by the limited use. However, this is not necessarily the case; the shelter can have daily use as a locker room, training room, meeting room, cafeteria, or employee lounge. There might also be a psychological value in having employees use the shelter area on a daily basis; the familiarity tends to minimize any depressing effect which use of the shelter might have in an actual emergency. Although most industrial employee shelters are designed

Fig. 18–6.—Civil Defense Fallout Shelter designation is used nationwide to mark shelters in both public and private buildings. Shelters are surveyed and approved by the Office of Civil Defense.

to give protection only from fallout, companies who have incorporated shelters in the construction of new buildings have found that with a small amount of additional expense they can also protect against blast.

Each employee shelter will have to have one person, with assistants, designated as the shelter manager. These managers will have to be carefully trained because they will be faced with all the psychological problems of life in close quarters, plus mass feeding, distribution of water, arranging sleeping accommodations, assigning duties, control of supplies, bolstering morale, and handling the personal problems that each individual will bring with him. The shelter manager may also be faced with emergency medical problems where training in first aid procedures is necessary.

Alarm systems

In most industrial operations a definite fire alarm system is set up, using existing signaling systems such as a plant whistle; however, to avoid confusion with the regularly used signals, some plants have special codes or other signaling devices. This type of signal also may indicate the location of the fire, or separate signaling devices may be used for the different building or working areas within the company property.

The alarm system that activates the emergency plan may or may not go through the communications center, but rather should be touched off in the emergency headquarters office. Alarm systems should be provided in all buildings (or fire-resistant sections of buildings).

Electric alarms are preferred to mechanical ones except in a shop having one large open area where there is only one alarm-summons station and one alarm-sending device, such as a manually operated gong. Manually operated alarms should supplement electric alarms. Closed circuit systems of the type specified by National Fire Protection Association standards are recommended; see References at the end of this chapter.

Companies in areas where municipal fire departments are available usually have a municipal alarm box close to the firm's entrance or in one of the buildings. Others may have auxiliary alarm box areas, connected to the municipal fire alarm system, at various points on the premises. Another system often used is a direct connection to the nearest fire station which may register by a water alarm on the sprinkler system or be set off manually. If possible, the fire alarm system should be connected with the local fire fighting alarm, and have an independent power supply.

In large cities private central station services are available and provide excellent protection. These central stations receive signals from plant fire alarm boxes, watchmen, sprinkler head operations, and other hazard control points in the plant. Being able to give undivided attention to matters of plant security, they can relay information to fire or police departments without delay. The signal received at the fire department or assistance agency should locate specifically the site of the fire, or at least the building or area, so the fire can be found quickly.

Automatic sending stations (thermostatic detectors) may be used, but should not interfere with the sending of the manual alarm.

Regular checks should be made on the alarm system. All stations should be inspected on a monthly basis by a responsible person, and the overall system should have a daily test to ensure it is in proper operating condition. These daily tests should be conducted at a prearranged time and under a variety of wind and weather conditions to determine whether the signal can be heard in all parts of the plant at all times.

Fire and emergency brigades

Because a fire can start from so many causes, fire prevention and fire protection must receive major attention in any emergency program.

The company fire chief must be able to command men as well as have special training in fire prevention and protection. A person who has had experience in city or volunteer fire department work, or a military service veteran with experience in fire fighting is a good choice. In a smaller company or plant, a master mechanic, maintenance department head, or other employee with mechanical experience can be a good part-time fire chief.

The fire chief should have one or more assistants who should have complete knowledge of the plant and equipment, command the respect and obedience of the men under them, and are qualified to perform the duties of the chief officer if he is absent.

The size of the plant and the fire potential presented by the occupancy determine the kind of plant fire fighting brigade. The majority of plants may require only first-aid fire fighters under the direction of departmental foremen or managers. In larger plants, the fire brigade organization, directed by a full-time fire chief, is composed of full-time and emergency members. The full-time members maintain fire brigade equipment and are responsible for the permanent fire protection of the plant. Emergency members report for fire duty when the alarm is sounded.

The fire brigade apparatus should be selected only after a study has been made of plant conditions, to thus make sure it will be adequate for any emergency. Advice and assistance can be obtained from the local fire department, the NFPA, Office of Civil Defense, insurance companies, or perhaps a neighboring plant. Two fire and rescue trucks are shown in Fig. 18–7.

The large plant fire brigade is usually organized into squads, each with specific duties.

Fig. 18–7a.—Many companies have found they can add a combination rescue-truck to their emergency equipment without exhorbitant cost by obtaining a used truck from a local fire department that may have "outgrown" it. The company equips the truck to handle its particular anticipated problems.

Courtesy Armstrong Cork Company R&D Center.

Fig. 18–7b.—A fire and rescue truck for in-plant use carries hose, fire extinguishers of various types, and miscellaneous emergency equipment.

Courtesy Victor Adding Machine Co.

The evacuation squad evacuates all employees from the emergency area as quickly and orderly as possible, without injury. They search closed areas, such as washrooms, to determine that everyone has been evacuated.

Utility control squad members are usually maintenance personnel, who are familiar with plant piping systems and the control of process gases, flammable liquids, and electricity.

Sprinkler control squad members must understand the automatic sprinkler system—the direction of rotation of the valve they are to operate, the use of sprinkler stops, and the replacement of sprinkler heads, if this is not a maintenance department function.

Fig. 18–8.—Employees can gain experience in the operation of fire extinguishers at the time extinguishers are due for recharge. Many companies schedule such drills twice a year.

Extinguisher squad. Portable fire extinguishers are frequently operated by designated employees who work in the vicinity. However, as the size of the plant increases, it is advisable that special squads be selected for handling fire extinguishers (Fig. 18–8).

Hose squad members are trained to operate fire hydrants and hoses. They should drill frequently with wet hose lines so they have the feel of a charged hose. After they become proficient in handling the hose, drills once or twice a year may be ample (Fig. 18–9).

The salvage squad is trained to protect as much stock and equipment as possible by controlling the directional flow of water and by covering stock with tarpaulins. Training should include proper methods of throwing tarpaulins and planned use of them in directing the flow of water. Members should also be familiar with the location of sawdust or other absorbent material and be aware of its value in controlling water on floors.

The brigade-at-large or rescue team consists

of maintenance personnel or specially trained men.

The main functions of this unit are to extricate casualties and eliminate hazards to other workers involved in the control of the emergency. Members respond to all alarms with a utility truck containing such rescue equipment as: ropes, chains, block and tackle, ladders, cutting torches, saws, axes, and jacks. The amount of equipment, of course, will depend upon the size of the plant and the hazards involved, but every effort should be made to anticipate possible problems. Under the direction of the brigade officer, men in this unit also control utilities, ventilating fans, and blowers; they close fire doors, windows, and other openings in division walls; they open windows and doors leading to fire exits; and, where escapes are the swinging section type, they should be the first to operate the escapes and to secure the steps to the ground.

Dependent upon the size and inherent hazards of the plant facility, it may be necessary to train squads in the handling and erection of ladders, use of foam lines, and recharging of foam generators.

Manning of fire pumps. There should be at

457

FIG. 18–9.—Fire brigade hose drill provides familiarity with equipment—makes the group more effective under actual emergency conditions.

Courtesy Firestone Tire & Rubber Company.

least two competent men for pump duty in the main pump room, and a man for each pump located elsewhere.

Training of fire fighters. The newly organized brigade should go through complete drills, preferably weekly. Later, less elaborate drills may be held at less frequent intervals. Drills should be held at unannounced times. They should be thorough in every respect, closely approximating fire conditions.

No matter how thoroughly the industrial fire brigade is trained, there still must be close cooperation between it and adjoining or nearby plants and the public fire department. As stated before, *it is recommended that the municipal fire department be called immediately upon alarm of every fire.*

Watchmen should be instructed in case of fire to open yard gates and be ready to direct fire apparatus. Where plants have railroad tracks, cars should not be permitted to block

crossings that may be needed in an emergency.

The brigade chief will be in full charge at a fire until the officer in charge of the public fire department takes over. He then serves as an advisor on plant processes and special hazards.

The fire station itself should be centrally located, but not exposed to possible fires. It should be of fire-resistant material or located in a sprinklered part of the plant and protected with portable extinguishers. A larger plant may require mobile units, such as light hand-drawn trucks outfitted for the special hazards of the plant.

Plant protection and security

Industrial security is management's responsibility. Government agencies can provide assistance and advice in establishing a policy. Since the basic problems of security are protection of property and control of persons, a company does not have to establish a new department to handle this function. A company's emergency security force could be built around the present security force.

The men need training in maintaining order, handling crowds, and coping with the threat of panic. They should also be prepared to prevent looting. They map emergency routes to shelters, both inside and outside the plant grounds.

A watchman of low mentality may be more of a liability than an asset. Fires have often been caused by watchmen smoking while on duty, or overlooking fire causes. Their failure to discover fires promptly, their shutting off of sprinklers without ascertaining whether the fires have been extinguished or their ignorance of the proper sprinkler valves to close after a fire is extinguished is often responsible for heavy water loss.

As a supplement to automatic alarm and signal systems, the employment of an intelligent and physically fit watchman can prevent or minimize fire loss. The watchman should be trained and made aware of his responsibilities. Because fires which start when the plant is idle produce more damage, the watchman becomes an important part of the fire prevention and detection organization (see Fig. 18–10).

The guard or watchman functions include protection against pilferage, burglary, vandalism, and espionage. The time of the inspection rounds should occur irregularly. The entire inspection should not create a detectable pattern.

The first round immediately after the plant closes is the most important. Most fires are likely to start just after employees have left, from machines or processes running unattended, or from careless smoking.

The guard or watchman should have enough time on his rounds to make a thorough inspection of the premises. His route should require no more than 40 minutes and should take him close to all hazardous occupancies. He should be provided with an approved flashlight or other illuminating device, and, where practicable, plant lights along his route should be left on.

The watchman can also look for violations of smoking rules, improper storage of flam-

Fig. 18–10.—Watchmen and guards should be acquainted with sprinkler system controls. The control panel behind this guard indicates if there is tampering with any master sprinkler control valve in building.

mable material, leaking oil, gasoline, gas, or other flammable materials, and he should report unsatisfactory conditions to the management. He should be physically capable of turning in a fire alarm, dealing immediately with small fires, and with such matters as shutting off gas and closing fire doors.

Failure of a guard to report on schedule at the end of his patrol calls for immediate investigation.

Additional protection and security measures call for closing off certain windows and other openings in plants which are not vital to operation and limiting the number of plant entrances and exits. (All measures must be consistent with good fire prevention practices.) Installation of protective wire mesh over windows along public thoroughfares is recommended. Floodlighting of critical parts of plants at night is necessary for good protection.

First aid and medical

A helpful publication in establishing a disaster medical service is the *Guide to Developing an Industrial Disaster Medical Service* compiled by the American Medical Association's Council on Occupational Health. This guide can also assist a company in evaluating its readiness and will reveal weak areas. Examples of forms and casualty tags are included. Although the guide is designed to combat peacetime disasters, it provides a basis on which to expand the medical services to meet the devastation following a nuclear attack.

First aid and medical service should be headed by the company doctor, if available, as discussed in Chapter 21, "Occupational Health Services."

In the organization of the medical phase of the emergency plan, those responsible must select and train personnel; decide what measures, equipment, and supplies are needed; and establish first aid stations and a treatment center.

All employees should be encouraged to enroll in a first aid course. People assigned to first aid and medical units should pass standard and advanced first aid courses. Local chapters of the American National Red Cross provide excellent training in this regard.

If a major disaster occurs, there may not be enough trained doctors and nurses available. In such an eventuality, care beyond the first aid level will have to be provided by nonmedical people who have received additional training. Such a medical team can be developed by recruiting volunteer medical aides, preferably with some experience. In addition to first aid training, more advanced instruction by regular medical personnel or local hospitals should be provided.

A major consideration to keep in mind when establishing an emergency medical program, is that while most of the medical aid will be given in a central medical station, some of it may be given on the job site, possibly under hazardous conditions.

Plans should be made for representatives of the medical team to check all personnel at the disaster scene for trauma and to provide a written clearance for them to leave the plant when they are able to leave. In this duty, representatives of the investigation team may interview personnel before they leave to be sure that any needed eyewitness information is recorded.

Welfare and medical service includes the investigation of needs for prevention of epidemics, food inspection, and sanitation inspection. In the planning stages, company trucks, if any, should be designated as ambulances and the necessary equipment for them supplied. Two-way radio communication for such ambulance service is essential. Provision should be made for nonperishable food and water rations. There should be close coordination of plant first aid measures with the local civil defense, health, and medical services.

The chemical service responsibility of this unit requires that gas masks be provided in case tear gas is used as part of the sabotage effort and that trained personnel, equipment, and supplies for chemical defense and decontamination be on hand. There should be a plan for priority sequence in decontamination of the plant—that is, water supply, power plant, machinery areas, warehouse areas, etc. Medical team personnel will also be responsible for any radiological monitoring thought to be necessary after a nuclear attack. The mere knowledge that such monitoring equipment is available is a morale-builder.

After disaster, no one should be permitted to drink water until it has been examined.

Warden service and evacuation

The warden service is responsible for maintaining employee control during emergencies, including (a) guiding employees to shelter, (b) directing employees away from hazardous areas, and (c) averting panic. In smaller companies, the wardens could also have the responsibility of taking charge of shelters. In some cases, the warden service may be responsible for seeing that shutdown of processes and equipment is carried out smoothly.

This type of service was devised primarily for areas of high population densities, such as commercial structures and factories with

Fig. 18–11.—Exit routes and escape passages properly identified expedite rapid evacuation. Especially important is provision for an alternate exit path if the primary route is blocked by flame, hot gases, or debris.

Courtesy Dunwoody Institute

a great many employees, and residential areas. Some plants have a high concentration of employees. For these, the use of warden teams is certainly a good idea. In some plants, however, the concentration is very low. In such plants or departments with low personnel concentration, warden service is not necessary. In these cases, the operators themselves will have to be trained in shutdown details.

Management, in checking and providing for safe exits and evacuation drills, should refer to the National Fire Protection Association's standards and to local codes. Smooth, safe functioning of an evacuation plan requires a thorough knowledge of all plant operations and types of employees, number and types of exits available, width of exits, proper location of exits, possible alternate exits, and location of hazards, as well as a knowledge of warning and evacuation facilities (see Fig. 18–11). The subject of building exits is covered on pages 354, 450–452, and 1325–1327.

Most plants have a rigid rule that only especially appointed people on the fire brigade shall go to the vicinity of the fire, and that everyone else shall proceed on signal to a refuge location in accordance with the organized evacuation plan.

Transportation

Disrupted transportation facilities or restrictive traffic regulations could make it impossible for many employees to get to work. The company may need to provide transportation with company trucks and cars. Advance planning for car pools and pick up stops will greatly facilitate such a step.

The transportation responsibility includes arrangements for ambulance service, preparation for transportation of employees to and from work, and movement of emergency service crewmen as needed.

The transportation unit should consist of a group of regularly assigned drivers. Station wagons, from which the seats can be removed, and company trucks can be used when it becomes necessary to handle stretcher cases in evacuating any injured. The unit will be the means of getting auxiliary firemen, first aid crewmen, and salvage and rescue workers to the scene of the disaster at the earliest possible instant. The unit will also be used to deliver needed equipment and material from outside suppliers.

Planning for adequate transportation service and traffic control requires liaison with the public police department, civil defense authorities, and possibly the military.

A source of motor fuel will have to be anticipated. One company uses oversized underground gasoline storage tanks for the company service pump. Emergency electric generating equipment also runs from gasoline-run engines because in this case, gasoline is more readily available.

Outside Help

A company's chance of survival and recovery is greater when knowledge, equipment, and personnel are pooled with its neighbors. Therefore, emergency plans should include a provision for exchanging aid with other plants in the industrial community.

Mutual aid plans

A number of industrial communities are organized to assist their members in the event of emergency or disaster. These organizations include manufacturing plants, large offices, stores, hotels, utility companies, chemical plants, law enforcement organizations, hospitals, newspapers, radio stations, and television stations. They operate independently of or as supplements to any civil defense groups.

One thing that has been learned from past disasters is that it is impossible to have adequate supplies available for a really large disaster. The best defense is to have adequate supplies in other areas committed for standby use, with communication channels and a plan for their rapid transportation to the stricken area. Especially important are adequate medical supplies and fire-fighting equipment. A plan for rapid and accurate communication is a necessity, as was discussed earlier in this chapter.

Planning with neighboring companies and community agencies for mutual aid during a disaster should include establishment of an organizational structure, standardization of an identification system, a communication system, standardization of procedures and equipment (such as fire hose couplings), formulation of a list of available equipment, stockpiling medical supplies, sharing facilities in an emergency, and cooperative test exercises and training.

Frequently these "cooperatives" establish a task force composed of men from each member company. Training is supplemented by detailed written instructions. Bulldozers, floodlights, and tools are marked by each plant for emergency use of crews. Training on a community basis might include instruction by members of the public fire department to plant fire brigade members and also some actual training by members of a construction or wrecking company to show the salvage and rescue crewmen how to handle heavy weights and to work safely among debris.

Contracting for disaster service

Some companies contract for disaster service. The service is paid for by a fixed annual retainer, plus additional pay for the actual hours worked. For example, a wrecking company can be engaged to supply the men and equipment necessary to clear debris created by a disaster. When a company contracts for such service, it can remove the burden of maintaining a great deal of idle equipment between emergencies and providing trained manpower, and reduces the possibility of emergency equipment being destroyed or damaged in the very emergency for which it was to have been used.

Municipal fire and police departments

Firemen from the station most likely to respond to an alarm should be fully acquainted with all fire hazards in the plant. Cooperation may be encouraged by inviting local fire officials to inspect the company area. As a result, they can become familiar with the location, construction, and arrangement of all buildings, as well as all special hazards, such as flammable gases, liquids, and materials. They can make sure that company equipment is compatible with the municipal equipment. Therefore, the local fire department can formulate an efficient plan of attack before a fire occurs. Such procedure is far better than waiting for fire to break out, and then running the risk of misunderstanding the situation and initiating improper methods of extinguishment.

Public fire department rescue equipment can supplement plant rescue units. Local fire and police departments can also help in training company forces, as discussed previously.

The public police force can aid in putting down large-scale disturbances and in assisting with evacuation from the plant premises in the event of a major disaster. Planning for this outside help should include arrangements for traffic control, particularly where a plant parking lot empties immediately onto a public highway.

Industry and medical agencies

• In 1970, the Manufacturing Chemists' Association created the Chemical Transportation Emergency Center (CHEMTREC). Under this program a national center located at 1825 Connecticut Ave., N.W., Washington, D.C. 20009 (MCA headquarters) can relay pertinent emergency information concerning specific chemicals upon request in particular information on the hazards, and information to take immediately to control the emergency. A phone is available for the 24 hour service. It is intended primarily for use by those who transport chemicals, but others may have need for the information.

• The Toxicology Information Conversation Online Network (Toxline) has been designed to provide current and prompt information requests on the toxicity of substances. It is intended to be used by health professionals and other scientists working with pollution, safety, health and other disciplines. The service is under the auspices of the National Library of Medicine, 8600 Rockville Pike, Bethesda, Md. 20014. The service is accessible via terminals on line through a national telephone-based network. A fee is required for use.

Governmental and community agencies

During a community-wide disaster, a large number of governmental and private agencies are available to assist industries; these include the Office of Civil Defense, the U.S. Army Corps of Engineers, the Salvation Army, the American Red Cross, the U.S. Public Health Service, and the U.S. Weather Bureau. To be effective in coping with an industrial community disaster, the efforts of all of these groups must be coordinated and directed toward a common end. Therefore each plant should have an up-to-date listing of all cooperating agencies; the administrator's name, address, and telephone number; and the task assignment of the agency. If possible, these people should meet periodically to discuss mutual problems and disaster control techniques.

The company's emergency planning director should become thoroughly familiar with the authority, organization, and emergency procedures that are established by law and which will become effective upon declaration of a civil defense emergency.

In wartime, the federal, state and local governments are responsible for relief measures to meet needs resulting from enemy attack. The Red Cross has offered to assist the government in providing food, clothing, and temporary shelter on a mass-care basis during the emergency period immediately following enemy attack. In many communities, local Civil Defense officials have requested Red Cross chapters to assume all or part of this responsibility, acting under Civil Defense authorities.

In natural disasters, the American Red Cross is responsible for assisting families and individuals to meet disaster-caused needs that cannot be met through their own resources. These relief operations are coordinated with the activities of the local, state, and federal governments. When a disaster occurs, the local chapter of the American Red Cross aids disaster sufferers. The resources of the national organization are available to supplement chapter assistance.

References

American Insurance Association, Engineering and Safety Service, 85 John St., New York, 10038.
Fire Hazards and Safeguards for Metalworking Industries,
Technical Survey No. 2.
Fire Safeguarding Warehouses, Technical Survey No. 1.

American Medical Association, Council on Occupational Health, 535 North Dearborn St., Chicago, Ill. 60610. *Guide to Developing an Industrial Disaster Medical Service.*

The Conference Board, 845 Third Ave., New York, N.Y. 10022.
Studies in Business Policy, No. 55, "Protecting Personnel in Wartime."

Factory Mutual System, 1151 Boston-Providence Turnpike, Norwood, Mass. 02062.
Handbook of Industrial Loss Prevention.
Loss Prevention Data.

National Fire Protection Association, 470 Atlantic Ave., Boston, Mass. 02210.
Auxiliary Protective Signaling Systems, Standard No. 72B.
Central Station Protective Signaling Systems, Standard No. 71.
Explosion Prevention Systems, Standard No. 69.
Fire Protection Handbook, latest ed.
Guard Operations in Fire Loss Prevention, Standard No. 601A.
Guard Service in Fire Loss Prevention, Standard No. 601.
Life Safety Code, Standard No. 101.
Local Protective Signaling Systems, No. 72A.
Management Control of Fire Emergencies, Standard No. 7.
Management Responsibility for Effects of Fire on Operations, Standard No. 8.
Private Fire Brigades, Standard No. 27.
Remote Station Protective Signaling Systems, Standard No. 72C.
Sprinkler System Installation, Standard No. 13.
Sprinkler System Care and Maintenance, Standard No. 13A.

National Petroleum Council, 1625 K St. NW., Washington, D.C. 20006.
Disaster Planning for the Oil and Gas Industries.
Security Principles for the Petroleum and Gas Industries.

National Safety Council, 425 North Michigan Ave., Chicago, Ill. 60611.
See other appropriate topics treated in this Manual, especially those pertaining to organization, training, medical and nursing services, fire extinguishment and control.

Underwriters Laboratories Inc., 207 East Ohio St., Chicago, Ill. 60611.
Classification of Fire-Resistance Record-Protection Equipment.
Gas Shutoff Valves—Earthquake.

Personal Protective Equipment

Chapter

19

Equipment Selection and Use   466
Selection of proper type . . . Proper use of equipment

Head Protection   469
Safety helmets . . . Bump caps . . . Hair protection . . .
Maintenance

Hearing Protection   476
Insert types . . . Muff types . . . Amount of attenuation . . .
Selection of hearing protectors

Face and Eye Protection   479
Background and supervision . . . Impact protection . . .
Selection of eye wear . . . Plastic *vs.* glass lenses . . .
Comfort and fit . . . Face protection . . . Acid hoods and
chemical goggles . . . Laser beam protection . . . Eye protection
for welding

Respiratory Equipment   489
Respirator selection

1. Air Purifying Respirators   493
Gas masks . . . Chemical cartridge respirators . . . Particulate
filter respirators . . . Combination respirators

19—Personal Protective Equipment

2. Atmosphere (Air) Supplied Respirators   **497**
Hose masks . . . Air line respirators . . . Abrasive blasting
respirators . . . Air supplied hoods . . . Air supplied suits . . .
Self-contained breathing devices . . . Care of respiratory equipment

Safety Belts   **509**
Selection . . . Belt care . . . Inspection and testing . . . Lifelines

Protective Footwear   **514**
Conductive shoes . . . Foundry shoes . . . Explosives operations
(nonsparking) shoes . . . Electrical hazard shoes . . . Special
shoes . . . Cleaning of rubber boots

Special Work Clothing   **517**
Protection against heat and hot metal . . . Flame-retardant work
clothes . . . Cleaning of clothing . . . Protection against impact
and cuts . . . Impervious clothing . . . Women's clothing and
protection . . . Cold weather clothing . . . Special clothing

References   **527**

The OSHAct makes it mandatory for each employer to furnish to each of his employees a place of employment that is free from recognized hazards that can cause (or are likely to cause) death or serious physical harm to his employees, and to comply with occupational safety and health standards promulgated under the Act.

This places a burden on the safety professional to take prompt steps to eliminate any hazardous situation he is faced with. Extensive engineering revision of processing or manufacturing methods may be involved, or only a simple change in materials handling methods may be necessary.

A machine so designed, for instance, that it effectively confines flying particles eliminates a cause of accidents. This is a more basic treatment of the problem than the use of goggles designed to prevent injury, for confinement stops particles at their source.

Reducing noise to acceptable levels by quieting down a machine or enclosing it is far superior than depending on personal hearing protective devices. (See Chapter 40, "Noise and Hearing Conservation.")

Likewise, dangerous solvents, chemicals and other vapor or fume hazard substances should be confined to a pipe or closed tank, or their vapors or fumes should be exhausted mechanically, instead of depending on a respirator to protect an operator required to work in a hazardous environment. Protection by mechanical means is generally more reliable than protection dependent upon human behavior.

If it is impractical to eliminate a cause of accidents by engineering revision or by safeguarding, or to limit exposure time to hazardous dusts, mists, vapors, or excessive noise to acceptable levels by administrative procedures, use of personal protective equipment is mandatory.

Equipment Selection and Use

Once the safety professional decides that personal protective equipment is needed, he must:

1. Select the proper type of equipment, and then

2. Make sure that the supervisor sees to it that the employee uses and maintains the equipment correctly.

Selection of proper type

After the need for personal protective equipment has been established, the safety professional is faced with the second problem, that of selecting the proper type. Two criteria should be used: the degree of protec-

Fig. 19–1.—If a firm does not have its own protective equipment store, it often arranges with nearby stores so that employees can be properly fitted.

tion which a particular piece of equipment affords under varying conditions, and the ease with which it may be used.

Unfortunately, with the exception of respiratory protective devices, few items of personal protective equipment available commercially are tested and approved by an impartial examiner according to published and generally accepted performance specifications. Satisfactory performance specifications exist for certain types of personal protective equipment, notably safety hats, devices to protect the eyes from impact and from harmful radiations, and rubber insulating gloves, but there are no approving laboratories to test equipment regularly according to these specifications.

The government is taking steps to correct this condition by proposing the accreditation of private testing laboratories that could test in accordance with criteria established by the National Institute for Safety and Health (NIOSH). Such equipment will bear an approval label.

In any event, unless the safety professional has ample testing facilities available to him, he must rely upon the equipment manufacturers' endorsement for devices that may fill his needs. Fortunately, the manufacturers have been aware of their responsibility in this respect, and have a remarkable record of reliability. Their representatives can usually be called in to demonstrate their products and discuss their conformance to safety standards.

The succeeding pages discuss seven major areas of personal protective equipment—head, eyes and face, ear, respiratory, hands, feet, and body. Each section will provide the latest information on the standards available or proposed, some details about the equipment available, suggestions for selecting equipment to meet the job hazard, and some case histories in the development and use of specific devices.

Proper use of equipment

The next problem is that of having workers wear personal protective equipment (if necessary), once it has been chosen. This is now required by law. Several factors influence the solution of this problem. Among them are: (a) the extent to which the men who must wear the equipment understand its necessity, (b) the ease and comfort with which it can be worn with a minimum of interference with normal work procedures, and (c) the available economic, social, and disciplinary sanctions which can be used to influence the attitudes of the men.

In an organization where workers are accustomed to wearing personal protective equipment as a condition of employment, this problem is minor. The men are simply issued equipment which meets the requirements of the job and is easy to wear, and are taught how and why it must be used. Thereafter, periodic checks are made until use of the issued equipment has become a matter of habit with the workers.

When a group of men are issued personal protective equipment for the first time or when new devices are introduced, the problem may be more difficult. The men will need to be given a clear and reasonable explanation as to why the equipment must be worn. Traditional work procedures may have to be changed. If such changes are required, a good deal of resistance, justifiable or not, may be generated. Also, workers may be reluctant to use the equipment because of bravado or vanity.

It should be pointed out to them that the Occupational Safety and Health Act (OSHAct) requires "each employee to comply with occupational safety and health standards, and all rules, regulations, and orders issued pursuant to this Act which are applicable to his own actions and conduct."

The practice of having supervisors and foremen try out new protective equipment and devices prior to actual adoption, and getting their comments and discussing the advantages, has been successfully used in many operations.

A good deal of the resistance to change can be overcome if the men are allowed to choose the particular style of equipment they will wear from a group of different styles which have been preselected to meet the job requirements. In some situations, it may be advisable to have a committee from the work force help select suitable devices. Management's desire to standardize on one style of equipment may not be realized immediately, and several styles may need to be stocked. In the latter case, the cost, though higher than the cost of stocking only one style, will be small compared to the potential cost of accidents resulting from failure to use the equipment.

For the convenience of their employees, some companies maintain equipment stores on the plant premises (Fig. 19–1).

Policies differ with respect to who pays for personal protective equipment. The OSHAct states that the employer is responsible for the employees' use of the equipment, but the question of "who pays for it?" depends upon individual company's employer-employee arrangements.

The cost of eye protection equipment that has prescription-ground lenses, such as safety spectacles, is often shared by worker and employer. For instance, a National Safety Council survey showed that 27 percent of the companies supplying employees with prescription goggles gave them free, and 60 percent assumed part of the cost or made them available at below retail.

Most companies do not furnish safety toe shoes free. Many industrial firms maintain shoe stores for the convenience of employees and make the shoes available at cost. In some instances, to encourage purchases, safety shoes are offered at below-cost prices.

In some areas, shoemobile service is available where a vendor comes into the plant with a trailer completely equipped and stocked to fit and sell safety shoes. These shoes are normally offered at an industrial price.

It is difficult to determine when management should pay the cost of personal protective equipment and when workers should. Factors are the probability and expected severity of injury, the willingness of workers to wear the equipment, its length of life, and the degree to which it may be depreciated by nonoccupational use, and provisions of collective bargaining agreements.

For example, the welder's helmet is almost universally supplied by management, for the job could not be performed at all without it. Work gloves sometimes must be purchased by the user. On the other hand, welder's or other special purpose gloves are usually considered as being a necessary part of the job work tools and are issued free.

No matter what decision is made, it remains management's responsibility under the law to develop and enforce the program.

As an added incentive to wearing equipment, several organizations sponsor "recognition awards" for those who have been spared injury by wearing personal protective equipment. The oldest one of these is the Wise Owl Club, sponsored by the National Society

Fig. 19–2.—Head protection is needed where there is a possibility of falling objects, or bumping into suspended or traveling stock.

for the Prevention of Blindness. The National Safety Council keeps an up-to-date list of U.S. and Canadian "clubs." It is available on request.

Head Protection

Although every worker should be encouraged to use his head to absorb knowledge—he should not use it to absorb blows. Men exposed to head hazards must be provided with head protection (Fig. 19–2). Particularly hazardous operations are: tree trimming, construction work, shipbuilding, logging, mining, overhead line construction or maintenance, and basic metal (steel or aluminum) or chemical production.

Safety professionals should be aware of changes in operations that may create a need for head protection. For example, a firm undergoing a slack season might transfer some

employees from relatively safe jobs to duties requiring safety hats or caps. In addition, construction, maintenance, and odd jobs requiring heat protection often occur in the normal operations of many companies.

Safety helmets

Safety helmets (hats or caps) are rigid headgear of varying materials designed to protect the workman's head—not only from impact but from flying particles and electric shock or any combination of the three. They also can protect the scalp, face, and neck from overhead spills of acid, other chemicals, or hot liquids. They can help shield the hair from entanglement in machinery or exposure to irritating dusts. Bump caps (for less-hazardous exposures) are discussed later in this chapter.

Helmets (see Fig. 19–3) have been classified into two types: (a) full brimmed, and

469

a. Protective cap molded of high dielectric plastic

b. Protective cap molded of glass fiber and plastic

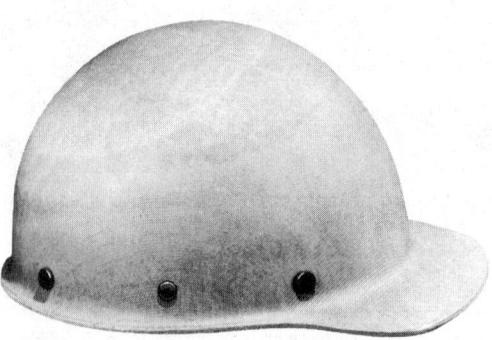

c. Protective cap molded of cloth and phenolic resin

d. Protective hat constructed of aluminum alloy

FIG. 19–3.—Various types of protective headgear.

(*b*) brimless with peak. The types have been further broken down into four classes: Class A, limited voltage resistance for general service; Class B, high voltage resistance; Class C, no voltage protection (metallic helmets); Class D, limited protection for fire fighters service.

All helmets that meet American National Standards (ANSI) Z89.1 or Z89.2, shall be identified on the inside of the helmet shell with the manufacturer's name, American National Standard designation, and class (A, B, C, or D).

Materials used in the construction of Class A and B helmet shells should be water resistant and slow burning. Materials in Class D helmets shall be fire resistant (self-extinguishing when tested in accordance with ASTM Standard D-635), and nonconductors of electricity.

Class B (electrical worker) headgear do not have holes in the shell or any metal parts at all.

The headpieces designed for use around electrical hazards meet voltage tests of 20,000 volts AC (rms), 60 Hz, for three minutes and leakage currents not exceeding nine ma.

The other classes have a less strict require-

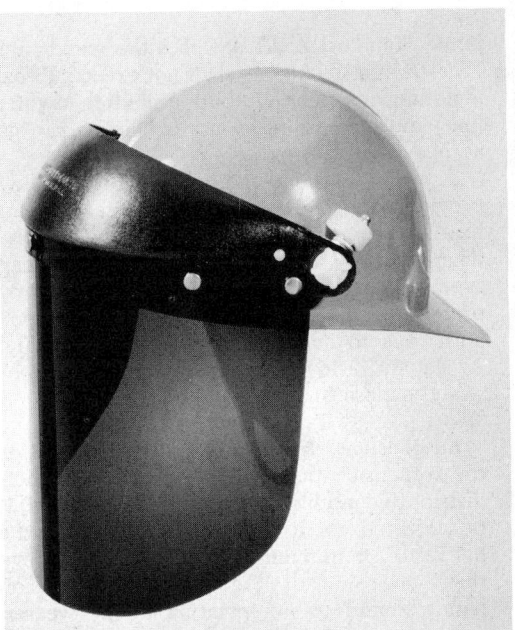

a. Protective cap with attachment for easy installation and removal of face shields, welding helmets, etc.

b. Man wearing chinstrap to secure protective headgear in place

c. Protective headgear with winter liner for cold weather wear

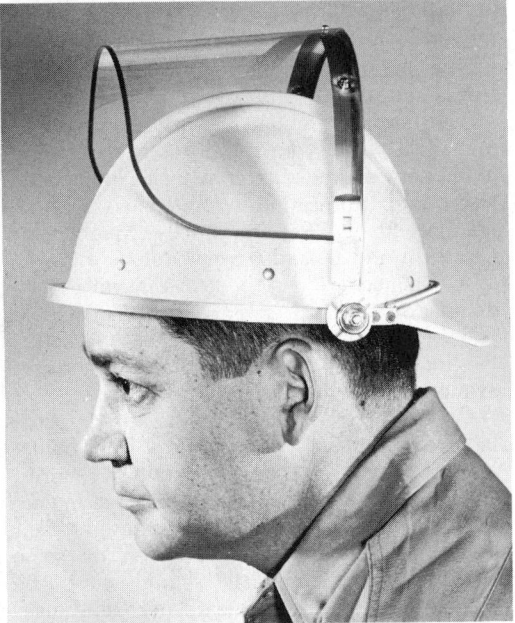

d. Protective headgear with removable-type attachment for face shields.

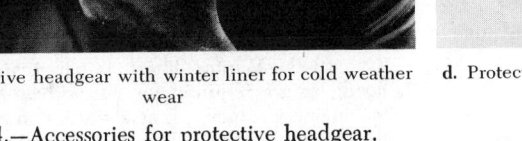

Fig. 19–4.—Accessories for protective headgear.

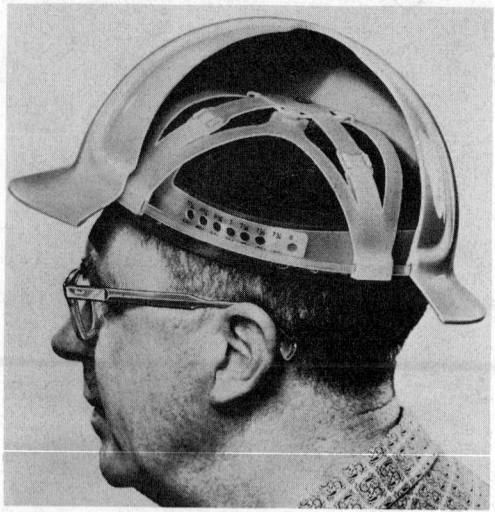

Fig. 19–5.—The crown straps of this safety hat are preset and unadjustable in order to guarantee clearance for impact protection. Underside of hat should be no less than 1¼ in. above head.

ment: 2200 volts AC (rms), at 60 Hz for one minute with no more than 9 milliamperes leakage.

The breakdown threshold for all helmets is the same: 30,000 volts.

The thinnest section of Class A and B helmet shells will not burn at a rate greater than 3 in. per minute. After a 24-hr immersion test, water absorption of the shell will be no more than 5.0 percent (by weight) for Classes A and D, and 0.5 percent for Class B.

All helmets are designed to transmit a maximum average force of not more than 850 lb when tested in accordance with ANSI Z89.1. No individual helmet shall transmit a force of more than 1000 lb.

When properly selected and used, these helmets will greatly reduce injury from falling objects as well as exposure to electric burns.

Metal helmets do not afford the high impact resistance of plastic ones, but because of their lighter weight, they are preferred by some workers. Metal helmets should never be used where electrical hazards or substances corrosive to aluminum are present (see ANSI Z89.1.)

Weight and shape of hat. American Na-

tional Standard Z89.1 specifies the weight for Class A and C helmets shall not exceed 15 oz, including suspension but excluding winter liner and chin strap. Class B helmets can go as heavy as 15½ oz.

A brim all around the helmet (Fig. 19–3d) provides the most complete protection for head, face, and back of the neck. For use where a brim may be in the way, the cap type (Fig. 19–2a, –3b, and –3c) is frequently preferred. This type may also be equipped with lugs to support a welding mask (Fig. 19–4a and –4d) if a welder works on jobs which expose him to head injury.

Suspensions, liners, and chinstraps. It is the suspension that gives a helmet its impact-distributing abilities. It is important that it be adjusted to fit the wearer and keep the hat itself a minimum distance of 1¼ in. above the wearer's head (Fig. 19–5). Suspension bands should be nonirritating to the wearer.

Liners are available so that protective helmets may be worn in comfort in cold weather (Figs. 19–4c and 19–6).

Various types of leather, fabric, and elastic chinstraps are available. Often hats are bumped or blown off, or fall off during a fall; a

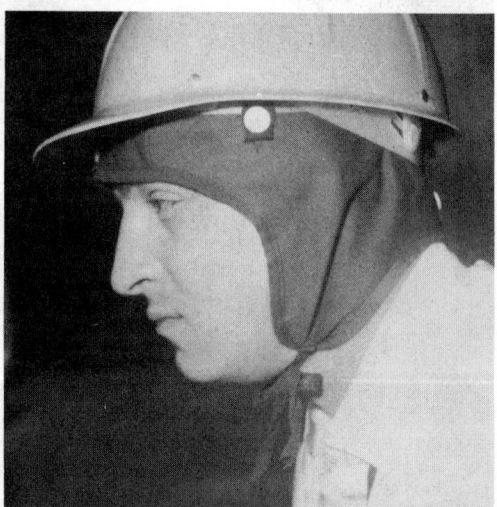

Fig. 19–6.—Winter liners rather than ordinary headgear are required for use with safety hats in inclement weather. Liners do not interfere with the shock-absorbing clearance of the hat.

FIG. 19–7.—Bump caps have limited application to protect against low head clearances.

chinstrap affords full protection (Fig. 19–4b). A nape strap (provided with most helmets) helps keep the headgear from falling off during normal use.

Bump caps

Another form of protective headgear is the bump cap, a thin-shelled, light-weight plastic affair (Fig. 19–7). It was originally conceived for use by aircraft workers, laboring in the close quarters of a fuselage—where a brim might get in the way.

Although some industries have adopted them, there are no specifications covering them. Authorities warn that, although they are fine for some applications, they are *not* a substitute for helmets. Use of such caps should be strictly limited.

Hair protection

It is important that women and men with long hair who work around chains, belts, or other machines protect their hair from contact with moving parts (Fig. 19–8). Besides the danger of direct contact with the machine, which may occur when they lean over, they are exposed to the hazard of having their hair lifted into moving belts or rolls that develop heavy charges of electricity. Since it is difficult to remove this hazard completely by mechanical means, people with long hair should be required to wear protective hair covering.

Hair nets, bandannas, and turbans are fre-

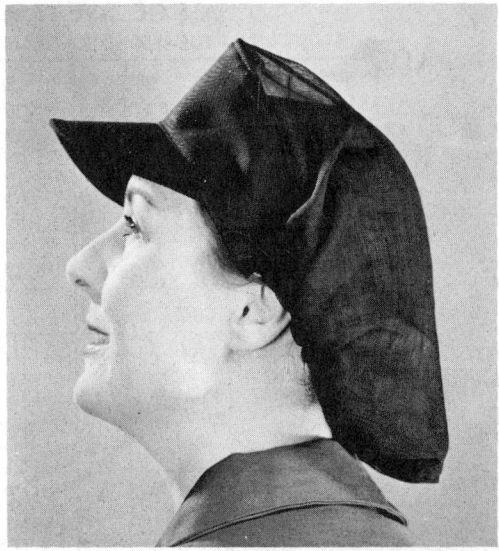

FIG. 19–8.—Cap protects hair of women in general factory work. The visor serves as a feeler guard to prevent injuries from contact with machines and objects. The netting construction on the top provides adequate comfort ventilation.

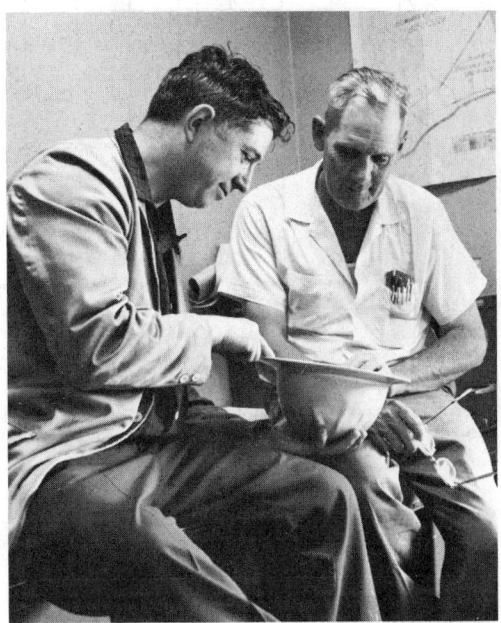

FIG. 19–9.—Periodic inspection of safety hats and suspensions is part of a maintenance program.

473

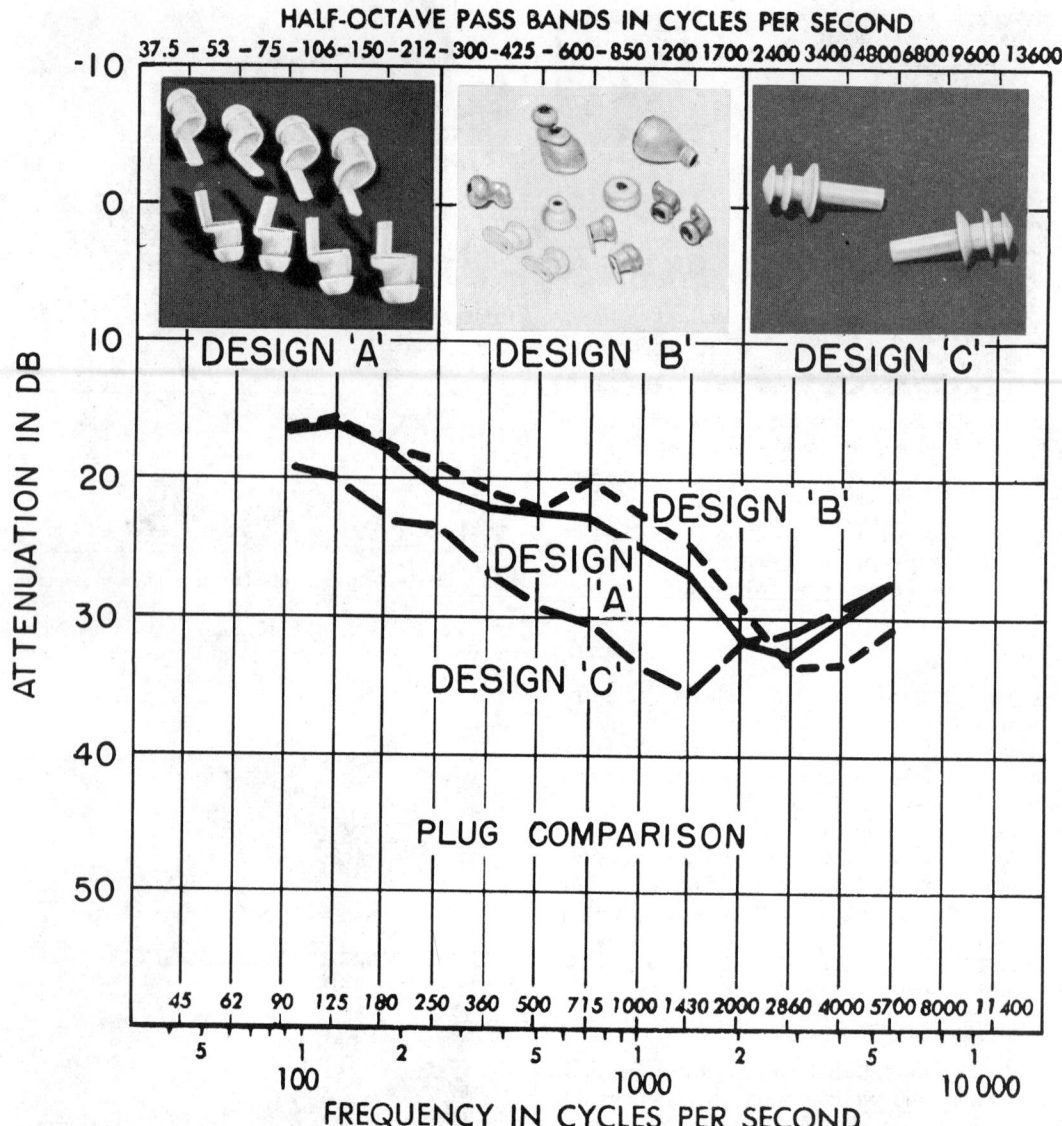

FIG. 19–10.—Attenuation (noise-reducing) characteristics of plug or insert type ear protection. Curves are for plugs of design shown in photographs above. (Cycles per second are now called hertz.)

quently unsatisfactory for hair protection because they do not cover the hair completely. Protective caps should completely cover the hair. If the wearer is exposed to sparks and hot metals, as in spot welding, the cap should be made of flame-resistant material. Disposable flame-proof caps are provided in some chemical plants.

No standards have been accepted for protective caps, but they should be made of a durable fabric to withstand regular laundering and disinfecting. Design should be simple so that they can be pressed or ironed by machine, and they should be available in a vari-

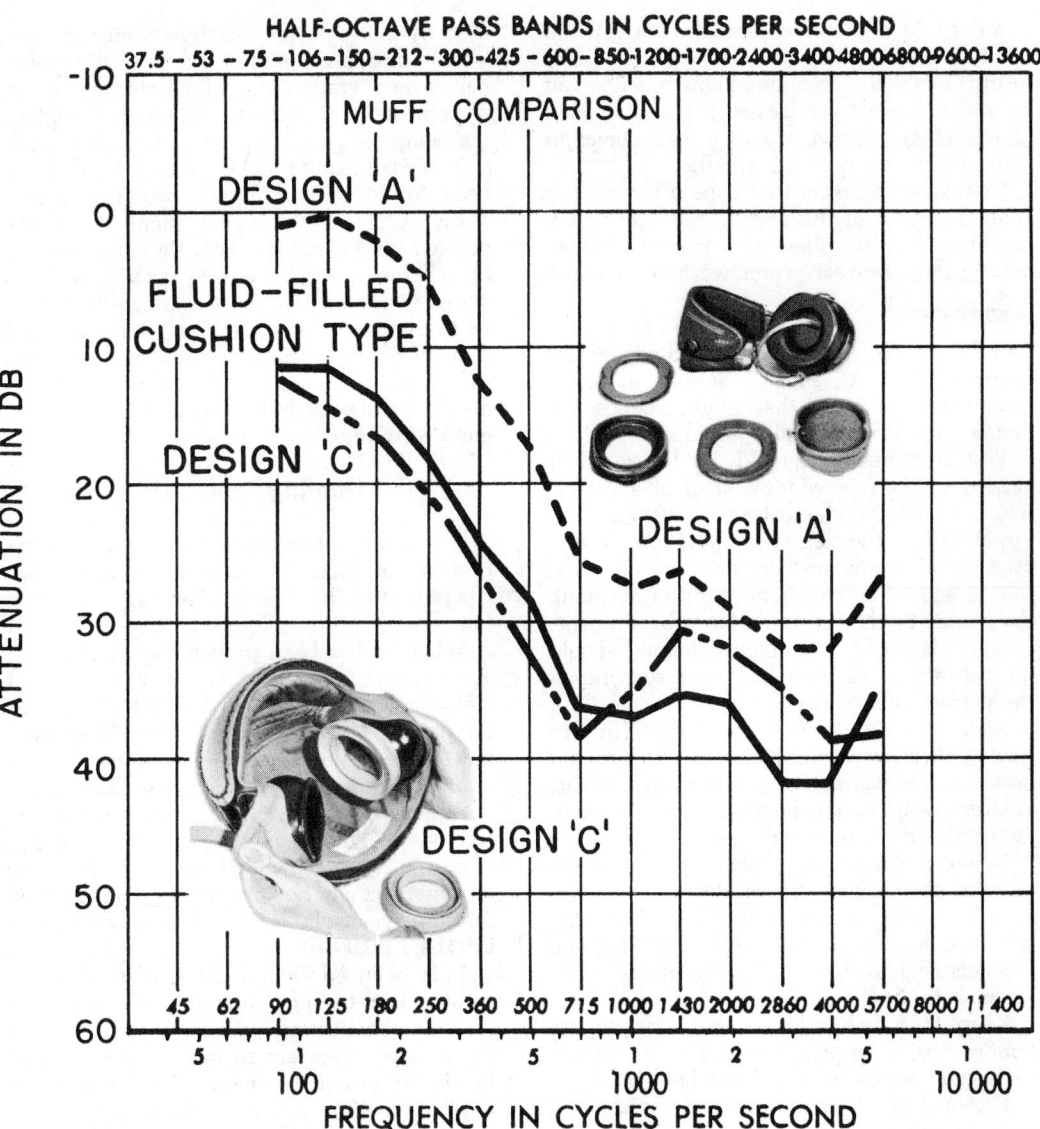

FIG. 19-11.—Attenuation characteristics of cup or muff type ear protectors.

Data from J. C. Webster and E. R. Rubin, U.S. Navy Electronics Laboratory.

ety of head sizes or should be adjustable to fit all wearers.

A cap should have a visor long enough and rigid enough to provide warning before the head itself comes into contact with a moving object, such as the spindle on a drill press.

In order to encourage its use, the cap should be as attractive as possible. It should be cool and lightweight. If dust protection is not required, the cap should be made of open weave material for better ventilation.

After a suitable cap has been chosen, its use should be required and enforced. It is common practice among workers, for reasons of vanity, to wear the cap on the back of the head so that part of the hair over the forehead is exposed. Sometimes this practice can be discouraged by a realistic demonstration of what may happen when the hair comes in contact with a revolving spindle.

Convincing workers that caps preserve hair from the effect of dusts, oils, and other shop conditions has resulted in some success in getting them to wear protective hair covering.

Maintenance

Before each use, helmets should be inspected for cracks, signs of impact or rough treatment, and wear that might reduce the degree of safety originally provided.

Protective helmets should not be stored or carried on the rear window shelf of a vehicle because sunlight and extreme heat may adversely affect the degree of protection. Another good reason not to carry hats there is that in case of an emergency stop or accident, the helmet might become a hazardous missile.

Once damaged, a protective helmet should be discarded. Alterations of any sort impair the performance of the headgear.

At least every 30 days, safety hats (in particular, their sweatbands and cradles) should be washed in warm, soapy water or a suitable detergent solution recommended by the manufacturer and then rinsed thoroughly.

Before reissuing used helmets to other employees, they should be scrubbed and disinfected. Solutions and powders are available which combine both cleaning and disinfecting. Helmets should be thoroughly rinsed with clean water and then dried.

Keep the wash solution and rinse water temperature at approximately 140 F; do not use steam, except on aluminum helmets.

Removal of tar, paint, oil, and other materials may require the use of a solvent. Because some solvents can damage the shell, the helmet manufacturer should be consulted as to what solvent should be used.

Pay particular attention to the condition of the suspension because of the important part it plays in absorbing the shock of a blow. Look for loose or torn cradle straps, broken sewing lines, loose rivets, defective lugs, and other defects (Fig. 19–9). Sweatbands are easily replaced. Disposable helmet liners made of plastic or paper are available for hats used by many people (such as visitors).

An adequate number of crowns, sweatbands, and cradles should be stocked as replacement parts. Some companies replace the complete suspension at least once a year.

Many companies make use of colored safety hats to identify different working crews. Many colors are available; some colors are painted on and others have the color molded in. Before painting a hat, consult with its manufacturer so that a coating can be chosen which will not only last but which will not reduce the dielectric properties or attack and soften the shell material. Lighter colored hats are cooler to wear in the sun or under infrared energy sources.

Hearing Protection

Increasing attention is being paid to the problem of excessive noise in industry. See Chapter 40, "Noise and Hearing Conservation."

Where it has been proven that engineering controls are not feasible as a permanent method of control, personal protective devices for noise control are acceptable. Their use, however, should be accompanied by an adequate hearing conservation program.

Some state regulations require audiometric testing of employees exposed to excessive noise. It is recommended that an audiometric testing program be initiated and maintained for employees who are exposed to noise levels in excess of 90 dBA.

It is believed that a properly carried out audiometric testing program will determine whether the hearing protective devices worn by the employees are in fact protecting their hearing from noise damage. More details are in Chapter 40.

Ear protectors in general use fall into two main groups—the plug or insert type and the cup or muff type (see Figs. 19–10 and –11).

Another type is the enclosure that completely surrounds the head; the astronaut helmet is of this type. Attenuation of sound is achieved through the acoustical properties of the helmet. Cost as well as bulk normally preclude use of this type for general use.

Insert type

The plug or insert type can be classified

into (*a*) aural, which can be inserted in the ear canal, and (*b*) superaural, which seals the external edges of the ear canal.

The aural type, placed into the ear canal, varies considerably both in design and material. Materials used are rubber, soft or hard plastic, wax, and cotton ("Swedish wool"). Rubber and plastic types are popular because they are easy to keep clean, inexpensive and give good performance.

Since ear canals vary in size, these inserts come in several sizes. In some cases an individual will require a different size plug for each ear. It is important that ear plugs be fitted individually by proper personnel (Fig. 19–14). Plugs must fit properly and remain correctly seated (Fig. 19–14 inset), as even the slightest leakage will lower the attenuation by as much as 15 dB (decibels) in some frequencies.

Custom-molded hearing protectors made of silicone rubber, if properly molded and correctly used, generally prove more comfortable than do prefabricated inserts. Soft plastic plugs give more comfort than hard types and hold their shape better than rubber types. Their attenuation properties are comparable with those of rubber or hard plastic.

Wax protectors vary in form from plugs of pure wax to impregnated cotton, cellular plastic foam, or paper. They have certain drawbacks, however, as they tend to lose their effectiveness during the workday. This is mainly caused by jaw movements which change the shape of the ear canal and break the acoustic seal between ear and insert. In dirty areas, wax protectors may be objectionable from a sanitary standpoint; as they must be shaped by hand, wax protectors should be used only once. While the initial cost may be low, the expense of frequent replacement can become an important consideration.

Cotton inserted in the ear is a poor choice as a noise suppressor, because of its low attenuating properties (2 to 12 dB depending on frequency) and the disadvantages associated with hand-formed protectors. The safety professional should discourage both workmen and employers from improvising any kind of inserts other than approved devices.

The superaural type depends on sealing the

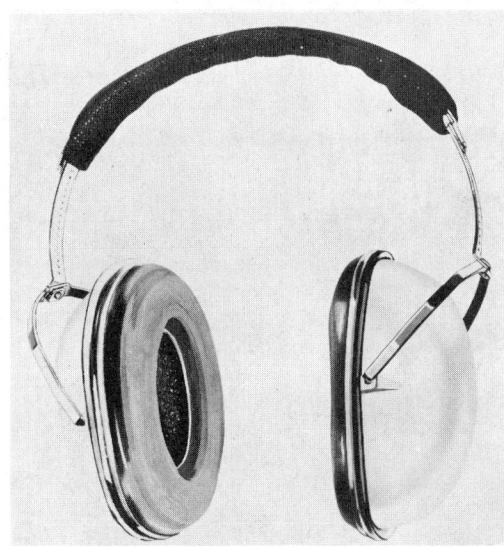

FIG. 19–12.—Fluid-filled, cushion-type muff.

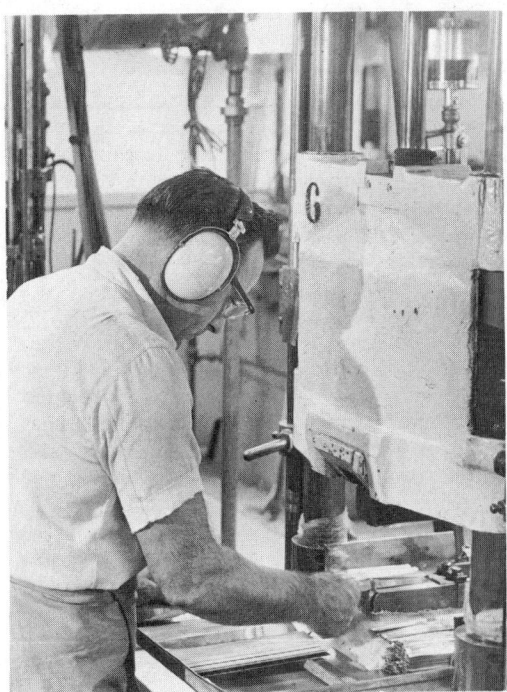

FIG. 19–13.—Earmuffs being worn around noisy equipment.

477

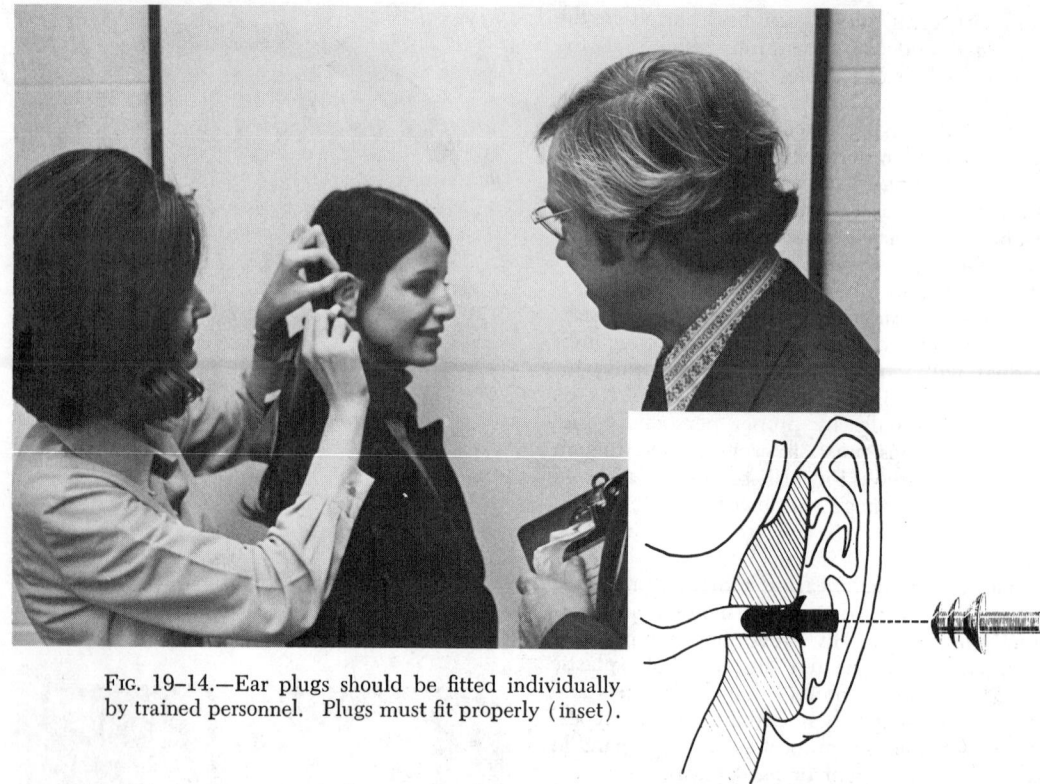

Fig. 19–14.—Ear plugs should be fitted individually by trained personnel. Plugs must fit properly (inset).

external edge of the ear canal in order to achieve sound reduction. A soft rubber-like material is used to make the caps, as they are sometimes called. They are held in place against the edges of the ear canal by a spring band or a head suspension.

Muff type

Cup (or muff) devices cover the external ear to provide an acoustic barrier (Figs. 19–12 and –13). The attenuation provided by ear muffs varies widely due to differences in size, shape, seal material, shell mass and type of suspension. Head size and shape also influence the attenuation characteristics of these protectors. The type of cushion used between the shell and the head has a great deal to do with attenuation efficiency. Liquid- or grease-filled cushions give better noise suppression than plastic or foam rubber types, but may present leakage problems.

Amount of attenuation

Commercially available earplugs if properly

fitted and used, generally reduced noise reaching the ear by 25–30 dB in the higher frequencies (Fig. 19–11a), which are conceded to be the most harmful. This will provide ample protection against sound levels of 115 to 120 dB. The better type of ear muffs may reduce noise an additional 10 to 15 decibels (Fig. 19–11b), making them effective against sound levels of 130 to 135 dB. Combinations of ear plugs and muffs give 3 to 5 more dB of protection. In no case will total attenuation be greater than about 50 dB because at this point, bone conduction becomes significant.

Variations from one model to another will provide varying degrees of noise reduction. Manufacturers supply attenuation data for their products so the safety professional can evaluate their effectiveness for use in a given situation. The NIOSH "Criteria Document on Noise" gives methods for determining the efficiency of a specific device for a given exposure. Called the "R factor," the figure takes into account the attenuation characteristics of the device, general physical characteristics of the wearer (such as hair and head structure), need for eye protection, and other factors.

Selection of hearing protectors

No matter how good a hearing protective device may be, its comfort has a great deal of influence on how well it will be accepted. Some people cannot wear plugs for physical or psychological reasons. Others may not be able to wear muffs. Consequently, many hearing conservation programs will include both types so the user may select the one most acceptable to him.

Face and Eye Protection

Protection of the eyes and face from injury by physical and chemical agents or by radiation is vital in any occupational safety program. In fact, this type of protection has the widest use and the widest range of styles, models, and types.

The eye and face protection standard, *Practice for Occupational and Educational Eye and Face Protection* (Z87.1), is a fairly comprehensive document. It sets performance standards, including detailed tests, for a broad area of hazards—excluding only X rays, gamma rays, high-energy particulate radiations, lasers, and masers.

Besides general requirements, applying to "all occupations and educational processes," the standard provides requirements on the following:

- Rigid welding helmets

- Welding hand shields

- Nonrigid welding helmets

- Attachments and auxiliary equipment—lift fronts, chin rests, snoods, aprons, magnifiers, etc.

- Flammability

- Face Shields

- Goggles—eyecup (chipper's), dust and splash, welder's, and cutter's

- Spectacles—metal, plastic, and combination.

Some of the requirements of the standard are stated generally. For example: "Aprons or bibs for helmets shall be of nonflammable, nonconducting material that is flexible and capable of withstanding disinfection."

Other requirements are more precise. As regards flammability, for example—"The thinnest section of the rigid helmet or hand shield shall not burn at a rate greater than 3 in. per minute when tested by inserting one end of a 5 by 1½ in. strip of the helmet material in a blue-flame Bunsen burner. . . ."

The most significant changes of Z87.1 from the previous Z2.1–1959 Standard are:

- A stringent requirement for the impact resistance of the lens when in frames. The revision illustrates the test block with two figures and prescribes that "A 1.00-in. diameter steel ball, weighing approximately 2.4 oz shall be dropped in free fall from a height of 50 in. onto the horizontal upper surface of the lens, impinging the lens within a circular area of a ⅝-in. diameter centered at the lens mechanical [geometric] center. The lens shall not be chipped and the lens shall not be displaced from the frame eye in this test." Supplementing this test for lens in frame—glass and plastic—are impact resistance tests for lens on block and breakage pattern. Also specified are tests for penetration resistance (plastic), flammability (plastic), haze (plastic), and ultraviolet, luminous, and infrared transmittance (plastic and glass).

- A new requirement for lenses in welder's and cutter's goggles to be hardened to meet impact resistance requirement. "All glass filter lenses or plates . . . shall be heat-treated and meet the impact-resistance requirements. . . ."

- A specific selection chart for eye and face protection, including recommendations for particular hazards and operations. The chart also includes line drawings of 11 types of goggles, spectacles, shields, and helmets (see Fig. 19–15).

- Disinfection requirements are more exact —"Completely immerse the protector for 10 minutes in a solution of modified phenol, hypochlorite, or quaternary ammonium compounds . . . at a room temperature of 68 F. . . ."

19—Personal Protective Equipment

SELECTION OF EYE AND FACE PROTECTIVE EQUIPMENT

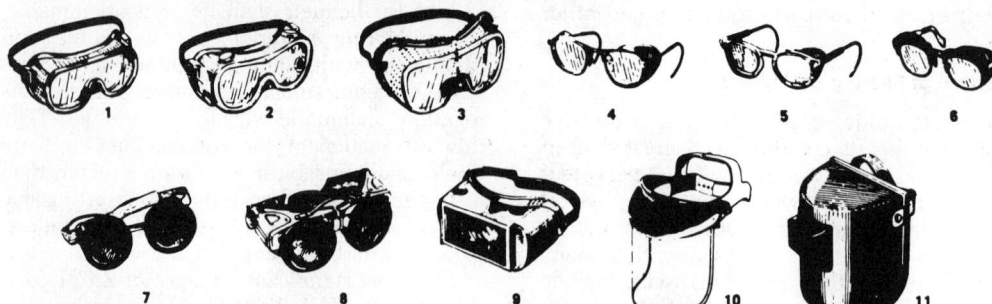

1. **GOGGLES,** Flexible Fitting, Regular Ventilation
2. **GOGGLES,** Flexible Fitting, Hooded Ventilation
3. **GOGGLES,** Cushioned Fitting, Rigid Body
*4. **SPECTACLES,** Metal Frame, with Sideshields
*5. **SPECTACLES,** Plastic Frame, with Sideshields
*6. **SPECTACLES,** Metal-Plastic Frame, with Sideshields

** 7. **WELDING GOGGLES,** Eyecup Type, Tinted Lenses (Illustrated)
  7A. **CHIPPING GOGGLES,** Eyecup Type, Clear Safety Lenses (Not Illustrated)
** 8. **WELDING GOGGLES,** Coverspec Type Tinted Lenses (Illustrated)
  8A. **CHIPPING GOGGLES,** Coverspec Type, Clear Safety Lenses (Not Illustrated)
** 9. **WELDING GOGGLES,** Coverspec Type, Tinted Plate Lens
10. **FACE SHIELD** (Available with Plastic or Mesh Window)
11. **WELDING HELMETS

*Non-sideshield spectacles are available for limited hazard use requiring only frontal protection.
**See appendix chart "Selection of Shade Numbers for Welding Filters."

APPLICATIONS

OPERATION	HAZARDS	RECOMMENDED PROTECTORS: Bold Type Numbers Signify Preferred Protection
ACETYLENE–BURNING ACETYLENE–CUTTING ACETYLENE–WELDING	SPARKS, HARMFUL RAYS, MOLTEN METAL, FLYING PARTICLES	7, **8**, 9
CHEMICAL HANDLING	SPLASH, ACID BURNS, FUMES	**2**, 10 (For severe exposure add 10 over 2)
CHIPPING	FLYING PARTICLES	1, 3, 4, 5, 6, 7A, **8A**
ELECTRIC (ARC) WELDING	SPARKS, INTENSE RAYS, MOLTEN METAL	9, 11 (11 in combination with 4, 5, 6, in tinted lenses, advisable)
FURNACE OPERATIONS	GLARE, HEAT, MOLTEN METAL	7, **8**, 9 (For severe exposure add 10)
GRINDING–LIGHT	FLYING PARTICLES	1, 3, 4, 5, **6**, 10
GRINDING–HEAVY	FLYING PARTICLES	**1**, 3, 7A, **8A** (For severe exposure add 10)
LABORATORY	CHEMICAL SPLASH, GLASS BREAKAGE	**2** (10 when in combination with 4, **5**, **6**)
MACHINING	FLYING PARTICLES	1, 3, 4, 5, **6**, 10
MOLTEN METALS	HEAT, GLARE, SPARKS, SPLASH	7, **8** (10 in combination with 4, **5**, **6**, in tinted lenses)
SPOT WELDING	FLYING PARTICLES, SPARKS	1, 3, 4, **5**, **6**, 10

Fig. 19–15.—Selection of eye and face protective equipment. Illustration and table are taken from American National Standard Z87.1, *Practice for Occupational and Educational Eye and Face Protection.*

Courtesy of the American National Standards Institute.

- Expansion and clarification of definitions.

The Z87.1 standard also contains editorial clarification, an informational appendix that covers visible light transmission and haze test for plastics, the selection of shade numbers for welding filters, and fitting details for goggles and spectacles.

Background and supervision

Eye protective devices must be considered as optical instruments, and they should be carefully selected, fitted, and used.

Contact lenses should never be considered as being a replacement for safe protective equipment for the eye. Workers have had their eyesight permanently impaired and have even been blinded by corrosive chemicals or small particles getting between their contact lenses and their eyes. The American Society for the Prevention of Blindness has issued a statement saying that contact lenses have no place in the industrial environment.

If corrective lenses are required, it is preferable to grind the correction into a goggle lens. Goggles, which cover ordinary spectacles, may be worn, but they require cups deep and wide enough to admit the whole spectacle.

The amount of money spent to acquire and fit eye protective devices is small when measured against the savings afforded by the protection given. For example, the purchase and fitting of a pair of impact-resistant spectacles may cost from 3 to 15 dollars; compensation costs for a lost eye may range from 5000 to 20,000 dollars.

If the eye protection program is supervised, directly or indirectly, by an industrial ophthalmologist, on either a full-time or consultation basis, there will be an additional return from proper utilization of the visual abilities of the work force.

Unfortunately, there is no impartial testing laboratory set up to approve eye protection devices according to generally known and accepted minimum performance specifications. Be sure that equipment is purchased from reputable manufacturers (Fig. 19–16).

Impact protection

Three general types of equipment are used to protect eyes from flying particles encountered in such jobs as chipping, and grinding—

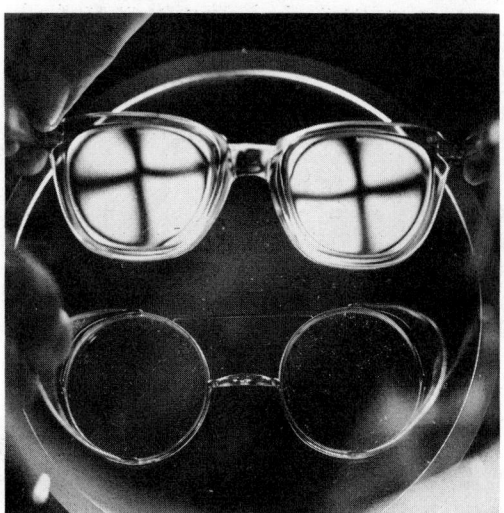

Fig. 19–16.—Photograph taken with light passing through two polarizing filters shows lack of heat treating in imported lenses (below). Standard safety glasses (at top) exhibit the proper cross-patterns.

spectacles with impact-resistant lenses, flexible or cushion-fitting goggles, and chipping goggles. (Refer to Fig. 19–17.)

Spectacles without sideshields should be used for limited hazards, requiring only frontal protection. Where side as well as frontal protection is required, the spectacles must have sideshields. Full-cup sideshields are designed to restrict the entry of flying particles from the side of the wearer. Semi- or flat-fold sideshields may be used where only lateral protection is required. Snap-on and clip-on sideshield types are not acceptable unless they are secured.

Flexible fitting goggles should have a wholly flexible frame forming the lens holder.

Cushion-fitting goggles should have a rigid plastic frame with a separate cushioned fitting surface on the facial contact area.

Both flexible and cushion goggles usually have a single plastic lens. These goggles are designed to give the eyes frontal and side protection from flying particles. Most models will fit over ordinary opthalmic spectacles.

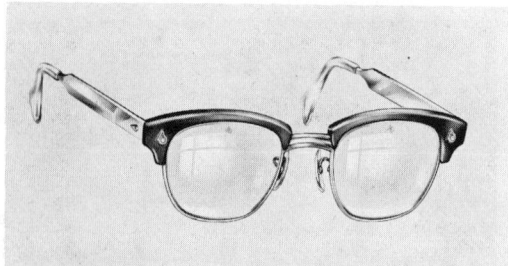

a. Plastic and metal frame safety spectacles, designed after regular streetwear glasses. The etched manufacturer's trademark on the lenses identify them as safety glass.

b. Plastic-frame safety spectacles with screened side-shields.

c. Rigid plastic chipper's goggles with standard round lenses designed to fit over regular glasses.

d. Light-weight, all-plastic visitor's spectacle.

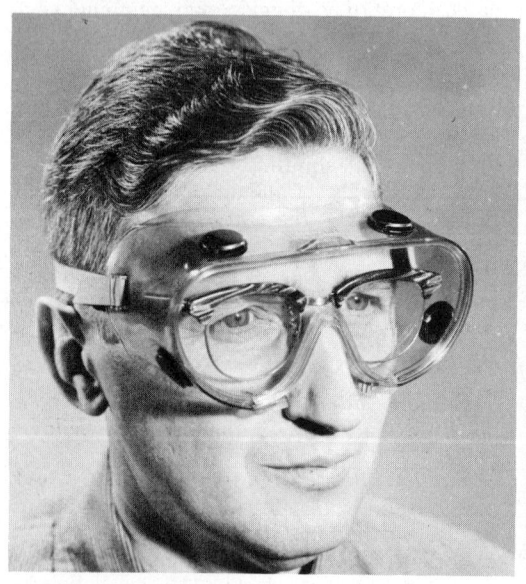

e. All-plastic, soft-sided cover goggle with shielded vents.

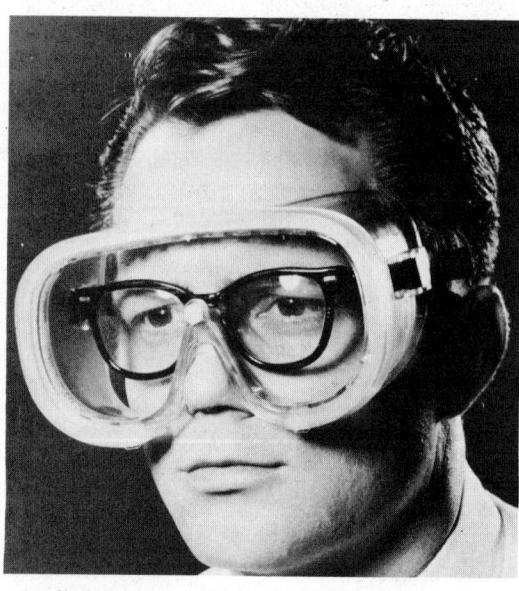

f. All-plastic, soft-sided chemical splash goggle with indirect and filtered ventilation for protection against heavy splash and driven mist.

Fig. 19–17.—Representative types of eye protection.

Chipping goggles, which have countour-shaped rigid plastic eyecups come in two styles—one for individuals who do not wear spectacles, and one to fit over corrective spectacles. Chipping goggles should be used where maximum protection from flying particles is indicated.

Selection of eye wear

Factors that should be considered in the selection of eyewear to protect from impact include the protection afforded, the comfort with which they can be worn, and the ease of keeping them in good repair. Metal frame spectacles with side shields and impact-resistant lenses rate high in all three. Styles now available (see Fig. 19–17a, 17b, and 17d) are similar to regular eyewear and are cosmetically pleasing. Flexible types are preferred by many because of their light weight and convenience. One drawback to the latter is they generally have a shorter wear-life than the more sturdier frame and glass lens type.

Proper eye protection devices should be selected, and their use should be impartially enforced so as to give maximum protection to the user for the degree of hazard involved. On certain jobs, 100 percent eye protection must be insisted upon.

Hardened glass opthalmic lenses are no substitute for safety lenses. They generally are 2 mm thick instead of 3 mm for safety lenses. The lens bezel which fits the groove of the safety frame is ground to an 80 degree angle as contrasted to the 113 degree angle of a safety frame and safety lens. See Fig. 19–18.

Face shields are not recommended by ANSI Z87.1 as basic eye protection against impact. To get impact protection, face shields must be used in combination with basic eye protection. Face shields have their purpose, and are discussed later in this section.

Plastic *vs.* glass lenses

When making a decision between plastic and glass lenses there are a number of things to consider:

- Both can pass impact tests when of certain formulation and thickness.

- Glass has a slightly lower resistance than plastic to breakage from sharp objects.

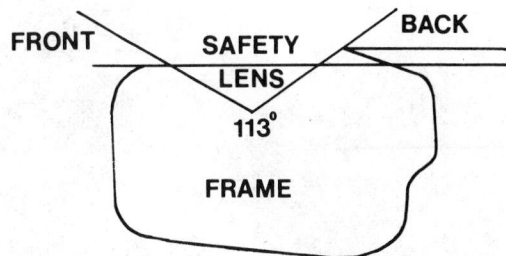

Fig. 19–18.—Cross section of safety lens in a safety frame.

- Tests show plastic lenses to have more favorable resistance to small objects moving at high rates of speed than glass.

- Abrasion resistance, while not good with plastic, is improved when plastics are coated.

- Plastics are resistant to hot materials. Hot metal invariably shatters glass but not plastic. Hot metal also tends to adhere to glass.

- Plastics generally show surface reaction to some chemicals but satisfactorily stop splashes and protect the eyes.

- Whereas fogging occurs on both glass and plastic, it usually takes longer for plastic to fog.

Plastic goggles are available with a hydrophylic coating which tends to prevent fogging. There are also double-lens plastic goggles which operate on the Thermopane principle and suppress fogging to a large extent except under conditions of extreme humidity or cold. For extreme conditions of humidity, wire screen lenses of face shields may be suitable.

Frames should be rigid enough to hold the lenses in the proper position in front of the eyes. Frames for safety glasses have the back of the groove slightly higher than the front to prevent the lenses from being pushed in; see Fig. 19–18. Frames should be constructed of corrosion-resistant material that will neither irritate nor color the skin. The simpler the design, the easier the goggle is to clean.

Cup goggles should have cups large enough to protect the eye socket and to distribute the impact over a wide area of the facial bones. Cups should be flame-resistant (as should be the frames), corrosion resistant, and nonirritating to the skin.

483

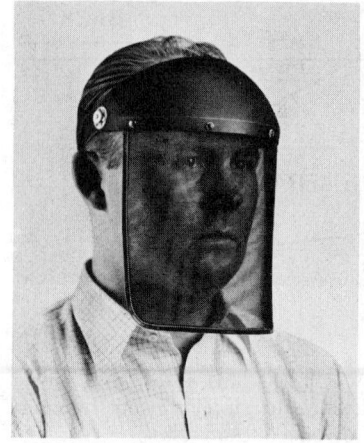 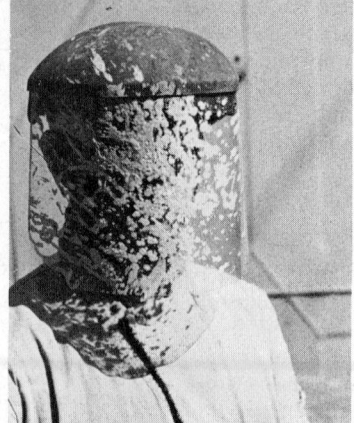 

FIG. 19–19.—Three types of face shields. *Left:* Metal-screen face shield affords protection to furnacemen and others exposed to radiant heat. *Center:* Chemical-resistant full-face shield stopped a splash of boiling caustic. *Right:* Sturdy, nonflammable helmet with specified cover and filter plates protects welder from sparks, flying particles, and injurious rays.

If lenses are to be exposed to pitting from grinding wheel sparks, a transparent and durable coating may be applied to them. Welding lenses should be protected by a cover lens of glass or plastic.

Lenses must not have appreciable distortion or prism effect. ANSI Z87.1 and Federal Specification GGG-G-50lb (*Federal Standard Stock Catalog*) limits the nonparallelism between the two faces to $\frac{1}{16}$ prism diopter (4 minutes of arc) and both the refraction in any meridian and the difference in refraction between any two meridians to $\frac{1}{16}$ diopter.

Both when supported in the eye cups of goggles and when supported on a rubber gasket on a wood tube, lenses are required to withstand the impact test described earlier in this section.

Comfort and fit

To be comfortable, eye protective equipment must be properly fitted. Corrective spectacles should be fitted only by members of the opthalmic profession. An employee can be trained to fit, adjust, and maintain eye protective equipment, however, and each employee can be taught the proper care of the device he uses.

To give the widest possible field of vision, goggles should be fitted as close to the eyes as possible, without bringing the eyelashes in contact with the lenses.

Various defogging materials are available. Before a selection is made, test to determine the most effective type for a specific application.

In areas where goggles or other types of eye protection are used extensively, goggle cleaning stations can be conveniently located. Defogging materials and wiping tissues can be provided there along with a receptacle for discarding them.

Use of sweatbands helps prevent eye irritation, aids visibility, and eliminates work interruptions for face mopping.

Sweatbands are usually made of a soft, light, highly absorbent cellulose sponge. An elastic band holds the sweatband in place on the wearer's forehead so that it does not interfere with glasses or goggles. Evaporation from the exposed surface produces a cooling effect which adds to comfort.

Face protection

Face shields are available in a wide variety of types to protect the face and neck from flying particles, sprays of hazardous liquids, splashes of molten metal, and from hot solutions—see Fig. 19–4a and –4d and Fig. 19–19. In addition, they may be used to provide antiglare protection where required. As a general rule, they should be worn over suitable basic eye protection.

Three basic styles of face shields include headgear without crown protectors, with crown protectors, and with crown and chin protectors. Each of the three is available with one of these replaceable window styles:

Clear transparent,

Tinted transparent,

Wire screen,

Combination of plastic and screen, and

Fiber window with a filter plate mounting.

The materials used in face shields should combine mechanical strength, light weight, nonirritation to skin, and capability of withstanding frequent disinfectings. Metals should be noncorrosive and plastics should be of the slow-burning type. Only optical grade (clear or tinted) plastic, which is free from flaws or distortions, should be used for the windows. And plastic windows should not be used in welding operations unless they conform to the standards of transmittance of absorptive lenses, filter lenses, and plates.

The metallic bindings on some plastic face shields help prevent the plastic from splitting and cracking. However, a slight bend in the binding will introduce an optical fault in the shield, making it unusable.

On some jobs, such as the pouring of low melting metals, protection against radiation is not necessary, but it is desirable to protect the head and face against splashes of metal.

A face shield similar to an arc welder's face shield, but made of wire screen (which provides much better ventilation than a solid shield), can be used. *Federal Standard Stock Catalog* GGG-H-171 *Specifications* require a screen of 28-, 30-, or 36-mesh with wire 0.012 to 0.017 in. in diameter.

This shield is commonly used without a window because the plain wire will not fog under high temperature and high humidity. A metallized plastic shield that reflects a substantial percentage of heat has been developed for use where there is exposure to radiant heat.

Acid hoods and chemical goggles

Head and face protection from splashes of acids, alkalis, or other hazardous liquids or

Fig. 19–20.—Acid hood of impervious plastic is designed for protection from acids and other corrosive chemical solutions. Rear-shielded ventilation ports can be used where ambient air is respirable. Where air is not, use air-fed models.

Fig. 19–21.—Laser protective spectacles. To designate specific wavelength protection, frames should be distinctively colored and optical density should be shown on the filter.

TABLE 19–A

TRANSMITTANCES AND TOLERANCES IN TRANSMITTANCE OF VARIOUS SHADES OF FILTER LENSES

Shade No.	Optical Density			Luminous Transmittance			Maximum Infrared Transmittance	Maximum Spectral Transmittance in the Ultraviolet and Violet for Four Wavelengths (millimicrons)			
	Min.	Std.	Max.	Max.	Std.	Min.		313	334	365	405
				per cent	per cent	per cent	per cent	per cent	per cent	per cent	per cent
1.5	0.17	0.214	0.26	67	61.1	55	25	0.2	0.8	25	65
1.7	0.26	0.300	0.36	55	50.1	43	20	0.2	0.7	20	50
2.0	0.36	0.429	0.54	43	37.3	29	15	0.2	0.5	14	35
2.5	0.54	0.643	0.75	29	22.8	18.0	12	0.2	0.3	5	15
3.0	0.75	0.857	1.07	18.0	13.9	8.50	9.0	0.2	0.2	0.5	6
4.0	1.07	1.286	1.50	8.50	5.18	3.16	5.0	0.2	0.2	0.5	1.0
5.0	1.50	1.714	1.93	3.16	1.93	1.18	2.5	0.2	0.2	0.2	0.5
6.0	1.93	2.143	2.36	1.18	0.72	0.44	1.5	0.1	0.1	0.1	0.5
7.0	2.36	2.571	2.79	0.44	0.27	0.164	1.3	0.1	0.1	0.1	0.5
8.0	2.79	3.000	3.21	0.164	0.100	0.061	1.0	0.1	0.1	0.1	0.5
9.0	3.21	3.429	3.64	0.061	0.037	0.023	0.8	0.1	0.1	0.1	0.5
10.0	3.64	3.857	4.07	0.023	0.0139	0.0085	0.6	0.1	0.1	0.1	0.5
11.0	4.07	4.286	4.50	0.0085	0.0052	0.0032	0.5	0.05	0.05	0.05	0.1
12.0	4.50	4.714	4.93	0.0032	0.0019	0.0012	0.5	0.05	0.05	0.05	0.1
13.0	4.93	5.143	5.36	0.0012	0.00072	0.00044	0.4	0.05	0.05	0.05	0.1
14.0	5.36	5.571	5.79	0.0004	0.00027	0.00016	0.3	0.05	0.05	0.05	0.1

NOTE: The values given apply to class I filter glass. For class II filter lenses, the transmittances and tolerances are the same, with the additional requirement that the transmittance of 589.3 millimicrons shall not exceed 15 per cent of the luminous transmittance. Some of the headings in this table have been changed to conform with NBS Letter Circular LC857.

Spectral-Transmissive Properties and Use of Eye-Protective Glasses. *Reprinted from National Bureau of Standards. Circular 471, 1948.*

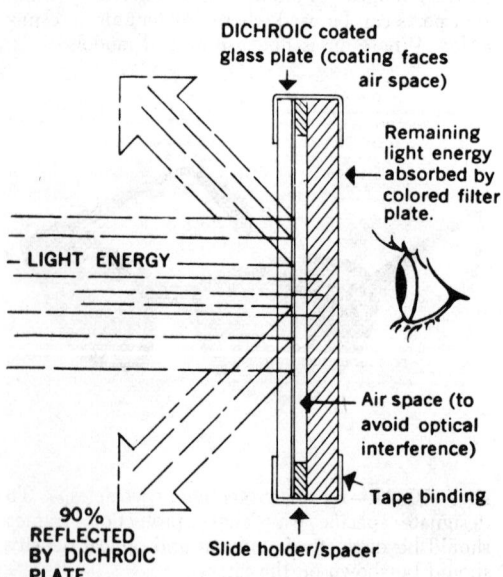

DICHROIC coated glass plate (coating faces air space)

Remaining light energy absorbed by colored filter plate.

– LIGHT ENERGY

Air space (to avoid optical interference)

Tape binding

90% REFLECTED BY DICHROIC PLATE

Slide holder/spacer

chemicals may be provided in a variety of ways, depending upon the hazard. Good protection is given by a hood made of chemical-resistant material with a glass or plastic window (Fig. 19–20). In all cases, there should be a secure joint between the window and the hood materials.

Hoods are extremely hot to wear, and can be obtained with air lines for the wearer's comfort. When a hood is so supplied, the wearer should have a harness or belt like that on an air line respirator (see Fig. 19–30 and –31) for support of the hose. A device based on a vortex principle has been developed to provide temperature-conditioned air. (See Fig. 19–36.)

If protection is necessary only from limited direct splashes, the wearer can don a face

◄FIG. 19–22.—Diagram shows how "antilaser eyeshield" deflects laser beam.

shield made of a material unaffected by the liquid or a flexible fitting chemical goggle with baffled ventilation, if the eyes are not exposed to irritating vapor.

For severe exposures a face shield should be worn in connection with the flexible fitting chemical goggles.

Face shields should be shaped to cover the whole face. They should be supported from a headband or harness, so they can be tipped back and clear the face easily. Any shield should be easily removed in case it becomes wet with corrosive liquid.

If goggles worn under the shield are of the nonventilated type for protection against vapor as well as against splashing, a nonfogging material should be provided, or they should be of the nonfogging type (described earlier).

Laser beam protection

No one type of glass offers protection from all laser wavelengths. Consequently, most laser-using firms don't depend on safety glasses to protect employee's eyes from laser burns. Some point out that laser goggles or glasses might give a false sense of security, tempting the wearer to expose himself to unnecessary hazards.

Nevertheless, researchers and laser technicians do frequently need eye protection.

Both spectacles and goggles are available— and glass for protection against nearly all the known lasers can be had on special order from eyewear manufacturers. Typically, the eyewear will have maximum attenuation at a specific laser wavelength—with protection falling off rather rapidly at other wavelengths.

Laser protective goggles or spectacles (Fig. 19–21) or an "anti-laser eyeshield" (Fig. 19–22) attenuate the He-Ne laser light (wavelength 6328 Å or 632.8 μ) by factors of 10 (O.D. = 1), 100 (O.D. = 2), 1000 (O.D. = 3), or more. An optical density (O.D.) of three or four still renders the beam visible in bright sunlight. The goggle-type of protective eyewear utilized in the laboratory is often unsuitable in the field because of fogging.

The American Conference of Governmental Industrial Hygienists cautions that safety glasses should be evaluated periodically to make sure maintenance of adequate optical density at the desired laser wavelength.

There should be assurance that laser goggles designed for protection from specific laser wavelengths are not mistakenly used with different wavelengths of laser radiation. The optical density values and wavelengths should be shown on the eyewear. Eyewear storage shelves can also carry this notation.

Laser safety glasses exposed to very intense energy or power density levels may lose effectiveness and should be discarded.

Technical details, uses, hazards, and exposure criteria for lasers are given in Chapter 37, "Industrial Hygiene."

Eye protection for welding

In addition to damage from physical and chemical agents, the eyes are subject to the effects of radiant energies. Ultraviolet, visible, and infrared bands of the spectrum are all able to produce harmful effects upon the eyes, and therefore require special attention.

Ultraviolet rays can produce cumulative destructive changes in the structure of the cornea and lens of the eye. Short exposures of intense ultraviolet radiation or prolonged exposures to ultraviolet radiations of low intensity will produce painful, but ordinarily self-repairing corneal damage.

Radiations in the visible light band, if too intense, can cause eyestrain and headache, and can destroy the tissue of the retina.

Infrared radiations transmit large amounts of heat energy to the eye, causing discomfort. The damage produced is superficial.

The filtering properties of filter lenses have been established by the National Bureau of Standards. The percentage transmittance of radiant energies in the three bands—ultraviolet, visible, and infrared—is established for 16 different filter lens shades (Table 19-A). Also see Table 34-B for type of operation for which each lens shade is recommended.

Photochromic lenses (they darken in sunlight and fade indoors in low light levels) should not be used as a substitute for established filter lens shades. They have high transmission in the near-ultraviolet and infrared.

Welding processes (see Chapter 34, "Welding and Cutting") emit radiations in three spectral bands. Depending upon the flux used and the size and temperature of the pool of melted metal, welding processes will emit more or less visible and infrared radiation—the proportion of the energy emitted in the visible range increases as the temperature rises. At least one manufacturer produces an aluminized cover for the usual black welding helmet. Its purpose is to reduce infrared absorption and the resulting heat stress to the wearer.

All welding presents problems, mostly in the control of infrared and visible radiations. Heavy gas welding and cutting operations, and arc cutting and welding exceeding 30 amperes, present additional problems in control of ultraviolet. Welding helmets must be used to provide head and face protection (Fig. 19–19, *right*).

Welders may choose the shade of lenses they prefer within one or two shade numbers. Following are shades commonly used:

SHADES No. 1.5 TO No. 3.0 are intended for glare from snow, ice, and reflecting surfaces; and for stray flashes and reflected radiation from cutting and welding operations in the immediate vicinity (for goggles or spectacles with side shields worn under helmets in arc welding operations, particularly gas-shielded arc welding operations).

SHADE No. 4, the same as shades 1.5 to 3.0, but for greater radiation intensity.

For welding, cutting, brazing, or soldering operations, the guide for the selection of proper shade numbers of filter lenses or windows in Chapter 34, "Welding and Cutting," should be used. (Recommendations are also in ANSI Z87.1, *Eye and Face Protection.*)

To protect the filter lenses against pitting, they should be worn with a replaceable plastic or glass cover plate.

Eye protection having mild filter shade lenses or polarizing lenses, and having opaque side shields, are adequate for protection against glare only. For conditions where hot metal may spatter and where visible glare must be reduced, a plastic face shield worn over mild filter shade spectacles with opaque side shields should be specified.

It is permissible to combine the shade of the plate in a welder's helmet with that of the shade of the goggle worn underneath to produce the desired total shade. This procedure has the added advantage of protecting the

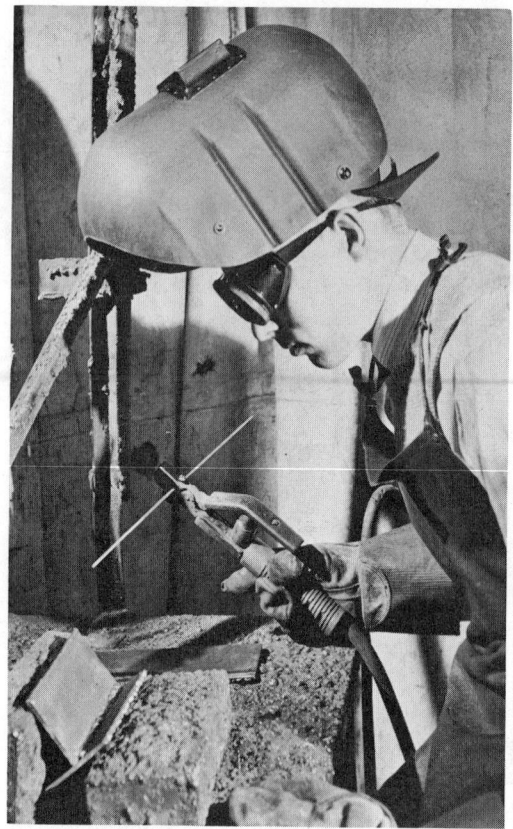

FIG. 19–23.—Double eye protection in electric welding. Flash goggles have No. 4 shade lens; they protect welder's eyes from the arc of other welders when he lifts his helmet to inspect work.

eyes from other welding operations or from an accidental arc when the helmet is raised (Fig. 19–23).

To protect against ultraviolet and infrared radiation as well as against visible glare in inspection operations, protective lenses should be installed in a hand shield or welder's helmet. The shield should be made of a nonflammable material, which is opaque to dangerous radiation and a poor conductor of heat. A metal shield is not desirable, because it heats under infrared radiation.

Some tinted lenses used in special work afford no protection from infrared and ultraviolet radiations. For instance, most melters' blue glass used in open-hearth furnaces and

FIG. 19–24.—The National Institute for Occupational Safety and Health and the Mining Enforcement and Safety Administration approval for respiratory equipment. Approval label must appear on the device.

the lenses used at Bessemer converters afford no protection against either type of harmful radiation. Probably no harm will come from continued use of these lenses if the exposures are of short duration. However, new men learning these flame reading skills should be provided with lenses that protect in these two portions of the spectrum.

The chemical composition of the lens rather than its color provides the filtering effect; this factor must be considered when selecting a filtering lens.

Respiratory Equipment

The National Institute for Occupational Safety and Health (NIOSH) of the U.S. Department of Health, Education, and Welfare, and the Mining Enforcement and Safety Administration (MESA) of the U.S. Department of the Interior have established minimum performance requirements for many respiratory protective devices. NIOSH maintains a testing laboratory and issues approval certificates (Fig. 19–24) to manufacturers for

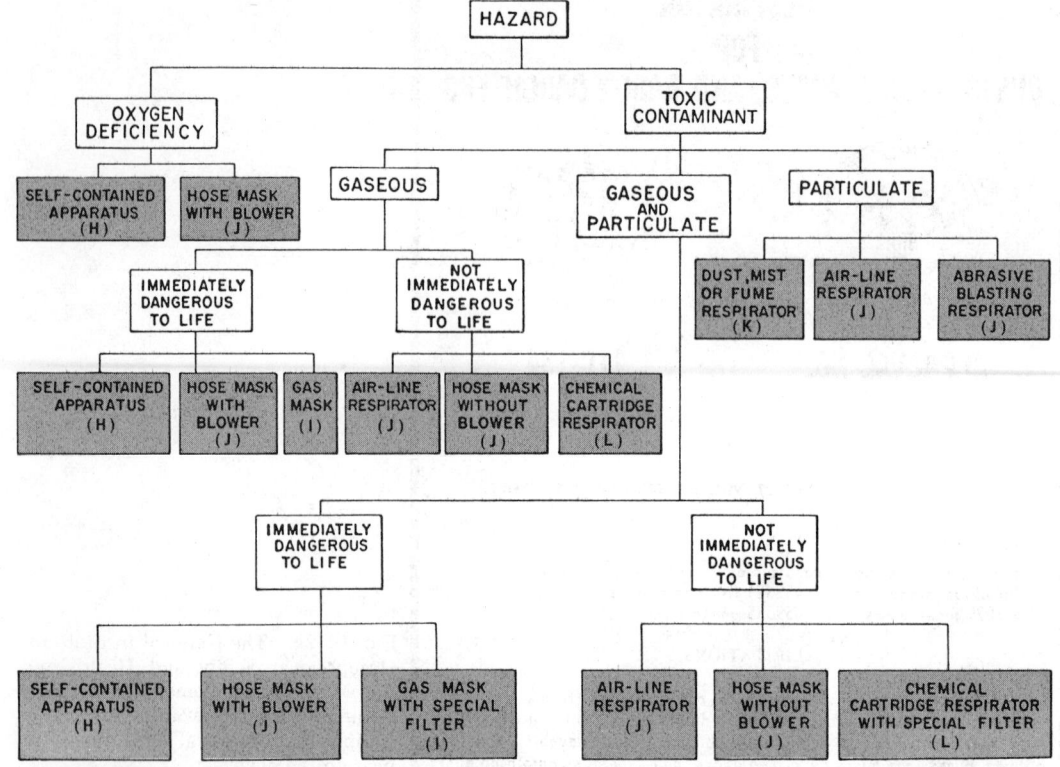

Fig. 19–25.—Suggested outline for selecting respiratory protective devices. Letters in parentheses (in shaded boxes) refer to Subparts of Title 30, C.F.R., Part 11, which discuss the items.

particular respiratory protective devices in their entirety. Individual components, as for instance face pieces, hoses, or canisters, are not approved separately. An approval number appears on each component of the certified assembly, where feasible.

The requirements for approval are contained in the Department of Interior's "Respiratory Protective Devices, Tests for Permissibility Fees," contained in the Title 30, *Code of Federal Regulations,* Chapter 1, Subchapter B, Parts 11, 12, 13, 14, and 14a. In some respects, these requirements are more stringent than the former Bureau of Mines requirements. The subparts of Part 11 that discuss various types of devices are shown in Fig. 19–25.

Although the NIOSH-MESA certification system does not extend to every type of hazard, whenever there is a choice between certified and uncertified devices, the certified devices should be selected.

Respirator selection

Among the many factors to be considered in the selection of the proper respiratory protection device for any given situation involving air contamination are the following. (The general principles for selecting respiratory protective devices are outlined in Fig. 19–25.)

1. The nature of the hazardous operation or process;

2. The type of air contaminant, including its physical properties, chemical properties, physiological effects on the body, and its concentration;

3. The period of time for which respiratory protection must be provided;

4. The location of the hazard area with respect to a source of uncontaminated respirable air;

5. The state of health of personnel involved;

6. The functional and physical characteristics of respiratory protective devices.

The safety professional should familiarize himself with the type of hazard for which a given type of respiratory equipment is approved, and should not permit its use for protection against hazards for which it is not designed. *For instance, particulate filter respirators are of no value as protection against solvent vapors, injurious gases, or lack of oxygen. Their use under these conditions is one of the most common and most dangerous abuses of respirators.*

Oher common misuses of respiratory equipment are to use chemical cartridge respirators where gas masks are required or to use chemical filtering types where atmosphere-supporting or self-contained units are necessary.

Some substances require protection against (a) damage to the respiratory system, and also (b) systemic injury through the skin. All substances should be investigated to learn whether this double hazard is involved. Some common substances which are hazardous through both the respiratory system and the skin are:

Aniline
Carbon disulfide
Cresol (cresylic acid)
Dichloroethylene-1,1
Decalin (decahydronaphthalene)
Diethylene oxide (dioxane-1,4)
Diethylphthalate
Dimethylaniline
Dinitrobenzene
Dinitrochlorobenzene
Hydrogen cyanide (hydrocyanic acid)
Iodoform
Lead tetraethyl
Mercury and its compounds
Nicotine
Nitrobenzene
Nitroglycerine
Organic phosphate insecticides
Phenyl hydrazine
Tetralin (tetrahydronaphthalene)

See Chapter 39, "Chemical Hazards," for a full list.

Some workers consider respiratory equipment a nuisance, not realizing that failure to wear it may endanger their lives. Usually, the attitude of such men can be changed if someone in authority will clearly explain why the equipment is necessary, show the men how to fit it in position, and explain its operation. If such educational efforts fail, the alternative is discipline, for the risk of injury is serious.

Excellent health and thorough training in the operation and maintenance of equipment are essential for persons who are to use respiratory protective devices. Anyone in questionable physical condition should be prevented from entering environments presenting respiratory hazards, and thus avoid having to wear any emergency breathing apparatus for protection.

From the standpoint of fire hazard, neither pure oxygen nor air containing more than 21 percent oxygen is to be preferred to ordinary air for use in atmosphere (air) supplied respirators or open-circuit (demand) type self-contained breathing apparatus, for ventilating, or for other purposes. A flammable substance in the presence of oxygen requires only a small fraction of the energy to ignite it as does the same substance in the presence of air, and an ensuing fire or explosion is much more violent.

Air or oxygen provided by or supplied to any respiratory equipment must be free from contaminants; quality as well as quantity is important. Respirable air should comply with the Compressed Gas Association *Commodity Specification for Air*, G-7.1-1966 (ANSI Z86.1), Grade D. This specifies that the maximum impurities be as follows: condensed hydrocarbons, 5 mg/m³ of gas at NTP; carbon monoxide, 10 mg/m³; and carbon dioxide, 1000 mg/m³. The presence of a pronounced odor renders the air unsatisfactory for breathing. The standard gives methods of verification and other details.

Oxygen should meet the requirements of the *United States Pharmacopoeia*.

Respiratory protective devices can be classified as follows.

1. Air purifying respirators,

TABLE 19-B. CHARACTERISTICS OF GAS MASKS AND THEIR CANISTERS

Mask	Type	Max. Con. of gas (% by vol.)°	Color of Canister (OSHA-NIOSH)†	Materials in Canister	Remarks
ACID GAS (for protection against gases such as hydrogen sulfide, sulfur dioxide, chlorine, hydrocyanic acid)	A	2%	White	Soda lime or soda lime and activated charcoal	The time of protection decreases rapidly as the concentration of gas increases. (See Notes 1, 2, 3, 5)
ORGANIC VAPOR (for protection against vapors such as aniline, benzene, ether, gasoline, carbon tetrachloride, chloropicrin)	B	2%	Black	Activated charcoal	(See Notes 1, 2, 3, 5)
AMMONIA GAS	C	3%	Green	Silica gel or porous granules impregnated with metallic salts such as those of copper or cobalt	(See Note 4)
CARBON MONOXIDE	D	2%	Blue	Hopcalite	The air becomes noticeably warmer as the percentage of carbon monoxide increases (See Notes 1, 2, 3.)
DUST FUMES, MISTS, FOGS AND SMOKES in combination with any of the above gases or vapors	AE, etc.	2%	Any of above, plus top grey stripe	Any of above plus mechanical filter	(See Note 6)
COMBINATION ACID GAS AND ORGANIC VAPOR	AB	2% acid gases 2% organic vapors	Yellow	Activated charcoal and soda lime	(See Notes 1, 2, 3, 4)
COMBINATION ACID GAS, ORGANIC VAPOR AND AMMONIA GAS	ABC	2% acid gas 2% organic vapor 2% ammonia	Brown	Soda lime, activated charcoal and silica gel or impregnated porous granules	As the number of gases increases, the service time of the canister decreases (See Notes 1, 2, 3.)
COMBINATION ACID GAS, AMMONIA GAS	AC	2% acid gases 3% ammonia	Green with white stripe at bottom	Soda lime, silica gel or porous granules impregnated with metal salts such as those of copper or cobalt	(See Notes 1, 2, 3.)

2. Atmosphere (air) supplied respirators, and

3. Self-contained breathing devices.

In turn, these are divided into a number of subclasses.

1. Air Purifying Respirators

Gas masks

The gas mask type of air purifying respirator consists of a facepiece connected by a flexible tube to a canister (Fig. 19–26). Contaminated air is purified by chemicals in the canister.

Because no one chemical has been found that will remove all gaseous contaminants, the canister must be carefully chosen to fit the specific need. A canister designed for a specific gas or vapor (or a single class of gases or vapors) will give longer protection than a same-sized canister designed for protection against a multitude of gases and vapors.

NIOSH-MESA tests and certifies gas masks for a number of gases and vapors and combinations of these hazards. Gas masks that protect against airborne particulates, in addition to gases and vapors, are available. Table 19-B lists the general characteristics of gas masks and their canisters.

Canister gas masks with full facepiece are for emergency protection in atmospheres immediately dangerous to life. They do *not* provide protection against oxygen deficiency; their effectiveness is limited to use in atmospheres containing at least 19.5 percent (by volume) oxygen, and not more than 2 percent of those toxic gases for which it is designed (except for ammonia, for which the limit is 3 percent, and phosphine, for which the limit is 0.5 percent).

Gas masks should not be used for fire fighting in accordance with the NFPA Standard No. 19B, *Respiratory Protective Equipment for Firefighters,* because of the possibility of oxygen deficiency. Self-contained breathing apparatus should be used; see the sections that follow on hose masks and self-contained breathing devices.

The period of protection that a gas mask provides depends upon (*a*) the type of canister, (*b*) the concentration of the gas or vapor, and (*c*) the activity of the user.

Each person who must use a gas mask

(for protection against combinations of acid gases, organic vapors, ammonia, carbon monoxide and smokes)

2% gases
3% ammonia
2% carbon monoxide
2% organic vapors
Not more than 2% total poisonous gases

charcoal, fused calcium chloride, silica gel or impregnated porous granules, hopcalite, and filter

Stripes to indicate filters are necessary on this type canister

NOTE 1: These gas masks are provided with a body harness to support the weight of the canister.

NOTE 2: These gas masks have check valves and exhalation valves.

NOTE 3: No gas mask will provide protection where the atmosphere is deficient in oxygen.

NOTE 4: If a dust, smoke or mist filter is combined with a gas filter, with the exception of the type-N canister, it is shown by a 1/2-in. black or white stripe on the canister.

NOTE 5: Canisters for a single gas or vapor other than ammonia or carbon monoxide shall have a 1/2-in. colored stripe around the canister near the bottom. The stripe for hydrocyanic acid gas is green; for chlorine it is yellow.

NOTE 6: Canisters with additional filters for dusts, fumes, mists, etc., shall have a 1/2-in. contrasting black or white stripe around the canister located near its top.

*These are high concentrations: In the case of acid gases and ammonia, they are as much as can be tolerated because of discomfort from skin irritation.

†The 1973 revision of ANSI K13.1, *Identification of Air-Purifying Respirator Canisters and Cartridges,* changes some of the identifying colors. Be sure to check latest OSHA-NIOSH standards.

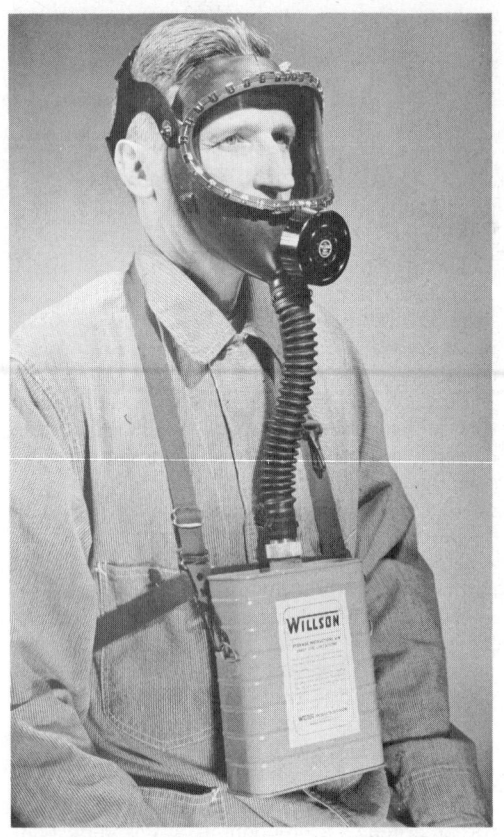 

FIG. 19–26.—Two different designs of canister-type gas masks. Note the speaking diaphragms and the face pieces which allow wide, panoramic vision. *Left:* Mask makes use of a rigid, adjustable head harness. *Right:* Mask has conventional strap-type head harness.

should first undergo a physical examination, especially of his heart and lungs.

Train carefully. Because the gas mask is ordinarily used in emergencies, where there is strain and excitement, those who use it should be carefully trained.

One of the first lessons should be how to make sure the mask will fit. To do this, the wearer should:

1. Remove the seal from the bottom of the canister and put on the head piece.

2. Adjust the head straps until the mask fits closely and comfortably.

3. See that there are no kinks in the tubes or the hose.

4. Breathe naturally and observe the amount of resistance to inhalation.

5. Grasp the hose tightly and close off the air intake. Breathe deeply until the facepiece collapses. If correctly adjusted, the facepiece will remain collapsed until the intake is opened.

In training practice the user should wear the mask long enough to become accustomed to the breathing resistance.

As a further check on the fit of the mask and the effectiveness of the canister, the user of a gas mask should enter the contaminated area cautiously. If the mask leaks or the canister is exhausted, the user will usually know by odor, taste, or irritation of eye, nose, or throat, and should immediately return to fresh air.

If the canister is used up, it should not be left attached, but removed, and a new one should be selected and fastened in place. When a respirator is worn in a gas or vapor that has little or no warning properties, like carbon monoxide, it is recommended that a fresh canister be used each time a man enters the toxic atmosphere.

Gas masks that offer respiratory protection against carbon monoxide must have an indicator that shows when the canister should be changed for protection against that gas. This is needed because water vapor renders useless the chemical (used for carbon monoxide removal), whether or not carbon monoxide is present. Nonwindow-type canisters should not be used; tests on nonwindow-type canisters that had been unsealed for some time showed they had lost their ability to remove carbon monoxide, even though the inhalation resistance of most canisters was the same as for fresh canisters.

Canisters are usually supplied with seals to prevent air from entering them until they are to be used. These seals should be removed when the canister is installed in the mask.

Storage. If a universal type N canister having an indicating window is used for protection against carbon monoxide, it should be discarded as soon as the color change in the window indicates that the canister will no longer convert carbon monoxide to carbon dioxide.

A card should be set up for each mask to indicate the date of the latest inspection and replacement of the canister, and the amount of use which the canister has had. If the mask is for emergencies only, it is wise to replace the canister after each use.

Gas masks should be kept easily available for emergencies. For instance, in an ice plant anticipating an ammonia leak, the mask should be placed just inside or outside one of the exit doors so that it can be reached quickly without disabling exposure to the gas. It should not be kept too near the ice machine or ammonia piping.

Masks should be stored away from moisture, heat, and direct sunlight, and should be regularly inspected.

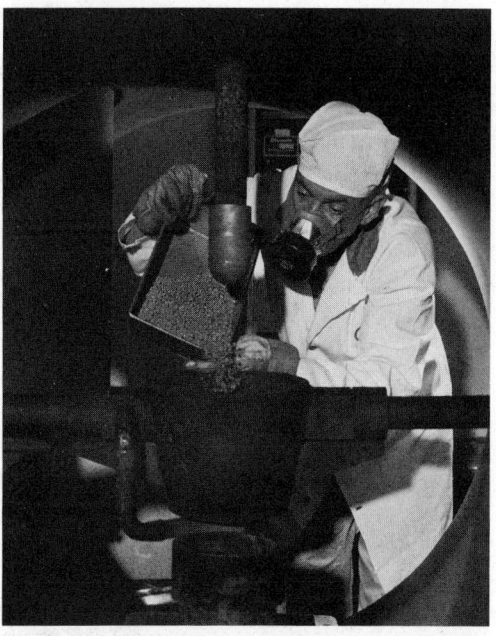

Fig. 19–27.—Single cartridge chemical respirator. Twin-cartridge units are also available and give longer life and less breathing resistance.

Chemical cartridge respirators

Chemical cartridge respirators consist of a half-mask facepiece connected directly to one or two small containers of chemicals (Fig. 19–27). Chemicals used are similar to those found in gas masks, but cartridge respirators are for use only in nonemergency situations; that is, for atmospheres which are harmful only after prolonged or repeated exposures.

Previously, Bureau of Mines approval has been given *only* for type B (organic vapor) and type BE (dusts, fumes, and mists in combination with organic vapors). These types are approved for concentrations not in excess of 1000 parts of organic vapor per one million parts of air. Now, Subpart K(23c) of Title 30, *Code of Federal Regulations*, Part 11, permits certification of respirators providing respiratory protection against certain other gases.

Particulate filter respirators

A particulate (mechanical) filter respirator can be designed to give satisfactory protection

495

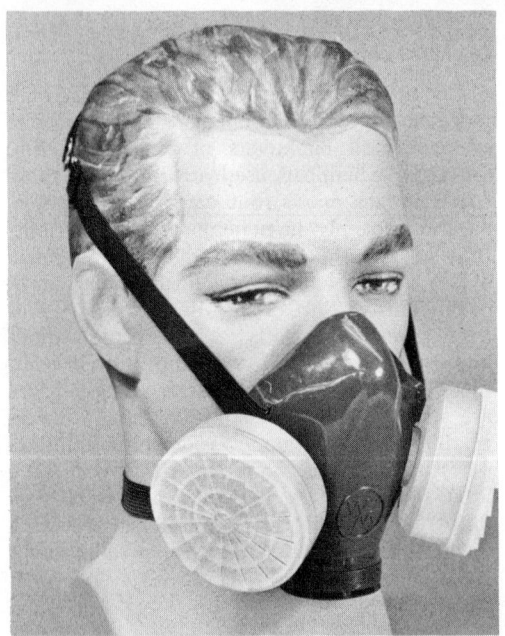

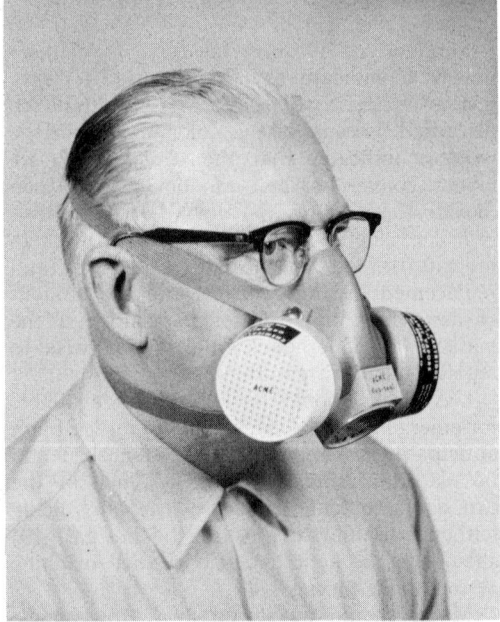

Fig. 19–28.—*Top:* Dust respirator and safety glasses protect employee during laboratory cleaning operation. *Lower left:* Twin-cartridge dust respirator. *Lower right:* Twin-cartridge organic vapor respirator.

against any kind of particle (Fig. 19–28). The major items to be considered are the resistance to breathing offered by the filtering element, the adaptation of the facepiece to faces of various sizes and shapes, and the fineness of the particles to be filtered out.

Low breathing resistance requires a fairly large area of filtering medium. Fineness of the filter is fixed by the permissible leakage for various dusts. It is important that breathing resistance be low even after the respirator has had considerable use under dusty conditions. Excessive breathing resistance wastes energy and may be the source of lung injury if long continued.

Filters of high porosity may give low breathing resistance with small filter area, but they will not stop fine dust. All such filters do is to improve comfort by catching the larger particles, which are usually caught in the nose and throat.

Breathing out against resistance does not engender the psychological effect that resistance to inhalation does. However, it can be detrimental in that it is physically tiring.

Filters should be replaced whenever breathing becomes difficult due to plugging of filters by retained particulates.

Combination respirators

Combination chemical and mechanical filter respirators utilize dust, mist, or fume filters with a chemical cartridge for dual or multiple exposure. Normally, the dust filter plugs up before the chemical cartridge is exhausted. It is therefore preferable to use respirators with independently replaceable filters. Carefully attach the filter to the chemical cartridge.

The combination respirator is well suited for spray painting and welding.

2. Atmosphere (Air) Supplied Respirators

Hose masks

Hose masks are available with blower (power driven or hand operated, Fig. 19–29), without a blower, or connected to a source of respirable air under pressure (air line hose mask, Fig. 19–30).

Of the three, only the hose mask with blower is to be used for work in atmospheres where the hazard is immediate. The hazard is considered immediate if, in the event of failure of the equipment, hasty escape from the dangerous (toxic, flammable, and/or oxygen deficient*) atmosphere would be impossible or could not be made without serious injury.

In case of failure of the air supply on a hose mask with blower, it is still possible to breathe through the hose while making an escape.

Hose masks with blowers are certified with a maximum of 300 ft of hose for each man. For hose masks without blowers, the certification is for a maximum of 75 ft of hose.

Most blowers for hose masks have two outlets so that two men, each with a maximum of 300 ft of hose, can be supplied. If the hose mask is used under conditions which may be immediately hazardous to life, a self-contained breathing apparatus should always be provided so that it will be available for an individual who may have to go in to assist the original workman.

A hose line is not intended for use as a life line, but as a breathing tube. When a hose mask is used, a helper should be assigned with no duties other than to observe the man in the hazardous atmosphere, tend the life line, and rescue the first man if necessary. The helper, too, should be equipped with a harness and life line, in addition to a self-contained breathing apparatus. When hose masks with blower are used, a blower operator also must be provided. Be sure the blower intake is located outside the contaminated area.

The body harness needed to pull the hose lines requires inspection prior to each use. The minimum requirement for certification is that the component parts of the harness shall withstand a pull of at least 250 pounds. Parts which are to be used subsequently should not be tested by being pulled, but only should be examined for signs of wear or deterioration. If a condition is found which appreciably reduces strength, the harness should be replaced.

NIOSH-MESA certification requires that 50

---

*Almost every flammable gas and vapor is toxic; however many toxic gases and vapors are nonflammable. See Chapter 39, "Chemical Hazards," for several examples.

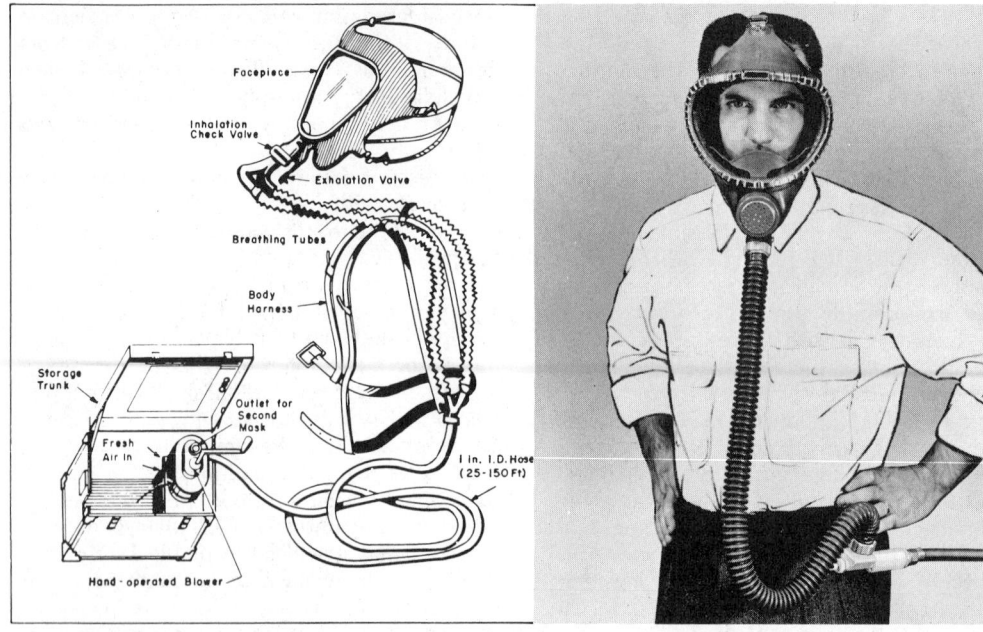

FIG. 19–29.—Diagram of hose mask with air supplied by hand-operated blower.

FIG. 19–30.—Air line respirator.

liters/min (106 cu ft/hr) be delivered to each facepiece through the maximum length of hose at not more than 50 rpm of the blower crank. Under most working conditions, this amount would furnish an excess of air to the wearers.

To eliminate dust and fumes that may have accumulated within a mask, air line and breather tubes should be blown out with air before a mask is put on. The couplings in the hose lines should be tested for tightness.

The hose mask should bear the certification number printed upon the container, and marked upon the facepiece and other parts of the equipment. This number establishes that the equipment was properly designed to do its job and will be safe to use if well maintained.

Where hose masks are supplied as emergency and rescue equipment, frequent and thorough training in their use should be given to all individuals who may use them.

The hose mask (Figs. 19–29 and –30) should be used by men entering tanks or pits where there may be dangerous concentrations of dust, mist, vapor, or gas, or insufficient oxygen. Inadequate oxygen should be suspected in any confined space, particularly in an iron or steel tank, unless a test shows that the supply will support life. Rusting of the walls of an otherwise clean tank may have depleted the oxygen supply.

No one should enter or remain in a tank or similar space that tests show has less than 16 percent oxygen in its atmosphere at any time unless he wears approved respiratory protective equipment, such as a fresh-air hose mask or self-contained breathing apparatus.

Insufficient oxygen and dangerous gases should also be suspected in a tank which has held organic materials such as grain or grain products, oil, or food material.

These materials, as well as a variety of others, may eventually react with the air to reduce its oxygen content and produce dangerous oxidization products.

Before a workman enters an enclosed space having an atmosphere that may contain toxic or flammable contaminants or that may be deficient in oxygen, the atmosphere should be tested with appropriate instruments. Instruments for determining the percentage of combustible gases are tested and approved by the U.S. Bureau of Mines. If the amount of any contaminant in the enclosure is more than its maximum acceptable concentration, or if the amount is within or close to the explosive limits, or if oxygen is deficient, the tank or enclosure should first be ventilated. This ventilation should never be done with oxygen because it could make an extremely hazardous atmosphere if flammable materials or gases were present.

After being purged of flammable or toxic concentrations and then reventilated, the tank or similar enclosure should be retested for contaminants in its atmosphere at intervals throughout the period that men are required to work in the space.

The hose mask with blower should be worn until the tank or similar enclosure has been thoroughly cleaned of scale and sludge and the flammable gas or vapor concentration has been reduced to 0.1 percent or less. Where there is a toxic gas or vapor and duration of stay in the enclosure is extended, the hose mask must be worn until the atmosphere is at or below the maximum acceptable concentration of the gas or vapor. It would be safer to use the device at all times in the confined space because the concentration of the hazardous gas or vapor can change suddenly.

The fit of the mask should be carefully checked before the worker enters the enclosure.

Air line respirators

The air line respirator type of atmosphere (air) supplied respirator (Fig. 19–31) has a hand-operated, quickly detachable coupling on the belt or body harness with which the operator can connect to a compressed air hose; it also contains a device to limit air flow.

There are two basic types: (a) the continuous flow, and (b) the demand flow.

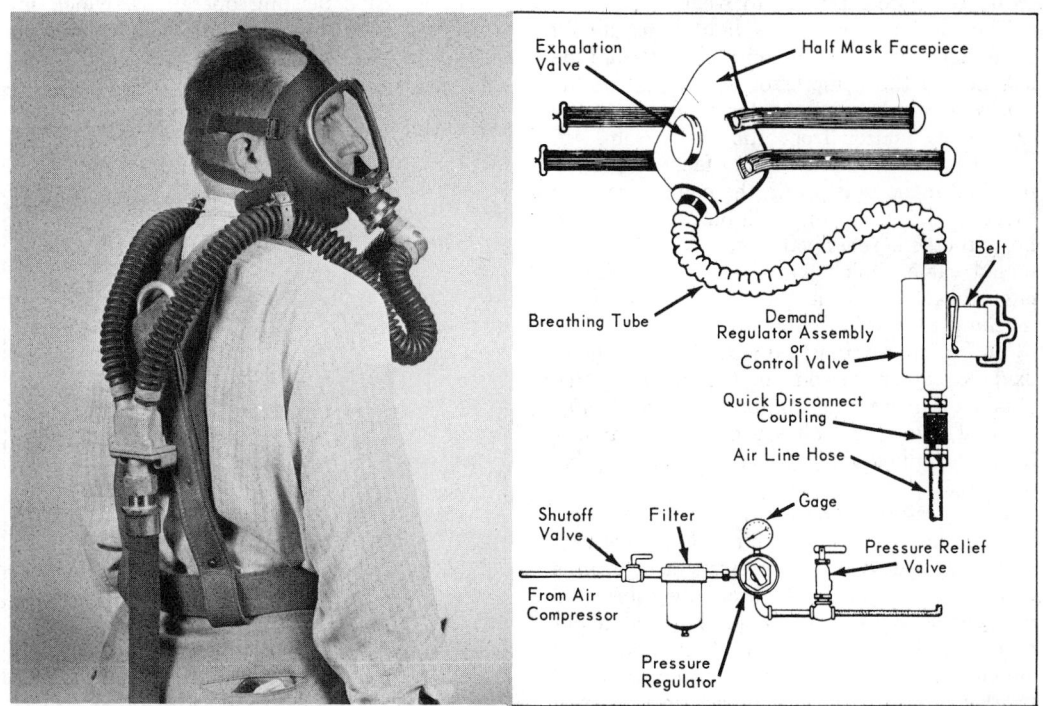

Fig. 19–31.—Air line respirator. Photograph at left shows full-face mask. Diagram (right) shows parts and connections.

• In the continuous flow type, a regulated amount of air is fed continuously to the face-piece.

• In the demand flow type, a diaphragm-actuated valve (demand valve) opens on inhalation of the user, and closes on exhalation. Exhaled air is discharged through an exhalation valve. The demand type is also available with a "pressure demand" capability. Maintaining a slight pressure at all times in the facepiece prevents any inward leakage of contaminated atmosphere. The demand type is commonly used when the air supply is limited, as from a compressed air cylinder.

The air line respirator furnishes complete protection against any atmosphere not immediately dangerous to life or health, and for such conditions is the most desirable protection for industrial operations requiring continuous use of a respirator. Although other types of respirators may give adequate protection, they offer a breathing resistance not present in the supplied-air type and are consequently more fatiguing to wear.

The air line respirator's limitations should be understood. A trap and filter should be installed in the compressor line ahead of the masks to separate oil, water, scale, or other extraneous matter from the air stream. A pressure regulator with an attached gage is required if the pressure in the compressor line exceeds 125 psi. In addition, there should be a pressure relief valve set at a predetermined value, which will operate if the regulator fails to prevent high-pressure air from reaching the user.

The air supply must be free of carbon monoxide or other gaseous contaminations. To obtain clean air, the compressor intake must be kept away from all sources of contamination, including internal combustion engine exhaust.

Low-pressure blowers which do not use internal lubricants are preferable to conventional high-pressure compressors as a source of respirable air. If an internally lubricated compressor is used, it should be well maintained and must not run hot; dangerous amounts of carbon monoxide can be produced by decomposition of the lubricating oil in a hot compressor.

To safeguard against the hazard of carbon monoxide in compressed air, there should be a temperature-actuated alarm on the compressor or a carbon monoxide alarm in the air line.

The air delivered to the mask should be of comfortable temperature and humidity. Compression and re-expansion often produce a temperature and a relative humidity widely different from that of the air entering the compressor. To lower its temperature, air can be passed through a cooler. Some states require both interstage and aftercoolers together with a high-temperature alarm.

On a job which requires a man to move from place to place, he will be hampered by dragging an air hose. This is usually a nuisance rather than a hazard, but it will have an effect upon efficiency which should be recognized in advance. Care must be exercised to prevent damage to the hose. For instance, it should not be permitted to lie in oil.

Abrasive blasting respirators

Abrasive blasting respirators are used to protect personnel engaged in shot, sand, or other abrasive blasting operations, which involve air contaminated with high concentra-

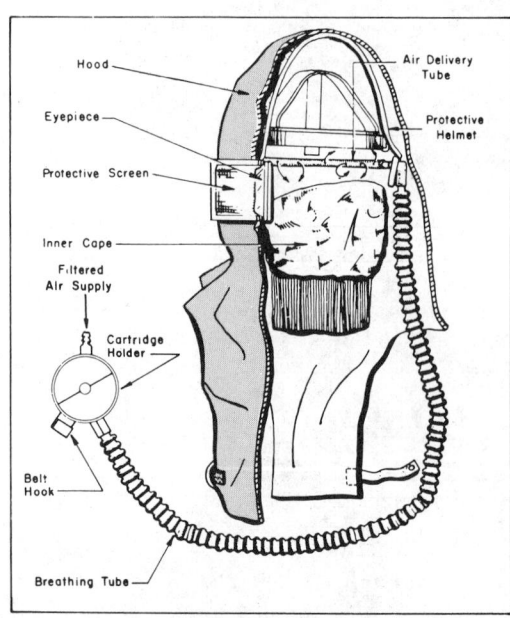

FIG. 19–32.—Diagram shows parts and connections for lightweight hood designed for use by persons doing abrasive blasting.

FIG. 19–33.—Solder grinders wear hoods to prevent foreign matter from entering their eyes, ears, noses, or mouths. Filtered air is supplied to the hoods at 12 to 15 psig and is exhausted around the bottoms.

tions of rapidly moving abrasive particles.

The requirements for abrasive blasting respirators are the same as those for an air line respirator of the continuous flow type, with the addition that mechanical protection from the abrasive particles is needed for the head and neck (Fig. 19–32). NIOSH-MESA tests and certifies such equipment.

There are two forms of abrasive blasting respirators that cover the head and neck, and even the shoulder and chest.

● One form has a full facepiece (made of rubber), and an eyepiece made of impact-resistant safety glass (or plastic covered by a metal screen). This unit is attached securely to a hood and cape made of tough, flexible rubber or rubber-covered fabric.

Air is supplied to the full facepiece and is exhausted through an exhalation valve located under the hood in order to expel the air at the edges of the hood.

● The other form uses a rigid helmet (generally made of metal) to encase the user's head. The eyepiece is of impact-resistant safety glass or of plastic covered by a metal screen. An adjustable knitted fabric collar,

covered with rubber- or plastic-coated fabric, fits over the metal helmet and down over the user's neck, shoulders, and chest in order to give additional protection. A flexible tube brings air to the helmet. Air, exhausted from the helmet, flows between the collar and the user's neck.

In both units, the eyepiece (window) and the protective screen should be easily replaceable, preferably without the use of tools.

Air supplied hoods

For some long-term operations where a completely enclosed suit is not necessary, an air supplied hood may be used. These are particularly useful in hot, dusty situations, Fig. 19–33.

Respirable air under suitable pressure should be delivered to a hood at a volume of at least 6 cfm.

Air supplied suits

The most extreme condition requiring respiratory equipment is that in which rescue or emergency repair work must be done in atmospheres extremely corrosive to the skin and mucous membranes, in addition to being acutely poisonous and immediately hazardous to life, such as atmospheres containing ammonia, or hydrofluoric or hydrochloric acid vapors.

For these conditions a complete suit of impervious clothing, with a respirable air supply, is available (Fig. 19–34). Complete units with self-contained breathing apparatus are discussed under "Self-contained breathing devices," which follows this section.

There are no generally accepted specifications for such suits, therefore considerable dependence must be placed upon the manufacturer. The material should have sufficient mechanical strength to resist rough handling and considerable abuse without tearing.

The hose line supplying the air should be connected to the suit itself, as well as to the helmet (Fig. 19–35), since it is not only extremely fatiguing but also dangerous to wear such a suit for a long period unless it is well ventilated.

Personal air conditioning devices utilizing a vortex tube are available for air supplied suits or hoods. These cooling devices are desirable to reduce fatigue where high am-

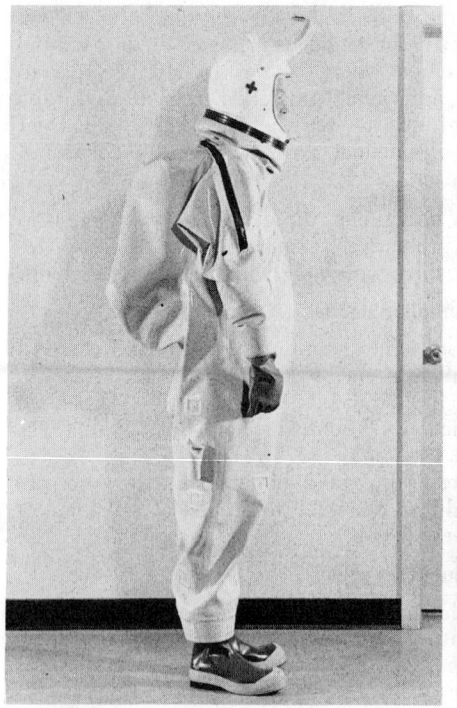

FIG. 19–34.—Environmental rocket fuel and corrosive chemical suits where complete protection from hazardous gases, mists, corrosive chemicals, or dusts is required. These suits include self-contained breathing apparatus. Suit at left has a liquid air system which provides a certain amount of pressure-regulated cool air for air conditioning. Suit at right is equipped with a standard self-contained gas mask with liquid air tank (in 25-lb back-pack unit) and pressure-regulated air supply. Unit maintains a constant internal temperature of 70 F, will enable a technician to live up to 2½ minutes in a flash fire (enough time to reach safety). Unit has a duration of 55 minutes, but could function as long as 1½ hours in emergency.

bient temperatures may be encountered (as in heat-protective clothing), or where body heat may build up (as under impermeable chemical protective clothing).

The vortex device (Fig. 19–36) works by taking an air stream under pressure and dividing it. One portion loses heat; the other gains heat. The cold portion passes into the suit or hood; the warm portion is vented to the atmosphere, or vice versa in cold weather.

Self-contained breathing devices

When a man must work in an atmosphere immediately hazardous to life, a self-contained breathing apparatus should be used. In an environment that contains a substance which is dangerously irritant or corrosive to the skin, a self-contained breathing device must be supplemented by impervious clothing.

Such devices afford complete respiratory protection in any toxic or oxygen-deficient atmosphere, regardless of the concentration of the contaminant. They also allow relative freedom of movement.

There are two main classes of self-contained breathing devices: (*a*) closed circuit (recirculating), and (*b*) open circuit (demand).

• There are two types of closed circuit apparatus generally available:

Oxygen-generating type, in which oxygen-generating chemicals in a container are activated by the moisture in the user's expired breath, and

Compressed air or liquid oxygen type, which employs a contained or compressed or liquid oxygen (Fig. 19–37).

• The open circuit (demand) type of apparatus uses a container or compressed or liquid air, or a container of compressed or liquid oxygen (Fig. 38–38).

When used in the conventional manner, no oxygen-supplying, self-contained breathing device should be used where the ambient pressure is more than two atmospheres, because of the danger of oxygen poisoning.

The oxygen-generating (recirculating) apparatus consists of a chemical canister, a breathing bag that acts as a reservoir, a facepiece with tube assembly, and a relief valve and check valves to regulate flow in accordance with respiratory requirements. The chemical in the canister evolves oxygen when contacted by the moisture and carbon dioxide in the exhaled breath, and also retains the carbon dioxide and moisture (moisture retention is important as it permits fog-free lenses). The apparatus is light in weight and contains no intricate parts.

The quick-start canister, however, might act

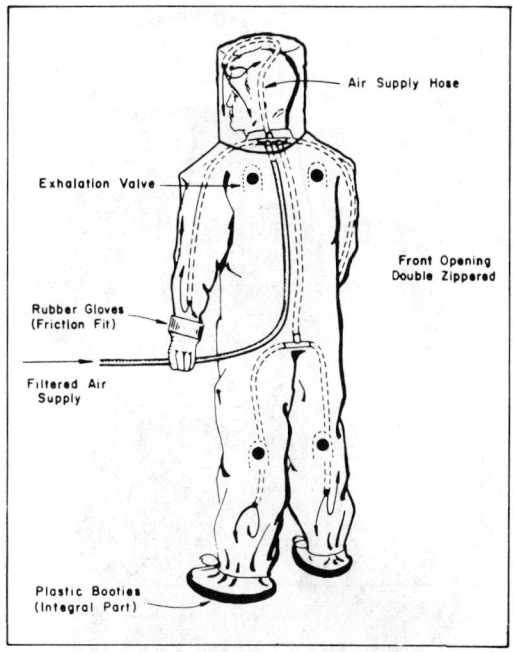

FIG. 19–35.—Diagram of air supplied suit for use in corrosive chemical atmospheres.

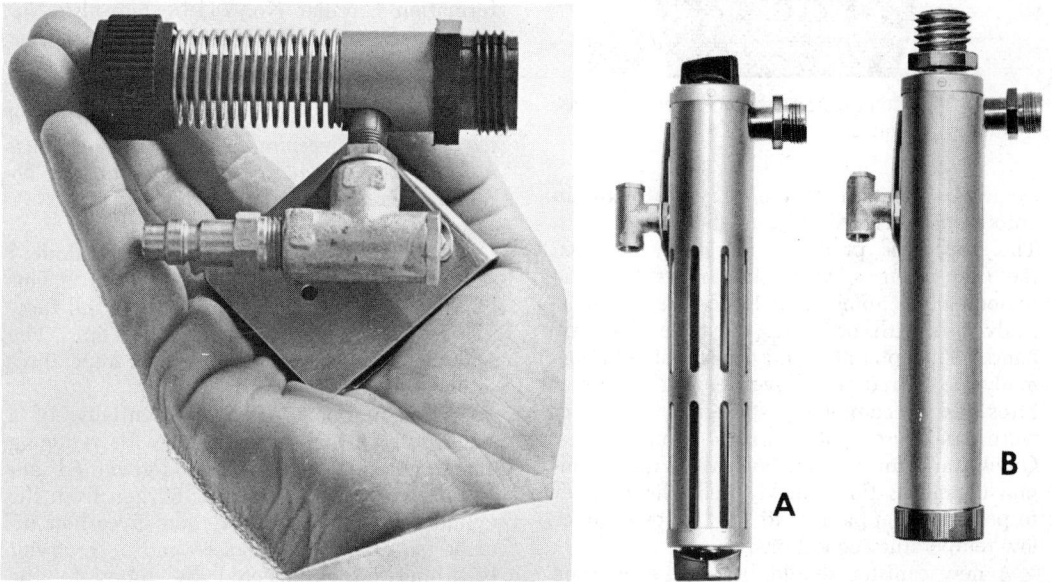

FIG. 19–36.—The vortex tube. Compressed air (100 psig, 100 F) enters tube from T-shaped fitting at side; it is fed to nozzles and accelerates to sonic speed, creating cyclone spinning at 500,000 rpm. Major portion of air spirals inward, expanding and cooling to 40 F before it is ejected. Smaller portion of air churns down the tube, heats up to 270 F. In photograph at right, tube A is capable of providing either cool or hot air; tube B provides cool air only.

Fig. 19–37.—Compressed oxygen (cylinder) rebreathing apparatus.

them. Used canisters should be kept away from combustibles so that residual oxygen will not have a chance to activate them. Only copious amounts of clean water should be used to destroy a canister.

It is important that no one wear self-contained breathing apparatus unless he is physically fit and well trained. Physical fitness and training are essential because poor physical condition of users and lack of proper training have been responsible for injuries and deaths. After the initial training, men should receive refresher training at least every six months in order to maintain efficiency.

The facepiece or mouthpiece and nose clip of a self-contained breathing apparatus should be carefully fitted to the wearer to ensure leak-proof protection against the hazardous atmosphere the man is to enter.

Half-hour and three-quarter hour apparatus are limited by the Bureau of Mines for use only as auxiliary equipment in mine rescue and recovery work. (See the U.S. Bureau of Mines, *Approval of Newly Developed Self-Contained Breathing Apparatus, Instruction in its Care and Use, and Training Procedure*, Information Circular No. 7413. See also National Safety Council, *Respiratory Protective Equipment*, Industrial Data Sheet 444.)

Because of the extreme hazard no one wearing self-contained breathing apparatus should work in an irrespirable atmosphere unless other persons similarly equipped are in attendance, ready to give assistance.

The compressed or liquid oxygen (cylinder) recirculating apparatus (Fig. 19–37). This equipment is supplied with either a full facepiece or a mouthpiece and nose clip. The seal around the facepiece must be kept absolutely tight.

Such equipment consists, essentially, of a high-pressure oxygen cylinder with reducing and regulating valves, a lung-governed admission valve which supplies oxygen from the cylinder only during inhalation, a carbon dioxide scrubber and cooler, and a reservoir breathing bag connected by tubes to the mouthpiece or facepiece. Check valves direct the flow through the circuit so that the exhaled oxygen is purified of carbon dioxide, and rebreathed from the bag, with replenishment from the cylinder as required. The

as an autoignition source to gases with an autoignition temperature of 600 F or less. The apparatus produces more oxygen than the user requires, so venting must be done periodically. Some models do this automatically but with others it must be done by hand. The apparatus and canister should normally be stored at above-freezing temperatures and be completely started before being worn and used at subzero temperatures. Quick-start canisters are available; these contain chemicals that quickly heat the canister to permit the apparatus to start quickly under low-temperature conditions.

A new canister should be used each time the apparatus is worn, for it is dangerous to enter a hazardous atmosphere (or even engage in a drill) with a previously opened canister. Used canisters should be promptly disposed of in accordance with instructions printed on liquid oxygen type operates similarly.

The rebreathing principle permits the most efficient utilization of the oxygen supply. The exhaled breath contains both oxygen and carbon dioxide because the body consumes only a small part of the inhaled oxygen.

Because this equipment is quite complicated to use and maintain, personnel should be thoroughly trained.

Open circuit apparatus. The compressed or liquid oxygen or compressed or liquid air (cylinder) open circuit (demand) apparatus (Fig. 19–38) consists of a high-pressure cylinder of oxygen or air, a cylinder valve, a demand regulator, and a facepiece and tube assembly with an exhalation valve. To use, the wearer turns on the cylinder valve after carefully putting on the facepiece, inhales to draw oxygen or air through the demand regulator at breathing pressure to the facepiece, and then exhales through the exhalation valve in the facepiece to the atmosphere. This makes the apparatus relatively inefficient when compared with rebreathing apparatus.

Pressure demand types are available. These maintain a slight pressure in the mask or facepiece to prevent leakage of ambient atmosphere into the mask or facepiece because of improper fit.

Once a cylinder has been used for compressed air, it should not be refilled with oxygen because of the possibility of fire if oxygen comes in contact with even a trace of oil or grease.

Care of respiratory equipment

Particulate filter and chemical cartridge types. Maintenance is simplified by selection of respirators that can be easily cleaned, disinfected, and repaired.

If possible, particulate filter respirators which use cotton facelets should be selected because the facelets protect the facepiece from skin oils. Facelets are not approved for use on chemical cartridge or metal fume respirators. If the filters are likely to clog rapidly, a respirator which allows convenient, inexpensive, and frequent filter replacement should be chosen.

Supervisors should be responsible for making daily inspection, particularly of functional parts such as exhalation valves and filter ele-

FIG. 19–38.—Compressed air (cylinder) nonrebreathing apparatus.

ments. They should see that the edges of the valves are not curled and that valve seats are smooth and clean. Inhalation and exhalation valves should be replaced periodically.

In addition to the daily check, respirators should be inspected weekly by trained persons. During the weekly inspection, rubber parts should be stretched slightly for detection of fine cracks. The rubber should be worked occasionally to prevent setting (one of the causes of cracking), and the headband should be checked to be sure that the wearer has not stretched it in an attempt to secure a snug fit.

Sometimes, in an effort to reduce resistance to breathing, workers will punch holes in the filter, the rubber facepiece, or other parts. Underlying causes of this mistreatment should be discovered and corrected. For instance, it may be found that in the interest of economy, filters are allowed to become completely plugged before they are replaced.

In cleaning respirators, dirt and dust should

first be blown from them by means of compressed air at not more than 10 to 20 psi of pressure, through a fixed nozzle directed towards an exhaust hood. The cleaners should wear dust-tight goggles. Dust filters should not be cleaned by brushing.

Then filters, screens, headbands, and cotton facelets should be removed. If the respirators are coated with paint or other foreign matter, they should be soaked for three hours in a cleaning solution of 1½ pounds of commercial alkaline base cleaner and 7 gal of water. Fresh paint can be wiped off with a clean rag moistened in alcohol. Thorough rinsing is always necessary before use.

Respirators having no visible accumulation of foreign matter should be scrubbed in warm, soapy water, rinsed, disinfected, then rinsed again and dried.

Knitted facelets should be washed in warm, soapy water, rinsed and dried before reuse. Dirty or oily elastic headbands should be washed in warm, soapy water and rinsed. The water should be warm to remove perspiration and hair oil from the elastic fabric.

Rubber parts should never be dried by direct application of heat or sunlight.

Employees should be instructed to wipe off oil, grease, and other harmful substances from headbands and other parts of the respirator as soon as they collect. They should be warned not to use solvents to clean plastic or rubber parts.

Most face and mouthpieces for respiratory protective devices are made from rubber or rubber-like compounds. Usually hand brushing or agitation in a washing machine, using detergent and warm water is sufficient to clean them.

Hypochlorite or quaternary ammonium compounds in the proper strength in aqueous solution can be used to disinfect the parts.

Ethylene oxide gas, handled under proper precautions, is sometimes used where large quantities of respirators must be disinfected.

All detergent should be rinsed from the device before disinfecting, except where combination detergent and quaternary ammonium compounds (that both clean and disinfect) are used.

After cleaning and disinfecting, the parts should be rinsed in clean water and dried quickly. It may also be desirable to rinse parts treated with quaternary ammonium compounds since their disinfecting properties continue and, except for rare cases, it will not produce a skin irritation.

Hot water, steam, solvents, and ultraviolet light should not be used to clean and disinfect rubber parts because they have a deteriorating effect.

Petroleum jelly should not be used to prevent skin irritation from rubber facepieces, for it is harmful to rubber. Disinfection and the use of clean cotton facelets will eliminate the need for a salve.

Respirators should be turned in at the end of each shift to be cleaned and repaired if necessary. They should be disinfected at least once a week when retained by the same employees.

In some plants, maintenance service for respirators, as well as for other kinds of personal protective equipment, can be effectively provided by traveling service carts (Fig. 19–39).

Where a number of respirators are in regular use, a central station is often set up for their care and maintenance, as well as for the care and maintenance of other items of personal protective equipment. Each employee is then provided with two respirators and either a locker or a hook at the central station.

Some plants have found that if two respirators are assigned to each man, equipment lasts more than twice as long. This plan is most desirable when cleaning cannot be done between shifts or before the next scheduled shift.

Under such a plan marked respirators are turned in daily or weekly, depending upon use, for cleaning, disinfection, inspection, and repair. The worker then uses the second respirator until the first can be serviced.

Another plan is to keep quantities of disinfected respirators on hand for groups. This plan works where individual needs vary, but in such a plan the user is not so easily charged with responsibility. If the same respirator is used by several persons, it is always well to clean and disinfect it after each use.

Respirators should be marked to indicate to whom they are assigned. The method of identification should be permanent enough so that the marking cannot be changed inadvertently or without effort.

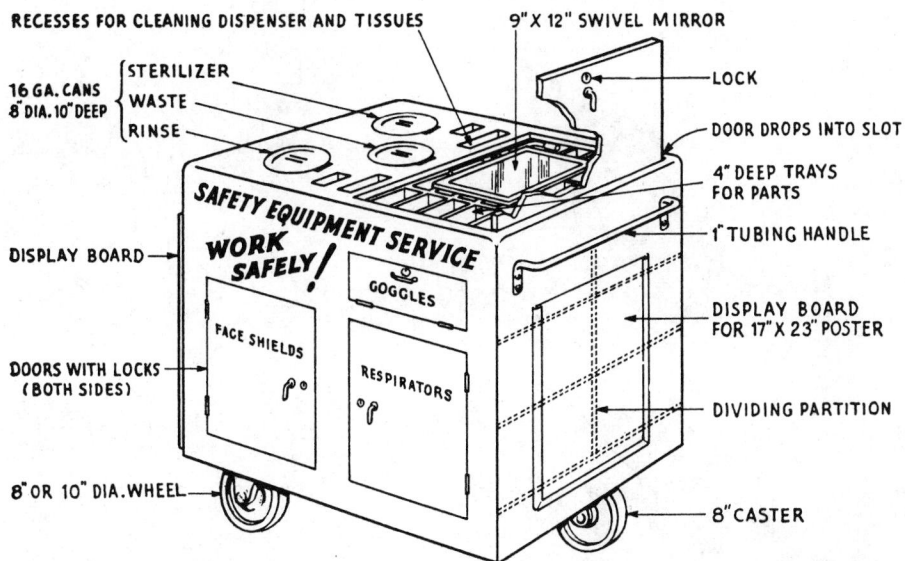

RECESSES FOR CLEANING DISPENSER AND TISSUES

9" X 12" SWIVEL MIRROR

16 GA. CANS
8" DIA. 10" DEEP
STERILIZER
WASTE
RINSE

LOCK

DOOR DROPS INTO SLOT

4" DEEP TRAYS
FOR PARTS

SAFETY EQUIPMENT SERVICE
WORK SAFELY!

DISPLAY BOARD

1" TUBING HANDLE

GOGGLES

FACE SHIELDS

RESPIRATORS

DISPLAY BOARD
FOR 17" X 23" POSTER

DOORS WITH LOCKS
(BOTH SIDES)

DIVIDING PARTITION

8" OR 10" DIA. WHEEL

8" CASTER

FIG. 19–39.—Maintenance service provided by this type of equipment improves employee attitude toward the use of personal protective devices. The service cart may be made of plywood or steel. Dimensions may be 36 in. long, 36 in. high, and 30 in. wide.

Before being stored, a respirator should be carefully wiped with a damp cloth and dried. It should be stored without sharp folds or creases. It should never be hung by the elastic headband or put down in a position which will stretch the facepiece.

Since heat, air, light, and oil cause rubber to deteriorate, respirators should be stored in a cool, dry place and protected from light and air as much as possible. Wood, fiber, or metal cases are provided with many respirators. Respirators should be sealed in clean plastic bags.

Respirators should not be thrown into tool boxes or left on work benches where they may be exposed to dust and damage by oil or other harmful materials.

Air line respirators. Maintenance suggestions for filter respirators apply also to air line respirators. In addition, the latter should be inspected at weekly or monthly intervals, depending on use. Pressure regulators, relief valves, air control valves, and demand (or pressure demand) regulators should be checked periodically for cleanliness and proper operation. Rubber (or other elastomer) and

plastic parts (such as hoses, facepieces, and exhalation valves) should be checked regularly for deterioration caused by age, heat, ozone, or other deleterious action.

The pressure regulator, filter, hose, and facepiece should be checked for deterioration.

The air hose and attachments should be cleaned regularly to prevent accumulation of paint, oil, grease, or solvents which might harm the rubber or connections. Oil and grease should be removed with steam at no more than 5 psig of pressure. The steam nozzle should not be held so close to the air hose as to "burn" it. Protective goggles and rubber gloves should be worn during cleaning operations.

Gas masks. Canisters should be stored and handled as directed by their manufacturer.

A canister with broken seals should not be kept in service for more than one year, regardless of how little it has been used. Although some depend on employees to record the length of time of canister use, an indicator is more reliable. A good rule to follow is that if there is any doubt, replace the canister.

When not in use, canisters should be pro-

507

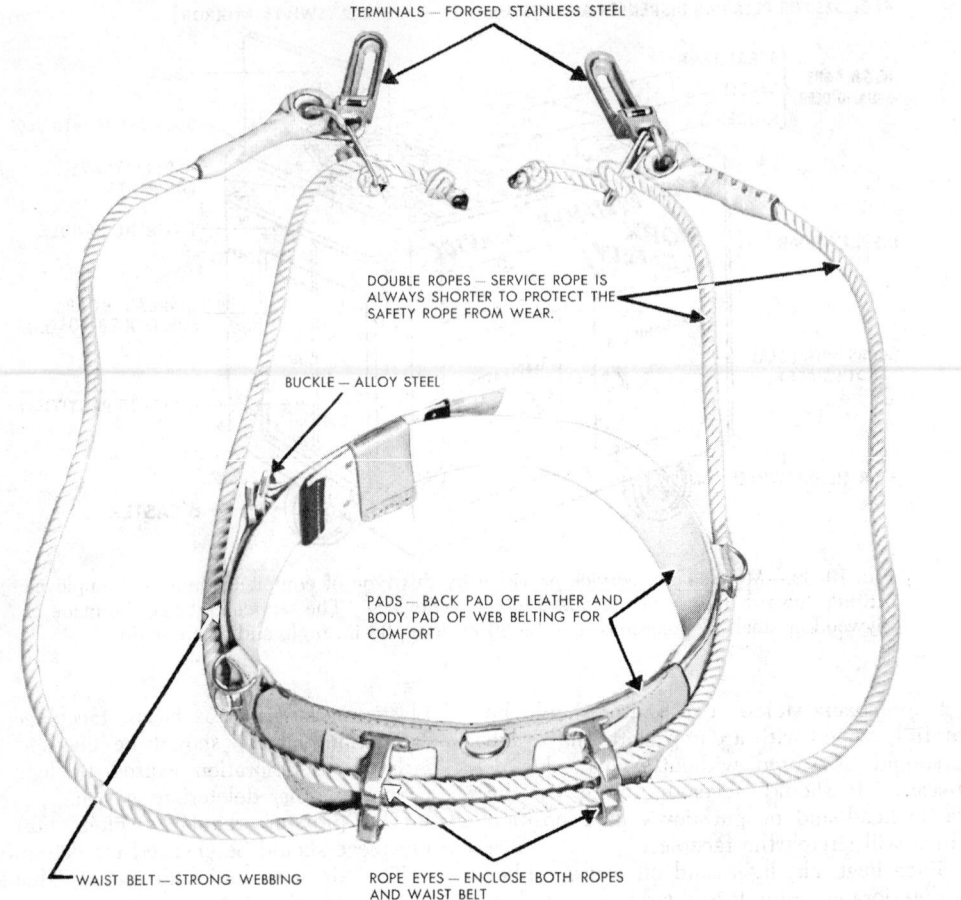

TERMINALS — FORGED STAINLESS STEEL

DOUBLE ROPES — SERVICE ROPE IS ALWAYS SHORTER TO PROTECT THE SAFETY ROPE FROM WEAR.

BUCKLE — ALLOY STEEL

PADS — BACK PAD OF LEATHER AND BODY PAD OF WEB BELTING FOR COMFORT

WAIST BELT — STRONG WEBBING

ROPE EYES — ENCLOSE BOTH ROPES AND WAIST BELT

FIG. 19–40.—Window cleaner's belt with separate nylon ropes—the service rope and the safety rope.

tected from moisture. When a mask with a canister is stored, the seal on the bottom opening of the canister should be unbroken. If the seal has been broken, the canister should be sealed again when stored.

After gas masks have been washed, disinfected, rinsed, and dried, they should be inspected for defects. Even if masks are not in use, valves, gaskets, and rubber parts should be checked every thirty to sixty days.

Gas masks should be stored in the carrying cases provided by the manufacturer or in other boxes or compartments which will permit the masks to retain their original shape. If the case or compartment is not equipped with a form, the facepiece should be stuffed with paper or cloth. Masks should be kept in a cool, dry place.

The lens and certain other parts of some gas masks may be replaced by trained personnel, but it is safest to return the mask to the manufacturer for all repairs. Employees should be cautioned not to attempt makeshift repairs which might damage the equipment and jeopardize the life of the wearer.

Hose masks. Ordinary maintenance consists of lubrication of the blowers, inspection and periodic replacement of the inhalation and exhalation valves, and regular cleaning of the rubber parts.

Self-contained breathing apparatus requires rigid inspection and maintenance because it is usually used under the most adverse circumstances. Periodic inspections should be made and a record kept.

All connectors, valves, and hoses should be carefully inspected. To assure proper function when needed, manufacturers' instructions should be followed to the letter.

A preventive maintenance program for respiratory protective equipment is the only assurance that the devices will operate properly when the need arises.

Safety Belts

This Manual refers specifically to window cleaners' belts and to belts used in general industrial work. Linemen's belts, motor vehicle and aviation seat belts are not included.

Selection

In the selection of a safety belt, two types of use must be considered: normal use and emergency use. Normal use refers to the comparatively mild stresses which will be more or less regularly applied to the belt during the usual course of work. These stresses will seldom, if ever, exceed the total static weight of the user, and they are usually far less than that. Examples are found in such operations as hoisting or lowering a man, or providing him with a steady support while he works.

Emergency use refers to stopping the man safely if he should fall. Such a use may occur with almost any belt at any time. It subjects every part of the belt to an impact loading which may, under certain conditions, amount to many times the weight of the user.

Belts with sparkproof hardware should be used where flammable dusts, vapors, or gases may be present and a spark might cause an explosion.

A window cleaner's belt (Fig. 19–40), for instance, is subject to a moderate static load most of the time it is in use, as the man leans back in the belt and is held in position by it while he works. It will, however, be subjected to a severe loading in case of a fall with only one terminal of the belt attached to the window anchor.

This is the most common kind of fall in window cleaning and also the most hazardous one with this type of belt, because the man will slide to the end of the safety line, trans-

Fig. 19–41.—Competent belt inspection before use is never excessive caution.

mitting impact to the single window anchor which may, particularly in older buildings, be broken off or pulled out.

The amount of impact force developed in arresting a fall depends chiefly on three elements: the weight of the man, the distance of his fall, and the suddenness of his stopping. Of these three elements, the suddenness of his stopping is of far greater importance than the other two factors.

The maximum possible fall in a window cleaner's belt is limited to the length of the safety rope or strap, usually 6 to 8 ft. In other types, the fall may be much longer than the rope length. For example, the maximum fall in a construction worker's belt may be as much as twice the length of his life line in case he falls from above his life line anchorage. In all cases where the life line is attached to the support on which he is standing, the free fall will be usually the full length of the life line plus the distance from his feet to the D-ring of his belt. If at all possible, the wearer of the safety belt should not "tie off" below waist level.

If a considerable free fall is possible in use of the belt, then some form of shock absorber or decelerating device should be used which will bring the fall to a more gradual stop. This will considerably lower the impact load on the equipment and the man. Shock absorbers and shock-absorbing belts are available which limit this impact force sufficiently to keep a normal workman from injury, as indicated by research. (See *Safety Belts, Harnesses and Accessories,* a joint project of the National Safety Council and the American Society of Safety Engineers.)

Belts should be scrutinized for weak points which might cause the belt to fail under a heavy impact (Fig. 19–41). For instance, the waist belt should always be inserted through the D-rings or other attaching devices and never riveted to them in such a way that the D-ring or life line could be separated from the belt through the failure of the rivets.

Hardware should have strength approximately equal to the full strength of the waist band webbing. All safety belt and lanyard hardware (except rivets) should be capable of withstanding a tensile loading of 4000 pounds without cracking, breaking, or taking a permanent deformation.

Buckles should hold securely without slippage or other failure, and this holding power should be achieved by only a single insertion of the strap through the buckle in the normal or natural way. If the buckle is of a type which requires that the free end of the webbing be turned back and inserted again through the buckle in order to achieve its full holding power, then management, through safety instruction and inspection, must make every effort to see that the men always use this method of fastening their belts; otherwise, the belt may slip under very minor strains. Where there is danger of falling into water or of being trapped by fire, a quick-release buckle should be used.

In general, webbing is superior to leather for any safety belt which may be called upon to take impact loads. Webbing has three to four times as much resistance to impact loading as leather of the same size, because webbing has both more strength and more stretch. Furthermore, a leather belt normally uses a tongue buckle which requires holes in the strap for adjusting the length. This permits the tongue to cut or pull through these straps at much less than the full strength of the strap. Webbing belts normally use friction buckles which avoid loss of strength at buckle holes.

An ordinary single tongue buckle will cut through ¼-in. thick leather strap of best commercial grade harness leather at a loading of 300 to 500 pounds; in contrast, web belts can be obtained that possess strengths of over 12,000 pounds.

The width of a leather strap does not materially affect its strength at the buckle. A leather belt ¼-in. thick and 1 in. wide has as much strength as a belt of the same strap 3 or 4 or even 10 in. wide. The only factors affecting the strength of such a belt are the quality of the leather, the thickness of the leather, the size of the metal tongue in the buckle, and the number of tongues in the buckle, a double tongue buckle having twice the strength of a single tongue buckle of similar size.

Leather requires special care and treatment to retain its strength; webbing does not. Leather stands ordinary abrasion well, but is easily cut. It is not easily attacked by chemicals, but is easily damaged by heat, dryness, or inadequate care and oiling.

No one but a leather expert can tell by visual inspection with any degree of accuracy the condition or strength of leather. Webbing, on the other hand, can be judged more accurately by visual inspection.

Cotton webbing is not seriously affected by dryness or moisture nor by heat, short of actual scorching or burning. Untreated cotton webbing should not, however, be long-subjected to moisture (which may promote mildew) or to corrosive chemicals unless the material has been prepared for such conditions.

There are several different materials and weaves used for webbing. The most common is cotton webbing, but synthetic fiber webbing such as nylon and Dacron are supplanting cotton because of their superior strength, mildew and moisture resistance. For some applications, namely chemical and oil work, webbing, coated or impregnated with plastic or neoprene rubber materials, which are impervious to oils and acids, are desirable. The most common weaves are the square or basket weave and the herringbone weave. For the same size and thickness, the herringbone weave usually has approximately twice the strength of the basket weave. One reason is that better yarn is used in the herringbone material, but most of the difference is due to the weave itself.

Under the stress of loading, herringbone weave webbing will elongate while basket weave webbing will not. However, the amount of elongation that occurs under loads common in this use will not significantly affect the fit of a safety belt.

Herringbone weave webbing is much more soft and pliable, and looks weaker than the stiffer, harder basket weave. For this reason, some concerns hesitate to use this material until better acquainted with its real superiority.

In the selection of a belt for general industrial use, the job requirements must be considered and matched against the types of belts available. The American National Standard A10.14, *Construction and Use of Individual Safety Belts, Harnesses, Lanyards, and Droplines,* classifies belts in this manner:

CLASS I—BODY BELTS. For limited movement and positioning to restrict the worker to a safe area to help prevent a fall.

CLASS II—CHEST HARNESS. Used where freedom of movement is most important and where only limited-fall hazards exist. Not recommended for use where vertical free-fall hazards exist.

CLASS III—BODY HARNESS. For use when the worker must move about at dangerous heights. In a fall, the harness distributes impact forces over a wider body area than does a belt; this further reduces the possibility of injury to the wearer.

CLASS IV—SUSPENSION BELT. For application where it is not possible to work from a fixed surface and the worker must be totally supported by a suspension harness. Some examples are shipboard painting, stack maintenance, and tree trimming.

If the only purpose of the belt is to stop a

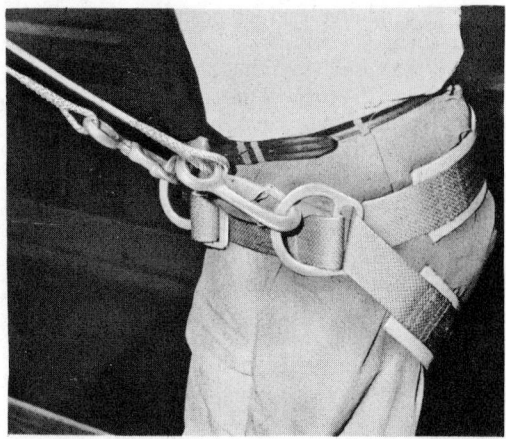

FIG. 19–42.—Multipurpose safety belt with two D-rings can be worn about the waist or slipped down around the buttocks for work on inclines.

workman in case of a fall, then a waist belt with a single D-ring may be satisfactory. This single D-ring may be attached for positioning at any point on the belt, for the belt can be turned to different positions after being buckled on the wearer.

Where possible, safety harness and chest-waist belts are preferable because they do a better job of absorbing impact and keeping a man upright in case of a fall.

If a belt is to furnish support to a man while he is working, it should have adequate means for such support. A belt for a man who will lean back in it as he works on a sloping roof or hillside should have two D-rings, one on each side of his belt, (Fig. 19–42), to which a suitable lanyard can be attached and connected to a suitable anchorage.

If the belt is to be used much for supporting the entire weight of a man vertically while he works, as in raising or lowering him along the wall of a building, a boatswain's suspension harness may be used. In this type of belt, one strap is used as a seat, sometimes with a board to make it more comfortable. Leg straps and a strap around the man's waist prevent him from falling out of the seat.

An industrial belt used to hoist a man out of a tank is usually subjected only to a static load. Such a belt should be of either the shoulder harness, chest-waist, or parachute type with the D-ring so placed as to hold a man in a relatively erect position.

In a window cleaner's belt, separate ropes should be used for the service and safety lines (Fig. 19–40). The service rope, which supports the man while he works, must be shorter than the safety rope so that the former will be taut and the safety rope slack. Thus, all wear will be placed upon the service rope, and the safety rope will retain its full strength for use in stopping a fall.

The safety rope should contain a shock absorber which will reduce the loading in the event of a fall, thus relieving strain on the building anchor, waist belt, and man. There should be a device in the belt to release the service rope in case of a fall and bring into play the shock-absorbing safety rope.

Belt care

Dirt adhering to leather safety belts should be brushed off carefully so as not to scratch the leather. The belt should then be washed with warm water and either saddle soap or castile soap. After a rinse in clean, warm water, it should be dried at room temperature.

Before becoming completely dry, a leather belt should be oiled with neatsfoot, castor, soy bean, or a compound oil, *never a mineral oil.* A leather belt which has been unusued and not oiled for a year or two is much weaker than one which has been regularly used and also adequately oiled over the same period, although the unused belt will look newer and stronger.

Leather belts should not be exposed to excessive heat, such as from a radiator, because they may be permanently damaged by a temperature as low as 150 F. Any heat painful to a man will damage leather.

Cotton or linen webbing belts should be washed in soapy water, rinsed, and dried by moderate heat. They are not damaged by temperatures up to 212 F. If a belt is to be subjected to unusual conditions of use, the manufacturer should be consulted as to its care.

Synthetic fiber belts should not be exposed to excessive heat that might soften or melt the fibers, or to chemicals that might affect the composition of the fiber.

Inspection and testing

Wearers of life belts should inspect them before each use (Fig. 19–41). At least every one to three months they should be examined by a trained inspector. Leather belts especially must be watched for cuts or scratches on either side of the strap. A deep cut of any considerable length, crosswise to the belt, warrants discard. However, cuts running lengthwise do not seriously affect strength.

Fabric belts should not be used if considerable portions of the outer fibers are cut or worn through.

Belt hardware should be examined and worn parts replaced. Each belt rivet should be examined to be certain that it is secure. No rivet should be used in a web belt which may be subjected to impact loading.

Safety belts in service should not be tested. Any service test to prove whether the belt could take the maximum impact loading which might be required of it would very likely so damage the belt as to render it unsafe for use. Therefore, only sample belts and worn or doubtful belts should be tested, and these should always be tested to destruction to determine their safety. They should be kept as samples and used only to help judge the safety of other belts. Belts subjected to the maximum impact in an accidental fall should not be reused because the fittings might have been overstressed and weakened.

In judging the safety of a belt, the following vital, and to some extent conflicting, factors must be considered:

1. Sufficient strength to stop the wearer after a maximum free fall. The safety belt lanyard should be a minimum of ½-in. nylon (or equivalent), with a maximum length to provide for a fall of no greater than 6 ft. The rope shall have a nominal breaking strength of 5400 pounds.

2. A shock absorber to limit the impact loading and prevent injury to the wearer or failure of the anchorage, life line, or belt.

3. Short enough stopping distance to prevent the wearer from striking some dangerous obstruction before he stops.

4. Sufficient margin on all these factors, so far as possible, to cover all unknowns such as weight of the wearer, distance of his fall, his physical fitness, the distance to any damaging obstruction, variations in the

strength or elasticity of materials, and deterioration of materials due to wear or other causes.

The primary caution regarding the use of safety belts, as with other personal protective equipment, is to see that they are worn and used correctly. A safety belt is worthless unless it is being worn at the time that a fall is possible. It should also be securely buckled and worn tight enough to prevent any possibility of the man's slipping out of it.

Lifelines

Lifelines should be secured above the point of operation to an anchorage or structural member capable of supporting a minimum dead weight of 5400 pounds.

For most lifelines, manila rope of ¾-in. diameter or nylon rope of ½-in. diameter is recommended. Nylon is more resistant to wear or abrasion than is manila and is also more resistant to some chemicals. Ropes made of other synthetic fibers such as Dacron, polyethylene, and polypropylene have characteristics that make them very good performers in certain applications. In a lifeline where shock loading strength and high energy absorption is paramount, nylon is superior.

Knots reduce the strength of all ropes. (See Table 26-C.) Tests have shown that manila lifelines of ½-in. diameter, when knotted around a bar or other anchorage, are not strong enough at the knots to be safe for general use unless a shock absorber of low load limit is used.

How much a knot reduces strength depends upon the type of knot or hitch used and the amount of moisture in the rope. Moisture increases the strength of a rope in a knot or around a sharp bend, and dryness greatly diminishes its strength at such points. In some cases, a knot will reduce the tensile strength of a rope to less than half of its breaking strength.

If rope is spliced into snaps and D-rings instead of knotted (Fig. 19–42), the splice will retain approximately 90 percent of the breaking strength of the rope. Breaking strength is always measured in a straight line pull on the rope. (See Chapter 26, "Ropes, Chains, and Slings.")

Manila rope of ½-in. diameter has a breaking strength of approximately 2650 pounds, or (dividing by a safety factor of 5) a safe load strength of 530 pounds. Spun nylon rope of ½-in. diameter is rated to have a breaking strength of 6400 pounds—a safe load strength of 710 pounds.

Manila rope of ¾-in. diameter rates a breaking strength of 5400 pounds—a safe load strength of 1080 pounds. When a low-load-limit shock absorber is used, ¾-in. manila rope or ½-in. nylon rope is considered strong enough for most lifelines. Without a shock absorber, even ¾-in. manila rope may not be strong enough to stop a long fall.

If no shock absorber is available, ¾-in. nylon rope is usually preferable to manila rope for impact use, because it stretches more and will act as a partial shock absorber. The chief disadvantage of nylon for this use is that much of its stretch occurs at a very low loading, thus the falling workman can fall a considerable distance closer to some possibly damaging obstruction before an adequate stopping force is exerted by the rope.

Thereafter, the loading increases very rapidly to a rather high force before the final stop. Also, a considerable amount of rebound is obtained from nylon at the higher loadings.

Wire ropes should not be used as lifelines where a free fall is possible unless some shock-absorbing device is also used, because their rigidity greatly magnifies the impact loading. Wire ropes are hazardous when used around electricity.

Lifelines should be tied to permit as little slack as possible, and thus stop a man with the minimum free fall. Special notice must be taken of the nearness of any beam or other obstruction which the workman might strike in case of a fall. Serious injury or death may result if the total free fall plus the total stretch or elongation of the lifeline and shock absorber will allow the workman to strike some damaging object before he is stopped.

If there is a long clear space in which a man may be stopped, then a low-limit shock absorber may be used with less discomfort to the workman. If this space is closely limited, a shorter stop is imperative, even though greater discomfort results.

Wire ropes should be kept clean and dry and should be frequently lubricated. Before use in acid atmospheres, they should be coated with oil. After such use, they should be thoroughly washed and again coated with oil.

Rope lines should be washed with mild soap and water and dried in circulating air. They should not be exposed to high temperatures.

The outer surface should be examined for cuts and for worn or broken fibers. Most of the fibers in a new manila rope are 3 to 6 ft long. The twist or "lay" of the rope brings each of these fibers to the surface many times along its length.

If there has been wear or abrasion sufficient to reduce the diameter of the rope only slightly and give it a "smooth" appearance, with the high ridges worn down and the "valleys" partly filled with the vertical ends of broken and worn fibers, the rope should be discarded. Each long fiber will have been cut or broken many times, and the rope will have very little remaining strength.

Inner fibers should be examined for breaks, discoloration, or deterioration. Rope should be kept in open coils and never bent sharply.

Protective Footwear

As a guide to the selection of protective footwear, the Office of Technical Services of the Division of Safety, U.S. Department of Labor, has classified safety shoes into five principal types:

Safety-toe shoes

Conductive shoes

Foundry (molders) shoes

Explosives-operations (nonsparking) shoes

Electrical hazard shoes

Specifications for various kinds of protective footwear (Fig. 19–44) have been standardized by ANSI Z41.1, *Men's Safety-Toe Footwear.* Safety-toe footwear has been divided into three classifications—75, 50, and 30—based on its ability to meet the minimum requirements for both compression and impact shown in Table 19-C.

So far only metal toe boxes have withstood these tests. The strength requirements are identical for both women's and men's shoes.

Metal toe boxes may be used in shoes of other types such as conductive but spark-

TABLE 19–C
MINIMUM REQUIREMENTS
OF AMERICAN STANDARD Z41

Classi-fication	Compression (pounds)	Impact (pounds)	Clearance (inches)
75	2500	75	16/32
50	1750	50	16/32
30	1000	30	16/32

resistant shoes, molders shoes, or nonconductive shoes. For work under wet conditions, rubber boots or rubber shoes may be obtained with a steel toe box to protect against impact.

OSHA requires that the safety-toe shoe be used for work requiring the handling of heavy materials. Safety shoes also afford good protection against rolling objects, such as barrels, heavy pipe, rolls, or truck wheels, and against the hazard of accidentally kicking sharp sheet metal.

The toe box does not add much either to the weight or to the cost of the shoe, and a well made and properly fitted safety shoe is as comfortable as any other. Comfort is an important factor in the wearing of any shoe, but particularly so when safety footwear is required. Great care should be exercised in selecting as well as fitting the correct type.

Many companies set up shoe departments, with the aid of shoe manufacturers, in their plants and provide trained employees to see that individuals are properly fitted with the correct types for the hazards involved (Fig. 19–1). Some retail organizations provide the same type of service.

To protect feet from the heaviest impacts, metatarsal (or over-foot) guards should be worn in addition to safety shoes (Fig. 19–43 and –44). Heavy-gage, flanged, and corrugated sheet metal footguards protect the feet from toes to ankles.

With the flanges resting upon a firm floor surface, they should resist an impact of at least 300 foot-pounds without sufficient deformation to damage the shoes underneath or injure the feet.

Conductive shoes

Safety shoes, boots, and rubber overshoes may also be obtained with conductive soles

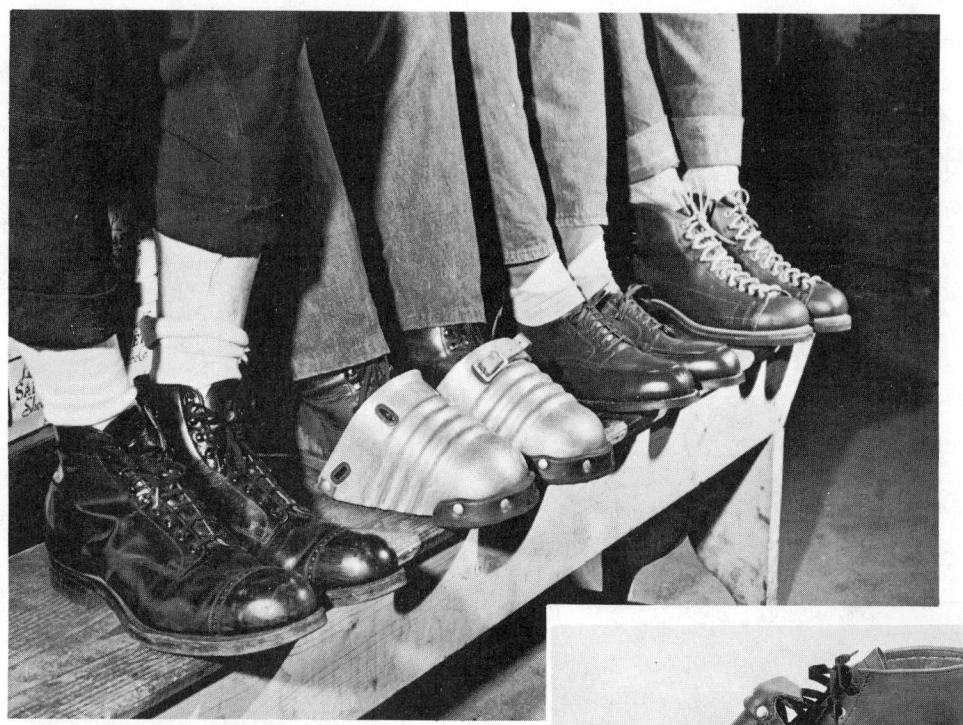

Fig. 19–43a.—*Above:* Typical safety shoes. Second from left illustrates metal instep guards which fit over shoes. Third from left are oxford-type shoes.

Fig. 19–43b.—*Right:* Shoe has built-in instep (metatarsal) protection. This style has proven especially valuable in the heavy metals industries.

to drain off static charges, and with nonferrous construction to reduce the possibility of friction sparks in locations with a fire or explosion hazard. Initial and subsequent periodic tests should be made on conductive footwear in order to make sure that the maximum allowable resistance of 450,000 ohms is not exceeded.

Foundry shoes

Safety shoes of the "congress" or gaiter type are used in some plants where employees are exposed to splashes of molten metal. Having no fasteners, such shoes are easily and rapidly removable in an emergency.

Some National Safety Council members engaged in foundry and steel mill operations have reported that serious burns have occurred to men unable to remove shoes of ordinary work type in an emergency. As in all such occupations, the tops of the shoes should be covered by the trouser leg, spats, or leggings, to keep out molten metal.

Explosives-operations (nonsparking) shoes

Explosives-operations shoes are used (*a*) in hazardous locations where the floors are nonconductive and grounded such as in the manufacture of certain explosive compounds, or (*b*) when cleaning tanks that have contained

FIG. 19–44.—Metal instep guards protect this demolition worker.

gasoline or other volatile hydrocarbons. These shoes do not have conductive soles. They have nonferrous eyelets and nails, and metal box toes are coated with a nonferrous metal.

Electrical hazard shoes

Electrical hazard shoes are intended to minimize hazards resulting from contacts with electric current where the path of the current would be from the point of contact to the ground. No metal is used in construction, except for the box toe which is insulated from the rest of the shoe. If damp or badly worn, they cannot be depended on for protection.

Special shoes

In some industries, such as construction, where there may exist an increased hazard from protruding nails, and where contact

with energized electric equipment is remote, shoes or boots are equipped with reinforced soles or inner soles of flexible metal (Fig. 19–45).

Special shoes made with a stitched and cemented construction are available for electricians. These shoes serve as good insulators, so when they are repaired, avoid the use of nails which will destroy their insulating value.

For wet conditions such as are found in dairies and breweries, leather shoes with wood soles, or wood-soled sandals worn over shoes, are effective.

Wood soles provide good foot protection on jobs which require walking upon hot surfaces which are not hot enough to char the wood. Wood soles have been so generally used by men handling hot asphalt that they are some-

FIG. 19–45.—This type of foot protection is worn extensively in the construction industry. A flexible steel insole prevents the nail from puncturing the foot.

times called paver's sandals or paver's shoes. They are, however, equally satisfactory for other work which requires soles that do not conduct heat.

Plastic shoe covers can be worn to protect a product from contamination (see Fig. 19–46).

Where shower baths are used, many organizations provide paper slippers or wooden sandals for each individual, to reduce the possibility of foot infection. Paper slippers are discarded after a single use, while the sandals are disinfected at frequent intervals, particularly before being assigned to others.

Cleaning of rubber boots

Some companies find it necessary for men on different shifts or jobs to wear the same pair of rubber boots. Where such conditions are encountered, great care should be exercised to disinfect boots after each shift or job. First, the boots are washed inside and outside with a hose under water pressure. Then they are dipped into a tub containing a solution of 1 part sodium hypochlorite and 19 parts water. The hose is again used for rinsing, after which the boots are ready for drying. Other disinfecting agents can be used, but this one has been satisfactory and is easily obtainable.

One company has a drying rack consisting of a tank with low-pressure steam coils having upright steel-pipe boot holders which permit circulation of hot air inside the boots. After the boots are washed throughly and dipped in the disinfecting solution, they are completely dried in about 12 minutes.

If much work of this type is necessary, the rack with water jets could be rearranged so that a number of boots could be cleaned and rinsed at one time.

Special Work Clothing

In our modern industrial environment, exposure to fire, extreme heat, molten metal, corrosive chemicals, cold temperature, body impact, cuts from materials which are handled, and other highly specialized hazards are often part of what is known as "job exposure."

Special protective clothing is available for all these hazards, however, to minimize their effect. Sample swatches of materials used can

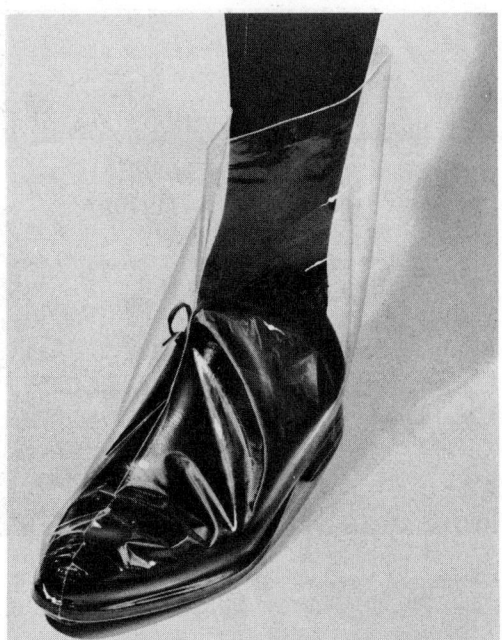

Fig. 19–46.—This polyethylene shoe cover is an example of footwear designed to protect the product—in the cleanroom. The cover fits either foot and is disposable.

usually be obtained from the manufacturers for testing.

Protection against heat and hot metal

Leather clothing is one of the more common forms of body protection against heat and splashes of hot metal. It also provides protection against limited impact forces and infrared and ultraviolet radiation.

The garments should be of good quality leather, solidly constructed, and provided with fastenings to prevent gaping during body movement. Fastenings should be so designed that the wearer can rapidly and easily remove the garment. There should be no turned-up cuffs or other projections to catch and hold hot metal. Pockets should have flaps which can be fastened shut.

For ordinary protection against hot metal, radiant heat, or flame hazards of somewhat more intensity than those represented by welding operations, asbestos and wool, as well as leather clothing is used. Specially

Fig. 19–47.—Open hearth worker opens tap hole. He is protected by long coat and hood made of asbestos-polyester cloth; gloves are sewn to cuffs of the coat.

Fig. 19–48.—Aluminum-coated, heat-protective suit is used in fighting fires without entering the burning area. Transparent face shield is metal coated to offer increased heat protection. Head fitting includes chin strap.

treated asbestos clothing has been developed which is impervious to metal splash up to 3000 F.

Asbestos and wool garment requirements are in general the same as those for leather, except that metal fastenings should be covered with flaps to keep them from becoming dangerously hot.

The most common types of asbestos clothing are the leggings and aprons usually worn by foundrymen working with molten metal (Fig. 19–47). Such leggings should completely encircle the leg from knee to ankle, with a flare at the bottom to cover the instep. The design of the leggings should permit rapid removal in emergencies. There is no evidence that wearing asbestos equipment presents a health hazard.

The front part of the legging may be reinforced to provide impact protection when it is required. The most common material for this reinforcement is fiber board.

Aluminized clothing. Where men must work in extremely high temperatures up to 2000 F, such as furnace and oven repair, coking, slagging, fire fighting and fire rescue work, aluminized fabrics are essential. The aluminized coating reflects much of the radiant heat and the underlying material insulates in the proximity of high temperature, such as slagging, coking, furnace repair work with hot ingots, and fire fighting where the flame area is not entered. These suits are seldom of 1-piece construction. They depend primarily on the reflective ability of an aluminized coating on a base cloth of asbestos, glass fiber, or synthetic fiber. *Never use fire proximity clothing where fire entry suits are required.*

Flame-retardant work clothes

Ordinary clothing can be protected against flame or small sparks by flameproofing. One

FIG. 19–49.—Demonstration of fire-entry suit. Spun glass material of suit is chemical resistant and will not burn, even in pure oxygen atmosphere.

against the remainder. Some of these suits consist of separate units of trousers, coats, gloves, boots, and hoods. Others are one-piece from head to foot. Some suits used in industrial operations are airfed to reduce heat and increase comfort.

Aluminized heat-resistant clothing generally falls into two classes:

- Emergency suits (Figs. 19–49 and –50) may be used where the temperatures may exceed 1000 F, as in a kiln or furnace, or where men must move through burning areas for fire fighting or rescue operations. These suits are constructed of aluminized asbestos or glass fiber, with layers of quilted glass fibers and a wool lining on the inside.

- Fire proximity suits (Fig. 19–48) are used

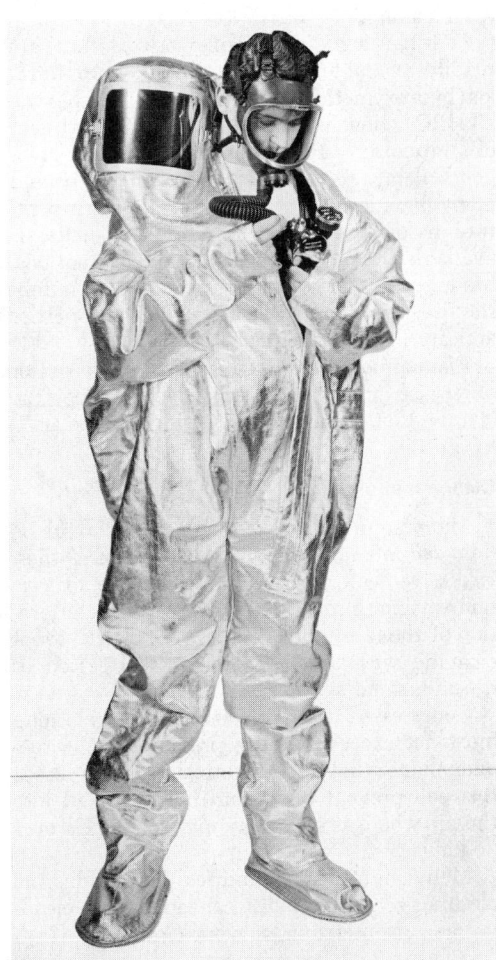

FIG. 15–50.—Fire-entry suit for use in entering a burning area. Note the self-contained breathing apparatus.

of the commercial preparations can be applied in ordinary laundry machinery after the garment is washed. Flameproofing will make the material highly flame-resistant and will not add much to the weight or stiffness of the cloth.

These compounds are usually water soluble, and must be replaced after each laundering. An effective flameproofing solution can be made by dissolving 8 oz of borax and 4 oz of boric acid in 1 gal of hot water. If stiffening of the fabric is not objectionable, more durable flameproofing methods can be applied, some of which are resistant to both laundering and dry cleaning.

Durable flame-retardant work clothes are readily available. Cotton treated with tetrakis(hydroxymethyl)-phosphonium chloride (THPC), developed by the U.S. Department of Agriculture Research Laboratories, gives good flame retardancy and will withstand many launderings. Nomex, a high-temperature nylon that chars rather than melts, is available for the most severe situations. Modacrylic fabrics that resemble cotton fabrics have permanent fire-retardant properties and are light in weight.

Flameproofed clothing should be marked or otherwise made distinctive to reduce the chance that untreated garments may be used by mistake.

Cleaning of clothing

Manufacturer's recommendations should be followed in laundering and cleaning of clothes. Excessive water temperatures or use of certain washing preparations can cause deterioration of the fabric or affect its properties. Spot cleaning with organic solvents will soften or dissolve some synthetics.

Compressed air used for dusting of clothing must not exceed 30 psig pressure. It is recommended that a vacuum system be used; this will prevent dust from being spread into the air where it could get into someone's eyes or lungs.

Many industrial laundries and industrial clothing rental agencies can advise on cleaning and maintenance of work clothing.

Protection against impact and cuts

It is necessary to protect the body from cuts, bruises, and abrasions on most jobs where heavy, sharp, or rough material is handled. Special protectors have been devised for almost all parts of the body and are available from suppliers of safety equipment.

Pads of cushioned or padded duck will protect the shoulders and back from bruises when men carry heavy loads or objects with rough edges.

Aprons of padded leather, fabric, plastic, hard fiber, or metal will protect the abdomen against blows. Similar devices of metal, hard fiber, or leather with metal reinforcements provide protection against sharp blows with edged tools. For jobs requiring ease of movement, aprons may be split and equipped with fasteners to draw them snugly around the legs.

Guards of hard fiber or metal are also widely used to protect the shins against impact.

Knee pads should be worn by mold loftsmen and others whose task requires continual kneeling.

Gloves. The material to be used for gloves depends largely upon what is being handled (Fig. 19–51). For most light work, a canvas glove is both satisfactory and cheap. For rough or abrasive material, leather or leather reinforced with metal stitching will be required. Leather reinforced by metal stitching or metal mesh also provides good protection from edged tools, as in butchering and similar occupations.

There are many plastic and plastic-coated gloves available. They are designed to give protection from a variety of hazards. Some surpass leather in wearing ability. Others have granules or rough materials incorporated in the plastic for better gripping ability. Some are disposable (Fig. 19–52).

Where the use of a complete glove is not necessary, finger stalls may be used. These are available in combinations of one or more fingers. Some of the more common materials used are asbestos, rubber, duck, leather, plastics, and metal mesh. The construction of the stall depends on the degree or type of hazard to be confronted.

Gloves should not be used while working on moving machinery such as drills, saws, grinders, or other rotating and moving equipment that might catch the glove and pull it and the

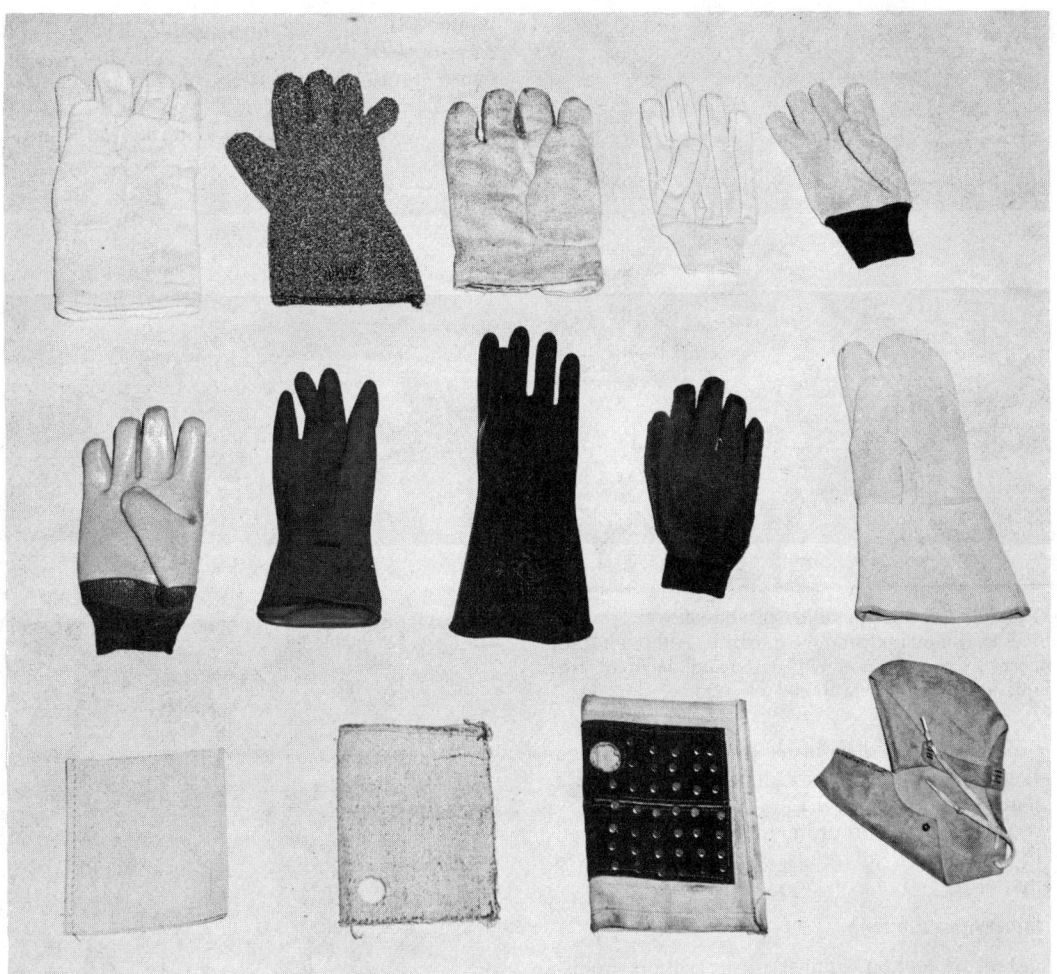

Fig. 19–51.—Pictured are various types of specialized hand protectors. **In the top row** (l. to r.) are asbestos, loop pile, and aluminized gloves, all heat resistant; 19-ounce cotton glove and loop pile glove are for general shop wear. **In the middle row** are plastic-dipped gloves for oily work, fine rubber gloves for protection against acids, etc., where finger dexterity is required, neoprene gauntlet for acid, caustic handling, neoprene and cork-dipped glove for slippery and oily jobs, and a chrome leather welder's glove for arc welding. **In the bottom row** are a neoprene sandwich palm pad for protection against sharp edges, asbestos palm pad for hot sharp edges, brass-studded palm pad for handling heavy material, and open-back leather palm pad for annealing operations.

worker's hand into hazardous areas.

Hand leathers and arm protectors. Where the problem is protection from heat or from extremely abrasive or splintery material, such as rough lumber, hand leathers or hand pads are likely to be more satisfactory than gloves, since they can be made heavier and less flexible without discomfort.

Since hand leathers or pads are primarily for heavy materials handling, they should not be used around moving machinery. They should at all times be sufficiently loose to release the hands and fingers if caught on a rough edge or nail.

For protection against heat, hand and arm

521

19—Personal Protective Equipment

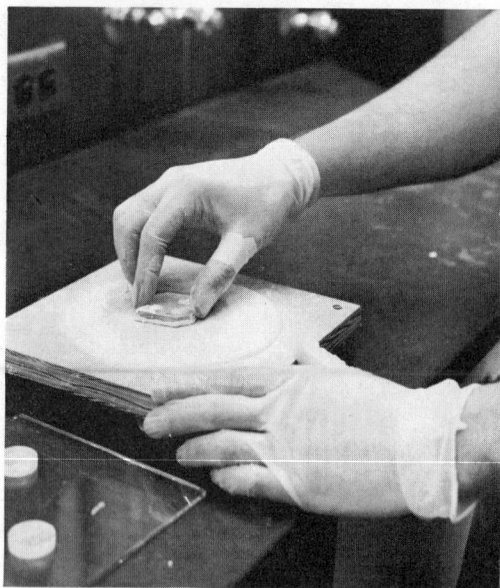

work clothing in chemical plants, where daily contact with acids and caustic solutions would cause rapid deterioration of regular cotton clothes, are not impervious and should not be used where impervious materials are indicated.

FIG. 19-52.—Disposable plastic gloves protect hands, maintain product purity. Although unit cost is low, replacement costs and accident frequency should be watched closely.

protectors should be of asbestos cloth or wool. Leather can be used, too, but will not stand a temperature over 150 F.

Wristlets or arm protectors may be obtained in any of the materials of which gloves are made.

Impervious clothing

For protection against dusts, vapors, moisture, and corrosive liquids, there are many types of impervious materials available. These are fabricated into clothing of all descriptions, depending on the hazards involved. They range from aprons and bibs of sheet plastic, to garments which completely enclose the body from head to foot and contain their own air supply.

Materials used include natural rubber, olefin, synthetic rubber, neoprene, vinyl, polypropylene, and polyethylene films and fabrics coated with them. Natural rubber is not suited for use with oils, greases, and many organic solvents and chemicals. Make sure that the clothing selected will protect against the hazards involved.

Some synthetic fabrics used for regular

FIG. 19-53.—Protective equipment worn by these linemen includes electrical-style safety hat, rubber sleeves, rubber gloves with leather glove protectors, and safety belt. In order for personal protective equipment to be effective, it must be put on before a man reaches a place where he could contact energized lines or equipment, even by accident. Rules should require equipment to be worn when the work area is reached.

Gloves coated with rubber, synthetic elastomers, polyvinyl chloride, or other plastics offer protection against all types of petroleum products, caustic soda, tannic acid, muriatic and hydrochloric acid. They are also recommended in the handling of sulfuric acid. Less deterioration takes place than with natural rubber. These gloves are available in varying degrees of strength to meet individual conditions.

Gloves should be long enough to come well above the wrists, leaving no gaps between the glove and the coat or shirt sleeve. Long, flaring gauntlets should be avoided unless they are equipped with locking devices to assure a snug fit about the wrist. Such gauntlets are especially desirable when acids and other chemicals are being poured.

In this operation, the chemicals may splash, and unless precautions are taken, harmful results may occur. When caustic substances and harmful solvents are being poured from large to small containers, sleeves should be worn outside gauntlets.

In many operations, rubber gloves with extra long cuffs have been used to advantage. The cuffs of these gloves are made with a heavy ridge near the top edge, which, when turned back, forms a trough to catch liquids running down the wrist or forearm.

Gloves or mittens having metal parts or reinforcements should never be used around electrical apparatus.

Work by such persons as linemen (Fig. 19–53) and electricians on energized or high-voltage electric equipment requires specially made and tested rubber gloves. The requirements for electrician's rubber gloves are detailed in ANSI J6.6, *Rubber Insulating Gloves*.

Over-gloves of leather must be worn to protect the rubber gloves against wire punctures and cuts and to protect the rubber in the event of flash. Frequent testing and inspection of linemen's rubber gloves are essential, and those gloves failing to meet original specifications should be discarded.

Where acid may splash, rubber boots or rubber shoes also should be worn. The tops of the boots should be high enough to come beneath the edge of the apron. If shoes are worn, the tops should come inside the legs of impervious trousers. These precautions keep the liquid from draining off apron or trousers

Fig. 19–54.—Laundering of gloves removes contaminants, prolongs gloves life, sanitizes, and permits reissue.

into the footwear.

Procedural setups. When personal protective equipment is used in a corrosive atmosphere, a rigid procedure should be set up for taking care of it after use to prevent contact with contaminated parts. Before the equipment is removed, whether or not it has come in contact with the corrosive chemical, it should be thoroughly washed with a hose stream. Gloves can be laundered to remove contaminants, prolong glove life, sanitize, and permit reissue (Fig. 19–54).

Boots, coats, aprons, and hats should then be removed, followed by removal of the gloves. This is the logical order of removal if the coat has been properly put on with the sleeves outside the cuffs of the gloves. Hands should be washed thoroughly before face shield and goggles are removed. The hands and face should then be thoroughly washed again, but complete shower and change of

523

Fig. 19–55.—Various types of thermal knit cotton materials used for regular style cold weather underwear. Note the air pockets which give materials their insulating properties.

clothing are much more desirable.

For protection against exposure to oil and the various other compounds which rapidly attack ordinary rubber, all the equipment discussed can be obtained in plastic and synthetic rubbers.

Women's clothing and protection

Women require practically the same safety features in protective clothing and equipment as men do, but styles of clothing may differ considerably in appearance.

It is desirable to have women wear the same type of dress, uniform, or smock. Skirts and loose, frilly clothing are easily caught in moving machinery. Where this danger is present, slacks and short-sleeved shirts are commonly worn. Tie strings for aprons should be of a type that can be easily broken.

The possibility of serious scalp injury is present when women work near moving machinery. Hair covering is desirable for cleanliness, particularly in food products industries.

Caps to be used near sparks or flame should be flame-resistant. A cap should have a stiff brim long enough and rigid enough to provide some warning before the head itself comes in contact with moving objects. Any fad of wearing cap visors turned up should be combatted. When string-tied caps are worn, the bow in the back should be tucked under the cap. Hair-nets and turbans do not give sufficient protection against moving machinery. Additional details on women's caps were given under Hair Protection.

Shoes with high or run-over heels, sneakers, and toeless shoes or sandals are not suitable for factory work, particularly where heavy materials or hot liquids are handled. Many companies require women employees to wear

524

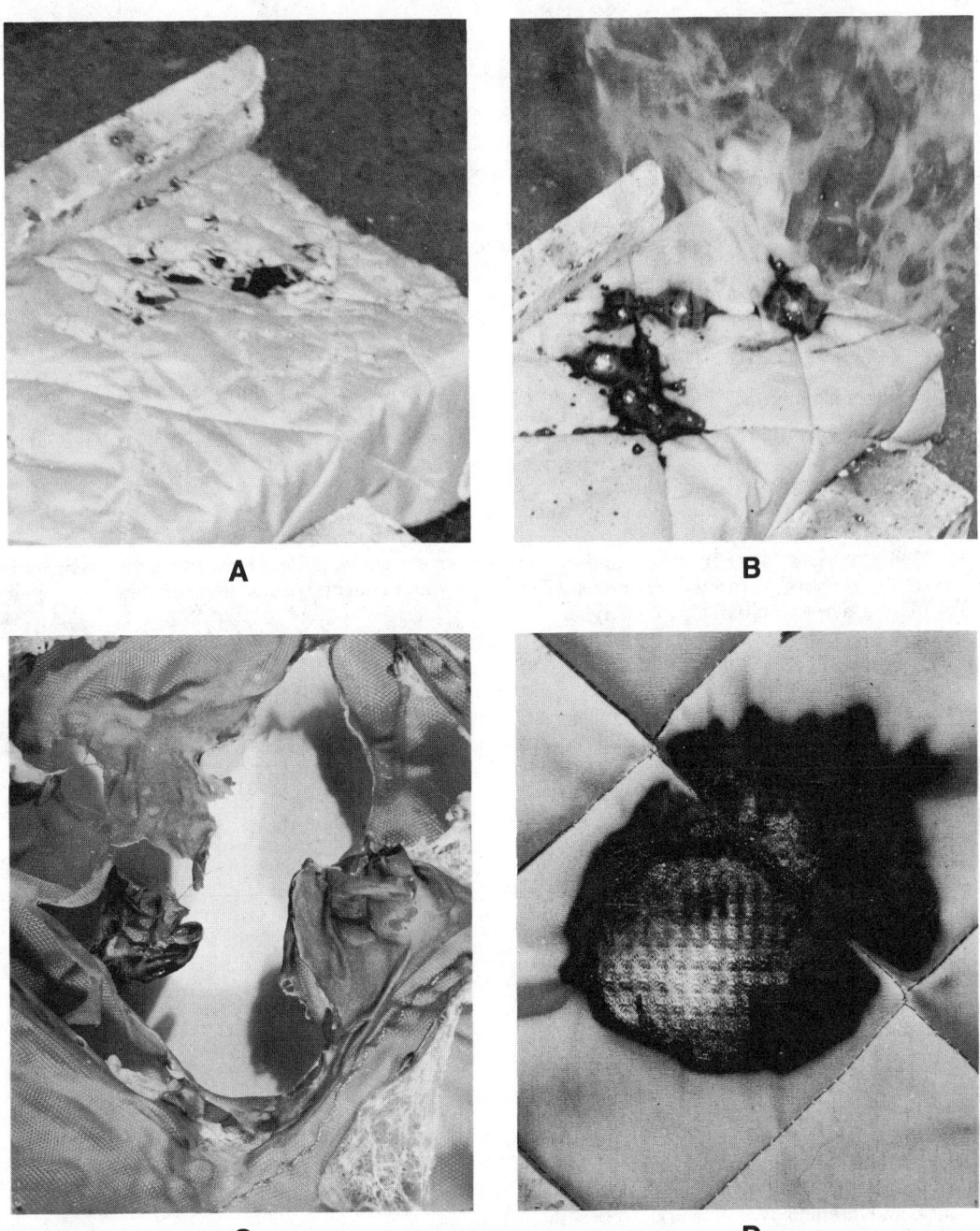

FIG. 19–56.—Demonstration of resistance to molten globules of welding metal by **A**—Regular thermal insulating underwear material; **B**—Fire-retardant material (note how fire-retardant material prevents burn-through, even though it smokes); **C**—Melting nylon and polyester materials form a hot, pitchy mass; **D**—Fire-resistive material merely scorches, as shown in this closeup photograph.

525

a medium or low heeled shoe. Safety shoes are also available in women's styles and their use should be encouraged where sharp or heavy objects are handled or where there is other danger of injury to the worker's toes.

Rings, bracelets, and earrings commonly cause accidents. Many companies require that no jewelry or ornaments be worn on or near jobs involving moving machinery.

Rules on the use of protective equipment, such as eye protective devices, face shields, and respiratory protective equipment, such as discussed previously in this chapter, should apply to women as well as to men.

Cold weather clothing

In recent years thermal insulating underwear has become popular among outdoor workers because of its lightweight protection from the cold. Two types are generally available. One is a thermal knit cotton (Fig. 19–55) patterned after regular underwear. The other consists of quilted materials (Fig. 19–56). Dacron quilted between nylon is a common type of construction.

While this material does not catch fire any easier than cotton, once it starts burning, the nylon and Dacron melt, forming a hot plastic mass, not unlike hot pitch, which will adhere to skin and cause serious burns. Quilted insulating underwear is now available that has been made fire retardant to combat this danger.

A nylon material (called "Nomex"), that chars at a relatively high temperature and does not melt, is available. Glass fiber material is also available for special uses (Fig. 19–49).

Special clothing

Safety experts have been very ingenious in developing many highly specialized types of clothing for protection against special hazards. A partial list includes such items as:

High visibility and night hazard clothing for construction, utility, maintenance workers, police and firemen whose work expose them to traffic hazards.

Disposable clothing made of plastic (Fig. 19–51) or reinforced paper is available for exposure to low level nuclear radiation, or for use in the drug and electronic industries where contamination may be a problem.

Leaded clothing of lead glass fiber cloth, leaded rubber or leaded plastic for laboratory workers and other personnel exposed to X rays or gamma radiation.

Electromagnetic radiation suit, which provides protection from the harmful biological effects of electromagnetic radiation found in high level radar fields and similar hazardous areas.

Conductive clothing, made of a conductive cloth, is available for use by linemen doing bare-hand work on extra-high voltage conductors. Such clothing keeps the worker at the proper potential.

For special applications, manufacturers have a vast number of materials they can draw upon to meet specific hazards.

Acknowledgment

The following companies and sources cooperated in furnishing illustrations of personal protective equipment for use in this chapter:

Acme Protective Equipment Co.
American Industrial Hygiene Assn.
American Optical Co.
Armco Steel Corp.
Bausch & Lomb, Inc.
E. D. Bullard Co.
Caterpillar Tractor Co.
Chicago Eyeshield Co.
David Clark Co.
Detroit, Mich., Division of Instruction
E. I. du Pont de Nemours and Co., Inc.
Eastern Safety Equipment Co., Inc.
Ellwood Safety Appliance Co.
Encon Manufacturing Co.
Factory Stores
Ford Motor Co.
Humble Oil & Refining Co.
International Shoe Co.
Jackson Products Co.
Kennedy Ingalls, Inc.
Midwest Glove Co.
Mine Safety Appliance Co.
NASA, Mississippi Test Facility
Owens-Corning Fiberglas
Raybestos-Manhattan, Inc.
Redwing Shoe Co.

Rohr Aircraft Corp.
Rose Manufacturing Co.
W. H. Salisbury & Co.
Scott Aviation Corp.
Sigma Engineering Co.
Standard Safety Equipment Co.
Universal Safety Equipment Co.

Welsh Manufacturing Co.
Western Electric Co., Inc.
Westinghouse Electric Corp.
Wheeler Protective Apparel, Inc.
Willson Products Div., ESB Inc.
U.S. Industrial Chemicals Co.
U.S. Navy Electronics Laboratory

References

American Conference of Governmental Industrial Hygienists, Cincinnati, Ohio 45202. *A Guide for Uniform Industrial Hygiene Codes or Regulations for Laser Installations*, 1968.

American Industrial Hygiene Association and American Conference of Governmental Industrial Hygienists. *Respiratory Protective Devices Manual.* 1963. Available from Committee on Respirators, P.O. Box 453, Lansing, Mich. 48902.

American National Standards Institute, 1430 Broadway, New York, N.Y. 10018.
Identification of Air-Purifying Respirator Canisters and Cartridges, K13.1.
Men's Safety-Toe Footwear, Z41.1.
Method for Measurement of Real-Ear Attenuation of Ear Protectors at Threshold, Z24.22.
Practice for Occupational and Educational Eye and Face Protection, Z87.1.
Practices for Respiratory Protection, Z88.2.
Requirements for Construction and Care of Industrial Safety Belts, Harnesses, Lanyards, and Droplines, A10.14.
Safety Guide for Respiratory Protection Against Radon Daughters, Z88.1.
Safety in Welding and Cutting, Z49.1.
Safety Requirements for Industrial Head Protection, Z89.1.
Safety Requirements for Industrial Protective Helmets for Electrical Workers, Z89.2.
Safety Requirements for Window Cleaning, A39.1.
Standard for the Safe Use of Lasers, Z136.1.

American Society for Testing and Materials, 1916 Race St., Philadelphia, Pa. 19103.
Standard Specification for Rubber Insulating Gloves, D 120–70, ANSI J6.6.
Standard Specification for Rubber Insulating Sleeves, D 1051–70, ANSI J6.5.

Appel, J. E. "Thermal Insulating Underwear." *Journal of the American Society of Safety Engineers* (January 1962).

Bureau of Mines, U.S. Department of Interior, Washington, D.C. *Respirators Approved by the Bureau of Mines as of May 24, 1972*, Information Circular 8559 (a revision of Information Circular 8436). (For mines only.)

Mack Publishing Co., 208 Northampton St., Easton, Pa. 18042. *U.S. Pharmacopoeia.*

National Safety Council, 425 N. Michigan Ave., Chicago, Ill. 60611.
Industrial Data Sheets
Flame-Retardant Treatment of Fabrics, 517.
Flexible Insulated Protective Equipment for Electrical Workers, 598.
Industrial Skin Diseases, 510.
Respiratory Protective Equipment, 444.
Safety With the Laser, NSNews Reprint No. 17.

U.S. Department of the Interior, Washington, D.C. 20240. 30 C.F.R. Subchapter B, Respiratory Protective Devices, Tests for Permissibility, Fees; Parts 11, 12, 13, 14, 14a.

Note. The *Code of Federal Regulations* is available through the U.S. Government Printing Office, Washington, D.C. 20402.

U.S. Department of Health Education and Welfare, Public Health Service Center for Disease Control, National Institute for Occupational Safety and Health, Morgantown, W.Va. 26505. *Approved or Certified Personal Protective Devices and Industrial Hazard Measuring Instruments* (with supplements), 1974.

Industrial Sanitation and Personnel Facilities

Chapter
20

Drinking Water   529
In-plant contamination . . . Plumbing . . . Private water supplies
. . . Water quality . . . Wells . . . Disinfecting the water system
. . . Water purification . . . Water storage

Sewage and Garbage Disposal   534
Building drains and sewers . . . Septic tanks . . . Garbage disposal
. . . Insect and rodent control . . . Refuse collection

Personal Service Facilities   536
Drinking fountains . . . Salt tablets . . . Washrooms and locker
rooms . . . Showers . . . Toilets . . . Janitorial service

Food Service   542
Nutrition . . . Types of service . . . Eating areas . . . Kitchens . . .
Controlling contamination of foods

References   546

A work environment should be kept clean and sanitary, and be well equipped for employee comfort and convenience. To achieve this, watch these five industrial health areas:

1. Potable water supply for drinking, washing, and food preparation;

2. Adequate disposal of sewage and garbage;

3. Adequate personal service facilities;

4. Sanitary food service; and

5. Satisfactory heating and ventilation.

These areas must be given the necessary attention if employees are to work efficiently, with the assurance that their health and welfare are well protected.

As with other industrial functions, maintaining a clean, sanitary work environment should rate a separately managed and comprehensive department if management and employees are to benefit fully. Sanitation, for example, must be properly managed and effectively integrated with production and maintenance if it is to be safe, efficient, orderly, and economical.

The general rules for sanitation include:

- Good housekeeping—as clean as nature of the work allows

- Personal cleanliness

- A good inspection system

Where wet processes are used, drainage must be maintained.

The director or supervisor responsible for maintaining the work environment must be at a level high enough in the organization to permit him to sustain his function against the pressures exerted by other departments, and to provide surveillance of the entire company or plant environment, in order to keep it at an appropriate and balanced level of cleanliness and order.

Some firms are appointing a manager of environment and safety, who has additional product safety responsibility.

Drinking Water

Most plants receive water for drinking, washing, and food preparation from a municipal supply. As delivered to the plant meter, this water most probably meets the requirements of the U.S. Public Health Service's *Drinking Water Standards* or an equivalent local ordinance. Since its purity is controlled by local health officers, the municipal water supply can usually be considered safe.

In-plant contamination

The fact that water is potable when delivered to the plant meter does not necessarily mean that it will be so when it is used, for there are many opportunities within a plant for water to become contaminated. If the requirements of American Standard and Sanitary Corporation's "National Plumbing Code," or applicable local ordinances are followed, the chances for in-plant contamination of the potable water supply will be minimized.

One of the most common causes of contamination of the water supply is direct or indirect cross-connection with a source of nonpotable water. Before the "National Plumbing Code," there were cases of typhoid fever contracted from drinking fountains with supply pipe lines connected to septic pipes. Other common causes are improper maintenance of drinking and cooking facilities and improper installation of plumbing facilities, permitting back-siphonage of used water.

The integrity of the drinking water system must be maintained throughout the plant. If there are piping systems containing water used for other purposes, such as sprinklers and fire hydrants or manufacturing processes, each should be clearly identified, particularly at outlets. There should be no direct connection between drinking water and other water systems. Long dead-end runs of pipe which cannot be flushed or drained and which might serve as a reservoir for contaminated water should also be eliminated. The location of, and piping for, drinking water should be easily identified.

Non-potable water may be used for cleaning work premises (other than food preparation and personal service rooms), provided it does not contain concentrations of chemicals or fecal coliform bacteria.

Where there is a possibility of misuse or cross connection of pipe lines, all non-potable water lines should be marked unsafe for drinking, washing the person or utensils, or food areas, personal service rooms, or clothes washing.

Plumbing

Fixtures and faucets should be installed to prevent back-siphonage of contaminated water if the pressure drops in the supply line. Faucets and similar outlets should be at least 1 in. above the floodrim of the receptacle below. To prevent backflow into the drinking water supply, surge tanks and air gaps may also be required in the drainage lines from process equipment.

Open joints in underground supply lines into which ground water or water from leaky sewers can seep are another common source of contamination. This condition may arise where pipes are subject to vibration or corrosion and the joints between pipes open mechanically or the pipe sections crack. Codes usually prohibit sewer and drinking water lines to be installed in the same trench, unless the sewer line is placed at a much greater depth and a certain horizontal offset is provided.

Frequently, contamination of the water supply, results when a system is opened for repair or the addition of new pipe and is not disinfected and properly flushed with clean water, before being put back into service.

If the supply for sprinklers and fire hydrants is the same as that for drinking water, hydrant drains or "weeps" connected directly to sewer lines may be a source of contamination. An open standpipe or reservoir may also permit contamination.

Plastic pipe can be considered, but be sure to check local code requirements. Unplasticized PVC (polyvinyl chloride) is good for cold water lines. Hot water up to 165 F and 100 psi can usually be handled in pipe made of chlorinated polyester or unplasticized PVC.

Private water supplies

Industrial establishments in outlying districts commonly supply and treat their own water from private sources. Such installations should be made and operated under the supervision of a thoroughly trained and experienced sanitary engineer. The information in the next few paragraphs is not meant to substitute for such supervision. The USPHS *Manual of Individual Water Supply Systems*

FIG. 20–1.—Periodic laboratory tests are needed to determine the quality of both source water and treated water.

Courtesy Olin Industries, Inc.

should also be consulted for methods of selecting, developing, and treating private supplies.

All underground and surface waters to be used for drinking purposes should be considered contaminated until proved otherwise. The water supplied from private sources for the personal use of plant personnel should meet the requirements of the USPHS *Drinking Water Standards,* and before it can meet these requirements will have to be treated according to the degree of its contamination. As a rule, ground water collected from deep-drilled wells will be free of biological contamination but may be contaminated by various mineral or chemical substances. In contrast, surface water usually will be reasonably free of mineral and chemical contamination but very likely to have biological contamination.

Where there are several sources of water,

TABLE 20–A

RECOMMENDED LIMITING CONCENTRATIONS OF CONTAMINANTS
IN DRINKING WATER

Undesirable Substance	Concentrations above which water should not be used if other sources are available (mg/L)	Dangerous Substance	Concentrations above which water supply should be rejected (mg/L)
Alkyl benzene		Arsenic	0.05
sulfonate	0.5	Barium	1.0
Arsenic	0.01	Cadmium	0.01
Chloride	250.	Chromium	
Copper	1.	(hexavalent)	0.05
Carbon chloroform		Cyanide	0.2
extract	0.2	Fluoride	See Table 39–B
Cyanide	0.01	Lead	0.05
Fluoride	See Table 39-B	Selenium	0.01
Iron	0.3	Silver	0.05
Manganese	0.05		
Nitrate	45.		
Phenols	0.001		
Sulfate	250.		
Total dissolved			
solids	500.		
Zinc	5.		

From Drinking Water Standards, *U.S. Public Health Service.*

the final choice will be influenced by (*a*) the daily water requirements, (*b*) the amount of treatment which water from each source will need to make it meet the purity standard, and (*c*) the potential each source has for additional contamination.

The daily per-person water requirements of an industrial plant can be estimated as follows: 15 to 20 gal for drinking, lavatory, and toilet usage; 20 to 25 gal per shower; and 5 to 10 gal per meal if food is prepared on the premises.

Water quality

The water supply source must be evaluated on the basis of the chemical and biological contaminants it may contain. Table 20-A lists the limiting concentrations of two classes of

TABLE 20–B

RECOMMENDED CONTROL LIMITS FOR FLUORIDE IN DRINKING WATER

Max Daily Air Temperature Five-Year Annual Average (deg F)	Concentration (mg/L)		
	Lower	Optimum	Upper
50.0-53.7	0.9	1.2	1.7
53.8-58.3	0.8	1.1	1.5
58.4-63.8	0.8	1.0	1.3
63.9-70.6	0.7	0.9	1.2
70.7-79.2	0.7	0.8	1.0
79.3-90.5	0.6	0.7	0.8

From Drinking Water Standards, *U.S. Public Health Service.*

contaminants: (a) those which usually have no toxicologic effect, but may give an undesirable taste or appearance to the water, and (b) those which constitute a health hazard and, if present, are grounds for rejecting the water supply. Standards under consideration for limiting mercury as a contaminant, propose values between 0.002 and 0.005 mg/L. (See EPA reference at end of chapter.) A limit of 0.5 mg/kg of fish is used currently.

Criteria for asbestos fibers, if present as a discharge from an industrial process in the surrounding area, may be less than those present in already treated or natural water. Information on proposed fibers testing and standards are available from the American Waterworks Assn.; see References.

Temperature criteria which affect the limiting concentrations of fluoride in drinking water are given in Table 20-B. Other factors influencing the limiting concentrations of these contaminants, particularly in combination with other substances, are given in the USPHS *Drinking Water Standards*, published in 42 C.F.R., part 72. These Standards also serve as a guide to radioactive substances in water supplies.

If the degree of contamination from the substances previously discussed is within recommended limits, the water supply source may be used, provided its bacteriological quality is acceptable. Standards for bacteriological quality are specified in the USPHS *Manual of Recommended Water-Sanitation Practice*.

The equipment necessary to treat water and make it potable depends on the degree of contamination and the likelihood that the source will become more heavily contaminated later. These factors can be evaluated only on the basis of a thorough sanitary survey of the water source. Such a survey will determine not only the type of treatment necessary, but also the nature and frequency of periodic laboratory tests of the source water and the treated water (Fig. 20–1).

Wells

The safest source of water is often a drilled well whose intake is well below the water table. Such wells show a reliable yield and are reasonably free from bacterial contamination and finely suspended fibers. If both well and city water are used, there should be no cross-connection between the two systems. Be sure to check the local code.

The wellhead should be carefully located away from sewage lines, septic tanks, and sewage drainage fields or process waste disposal systems. The following distances are often considered adequate for separation: sewers, pit privies, and septic tanks, 50 ft; seepage pits and disposal fields, 100 ft; cesspools, 150 ft. Process waste disposal systems require special consideration. As soil and drainage conditions vary from one location to another, approval by local health authorities is recommended.

The USPHS *Manual of Individual Water Supply Systems* recommends that the casing for such wells be made of wrought iron or steel with threaded couplings or welded joints. "Stovepipe" or sheet metal casings are not recommended.

To prevent contamination of the underground water by seepage of surface waters, the space between the casing and the surrounding area should be sealed with a cement grout to a minimum depth of 10 ft below the finished ground level or floor. As a further precaution, the casing should be grout-sealed to the lowest impervious stratum it passes through.

The well casing should extend 6 in. above the pump platform, which should be of reinforced, waterproof concrete at least 4 in. thick and continuous with the grout seal which surrounds the well casing. The platform should be designed so that water spilled at the wellhead will drain away from the casing. The joint between the casing and the concrete pump platform should be sealed with an asphalt caulking compound.

The top of the well casing should be at least 2 ft above the highest known floodwater mark. The pump should be self-priming and designed so that it makes a watertight seal with the well casing. The wellhead should not be covered over by paving or other material which would make access difficult.

Both submersible and turbine pumps must be considered. Submersibles are located in the well and do not require a pumphouse.

A safety factor is provided by two wells and two pumps. An automatic alternator can take effect if either unit fails.

Disinfecting the water system

The pipes, reservoirs, standpipes, pump, and well casing of a new system should be thoroughly disinfected before being put into service. An old system carrying treated water for the first time following an extended outage should also be disinfected on the discharge side of the treatment plant, and a system which has been opened for repairs should be disinfected before being put back into service.

A drinking water system can be disinfected most easily by filling with water containing not less than 100 mg/L of available chlorine. The solution should be allowed to remain for 24 hr in a new system or one which has not previously held treated waters. If the system has previously held treated water and is being put back into service following minor repairs, 12 hr will probably be sufficient.

To determine the success of the disinfecting job, the residual chlorine in the solution is measured at the end of the required time. Test kits for this purpose are available commercially and are easy to operate. If tests show residual chlorine, the biological chlorine demand of the system has been met and it can be connected to the drinking water supply, flushed out, and put into service. If no residual chlorine is present, the system should be drained and recharged with new disinfectant solution and the procedure repeated.

If the system contains a standpipe or reservoir, the disinfectant solution can be added through it. Otherwise, the solution can be supplied in a temporary reservoir on the supply side of the system pump and injected through it. A solution containing 500 mg/L available chlorine, applied with a fog nozzle, will disinfect standpipes and covered reservoirs.

Underground water supplies may become contaminated while being developed. If so, they too will have to be disinfected, as follows. After the 24-hr yield of the well has been determined, the test pump should be run to clear the well of turbidity. A chlorine solution should then be added to the well to make, with the 24-hr yield, a solution of 50 mg/L. Connect the permanent pumping equipment to the wellhead and operate it until the discharge has a distinct odor of chlorine.

There are several methods for uniformly distributing the disinfecting solution. The well casing can be sealed and the solution injected under pressure, or the solution can be added from a hose or a small pipe at several levels beneath the surface of water in the well. The chlorine solution should remain in the well for 24-hr, as mentioned.

Water purification

Of the several methods of water purification available, filtration and chemical disinfection are the most practical for industrial private water supplies.

Filtration. This method of water purification is used primarily to clarify turbid waters, but it may also serve to remove some bacterial contamination. Filtering plants are of two types: slow filters and rapid filters.

Slow sand filters will clarify turbid waters when operated at a rate of 25 to 50 gal per day per square foot of filter area. Such filters should be made with 0.25 to 0.35 mm sand and should be at least 20 in., but preferably 36 to 40 in., deep. During the operation of the filters, the film that accumulates on the surface of the sand must be kept below water level as this film increases the effectiveness of the filter.

Rapid sand filters, made with a uniform 0.4 to 0.5 mm sand with a depth of 30 in., will handle about 3000 gal of turbid water a day per square foot of filter surface.

Both types of filters should be made under competent engineering supervision and operated under continuous inspection. Depending upon the water source, filters may require a presedimentation basin for preliminary treatment of the water. Filters should be provided in pairs so that one can be removed for cleaning and maintenance without disrupting the supply of filtered water.

Disinfection. Chlorine is the best available disinfecting agent for drinking water. It can be added to the water directly as a gas or as a soluble salt (calcium hypochlorite or chlorinated lime—refer to Chapter 38, "Industrial Toxicology," for hazards of these chemicals.) The free chlorine available from any of these materials makes them preferable to chloramine-B, which is frequently used as a disinfecting agent.

Small-capacity chlorinators which inject gaseous chlorine into a water system are available and easy to operate. Injection pumps which supply high concentration chlorine solutions to the system at a proper rate are preferable because of the ease and safety of operation.

Standby equipment should be maintained at all chlorinating stations, together with an adequate supply of spare parts. Gas masks which are effective against chlorine and a small bottle of ammonia to test for leaks should be kept just outside of areas in which chlorine is stored or used. Masks should be inspected at regular intervals, and authorized employees should be trained in emergency procedures.

The chlorinator should be adjusted to leave a chlorine residue of about 0.2 mg/L in the water after 20 minutes of contact between chlorine and the untreated water. Test kits are available which will measure residual chlorine rapidly.

Small quantities of water for emergency use may be disinfected in one of several ways. Commercial preparations should be used according to the manufacturer's instructions. Boiling water for five minutes, or adding four drops of household bleach (hypochlorite solution, 4 percent available chlorine) or two or three drops of common tincture of iodine to a quart of water and allowing it to stand for 30 minutes will also produce safe drinking water. Its flatness and medicinal taste can be partially removed if it is aerated after disinfection by being poured from one container to another.

Water storage

Reservoirs or standpipes for treated water should be completely enclosed and located so that accidental contamination is impossible. The reservoir should be large enough to hold a 48-hr reserve supply of treated water. Vents should be fitted with screened downspouts well above flood level. Entrance manholes should be enclosed by watertight frames at least 6 in. higher than the surrounding surface and fitted with watertight covers extending at least 2 in. down the outside of the frames. When not in use, the cover should be closed and locked.

A reservoir permits full utilization of a smaller well and pump and still provides a buildup for peak demand. Quality of water is generally improved by aeration.

Sewage and Garbage Disposal

In some outlying districts and rural areas, industrial plants must provide their own sewage disposal systems. If there are state or local ordinances governing disposal, they must be adhered to. In addition, the recommendations given in the USPHS *Individual Sewage Disposal Systems* can be used as a guide for providing safe disposal of sanitary sewage from toilets, washrooms, showers, and kitchens. The disposal of process waste requires separate facilities. Many states and health departments require prior examination of all plans concerning disposal treatment.

Provisions should also be made for the storage and collection or disposal of garbage and refuse.

Building drains and sewers

The in-plant sanitary sewage collection system should conform to local codes, and the recommendations of American Standard and Sanitary Corporation's "National Plumbing Code" should also be considered.

Every fixture should be properly trapped and vented by means of drain(s) and stack(s) serving it, to prevent the discharge of sewer gases into the building and to assure proper draining. Traps and especially grease interceptors (such as those placed in waste pipes serving the plant cafeteria and kitchen) and interceptors designed to collect other particular foreign materials should be of adequate size, located for easy access, and should be cleaned periodically. Be sure not to (a) install a type trap that is prohibited by the local code, or (b) place a trap in a prohibited location.

The building drain and sewer should be constructed of extra-heavy cast iron, bell-and-spigot pipe with drainage fittings. This material is less susceptible to clogging and much easier to clean than other materials, and it provides good "insurance" when installed under floors which would be expensive to tear up. Lead or other suitable material should be used for joints. The sewer should be tight under a 10-ft head of water.

A cleanout should be provided where the building drain passes through the building wall, and at other selected places as the code requires. Check local codes for trap size, permitted locations, and whether a strainer is required. In some areas, codes call for installation of backwater valves to prevent backup in the sewer line. These should be located where they are accessible for inspection and cleaning.

Septic tanks

The main function of a septic tank is to separate solid from liquid wastes. It serves a secondary purpose of permitting aerobic bacteria to convert some solid waste matter into liquid waste.

The effluent from the dosing chamber of the septic tank should pass through a tight sewer over as short a distance as possible to the disposal field. The distribution box or boxes, which are the first unit of the disposal field, equalize the flow of liquid waste through the disposal field lines and serve as an inspection point where the quality of the effluent may be checked. Sludge must not be carried into the disposal field because it will clog the absorption system.

The disposal field should have enough discharge lines to permit the daily liquid waste to be absorbed and disposed of by aerobic bacteria. The number of lines and the length of each are based on the permeability of the soil to the liquid.

A septic tank should be located at least 50 ft from any drinking water source. Surface drainage from the area around the septic tank should not be permitted to reach the water sources. The tank itself should be located well below neighboring water sources.

Details of construction and maintenance of septic tanks and disposal fields are given in two USPHS publications: *Individual Sewage Disposal Systems* and *Manual of Septic Tank Practice*. Compact sewage treatment plants for populations of 50 to 500 persons are available commercially. If required, they can be specified with a grinder and chlorination apparatus.

The disposal field should be at least 100 ft from any water supply, 50 ft from any stream, and 10 ft from any building or property line. A minimum distance of 50 ft between a disposal field and the head of a deep well is acceptable if the well has a watertight casing extending downward at least 50 ft.

A detailed map should be made of the sewage disposal system, showing the location of septic tanks, distribution boxes, and tile fields. The area covering the disposal field should be kept free from vehicular traffic. Sodium and potassium hydroxides and similar "conditioning" agents, as well as brine discharges, should not be emptied into the sewage disposal system or building drains. If they are, a clogged and useless disposal field may be the result. Frequent monitoring of the effluent of the absorption field to assure the proper function of the system should be done.

Garbage disposal

Companies and plants that have food services for their employees must provide for proper disposal of food wastes and refuse. There are several methods of disposal—local ordinances should be followed in each case.

Many plants collect garbage and store it for later pickup and disposal by municipal or private collection services. Large outside metal storage receptacles and compactors should be fenced in or covered and locked to prevent children from entering and playing in them.

Garbage and refuse containers should be metal or plastic with tight-fitting covers to prevent the entry of insects or rodents. Containers should be easy to clean and handle, and washed with detergent-deodorant solutions. The use of plastic or polyethelene bags, or liners is desirable. The liner and contents can be removed easily and will help keep garbage containers clean. Garbage containers in the plant should be located so that employees will not throw food waste in waste baskets or other unsuitable receptacles. Collections should be frequent enough to prevent undue accumulation of garbage.

Discharge of ground food waste into the municipal sewage system is an acceptable method of disposal in many localities. Installation of food waste disposers in the plant kitchen can provide a convenient, efficient way of eliminating the need for garbage stor-

535

age and collection. Disposers should be located, grounded, and installed according to approved plumbing practices and local code requirements.

Kitchen employees should be instructed to use nonmetallic tampers, keep silverware out of the disposer, and clean the trap metal catcher daily. The disposer should be stopped and the power disconnected before any attempt to clean or clear it is made.

Insect and rodent control

The bacterial problem is often the limiting factor for securing good sanitary procedures. We must recognize other potential biological hazards, the flying, leaping, crawling vermin, such as insects and rodents, which create esthetic and microbial problems.

In plants where insect or rodent infestation is a problem, it is always best to employ a professional exterminator. The hazard of poisonous chemicals and preparations should not be risked by personnel with little or no previous experience in exterminating.

Communication with all plant departments, food vendors, and others concerned should be set up well in advance of the exterminator's visit. Outside signs may provide an extra warning to neighboring plants or residents, to protect pets from exposure.

To protect the workplace against entrance of vermin, wire mesh and metal screens can be used near base of foundations. See Hopkins and Schulze (References) for ideas.

Refuse collection

Hazards in refuse collection vary with the type of equipment used and the various conditions surrounding the operation. A frequent cause of accidents involve packing blades which cause partial loss of fingers, hands, arms, and feet. Other hazards arise from "booby traps" unwittingly laid by companies whom they serve—loose broken glass in a refuse container, lightweight trash cans filled with heavy objects (like chunks of concrete), heavy objects concealed by paper or other trash, hose or other obstacles strewn along the pathway to a rubbish can. Containers that are rusted through or that have unserviceable handles increase the risk of job injury.

The injury frequency rates for refuse collection and disposal, for those units reporting injuries for 1972 to the National Safety Council, are 110 and 88, respectively. This compares to about 10 for the 1972 all-industry average.

Cuts, lacerations, and punctures accounted for about 14 percent of the lost-time injuries; this compares favorably to industry in general where 17 percent of the workers suffered such injuries. Wearing heavy work gloves minimizes these types of injuries.

Refer to the booklet, "Operation Responsible—Safe Refuse Collection," published by Environmental Protection Agency (see References).

Personal Service Facilities

Drinking fountains, washrooms, locker rooms, showers, and toilets—personal service facilities that contribute to employee comfort —should be conveniently located. These facilities make up an essential part of the occupational health program in most industries.

Drinking fountains

Sanitary drinking fountains, one to about every 50 people, should be installed at convenient places throughout an industrial plant, in accordance with American National Standard A112, *Specifications for Drinking Fountains,* and local code requirements. The fountain should have an angle jet and a lip guard (Fig. 20–2). A waste can may be located at each fountain. It is important that the stream projector cannot be flooded or submerged in the event of stoppage, and that the stream be directed and projected so that it cannot be contaminated by the user. In dusty areas, fountains should be covered.

The water temperature should be 50 to 55 F for heavy manual labor, or 45 F for less active office work. If ice is used, it should be in a separate compartment, without direct contact between the water and the ice.

Where city water is available on construction work, a water line can be extended to upper floors as the building is erected. A standard drinking fountain can be installed on each floor.

On some types of work, such as highway, pipeline, power line construction, and timber clearing, the drinking water source is so re-

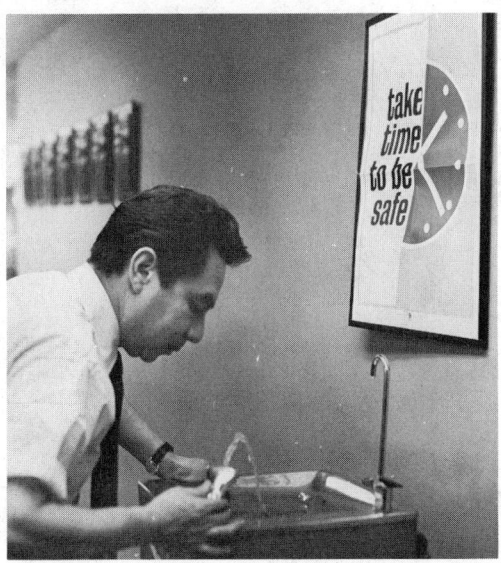

Fig. 20–2.—Fountains should be conveniently located so that employees can maintain their daily water intake. A safety poster can be located nearby the fountain.

mote that it is impractical to pipe water to the job. Some companies have successfully solved this problem by using portable drinking fountains. These fountains have an insulated tank equipped with an angle jet drinking nozzle. The tank has an air pump and pressure release valve so that it can be pumped up to the necessary operating pressure. Containers should be kept scrupulously clean and should be sterilized daily with steam, boiling water, or chlorine solution.

Under no circumstances should use of a common drinking cup or ladle be permitted. If drinking cups are required, they should be single-service paper cups kept in a sanitary container at the drinking faucet, with a receptacle provided for disposal.

American National Standard A117.1, section 5.7 *Making Buildings and Facilities Accessible to, and Usable by, the Physically Handicapped* mentions modifications necessary for use by the physically handicapped.

Drinking fountains are usually not permitted to be installed in any toilet room. Bubblers are usually not permitted by codes to be installed as an integral part of—or connected to—another fixture, such as a lavatory or sink.

Carafes (vacuum-type bottles) that are frequently used in private offices are a potential source of bacterial contamination. They should be rinsed and refilled daily and cleaned periodically, using a sanitizer such as a cationic quaternary ammonium germicide.

Salt tablets

On extremely hot, heavy jobs, it may be advisable to provide extra salt in the form of salt tablets. In any case, the indiscriminate use of tablets should be avoided. A doctor should supervise the providing and use of salt tablets. He may recommend use of buffered soft drinks to aid in maintaining body fluids in balance.

Some workers may be on salt-restricted diets, but they will have already been warned about their own salt requirements and are not likely to be doing heavy work. Other persons are nauseated by fairly high salt concentrations. This condition can be avoided by the use of enteric-coated salt tablets which prevent dissolving of the salt in the stomach, by the use of impregnated tablets which dissolve slowly in the stomach, or to some extent by the use of salt tablets containing dextrose. Diabetics who may be working in the plant should be warned about tablets containing dextrose and, if necessary, should be provided with tablets without the sugar.

Washrooms and locker rooms

ANSI Z4.1, *Minimum Requirements for Sanitation in Places of Employment*, serves as a guide to the types and sizes of washroom, locker rooms, and accessories.

A large, single washroom and locker room may be sufficient for a compact plant employing fewer than 500 people. Washrooms in a large one-story plant generally are scattered throughout the building. If the plant consists of a series of separate buildings with only a few people working in each, all the facilities may be placed in a centralized building. This arrangement has been successful in such establishments as chemical plants, oil refineries, and railroad yards.

If the plant is relatively small, it is advisable to have the dressing rooms, lockers, and washrooms near the entrance. In a larger plant it

537

TABLE 20–C

LAVATORIES PER EMPLOYEE

Type of Employment	Number of Employees	Minimum Number of Lavatories
Nonindustrial—Office buildings, public buildings, and similar establishments	1—15	1
	16—35	2
	36—60	3
	61—90	4
	91—125	5
	Over 125	1 additional fixture for each additional 45 employees
Industrial—factories, warehouses, loft buildings, and similar establishments.	1—100	1 fixture for each 10 employees
	more than 100	1 fixture for each additional 15 employees

From OSHA Regulations, § 1910.141(d)(2).

is better to have these facilities in a single building centrally located or in several buildings near the work areas.

All washing facilities should be maintained in a sanitary condition. Each lavatory should have hot and cold water, or at least must have tepid running water, and hand soap or similar cleansing agent.

Washrooms and locker rooms should be well ventilated, kept warm and comfortable, and at 50 percent relative humidity. The heating equipment should be so installed as to protect against burns and it should comply with state codes.

The floors of washrooms and locker rooms should be of nonabsorbent material such as glazed brick, tile, or concrete. The floor material should be continued up into the walls as a cove for at least 6 in. before there is a joint. The walls should then be connected to the floor cove with a tight joint and should be impervious to water to a height of at least 5 ft.

Flooring material should be selected for durability and sanitation and to minimize the hazard of slipping and falling. Terrazzo, tile, marble, and polished concrete floor surfaces are particularly hazardous when wet. For safety with such floors, a rigid cleaning and mopping schedule must be maintained to keep them dry when they are in use.

Concrete floors can be made much less hazardous by covering the surface with a finishing layer of abrasive grain concrete. Abrasive strips may be helpful on old concrete floors, which have been worn smooth. Ceramic tiles with a nonskid, nonabsorbent, and watertight surface, are also available.

A floor should frequently be inspected for watertightness. Leaky floors cause damage to joists and other structural members of the building, and, if organic materials collect in them, may attract vermin. Worn wood or concrete floors may be covered with a plastic material to obtain a watertight surface.

Lavatories should be available in accordance with Table 20-C. A lavatory (wash basin), 24 in. of trough (wash sink), and 20 in. of circular basin are considered equivalent fixtures if there are water outlets (preferably a mixing faucet) at each space. In industries where workers need additional

FIG. 20–3.—Wash fountains should be supplied with running water at a controlled temperature. A sufficient number of wash-up facilities should be available.

Courtesy E. I. du Pont de Nemours and Co., Inc.

washing time, one lavatory for every five employees is recommended.

To eliminate standing water, which can transmit disease from one employee to another, lavatories should have no stoppers. A mixing faucet or a spray will permit employees to wash in a flowing stream with controlled temperature. Knee-actuated water controls are available.

Circular wash fountains (Fig. 20–3) of stainless steel, stoneware, enameled iron, and other materials impervious to moisture permit a number of persons to wash at the same time by means of center water sprays which are continuous or are controlled by a treadle. These fountains are easily kept clean and sterile. Their construction prevents splashing and spilling of water.

Wherever practicable, a thermostatic control should be installed in the hot water supply system in order to keep temperature below 140 F. Injecting live steam into tanks or lines of a cold water system (to make warm water) is dangerous, since failure of pressure in such a system could release steam through the taps.

A regular maintenance program for equipment should be in effect, and employees should be requested to report defective equipment. Broken faucets and valve handles may cause serious cuts or lacerations. Handles should be made of metal, not a breakable

material, such as porcelain. If leaky faucets are repaired at once, employees will not develop the bad habit of turning valves off too tightly.

The proper type of soap is important, not only for ordinary hygiene, but as a protection against dermatitis caused by the cleaning agent. The soap used should have no free alkali and should have a pH less than 10.5. It should be free of mineral abrasives. Individually dispensed paste, liquid, or powder (not bar soap) for common use should be provided. Liquid or powdered soaps are preferable, because they lend themselves to ready dispensing and are also an aid to housekeeping.

The practice of removing paint, dye, and other stains with solvents or other chemicals, and especially the practice of removing grease from the hands with naphthas, should be strongly discouraged. Solvents may cause a severe skin irritation.

Waterless skin cleansers are not substitutes for soap and water, but are convenient for special use or where water is scarce.

Common-use towels should not be permitted. Paper towels should be soft enough not to cause irritation. They should be kept in a covered container with a self-closing disposal receptacle nearby.

Hot air hand driers should be well secured either to the floor or the wall to prevent loosening of the fixtures or the electric element. The equipment must be grounded and permanently installed without extension cords or plugs. Blowers must provide air at not less than 90 F, nor more than 140 F.

Lockers should be perforated for ventilation and be large enough to permit clothing to be hung up to dry. If the clothing may be heavy or wet, it is highly desirable to provide forced circulation of hot air through the base of the lockers and out through the top, or to provide hangers on elevating chains so that the work clothing may be dried between shifts.

Lockers should have sloped tops to prevent material from being stored on the tops. The multiple legs of lockers are serious impediments to floor cleaning; lockers should be placed on metal frames with a minimum of floor supports. They should be anchored together to prevent their being overturned.

Men working with highly toxic materials which are dusty or which may otherwise contaminate the clothing should have separate lockers for work clothing and street clothing. These lockers should preferably be in rooms on opposite sides of the shower room so that the men will have to pass through the shower room when changing from work clothing to street clothing, and back.

Benches in front of the lockers should be permanently fastened to the locker base, preferably on a hinged support so that they can be turned up against the faces of the lockers while the aisles are swept. The benches should be checked at regular intervals and kept in repair, free from splinters, breaks, and other imperfections.

Showers

Showers should be installed in plants in which workmen become dirty or wet with perspiration, or are exposed to dust or vapors. The showers should be as close to the job as possible, preferably in a separate room adjacent to the dressing and locker rooms. Men exposed to high temperature who come off the job wet with perspiration should not be exposed to cold weather in going to the shower room and change house.

One shower should be provided for each 10 employees of each sex, who are required to shower during the same shift. Each shower should be supplied with hot and cold water through a mixing fixture that the user can regulate. The maximum temperature of the hot water should be automatically maintained at 140 F.

Flood showers, eyewash fountains, and similar installations for emergency use are discussed in Chapter 21, "Occupational Health Services."

Body soap or other appropriate cleaning agents, convenient to the showers, should be provided, and also hot and cold water feeding a common discharge line. Employees who use showers should be provided with individual clean towels.

When employees are required by a particular standard to wear protective clothing because of possible contamination with toxic material, change rooms should be equipped with storage facilities for street clothes. Separate storage should be provided for protective clothing.

Where clothes are provided by the employer, and become wet, or are washed between shifts, provision should be made to make sure that the clothing is dry before re-use.

The floor of the shower room and of the individual compartments should be made of nonskid material to provide good footing when it is wet. Either abrasive grain concrete or concrete with a wood-float finish is a satisfactory surface.

Existing floors which were made smooth or which have become smooth through long wear can be given a nonskid surface. Concrete floors should be scrubbed thoroughly with an abrasive pad using a synthetic detergent. Strips of abrasive material can be applied to other types of floors to provide a nonslip surface. The floor throughout the shower room area should slope toward drains, preferably at the back of the shower stalls. Curbs around the individual shower stalls are not necessary if the floor is properly sloped. They are a tripping hazard and, if used, should be dyed or painted a contrasting color.

Wood mats should not be used on shower room floors because of the tripping hazard and the probable exposure to splinters and loose joining members.

The pans of antiseptic solution commonly seen at the entrance of shower stalls or shower rooms are useless for killing organisms and a nuisance to keep clean.

As an item of general sanitation, shower rooms and stalls should be well ventilated and adequately lighted to prevent the formation of mold. The floor of the shower should be mopped daily with detergent, hot water and disinfectant to combat athlete's foot (fungus and ringworm infection). A foot-actuated spray can aid in controlling athlete's foot.

Toilets

Wall-hung, elongated-bowl flush toilets with open-front seats should be provided according to the number of employees (Table 20-D). If persons other than employees are allowed to use toilet facilities, their number should be increased accordingly. Paper holders must be provided for every water closet. For each three toilet facilities, there must be at least one lavatory in the toilet room or adjacent to it.

Wall-hung units are easier to keep sanitary and to clean under. Codes prohibit any type that is not thoroughly washed at each discharge or that might permit siphonage of bowl contents back into the tank. Water supplied to tanks must have vacuum breakers or have a positive air gap between top of water in the tank and the water supply inlet.

Toilets should be placed not more than 200 ft from any work place. In multistory buildings, toilets should be not more than one floor above or below the work area. With toilets and lavatories at various points throughout the plant, the main locker room and shower room can be closed for cleaning during the work period, an advantage for the janitorial crew.

Toilet rooms and washrooms for women sometimes have an attendant on duty during use. Washroom attendants should not attempt to give first aid to women who become ill at work. Such aid is best given by the plant nurse or physician. Some states also require that women work no more than a certain distance from a woman's rest room. This should be checked during design stages. Some states also require cots to be installed.

Ventilation is required for toilet rooms. If natural ventilation is relied upon, there should be windows or skylights having a ventilation area of 6 sq ft for a room with one toilet,

TABLE 20-D

MINIMUM TOILET FACILITIES

Number of Employees	Minimum Number of Water Closets*
1 to 15	1
16 to 35	2
36 to 55	3
56 to 80	4
81 to 110	5
111 to 150	6
More than 150	One additional fixture for each additional 40 employees.

* Where toilet facilities will not be used by women, urinals may be provided instead of water closets, except that the number of water closets in such case shall not be reduced to less than ⅔ of the minimum specified.

From OSHA Regulations, § 1910.141(c)(1).

with an additional square foot of window ventilation space for each additional toilet. If this amount of window space cannot be provided, forced ventilation should be supplied at a rate of three to four air changes per hour in the room.

Since windows and skylights generally do not afford sufficient light, light fixtures should be installed in all toilet rooms and washrooms. Switches, for the lights or for electric driers or other equipment, should be located so that they cannot be operated by persons who are at the same time in contact with piping or other grounded conductors.

Individual wall-hung urinals should be provided in the men's room. These may be substituted to the extent of one-third or less of the number of stools specified. Trough urinals are poor substitutes for individual fixtures and are prohibited in many states. Approved urinals must have all surfaces which are subject to soiling accessible for cleaning. Integral screens over the discharge openings are the major cause of chronic toilet room odors because the decomposing soil under the screen cannot be removed by any practicable method. Blow-down washout urinals are the only acceptable type. Floor-type urinals, in which the drain pipe becomes chronically offensive, and wall-hung urinals with integral screens, should be replaced by the approved sanitary type (blow-down wash-out), thus making room deodorants unnecessary.

Employees should be prohibited from lunching in toilet rooms, or in process areas where toxic or noxious materials are present. The habit of some workers to heat foods in molten lead reservoirs or other process heating equipment can be dangerous to their health and should be prohibited. This prohibition naturally implies the provision of proper lunchrooms or other eating facilities outside the toilet rooms or process areas.

Covered receptacles should be provided in plant lunchrooms for disposal of waste food and papers, and employees should be prohibited from disposing of such refuse in the toilet rooms. If cups of coffee or other drinks are carried from the lunchroom, they should be in covered containers or on trays to prevent spillage which might create unsanitary or slippery conditions.

Privies are unsatisfactory, but where no other method is feasible, privies and chemical closets should be approved by health authorities having jurisdiction or conform to USPHS *Individual Sewage Disposal Systems*. Portable toilets are also available. These are often necessary on construction jobs. The supplier can provide waste removal and maintenance.

Janitorial service

As a part of an overall, managed plant sanitation function, a minimum, daily janitorial service should be provided for all personal service facilities. When properly designed, washrooms, shower rooms, and toilets can be thoroughly cleaned with little personal involvement in the process. Floors and fixtures should be mopped and cleaned with detergent and hot water, at least once daily. A sanitizing cleaner should be used as often as necessary. The occasional use of an acid-type cleaner may be necessary on toilet bowls and urinals.

Rubber gloves and goggles should be worn and the fixtures thoroughly flushed following use.

When floors are being mopped, the area should be blocked off by signs reading CAUTION WET FLOOR, as a precaution against possible slipping accidents.

Food Service

Nutrition

Nutrition, another factor in industrial health and safety, concerns the medical and safety departments of any plant. If a survey of the food service establishments in the neighborhood shows that they cannot supply the nutritional needs of employees, then the plant is justified in establishing its own food service. With care and thought, adequate, balanced in-plant meals can be provided. Food must also be properly prepared and attractively served, with strict adherence to sanitary practices.

The company nurse, working with the company or visiting physician, can provide employees with leaflets on better nutrition. The Council on Foods of the American Medical Association, and the American Dietetic Association (see References) have many excellent articles and materials available.

The importance of a good breakfast, high in protein for "timed energy release" during the day, contributes to less fatigue and, consequently, less chance of accidents.

Workers usually need additional nutrition during the first half of a shift. Low blood sugar tends to be a health and accident hazard. A survey of vending machines will show that about two out of three food snack items are overbalanced in sweets. Overeating of sweets can contribute to low blood sugar. Snacks that contain higher protein, such as peanuts, meat sticks, and peanutbutter foods, are desirable. Brown sugar rolls and buns, raisins, and sandwiches made of protein breads are helpful.

Some organizations have dietitians review their food service menus and even talk to employees.

Types of service

There are five main types of industrial food service:

1. Cafeterias preparing and serving hot meals

2. Canteens or lunchrooms serving sandwiches, other packaged foods, hot and cold beverages, and a few hot foods

3. Mobile canteens which move through the work areas, dispensing hot and cold foods

Fig. 20–4.—A spacious, well-lighted cafeteria lends a pleasant atmosphere to an in-plant meal. With care and thought, adequate, balanced meals can be provided.

Courtesy The Austin Company.

and beverages from insulated containers

4. Box lunch service

5. Vending machines

Even using a mobile canteen to provide a midshift snack adds considerably to the nutrition of the average worker. If lunches are also served, the mobile canteen should carry both hot and cold foods and beverages.

The central cafeteria with a kitchen where full meals can be prepared and served is often the most satisfactory form of food service. In large plants, it may be economical to supply several cafeterias from a central kitchen.

Vending machines and microwave ovens. Self-service vending machines offer a wide variety of packaged, ready-to-eat foods. Some

machines have ovens that let the user quick-cook his meal. Two important safety and health precautions should be followed in the use of microwave ovens.

1. All repairs should be made by a manufacturer's trained repairman.

2. Persons with pacemaker heart units should be warned against coming too close to microwave ovens.

Proper installation and maintenance of food heating and refrigeration systems is important. Normal sanitary precautions of course apply to vending machines. Can openers in safe working order, sufficient utensils, and adequate waste disposal facilities, in both kitchens and eating areas, are other necessary provisions of a self-service operation.

543

TABLE 20–E

MINIMUM FLOOR SPACE IN EATING AREAS

Number of People	Sq Ft per Person
25 or less 13	
26 through 74. .12	
75 through 149. 11	
150 and over. .10	

From American National Standard Z4.1

Eating areas

The cafeteria or lunchroom should be clean and attractive to encourage employees to eat away from their work area (Fig. 20–4). Refrigerators for the storage of lunches will also help convince employees that they should eat in the proper area.

Minimum floor spaces for the number of people using the eating area at one time are given in Table 20-E. Where space is limited, lunch periods should be staggered so that employees do not have to eat on the job.

Kitchens

When a cafeteria kitchen is set up, the same attention should be paid to proper equipment and working conditions as would be in any other part of the plant. Food equipment should be of types approved by the National Sanitation Foundation. The layout should conform with public health food service codes, and be such that the various operations are segregated, with adequate walkways from point to point about the kitchen.

Floors should be made of impervious, water-resistant, nonskid material to minimize the hazard of slips and falls if water or grease should be spilled on them, as often happens in a busy kitchen.

Ranges and other heat-producing equipment should be hooded and ventilated to carry away heat, combustion products, and vapors. Since these ventilating systems get very greasy, the duct work should be easily available for cleaning. They should be made of heavy gage steel so any fire can be self-contained.

Sprinkler systems and portable extinguishers should be installed for fire protection (Fig. 20–5). Extinguishers containing carbon tetrachloride should be prohibited. A fire blanket should be located near the ranges or in areas where clothing may be ignited.

Easily changeable racks for handtools, such as knives, cleavers, and saws, and storage racks or cabinets for utensils should be provided in convenient places.

Controlling contamination of foods

Incorporated communities have detailed sanitary regulations for the installation and operation of industrial food service facilities. Regulations of the local authority having jurisdiction should be followed in detail. In unincorporated areas with no local authority, the state code or the recommendations of the USPHS *Food Service Sanitation Manual* should be followed.

Perishable food and drinks, particularly custard-filled and cream-filled pastry, milk and milk products, egg products, fish, meat, shellfish, gravies, poultry, stuffing, sauces, dressings, and salads containing meat, should be kept at or below 40 F except while being prepared or served. If foods of this kind have been permitted to stand for some time at room temperature after preparation, reheating them is not a sufficient protection against bacterial poisoning.

The types of bacteria (staphylococcus) which may infect these foods produce a toxin that is not destroyed by normal cooking temperatures. If the bacteria have grown in the food during room temperature storage, they will be killed by reheating, but the toxin will remain and food poisoning will result.

To prevent bacterial food poisoning, the following suggestions are made.

1. Keep perishable foods under refrigeration until they are to be used.

2. Keep hot foods hot (160 F or above) and cold foods chilled (40 F or below).

3. Remove leftover foods from food-warming devices immediately after the last feeding period. Never hold hot foods in warmers from one meal to the next, or for several hours before a dinner is served.

4. Place leftover food under refrigeration as quickly as possible.

FIG. 20–5.—The modern industrial kitchen should include fire protection equipment, such as the carbon dioxide nozzles over the deep fat friers (at left) in this installation. Also note racks for hand tools.

Courtesy Walter Kiddle and Co.

5. Instruct employees who handle food and utensils that they must wash their hands thoroughly with soap and water after using the restrooms and before handling foods.

6. Eliminate flies, roaches, rodents, and other pests that may transmit disease. Consult a professional exterminator if necessary.

7. Never use galvanized or cadmium-plated containers for storage of moist or acid foods.

8. Consult your local health authorities if you have questions on sanitation.

9. Employees with open infections or communicable diseases, should not handle food.

As a further precaution against the transfer of infections, no first aid material should be permitted in the kitchen. All conditions requiring first aid should be seen immediately by the plant nurse or the physician, and the individual should continue on the job only at the doctor's discretion.

To prevent cross infections in large dining rooms, the proper cleaning, sanitizing, and storing of containers, utensils, glassware, dishes, and silverware are highly important.

Use of single-service containers and utensils, however, can eliminate washing and handling.

It is generally easier to sanitize utensils by machine washing than it is by hand washing if the machine is kept clean and in top operating condition. In either case, one of the main requirements in maintaining adequate sanitation is proper training of employees.

Utensils should be carefully scraped and preferably pre-rinsed before being put into the detergent solution. They should be thoroughly washed with soap or a detergent, and well rinsed by a method which will destroy bacteria.

For thorough machine washing, the utensils must be stacked in the trays loosely enough so that the cleansing agent gets to every part, the concentration of detergent in the wash water and the wash water temperature must be maintained. The wash water must be changed before it becomes excessively dirty. Spray nozzles in the dish washing machine must be cleaned daily to maintain proper flow and distribution.

Requirements for hand washing are the same, except that each utensil must be individually scrubbed in all parts rather than simply stacked in a tray.

For rinsing, the cleaned utensils may be immersed in clean hot water at 170 F for one-half minute. One problem is that of maintaining the temperature of the rinse water over long sessions of dishwashing, since a large volume of fresh hot water is required for this method. Water heaters should be of adequate size. Less water, however, will be needed if a chemical sterilizing agent is used. Hypochlorite solutions at a concentration of at least 50 ppm of available chlorine for an immersion time of one minute at 75 F will provide adequate sterilization. Cationic quaternary ammonium germicides are also suitable.

When the utensils have been properly cleaned, they should be stored and handled so as to prevent contamination by the handler's fingers and from ordinary dust and dirt or from leakage from overhead pipes.

References

American Dietetic Association, 620 N. Michigan Ave., Chicago, Ill. 60611.

American Medical Association, Council on Foods, 535 N. Dearborn St., Chicago, Ill. 60610.

American Public Health Assn., 1790 Broadway, New York City 10019.
  Standard Methods for the Examination of Water and Wastewater. 12th ed. 1965.

American National Standards Institute, 1430 Broadway, New York, N.Y. 10018.
  Air Gaps in Plumbing Systems, A112.1.2.
  "Cast Iron Pipe and Fittings," A21 Series.
  Drinking Fountains and Mechanically Refrigerated Drinking Water Coolers, A112.11.1.
  "Gas-Burning Appliances," Z21 Series.
  "Iron and Steel Pipe," B36 Series.
  Minimum Requirements for Non-Water Carriage Disposal Systems, Z4.3.
  Minimum Requirements for Sanitation in Places of Employment, Z4.1.
  Minimum Requirements for Sanitation in Temporary Labor Camps, Z4.4.
  National Electrical Code, C1.
  "Pipe Flanges and Fittings," B16 Series.
  "Plastic Pipe," B72 Series.
  Specifications for Making Buildings and Facilities Accessible to, and Usable by, the Physically Handicapped, A117.1.
  Threaded Cast-Iron Pipe for Drainage, Vent, and Waste Services, A40.5.

American Standard and Sanitary Corp., 40 W. 40th St., New York, N.Y. 10018. *National Plumbing Code.*

American Waterworks Association, 2 Park Ave., New York, N.Y. 10016.

Burton, B. T. (ed.). *The Heinz Handbook of Nutrition.* New York, McGraw-Hill Book Co., 1959.

Environmental Management Association, 1710 Drew St., Clearwater, Fla. 33515.

Hopkins and Schulze, *The Practice of Sanitation,* 4th ed. Baltimore, Md., Williams & Wilkins, 1970.

McKee, Jack E., and Harold W. Wolf, eds. *Water Quality Criteria*, 2nd ed. Sacramento, Calif., Dept. of General Services, Office of Procurement—Stores, Documents Section, PO Box 1612.

National Restaurant Association, One IBM Plaza, Chicago, Ill. 60611.
"A Safety Self-Inspection Program for Food-Service Operators."

National Safety Council, 425 N. Michigan Ave., Chicago 60611.
Industrial Data Sheets
Dusts, Fumes, and Mists in Industry, No. 531.
Industrial Skin Diseases, No. 510.
Refuse Collection in Municipalities, No. 618.
Public Employee Safety Guides
Refuse Collection.
Water Department.

National Sanitation Foundation, 3475 Plymouth Rd., Ann Arbor, Mich. 38106.

Public Health Service, U.S. Department of Health, Education, and Welfare, Washington, D.C. 20234.
Drinking Water Standards, Publication No. 956.
Food Service Sanitation Manual, Publication No. 934.
Individual Sewage Disposal Systems, Reprint No. 2461.
Inspection Report for Food-Service Establishments, Publication No. 4006.
Manual of Individual Water Supply Systems, Publication No. 24.
Manual of Recommended Water-Sanitation Practice, Publication No. 525.
Manual of Septic Tank Practice, Publication No. 526.
Vending of Foods and Beverages—A Sanitation Ordinance and Code, Publication No. 546.

U.S. Environmental Protection Agency, Water Supply Program Div., Washington, D.C. 20460.
Manual for Evaluating Public Drinking Water Supply.
"Operation Responsible—Safe Refuse Collection." Available through National Audiovisual Center, General Services Administration, Washington, D.C. 20409.

Water Treatment for Industrial and Public Supply. Department of Trade and Industry, Control office of Information, London, England. 1971.

Williams, Roger J. *Nutrition Against Disease.* New York, Pitman Publishing Co., 1971; reprint ed., New York, Bantam Books, Inc., 1973.

547

Occupational
Health
Services

Chapter
21

Occupational Health   549

Medical Service   550
The industrial physician . . . Duties of the industrial physician
. . . Emergency medical planning . . . Duties of the occupational
health nurse . . . Physical examination program . . . Health problems
of women . . . Placement of women . . . First aid . . . Employee
health services . . . Medical records

Alcohol and Drug Control   563

Selective Placement   563
Placement of the handicapped

References   568

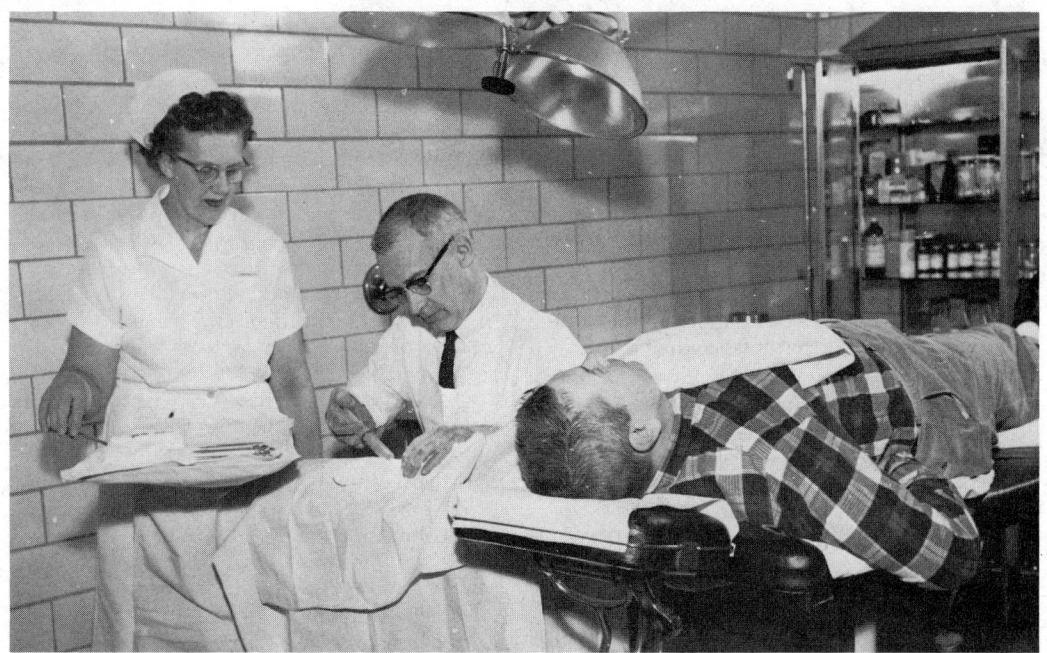

Fig. 21–1.—Operating room in a well staffed and fully equipped medical department.

Courtesy George A. Hormel Co.

Occupational Health

Occupational health programs are concerned with all aspects of a worker's health and his relationship with his environment.

The American Medical Association's Council on Occupational Health's official guides to occupational health programs, *Scope, Objectives and Functions of Occupational Health Programs* and *Guide to Small Plant Occupational Health Programs,* state that the basic objectives of a good occupational health program should be:

1. "To protect employees against health hazards in their work environment

2. "To facilitate placement and ensure the suitability of individuals according to their physical capacities, mental abilities, and emotional makeup in work that they can perform with an acceptable degree of efficiency and without endangering their own health and safety or that of their fellow employees

3. "To assure adequate medical care and rehabilitation of the occupationally injured

4. "To encourage personal health maintenance

"The achievement of these objectives benefits both employees and employers by improving health, morale, and productivity."

If an industrial worker is to perform his tasks safely and efficiently, he must be in good health. It is not uncommon for a worker to have lower personal efficiency and increased accident susceptibility when feeling below par because of a nonoccupational illness, sometimes aggravated by self-diagnosis and self-medication.

Application of occupational health principles also helps to assure that workers are placed in jobs according to their physical capacities, mental abilities, and emotional makeup. This phase of medical-safety teamwork is also effective in the proper placement of severely handicapped workers. It also assures continuing medical care and rehabilitation of occupationally ill and injured workers.

549

There is a relationship between accident prevention and occupational health. For example, some industrial chemicals present a variety of serious hazards to health and property when improperly handled. That is, depending on conditions, the vapor from a chemical can ignite or explode, it can cause dizziness or death when inhaled, or dermatitis when touched. For details on the effects of specific chemicals see Chapter 38, "Industrial Toxicology."

The safety professional has ably demonstrated his ability to reduce accidental injury frequency by control of many phases of the industrial environment, through education of the worker, and by improved supervisory techniques. A large part of the remainder of the problem rests within the physical and emotional characteristics of individual workers. Here lies the key to variations in job attitude, productivity, safety, and absence for personal health reasons.

The services and skills of several additional professions are also frequently needed if maximum results are to be attained.

● Medicine plays an important role, with physicians being specially trained in industrial and preventive medicine. They are assisted by specialists in orthopedic surgery, ophthalmology, radiology, surgery, dermatology, and psychiatry, to name a few.

● Nursing in an occupational health service requires a specialized knowledge not only of good basic nursing procedures and health maintenance, but also of the legal, economic, social, and labor laws that form the parameters within which the nurse practices. Often, the nurse is the only full-time medically oriented employee in a company or plant.

● The industrial hygienist, a professional development of this technological age, serves as the analytical preventive engineering arm of occupational medicine by applying specialized knowledge to the recognition, evaluation, and control of health hazards in the work environment. (More details in Chapter 37, "Industrial Hygiene.")

● Other phases of occupational health may utilize the industrial dentist, the sanitary engineer, the public health expert, the psychiatrist, the podiatrist, and the psychologist.

Working individually and collectively, these specialists have helped to improve the occupational health and safety records of many industries. In some companies, the application of these talents is so well organized and effective in the anticipation and correction of hazards that the employee is unquestionably safer and healthier at work than he is at home. In fact, the work environment is so well controlled, in many cases, that the problem now is how to get the worker to avoid the hazards of home life, recreation, and travel so that he will be able to return to work each day safe and sound.

Occupational health services of a company should be involved in the off-the-job safety program. The personnel of the medical department can be influential in extending this program beyond the facility to include the employee's and his family's off-the-job activities. The health and well-being of the employee's family has a direct bearing on his efficiency and safety on the job. If the employee is injured off-the-job, he is as much a loss to the operation as if he were injured on the job.

Some insurance carriers offer a consulting service which will help organizations set up an occupational health program suitable for their needs. They usually know who the doctors and clinics are that would be available for this kind of service. The basic work, however, has to be done by the organization desiring to set up a program.

The justification for these services lies in their accomplishments. Prevention is not only better than cure—it is easier and less expensive. Off-the-job safety programs and activities can benefit a plant or shop of any size.

Medical Service

Occupational health services may range from the truly elaborate to the bare minimum required by OSHA. One establishment may have a full-time staff of physicians, nurses, and technicians, housed in a model dispensary; another may have only the required first aid kit with an adequately trained person to render first aid. There are, of course, many intermediate stages. The operating and treatment room in a well staffed and fully equipped medical department is shown in Fig. 21–1.

Modern occupational health programs, regardless of size, ideally are composed of elements and services designed to maintain the health of the work force, to prevent or control occupational and non-occupational diseases and accidents, and to prevent and lower disability and the resulting lost time. They should provide for:

1. Maintenance of healthful environment

2. Health examinations

3. Diagnosis and treatment

4. Immunization programs

5. Medical records

6. Health education and counseling

7. Open communication between plant physician and personal physician

Any treatment of ill or injured persons has some bearing on the practice of medicine. All states, therefore, regulate medical and nursing practice to curb the activities of persons who are not properly trained and licensed. Ideally, all the services related to health, injuries, first aid, and medication of any kind should be under the *supervision* of a licensed physician.

The industrial physician

Physicians in industry may be employed on a number of different bases, dependent upon such considerations as the number of employees, the hazards in the operations, and the type and extent of the occupational health program. Arrangements may be made for full-time, part-time, on-call, or consulting service.

● Where the medical service program requires a full-time physician, one should be employed by the company. Some large organizations employ a full-time medical director on their headquarters staff, with part-time physicians serving their decentralized operations.

● Some part-time physicians devote a scheduled number of hours, either daily or weekly, to the medical service needs of a company, yet they are available at other times for emergencies. Others arrange their service by telephone on the basis of current needs of the company—when applicants require examina-

tion, injured employees need medical care, or other medical problems arise.

● The on-call physician arrangement is most often used by companies (or establishments) having fewer than 500 employees, a low frequency of accidental injuries, or a minimum program of health services for employees. The on-call physician is usually located nearby and is available in emergencies. He cares for most injuries not requiring hospitalization and for those cases requiring hospitalization that fall within his competence. He often uses specialists as consultants.

Frequently, the on-call physician has similar arrangements to serve a number of companies in the vicinity—a particularly convenient system with the small-plant or small-company clusters, such as in industrial parks.

It is not unusual to find physicians specializing in the care of industrial injuries and diseases, and the provision of occupational medical services in some manufacturing centers.

● Consulting service is usually reserved for diagnosis and treatment of serious injuries, illnesses, toxicological problems, and special kinds of injuries or disorders (such as eye injuries) that require the services of a specialist.

State and local health departments often supply, without charge, medical, nursing, and engineering consultation, as well as make industrial hygiene and radiological surveys to industries within their areas. Many insurance companies also provide this type of service to their clients.

Duties of the industrial physician

An effective medical service program should be planned by the medical director, with the approval of management. Full support of management can only be obtained when management is sympathetic to a cause that reflects management's policies. Only then is it possible to establish and maintain an adequate professional staff and facilities for examinations, emergency cases, and storage of records. Also, the medical director must be given enough authority so that workers will respect his judgment and follow his instructions on personal health and safety.

The industrial physician should be familiar

with all jobs, materials, and processes that are used. An occasional inspection trip will help him keep abreast of what is going on in these areas. This inventory of hazards will help the physician suggest to the safety professional ways to protect employees from actually or potentially harmful environments.

This basic knowledge will also aid the industrial physician when he recommends placement of employees in jobs that are suited to their mental, emotional, and physical fitness. This knowledge is especially useful in placing handicapped workers.

The physician should be involved in other company services that relate to health of workers, such as food service, welfare service, safety program, sanitation, and mental health. He can also initiate and be responsible for sponsorship of company-wide immunization programs (against tetanus, polio, or flu) as well as for blood donor and chest X-ray programs.

Maintenance of the true physician-patient relationship (with fairness to both employee and employer) is essential to the success of any occupational health program—workmen should receive the same courtesy and professional honesty as do private patients. The first meeting of physician and employee usually occurs at the preplacement examination and this meeting may be followed up by subsequent examinations. The examining physician in his professional discretion should acquaint the examinee with the results of these examinations and, if necessary, refer him to his own personal physician for correction of defects.

The industrial physician should provide emergency medical care for all employees who are injured or become ill on the job, and he should arrange for the necessary followup treatment of employees suffering from occupational disease or injuries. The treatment of employee injuries or diseases not industrially induced is the function of private medical practice. Therefore, the industrial physician should abstain from rendering such services except where independent facilities are not readily available, or where the ailment or discomfort is so minor that the employee would ordinarily not seek medical attention, or where the rendering of such service would enable an employee to complete a shift.

The industrial physician should not devote time or facilities to diagnose or treat dependents of all employees. He should, however, promote health education programs for employees and their families so that they will be encouraged to seek proper private consultation.

Medical and surgical management in every case of industrial injury or disease should aim to restore the disabled worker to his former earning power and occupation as completely as possible and without unnecessary delay. Concise, dependable medical reports that are promptly submitted to those agencies entitled to them are a part of this obligation. In the same way, equitable administration of workmen's compensation rests on medical testimony which adheres closely to reasonable scientific deductions regarding the injury or its possible consequences.

The industrial physician is responsible for maintenance and completeness of medical records. Records and reports are a necessary part of an occupational medical program. Comprehensive records must be kept and reports rendered in order that management be guided in their continuing responsibility for health and safety, and the employees kept informed of the degree of success of the safety program afforded by their cooperation. Records and reports are necessary to direct and evaluate preventive medical and safety engineering techniques. They are necessary to chart progress in the reduction of accidents and to meet the recordkeeping requirements of the Williams-Steiger Occupational Safety and Health Act of 1970 (see Chapter 2).

The physician is also responsible for properly instructing nurses and other paramedical personnel (such as first-aiders) and directing their activities. Their duties, therefore, should be described in clear, concisely written directives, a copy of which should be posted in the medical department. There must be no delegation of services requiring expert medical attention.

Emergency medical planning

The industrial physician, nurse, and safety professional should confer with management to plan for emergency handling of large numbers of seriously injured employees in the event of disaster, such as explosion, fire, or other catastrophe. These plans should be coordinated with community plans for such events. See "Planning for Emergencies," Chapter 18.

Procedures should include selecting, training, and supervising auxiliary personnel, and should provide for:

1. Selecting and training auxiliary nursing personnel.

2. Selecting and training first aid workers.

3. Transporting and caring for the injured.

4. Transferring seriously injured to hospitals.

5. Coordinating these plans with the safety department, guards, police, road patrols, fire departments, and other interested community groups.

Duties of the occupational health nurse

Occupational health nursing is a specialized branch of the nursing profession. The position requires a registered professional nurse, who is licensed in the state where employed. In addition, it is desirable that the occupational health nurse have some knowledge of workmen's compensation laws, insurance, health and safety laws, occupational diseases, sanitation, first aid, and record keeping.

The occupational health nurse contributes the most when she works with the company physician who provides her with written directives that have been discussed between them, so they are mutually agreed upon and understood.

Working with the physician, the occupational health nurse can provide a variety of nursing services, such as initial care for injuries or illnesses, counseling, health education, consultation about sanitary standards, and referral to community health agencies. She also can participate in programs to evaluate employee health, such as health examinations, or to prevent disease, such as immunizations.

Working with the physician, the plant nurse can perform excellent employee health education services in the distribution of literature on heart care, weight control, cancer and tuberculosis prevention, and the prevention and treatment of venereal disease.

When the physician is only at the establishment part time, the nurse is responsible to

him on professional matters, and to a member of management on administrative matters. In this case, the nurse works with the safety director in planning and conducting accident prevention programs.

The occupational health nurse must maintain a confidential professional relationship with the employees in conformity with legal and ethical codes. She may not divulge information contained in individual employee health records unless the employee gives his signed permission; the medical files are only accessible to medical personnel.

It must be clear that establishment of medical diagnosis and definition of treatment are the functions of the physician. The nurse is not a substitute for him. Each has a legally defined area of practice and responsibility. An effective company medical program requires the services of both a physician and a nurse.

According to a statement published by the American Medical Association in 1959: "Courts have held that professional nurses have a legal duty to interpret evidence presented by the patient possibly indicating the need for medical attention, and to proceed in the light of that interpretation to do what is required for the patient, as for example the need to call a physician; to discontinue a treatment where there is evidence of its harmful effect, for example on an unconscious patient; or to determine the patient's need for special medication, for example, sedatives. There can be little doubt that by custom and usage the relationship between doctor, patient, and nurse is one in which the parties recognize that the nurse as well as the physician has the function and responsibility to observe and interpret the patient's reactions.

"In emergency cases, industrial nurses have a duty to determine the need for prompt medical attention, to make a tentative diagnosis of the patient's condition, and to employ necessary resuscitative and first aid measures."

Physical examination program

Supervision of the health status of workers by qualified medical personnel is essential if any occupational health program is to obtain maximum benefits for both employer and employee. Therefore, a program of physical examinations should be established. The examining physician should discuss all significant findings with the worker. He should use good judgment to prevent the raising of unnecessary fears, while emphasizing the importance of obtaining adequate personal medical care. A transcript of the data may be supplied to the worker's personal physician or for insurance purposes, with the consent of the employee. Courts, workmen's compensation commissions, or health authorities may request this information by legal means, but employee consent is a more agreeable method. Certainly the confidential character of health examination records should be observed.

The employer should be informed of a potentially harmful work environment detected through examination of persons subjected to it, because OSHA requires employers to keep accurate records of employee exposure to potentially toxic materials or harmful physical agents.

Scope of the examination. It is impossible to set forth what constitutes a complete examination, or even a suitable examination. Physicians have different opinions regarding the relative values of various test procedures, based on their own training and experience. Therefore, the scope of physical examinations should be determined by the company physician, who is familiar with the operations involved. The nature of the industry, its inherent hazards, the variations in jobs, in physical demands and in health exposures are determinants. The values of different test procedures and their cost in time and dollars must be assayed. Perhaps examinations should be different in scope for different jobs. For example, the physical condition of the ironworker to be engaged in construction of the skyscraper is a far different problem from that of the sedentary seamstress, yet there are basic physical examination considerations applicable to each.

The various kinds of examinations may be classified as follows—preplacement, periodic, transfer, promotion, special, and termination. the nurse works with the physician in all these examinations.

Preplacement examinations. The preplacement examination is made to determine and record the physical condition of the prospec-tive worker so that he can be assigned to a suitable job in accordance with his mental ability and physical capacity, and in which his disabilities, if any, will not affect his personal efficiency, safety, and health, or the safety of others. The applicant (or his personal physician, on the applicant's approval) is advised of conditions needing attention. Medical department followup may or may not be necessary.

It must be a paramount principle that the purpose of the program is selection *and* placement—not merely selection of the physically perfect and rejection of all others. For further information, see Selective Placement later in this chapter.

From the public and occupational health standpoints, the only bars to immediate employment in nonhazardous occupations should be communicable disease, or progressively incapacitating injury or disease, or incapacitating mental illness. It is obvious that communicable diseases must be controlled, and this may involve the assistance of public health officials. One of the values of an examination program is the detection of disease in its early stages, when it is most amenable to treatment. Applicants with incipient but still nondisabling disorders can often be employed while being treated by their personal physicians. Many applicants with incapacitating injury or disease can improve their employability and job opportunities after being assisted by rehabilitation agencies. (See the section on Placement of Handicapped later in this chapter.)

A significant percentage of persons have mental illness or emotional disturbance that impair judgment or prevent them from performing normal work. These aberrations vary in degree and can be serious enough to bar employment. The trained physician can frequently detect them during the preplacement examination.

It is not the function, however, of the physician to inform the applicant whether he is to be employed. This is the prerogative and duty of management, as there are other factors in addition to physical qualifications that bear upon suitability for employment.

Periodic examinations of all employees may be on a required or voluntary basis.

A required program should be applied to

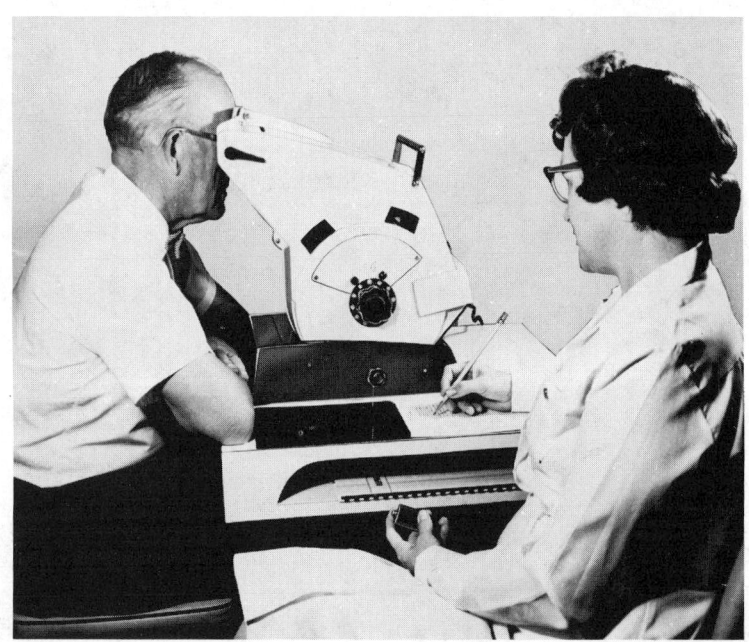

Fig. 21–2.—Typical device used to conduct screening tests of vision.

Courtesy Titmus Optical Co., Inc.

workers who are exposed to health-hazardous processes or materials, or whose work involves responsibility for the safety of others, such as vehicle operators. Substances like lead or carbon tetrachloride that are capable of causing occupational disease are usually subjected to process controls that will keep the workers safe from poisoning; however, caution dictates the advisability of periodic examination of such workers, to be certain that the engineering and hygiene controls are effective and continuing to be so. This procedure also enables early detection of the hypersusceptible individual, and the worker whose personal unsafe practices defeat the control measures.

Frequency of the examination must vary in accordance with the quality of the engineering control, the nature of the exposure (this is influenced by the rapidity of the action of the hazardous substance on the human body), and the findings on each examination. Thus, exposures to some substances might justify examinations or laboratory tests on a weekly basis, others monthly or quarterly, whereas annually or biennially may be adequate in some dust exposures.

In many cases, laboratory tests of blood or urine will suffice as the major portion of a periodic examination program, with complete examinations being made less frequently. The type of special examination (laboratory, X ray, etc.) necessary for any exposure and the interpretation of the results are decisions requiring the most expert medical personnel.

Special examinations. Employees having on-the-job difficulties that may be health related are often benefited by special examinations. Job transfer may also require medical evaluation.

Many organizations find it worthwhile to make "return to work" examinations of employees who have been absent more than a specified number of days as a result of a non-occupational illness or injury. This is done to control communicable disease, as well as to determine suitability for return to work. There is a wide difference in the actual effects of the same disease on different persons, and an even greater divergence of personality reaction to the disorder. One person will go to bed at the first sign of discomfort, whereas another must be truly overwhelmed before he will fail to report for work.

555

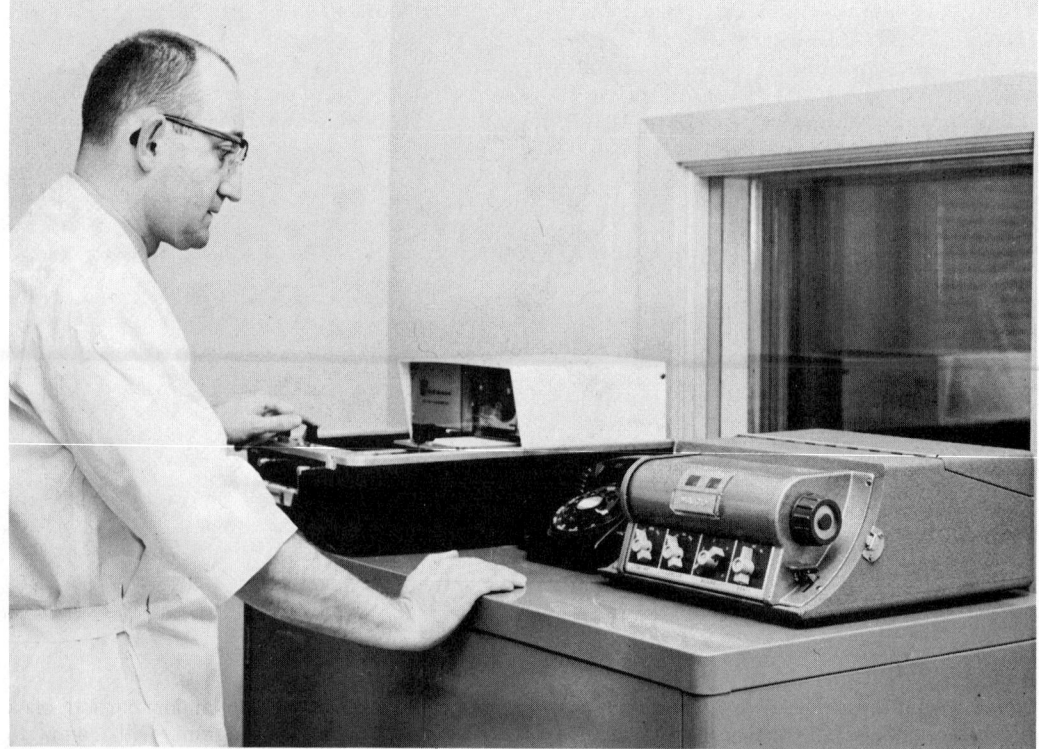

FIG. 21–3.—Automatic audiometer (left) and manual audiometer (right) can be used for accurate hearing evaluations.

Courtesy Erie Mining Co., Hoyt Lakes, Minn.

When an employee returns to work following serious injury, occupational or nonoccupational, a new evaluation of physical capacities may be necessary. Also the application of rehabilitation procedures may reduce the disability and improve the range of employability.

Exit examinations. Upon termination of employment of workers, some organizations make examinations and record the findings, particularly where operations involve definite exposure to health hazardous substances, such as lead, benzene (benzol), silica, and asbestos dust, or to harmful noise.

Laboratory tests. Urine and blood tests are a good investment in detecting liver and kidney disease, diabetes, anemia, etc. Where

there is to be an exposure to toxic substances, appropriate laboratory tests may be indispensable.

Chest X ray. A record of the condition of the lungs as shown by X ray is desirable for every applicant. Those who will be engaged in a dusty trade should have periodic chest X rays. Every X ray should be carefully identified with date and name and each should be compared with previous X-ray photographs.

Vision tests. In recent years, special devices have been developed for routine testing of several aspects of vision (Fig. 21–2). Near, as well as far, vision should be recorded. The old method of testing only distance vision at 20 ft is inadequate. Failure to compare the visual requirements of a job with the visual abilities of employees may result in employees becoming easily fatigued, inefficient, and involved in an accident.

If facilities are not available at the establishment, arrangements can be made with outside doctors for annual eye examinations and fitting prescription safety glasses if needed.

If the job demands it, color vision should also be tested.

Hearing tests. Workers who are to be exposed to hazardous noise levels should be examined for hearing acuity before placement to determine prior hearing loss, if any, and periodically thereafter to detect early loss due to noise. Personal protective devices can reduce noise reaching the auditory nerve, but wisdom dictates the value of tests and records. The audiometer is the accepted method of testing (Fig. 21–3). A special booth will usually be required to administer these tests. (See Chapter 40, "Noise and Hearing Conservation.")

Health history. Some doctors think that a carefully taken personal health and occupational history gives clues to the areas of examination which warrant extra-careful study. A nurse or other specially trained person can secure basic data from the examinee, such as height, weight, age, and the history. Nurses may also assist in other portions of the examination when specially trained and authorized to do so. This relief from minor details will give the physician more time to devote to an analysis of the whole person.

Health problems of women

There is no reliable evidence to support the view that women are more susceptible than men to occupational health hazards, except during pregnancy.

Some companies used to dismiss women on learning of their pregnancy. As a result, many women concealed their condition until it became obvious, frequently until the last half of the pregnancy. This policy can work to the disadvantage of both employee and company inasmuch as the early months of the pregnancy present the greatest danger. It is desirable that women who work during this time be under medical department supervision. The woman should consult her own doctor who should,

in turn, report her condition to the plant doctor so that he may determine the advisability of her continuing to work.

Chemical substances, such as carbon monoxide, chloroform, phosphorus, and mercury, may produce harmful effects on the fetus and lead to abortion.

Pregnancy limits the ability of women to do physical work, since a pregnant woman tires more readily, has poorer balance, and is unable to respond normally to the physiological demands of strenuous physical work.

Placement of women

State and federal laws and regulations governing employment of women in industry should be studied carefully before female employees are placed, and should be rechecked before women are transferred to other types of work or are placed on different shifts.

The examining physician, at the time of employment, should furnish enough information to help place the applicant most advantageously from the standpoint of her health and safety. Periodic examinations and medical histories have indicated that properly trained women are capable of performing safely most types of work except those involving heavy muscular effort.

Methods for the prevention of industrial accidents among women are the same as those for men. However, when women are placed on machine jobs ordinarily done by men, it is important that adjustments be made at all points of operation because improper adjustment for women can result in accidents. For example, machine guards may have to be set so that women's smaller hands cannot enter the openings. Height of benches, distance away from parts, and foot or hand controls should be adjusted to conform to the generally shorter stature and reach of women. For some jobs, low platforms may be provided. Smaller hand tools may also be advisable. Because women have less muscular strength, some jobs may need to be broken down into simpler operations or mechanical aids may need to be provided.

First aid

Good administration of first aid is an important part of every safety program. It is

557

recommended that a first aid facility of some sort be set up in all establishments regardless of size. This may range from a deluxe first aid kit to a well staffed first aid and medical facility, depending on the size of the establishment.

First aid kits and supplies should be kept in a central location readily accessible to the establishment or department. Under no circumstances should medical supplies be spread about the plant for self-administration by employees. Where there is no nurse in charge, a supervisory employee for each shift should be delegated the responsibility for all medical supplies.

A careful record should be kept of each administration of first aid and an injury investigation report sent to the injured person's supervisor at the time first aid is administered.

Establishments that do not have a full-time medical doctor should maintain good liaison with a doctor, or doctors, designated to handle plant injuries. The doctor should be invited to the plant occasionally to evaluate the quality of first aid procedure, to make recommendations for improvement, and occasionally to tour the establishment, as discussed earlier in this chapter. The company-designated doctor (or doctors) should be well aware of the type of work done so he may better evaluate information given him by patients. The doctor or nurse or supervisor designated should routinely inspect first aid supplies, stretchers, and stretcher locations.

Definition and limitations. In many small organizations and in field operations, it is neither practical nor justifiable to have qualified professional medical personnel available. In such cases, the best arrangement is to use suitable first aid attendants who follow procedures and treatments outlined by a doctor. The doctor should be available on an on-call or referral basis to take care of serious injuries.

It should be noted that in some jurisdictions injured employees have their choice of a physician. In such cases, the employer should comply with this request, if possible.

There are two kinds of first aid.

• One is emergency treatment—according to the American Red Cross first aid textbook, "First aid is the immediate, temporary treatment given in the case of accident or sudden illness before the services of a physician can be secured." Proper first aid measures reduce suffering and place the injured person in a physician's hands in a better condition to receive subsequent treatment.

• The other kind of first aid is the prompt attention given to injuries, such as cuts, scratches, bruises, and burns, which are usually so minor that the injured person would not ordinarily seek medical attention.

Under OSHA recordkeeping procedures, first aid is defined as a one-time treatment plus any followup visit for observation of minor scratches, cuts, burns, splinters, and the like that do not ordinarily require medical care.

The requirement that all employees report for treatment immediately upon being injured, regardless of the extent of the injury, has resulted in much headway in the reduction of infection and disability, and also in the avoidance of false claims of injury and disability.

A first aid program should include:*

1. Properly trained and designated first aiders on every shift

2. A first aid unit and supplies, or first aid kit

3. A first aid manual

4. Posted instructions for calling a physician and notifying the hospital that the patient is en route

5. Posted method for transporting ill or injured employees and instructions for calling an ambulance or rescue squad

6. An adequate first aid record system

First aid procedures, approved by the consulting physician, should embrace the type of medication, if any, to be used on minor injuries, such as cuts and burns. These are two examples where there is often considerable difference of opinion regarding proper treatment. In areas where chemicals are stored, handled, or used, emergency flood showers and eye-wash fountains (Fig. 21–4) should be available. If possible, they should

---

* Small Plant Health and First Aid Services. Division of Occupational Health, Pennsylvania Department of Health, Harrisburg, Pa.

Fig. 21–4a.—Checking a safety shower. Note deluge quantity of water.

Fig. 21–4b (*upper right*).—Portable shower/ eye-wash fountain can be taken to location of temporary operations. Be sure connecting hose is of large enough diameter to supply adequate amount of water. Make sure water and connections are clean.

Fig. 21–4c.—Emergency eye-wash fountain.

be well identified (Fig. 21–5).

The equipment and supplies should be in accordance with the recommendations of the physician, and service should be rendered only as covered by written standard procedures, signed and dated by him. If it is intended to furnish temporary relief for minor nonoccupational ailments, such as colds and headache, the physician should specify the procedures to be followed. The limitations of first aid must be thoroughly understood.

The majority of states have medical practice acts under which a person is limited to a certain definite procedure when attending anyone who is sick or injured—except, of course, under the direct supervision of a physician. It is important, therefore, that anyone who is responsible for first aid have a full understanding of the limits which restrict the work. Since improper treatment might involve the company in serious legal problems, the first aid attendant should be duly qualified and certified by the Bureau of Mines or American Red Cross. These certificates must be renewed at specified intervals.

First aid training. The American Red Cross first aid textbook and the United States Bureau of Mines manual of first aid instruction are recommended for the teaching of first aid. It is often found that accidents occur less frequently and as a rule are less severe among persons trained in first aid work. It is therefore advisable that as many industrial workers as possible be given this training.

The National Safety Council publishes posters and pamphlets that can be used for training employees.

Another valuable reference is *Emergency Care of the Sick and Injured*, by the Committee on Trauma, American College of Surgeons. (See References.)

A universal symbol for emergency medical identification has been developed by the American Medical Association (Fig. 21–6). The object of the symbol is to identify its wearer immediately as a person with a physical condition requiring special attention. If the wearer is unconscious or otherwise unable to communicate, the symbol will indicate that there are vital medical facts to be found on a health information card in the bearer's purse, wallet, or elsewhere. These should be known by anyone helping the individual struck down by an accident or sudden illness.

FIG. 21–6.—Emergency identification symbol.

FIG. 21–5.—Eye-wash fountain is well identified.
Courtesy City of Chicago, Dept. of Water & Sewers.

Employee health services

Although good industrial medicine can be practiced in any logical location in the plant that is clean and private, experience indicates that the medical unit commands respect only if careful attention is paid to suitable and efficient housing, appearance, and equipment. The entire unit should be painted in light colors and be kept spotlessly clean. The dispensary should have hot and cold running water and be adequately heated, ventilated, and illuminated. Proximity to toilet facilities is necessary. Suitable provision should be made for men and women if both are employed.

The medical department should be easily accessible and near the greatest number of employees so that distance does not become a barrier against the immediate reporting of all minor injuries for treatment. If possible, it should also be connected with the employment and safety departments; this facilitates prompt physical examinations of applicants, the mutual use of clerical service, and the interchange of ideas and plans relative to employment, accident, and health problems. Another location consideration is to have the facilities near the entrance so that an ambulance can be brought to the door, if necessary; or so injured workers, who are off duty but under treatment, may come and go through a separate entrance.

The medical department should be in a place of greatest safety in case of a major company or plant disaster which might otherwise destroy first aid or dispensary facilities.

Health service office (dispensary). A minimum of three rooms, consisting of a waiting room, a treatment room, and a room for consultation or for making physical examinations, is recommended (Fig. 21–7). Rooms for special purposes can be added according to the needs and size of the company. Applicants for employment waiting for physical examinations should not mingle with the injured workmen.

The surgical treatment room should be large enough to treat more than one person at a time—small dressing booths can be arranged to give some degree of privacy.

First aid room. It is always advisable to set aside a room at a convenient location for the sole purpose of administering first aid. It upsets morale to administer treatments in public as injured persons prefer privacy. Furthermore, the person administering first aid should have a proper place to do his work.

The environment of a good first aid room should be similar to that of the dispensary. The room should be equipped with the following items:

1. Examining table
2. Cot for emergency cases, enclosed by movable curtain
3. Dustproof cabinet for supplies
4. Covered waste receptacle operated by foot pedal
5. Small table
6. Chair with arms, and one without arms
7. Magnifying light on a stand
8. Dispensers for soap, towels, cleansing tissues and paper cups

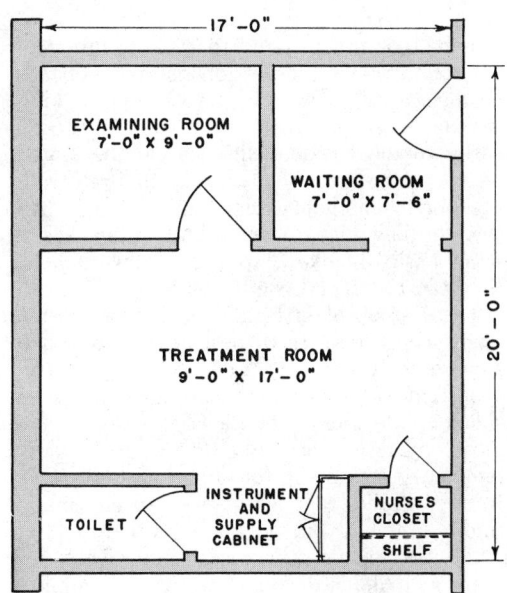

FIG. 21–7.—Floor plan for a health service office.

9. Wheel chair
10. Stretcher
11. Blankets
12. Bulletin board on which are posted all important telephone numbers for emergencies

Emergency oxygen is of great benefit in the treatment of many cases requiring first aid. Where oxygen is being administered, smoking should be prohibited because of the danger of fire or explosion. Any type of resuscitating device should only be used by trained persons.

First aid kits. There are many types of emergency first aid kits; they are designed to fill every need, depending on the special hazards that might occur. Commercial or cabinet type first aid kits, as well as unit-type kits, must meet OSHA requirements.

Kits vary in size from the pocket model to what amounts to a portable first aid room. The size and contents depend on the intended use and the types of injuries that could occur. For example, "personal kits" contain only essential articles for the immediate treatment of injuries. "Departmental kits" are

561

larger—they are planned to cover a group of workers and the quantity of material depends on the size of the working force. "Trunk kits" are the most complete—they can be carried easily to the accident site, or can be stored near working areas that are distant from well-equipped emergency first aid rooms. Although they include such bulky items as a wash basin, blankets, splints, and stretchers, they can be carried conveniently by two men.

Distribution of first aid kits throughout the plant seems to work to best advantage when supervised properly. Each kit is the responsibility of a single, trained individual who should understand that he is to care for the most trivial injuries only or to render only temporary treatment for more serious cases. It is believed that with this system many slight scratches and cuts are given attention which they would otherwise not receive.

In such industrial organizations as mining companies and public utilities, where activities are widely scattered, the use of first aid kits and some self-medication by employees may be necessary. In these instances, however, it is better if first aid service can be controlled by having the attendant in charge properly instructed in first aid and by seeing that the service as a whole has proper medical supervision.

The maintenance and use of all of these first aid kits should be under the supervision of medical personnel, and they should contain the materials approved by the consultant physician. A member of the medical services group should be assigned to inspect all first aid materials regularly, and to submit a report of their content level and serviceability.

Maintenance of quantities of materials in the first aid kit is facilitated if each kit contains a list showing the original contents and the minimum quantities below which new materials should be ordered. All bottles or other containers should be clearly labeled.

Recommended materials for first aid kits are listed in the American Red Cross and the U.S. Bureau of Mines first aid textbooks. Suggestions are also available from the American Medical Association and from manufacturers of first aid materials.

Stretchers. Inadequate facilities for trans-

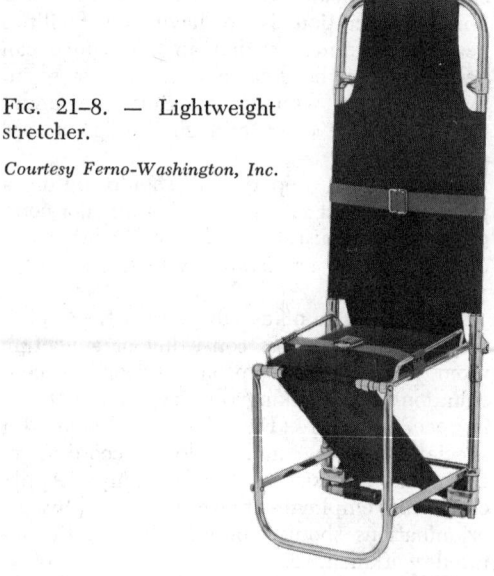

Fig. 21–8. — Lightweight stretcher.

Courtesy Ferno-Washington, Inc.

porting a seriously injured person from the scene of the accident to a first aid room or a hospital can add to the seriousness of the injury and may be the determining factor between life and death.

The stretcher provides the most acceptable method of hand transportation and it can be used as a temporary cot at the scene of the accident, during transit in a vehicle, and at the first aid room or dispensary.

There are several types of stretchers. The ordinary army type is commonly used, and is satisfactory. However, when it is necessary to hoist or lower the injured person out of awkward places, specially shaped stretchers to which the patient may be strapped and thus kept in an immovable position are better. A lightweight stretcher is shown in Fig. 21–8.

Stretchers should be located conveniently near all places where employees are exposed to serious hazards. It is customary to keep stretchers, blankets, and splints in cabinets that are marked and placed to stand out among everything surrounding them. Stretchers should be kept clean; should not be exposed to destructive fumes, dust, or other substances; should be protected against mechanical damage; and should be ready for use at all times. If the stretcher is of a type that will deteriorate, it should be tested periodically for durability and for strength.

Medical records

Employers are required under OSHA to maintain accurate records and make periodic reports of work-related deaths, injuries, and illnesses. See Chapter 6, "Accident Records and Injury Rates."

Industrial medical records also provide data for use in job placement, in establishing health standards, in health maintenance, in treatment and rehabilitation, in workmen's compensation cases, in epidemiologic studies, and in helping management with program evaluation and improvement. These data are collected in the history interview, from the preplacement examination and any subsequent examinations, and from all visits the worker makes to the dispensary or first aid room—they establish a medical profile of each worker.

The key to accurate diagnosis and treatment often lies in the adequacy and completeness of this medical profile; therefore, its maintenance is a professional responsibility. Further, to make a complete history, medical records should include absences caused by illness or off-the-job injury. Thus record keeping, nonoccupational as well as occupational, often uncovers chronic or recurrent conditions where treatment (referral to family doctors) and preventive measures pay dividends, assist in absenteeism control, and reduce accident rates.

Maintenance of health records, however, should not be so laborious a task that the occupational health nurse (or first aid attendant) becomes a file clerk. It is important that the recording forms and filing systems be simple so that they are usable by and interpretable by a physician, nurse, or first aid attendant.

Descriptions and illustrations of medical record forms are shown in Parts 1, 2, and 3 of AMA's *Guide to Development of an Industrial Medical Records System.* These forms can be modified by the medical director to suit the specific industry. Basically, the information should be important and useful, easily and accurately obtained, and the yield should justify the cost of obtaining.

Although the employer should know the individual's limitations from a placement standpoint, only persons who are in positions of responsibility should look at the records. These records are confidential.

If a company does not have a resident medical director or nurse, its medical records are usually filed in the personnel department, under the direction of the part-time physician.

Reports to the insurance company and to the state authorities, as well as compensation payments, should, of course, be attended to promptly.

Alcohol and Drug Control

Alcoholism and drug abuse among employees are becoming serious problems in industry. The occupational health services department should have a program that helps recognize the employees who have this problem. These employees should be counseled about their problem and treatment initiated as may be most appropriate from the standpoint of organization policy.

The *combination of alcohol and drugs* can produce a variety of effects that may severely impair a worker. Concentrations of alcohol and drugs remain in the blood stream much longer than most users realize, and the effects of this combination may arise unexpectedly. Little scientific study has been conducted on the interaction of alcohol and drugs, but there is sufficient evidence to conclude that such a combination can lead to increased impairment of judgment and skill.

Managers and supervisors should be alert to employees whose work and performance are deteriorating because of an alcohol problem. These employees should be referred as speedily as possible for medical care.

The extent of the problem is reflected in one large printing establishment of over 3000 employees where 400 persons are regularly enrolled in alcohol and drug control programs.

Selective Placement

The technique of job surveys is applicable to the placement, reassignment, and transfer of all workers—male or female, young or old, able-bodied or handicapped. Therefore, to use manpower most efficiently, the job should be evaluated for the physical demands it imposes and the working conditions under which it is performed; moreover, in parallel with this, the person should be evaluated for his abilities and limitations. These analyses are applicable to all workers and all jobs; however, they are particularly helpful in the selective placement of women and anyone with physical or mental limitations.

Placement of handicapped

To place a handicapped worker properly, properly, the following requirements should be observed:

1. Worker should meet physical demands of the job. For example, a man on crutches should not be placed in a job requiring heavy lifting, carrying heavy weights, or extensive walking.

2. Worker should not be a hazard to himself. For example, a person subject to dizzy spells should not work on a ladder or scaffold or around moving machinery, where he may be seriously injured or killed if he becomes dizzy on the job.

3. Worker should not be a hazard to others. For example, an epileptic should not drive a bus or operate an overhead crane, because he may have an attack while working, lose control, and cause injury to others.

4. Task should not aggravate the degree of disability. For example, a person with heart trouble should not be placed on a job that requires considerable stair climbing, running, heavy lifting, or other strenuous tasks. A person with skin disease should not be exposed to skin irritants.

What proper placement does, then, is match the worker to the job on the basis of his ability to meet the demands of the job. When this is done, the impairment disappears as a job factor. Moreover, it should be realized that most disabled persons have more ability than disability, because few jobs actually require all of a nonhandicapped worker's ability.

It is important to remember that each impairment may impose limitations on the activities in which the individual may engage and the working conditions and accident and health hazards to which he may be exposed. Although a majority of handicapped workers are employable, a small percentage can only be employed in special workshops, or at home.

When properly placed and trained, and competing on an equal basis, the handicapped equal or excel able-bodied workers in production and safety; and their attendance and labor turnover records are usually superior to those of the able-bodied.

Effect on insurance rates. There has been a general misconception of the effect of employment of handicapped workers upon insurance rates. The following statement[*] clarifies this point. "It has been thought by many that companies issuing workmen's compensation insurance coverage increase the premium where the physically handicapped are employed. This is not true. Rates are based on experience by the class of industry and modified in most cases by the individual plant experience. There is no indication that losses are increased when the physically handicapped are properly placed."

Rehabilitation. Through federal-state programs, the Vocational Rehabilitation Administration of the Department of Health, Education, and Welfare continues to rehabilitate handicapped persons. Its stated purpose is to develop, preserve, or restore the ability of mentally and physically disabled men and women to perform useful work. As the final step in their rehabilitation process, these persons seek employment. Inevitably, some will be rejected, but rejection rates are higher in some industries than in others. Whether or not an industry can use any or many of the handicapped depends on the nature of the work, the variety of jobs available, and other factors, such as workmen's compensation laws. The greatest need is for an understanding that workers can be rehabilitated and employed with benefit and profit to both the employer and employee. Often supervisors can revise methods to permit the efficient employment of greater numbers of the partially disabled.

---

[*] "The Pre-placement Physical Examination Program" and "Periodic Examinations in Industry," Robert B. O'Connor, *Health at Work.* Liberty Mutual Insurance Co.

JOB ANALYSIS
FOR PHYSICAL FITNESS REQUIREMENTS

TITLE OF POSITION	GRADE

NAME AND LOCATION OF ESTABLISHMENT	AGENCY

Does establishment have medical supervision ☐ Yes ☐ No
Is there an industrial safety branch ☐ Yes ☐ No

Refer to the manual for job analyses before using this form. Check all functional and working condition factors as well as acceptable disabilities whenever appropriate.

I. FUNCTIONAL FACTORS
L - *Little*   **M** - *Moderate*   **G** - *Great*   **O** - *None*

Hands - Fingers	L	M	G	O	Arms	L	M	G	O	Legs - Feet	L	M	G	O	Body - Trunk	L	M	G	O
1. Reaching					8. Reaching					14. Walking or running					22. Sitting				
2. Pushing or pulling					9. Lifting					15. Standing					23. Bending				
3. Handling					10. Pushing or pulling					16. Sitting					24. Reaching				
4. Fingering					11. Carrying					17. Carrying					25. Lifting				
5. Climbing					12. Climbing					18. Climbing					26. Carrying				
6. Throwing					13. Throwing					19. Jumping					27. Jumping				
7. Touching					**Eyes**					20. Turning					28. Turning				
					30. Near vision					21. Lifting									
Voice					31. Far vision					**Ears**									
29. Talking					32. Color vision					33. Hearing									

II. WORKING CONDITION FACTORS

34. Inside				41. High humidity				48. Odors				55. Toxic conditions	
35. Outside				42. Low humidity				49. Body injuries				56. Infections	
36. High elevations				43. Wetness				50. Burns				57. Dust	
37. Cramped body positions				44. Air pressure				51. Electrical hazards				58. Silica dust	
38. High temperature				45. Noise				52. Explosives				59. Moving objects	
39. Low temperature				46. Vibration				53. Slippery surfaces				60. Working with others	
40. Sudden temperature changes				47. Oily				54. Radiant energy					

III. ACCEPTABLE DISABILITIES   *Check appropriate square if acceptable*
A - *Amputation*   **D** - *Disability*   **Y** - *Yes*   **N** - *No*

Hands - Fingers	A	D	Arms	A	D	Legs - Feet	A	D	Body - Trunk	D
1 or 2 on primary hand			1 Arm			1 Leg (high)			1 Hip	
1 or 2 on secondary hand			2 Arms			2 Legs (high)			2 Hips	
More than 2 on primary hand			None ☐			1 Leg (low)			1 Shoulder	
More than 2 on secondary hand						2 Legs (low)			2 Shoulders	
1 Hand						1 Foot			Back	
2 Hands						2 Feet			None ☐	
None ☐						None ☐				

Eyes	Y	N	Ears	Y	N	Cardio - Vascular	Y	N	Tuberculosis	Y	N
Blind			Deaf			Moderate tension			Minimal (healed, stable or arrested)		
Industrially blind			Hard of hearing, 1 ear			High tension			Moderate (healed, stable or arrested)		
Blind one eye			Hard of hearing, 2 ears			Organic heart disease compensated			Far advanced (healed, stable or arrested)		
Color blind			Hearing aid acceptable						Collapse therapy		
Color blind for shades											

FIG. 21–9.—Form used by U.S. Civil Service Commission to facilitate placement of the handicapped worker in a productive environment.

PHYSICAL CAPACITIES FORM

Leg Amputation 5" below knee
Artificial leg—good fitting

Name.. Sex ..M.. Age ..31.. Height 5'9½"Weight .155.

PHYSICAL ACTIVITIES			WORKING CONDITIONS		
√ 1 Walking	16 Throwing		51 Inside	√66 Mechanical Hazards	
0 2 Jumping	√17 Pushing		52 Outside	√67 Moving Objects	
0 3 Running	√18 Pulling		53 Hot	0.68 Cramped Quarters	
√ 4 Balancing	19 Handling		54 Cold	69 High Places	
√ 5 Climbing	20 Fingering		55 Sudden Temp. Changes	70 Exposure to Burns	
0 6 Crawling	21 Feeling		√56 Humid	71 Electrical Hazards	
√ 7 Standing	22 Talking		57 Dry	72 Explosives	
8 Turning	23 Hearing		√58 Wet	73 Radiant Energy	
9 Stooping	24 Seeing		59 Dusty	74 Toxic Conditions	
010 Crouching	25 Color Vision		60 Dirty	75 Working With Others	
√11 Kneeling	26 Depth Perception		61 Odors	76 Working Around Others	
12 Sitting	27 Working Speed		62 Noisy	77 Working Alone	
13 Reaching	28		63 Adequate Lighting	78	
√14 Lifting	29		64 Adequate Ventilation	79	
√15 Carrying	30		65 Vibration	80	

Blank Space = Full Capacity   √ = Partial Capacity;   0 = No Capacity

May work............hours per day..........days per week.   (If TB, cardiac or other disability requiring limited working hours.)

May lift or carry up to............**............pounds.

Details of limitations for specific physical activities Should not be required to walk, balance, climb, stand, kneel for prolonged periods of time.

**Should not lift heavy weights continuously.

Should not carry long distances.

PHYSICAL DEMANDS FORM

Job Title ....Linotype Operator.......... Occupational Code ..4-44.110.

Dictionary Title..LINOTYPE OPERATOR.

Firm Name & Address..

Industry.................................. Industrial Code............

Branch...................................... Department............

Company Officer............................ Analyst ..Wetzel.. Date............

PHYSICAL ACTIVITIES			WORKING CONDITIONS		
x 1 Walking	61 Throwing		x51 Inside	66 Mechanical Hazards	
2 Jumping	x17 Pushing		52 Outside	67 Moving Objects	
3 Running	x18 Pulling		53 Hot	68 Cramped Quarters	
4 Balancing	x19 Handling		54 Cold	69 High Places	
5 Climbing	x20 Fingering		55 Sudden Temp. Changes	70 Exposure to Burns	
6 Crawling	21 Feeling		56 Humid	71 Electrical Hazards	
7 Standing	22 Talking		57 Dry	72 Explosives	
8 Turning	23 Hearing		58 Wet	73 Radiant Energy	
9 Stooping	x24 Seeing		59 Dusty	74 Toxic Conditions	
10 Crouching	25 Color Vision		60 Dirty	75 Working With Others	
11 Kneeling	26 Depth Perception		61 Odors	x76 Working Around Others	
12 Sitting	27 Working Speed		x62 Noisy	77 Working Alone	
x13 Reaching	28		63 Adequate Lighting	78	
x14 Lifting	29		64 Adequate Ventilation	79	
x15 Carrying	30		65 Vibration	80	

DETAILS OF PHYSICAL ACTIVITIES:

Sits at linotype machine most of the day, reads copy, and fingers keyboard to set lines of type. Periodically walks short distances, reaches for, lifts and carries matrices, galleys of type, and pigs of type metal. Reaches for, handles, pushes, and pulls handwheels and levers and fingers gages, stops, and micrometer in setting up machine.

Fig. 21-10.—State employment service forms used in matching workers to jobs may be obtained from state employment service offices.

The majority of physically handicapped persons can be employed without preliminary rehabilitation processes. They require only selective placement in suitable jobs; however, they could enhance their skills, enlarge their usefulness to industry, and increase their earning power by taking advantage of rehabilitation services.

Information regarding available services can be obtained from the state rehabilitation agency or the Vocational Rehabilitation Administration, Department of Health, Education, and Welfare, Washington, D.C. This federal department coordinates efforts with the Secretary of Labor and both departments work with the President's Committee on Employment of the Handicapped.

Physical classification. The labor market simply cannot supply all physically perfect workers. In fact, the percentage of workers in perfect health is extremely low. The working population now includes more persons with disabilities than ever before. Advances in medical science now prolong the lives of many who would have died of war injuries, or of illnesses, such as tuberculosis, diabetes, and heart disease. Accidents in industry, in traffic, and in the home continue to increase the number of handicapped persons.

Because of his proximity to the problem and his regular plant inspections, the company's physician should have a better understanding of the job requirements than other physicians. Therefore, it is his responsibility to provide management with clear recommendations on the employability, limited employability, or nonemployability of applicants for jobs.

Many systems of classification are now in use. Generally, however, these systems use broad statements, such as "physically fit for any work"; "defect that limits applicant to certain jobs" (the defect may or may not be correctible, but may require medical supervision); and "defect that requires medical attention and is presently handicapping." This disqualifies a person for any type of employment.

Another method provides a greater range in expression of limiting factors, allowing more alternatives for the individual case.

Yet another method, which approaches the ultimate in functional evaluation of the individual, appraises capacities on a form with the identical terminology used in evaluating the physical or functional factors and working conditions of jobs (Fig. 21–9). This effective method of presenting information from a physical examination clearly indicates the specific work capability and limitations of the individual. Thus the medical report is more meaningful to the placement officer because the examining physician indicates ability or limitations with respect to each job factor. The examining physician, then, is responsible for determining the occupational significance of physical disorders.

This now makes proper job evaluation important. That is, employers must be aware of the physical requirements of jobs and the accident and health hazards involved. Each job-appraisal factor has a direct relationship which makes it either definitely unsuitable or potentially undesirable for one or more types of disability. The factors to be considered in job appraisal are physical demands, working conditions, health hazards, and accident hazards.

Physical demands include agility, strength, exertion, vision, hearing, talking, sitting, standing, walking, running, climbing, crawling, kneeling, squatting, stooping, twisting, lifting, and handling. They should be evaluated according to quality of ability and duration of activity. For example, a job involving a considerable amount of stair climbing is unsuitable for workers with heart disease, respiratory diseases, obesity, or lower limb orthopedic disorders, but a small amount of stair climbing may be tolerated.

Working conditions include indoors, outdoors, excessive heat or cold, excessive humidity or dryness, wetness, sudden temperature changes, ventilation, lighting, noise, whether the work is performed alone, near others, with others, or as shift work or piece work. Some of these conditions could have a harmful influence upon certain disabilities. For example, work in excessive heat is generally unsuitable for persons who have had malaria, or for those with high blood pressure, heart disease, skin disease, the aged, and the obese.

567

21—Occupational Health Services

Health hazards include air pressure extremes; radiant energy (ultraviolet, infrared, radium emanations, and X rays); silica, asbestos, dusts, and skin irritants; respiratory irritants; systemic poisons; and asphyxiants. These hazards have serious effects and can aggravate a preexisting bodily defect. For example, a job might involve exposure to respiratory irritants of insignificant quantities to a normal person; yet this condition might aggravate the disability of a person who has chronic bronchitis.

Accident hazards include danger of falls from elevations, work on moving surfaces, slipping and tripping hazards, exposure to vehicles or moving objects, falling objects from overhead, foot injuries, eye injuries, cuts and abrasions, burns, mechanical and electrical hazards, and fire and explosion hazards. These hazards could have an unfavorable relationship to the disability of the handicapped person. For example, a job that may involve foot injury hazards is unsuitable for the diabetic because of his susceptibility to gangrene

of the feet and slow healing of wounds or union of fractures.

Will it work? Many industries have perfected their own special plans for selective placement, based on an adaptation of the principles suggested here. The United States Civil Service Commission has made great progress and had outstanding success in matching disabled workers to suitable jobs through the use of a similar method. (See form reproduced in Fig. 21–9. Also see the state forms reproduced in Fig. 21–10.)

Details of the technique for using the form shown in Fig. 21–9 are obtainable from the U.S. Civil Service Commission, Washington, D.C. It is recommended as well worth study by those who wish to develop improved methods of employee placement. Other volumes suggesting different methods which warrant review and study are shown in the references. Particularly recommended is the work by Bert Hanman, *Physical Abilities to Fit the Job,* published by American Mutual Liability Insurance Company, Boston.

References

American Association of Industrial Nurses, Inc., 79 Madison Ave., New York, N.Y. 10016. *Standards and Criteria for Evaluating an Occupational Nursing Service.* 1965.

American College of Surgeons, Committee on Trauma. *Emergency Care of the Sick and Injured.* Available from W. B. Saunders Company, West Washington Square, Philadelphia, Pa. 19105.

American Conference of Governmental Industrial Hygienists, 1014 Broadway, Cincinnati, Ohio. *Guide to Health Records for Health Services in Small Industries.*

American Medical Association, Dept. of Occupational Medicine, 535 North Dearborn St., Chicago, Ill. 60610.
Guide to Developing an Industrial Disaster Medical Service.
Guide to the Development of an Industrial Medical Records System.
Guide to Small Plant Occupational Health Programs.
Guiding Principles of Medical Examinations in Industry.
The Legal Scope of Industrial Nursing Practice.
Occupational Health Services for Women Employees.
Scope, Objectives, and Functions of Occupational Health Programs.

American Mutual Liability Insurance Company, Engineering Department, Boston, Mass. *Physical Abilities to Fit the Job.* 1956.

American National Red Cross, 17th and E Sts. NW., Washington, D.C. 20006.
"Advanced First Aid and Emergency Care."
"Standard First Aid and Personal Safety."

American Nurses' Association, Inc., 10 Columbus Circle, New York, N.Y. 10019. *Functions, Standards, and Qualifications for Occupational Health Nurses. Guide for the Development of a Manual for an Employee Health Program.*

Bridges, Clark D. *Organization and Operation of an Occupational Health Program.* Chicago, Industrial Medical Association.

Hess, Gaylord R. *Medical Service in Industry and Workmen's Compensation Laws.* Chicago, American College of Surgeons.

Medic Alert Foundation, Turlock, Calif. 95380. "Why Medic Alert?"

Metropolitan Life Insurance Company, 1 Madison Ave., New York, N.Y. 10010. *Correlated Activities in an Employee Health Program, A Guide for Management, the Physician, and the Nurse.*

North American Association of Alcoholism Programs, 1611 Devenshire Dr., Columbia, S.C. 29204. General information.

O'Connor, Robert B. "The Preplacement Physical Examination Program," and "Periodic Examinations in Industry." *Health at Work,* Boston, Mass., Libery Mutual Insurance Company, 175 Berkeley St.

Occupational Health Institute, Inc., 150 N. Wacker Dr., Chicago, Ill. 60606. *Minimal Standards for Accreditation of Industrial Medical Services. Organization and Operation of an Occupational Health Program.*

Olishifski, Julian B. "Guidelines for Alcohol and Drug Abuse Programs," NSNews Reprint No. 111.17–93. Chicago, Ill., National Safety Council.

Pennsylvania Department of Health, Division of Occupational Health, P.O. Box 90, Harrisburg, Pa. *Guide for First Aid Services. Small Plant Health and First Aid Services.*

President's Committee on Employment of the Handicapped, Dept. of Labor Building, Washington, D.C. 20210. "Reports on Employment of the Handicapped."

Sutter, Richard A. "A Modern Medical Program for the Small Plant: Its Contribution to Society." *Transactions of the National Safety Congress, 1963.*

U.S. Department of Health, Education, and Welfare, Washington, D.C. 20201. *International Classification of Diseases, Adapted for Use in the United States.*

U.S. Department of the Interior, Bureau of Mines, Washington, D.C. 20240. *A Manual of First Aid Instructions.* Washington, D.C., Government Printing Office.

U.S. Public Health Service, Division of Industrial Hygiene, Washington, D.C. *Outline of an Industrial Hygiene Program.* Washington, D.C. 20402. Government Printing Office, Supplement No. 171.

Young, Carl B. *First Aid and Resuscitation—Emergency Procedures for Rescue Squads, Ambulance Crews, Interns, and Industrial Nurses.* Springfield, Ill., Charles C Thomas.

Zurich-American Insurance Companies, 135 South LaSalle St., Chicago, Ill. 60603. *Ability, Disability, Employability.*

Non-Employee Accident Prevention

Chapter 22

The Legal Side of Non-Occupational Injuries   572

Hazard Recognition   573
Glazing . . . Parking lots . . . Walking surfaces . . . Displays
Food preparation . . . Handicapped patrons . . . Fire losses

Measuring Non-Employee Losses   575

Identifying the Hazard   576

Reducing or Eliminating the Hazards   579
Slips and falls . . . Types of floor surfaces . . . Good housekeeping
prevents falls . . . Floor coverings and mats

Protection Against Fire, Explosion, and Smoke   582
Fire detection . . . Response methods . . . High-rise building fire
and evacuation controls

Contamination   585

Protect Attractive Nuisances   585

Transportation   586

Escalators   586
Escalator standards and regulations

Product Safety   588
Major provisions of the Act . . . Claims, defenses, and precautions

References   590

Fig. 22–1.—Well marked parking stalls, entrances, and exits encourage safe parking and traffic flow.
Courtesy 3M Company.

So-called "Murphy's Law" states, "If something *can* go wrong, it will." Actually something is more likely to go wrong to a stranger (customer, visitor, deliveryman, contractor, or the like) on business premises than it is to someone who is more familiar with the area.

The non-employee accident exposure affects not only non-employees, but also the employees and the quality of the product or service they sell. Operations stop (or at least slow down) whenever there is a bad accident, no matter who had it. With the increasing legislation concerning consumer product safety (discussed later in this chapter) and greater cost of claims, more attention must be given to this area of business loss.

In today's society, there are increasingly more businesses and people involved in serving and selling to the public than there are those engaged in making the items that are sold. Just look at the types of businesses—specialty shops, department stores, shopping center complexes, restaurants and fast-food service operations, hotels and motels, automotive service and dealerships, and hardware building supply stores. Amusement parks and banks also serve the public. Patrons, guests, or visitors are a source of non-employee injuries.

Most of these operations, unfortunately, cannot provide complete and constant surveillance. Neither can they control much of the public's personal activity or habits (such as smoking). In some areas, such as automotive service stations and lumber yards, smoking and open flames must be prohibited and cus-

571

tomers should be watched for violations. The inebriated guest or patron of a hotel or motel can be a threat when he smokes in bed or his instability contributes to slips, falls, or other problems. Often, such individuals do not heed warning signs.

Loss control plans should recognize the possibility of non-employee mishaps. These unplanned events can be minor or catastrophic; businesses that invite customers on their premises must provide extensive plans to protect non-employees. A firm can be liable for damage or for injuries from the minute someone enters the property (including the parking lot), especially if these result from any hazardous condition that could have been eliminated. Also, a firm can be liable for the actions of its employees that result in damage or personal injury.

The Legal Side of Non-Occupational Injuries

Customer or product claim cases often wind up in a courtroom. The legal terminology and aspects of the law that deal with accident claims, in situations in which one is likely to become involved, include the following definitions and explanations. They are not intended to be all-inclusive but to provide a quick review of some of the principal considerations.

TORT. A private or civil wrong or injury—a violation of a right not arising out of a contract. It may be either (a) a direct invasion of some legal right of the individual, (b) the infraction of some public duty by which special damage accrues to the individual, or (c) the violation of some private obligation by which like damage accrues to the individual.

Torts deal with negligence, accidents, trespass, assault, battery, seduction, deceit, conspiracy, malicious persecution, and many others.

NEGLIGENCE. Failure to exercise that degree of care which an ordinarily careful and prudent person ("reasonable man") would exercise under similar circumstances. To establish a proper claim of negligence, however, there must be (a) a legal duty to use care, (b) a breach of that duty, and (c) injury or damage.

DEGREE OF CARE. The degree of attention, caution, concern, diligence, discretion, prudence, or watchfulness depends upon the circumstances. For example, a high degree of care is demanded from people who invite others onto their premises, by formal, verbal, or implied invitation. All sales and services enterprises must exercise a high degree of care for the safety of their patrons. As long as a business is open, it assumes a responsibility to its clientele.

INVITEE. One whose presence on the premises is upon the invitation of another, such as a patron at a sports stadium or a person who visits an exhibition hall even though no admission is charged.

LICENSEES. Licensees are neither "invitees" nor "trespassers." They have not been invited to enter upon the property but they have a reasonable excuse (by permission or by operation of the law) for being there. These could be vendors, deliverymen, people visiting executives for business purposes, and the like. Note that policemen and firemen who enter property in the course of their duties have sometimes been held by the courts to be invitees (patrons) and sometimes licensees (non-patrons).

CONTRIBUTORY NEGLIGENCE. Not every injury gives rise to a claim for damages. If an insurance company can prove that the plaintiff (claimant) contributed to the injury by not exercising ordinary care, the resultant damages may be considerably reduced or negated. If it can be proven that the plaintiff was even only slightly negligent, he might not be allowed to collect damages.

This points up the importance of a proper accident investigation procedure; this can strengthen a company's ability to reduce the overall cost of doing business.

ASSUMPTION OF RISK. The claimant cannot collect damages when the law presumes he was aware of peril or danger, yet was willing to proceed with his original intention and undertake his action. "That to which a person assents is not regarded by law as an injury." For example, a skier who falls while descending a slope is said to assume the normal risks that can happen when participating in this sport. However, an injury involving a chair lift or tow rope could be of a mechanical nature and could be costly.

HOLD-HARMLESS. A clause in a contract agreeing for one party to assume all liabilities or losses involved, thus holding the other party not liable. For example, a department store may have a hold-harmless agreement with a manufacturer who supplies it with a particular type of merchandise. Should a claim for injury arise out of the use of that product by a consumer, the manufacturer and not the store would be liable. Even though the consumer purchased the item from the department store, the hold-harmless agreement may relieve the store of direct liability.

ATTRACTIVE NUISANCE. Liability growing out of a dangerous condition, generally to children. It excuses trespassing and penalizes for failure to keep children away or for failure to protect or eliminate a hazard that may reasonably be expected to attract them to premises.

BURDEN OF PROOF. The injured party must prove his injury or damage and its relation to the accident. The defense, on the other hand, must prove it is without fault. Proof must be established by facts, not opinion, suspicion, rumor, hearsay, gossip, or emotional reaction. Proof is the conclusion drawn from the evidence.

Honest and sincere witnesses convey different impressions from the same evidence attested to by dishonest witnesses. Thus we see how important promptness is when assembling and preserving the evidence. Signed statements taken shortly after the accident or an all-important photograph can often make the big difference. In public liability claims, the onus of proof rests upon the plaintiff (claimant).

Furthermore, those responsible for property may be sued for alleged injuries that may not have occurred on the premises. Investigation to prove innocence can be expensive and time consuming. Supervisors and employees should be forewarned and trained to report any occurrence.

This chapter cannot begin to cover all possible non-employee involvements. Instead it will describe prominent ways that non-employees can get hurt. Accident prevention techniques will necessarily be broad and varied. It will be up to each safety professional, regardless of title or business, to examine his own operations in light of these guidelines.

Hazard Recognition

Past accident records are a jumping-off place for possible areas of concern and action. However these are not always complete. They cannot reflect day-to-day conditions that include the myriad of accident potentials that exist but are innocent looking and previously accident-free.

Some types of hazards have been isolated and have received a goodly share of publicity. Here are a few that may trigger exploration into other and similar areas of concern.

Glazing

Modern design of office buildings, stores, and manufacturing operations often includes the extensive use of glass in doors, show windows, panels, enclosures, to mention a few. Often such areas provide a confusing pattern to the stranger. Most likely the non-employee with whom we are concerned will be a first-time visitor to the property, and therefore unfamiliar with "hard-to-see" glass doors.

Unmarked glass panels and doors invite severe injuries and cuts. Be sure to follow recommended practices and standards for glazing, strength of glass, and marked identification of doors and panels. Marked identification is required by law in some states.

Parking lots

Private parking facilities involve hazards of collision and falls and when drivers become pedestrians this added problem exists. Driving skill, condition of pavement, limited sight distance and visibility, and preoccupied drivers looking for a parking space can pose a real threat to safety. When adverse weather strikes, with rain, snow, and ice, such problems are multiplied. Snow removal and deicing must be provided to reduce these hazards as much as possible.

Traffic control, parking lanes, lighting, pavement maintenance, directional signs and law enforcement problems must also be considered. (See Fig. 22–1.)

Fig. 22–2.—Cotton and other fabric mats used in entrances are useful for absorbing moisture and dust, but require constant watching to minimize tripping hazards.

Above: The first person's toe or heel will catch in the mat.

Right: The next person will either trip on the raised portion or dislodge it further, building up the hazard. The remedy is to lay the mat flat and fasten it securely.

Walking surfaces

Slips and falls are perhaps the most likely source of non-employee injuries. These injuries can occur almost everywhere at any time. There are few surfaces that can be ignored and the dangerous ones include everything from asphalt roads, concrete walks, wooden, tiled or rug covered floors and special surfaces on stairs, conveyances (treadmills, escalators, elevators) to bridges and catwalks.

The natural properties of any surface can be changed substantially when people track in mud, snow and moisture on their shoes, boots and galoshes and such surface can be rendered extremely slippery. Mats, runners, or rugs designed to reduce such hazards and

floor maintenance themselves require special attention to eliminate the hazard of torn or curled-up floor coverings. (See Fig. 22–2.)

Displays

Food and merchandise displays must be constructed in such a manner as to prevent the dislodging of breakables and other articles that could strike and injure the customer. Sharp or broken edges of displays and counters should be made less hazardous.

Food preparation

Restaurants, hotels, motels, hospitals, rest homes, and the like must constantly monitor food preparation and service to prevent serving spoiled food or food contaminated by

TABLE 22-A

ANNUAL FIRE LOSSES

Types of Building	1971		1972		Percent Increase
	Number	Losses*	Number	Losses*	
Industrial plants	41,300	390,000	48,000	470,600	14%
Stores and offices	71,000	332,200	76,900	337,700	8½
Schools and colleges	20,500	87,000	22,400	90,900	9
Churches	3,400	22,300	4,300	28,100	26
Hospitals, nursing homes, et cetera	18,200	22,400	21,200	24,800	16
Hotel/Motel	15,200	37,900	16,400	43,600	8
Restaurants	18,200	50,900	21,700	54,300	20

* Losses in thousands of dollars.

Source: National Fire Protection Association.

broken glassware, insecticides, and other harmful materials.

Handicapped patrons

The handicapped patron demands special precautions. Special notice and care must be used in handling them and providing special services. Such persons pose problems in the event of an emergency through early warning and special evacuation procedures, but special ramps, rails or other devices may be necessary for routine service to such patrons.

Fire losses

A constant threat to every business, in which both employees and non-employees can be hurt, is fire or the residue of fire such as smoke, toxic gases, and heat. Buildings containing heavy concentrations of people are especially vulnerable to catastrophic fires. Hotels, department stores, and high-rise buildings have had severe loss of life when fire breaks out. Cooking and heating systems, defective and misused electrical equipment and wiring, and careless use of smoking materials are listed as the three leading causes of fires. Combustible contents and building materials must also be considered.

According to the National Fire Protection

Association, fires and fire losses are on the increase. Table 22-A indicates the comparison between 1971 and 1972.

Measuring Non-Employee Losses

In order to provide an orderly and uniform method of comparing patron losses, the ANSI Standard Z108.1, *Method of Measuring and Recording Patron and Non-Employee Injury Statistics,* was established. As explained in Chapter 6, "Accident Records and Injury Rates," this standard provides a variety of plans on which the frequency of non-employee injuries can be measured. A variety of exposure methods is suggested due to the very difficult accumulation of exposure hours within sales and service operations. Non-employee exposure time is not "clocked" in public places such as restaurants, stores and offices and averages have little in common—even within single businesses.

However, within the system suggested, a degree of comparative information can be found and it can be determined whether or not a particular segment of the business is extremely high. Sales volume or number of transactions, for example, can be used as a common denominator to measure frequency.

FIG. 22–3.—Permanent aisles and passageways should be appropriately marked. Stock must be kept out of aisles and fire lanes.

Courtesy Macy's, Jamaica.

Identifying the Hazard

Some accident prevention measures may be demanding of time, extensive in execution, and costly. Some problems even require surveillance with sophisticated electronic monitoring equipment such as closed circuit television. Some operations have recorded computerized print-outs of entry into restricted areas. These report and pinpoint danger areas of unauthorized entry in which the party could be injured by virtue of contact with high voltage, construction in progress, and areas of restricted passage.

When the hazard exists because of poor illumination, poor identification, or bad visibility or possible camouflage, highlight the hazard by placing prominent designs at eye level on glass doors and panels or by highlighting areas of poor contrast. Doors should be provided with handle and kick plate protection panels. Glass panels should be highlighted for best possible visibility. Make sure that there are no obstacles in the reception areas such as furniture or planters too close to the normal traffic pattern, telephone cords or

lamp cords that could be tripped over, and sharp objects on or near the reception desk. Avoid the use of glass (and remove any broken glass) to cover the top of reception desks, tables, et cetera.

Displays, counters, and exhibit areas offer their own brand of hazard. For example, unprotected display hooks at eye level or below can be extremely dangerous. Make sure that display cases, racks, and cabinets are in good repair and do not have projections. Make sure all stock is safely contained and safely displayed (such as not being stacked too high).

Demonstrations of do-it-yourself shop or kitchen equipment require careful attention. Electrical and mechanical equipment should be safeguarded to protect employees and customers.

Buildings, furniture, and fixtures can be sources of injury. They should be inspected regularly to find hazards that may have developed since the previous inspection. Aisles, pedestrian lanes, exits, and stairs should be unobstructed and free of tripping and stum-

FIG. 22–4.—An effective warning sign like this helps prevent public and employee falls due to spillage.

Fig. 22–5.—Slipometers, ranging from the motorized type (shown here) to a simple spring scale and heavy block pulled by hand, can be used to gage the slipperiness of floors.

Courtesy Liberty Mutual Insurance Company.

TABLE 22-B

PHYSICAL PROPERTIES OF FLOOR FINISHES

Types of Finish	Resistance to			Quality of			
	Abrasion	*Impact*	*Indentation*	*Slipperiness*	*Warmth*	*Quietness*	*Ease of Cleaning*
Portland cement concrete *in situ*	VG - P	G - P	VG	G - F	P	P	F
Portland cement concrete precast	VG - G	G - F	VG	G - F	P	P	F
High-alumina cement concrete *in situ*	VG - P	G - P	VG	G - F	P	P	F
Magnesite	G - F	G - F	G	F	F	F	G
Latex-cement	G - F	G - F	F	G	F	F	G - F
Resin emulsion cement	G - F	G - F	F	G	F	F	G - F
Bitumen emulsion cement	G - F	G - F	F - P	G	F	F	F
Pitch mastic	G - F	G - F	F - P	G - F	F	F	G
Wood block (hardwood)	VG - F	VG - F	F - P	G - F	F	F	G
Mastic asphalt	VG - F	VG - F	VG - F	VG	G	G	G - F
Wood block (softwood)	F - P	F - P	F	VG	G	G	G - P
Metal tiles	VG	VG	VG	F	P	P	G - F
Clay tiles and bricks	VG - G	VG - F	VG	G - F	P	P	VG
Epoxy resin compositions	VG	VG	VG	VG	F	F	VG

Code: VG—Very Good; G—Good; F—Fair; P—Poor; VP—Very Poor.

577

TABLE 22-C

GUIDE TO FLOOR MATERIALS AND SURFACINGS

Floor Types*	Characteristics	Use of Abrasives	Dressing Materials
Asphalt tile	Composed of blended asphaltic and/or resinous thermoplastic binders, asbestos fibers and/or other inert filler materials, and pigments.	Abrasive materials of various types may be used to reduce slipperiness of floors. Colloidal silica** can be incorporated in wax and synthetic resin floor coatings.	Wax or wax-base products—For most purposes, wax has several advantages. This is especially true of Carnauba wax, an ingredient generally used in so-called wax products. This wax, a Brazilian palm tree product, dries in place with a very hard and glossy finish, but with a characteristically slippery surface. Because of its many good qualities, it is widely used as a base for floor surface preparations, both in paste and emulsion forms. Other waxes, notably petroleum wax and beeswax, have their place in floor dressing formulas; they are softer and less slippery than Carnauba, but are still slippery to a degree depending on the formulation.
Linoleum	Cork dust, wood flour, or both, held together by binders consisting of linseed oil or resins and gum. Pigments are added for color.		
Rubber	Vulcanized, natural, synthetic, or combination rubber compound cured to a sufficient density to prevent creeping under heavy foot traffic.	Slip-resistant except when wet.	
Vinyls	Composed of inert, nonflammable, nontoxic resins compounded with other filler and stabilizing ingredients.	Adhesive fabric with ingrained abrasives can be used. They are patterned in strips, tiles, and cleats.	
Terrazzo	Consists of marble or granite chips mixed with a cement matrix.	Silicon carbide or aluminum oxide can be included in mix when floor is laid. Also an abrasive-reinforced plastic coating can be painted on.	
Concrete	Made of portland cement mixed with sand, gravel, and water and then poured.		
Mastic	Like asphalt tile in composition but is heated on the job and troweled onto the floor to form a seamless flooring. Such floors are often used over concrete to give a new durable, resilient surface.	(Same as asphalt tile)	Synthetic resins—These preparations, known as "synthetics," "resins," "synthetic resins," or "polishes," are intended to supply the desirable characteristics of wax without producing the same degree of surface slipperiness. They include soaps, oils, resins, gums, and other ingredients, compounded in the various ways to produce the desired result.
Wood	May be either soft or hard, in a variety of thicknesses and designs.	Metallic particles and artificial abrasives in varnish or paint give good nonslip qualities to various floors.	

Other materials	**Other materials**—Paint products (paint, enamel, shellac, varnish, plastic) are semipermanent finishes used principally on wood and concrete floors. They do not materially increase the slipperiness of the base.
Cork tile	Made of molded and compressed ground cork bark with natural resins of the cork to bind the mass together when heat cured under pressure. (Same as asphalt tile)
Steel	Iron containing carbon in any amount up to about 1.7 percent as an alloying constituent, and malleable when used under suitable conditions. Surface can be touched up with an arc welding electrode so the shape of raised places on the surface resembles angle worms. Also an abrasive-reinforced plastic coating can be painted on to any desired thickness, dries hard as cement and has a sandpaper like finish. If a temporary non-skid surface is needed two uses of mats can be employed: (a) flexible rubber mats made from old automobile tires; (b) rubber or vinyl runners.

* Floors and stairways should be designed to have slip-resistant surfaces insofar as possible; adhesive carborundum strips may be used on stair treads or ramps and at critical concrete areas. Etching with mild hydrochloric (muriatic) acid solution will lessen slip problems.

** Colloidal silica is an opalescent, acqueous solution containing 30 percent amorphous silicon dioxide and a small amount of alkali as a stabilizer.

---

bling hazards. Spilled materials must be cleaned up promptly. (See Fig. 22–4.)

Reducing or Eliminating the Hazards

Slips and falls

Floors, stairs, and other walking surfaces are hazardous enough when they are clean, clear, and free of defects—they are a major source of slips and falls. Since the footwear worn by non-employees cannot be controlled, the safety professional must make sure that walking surfaces offer no additional problems. Walking surfaces should be non-slippery, clear and in good repair, and adequately illuminated.

Slipmeters developed by testing agencies and insurance companies measure the slipperiness of floors. One type of instrument (see Fig. 22–5) is mounted on three leather "feet." It is pulled across the floor by a motorized winch; the dial on top measures the intensity of the pull required to start moving it. This is converted into a "slip index." Floor slipperiness, however, may increase because of moisture, oil, grease, foreign or waste materials, and incorrect cleaning or waxing.

Falls on floors occur in various ways and from various causes. A person may slip and thus lose traction, or he may trip over an open drawer, box in the aisle, or other object. In either case, he may suffer an injurious fall.

In nearly every floor fall accident claim, this condition is alleged to have existed. However, of the actual unsafe conditions the primary mechanical causes of falls on floors are, in reality, unobserved, misplaced, or poorly designed movable equipment, fixtures, or displays; poor housekeeping; and defective equipment. Obviously, the condition of shoes or type of footwear soles and heels might be major contributing factors.

Inadequate illumination can also be a cause of falls. Light values at floor level should be uniform with no glare or shadows. Also, there should be no violent contrasts in light levels between floor areas, such as from bright sunlight outside the entrance to a dimly lit lounge or restaurant.

Undoubtedly, a large percentage of falls stem from unsafe acts and from purely personal causes such as age, illness, emotional disturbances, fatigue, lack of familiarity with the environment, and poor vision, which cannot be readily identified or controlled. It thus becomes doubly important to eliminate unsafe conditions and unsafe employee practices to which the blame for an accident can be shifted or which might contribute to the personal cause. As an example, mirrors and other distracting decorations should not be placed in areas visible from steps or from approaches to steps or escalators.

Types of floor surfaces

A wide variety of floor surfaces are available. In office buildings, hotels, mercantile, and similar establishments, it is common to find masonry (terrazzo, cement, or quarry tile) floors at entrances, in lobbies, on stairways, and sometimes extensively throughout the ground floor and in upper floor corridors. Decorative materials such as terrazzo, marble, and ceramic tile are most often used for interiors while concrete and granite are generally considered more practical for exterior use. Details are in Tables 22-B and -C.

In other public areas in these buildings, the base floor, usually of concrete or wood, is generally surfaced with one or more of the popular resilient floor covering materials. Carpeting is commonly used on limited areas in the department, furniture, specialty, and similar stores and in hotels. Elsewhere, asphalt, linoleum, rubber, or plastic in either sheet or tile form, will usually be found. Obviously, safety, initial cost, durability and maintenance costs are some factors which govern the choice of floor covering.

Most flooring materials, whether wood, masonry, or the resilient types, are reasonably slip resistant in their original untreated condition. Exceptions will be found among some of the masonry materials. A highly polished marble, terrazzo, or ceramic tile may be used to achieve an ornamental effect. These highly polished surfaces can be slippery when dry. Their slipperiness will be greatly increased by moisture, by improper surface treating preparations, and by improper cleaning materials and methods. Unless a non-slip material is added to the aggregate during construction, the only preparation which should be used on such floors is a penetrating sealer of the slip-resistant type.

Good housekeeping prevents falls

In food retail operations, stores, and mercantile establishments, it is estimated that fixtures, displays, and other portable equipment are involved in over 40 percent of customer falls. Therefore, it is essential that management provide safe equipment and that the accident control program place particular emphasis on safe placement and use of that equipment.

Dress racks and stock trucks should be removed from the sales area and returned to the stock room as soon as they have been emptied.

All electrical wiring and extension cords for store machines, displays, special decorations, and the like should be designed so as not to lie on the floor. Where necessary, wires or cords may be installed in low-profile channels.

A high percentage of falls is caused by low-profile stock left in aisles. Stock should be kept in large piles while employees are filling shelves or stock drawers, or dressing a display.

Stock containers, such as baskets, boxes, bags, trays, and cartons, must be removed from the aisles immediately after they have been emptied.

Poor housekeeping accounts for one-third or more of all customer falls. Each employee should be made to realize that it is part of his or her responsibility to maintain good housekeeping in his work area, to report promptly unsafe floor conditions, such as tears in carpets and holes in the floors. He should wipe up spills immediately, or else barricade them until the hazard can be removed. A special warning sign can be used (see Fig. 22–4).

Floor coverings and mats

Reduce the possibility of slips and falls by using good carpeting, bound edging, and flush floor-level mats and runners (see Fig. 22–6). If material such as an extruded metal runner is used to provide self-cleaning removal of snow, ice, or mud at entryways, it should be flush and not present a tripping hazard. Care also in the use of rubber mats, rug runners, and the like must be taken to prevent them from becoming tripping hazards. Oftentimes the edges become rumpled (Fig. 22–2), corners and ends are torn or do not lie flat and excess wear can cause tears. Mats and

FIG. 22–6a.—Recessed coconut-fiber mat reduces slipping, keeps inside of building clean. Mat is kept in place year-round and is vacuumed about twice a week. Each year it is taken up and thoroughly cleaned.

Courtesy Kingsport Press, Inc.

FIG. 22–6b.—A heavyweight rubber or plastic mat with a nubby finish or raised design and beveled edges tends to lie flat and in place. Rotating the mat distributes the wear and minimizes "bald" spots in locations such as this, where persons step off elevators.

runners should be replaced at the first sign of such unsafe condition.

Floor mats, runners and carpeting are often used in places other than building and store entrances such as around swimming pools, in shower stalls, around drinking fountains and vending machines, on boat decks, and in garages, factories, and other areas where water, oil, food, waste, and other material on the floor might make it slippery.

Definite procedures should be set up for the placing, cleaning, removing, and storing mats. Those who put mats in place during inclement weather should have clear-cut instructions as to where and when mats should be put down and removed. Failure to get the mats down promptly and close enough to the door may result in slippery entrance-ways and in water and dirt being carried beyond the entrance-way to create a hazard and maintenance problem in another location.

Definite procedures for inspecting and checking the condition of mats and for maintaining them in safe condition should be followed.

Stair rails, treads, and surfaces should be in good repair and checked frequently for defects. Nothing should be stored on the stairways and landings that can contribute to falls. See also pages 353–354 and 390–392.

Protection Against Fire, Explosion, and Smoke

If fire breaks out or an explosion occurs, the quick and orderly evacuation of the premises will protect all persons including visitors. All evacuation plans should be based on the premise that visitors will be on the property for the first time. Such persons should be protected by ample and special direction signs, even though location of exits might be a well established and known fact to regular employees.

Panic is one of the most serious situations that can grip people; through panic, they may take irrational means for self-preservation. Often panic leads to injury of many persons who might otherwise have been saved. An evacuation plan, well rehearsed with supervisors and employees, is needed for every business. (A good discussion of panic and techniques for handling it is given in NSC publication *Supervisors Guide to Human Relations.*)

Businesses that generally attract large numbers of customers, guests, or patrons should provide well marked exits, emergency lighting (see Fig. 22–7), and ample direction signs inscribed to direct people to safe exits.

Dense, penetrating smoke can be as deadly as the heat or flame of a fire. Lungs can be seared quickly.

Enclosed stair wells provide the best fire escape routes. Doors to such stair wells must never be obstructed, locked, or propped open.

To help evacuate a building, a public address system, manned by a qualified and trained person, can be used to direct the evacuation and issue life-saving instructions.

Usually the early detection of a fire and the use of a good evacuation plan can prevent panic and personal injuries. More details are in Chapter 18, "Planning for Emergencies."

Fire detection

Properly engineered fire detection systems are sound investments. But the best installation is useless if there is no response to the alarm. Systems should have a direct connection to the local fire department or to some alarm center.

The four stages of fire. Fire is a chemical combustion process created by the rapid combination of fuel, oxygen, and heat. A full discussion is found in Chapter 43, "Fire Protection."

Most fires develop in four distinct stages and detectors are available for each.

• INCIPIENT STAGE. No visible smoke, flame, or significant heat is developed, but a significant amount of combustion particles are generated over a period of time. These particles, created by chemical decomposition, have weight and mass, but are too small to be visible to the human eye. They behave according to gas laws and quickly rise to the ceiling. Ionization detectors respond to these particles.

• SMOLDERING STAGE. As the incipient condition continues, the quantity of combustion particles increases to the point where they become visible—this is called "smoke." There is still no flame or significant heat developed. Photoelectric detectors "see" visible smoke.

Fig. 22–7.—This unit is positioned at a stair well to afford proper illumination during a power failure.

- FLAME STAGE. As the fire condition develops further, the point of ignition occurs and flames start. The level of visible smoke decreases and the heat level increases. Infrared energy is given off; this can be picked up by infrared detectors.

- HEAT STAGE. At this point, large amounts of heat, flame, smoke, and toxic gases are produced. This stage develops very quickly, usually in seconds. Thermal detectors respond to heat energy.

Engineering and control procedures. The best fire detection system is only as good as its weakest component. The services of a fire protection consultant should be obtained in engineering the system and establishing the control procedures. Here are four steps to consider.

1. Select the proper type detector(s) for the hazard areas. For example, a computer area may involve ionization or combination detectors. A warehouse may have infrared and ionization detectors. In low-risk areas, thermal detectors or combinations of detectors may be used.

2. Determine the spacing and locations of detectors to provide the earliest possible warning.

3. Select the best control system arrangement to provide fast identification of the exact source of alarm initiation.

4. Assure notification of responsible authorities who can immediately respond to the alarm and can take appropriate action. Every detection system must have an alarm signal transmitted to a point of constant supervision. If this cannot be assured at the premises, the signal must be transmitted to a central station, fire department, or other reporting source to assure prompt response.

Response methods

Early warning systems are as important during hours of occupancy as they are when the premises are vacant.

When detection pinpoints a trouble spot, immediate response by a responsible trained company representative is all important. Shaving seconds and minutes can mean the difference between lives lost or saved, and fire confined or allowed to spread out of control.

Here are some systems that are used.

Twenty-four hour supervisory service. If the installation has 24-hours-a-day, seven-days-a-week supervision at some point in the building, or complex of buildings, then it is recommended that alarm, trouble, and zone signals be indicated at this location.

Less than 24 hour supervisory service. For periods when the installation does not have responsible personnel to respond to the alarm, then a backup annunciation system should be provided. The National Fire Protection Association recommends that this be done with a connection to a central station supervisor's service or other service.

Such systems should include the means of initiating fire and trouble systems to the central station transmitter equipment. In the case of a trouble signal, the station can immediately dispatch a representative to investigate the trouble and notify proper representatives of the property under surveillance.

If central station tie-in is not available. In areas where no central station supervision is available, the local fire or police department may accept a remote fire alarm panel to be installed at their headquarters or firehouse.

If central station or telephone-leased line tie-in is not available. If none of the foregoing possibilities is available, then consideration should be given to the use of qualified and licensed telephone answering services. Automatic dialing units connected to responsible officials of the property is another alternative.

Burning plastics. It should be noted at this point that some fuels such as plastic waste receptacles can produce a great deal of toxic smoke when they burn. Therefore non-toxic and non-combustible materials (such as metal cans) might be used. For example, polyvinyl chloride (PVC) in a single foot of one-inch size PVC rigid non-metallic conduit involved in a fire:

1. Can produce a sufficiently heavy, dense smoke to obscure 3500 cu ft of space, and

2. Can generate enough hydrogen chloride to provide a lethal concentration of HCl in approximately 1650 cu ft of space.

High-rise building fire and evacuation controls

Just what is a high-rise building? The General Services Administration, at a conference in Warrenton, Virginia, in April of 1971, established four basic criteria to designate a high-rise. First, the size of the building made personnel evacuation impossible or not practical. Second, part or most of the building was beyond the reach of fire department aerial equipment. Third, any fire within the building must be attacked from within because of building height. Fourth, the building had the potential for "stack effect."

The huge high-rise megastructures now building or on the drafting boards all have one thing in common—they are intended to house people. As buildings go higher, the population density per square foot of ground areas increases, which poses a whole new set of problems concerning the health, safety, and welfare of their occupants. Actually, each building is a sealed life-support system. Present engineering approaches facilitate heating and cooling but are extremely wasteful of energy. They are more airtight than ever, thus there is increasing danger from smoke and toxic vapors.

All businesses to some extent must be concerned about panic and crowd control during those times when an emergency and evacuation takes place. The threat of unusual occurrence (such as riot, bomb threat, and the like) cannot be overlooked and plans must be made to handle such a possibility. High-rise buildings pose new and special problems but other public places such as theaters and other amusement and recreation facilities must also provide well planned emergency procedures. Panic and the press of frantic, hysterical people have caused wholesale destruction of life in many emergencies. Often times, such losses could have been prevented. Strict observance to building and fire codes can do much to eliminate physical hazards and "death traps" through improper design and installation of facilities.

Design and installation of the suppression system for life safety should be considered by every business. Architects should be held accountable for proper design for new construction, especially in the high-rise buildings.

Stack effect. Every building has its own peculiarities for creation of a "stack effect," among them being structure configuration, height, number and size of openings, wind velocities, temperature extremes, number and location of mail chute openings, and elevator shafts. All of these factors create varying air flows which tend to accelerate and intensify an interior fire. Unprotected air conditioning systems are an open invitation to catastrophe. If there is no automatic smoke and heat detection, no automatic fan shutdown, and no automatic fire dampers, smoke and toxic fumes would be quickly drawn into the exhaust or return air duct system and promptly distributed to all other floors and areas of the building served by the air conditioning system.

The National Fire Protection Association

recognizes this potential and in its Standard No. 90A, *Installation of Air Conditioning and Ventilating Systems,* states that in systems of over 15,000 cfm capacity, smoke detectors *shall* be installed in the return air steam prior to exhaust and *shall* be installed in the main supply duct downstream of the filters. These detectors shall automatically shut down fans and close smoke dampers to stop the recirculation of the smoke—or they may incorporate automatic exhaust.

Considering evacuation, it must be assumed that elderly people will be involved and cannot move promptly. In addition, some people are psychologically prone to panic in a fire situation. The quantity and size of staircases will undoubtedly prohibit complete evacuation. Tests conducted in Canada indicate that based on an occupant load of 240 persons per floor, total evacuation of an 11-story building can take up to 6½ minutes, while an 18-story building can take up to 7½ minutes. Exits are just not designed to handle all occupants simultaneously.

Elevators cannot be relied upon. Many elevators use capacitance-type call buttons which may bring them to a stop on the fire-involved floor, from which they will not move because smoke interrupts the light beam that keeps the doors open. Other possibilities include the inadvertent arrival of the elevator at the fire-involved floor by a passenger not knowing the fire exists and wishing to get off, and also the possibility of a person pushing the call button in the elevator lobby and then, in panic, using the staircase for exit. With problems of this magnitude, it can be assumed that evacuation is impossible.

Regarding most fire department aerial equipment, most ladders are limited to a height of approximately 85 feet. This means that a building higher than about 8 to 10 stories cannot be served by this equipment.

From a height standpoint and also due to increased floor areas, fires must be fought from within the building. This can be accomplished by automatic sprinklers, hose standpipes, and portable extinguishers, as well as hose lines from the building exterior.

Contamination

With emphasis given to environment control, many business operations can risk heavy losses and possible injury to non-employees through airborne pollutants or contaminants discharged into water or sewer systems.

Despite environmental laws and government controls together with scientific detection devices, some operations continue to use methods that pollute air, streams, and sewer systems with flammable or toxic materials. Cases on record reveal large areas of contamination when unauthorized chemical mixtures (such as chlorine) or accidental dangerous emissions spread out over a community and affect large numbers of people. Even the burning of rubbish is now prohibited in many areas and other means of disposal must be sought. See Chapter 20, "Industrial Sanitation and Personnel Facilities."

Adhere strictly to proper procedures and internal policing of any process that could contaminate. Engineering studies and analysis of doubtful procedures can reduce the possibility of or eliminate the hazard.

Protect Attractive Nuisances

The natural curiosity of human beings often brings grief to themselves and trouble to others. Building sites and excavations are "attractive nuisances"—see the definitions of legal terms earlier in this chapter. There are many opportunities for every business to be involved in losses caused by the public's curiosity.

Any unattended vehicle and machine that is left in an operable condition is attractive to the young (and the so-called "young in heart"). Unauthorized use of vehicles or machines should be discouraged by use of "tamper-proof" locks. Watchmen and security guards should be employed if necessary.

Frequently, partially finished road repairs or other construction can "booby trap" a vehicle. So, also, can unprotected and unbarricaded hazards, especially in adverse weather or in storms. American National Standard D6.1, *Manual on Uniform Traffic Control Devices for Streets and Highways,* should be referred to.

Many contractors or builders provide special, but safe, observation facilities for public "sidewalk superintendents." Local authorities and insurance engineers should be consulted

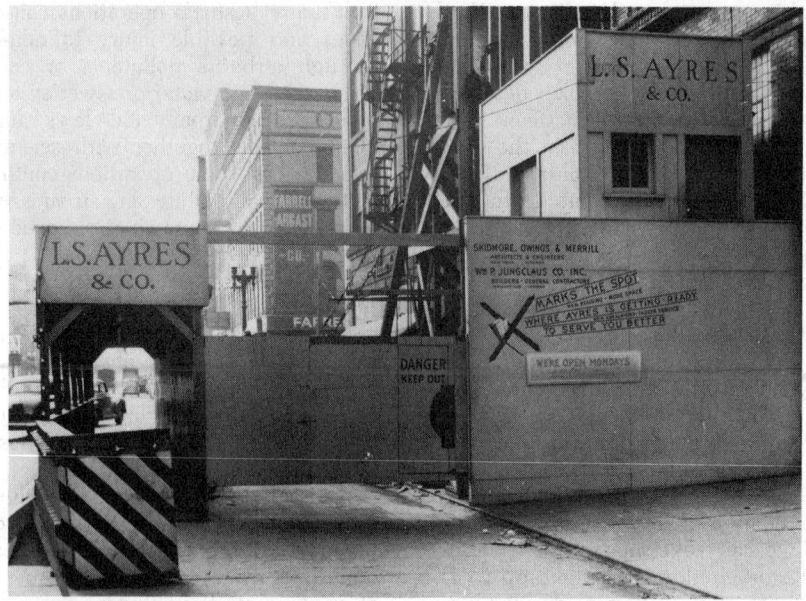

FIG. 22–8.—Place red warning lights on the street side of a temporary structure; and use diagonal striping on the end and interior lights.

for regulations and control measures. (See Fig. 22–8.)

Transportation

Some businesses by their very nature have special, and often prominent accident control problems. A majority of their loss control efforts must be directed at protecting the non-employee from harm.

All types of transportation—from commercial airlines, railroads, marine, and buses to local transit, taxi, and school bus operations—must be vitally concerned with prevention of injury through accidents. Not only is maintenance and operation of the vehicle or unit itself involved, but the problem also extends into areas around it, such as terminals, stations, school bus loading areas, and the like.

Courtesy cars. When transportation is provided by a business for customer courtesy and convenience (such as a hotel or motel courtesy car or bus), the same concern and precautions used in commercial operations should be considered. These include providing the

safest vehicle, maintained in proper working condition (meeting all local, state, and federal requirements), and operated by a professional who is trained and skilled in all elements of operating the vehicle. This also holds true for boats, airplanes, and other means of transportation.

Company-owned vehicles. Another source of damage and injury claims arises out of the operation of company motor vehicles by employees. Chances are that one out of every four drivers will have a collision in any given year. This probability makes the occurrence of such liability accidents a constant threat and often a very real dollar drain for insurance protection and claim settlement. The question of liability resulting from an employee's use of his own car on company business must also be considered.

Escalators

A major concern in most businesses is to prevent falls, bumping into objects, having objects strike a customer, and also protecting

FIG. 22–9.—Close-up of escalator entrance shows PLEASE HOLD HANDRAIL sign that is placed conspicuously on the balustrade.

him on stairs, elevators, escalators, and moving ramps.

Many trades and service organizations move people vertically by escalators, elevators, and ramps. Escalators can be operated from a low of 70 feet per minute (fpm) to a speed of 125 fpm. The average recommended speed is 90 fpm. Faster or slower operation often is a source of injury to the young or the elderly.

Escalator standards and regulations

When installing or modifying escalators, check local and state ordinances. Also refer to NSC Data Sheet No. 561, *Escalators*. The American National Standard A17.1, *Safety Code for Elevators, Dumbwaiters, Escalators, and Moving Walks,* is known as the Elevator Code. Some of its important points are:

● Hand and finger guards are to be protected at the point where the handrail enters the balustrade.

● Prominent signs recommending PLEASE

HOLD HANDRAIL (or equivalent message) should be displayed (see Fig. 22–9).

● Comb plates with broken teeth should be replaced immediately.

● Strollers, carts, and the like must be prohibited on escalators.

The Elevator Code also states that balustrades must be provided with handrails moving in the same direction and substantially at the same speed as the steps.

Daily inspection. Examine all escalators from landing to landing every day. This includes riding them before store opening to find out if any visual or sound defects exist.

Start and stop controls. Escalators should have an emergency STOP button or switch

FIG. 22–10.—Those who are authorized to stop escalators should know the location of STOP buttons and the safe procedures for handling emergencies. Note the close-fitting guards where the handrail enters the balustrade.

accessibly located at the top and the bottom landing. These must stop, but not start the escalator. Placement of a STOP button at the base of a newel (Fig. 22–10), with either recessed design or provision of a cover, will help protect the button from being activated unnecessarily. Selected employees should be trained in the location and use of emergency switches.

Product Safety

Today, attention is being given to the subject of "total safety engineering." To this end, federal legislation is becoming more and more encompassing. Liability now extends to non-employees beyond the boundaries of property lines. "Caveat emptor" (let the buyer beware) has been changed to "protect the buyer" by providing safe-operating products. The Consumer Product Safety Act has been enacted to reach out and protect almost everyone, everywhere, all the time.

Major provisions of the Act

Congress passed the Consumer Product Safety Act on October 12, 1972. Provisions of this act provide the authority to implement it in an independent five-member regulatory agency known as the Consumer Product Safety Commission.

Purpose. The fundamental purposes of the act are fourfold:

1. To protect the public against unreasonable risks of injury associated with consumer products;

2. To assist consumers in evaluating the comparative safety of consumer products;

3. To develop uniform safety standards for consumer products;

4. To promote research and investigation into the causes and prevention of product-related injuries and illnesses.

Consumer product. A consumer product is defined as any article or component part that is produced or distributed for sale to a consumer for use in or around a household or residence, a school, in recreation or otherwise, or for the personal use, consumption, or enjoyment of a consumer in the same places. The act specifically excludes from coverage tobacco products, motor vehicles, pesticides, aircraft, boats, food, drugs, medical devices or cosmetics, and firearms and ammunition.

The Consumer Product Safety Commission has, among other things, the authority to:

1. Maintain an injury information clearinghouse

2. Conduct studies and investigations resulting from accidents

3. Conduct studies on improving safety of products

4. Test consumer products

5. Develop product safety test methods and testing devices

6. Develop and promulgate product safety standards

7. Declare a product a "banned hazardous product"

8. Inspect areas such as factories and warehouses

9. Require manufacturers to establish and maintain certain records

10. Develop product defect notification procedures for manufacturers, distributors, and retailers

In addition, the commission is to administer the existing federal Hazardous Substances Act, the Poison Prevention Packaging Act, the Flammable Fabrics Act, and the Refrigerator Safety Act.

Notable in the Flammable Fabrics Act is the establishment of standards for carpets and rugs, children's sleepwear, and mattresses. Also considered among other items are mattresses.

Some of these standards are extremely important to people in the hotel and motel business since the fire rating of a building is the extent of its fire-sustaining load. By this is meant the quantity and quality of flammable items such as bed clothing, linens, rugs, mattresses, and the like.

The Product Safety Advisory Council. A Product Safety Advisory Council composed of

15 persons who are qualified by training and experience in fields applicable to product safety is part of the act.

Five Advisory Council members are to be selected from government agencies, five from consumer product industries, and five from consumer organizations, community organizations, and recognized consumer leaders. The Advisory Council may propose consumer product safety standards or ban orders to the commission.

Banned hazardous products. The commission can take steps to ban a product as a hazardous product if it finds that a specified consumer product presents an unreasonable risk of injury and that no feasible standard promulgated under the act will adequately protect the public from the unreasonable risk associated with the product. In a situation where an imminent hazard exists (a hazard that presents an imminent and unreasonable risk of death, serious illness, or severe personal injury), the commission has the authority to go directly to the U.S. District Court to seek relief.

Certification and labeling. Every manufacturer or private labeler of a product subject to a standard promulgated under the act is required to issue a certificate that such product conforms to all applicable safety standards and specify such standards. The certificate must accompany the product or be furnished to any distributor or retailer to whom the product is delivered. The certificate is to include the date and place of manufacture. The commission has the authority to prescribe reasonable testing programs under which such certificates may be issued and specify the form and content of product labels.

Inspection and recordkeeping. For the purposes of implementing the law, designated employees of the commission are authorized to enter an establishment and inspect those things that are related to the safety of the consumer product. Further, every manufacturer, private labeler or distributor of a consumer product is required to establish and maintain records, make reports, and provide such information as the commission may by rule require

and make such data available upon request of the commission.

Prohibited acts. Among other things, the new law makes it illegal to manufacture, sell, distribute, or import any consumer product that does not meet the requirements of applicable standards or any product that has been banned as hazardous. Private labelers and importers are also subject to such provisions. It is also illegal to fail to comply with the inspection and recordkeeping requirements.

The Consumer Product Safety Act and the Consumer Product Safety Commission will ultimately have a major economic and safety impact on consumer products covered by the act. The consumer product industry as well as the consumer can expect accelerated action by the commission in fulfilling its obligations under the law.

Claims, defenses, and precautions

Here are some of the reasons claims have been instituted against the seller and manufacturer of a product:

Faulty design

Faulty manufacture

Faulty components or materials used

Faulty specifications

Incomplete instructions

Lack of warning about use or misuse of product

Inadequate warning of hazards that might arise from normal use

Instructions not clear due to faulty translation or poor copy

The seller has some of the following defenses against such claims:

Proof that the product was not defective when it left his hands.

Proof that a product has a reasonable durability span.

Proof that the seller is entitled to expect his product to be utilized for the use intended.

Proof that the claimant contributed to the

cause of injury (contributory negligence); For example, when opening a glass toothbrush container, claimant used such undue force that he cracked the glass and suffered an injury as a result.

The retailer or other middleman can subrogate the claim by instituting an action against the manufacturer claiming the latter bears the primary responsibility for putting defective products in the "stream of trade."

Precautions the seller can take to avoid claims:

Appropriate warning labels should be placed on the product if necessary, or require the manufacturer to do so.

Imported products must have warning labels and directions printed in understandable English.

Where necessary, an appropriate instruction manual should be attached or sent along with the package. The "proper" use of the item must be stressed.

Check the phrases used in the product advertising copy for possible implied misrepresentation, such as "fire resistant," "absolutely safe," "foolproof," "nonflammable."

Anytime any sketches, photos, or diagrams accompany products they should emphasize accident prevention features and safe operating devices such as safety switches and guards.

Teach and train sales and service personnel in the use and dangers of products and how to advise customer accordingly.

Instruct employees not to contradict or change any of the manufacturer's instructions.

References

American National Standards Institute, 1430 Broadway, New York, N.Y. 10018.
  Manual on Uniform Traffic Control Devices for Streets and Highways, D6.1.
  Method of Measuring and Recording Patron and Non-Employee Injury Statistics, Z108.1.
  Safety Code for Elevators, Dumbwaiters, Escalators, and Moving Walks, A17.1.

Matwes, George and Helen. *Loss Control: A Safety Guidebook for Trades and Services.* New York, N.Y., Van Nostrand Reinhold Co. 1973.

National Fire Protection Association, 470 Atlantic Ave., Boston, Mass. 02210.
  Auxiliary Protective Signaling Systems, Standard No. 72B.
  Installation of Air Conditioning and Ventilating Systems, Standard No. 90A.
  Life Safety Code, Standard No. 100.
  Local Protective Signaling Systems, Standard No. 72A.
  Proprietary Protective Signaling Systems, Standard No. 72D
  Remote Station Protective Signaling Systems, Standard No. 72C
  Central Station Protective Signaling Systems, Standard No. 71.

National Safety Council, 425 North Michigan Ave., Chicago, Ill. 60611
  Industrial Data Sheets
    Carbon Monoxide, 415.
    Escalators, 561.
    Falls on Floors, 495.
    Fire Prevention in Stores, 549.
    Flame-Retardant Treatment of Fabrics, 517.
    Floor Mats and Runners, 595.
    Parking Lots—Self-Service, 597.
    Sidewalk Sheds, 368.
  National Safety News Reprints
    "Slips and Falls—Part I, Factors in Detecting and Correcting Floor Slipperiness."
    "Slips and Falls—Part II, Elusive Factor in Falls: The Shoe Sole."
  Supervisors Guide to Human Relations.

"Property Conservation Engineering and Management." *Record,* 50:3 (May–June 1973). Factory Mutual System, Norwood, Mass. 02062.

Sources
of Help

Chapter
23

Service Organizations   592
Community safety councils

Standards and Specifications Groups   597
Other standards groups

Fire Protection Organizations   599

Insurance Associations   600
Insurance service

Professional Societies   603

Trade Associations   608

U.S. Government Agencies   618

Departments and Bureaus in the States and Possessions   618
Labor offices . . . Health and hygiene services

Canadian Departments, Associations, and Boards   624
Governmental agencies . . . Private provincial agencies

International Safety Organizations   626

Educational Institutions   628
Correspondence courses

Bibliography of Safety and Related Periodicals   629
Safety . . . Industrial hygiene and medicine . . . Fire

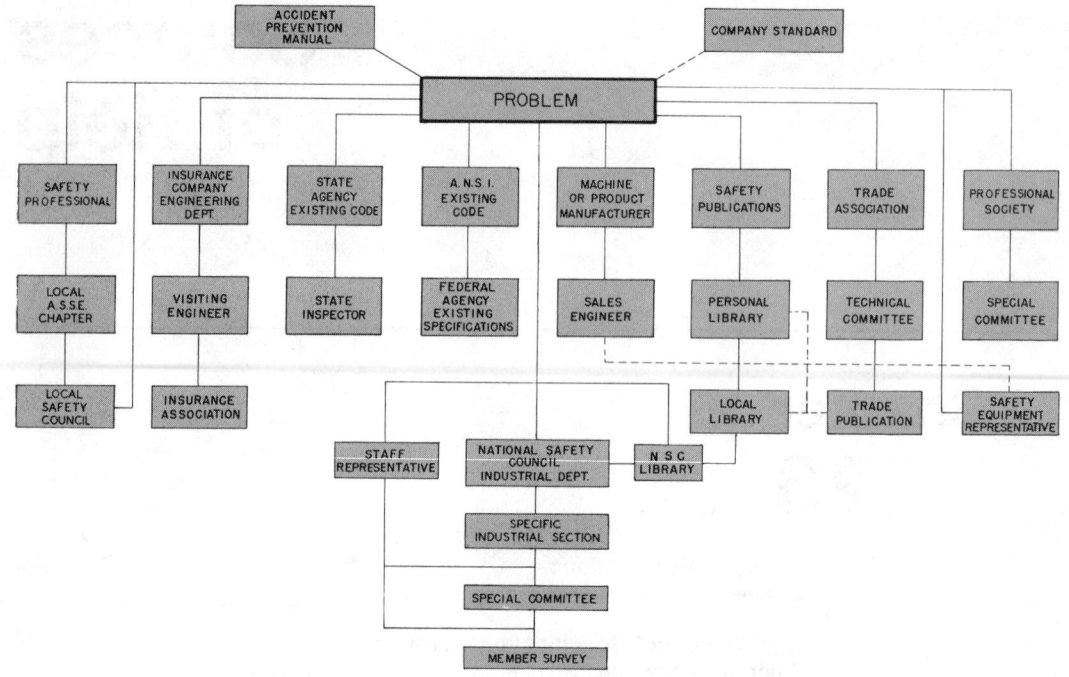

Fɪɢ. 23–1.—Method of finding information necessary to solve a safety problem. Try sources in order presented, starting with this Manual.

The safety professional frequently needs highly specialized or up-to-the-minute, yet-unpublished information, as well as knowledge of the work and publications of trade associations, public agencies, and private service organizations. These sources are numerous and their charters of responsibility are varied. As an aid in the safety professional's search for information, this chapter classifies these sources and defines their functions. A chart showing how the various sources of information can be tapped in order to find a solution to a safety problem is exhibited in Fig. 23–1.

On a particularly difficult problem, such as the guarding of a special machine for example, it may be necessary to write to a number of sources before the best solution can be obtained. Federal and state authorities can furnish the minimum requirements under the law. Standards from a number of sources must be consulted to see whether or not the problem has already been answered.

After a review of the laws and the standards, it may be found that requirements for safety set up in the establishment are still not met and further advice is necessary. Insurance companies or their associations may shed light through their knowledge of a similar problem solved in another plant. The trade association in the industry that is the prime user of the equipment may have a safety committee that can be contacted for specific information.

If no solution is found, it may be necessary to consult a service organization, such as the National Safety Council, and present the problem to the unified thinking of one of its sectional committees, its industrial staff, or both.

Service Organizations

National Safety Council
425 North Michigan Ave., Chicago, Ill. 60611

The National Safety Council is the largest organization in the world devoting its entire effort to the prevention of injury. It is nonprofit and nonpolitical. Its staff members work as a team with more than 2000 volunteer officers, directors, and members of various conferences and committees to develop and maintain accident prevention material and programs in specific areas of safety. These areas include industrial, traffic, home, recreational, and public. There are also volunteer-staff teams in such areas as public information, publications, membership extension, and field organization. Council headquarters facilities include the largest safety library in the world.

The bulk of this work is done at Chicago headquarters by a staff of more than 450, about half of whom are engineers, editors, statisticians, writers, educators, librarians, and other specialists. In addition to the main office at 425 N. Michigan Ave., Chicago, the council has a regional office at the Lincoln Bldg., 60 E. 42nd St., New York, N.Y. 10017.

Recognizing that industry's safety problems often require specialized treatment, the Council has divided its industrial effort into sections; each is administered by its own executive committee, nominated and elected from the membership within that industry. Each executive committee consists of a general chairman, a vice-chairman, secretary, a newsletter editor, and such other officers as may be needed to head the functional committees of the section.

The industrial membership of the Council is organized according to the following industries. These sections are designed to provide special help for all facets of the industrial section, as shown below.

AEROSPACE
Missile and aircraft manufacture, related components

AIR TRANSPORT
Ground safety; personnel and equipment

AUTOMOTIVE, TOOLING, METALWORKING, AND ASSOCIATED INDUSTRIES
Machining, fabrication, assembly, general manufacturing

CEMENT, QUARRY AND MINERAL AGGREGATES
Quarrying, processing, manufacturing, production

CHEMICAL
Manufacturing compounds and substances

COAL MINING
Underground and open pit

CONSTRUCTION
Highway, buildings, heavy, home, specialty

ELECTRONIC AND ELECTRICAL EQUIPMENT
Manufacturing and assembly

FERTILIZER
Manufacturing, storage, transportation, use

FOOD AND BEVERAGE
Production and processing; bakers, dairies, brewers, confectioners, distillers, poultry

GLASS AND CERAMICS
Manufacturing; flat, containers, miscellaneous products, fiber, refractories, molds

HOSPITAL—HEALTH CARE
Patient, employee, visitor safety, security

MARINE
Deep water and inland waterway; crew, passenger, vessel safety, stevedoring, shipbuilding, repair

MEAT AND LEATHER INDUSTRIES
Meat packing, processing, leather products

METALS
Foundries, manufacturing ferrous and nonferrous, fabricating, steel service centers

PETROLEUM
Exploration, drilling, production, pipeline, marketing, retail

POWER PRESS AND FORGING
Metal stamping and forming, forging

PRINTING AND PUBLISHING
Letterpress, offset, newspaper, bindery

PUBLIC EMPLOYEE
City, state, county

PUBLIC UTILITIES
Communications, electric, gas, water, constructor

PULP, PAPER AND RELATED PRODUCTS
Logging, manufacturing, converting, related products

RAILROAD
Employee, passenger, public safety

RESEARCH AND DEVELOPMENT
Safety; laboratory, physical, fire, health

RUBBER AND PLASTICS
Tires, molded products, belts, footwear, synthetics

TEXTILE
Manufacturing and fabrication; natural and synthetic fibers, ginning

TRADES AND SERVICES
Food service, retailers, hotels, motels, mercantile, warehouses, offices

WOOD PRODUCTS
Logging, sawmills, veneer, plywood, furniture, millwork, cooperage

The Council assigns a staff engineer to each section to help with programs, membership, organization, and the preparation of informational materials. In this way, each major industry group is assured of representation in the affairs of the Council and has the means to develop material and services to meet its needs.

Section executive committees usually meet two or three times a year in cities designated as most convenient to the members. Improvement of Council services through new technical materials and visual aids takes a major portion of each committee's meeting time. Since increased membership can improve these services, membership solicitation is encouraged. Planning of National Safety Congress programs is given careful consideration in the committee agenda. Special committees are often assigned to work on problems peculiar to an industry and on which there is no ready reference.

To coordinate the entire Council industrial program, a committee representing all industries, the Industrial Conference, meets three times a year. The function of this Conference is to review current industrial safety problems and to determine on a national scale the best procedures to follow in providing increasingly beneficial programs to industry.

The Industrial Conference is made up of industrial members of the Council and representatives of cooperating groups. It comprises 28 sectional general chairmen, 28 sectional vice-general chairmen, and 125 members at large, drawn from member organizations, governmental agencies, insurance organizations, professional and trade associations, and other groups.

The Conference is divided into subcommittees that cover various segments of the occupational safety area to assist the Industrial Department in carrying out its responsibilities to industry.

The entire industrial safety program is under the leadership of the Vice President for Industry who serves as a member of the Board of Directors of the National Safety Council in order to help formulate policies.

The following publications for industrial safety programs are available through the Council. (Unless otherwise stated, they are monthly.)

Family Safety (quarterly)
Farm Safety Review (six issues a year)
Industrial Supervisor
Journal of Safety Research (quarterly)
National Safety News
"OSHA Up-To-Date"
"Product Safety Up-To-Date"
Sectional "Newsletters"
Safe Driver (issued in three editions—Truck, Passenger Car, and Bus)
Safe Worker
Traffic Safety

In addition, the following administrative, supervisory, and employee training materials can be obtained.

Statistical information:
Accident Facts (annually)
"Motor Fleet Accident Rates" (annually)
"Work Injury Rates" (annually)
Sectional Contest Bulletins (monthly)
Reporting forms

Technical materials (see current NSC *Catalog* for a complete listing):
Accident Prevention Manual for Industrial Operations (this book)
Aviation Ground Operations Handbook
"Emergency Removal of Patients and First Aid Firefighting in Hospitals"
Forging Safety Manual
Fundamentals of Industrial Hygiene
Guards Illustrated
"Guide for Operating Ambulance Fleets"
"Guide to Traffic Safety Literature"
Handbook of Accident Prevention
"How To Make a Safety Speech"
"Industrial Data Sheets" (a series)
Meat Industry Guidelines
Motor Fleet Safety Manual
"National Directory of Safety Films"
"Power Press Safety Manual"

Safety Guide for Health Care Institutions
"Safety Handbook for Office Supervisors"
Safety Manual for the Graphic Arts Industry
"Safety Reprints" (a series)
"School Transportation: A Guide for Supervisors"
"Small Fleet Guide"
"Successful Supervision"
Supervisors Guide to Human Relations
Supervisors Safety Manual

Training materials:
Banners
Booklets
Calendars
Films
Posters
Safety slides
Supervisory training pamphlets

The Council sponsors the National Safety Congress in Chicago in the fall of each year. In terms of program, it is one of the largest conventions held anywhere. Proceedings of each Congress are published in *Transactions of the National Safety Congress.*

Special services available through the Council are:

Consultation
Employee publication
Library
Statistical

Safety training. The National Safety Council's Safety Training Institute conducts basic and advanced training courses in safety engineering and management, industrial hygiene, and industrial noise and hearing conservation. There are also courses on safety fundamentals for public utilities, hospitals, chemical plants, and laboratories. These courses are offered to employees of member companies and to representatives of city, state, and federal departments and foreign governments.

Special courses covering some of these subjects are being sponsored by local safety councils under the auspices of the Safety Training Institute.

Courses are also sponsored by other educational institutions. See the discussion toward the end of this chapter.

The Labor Conference. Members of the Council's Labor Conference represent unions and governmental labor agencies. The Conference and its members represent labor and its safety and health viewpoints in many of the Council's areas of work, such as construction, public utilities, coal mining, legislation, defensive driving, and vocational education.

Community safety councils

Some eighty community and state safety councils located throughout the United States have been chartered by the National Safety Council. These councils work under the leadership of public-spirited citizens, commercial and industrial interests, responsible official agencies, and other important groups. They are nonprofit, nonpolitical, self-supporting organizations whose purpose is to reduce accidents.

Accredited community councils operate under the guidance of a full-time manager and staff and receive a continuous service from the National Safety Council. (A list of accredited councils can be obtained from the Field Service Dept., National Safety Council.)

These organizations give assistance to local safety engineers and others concerned with occupational safety. Safety professionals, in turn, render substantial service through participation in the local council's work.

Many of the local councils offer the following services:

• Act as a clearinghouse of information on occupational safety problems, and maintain a library of films and other aids,

• Provide a forum for exchange of experience through regularly scheduled meetings of supervisory personnel,

• Sponsor courses of safety instruction, in some cases with the aid of educational institutions (many now present approved National Safety Council training courses),

• Conduct annual, area-wide occupational safety conferences,

• On request, give advice on safety problems and programs,

• Stimulate and assist in the development of accident prevention programs for all employers,

• Conduct safety contests with awards for outstanding safety records.

The scope and extent of activities of each group depend, of course, upon local conditions and available resources.

American National Red Cross
Safety Services
17th and "D" Sts. NW, Washington, D.C. 20006

The American Red Cross, through its more than 3300 chapters, offers free courses in first aid, swimming, lifesaving, and small craft handling.

Experience in industry shows that first aid and safety training contributes to the reduction of accidents—both on and off the job—by creating an understanding of accident causes and effects and by improving attitudes toward safety. In addition, this training prepares individuals to give proper emergency care to accident victims. In some situations, immediate action may mean the difference between life and death.

Arrangements for first aid training can be made through local Red Cross chapters. Most industries prefer to select key personnel to receive training as volunteer instructors who, in turn, can conduct classes for fellow employees. Others may wish to arrange for employee training by instructors provided through a local chapter.

Texts, instructor's manuals, charts and other teaching materials and visual aids, such as films and posters, are available through local chapters.

Industrial Health Foundation, Inc.
5231 Centre Ave., Pittsburgh, Pa. 15232

The foundation, a nonprofit research association of industries, advocates industrial health programs, improved working conditions, and bettered human relations.

The foundation maintains a staff of physicians, chemists, engineers, biochemists, and medical technicians at Mellon Institute, the foundation headquarters.

Activities fall into three major categories:

● To give direct professional assistance to member companies in the study of industrial health hazards and their control.

● To assist companies in the development of health programs as an essential part of industrial organization.

● To contribute to the technical advancement of industrial medicine and hygiene through investigation, research, and allied activities.

Activities are classified as follows:

Medical.
Organization and administrative practices.
Opinions on doubtful X rays.
Surveys of health problems.
Specific industrial medical problems.

Chemistry, toxicology, industrial hygiene.
Field studies—plant or industry basis.
Toxicity of chemicals, physical agents, or processes.
Sampling and analytical procedures.
Training industrial hygiene personnel.

Engineering.
Design and control of new hazardous processes.
Consultation on new plant design.
Ventilating systems.
Exhaust hoods.
Air pollution.
Engineering design programs.

Research.
Toxicity studies on new chemicals.
Hazardous dusts, fumes, vapors, and gases.
Physiological response to toxic agents.

Legal.
Legislative activities pertaining to occupational disease.
Review of new compensation laws.
Analysis of new health codes.

The foundation holds an annual meeting of members, conferences of member company specialists, and special conferences on problems common in a particular industry.

The following publications are issued:

Industrial Hygiene Digest, monthly.
Transactions of annual meetings.
Special bulletins prepared by advisory committees.
Pamphlets on current interest subjects.

National Society for the Prevention of Blindness, Inc.
79 Madison Avenue, New York, N.Y. 10016

The NSPB is the oldest voluntary health agency nationally engaged in the prevention of blindness through a comprehensive program of community services, public and professional education, and research. The Society's industrial service program is guided by its Advisory Committee on Industrial Eye Health and Safety, comprised of experts in the fields of industry, education, medicine, nursing, and accident prevention.

Activities and programs of the department are:

• Promotes and administers the Wise Owl Club of America, eye-safety incentive program, among industrial, military, municipal and educational organizations. Membership in the Club is restricted to those who save their vision from being damaged or destroyed by wearing proper eye-protection both on and off-the-job. Junior Wise Owl Club membership recognizes sight saved by children and teenagers through wearing safety glasses or other forms of protectors.

• Promotes state-wide eye safety for all school and college laboratory and shop students, and their teachers and visitors. Both voluntary and legislative means are used. Provides counseling to school administrators and teachers in establishing and implementing eye safety programs. Encourages amendments to state school eye safety laws which presently exclude compliance by private schools.

• Provides secretarial, counseling, and liaison services for the American National Standards Institute Z87 Committee, engaged in upgrading current eye and face protector code Z2.1. Participates as member of other American National Standards Institute studies dealing with topics related to illumination, vision, and eye protection.

• Promotes universal usage of nonflammable prescription and plano safety eyeglasses and sunglasses by the general public through voluntary and legislative means, and encourages cooperation by the eye care professions and the ophthalmic industry.

• Promotes periodic vision testing in industry and technical schools to determine visual defects and their relationship to visual requirements for various tasks and occupations. Provides counsel on improvement of visual working conditions through proper use of illumination and color.

• Provides exhibits on eye health and safety topics at industrial and educational meetings. Distributes literature and films on eye safety.

• Stimulates relationships with organizations, groups, and individuals with a related interest in sight conservation. Addresses industry, safety and educational meetings to project NSPB recommendations and program aims.

Standards and Specifications Groups

American National Standards Institute
1430 Broadway, New York, N.Y. 10018

The American National Standards Institute coordinates and administers the federated voluntary standardization system in the United States, which provides all segments of the economy with national consensus standards required for their operations and for protection of the consumer and industrial worker. It also represents the nation in international standardization efforts through the International Organization for Standardization (ISO), the International Electrotechnical Commission (IEC), and the Pacific Area Standards Congress (PASC).

ANSI is a federation of some 1000 national trade, technical, professional, labor, and consumer organizations, government agencies, and individual companies. It coordinates the standards development efforts of these groups and approves the standards they produce as American National Standards when its Board of Standards Review determines that a national consensus exists in their favor.

Many American National Standards, as well as other national consensus standards, have taken on additional importance since the passage of the Williams-Steiger Occupational Safety and Health Act of 1970. In promulgating standards under the act, the Occupational Safety and Health Administration has stated a definite preference for basing its regulations on consensus standards that have proved their value and practicality by use.

Under ANSI procedures, the responsibility for the management of specific standards projects is divided according to subject matter and assigned to an ANSI Technical Advisory Board. Standards dealing with safety fall under the jurisdiction of either the Safety

597

Technical Advisory Board or the Highway Traffic Safety Technical Advisory Board. Many American National Standards on safety and health are developed by American National Standards Committees formed under ANSI procedures.

American Society for Testing and Materials
1916 Race Street, Philadelphia, Pa. 19103

ASTM is the world's largest source of voluntary consensus standards for materials, products, systems and services. There are currently more than 4800 ASTM standards.

ASTM membership is drawn from a broad spectrum of individuals, agencies, and industries concerned with materials. The 22,000 members include engineers, scientists, researchers, educators, testing experts, companies, associations and research institutes, governmental agencies and departments (federal, state, and municipal), educational institutions, and libraries.

ASTM standards are published under the following categories:

Nonferrous metals
Ferrous metals
Cementitious, ceramic and masonry materials, soils, roads, and waterproofing.
Paint, wood, adhesives, paper, packaging
Fuels, petroleum, aromatic hydrocarbons, soap, water, textiles, leather
Electrical insulations, electronics, plastics, rubber, paper
Space simulation, nuclear, aerospace, surgical implants, environmental acoustics
Forensic sciences, occupational health and safety aspects of materials, physical and biological agents, and hazard potential of chemicals
Protective equipment for sports, tires, meat and meat products, vacuum cleaners, fences, security systems, aerosols and closures, safety and traction for footwear, and pesticides

ASTM recently formed a committee on consumer product safety to develop standards that will help protect the public by reducing the risk of injury associated with the use of consumer products such as matches, cigarette lighters, bathtubs and shower structures, children's high chairs, trampolines, and football helmets.

ASTM standards are of interest to the safety professional since they identify areas of hazard and establish guidelines for safe performance. The society also publishes standards for atmospheric sampling and analysis, fire tests of materials and construction, methods of testing building construction, nondestructive testing, fatigue testing, radiation effects, pavement skid resistance, protective equipment for electrical workers, and others.

These constitute basic reference materials for the safety professional who will frequently be confronted with ASTM standards in processes, as with power plant installations in which vessels, piping, valves, and other component parts are designed and fabricated according to these standards. ASTM standards may also be factors in the raw materials used in protective equipment or other devices.

Other standards groups

In addition to the American National Standards Institute, many governmental and other agencies have established specifications used by safety professionals. Many industries through their trade associations have also established either (a) codes covering operations in their own plants or (b) safe practices to be followed in the use of their products. Some of these groups are:

AMERICAN SOCIETY OF MECHANICAL ENGINEERS

AMERICAN WELDING SOCIETY

BUREAU OF MINES

COMPRESSED GAS ASSOCIATION

DEPARTMENT OF TRANSPORTATION

GENERAL SERVICES ADMINISTRATION

INDUSTRIAL SAFETY EQUIPMENT ASSOCIATION

INTERSTATE COMMERCE COMMISSION

NATIONAL ASSOCIATION OF PLUMBING AND MECHANICAL OFFICIALS, 5032 Alhambra Ave., Los Angeles, Calif. 90032

NATIONAL BOARD OF BOILER AND PRESSURE VESSEL INSPECTORS, 1155 North High St., Columbus, Ohio 43201

NATIONAL BUREAU OF STANDARDS

NATIONAL FIRE PROTECTION ASSOCIATION

OCCUPATIONAL SAFETY AND HEALTH ADMINISTRATION

Fire Protection Organizations

Surveys made of members attending National Safety Council member training classes have shown that more than 50 percent were charged with the duties of fire prevention and fire control in addition to other aspects of safety in their companies. The following organizations are among those with which engineers would have most frequent contact on technical matters. (Also see Insurance Associations, the next section in this chapter.)

Factory Insurance Association
85 Woodland St., Hartford, Conn. 06105

The association is composed of a group of capital stock insurance companies to provide engineering, inspection, and loss adjustment service to industry. It maintains a staff of engineers with representatives in key industrial centers.

The fire safety laboratory in Hartford contains many types of fire protection equipment for examination and demonstration under working conditions. This laboratory is used for the basic and advanced training of F.I.A. fieldmen, and training plant protection personnel of F.I.A. policyholders and of other insurance organizations who are given short courses in the proper use of fire protection devices.

Factory Mutual System
1151 Boston-Providence Turnpike, Norwood, Mass. 02062

The Factory Mutual System is an industrial fire protection engineering and inspection bureau established and maintained by mutual fire insurance companies. Its staff of trained fire protection engineers minimizes the likelihood of heavy property losses and safeguards production in industry.

The Factory Mutual laboratories test and list fire protection equipment for approval, assist in the development of better standards, and conduct research in industrial fire protection. The testing station at West Gloucester, Rhode Island, the largest such facility in the world, enables industrial conditions to be duplicated in fire and explosion tests.

Its publications include the *Factory Mutual Record*, an illustrated bimonthly digest of current fire protection information and ex-perience.

The *Factory Mutual Handbook of Industrial Loss Prevention* and *Loss Prevention Data* cover recommended practices for protection against fire and related hazards.

A manual which lists industrial fire protection equipment approved by the laboratories is available. Reputable manufacturers of fire prevention and control equipment carry both Factory Mutual and Underwriters' listings.

Factory Mutual has inspectors in major industrial centers.

National Fire Protection Association
470 Atlantic Ave., Boston, Mass. 02210

The National Fire Protection Association is the clearinghouse for information on the subject of fire protection and fire prevention. It is a nonprofit technical and educational organization with a membership of some 28,500 companies and individuals.

The technical standards issued as a result of NFPA committee work are widely accepted by federal, state, and municipal governments as the basis of legislation, and widely used as the basis of good practice. More than 50 are used as OSHA regulations. Constantly revised and updated, 225 standards have to date been issued by NFPA, which are available in separate pamphlet form.

Many of them supply authoritative guidance to safety engineers. Representative subjects include:

Industrial Fire Loss Prevention
Installation, Maintenance and Use of Portable Fire Extinguishers
Care and Maintenance of Sprinkler Systems
Organization and Training of Private Fire Brigades
Flammable and Combustible Liquids Code
Hazardous Chemicals Data
Cutting and Welding Processes
Storage and Handling of Liquefied Petroleum Gases
Prevention of Dust Explosions in Industrial Plants
National Electrical Code
Lightning Protection Code
Air Conditioning and Ventilating Systems
Life Safety Code
Safeguarding Building Construction Operations

Protection of Records
Truck Fire Protection
Use, Maintenance and Operation of Industrial Trucks

The standards are also published as the "National Fire Codes" in 10 volumes totaling 9500 pages, each volume covering a particular subject area.

Other books and pamphlets of interest, such as the manual on industrial fire brigades and the NFPA *Inspection Manual,* are available to safety professionals from the association.

The *Fire Protection Handbook* is also published by NFPA. An authoritative encyclopedia on fire and its control, the 2200-page handbook is divided into 21 sections and 130 chapters.

Underwriters Laboratories Inc.

207 East Ohio St., Chicago, Ill. 60611

The Underwriters Laboratories, a not-for-profit organization, maintains laboratories for the examination and testing of devices, systems, and materials, to determine their safety.

The laboratories publish annual lists of manufacturers whose products have proved acceptable under appropriate standards and who continue to pass their follow-up checking service.

These lists are:

"Accident, Automotive, and Burglary Protection Equipment"
"Building Materials"
"Classified Products Index"
"Electrical Appliance and Utilization Equipment"
"Electrical Construction Materials"
"Fire Protection Equipment"
"Fire Resistance Index"
"Gas and Oil Equipment"
"Hazardous Location Equipment"
"Marine Products"

The laboratories follow up on listed material at annual or more frequent intervals.

Safety professionals have come to regard the Underwriters' label as a requisite when they purchase fire, electrical, and other equipment which falls in categories tested in the laboratories.

Engineers should be aware, however, that UL listings apply only within the scope of the tests made, and may have no bearing on performance or other factors not involved in the examination procedure. If the function of the device or material tested and listed is a safety function then, of course, the test tells the safety professional what he wants to know. UL tests are made under conditions of installation and use which conform to the appropriate standards of the NFPA or other applicable codes. Any departure from these standards by the user himself may affect the performance qualifications found by the Underwriters Laboratories.

Insurance Associations

In addition to insurance associations listed under "Fire Protection Organizations," there are a number of insurance federations with accident prevention departments that produce technical information available to safety engineers.

The American Association of State Compensation Insurance Funds

P.O. Box 5922, San Francisco, Calif. 94101

The American Association of State Compensation Insurance Funds is an International organization, the members of which are the state insurance funds of 17 states, Puerto Rico, Virgin Islands, and many of the provinces of Canada. It has a permanent Safety Committee and several times a year issues bulletins to members relating to workmen's compensation.

American Mutual Insurance Alliance

20 North Wacker Dr., Chicago, Ill. 60606

The American Mutual Insurance Alliance is a national organization of leading mutual property-casualty insurance companies. Its membership includes more than 100 companies that safeguard the value of lives and property by providing protection against mishaps in workplaces and losses from fires, traffic accidents and other perils. Alliance member companies have a tradition of loss prevention which dates from the organization of the first American mutual insurance company in 1752.

Through its Accident and Fire Prevention Department, the Alliance makes a concerted effort to reduce accidents, fires, and other loss-producing incidents. Under the guidance of its loss control advisory committee, sound safety engineering, industrial hygiene, fire protection engineering, and other loss prevention principles are promoted. Major activities include the dissemination of information on safety subjects, conduct of specialized training courses for member company personnel, sponsorship of research, cooperation in the development of safety standards, development of visual aids, and the publication of technical and promotional safety literature. A catalog of safety materials is available without charge.

Notable among the publications issued by the department are:

Safe Openings for Some Point-of-Operation Guards
Wood Working Circular Saws, Protection for Variety and Universal Types
Spreaders for Variety and Universal Saws
Nip Hazards on Paper Machines
Handbook of Organic Industrial Solvents
Judging the Fire Risk
Tested Activities for Fire Prevention Committees
Exit Drills in the Home
Handbook of Hazardous Material
Commercial Driver's Code
Safety Memos for Fleet Supervisors

Other Alliance departments also regularly issue a variety of bulletins, reports, research reports and similar materials that often may relate to and support work safety and accident prevention. The communications department, for example, publishes the general interest magazine, *Journal of American Insurance,* as well as leaflets, brochures, and other informational and educational materials. Inquiry is invited.

The Alliance cooperates extensively with trade associations, professional societies, and other organizations with similar interests, such as the National Safety Council, the National Fire Protection Association, the American National Standards Institute, Inc., the American Society of Safety Engineers, and the American Industrial Hygiene Association.

American Insurance Association Engineering and Safety Service
85 John Street, New York, N.Y. 10038

The American Insurance Association, or-

ganized January 1, 1965, is an advisory organization serving a large number of companies in the property-liability insurance field. It is a multi-line organization designed to help its members' and subscribers' companies meet many of the diverse problems confronting the insurance business today. Embodied in the Association are the traditions, experience, and accomplishments of the National Board of Fire Underwriters, the Association of Casualty and Surety Companies, and the former American Insurance Association.

The Association's Engineering and Safety Service develops and publishes recommended safety programs and procedures, codes and ordinances for buildings, and the control of fire hazards. It prepares and issues bulletins to Association member companies and to public officials on matters of special interest in all manner of accident and fire prevention. It studies special hazards in specific industries and recommends safety measures.

The Association, through staff and subscriber representatives, has working representation on more than 175 committees of the National Safety Council, American National Standards Institute, and the National Fire Protection Association. In addition, the Association has worked closely with other trade, industry and professional groups with similar interests.

Typical of the publications issued are the following:

Accident Prevention (partial listing).

Commercial and industrial
An Emergency First Aid Guide
Office Building Safety Instructions
Your Guide to Safety as a Hospital Employee
Women in the Factory
General Safety Instructions
Supervisor's Safety Guide Book

Commercial vehicles fleet
Commercial Vehicle Driver's Guide Book
How To Hire Better Drivers

Construction
Your Guide to Safety in Demolition Operations
Your Guide to Safety on Construction Projects

Construction management bulletins
Cranes, Derricks, Draglines, Power Shovels (4 bulletins)
Organizing for Loss Control (10 bulletins)

Home, traffic, and off-the-job
Bicycle Safety
Too Fast for Conditions (deals with high speeds on the highway)

Physically impaired workers
Supervising the Physically Impaired

Public and semi-public
A Safety Guide for Hotels
Safe Hospitals
Safe Schools
Your Guide to Safety as a Hospital Employee

Special and chemical hazards bulletins.
Research constitutes an important phase of the department's program. A continuing study of hazardous chemicals and processes has resulted in an encyclopedia of chemical data related to industrial exposures in the form of *Chemical Hazards Information Series* Bulletins. A companion service known as the *Special Hazards Bulletin Series* provides related coverage of many hazards on which safety information is needed. Representative chemical hazards bulletins cover such subjects as: aluminum alkyls, titanium, zirconium, phthalic anhydride, epoxy resins, and acrolein. Representative special hazards bulletins cover swimming pool purification, lasers and masers, insecticides, ultrasonics, and microwave radiation.

Fire and natural hazards.

Codes and ordinances (suggested)
National Building Code
Fire Prevention Code

Fire resistance ratings
Fire Resistance Ratings (looseleaf)

Fire prevention education
"Self-Inspection Blanks" (churches, farms, hospitals, hotels, industrial plants, mercantiles, places of assembly, and schools)

Fire and natural hazards
Building Loss Possibilities from Fire and Natural Hazards
Water in Modern Fire Control

Highway transportation of extra-hazardous commodities
Suggested Guide for State Action on Safety from Fire, Explosion and Health Hazards

Standards and recommended safeguards
Safe Handling and Use of LP-Gas

Research publications
Fire, Explosion, and Health Hazards of Organic Peroxides

Special-interest bulletins
The Association has over 300 special-interest bulletins, covering subjects like salamanders, fire-flow tests, small hose, and other items that do not require the full treatment given a standard. They contain information of special interest to fire fighting services, municipal building inspectors and fire protection people generally. An index to the bulletins is available on request.

Highway emergency bulletins, police and fire
These bulletins promote public safety in the highway transportation of extra-hazardous commodities and furnish fire and police officials with educational and training material.

Technical surveys
Hazard Survey of the Chemical and Allied Industries, 1968

Card data service
These 4 x 6-inch cards contain brief technical information on special hazards and related items of interest in underwriting. Issued jointly with Underwriters' Laboratories, Inc. Examples of chemicals covered are dinitrotoluene, phosphorus, and sodium nitrate.

Insurance service

The safety professional from the insurance company should not be overlooked as being a good source of information. His work requires him to visit many plants, and his experience and acquaintance with other safety professionals can be extremely helpful.

Private insurance carriers may be called upon for assistance with regard to:

Planned accident prevention programs
Inspections
Special bulletins on processes, machines, or

hazardous substances
Contest and awards
Safety meeting subject material
Speakers for safety meetings
Posters or other promotional material
Industrial hygiene services, often in their own
    laboratories
Statistical information
Compensation and claims information
Consultation

Professional Societies

American Society of Safety Engineers
850 Busse Highway, Park Ridge, Ill., 60068

The American Society of Safety Engineers is the only organization of individual safety professionals dedicated to the advancement of the safety profession and to foster the well-being and professional development of its members.

In fulfilling its purpose, the Society has the following objectives:

Promote the growth and development of the profession

Establish and maintain standards for the profession

Develop and disseminate material which will carry out the purpose of the Society

Promote and develop educational programs for obtaining the knowledge required to perform the functions of a safety professional

Promote and conduct research in areas which further the purpose and objectives of the Society

Provide forums for the interchange of professional knowledge among its members

Provide for liaison with related disciplines

The Society has vice presidents elected by and representing each of its twelve regions. In addition, there are appointed vice presidents responsible for professional development, conferences, professional affairs, communications, and research and technical development. Special committees are established each year as the need arises. An annual professional conference is conducted for members.

The Society is actively pursuing its Professional Development Programs including the development of curricula for safety professionals, accreditation of degree programs, member education courses, additional publications, and defining research needs and communications to keep safety practitioners current. In addition, the Society has increased its participation and activity in national government affairs.

The Society has established a separate research corporation, the American Society for Safety Research, because of its concern for the need for greater research efforts in the prevention of accidents and injuries. The Society believes that increased research will improve accident and injury control techniques, thus will serve the interests of its members as well as contributing to the economy of our nation and to the health and welfare of all persons.

The Society was also instrumental in establishing a separate corporation to develop a certification program for safety professionals. This corporation, known as Board of Certified Safety Professionals of America (BCSP) is now accepting applications. (See BCSP later in this chapter.)

The members of the Society receive the monthly *ASSE Journal* as part of their membership services. (Non-members may also subscribe.) Articles on new developments in the technology of accident prevention are included, as well as information on the activities of the Society, its chapters and members. The Society also publishes other technical or specialized information, such as "A Selected Bibliography of Reference Materials in Safety Engineering and Related Fields," a glossary of terms used in the safety profession, and a series of Monographs.

In 1971, the Society observed its 60th Anniversary as the oldest National Safety organization in the United States. Founded in 1911 as the United Association of Casualty Inspectors, it grew from the original enrollment of 35 to more than 1000 members by 1930. In 1947, the membership reached 4500; in 1973, there were more than 12,000 members internationally.

The Society was originally chartered in New York in 1915 and incorporated in Illinois in 1962. In 1924, the Society merged with the National Safety Council, becoming the Engineering Section of the Council. Twenty-three years later (1947), the membership voted to re-establish itself as an independent organization. Independent offices were established in 1959 and the present headquarters office building was constructed in 1967.

In 1973, there were 93 Chapters in the United States and Canada. Chapters engage in a number of activities designed to enhance the professional competence of their members. Most hold monthly meetings featuring speakers, demonstrations, workshops and discussions designed to help members keep abreast of developments in their professional field. A list of chapters grouped by states and listed by name and by location can be obtained from the national office.

**American Association of Industrial Dentists
Office of the Secretary**

14 Hunter Lane, Camp Hill, Pa. 17011

This national organization is fostered by the American Dental Association and co-operates closely with it as well as with other national industrial health organizations, such as the Industrial Medical Association, the American Industrial Hygiene Association, the American Association of Industrial Nurses, and the American Conference of Governmental Industrial Hygienists.

The AAID sponsors local meetings of industrial dentists, holds an annual meeting wherever the Industrial Health Conference is held, conducts tours of plants, publishes scientific papers, and issues an annual report of transactions.

American Association of Industrial Nurses, Inc.

79 Madison Ave., New York, N.Y. 10016

As the national professional organization for the registered nurse working in industry, the association strives to raise the qualifications for industrial nurses, improve nursing services and standards, and provide educational programs for nurses in this special field. *Occupational Health Nursing* is published monthly.

The annual meeting is held in conjunction with that of the Industrial Medical Association at the American Industrial Health Conference.

American Chemical Society

1155 16th St. NW., Washington, D.C. 20036

This society is devoted to the science of chemistry in all its branches, the promotion of research, the improvement of the qualifications and usefulness of chemists, and the distribution of chemical knowledge.

Articles on safety appear in the monthly publication *Industrial and Engineering Chemistry* and the weekly publication *Chemical and Engineering News*. Digests of papers dealing with aspects of industrial hygiene appear monthly in *Chemical Abstracts*. The environment is discussed in *Environmental Sciences and Technology*.

The society has a committee on Chemical Safety. In addition, the ACS is represented on the committee on Chemicals and Explosives of the National Fire Protection Association.

American College of Surgeons

40 East Erie St., Chicago, Ill. 60611

The American College of Surgeons and the American Association for the Surgery of Trauma are working on a major joint-action program with the National Safety Council. The college has a committee on Trauma in all states and provinces of Canada and local committees on Trauma in most cities. These committees and the central committee on Trauma of the American College of Surgeons are continuing their efforts on acceptable plans for the care of injured persons.

**American Conference of Governmental
Industrial Hygienists**

P.O. Box 1937, Cincinnati, Ohio 45201

A professional association composed of industrial hygiene personnel in government (federal, state, county, or municipal government), or working under a government grant. ACGIH was organized in 1938 by a group of governmental industrial hygienists as a medium for the exchange of ideas, experiences, and the promotion of standards and techniques in occupational health.

ACGIH's wide scope of activities are accomplished through the work of its standing committees, of which there are now 15, plus five ad hoc committees, and six joint committees with the American Industrial Hygiene

Association. Particularly valuable to safety professionals is *Industrial Ventilation—A Manual of Recommended Practice,* by the committee on industrial ventilation. ACGIH annually publishes a table of recommended threshold limits.

A list of publications will be sent on request to ACGIH.

American Industrial Hygiene Association
66 S. Miller Rd., Akron, Ohio 44313

The AIHA is a professional association composed of industrial hygiene personnel. Established in 1939 by leading industrial hygienists as a result of a need for an association devoted exclusively to industrial hygiene, its purpose is to disseminate knowledge of the field and to promote the study and control of environmental factors effecting the health of industrial workers. Requirements for membership are a college degree and three years of industrial hygiene experience.

AIHA publishes "Hygienic Guides" (summarizing current information on physiological effects of specific chemicals, and methods of control), the *American Industrial Hygiene Association Journal,* the *Air Pollution Manual* and the *Industrial Noise Manual,* and reports and monographs. A list of publications will be sent on request to AIHA.

The AIHA, in cooperation with the American Conference of Governmental Industrial Hygienists, sponsors the annual American Industrial Hygiene Conference, the largest national assembly for the presentation and exchange of industrial hygiene information.

American Institute of Mining, Metallurgical, and Petroleum Engineers
345 East 47th St., New York, N.Y. 10017

The AIME promotes the arts and sciences connected with the economic production of useful minerals and metals and the welfare of those employed in this work.

Publications are *Mining Engineering, Journal of Metals,* and *Journal of Petroleum Technology,* all issued monthly; and *Transactions of the Society of Mining Engineers* (quarterly), *Transactions of the Metallurgical Society* (bimonthly), and *Society of Petroleum Engineers Journal* (quarterly).

American Medical Association
535 North Dearborn St., Chicago, Ill. 60610

The American Medical Association, the professional society for physicians, has a membership over 213,000.

Its Council on Occupational Health is concerned with the protection and improvement of the health of the nation's working population through promotion of occupational health programs. Through its committees and the staff of the Department of Occupational Health, the Council encourages (*a*) the application of protective measures against health hazards in their working environment, (*b*) the placement of individuals according to physical, mental, and emotional make-up, (*c*) adequate medical care and rehabilitation of the occupationally ill and injured, and (*d*) personal health maintenance.

The three committees of the Council are the Committee on Aerospace Medicine, the Committee on Occupational Toxicology, and the Joint Committee on Mental Health in Industry (Council on Occupational Health and the Council on Mental Health).

More than 70 statements and pamphlets prepared by the Council and its committees on various occupational health subjects are available from the Department of Occupational Health. Although these are primarily directed to the physician or nurse, many are of interest to the safety professional. Some of the subjects covered are occupational diseases, occupational dermatoses, hazard prevention, workmen's compensation, industrial nursing, and management relations.

The Council on Environmental and Public Health is concerned with problems closely allied to occupational health. Its areas of interest are those which pertain to the control of transmissible diseases; air, water and soil pollution; ionizing and other radiation affecting the public; traffic and public safety; health problems relating to population growth; urbanization; and technological changes bearing on ecology of man. Publications on these subjects are available from the Department of Environmental Health.

The Committee on Medical Aspects of Automotive Safety over the past years has had a close working relationship with the various committees on traffic safety of the National Safety Council.

605

The *Journal of the American Medical Association (JAMA)* is published weekly, and frequently contains articles on some aspect of occupational health.

The AMA *Archives of Environmental Health* is one of the ten specialty journals of the American Medical Association. This monthly publication contains articles on the socio-economic aspects of occupational and environmental health, toxicology, industrial hygiene, air pollution, radiation, and numerous other subjects of interest to the safety engineer.

American Nurses' Association, Inc.
2420 Pershing Rd., Kansas City, Mo. 64108

The American Nurses' Association is the voluntary membership organization for all registered nurses. The Division of Community Health Nursing Section is the component of the ANA which provides authoritative information about the practices of occupational health nursing. The objectives of the division are to improve community health nursing practice including occupational health nursing practice for better employee health care and also to improve the economic and general welfare of members.

Publications of interest are: "Functions, Standards, and Qualifications for Occupational Health Nurses," "Guide for Development of a Manual for an Employee Health Program," and "Selected Areas of Knowledge or Skill Basic to Effective Practice of Occupational Health Nursing." Titles of other brochures, pamphlets, guides, and articles are included in the association's publications list which will be sent on request.

American Psychiatric Association
1700 18th St. N.W., Washington, D.C. 20009

This association is concerned with research in all phases of mental disorders, standards of psychiatric education, and the medico-legal aspects of psychiatric practice. It publishes the *American Journal of Psychiatry* bimonthly.

American Public Health Association
1015 18th St. N.W., Washington, D.C. 20036

This multidisciplinary, professional association for health workers, through its monthly *American Journal of Public Health* and other publications, distributes safety information to those responsible for state and community health-service programs. One of the program area committees is devoted to injury control and emergency services.

American Society for Industrial Security
2000 K St. N.W., Washington, D.C. 20006

The American Society for Industrial Security, a professional Society of industrial security executives and supervisors, has more than 50 chapters in seven regions. Committee activities that would be of interest to the safety professionals are safeguarding, classified information, classification management, safeguarding proprietary data, physical security, subversive activities in industry, emergency planning, investigations, fire protection, and public relations.

The society publishes the magazine *Industrial Security* bimonthly and issues a newsletter to its members monthly.

American Society of Mechanical Engineers
345 East 47th St., New York, N.Y. 10017

This Society, the professional mechanical engineers' organization, encourages research, prepares papers and publications, sponsors meetings for the dissemination of information, and develops standards and codes necessary in this phase of engineering.

The Society sponsors the following code under the procedures of the American National Standards Institute:

Safety Code for Elevators,
Safety Code for Mechanical Power-Transmission Apparatus,
Safety Code for Conveyors, Cableways, and Related Equipment,
Safety Code for Cranes, Derricks, and Hoists,
Safety Code for Manlifts,
Safety Code for Powered Industrial Trucks,
Safety Code for Aerial Passenger Tramways,
Safety Code for Mechanical Packing,
Safety Standards for Compressor Systems, and
Safety Code for Garage Equipment.

The society also sponsors a committee on the *Code for Pressure Piping and Scheme for Identification of Piping Systems.* In addition, ASME is represented on 30 other safety code committees under the American National

Standards Institute.

The ASME Boiler and Pressure Vessel Committee, under the direction of the Policy Board on Codes and Standards, is responsible for the formation and revision of the *ASME Boiler and Pressure Vessel Code*.

The society publishes the *Transactions of the American Society of Mechanical Engineers* and the monthly publication *Mechanical Engineering and Applied Mechanics Review*.

Board of Certified Safety Professionals of America
501 S. 6th St., Champaign, Ill. 61820

This organization was incorporated in the state of Illinois in 1969 to grant the designation of Certified Safety Professional to any individual whose academic training and professional safety experience meet the established criteria. Contact the organization for details on certification.

Flight Safety Foundation, Inc.
1800 N. Kent St., Arlington, Va. 22209

The Flight Safety Foundation works to improve standards and techniques in all aircraft operations. Its specific objectives are to promote research, to provide a forum where controversial safety issues may be resolved, to disseminate accident prevention information (through such means as articles, bulletins, reports, books, and lectures), to encourage the adoption of proven safety devices or procedures, and to try to foresee hazards and to press for corrective action.

To further these objectives, the foundation acts as a clearinghouse for the collection, analysis, and dissemination of safety information. Its personnel participate in safety discussions and safety studies and conduct safety seminars for aviation personnel. It presents awards and otherwise encourages the growth of safety programs. It acts as a catalytic agent in drawing attention to needed improvements and changes in safety techniques. Its publications include:

Accident Prevention Bulletin (monthly)
Pilots Safety Exchange Bulletin (monthly)
Business Pilots Safety Bulletin (monthly)
Aviation Mechanics Bulletin (bimonthly)
Cabin Crew Safety Exchange (bimonthly)

Among the foundation's reports and special studies are:

Aircraft Fueling
Ramp Service
Problem of Bogus Parts
Collision Prevention
Accident Reports and Accident Prevention

Health Physics Society
P.O. Box 156, E. Weymouth, Mass. 02189

Organized in 1955 and incorporated in 1961, the HPS has as its objectives: (*a*) to aid and advance health physics research and applied activities, (*b*) to encourage dissemination of information between individuals in this and related fields, (*c*) to improve public understanding of the problems and needs in radiation protection, (*d*) to initiate and develop programs for training of health physicists, and (*e*) to promote the health physics profession.

Health physics is devoted to the protection of man and his environment from unwarranted radiation exposure. The health physicist is a person engaged in the study of problems and practices of providing radiation protection. He is concerned with an understanding of the mechanisms of radiation damage, with the developing and implementing methods and procedures necessary to evaluate hazards, and with providing protection to man and his environment from unwarranted radiation exposure.

Health Physics is the official journal of the society.

Illuminating Engineering Society
345 East 47th St., New York, N.Y. 10017

The society is the scientific and engineering stimulus in the field of lighting. The work of the Industrial Lighting Committee and its numerous subcommittees for various specific industries should be of particular interest to industrial safety professionals. Many other projects, such as street and highway, aviation, and office lighting, may also be of interest.

Through the society, safety engineers can obtain reference material on all phases of lighting, including authoritative treatise on nomenclature, testing, and measurement procedures.

The society publishes a monthly journal, *Illuminating Engineering,* often of interest to the safety professional, and the *IES Lighting Handbook,* a reference guide.

The society has surveyed a number of industries and has reported its recommendations on light requirements at work areas on each operation, in accordance with its engineering studies.

Industrial Medical Association
150 N. Wacker Dr., Chicago, Ill. 60606

This association fosters the study and discussion of problems peculiar to the practice of industrial medicine and surgery, encourages the development of methods adapted to the conservation and improvement of health among workers, and promotes a more general understanding of the purpose and results of employee medical care.

The association sponsors committees on:

Standards for Medical Service in Industry
Industrial Hygiene and Clinical Toxicology
Problem Drinking
Radiation
Education and Training
Medical Aspects of Driver Safety
Recommended Publications in Occupational
    Medicine
Careers in Industrial Medicine
Industrial Medical Practice
Medical Problems in Overseas Operations
Surgery of Trauma

The official publication of the association is the monthly *Journal of Occupational Medicine.*

An annual American Industrial Health Conference is held, usually in April, in collaboration with the American Association of Industrial Nurses and related groups.

National Association of Suggestion Systems
435 N. Michigan Ave., Chicago, Ill. 60611

The National Association of Suggestion Systems, incorporated as a nonprofit organization, encourages suggestion system activity in industry, commerce, finance, and government. Specific objectives are:

● To increase appreciation of the usefulness of employee suggestion systems.

● To encourage study of the elements nec-essary to successful utilization of employee thinking.

● To provide an opportunity for the personal development of those who represent member institutions.

● To gather and disseminate useful information through meetings, publications, factual surveys, etc.

● To promote personal contacts between suggestion system administrators and the leaders of various industries.

Trade Associations

American Foundrymen's Society
Golf and Wolf Rds., Des Plaines, Ill. 60016

This association serves as a clearinghouse for information in the foundry industry, including safety, hygiene and air pollution control. The following publications available through the society are of interest to safety professionals:

*Control of Emissions from Metal Working
    Operations,*
*Engineering Manual for Control of In-Plant
    Environment in Foundries,*
Foundry Noise Manual,
Foundry Radiation Protection Manual,
Grinding, Polishing, and Buffing Operations,
Safety in Metal Casting,
Water Pollution from Foundry Wastes.

American Gas Association
1515 Wilson Blvd., Arlington, Va. 22209

The association, through its Accident Prevention Committee, serves as a clearinghouse and in an advisory capacity to persons responsible for employee safety and to safety departments of its member companies. Its purposes are to study accident causes, recommend corrective measures, prepare manuals, and disseminate information to the gas industry that will help reduce employee injuries, motor vehicle accidents, and accidents involving the public. The committee meets several times a year and conducts an accident prevention conference each year. It also provides speakers for regional gas associations and other gas industry meetings.

The committee has subcommittees on each

of the following aspects of employee safety: awards and contests, distribution and utilization, education, motor vehicles, posters, program and speakers, publications, gas transmission, research legislative, off-the-job statistics, and promotional and advisory.

Published material includes suggested safe practices manuals, quarterly and annual reports on the industry's accident experience, annual summaries and analyses of employee fatalities, safety flipcharts and posters, and an annual analysis of disabling injuries occurring to employees of more than 100 gas companies. One or more 35mm sound slide films are produced each year.

The association also maintains laboratories in Cleveland and Los Angeles, where gas appliances are tested and certified.

American Iron and Steel Institute
1000 16th St. N.W., Washington, D.C. 20036

This Institute represents 95% of the steel producers in the United States. Special committees on Safety, Industrial Hygiene and Industrial Health meet quarterly. Regional safety committees in the eastern, midwest and western sections of the United States also meet quarterly to provide informational forums for steel plant safety personnel.

Publications available to the public are "Safety Firsts in Steel," "Steel Mill Ventilation" and "Occupational Health Practices—Iron and Steel Industry."

American Meat Institute
1600 N. Wilson Blvd., Arlington, Va. 20007

The institute is the trade research and educational association of the meat packing industry. It has a Safety Committee composed of safety engineers from member companies. Distribution of information is limited to bulletins on special subjects sent to members.

American Mining Congress
1200 18th St. NW., Washington, D.C. 20036

Membership of the congress is from coal, metal, and nonmetal mining companies. This association has a Safety Committee organized within its Coal Division. The monthly *Mining Congress Journal* regularly carries articles and news items concerned with safety.

American Paper Institute
260 Madison Ave., New York, N.Y. 10016

The American Paper Institute, through its Safety Task Force, conducts a broad safety education and informational service for the paper and pulp industry by means of a monthly safety letter covering topics of current interest, and by reports on selected fatalities and disabilities which have occurred during the month. Each report includes the department in which the accident occurred, the machine, tool or other agency involved, and the corrective action taken by the particular mill reporting the accident.

The institute also issues, to all participating companies, a "Safety Reference Desk Book" which contains information on subjects ranging from the organization of a safety department to technical releases on advances in noise control.

Other facets of the safety work of the institute include a subcommittee which is developing standardized machine guard designs, and the P1 Code Committee, which, in conjunction with the National Safety Council, has updated the P1 Code, *Safety in Pulp and Paper Mills.*

The institute compiles an "Annual Injury Frequency Rate Report." It also issues safety awards to those mills or companies having the best annual and five-year cumulative performance on a national and regional basis. Other special awards are given for other outstanding accomplishments in accident prevention.

American Petroleum Institute, Safety and Fire Protection Services
1801 K St. N.W., Washington, D.C. 20006

The objective of the Safety and Fire Protection Services of the American Petroleum Institute is to reduce the incidence of accidental occurrences, such as injuries to employees and the public, damage to property, motor vehicle accidents, and fires. To attain this objective, the Safety and Fire Protection Services:

● Provide statistical reports, pamphlets, data sheets, and other publications to assist the industry in the prevention of accidents and the prevention, control, and extinguishment of fires.

609

• Provide a means for the development and exchange of information on accident prevention and fire protection to be used for education and training in the industry.

• Provide a forum for discussion and exchange of information concerning safe practices and the science and technology of fire protection and safety engineering.

• Promote research and development in the fields of accident prevention and fire protection for the benefit of the petroleum industry as a whole.

Safety and fire protection manuals have been published on such subjects as:

"Cleaning Petroleum Storage Tanks"
"Gas and Electric Cutting and Welding"
"Service Station Safety"
"Safe Practices in Well-Pulling Operations"
"Safe Practices in Bulk-Plant Operations"
"Safe Practices in Drilling Operations"
"Cleaning Mobile Tanks Used for Transportation of Flammable Liquids"
"Guide for Tank Venting"
"Fire Protection in Refineries"
"Fire Protection in Natural-Gasoline Plants"
"Protection Against Ignitions Arising out of Static, Lightning, and Stray Currents"
"Safe Maintenance Practices in Refineries"

Also published are "Annual Summary of Injuries in the Petroleum Industry," "Fire Losses in the Petroleum Industry," "Summary of Motor Vehicle Accidents in the Petroleum Industry," and "Review of Fatal Injuries in the Petroleum Industry."

American Pulpwood Association
605 Third Ave., New York, N.Y. 10016

This association fosters study, discussion, and action programs to guide and help the pulpwood industry in growing and harvesting pulpwood raw material for the pulp and paper industry. The safety and training program of the Association is served through six regional Technical Divisions, each one of which has a safety and training committee.

Available literature includes training guides, notebooks, and technical releases which describe items of personal protective equipment, safe working procedures, and other pertinent accident control items.

American Road Builders Association
525 School St. SW., Washington, D.C. 20024

Accident prevention in the construction of highways is the objective of a three-phase program:

• A continuing promotional program,

• Development of statistical data on accident frequency and severity in the industry,

• Development and utilization of safety standards in the industry.

The program is carried on cooperatively with private and governmental agencies concerned.

Information is disseminated to members through special bulletins, newsletters, and the *American Road Builder* (monthly magazine).

The organization is also interested in highway design, signing, lighting, and other factors relating to highway safety.

American Trucking Associations, Inc.
1616 P St. NW., Washington, D.C. 20036

American Trucking Associations is the national federation representing the trucking industry. Through its 3000-member Council of Safety Supervisors, standards for the selection, training, and supervision of truck fleet personnel have been developed—these form the foundation for the ATA Safety Service, a basic safety program for truck fleets.

Guidebooks, forms, and a driver safety program are available through the ATA Department of Safety as are monthly mailings of safety bulletins, driver letters and safety posters. In addition to providing these services and materials, and acting as secretariat for the Council of Safety Supervisors, the Department of Safety is the trucking industry's liaison with federal agencies and national organizations concerned with safety of highway truck operations.

Membership in the Council of Safety Supervisors is available to any person concerned with truck safety. In addition to regional and national meetings, committees work on such problems as employee selection, training and supervision, accident investigation and reporting, transportation of hazardous materials, physical qualifications, and injury control.

There are 39 state councils of safety supervisors and two councils concerned with safety of tank truck operations and of automobile transporters. Such councils conduct monthly and quarterly meetings and engage in safety engineering activities that relate to their particular interests according to geographic locations or type of operations.

The American Waterways Operators, Inc.
1250 Connecticut Ave. NW., Suite 502, Washington, D.C. 20036

The American Waterways Operators, Inc., is a trade association representing the national interests of operators of towboats, tugboats, barges and shallow-draft self-propelled freighters and tankers engaged in domestic trade primarily over the inland waterways of the United States. AWO has a safety committee which pursues three primary objectives:

• Promotion of individual member company's safety programs,

• More participation by members in the Barge and Towing Vessel Industry Safety Contest which the Association co-sponsors with the National Safety Council, and

• Preparation and distribution of AWO safety posters which the Association issues to its members each month.

The association's manual, *Basic Safety Program for the Barge and Towing Vessel Industry,* is designed either (*a*) to be adopted as a complete company program, or (*b*) be used as a guide to develop or supplement an individual company's program. The manual covers methods and techniques for accident prevention, and contains valuable guidelines on all aspects of personnel safety in the barge and towing vessel industry.

American Water Works Association
2 Park Ave., New York, N.Y. 10016

The AWWA Safety Practices Committee (*a*) collects and compiles annual injury statistics on more than 2500 water utilities employing 80,000 workers, (*b*) issues awards to utilities with outstanding or improved performance in accident prevention, and (*c*) sponsors accident prevention schools for water utility personnel. The committee has prepared a manual entitled *Safety Practice for Water Utilities,* which may be purchased from the Association. Another committee manual, *Tailgate Lectures for Water Utilities,* was issued in 1967.

American Welding Society
2501 NW. 7th St., Miami, Fla. 33125

The society is devoted to the proper and safe use of welding by industry. Through its Committee on Safety Recommendations the society coordinates safe practices in welding by cooperating with committees on promoting new and revising existing standards.

"Safe Practices for Welding and Cutting Containers that Have Held Combustibles" is one booklet produced by this committee and available to industry. Another is "Recommended Safe Practices for Gas Shielded Arc Welding," which covers such subjects as ventilation, clothing, eye protection, and radioactivity. The necessary precautions are indicated, and unfounded fears are dispelled.

The society also sponsors American National Standards Institute Sectional Committee Z49, whose publication, *Safety in Welding and Cutting,* is the authoritative standard in this field. It deals with the protection of workers from accidents, occupational diseases, and fires arising out of the installation, operation, and maintenance of electric and gas welding and cutting equipment.

Frequent articles on safety in welding appear in the official publication of the society, *The Welding Journal.*

Associated General Contractors of America, Inc.
1957 E St. NW., Washington, D.C. 20006

This association of contractors specializes in the building, highway, railroad, and heavy construction fields. All areas of the United States are served by the association's 131 chapters, which carry on their own programs and render assistance to their members.

The association has had about 50 years of continuing interest in the activities of its Accident Prevention Committee, to which member contractors have freely contributed their time and effort. The committee's *Manual of Accident Prevention in Construction* is revised periodically. Each year, the national organization presents awards to members and chapters for significant achievement in accident prevention.

611

23—Sources of Help

Association of American Railroads
1920 L St. NW., Washington, D.C. 20006

All Class I railroads (those with an annual revenue in excess of $5 million) are members of the AAR. The following divisions and committees of the association are concerned with safety:

Safety Section
Communication and Signal Section
Medical Section
Committee on Grade Crossings
Mechanical Division
Engineering Division

The Safety Section holds an annual meeting and sponsors several regional meetings during the year. It issues a monthly newsletter, produces special posters, and publishes pamphlets on railroad safety.

The Bureau for the Safe Transportation of Explosives and Other Dangerous Articles is also located at 1920 L St. NW., Washington, D.C. 20036.

Bituminous Coal Operators' Association
303 World Center Bldg., Washington, D.C. 20006

The Bituminous Coal Operators' Association has a Safety Department to furnish health and accident services to its members.

Can Manufacturers Institute, Inc.
1625 Massachusetts Ave. NW., Washington, D.C. 20036

Can Manufacturers Institute (CMI), the manufacturing trade association for companies who make all-metal cans out of tin plate, black plate, terne plate, aluminum, or a combination thereof, acts as a clearinghouse of information on the metal can industry in the United States.

Most of the operations are carried on through standing committees, one of which is the Safety and Industrial Hygiene Committee. This committee administers a Safety Program for the members that is designed to cover operations peculiar to can plants. In other safety matters, the committee provides referral service.

The program consists of committee-planned material on metal can safety, a monthly *Safety Report,* statistics, safety contest, *Safety Manual,* and articles of special interest.

The Chlorine Institute
342 Madison Ave., New York, N.Y. 10017

Founded in 1924, the institute provides "a means for chlorine producers and firms with related interests to deal constructively with common industry problems—especially in safety, transportation, regulations and legislation, and community relations."

Safety, transportation, medical, and related experiences and problems of all members and their customers are shared through committees, principally the Technical Committee on Container Specifications and Safety, the Transportation Committee, and the Committee on Environmental Health.

Results of committee deliberations are distributed worldwide to chlorine producers, consumers, and other interested groups and persons. In addition to the *Chlorine Manual,* some 70 engineering and design recommendations, specifications, and drawings are continually reviewed and revised.

Compressed Gas Association, Inc.
500 Fifth Ave., New York, N.Y. 10036

The major purpose of the Compressed Gas Association is to provide, develop, and coordinate technical activities in the compressed gas industries, in the interest of safety and efficiency.

Most of the work of the association is done by more than 30 technical committees, made up of representatives of member companies who are highly qualified technically in their respective areas:

Atmospheric gases: nitrogen, oxygen, argon, and the rare gases,
Cryogenic and low temperature gases,
Hydrocarbons: propane, butane, etc.,
Refrigerants: ammonia, fluorocarbons,
Aerosol propellants,
Poisonous gases: hydrogen cyanide, phosgene, etc.
Medical gases: nitrous oxide, cyclopropane, ethylene, etc.,
Other gases: anhydrous ammonia, chlorine, methylamines, sulfur dioxide, carbon monoxide, fluorine.

The association publishes the *Handbook of Compressed Gases,* which contains complete descriptions of 49 widely used gases, and gives the safest recognized methods for handling and storing them. More than 50 standards and bulletins are published. A list is available on request to CGA.

Edison Electric Institute
90 Park Ave., New York, N.Y. 10016

The Edison Electric Institute, a trade organization, is composed of 85 electric utility companies who are dedicated to improving working conditions through development of safe work practices.

The Accident Prevention Committee, which meets twice a year, publishes engineering reports on subjects of interest to safety engineers in the industry. Some representative titles are:

"Use and Care of Pole Climbing Equipment,"
"Resuscitation Manual,"
"Specifications for Linemen's Climbing Equipment," and
"Specifications for Electrical Workers' Insulating Safety Headgear."

Also prepared by the committee are sound slide films on such subjects as pole top safety, electric shock facts, and meter testing.

Graphic Arts Technical Foundation
4615 Forbes Ave., Pittsburgh, Pa. 15213

The education council is a coordinating organization, the membership of which is made up of large national and various local printing and allied trade associations. Individual companies are also members of the council.

In cooperation with the National Safety Council, the Education Council published the *Safety Manual for the Graphic Arts Industry.*

Gray and Ductile Iron Founders' Society, Inc.
Cast Metals Federation Bldg., Rocky River, Ohio 44126

The Gray and Ductile Iron Founders' Society is a trade association which represents gray and ductile iron foundries in the United States, Canada, and Mexico.

The Society's Safety Committee convenes at least once a year to discuss in-plant safety programs for foundries. The committee has developed a manual of safe working conditions for gray and ductile iron foundries, titled "How You Can Work Safely," and a slide film series and numerous inspections, check

lists, medical and accident reporting forms.

The Committee also sponsors a National Safety Recognition Award in the foundry industry and has cooperated with other organizations to provide standard recommendations on lighting in foundries and eye protection.

Industrial Safety Equipment Association, Inc.
2425 Wilson Blvd., Arlington, Va. 22201

This association has represented manufacturers and distributors of industrial safety equipment since 1934. It is devoted to the promotion of public interest in safety and encourages development of efficient and practical devices and personal protective equipment for industry.

It is umbrella-like in nature, providing technical improvement through the constant activities of its 15 separate product groups. Of outstanding importance is the broad representation of its members on numerous American National Standards Institute standards committees engaged in promulgation of industry-wide standards of performance for specific types of personal protective equipment. Safety engineers can be guided by these codes and can anticipate that in the future certified products will conform to published standards.

Institute of Makers of Explosives
420 Lexington Ave., New York, N.Y. 10017

The institute functions through committees composed of qualified representatives of member companies experienced in the activities assigned to the various groups. A technical committee and a committee on traffic and storage conditions are responsible for the institute's booklets on the safe transportation, handling, and use of explosives. Included among such booklets are:

"Explosives in Agriculture,"
"Safety in the Handling and Use of Explosives,"
"Rules for Storing, Transporting and Shipping Explosives," and
"Radio Frequency Energy—a Potential Hazard in the Use and Transportation of Electric Blasting Caps."

Blasting cap safety posters are circulated, especially in schools and to youth groups; related material, available in mat form for newspaper reproduction, is furnished on request.

International Association of Drilling Contractors
211 N. Ervay, Suite 505, Dallas, Texas 75201

This association works to improve oil well-drilling contracting operations as a whole and to increase the value of oil well drilling as an integral part of the petroleum industry.

The association holds an annual safety clinic and has standing and special safety committees of contractor representatives to study current problems. *Safety Hints on Drilling*, an eight-page monthly periodical, gives pertinent safety information, accident case histories, and statistical data.

Safety clinics for tool pushers, drillers, and crewmen, sponsored by the association, are conducted throughout the country in locations where these men normally reside.

A Supervisory Accident Prevention Training Program has been developed to instruct supervisors and tool pushers in how to establish and maintain effective accident prevention programs. A professional safety instructor personally conducts these programs anywhere in the world where 18 to 25 people wish to enroll. Each enrollee receives a specially prepared drilling safety manual. Schools of either two-day or five-day duration are available.

Safety award certificates, cards, safety hat decals, and plaques are given to member personnel and rigs which have completed one or more years without a disabling injury.

The group has produced safety manuals for the industry, inspection reports, color codes, safety signs, studies on protective clothing, and other publications. They have produced color films, film strips, and slides on specific drilling rig safety practices. The association also produces safety posters keyed to the hazards of the drilling industry.

International Association of Refrigerated Warehouses
7315 Wisconsin Ave. NW., Washington, D. C. 20014

The association's Safety Committee conducts a program specifically aimed at reducing accidents and injuries in refrigerated warehouses. The program includes periodic industry surveys to determine types of injuries being experienced, their causes, frequency and severity, safety bulletins, posters, awards, and information on how to establish and operate a safety program. Members are encouraged to submit problems to the Safety Committee for study and suggested solutions.

Linen Supply Association of America
975 Arthur Godfrey Rd., Miami Beach, Fla. 33140

The association represents approximately 95 percent of all the linen supply volume in the United States and Canada. In addition, the Association has members in 18 other countries throughout the world.

The safety program, which was started in 1960, now includes as its major projects and activities:

Accident Prevention Manual
Quarterly Accident Frequency Report
Safe Driver Award Program
Traffic Safety Program
Traffic Safety Posters
Safety Posters for Training Route Salesmen

The safety committee meets regularly (at least twice a year) to review existing programs and to discuss new projects.

Manufacturing Chemists' Association, Inc.
1825 Connecticut Ave. NW., Washington, D.C. 20009

One of the most important of this association's many services is dissemination of information on safe handling, transportation, and use of chemicals.

The association supports a Safety and Fire Protection Committee composed of safety directors selected from its member companies. This committee meets four to six times a year and develops chemical safety information for use by member companies, state and federal health organizations, and the public. It also holds an annual safety workshop for those interested in chemical safety.

The Labels and Precautionary Information Committee establishes guidelines for the preparation of labels for containers in which hazardous chemicals are shipped. The Medical Advisory Committee, composed of medical directors of member companies, provides medical advice on prevention and treatment of chemical injuries.

Other committees of the association which include safety in their programs are:

Technical Committee on Rocket Propellant Safety,

Chemical Packaging Committee,

Committee on Tank Cars,

Tank Trucks, and Portable Tanks, and

Transportation and Distribution Committee.

No regular periodical is published, but manual sheets and "Chemical Safety Data Sheets" covering hazards, both within the industry and in the handling and transporting of its products by the consumer, are issued from time to time. These publications embody the best thinking of chemical safety engineers on the prevention of accidents in the handling of flammable, corrosive, or toxic materials. "Chem-Card" sheets cover recommendations for shipping containers, methods of storage, and safe unloading procedures.

Mine Inspectors' Institute of America
1900 Grant Bldg., Pittsburgh, Pa. 15219

The institute is composed of "all men commissioned by a state, commonwealth, province, county, or by a foreign nation to act as mine inspectors, or chief of a department of mines or minerals; all persons commissioned by the United States Government as director, assistant director, chief of any division of the Bureau of Mines, or as mine inspectors of any grade within the bureau of mines; all persons engaged in safety or responsible supervisory work in or around mines or in teaching safety pertaining to mines; all persons in a responsible position in the manufacture or distribution of equipment for the promotion and the preservation of safety for mines or mining employees; all persons in responsible positions with labor organizations of mining employees; and all mining safety and engineering employees of Mining and Workmen's Compensation Liability insurors."

The institute advocates uniformity of mining legislation and better protection of the lives and health of mine employees. Annual meetings discuss such subjects as mine ventilation, electrical hazards, mine safety, and causes of accidents.

Its annual publication is the *Proceedings of the MIIA.*

National Association of Manufacturers
1776 F St. NW., Washington, D.C. 20006

The NAM safety activities are carried on under the aegis of its Employee Health and Safety Committee which has a dual function: (*a*) promoting sound health and safety policies and programs in industry; and (*b*) working with the federal government to assure that present regulation of health and safety practices in industry and proposals for new legislation are realistic from industry's viewpoint.

National Coal Association
1130 17th St. NW., Washington, D.C. 20036

The association has a Department of Safety, which assists its members in the promotion of safety and accident prevention through cooperative, supervisory, educational, and research endeavors. Cooperation is extended to federal and state governments and private safety organizations.

National Constructor's Association
1133 15th St. NW., Washington, D.C. 20005

This association composed of engineering and construction contractors engaged primarily in designing and building chemical plants, steel mills, power-generating facilities and oil refineries carries out concerted programs to improve and stabilize field labor conditions. Its accident prevention committee carries out a number of programs to enhance the safety of workmen and the public.

National LP-Gas Association
79 W. Monroe St., Chicago, Ill. 60603

Founded in 1931, the association is a nonprofit, cooperative group of producers and distributors of liquefied petroleum gas (LP-gas), manufacturers of LP-gas equipment, and manufacturers and marketers of LP-gas appliances. NLPGA promotes technical information and industry standards in its special field.

Its Safety Committee develops and maintains educational programs to train the public and industry in the safe handling and use of LP-gas and in safe practices for the installation and maintenance of equipment and appliances.

An educational committee working closely with the safety committee arranges training school and conferences for dealers and distributors.

The association distributes informational, technical and legislative bulletins and publishes a magazine, *NLPGA Times,* for members. It holds an annual meeting and sectional meetings with a definite portion of each program devoted to safety.

National Petroleum Refiners Association
1725 De Sales St. NW., Washington, D.C. 20036

Primarily a service organization, the association services the petroleum refining industry and at the same time, serves as a clearinghouse for new ideas and developments in refining technology.

Among the meetings sponsored by the NPRA are those of its Fire and Accident Prevention Group. These are one-day, regional meetings held at different refinery locations. Held throughout the year, these meetings, primarily for first-line supervisors and safety engineers, were established to promote the exchange of information and experiences pertaining to fire and accident prevention in refining operations.

In carrying out the association's efforts in fire and accident prevention, it prepares reports, bulletins, and statistics. A *Newsletter,* published quarterly, keeps members informed of regional group activities. Lost-time injury statistics on a group basis, compiled and reported twice annually, provide the basis upon which association certificate awards are made to individual refineries for exceptional safety achievement. Copies of any of the reports of surveys, meetings, statistics, and the quarterly *Newsletter,* are available without cost upon request.

National Restaurant Association
No. 1 IBM Plaza, Chicago, Ill. 60611

The National Restaurant Association, through its Public Health and Safety Committee, carries on a program to reduce accidents and hazards which affect the safety of food service employees and patrons. The association works with governmental and private agencies and member establishments to identify hazards in food service operations and establish safeguards and procedures to prevent accidents. They prepare or collaborate in preparation of safety bulletins and posters directed specifically to food service safety problems.

Typical of the association's publications are a series of case histories, management-level filmstrip programs, helpful safety articles and reprints which appear periodically in the *Food Service Research Digest,* published three times a year.

National Rural Electric Cooperative Association
2000 Florida Ave. NW., Washington, D.C. 20009

The association, through its Retirement, Safety and Insurance Department, promotes a vigorous and diversified program of accident prevention among rural electric cooperatives. The following is a brief résumé of safety activities:

Job Training and Safety Fund. A fund is distributed to state safety committees based on the proportionate amount of premium developed within each state in the casualty dividend pool. It is required that the fund be used for the benefit of all NRECA member systems which participate in the state safety program.

Publications and Film. Safety articles are published each month in the association's magazine, *Rural Electrification.* Safety releases are issued several times a year to job training and safety instructors and state safety committees. Farm safety folders and electrical safety "comic" booklets are designed and made available to the membership at cost. Safety films have been produced and made available to member systems.

Meetings. NRECA publicizes the National Job Training and Safety Conference; staff members participate each year. NRECA also arranges for exhibits and panels on safety at both annual and regional meetings.

Safety Accreditation Program. NRECA assisted in the development of Safety Accreditation Program which sets criteria to measure system operating practices and provides a procedure to recognize systems for effective and safe operating practices and procedures.

National Soft Drink Association
1101 16th St. NW., Washington, D.C. 20036

The association's safety committee prepares

safety posters for in-plant personnel and driver-salesmen. These posters are mailed to soft drink manufacturers monthly. Through a safety idea award program, the association periodically releases safety suggestions to managers of soft drink plants.

New York Shipping Association, Inc.
80 Broad St., New York, N.Y. 10004

The association consists of American and foreign steamship lines, steamship agencies, contracting stevedores, contracting checkers and clerks, contracting watching agencies, and other employers of waterfront labor within the Port of New York.

Its Safety Bureau inspects all stevedoring operations being performed within the Port. This includes operations on board vessels and at piers and terminals.

The Safety Bureau has several models designed for demonstrating safe operating methods, including: (a) A ship's boom model, (b) guy stress model, (c) lifting model, and (d) safety shoe demonstration. In addition, the Bureau has developed the following safety educational films: *Hand Hooks, Gangways and Ladders, Deck Conditions,* and *Mobile Equipment.*

Portland Cement Association
5420 Old Orchard Rd., Skokie, Ill. 60078

This organization, devoted to research, educational, and promotional activities to extend and improve the use of portland cement, is supported by more than 70 member companies.

The Accident Prevention Department of the association develops and administers the safety program. This work is based upon the safety needs of members as revealed in studies of disabling injuries (reported on a standard form to the association), in committee and regional conference discussions, and in plant visits. Loose-leaf accident prevention manuals and other publications and materials transmit findings of studies to all plants, impart accident prevention technology, publicize the successful features of company safety programs, and stimulate safety work at all levels.

The association holds an annual series of regional safety conferences throughout the country during which mutual safety problems are discussed and ideas exchanged. A summer safety campaign for all-out accident prevention is conducted annually, and special program materials are provided for the fall and holiday seasons.

A safety trophy is awarded annually by the association to cement plants that operate a full calendar year without a disabling injury. More than 150 plants have earned membership in the association's Thousand-Day Club, which comprises plants credited with more than 1000 successive days of safe operation.

Since the inception of the association in 1916, its safety work has been sponsored by the Accident Prevention Committee. The committee meets twice a year to review activities, to consider the effectiveness of the accident prevention program, and to make plans.

Printing Industries of America, Inc.
1730 N. Lynn St., Arlington, Va. 22209

Printing Industries of America, an association of local printers' organizations, actively sponsors the development of safety in the graphic arts through its affiliated local organizations and through its participation in the Graphic Arts Technical Foundation.

Scaffolding and Shoring Institute
2130 Keith Bldg., Cleveland, Ohio 44115

Institute has a deep interest in safety and members try to do everything possible to improve this situation in the construction industry. Listings of "Steel Scaffolding Safety Rules," "Steel Frame Shoring Safety Rules," "Single Post Shore Safety Rules" are available, as are booklets "Recommended Standard Safety Code for Vertical Shoring" and "Recommended Steel Frame Shoring Erection Procedure."

The Society of the Plastics Industry, Inc.
250 Park Ave., New York, N.Y. 10017

The Society of the Plastics Industry, Inc., is the national commercial trade association serving all segments of the plastics industry.

SPI functions through divisions representing functional or product interest and committees representing comprehensive interests of the entire industry. Among its committees concerned with safety matters are:

The Safety and Loss Prevention Committee

Machinery Safety and OSHA Committee
Coordinating Committee on Consumer Safety
Urethane Safety Group

The Safety and Loss Prevention Committee conducts semiannual survey of disabling injury experiences among plastic processors. The recently published *Plastics Industry Safety Handbook* is designed to assist supervisors and other management personnel. Safety posters and other visual aids have been developed to supplement total management safety services.

Machinery Safety and OSHA Committee is made up of representatives from members of the Society's machine builders. This committee is developing machinery standards for promulgation through the American National Standards Institute. These standards will encompass all the machinery that is prominently used in the plastics-processing industry.

The Coordinating Committee on Consumer Safety is responsible for the total consumer fire safety program. It works closely with research organizations and other trade associations to help the general public to better understand how synthetic and natural materials perform in a fire environment. This group also coordinates the activities with similar existing federal programs.

The Urethane Safety Group represents the urethane industry, fosters intelligent regulations, meaningful standards, and truthful communications looking toward safety for the consumer. It will assure that the urethane industry is represented on matters before governmental agencies, industry, and other concerned groups as regards questions of urethane flammability and the safe use of urethane products. It will collect statistics and other data concerning the use of urethane, as may be necessary or appropriate, and will support and conduct research and testing programs related to the safe use of urethane.

Steel Plate Fabricators Association, Inc.

15 Spinning Wheel Rd., Hinsdale, Ill. 60521

The association has an active safety committee which prepares publications on safety for member companies and their employees. Some of these are:

"Supervisor's Accident Prevention Manual for Field Erection and Construction,"

"Basic Safety Rules for Fabricating Shops," and

"Basic Safety Rules for Field Erection and Construction."

The Association sponsors jointly with the National Safety Council the Steel Plate Fabricators Safety Contest.

U. S. Government Agencies

There is an overwhelming amount of safety information available from the federal government concerning all aspects of safety and health, environmental problems, pollution, statistical data, and other industry problems.

Because of the constant change in government agency activities and frequent reorganizations, it is recommended that the reader consult the *United States Government Organization Manual,* published by the Government Printing Office, Washington, D.C. 20402. It can be found in most libraries.

Information on the Occupational Safety and Health Administration, the National Institute for Occupational Safety and Health, along with the Consumer Product Safety Commission can be found in Chapter 2 of this Manual.

Departments and Bureaus in the States and Possessions

It is most important that a safety professional has a good working knowledge of the state agencies responsible for the enforcement of safety and health laws. He should therefore contact the proper groups in his specific labor department or other agency and find out how the various boards, divisions, and services function, so that he can properly direct inquiries when the occasion arises.

Much difficulty can be avoided if the safety professional is familiar with the labor legislation and safety and health codes under which he is working. Codes and laws vary widely in the different states and provinces, and men who have safety jurisdiction in plants in a number of places should understand these differences.

In many cases, the standards set up by the code may serve only as a minimum, and the safety professional will want to compare them with American National Standards or other regulations to establish more rigid rules for his own plant. He should also know the jurisdiction rights of the factory inspectors, so that he can better understand the job they have to do and how he can help them in the performance of their duties.

Labor offices

Labor offices in the several states perform many functions, generally depending on the number and kind of labor problems requiring state attention or supervision.

In the following list, the term "labor offices" interprets the term broadly and includes state and provincial agencies dealing with such matters as safety in general; inspection of boilers, buildings, elevators, mines, and the like; and industrial safety and accidents.

Workmen's compensation laws in the various places may be enforced by the labor department or by the industrial accident commission or board. The safety professional should understand the basic regulations of the compensation law in his area so that he will know how the liability of an employer is determined, what procedure should be followed in case of an accident, and how disputed cases may be appealed. Some compensation cases are complicated and require legal or medical advice.

Every state or province has compensation for industrial injury, but there is little uniformity in the benefits. A safety professional can obtain copies from the state labor department, industrial commission, or board of appeals. (See discussions in Chapter 8, "Workmen's Compensation Insurance.")

The following list of labor offices gives the title of the chief executive of each agency or subdivision to whom inquiries should be addressed. (An up-to-date listing is given in the U.S. Department of Labor Bulletin 177, "Labor Offices in the United States and Canada." The bulletin, revised periodically, is available from the Bureau of Labor Standards, Washington, D.C. 20210.)

Alabama

Department of Industrial Relations (Director), Industrial Relations Building, Montgomery 36104

Division of Workmen's Compensation (Supervisor)

Division of Safety and Inspection, 1816 Eighth Avenue North, Birmingham 35203 (Chief)

Department of Labor (Director), administers occupational safety and health act, State Administrative Bldg. Montgomery 36104

Board of Mine Examiners (Chairman), 1816 Eighth Avenue North, Birmingham 35203

Alaska

Department of Labor (Commissioner), P.O. Box 1149, Juneau 99801

Workmen's Compensation Division (Director)

Industrial Safety Division (Director)

Arizona

Industrial Commission (Chairman), Administers Workmen's Compensation, 1601 W. Jefferson St., Phoenix 85005

Labor Department (Director)

Office of State Mine Inspector (State Mine Inspector), 431 Capitol Bldg., Phoenix 85007

Arkansas

Department of Labor (Commissioner), Capitol Hill Bldg., Little Rock 72201

Safety Division (Director)

Boiler Division (Chief Inspector)

Workmen's Compensation Commission (Chairman), State Capitol Grounds, Little Rock 72201

Mine Inspection Department (State Mine Inspector), 700 First National Bank Bldg., Fort Smith 72901

California

Department of Industrial Relations (Director), 455 Golden Gate Ave., P.O. Box 603, San Francisco 94101

Division of Industrial Safety (Chief)

Division of Industrial Accidents (Director), administers workmen's compensation

State Compensation Insurance Fund (General Manager), 525 Golden Gate Ave., San Francisco 94102

Colorado

Department of Labor and Employment, 200 East Ninth Avenue, Denver 80203

Division of Labor (Director)

Safety Section (Supervisor)

Safety Education (Supervisor)

Boiler Inspection (Supervisor)

Safety Inspection (Supervisor)

Workmen's Compensation Benefit Section (Supervisor)

State Compensation Insurance Fund (Director)

Coal Mine Inspection Department (Chief Inspector), 1845 Sherman St., Denver 80203

Bureau of Mines (Deputy Commissioner), mining other than coal

Connecticut

Labor Department (Commissioner), 200 Folly Brook Blvd., Wethersfield, Hartford 06115
  Bureau of Labor Statistics (Director)
  Division of Factory Inspection (Deputy Commissioner of Factory Inspection)
  Boiler Safety Board (Chairman)
Workmen's Compensation Commission (Chairman), 110 Broadway, Norwich 06360

Delaware

Department of Labor (Secretary), 801 West St., Wilmington 19899
Division of Industrial Affairs (Director), administers workmen's compensation, 1102 West St., Wilmington 19801
  Occupational Safety and Health Section (Safety Engineer)

District of Columbia

Minimum Wage and Industrial Safety Board (Chairman), 614 H St. NW., Washington 20001
  Industrial Safety Division (Director), 615 Eye St. NW., Washington 20001
Office of Workmen's Compensation Programs (Deputy Commissioner), Vanguard Bldg., Washington 20211

Florida

Department of Commerce, Division of Labor and Employment Opportunities, Caldwell Bldg., Tallahassee 32304
  Bureau of Workmen's Compensation (Chief)
  Industrial Relations Commission (Chairman)

Georgia

Department of Labor (Commissioner), State Labor Bldg., Atlanta 30334
  Inspection Division (Chief)
Board of Workmen's Compensation (chairman), 494 Labor Bldg., 254 Washington St. SW., Atlanta 30334

Guam

Department of Labor (Director), P.O. Box 884, Agana 96910

Hawaii

Department of Labor and Industrial Relations (Director), 825 Mililani St., Honolulu 96813
  Workmen's Compensation Division (Administrator)
  Industrial Safety Division (Administrator)

Idaho

Department of Labor (Commissioner), Industrial Administration Bldg., 317 Main St., Boise 83702
  Industrial Commission (Chairman), administers workmen's compensation
  State Insurance Fund (Manager), 317 Main St.,

P. O. Box 1038, Boise
Office of Inspector of Mines (Inspector), 117 State House, Boise 83707

Illinois

Department of Labor (Director), 160 N. LaSalle St., Chicago 60601
  Division of Safety Inspection and Education (Chief)
Industrial Commission (Chairman), administers workmen's compensation, 160 N. LaSalle St., Chicago 60601
Department of Mines and Minerals (Director), 704 State Office Bldg., Springfield, 62706

Indiana

Division of Labor (Commissioner), Indiana State Office Bldg., Room 1013, 100 N. Senate Ave., Indianapolis 46204
  Bureau of Building and Factory Inspection (Director)
  Bureau of Safety Training and Education (Director)
  Elevator Safety Subdivision (Director)
  Bureau of Mines and Mining (Director), Terre Haute 47802
Industrial Board (Chairman), administers workmen's compensation, State Office Bldg., Room 601, Indianapolis 46204

Iowa

Bureau of Labor (Commissioner), State House, East Seventh and Court Aves., Des Moines 50319
  Boiler Inspection Department (Boiler Inspector)
Workmen's Compensation Service (Industrial Commissioner), State Office Bldg., Des Moines 50319
Department of Mines and Minerals (State Mine Inspector), 812 East Grand Ave., Des Moines 50319

Kansas

Department of Labor (Commissioner), 401 Topeka Ave., Topeka 66603
  Industrial Safety Division (Director)
Workmen's Compensation Director (Director), State Office Bldg., Room 1007N, Topeka 66612

Kentucky

Department of Labor (Commissioner), State Office Bldg. Annex, Frankfort 40601
  Workmen's Compensation Board (Chairman)
  Industrial Safety Board (Chairman)
  Division of Occupational Safety (Director)
Department of Mines and Minerals (Commissioner), P.O. Box 680, Lexington 40501

Louisiana

Department of Labor (Commissioner), 205 Capi-

620

tol Annex, P.O. Box 44063, Baton Rouge 70804
 Division of Occupational Safety and Health (Director)
 Division of Boiler Inspection (Director)
Workmen's Compensation (court administered)

Maine
Department of Labor and Industry (Commissioner), State Office Bldg., Augusta 04330
 Division of Boiler Inspection (Chief Inspector)
 Division of Industrial Safety (Director)
 Division of Elevator Inspection (Supervising Inspector)
Industrial Accident Commission (Chairman), administers workmen's compensation, State Office Bldg., Augusta 04330

Maryland
Department of Labor and Industry (Commissioner), 301 West Preston St., Baltimore 21201
 Division of Boiler Inspection (Chief Boiler Inspector)
 Division of Industrial Safety (Director)
 Division of Safety Inspection (Chief)
 Occupational Health and Safety Advisory Board (Chairman)
Workmen's Compensation Commission (Chairman), administers Workmen's Compensation, 108 E. Lexington St., Baltimore 21202
State Accident Fund (Chairman), 301 West Preston St., Baltimore 21201

Massachusetts
Department of Labor and Industries (Commissioner), State Office Bldg., Government Center, 100 Cambridge St., Boston 02202
 Division of Industrial Safety (Director)
Industrial Accident Board (Chairman), administers workmen's compensation (Division in the Department of Labor and Industries but not under its supervision or control; contact Board directly)
Department of Public Safety (Commissioner), 1010 Commonwealth Ave., Boston 02115
 Division of Inspections (Chief of Inspections)
 Board of Boiler Rules (Chairman)
 Division of Fire Prevention (State Fire Marshal)

Michigan
Department of Labor (Director), 300 E. Michigan Ave., Lansing 48913
 Bureau of Safety and Regulation (Director)
 Bureau of Workmen's Compensation (Director)
 Ocupational Safety Division (Chief)
 Workmen's Compensation Appeal Board (Chairman)

Minnesota
Department of Labor and Industry (Commis-

sioner), 110 State Office Bldg., St. Paul 55101
 Division of Accident Prevention (Chief)
 Division of Boiler Inspection (Chief)
Workmen's Compensation Commission (Chairman), 110 State Office Bldg., St. Paul 55101

Mississippi
Workmen's Compensation Commission (Chairman), Barnett Bldg., P.O. Box 651, Jackson 39205
Division of Occupational Health and Factory Inspection (Board of Health) (Director), 2423 N. State St., Jackson 39205

Missouri
Department of Labor and Industrial Relations (Chairman, Industrial Commission), 1904 Missouri Blvd., Jefferson City 65101
 Division of Workmen's Compensation (Director), P.O. Box 58
 Division of Industrial Inspection (Director), P.O. Box 449
 Division of Mine Inspection (Director)

Montana
Department of Labor and Industry (Commissioner), Mitchell Bldg., 1331 Helena Ave., Helena 59601
Division of Workmen's Compensation (Administrator), administers workmen's compensation and industrial safety program, 815 Front St., Helena 59601

Nebraska
Department of Labor (Commissioner), State Capitol, Lincoln 68509
 Division of Safety (Chief Boiler Inspector) (State Safety Engineer)
Workmen's Compensation Court (Presiding Judge), State Capitol, Lincoln 68509

Nevada
Department of Labor (Commissioner), 111 W. Telegraph St., Carson City 89701
Industrial Commission (Chairman), administers workmen's compensation and safety functions, 515 E. Musser St., Carson City 89701
 Department of Industrial Safety (Director)
Inspector of Mines (Inspector), 6 Capitol Bldg., Carson City 89701

New Hampshire
Department of Labor (Commissioner), 1 Pillsbury St., Concord 03301
 Inspection Division (Chief Inspector)
 Workmen's Compensation Division (Claims Supervisor)

New Jersey
Department of Labor and Industry (Commissioner), John Fitch Plaza, P.O. Box V, Trenton, 08625

Division of Labor, Bureau of Engineering and Safety (Deputy Director)
Division of Workmen's Compensation (Director), P.O. Box W,

New Mexico

Labor and Industrial Commission (Labor Commissioner), 137 E. DeVargas St., Santa Fe 87501
State Inspector of Mines (Inspector), 505 Marquette N.W., Albuquerque 87101
Environmental Improvement Agency (Director) administers safety act, P.O. Box 2348, Santa Fe 87501

New York

Department of Labor (Industrial Commissioner), State Campus, Albany 12226
Division of Industrial Safety Service (Director), 80 Centre St., New York, N.Y. 10013
Division of Industrial Hygiene (Director), 80 Centre St., New York, N.Y. 10013
Workmen's Compensation Board (Chairman), 50 Park Pl., New York, N.Y. 10007
State Insurance Fund (Executive Director), 199 Church St., New York, N.Y. 10007

North Carolina

Department of Labor (Commissioner), P.O. Box 1151, Raleigh 27602
Bureau of Boiler Inspection (Chief Boiler Inspector)
Bureau of Elevator Inspection (Elevator Inspector)
Bureau of Mine Inspection (Chief Inspector)
Industrial Commission (Chairman), administers workmen's compensation, Albemarle Bldg., P.O. Box 27546, Raleigh 27603
Safety Director

North Dakota

Department of Labor (Commissioner), State Capitol, Bismarck 58501
Workmen's Compensation Bureau (Chairman), State Capitol, Bismarck 58501
Safety Inspector
State Boiler Inspector
State Mine Inspector

Ohio

Department of Industrial Relations (Director), 220 Parsons Ave., Columbus 43215
Division of Factory and Building Inspection (Chief)
Division of Boiler Inspection (Chief)
Division of Mines (Chief)
Division of Elevator Inspection (Chief)
Division of Pressure Piping Inspection (Chief)
Industrial Commission (Chairman), Ohio Department Bldg., Columbus 43215
Division of Safety and Hygiene (Superintendent)

Bureau of Workmen's Compensation (Administrator), Ohio Department Bldg., Columbus 43215

Oklahoma

Department of Labor (Commissioner), State Capitol, Oklahoma City 73105
Chief Safety Engineer
State Industrial Court (Presiding Judge), administers workmen's compensation, State Capitol, Oklahoma City 73105
Department of Mines (Chief Inspector), State Capitol Building, Oklahoma City 73105

Oregon

Bureau of Labor (Commissioner), 115 Labor and Industries Bldg., Salem 97310
Workmen's Compensation Board (Chairman), Labor and Industries Bldg., Salem 97310
Accident Prevention Division (Director)

Pennsylvania

Department of Labor and Industry (Secretary), Labor and Industry Bldg., Harrisburg 17120
Bureau of Occupational and Industrial Safety (Director)
Bureau of Industrial Standards (Director)
Workmen's Compensation Board (Chairman)

Puerto Rico

Department of Labor (Secretary of Labor), 414 Barbosa Ave., Hato Rey 00917
Bureau of Work Accident Prevention (Director)
Bureau of Labor Standards (Director)
Inspection Office (Executive Assistant to the Secretary of Labor)
State Insurance Fund (Administrator), G.P.O. Box 528, San Juan 00936
Industrial Commission (President), G.P.O. Box 4416, San Juan 00936

Rhode Island

Department of Labor (Director), 235 Promenade St., Providence 02908
Division of Industrial Inspection (Chief)
Division of Workmen's Compensation (Chief)
Industrial Code Commission for Safety and Health (Chairman)
Workmen's Compensation Commission (Chairman), handles disputed cases, 25 Canal St., Providence 02903

South Carolina

Department of Labor (Commissioner), Rutledge Bldg., 1710 Gervais St., P.O. Box 11329, Columbia 29201
Division of Inspection (Director)
Division of Safety (Director)
Industrial Commission (Chairman), administers workmen's compensation, 1429 Senate St., Columbia 29201

South Dakota
Department of Labor and Management Relations, State Capitol Bldg., Pierre 57501
State Inspector of Mines (Inspector), P.O. Box 101, Whitewood 57793

Tennessee
Department of Labor (Commissioner), Cordell Hull Bldg., Nashville 37219
  Workmen's Compensation Division (Director)
  Workshops, Factories, and Elevators Division (Director)
  Division of Mines (Director), Knoxville 37902
  Division of Boiler Inspection (Director)
  Construction Safety (Director)
  Elevator Safety Board (Chairman)

Texas
Industrial Accident Board (Chairman), State Insurance Bldg., Austin 78714
Bureau of Labor Statistics (Commissioner), Box T, Capitol Station, Austin 78711
  Boiler Inspection Division (Chief Boiler Inspector)
  Safety Division (Director)

Utah
Industrial Commission (Chairman), 350 East Fifth South, Salt Lake City 84111
  Industrial Safety (including mine inspection) (Commissioner)
  Workmen's Compensation (Commissioner)

Vermont
Department of Labor and Industry (Commissioner), administers Workmen's Compensation, Montpelier 05602
  Industrial Safety (Deputy Commissioner)

Virgin Islands
Department of Labor (Commissioner), P.O. Box 708, Christiansted, St. Croix 00820
  Division of Workmen's Compensation (Deputy Commissioner), St. Thomas 00801

Virginia
Department of Labor and Industry (Commissioner), P.O. Box 1814, Ninth Street Office Bldg., Richmond 23214
  Division of Industrial Safety (Director)
  Division of Mines and Quarries (Chief)
Industrial Commission, Blanton Bldg., P.O. Box 1794, Richmond 23214
  Department of Workmen's Compensation (Chairman)

Washington
Department of Labor and Industries (Director), General Administration Bldg., Olympia 98501
  Safety Division (Supervisor)
  Electrical Division (Chief Electrical Inspector)
  Industrial Insurance Division (Supervisor)

West Virginia
Department of Labor (Commissioner), 1900 Washington St., East, Charleston 25305
  Division of Safety (Director)
Workmen's Compensation Fund (Commissioner), 112 California Ave., Charleston 25305
Department of Mines (Director), Capitol Bldg., Charleston 25305

Wisconsin
Department of Industry, Labor and Human Relations (Chairman), Hill Farms State Office Bldg., P.O. Box 2209, Madison 53701
  Industrial Safety and Buildings Division (Director)
  Division of Labor Standards (Director)
  Workmen's Compensation Division (Administrator)

Wyoming
Department of Labor and Statistics (Commissioner), 304 Capitol Bldg., Cheyenne 82001
Workmen's Compensation Department (Director), 2305 Carey Ave., P.O. Box 408, Cheyenne 82001
State Mine Inspection Department (State Inspector of Mines), P.O. Box 1094, Rock Springs 82901

Health and hygiene services

Departments or boards of health and industrial hygiene services are integral parts of the organization of each of the states, the District of Columbia, and the autonomous territories of the United States.

It is to these organizations that the safety professional must look for his state's specific standards and recommendations on such points as occupational health, food and health engineering, disease control, water pollution, and other facets of the overall field of plant hygiene.

Industrial hygiene units usually function full time or, in several states, on a limited basis. In addition to the units that operate under state health departments, a number of other industrial hygiene units are run by municipalities or other local authorities.

In addition to direct industrial hygiene services, these state units are able to bring to industry a more or less complete health program by integrating their work with that of other divisions in the state government, such as sanitation and infectious disease control.

State and local programs coordinate their

efforts with the U.S. Department of Health, Education, and Welfare, Public Health Service, and the U.S. Department of Labor. They also cooperate with medical societies and nurses' associations. (See listings earlier in this chapter.)

The names and addresses of such state, commonwealth, or territorial agencies with which the safety professional may need to communicate, are available in an up-to-date listing of health authorities in Public Health Publication No. 75, "Directory of State, Territorial, and Regional Health Authorities," for sale by the U.S. Government Printing Office. Occupational health personnel are listed in the annual "Directory of Governmental Occupational Health Personnel," available from the Bureau of Occupational Safety and Health, Environmental Control Administration, Consumer Protection and Environmental Health Service, Department of Health, Education, and Welfare, 1014 Broadway, Cincinnati, Ohio 45202.

Canadian Departments, Associations, and Boards

In all provinces of Canada there is a Workmen's Compensation Board or Commission. Some of these handle accident prevention directly. In other provinces there are provisions similar to Section 110 of the Quebec Workmen's Compensation Act, which stipulates "that industries included in any of the classes under Schedule I may form themselves into an Association for accident prevention and formulate rules for that purpose. Further, the Workmen's Compensation Commission, if satisfied that an Association so formed sufficiently represents the employers in the industries included in the class, may make a special grant toward the expense of any such Association."

It is under these provisions that the various safety associations were organized and are functioning. In some provinces, accident prevention is directly assumed by the board itself by establishing a safety department.

Furthermore, all provinces have legal safety requirements which are administered by the Department of Highways, Department of Labor, and the Department of Mines. These sources can be contacted by writing to the deputy minister of the department located in the capital of each province.

Governmental agencies

The following list, arranged alphabetically by provinces, shows the departments of labor and workmen's compensation boards. For the convenience of safety professionals who want to communicate with such agencies, titles of top executives are indicated.

Federal
Canada Department of Labour, Accident Prevention and Compensation Branch (Director), 340 Laurier Avenue West, Ottawa, Ontario
This branch is responsible for the implementation and administration of the Canada Labour (Safety) Code, which became effective January 1, 1968. It is also responsible for the development of occupational safety regulations and standards, for the inspection of work places under federal jurisdiction, and for the enforcement of the Safety Code and all regulations prescribed under its authority.

Alberta
Department of Labour (Minister), 10808 99th Ave., Edmonton
Workmen's Compensation Board (Chairman), 9912 107th St., Edmonton

British Columbia
Department of Labour (Minister), Parliament Buildings, Victoria
Chief Factory Inspector, 707 W. 37th Ave., Vancouver 13
Workmen's Compensation Board (Chairman), 707 W. 37th Ave., Vancouver 13

Manitoba
Department of Labour (Minister), Legislative Building, Winnipeg 1
Mechanical and Engineering Division (Chief Inspector), Legislative Buildings, Winnipeg 1
Workmen's Compensation Board (Chairman), Legislative Buildings, Winnipeg 1

New Brunswick
Department of Labour (Minister), P.O. Box 580, Fredericton
Workmen's Compensation Board (Chairman)

Newfoundland
Department of Labour (Deputy Minister), Confederation Building, St. John's
Boiler Inspection Branch (Chief Inspector)
Industrial Standards (Industrial Standards Officer)
Workmen's Compensation Board (Chairman)

Nova Scotia

Department of Labour (Minister), Sir John Thompson Bldg., 1256 Barrington St., Halifax

Fire Marshal

Director of Industrial Safety

Workmen's Compensation Board (Chairman), P.O. Box 1150, Halifax

Ontario

Department of Labour (Minister), 8 York Street, Toronto

Safety and Technical Services (Director), 44 Victoria St., Toronto

Boiler Inspection, Construction Safety, Elevator Inspection, Industrial Safety Branch

Prince Edward Island

Workmen's Compensation Board (Chairman), P.O. Box 757, Charlottetown

Department of Labour, Industry and Commerce (Minister), P.O. Box 2000, Charlottetown

Quebec

Department of Labour (Minister), 2875 Laurier Blvd., Quebec 10

Inspection Service of Industrial and Commercial Establishments and Public Buildings (Provincial Director)

Workmen's Compensation Board (Chairman), 524 Bourdageg St., Quebec

Saskatchewan

Department of Labour (Minister), 2350 Alberta St., Regina

Electrical and Elevator Inspection (Chief)

Boilers and Pressure Vessels (Chief Inspector)

Fire Commissioner

Labor Standards (Director)

Workmen's Compensation Board (Chairman), 1840 Lorne St., Regina

Private provincial agencies

In many provinces of Canada, accident prevention associations exist under the law. Some are concerned with the safety problems of a single industry; others comprise associations representing various industries; still others encompass all industry within the province. Employers assessed under the Workmen's Compensation Act of the province become members.

Industrial accident prevention in Canada is largely handled by the following groups:

British Columbia

British Columbia Loggers' Association (Manager), 550 Burrard St., Vancouver

British Columbia Lumber Manufacturers Association (Safety Director), 302 Forest Industries Bldg., 550 Burrard St., Vancouver

New Brunswick

New Brunswick Accident Prevention Association (Secretary-Manager), P.O. Box 755, 55 Canterbury St., Saint John

Nova Scotia

Nova Scotia Accident Prevention Association (Manager), 231 Hollis St., Halifax

Ontario

Transportation Safety Association (Trucking) (General Manager)

Industrial Accident Prevention Associations (General Manager), 74 Victoria St., Toronto. Included are the following Accident Prevention Associations:

Ceramics and Stone
Chemical Industries
Food Products
Forest Products
Metal Trades
Miller's Feed Manufacturers
Leather, Rubber, and Tanners
Ontario Retail
Printing Trades
Textile and Allied Industries
Woodworker's

Ontario Pulp and Paper Manufacturers' Association (Manager), 74 Victoria St., Toronto

Ontario Safety League (General Manager), 409 King St., Toronto

Mines Accident Prevention Association of Ontario (Executive Director), 199 Bay St., Toronto

Electric Utilities Safety Association of Toronto (Manager), 81 Kelfield St., Rexdale

Forest Products Accident Prevention Association (General Superintendent), 183 First Ave., West, North Bay

Ontario Highway Construction Safety Association

Ontario Pulp and Papermakers Safety Association (Secretary-Engineer)

Steel Erectors Accident Prevention Association. Above addresses: 90 Harbour St., Toronto

Mines Accident Prevention Association of Ontario (Executive Director), 320 Bay St., Toronto

Quebec

Industrial Accident Prevention Association of Quebec (General Manager), 50 Place Cremazie, Montreal

The Lumberman's Accident Prevention Association, Inc. (President, Secretary-Manager), 65 Ste. Anne St., Quebec

Quebec Asbestos Mining Association (Safety Director), Suite 102, 1510 Drummond St., Montreal

The Quebec Metal Mines Accident Prevention

Association (Manager), 46 St. Louis St., Quebec

The Quebec Public Utilities Safety Association (Manager), 600 Dorchester St., W., Montreal

The Quebec Pulp and Paper Safety Association, Inc. (President, Secretary-Treasurer), 65 Ste. Anne St., Quebec

Western Quebec Mines Accident Prevention Association, 46 St. Louis Street, Quebec

Saskatchewan

Miller's Accident Prevention Association, Flour Mill, Saskatchewan

Motor Transport Accident Prevention Society, 6 Victoria Park Bldg., Regina

In addition to the foregoing groups, there is also the Canadian Industrial Safety Conference, with officers elected annually. The organization operates in a manner not dissimilar to that of the Industrial Conference conducted in the United States under the auspices of the National Safety Council. Further information may be obtained from the President, 550 Burrard Street, Vancouver, B.C.

International Safety Organizations

Inter-American Safety Council
(Consejo Interamericano de Seguridad)
33 Park Place, Englewood, N.J. 07631

The Inter-American Safety Council was founded and incorporated in 1938, as a non-commercial, nonpolitical, and nonprofit educational association for the prevention of accidents. It is the Spanish and Portuguese language counterpart of the National Safety Council.

The Council is the first and only association of its kind rendering services to all industries and agencies in the Latin American countries and Spain. The objectives are to prevent accidents—to reduce the number and severity of accidents in every activity, both on the job and off the job. The services which the Council provides for its members are paid by membership dues and sales of the Council's monthly publications and other educational materials. All of its work is done from the headquarters in New Jersey.

Membership is open to all industries, organizations, institutions, or other groups with two or more employees, interested in accident prevention in Latin America and Spain. More than 1800 plants or work locations in 22 countries are members or are using the materials and services of the Council. In addition, over 300 universities, technical schools, public libraries, and the like, receive the monthly publications free of charge.

Among the services available to members are: monthly publications, annual contest, special awards, consultation, statistical service, reproduction and translation rights and participation in the election of Council officers. In addition, the Council acts as a clearing house of accident prevention materials available in the United States. A catalog is available from the organization.

The Council's monthly publications include two magazines and safety posters:

Noticias de Seguridad (Safety News)
El Supervisor (The Supervisor)
Safety posters in sizes 8½ by 11 and 17 by 22 inches

In addition, the Council publishes translations of publications, films, safety slides, training programs and other materials of the National Safety Council and other accident prevention organizations.

Among the materials available in Spanish are:

Training Courses.
"S-T-E-P Program" (programmed instruction)
"Defensive Driving," a six-film series
"Safety and the Foreman," a four-film series, also available in Portuguese
"Psychology of Safety in Supervision," a six-lesson program
"Defensive Driving Program" (DIP)

Training materials.
Films—more than 40 films are available in Spanish and Portuguese
Safety slides—series of 30–35 mm color slides
Supervisory pamphlets
Booklets
Posters
Miniature posters
"Five-Minute Safety Talks"
Conversation topics
Miniguias (pocket-size guides)

Technical materials.

Accident Prevention Manual for Industrial Operations (National Safety Council's 4th edition, 14 chapters available)
Construction Accident Prevention Manual
Fleet Safety Manual
Fundamentals of Industrial Hygiene
Industrial Accident Prevention Manual, by H. W. Heinrich
Introduction to Industrial Hygiene, by J. J. Bloomfield
National Electrical Code Handbook (NFPA)
Power Line Safety Manual
"Safety Subjects" (U.S. Department of Labor)
Practiguias ("Safe Practices" pamphlets)
First Aid Manual (American Red Cross)
Fire Prevention and Control
Portable Fire Extinguishers
Respiratory Protection
The Supervisor and Accident Prevention
Explosives Manual
U.S. Bureau of Mines Publications
Recording and Measuring Work Injury Experience and Method of Recording Basic Facts Relating to Nature and Occurrence of Work Injuries (American National Standards Institute)
Construction Safety Manual

International Association of Governmental Labor Officials

401 Railway Labor Bldg., First and "D" Sts. NW., Washington, D.C. 20210

This association, made up both of state labor commissioners and of ministers of labor from Canadian provinces, holds annual meetings to consider problems in industrial safety and health.

International Association of Industrial Accident Boards and Commissions

Justice Building, State Capitol Grounds, Little Rock, Ark. 72201

This group, composed of American, Canadian, New Zealand, and Philippine members, is concerned with workman's compensation and safety.

International Labor Organization

International Labor Office, CH 1211, Geneva 22, Switzerland

The International Labor Organization (ILO), a specialized agency associated with the United Nations, was created by the Treaty of Versailles in 1919 as part of the League of Nations. Its purpose is to improve labor conditions, raise living standards, and promote economic and social stability as the foundation for lasting peace throughout the world. To this purpose, one of ILO's functions is "the protection of the worker against sickness, disease, and injury arising out of his employment."

The organization consists of about 120 member countries, including the United States (which joined in 1934). ILO functions through an annual conference of member states, a governing body, advisory committees, and a permanent office, the International Labor Office. ILO is distinctive from all other international agencies in that it is tripartite in character—that is, the conference, the governing body, and some of the committees are composed of representatives of governments, employers, and workers.

In the field of safety and health, the International Labor Office maintains a permanent international staff of medical doctors, engineers, and industrial hygienists. Assistance in specific fields is given by panels of consultants, drawn from all parts of the world to act in an advisory capacity and to discuss problems, draft regulations, or render help in emergencies. The office also maintains an Occupational Safety and Health Information Centre (CIS) which analyses and provides abstracts of relevant articles appearing in official publications and journals throughout the world. (See bibliography of "Industrial hygiene and medicine" periodicals later in this chapter.)

The office is also assisted by temporary committees of experts, which have discussed safety in coal mines, the prevention and suppression of dust in mining, tunneling, and quarrying, safety and health in dock work, protection of workers against ionizing radiation, industrial medical services, pneumoconiosis and other matters relating to occupational safety and health. Meetings of experts and symposia have been held on such subjects as: the medical inspection of labor, electrical accidents, the international classification of radiographs of pneumoconiosis, and radiological health and safety in mining and milling of unclean materials.

The United States has several members on the panels and has been represented on all the temporary expert committees and special conferences.

The main tasks of the ILO in the field of occupational safety and health are:

International instruments. These include conventions and recommendations, and also model safety codes and codes of practice.

An *Encyclopedia of Occupational Health and Safety* has been prepared in English and French, to succeed *Occupation and Health,* which was published in 1930. This is designed to provide guidance to a wide range of people concerned with health, safety, and welfare at work. Although problems are reviewed from an international angle, special account is taken of the needs of developing countries.

The compilation of technical studies.

The publication of medical and technical studies.

Direct assistance to governments, by furnishing experts, drafting regulations, supplying information, etc.

Collaboration with other international organizations, the World Health Organization, and the International Organization for Standardization.

Assistance to national safety and health organizations, research centers, employers' associations, trade unions, etc., in different countries.

In general, keeping in touch with the safety and health movement throughout the world and assisting the movement by all the means in its power.

Under the expanded technical assistance program of the United Nations, the ILO office has sent safety and health and labor inspection experts to Argentina, Burma, Chile, Egypt, Guatemala, Indonesia, Iraq, Iran, Nicaragua, Pakistan, the Philippines, Thailand, and Turkey. Under the United Nations Development Program, the ILO has sent safety and health and labour inspection experts to more than 30 countries throughout the world. In many cases it has supplied technical equipment for laboratory and demonstration purposes.

Fellowships under the program have been awarded to safety and health officials in several countries to enable them to study in highly industrialized countries.

Pan American Health Organization
Pan American Sanitary Bureau, 525 23rd St. NW., Washington, D.C. 20037

Originally established as the International Sanitary Bureau in 1902, the Pan American Health Organization serves as the regional office for the World Health Organization for the Americas. The purposes of the PAHO are to promote and coordinate the efforts of the countries of the Western Hemisphere to combat disease, lengthen life, and promote the physical and mental health of the people.

Programs encompass technical collaboration with governments in the field of public health, including such subjects as sanitary engineering and environmental sanitation, eradication or control of communicable diseases, and maternal and child health.

World Health Organization
Avenue Appia, Geneva, Switzerland

The United States became a member of the World Health Organization on June 21, 1948, by joint resolution of Congress. There are over 125 member nations and 3 associate member nations in WHO.

WHO's objective is the attainment by all peoples of the highest possible level of health —physical, mental, and social. The organization recognizes health as fundamental to the attainment of peace and security, and as being dependent upon the fullest cooperation of individuals and states.

WHO assists countries to strengthen their public health services, including environmental sanitation, mental health, communicable disease control, and health aspects of the peaceful uses of atomic energy. Advisory and demonstration teams are sent to countries requesting assistance.

Educational Institutions

Many colleges and universities offer formal courses in industrial safety. In a publication "Educational Opportunities in Occupational Safety and Health," compiled by the American Society of Safety Engineers, accredited four-year colleges and universities are grouped by those that offer a degree program with concentration on industrial safety and those that offer one or more credit courses as an elective within engineering or education curricula. About one-half of the 1200 four-year colleges and universities in the U.S. provided catalogues for a recent study in occupational safety and health offered by post-secondary educational institutions.

Four-year and two-year degree programs and credit courses are listed in the ASSE report. Courses in water safety, first aid and safety, fire fighting, and driver or traffic education are not listed. The report shows almost 200 four-year institutions offer one or more courses in the five main occupational categories. There are 177 with specialized courses. Write to the ASSE for a copy of this publication. (Address is listed under Professional Societies earlier in this chapter.)

This survey showed that no clear-cut curriculum in safety has been developed. Because of the broad scope, overlap, and interdisciplinary relationships in the field, degree programs (primarily at the Masters level) have been established in engineering and education departments of institutions, mostly as a result of an individual at that institution who saw a need.

As a result of the increasing national interest in product, traffic, and occupational safety, it is ASSE's intent to carry out a vigorous role in working with institutions that presently have, who are actively planning, or who are interested in developing such programs.

Correspondence courses

Numerous courses in industrial safety are available through noncredit seminars and workshops, usually sponsored by local organizations and insurance companies, or through credit and noncredit correspondence courses. Specific data on correspondence courses may be obtained from the National Home Study Council, or the National University Extension Association, and from the following sponsors:

International Correspondence Schools
  Scranton, Pa. 18515 (also Montreal, Quebec and Honolulu, Hawaii)

Massachusetts Department of Education, Division of University Extension, Department of Education Bldg.

Newbury and Exeter Sts.
Boston, Mass. 02116

National Safety Council
425 North Michigan Ave.
Chicago, Ill. 60611

Pennsylvania State University, College of Agriculture, Extension Service
University Park, Pa. 16802

Personnel Training and Development Co.
10 Rodgers Drive
Pittsburgh, Pa. 15238

U.S. Department of Agriculture Graduate School
Fourteenth St. and Independence Ave. SW.
Washington, D.C. 20250

University of Florida, Division of General Extension
Seagle Building
Gainesville, Fla. 32601

University of Tennessee, University of Correspondence Study
Knoxville, Tenn. 37916

University of Wisconsin, Extension Division
Madison, Wis. 53706

Bibliography of Safety and Related Periodicals

Safety

National Safety Council
425 N. Michigan Ave.
Chicago, Ill. 60611

See details of publications in descriptive listing earlier in this chapter.

Journal of the American Society of Safety Engineers (monthly)
American Society of Safety Engineers
850 Busse Highway
Park Ridge, Ill. 60068

Occupational Hazards (bimonthly)
Industrial Publishing Corporation
614 Superior Ave., W.
Cleveland, Ohio 44113

Protection (monthly)
Institution of Industrial Safety Officers of Great Britain
113 Blackheath Park
London S.E.3, England

Safety (quarterly)
Greater New York Safety Council
302 Fifth Ave.
New York, N.Y. 10001

Job Safety and Health (monthly)
U.S. Department of Labor, Occupational Safety and Health Administration
Available from: U.S. Government Printing Office, Washington, D.C. 20402

Occupational Safety and Health (monthly)
Royal Society for the Prevention of Accidents
6 Buckingham Place
London S.W.1, England

Industrial hygiene and medicine

A.M.A. Archives of Environmental Health (monthly)
American Medical Association
535 North Dearborn St.
Chicago, Ill. 60610

American Industrial Hygiene Association Journal (monthly)
American Industrial Hygiene Association
66 S. Miller Rd.
Akron, Ohio 44313

American Journal of Nursing (monthly)
American Nurses Association
2420 Pershing Rd.
Kansas City, Mo. 64108

Industrial Hygiene Digest (monthly)
Industrial Health Foundation
5231 Centre Ave.
Pittsburgh, Pa. 15232

Industrial Hygiene News Report (monthly)
Flournoy & Associates
1845 W. Morse Ave.
Chicago, Ill. 60626

Journal of Occupational Medicine (monthly)
Industrial Medical Association

150 N. Wacker Dr.
Chicago, Ill. 60606

Occupational Health Nursing (monthly)
American Association of Industrial Nurses
79 Madison Ave.
New York, N.Y. 10016

British Journal of Industrial Medicine (quarterly)
British Medical Association
Tavistock Square
London W.C.1, England

CIS Abstracts (8 a year)
International Occupational Safety and Health Information Centre (CIS)
International Labor Office
1211 Geneva 22, Switzerland

Fire

Fire Engineering (monthly)
Dun-Donnelley Publishing Corporation
666 Fifth Ave.
New York, N.Y. 10019

Fire Command! (monthly)

Fire Journal (bimonthly)

Fire News (monthly)

Fire Technology (quarterly)
National Fire Protection Association
470 Atlantic Ave.
Boston, Mass. 02210

Fire Prevention (quarterly)
Fire Protection Association
Aldermary House
Queen St.
London E.C.4, England

Many of the state departments of labor and federal agencies publish periodicals which are available upon request. With the exception of the state agencies, most of the others are mentioned under the subject headings in this section.

A number of trade journals have sections devoted to industrial safety. The safety professional should become acquainted with those servicing the industry in which he is primarily interested.

Materials Handling and Storage

Chapter
24

Preventing Common Injuries   632
Materials handling problems . . . Lifting by hand . . . Team lifting
and carrying . . . Handling specific shapes . . . Machines and
other heavy objects

Accessories for Manual Handling   640
Hand tools . . . Jacks . . . Handtrucks, dollies, and wheelbarrows

Storage of Specific Materials   644
Rigid containers . . . Uncrated stock

Problems With Hazardous Materials   648
Liquids . . . Solids

Shipping and Receiving   655
Floors, ramps, and aisles . . . Lighting . . . Stock picking . . .
Dockboards (bridge plates) . . . Machines and tools . . . Steel
strapping and sacking . . . Glass and nails . . . Pitch and glue
. . . Barrels and kegs . . . Boxes and cartons . . . Car loading

Personal Protection   660

References   661

Materials are handled between operations in every department, division, or plant of a company. It is a job that almost every worker in industry performs—either as his sole duty or as part of his regular work, either by hand or with mechanical help.

Mechanized materials handling equipment is being used more and more. In many industries, materials could not be processed at low cost if it were not for efficient mechanical handling. Although mechanical handling creates a new set of hazards, the net result (entirely aside from increased efficiency) is fewer injuries.

The problems and safe handling techniques involved in the manual and mechanical handling of materials are discussed in this and the next four chapters.

Just as all Job Safety Analyses (JSA's) should contain the amount and extent of manual lifting involved in a particular position of employment, so should this requirement be considered in job selection and placement of employees. Preemployment physical examinations can show those employees who are most likely to incur serious back injuries or hernias.

Areas to be considered when looking at materials handling are the work environment, the need for specific training, and proper materials handling engineering.

Preventing Common Injuries

Materials handling problems

Handling of materials accounts for 20 to 25 percent of all occupational injuries—these injuries are from every part of an operation, not just the stock room or warehouse. As an average, industry moves about 50 tons of material for each ton of product produced. Some industries move 180 tons for each ton of product.

Strains and sprains, fractures and bruises are the common injuries. They are caused, primarily, by unsafe work practices—improper lifting, carrying too heavy a load, incorrect gripping, failing to observe proper foot or hand clearances, and failing to wear personal protective equipment.

To gain insight on his materials handling injury problem, the safety professional should ask these questions about his company's present operating practices and management policies:

1. Can the job be engineered so that manual handling will not be necessary?

2. How do the materials being handled (such as chemicals, dusts, rough and sharp objects) hurt the people doing the handling?

3. Can employees be given handling aids, such as properly sized boxes, adequate trucks, or hooks, that will make their jobs safer?

4. Can the material be conveyed or moved mechanically to eliminate manual handling?

5. Will protective clothing, or other personal equipment, help prevent injuries?

These are by no means the only questions that might be asked, but they serve as a start toward overall appraisal and detailed inquiry. The largest number of injuries occur to fingers and hands. People need instruction if they are to avoid the "natural" ways of picking objects up and putting them down. Training in safe work habits, breakdown and study of even the simplest job operations, and adequate supervision can help minimize these accidents.

General pointers that can be given to those who handle materials include:

1. Inspect materials for slivers, jagged edges, burrs, rough or slippery surfaces

2. Get a firm grip on the object

3. Keep fingers away from pinch points, especially when setting down materials

4. When handling lumber, pipe, or other long objects, keep hands away from the ends to prevent them from being pinched

5. Wipe off greasy, wet, slippery, or dirty objects before trying to handle them

6. Keep hands free of oil and grease

In most cases, gloves, hand leathers, or other hand protectors have to be used to prevent hand injuries. Their use must be controlled, if they are to be worn around moving machinery. (See details in Chapter 19, "Personal Protective Equipment.")

In other cases, handles or holders can be attached to objects themselves, such as handles for moving auto batteries, tongs for feeding material to metal-forming machinery, or wicker baskets for carrying control laboratory samples.

Feet and legs sustain a share of materials handling injuries, the greatest share being to the toes. One of the best ways to avoid injuries is to have people wear foot protection —safety shoes, instep protectors, and ankle guards.

Eyes, head, and trunk of body can also be injured. When opening a wire-bound or metal-bound bale or box, a person should wear eye protection, as well as stout gloves, and take special care to prevent the ends of the bindings from flying loose and striking his face or body. The same precaution applies to coils or wire, strapping, or cable. In many cases, special tools are available to safely cut bonds, strapping, etc.

If material is dusty or is toxic, the person handling it should wear a respirator or other suitable personal protective equipment. (See Chapter 19.)

Manual handling of materials increases the possibility of injuries and adds to the cost of a product. To reduce the number of materials handling injuries and to increase efficiency, materials handling should be minimized by combining or eliminating operations. What materials handling is left should be done mechanically, insofar as possible. For those jobs that cannot be mechanized, here are some suggestions.

Lifting by hand

Physical differences make it impractical to set up safe lifting limits for all workers. Height and weight do not necessarily indicate lifting ability. Some small, wiry men can handle heavier loads than can tall, heavy men.

Codes limiting the load a worker may be required to lift in repetitive operations differ widely. Each employer should know the provisions of his local codes.

When a worker is to lift a heavy or bulky object and carry it to another point, he should first inspect the ground around the object and the route over where it is to be carried, making sure that there is no obstruction or spillage on the floor on which he might trip or slip. He should make sure that clearances are sufficient. If there are obstructions that

will interfere, he should determine a safe route around them.

He should next inspect the object to decide how it should be grasped and how he can avoid sharp edges, slivers, or other things that might cause injury. He may have to turn the object over before he attempts to lift it from the ground. If the object is wet or greasy, he should wipe it dry so that it will not slip from his grasp. If this is not practical, he could use a rope sling or other device that will give him a positive grip.

Most strains and back injuries occur when lifting and setting down objects by hand. It is important that those who do this work be trained in the proper lifting techniques, if these injuries are to be reduced.

The correct application of six basic factors is essential. In practice, their order of application is as shown in Fig. 24–1:

1. Correct position of feet

2. Straight back

3. Arms close to the body (for lifting and carrying)

4. Correct hold

5. Chin in

6. Use of bodyweight

Correct position of feet. One of the causes of muscle injury, particularly to the back, is loss of balance due to working with the feet too close together. Lifting off the ground, pushing and pulling, or reaching (and in many instances overreaching) may throw the weight of the body off balance. To counteract this, the muscles of the lower limbs and back "stiffen." In the kinetic method, however, the feet are correctly positioned with one placed in the proposed direction of movement and the other in a position where it can give thrust to the body. The worker can decide whether he prefers the forward foot to be his left or right. Following is a description of the kinetic method of lifting.

Straight back. A straight back is not necessarily a vertical back. In the kinetic method, the back is often inclined, particularly when lifting weights from the ground, but the inclination should be from the hips so that

Proper Way To Lift

Lifting is so much a part of everyday jobs that most of us don't think about it. But it is often done wrong, with bad results: pulled muscles, disk lesions, or painful hernia.

Here are six steps to safe lifting.
1. Keep feet parted—one alongside, one behind the object.
2. Keep back straight, nearly vertical.
3. Tuck your chin in.
4. Grip the object with the whole hand.
5. Tuck elbows and arms in.
6. Keep body weight directly over feet.

FEET should be parted, with one foot alongside the object being lifted and one behind. Feet comfortably spread give greater stability; the rear foot is in position for the upward thrust of the lift.

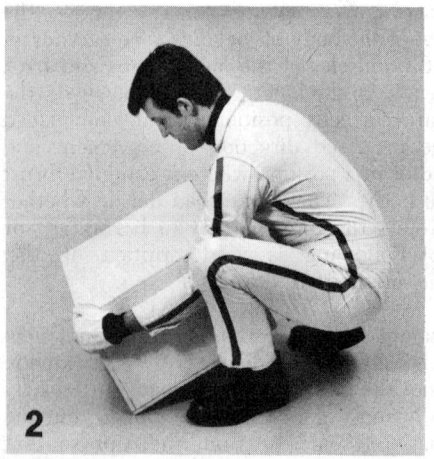

BACK. Use the sit-down position and keep the back straight —but remember that "straight" does not mean "vertical." A straight back keeps the spine, back muscles, and organs of the body in correct alignment. It minimizes the compression of the guts that can cause hernia.

FIG. 24–1.

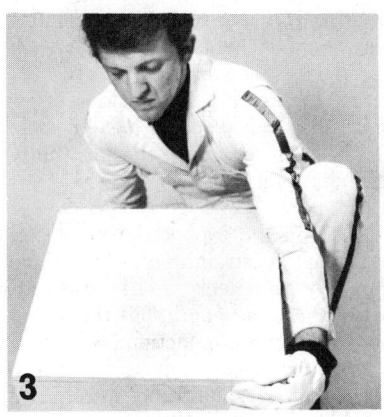

ARMS AND ELBOWS. The load should be drawn close, and the arms and elbows should be tucked into the side of the body. When the arms are held away from the body, they lose much of their strength and power. Keeping the arms tucked in also helps keep body weight centered.

PALM. The palmer grip is one of the most important elements of correct lifting. The fingers and the hand are extended around the object you're going to lift. Use the full palm; fingers alone have very little power. Glove has been removed to show finger positions better.

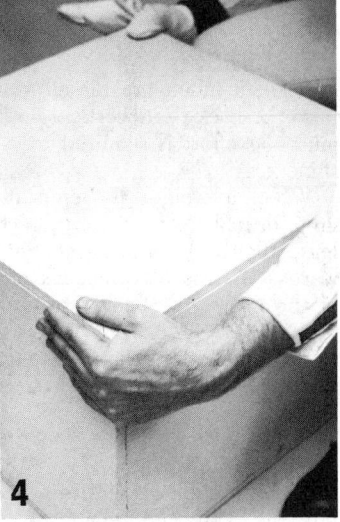

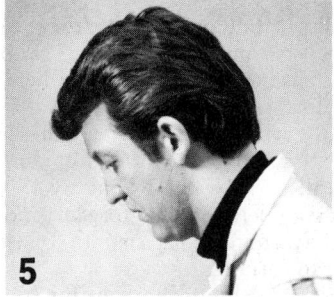

CHIN. Tuck in the chin so your neck and head continue the straight back line and keep your spine straight and firm.

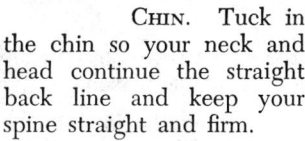

BODY WEIGHT. Position body so its weight is centered over the feet. This provides a more powerful line of thrust and ensures better balance. Start the lift with a thrust of the rear foot.

635

24—Materials Handling and Storage

the normal curvatures are maintained. This normally curved spine is termed a "straight back."

With "straight back lifting" the spine is fairly rigid (nonmedically speaking) and the pressure on the lumbar intervertebral disks is evenly distributed. When lifting with the back bent, the spine forms an arc with the result that the lower back muscles are subject to strain and there is uneven pressure on the disks.

In addition to the risk of intervertebral disk lesions, lifting an object with the back bent and the legs straight imposes excessive stress on the muscles of the back for two reasons. First, the back must be inclined at a greater angle to the vertical for the hands to reach the object. Since the "effective weight" (of the object plus the upper part of the worker's body) increases rapidly as this angle is increased, a much greater effort is required to raise the back to its vertical position. Second, muscular effort is required to "straighten" the spine.

When a weight is being lifted from the ground, making maximum effective use of the legs, the back is straight but inclined forwards. As the lift proceeds by the extension of the knees, the back returns to the vertical position.

The position of the feet and the flexion of the knees are the key factors for maintaining a straight back.

Arms close to the body. When lifting and carrying weights, the arms should be close into the body and remain straight whenever possible. This is because flexing the elbows and raising the shoulders imposes unnecessary strain on the muscles of the upper arms and chest.

Carrying involves a static posture of the arms, and particularly in the case of long distances, any assistance given by the body in supporting the weight will lessen the tension in the muscles. Carrying with the arms straight enables the weight to rest against the thighs.

Correct hold. An insecure grip may be due to taking the load on the finger tips, thus creating undue pressure at the ends of the digits and strain to certain muscles and tendons of the arm. Greasy surfaces often prevent a secure hold—whenever possible such surfaces should be wiped clean. The use of suitable gloves should also be considered.

A full palm grip will reduce local muscle stress in the arms and decrease the possibility of the weight slipping.

Chin in. Raising the top of the head and tucking the chin in straightens the whole spine, not merely the neck. This automatically raises the chest and conditions the shoulders for more efficient arm action. See illustrations on previous page.

This chin-in action should be introduced immediately before lifting and maintained throughout the movement. The worker will be looking down at the early stages of the lift and this may conflict with his desire to raise his head to see where he is going. However, as he returns to the upright position, his head will automatically be raised at the same time.

Use of bodyweight. With the correct positioning of the feet and the flexion and extension of the knees, the weight of the body can be effectively utilized to push and pull objects and to initiate a forward movement such as placing an object on a shelf or walking.

When lifting an object from the ground, the thrust from the back foot, combined with the extension of the knee joints, will move the body forwards and upwards and for a brief period it will be off balance. This is immediately countered by bringing the back leg forward, as in walking, but by this time the lift is completed. This forward movement of the body results in a smooth transition from lifting to carrying.

Here are some techniques for specific situations.

1. If the object is too large or too heavy to be handled by one person, he should get help.

2. Before lifting the load to be carried, the worker should consider the distance to be traveled and the length of time he will have to maintain the grip. He should recognize the fact that his gripping power may tire if he has to carry the load a long distance, especially if he has to climb stairs or ramps.

3. *To place an object on a bench or table,* first set it on edge and push it far enough onto the support to be sure it will not fall. Release it gradually as you set it down. Move it in place by pushing with the hands and body from in front of the object. This method prevents fingers from getting pinched.

4. It is especially important that an object, placed on a bench or other support, be securely set so that it will not fall, tip over, or roll off. Supports should be correctly placed and strong enough to carry the load. Heavy objects, like lathe chucks, dies, and other jigs and fixtures, should be stored at approximately waist height.

5. *To raise an object above shoulder height,* lift it first to waist height. Rest the edge of the object on a ledge, stand, or hip. Shift hand position, so object can be boosted after knees are bent. Straighten out knees as the object is lifted or shifted to the shoulders.

6. *To change direction,* lift the object to carrying position and turn the entire body, including the feet. Do not twist your body. In repetitive work, the person and the material should both be positioned, so that the person will not have to twist his body when moving the material.

7. To deposit an object manually in a tight space, it is safest to slide it into place with the hands in the clear, rather than to lift it.

Team lifting and carrying

When two or more men must carry a single object, they should adjust the load so that it rides level and so that each carries an equal part of the load. Test lifts can be made before proceeding.

When two men carry long sections of pipe or lumber, they should carry them on the same shoulder and walk in step. Shoulder pads will prevent cutting of the shoulders and help to reduce fatigue.

When a gang of men carries a heavy object like a rail, the foreman should direct the work and special tools, such as tongs, should be used. In some companies, on such jobs the foreman uses a whistle to signal for "lift," "walk," or "down." New employees and men

Fig. 24–2.—Two-handled lifter minimizes strain and provides safe control when lifting or lowering a drum.

Courtesy Pickands Mather & Co., Hoyt Lakes, Minn.

who move slowly need special attention.

Handling specific shapes

Boxes, cartons, and sacks. The best way to handle boxes and cartons is to grasp the alternate top and bottom corners and to draw a corner between the legs.

Sacked materials are also grasped at opposite corners. When he has reached an erect position, the worker should let the sack rest against his hip and belly and then swing the sack to one shoulder.

As the sack reaches his shoulder, he should stoop slightly and put his hand on his hip, so that the sack rests partly on his shoulder and partly on his arm and back. His other hand should be holding the sack at the front corner. When the sack is to be put down, he should swing it slowly from his shoulder until it rests

If it is necessary to roll a barrel or drum, a man should push against the sides with his hands. To change direction of the roll, he should grip the chime, not kick the drum with his feet. A clamp device for carrying a drum is shown in Fig. 24–3.

To lower a drum or barrel down a skid, the drum should be turned and slid endwise. Rolling a drum or barrel up a skid takes two men, who should stand outside the skid, not inside the rails nor below the drum or barrel being raised or lowered.

If drums or barrels are to be handled on an incline or skid, it is safer to use ropes or other tackle to control their motion. The drum or barrel should be snubbed with a rope, one end of which is securely fastened to the platform from which the drum or barrel is to be lowered. The rope should then be passed around the barrel or drum, and the operator, keeping a firm grip on the free end, can gradually lower the load.

Sheet metal usually has sharp edges and corners and should be handled with leather gloves, hand leathers, or gloves with metal inserts. Gauntlet-type gloves or wristlets will give added protection to wrist and forearm. Bundles of sheet metal should be handled

FIG. 24–3.—Truck-clamp device provides handles for carrying grease barrel.

Courtesy Pickands Mather & Co., Hoyt Lakes, Minn.

against his hip and belly. If it must be placed on the ground, he should then bend his legs and lower the sack, keeping his back straight.

Barrels and drums. Men handling heavy barrels and drums need special training. Barrels are generally less hazardous to handle than drums because the shape of the barrels aids in upending them. Since the weight and contents of a barrel or drum may vary greatly, special attention should be given to these factors.

If one man must handle a drum, he should have a lifter bar which hooks over the chime and gives him powerful leverage and excellent control. Commercial barrel and drum lifters are available (Fig. 24–2).

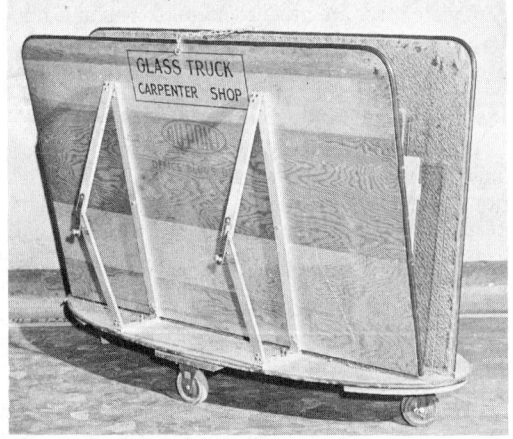

FIG. 24–4.—Service truck holds panes of glass between fixed and hinged panels to ensure safe transport.

Courtesy E. I. du Pont de Nemours & Co., Inc.

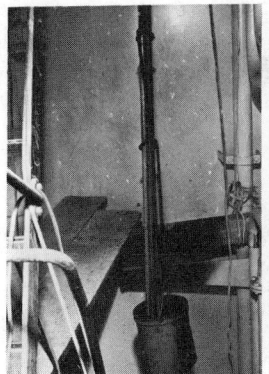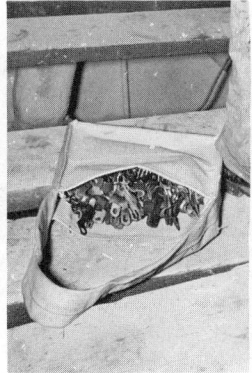

Fig. 24–5.—Special slings, buckets, and bags are used to handle various materials around platforms and staging. Shown, from left: (**A**) Straight pipe without flanges can be lifted with a bucket and single-rope sling, if bail is strong enough. (**B**) Barrel hitch prevents bail from pulling out when heavily loaded; material in buckets is not to be handled over employees. (**C**) Small material should be carried and laid on staging in canvas bag. (**D**) Plates must be well secured to prevent dropping; each side must be supported by a sling made of heavy manila rope or wire.

Courtesy Newport News Shipbuilding and Dry Dock Co.

with power equipment. A sheet metal "grab" can be purchased.

Window glass or other sheet glass should be handled with gloves or hand leathers, and the wrists and forearms should be covered with long leather sleeves. The worker should wear a leather or canvas apron and shoes that protect the feet and ankles completely.

Unless the panes are small, the worker should carry only one at a time and should walk carefully.

He should pick the pane up carefully and carry it with the bottom edge resting in his palm turned outward and with the other hand holding the top edge to steady it. Glass panes should never be carried under the arm because a fall might sever an artery. To carry large panes a considerable distance, an A-frame truck (Fig. 24–4) should be used.

Plate glass should not be carried in such a way that it will bend. Large panes of plate glass should be marked with labels, tape, or grease pencil, so that they can be seen, and should be handled by two or more men, using canvas slings and padding to protect head, neck, and shoulders. Rubber suction cups may also be used, as can an A-frame truck.

Long objects. Long pieces of pipe, bar stock, or lumber should be carried over the shoulder, with the front end held as high as possible to prevent striking other employees, especially at corners. Workers should wear shoulder pads for this operation.

Irregular objects present special problems. Often the object must be turned over or up on end, so that the best possible grip can be secured. If the worker questions his ability to handle the object, because of either its weight or shape, he should get help.

Miscellaneous objects. Fig. 24–5 shows four safe methods of securing miscellaneous material and small objects on overhead platforms or when raising or lowering to different elevations.

Scrap metals. In a scrap storage area, the best possible housekeeping practices should be observed, but even then irregularly shaped, jagged pieces may be mingled in such a way that strips or pieces may fly when a piece is removed from a pile. Therefore workers should be provided with goggles, leather gloves or mittens, safety shoes, safety hats, and protection for the legs and body.

Workers should be cautioned against stepping on objects which may roll or slide.

639

Heavy, round, flat objects (such as car wheels or tank covers) are rolled by hand only with difficulty, and with considerable risk even to skilled men. The operation requires careful training and exacting precautions. It is preferable to use hand trucks or power equipment designed for the purpose.

Heavy rolls can be safely secured and handled by specially designed devices.

Machines and other heavy objects

Manual handling of heavy machinery and like objects requires special skill and knowledge. Sometimes machines or castings weighing 100 tons or more must be moved from freight cars to ground level and into permanent position without use of heavy-duty cranes or similar equipment.

Only general safety principles for such jobs can be suggested. Each one presents its own problems and requires careful study and thorough planning. Some companies build scale models of the machines and the blocking, jacks, rollers, and other equipment to be used and then work out the procedure in miniature.

In all cases, the safe floor load limits for areas over which the machine or part will move, as well as for the place at which it is to be installed or stored, should be determined.

Blocking and timbers should be selected with great care. They should be of hardwood, preferably oak, and of the proper sizes to allow the machine to be safely blocked or cribbed as it is raised or lowered. Blocking that has rounded corners or evidence of dry rot should be rejected.

One company has found through many years of experience that the best sizes for cribbing timber are as follows:

8 in × 12 in. × 10 ft
8 in. × 12 in. × 8 ft

Smaller blocking may be as follows:

8 in. × 12 in. × 24 in.
6 in. × 12 in. × 24 in.
3 in. × 12 in. × 24 in.
4 in. × 4 in. × 18 in.
2 in. × 10 in. × 18 in.

For sufficient strength, cribbing should have a safety factor of at least 4. The natural tendency to underrate the load should be guarded against.

Cribbing must be placed on a foundation, so that it can be removed readily as the machine is lowered.

Accessories for Manual Handling

In handling materials, a variety of hand-operated accessories can be used. Each tool, jig, or other device should be kept in good repair and used only for the job for which it is designed.

Hand tools

Hooks. The worker should be trained to use hand or packing hooks, so that they will not glance off hard objects, with possible injury to himself or others. If the hook is to be carried in the belt, the point should be covered.

Hook handles should be of hardwood and in good condition. Hooks for handling logs, lumber, crates, boxes, and barrels should be kept sharp and inspected daily.

Bars. The principal hazard in the use of a crowbar is that it may slip. A dull, broken crowbar is more likely to cause injury than a sharp one. The point or edge should have a good "bite." The workman should position himself to avoid falls or pinched hands, if the bar slips or the object moves suddenly—he should never work astride it. His hands or gloves should be dry and free of grease or oil.

Crowbars not in use should be stored so that they will not fall or cause a tripping hazard.

Ordinary crowbars should not be used to move cars on steel rails. Car movers which do not readily slip are available. When two men are needed to move a car, two car movers should be used, as two men should not try to work with the same mover.

Rollers. Heavy, bulky objects must often be moved by means of rollers. The principal hazard is that the fingers or toes may be pinched or crushed between a roller and the floor. Rollers should extend beyond the load to be moved and be sufficiently strong. Rollers under a load should be moved with a sledge or bar, not the hand or foot.

WOODEN SHIM

STANDS STRAIGHT

SOLID BASE

Fig. 24-6.—Three elements that help make the operation of a pneumatic jack safe are pointed out here.

Jacks

When using a jack, a man should check the capacity plate or other marking on the jack to make sure the jack can support the load. Where there is no plate, capacity should be determined and painted on the side of the jack. If a properly rated jack is used, it should not collapse under load.

Jacks should be inspected before and after each use. Any signs of hydraulic fluid leakage should eliminate the jack from use.

A heavy jack is best moved from one location to another on a dolly or special hand truck. If it has to be manually transported, it should have carrying handles and, at least, two workmen should form a team to move it. The operating handle should never be left in the socket while a jack is being carried, because it may strike other workers.

The workmen should see that jacks are well lubricated, but only at points where lubrication is specified, and should inspect them for broken teeth or faulty holding fixtures. A jack should never be thrown or dropped upon the floor; such treatment may crack or distort the metal, thus causing the jack to break when a load is lifted.

It is important that the floor or ground surface, upon which the jack is placed, be level and clean and that the safe limit of floor loading is not exceeded. If the surface is earth, the jack base should be set on substantial hardwood blocking (at least twice the size of the jack), so that the blocking will not turn over, shift, or sink. If the surface is not perfectly level, the jack may be set on blocking, which should be leveled by substantial shims or wedges, placed so securely that they cannot be crushed or forced out of place.

To prevent the load from slipping, no metal-to-metal contact should be permitted between the jack head and the load. A hardwood shim, longer and wider than the face of the jack head, should be placed between the jack head and the contact surface of the load. Two-inch wood stock is suitable for this purpose (Fig. 24-6).

641

"Extenders" of wood or metal should never be used. Instead, either a larger jack should be obtained or higher blocking which is correspondingly wider and longer should be placed under the jack.

All lifts should be vertical, with the jack correctly centered for the lift, the base on a perfectly level surface, and the head with its shim bearing against a perfectly level meeting surface. When an emergency requires that the lifting force be applied at an angle, extra precautions must be taken which include:

1. A base of blocking, securely fastened together and to the ground, to make an immovable surface at right angles to the lift for the jack base to rest on

2. Cleats on the blocking to prevent shifting of the jack base

3. A meeting surface at right angles to the direction of the lift for the jack head with its shim to bear against

4. Props or guys to the load to prevent its swinging sidewise when lifting begins

When a jack handle is placed in the socket, the workman should make sure that the area is clear and that there is ample room for an unobstructed swing of the handle before he applies pressure. A faulty movement in the load may cause the handle to pop up and strike another worker. The man operating the handle should stand to one side so that, if the handle kicks, it will not catch him in the body or face.

After the load is raised, metal horses or substantial heavy wooden horses or blocking should be placed under it for support in case the jack should let go. A raised load should never be allowed to remain supported only by jacks. The handles of the jacks should be removed immediately and placed out of the way to prevent workers from walking into them.

When a jack is released, the worker should keep all parts of his body clear of the movement of the handle.

Hydraulic jacks may settle after raising a load. It is, therefore, especially important to place blocking under a load that has been raised by such jacks.

Screw jacks have a tendency to twist when a heavy load causes the floating head of the jack to bind. The base of a screw jack should be anchored as securely as possible, so that the jack base will not twist and slip out from under the load when force is applied on the bars to raise the screw.

To raise a large piece of equipment with screw jacks, two or more jacks should be used. The load should be equally distributed on each jack. Each jack should be raised a little at a time to keep the load level and the strain equal on each screw jack head. Special signals are required to verify that all jacks are rising uniformly.

Workmen using jacks should wear safety shoes and instep-guard protection, because jack handles may slip and fall or parts of machinery or equipment may come loose and drop, as the load is being lifted or shifted. Waste or other wiping material should be furnished to jack operators who should use it to remove oil from their hands and from the jack handles, so that they will always have a firm grip.

Oil that has collected in the bases of equipment or machines to be jacked should be removed before the operation is begun, to prevent spillage when the equipment or machines are tilted. Spillage of residual oil should be wiped up immediately.

When a jack develops any defect, it should be removed from service, repaired, and tested under load before being returned to service.

Handtrucks, dollies, and wheelbarrows

Many special types of handtrucks and dollies are available, such as two-wheeled, flat, platform, refrigerator, appliance, fox, special racks or dollies, and lift trucks. Trucks can be purchased or designed for objects of various sizes and kinds. Operators should wear gloves and safety shoes.

Two-wheeled trucks and wheelbarrows should be equipped with knuckle guards to protect against the jamming of hands against door frames or other obstructions.

Two-wheeled trucks should have brakes (Fig. 24–7), so that the worker need not hold the truck with a foot on the wheel or axle. The latter practice causes many injuries.

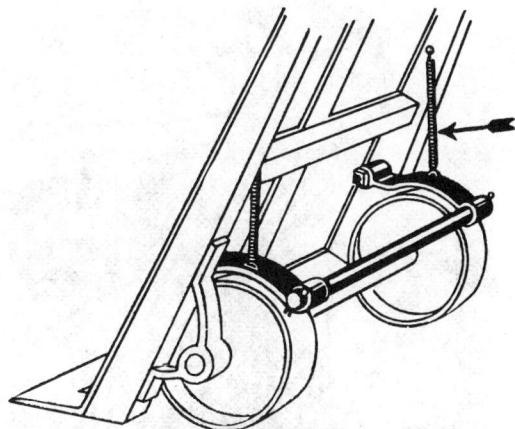

Fig. 24–7.—Foot brake keeps wheels from slipping while truck is being loaded. Springs (arrow) hold brake away from wheels when it is not needed.

To decrease hazards to toes and feet, wheels should be as far under the truck as possible. Wheel guards can be installed on many types of trucks.

Tongues of flat trucks should be provided with counterweights, springs, or hooks to hold them vertical when not in use. If this is not possible, workers should be trained to leave handles in such a position that they will not be a tripping hazard.

Equipment should be inspected daily and kept in good repair. Repair and maintenance records should be kept to show the condition of each piece of equipment. Axles should be kept well-greased.

The type of truck most suitable for the work at hand should be used. No one truck is right for handling all types of material.

Two-wheeled trucks look as though they should be easy to handle, but there are safe procedures that must be followed.

1. Tip the load to be lifted forward slightly, so that the tongue of the truck goes under the load.

2. Push the truck all the way under the load to be moved.

3. Keep the center of gravity of the load as low as possible. Place heavy objects below lighter objects. When loading trucks, truckers and loaders should keep their feet clear of the wheels.

4. Place the load well forward so the weight will be carried by the axle, not by the handles.

5. Place the load so it will not slip, shift, or fall. Load only to a height that will allow a clear view ahead.

6. When a two-wheeled truck or wheelbarrow is loaded in a horizontal position, raise it to traveling position by lifting with the leg muscles and keeping the back straight. Observe the same principle in setting a loaded truck or wheelbarrow down—the leg muscles should do the work.

7. Let the truck carry the load. The operator should only balance and push.

8. Never walk backwards with a hand truck.

9. For extremely bulky items or pressurized items, such as gas cylinders, strap or chain the item to the truck.

10. When going down an incline, keep truck ahead. When going up, keep truck behind. (This advice applies to four-wheeled as well as two-wheeled trucks.)

11. Move trucks at a safe speed. Do not run. Keep truck constantly under control.

Four-wheeled truck operation follows rules similar to those for two-wheeled trucks. Extra emphasis should be placed on proper loading, however. Four-wheeled trucks should be evenly loaded to prevent their tipping. Four-wheeled trucks should be pushed rather than pulled, except for a truck that has a fifth wheel and a handle for pulling.

Trucks should not be loaded so high that operators cannot see where they are going. If there are high racks on the truck, two men should move the vehicle—one to guide the front end, the other to guide the back end. Handles should be placed at protected places on the racks or truck body, so that passing traffic, walls, or other objects will not crush or scrape the operators' hands.

Truck contents should be arranged, so that they will not fall or be damaged in case the truck or the load is bumped.

Pallets and palletances are now constructed of wood, glass fiber, and plastic. They should always be kept in good condition and repair. Systematic inspection and the removal of defective pallets is essential. Those that can be repaired should be.

Idle pallets constitute a falling and tripping hazard as well as a fire hazard. Idle pallets should never be stacked on end or over 12 pallets high.

General precautions. Truckers should be warned of three main hazards: (*a*) running wheels off bridge plates or platforms, (*b*) colliding with other trucks or obstructions, and (*c*) jamming their hands between the truck and other objects (Figure 24–8).

Workers should operate trucks at a safe speed and keep them constantly under control. Special care is required at blind corners and doorways. Properly placed mirrors can aid visibility at these places.

When not in use, trucks should be stored in a designated area, not parked in aisles or other places where they would be a tripping hazard or traffic obstruction. Trucks with drawbar handles should be parked with handles up and out of the way. Two-wheeled trucks should be stored on the chisel with handles leaning against a wall or the next truck. Wheels of trucks not being used should be blocked.

Powered hand trucks and power trucks are discussed in Chapter 27.

Conveyors, cranes, and hoists. See Chapter 25, "Hoisting Apparatus and Conveyors," for description of many types of lifting and transporting devices used to reduce physical strain and facilitate material handling.

Storage of Specific Materials

Both temporary and permanent storage must be neat and orderly. Materials piled haphazardly or strewn about increase the possibility of accidents to employees and of damage to materials.

The warehouse supervisor must direct the storage of raw materials (and sometimes processed stock) that are kept in quantity lots for some time. The production supervisor is usually responsible for storage of limited amounts of materials and stock for short

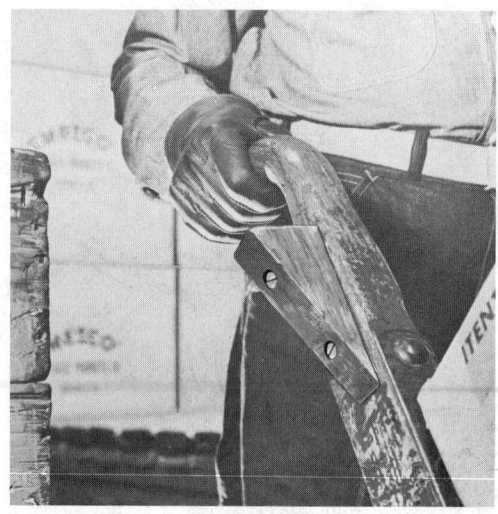

FIG. 24–8.—This man's hands are protected by sturdy gloves and by knuckle guards on the handles of this two-wheeled truck. Such guards can be made and installed by maintenance personnel.

periods near the processing operations.

Planned materials storage minimizes the handling necessary to bring materials into production, as well as remove finished products from production to shipping.

When planning materials storage, allow 18 to 36 in. for clearance under sprinklers. Be sure automatic sprinkler system controls and electrical panel boxes are free and clear. Unobstructed access to fire hoses and extinguishers must be maintained. All exits and aisles must be kept clear at all times (Fig. 24–9). The amount of clearance between storage and automatic sprinkler heads will vary with the commodity being stored and the height of storage. (See NFPA Std. No. 13, *Installation of Sprinkler Systems.*)

Aisles that carry one-way traffic should not be less than 3 ft wider than the widest vehicle when loaded. If materials are to be handled from the aisles, the turning radius of the power truck may also need to be considered. Employees should keep materials out of the aisles and out of loading and unloading areas, which should be marked with painted or taped lines.

Storage is facilitated and hazards are reduced by use of bins and racks. Material

Fig. 24-9.—Various methods of stocking and piling materials. Palletized materials are speedily and efficiently handled by lift trucks; however, pallets must be securely and evenly loaded. Note cartons are overlapped. Barrels and kegs must be adequately supported by planking or channels between layers, or else stored on end.

Courtesy General Electric Co.

stored on racks, pallets, or skids can be moved easily and quickly from one work station to another with less material damage and fewer employee injuries. When possible, material piled on skids or pallets should be cross-tied.

Storage racks should be secured to floor, wall, and each other. If flue spaces are provided, stock should never block the flue. If in the racks automatic sprinklers or fire preventive devices are provided, special care should be exercised to avoid damage to them. Racks, when damaged, should be repaired. Employees should never be allowed to climb racks.

In an area where the same type of material is stored continuously, it is a good idea to paint a horizontal line on the wall to indicate the maximum height to which material may be piled. This will help keep the floor load within the proper limits and the sprinkler heads in the clear.

High rack storage. The modern trend in storing uniform-sized containers or stock is high bay storage. Some European storage facilities now have approximately 100-ft high automated storage, and more of these are being constructed in the U.S.

High rack facilities require unique, specially designed, high-lift materials handling equipment. Some, up to 30 ft, are operated manually. Others are operated under computer automation or control. Applicable standards include the ANSI B56 series, "Powered Industrial Trucks," and the Crawler Cranes section of ANSI B30.5, *Safety Code for Crawler, Locomotive, and Truck Cranes.* Not only do these represent unique materials handling problems, but also special fire protection problems. (See NFPA Std. No. 231C, *Rack Storage of Materials.*)

The protection of personnel who operate and maintain such facilities must include disaster and emergency planning, physical protection at the operating position, and visual and audible warnings on moving equipment.

Special procedures and equipment are also required for maintenance and physical inventory.

Rigid containers

Boxes and cartons. Box containers are of three general types: wire-bound shooks, box shooks, and cardboard cartons. Wire-bound shooks should be laid, so that the sharp ends of wires do not protrude into passageways.

645

Piles of both types of shooks should be tied, so that they will not topple.

A good method is to store shooks on skids, so that they may be easily handled when they are to be assembled. Piles of wire-bound shooks should be not more than 5 ft high.

Loaded cartons should be stored on platforms as protection against moisture since even low piles, when wet, will collapse.

Since the height of piles of cartons is regulated by the kind of materials in the cartons, it is not possible to set a standard height. An important factor to consider is that the sides of the cartons will not support much load. Sheets of heavy wrapping paper placed between layers of cartons help prevent the shifting of the pile. Interlocking the cartons increases the stability of the piles.

Preferably, cartons should be stored on pallets or racks. Any sign that the lower cartons are being crushed requires a restacking of the pile.

Certain bulky materials, such as skids of paper, should not be stacked to maximum allowable heights in the rows bordering on aisles, especially those which carry hand- or power-truck traffic. A good rule is to make the first row one item high unless the material can be tightly interlocked.

Barrels and kegs. Piles of barrels should be symmetrical and stable, preferably in the shape of a pyramid. The first or bottom row should be blocked to prevent the barrels from rolling. If barrels or kegs are piled on end, place planks between rows.

When barrels or kegs are piled on their sides other than in pyramid form, they should be laid on specially constructed racks. Otherwise, planks should be laid between rows and the ends of the rows should be blocked.

Drums are discussed under Portable Containers in the Liquids section that follows.

Rolled paper and reels. Clamp-type trucks stack these items three or four high. Extreme care is required, so that such stacks are even. Physical damage to lower rolls or reels requires that there be a restacking.

Compressed gas cylinders. For their safe handling and storage refer to Chapter 34, "Welding and Cutting."

Uncrated stock

Lumber. Except for the amount for immediate need, lumber is best stored outdoors or in a building separate from the general warehouse. Lumber should be sorted by size and length and stored in separate piles. It is important for outdoor lumber piling that firm ground be selected. The area should be well drained to remove surface water and prevent softening of the ground. A periodic check should be made to determine if there is any shifting of material.

For long-time piling, substantial bearings or dunnage is recommended. Concrete with spread footing extending below the frost line is a good method.

For temporary piling, heavy timbers may be used to support the crosspieces. This type of support should be inspected periodically for deterioration, which may cause the pile to list dangerously.

If lumber must be removed by hand, it is best stored in low piles or in racks, with galleries provided, if necessary, to permit workmen to reach the top of the piles.

If lumber must be handled manually to or from a higher pile, the pile should be not more than 16 ft high, and a safe means of access to the top should be provided.

Twenty feet is generally considered the maximum safe height for lumber piles when lumber is piled and removed mechanically, as by lift trucks.

Tie pieces are needed not only to stabilize the pile, but also to provide air circulation. Tie pieces should not extend into walkways, but should be cut flush with the pile. Green lumber should have tie pieces on every layer, whether stored indoors or outdoors.

Lumber stored outdoors should be covered to prevent checking or twisting. Lumber stored indoors should be in a well-ventilated building.

Bagged material should be cross-tied with the mouths of the bags toward the inside of the pile. When the pile is 5 ft high, it should be stepped back one row. It should then be stepped back one row for each additional 3 ft of height. A pile of sacks should never be undermined by the removal of sacks from lower rows first.

Pipe and bar stock places a heavy load on the floor. Therefore, the floor area must be selected with bearing strength in mind. Because of the hazard to passersby while stock is being withdrawn from the racks, fronts of pipe and bar stock storage racks should not be located on main aisles.

Pipe and other round materials should be piled in layers with strips of wood or iron between the layers. Either the strip should have blocks at one end, or the end should be turned up.

Material, such as lumber or pipe, is particularly dangerous because of its tendency to roll or slide. Dropping instead of placing such objects on a pile frequently causes them to slide or bounce. Employees are likely to be injured if they attempt to stop rolling or sliding objects with their hands or feet.

Bar steel stock in the larger sizes should be stored in racks that are designed to rest the bars on rollers. The center distance between rollers will be governed by the sizes of the bars and should permit easy withdrawal of them. Rollers with multiple sections make withdrawal easier.

Racks should incline toward the back so that bars cannot roll out. Light bar stock may be stored vertically in special racks.

Special A-frame racks of metal can hold a variety of pipes and bars safely in quantity, if loaded evenly and supported properly.

Sheet metal. Racks, similar to those for bar stock, may be provided for plate and sheet stock except that rollers are not always applicable. Oiled sheets require additional caution in handling.

Sheet metal usually has sharp edges. It should be handled with hand leathers or leather gloves, or gloves with metal inserts. Large quantities should be handled in bundles by power equipment. These should be separated by strips of wood to faciliate handling, when the material is needed for production, and to lessen chances of shifting or sliding of the piles of material.

Tin plate strip stock is heavy and razor sharp. Should a load or partial load fall, it could badly injure anyone in the way. Two measures can be taken to prevent spillage and injuries: (*a*) band the stock after shearing, and (*b*) use wooden or metal stakes around

Fig. 24–10.—Pickup and drop stations should be convenient and well marked.

the stock tables and pallets that hold the loads. It is the responsibility of the supervisor and all who handle the bundles to make sure the load is banded properly and that the stakes are in place when the load is on the table. (When removing bands, special cutting tools are required.)

Burlap sacking. Large reserves of burlap sacking must be protected against spontaneous ignition. The room should be of fire-resistant construction and provided with sprinklers and dust-tight lights. The sacks should not be piled too high, and the interior of the pile should be ventilated by air vents, permitting air flow from the outside of the pile to the center.

Straw, excelsior, and other packing materials are usually received baled and should preferably be stored that way, either in a separate building or in a fire-resistant room provided with sprinklers and dust-proof electric equipment.

Because they are a fire hazard, only the amount necessary for immediate use should be taken into the packing room. For storing enough for immediate use, the best bins are made entirely of metal, or of wood lined with metal, and are provided with covers that are normally closed.

Large bins may have several compartments with counterweighted covers. The counter-

647

weight ropes should have fusible links to ensure automatic closing of covers in case of fire. Counterweights should be boxed in to prevent injury if the ropes break.

No SMOKING signs must be posted.

Problems With Hazardous Materials

Storage and handling of specific hazardous materials are discussed in other chapters: gases in Chapter 34, "Welding and Cutting"; flammable liquids (including refrigerants) and tank car and tank truck loading and unloading, in Chapter 42, "Flammable and Combustible Liquids"; NFPA hazard symbols in Chapter 43, "Fire Protection"; and incompatible materials in Chapter 46, Table 46-N.

Liquids

Tanks. The structure of a new building should be designed with an ample safety factor to permit supporting the weight of storage tanks. However, inspection by a competent structural engineer should be made before a tank is installed in an old building. Storage tanks for hazardous liquids are preferably installed outdoors either above or below ground.

There are many advantages in underground installation of outdoor storage tanks. However, the danger of undetected leaks in tanks containing corrosive or toxic materials probably outweighs the advantage of freedom from drips and sprays. When an outdoor tank is located in a pit, the pit should be large enough to permit easy access to all parts of the tank. A permanent ladder and an access door which can be fastened shut should be provided.

No one should be permitted to enter a pit without an approved type of a supplied air mask or hood or oxygen breathing apparatus, unless tests have established the presence of enough oxygen and the absence of dangerous amounts of toxic vapors. In any case, a worker who enters a pit should wear a safety belt with attached lifeline and a similarly equipped observer should be stationed to watch him.

Process tanks that will contain volatile or corrosive liquids should be installed only at or above grade, in areas having adequate drainage, and should be separated from the processing area by construction having a fire-resistive rating of at least two hours.

Tanks should be installed where traffic cannot pass under them. If people must walk beneath them, drip pans should be installed and provided with drainage to a safe disposal or recovery location.

Tanks should be provided with permanent stairs or ladders and walkways with standard guardrails and toeboards. Tanks should be emptied, cleaned, and inspected for structural weaknesses at regular intervals, and records kept of each inspection.

Tanks for holding volatile materials should be bonded, grounded, and provided with emergency venting devices. Venting should be in accordance with provisions of the *Flammable and Combustible Liquids Code*, NFPA Standard No. 30. If tanks are inside buildings, vents should discharge outside the building, at a location free from any ignition source, and from contact with personnel. Be sure to consider the effects of corrosion on venting devices for tanks that will contain corrosive liquids.

Connections for filling and emptying tanks are preferably made through the top, to minimize liquid loss and the possibility of injury from a broken fitting. Fill lines should be plainly labeled and be equipped with a drain. Use of compressed air is not permitted for the transfer of flammable or corrosive liquids, according to NFPA Standard No. 30, paragraph 8331.

Cleaning tanks is an exceedingly dangerous operation. An exact and specific procedure should be set up, preferably in written form, and strictly adhered to. Specifications for tank-cleaning procedure are set forth in NFPA Standard No. 327, and API publications RP 2015. (See References.)

The procedure should be modified for toxic compounds only to the extent that more complete protective equipment may be required. Liquid aromatic nitro-compounds and amines are solvents for rubber and are absorbed through it gradually as well as through the skin. For such exposures, personal protective equipment should be made of one of the inert synthetic rubbers or plastics. (This includes gloves, aprons, boots, respiratory- and eye-protective equipment.)

The handling of phenolic compounds, such as carbolic acid, cresylic acid, and the cresols,

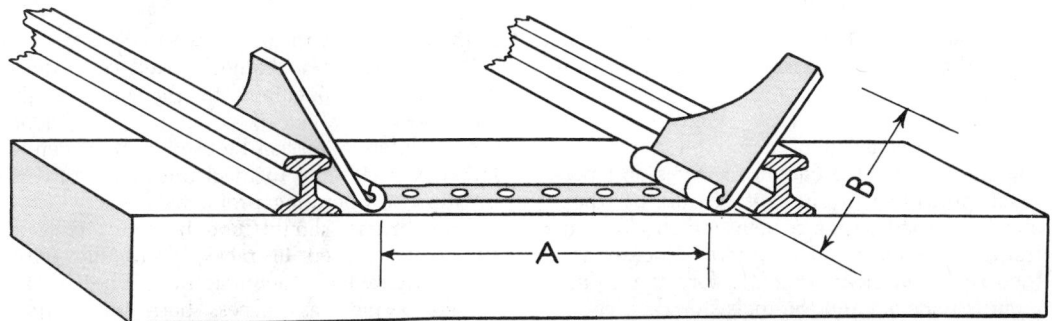

FIG. 24–11a.—Automatic spring-closed barrel stops on rails in storage racks help in removing and inserting barrels. While dimensions A and B are not critical, B should be less than half of A so the leaves can fold between tracks.

requires precautions against skin absorption, although these compounds are not such effective rubber solvents as the nitro and amino compounds.

To prevent pumps from being primed with dangerous liquids where connection is made through the top of a tank, self-priming pumps or pumps that generate enough suction to lift the liquid from the bottom of the tank should be used.

Pipelines for carrying chemicals are preferably installed in trenches or tunnels. If they must be installed overhead, they should be isolated so that they will not drip on men working underneath. Pipelines for carrying flammables should not be installed in tunnels. All pipelines must be identified as to content.

There are three major sources of chemical injury in pipeline work:

FIG. 24–11b.—Illustrates safety device to prevent barrels from rolling out of storage racks. Stops will push in but not push out, as man shows here.

● FAILURE OF PACKING IN VALVE STEMS OR OF GASKETS IN BOLTED FLANGES. To minimize injuries from valve packing failure, the valve stem can be surrounded by a sheet metal box or hood which will deflect spray away from the man operating the valve. So far as possible, packing should be renewed without pressure on the valve.

● OPENING THE WRONG VALVE. To prevent injuries and accidents from this source, pipelines and valves should be identified by tags, lettered markings, and distinctive colors. (See the next section, Identification of Piping.) It is also desirable to have valves

well separated and the immediate area well lit to assure quick and easy identification.

● FAILURE TO CHECK THAT VALVES ARE CLOSED AND LOCKED AND THE LINES DRAINED BEFORE TENSION IS RELEASED ON FLANGE BOLTS. The opening between the faces of the flange may be temporarily covered with a piece of sheet lead, while the flange bolts are being loosened and the faces separated.

The bolts farthest away should be loosened first so that drainage will tend to go away from the workman. Blinds should be inserted in the flanges as soon as they are opened.

For lines which are opened often, blinds

permanently pivoted on a flange bolt, with one end acting as a gasket and the other as a blind, can be used.

At the conclusion of a job on a pipeline containing corrosive chemicals, tools and personal protective equipment should be thoroughly washed with a reagent which will neutralize or remove the corrosive material and then rinsed in clean water, before the equipment is removed and the tools stored.

Identification of piping. Distinctive colors for identifying piping have been standardized in American National Standard A13.1, *Scheme for the Identification of Piping Systems,* and are described in Chapter 16, "Industrial Buildings and Plant Layout," in the section titled Use of Color, pages 385–386.

Specific identification of piping should be provided by a lettered legend which names the material being piped, summarizes the hazards involved, and gives directions for safe use. Stencils or decals may be used to apply legends.

Legends should be made in accordance with MCA Manual L-1, listed in References, be moisture-resistant, and contain pigments that are colorfast.

Portable containers. Where liquid chemicals are used in quantity, it is generally better to install pipelines and outside storage tanks than to use portable containers. Spillage is then reduced and localized, so that it is easier to take necessary precautions.

Portable containers, such as drums, barrels, and carboys, should be correctly stored. The minimum of liquid should be kept at the point of operation; only enough for one shift is a common rule. The main stock should be stored in a safe, isolated place.

If the liquid is corrosive or highly toxic, the storage area should be isolated from the rest of the plant by impervious walls and floor, with provision for safe disposal of spillage, or a separate building should be used.

Floors in the storage area for corrosive liquids should be made of cinders, concrete treated to decrease its solubility, or other resistant material. Concrete is also satisfactory for flammable liquids. Good drainage will permit easy cleaning in case a container

in the storage area leaks or breaks.

The storage area should be well ventilated, preferably by openings to the outside air equivalent to about 5 percent of the floor area. Whenever it can be used, natural ventilation is preferable to mechanical, because it involves no operating problems.

Full drums should not be stacked, but should be placed in racks, preferably with a separate rack for each material. These racks should permit easy access both for getting the drums in and out and for ready inspection of the stock (Fig. 24–11a and b).

Barrels may be stacked vertically with dunnage between the tiers, but are more conveniently handled if they are kept in racks similar to those used for drums. There should be equipment for moving barrels or drums in and out of racks and bracing to keep them in the racks.

Be sure grounding is adequate, if flammable liquids are involved. See Chapter 42, "Flammable and Combustible Liquids," for details.

Different materials should be stored in different designated areas, separated by wide aisles. Boxed carboys should generally be stacked not higher than two tiers, never higher than three. Not more than two tiers should be used for carboys of strong oxidizing agents, such as concentrated nitric acid or concentrated hydrogen peroxide.

Before acid carboy boxes are handled, they should be inspected for corrosion of nails or weakening of the wood by acid.

Before empty carboys are piled, they should be thoroughly drained and the stoppers replaced.

Special equipment should include a long-handled truck (Fig. 24–12) which picks up the boxed carboys under the handling cleats or between the bottom cleats provided on all standard 12- and 13-gallon boxed carboys. These trucks have handles long enough to keep the men handling them away from splashes, in case carboys are dropped.

There is also less danger of dropping a carboy with this kind of truck than with the standard two-wheeled truck. The two-wheeled truck becomes a much safer device for handling drums and barrels, if it has a bed curved to fit the drum and a hook to catch the chime.

The safest way to empty a carboy is to move

the liquid by suction from a vacuum pump or aspirator or start a siphon with a rubber bulb or ejector. The carboy inclinator, if it holds the carboy firmly by the top, as well as the sides, and automatically returns to the neutral position on being released, is satisfactory.

Compressed air, even from a hand pump, should not be used on a carboy unless the carboy is enclosed by another container so that the pressures inside and outside remain the same (Fig. 24–13). *Pouring by hand or starting pipettes or siphons by mouth suction should never be permitted.* Mechanical pumps are not desirable unless they are self-priming or have sufficient suction pressure to start themselves.

A corrosive or poisonous liquid sometimes requires a specially identified container. The simplest is a glass or plastic jug or jar, with a good closure, set into a metal can. A container resistant to shock can be improvised by putting a jug into a metal pail and filling the space between the pail and the jug with pitch or foamed plastic. A container may be made to fit the jug with only a thin layer of padding, such as a layer of gasket rubber. Containers are also available commercially.

Highly toxic substances, such as cyanides and soluble oxalates, should be kept in containers of a distinctive shape if they must be handled manually. The containers should be plainly labeled or otherwise identified. Materials should be locked up at all times and dispensed only by authorized personnel.

Where caustics or acids are stored, handled, or used, emergency flood showers or eyewash fountains must be available. Men should be provided with chemical goggles, rubber aprons, boots, gloves, and other protective equipment necessary to handle the particular liquid. (See Chapter 19.)

Tank cars should be isolated on sidings by derails and by blue stop flags or blue lights before they are loaded or unloaded. Hand brakes should be set and wheels chocked. Before the car is opened, it should be bonded to the loading line. The track and the loading or unloading rack should be grounded, and all connections checked regularly. (Also see Chapter 42, "Flammable and Combustible Liquids.")

Chemical tank cars should be unloaded

Fig. 24–12.—Carboy truck is safe and convenient.

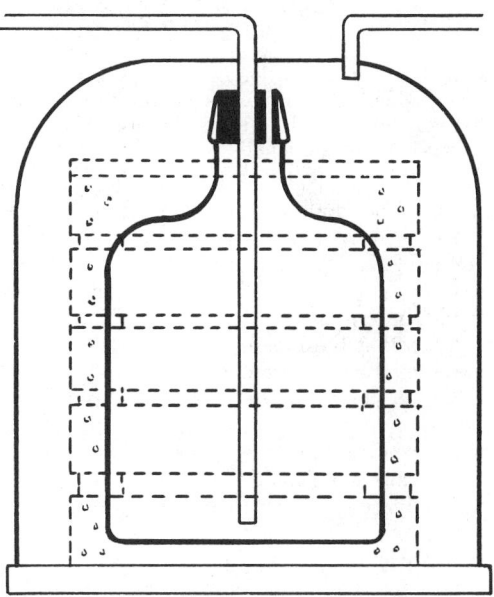

Fig. 24–13.—Air pressure admitted through the short pipe forces liquid from the carboy through the long pipe. Pressure is exerted on the bell cover, not the carboy.

651

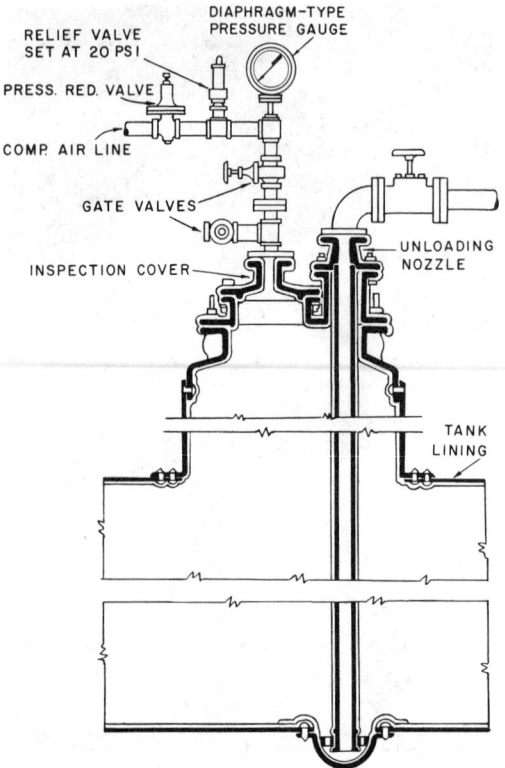

RELIEF VALVE
SET AT 20 PSI

DIAPHRAGM-TYPE
PRESSURE GAUGE

PRESS. RED. VALVE

COMP. AIR LINE

GATE VALVES

INSPECTION COVER

UNLOADING
NOZZLE

TANK
LINING

FIG. 24–14.—Cross section of one type of tank car designed for unloading under air pressure. Note that air pressure should not be used for discharging flammable liquids.

through the dome rather than through the bottom connection. If the contents are non-flammable, air pressure not to exceed 25 psi may be used for unloading. The connections should be equipped with a safety valve and gage, so that the pressure on the tank can be determined at any time (Fig. 24–14).

Before the car is opened, the cap of the unloading pipe should be gently backed off without being completely removed. Pressure in the tank car should be allowed to escape gradually before the cap is entirely removed.

The unloading dock should be equipped with a walkway at the height of the tank car domes and with drawbridges which can be lowered to make a firm walkway directly to the domes of the cars. Standard handrails and nonslip surfaces are required. If corrosive materials are handled, emergency showers should be provided along the walkway.

Some materials normally shipped in tank cars solidify at temperatures reached during shipment. Tank cars used for such materials are ordinarily equipped with steam lines for melting the contents. The eduction lines and valves should be thoroughly thawed and clear before unloading is begun.

The plant line to the unloading dock must be completely drained after unloading. This line should be installed with a slope toward the storage tank so it drains itself.

Solids

Bins. When new bins are to be installed in an old structure or when new materials are to be stored in old bins, the mechanical strength of the structure should be checked. Solids vary in unit weight, and the more dense materials produce high unit loads.

A fundamental factor in the design of equipment for the handling of bulk solids is sufficient slope in the cone bottom of a tank or bin to permit the solid to run freely and to prevent its arching over. Where arching takes place, means should be provided so that the flow can be started again without it being necessary that a workman enter the bin either above or below the solid material.

A vibrator to shake the bottom of a small metal bin, or an agitator bottom, is a simple means to start the flow. Both are standard equipment, and one or the other can be applied to bins of almost any size or shape.

It is sometimes possible to work either from the bottom or from the top of a bin. If a man can break up the arch from the top with long tools without entering the bin, the job is reasonably safe; it is dangerous when done from the bottom. There is also the ever-present temptation to step inside and work with a little more convenience until the material starts to flow.

Falls into open storage bins often result in injuries. Bin openings at floor level or within 2 ft of it should be surrounded with standard guardrails and toeboards.

If guardrails are impracticable or if the opening is not easily accessible, like the fill opening of a high bin, the opening can be covered with a grating, which will not materially obstruct the opening but will prevent a man from falling in.

Many bin openings can be covered with a 2-in. mesh, and most of them can be covered by a 6-in. grating or by parallel bars on 6-in. centers.

Before repair and maintenance work is started, a test should be made for oxygen content of the air and for the presence of toxic materials, particularly carbon monoxide. If the bin has contained an organic material, there may have been a reaction that dangerously depleted the oxygen content of the air without noticeable rise in temperature or other warning signs.

Before a workman enters a bin, the filling equipment should be made inoperative so that it cannot be started again, except by the workman after he leaves the bin or by his immediate supervisor after he has checked the man out of the bin.

The workman entering the bin should be provided with a safety belt and lifeline, and a similarly equipped workman should be stationed with the sole duty of tending the line and observing the man inside the bin.

Where bins are filled and emptied by continuous conveyors, the control of dust is likely to be a serious problem. In filling a bin, material is generally dropped from a belt conveyor. If the material is dropped through a chute from an elevating conveyor, even more dust may be produced.

Escape of dust into the rest of the plant can generally be prevented by enclosing the bin, except for the fill opening, with a skirt of either metal or fabric, and by taking an exhaust up through the filling chute.

If the material is scraped from a belt conveyor, it is usually enough to cover the conveyor at the point of discharge with an exhaust hood and provide a closed chute from the discharge point to the bin.

Since the dust in these cases is released at a low velocity, it is sufficient to provide an inward air velocity of about 50 fpm through all the openings, provided that there are no seriously disturbing air currents to blow dust out of the openings and that enough velocity is provided through the rest of the system.

The same general principles apply to the discharge of bins onto conveyors. There is seldom a serious dust problem except at the loading and discharge points of the conveyors. These points must generally be ventilated and covered with hoods, because it is seldom feasible to provide them with dust-tight enclosures or to reduce the dust by wetting.

Combustible solids. Where combustible materials are handled, the dust content of the air must be kept below the lower explosive limit. In addition to tight enclosures and dust collection systems, good plant housekeeping will go far toward preventing disaster.

Dust explosions commonly occur as a series, not as a single shock. The first explosion uses up the dust in the air, but the shock stirs up more dust from the building members, which is in turn set off. If the building is kept clean, this sequence cannot occur and damage will be minimized.

All sources of ignition should be excluded from the area of a potentially explosive dust. Wiring, lights, and switches should be in compliance with the NFPA Standard 63 for hazardous locations. Electric motors should be of the totally enclosed, explosion-proof type or should be in a tight enclosure, which is independently ventilated from a nonhazardous area and kept under positive pressure.

Bearings should be large and well protected; a hot bearing may ignite many types of dust. Heating systems should be of the indirect type only, and radiators should be constructed to permit easy cleaning. The no smoking rule must be rigidly enforced.

Static electricity is the source of ignition in many fires. It can be prevented from accumulating on most surfaces, if relative humidity is maintained at 60 percent to 70 percent. If this cannot be maintained, use a ground to minimize static buildup.

Static electricity can be removed from moving parts, such as conveyor belts and shafts, by methods that include static collectors, grounding brushes, and conductive V-belts and conductive leather belt dressings.

Metal bins should be grounded to the conveyor frame. Electrical interconnection at loading and discharge points of conveyors should be checked especially. Static voltmeters will quantitatively measure the effectiveness of the grounding system, or a rough estimate can be made with a simple electroscope.

Automatic sprinkler protection should be

installed inside bins and processing equipment containing combustible materials. Where water is undesirable, either because of reaction with or damage to the material, protection can be provided by inert gas extinguishers.

If a particularly hazardous material like metal powder is being handled, the apparatus may be completely enclosed and flooded with an inert gas like carbon dioxide, nitrogen, or helium to remove or reduce the oxygen content and, thus, prevent ignition.

The area should be explosion-vented, preferably by windows and skylights which swing out on friction catches. If other means are not available, the windows may be scored with a diamond so that they will easily break outward in case of explosion.

Portable containers. The same general rules that apply to the storage and handling of containers of liquids apply also to the storage and handling of containers of solid materials. The most popular containers for solids are 50- and 100-pound paper bags. These bags are free from sifting or leaking, but they must be handled with some care to prevent mechanical damage. A few slipover bags should be available to cover the occasional broken or leaking container, both to save material and to prevent skin contact with the dust.

Full bags should be stacked on pallets or staging to prevent water damage. Interlocked stacking on pallets generally leads to better piling and less mechanical hazard in moving material. Bags should be protected from the weather, although some of the laminated bags now made are remarkably weather resistant.

Large quantities of solid chemicals are handled and shipped in bulk, cloth bags, barrels, and barrels with paper liners.

The filling of bags or barrels with solids is always a potentially dusty operation and, if the material being handled is finely divided or dangerous, the health and fire hazards may be serious. The simplest solution to the problem is to moisten the material, so that it does not produce fine dust.

This solution, however, cannot be used with some materials, and some methods of handling and hoods and exhaust ventilation in quantity must be provided. (See Chapter 36, "Exhaust and Ventilation.")

The common way to open bags, both cloth and paper, is to slit one side crosswise and fold back the top and bottom. The hazard of knife cuts can be minimized by keeping knives in scabbards where bags are opened and by using knives with hilts or guards.

The emptying of bags and barrels involves health hazards similar to those of filling. However, prevention of dust and of skin contact is somewhat more difficult where such containers are emptied than where they are filled because of the tendency to dump them suddenly, with consequent rapid dispersion of dust which may not be trapped by the collecting system.

Exhaust hoods that are larger than the containers and careful supervision can help solve this problem. In some instances, toxic materials can be handled in complete enclosures. The head of the barrel is broken in after the enclosure is sealed, and all dust is removed before the next barrel is put in.

Explosives should be stored in magazines of approved fireproof and bulletproof construction, located at a safe distance from railroads and other buildings (Table 16-A). Federal, state, and municipal codes regarding storage of explosives should be consulted and followed closely. NFPA Standard 495 gives detailed specifications for handling and storage, as does the Institute of Makers of Explosives and the duPont *Blasters Handbook*.

Explosives should be stored under lock and key, and records should be maintained of all explosives issued.

No matches, flammable materials, metal, or metal tools should be brought into an explosives magazine. Floors must be kept clean and free from loose explosives. The floors, which are usually of wood, should be blind-nailed; no nail or bolthead should be exposed.

Magazines should be clean, dry, and well ventilated. Ventilation openings should not exceed 110 sq in. in area and should be screened to prevent the entrance of sparks and rodents.

Only portable lights, approved for such use, are to be permitted in a magazine. Fire or sparks should not be allowed near a magazine, and the surrounding ground should be kept clear of brush, leaves, grass, debris, and other flammable material. Explosives should not be exposed to the direct rays of the sun.

Storage should be arranged so that the oldest explosives will be used first.

Ammonium nitrate requires special precautions including stacking limitations, air space, and ventilation. No oils or hydrocarbons are to be permitted near ammonium nitrate.

Packages of explosives should always be opened at least 50 ft from the magazine. Only wood wedges and wood, fiber, rawhide, zinc, babbitt metal, or rubber mallets should be used to open cases of explosives.

Blasting caps or detonators of any kind should never be kept in the same magazine with other explosives.

Explosives and blasting supplies should always be kept in the magazines, never in any place where animals, unauthorized persons, or children can get at them. Many children have been killed or crippled because they have obtained detonators from unwatched or unguarded sources.

Shipping and Receiving

The supervisor of the shipping and receiving room must be aware of Department of Transportation (DOT) Regulations and labels. Hazardous or flammable materials must be properly labeled and identified. Bills of lading or shipping must identify the item.

Floors, ramps, and aisles

Floors in warehouses, storerooms, and shipping rooms must be level. Unevenness of floors may lead to the toppling of piles of stored materials.

Safe floor load capacities and maximum heights to which specific materials may be piled should be posted conspicuously. Where bulk material, boxes, or cartons of the same weight are regularly stored, it is good practice to paint a horizontal line on the wall indicating the maximum height to which the material may be piled.

The strength of floors should be checked before the use of power trucks is adopted. Floor load capacity should be determined by a structural expert from the architectural data, the age and condition of the floor members, the type of floor, and other pertinent factors. Wherever materials are stored or trans-

ported, the surface of floors, platforms, and ramps should be kept in good condition. Breaks should be repaired immediately, and the area around doorways and elevator entrances should particularly be watched.

Ramps should have nonskid surfaces. When ramps are used for hand trucking, a nonskid foot strip may be laid in the center, or in the center of each lane for two-way traffic. Ramps should have handrails and, where there is heavy trucking, substantial curbs. A separate pedestrian lane, divided from the truck lane by a handrail, is a good idea for ramps used by both pedestrians and trucks.

Good housekeeping contributes to safety in handling of material, whether by hand or mechanically. A fall by a person carrying an object might result in more serious injury than it would if his hands were free.

Mirrors, placed at blind intersections, help to prevent collisions (Fig. 24–15). Warning signs and signals at such locations also serve as useful reminders, particularly to operators of power equipment. Doorways and entrances to tunnels and elevators may be similarly protected.

Mobile equipment used in storage areas should be equipped with backup warning devices.

Aisles should be wide enough to enable employees to move about freely while handling material or removing it from bins, racks, or piles, and to allow safe passage of loaded equipment. Aisleways and unloading areas should be clearly marked with white paint or black and white stripes. (See American National Standard Z53.1.) Trucks not in use, material, and other objects should not be allowed to stand in or extend into aisles.

Aisles leading to sprinkler valves and fire extinguishing equipment should be kept clear. Materials should not be piled closer than 18 in. to sprinkler heads. Closer spacing may reduce the effectiveness of heads in event of fire. For overly large, closely packed piles of combustible cases, bales, cartons, and similar stock, up to 36 in. of clearance should be provided. (See NFPA Standard No. 13.)

Lighting

General illumination of warehouses and storage rooms should follow American Na-

655

Fig. 24–15.—Blind-corner mirrors minimize collison hazard at doorways, 90-degree turns, and cross-aisles.

tional Standard *Practice for Industrial Lighting*, A11.1, published by the Illuminating Engineering Society. (See Table 46-G in Chapter 46.) Also see the *IES Handbook*.

Special lighting should be provided for operations requiring greater illumination. All lighting fixtures and wiring should be in accordance with the requirements of the NFPA *National Electrical Code*, Standard No. 70.

Stock picking

The movement of full pallet loads is most readily accomplished with the modern powered industrial truck. However, many industrial operations today require sequence picking of component parts for final assembly and fabrication.

With the variety of smaller finished products manufactured in industrial plants, shipments are typically made up by truck and/or rail carloads of mixed lots.

In both operations, such stock is usually found in racks or bins and, with the increase in high bay storage, special order picking equipment is required. Such operations lend new efficiency to material handling as well as new hazards in the areas of traffic, personal injury, and fire protection. The worker operating from a mobile order picking truck is exposed to falls from a height, as well as falling objects and material handling accidents. (See Chapter 22.)

Often, a worker is required to climb a ladder to get small parts or stock. Only heavy-duty material handling ladders should be used. These may be on rollers, but will have a braking mechanism with rubber feet which contact the surface as the worker's weight is imposed on the ladder. A working platform is provided, and standard guardrails protect the worker from falls as he reaches for stock or parts.

Under no circumstances should employees be allowed to use ordinary stepladders (particularly the short 2- or 3-step stools). Heavy-duty material handling equipment is required. Under no circumstances should employees climb racks or shelves.

Dockboards (bridge plates)

Dockboards, used in trailer and rail car loading and unloading, should be designed to carry four times the heaviest expected load and be wide enough to permit easy maneuvering of hand or power trucks. Dockboards are also known as bridge plates, dock plates, gangplanks, and bridge ramps. (Information given here applies to both hand and power truck operations.)

Many modern facilities today utilize automatic dock levelers and fixed-position hydraulic dockboards on both truck and rail docks. Dock shelters are also found in inclement weather zones, which effectively keep moisture from the dockboard and give the shipping and receiving department personnel a safer work environment.

These units require regularly scheduled maintenance. Most operate hydraulically and will provide a solid working and walking surface for heavy industrial trucks as well as hand truck operations. The design must be considered. All shear points must be guarded. The edges of movable sections should be painted yellow to denote a possible tripping hazard.

Trailers onto which material handling equipment operate must have wheel chocks at each wheel and the nose provided with a special jack, when the equipment operates in the forward portion of the trailer.

Dockboards should be designed and maintained so that, when they are in a secure position, the ends will have substantial contact with the dock (or loading platform) and the carrier to prevent the boards from rocking or sliding, when they are being used.

The sides of dockboards should be turned up at right angles, or otherwise designed, to prevent trucks from running over the edge.

Dockboards should have a nonskid surface to prevent employees and trucks from slipping. They should be kept clean and free of oil, grease, water, ice, and snow.

Handholds, or similar means, should be provided on dockboards to facilitate safe handling. Where practicable, fork loops or lugs should be fitted to the plates, so that they can be handled by fork trucks. Another method for lifting steel dockboards uses a low-voltage magnet, which is hung from the forks of a forklift truck and powered by the truck's battery.

To prevent foot injuries when dockboards are handled manually, they should be lowered or slid into place and not dropped. Enough men should be assigned to the job to permit safe, easy handling.

Dockboards must be secured in position. They should be either anchored down or equipped with devices to prevent slippage from the platform or the car threshold.

Dockboards, when not in use, should be stored in a safe place provided for the purpose. In some cases, heavy boards may be stacked horizontally by power trucks.

Positive protection should be provided to prevent railroad cars from being moved, by engines or car pullers, while dockboards are in position and men are working across them. Standard blue flags for daytime and blue lights for night work should be used to warn train crews. Local railroad authorities should be consulted on the specific warning device.

A regular program of dockboard inspection and maintenance should provide for necessary repairs or replacement.

Machines and tools

Machines which may be used in receiving and shipping, such as shears, saws, and nailing machines, should be guarded. Workmen should wear the necessary protective clothing—goggles, for instance—when operating a nailing or banding machine.

Tools should be of good quality and kept in good condition. Edged tools should be sharp, and holsters provided for workmen to carry them. Files should be provided that have a handle on the tang.

When the worker uses a drawknife instead of a scraper to remove markings from cases, boxes, and barrels, he should never brace the work with his knees. Should the drawknife slip, injury would almost certainly occur.

Steel strapping and sacking

Steel strapping, which may be flat or round,

requires that the worker be trained in both its application and removal. In all cases, the worker should wear safety goggles and leather-palm gloves (Fig. 24–16). Heavy strap may require steel-studded gloves.

Equipment designed for applying and re-

Fig. 24–16.—Note gloves and eye protection used by packers of electronic crossbar frame. Packers stand slightly to one side, so they are out of direct line of the strap in case it breaks.

Courtesy Western Electric Co., Omaha Works.

moving strapping should be used. When he operates a strapping tool, the workman should face in the direction of pull, one foot ahead of the other. Then, if the strap breaks or the tool slips, he will be in a position to protect himself.

When final tension is being attained, the operator should get out of direct line of the strap so that, in case of breakage, ends of strap will not strike his face or body. Excess strap beyond the tension-holding seal should be broken or cut off before a bound shipment is considered safe for further handling or shipment.

Before attempting to move bound merchandise or material, the workman should examine it for broken bands or loose ends. Broken bands should be removed and safely disposed of and, if possible, replaced to keep the ship-

ment from coming apart.

A box, carton, or package should never be handled by the steel bands, either manually or with a lift truck or other mechanical handling device. Stored packages, boxes, cartons, or other material should be checked for loose or protruding banding ends that might cut or otherwise injure passersby.

To remove strapping from bound containers, a cutter designed for the work should be used (Fig. 24–17). Workmen should never break the steel strap by applying leverage with a claw hammer, chisel, crowbar, or other tool.

Before cutting strapping, workmen should make sure that no one is standing where he might be hit by loose ends of strap.

To cut bands safely, the workman should place one gloved hand on the portion of the strap nearer him. Then, if the strap springs, it will be held to one side and fly away from the workman's face. In addition, the face should be held out of direct line of the strap. Eye protection should be worn.

Strap should be cut square, never at an angle. Strap cut at an angle has much sharper ends. A container for scrap steel should be close at hand, so that each piece can be safely disposed of as it is cut off.

Burlap and sacking are often received baled. Opening these bales is a job requiring some skill and experience. The foreman

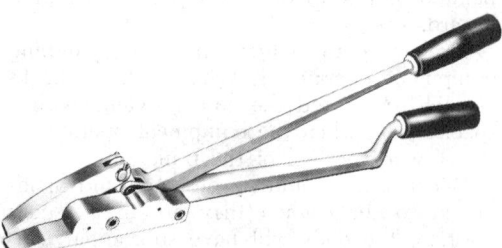

Fig. 24–17.—Special tool for cutting steel strap leaves only rounded edges.

Courtesy A. J. Gerrard & Co.

of the department should thoroughly instruct employees in the exact procedure to be used. Burred ends of wire used to tie sacks may cause many cuts. The workman should hold down one end of the wire, when he makes the cut, and should stand clear of the free end.

Glass and nails

Broken glass, often found in containers being unpacked, is a serious hazard in the shipping room. When unpacking glass or crockery materials, the employee should assume that broken material may be present. If possible, he should wear gloves.

Companies operating large shipping rooms report many injuries from flying nails. A nail should be started with a few light taps, so that it is given a good hold and will not fly. The face of the hammer should strike the nail at right angles. Eye and face protection should be worn.

Employees should be instructed to pull out and drive over nails which have been started at the wrong angle. Not only are fellow workers likely to be injured by the projecting nails, as they handle the cases, but also employees of the customer or railroad may cut their hands or fingers on the nails. Poorly driven nails may cause merchandise loss, if the package should come apart while being shipped.

Loose nails should not be left on the floors. Loaded trucks passing over the nails may drive them into the floor with the points up. In this position, they are a serious hazard since they are easily driven through workers' shoes and can also puncture pneumatic tires. The best practice is to have dropped nails removed from the floor at regular intervals during the day.

Employees who are opening cases should be instructed to bend nails over or remove them from the boards. If the box wood is to be used to make other or smaller boxes, removal of nails is important.

More often than not, crating lumber is discarded. Where quantities of such lumber are found, special carts should be provided for prompt removal.

If nails are used directly from kegs, supervisors should make certain that the nails holding the keg head in are withdrawn. Kegs, used for storage of nails at the work station, should be placed in an inclined rack so that the nails will feed out of the kegs.

Pitch and glue

To protect export shipments, parts are wrapped first with paper, plastic, or cheese-cloth and often covered with burlap. Pitch or other material is used for sealing. Hot pitch, however, can severely burn the skin not only because it is hot, but also because it is difficult to remove.

This work should be given to skilled men who have been instructed in the hazards and know how to avoid them. Goggles, face masks, gloves with the sleeves rolled over the gauntlets, and aprons should be worn. Wherever possible, it is better to use cold mastic asphalt instead of hot pitch.

Labeling glue, which often contains silicate of soda, causes discomfort when it splashes into the eyes. Workmen should wear eye protection and use the glue brush carefully.

Barrels and kegs

Projecting nails, jagged hoops and metal bands, ends of wire, splinters, and slivers cause many barrel-handling injuries, some of which lead to infection. Before handling barrels and kegs, employees should inspect them and take precautions against these hazards.

One method of opening a barrel is to use a lather's hatchet or a crate-opening tool to remove the nails. Then, when the top hoop is removed and the second hoop loosened, the head can easily be removed intact. Loosening nails on a barrel with wood hoops is simple, if the hoop is struck sharply with a hammer or hatchet near the point where the nail is inserted. The nail can then easily be pulled. This method of opening barrels not only preserves the barrel for future use, but also prevents contamination.

The opening of single-trip drums with a hammer and chisel is a frequent source of cuts and scratches. A commercial drum opener will open these drums without hazard and leave a smooth rolled edge.

Boxes and cartons

Employees who open boxes or cartons may incur wire punctures, nail punctures, or cuts from the device used for opening pasteboard cartons. When wirebound boxes or nailed boxes are opened, wires should be bent back and nails should be turned over or removed. Eye protection should be worn.

The boxes and covers should be piled neatly out of passageways. Safety carton openers, made of a protected sharpened blade, are available. The blade is slid along the edge of the carton.

These tools are useful only, of course, where the carton will not be reused. Cartons which are to be reused should be pried open with a flat steel pry bar, so that the flaps will not be damaged.

Corners of boxes and crates receive more blows than other parts and, therefore, should be constructed strongly. The interlocking-corner crate (Fig. 24–18) is stronger than any

Fig. 24–18.—The interlocking corner method (right) of crate construction is stronger than the common method (left).

other type, requires less lumber, and should be used wherever possible.

Diagonal braces help greatly in making crate sections rigid. One diagonal brace in the section will give more stiffening than several parallel slats. The diagonal should extend from corner to corner and should be placed so that it does not project beyond the other members of the crate.

Skids constructed as an integral part of a box should be made of sound lumber, free from knots, and of sufficient size to support the box without breaking. The skids should be firmly attached to withstand dragging across the floor.

The principal reason for failure of boxes and crates is poor nailing. Correct nailing is very essential to safe shipment. Where trouble is experienced with the nails pulling out, cement-coated nails, which hold better than uncoated nails, should be used.

Broken or damaged containers of consumables. Handlers should not attempt to sample or distribute food or other commodities which are available when containers are damaged

or broken. These commodities could have become contaminated, tainted, or otherwise unsafe while en route. In Columbia, South America, contents of insecticide and food bags became mixed during rough shipment and use of the contaminated flour caused the deaths of many people. Careless handling of samples could also endanger health.

Car loading

Heavy machinery shipped on skids should be braced inside the car to prevent shifting. Lag screws should not be used to fasten the skid to the car floor, since men may drive them in with a hammer rather than use a wrench. Using a hammer damages the wood, thus reducing the holding power of the lag screws.

Skids with large knots are hazardous when used on shipments of heavy machinery. When rollers are used to move the object, the skid is likely to break when the roller comes under a knot. (Refer to Association of American Railroads Pamphlet No. 21 in References.)

Before opening rail cars, the doors should be carefully inspected. If damaged, runners should be repaired and special precautions should be taken. Door openers should be used and employees should stand clear, in case an improperly loaded car is received.

Workers, who are opening and closing car doors, may catch their hands between the doors and the car doorposts. Workers should be instructed never to grasp the leading end of the door which might cause their fingers to catch between the car door and the side of the car. Likewise, they should keep their hands and fingers away from the doorpost when they are closing the door.

To avoid leaving hazards for railroad or other employees, as well as to avoid contamination of future lading or damage to it, consignees should clean cars after they have been unloaded.

Personal Protection

Certain items of protective equipment are desirable for the prevention of various types of materials handling injuries. Since toe and finger injuries are among the most common types, handlers should wear safety shoes and stout gloves, preferably with leather faces.

Other special protective clothing, such as eye protection, aprons, and leggings, should be required for the handling of certain types of materials.

Gloves should be dry and free of grease and oil. Hand protection to prevent injury from splinters should be worn when wooden crates are handled. Clean leather-palm gloves give better holding power on smooth metal objects than do cotton or other types. However, it may be unsafe for workers to wear gloves near conveyors, or elsewhere where the risk of catching exists. Care should be taken not to bruise or squeeze the hands at doorways or other points where clearance is close. During inclement weather, special foul weather apparel may be necessary.

Where toxic or irritating solids are handled, workmen should take daily showers to remove the material from their persons before they leave the plant. Even though the exposure does not necessitate showers, workers should be encouraged to wash thoroughly at the end of their shifts. Cleansing materials, shower stalls, and wash basins should be provided.

Washable suits of tightly woven fabric, preferably full-length coveralls, and washable caps should be worn. Suits, caps, socks and underwear should be laundered daily at the plant. Clothing should be laundered less frequently only at the express direction of the plant medical department.

Skin contact with chemicals should be avoided. See Chapters 37 and 39 for details.

References

American Petroleum Institute, 1801 K St., N.W., Washington, D.C. 20006. *Cleaning Petroleum Storage Tanks*, RP 2015.

American National Standards Institute, 1430 Broadway, New York, N.Y. 10018.
Gray Finishes for Industrial Apparatus and Equipment, Z55.1.
Practice For Industrial Lighting, A11.1.
Safety Code for Crawler, Locomotive, and Truck Cranes, B30.5.
Safety Code for Fixed Ladders, A14.3.
Safety Codes for Powered Industrial Trucks, B56.1–4.
Safety Color Code for Marking Physical Hazards, Z53.1.
Scheme for the Identification of Piping Systems, A13.1.

Association of American Railroads, Operating-Transportation Div., 1920 L St. NW., Washington, D.C. 20006. *Recommended Methods for Loading, Bracing and Blocking Carload Shipments of Machinery in Closed Cars*, Pamphlet No. 21.

E. I. du Pont de Nemours & Co. (Inc.), Explosives Dept., Wilmington, Del. 19898. *Blasters Handbook*, 15th ed., 1967.

Illuminating Engineering Society, 345 East 47th St., New York, N.Y. 10017. *IES Handbook*.

Institute of Makers of Explosives, 420 Lexington Ave., New York, N.Y. 10017. *Safety in the Handling and Use of Explosives*.

Manufacturing Chemists' Association, 1825 Connecticut Ave. NW., Washington, D.C. 20009. Manual Sheets, TC Series (on unloading various hazardous liquids from tank cars). *Guide to Precautionary Labeling of Hazardous Materials*. Manual L-1.

National Fire Protection Association, 470 Atlantic Ave., Boston, Mass. 02210.
Cleaning or Safeguarding Small Tanks and Containers, No. 327.
Code for Explosives and Blasting Agents, No. 495.
Flammable and Combustible Liquids Code, No. 30.
Fundamental Principles for the Prevention of Dust Explosions in Industrial Plants, No. 63.
National Electrical Code, No. 70.
National Fire Codes, Vol. 4, "Flammable Liquids and Gases."
Rack Storage of Materials, 231C.

24—Materials Handling and Storage

National Safety Council, 425 North Michigan Ave., Chicago, Ill. 60611.
    Industrial Data Sheets

        Automotive Hoisting Equipment, 437.
        Baling Presses, 421.
        Bottles and Broken Glass, 355.
        Boxcars, Unloading Bulk Grain from, 521.
        Calender Rolls, Handling, 403.
        Chains (Alloy Steel) for Overhead Lifting, 478.
        Conveyors, Belt (Equipment), 569.
        Conveyors, Belt (Operation), 570.
        Conveyors, Roller, 528.
        Conveyors, Underground Belt, 447.
        Cranes, Pendant-Operated and Console-Controlled, 558.
        Dock Plates and Gangplanks, 318.
        Electromagnets Used with Crane Hoist, 359.
        Flammable Liquids in Small Containers, 532.
        Food Bins and Tanks, 524.
        Forging Industry, Handling Materials in, 521.
        Front-End Loaders, 589.
        Fusees and Torpedoes—Used in Railroad Operations, Handling and Storage of, 639.
        Hand Trucks, Powered, 317.
        Hauler-Loaders, 576.
        Hoisting Equipment, Automotive, 437.
        Hoists, Construction Material, 511.
        Liquefied Petroleum Gases for Industrial Trucks, 479.
        Log Skidding by Tractor, 377.
        Lumber Handling and Piling, 345.
        Molds (Heavy) in the Rubber Industry, Handling and Storing of, 507.
        Motor Trucks for Mines, Quarries and Construction, 330.
        Oil Field Pipe, Handling Large-Diameter, 463.
        Paper Rolls, Handling and Storing, 596.
        Power Presses, Handling Finished Pieces at, 534.
        Pulpwood, Unloading at the Mill, 274.
        Scrap Ballers, 611.
        Sheet Metal, Handling and Storage of, 434.
        Skids, 260.
        Steel Plate, Handling for Fabrication, 565.
        Stores, Fire Prevention in, 549.
        Sulfur, Handling and Storage of Solid, 612.
        Sulfur, Handling Liquid, 592.
        Thawing Frozen Ladings, 233.
        Tractor Operation and Protective Frames, 622.
        Winches, Truck Mounted Power, 441.
        Wire Rope Slings, Recommended Loads for, 380.
        Vehicular Traffic, Airport, 539.

"R. M. Graziano's Tariff," Bureau of Explosives, 1920 L St., N.W., Washington, D.C. 20036. *Hazardous Materials Regulations of the Department of Transportation*, Tariff No. 25.

U.S. Department of Transportation, Office of Hazardous Materials Washington, D.C. 20590. "Newly Authorized Hazardous Materials Warning Labels." Rev. January 1974. (Based on Title 49, C.F.R., sections 173.402, -403, and -404; import or export shipments are covered in Title 14, C.F.R., Section 103.13.)

Hoisting Apparatus and Conveyors

Chapter
25

Hoisting Apparatus 664

Hoists—general . . . Electric hoists . . . Air hoists . . . Hand-operated chain hoists . . . Cranes—design and construction . . . Cranes—guards and limit devices . . . Cranes—ropes and sheaves . . . Crane and hoist signals . . . Selection and training of operators . . . Inspection . . . Safety rules . . . Overhead cranes . . . Storage bridge and gantry cranes . . . Monorails . . . Jib cranes . . . Derricks . . . Mobile cranes . . . Aerial baskets . . . Crabs and winches . . . Blocks and tackles . . . Portable floor cranes or hoists . . . Tiering hoists

Conveyors 696

General precautions . . . Slat and apron conveyors . . . Flight conveyors . . . Chain conveyors . . . Shackle conveyors . . . Screw conveyors . . . Bucket conveyors . . . Pneumatic conveyors . . . Aerial conveyors . . . Portable conveyors . . . Gravity conveyors . . . Live roll conveyors . . . Vertical conveyors

References 709

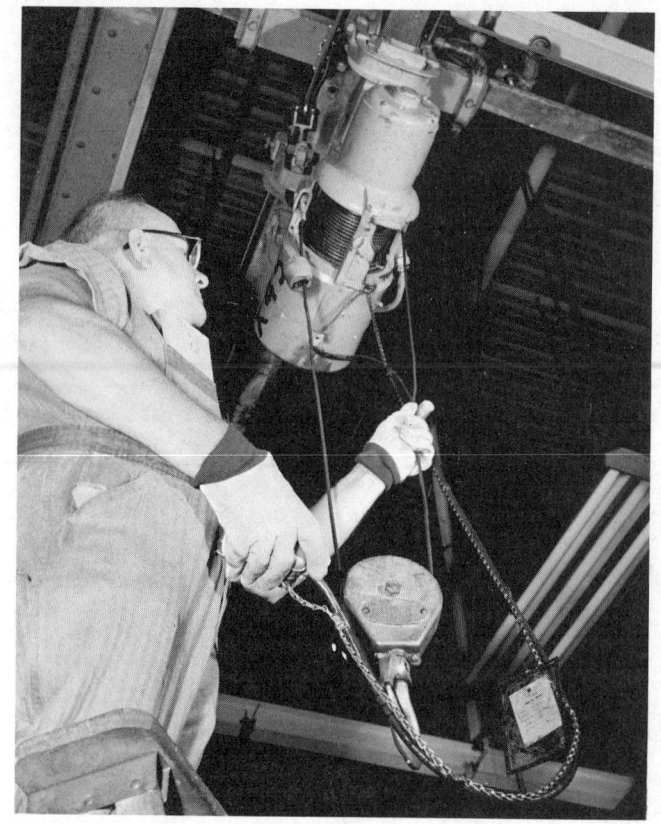

FIG. 25–1.—To check the rope of this 1000-lb electric hoist, man inspects 2- to 3-ft increments after rope has come to complete stop. To check for frays, he wipes a rag up and down rope.

Courtesy Detroit Diesel Engineering Div., General Motors Corp.

Hoisting Apparatus

To raise, lower, and transport heavy loads for limited distances, hoisting apparatus has been in use for centuries. Today there are traveling cranes in steel mills and power plants and hammerhead cranes in naval shipyards capable of lifting hundreds of tons.

Also see the discussion of material and passenger hoists in Chapter 17, "Building Construction and Maintenance."

Hoists – general

Many thousands of hoists (electric, air, and hand powered) are in use in industry. Typically, they range from ¼ to 2 tons in capacity, but they can go larger.

During a 20-year period, one division of an automobile manufacturing company made 72,000 formal hoist inspections, involving over 600 electric hoists and cranes. The inspections disclosed approximately 3000 unsafe cables, 1000 faulty brakes, and 1000 defective limit switches.

Scheduled detailed inspection of all hoists, with special attention to load hooks, ropes, brakes, and limit switches, is important (Fig. 25–1).

Safety factors, similar to those indicated for cranes, should be allowed for hoists. The safe load capacity of each hoist should be shown in conspicuous figures on the hoist body of the machine.

Flanges on hoist drums with single-layer spiral grooves should be free of projections that could damage a cable.

All hoists should be attached to their supports (fixed member or trolley) with shackles, or the support hooks should be moused or have safety latches. Latches are recommended also for load hooks. Hoist supports should also have an adequate safety factor for the maximum loads to be imposed.

Material hoists operating on rails, tracks, or trolleys should have positive stops or limiting devices on the equipment, rails, tracks, or trolleys to prevent the overrunning of safe limits, and hoists should also be equipped with overspeed devices.

Extra protection against failure of the supporting hook, shackle, or block can be provided by a retaining cable or chain looped around the body of the hoist and the support (jib, beam, or carrying pin of the trolley).

A load should be picked up only when it is directly under the hoist. Otherwise, stresses for which the hoist was not designed may be imposed upon it. If the load is not properly centered, it may swing (upon being hoisted), and injury could result. Everyone must stay out from under raised loads (Fig. 25–2).

Caution: Unless used in combination with other safety devices approved by the hoist manufacturer, the simple floor-operated electric-, air-, or hand-powered hoist should not be used to lift, support, or otherwise transport people. The standard commercial hoist or crane does not provide a secondary means of supporting the load should the wire rope or other suspension element fail.

Electric hoists

Electric hoists should have nonconducting control cords unless they are grounded. Control cords should have handles of distinctly different contours so that, even without looking, the operator will know which is the hoisting and which is the lowering handle. (See discussion in Chapter 10, "Human Factors Engineering.")

Each control cord should be clearly marked "hoist" or "lower." Some companies attach an arrow to each control cord, pointing in the direction in which the load will move when the rope is pulled. Also, it may be advisable to pass the control cords through a spreader to keep them from becoming tangled. The spreader can be a 1- by 3-in. board with equally spaced holes, resting upon the pull handles.

Control cords, usually made of fiber or

Fig. 25–2.—Company safety rule states: "Be watchful of loads suspended in air; keep out from under them." Fortunately everyone was in the clear when this 150-lb crane hook failed.

Courtesy Colorado Fuel & Iron Corp.

light wire rope, should be inspected weekly for wear and other defects.

On pendant-controlled electric hoists, means for effecting automatic return to the OFF position should be provided on the control, so that a constant pull on the control rope or push on the control button must be maintained to raise or lower the load.

Pushbutton control circuits should be limited to 110–120 volts.

A limit stop should be installed on the hoist motion, and at least two turns of rope should remain on the drum when the load block is on the floor.

Air hoists

After a piston-type air hoist has been in operation for a time, the locknut that holds the piston on its rod may become loose so
(*Text continues on page 668.*)

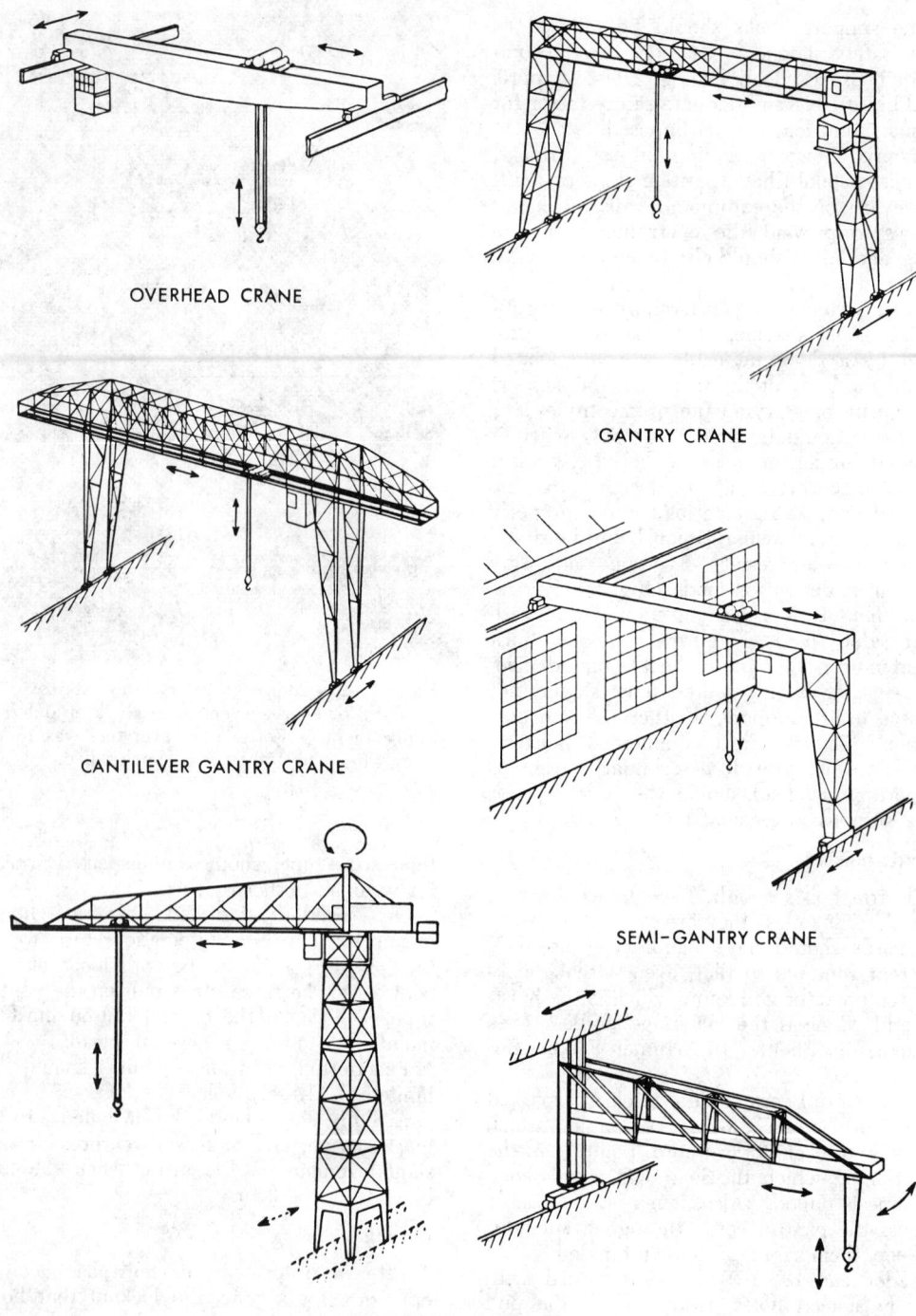

OVERHEAD CRANE

GANTRY CRANE

CANTILEVER GANTRY CRANE

SEMI-GANTRY CRANE

HAMMERHEAD CRANE

WALL CRANE

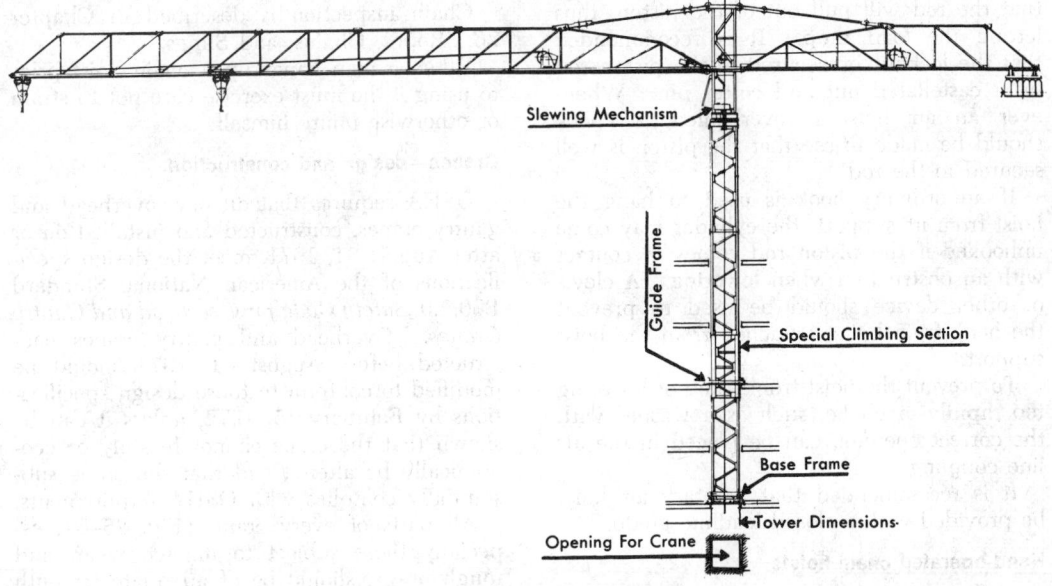

CLIMBING-TYPE CRANE

Labels: Slewing Mechanism, Guide Frame, Special Climbing Section, Base Frame, Tower Dimensions, Opening For Crane

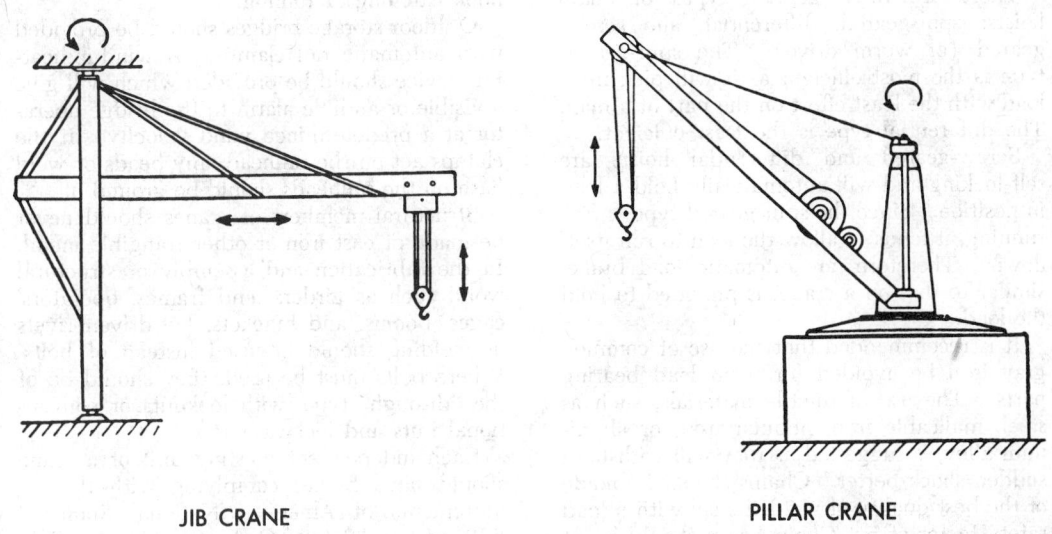

JIB CRANE

PILLAR CRANE

FIG. 25–3.—Diagrammatic sketches of various types of cranes.

From ANSI Standard Series B30, "Safety Code for Cranes, Derricks, Hoists, Jacks and Slings" (except for climbing crane).

that the rod will pull out of the piston, thus letting the load drop. It is recommended that the locknut be secured to the piston rod by a castellated nut and cotter pin. Whenever an air hoist is overhauled, a check should be made to see that the piston is well secured to the rod.

If an ordinary hook is used to hang the hoist from its support, the cylinder may come unhooked if the piston rod comes in contact with an obstruction when lowering. A clevis or other device should be used to prevent the hook from being detached from the hoist support.

To prevent the hoist from rising or lowering too rapidly, a choke, such as a washer with the correct opening, can be placed in the air line coupling.

It is recommended that a rotary air hoist be provided with a closed loadline guide.

Hand-operated chain hoists

Chain hoists may be portable, but should be either permanently hooked onto a monorail trolley, or built into the trolley as an integral part. They are suitable for many operations on which a block and tackle, fitted with fiber rope, is used and are stronger, more dependable, and more durable than fiber-rope tackle.

There are three general types of chain hoists: spur-geared, differential, and screw-geared (or worm drive). The spur-geared type is the most efficient as it will pick up a load with the least effort on the part of a man. The differential type is the least efficient.

Screw-geared and differential hoists are self-locking and will automatically hold a load in position. Since the spur-geared type is free running, it tends to allow the load to run itself down. Therefore, an automatic load brake, similar to that on a crane, is provided to hold the load.

It is recommended that the use of common gray iron be avoided for main load bearing parts. The use of ductile materials, such as steel, malleable iron, nodular iron, or aluminum alloy, is suggested as they will withstand sudden shock better. Chains should be made of the best quality of welded steel with a load safety factor of 5. Chain hoists should be of larger capacity than the regular work requires. Supports for the hoists must be strong enough to carry the load imposed on them.

Chain inspection is described in Chapter 26, "Ropes, Chains, and Slings."

When a man hangs up a chain hoist prior to using it, he must exercise care not to strain or otherwise injure himself.

Cranes—design and construction

OSHA requires that all new overhead and gantry cranes, constructed and installed on or after August 31, 1971, meet the design specifications of the American National Standard B30.2.0, *Safety Code for Overhead and Gantry Cranes.* Overhead and gantry cranes constructed before August 31, 1971 should be modified to conform to those design specifications by February 15, 1972, unless it can be shown that the crane cannot feasibly or economically be altered and that the crane substantially complies with OSHA requirements.

All parts of every crane (Fig. 25–3), especially those subject to impact, wear, and rough usage, should be of adequate strength for maximum service. Journals and shafts should be of sufficient strength and size to bring the bearing pressure within safe limits.

Open hooks should not be used to support human loads, loads that pass over workmen, or loads where there is danger of relieving the tension on the hook due to the load or hook catching or fouling.

Outdoor storage bridges should be provided with automatic rail clamps. A wind-indicating device should be provided which will give a visible or audible alarm to the bridge operator at a predetermined wind velocity. If the clamps act on the railhead, any beads or weld flash on the railheads should be ground off.

Structural members of cranes should never be made of cast iron or other frangible metal. In the fabrication and assembly of structural work, such as girders, end frames, operators' cages, booms, and brackets, hot-driven rivets or welding should be used instead of bolts. Where bolts must be used, they should be of the "through" type, with locknuts or conventional nuts and lockwashers.

Each independent hoisting unit of a crane should have brakes complying with the requirements of American National Standard B30 Series, "Safety Code for Cranes, Derricks, Hoists, Jacks, and Slings."

The rated load of the crane should be plainly marked on each side of the crane. If the crane has more than one hoisting unit, each hoist should have its rated load marked on it or its load block, and this marking should be clearly legible from the ground or floor. The crane should not be loaded beyond its rated capacity, except for testing.

The general arrangement of the cab and the location of control and protective equipment should be such that all operating handles are within convenient reach of the operator, when facing the area to be served by the load hook or when facing the direction of travel of the cab. The arrangement should allow the operator a full view of the load hook in all positions.

Each controller and operating lever should be marked with the motion it controls and its direction. These levers should have spring returns, so that they will move automatically into the OFF position and latch themselves there if the operator releases the handle.

Other factors that should be considered:

1. Platforms, footwalks, steps, handholds, guardrails, and toeboards should be provided for safe footing and accessways.

2. All machinery, equipment, and material hoists operating on rails, tracks, or trolleys should have positive stops or limiting devices on the equipment, rails, tracks, or trolleys to prevent overrunning safe limits and should be equipped with overspeed devices.

3. All joints requiring lubrication during operation should have fittings so located or guarded as to be accessible without hazardous exposure.

4. Platforms, footwalks, steps, handholds, guardrails with intermediate rails, and toeboards should be provided on machinery and equipment to provide safe footing and accessways. Platforms and steps should be of nonskid material.

5. Access to the cab and/or bridge walkway should be provided by conveniently placed fixed ladders, stairs, or platforms whose steps leave no gap exceeding 12 inches. Fixed ladders should be in conformance with the American National Standard A14.3, *Safety Code for Fixed Ladders.*

6. A carbon-dioxide, dry-chemical, or equiva-

FIG. 25–4.—Load blocks are safer when guarded.

lent fire extinguisher should be kept in the cab.

Cranes—guards and limit devices

Gears and other moving parts should be totally enclosed, covered by screen guards, or out of reach.

No overhung gears should be used, unless means are provided to prevent their falling if they should break or work loose.

The bolts in shaft couplings should be recessed, so that the tops of the nuts do not project.

To prevent fingers from being crushed, load hooks on cranes should have handles, so that the men can hold or guide the hooks when slings are being placed on them. Also, on small cranes, the pinch points where cables pass over sheaves in the load block should be guarded (Fig. 25–4).

The hoist motion of every crane, with the exception of boom-type cranes and derricks, should be provided with an approved limit device, which is designed to be operated by the ascending load block as it approaches its upper limit of travel. The device may be paddle or weight operated and may also be of the geared type. In the latter case, the limit device should be reset each time a new cable is installed or a new attachment of the cable is made to the hoist drum.

Limit devices always should operate on a normally closed circuit and should be tested daily. To make this test, the unloaded block should be carefully run up so as to actuate the device. Here is a suggested testing procedure:

1. Inch block into limit switch.

2. Drop block 4 to 6 ft, and run into limit at half speed.

3. Drop hoist approximately 10 ft; stop and run at full speed into limit switch.

4. Test should always be made clear of equipment and employees.

Hoist limit switches are normally a safety device to prevent accidental overtravel of the load block and are not intended for constant-duty service. Requirements for hoists with constant-duty limit switches should be referred to a hoist manufacturer.

The hook block should be designed so that it will lift vertically without twisting. The hook should be of solid forged steel or built-up steel plates. Large hooks should swivel on roller or ball bearings.

Electric wiring should be installed in accordance with Article 610, "Cranes and Hoists," of the *National Electrical Code* (American National Standard C1 and NFPA Standard No. 70).

On electric-power operated cranes, the power supply to the runway conductors should be controlled by a switch or circuit breaker, which is located on a fixed structure accessible from the floor and arranged to be locked in the OPEN position.

Other precautions include:

1. No guard, safety appliance, or device should be removed or otherwise be made ineffective in machinery or equipment, ex-

cept for the purpose of making immediate repairs, lubrications, or adjustments and, then, only after the power has been shut off.

2. All guards and devices should be replaced immediately after completion of repairs and adjustments.

3. Traveling cranes should be equipped with a warning device that will sound continuously while the crane is traveling.

4. Cranes should be equipped with a load limiting device which will effectively prevent overloading beyond the manufacturers' ratings at any load, boom radius, and counterweight position.

Before starting maintenance or repair work on a crane, workmen should apply their personal padlocks to the main switch while it is in OFF position. When working on a multiple-crane runway, positive provision should be made to prevent other cranes from running into the crane being worked on.

Cranes—ropes and sheaves

The hoisting ropes should be of a recommended construction for crane service. The crane or rope manufacturer should be consulted whenever a change is contemplated. The rated load divided by the number of parts of rope should not exceed 20 percent of the normal breaking strength of the rope.

Sheaves and drums should be watched for wear. If the grooves become enlarged by wear or corrugated from excessive rope pressure, they should be turned down or the sheaves should be replaced at once, for these conditions will cause rapid wear and loss of strength of the rope. In cases where considerable material must be removed to regroove the drum or sheave, the strength of these parts may be impaired. In such cases, it is recommended that the hoist or crane manufacturer be consulted prior to regrooving. (Further information is in Chapter 26, "Ropes, Chains, and Slings.")

The ratio of drum diameter to rope diameter should not be less than 20 to 1 (not less than 10 to 1 for equalizing sheaves). Ratio for rotating sheaves should be 16:1. It is recommended that 6 x 37 construction wire rope

(*Text continues on page 674.*)

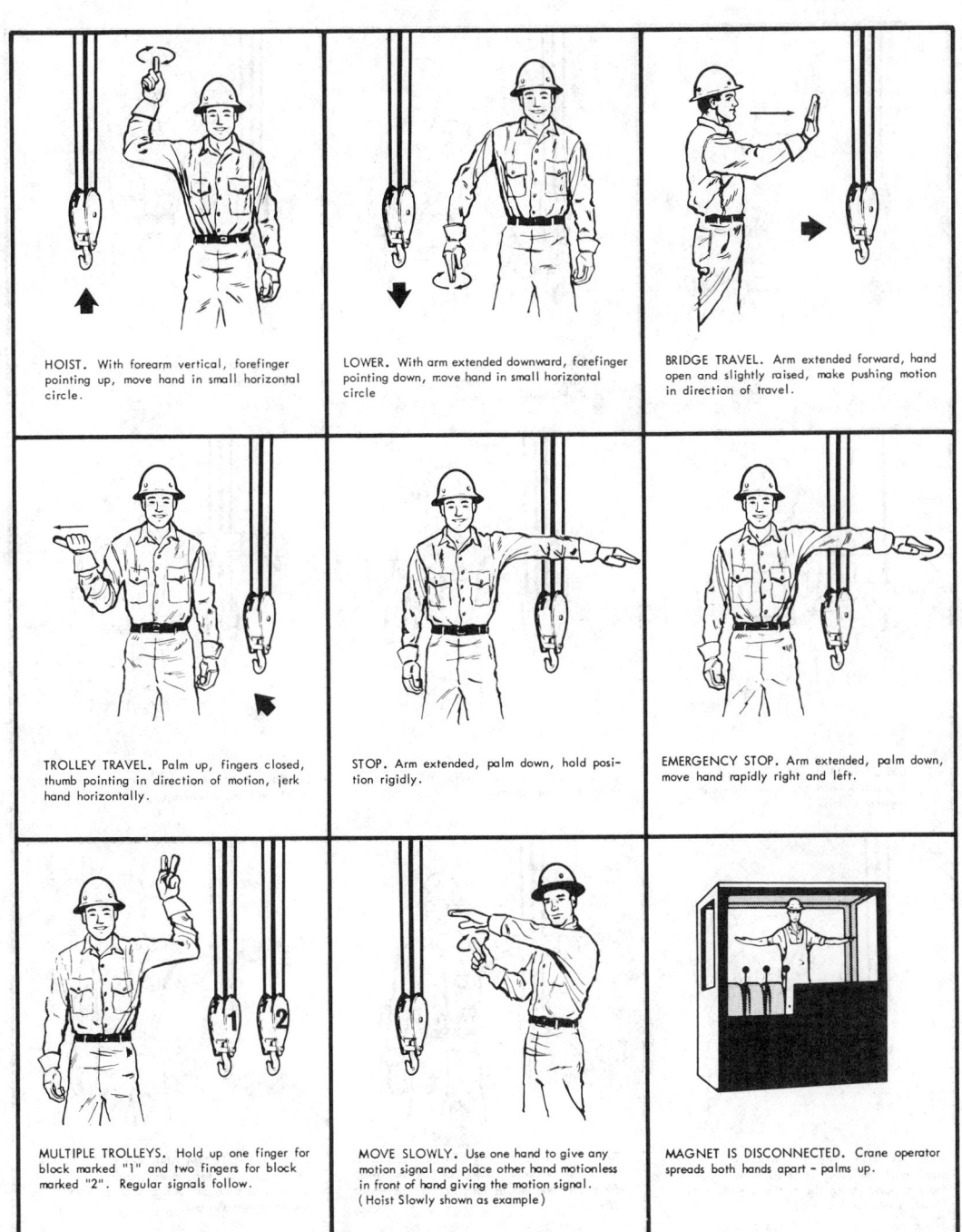

HOIST. With forearm vertical, forefinger pointing up, move hand in small horizontal circle.

LOWER. With arm extended downward, forefinger pointing down, move hand in small horizontal circle

BRIDGE TRAVEL. Arm extended forward, hand open and slightly raised, make pushing motion in direction of travel.

TROLLEY TRAVEL. Palm up, fingers closed, thumb pointing in direction of motion, jerk hand horizontally.

STOP. Arm extended, palm down, hold position rigidly.

EMERGENCY STOP. Arm extended, palm down, move hand rapidly right and left.

MULTIPLE TROLLEYS. Hold up one finger for block marked "1" and two fingers for block marked "2". Regular signals follow.

MOVE SLOWLY. Use one hand to give any motion signal and place other hand motionless in front of hand giving the motion signal. (Hoist Slowly shown as example)

MAGNET IS DISCONNECTED. Crane operator spreads both hands apart – palms up.

FIG. 25–5.—Standard hand signals for overhead and gantry cranes.

From ANSI Standard Series B30, "Safety Code for Cranes, Derricks, Hoists, Jacks, and Slings," with permission of the publishers, The American Society of Mechanical Engineers, New York, N.Y. 10017.

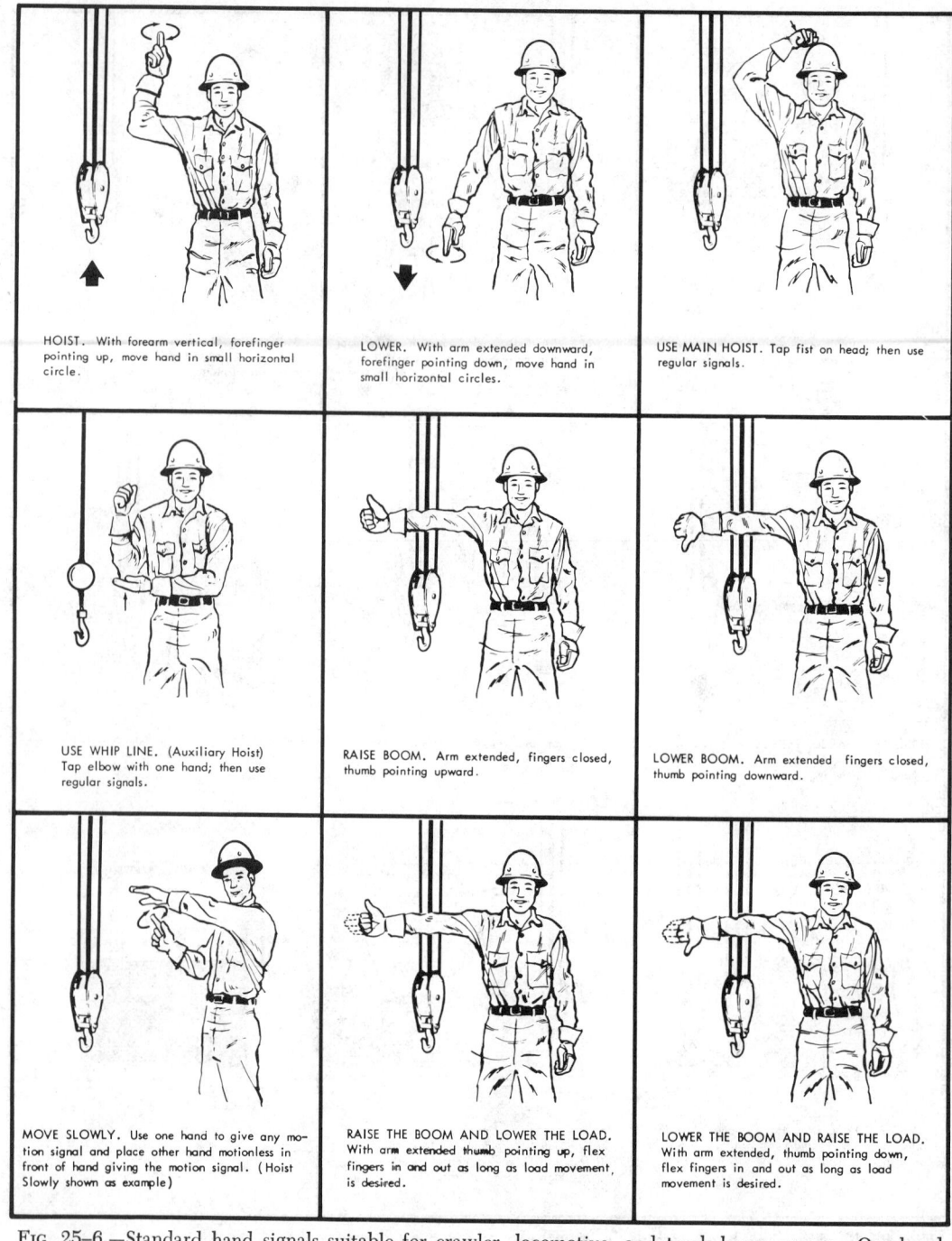

HOIST. With forearm vertical, forefinger pointing up, move hand in small horizontal circle.

LOWER. With arm extended downward, forefinger pointing down, move hand in small horizontal circles.

USE MAIN HOIST. Tap fist on head; then use regular signals.

USE WHIP LINE. (Auxiliary Hoist) Tap elbow with one hand; then use regular signals.

RAISE BOOM. Arm extended, fingers closed, thumb pointing upward.

LOWER BOOM. Arm extended fingers closed, thumb pointing downward.

MOVE SLOWLY. Use one hand to give any motion signal and place other hand motionless in front of hand giving the motion signal. (Hoist Slowly shown as example)

RAISE THE BOOM AND LOWER THE LOAD. With arm extended thumb pointing up, flex fingers in and out as long as load movement, is desired.

LOWER THE BOOM AND RAISE THE LOAD. With arm extended, thumb pointing down, flex fingers in and out as long as load movement is desired.

Fig. 25–6.—Standard hand signals suitable for crawler, locomotive, and truck boom cranes. One-hand signal for extending or retracting telescoping boom (not shown above): Extend boom—one fist in front of chest with thumb topping chest. Retract boom—one fist in front of chest, thumb pointing out-

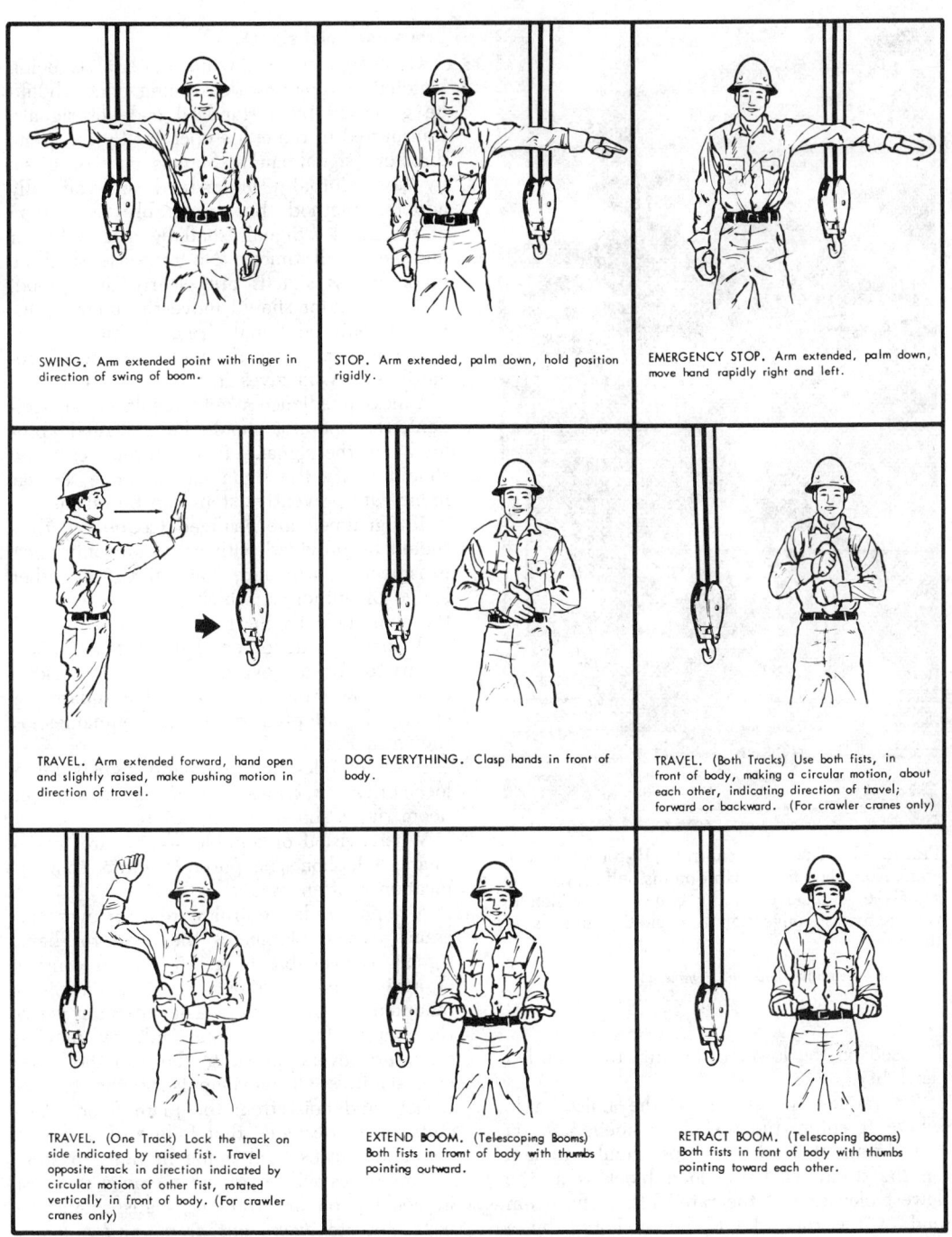

SWING. Arm extended point with finger in direction of swing of boom.

STOP. Arm extended, palm down, hold position rigidly.

EMERGENCY STOP. Arm extended, palm down, move hand rapidly right and left.

TRAVEL. Arm extended forward, hand open and slightly raised, make pushing motion in direction of travel.

DOG EVERYTHING. Clasp hands in front of body.

TRAVEL. (Both Tracks) Use both fists, in front of body, making a circular motion, about each other, indicating direction of travel; forward or backward. (For crawler cranes only)

TRAVEL. (One Track) Lock the track on side indicated by raised fist. Travel opposite track in direction indicated by circular motion of other fist, rotated vertically in front of body. (For crawler cranes only)

EXTEND BOOM. (Telescoping Booms) Both fists in front of body with thumbs pointing outward.

RETRACT BOOM. (Telescoping Booms) Both fists in front of body with thumbs pointing toward each other.

ward and heel of fist tapping chest.

From ANSI Standard Series B30, "Safety Code for Cranes, Derricks, Hoists, Jacks, and Slings," with permission of the publishers, The American Society of Mechanical Engineers, New York, N.Y. 10017.

FIG. 25–7.—Operator controls 10-ton overhead crane from shop floor. Unit on his belt sends radio signals to a receiver in the crane cab, which in turn actuates conventional magnetic controls of the crane.

Courtesy Telemotive Division, Dynascan Corp.

be used on the lower ratios due to its added flexibility.

To reduce the strain on the hoist cable where it enters the socket or anchorage on the drum, at least 2 wraps should remain on the drum when the load block is at the lowest elevation of the rated lift. The drum end of the rope should be anchored by a clamp securely attached to the drum or by a socket arrangement approved by the crane or rope manufacturer.

Crane and hoist signals

Crane movements, while material is being handled or repair work is being done, should be governed by a standard code of signals, transmitted to the crane operator by the crane director (signalman). Signals may be given by any mutually understood and officially adopted method, but preferably by motion of the hand. Signals shall be discernible or audible at all times. No response shall be made unless signals are clearly understood.

The operator should move the hoisting apparatus only on signals from the proper person, but a stop signal should be obeyed regardless of who gives it.

Unless obedience would result in an accident, the operator should be governed absolutely by the signal. In the former case, he should notify the signalman at once so that immediate preventive steps can be taken.

If signalmen are changed frequently, they should be provided with one (and only one) conspicuous armband, hat, glove, or other badge of authority, which must be worn by the man currently in charge.

A simple code of one-hand signals is appropriate for an overhead crane or bridge crane. The American National Standard set of signals, adopted by many companies, is shown in Fig. 25–5.

A set of one- and two-hand signals for a locomotive or crawler crane, or any other boom rig, is shown in Fig. 25–6.

Where visual or audible signals are inadequate, telephone or portable radio communication is often used.

A remote radio-control system for overhead cranes, which eliminates the need for hand signals, is available (Fig. 25–7). It consists of a cigarbox-size transmitter and a receiver mounted on the crane. The operator wears the transmitter on a belt and, by moving miniature levers similar to those in the crane cab, controls all movements of the bridge, trolley, and hoist from the plant floor. Circuits are so designed that failure of a system component causes all crane motions to stop.

Employees who work near cranes or assist in hooking on or arranging loads should be instructed to *keep out from under loads*. Supervisors should watch closely to see that this rule is strictly followed.

One manufacturing company publishes this

warning: "From a safety standpoint, one factor is paramount: Conduct all lifting operations in such a manner that, if there were an equipment failure, no personnel would be injured. This means *keep out from under raised loads!*" (See Fig. 25–2.)

Selection and training of operators

Cranes should be operated only by employees who are at least 21 years of age; are physically qualified, particularly with regard to acuity of vision, depth and color perception, hearing, muscular coordination, and reaction time; and are not subject to epilepsy, heart, or similar ailments that would be detrimental to safe operation of equipment. A physical examination should be required prior to employment and annually thereafter.

Operators should undergo a course of training and be certified to operate cranes. Equally important are training and authorization of hook-on men. Some companies require both operators and hook-on men to have permits, renewable at intervals of a year or two upon reexamination.

Inspection

All crane machinery, apparatus, and appliances (including ropes, chains, and slings) should be inspected daily by a qualified person assigned to this task; and the date, findings, and action taken should be recorded on a special report form. The crane director also should make such daily inspections. In addition, detailed weekly inspection of the entire crane is recommended.

Hooks should be tested by X ray or similar means. Under no circumstances should a hook be heated to reduce a permanent set; rather, the hook should be replaced. Hooks that have cracks, that have more than 15 percent in excess of normal throat opening (Fig. 25–8), or that have more than a 10-degree twist from the plane of the unbent hook should be discarded. Repairs by welding or reshaping are generally not recommended.

A crane operator never should attempt to make repairs himself, but should report to his foreman any condition that might make the crane unsafe to operate. Certain faults may be so dangerous that the crane should be shut down at once and not operated until the faults are corrected.

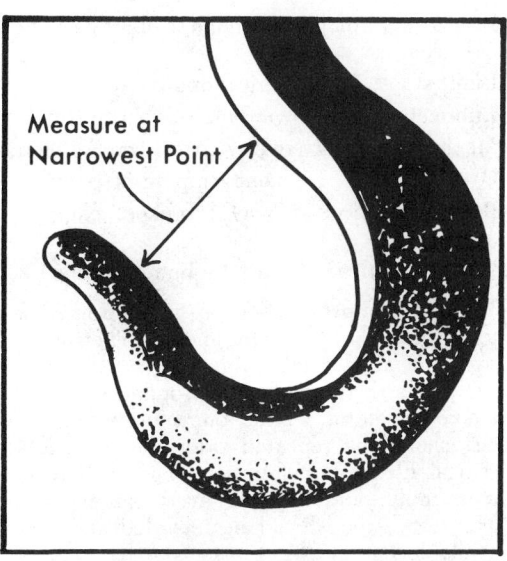

Measure at Narrowest Point

Fig. 25–8.—When permanent set of hook is greater than 15 percent in excess of normal throat opening, the hook should be replaced.

Following is a list of unsafe conditions to be checked mainly by operators of overhead traveling cranes. (Also see next section.)

Bearings: loose, worn

Brakes: shoe wear

Bridge: alignment out of true (indicated by screeching or squealing wheels)

Bumpers on bridge: loose, missing, improper placement

Collector shoes or bars: worn, pitted, loose, broken

Controllers: faulty operation because of electrical or mechanical defects

Couplings: loose, worn

Drum: rough edges on cable grooves

End stops on trolley: loose, missing, improper placement

Footwalk: condition

Gears: lack of lubrication of foreign material in gear teeth (indicated by grinding or squealing)

Guards: bent, broken, lost

Hoisting cable: broken strands

Hook block: chipped sheave wheels

Hooks: straightening

Lights (warning or signal): burned out, broken

Limit stops: functioning properly

Lubrication: overflowing on rails, dirty cups

Mechanical parts (rivets, covers, etc.): loose

Overload relay: frequent tripping of power

Rails (trolley or runway): broken, chipped, cracked

Wheels: worn (indicated by bumpy riding)

Many companies believe in performance tests for all hoisting equipment. They make sure that all hoisting equipment satisfactorily completes a performance (operating) test when placed in service on a project. The test should be repeated prior to unusual or critical lifts; after alteration, modification, or reassembly; and at least every six months. The test results should be recorded and kept available for review.

Safety rules

Following is a set of safety rules for electric cranemen.*

1. It is important that no one but an authorized operator be allowed to use any crane.

2. When on duty, remain in the crane cab ready for prompt service.

3. Never go on top of the crane, or permit any one else to do so, without first opening the main power disconnect switch and locking it "off" with a padlock.

4. Before traveling the trolley or the crane bridge, be sure that the hook is high enough to clear obstacles.

5. Never permit your crane to bump into another crane.

6. Examine the crane at the start of every shift for loose or defective gears, keys, runways, railings, warning bells, signs, switches, sweep-brushes, cables, etc., and report defects. Make sure the crane is kept clean and well lubricated.

7. While hoisting equipment is in operation, the operator should not be permitted to perform any other work, and he should not leave his position at the controls until the load has been safely landed or returned to ground level.

8. Do not carry a load over men on the floor; sound the gong or siren when necessary.

9. Do not allow men to ride on a load or on crane hooks.

10. If the power goes off, move the controller to OFF position until power is available again.

11. See that the fire extinguisher on the crane is kept filled and in good condition.

12. Do not operate a crane if you are not physically fit to do so. If you are ill, report to your foreman.

13. Do not drag slings, chains, or load block. After the load is taken off, do not move the crane until you have lowered the hook and the hook-on man has hooked up the chain or sling.

14. If you are asked to do something that seems unsafe, call your foreman or the repairman in charge for advice.

15. Before leaving the cab, open the main switch. Make sure the magnet or hook is empty and the magnet-controller (if any) is off. Lock, or otherwise secure equipment, to prevent starting by unauthorized persons.

16. When parking an outside crane at the end of the shift, always set the brake or chain the crane to the track. Lower booms to ground level or secure them against displacement by wind or other outside forces.

17. Stop operation and open the power switch if your crane fails to respond correctly. Then call your foreman. Attempting to get out of difficulty by repeated operation may make the condition worse instead of better.

18. Whenever a slack line condition occurs, prior to further operations, check the proper seating of the rope in the sheaves and on the drum.

---

* Based on *How To Operate a Crane*, Booklet 920 of the Electric Controller and Manufacturing Co., and *General Safety Requirements* (EM 385–1–1) of the U.S. Army Corps of Engineers.

Additional safety rules for crane operators are:

1. Never pick up a load beyond the rated load capacity of the crane. In case of doubt, call the foreman.

2. Never move the load or the crane unless you are sure that you understand the floor signal.

3. When there are several hook-on men, obey the signals of the head hooker only. (Obey an emergency stop signal given by anyone.)

4. Do not allow the load to swing against the hook-on man or other floormen; make certain they are in the clear.

5. When raising or lowering the load, see that it safely clears adjacent stockpiles or machinery.

6. Never leave a load suspended.

Following are general safety rules for the hook-on man:*

1. At the beginning of each shift, check slings and hooks for defects, dirt, and grit. Give special attention to slings to which hooks or rings have been added to make certain they are secure.

2. Never use a sling that has stretched or a hook that has begun to straighten (Fig. 25–8). Take defective equipment out of service at once, and notify the supervisor so he can have it tagged.

3. If one sling in a set needs repairs, turn in the entire set.

4. Use a sling heavy enough to carry the load safely. Never overload a sling.

5. Distribute the load equally on the legs of a sling. Have as small an angle as possible between the legs and the vertical; the stress on a leg increases as the angle increases.

6. See that slings are not kinked, twisted, or knotted. Use an adjuster or wooden wedge if necessary.

7. Make sure that the load does not exceed the rated load capacity of the crane.

8. Load compressed gas cylinders or acid carboys *only* in a cradle or similar device.

9. Center the crane block directly over the load to prevent the load from swinging when it is raised. Be sure load is balanced.

10. Set the load in the bowl of the hook, never on or toward the point (unless the hook has been specially designed for such use).

11. While hooking or unhooking a load, keep hands out of pinch points. Use hand hooks whenever possible.

12. Do not leave separator blocks or short lengths of steel loose on top of a load.

13. Before giving the hoisting signal, make sure the load is safe and warn everyone to stay well out of the way in case the load should swing or fall.

14. Use only standard signals to direct the operator. Stand where he can see you clearly.

15. Whenever possible, walk ahead of the moving load. See that it is carried high enough to clear all obstructions. Use tag or restraint lines where needed.

16. *Never walk under a load or permit others to do so.*

17. When releasing slings, make sure that they are entirely free from the load.

18. Hook both ends of an empty sling to the block before signaling the operator to move the crane.

19. Do not drag, throw, or drop slings.

20. Pile and block material so that it will not slip or overbalance.

21. Use hardwood separator blocks in piles of steel stock or dies.

General safety rules for maintenance:

1. All machinery and equipment should be shut down and positive means taken to

---

* Adapted from rules prepared by the Chrysler Corporation, Space Division, P.O. Box 26018, New Orleans, La. 70126.

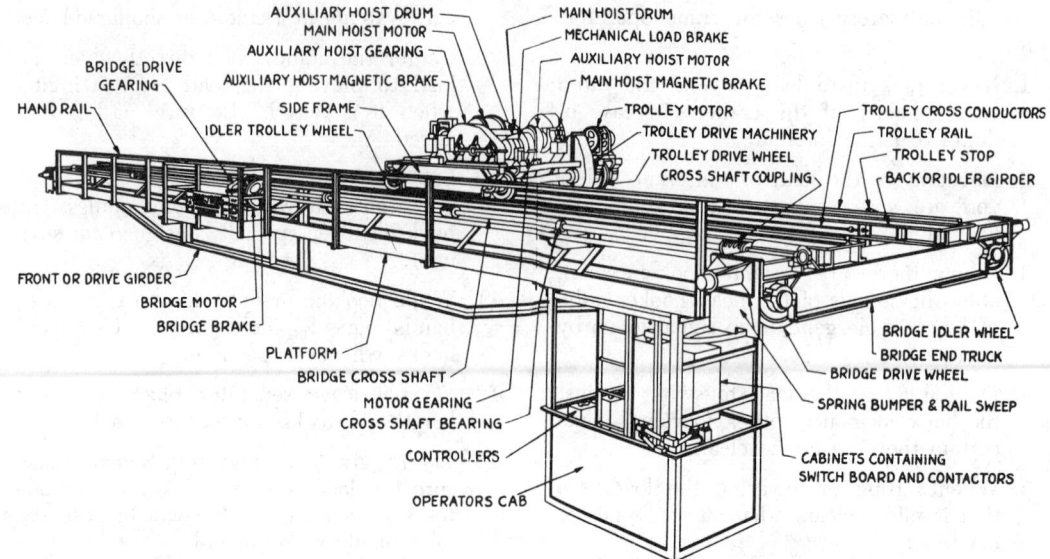

FIG. 25–9.—Essential parts of a typical traveling crane.

Courtesy Shaw Box Crane and Hoist Division of Manning, Maxwell & Moore, Inc.

prevent its operation while repairs or manual lubrications are being done.

2. All repairs on machinery or equipment should be made at a location which will provide protection from traffic for repairmen.

3. Heavy machinery, equipment, or parts thereof that are suspended or held apart by use of slings, hoists, or jacks should also be substantially blocked or cribbed before men are permitted to work underneath or between them.

4. All modifications, extensions, replacement parts, or repairs of equipment should maintain at least the same factor of safety as the original designed equipment.

5. After completion of a repair job, make sure that bolts, tools, and other materials have been removed so that no damage to machinery will result when the crane is started and so that nothing can fall off the crane. Keep tools, oilcans, and other loose objects in a box provided for them.

Overhead cranes

An overhead crane (Fig. 25–9) may be operated either from a cab or from the floor. In the latter case, control devices may be either pendant pushbuttons or pull ropes. (In some cases they are radio controlled.) The control handles should be clearly identified by signs and by shape or position so that the operator, without taking his eyes off the signalman, can tell by the "feel" which motion he is about to control. Controls on all floor-operated overhead traveling cranes should be identified. Likewise in cab-operated cranes the controls should be identified.

If there are several cranes on the same runway or in the same building, all should have the controls in identical positions so that a substitute operator will not be confused.

Safe means should be provided for the operator to pass from the cab to the footwalk. Stairs are preferable to ladders. If ladders are used, they should meet safety standards.

The space that the operator must step across in going from the landing or the runway girder to the crane should not exceed 12 in.

In case of fire, the operator must be able to escape from the crane regardless of its location on the runway. He would be in particular danger if a fire should occur when the crane was attached to a load which could not be picked up quickly so that the crane could travel at once to the access landing. Unless a means of escape via the bridge and runway is provided, an emergency rope ladder or other means of escape should be installed in the cab.

Safe access must be provided also to the bridge motor and brake and to the equipment on the crane trolley. OSHA requires that a clearance of not less than 3 in. overhead and 2 in. laterally be provided and maintained between the crane and obstructions. A small CO_2, dry chemical, or equivalent fire extinguisher should be installed in the cab.

When less than 30 in. is provided between the highest part of the trolley and overhead obstructions, precautions should be taken to restrict personnel from servicing or otherwise riding the crane while it is in motion. This will prevent a man from being brushed off the crane by low beams or trusses. The 30 in. clearance, although desirable, is often impractical because of head room restrictions of many industrial plants. However, the movement of the crane must *never* jeopardize the safety of personnel. This is one reason the safety professional should be active during the building design stage.

Footwalks and platforms should be substantial, rigidly braced, and protected on open sides with standard railings and toeboards.

A footwalk with handrails should be installed along the entire length of the crane runway. The walk should be reached by one or more fixed ladders not less than 16 in. wide. The outside edge of the walk should be not less than 30 in. from the nearest part of the trolley.

The bridge walkway should have a 42-in. high handrail, an intermediate rail, and a 4-in. toeboard and the space at the squaring shaft should be guarded so that a man cannot fall between the walkway and the crane girder.

A footwalk should be provided also, if headroom permits, along the entire length of the bridge of any crane having the trolley running on the tops of the girders. This footwalk should be on the drive side of the bridge and should have toeboards and metal handrails. Safe access to the opposite side of the trolley also should be provided.

Floors of walks should be neatly fitted so as to leave no openings and should have a nonslip surface. Vertical clearance between the floor of the walk and overhead trusses, structural parts, or other permanent fixtures should be at least 6½ ft. Where such clearance is structurally impossible, built-in members should be distinctively painted or striped and padded where necessary.

Toeboards should be provided at the edges of flooring on the trolley to prevent tools from falling to the floor below.

Stanchions or grab irons should be installed to enable a man to climb onto the trolley in safety.

To guard against electric shock, a heavy rubber mat should be provided at the control panel in the cab. The operator should have an unobstructed view of the load hook in any possible position.

The bridge truck wheels and the trolley wheels should have sweeps to push away a man's foot or hand. To prevent a serious impact on the crane, if a bridge wheel should fail, the end frames or trucks should have safety lugs only 1 in. above the top of the rails.

To prevent clothing from being caught on revolving equipment, the cross-shaft couplings should be guarded.

Runway rails, after years of service, may become distorted or the span between them may be altered because of settling of the column footings, so that wear on the flanges of the bridge wheels results. Rail alignment, therefore, should be checked every few years. If the tread of bridge truck wheels is tapered, the crane will run square with the runway constantly.

Rail stops or bumpers must be so located that, when the wheels are against them, the crane bridge is square with the runway.

If two or more cranes operate on the same runway, limit stop switches should be installed to prevent collision.

An automatic alarm should sound continuously from the time the travel controller handle is first moved from OFF position until it is returned to OFF position. Warning bells or horns and flashing lights, arranged to operate automatically when a crane is approaching, should be placed at aisles and walkways. An electric or electronic device which sounds a distinct alarm at a predetermined distance should be used to warn operators, when two or more cranes are operating on the same runway.

Operation. When not in use, the crane should be parked with the load hook (and the slings if they remain on the hook) raised high enough to clear the heads of men at work on the floor below, and the operator should throw all controls into OFF position and open the main switch.

A light should be visible from the floor to indicate when the main switch is on. Controllers should be the spring-return type or momentary contact pushbutton.

The operator should center the trolley over the load, when about to make a lift, and should accelerate and brake the lifting and lowering motions slowly to minimize stresses on the crane.

He should check frequently to see that the crane is square with the runway rails, by carefully running up against the rail stops. If the crane is out of square, he should apply power for travel motion (without a load on the crane and with the trolley at the end of the bridge where the wheel is away from the rail stop). The drive wheel at one end of the crane then will slip on the rail while the other drive wheel rolls to the rail stop.

Two or more cranes operated in the same bay or runway should be kept at least 30 ft apart where possible. If they operate closer, operators and foremen should take special care to avoid collision.

When repairs to one crane are necessary, every precaution must be taken to prevent other cranes from colliding with it. Safety stops should be installed.

Electromagnets often are used with electric overhead cranes, as well as with gantry cranes and several other types, to handle ferrous scrap and hot or cold ingots and to move iron and steel products (Fig. 25–10).

Magnets should be used neither close to steel machines or parts nor near ferrous materials in process, to prevent them from becoming magnetized and thereby causing trouble. Watches and other delicate instruments should be kept out of the electromagnetic field.

Switches or switchboxes controlling power

FIG. 25–10.—A six-magnet "spider" is shown positioning an open coil. Coil is being placed on cooling base following open coil annealing in cold strip mill.

Courtesy Inland Steel Co.

to the magnet should be labeled DANGER— DO NOT OPEN SWITCH—POWER TO ELECTROMAGNET.

The metal body of an electromagnet should be grounded. The magnet's power supply circuit should be tapped directly from the feed lines, entirely separate from other crane electric circuits. The switchboard, wiring, and all other electrical equipment should comply with the American National Standards C2, *National Electrical Safety Code,* and C1, *National Electrical Code* (NFPA Standard No. 70).

A load suspended from a magnet *never* should be moved over personnel.

Further information is in NSC Data Sheet 359, *Electromagnets Used With Crane Hoists.*

Special hook-on devices can be designed and made for handling special shapes. A

Fig. 25–11.—Custom-made clamping device for lifting rolls of paper. As chain is pulled upward, clamp (center) tightens on roll (right).

custom-made clamping device for lifting rolls of paper is shown in Fig. 25–11.

Storage bridge and gantry cranes

Storage bridge and gantry cranes are similar to traveling cranes, but run on rails at ground level instead of on elevated runway girders. Gantry cranes have relatively short spans, whereas storage bridge cranes may have spans up to 300 ft or more, sometimes with a cantilever on one or both ends. Storage bridge cranes usually are used for handling coal or ore (Fig. 25–25). Ordinarily, a caged ladder on one of the legs of the crane provides access to the cab.

Track riding cranes should be provided with substantial rail scrapers or track clearers at each end of the trucks and which are effective in both directions of travel (Fig. 25–12).

Since there may be a serious shearing or crushing hazard in the area between the crane trucks and adjacent structures or stored material (Fig. 25–12), an automatic alarm should sound continuously from the time the travel controller handle is first moved from Off position until it is returned to Off position.

The wheel truck of gantry cranes should have side clearance of at least 30 in.

Bumpers, made of cast steel plates and angles, should be provided on storage bridge and gantry cranes and should be at least one-half the diameter of the truck wheels in height. Both truck wheel and trolley bumpers should be fastened to the girder and not to the rail.

Spring bumpers usually are provided where bridge axles have antifriction bearings. If compression springs are used, they should be at least 5 in. in diameter at the point of contact and so arranged that, if a spring or guide pin breaks, no part can fall on the crane.

To prevent the crane from being moved down the track by a strong wind, the operator should apply rail clamps before he leaves the cab, even for a short time. The holding power of the clamps should be sufficient to withstand wind pressure of 30 pounds per square foot of projected area of the crane.

The electric contact rails or wires should be high enough not to create a hazard to persons on the ground.

So that a bridge crane can be squared to the track, the squaring shaft should have a clutch. One end of the bridge can then be moved, while the other end remains stationary, to bring the crane into proper position.

All outside cranes should have the following features:

1. Floors of the footwalk constructed to provide drainage.

2. The operator's cab constructed of fire-resistant material; weatherproof; with pro-

Fig. 25–12.—This man should not be riding on this part of the crane; he could be crushed between the moving truck and a nearby object. Note sweep guard (lower left).

vision for heating, with ample space for control equipment, and with the operator located so that signals can be clearly seen.

3. The floor of the cab extended to an entrance landing and equipped with a handrail and toeboard of standard construction.

4. A rope ladder or other means of emergency escape from the cab.

5. Locking ratchets on wheel locks, rail clamps, and brakes so that the crane will not move in a high wind.

6. Skew switches to prevent excessive distortion of the bridge.

7. A screen of wire or expanded metal between the contact bars and the bridge walkways to prevent accidental contact with the current conductor.

The main line switch should be so constructed that it can be locked in OPEN position and should be mounted above the cab where it can be reached conveniently from the footwalk.

Monorails

A monorail system consists of one or more independent trolleys, supported from or within an overhead track, from which hoists are suspended. All applicable safety features should be incorporated in monorail equipment.

Monorail hoists are used extensively in many industries to raise, lower, and transport materials. They can be classified in three major groups: hand-operated, semi-hand-operated, and power-operated.

On the hand-operated monorail, the material is raised with a hand-powered hoist, and the trolley is propelled by hand. The semi-hand-operated monorail has a power hoist and is moved horizontally by hand. The power-operated monorail is fully power actuated for both vertical and horizontal movements.

No attempt should be made with a monorail hoist to lift or otherwise move an object by a side pull, unless the hoist has been designed for such use. Monorail hoists, operated in swivels, should have one or more safety catches that will support the load should a suspension pin fail. All trolley frames should be safeguarded against spreading.

Monorail track supports and track should be designed to carry safely the intended loads and erected according to good engineering practice. Both the track and its support should be inspected frequently for signs of weakening and wear, and necessary repairs should be made as soon as possible.

Rail stops should be provided at the ends of monorail track. The stops should extend at least as high as the centers of the trolley wheels.

At switches, turntables, and transfer tables, automatic bumpers should drop into position to prevent the trolley from running off the open ends of the fixed and movable track when they are not properly lined up with each other. The movable track should be interlocked with the bumpers so that it cannot move until the rail stops are in position.

If an electric monorail carrier is run from a cab attached like a trailer to the hoist trolley, an access platform should be provided as well as means of emergency escape for operator. Pull ropes should be so located that he can safely operate them from the cab. The electric contact wires and the current collectors should be so located that he cannot reach them when he is entering or leaving the cab or inside it.

Jib cranes

A jib crane is a crane capable of lifting, lowering, and rotating a load within the circle covered by a rotating arm or jib upon which runs a trolley. The jib usually is supported from a building wall or column or from a pillar. A hoist (chain, air, or electric), with which the loads are lifted, is suspended from the trolley.

Before a jib crane is installed in a building, the strength of the structure should be checked by a qualified engineer to determine whether or not the column or wall to be used is strong enough to support the jib, hoist, and load.

A safety band, fastened to the building structure, should be installed at the top of the vertical member of the jib frame, just below the angle brace to prevent the crane from falling, if the top bracket should fail.

The jib should be braced or guyed to withstand the loads it is expected to carry.

A bumper should be installed at the outboard end of the jib to prevent the load trolley from running off the beam. The end stop requires frequent inspection to see that it is not becoming loose or rusting off.

Wherever possible, a bail, fastened to each side of the trolley, should be provided over the jib.

Derricks

The principal types of derricks are the A-frame derrick, the stiff-leg derrick, and the guy derrick. There are other types, but these are the most common. All derricks must have every part firmly anchored.

The A-frame derrick (Fig. 25–13), as the name implies, has a frame of steel or timber shaped like the letter "A" and erected in a vertical plane, with a brace or leg extending from the top of the A at a 45 degree angle to

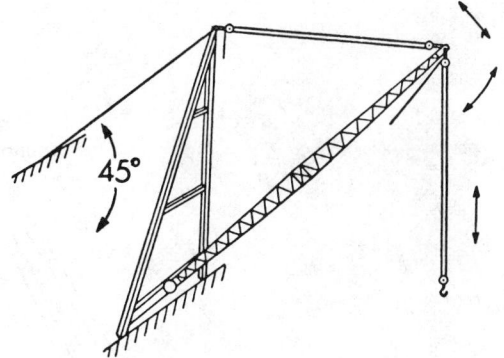

FIG. 25–13.—A-frame derrick. Brace is set at 45 degrees to the ground.

the ground. The ground member(s) ties this brace to the bottom of the A-frame. The boom is hinged at the horizontal member of the A. The base of the A-frame and the rear brace must be firmly weighted down.

The stiff-leg derrick (Fig. 25–14) has a mast with two braces at a 90 degree angle to each other and at a 45 degree angle to the ground. Usually steel or timber sills tie the mast and the braces together at the ground. To withstand the uplift caused by a heavy load on the boom, sandbags, cast iron weights, or concrete blocks are used to hold down the stiff-legs.

The hoist engines for both stiff-leg and A-frame derricks usually are bolted to the sills or ground members. On smaller derricks, the suspended loads may be slewed by being pushed manually. The boom of a large derrick may be swung by a "bull-wheel" to which cables from another drum on the hoist engine are attached.

Since a loaded cable may whip considerably and cause severe injury, the horizontal cables between the hoist engine and the boom hinge should be barricaded, and workmen should be prohibited from crossing over or under them.

The guy derrick (Fig. 25–15) is used largely for erection of the structural steel of tall buildings, especially those over 10 stories high that cannot be reached by the boom of a crawler crane operating on the ground. Such derricks usually are of latticed steel and have an odd number of equally spaced wire rope guys, each equipped with a turnbuckle and attached to the steel beams

683

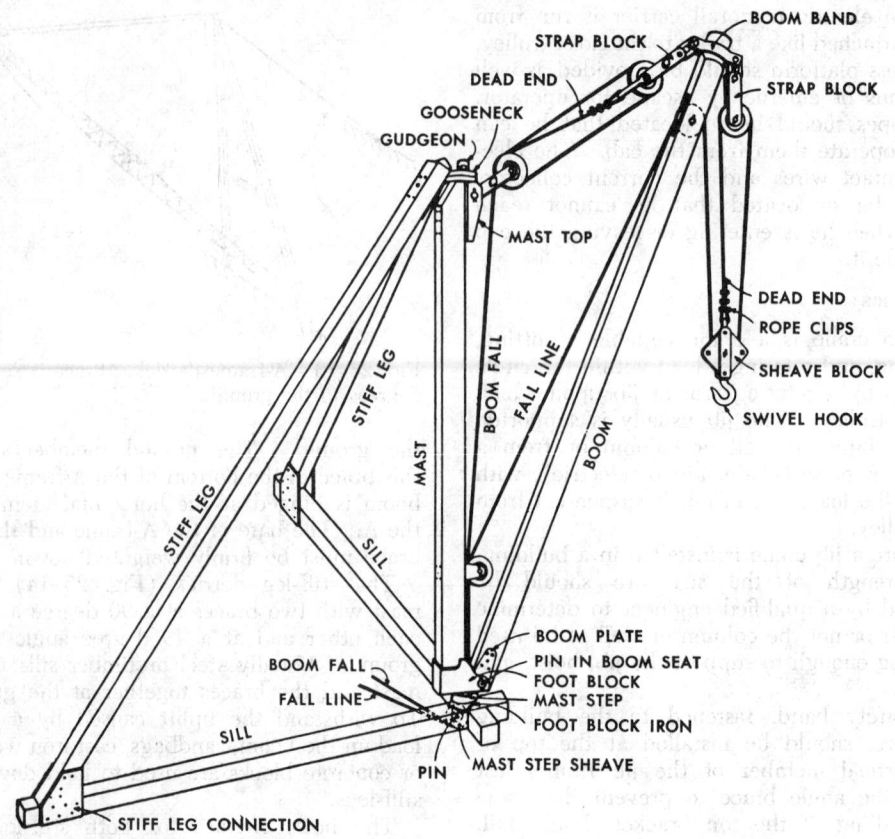

FIG. 25–14.—Diagram of stiff-leg derrick shows names of various parts.

Courtesy Travelers Insurance Co.

or columns on the erection floor (Fig. 25–16).

If the derrick is erected on the ground, the guys should be secured to heavy steel anchors buried deep in the ground, with additional weights placed on the anchorages.

Heavy timbers, 12 by 12 in. or 12 by 16 in., should be placed on the floor beams to support the foot of the mast. These foot blocks must be braced against the stubs of the building columns to prevent their being "kicked" out of position when a heavy load is picked up with the boom at a low angle. Wire rope and turnbuckles may be used in place of 8 by 8-in. timbers to secure the base of the mast.

The hoist engine, whether on the same level with the derrick base or on the ground many floors below, should be securely anchored by steel cables or shoring timbers, to prevent it

from being pulled towards the base of the mast by the tension on the cables.

A unique feature of the lattice-type steel guy derrick is its ability to lift itself from one erection floor to the next. First, the hinge pin is removed to disconnect the boom (which is shorter than the mast and has four spare guy ropes near its upper end) from the mast. The boom hoist cable (or topping lift) then lifts the boom in a vertical position and sets it on the foot blocks close to the mast. The boom is rotated 180 degrees, the normally unused guys are secured, and the boom then stands as a guyed gin pole.

The load hoist of the boom then is used to pick up the mast, the mast guys being slackened off a few at a time and reattached at the upper level. When the mast is secured at

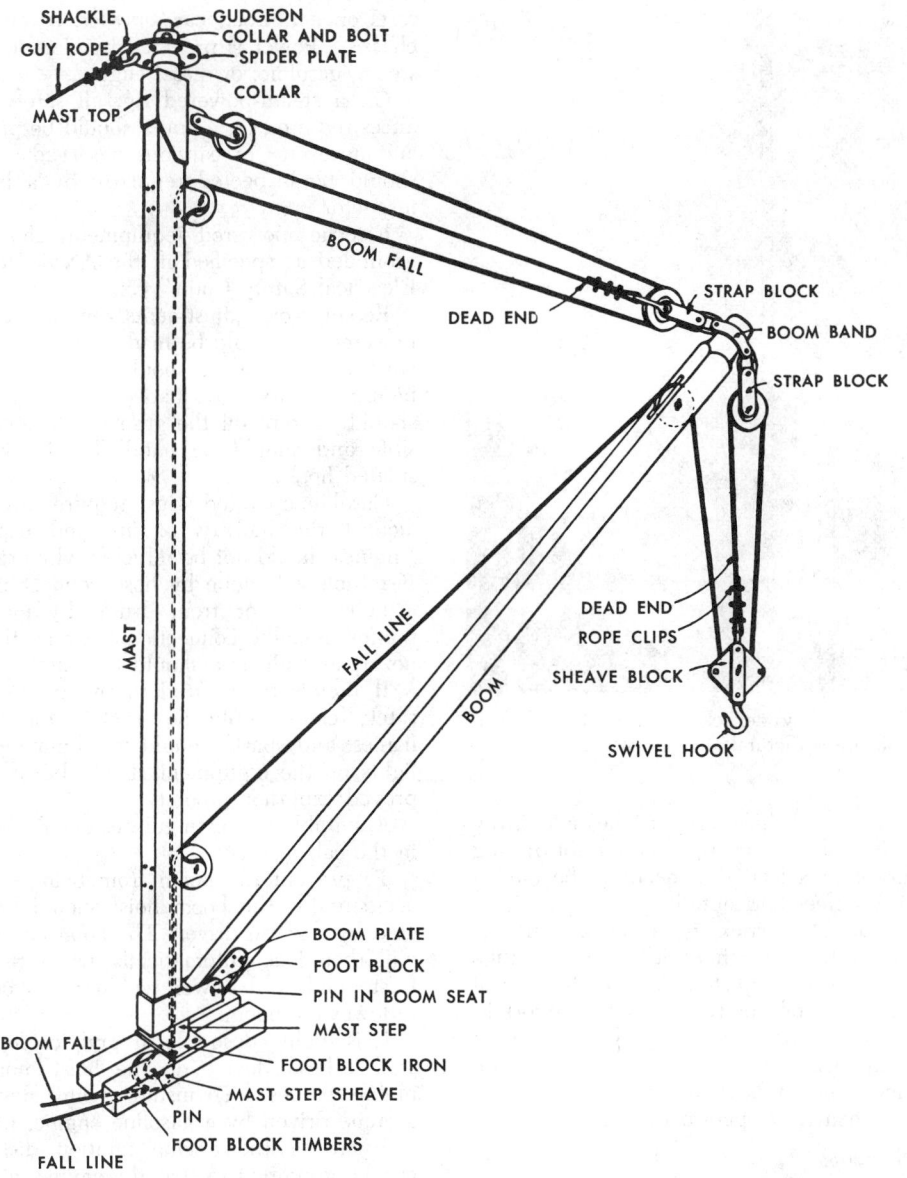

SHACKLE
GUY ROPE
GUDGEON
COLLAR AND BOLT
SPIDER PLATE
COLLAR
MAST TOP
BOOM FALL
DEAD END
STRAP BLOCK
BOOM BAND
STRAP BLOCK
DEAD END
ROPE CLIPS
SHEAVE BLOCK
SWIVEL HOOK
MAST
FALL LINE
BOOM
BOOM PLATE
FOOT BLOCK
PIN IN BOOM SEAT
MAST STEP
BOOM FALL
FOOT BLOCK IRON
MAST STEP SHEAVE
PIN
FOOT BLOCK TIMBERS
FALL LINE

FIG. 25–15.—Diagram of guy derrick. (For simplicity, only one guy is shown.)

Courtesy Travelers Insurance Co.

the new erection floor, the boom is raised and again connected at the hinge.

To swing guy derricks, the men push either on the suspended load or on a pipe "bull stick" attached near the base of the mast.

Other types of derricks are the gin pole and the breast derrick. The gin pole is merely a mast slightly out of plumb, with a hoisting tackle suspended from its upper end. The gin pole is supported by a number of

Fig. 25–16.—Guy derrick of latticed steel construction lifts structural steel in place.

rests on a railroad car, crawler, or autotruck chassis. Power is provided by electric motor, steam, gasoline, or diesel engine.

On a steam-powered rig, all safety appliances required for a boiler should be provided and necessary precautions observed. Boilers should be inspected regularly by a licensed inspector.

Electric powered equipment should be grounded as specified in the ANSI "National Electrical Safety Code," C2.

Repairs or adjustments on an electric-powered rig should be made only by qualified electricians. Power should be disconnected before repairs are made. Trailing cables should be kept off the ground whenever possible and should be handled only with insulated hooks.

Gasoline-operated rigs require protection against the hazards of fire and explosion. Engines should not be refueled while running. If refueling is done by hose connection from a tank truck or from drums by means of pumps, metallic connection between the hose nozzle and fill pipe should be maintained.

If fuel is transported to the rig by hand, safety cans should be used. Open lights, flames, and sparks should be eliminated, and lights on the equipment should be of an approved explosion-proof type.

A suitable fire extinguisher should be kept in the cab of a rig.

To prevent the boom from being dropped accidentally, the boom hoist should be operated by a worm drive. The boom hoist drum will then lock automatically when the hoist is stopped, and power will have to be used to lower the boom.

It is highly desirable, if practicable, to install a limit device on the hoist motion of mobile cranes. To install a limit device on a crane driven by a gasoline engine, the center wire of the engine ignition distributor can be grounded so that the engine will stall.

Even if the operator fails to step on the brake quickly, the worst that can happen is that the load will turn the engine over backward while it slowly sinks to the ground. Without this limit device, the load block may strike the boom and break a hoist cable.

No attempt should be made to lift the boom by means of the load hoist cable. A load should not be lifted by the boom hoist

guys, most of which are on the side away from the load. This rig is used for raising and lowering a load that needs to be moved only a few feet horizontally.

The breast derrick is a small portable A-frame with a winch attached to it. Like the gin pole, it is erected in a nearly vertical position, with one or two guys to support it. Care must be exercised to prevent the base from slipping and causing the load to fall. The men should be warned to watch their fingers when they operate the winch.

Mobile cranes

Mobile cranes include locomotive cranes, crawler cranes, truck cranes (Fig. 25–17), and industrial truck cranes. The first three types are well standardized in design, but there is a great variety of industrial truck cranes.

All mobile cranes have booms with load hoists and boom hoists. In most instances, the rig swings or rotates on a turntable, which

FIG. 25–17.—Truck-mounted hydraulic crane.

Courtesy Grove Manufacturing Co.

line unless the rig is designed for this purpose.

Every crane should have on it a capacity plate or a sign plainly legible to the crane operator, signalman, and rigger, stating the safe loads at various radii from the centerpin of the turntable. A plate can be mounted on the side of the boom near the hinge, with a pointer actuated by gravity suspended freely in front of it.

The safe loads are painted on the plate at the proper places, so that the pointer will directly indicate the safe load for any angle of the boom (Fig. 25–18).

It must be understood that the safe loads posted on a boom-indicating device are valid for only one particular boom length and only one particular configuration of crane. Should sections be added or removed from the boom, the calibrated safe load chart would be invalid as the safe load indicated is dependent on the boom angle.

The determination of load ratings, with booms of stipulated lengths at stipulated working radii for truck- and wheel-mounted cranes, is established by taking 85 percent or less of the load that will produce a condition of tipping or balance, with the boom in the least stable direction relative to the mounting.

Often the weight of the load to be lifted is not accurately known, and operators will make a test lift over the end of the machine to determine the safe lifting boom angle or boom radius. If the load is critical, this can lead

to tipping when the load is shifted across the side.

Boom stops of the shock-absorbing type should be installed on all cranes. These stops limit the travel of the boom, when beyond an angle of 80 degrees above the horizontal plane, and prevent the boom from being pulled over the top of the machine by the boom hoisting mechanism or by the sudden release of a heavy load suspended at a short radius. Either of these occurrences usually results in serious damage to the equipment and injuries to the operator or other workmen.

Boom stops are best suited to medium-size cranes in the 5- to 60-ton range.

The operator must have safe access to and egress from his cab or seat, regardless of the position of the crane boom. To do his work in safety, he must have an unobstructed view of the load hook and the point of operation at all times. He must also be able to see ahead of the crane when it is traveling on the ground, whether the chassis is moving forward or backward. On some cranes, visibility ahead is obstructed, and the operator must use extreme caution.

A crane operated after dark should have clearance lights. Floodlights should illuminate the area beneath the boom, and lights mounted on the underside of the boom are recommended.

A warning bell or horn and an automatic backup alarm are necessary equipment for a

687

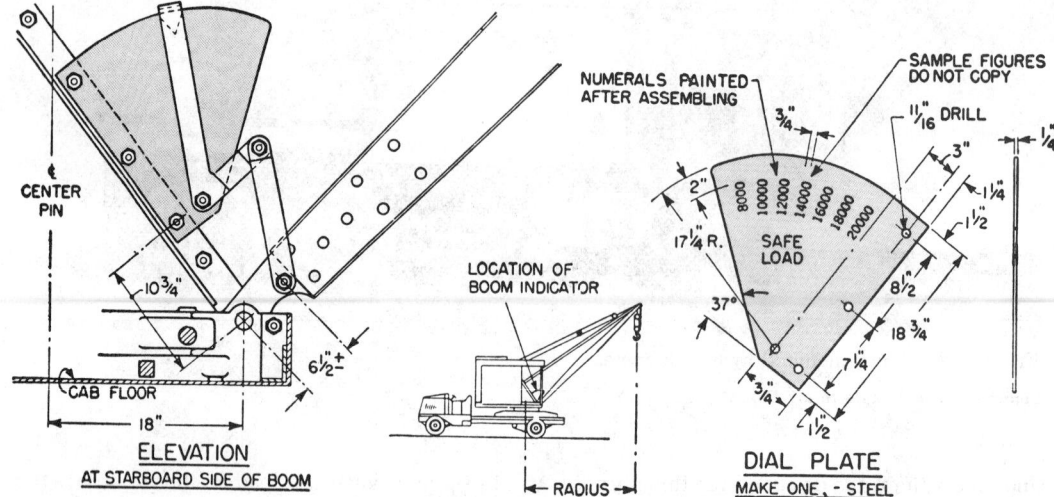

ELEVATION
AT STARBOARD SIDE OF BOOM

DIAL PLATE
MAKE ONE, - STEEL

Fig. 25–18.—Some details of boom-indicating device that shows the safe load with the boom at any angle. Notes: (*a*) Verify all dimensions of the crane before fabricating the indicator. (*b*) After assembling the indicator, ascertain from the manufacturer of the crane the radii at which each additional 1000 pounds may be lifted safely. Raise or lower the boom to give the desired radius and paint the figures on the dial at the proper point. (*c*) Note that the boom radius is measured from the center pin of the crane, and not from the boom hinge pin.

From Consolidated Edison Company of New York and from Handbook of Rigging, W. E. Rossnagel (© 1950, *McGraw-Hill Book Company*)

Fig. 25–19.—Guard on locomotive crane prevents a man from being caught in the shear between the cab and the car body. The guard is pivoted under the corner of the cab, and the slide at the farther corner follows the continuous guide track. When the cab is parallel to length of the bed, guard is entirely under the cab; as cab swings out, the guard is extended.

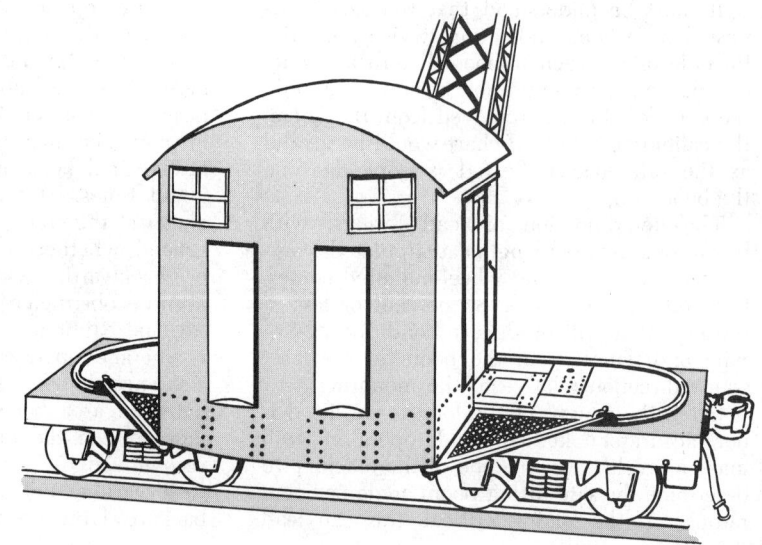

688

mobile crane.

Men should *always* be kept beyond the range of the cab swing and out from under the boom and the load.

Locomotive and truck cranes frequently present a serious shear hazard between the side of the operator's cab near the counterweight and the side of the car body or chassis. Many a man has been killed by being caught in this shear point. Fig. 25–19 shows one way of guarding this location on a locomotive crane.

A guard consisting of two segments of metal, attached one to each side of the cab and suspended in a horizontal position by chains, is available commercially. This guard completely covers the shear points in all positions of the cab.

Since legal restrictions on the width of vehicles traveling on highways prevent use of such a guard on a truck crane, other methods must be employed. An alarm can be installed to sound when an operator first touches the swing lever and before the crane begins to move, so a man has time to jump clear of the hazard.

Also, a bar gate can be installed, so placed that a man must move it to reach the box of tools, shackles, clips, and similar items that usually is located on the chassis. Movement of the gate causes an alarm to sound in the crane cab, warning the operator not to swing the crane.

Mirrors can be placed so that the operator can follow the swing of the rear of the cab and, thus, make sure that no one will be caught. On heavy, slow-moving cranes, "feeler" bars, electrically connected to stop the swing, will guard against the shear hazard and help prevent accidents from insufficient clearance.

Additional information is given in Chapter 45, "Motorized Equipment."

Operation. A load never should be picked up when the weight, supported by the chassis, rests on springs over the axles. It would be difficult to control the load, and its movement could be dangerous. For instance, the boom on a truck crane may be swung high to one side of the chassis when it picks up the load. As the strain is taken on the boom, the springs on that side of the truck will compress, while those on the opposite side will ease up somewhat. As a result, the top of the boom will move outward and the load, as it is picked up, will swing.

Then the boom may be swung to the other side of the chassis. When the load passes the center of the chassis, the action of the springs on each side will be reversed, the crane will tilt the other way, and the load will swing. To prevent this hazard, all loads should be taken off the springs of the vehicle by means of built-in jacks, outriggers, or blocks and wedges.

A boom must never be swung too rapidly. It it is, the suspended load will be swung outward by centrifugal force, so that the crane may rock or even be upset and the load may swing and strike a person or object.

Operating a crane on soft or sloping ground or close to the sides of trenches or excavations is dangerous. The crane should always be level before it is put into operation. Outriggers can be relied upon to give stability only when used on solid ground. Heavy timber mats should be used whenever there is doubt as to the stability of the soil on which a crane is to be operated.

The use of any makeshift methods to increase the capacity of a crane, such as timbers with blocking or adding counterweight, should not be permitted.

If the crane tips when hoisting or lowering a load, the operator should lower the load as quickly as possible by snubbing it lightly with the brakes. Workers, therefore, should never be allowed to ride a load that is being hoisted, swung, or transported.

When operating a crane with the boom at a high angle, the operator should take care that the suspended load does not strike the boom and bend the steel lattice bars on its underside. Bending these members will weaken the boom; and, when it picks up the next heavy load, it may collapse. Likewise, if the main members of the boom are bent even slightly, the strength of the boom may be materially reduced.

When an extension boom is used on a crane, as for structural steel erection, the operator must use extreme care in lowering it to the ground at the completion of the job. An extension boom never should be lowered to one side of the chassis or crawler, for the

689

stability of the rig is much reduced in that position and the crane may upset.

When using a boom tip extension or jib, the allowable load on the jib is limited. Its capacity must be known by the operator, and he must use care when swinging with a load, especially when the jib is lowered at an angle to the main boom.

The hook must be centered over the load to keep it from swinging while it is being lifted. Employees should keep their hands out of pinch points, when holding the hook or slings in place, while the slack is taken up. A hook, or even a small piece of board, may be used for the purpose. If a man must use his hand, he should place it flat against the sling to hold it. The hook-on man, the rigger, and everyone else must be in the clear before the load is lifted. A tag line should be used for guiding loads.

On a crane operated by two men, such as a truck crane (with one man in the crane cab and the other in the truck cab), a code of signals on a loud bell or gong should be used to tell the truck driver when to move the crane forward or backward.

A heavy load should never be removed from a truck by hooking a crane to the load and, then, having the truck pull out from under it. If the load should prove too heavy for the crane, the crane will upset before the operator can lower the load to the ground. The load should be lifted clear of the truck body, and the operator should make sure that the crane can handle it safely before the truck is moved out from under it.

When handling unknown loads as those encountered in pile extraction and underwater work of obstruction removal, the boom of the crane should be in the most stable position and should be at an angle not greater than 60 degrees above horizontal. This procedure limits the load on the boom to that permitted by the stability of the entire crane unit and greatly reduces the possibility of a buckled boom.

A crane should never be used to jerk piling. If piling cannot be pulled by a straight steady pull, a pile extractor should be used. When a pile extractor is used, the boom angle should be kept at or less than 60 deg. above horizontal.

Before leaving the crane at the end of the work day, the operator should lower the load block or bucket onto the ground in such a manner that it cannot be upset. Unless the boom hoist is driven by a worm, which will positively prevent anyone from lowering the boom, it should be rested on a steel oil drum or other support. In one case on record, the cab of a truck crane was locked for the night, but some boys mischievously pried the pawl-actuating lever below the cab with a bar, and the boom crashed to the ground.

All modifications, extensions, replacement parts, or repairs of equipment shall maintain at least the same factor of safety as the original designed equipment.

Travel of cranes. Except for very short distances, a crane should not travel with a load suspended from the boom. When a crawler crane must travel on a public thoroughfare, the boom should point forward and a flagman should walk ahead of it.

A truck crane on a truck chassis and a crawler crane on a semitrailer should be transported with the boom pointing toward the rear and high enough to clear an automobile. Truck cranes with short booms may travel with the boom forward in the boom rest. One of the work crew should follow the rig in a car or truck to keep other vehicles from traveling beneath the boom. Otherwise, if the truck wheels should roll into a low spot in the pavement, the boom might suddenly crash through a car roof.

If not disassembled, a crane being transported should have the crane engine running and the operator in the cab swinging the boom, when necessary, to avoid fouling trees, poles, or buildings when the vehicle turns corners.

Before heavy, slow-moving equipment or heavy equipment on a low-slung trailer is moved over any public *or private* railroad grade crossing, a responsible representative of the railroad company should be notified. The railroad then can provide flag protection to guard against a train's striking the equipment, while it is moving over the crossing or if the equipment becomes stalled or "hung up" on the crossing.

This is an important precaution for the benefit of both the equipment owner and the railroad company.

Electric wires. The boom and cables of a crane should be kept away from all electric wires, regardless of their voltage. OSHA requires that, except where the electrical distribution and transmission lines have been de-energized and visibly grounded at the point of work, or where insulating barriers, not a part of or an attachment to the crane, have been erected to prevent physical contact with the lines, cranes should be operated near power lines only in accordance with the following:

- For lines rated 50 kv or below, minimum clearance between the lines and any part of the crane must be 10 feet.

- For lines rated over 50 kv, minimum clearance between the lines and any part of the crane must be either 10 feet plus 0.4 inch for each 1 kv over 50 kv, or twice the length of the line insulator but never less than 10 feet.

- In transit and with no load and boom lowered, the clearance should be a minimum of 4 feet.

Cage-type boom guards, insulating links, or proximity warning devices may be used on cranes, but the use of such devices should not operate to alter the requirements as spelled out above.

Any overhead wire should be considered an energized line until either the person who owns the line, or the electric utility authorities indicate that it is not energized.

If a crane boom contacts a conductor, the hazard is greatest to the hook-on man and others who may touch the load or the sling. To protect these men, a commercially available load hook with an insulated link can be used.

If the boom or cables accidentally come into contact with a wire, the operator should swing the crane to get clear. If the wire has been broken and the boom cannot be cleared from it, the operator should stay on the crane and remain calm.

A crawler crane, if the ground is wet or damp, will be electrically grounded and, when the boom touches a power line, the wire will, in turn, be grounded and the power company circuit breaker will open. Some arcing may occur. After a few seconds, however, the circuit breaker will automatically close and reenergize the wire. Again the circuit breaker may open, and again it will close. Thus the wire may be "dead" at one instant, but live a few seconds later.

On the other hand, if the boom of an automotive crane on rubber tires should become tangled with a "hot" electric wire, the entire crane will be energized, for the rubber tires will insulate the crane and truck from the ground. Hence, the circuit breaker may not open, and the wire *and* the crane may remain energized.

Stepping from the crane to the ground is often fatal, for one hand and one foot may be in contact with the crane when the other foot touches the ground. Therefore, the operator should remain on the crane until the emergency crew from the electric company arrives and frees the crane from the live wire.

If the gasoline tank should become ignited, or if for any other reason it is impossible for the operator to remain on the crane, he should *jump,* after making sure that all parts of his body are clear of the crane before his feet touch the ground.

Crawler and truck cranes are frequently used in structural steel erection for buildings less than 100 ft high. Extension sections may be inserted to lengthen the booms of these cranes. When the boom is reassembled, all the lattice bars or braces must be installed properly. In one documented case, a lattice bar was accidentally omitted. Later, when a heavy load was picked up, the boom buckled at this point and several men were killed.

Locomotive cranes. In the cab of a locomotive crane (Fig. 25–20), a clear passageway should be provided from the operator's platform to an exit door on the nearer side of

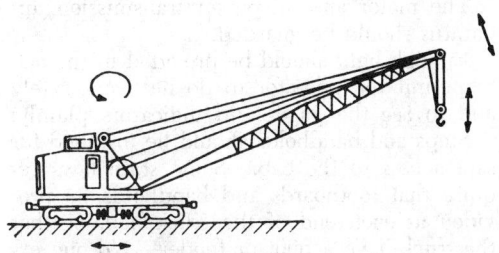

FIG. 25–20.—Locomotive crane.

Point of contact is 29'-2" above ground.

2.

Jumper to be removed

Contact →

N

S

Victim contacted energized truck.

Fig. 25–21.—Employee standing on ground (stick figure drawn in at arrow) was fatally shocked when metal parts of boom contacted one phase of 25 kv circuit. This photograph of the accident scene was used as part of the report of the accident. Note how the details were written with a felt-tip pen right on the photographic enlargement.

Courtesy Idaho Power Co.

the cab. Doors should be hinged at the rear edge and should open outward. Sliding doors should slide to the rear to open.

The motor and all power transmission apparatus should be guarded.

Enough light should be provided in the cab to permit the operator to do his work safely and to see the gages and indicators plainly.

Steps and handholds should be installed for safe access to the cab. Some state laws require that footboards and handholds be provided at each end of the truck bed, or that the truck have a pilot or fender. No one except necessary operating personnel should be

permitted on a rig while it is operating.

An on-track crane should have standard automatic couplers and uncoupling levers, and air brakes as well as hand brakes. It should have rail clamps at each corner to hold it in position while at work. A guard should be provided at the end of the boom to prevent the thimble on the cable from coming into contact with the sheave.

An on-track crane should be moved only on signal from an authorized signalman or switchman, who should walk ahead of the crane to warn others and to see that switches are properly set and the track is free of obstructions. When no signalman or switchman is employed, the operator should move the crane only on orders from the foreman of the department in which the crane is working.

The crane should not be swung across another track until the craneman and signalman have made sure that cars are not on that track and will not be moved on to it.

When he moves the crane about the yard or working space, the operator should keep the crane and boom parallel to the track to avoid striking buildings or other structures and should carry the boom low enough to clear overhead wires. Buckets and magnets should not be carried on the boom when the crane is going from one location to another.

Industrial truck cranes usually have four wheels, although some have three. In one model, the operator sits behind a small pillar-type jib crane mounted on a chassis, while, in another, the operator stands on a platform and operates a fully or partly rotating crane. Still another type has a fixed boom (which cannot be swung), so that to make side motions the entire rig must be moved from one position to another.

A crane carrying a load should be driven at the lowest possible speed, and the load should be carried as low as possible.

The operator should have a helper to act as a hook-on man and to give signals. When a long load is being carried, the helper should walk alongside and, by means of a tag line, keep the load from swinging and striking against objects along the way.

When the crane is traveling without a load, the hook should be fastened to the lower end of the boom to prevent the hook block from swinging.

Aerial baskets

A study of aerial basket accidents reported for the years 1967–1971 (see NSC publication, *Study of Aerial Basket Accidents*, Volume II) shows that approximately 5 out of 6 accidents can be attributed to the acts of individuals, and that many of the accidents due to equipment failure were probably the result of abuse or misuse of the equipment.

In the period from 1962 to 1971, almost 18 percent of the accidents reported involved electrical contact (Fig. 25–21). The same

precautions used on live-line work should be followed. (See NSC Data Sheet 598, *Flexible Insulated Protective Equipment for Electrical Workers.*) Use of baskets for "bare-handed" work requires extra caution to maintain adequate clearances between crews and equipment, where work involves one crew working from a grounded structure and the other doing "bare-handed" work from an insulated basket.

Many accidents were caused by control levers catching on obstructions, which re-emphasizes the need for guarded controls.

To reduce accidents due to equipment failures, the following are *musts*:

1. Aerial baskets must be of the proper design and construction for the intended work.

2. The design limits of the equipment must be thoroughly understood and operated within the limits of their capabilities.

3. Daily and periodic inspections are necessary to uncover defects before they become serious in nature.

4. All maintenance, both preventive and corrective, must be performed by qualified personnel.

From a realistic viewpoint, it would be naive to say that all aerial basket accidents due to actions of individuals can be prevented. However, the lessons that can be learned from this accident study are:

1. Adequate instruction and training of personnel is a must. The capabilities of the equipment must be thoroughly understood and not exceeded.

2. It is not safe to assume that an operator familiar with one type of aerial basket or equipment can operate other types.

3. Adequate clearances must be observed and ability to judge distances is essential.

4. Sufficient rubber protective equipment is as necessary in working from aerial baskets as in working from a pole.

5. When jobs involving both "bare-hand" work from a basket and work from a structure are performed, coordination and teamwork between the two methods is of primary importance.

693

6. Job briefing and followup on training are essential for safe operation.

Crabs and winches

Crabs and winches (Fig. 25–22) may be either hand operated or electrically driven. Some form of brake or safety lowering device should be installed, and portable units should be anchored securely against the pull of the hoisting rope or chain.

Barricade guards should be installed to protect the operator against flying strands of wire and the recoil of broken cables.

The locking pawl on the ratchet of a winch frequently presents a serious finger hazard, particularly when the operator attempts to disengage it. To reduce this hazard, a small lever can be welded to the pawl so that it can be grasped safely.

A major danger with hand-operated equipment (that has a crank handle instead of a hand wheel) is that the operator may be struck by the revolving crank handle, if he loses control while lowering a load. A dog should be provided to lock the gears.

To lower loads rapidly, a strap brake is practicable. Before using the brake, the crank should be removed, or other steps taken, to prevent the crank handle from flying around, such as replacing a spur gear and dog with a worm gear.

A pin through the end of a crank will keep it in the socket during hoisting operations.

Be sure gears are fully guarded. Power-driven crabs and winches should have their

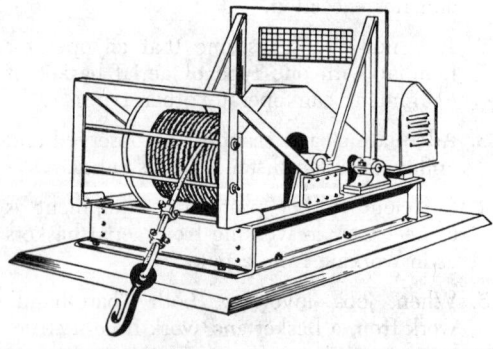

Fig. 25–22.—Operator of this car puller winch stands behind the shield, which protects him if the rope breaks.

moving parts encased and should be electrically grounded.

Blocks and tackles

A factor of safety of 10 is recommended for determining the safe working load of manila rope falls in a block and tackle assembly. The purpose of this large safety factor is to allow for error in estimating the weight of the load, for vibration or shock in handling the load on the tackle, for loss of strength at knots and bends, and for deterioration of the rope due to wear (if it has been in service more than six months) or other causes.

The governing value usually is the safe working load limit of the blocks, rather than of the falls (rope). The reason is that multiplying the number of sheaves and rope parts multiplies the weight of the load that can be handled by the rope, but does not correspondingly increase the strength of the blocks. Calculation will show that, in most instances, using a safety factor of 10 for the rope will automatically keep the load on blocks corresponding to the rope size within safe working load limits.

Blocks should be plainly marked with their safe working loads, as specified by their manufacturers, and the total weight on the tackle never should exceed this.

Breaking strengths of fiber ropes are shown in Chapter 26, "Ropes, Chains, and Slings." Safe working loads for rope used in block and tackle assemblies are conversely 1/10 of the block's breaking strength, based on a safety factor of 10 as before.

For new rope, to find the required breaking strength, proceed as follows: For each sheave 3 in. in diameter or larger, add 10 percent to the weight of the load to compensate for friction loss. Divide this figure by the number of ropes or parts running from the movable block, and multiply the resultant figure by a safety factor of 10.

For example, a load to be lifted weighs 2000 pounds and the tackle consists of two double blocks—four sheaves, four rope parts at the movable block. Friction loss (10 percent for each sheave) = 40 percent, or 800 pounds; 2000 + 800 = 2800 pounds, which divided by 4 (the number of parts at the movable block) = 700. Applying the safety factor of 10 (10 × 700) gives 7000 pounds,

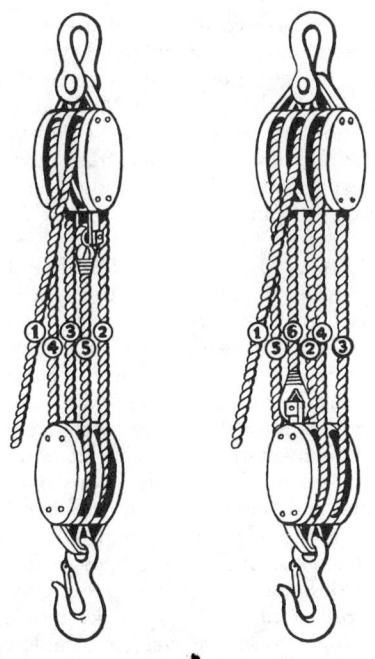

Double and Double   Triple and Double

Fig. 25–23.—Reeving tackle blocks. Lead line and becket line should come off a middle sheave when blocks contain more than two sheaves. The upper and lower blocks then will be at right angles to each other. This eliminates tipping and the accompanying losses in efficiency.

the required breaking strength of the rope. New manila rope of ⅞-inch diameter has a breaking strength of 7700 pounds and, therefore, is the proper size for the load. (Synthetic fibers would have greater tensile strength. See Table 26-A.)

The safe working load for two double blocks made for rope of ⅞-inch diameter, as given by a prominent manufacturer for one series of its standard blocks (regular mortise, inside iron strapped blocks, with loose side hooks, intended for use with manila rope), is 2000 pounds—the equivalent of the total load in the example.

The rope should be attached to the block with a thimble and a proper eye splice.

A mousing of yarn or small rope should be placed on the upper hook of a set of falls as a precaution against its accidental detachment.

Blocks should be inspected thoroughly and frequently, with particular attention to all parts subject to wear.

Fig. 25–23 shows how tackle blocks should be reeved.

If the sheave holes in blocks are too small to permit sufficient clearance, excessive surface wear of the rope will occur. Likewise, excessive internal frictional wear on the fibers will occur if the diameter of the sheave is too small for the rope.

When men are using block and tackle in confined spaces, the pulley block must be guarded so that their hands cannot be caught between the pulley and the rope.

When blocks and falls are used to lift heavy materials or to hold heavy loads in suspension, as on heavy duty scaffolds, wire rope is more serviceable than fiber rope.

Portable floor cranes or hoists

Portable floor cranes or hoists are hoists mounted on wheels. They can be moved from place to place, either by hand power or under their own power. They can raise and lower loads in a vertical line, but cannot rotate around a fixed point.

Portable floor hoists are useful where overhead construction, belting, or shafting prevents the use of overhead hoists or cranes and where more expensive equipment is not justified because of infrequent use. These hoists are handy for placing work on machines, loading heavy material on trucks, and transporting material from one location in the shop to another.

Portable floor hoists usually are operated by hand or by electric power. The lifting mechanism ordinarily is either a winch with wire rope and block or a chain hoist.

Hoists operated by electric power should be effectively grounded to prevent shock in case of short circuit. A special ground wire can be used. One end is fixed permanently to the frame of the hoist, and the other end is equipped with a device that can be attached to a grounded building column, water line, or other direct-to-ground connection.

Where conditions permit, it may be advisable to install sweep guards on the truck wheels to prevent foot injuries when the hoist is moved.

Truck handles on hoists should be designed

to stand upright when not in use. If permitted to project horizontally from the hoist or to lie on the floor, these handles present a serious tripping hazard.

Tiering hoists

The tiering hoist—sometimes called a stacking elevator, portable elevator, tiering machine, or platform hoist—is designed to raise material in a vertical line on a moving platform. This hoist is portable and is used extensively in warehouses for piling and storing materials. It is operated either manually or electrically. The large capacity machines usually are power driven.

Tiering hoists should have a braking device to permit safe lowering of the platform, and a ratchet lock (or dog) to lock the platform in position for loading and unloading operations. Workmen should not be permitted to ride the platform because they may be crushed should the platform meet an obstruction.

Tiering machines, especially the revolving type, should be operated so that they will not tip over. Essential precautions include having the machine solidly on the floor, making sure that its safe capacity is not exceeded, and placing the load properly on the platform.

Before the machine is used, the casters should be lifted off the floor. One type of hand-operated hoist is arranged so that the platform cannot be moved unless the machine stands solidly on the floor—on the frame and not on the casters.

Material in the form of rolls and other round objects should be blocked to prevent their rolling off the platform.

Power-operated tiering hoists should be grounded, and the wiring preferably should be safeguarded by armored cable. On two-section machines, locks should be provided to prevent release of the upper section. Gears and channel iron guides should be guarded to prevent shearing of fingers.

Some tiering machines have practically all the safeguards that a freight elevator has, including limit stops for top and bottom travel on the hoisting cable drum as well as for the shipper rope, if one is provided.

Conveyors

Because there has been much confusion as to what is a conveyor and what is not, and as to correct names for different types of conveyors, the general definition given in Sec. 4, Par. 4.01, of the American National Standard B20.1, *Safety Standards for Conveyors and Related Equipment* (called, from now on, the Conveyor Code), is here quoted in full, including exceptions:

● CONVEYOR. A horizontal, inclined, or vertical device for moving or transporting bulk material, packages, or objects, in a path predetermined by the design of the device, and having points of loading or discharge, fixed or selective; included are skip hoists, and vertical reciprocating and inclined reciprocating conveyors. Typical exceptions are those devices known as industrial trucks, tractors, and trailers; tiering machines; cranes, hoists, monorails, power and hand shovels; power scoops, bucket drag lines; platform elevators designated to carry passengers or the operator; manlifts, moving walks; moving stairways; highway or rail vehicles; cableways; or tramways.

By industry agreement, nomenclature and definitions have been standardized and published as American National Standard MH4.1, *Conveyor Terms and Definitions*. Terms and definitions in this Manual and in the Conveyor Code follow this standard.

General precautions

The most common conveyors are of the belt, slat, apron, chain, screw, bucket, pneumatic, aerial, portable, gravity, live roll, and vertical types. All should be designed and constructed to conform with applicable codes and regulations. Fig. 25–24 shows some suggested guards which can be used along the conveyor's entire length.

Manually loaded conveyors, traveling partially or entirely in a vertical path, should have a conspicuous sign at each loading point showing the safe load that they can raise or lower.

Gears, sprockets, sheaves, and other moving parts must be protected either by standard guards or must be positioned in such a way to ensure against personal injuries.

The entire conveyor mechanism should be inspected periodically, and any part showing signs of excessive wear should be replaced

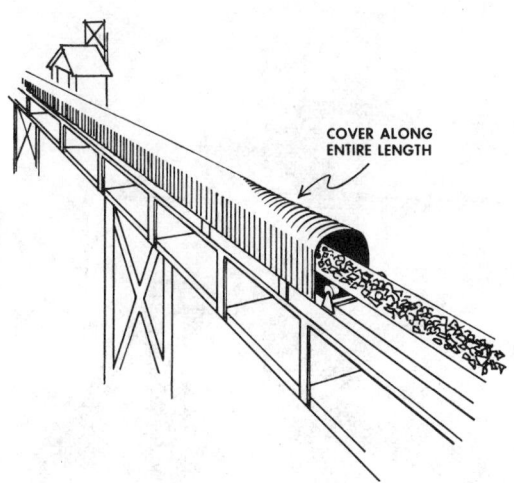

Fig. 25–24a.—Covered conveyor.

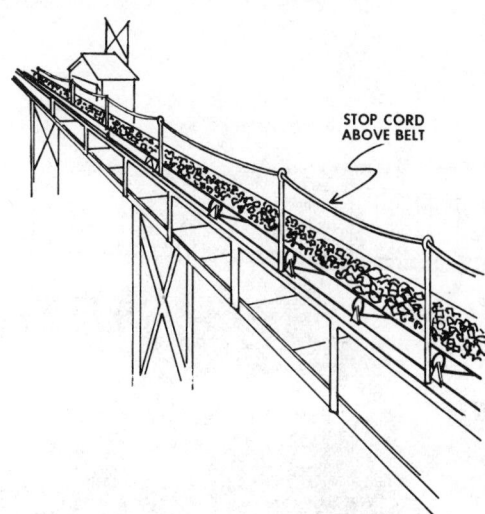

Fig. 25–24b.—Emergency stop cord along along entire length of conveyor.

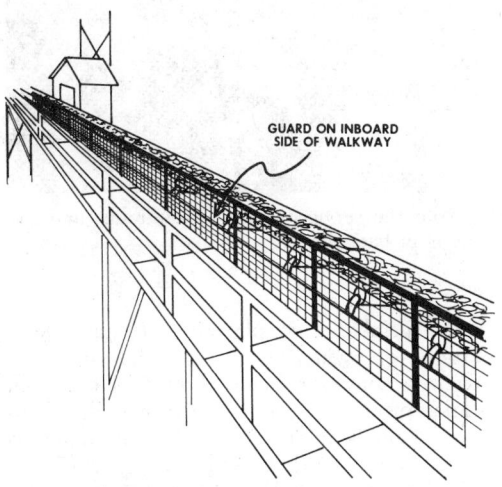

Fig. 25–24c.—Expanded metal along entire length.

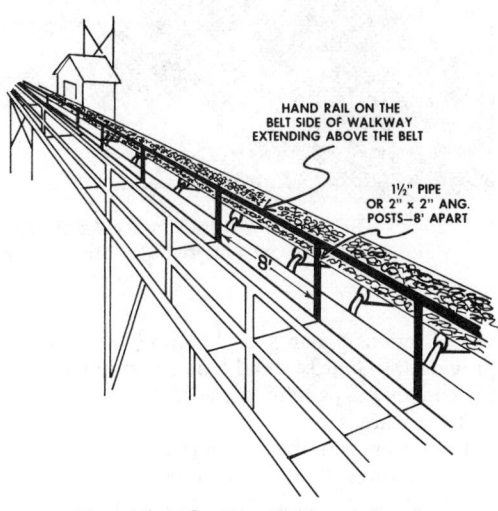

Fig. 25–24d.—Detail of guard rail.

immediately. Particular attention should be paid to brakes, backstops, antirunaway devices, overload releases, and other safety devices to ensure that all are operative and in good repair.

All machine parts should be lubricated according to manufacturer's instructions. If grease nipples are installed on long tubes or pipes that permit oilers to keep a safe distance from moving parts, then the conveyor need not be shut down for greasing.

All conveyors within 6 ft-8 in. of a floor or walkway surface should have crossovers or other passages that comply with NFPA Standard 101 requirements if the walkway is a means of egress. Frequently, a work platform on a movable conveyor tripper can be used as a crossover, if properly railed.

Underpasses should have sheet metal ceilings. Where overhead conveyors dip down at

697

Fig. 25–25.—Belt conveyor used in a coke storage yard. Note the service walkway, handrails, and the lights along the walkway (far right). Storage bridge crane is in background.

work stations, guards or handrails should be provided. Guards should be provided below all conveyors passing over roads, walkways, and work areas.

Conveyors running in tunnels, pits, and similar enclosures should be provided with adequate drainage, lighting, ventilation, guards, and escapeways wherever it is necessary for persons to work in or enter such areas. Sufficient side clearance should be provided to allow safe accessway and operating space for essential inspection, lubrication, repair, and maintenance operations.

Where conveyors pass through building floors, the openings should be guarded by standard handrails and toeboards. As a fire precaution, each opening should be protected against the passage of flame or super-heated gases, from one floor to the next, by doors that close automatically or by fog-type auto-

matic sprinklers, so placed as to provide a curtain of waterfog across the opening. Where a conveyor passes through a fire wall, similar protection should be provided. Conveyor tunnels under stock piles of materials should be open at both ends.

If the top of a loading hopper is at or near the level of a floor or platform, the hopper should be protected by a bar guard with openings not greater than 2 in. in one dimension (such as 2 x 12, 14 x 2, 2 x 16 in.) or by standard railings and toeboards.

Elevated conveyors should have access platforms or walkways on one or both sides (Figs. 25–24a and 25–25). Handrails should be 42-in. high with an intermediate rail, and platforms should have 4-in. toeplates. Flooring should be of checkered plate or other nonslip surface, particularly on sloping walkways.

Both sideboards, along edges and at corners and turns of overhead conveyors, and screen guards, underneath high runs, will protect workers from falling material.

Crossovers or underpasses with proper safeguards should be provided for passage over or under all conveyors. Crossing over or under conveyors, except where safe passageways are provided, should be prohibited. Riding on a conveyor should be absolutely forbidden.

Operating precautions. The starter button or switch for a conveyor should be so located that the operator can see as much of the conveyor as possible. If the conveyor passes through a floor or wall, then each area should be equipped with starting and stopping devices, and the simultaneous operation of all starting buttons or switches should be required to start the conveyor. These start-stop devices should be marked clearly, and the area about them must be kept free of obstructions so they can be seen and reached easily. All personnel working on or about the conveyor should be instructed in the location and operation of all stopping devices.

Electrical or mechanical interlocking devices (or both types) should be provided which will automatically stop a conveyor when the unit it feeds (another conveyor, bin, hopper, or chute) has been stopped or is blocked, so that it cannot receive additional loads.

If two or more conveyors operate in series, controls should be designed so that, if one conveyor is stopped, all conveyors feeding it are stopped also.

Emergency stop devices should be located not more than 40 ft apart along walkways by the conveyor. For some installations, a good solution is to have a lever-operated emergency stop device at the tail end of the conveyor, with a strong cord or wire strung on each side of the conveyor for its entire length. A pull on the cord or wire will stop the conveyor (Fig. 25–24c).

On conveyors, where there is a possibility of reversing or running away, antirunaway and backstop devices can be provided or the conveyor track should be designed, so that the load (or conveyor parts) cannot slide or fall in event of mechanical or electrical failure. If such design is not practicable, guards capable of withstanding the shock and of holding the falling load should be installed.

Electric machines, with brakes that are mechanically applied and released by the movement of operating devices, must be so designed that, if the power is interrupted with the brakes in the OFF position, the load can descend only at a controlled speed.

In addition to overload protection customarily provided for electric motors, there should be an overload device designed to protect the conveyor and mechanical drive parts. In the event of overload, the device must shut off the electric power quickly, disconnect the conveyor or drive parts from the motive power, or limit the applied torque. Shear pins and slip or fluid couplings are examples of overload protective devices.

When a conveyor has stopped because of an overload, all starting devices should be locked out and the cause of the overload removed. The entire conveyor should be inspected before it is restarted.

The loading and discharge points of a conveyor carrying material in fine or powdered form should be covered with exhaust hoods and should have good general ventilation, to prevent the formation of dust clouds.

If the material is combustible, the concentration of dust must be kept below the lower explosive limit. Only approved explosion-proof electrical fixtures should be used, in accordance with the American National Standard Z12 Series, "Prevention of Dust Explosions," and all sources of ignition should be excluded from the dusty area. The conveyor should be grounded and its parts bonded electrically. The container into or from which the material is conveyed also should be grounded and its parts bonded electrically.

Men working near or on conveyors should wear close-fitting clothing that cannot become caught in moving parts. Safety shoes are recommended. If the conveyor galleries are dusty, approved dust goggles and, if necessary, dust respirators should be worn.

One insurance company found that, in a large percentage of conveyor accidents reported to them, material had fallen from a conveyor and struck a person. Workmen should always be warned to place material carefully on a conveyor so that it will ride safely.

699

Maintenance precautions. Before maintenance men commence working on a conveyor, they should padlock the main power control in the OFF position. The maintenance foreman should carry the only key to this lock. If two gangs work on one conveyor, the foremen of both gangs should each padlock the master switch.

Areas should be readily accessible so that maintenance men can change the position of pulleys, sprockets, or sheaves to compensate for normal working conveyor stretch and wear. If adjustment is required while equipment is operating, moving parts should be thoroughly guarded.

The on-running belt should be guarded for at least 18 in. from the points of tangency between the belt and the head and tail pulleys and between the belt and the tripper and hump pulleys. If the hazard is out of reach (more than 8 ft above the floor or platform) or close to a wall or other obstruction, no one in the normal course of his duties should be exposed to it.

The fact that the skirtboards of the loading boot are close to the upper surface of the belt means that, if a man's arm should be caught, the belt could not be raised sufficiently to allow the arm to ride over an idler pulley under the belt where the arm undoubtedly would be badly mangled. Therefore, guards should be installed at the sides of the conveyor at the loading boot.

The points of contact where the wheels of movable trippers roll on the rails also must be guarded. Where the operator travels on the tripper, a platform should be provided for him. It should be so located and constructed as to protect him from slipping and falling and to prevent contact between him or his clothing and moving parts. To help prevent his falling into the hopper, the platform should have handholds and railings.

At the conveyor floor above the coal bunkers in power plants, the slot through which coal from the tripper chute is discharged into the bunker is protected by bars placed across it about 12 in. apart at floor level. A piece of discarded belting of the required width and length can be placed to cover the slot, and its ends are securely anchored. At the tripper platform are four pulleys that raise this belt vertically so that it passes over the access to

the tripper platform. This device not only provides safety, but also seals the slot and thus keeps the dust in the bunker.

Whenever a man might fall onto a conveyor, a gate or paddle can be suspended as low as practicable above the belt near the head pulley, so that a man's body riding on the belt would automatically pull a stop rope and quickly stop the conveyor.

On belt conveyors at floor level or on balconies or galleries, a shield guard or housing should completely enclose each end. Guardrails and toeboards should extend the length of the conveyor.

To help remove static from belt conveyors, tinsel or needlepoint static collectors can be placed close to the outrunning sides of the drive pulleys and idlers, which, along with the shafting, can be grounded through carbon or bronze brushes running on the shaft.

A belt that does not run too fast can be grounded to the drive gear by a continuous strip of copper foil on the pulley side. Other belt conveyors can be grounded by being treated with conductive belt dressings.

One of the outstanding dangers of belt conveyors is that workmen are tempted to clean off material that sticks to the tail drums or pulleys while they are in motion. Fixed scrapers and revolving brushes eliminate the need for a hand operation.

Barrier guards can be placed directly in front of the pinch points of the belts and drums to protect workmen should they attempt to clean or dress the belts while they are moving. The belt and drum should be guarded on the side also at a sufficient distance from the drum to prevent contact.

Slat and apron conveyors

The slat conveyor has one or two endless chains operating on sprockets and usually runs horizontally or at a slight incline (Fig. 25–26). Attached to the chain or chains are non-overlapping, noninterlocking slats that are spaced closely.

Apron conveyors (Fig. 25–27) have overlapping or interlocking plates that form a continuous moving bed. They vary greatly in size: one may be part of a bottling machine, and another may handle billets and castings in a steel mill.

Pinch points between slats or plates, and

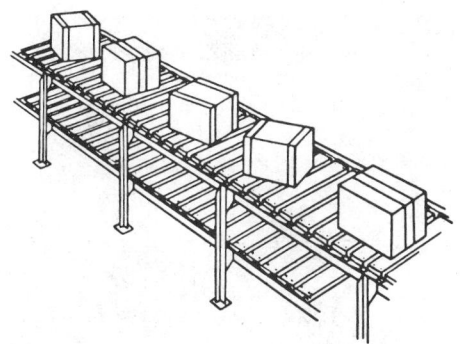

FIG. 25–26.—Slat conveyor.

between them and the chain, sprockets, and guides, should be guarded. Where slats are spaced farther apart than 1 in., a serious shearing hazard exists between the slats and the conveyor substructure. When a slat conveyor is located at floor level or in working areas, the space under the top run of the slats should be filled in solid.

When designed for handling heavy material, slat and apron conveyors usually are installed flush with the floor to facilitate load-

FIG. 25–27.—Two small apron conveyors used in conjunction with a bottling machine.

ing and unloading. When so located, conveyors should be guarded by handrails (except at loading stations) so that workmen will not step onto the moving conveyor. Openings in the floor at the loading platforms should be guarded or covered.

Flight conveyors

The flight conveyor (Fig. 25–28) is similar to the slat conveyor except that the metal flights are attached at right angles to the chain rather than parallel to it. These equally spaced flights act as scrapers as they are pulled through a trough containing stone, coal, or other bulk material.

The flight conveyor uses one or two chains

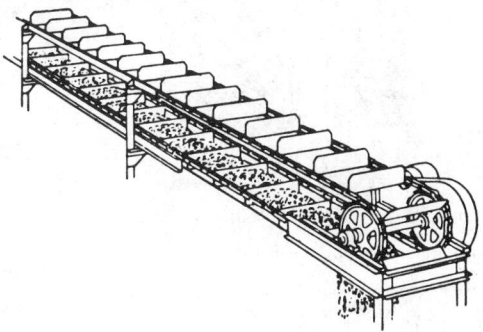

FIG. 25–28.—Flight conveyor.

and may be operated at an incline. Shearing and pinching hazards are similar to those on the slat conveyor and should be guarded.

Chain conveyors

Chain conveyors take many forms, but they all carry, pull, push, haul, or tow the load either directly by the chain or by means of attachments, pushers, cars, or similar devices.

Types in which the chain itself directly moves the load are *drag, rolling,* and *sliding chain conveyors.*

Those consisting of an endless chain that is supported by trolleys from an overhead track or that runs in a track above, flush with, or under the floor, with attachments for towing trucks, dollies, or cars, are known as *tow conveyors* (Fig. 25–29).

Trolley conveyors. Where an endless chain

FIG. 25–29.—*Left:* Overhead chain tow conveyor. *Right:* Underfloor chain tow conveyor.

Courtesy Jervis B. Webb Co.

propels a series of trolleys (Fig. 25–30) supported from or within an overhead track, with the loads suspended from the trolleys, the assembly is known as a trolley conveyor.

In modern manufacturing plants, various devices are used on chain conveyors operated on overhead tracks: equally spaced hooks, suspended directly from the endless chain or from trolleys, may carry, for example, in automobile plants, gears, fenders, or other parts. Smaller parts are carried in suspended wire baskets. Suspended tongs often are used to carry boxes and similar objects.

If the parts are to be washed in a solvent or immersed in paint, the track is bent downward as required immediately above the tank.

In the paper manufacturing industry, a chain with equally spaced vertical lugs is located below a slot at the bottom of a V-trough. Short logs are thrown into the trough and are carried forward by the lugs. This conveyor may be operated on the level or on an incline.

The return portion of the chain should be well guarded. Whenever possible, guardrails should be installed along both sides of the trough to prevent a man from stepping or falling into it.

Shackle conveyors

A shackle conveyor consists normally of chain-type conveyors, with suspended shackles evenly spaced along the line for conveying poultry or other meat products through a processing plant. In the case of a poultry processing plant, the shackle conveyor or conveyors carry the poultry from the beginning of the line where live poultry is hung on the line, from each shackle, head down, through all processing to the packing department where the finished product is removed for packaging and shipment. In some plants, one continuous conveyor travels through all departments and processes, while in other plants two or more similar conveyors carrying suspended shackles are used.

The speed of such conveyors varies from plant to plant dependent on the number of workers on any given conveyor line, with the speed generally increased with more workers on the line. Poultry is normally processed by

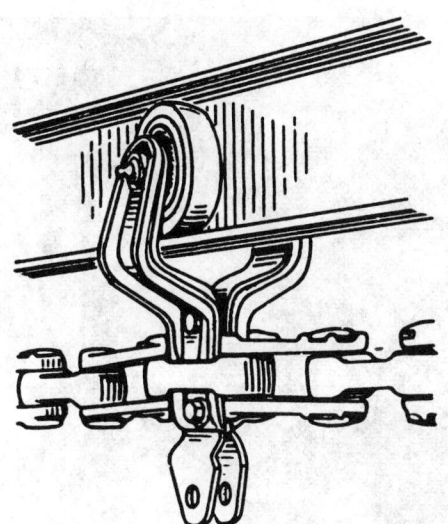

hanging head down and, after cutting and bleeding, it is then repositioned feet down before going through scalders and beaters and then to viscerating and cleaning.

The unique hazard involves workers accidentally placing a thumb or finger in one of the shackles and being dragged through a beater or scalder or to the end of a platform where a worker would be suspended by a finger resulting in loss of life, a finger, or a thumb. The hazard should be controlled by placing emergency stop switches, actuated by pull wires positioned below and perpendicular to the line of travel, at critical points such as at the end of a platform or before entry into an area hazardous to workers.

Screw conveyors

Screw conveyors usually are of limited length and consist of high-pitch spiral or twisted plates on a longitudinal shaft, operating in a semicircular trough with a flat top (Fig. 25–31). As the screw revolves, the material is carried forward in the trough. Since the friction in this type of conveyor is great, more power is required to operate it than is required by other types of conveyors.

Some screw conveyors have spirally placed paddles rather than a continuous screw. Thus, when used for conveying cement-sand mixtures or paint pigments, the conveyor acts as a mixer while it is transporting the material to its destination.

Screw conveyors present a hazard to feet and hands because these can get caught and mangled in the rotating screw. Therefore, conveyor troughs should be completely covered. For inspection and for convenience in dislodging choked material, the covers should have hinged or removable sections interlocked so that, when one is lifted, the screw stops. A heavy wire mesh screen, which will permit inspection, can be placed beneath the solid cover to guard the screw when the cover is not in place.

Fig. 25–32 illustrates the hazard of an open, hinged cover. Sections which do not require opening can have tack-welded covers to prevent their opening.

Screw conveyors that transport powdered materials, having an explosion hazard, should be provided with choke seals to prevent propagation of an explosion from one building to another.

Bucket conveyors

Bucket conveyors are of three general types, in all of which an endless belt, chain, or chains carry elevator buckets, either fixed or pivoted (Fig. 25–33).

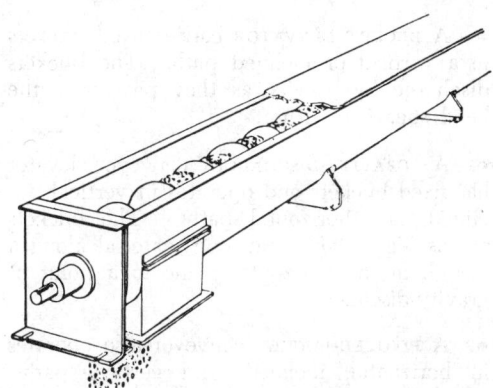

FIG. 25–31.—Screw conveyor. (Cover has been cut away to show interior.)

Fig. 25–32.—Failure to replace hinged cover or any other safeguards could lead to serious injury. Although this photograph is posed, it is a reconstruction of the facts of one lost-time accident. One worker failed to replace this cover. Another, taking a shortcut, missed his footing and stepped into the open conveyor instead of on to the adjacent grating. The resulting injury to his right foot disabled him for three months.

• A BUCKET ELEVATOR carries fixed buckets in a vertical or inclined path. The buckets discharge by gravity as they pass over the head sheave or drum.

• A GRAVITY-DISCHARGE conveyor-elevator has fixed buckets and operates in vertical, inclined, and horizontal paths. The buckets act as flights while carrying material along a trough in the horizontal plane to a point of gravity-discharge.

• A PIVOTED-BUCKET conveyor also operates in horizontal, inclined, and vertical paths. The buckets remain in carrying position until they are tipped or inverted to discharge.

Bucket conveyors should be totally enclosed in a housing to protect operating personnel.

No attempt should be made to take samples when the conveyor is in motion.

The pivoted-bucket conveyor has tripping devices for emptying the buckets at desired locations. These tripping devices frequently are movable so that the material can be distributed evenly in the storage bins. If the control lever mechanism for shifting and locking the trippers is arranged for remote control, workmen will not have to go on the conveyorway to change the position of the trippers.

For the safety and convenience of repairmen and operators, it is advisable to provide a permanent footwalk alongside a conveyor that hoists material and carries it over stokers or bins. The footwalk should be well lit and

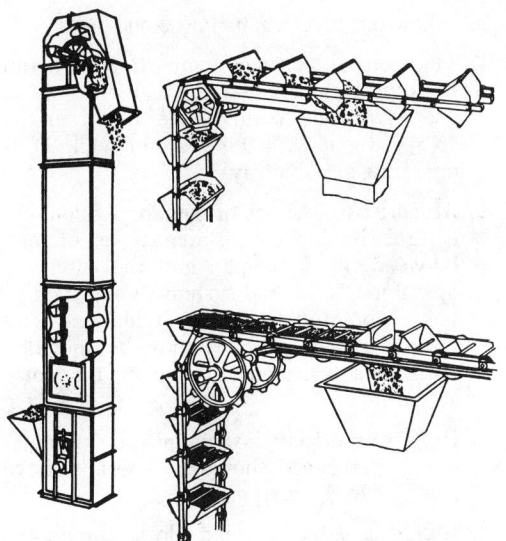

FIG. 25–33.—Three types of bucket conveyors. *Left:* bucket elevator. *Upper right:* gravity-discharge conveyor-elevator. Lower right: pivoted-bucket conveyor.

Courtesy Conveyor Equipment Manufacturer's Assn.

equipped with standard handrails and toe-boards.

Pneumatic conveyors

A pneumatic conveyor consists of an arrangement of tubes or ducts through which solid objects, such as cash, mail, and other small items or grain, dust, and similar bulk materials, are transported by means of compressed air or vacuum.

Solid objects to be conveyed are placed inside a cylindrical cartridge having packing on its outer surface to seal it inside the smooth tubing. Compressed air injected into the tubing system pushes the cylinder forward at a relatively high velocity.

To convey bulk material, compressed air is injected into the piping through a nozzle below the loading chute. The air mixes with the material and makes it flow rather like a fluid through the piping.

A constant volume, variable pressure (positive type) blower should have a relief valve on or adjacent to it. Doors of blower conveyors should be interlocked, so that they cannot be opened if there is positive internal pres-sure. Gaskets holding the line pressures should be shielded so that, if a gasket leaks, the conveyed material will not be thrown against workmen or into the working area.

Where suction lines are large enough to draw in a man, bar guards or screening should be provided over the intake and employees should be instructed to stay at a safe distance.

Receivers and storage bins should have full-bin indicators or controls to prevent over-filling.

A pneumatic conveyor serving an area containing contaminated air must be so arranged that none of the contaminated air can enter the conveyor tube and, through it, be carried to other areas.

When material that has a dust explosion hazard is being transported, the air velocity should exceed the critical velocity for flame propagation of the material to prevent propagation of an explosion from one point to another within the system.

Equipment should be bonded electrically and grounded to prevent static electricity being an ignition source. Details are given in General Precautions section, earlier in this chapter.

Aerial conveyors*

There are two types of aerial cable systems:

● A CABLEWAY is a wire rope-supported system in which the materials handling carriers are not detached from the operating span and the travel is wholly within the span;

● A TRAMWAY is a wire rope-supported system in which the travel of the materials handling carriers is continuous over the supports of one or more spans.

Aerial conveyors frequently are used in industrial plants, and particularly on large construction work, to carry material from point to point. They are used also to some extent for handling coal and ore. The principal hazards in their use are falling material, injury to workmen inspecting and oiling the cables and

---

* The Conveyor Code no longer lists these as conveyors in their definition. However, because of their similarities with the conveyors, this subject is discussed here.

carriages, and misunderstanding of signals.

No attempt should be made to carry on work directly underneath the conveyor except, of course, at the loading and unloading stations where exceptional precautions should be taken. Wherever workmen must pass underneath the conveyor, a covered passageway should be provided.

Another form of protection against material that falls from buckets over roadways and working spaces consists of heavy wire screens suspended under the conveyor from pole to pole and wide enough to catch the material.

The equipment should be inspected regularly and frequently with special attention to the sheave wheels and bearings, the rope fastening, the bucket latch and trunnions, and all load-sustaining parts.

All ropes should be kept well oiled, both for efficiency of operation and for protection against the weather. Lubrication of the hauling rope should be done continuously by means of a controlled drop feed from an oil reservoir at the point (one or both ends of the line) where the rope leaves the drive sheave and passes over a support sheave.

In one plant, to oil a tramway rope used for conveying coal, the workman had to ride a trolley carriage. An automatic lubrication system was devised to eliminate this dangerous practice.

Suitable lighting should be provided at critical points for night operation and repairs.

Every aerial cableway must have a suitable signaling system. A telephone or electric pushbutton system is advisable. On large construction jobs, a portable telephone system is often needed so that a signalman can direct the operation of raising and lowering loads that are out of the operator's sight.

For a tramway, at least three control systems are recommended:

1. Pushbutton stations and a bell signal code that indicates stop, start, slow speed, high speed, and reverse.

2. An all-metallic aerial wire circuit telephone with instruments at certain points along the line in addition to terminal sets.

3. A second telephone circuit, which may be grounded if desired.

The U.S. Army Corps of Engineers suggests the following precautions be taken.

1. The control console compartment should be kept locked when the cableway is in use. No one should be permitted in the compartment with the operator while he is operating a cableway.

2. At least two systems of communication and control should be continuously maintained between the signalman and the cableway operator. This dual system should include voice communication by telephone and radio. Lights or bells may be included with or substituted for one of the voice systems.

3. Only authorized inspection and maintenance personnel should be permitted to ride cableway carriages.

4. Riding of cableway load blocks should not be permitted.

Portable conveyors

Belt, flight, apron, and fixed bucket conveyors are made also as inclined portable units on a pair of large wheels, for loading railroad cars and trucks with bulk materials and for raising construction material from one elevation to another (Fig. 25–34).

The various types of portable conveyors should be provided with guards as specified for the corresponding types of fixed conveyors.

Weatherproof electrical equipment should be used. With three-phase power, the flexible cord that is connected to the power outlet should be a four-conductor, the fourth wire being grounded in all plugs and receptacles. The cable should be so arranged that it cannot be run over by trucks or other machines. If the cable must cross a driveway, it should be hung on poles at a minimum height of 14 ft. If two or more sections of cable are required, the connectors should be kept above the ground, since these machines are used most in the worst weather.

Portable conveyors should have skirtboards or sideboards (not less than 10 in. high) to keep heavy material from falling over the sides and light or loose material from blowing off. This safety feature applies to belt conveyors, as well as to the other types, because troughing of the belt gives insufficient

FIG. 25–34.—Portable conveyor.

protection against such spillage.

The conveyor should be stable and should be fitted with a locking device to hold the conveying unit at various fixed elevations.

The mechanism for raising and lowering the boom on all types of portable conveyors should be closely checked periodically—it constitutes one of the greatest sources of hazard in the use of such machinery. Any positive type may be used, but the self-walking worm or jack-screw type is preferable. Gear drives should be completely housed and run in oil. All chains within easy reach of workmen should be thoroughly guarded, and a system should be provided for convenient oiling of chains. In warehouses, the boom of portable conveyors usually consists of two sections to permit their use in a small space. These conveyors should be so designed that material will not roll back from one section to the other at the transfer point.

When portable conveyors are used for loading or unloading railroad cars, a suitable safety device should be provided to prevent the car from being shifted during the operation. The device should not be removed until the crew is sure that no one is in the car. A red banner or a standard blue warning sign at each end of the car is recommended as a warning to switchmen. (See the section, Plant Railways, in Chapter 28.)

When portable conveyors are used in raising or lowering construction materials, ample stairs or ladders should be provided in the immediate vicinity of the conveyor. Walking on idle or moving conveyors should be prohibited.

Gravity conveyors

The fact that gravity conveyors (Fig. 25–35) depend wholly upon the natural force of gravity for their operation often leads to disregard of some necessary safe practices. Even though it is not power-driven, this equipment can be the source of serious accidents.

There is real danger if an employee climbs upon a conveyor to release a blockade. He may either slip on the rollers or be knocked

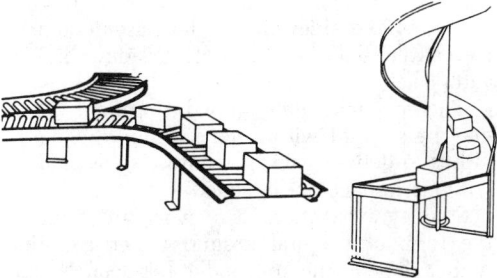

Courtesy Conveyor Equipment Manufacturer's Assn.

FIG. 25–35.—Types of gravity conveyors. *Left and below:* roller conveyor. *Right:* spiral chute.

Courtesy Dow Corning Corp.

707

down should the jam suddenly be released and strike him. To guard against this hazard, workers should be prohibited from climbing onto conveyors. In addition, steel or wood plates should be installed between rolls, so that neither a worker's body nor limb can fit between the rolls.

Gravity conveyors are divided into two general classes—chute conveyors and roller or wheel conveyors.

Chute conveyors. Chutes of polished metal sheets or bars are used to lower packing cases, cartons, and crates from one floor to another and from the sidewalk to the basement of a mercantile building through a sidewalk elevator shaft or other opening.

The inclined chute may be straight, or it may be straight with a vertical curve of large radius to deliver packages onto the lower floor without impact or damage.

Some gravity-chute conveyors are built in the form of a spiral around a vertical pipe with a slope at the outer edge between 18 and 30 degrees.

In removing packages from the delivery end of spiral chutes, workmen frequently injure their hands, which become caught by descending packages or are mashed against others on the delivery table. Where the chute is enclosed, a warning sign should be placed over the delivery end. A simple mechanical or electrical device that will give warning when a package is about to be delivered from the chute is advisable, especially where the packages cannot be plainly seen in descent.

Spiral chutes present a serious fire hazard because they form flues from lower floors to upper floors through which fire will spread quickly. Two methods of eliminating this hazard are in general use: one is to enclose the chute in a tower made of fire-resistant material, such as steel, concrete, or masonry; the other is to provide automatic fire doors (draft checks) where the chute passes through floors.

The enclosed tower has doors at each charging station and a door at the delivery end. The charging station doors should be kept closed, except when charging is being done. The door at the delivery end should close automatically in case of fire.

Shutoff doors (draft checks) are of two kinds—the vertical sliding and the shutter type. Both types should have fusible links so that they will close automatically in case of fire.

Where an open chute is used, a guardrail and toeboard should be provided at each floor. The charging stations should be guarded either by a movable railing or by a hinged door or gate.

Roller or wheel conveyors. Gravity roller or wheel conveyors are similar to chute conveyors, except that the angle of slope is much less (2 to 4 percent). The conveyors, therefore, can be used to convey packages for considerable distances on one floor. If the rollers or wheels are placed radially instead of parallel, the course of travel can be changed from a straight line to a curve.

The principal hazards in the use of roller conveyors are that material may run off the edge of the rollway and fall to the floor, and that loads may run away. A guard railing often is provided on each side of the roller conveyorway, to guide the material and prevent it from running off. Such guardrails are especially advisable at corners and turns and on elevated conveyors under which men must work or pass.

Where heavy loads are conveyed, retarders, brakes, or similar means should be provided to prevent the loads from running away, or a power conveyor on which speed can be controlled should be substituted.

A vertically swinging hinged section of a roller or wheel conveyor should be hinged to that end of a stationary section of the conveyor from which the material is flowing to help block the oncoming material. The open end of the conveyor line should have a stop that automatically projects above the level of the rollers or wheels when the hinged section is opened and automatically retracts when the hinged section is closed.

Where a horizontally swinging hinged section occurs in a conveyor, the open ends of stationary sections (the two ends adjacent to the hinged section) should be equipped with retractable stops to prevent loads from dropping off when the hinged section is open.

Live roll conveyors

A live roll (or roller) conveyor consists of a

series of rolls over which objects are moved by power applied to all or some of the rolls through belts or chains.

Where installed at floor level or used in working areas, live roll conveyors should be designed to eliminate hazards from pinch points and moving parts, unless other provisions are made to prevent personnel from coming in contact with or crossing the conveyor.

Vertical conveyors

Vertical conveyors handle packages or other objects in a vertical or substantially vertical direction. The Conveyor Code designates three types:

● VERTICAL RECIPROCATING CONVEYOR: a power- or gravity-actuated unit that receives objects on a carrier or car bed, usually constructed of a power or roller conveyor, and elevates or lowers them to other locations.

● SUSPENDED TRAY CONVEYOR: a vertical conveyor having one or more endless chains with pendant trays, cars, or carriers that receive objects at one or more elevations and deliver them to another or several elevations.

● VERTICAL CHAIN CONVEYOR (opposed shelf type): two or more vertical elevating conveying units opposed to each other, each unit consisting of one or more endless chains whose adjacent facing runs operate in parallel paths. Thus, each pair of opposing shelves or brackets receives objects (usually dish trays) and delivers them to any number of elevations.

Where vertical conveyors are loaded and unloaded automatically, guards should be provided to protect personnel from contact with moving parts. Where they are loaded and unloaded manually, guards and safety devices, such as lintel and sill switches and deflectors, should be installed.

Carriages of vertical reciprocating conveyors designed to register at a floor, balcony, gallery, or mezzanine level never should have a solid bed. This type of conveyor is not intended to carry passengers or operators, or to have its car or carriage called to a station by a manually-operated pushbutton.

References

American National Standards Institute, 1430 Broadway, New York, N.Y. 10018.
  Conveyor Terms and Definitions, MH4.1.
  "National Electrical Safety Code," C2.
  "Prevention of Dust Explosions," Z12 Series.
  Safety Code for Building Construction, A10.2.
  "Safety Code for Cranes, Derricks, Hoists, Jacks, and Slings," B30 Series.
  Safety Requirements for Floor and Wall Openings, Railings, and Toeboards, A12.1.
  Safety Requirements for Personnel Hoists, A10.4.
  Safety Standards for Conveyors and Related Equipment, B20.1.
  Safety Standards for Mechanical Power-Transmission Apparatus, B15.1.

Conveyor Equipment Manufacturer's Association, 100 Vermont Ave., NW., Washington, D.C. 20005. (General.)

Electric Controller and Manufacturing Co., 2704 East 79th St., Cleveland, Ohio 44104.
  How to Operate a Crane, Booklet 920.

International Union of Operating Engineers, 1125 17th St., NW., Washington, D.C. (General.)

National Fire Protection Association, 470 Atlantic Ave., Boston, Mass. 02210.
  Life Safety Code, Std. No. 101.
  National Electrical Code, Std. No. 70 (ANSI C1).

National Safety Council, 425 N. Michigan Ave., Chicago, Ill. 60611.
  Industrial Data Sheets
    Aerial Baskets, 572.
    Automotive Hoisting Equipment, 437.

709

25—Hoisting Apparatus and Conveyors

Belt Conveyors for Bulk Material, 569 and 570.
Electromagnets Used with Crane Hoists, 359.
Flexible Insulated Protective Equipment for Electrical Workers, 598.
Pendant-Operated and Radio-Controlled Cranes, 558.
Roller Conveyors, 528.
Tower Cranes, 630.
Truck-Mounted Power Winches, 441.
Underground Belt Conveyors, 447.
"Study of Aerial Basket Accidents, Vol. I—1962 to 1967."
"Study of Aerial Basket Accidents, Vol. II—1967 to 1971."

Ropes, Chains, and Slings

Chapter
26

Fiber Rope   712
Natural fibers . . . Synthetic fibers . . . Inspection . . . Care of
rope in use . . . Care of rope in storage

Wire Rope   717
Construction . . . Safety factor for hoisting ropes . . . Causes of
deterioration . . . Sheaves and drums . . . Lubrication . . .
Overloading . . . Rope fittings . . . Inspection and replacement

Rope Slings   726
Materials . . . Methods of attachment . . . Safe load limits . . .
Inspection

Chains and Chain Slings   730
Materials . . . Properties of alloy chains . . . Hooks and attachments
. . . Inspection . . . Safe practices

Metal and Nylon Mesh Slings   734
Materials . . . Safe practices . . . Inspection

References   737

Special safety precautions apply to using and storing ropes, rope slings, wire rope, chains, and chain slings. The safety professional should know the properties of the various types used and the precautions both in use and maintenance.

Fiber Rope

Fiber rope is used extensively in handling and moving materials. The rope is generally made from manila (abaca), sisal, hemp, or nylon. Manila or nylon ropes give the best uniform strength and service. Other types of rope on the market today include those made from polyester and polypropylene; these are adaptable to special uses.

Natural fibers

The properties of abaca (commonly known as manila) fiber make it the best-suited natural fiber for cordage. Manila rope is often recommended for capstan work because of its ability to render or pay out evenly when so used. High-grade manila rope, when new, is firm but pliant, ivory to light yellow in color, and with considerable luster. Its good reputation in fresh and salt water has been established for many years.

The properties of the several agave fibers (commonly known as sisals) do not give these ropes the high general acceptance of manila. The sisals are confined mostly to the smaller size ropes. The agaves are not as satisfactory for general use partly because their breaking strengths are generally lower than those of manila. Sisal rope varies in color from white to yellowish white and lacks the gloss of high-grade manila. The fibers are stiff and harsh and tend to splinter. This makes the ropes uncomfortable to handle.

Both sisal and manila fibers deteriorate when in contact with acids and caustics (and their mists or vapors). This deterioration is accelerated by hot, humid conditions.

Sisal and hemp are not as satisfactory as manila because their strength varies in different grades. Sisal rope is about 80 percent as strong as manila; hemp is about 50 percent as strong, but is more resistant to atmospheric deterioration.

Other natural fibers are also used in ropes but to a lesser or negligible degree or for very special reasons. These fibers include: cotton, flax, coir, straw, asbestos, istle, jute, kenaf, silk, rawhide, and sanseveria.

Synthetic fibers

The popularity of synthetic fiber ropes now rivals that of natural fiber ropes. There are several reasons.

• Greater knowledge of the properties of various synthetics is an important reason for their increased use. Successful use of synthetics depends largely on the selection of the synthetic with the physical properties and characteristics that most closely match the requirements of the job.

• Splices can be made readily and can develop nearly the full strength of the rope. Tapered splices are highly recommended for rope sizes with a 1-in. diameter and larger.

• It is recognized that synthetics have "magic-sounding" names; caution dictates, however, that no more than the true attributes be assumed for these materials.

Nylon rope has over 2½ times the breaking strength of manila rope and about 4 times its working elasticity. It is, therefore, particularly suitable for shock loading, such as required for safety lines. Its resistance to abrasion is remarkably high in comparison to other ropes. While nylon rope is wet or frozen, its breaking strength is reduced by 10 to 15 percent.

It also is highly resistant to organisms that cause mildew and rotting and to attack by marine borers in sea water. Atmospheric exposure produces little loss of strength over a considerable period. Wet nylon rope runs through blocks as easily as dry nylon rope, since there is no swelling. Although resistant to petroleum oils and most common solvents and chemicals, nylon strength is affected by drying oils, such as linseed oil or the phenols, and is quickly deteriorated by strong mineral acids, such as sulfuric and muriatic.

Whereas manila rope begins to char at 300 F, nylon has lost some of its strength at this temperature and all of it at 482 F, its melting point. Short of melting, most strength is regained on cooling to normal temperature. At an increased initial cost, nylon of higher

ROPE CHECKLIST

Strength	Ruggedness in shape
Stretch with load	Temperature
Permanent stretch	resistance
Recovery from stretch	Friction melting
Length	Combustibility
Size	Sunlight resistance
Yardage	Marine growth
Floatability	resistance
Flexibility	Rot resistance
Twist direction and	Chemical resistance
torque	Color
Flex life in bending	Aging
Slipperiness	Contamination
Texture	Uniformity
Water repellency	Service Cost
Hygroscopicity	Toughness against
	wear

FIG. 26–1.—Factors that may be of significance when specifying rope for a specific use.

melting point can be secured.

Nylon, more than any other rope material, will absorb and store energy in the same manner as a spring. This energy, released at break, will make the moving ends as dangerous as a projectile. Caution must therefore be exercised when working lines around corners, capstans, timber heads, and the like.

Polyester makes probably the best general-purpose rope available, especially for critical uses. Polyesters stretch about half that of nylon, so energy absorption is also about half. It is not weakened by rot, mildew, or prolonged exposure to sea water. It retains its full strength when wet and shows little deterioration from long exposure to sunlight. It has good resistance to abrasive wear and to most alkalis and acids, except benzoic acid.

Polyolefin ropes, in general, are strong and inexpensive; they float and are unaffected by water. Polyolefins, like the polyesters, are not hygroscopic; therefore, they do not shrink or swell with water. The movement of crossed ropes, as well as other types of abrasion, must be avoided because of a very rapid friction sawing which even modest loads will cause. They are unaffected by rot, mildew, and fungus growth and are highly re-

sistant to a wide variety of acids (except nitric) and alkalis, as well as to alcohol-type solvents and bleaching solutions. They swell and soften with hydrocarbons, particularly at temperatures above 150 F. There are two types:

● POLYPROPYLENE, with a specific gravity of 0.91 and a softening point of 300 F, is made in several different size filaments and from film with or without longitudinal fracturing. Polypropylene rope is about 50 percent stronger than manila, size for size. Pure polypropylene ropes have relatively poor rendering properties.

● POLYETHYLENE, with a specific gravity of 0.95 and a softening point of 250 F, is characteristically slippery and has very little springiness. Stretch is low and strength is good. It has a comparative low softening point and low coefficient of friction.

Composite ropes (combining several types of synthetic fibers or synthetic and natural fibers) are available in all sizes. They result from attempts to give the surface of the rope or strand more wear resistance, internal tensile strength, or structural strength to resist deformation. Many of these ropes are made to match the requirements of specific jobs. These ropes can be expected to perform to the ends their makers declare.

Other types of rope are available, but enjoy only a small percentage of the market for reasons of cost, limited use, or short supply. These ropes include in various modifications: paper, glass, acrylic, rayon, polyvinyl chloride, fluorocarbon, rubber, cellulose acetate, and polyurethane.

Table 26–A lists breaking strengths and safe loads (working load limits) for manila and synthetic ropes. Fig. 26–1 gives a checklist of factors that may be of significance in a specific application.

Inspection

New rope should be inspected along its entire length thoroughly, to determine that no part of it is damaged or defective, before being placed in service. Any irregularity in the uniformity of appearance is evidence of possible degradation.

TABLE 26-A. TENSILE STRENGTH AND WEIGHT OF SYNTHETIC AND NATURAL FIBER ROPE

SIZE IN INCHES		NYLON		POLYESTER		POLYPROPYLENE		3-STRAND NATURAL FIBER		
Diameter	Circumference	Pounds per 100 feet	Tensile Strength (lb)	Pounds per 100 feet	Tensile Strength (lb)	Pounds per 100 feet	Tensile Strength (lb)	Pounds per 100 feet	Tensile Strength (lb) Manila	Tensile Strength (lb) Sisal
3/16	5/8	1.0	1,000	1.2	1,000	0.70	800	–	–	–
1/4	3/4	1.5	1,650	2.0	1,650	1.2	1,250	2.0	600	480
5/16	1	2.5	2,550	3.1	2,550	1.8	1,900	–	–	–
3/8	1 1/8	3.5	3,700	4.5	3,700	2.8	2,700	4.1	1,350	1,080
7/16	1 1/4	5.0	5,000	6.2	5,000	3.8	3,500	–	–	–
1/2	1 1/2	6.5	6,400	8.0	6,400	4.7	4,200	7.5	2,650	2,120
9/16	1 3/4	8.3	8,000	10.2	8,000	6.1	5,100	–	–	–
5/8	2	10.5	10,400	13.0	10,000	7.5	6,200	13.3	4,400	3,520
3/4	2 1/4	14.5	14,200	17.5	12,500	10.7	8,500	16.7	5,400	4,320
13/16	2 1/2	17.0	17,000	21.0	15,500	12.7	9,900	–	–	–
7/8	2 3/4	20.0	20,000	25.0	18,000	15.0	11,500	22.5	7,700	1,540
1	3	26.0	25,000	30.5	22,000	18.0	14,000	27.0	9,000	7,200
1 1/16	3 1/4	29.0	28,800	34.5	25,500	20.4	16,000	36.0	12,000	9,600
1 1/8	3 1/2	34.0	33,000	40.0	29,500	23.7	18,300	41.8	13,500	10,800
1 1/4	3 3/4	40.0	37,500	46.3	33,200	27.0	21,000	–	–	–
1 5/16	4	45.0	43,000	52.5	37,500	30.5	23,500	–	–	–
1 1/2	4 1/2	55.0	53,000	66.8	46,800	38.5	29,700	60.0	18,500	14,800
1 5/8	5	68.0	65,000	82.0	57,000	47.5	36,000	74.4	22,500	18,000
1 3/4	5 1/2	83.0	78,000	98.0	67,800	57.0	43,000	89.5	26,500	21,200
2	6	95.0	92,000	118	80,000	69.0	52,000	108	31,000	24,800
2 1/8	6 1/2	109	106,000	135	92,000	80.0	61,000	125	36,000	28,800
2 1/4	7	129	125,000	157	107,000	92.0	69,000	146	41,000	32,800
2 1/2	7 1/2	149	140,000	181	122,000	107	80,000	167	46,500	37,200
2 5/8	8	168	162,000	205	137,000	120	90,000	191	52,000	41,600
2 7/8	8 1/2	189	180,000	230	154,000	137	101,000	215	58,000	46,400
3	9	210	200,000	258	174,000	153	114,000	242	64,000	51,200
3 1/4	10	263	250,000	318	210,000	190	137,000	299	77,000	61,600
3 1/2	11	316	300,000	384	254,000	232	162,000	–	–	–
3 5/8	11 1/2	–	–	–	–	–	–	367	91,000	72,800
4	12	379	360,000	460	300,000	275	190,000	436	105,000	84,000

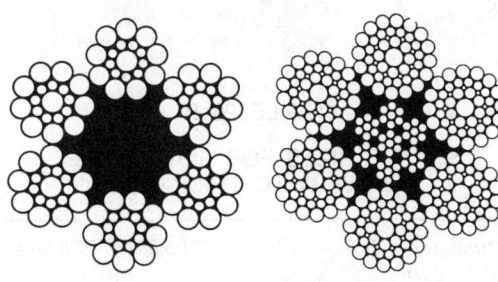

FIG. 26–2.—Wire rope is made from large numbers of individual wires, most of which are not exposed. This, plus the fact that wires are continuous in any given length of rope, accounts for the high degree of dependability and uniformity. Cross sections show usual 6 × 19 rope containing 114 wires (*left*), and 6 × 37 classification rope made up of 313 separate wires (*at right*).

Courtesy John A. Roebling's Sons Corp.,
Subsidiary of The Colorado Fuel and Iron Corp.

NOTES to Table 26–A.

Recommended Factors of Safety: Nylon 9. Polyester 9. Polypropylene 6. Manila 5. Sisal 5.

These specifications of weight, circumference and strength for the synthetic ropes have been derived from tests made in accordance with the "Cordage, Institute Standard Test Methods for Synthetic Fiber Ropes" dated November 10, 1960; revised December 8, 1965. They represent the Cordage Institute specifications (adopted January 13, 1966).

Tensile Strength figures are "Average." Minimum will be 5% below "Average." Pounds per 100 feet are "Average." Maximum weight 5% above these figures.

In-service rope should be inspected every 30 days under ordinary conditions and much oftener if used in critical applications, such as to support scaffolding on which men work. Inspection consists of an examination of the entire length of the rope, inch by inch, for wear, abrasions, powdered fiber between strands, broken or cut fibers, displacement of yarns or strands, variation in size or roundness of strands, discoloration, and rotting. To inspect the inner fibers, the rope should be untwisted in several places to see whether the inner yarns are bright, clear, and unspotted. If exposed to acids, natural fiber ropes, such as manila, should be scrapped or retired from critical operations, as visual inspection will not always reveal acid damage. A rope, like a chain, "is only as strong as its weakest link" (or with rope, its cross section).

Natural fiber rope loaded to over 50 percent of its breaking strength will be permanently damaged; synthetics loaded to over 65 percent may be damaged. Damage from this cause may be detected by examining the inside fibers. These will be broken into short lengths in proportion to the degree of overload. A good estimate of the strength of fibers can be made by scratching the fibers with a fingernail—fibers of poor strength will readily part. This "fingernail test" is a quick test for chemical damage.

In small ropes (up to ¾-in. diameter), surface wear that has progressed to the center of the twisted element (yarn) may account for more than an 80 percent loss of rope strength. In ropes with a ¾-in. diameter or more, surface wear may destroy the strength of the cover yarns, yet not affect the original strength of the core yarns; the remaining strength of the rope will be in the proportion that the core yarns are to the original total of yarns.

If sample fibers can be secured from the rope, an approximation of rope strength can be made. Estimate the fiber strength by manually breaking fiber samples, and estimate the distribution of fibers in a cross section, quartered to allow for twist configuration.

Due to the motion or slippage on a supporting surface when under high tension, synthetics sometimes melt on the surface and form a skin. This skin may be evidence of degradation.

Ropes, using multifilament synthetic fiber

715

26—Ropes, Chains, and Slings

on the surface, will often "fuzz," which is due to the minute fiber breakage. This condition can be beneficial because it provides a buffer or cushion which helps protect the rope from further rapid wear.

Care of rope in use

The factors of safety in Table 26–A are for average use; unusual or erratic loading requires higher factors. The factors recommended may be too low in many instances and, therefore, they can be used only with an expert knowledge of conditions or professional estimates of risk.

The factors of safety variations among different types of rope are derived from elements, such as chaffing, cutting, elasticity, diameter-strength ratio, and general anticipated mishandling.

If possible, a rope should not be dragged as this abrades the outer fibers. If the rope picks up dirt and sand, abrasion within the lay of the rope will rapidly wear it out.

Precautions should be taken to keep rope in good condition. Kinking, for example, strains the rope and may overstress the fibers. It may be difficult to detect a weak spot made by a kink. To prevent a new rope from kinking while it is being uncoiled, lay the rope coil on the floor with the bottom end down. Pull the bottom end up through the coil, and unwind the rope counterclockwise. If it uncoils in the other direction, turn the coil of rope over, and pull the end out on the other side.

Twisted rope should be handled so as to retain the amount of twist (called balance) that the rope seeks when free and relaxed. If rotating loads and improper coiling and uncoiling change the balance, it can be restored by proper twisting of either end. Severe unbalance can cause permanent damage; localized overtwisting causes kinking or hocking.

Sharp bends over an unyielding surface cause extreme tension on the fibers. To make a rope fast, an object with a smooth round surface of sufficient diameter should be selected. If the object does have sharp corners, pads should be used. To avoid excessive bending, sheaves or surface curvatures should be of suitable size for the diameter of the rope, as shown in Table 26–B.

TABLE 26–B

SHEAVE SIZES FOR FIBER ROPES OF VARYING THICKNESS

Diameter of Rope (in.)	Diameter of Sheave (in.)
3/4	6
7/8	7
1	8
1 1/4	10
1 3/8	11

When lengths of rope must be joined, they should be spliced and not knotted. A well made splice will retain up to 100 percent of the strength of the rope, but a knot, only half (Table 26–C).

Use of wet rope, or of rope reinforced with metallic strands, near power lines and other electrical equipment is extremely dangerous.

Rope must be thoroughly dried out after it becomes wet, otherwise it will deteriorate quickly. A wet rope should be hung up, or laid in a loose coil, in a dry place until thoroughly dry. Rope will deteriorate more rapidly, if it is alternately wet and dry than if it remains wet. Wet rope should not be allowed to freeze.

Care of rope in storage

To maintain the existing strength of any rope that is properly prepared, it should be stored safe from deleterious fumes, heat,

TABLE 26–C

EFFICIENCY OF MANILA FIBER ROPE WITH SPLICES, HITCHES, AND KNOTS

Jointure	Per Cent Efficiency
Full strength of dry rope	100
Eye splice over metal thimble	90
Short splice in rope	80
Timber hitch, round turn, half hitch	65
Bowline, slip knot, clove hitch	60
Square knot, weaver's knot, sheet bend	50
Flemish eye, overhand knot	45

TABLE 26–D

BREAKING STRENGTH IN TONS OF IMPROVED PLOW STEEL (IPS) AND EXTRA IMPROVED PLOW STEEL (EIPS) ROPES

Diameter (in.)	6 × 19 CLASSIFICATION			6 × 37 CLASSIFICATION			8 × 19 CLASSIFICATION
	IPS		EIPS	IPS		EIPS	IPS
	FC	IWRC	IWRC	FC	IWRC	IWRC	FC
3/8	6.10	6.56	7.55	5.77	6.20	7.14	5.24
7/16	8.27	8.89	10.20	7.82	8.41	9.67	7.09
1/2	10.70	11.50	13.30	10.20	11.00	12.60	9.23
9/16	13.50	14.50	16.80	12.90	13.90	15.90	11.60
5/8	16.70	17.90	20.60	15.80	17.00	19.60	14.30
3/4	23.80	25.60	29.40	22.60	24.30	27.90	20.50
7/8	32.20	34.60	39.80	30.60	32.90	37.80	27.70
1	41.80	44.90	51.70	39.80	42.80	49.10	36.00
1 1/8	52.60	56.50	65.00	50.10	53.90	61.90	45.30
1 1/4	64.60	69.40	79.90	61.50	66.10	76.10	55.70
1 3/8	77.70	83.50	96.00	74.10	79.70	91.70	67.10
1 1/2	92.00	98.90	114.00	87.90	94.50	109.00	79.40
1 5/8	107.00	115.00	132.00	103.00	111.00	127.00	. . . .
1 3/4	124.00	133.00	153.00	119.00	128.00	146.00	. . . .
1 7/8	141.00	152.00	174.00	136.00	146.00	168.00	. . . .
2	160.00	172.00	198.00	154.00	165.00	190.00	. . . .
2 1/8	179.00	192.00	221.00	173.00	186.00	214.00	. . . .
2 1/4	200.00	215.00	247.00	193.00	207.00	239.00	. . . .

Courtesy Armco Steel Corp.

chemicals, moisture, sunlight, rodents, and biological attack.

Rope should be stored in a dry place where air circulates freely about it. Air should not be extremely dry however. Small ropes can be hung up and larger ropes can be laid on gratings, so air can get underneath and around them.

Rope should not be stored or used in an atmosphere containing acid or acid fumes, as it will quickly deteriorate. Signs of deterioration from this cause are dark brown or black spots on the rope.

Rope should not be stored unless it has been cleaned. Dirty rope can be hung in loops over a bar or beam and, then, sprayed with water to remove the dirt. The spray should not be so powerful that it forces the dirt into the fibers. After washing, the rope should be allowed to dry and, then, be shaken

to remove the rest of the dirt.

Wire Rope

Wire rope is used widely instead of fiber rope because:

● It has greater strength for the same diameter and weight;

● Its strength is constant when wet or dry;

● It has constant length under varying weather conditions; and

● It has greater durability.

Construction

Wire rope is composed of wires, strands, and core. The wires are drawn to a predetermined size and laid together in various

(*Text continues on page 720.*)

WIRE ROPE WEAR AND DAMAGE

The evidence in these illustrations will aid the inspector in determining the actual cause of wear or damage that he may find in any wire rope.

A wire rope which has been kinked. A kink is caused by pulling down a loop in a slack line during improper handling, installation, or operation. Note the distortion of the strands and individual wires. Early rope failure will undoubtedly occur at this point.

Localized wear over an equalizing sheave. The danger of this type wear is that it is not visible during operation of the rope. This emphasizes the need of regular inspection of this portion of an operating rope.

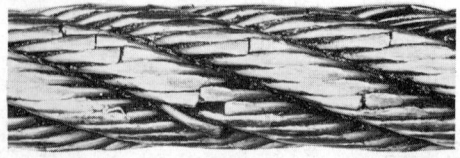

A typical failure of a rotary drill line with a poor cut-off practice. These wires have been subjected to excessive peening causing fatigue type failures. A predetermined, regularly scheduled, cut-off practice will go far toward eliminating this type of break.

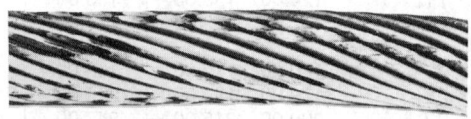

A single strand removed from a wire rope subjected to "strand nicking". This condition is the result of adjacent strands rubbing against one another and is usually caused by core failure due to continued operation of a rope under high tensile load. The ultimate result will be individual wire breaks in the valleys of the strands.

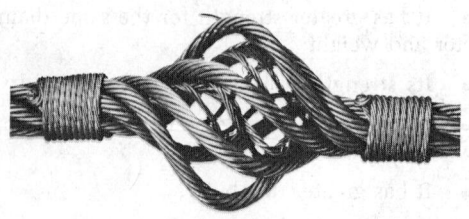

A "bird cage". Caused by sudden release of tension and resultant rebound of rope from overloaded condition. These strands and wires will not return to their original positions.

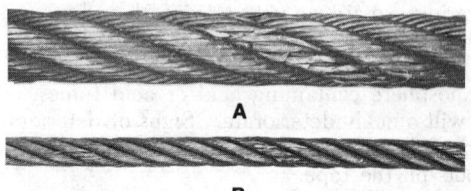

An example of a wire rope with a high strand—a condition in which one or two strands are worn before adjoining strands. This is caused by improper socketing or seizing, kinks or dog legs. Picture A is a close-up of the concentration of wear and B shows how it recurs in every sixth strand (in a six strand rope).

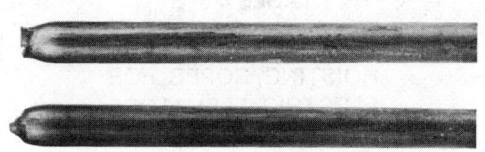

An illustration of a wire which has broken under tensile load in excess of its strength. It is typically recognized by the "cup and cone" appearance at the point of fracture. The necking down of the wire at the point of failure to form the cup and cone indicates that failure occurred while the wire retained its ductility.

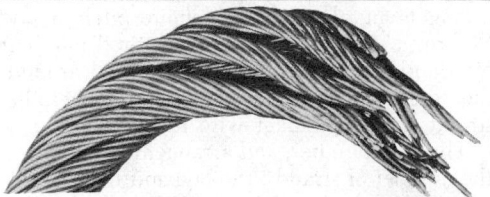

A wire rope which has jumped a sheave. The rope itself is deformed into a "curl" as if bent around a round shaft. Close examination of the wires show two types of breaks—normal tensile "cup and cone" breaks and shear breaks which give the appearance of having been cut on an angle with a cold chisel.

A wire rope which has been subjected to repeated bending over sheaves under normal loads. This results in "fatigue" breaks in individual wires—these breaks are square and usually in the crown of the strands.

An illustration of a wire which shows a fatigue break. It is recognized by the squared off ends perpendicular to the wire. This break was produced by a torsion machine, which is used to measure the ductility. This break is similar to wire failures in the field caused by excessive bending.

An example of "fatigue" failure of a wire rope which has been subjected to heavy loads over small sheaves. The usual crown breaks are accompanied by breaks in the valleys of the strands—these breaks are caused by "strand nicking" resulting from the heavy loads.

An example of a wire rope that has provided maximum service and is ready for replacement.

A close-up of a rope subjected to drum crushing. Note the distortion of the individual wires and displacement from their normal position. This is usually caused by the rope scrubbing on itself.

A fatigue break in a cable tool drill line caused by a tight kink developed in the rope during operation.

FIG. 26–3.—Typical characteristics and causes of broken wires in wire ropes.

Courtesy Wire Rope Corp. of America, Inc.

arrangements that have a definite pitch or lay to form a strand. Then, the required number of strands are helically laid or formed around the core, which may be sisal rope, a metallic strand, or independent wire rope core.

The size, number, and arrangement of wires, the number of strands, the lay, and the type of core in a rope are determined largely by the service for which the rope is to be used.

In general, the greater the number of wires in a strand and the greater the number of strands, the more flexible the rope. Hoisting ropes require flexibility and are usually made up of six strands of 19 (16 to 26 wires per strand) and six strands of 37 (27 to 49 wires per strand) wire rope classifications (Fig. 26–2).

Depending upon the service conditions, six-strand ropes may have a fiber core (FC), a wire strand core (WSC), or an independent wire rope core (IWRC) and may be either regular lay (wires in the strands are laid in the opposite direction from that of the strands in the rope) or lang lay (wires in the strands are laid in the same direction as that of the strands in the rope).

Eight-strand hoisting ropes can also be found. They are usually 19-wire rope classification with regular lay and fiber core.

Flexibility is not a requirement for guy wires, highway guards, and similar services. Therefore, wire rope of six by seven construction (six strands with seven or eight wires per strand) is suitable. When rope for a particular service is to be selected, it is recommended that engineers of reliable wire rope manufacturers be consulted for the most suitable type.

Some service conditions require rope with special qualities. Fiber cores are affected by temperatures above 212 F. Under such conditions, a metallic core provides greater efficiency and safety. A zinc-coated or stainless steel wire rope effectively resists some types of corrosion. Specific corrosion problems should be referred to a wire rope manufacturer.

Since preformed wire rope does not unravel, it has advantages for certain services, such as for slings and on construction and similar heavy equipment. Preformed rope is less likely to set or kink, and broken wires do not protrude to create a hazard to the

TABLE 26–E

SAFETY FACTORS FOR WIRE HOISTING ROPES FOR ELECTRIC ELEVATORS

Rope Speed (fpm)	Minimum Factor of Safety	
	Passenger	Freight
50	7.60	6.65
250	8.90	7.90
700	11.00	9.80
1100	11.70	10.40
1500	11.90	10.55

Excerpted from Table 212.3 of American National Standard *Safety Code for Elevators, Dumbwaiters, Escalators, and Moving Walks,* A17.1–1971.

hands of workmen. However, closer inspection is necessary to detect broken wires.

It is recommended that hoisting rope have at least the strength of the "improved plow steel" grade. For many applications the "extra improved plow steel" grade, which is the greatest in strength, should be used to provide an adequate factor of safety and better service. It resists abrasion, shock, vibration, and fatigue (Table 26–D).

Safety factor for hoisting ropes

The safety factor for hoisting ropes is calculated by dividing the breaking strength of the rope (rated by the manufacturer or determined by testing) by the sum of the maximum weights to be hoisted (static load).

In the case of mine hoisting rope, the maximum weights to be hoisted include the weight of the skip or car and cage, plus the weight

TABLE 26–F

TREAD DIAMETERS OF SHEAVES AND DRUMS FOR WIRE ROPE

Rope Classification	Average Recommended (Times rope diameter)	Minimum
6 × 7	72	42
6 × 19	45	30
6 × 37	27	18
8 × 19	31	21

of the material, plus the weight of the rope when it is extended to the bottom of the shaft.

The minimum safety factor for wire hoisting rope should not be less than that shown in Table 26-E.

Recommended minimum safety factors for other types of services are 6 for haulage ropes and overhead cranes and derricks, 7 for small electric and air hoists, and 8 for hot-ladle cranes.

The safety factors for other types of elevators and other rope requirements are specified in American National Standard A17.1, *Safety Code for Elevators, Dumbwaiters, Escalators, and Moving Walks.*

The factor of safety for wire rope should be determined by consideration of all pertinent data, which includes the type of load, acceleration, deceleration, rope speed, rope attachments, the number, size, and arrangement of sheaves and drums, and possible exposure to moisture and corrosives. If difficulty is encountered in determining the factor of safety, the advice of a reliable wire rope manufacturer should be obtained.

Causes of deterioration

Deterioration of wire ropes is due largely to the following factors, which vary considerably in importance, depending on the conditions of service (Fig. 26–3). For example, corrosion often is the principal cause of deterioration of mine hoisting rope in wet mine shafts because of moisture and the presence of acid in the water. Among other factors are:

● WEAR, particularly on the crown or outside wires, from contact with sheaves and drums.

● CORROSION, particularly of the interior wires, indicated by pitting. This condition is difficult to detect and highly dangerous. Wear is accelerated by corrosion.

● KINKS, acquired in improper installation of a new rope, hoisting with slack in the rope, and so forth. A kink cannot be removed without creating a weak place.

● FATIGUE, indicated by a square break of a wire—a break showing granular structure—and particularly due to excessive bending

stresses from sheaves and drums with small radii, whipping, vibration, pounding, and torsional stresses.

● DRYING OUT of lubrication, often hastened by heat and operating pressure.

● OVERLOADING, including dynamic overloading, if acceleration and deceleration are factors of importance.

● OVERWINDING, which is drum crushing caused by uncontrolled multiple wrapping.

● MECHANICAL ABUSE, such as pinching down and cutting wires or dragging ropes.

The safety and efficiency of hoisting rope installations can be greatly increased by the use of sheaves and drums of suitable size and design, proper lubrication, and good maintenance of the rope and the hoisting equipment.

Sheaves and drums

Fatigue resulting from bending stresses is greatly lessened when drum and sheave diameters are as large as practicable, such as 80 to 120 rope diameters for some mine hoisting service. However, such ratios are impractical on cranes, construction, and other types of equipment. In any case, sheave and drum diameters should not be less than the minimum sizes shown in Table 26-F.

It is essential that head, idler, knuckle, curved sheaves, and grooved drums have grooves which support the rope properly. Before a new rope is installed, the grooves should be inspected and, where necessary, machined to proper contour and groove diameter, which should exceed the nominal rope diameter by the amount shown in Table 26-G.

The recommended grooving for wire rope drums is as follows:

1. On drums designed for multiple-layer winding, the distance between groove centerlines should equal the nominal diameter of the rope plus the oversize tolerance, suggested for new and remachined grooves, as shown in Table 26-G.

2. The radius of curvature of the groove profile should equal one-half the nominal diameter of the rope plus the tolerance,

TABLE 26-G

GROOVE DIAMETER TOLERANCE IN RELATION TO WIRE ROPE DIAMETER

Rope Size (in.)	Amount that groove diameter should be larger than nominal rope diameter (in.)	
	Used	*New*
¼ and ⁵⁄₁₆	¹⁄₆₄	¹⁄₃₂
³⁄₈ to ¾ incl.	¹⁄₃₂	¹⁄₁₆
¹³⁄₁₆ to 1⅛ incl.	³⁄₆₄	³⁄₃₂
1³⁄₁₆ to 1½ incl.	¹⁄₁₆	⅛
1⁹⁄₁₆ to 2¼ incl.	³⁄₃₂	³⁄₁₆
2⁵⁄₁₆ and larger	⅛	¼

Source: A.P.I. Spec. RP-9B, March 1966, Use and Care of Wire Rope.

shown in Table 26-G, for new and re-machined grooves.

3. The depth of the groove should be approximately 20 percent of the nominal diameter of the wire rope. The crests between the grooves should be rounded off to the circular contour, which will provide the recommended groove depth.

So far as the life of wire rope is concerned, the condition and contour of sheave grooves are important. Sheave grooves should be checked periodically and should not be allowed to wear to a smaller diameter than those shown for used grooves in Table 26-G. If they become worn more than this, a reduction in rope life can be expected. Reconditioned sheave grooves should conform to the tolerance, shown in Table 26-G, for new (or remachined) grooves.

On all new sheaves, the grooves should be made for the size of rope specified. The bottom of the groove should have a 150-degree arc of support, and the sides of the groove should be tangent to the ends of the bottom arc. The total depth of the groove should be 1¾ times the nominal diameter of the rope, and the radius of the arc should be one-half the nominal rope diameter plus the tolerance, shown in Table 26-G, for new (or remachined) grooves.

Multiple layer winding on drums should be avoided, if possible, because the rope will mash and jam and its life will be shortened materially, particularly at the point where the rope rises to the next layer. Where practicable, the drum should be of such diameter and length as to take all the rope in a single layer.

Crushing and excessive wear of wire rope are minimized by use of helically grooved drums with capacity for one layer of rope. In any case, the number of layers should be limited to three. Rope lifters are recommended when two or more layers are wound on drums. To distribute wear at crossover points uniformly, 1¼ wraps can be cut off every six months or three or four times during the life of the rope. In no case should there be fewer than three full wraps on a drum.

In general, wire rope bending—first in one direction and then in the opposite—over sheaves or drums wears out rope faster than any other known cause and should be avoided.

It is desirable that the fleet angle* not exceed 1 degree, 30 minutes, to reduce any tendency for the rope to open-wind. Also, a minimum limit of 0 degree, 30 minutes, to assure the rope's starting back on the next layer, has been found desirable. Adherence to these specifications will help ensure uniform winding on smooth-faced drums and will

---

* The fleet angle is the included angle between the rope winding on the drum and a line perpendicular to the drum shaft and running through the head or lead sheave.

also increase the efficiency of grooved drums. For smooth-faced drums, proper direction of lay of rope for specified winding conditions further helps ensure uniform winding.

The head sheave groove should be checked for poor alignment, worn bearings, broken flanges, and size and shape of the groove at intervals of two or three months, depending on the amount of usage. Proper groove size and shape are especially necessary when a new rope is installed.

Excessive sheave wear and its bad effects on the life of a rope often result from use of sheave or liner material unsuitable for the pressure of the rope. If the radial pressure of a 6 x 19 rope exceeds 900 psi, a harder material than cast steel, such as manganese steel, is more suitable.

Lubrication

Regular application of a suitable lubricant to wire hoisting rope prevents corrosion, wear from friction, and drying out of the core. Thin lubricants, which are fluid at normal temperatures and are fortified with polar additives and rust inhibitors, can penetrate the rope and also afford good protection even under wet operating conditions. Ropes should be dry, when lubricant is applied, so that moisture will not be entrapped by the lubricant. Thin lubricants can be applied by hand, but the best arrangement is to provide some means of dripping them on the rope, or using a spray device to apply the proper quantity automatically.

Periodic monthly cleaning, done in mine shafts, eliminates dirt, abrasive particles, and corrosion-producing moisture. Cleaning fluids should not be used because of their detrimental effect on the core lubricant. Light oils are sometimes used to loosen the coating of lubricant and harmful materials. A compressed air or steam jet (or other mechanical method) cleans a rope effectively and thoroughly.

Overloading

Overloading may occur in the operation of most types of hoisting equipment. For example, it is difficult to estimate the load within 25 percent when making a lift with a crane, and in many instances a load is under-

TABLE 26–H

NUMBER AND SPACING OF CLIPS FOR ROPES OF VARIOUS SIZES

Diameter of rope (in.)	Number of clips	Center-to-center space between clips (in.)	Length of rope turned back exclusive of eye (in.)	Length of wrench (in.)
1/2	3	3	9	12
5/8	3	3 3/4	12	12
3/4	4	4 1/2	18	18
7/8	4	5 1/4	21	18
1	4	6	24	24
1 1/8	5	6 3/4	34	24
1 1/4	5	7 1/2	38	24
1 3/8	6	8 1/4	50	24
1 1/2	6	9	54	24
1 5/8	6	9 3/4	60	30
1 3/4	7	10 1/2	74	30
1 7/8	8	11 1/4	90	30
2	8	12	96	30
2 1/8	8	13	104	30
2 1/4	8	14	112	30

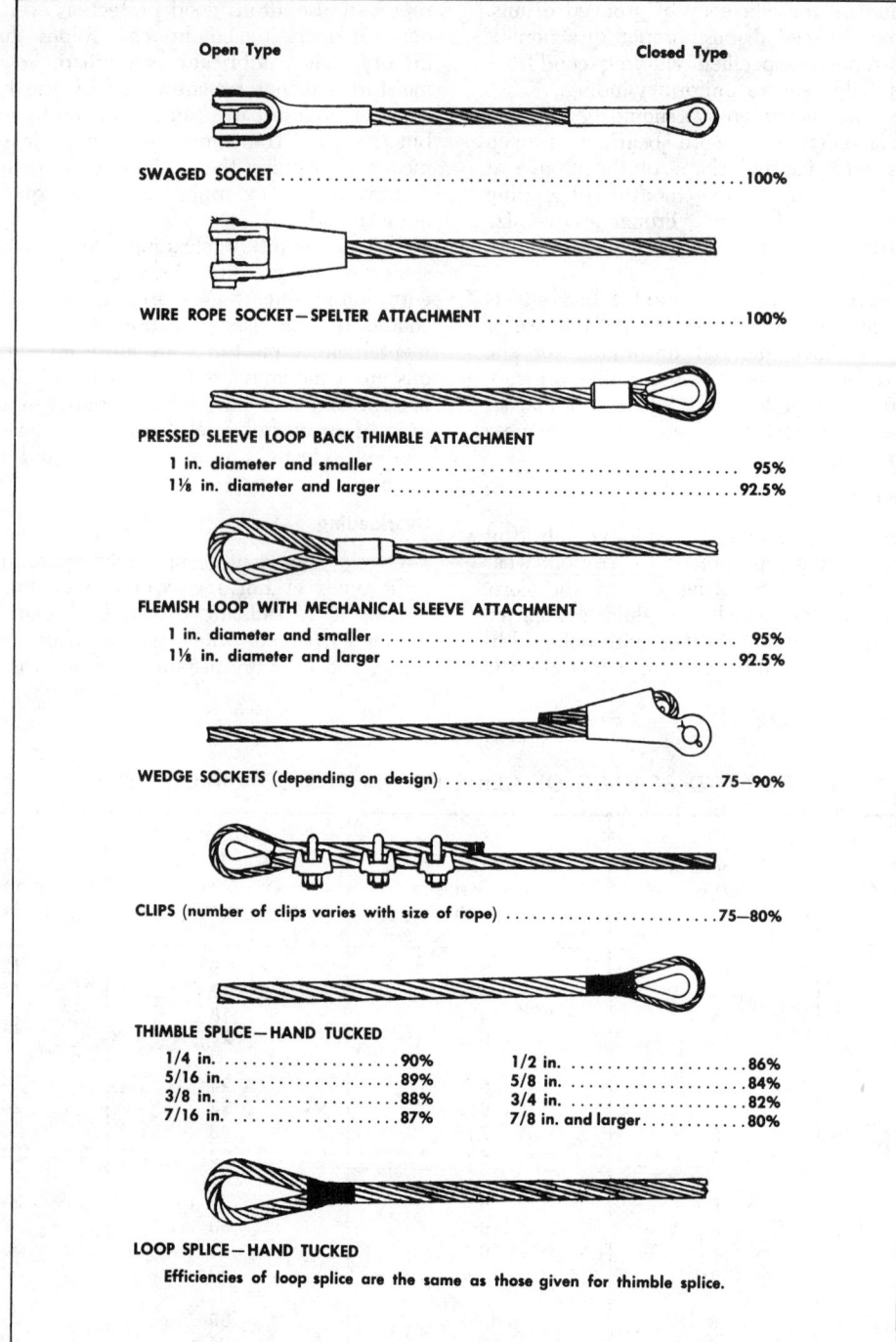

FIG. 26–4.—Efficiencies of attaching wire rope to fittings in percentages of strength of rope.

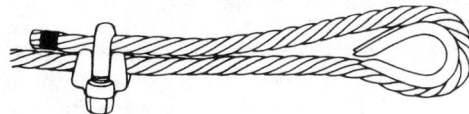

APPLY FIRST CLIP—one base width from dead end of wire rope—U-Bolt over dead end—live end rests in clip saddle. Tighten nuts evenly to recommended torque.

STEP 2.

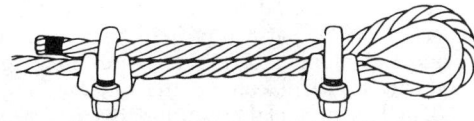

APPLY SECOND CLIP—nearest loop as possible— U-Bolt over dead end—turn on nuts firm but DO NOT TIGHTEN.

STEP 3.

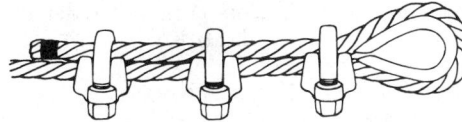

ALL OTHER CLIPS—Space equally between first two—this should be no more than one clip base apart—turn on nuts—take up rope slack—TIGHTEN ALL NUTS EVENLY ON ALL CLIPS to recommended torque.

FIG. 26–5.—Method of clip installation.

estimated by two or three times its weight.

Overwinding likewise may stress a rope excessively and, if a conveyance is caught by obstructions, excessive elongation of the rope may result.

Rope fittings

Methods of attaching wire rope to fittings

are important for safety because they develop from 75 to 100 percent of the breaking strength of the rope. Manufacturers also specify fittings of suitable size and design for ropes of different sizes. The strength of an attachment is attained only when the connection is made exactly according to the manufacturer's instructions (Fig. 26–4).

Some types of attachments, such as pressed fittings or mechanical sleeve splices which are used in making slings, are usually made at the wire rope manufacturer's plant.

A hand-tucked splice develops approximately 90 percent of the breaking strength of the rope for ropes ¼ in. in diameter, and 80 percent of the breaking strength for ropes ⅞ in. in diameter.

The various kinds of pressed fittings and mechanical sleeve splices will develop 92.5 to 100 percent of the breaking strength of the rope.

Rope often is connected to the fittings of conveyances by means of clips and clamps. They are rated to develop 75 to 80 percent of the breaking strength of the rope. Fig. 26–5 shows how the clips should be attached, and Table 26-H gives the number of clips and the spacing required for ropes of different sizes. It is important to retighten the nuts on all clips after the initial load-carrying use of

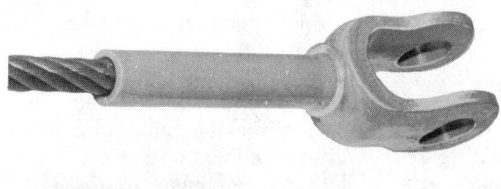

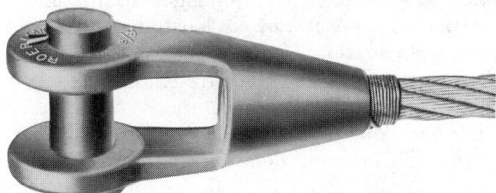

FIG. 26–6.—End fittings should be of best possible type for specific use. Zinc-poured sockets (*bottom*) are efficient in straight tension but are not as fatigue-resistant as swaged sockets (*above*).

Courtesy John A. Roebling's Sons Corp., Subsidiary of The Colorado Fuel and Iron Corp.

the rope, as well as at all subsequent regular inspection periods.

Socketing with zinc will develop 100 percent of the breaking strength of the rope. Since there is no ready way to detect flaws in the finished job, the recommended procedure must be followed exactly. In high-speed hoisting, fatigue is especially likely to develop with this type of attachment. For this reason, the section adjacent to the conveyance should be cut off and discarded at frequent intervals. The required interval in some state mining laws is every 6 months.

Fig. 26–6 shows zinc-poured and swaged sockets.

Square and other types of knots have low efficiency, 50 percent of the strength of the rope; their use is likely to result in failure of a rope assembly and, under certain conditions, in a serious accident.

Inspection and replacement

The frequency of inspections and replacement of a rope depends considerably on service conditions. OSHA requires wire rope or cable to be inspected when installed and weekly during use. In any case, at regular intervals, a specially trained inspector should examine the ropes on which human life depends. Some mines, for instance, make a daily inspection for readily observable defects, such as kinking and loose wires, and a thorough inspection weekly. For the latter inspection, the rope speed is generally less than 60 fpm.

The inspector specifically checks for wear of the crown wires, broken wires, kinking, high strands, loose wires, nicking, and lubrication (Fig. 26–3). Rope calipers and micrometers are used to determine changes in the cross section of rope at various locations. In some cases, elongation is an indication of fatigue. That is, except for initial stretch, elongation is generally the result of internal corrosion or deterioration of the core. It is a warning signal of internal deterioration and is generally accompanied by a noticeable reduction in rope diameter. Therefore, a rope showing excessive elongation or reduction in diameter should be removed from operation.

The number of broken wires per lay is one of the principal bases for judging the condition of a rope. If most of the broken wires in a lay are concentrated in several strands, that section of the rope is weaker than it would be if the broken wires were uniformly distributed throughout all strands and along the length of the rope. If, however, the number of broken wires along the length of a rope increases rapidly between inspections, the rope is becoming fatigued and nearing the end of its useful life.

Governmental regulations covering the number of broken wires permitted in a given section of mine hoist rope vary considerably. OSHA specifies that hoisting ropes have no more than three broken wires in one strand and no more than six randomly distributed broken wires in one lay. A Canadian mining province specifies no more than six broken wires in one strand per lay. A mining state once specified that a rope must be discarded when the number of broken wires in a foot amounts to 10 percent of the total.

The loss of metal, resulting from wear of the crown wires, is generally limited to 35 to 40 percent. However, the number of broken wires and the wear in a given section have major significance only when there is no corrosion.

Experience and judgment of all factors, combined with the length of time in service and the tonnage hoisted or other unit for judging the work done by the rope, determine when it should be discarded.

Rope Slings

The safety of a sling assembly for special or ordinary uses is determined particularly by these factors: use of rope or chain and fittings of suitable strength for the load; the method of fastening the rope or chain to the fittings; the type of sling, such as single or three-leg; the kind of hitch; and regular inspection and maintenance.

Materials

Fiber rope slings should be made from first grade manila rope. Since the strength of fiber rope is affected by chemicals, freezing, high temperatures, and sharp bends, these factors should be considered when such rope is used for slings. Fiber rope is particularly suitable for the handling of loads that might be damaged by contact with metal slings.

FIG. 26-7.—Braided slings are resistant to kinking. Be sure loads are hoisted uniformly and that all slings have a minimum safety factor of five.

Manila rope also has the advantage of flexibility.

Wire rope slings are preferably made from improved plow steel and from preformed wire rope with an independent wire rope core. Wire rope will withstand temperatures up to 400 F for a limited time, but corrodes when exposed to acids and other chemicals. Since flexibility is essential, the 6 by 19 classification is used when the service requires a rope 1⅛ in. or smaller. The 6 by 37 classification is used for slings made from larger rope. Braided rope is available for service where flexibility combined with great strength is required, as in railroad shops.

Braided slings (Fig. 26-7) are popular in numerous industries as a safe, economical, and versatile type of sling. Because of the nature of its construction, the braided sling is flexible and resistant to kinking. These characteristics make it easy to handle, especially in the larger sizes.

The construction of the braided sling permits easy and thorough inspection of its parts, so that its condition can be readily known with greater assurance of safety. Another feature of these slings is that they have little tendency to twist when free-end loads are being handled.

Methods of attachment

The hook and ring method of attaching wire rope fitting develops from 70 to 100 percent of the breaking strength of the rope. Socketing with zinc and the use of compression fittings are compact methods which develop 100 percent of the strength of the rope when properly used. Swaged sleeve attachments develop about 90 to 95 percent of the efficiency of the rope. Compression and swaged sleeve fittings are usually made only by the manufacturer.

727

TABLE 26-I

SAFE LOAD LIMITS (IN TONS) OF WIRE ROPE SLINGS, USING PREFORMED IMPROVED PLOW STEEL IWRC ROPE
(Depending on method of attaching the rope to the fittings)

Rope Diameter (in.)	Single Leg						Two-Leg Bridle or Basket Hitch											
	Vertical			Choker			Vertical°			60 degrees			45 degrees			30 degrees		
	A	B	C	A	B	C	A	B	C	A	B	C	A	B	C	A	B	C
6 × 19 Classification Construction																		
3/8	1.3	1.2	1.1	0.98	0.93	0.86	2.6	2.5	2.3	2.3	2.1	2.0	1.8	1.8	1.6	1.3	1.2	1.1
1/2	2.3	2.2	2.0	1.70	1.60	1.50	4.6	4.4	3.9	4.0	3.8	3.4	3.2	3.1	2.8	2.3	2.2	2.0
5/8	3.6	3.4	3.0	2.70	2.50	2.20	7.2	6.8	6.0	6.2	5.9	5.2	5.1	4.8	4.2	3.6	3.4	3.0
3/4	5.1	4.9	4.2	3.80	3.60	3.10	10.0	9.7	8.4	8.9	8.4	7.3	7.2	6.9	5.9	5.1	4.9	4.2
7/8	6.9	6.6	5.5	5.20	4.90	4.10	14.0	13.0	11.0	12.0	11.0	9.6	9.8	9.3	7.8	6.9	6.6	5.5
1	9.0	8.5	7.2	6.70	6.40	5.40	18.0	17.0	14.0	15.0	15.0	12.0	13.0	12.0	10.0	9.0	8.5	7.2
1 1/8	11.0	10.0	9.0	8.50	7.80	6.80	23.0	21.0	18.0	19.0	18.0	16.0	16.0	15.0	13.0	11.0	10.0	9.0
6 × 37 Classification Construction																		
1 1/4	13.0	12.0	10.0	9.9	9.2	7.9	26.0	24.0	21.0	23.0	21.0	18.0	19.0	17.0	15.0	13.0	12.0	10.0
1 3/8	16.0	15.0	13.0	12.0	11.0	9.6	32.0	29.0	25.0	28.0	25.0	22.0	22.0	21.0	18.0	16.0	15.0	13.0
1 1/2	19.0	17.0	15.0	14.0	13.0	11.0	38.0	35.0	30.0	33.0	30.0	26.0	27.0	25.0	21.0	19.0	17.0	15.0
1 3/4	26.0	24.0	20.0	19.0	18.0	15.0	51.0	47.0	41.0	44.0	41.0	35.0	36.0	33.0	29.0	26.0	24.0	20.0
2	33.0	30.0	26.0	25.0	23.0	20.0	66.0	61.0	53.0	57.0	53.0	46.0	47.0	43.0	37.0	33.0	30.0	26.0
2 1/4	41.0	38.0	33.0	31.0	29.0	25.0	83.0	76.0	66.0	72.0	66.0	57.0	58.0	54.0	47.0	41.0	38.0	33.0

Courtesy Armco Steel Corp.

A = **Socket or swaged terminal attachment**
B = **Mechanical sleeve attachment**
C = **Hand-tucked splice attachment**

*If slings are used to handle loads with sharp corners, pads or saddles should be used to protect the rope. The radius of bend should not be smaller than five times the diameter of the rope. If the radius of bend is smaller, a choker hitch rating should be used.

Note 1. Table is based on: a safety factor of 5, sling angles formed by one leg and a horizontal line through the crane hook, and uniform loading.

Note 2. For 3-leg bridle slings, multiply safe load limits for 2-leg bridle slings by 1.5; and for 4-leg bridle slings, multiply by 2.0.

Note 3. For fiber core slings having Type A or Type C attachments, multiply the above values by 0.93; for fiber core slings with Type B attachments, multiply the above values by 0.91.

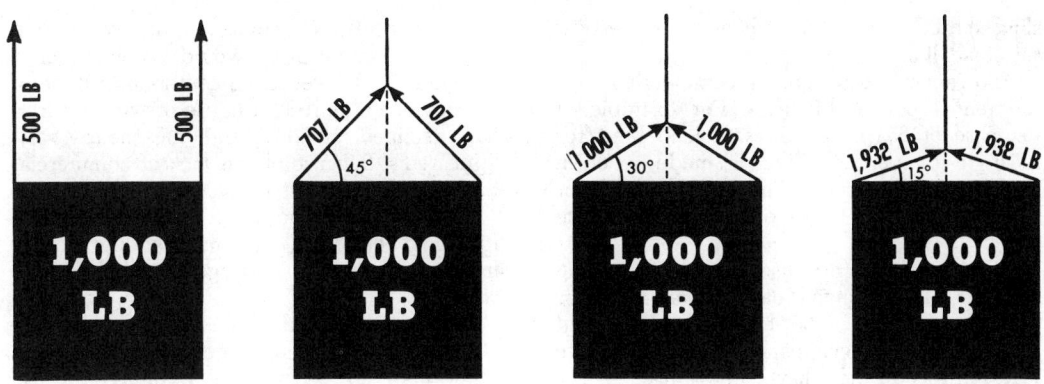

FIG. 26–8.—Decreasing the angle between the sling leg and horizontal increases the stress on each leg of the sling, even though the load remains constant. Leg stress varies as the cosecant of the included angle.

Hand-tucked splices develop about 90 percent of the strength of rope less than ½ in. in diameter. This efficiency decreases to 80 percent for ropes ⅞ in. and larger in diameter. Fittings should be specified according to rope sizes. The recommended load for a sling assembly is not more than one-fifth the strength of the assembly, taking into account the efficiency of the rope connections.

Safe load limits

The effects of unforeseen jerks, of acceleration or deceleration that multiplies load weights, of aging of the wire rope, and of abrasion, nicking, and other damaging contacts are among the matters that must be taken into account when determining safety factors for slings.

Because slings can be used at various angles and because strain builds up rapidly on angles greater than 60 degrees, it is essential that all slings be ordered with a minimum safety factor of 5, after calculating the load and strains at the anticipated lift angle to be used, to prevent injury to personnel and to prevent damage to loads and hoisting equipment.

For special applications where danger to personnel or loads is high, a safety factor of at least 8 should be specified. For use in circumstances involving heat, wire rope slings should always be made of rope having a wire rope core.

The human element is the most important reason for maintaining substantial safety factors in slings. Too many workmen assume that, if a sling is long enough to go around a load, the sling, regardless of its actual strength, can lift the load. Such thinking frequently results in accidents and loss of life.

A suitable load for a two-leg sling vertical is twice the load for a single-leg sling vertical; the three-leg sling vertical will have a capacity equal to 1½ times the capacity of a two-leg sling vertical; and the capacity of a four-leg sling vertical is twice the capacity of a two-leg sling vertical.

When the rope is made into a sling and placed in position on a load, the angle formed by the ropes and the horizontal should always exceed 45 degrees to avoid excessive stresses. The safe load limit (working load limit) of the rope decreases sharply as the angle formed by a leg and the horizontal becomes smaller. When this angle is 45 degrees, the safe load limit has decreased to 71 percent of the load that can be lifted when the legs are vertical (Table 26-I).

Fig. 26–8 shows how the tension on a leg of a sling increases as the angle decreases from the horizontal. When the angle formed by a leg and the horizontal is 60 degrees, the safe load limit is only 87 percent of that for both legs vertical. For an angle of 30 degrees, the safe load limit is only 50 percent of that for both legs vertical. Excessive angles can be avoided by the use of longer slings, if head room permits.

Tables showing safe loads should be posted conspicuously around the shop. Also, each sling should bear a tag indicating its vertical safe load limit.

Allowances also are recommended for different types of hitches. For example, a decrease of 25 percent in safe load limit for a single-leg vertical sling is made when a choker hitch is used. The suitable load for a basket hitch is based on the angle of the legs.

Special clamps are used in the handling of steel plate, flanged castings, and similar shaped products. The horizontal type and the vertical type clamps have the same safe load limit as do the other bridle slings.

If loads that have sharp edges or sharp corners must be lifted, pads or saddles should be used to protect the ropes or chains.

For wire rope, the radius of bend should not be smaller than ten times the diameter of the rope. For braided, cable-laid, and multi-part slings, the radius should not be smaller than twenty times the diameter of the rope. If the radius of bend is smaller, a choker hitch rating should be used.

Thimbles spliced in the ends of slings will materially reduce wear.

Inspection

Employees should be trained to check slings daily and whenever they may suspect damage after a lift, to report promptly any questionable conditions of the equipment or assembly. A thorough inspection by a trained man should be made at least every 6 months.

Slings that do not meet standard requirements should be promptly withdrawn from service and either discarded or reconditioned in accordance with the recommendations of the manufacturer.

Chains and Chain Slings

Materials

Alloy steel chain has become the standard material for slings because it is at least twice as strong, size for size, as wrought iron chain. Alloy steel chain has high resistance to abrasion and is practically immune to failure from the cold-working of the metal.

Wrought iron chain (commonly called iron crane or dredge chain) used to be the accepted material for slings, hoists, cranes,

power shovels, and marine purposes where human life or property would be endangered by failure. However, since the introduction of alloy chain in 1933, the use of wrought iron has declined steadily and has nearly disappeared. Because of this lack of commercial importance, it is not considered here.

Heat-treated carbon steel is sometimes used for sling chains. It has high strength, high impact resistance, and good abrasion resistance.

Special-purpose alloy chains are made from stainless steel, monel metal, bronze, and other materials. They are designed for use where resistance to corrosive substances is required, or where other special properties are desirable.

Proof coil chain (also known as common or hardware chain) is used for miscellaneous purposes where failure of the chain would not endanger human life or result in serious damage to property or equipment. Proof coil chain should never be used for slings.

Properties of alloy chain

Alloy chain is produced from heat-treatable alloy steel in conformance with ASTM specifications A391–65. Types and amounts of the alloying elements may vary according to requirements of the individual chain manufacturer. Chain after heat treatment has typical mechanical properties as follows:

Tensile Strength	125,000 psi minimum
Elongation	15 percent minimum

Table 26-J shows the recommended working load limits, proof test loads, and the minimum breaking strengths of alloy steel chain.

The working load limit (safe load strength) is arrived at by dividing the breaking strength (ultimate strength) by a specified safety factor.

The values, shown in Table 26-J, represent the maximum loads that should ever be applied in direct tension to a length of alloy chain. All alloy chain is tested in direct tension under the proof test loads shown, prior to final inspection and shipment. The data for other types of chain may be obtained from the National Association of Chain Manufacturers' Specification No. 3001.

Stress on each leg of a multi-branch sling is increased appreciably when the angle between the chain and the horizontal is decreased. The effects of angle of loading on the working load limit of a two-leg alloy sling are shown in Fig. 26–9.

The tensile strength of alloy steel chain increases in proportion to its hardness (produced by heat treating). Resistance to abrasion also increases proportionately with hardness.

Impact conditions caused by faulty hitches, bumpy crane tracks, and slipping hookups can materially add to the stress in the chain. If severe impact loading may be encountered, a lower working load limit should be used, regardless of the chain type.

Alloy steel chains are suitable for high temperature operations. However, continuous operation at a temperature of 800 F (the highest temperature for which continuous operation is recommended) requires a reduction of 30 percent in the regular working load limit. These chains may be used at temperatures up to 1000 F at 50 percent of the regular working load limit, but only for intermittent service.

TABLE 26–J

WORKING LOAD LIMITS, PROOF TEST LOADS AND MINIMUM BREAKING LOADS FOR ALLOY STEEL CHAIN

Nominal Size of chain (in.)	Working Load limit (lb)	Proof Test (lb)	Minimum Break (lb)
1/4	3,250	6,500	10,000
3/8	6,600	13,200	19,000
1/2	11,250	22,500	32,500
5/8	16,500	33,000	50,000
3/4	23,000	46,000	69,500
7/8	28,750	57,500	93,500
1	38,750	77,500	122,000
1 1/8	44,500	89,000	143,000
1 1/4	57,500	115,000	180,000
1 3/8	67,000	134,000	207,000
1 1/2	80,000	160,000	244,000
1 3/4	100,000	200,000	325,000

SOURCE: *Specification for Alloy Chain*, American Society for Testing and Materials, A-391-65. *Alloy Steel Chain Specifications*, No. 3001, National Association of Chain Manufacturers.

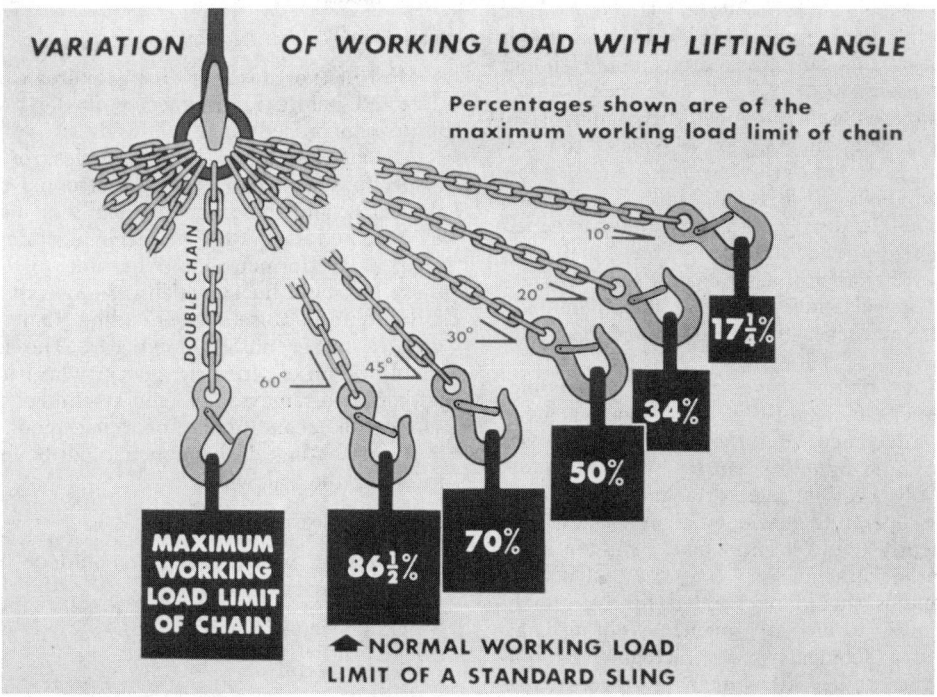

Fig. 26–9.—Diagram shows how the angle of the sling affects the working load limit.

The general strength and working load limits of alloy steel chain are not altered appreciably by low atmospheric temperatures.

Chain slings should preferably be purchased complete from the manufacturer and, whenever repairs are required, should be sent back to him.

Hooks and attachments

As a general rule, hooks, rings, oblong links, pear-shaped links, coupling links, and other attachments should be made of the same, or equivalent, heat-treatable alloy steel as the chain itself. In most cases, attachments will be installed on the chain by the chain manufacturer, who will then heat-treat and proof-test the assembly.

If emergency conditions make it necessary for the user to replace an attachment, he should select the grade and size with extreme care. High-strength, heat-treatable alloy connecting links of the same analysis type as that used by the chain manufacturer should be employed.

Unalloyed carbon-steel hooks, repair links, rings, pear-shaped links, and other attachments should not be used. Homemade or makeshift bolts, rods, shackles, hooks without safety catches, or other attachments should never be used.

Standard items produced from alloy steel include sling hooks, grab hooks, foundry hooks, grab links, rings, oblong links, pear-shaped links, and repair links. All such attachments used with the recommended chain size provide a safety factor equal to or greater than that of alloy chain itself. Dimensional specifications for these attachments will vary somewhat with the individual manufacturer.

Many injuries have resulted when employees have caught their fingers between the hook attachment and the load. To prevent such injuries, handles can be attached to the assembly hook or end attachment. To increase operating efficiency, handles are also frequently used on large hooks, master links, and other attachments.

Handles should be welded to the attachment prior to heat treatment. Welding heat-treated attachments is not recommended unless the *entire* attachment is heat-treated again *after* welding. The weldments should

TABLE 26–K

MAXIMUM ALLOWABLE WEAR AT ANY POINT OF LINK

Chain size (in.)	Maximum allowable wear (in.)
1/4	3/64
3/8	5/64
1/2	7/64
5/8	9/64
3/4	5/32
7/8	11/64
1	3/16
1 1/8	7/32
1 1/4	1/4
1 3/8	9/32
1 1/2	5/16
1 3/4	11/32

be inspected critically for possible zones of high-stress concentration, caused by welding, that would lower the strength of the entire attachment.

Inspection

Most of the causes of chain failures can be detected before failure occurs if the proper inspection procedure is followed.

A good inspection plan provides for two inspections; one daily by the personnel using the chain, and the other biannually or oftener by a trained man from the mechanical department. The former should be able to detect those links and hooks which have become visibly unsafe because of overloading, faulty rigging, or other unsafe practices. The latter should be an experienced person who has the authority to remove damaged assemblies from service for reconditioning or replacement.

A link-by-link inspection should be made to detect the following:

1. Bent links.

2. Cracks in weld areas, in shoulders, or in any other section of link.

3. Tranverse nicks and gouges.

4. Corrosion pits.

5. Elongation caused by stretching can only

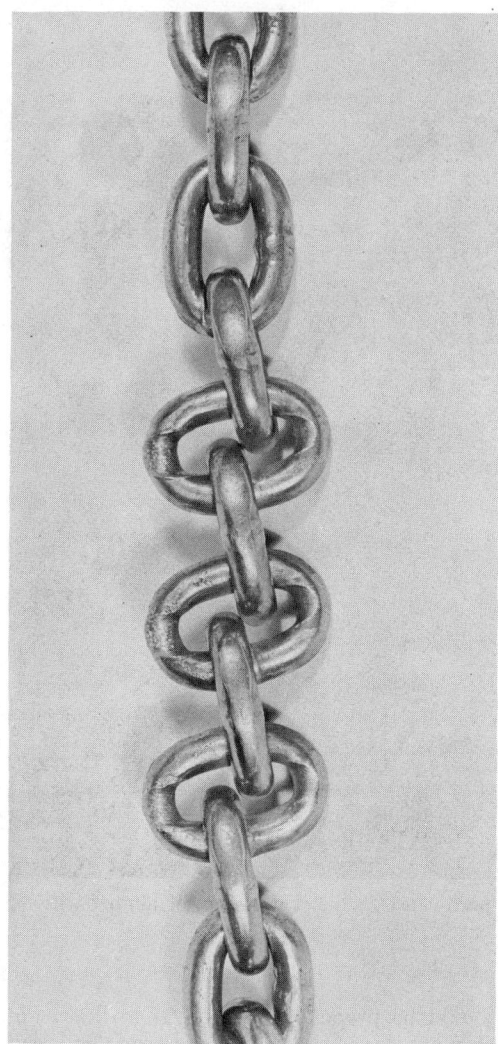

FIG. 26–10.—Extreme wear at bearing surfaces. Links are turned to detect wear.

be caused by overloading. The best way to detect such elongation and also link wear is by a visual link-by-link inspection. Overall sling length measurements and even measurements of 1- to 3-ft lengths are inadequate because not all links are affected uniformly. Likewise caliper readings only over several links can miss such a condition (Table 26-K and Fig. 26–10).

When inspecting the hook, measurement must be made between the shank and the narrowest point of the hook opening. (See Fig. 25–8 of the previous chapter.) Whenever the throat opening exceeds 15 percent of the normal opening, the hook should be replaced. Special attention should be given to slings to which hooks have been added, in order to make sure that the hooks are secure.

Unless full and adequate facilities for repair are available, chain showing faults by inspection should be returned to the manufacturer for reconditioning.

Alloy steel chains and heat-treated carbon steel sling chains and hooks should never be annealed or normalized because these processes reduce the hardness and, therefore, greatly reduce the strength.

Safe practices

In the use of chains, recognized safe practices will do much to prevent failures:

1. Never splice a chain by inserting a bolt between two links.

2. Never put a strain on a kinked chain. Workmen should be trained to take up the slack slowly and see that every link in the chain seats properly.

3. Do not use a hammer to force a hook over a chain link.

4. Permanent identification tags are usually attached to chain slings by the manufacturer. Tags should never be removed.

5. Remember that decreasing the angle between the legs of a chain sling and the horizontal increases the load in the legs.

6. Use chain attachments (rings, shackles, couplings, and end links) designed for use with the chain to which they are fastened.

7. See that the load is always properly set in the bowl of the hook. Loading on or toward the point (except in the case of grab hooks or others especially designed for the purpose) overloads the hook and leads to spreading and possible failure.

8. Chains not in use should be stored in a suitable rack. Do not let them lie on the ground or floor for extended periods.

Fig. 26–11.—A four-legged, basket-hitch sling of steel mesh can take the sharp edges of lumber without failure.

Metal and Nylon Mesh Slings

Two widely used types of slings today are metal mesh and nylon. They are strong and dependable types of slings, if used in the proper manner. However, even these two categories should be selected according to the needs of the particular operation.

Materials

Safety in the use of metal mesh slings is primarily determined by two factors: the correct sling for the load and the construction of the sling. With metal mesh you are certain of safe lifting within the limits stamped on the handles of the sling. All metal mesh slings should be properly identified as to their safe working load capacities and can be used efficiently with all weights that fall within that limit.

A great safety factor in metal mesh slings is that they are all pretested and are classified either heavy-duty, medium-duty, or light-duty. The safety factor of metal mesh is five-to-one or five times the amount stated on the sling.

With metal mesh, the handling of sharp-edged metals, concrete in its many prestressed forms, and high temperature materials up to 500 F is a safe operation (Fig. 26–11). Danger in the use of metal mesh stems mainly from improper use. Damaging the spirals at the edges by faulty loading or dragging a sling out from under a load may cause wear to the spiral, thus reducing the wire diameter and requiring the sling to be taken out of service.

Nylon has definite advantages for some material handling applications. It is light-weight and inexpensive. However, it also has limits in safe material handling operations.

Nylon slings are not pretested, but they are rated by the manufacturer as to the slings' lifting capability. This capability should al-ways be checked before use. Many nylon slings have their weight limits printed on their surface, and there is a tendency for the ink to fade with wear. A safety professional should make the renewal of this stamping a part of his continuing inspection of materials handling equipment.

Applications for which nylon mesh is safe will vary from plant to plant. Nylon can be dangerous when used near machinery or steel racks and in the lifting of sharp-edged metals. Yet for safely transporting polished metal it is widely accepted (Fig. 26–12).

Wear pads that can be attached to nylon can help alleviate the cutting problem at the lifting edge of the sling but, when taut, nylon can still be sliced easily. Nylon should never be used in temperatures above 180 F or in lifting materials that have been heated beyond that point.

Fig. 26–12.—Web sling protects roller's polished surface.

Courtesy The Wear-Flex Corp.

Nylon has a certain degree of permanent stretch, but through constant use, fibers will become more elongated and cause an even greater degree of elasticity. This stretch factor should be taken into consideration when using nylon, especially with heavy materials.

Safe practices

Specific safety guidelines, which should be followed in the use of either nylon or metal mesh, will result in safe materials handling operations.

Metal mesh or nylon should not be used before first determining the weight of the load. Either a safety professional or the materials handling department should also check that the sling is suitable for the environment in which it is being used.

With metal mesh slings, the length of the sling should be long enough to provide the maximum practical angle between the sling leg and the horizontal. This would also mean being sure there is a minimum practical angle at the crane hook, if vertical angles are used.

Metal mesh slings should never be shortened by riggers or crane operators with knots, bolts, or other unapproved methods. Consult with the sling manufacturer if shortening becomes necessary. Tampering with the surface of any sling can weaken it and make it highly dangerous.

Precautions should be taken never to twist or kink the legs of a metal mesh sling or use one when the spirals are locked. A sudden jolt could break the spirals, cause the load to shift, and create havoc on the floor below. One method *never* to be used to straighten a spiral or cross rod on a metal mesh sling is hammering. It causes immediate damage to the sling construction and eventual damage to its capability.

With metal mesh, as well as all other slings,

adhering to the rules for certain hitches also insures the safe handling of a load.

When using nylon, many of the same precautions that pertain to metal mesh should be considered. Avoid knotting, provide sufficient length, and balance the load. Of special importance are the heat limitations of nylon as already mentioned.

Environmental considerations also exist in the form of acidic or caustic materials. With metal mesh, the chemicals can be washed away or drained through the open face of the sling. Nylon and dacron are not as easily protected or kept clean, and deterioration can occur at a rapid pace. Check with the manufacturer before bringing your nylon sling into contact with corrosive chemicals.

Inspection

One of the most important precautions in the use of slings is regular inspection by a safety professional. Inspection frequency should be based on the amount of sling use and the severity of conditions. It would be wise to keep written inspection records, with the identification of each sling recorded as is provided on the sling by the manufacturer.

In the case of metal mesh slings, they should be removed from service, if a broken weld or brazed joint is discovered along the sling edge. Other breakdowns in construction are broken wires in any part of the mesh, a reduction of 25 percent in wire diameter due to abrasion, and a lack of sling flexibility. Any one of these conditions, or a combination of them, if ignored, could eventually result in sling breakdown.

With nylon slings, inspection should include checks for caustic or acid burns, for melting or charring on any part of the surface, and for snags, punctures, tears, and broken or worn stitches.

References

American Petroleum Institute, 1801 K St., N.W., Washington, D.C. 20006. *Recommended Practice on Application, Care, and Use of Wire Rope for Oil-Field Service*, Code No. API-RP-9B.

American Society for Testing and Materials, 1916 Race St., Philadelphia, Pa. 19103.
Specification for Alloy Steel Chain, A391-65.
Specification for Carbon Steel Chain, A413-65.

American National Standards Institute, 1430 Broadway, New York, N.Y. 10018.
"Safety Code for Cranes, Derricks, Hoists, Jacks, and Slings," B30 Series.
Safety Code for Elevators, Dumbwaiters, Escalators, and Moving Walks, A17.1.
Safety Color Code for Marking Physical Hazards, Z53.1.
Specifications for and Use of Wire Ropes for Mines, M11.1.

Cordage Institute, 2300 Calvert St., N.W., Washington, D.C. 20008.

Electric Controller and Manufacturing Company. 2704 East 79th St., Cleveland, Ohio 44104.
How to Operate a Crane, Booklet 920.

Hess, Owen. "Metal Mesh and Nylon Slings." *National Safety News*, 103:78–79, (June 1971).

National Association of Chain Manufacturers, 111 West Washington St., Chicago, Ill. 60602.
Alloy Steel Chain Specifications, No. 3001.

National Fire Protection Association, 470 Atlantic Ave., Boston, Mass. 02210.
National Electrical Code, Standard No. 70.
Flammable and Combustible Liquids Code, Standard No. 30.

National Safety Council, 425 North Michigan Ave., Chicago, Ill. 60611.
Industrial Data Sheets
Alloy Steel Chains for Overhead Lifting, 478.
Recommended Loads for Wire Rope Slings, 380.
Safe Use of Manila Rope Slings, 259.

U.S. Department of the Interior, Bureau of Mines, Washington, D.C. 20240.
Recommended Procedures for Mine Hoists and Shaft Installation, Inspection, and Maintenance, Information Circular 8031.

Powered
Industrial
Trucks

Chapter
27

Types of Trucks   739

Guards and Safety Devices   740
Lift trucks . . . Vacuum-handling trucks . . . Straddle carriers
. . . Crane trucks . . . Tractors and trailers . . . Motorized
handtrucks . . . Remotely controlled trucks

Safe Operations   744
Lift trucks . . . Vacuum-handling trucks . . . Straddle carriers
. . . Motorized handtrucks . . . Remotely controlled trucks

Pallets   752

Operators   752
Selection . . . Training . . . Five-day training program

Inspection and Maintenance   757
Electric trucks . . . Gasoline-operated trucks . . . Liquefied
petroleum gas trucks

References   761

Factories, warehouses, docks, and railroad terminals use powered industrial trucks to carry, push, pull, lift, stack, and tier material. Each of the many types of trucks used in these operations requires guarding for the operator's protection and for the safety of other workers. Equally important is the establishment of safe practices for the operation, maintenance, and inspection of powered industrial trucks.

This chapter, however, does not discuss powered industrial trucks employed in airport and air terminal areas. For discussion of trucks and operators in these areas, see *Aviation Ground Operations Safety Handbook,* and Industrial Data Sheet 539, *Airport Vehicular Traffic,* and Data Sheet 540, *Airport Motor Vehicle Operators,* published by the National Safety Council. See References.

Types of Trucks

Powered industrial trucks may be classified by power source, operator position, and means of engaging the load. Electric motors energized by storage batteries, engines using gasoline, LP-gas, diesel fuel, or trucks using a combination of gas or diesel and electric are the power sources. Provisions for safe operation, maintenance and design of powered industrial trucks are given in ANSI *Safety Standard for Powered Industrial Trucks,* B56.1.

● *One class of powered industrial truck is designed to be controlled by a riding operator* (Fig. 27–1). The widely used lift truck, with its cantilever load engaging means, vertical uprights, and elevating mechanism, is usually a rider truck. Some rider trucks use a platform to engage the load. Both types may be either high-lift—with an elevating mechanism that permits the tiering of one load on another—or low-lift—with a mechanism that raises the load only enough to permit horizontal movement.

Variations in rider-operated trucks occur mainly in the means of engaging the load. Scoop trucks shovel loose materials and dump them into hoppers and bins. Ram trucks may use a single horizontal ram or arm to pick up coils, rolls, and other open center type loads. Roll-handling trucks may have a cable drum and two cables with hooks to engage a rod in-

Fig. 27–1.—Lift truck with driver's overhead guard and load backrest extension. Trucks capable of lifting loads higher than the operator's head should have guards of this type. It is important that guards do not obstruct operator's vision.

Courtesy Hyster Co.

serted in a roll of paper or a spool of cable, while others have a roll clamp or a vacuum system for engaging and holding the roll. Industrial crane trucks are equipped with booms, cables, and drums for moving heavy, large loads. Straddle carriers carry long material, such as pipe or lumber, under the truck body, which rides on four high legs. Powered industrial tractors draw trailers, nonpowered trucks, and other mobile loads.

● *A second category of powered industrial trucks is the motorized handtruck controlled by a walking operator.* They also have a platform or lifting forks to engage the load and may be either high-lift for tiering or low-lift to raise the load only enough for horizontal movement.

A unique powered industrial truck is an electronically controlled vehicle without an operator. It travels a prearranged route, outlined on or under the floor, and is controlled by a light beam or induction tape.

27—Powered Industrial Trucks

Guards and Safety Devices

Powered industrial trucks capable of lifting loads higher than the operator's head or operated in areas where there is a hazard from falling objects should be equipped with an overhead guard (Fig. 27–1). This guard should not interfere with good visibility. Openings in the guard should be small enough to protect the operator from being struck by objects or material falling from an overhead load or stack, and should conform to American National Standard B56.1. A load backrest extension should always be used when operating conditions permit.

Trucks should be equipped with platforms extending beyond the operator's position, strong enough to withstand a compression load equal to the weight of the loaded vehicle applied along the longitudinal axis of the truck with the outermost projection of the platform axis of the truck against a flat vertical surface (Fig. 27–2).

Additional operator enclosures are not recommended because rapid and unobstructed ingress/egress for the operator is considered more desirable. However, should additional enclosures be provided in conjunction with the platform, they should not prevent easy ingress or egress from the platform.

Exposed tires should have guards which will stop particles from being thrown at the operator. Hazardous moving parts, such as chain and sprocket drives and exposed gears, should be guarded if possible.

Although many lift trucks may come equipped with steering hand wheel knobs, their use is prohibited by many companies. If knobs are used, they should be of the mushroom type to engage the palm of the operator's hand in the horizontal position, and should be mounted within the periphery of the wheel. Spinner knobs must not be attached to steering handwheels of trucks not originally equipped with them (OSHA regulations, section 1910.178). The steering mechanism should minimize transmission of road shock to the steering hand wheel.

Every powered industrial truck should carry a name plate showing the weight of the truck and its rated capacity as specified by the ANSI B56.1, *Safety Standard for Powered Industrial Trucks,* previously mentioned.

Fig. 27–2.—Enclosure on this stand-up lift truck protects the operator from cruching injury in the event of a collision with other objects.

Courtesy Raymond Corp.

Powered industrial trucks should be painted a distinctive color to make them readily visible. Color should comply with the ANSI Z53.1, *Safety Color Code for Marking Physical Hazards.* See page 386.

Powered industrial trucks should have horns or other warning devices with a distinctive sound, loud enough to be heard clearly above other local noises. The warning device should be under the control of the operator.

Specifications of steering, braking and control arrangements should conform in detail to the ANSI B56.1.

Powered industrial trucks should also be constructed and equipped to comply with Underwriters Laboratories "Standards for Safety," UL 558 and UL 583; and in hazardous locations, trucks should comply with NFPA

Fig. 27–3.—Vacuum-handling truck, carrying newsprint roll in this photo, has canopy guard for operator's protection. Truck engine or separate source provides vacuum.

Courtesy Industrial Truck Div., Clark Equipment Co.

Standard No. 505, *Type Designations, Areas of Use, Maintenance and Operation of Powered Industrial Trucks.*

The definitions of hazardous locations are given in the NFPA *National Electrical Code.* Class I locations are those in which flammable gases or vapors are or may be present; Class II, combustible dust; and Class III, easily ignitible fibers or flyings. The B56.1 *Safety Standard for Powered Industrial Trucks* stipulates that trucks, electric- or gasoline-powered, shall not be used in certain hazardous locations unless they either comply with NFPA requirements or, in some cases, are specifically approved by the inspection authority for the locations involved. This stip-

ulation should apply to any internal combustion engine truck, although a diesel-powered lift truck has been approved by Underwriters Laboratories and Factory Mutual for use in hazardous areas. (For definitions of hazardous locations, see Fig. 41–6, Chapter 41, "Electrical Hazards," pages 1269–1274.)

Powered industrial trucks should be equipped with a carbon dioxide fire extinguisher or a gas-pressured dry chemical fire extinguisher, and all operators should be trained in its use. Also, an important incidental advantage is that the truck-mounted fire extinguisher and trained operator are quickly available to combat other fires on the premises.

Fig. 27–4.—Straddle trucks should have many safety features: drive chain guards, access ladder and hand-holds, horn, lights, and weather protection for the operator.

Courtesy Towmotor Corp.

Lift trucks

Lift trucks that can elevate loads higher than the operator's head, or that operate in areas where stacks are higher than the operator's head, should have substantial overhead guards and load backrest extension (Fig. 27–1). If an overhead guard is attached to the rear of the truck body, it should also be attached to the body in front and not the mast. An exception is made for those trucks that have the tilt cylinder as an integral part of the overhead guard construction.

Fork trucks equipped with a single tilt cylinder should have provisions to avoid injury to the operator resulting from failure of the cylinder or associated parts.

Forks should be locked to the carriage, and the fork extension, if used, should be designed to prevent unintentional lifting of the toe or displacement of the fork extension.

Lift trucks should have means or be equipped with mechanical hoist and tilt mechanisms to prevent overtravel of hoist and tilt motions. If the lifting systems are hydraulically driven, an overload valve should be installed in the system and suitable stops provided to prevent overtravel.

Vacuum-handling trucks

In recent years, another type of powered industrial truck has come into use—the vacuum-handling truck (Fig. 27–3). It uses a suction or vacuum system to hold and carry loads such as boxes, cased goods, board sheets, bags, and cylindrical shapes such as drums, tiles, and rolls of paper and textiles. Vacuum-handling trucks should conform to the ANSI B56.1.

One such vehicle uses the engine vacuum to operate the load-engaging device. Other vacuum-handling trucks have an independent primary vacuum engine mounted in a pod on the rear of the truck. The truck engine is used as a secondary vacuum source. These trucks should be provided with controls that cut in the secondary vacuum system if the primary system fails.

Straddle carriers

Straddle trucks (Fig. 27–4) should be provided with horns or other warning devices, and with headlights and tail lamps for night operation. Safe access ladders, wheel guards, and chain drive guards should be provided. Certain types of operation may require a rigid overhead guard for the operator.

Overhead clearances can be determined by telltales installed in advance of overhead obstructions across roadways and other passageways where straddle trucks operate. Gage rods may be mounted on the truck at front and rear.

Crane trucks

Crane trucks usually have a capacity of less than 10,000 pounds and are powered by LP-gas, diesel, or gasoline engines. Unless the truck crane has a fixed boom, outriggers should be used if the boom is to be swung. Specific details of safe design and operation are given in Chapter 25, "Hoisting Apparatus and Conveyors," under section, Mobile Cranes.

Tractors and trailers

In addition to conforming to the ANSI B56.1, tractors and trailers should incorporate the necessary safety features in the coupling used to make up the tractor-trailer train. The type of coupling used depends on the con-

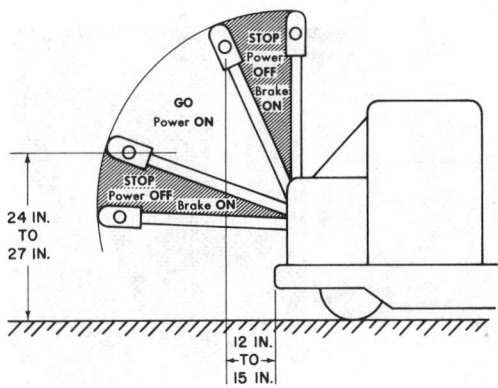

Fig. 27–5.—By modifying the controls to reduce the "power-on" area to the dimensions shown, the hazard of contact with frame or wheels of a powered industrial handtruck can be minimized.

struction of the trailer, the loads carried, and whether or not the route traveled includes sharp curves, ramps, or inclines. The coupling must be one that will not come unhitched on grades nor permit the trailer to whip or cut in on curves.

Motorized handtrucks

A powered handtruck should be equipped so that its brakes will be applied when the handle is in either the fully raised or the fully lowered position. When the handle is released, it should automatically apply the brakes. By modifying the controls on trucks to limit the "power-on" zone to an area between the fully raised and fully lowered positions of the handle, the hazard of contact with frame or wheels of a powered hand truck can be minimized (Fig. 27–5).

Unless controls are designed to keep the operator and his extremities within the outline of the trucks, guards should be installed to prevent the operator's hands or the controls from coming into contact with obstacles when the truck must be maneuvered in close quarters.

The wheels of many powered handtrucks can be considered guarded by their position under the frame or lift platform. However, where the wheels might injure the operator or others, guards should be installed. High-lift platform rollers and chain sprockets should also be guarded.

743

FIG. 27–6.—Triangular guard prevents the operator from riding a powered handtruck.

Many trucks, however, are being manufactured with a platform for the operator to ride in a standing position. If this practice is permitted, the operator should be protected against foot or leg injuries. Prism-shaped covers can be installed over the battery box to discourage operators from sitting on the truck during operation (Fig. 27–6).

Remotely controlled trucks

Since they are operated electronically without the guidance of an operator, remote control (programed) trucks (Fig. 27–7) must be provided with some means of bringing them to a complete stop if someone steps in front of the truck. These trucks should be equipped with a lightweight, flexible bumper which, when contacted, shuts off the power and applies the brakes. There should be sufficient clearance between the bumper and the front of the truck to permit the truck to come to a full stop without contacting anything in its path.

Safe Operation

Powered industrial truck traffic accidents can be prevented by putting into operation the same safe practices that apply to highway traffic. For example, a powered industrial truck operated at excessive speed is being run at a speed too high for the conditions at a

given moment. Safe speed is the rate of travel which will permit stopping the truck, well within the *certain clear distance ahead,* and making a turn without hazard of overturn. Wet or slippery floors require a slower-than-ordinary speed.

Specific speed limits for trucks should be established. A favored maximum speed for loaded power trucks moving forward is 6 mph, although this speed may be too high in congested areas. Many companies have installed governors, locking them in some instances.

Collisions between trucks and stationary objects often occur while trucks are backing, usually while turning and maneuvering. In such cases, the operator may be so intent on handling the load that he forgets where the rear of his truck is going. Backing accidents usually result from failure to look. Operators should be required to look in the direction of travel and keep a clear view of it. Some trucks permit the operator to sit sideways and look backward and forward more easily. (Stunt driving and horseplay must not be permitted.)

FIG. 27–7.—Electronically controlled robot truck (guided by floor strip shown at lower right) has bumper which shuts off power when contacted.

Courtesy Barrett Electronics Corp.

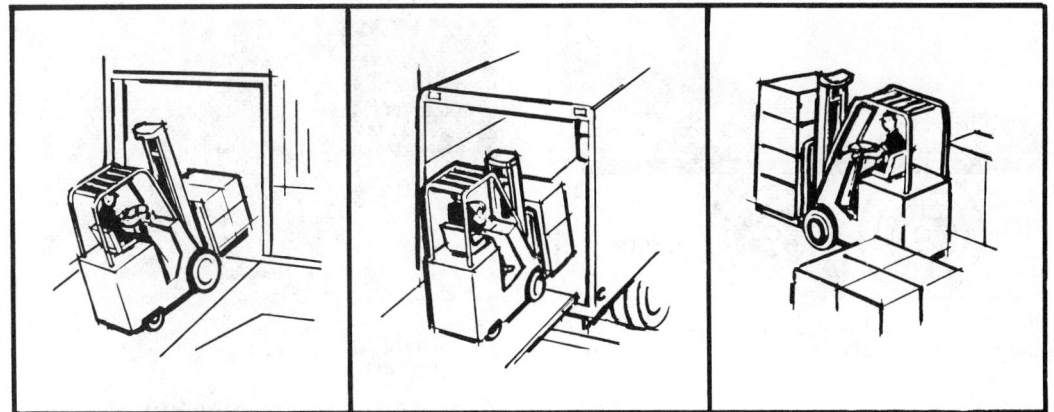

Fig. 27–8.—Driving with load pointing upgrade permits better control. When forward vision is obscured by big or bulky loads, driving in reverse permits better visibility (right).

The operator should come to a stop at blind corners and before passing through doorways and go ahead only when he can see the way is clear. Many companies have installed large convex mirrors at blind corners, so that operators and pedestrians can see each other approaching. (See Fig. 24–14 in Chapter 24, "Materials Handling and Storage.") The operator should keep his truck a safe distance behind another moving truck—some companies specify 3 truck lengths. At intersections, blind spots, or other dangerous locations, he should not pass another truck traveling in the same direction. He should keep to the right, if aisle width permits him to do so, without passing dangerously close to machine operators and others.

Operators should avoid making quick starts, jerky stops, or quick turns at excessive speed. Operators should use extreme caution when operating on turns, ramps, grades, or inclines. On descending grades, the truck should be kept under control, so that it can be brought to an emergency stop in the clear space in front of the truck (Fig. 27–8). The reverse control should not be used for braking, and batteries should not be operated beyond their rated capacity.

No powered industrial truck should be used for any purpose other than the one for which it is designed. Common dangerous misuses of trucks include bumping skids, pushing piles of material out of the way, moving heavy objects by means of makeshift connections, and pushing other trucks. Disabled trucks should not be pushed or carried by another lift truck. They should be moved by towing with tow bar and safety chain or by transporting them on a lowboy trailer. Powered industrial trucks, operated on elevated docks or platforms above the track level, should never be used to tow or push freight cars or to close freight car or truck doors.

Truckers should approach elevators at right angles to the gate, stop at least 5 ft from the gate, and wait for a signal from the elevator operator before entering. They should get off the truck, when it is on the elevator, only after making sure that the brakes are set, the power is shut off, and the controls are in neutral.

Powered industrial trucks should be driven carefully and slowly over bridge plates that are properly secured. Trucks should cross railroad tracks diagonally wherever possible and park at least 8 ft from the nearest rail.

Highway trucks, trailers, and railroad cars should have their brakes set and their wheels securely blocked while they are being loaded or unloaded by powered industrial trucks (Fig. 27–9).

The operator should keep his feet and legs inside the guard or the operating station of his truck. Driving with a foot or a leg outside is obviously unsafe, as is placing a hand, arm, or leg on or between the uprights of a truck. When in close quarters, operators should keep their hands where they cannot be pinched

745

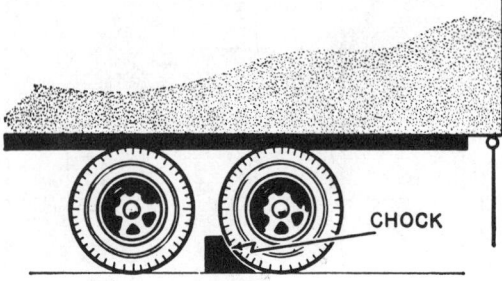

Fig. 27-9a.—Preferred placing of chocks is under the rear set of trailer wheels.

Fig. 9b.—Truck and trailer chocks connected to the dock by a chain or to electrical warning devices provide additional protection, even though warning signs are posted.

Courtesy Rite-Hite Corporation.

between the steering or control levers and projecting stationary objects. Steering handles on motorized hand trucks should have guards to protect against injuries of this kind.

The operator should leave his truck unattended only after he has put the controls in neutral, shut off the power, set the brakes, removed the key or pulled the connector plug, and placed the load-engaging means in a lowered and inoperative position. (OSHA defines an "unattended truck" as one where the operator is more than 25 ft from it or cannot see it.) Although it is not usually good practice to leave a truck on an incline, when such action is necessary, the wheels should be blocked as an added precaution.

It is the operator's responsibility never to park his truck in an aisle or doorway, nor to obstruct material or equipment to which another worker may need access. Accidents often happen because an operator leaves a truck blocking a passageway, and an unauthorized workman tries to move it out of his way.

Looking out for pedestrians should also be the truck operator's responsibility. He should sound the horn when approaching pedestrians. Excessive hornblowing, however, is to be discouraged. Having sounded the warning, the operator should go ahead with caution and pass only when he knows the pedestrians are aware of him and in the clear. The operator should not use the horn to "blast" his way through. He should not drive his truck directly toward anyone who is standing in front of a bench or other fixed object.

Pedestrians also have a responsibility to watch out for trucks and to get out of the way with reasonable promptness. Ill feeling and accidents can result from their refusal to move out of the way when a truck approaches—consideration on both sides is needed.

Passengers should not be permitted to ride on a truck, fork, coupling, or trailer. It is the operator's responsibility to keep riders off.

Loads, whether on trucks or trailers or on skids or pallets, should be stable. Objects should be neatly piled and crosstied, if the shapes permit (Fig. 27-10). Irregularly shaped objects should be loaded so that they cannot roll or fall off. Heavy, odd-shaped

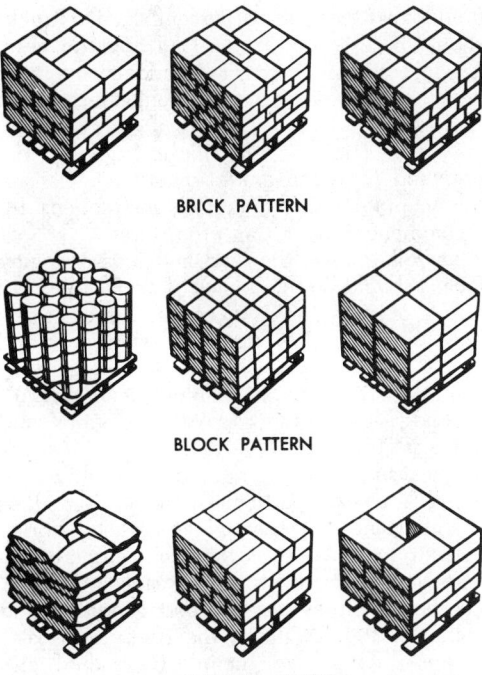

BRICK PATTERN

BLOCK PATTERN

PINWHEEL PATTERN

Fig. 27–10.—Typical pallet-loading patterns.

Courtesy The Industrial Truck Assn.

objects should be placed with the weight as low as possible. Round objects, like pipe or shafting, should be blocked and, if necessary, tied so that they cannot roll. Loading to an excessive height not only blocks the view ahead, but makes it likely that part of the load may fall.

Operators should not permit gasoline engines to idle for long periods in enclosed or semi-enclosed areas because of the accumulation of exhaust fumes and gases. Concentration of carbon monoxide (CO) gas in areas where powered industrial trucks are operated should not exceed the levels established by OSHA or specified by local, state, or municipal codes. Sampling should be done by a qualified industrial hygienist. See Chapter 36, "Exhaust and Ventilation," for additional information.

Catalytic exhaust purifiers are available which are reported to reduce considerably the level of carbon monoxide and other noxious gases in the engine exhaust. Even though exhaust purifiers are installed on lift trucks, the user must provide adequate ventilation for enclosed areas, in order to maintain a clean atmosphere.

Lift trucks

The operation of a lift truck differs basically from that of an automobile or a highway truck. The prospective operator should be impressed with the fact that a lift truck:

1. Is generally steered by the *rear* wheels

2. Steers more easily loaded than empty

3. Is often driven in reverse direction as much as in forward

4. Is often steered with one hand—the other hand being used to operate the controls.

Because a lift truck is generally steered by the rear wheels, the swing of the rear of the truck must always be carefully watched (Fig. 27–11). Beginners usually try to turn too sharply. Some lift trucks when traveling forward have a peculiarity known as "free turning," that is, once the turn is started, the truck tends to turn more and more sharply in a smaller and smaller circle. To counteract this tendency and to slow down the sharpness of the turn, the operator must apply force on the steering wheel in the opposite direction. When such a truck is traveling in reverse, the opposite holds true—the operator must apply force in the direction of the turn.

Operators should learn to judge the correct aisle width for the truck size and load. They should also observe the general operating safety rules, given earlier in this chapter, and also the following rules which apply specifically to lift trucks.

All starts and stops should be easy and gradual to prevent the load from shifting. Turns should be made at a safe speed, smoothly and gradually.

The operator should take particular care, while either traveling or maneuvering, to avoid striking overhead structures and nearby objects, such as sprinkler piping, electrical conduit, or fixed structures. Critical equipment and materials, such as electrical panels, fire equipment, fire doors, and load supporting

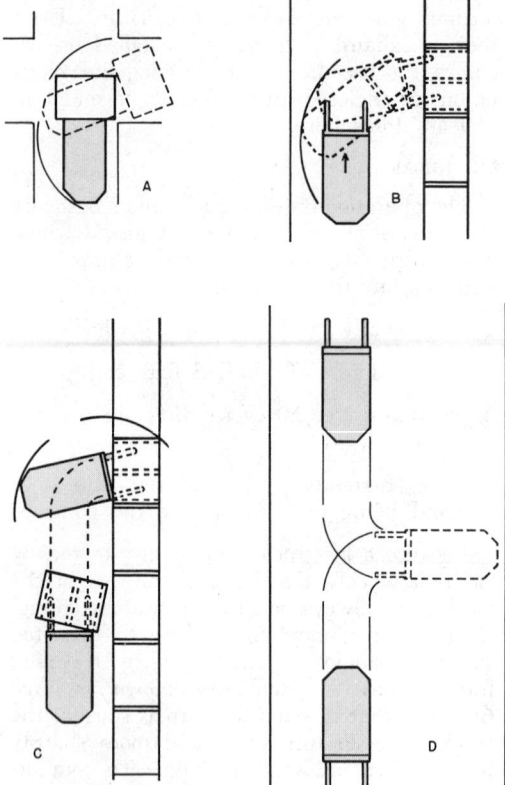

Fig. 27-11.—Lift truck maneuvers: **A**—Turning a sharp corner; **B**—Turning across an isle; **C**—Turning on exceptionally narrow aisle; and **D**—Turning around in a narrow passage. Turns should be gradual. Driver should allow ample space for rear-end swing.

structures, should be protected by strong barriers, posts, or curbing to prevent damage (Fig. 27–12).

Loads should not be raised or lowered en route. Loaded or empty, the forks should be carried as low as possible, but high enough not to strike any raised or uneven surface. Tilting back the upright keeps the load steady and secure (Fig. 27–13).

If a bulky load is to be carried which cannot be lowered enough to prevent its obstructing the view, the operator should drive the truck backward so that he can see where he is going.

Trucks should ascend or descend grades slowly. When ascending or descending grades in excess of 10 percent, loaded trucks should be driven with the load upgrade (Fig. 27–8). Unloaded trucks should be operated on all grades with the load engaging means downgrade.

NOTE: *High-lift order-picker trucks are not designed for steep grade operation.* Consult the manufacturer's operating instructions for recommended operating procedures.

On all grades, the load and load-engaging means should be tilted back, if applicable, and raised only as far as necessary to clear the road surface. Low gear or slowest speed should be used when the truck is descending a grade. The operator should keep clear of the edge of loading docks and ramps and never make a turn on a ramp.

Because of the comparative instability of high-lift trucks, particularly when loaded, it is important that the operator check bridge plates (to make sure they are properly secured) and the floors of boxcars and trucks (to be sure they are in good condition and will bear the weight of the truck and load) before driving over them. He should also

Fig. 27-12.—This switch box was damaged when it was struck by a loaded forklift truck.

Fig. 27–13.—Note how the mast is tilted back and the load carried as low as safely possible.

Courtesy British Columbia Cement Co., Ltd., Bamberton, B.C.

check the truck or trailer to see that it is properly chocked (Fig. 27–9a). Failure to do this may result in the bridge plate shifting or the trailer or truck moving away from the dock (Fig. 27–14).

Lift trucks have a rated capacity, which is stated in pounds at a 24 in. load center distance from the vertical face of the forks. Thus, the rated capacity may be "2000 pounds at 24 in." A wide, flat load of 2000 pounds with the center of gravity 30 in. from the forks is, therefore, a considerable overload. Every operator should be familiar with the maximum load limits of the truck he operates and be required to observe them.

Placing extra weight on the rear of a lift truck to counter-balance an overload should not be permitted as it may strain chains, forks, tires, axles, and motor and also may cause accidents.

The stability of lift trucks is covered in the ANSI B56.1. Most lift truck manufacturers use the stability values in the standard as a criteria for design and to determine the rated capacity for various truck models.

Operators should never, under any circumstances, attempt to operate a truck with an overload that removes enough weight from the rear wheels to tip the truck or make steering uncertain. Such a load is dangerous. Standing on a truck or adding counterweights to compensate for an overload should never be permitted.

Side stability is a critical factor in making turns at speed or on a slope or ramp. Back tilt of uprights reduces side stability on high lifts, and allowance should be made for this factor.

Particular care should be taken not to exceed floor load limits. The force exerted by a truck on a floor varies with the speed, load, and total weight distribution. It is also

749

Fig. 27–14.—This is what happened when a lift truck tried to unload a trailer whose wheels were not chocked.

Courtesy Printing and Publishing Section Newsletter.

affected by the number of wheels, wheelbase, and other variable factors. All questions about floor capacities should be referred to a qualified building architect or structural engineer.

When standard forks are used to pick up round objects, such as rolls and drums, care must be taken to see that the tips do not damage the load or push it against workers. The uprights should first be tilted, so that the tips of the forks touch the floor, and then moved forward so that the forks can slide under the object. Tilting the uprights backward will then cause the load to roll back against the vertical face of the forks and/or carriage and

the load backrest extension—a secure carrying position (Fig. 27–15).

To unload a large case or similar object without a pallet, the operator should first drive into position for stacking. He should then lower the load onto a base having a block near the edge, withdraw the forks so that only their tips hold up the end of the load, withdraw the block, tilt the uprights forward, and back away (Fig. 27–16).

In attempting to pick up a palletized load, the forks should be fully and squarely seated in the pallet, an equal distance from the center stringers and well out toward the sides. Forks to be inserted in a pallet should be level, not tilted forward or back. If the forks are placed close together, the pallet tends to drop at the sides and seesaw, causing strain and instability (Fig. 27–17).

When raising or lowering loads while standing still, the operator should not leave the truck in gear with the clutch depressed; he should return the shift to neutral and disengage the clutch.

Operators of lift trucks should be told to re-

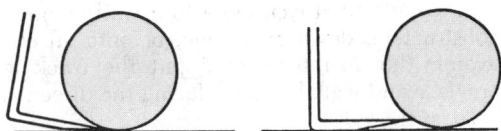

Fig. 27–15.—To pick up round objects, forks should be tilted (left), not used flat (right).

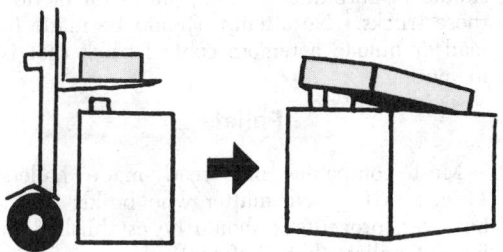

Fig. 27–16.—Unloading a large case or similar object without a pallet requires a base and a movable block.

Fig. 27–17.—With forks spread wide (left), load is well distributed and tends to bind itself together. Effect of placing forks too close is shown on the right.

Fig. 27–18.—Guarded platform attaches to lift truck for overhead maintenance work. Truck operator is protected by overhead guard and side shields.

Courtesy Eaton, Yale & Towne, Inc.

fuse improperly loaded skids or pallets, or loads too heavy for the truck.

The forks should be placed flat on the floor when a lift truck is parked. No one should be allowed to stand or walk under elevated forks.

Using a lift truck as an elevator for employees (for example, to service light fixtures) should be done only if a safe platform that is securely seated on the forks, fastened to the vertical face, and provided with handrails and toeboards is used. The truck should also have

Fig. 27–19.—Mobile personnel work platform.

Courtesy Lift-A-Loft.

an overhead guard for the operator's protection (Fig. 27–18). There are special trucks made for this purpose (Fig. 27–19).

Vacuum-handling trucks

Vacuum-handling, and other truck types that are similar to lift trucks except in the way they engage a load, should be operated in accordance with the general principles of safe operation previously discussed. In addition, operators should follow those principles used for lift trucks—turning, maneuvering, and load-carrying procedures—to operate trucks with same steering and power-transmission design as lifts.

Straddle carriers

Because some early straddle truck models have a blind spot either to the front or the rear, depending on the location of the operator's seat, precautions must be taken to avoid striking pedestrians, especially when these trucks are carrying long loads. Red

flags may be attached to the ends of the load or flagmen may be stationed in congested areas, particularly if the truck is used after dark.

Motorized handtrucks

The principal hazards in the operation of a motorized handtruck include the operator getting pinned between the truck and a fixed object and the truck running up on his heels. Operators should walk ahead of the truck, leading it from either side of the handle and facing the direction of travel. Where the truck must be driven close to a wall or other obstruction, down an incline, or onto an elevator, the operator should put the truck in reverse and walk behind it facing the direction of travel.

Remotely controlled trucks

The use of such "robot" trucks requires that aisles where the truck operates be clearly marked and clear of material. Employees should be forbidden to jump on or off or ride these trucks. No attempt should be made to load or unload a remote control truck that is in motion.

Pallets

Most companies buy ready-made pallets (Fig. 27–20). No matter who builds them, however, procedures should be established to inspect pallets, both before they are put into service and at regular intervals, to be sure that they are in a safe condition. Top deckboards should be sound and securely fastened to runners. Splintered, broken, or loose parts should call for repair or replacement. Loose nails or chunks of wood can cause injury to workers and damage to trucks.

A safe place, out of the way of traffic and work areas, should be provided for the storage of pallets. Pallets should be neatly piled (not over 4 ft high), and they should not be left in a leaning position from which they might slide or topple onto persons or damage other objects.

Operators

Selection

Physical examination of prospective operators is desirable, with special emphasis on

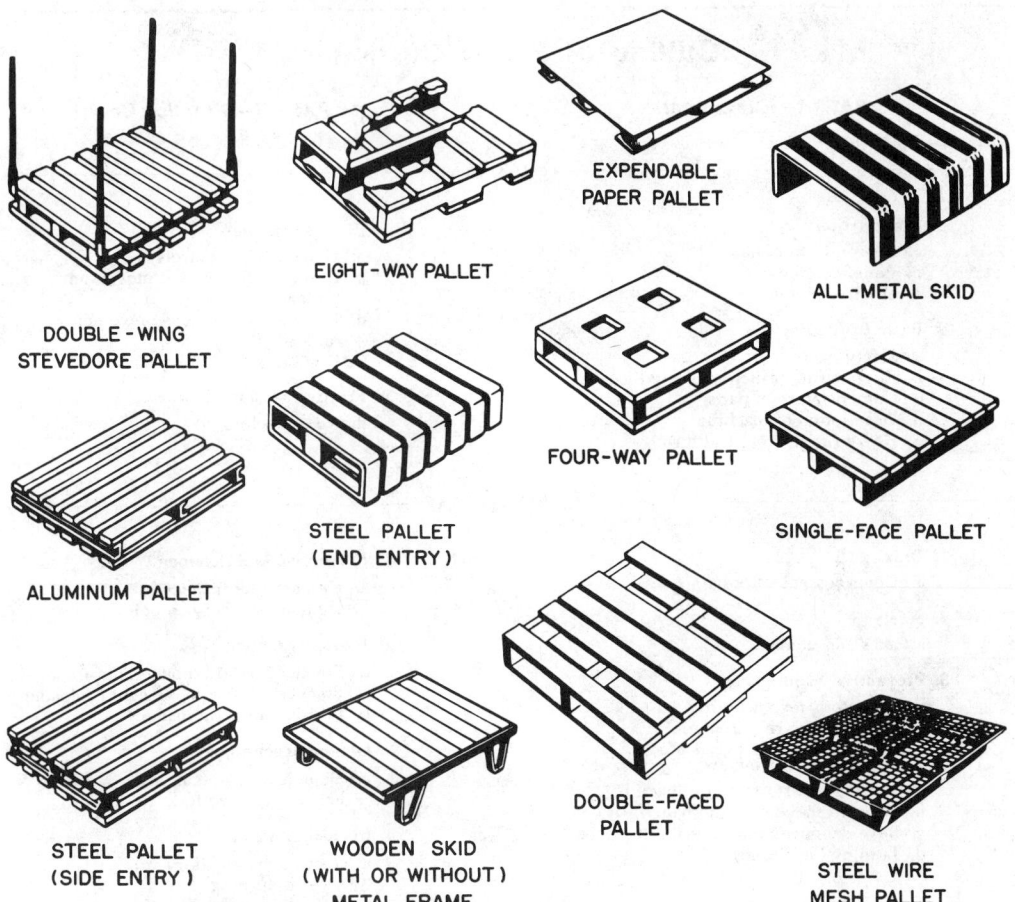

DOUBLE-WING STEVEDORE PALLET

EIGHT-WAY PALLET

EXPENDABLE PAPER PALLET

ALL-METAL SKID

ALUMINUM PALLET

STEEL PALLET (END ENTRY)

FOUR-WAY PALLET

SINGLE-FACE PALLET

STEEL PALLET (SIDE ENTRY)

WOODEN SKID (WITH OR WITHOUT) METAL FRAME

DOUBLE-FACED PALLET

STEEL WIRE MESH PALLET

FIG. 27–20.—Lift truck operators should be familiar with pallet and skid types.

Courtesy Industrial Truck Assn.

acuity of vision, depth and color perception, hearing, muscular coordination, and reaction time.

Physical reexaminations should be given to operators at the time of permit or license renewal, after each accident, injury, or extended illness, or at other appropriate intervals.

A fair amount of intelligence is required, together with respect for the safety of personnel and property. It should be made clear to operators that they carry considerable responsibility and that reckless or careless operation will not be tolerated. They should be checked on the job from time to time for observance of this rule.

Each plant should adopt a set of rules governing the operation of powered industrial trucks. Because plant conditions and equipment vary widely, rules which cover specific conditions should be set up.

For ready identification, many companies issue badges to authorized truck operators. Badges also tend to remind operators of their responsibilities and give them pride in their job.

Training

A truck incompetently operated may cause severe injury or substantial property damage; no company can afford even one improperly

Outline of 5-Day Curriculum

PART I — *(Classroom)* **PART II — *(Practice Court)***

FIRST DAY

1. **Introduction**
   a. Reason for program
   b. Benefits of program
   c. Outline of course

2. **Basic Instruction**
   a. Safety
   b. Engineering principle of forklift truck
   c. Nomenclature of parts
   d. Description of machine
   e. How to start, stop, and turn

1. **Basic Instruction**
   a. Demonstrate controls
   b. Demonstrate forward movement (unloaded)
   c. Demonstrate reverse movement (unloaded)
   d. Demonstrate turning (unloaded)

2. **Practice Session**
   a. No obstacles

SECOND DAY

1. **Quiz**
   a. Cover lessons of previous day

2. **Safety**
   a. Do's and don'ts

3. **Preventive Maintenance**
   a. Operator's responsibility
   b. Basic service requirements

4. **Basic Instruction (Review)**
   a. Starting and stopping (unloaded)
   b. Forward movement (unloaded)
   c. Reverse movement (unloaded)
   d. Turning (unloaded)

1. **Preventive Maintenance**
   a. Demonstrate inspection
   b. Demonstrate service points

2. **Basic Instruction**
   a. Practice forward movement (unloaded)
   b. Practice reverse movement (unloaded)
   c. Practice turns (unloaded)

3. **General Techniques**
   a. Demonstrate operation for hoisting, lowering, tilting forks (unloaded)

4. **Practice Session**
   a. Obstacle course (forward, reverse, turning)
   b. Operation of forks

THIRD DAY

1. **Quiz**
   a. Cover lessons of 2nd day

2. **Safety**
   a. Do's and don'ts

3. **Preventive Maintenance**
   a. Review
   b. New material

4. **Basic Instruction**
   a. Starting and stopping (loaded)
   b. Forward movement (loaded)
   c. Reverse movement (loaded)
   d. Turning (loaded)

5. **General Operating Techniques**
   a. Operating forks (loaded)
   b. Maneuvering (unloaded)

1. **Preventive Maintenance**
   a. Review by actual demonstration

2. **Basic Instruction**
   a. Practice forward movement (loaded)
   b. Practice reverse (loaded)
   c. Practice turns (loaded)
   d. Practice operating forks (unloaded)

3. **General Techniques**
   a. Operating forks (loaded)
   b. Maneuvering (unloaded)

4. **Practice Session**
   a. Obstacle course

1. **Quiz**
   a. Cover lessons of 3rd day

2. **Basic Materials Handling Techniques**
   a. Types of pallets
   b. Methods of loading pallets
   c. Aisle widths
   d. Floor loading

3. **Safety**
   a. Do's and don'ts

4. **Preventive Maintenance**
   a. Review of all previous material

5. **Basic Instruction**
   a. Operating forks (loaded)
   b. Maneuvering (unloaded)

6. **General Operating Techniques**
   a. Basic maneuvers (loaded)
   b. Advanced maneuvering (unloaded)
   c. Operating under adverse conditions (unloaded)
   d. Operating attachments (optional)

1. **Basic Instruction**
   a. Practice maneuvers (unloaded)
   b. Practice operating forks (loaded)

2. **General Techniques**
   a. Demonstrate advance maneuvers (unloaded)
   b. Demonstrate operating under adverse conditions (unloaded)
   c. Demonstrate attachments (optional)

3. **Practice Session**
   a. Obstacle course

1. **Quiz**
   a. Cover lessons of 4th day

2. **Basic Materials Handling**
   a. Review of previous material

3. **Basic Instruction**
   a. Review of maneuvering (unloaded)
   b. Maneuvering (loaded)

4. **General Operating Techniques**
   a. Advanced maneuvers (loaded)
   b. Review of advanced maneuvering (unloaded)
   c. Review of operating under adverse conditions
   d. Operating under adverse conditions (loaded)
   e. Review of attachments

1. **Basic Instruction**
   a. Review maneuvering (loaded)
   b. Review of operating under adverse conditions (unloaded)

2. **General Techniques**
   a. Review of adverse maneuvering (loaded)
   b. Demonstrate operating under adverse conditions (loaded)
   c. Review of attachments

3. **Practice Session**
   a. Obstacle course
   b. Qualification tests

SUPPLEMENTARY

I. Presentation of certificate of eligibility (Operating Permit)
II. Forklift Truck Rodeo
III. Periodic qualification tests
IV. OPERATING MANUAL for eligible operators.

FIG. 27–21.—Sample outline of 5-day driver training course.

trained truck operator. Many companies have set up operator training programs. Truck manufacturers can often supply course outlines, instructors' manuals, study text, films, and booklets for training in correct operation and safety.

Five-day training program

A brief discussion of a five-day training course for lift truck operators follows. A detailed description and discussion of this course is available from the manufacturer who supplied the information.* Courses for training operators of other types of powered industrial trucks are also available.

A classroom and a paved practice area, at least 50 ft square, should be provided. Each class should have no more than five trainees. The course is based on four principles: tell, show, practice, correct. As indicated in the sample outline of curriculum (Fig. 27–21), each day's session should be divided between classroom work and practical instruction in the practice area with the trucks.

On the second and each following day, the session should open with a quiz covering the previous day's work. Each trainee should have a record card showing the day's progress and his own weak points. He and his instructor should try to eliminate these weak points before the next step is undertaken.

The final examination should take the form of qualification tests, which require the trainee to demonstrate his skill in handling the truck, his judgment, and his knowledge of safety, basic materials-handling, and preventive maintenance. A certificate of graduation, which may be an official operator's permit, should be issued upon successful completion of the course.

Practical training and testing are conducted on obstacle courses consisting of pallets set on end in patterns designed to simulate boxcars, aisles, material piles, and other obstructions (Fig. 27–22). These obstacle courses and the maneuvers required to negotiate them have been largely standardized in operator training programs all over the country. In many training courses, the final qualifying examination consists of one or more test problems presented on the obstacle course. The instructor may be provided with a checklist and test sheet (Fig. 27–23) upon which he

FIG. 27–22.—Judges watch a driver maneuver his lift truck through a mockup of a crowded aisle. One-inch clearance on both sides of truck tests operator's judgment.

Courtesy Crouse-Hinds Co.

grades the prospective operator as he observes the performance. Many other points may be included in the checklist, such as failure to look behind when backing, failure to inspect the load before starting, failure to center the forks properly under the load, and traveling with forks raised too high.

Although other training programs may vary in details from the example given, it demonstrates the essentials of good training procedures which include:

1. Physical qualification

2. Aptitude selection

3. Classroom instruction and testing

4. Field instruction and testing

---

* Donald R. Holm, "Selecting and Training Fork Truck Operators." Reprinted from *Space Magazine* by the Hyster Company, Portland, Ore.

PERFORMANCE TEST FOR LIFT TRUCK OPERATOR EXAMINER'S TEST SHEET

NAME OF TEST _____

ERROR SCORE _____

FINAL RATING _____

NAME _____ DATE _____

Operating Errors

_____ 1. Started jerkily, overcautious, unfamiliar with controls.

_____ 2. Did not give proper signals when turning.

_____ 3. Did not slow down at intersections.

_____ 4. Did not sound horn at intersections.

_____ 5. Did not obey signs.

_____ 6. Did not look where he was going.

_____ 7. Made too wide a turn on corners.

_____ 8. Cut corners too sharply.

_____ 9. Drove recklessly.

_____ 10. Approached load improperly.

_____ 11. Lifted load improperly.

_____ 12. Maneuvered unnecessarily.

_____ 13. Traveled with load too high.

_____ 14. Lowered load too fast.

_____ 15. Stopped too suddenly.

_____ 16. Load not balanced properly.

_____ 17. Forks not under load all the way.

_____ 18. Struck pallet or floor with fork while loading.

_____ 19. Failed to check bridge plates.

_____ 20. Did not place load within marked area.

_____ 21. Did not drive backward when required.

_____ 22. Number of pallets displaced (or lines crossed).

_____ 23. Number of unnecessary stops made.

_____ 24. Did not follow instructions.

_____ 25. Did not complete problem.

FIG. 27–23.—Examiner's checklist and test sheet can be used to evaluate performance of lift truck operators. Similar tests can be given operators of other types of trucks.

5. Authorization only upon satisfactory completion of training.

In some training courses, the candidate is issued a student badge upon successful completion of the training course and can qualify for an operator's badge only after a specified period of driving on the job under close supervision, followed by an additional final examination. In any case, close supervision is essential, especially immediately following the training period when the newly qualified driver is applying his lessons on the job. In fact, as long as an operator remains on the job, his performance should be periodically checked to stop any tendency to lapse into unsafe practices.

There are a number of Safe Driver Award programs available. These programs have been very effective in achieving extended periods of safe and productive operation.

Inspection and Maintenance

Powered industrial trucks should be inspected on a regular schedule. Operators should make daily inspection of controls, brakes, tires, and other moving parts. Checklists (Fig. 27–24) should be used to record conditions requiring correction. In addition, competent maintenance men should be assigned to keep the equipment in safe operating condition. Defective brakes, controls,

757

```
                    GAS
        Daily Operator Check Off List
Truck No._____ Operator_____
Truck Model_____ Dept._____
Date_____ Running Time_____Hours
Shift_____
```

Check	O.K.	Adjust	Add (Amount)
1. Gas			
2. Water or Anti-Freeze			
3. Engine Oil			
4. Hydraulic Oil			
5. Steering and Horn			
6. Brakes			
7. Tires			
8. Hoist Cylinder			
9. Tilt Cylinders			
10. Air Cleaner			
11. Oil Pressure			
12. Forks			
13. Battery			
14. Fire Extinguisher			
15.			

P1639   Operators Signature

```
                  ELECTRIC
        Daily Operator Check Off List
Truck No._____ Operator_____
Truck Model_____ Dept._____
Date_____ Running Time_____Hours
Shift_____
```

Check	O.K.	Adjust	Add (Amount)
1. Steering and Horn			
2. Brakes			
3. Safety Seat Brake			
4. Tires			
5. Hoist Cylinder			
6. Tilt Cylinders			
7. Hydraulic Oil			
8. Controller			
9. Forks			
10. Battery (Specific Gravity)			
11.			
12.			

P1641   Operators Signature

FIG. 27–24.—Two daily inspection checklists.

Courtesy Eaton, Yale & Towne, Inc.

tires, lights, power supply, load-engaging mechanism, lift system, steering mechanism, and signal equipment should be repaired before trucks are allowed to go back into service. Operators should be prohibited from making repairs to trucks.

The operating mechanism should be locked "off" before repairs are made to any part of a powered industrial truck.

It is advisable to keep a detailed inspection and repair record for each truck. The maintenance force should give each truck a thorough inspection at stated intervals and a complete overhaul after regular extended periods of operation.

Electric trucks

Handling and charging of storage batteries for electric trucks introduce several hazards. By wearing chemical goggles, rubber gloves, aprons, and rubber boots, operators of charging equipment can be protected against acid burns during refilling or handling operations. Wood slat mats, rubber mats, or clean floorboards will help prevent slips and falls and will protect against electric shock from the charging equipment. See NFPA Standard 505, *Type Designations, Areas of Use, Maintenance & Operation of Powered Industrial Trucks,* and NSC Data Sheet No. 635, *Lead-Acid Storage Batteries.*

Battery changing and charging operations should be performed by trained and authorized personnel. Truck operators may or may not be so authorized depending upon individual plant setups.

Battery charging installations should be located in areas designated for that purpose. Facilities should be provided for flushing and neutralizing spilled electrolyte, for fire protection, for protecting charging apparatus from damage by trucks, and for adequate ventilation to disperse gases and fumes from batteries.

When racks are used for support of batteries, they should be made of materials nonconductive to spark generation or be coated or covered to achieve this objective.

When charging batteries, acid should be poured into water; *never* the reverse. A carboy tilter or siphon should be provided for handling electrolyte. If acid or electrolyte is spilled on the worker's skin or clothing, it should be washed off immediately with plenty of water.

Trucks should be properly positioned and the brake applied before attempting to change or charge batteries. Reinstalled batteries should be properly positioned and secured in the truck.

When charging batteries, the vent caps should be kept in place to avoid electrolyte spray. Care should be taken to assure that vent caps are functioning. The battery (or compartment) cover(s) should be open to dissipate heat.

Smoking should be prohibited in the charging area. Sulfuric acid used for refilling should not be allowed to run into ordinary cast iron, lead, steel, or brass drains.

Precautions should be taken to prevent open flames, sparks, or electric arcs in battery charging areas. Electrical installations should conform to the local codes and the *National Electrical Code.*

Battery terminals should be clean, connections tight, and the battery securely locked in place in the truck. Tools or metal parts should never be laid on a battery.

To prevent operator strains from manual handling of heavy or awkward loads, a roller conveyor, an overhead hoist, or equivalent materials-handling equipment should be provided. Chains, hooks, yokes, and other parts of the hoisting mechanism should be insulated to prevent short-circuiting.

Gasoline-operated trucks

Gasoline for trucks should be handled and stored in accordance with the provisions of the NFPA Standard No. 30, *Flammable and Combustible Liquids Code.*

Fuel tanks on gasoline-operated trucks and tractors should be filled at designated locations, preferably in the open air, with the filling hose and equipment properly grounded and bonded. Locations outside main buildings should be selected to minimize the chances of involving combustible material in a fire.

Safety cans used for fuel handling should be tested and approved by Factory Mutual or listed by Underwriters Laboratories. (See Chapter 42, "Flammable and Combustible Liquids".)

Engines should be stopped, and operators must be off trucks, before trucks are refueled. *No smoking should be permitted during refueling.*

Workers should avoid spilling gasoline or overflowing the gasoline tank during refueling. Before an attempt is made to start the engine, the gasoline tank cap should be replaced and spilled fuel should be flushed down or allowed to vaporize.

Gasoline tanks should be drained at safe locations into grounded self-closing cans.

Liquefied petroleum gas trucks

The use of liquefied petroleum gas (LP-gas) as a fuel for industrial power trucks is increasing, largely through conversion of gasoline-powered units.

A properly adjusted engine burning LP-gas will generally produce a substantially lower concentration of carbon monoxide in the exhaust than a similar engine which uses gasoline as fuel. Only air sampling, however, can prove whether or not the concentration in an area is below the maximum allowable concentration. Fifty ppm for an 8-hour exposure is the maximum exposure allowable, unless local codes set a lower limit.

The principal hazard in the use of LP-gas is that of fire. Fittings, not listed by a nationally recognized testing agency or incorrectly installed, or connections, not properly tightened before refueling is begun, may fail and release combustible gas into the air. Only conversion units and fittings listed by an agency, such as Underwriters Laboratories or Factory Mutual, should be used. They should include all safety features that are incorporated in LP-gas-fueled trucks listed by the testing agency. The units and fittings should be installed in strict conformity with requirements specified in NFPA Standard No. 58, *Storage and Handling of Liquefied Petroleum Gases,* and Underwriters Laboratories No. 558, *Standards for Safety—Internal Combustion Engine-Powered Industrial Trucks* to provide maximum protection against damage to the system by vibration, shock, or objects striking against it, and against failure from other causes.

The manufacturer of a truck which is to be converted to LP-gas operation should be consulted. The manufacturer may be able to supply listed conversion units and assign a

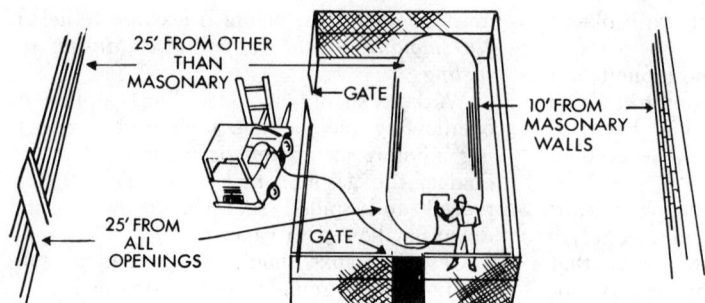

25' FROM OTHER THAN MASONARY

GATE

10' FROM MASONARY WALLS

25' FROM ALL OPENINGS

GATE

Fig. 27–25.—Schematic shows minimum distances allowed for industrial truck refueling operations by LP-Gas Assn.

qualified representative to supervise the installation. Conversions should be attempted only by qualified mechanics who are familiar with handling LP-gas equipment and who are using listed parts.

Rather than convert gasoline-powered equipment, it is preferable to purchase new trucks or tractors designed for LP-gas operation, which have been tested and listed by a nationally recognized testing agency.

Only listed fuel containers, designed in accordance with NFPA standards, should be used. Permanently mounted fuel containers should be charged outdoors, and storage facilities and charging equipment should meet standards of NFPA No. 58, *Storage and Handling of Liquefied Petroleum Gases.*

For fuel containers used on trucks, the standards require a device, such as an excess flow check valve, be provided in the fuel system to prevent the escape of fuel in the event that a pressure fuel line or fitting breaks. Exchange of removable fuel containers preferably should be done outdoors, but may be done indoors if safety devices are installed as specified in the NFPA standards.

A special building or outside storage is recommended for the storage of fuel containers. Where the cylinders must be stored inside of the building, they must be kept in a special room or designated safe area in accordance with NFPA Standard 58. NFPA standards permit no more than two containers for LP-gas motor fuel on each industrial truck and require that storage inside a building not frequented by the public be limited to a total gas capacity of 300 pounds. Containers should be enclosed in a separate room that is well ventilated and of ample size. The walls, floor, and ceiling of the room are required

to be of specified fire-resistive construction. Openings to other parts of the building should be protected by specified fire doors.

Proper container filling is of the utmost importance, and the person filling the containers must be trained in the safe handling of LP-gas.

Container filling from bulk storage must be done at least 10 ft from the nearest important masonry-walled building and at least 25 ft from important non-masonry buildings and openings in masonry and nonmasonry buildings (Fig. 27–25).

The filling plant must conform to NFPA Standard 58 and applicable state or local insurance regulations. Trucks themselves must comply with NFPA Standard 505.

LP-gas-fueled trucks should be garaged in a well ventilated special building or fire-resistive enclosure. Ventilation should be provided at floor level, since fuel gas is heavier than air. The trucks should not be garaged in the same room with stored cylinders because the hazard of leakage from a truck's fuel container during downtime—overnight or over a weekend—would exist, and such a risk should be minimized.

A rigid and thorough inspection and maintenance procedure for LP-gas-fueled trucks should be followed. No repair work on LP-gas-fueled industrial trucks should be permitted indoors without first removing the fuel container.

If the fuel container is permanently mounted, major repairs should be made outdoors or in a well ventilated, fire-resistive area provided for this purpose.

Subparts of the OSHA regulations that apply to this chapter must be strictly adhered to.

References

American Insurance Association, 85 John St., New York, N.Y. 10038.
Safe Handling and Use of LP-Gas.

American National Standards Institute, 1430 Broadway, New York, N.Y. 10018.
Allowable Concentration of Carbon Monoxide, Z37.1.
Safety Standard for Powered Industrial Trucks, B56.1.
Safety Color Code for Marking Physical Hazards, Z53.1.

Industrial Truck Association, 250 Gateway Towers, Gateway Center, Pittsburgh, Pa. 15222.
ITA Recommended Practices.

National Fire Protection Association, 470 Atlantic Ave., Boston, Mass. 02210.
Flammable and Combustible Liquids Code, Standard No. 30.
Liquefied Petroleum Gases, Standard No. 58.
National Electrical Code, Standard No. 70.
Self-Service Gasoline Stations, Standard No. 30E.
Type Designations, Areas of Use, Maintenance and Operation of Powered Industrial Trucks, Standard No. 505 (ANSI B-56.2).

National LP-Gas Association, 11 S. LaSalle St., Chicago, Ill. 60603.

National Safety Council, 425 North Michigan Ave., Chicago, Ill. 60611.
Aviation Ground Operations Safety Handbook.
Industrial Data Sheets
Airport Motor Vehicle Operators, 540.
Airport Vehicular Traffic, 539.
Carbon Monoxide, 415.
Lead-Acid Storage Batteries, 635.
Liquefied Petroleum Gases for Industrial Trucks, 479.
Powered Hand Trucks, 317.

Underwriters Laboratories Inc., 207 East Ohio St., Chicago, Ill. 60611.
"Standards for Safety," No. 558 and No. 583.

761

Elevators and Plant Railways

Power Elevators   763
Use of codes . . . Types of drives . . . New elevator equipment . . .
Hoistways and machine rooms . . . Landings and doors . . . Cars
. . . Hoisting ropes . . . Operating controls . . . Inspection and
maintenance . . . Operation . . . Emergency procedures

Sidewalk Elevators   778
Operation . . . Hatch covers

Hand Elevators   779
Hoistway doors . . . Safety devices and brakes

Dumbwaiters   780
Hoistways and openings . . . Safety devices and brakes

Escalators   781
Safety devices and brakes . . . Machinery . . . Protection of riders
. . . Fire protection

Moving Walks   783

Manlifts   783
Construction . . . Brakes, safety devices, and ladders . . .
Inspections . . . General precautions

Plant Railways   786
Clearances and warning methods . . . Track . . . Loading and
unloading . . . Overhead crane runways . . . Types of motive
power . . . Tools and appliances . . . Moving cars . . . Safe
practices

References   792

This chapter, in addition to dealing with passenger and freight elevators, also deals with sidewalk elevators, hand elevators, dumbwaiters, escalators, and manlifts.

Power Elevators

Information essential to the safe operation and maintenance of elevators is provided. This discussion is generally limited to electric-drive passenger and freight elevators (Fig. 28–1). It is not intended as a guide for design or specification of new equipment, nor is it a manual for skilled elevator inspectors and mechanics. Refer also to Chapter 22, "Non-Employee Accident Prevention."

Use of codes

Before new elevators are specified or major alterations scheduled for existing installations, reference should be made to the American National Standard A17.1, *Safety Code for Elevators, Dumbwaiters, Escalators, and Moving Walks.* For the rest of this section, this will be referred to as the Elevator Code. This Standard is the basis for much of the material presented in this section.

To make sure operations are safe, the Elevator Code requires inspections. These should be made as recommended by the *Practice for the Inspection of Elevators (Inspectors' Manual)*, ANSI A17.2, including addenda. This is discussed later under Inspection and Maintenance.

New equipment should be specified so that it will conform to Elevator Code requirements, unless federal, state, or local codes are more stringent, in which case these should be observed as the minimum requirements. Usually, the Elevator Code will be the more strict and, in this case, it should be referred to in the specifications. Two commonly used wordings are:

"The elevator and associated equipment shall meet the requirements of American National Standard A17.1, *Safety Code for Elevators, Dumbwaiters, Escalators, and Moving Walks,* latest edition, except as hereinafter specifically exempted or modified."

The second typical paragraph is: "Except as changed or modified herein, the elevator and associated equipment shall meet the requirements of American National Standard

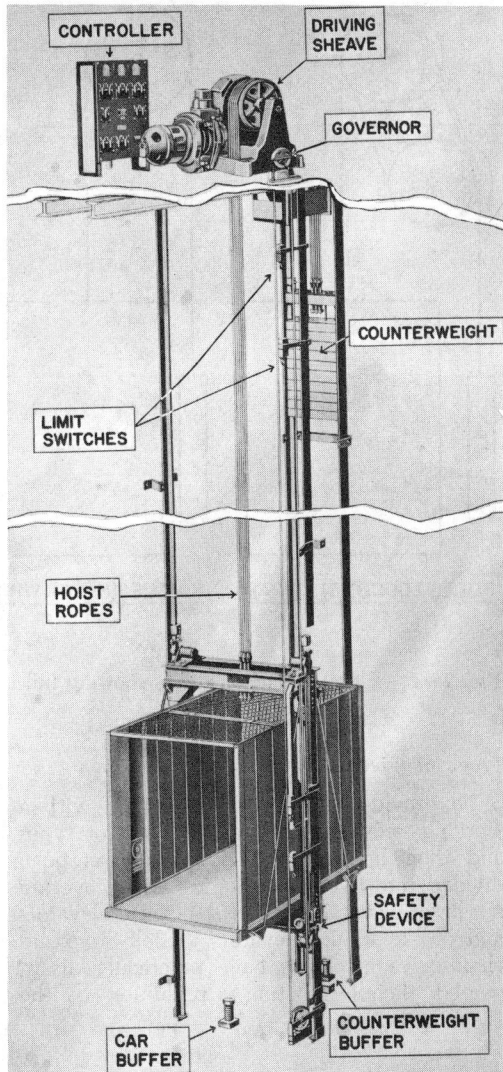

Fig. 28–1.—Typical worm-geared traction freight elevator with electric drive.

Courtesy Westinghouse Electric Corp.

A17.1, *Safety Code for Elevators, Dumbwaiters, Escalators, and Moving Walks,* latest edition, and also shall comply with all applicable local laws and ordinances."

Such paragraphs will generally take care of items not specifically spelled out in the specifications and drawings.

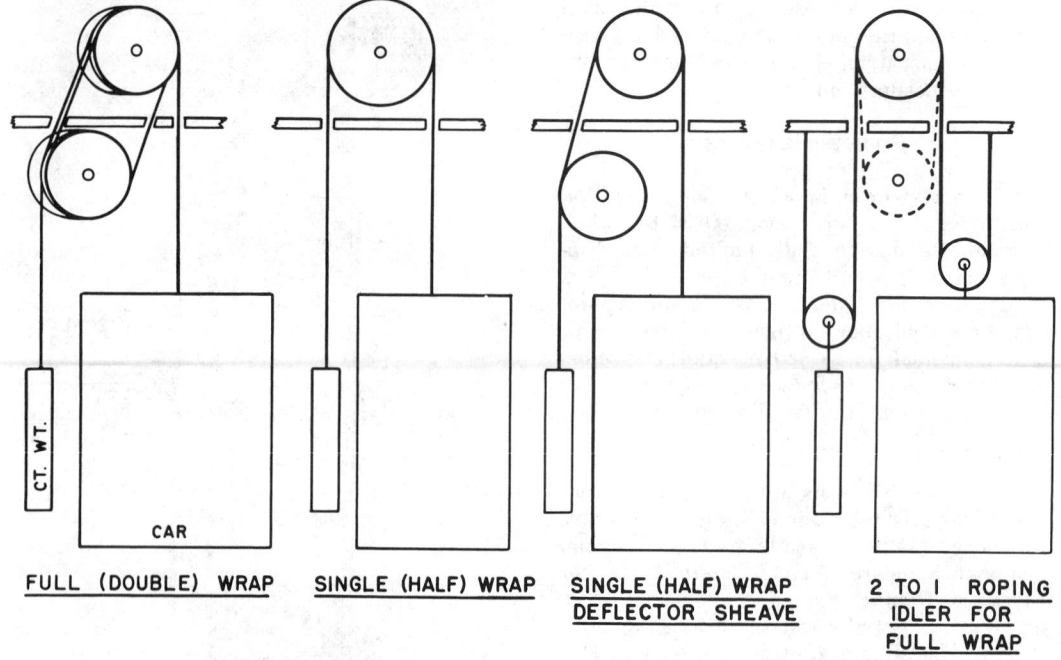

FULL (DOUBLE) WRAP   SINGLE (HALF) WRAP   SINGLE (HALF) WRAP DEFLECTOR SHEAVE   2 TO 1 ROPING IDLER FOR FULL WRAP

FIG. 28–2.—Elevator traction drives showing full (double) and single (half) wraps.

Types of drives

To provide the safety professional with a working knowledge of various elevator types and their limitations, this section has been divided into two major topics, each covering a type of elevator drive: (*a*) electric elevators and (*b*) hydraulic elevators. (Belt-drive and chain-drive machines have practically disappeared; their installation is prohibited by the Elevator Code.)

Electric elevators. The two general types of drives for electric elevators are traction drive and winding drum. The winding drum type, however, is now almost obsolete and is presently being used only in new dumbwaiter construction.

The Elevator Code requires that all driving machines be of the traction type, except that it permits the use of winding-drum machines on freight elevators which travel not more than 40 ft at speeds not exceeding 50 ft per minute (fpm) and which are not provided with counterweights.

In the winding-drum type, the hoisting rope is anchored in and winds on a spirally grooved drum. This is a positive drive. If the machine is not stopped at the limits of travel, the car may be pulled into the overhead structure and, if the motor is powerful enough, the ropes may be pulled from their anchorage.

In the traction-drive type (Fig. 28–2), the hoisting rope is not attached to the machine. The elevator is moved by the traction (friction) of the ropes in grooves on the drive sheave. The grooves may be semicircular (U-groove) or undercut U-groove. Some V-groove sheaves are still in use.

With the U-groove, the ropes run through one set of grooves on the drive sheave, over an idler sheave, and then through a second set of grooves on the drive sheave—effecting a nearly full (double) wrap.

A simple rule is that when the drive sheave has twice as many grooves as there are hoisting ropes, the machine is a full (double) wrap. The total angle of contact of each rope with the drive sheave generally is between 300 and 360 degrees, and the traction relation changes very little with wear.

Most companies use some form of undercut U-groove for their single-wrap machines. The friction is higher than an ordinary U-groove because of the pinching action. The width of the undercut is varied to suit traction needs. The traction remains substantially constant until the groove has worn to near the bottom of the undercut. The groove should be checked periodically to make sure that it is not worn so much that the rope bottoms.

The V-groove likewise has higher friction when new but, as the V-groove wears, the rope seats deeper and deeper, so that the area of contact of rope with groove is increased and the unit pressure is decreased. When the rope reaches the bottom of the groove, much of the driving traction probably will be lost, and a loaded (or empty) car may "break traction" and slide with the brake locked. For this reason, such grooves should be checked frequently for wear and should be remachined before the rope wears the groove down to the bottom.

Generally, the V-groove or undercut U-groove is used on geared machines and the U-groove on gearless machines, but there are exceptions.

Traction machines have an inherent safety feature: when the descending member (counterweight or car) bottoms, driving traction is lost and, in most cases, the ascending member cannot be pulled into the overhead. With extremely high rises, the weight of rope hanging on the downrun side may be sufficient to maintain traction after the car or counterweight has landed but, since the compensative sheave is tied down, there is no danger.

For safe and successful operation, rope tension on the car side must bear a definite relation to rope tension on the counterweight side. Normally, this relation ranges from 2-to-1 to 1-to-2. See discussion of overloads in "Operation" later in this chapter.

Hydraulic elevators. Except for the electrohydraulic type (now required in new installations by the Elevator Code), hydraulic elevators, once widely used, have been largely replaced by electric machines. Their efficiency generally is lower, and the difficulty of keeping the valves and stuffing boxes tight creates maintenance problems.

Leakage may permit the car to drift away from a landing and out of range of a mechanical hoistway door interlock. Many accidents have been attributed to such drifting; in most cases, a hoistway door has been left open.

The Elevator Code requires that electrohydraulic elevators be equipped with anticreep devices and with hoistway-door locking devices, electric car-door or gate contacts, hoistway access, and parking devices the same as those required for electric-drive elevators.

Because most of these elevators do not have a counterweight, the motor must supply pressure to lift the entire weight of the car and the load, and it must be considerably larger than the motor of a traction-drive electric machine, on which the weight of the car and part of the load is compensated for by the counterweight, to maintain the same speed. As built, the electrohydraulic type has all the electrical protective devices (interlocks, car gate contacts, limit switches, and similar devices) found on an electric elevator.

No safety is required on electrohydraulic elevators, since they can come down no faster than the fluid can be forced out of the cylinder by the descending plunger.

Belted machines. The Elevator Code prohibits the installation of belt-driven and chain-driven machines. Existing installations should be provided with electrically released brakes, terminal stopping, and safety devices as required for electric elevators. Few, if any, of these elevators remain in operation.

New elevator equipment

Requirements. When new elevators are under consideration, care should be taken to be sure that all requirements of the Elevator Code are met. Examples of the requirements to be checked are: safe and convenient access to the machine room and the pit, adequate lighting in the machine room and overhead spaces, and convenient electric outlets on the crosshead and in the pit.

An inspection station (with slow-speed UP and DOWN operating buttons and an emergency STOP switch) should be provided on the top of the car for the use of maintenance men and inspectors.

The elevator should have normal and final limit stops, interlocks on all hoistway doors, a contact on the car door, and a car emergency exit or exits. Unless the elevator is the hydraulic-plunger type, it should have a governor-actuated safety that meets the latest code requirements (city, state, federal, and/or ANSI standards).

Buffers are required to absorb the energy of descending cars and counterweights at the limits of travel. The two types in use are spring buffers and oil buffers.

Spring buffers are permissible under the Elevator Code only for cars traveling up to 200 fpm, but a few states and cities permit them for higher speeds. A spring at best is a poor absorber of energy. A good spring will return approximately 95 percent of the stored energy, so that a spring buffer stop affords only a series of decreasing surges.

Capacity ratings. The size, capacity, and speed of new freight elevators will depend upon the purposes for which they are to be used. The safety professional should, therefore, find out from production personnel what size units (both freight "package" and freight carrier) will probably be in use twenty years hence.

In specifying size, capacity, and speed of new freight elevators, it is important to anticipate future load requirements as well as present ones.

The Elevator Code classifies freight elevators as follows:

CLASS A—Elevators loaded by hand or by handtrucks. Here, the weight of any single piece of freight or of any single handtruck and its load is limited to a maximum of one-fourth the rated load capacity.

CLASS B—Elevators handling motor vehicles.

CLASS C1—Industrial truck loading where the truck is carried by the elevator.

CLASS C2—Industrial truck loading where the truck is *not* carried by the elevator but used only for loading and unloading.

CLASS C3—Other loading with heavy concentrations where a truck is *not* used.

For Class C1, Class C2, and Class C3 loadings, the Elevator Code requires that the rated load of the elevator be not less than the load (including any truck) to be carried, and

that the elevator be provided with a two-way automatic leveling device.

With palletized loads increasing in size year by year and with lift trucks being made larger to handle them, cases have been reported in which old elevators were overloaded and started downward when the last loaded lift truck was run onto the car. In some cases, the brakes would not hold. In others, the traction relation was broken, and the hoisting ropes slipped through the drive sheave.

Adherence to the Elevator Code should prevent such cases in new installations. Where old elevators are being used to handle heavy palletized loads, however, safe load limits and safe operating procedures must be determined and strictly enforced, if serious accidents are to be prevented.

The rated capacity of the passenger elevators should conform to the Elevator Code.

Hoistways and machine rooms

Hoistways. Most building exit and elevator codes require that new elevators be installed in fire-resistant hoistways, with fire doors that fill the entire opening, in order to prevent the rapid spread of fire from floor to floor.

Hoistways, guide rails, and all other appurtenances should be thoroughly cleaned of grease and dirt accumulations, at frequent intervals, to eliminate fire hazards. Projections, windows, and stairways adjacent to hoistways should be treated as recommended in the Elevator Code.

It is particularly important that no pipe conveying gas or liquids, which—if discharged into the hoistway—would endanger lives, be installed in or under any elevator or counterweight hoistway. However, low-pressure steam (5 psig or less) or hot water pipes only for heating the hoistway and the machine room (or penthouse) are permitted.

The entire conduit system should be effectively grounded. All electrical installations, including grounding, should conform to the *National Electrical Code* (NFPA Standard No. 70; ANSI C1).

Pits. To protect persons working in elevator pits, a minimum clearance of 2 ft is necessary between the lowest projection on the underside of the car platform (not including

FIG. 28–3.—View from pit looking up hoistway, showing hoistway illumination (arrows).

Courtesy Architect U.S. Capitol.

guide shoes and aprons attached to the sill) and any obstruction in the pit (exclusive of compensating devices, buffers, buffer supports, and similar devices). Measurements should be made when the car is resting on fully compressed buffers.

Counterweight runways should be enclosed from a point not more than 1 ft above the pit floor to a point at least 7 ft above the pit floor and adjacent pit floors, except where compensating chains or cables are used.

Screen partitions, at least 7 ft high between adjacent pits, will protect men in one pit from cars and counterweights in adjacent pits and will protect employees from hazards, when adjacent pits are at different levels.

An elevator pit should never be used as a thoroughfare or storage space. It should be fully enclosed, and the entrances kept locked. A drain should be provided for removal of water.

Pits should be kept clean and free of debris. Rubbish never should be swept into pits.

Lighting of at least 5 foot-candles at the pit floor level should be provided.

Machine rooms should have safe and convenient access, as specified in the Elevator Code. So that persons repairing or inspecting elevator hoisting machinery have sufficient room and are safe, there should be at least 7 ft of headroom between the machinery plat-

767

form and the machine room roof.

Like pits, elevator machine rooms should never be used as thoroughfares. The one exception would be if the elevator equipment were in a separate locked enclosure. Rooms should be well-ventilated and lighted with not less than 10 foot-candles at floor level. Doors should be kept locked, to prevent entry of unauthorized persons, with a sign affixed to the door stating such.

Machine rooms should be kept clean and should not be used for storage. Small quantities of ordinary maintenance supplies should be placed in a wall cabinet. A portable carbon dioxide or dry-powder fire extinguisher should be kept within reach of someone standing at the door.

Overhead protection. If the elevator machine is located in the penthouse, a substantial grating or floor of fire-resistant construction should be provided under the machine. For detailed information concerning construction, the Elevator Code should be consulted.

On all overhead machine installations there should be a cradle below the secondary sheaves if they extend below the floor or grating.

No elevator machinery, except the idler or deflecting sheaves, should be hung underneath the supporting beams at the top of the hoistway. When the governor or other devices (other than terminal stopping switches) must be installed below the machine floor, they should be set on a substantial secondary floor.

For winding-drum machines, a substantial beam or bar should be placed at the top of the counterweight guide rails and beneath the counterweight sheaves, to prevent the counterweights from being drawn into the sheaves.

Landings and doors

The Elevator Code specifies that hoistway landing openings of all power-driven elevators must be provided with hoistway doors that guard the full height and width of the openings.

Hoistway doors (passenger elevators). Records show that most elevator accidents occur at the hoistway door. Most are "tripping" accidents. Some are the "caught and crushed" or "fall down hoistway" type. These are on old elevators, because the interlocks required by the Elevator Code practically eliminate these two types of accident, if the interlocks are properly installed and maintained.

Many serious accidents and fatalities have occurred on older elevators equipped with door-locking devices which cannot be unlocked from the corridor side without the use of a special key, even when the elevator is at the landing. Frequently, the victim opened the door and stepped into the shaft under the mistaken impression that the car was at that floor.

The Elevator Code permits the use of unlocking devices and other means for emergency access to elevators equipped with doors that are unlocked, when the car is at the landing or can be unlocked from the corridor side without tools if the car is in the landing zone.

All emergency keys should be kept where they are accessible only to qualified personnel who are aware that they should be carefully guarded and used.

To help prevent hoistway accidents in older elevators, conduit can be installed in each hoistway so as to provide an electric light opposite each opening. If the door is opened with an emergency key and the car is not at the landing, the light, which is kept burning continuously, will help a person see that the shaft is empty in time to prevent him from stepping into it (Fig. 28–3).

These requirements can be met in either of the following two ways:

1. The provision of only two means of access to the hoistway—one at an upper landing to permit access to the top of the car, and one at the lowest landing if this landing is the normal point of access to the pit.

2. Where elevators operate in a single hoistway, hoistway doors should be provided that can be unlocked when closed with the car at the floor, or locked but openable from the landing by means that are effective only when the car is in the landing zone.

In general, three types of doors at landings are used: vertically or horizontally sliding

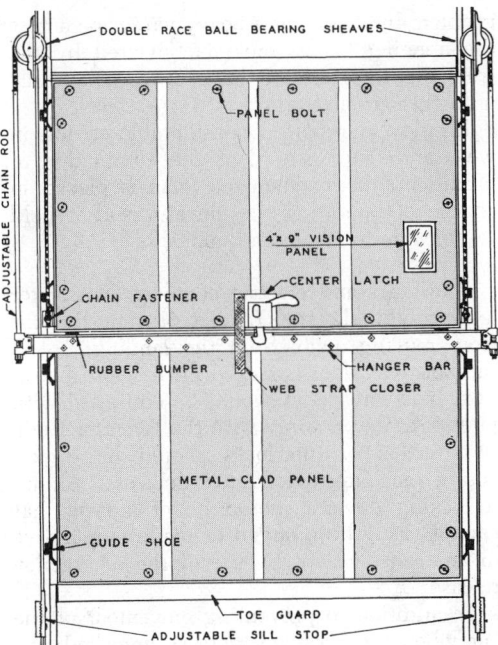

Double race ball bearing sheaves

Adjustable chain rod

Panel bolt

4" x 9" vision panel

Center latch

Chain fastener

Rubber bumper

Hanger bar

Web strap closer

Metal-clad panel

Guide shoe

Toe guard

Adjustable sill stop

Fig. 28–4.—Design of a typical vertical sliding biparting steel hoistway door for freight elevators, as seen from inside the car.

Courtesy The Peelle Co.

doors, combination sliding and swinging doors, and swinging doors. All three types should have direct-acting interlocks. Nothing less than direct-acting mechanical interlocks should be used.

Power-operated vertical-slide doors and gates must operate in sequence, if the elevator is used for passengers and, in any case, if the doors close automatically.

For new installations, according to the Elevator Code, the distance between the hoistway side of the hoistway door opposite the car opening and the hoistway edge of the landing threshold should be not more than ½ in. for swinging doors and 2¼ in. for sliding doors. The face of the hoistway door should not project into the hoistway beyond the edge of the landing sill. On existing installations, if this distance exceeds 1½ in. for swinging doors and 2½ in. for sliding doors, it is advisable to fill in the excess space.

No automatic fire door, whose functioning depends on the action of heat, should be designed to lock any landing opening in the hoistway of any elevator or any exit leading to the outside of the building.

The loading platform for at least 2 ft back from the door should be so constructed and maintained that persons will not readily slip. Many members of the National Safety Council use rubber mats, firmly secured, or adhesive abrasive strips, or abrasive-surfaced concrete directly in front of all hoistway doors.[*] To eliminate tripping hazards, such surfaces should be made flush with the surrounding floor.

Hoistway doors and gates (freight elevators). Like those for passengers, most freight elevators are now installed in fire-resistant hoistways with fire doors that fill the entire opening, as required by practically all codes. However, there are some older elevators—mostly freight—which have hoistways enclosed only to a 6- or 7-ft height and hoistway gates of hardwood slats which are 5 or 6 ft high and have a clearance under them of as much as 8 to 10 in. Many of these elevators are shipper-rope operated, so that it is necessary to reach into the hoistway to bring the car to a landing—a procedure which has been prohibited by the Elevator Code for many years.

Hoistway doors (Fig. 28–4) should comply with applicable state and municipal requirements for fire resistance. Doors closed by hand should be so arranged that it is not necessary to reach back of any panel, jamb, or sash to operate them.

To facilitate movement of trucks, a door sill is recommended to fill the gap between the landing and the car (Fig. 28–5).

Hoistway gates in older installations should be replaced with doors as soon as practicable. Where gates are used, special attention must be given to maintain them at the best possible standard of safety. The openings in gates made of grille, lattice, or other open-work should reject a ball 2 in. in diameter.

The bottom of the gate should come down to the threshold to prevent objects from sliding under the gate into the hoistway.

---

[*] See NSC Data Sheet 595, *Floor Mats and Runners*. See also pages 580–582.

FIG. 28-5.—A truckable sill fills the gap between the landing and the elevator car.

Where lack of headroom precludes a standard gate at the lowest landing, the gate should be made in two sections.

Gates should have convenient handles or straps for manual operation. However, an attachment for closing the gate by power, controlled from the landing buttons, saves time and labor and decreases the possibility of leaving or propping the gate open.

Power-operated doors and gates are recommended. For freight elevators, these are usually of the continuous-pressure-operation type.

The growing weight and speed of forklift trucks presents the problem of protecting the hoistway doors and gates of freight elevators. A 5-ton truck with a 5-ton load moving only 2 mph (2.93 fps) has an impact of 2690 foot-pounds ($F = \frac{1}{2} MV^2$). If it is assumed that the door or gate can deflect 1 in., a force of 32,300 pounds results—certainly more than any hoistway door can stand.

One large industrial concern has solved this problem by installing a heavy wire rope across the opening. This must be lowered by the truck operator before he can run the truck onto the car platform.

Employers report a considerable saving in wear and tear, as well as the elimination of unsafe practices, when one man is placed in charge of operating the elevator and is held responsible for its operation.

Interlocks and electric contacts (passenger and freight elevators). To prevent the car from moving away from the landing, unless the hoistway door is locked in the closed position, hoistway doors should be equipped with interlocks that comply with the Elevator Code (Fig. 28-6). Interlocks should be direct-acting mechanical-activated devices. All interlocking devices should be of a type that cannot easily be plugged or made inoperative in any way, except by use of the emergency release.

In addition to preventing movement of the car when the hoistway door is unlocked, the interlock also prevents opening of the hoistway door from the landing side, except by emergency key, unless the car is within the landing zone and is either stopped or being stopped.

Locks and contacts (not interlocks) are permitted in a very few cases. Contacts are required for the car door or gate, which is considered closed, if within 2 in. of the nearest face of the jamb.

The Elevator Code defines the "closed position" of hoistway doors as being ⅜ in. from the jamb or between panels of center-opening doors, except that, for vertical-slide bi-parting doors, the dimension is ¾ in. Under certain conditions, the elevator may be started when the doors are 4 in. from full closure.

The Elevator Code requires that all interlocks be so designed that the door must be locked in the closed position, as defined here, before the car can be operated.

Like any other safety device, an elevator interlock or electric contact will be useless if it easily gets out of order. It is advisable, therefore, to use only those devices that comply with the Elevator Code and that have been either tested and listed by competent, designated testing laboratories, or approved by the city or state authorities that have jurisdiction. Also, it is essential that interlocking devices be inspected frequently and maintained in proper working order.

The car-leveling device automatically brings the car to a stop when the platform is level with the desired landing. A tripping hazard from uneven surfaces thus is eliminated. Moreover, in the case of freight elevators, the level surface automatically provided for the passage of trucks saves wear and damage to sills, and reduces the possibility of material being jarred off trucks.

Cars

Enclosure. All elevator cars should be enclosed on the sides and top, except for the side or sides used for exit and entrance. Openings for ventilation, emergency exit, signal, operating, or communication equipment are allowed.

The sides and top of every passenger elevator car should be made of metal or of an approved, fire-retardant treated material.

Although wood and openwork have been used in the past for freight elevator cars, the Elevator Code now specifies for them a solid metal enclosure to a height of at least 6 ft. The enclosure may have perforations above the 6-ft level, except in the area where the counterweight passes the car. That portion of the enclosure, for 6 in. on each side of the counterweight and up to the crosshead or car top, should be solid.

The enclosure should be fastened securely to the floor and to the suspension sling. If the enclosure is cut away at the front to provide access to the hand rope, the opening should be low enough to prevent injury to the operator's hand.

Enclosures should be kept in good condition to prevent injuries to persons loading and unloading the car. Broken wooden wainscoting or torn sheet metal may cause severe lacerations.

One way to prevent such damage is to install bumper strips at the level of truck platforms and truck push bars (Fig. 28–7). Strips should be made of oak, ash, hickory, or similar tough, resilient wood and should be at least 1½ in. thick (2 in. thick where heavy trucks are used). Steel channels are also effective.

FIG. 28–6.—An electromechanical interlock designed for use on biparting, manually operated steel doors on pushbutton-controlled freight elevators.

Courtesy The Peelle Co.

Strips should be wide enough to match all heights of platforms and push bars of trucks that are transported in the elevator.

Painting the inside of the car a flat black, from the floor to the top of the upper guard strip, will make bumps and scrapes less conspicuous. If the sides above the top strip are painted aluminum, the high reflectance value of the paint will help overall illumination.

Top covers. Freight elevator cars may have either solid or openwork top covers. If openwork is used, the Elevator Code requires that the openings be small enough to reject a ball

1½ in. in diameter; some companies, however, recommend ¾-in. openings.

Wire mesh is recommended for openwork covers, because it combines maximum strength with minimum weight and offers little interference with light and air. Not less than No. 10 gage steel wire should be used.

The cover should be strong enough to sustain a load of 300 pounds on any square area, 2 ft on a side, or of 100 pounds applied at any one point.

In the top of every elevator car, there should be an emergency exit of at least 400 sq in. in area, providing an unobstructed passageway to permit easy escape if the car becomes stalled between floors or some other emergency arises.

The exit cover should open outward and be so hinged or attached that it can be opened easily from the top of the car only. A heavy cover should be counterweighted or divided into several hinged sections, which should be kept down except when being used.

Doors or gates. To conform to Elevator Code requirements for new installations, a car door or gate must be provided at each entrance to the car. If the elevator is electrically operated, the car door or gate should have electric contacts that will prevent movement of the car unless the door or gate is within 2 in. of the fully closed position.

On existing elevator cars having more than one opening, it is recommended that each entrance be equipped with a car gate. The gate should be provided with a contact to prevent its being opened when the car is in motion. When closed, the gate should guard the full width and height of the opening (except for vertically sliding gates, which should extend from a point not more than 1 in. above the car floor to a height of at least 6 ft). Collapsible gates, when fully expanded, should reject a ball 3 in. in diameter. They should have convenient handles with guards for protection of the operator's fingers.

Floors of cars should be kept in good condition. Ordinary metal sheets should not be used for surfacing or repairing floors because they soon become smooth and slick. In time, even sheet metal with raised surface markings may wear smooth and then should be replaced.

Fig. 28–7.—Elevator car showing hardwood buffer strips for hand trucks. Note the black paint on lower elevator car section, which is subject to truck impact, and aluminum paint for high-reflectance above.

Courtesy National Bureau of Standards

Neither the edge of the car platform nor the edge of the landing should be permitted to become slippery or badly worn. Cast iron and steel sills at such points are slipping hazards unless provided with antislip surfaces. Abrasive strips or welded beading may be applied to eliminate slipperiness.

Loads. The safe working load (or rated capacity) of an elevator includes a safety factor, but nevertheless it should not be exceeded. The rated safe load should be indicated by a conspicuous sign inside the car. Metal signs with stamped, etched, or raised letters and figures, not less than ¼-in. high, are satisfactory for passenger elevators and a minimum of 1-in. high letters and figures for freight elevators.

To find the maximum number of passengers allowable under emergency conditions, common practice is to divide the rated safe capacity load in pounds by 150.

Lighting. The Elevator Code requires that cars and landings be well-lit at all times when in use. Code requirements are minimums and should be exceeded if necessary.

A car should have at least two lights that provide a minimum illumination of 5 foot-candles for passenger elevators and 2½ foot-candles for freight elevators at the landing edge of the car platform, when the car and loading doors are open.

On freight elevators with perforated car tops and that travel not more than 15 ft, car lights may be omitted, if at least two electric lights are installed at the top of the hoistway and furnish the minimum illumination specified.

Illumination on landing thresholds should be at least 5 foot-candles.

Hoisting ropes

Car and counterweight ropes must be of iron (low-carbon steel) or steel. Ropes less than ½ in. in diameter may not be used on passenger elevators. Traction elevators should have not fewer than three hoisting ropes; winding-drum elevators should have not fewer than two hoisting ropes and two ropes for each counterweight used.

The Elevator Code requires that the diameter of sheaves or drums for hoisting or counterweight ropes be not less than 40 times the diameter of the ropes. In practical application, however, this ratio is usually higher than 40.

On winding-drum machines, rope should be long enough so there will not be less than one full turn of rope on the drum, when the car is at the extreme limit of its overtravel. Drum ends of ropes are usually secured by tapered babbitted sockets or by clamps on the inside of the drum.

Rope fastenings. Ends of car and counterweight suspension ropes are usually fastened by individual babbitt-filled tapered sockets (Fig. 28-8). The Elevator Code gives detailed specifications on proper application.

Rope sockets must develop at least 80 percent of the breaking strength of the rope used. Shackle rods, eyebolts, and other means used to connect sockets to the car or counterweight must have a breaking strength at least equal to that of the rope.

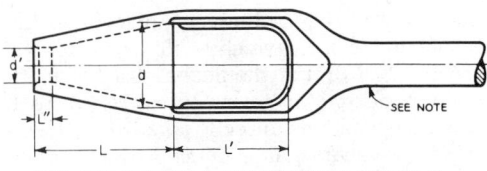

NOTE – ROPE SOCKET AND SHACKLE ROD MAY BE IN ONE PIECE, AS SHOWN, (UNIT CONSTRUCTION) OR THE SOCKET AND ROD MAY BE SEPARATE. (SEE RULE 212.9C)

Fig. 28-8.—Tapered babbitted rope sockets. Diameter (d') of the hole at the small end of the tapered portion of the socket shall not be more than as follows:

Rope Diameter (in., nom.)	Max. Diameter of Hole
½ to ¾ (inc.)	⅛ in. larger than nominal diameter
⅞ to 1⅛ (inc.)	5/32 in. larger
1¼ to 1½ (inc.)	3/16 in. larger

Courtesy American National Standards Institute.

Governor ropes should be of iron, steel, Monel metal, phosphor bronze, or stainless steel and be of regular lay construction not less than ⅜ in. in diameter. Tiller rope is prohibited by the Elevator Code. Rope that winds on the safety drum must be corrosion-resistant.

Operating controls

Stopping devices. Winding-drum elevators are required by the Elevator Code to have an adjustable, automatic, machine-limit stop mechanism, which will stop the car, if it overruns the highest and lowest landings. In addition, limit switches should be placed either in the hoistway or on the car.

Electric elevators with traction machines are required to have stopping switches in the hoistway, on the car, or in the machine room, if the switches are operated by the motion of the car.

Drum-type elevator machines are required to have a slack rope device to shut off power to the machine and brake in case the rope becomes slack. This device must be such that it won't reset automatically when slack in the rope is removed.

Every electric elevator must have an emergency stop switch in the car, adjacent to the operating device, which will cut off the power. The button should be clearly identified and be red in color. Contacts of such switches should be directly opened mechanically and should not depend solely upon springs for opening contacts.

Grounding. The motor frame (and the operating cable, if insulated from the motor frame) should be grounded. All switches and wiring should conform to the National Bureau of Standards' *National Electrical Safety Code* and the NFPA *National Electrical Code*.

Safety devices. Every elevator car suspended by wire ropes is required by the Elevator Code to have one or more car safety devices that will catch and stop the car in the event of overspeed or failure of the hoisting ropes. The safeties must be attached to the car frame, and one safety must be located within or below the lower members of the car frame (safety plank).

The Elevator Code classifies three types of safeties:

TYPE A (instantaneous) safeties rapidly increase their pressure on the guide rails to give a very short stopping distance. They are permissible on elevators having a rated speed of not more than 150 fpm.

TYPE B safeties apply limited pressure to the guide rails, and the retarding forces are reasonably uniform after full application. They are permissible on elevators of any speed and may be used in multiple.

TYPE C safeties (Type A with oil buffers) are permissible on elevators with a rated speed of not more than 500 fpm.

Counterweight safeties should conform to the car safety requirements.

Car switches. The handle of the car switch (operating control) should be designed to return to the "stop" position and lock, when the hand of the operator is removed.

Signal system. Every elevator (except automatic-operation and continuous-pressure-operation elevators) should have a signal system that can be operated from any landing

to signal the car when it is wanted at that landing.

Inspection and maintenance

Inspections and tests must conform to requirements and regulations set forth by the particular municipality or regulatory agency. Much of the discussion in this section is based on ANSI A17.2, *Practice for the Inspection of Elevators (Inspectors' Manual)*, published by the American Society of Mechanical Engineers and also available from the American National Standards Institute. It is important that anyone concerned with maintenance and inspection should have and use a copy of the Elevator Code and the Inspectors' Manual.

A guide for inspection is given in the Inspectors' Manual. It contains these requirements:

1. Acceptance inspections and tests of all new installations and alterations of elevators, dumbwaiters, and escalators should be made by an inspector employed by the enforcing authority.

2. Periodic inspections and tests of all installations should be made by a person qualified to perform such services, in the presence of an inspector employed or authorized by the enforcing authorities. It is recommended that periodic inspections and tests be made at intervals not longer than:

    THREE MONTHS for power passenger elevators.

    SIX MONTHS for escalators and power freight elevators.

    TWELVE MONTHS for hand elevators and power and hand dumbwaiters.

3. Maintenance inspections and tests should be made by a person qualified to perform such service in the presence of an inspector employed or authorized by the enforcing authority. Car and counterweight safeties, governors, and oil buffers should be given maintenance inspections and tests at least every five years.

Inspection program. Careful maintenance is essential to the conservation and safe operation of elevators and their appurtenances. It also reduces the need for repairs. Frequent and thorough inspections by qualified personnel are the first requisite of an efficient maintenance program. However, minor day-to-day inspection and maintenance can be performed by plant personnel.

Items of regular maintenance that require special consideration include: lubrication of hoisting and counterweight wire ropes and of oil buffers; cleaning and lubrication of guide rails, controller contactors and relays, and car safety mechanisms; cleaning of hoistways and pits, machine rooms, and tops of cars.

Hoistways and landings. Accidents occurring or originating at landings usually are due to: (*a*) tripping or slipping at the car entrance or landing, (*b*) being caught by the car, (*c*) falling down the hoistway, and (*d*) being caught by the doors.

To help prevent tripping and slipping accidents at hoistway landings, the following points should receive attention:

1. The leveling of the car should be checked. Improper leveling may be due to careless operation or, where cars are leveled automatically, to improper adjustment. Operation should be observed and necessary instructions given to the operator. With an automatic car-leveling device, the operator has no control over the final stop. It should be adjusted by a competent mechanic, preferably one trained or employed by the elevator manufacturer.

2. The condition of landing sills and floors should be watched. Landing sills should be of nonslip material; if they are not, they should be replaced or roughened when worn smooth. Broken sills, holes in flooring, worn floor coverings, and other conditions that create tripping hazards should be repaired. The finished surface should be flush with the surrounding floor.

3. Illumination of landings should be checked, with particular attention to those near building entrances where the difference between outdoor and indoor light intensity is noticeable. Globes and reflectors should be clean, and lamps should be of adequate size.

To provide the required protection at landings, interlocks and contacts must be well maintained and correctly adjusted.

Track grooves for hoistway doors should be clean, so that the doors will move freely. There should be no excessive play. Vision panels should be clean and unbroken. Counterweights should operate freely, and sheaves should be properly aligned.

Because the entire load is transferred to the guide rails when the safety operates, the guide rails must be properly aligned at the joints and securely attached to the brackets, and the bolts with which the brackets are attached to the walls must remain tight. Alignment of rails can be checked easily by sighting along the faces; bracket bolts must be tested individually.

Except where roller guide shoes, which run on dry rails, are provided, guide rails should be kept properly lubricated to help reduce wear between the rails and the guide shoes.

Roller guide shoes should be clean and, if necessary, adjusted for pressure against the rails.

Elevator cars should be checked for structural defects, such as loose bolts and other fastenings, excessive play in guide shoes, and worn or damaged flooring.

Each emergency exit should be tested by being opened. If panels are held in place by locks, the key should be in the car; if by thumbscrews, they should be removable without the use of pliers.

Car doors or gates should be subjected to the same examination as hoistway doors, and the same standards of maintenance should be followed. Contacts should be checked for adjustment in the same manner as hoistway door contacts.

The car operating switch should be tested to see that it returns to neutral position, when released by the operator, and locks there. On a cable-operated car, the cable lock should be checked to make sure that, when it is locked, the cable cannot be operated. All switch contacts in the car-operating device should be examined, and the glass cover should be in place over the emergency release switch.

Car lights should be checked for operation and for loose or missing screws and broken or cracked glassware. Glassware or bulbs should be clean.

Safety devices. Maintenance of safety devices often is neglected because they do not affect normal operation of the car. However, in case of emergency, the safety of passengers depends entirely upon proper performance of these devices. It is of utmost importance, therefore, that they be maintained in proper working condition.

At frequent intervals, all types of safeties should be cleaned and lubricated; inspected for worn, cracked, broken, or loose parts; and tested to determine their ability to stop and hold the car. Safety devices should be tested and adjusted only by a person qualified to perform such services, in the presence of an inspector employed or authorized by the enforcing authorities.

Limit switches. The inspector should check the switches for proper alignment and the mounting of switches and cams for rigidity.

Buffers. Since oil buffers lose some oil during normal usage, the oil levels should be checked at least once every month and each time the buffer is known to have been compressed.

For refilling, an oil of the type specified by the manufacturer should be used. Buffers that have been submerged by floods or pit leakage should be emptied, cleaned, and refilled with fresh oil. The alignment and the tightness of bolts in the anchorage should be checked.

Spring buffers should be checked for alignment and for proper seating in the cups or mountings. Springs should be examined for deformation and permanent set.

Hoisting machines should be examined carefully at each inspection. The machine base should be checked for misalignment and cracks, and defects should be repaired immediately by the manufacturer. The inspection should also include the following details:

1. The oil level in the motor

2. Brake operation

3. The oil level in the gear housing

4. Sheaves and drums for cracks and wear in the grooves

Belted machines. At frequent intervals, belts should be examined for proper tension, wear, burns, condition of splices, and cuts and breaks in the surface. Chains should be checked for excessive wear. Machine fastening bolts, belt guards, and the fastenings of platforms under any ceiling machinery also should be checked regularly.

Hoisting ropes. Records show that it is relatively unusual that the parting of hoisting ropes causes an elevator accident. When it does happen, though, it is more likely to occur with winding-drum machines than with traction machines. Although ropes usually are installed to give the high safety factors specified in the Elevator Code, they should be inspected closely to avoid the possibility of accident.

Rope life is shortened by improper brake action. Unduly sudden stops may result from brake defects, such as heavy spring pressure or brakeshoe wear.

Unnecessary starting and stopping also shorten rope life. Inching for landings should be avoided. In some cases, it can be eliminated by proper adjustment of the stopping devices; in most cases, it is due to faulty operation and the operator should be reinstructed.

Inspection routine. Persons making elevator inspections should take all necessary precautions for their own safety. American National Standard A17.2, *Practice for the Inspection of Elevators (Inspectors' Manual)* contains not only detailed instructions for the conduct of all tests, but also comprehensive information on personal safety for inspectors working in machine rooms, in and on top of cars, and in pits.

Close-fitting clothing, preferably one-piece overalls with all buttons fastened and without cuffs, is recommended. Gloves should not be worn.

Inspectors must pay close attention to moving objects, such as counterweights, to hoistway projections, and to limited overhead and pit clearance. Before electrical parts are inspected, switches should be locked in the open position, so that current-carrying parts cannot be energized. When a man is on top of a moving car, he should keep one hand free to hold onto the crosshead or other part of the car frame on top. He should not hold the hoisting ropes. Safety belts should not be attached to ropes, but to a fixed structure of the car or frame.

If controls are not provided on top of car for inspector's use, specific instructions should be given to another person who operates the elevator car, while the inspector is inspecting the top of car and other appurtenances.

The order in which the various parts are inspected depends upon the type of installation and the preference of the inspector. However, inspection time and interference with operation of the elevator can be reduced by planning—by determining which parts of the job can be done from each location.

A written report of each inspection should be made and kept on file. Such a record is particularly valuable for checking the progress of defects in ropes. Each report, therefore, should give definite details concerning rope condition—diameter, number of broken wires per unit length, and estimated percentage of wear. Most insurance carriers can supply the most recent elevator inspection reports.

A report of each service interruption should be made by the mechanic who corrects the trouble. The report should be entered on the log-sheet for the particular elevator, so that the maintenance engineer can spot defective equipment that causes repeated interruptions and correct the basic fault.

Operation

Training and placing operators. Many common causes of elevator accidents can be eliminated by safe operating practices. Best results are obtained where a properly instructed operator can be assigned to full-time duty. In any case, only specified employees who have been properly instructed should be permitted to act as elevator operators.

Elevator operators should be selected with extreme care because of the risks involved. In many cases, the lives of others depend upon the operator's efficiency. Moreover, an incompetent or poorly trained operator may damage valuable and indispensable equipment.

Faulty practices, such as starting a car be-fore the doors and gates are closed, blocking gates open, and permitting crowding on cars not provided with gates, have caused many serious injuries. Actions, such as improper loading, unnecessary starting and stopping, and reversing of the controller, have caused damage to equipment.

Persons selected to operate elevators should be free from physical defects that might interfere with their duties. They should be mentally alert, not easily excited, and capable of carrying out instructions and of insisting upon compliance with rules.

Some companies have found that elevator operators work more safely, if they are given cards which state that they have completed the training course and are authorized to operate the equipment. In some cases, company rules require that operators have such cards.

Operating rules. Definite operating rules should be adopted. In the case of industrial freight elevators, the rules should be posted in the car, and persons serving as operators should be required to know and observe them. Many concerns that employ a number of elevator operators prepare pocket-size rule books for their guidance. In plants with a number of elevators that are used under variable conditions, specific rules for specific elevators or groups of elevators are prepared.

Accident investigations. Careful investigation of minor elevator accidents often will disclose conditions or practices that, if left uncorrected, later may cause serious accidents. Every accident, therefore, should be carefully investigated to determine the responsible conditions or practices, and prompt corrective action should be taken. See Chapter 7 for details.

Overloads. The overloading of elevators may cause injury to personnel, mechanical failure to the machine or the car, or both. Many machines, still in service, were built with much lower safety factors than are now required, and with them the hazard is particularly great. However, even though machines meet current requirements for factors of safety, the danger of overloads cannot be overstressed, as the consideration of a few engineering factors will show.

In a traction elevator, the ratio between rope tension on the driving sheave (car side) and rope tension on the counterweight side must be kept within certain limits. The counterweight normally is equal to the weight of the car plus 40 percent of the rated safe load. The motor torque and the brake are designed to handle this difference in weight.

In the case of a freight traction elevator with a car weighing 8000 pounds and a rated safe load of 10,000 pounds, the counterweight would equal 12,000 pounds—the weight of the car, or 8000 pounds, plus 40 percent of the rated safe load, or 4000 pounds. The motor would be designed to lift, and the brake to hold, a 6000-pound load.

If the elevator is overloaded by 50 percent of its rated safe load (for this example 5000 pounds), the total platform load will be 15,000 pounds. If 4000 pounds (over-balance of the counterweight) are subtracted from the 15,000 pounds on the platform, a net weight of 11,000 pounds must be handled by a motor and brake designed to handle only 6000 pounds. This is an 83 percent overload.

In such a case, the traction relation may be broken, and the motor may not even pick up the load. As the brake is lifted, the car may start to move downward, probably a fuse will blow, and other mechanical failures, as well as injury to personnel, may result.

In no case should an elevator be overloaded unless someone, familiar with the design of the particular equipment, has checked the entire installation for its ability to handle the load. Failure to do so may result in serious injury or death to employees and in serious damage to valuable equipment.

If an overload or a one-piece load (a transformer, for instance) as heavy as or heavier than the rated safe load of the car must be handled, the company that installed the elevator or another reputable elevator company should be called in to see that:

1. The machine is strong enough to handle the load

2. The elevator structure, including the car frame (sling), platform, and undercar safeties, is adequate

3. The traction relation will not be exceeded

If the machine is otherwise strong enough

ELEVATOR EMERGENCY INSTRUCTIONS

When the car stops:

1. Throw emergency switch to STOP

2. Press alarm button until help arrives, or use emergency phone if so equipped

3. Do not try to exit car until told to do so by rescuers

4. Be patient

Fig. 28–9.—Typical

for the overload, the elevator company's crew may increase the counterweight to maintain the traction relation while the overload is being lifted.

Emergency procedures

It is recommended that an emergency procedure be implemented, similar to the one listed in Fig. 28–9, for the safe removal of persons from elevators stalled between floors. As is illustrated, emergency instructions for persons involved in such incidents are spelled out in two printed self-adhesive stickers. It is suggested that these stickers be mounted behind a protective transparent material to preserve legibility. In addition, foreign language stickers should be posted, if necessary.

In gaseous or toxic environments, tests should be taken to determine accident/injury potential and, if necessary, other emergency procedures should be implemented.

It is also recommended that a telephone or other means of communication be installed in the car. This may prevent panic by the occupants and may assist coordination of emergency procedures.

Elevator emergency operation and signal devices are discussed in the 1973 Supplement to the Elevator Code.

Sidewalk Elevators

A sidewalk elevator is defined in the Elevator Code as an elevator of the freight type for carrying material, exclusive of automobiles, and operating between a landing in a sidewalk (or other area exterior to a building) and the

Emergency stickers.

floors below the sidewalk or grade level.

Because sidewalk elevators present hazards not easily eliminated, it is always preferable to locate them inside the building line or in an area not accessible to the public.

Sidewalk elevators should conform to the requirements of the Elevator Code for power freight elevators, with certain minor exceptions.

Except by permission of the administrative authority, the maximum dimensions of sidewalk openings should be 5 ft at right angles to and 7 ft parallel with the building line, and the side of the opening nearest the building should be not more than 4 in. from the building wall.

Where hinged doors or vertically lifting covers are provided at the sidewalk or at other areas exterior to the building, bow-irons or stanchions shall be provided on the car to operate such doors or covers.

A sidewalk elevator with winding-drum machinery should have a normal terminal stopping device on the machine and one either in the hoistway or on the operating device.

Operation

A sidewalk elevator is required by the Elevator Code to be operable in both the up and the down directions through the opening, only from the sidewalk or other exterior area.

Either a key-operated, continuous-pressure up-and-down switch, or continuous-pressure up-and-down buttons on the free end of a detachable flexible cord, 5 ft or less in length, should be used.

Hatch covers

Automatic hatch covers, when closed, should be capable of sustaining a uniformly distributed static load of 300 pounds per sq ft.

Hatch covers are required by the Elevator Code to be self-closing. Fastening or holding them open when the car is away from the top landing is forbidden by the Code.

The covers should be made of metal and should be vertically lifting or hinged with the line of the hinges at right angles to the building wall. When the covers are fully open, there should be minimum clearance of 18 in. between them and any obstruction.

Hatch covers should have a coefficient of friction not less than that of the surrounding walkway surface. No hinges, locks, or flanges should project above the closed covers to constitute tripping hazards to passers-by.

Hand Elevators

Hand-powered elevators once were used extensively, but today they are found occasionally in storage rooms or warehouses.

In no case is it advisable to install a hand elevator where it is to be used constantly during the working day. Instead some form of power equipment should be installed wherever an elevator is a basic part of the manufacturing process or service function.

Mechanical power never should be applied to hand elevators by means of rope-grip or similar attachments. If power is to be used, the entire installation should be changed and all the protective devices, called for by the Elevator Code for power elevators, are required.

A hand elevator having a travel of more than 15 ft is required by the Elevator Code to have a safety, attached to the underside of the car frame, which is capable of stopping and sustaining the car and its rated safe load.

The construction of hand elevators is specified in the Elevator Code. No hand elevator car upon which persons are permitted to ride should have more than one compartment or be arranged to counterbalance another car.

With regard to hoistways, hoistway openings, pits, machinery spaces, supports, and foundations, hand elevators should conform to the requirements of the Elevator Code for power elevators.

779

Hoistway doors

Every hoistway door should have conspicuously displayed on the landing side, in letters not less than 2 in. high: DANGER—ELEVATOR—KEEP CLOSED.

Hoistway openings may have self-closing doors that extend to the floor; doors made in two parts, one above the other, and so arranged that the lower part can be opened only after the upper part has been opened; or doors equipped with two spring locks or latches, one located 6 ft above the floor.

Safety devices and brakes

Hand elevators should have hand or automatc brakes that operate in either direction of motion and, when applied, remain locked in the ON position until released.

Where the travel exceeds 40 ft, a driving machine with a hand-operated brake also should have an automatic speed retarder.

Dumbwaiters

In the Elevator Code, a dumbwaiter is defined as a hoisting and lowering mechanism equipped with a car that moves in guides in a substantially vertical direction, that has a floor area not exceeding 9 sq ft, a compartment height not exceeding 4 ft, a rated safe load not greater than 500 pounds, and which is used exclusively for carrying materials. It may be hand- or power-operated.

Hoistways and openings

The Elevator Code requirements for dumbwaiter hoistways are substantially the same as those for elevator hoistways, with certain exceptions. The requirements for dumbwaiter landing openings and doors are designed to protect persons from falling into the hoistways and from being struck by the car as it rises or descends.

The Elevator Code specifies that the hoistway landing openings of power-driven dumbwaiters must be provided with hoistway doors that guard the full height and width of the openings.

With certain specified exceptions, the Code requires that power dumbwaiter doors be equipped with hoistway-unit-system hoistway-door interlocks that will prevent operation of the machine while any hoistway door or

FIG. 28–10.—A well guarded dumbwaiter has car gates at both openings. Gates and biparting hoistway doors are interlocked (arrow, upper right, points to gate interlock) to prevent operation while open. Car interior should be well lighted.

Courtesy Sheraton Hotels.

gate is open (Fig. 28–10).

Every hoistway door of hand dumbwaiters should have conspicuously displayed on the landing side, in letters not less than 2 in. high, the words, DANGER—DUMBWAITER—KEEP CLOSED.

Safety devices and brakes

The Elevator Code specifies that power dumbwaiters (except hydraulic dumbwaiters) have brakes that are automatically applied when the power is cut off or fails, and that they also have an automatic means to stop the car within the limit of overtravel at each terminal. Hand-driven machines should have hand-operated or automatic brakes capable of sustaining the weight of the car and its load.

A power dumbwaiter having winding-drum machinery, a travel greater than 30 ft, and a capacity in excess of 100 pounds requires a slack-rope device that will cut off the power from the motor and stop the car, if it is obstructed in its descent.

Escalators

An escalator is defined as a power-driven, inclined, continuous stairway for raising or lowering passengers. See Chapter 22, "Non-employee Accident Prevention."

Safety devices and brakes

Emergency STOP buttons (or other type of manually operated switches having red buttons or handles) shall be accessibly located at or near the top and bottom landings of each escalator and shall be protected against accidental operation. An escalator STOP button with an unlocked cover over it, which can readily be lifted or pushed aside, shall be considered accessible. The operation of either of these buttons or switches shall interrupt the power to the driving machine. It shall not be possible to start the driving machine by these buttons or switches (Fig. 22–10 on page 587).

Buttons or switches used for starting the units should be key operated and located within sight of the escalator steps. Means should be provided to cut the power in case an ascending escalator accidentally reverses its travel.

Each escalator should have a speed governor that will interrupt the power if the speed exceeds a predetermined value (not more than 40 percent greater than the rated speed). The overspeed governor is not required where a low-slip, alternating current, squirrel cage induction motor is used and the motor is directly connected to the driving machine.

If a tread chain breaks, a broken-chain sensing device should cut the power. Where an escalator has tightening devices that are operated by tension weights, provision should be made to retain these weights in the escalator truss if they should fall.

Each escalator should have an electrically released and mechanically applied brake capable of stopping the fully loaded escalator when it is traveling either up or down. The brake should stop the escalator automatically if any safety devices function.

Machinery

Every escalator machine room should have a light that can be lit without having to pass over or reach over any part of the machinery. Reasonable access to the interior of the escalator should be provided for inspection and maintenance. For the protection of maintenance personnel, all chains in escalator machinery compartments should be guarded.

While bearings on modern installations are sealed and require no oiling, the chains do require lubrication. Care should be taken not to overlubricate. An oil pan should be provided in the bottom of the truss to catch oil or grease that may drip from moving parts, as well as dust and dirt dropping between the treads.

The oil pan should be cleaned periodically to eliminate any fire hazard from the accumulated oil-soaked dust and dirt. A brush available for this purpose can be attached to a step axle and drawn down over the drip pan to brush all the foreign matter to the lower end of the truss. The sweepings then can readily be removed. Reversing the unit returns the brush to the top, where it can be disconnected.

The moving handrail, if of the common canvas duck and rubber construction, will show some stretch over a period of time. It should be checked at intervals, and the handrail drive adjustment should be used to take up the slack.

All parts of escalators and their driving machinery should be inspected at regular intervals and be well maintained. All safety devices should be tested for proper functioning.

Protection of riders

Most escalators are installed in public places, and their principal hazards arise from misuse by the public. Escalators have been developed to a high degree of safety, and their accident record, in general, is good.

Most escalator accidents occur when shoe heels, fingers, or toes are caught between the surface grooves or slots on the treads and the combplate. The Elevator Code limits the width of each slot to not more than 1/4 in. and the depth to not less than 3/8 in., with a center-to-center spacing of not more than 3/8 in. between adjoining slots.

In some accidents, edges of shoe soles have been caught between the step and the vertical side member (skirt guard). The Elevator Code limits this clearance to 3/16 in. on each side and limits the sum of the clearances on the two sides to ¼ in., unless skirt switches are installed.

The Elevator Code requires that each balustrade have a handrail moving in the same direction and at substantially the same speed as the steps. Each moving handrail is to extend at normal handrail height (not less than 12 in. beyond the points of the comb-plate teeth at the upper and lower landings). Hand or finger guards are to be provided at the point where the handrail enters the balustrade.

In some localities, barefoot passengers have been injured, and signs warning barefoot persons not to ride the escalator have been posted. Umbrella tips frequently are caught between the grooves and the comb-plate, and this type of accident occasionally results in minor injuries and in damage to equipment.

Many accidents have resulted from attempts to handle baggage on escalators. Suitcases and handbags should not be placed on the steps and, if carried by hand, always should be held parallel to the run of the escalator.

Escalator treads and landings should be of incombustible material affording secure foothold.

Some riders, through inexperience or infirmity, have trouble seeing the parting point where treads rise or descend and the point where they level off. To aid them especially, the Elevator Code calls for illumination of at least 2 foot-candles on all tread surfaces, but some firms provide additional warnings. In one case, all risers were painted a bright yellow, so that the rising face would contrast with the tread. In another application, green lights (red is also a good color) were mounted inside the truss at top and bottom to shine through the treads where they break away and come together—the trouble points.

Some manufacturers mark the edges of the steps to emphasize the lines between adjacent steps. One manufacturer, for example, adds a distinctive color strip to the edge of the step adjacent to the riser of the next step.

Signs reading PLEASE HOLD HANDRAIL are often posted at the top and bottom and are found to be helpful. It is important that no overly conspicuous or distracting signs (such as advertising) be placed near these critical points. (See Fig. 22–9 on page 587.)

At the various floor levels, directional signs and floor number markings are recommended in order to improve traffic flow from the escalators.

When traffic on escalators is extremely heavy, as in department stores during the Christmas shopping season, proprietors often station uniformed employees at each flight to repeat warnings and, if necessary, help riders on and off.

It is extremely important that no object or construction of any kind obstruct the free flow of passengers from the area at the exit of an escalator. (This area is not a part of the escalator.) Serious accidents have occurred where the flow of traffic from the exit was restricted by a fence or barrier placed at some distance from the escalator.

Fire protection

Protection of escalator floor openings against the passage of fire and smoke is required by the Elevator Code. One of the best safeguards against burnout is to divide buildings of fire-resistive construction into limited areas in which fire can be readily controlled. Vertical openings need to be protected against the passage of fire from story to story. This principle often is disregarded where escalators are installed.

Escalators, accredited as a required means of egress, must be fully enclosed in accordance with the requirements of local laws and ordinances pertaining to interior stairways.

Escalators, not accredited as a required means of egress, must have the floor openings protected by one of the following methods, in accordance with national and local standards and regulations:

1. Full enclosures

2. Kiosks

3. Automatic rolling shutters

4. Sprinklers so spaced as to protect the exposed sides of the opening

5. Spray nozzles (only where the building area is fully protected by a supervised automatic sprinkler system)

For detailed specifications, see "National Fire Codes," Vol. 4, *Building Construction and Facilities* (National Fire Protection Association).

Moving Walks

A moving passenger-carrying device, on which persons stand or walk and in which the passenger-carrying surface remains parallel to its direction of motion and is uninterrupted, is called a moving walk.

Criteria for the design, construction, installation, operation, inspection and testing of moving walks installed for the purpose of transporting passengers are set forth in the Elevator Code, part IX.

Moving walks may operate in a horizontal plane or in a slope up to a maximum of 15 degrees. Operating speed and treadway width are governed by the slope.

Comments as to protection of passengers, found in the previous Escalators section of this chapter, apply also to moving walks.

Manlifts

The following are the principal hazards in the use of manlifts:

1. The rider may be carried over the top

2. He may be unable to make an emergency stop

3. He may jump on or off after the step has passed the floor

4. His head or shoulders may strike the edges of floor openings

5. He may be unable to reach the landing because of power failure and belt stoppage

6. Parts of the manlift may fail or operate unsafely

Construction

Manlifts should be constructed, maintained, and operated in strict compliance with the recommendations of American National Standard A90.1, *Safety Code for Manlifts*. A safety factor of 6 (based on a 200-pound load, on each step, on both the up and down runs) should be used, and all equipment should be braced securely at top, bottom, and intermediate landings. Steps should be nonslip. The entire manlift should be suspended from the top to prevent bending or buckling of the rails.

Handholds (either open or closed type) should be painted a conspicuous color, such as orange or yellow. They should be not less than 48 in. nor more than 56 in. above each step, and steps should not be less than 16 ft apart.

Floor landings or emergency landings should be provided for each 25 ft of manlift travel. Clearance between the surface of any landing and the lower edge of the conical guard suspended from the ceiling should be at least 7½ ft. The minimum clearance between the center of the head pulley and the roof or other construction should be 5 ft.

The bottom landing on the up side should have steps to a platform, level with the manlift step as it rises to a horizontal position.

Floor openings and shear guards. At floor landings, standard 42-in. guard rails and 4-in. toeboards should be provided around floor openings in such a way as to permit a landing space at least 2 ft wide. Rails or guards should have maze or staggered openings or self-closing gates that open away from the manlift (Fig. 28–11). At each floor opening on the upside, funnel-shaped (conical) guards should be installed (Fig. 28–12).

Brakes, safety devices, and ladders

A brake, designed to be applied automatically when the power is shut off, should be installed on the motor shaft for direct-connected units and on the input shaft for belt-driven units. The brake should get its power or force from an external source. The brake must be capable of quickly stopping the manlift and holding it when the down side is loaded with 250 pounds on each step. It should be electrically released.

A control rope, not less than ⅜ in., or a rod should be provided within easy reach of both the up and down runs. When pulled in the direction of belt travel, the rope or rod should cut off the power and apply the brake. A mechanical-electrical device that will automatically shut off the motor power supply and apply an electric brake, if the rider fails to alight at the top landing, should be installed.

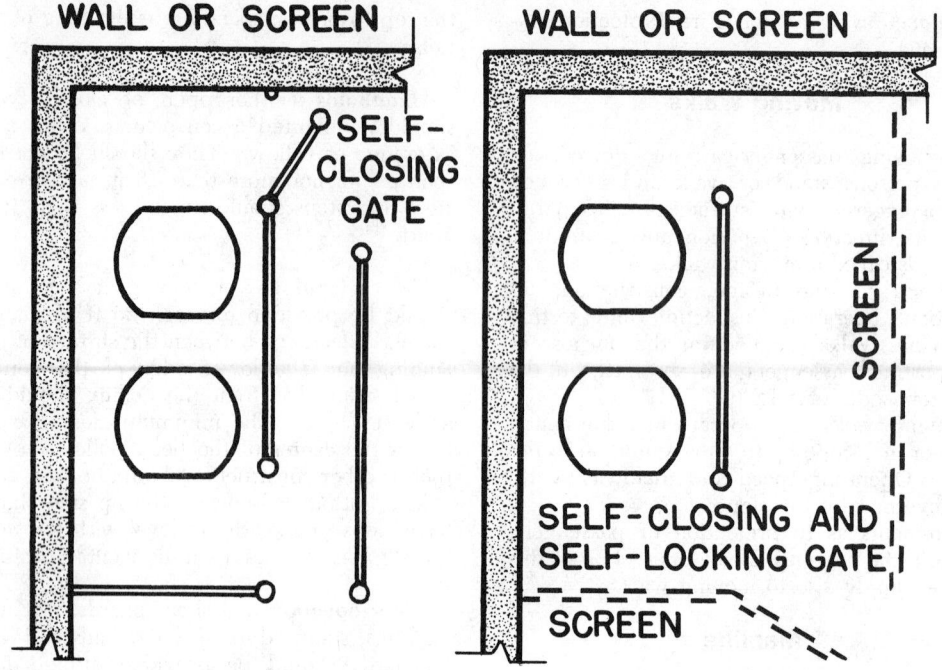

FIG. 28–11.—Guardrails (left) and screen enclosures (right) for manlift floor openings.

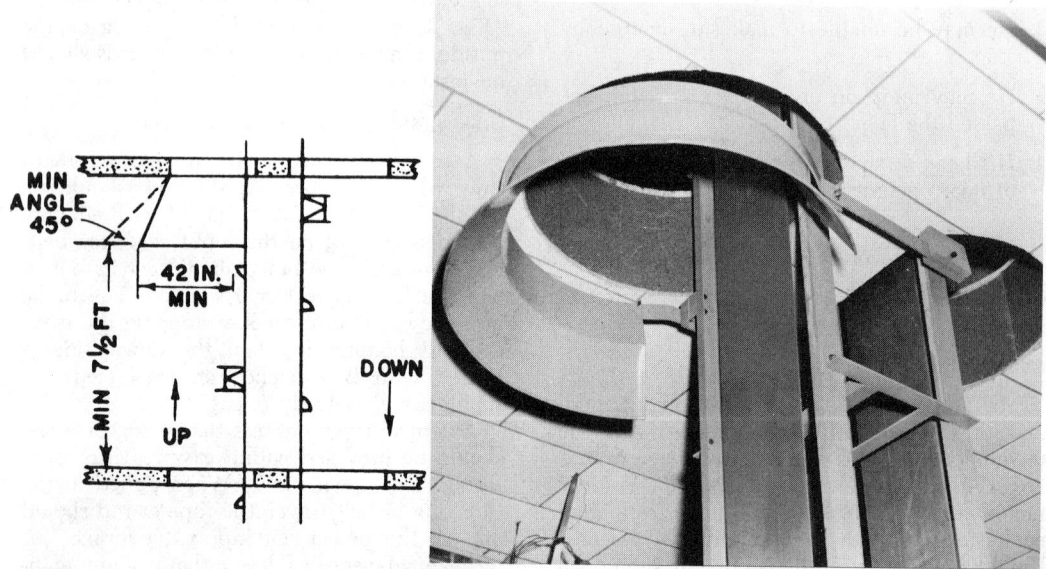

FIG. 28–12.—Drawing shows dimensions for flared-opening guards. In installation (shown at right), guard is counterbalanced and will yield slightly if hit.

Photo courtesy Humphrey Elevator Co.

Fig. 28-13.—A bar mounted vertically on head pulley trips switch to shut off power and prevent a rider from being carried over the top of the manlift.

Courtesy Humphrey Elevator Company.

It should be arranged so that the manlift cannot be started again except from above the upper landing.

A secondary safety control located on the top operating floor is required. It should be set to operate when the belt has traveled 6 in. beyond the point of operation of the primary safety switch, in case the latter fails. The device should stop the manlift before the loaded step reaches a point 24 in. above the top landing (Fig. 28-13).

A fixed metal ladder, accessible from both the up and down runs, should be provided for emergency exit where the vertical distance between landings exceeds 20 ft. It should meet the requirements of the governing agency for ladders, except that an enclosing cage should not be provided, since the ladder should be accessible from either side throughout its entire run.

Inspections

Every manlift should be tested and inspected at least every 30 days. Indicators of a defect are such things as unusual or excessive vibrations, continual misalignments, or "skips" when mounting steps (which indicates worn gears). On discovery of a defect, the manlift should be put out of operation *immediately* and not used until repaired. Each periodic inspection should cover, but not necessarily be limited to, the following items:

Steps	Lubrication
Step fastenings	Illumination
Rails	Warning signs and lights
Rail supports and fastenings	Signal equipment
Rollers and slides	Drive pulley
Belt and belt tension	Bottom (boot) pulley and clearance
Handholds and fastenings	Pulley supports
Floor landings	Motor
Guard rails	Driving mechanism
Limit switches	Gears
Electric switches	Brake

The safety mechanism of a manlift should be inspected daily for free operation and for accumulations of dirt and grease. A manlift should be completely dismantled once every three years, and defective or excessively worn parts should be replaced.

A written record should be kept of findings at each inspection. Records of inspections should be available, in case a request to see such records is made by OSHA or state compliance officers.

General precautions

The maximum speed of a manlift belt should not exceed 80 fpm and should be uniform on all manlifts throughout the plant. If the lift carries a great deal of traffic, a maximum speed of 60 fpm is recommended.

Floors should be numbered with large figures in full view of both ascending and descending riders. A constantly illuminated sign, TOP FLOOR—GET OFF, in block letters at least 2 in. high, should be conspicuously placed not more than 2 ft above the top landing, in full view of an ascending passenger.

In addition, at least a 40-watt red warning light should be located immediately below the top landing to shine in an ascending rider's face.

Signs carrying instructions for use of the manlift should be displayed prominently at each landing. Only authorized employees should be permitted to ride manlifts, and signs so stating should be displayed at each landing.

The entire manlift should be illuminated at all times while it is in operation, with at least 1 foot-candle at all points and at least 5 foot-candles at landings.

Nothing which cannot be placed entirely inside a pocket, a sling, or a pouch should be carried by a rider on a manlift. Tools may be carried in a leather-bottom canvas bag, not larger than 11 by 13 in., and provided with carrying loops or handles. While riding the manlift, the passenger should hold the bag in his hand.

Employees, particularly new ones, should be carefully instructed in safe use of the manlift and should be instructed to report immediately any defects or irregularity in operation of the manlift or the safety devices. Corrective action should be initiated immediately by supervisors.

Plant Railways

Clearances and warning methods

Insofar as possible, plant railway hazards should be eliminated in the design of a new plant. Horizontal and vertical track clearances are primary considerations.

It is recommended that in this and all other phases of *track construction*, AREA standards[*] be adopted, provided municipal, state or federal regulations do not conflict. For example, some state regulations are even more stringent than the AREA recommendations for overhead clearances. Fig. 28–14 shows the AREA recommended clearances.

Where platforms, building entrances, or structures are located along curved track, additional clearance must be allowed on both sides of the curve for lateral movement of cars and loaded flatcars at center and at ends.

Additional track clearance must be allowed also when awning or jalousie windows are installed in adjacent buildings.

CLOSE CLEARANCE warning signs should be placed at points where buildings or other obstructions are so close to a track that they will not clear a man riding on the side of a car. In addition, a red light should be installed and *continuously* burned. Side or overhead telltales sometimes are installed to give positive warning of approach to structures with close clearance.

Standard clearances may not give enough protection where tracks pass doorways or corners of buildings or other places where workers may walk directly onto tracks in front of moving cars. These locations may be safeguarded with fixed railings that force pedestrians to detour a short distance before stepping onto the track. If a barrier railing is impractical, hinged bars or gates swinging horizontally through an angle of not more than 90 degrees are effective.

Still another means is a convex mirror located at the intersection of the passageway and the track at a 45-degree angle, in which approaching railway equipment can be seen

---

[*] Association of American Railroads, Engineering Division (American Railway Engineering Association), *Manual of Recommended Practice.*

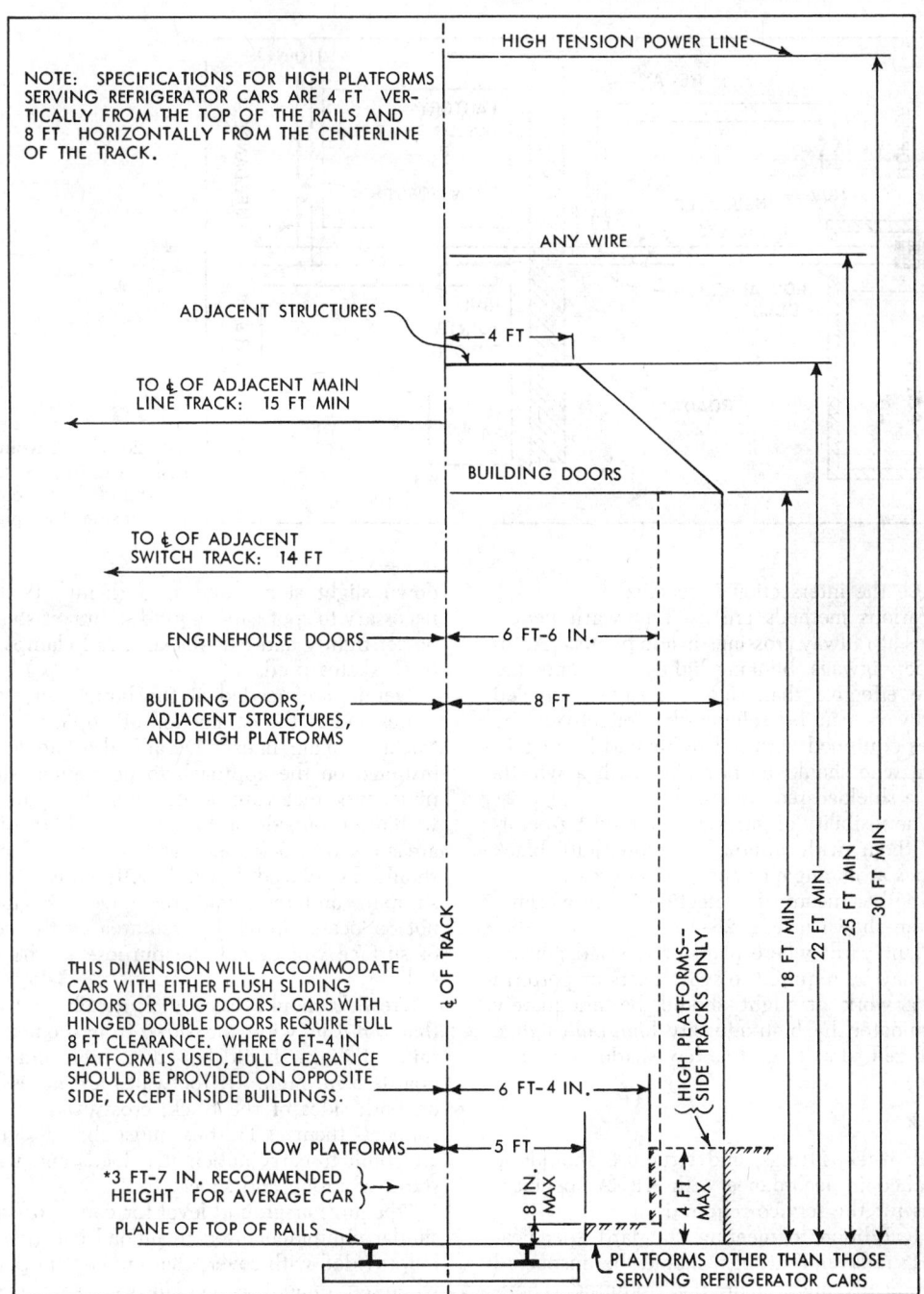

FIG. 28–14.—Clearances for standard gage tracks. Specific requirements in each state should be checked. (Drawing is not to scale.)

Based on information supplied by American Railway Engineering Assn. (Engineering Division, Association of American Railroads).

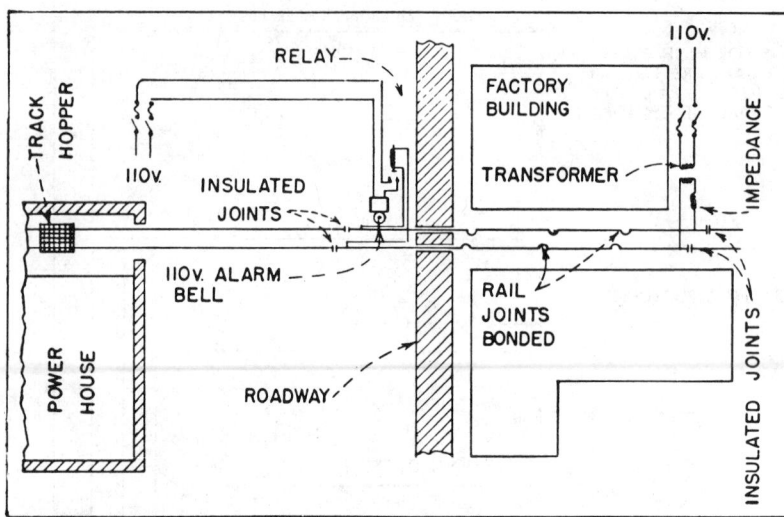

Fig. 28–15.—Layout of bell warning system for blind level crossings inside the plant.

before the intersection is reached.

Various methods are used to warn pedestrians at railway crossings inside plants. Automatic wigwags, blinking lights, and gongs are more effective than signs. Heavily traveled roadways can be safeguarded effectively by gates equipped with red lights and by watchmen, who should be provided with a whistle and a shielded red lantern.

The visibility of gates is increased by painting them with alternate yellow and black stripes at an angle of about 45 degrees.

Another means of protection is the warning system shown in Fig. 28–15.

Plant yards where plant or railroad personnel may be required to switch cars or perform other work at night should be adequately illuminated by high-intensity high-pole lights, arranged so as to cast as few shadows as possible.

Track

All track, fittings, and structure should be installed in accordance with AREA specifications for the service required.

In addition to meeting standard specifications, rails and fittings should be inspected periodically and repaired as required. Serious accidents may be caused if defects are permitted to go unremedied.

Tracks should be level at loading points. Workmen have been killed when cars roll down slight slopes and trap them. If it is necessary to spot cars on grades, brakes should be set tightly and car blockers, rail clamps, or track skates used.

Derails are needed at the bottom of steep slopes and at the connection of sloping switch tracks to main lines. Derails also should be installed on the approach to permanent shipping and receiving areas, whether located within or outside a building. Where such areas are on tracks open at both ends, derails should be placed beyond both ends of the shipping and receiving area. Derails should not be located in hard-paved area, as this type of surface will defeat the purpose of the derail.

Trestles should have a footwalk not less than 5 ft-1 in. wide, measured from the nearest rail. Railings should be 42 in. high and toeboards 6 in. high. If footwalks are necessary on both sides of the track, crosswalks should connect them. Trestles must be designed and built to carry anticipated loads and withstand vibration and shock.

Openings at ground level for conveyors and similar equipment used to unload cars should be provided with covers that are kept in place, when equipment is not in use, to prevent persons from falling into them.

Hoppers or trackside bins into which material is dumped should also be covered. Heavy steel bars, 12 in. apart, will prevent a man from falling or being carried through the openings. It may also be necessary to cover walkways that are under trestles, in order to protect those who use them from falling materials.

Where tracks are at a dead end, a standard bumping post should be installed.

A switch with a handle swinging parallel to the rails is safest. Targets should have rounded corners to lessen the possibility of cuts, scratches, and torn clothing. Switch lamps are needed if tracks are to be used at night. Blocking should be installed in switch points and frogs to prevent employees from getting their feet caught.

Loading and unloading

Tracks at loading and unloading areas warrant special attention because numerous accidents occur when loads are trucked into and out of cars. Excessive clearance requires heavy, large bridge plates or dockboards. If the dock level is considerably above or below the level of car door openings, trucking is more hazardous.

The dock should be wide enough to provide a temporary storage area without interference with truck movements. A narrow dock may force trucks to turn onto dockboards at a dangerous angle. The hazard may be reduced somewhat by making dockboards wider at the dock side, with flanges on the plates to turn wheels away from the edge. Portable dockboards must be securely anchored and be strong enough to carry the load imposed on them. Handholds or other means must be provided to permit safe handling.

Cars spotted for loading or unloading should be protected against being moved by switching crews. Standard railroad blue flags for daytime, and blue lights for night use, furnish warnings to train crews. Signals should be placed between the rails at both ends of a car that is accessible from either direction. Train crews should be strictly prohibited from coupling engines or cars to any cars so protected. The blue signals should be removed only by the employees, engaged in the loading or unloading operations, and only when they are ready to release the cars.

Bells and oscillating warning lights can be installed along the tracks in working areas to warn personnel that switching operations are

going on. The warning lights and bells should be turned on by plant supervision before switching operations are begun.

It is the responsibility of the local plant management to clear employees from cars before releasing the track to railroad employees for switching. Therefore, before derails and blue flags are removed, plant supervision must make sure that:

- Building railroad doors are opened and other obstructions cleared six feet from the track area

- All plant personnel are cleared from railroad cars and track area

- All overhead building cranes in the area being switched have stopped operations and are clear of the tracks

- All dockboards (bridge plates) are removed from cars

- All counterweighted, retractable, service platforms are retracted and secured

- All car moving equipment (cables, hooks, etc.) is removed from cars

It is preferable that tracks for loading or unloading flammable liquids or other dangerous materials are used only for those purposes. For additional protection, switches should be provided with locks. Specific recommendations in regard to tank cars, grounding, rail bonding, etc., are given in Chapter 42, "Flammable and Combustible Liquids."

The consignee of a carload shipment of material or merchandise, after unloading the car, should clean it before releasing it to the railroad that serves the establishment. In addition, they should make sure that the doors are properly closed and secured. Plug doors on box cars create a distinct hazard when the doors are not properly secured (Fig. 28–16). Pieces of crating and dunnage, nails, and strapping left loose in the car are serious hazards to railroad employees and others who may have to enter it later.

Overhead crane runways

A very serious hazard exists at points where an overhead crane runway crosses above a railroad track, either inside or outside. The hazard lies in the possibility of crane loads or hook blocks striking locomotives or cars while

Fig. 28–16.—Plug doors should be locked before car is moved.

Courtesy Inland Steel Co.

switching is going on.

A system of interlocked signal lights, with one set visible to the crane operator and the other to the switch crew, can be installed to guard against movement of the crane near the track while it is occupied. All personnel involved, especially crane operators, must be trained to respect the signals without exception. The signals can be actuated manually by a key switch, whose key is kept by the area foreman. The signals can also be interlocked with a derail.

Another method is to provide a zone power cutoff for the crane runways in the vicinity of the track, with the cutoff being actuated by a key switch under the control of the area supervisor. Other methods that assure positive control of the crane movements may be used.

Switch crews should be required to get clearance from the area foreman before moving into the area, and the foreman should be responsible for keeping cranes clear until the switch engine and cars move out.

Overhead cranes are also discussed in Chapter 25, "Hoisting Apparatus and Conveyors," pages 678–681.

Types of motive power

All plant locomotives (and cars as well) should be equipped with all safety appliances and with standard automatic couplers and air brakes, as required for common carrier railroads by federal law. All equipment and appliances should be maintained in sound, safe operating condition.

The type of motive power for a plant railway is important from the standpoint of accident and fire prevention. Explosive gases are easily ignited by flames or sparks from fuel-fired locomotives. Where such gases may be present, electric, compressed air, or storage battery locomotives should be used.

Where ventilation is insufficient to keep the concentration of noxious and even toxic exhaust gases at a safe level, as in mines, the use of fuel-fired locomotives should be prohibited. However, since diesel engines can be equipped with devices to eliminate toxic gases from the exhaust, this type of locomotive is being used in some adequately ventilated mines (other than coal).

A major hazard with electric locomotives, in addition to that of sparking in explosive atmospheres, is the possibility of employee contact with the trolley. The trolley should be guarded at all points where employees may pass under it and should be high enough to prevent contact anywhere along it.

Boilers of steam locomotives are constructed in accordance with the ASME boiler code and inspected in accordance with Federal Railroad Administration requirements.* Although the regulations of the administration are not mandatory for most industrial plant railways, the standards are an excellent guide.

Diesel locomotives, which operate more quietly than steam locomotives, should be equipped with bells. Some companies paint the front and rear of diesel locomotives in contrasting colors, such as yellow and black stripes. Deck walks should have handrails around the outside. Each locomotive should carry an extinguisher for oil fires.

The tanks of compressed air locomotives should be constructed in accordance with ASME standards.** Each air receiver requires an air pressure gage, a pop safety valve, and drainpipe with valve at bottom.

A battery locomotive should have a dead-man control, so that the operating lever will return automatically to the "off" position when released, or so that the circuit is broken when the operator leaves his position. Safety provisions for battery charging rooms are in Chapter 27, "Powered Industrial Trucks."

Tools and appliances

Devices, tools, and appliances are important in safe operation of a plant railway.

Automatic couplers eliminate the serious hazards involved when men go between either standard or narrow gage cars to insert or withdraw pins manually.

Hopper bottom cars should be opened only with standard car wrenches. A special tool for closing the latches on bulk cars eliminates the need for a man to go on top of the car, from which he might fall. Rerailers should be provided for narrow gage cars; other means of rerailing cars generally are dangerous.

Moving cars

Moving cars by hand methods often results in accidents. The safest procedure is to use a switch engine.

If the ordinary hand car mover is used, there should be a shield around the bar, so that the employee will not strike his hands or otherwise injure himself, if the tool should slip. Crowbars and other makeshift tools should not be used to move cars.

When a car is on a grade, a workman should test the handbrake to make sure that it takes hold and that excess slack in the brake chain is taken up. The man should remain on the brake platform to stop the car with the brake at the required point.

Winch-type car pullers are used extensively. If the cable breaks, the operator may be killed, since he is in line with it while operating the equipment. A shield of steel plate or of expanded metal should be installed to protect him. A forged steel hook should fasten the cable to the car.

A self-propelled, rider-operated car mover is used in many industries (Fig. 28–17). This vehicle has rubber-tired wheels and steel rail wheels, both retractable, so that it can run either on plant grounds or on rails. It is fitted with standard couplers and, depending upon the model, can develop a drawbar pull of from 8400 to 18,000 pounds. Its use eliminates the need to use hand car movers, winches, capstans, or power equipment, which are not designed for the purpose and which are often hazardous.

It should be understod that use of a push-pole to move cars by a locomotive is dangerous under any circumstances and is not recommended.

Safe practices

It is vital that transportation personnel—as well as all other employees—observe safe practices. They can be guided by the very same safety rules observed by all large railroad systems. When plant railway safety problems arise, help and suggestions can be sought from safety professionals and operating officers of the plant's connecting railway line.

---

* See the ASME *Boiler and Pressure Vessel Code*, Section III, "Boilers of Locomotives" (American Society of Mechanical Engineers); and *Laws, Rules and Instructions for Inspection and Testing in Steam Locomotives and Tenders and Their Appurtenances* (Federal Railroad Administration, U.S. Dept. of Transportation).

** *ASME Boiler and Pressure Vessel Code*, Section VIII, "Unfired Pressure Vessels." Also see Chapter 44, "Boilers and Unfired Pressure Vessels."

Fig. 28–17.—Self-propelled, rider-operated car mover. This model has special coupler which transfers part of freight car's weight to the power unit to help increase traction.

Courtesy Whiting Corp.

Safety meetings and other educational activities should be used to inform plant personnel about railway hazards.

To include all the safe practices required for the safe operation of railway equipment would require a book in itself. However, a few of the more important ones are included here:

1. Stop and look both ways before crossing any track.

2. Expect trains or cars to move at any time, on any track, in either direction.

3. *Step over* rails when crossing tracks. Never step, walk, or sit on any rail.

4. Never go between moving cars (or cars that may move) for the purpose of adjusting couplers or for any other purpose. (The once common practice of kicking couplers to align them is particularly dangerous.)

5. Give a hand or lamp stop signal, and receive an acknowledgment of the signal, before going between standing engines or cars.

6. To close a boxcar door, place one hand on the door handle and the other on the back end of the door.

7. Step down from cars—do not jump.

8. *Never* attempt to ride the leading footboard of an engine. (This rule applies to *all* employees, including switchmen.)

References

American Society of Mechanical Engineers, 345 East 47th St., New York, N.Y. 10017.
    ASME Boiler and Pressure Vessel Code, Section III, "Boilers of Locomotives," and Section VIII, "Unfired Pressure Vessels."

American National Standards Institute, 1430 Broadway, New York, N.Y. 10018.
  Practice for the Inspection of Elevators, Escalators, and Moving Walks (Inspectors' Manual), A17.2.
  Safety Code for Elevators, Dumbwaiters, Escalators, and Moving Walks, A17.1 (Also published by American Society of Mechanical Engineers.)
  Safety Standard for Manlifts, A90.1.

Association of American Railroads, 59 East Van Buren Street, Chicago, Ill. 60605.
  Engineering Division (American Railway Engineering Association)
    Manual of Recommended Practice

Federal Railroad Administration, U.S. Dept. of Transportation, Washington, D.C. 20590.
  Laws, Rules and Instructions for Inspection and Testing in Steam Locomotives and Tenders and Their Appurtenances.

National Bureau of Standards, U.S. Department of Commerce, Washington, D.C. 20234.
  National Electrical Safety Code, NBS Handbook H30. (Also adopted as American National Standard C2.)

National Fire Protection Association, 470 Atlantic Ave., Boston, Mass. 02210.
  National Electrical Code. Standard No. 70. (Also adopted as American National Standard C1.)
  "National Fire Codes," Vol. 4, *Building Construction and Facilities.*

National Safety Council, 425 North Michigan Ave., Chicago, Ill. 60611.
  Industrial Data Sheets
    Escalators, Data Sheet 516.
    Floor Mats and Runners, Data Sheet 595.
    Manlifts, Data Sheet 401.

Point-of-Operation and Transmission Guards

Guarding Under OSHA   795
Guarding terminology . . . Built-in guards . . . Checklists for
guarding . . . Types of guards

Point-of-Operation Guards   800
Guard openings . . . Test results . . . Application of test findings
. . . Guard construction

Power Transmission Guards   806
Guarding materials

Guards and Noise Control   808

Summary   809

References   809

Guarding Under OSHA

One of the major goals of OSHAct is the guarding of all machinery and equipment to eliminate personnel hazards created by points of operation, ingoing nip points, rotating parts, flying chips and sparks. These hazards have been responsible for countless numbers of serious injuries and fatalities of personnel. If the now-required guarding had been more widely used then, many if not most of these accidents might never have occurred and even OSHAct would probably not be the law of the land.

In the past, the reasons for not guarding were almost as many as there were accidents. The commonest reason, and one still heard today, probably was that the guard either interfered with or made it impossible to achieve production goals. Thus points of operation and/or power transmission were either never guarded, or the guards were removed and never replaced. In some cases, removal of guards was justified for they were poorly designed and created additional hazards to personnel, but in most instances guarding was not installed, or was removed and not replaced simply because "it really wasn't necessary" or "it was inconvenient to take it off and put it back each time the machine had to be adjusted, serviced, or repaired."

Today, most of these reasons and the attitudes which went along with them, are not acceptable excuses to OSHA. The regulations clearly state that points of operation and power transmission *shall* be guarded. Exceptions to this can be obtained only by applying for a variance from the regulations. Variances may be granted if it can be proved that the current "state of the art" (in other words —technology) is not capable of guarding the hazard without creating an additional hazard or without seriously degrading the production process. Even then, those applying for a variance must develop alternate safeguarding methods which in effect will be equivalent or better than the protection of a guard. Details are in Chapter 2.

It should be obvious that the words "shall be guarded" apply to most, if not all, machines and equipment that a company uses. How they apply may not be very clear, particularly when each machine is considered.

Some machines require specific methods of guarding; all machines are regulated by the general requirements.

Guarding terminology

As in many problems, a lack of understanding or agreement on the meanings of terms used causes much confusion and even disagreement. To avoid this, some of the most common terms are defined as used both in standards and in this chapter.

GUARDING. Any means of effectively preventing personnel from coming in contact with the moving parts of machinery or equipment which could cause physical harm to the personnel. When discussing power press guarding, the word "guard" is used exclusively for referring to barriers designed for safeguarding at the point of operation. Therefore, the word "safeguard" should be used for a barrier or cover that protects other danger zones.

ENCLOSURES. Guarding by fixed physical barriers that are mounted on a machine to prevent access to the moving parts. They are most effective when they are designed as an integral part of the machine, but they can be bolted or welded to the frame.

FENCING.° Guarding by means of a fence or rail enclosure which restricts access to a machine except by authorized personnel. Such enclosures must be a minimum of 42 in. away from the dangerous part of a machine.

LOCATION.° Guarding which is the result of the physical inaccessibility of a particular hazard under normal operating conditions or use.

POINT OF OPERATION. That area on a machine where material is positioned for processing or change by the machine.

POWER TRANSMISSION. All mechanical components including gears, cams, shafts, pulleys, belts, and rods which transmit energy and motion from the source of power to the point of operation.

INGOING NIP POINTS OR BITES. A hazard

° Both fencing and location are very limited as guarding techniques and are permitted only if restrictions can be met.

area created by two or more mechanical components rotating in opposite directions in the same plane and in close conjunction or interaction.

SHEAR POINTS. A hazard area created by a reciprocal (sliding) movement of a mechanical component past a stationary point on a machine.

Built-in guards

Most machine manufacturers make point-of-operation and power transmission guarding "standard equipment" on all stock type machines, meaning that the machine cannot be purchased without guards. In some cases, when they know the end use of the machine, some manufacturers have refused to sell the machine without a point-of-operation guard. This highly desirable state of affairs has in the main resulted from legal actions and legislation holding the manufacturer liable for accidents arising from "unsafe design" or manufacturing of his machines. However, long before OSHA, many reliable machine manufacturers produced machines which met or exceeded safety standards and they tried to sell only well guarded machines.

Regardless of these reasons for guarding, it has long been agreed that guards designed and installed at the point of manufacture offer three advantages to the user.

1. Guards designed and installed by the manufacturer are usually less expensive than guards purchased or built and installed by the user.

2. Manufacturer-built guards usually conform better to the design and operation of the machine.

3. Manufacturer-built guards can be designed to strengthen the machine or serve other functional purposes.

All of these advantages can be seen in the newer machines available today, but these machines account for probably less than 20 percent of all machines currently in use. Of the remaining machines, perhaps less than 20 percent are in compliance with the OSHA guarding regulations. The inescapable conclusion is that there are many machines being operated with little to no guarding of hazardous points of operation and/or power transmission.

Checklists for guarding

These conditions are inviting two consequences to owners of such machines—accidents and penalties for OSHA violations. Most managements would like to avoid these consequences and they can by initiating and following up an OSHA compliance checklist for all their machines and equipment.

Basically the checklist will list individually all machines and equipment used. For each unit, it will then list the point(s) of operation and power transmission components. It should then indicate if they are guarded or not. If they are guarded, the kind of guarding should be noted—enclosure, fencing, or location, and whether the guard meets OSHA or other requirements. In some machines these facts may be readily apparent, but most machines are complex and may require a detailed analysis of the design, the interaction of all components, and the methods of servicing, including loading and unloading of materials as well as maintenance and repair. In other words, the total operation or process should be looked at as a complete system of interrelated parts and actions. A hazard point may appear well guarded when the machine is at rest, but when the machine is in operation, portions of it may be exposed or new hazards created when a component interacts with or passes by another component.

A poorly designed or maintained guard or safety device can interact with parts in motion and produce a hazard point. Newer machines may appear very well guarded, and may well be, but sometimes their unitized, streamlined designs are more for appearance than safety of operation. A good checkout may take some time and effort to complete, but it will be worth it for it will pinpoint guarding problems and indicate a general course of action. In the main, high-hazard problems should be dealt with first, particularly where there is no guarding at all. As previously indicated, OSHA does not specify whether guarding is accomplished by enclosure, fencing, or location of the hazardous part or operation. OSHA standards do require that the method chosen must be adequate to eliminate the hazard under normal operating conditions. Since fencing and location methods can be used only when a machine can be isolated, these methods will have limited use on most production setups. Most production machines will require some kind of enclosure of the hazard points to comply with the OSHA regulations.

Types of guards

There are several types of enclosures that are suitable for guarding hazard points depending on how and where they are used.

The fixed guard is considered preferable to all other types and should be used in every case unless it has been definitely determined that this type is not at all practicable. The fixed guard at all times prevents access to the dangerous parts of the machine.

Fixed guards may be adjustable to accommodate different sets of tools or various kinds of work. Once adjusted, they should remain "fixed," and there should be no movement nor detachment of them.

Some fixed guards are installed at a distance from the danger point in association with remote feeding arrangements which make it unnecessary for the operator to approach the danger point.

A table of the safe distance of a guard from the danger point and the permissible size of openings in a fixed guard is given in the Point-of-Operation Guards section of this chapter.

Examples of fixed guards are found on power presses, sheet leveling or flattening machines, milling machines, gear trains, drilling machines, and guillotine cutters.

Interlocking guards. If a fixed guard cannot be used, an interlocking guard should be fitted onto the machine as the first alternative. Interlocking guards may be mechanical, electrical, pneumatic, or a combination of types.

The interlocking guard prevents operation of the control that sets the machine in motion until the guard is moved into position so that the operator cannot reach the point of operation or the point of danger.

When the guard is open, permitting access to dangerous parts, the starting mechanism is locked, and a locking pin or other safety device prevents the main shaft from turning or other basic mechanisms from operating.

When the machine is in motion, the guard cannot be opened. It can be opened only when the machine has come to rest or has reached a fixed position in its travel.

An effective interlocking guard must satisfy three requirements; it must:

1. Guard the dangerous part before the machine can be operated

2. Stay closed until the dangerous part is at rest

3. Prevent operation of the machine if the interlocking device fails

Where neither a fixed guard nor an interlocking guard is practicable, mechanical interlocks may be used.

Automatic guards. An automatic guard may be used, subject to limitations outlined in Table 29-A, where neither a fixed guard nor an interlocking guard is practicable. Such a guard must prevent the operator from coming in contact with the dangerous part of the machine while it is in motion, or must be able to stop the machine in case of danger. (Table 29-A is on the next three pages.)

An automatic guard functions independently of the operator, and its action is repeated as long as the machine is in motion.

An automatic guard is usually operated by the machine itself through a system of linkage or levers.

Whenever an automatic guard is used on a machine which is loaded and unloaded by hand, the operator always should use hand tools.

All of the foregoing types of guard are suitable for guarding points of operation, but only the fixed guard can be used for guarding power transmission components (unless they are guarded by fencing or location as previously noted).

Safeguarding devices can be used in lieu of or in conjunction with barrier guards at the point of operation. Presence-sensing devices (either radio or electronic beam), pullbacks, restraints, and two-hand trips may be used under specified conditions and constraints. (See Chapter 32, "Cold Forming of Metals," for a fuller discussion of these devices.)

(*Text continues on page 800.*)

TABLE 29-A. POINT-OF-OPERATION PROTECTION

Type of Guarding Method	Action of Guard	Advantages	Limitations	Typical Machines on Which Used
ENCLOSURES OR BARRIERS				
Complete, simple fixed enclosure	Barrier or enclosure which admits the stock but which will not admit hands into danger zone because of feed opening size, remote location, or unusual shape.	Provides complete enclosure if kept in place. Both hands free. Generally permits increased production. Easy to install. Ideal for blanking on power presses. Can be combined with automatic or semiautomatic feeds.	Limited to specific operations. May require special tools to remove jammed stock. May interfere with visibility.	Bread slicers Embossing presses Meat grinders Metal square shears Nip points of inrunning rubber, paper, textile rolls Paper corner cutters Power presses
Warning enclosures (usually adjustable to stock being fed)	Barrier or enclosure admits the operator's hand but warns him before danger zone is reached.	Makes "hard to guard" machines safer. Generally does not interfere with production. Easy to install. Admits varying sizes of stock.	Hands may enter danger zone—enclosure not complete at all times. Danger of operator not using guard. Often requires frequent adjustment and careful maintenance.	Band saws Circular saws Cloth cutters Dough brakes Ice Crushers Jointers Leather strippers Rock crushers Wood shapers
Barrier with electric contact or mechanical stop activating mechanical or electric brake	Barrier quickly stops machine or prevents application of injurious pressure when any part of operator's body contacts it or approaches danger zone.	Makes "hard to guard" machines safer. Does not interfere with production.	Requires careful adjustment and maintenance. Possibility of minor injury before guard operates. Operator can make guard inoperative.	Calenders Dough brakes Flat roll ironers Paper box corner stayers Paper box enders Power presses Rubber mills
Enclosure with electrical or mechanical interlock	Enclosure or barrier shuts off or disengages power and prevents starting of machine when guard is open; prevents opening of the guard while machine is under power or coasting. (Interlocks should not prevent manual operation or "inching" by remote control.)	Does not interfere with production. Hands are free; operation of guard is automatic. Provides complete and positive enclosure.	Requires careful adjustment and maintenance. Operator may be able to make guard inoperative. Does not protect in event of mechanical repeat.	Dough brakes and mixers Foundry tumblers Laundry extractors, driers, and tumblers Power presses Tanning drums Textile pickers, cards

AUTOMATIC OR SEMIAUTOMATIC FEED

AUTOMATIC OR SEMIAUTOMATIC FEED				
Nonmanual or partly manual loading of feed mechanism, with point of operation enclosed	Stock fed by chutes, hoppers, conveyors, movable dies, dial feed, rolls, etc. Enclosure will not admit any part of body.	Generally increases production Operator cannot place hands in danger zone.	Excessive installation cost for short run. Requires skilled maintenance. Not adaptable to variations in stock.	Baking and candy machines Circular saws Power presses Textile pickers Wood planers Wood shapers
HAND REMOVAL DEVICES				
Hand restraints	A fixed bar and cord or strap with hand attachments which, when worn and adjusted, do not permit an operator to reach into the point of operation.	Operator cannot place hands in danger zone. Permits maximum hand feeding; can be used on higher-speed machines. No obstruction to feeding a variety of stock. Easy to install.	Requires frequent inspection, maintenance, and adjustment to each operator. Limits movement of operator. May obstruct space around operator. Does not permit blanking from hand-fed strip.	Embossing presses Power presses
Hand pullaway device	A cable-operated attachment on slide, connected to the operator's hands or arms to pull the hands back only if they remain in the danger zone; otherwise it does not interfere with normal operation.	Acts even in event of repeat. Permits maximum hand feeding; can be used on higher speed machines. No obstruction to feeding a variety of stock. Easy to install.	Requires unusually good maintenance and adjustment to each operator. Frequent inspection necessary. Limits movement of operator. May obstruct work space around operator. Does not permit blanking from hand-fed strip stock.	Embossing presses Power presses
TWO-HAND TRIP				
Electric	Simultaneous pressure of two hands on switch buttons in series actuates machine.	Can be adapted to multiple operation. Operator's hands away from danger zone. No obstruction to hand feeding. Does not require adjustment. Can be equipped with continuous pressure remote controls to permit "inching." Generally easy to install.	Operator may try to reach into danger zone after tripping machine. Does not protect against mechanical repeat unless blocks or stops are used. Some trips can be rendered unsafe by holding with the arm, blocking or tying down one control, thereby permitting one-hand operation. Not used for some blanking operations.	Dough mixers Embossing presses Paper cutters Pressing machines Power presses Washing tumblers
Mechanical	Simultaneous pressure of two hands on air control valves, mechanical levers, controls interlocked with foot control, or the removal of solid blocks or stops permits normal operation of machine.			

TABLE 29-A. POINT-OF-OPERATION PROTECTION—*Concluded*

Type of Guarding Method	Action of Guard	Advantages	Limitations	Typical Machines on Which Used
		MISCELLANEOUS		
Limited slide travel	Slide travel limited to ¼ in. or less; fingers cannot enter between pressure points.	Provides positive protection. Requires no maintenance or adjustment.	Small opening limits size of stock.	Foot power (kick) presses Power presses
Electric eye	Electric eye beam and brake quickly stop machine or prevent its starting if the hands are in the danger zone.	Does not interfere with normal feeding or production. No obstruction on machine or around operator.	Expensive to install. Does not protect against mechanical repeat. Generally limited to use on slow speed machines with friction clutches or other means to stop the machine during the operating cycle. Can be circumvented	Embossing presses Power presses Rubber mills Squaring shears Press brakes
Special tools or handles on dies	Long-handled tongs, vacuum lifters, or hand die holders which avoid need for operator's putting his hand in the danger zone.	Inexpensive and adaptable to different types of stock. Sometimes increases protection of other guards.	Operator must keep his hands out of danger zone. Requires unusually good employee training and close supervision.	Dough brakes Leather die cutters Power presses Forging hammers
Special jigs or feeding devices	Hand-operated feeding devices of metal or wood which keep the operator's hands at a safe distance from the danger zone.	May speed production as well as safeguard machines. Generally economical for long jobs.	Machine itself not guarded; safe operation depends upon correct use of device. Requires good employee training, close supervision. Suitable for limited types of work.	Circular saws Dough brakes Jointers Meat grinders Paper cutters Power presses Drill presses

Point-of-Operation Guards

If all machines were alike, it would be simple to design a universal point-of-operation guard and install it at the factory. But they are not all alike and to complicate the problem further, different users of the same machine may use it in different ways and even for different purposes. These differences constitute the main reason why more point-of-operation guarding is not done by the manufacturer. In many cases a point-of-operation guard can be made and installed by the user only after his operational data is clearly defined and understood.

Whenever point-of-operation guards are needed, certain principles and conditions must be kept in mind and applied to the design and construction of the guard.

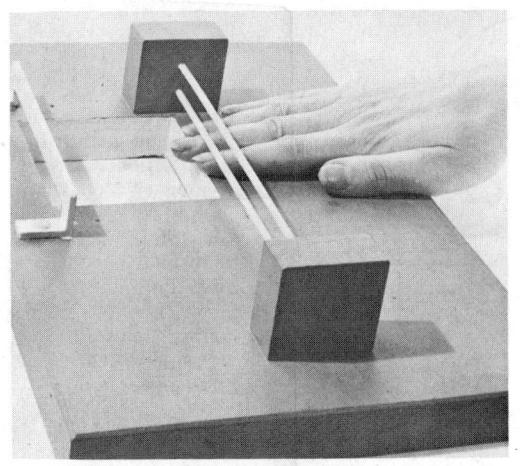

FIG. 29–1.—A ⅜-in. opening still permits a small part of the fingers to slip past the guard. A ¼-in. opening would keep out the fingers if the guard must be close to the danger zone.

Courtesy American Mutual Insurance Alliance; Liberty Mutual Insurance Company.

Guard openings

The following is a summary of these principles and other important data as found in *Safe Openings for Some Point-of-Operation Guards,* a publication based on research material developed by Liberty Mutual Insurance Company and the Engineering Committee of the American Mutual Insurance Alliance.

It has been common practice in the design of point-of-operation guards to work on the basis that any opening not over ⅜ in. was relatively safe, as it would not permit entrance of any considerable part of a hand inside the guard (Fig. 29–1). In many instances, however, a ⅜-in. opening is not a sufficient space for passing material to be processed through or under the guard. But as the width (or height) of the opening is increased to allow for entrance of material, the additional space permits the operator to reach farther inside the guard. Under such conditions, it is no longer possible to prevent entrance of some part of the hand within the guard. The problem is to stop movement of the hand *inside* the guard at a safe distance from the danger zone.

Fig. 29–2 shows diagrammatic sketches of the conditions found in connection with an inrunning roll hazard where a feed table is used. There are many other types of point-of-operation hazards, however, where the same or similar conditions exist where an opening more than ⅜ in. may be required in a guard design.

In Fig. 29–3 the proper location of the guard distance X for the use of required opening Y must be determined.

If the dimensions of the opening and its location (distance from the hazard) are properly selected, adequate safety for the operator can be established.

Some guard designers have made use of the formula:[*]

$$\text{Maximum safe opening} = \tfrac{1}{4} \text{ in.} +$$
$$\tfrac{1}{8} \text{ distance to guard from danger zone (in.)}$$

This formula is not intended for use where the distance from the guard to the danger zone exceeds 12 in.

Fig. 29–4 illustrates a situation where it may be essential to have part of the hand and fingers extend through the guard to permit manipulation of the material inside the guard. This condition may exist where it is impossible to use hand tools or mechanical devices to carry on the operation.

Test results

For their design analysis, the Industrial Plant Services Department of the Liberty Mutual Insurance Company constructed a number of test fixtures which approximated types of guards commonly used. By testing different-size openings with various hands (men's and women's) in these fixtures, information about allowable openings and location of guards for guidance of guard designers was developed. These design data are shown in Fig. 29–7, with a feed table, and Fig. 29–8 without a feed table.

One set of guide data was established for general use. Therefore, because of variations in hands and inaccuracies to be expected in maintaining a ⅜-in. opening, it was decided that for the first 1½ in. away from the danger line, in Fig. 29–5, no openings over ¼ in. could

[*] Royal Society for the Prevention of Accidents, London. *Industrial Accident Prevention Bulletin,* Vol. II, No. 118.

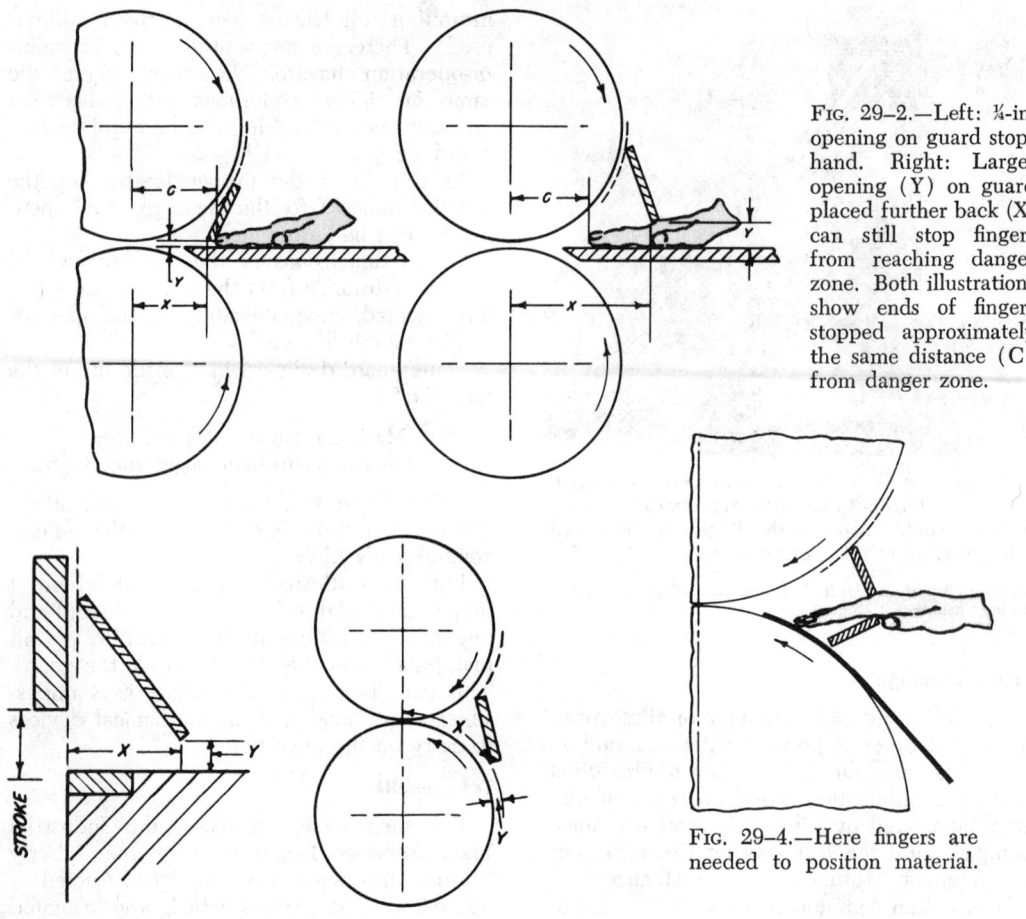

FIG. 29–2.—Left: ¼-in. opening on guard stops hand. Right: Larger opening (Y) on guard placed further back (X) can still stop fingers from reaching danger zone. Both illustrations show ends of fingers stopped approximately the same distance (C) from danger zone.

FIG. 29–4.—Here fingers are needed to position material.

FIG. 29–3.—Left: Vertical shear hazard. Right: Inrunning roll hazard (no feed table used).

Figures courtesy American Mutual Insurance Alliance; Liberty Mutual Insurance Company.

be considered safe.

Most men and women have finger tips which will not travel any considerable distance through a ⅜-in. opening. However, if a designer wishes to maintain a definitely safe zone beyond a ⅜-in. opening, he will not use it within 1½ in. of the danger point, as shown in Fig. 29–5.

Application of test findings

Vertical shear. The findings of the tests, Fig. 29–6, when applied to vertical shear design apply to:

1. All vertical openings in the guard proper, such as visibility slots, clearance slots for ejection devices and stop gages.

2. All horizontal openings in the guard proper as are necessary for feeding stock (front or sides) and for ejection of finished parts or scrap.

3. The installation or adjustment of a guard.

Inrunning rolls with feed table. In applying findings of tests to guards for use on inrunning rolls, it is necessary to take into ac-

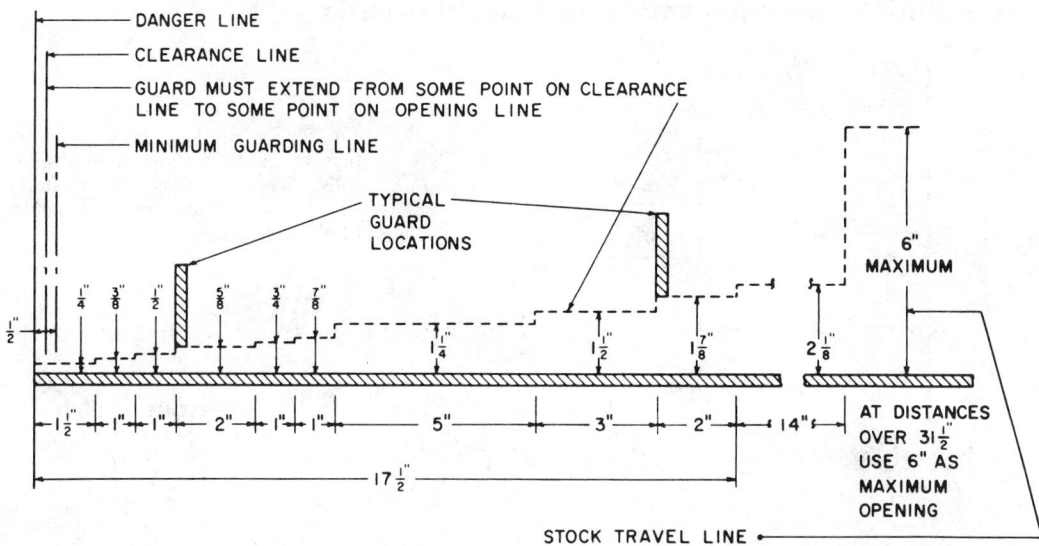

DANGER LINE
CLEARANCE LINE
GUARD MUST EXTEND FROM SOME POINT ON CLEARANCE
LINE TO SOME POINT ON OPENING LINE
MINIMUM GUARDING LINE

TYPICAL
GUARD
LOCATIONS

6"
MAXIMUM

AT DISTANCES
OVER $31\frac{1}{2}$
USE 6" AS
MAXIMUM
OPENING

STOCK TRAVEL LINE

Fig. 29–5.—Point-of-operation guard locations. Barriers placed to touch dashed line will wedge hand or forearm. The danger line is the point of operation. The clearance line marks the distance required to prevent contact between the guard and the moving parts. The minimum guarding line is ½ in.

From American National Standard B11.1, Safety Requirements for the Construction, Care, and Use of Mechanical Power Presses.

count the characteristics of a nip point. In Fig. 29–5, the "danger" line represents a hazardous contact; for vertical shear exposures, it is equivalent to the shear line.

On rolls, the hazard (nip or pinch zone) is not defined by a straight line. Therefore a ⅜-in. width of nip zone is considered the actual nip point through which the danger line *DE* is drawn (see Fig. 29–6). The distance of the ⅜-in. width of nip zone from the contact point between the rolls is designated as dimension S. It is recommended that rolls held less than ⅜-in. apart be considered as rolls in contact.

Fig. 29–7 shows an inrunning roll nip where a feed table is used. To design a properly placed barrier guard, it is suggested that the following procedure be used:

1. Draw a full-scale outline of the nip zone with the top surface of the feed table accurately shown. Indicate the clearance line on the top roll. If more than a ⅜-in. clearance is required, the top edge of the guard should be located in accordance with the safe opening layout shown in Fig. 29–9 for layout on a roll surface.

2. Determine distance S—the distance from the center line of the rolls to a point where a ⅜-in. vertical space exists between the top of the feed table and the surface of the upper roll.

3. At this distance, begin the layout of the safe opening dimensions, as shown in Fig. 29–5, up to the opening necessary for the particular guard being designed. Outline the guard section (top edge on clearance line on upper roll and bottom edge at proper point on safe opening layout), and determine the necessary dimensions for installing the guard. (Width of guard can be determined in addition to locating distances.)

4. Before the guard is put in operation, check carefully for hand travel under the guard, stability of mounting, and rigidity of construction.

Inrunning rolls with central feed (no feed table). Fig. 29–8 shows an inrunning roll nip where no feed table is used and the stock processed runs into the nip at right angles to the center of the rolls. (If the run of the stock is slightly above or below the horizontal, the center of the guard opening should be

803

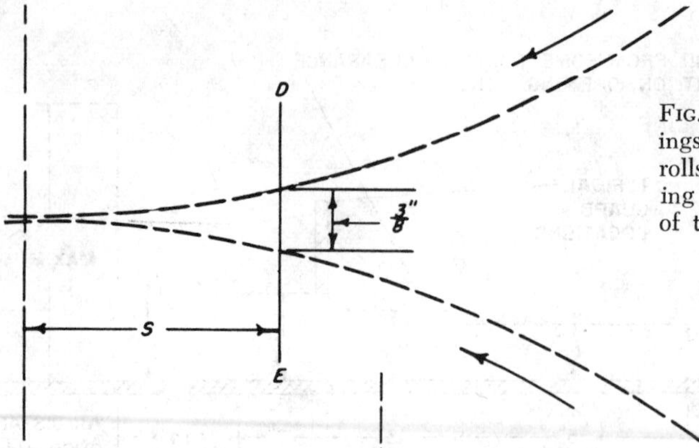

FIG. 29–6.—Application of test findings to guard design on inrunning rolls. See accompanying text (starting on page 800) for explanation of these figures.

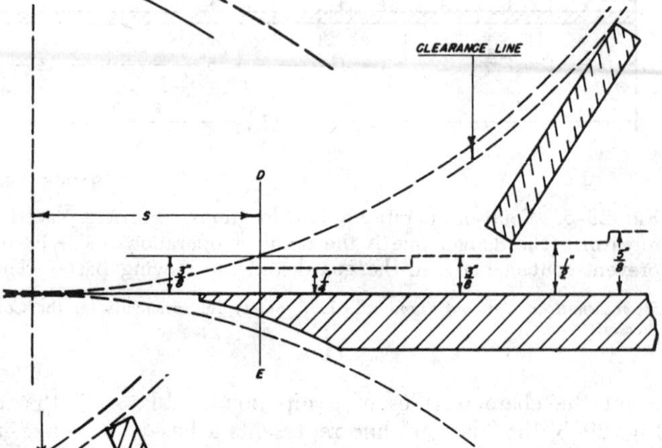

FIG. 29–7.—Inrunning roll nip where a feed table is used. "D-E" is the "stop" line. "S" is distance of ⅜-in. wide nip zone form contact point between rolls.

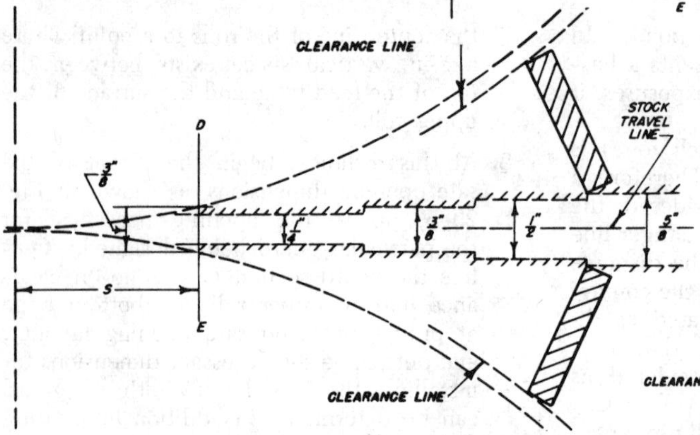

FIG. 29–8.—Inrunning rolls with central feed and no feed table.

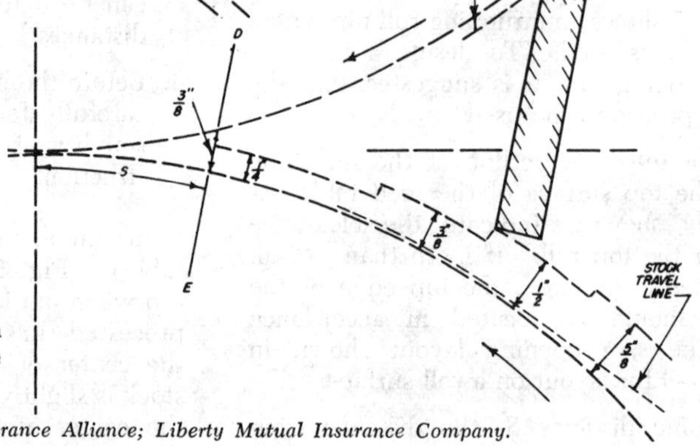

FIG. 29–9.—Inrunning rolls with stock traveling over one roll before entering nip zone.

Figures courtesy American Mutual Insurance Alliance; Liberty Mutual Insurance Company.

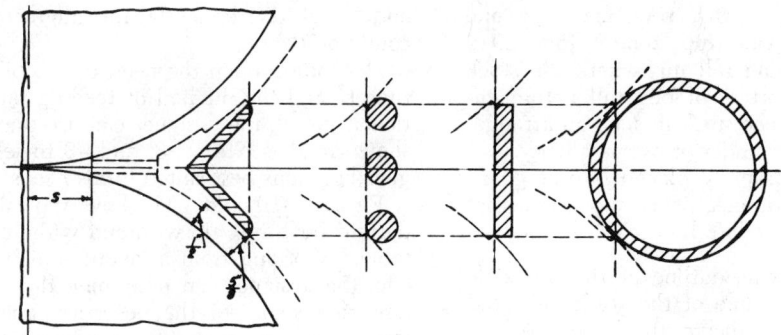

FIG. 29–10.—Different designs for a roll nip guard where feeding over or under a guard with ⅜-in. opening is desired.

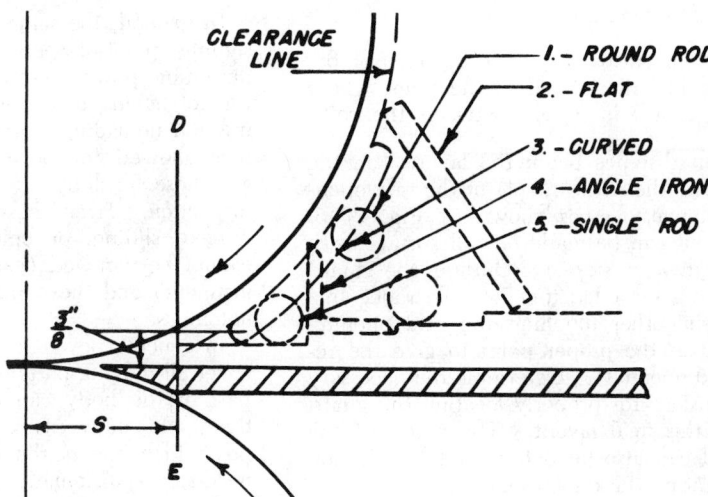

CLEARANCE LINE

D

1.– ROUND ROD
2.– FLAT
3.– CURVED
4.—ANGLE IRON
5.—SINGLE ROD

FIG. 29–11.—Different designs (shown in outline) for a good guard on a roll nip with a feed table.

3/8"

S

E

shifted accordingly.)

To design a properly placed barrier guard fitted with an opening for passage of stock, the following procedure is suggested:

1. Draw a full-scale outline of the nip zone with the stock travel line accurately shown. Indicate the clearance line on both rolls. If more than a ⅜-in. clearance is required between the edges of the guard and the rolls, the edges of the guard should be located in accordance with the safe opening layout shown in Fig. 29–9 for layout on a roll surface.

2. Determine distance S—the distance from the vertical line of the rolls to a point where a ³⁄₁₆-in. vertical space exists on each side of the travel line of the stock (total nip width of ⅜ in.), as shown in Fig. 29–8.

3. At this point, begin the layout of the dimensions shown centered on the travel line in Fig. 29–5. Outline the guard sections, giving the required opening between sections. (One edge of each section will touch a clearance line; the other will touch the safe opening layout at the proper point to give the required opening.) Determine the necessary dimensions for properly locating the guard from this final layout. The width of the guard sections can also be determined.

4. Before the guard is put in operation, check carefully for hand travel under the guard, stability of mounting, and rigidity of construction.

805

Inrunning rolls — stock traveling over one roll before entering nip zone. Fig. 29–9 shows an inrunning roll nip where the stock travels over a portion of one roll before entering the nip. The stock in such an arrangement is fed either under or over a barrier.

To design a properly placed barrier guard under such conditions, it is suggested that the following procedure be used:

1. Draw a full-scale outline of the nip zone with the travel line of the stock indicated on the roll. Indicate the clearance line on the top roll. If more than a ⅜-in. clearance is required, the top edge of the guard should be located in accordance with the safe opening layout.

2. Determine distance S — the distance of the center line of the roll to the point where there is a ⅜-in. space between the rolls.

3. At this distance begin the layout (on the roll with the stock travel) of the safe opening dimensions, as shown in Fig. 29–5. (Layout can be made on roll surface with ½-in. divider steps.) Outline the guard section (one edge touching clearance line and the other touching the safe opening layout at the proper point to give the required opening). Determine the necessary dimensions for properly locating the guard from this final layout. The width of the guard can also be determined in addition to locating the dimensions.

4. Before the guard is put in operation, check carefully for hand travel under the guard, stability of mounting, and rigidity of construction.

Guard construction

To assure maintenance of the dimensions established for safe openings, it is important that the guards be constructed to minimize the possibility of distortion or movement which would destroy the effectiveness of the guard. All guard parts should be strong enough to withstand expected stress. Fastenings should be secure and of such design as to prevent the guard proper from shifting and its unauthorized removal or movement. Any guard with openings over ¼ in. should be considered more or less a "precision" construction

and checked frequently for alignment and condition.

Depending upon the need for visibility and rigidity and the method of feeding for a particular guard, the designer can use the layouts shown in Figs. 29–7, –8, and –9 to select the guard sections best suited to his needs.

Figs. 29–10 and –11 show typical guard designs for a roll nip with and without a feed table. Working from a layout similar to this one, the designer can determine the location, size, and shape of the necessary guard section to meet his particular requirements.

Power Transmission Guards

In general, the same principles used in designing point-of-operation guards apply to designing power transmission guards except for not having to consider openings for loading and unloading materials. The only openings allowed for power transmission guards are those for lubrication, adjustment, and/or inspection. Even these openings must have hinged, sliding, or bolted cover plates that cannot be removed (except for service or adjustment) and they must remain closed when not in use.

In general, power transmission guards must cover all moving parts in such manner that no part of the body can come in contact with them. In many cases, a simple flat plate or box which covers the opening is all that is necessary, particularly if the parts are flush with or recessed within the frame of the machine. Where parts protrude beyond the frame, it may be necessary to "slipcover" the part; that is, build a guard which conforms to the dimensions and forms of the parts being guarded. In such cases, any openings to permit shafts or other components to pass into the machine must follow the requirements of the maximum size of permissible opening as related to the distance from the moving part. (See Table 29-B.)

Guarding materials

The preferable material for guards under most circumstances is metal. Framework of guards is usually made from structural shapes, pipe, strapping, bar, or rod stock. Filler material generally is expanded or perforated or solid sheet metal or wire mesh (Table 29-B).

TABLE 29-B

STANDARD MATERIALS AND DIMENSIONS FOR MACHINERY GUARDS

SIZE OF FILLER MATERIALS

Material	Clearance from Moving Part at All Points (inches)	Largest Mesh or Opening Allowable B (inches)	Minimum Gage (U.S. Standard) or Thickness	Min. Height of Guard from Floor or Platform Level (ft, in.)
Woven Wire	Under 2 2-4 4-15	3/8 1/2 2	No. 16-3/8 in. No. 16-1/2 No. 12-2	8-0* 8-0 8-0
Expanded Metal	Under 4 4-15	1/2 2	No. 18-1/2 in. No. 13-2	8-0 8-0
Perforated Metal	Under 4 4-15	1/2 2	No. 20-1/2 in. No. 14-2	8-0 8-0
Sheet Metal	Under 4 4-15	. . .	No. 22 No. 22	8-0 8-0
Wood or Metal Strips Crossed	Under 4 4-15	3/8 2	3/4 in. wood or No. 16 metal	8-0
Wood or Metal Strips Not Crossed	Under 4 4-15	1/2 the width One width	3/4 in. wood or No. 16 metal	8-0
Plywood, Plastic or Equivalent	Under 4 4-15	. . .	1/4 in. 1/4 in.	8-0
Standard Railing	Min. 15 Max. 20	. . .	. . .	3-6

* Guards for rotating protruding objects should extend to a minimum height of 9 ft from the floor or platform.

From American National Standard B15.1, Safety Code for Mechanical Power-Transmission Apparatus.

The use of plastic or safety glass where visibility is required is widely practiced.

Guards made of wood have limited application. Their lack of durability and strength, relatively high maintenance cost, and flammability are objectionable. Wood guards, particularly when they become oil-soaked, can be ignited by nearby welding operations, by overheated bearings, by rubbing belts and defective wiring, and by other sources of heat. Wood is also subject to splintering which may contaminate products or cause injury.

Where resistance to rust or possible damage to tools and machinery is an important factor, guards of aluminum or other soft metal, or plastic are sometimes used. Plastic guards

807

are coming into increased use where inspection of the moving parts is necessary. Shatterproof glass is similarly used, particularly where visibility of guarded parts is a problem and where the flexibility of plastic is not required. Safety glass and plastic used where chips or other flying particles are likely to mar the surface may be protected by inexpensive and easily replaced cover glasses.

When a guard cannot be made to exclude lint, ample ventilation should be provided. Vents, too small to admit a hand, likewise should be built into the bottom of larger guards to let lint or dust drop through. Larger guards also should have self-closing access doors for cleaning by brush, vacuum hose, or compressed air. Consideration should be given to latches interlocked with the power source to prevent operation of the machine while door is open.

Whatever material(s) is selected for a guard, it should also be substantial enough to withstand internal as well as external impacts of materials, parts under stress, or passing pedestrians or vehicles. Parts or other materials in process can become lethal projectiles. Machines located close to heavy traffic aisles can be very vulnerable to damage, particularly if parts of them overhang the aisle. If the machine cannot be relocated or traffic rerouted, a regular enclosure guard may not be sufficient for all the hazards involved to protect both people and the machine. Loaded lift trucks could snag on the overhang and damage the guard, machine, and even nearby personnel if the load topples. Such cases require a floor guard in addition to the enclosure guard to prevent vehicles from straying into the hazard area.

A major problem in designing enclosure guards is a lack of planning for routine maintenance. Failure to establish, enforce, and facilitate safe maintenance procedures is probably the major cause of failure to replace a guard, particularly if the maintenance has to be done frequently. Then, it becomes easier to leave the guard off permanently. This condition now allows (and even promotes) an additional problem—maintenance of the machine while it is in operation. These conditions are highly hazardous for all personnel concerned, and can be solved in any one or combination of three ways.

• The first would be to apply engineering techniques which would reduce or eliminate the frequency of doing the job. If, for example, the fittings for those parts needing service were relocated to the outside of the guard, this would make it unnecessary to remove the guard and it would permit maintenance during the machine operation. For example, an oil or grease fitting might be lubricated by an extension through the guard. This procedure is highly recommended for operations which cannot be shut down for adjustment or maintenance.

• A second way, which is similar to the first, involves equipping the machine with automatic controls for lubrication, adjustment or service. Sophisticated equipment like this may be costly, but for some machines the cost is offset by real savings effected by better and more reliable adjustment and maintenance procedures.

• The third method requires interlocking all guards with the source of power so that the machine cannot be operated without all guards in place. This method ensures the replacement of the guard each time it is removed. Interlocking is effective by itself; and if used in conjunction with the other methods, makes the total guarding system almost foolproof.

Consideration should also be given to the installation or mounting of guards. They should, of course, be mounted securely, and the best ways are either tack welding or special bolting. Special bolting would require special tools for removal, and thus either method of mounting would eliminate unauthorized removal of a guard. They should be the only ways of mounting guards if they are not interlocked.

Guards and Noise Control

Earlier in this chapter it was pointed out that guards and safeguards can be designed to strengthen machines and serve other functional purposes. One of the best of these other functional purposes is the control of noise. A well designed, well made, and well mounted guard can be very effective in reducing unwanted noise, instead of contributing to it. In too many cases, the guard is an afterthought and it is flimsily made and tacked on to the machine as cheaply as possible. Such guards may even aggravate noise from a machine. This is one reason why manufacturer-built guards are recommended when possible over the "home made" kind. However, for noise reduction, many manufacturer-built guards are no better than "home made" guards.

Noise travels primarily by conduction and vibration through air. It can be attenuated by barriers which effectively stop or restrict further air activity either by absorption or by reflection and confinement of the sound waves. Since guards are usually positioned at either the point of operation or power transmission (whence noise originates), they can be designed as a barrier for noise as well as a barrier against personal injury.

A guard can be designed to be either absorbent or reflective of sound waves. A common way to absorb sound is to either line or cover the surrounding frame with a soft cellular material which soaks up the sound with dead-air spaces. Sometimes a thin layer of lead is sandwiched between two layers of such soft material to further reduce sound transmission. However, if oil is used in the process, this could create a serious fire hazard if the area has a heat buildup, for some of these soft materials tend to soak up oil. It

has also been found that if the guard is made of heavier metal stock (16 gage or better), it provides a dense, vibrationless surface when it is properly mounted. Proper mounting is essential for the success of either method, because both methods (absorptive and reflective barriers) require that the guard have an identical configuration with the surface on which it is attached. Some sort of gasket material should be used around the edges and any other place that metal-to-metal contact can occur, and the whole guard secured with shakeproof fittings. In effect, this unitizes the guard with the machine and contains the sound within, thus reducing the conducted sound. More details are in Chapter 40, "Noise and Hearing Conservation."

Summary

In summary, a complete guarding program is essential for any company. It prevents accidents, whether to people or machines, and thus facilitates production goals. It can be achieved only by studying and understanding the relationships between people and the machines they operate. Machines can produce only if they are used correctly—and they are used correctly only when all guards and safeguards are in place.

References

American Mutual Insurance Alliance, 20 N. Wacker Drive, Chicago, Ill. 60606. *Safe Openings for Some Point of Operation Guards,* Technical Guide No. 2, 3rd ed., 1966.

American National Standards Institute, 1430 Broadway, New York, N.Y. 10018.
"How To Operate a Power Press Safely," 1975.
Safety Requirements for Floor and Wall Openings, Railings, and Toe Boards, A12.1.
Safety Specifications for Mills and Calenders in the Rubber and Plastics Industries, B28.1.
Safety Standard for Construction, Care and Use of Mechanical Power Presses, B11.1.
Safety Standard for Forging, B24.1.
Safety Standard for Mechanical Power-Transmission Apparatus, B15.1.

National Fire Protection Association, 470 Atlantic Ave., Boston, Mass. 02210.
Electrical Metalworking Machine Tools, Standard No. 79.
National Electrical Code, Standard No. 70.

National Safety Council, 425 N. Michigan Ave., Chicago, Ill. 60611. *Guards Illustrated,* 2nd ed.

U.S. Department of Labor, Occupational Safety and Health Administration, Washington, D.C. 20210. *Principles and Techniques of Mechanical Guarding,* Bulletin 2057, 1972.

Woodworking Machinery

General Safety Principles   811
Electrical . . . Guards . . . Environment . . . Materials handling
. . . Inspection . . . Health . . . Personal protective equipment
. . . Sources

Specific Safety Principles   813
Circular saws . . . Circular saw blade maintenance . . . Overhead
swing saws and straight line pull cutoff saws . . . Underslung cutoff
saws . . . Radial saws . . . Power-feed ripsaws . . . Band saws
. . . Jig saws

Other Woodworking Equipment   822
Jointers . . . Shapers . . . Power-feed planers . . . Sanding
machines

References   826

The operation of powered woodworking equipment can cause a variety of serious accidents. Each machine generates its own characteristic type of accident problems. To reduce unsafe conditions, the worker should be provided with the right type of equipment that is fully guarded to do the job. Adequate jigs or fixtures must be available to afford maximum protection. It is important that the hands be kept as far away as possible from the point of operation.

The worker must be taught proper and safe procedures. He needs to be observed frequently to see that the established procedures are being followed. He needs to be trained to recognize potential accident situations and to know what to do when they are seen. Because he is working with mechanical equipment, he should know what to do when there is a change in noise, pitch, or any operating characteristic.

Because woodworking equipment is used in many industries, this chapter will be concerned with the equipment and not with the specific types of operations in the wood industry. Suppliers of new equipment should meet standard and code requirements of mechanical and electrical safeguarding.

General Safety Principles

All machines should be constructed and maintained, so that, while running at full or idle speed and with the largest cutting tool attached, they are free of excessive noise and harmful vibration.

The machines should be designed, so that a tool larger than those for which the machine was designed cannot be mounted on it.

All arbors and mandrels should be constructed so that they have a firm and secure bearings and are free from slip or play.

Electrical

All of the metal framework on electrical machines should be grounded (including the motor itself).

Each power-driven woodworking machine should be equipped with an electrical disconnect or valve which can be locked in the OFF position for maintenance, repair, or security. The switch or valve should automatically return to the OFF position, if there is any interruption to the main power supply and, on the return of the power, the switch should be reset manually to the RUN position.

Every machine should have a STOP switch conveniently located for the operator's use. Sometimes a foot control can be provided as an emergency control. On some machines, it is desirable (and sometimes required) to have a brake, often an electric one, to stop the mechanism after the power is shut off.

Guards

All belts, shafts, gears, and other moving parts should be fully enclosed or safeguarded, so that the worker cannot touch them. If there are moving parts in back (*i.e.*, the side away from the workers), these parts should be covered or the area closed to prevent entry.

Because most woodworking operations involve cutting, it is often difficult, although necessary, to provide guards at the point of operation. On most machines, the point-of-operation guard must be moveable to accommodate the wood, balanced so as not to impede the operation, and strong enough to provide protection to the operator.

Environment

All machines, except portable or mobile ones, should be securely fastened to the floor or other suitable foundation to eliminate all movement or "walking."

There should be ample work space around the machine, as required by the type of operation (Table 30-A).

Floors should be well maintained to prevent splintering conditions and protruding nails. Floors should be kept even and free from holes and irregularities. The work area floor near the machines should have a nonslip surface. Aisleways should be marked by paint and other markings.

A rule that should be followed is to forbid the making of adjustments while the machine is running. When possible, work should be scheduled to avoid frequent adjustments of machines and alterations of the position of the guards.

There should be a periodic inspection (about every six months, depending on usage) of each machine. This includes the infeed and outfeed, the temporary storage, and the activity areas.

The machines should be located to receive natural and artificial illumination. Generally 50 foot-candles will be needed for work, but fine work may require 100 or more foot-candles. There should be no shadows or reflected glare.

The working surfaces of the machine should be at a height that will contribute a minimum of fatigue for the operation (Table 30-A). Adjustments should be made if the worker is taller or shorter than average. All accessory or feed tables should be at the same height as the working surface.

Materials handling

The machine layout should encourage an even flow of materials and keep to a minimum back-tracking and crisscrossing. Operators should not have to stand in or near aisles.

The machines should be arranged, so that the material handled by the operator and others requires a minimum of movement and change of heights. This applies to both incoming supply and outgoing stock.

Provision should be made for the removal of sawdust and scrap, so that they do not accumulate. Automatic vacuum systems are desirable and efficient.

The working surface should be kept entirely free from scrap and waste.

Inspection

The operator should make an inspection of the machine prior to each start. This would include a check of operating controls, safety devices, power drives, sharpness of cutting edges and other parts which are to be used.

All cutting edges and tools must be kept sharp at all times. They must be properly adjusted and firmly secured.

Health

If the operation tends to be noisy, sound level measurements should be taken. It is generally accepted that a sound level of 85 to 90 dBA (or greater) for an eight-hour workday requires some attention. Ear protection for the operator, shorter work hours, and/or machine enclosures will reduce the noise level to the operator. See Chapter 40, "Noise and Hearing Conservation."

The amount of finely divided dust which oc-

TABLE 30-A

TABLE HEIGHTS AND WORK SPACE

Machine	Table Heights (in.)	Work Area
Band Saws	46	On three sides—a radii equal to twice the band saw diameter (as measured from the point cut).
Circular Saws	36 (Hand feed) 33 (Power feed)	Clearance on the working side should be 3 ft plus the length of the stock.
Jointers	39	3 ft plus the length of stock.
Lathes	41	Clearance of at least 30 in. from stand, with smaller distances on ends and backside allowable.
Radial Saws	39	Ripping—saw table equal to twice the length of the stock. Crosscutting—saw table equal to length of the stock plus 3 ft.
Sanders	36	3 ft plus the length of stock.
Shapers	36	3 ft plus the length of stock.

curs in many operations should be measured. The threshold limit values have been established for many materials, and these should be followed. Fine dust may be a health problem and can also be a cause of an explosion or fire. (See Chapter 38, "Industrial Toxicology.")

Personal protective equipment

All individuals in machine areas should wear safety glasses or face shields.

All workers should wear close-fitting apparel without rings, bracelets, or other jewelry which may become entangled in moving machinery. Loose sleeves or other loose clothing also possess the same danger.

Hair nets or caps should be worn to keep long hair out of moving parts, and gloves or hand pads can be worn to protect hands from splinters and rough lumber. However, gloves should not be worn where there is any chance of being caught by the moving parts of machines, the rotating stock, etc.

Safety shoes should be worn when handling heavy material or when there is danger of foot injury.

See Chapter 19, "Personal Protective Equipment," for details.

Sources

There are a number of specific OSHA standards which require safety features on some woodworking machines. Most of these standards are based on American National Standard O1.1–1971, *Safety Requirements for Woodworking Machinery*. In addition, some states and other jurisdictions have codes which specify certain requirements. All of these sources should be consulted. The National Safety Council publishes data sheets on a number of woodworking machines. See References.

Specific Safety Principles

Circular saws

There are two principal types of accidents involving circular power saws: (*a*) blade cuts or abrasions, and (*b*) kickbacks. These two kinds of accidents can be minimized by proper guarding and by establishing and enforcing safe work procedures.

Power saw operators are most frequently injured when their hands slip off the stock, while pushing it into the saw, or when holding the hands too close to the blade during the cutting operations.

Helpers and take-away men are injured by coming into contact with the blade while removing scrap or the finished pieces from the table.

Poor housekeeping practices and slippery floors are another source of accidents involving circular saws.

Circular saws are designed to permit a wide range of operations. Therefore, as is the case with most multiple-use equipment, it is difficult to design one type of guard for maximum protection for all types of operations. Contact with the saw blade can be prevented through use of the proper type of spreader, hood guard, jigs, fixtures, combs, or other devices.

Kickbacks. Kickbacks can cause serious injuries or death. Therefore, they must be prevented by *every* means possible.

Kickbacks on circular saws are usually caused by one of the following conditions.

1. No spreader to prevent work from binding on blade.

2. An improperly conditioned saw that allows the material to pinch on the outfeed edge of the saw and rise from the table. A dull saw blade is often a contributing factor.

3. Improper alignment of the gage or rip fence. It should be parallel to the saw blade to prevent the work from pinching between the rear of the blade and the fence.

4. Lumber that has not been properly planed, is not square, or has a twisted grain.

5. Attempting to ripsaw or crosscut stock too large to control on the table.

6. Confining the cutoff piece whether ripping or crosscutting, such as between the blade and the fence when ripping or being caught by the length stop when crosscutting.

7. Ripping by applying feed force on the cutoff piece. Feed force for hand or power rip feeding should always be applied on the stock between the saw blade and the fence.

8. A sudden encounter with a knot or other hard material by the saw blade.

Kickbacks can be minimized by avoiding the causes mentioned. The proper use of the spreader and the antikickback fingers (dogs)

FIG. 30–1.—A hood guard fabricated of ¼-in. clear plastic can be mounted on the spreader. Anti-kickback dogs help control kickbacks.

Courtesy U.S. Naval Civil Engineering Laboratory, Port Hueneme, Calif.

and the following of a safe procedure will keep kickbacks to a low level. Careful selection of the kind of lumber stock being cut can reduce kickbacks.

Guards. Guards that greatly reduce the likelihood of injury are available and in common use. If they are not furnished with a saw, it is necessary to provide them when the saw is installed. The majority of work done on table saws can be done with standard guards in place. The protection gained by using guards on most work makes them essential. It is, however, important that the guards be practical and correct for the job being done, or else they may be removed.

A table saw should be provided with a spreader, which prevents wood with internal stresses from clamping down or binding at the outfeed edge of the saw blade. A spreader thus helps prevent kickbacks. It also keeps chips and slivers away from the back of the saw where they might be caught by the saw teeth and thrown.

It should be mounted rigidly, not more than ½ in. in back of the saw blade, and should be at least 2 in. wide at table level. The spreader should conform to the radius of the saw as nearly as practicable and be high enough above the table to penetrate the full thickness of the stock. The spreader shall be attached so it will remain in true alignment with the saw blade, even when the table or arbor is tilted.

A circular table saw should be guarded by a hood that covers the part of the saw projecting above the stock, adjusts itself to the thickness of the stock, and rides on the stock.

The hood shall be of adequate strength to resist blows incidental to reasonable operation, adjusting, and handling. It should be made of material that will not shatter when broken, be nonexplosive, and no more flammable than wood. The hood must remain in true alignment with the saw blade to be effective, even if the table or arbor is tilted.

The hood may be suspended from above, either from the ceiling or, more commonly, from a post attached to the side of the machine, or supported on the spreader (Fig. 30–1). The mounting must be secured and supported, so that it will not wobble and strike against the saw blade. In both strength and design, it must protect the operator against flying slivers or broken saw teeth, and keep his hands away from the blade.

Tilt table and tilt arbor. The part of the saw blade underneath the table should be completely enclosed. The enclosure, which may be an exhaust hood, should be so constructed that saw blades can be easily changed. Preferably, it should have a hinged cover, so that the blades can be changed without removal of the enclosure.

A circular table saw used for ripping should be provided with antikickback fingers (dogs), with points riding on the stock to oppose the tendency of the saw to force the stock upward and toward the operator (Fig. 30–2). Anti-kickback fingers should be inspected frequently and kept sharp.

On rabbeting and dadoing jobs, it is impossible to use a spreader and often impracticable to use the standard hood guard. These operations can be effectively guarded by a jig that slides in the grooves of the transverse guide. The work is locked in the jig, and the operator's hands are kept well away from the saws or cutting head.

Because of variation in rabbeting and dadoing jobs, special jigs may have to be made. The hazard in these operations, particularly when work is being done on small stock, justifies the effort. If a shop does much dadoing and rabbeting, one or more machines ought to be set aside for this work to eliminate the need for frequent removal of the standard guards from machines normally used for cutting and ripping.

Feather boards can often be used to guard operations where standard guards cannot be used. They are suitable for short runs because they can be set up quickly and are inexpensive. A feather board should be made from straight-grained stock, either softwood or hardwood, and the parallel saw cuts (the "comb") should be in the direction of the grain. The feather board should bear against the stock at an angle of 45 to 60 degrees. On ripping operations, a feather board should press against the stock at a point between the saw and the operator so that the stock will not be forced against the saw. On dadoing or shaping jobs, a feather board should be placed opposite the cutting head.

Operating methods. The hands should be kept out of the line of the cut in feeding a table saw. Even the best guard is primarily a warning device and will permit the hands to follow the stock into the saw.

When the operator is ripping with the fence gage close to the saw, he should use a push stick to keep his fingers away from the saw. Push sticks or blocks of various sizes and shapes should be kept near the machine.

Stock should be held against a gage, never sawed freehand. Freehand sawing endangers the hands and may cause work to get out of line and bind on the saw. When ripping stock with narrow clearance on the gage side, the operator can gain more clearance by clamping a filler board to the table between the gage and the saw and guiding the stock against it. Use of a filler makes unnecessary the hazardous practice of removing the hood guard because of lack of clearance.

Because of kickbacks, the operator should stand out of the line of the stock he is ripping. A heavy leather or plastic apron or abdomen guard gives additional protection.

The best height of the saw blade for cutting involves two major considerations. For maximum safety to the operator, the blade should clear the stock by about ⅛ in. To provide the

Fig. 30–2.—Close-up of an anti-kickback arrangement attached to the hood.

maximum down force to hold the stock on the table and, thus, reduce the chance of a kickback, the saw blade should be as high as possible. With more of the blade exposed, there is a greater danger of deep cuts and amputations.

A crosscut saw should not be used for ripping nor a ripsaw for crosscutting. Using the wrong saw for the job makes the work harder and requires additional force to feed the stock.

Use of a general-purpose table saw for work that should be done on special machines is bad practice. For example, a table saw is often used for ripping operations. If such work is done on a power-feed ripsaw instead, the danger of getting hands against the saw and having the work kick back is virtually eliminated.

Long stock is sometimes crosscut on a table saw. This is another dangerous practice, because stock extending beyond one or both ends of the table interferes with other operations and may be struck by persons or trucks. Also, it is difficult to guide long pieces, and the operator must exert considerable pressure with his hands close to the saw. Such stock should be cut on a swing or pull saw.

Work that can be done on special or power-feed machines should not be done on hand-

feed, general-purpose machines.

A circular saw should be stopped when the operator leaves it. It is not sufficient to cut the switch and walk away, since amputations have been caused by saws still coasting with the power off. An electric brake attached to the motor arbor offers fast, positive stopping action. Another effective method is shown in Fig. 30–3. Sawdust slivers should be cleared away from a saw with a brush or a stick, never with the hands.

Under no circumstances should either the saw guard or the fence gage be adjusted while the saw is running. Parallel setting of the gage is of particular importance.

To enable the operator to set the gage without removing the guards, it is good practice to mark the top of the saw table with a permanent, distinctive line or other suitable device, directly in front of and in line with the saw.

Circular saw blade maintenance

The characteristics and condition of circular saws have an important bearing on the safety of the men who use them. Saw manufacturers have published valuable information on the selection, use, and care of saws.

In the designing, building, and tensioning of a saw, the maker gives it enough rigidity and tensile strength to cut without harmful distortion. Altering a saw from its original design, operating it at other than its rated speed, or changing its balance or tension seriously affects its efficiency and safety.

Some faults of circular saws that may cause difficult and unsatisfactory operation are:

● SAW OUT OF ROUND. If some teeth are longer than others, the long teeth do most of the work. An unequal strain is thus imposed on the saw, which may cause it to run out of line, to heat up, and to warp.

● SAW NOT STRAIGHT (OUT OF PLANE). Lumps or warps can be checked with a straightedge across the length of the diameter of the saw.

● IMPROPER HOOK OR PITCH OF TEETH. Ripsaw and cutoff teeth differ in design for different kinds of wood and for different pur-

FIG. 30–3.—When the hand lever is pushed down, it operates the STOP switch and engages the brake on the collar on the mandrel, stopping the saw blade. The foot kick switch shuts off the power.

Courtesy Western Electric Company

poses. Combination saws may be used for both crosscutting and ripping.

● IMPROPER OR UNEVEN SET. A circular saw has to cut a kerf thicker than the blade to give clearance which permits the saw to pass through the wood. Set or swage is given the teeth by bending alternate teeth right and left or by spreading the point of every tooth, so that each is slightly wider than the saw.

● CRACKED BLADES. Saw blades should be inspected for cracks each time the teeth are filed or set. Some cracks are so small that they may be invisible to the naked eye. The Magnaflux method for testing saw blades is extensively used. If cracked saw blades are continued in service, the crack frequently grows larger and eventually may cause partial fragmentation. Most cracks start in the gullets. As soon as a crack is visible to the naked eye, the blade should be removed from service.

Whatever method is used, the saw blade must be retensioned after repairs have been made. This is a job for a sawsmith and, unless the company has the services of such a man, the blade should be repaired by the manufacturer.

Excessive heat and vibration cause saw blades to crack. To prevent cracking, these precautions should be followed:

1. The saw blade should be tensioned for the speed at which it will operate. Otherwise, the blade will wobble and vibrate and, as a result, will heat, expand, and crack.

2. The teeth must have sufficient clearance (set or hollow grinding) to prevent burning, with consequent heating and cracking.

3. The blade should be in perfect round and balance.

4. The blade must be kept sharp at all times. A dull blade will not cut; rather, it will pound itself through the wood, so that vibration, heating, and then cracking result.

Proper operation. A saw in good condition and running at correct speed should cut easily. The operator should not crowd the saw, that is, force the stock faster than it can be easily cut. If the saw does not cut as fast as it should, or if it does not saw a clean, straight line, it is probable that something is wrong with the saw or the running speed. These conditions, potential sources of accidents, should be checked and corrected before proceding with the job.

Most production hand and power-feed saws run at 3600 rpm (actually about 3450 rpm). This will give a 12-in. blade a rim speed of 10,839 sfm; a 16-in. blade, 14,451 sfm; and an 18-in. blade, 16,258 sfm. (The manufacturer's instructions and specifications should always be followed.)

Only the outer edge of the collar should come in contact with the saw when it is tightly clamped in position. If the inside of the collar has not been machined out properly, it will force the rim of the saw out of line. When the saw blade comes in contact with the stock being cut, a buckling effect on the saw is produced. After the loose collar is fastened securely in place, it is well to test the saw with a straightedge. This test is of considerable importance on head rig circular saws as well as on edgers and trimmers.

Some means should be provided to prevent men from placing on the mandrel a saw larger than allowable for the speed of the mandrel.

Overhead swing saws and straight line pull cutoff saws

Overhead swing saws and cutoff saws cause hand injuries because of several of their characteristics. Hands are cut with the saw, while it coasts or idles, when the operator reaches to remove a sawed section of board or a piece of scrap, or when he measures a board or places it in position for the cut.

The operator's hands may likewise be struck by the saw, as either it bounces forward from the idle position, or swings or drifts forward when the spring or counterweight fails. The operator may pull the saw against his hands or may suffer body cuts from a saw that swings beyond its safe limits.

FIG. 30–4.—A cutoff saw guarded with a hinged (floating) hood guard.

Courtesy Western Electric Company

817

30—Woodworking Machinery

Guards. Cutoff saws should be guarded with a hood guard. Some guards cover the lower half of the saw, when the saw is not cutting, and ride on top of the stock as the saw cuts (Fig. 30–4).

A counterweight or other device that will automatically return the swing saw to the back of the table when released should be used. The counterweight should be secured, so that it cannot fall.

There should be a limit chain or other device to prevent the saw from swinging beyond the front edge of the table. A device should likewise keep the saw from rebounding from its idling position. A latch with a ratchet release on the handle is positive, but in some instances a nonrecoil spring or bumper is adequate. A magnetic ratch provides one method.

STOP and START buttons should be located for quick and easy access. The saw table can be provided with a wood bumper or a pipe guard to prevent bodily contact with the saw blade when it is extended the full length of the support arm.

Operating methods. If the saw is pulled by a handle, the handle should be attached either to the right or left of the saw rather than in line with it. The operator should stand to the handle side and pull the saw with the hand nearer it.

Thus, if the handle is on the right side of the saw, boards should be pulled from the right with the right hand, and the saw should be pulled with the left hand. This method makes it unnecessary for the operator to bring his hand near the saw while it is cutting, and keeps his body out of line of the saw.

Saws may be ordered with either right or left handles. On a new installation, a saw should be ordered with the handle on the side from which the stock is to be pulled. If it is necessary to pull stock from the other side, a handle should be built on that side, so that the operator can stand in the correct position.

To measure boards, the ends should be placed against a gage stop. When it is necessary to measure the board with a scale while it is on the table, move the board away from the saw blade.

At the completion of each cut, the operator should ease the saw back to the idling position and make sure that all bounce has stopped before putting his hand on the table.

No automatic or constant-stroking saws should be used because, if the operator does not maintain his exact rhythm, he may suffer an amputation.

Underslung cutoff saws

An underslung cutoff saw is usually operated by a treadle, and its forward movement is fast. It should be completely enclosed in the idling position. For general work, it should also be covered by a movable hood guard that slides forward or drops to rest on the stock when the saw is cutting. A treadle guard ensures that application of the foot on the treadle is intentional.

Underslung cutoff saws are commonly used to cut knots out of such narrow pieces as flooring and molding. The stock is placed by hand, and the hands are customarily held close to the line of the cut on either side. The movable guard gives little protection, because the action of the saw is so fast that the guard can ride over the top of the hands.

On either side of the line of travel a barrier guard can be constructed, with enough clearance between the guard and the table top to admit the stock, but not the hands or fingers of the operator. With practice an operator can feed stock rapidly under this type of guard.

Radial saws

Radial saws cut downward and pull the wood away from the operator and against a fence. These saws, like straight line pull cutoff saws, require many adjustments to permit their full use. These adjustments may create additional hazards for the user.

The head of a radial saw may be tilted to cut a bevel, or the supporting beam and track may be swung at an angle to make a diagonal cut. Both adjustments may be used to cut a compound bevel or miter. Likewise, the head may be turned parallel to the length of the table and the saw used for ripping. In this case it is an overhead, stationary saw against which the stock is fed by hand.

Obviously, a saw with so many features should be operated only by a skilled woodworker who has had good training and wide experience.

Fig. 30–5.—A power ripsaw operation. Note that there are two emergency Stop buttons, one for the operator and one for the off-bearer. Floor mats are provided to decrease slipping hazards and reduce fatigue. Housekeeping is good. The operator stands to one side to avoid possible injury caused by kickback.

The principal sources of injury connected with the operation of the radial saw are those common to other power-driven saw operations. They include cutting injuries to the arms and hands that are caused by the saw blade; being hit by flying wood and chips; and injuries caused by handling materials.

As with most power saws, prevention of injuries requires using equipment properly. The operator should be trained well, and he should be aware of likely hazards in the working environment. He needs to know what to do when his machine is performing below standards.

The upper half of the saw, including the arbor end, should always be guarded. The lower half of the saw should have a floating guard.

The radial saw should be installed, so that its arm or beam is raised slightly at the free (or operator) end. This slight inclination of the arm is to prevent the cutting head from "creeping" toward the operator during cross-cut operations. However, the tilt should not be enough to cause rebound.

Operators should always wear safety glasses or a face shield. They should never wear gloves, rings, or loose clothing—all of which might accidentally become caught by the saw.

Ripping. In using a radial saw for ripping purposes, special precautions are required. A spreader should be provided, used, and properly aligned with the saw blade. Antikickback fingers should be used on both sides of the blade and should be adjusted to hold the material. The direction of saw blade rotation must be upward toward the operator. (It is helpful to have the direction of the saw blade rotation conspicuously marked on the hood.)

819

FIG. 30–6.—Anti-kickback fingers should spread at least the full width of the feed rolls between the operator and the saw blade on a power-feed ripsaw. Shown here is a double set of anti-kickback fingers.

Courtesy Western Electric Company.

Some automatic feed rolls that help reduce kickbacks when ripping are available.

The saw blade should extend slightly below the table top for ripping operations.

When ripping, the operator should wear a protective apron.

Crosscutting. In this operation, the radial saw is pulled across the cutting area by a handle located on one side of the saw. As with the cutoff saw, the operator should stand to the side on which the handle is located and pull the saw with the hand nearest the handle, while holding the stock with the other hand. In this position, the operator keeps his body out of line of the saw blade, while, at the same time, making it unnecessary for him to bring his hands near the cutting zone of the blade. The operator should never place his hands in the path of the saw blade.

The stock should be removed from the table after the saw has been returned to the rear position. The stock must be placed solidly on the table and tightly against the back guide.

Usually the measuring of the stock should be completed by placing the board against a stop gage. When this is not possible, the lumber should be measured and marked before placing it on the saw table.

Scrap and sawdust should be brushed and/or vacuumed from the table.

Prospective purchasers of the radial saw, perhaps intrigued by the apparent advantages of a single saw to do many kinds of work, should give careful consideration to its hazards. In most shops, two or three saws, each intended for a special kind of work, will be safer and more efficient.

Power-feed ripsaws

A power-feed ripsaw should be guarded with a hood guard that covers the top and sides of the saw and the rolls (Fig. 30–5). Such a guard prevents the hands from coming in contact with the saw or from being pinched between the stock and the inrunning rolls. At the front, the guard should extend to within ⅜ in. of the lowest part of the rolls. There should also be sectional antikickback fingers for the full width of the feed rolls between the operator and the saw (Fig. 30–6).

820

The antikickback fingers should be checked regularly to make sure they are sharp and that none of the fingers is bent.

Because long stock is often ripped on power-feed ripsaws, the clearance at each working end of the saw table should be at least 3 ft longer than the length of the longest material handled. Feed rolls should be adjusted to the thickness of the stock being ripped. Insufficient pressure on the stock can contribute to kickbacks.

Where multiple-cut power ripsaws are used on a production basis, it is advisable to have a dado head attachment installed alongside the last saw blade. This head disposes of the edging. The offbearer then does not have to handle any scrap pieces and can pay more attention to material coming from the saw.

A common accident occurs in the use of power-feed chain rip saws as pictured in Fig. 30-5. One type of saw employs an overhead cutting saw with a solid chain, which has a babbitted center trough. Unless care is taken to maintain the babbitt in the trough, very thin slivers may drop from a ripped edge into the center trough of the chain—these can come flying out like bullets toward the operator. Regular periodic inspection is necessary.

Band saws

Although injuries from band saws are less frequent and less severe than those from circular saws, they are not uncommon. The usual cause of band saw injuries is that hands come in contact with the saw blade. Feeding is by hand, and the operator's hands must come close to the blades. Therefore, it is particularly important that the saw table be well lit, but free from glare.

The band saw point of operation cannot be completely covered, but an adjustable guard, which protects the saw at the front and on both sides, should be set as close as possible to the table, without interfering with movement of stock and clear vision by the operator (Fig. 30-7).

The wheels and all unused parts of the blade should be encased, and the periphery of the enclosure should be of solid metal (Fig. 30-8). The front and back of the enclosure should be of solid or mesh material.

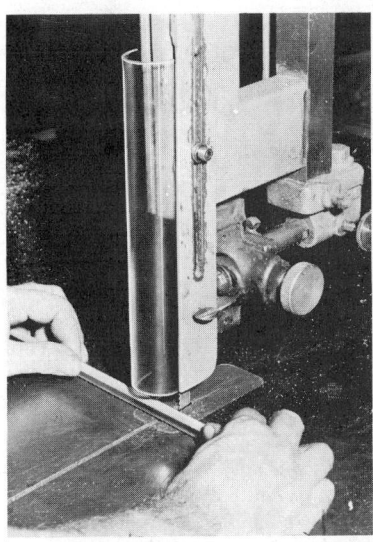

Fig. 30-7.—A clear plastic shield formed into a guard for a band saw. Note its good transparency.

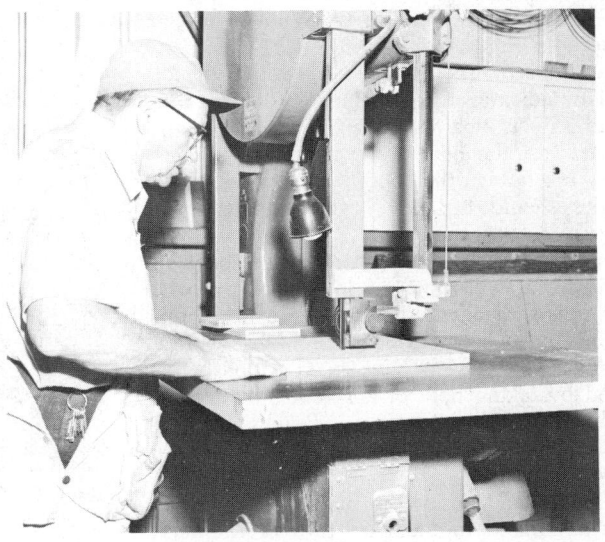

Fig. 30-8.—Band saw has metal enclosure over wheels and saw blade. A light fixture with flexible connection provides essential lighting.

Courtesy U.S. Naval Civil Engineering Laboratory, Port Hueneme, Calif.

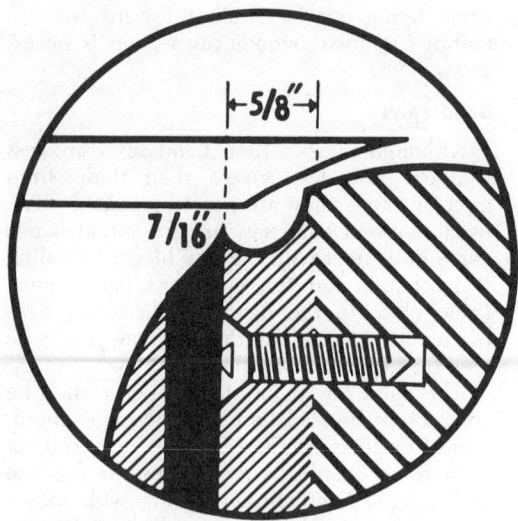

Fig. 30–9.—Detail of jointer head and knife (circular) throat not deeper than ⁷/₁₆ in., not wider than ⅝ in.

Courtesy U.S. Dept. of Labor, Bureau of Labor Standards.

An automatic tension control device will help to prevent breakage of blades. One device also prevents the motor from starting, if the tension on the blades is too much or too little.

A bandsaw should have a control device which indicates the proper tension. If the band saw is not so equipped, the operator should test the blade for correct tension before beginning the operation. Band saw blades should be periodically examined for cracks and broken teeth.

Inasmuch as a modern band saw, especially a large one, will run for a long time after the power is shut off, it should have a brake operating on one or both wheels to bring it to a quick stop. Men have been seriously injured by taking hold of a running blade, not realizing it was in motion. Brakes are also desirable to stop the wheels in case the blade should break.

On a band resaw, a heavy sheet metal guard, formed to the curvature of the feed rolls, should cover the nip point. It should be installed, so that the edge is ⅜ in. from the plane that is formed by the inside face of the feed roll in contact with the stock.

When small pieces of stock are to be cut, a special jig or fixture should be used.

Jig saws

Jig saws are not normally considered hazardous, but they occasionally cause disabling injuries.

Safe operating procedure requires that the blade be properly attached and secure, that the threshold rest on the stock, that the guard be in an effective position, and that the operator keep his hands a safe distance from the blade.

In addition, turn cuts should be made slowly, with no sharp or small radii turns if working with a large blade, and cuts should be planned to eliminate the need to back out of curves. The table should be cleaned with a long-handled brush after the blade has stopped.

Other Woodworking Equipment

Jointers

Hand-feed jointers or surface planers are, second to circular saws, the most dangerous woodworking machines. Injuries come al-

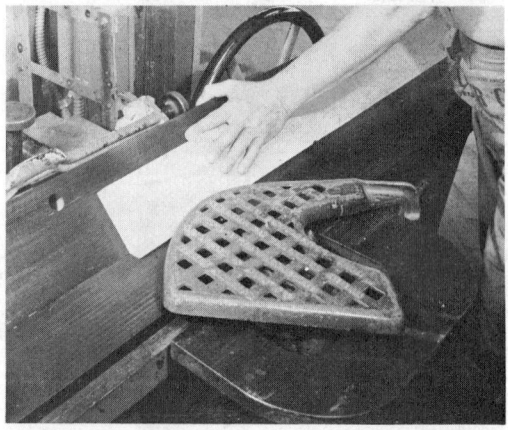

Fig. 30–10.—"Leg of mutton" guard provides good protection during edging, but does not give full protection for surfacing—much of the cutting head is exposed as the read end of the stock passes over it. Note how operator holds thumb over fence to prevent his hand from slipping when fence is not perpendicular to the jointed table. A push block should be used.

FIG. 30–11.—Jointer guard has been modified by user, allowing the material to pass over the cutting head. During edging operation, guard moves horizontally, After stock has passed over the knives, the guard automatically returns to the fence.

Courtesy Western Electric Company.

most entirely from getting the hands and fingers against the knives, and many accidents occur when short lengths of stock are being jointed.

Jointers should have cylindrical heads. The throat in the cylinder should not be deeper than 7/16 in. nor wider than 5/8 in. (Fig. 30–9).

The openings between the table and the head should be just large enough to clear the knife. Deeper cuts should be avoided because of the danger of kickbacks and also because a deep cut requires a larger table opening.

A guard which adjusts itself as the stock rides against it or under it should cover the table opening on the working side of the gage. For edge jointing, a swinging leg-of-mutton guard (Fig. 30–10) gives good protection, but, on surface jointing, this guard leaves much of the cutter head exposed as the rear end of the stock passes over it. Unless a push stick (or block) is used, the heel of the operator's hand is exposed to the revolving cutter head.

For surfacing work, a guard that rises and rides on top of the work gives much better protection. Some guards are built to move horizontally when edging, vertically when surfacing (Fig. 30–11).

The unused end of the head (the part back of the gage) should be enclosed at all times. A sheet metal telescoping guard is good.

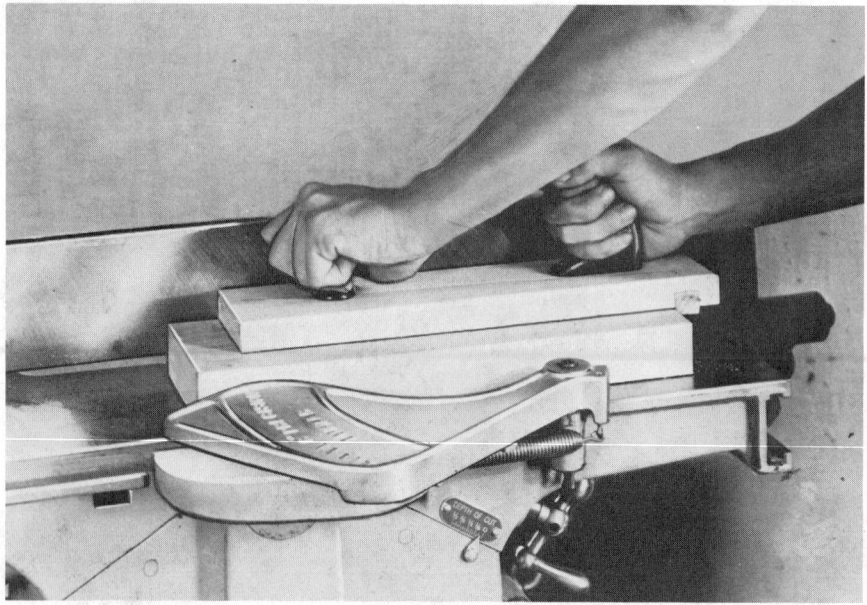

FIG. 30–12.—Notched push sticks, or, preferably, push blocks with handles (such as shown here) help keep the operator's hands away from the cutting head of the jointer.

A narrow jointer is safer than a wide one, and there is comparatively little work that requires a wide jointer.

Jointers are sometimes used for planing off cupped or warped stock to make it flat. To do this job, and still have the advantage of a power-feed machine, power-feed attachments are available with resilient holddown devices that simulate the pressure of the hands. There should be comparatively little work, however, which has to be straightened in this manner. If stock is properly conditioned, it will be straight enough so that planing on a roll-feed surface planer will give a flat surface.

For surfacing work on the jointer, the operator should have both hands on top of the stock, never over the front or back edge where they can easily come in contact with the head. If surfacing is done on a jointer equipped with a leg-of-mutton guard, the operator should be provided with push blocks of various sizes to accommodate the widths and thicknesses of stock planed. Push sticks or push blocks should be available, when short pieces of stock are jointed (Fig. 30–12).

Shapers

The principal danger in the use of the wood shaper is that hands may strike against the revolving knives. Severe accidents also result from broken knives thrown by the machine. When a shaper knife breaks or is thrown from the collar, the other knife is usually thrown too, so that four or five pieces of heavy, sharp steel are thrown about the shop with sufficient speed to kill a man.

The danger from broken or thrown knives can be eliminated by the use of solid cutters which fit over the spindle. The initial cost is greater for cutters than for knives, but, on moderately long runs, cutters are less expensive. In all cases they are safer.

When knives are used, several precautions can be taken to keep them from breaking or flying:

1. Knives must be of the best shaper steel obtainable, purchased according to rigid specifications.

2. Knives must be set only by a fully qualified shaper man.

3. Knives and the grooves in the collars must fit perfectly and be free of dust.

4. The two knives must balance perfectly. They must be weighed against each other in a beam balance each time they are set.

5. A knife must not be used after it has become so short that the butt end does not extend beyond the middle point of the collar.

6. Deep cuts should be avoided. It is safer and more efficient to take two light cuts than one heavy cut.

7. When he starts a shaper, the operator should apply the power in a series of short starts and stops to bring the spindle up to operating speed slowly. He should listen carefully for chatter and watch for other evidence that knives are out of balance.

Various types of "safety collars" are in use, and it is claimed that they prevent shaper knives from flying. Nothing should be done to discourage the use of such collars, but they should not be considered substitutes for knives of adequate length, perfectly balanced and fitted.

There should be some type of braking device to stop the spindle after the power is shut off. With double spindle shapers, there should be starting and stopping devices for both spindles, and the spindles should be started one at a time.

Only a long-handled brush should be used

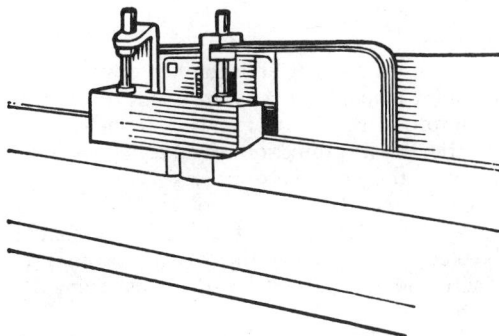

Fig. 30–14.—A shaper guarded by a fence and hold-down.

to remove chips and scraps from the work table.

A number of guards are available to protect the fingers, all of which cover the top of the spindle and the knives and surround the knives just enough to clear the stock (Fig. 30–13). The greatest number of accidents comes from the use of narrow stock which, held in the operator's hand, brings the hand close to the knives.

A jig keeps the hands a safe distance from the cutter and also permits the use of a guard, which cannot be kept in position if narrow stock is held by hand. Each shop should have a well-understood rule that stock narrower than a specified width must be held in a jig. Some shops put the limit at 6 in., others as high as 12 in.

Shaper work should be held against guide pins or a fence (Fig. 30–14).

When feeding the shaper, the operator must never forget that the direction of the cut must be made in the opposite direction of the rotation of the cutting head. It is important to instruct a new operator well.

Power-feed planers

The planer is so powerful and fast running that it tends to vibrate excessively. Vibration can be reduced by anchoring the planer on a solid foundation and by insulating it from the foundation with cork or other vibration-absorbing material. If the planer is bolted down, it should have a three-point bearing and *must not* be bolted into a twist. Distorting any woodworking machine will ultimately cause it to malfunction and, perhaps,

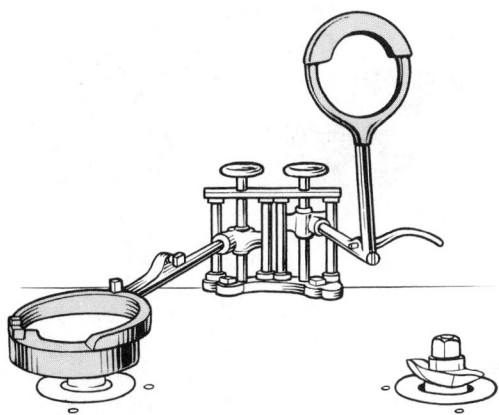

Fig. 30–13.—Ring guards for shaper heads.

to become unsafe.

Because of noise, it is highly desirable to isolate a planer in a separate room or build a soundproof enclosure for it. If neither is practical, ear protection should be worn by those in the immediate area.

Cutter heads should be completely enclosed in solid metal guards which should be kept closed when the planer is running. There should be good local exhaust from the cutting heads.

Feed rolls, cutter heads, and cylinders should be stopped before reaching into the bed plate to remove wood fragments, to make adjustments, or for any other reason. If planer parts are driven by belts running on the back side of the planer, belts and pulleys should be completely enclosed by sheet metal or heavy mesh guards, even though the planer is fenced at the back or is next to the wall.

Feed rolls should be guarded by a wide metal strip or bar which will allow boards to pass, but will keep the operator's fingers out of the rolls.

Danger of kickbacks cannot be entirely overcome by mechanical means. Therefore, the operator should always stand out of the line of board travel. Other men should not work or walk directly behind the feeding end of the planer. Operators must avoid feeding boards of different thicknesses at the same time because a thinner board is not held by the feed rolls and can be kicked back from the heads.

The operator should wear a face shield or eye protection.

The space at the outrunning end should be fenced or marked off to keep workers out of the area where they might be struck by long, fast-moving boards.

Sanding machines

Drum, disk, or belt sanding machines should be enclosed by an exhausting dust hood which encloses all portions of the machine except the portion designed for the work feed. Personnel who operate sanders should wear goggles and dust respirators during sanding operations and cleanup.

On a belt sanding machine, a guard should be placed at each inrunning nip point on both power transmission and feed roll parts. The unused run of the abrasive belt should be guarded to prevent human contact.

All manually fed sanders should have a work rest and be properly adjusted (a) to provide minimum clearance between the belt and the rest, and (b) to secure support for the work. Small pieces should be held in a jig or holding device.

Abrasive belts used on sanders should be the same width as the pulley-drum. The drums should be adjusted to keep the abrasive belt taut enough to turn at the same speed as the pulley-drum, yet not slip on the drum when material is brought into contact with the moving abrasive belt.

Inspect abrasive belts before using them. Those found to be cracked, frayed, or excessively worn in spots should be replaced.

References

American National Standards Institute, 1430 Broadway, New York, N.Y. 10018. *Safety Requirements for Woodworking Machinery,* O1.1.

National Safety Council, 425 North Michigan Ave., Chicago, Ill. 60611.
Guards Illustrated, 2nd. ed.
Industrial Data Sheets
Overhead Swing Cut-Off Saws, 277.
Power Feed Wood Planers, 225.
Sawmill Edgers, 571.
Sawmill Head Rigs, 460.
The Radial Saw, 353.
Tilting-Arbor and Tilting Table Saws, 605.
Wood Jointers, 280.
Wood Shapers, 333.
Wood Stickers, Molders and Matchers, 384.
Wood Turning Lathes, 253.
Woodworking Band Saws, 235.

Machine
Tools

Chapter
31

General Safety Rules   828
Personal protection

Turning Machines   830
Engine lathes . . . Turret lathes and screw machines

Boring Machines   832
Drills . . . Boring mills

Milling Machines   835
Basic milling machines . . . Metal saws . . . Spinning lathes . . .
Gear cutters

Planing Machines   839
Planers and shapers . . . Slotters . . . Broaches

Grinding Machines   841
Abrasive disks and wheels . . . Inspection, handling, and storage
. . . Wheel mountings . . . Guarding disks and wheels . . .
Adjusting safety guards . . . Safe speeds . . . Work rests . . .
Abrasive wheel dressing . . . Surface grinders and internal grinders
. . . Grindstones . . . Polishing and buffing wheels . . . Wire brush
wheels

References   855

31—Machine Tools

Injuries on machine tools occur most often because of unsafe work practices or incorrect procedures—basically problems of training and supervision. If equipment were kept in top shape and operated correctly, injuries from machine tools would be rare. Better use of safeguards would help too. There are many effective guarding devices that can be used without hampering operation or reducing production. Of course, certain safeguards are required by OSHA, state, or local regulations.

More rarely, but still too often, injuries result when a machine fails mechanically or is operated after an unsafe condition develops.

Good housekeeping, too, contributes to safe operation. Good habits in maintaining a ship-shape work area can carry over to establishing good habits in machine operation, with a resulting prevention of accidents (Fig. 31–1).

Definition. Machine tools include all power-driven machines, not portable by hand, used to shape or form metal by cutting, impact, pressure, electrical techniques, or a combination of these processes. Grinders, buffers, and similar machines are included in this definition.

The National Machine Tool Builders Association has classified some 200 types of machine tools into five basic groups—*turning, boring, milling, planing,* and *grinding.* Some machines combine the functions of two or more basic groups.

Portable power tools, normally held in the operator's hands while in use, are not considered to be machine tools. They are covered in Chapter 35, "Hand and Portable Power Tools."

There is insufficient statistical data to make a conclusive evaluation of the hazards of machine tools. Injuries resulting from machine tool operations are reported less frequently than injuries from power presses or metal shears, but still present some of the more serious problems in the metal products industries.

General Safety Rules

The greatest emphasis and dependence must be placed upon safe operation. A policy to eliminate unsafe practices which result in operator injury should be established and should include these provisions:

1. Operation, adjustment, and repair of any machine tool must be restricted to experienced and trained personnel or apprentices under *close* supervision.

2. Safe work procedures must be established, with short cuts and chance-taking prohibited.

3. Supervisors must be responsible for the enforcement of this policy and for making certain that no deviation from points 1 and 2 is permitted.

4. When purchasing new equipment, specify that it should conform to all applicable regulations concerning guarding, electrical safety, etc.

5. New equipment should be inspected and safety innovations made before allowing operator(s) to use the equipment.

Maintenance personnel and repairmen should comply with this five-point policy.

A tool rack should be provided for the convenience of the operator, repairmen, and maintenance personnel. All wrenches and tools needed for operation or adjustment should be included as standard equipment.

The *National Electrical Code* (NFPA Standard No. 70) and the *National Electrical Safety Code* (American National Standard Series C2) should govern the installation of electrical circuits and switches.

In addition to the electrical controls installed on machine tools by the manufacturer, each machine should have a disconnect switch, which can be locked in the OFF position, to isolate the machine from the power source.

Maintenance or repair should be permitted on any machine only after its disconnect switch has been shut off, padlocked in the OFF position, and tagged.

The following general rules apply to the safe operation of any machine tool:

1. Machine tools should never be left running unattended.

2. Operators should not wear jewelry or loose-fitting clothing, especially loose sleeves or cuffs of shirts or jackets and neckties. Long hair, which could be caught by moving parts, should be covered.

3. All operators should wear eye protection,

Fig. 31–1.—Operator of this engine lathe is safely attired. Workplace shows good housekeeping.

Courtesy General Electric Company.

as should others in the area (such as inspectors, stock handlers, and supervisors).

4. Throwing refuse or spitting in the machine tool coolant should not be allowed—such actions foul the coolant and may spread disease.

5. Manual adjusting and gaging (calipering) of work should not be permitted while the machine is running.

6. Operators should use brushes, vacuum equipment, or special tools for removing chips.

7. Operators should use the proper hand tools.

One of the major causes of eye accidents on all machine tools, especially on drilling equipment, is indiscriminate use of high-pressure compressed air to blow chips from machines or workers' clothing. Brushes provide a less dangerous method.

In cases where neither a brush nor a vacuum system proves practicable, it may be necessary to continue the use of air. The line pressure should be as low as possible—many companies have found a nozzle pressure of 10 to 15 psi sufficient for most operations. (OSHA specifies that the pressure be less than 30 psig.)

The operation should be isolated so that other employees are not endangered. Baffles should be placed around the machine. Only reliable and trained employees should be permitted to use compressed air, and they should wear cup-type goggles and other personal protective equipment while using it. Nozzles that meet OSHA requirements are readily available.

Employees should be prohibited from using compressed air to blow dust or dirt from their clothing or out of their hair because damage to ears and eyes can result.

Fig. 31–2.—This transparent lathe guard protects against chips and permits the operator to see the work.

Personal protection

The safety of a machine tool operator depends largely upon his following established safe work procedures and his wearing of proper protective clothing.

Obviously, all machine tool operators should wear eye protection. Without some effort to confine or control the chips and coils removed from the stock being machined, there is no way for an operator to determine or control the direction of flight of these metal particles.

Closely fitting clothing is of vital importance to the operator's safety. Many serious injuries and fatalities have resulted when neckties, loose shirt sleeves, or other clothing were caught in a belt and pulley, between gears, in a revolving shaft, or in the revolving work being held in the chuck (Fig. 31–1).

Both male and female operators who wear long hair should wear caps, snoods, hairnets, or other protection to cover their hair completely. There have been many instances of people being partially (or entirely) scalped when their hair became entangled in the moving parts of a machine. Operators should also not wear gloves, rings, necklaces or neckties, or jewelry.

Because most machine operations involve the handling of heavy stock or heavy machine parts, such as face plates, chucks, etc., every operator should wear safety shoes.

Splash guards, shields, and other means may be practical to minimize exposure of workers to cutting oils which may cause skin irritations. Personal hygienic measures by the employee will also tend to minimize skin irritations.

Turning Machines

Turning—shaping a rotating piece with a

Fɪɢ. 31–3.—Swinging welded pipe fixture (right) supports lathe chucks and face plates in order to take the strain away from the operator when he makes changes.

cutting tool, usually to give a circular cross section—is done on machines such as engine lathes, turret lathes, chuckers, semiautomatic lathes, and automatic screw machines.

Engine lathes

An engine lathe must be operated intelligently, if accidents are to be prevented. Injuries are likely to result from:

1. Contact with projections on work or stock, face plates, chucks, or lathe dogs, especially those with projecting set-screws.

2. Flying metal chips.

3. Hand braking of the machine.

4. Filing right-handed, using file with unprotected tang, or using the hand instead of a stick to hold emery cloth against the work.

5. Calipering or gaging the job while the machine is in operation.

6. Attempting to remove chips when machine is in operation.

7. Contact with rotating stock projecting from turret lathes or screw machines.

8. Leaving chuck wrench in chuck. Using a spring-loaded wrench can eliminate this hazard.

9. Catching loose clothing or wiping rags on revolving parts.

Injuries may also occur on lathes because operators fail to keep the center holes of taper work clean and true and the lathe centers true and sharp. Furthermore, it is hazardous for operators to leave the machines running unattended, or to handle chips by hand instead

Fig. 31–4.—An automatic screw machine with its enclosure guard that prevents splashing of cutting oil.

Courtesy Delco-Remy, Division of General Motors.

of using a hooked rod.

Face plates and chucks without projections should be used whenever possible. Otherwise, a simple shield formed to the contour of the chuck or plate and hinged at the back should be installed to prevent contact with the revolving plate or chuck.

Safety-type lathe dogs are relatively inexpensive to install and should be substituted for those with projecting setscrews. Chip guards, particularly on high-speed operations, will control flying chips. Plastic or small-mesh screen chip guards have been successfully used because they allow visibility and confine the flying chips (Fig. 31–2). Use of these guards, however, does not eliminate the need for protective eye equipment.

Mechanical means, such as an overhead hoist or a special device (Fig. 31–3), should be provided to lift heavy face plates, chucks, and stock on both lathes and screw machines and eliminate operator strain.

Turret lathes and screw machines

Injuries from turret lathes and screw machines (Fig. 31–4) result from the same causes as those listed for other lathes. In addition, injuries frequently occur when the operator fails to move the turret back as far as possible when changing or gaging work, or uses the power of the machine to start the face plate or chuck onto the spindle.

Other injuries result when the operator fails to keep his hand clear of the turret slide or permits his hand, arm, or elbow to strike the cutter while adjusting or setting up.

Splash guards, especially on automatic machines, should be installed and kept in good condition. Enclosure guards over the chuck confine hot metal chips and oil splashes, and also act as exhaust hoods for removal of fumes.

When steel and some other materials are turned on lathes, the chip produced is in a continuous spiral which frequently causes hand and arm injuries. Proper chip breakers will provide protection against this. Fig. 31–5

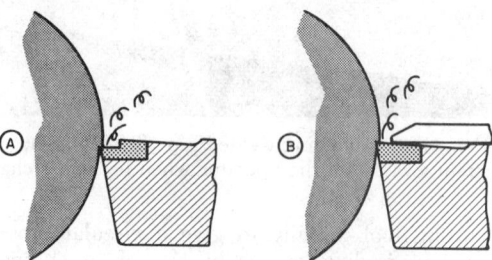

Fig. 31–5.—Chip breakers may be ground into the tip of the tool (A) or a breaker in the form of a piece of metal may be clamped or brazed onto the tool itself (B).

illustrates two types of chip breakers.

Screw machines may create noise which exceeds OSHA standards. Consideration should be given to the reduction of the noise level by engineering methods, including commercially available sleeves for rotating bar stock. See details in Chapter 40, "Noise and Hearing Conservation."

Boring Machines

Boring consists of cutting a round hole by means of drills, boring cutters, or reamers.

Fig. 31–6.—Guards that afford protection against chips and broken drills and yet preserve visibility. The filler on these guards is wire mesh.

The drill press, in its various forms, is probably the best known machine tool because of its frequent use in home workshops. Drilling machines are equipped with rotating spindles, handles, and chucks carrying pointed or fluted cutting tools. Operations performed with drilling machines include countersinking, reaming, tapping, facing, spot facing, and routing.

Boring mills use a cutter, either single or multiple edged, and mounted on a supporting shaft to true up or enlarge a hole that has already been roughed out by a drill or formed in a rough casting or forging.

Drills

The most common causes of injury in drilling operations are:

1. Contacting the spindle or the tool (Do not touch the tool while using a quick-change clutch.)

2. Breaking a drill

3. Using dull drills

4. Being struck by insecurely clamped work

5. Catching hair or clothing in the revolving parts

6. Sweeping chips or trying to remove long, spiral chips by hand

7. Leaving key or drift in chuck

8. Being struck by flying metal chips

9. Failing to replace guard over speed change pulley or gears

To guard a spindle, a simple sleeve guard or other barrier that protects the operator from contact with the spindle can be used (Fig. 31–6).

The tool itself can be guarded by a telescoping guard that covers the end of the tool,

833

leaving only enough of it exposed to allow easy placement into the piece being worked. A spring safety guard can be attached to a small drill press and, as the head descends, the spring compresses so that it can contain metal slivers and chips. Brackets of various lengths can be used, depending upon the depth of the hole to be drilled (Fig. 31–7).

Breakage of tools results most frequently from use of dull tools. A drill, smaller than ⅛-in. diameter will often break and cause injury. A larger drill may "fire up" to the extent that it freezes in the hole and then breaks. Furthermore, a frozen tool may spin unclamped or insecurely clamped work, with possible injury to the operator.

To avoid having a drill catch in thin material and spin it, the work can be clamped between two pieces of metal or wood before drilling. Generally, when drilling thin stock, it is advisable to grind the drill point to an included angle of about 160 degrees, and thin the point of the drill by grinding the flutes.

Drill press accidents are more likely to occur during odd jobs, because special jigs or vises for holding the work are not usually provided as they are in production work.

When deep holes are being drilled beyond the flutes of the drill, the drill should be removed frequently and the chips cleaned out. If chips are allowed to pile up in such an operation, the tool may jam, with results similar to those of freezing.

Counterweight chains should be maintained in good condition, and a shield should be installed around the counterweight.

Radial drill accidents are frequently caused by incorrect manipulation of controls. The drill head and arm, as well as the workpiece, should be properly clamped prior to cutting metal.

Boring mills

Among common causes of injury in boring mill operations are:

1. Being struck by insecurely clamped work or by tools left on or near revolving table

2. Catching clothing, wiping rags, etc., in revolving parts

3. Falling against revolving work

4. Calipering or checking work while ma-

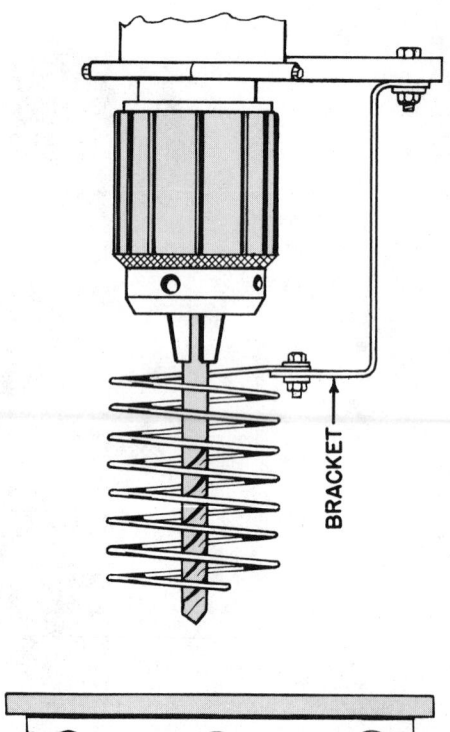

FIG. 31–7.—Spring safety guard compresses as drill cuts into metal in order to contain metal slivers and chips.

chine is in motion

5. Allowing turnings to build up on table

6. Removing turnings by hand

The same accident prevention measures are effective on both the table type and the floor type of horizontal boring mills. The operator should never attempt to make measurements near the tool, reach across the table, or adjust the machine or work while the machine is in motion.

Clamps and blocking should be inspected periodically to make sure clamping is positive. Makeshift setups should always be avoided.

Before attempting to raise or lower the head of the boring mill, the operator should make sure that the clamps on the column have been loosened. Otherwise, the boring bar may be bent or the clamps or bolts broken, with possible damage to the machine and injury to the operator.

Fɪɢ. 31–8.—A sheet metal guard for a vertical boring mill is made in two sections which are hinged to the machine. Left view shows guards closed. Right view shows guards when opened to permit setup or machine adjustment.

Before the boring bar is inserted into the spindle, the operator should make certain that the spindle hole and the bar are clean and free from nicks. No attempt should be made to drive the bar through the tail stock bearing with a hammer or other heavy tool.

A soft metal hammer should be used to drive the bar into the spindle. If a steel hammer or piece of steel must be used, the operator should hold a piece of soft copper or brass against the bar while driving it into the spindle.

The same procedures apply to the safe operation of a vertical boring mill. The table of each mill, particularly those tables 100 in. or less in diameter, should have the rim enclosed in a metal band guard to protect the operator from being struck by the revolving table or by projecting work. Such guards should be hinged so that they may be easily opened during setting up and adjustment (Fig. 31–8).

If the table is flush with the floor, a portable fence, usually of iron pipe sections, should be installed. Such fencing should conform to the state code or the specifications of the American National Standard A12.1, *Safety Requirements for Floor and Wall Openings, Railings, and Toe Boards.*

The operator should never attempt to tighten the work or the tool, caliper or measure the work, feel the edges of the cutting tool, or oil the mill while it is in operation. Further, he should never ride the table while it is in motion, except that on some large mills, like those used for boring turbine castings, the operator may have to ride the table in order to observe the progress of the work. In such cases, he should always make certain that no portion of his body will come in contact with a stationary part of the mill.

Steps or stairs which provide access to the machine or to the work should have a pitch of not more than 50 degrees and should have nonslip treads. Stairs with four or more risers must have a handrail.

Milling Machines

Milling—machining a piece of metal by bringing it into contact with a rotating multi-edged cutter—is done by horizontal and vertical milling machines, by gear hobbers, profiling machines, circular and band saws,

FIG. 31–9.—All the power transmission apparatus on this milling machine is guarded.

Courtesy Sundstrand Corporation.

and a number of other types of related machines (Fig. 31–9).

About two of every three milling machine accidents occur when operators unload or make adjustments. Other causes of injuries include:

1. Failure to draw the job back to a safe distance when loading or unloading

2. Using a jig or vise which prevents close adjustment of the guard

3. Placing the jig or vise-locking arrangement in such a position that force must be exerted toward the cutter

4. Leaving the cutter exposed after the job has been withdrawn

5. Leaving hand tools on the worktable

6. Failure to clamp the work securely

7. Reaching around the cutter or hob to remove chips while the machine is in motion

8. Removing swarf by hand instead of with a brush

9. Adjusting the coolant flow while the cutter is turning

10. Calipering or measuring the work while the machine is operating

11. Using a rag to clean excess oil off the table while the cutter is turning

12. Wearing gloves, ties, or loose clothing

13. Using cutters incorrectly dressed

14. Storing cutters incorrectly

15. Attempting to remove a nut from machine arbor by applying power to the machine

16. Striking cutter with hand or arm while setting up or adjusting stopped machine

Basic milling machines

Regardless of the classification, direction of movement, or special attachments which make varied operations possible on a milling machine, the safeguarding requirements are basically the same. To guard the cutter, one of several methods should be employed (Fig. 31–10).

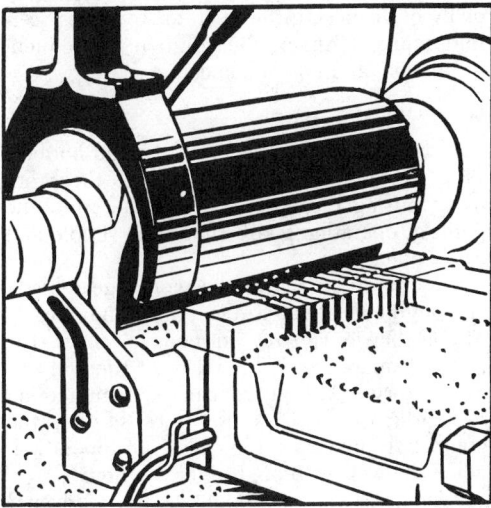

FIG. 31–10.—Self-closing guard for milling machine cutter. Cutter is entirely enclosed when the table is withdrawn (top), and the guard opens automatically as the table moves forward.

The hand-adjusting wheels, for quick or automatic traverse on some models, should be mounted on the shaft by either clutches or ratchet devices so that the wheels do not revolve when the automatic feed is used; or else the wheels should be provided with removable handles with compression springs so that the handles cannot remain in the wheels, unless held in place by the operator.

The horizontal milling machine should have a splash guard and pans for catching thrown cutting lubricant and lubricant running from the tools. The lubricant should be directed on the work in such a way that the distribution setup will not be drawn into the cut by means of the cutter rotation. When possible, all cuts should be made into the travel of the table, rather than away from the direction of travel.

Metal saws

A circular saw for cutting cold metal stock should have a hood guard at least as deep as the roots of the teeth, and the guard should automatically adjust itself to the thickness of the stock being cut.

A sliding stock guard should be used when tube or bar stock is cut. The portion of the saw under the table should be guarded with a complete enclosure that provides for disposal of scrap metal. A plastic or metal guard placed in front of and over the saw will provide protection against flying pieces of metal. This guard should not, however, be considered a substitute for eye protection.

In swing-type saws, the length of the stroke should be adjusted, so that the blade will not pass the table at its most-forward point. The control should be located so that the saw can be operated with the left hand when fed from the left, or with the right hand if fed from the right. This will position the operator to the side away from the whirling blade.

Band saws. The upper and lower wheels of band saws for cutting metal should be completely enclosed with sheet metal or heavy, small mesh screen mounted on angle iron frames. Access doors equipped with latches should be provided for changing of blades. The portion of the saw blade between the upper wheel and the saw table should be completely enclosed with a sliding fixture attached to the slide, except for the point at which the cut is made.

The length of blade exposed should be not more than the thickness of the stock being cut plus ⅜ in. Flying particles of metal can be confined by a metal or transparent plastic guard installed in front of the saw. On a hand-fed operation, care should be taken at the end of a cut. Use a push block, not hands.

Spinning lathes

A spinning lathe is a forming tool rather than a machine tool and usually requires a specially skilled and qualified operator.

Unsafe practices which should be prohibited in operation of a spinning lathe include:

1. Inserting blanks and removing the processed part without first stopping the machine

2. Failing to tighten the handle of the tailstock sufficiently, so that the blank may work loose or ruin the stock or tool

3. In trimming copper and certain grades of steel, allowing the swarf to build up into a long coil

This last practice has caused hands to be severed and arms severely cut. One fatality has been reported from a coil which became snarled around the operator's neck. Operators should remove the tool, when necessary to allow the swarf to break off.

The chuck of a spinning lathe is usually a form built up of hardwood, shaped exactly like the finished part. If the piece being worked must be "necked-down," so that after the piece is formed the chuck cannot be taken out, the chuck is made in sections held together by locking rings or locking grooves.

There is a great hazard from the chuck's flying apart if the grooves should become worn or the rings should break. In view of the speed of spinning lathes (500 to 2000 rpm), the danger of flying sections of the chuck is obvious. Prevention of injuries from this cause lies in periodic inspection and maintenance of chucks.

Similarly, lathe tools should be inspected frequently for cracks in the handles or in the tools themselves.

Fig. 31–11.—A gear hobbing machine with chip and splash guards.

Courtesy International Harvester Company.

On some older equipment, vibration or worn parts may cause the tailstock to loosen during operation and the piece to work loose or fly off, exposing the operator to almost certain injury. Again, the solution is frequent inspection and maintenance.

Gear cutters

In the operation of gear cutters and hobbers (Fig. 31–11), both the tool and the work move during the operation. As a result, the point of operation guard should be simple and easily adjusted.

On operations where the work (gear blank or rough-cut gear) is moved to the tool, a simple barrier guard, formed to cover the point of operation and sized to fit the work, is satisfactory. The guard can be mounted on a spindle that carries the work so that the guard will fit over the point of operation, when the work is brought into position.

When the tool is brought to the work and when the tool and work are adjustable, an encircling type of guard may be attached to the tool head. Such a guard can be an automatic dropgate device, which can be equipped with a release latch to open the guard enclosure and a spring release to return the guard to a position clearing the work. Each guard should have an automatic interlock, so that the machine will not operate except when the guard is in place.

On some makes of machines still in use, the lever which controls the direction of operation of the spindle is located so that the operator's hand may be caught on the back gears driving the spindle. An auxiliary lever should be installed which can be operated at a point outside the danger zone created by these gears.

On large machines where the operator may not be close to the regular control switch, a pendant switch, mounted on an arm or sweep, should be installed to operate a magnetic brake so that the machine can be stopped instantly.

When the operator inserts an arbor into the spindle, he should be sure that both arbor and spindle holes are clean and free from nicks. He should draw the arbor firmly into place by a sleeve nut and securely tighten the nut. Before he removes the arbor from the spindle, he should make certain that the machine is at a standstill.

Planing Machines

Similar to a carpenter's hand plane, a planer machines a metal surface with the cutting tool held stationary while the work is moved back and forth underneath it. Shapers are generally classified as planing machines, but the process is reversed—the work is held stationary while the cutting tool is moved back and forth. Other machine tools coming within this classification are slotters, broaches, and keyseaters.

Planers and shapers

Planer accidents result frequently from unsafe practices which are basically problems of training and supervision:

1. Placing the hand or fingers between the

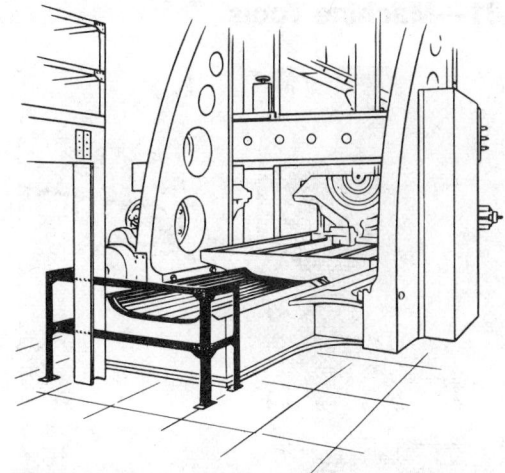

Fig. 31–12a.—A guardrail or similar barrier should close off any space 18 in. or less between a fixed object and parts of a fully extended planer or its stock. Openings in the bed of the planer should be filled to eliminate shear hazards.

tool and the work.

2. Running the bare hand over sharp metal edges.

3. Measuring the job while the machine is running.

4. Failing to clamp the work or tool securely before starting the cut.

5. Riding the job.

6. Having insufficient clearance for work.

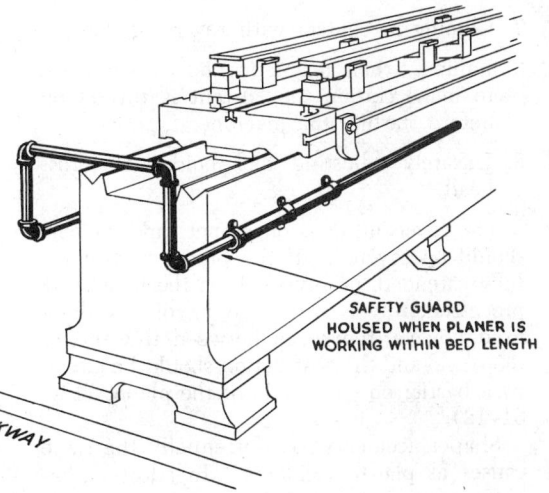

SAFETY GUARD HOUSED WHEN PLANER IS WORKING WITHIN BED LENGTH

WALKWAY

Fig. 31–12b.—Self-adjusting planer table guard moves out with the table and is retained in position by friction sleeves.

Fɪɢ. 31–13.—Foot-operated emergency stop bar on broaching machine has a wide surface plate to facilitate its use. The platform on which the operator stands should be anchored to avoid movement under the stop bar.

Courtesy Detroit Diesel Engine Div. General Motors Corp.

7. Coming in contact with reversing dogs.

8. Failing, when magnetic chucks are used, to make certain the current is turned on before starting the machine.

9. Unsafely adjusting tool holder on cross head.

The reversing dogs on planers and shapers should be covered. If the planer bed, when fully extended, or any stock on the bed being processed, travels within 18 in. of a wall or fixed objects, the space between the end of the travel and the obstruction should be closed by a barrier on either side of the planer (Fig. 31–12).

Shaper accidents have essentially the same causes as planer accidents. In addition, in-

juries frequently result from contact with projections on the work or with projecting bolts or brackets, especially when the table is being adjusted vertically. The ram of the shaper should be left projecting over the table, so that the operator will know when the table is high enough.

Failure to locate the stops or dogs properly may cause injury to shaper operators. The stops should be rigidly bolted to the table, especially on heavy jobs.

The shaper operator should make sure that the tool is set so that, if it shifts away from the cut, it will rise away from the cut and not dig into the work. The handle of the stroke change screw should be removed before the shaper is started. Flying chips must be controlled in order to prevent injury to the operator and to adjacent workers.

Slotters

The most serious accident which occurs in the operation of slotters is catching the fingers between the tool and the work, or between the ram and the table when the ram is at the end of the downstroke. Since the ram works at slow speed and the platen or machine table is small, the operator may reach across the table and under the ram to pick up a tool or other object. The ram eccentric should be enclosed, preferably with a hinged guard of sheet metal or cast iron.

Broaches

Broaches, like heavy production machine tools, may be safeguarded during normal operations by the use of supplementary controls. The most widely used safeguard for broaches is a standard two-hand, constant-pressure control. An emergency stop button, preferably of the mushroom type, should be installed adjacent to one of the two-hand controls. Another type of safeguard for broaches is a foot-operated emergency stop bar with a wide surface plate (Fig. 31–13).

All pneumatically or hydraulically powered clamping equipment should be actuated by two-hand controls, so located that the operator's hands cannot reach the pinch area before the clamps close. Controls should be shielded, if they could be tripped by parts of the body other than hands.

Tongs should be used for loading and unloading, if the hands are exposed in the clamping area.

The rated capacity of the broach should be equal to or greater than the force required for the job. The centerlines of the work, the ram head puller, and the follow rest, if used, should all line up.

Fixtures should be checked to make sure that the work is securely held. Trial runs at slow speeds are advisable to make sure that the chips do not pack tightly between the teeth.

Grinding Machines

Grinding—shaping material by bringing it into contact with a rotating abrasive wheel or disk—includes surface, internal, external cylindrical, and centerless operations. Polishing, buffing, honing, and wire brushing are also classed as grinding operations. Portable machines that use small, high-speed grinding wheels are discussed in Chapter 35, "Hand and Portable Power Tools."

The text and illustrations in this section have been adapted with permission from American National Standard B7.1, *Safety Code for the Use, Care, and Protection of Abrasive Wheels.* Specifications for the operation of grinding machines and for the construction of guards and safety devices are discussed in this code.

Abrasive disks and wheels

- **An abrasive disk** is made of bonded abrasive, with inserted nuts or washers, projecting studs, or tapped plate holes on one side of the disk, which is mounted on the machine face plate of a grinding machine. Only the exposed flat side of an abrasive disk is designed for grinding.
- **An abrasive wheel** is made of bonded abrasive and is designed to be mounted, either directly or with adapters, on the spindle or arbor of a grinding machine (Fig. 31–14). Only the periphery or circumference of an abrasive wheel is designed for grinding.

The causes of personal injury involving abrasive wheels and disks include:

1. Failure to use eye protection in addition to the eye shield mounted on the grinder

2. Holding the work incorrectly

3. Incorrect adjustment or lack of work rest

4. Using the wrong type of wheel or disk or a poorly maintained or imbalanced one

5. Grinding on the side of a wheel

6. Taking too heavy a cut

7. Applying work too quickly to a cold wheel or disk

8. Grinding too high above the center of a wheel

9. Failure to use wheel washers (blotters)

10. Vibration and excessive speed which lead to bursting of wheel or disk

11. Use of bearing boxes with insufficient bearing surface

12. Using spindle with incorrect diameter or with the threads cut so that nut loosens as spindle revolves

13. Installing flanges of the wrong size, with unequal diameters, or unrelieved centers

14. Incorrect dressing of wheel

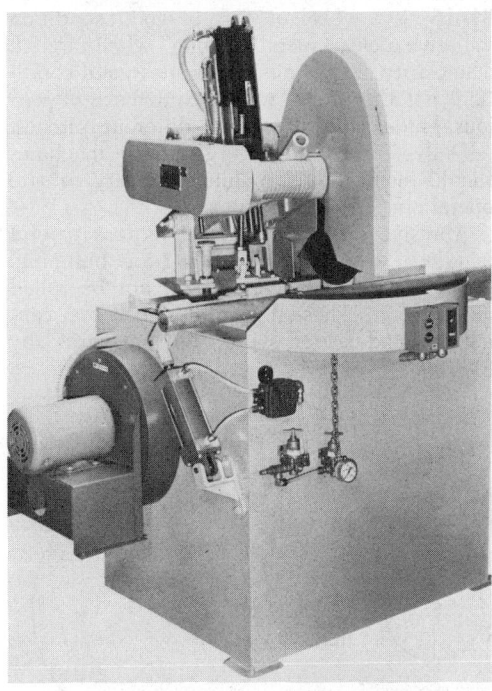

Fig. 31–14.—An abrasive wheel cutoff machine with wheel and motor guards.

Courtesy Everett Industries Inc.

15. Contacting unguarded moving parts

16. Using controls that are out of operator's reach

17. Using an abrasive saw blade instead of a grinder disk

Inspection, handling, and storage

While they are being unpacked, abrasive disks and wheels should be inspected for damage received in shipment and, then, given the "ring" test by a qualified person. This test can be used for both light and heavy disks or wheels that are dry and free of foreign material. To conduct the test, a light disk or wheel should be suspended from its hole on a small pin or the finger, and a heavy one should be placed vertically on a hard floor. Then the wheel or disk should be tapped gently with a light tool, such as a wooden screwdriver handle; a mallet may be used for heavy wheels or disks. The tap should be made at a point 45 degrees from the vertical centerline and about 1 or 2 in. from the periphery (Fig. 31–15). A wheel or disk in good condition will give a clear, metallic "ping" when tapped. The clarity of the "ping" indicates good condition, not the pitch—wheels and disks of various grades and sizes give different pitches.

Daily inspection of grinding machines should include those points necessary to safe operation (Fig. 31–16).

Abrasive disks and wheels require careful handling to prevent dropping or bumping. Large disks and wheels should not be rolled on the floor. Those disks and wheels that cannot be hand-carried should be transported by

GRINDER CHECKLIST	
TYPE_____RPM_____	
SIZE_____PERIPHERAL SPEED_____	
Item	**OK**
HOOD: securely fastened............................	☐
properly aligned	☐
GLASS SHIELD: clean	☐
unscored	☐
in place	☐
WORK REST: within ⅛ inch of wheel	☐
securely clamped.................................	☐
FRAME: securely mounted	☐
no vibration	☐
WHEEL FACE: well lighted	☐
dressed evenly	☐
FLANGES: equal size	☐
correct diameter (½ wheel diam.)........	☐
SPEED: correct for wheel mounted	☐
GUARD FOR POWER BELT OR DRIVE:	
in place..	☐
DATE_____DEPARTMENT_____	
INSPECTED BY_____	

FIG. 31–16.—A summary of checkpoints for safe grinder operation.

truck or other conveyance which provides support.

Abrasive disks and wheels should be stored in a dry area which is not subject to extreme temperature changes, especially below-freezing temperatures. Wet wheels may break or crack if stored below 32 F. Breakage may also occur, if a wheel or disk is taken from a cold room and work is applied to it before it has warmed up.

Abrasive disks and wheels should be stored in racks in a central storage area (Fig. 31–17) under the control of a specially trained person. The storage area should be as close as possible to grinding operations to minimize handling and transportation.

The length of time abrasive disks and wheels may be stored and still be safely used should be in accordance with manufacturers' recommendations. Disks and wheels taken out of long storage should be given the "ring" test, followed by a check for recommended

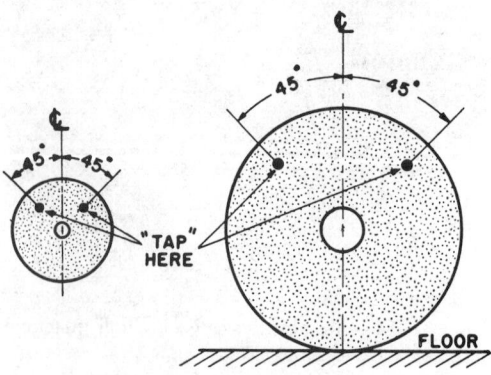

FIG. 31–15.—Tap points for the "ring" test.

speed and a speed test on the machine on which they are to be mounted. Checking the speed of vitrified wheels is especially important—some are designed only for low-speed use.

Wheel mountings

All abrasive wheels should be mounted between flanges; exceptions include: mounted wheels, threaded wheels (plugs and cones), plate-mounted wheels, and cylinder, cup, or segmental wheels mounted in chucks.

Flanges should have a diameter not less than one-third of the wheel diameter and should preferably be made of mild steel. However, cast iron or other material of equal strength is sometimes used for flanges less than 10 in. in diameter. Flanges for the same wheel, whether straight or tapered, should be of the same diameter and thickness, accurately turned to correct dimensions, and in balance. The requirement for balance does not apply to those flanges that are made out of balance to counteract an unbalanced wheel.

The inner or driving flange should be keyed, screwed, shrunk, or pressed onto the spindle, and the bearing surface of the flange should run true with the spindle. The bore of the outer flange should have an easy sliding fit onto the spindle.

Flanges should be frequently inspected at regular intervals. A flange found to be sprung, not bearing evenly on the wheel, or defective in any other way should be removed from the spindle at once and replaced with a flange in good condition.

There are two classes of flanges used to mount abrasive wheels: straight flanges and safety flanges.

Straight flanges serve only as a support and driving medium for the wheel and should only be used where a safety guard (protection hood) is installed on the wheel. They may be simple collar type flanges or integral parts of the wheel sleeves and adapters.

Safety flanges are used on abrasive wheels of special shapes and those, 6 in. or more in diameter, that are not provided with safety guards, chucks, or bands. These flanges are designed so that, in addition to clamping the wheel to the spindle, they tend to retain the

Fig. 31–17.—Wooden storage racks for abrasive wheels designed for storage of heavier wheels at the bottom and smaller wheels in the upper tiers of shelves.

Courtesy General Motors Corporation.

pieces of wheel in case of breakage during operation. Tapered, hub, and ring are the three types of safety flanges most commonly used.

Incorrect mounting of an abrasive wheel is responsible for much wheel breakage. Since rotational forces and grinding heat cause high stresses around the central hole of the wheel, it is most important that safety regulations concerning size and design of mounting flanges and mounting techniques be followed.

Before a wheel is mounted, it should be given the same inspection and "ring" test as was given when it was originally received and stored. The bushings, particularly on wheels which have been rebushed by the user, should be checked for shifting or looseness.

Compression washers should be used to compensate for unevenness of the wheel or flanges. Blotting paper (not more than 0.025 in. thick) or rubber or leather compression washers (not more than 0.125 in. thick) may be used for this purpose. If flanges with babbitt or lead facings are used, the thickness of

843

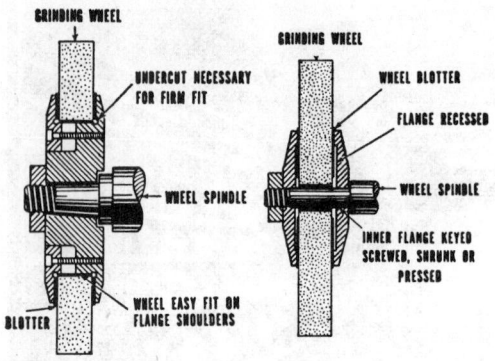

Fig. 31–18.—Correct methods of mounting wheels with small holes (*right*) and wheels with large holes (*left*).

the facing should not exceed 0.125 in. and the diameter of washers, regardless of the material, should not be less than the diameter of the flange.

Allowance for the mounting fit of the wheel should be made in the wheel hole rather than in the arbor or wheel mount (Fig. 31–18). The wheel should not be forced on the spindle, because such forcing may loosen, or otherwise damage, the wheel bushing or crack the wheel. A wheel that is too loose on the spindle will run off-center. Spindle end nuts should hold the wheel firmly but not too tightly. Too much pressure may spring or distort the flange or may even break the wheel.

If rebushing is necessary to make the wheel fit the spindle, the job should be done by the manufacturer or, in the plant, by an experienced man with suitable equipment.

Immediately after mounting the wheel and before turning on the power, the operator should turn the wheel by hand for a few revolutions to see that it clears both the work rest and the hood guard.

Guarding disks and wheels

A safety guard is an enclosure for an abrasive disk or wheel consisting of a peripheral member and usually two side members. *Band-type guards do not have side members.* The main function of a safety guard is to re-

TABLE 31-A

MINIMUM THICKNESSES OF PERIPHERAL AND SIDE MEMBERS FOR SAFETY GUARDS ON CUTTING-OFF AND GRINDING WHEELS

Max. Peripheral Speed (sfpm)	Material	Cutting-Off Wheel Diameters (Max. Thickness ½ in.)									
		6 to 11 in.		11+ to 20 in		20+ to 30 in.		30+ to 48 in.		48+ to 72 in.	
		A (in.)	B (in.)	A (in.)	B (in.)	A (in.)	B (in.)	A (in.)	B (in.)	A (in.)	B (in.)
16,000	Structural Steel (min. tensile strength 60,000 psi)	3/32	1/8	1/8	1/8	3/16	1/8	1/4	3/16	5/16	1/4

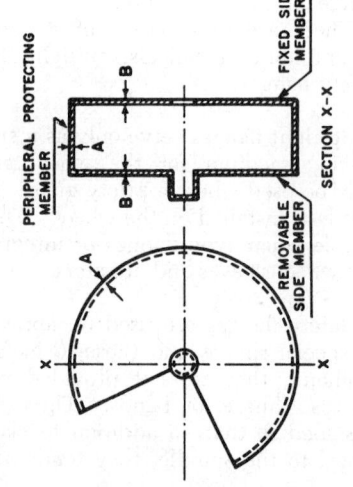

844

GRINDING WHEEL DIAMETERS

Maximum Peripheral Speed (sfpm)	Material	Maximum Thickness of Grinding Wheel (in.)	3 to 6 in. A	3 to 6 in. B	6+ to 12 in. A	6+ to 12 in. B	12+ to 16 in. A	12+ to 16 in. B	16+ to 20 in. A	16+ to 20 in. B	20+ to 24 in. A	20+ to 24 in. B	24+ to 30 in. A	24+ to 30 in. B	30+ to 48 in. A	30+ to 48 in. B
8,000	Cast Iron (min tensile strength 20,000 psi) Class 20	2	1/4	1/4	3/8	5/16	1/2	3/8	5/8	1/2	7/8	5/8	1	3/4	1 1/4	1
		4	5/16	5/16	3/8	5/16	1/2	3/8	3/4	5/8	1	5/8	1 1/8	3/4	1 3/8	1
		6	3/8	5/16	1/2	7/16	5/8	1/2	1	5/8	1 1/8	3/4	1 1/4	7/8	1 1/2	1 1/8
		8	…	…	5/8	9/16	7/8	3/4	1	3/4	1 1/8	3/4	1 1/4	7/8	1 1/2	1 1/8
		10	…	…	3/4	11/16	7/8	3/4	1 1/4	1	1 5/16	1	1 1/4	7/8	1 1/2	1 1/8
		16	…	…	…	…	1 1/8	1	1 3/8	1 1/8	1 3/8	1 1/8	1 7/16	1 1/16	1 3/4	1 3/8
		20	…	…	…	…	…	…	…	…	…	…	1 1/2	1 3/8	2	1 5/8
9,000	Malleable Iron (min tensile strength 50,000 psi) Grade 32510	2	1/4	1/4	3/8	5/16	1/2	3/8	5/8	1/2	3/4	5/8	7/8	3/4	1	7/8
		4	5/16	5/16	3/8	5/16	1/2	3/8	5/8	1/2	3/4	5/8	7/8	3/4	1 1/8	7/8
		6	3/8	5/16	1/2	7/16	5/8	1/2	3/4	5/8	7/8	5/8	1	3/4	1 1/4	7/8
		8	…	…	1/2	7/16	5/8	1/2	3/4	5/8	7/8	5/8	1	3/4	1 1/4	7/8
		10	…	…	1/2	7/16	13/16	11/16	13/16	11/16	1	3/4	1 1/8	7/8	1 1/4	7/8
		16	…	…	…	…	…	…	7/8	3/4	1	3/4	1 1/8	7/8	1 3/8	1
		20	…	…	…	…	…	…	…	…	…	…	…	…	1 1/2	1 1/8
16,000	Steel Castings (min tensile strength 60,000 psi) Grade V60-30	2	1/4	1/4	5/16	5/16	3/8	3/8	1/2	7/16	5/8	1/2	3/4	5/8	3/4	3/4
		4	1/4	1/4	1/2	1/2	1/2	1/2	9/16	1/2	5/8	1/2	3/4	5/8	1	3/4
		6	3/8	1/4	3/4	5/8	3/4	5/8	3/4	5/8	13/16	11/16	13/16	11/16	1 1/8	1
		8	…	…	7/8	3/4	7/8	3/4	7/8	7/8	1 1/8	3/4	15/16	13/16	1 13/16	1 1/16
		10	…	…	1	7/8	1	7/8	1 1/4	1 1/8	1 1/4	1 1/8	1 1/8	1	1 13/16	1 11/16
		16	…	…	…	…	1 1/4	1 1/8	1 3/8	1 1/4	1 3/8	1 1/4	1 7/16	1 5/16	2 1/16	1 11/16
		20	…	…	…	…	…	…	…	…	…	…	…	…	…	…
16,000	Structural Steel (min tensile strength 60,000 psi)	2	1/8	1/16	5/16	1/4	5/16	1/4	5/16	1/4	5/16	1/4	3/8	5/16	1/2	3/8
		4	1/8	1/16	3/8	5/16	3/8	5/16	3/8	5/16	3/8	5/16	3/8	5/16	1/2	3/8
		6	3/16	1/16	1/2	3/8	7/16	3/8	7/16	3/8	7/16	3/8	7/16	3/8	3/4	1/2
		8	…	…	1/2	3/8	9/16	7/16	9/16	1/2	5/8	7/16	5/8	1/2	3/4	1/2
		10	…	…	9/16	7/16	5/8	1/2	5/8	5/8	3/4	5/8	5/8	1/2	7/8	5/8
		16	…	…	5/8	…	…	9/16	3/4	11/16	13/16	11/16	13/16	11/16	1 1/16	13/16
		20	…	…	…	…	…	…	13/16	…	…	…	7/8	3/4	1 13/16	15/16

TABLE 31-B

DIMENSIONS FOR BAND-TYPE GUARDS FOR ABRASIVE DISKS

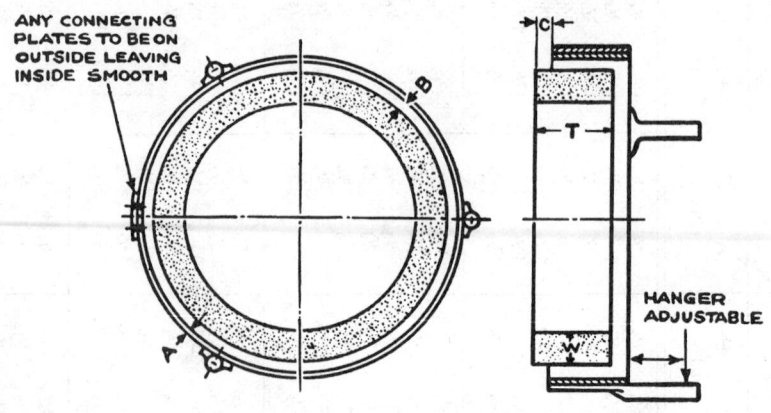

Disk Diameter (in.)	A Min Thickness of Band (in.)	Min Diameter of Rivets (in.)	Max Distance Between Centers of Rivets (in.)	T Overall Thick. of Disk (in.)	C Max Exposure of Disk (in.)
Less than 8	1/16	3/16	3/4	1/2	1/4
				1	1/2
8 to 24	1/8	1/4	1	2	3/4
				3	1
25 to 30	1/4	3/8	1 1/4	4	1 1/2
				5 and over	2

tain the pieces of the disk or wheel, if it should be broken in operation. Safety guards (protection hoods) should be constructed so that, in changing wheels, there is no need for removing the peripheral member which is connected to the machine. Guards on machines used for dry grinding should have provisions for connecting to an exhaust system. Table 31-A gives the material specifications for safety guards.

Safety guard is a general term and includes all of the following specific types:

Band-type guard

Cast guard

Drawn steel guard

Fabricated guard

Revolving cup-type guard

A **band-type safety guard** is designed especially for an abrasive disk where grinding is done only on the exposed side of the disk and where its diameter remains constant. The band-type guard differs from other safety guards in that it consists only of a properly supported peripheral member or band. This band is continuous and completely surrounds the disk and, though its diameter is only slightly larger than the disk diameter, it need not be as heavy as other types of guards that are used on abrasive wheels.

A band-type guard should be made of wrought iron or steel plate or other material of equal or greater strength. It should be continuous and should conform as closely as practicable to the periphery of the disk. The ends should be riveted, bolted, or welded together to leave no projections inside the band.

TABLE 31-C

MINIMUM DIMENSIONS FOR DRAWN STEEL GUARDS ON GRINDING WHEELS 8 INCHES OR LESS IN DIAMETER

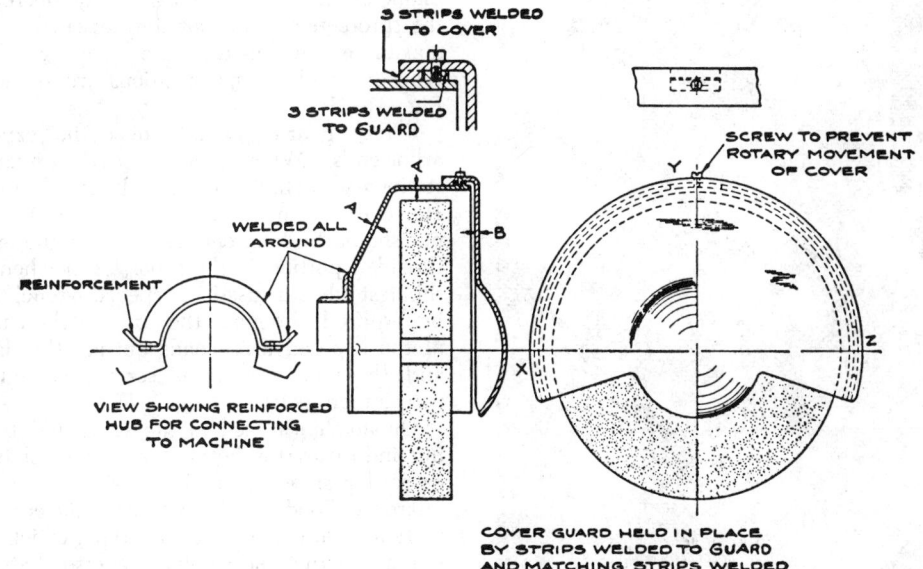

VIEW SHOWING REINFORCED HUB FOR CONNECTING TO MACHINE

COVER GUARD HELD IN PLACE BY STRIPS WELDED TO GUARD AND MATCHING STRIPS WELDED TO COVER. THREE PAIRS LOCATED AT X-Y-Z

Material Used in Construction of Guard Hot Rolled Steel SAE 1008 Min. Tensile Strength 60,000 PSI For Speeds Up to (SFPM)	Maximum Thickness of Wheel, Inches	2 to 5 Inches		Above 5 to 8 Inches	
		A	B	A	B
		Inches		Inches	
9,500	2	$\frac{1}{16}$	$\frac{1}{16}$	$\frac{3}{32}$	$\frac{1}{16}$
12,500	2	$\frac{3}{32}$	$\frac{1}{16}$	$\frac{3}{32}$	$\frac{3}{32}$
17,000	1	$\frac{3}{32}$	$\frac{1}{16}$	$\frac{1}{8}$	$\frac{3}{32}$

The band should be wide enough and kept in position, so that at no time will the disk protrude beyond the edge of the band farther than maximum exposure (C), indicated in Table 31-B, or at a distance greater than the thickness of the wall (W) whichever is less.

A cast safety guard is one which has the peripheral protecting member cast integral with one side member and may be of iron, steel, or other suitable material. One member, the peripheral member, or both may also be an integral part of the base casting or wheel head casting of the grinding machine.

The guard should be no more than ½ in. from the wheel (B).

A drawn steel guard is a safety guard with the peripheral member and the fixed side member formed from a single plate or sheet of steel. A cover, consisting of a side plate and a relatively narrow peripheral member, is similarly formed from a single plate or sheet of steel. The cover fits over and outside the peripheral member of the guard. The cover, therefore, adds strength to the guard proper, in addition to acting as a side guard. Construction details are shown in Table 31-C.

847

Fɪɢ. 31–19.—A good example of a fabricated safety guard.

A fabricated guard is a safety guard, which is built up or constructed by bolting, pinning, riveting, or welding the peripheral protecting member to the side members, and may be made of structural steel plate, in combination with iron or steel castings, or a material possessing an equivalent tensile strength (Fig. 31–19). Table 31-A indicates the material specifications, and Table 31-D shows dimensions and construction details for fabricated guards.

Adjusting safety guards

The peripheral protecting member can be adjusted to the constantly decreasing diameter of the wheel by means of an adjustable tongue or similar device. Then the angular protection specified for bench and stand grinders will be maintained throughout the life of the wheel, and the maximum distance between the wheel periphery and the tongue

or end of the peripheral band at the top of the opening will not exceed ¼ in. (Fig. 31–20).

The guard should enclose the wheel as completely as the nature of the work will permit and should be adjustable so that, as the diameter of the wheel constantly decreases, the protection will not be lessened. The maximum angular exposure varies with the type of grinding; specifications are shown in Fig. 31–21.

Safety guards should cover the exposed arbor ends. Where there is no safety guard, a stationary cylindrical guard is desirable. However, a smooth nut may be used. The nut should be long enough to cover the entire threaded portion of the arbor, even when the thinnest wheel is used. A special wrench may be required to remove the nut, but the hazard of the exposed arbor ends outside the flange and.threaded spindle is serious enough to warrant protection.

On machines used in the stone and building industries for cutting, grooving, slotting, or coping stone or other materials, the safety guard or hood seldom affords sufficient protection. On machines of this type, which permit a relative horizontal traverse between wheel and work greater than 10 in. and which use solid cutting wheels 10 in. or more in diameter, an auxiliary enclosure should be provided in addition to the guard.

This auxiliary enclosure may consist of a set of heavy screen panels, suspended from approximately 8 ft above the floor to or below the table on which the work is placed. The screens for such panels should be ½-in. mesh or smaller, and the wire should be ⅛ in. in diameter or more. The framework of the panels should be made of 1- by 1¼-in. or heavier structural steel angles or channels.

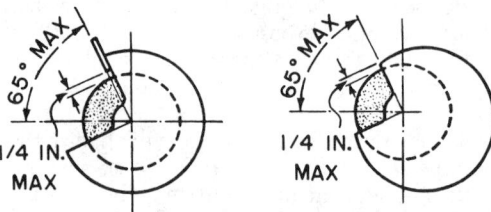

Fɪɢ. 31–20.—The correct wheel exposure can be maintained with an adjustable tongue (left) or a movable guard (right).

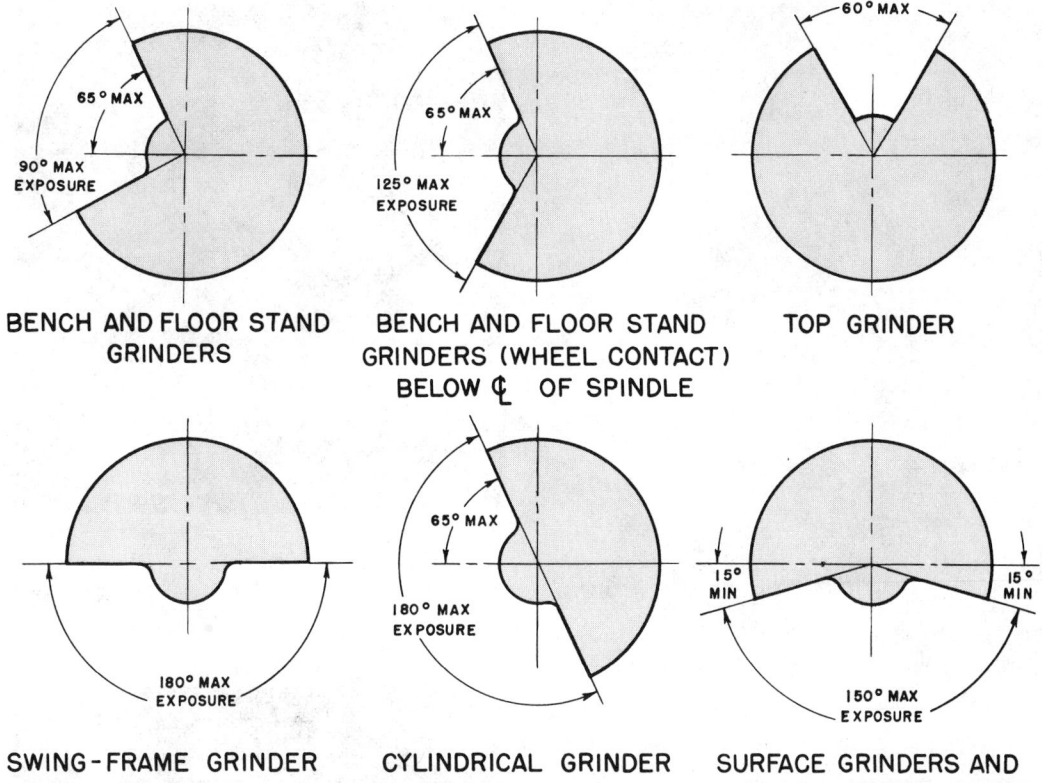

FIG. 31–21.—Maximum exposure angles for various grinding applications.

Safe speeds

The various types of abrasive wheels and disks should be operated at speeds not exceeding those recommended by the manufacturer. In particular, unmarked wheels of unusual shape, such as deep cups with thin walls or backs or with long drums, should be operated according to the manufacturer's recommendations.

All wheels should be tested by the manufacturer (Fig. 31–22) at specific testing speeds shown in Table 31-E. Wheels which need not be tested include those less than 6 in. in diameter, diamond wheels, mounted wheels, and other special types. A letter of certification should accompany each speed-tested wheel, and the maximum operating speed should be clearly marked on the wheel.

Before a wheel is mounted, the machine spindle should be checked for the correct size. Spindles, including those with adjustable speeds, should be changed only by authorized persons.

Because most defective wheels break when first started, new wheels should be run at full operating speed for at least one minute before work is applied. During this time, the operator should stand away from the machine.

As the wheel wears down, the spindle speed (rpm) is sometimes increased to maintain the surface speed (sfpm). When the wheel is nearly worn down, the spindle is running at the highest rpm. Therefore, when the worn wheel is replaced, spindle speed must be adjusted or the new wheel may break at a surface speed that exceeds manufacturer's recommendations.

The cause of grinding wheel or disk failure should be thoroughly investigated, preferably with the manufacturer's representative.

849

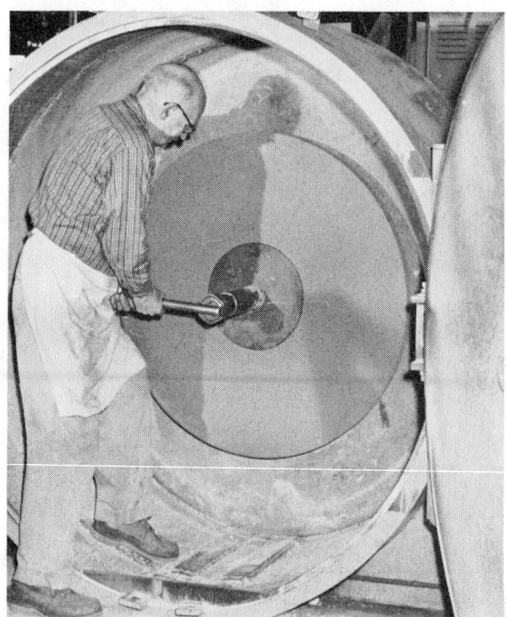

FIG. 31–22.—Speed testing of grinding wheels is done in totally enclosed variable speed machines.

FIG. 31–23a.—With a properly adjusted work rest, operator can keep his hands away from the wheel and also hold the work firmly. Note ⅛-in. clearance between wheel and work rest.

This type of investigation, with immediate corrective action, greatly reduces the possibility of recurrent failures.

Grinding equipment for high-speed operation should be specially designed, with particular reference to spindle bearings for vibration control, wheel strength, guards, and flanges for eliminating mounting stresses. The manufacturer's approval should be obtained for high-speed operation of wheels and disks. Such things as pressure from side grinding and the shape of the wheel must also be considered. Approval of special high-speed operation should apply only to the particular machine investigated. Also important to safe high-speed operation is the maintenance of equipment and protective devices.

Work rests

Many wheels have broken, causing injury to operators, because work has become wedged between the work rest and the wheel. The work rest should be substantially constructed and securely clamped not more than ⅛ in. from the wheel (Fig. 31–23). The position of the work rest should be checked fre-

FIG. 31–23b.—Tool rest should leave a safe space (⅛-in.) between it and the wheel.

TABLE 31-D

DIMENSIONS FOR CONSTRUCTION OF FABRICATED GUARDS OF STRUCTURAL STEEL

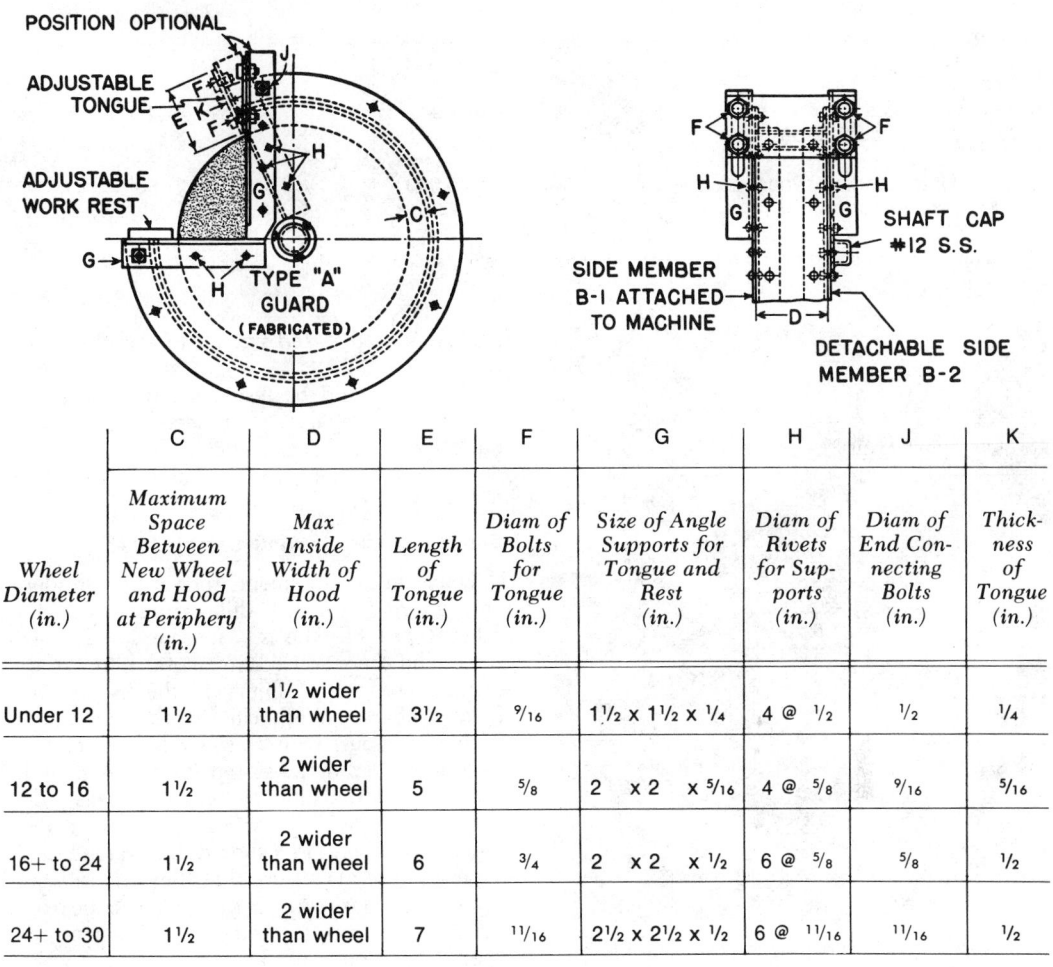

	C	D	E	F	G	H	J	K
Wheel Diameter (in.)	Maximum Space Between New Wheel and Hood at Periphery (in.)	Max Inside Width of Hood (in.)	Length of Tongue (in.)	Diam of Bolts for Tongue (in.)	Size of Angle Supports for Tongue and Rest (in.)	Diam of Rivets for Supports (in.)	Diam of End Connecting Bolts (in.)	Thickness of Tongue (in.)
Under 12	1½	1½ wider than wheel	3½	9/16	1½ x 1½ x ¼	4 @ ½	½	¼
12 to 16	1½	2 wider than wheel	5	5/8	2 x 2 x 5/16	4 @ 5/8	9/16	5/16
16+ to 24	1½	2 wider than wheel	6	¾	2 x 2 x ½	6 @ 5/8	5/8	½
24+ to 30	1½	2 wider than wheel	7	11/16	2½ x 2½ x ½	6 @ 11/16	11/16	½

quently. The rest should never be adjusted while the wheel is in motion—the rest may slip and strike the wheel and break it, or the operator may catch his finger between the wheel and the rest.

To prevent twisting of the work and the imposition of a bending stress on the wheel, operators should use guides to hold the work in position when slot grinding or performing similar operations.

Abrasive wheel dressing

Abrasive wheels that are not true or not in balance (Fig. 31-24) will produce poor work and may damage the machine and injure the operator. Keeping the wheels in good condition eliminates these possibilities, decreases wheel wastage, and lengthens wheel life.

To put a rutted or excessively rough wheel in good condition, it is often necessary to dress it by removing a large area of the face.

TABLE 31-E

WHEEL MANUFACTURERS' TEST FACTORS

Class of Wheel	Operating Speed (sfpm)	Test Factor°
Cutting-off	All speeds	1.20
All Bonds and Types (except cutting-off)	Less than 5,000	1.25
All Bonds and Types (except cutting-off)	More than 5,000	1.50

*Multiply operating speed by test factor to obtain minimum testing speed.

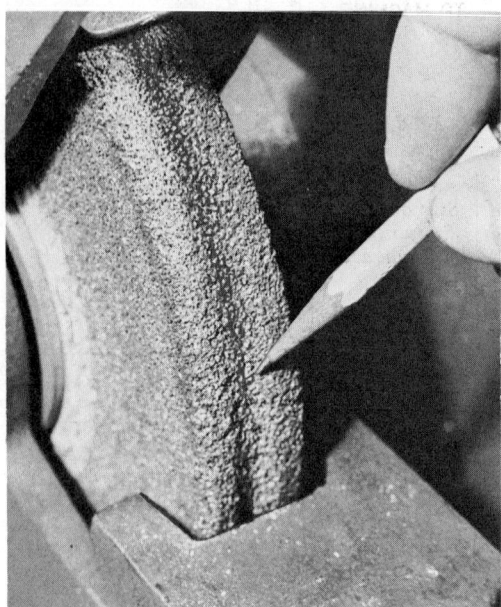

Fig. 31–24.—Badly rutted or out-of-balance grinding wheel requires dressing.

Wheel dressing tools should be equipped with hood guards over the tops of the cutters to protect the operator from flying wheel particles or pieces of broken cutters. The operator of a wheel dresser should use a rigid work rest set close to the wheel. The wheel dresser should be moved back and forth across the face of the wheel, with the heel or lug on the underside of the dresser head held firmly against the edge and not on top of the work rest (Fig. 31–25).

Wheels should be tested for balance occasionally and rebalanced if necessary. Worn, out-of-balance wheels, which cannot be balanced by truing or dressing, should be taken out of service.

Surface grinders and internal grinders

Operating requirements for surface grinders and internal grinders differ from those for other types of wheels. Insecurely clamped work and unenergized magnetic chucks are common sources of injury in the use of surface grinders. Work can be thrown, under such conditions, with considerable force.

If the operator takes too deep a cut, or if he traverses the table or wheel too fast, the wheel may overheat at the rim and crack. Operators, therefore, should be trained and supervised to make sure that they clamp work tightly, always turn a magnetic chuck on before applying the wheel, and control the speed and depth of the work.

Baffle plates on each end of a surface grinder are generally standard equipment, but should include some provision for exhausting the grinding dust (Fig. 31–26).

Internal grinders can be guarded with an automatic positioning hood that covers the grinding wheel when it is in the retracted or idling position.

Grindstones

In using grindstones, the manufacturer's suggested running speeds and operational pro-

BENCH AND STAND GRINDERS DRESSING

1. Wear a face shield over your safety glasses for protection against heavy particles.
2. Use a dressing tool approved for the job. Never use a lathe cutting tool.
3. Inspect star dressers for loose shaft and worn disks.
4. Round off the wheel edges with a hand stone before and after dressing to prevent the edges from chipping.
5. Use the work rest to support and guide the tool. Use a tool holder if one is available.
6. Apply moderate pressure slowly and evenly.
7. Always apply diamond dressers at the center or slightly below the center, never above.

FIG. 31–25.—Major safety measures to be taken in a wheel dressing operation.

cedures should be followed. Stones of unknown composition or manufacture should never be run more than 2500 sfpm and ordinarily not more than 2000 sfpm.

The size and weight of grindstones require a stand that is rigidly constructed, that is heavy enough to hold the stone securely, and that is mounted on a solid foundation to withstand vibration.

Since grindstones are run wet, all possible precautions should be taken to prevent slipping accidents near the stones. Rough concrete or other nonskid floor material is recommended.

Grindstones should be carefully inspected for cracks and other defects as soon as they are received from the manufacturer. Those not to be used at once should be stored in a dry, uniformly heated room. In storage they should be placed so that they will not be damaged.

Many grindstone failures result from faulty handling and from incorrect methods of mounting. Grindstones should not be left partially submerged in water. This practice causes an unbalanced stone that may break when rotated. Wooden wedges should not be used on power-driven stones. Often, such wedges are driven excessively tight or become wet and swell. In either case, cracks start in the corners of a square center hole, radiate outward, and so weaken the stone that ruptures occur at normal speed.

After the stone has been centered, the central space about the arbor should be filled with lead or cement. Double thicknesses of leather or rubber gaskets, rather than wood washers, should be used wherever possible. If wood washers are used between the flanges and the stone, the washers should be ½- to 1-in. thick and the flanges should be clamped in place by heavy nuts.

Work rests should comply with the same conditions as those for grinding wheels.

For dressing or operating power-driven grindstones (either wet or dry), exhaust systems should be provided to remove dust and wet spray or mist.

Polishing and buffing wheels

Polishing wheels are either wood wheels faced with leather or wheels made of disks of canvas or similar material stitched together, with a coat of emery or other abrasive glued to the periphery of the wheels.

Buffing wheels are made of disks of felt, linen, or canvas, and the periphery is given a

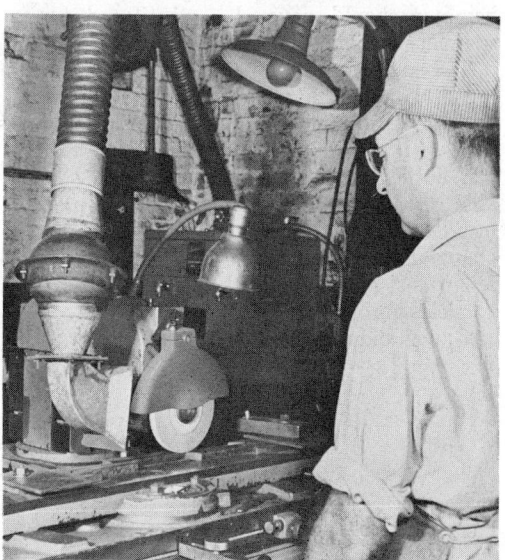

FIG. 31–26.—Guarding and exhaust provisions for a surface grinding operation.

coat of rouge, tripoli, or other mild abrasive.

The softness of the wheel built up of linen, canvas, felt, or leather is determined by the size of the flanges used—the larger the flange, the harder the surface. When large flanges are used, it is often necessary to soften the working surface of the built-up wheels to conform to the contour of the object being polished. A safe procedure for softening is to place the wheel on the floor or other flat surface and pound the edges of the wheel with a hammer or mallet. The wheel should not be placed on the spindle with a file or other object held against it. Such practice is almost certain to result in the object's catching in the wheel and being thrown with such force that the operator or nearby workers may be injured.

Polishing and buffing wheels should be mounted on rigidly constructed substantial stands, heavy enough for the wheels used. Mounting procedures are the same as for grinding wheels. Hood guards should be designed to prevent the operator's hands or clothing from catching on protruding nuts or the ends of spindles. If working conditions require a hood which does not give such protection, then smooth nuts which cover the spindle ends and are installed with a spanner wrench should be used. A prick punch and hammer should never be substituted for a spanner wrench.

The range of peripheral speed in polishing and buffing is from 3000 to 7000 sfpm, with 4000 sfpm in general use for most purposes. If the motors that drive the polishing and buffing wheels are equipped with adjustable speed controls, the controls should be installed in a locked case and the speed changed only by an authorized person.

Exhaust hoods should be designed to catch particles thrown off by the wheels.

Gloves should not be worn by polishers and buffers because a glove may catch and drag the operator's hand against the wheel. Small pieces being polished or buffed can frequently be held in a simple jig or fixture. Some operators use a piece of an old linen or canvas wheel for holding small pieces. The operator should not attempt to hold a small piece against the wheel with his bare hands.

When rouge or tripoli is applied to a revolving wheel, the side of the cake should be held lightly against the periphery of the wheel. If a stick is used, the side of the stick should be applied on the "off" side so that, if thrown, it will fly away from the wheel.

Wire brush wheels

Brushing or, more commonly, scratch wheel operation, is used to remove burrs, scale, sand, and other materials.

Wheels are made of various kinds of protruding wires and with different thicknesses. Scratch wheels should be rigidly held in place by flanges or nuts.

The same machine setup and conditions that apply to polishing and buffing wheels apply to brushes. The speed recommended by the manufacturer should be followed. The hood on scratch wheels should enclose the wheel as completely as the nature of the work allows and should be adjustable, so that the protection will not be lessened as the diameter of the wheel decreases. The hood should cover the exposed arbor ends, or a smooth nut should be installed on them. The work rest should be adjusted so that it is about ⅛ in. from the brush wheel.

Personal protective equipment is especially important in the operation of scratch wheels because of the tendency of the wires to break off. The wearing of aprons of leather, heavy canvas, or other heavy material, leather gloves, face shields, and goggles should be mandatory.

References

American National Standards Institute, 1430 Broadway, New York, N.Y. 10018.
  National Electrical Code, C1.   (Also available from National Fire Protection Association.)
  National Electrical Safety Code, C2.   (Also available from National Bureau of Standards.)
  Safety Code for Use, Care, and Protection of Abrasive Wheels, B7.1.   (Also available from Grinding Wheel Institute.)
  Safety Requirements for Floor and Wall Openings, Railings, and Toeboards, A12.1.
  Specifications for Shapes and Sizes of Grinding Wheels, B74.2.
  Ventilation Control of Grinding, Polishing, and Buffing Operations, Z43.1.

Grinding Wheel Institute, 2130 Keith Bldg., Cleveland, Ohio 44115. *Safety Recommendations for Grinding Wheel Operation.*

National Machine Tool Builders Assn., 2139 Wisconsin Ave. NW., Washington, D.C. 20007.

National Safety Council, 425 North Michigan Ave., Chicago, Ill. 60611.
  Guards Illustrated.
  Industrial Data Sheets
    Coated Abrasives, 452.
    Engine Lathes, 264.
    Gear-Hobbing Machines, 362.
    Horizontal Metal Boring Mills, 269.
    Metal Planers, 383.
    Metal Saws (Cold), 322.
    Metal Shapers, 216.
    Metal Slotters, 263.
    Metal-Working Drill Presses, 335.
    Metal-Working Milling Machines, 364.
    Portable Grinders, 583.
    Vertical Metal Boring Mills, 347.

Cold Forming
of Metals

Power Presses   857
Mechanical power presses . . . Hydraulic and pneumatic presses . . .
Press brakes . . . Power shears . . . Hand or foot operated presses
. . . Foot shears and rollers

Power Press Electric Controls   865
Installation . . . Part revolution clutch presses . . . Full revolution
clutch presses . . . Controls for hydraulic or pneumatic presses

Point-of-Operation Guarding   869
Die-enclosure guards . . . Fixed barrier guards . . . Adjustable barrier
guards . . . Gate or movable barriers . . . Point-of-operation devices
. . . Two-hand tripping devices . . . Pull-back devices . . .
Automatic electrical devices . . . Auxiliary devices . . . Sweep devices

Feeding and Ejection of Parts   883
Ejecting devices

Die Handling   890
Installation of dies . . . Removing dies

Inspection and Maintenance   895

Power Press Noise   896

References   897

Of all the machines used by industry, probably none are more controversial than those used for cold forming metal—namely power presses and other related equipment. The controversy arises primarily because of the many serious accidents associated with press operations, particularly at the point of operation.

For many years, concerned and knowledgeable safety professionals have advocated a "hands out of dies" policy for press operations, reasoning that if the operator could not put his hands or other parts of his body between the dies, there could be no injury. This policy was incorporated into the 1971 revision of the voluntary American National Standard B11.1, *Safety Requirements for Construction, Care, and Use of Mechanical Power Presses,* which in turn in 1972 was made an OSH standard, section 1910.217. The policy thus passed from voluntary to mandatory—with wide-sweeping consequences for all power press users.

There is no useful purpose served in debating the merits or faults of such a policy. Both sides of the issue have presented many arguments supporting their position. Neither side disputes the goals of the "no hands in dies" policy, which is the elimination of press injuries in the die area. It is the means of accomplishing the goal which concerns everyone.

For many years the Power Press and Forging section of the National Safety Council has also advocated the "no hands in dies" policy, and many NSC members have adopted the policy for their operations. Others have adopted modified versions of the policy. It really matters little what the final policy or means are as long as the goal remains "no point-of-operation press accidents."

The problem is not easy to solve for many reasons. At this time, one of the largest reasons is that no one has objectively studied press operations systematically in all its components. This is particularly true with regard to the interaction of the operator in the process at the interface. (See Chapter 10, "Human Factors Engineering.") Be sure to obtain the American National Standards that pertain to your operations. These standards contain much detail on the operation and functioning of such machines which comple-

ments the discussion in this Manual. The B11 standards are particularly helpful for they contain explanatory notes which facilitate better production with safety.

Power Presses

Probably the easiest and most common way of creating a useful object from a sheet of metal is to cold form it in one, or more, of a unique group of machines called power presses. Such machines usually consist of a frame within which a powered slide or ram is moved, in a reciprocating motion, at right angles to a stationary bed. Attached to the slide and the stationary bed there are mated dies which cut or form material placed between them when tremendous pressure is applied by the slide to close the dies. This action of closing the dies creates particular hazards for an operator, unless safeguarding is used to keep the operator out of the die area, and to confine or deflect the effects of the exerted force of the dies.

There are four kinds of power presses in wide use today: the mechanical power press, the hydraulic power press, the power press brake, and the shear. As was just pointed out, each operates on similar principles and, at times (with the exception of the shear), the same job might be run on all machines. However each machine is best suited for use in a particular way by reason of its construction differences.

Mechanical power presses

Mechanical power presses are of two basic types of construction with variations of each; the gap frame (OBI), which is like a "C," and the straight side frame, which is like an "H." (See Fig. 32–1.) The gap frame is the more common and is generally lighter and with a lower tonnage capacity. It is used extensively for small- to medium-sized parts production. It is very versatile and can be used for both primary and secondary operations, as well as for specialized operations such as coining, riveting or staking, and other assembly operations.

In the past, many of these operations were accomplished by manually loading the material to be processed directly into the lower

(*Text continues on page 860.*)

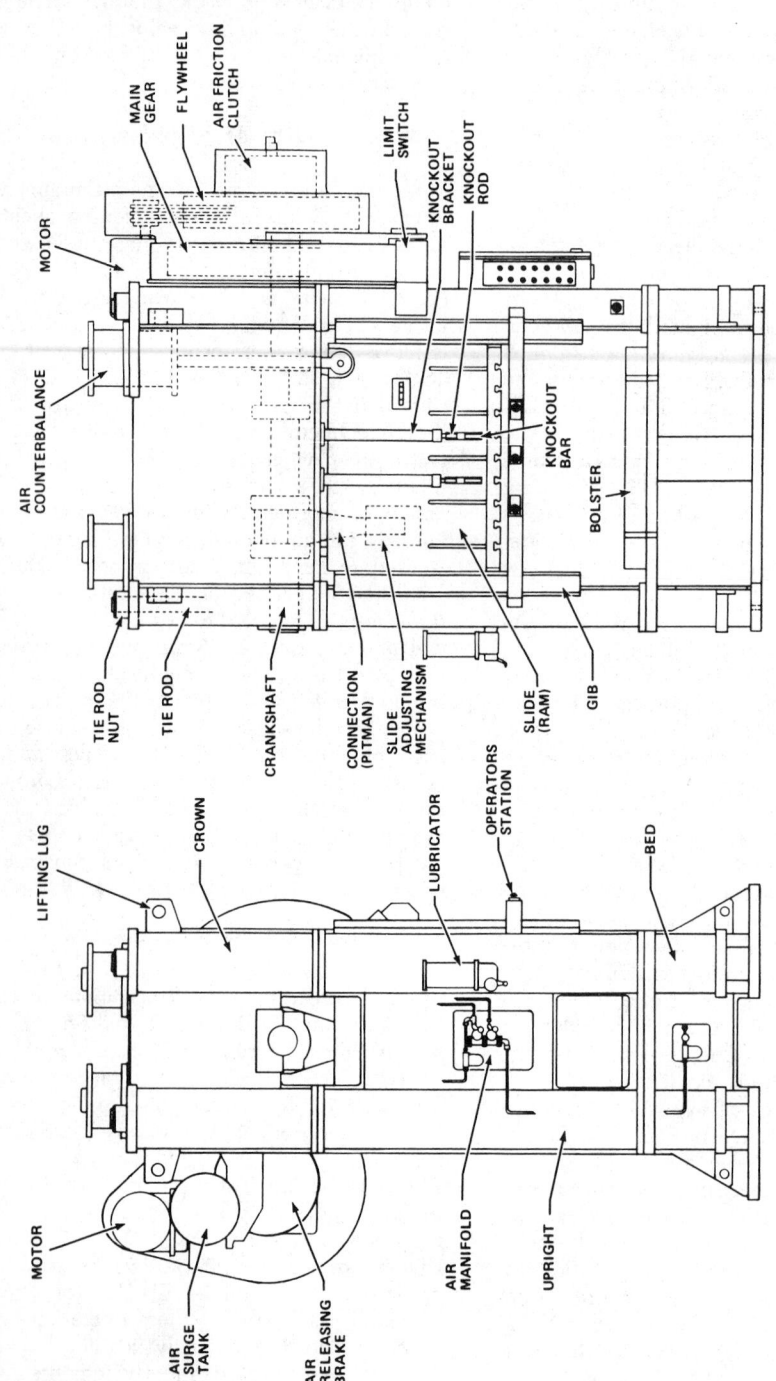

Fig. 32-1a.—Standard terminology for typical straight side power press.

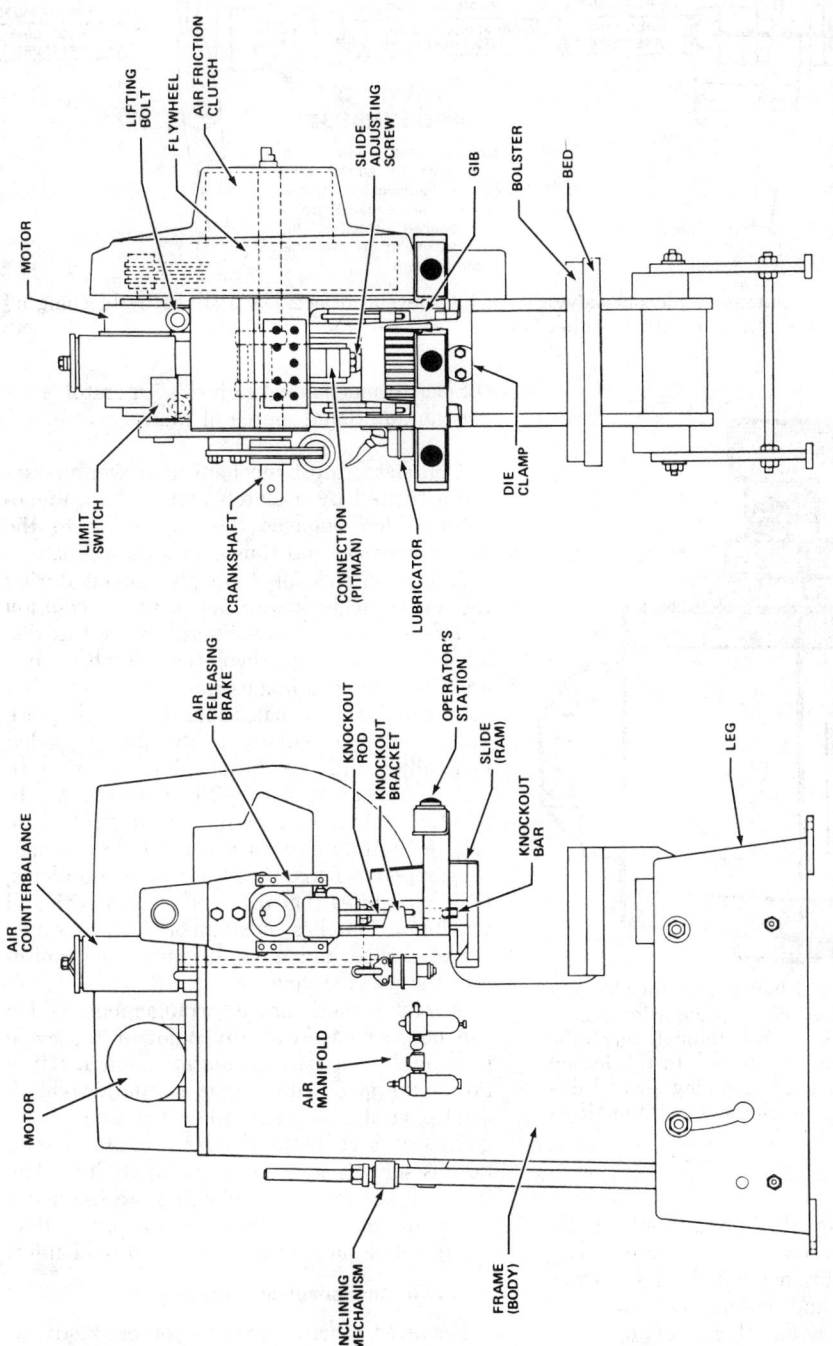

MOTOR

LIFTING BOLT

FLYWHEEL

AIR FRICTION CLUTCH

SLIDE ADJUSTING SCREW

GIB

BOLSTER

BED

LIMIT SWITCH

CRANKSHAFT

CONNECTION (PITMAN)

LUBRICATOR

DIE CLAMP

AIR COUNTERBALANCE

AIR RELEASING BRAKE

KNOCKOUT ROD

KNOCKOUT BRACKET

OPERATOR'S STATION

SLIDE (RAM)

KNOCKOUT BAR

MOTOR

AIR MANIFOLD

INCLINING MECHANISM

FRAME (BODY)

LEG

Fig. 32–1b.—Standard terminology for a typical OBI (open back inclinable) press.

859

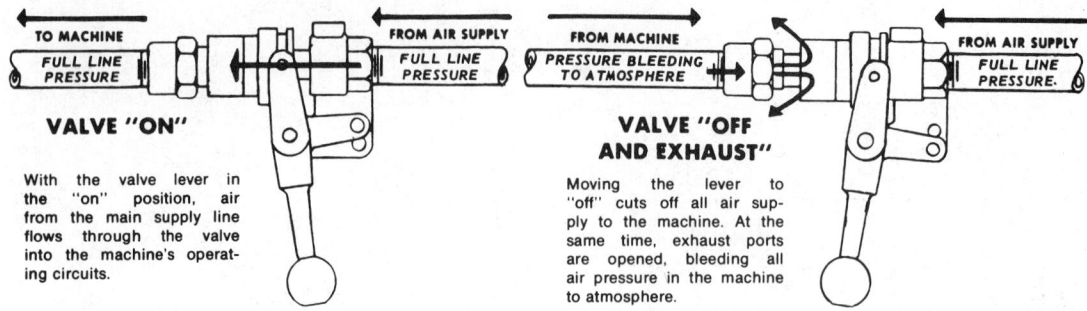

VALVE "ON"

With the valve lever in the "on" position, air from the main supply line flows through the valve into the machine's operating circuits.

VALVE "OFF AND EXHAUST"

Moving the lever to "off" cuts off all air supply to the machine. At the same time, exhaust ports are opened, bleeding all air pressure in the machine to atmosphere.

FIG. 32–2.—One type of automatic bleeder valve that helps prevent an accidental stroke by bleeding off residual pressure. Valve can be locked in either "on" or "off" position.

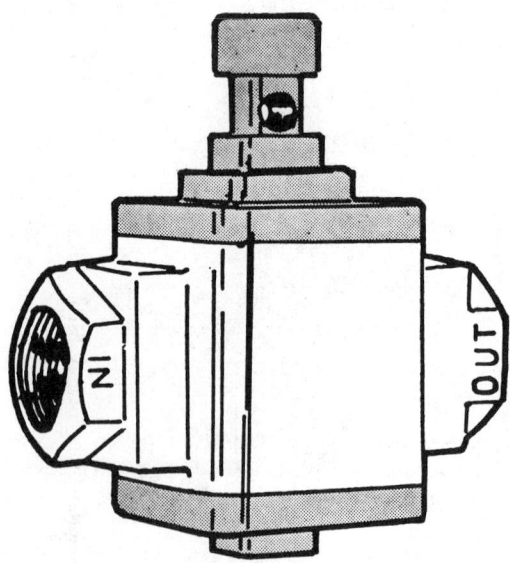

FIG. 32–3.—Shut-off and bleed valve for air-operated equipment. In the "off" position, the stem is extended so as to disclose a hole through which the hasp of a padlock may be passed. In this locked position, the valve shuts off incoming air and dissipates all trapped air in the system through its bottom port.

die, then activating the press, and finally removing the processed part by hand. The safety of the operator depended on his strict attention to the job and on how well the press was maintained. Even then, an operator could not always be sure of his safety, for presses were known to malfunction or were subject to mechanical failure. Operator error was an additional source of injury.

Clutches. Most mechanical power presses are activated by a clutch and brake arrangement which engages the crankshaft to the heavy flywheel and thus cycles the press.

If the clutch cannot be disengaged during the cycle, it is known as a *full revolution clutch*. There are advantages as well as disadvantages to a full revolution clutch. They are fine for continuous-run production but they can be very hazardous for single-piece production, particularly if the die is loaded manually. Full revolution clutches are subject to repeat strokes without warning. In addition, unintended or premature strokes are easy to initiate and once started they cannot be stopped. To correct both these problems, OSHA requires that users of presses with full revolution clutches must incorporate a single stroke device and an anti-repeat mechanism into the press system.

A clutch and brake arrangement which can be disengaged at any point of the cycle is known as a *part revolution clutch*. If a press equipped with a part revolution clutch is also equipped with other necessary safeguards, it is generally a safer press to operate because of this start and stop capability. The full and part revolution clutches are the major types of drive mechanisms in use today. More details are given later in this chapter.

Hydraulic and pneumatic presses

However, there are two other kinds of presses that have the same operating characteristics as the part revolution clutch press—

the hydraulic and the pneumatic-powered presses. Each is driven by a gear train which can be started or stopped by simply cutting the power supply.

All that has been said so far about mechanical power presses (except for the source of power) applies to hydraulic or pneumatic powered presses. These are used for the same kind of work and have similar hazards, such as sudden dropping of the slide because of power failure or breaking of a pressure line, defective dual controls, unexpected energizing of one or more electric circuits because of oil or dirt, defects in wiring which cause the press to close, leaking valves, residual or back pressure, and air pressure built up between dual controls.

Where air or hydraulic power is used either alone or in combination, there is the hazard of residual or back pressure. Even though the control valve be turned off and locked in place, some pressure may be locked in the system past the valve point.

This residual pressure has produced accidental cycling of equipment even though persons involved had thought all the necessary precautions had been taken by shutting off the power source valve. As a result, when the tripping device for the machine was next touched or when the power source valve was again turned on, the equipment accidentally cycled, producing a dangerous situation. Figs. 32–2 and –3 show methods used to prevent this.

Press brakes

The press brake is classed as a power press because it utilizes the same construction and operational features that characterize the mechanical and hydraulic power presses. It has a bed and a descending slide, called the ram, and can be powered either mechanically or hydraulically—but there the similarity ends. The ram and bed are long and narrow, matched to each other, and usually set in front of the gapped frame (Fig. 32–4). In some cases the bed and ram are located between uprights, but this arrangement limits the length of material that can be processed.

In operation, both types of press brakes utilize the part revolution clutch type of drive, and thus can be started and stopped at will. Most press brakes operate only in what

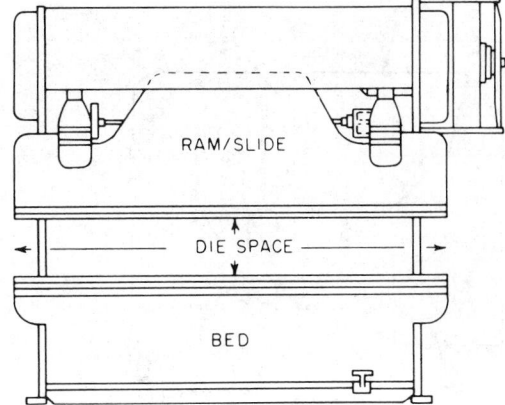

FIG. 32–4a.—Typical mechanical press brake.

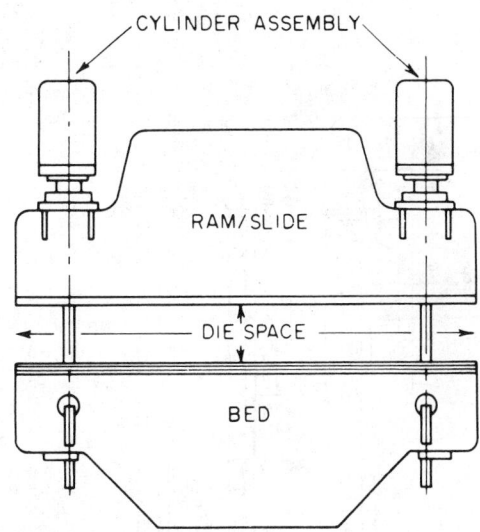

FIG. 32–4b.—Typical hydraulic press brake.

From ANSI Standard B11.3–1973.

would be termed "low gear," but some special purpose brakes are equipped with two speeds which can be used either individually or in combination.

Most press brakes are used to form or bend large or small pieces of sheet metal to a desired shape. The open end construction of the ram and bed even permits the continuous forming or bending of pieces that are much longer than the bed and ram, when appropriate accessories are used.

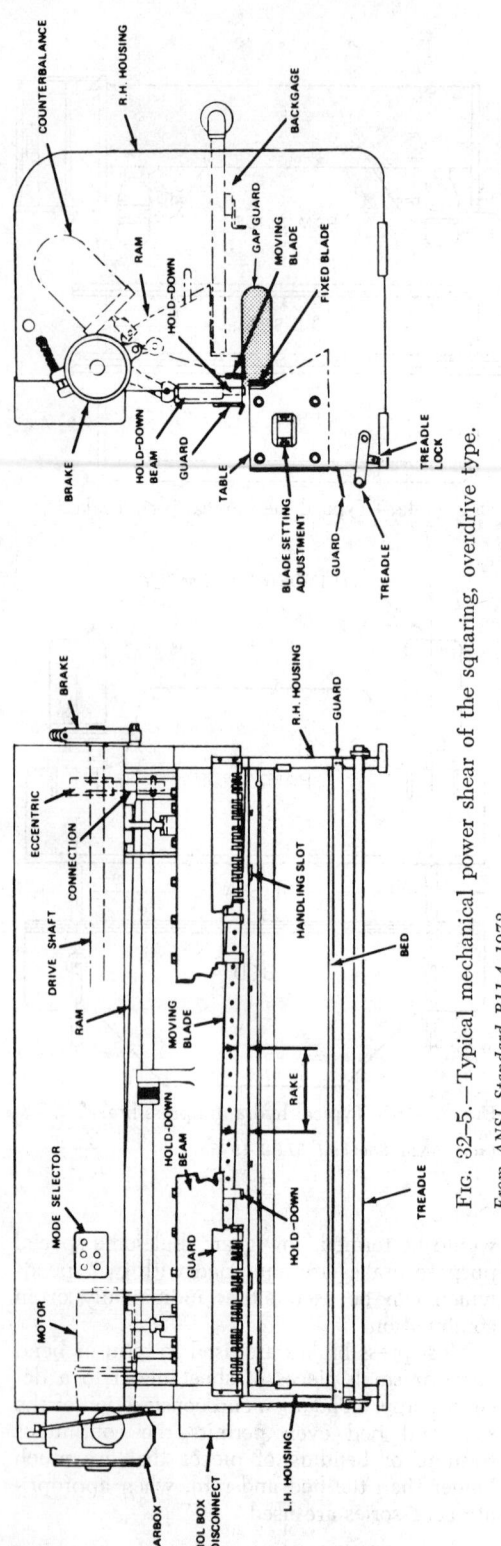

Fig. 32–5.—Typical mechanical power shear of the squaring, overdrive type.

From ANSI Standard B11.4–1973.

TABLE 32-A

PERMISSIBLE GUARD OPENINGS

Distance of opening from pinch point	Maximum width of opening
0 in. to 1½ in.	¼ in.
1½ in. to 2½ in.	⅜ in.
2½ in. to 3½ in.	½ in.
3½ in. to 5½ in.	⅝ in.
5½ in. to 6½ in.	¾ in.
6½ in. to 7½ in.	⅞ in.
7½ in. to 8½ in.	1¼ in.

Also see topic Point-of-Operation Guards in Chapter 29, "Point-of-Operation and Transmission Guards."

Press brakes are generally used for custom work. An operator can work all day, using the same die, and never produce the same item twice. Yet it is quite capable of turning out small production runs of work that would be much too costly to assign to other machines such as a regular power press.

Many of the hazards of press brake operation arise out of this "custom" operation. A press brake will accept and process almost any size work and because of this, there is no such thing as "normal" guarding for press brake operations. In fact, there are times when a guard would seriously interfere with the production process and constitute a hazard in itself.

Good guarding of a press brake starts with a good job analysis of each job to be run, with particular attention to the handling of the work and the closeness of the operator to the point of operation. Any part of the job, or mode of operation, that allows or requires any part of the operator to be in close proximity to the point of operation must be looked at very carefully with respect to the safeguarding applied. Most shops will probably find that two or three safeguarding methods, used either singly or in combination with each other, will enable them to operate efficiently and with safety in all jobs.

Guarding press brakes is not difficult if the foregoing procedures are followed. If a press brake is ever used as a regular power press, however, such operations must be guarded as required for power presses.

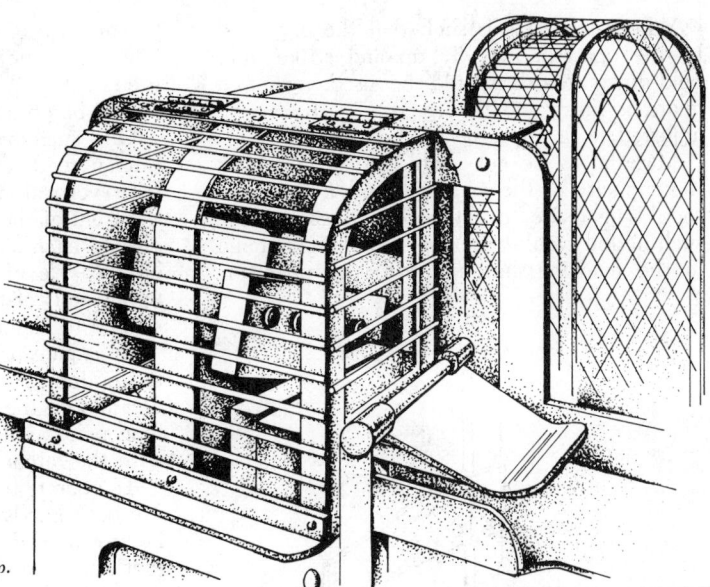

F‌ɪɢ. 32–6.—A well designed guard for an alligator shear should permit easy maintenance and adjustment. Hinged section of the guard should be interlocked electrically to prevent shear operation if the hinged section is not in place.

Courtesy Jones & Laughlin Steel Corp.

Power shears

The fourth of the cold forming metal presses is the shear (Fig. 32–5). It is available with either full or part revolution clutch/ brake drives as a mechanical shear, and also with hydraulic or pneumatic drives. Fortunately they are used for one purpose only— shearing metal.

Aside from the usual precautions a user must take as outlined in American National Standard B11.4, guarding a shear is a fairly simple and straightforward thing to do. A fixed barrier which conforms to the table of safe distances is generally acceptable, especially if the user is working with ⅜ gage stock or less. A plain barrier across the shear which admits this thickness of metal, but not the user's fingers, affords positive protection. Larger gages should utilize semi-adjustable "awareness barriers" that will admit heavier stock, but will continue to curtain the areas to each side of the stock being sheared and thus warn the operator against farther hand movement into the danger zone. (See Table 32-A.)

Some companies have developed a system of feeding from the rear of the shear. This practice should be eliminated unless equal guarding is used both front and back to protect the operator and/or his helper.

There is one further type of shear which, while not a power shear in the foregoing sense, needs mentioning because of its wide use in forging and salvage yard operations. Alligator shears chop rather than cut and they are used for cutting heavy bar or rod stock to more convenient lengths.

The alligator shear operates continuously, and the operator must be trained to time his movements with the opening and closing of the cutter. Because the machine is relatively simple and comparatively slow in its movement, management and operators are often willing to disregard the hazards created. Consequently, alligator shears are responsible for far more injuries than their inherent hazards or frequency of use warrants.

If possible, a long bench should be built to the right or left of the shear, depending upon the type of machine, and the material slid along and through the cutter.

Most finger and hand injuries can be avoided by the installation of an adjustable guard. The wide variety of sizes and shapes of material to be cut makes it difficult at times to guard closely the point of operation. However, such a guard can often be used. When it is, it should be set close enough to prevent the fingers from entering the danger zone (Fig. 32–6).

When stock size is such that the end held by the operator may fly up and strike him, hold-down guards should be used. Such a guard is illustrated in Fig. 32–7. It can be adjusted to fit any type of shear.

Material to be cut should be kept within the capacity of the machine, and no attempt should be made to cut hardened steel since such action would probably result in damage to the machine and injury to the operator.

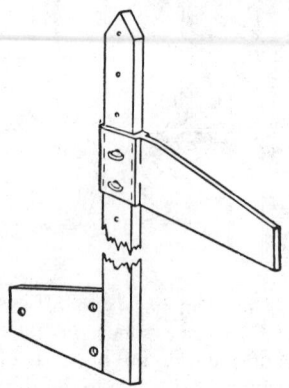

Fig. 32–7.—Adjustable hold-down bar for use on an alligator shear, for cutting heavy or long material.

Because of their operational differences and the wide variety of work they can do, power presses need close attention by the user and operator to the safeguarding applied to each press. This is particularly true if a fixed barrier guard cannot be used for the job, and the user must depend on auxiliary safety devices to achieve the necessary degree of safety.

Hand or foot operated presses

All power presses have counterparts in presses which are hand or foot powered—there are still many shops using kick presses, foot shears, hand folders, and hand rollers. Unfortunately, serious injuries can occur even though the operator is the source of power. Fortunately such presses are easy to guard and without much expense.

Foot press operations are most effectively safeguarded by some type of interlocking tripping mechanism which requires the simultaneous use of both hands of the operator to release the slide head before the pedal can be used. At all other times, the slide head should be positively locked in the top position.

One such device using a steel plate with ratchet teeth cut into the back side, next to both vertical edges, is securely fastened to the front of the slide head. Two hand levers, one on each side, are anchored to pivot pins on the side of the slide head casting. The upper ends of the levers, or locking dogs, are shaped to fit into the ratchet teeth.

An expansion spring fastened on one end to the bottom of the levers, just above the handle, and on the other end to the casting of the foot press, holds the locking dogs engaged with the ratchet teeth.

To operate this foot press, the operator must use both hands to disengage the locking dogs before the slide head may be forced down by foot pressure on the treadle bar.

A secondary advantage of this type of tripping mechanism is that it provides means for the operator to brace himself when applying pressure on the treadle.

Another simple, yet effective, type of two-hand device can be installed with little expense. Half-inch holes are drilled through the sides of the frame and into the swinging arm of the slide. Then, when the slide is in the top position, automatic locking pins that can be withdrawn by the operator are inserted into the holes. Conveniently located levers with springs attached to the pins can be used by the operator to lock the slide when it is in top position.

Back-to-front and side-to-side guards also have proved to be practical methods of guarding foot or kick press operations.

Fig. 32–8 details a moving ring guard, applicable to small bench-type pneumatic presses.

Foot shears and rollers

These operations are even easier to guard than foot (kick) presses. All that is needed is some type of barrier placed in front of the shear or roller which will effectively bar the operator's fingers from entering the point of operation. It can be angle iron, expanded metal, or even a rod if it is positioned with due regard for the gage of the material and the distance from the point of operation.

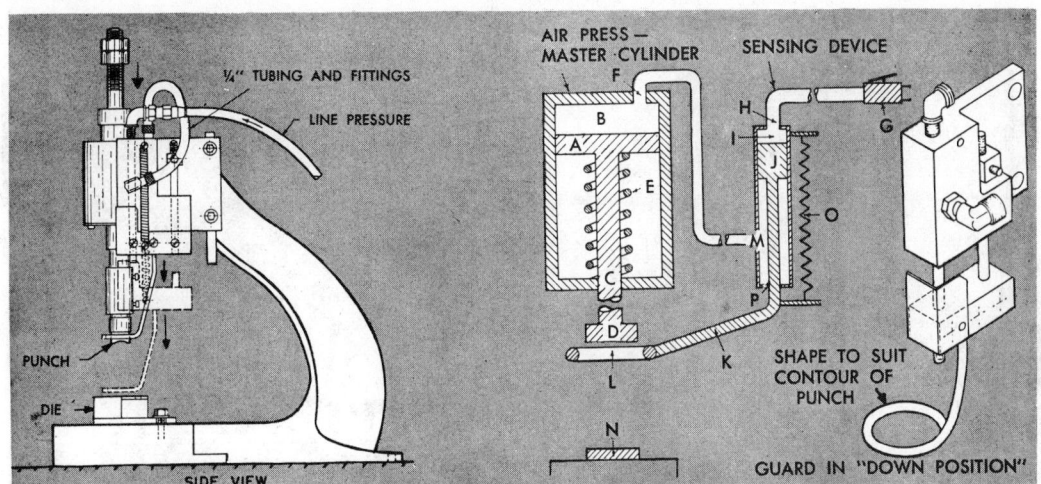

Fig. 32–8.—This sketch represents a standard pneumatic bench press in combination with a specially developed safeguarding device.

Operation: Piston A, sliding in cylinder B, is connected by means of rod C with punch D. Spring E returns piston A to its highest position after the air pressure is released. Air enters cylinder B through port F. Without the safety device, the foot-operated valve G admits air from the storage tank directly to inlet F.

With the sensing device installed, the air admitted through valve G, now enters port H into the auxiliary small-diameter cylinder I. Piston J in cylinder I is directly connected by means of rod K with the "ring" guard L which fits snugly around punch D. The air pressure will depress piston J, allowing air to enter main cylinder B through the orifice M in cylinder I.

Air can enter the main cylinder only under the condition that piston J has passed beyond orifice M. Any obstruction between punch D and die N will prevent the ring guard L and its attached piston J from traveling downward sufficiently to open port M. After releasing the pressure, piston J is returned to its highest position by spring O. The air in main cylinder B exhausts through port P, by way of inlet F and orifice M, allowing spring E to return piston A to its original position.

From American Machinist, © McGraw-Hill, Inc. Used with permission.

Power Press Electric Controls

The best safeguarding for all presses starts with the electric control system. Much of the electrical safeguarding discussed in this section is required by OSHA. Those items not required are recommended to users either as supplemental to the required items or as items which offer additional safeguards to increase the reliability of the system.

Typical press control systems—whether on an air friction clutch/brake control or on an electrical device for actuation of a full revolution (mechanical) clutch—provide:

1. A main disconnect switch or circuit breaker for isolating the machine from power. (See details in Chapter 41, "Electrical Hazards.")

2. Starters for all motors including main drive, slide-adjustment, inclining, lubrication, and auxiliary motors.

3. A transformer to step down line voltage to a value not exceeding 120 volts for control circuits, operator controls, relays, and solenoids.

4. Control relays for clutch, brake, air blow-off, and auxiliary devices.

5. Operator controls, flywheel brake, rotary limit switches, air pressure switches for clutch-brake, counterbalances, and other control items.

6. Air valves for clutch, flywheel brake, air blow-off, and auxiliary devices.

FIG. 32–9.—Presses must also be made safe for electricians, millwrights, pipe fitters, and others who must maintain them in a safe condition. Automotive companies have requirements similar to the following for electrical installations on equipment.

1. All motor starters, relays, etc., required for each machine must be built into a single oil-tight (NEMA Type XII) control panel, with externally operated main power disconnect mechanically interlocked with the enclosure doors so that no door can be opened until after the disconnect has been opened.

2. The disconnect operating handle must be permanently attached to the disconnecting device and must be located not more than 6½ ft and not less than 3 ft above the floor. The handle must be arranged so that it can be locked only in the off position with a minimum of three locks.

3. Enclosure doors must be hinged vertically, be not more than 36 in. wide, and be painted "safety orange" on the inside so that an open door will be readily seen.

4. The inside of the enclosure should be painted with white enamel to increase visibility for maintenance and large panels must have interior lights that are turned on by door switches.

5. The control circuit voltage must be 115 or 120, obtained from a transformer with isolated primary and secondary windings.

6. Equipment on the control panel must be located at least 18 in. above the floor and not more than 84 in. above the floor. Devices carrying line voltage must be grouped above or to the side and be segregated from devices such as relays and timers carrying only the control voltage. Nothing may be mounted above the main disconnect.

7. All wires must be color coded as follows for easy identification of their voltage level:
a) AC and DC power circuits (line voltage) – Black,
b) AC 115/120 v control circuits – Red,
c) DC control circuits – Blue,
d) 115/120 v circuits still energized when main disconnect is open, such as enclosure lighting circuit – Yellow,
e) Equipment grounding conductors – Green,
f) Current-carrying grounded conductors – White.

Courtesy Clark Control Div., A. O. Smith Corp.

Installation

Controls for new machines should be built and installed in accordance with the *National Electrical Code* (ANSI Standard C1 and NFPA Standard No. 70), applicable local codes, and recognized industry standards. The purpose of these codes and standards is to safeguard personnel and equipment.

The machine control panel should contain the main disconnect switch, motor starters, clutch control relays, and control transformer. This panel should be on or immediately adjacent to the machine it serves, but out of the operator's way during normal operation. It should be properly identified and accessible for maintenance (Fig. 32–9).

The main disconnect switch handle should be on the exterior of the control enclosure and operable from floor level. The recommended maximum height of the centerline for the switch handle is 6½ ft above the floor. The switch should be provided with a means for positive lockout.

Selector switches used for setting the mode of operation—such as single-stroke, run, inch, two-hand, or foot—should be of the key-lock type to prevent unauthorized change of operation.

Operating controls should be located or guarded to minimize the possibility of unintentional operation. Foot switches should be guarded to prevent activation by falling objects or by accidentally stepping on them. A foot switch should only be used where satisfactory operator protection is provided. Hand buttons should be physically located or guarded so that operation by a single hand and arm or elbow is impossible. Stop buttons should be conveniently and conspicuously located and left unguarded. Emergency stop buttons should be mushroom shaped and colored red.

All operating buttons should have metal cases and covers. Where mushroom heads of run buttons are used, they should be of metal or plastic (metal preferred) and grounded to protect the operator from shock in case of electrical defect.

Cases or frames of all control devices should be grounded to the press frame. Mounting bolts are satisfactory for grounding provided all paint and dirt are removed from joint surfaces before assembly. Movable control devices not secured to the press frame should be grounded by means of a grounding conductor in the connecting cable or conduit.

The rotary limit switch is a vital element in the control system. Mechanical or electrical failure of this switch may cause the press to repeat. The switch driving mechanism—including all couplings, gears, sprockets, and associated parts—should be securely assembled and of rugged design. A positive, keyed connection is preferred to set screws, as these have a tendency to vibrate loose.

The use of two separately driven switches or a drive failure detector is recommended. All cam switches should be of a type in which the switch contacts are forced open by the actuating cam.

1. Where a flywheel brake is used, it should be electrically interlocked with the main drive starter to prevent simultaneous operation.

2. Where safety blocks are used, they should be provided with interlock plugs which disconnect the clutch and motor control circuits when blocks are in place between the dies. (See Fig. 32–10.)

Part revolution clutch presses

On a press with air friction clutch/brake, the control performs the following functions:

1. Provides that the press operator must hold both run buttons during the down stroke until the slide has descended to a point where there is no longer a hazard. Release of any button during "holding time" should stop the slide.

2. Allows the operator to release buttons after completion of "holding time" and press will complete stroke.

3. On "single cycle" operation, it stops the slide at end of a stroke even though operator holds control buttons depressed.

4. Actuates clutch/brake air valve(s) so that friction brake will stop machine at top of stroke.

5. Provides anti-repeat protection and prevents repeat operation with any buttons tied down.

Starters for main drive motor should be the

Fig. 32–10.—When making adjustments on press, the upper die should be propped to eliminate any unexpected motion.

Courtesy McDonnell Aircraft Corp.

magnetic, reversing type. This feature permits backing up of the press in the event of a jam without the need for an electrician to change wiring. Selector switches should require keys so that backing up can be done by authorized personnel only.

The clutch control should be provided with a main drive motor "forward interlock" so that the press can be run only when the motor is turning in the forward direction. "Backward" operation of the press is unsafe because there is no run button "holding time." This interlock also prevents "run" actuation of the clutch air valve before the motor is started or when the flywheel is coasting. Starting the motor with the clutch engaged is unsafe since it may result in unexpected downward movement of the slide. Inch or jog operation

should normally bypass the interlock to allow power-off "inching."

Pressure switches should be provided on the compressed air system downstream from pressure regulators and shut-off valves to assure that air is present for proper actuation of clutch-brake mechanisms and counterbalance cylinders. These switches should be normally open and wired in series into the control circuit ahead of the main-drive-motor starter and clutch/brake control.

1. The clutch pressure switch assures that the correct air pressure is available before the machine is started. This assures proper movement of the slide and protects clutch plates from excessive wear resulting from slippage.

2. The counterbalance pressure switch prevents a sudden drop of the slide during the instant between release of the brake and engagement of the clutch. Proper counterbalance pressure is also required for quick stopping of the slide during the downstroke, or prevention of drop back if the slide is stopped during the up stroke.

Jogging or inching of the slide is often required when setting up or trying out dies. The use of two-hand inch buttons is preferred for these operations so that both hands must be out of the danger area. Where a single button is used, it should be guarded so that it cannot be operated unintentionally.

The clutch/brake solenoid valve(s) should not be energized directly across the line through the contacts of a single relay, in such a way that faulty operation of this single relay will result in unintended energization of the valve(s). Protection can be accomplished through the use of multiple relays and/or energization of the valve(s) through push button switches and rotary limit switches.

The previous paragraph makes stopping a machine independent of operation of the valve or run relay. Normally closed contacts of this relay should be incorporated into the circuit to check actual dropout of the relay.

The use of a multiple clutch control valve is recommended. This is a relatively inexpensive means for reducing the possibility of press repeats due to malfunctioning of a single valve.

Press clutch control circuits can be either of two types:

1. GROUNDED, with all coils connected to one side of the isolated secondary of a control transformer, and this side grounded to the machine frame, or

2. UNGROUNDED (double ended) with ground detector lights connected to each side of the line and a double break arrangement.

On multiple operator presses, complete operating stations not required should be bypassed. This can be done by means of dummy plugs or terminal board jumpers. The control circuit should be arranged so that the clutch cannot be actuated when all operating stations are bypassed. This arrangement is preferable to individual button lockouts which may result in one button being locked out unintentionally.

Maintenance electricians should be made to realize the importance of anti-repeat and two-hand control features so they are not removed or modified when incorporating tooling protective switches or to aid production.

The "holding time" during which the run buttons must be depressed will vary with different types of presses. It is frequently 150 degrees of crank or eccentric shaft rotation. It may be less on multiple action presses where blank-holder dies are used. It should be considerably longer on slow speed, single-action presses. In all cases, the "holding time" should be sufficiently long so that the slide has descended to a point where the operator cannot reach into the point of operation.

When a continuous operation position is provided on the machine, the circuit should be arranged so that a locking selector switch must be operated or terminal board jumpers reconnected to prevent unintentional or unauthorized selection of continuous operation.

Full revolution clutch presses

Presses with full revolution clutches should be provided with two run buttons, which are arranged to give positive anti-repeat protection even if the clutch is not set for single stroke operation. Safety is limited because the clutch cannot be disengaged to stop the slide during a stroke. The press control should not be simply a clutch operator without anti-repeat protection.

Motor starter interlocks and 110/120 volt control circuit to the air clutch-brake control should be used.

Run buttons should be located as far as possible from the work point since no "holding time" capabilities exist. The operation speed of the press should be considered when locating the buttons to maximize operator safety.

Controls for hydraulic or pneumatic presses

One method recommended for prevention of accidents in case of power failure, which would allow the slide to drop, is to connect an electrically controlled solenoid valve to all sources of electrical power supplied to the press. In case of power failure, the normally closed springloaded solenoid valve opens. Air then travels through and instantly actuates the slide hold-back cylinder, which holds the slide in an up position.

If the coil on the solenoid valve should burn out and air pressure should open the control, the circuit is so interlocked that the main pump drive motor will shut off to prevent damage to the pump.

Valve controls for hydraulic or air presses should be located within easy reach of the operator where he has a clear and unobstructed view of the press. Otherwise, mirrors should be so placed that the operator can see the point-of-operation area.

Point-of-Operation Guarding

In the past, many point-of-operation press accidents occurred because of either electrical or mechanical malfunctioning of the press. Many point-of-operation accidents occurred and still occur, however, simply because the point of operations is inadequately guarded or not guarded at all. From a safety viewpoint, there really is very little difference between an inadequately guarded point of operation and one which is unguarded. In effect, both situations are unguarded and are potentially hazardous to an operator. There is a very large difference, however, from a psychological viewpoint. Most press operators have a healthy respect for any open die. In fact, many older operators may say that they prefer open die work over guarded die work.

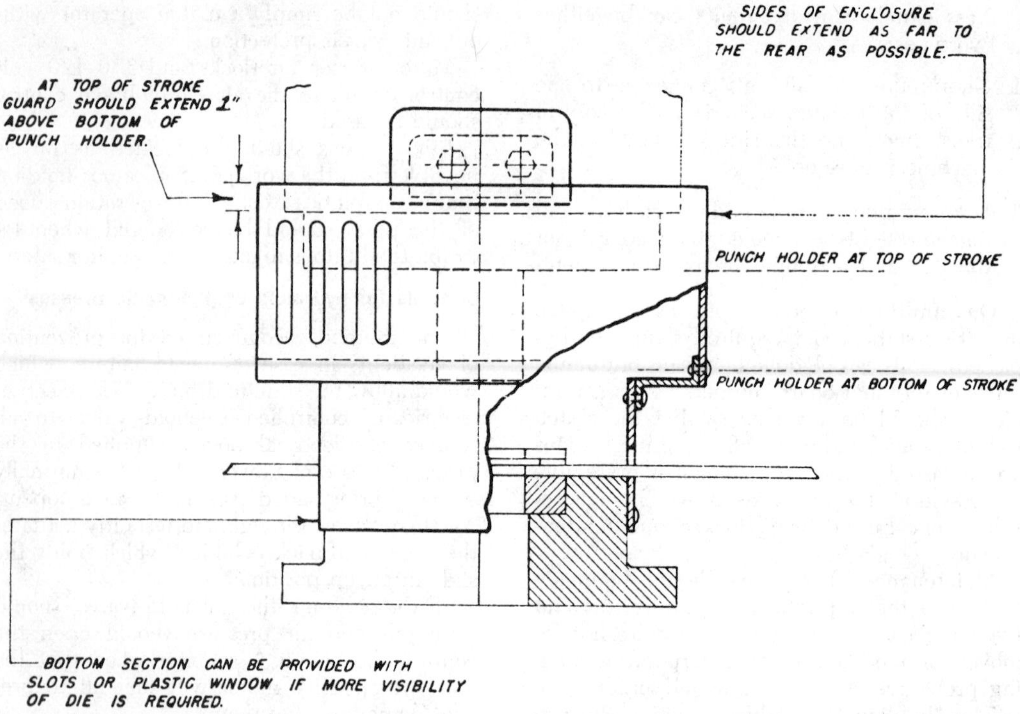

SIDES OF ENCLOSURE
SHOULD EXTEND AS FAR TO
THE REAR AS POSSIBLE.

AT TOP OF STROKE
GUARD SHOULD EXTEND *1"*
ABOVE BOTTOM OF
PUNCH HOLDER.

PUNCH HOLDER AT TOP OF STROKE

PUNCH HOLDER AT BOTTOM OF STROKE

BOTTOM SECTION CAN BE PROVIDED WITH
SLOTS OR PLASTIC WINDOW IF MORE VISIBILITY
OF DIE IS REQUIRED.

Fɪɢ. 32–11.—Die enclosure guard for strip or coil stock showing the vertical clearances required.

Courtesy Liberty Mutual Insurance Co.

Fɪɢ. 32–12.—Die guard made of preslotted material can be easily fabricated in the shop.

Many of them will say that they simply distrust guards. What they are really saying is —that they distrust guards that do not guard.

There have been a number of studies of well guarded versus unguarded press operations. In almost all cases, the guarded situations outscored the unguarded ones; aside from the added safety, there were fewer rejects, less operator fatigue, and (perhaps most significant) total production equaled or exceeded what was produced in the unguarded situations.

It follows, therefore, that good point-of-operation guarding is beneficial to all concerned. It fulfills the law, it allows operators to work without fear of injury, and it results in better production.

What remains now is to decide what kind of guarding applies to which operation, and how shall it be installed.

Fig. 32–13.—Fully enclosed dies can be operated with complete safety to personnel. Note the various materials that can be used for guards in order to achieve production flexibility.

Courtesy Allis-Chalmers Mfg. Co.

Several kinds of guards, devices, and combinations of devices can be used for safeguarding. Actual approval depends on the suitability of the guard or device to the actual operation. Ideally, the best guard for all operations would be a fixed barrier guard (mounted on either the press frame or the die), which would keep the operator's hands out of the danger zone. At the present time, however, this would be less than ideal for many operations. It can and should be used on most primary operations, but secondary op-

erations will require more thought and study. This process will be simplified if the thought and study is started at the initial planning sessions. Too many times an idea is carried right into the production stage with little or no thought given to the production problems involved. Too many times the result is a "jerry-built" system in which safety is an afterthought.

Good practice for both production and safety is to operate power presses at minimum slide travel consistent with efficiency. If the

871

32—Cold Forming of Metals

Fig. 32–14.— Slide feed allows loading of die outside the danger zone. Permanent plastic barrier guard permits full visibility of operation. Note separation of guard at top, to permit die maintenance. Overlap of guard at separation eliminates shear hazard during travel of slide.

Courtesy Allis-Chalmers Mfg. Co.

nearest pinch point between the slide and the bottom die can be limited to ¼ in., guarding is unnecessary. Slide travel can be limited by means of a cylinder designed for short travel, by die or jig and fixture design, or by auxiliary stops.

Following is a discussion of various guards and devices. Their selection and use will depend on thoughtful analysis of the feeding, locating, and ejecting requirements of the stock, formed part, and scrap.

Die-enclosure guards

A fixed die-enclosure guard, such as the individual die guard, provides the most positive protection for the operator since the die is completely enclosed and the guard is a permanent part of it. Die enclosures can be used on many types of press operations to prevent operators from placing their hands between the punch and the die.

Die enclosures are attached to the die shoe or stripper in a fixed position and are designed so that hands cannot reach over, under, or around the guard into the point of operation.

It is important that this type of guard be constructed to permit easy feeding of the stock, ejection of the part, and scrap removal. It should also afford good visibility at all times.

A minimum clearance of 1 in. must be provided between the top edge of the guard and the slide or any projection on the slide (Fig. 32–24). In addition, the guard should extend at least 1 in. above the bottom of the punch holder to prevent a shearing hazard caused by travel of the slide (Fig. 32–11).

Enclosure guards can be built of various types of material. Prefabricated, slotted material is often used (Fig. 32–12). A metal frame can be built, and rod stock can be welded or otherwise fastened to it. Openings should run vertically to lessen eye fatigue.

Transparent plastic also may be used (Figs. 32–13 and –14). When properly maintained, it has the advantage of affording good visibility. However, this material scratches easily and is damaged by oil and grease. Plastic material should be at least ¼-in. thick and will give longer service if mounted in a metal frame. Thinner plastic sheeting is often too brittle (or else becomes that way) and cracks or breaks easily.

The expense of designing and installing permanent die enclosures is very small and yields positive safety results. In many plants in which this type of guarding is applicable, each die is so guarded before it leaves the tool room.

Fixed barrier guards

Fixed barrier guards may be designed with a pivoting, sliding, or removable section to allow ready access to the die (Figs. 32–15 and –16). The pivoting or sliding section must be interlocked with the press control if operation of the machine is to be prevented when the section is open.

This type of interlocked fixed barrier guard is used successfully on automatic presses where the point of operation must occasionally be exposed so that jams can be relieved. As a further safety measure, protecting against failure of interlocked press controls, hand tools or picks should be used to relieve jams, or the slide should be blocked.

This type of guard should be secured to the press frame with fasteners which require

Fɪɢ. 32–15.—Fixed barrier guard on automatic press is equipped with a hinged plastic front with interlocking handle. Note take-up reel for coil scrap.

Courtesy Automatic Electric Co.

Fig. 32–16.—There are advantages to inclining a press. The guard is simple and economical to make. This operation (forming pail covers) also permits automatic gravity feed direct from blanking operation (via the belt conveyor). This job could also be hand fed.

Courtesy United States Steel Corp.

a special tool, retained only by the foreman or the job setter, for removal.

Fixed barrier guards are also applicable on hydraulic presses.

A universal barrier guard that completely encloses the danger zone is shown in Fig. 32–17. The lower guard conforms to the universal guard, the feed device, and the lower die contour.

Adjustable barrier guards

Where a die enclosure guard or fixed barrier guard would take some time to complete or its provision is considered functionally impracticable, or both, an adjustable barrier device can be provided on each press. This type of guard can be used on many operations to prevent the operator's hands from entering the point-of-operation zone (Fig. 32–18).

Adjustable barrier guards are available commercially or may be made in the plant. They are attached to the frame of the press and have front and side sections which can be adjusted for dies of almost any size.

This type of device is usually constructed of rod stock, set vertically to minimize eye strain, or perforated metal. The pivoting or sliding sections should be interlocked with the press control for maximum safety.

Unless feeding or ejection is automatic, it may be necessary to leave an opening in the barrier for insertion of a tool to remove the piece from the die. This opening should not be wide enough for a hand or finger to extend into the danger zone.

The adjustable barrier guard requires adjustment for each job or die setup and is not usually interlocked with the press control. When adjusting the sections of this type of guarding device, die setters should be instructed to follow the dimensions for permissible openings given in Table 32-A. The operator should never be allowed to make changes in the adjustments without the supervisor's approval. (Table 32-A is on page 862.)

Gate or movable barriers

These devices are designed to enclose the point of operation completely before the clutch can be engaged.

The clutch is tripped by part of the gate or barrier mechanism during the last ½ in. of the guard's downward travel, or at whatever predetermined height is set. If the fingers or hands or any other obstruction ¼ in. or more in height is in the point-of-operation area, the gate will strike against the object and the press cannot be tripped until the hand or the object is removed.

The clutch tripping mechanism is usually linked to the gate so that the gate trips the press by mechanical or other means. Counterweights are usually provided to return the gate to the raised position, or it may be lifted by mechanical or other means.

Gate or movable barrier devices must be

Low-Cost Universal Barrier Guard

Press guarding is simplified by designing universal barrier guards. The finished guard is made in two sections that completely enclose the danger zone. The upper guard will enclose any operation done on the press; the lower guard conforms to the upper guard, the feed device, and the lower die contour.

¹/₄-in. diameter steel rods are spot welded (⁷/₈-in. center-to-center) on a frame made of ³/₈-in. diameter steel rods. Height should be sufficient to cover the ram at the top of the stroke.

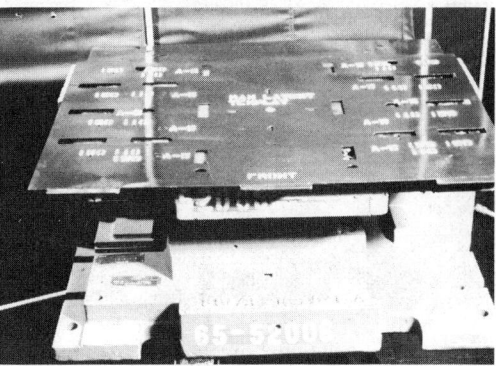

The upper guard is positioned on the die with the aid of a ram template.

An opening, standardized to the widest feeding arrangement anticipated, is now made in the upper guard. This will allow flexibility of the upper guard in all press operations of similar size.

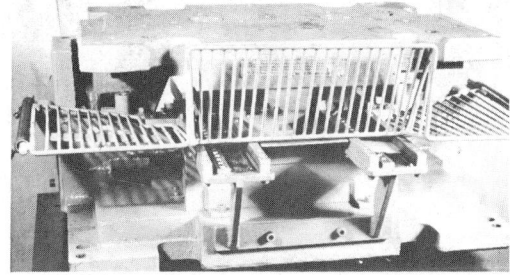

The design of the lower guard is made to conform to the upper guard, the feed device, and the lower die contour. The framing and webbing are angled or bent to permit easy loading and eliminate finger entry into danger zone.

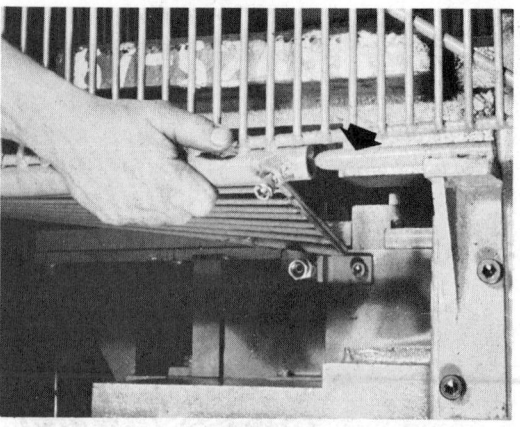

The guard is completed by clip welding the lower portion to the upper portion. Clip welding prevents employees from making their own adjustments. If the upper guard is to be used with other lower guards and dies, the clips are cut and a new lower guard installed.

Mounting pins on the die block make positioning easy when guard is removed and replaced. Dies and guards should be given corresponding numbers to ease matching during setup.

To prevent damage when not in use, guards are hung securely in unused overhead space. Average weight of guard is about 50 lb. Cost is about $18 for materials, plus 15 hr of labor.

Fig. 32–17.—Low-cost universal barrier guard (Picture sequence begins on previous page.)

Courtesy Fisher Body Div., General Motors Corp.

maintained and supervised to prevent their being thrown out of adjustment intentionally or otherwise. They should be constructed so operators cannot reach around, over, or under them. Table 32-A (earlier in this chapter)

shows recommended tripping areas between the bolster plate and the bottom of the gate guard.

A gate or movable barrier device, connected to the tripping mechanism only, does not prevent accidents caused by repeating of the press or by mechanical failure. A single-stroke attachment should therefore be installed. This device should be designed so that it does not return to its up position before the slide has completed the downstroke.

Movable barrier guards on hydraulic presses are operated by single or dual switches or valves. This guard actuates the controls in the last ½ in. of travel, to completely enclose the point of operation.

Dual or multiple control may be obtained by means of air or hydraulic valves or by direct manual operation of valves. The valves should be so designed that pressure cannot be built up between them and should be self-closing.

Switches should be positioned or guarded so that the operator must use his hand and cannot use his elbow, knee, or any other part of his body to trip the press.

Point-of-operation devices

Auxiliary point-of-operation devices differ from guards in that they permit access to the point of operation for loading and unloading the die. Most such devices provide limited or no protection against press failure. Moreover, since the press can be operated without protection of the device, use of the device must be closely supervised.

Malfunction of the device or of the press, causing an unintentional press cycle or repeat, could result in injury to the operator. Unless mechanical loading and unloading facilities are provided, hand-feeding tools for this purpose must be mandatory if maximum safety is to be provided for the operator.

Two-hand tripping devices

There are several types of two-hand tripping devices which require simultaneous application of both hands to operate the controls, so that the hands of the operator are kept out of the point-of-operation area while the slide is descending. These controls can be electrical, pneumatic, or mechanical. (Figs. 32–19 and –20.)

For electric controls, a solenoid should be used in the point-of-operation control circuit so the press cannot be actuated unless its motor is running.

For pneumatic tripping devices, dual sole-

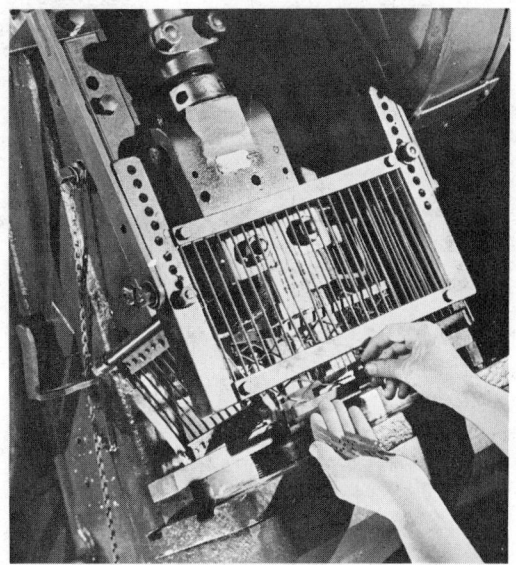

FIG. 32–18.—Adjustable barrier fits any size die used. This barrier is especially practical for short-run jobs.

FIG. 32–19.—Two-hand controls. Center button is STOP switch.

877

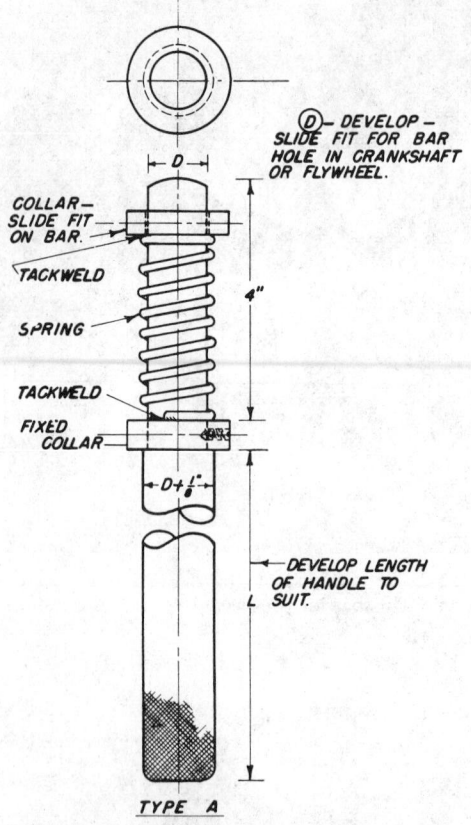

(D) – DEVELOP –
SLIDE FIT FOR BAR
HOLE IN CRANKSHAFT
OR FLYWHEEL.

COLLAR–
SLIDE FIT
ON BAR.

TACKWELD

SPRING

TACKWELD

FIXED
COLLAR

DEVELOP LENGTH
OF HANDLE TO
SUIT.

TYPE A

SPRING ACTION ON END OF BAR
MAKES IT IMPOSSIBLE TO LEAVE BAR
IN BAR HOLE. TYPE (A) SHOWS SPRING
WELDED TO COLLARS.

FIG. 32–20.—Spring-action safety bar for turning power press flywheels during setups.

Courtesy Liberty Mutual Insurance Co.

noid-controlled air valves should be incorporated into the system to prevent accidental clutch release resulting from air leakage and pressure build-up in the air cylinder. The dual solenoid-controlled air valves are recommended as a back-up safety factor for the system. Single solenoid-controlled air valves have been known to fail when foreign particles are caught between the valve and the valve seat. This condition can cause leakage and sufficient pressure build-up in the air-cylinder to actuate the clutch even though the motor is off. A simultaneous failure of both air valves due to this condition is very unlikely when dual solenoid-controlled air valves are used.

Certain other precautions are essential to prevent the operator from making these devices inoperative. Two-hand tripping devices should be used in conjunction with single stroke attachments. When electrically operated tripping devices are used, the circuit should contain the necessary relays to permit only one stroke of the press for each operation of the tripping device.

On pneumatically operated controls, the system should contain a measured chamber that will permit only one stroke of the press for each operation of the hand valve.

On mechanical controls, the mechanism should be such that both hand levers are released and depressed for each stroke of the press. Where the press itself is equipped with a single-stroke device, it is important to utilize the added protection.

On presses equipped with two-hand tripping devices, it is sometimes convenient to provide a selector switch for the selection of hand or foot controls. The selector switch should be equipped with a lock, and the setup man or supervisor should be responsible for locking the selector in the proper position. See the section on Foot Control Guards later in this chapter.

All two-hand controls should be installed and supervised so as to prevent operation by one hand only, and the buttons or levers should be located far enough apart (21 in.) to prevent the use of the hand and elbow of one arm to bridge the controls. Some companies place a baffle between the controls when it is impossible to separate them by at least 21 in.

Where more than one operator is required on a press, a duplicate set of controls should be provided for each operator, and the press should not cycle until all controls are energized.

On presses having positive clutches, the operator can release his hand from the controls immediately after he has energized the circuit. On long stroke, slow speed presses, therefore, two-hand trips are particularly hazardous unless they are located far enough from the die area that it is impossible for the operator to energize the controls and then reach into the point-of-operation area before the press slide has reached the bottom of the stroke.

Two-hand devices installed on presses with friction clutches must be the constant pressure type which requires the operator to maintain pressure on the controls until the slide has bottomed or reached a point at which the operator can no longer reach into the area between the dies.

Pull-back devices

Also called pull-out or pull-away devices, these have the advantage that, when properly used, supervised, and maintained, they always remove the hands from the point-of-operation zone as the slide descends. This type of device is usually limited to secondary operations and jobs where the operator can remain at the feeding position. Pull-back devices should be attached to the slide or upper shoe to attain maximum safety. *It is very important that the pull-back device be adjusted to fit each operator and after each die change.*

One manufacturer reports that "the length of the arm is not a factor. The length of the hand from the crutch of the thumb and forefinger to the tip of the middle finger is the critical measurement. The wrist band is fastened around the wrist with the strap looped around the thumb, therefore the finger length is really the measurement to be considered."

Where more than one person is working on a press, pull-back devices must be provided for each.

Automatic electrical devices

Regardless of the energy source, all electrical or electronic devices sold for use on power presses perform the same end function; when their field or frequency is blocked, they act to interrupt the electric current to the press. This has the same effect as if the "stop" button had been pushed.

To be effective, this type of device should be operated from a closed electric circuit so that interruption of the current will automatically prevent the press from cycling.

One advantage claimed for this device is the absence of mechanism in front of the operator. It is particularly applicable on large presses. The photoelectric cell or electric eye device should be installed far enough from the point-of-operation area that it will stop the ram before the operator's hand can get under it, and enough light beams must be used to cover the open area with a curtain of light.

These devices cannot be used on presses with positive clutches but rather on presses having "friction," air, or other clutches which can be held at any position of the stroke (cycle), and with adequate brakes for stopping the press at any point of slide travel. Automatic electrical control devices are *not* effective on punch presses with positive clutches because once the operating cycle of a power press starts, no device can prevent completion of the cycle.

These devices should always be supplemented by hand tools or feeding and ejection devices so that the operator need not place his hands in the point-of-operation zone.

Auxiliary devices

Feeding-extracting tools. A variety of special tools has been developed for use with automatic feeds or enclosure guards. These tools are made of soft metal, aluminum or magnesium (some are magnetized) and include pushers, pickers, pliers, tweezers of various types, forks, and suction disks (Fig. 32–21).

Tools provide protection only if they are always used by the operator. They should never be permitted to substitute for proper guarding.

One way to make sure tools are used properly is to provide storage convenient to the work place. Pegboard can be used to mount various tools; a silhouette of each tool can be painted on the board to indicate where it belongs. This encourages keeping tools in their correct place.

Foot control guards. All presses operated by foot pressure should have a guard over the pedal or switch button. The guard shown in Fig. 32–22 may be equipped with a spring-closed shield so that when the operator's foot is removed, the shield snaps shut, enclosing the switch button.

Guards, such as shown in Fig. 32–23, should be large enough to allow room for the operator to place his foot in the operating position

(*Text continues on page 882.*)

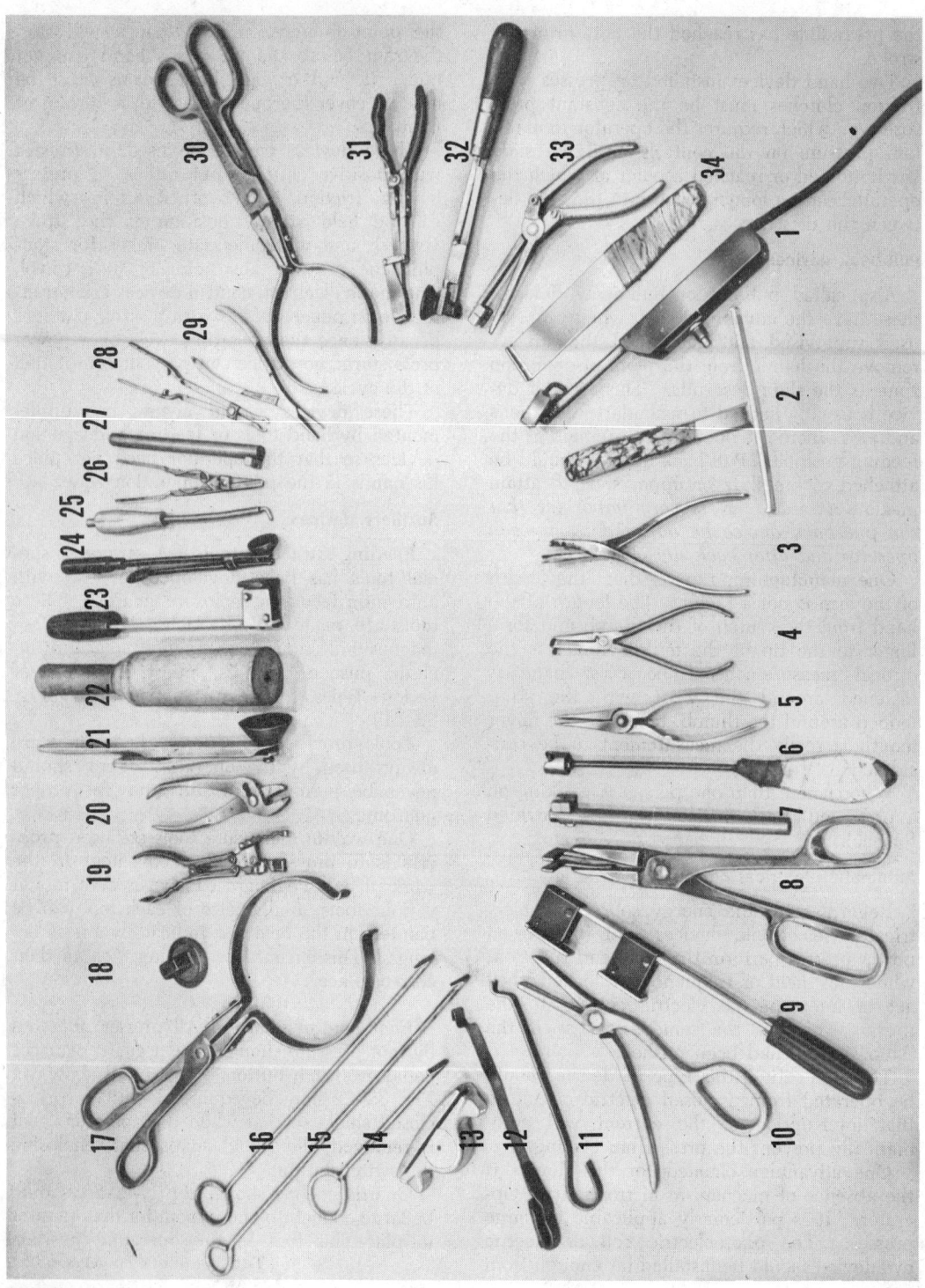

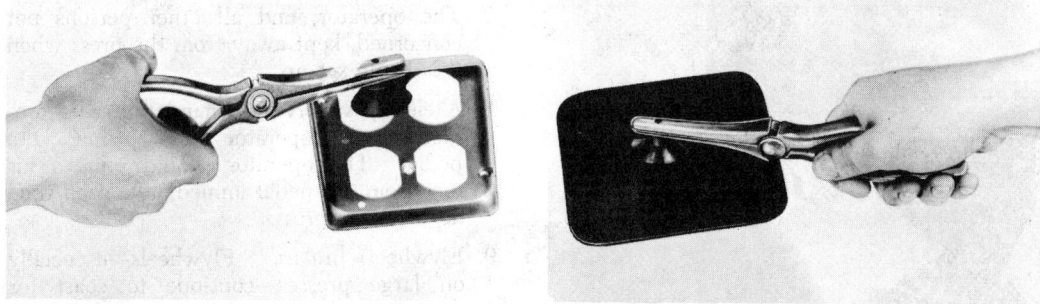

F<small>IG</small>. 32–21b.—Closeup views of one versatile hand-feeding tool in use. Tools may be purchased or "custom made" in company shop.

TOOL IDENTIFICATION

Shown in the photograph at left are 34 simple, safe hand tools – all of which can be made in a shop, and all of which can save the hands and fingers of power press operators.

The tools shown include pliers up to 12-in. long which permit a hand to be kept out of the danger zone; other pliers are designed to grasp a vertical flange by means of bent jaws; others have sharp lower jaws for picking up thin pieces; others have a claw-like grip for holding material in work. Tools also include vacuum cups for handling sheet metal at slitting and shearing machines, permanent and electromagnets, pliers with magnets, steel hooks, and steel or brass pusher sticks.

Not only do tools save workers' fingers and hands but they also contribute to speed of operation. Studies show that it takes 1.4 seconds to load a press with a 12-in. pliers compared to 1.8 seconds by hand.

Key to numbers on the accompanying photograph:

1. 110-v electric magnet for picking up sheet metal.
2. Steel or brass pusher.
3. Pliers with extra-long handles.
4. Pliers with adapters for grasping vertical edges.
5. Pliers with long handles and long grip.
6. Alnico magnet on a stick.
7. Fiber stick with Alnico magnet.
8. Pliers with adapters for grasping vertical edges.
9. Fiber stick with Alnico magnet.
10. Pistol-grip pliers.
11. Push stick.
12. Push stick.
13. Sheet-edge gripper.
14–16. Hook with 90-degree bend.
17. Pliers with adapter for grasping vertical edges.
18. Vacuum cup.
19–20. Pliers with high-pressure grip.
21. Releasable vacuum gripper.
22. Cylindrical holding tool.
23. Fiber stick with magnet.
24. Releasable vacuum gripper.
25. Hook with 90-degree bend.
26. Normally closed pliers.
27. Fiber stick with Alnico magnet.
28. Pliers with adapter for grasping vertical edges.
29. Long-nosed pliers.
30. Pliers with adapter for grasping vertical edges.
31. High-pressure pliers.
32. Vacuum cup with handle.
33. Adjustable-handled pliers with Alnico magnet.
34. Push stick.

Information courtesy Hersche Smith, Fort Wayne, Ind.

F<small>IG</small>. 32–21a.—Thirty-four "mechanical hands" show what can be accomplished through determination and ingenuity. These hand tools have been used for loading and unloading dies and work in 20 presses of one large plant, and have contributed to making a "No Accident Month" last for more than 25 years. Tool description is above.

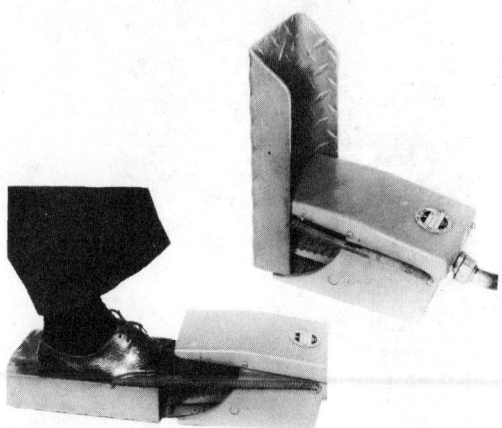

Fig. 32–22.—Two views of a foot switch to which a spring-loaded shield has been attached by the shop in such a way that it closes automatically when foot is removed.

without undue fatigue and without striking or scraping his leg against sharp edges of the guard. A split length of ordinary garden hose around the open edge of the pedal guard will protect against such injury.

The tripping pedal should travel about 3 in. The clutch rod should be connected to the pedal lever so that the distance between the clutch rod and the rear pivot on the pedal will be approximately one-third the length of the pedal. The travel of the connecting point will thus be about 1 in. when the pedal travel is 3 in.

The pedal lever should have no side play, and the tension of the pedal should be maintained as recommended by the manufacturer. No spring or counterweights should be added to the pedal shaft or to the pedal shaft lever.

Measures to prevent accidental tripping of the press in addition to placing a guard over the pedal include:

1. A safety spring on the trip rod.

2. Aisles of ample width for trucking material to the presses.

3. Aisles adjacent to the press used only for necessary trucking of dies and stock.

4. Enough working space between adjacent machines in the same line.

5. Unauthorized persons kept away.

6. Operators protected from interruption.

7. The operator, and all other persons not concerned, kept away from the press when it is being set up.

8. A closely supervised, mandatory rule prohibiting the operator from "riding" the pedal. The operator should remove his foot from the pedal immediately each time he trips the press.

9. Flywheel brakes. Flywheels—especially on large presses—continue to coast for some time after the power source has been turned off, and they have been known to cycle a press when the actuating control (or controls) was tripped. A flywheel brake will stop the flywheel within 10 to 20 seconds.

On all squaring shears the entire length of the treadles should be provided with fixed guards. Such guards should be placed above the treadle, with room between the guard and the treadle only for the operator's foot. To make accidental tripping of the shear impossible, an extra foot latch can be attached to the treadle.

Single-stroke attachments. Through common usage the term "nonrepeat" has more or less come to mean both the nonrepeat device which is installed in the clutch mechanism

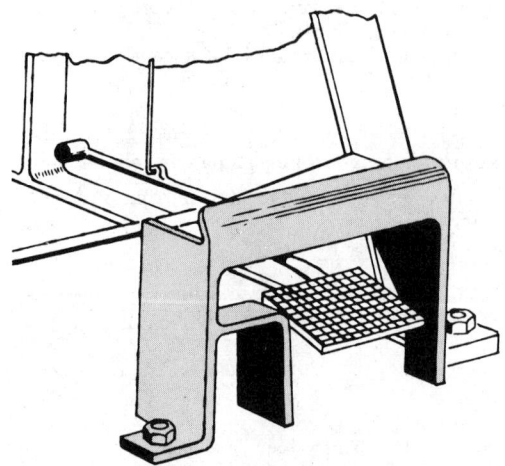

Fig. 32–23.—This foot pedal guard has a foot rest at the left to prevent fatigue.

by the manufacturer of the clutch and the single-stroke device which is attached to the trip rod and which disconnects the pedal after each stroke of the press.

A press with a positive clutch should have a single-stroke attachment which disconnects the pedal or operating lever after each stroke.

A single-stroke spring device should depend on spring action only if it is a compression spring encased in a close-fitting tube or closely wound on a rod. The type wound on a rod is preferable because it permits easy detection of a broken spring. The space between the coils should be less than the diameter of the wire.

Of necessity, a single-stroke device is made inoperative when the press is used on continuous operation. In this case the die should be completely enclosed, regardless of the method of feeding. When the press is set up for other than continuous work, it is of the utmost importance that the single-stroke device be reconnected.

If the press is controlled by means of a hand lever, the lever should have a spring latch to prevent accidental or premature tripping. If a hand operated press requires more than one operator, the hand lever should be interlocked.

The foregoing guards and safeguarding devices are satisfactory for use on power presses depending on their suitability to the operation and their condition of maintenance. The approval of any safeguarding by OSHA is dependent on these criteria—is it suitable to the operation and is the safeguarding maintained as required?

Sweep devices

There is one device not covered in the listing for a number of good reasons—the sweep device. Their action is described by their name—they are supposed to "sweep" the operator's hand out of the way. They have serious disadvantages, however: their action can be violent and, more importantly, they can be "beat" or avoided in too many cases. They are activated by the press, which in most cases is too late for positive protective, sweeping action. In short they are deemed ineffective and, at times, hazardous by press safety experts. The B11.1 Standard Committee is considering dropping the "sweep device" as an approved safeguarding, and OSHA regards them in the same light.

All of this applies only to those sweep devices which are activated by the action of the press. There are some new sweep devices that are interlocked with the control system so that the press cannot be operated unless the sweep is in the closed position. This requirement is similar to the other approved devices and so they will probably be allowed if suitable to the operation and well maintained. Current regulations should be checked and followed.

Feeding and Ejection of Parts

As was mentioned in the preceding section on guards and devices, the final selection of a guard must be correlated with the feed and ejection requirements of the processed part. Feed and ejection must also be correlated with the processing die. In the case of existing dies, this may require modification of the die to facilitate the coordination of the various mechanisms or devices used. New dies can be built with these functions as an integral part of the whole die. This approach should definitely be considered if the part being processed has any complexity of size, shape, or form which should make it difficult to feed or eject it from the die. Following are some useful feeding and ejection methods which have been used with success for many years.

Automatic feed operations are normally conducted with random length strip stock or coiled stock. Strip stock is usually manually fed. Coiled stock can be fed either manually or by means of a roll feed or hitch feed.

Automatic roll feeds are ideal for continuous operations of blanking from strip stock. Small gears on the roll feed should be enclosed and the run-in nip point of the feed rolls should also be guarded.

Automatic push or pull feeds are similar to roll feeds and are used mostly in blanking larger pieces.

When coiled stock is used with a reel and roll feed or hitch feed, a feed table is generally unnecessary. Where coiled stock is used with a stock reel but not with a roll feed, a feed table helps the operator backgage the stock efficiently and feed it to the die with

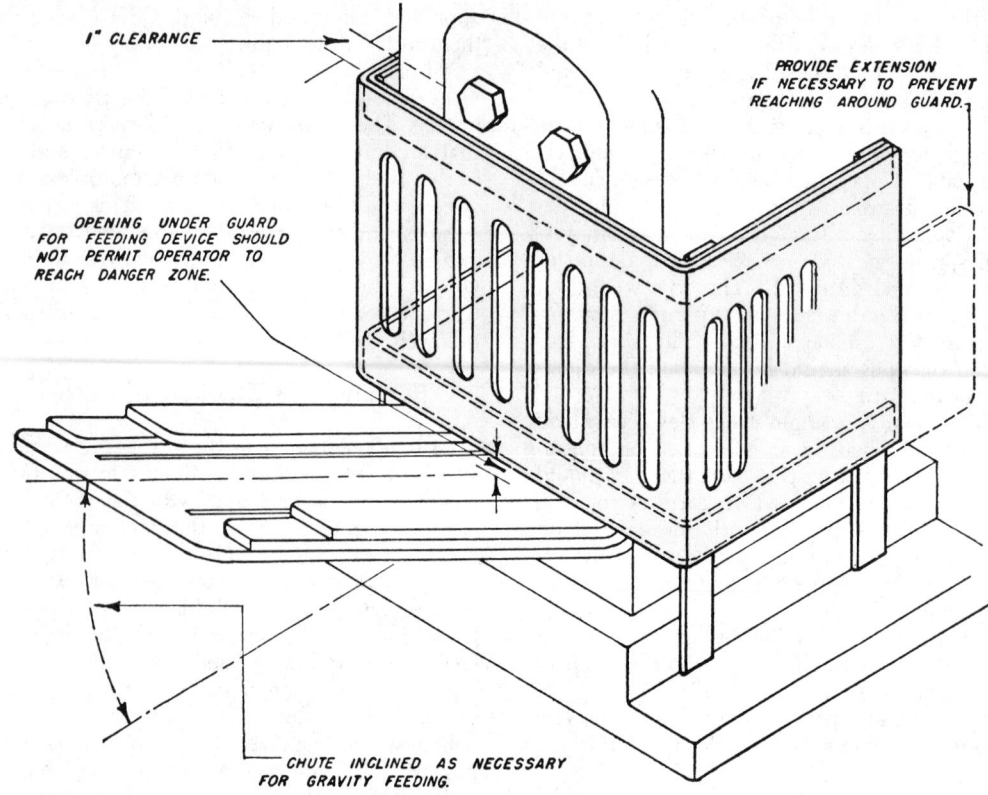

I" CLEARANCE

PROVIDE EXTENSION IF NECESSARY TO PREVENT REACHING AROUND GUARD.

OPENING UNDER GUARD FOR FEEDING DEVICE SHOULD NOT PERMIT OPERATOR TO REACH DANGER ZONE.

CHUTE INCLINED AS NECESSARY FOR GRAVITY FEEDING.

Fig. 32–24.—A die enclosure guard with an inclined chute for gravity feeding. This guard may also be used on an inclined press but with a straight chute.

Courtesy Liberty Mutual Insurance Co.

minimum effort.

In manual feeding of strip stock, a feed table eliminates unnecessary motion and reduces operator fatigue. The feed table should be adjusted to the height at which the operator can work with minimum effort.

Oiling rolls or pressure guns can be used instead of a paintbrush system to lubricate strip or roll stock. Automatic or manual control pressure guns can be provided to lubricate the punch and the die.

When automatic feeding methods are used, it is still necessary to enclose the slide, to limit its stroke to ¼ in. or less, or to provide a gate guard device, according to ANSI Standard B11.1, *Safety Requirements for the Construction, Care, and Use of Mechanical Power Presses.*

Semiautomatic feeding devices place the piece being processed under the slide by a mechanical device that requires the attention of the operator at each stroke of the press. Such feeds have a distinct advantage in that the operator is not required to reach into the point-of-operation area to feed the press, and the feeding method permits complete enclosure of the die. This type of feeding is not adaptable for certain blanking operations nor for nesting of odd-shaped pieces.

The six principal types of semiautomatic feeds are chute (both gravity and follow), plunger, slide or push, sliding dies, dial, and revolving dies.

Chute feed. Of the six semiautomatic methods, the chute feed is probably the most

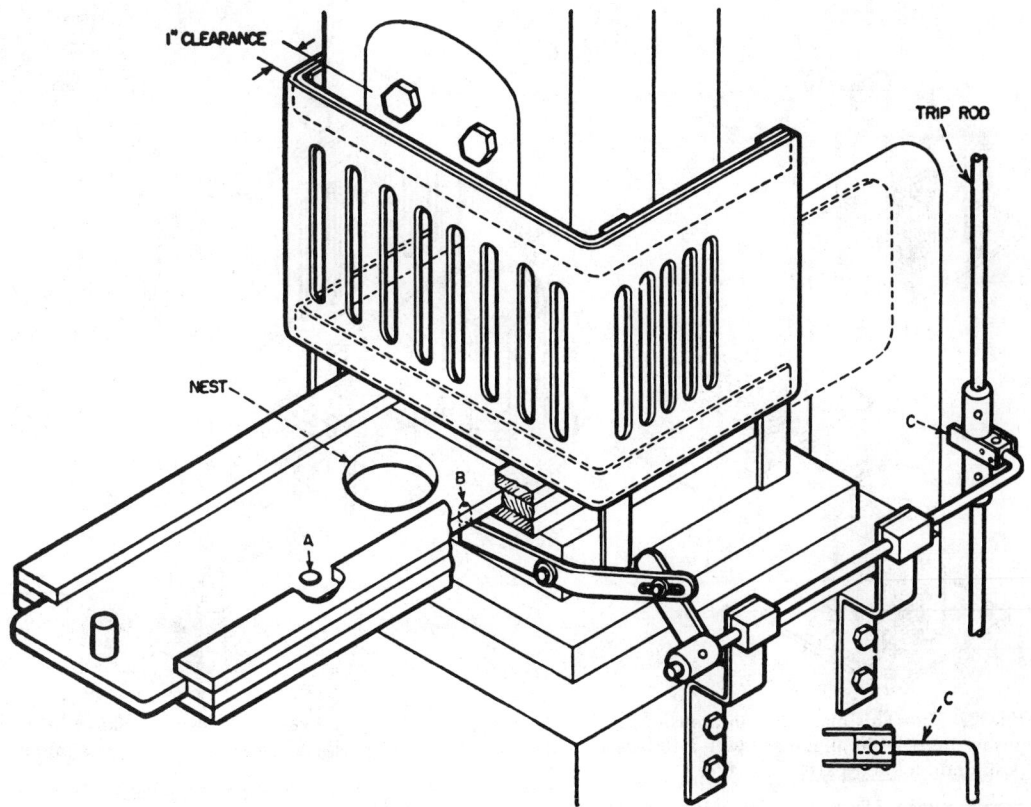

Fig. 32–25.—A manually operated plunger feed. Note that the press cannot be tripped until the part is pushed to the next location. At this point, hole A is directly over tapered pin B; it can rise and release yoke C so that the press can be tripped.

widely used. It is a horizontal or inclined chute into which each piece is placed by hand. The pieces then slide or are pushed one at a time into position in the lower die. The entire die may be enclosed, since it is unnecessary for the operator to place his hands in the point-of-operation area if nonmanual ejection is also provided (Fig. 32–24).

It is customary to use a soft metal pick or rod or a wood stick to remove pieces that jam in the die. *A steel rod should never be used. If caught in the die, it could shatter, with the result that the die might be ruined and the operator injured.*

Many hand-fed dies can be changed to chute feed by reversing the dies and inclining the press. A chute feed on an inclined press not only helps center the piece as it slides

into the die but also simplifies the problem of ejection.

Plunger feeds are a variation of the push feed and may be semiautomatic or manual in operation. The semiautomatic plunger feed is a magazine or chute in which blanks or partly formed pieces are placed. The blanks or pieces are fed, one at a time, by a mechanical plunger or other device which pushes them under the slide.

Manually operated plunger feeds are used for individual pieces which, because of their irregular shape, will not stack in a magazine or will not slide easily down a gravity chute. Each piece is placed in the nest in the pusher and moved to the die by manual operation of the pusher. To get correct location of the part

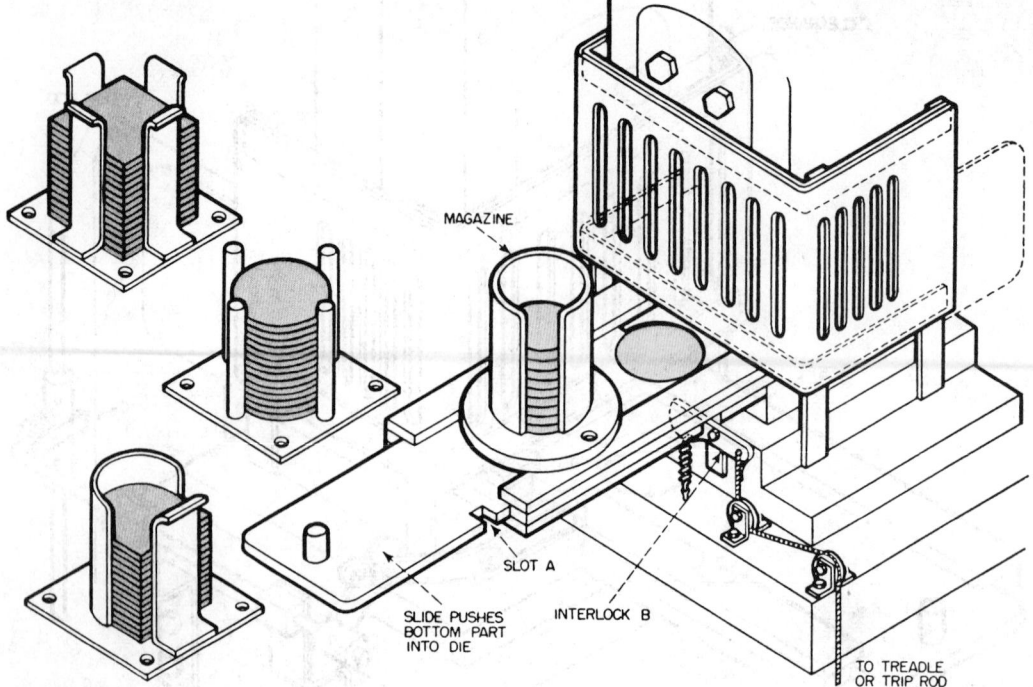

MAGAZINE

SLOT A

SLIDE PUSHES
BOTTOM PART
INTO DIE

INTERLOCK B

TO TREADLE
OR TRIP ROD

Fig. 32–26.—This magazine on a push feed enables the operator to catch every press stroke. Slot A in the pusher must be in alignment with interlock B before the press can be tripped. This feature assures proper positioning of the part in the die.

Figures courtesy Liberty Mutual Insurance Co.

in the die, an interlock is sometimes necessary so that the press cannot be operated until the pusher has spotted the part accurately (Fig. 32–25).

Slide or push feeds are a variation of the chute feed, combined with magazines and plungers. The pieces are stacked in the magazine. As each piece reaches the bottom, it is pushed into the die by means of a hand operated plunger (Fig. 32–26).

Sliding dies are pulled toward the operator for safe feeding and then pushed into position under the slide (ram) for the downward stroke. The die may be moved in and out by hand or by means of a foot lever. Regardless of how the die is actuated, it should be interlocked with the press to prevent tripping when the die is out of alignment with the slide (Fig. 32–27), and "stops" should be provided to

prevent the die from being inadvertently pulled out of the slides.

Sliding bolsters are a new variation on the sliding die. Here the press bed is modified with a hydraulically or pneumatically controlled bolster that slides in when the palm buttons are depressed, and out when the stroke is complete. It is completely aligned and interlocked with the press. In reality, it is an automated sliding die.

Dial feeds consist of two or more nests arranged in dial form. The dial revolves with each stroke of the press, so that the operator can feed the machine safely. The part to be processed is placed in a nest on the dial which is positioned in front of the die. The dial is indexed with each upstroke of the press to deliver the next nested part into the die (Fig. 32–28).

886

FIG. 32-27.—An adjustable barrier with a sliding die. A locating device is provided on the slide bolster to locate and lock the die slide in alignment with the punch holder. An interlock should be provided to prevent tripping of the press until the die slide is in the proper location.

The best method of ejection is usually pick-up fingers or compressed air. However, in many installations the operator both nests the part on the outside and removes it when it is returned on the dial from under the slide.

Two operators are sometimes used on a dial fed press, one to feed the press and the other to remove the processed parts.

Revolving dies operate on the same principle as dial feeds except that they may consist of multiple dies, sometimes only two.

When a dial feed is used, the point-of-operation area should be enclosed.

Ejecting devices

Properly designed and installed ejector mechanisms will eliminate many common hazards and also increase the cycle of production. Ejector mechanisms can automatically clear the press faster than human hands and with greater safety.

In the development of ejection facilities, two problems must be solved: how to strip the piece from the punch or the die, and how to eject the piece from the die and the press to a container or conveyor.

This process can be accomplished in many ways. In some cases, a single mechanical means performs both stages. In other cases, a separate method is used for each stage.

Some of the more common methods which are used singly or in combination to strip pieces from the die are shown in Fig. 32-29:

A—Positive stripper plates

B—Spring pressure pads or pins

C—Latch-type mechanical lift dogs

D—Compressed air jets

Pneumatic or hydraulic lift pins or pickup fingers are shown in Fig. 32-30.

Single or multiple air jets can be used for effective removal of small pieces. Air ejection can be combined with other mechanical release means or with gravity removal. All jets should be anchored securely to direct the air stream effectively and to prevent jet tubes from shifting into the die working area.

Both the consumption of air and the production of noise can be reduced considerably by incorporation of one or more of the following principles:

1. Locate the jet discharge as close as possible to the piece to be removed. Optimum positioning is essential for maximum effectiveness and frequently can be accomplished through die design.

2. Limit the duration of air discharge to the minimum period required to remove the piece. A simple cam-actuated air valve will operate only for the ejection time.

3. By means of flow valves or pressure regulators, reduce the discharge pressure to the minimum required. Reducing the discharge pressure will also result in greater operator safety from flying particles. However, this statement should not be taken to mean that the operator need not wear eye protection. There are always hazards from flying particles in press operations, and the addition of air ejection increases the hazards. Be sure all operators wear eye protection at all times.

4. Use an air-ejector nozzle with several orifices (Fig. 32-31).

(*Text continues on page 890.*)

FIG. 32–28.—Well guarded double-dial feed. Note plastic coating (arrow) on lower part of metal guard, which eliminates sharp edges and reduces glare.

Courtesy Automatic Electric Co.

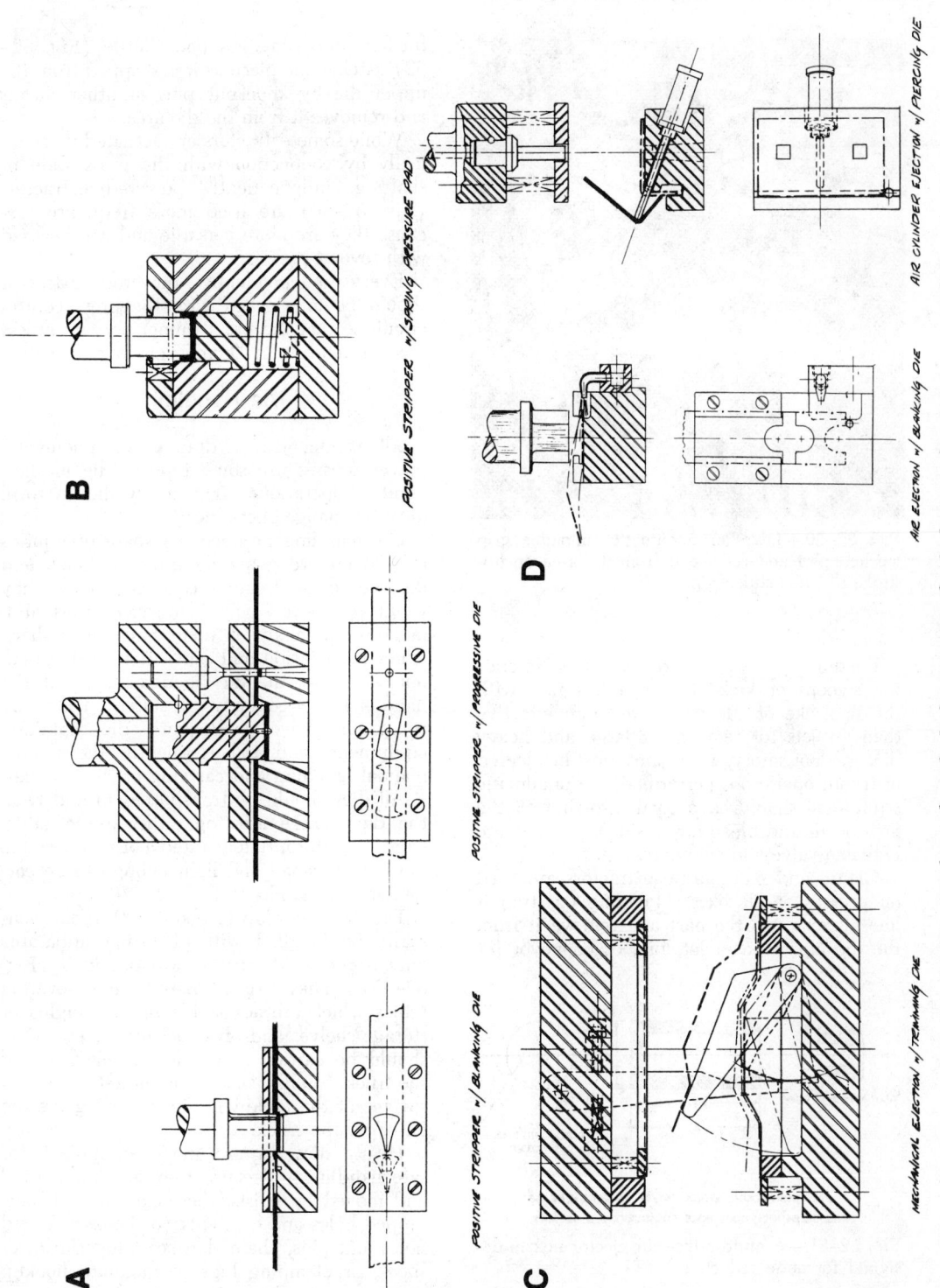

A

POSITIVE STRIPPER w/BLANKING DIE

POSITIVE STRIPPER w/PROGRESSIVE DIE

B

POSITIVE STRIPPER w/SPRING PRESSURE PAD

C

MECHANICAL EJECTION w/TRIMMING DIE

D

AIR EJECTION w/BLANKING DIE

AIR CYLINDER EJECTION w/PIERCING DIE

FIG. 32–29.—Mechanical methods for stripping finished pieces from power press dies.

FIG. 32–30.—Jaws on pneumatic unloader grip finished part and remove it from die area on upstroke of press slide.

Pneumatically powered cylinders operating sweeps or kickout pins and timed with the upstroke of the press are more effective than air jets for removal of large and heavy pieces. For safety, when jams must be cleared or tryout operations performed, the pneumatic equipment should have a valve to shut off the flow of air and bleed any residual air pressure between valve and cylinders.

Clamp and pan shuttle extractors are used on presses of all sizes. Pivoting or straight line clamps grip the part and remove it from the die area to a pallet, bin, or conveyor for

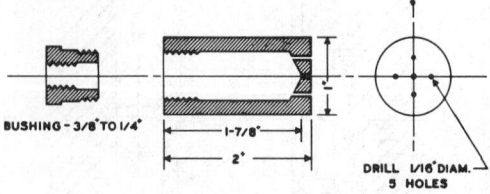

BUSHING - 3/8" TO 1/4" |— 1-7/8" —|
|— 2" —|
DRILL 1/16" DIAM.
5 HOLES

USE 1-INCH ROUND STOCK - STEEL, ALUMINUM, BRASS

DRILL 37/64 INCH DIAM. HOLE TO ACCEPT 3/8"-18 N.P.T. TAP

FIG. 32–31.—A multi-orifice air ejector nozzle designed for noise reduction.

further processing. A pan shuttle (Fig. 32–32) catches the piece as it is stripped from the upper die by knockout pins or other means and removes it from the die area.

While some extractors are actuated mechanically by connection with the press slide or shafting, independently powered extractors (Fig. 32–30) are used more frequently because they are more versatile and can be used with several presses.

Every independently powered extractor should be interlocked with the press control circuit so that the press cannot operate unless the extractor is in "home" or "out" position.

Die Handling

All mechanical, hydraulic, or pneumatic power presses use some type of die in their point of operations. The heavy die, in turn, must be changed periodically.

Die handling produces its share of injuries too. Dies are generally precision tools and deserve to be treated as such. They may weigh from a few pounds to several tons, and, in addition, many of them have razor-sharp edges both inside and out. If it is dropped, not only may the die be damaged, but the hand or foot it contacts will suffer. Too many companies have inadequate die handling equipment, and too frequently they allow manual handling and carrying of smaller dies.

All dies should be transported to and from the die storage area on either a die table, truck, or appropriately rated hoist, unless the die is truly small and light enough to be carried in one hand.

Dies weighing up to about 100 lb can generally be handled without lifting apparatus only if proper die trucks are provided (Fig. 32–33). These trucks should have elevating tables which are adjustable to the heights of storage shelves and press bolster plates. Dies should be carried at the lowest elevation of the truck. Rollers or balls mounted in the top surface of the table will aid in sliding the die on or off the truck.

Heavy dies require more equipment for safe handling. Because they are often lifted and moved by hoists, these dies should have tapped holes and eye bolts (or hooks), drilled holes and pins, chain slots, cast lugs in lower shoes, or clamping lugs to facilitate hookup

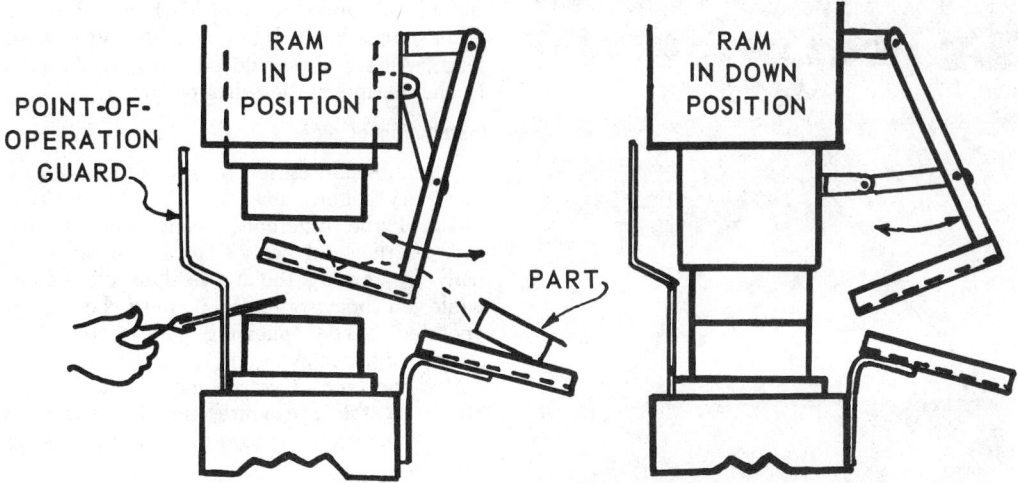

Fig. 32–32.—Pan shuttle mechanism moves under finished part on upside of press slide. Shuttle then catches part stripped from slide by knockout pins and deflects part into chute. When press slide moves down toward next blank, pan shuttle moves away from die area.

Fig. 32–33.—Die truck (left), with adjustable-height and adjustable-angle die table, can match angle of inclined presses. Truck is secured to press before moving die. Holes in upper die shoe indicate drilled-hole-and-pin method of lifting.

Fig. 32–34.—Drilled-hole-and-pin method of lifting a forging die. For easy fit, holes in die are ⅟₁₆ in. greater diameter than pin diameter. Depth of hole, 1½ times pin diameter, should be equal to pin length from its point to the collar, so that the collar will be held against the die and pressure will keep the pin from slipping out.

Courtesy Tractor Works, International Harvester Co.

and transfer (Fig. 32–34). When tapped holes are used for securing lifting hooks or eye bolts, they should be either ¾ or 1 in. in diameter. (The use of ⅞-in. holes is not recommended because ¾-in. bolts might be used in them and appear to fit, only to pull out during the lift.) The depth of the hole should be 1½ times the diameter.

For moving dies that are about 1000 lb or heavier, special die-handling power trucks are recommended. On these trucks, special equipment for pushing or pulling the die includes power winches, roller tables, and hydraulic or pneumatic clamps.

Where lifting is needed to effect the transfer, use a hoist and never lift higher than is necessary for minimum clearance. At no time should an employee have his hands, feet, or other part of his body underneath a suspended load. All signals should be given by the man in charge and by no other person.

Installation of dies

Safe procedures for setting and removing dies vary slightly for large or small presses. Most of the difference occurs because the slide on the light presses can be moved manually by turning the flywheel or crankshaft, while on heavy presses it must be power operated. The following safe method for setting and removing power press dies may be followed *for all presses,* except where variations in the procedure are designated as being specifically for *light presses* or for *heavy presses.*

First, clear away all stock, containers, tools, and other tripping hazards from the work area.

For light presses. Disconnect or shut off the power and lock the switch. Bring the slide down to its lowest position by turning the flywheel or crankshaft. (If a safety bar is used as a lever to turn the crankshaft, it should have a spring and collar arrangement that will prevent its being accidentally left inserted in the crankshaft. See Fig. 32–20.) Measure the clear height between the bottom of the slide and the bolster plate. This distance should be slightly greater than, or at least equal to, the height of the closed die. If not, raise the slide by means of the adjusting screw until this is assured. When physical recessing is necessary, block up the slide with timber, metal blocks or posts provided for this purpose. Blocks should be equipped with an electric receptacle plug to hold open the circuit when they are in the die. Thus, they will not be left in the press when power is applied.

For heavy presses. Jog or inch the press to bring the slide down to its lowest position and measure the clear height between it and the bolster plate. Again, if this distance is not slightly greater than, or at least equal to, the height of the die in the closed position, adjust the slide until such clearance is assured. Raise the slide, block it in position, shut off the power, and lock the switch.

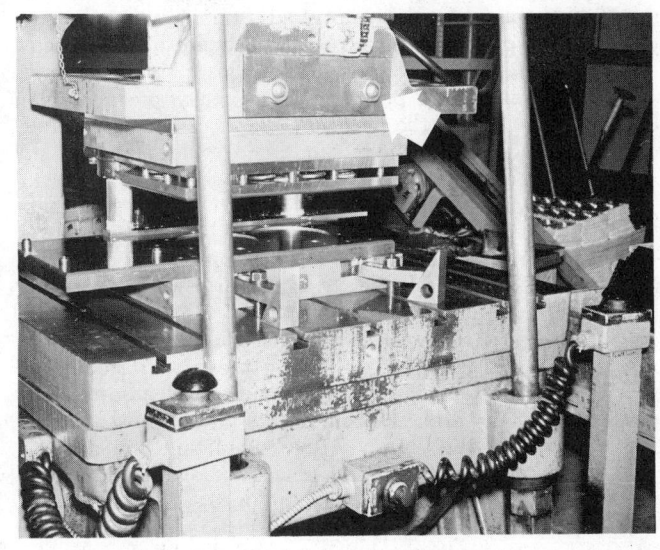

Fig. 32–35.—Mounting bolts (arrow) pass through the ram to fasten upper die securely. Lower die is clamped to bolster. Note two-hand control.

Dismantle or disconnect a point-of-operation safety device only if it is absolutely necessary. For example, enclosure guards fastened to the press, gate or barrier guards, and some types of sweep guards will have to be removed, but not, in general, pull-back guards or photoelectric cells. All parts of any dismantled safety devices should be carefully put aside so that they may be reinstalled in good condition as soon as the new dies have been placed.

Clean off the bolster plate, preferably with a brush. Check that bolt holes are clear of slugs. (A pencil-shaped magnet is valuable for removing iron or steel slugs.) Burrs can be removed with a flat file.

Check the die to make sure it contains no overlooked chips, tools, or finished parts, and that it is in good operating order. Then transfer the die from the truck to the press, as discussed earlier in this section.

Line up the die in the correct operating position and remove the posts or blocks from under the slide. (For heavy presses—reconnect the power to operate the slide. For light presses—manually turn the flywheel to move the slide, using a safety bar if necessary—Fig. 32–20.) Then, carefully lower the slide until it fits firmly against the top die. It is extremely important not to put too much pressure on pierce or trim dies. Tighten all bolts

and clamps necessary to secure the top half of the die to the slide. Bolting the die to the slide with bolts through holes in the upper die shoe is recommended (Fig. 32–35). On heavy presses—if air cushion pads are used, keep the slide close to the top of the die until the die is properly seated on the cushion pins.

Shim or block up the lower half of the die to the power level; bolt and clamp it to the bolster plate. (Make sure that the bolts, bolt heads, and wrenches are of proper size and in good condition.) Bolting the die shoe to the bolster produces the most secure die setup. If clamps must be used (Fig. 32–35), their outer ends should be blocked up slightly higher than the die surface on which their inner ends will rest. Clamps should be of minimum length. Clamp fastening bolts should be closer to the die than to the block end of the clamp. See Fig. 32–36.

Check all bolts and clamps to see that they are tight and that dies are securely fastened in the press. Remove all tools and equipment from the dies, bolster plates, or other ledges on the press. For forming or drawing operations, adjust the slide down to almost its proper depth. For a pierce or trim die, raise the slide slightly so that the punch will not shear when the crank is again brought over and the die entered. Turn light presses several times by hand, checking to see that they

893

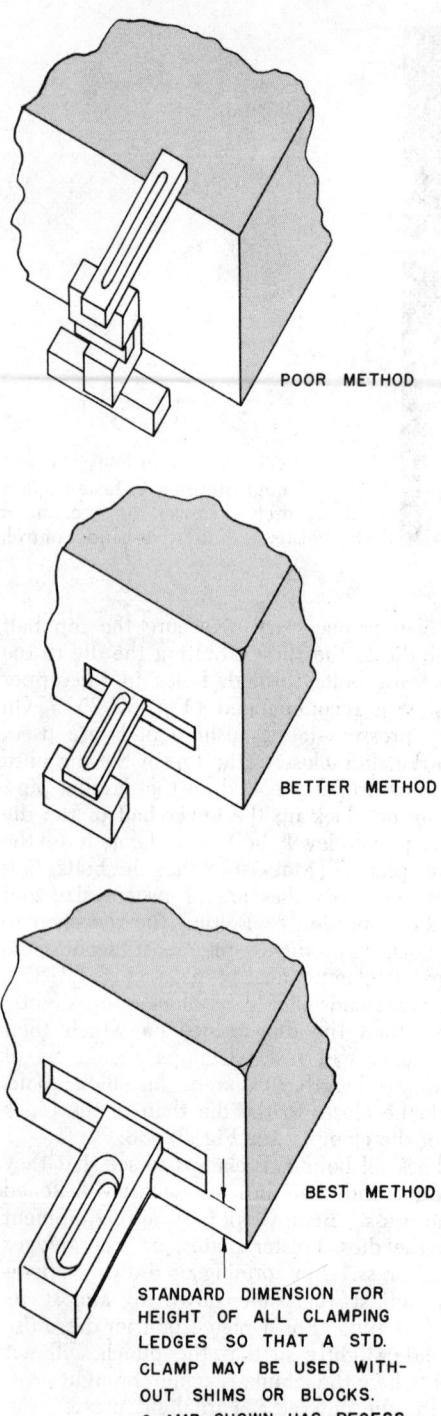

POOR METHOD

BETTER METHOD

BEST METHOD

STANDARD DIMENSION FOR
HEIGHT OF ALL CLAMPING
LEDGES SO THAT A STD.
CLAMP MAY BE USED WITH-
OUT SHIMS OR BLOCKS.
CLAMP SHOWN HAS RECESS
FOR CLAMP SCREW.

are in satisfactory adjustment. On heavy presses do not check adjustment at this point.

Raise the slide to its highest point and block it in this position. (On heavy presses—disconnect the power before proceeding further.) Wipe out the die and remove the safety blocks or posts. Replace the safety device and check it for adjustment and operation. Be sure that pull-out devices are adjusted properly so that an accident could not occur to an operator who has longer hands or arms than the previous operator.

Reconnect the power and try out several actual operations, using the proper stock, on the press. Make any necessary adjustments only after shutting off the power and blocking up the slide. After completing the adjustments, turn on the power and again try several actual operations on the press. If satisfactory, the operator should be given instructions by an authorized person who should observe his methods for a few minutes to see that he is operating the press correctly.

Removing dies

Dies should be removed from the press with the same care used in setting them up. Although modifications may be necessary in special cases, the safe procedure is as follows:

1. Make sure that the working space is cleared of all stock, containers, tools, and other items.

2. Disconnect or shut off the power and lock the switch. Turn the flywheel by hand or by means of the bar until the slide is at the bottom of the stroke. If the press cannot be turned over manually, jog it under power, then shut off the power and lock the switch.

3. Dismantle or disconnect the point-of-operation safety device only if absolutely necessary. Put aside the parts of a dismantled safety device so that it can be reinstalled in good condition as soon as the new die is in place.

← Fig. 32–36.—Three methods of clamping dies are compared. It is always important to use correct-size clamps. The method illustrated at the top shows a practice that is seriously hazardous.

4. Clean off the bolster plate, preferably with a brush.

5. If the die is to be operated with an air pad, shut off the air supply and open the release valve to permit the pins to go down. Also shut off the air supply to the automatic blowout system used in the die.

6. Loosen and remove bolts and clamps bolting the die to the slide and to the bolster plate. Place bolts, nuts, and clamps on a bench or in a special place on the die truck as soon as they have been removed from the press.

7. Make certain that the die is loose and that bolts, nuts, clamps, and other obstructions have been removed.

8. Raise the slide slowly, manually on light presses, and by jogging or inching under power on large presses, and make sure that the die does not hang in the slide.

9. Block the slide in its highest position, and if power was used, shut off the power and lock the switch.

Inspection and Maintenance

Many serious press accidents occur during what should be routine inspection and maintenance procedures. In too many instances, these procedures are carried out on either a haphazard or a "rush-rush" demand basis. The problem is further aggravated by the ignoring (or not knowing of) the manufacturer's recommendations for these procedures.

It is not practical to detail all of the things that personnel should check when inspecting and/or maintaining their presses. These items are well outlined in the manufacturers' manuals and the ANSI B11 standards. If problems develop or if some procedures are unclear, consult the manufacturer or the B11 interpretations committee for clarification. This volunteer committee was formed for the express purpose of solving difficult press problems.

However, none of this information will be of value if the user does not set up and follow a scheduling and recordkeeping system. The advantages of scheduling inspection and maintenance procedures are: it allows planning

Fig. 32–37.—Auto safety glass shield placed between the power press and the operator serves as sound barrier.

ahead, it allows better integration with production scheduling, and it avoids or minimizes downtime due to mechanical failure.

Scheduling inspections and maintenance can be very simple if a computer is available. Essentially all that is involved is listing all inspection and maintenance work required for each machine and the frequency with which it must be done. This information is then entered into a storage bank for retrieval as required. This system makes it very easy to know exactly what must be done and when.

Another system is to enter the same information on punch-sort cards which can then be picked for scheduling. A third method (which might be best for small companies) is simply a list of all machines and their jobs posted vertically on a large chart. These are then plotted against the months of the year with check marks indicating when what job is to be done.

By and large, whatever scheduling method is used, it should be followed as closely as possible. Adherence to the schedule is easy to check if job cards are used with whichever system is used. The computer can be programmed to issue job cards, and the punch-sort cards can sometimes be used as job cards. Job cards would have to be made out for the chart system, but usually this would not be

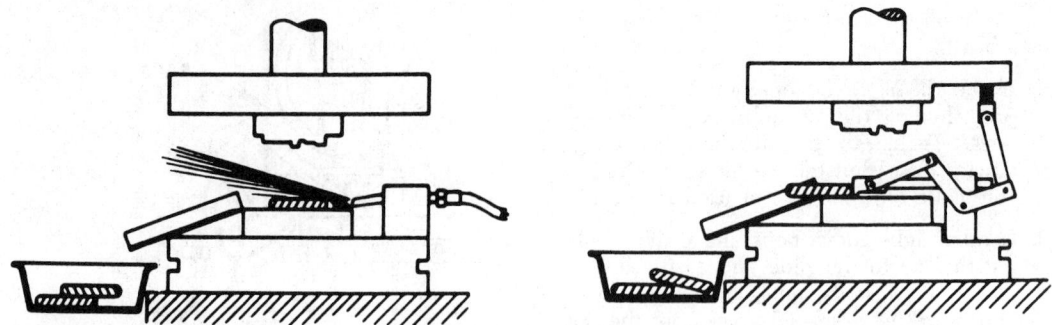

Fig. 32–38.—Quieter method cuts noise. At left is noisy air-ejection method; at right, quieter mechanical ejector.

too large a job to be burdensome.

The caution when using job cards is to have a feedback system to check on their effectiveness. Entering the completed job card back into the file keeps it up-to-date and ready for coordination with the work or production schedule.

Good scheduling involves recordkeeping, and the benefits of recordkeeping are many. It eliminates guesswork which ultimately may create breakdowns in the production process. However if a better reason for recordkeeping is needed, it can be found in the fact that OSHA inspectors frequently will ask to see such records so that they may better assess a company's safety efforts. There really can be no better proof of a company's sincere intention of "providing a safe and healthful workplace" for its employees than complete and current records for all equipment.

Power Press Noise

Press rooms provide an additional hazard to workers which has only briefly been mentioned so far under Air Ejection Methods. For years, noise has been an accepted way of life in the pressroom. Readings of 100 plus dBA are the rule in many cases. Most pressroom noise originates from two sources—the drive mechanism and the point of operations.

Noise needs mentioning here because good die design and good guarding can do much to reduce the overall noise level. There is a great deal of noise generated when the upper die smashes against the work in process and the lower die. The noise is further augmented if punching various holes is done at the same time. Impact noise of this type can be lessened considerably if the die and punches are designed to have a shearing action rather than the flat punching action. It also helps to stagger the entry times of the punches. Some manufacturers have found that polyurethanes are excellent for use in place of springs or wherever cushioning is desired. In fact, some of them use polyurethane pads as female dies for some forming/drawing operations.

Further reductions in noise can be accomplished by making the point-of-operations guard out of solid materials which enclose all of the point of operations. This acts as a barrier to the transmission of the die noise. If visibility is needed, safety glass or plastic can be inset where needed (Fig. 32–37).

Noise is easily generated wherever there is metal-to-metal contact. This means that the guards and any other auxiliary equipment attached to (or used in conjunction with) the press should be firmly attached to the press and isolated from it by some kind of insulating gasket. If metal chutes, tables, or conveyors are used for stock handling or disposal, lining or padding them with bonded rubberlike plastics will also help bring down the decibels. Check Chapter 40, "Noise and Hearing Conservation," for additional ways of reducing pressroom noise hazards.

References

American National Standards Institute, 1430 Broadway, New York, N.Y. 10016.
Safety Requirements for the Construction, Care, and Use of Power Press Brakes, B11.3.
Safety Requirements for the Construction, Care, and Use of Mechanical Power Presses,
B11.1.
Safety Requirements for the Construction, Care, and Use of Shears, B11.4.

National Safety Council, 425 N. Michigan Ave., Chicago, Ill. 60611.
Forging Safety Manual.
Guards Illustrated.
Industrial Data Sheets
Alligator Shears, 213.
Cold Shearing Billets and Bars in the Forging Industry, 557.
Cutting Oils, Emulsions, and Drawing Compounds, 501.
Electrical Controls for Power Presses, 624.
Handling Finished Pieces at Power Presses, 534.
Handling Steel Plate for Fabrication, 565.
Individual Die Guards and Adjustable Press Barriers, 365.
Inspection and Maintenance of Mechanical Power Presses, 603.
Kick (Foot) Presses, 363.
Metal Squaring Shears, 328.
Press Brakes, 419.
Scrap Ballers, 611.
Setting Up and Removing Power Press Dies, 211.
Veneer Clippers, 542.
Power Press Safety Manual.

United Auto Workers, 8000 E. Jefferson Ave., Detroit, Mich. 48214. (General.)

Wilson, Frank W. (ed.). *Handbook of Fixture Design.* New York, N.Y., McGraw-Hill
Book Co., 1962.

Hot Working of Metals

Foundry Health Hazards   899
Hazardous materials . . . Medical program . . . Personnel facilities

Foundry Work Environment   902
Housekeeping . . . Ventilation and heating . . . Illumination . . .
Inspection and maintenance . . . Plant structures . . . Compressed
air hose . . . Fire protection

Materials Handling in Foundries   904
Handling, sand, coal, and coke . . . Ladles . . . Hoists and cranes
. . . Conveyors . . . Elevators and monorails . . . Scrap breakers
. . . Storage . . . Slag disposal

Cupolas   909
Charging . . . Charging floor . . . Carbon monoxide . . . Blast gates
. . . Tapping out . . . Dropping cupola bottom . . . Repairing linings

Open Hearth Furnaces   913
Charging machines . . . Charging boxes . . . Tapping out

Crucibles   914
Charging and handling . . . Crucible furnaces

Ovens   916
Gas-fired ovens . . . Ventilation . . . Inspection

Foundry Production Equipment   918
Sand mills . . . Dough-type mixers . . . Sand cutters . . . Sifters . . . Molds and cores . . . Molding machines . . . Core blowing machines . . . Flasks . . . Sandblast rooms . . . Tumbling barrels . . . Shakeouts

Cleaning and Finishing Foundry Products   923
Magnesium grinding . . . Chipping . . . Welding . . . Power presses

Nondestructive testing   925
Magnetic particle methods . . . Penetrant inspection . . . Ultrasonics . . . Triboelectric method . . . Electromagnetic methods . . . Radiography

Forging Hammers   928
Open-frame hammers . . . Gravity drop hammers . . . Steam hammers . . . Hammer hazards . . . Guarding . . . Props . . . Hand tools . . . Die keys . . . Safe operating practices . . . Personal protection . . . Setup and removal of dies . . . Design of dies . . . Maintenance and inspection . . . lead casts

Forging Upsetters   940
Design of dies . . . Setup and removal of dies . . . Inspection and maintenance . . . Auxiliary equipment

Forging Presses   942
Basic precautions . . . Die setting . . . Maintenance

Other Forging Equipment   944

References   945

The control of materials handling hazards and environmental stresses (dust, fumes, gases, heat, and noise) that are present in foundries and permanent mold and die-casting plants is discussed in this chapter. Also discussed are methods of safeguarding and safe operating practices for forging and hot metal stamping operations. The discussion of these industrial operations is supplemented by a survey of the use of nondestructive testing methods.

Foundry Health Hazards

Metals are formed into finished castings in foundries and permanent mold and die-casting plants. The overall foundry operation usually includes a pattern shop and machine shop.

The principles and practices used for the recognition, evaluation, and control of health hazards are discussed in Chapter 37, "Industrial Hygiene." A technically competent person familiar with these subjects should survey the foundry operations; his recommendations should be followed in the design and installation of control equipment. He should also be consulted for test procedures, analysis of new processes, and similar problems that arise in the course of the hygiene program. Other sources of professional help include the National Safety Council, the American Foundrymen's Society, the U.S. Public Health Service, insurance carriers, and state and local departments of industrial health.

Hazardous materials

The types of materials that present a health hazard and their mode of production in foundry operations are discussed below. See Chapter 38, "Industrial Toxicology," for threshold limit values, maximal acceptable concentrations, and physiological effects of these materials. Also see Chapter 39, "Chemical Hazards," for summaries of their physical properties.

899

Fig. 33–1.—Fume-diverting baffles for barrel furnaces in a foundry. Hoods roll on trolley so that correct positioning is made easy.

Courtesy American Brake Shoe Co.

Dust is generated in many foundry processes and presents a two-fold problem: cleaning to remove deposits and control at the point of origin to prevent further accumulation.

Some foundries have found water under pressure or mixed with steam to be a good cleaning agent. A combination water and compressed air hose with a special nozzle also does a good job. With this method, however, workers should wear approved respirators; see Chapter 19, "Personal Protective Equipment." In any case, water cleaning should not be done during or immediately before melting and pouring. These operations should be started only after all equipment has been allowed to dry thoroughly.

Vacuum cleaning is probably the most satisfactory method used for dust removal in foundries, and the special equipment needed is well worth the investment.

Once dust has been removed, further accumulation can be prevented by local exhaust systems which remove it at the point of origin. See the discussion in Chapter 36, "Exhaust and Ventilation."

Solvents include many different substances, each of which must be evaluated on the basis of its chemical ingredients. Proper labeling, limitations on quantities in use, and other methods of control as detailed in Chapter 37, "Industrial Hygiene," can help minimize the toxic and flammable hazards involved in solvent use.

Other materials.

ACROLEIN occurs in foundry operations as a result of the thermal decomposition of core oil.

ALUMINUM is not usually a toxic hazard in casting processes, but does present a fire and explosion hazard in dust-collecting systems.

ANTIMONY is usually an unimportant contaminant in foundry operations.

BERYLLIUM may produce a typical pulmonary disease, such as reported in one plant casting a 1 percent beryllium-copper alloy.

CARBON, as sea coal, is a common ingredient of molding sand used for facing. Carbon dust may cause anthracosis, which produces characteristic lung shadows in an X ray, but is a relatively harmless condition.

CARBON MONOXIDE is generated during some cycles in the operation of a cupola.

CHROMIUM is encountered in stainless steel casting as the element or the oxide. Exposures occur during melting, gate and head burning, and grinding.

FLUORIDES, sometimes in the form of cryolite (sodium aluminum fluoride), are used in the manufacture of ductile iron and magnesium castings.

IRON OXIDE fumes and dust are created during melting, burning, pouring, grinding, welding, and machining of ferrous castings. Exposure may be particularly high where manganese steel castings or oxygen-lancing of the furnace is involved. Local exhaust can be used (Fig. 33–1).

LEAD is the greatest health hazard in nonferrous foundries. It forms the oxide in melting, pouring, and welding operations. Elemental lead dust is produced in cleaning and machining operations.

MAGNESIUM dust or chips create serious fire and explosion hazards. Physiological effects are confined to a form of "metal fume fever" from the inhalation of finely divided magnesium.

MANGANESE is usually associated with steel castings and bronze alloys in foundry work and presents no special control problem.

PHOSPHORUS is used in the production of phosphor-copper. Acute cases of poisoning have not been reported and chronic cases are rare. The drying of phosphor-copper shot may produce phosphine gas.

RESINS—phenol-formaldehyde and urea-formaldehyde—are used in shell molding and create several hazards. The phenol-formaldehyde type contain hexamethylenetetramine ("hex") which is a skin irritant and highly explosive. This type of resin also decomposes on heating to give a mixture of phenol and formaldehyde vapors. Urea decomposes to give ammonia and carbon dioxide. In practice, however, vapors from resins are a nuisance, because the concentrations needed to produce toxic effects cannot usually be tolerated by man.

Resin dust, especially "hex," is highly explosive when suspended in air and therefore requires wet-type dust collectors.

Alcohol, sometimes used for cold coating sand with resin, must also be controlled to keep its concentration well below the lower flammable limit.

SILICA is usually encountered in the use of silica flour in molding sand or in core washes and sprays. This material is often 100 percent free-silica. Zircon, which is more dense and therefore settles more rapidly, is an effective substitute for silica flour in some applications.

Sand handling and conditioning systems, shake-out operations, and sand-slinging constitute other sources of exposure to silica dust.

Chapter 37, "Industrial Hygiene," discusses the mechanism and effects of silicosis.

SILICONES are used as mold-release agents in shell-molding. The hydrolyzing types are highly corrosive and hazardous on skin contact and inhalation. The concentrations necessary to produce these effects, however, cannot usually be tolerated. Care in handling can eliminate the dangers of skin or eye contact. The nonhydrolyzing types (methyl, mixed methyl, and phenylpolysiloxane) can be just as effective as mold-release agents, are of a very low order of toxicity, and are thus the better choice.

SULFUR DIOXIDE is the result of the oxidation of sulfur used in magnesium castings, and, in concentrations normally present in foundries, is an irritant nuisance.

Medical program

Health protection of foundry workers should be the result of a program based on the principles outlined in Chapter 21, "Occupational Health Services." Such a program includes these integral functions:

1. Preplacement physical examinations and chest X rays to place applicants on jobs most suitable to both their safety and health and that of other employees.

2. Periodic examinations and chest X rays to keep track of the employees' health, detect incipient disease, and reclassify workers as necessary.

3. Adequate first aid facilities and employee training in first aid work.

Personnel facilities

Dermatitis among foundry employees occurs in coreroom workers whose hands and arms are exposed to sand and core oil mixtures. Prolonged contact with oil, grease, acids, alkalis, and dirt can also produce dermatitis.

For these reasons, frequent washing with soap and water should be encouraged and adequate facilities should be installed.

Recommendations for toilets, washrooms, shower and locker rooms, and food service are detailed in Chapter 20, "Industrial Sanitation and Personnel Facilities." Sanitary food preparation and service is important especially in nonferrous foundries where eating should be prohibited in work areas where toxic materials are handled.

Foundry Work Environment

Housekeeping

The best housekeeping results when each individual is held responsible for maintaining order in his own work area. Time should be set aside for housekeeping when work is scheduled. Necessary housekeeping equipment should be available, and trash cans and special disposal bins should be handy and emptied regularly.

Each worker should:

1. Clean his machines and equipment after each shift and keep them reasonably clean while working.

2. Put all scrap and trash in the proper trash bins for easy removal.

3. Keep the floors and aisles in his work area free and unobstructed.

4. Properly stack and store materials he uses.

The foundry floor should be arranged for efficient operating economy and to prevent accidents, especially those caused by spills and "run-outs" of molten metal.

Floors should be cleaned frequently and kept in good condition, firm, and level. Worn spots, holes, or other defects should be reported and repaired immediately.

Special types of flooring are necessary where fires or explosions may occur or other serious hazards may exist.

Floor loading should be controlled. Many buildings are being used for purposes for which they were not designed. Mechanized materials handling introduces floor load problems because of the heavy dead weight of platform and lift trucks. The suspension of overhead cranes and hoists from wood ceiling joists severely taxes roof and floor members. Insurance engineers, local building inspectors, or private consulting engineers can help determine safe load limits.

Ventilation and heating

Control of air contaminants is the primary purpose of ventilation in foundries. The degree to which air contaminants should be controlled by a ventilating system is determined by applicable standards and codes and need for employee comfort.

The need for ventilation control may be determined by one or more of the following:

1. Reference to applicable standards, codes, and recommendations. (See Chapter 37, "Industrial Hygiene" and Chapter 38, "Industrial Toxicology.")

2. Comparison with similar operations in a like environment.

3. Collection and analysis of representative air samples taken by qualified personnel in the breathing zone of the workers.

A comfortable working temperature should be maintained in all parts of the foundry where men are at work. Natural gas, propane, or electric infrared heaters should be used for local heating.

Illumination

Good illumination is difficult to achieve in foundries because of the nature of the operations. Where craneways are used, lighting fixtures must be placed high and at considerable distances from the work areas. Nevertheless, good lighting should be provided for each work area. See in Chapter 46, "Safety Engineering Tables," for recommended levels of illumination.

Foundries having difficulty in maintaining recommended light intensities can call on their local power companies or illuminating consultants for expert information.

Inspection and maintenance

Standard inspection and maintenance procedures should be followed in foundries. Maintenance men should be carefully selected and trained in safe practices, particularly in procedures for locking out controls.

Plant structures

Entrances and exits. In some cases, it is desirable to provide entrances to heated buildings with vestibules or enclosures constructed or located to prevent harmful drafts from reaching employees and sized to permit the passage of conveyances regularly used inside the plant. This provision does not apply to entrances used for railroad or industrial cars handled by locomotives or for traveling cranes, trucks, and automobiles.

All doors, particularly double-acting swinging doors, should have a window opening not less than 8 by 8 in., located at normal eye level, to permit a view beyond the door.

Stairways. All permanent and all portable stairways having four or more risers should be provided with substantial handrails, standard guard rails, and toeboards. See American National Standard A12.1, *Safety Requirements for Floor and Wall Openings, Railings, and Toe Boards.*

Floors and pits. The floor beneath and immediately surrounding foundry melting units should be pitched away from the melting units to provide drainage. The floor should be kept free from pools of water to prevent an explosion hazard. Where water is needed on dusty operations, only enough water to hold down the dust should be used.

Pits in which molten metal is handled should be free from dampness because of the danger of explosion.

Pig molds and receiving stations for excess molten metal from ladles should be located clear of passageways and at least 1 ft above floor level. The practice of having pig holes in the floor near pouring areas is unsafe and should never be allowed.

Pits connected with ovens or furnaces and floor openings should be protected with either a cover or a standard guard rail when not in use.

Where tram or standard gage railroad tracks

903

run into or through a foundry, the top of the rails should be flush with the foundry floor, which should be kept built up to maintain this level.

Galleries where molten metal is poured into molds should be provided with solid, leakproof floors (concrete or sheet steel covered with sand) and with partitions of sheet steel. The partitions should be not less than 42 in. high and should be installed on the open sides of such galleries.

Where floor space is cramped, it is sometimes desirable to construct galleries on which to store ladles, flasks, flask boards, and other equipment. These galleries should be equipped with standard handrails and toeboards and should have sturdy stairways instead of ladders.

Gangways other than those for carrying molten metal should be at least 3 ft wide and should be kept in good condition, sufficiently firm to withstand the travel for which they are intended, uniformly smooth, and without obstructions. See American Foundrymen's Society, *Recommended Safety Practice for Protection of Workers in Foundries*.

Every gangway employed for the handling of molten metal should be kept uniformly smooth, clear of obstructions, and free from pools of water.

Concrete pavements around pouring floors should be kept coated with sand during pouring operations to reduce splattering of hot metal in case of a spill.

Gangways should have the following widths:

• For parallel travel of truck ladles or manually operated monorail ladles—not less than twice the width required for one ladle operation.

• For distribution of molten metal in trucks or manually operated monorail ladles exclusively—not less than 24 in. wider than the widest span of the ladle equipment.

• For distribution of molten metal in hand shank ladles or crucibles carried by not more than two workmen—not less than 3 ft wide. Where more than two workmen are required for this operation, gangways should be not less than 4 ft wide.

Aisles in which molten metal is handled should be kept in good condition, clear of obstructions, firm, uniformly smooth, and free from pools of water at all times.

Aisles should have the following widths:

• Where molten metal is carried in hand or bull ladles or crucibles by not more than two workmen per ladle or crucible—not less than 15 in. wide.

• Where molds alongside the aisle are more than 20 in. above aisle level—not less than 24 in. wide.

• Where there are more than two workmen per ladle of crucible—not less than 36 in. wide.

Compressed air hose

The compressed air hose presents one of the most serious foundry hazards. Improper use of the hose and "horseplay" have caused severe injuries to internal organs and eardrums.

Such unsafe practices as blowing and brushing sand from new castings without regard for the cloud of dust produced, blowing dust off patterns, and using compressed air to remove parting compounds and other light materials should be prohibited.

Vacuum methods can be substituted for compressed air cleaning of molds. Workers should be carefully instructed in safe use of air hoses where the latter method is employed. Use of nonsilica parting eliminates the possibility of silicosis from this source.

Fire protection

Foundries should organize brigades to make periodic fire inspections and to perform emergency fire-fighting before the arrival of the local fire department. A fire brigade will also aid the safety program by keeping its members, as well as other employees in the foundry, safety-conscious. See Chapter 43, "Fire Protection," and Chapter 18, "Planning for Emergencies."

Materials Handling in Foundries

Materials handling produces a wide variety of injuries in foundries. Cuts and crushing injuries involving fingers, hands, toes, and feet, fractures of legs and arms, back strains, ruptures, and burns can occur during the manual handling of scrap metals, pig iron, and similar materials.

Some of the precautions that can be taken to prevent materials handling accidents include:

1. Instructing workers in the safe methods of manual and mechanical handling of material.

2. Providing personal protective equipment —eye protection, face shields, leather mitts or gloves, preferably studded with steel (unless hot metal is to be handled), hand pads, aprons, safety shoes, and other items —to be worn as required.

3. Planning the sequence and method of handling materials to eliminate unnecessary handling.

4. Safeguarding mechanical devices and setting up inspection procedures to assure their proper maintenance.

5. Keeping good order at storage piles and bins, and piling materials properly.

6. Keeping ground and floor surfaces level so that men handling materials will have good footing.

7. Installing side stakes or side boards on tramway or railroad cars to prevent materials from falling off.

Many foundries have replaced hand methods with materials handling machinery and thus reduced exposure to manual hazards. However, the machinery usually involves hazards of its own. For instance, lifting magnets are dangerous if swung over areas where men are at work. A break in the magnet circuit could cause the magnet to release its load without warning or pieces dangling from the magnet could be jarred loose. In no case, of course, should suspended loads be carried over workmen's heads.

Handling sand, coal, and coke

Certain hazards in the handling of materials such as sand, coal, and coke can be remedied as follows:

1. Falls through hoppers while unloading bottom dump-type railroad cars can be prevented by requiring the use of safety belts and lifelines. Men should never do this work alone. Observers should always be on the scene and should be prepared to effect rescues in emergencies.

2. Using hopper car door safety ratchet wrenches can keep doors from swinging and striking workers.

3. Hand and foot injuries can be prevented by using safety car movers instead of ordinary pinch bars to spot cars by hand. (Where a locomotive is available, it should, of course, be used.)

4. By using warning targets, derails, and red lanterns at night, and by locking switches, workers can keep cars they are working on from being moved.

5. Prohibiting the undermining of piles and avoiding overhangs can reduce the danger of cave-ins of loose material.

6. Electric shock can be prevented by grounding portable belt conveyor loaders.

Some foundries have eliminated double handling by having the raw materials put directly from the cars, storage piles, or bins into unit charging trays or boxes, which are taken to the point of use and dumped mechanically. Trays or boxes to be carried overhead must be properly trimmed.

Ladles

Ladles for distributing molten metal or reservoir and mixing ladles, which are mounted on stationary supports or trucks or handled by overhead crane or monorail and which have a capacity of not more than 2000 pounds may be of the hand-shank type and should be provided with a manually operated safety lock (Fig. 33–2). Shanks should be made from solid material, and shields should be installed on them.

Ladles which have a capacity over 2000 pounds should be of the gear-operated type. Such ladles and those which are mechanically or electrically operated should be equipped with an automatic safety lock or brake to prevent overturning or uncontrolled sway.

Suitable covers should be provided on portable ladles.

The rim or lip on hand or bull ladles should be built up above the top of the metal shell

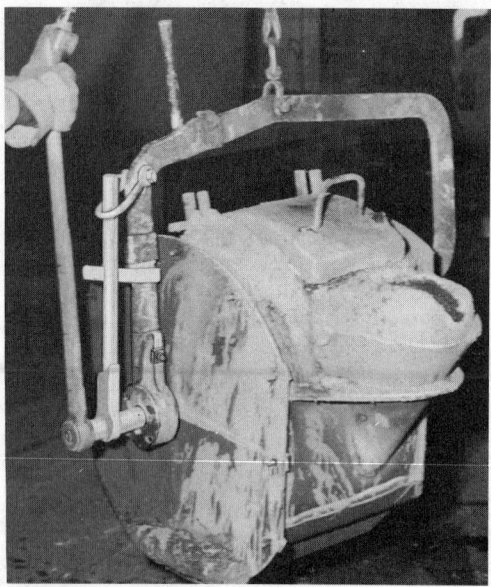

FIG. 33–2.—Tilting ladle equipped with manually operated anti-tilt level shown in white. Hoist control switch cable is at left.

with fire clay no more than ½ in. if the refractory ladle lining is less than 1½ in. thick at the rim. In any case, the maximum height of the rim or lip should be not greater than 1 in.

Ladles should be thoroughly dried out before being used. Local exhaust should be provided to control vapors or fumes produced during ladle drying. Some foundries control both vapors and excessive noise produced by gas torches used for drying by performing all operations in a ladle-drying shed located outside the foundry.

Monorail ladles and trucks used to transport molten metal ladles should be equipped with warning devices (bells or sirens) to be used whenever molten metal is being transported.

Trunnions and the devices used to attach them to flasks, buckets, ladles, and other equipment should be constructed with a factor of safety of at least 10. The diameter of the head on the outside end of the trunnion shaft should be not less than one and one-half times the diameter of the trunnion shaft. The inside corners where the trunnion shaft joins the base and the head should be filleted

to prevent the sling or hook from riding the trunnion base or head.

Inoculation or treatment of molten metal to desulfurize it or to change its composition or type, as in the making of an alloy or ductile iron, is done in the reservoir or in a pouring ladle. A hood can be made to cover this operation so that the workmen are effectively shielded from possible spatters of metal caused by the violence of the reaction. The resulting fumes are drawn off and exhausted through a stack. (Details for local exhaust removal of contaminants generated in this operation may be found in *Engineering Manual for Control of In-Plant Environment in Foundries,* published by American Foundrymen's Society.)

Hoists and cranes

Hoists and cranes that handle molten metal require a preventive maintenance program, conducted by men trained and thoroughly familiar with the equipment. (This is in addition to observations and inspections made by supervisors and operators.) The degree to which the program should be carried out depends on both the equipment being used and the tonnage moved. The program should be geared to making sure the operation is safe, not to a minimum compliance with existing regulations.

For example, a program for a 300-employee, ordinary gray iron foundry could require monthly inspection of crane and hoist structures, and an inspection of wire ropes and hooks before every shift.

Because of the severe stresses and demanding service required in some high-tonnage operations, inspection programs become more elaborate. Some programs regularly schedule nondestructive testing—ultrasonic testing of crane hoist shafts and parts, and dye penetrant inspection for surface cracks on bales, dumping chains, clevises, and pins. (See the discussion under Nondestructive Testing later in this chapter.)

An effective program of preventive maintenance can be arrived at by consulting recommended safe practices. (See Chapter 25, "Hoisting Apparatus and Conveyors," and References.)

Conveyors

Conveyorized systems are now being used

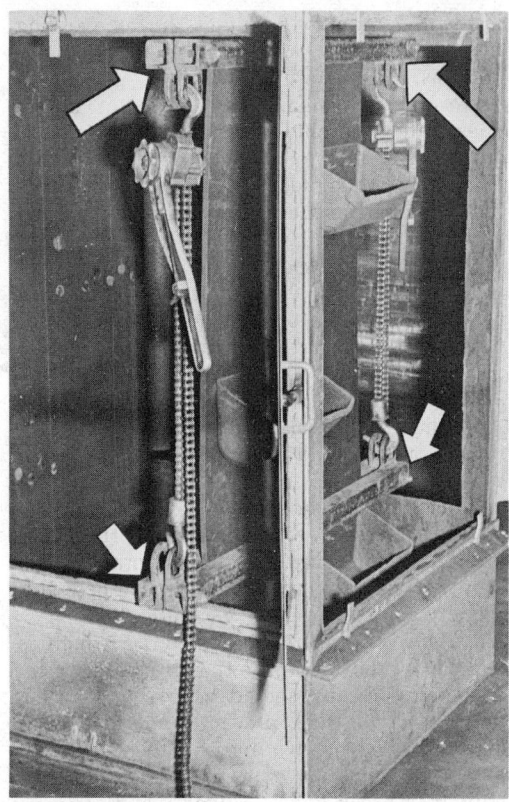

FIG. 33–3.—Bucket conveyor belt locked in position by squeeze clamps at top and bottom of elevator shaft opening. Clamps prevent movement of the belt during maintenance or repairs.

Courtesy McKinnon Industries, Ltd.

in a number of foundries. Sand mixed in the mixing room is carried by belt conveyor to hoppers at molding stations, where each hopper is filled by means of a movable plow. Surplus sand is carried to the end of the belt and returned by bucket to the storage bin (Fig. 33–3).

An endless conveyor is used to handle molds. Empty flasks come from the shake-out to the molders, who remove them and make their molds on molding machines, taking sand from the overhead hoppers as required. Spilled sand goes through a grating onto a belt conveyor which returns it to the mixing room. Thus all shoveling operations are eliminated.

The molds are then placed on a conveyor

(Fig. 33–4) and passed into the pouring room. The pourers get their metal from the cupola and then step onto a moving platform geared to an endless single-rail conveyor moving at the same rate of speed, on which hand ladles can be supported. Pouring is done as the men move along, and they are as safe as they would be if they were standing still.

An electric switch should be installed near the end of the conveyor, so that if a worker rides that far, his foot will come in contact with the switch and the conveyor will stop.

The mold conveyors then pass into a cooling zone. Weights can be removed from the molds by a mechanical device and returned by another conveyor to the place where they are originally used.

The molds then move to the shake-out, where sand is dumped from flasks onto a vibrating grating. The sand falls to a belt conveyor, which returns it to the mixing room. Using a hook, a workman pulls the

FIG. 33–4.—Safe, production-line pouring of brass castings is done from platform that moves at same rate as molds passing by.

Courtesy Link-Belt Co.

907

castings onto another conveyor, which takes them to the tumbling barrels. Empty flasks are brought back to the molding section by still another conveyor.

This is a complete system for mass production where each man has one job to do rather than several. In the installation of such a system, shear points, crush points, and moving parts must be effectively guarded.

Where conveyor systems run over passageways and working areas, the employees beneath should be protected by screens, grilles, or guards strong enough to resist the impact of the heaviest piece handled by the conveyor.

Where chain conveyors operate at various levels, other than in a fixed horizontal plane, a mechanism of safety dogs should be installed in accordance with applicable standards on both the upgrade and the downgrade. In case of chain failure, the safety dogs will hold the chain and prevent the load from piling up at the bottom of the incline.

Elevators and monorails

Elevator openings at floor levels should be protected by manually or mechanically operated elevator gates at least 6 ft high. Slots should be no more than 1 in. wide, and the gates should extend as close to the floor as practicable.

The drive mechanism should be electrically or mechanically interlocked with the gates so that the elevator cannot be started until the gate is closed and so that the gate cannot be opened until the elevator is at floor level.

Elevators, except for those that are hydraulically operated, should be equipped with safety blocks controlled by a speed governor that will hold the elevator in case of cable failure or overspeeding.

All types of elevators should have upper and lower travel limit devices.

Monorail systems should be provided with stops or interlocks that will prevent the trolley equipment from running off at open switches. Permanent rail stops should be installed at the ends of all such equipment.

Scrap breakers

Shears should be guarded to protect operators and passersby from flying particles. The working floor should be clear and level.

Use of a drop to break castings or scrap

inside the foundry buildings during working hours should be prohibited unless such operations are performed within a permanent enclosure of planking or equivalent strong enough to withstand the most severe impacts from the demolition weight or flying scrap. The enclosure should be high enough to protect workmen in the vicinity from flying fragments of metal.

If a rope is used, it should extend over pulleys to a point clear of the breaking area to assure that the operator will be at a safe distance and to prevent entanglement.

Storage

Foundry materials and equipment not in regular use should be stored in the space provided in a safe, orderly manner on level and substantial foundations.

When workmen remove materials from bins located at floor level or from storage piles, they should not undermine the piles and thereby cause cave-ins.

Hopper bins containing material which is fed out at the bottom, either by hand or by mechanical means, should be covered with a grating which will prevent workmen from entering the bin. No one should be allowed to get on the rails of a bin nor to enter a hopper to break down bridged material.

A worker who must enter a bin should wear a safety belt with attached lifeline, tended by a man similarly equipped.

Pattern storage buildings should have racks and shelves substantial enough to hold the loads placed upon them. The floors and stairways should be well designed and kept in good condition. The storage area should be well lighted. The pattern keeper should be provided with a sound ladder so that he can safely reach the patterns.

The pattern keeper, who is likely to be alone in the pattern storage area, should report at regular intervals to the molding floor supervisor.

Flammable liquids should be stored in accordance with National Fire Protection Association Standard No. 30, *Flammable & Combustible Liquids Code.* See Chapter 42, "Flammable and Combustible Liquids."

Slag disposal

Furnaces and pits should be designed to

have removable receptacles into which slag and kish (separated graphite) may flow or be dumped. Enough of these receptacles should be provided so that slag can solidify before it is dumped, unless it is disposed of in the molten state.

To decrease the amount of slag that goes into the slag pits, slag or cinder pots should be used. The pots can be set aside and allowed to cool, to eliminate the danger of explosion when they are emptied.

Open hearth slag ladles should not be dumped until 20 hr after they have been filled. The slag should be dumped where there is absolutely no water or dampness, which might cause an explosion if some of the slag is still molten. The slag should then be allowed to stand for several hours before being broken up so that there will be no molten slag in the center.

Cupolas

Charging

The dangers in the charging of cupolas are principally confined to the handling of material. Barrows or buggies should never be unevenly loaded or overloaded. "Tip-up" barrows used for charging coke are sometimes so poorly balanced that they will not stay in the tipped-up position after being emptied, but fall back on the chargers' feet at the slightest touch. The preventive measure is to lower the center of gravity.

The use of mechanical devices for charging cupolas not only saves labor but also reduces the number of materials handling accidents. Most foundry cupolas are now charged by either fully automatic charging machines equipped with crane and cone bottom buckets or by lift trucks equipped with tilting boxes.

The charging opening on some cupolas is covered by a door or chain curtain, which should be kept closed except for charging.

The space underneath cupola charging elevators, machines, lift hoists, skip hoists, and cranes should be railed off or guarded to prevent material from dropping onto workers during charging operations.

Occasionally, during idle periods, men may rest under the charging platforms, close to the warm chambers and flues. This practice

should be prohibited because of the danger of objects falling from the platform and the possibility of carbon monoxide escaping from the flues. Clothes lockers should not be located under these platforms.

Scrap cylinders, tanks, drums, and the like should be broken open before charging to prevent explosion in the cupola.

Charging floor

Charging floors should be kept free from loose materials and storage racks should be provided for equipment not in use.

Steel floor plates should be substantial enough not to turn up and should be securely riveted in place. Steel plate flooring in the immediate vicinity of the furnace becomes extremely hot. Therefore, brick flooring laid on substantial steel framework is more satisfactory.

Standard railings 42 in. high and 4-in. toeboards should be provided around all floor openings.

Since railings on the charging floor are likely to be subjected to considerable abuse, it is usually best to construct them of angle iron, which is more easily repaired than pipe railings. At the tapping platforms, hinged gates or chains which may be hooked in place must often be provided.

Where cupolas are manually charged, a guard rail should be placed across the charging opening.

Where cupolas are charged with wheelbarrows or cars, a curb of a height equal to the radius of the wheel of the barrow or car should be provided to prevent the barrow or car from pitching over and falling into it.

Carbon monoxide

Carbon monoxide (CO) is generated during some cycles in the operation of a cupola. It is an explosion hazard if it gets into the wind boxes and blast pipes when the blowers are shut down. To eliminate this hazard, adequate natural or mechanical ventilation should be supplied in back of the cupola, and two or more tuyeres should be opened after shutdown.

The large amount of blast air in the cupola generally carries the carbon monoxide out the stack. In some cases, it is burned in the stack before it can be discharged. Sometimes, however, the gas may escape. It gives no warning, so CO indicators that light and give a loud sound should be located about the cupola.

Approved gas masks (see Chapter 19, "Personal Protective Equipment") should be at hand, in good condition, and workers trained in their use. If concentration of CO is more than 2 percent, self-contained breathing apparatus or a hosemask (with blower) should be provided.

Blast gates

Blast gates and explosion doors are used successfully to prevent damage from gas explosions. They are sometimes placed in front of the tuyeres, so that they can be opened to admit fresh air when the blowers are shut down. They should never be closed until the blast has entered the wind box and driven out all gas.

Blast gates should be provided in the air blast pipe that supplies air to the melting equipment. They should be closed when the air supply fails or when the melting equipment is shut down, in order to prevent the accumulation of combustible gases in the air supply system. In the cupola, the blast gate may be omitted if alternate tuyeres are opened to permit air circulation.

Blast gates should be so located in relation to the cupola wind box that the duct volume will be kept to a minimum. Motorized dampers may be installed at centrifugal blowers so that they will close automatically when the air supply fails.

Positive-pressure blowers should be equipped with safety valves having liberal discharge areas. If these are not provided, clogging of the cupola with slag or quick closing of the gate or damper may produce sufficient pressure in the blast pipe to burst it.

Every cupola should have at least one safety-type tuyere, with a small channel 1½ or 2 in. below the normal tuyere level. This channel has a fusible plate that will melt through should the slag and iron rise to an unsafe level.

Tapping out

Tapping out with safety requires skill and should be done only by experienced and dependable men. In "botting-up" the hole, the bott should not be thrust directly into the stream of molten metal or it will cause spattering. To eliminate this, place the bott immediately over the stream of metal close to the hole and aim it down toward the hole at a sharp angle.

A supply of botts ready for use should be kept within convenient reach of the man who does the tapping.

When the cupola is tapped, the back end of the tapping bar should be held below the level of the hole, to prevent puncturing the sand bed and causing run-out through the bottom.

A tilting spout placed with one end directly beneath the stationary cupola spout and mounted on trunnions on a stand adds to safety and efficiency. It can be tilted back and forth by a foot lever. The rear end of the tilting spout is closed so that when that end is tilted down, it forms a reservoir to receive the molten metal from the cupola. When the supplementary spout is tilted forward, the metal runs from it into the waiting ladle. At the same time more metal continues to run into the spout from the cupola. Thus, the stream of metal runs from the cupola continuously, and the tilting spout acts as a reservoir between ladle loads.

The slag spout of the cupola should be equipped with a shield or guard to protect workers from sprays of molten slag and to form a hood to collect slag wool.

The slag wool is sometimes collected through a wet slagging system wherein slag is thrown off into a water-filled container or trough and flushed away.

Dropping cupola bottom

When the cupola is in operation, its bottom doors should be supported by one solid prop and two adjustable screw props (of the required structural strength) on a metal prop base set on a concrete or other fabricated footing of equivalent strength (Fig. 33–5).

Temporary supports (timbers, blocking, etc.) should be placed under the cupola bottom doors to prevent their falling on employees while the metal props are being adjusted to proper height. Mechanical means for raising the bottom doors of the cupola should be provided (Fig. 33–6).

Dropping the bottom doors of a cupola requires extraordinary care. One of the best methods is to use a block and tackle with a wire rope and chain leader attached to the props which support the doors. The props can then be pulled out by means of the block and tackle from a safe distance or from behind a suitable barrier (Fig. 33–7).

Before the bottom is dropped, the area

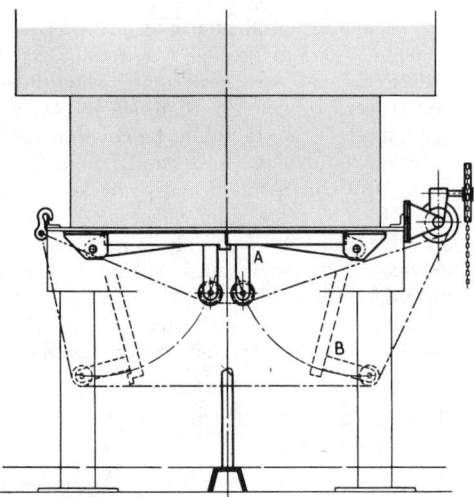

FIG. 33–6.—Suggested method of raising bottom doors of cupola by mechanical means.

underneath the cupola should be carefully inspected to see that no water has seeped under the sand. One man should make sure that no one is in the danger zone and that workers stay away during the operation. Employees should be warned by means of a whistle or other signal before the bottom is dropped.

If the cupola bottom doors fail to drop or if the remaining charge inside the cupola bridges over, employees should never be permitted to enter the danger zone to force the doors or relieve the bridging.

The bridging may be relieved by turning on the blast fan. The vibration produced usually corrects the condition. A mechanical vibrator attached to the bottom doors is also effective. Another method is to drop a demolition ball from the charging door. If these methods fail, the doors must be flame cut with a lance, but only after the cupola has cooled to a safe temperature.

Special bottom door locking devices may be used if the cupola drop is to be caught in a container, car, or skid.

Repairing linings

Only careful and experienced men should be allowed to make repairs inside cupolas. The precautions to be followed are:

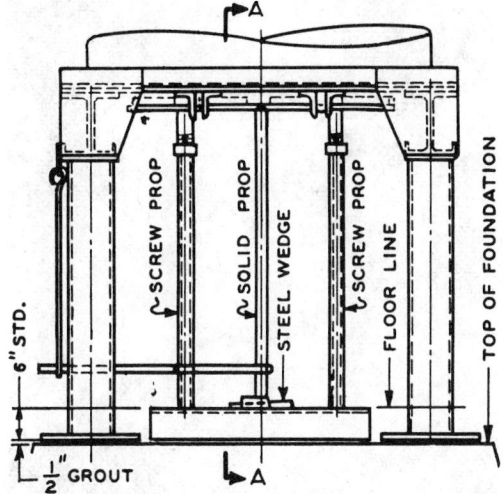

NOTE :– SCREW PROPS MUST BE REMOVED BEFORE REMOVING SOLID PROP

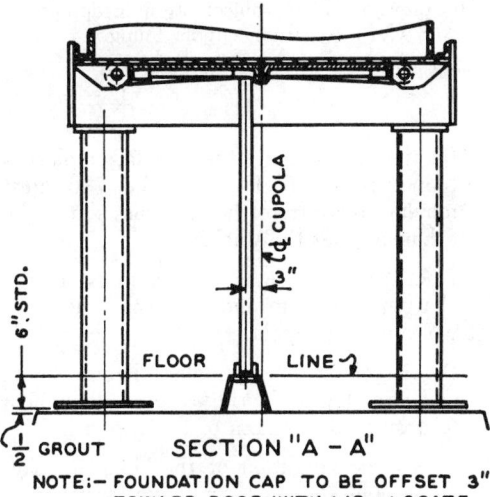

NOTE:– FOUNDATION CAP TO BE OFFSET 3" TOWARD DOOR WITH LIP – LOCATE AFTER CUPOLA IS IN PLACE

FIG. 33–5.—Proper method of supporting cupola bottom doors.

1. Install a substantial guard over the cupola charging door to protect workmen against falling objects. Such a guard should be constructed of not less than 1½ by 1½ by ¼-in. angle iron. It should be covered with a screen equivalent in strength to a 1-in. mesh of ³⁄₁₆-in. wire or with not less than No. 16 U.S. gage solid sheet steel. The guard should be securely supported by means of overhead slings or underpinning to resist falling objects (Fig. 33–8). An alternate method is to place a solid steel plate on the bucket ring in the cupola at charging door level.

2. Require all workers who enter a cupola to wear hard hats.

3. Provide approved respiratory equipment

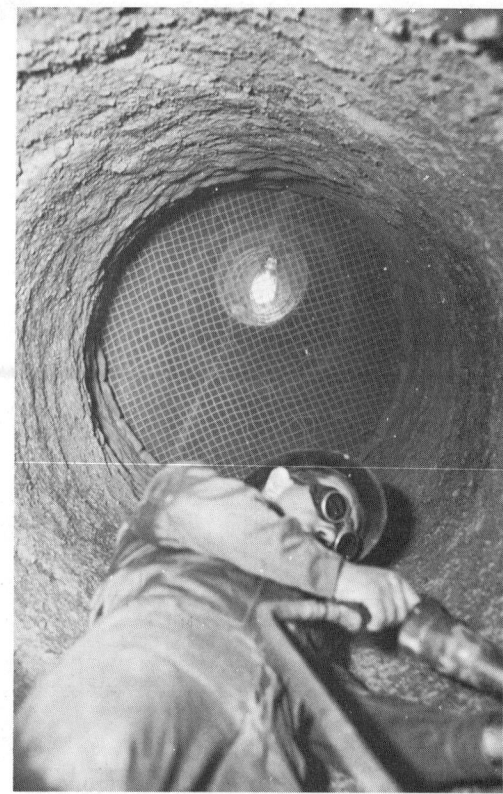

Fig. 33–8.—A screen placed over the charging door prevents falling objects from dropping on worker who is repairing cupola lining.

Courtesy Hamilton Foundry & Machinery Co.

Fig. 33–7.—Wire rope with hook being secured to solid prop beneath cupola doors.

Courtesy International Harvester Co.

for the men relining the cupola, or place a blower fan in the bottom to keep the dust moving away from them. Dust should be exhausted out the stack.

4. Place warning signs or crossbars at the charging door to indicate that men are working in a cupola.

5. Before relining of a cupola is begun, break all loose slag and bridges and allow them to drop to the bottom of the cupola.

6. Check the condition of the shell and riveting while the cupola is down for relining. A weak shell is likely to increase the risk of a gas explosion.

7. Leave ample clearance (at least ¾ in.)

between the new brick lining and the shell to allow for expansion. This space should be filled with dry sand to serve as a cushion to protect the shell against severe stresses.

8. Before the cupola is started up, check to make sure that all personnel have made their exit from within the cupola and the area beneath it, that the lining is thoroughly dry, and that all tools and other equipment have been removed.

9. Request all cupola tenders and other individuals working apart from a group to report to their foremen at regular intervals as a safety check practice.

Open Hearth Furnaces

The hazards of open hearth furnaces are somewhat different from those presented by other types of melting equipment because of the larger quantities of metal which are usually handled by mechanical methods.

Furnace men are exposed to several types of eye hazards. For protection against the frequent splashes of molten metal, the men should wear special glasses to enable them to look into the furnaces. An aluminum shield installed between the furnace and operator will protect against infrared radiation.

Another possibility for eye injury occurs in the breaking of test bars. Use of a special test bar breaking device will eliminate this hazard.

Charging door cables and sheave supports should be well designed and kept in the best condition, to prevent them from giving way. Proper ventilation on the charging and tapping platforms without undue draft is important. Tapping-hole shields of sheet metal or chains that can be swung out of the way when necessary should be provided.

Charging machines

The average open hearth charging machine is somewhat like an electric traveling crane in its fundamental arrangement and many of the safety engineering features used in these cranes are applicable.

Since the noise of open hearth furnaces, supplemented by the noise of nearby gas producers, is often sufficient to drown out the noise of the charging machine as it travels back and forth, automatic warning gongs which ring continuously while the charging machine is in motion should be installed. Wheel guards should also be provided in front of the track wheels to clear the rails ahead of the machine.

During charging operations, splashes of molten metal may expose both the charging machine operator and his helpers to burn and clothing fires. To protect against this danger, a screen may be installed which drops down between the operator and the furnace as the machine goes through the charging operation.

Also, fire-resistant coats and hoods should be worn by operator and helpers.

There should be at least 3 ft of clearance between the charging machine and permanent objects. On some charging machines, plate guards at least 3 in. high must be provided along both sides of the traversing carriage runway, to keep men from placing their feet on the track. Bumpers installed 2 in. from the endplate on each traversing carriage track will prevent the carriage from running into the endplate.

Charging boxes

Charging cars should have angle irons or other equally effective stops riveted along the edges to prevent the boxes from slipping off.

Proper scheduling of the charging operations and good supervision will prevent overloading of the boxes and falls of material from them.

To prevent water from accumulating in charging boxes, a number of holes should be drilled in the bottom of each box. This will also keep molten metal from splashing when the furnace is charged.

Charging boxes should be inspected periodically for weakened ends which may break.

Lifting chains should not be looped around one end of the box, because the box may tilt when hoisted and drop material.

The charging-box cars should be equipped with automatic couplers; link-and-pin couplers are a source of many hand injuries. When automatic couplers are used, a heavy piece of angle iron which extends out over the lift pin should be riveted to the car frame to protect the pin from damage if heavy pieces should fall from charging boxes.

Tapping out

A tapping platform on pivot-type hinges may be installed which can be swung out in front of the furnace after the ladle has been placed in position so that men can tap the furnace with a long bar.

A rope should be attached to one outer corner of the platform so when the metal starts flowing, a man on the floor can swing the platform back against the side of the furnace, allowing the men who have done the tapping to get back safely without having to jump down.

For tapping acid furnaces, substantial tapping platforms, either permanent or portable, should be provided.

In jet tapping, safe procedures must be followed. The manufacturer of the equipment may be consulted for specific information. (Also see NSC Data Sheet 394, *Jet Tapping of Open Hearth Furnaces.*)

A whistle or gong installed in back of the furnace can be sounded several times to allow ample time for all men to get out of the pit. This is a good way to warn men the furnace is about to be tapped.

Ladle additions should not be thrown into the ladle from the tapping platform with shovels or in bags. This is a hazardous practice if the platform has no railings. A chute to add materials to the ladle is preferable. Such a chute may be constructed to feed material into the flowing stream of molten metal from the furnace spout or to swing out over the ladle and feed directly into it. Ladle additions, particularly charcoal, should not be made until there is some molten metal in the ladle.

An oxygen torch can be used to meet a number of emergency conditions in furnace work. For instance, it can open the way for a stuck tapping bar, cut away the dam formed when the tapping hole is a little above the furnace bottom, or open frozen tuyeres.

When tap-out holes are to be lanced or burned out with oxygen, a guard should be placed in front of the tap hole to prevent hot metal from being blown onto the workmen performing the operations.

Crucibles

The principal danger in handling refractory clay crucibles is that one may break when full of molten metal.

All new crucibles should be inspected for cracks, thin spots, and other flaws by a competent inspector. Those showing signs of dampness should be put aside to be returned to the manufacturer. Examination of the packages and car in which they were shipped may reveal whether or not they were exposed to moisture in transit.

Crucibles should be stored in a warm (about 250 F), dry place and protected from moist air as much as possible. It is generally best to place them in an oven built on top of a core oven or at some other point where waste heat can be utilized. If not all crucibles in stock can be kept in ovens, those stored elsewhere should be accurately dated and the oldest and best-seasoned crucibles should be selected first.

In the annealing process, the crucibles are brought up to red heat very slowly and uniformly, usually over a period of eight to ten hours. Crucibles should not be allowed to cool before being charged, because as they cool they may again absorb moisture. Moisture in the walls of crucibles that are heated quickly is converted into steam, which expands and causes cracks or ruptures and may also cause pinholes or "skelping." Neither damp or high-sulfur coke or coal nor fuel oil containing excessive moisture should be used to heat crucibles.

Too high a percentage of sulfur in the fuel used in the drying or annealing process is likely to cause fine cracks, sometimes called "alligator cracks." Too little oil or too much air or steam used at the burners of oil furnaces tends to oxidize a portion of the graphite in the crucible wall, leaving the binding material somewhat porous.

Charging and handling

Proper care of crucibles is good economy as well as good safety. Since crucibles are costly, they should be made to last through as many heats as possible.

Crucibles should be charged carefully. Ingots should not be thrown in with such force that they may bend the bottom or walls of the crucible out of shape. Neither should the ingots be forced into the crucibles so as to become wedged or jammed.

New crucibles should be heated very slowly for the first few runs, especially the first. Because crucibles are soft at white heat and easily forced out of shape, they must be handled with great care.

In hoisting larger crucibles with air or electric hoists, it is preferable to have one man at each sling and one man controlling the hoist.

Tongs of the proper size and shape for the particular crucible should be selected to prevent damage to it. Tongs should fit well around the bilge or belly of the crucible and should extend to within a few inches of the bottom.

Before tongs are applied, the sides of the crucible should be checked to see that no clinkers are adhering to them. Tong rings should never be driven down tight with a skimmer or other tool, because this practice is almost certain to squeeze the crucible out of shape and produce cracks and fissures.

At least two pairs of tongs should be provided for each size of crucible so that if one pair becomes bent, the other will be available.

The blacksmith should have a complete set of cast iron forms in the exact shapes of the crucibles used. Then he will have only to heat the tongs to red heat, clamp them onto the forms, and bring them into exact shape with a heavy hammer.

Ramming of the fuel bed around a crucible should be avoided. If it becomes necessary, it should be done cautiously and only by experienced workmen. Crucibles should be supported on foundations or pedestals of firebrick, graphite, or other infusible material.

The removal of heavy crucibles from furnaces calls not only for special skill but also for physical strength. If the men are overstrained, serious accidents are likely to result. Where possible, therefore, a mechanical device should be employed for removing heavy crucibles—those exceeding 100 lb in combined weight of crucible, tongs, and metal (Fig. 33–9).

Crucible furnaces

The operation of crucible furnaces is a relatively hazard-free operation if suitable exhaust hoods are installed on all furnaces used for melting metals which give off harmful fumes.

Upright furnaces having crown plates more than 12 in. above the surrounding floor should

Fig. 33–9.—Mechanical lifting device is used to move crucible, which has been covered to prevent escape of fumes.

Courtesy American Brake Shoe Co.

be equipped with metal platforms having standard rails. Such platforms should extend along the front and sides of the furnace flush with the crown plates and clear of all obstructions. Crucibles containing molten metal should be lowered from these platforms mechanically.

Many crucible furnaces are oil fired, and unless the motors driving the oil pump and the air supply for the furnaces are connected to the same source of power, a considerable quantity of oil may flow onto the floor if the power goes off the air line.

One remedy is to put a gate valve in the oil supply line so that in case of an air failure, oil can be shut off from the entire battery of crucible furnaces in one operation. If the pilot and burner are electrically controlled, this valve can be so arranged that it will be similar to an interlock.

A mechanical shutoff can be made by the installation in the oil line of a gate lever-operated valve which is closed by the release of a weight. Since the weight is held up normally by an air cylinder, the oil supply is stopped at once when the air goes off. A small hole may be drilled in the cylinder so that air in it may be released promptly when the oil pressure goes off.

Ovens

The principal hazards in the construction and operation of core ovens and mold drying ovens are excess smoke, gases, and fumes given off into the foundry atmosphere. Other unsafe conditions are unprotected firing pits, unguarded vertical sliding doors or their counterweights (which may drop on workers), and flashbacks from fire boxes.

Firing pits into which men may fall should be guarded with substantial railings or with grating covers having trap doors to give access to the steps leading down into the pits.

Safe types of vertical sliding doors should be installed. Cables and chains having a high factor of safety, substantial cable and chain fastenings, large sheaves to prevent undue wear of cables, and guarded counterweights are also necessary. All sliding doors should be thoroughly inspected at frequent intervals. Safety dogs may be used to hold the doors in raised position.

Many vertical core ovens are 60 to 70 ft high and have a driving mechanism located at the top. Some foundries have installed steel stairways with standard treads wherever possible. Where metal ladders must be used, not less than 6 in. of clearance should be maintained from the center of each ladder rung to the side of the oven.

A caution sign should be placed on the wall near core ovens, giving necessary precautions and manufacturer's instructions to be observed when ovens are being lighted.

Gas-fired ovens

Gas-fired ovens should, whenever possible, be located in a room separate from the molding floors and also from the coremaking room by a partition. This measure will help prevent equipment failures caused by sand in the controls.

Blast tip pipe burners should be equipped with baffles to keep sand out of the tip and also to spread the flame. Tips should also be horizontal to protect them from sand.

Safety pilot valves will prevent the flow of unburned gas into the oven combustion chamber should the burner pilot light go out, or should a cock or burner be accidentally opened. A valve of this nature should be installed on every gas-burning furnace or oven.

A bleeder valve in the line between two control valves close to the burner is an additional safety device. The operator can then allow gas to escape safely to the atmosphere instead of to the fire box, should there be leakage past the main control valve when the burner is not in use.

Ventilation

Where fumes, gases, and smoke are emitted from drying ovens, hoods and ducts, exhaust fans, or other means for removal should be provided near the oven doors to keep concentrations below toxic and irritant levels.

To prevent flashbacks, flues should be adequate in size and kept free from soot. Where oil burners are used, the type of equipment and the arrangement and control of drafts should ensure as perfect combustion as possible. In some installations, forced draft equipment may be needed.

Core ovens should be equipped with explosion vents in the ratio of 1 sq ft of vent area to each 15 cu ft of oven volume. Lightweight panels may be installed on the top of the oven, or the oven may have hinged doors with explosion latches.

Natural draft ventilation is usually considered adequate for ovens under 500 cu ft in volume, but the doors must be arranged so that they have to be wide open when burners are being lighted.

Larger ovens, particularly those with vertically sliding doors and other heavy construction, should have forced draft ventilation. The ventilation system should be interlocked with the gas supply through a time relay arranged to allow for at least three complete changes of air in the oven before the burners are lighted.

Inspection

Before foundry core ovens are lighted, the

ovens and burners should be given a thorough-going inspection. Only trained and qualified personnel should do this work. This equipment should already be covered by a preventive maintenance and inspection program.

The following checklists represent an example procedure, and should be helpful as a guide that includes many of the points of good inspection practice. First, the main valve controlling the fuel supply should be shut off (which automatically turns off the gas line to the burner pilots), and the following items should then be checked:

1. Do the safety cutoff valves for gas and oil close when the pilot burners fail? If not, check for stuck valves or unusually long flame electrodes.

2. Do the red signal lights on the flame-detecting device light up? If not, check first for burned-out signal lamps and then for defective relays on the flame-detecting device.

3. Does the oven-tender warning light on the back of the control panel light up? (This light is mounted so that it can readily be seen throughout the core room.)

4. Does the warning horn blow? Can it be shut off by manipulating the safety cutoff valve handle? (It should not go off.)

Other points to be checked with the main fuel valve and burner pilots turned off include:

1. Are spark plugs clean and ignition wires in good condition?

2. Are OPEN and SHUT markings on gas and oil safety shutoff valves legible?

3. Are there core oil resin deposits on safety shutoff valves or the pilot gas solenoid valve? (Extensive deposits may cause valves to stick.)

4. Are all cover retaining screws in place on flame-detecting devices and flame electrode holders to prevent entry of dust or core oil resins?

5. Are all motor controller, relay box, and wiring junction box covers tightly in place?

6. Is there excess fuel oil or an oil-soaked accumulation of dirt or rags on the burner deck?

7. Are fuel oil atomizing heads removed from the burners, cleaned if necessary, and checked for burned tips?

8. Is atomizing air tube removed and are swirl vanes and the main burner nozzle checked for burned parts or accumulations?

9. If the oven is being operated on oil, is the three-way cock between the gas control valve and the burners turned to permit outside air to be pulled into the normal gas inlet to the burners?

10. Are there any broken refractory burner blocks?

11. Are gas, oil, and air control valves, valve control motor, and valve control linkage checked for loose valve adjustment screws or broken valve cam springs?

12. Is the combustion chamber checked for defective or burned-out arches, side walls, combustion wall, or metal tie bars which may span the combustion chamber above the arches?

13. Are all electrical contacts in disconnect switches, motor controllers, and relays inspected? (After inspection, replace all enclosure covers securely.)

14. Are temperature controller, controller relays, and control resistance lamp (if one is used) checked? Make sure that the burner control motor returns to low burner position when the controller disconnect switch is in OFF position.

15. Are covers from all air flow switches removed and gummy or worn parts, damaged wires, or damaged mercury tube elements removed?

16. Has performance of flow switches been checked by starting the proper fans and observing the flow switch operation?

17. As fans are started, have they been checked for V-belt squeal or fan vibration?

18. Is it certain that fans can be started only

in proper sequence as follows:

a) Main circulation fan,

b) Power exhaust fan for drying zone,

c) Combustion air blower,

d) Similar fans for additional burner units, if used, and

e) Cooling zone supply and exhaust fans?

19. Has the operation of the purge duct damper motor and linkage at the start and end of the purge cycle been checked?

20. The purge cycle time should be 10 minutes on all ovens (except horizontal drying oven burner units, which have 5-minute cycles). Have red signal lights of the flame-detecting device come on at the end of the purge cycle?

21. Do spark plugs and flame electrodes project approximately 2 in. beyond the burner-pilot casting? (The flame electrode should be centered in the pilot orifice and set to one side about ¼ in. at its tip.)

22. Have all explosion doors been checked for proper operation using a light to check for evidence of damage to internal structures or ductwork? (Make sure that the asbestos paper explosion panel in the roof is intact.)

Next, the main fuel valve should be turned on (which opens the pilot gas valve), the ignition button pressed, and the following points checked:

1. Do the pilot lights ignite readily?

2. Are there fuel oil leaks in the piping around the burners?

3. Is there an odor of leaking gas around the burners, the gas regulators, or the gas valves? (The various gas valves should be tested for leaks periodically with a soap solution.)

4. Is there any indication of flame flashback to the outside from either the pilot burner or the main burners?

All fans should be inspected for these items:

1. Using a tachometer, check the rmp against the proper speed shown on the metal tag on the fan drive guard. If fan speed is too low, tighten the belts and recheck the speed.

2. Check all V-belt drives for proper belt tension. (A tag showing the V-belt size should be mounted on each belt guard.)

3. All belt guards should be in good condition and securely fastened in place.

4. Fans should be checked for unnecessary vibration.

5. Air leakage at fans or attached ductwork should be overcome.

6. Fan motors should be clean and securely bolted in place.

7. Damper motors, control valves, and control linkages must be securely in place.

Foundry Production Equipment

On production-line equipment, moving parts (such as belts, pulleys, gears, chains, and sprockets) and other common machine hazards (such as projecting setscrews) should be fully guarded in accordance with standard practice (see Chapter 29, "Point-of-Operation and Transmission Guards").

Repairs should only be done on equipment that is locked in the OFF position. Electrical equipment should be grounded to eliminate shock hazards.

Some operations require mills, mixers, and cutters of such size that an employee can enter the machine for cleaning or repair. In these cases, a lockout procedure must be set up and enforced. See section titled Lockouts in Chapter 41. "Electrical Hazards," and NSC Data Sheet 237, *Method of Locking Out Electric Switches.*

Sand mills

The principal danger of sand mills (mullers) exists when operators reach in for samples of sand or attempt to shovel out sand while the mill is running. In doing so, they may be caught and pulled into the mill. To protect against this hazard, one or more of the following measures may be used:

1. Provide screen enclosures for charging and discharging openings of mills.

2. Install self-discharging mills, or equip mills with discharge gates or scoops.

3. Provide sampling cones for taking samples of the sand during the mixing operation.

4. Prohibit the shoveling of sand out of mills while they are running.

5. Install an interlocking device so that the mill cannot be operated until the doors are closed.

Dough-type mixers

To prevent men from reaching into a dough-type mixer while the blades are in motion, the top of the mixer should be covered with a substantial grating made of ⅜-in.-round bars or their equivalent.

An interlocking device so arranged that the cover cannot be opened nor the bowl tilted until the blade drive mechanism has been shut off and so that the blades cannot be set in motion again until the cover is in place is advisable.

Where the mixer is driven by an individual motor, a small steel cable can be attached to the cover and extended over a pulley to a counterweight. This cable is attached to a ring on the motor control switch handle so that when the cover of the mixer is lifted a predetermined distance, the switch is pulled open and cannot be closed again until the cover is back in place.

Sand cutters

Sand cutters throw sand and pieces of tramp metal with bullet-like force, sometimes causing serious puncture wounds. If a guard that would not seriously impair the efficiency of the operation cannot be devised, then suitable personal protective equipment should be worn by the operator.

Since it is often difficult to operate a power-driven cutter on a sand floor, parallel concrete strips can be installed to act as runways for the cutter wheels.

Sifters

Rotary sand sifters should be guarded by enclosures or by angle iron or pipe railings 42 in. high placed from 15 to 20 in. from the sifter. Belt shifters and motor control switches should be placed within convenient

FIG. 33–10.—To avoid crawling under heavy sand core, inspector uses illuminated stainless steel mirror to check the bottom of the sand core.

Courtesy American Brake Shoe Co.

reach. The control switches should be so designed that they cannot be actuated accidentally.

Portable sand sifters equipped with pneumatic vibrators are usually slower moving than those equipped with electric vibrators, and oscillation of heavy parts causes the entire machine to move around the floor in jerky fashion. Often, machine travel is limited only by the air hose. If the hose coupling breaks, the hose flails around and blows sand in every direction, presenting a particular hazard to workers' eyes. Anchoring the sifter with a rope a little shorter than the hose helps to prevent such accidents.

Molds and cores

Letting flasks down on feet, pinching fingers between flasks, dropping heavy core boxes on feet, cutting hands on nails and other sharp

919

Fɪɢ. 33–11.—Each operator must depress his own dual controls to actuate this straight molding machine.
Courtesy General Motors Corp.

pieces of metal in the sand, and stepping on nails constitute the principal hazards in hand molding and core making. Most of the remedies are obvious.

Hand and foot injuries can be minimized if workers use proper methods of handling flasks and core boxes and wear good safety shoes with stout soles. Screening or magnetic separation to remove nails and other sharp metal from the sand is also essential to safety.

In general molding and core making, gagger rods and core wires are cut, straightened, and bent by means of hammers and cutting sets. This operation presents danger from flying pieces of metal and dirt. Machines are available for performing this work, but many of their hazards are similar to those found in the use of hand tools.

Heavy cores in large molds must be carefully braced as the work progresses to keep the cores from toppling over.

Working underneath molds suspended from cranes should be *prohibited*. Substantial tripod supports or wooden or steel horses will provide greater safety and efficiency (Fig. 33–10).

Proper venting of molds is essential if explosions are to be averted during pouring. When the sand in an undried mold is too wet, metal can boil and explosions may occur even though the molds are well vented.

In ramming a mold, the peen of the rammer should not be placed too close to the pattern. Otherwise a hard spot in the sand will be made and molten metal coming in contact with it will boil and tear the sand away to the depth of the hard spot. If a gagger iron is rammed against a pattern and the sand between is pressed into a hard spot, this same thing occurs, and when the molten metal reaches the wet gagger iron, an explosion usually results.

920

Molding machines

Three types of molding machines are used in modern foundries: straight, semiautomatic, and automatic. All molding machines should be equipped with two-hand controls for each man assigned to the machine (Fig. 33–11).

The carry-out man should stand so that he is clear of the squeeze at the back of the machine. Operators should never touch the frame while it is moving.

When patterns are to be changed, the frame should be blocked to prevent the table from falling and trapping pattern setup men, should the dog fail or the stripping frame operate accidentally.

On automatic molding machines, shields or apron-type metal guards should be used to protect pinch points.

Where molding machines have sand delivered by elevator buckets, the side on which the buckets return should be enclosed or railed off with 42-in. double railings placed 15 to 20 in. from the moving parts.

The jolt squeezer machine is used as an accessory to molding machines. The main hazard of this machine is that the worker may keep his hand on the edge of the flask or get his hand between the head of the machine and the flask. Dual safety squeeze controls and a knee valve jolt control will eliminate this hazard.

Core blowing machines

Straight, semiautomatic, and automatic core blowing machines are used in foundries.

To prevent sand blows, parting lines should be maintained in good condition. Also, the parting line of the core box can be guarded by a dike seal (Fig. 33–12).

Vents should be replaced when ruptured. Good feeding systems with vibrators should be installed to keep the magazine full. Core box and cover seals should be kept free of sand.

When a core machine is to be repaired, the air lines should be shut off, the controls should be locked in the OFF position, and the air should be bled from the machine.

Where practical, two-hand operating controls should be provided to prevent the operator from placing his hand or fingers between the top of the core box and the ram. Where

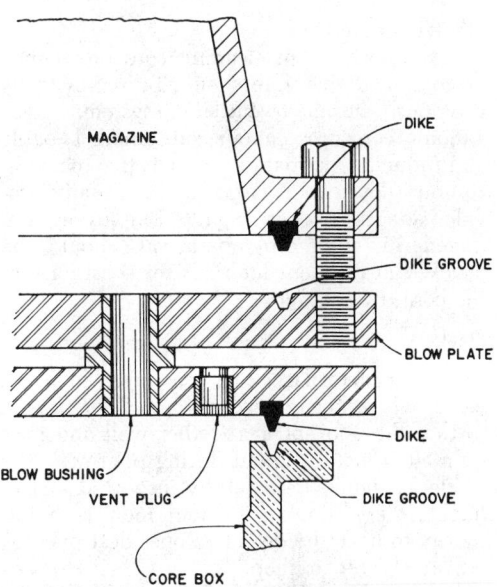

Fig. 33–12.—Section of core box shows rubber dike seal, which prevents sand blows and abrasion of the box.

Courtesy Dike-O-Seal Corp.

two operators are employed on a core blowing machine, four-hand control buttons should be provided.

All core boxes should be equipped with handles so that employees can move the boxes without placing their hands on top of them.

If driers are located above the rollover area for each core, they should be placed high enough so that they will not become entangled during the rollover process.

On semiautomatic and automatic machines, core box push cylinders, counterweight cable pulleys, wheel guides, and table adjusting footpads should be guarded, and an automatic barrier guard should be installed between the operator and the machine. If the drier is lowered automatically from the rollover and then pushed and raised toward the operator, there can be a pinch point between the lowering table and the raise table. This hazard is eliminated by an automatic barrier guard.

Automatic and semiautomatic core machines should be equipped with double solenoid valves. The slide valve should be well maintained and lubricated to prevent recycling or

921

other malfunction.

Materials used in cleaning core boxes may be toxic and therefore should be removed by a properly installed ventilation system.

Some core dips contain substances capable of producing dermatitis on sensitive persons. Rubber gloves and plastic sleeves usually provide adequate protection. Employees engaged in core dip operations should be checked at frequent intervals for sensitivity to the core dip solution.

Flasks

Iron or steel flasks are preferable to wood flasks, which become worn, burned, or broken so that they do not fit together well and may let molten metal run out during pouring.

Flasks should be carefully inspected at frequent intervals by competent men with authority to have the defective ones destroyed or sent to the repair shop.

It is unwise to leave defective flasks in the foundry building or in yard storage piles because they may be put back into service without having been repaired.

Flask trunnions should have end flanges at least twice the diameter of the trunnions to minimize the danger of hooks slipping or jumping off. Trunnions should preferably be turned, or otherwise be smooth castings.

It is sometimes best to cast the trunnions separately and rivet them in place. This procedure facilitates the machining operations and permits reuse of trunnions recovered from broken flasks. Trunnions cast separately should be of steel.

When trunnions are bolted in place, the nuts should be inside the flask. If they project on the outside, slings are likely to catch on them and slip off with a jerk, which subjects both the sling and the trunnion to severe strain.

Large flasks should have loop handles of wrought iron. On steel flasks, handles should be cast in place at frequent intervals to facilitate chaining.

Trunnions and handles should be designed for the loads they are to carry and constructed with a factor of safety of at least 10. The bolts with which they are fastened to the flasks should be of sufficiently heavy construction also.

The diameter of the button should be equal to the diameter of the groove plus one and one-half times the diameter of the sling used to handle the flask. Inside corners shall be well filleted. To prevent the sling from sliding off and riding the button, the radius of the corner between groove and button should be approximately equal to the radius of the sling used and the remainder of the inside edge of the button should be straight.

Sandblast rooms

Each foundry should have dust-tight sandblast rooms. The doors to these sandblast rooms should be kept closed, and castings should be dusted before they are removed from the rooms. Even small cracks in the walls or under doors will allow fine dust to escape and contaminate air in the foundry.

Tumbling barrels

Tumbling barrels need frequent care to keep them dust-tight. Barrels that cannot be maintained dust-tight should be enclosed in booths connected to an exhaust system. Modern barrels are equipped with exhaust ducts through the trunnions. A removable guard rail should be placed around the machine.

While barrels are being loaded or unloaded, they should be locked in stationary position.

Shake-outs

The operation of shaking out castings presents the danger that hands and feet may be crushed or arms and legs broken. If steel hooks or rakes are used to pull castings from the screen, men should be instructed to stand so that one foot is kept behind, so they can keep their balance in case the hook slips from the casting while they are pulling. Workers must wear foot protection.

Because this operation is also often a source of dust, it should be hooded, and local exhaust should be provided to draw the dust to a collector. In fact, many foundries perform shake-out operations at night so that as few people as possible will be exposed to dust.

Shake-out machines should be designed so that the flasks cannot fall off the plunger. Foundrymen should not attempt to retrieve gagger irons while these machines are in operation.

The hazards of sand conveyors are found at

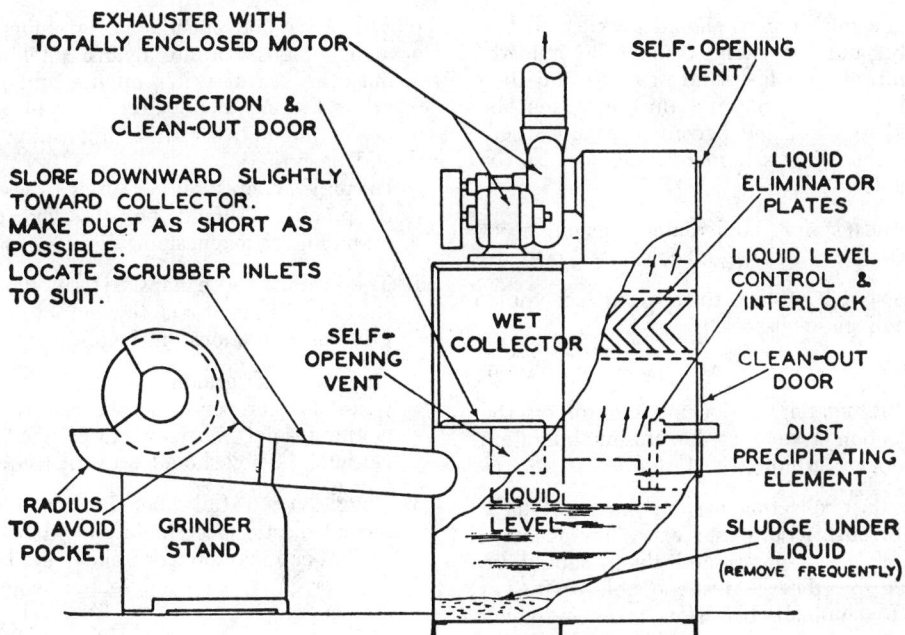

FIG. 33–13.—Magnesium dust-collection system converts dust into wet sludge for later removal.

Courtesy Dow Chemical Co.

Labels in figure:

EXHAUSTER WITH TOTALLY ENCLOSED MOTOR

SELF-OPENING VENT

INSPECTION & CLEAN-OUT DOOR

LIQUID ELIMINATOR PLATES

SLORE DOWNWARD SLIGHTLY TOWARD COLLECTOR. MAKE DUCT AS SHORT AS POSSIBLE. LOCATE SCRUBBER INLETS TO SUIT.

LIQUID LEVEL CONTROL & INTERLOCK

SELF-OPENING VENT

WET COLLECTOR

CLEAN-OUT DOOR

DUST PRECIPITATING ELEMENT

RADIUS TO AVOID POCKET

GRINDER STAND

LIQUID LEVEL

SLUDGE UNDER LIQUID (REMOVE FREQUENTLY)

shake-outs since the sand is collected under the shake-out on a conveyor belt which conveys it to storage for reclaiming and reuse. The area about the shake-out should be kept free of sand and scrap, and the conveyor belt opening should be guarded at sides.

Cleaning and Finishing Foundry Products

Abrasive, polishing, and buffing equipment for foundry use should be installed and operated as recommended in Chapter 31, "Machine Tools."

Operators should be required to wear full eye and face protection specified for the operations.

Excessive dust generated in the use of dry abrasive wheels may be a health hazard and should be effectively removed at the point of origin by an exhaust system. If castings are precleaned in a barrel, mill, or abrasive chamber, dust from grinding can also be minimized. The space about the machines should be kept dry, clean, and as free as possible of castings and other obstructions.

Abrasive grinding wheels should be mounted and changed by qualified personnel. Use of correct washers and wheel-mounting procedures must be closely supervised. (See details in Chapter 31, "Machine Tools.")

Required wheel guarding must be kept intact. Eye, hand, and foot protection are necessary. Some companies speed-test new wheels before allowing them to be used on the job.

Magnesium grinding

The fundamental hazard in the grinding of magnesium lies in the possibility of fire or explosion. To eliminate this hazard, a proper dust-collection system must be used (Fig. 33–13).

In a magnesium dust-collecting system, the dust should be precipitated by a heavy spray of water and immediately washed into a sludge pit in which it is collected under water.

Sludge pits or pans must be well ventilated, since reaction of the collected dust with water evolves hydrogen.

923

Sludge pits or pans should be cleaned frequently, and the sludge should be immediately mixed with fine sand or earth and then buried. Wet magnesium dust must not be allowed to stand and become partially dried.

The following safeguards should also be provided:

1. A means for immediate quenching of sparks from grinding wheels, disks, or belts.

2. Dust-proof motors to prevent accumulation of static charges.

3. Explosion doors on the collection system.

4. An automatic interlocking control on the collection system to ensure its operation whenever grinding is started.

The dust-collecting system must not have filters or obstructions that will allow accumulation of dust. Pipes and ducts should be installed by the shortest possible route, in order to eliminate bends or turns in which magnesium dust or fines could collect. Pipes and ducts connecting the grinder and the collecting device should be cleaned daily and should be disconnected while wheels are being dressed.

Good housekeeping is essential for safe handling of magnesium. Accumulations of magnesium dust on benches, floors, window ledges, overhead beams and pipes, and other equipment should be prevented. Vacuum cleaners should not be used to collect the dust. It should be swept up and placed in covered, plainly labeled iron containers. It should not be mixed with regular floor sweepings. It should then be mixed with fine dry sand and buried.

Grinding of other metals on equipment used for magnesium is dangerous because of the possibility of producing sparks. Equipment for magnesium grinding should be marked FOR MAGNESIUM ONLY.

Benches of wood grating are recommended for rough finishing operations.

An ample supply of powdered graphite in plainly labeled, covered metal containers should be kept close to each grinding unit. A scoop should be kept inside each container, and the lid should be kept loose for easy access.

Warning signs should be prominently displayed inside and outside the grinding rooms or areas. Signs warning against smoking and against the use of water on fire and recommending the use of powdered graphite, limestone, or dolomite as an extinguishing agent should be posted.

To prevent fires and injuries, these safe work practices should be observed during the grinding of magnesium alloys:

1. The grinder and exhaust system should be started and run for a few minutes before grinding operations are begun.

2. Operators of grinding equipment should wear leather or smooth, fire-retardant clothing, not coarse-textured or fuzzy clothing. They should brush it frequently.

3. Goggles or a full fiber helmet with attached plastic face shield should be worn. A skull cap may be worn under the helmet. Leather gloves with long gauntlets are recommended.

4. Machine tools must be sharp and properly ground for magnesium alloys, or they may cause fire from friction.

5. Use only neutral mineral oils and greases for cooling and lubrication. Animal or vegetable oils, acid-containing mineral oils, or oil-water emulsions can be hazardous.

Chipping

Where castings are cleaned or chipped, tables, benches, and jigs or fixtures specifically designed and shaped to hold the particular casting should be provided. Screens or partitions should be installed to protect other employees from flying chips. These areas should be provided with hoods and exhaust systems to remove dust and workers should wear eye and face protection.

Welding

Welding is done in a foundry to a considerable extent in the cleaning or reclaiming of castings. The safe practices given in Chapter 34, "Welding and Cutting," should be followed.

Sand is plentiful in all foundries and is one of the best noncombustible materials. It can be spread on the floor to a depth of 2 in. in areas where welding operations are conducted, to help prevent fire.

Powder washing, a method in which a stream of powdered iron oxide is introduced into the gas flame to intensify the heat produced, should be done according to the same safe practices as other welding and cutting of carbon steel or cast iron. However, when this method is used to clean or cut sprues, gates, and risers from alloyed castings, exhaust ventilation is recommended.

Power presses

Power presses have wide application in finishing departments of foundries. Sufficient aisle space, good housekeeping, and effective lighting should be provided for safety in power press operation. Machines should be fully guarded and maintained in good working order. Operators should be carefully selected and should be trained in efficient and safe operation of the machine. Mechanical feed and ejection equipment should be used where practical.

These topics are fully discussed in Chapter 32, "Cold Forming of Metals."

Nondestructive Testing

Visual observation, even with magnification, cannot locate all small, below-the-surface defects in cast and forged metals. Proper nondestructive testing will, however, reveal all such defects positively without damage to the parts being tested. Nondestructive testing methods will locate defects inherent in the metal (nonmetallic inclusions, shrinkage, porosity), defects that result from processing (high residual stresses, cracks and checks caused by handling, spruing, or grinding of castings and forgings), and in-service defects (sharp changes in section, corrosion, erosion).

The types of testing most commonly used for forged and cast metals are:

1. Magnetic particle
2. Penetrant
3. Ultrasonic
4. Triboelectric
5. Electromagnetic
6. Radiographic

These methods, as well as others applicable to nonmetallic substances, are fully discussed in NSC Data Sheet 488, *Nondestructive Testing of Materials*.

Recommendations for installation, inspection, and maintenance of the electrical equipment used in many of these testing procedures are given in Chapter 41, "Electrical Hazards."

Magnetic particle methods

Magnetic particle inspection is the most widely used test method. It utilizes magnetism to attract and hold very fine magnetic particles right on the part itself. If a defect is present, it interrupts the magnetic field and is clearly shown by the pattern made by the particles. The part is magnetized in suitable directions by dc line voltages transformed to low-voltage (4 to 18 volts), high-amperage ac, half-wave current, or three-phase full-wave current.

Inspection materials are finely divided ferromagnetic particles selected, ground, and controlled to provide mobility and sensitivity. Materials are available in several forms and colors. The type of defect to be located and the condition of the surface to be inspected determine which form of material and which method—dry, wet, or fluorescent—should be used. Color is selected to provide maximum contrast with the surface of the part.

All electric circuits should be installed and grounded according to standard procedures.

Local exhaust is required to control the dust particles used for testing. If local exhaust is not feasible, operators should wear respiratory protective equipment. Eye protection should also be worn to guard against the irritating effects of the dust particles.

Since the magnetic particle magnetizing equipment may produce arcing, it should not be used in an area where combustible gases or vapors may be present.

Penetrant inspection

Penetrant inspection methods are useful for revealing cracks, pores, leaks, and similar defects which are open to the surface in a metal or other solid material. First, the part to be inspected is cleaned. A penetrant is then applied to the surface and within a few minutes is drawn into defects by capillary action. The penetrant is removed from the surface but remains in the defects.

A developer which acts like a blotter and brings the penetrant in the defects back to the surface is then applied, and the surface is inspected.

Depending upon the sensitivity of the material, the penetrant is removed by water wash, a solvent cleaner, or an emulsifier followed by water wash.

Fluorescent penetrants may be used to reveal defects under ultraviolet ("black light"). Defects may also be detected by a dye penetrant which contrasts with white.

Most penetrants are organic compounds that may cause dermatitis. Skin contact should be avoided and personal hygiene strictly followed.

Ultraviolet equipment should be effectively shielded or filter lenses of the correct shade should be worn (see Chapter 19, "Personal Protective Equipment").

Ultrasonics

Ultrasonic waves (above the audible range of 20,000 Hz—cycles per second) are created by an electronic generator which supplies high-frequency voltage to a piezoelectric crystal.

Three basic ultrasonic methods have been developed: the reflection method, the through-transmission method, and the resonant-frequency method.

● In the **reflection method,** that portion of the ultrasonic beam which strikes a flaw or discontinuity in the material is reflected; the rest of the beam goes on. The piezoelectric crystal transducer radiates these waves through a coupling medium into the material and also acts as a receiver to detect reflections, which are then picked up by an electronic amplifier and applied to a cathode ray oscilloscope on which the time intervals between the outgoing and the incoming waves can be measured.

● In the **through-transmission method,** a beam or wave is directed through a piece of material. If a flaw or discontinuity is encountered, the energy is absorbed, and the beam or wave does not get through. Since fluids such as water, oil, and glycerin give better coupling than air, they are generally used as the coupling medium between the transmitter, material, and receiver. In some applications, however, air or other gases can be used.

● The **resonant-frequency method** is used primarily to measure the thickness of material.

The equipment consists of an electronic oscillator which supplies voltage of ultrasonic frequencies to a piezoelectric transducer. The transducer is pressed into contact with the part to be tested and induces longitudinal vibrations in the test piece under the area of contact.

One type of resonance instrument displays the thickness reading on a cathode ray tube as a pip on a calibrated scale. Equipment should be disconnected from the power supply and the condensers discharged whenever a cathode ray tube must be adjusted or removed.

Triboelectric method

The triboelectric method is used to detect minute quantities of current generated when two metallurgically or chemically unlike conductors are moved into frictional contact. If the conductors are alike, no current will be generated.

The equipment consists of a control unit and a portable sorting head connected by means of a cable. The sorting head contains all the main controls for actuating the test and is designed for one-handed operation. This method is designed to sort and identify metal parts of not more than four alloy types.

Electromagnetic methods

Two types of electromagnetic tests are currently being used in industry: magneto-inductive and eddy current. A third type, employing radar frequency, is also being used, but only to a limited extent.

● The **magneto-inductive method** utilizes variations in the permeability of magnetic materials to create variations in a pickup coil or probe.

● **Eddy current.** The second type of electromagnetic testing utilizes ac in a coil or probe to induce eddy current into the part being tested. Defects and variations in properties or geometry cause changes in the strength and distribution of the eddy current. Readout is presented on a cathode ray tube, on a meter, by audible or visible alarm, or by a combination of these methods.

● **Radar frequency** methods use high-fre-

Fig. 33–14.—"Pill" of radioactive cobalt-60 is placed inside newly cast pump housings. Radiation exposes X-ray film strapped to outside of castings and shows up defects. Aluminum rod is long enough to keep operators and radiation source safely separated. Second man monitors exposure during inspection.

quency radar waves to measure the electromagnetic properties of thin coatings and surface layers of material. To make such tests, a wave guide or cavity oscillator is coupled to the test object; high-frequency waves are then reflected from the object, providing an indication of the surface electrical resistance and of the thickness of the nonconducting coatings.

In some radar-frequency testing installations, operators have been burned internally when they passed between the object being tested and the testing device. Special regulations should be formulated and enforced, and barriers should be set up to prevent operators and other workers from entering such areas. Recommendations of the manufacturer of the equipment should be followed explicitly.

Radiography

Radiography uses X rays, gamma rays, or beta rays. X rays are unidirectional, and their wavelengths can be varied (within certain limits) to suit the condition. Gamma and beta radiography differ from X-ray radiography in that the gamma and beta rays are multidirectional and their wave lengths, being characteristic of the source, cannot be regulated. Gamma rays for radiography usually are obtained from isotopes of cobalt-60, cesium-137, iridium-192 (Fig. 33–14). Important pure beta-emitting sources are: strontium-90, yttrium-90, technetium-99, and promethium-147.

Exposures made by gamma rays usually take longer to complete than do X-ray exposures. Also, in some instances, gamma exposures are inferior to X-ray exposures in

927

sensitivity and contrast.

Gamma radiography, however, has several advantages. Because of the nature of isotopes, it frequently is possible to make a number of tests simultaneously, provided that specimens can be suitably located. Moreover, isotopes are independent of electrical power, their sources are portable, and the small size of the sources makes it possible to obtain radiographs in tight quarters.

Beta ray emitters are provided in the form of sealed sources and are used primarily for measuring sheet metal thickness.

Devices used to transform differences in intensity of the penetrating radiation into visible images are X-ray film fluorescent screens, geiger proportional end window scintillation counters, and ionization gages.

All sources of ionizing radiation are potentially dangerous. X-ray and gamma ray sources may also produce hazardous secondary radiation. In addition, X-ray units involve both low- and high-potential electrical hazards.

Radiation safety is discussed in Chapter 37, "Industrial Hygiene."

Forging Hammers

Open-frame hammers

Open-frame or Smith forging hammers are constructed so that the anvil assembly is separate from the foundation of the frame and operating mechanism of the hammer. They may be single or double frames.

Flat dies are generally used in Smith hammers, and the work done allows for more material to be machined off. Smith hammers are used where the quantity of forgings to be run is too small to warrant the expense of impression dies or where the forgings are too large or too irregular to be contained in the usual impression dies.

Gravity drop hammers

Drop forgings in closed impression dies are produced on gravity drop hammers (both board drop hammers and steam- or air-lift drop hammers) and steam hammers.

Both types of gravity drop hammers shape the hot metal in closed impression dies. The impact of the hammer blows shapes the forging through one or more stages to the finished shape.

Forgings on gravity drop hammers may range in weight from less than an ounce to 100 pounds and may be made from any type of malleable metal, such as steel, brass, bronze, aluminum, or magnesium alloys.

Gravity drop hammers are designed so that alignment between the dies can be maintained by the use of guides, die pins, die locks, or a combination of these devices.

On gravity drop hammers, the ram and the upper die are raised to the top of the hammer stroke, and the impact blow is given by the free fall of the ram and the die.

The board drop hammer, the more common type, utilizes hardwood boards which are secured to the ram and held in it by wedges. The boards pass between rotating rolls which grip the boards and raise the ram and the upper die for the successive blows.

At the top of the stroke, the rolls release and the ram is held in this position by clamps until released by depression of the treadle. The impact of a board drop hammer cannot be varied while the hammer is in operation.

A steam- or air-lift gravity drop hammer is controlled by a valve which admits steam or air under the piston into the cylinder in the head of the hammer to raise the ram and the die. At the top point, the air or steam is exhausted and has no effect upon the hammer blow, which is controlled entirely by the weight of the freely falling ram and die. Like the board drop hammer, the steam- or air-lift hammer is operated by treadles or pedals, levers, or air valves.

The height to which the ram is raised and consequently the impact can be varied during the operation of an air- or steam-lift hammer.

A third type of gravity drop hammer has the ram suspended by ropes or belts lifted by cranes, by a drum on which the rope or belt is wound, or by similar means.

The falling weight of the ram assembly and upper die of gravity drop hammers may range from 400 to 7500 pounds.

Steam hammers

Steam hammers are also classified as drop hammers. Most steam hammers are double-acting and use steam pressure (or air pressure) through the medium of a piston and

Fig. 33–15.—Safety basket, made of 2 by ½ in. welded steel, holds the piston rod of a forging hammer when the steam is shut off for cleaning ram bore. The ends of the bars are bent around the chain eyes.

Courtesy Tractor Works, International Harvester Co.

cylinder to raise the ram and the die and to assist in striking the impact blow.

Since steam or air power is used in addition to the weight of the falling ram and die, the steam hammer will strike a heavier blow than will a gravity drop hammer using an equivalent falling weight.

The falling weight of the ram assembly and upper die of double-acting steam hammers ranges from 1000 to about 50,000 pounds. They commonly produce forgings ranging from a few ounces to above 500 pounds, and sometimes up to 2000 pounds.

Steam and air operated hammers are manufactured with many built-in safety features. Some of the most outstanding are:

1. Hammers are designed so that the distance between the floor and the die seat is approximately 36 in.

2. Safety latches prevent the ram from dropping when a job is being set up or when work is being done on the die. However, for complete protection, the steam or air should be shut off from the press and the ram should be blocked up.

3. A safety cylinder head protects against piston overtravel.

4. Many of the working parts are safely enclosed.

5. Operating levers are placed in a safe position.

Hammer hazards

For the most part, all types of forging hammers have identical hazards. The most frequent causes of injury are:

1. Being struck by flying drift and key fragments or by flash or slugs.

2. Using feeler gages to check guide wear or the matching of dies.

3. Using materials handling equipment improperly, such as tong lifts.

4. Crushing fingers, hands, or arms between the dies.

Fig. 33–16.—Permanent catwalk installed along row of board drop hammers makes repair and servicing of hammers easy and safe.

929

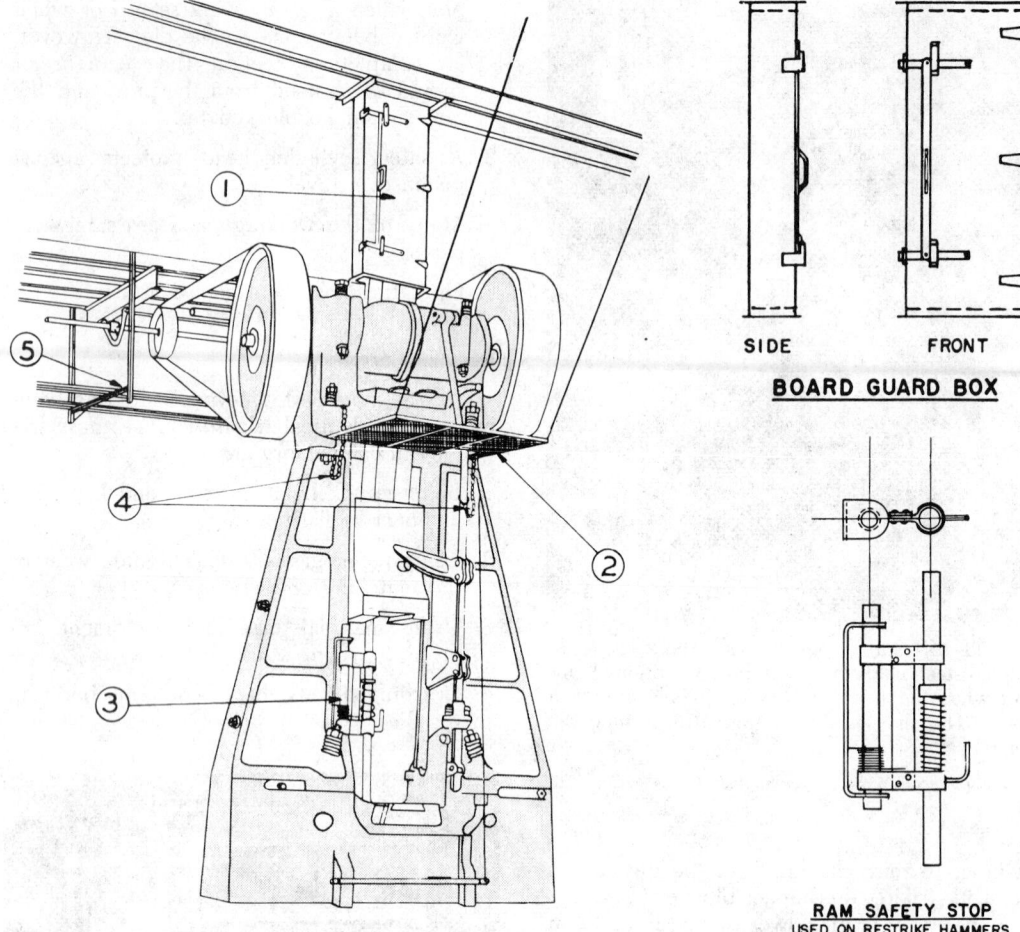

SIDE   FRONT

BOARD GUARD BOX

RAM SAFETY STOP
USED ON RESTRIKE HAMMERS

Fig. 33–17.—Well-guarded board drop hammer features: (*1*) sheet steel board guard box, (*2*) screen-platform made from No. 9 expanded metal, (*3*) steel ram safety stop that swivels on left column, (*4*) safety chain to restrain tiebolt and nut, and (*5*) catwalk and belt catcher. Details of board guard box and ram safety stop are shown in drawings at right.

Courtesy American Brake Shoe Co.

5. Crushing fingers between tong reins.

6. Receiving kickbacks from tongs.

7. Using swabs or scale blowing pipes with short handles.

8. Being burned by hot scale.

9. Dropping stock on the feet.

Injury may be incurred on a steam drop hammer when the ram pulls off a new piston rod (Fig. 33–15). Sometimes the rod must be set in the ram several times before it holds. If the piston rod breaks, the ram will fall. This common hazard emphasizes the importance of the operator's using a safety prop to support the ram before he reaches under it.

Operating a hammer with a worn cylinder sleeve is hazardous. When the sleeve is so worn that the swing of the ram cannot be controlled at the throttle control, the hammer should be shut down and repaired.

Operating a hammer with broken piston rings is also dangerous. A piece of broken piston ring passing through the steam ports and lodging in the throttle valve can cause the ram to drop out of control, often catching the operator's tongs or the transfer tool and causing serious injury.

Repairmen, particularly, are exposed to crushing injuries when they remove and install parts on the top of the hammer and when they remove sow blocks, anvils, and columns. Means for locking out the power should be provided and used. Catwalks and guard rails should be installed on all hammers to provide safe footing for repairmen (Fig. 33–16).

Guarding

On gravity drop steam- or air-lift hammers, a hand lever is preferable to a treadle on cold restrike operations. Also, a positive sweep device, which moves from back to front and is actuated by the hammer ram, should be used.

Two-hand tripping controls should be provided where the material being forged is not held by the hands or by hand tools or where a sweep device, safety stop, or tripping lever cannot be installed.

On board drop hammers, a substantial guard should be provided around the boards above the rolls to prevent the boards from falling if they should break or come loose from

FIG. 33–20.—Hinged scale guard includes forging chute that can be lifted out of the way for die setting and key driving.

the ram (Fig. 33–17).

Other standard protective features for a board drop hammer include the ram stop and safety chain for tie bolt and nut.

A pneumatic key driver is superior to a manually operated driver and offers a far greater margin of safety (Fig. 33–18). The driver should be made of steel of proper hardness so that it will not chip on impact.

A manually operated key driver should be sturdy and well balanced (Fig. 33–19). The driving face should not be allowed to mushroom, but should be kept ground down—not burned off with a cutting torch.

A scale guard to confine pieces of flying scale should be standard equipment on the back of every hammer. The guard should allow ample clearance for the ram and easy access to the dies. It may be installed in one of the following ways:

1. Hinged on one side to an upright post so that the guard can be swung closed or open—out of position when access to the die area is required (Fig. 33–20). This installation is considered the most efficient and is widely used throughout the forging industry.

2. Supported on a floor standard.

3. Suspended from the ceiling or anchored to a rail.

Treadles and pedals should have ample clearance and should be guarded to prevent accidental tripping by a falling object. Any portion of a treadle or pedal at the rear of the hammer should also be guarded so that scrap or other material cannot interfere with treadle action.

Several methods for interlocking the treadle have been developed. A simple mechanical interlock (Fig. 33–21) is effective. To lock the treadle when it is necessary to use a pry bar to remove stuck forgings from die cavities, the pry bar is kept in a sleeve-type holder with its weight resting on the actuator of an air valve. If the pry bar is removed, the air valve opens and the air cylinder drives a wedge under the treadle arm to prevent its movement.

Flywheels or drive pulleys should be enclosed by a guard strong enough to prevent the pulley from falling to the floor if the shaft

FIG. 33–21.—An easy-to-install treadle lock for a steam hammer. Safety cables are attached to the base bolts and springs.

Courtesy Wyman-Gordon Co.

should break. In this installation, the strength and location of the guard bracket or the frame are more important safety factors than is the gage of the sheet metal used for the enclosure. Brackets should be bolted to the column of the hammer. In some instances, the guard enclosure is supported from the floor by an I-beam.

All cylinder bolts, gland bolts, guide bolts and liners, and the head assembly over the operator's working position should be restrained by cables or chains (Fig. 33–22).

Steam or air drop hammers should have a stop valve or quick opening and closing valve. Also, a safety head in the form of a steam or air cushion, if not already standard on the hammer, should be provided to prevent the piston from striking the top of its cylinder.

TABLE 33-A

STRENGTH AND DIMENSIONS FOR RAM PROPS MADE OF TIMBER

Actual Dimensions of Timber (in.)	Cross Sectional Area (sq in.)	Min Allowable Crushing Strength[1] Parallel to Grain (psi)	Max Static Load within Short Column Range (lb)	Safety Factor	Max Recommended Weight of Forging Hammer for Timber Used (lb)	Maximum Allowable Length[2] of Timber (in.)
4×4	16	5,000	80,000	10	8,000	44
6×6	36	5,000	180,000	10	18,000	66
8×8	64	5,000	320,000	10	32,000	88
10×10	100	5,000	500,000	10	50,000	100
12×12	144	5,000	720,000	10	72,000	132

[1]Adapted from U.S. Department of Agriculture Technical Bulletin 479. Hardwoods recommended are those whose ultimate crushing strengths in compression parallel to grain are 5000 psi or greater.

[2]Slenderness ratio formula for short columns is L/D = 11 where L = length of timber in inches, D = least dimension in inches. This ratio should not exceed 11.

The cylinder head and safety head bolts should be connected to an anchored wire or cable.

If the hammer has no self-draining arrangement, a drain cock, preferably the quick acting type, should be installed in the lower part of the cylinder at the back of the hammer. This cock should be arranged so that it can be opened without danger to anyone, or it should be piped to discharge at a safe place. Steam lines should be well trapped.

If air or steam is used to remove scale, a quick shutoff valve should be provided so that the pressure can be regulated. The operator should adjust the scale guard to protect other employees from flying scale.

Props

Safety props equipped with handles at the middle should be provided and their use required when repairs or adjustments are to be made on the dies or when dies are to be changed. The props should be held in place while the power is released to permit the weight of the upper die and the ram to rest on the props. Operators should never place their hands on the top of a prop.

The props can either be chained to the hammer so that they cannot slip out of position or be hinged to the side of the hammer so that they are readily available and easily moved into and out of their blocking position

(Fig. 33–17).

The material most commonly used for ram props is a hardwood timber not less than 4 by 4 in. in cross section with a ferrule on each end. Specifications for hardwood props are given in Table 33-A. They may also be made of a section of structural shape carefully squared at the ends or a section of steel tubing not less than 2½ in. outside diameter with a wall thickness of ½ in.

Hand tools

Pliers, tongs, and other devices specially designed for the work to be handled should be used for feeding the material so that the operator's hands need not be placed under the hammer at any time. Tongs should be long enough that they can be held at the side of the body rather than in front. Tongs should fit the shape of the material being held for forging.

Oil swabs and scale brushes or pipes should have handles long enough to make it unnecessary for the operator to place his hands or arms underneath the die.

Die keys

Die keys should be made of a suitable grade of medium carbon alloy steel that has been properly heat treated so that it will not crack or splinter. Both ends of the key should be tapered for clearance in driving and removing

33—Hot Working of Metals

the key. Mushroomed keys should never be used.

Die keys must be the correct length, so they do not project more than 2 in. in front and 4 in. at the back of the hammer. If necessary, shims should be used. If keys project farther, they become a hazard to the operator working in front and may break off while the hammer is operating and fall between the dies in back.

An adequate supply of die keys should be stocked so that drifts will be needed only when the end of a key becomes distorted and must be cut off before the key can be driven out. The drift must be blocked or held securely with a drift holder.

Safe operating practices

The hammer man who directs the activities of the men in the hammer crew should be made responsible for their following safe work practices.

All guards should be in place when the hammer is in operation and should be kept in good repair.

Material and tools should be moved from the aisles and the operator's working space and stored in the proper place. The floor area around hammers should be free of scale, oil, water, and other material, to ensure safe footing.

Before starting work, the hammer crew should make its own inspection to see that the equipment and working conditions are in good order. A frequent check should be made for breakage at all critical points which are subject to severe strain and can therefore often fail. If an unsafe condition is found, the foreman should be informed of it at once.

Drop hammers should never be operated when the dies are cold. Dies should always be preheated by hot steel placed between them.

No adjustments nor repairs should be allowed until the power has been turned off and the master switch locked out, the treadle blocked to prevent accidental tripping, and the ram propped.

When dies are being set on a board or steam- or air-lift gravity drop hammer, it is best to fit the dowel in the upper die and the ram with as few shims as possible. The bottom die, which should have enough shims

Fig. 33–22.—Safety cables strung through and around packing gland bolts, cylinder bolts, and guide bolts and liners of a steam hammer restrain these parts should any of them fail.

Courtesy Oldsmobile Div., General Motors Corp.

so it can be lifted easily, should be moved first. (This procedure is the opposite of that followed in setting dies in a double-acting steam drop hammer.)

On steam hammers a prop should be placed between the ram and the shank of the top die before the die is moved. When it is necessary to move the bottom die, the prop should be placed between the sow block and the ram on the side containing the dowels, so that the die can be moved.

On steam drop hammers, the spool bolt should not be adjusted until the main steam valve has been turned off and the treadle blocked, to prevent the ram from accidentally picking up while the operator is adjusting the bolt.

Laying liner stock between the dies to jar loose a stuck forging is dangerous. When a forging sticks, the safe method is to stop the hammer, remove the forging, relieve the die, and then continue the operation. On some operations where this method is not practicable, a safety liner made of soft steel may be used.

Whenever an operator leaves the hammer, even if only for a few moments, the upper die should be left resting on the lower die, to prevent accidental tripping.

Flywheel speeds must be carefully observed and should not as a rule be permitted to exceed the number of rpm's given on the specification sheets since this is the speed upon which proper operation of the press is based.

Personal protection

Operators of forging hammers and other employees working in the vicinity of this equipment should wear suitable personal protective equipment—full eye protection, hard hats, safety shoes, and leather leggings and aprons. Light wool clothing is preferable since it does not burn as readily as cotton and provides better protection against skin burns.

Cotton fabric gloves should be worn and when wet, should be removed and allowed to dry. Leather gloves should not be worn since perspiration may cause steam burns.

Setup and removal of dies

Dies are usually heavy and hazardous to handle without proper equipment. Uniform holes may be drilled in both sides of each heavy die block so that pins can be inserted to make lifting and moving easy. The diameter of the pin and hole will depend upon the weight of the die. Standard practice in many companies is to have the pin $\frac{1}{16}$ in. smaller in diameter than the hole. The depth of the hole and the diameter of the hole and the pin should be uniform in dies of a certain weight group to assure that there will be sufficient pressure to prevent the pin from falling out.

Transfer boards should not be used to transfer dies between the work bench and the machine. Transfer trucks, preferably the elevating type, are safer and more efficient. The top or table of the truck should be covered with sheet metal at least $\frac{1}{4}$ in. thick and should be securely fastened in place.

Power lift trucks should be used for transporting and installing dies. Lift trucks should be well blocked or secured to the base of the hammer when dies are to be set or removed. Otherwise, the truck may slip away from the hammer, causing the die to slip and fall.

Where die trucks are still used, the frames of the hammers should have eyes to receive hooks attached to the trucks. When a die is to be moved from the die truck onto the hammer bed, or vice versa, the hooks are first engaged in the eyes or swivels to prevent the truck from being pushed away from the hammer.

When forge dies are set up or removed, the hammer man should act as leader of the group and should see that all efforts are coordinated and all safety rules are observed so that the work will be done efficiently and safely.

Setting up dies. The immediate area around the hammer must be clean and clear of obstructions. Maintenance work should not be performed on the equipment during a die setup.

The hammer crew should make a complete check of the equipment between setups: the hammer should be in good working order; the seats of both the sow block and the ram should be flat and clean; dowels and die keys should be inspected for galls and burrs; dies should be checked for burrs and other defects, such as cracks or sharp corners.

It is important that the overall height of the dies be greater than the shut height of the hammer.

Good lighting is essential for accurate setting of dies and for the safety of the crew. Portable lights may be used, and should have heavy-duty cords, with bulbs protected by heavy screen guards.

If lift trucks are used, the floor should be level, in good condition, and free of obstructions. If cranes are used, the lift chains should be in good condition and the die pins should have a snug, but free fit.

Many methods are used in setting dies in hammers. The type and size of the dies and the type of hammer determine the method to be selected. The first action in setting up dies should be to prop the ram securely and to shut off and lock out the power, whether steam, electricity, or air.

Die dowels made of a suitable grade of steel which will not splinter or crack should then be driven into the dowel holes in the die shank. Dimensions should be accurate to ensure a tight driven fit.

After the bottom die of a steam hammer has been set in place, the bottom key should be driven to help line up the die and partially tighten it. The top die should then be inverted and set in position so that the dies are face to face with the match lines lined up. This procedure should be reversed for a gravity drop hammer: the top die is set and keyed first. Sometimes both dies can be set at once.

The safety prop should be removed from between the ram and sow block at this time, and the ram allowed to descend slowly until it engages the top die.

If shims are used on the dowel in the top die, an extra hazard is created. Normalized spring steel is used to shim dowels, and they must be set so that they will fall into place when the ram engages the die. The number of shims and their location (whether front or back) should be recorded so that succeeding shifts or different hammer crews can refer to the record of the setup for that specific set of dies.

If a die must be moved to match, a prop should be used after the ram is raised and before the operator reaches under the hammer to reset the shims. This prop must be strong enough to support the ram and long enough to extend from the top of the die to the ram.

If allowance is made for moving the dies, the allowance on a steam hammer should be made in the top die only and the bottom die should have a tight fit. On gravity drop hammers, the general safe practice is to have the top die tight, allowing for movement in the bottom die.

At this stage of the setup, the die keys should be driven with a hand sledge only. Common practice is to drive the keys up tight with a sledge or light ram, and then "bounce" the dies. The safe procedure for bouncing is as follows: bounce, shut off the power, ram the key, bounce, shut off the power, ram the key, etc. The impact helps align the dies, but creates an additional hazard if lock dies are being set.

Extra precautions and special equipment may be required for abnormally large or long dies. In setting such dies, the regular safety procedures for propping and handling may have to be changed. Any change should have approval of management.

After the die keys have been driven and before adjustments are made to the gibs or column wedges, heaters should be applied to the dies if they have not been preheated. On deep impression jobs, it is good practice to preheat dies in special low-temperature furnaces, in hot water baths, or with hot scrap steel before they are set up. Driving die keys too tight when the dies are cold may crack the shanks, sow blocks, or rams.

The hammer crew should use any waiting time to make a final check and get ready for production. The dies should be checked for proper alignment and for proper wing clearance. (Tight wings may cause breakage.) A check can be made for necessary tools, and they can be put in their proper work stations. Scale guards can be moved into position, and final adjustments can be made to the billet heating furnace.

After the dies have been heated to approximately 300 F, die keys should be driven tight again by means of either a pneumatic ram or a light suspended ram (Figs. 33–18 and 33–19). If any further adjustment to the hammer is required, it can be done after a tryout forging has been made.

Removing dies. Before dies are removed, the immediate area around the hammer should be cleared of overhead trolleys, suspended tongs, portable conveyors, tool and billet stands, and other equipment.

Overhead trolleys should be tied down so that they will not creep back into the working area. The scale guard should be moved back, and accumulated scale that would interfere with safe footing should be removed. Forgings should be moved away from the unit immediately and sent to the next work station.

If another set of forging equipment has been delivered, it should be placed nearby but not directly in the area where the hammer crew will work. Service personnel, such as truckers, crane men, and hookers, should be familiar with the proper procedure so that unnecessary handling can be eliminated.

The hammer's main power switch or control must be shut off and locked before the die keys are loosened. The top key is generally loosened first, usually with a mounted pneumatic ram (Fig. 33–18). A light, well balanced ram suspended from a cross beam or from an overhead crane or chain fall can also be used successfully.

Using a manually held drift pin or a knock-

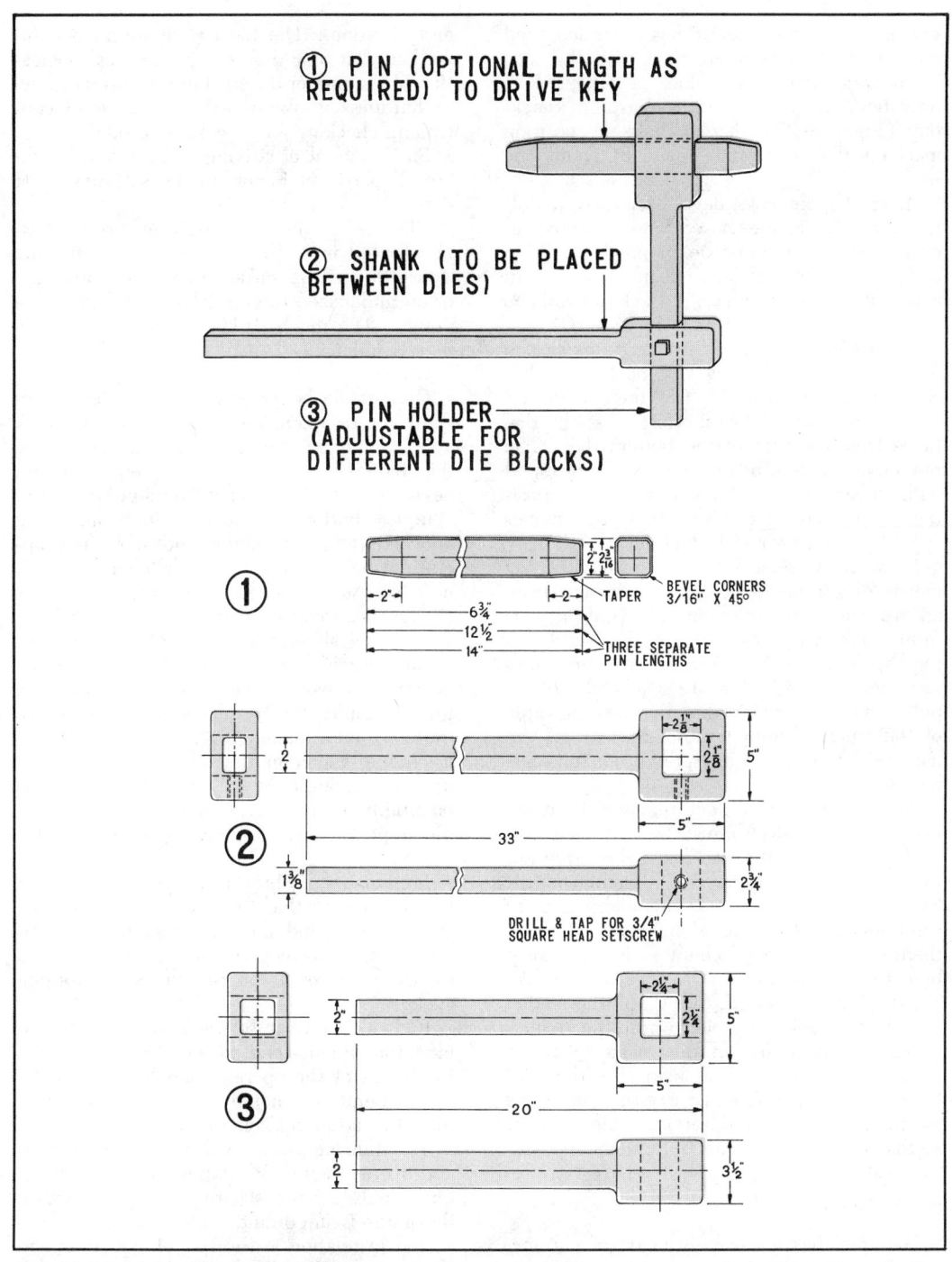

Fig. 33–23.—A key knockout, which can be made easily, is held in position mechanically. The only machining required is to drill and tap the ¾-in. hole for the setscrew.

out on a die key after it has been loosened and driven to a position even with the face of the ram or the sow block is a hazardous operation. A special type of adjustable knock-out (Fig. 33–23) that is held in position mechanically rather than manually should be used.

After the die keys have been driven out, the ram should be raised and propped at once. The prop must be in good condition and must be placed on a clean surface. On a gravity drop hammer, a jack should be used and a special prop may be required.

No attempt should be made to raise the hammer to propping level if the top die has a tendency to "hang." The die should be freed first within the shortest possible distance from the face of the bottom die. The ram on an air drop hammer should be propped with special care. After the prop is positioned securely under the ram, the power should be shut off and locked out.

Dies can be removed from the hammer by means of special die chains suspended from an overhead crane or chain fall. Special platform trucks with winches are also employed for this operation. They are practical and safe because the dies are winched out or pulled out horizontally directly onto the table of the truck. Dumping the dies out of the hammer onto the floor should not be permitted.

After the dies have been removed from the hammer, the dowels must be extracted. It is important that the two men who drive out the dowels have proper tools, usually a drift and a sledge. These tools should be in good condition and have sound handles. Because there is metal-to-metal contact, the men must be extremely careful.

After being removed, dies should be loaded on low steel pallets and taken from the vicinity as soon as possible. If dies need repair or modification, the hammer man should notify his foreman. The foreman can then send this information to the die servicing department so that the condition can be remedied before the next run.

Design of dies

Hammer dies are usually made of chrome, nickel, or molybdenum stellite—materials which have high resistance to heat, shock, and abrasion. Die blocks are supplied commercially in four different hardnesses. Selection of the proper die steel in the correct range of hardness is of utmost importance in controlling checking and breakage of dies.

Size, amount of striking surface, and height are all pertinent factors in the safe design of dies.

The correct amount of striking surface must be allowed in relation to the size of the die. Too little striking surface may cause breakage or an undersized forging when the dies pound down. Too much striking surface, especially if it is unbalanced, may cause a pull or misalignment.

Correct die height must be specified, especially for resinking forge dies. Normal work level should be maintained for the benefit of the hammer man, and the height should never be below a specified minimum, to prevent the bottom of the cylinder from being knocked out. Maximum and minimum operating heights should be labeled on each unit for use by the design engineer.

The dies must be made so that they meet in precise alignment. Layout of the dies should provide for the major portion of the heavy forge work to be done in the center of the die under the center of the ram, where the maximum hammer force is transmitted. Each impression in the die must be backed up with enough die material to reduce the possibility of breakage, especially where multiple impressions or nesting methods are employed.

Preliminary or breakdown operations must be designed so that they accomplish the specific purpose and must be systematically arranged so that they do not create a hazard for the operator as he completes the forging cycle.

Radical bends or severe reductions in volume that might tend to jerk the tongs from the hands of the operator should be avoided. Such operations should be modified or completed in additional operations.

The thickness and width of flash, gutter, and sprue should be ample so that flash or tong holds are not sheared off. The size of the gates is important. They should be designed in relation to the size of the stock and the tongs used. Gates should have enough width, depth, and clearance to allow safe

BOARD HAMMER MAINTENANCE CHECK			
Date_____Hammer No._____Location_____			
ITEM CHECKED	CONDITION	TYPE OF REPAIR	EST. HRS. TO REPAIR
TIE BARS & SPRINGS			
FRAME STUDS & SPRINGS			
DIE KEYS & SHIMS			
SOW BLOCK KEY & SHIMS			
RAM CLEARANCE			
GUIDE BOLTS & ADJ.			
MOTOR MOUNTS			
MOTOR COUPLING			
ROLLSHAFT BEARINGS			
WIRING & CONTROLS			
FLYWHEEL & BEARINGS			
DRIVE GEARS			
ROLL ADJUSTMENT			
LUBRICATION			
STEAM LINES			
BOARDS & WEDGES			
AIR LINES			
AIR FOOT SWITCH			
AIR CYL. & LINKAGE			
TREADLE & LINKAGE			
BOARD CLAMPS & LINKAGE			
DOGS & STOPS			
KNOCKOUT ARM			
FRICTION RODS			
SAFETY RODS			
REMARKS:			

FIG. 33–24.—Maintenance checklist for board drop hammers.

handling.

Some dies, especially for smaller hammers, are designed with cutoffs which shear the completed forging from the end of the bar. If possible, such cutoffs should be placed on one of the rear corners of the die for the operator's safety. If cutoffs are placed on the front of the die and are used by placing the stock across the knife portion at an angle, enough clearance must be provided between the die and the hammer frame or gib.

Hammer die maintenance is important because of the nature of forging work and the abnormal abuse to which the dies are subject. For example, fillet and corner radii in the impressions may be enlarged or sharpened by impact pressures, scale, abrasion, and wear. It is therefore necessary to inspect these radii and correct any alteration that may be hazardous.

Sometimes the face of a forging die may be welded either to correct some defect or to maintain specifications. A hard, brittle, or thin-skinned weld can become a flying hazard under impact. The rod and the preheating

and postheating methods should be selected with extreme caution.

Provisions for storing dies, such as racks and rails, are essential to safety, good housekeeping, and efficiency.

Dies preferably should be stored in an area separate from the forge shop and away from vibration.

Maintenance and inspection

A well planned preventive maintenance program for forging hammers will help to reduce the number and severity of accidents by minimizing part breakage and wear. Regular inspections will disclose production units which are not operating up to standard so that repairs or adjustments can be made.

The results of a good maintenance program can be measured in reduced operating costs which include:

1. Cost of machine downtime, breakage, and lost production

2. Cost of replacement parts and labor

3. Cost of accidents due to faulty equipment

Maintenance checklists for hammers (Figs. 33–24 and 33–25) can be the basis for formulating a definite, planned inspection program. A written checklist avoids the errors resulting from verbal reports that are often forgotten or misunderstood.

A work schedule for repairs may be set up on the basis of the data recorded on the checklists. The data may be transferred to the permanent records of the equipment and used for future planning in the maintenance program as well as for comparison of costs.

Steam hammers constitute a considerable portion of the forging equipment. Many hammers are not kept in as efficient condition as possible, usually because management fails to realize that operating and upkeep costs are higher for units in poor condition. Waste of steam usually results from worn piston rings or sleeves, loose heads, blown head gaskets, and leaky glands.

Replacement of worn rings is good economy. Worn piston sleeves and sloppy linkage make the hammer hard to control and create a hazard. Loose cylinder heads also are dangerous.

Frequent periodic inspection of every forging hammer will help ensure proper condition of all bolts, screws, keys, valves, and other parts which may be loosened by vibration. Similarly, thorough periodic inspection and adjustment should be made of all parts of the treadle or pedal, clutch, and other operating mechanisms.

Worn or loose treadle linkage, motion arm, crank arm, and treadle can cause the hammer to go out of control. These parts especially should be kept in good repair.

The clutch is a vital part of the forging press and must be kept in good condition if the press is to operate satisfactorily. A broken spring or part which shows wear should be replaced at once.

Lead casts

Lead casts are taken in practically every conceivable manner in the forging industry. Casts should be taken only in an isolated area, if possible, where there is no likelihood of interference from or injury to other workers. The die impressions should be dry since hot metal that contacts water produces flying particles of molten metal.

Lead pots should be ventilated to the outside of the building.

Forging Upsetters

The upsetter is a horizontal forging machine which forges hot bar stock, usually round, into a great many forms by a squeezing action instead of impact blows, as in the case of forging hammers. Although numerous hazards are involved in the operation of an upsetter, the most serious problems are encountered in changing the dies.

The entire machine should be completely enclosed, except for the feed area. Heavy wire mesh or expanded or sheet metal reinforced with structural steel should be used. Doors may be cut into the enclosure for servicing flywheel, brake, and other moving parts. A guard should be installed over the operating pedal.

Safe operating conditions call for the area around the machine to be clean and clear of obstructions and litter. It is especially important that the top of the machine be clear of any objects, such as loose bolts, bars, nuts, or shims, which might fall into it or from it.

STEAM HAMMER MAINTENANCE CHECK			
Date_____ Hammer No._____ Location_____			
ITEM CHECKED	CONDITION	TYPE OF REPAIR	EST. HRS. TO REPAIR
CYL. HEAD BOLTS			
MOTION VALVE STEM			
MOTION VALVE CRANK			
MOTION VALVE CONNECT.			
WIPER BAR & CRANK			
THROTTLE CRANK			
THROTTLE LINKAGE			
RAM & SOW BLOCK			
DIE KEYS & SHIMS			
SOW BLOCK KEYS & SHIMS			
GUIDE BOLTS			
GUIDE WEAR			
GUIDE ADJUSTING BOLTS			
GUIDE WEDGE POSITION			
HOUSING BOLTS & SPRINGS			
TREADLE			
TIE PLATE LINER			
PISTON ROD GLAND PLATE			
GLAND BOLTS			
TREADLE PLATFORM			
STEAM CUSHION LINE			
SCALE HOSE			
SAFETY LINER			
SAFETY PROP			
STEAM LINES			
STEAM SHUTOFF VALVES			
REACH ROD			
COLUMNS FLAT ON BASE			
COLUMN WEDGE & BOLTS			
SPOOL BOLT & PIN			
SAFETY CABLES			
CRACKS IN BASE			
REMARKS:			

Fig. 33–25.—Maintenance checklist for steam hammers.

The operator should shut off the power, lock the main power switch, and immobilize the flywheel before attempting to adjust dies, heading tools, stock gages, or backstops.

Design of dies

Dies and heading tools used in an upsetter or horizontal forging machine are not usually subjected to such severe abuse as are hammer dies. A good grade of chrome, nickel, or molybdenum steel of the correct hardness is recommended for gripper dies.

It is common practice in die shops when resinking upset dies to use inserts rather than to cut down the dies. The inserts most commonly used are made of a high chrome, nickel, or molybdenum steel with silicon added.

Tool steels with a vanadium and tungsten analysis are also used for inserts and headers, but they must be selected to satisfy operating conditions because of their limited use where coolants are employed.

Front or back stock gages or backstops should be designed and located to serve the specific purpose and should not be a handicap to the operator. Special care should be taken to eliminate all hazards, especially pinch points.

Jobs which are abnormally heavy or which would create an unbalanced condition when running should be designed to employ balancing equipment to facilitate handling and reduce operator fatigue. The grip sometimes provided on upset die impressions should be sufficient to hold the stock securely, and should be checked after every run. This precaution is important for the operator's safety, especially on jobs where the heading tool could push the stock bar out of the impression toward the operator.

Setup and removal of dies

At the end of a run and before further work is done, all skids of stock or forgings should be moved out of the area to allow as much room as possible when the dies are changed.

Die setters should not try to make a complete setup dimensionally by measuring headers, strokes, and dies, without inching the header slide forward. This practice should be prohibited particularly on a worn machine where special shimming is required for proper alignment.

The proper sequence and procedure are as follows:

1. Check to make sure that all tools of correct size and other equipment are at hand.

2. Check dies and headers for defects which could develop into hazards.

3. Shut off all power and lock the master switch off.

4. Turn off the water or header lubricant and set the air brake. If the upsetter does not have an air brake, wait until the flywheel has stopped completely.

5. Loosen all setscrews, lock nuts, and hold-down bolts on the die clamps.

6. Remove the dies by means of a swivel or arm crane. (Eyebolt holes 2 in. deep should be drilled into all dies. Eyebolts or swivels should be stocked only in ¼-in.

graduations to a minimum of ¾ in. in diameter.)

7. Remove all die packing and thoroughly clean the die set seat; inspect for burrs, especially along the die keys.

8. Remove headers and dummies and any shims which may have been used.

9. Measure the new dies to determine the amount of packing needed. At least an ⅛-in. liner should be used to help protect the die seats.

10. Set in the new stationary die. Check to make sure that it is seated properly and packed correctly.

11. Tighten the hold-downs by hand.

12. Turn on the power, open the safety, release the brake and the flywheel. Set the machine on "inch" and slowly inch the header slide forward to bring the tool holders into correct position for the headers.

13. Disconnect the power and lock out the master switch. Make certain that the flywheel has completely stopped.

14. Assemble the new headers, tool dummies, and tool holders. Be sure that the headers to the die cavity are correctly matched and on correct center. It is good practice to set up according to a die layout which shows all principal dimensions on the equipment.

15. Check for the need to use shims behind the headers. Shims should be the washer type, not the horseshoe type, which fit around the shank of the header and cannot fall out.

16. Make a complete check of the assembly before finally tightening the header and dummy setscrews.

17. Insert the moving die, make sure that it is properly located, and then tighten the hold-down bolts.

18. Turn on the power. Set the machine on "inch" and slowly close the dies.

19. Check for match alignment and proper amount of packing. If too much packing

is used, the safety pin should open the dies.

20. Allow for expansion of the dies when they become hot. It may be necessary to remove a shim from back of the die.

21. If everything checks out, shut off the power, lock out the master switch, and allow the flywheel to come to a complete stop.

22. Tighten hold-down bolts, lock nuts, and setscrews.

23. If needed, attach a front gage or back-stop gage.

24. Turn on the power and try out one forging. Check dimensionally. If dies or headers need further adjustment, again turn off the power, lock out the master switch, and wait until the flywheel comes to a complete stop.

Inspection and maintenance

Since worn or defective upsetters can be dangerous to operate, these machines must be kept in excellent working order. A definite program of inspection and maintenance is essential.

The maintenance crew should check all working parts for wear and proper adjustment at least once each week. *The air clutch and brake should be inspected daily.*

The operator should also make daily inspection of air gages, air lines, water lines, water valves, belts, pulleys, and tools. He should check daily on the performance of the machine and should report any abnormal functioning immediately.

All equipment for handling dies, such as chains, cables, and eyes or swivels, should be inspected periodically as well as at each use.

Upsetters require daily lubrication. If possible, automatic lubrication should be installed.

Auxiliary equipment

All devices essential for safe operation, such as stock gages, should be designed for the particular job. There are three basic types of stock gages: the front gage (locates and swings away), the backstop gage (locates and helps hold the stock fixed), and the special tong gage or finger gage (locates and helps control the stock).

Tools should be in good condition. A complete set of wrenches to fit all sizes of nuts or bolts on the machine should be provided.

The jaws of tongs should conform to the shape of the stock being handled. Tongs should be made of tough, low-carbon steel so that they will not harden from repeated quenching in water.

Oil swabs and scale removers should have handles long enough to enable the operator to reach the full length of the dies without having to put his arm or hand between the dies.

Air, electrical, water, and oil lines should have distinctly marked shutoffs. Safety valves or switches should be located in a spot convenient for the operator.

Forging Presses

The forging press, because of its basic design, is similar to the power presses discussed in Chapter 32, "Cold Forming of Metals." However, since the work it performs is quite different from the conventional cold stamping operation, it has its own operating technique, die setup, and maintenance problems.

The increase in use of the forging press has been greater than that of any other type of forging equipment. Presses range from 500-ton capacity to 6000-ton capacity.

Compared to the cold stamping press, the forging press is designed with a faster-acting slide. The speed of the downward motion and the pick-up of the slide is one of the factors which determine the life of the forging dies. Another is control of the temperature of the dies. It is therefore important that the action of the press be fast enough to minimize the length of time that the dies are exposed to the forging temperature of the billet.

Basic precautions

This rapid action of the slide creates certain hazards which must be recognized and controlled by the operator. The size and shape of the forgings and the method of moving them into and out of the dies limit application of point-of-operation guarding. The most important single factor in the prevention of accidents, therefore, is the operator's control of the tools and methods he uses.

Because of the temperature of the forging dies and the forging temperature of the billet being forged, the operator must move the billet into the dies by means of tongs or special handling tools.

Tong handles should not be held in front of the body, and the fingers should not be placed between the handles. Die swabs must have handles long enough to keep hands and arms from between the dies.

Tongs, special handling tools, and die swabs are ordinarily the only tools used by an operator during the forging operation. It is important that they be kept in good condition at all times.

Proper clearance must be maintained at the front of the press. The operator must have enough hand room to allow for upward or downward motion of his tongs, which can be caused by improper spotting of the billets or tongs at the striking surfaces of the forging dies.

Steam lines, air lines, water headers, and scale and oil splash aprons must be properly located. If too near the working area of the operator's hands, they create pinch and shear points when tong handles are forced against them.

Unless properly confined, hot scale produced during the forging operation can cause serious burns. Air or steam curtains at the front of the die and directed onto the die facing will prevent scale from coming out the front of the press. Combination scale and smoke exhaust hoods should be installed at the back of the press to confine the scale and exhaust the smoke created by die lubricant.

Safety hats, eye protection, and safety shoes should be worn by operators at all forging press operations.

All forging presses should be equipped with pedal guards and nonrepeat devices. A forging press should never be operated on continuous stroke. The operating controls should require depression of the pedal for every cycle of the slide. The press controls should also permit inching of the slide for die setting or other press adjustments.

Die setting

Since the setting of dies in forging presses is very different from the setting of dies in cold stamping presses, certain extra precautions not required on the latter type of press must be taken.

Forge press dies are set in die holders designed in sets, each set consisting of an upper and a lower section. The upper section of the holder is secured to the press slide by threaded bolts, and the lower section to the press bed by threaded bolts. The forging dies are set or nested in die pockets recessed in both the upper and lower sections of the holder.

The problem of setting the dies in the recessed pockets and of removing them can be solved by the use of special equipment consisting of an eyebolt attachment at the face of the die and a die truck with boom attachment. Pry bars and blocking should be carefully used since the pry bar may easily slip from the die and injure the operator's hands or fingers.

The dies are secured in the pockets of the holder by sectional flat clamps and a series of cap screws through the holder at all four sides of the die. These screws are used primarily to shift the die for matching and to prevent the die from floating after it has been properly aligned.

Suitable blocking should be installed to hold the top die in the pocket before the clamps which secure the top die in the holder are removed. The adjustment cap screw should not be relied upon to hold the die after the clamps have been removed.

The press should be equipped with safety props interlocked with the press motor circuit so that the motor cannot be started when the props are in use.

All die clamps, bolts, washers, blocking, wrenches, and handling equipment should be in good condition at all times.

The table wedges which are used for making vertical adjustment of the dies must be kept free of scale so that the wedge can be raised and lowered with minimum pressure. Trucks or driving rams should not be used for this operation. If wedges are kept free of scale, they can be raised or lowered by hand or by air motor wrench.

Maintenance

Maintenance of forging presses requires the same precautions used in maintenance work on power presses used in cold stamping. The operating power source should be locked out so that the equipment cannot be energized while it is being worked on. If adjustments have to be made while the press is energized, the work must be carried out under the direct supervision of the maintenance supervisor.

Suitable permanent work platforms should be installed for making brake adjustments and doing repair work at the surge tank and booster cylinders. To prevent falls, maintenance men should not use portable straight ladders nor stand on parts of the press, such as the press crown or the backshaft.

Major repair work on forging presses usually requires removal of bulky, heavy parts. Tearing down of the rolling type clutch, flywheel, slide, pitman, and crankshaft requires special heavy-duty rigging used by skilled workmen trained in this type of work.

The brake should be adjusted properly and kept in good repair. Precautions must be taken to prevent the brake lining from becoming flooded with oil. A sheet steel disk about 2 in. larger than the brake wheel, placed on the eccentric shaft back of the brake wheel, will help reduce the amount of lubricant on the brake.

Pitman bolts should be securely tightened. It may be necessary to design and make a socket or box end wrench with a strong heavy handle for the specific press.

Flywheel hubs, clutch spline hubs, pinion gears, and brake wheels should not be permitted to become loose on the pinion shaft. If they do become loose and run that way very long, keys and keyways will be ruined and the keys will break out a portion of the pinion shaft on one side of the keyway. The keys on each side of the pinion shaft should be inspected periodically and driven tight if they are loose. When a key will no longer stay tight, it should be replaced with a carefully fitted new key.

When the hub or gear on the shaft becomes too loose, it will be impossible to hold the keys tight and the shaft will have to be turned down. The inside of a steel hub should be welded and bored to a shrink fit on the new shaft size. The keyways should be remachined and new keys made. If the hub is cast iron, a tapered sleeve can be made or a new hub fitted to the shaft.

Experienced repairmen warn against welding the flywheel end of the pinion shaft. If normalizing of the welded section is not complete, a crack may start and allow the shaft to break and drop the rotating flywheel. Crankshafts and backshafts on forging presses likewise should not be welded.

The friction slip on a press should be tightened so that it will not creep even on heavy jobs. If the friction slip is allowed to move a small amount with every forging, this movement will polish or glaze the friction surfaces and soon the friction slip will move more with each forging. This movement will cause the friction surface to score the hub and clamp surfaces.

Rotation of the flywheel on the friction hub will frequently wear or cut the inside diameter of the flywheel so that it runs off center. To correct this condition, the inside diameter of the flywheel can be bored and fitted with a bronze bushing.

Sometimes the eccentric shaft becomes cracked in the fillet on each side of the eccentric. This defect is usually noticed when the shaft is taken out so that the main bearings can be checked or the shaft can be machined. This crack should be examined carefully to determine its length, depth, and direction of travel. If the crack is in a critical area where failure would permit the rolling clutch to drop off, then the shaft must be replaced.

Other Forging Equipment

Hot trimming dies can be made from hardened chrome, nickel, carbon, and vanadium steels or from medium carbon steels with cutting edges which have been hard faced with a rod similar to a Haynes Stellite No. 6 rod.

Stellite should be used on the trimmer only, and not on both the trimmer and the punch. A stellite punch should never work against a stellited punch die. One or the other, preferably the punch, should be softer.

Cold trim dies and punches are usually made from a high carbon, high chrome, molybdenum steel hardened and drawn to a 60 to 62 Rockwell C hardness. In both hot and cold trimmers it is important for the designer to equalize the trim so that stresses are equally distributed.

Proper working clearance must be maintained to facilitate correct location for unloading. If possible, cold trimmers should be designed to work on guide pins for proper alignment.

Padding, bending, or straightening equipment is sometimes required to complete a forging. This equipment can be made from a good grade of wear-resistant carbon, nickel steel. Hot pad dies are usually made with an opening between the die faces so that the hot forging acts as a cushion. Additional clearance should be provided in a hot pad or restrike die because the forging is constantly cooling and shrinking.

Cold coin dies are made from high carbon, high chrome steels hardened to 62 to 64 Rockwell C hardness. The forging is coined cold. For best result, either interlocking dies or dies with guide pins are used. Additional safety measures such as magazine type loaders are sometimes made an integral part of the die for the operator's convenience.

The faces of the coining dies should be ground as smooth as possible. No lubricant should be used because it may cause the forging to stick to one die face. The opening between the dies is sometimes limited mechanically so that no more than one forging can be loaded at a time.

Bulldozers. The greatest danger in the operation of a bulldozer is the possibility of a workman's being caught between the dies. There are several ways to decrease this hazard. A guard may be attached to the side of the moving head and travel with it past the stationary head, or telescoping rods or rails may be used to serve the same purpose. The base plate can be notched out to leave room for the operator's leg. The clutch should be kept in good order so that the machine will not repeat. The power transmitting mechanisms should be guarded.

Cold heading and similar machines should have screen shields to protect workers from flying pieces. Relief springs should be guarded to prevent the bolts and nuts from being thrown out in case of breakage.

Bolt headers and riveting machines should have treadle guards to prevent accidental operation. The machines should be stopped and blocked before dies are changed or adjustments are made.

Hot saws should be provided with tanks of water placed below the saws and 8-in. sheet metal guards positioned to stop the flying sparks.

References

Forging

American National Standards Institute, 1430 Broadway, New York, N.Y. 10018.
> *Safety Requirements for Forging*, B24.1.

National Safety Council, 425 North Michigan Ave., Chicago, Ill. 60611.
> *Forging Safety Manual.*
> *Guards Illustrated.*
> Industrial Data Sheets
> > *Cold Shearing Billets and Bars*, 557.
> > *Handling Materials in the Forging Industry*, 551.
> > *Handling Steel Plate for Fabrication*, 565.
> > *Mechanical Forging Presses*, 560.
> > *Method of Locking Out Electric Switches*, 237.
> > *Setup and Removal of Forging Hammer Dies*, 467.
> > *Steam Drop Hammers*, 450.
> > *Upsetters*, 466.

United Auto Workers, 8000 E. Jefferson Ave., Detroit, Mich. 48214. (General.)

33—Hot Working of Metals

Foundries

American Foundrymen's Society, Golf and Wolf Rds., Des Plaines, Ill. 60016.
Engineering Manual for Control of In-Plant Environment in Foundries.
Health Protection in Foundry Practice.
Safety in Metal Casting.
Recommended Practices for Grinding, Polishing, and Buffing Equipment.

American National Standards Institute, 1430 Broadway, New York, N.Y. 10016.
Safety Code for Floor and Wall Openings, Railings, and Toe Boards, A12.1.
Ventilation Control of Grinding, Polishing, and Buffing Operations, Z43.1.

National Safety Council, 425 North Michigan Ave., Chicago, Ill. 60611.
Fundamentals of Industrial Hygiene.
Industrial Data Sheets
  Coated Abrasives, 452.
  Magnesium, 426.
  Mullers and Other Pan-Type Mixers, 526.

U.S. Department of Health, Education, and Welfare, Public Health Service, 404 N. Fairfax Dr., Arlington, Va. 22203.

Nondestructive testing

McMaster, Robert. *Nondestructive Testing Handbook.* New York, Ronald Press.

National Safety Council, 425 North Michigan Ave., Chicago, Ill. 60611.
Industrial Data Sheets
  Beta Ray Sealed Sources, 461.
  Nondestructive Testing of Materials, 488.

Welding
and
Cutting

Chapter
34

Gas Welding and Oxygen Cutting   **948**
Welding and cutting gases . . . Compressed gas cylinders . . .
Handling cylinders . . . Storing cylinders . . . Using cylinders
. . . Manifolds . . . Distribution piping . . . Portable outlet headers
. . . Regulators . . . Hose and hose connections . . . Torches . . .
Powder cutting

Resistance Welding   **957**
Type . . . Power supply . . . Cables . . . Electrodes and holders
. . . Machine installation

Arc Welding and Cutting   **960**
Power supply . . . Voltages . . . Cables . . . Electrodes and holders
. . . Automatic and semiautomatic welding . . . Protection against
electric shock . . . Gas-shielded arc welding

Other Welding and Cutting Processes   **966**

Common Hazards   **966**
Light rays . . . Fire protection . . . Floors and combustible materials
. . . Hazardous locations . . . Drums, tanks, and closed containers

Personal Protection   **969**
Respiratory protection . . . Eye protection . . . Protective clothing

Training in Safe Practices   **974**

References   **974**

Most welding and cutting processes—for construction, demolition, maintenance, and repair—use portable, manually operated equipment. The major portion of this chapter discusses ways of minimizing the hazards encountered in this type of intermittent operation. Since production line welding and cutting equipment is usually permanently installed, the source of hazards can be minimized through safe design. The safety professional, however, can use this chapter to help him check the operation and maintenance of production line equipment.

The detailed specifications of ANSI Standard Z49.1, *Safety in Welding and Cutting*, should be followed. In addition, city, state, and Federal codes and regulations, where applicable, should be consulted.

This chapter uses the American Welding Society's welding and cutting definitions:

A WELDER is "one who is capable of performing a manual or semiautomatic welding operation."

A WELDING OPERATOR is "one who operates machine or automatic welding equipment."

Gas Welding and Oxygen Cutting

A gas-welding process unites metals by heating them with the flame from the combustion of a fuel gas or gases and sometimes includes the use of pressure and a filler metal.

An oxygen-cutting process severs or removes metal by the chemical reaction of the metal with oxygen at an elevated temperature maintained with heat from the combustion of fuel gases. In the powder-cutting process, a finely divided material, such as iron powder, is added to the cutting oxygen stream. The powder bursts into flame in the oxygen stream and starts cutting without preheating the material to be cut. Powder cutting is used on stainless and other steels, on many nonferrous metals, and on concrete in construction and demolition jobs. Plasma arc cutting is now replacing powder oxy-fuel cutting.

Welding and cutting gases

Oxygen is furnished to the consumer in steel cylinders, usually under a pressure of about 2200 psi at 70 F, or as a liquid to be gasified on the consumer's premises.

Pure oxygen will not burn or explode. It supports combustion; that is, it causes other substances to burn when they are raised to the kindling temperature. Combustible materials burn much more rapidly in oxygen than in air. Oxygen forms explosive mixtures in certain proportions with acetylene, hydrogen, and other combustible gases.

Acetylene (C_2H_2) consists of 92.3 percent by weight of carbon and 7.7 percent by weight of hydrogen in chemical combination. It contains stored-up energy which is released as heat when it burns, as in the welding flame. This heat is in addition to that which would be obtained by combustion of equivalent amounts of elemental carbon and hydrogen.

Acetylene burned with oxygen can produce a higher flame temperature (approximately 6000 F) than any other gas used commercially. Acetylene, like other combustible gases, ignites readily, and in certain proportions forms a flammable mixture with air or oxygen. The range of flammable limits of acetylene (2.5 to 81 percent acetylene in air) is greater than that of other commonly used gases, with consequently greater hazard.

Acetylene is either supplied in cylinders or generated as needed. It is a product of the reaction between water and calcium carbide, a gray crystalline substance made commercially by fusing lime and coke in an electric furnace. Calcium carbide itself is neither flammable nor explosive. It is stored and sold in airtight and watertight cans or drums. If the drums are damaged in handling and if water comes in contact with carbide, acetylene will be generated and there is then danger of ignition and explosion.

Hydrogen is furnished in cylinders under a pressure of about 2000 psi at 70 F. It may ignite in the presence of air or oxygen when in contact with a spark, open flame, or other source of ignition. Hydrogen-air mixtures in the range from 4.1 to 74.2 percent hydrogen are flammable.

Other fuel gases are used with oxygen in torches, primarily for oxygen cutting. For example, propane, butane, and their mixtures are supplied in cylinders in liquid form, generally under various trade names. These gases are discussed in Chapter 42, "Flammable and Combustible Liquids."

Compressed gas cylinders

Most of the gas used for welding and cutting is purchased in cylinders. These cylinders should be constructed and maintained in accordance with regulations of the Department of Transportation. The purchaser should make sure that all cylinders bear DOT markings. The contents should be legibly marked on each cylinder in large letters.

Oxygen is supplied in cylinders; the usual size for welding contains 244 cu ft of oxygen under pressure of 2200 psi at 70 F. A cap should be provided to protect the outlet valve when the cylinder is not connected for use.

Acetylene for welding and cutting is usually supplied in cylinders having a capacity up to about 300 cu ft of dissolved acetylene under pressure of 250 psi at 70 F.

Acetylene cylinders should be completely filled with an approved porous material impregnated with acetone, the solvent for acetylene. The porous material should have no voids of appreciable size so that acetylene can be safely stored at the prescribed full cylinder pressure. Since acetylene is highly soluble in acetone at cylinder filling pressure, large quantities of acetylene can be stored in comparatively small cylinders at relatively low pressures.

LP-gas. Handling, storage, and use of LP-gas are discussed in Chapter 42, "Flammable and Combustible Liquids."

Handling cylinders

Serious accidents may result from the misuse, abuse, or mishandling of compressed gas cylinders. Workers assigned to the handling of cylinders under pressure should be carefully trained and should work only under competent supervision. Observance of the following rules will help control hazards in the handling of compressed gas cylinders.

1. Accept only cylinders approved for use in interstate commerce for transportation of compressed gases.

2. Do not remove or change numbers or marks stamped on cylinders.

3. Because of their shape, smooth surface, and weight, cylinders are difficult to carry by hand. Cylinders may be rolled on bottom edge but never dragged. Cylinders weighing more than 40 pounds (total) shall be transported on a hand or motorized truck.

4. Protect cylinders from cuts or abrasions.

5. Do not lift compressed gas cylinders with an electromagnet. Where cylinders must be handled by a crane or derrick, as on construction jobs, carry them in a cradle or suitable platform and take extreme care that they are not dropped. Do not use slings.

6. Do not drop cylinders or let them strike each other violently.

7. Do not use cylinders for rollers, supports, or any purpose other than to contain gas.

8. Do not tamper with safety devices in valves on cylinders.

9. When in doubt about the proper handling of a compressed gas cylinder or its contents, consult the supplier of the gas.

10. When empty cylinders are to be returned to the vendor, mark them EMPTY or MT with chalk. Close the valves and replace the valve protection caps.

11. Load cylinders to be transported to allow as little movement as possible. Secure them to prevent violent contact or upsetting.

12. Always consider cylinders as full and handle them with corresponding care. Accidents have resulted when containers under partial pressure were thought to be empty.

The fusible safety plugs on acetylene cylinders melt at about the boiling point of water. If an outlet valve becomes clogged with ice or frozen, it should be thawed with warm (not boiling) water, applied only to the valve. A flame should never be used.

Storing cylinders

Cylinders should be stored in a safe, dry, well ventilated place prepared and reserved for the purpose. Flammable substances, such as oil and volatile liquids, should not be stored in the same area. Cylinders should not be stored near elevators, gangways, stair wells, or other places where they can be knocked down or damaged.

Cylinders of oxygen should not be stored indoors within 20 ft of cylinders containing flammable gases or to highly combustible materials. If closer, cylinders should be separated by a fire-resistive partition (½-hr minimum rating).

Acetylene and liquefied fuel gas cylinders should be stored with the valve end up. The total capacity of acetylene cylinders stored and used inside a building should be limited to 2000 cu ft of gas, exclusive of cylinders in use or connected for use. Quantities exceeding this total should be stored in a special room built in accordance with the specifications of NFPA Standard No. 51, in a separate building, or outdoors. Acetylene storage rooms and buildings must be well ventilated, and open flames must be prohibited. Storage rooms should have no other occupancy.

Cylinders should be stored on a level, fireproof floor. One common type of storage house consists of a shed roof with side walls extending approximately halfway down from the roof and a dividing wall between one kind of gas and another.

To prevent rusting, cylinders stored in the open should be protected from contact with the ground and against extremes of weather—accumulations of ice and snow in winter and continuous direct rays of the sun in summer.

Cylinders are not designed for temperatures in excess of 130 F. Accordingly, they should not be stored near sources of heat, such as radiators or furnaces, or near highly flammable substances like gasoline.

Cylinder storage should be planned so that cylinders will be used in the order in which they are received from the supplier. Empty and full cylinders should be stored separately, with empty cylinders being plainly identified as such to avoid confusion. Group together cylinders which have held the same contents.

A direct flame or electric arc should never be permitted to contact any part of a compressed gas cylinder.

Storage rooms for cylinders containing flammable gases should be well ventilated to prevent the accumulation of explosive concentrations of gas. No source of ignition should be permitted. Smoking should be prohibited. Wiring should be in conduit. Electric lights should be in fixed position and enclosed in glass or other transparent material to prevent gas from contacting lighted sockets or lamps and should be equipped with guards to prevent breakage. Electric switches should be located outside the room.

Using cylinders

Safe procedures for the use of compressed gas cylinders include:

1. Use cylinders, particularly those containing liquefied gases and acetylene, in an upright position and secure them against accidentally being knocked over.

2. Unless the cylinder valve is protected by a recess in the head, keep the metal cap in place to protect the valve when the cylinder is not connected for use. A blow on an unprotected valve might cause gas under high pressure to escape.

3. Make sure the threads on a regulator or union correspond to those on the cylinder valve outlet. Do not force connections that do not fit.

4. Open cylinder valves slowly. A cylinder not provided with a handwheel valve should be opened with a spindle key or a special wrench or other tool provided or approved by the gas supplier.

5. Do not use a cylinder of compressed gas without a pressure-reducing regulator attached to the cylinder valve, except where cylinders are attached to a manifold, in which case the regulator will be attached to the manifold header.

6. Before making connection to a cylinder valve outlet, "crack" the valve for an instant to clear the opening of particles of dust or dirt. Always point the valve and opening away from the body and not toward anyone else. Never crack a fuel gas cylinder valve near other welding work or near sparks, open flames, or other possible sources of ignition.

7. Use regulators and pressure gages only

FIG. 34–1.—Well-designed manifold system for acetylene cylinders.

Courtesy Linde Co., Division of Union Carbide Corp.

with gases for which they are designed and intended. Do not attempt to repair or alter cylinders, valves, or attachments. This work should be done only by the manufacturer.

8. Unless the cylinder valve has first been closed tightly, do not attempt to stop a leak between the cylinder and the regulator by tightening the union nut.

9. Fuel gas cylinders in which leaks occur should be taken out of use immediately and handled as follows:

Close the valve, and take the cylinder outdoors well away from any source of ignition. Properly tag the cylinder, and notify the supplier. A regulator attached to the valve may be used temporarily to stop a leak through the valve seat.

If the leak occurs at a fuse plug or other safety device, take the cylinder outdoors well away from any source of ignition, open the cylinder valve slightly, and permit the fuel gas to escape slowly. Tag the cylinder plainly. Post warnings against approaching with lighted cigarettes or other sources of ignition. A responsible person should stay in the area until the cylinder is depressured to make sure that no fire occurs. Promptly notify the supplier, and follow his instructions for returning the cylinder.

10. Do not permit sparks, molten metal, electric currents, excessive heat, or flames to come in contact with the cylinder or attachments.

11. Never use oil or grease as a lubricant on valves or attachments of oxygen cylinders. *Keep oxygen cylinders and fittings away from oil and grease, and do not handle such cylinders or apparatus with oily hands, gloves, or clothing.*

12. Never use oxygen as a substitute for compressed air in pneumatic tools, in oil preheating burners, to start internal combustion engines, or to dust clothing. Use it only for the purpose for which it is intended.

13. Never bring cylinders into tanks or unventilated rooms or other closed quarters.

14. Do not fill cylinders except with the consent of the owner and then only in accordance with DOT regulations. Do not attempt to mix gases in a compressed gas cylinder or to use it for purposes other than those intended by the supplier.

15. Before a regulator is removed from a cylinder valve, close the cylinder valve and release the gas from the regulator.

Manifolds

Cylinders are manifolded to centralize the gas supply and to provide gas continuously and at a rate in excess of that which may be obtained from a single cylinder. Manifolds must be of substantial construction and of a design and material suitable for the particular gas and service for which they are to be used. Manifolds should be obtained from, and installed under the supervision of, a reliable manufacturer familiar with safe practices in construction and use of manifolds.

Portable manifolds connect a small number (usually not over five) of cylinders for direct supply to a consuming device. The cylinders may be connected by individual leads to a single, common coupler block or individual cylinders may be connected to a common line with coupler tees attached to the cylinder valves. A properly supported regulator serves the group of connected cylinders.

Stationary manifolds connect a larger number of cylinders for supply through piped distribution systems. This type of manifold consists of a substantially supported stationary pipe header to which the cylinders are connected by individual leads (Fig. 34–1). One or more permanently mounted regulators serve to reduce and regulate the pressure of the gas flowing from the manifold.

Oxygen manifolds should be located away from highly flammable material, especially oil, grease, and the like. They should not be located in acetylene generator rooms, or in close proximity to cylinders of combustible gases. There should be a fire-resistant partition between an oxygen manifold and combustible gas cylinders, unless the manifold and such cylinders are separated at least 50 ft. Regulations of NFPA Standard No. 51, *Oxygen-Fuel Gas Systems for Welding*, should be followed.

Distribution piping

Distribution piping carrying oxygen from a manifold or other centralized supply should be of steel, wrought iron, brass, or copper. For pressures over 150 psi, extra-heavy pipe and fittings should be used. For lower pressures, standard weight pipe and fittings or seamless nonferrous gas tubing and fittings may be used.

All pipe and fittings for oxygen service lines should be examined before use, and, if necessary, tapped with a hammer to free them from dirt and scale. They should, in every case, be washed out with a suitable non-flammable cleaner—hot water solutions of caustic soda and trisodium phosphate are effective.

Only steel or wrought iron piping should be used for acetylene distribution systems. Under no circumstances should acetylene gas be brought into contact with unalloyed copper except in a torch treated to prevent chemical reaction. Joints in steel or wrought iron pipe should be welded or made up with threaded or flanged fittings. Flanged connections in acetylene lines should be electrically bonded. Cast iron fittings should not be used.

Joints in brass or copper pipe may be welded, threaded, or flanged. A socket joint may be brazed with silver solder or similar high melting point material. Threaded connections in oxygen piping should be tinned or made up with litharge and glycerine (litharge and water for service pressures over 300 psi) or other joint compound approved for oxygen service.

In fuel gas distribution systems, a pressure-reducing regulator, back-flow check valve, or hydraulic seal should be used to prevent back flow at every point where gas is withdrawn from the piping system to supply a torch or machine. Such devices should be listed (or approved) by an agency such as Factory Mutual or Underwriters Laboratories Inc.

Portable outlet headers

Portable outlet headers are assemblies of valves and connections used for service outlet purposes and are connected to a permanent service piping system by means of hose or other nonrigid conductors. Devices of this nature are commonly used at piers and dry docks in shipyards where the service piping cannot be located close enough to the work to provide a direct supply. Their use should be restricted to outdoor locations and to temporary service where conditions preclude a direct supply, and they should be used in accordance with regulations in NFPA Standard No. 51.

Regulators

Regulators or reducing valves must be used on both oxygen and fuel gas cylinders to maintain a uniform gas supply to the torches at the correct pressure. The oxygen regulator should be equipped with a safety relief valve or be so designed that, should the diaphragm rupture, broken parts will not fly. Workers should stand to one side and away from regulator gage faces when opening cylinder valves.

Only regulators listed by agencies such as Underwriters Laboratories Inc., or Factory Mutual should be used on cylinders of compressed gas. If unlisted regulators are used, they should be fully checked by a competent welding engineer. Each regulator (oxygen or fuel gas) should be equipped with both a high pressure (contents) gage and a low pressure (working) gage.

High pressure oxygen gages should have safety vent covers to protect the operator from broken glass in case of an internal explosion. Each oxygen gage should be marked OXYGEN —USE NO OIL.

Serious, even fatal, accidents have resulted when oxygen regulators have been attached to cylinders containing fuel gas, or vice versa. To guard against this hazard, it has been customary to make connections for oxygen regulators with right-hand threads and those for acetylene regulators with left-hand threads, to mark the gas service on the regulator case, and to paint the two types of regulators, different colors. Cylinder valve outlet threads have been standardized for most industrial and medical gases; see ANSI B57.1, *Compressed Gas Cylinder Valve Outlet and Inlet Connections* (CGA V-1). Different combinations of right-hand and left-hand threads, internal and external threads, and different diameters to guard against wrong connections are now standard. (Standards are being rapidly adopted whenever gas manufacturers and industrial users reach agreement to change both valve outlets and regulator connections.)

The regulator is a delicate apparatus and should be handled carefully. It should not be dropped or pounded on. Regulators should be repaired or tested only by skilled workmen or sent to manufacturer for repairs.

Leaky or "creeping" regulators are a source of danger and should be withdrawn from service at once for repairs. If a regulator shows a continuous creep, indicated on the low pressure (delivery) gage by a steady buildup of pressure when the torch valves are closed, the cylinder valve should be closed and the regulator removed for repairs.

If the regulator pressure gages have been strained so that the hands do not register properly, the regulator must be replaced or repaired before it is used again.

When regulators are connected but are not in use, the pressure-adjusting device should be released. Cylinder valves should never be opened until the regulator is drained of gas and the pressure-adjusting device on the regulator is fully released.

These procedures should be followed in detail when regulators or reducing valves are being attached to a gas cylinder.

1. To blow out dust or dirt that otherwise might enter the regulator, "crack" the discharge valve on the cylinder by opening it slightly for an instant and then close it. On a fuel gas cylinder, first see that no open flame or other source of ignition is near; otherwise, the gas may ignite at the valve.

2. Connect the regulator to the outlet valve on the cylinder. Be sure the regulator inlet threads match the cylinder valve outlet threads. Never connect an oxygen regulator to a cylinder containing fuel gas, or vice versa. Don't force connections which do not fit easily. Be sure that the connections between the regulators and cylinder valves are gas-tight.

3. Release the pressure-adjusting screw on the regulator to its limit—turn it counterclockwise until it is loose.

4. Open the cylinder valve slightly to let the hand on the high pressure gage move up *slowly*. On an oxygen cylinder gradually open the cylinder valve to its full limit, but on an acetylene cylinder make no more than 1½ turns of the valve spindle.

5. Attach oxygen hose to outlet of oxygen regulator and to oxygen inlet valve on torch. Attach acetylene hose to outlet of acetylene regulator and to acetylene inlet on torch.

6. Test oxygen connections for leaks. Be sure torch oxygen valve is closed; then turn oxygen regulator pressure-adjusting screw clockwise to give about normal working pressure. Using soapy water

953

FIG. 34-2.—Testing for leaks. With the pressure on and the torch valves closed, hold the hose (*upper left*) and the torch tip under water. Bubbles indicate leaks. Use soapsuds to test for leaks on the torch valves and hose-to-torch connections (*upper right*) as shown by arrows. Separately test the oxygen and acetylene cylinder and regulator connections (*lower left and right*) for leaks at points marked by arrows.

(nonfat soap) or approved leaktest solution, check connections for leaks as in Fig. 34–2. At the same time, check regulator for creeping indicated by an increase in the reading on the low pressure (delivery) gage. If the regulator creeps, have it replaced or repaired before it is used.

7. Test acetylene connections for leaks. Be sure torch acetylene valve is closed and proceed in manner similar to No. 6 above —except that acetylene regulator pressure-adjusting screw should be set to produce a pressure of about 10 psi (see Fig. 34–2).

8. If torch is to be used immediately, proceed as in No. 9. If not, close cylinder valves, open torch valves to release pressure on regulator, etc., close torch valves, and release pressure-adjusting screws on regulators.

9. To adjust pressures of oxygen and fuel gas prior to using torch, proceed as follows: with all torch valves closed, slowly open oxygen cylinder valve, open torch oxygen valve, turn in pressure-adjusting screw on oxygen regulator to desired pressure, then close torch oxygen valve. Open acetylene cylinder valve (1½ turns only), and with torch acetylene valve

closed, turn in pressure-adjusting screw on acetylene regulator to desired pressure.

10. Purge each line individually. Open oxygen torch valve and release oxygen to the atmosphere for a few seconds before closing the valve; then open acetylene torch valve and release acetylene to the atmosphere for a few seconds and close the valve.

11. Open torch acetylene valve, light flame, and readjust regulator. Then close torch acetylene valve. (Acetylene pressure should first be adjusted with torch valve closed to prevent release of acetylene to air.)

12. Open torch valves and light torch according to procedure described in instructions provided with the equipment. The procedure for operating one torch is not necessarily best or even satisfactory for another.

Hose and hose connections

The oxygen and acetylene hoses should be of different colors or otherwise identified and distinguished from each other. Red is the generally recognized color for fuel gas hose and green for oxygen hose (Fig. 34–3). The hose connections are usually marked STD-OXY for oxygen and STD-ACET for acetylene. The acetylene union nut has a groove cut around the center to indicate left-hand threads.

Black is used for inert gas and air hoses.

Connections for joining the hose to the hose

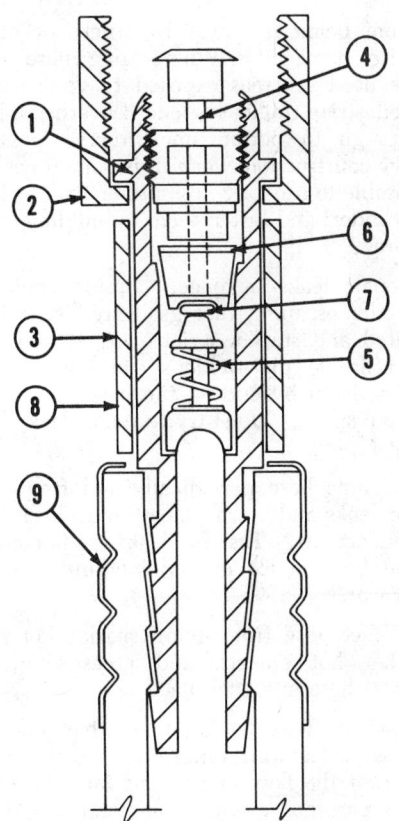

Fig. 34–4.—This type connector can provide positive gas shutoff and eliminate need for purging hose when torches are changed. Numbers indicate: 1—valve body, 2—connecting nut, 3—retainer sleeve, 4—valve stem, 5—spring, 6—seat, 7—valve body seat, 8—sleeve, and 9—hose collar. Overall length is 2⅝ in.

Courtesy Anderson Equipment Co.

nipple on the torches and regulators may be either the ferrule or clamp type. Gaskets should not be used on these connections. Special torch connectors with built-in shutoff valves are available (Fig. 34–4 shows one type).

Following are suggestions for safe use of hose in welding and cutting operations:

1. Do not use unnecessarily long hose—it's hard to purge properly. When long hose must be used, see that it does not become kinked or tangled and that it is protected

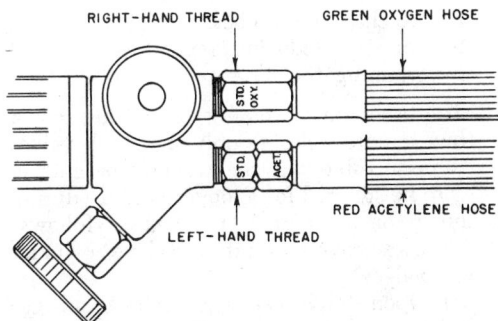

RIGHT-HAND THREAD   GREEN OXYGEN HOSE

STD OXY

STD ACET

RED ACETYLENE HOSE

LEFT-HAND THREAD

Fig. 34–3.—When attaching hoses to a welding or cutting torch, use the red hose for acetylene and the green hose for oxygen; then test connections for leaks.

from being run over by trucks or otherwise damaged. Where long hose must be used in areas exposed to vehicular or pedestrian traffic, suspend it overhead, high enough to permit unobstructed passage. On construction work it is sometimes advisable to use long hose rather than to hoist cylinders and fasten them to building structures.

2. Repair leaks at once. Besides being a waste, escaping fuel gas may become ignited and start a serious fire; it may also set fire to the welder's clothing. Repair hose leaks by cutting the hose and inserting a splice. Don't try to repair leaky hose by taping.

3. Examine hose periodically and frequently for leaks and worn places, and check hose connections. Test for leaks by immersing the hose under normal working pressure in water.

4. Protect hose from flying sparks, hot slag, other hot objects, and grease and oil. Store hose in a cool place.

5. A single hose having more than one gas passage, a wall failure of which would permit the flow of one gas into the other gas passage, is not recommended. When parallel links of oxygen and acetylene hose are taped together for convenience and to prevent tangling, not more than 4 in. of each 12 in. of hose should be taped.

6. The use of hose with an external metallic covering is not recommended. In some machine processes and in certain types of operations, hose with an inner metallic reinforcement, which is exposed neither to the gas passage nor to the outside atmosphere, is acceptable.

7. If a flashback occurs and burns the hose, discard the burned section. Purge new hose before connecting it to the torch and regulator.

Torches

Torches are constructed of metal castings, forgings, and tubing. Usually, they are made of brass or bronze, but stainless steel may also be used. They should be of substantial design to withstand the rough handling they sometimes receive. It is best to use only those torches listed (or approved) by an agency such as Underwriters Laboratories Inc., or Factory Mutual.

The gases enter the torch by separate inlets, go through valves to the mixing chamber, and then to the outlet orifice, located in the torch tip. Several interchangeable tips are provided with each torch and have orifices of various sizes according to the work to be done.

The cutting torch, unlike the welding torch, uses a separate jet of oxygen in addition to the jet or jets of mixed oxygen and fuel gas. The jets of mixed gases are for preheating the metal, and the pure oxygen jet is for cutting. The flow of oxygen to the cutting jet is controlled by a separate valve.

There are two types of torches in general use: the "injector" or low-pressure type, and the "pressure" or medium-pressure type. In the injector torch, the acetylene is drawn into the mixing chamber by the velocity of the oxygen. The acetylene may be supplied either from a low-pressure generator or a medium-pressure generator or from cylinders. In the medium-pressure torch, the gases enter under pressure; therefore, the acetylene is supplied from cylinders or from a medium-pressure generator.

In the operation of torches several precautions should be observed.

1. Select the proper welding head or mixer, tip or cutting nozzle (according to charts supplied by the manufacturer), and screw it firmly into the torch.

2. Before changing torches, shut off the gas at the pressure-reducing regulators and not by crimping the hose.

3. To discontinue welding or cutting for a few minutes, closing only the torch valves is permissible. If the welding or cutting is to be stopped for a longer period (during lunch or overnight), proceed as follows:
   a) Close oxygen and acetylene cylinder valves.
   b) Open torch valves to relieve all gas pressure from hose and regulator.
   c) Close torch valves and release regulator pressure-adjusting screws.

4. Do not use matches to light torches. Use

a friction lighter, stationary pilot flame, or other suitable source of ignition. When lighting, point the torch tip so no one will be burned when the gas ignites.

5. Never put down a torch until the gases have been completely shut off. Do not hang torches from a regulator or other equipment so that they come in contact with the sides of gas cylinders. If the flame has not been completely extinguished, it may heat the cylinder or even burn a hole through it.

6. When extinguishing the flame, close the acetylene and oxygen valves in the order recommended by the torch manufacturer. However, if the oxygen valve is closed first, the acetylene flame enlarges appreciably and could burn the welder. Unburned carbon "feathers" will also be deposited in the area. If the acetylene is turned off first, the loud report or "bang" which results can distract nearby workers.

Powder cutting

Powder-cutting processes for metal and concrete use similar equipment and gas supplies as do oxygen-cutting operations. The precautions previously discussed for safe handling and use of compressed gas equipment and cutting torches therefore apply. Manufacturers' recommendations for the operation and maintenance of the powder-dispensing apparatus—both pneumatic and vibratory—should be followed.

Resistance Welding

Since resistance welding equipment is normally permanently installed, the hazards are usually minimized if the equipment has been properly designed and safe operating practices have been established.

Certain hazards in the operation of this equipment—lack of point-of-operation guards, flying hot metallic particles, improper handling of materials, unauthorized adjustments and repairs—may cause eye injuries, burns, and electrical shock. Most of these hazards can be eliminated by safeguarding the equipment, by the wearing of protective clothing, and by strict control of operating practices.

Resistance welding is a metal-joining proc-

FIG. 34–5.—An automatic resistance welding machine with a dial feed. The operater removes and places the work when the proper dial fixture comes to the front.

Courtesy General Electric Co.

ess in which the welding heat is generated at the joint by the resistance to the flow of electric current. The three fundamental parameters of resistance welding are current magnitude, current time, and tip pressure, each of which must be accurately controlled.

Type

A SPOT WELD is made by applying heat and

pressure at one point, usually on overlapped parts.

A SEAM WELD is a series of overlapping spot welds, spaced close enough to make a single, continuous joint.

FLASH WELDING. A resistance-welding process wherein coalescence is produced, simultaneously over the entire area of abutting surfaces, by the heat obtained from resistance to the flow of electric current between the two surfaces, and by the application of pressure after heating is substantially completed. Flashing and upsetting are accompanied by expulsion of metal from the joint.

PERCUSSION WELDING. A welding process wherein coalescence is produced simultaneously over the entire abutting surfaces by the heat obtained from an arc. This arc is produced by a rapid discharge of electrical energy. It is extinguished by pressure percussively applied during the discharge.

PROJECTION WELDING. A resistance-welding process wherein coalescence is produced by the heat obtained from resistance to the flow of electric current through the work parts held together under pressure by electrodes. The resulting welds are localized at predetermined points by the design of the parts to be welded. The localization is usually accomplished by projections, embossments or intersections.

UPSET WELDING. A resistance-welding process wherein coalescence is produced, simultaneously over the entire area of abutting surfaces or progressively along a joint, by the heat obtained from resistance to the flow of electric current through the area of contact of those surfaces. Pressure is applied before heating is started and is maintained throughout the heating period.

Resistance welding machines are operated manually, semiautomatically, and automatically (Fig. 34–5). Fixed installation is most common, but portable machines are also available. They use manual, electrical, mechanical, pneumatic, or hydraulic power to apply pressure. They are often equipped with intricate control and timing devices. Modern mechanical handling devices for the work add to their efficiency. Expert maintenance is

required for these welding machines, even though they may be operated by unskilled labor.

Power supply

Resistance welding usually employs 60-hertz alternating current which is fed to the primary of the water-cooled welding transformer. The primary can vary from 150 to 10,000 amp, at 240, 440, or 550 volts. The output, at the secondary of the transformer is a low voltage (max 30 volts) and high current (up to 200,000 amp) used for welding.

The welding current is sometimes furnished by the "stored energy" type of equipment, in which energy is built up and stored either in capacitors or in a combination transformer-reactor during the nonwelding period, and then is discharged to form the weld. This process involves low primary currents and high voltages, which must be guarded against.

To facilitate servicing the equipment, a safety type disconnecting switch or a circuit breaker of the correct rating for opening supply circuits should be installed near the welding machine. Permanent injuries, and several fatalities have been caused by neglecting to use the line-disconnecting switch before making adjustments. The use of single-pole primary circuit breakers and electronic contactors which leave one line to the welder "hot" makes this precaution imperative.

Current and timing controls are of two types, mechanical and electronic. If worn parts throw the mechanical control out of synchronism with the current and voltage waves, transients may cause severe damage to the equipment. Synchronism should be carefully maintained, particularly on high-power equipment. These mechanical controls are becoming obsolete and replaced by electronic units. The electronic controls are placed in the primary circuit and involve high voltages which must be guarded by proper insulation, lockouts, and safety switches. The latest trend in controls is to the use of solid state units.

All electrical installation work should conform to applicable codes and federal regulations.

Cables

The high current of the primary circuits

(up to 10,000 amp) in resistance welding does not permit the use of plugs except for control circuits, usually 120 volts or less. Power lead terminals should be securely and permanently attached with screws or bolts for the duration of service.

The magnitude of the secondary welding current is such that the cables for a portable welding machine must be heavy. Even with sizes from 200,000 to 1,000,000 circular mils, power losses are high and in many instances water cooling is required, especially for high production portable, as well as multiple-spot station, welding machines. The length of these cables varies from 1 to 14 ft, the latter being the maximum considered reasonably economical. The covering for the cable used with a portable welding machine is thick, high tensile rubber, or now, more commonly, neoprene. These are now commonly concentric—two cables in one with water cooling.

Abuse of the cables for resistance welding is severe. The production requirements demand the utmost of the cable materials used, and even the best cables need frequent replacement during a production year. In use the cables are subjected to electrical pulsation, bending, and twisting. Electrical pulses at the rate of 3600 per minute cause a continuous whipping of the two cables which leads to fatigue and eventual breakdown. This condition is minimized by the use of concentric cables.

The secondary voltage presents little shock hazard, since the maximum voltage is about 30 volts; but the operator can be hurt by a cable blowout caused by steam pressure due to overheating from faulty water cooling circulation.

A periodic check for weak spots in the cable covering is good practice. The use of concentric welding cables is now common because they do not have the undesirable features of the pulsating cables. Portable welding machines, including the cables, should have proper weight balance to permit operation without undue strain to the operator.

Electrodes and holders

For resistance welding, copper or copper alloy electrodes are in general use. Made in various shapes for particular applications, they are, with few exceptions, water cooled. Resistance-welding electrodes, unlike those used in arc welding, are not deposited on the work. However, the high current density and continuous application of mechanical force finally cause wear and necessitate redressing and replacement.

Electrodes are placed in holders connected to the secondary circuit. The welding pressure is also applied through the holders. Their designs cover a multitude of applications and vary considerably with the welding methods used. Electrode holders and electrodes should be water cooled to minimize softening of the metal from heat and to reduce the chance of heat burns to the operator.

Machine installation

Installation of resistance welding equipment should conform to the NFPA *National Electrical Code*, Standard No. 70. Some items worthy of special attention are listed below.

1. Control circuits should operate on low voltage, not exceeding 24/36 volts maximum for portable spot welders.

2. Stored energy equipment (capacitor discharge or resistance welding) having control panels involving high voltage (more than 550 volts) should be completely enclosed. Doors should have locks and contacts wired in the control circuit to short circuit the capacitors when door or panel is opened. A manually operated switch will serve as an additional safety measure, assuring complete discharge of the capacitors.

3. Back doors of machines and panels should be kept locked or interlocked to prevent tampering.

4. A fused safety switch or circuit breaker should be located conveniently near the welding machine so that power supply circuits may be opened before a man services the machine and its controls.

5. The point-of-operation hazard should be eliminated by suitable guards. Enclosure guards, gate guards, two-hand controls, and similar standard guards as designed for punch press operations are applicable.

6. A flash welding machine should have a

shield or hood to control flash and fumes, and a ventilating system to carry off the metallic dust and oil fumes.

7. Where flying sparks are not confined, the operators and nearby persons should be protected by shields of safety glass or other transparent material.

8. Foot switches, air or electrical, should be guarded to prevent accidental operation.

Arc Welding and Cutting

Arc welding is a process for joining metals by heating with an electric arc or arcs with or without the application of pressure and with or without the use of filler metal. The process includes shielded welding which uses gas or a solid flux to blanket the weld. Arc welding is used to fabricate nearly all types of carbon or alloy steels, the common non-ferrous metals, and is indispensable in the repair and reclamation of metallic machine parts.

Arc cutting is being used only for rough cuts or for scrapping because of the unevenness of the cut obtained. It has been used for underwater cutting in salvaging operations. Oxy-arc or arc-oxygen cutting has been used for finer cutting and is especially useful for cutting metals that do not oxidize readily. Plasma arc cutting is now being used widely for quality cuts. Also being used is carbon arc-air cutting which leaves a smooth cut.

For arc welding or cutting, two welding leads, the electrode lead and the work lead, are required from the source of current supply. Usually, one lead is connected to the work and the other to the electrode holder. The work lead (cable) is the most satisfactory means of providing the return (ground) circuit to the welding machine (Fig. 34–6), but in some cases operating conditions may require the use of a grounded steel structure.

Power supply

Either AC or DC may be used for arc welding or cutting of any kind, including carbon arc cutting and cutting with heavily coated electrodes. With small diameter electrodes used on thin sheets for manual arc welding, current values vary from 10 to 50 amp. For most manual welding, because the welder

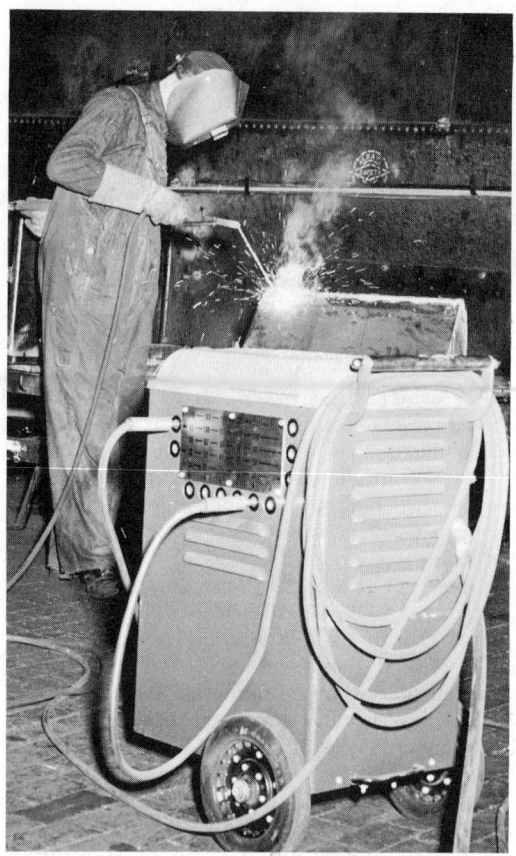

FIG. 34–6.—One conductor of a three conductor primary cable is permanently connected to the case of this AC welding unit and to the ground prong of a three-prong polarized plug, thus making grounding of the case both sure and easy.

Courtesy Illinois Central Railroad.

must withstand the heat, current values should not exceed 500 or 600 amp.

Automatic machine arc welding may use current values up to 200 amp or even higher on special applications.

Commercial AC and DC power supplies are not suitable for arc welding. Consequently, to convert commercial power for arc welding, a transformer and rectifier must be used for DC welding, and a transformer for AC welding. A motor generator may also be used to produce alternating or direct current. Transformer, rectifier or motor generator welding may be either single or 3 phase.

Since these welding converters are subject to abuse and are given little expert attention as compared to ordinary electrical equipment, they should be well protected, both mechanically and electrically. Often the machines are moved about, so convenient primary power receptacles should be provided about the shop or construction operation. Plugs and receptacles should meet the requirements of Underwriters Laboratories Inc.

AC transformers for welding are usually air-cooled. The transformers should be built in accordance with the high standards of electrical construction used on power or distribution transformers.

A typical industrial-rated AC transformer-type welding unit should be equipped with a voltage-reducing control which automatically reduces the open-circuit secondary voltage to 38 volts during idling. This reduction of the voltage is delayed two to three seconds from the instant the arc is broken to prevent interference with normal welding operations. Contact of the electrode to the work automatically restores normal welding voltage. Other safety features are (a) circuit breaker, integrally mounted on the unit, for providing thermal overload protection and serving as a convenient disconnect switch; (b) circuit breaker of the automatic type or a fused safety switch mounted on the wall and rated for use with the unit and the power supply circuit; (c) three-conductor power supply cable, one conductor of which is for grounding purposes, between the wall mounted breaker and the unit.

Large welding machines are usually installed as stationary units. Since primary line switches are not normally provided as part of the equipment, they should be installed by the person making the installation.

The smaller machines, which may be used as portable units, customarily have primary switches built into them and are provided with primary cables and attachment plugs of proper capacity. Welding transformers should not be attached to lighting circuits under any circumstance.

If a gasoline-powered welding generator is used inside a building or in a confined area, the engine exhaust should lead to the outside atmosphere. Otherwise, carbon monoxide and other toxic gases may accumulate.

Voltages

The voltage across the welding arc varies from 15 to 40 volts, depending on the type and size of electrode use. The welding circuit must supply somewhat higher voltage to strike the arc. This voltage is called the open circuit or "no load" voltage. After the arc is established, the open circuit voltage drops to a value about equal to the arc voltage plus the lead voltage drop. The open circuit voltages on DC welding machines should be less than 100 volts. Constant voltage power supplies (welder or converter) are now also being widely used.

For AC transformer welding machines, ANSI Z49.1, *Safety in Welding and Cutting,* prescribes a maximum open circuit voltage of 80 volts on manuals, 100 volts on automatics.

Some standard heavy duty industrial AC welding machines have open circuit voltages of 65 volts. Reliable control equipment is available which automatically reduces the open circuit voltage to about 30 volts during the "off" arc period, at which time the welder may be exposed to the open circuit voltage. This control equipment is commonly used where welders are working outdoors on wet steel or in confined quarters or under other severe conditions where the probability of contact with the open circuit voltage may be greater.

Heavy duty AC welding machines (ratings usually over 500 amp for automatic machine welding) are also built with open circuit voltages of 75 to 80 volts with a special tap to provide 100 volts where necessary. The tap may be needed to obtain rated output from the machine if the line voltage is low or if the voltage of the secondary circuit drops. The tap should not be accessible to the welder for current adjustment but should be under the control of a responsible electrician or supervisor.

Open circuit voltages should be as low as 50 volts on small AC welding machines used without expert supervision. Since these machines are often used with low arc voltage electrodes and with leads not more than 20 ft long, the probability of a large voltage drop in the welding circuit is small.

For other manual and automatic welding and cutting processes in which the work metal is connected electrically to one side of the circuit, open circuit voltage of 150 volts may be allowed provided the following conditions are present:

1. All equipment and circuiting are fully insulated and the operator cannot make electrical contact other than through the arc itself, while the arc is maintained.

2. Disconnecting or voltage reducing devices operate within a time limit not exceeding one second after breaking the arc.

Where neither side of the circuit is electrically connected to the work, open circuit voltage of 300 volts is allowed if controls are present to prevent the operator from touching both sides of the circuit. One hand should be used to operate control devices. Also, the voltage should be disconnected automatically by a reliable switch instantly upon breaking the arc.

For AC or DC welding under wet or humid atmospheric conditions where perspiration is a factor, a reliable automatic control device for reducing no-load voltage is recommended.

Cables

Welding cable is purchased in 50-ft lengths, and several lengths may be used in one circuit. Substantially insulated connectors, of a capacity at least equivalent to that of the cable, should be used to splice or connect cables. Cable lugs used for connections should be securely fastened together to give good electric contact. The exposed metal parts of the lugs should be completely insulated.

Welding cable is subjected to severe abuse if it is dragged over work under construction and across sharp corners, or run over by shop trucks. Special cable with high quality insulation should be used. The fact that welding circuit voltages are low may lead to laxity in keeping the welding cable in good repair. Operators and maintenance men should be instructed to see that defective cable is immediately replaced or repaired.

On large jobs, there is likely to be much loose cable lying around. Welders should keep this cable orderly and out of the way, preferably strung overhead to permit the passage of persons and vehicles. Welding cables should not lie in water or oil, in ditches or bottoms of tanks. Rooms in which arc welding is to be done regularly should be permanently wired with enough outlets so that extension cables will not have to be strewn about.

Electrodes and holders

Arc welding is done with either a metallic or a carbon electrode. The carbon electrode is usually a solid carbon or a graphite pencil, ¼ in. in diameter or larger, depending upon the amount of current used.

For gas-shielded metal arc welding, the electrode is a solid or flux-cored wire. For shielded metal arc welding, the electrode is a covered wire.

Electrode holders as used for shielded-metal arc welding (SMAW) are used to connect the electrode to the welding cable supplying secondary current. Fully insulated holders (Fig. 34-7) are preferred because there is less likelihood of accidentally striking an arc with such holders, particularly in close quarters.

Electrode holders will become hot during welding operations if holders designed for light work are used on heavy welding or if connections between the cable and the holder are loose.

If a holder of the correct size for the electrode cannot be used, an extra holder should be provided so that one can cool while the other is in use. *Dipping hot electrode holders in water should be prohibited.*

On light or medium heavy work, where light, extremely flexible cables are used, holders may be attached directly to the work lead running to the machine. On heavier work, welders generally prefer a short length attached to the holder, which is more flexible than the main work lead. Properly insulated cables of weight and flexibility which will not inconvenience the welder are available.

Automatic and semiautomatic welding

Arc welding may also be done automatically or semiautomatically to increase the speed of welding and provide more uniform weld quality. Machine welding is especially efficient for long welds or where a high rate of production is involved and the work can be jigged. The rate of electrode feed and arc

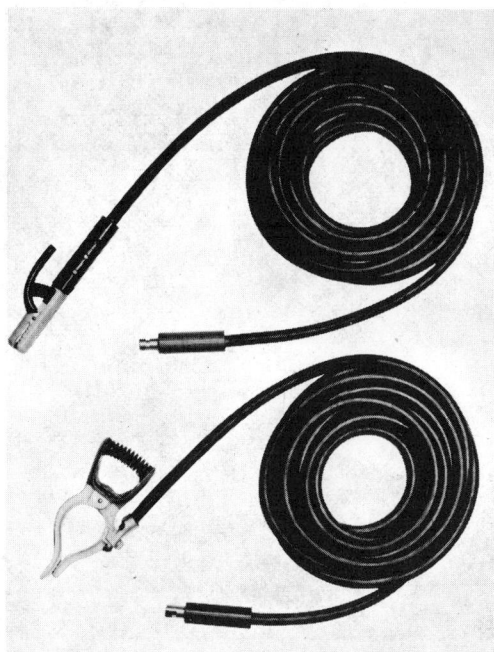

Fig. 34–7.—Fully insulated electrode holder on the electrode lead and the ground clamp on the work lead have insulated locking-type plugs for connection to receptacles on the welding machine.

Courtesy Westinghouse Electric Corp.

length are controlled by mechanical or electrical means.

If speed of travel or feed of the work is controlled manually, the operation is semi-automatic. If the work is fed or the arc traverses the work automatically, the operation is automatic.

Insulation of current-carrying parts, cables, etc., applies to machine welding as well as to manual welding.

Mechanical power-transmission apparatus, such as gears, shaftings, couplings, and clutches, and other moving parts to which operators may be exposed should be guarded. (See Chapter 29, "Point-of-Operation and Transmission Guards.")

Protection against electric shock

Although open-circuit voltages on standard arc-welding units are not high compared to those of other processes, they cannot be neg-

lected as a potential hazard. Normally, the work setup is such that the work is grounded, and unless care is exercised, the welder or operator can easily become grounded.

The voltage between the electrode holder and the ground, during the "off" arc or "no-load" period, is the open circuit voltage. Unless the welder or operator is properly instructed and uses the equipment provided for his protection, he can become exposed to this voltage while changing electrodes, setting up work, or changing working position. The danger is particularly great in hot weather when he is sweaty.

He should develop the habit of keeping his body insulated from both the work and the metal electrode and holder. He should never permit the bare metal part of an electrode, the electrode insulation, or any metal part of the electrode holder to touch either his bare skin or any wet covering on his body.

Consistent use of well insulated electrode holders and cables, dry clothing on the hands and body, and insulation from ground will be helpful in preventing contact.

Some specific precautions for prevention of electric shock are:

1. In confined places, cover or arrange cables to prevent contact with falling sparks.

2. Never change electrodes with bare hands or wet gloves, or when standing on wet floors or grounded surfaces.

3. Ground the frames of welding units, portable or stationary, in accordance with the *National Electrical Code*. With a small welding unit, a primary cable containing an extra conductor, one end of which is attached to the frame of the welding unit, can be used. By means of the proper polarized plug, this ground connection can be carried back to the permanently grounded connection in the receptacle of the power supply.

4. Arrange receptacles of power cables for portable welding units so that it is impossible to remove the plug without opening the power supply switch, or use plugs and receptacles which have been approved to break full load circuits of the unit.

5. If a cable (either work lead or electrode lead) becomes worn, exposing bare con-

Fig. 34–8.—Cable and hose are suspended to keep them clear of work area on this gas-shielded arc welding operation.

6. Keep welding cables dry and free of grease and oil to prevent premature breakdown of the insulation.

7. Suspend cables on substantial overhead supports if the cables must be run some distance from the welding unit. Protect cables that must be laid on the floor or ground so that they will not interfere with safe passage or become damaged or entangled.

8. Take special care to keep welding cables away from power supply cables or high-tension wires.

Gas-shielded arc welding

Gas-shielded metal arc welding (Fig. 34–8) is used for joining all types of metals and castings. In gas-shielded methods, only one electrode is used. A gas or gas mixture is introduced through the torch or gun to surround the electrode and the welding point

ductors, cover the exposed portion with rubber, plastic, or friction tape equivalent in insulation to the cable covering.

like a shield. The gas cone protects the molten weld puddle from the atmosphere and stops oxidation of the base metal. The electrode conducts either alternating or direct current to provide heat.

Tungsten arc (GTAW). In gas-shielded tungsten arc welding, the electrode does not melt and is not used for filler metal. The electrode is tungsten, which is highly resistant to heat and non-consumable in the welding process. Weld metal may be added by using a cold (non-electrical) welding rod which is introduced into the arc or molten weld puddle.

Metal arc (GMAW). In gas-shielded metal arc welding, the electrode melts and provides the filler metal (the word "metal" used in the title of this type of welding signifies that the electrode is consumable as filler). The need for filler rods on light gage metals is eliminated if the pieces to be joined are fitted up closely. The pieces are generally cut in a die and held in place with clamps or are set up in

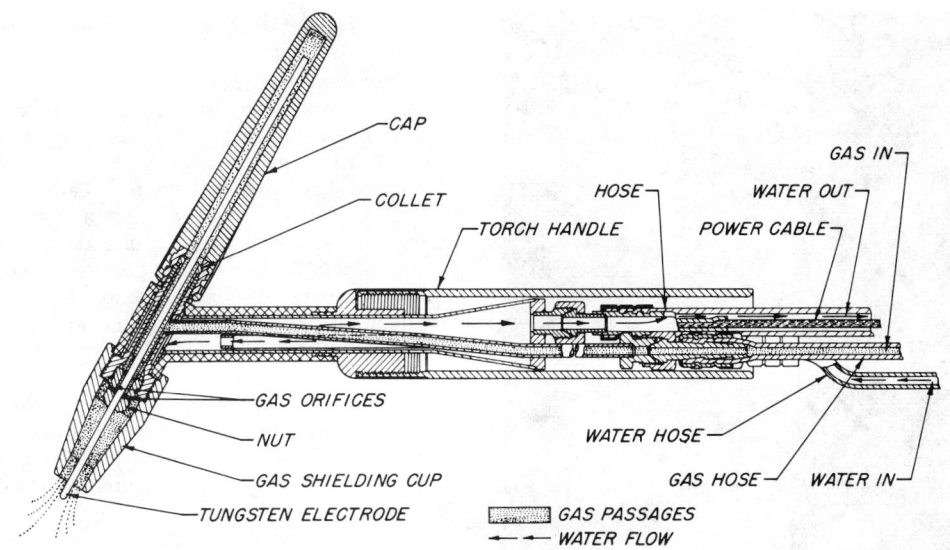

Labels on figure:
- CAP
- COLLET
- TORCH HANDLE
- HOSE
- GAS IN
- WATER OUT
- POWER CABLE
- GAS ORIFICES
- NUT
- GAS SHIELDING CUP
- TUNGSTEN ELECTRODE
- WATER HOSE
- GAS HOSE
- WATER IN
- GAS PASSAGES
- WATER FLOW

Fig. 34–9.—Gas-shielded tungsten arc welding torch, in which the electrode is nonconsumable.

From the Welding Handbook, *American Welding Society.*

jigs for maintenance of close contact. The clamps also prevent distortion of the metal from heat. Because the gas shields the weld area, flux is not necessary and the spattering of molten metal associated with other types of welding generally does not occur.

Argon, helium, carbon dioxide, and mixtures of these gases are used to shield the welding zone from the atmosphere. Argon, having greater density, does not mix with the surrounding air as rapidly and is more widely used than helium in the welding of light-gage metals. The greater penetration obtained with helium makes it useful on heavier metals and castings. It is also cheaper and more available. The bare rod is fed into the weld puddle, where it melts and mixes with the weld metal.

Equipment. Either AC or DC welding units can be used for gas-shielded tungsten arc welding, depending on whether the weld is wide, deep, or narrow, whether the job is to be performed on a permanent fixture or a portable machine, and whether it is to be a manual or machine welding operation.

For gas-shielded metal arc, DC is used with a reverse polarity hookup, with current sup-plied by a generator or rectifier. The AC supply may be obtained through a transformer or high-frequency generator.

The manufacturer of the welding equipment should be consulted as to the specific job before the equipment is installed, especially when high-frequency AC is used; otherwise, there may be interference with radio transmission.

Argon, helium, and gas mixtures are supplied by manufacturers in cylinders similar to oxygen cylinders. Since cylinder pressures range from 2200 to 2640 psi, argon and helium cylinders should be stored and handled like other high-pressure gas cylinders.

Carbon dioxide, although not strictly an inert gas, is sometimes used as a shield gas when steel is welded by the gas-shielded metal arc welding process. It is usually supplied in partially liquid and partially gaseous form in cylinders at approximately 835 psi pressure. These cylinders should therefore be handled like other high-pressure gas cylinders.

To supply gas to the welding torch, a regulator must be used to lower the pressure to 25 psig or less, and a flowmeter should measure the volume of gas. If more than one torch is used from the same gas line, a flow-

965

FIG. 34–10.—Self-standing safety shield provides close-quarter protection, rolls up for storage.

meter should be installed at each torch connection.

Air is used to cool the torch and electric current cables. Water is also used for cooling, generally where the welding current is more than 250 amp. The water supply line, even if city water is used, should be equipped with a strainer to keep out impurities which might get into the torch and plug the water cooling passages.

In the gas-shielded tungsten arc torch (Fig. 34–9), gas is conducted to the welding point through orifices in the torch around the electrode holder. Cooling water goes through passages through the torch handle and about the holder. In the smaller torches, ceramic cups are used. A torch for heavier work generally has a water-cooled gas cup.

Other Welding and Cutting Processes

There are several heat sources for welding and cutting which are presently in the experimental stage, such as friction, ultrasonics, and lasers. During each stage of their development, these heat sources require guarding and the evolution of a set of safe practices.

For example, the *laser* (Light Amplification by Stimulated Emission of Radiation) presents the hazard of eye damage from the optically amplified light beam, which because of its intensity, can do damage even at great distances. All employees should therefore be given preemployment and periodic followup eye examinations. Most companies require that employees work in pairs when using laser equipment.

To confine the laser beam, suitable shields should be developed and installed. Since a reflected laser beam is also hazardous, the work area should contain no glossy surfaces.

Power supplies for lasers are high-voltage equipment, which should be operated with the precautions developed for this type of equipment.

See discussion in Chapter 19, "Personal Protective Equipment" and Chapter 37, "Industrial Hygiene."

Common Hazards

Light rays

Electric arcs and gas flames both produce ultraviolet and infrared rays which have a harmful effect on the eyes and skin upon continued or repeated exposure. The usual effect of ultraviolet is to "sunburn" the surface of the eye, which is painful and disabling but temporary in most instances. However, permanent eye injury may result from looking directly into a very powerful arc without eye protection. Ultraviolet may also produce the same effects on the skin as a severe sunburn.

Production of ultraviolet radiation is high in gas-shielded arc welding. For example, a shield of argon gas around the arc doubles the intensity of the ultraviolet radiation, and, with the greater current densities required (particularly with a consumable electrode), the intensity may be five to thirty times as great as with non-shielded welding such as covered electrode or gas-shielded metal arc welding.

Infrared has only the effect of heating the tissue with which it comes in contact. If the heat is not enough to cause an ordinary thermal burn, there is no harm.

Whenever possible, arc welding operations should be isolated so that other workers will not be exposed to either direct or reflected rays (Fig. 34–10). Walls, ceilings, and other exposed inner surfaces of such rooms should have a dull finish produced by a dark non-reflective paint, such as zinc oxide and lamp-black.

Arc welding stations for regular production work can be enclosed in booths if the size of the work permits. The inside of the booth should be coated with a dark non-reflective paint and provided with portable flameproof screens similarly painted or with flameproof curtains. Booths should be designed to permit circulation of air at the floor level.

Fire protection

Because portable welding and cutting equipment creates special fire hazards, it should be used in a permanent welding and cutting location which can be designed to provide maximum safety and fire protection. Otherwise, the welding and cutting site should be inspected to determine what fire protection equipment is necessary. See ANSI Z49.1 and NFPA 51B.

It is advisable, particularly in hazardous locations, to require written "hot work" permits issued by the welding supervisor, a member of the plant fire department, or some other qualified person before welding or cutting operations are started. In a small plant, this responsibility may be delegated to a competent welder. Specifications for hot work permits are outlined in Chapter 43, "Fire Protection."

Floors and combustible materials

Where welding or cutting must be done near combustible materials, special precautions are necessary to prevent sparks or hot slag from reaching such material and starting fires. If the work itself cannot be moved, the exposed combustible material should, if possible, be moved a safe distance away. Otherwise, it should be covered with asbestos curtains or sheet metal. Spray booths and ducts should be cleaned to remove combustible deposits. Before welding or cutting is started, wood floors should be swept clean and, preferably, covered with metal or other noncombustible material where sparks or hot metal may fall. In some cases, it is advisable to wet down the floor, though the wet floor increases the shock hazard to electric (arc and resistance) welders and necessitates special protection for them.

If gas welding or oxygen cutting is done inside a booth provided for arc welding, the gas cylinders should be placed in an upright and secured position away from sparks to prevent contact with the flame or heat.

Hot metal or slag should not be allowed to fall through cracks in the floor or other openings, nor into machine tool pits. Cracks or holes in walls, open doorways, and open or broken windows should be covered with sheet metal guards or asbestos curtains. Because hot slag may roll along the floor, it is important that no openings exist between the asbestos curtain and the floor. Similar protection should be installed for wall openings through which hot metal or slag may enter when welding or cutting operations are conducted on the outside of the building.

If it is necessary to weld or cut close to wood construction or near combustible material which cannot be removed or protected, a small fire hose, water pump tank extinguisher, or fire pails should be conveniently located. Portable extinguishers for specific protection against Class B and C fires should also be provided (see Chapter 43, "Fire Protection"). Pails of limestone dust or sand may be useful. It is good practice to provide a fire extinguisher, either dry chemical, multi-purpose chemical, or carbon dioxide, for each welder, as part of his kit.

A fire watcher equipped with a suitable fire extinguisher should be stationed at or near welding or cutting operations conducted in hazardous locations to see that sparks do not lodge in floor cracks or pass through floor or wall openings. The fire watch should be continued for at least 30 minutes after the job is completed, to make sure that smoldering fires have not been started.

Hazardous locations

Welding and cutting operations should not be permitted in or near rooms containing flammable or combustible vapors, liquids, or dusts, or on or inside closed tanks or other containers which have held such materials, until all fire and explosion hazards have been eliminated. All of the surrounding premises should be thoroughly ventilated, and frequent gas testing provided. Sufficient draft should be maintained to prevent accumulation of explosive concentrations. Local exhaust equipment should be provided for removal of hazardous gases, vapors, and fumes (present in the surroundings or generated by the welding or cutting operations) that ventilation fails to dispel.

Drums, tanks, and closed containers

Closed containers that have held flammable liquids or other combustibles should be thoroughly cleaned before welding or cutting. Sometimes containers which cannot be removed and handled properly for standard cleaning procedures are purged with an inert gas (Fig. 34–11) or filled with water to within an inch or two of the place where the work is to be done, with a vent left open. Either of these two measures may also be employed as an added precaution after cleaning according to recommended methods. (See American Welding Society. *Safe Practices for Welding and Cutting Containers That Have Held Combustibles,* A6.1. See also National Fire Protection Association, *Cleaning or Safeguarding Small Tanks and Containers,* Standard No. 327. Also see Chapter 42, "Flammable and Combustible Liquids," in this Manual.)

The accepted method for preparing tanks and drums for welding is:

1. Remove all sources of ignition (open flames, unguarded electric lights, etc.) from the vicinity of the drums to be cleaned.

2. Remove the bung with a special long-handled wrench.

3. Examine the inside for rags, waste, or other debris which might interfere with free draining. Use a portable electric hand lamp that is listed for hazardous locations, or an electric extension lamp protected by a guard of spark-resistant material.

4. Place the drums on a steam rack with the bung holes at the lowest possible point, and let the drums drain for 5 minutes.

5. Steam the drums for at least 10 minutes.

Drums that have contained shellac, turpentine, or similar materials require longer steaming.

6. Remove the drums from the steaming rack, and fill them part way with caustic soda or soda ash solution. Rotate the drums for at least 5 minutes. Light hammering with a wood mallet will help to loosen scale.

7. Thoroughly flush the drums for at least 5 minutes with boiling water. A water spray nozzle placed 6 or 8 in. inside the drum can be used. Drums should be so placed that water can drain out the bung openings during this operation.

8. Wash down the outside of the drum with a hose stream of hot water.

9. Dry the drum thoroughly by circulating warm air throughout the inside.

10. Thoroughly inspect the interior of the drum, using a light that is listed for hazardous locations, and a small mirror. If it is not clean, repeat the cleaning process.

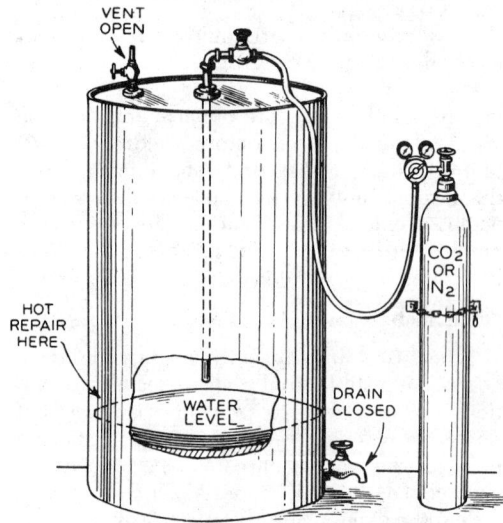

FIG. 34–11.—As an added precaution after cleaning, a container to be welded or cut may be filled with either carbon dioxide or nitrogen to dilute any combustible gas or vapor remaining—dilute it enough to render it nonhazardous.

Courtesy American Welding Society.

11. Test the container for the presence of flammable vapors, preferably with a combustible gas indicator. Test for toxic contaminants and for oxygen sufficiency if personnel are to enter.

12. Make similar tests just before welding repair operations are performed. If the operations extend over an appreciable period of time, make repeated tests.

Precautions for employee protection during container cleaning operations include:

1. Wear head and eye protection, rubber gloves, boots, and aprons when handling steam, hot water, and caustic solutions. When handling dry caustic soda or soda ash, wear approved respiratory protective equipment, long sleeves, and gloves.

2. To handle hot drums, wear asbestos hand pads or gloves. Steam irons or other hot surfaces which may be touched should be insulated or otherwise guarded.

3. Dispose of residue in a safe manner. In each instance, the method of disposal should be checked for hazards.

4. If a vessel must be entered, wear respiratory protective equipment approved for the exposure and a safety harness with attached lifeline tended by a helper who is similarly equipped and stationed outside the vessel.

Many containers which have held combustible or explosive material present special problems. Detailed information can be secured from the manufacturer of such materials. For gasometers or gasholders for natural or manufactured gas, the American Gas Association should be consulted.

For cleaning and gas-freeing of tanks, bunkers, or compartments on board ship, refer to NFPA Standard No. 306, *Control of Gas Hazards on Vessels To Be Repaired.*

Personal Protection

Respiratory protection

During welding and cutting operations, toxic gases, fumes, and dusts may be evolved, depending on the type of electrode used, the base metal being welded or cut, and whether or not the base metal is coated with such materials as tar, paint, lead, or zinc. Gases—the oxides of nitrogen, carbon monoxide, ozone—and metallic dusts and fumes may, in varying degrees, present inhalation hazards if adequate respiratory protection is not provided.

American National Standard Z49.1, *Safety in Welding and Cutting,* gives specifications for local exhaust systems, general ventilation, and health protection of workers on various types of welding and cutting. The following discussion is based on this standard, which should be consulted for further information.

In open-air welding or cutting or in large, well-ventilated areas, where clean carbon steel is welded or cut with bare or coated carbon-steel electrodes and without inert-gas shielding, minimum health hazards exist.

In confined areas, such as tanks, pressure vessels, and holds of ships, local exhaust or general ventilation systems should be provided to maintain concentrations of toxic gases, fumes, or dusts below maximum allowable concentrations (MAC) as defined in American National Standard Z37 Series "Maximal Acceptable Concentrations of Toxic Dust and Gases," or below threshold limit values (TLV) specified by the American Conference of Governmental Industrial Hygienists. For a complete discussion, see Chapters 37 and 38, "Industrial Hygiene" and "Industrial Toxicology."

If gases, dusts, and fumes cannot be kept below the TLV because work is intermittent or for other reasons, welders should wear respiratory protective equipment approved for the exposure by NIOSH. Where oxygen is also deficient, self-contained breathing apparatus or hose masks with blowers are necessary.

Coatings, fluxes, and base metals. When welding or cutting involves coatings or fluxes containing base metals that contain elements such as zinc, fluorine, beryllium, lead, or cadmium, and their compounds, local exhaust removal or general ventilation should be provided to maintain the concentration of any toxic fume generated below the MAC or TLV. Outdoor welding and cutting involving lead,

Fig. 34–12.—Local exhaust for arc welding.

mercury, and cadmium require that the worker wear respiratory protective equipment.

Generation of gases and fumes. Oxides of nitrogen from the high-temperature combination of nitrogen and oxygen in the air are always generated near the welding or cutting arc. An inert gas shield, however, minimizes the introduction of air to the arc. Concentrations of these oxides are generally above their MAC or TLV within a few inches of the arc, but are diluted rapidly by air movements. Local exhaust or general ventilation should be used to keep the concentrations of oxides of nitrogen within safe limits.

Ozone—a triatomic form of gaseous oxygen, chemically designated by the symbol O_3—is formed by ultraviolet radiation acting on oxygen in air. It is a highly toxic and irritating gas. Since ultraviolet radiation passes through the air, ozone may be formed several feet from the welding or cutting arc. The amount of ozone formed depends on the metal and the gas shield being used and the arc temperature. A greater amount of ultraviolet is produced with argon gas shield than with helium. General ventilation will usually control the production of ozone in the welding or cutting area.

Ultraviolet radiation from the welding or cutting arc can also decompose chlorinated hydrocarbons, such as trichloroethylene and perchloroethylene, to form highly toxic substances. Since this decomposition can occur even at a considerable distance from the arc, degreasing operations and other work using these chlorinated solvents should be located so that no solvent vapor will reach the welding or cutting area.

Inert-gas shielded arc welding requires that precautions be taken to provide proper respiratory protection. Depending upon a number of factors, including the particular variety of gas-shielded arc welding to be done, the nature of the materials to be welded, and whether or not the work must be done in a confined space, there will be a need for provision of positive ventilation, local exhaust removal, or approved respiratory equipment, or a combination of them.

Indoor exhaust and ventilation. In spaces of 50,000 cu ft and over, where welding is an essential part of the work, local exhaust or positive ventilation is not required for the protection of welders on uncoated ferrous metals, provided that (a) welding bays are not structurally blocked so as to obstruct cross ventilation, (b) the work is not done inside tanks, boilers, or other closed iron or steel containers, (c) space allowance of 10,000 cu ft is assured each welder, (d) ceiling heights are greater than 16 ft, and (e) the process involved is other than inert-gas shielded arc welding.

Where these requirements are not met, the Code specifies mechanical ventilation at a minimum rate of 2000 cfm of air per welder, or four air changes per hour, whichever is the greater, except when local exhaust hoods and booths or respiratory equipment approved by the NIOSH are used (Fig. 34–12).

When welding must be performed in a space screened on all sides, the screens should be so arranged that they do not seriously restrict ventilation. They can be mounted about 2 ft above the floor, unless the work is performed at so low a level that they must be nearer the floor to protect nearby workers from welding glare.

Local exhaust removal may be by means of movable hoods placed by the welder as near as practicable to the work being welded and provided with a rate of air flow sufficient to maintain a velocity in the direction of the hood of 100 fpm at the point of welding when the hood is at its most remote distance from the point of welding. The duct diameters and air-flow volumes that will produce this control velocity using a 3-in.-wide flanged suction opening are shown in Table 34-A. Exhaust air should be discharged outdoors.

Local exhaust may also be by a fixed enclosure with a top and not less than two sides, which surround the welding or cutting op-

TABLE 34-A

SPECIFICATIONS FOR LOCAL EXHAUST DUCT DIAMETERS AND FLOW RATES

Distance from Arc or Torch (in.)	Min. Air Flow[1] (cfm)	Duct Diameter[2] (in.)
4 to 6	150	3
6 to 8	275	3½
8 to 10	425	4½
10 to 12	600	5½

[1] Increase by 20 per cent for hoods without flanges.
[2] To nearest ½ in. based on velocity of 4000 fpm in duct.

erations, and with an air flow sufficient to maintain a velocity away from the welder of not less than 100 fpm. See ANSI Z49.1, *Safety in Welding and Cutting.* Local exhausts are further discussed in Chapter 36, "Exhaust and Ventilation."

Medical control. Preplacement physical, including chest X ray, examinations are recommended for all persons engaged in welding. Periodic re-examinations should be made as recommended by the plant physician.

Eye protection

Goggles, helmets, and shields that give maximum eye protection for each welding and cutting process should be worn by operators, welders, and their helpers. These items should conform to ANSI Z87.1, *Practice for Occupational and Educational Eye Protection,* and Z89.1, *Safety Requirements for Industrial Head Protection.* Table 34-B is a guide for selecting the correct filter lens for various welding and cutting operations. Goggles or spectacles should have side shields. Guidance for lens care is given in Chapter 19, "Personal Protective Equipment."

Protective clothing

Some of the items of protective clothing needed by welders (Fig. 34–13) are:

1. Flame-resistant gauntlet gloves, except on very light work.

2. Aprons of leather, asbestos, or other flame-

TABLE 34-B

FILTER LENS SHADE NUMBERS FOR
VARIOUS WELDING AND CUTTING
OPERATIONS (WELDER AND HELPERS)

Welding Operation	Suggested Shade Number*
Shielded metal-arc welding, up to 5/32 in. (4 mm) electrodes	10
Shielded metal-arc welding, 3/16 to 1/4 in. (4.8 to 6.4 mm) electrodes	12
Shielded metal-arc welding, over 1/4 in. (6.4 mm) electrodes	14
Gas metal-arc welding (nonferrous)	11
Gas metal-arc welding (ferrous)	12
Gas tungsten-arc welding	12
Atomic hydrogen welding	12
Carbon arc welding	14
Torch soldering	2
Torch brazing	3 or 4
Light cutting, up to 1 in. (25 mm)	3 or 4
Medium cutting, 1 to 6 in. (25 to 150 mm)	4 or 5
Heavy cutting, over 6 in. (150 mm)	5 or 6
Gas welding (light) up to 1/8 in. (3.2 mm)	4 or 5
Gas welding (medium) 1/8 to 1/2 in. (3.2 to 12.7 mm)	5 or 6
Gas welding (heavy) over 1/2 in. (12.7 mm)	6 or 8

* The choice of a filter shade may be made on the basis of visual acuity and may therefore vary widely from one individual to another, particularly under different current densities, materials, and welding processes. However, the degree of protection from radiant energy afforded by the filter plate or lens when chosen to allow visual acuity will still remain in excess of the needs of eye filter protection. Filter plate shades as low as shade 8 have proven suitably radiation-absorbent for protection from the arc welding processes.

Note 1. *In gas welding or oxygen cutting where the torch produces a high yellow light, it is desirable to use a filter lens that absorbs the yellow or sodium line in the visible light of the operation (spectrum).*

Note 2. *See Table 19-A for transmittance characteristics.*

From American National Standard Z49.1, Safety in Welding and Cutting.

resistant material to withstand radiated heat and sparks.

3. For heavy work, fire-resistant leggings, high boots, or similar protection.

4. Safety shoes, wherever heavy objects are handled. Low-cut shoes with unprotected tops should not be used because of the spark hazard.

5. For overhead work, capes or shoulder covers of leather or other suitable material. Skull caps of leather or flame-resistant fabric may be worn under helmets to prevent head burns. Also, for overhead welding, ear protection (wool or rubber

plugs or wire screen protectors) is sometimes desirable.

6. Safety hats or other head protection against sharp or heavy falling objects.

Operators and other persons working with inert-gas shielded arc welding should keep all parts of the body which could be exposed to the ultraviolet and infrared radiation covered to protect against skin burns and other types of injuries. Dark clothing, particularly a dark shirt, is preferable to light-colored clothing in order to reduce reflection to the operator's face underneath the helmet.

Cotton clothing disintegrates in from one

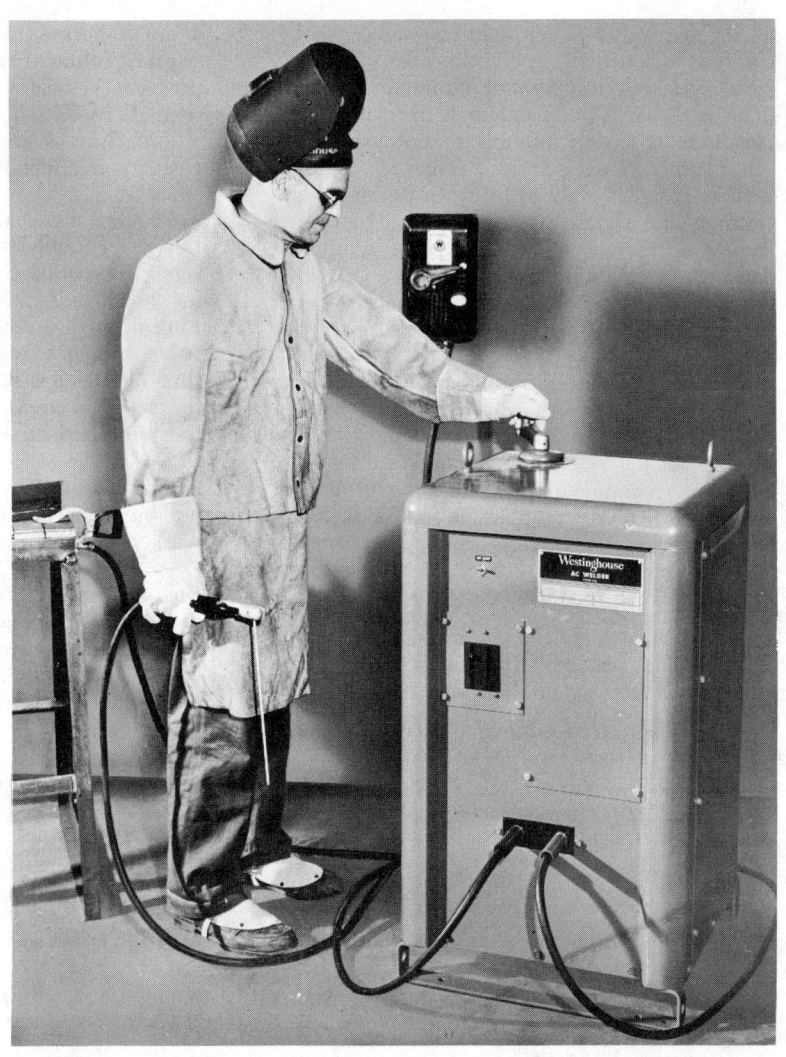

FIG. 34–13.—Proper personal protection for a welder—flash goggles worn under helmet, chrome leather jacket, apron, gauntlet gloves, and leggings.

Courtesy Westinghouse Electric Corp.

day to two weeks, presumably because of the high ultraviolet radiation from arc welding and cutting. Wool or leather clothing, therefore, is preferable to cotton because it is more resistant to deterioration.

For gas-shielded arc welding, woolen clothing is also preferable to cotton. It is not readily ignited and protects the welder from changes in temperature. Cotton clothing, if used, should be chemically treated to reduce flammability. In either case, clothing should be thick enough to keep radiation from penetrating it.

Outer clothing should be reasonably free from oil and grease. Sleeves and collars should be kept buttoned. Aprons and overalls should have no pockets, in which sparks could be caught, on the front of them. For the same reason, trousers or overalls should not have turned-up cuffs.

Thermal insulated underwear for cold-weather protection should be made of down-filled or waffle-weave cotton and wool. Quilted nylon-shell, polyester-filled underwear, although it does not ignite any easier than cotton, will burn and melt to form a hot plastic mass which adheres to the skin and can cause serious injuries. Thermal underwear is designed only to be worn under other clothing and should not be exposed to open flames, sparks, or other sources of ignition. See Fig. 19–56 on page 525.

Training in Safe Practices

Welders and cutters should be well trained in the safe practices that apply to their work. The standards for training and qualification of welders set up by the American Welding Society are recommended. A training program should particularly emphasize that a welder or cutter can best provide for his own safety and that of his coworkers by observing safe practices which include:

1. For work at more than 5 ft above the floor or ground, use a platform with railings, or a safety belt and lifeline.

2. Wear respiratory protection as needed and a safety harness with attached lifeline for work in confined spaces, such as tanks and pressure vessels. The lifeline should be tended by a similarly equipped helper whose duty is to observe the welder or cutter and effect rescue in an emergency.

3. Take special precautions if welding or cutting in a confined space is stopped for some time. Disconnect the power on arc welding or cutting units and remove the electrode from the holder. Turn off the torch valves on gas welding or cutting units, shut off the gas supply at a point outside the confined area, and, if possible, remove the torch and hose from the area.

4. After welding or cutting is completed, mark hot metal or post a warning sign to keep workers away from heated surfaces.

5. Follow safe housekeeping principles. Don't throw electrode or rod stubs on the floor—discard them in the proper waste container. Keep tools and other tripping hazards off the floor—put them in a safe storage area.

References

American Gas Association, 605 Third Ave., New York, N.Y. 10016.

American Insurance Association, 85 John St., New York, N.Y. 10038. *Lasers and Masers,* Special Hazards Bulletin Z-125.

American Petroleum Institute, 1801 K St. NW., Washington, D.C. 20006. *Cleaning Mobile Tanks Used for Transportation of Flammable Liquids.* Accident Prevention Manual No. 13. 1958.
Cleaning Petroleum Storage Tanks. Accident Prevention Manual No. 1.
Gas and Electric Cutting and Welding. Accident Prevention Manual No. 3.

American Society of Mechanical Engineers, 345 East 47th St., New York, N.Y. 10017. *ASME Boiler and Pressure Vessel Code,* Section IX, "Qualification Standard for Welding and Brazing Procedures, Welders, Brazers, and Welding and Brazing Operators." 1965.

American National Standards Institute, 1430 Broadway, New York, N.Y. 10018.
"Acceptable Concentrations of Toxic Dusts and Gases," Z37 Series.
Compressed Gas Cylinder Valve Outlet and Inlet Connections, B57.1.
Method of Marking Portable Compressed Gas Containers To Identify the Material Contained, Z48.1.
Practice for Occupational and Educational Eye and Face Protection, Z87.1.
Practices for Respiratory Protection, Z88.2
Safety in Welding and Cutting, Z49.1.
Safety Requirements for Industrial Head Protection, Z89.1.
Safety Standard for Mechanical Power-Transmission Apparatus, B15.1.

American Welding Society, 2501 NW 7th St., Miami, Fla. 33135.
    Recommended Safe Practices for Inert-Gas Metal-Arc Welding. Z49.1–58.
    Safe Practices for Welding and Cutting Containers That Have Held Combustibles. A6.1–55.
    Standard Qualification Procedure. B3.0–41.
    Welding Handbook, 5th Edition. Section I. 1962.
    Welding in Building Construction, D1.0–1969.

Compressed Gas Association, 500 Fifth Ave., New York, N.Y. 10036.
    Acetylene.
    Publications of the International Acetylene Association.
    Safe Handling of Compressed Gases.

Linde Division, Union Carbide Corp., 270 Park Ave., New York, N.Y. 10017. "Plasma-Arc Process" Bulletins.

National Bureau of Standards, U.S. Department of Commerce, Washington, D.C. 20234.
    National Electrical Safety Code. Handbook H30.

National Fire Protection Association, 470 Atlantic Ave., Boston, Mass. 02210.
    Bulk Oxygen Systems at Consumer Sites, Standard No. 50.
    Cleaning or Safeguarding Small Tanks and Containers, Standard No. 327.
    Control of Gas Hazards on Vessels To Be Repaired, Standard No. 306.
    Cutting and Welding Processes, Standard No. 51B.
    National Electrical Code, Standard No. 70.
    Oxygen Fuel-Gas Systems for Welding and Cutting, Standard No. 51.

National Safety Council, 425 North Michigan Ave., Chicago, Ill. 60611.
    Industrial Data Sheets
        Acetylene, 494.
        Cleaning Small Containers That Have Held Combustibles, 432.
        Gas-Shielded Arcs and Plasma Jet Torches, 552.
        Gaseous Oxygen, 472.
        Hand Soldering and Brazing, 445.
        Hot Work Permits (Flame or Spark), 522.

Solon, Leonard R. "Occupational Safety with Laser (Optical Maser) Beams." *Archives of Environmental Health,* 6:414–17 (March 1963).

U.S. Navy, The Pentagon. Washington, D.C. 20350.
    Bureau of Ships Manual. Chapter 92, "Welding and Allied Processes." NAVSHIPS 250–00–92.
    Underwater Cutting and Welding Manual. NAVSHIPS 250–692–9.

Hand and Portable Power Tools

Chapter

35

Control of Tool Accidents 977
Centralized tool control . . . Carrying tools

Maintenance and Repair 979
Tempering tools . . . Safe-ending tools . . . Dressing tools . . .
Handles

Use of Hand Tools 982
Metal-cutting tools . . . Wood-cutting tools . . . Miscellaneous cutting
tools . . . Materials handling tools . . . Torsion tools . . . Hammers
. . . Spark-resistant tools

Portable Power Tools 994
Selection of tools . . . Inspection and repair . . . Electric tools . . .
Air power tools . . . Special power tools

Personal Protective Equipment 1004
Eye protection

References 1005

Fig. 35–1.—Tool boxes containing personal tools can be neatly and safely stored in a rack like this. Shelves are movable so employees can reach all sections to get their tool boxes out.

Courtesy Lackawanna Plant, The Bethlehem Steel Co.

Control of Tool Accidents

Because of the widespread use and abuse of hand and powered hand tools and the severity of many tool injuries, it is important that control of tool accidents be made a part of every safety program.

Each year hand tools are the source of about 7 or 8 percent of all compensable injuries. Disabilities resulting from misuse of tools or using defective tools include loss of eyes and vision, puncture wounds from flying chips, slivers from concussion tools, severed fingers, tendons and arteries from cutting tools, broken bones, contusions from slipping wrenches, infections from puncture wounds, and other injuries too numerous to mention.

Failure to observe one or more of the following four safe practices accounts for most hand and powered hand tool accidents:

I. Select the right tool for the job

Examples of unsafe practices are: Striking hardened faces of hand tools together (such as using a carpenter's hammer to strike another hammer, hatchet, or metal chisel), using a file or a screw driver for a pry, a wrench for a hammer, and pliers instead of the proper wrench.

II. Keep tools in good condition

Unsafe tools include wrenches with cracked or worn jaws; screw drivers with broken points, or split or broken handles; hammers with loose heads, broken or split handles; mushroomed heads on chisels (Fig. 35–2); dull saws; and extension cords or electric tools with broken plugs, improper or removed grounding system, or split insulation.

III. Use tools correctly

Screw drivers applied to objects held

977

FIG. 35–2.—Poorly maintained tools—split hammer handle, mushroomed chisel heads.

in the hand; knives pulled toward the body, and failure to ground electrical equipment are common causes of accidents.

IV. KEEP TOOLS IN A SAFE PLACE

Many accidents have been caused by tools falling from overhead and by knives, chisels, and other sharp tools carried in pockets or left in tool boxes with cutting edges exposed.

A good safety program to control tool accidents should include the following activities:

1. Train employees to select the right tools for each job and see that they are available.

2. Establish regular tool inspection procedures (including inspection of employee-owned tools), and provide good repair facilities to ensure that tools will be maintained in safe condition.

3. Train and supervise employees in the correct use of tools for each job.

4. Establish a procedure for control of company tools, such as a check-out system at tool cribs.

5. Provide proper storage facilities in the tool room and on the job (Fig. 35–1 and –3).

Each supervisor should also make a complete check of his operations to determine the need for special tools that will do the work more safely than ordinary tools. Special tools should be kept readily available.

Centralized tool control

The principal advantage of centralized tool control from the standpoint of accident prevention is that it assures uniform inspection and maintenance of tools by a trained man.

Special fixtures and equipment, available at a central location, permit uniformly good maintenance. The correct type of personal protective equipment, such as safety goggles, can be issued when the tool is distributed.

Centralized control facilities and the keeping of effective records on tool failure and other accident causes will help locate hazardous conditions and unsafe acts. Central storage facilities will also assure more positive control than will scattered storage. Tools that are exposed to less damage and deterioration are less likely to fall or create other hazards.

The tool room attendant can help promote safety by recommending or issuing the right type of tool, by encouraging employees to turn in defective or worn tools, and by en-

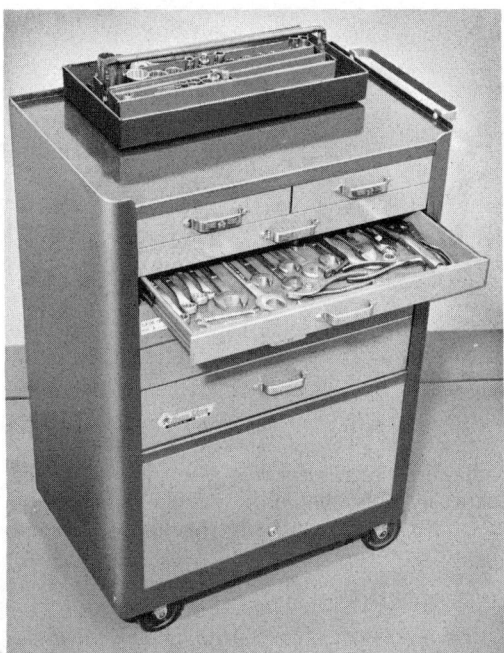

Fig. 35-3.—Cabinet and shelves provide tool storage and availability at work site.

Courtesy Buick Div., General Motors Corp.

couraging the safe use of tools.

A procedure should be set up so that the tool supply room attendant can send tools in need of repair to a department thoroughly familiar with methods of repair and reconditioning.

Some companies issue to each employee a set of numbered checks which are exchanged for tools from the supply room. With this system the attendant knows the location of each tool and can recall it for inspection at regular intervals.

Companies performing work at scattered locations may find that it is not always practicable to maintain a tool supply room. In such cases, the supervisor should inspect all tools frequently and remove from service those found to be defective. Many companies have each supervisor check all tools each week. A checklist for hand tools considered most hazardous can be helpful in systematizing inspection.

Some workmen prefer to use their own tools even though tools are furnished by the company. In this case, supervisors should examine the tools frequently to prevent the use of those that are unsafe. If privately owned tools are found to be defective, supervisors should insist that they be replaced.

The Federal Occupational Safety and Health rules and regulations state that the employer is responsible for seeing that nondefective tools are used.

Carrying tools

The workman should never carry tools which in any way might interfere with his using both hands freely on a ladder or while he is climbing on a structure. A strong bag, bucket, or similar container should be used to hoist tools from the ground to the job. Tools should be returned in the same manner, that is, not brought down by hand, carried in pockets, or dropped to the ground.

Mislaid and loose tools cause a substantial portion of the hand tool injuries. Tools are laid down on scaffolds, on overhead piping, on top of step ladders, and in other locations from which they can fall on persons below. Leaving tools overhead is especially hazardous where there is vibration or where men are moving about.

Chisels, screw drivers, and pointed tools should never be carried edge or point up in a workman's pocket. They should be carried in a tool box, a cart, or in a carrying belt like that used by electricians and steelworkers, in a pocket tool pouch, or in the hand with points and cutting edges away from the body.

Tools should be handed from one workman to another, never thrown. Edged or pointed tools should be passed, preferably in their carrying case, with the handle toward the receiver.

Workmen carrying tools on their shoulders should pay close attention to clearances when turning around and should handle the tools so that they will not strike other workmen.

Maintenance and Repair

The tool room attendant or tool inspector should be qualified by training and experience to pass judgment on the condition of tools for further use. No dull or damaged tool should be returned to stock. Enough tools of

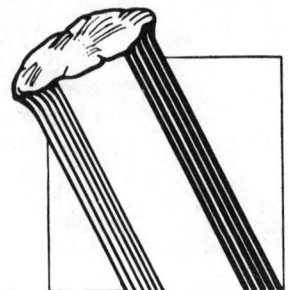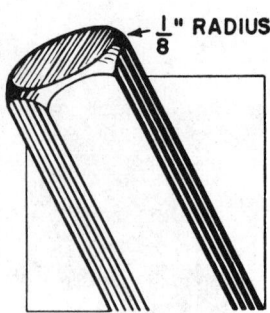

⟵ $\frac{1}{8}$" RADIUS

Fig. 35–4.—Chisel at left has mushroomed dangerously. The one in the center has started to mushroom and should be ground until it looks like the chisel at right.

Courtesy Div. of Safety and Hygiene, The Industrial Commission of Ohio.

each kind should be on hand so that when a defective or worn tool is removed from service, it can be replaced immediately with a safe tool.

Efficient tool control requires periodic inspections of all tool operations. These inspections should cover housekeeping in the tool supply room, tool maintenance, service, number of tools in the inventory, handling routine, and conditions of tools.

Responsibility for such periodic inspections is usually placed with the department head and should not be delegated by him.

When metal tools break in normal use, there are usually detectable causes: overheating or underheating of the forging or steel when it was being hardened, cracks from improper forging, improper tempering, failure to relieve stresses in forging, improper quenching, incorrect angle of cutting edge, or steel of poor quality.

Defects such as these will usually be found in tools of inferior construction, which, because of breakage and inefficiency, are more expensive in the long run than are tools of good quality. Therefore, it pays to purchase tools of the best quality obtainable.

Hand tools receiving the heaviest wear, such as chisels, wrenches, sledges, star-drills, blacksmith's tools, and cold cutters, require frequent maintenance on a regular schedule.

Proper maintenance and repair of tools require adequate facilities and equipment—workbenches, vises, a forge or furnace for hardening and tempering, tempering baths, safety goggles, repair tools, grinders, and good

lighting. Workmen especially trained in the care of tools should be in charge of these facilities, otherwise tools should be sent out for repairs.

Tempering tools

Hammer-struck and striking tools (chisels, stamps, punches, cutters, hammers, sledges, and rock drills) should be made of carefully selected steel and heat-treated so that they are hard enough to withstand blows without mushrooming excessively (Fig. 35–4), and yet not be so hard that they chip or crack.

For safety, it is better that percussion tools, some of which can be dressed frequently, be a little soft rather than too hard, because a chip may fly from an excessively hard tool without warning when the tool is struck with a hammer or sledge.

Before a chisel is heat-treated, the proper heat treatment should be obtained from the manufacturer, because different grades of steel are used for chisels. Usually, each requires a different heat treatment, from the quenching heat to the quenching bath. Therefore, the forming and tempering of tools is a skilled operation to which only experienced workmen should be assigned. It is possible to temper a tool so that the surface is hard enough for cutting, yet the body is soft enough not to fracture.

In very cold weather, some tools may break when they are struck by a hammer or sledge hammer unless they are first warmed up to about 50 F. Many firms keep such tools on a steel bench equipped with a heating element to eliminate this danger. Non-ferrous metals, such as aluminum and aluminum alloys, nickel, copper and copper alloys, chromium, zinc and zinc alloys, magnesium alloys, and lead, are more resistant to low temperatures and are widely used in areas of extreme cold.

Safe-ending tools

Hammer-struck tools, such as chisels, rock drills, flatters, wedges, punches, cold cutters, and number dies, should have heads properly hardened by a qualified workman. The hazard of burred heads can be reduced by use of the following quick and economical procedure for safe-ending tools.

In this procedure, a bronze band is welded into a recess about ⅛ in. wide by ¼ in. deep (Fig. 35–5) ground into the tool at the head.

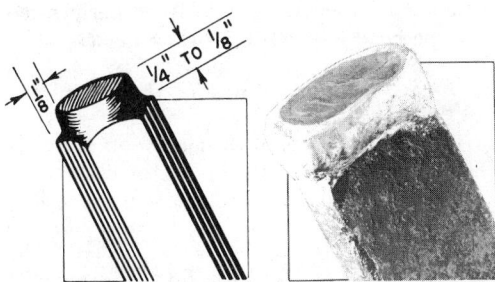

FIG. 35–5.—A more-lasting job of safe-ending results from grinding or flame cutting a shoulder on the chisel and then bronze-welding it.

This band retards checking of the tool head and reduces the frequency of tool maintenance. Safe-ending is applicable to both new and old tools.

The proper base-metal temperature for bronze-welding is 1600 to 1700 F. The correct temperature is indicated by a bright red color when the tool is viewed in the dark, but, when it is looked at through dark glasses in the light of the oxyacetylene flame, the correct temperature has been reached as soon as a red coloring can be noted.

Bronze-welding can best be done by using as small a flame as possible, and by speeding up the welding operation as it nears the finish. In welding around the ends of a tool, there is a build-up of heat as the operator proceeds,

so that as the finish is neared, less heat must be supplied to the base metal to allow complete tinning.

Short sections of tight-fitting rubber hose can be set flush with the striking ends of chisels, hand drills, mauls, and blacksmith's tools to keep chips from flying, since they usually imbed themselves in the rubber sleeve.

Safety chisels, drift punches, cutters, and marking tools are available that are warranted not to spall or mushroom. This feature is attributed to a combination of alloys, plus scientific heat treatment. Blank steel bars are also available for tool manufacture.

Dressing tools

Shock, cutting, and pointed tools require regular maintenance of their edges or striking surfaces. In most cases, once the cutting or striking surfaces have been properly hardened and tempered, only an emery wheel, grindstone, file, or oilstone need be used to keep the head in shape and the edges clean and sharp.

Never grind hardened tools until after they have been drawn or tempered. Grind in easy stages with no attempt to take off much metal at one time. While grinding, keep the tool as cool as possible with water or other cooling medium.

Shock tools that require a soft or medium-soft edge should be dressed as soon as they begin to mushroom (Fig. 35–4). A slight radius ground on the edge of the head when it is dressed will enable the tool to stand up better under pounding and will reduce the danger of chips being knocked off. Tests have shown that a radius of ⅛ to 3⁄16 in. is satisfactory on ordinary hand tools.

A file, rather than an abrasive wheel, is recommended for re-forming screw driver tips and for sharpening pike poles and axes. With a file, there is less danger of drawing the temper.

A wood-cutting tool, because of its fine cutting edge, should be shaped on a wet grindstone with plenty of water, or on a grinder having a wheel recommended by the manufacturer for that type of tool. To prevent overheating or burning, the tool should not be pressed too hard against the grinder. Frequent dipping in water will help to keep it cool.

FIG. 35–6.—The correct way to whet a wood chisel or plane iron on a stone to produce a sharp cutting edge. The bevel should be placed on the stone with the back edge slightly raised.

To shape wood chisels, an oilstone set securely in a wood block placed on a bench can be used. The oilstone should never be held in the hand because a slip off the face of the stone could cause a severe hand injury.

After a wood-cutting tool has been shaped, it should be whetted on a clean stone to produce a sharp cutting edge (Fig. 35–6). Often a few finishing strokes on a leather strop will produce a keener edge.

Metal cutting tools and hammer-struck surfaces, because they have greater body, can be dressed and sharpened on an emery wheel. Care should be taken that the tool does not overheat from too much pressure against the grinder. The manufacturer's recommendations for type and kind of grinding wheel should be followed. Each cutting edge should have the correct angle according to its use and be finished off with a file.

Handles

The handles of hand tools used for striking, such as hammers and sledges, should be of the best straight-grained material, preferably hickory, ash, or maple, and should be free from slivers. Screw drivers and wood chisels are normally provided with plastic handles. To make sure that they are properly attached, handles should be fitted to tools only by an experienced person. Poorly fitted handles make it difficult for the worker to control the tool.

Loose wood handles in sledges, axes, hammers, cold cutters, and similar tools create a hazard. No matter how tightly a handle may be wedged at the factory, both use and shrinkage will loosen it. In some cases, tapping the wedge will take up the shrinkage. In others, the head of the tool can be driven back on the handle and the protruding end of the handle cut off.

Eventually, any tool will have to be rewedged. The wedge is selected and the handle is driven off. If the wood of the handle does not bear against the head eye at all points, shave the handle until it fits snugly. Then replace the handle in the tool and sight along it to be sure that the head is properly centered. After the handle is properly fitted into the head eye, replace the wedge.

Use of Hand Tools

The misuse of common hand tools is a prolific source of injury to the industrial worker. In many instances, injury results because it is assumed that "anybody knows how" to use common hand tools. Observation and the record of injury show that this is not the case.

A part of every job instruction program should, therefore, be detailed training in the proper use of hand tools. So important is this training that considerable attention is given, in the following pages, to those safe practices that characterize the competent and safe worker.

Metal-cutting tools

Chisels. Factors determining the selection of a cold chisel are the materials to be cut, the size and shape of the tool, and the depth of the cut to be made. The chisel should be made heavy enough so that it will not buckle or spring when struck. It should be kept sharp and ground to a 60-degree angle, and by the Rockwell test should show 57 at the point and 50 at the head (Table 35-A).

A chisel only large enough for the job should be selected so that the blade is used rather than the point or corner. It is good

TABLE 35-A

DIMENSIONS OF FLAT HAND COLD CHISELS

Width of Cutting Edge "A" (in.)	Nominal Dimensions		Minimum Overall Length "L" (in.)
	Thickness of Cutting Edge "D" (in.)	Length of Taper "E" (in.)	
1/4	1/16	1 3/4	4
3/8	1/16	2	5 1/2
1/2	3/32	2	6
3/4	3/32	2 1/2	8
1	5/32	3 1/4	8 1/2

practice to round the corners of the cutting edges slightly to prevent their breaking if used in cutting. Also, a hammer heavy enough to do the job should be used.

Some workmen prefer to hold the chisel lightly in the hollow of the hand with the palm up, supporting the chisel by the thumb and first and second fingers. They claim that if the hammer glances from the chisel, it will strike the soft palm rather than the knuckles. Other workmen think that a grip with the fist holds the chisel more steady and minimizes the chances of glancing blows. Moreover, in some positions, this is the only grip that is natural or even possible. For regular use, a sponge rubber pad, forced down over the chisel, provides a protective cushion for the hand (Fig. 35-7).

When shearing with a cold chisel, the work-man should hold the tool at the vertical angle that permits one bevel of the cutting edge to be flat against the shearing plane. Protective holders are also commercially available.

Workmen should wear safety goggles when chipping and should set up a shield or screen to prevent injury to other workmen from flying chips. If a shield does not afford positive protection to all exposed employees, they should wear safety goggles.

Bull chisels held by one man and struck by another require the use of tongs or a chisel holder to guide the chisel so that the work-men will not be exposed to injury. Both workmen should wear safety goggles.

Stamping and marking tools of special alloy and design are available. If possible, marking tool holders should be used so that the workman does not have to hold his fingers close to the face of the tool being struck.

Tap and die work requires certain precautions. The work should be firmly mounted in the vise. A tap wrench of the proper size should be secured. The hands should be kept away from broken tap ends. If a broken tap is removed by using a tap extractor or a punch and hammer, the worker should wear safety goggles. When a long thread is being cut with a hand die, the hands and arms should be kept clear of the sharp threads coming through the die.

Hack saws should be adjusted in the frame to prevent buckling and breaking, but should

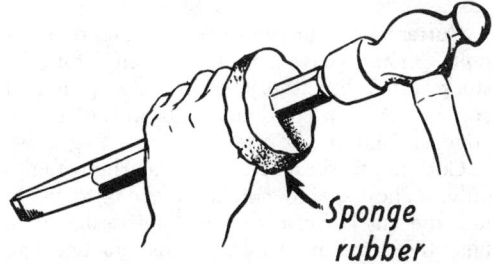

Sponge rubber

FIG. 35-7.—Inexpensive and effective hand protection is provided by a sponge rubber shield forced onto hammer-struck tools. Combination rubber hand grips and shields are available for some hammer-struck tools.

983

not be tight enough to break off the pins that support the blade. Install blade with teeth pointing forward.

Blades with 14 teeth to the inch should be used for cutting soft solid metal; 18 teeth for tool steel, iron pipe, hard metal, and general shop use; 24 teeth for drill rods, sheet metal, and tubing; and 32 teeth for thin sheet metal (less than 18 gage) and tubing.

Pressure should be applied on the forward stroke only. Lift the saw slightly and pull back in the cut lightly to protect the teeth. If the blade is twisted or too much pressure is applied, the blade may break and cause injury to the hands or arms of the user. Do not continue an old cut after changing to a new blade; it may bend and break.

Files. Selection of the right kind of file for the job will prevent injuries, lengthen the life of the file, and increase production. Files should not be used without a handle. Inasmuch as the extremely hard and brittle steel of the file chips easily, the file should never be cleaned by being struck against a vise or other metal object—a file-cleaning card should be used.

For the same reason a file should not be hammered or used as a pry. Such abuse frequently results in the file's chipping or breaking, causing injury to the user. A file should not be made into a center punch, chisel, or any other type of tool because the hardened steel may fracture in use.

The correct way to hold a file is to grasp the handle firmly in one hand and use the thumb and forefinger of the other to guide the point. This technique will give good control and ensure better and safer work.

A file should never be used without a smooth, crack-free handle; otherwise, if the file should slip or be struck by a revolving part of a machine, the tang may puncture the palm of the hand, the wrist, or other part of the body. Under some conditions, a clamp-on, raised offset handle may be useful to give extra clearance for the hands.

When work to be filed is placed in a lathe, the job should be done left-handedly, with the file and hands clear of the chuck jaws or the dog. See Fig. 35–8.

Tin snips should be heavy enough to cut

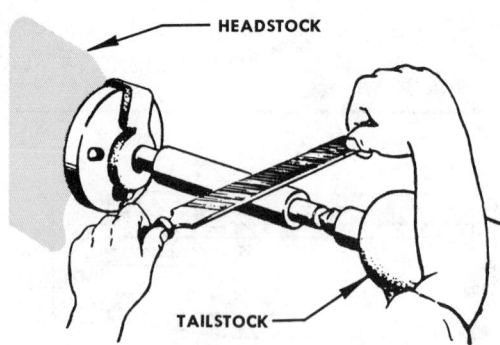

Fig. 35–8.—Although filing on a lathe is not recommended (an accurately ground tool bit and careful workmanship will keep the amount of filing to a minimum), the recommended procedure is to grip the file handle in the left hand and use the right hand to hold the top of the file. Be sure arm is kept out of range of driver plate and lathe dog or chuck.

the material so easily that the workman needs only one hand on the snips and can use the other to keep the edges of the cut material pulled aside. The material should be well supported before the last cut is made so that the cut edges do not press against the hands.

Jaws of snips should be kept tight and well lubricated.

Workmen should wear safety goggles when trimming corners or slivers of sheet metal, because small particles often fly with considerable force. Leather or heavy canvas work gloves will help prevent hand cuts or scratches due to handling sharp edges of the sheet metal. When cutting long sheet metal pieces, push down the sharp ends next to the hand holding the snips.

Cutters used on wire, reinforcing rods, or bolts should have ample capacity for the stock; otherwise, the jaws may be sprung or spread. Also, a chip may fly from the cutting edge and injure the user.

Cutters are designed to cut at right angles only. They should not be "rocked" to facilitate the cut because they are not designed to take the resulting strain. This practice can also cause the knives to chip.

Cutters require frequent lubrication. To keep cutting edges from becoming nicked or chipped, cutters should not be used as nail pullers or pry bars.

Cutter jaws should have the hardness specified by the manufacturer for the particular kind of material to be cut. By adjustment of the bumper stop behind the jaws, cutting edges should be set to have a clearance of 0.003 in. when closed.

Wood-cutting tools

Edged tools should be used so that, if a slip should occur, the direction of force will be away from the body. For efficient and safe work, edged tools should be kept sharp and ground to the proper angle. A dull tool does a poor job and may stick or bind. A sudden release may throw the user off balance or cause his hand to strike an obstruction. Dressing of wood-cutting tools is covered earlier in this chapter.

Wood chisels. Inexperienced employees should be instructed in the proper method of holding and using chisels. Handles should be free of splinters. The wood handle of a chisel struck by a hammer or mallet should be protected by a metal or leather band to prevent its splitting.

The work to be cut should be free of nails to avoid damage to the blade of the chisel. Should metal be struck by the cutting edge, a chip from the chisel might fly.

The wood chisel should not be used as a pry or a wedge. The steel in a chisel is hard so that the cutting edge will hold, and is therefore brittle enough to break if the chisel is used as a pry.

When not in use, the chisel should be kept in a rack, on a workbench, or in a slotted section of the tool box so that the sharp edges will be out of the way.

Saws should be selected for the work they must do. For cutting across the grain of the wood, use a crosscut saw; for cutting with the grain, a ripping saw. The difference between them is their teeth angle. For fast crosscut work on green wood, use a coarse saw (4 to 5 points per inch); for smooth, accurate cutting of dry wood, use a fine saw (7 to 8 points per inch). Points per inch is stamped on the heel of a saw.

Saws should be kept sharp and the teeth kept well set to prevent binding. When not in use, saws should be kept in racks or hung by the handle to prevent the teeth from being dulled. A Teflon-coated saw is available that minimizes binding.

Axes. To use an axe safely, workmen must be taught to lift it properly, to swing correctly, and to place the stroke accurately. The proper grip for a right-handed person is to have the left hand about 3 in. from the end of the handle, and the right hand about three-fourths of the way up. A left-handed person should reverse the position of the hands.

A narrow axe with a thin blade should be used for hard wood, and a wide axe with a thick blade for soft wood. A sharp, well-honed axe yields better chopping speed and is much safer to use because it bites into the wood (Fig. 35–9). A dull axe will often glance off the wood being cut and may strike the user in the foot or leg.

An axeman should make sure that he has a clear circle in which to swing his axe before he starts chipping. Also, he should remove all vines, brush, and shrubbery within the

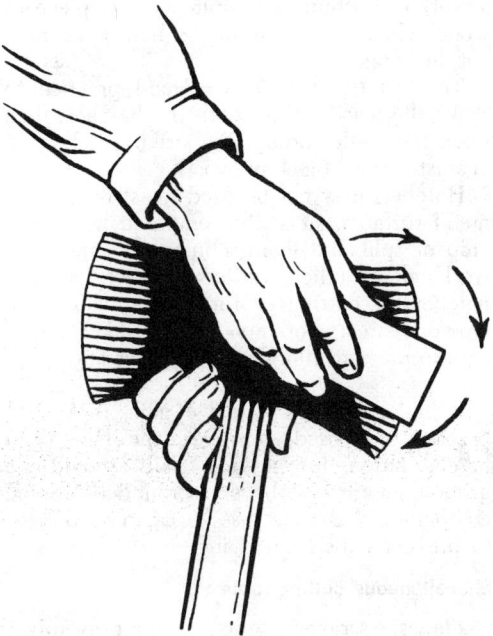

Fig. 35–9.—Good honing saves labor and makes an axe safer to use. The axe should be honed after each sharpening and each use. Correct honing motion is shown here.

range, especially overhead vines that may catch or deflect his axe. It is advisable to wear safety shoes when using an axe.

Axe blades should be protected with a sheath or metal guard wherever possible. When the blade cannot be guarded, it is safer to carry the axe at one's side. The blade on a single-edged axe should be pointed down.

Adzes are hazardous tools in the hands of inexperienced men and should be handled only by those skilled in their use. Workmen should straddle the work or, as is necessary at times, stand at the side of the material. They should see that the adze is sharp and that the handle is sound and fitted securely in the head. Safety shoes and shin guards should be worn. When not in use, the adze should be set aside in a safe place with cutting edge covered or secured to protect passers-by or, temporarily, left stuck in the timber.

Hatchets are used for many purposes and frequently cause injury. For example, when a workman attempts to split a small piece of wood while holding it in his hands, he may cut his fingers.

To start the cut, it is a good practice to strike the wood lightly with the hatchet, then force the blade through by striking the wood against a solid block of wood.

Hatchets may not be used for striking hard metal surfaces, since the tempered head may chip or split and flying chips may injure the workman or other workers. When using a hatchet for cutting or for driving nails in a crowded area, workmen should take special care to prevent injury to themselves and other workers.

Using a hatchet to drive nails is a poor practice. If used, however, the face of a hatchet used for driving nails should be ground square if it becomes rounded. Some companies prefer to use a corrugated face to prevent nails from flying.

Miscellaneous cutting tools

Planes, scrapers, bits, and drawknives should be used only by experienced men. These tools should be kept sharp and in good condition. When not in use, they should be placed in a rack on the bench, or in a tool box in such a way that will protect the user and prevent damage to the cutting edge.

Knives are more frequently the source of disabling injuries than any other hand tool. In the meatpacking industry, hand knives caused more than 15 percent of all disabling injuries, which is more than caused by any other agency.

The principal hazard in the use of knives is that the hands may slip from the handle onto the blade or that the knife may strike the body or the free hand. A handle guard or a finger ring (and swivel) on the handle eliminates these hazards. Adequate guarding is important.

The cutting stroke should be away from the body. If that is not possible, then the hands and body should be in the clear, a heavy leather apron or other protective clothing should be worn, and, where possible, a rack or holder should be used for the material to be cut. Jerky motions should be avoided to help maintain balance. Be sure employees are trained and supervised.

Belt repairmen and other workmen who must carry knives with them on the job should keep them in sheaths or holders. Never carry a sheathed knife on the front part of a belt—always carry it over the right or left hip, toward the back. This will prevent severing a leg artery or vein in case of a fall.

Knives should never be left lying on benches or in other places where they may cause hand injuries. When not in use, they should be kept in racks with the edges guarded. Safe placing and storing of knives is important to knife safety. (See Fig. 35–10.)

Ring knives—small, hooked knives attached to a finger ring—are used where string or twine must be cut frequently. Supervisors should make sure that the cutting edge is kept outside the hand, not pointed inside. A wall-mounted cutter or blunt-nose scissors would be safer.

Carton cutters are safer than hooked or pocket knives for opening cartons. They not only protect the user, but eliminate deep cuts that could damage carton contents. Frequently, damage to contents of soft plastic bottles may not be detected immediately; subsequent leakage may cause chemical burns, damage other products, or start a fire.

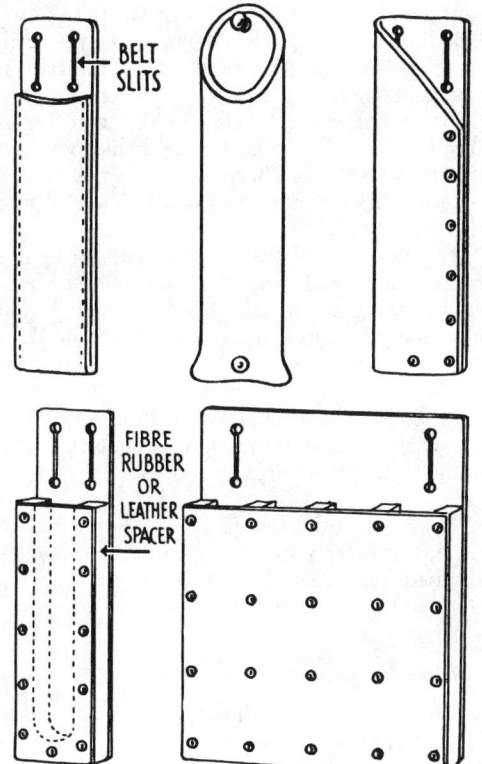

BELT SLITS

FIBRE RUBBER OR LEATHER SPACER

Fig. 35–10.—Sheaths for sharp-pointed tools can be made from leather scrap, canvas belting, rubber, plastic, or sheet metal.

To cut corrugated paper, a hooked linoleum knife permits good control of pressure on the cutting edge and eliminates the danger of the blade suddenly collapsing as in the case of a pocket knife. Be sure hooked knives are carried in a pouch or heavy leather (or plastic) holder. The sharp tip must not stick out.

Supervisors should make certain that employees who handle knives have ample room in which to work so they are not in danger of being bumped by trucks, the product, overhead equipment, or other employees. For instance, a left-handed worker should not stand close to a right-handed person; the left-handed person might be placed at the end of the bench or otherwise given more room. Workers should be trained to cut away from or out of line with their bodies.

Supervisors should be particularly careful

about the hazard of employees leaving knives hidden under product, under scrap paper or wiping rags, or among other tools in work boxes or drawers. Knives should be kept separate from other tools to protect the cutting edge of the knife as well as the employee.

Work tables should be smooth and free of slivers. Floors and working platforms should have slip-resistant surfaces and should be kept unobstructed and clean. If sanitary requirements permit mats or wooden duck boards, they should be in good repair, so workers do not trip or stumble. Conditions which cause slippery floors should be controlled as much as possible by good housekeeping and frequent cleaning.

Careful job and accident analysis may suggest some slight change in the operating procedure which will make knives safer to use. For instance, on some jobs special jigs, racks, or holders may be provided so it is not necessary for the operator to stand too close to the piece being cut.

The practice of wiping a dirty or oily knife on the apron or clothing should be discouraged. The blade should be wiped with a towel or cloth with the sharp edge turned away from the wiping hand. Sharp knives should be washed separately from other utensils and in such a way that they will not be hidden under soapy wash water.

Horseplay should be prohibited around knife operations. Throwing, "fencing," trying to cut objects into smaller and smaller pieces, and similar practices are not only dangerous but reflect inadequate supervision.

Supervisors should make sure that nothing is cut that requires excessive pressure on the knife—such as frozen meat. Food should be thawed before it is cut or else it should be sawed. Knives are not a substitute for can openers, screw drivers, or ice picks.

A brad awl should be started with the edges across the grain to keep the wood from splitting. Then it should be turned back and forth as it is pressed down. It should be held at right angles to the surface to prevent its slipping.

Materials handling tools

Crowbars. Whenever a crowbar is needed, the proper size and kind of bar for the job should be used. Makeshifts such as a piece of pipe or an iron bar should never be substituted for a crowbar, since they may slip and cause injury.

The crowbar should have a point or toe of such shape that it will grip the object to be moved, and a heel to act as a pivot or fulcrum. In some cases, a block of wood under the heel will prevent the crowbar from slipping and injuring the hand.

Crowbars not in use should be secured if stood on end so that they will not fall; or if laid on the ground, placed where they will not create a stumbling hazard.

Hooks. Hand hooks used for handling material should be kept sharp so that they will not slip when applied to a box or other object.

The handle should be strong and securely attached and shaped to fit the hand. The handle and the point of long hooks should be bent on the same plane so that the bar will lie flat when not in use and not constitute

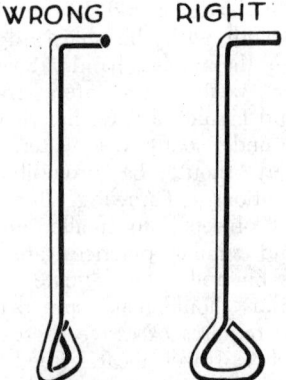

WRONG   RIGHT

Fig. 35–11.—Hand hooks can be made safer if the hook is bent so that it will lie flat when set down on a flat surface.

a tripping hazard (Fig. 35–11). The hook point should be shielded when not in use.

Shovels. Shovel edges should be kept trimmed, and handles should be checked for splinters. Workers should wear good shoes with sturdy soles, preferably safety shoes. They should have their feet well separated to

get good balance and spring in the knees. The leg muscles should take much of the load.

To reduce the chance of injury, the ball of the foot—not the arch—should be used to press the shovel into clay or other stiff material. If the instep is used and the foot slips off the shovel, the sharp corner of the shovel may cut through the workman's shoe and into the foot.

Dipping the shovel into a pail of water occasionally will help to keep it free from sticky material, making it easier to use and less likely to cause strain. Greasing or waxing the shovel blade will also prevent some kinds of material from sticking. Teflon-coated shovels prevent sticking of certain materials. When not in use, stand shovels against a wall or keep them in racks or boxes.

Rakes. A rake should not be left with the prongs turned upward where they may be stepped on, causing foot injury or causing the handle to fly up. Rakes should be racked when not in use.

Torsion tools

The great variety of wrenches and socket wrenches used for turning nuts, bolts, and fittings makes it important that workers know the purpose and limitations of each type and size.

Box and socket wrenches are indicated where a heavy pull is necessary and safety is a consideration. Box and socket wrenches completely encircle the nut, bolt, or fitting and grip it at all corners, as opposed to the two corners gripped by an open-end wrench, Fig. 35–12. They will not slip off laterally

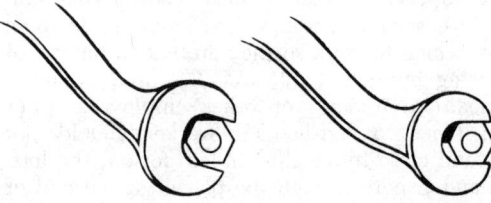

Fig. 35–12.—Wrench jaws that fit (and are not sprung or cracked) prevent damage to the head of the nut, and are not likely to slip and cause injury to the user.

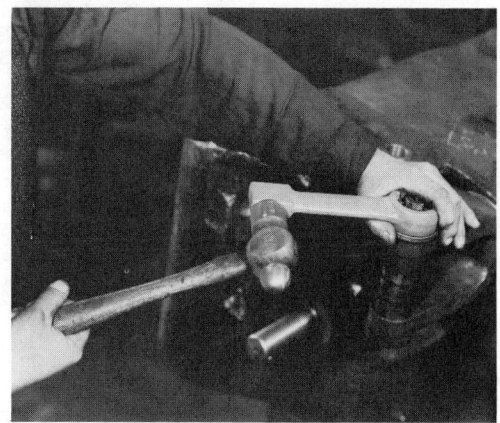

FIG. 35–13.—Where handle length does not offer the necessary leverage, use a striking-face wrench. This is a safe approach where large nuts must be set up tight or frozen nuts loosened.

and they eliminate the dangers of sprung jaws.

Socket and box wrenches normally come in three styles of openings—single hex, double hex, or double square. On square-headed bolts, nuts, and fittings, either a single or double square design should be used. Single or double hexagonal wrenches are designed for hexagonal-shaped fittings, bolts, and nuts. Ratchet handles and universal joint fittings for socket wrenches allow them to be used where space is limited.

Never overload the capacity of a wrench by using a pipe extension on the handle or strike the handle of a wrench with a hammer. Hammer abuse weakens the metal of a wrench and can cause the tool to break. Special heavy-duty wrenches are available which can be used with handles as long as three feet. For extra stubborn bolts and nuts, heavy-duty, sledge type box wrenches are available. These are of a heavy design, properly tempered, and have a striking surface for the hammer or sledge, Fig. 35–13. Where possible, special penetrating oil should be used to first loosen tight nuts.

There is a correct wrench for every nut and bolt. Oversize openings will not grip the corners securely and shims should not be used to compensate for an oversize opening. The use of the wrong size can round the corners of the bolt, or cause slippage, as well as make it difficult to then apply the proper size. Be sure to use the proper tool and not a makeshift approach (Figs. 35–14 and –15).

Sockets should be kept clean of caked dirt and grime inside the socket as it will prevent the socket from seating fully and the concentration of the pull force at the end of the socket opening even with a moderate pull can easily damage the socket.

A common cause of socket and box wrench breakage is "cocking." Cocking is when the

FIG. 35–14.—This makeshift is made up from the nearest tools at hand. One slip could cause injury.

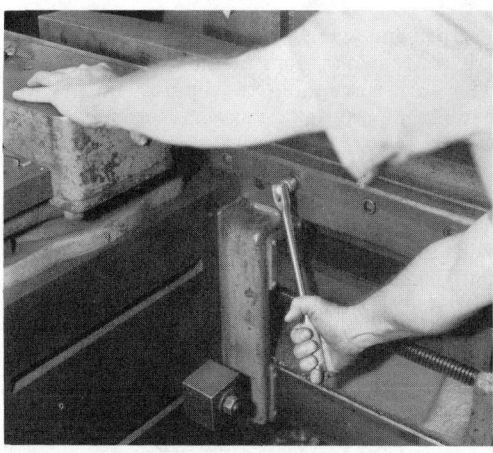

FIG. 35–15.—The correct tool at the right time makes an otherwise hard job safe and easy.

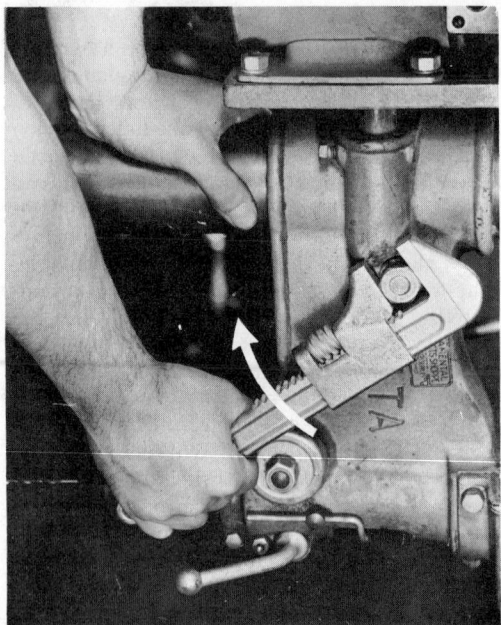

Fig. 35–16.—Correct use of a wrench. The wrench is in good condition and securely gripped. Hand of workman is braced and clear in the event that the nut should suddenly turn. To protect workman's hand, wrench is pulled, not pushed.

tool does not fit securely on the bolt or nut but fits at an angle. This concentrates the entire strain at one point, making the tool vulnerable to fracture.

Open-end wrenches have strong jaws and are satisfactory for medium-duty turning. They are susceptible to slipping if they do not fit properly or are used incorrectly.

Combination wrenches have a box opening at one end and an open end at the other end. They are very handy for speeding the turning with the open end and using the box end for initial loosening or final tightening.

Torque wrenches measure the amount of twisting force being applied by means of a dial or calibrated arm. A torque wrench is safer to use than an end or box wrench of similar size because on a torque wrench an excessive amount of twisting force will be indicated. The worker can then reduce the

force and thus prevent slips and falls that might occur if the wrench slips, nut twists off, or stud breaks.

A machine wrench is often misused as a hammer. As a result, it soon becomes distorted and, therefore, unsafe. A wrench of the proper size for the job should be selected. It is an unsafe practice to try to make the wrong wrench fit by using shims.

Adjustable wrenches are generally recommended for light-duty jobs or when the proper-size fixed-opening wrench is not available. They are likely to slip because of the difficulty in setting the correct size and the tendency for the jaws to "work" as the wrench is being used.

Unless the space in which the job is being done makes the method impracticable, an adjustable wrench should be placed on the nut with the open jaws facing the user. With the open jaw facing the user, the *pulling force* applied to the handle tends to force the movable jaw onto the nut. For that reason, and for reasons of safety, wrenches should be pulled, not pushed (Fig. 35–16).

According to the manufacturers, the movable jaws on adjustable wrenches are weaker than the fixed jaws. One manufacturer of forged wrenches tested a 10-in. adjustable wrench and found that when the wrench was correctly used, as described in the preceding paragraph, it easily took a 250-pound direct pull on the extreme end of the handle. Turned around with the pull against the movable jaw, the wrench began to slip at between 75 and 100 pounds of pull at the end of the handle.

Pipe wrenches. Workmen, especially those on overhead jobs, have been seriously injured when pipe wrenches slipped on pipes or fittings, causing the men to lose their balance and fall. Pipe wrenches, both straight and chain tong, should have sharp jaws and be kept clean to prevent their slipping.

The adjusting nut of the wrench should be inspected frequently. If it is cracked, the wrench should be taken out of service. A cracked nut may break under strain, causing complete failure of the wrench and possible injury to the user.

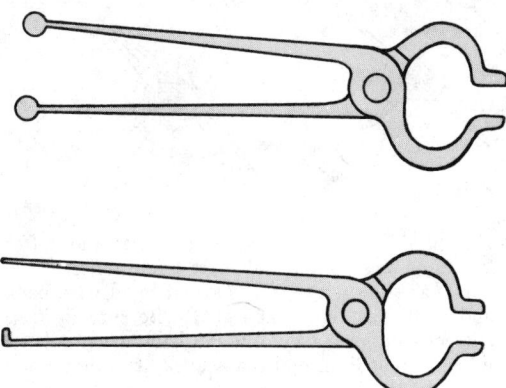

FIG. 35–17.—Ends of tongs can be upended, or projections can be welded on them to give clearance for user's hands.

Using a wrench of the wrong length is also a source of accidents. A wrench handle too small for the job does not give proper grip or leverage. An oversized wrench handle may strip the threads or break the work suddenly, causing a slip or fall.

A piece of pipe slipped over the handle to give added leverage also can strain a pipe wrench or the work to the breaking point. Using a makeshift as an extension to secure greater leverage may easily cause the wrench head to break. The handle of every wrench is designed to be long enough for the maximum allowable safe pressure.

A pipe wrench should never be used on nuts or bolts, the corners of which will break the teeth of the wrench, thereby making it unsafe to use on pipe and fittings. A pipe wrench should not be used on valves or small brass, copper, or other soft fittings which may be crushed or bent out of shape. Also, a pipe wrench used on nuts and bolts ruins their heads. A wrench should not be struck with a hammer nor used as a hammer unless, as with specialized types, it is specifically designed for such use.

Pipe tongs should be placed on the pipe only after the pipe has been lined up and is ready to be made up. A 3- or 4-in block of wood should be placed near the end of the travel of the tong handle and parallel to the pipe to prevent injury to the hands or feet in the event the tongs slip.

Workmen should neither stand nor jump on the tongs nor place extensions on the handles to obtain more leverage. They should use larger tongs, if necessary, to do the job.

Tongs usually are bought, but some companies make their own to perform specific jobs. Often they are designed in such a way that the hands are pinched when the tongs are closed.

To prevent pinching, the end of one handle should be up-ended toward the other handle, to act as a stop. It is also possible to braze, weld, or bolt bumpers on the handles a short

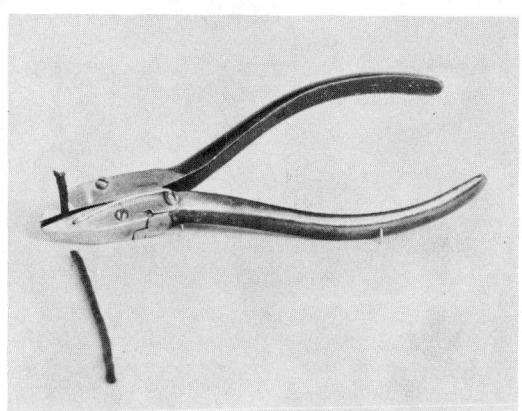
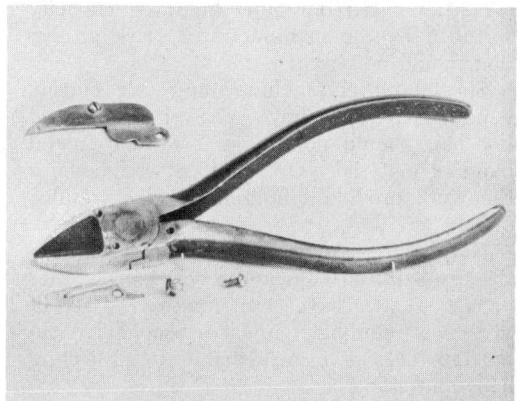

FIG. 35–18.—*At left,* the guard on a pair of pliers holds a wire end after it has been cut, thus preventing it from flying. *At right,* a disassembled view of the same pliers.

Courtesy Capco Co., Ottumwa, Iowa.

distance behind the pivot point so that the handles cannot close against the fingers (Fig. 35–17).

Pliers are often considered a general-purpose tool and are often misused for purposes for which they were not designed. Pliers are meant for gripping and cutting operations; they are not recommended as a substitute for wrenches because their jaws are flexible and frequently slip when used for this work. Pliers also tend to round the corners of bolt heads and nuts and leave jaw marks on the surface; this makes it difficult to use a wrench at some future time.

Side-cutting pliers sometimes cause injuries when short ends of wire are cut. A guard over the cutting edge (Fig. 32–18) and the use of safety goggles will prevent flying short ends from causing injuries.

The handles of electricians' pliers should be insulated. In addition, the men should wear electricians' gloves, if company policy requires them to.

Because pliers do not hold the work securely, they should not be used as a substitute for a wrench.

Special cutters for heavy wire, reinforcing wire, and bolts are safer than makeshift tools. It is important that the cutting edges apply force at right angles to the wire or other work being cut (Fig. 35–19). The cutter should not be used near live electrical circuits and should be used only for the rated capacity specified by the manufacturer. Eye protection should be worn.

Special cutters include those for cutting banding wire and strap. Claw hammers and pry bars should not be used to snap metal banding material. Only cutters designed for the work provide safe and effective results.

Nail band crimpers make it possible to keep the top band on kegs and wood barrels after nails or staples have been removed. Use of these tools eliminates injury caused by reaching into kegs or barrels that have projecting nails and staples.

Pullers are the only quick, safe, and easy way for pulling a gear, wheel, pulley, or bearing from a shaft. Prybars and chisels should

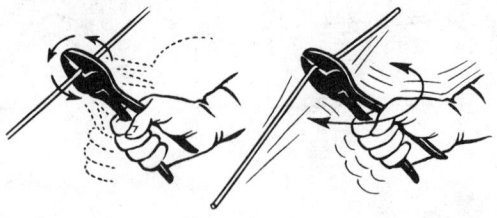

Fig. 35–19.—Wire should be cut with the cutter knives swung up and down, at right angles to the wire (as shown at left). Do not bend wire back and forth (as shown at right). Be sure to keep cut ends under control. Stand back when cutting wires so that if an end snaps back, it cannot reach you.

not be used as they concentrate the force at one point and tend to cock the part on the shaft. Select the correct-sized puller. The jaw capacity should be such that the jaws press tightly against the part being pulled. Use a puller with as large a pressure screw as possible.

Screw drivers. The screw driver is probably the most commonly used and abused tool. The practice of using screw drivers for punches, wedges, pinch bars, or pries should be discouraged. If used in such manner, they become unfit for the work they are intended to do. Furthermore, a broken handle, bent blade, dull or twisted tip may cause a screw driver to slip out of the slot and cause a hand injury.

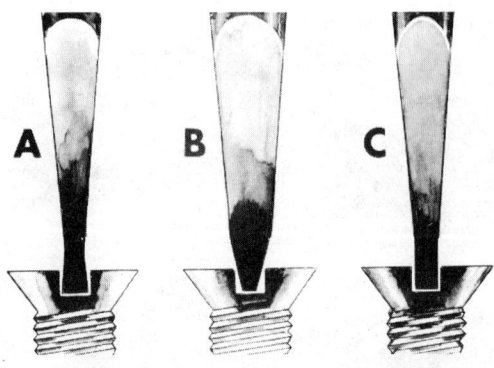

Fig. 35–20.—Screw driver bits with parallel sides shaped like an illustration **C**, or undercut slightly like in **A**, will grip the slot more securely than the bit shown in **B**.

Cross-slot (Phillips-head) screw drivers are safer than the square bit type, because they have less tendency to slip. The pressure on the tip is more evenly distributed and results in less wear on the tip.

A screw driver tip should be filed to fit the screw. A sharp square-edged bit will not slip as easily as a dull, rounded one, and requires less pressure. The tip must be kept clean and sharp, however, to permit a good grip on the head of the screw (Fig. 35–20).

The part to be worked upon should never be held in the hands; it should be laid on a bench or flat surface, or held in a vise. This practice will lessen the chance of injury to the hands if the screw driver should slip from the work.

No screw driver used for electrical work should have the blade or rivet extending through the handle. Also, both blade and handle should be insulated except at the tip.

Hammers

A hammer should have a securely wedged handle suited to the type of head used. The handle should be smooth, free of oil, shaped to fit the hand, and of the specified size and length. Employees should be warned against using a steel hammer on hardened steel surfaces. Instead, a soft metal hammer or one with a plastic, wood, or rawhide head should be used. Safety goggles should be furnished and worn to protect against flying chips, nail heads, or scale.

Federal specifications for the hardness of ball peen hammers require that the Rockwell number on the base and peen be no less than 50 nor more than 60. U.S. Government Printing Office, *Federal Specifications for Hammers, Mauls and Sledges,* GGG-H-86.

Selection of the proper hammer is important. One that is too light is as unsafe and inefficient as one that is too heavy. Either is hard to control.

Sledge hammers can have two common unsafe conditions: split handles and loose or chipped heads. Because these tools are used infrequently in some industries, the heads may become loose or chipped and the defect not noticed.

Some companies place a steel band around the head and bolt it to the handle to prevent the head from flying off. The heads should be dressed whenever they start to check or mushroom.

A hammer so light that it bounces off the work is hazardous; likewise, one too heavy and hard to control may cause body strain.

Riveting hammers, often used by sheet metal workers, should have the same kind of use and care as ball peen hammers and should be watched closely for checked or chipped faces.

Carpenter's or claw hammers are designed primarily for driving and drawing nails. Their shape, depth of face, and balance make them unsuitable for striking heavier objects, such as cold chisels. The corners of the claw should not be used as pries because they are susceptible to chipping and breaking.

The faces should be kept well dressed at all times to reduce the hazard of nails flying while they are being started into a piece of wood. A checker-faced head is sometimes used to reduce this hazard.

Some companies have found eye protection advisable for all men working close together, as in a shipping room, when some are using these hammers.

When a nail is to be drawn from a piece of wood, a block of wood may be used under the head to increase the leverage and reduce the strain on the handle.

Spark-resistant tools

So-called spark-resistant tools made of non-ferrous materials (such as a beryllium copper alloy) are sometimes advised for use where flammable gases, highly volatile liquids, and other explosive substances are stored or used. The intensified sparks from steel tools are capable of igniting substances such as gunpowder, lint, TNT, carbon disulfide, and ethyl ether. (See U.S. Department of Commerce, Office of Technical Services. *Sparking Characteristics and Safety Hazards of Metallic Materials,* Technical Report No. NGF-T-1-57, PB 131131.)

In certain circumstances, steel coated with aluminum paint can emit sparks when struck with a metal striker (steel, brass, or spark-resistant alloys), and such sparks may ignite certain mixtures of flammable gases or vapors

993

in air. There is some question about the hazard of friction sparks igniting gasoline vapors and petroleum products. See Chapter 42, "Flammable and Combustible Liquids."

Nonferrous tools reduce the hazard from sparking but do not eliminate it. They need inspection before each use to be certain that they have not picked up foreign particles which could produce friction sparks, thereby obviating the value of these special tools.

Portable Power Tools

Portable power tools are divided into five primary groups according to the power source: electric, pneumatic, gasoline, hydraulic, and explosive (powder actuated). Several types of tools, such as saws, drills, and grinders, are common to the first three groups; hydraulic tools are used mainly for compression work; explosive tools are used exclusively for penetration work, cutting, and compression.

A portable power tool presents similar hazards as a stationary machine of the same kind, in addition to the risks of handling. Typical injuries caused by portable power tools are burns, cuts, and strains. Sources of injury include electric shock, particles in the eyes, fires, falls, explosion of gases, and falling tools.

The source of power should always be disconnected before accessories on a portable tool are changed, and guards should be replaced or put in correct adjustment before the tool is used again.

A tool should not be left in an overhead place where there is a chance that the cord or hose, if pulled, will cause the tool to fall. The cord or hose and the tool may be suspended by counterweighted rope or string which keeps them out of the operator's way and also counterbalances some of the weight of the tool and the power source. Cords and hoses on the floor create a stumbling or tripping hazard. They should be suspended over aisles or work areas or, if laid across the floor, protected by wooden strips or special raceways. Suspend them in such a way that they will not be struck by other objects or by material being handled or moved. An unexpected pull might cause the tool to jam or otherwise expose the operator to injury.

Do not hang cords or hoses over nails, bolts, or sharp edges. They should also be kept away from oil, hot surfaces, and chemicals.

Because of the extreme mobility of power-driven tools, they can easily come in contact with the operator's body. At the same time, it is difficult to guard such equipment completely. There is the possibility of breakage because the tool may be dropped or roughly handled. Furthermore, the source of power (electrical, mechanical, air, hydraulic, or explosive cartridge) is brought close to the operator, thus creating additional hazards.

When using explosive powder load equipment for drilling anchors into concrete, or when using air-driven hammers or jacks, it is recommended that proper hearing protection equipment be used. All companies and manufacturers of portable power tools attach to each tool a set of operating rules or safe practices. Their use will supplement the thorough training which each powered tool operator should receive.

Power-driven tools should be set in safe places and not left in areas where they may be struck by passers-by, and activated.

Selection of tools

Replacement of a hand tool by a power tool designed for the same purpose may mean merely a substitution of electrical or mechanical hazards for comparatively less serious manual hazards. Therefore, the safety professional should anticipate the new problems and avoid as many as possible through insistence upon safe design and proper training.

The tool source should be given complete information about the job on which a tool is to be used so that the most appropriate tool can be recommended. Factors to be considered include clearance in the working quarters, depth of hole to be drilled, nature and thickness of materials.

Portable power tools designed to be used intermittently or on light work are generally designated as "home owners grade." Those intended for continuous operation and production service or for heavy work are usually identified as "industrial duty."

For safe operation of portable power tools, workers should be trained in the proper selection and limitations of the tools used, and they should be taught never to tackle a job with

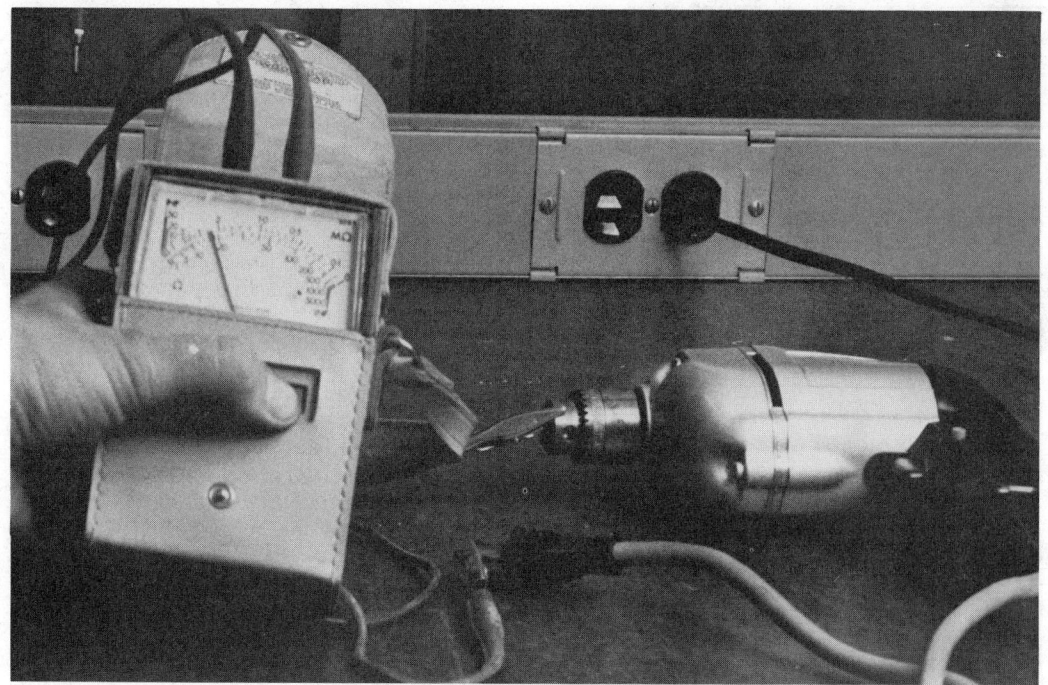

FIG. 35–21.—Insulation resistance tester has test leads attached to grounding pin of attachment plug and chuck of portable tool. Meter indicates condition of insulation of portable tool and permits identification of impending failures.

Courtesy Daniel Woodhead Company.

an undersized tool. A tool that is too light may not only fail, but may cause undue fatigue to the operator.

Inspection and repair

Periodic inspections are essential to the maintenance of power tools. Thorough inspections will uncover operating defects that can be corrected, and prevent breakdowns and costly repair charges. A schedule for systematic inspection and maintenance records for each tool should be set up. Defective tools should be tagged and withdrawn from service until repaired.

An insulation resistance meter (Fig. 35–21) will identify a short circuit or destroyed insulation. Electrical tools should be checked periodically. Some companies provide for a visual or external inspection at the tool room each time a tool is returned and for a thorough "knock down" inspection at specified intervals.

A colored tag can be used to identify when the equipment was last inspected. The important thing is to record the condition of the tool and to correct any unsafe conditions.

Employees should be instructed and trained to inspect tools and to recognize and report (and, if authorized, to correct) defects. The extent of this inspection and of the responsibility for correcting defects should be clearly outlined so that there is neither unnecessary duplication of effort nor misunderstanding about the responsibility for maintenance. A convenient reminder of points to check is a card similar to that shown in Fig. 35–22. Employees must be warned not to make makeshift repairs and to do no repair work unless authorized.

Power tools should be cleaned with a recommended nonflammable and nontoxic solvent. Air drying should be used in place of compressed air.

PORTABLE ELECTRIC TOOLS

Inspection Checklist

GENERAL
- Low voltage equipment used in tanks and wet areas? ☐
- Tools well maintained? ☐
- Motors in good condition? ☐
- Approved tools used in explosive atmospheres? ☐
- Tools left where they cannot fall? ☐

CORDS
- Insulation and plugs unbroken? ☐
- Cords protected against trucks and oil? ☐
- Cords not in aisles? ☐

GROUNDING
- Ground wire fastener in safe condition? ☐
- 3-wire plug extension cord (if a 3-wire tool)? ☐
- Ground wire used? ☐
- Defects or minor shocks reported? ☐

GUARDING
- Guards used on grinders and saws? ☐
- Movable guards operate freely? ☐
- Eye or face protection worn? ☐

FIG. 35–22.—An inspection checklist card can be used for portable electric tools. Such a card encourages workers to inspect equipment before and after use. More specific cards or tags simplify the prompt recording of defects and result in better maintenance records.

Electric tools

Electric shock is the chief hazard from electrically powered tools. Types of injuries are electric flash burns, minor shock that may cause falls, and shock resulting in death.

Serious electric shock is not entirely dependent on the voltage of the power input. The ratio of the voltage to the resistance determines the amount of current that will flow and the resultant degree of hazard. The current is regulated by the resistance to the ground of the body of the operator and by the conditions under which he is working. It is possible for a tool to operate with a defect or short in the wiring. The use of a ground wire protects the operator under all conditions and is mandatory for all but double insulated electric power tools.

Insulating platforms, rubber mats, and rubber gloves provide an additional factor of safety when tools are used in wet locations, such as in tanks or boilers and on wet floors.

Low voltage of 6, 12, 24, or 32 volts through portable transformers will reduce the shock hazard in wet locations. Standing orders should be issued to supervisors and employees to use such low-voltage equipment on work in such locations.

Electric tools used in wet areas or in metal tanks expose the operator to conditions favorable to the flow of current through his body, particularly if he is wet with perspiration. Most electric shocks from tools have been caused by the failure of insulation between the current-carrying parts and the metal frames of the tools. Only tools listed by Underwriters' Laboratories, Inc., should be used.

Double insulated tools. Protection from electric shock, while using portable power tools, has been described as depending upon third wire protective grounding. "Double insulated" tools, however, are available which generally provide more reliable shock protection without third-wire grounding. Paragraph 250–45 of the 1971 *National Electrical Code* permits "double insulation" for portable tools and appliances. Tools in this category are permanently marked by the words "double insulation" or "double insulated." Units designed to this category which have been tested and listed by Underwriters' Laboratories, Inc., will also employ the use of the UL symbol. Many U.S. manufacturers are also using the symbol

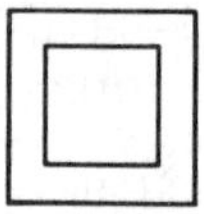

to denote "double insulation." This symbol has been widely used in most European countries.

Conventional electric tools have a single layer of functional insulation and are metal encased. For small-capacity tools, "double insulation" can be provided by encasing the entire tool in a nonconductive material which is also shatterproof. The switch is also non-

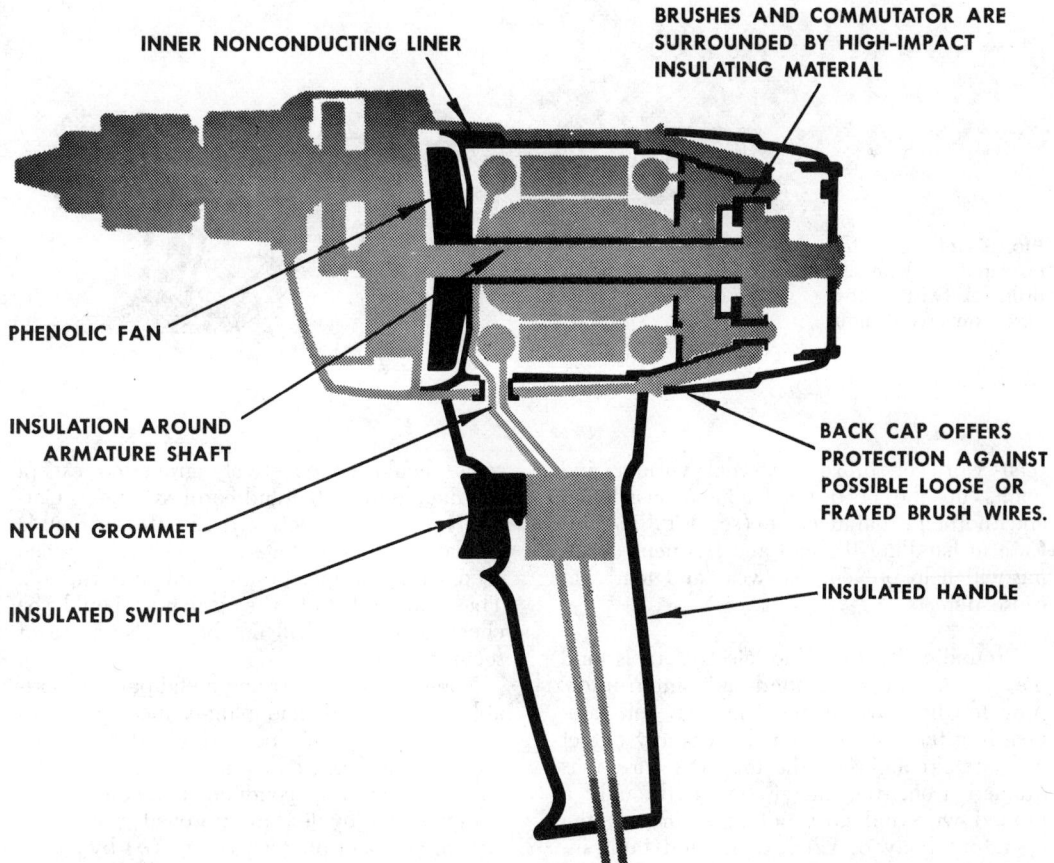

INNER NONCONDUCTING LINER

BRUSHES AND COMMUTATOR ARE
SURROUNDED BY HIGH-IMPACT
INSULATING MATERIAL

PHENOLIC FAN

INSULATION AROUND
ARMATURE SHAFT

NYLON GROMMET

INSULATED SWITCH

BACK CAP OFFERS
PROTECTION AGAINST
POSSIBLE LOOSE OR
FRAYED BRUSH WIRES.

INSULATED HANDLE

FIG. 35–23.—On a double-insulated, shock-proof electric tool, an internal layer of protective insulation completely isolates the electrical components from the outer metal housing.

Courtesy Millers Falls Co.

conductive, so that no metal part comes in contact with the operator.

Large capacity electric tools require a more rigid design in order to provide for greater stress requirements where more power and high-torque gearing are involved. Double-insulated tools with metal housings have an internal layer of protecting insulation completely isolating the electrical components from the outer metal housing (Fig. 35–23). This is in *addition* to the functional insulation found in conventional tools. Preferably, these tools contain a nonconductive handle, an insulated armature shaft, and a completely insulated motor. Brushes, commutators, and built-in switches are designed and utilized

under the concept of "reinforced insulation." This means that in addition to the functional insulation, a reinforced or protecting insulation is also incorporated into the tool. This extra *or* reinforced insulation is physically separated from the functional insulation and is arranged so that deteriorating influences such as temperature, contaminants, and wear will not affect both insulations at the same time. Unless subject to immersion or extensive moisture which might nullify the double insulation, a double insulated or all-insulated tool does not require separate ground connections; the third wire or ground wire is not needed. See Fig. 35–24.

Failure of insulation is harder to detect

997

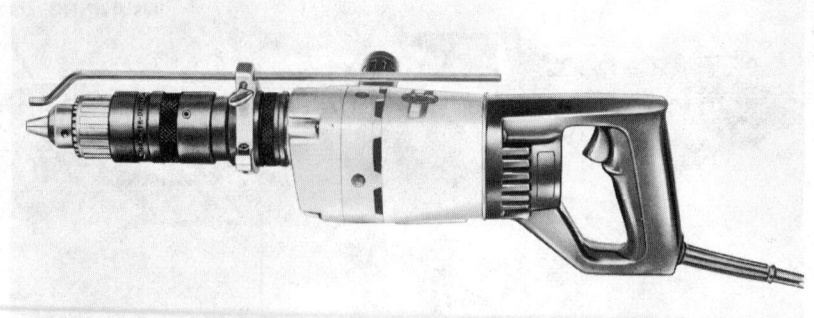

Fig. 35–24.—Double-insulated hammer drill. A twist of the nose converts it into a regular drill.

Courtesy Millers Falls Co.

than worn or broken external wiring, and points up the need for frequent inspection and thorough maintenance (see Fig. 35–21). Care in handling the tool and frequent cleaning will help prevent the wear and tear that cause defects.

Grounding of portable electric tools and the use of proper ground-fault interruptors provide the most convenient way of safeguarding the operator. If there is any defect or short circuit inside the tool, the current is drained from the metal frame through a ground wire and does not pass through the operator's body or where a ground fault interruptor is used the current is shut off before a serious shock can occur. All electric power tools should be effectively grounded except the double insulated and cordless types. Correctly grounded tools are as safe as double insulated or low voltage tools, especially when used with a proper ground fault interruption. The continuity of the ground should be checked so there will not be a false sense of security.

The noncurrent-carrying metal parts of portable and/or cord- and plug-connected equipment required to be grounded may be grounded either (*a*) by means of the metal enclosure of the conductors feeding such equipment, provided an approved grounding-type attachment plug is used, or (*b*) by means of a grounding conductor run with the power supply conductors in a cable assembly or

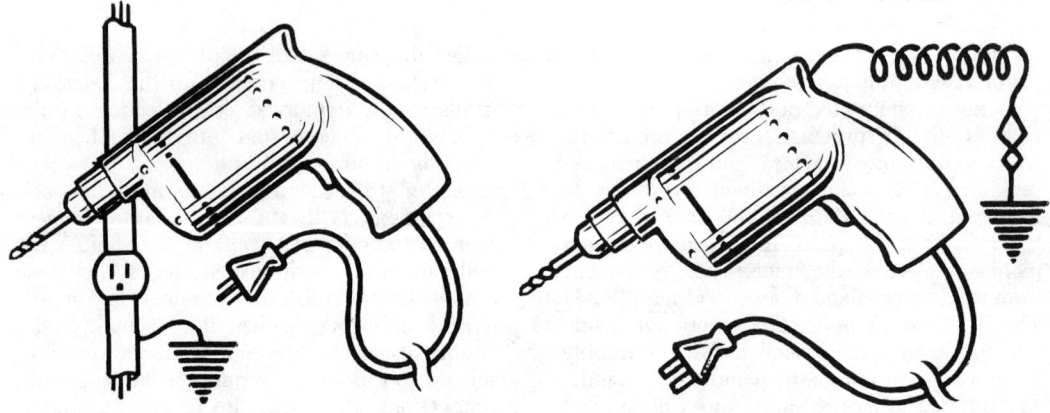

Fig. 35–25.—Electric drill at left is automatically grounded by means of built-in ground wire connected to grounded receptacle by means of three-prong plug. Drill at right is unacceptably grounded.

flexible cord that is properly terminated in an approved grounding-type attachment plug having a fixed grounding contacting member. The grounding conductor may be uninsulated; if individually covered, however, it must be finished a continuous green color or a continuous green color with one or more yellow stripes.

There is an exception to (*a*) and (*b*) in that the grounding contacting member of grounding-type attachment plugs on the power-supply cord of portable, hand-held, hand-guided or hand-supported tools or appliances may be of the movable self-restoring type.

(*c*) By special permission, nonportable cord- and plug-connected equipment can be grounded by a separate flexible wire or strap, insulated or bare, that has been protected (as well as practicable) against physical damage.

For more information on grounding refer to Chapter 41, "Electrical Hazards."

Electric cords should be inspected periodically and kept in good condition. Heavy-duty plugs that clamp to the cord should be used to prevent strain on the current-carrying parts if the cord is accidentally pulled. Terminal screws or connections on plugs and connectors should be covered with proper insulation. Employees should be trained not to jerk cords and to protect them from sharp objects, heat, and oil or solvents that might damage or soften the insulation. To ensure the continuity of grounding, extension cords used with tools and equipment that require grounding should also be of the three-wire, grounded-connection type.

Electric drills cause injuries in several ways: a part of the drill may be pushed into the hand, the leg, or other parts of the body, the drill may be dropped when the operator is not actually drilling, and the eyes may be hit either by material being drilled or by parts of a broken drill. Although no guards are available for drill bits, some protection is afforded if drill bits are carefully chosen for the work to be done, such as being no longer than necessary to do the work.

Where the operator must guide the drill with his hand, the drill should be equipped with a sleeve that fits over the drill bit. The

Fig. 35–26.—Lower movable guard of this portable electric saw always returns to the guarded position. Operator should keep his fingers away from the trigger when saw is not being used.

Courtesy Black & Decker Manufacturing Company.

sleeve protects the operator's hands and also serves as a limit stop if the drill should plunge through the material.

Oversized bits should not be ground down to fit small electric drills; instead, an adapter should be used that will fit the large bit and provide extra power through a speed reduction gear. When large, powerful drills are used, small pieces of work should be clamped or anchored to prevent whipping.

Electric saws must be equipped with guards above and below the face plate—the lower guard must automatically retract to cover the exposed saw teeth. Employees must be trained to use the guard as intended. The guard should be checked frequently to be sure that it operates freely and encloses the teeth completely when it is not cutting. (Fig. 35–26.)

OSHA standards require that frames and exposed metal parts of portable saws and

999

other portable, electric woodworking tools operating at over 90 volts should be grounded.

Circular saws should not be jammed or crowded into the work. The saw should be started and stopped outside the work. At the beginning and end of the stroke, or when the teeth are exposed, the operator must use extra care to keep his body out of the line-of-cut. The saws must be equipped with "dead-man" controls or a trigger switch that shuts off the power when pressure is released. Such a saw cannot run when not in use.

Grinding wheels, buffers, and scratch brushes should be guarded as completely as possible. For portable grinding (unless the diameter of the wheel is less than 2 in.), the maximum angular exposure of the periphery and sides should not exceed 180 degrees and the top half of the wheel (or the portion facing the operator) should always be enclosed. Grinding wheels shall never be used at greater than their rated speed which is shown on the wheel. To do so may result in the wheel breaking apart due to excessive centrifugal force.

Guards should be adjustable so that operators will be inclined to make the correct adjustment rather than remove the guard. However, the guard should be easily removable to facilitate replacement of the wheel. In addition to mechanical guarding the operator should wear safety goggles at all times in case the wheel disintegrates.

The portability of a grinding wheel exposes it to more abuse than that given a stationary grinder. The wheel should be kept away from water and oil, which might affect its balance; the wheel should be protected against blows from other tools; and care should be exercised not to strike the sides of a wheel against objects or to drop the wheel. Cabinets or racks will help protect the wheel against damage.

The speed and weight of a grinding wheel, particularly a larger one, make it more difficult to handle than some other power tools. Since part of the wheel must necessarily be exposed, it is important that employees be trained in the correct way to hold and use the wheel so that it does not touch the clothes or the body. Eye protection should always be worn when grinding.

The wheels should be mounted only by trained workmen, with the wheels and safety guards conforming to ANSI Standard B7.1, *Safety Code for the Use, Care, and Protection of Abrasive Wheels.* Grinders should be marked to show the maximum wheel size and speed. Wheels should be sound-tested before being mounted. Wheels should be examined before each use for gouges, cracks, and general condition. See Chapter 31, "Machine Tools."

Sanders of the belt and disk type cause serious skin "burns" when the rapidly moving abrasive touches the body. It is impossible to guard sanders completely; therefore, employees require thorough training in their use. The motion of the sander should be away from the body, and all clothing should be kept clear of the moving parts. Dust-type safety goggles or plastic face shields should be worn and, if harmful dusts are created, a respirator approved for the exposure should be worn.

Sanders require especially careful cleaning because of the dusty nature of the work. If a sander is used steadily, it should be dismantled periodically, as well as thoroughly cleaned every day by being blown out with low-pressure air. If compressed air is used, the operator should wear safety goggles or work with a transparent shield between his body and the air blast.

The fire and explosion hazard of wood sanding is considerable, and precautions should be taken to keep the dust to a minimum through adequate ventilation or, if the machine is so designed, with a dust collector or vacuum bag. Because of the extreme combustibility of wood dust, or woodwork finishes, such dust should not be disposed of in enclosed burners, furnaces, or incinerators where it would burn with almost explosive force.

If much wood sanding must be done, electrical equipment designed for this exposure should be used to minimize the explosion hazard. Fire extinguishers approved for electrical or Class C fires should be available, and employees should be told what to do in case of fire.

Soldering irons are the source of burns and of illness resulting from inhalation of fumes. Insulated, noncombustible holders will practically eliminate the fire hazard and the danger of burns from accidental contact. Ordinary metal covering on wood tables is not sufficient because the metal conducts heat and may ignite the wood below.

Holders should be designed so that employees cannot accidentally touch the hot irons if they reach for them without looking. The best holder completely encloses the heated surface and is inclined so that the weight of the iron prevents it from falling out. Also see NSC Data Sheet 445, *Hand Soldering and Brazing*.

Harmful quantities of fumes from lead soldering should not be allowed to accumulate. Local regulations may require exhaust facilities if much lead soldering is done. Even if lead fumes are not present in harmful quantities, it may be desirable to exhaust the nuisance fumes and smoke. Air samples should be taken to verify that the amount of lead in the air is not harmful.

Lead solder particles should not be allowed to accumulate on the floor and on work tables. If the operation is such that the solder or flux may spatter, employees should wear face shields or do the work under a transparent shield.

Air power tools

Air hose. An air hose presents the same tripping or stumbling hazard as do cords on electric tools. Persons or material accidentally hitting the hose may unbalance the operator or cause the tool to fall from an overhead place. An air hose on the floor should be protected against trucks and pedestrians by two planks laid on either side of it or by a runway built over it. It is preferable to suspend hoses over aisles and work areas.

Workmen should be warned against disconnecting the air hose from the tool and using it for cleaning machines or removing dust from clothing. The federal safety and health regulations mandate that air pressure in excess of 30 psig must not be used for cleaning machines, and then only with effective chip guarding and personal protective equipment such as safety glasses. Brushing or vacuum equipment is recommended for removing dust from clothing.

Accidents sometimes occur when the air hose becomes disconnected and whips about.

A short chain attached to the hose and to the tool housing will keep the hose from whipping about if the coupling should break. In some cases, couplings should also have such chains between the sections of hose.

Air should be cut off before attempting to disconnect the air hose from the air line. Air pressure inside the line should be released before disconnecting.

A safety check valve installed in the air line at the manifold will shut off the air supply automatically if a fracture occurs anywhere in the line.

If kinking or excessive wear of the hose is a problem, it can be protected by a wrapping of strip metal or wire. One objection to armored hose is that it may become dented and thus restrict the flow of air.

Air power grinders require the same type of guarding as electric grinders. Maintenance of the speed regulator or governor on these machines is of particular importance in order to avoid over-speeding the wheel.

Regular inspection by qualified personnel at each wheel change is recommended.

Pneumatic impact tools, such as riveting guns and jackhammers, are essentially the same in that the tool proper is fitted into the gun and receives its impact from a rapidly moving reciprocating piston driven by compressed air at about 90 psig pressure.

Noise levels from pneumatic impact tools should be determined to see if hearing protective devices for workers are necessary to comply with the 90 dBA requirement of federal standards.

Two safety devices are required. The first is an automatically closing valve actuated by a trigger located inside the handle where it is reasonably safe from accidental operation. The machine can operate only when the trigger is depressed.

The second is a retaining device that holds the tool in place so that it cannot be fired accidentally from the barrel (Fig. 35–27).

A good safety rule to impress on all operators of small air hammers is—Do not squeeze the trigger until the tool is on the work.

In the use of all pneumatic impact tools there is, of course, a hazard from flying chips.

1001

Operators should wear safety goggles, and, if other employees must be in the vicinity, they should be similarly protected. Where possible, screens should be set up to shield persons nearby where chippers, riveting guns, or air drills are being used.

Because excessive noise is a definite possibility, consideration should be given to isolating the noisy operation, substituting quieter methods, or providing hearing protection to all exposed workers.

Two chippers should work away from each other, that is, back to back, to prevent face cuts from flying chips. Workmen should not point a pneumatic hammer at anyone, nor should they stand in front of operators handling pneumatic hammers.

Handling of heavy jackhammers causes fatigue and may even cause strains. Jackhammer handles should be provided with heavy rubber grips to reduce vibration and fatigue, and operators should wear safety shoes to reduce the possibility of injury should the hammer fall.

Many accidents are caused by breaking of the steel drill because the operator loses his balance and falls. Also, if the steel is too hard, a particle of metal may break off and strike him. The manufacturer's instructions for sharpening and tempering the steel should be followed.

Special power tools

Flexible shaft tools require the same type of personal protective equipment as do direct power tools of the same type. Abrasive wheels should be installed and operated in conformance with the American National Standard B7.1 and federal regulations. The flexible shaft must be protected against denting and kinking, which may damage the inner core and shaft.

It is important that the power be shut off whenever the tool is not in use. When the motor is being started, the tool end should be held with a firm grip to prevent injury from sudden whipping. The abrasive wheel or buffer of the tool is difficult to guard and, because it is more exposed than the wheel or buffer on a stationary grinder, extra care should be exercised to avoid damage. Wheels should be placed on the machine or put on a rack, not on the floor.

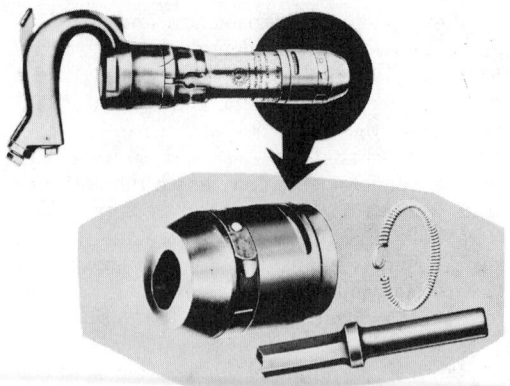

FIG. 35–27.—Chipping hammer safety retainer prevents the discharge of the tool.

Courtesy Chicago Pneumatic Tool Co.

Hydraulic power tools are used in some industries, notably the electric utility industry where workers work aloft from a hydraulically powered aerial lift device. The power is obtained from the source used to operate devices such as hydraulic chain saws and compression devices. Some compression devices have a small hydraulic press incorporated in the device that is pumped by the operator. A hazard in the use of such equipment is small leaks in the hydraulic hose or around fittings. There have been instances where workmen have put a hand over a pinhole leak and had oil forced into a finger by the high pressure. Care must also be taken to always use hose built for the pressure involved because a rupture can cause serious consequences.

Gasoline power tools are widely used in logging, construction, and other heavy industry. Probably the best known and most prevalent is the chain saw. (See NSC Data Sheet 320, *Portable Power Chain Saws.*)

Operators of gasoline power tools must be trained in their proper operation according to the manufacturer's instructions. They must also be familiar with the fuel hazards.

Explosive-actuated tools (see NSC Data Sheet 236, *Powder-Actuated Hand Tools*) are used for fastening fixtures and materials to

QUALIFIED OPERATOR OF POWDER-ACTUATED TOOLS

Make(s)________________ Model(s)________________

This certifies that ________________________
(NAME OF OPERATOR)

Card No.__________ Soc. Sec. No.______________

Has received the prescribed training in the operation of powder actuated tools manufactured by

____
(NAME OF MANUFACTURER)

Trained and issued by ________________________
(SIGNATURE OF AUTHORIZED INSTRUCTOR)

I have received instruction in the safe operation and maintenance of powder actuated fastening tools of the makes and models specified and agree to conform to all rules and regulations governing their use. Failure to comply shall be cause for immediate revocation of this card.

____________________ __________
(SIGNATURE) (DATE)

FIG. 35–28.—Wallet card for qualified powder-actuated tool operators are available from tool manufacturers. List of instructors should be maintained by each manufacturer.

Based on American National Standard A10.3–1972.

metal, pre-cast or pre-stressed concrete, masonry block, brick, stone, and wood surfaces, tightening rivets, and punching holes. Cased (cartridge) or uncased (pellet) power loads provide the energy and are ignited by means of a conventional percussion primer.

The hazards encountered in the use of these tools are similar to those of a firearm. The handling, storing, and control of explosive power loads present additional hazards. Therefore, instructions for the use, handling, and storage of both tools and power loads should be just as rigid as those governing blasting caps and firearms.

Specific hazards are accidental discharge, ricochets, ignition of explosive or combustible atmospheres, projectiles penetrating the work, and flying dirt, scale, and other particles.

Tools are divided into two types. A direct-acting tool is one in which the expanding gases of the power load acts directly on the fastener to be driven. An indirect-acting tool has the expanding gases of the power load acting on a captive piston, which in turn drives the fastener.

The two types are divided into three classes —low, medium, and high velocity of the lightest fastener driven by the strongest available commercial power load that will properly chamber in the tool.

In case of misfire, the operator should hold the tool in operating position for at least 30 seconds. He should try to fire the tool a second time. If it still misfires, he must hold the tool in operating position for another 30 seconds, then follow the manufacturer's instructions to remove the load.

Powder-actuated tools can be used safely if special training and proper supervision are provided for the operator. Manufacturers of the tools will aid in this training. Only trained and properly qualified personnel holding a "qualified operator's card," Fig. 35–28, should be permitted to operate or handle the tools. Qualification requirements for both instructor and operator are given in ANSI A10.3, *Safety Requirements for Powder Actuated Fastening Systems.*"

Power-assisted, hammer-driven tools are used for the same purposes as powder-actuated tools and generally the same precautions should be followed (Fig. 35–29).

Fig. 35–29.—Workman demonstrates correct use of power-assisted hammer-driven tool to fasten conduit clips to masonry blocks.

Courtesy Hilti Inc.

Personnel should be trained to use powder charges of the correct size to drive studs into specific surfaces and should be made responsible for safe handling and storing of the cartridges and the tools.

Powder-actuated tools should not be used on concrete less than 2 in. thick nor on steel less than ¼ in. thick unless suitable material, such as timber or sandbags, that will contain the flying stud is placed directly behind the work. As an added precaution, the tool should be discharged only when personnel are not in line with the path of the projectile. This material should be left in place until the job is completed. Fasteners should not be driven closer than 3 in. from an unsupported edge or corner. A low-velocity tool may drive no closer than 2 in. from an edge in concrete or ¼ in. in steel.

Operators should wear adequate eye pro-

tection when firing the tool. Where the standard shield cannot be used for a particular operation, special shields can be obtained from the tool manufacturer.

Personal Protective Equipment

Gloves, ties, loose clothing, and jewelry should not be worn by workers using revolving tools such as drills, saws, and grinders. *The weight of most power tools makes it advisable for users to wear safety shoes to reduce chances of injury should the tools fall or be dropped.* (See Chapter 19, "Personal Protective Equipment," for more details.)

When power tools are used in overhead places, the operator should wear a safety belt or shoulder harness to minimize the danger of falling should the tool break suddenly or shock the operator.

Approved dust-type respirators should be worn on buffing, grinding, or sanding jobs which produce harmful dusts.

Safety goggles or face shields should be worn for work on grinders, buffing wheels, and scratch brushes because the unusual positions in which the wheel operates may cause particles to be thrown off in all directions. For this reason, protective equipment is even more important than it is for work on stationary grinders.

As with all revolving tools, loose clothing and gloves that will catch in the wheel should not be worn. Clothing should be free of oil, solvents, or frayed edges to minimize the fire hazard from sparks.

For powder-actuated tools or jack hammers, hearing protection equipment should be used if more positive noise controls are not possible.

Eye protection

Usually, in all operations where hardened metal tools are struck together, where equipment or material is struck by a metal hand tool, or where the cutting action of a tool causes particles to fly, eye protection is needed by the user of the tool and by other workmen who may be exposed to flying particles.

The hazard can be minimized through the use of nonferrous, "soft" striking tools and through shielding the job by metal, wood, or canvas.

Safety goggles or face shields should be

worn when woodworking or cutting tools, such as chisels, brace and bits, planes, scrapers, and saws, are used head-high or overhead, with the chance of particles falling or flying into the eyes.

Occasionally, the need for eye protection is overlooked on such jobs as cutting wire and cable, striking wrenches, using hand drills, chipping concrete, removing nails from scrap lumber, shoveling material head-high or working on the leeward side of a job, and using wrenches and hammers overhead, and on other jobs where particles of materials or debris may fall.

References

American National Standards Institute, 1430 Broadway, New York, N.Y. 10018.
    Safety Code for the Use, Care, and Protection of Abrasive Wheels, B7.1.
    Safety Requirements for Powder Actuated Fastening Systems, A10.3.

Grinding Wheel Institute, 2130 Keith Bldg., Cleveland, Ohio 44115.

McMahan, E. L. and R. T. Wray. *Aluminum Paint on Hot Surfaces.* Aluminum Research Laboratories, New Kensington, Pa.

National Safety Council, 425 North Michigan Ave., Chicago, Ill. 60611
    Industrial Data Sheets
        Air-Powered Hand Tools, 392.
        Brush Cutting Tools, 427.
        Electric Hand Saws (Circular Blade Type), 344.
        Ground Fault Circuit Interrupters for Personnel Protection, 636.
        Grounding Portable Electric Equipment, 299.
        Hand Knives, 369.
        Hand Soldering and Brazing, 445.
        Portable Power Chain Saws, 320.
        Powder-Actuated Hand Tools, 236.

    Meat Industry Safety Guidelines.

Powder Actuated Tool Manufacturer's Institute, 331 Madison Ave., New York, N.Y. 10017.
    Basic Training Manual.

Underwriters Laboratories Inc., 207 East Ohio St., Chicago, Ill. 60611.
    Underwriters' Reports. UL 45, *Electric Tools.*

U.S. Department of Commerce, Office of Technical Services, Washington, D.C. 20234.
    Sparking Characteristics and Safety Hazards of Metallic Materials, Technical Report No. NGF-T-1-57, PB 131131.

U.S. Department of Labor, Washington, D.C. 20210. *Safe Use of Hand and Portable Power Tools,* Bulletin 293.

U.S. Government Printing Office, Superintendent of Documents, Washington, D.C. 20402.
    Maintenance and Care of Hand Tools. TM-9-867
    Federal Specifications for Hammers, Mauls, and Sledges. GGG-H-86.

Exhaust and Ventilation

Chapter
36

Air Supply Systems   1007
Dilution ventilation . . . Summer cooling . . . Makeup air . . . Spot
cooling

Exhaust Systems   1009
General exhaust . . . Local exhaust . . . Hoods . . . Hood air flow
. . . Ducts

Other Design Considerations   1014
Weather problems . . . Air movers . . . Testing

References   1027

In most companies and plants, safety professionals have the responsibility for ventilation, because most systems are (or will be) installed to control air-borne material that may either cause illness or discomfort to the employees.

Although many people associate the word *ventilation* only with exhaust equipment, the term refers to both the simultaneous supply and exhausting of air from a space. Let us first discuss the air-supply system. Exhaust and construction will follow later in the chapter.

Air Supply Systems

When properly designed, installed and operated, air-supply systems can:

1. Replace the air that is exhausted, thereby increasing the efficiency of the exhaust system

2. Dilute air-borne dust, fumes, and heat sufficiently so that exhaust volumes can be reduced

3. Provide pressurization to prevent cross-contamination between areas

4. Serve as a carrying vehicle to allow heating, humidification, and air cleaning to be provided

5. In winter, to reduce or eliminate objectionable cold drafts

6. In summer, provide air motion over people to lower the effective temperature.

Dilution ventilation

Whenever air enters a space and is exhausted, some dilution ventilation takes place. Often, this alone is sufficient to provide satisfactory conditions in an area. If the air is reduced under controlled conditions, maximum dilution benefits will occur. Ideally, the best air available is always introduced to the clean part of the building or area and is allowed to flow across individuals to the contaminated areas where it is exhausted. See Fig. 36–1.

Summer cooling

One of the best methods of improving the environment for personnel during hot weather

is to direct high-velocity air over their bodies. If properly installed with suitable grilles, air supply systems can provide such air motion. This air must be delivered to the lower 10 ft of the work area in order to be effective for this purpose.

When air is mechanically supplied to a space, it will provide pressurization if sufficient volumes are introduced. Usually, it is easier and more economical to control the environment within a space that is balanced or pressurized, because when air enters a space by infiltration there is no control of temperature or direction. This results in discomfort to occupants. The most practical solution is to provide properly treated and sufficient distribution of air.

Makeup air

Safety professionals should look for these basic items when evaluating makeup air systems or when discussing them with engineering personnel.

• Air must be supplied if the exhaust equipment is to function properly. The usual recommendation is for 10 percent more air to be supplied than is exhausted, unless it is desired to operate under a negative pressure condition. An industrial exhaust system can fail because it is impossible for it to work against the added resistance required to draw large volumes of air through cracks in windows and doors when mechanical air supply is lacking.

• Whenever a door is opened in cold weather, large amounts of cold air enter the building. People located in these indrafts are chilled and cannot be made comfortable regardless of how much heat is added to the building itself. Similarly, when the building is being cooled, the indraft of untreated air can be detrimental. A high-velocity indraft can be controlled by providing slots at each side of the door. Through these slots, treated air is projected horizontally, at about 15 degrees toward the opening. The velocity is about 3000 fpm at a rate of 100 cfm per square foot of opening.

• Sometimes, facilities are cooled during warm weather. Filtered (cleaned) air provides a cleaner environment for the employee

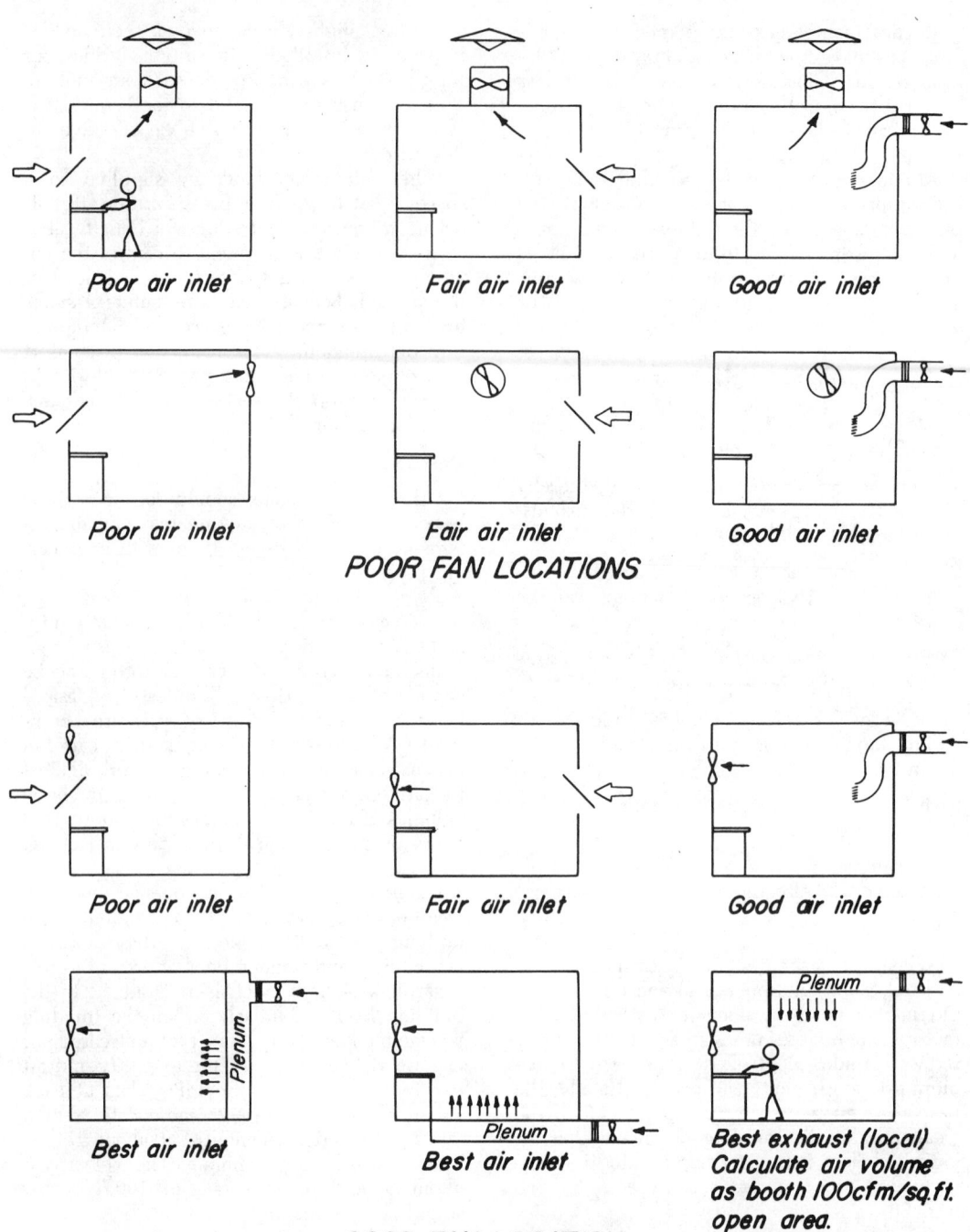

Poor air inlet Fair air inlet Good air inlet

Poor air inlet Fair air inlet Good air inlet

POOR FAN LOCATIONS

Poor air inlet Fair air inlet Good air inlet

Best air inlet Best air inlet Best exhaust (local)
Calculate air volume
as booth 100cfm/sq.ft.
open area.
Best air inlet

GOOD FAN LOCATION

FIG. 36–1.—Principles of dilution ventilation. During winter months, inlet air requires tempering.

From *American Conference of Governmental Industrial Hygienists* Industrial Ventilation.

and also protects the equipment. Properly designed air supply systems can fulfill all of these requirements.

Spot cooling

Local or spot cooling is cooling one or more individuals at specific locations, rather than controlling the temperature of the entire space. Most of the cooling is accomplished by convection, the quantity of heat removed being a function of air temperature and velocity. Air must be below skin temperature to be effective.

One method is to provide air vertically through a floor grating. The air surrounds the individual and flows between his body and his outer clothing. Grille velocities vary between 3000 and 4000 fpm; provision should be made for simple control by each individual.

Where floor gratings are not possible or where excessive dust might result in eye injuries or respiratory problems, air may be delivered at or near the floor through double-deflecting grilles that are controlled by the worker. Air supplies that strike the head or neck should only be used when other methods of distribution are not feasible.

Outside air can be passed through an evaporative cooler before being delivered to the worker. When the work room air temperature is as high as 110 to 115 F, a 10 to 15 F drop in the air temperature provides extremely welcome relief. An evaporative cooler operating at an efficiency of 80 percent or greater will cool the air passing through it to a temperature below that of a man's skin.

More details can be found in the American Industrial Hygiene Association's book, *Heating and Cooling of Man in Industry*. See References.

Exhaust Systems

The method of exhausting air can be general or local. General exhaust is the venting of air from a building through gravity or mechanical flow through openings usually located in the roof, or high on the side walls, with no specific connection to the operation. In contrast, local ventilation is as close as possible, and often encloses a specific operation. Hoods on plating tanks or grinding

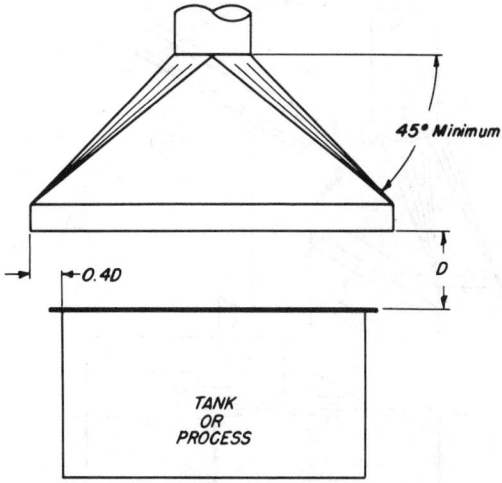

FIG. 36–2.—Canopy hood. Side curtains are necessary when extreme cross-drafts are present. Hood is not to be used where material is toxic and worker must bend over the tank or process.

From ACGIH Industrial Ventilation.

wheels are examples of local exhaust ventilation.

General exhaust

General exhaust can provide satisfactory control for certain limited operations. They do require larger air volumes, however, and, although they are often more economical to install, they are expensive to operate because of their low efficiency in controlling air-borne contaminants—large volumes of air must be replaced. Even then, general exhaust cannot usually provide adequate control of contaminants having a low TLV or low ceiling value or both. See the next chapter, "Industrial Hygiene." Mechanical exhausting of flammables is discussed in Chapter 42, "Flammable and Combustible Liquids."

Most general exhaust units, even if mechanically driven, will not operate against more than about ½-in. (water gage) resistance. A good air supply is necessary, therefore, when the building is closed. If doors and windows are open, the exhaust may work satisfactorily if it is located high enough to benefit from a chimney effect. In winter, this will require

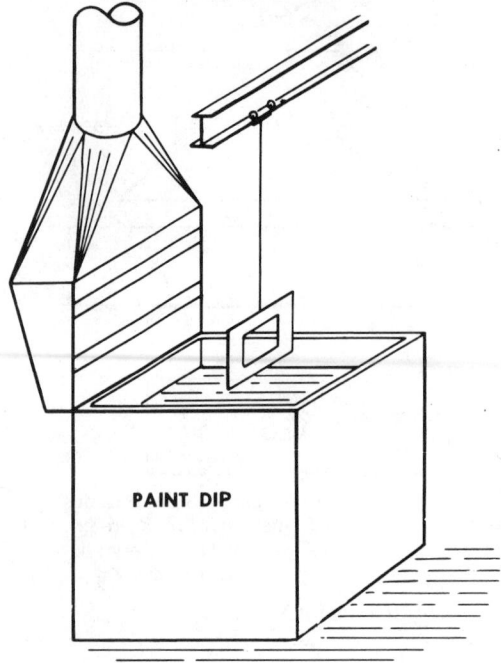

PAINT DIP

FIG. 36–3.—Side draft hood.

From ACGIH Industrial Ventilation.

large quantities of heat, such as will be found in a heat-treating area, to provide the thermal lift necessary. Modern, low, wide buildings cannot, in most cases, be adequately ventilated by general exhaust systems.

Be sure that roof-mounted units can be closed tightly in order to eliminate water leakage and indrafts in winter. They should not be located close to any supply system, otherwise contaminated exhaust could be recycled.

Local exhaust

Local exhaust systems are the backbone of air control procedures for most industrial operations. With the increasing emphasis on air pollution, they are even more important because they use air more effectively and economically. Federal and state codes, and most local codes, require the use of local exhaust on most operations.

A local exhaust system consists basically of hoods, ductwork, air-moving devices, and usually an air-cleaning device.

Hoods

Although there are many variations in hoods, there are only a few basic types.

The canopy hood is probably the oldest form of hood known. As the name implies, it is merely a canopy placed above an operation so contaminants can be evacuated through the exhaust duct. See Fig. 36–2. This type of hood is not to be used where material is toxic and the worker must bend over the tank or process. Side curtains may be necessary when extreme cross-drafts are present.

The booth-type hood, such as a paint booth, usually (but not always) has a large face opening so the part or operation that creates the contaminant can be placed in the hood itself. Air flows horizontally rather than vertically as it does in the canopy hood. The booth hood, therefore, could be thought of as a canopy hood laid on its side. Filters or other air-cleaning devices may be needed to meet air pollution or other regulations.

The side draft hood is similar to the booth hood, but usually smaller in face area. Usually the operation takes place in front of the

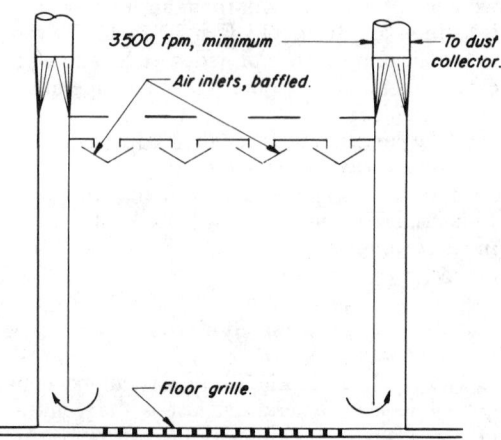

3500 fpm, mimimum

To dust collector.

Air inlets, baffled.

Floor grille.

SECTION THRU TYPICAL ROOM

FIG. 36–4.—Downdraft hood, such as used for abrasive blasting ventilation. Operator still requires an approved abrasive blasting helmet. Downdraft is usually 80 fpm.

From ACGIH Industrial Ventilation.

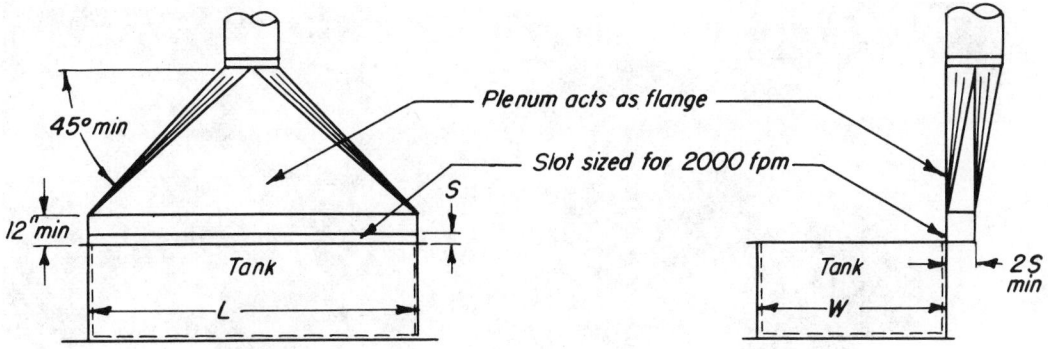

FIG. 36–5.—Open surface tank with upward plenum slot exhaust. Downward plenum and central slot exhausts can also be used.

From ACGIH Industrial Ventilation.

hood and is arranged so the air entering the hood flows over the operation, thereby providing the necessary control; see Fig. 36–3.

This type of hood must handle more air to control the operation than the booth type. It is usually desirable to install baffles at the face to provide an even air flow into the hood.

The downdraft hood, Fig. 36–4, causes air to flow downward. Although it is out of the way of the operation, use of this type hood is limited. The flow of air through such a hood is often ineffective in controlling the contaminant, especially if it is from a hot operation. Any thermal lift or strong cross draft has an adverse effect on the pattern of air entering the exhaust openings, so that control of the contaminant is lost.

The slot exhaust hood, such as installed on the upper edge of a plating tank (Fig. 36–5), is merely a miniature side draft hood with a less efficient aspect ratio (ratio of width to length).

Hood air flow

The safety professional should understand how air flows into a hood. This is often misunderstood, even by designers, yet it is critical to effective local exhaust.

Some people erroneously assume that because air motion is felt many feet in front of a supply outlet or circulating fan, that air drawn *into* a hood will create measurable air movement many feet in front of the hood.

Actually, air flowing into a hood does not create motion more than a few inches in front, Fig. 36–6. Air under pressure has a "throw" of about thirty times farther than the "pull" on the suction side of the fan or blower.

To understand this, consider how air flows into a system. Air flows into an intake from all directions. In contrast, air flows out of the open end of a duct in almost a straight line. Another influence on this difference is that energy cannot be put into air flowing into an intake. All that can be done is to reduce the pressure in the duct until the atmospheric pressure in the room pushes the air into the duct itself. In contrast, an almost unlimited amount of energy can be put into air being expelled from a duct.

Fig. 36–7 shows this difference in velocities in front of an exhaust and at the discharge of a supply duct of the same diameter. The velocity at the face of both is 4000 fpm. One diameter (d) from the intake, the air velocity is approximately 400 fpm, 10 percent of the entrance velocity. In contrast, the air in front of the outlet will travel about thirty diameters (30d) before the velocity is reduced to about 400 fpm.

The air flow into an open duct is shown in Fig. 36–8. Note that at a distance of one duct diameter away from the duct entrance, the air velocity is about 7.5 percent that of the entrance velocity. If the duct is brought flush with the surface, as shown at the right, its "reach" is extended slightly.

The velocity in front of the duct opening

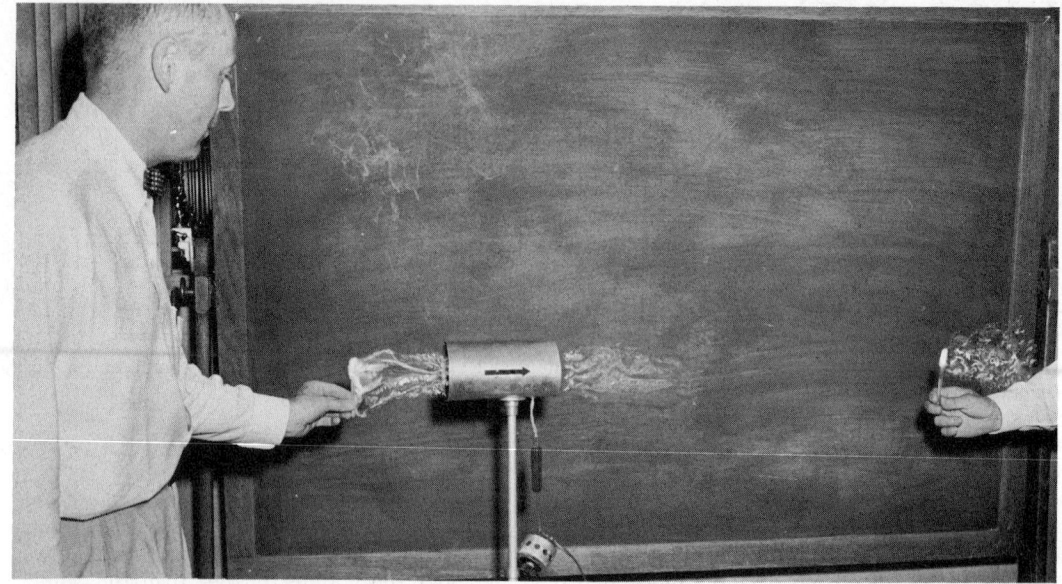

FIG. 36–6.—"Blowing" effect of a fan is felt at a much greater distance than the "sucking" effect. Illustration shows a short length of duct, containing a propeller-type fan. Smoke tube held in hand at left has to be within 8 in. of duct inlet for smoke to be drawn in, but smoke tube in hand at right is affected by the blowing air, even though it is several feet away.

Courtesy Owens-Illinois, Inc.

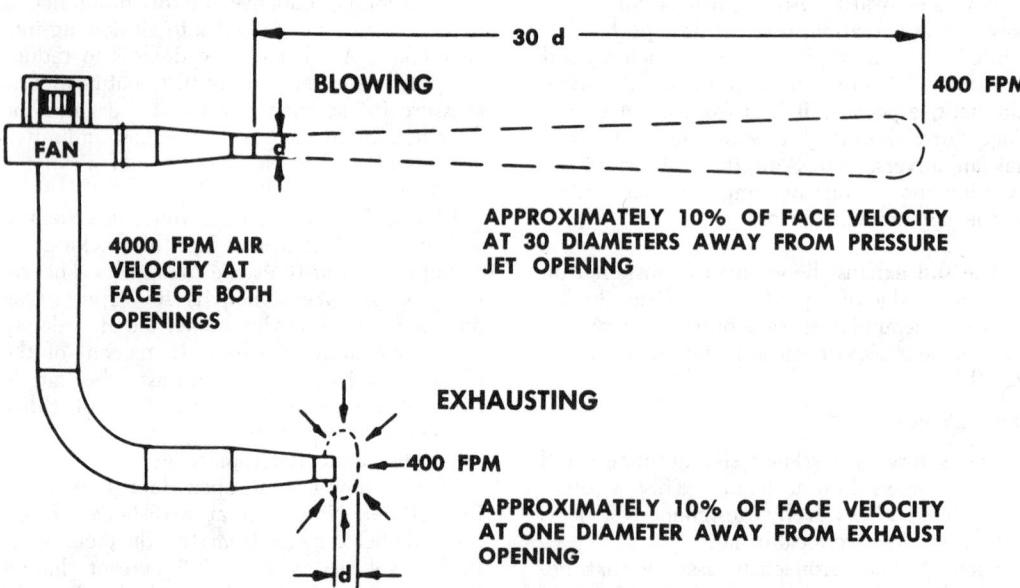

FIG. 36–7.—"Throw" of blower is quite different, depending on whether it is exhausting or blowing. (See photo above.)

From ACGIH Industrial Ventilation.

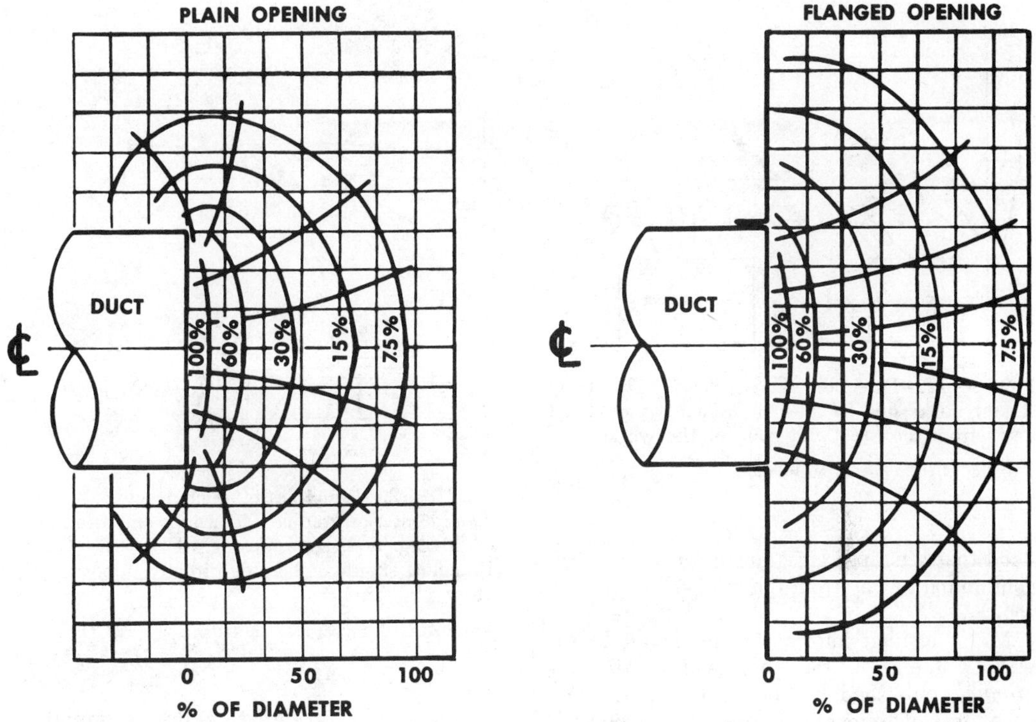

PLAIN OPENING

DUCT

¢

100% 60% 30% 15% 7.5%

0   50   100
% OF DIAMETER

FLANGED OPENING

DUCT

¢

100% 60% 30% 15% 7.5%

0   50   100
% OF DIAMETER

Fig. 36–8.—Velocity contours (expressed as percentages of velocity at the opening) and stream lines for both plain and flanged circular openings.

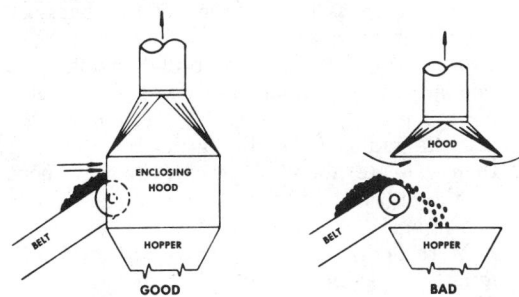

Fig. 36–9.—Enclosure. The more completely the hood encloses the source, the less air is required for control.

may be increased by exhausting larger volumes of air. It is more economical, however, to add baffles to prevent the exhausted air from coming in front of the duct. A simple baffle will improve the velocity equal to that obtained by increasing the air volume into the hood by 25 percent. Further baffling of unwanted air flow will provide even more efficient control, Fig. 36–9. The volume of air required for control may also be reduced by placing the hood closer to the operation, Fig. 36–10.

Fig. 36–11 shows how a canopy hood only captures material that rises directly into it; in this case, the contaminant rises in the worker's breathing zone. A better design is the slot

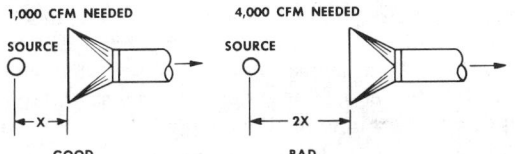

Fig. 36–10.—Place hood as close to the source of contaminant as possible. The required volume varies with the square of the distance from the source.

Illustrations from ACGIH Industrial Ventilation.

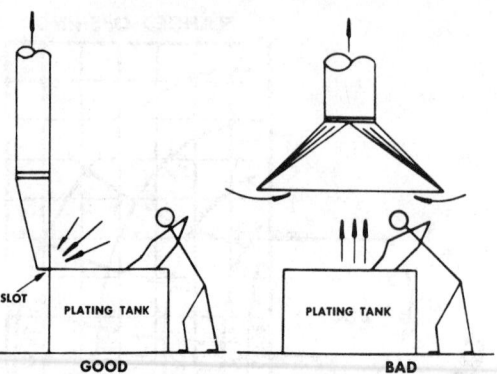

FIG. 36–11.—Direction of air flow. The hood should be located so the contaminant is removed away from the breathing zone of the worker.

From ACGIH Industrial Ventilation.

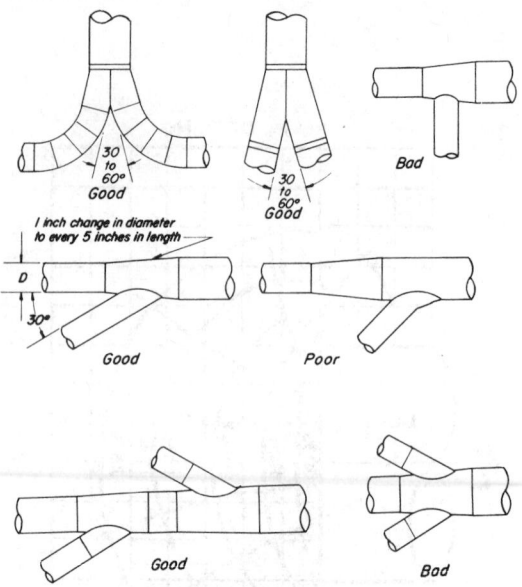

FIG. 36–12.—Principles of duct design. Branches should enter at gradual expansions and at an angle of 30 deg. or less, to 45 deg., if necessary (top). Branches should not enter directly opposite each other (bottom).

From ACGIH Industrial Ventilation.

hood that sets up a cross draft that carries the contaminant away from the worker's breathing zone.

More detailed information on hood design will be found in the *Industrial Ventilation Manual,* published by the American Conference of Governmental Industrial Hygienists. See References.

Ducts

No matter how well designed a hood is, it is of no value unless adequate air volumes are passed through it. This requires well designed ductwork and properly selected fans.

Air handling can be compared to traffic handling. If you are driving on a highway that enters another at a right angle, you will find it difficult to turn the corner without slowing down first. On the other hand, if there is a gradual "merge" from one highway to the other, like at an expressway cloverleaf, you can enter the new line of traffic without difficulty.

Air has the same problem as traffic. If the ductwork looks like a miniature superhighway, it will probably handle air with minimum problems and minimum energy loss. If, on the other hand, the ducts look like right-angle city streets, then air flow will be reduced and material being carried in the air stream may drop out at intersections because of reduced speeds and increased turbulence. Good and bad duct design is shown in Fig. 36–12.

The inside, or throat, radius is important. Some people mistakenly believe that the inside of a rectangular elbow requires little or no consideration as long as the heel, or back, of the elbow has a large radius; just the reverse is true. Again, look at the traffic analogy. If you are driving along and come to a right-angle turn, you are more interested in how big the radius is on the inside of the turn than you are about a lot of extra pavement on the outside, where you have no intentions of driving.

More details are given in *Industrial Ventilation* and Tables 36-A through E.

Other Design Considerations

Weather problems

The "Dutch cap" that is often used on the top of an exhaust stack has been shown to be ineffective in keeping rain out unless it is placed very close to the end of the stack. When located less than one diameter from the end of the stack, however, this type cap gives high resistance to air flow. For example, velocity pressure loss is about 75 percent

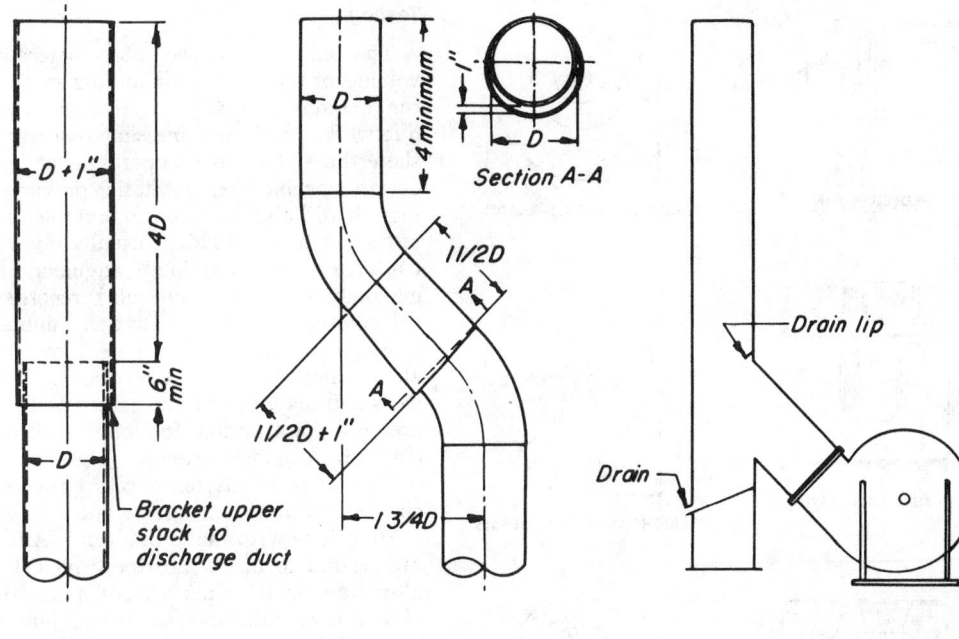

VERTICAL DISCHARGE   OFFSET ELBOWS   OFFSET STACK

No loss   Calculate losses due to elbows

Fig. 36–13.—Stackhead designs. Rain protection characteristics of these caps are superior to a deflecting (Dutch) cap located 0.75 D from top of stack. The length of the upper stack is related to rain protection.

From ACGIH Industrial Ventilation.

when the cap is only 50 percent of the duct diameter away from the end of the duct. In addition, the cap causes discharge air to be directed horizontally, which increases the possibility of its reentry into the building.

Fig. 36–13 shows three different stackhead designs, all with little or no resistance to flow and all which discharge vertically.

Air movers

Air is moved through ductwork because of a difference in pressure created by fans or blowers. Fans are of axial flow or centrifugal types (Fig. 36–14). The proper sizing and selection is important if an efficient, low-maintenance system is to be installed. The *Industrial Ventilation Manual* and NSC's Data Sheet 428, *Checking Performance of Local Exhaust Systems*, give details. A brief summary follows.

Centrifugal fans are most often used in industry. If field tests show a fan is not delivering its rated volume of air, remember that fan rating tables were developed from tests carried out under ideal conditions with several diameters of straight pipe installed on the inlet and outlet. Here are some things that can go wrong.

Ductwork problems. Sharp bends in the ductwork, or even having no duct connected to the outlet, will reduce air flow. If the entrance box is too small, high turbulence will reduce the quantity of air passing through the fan. If air entering the fan spins with the rotation of the fan wheel, the volume of air will be reduced and the power required will be reduced. If, on the other hand, the spinning action of the air entering the fan is opposite to the rotation of the fan wheel, the

1015

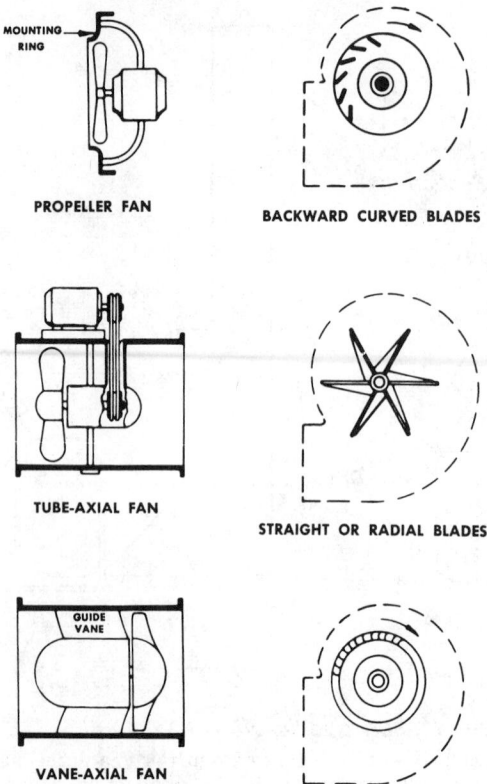

FIG. 36–14.—Types of exhaust fans.

From ACGIH Industrial Ventilation.

volume of the air will be reduced and the power required will be increased.

Wheel turning in wrong direction. Centrifugal fans will move from 25 to 50 percent of their capacity even if the wheel is rotating in the wrong direction. Inasmuch as there is a 50 percent chance that a motor will run in the wrong direction when first connected to the electric supply, each fan installation should have its rotation carefully determined. One cannot determine, therefore, the quantity of air the fan is handling, or determine the horsepower of the fan automatically because both are functions of the fan or pressure at the outlet or both.

Testing

The only way to accurately determine the volume of air in an air-handling system is to run a pitot traverse on a section of straight ductwork. It is not unusual that such tests show the system to be operating at less than design specifications. Static pressure readings should also be taken in suitable locations in the ductwork. This is usually done by the contractor or a trained engineer, but if manometer readings are also recorded, the safety professional can retest in minutes with a simple manometer, if there have been no alterations. This should be done on a routine basis and may well be the safety professional's greatest contribution to control of contaminants by exhaust systems. Figs. 36–15 and –16 show positions for a pitot traverse in a circular and in a rectangular duct.

Here is how to make this test. After looking around to make sure there have been no alterations in the duct system, place the end of a rubber tube over a drilled hole in the sidewall of the duct. When connected to a manometer, this will provide the static pres-

(*Text continues on page 1023.*)

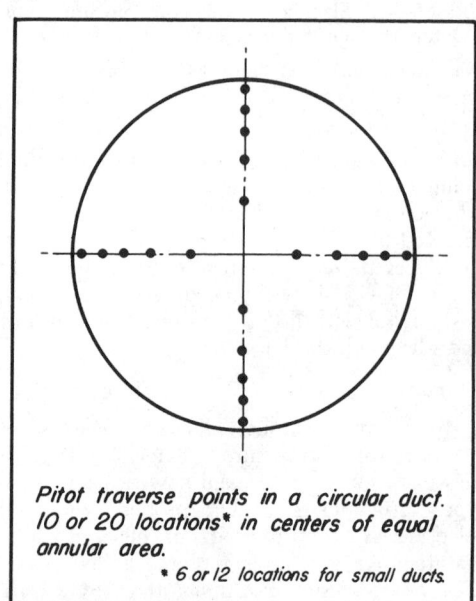

Pitot traverse points in a circular duct. 10 or 20 locations* in centers of equal annular area.

* *6 or 12 locations for small ducts.*

FIG. 36–15.

From ACGIH Industrial Ventilation.

TABLE 36-A

A GUIDE TO VENTILATION RATES FOR TYPICAL INDUSTRIAL EQUIPMENT
Federal, State, or Local Regulations Should Be Followed Where Higher Ventilation Rates Are Required

Operation	Ventilation		Usual Transport Velocity (fpm)	Remarks
	Type of Hood	Air Flow		
Abrasive blast rooms (sand, grit, or shot)	Tight enclosure with air inlets (usually in roof)	60–100 fpm downdraft (long rooms of tunnel proportions 100 fpm crossdraft)	3,500	Many codes specify minimum down draft of 80 fpm
Abrasive blast cabinets	Tight enclosure with access openings	20 air changes per minute but not less than 500 fpm through all openings	3,500	
Asbestos Carding Spool winding	Enclosure Local hoods	800 cfm per machine 50 cfm per spool	3,000 3,000	See references for details and other operations
Bagging Open bag top	Booth or enclosure (provide spillage hopper)	Paper bags—100 cfm per sq. ft. open area Cloth bags—200 cfm per sq. ft. open area	3,500 3,500	
Barrels—Drums (filling or removing material by scoop)	Local hood	100 cfm/sq. ft. of container cross section	3,500	Hood with 1-in. slot extending 120 to 180 deg container. Does not confine spillage.
	Booth	100 fpm at face	3,500	Confine spillage—also recommended for container up-setting.
Belt conveyors	Hood at transfer point	Belt speeds less than 200 fpm—350 cfm per foot of belt width, but not less than 150 fpm through open area. Belt speeds over 200 fpm—500 cfm per foot of belt width but not less than 200 fpm through open area	3,500	

TABLE 36-A. A GUIDE TO VENTILATION RATES FOR TYPICAL INDUSTRIAL EQUIPMENT *(Continued)*
Federal, State, or Local Regulations Should Be Followed Where Higher Ventilation Rates Are Required

Operation	Ventilation		Usual Transport Velocity (fpm)	Remarks
	Type of Hood	Air Flow		
Bins (closed top)	Connect to bin top away from feed point	150–200 fpm through open area at feed points	3,500	
Bucket elevators	Tight casing required	10 cfm per sq. ft. of elevator casing cross-section	3,500	
Brick cutting and sizing (abrasive cut-off wheel used dry)	Local hood	500 cfm	3,500	For portable operations. Control not as effective as booth
	Booth with saw at face of booth	150 fpm at face	3,500	
Ceramics				
Dry pan	Enclosure	200 fpm through all openings	3,500	
Dry press	Local at die	500 cfm	3,500	Automatic feed
	Local at die	500 cfm	3,500	Manual feed
	At supply bin	500 cfm	3,500	Manual feed
Aerographing	Booth	100 fpm (face)		
Fettling, brushing, sagger filling, and unload	Downdraft or side hood	100–150 cfm per sq. ft. of plan area of dust producing operation	3,500	
Cooling tunnels (foundry molds)	Enclosure	75–100 cfm per running foot of enclosure		
Crushers and grinders	Enclosure	200 fpm through openings	3,500	
Furnaces				
Stationary melting pots for nonferrous	Enclosure	100–200 fpm at hood opening	1,500–2,000	Use higher ventilation rate for toxic fumes
Tilting or rocking melting for nonferrous	Canopy	3,000–6,000 cfm	1,500–2,000	Ventilation rate can be materially reduced with more complete enclosures
Electric arc for steel	Hood attached to roof ring	2,500 cfm per ton charged	2,500–3,500	General ventilation of melting room may be substituted but involves greatly increased ventilation rate
Forge (hand)	Canopy	200 fpm at face	1,500	

Operation	Type of hood	Air volume / velocity	cfm	Remarks
Garage (tail pipe at servicing location)	Local hood slipped over tailpipe	100 cfm through 3-in. flexible duct for autos up to 200 hp; 200 cfm through 4-in. flexible duct for trucks and autos above 200 hp; 400 cfm through 4½-in. flexible duct for Diesel engines	2,000	
Granite cutting, finishing Pneumatic hand tools	Local hood	500 cfm	5,500–6,000	Typical hood 3 × 8 in. opening with 3 in. flange. Hood surrounds tool
Surfacing machine	Local hood	500 cfm for tools up to 2⅜ in. diameter; 1000 cfm for 2⅜ to 2⅞ in. diameter	5,500–6,000	
Grinders Polishers, buffers, etc. Portable	Standard wheel hood Downdraft bench	(See Table 45-I) Bench type, 2–400 cfm per sq. ft. of exhaust grille but not less than 150 cfm per sq. ft. of plan working area	3,500	
Swing frame	Booth / Booth	100 fpm at face; 100–200 fpm indraft through opening in booth face	3,500 / 3,000	Recommended for larger parts (Usual cfm per grinder 2,000–4,000)
Kitchen range	Canopy	100 fpm at hood face	1,500–1,800	Provide drip gutter on all interior vertical walls to catch condensed grease. Ref. 30
Laboratory hood (provide with door)	Booth type	50–100 fpm		Air supply location very critical
Metalizing	Local hood / Booth	200 fpm at hood face; 125–200 fpm at booth face	3,500 / 3,000	Not recommended for toxic materials. Use higher ventilation rates for toxic materials
Mixers	Enclosure	100–200 fpm through feed and inspection openings	3,000–3,500	
Motion picture projector (carbon arc)	Enclosure	12–100 cfm (for fume and gas removal) directly exhausted from projector housing		1,000–1,500 cfm or 20–30 air changes per hour for heat removal

TABLE 36-A. GUIDE TO VENTILATION RATES FOR TYPICAL INDUSTRIAL EQUIPMENT *(Concluded)*
Federal, State, or Local Regulations Should Be Followed Where Higher Ventilation Rates Are Required

Operation	Ventilation		Usual Transport Velocity (fpm)	Remarks and References
	Type of Hood	Air Flow		
Pharmaceuticals				
Blenders	Fully enclosed	100 to 200 fpm through opening	2,500–3,500	Exhaust required only when filling or emptying
Coating pans	Hood	120 cfm exh 24-in. diam. pan 60 cfm supply direct to pan. Differential 60 cfm	3,000	Flexible exhaust hood connection extended into pans. Feed opening covered to greatest possible extent.
Centrifuges	Enclosure	250–300 fpm through opening	1,500–2,000	Centrifuge to be provided with cover for 50-75 per cent of opening
Hammer mills Oscillators Shakers	Local hood	200 fpm but not less than 50 cfm		
Mixers	Hood	Not less than 75 cfm per sq. ft. of plan area	2,500–3,500	Hood with slot at each end of mixer
Process kettles and tanks	Enclosure	100–250 fpm through opening or manhole	1,500–2,000	Use higher ventilation rate when contents are being heated
Pouring hoods Foundry	Side hood	200 to 300 cfm per linear ft. of hood with slot velocities of 1500 fpm. Exhaust take-off every 8 to 10 ft.	3,500	
Rock drilling Dry drilling (rock)	Special trap (see references)	60 cfm—vertical (downward) work 200 cfm—horizontal work	3,500	May vary with size and speed of drill. Wet drilling offers alternate control
Rubber calender rolls	Canopy—side panels	75–100 fpm indraft	3,500	
Quartz fusing	Booth on bench	150–200 fpm at face	3,500	
Screens Vibrating Flat deck	Enclosure	150–200 fpm indraft through hood openings but not less than 25–50 cfm per sq. ft. of screen area	3,500	

Operation	Hood Type	Air Flow	Control Velocity	Remarks
Shakeouts Foundry	Enclosure	200 fpm through all openings in enclosure, but not less than 200 cfm per sq. ft. of grate area	3,500	
Spray coating	Booth—operator inside	100–200 fpm at booth cross section	1,500–2,000	Use higher ventilation rate for small booths 4 sq ft or less
	Booth—operator outside	150–200 fpm at booth cross section	1,500–2,000	
	Booth—downdraft	100–200 fpm downdraft	1,500–2,000	
Tanks, open surface	(See Table 47-L)			
Tumbling mills Hollow trunnion type	Exhaust connection by manufacturer	Use branch diameter same size as exhaust outlet. For round mills branch diam should be 1/6 diam of mill; for square mills branch diam should be 1 in. plus 1/6 side dimension of mill.	3,500–5,000	
Tumbling mills, drums, cages, barrels	Enclosure	400 fpm through openings but not less than 75 cfm sq ft plan area.	3,500	Where equipment is enclosed and dust tight during rotation, enclosure may not be needed if feed and discharge operations can be otherwise run.
Welding	Local hood with flange	6 in. from arc—150 cfm 6–9 in. from arc—275 cfm 8–10 in. from arc—425 cfm 10–12 in. from arc—600 cfm	2,000–4,000	
	Downdraft bench	150–250 cfm per sq ft grille area	2,000	
	Booth	100 fpm at booth face	2,000	
Woodworking	(See Table 47-K)			
Miscellaneous Packaging machines, granulators, enclosed dust producing units. Packaging, weighing, container filling, inspection	Complete enclosure	100–400 fpm indraft through inspection or working openings, but not less than 25 cfm per sq ft of enclosed plan area	3,000	
	Booth	50–150 cfm per sq ft of open face area	3,000	
	Downdraft	75–150 cfm per sq ft of dust producing plan area.	3,500	

From ASHRAE Guide and Data Book, 1970, Chapter 20. Used by permission.

1021

TABLE 36-B

HOOD EXHAUST REQUIREMENTS FOR GRINDING, POLISHING, BUFFING, SCRATCH BRUSHING, ABRASIVE CUT-OFF WHEELS, GRINDING AND POLISHING BELTS

(Air velocity: 4500 fpm in all branches)

Consult federal, state, or local regulations to see if higher ventilation rates are required

Grinding and Abrasive Cutting-off Wheels			Buffing, Polishing, and Scratch Brushing Wheels *(Note 1)*		
Wheel Diam. *(In.)*	Max. Width *(In.)*	Minimum CFM *(Note 4)*	Wheel Diam. *(In.)*	Max. Width *(In.)*	Minimum CFM *(Note 4)*
Up to 9	Not over 1½	220	Up to 9	Not over 2	300
Over 9 to 16	2	390	Over 9 to 16	3	500
Over 16 to 19	3	500	Over 16 to 19	4	610
Over 19 to 24	4	610	Over 19 to 24	5	740
Over 24 to 30	5	880	Over 24 to 30	6	1040
Over 30 to 36	6	1200			

For abrasive wheels smaller than 9 in., use the minimum flow of 220 cfm. If the smaller wheel is also wider, increase the flow by the ratio of the width to 1½ in. For example, a 3-in. wide, 9-in. wheel should have a minimum flow of 440 cfm.

Horizontal Single Spindle Disk Grinders		Horizontal Double Spindle Disk Grinders			Vertical Single Spindle Disk Grinders *(not covered)*		
Wheel Diam. *(In.)*	Minimum CFM *(Note 4)*	Wheel Diam. *(In.)*	Pipes *(Note 2)*	Minimum CFM *(Note 4)*	Wheel Diam. *(In.)*	Pipes *(Note 2)*	Minimum CFM *(Note 4)*
Up to 12	220	Up to 19	1	610	Up to 20	2	610
Over 12 to 19	390	Over 19 to 25	1	880	Over 20 to 30	2	1490
Over 19 to 30	610	Over 25 to 30	1	1200	Over 30 to 53	4	3530
Over 30 to 36	880	Over 30 to 53	2	1770	Over 53 to 72	5	6010
		Over 53 to 72	4	6240			

Vertical Single Spindle Disk Grinders *(More than half covered)*			Grinding and Polishing Belts and Straps *(Note 5)*	
Wheel Diam. *(In.)*	Pipes	Minimum CFM *(Note 4)*	Max. Width *(In.)*	Minimum CFM *(Note 4)*
Up to 20	1 (Note 3)	500	Up to 3	220
Over 20 to 30	2 (Note 2)	790	Over 3 to 5	300
Over 30 to 53	2 (Note 2)	1770	Over 5 to 7	390
Over 53 to 72	2 (Note 2)	3140	Over 7 to 9	500
			Over 9 to 11	610
			Over 11 to 13	740

TABLE 36-C

EXHAUST REQUIREMENTS FOR WOODWORKING AND MISCELLANEOUS OPERATIONS

Consult federal, state, or local regulations to see if higher ventilation rates are required

Pipe Sizes for Indicated Air Volume — Air Velocity is 4000 fpm

Branch pipe diam (in.)	3	4	4½	5	5½	6	7	8	9	10	12
Air volume (cfm)	200	350	440	550	660	800	1,100	1,400	1,800	2,200	3,100

From Exhaust Systems. Industrial Code Rule 18 (April 1961). New York State Department of Labor.

(Text continued from page 1016.)

sure at that point. If some readings are higher or lower than originally, then somewhere in the system the air flow has been restricted. If all the readings taken before the collector or fan inlet are low, then either the air collector is dirty or the fan is not doing its job because of worn blades or slipping belts. Readings taken on each side of the collector will tell if it is clean or plugged.

It is suggested that all new installations be tested in accordance with the *Test Manual* published by the American Conference of Governmental Industrial Hygienists (see References). This small publication tells how a system should be tested and supplies forms that are to be filled in by test personnel. When signed, these forms certify that the system has been tested in accordance with approved practice. The forms are kept by the owner of the installation and serve as a good reference.

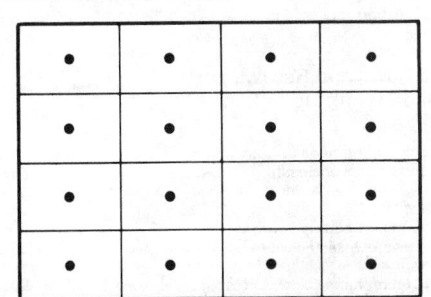

Pitot traverse points in a rectangular duct. Centers of 16 to 64 equal areas. Locations not more than 6" apart.

FIG. 36–16.

From ACGIH Industrial Ventilation.

Notes to Table 36-B.

NOTE 1: "Polishing wheels" means "made-up" hard wheels faced with abrasive.

NOTE 2: One branch pipe may be used wherever two or more branch pipes are required, provided the cross-sectional area of the single branch pipe is equal to the combined cross-sectional area of the required number of branch pipes and the air flow into the encircling hood is substantially uniform at all points.

NOTE 3: "Half covered" means a minimum covering of not less than 50 percent of the open area over the top of the grinding wheel.

NOTE 4: Local exhaust ventilation of high speed machinery shall provide such higher rates of air flow as may be necessary to control the contaminants.

NOTE 5: The listed air flow rates are for short belts and straps (up to six feet on centers) which are completely covered except for the necessary working openings. More than one pipe and higher air flow rate for longer belts or straps or for belts or straps which are essentially uncovered shall be provided where necessary to effect control.

TABLE 36-D

EXHAUST REQUIREMENTS FOR WOODWORKING OPERATIONS
Air Volume for Various Operations—Air Velocity is 4000 fpm
(Use in Conjunction with Table 36-C to Find Branch Pipe Size)
Consult federal, state, or local regulations to see if higher ventilation rates are required

Self-Feed Table Rip Saw

	cfm
Top hood	350
Bottom hood	350

NOTE: Where saw does not cut through table, use 550 cfm and top hood.

All Other Rip, Table, Mitre, Swing, and Variety Saws

	cfm
Up to 12 in. diam.	350
Over 12 to 16 in. diam.	440
Over 16 in. diam.	550
Dado heads, any size	550

Gang Circular Saws

Ventilation rate shall be such as to collect and remove the sawdust.

Band Saws and Resaws
Blade Width Up to 2 in.

	cfm
Each cutting hood	350
Lower wheel enclosure	350
Single hood with wheel diam. less than 16 in.	350
Over 2 to 3 in.	
Cutting hood	550
Lower wheel enclosure	350
Over 3 to 4 in.	
Cutting hood	800
Lower wheel enclosure	550
Over 4 to 5 in.	
Cutting hood	1,100
Lower wheel enclosure	550
Over 5 in.	
Cutting hood	1,400
Lower wheel enclosure	550

Disk Sanders

Disk diameter	cfm per branch
Up to 12 in.	One branch, 350
Over 12 to 18 in.	One branch, 440
Over 18 to 26 in.	One branch, 550
Over 26 to 32 in.	Two branches, 350 each
Over 32 to 38 in.	One branch, 550 and one branch, 350
Over 38 in.	One branch, 550 and two branches, 350 each

Single Drum and Spindle Sanders

Sq in. of sanding surface	cfm
Up to 50	200
Over 50 to 200	350
Over 200 to 400	440
Over 400 to 700	550
Over 700 to 1,400	800
Over 1,400 to 2,400	1,100
Over 2,400 sq in. sanding surface: Use formula	

$d = 0.133 \sqrt{a}$, where d = branch diam in nearest standard size, and a = sq in. of sanding surface.

NOTE: On long drums of small diameter, above air volumes should be divided into two or more branches at least equalling in volume that of the single volume specified.

Horizontal Belt or Edge Sanders

Belt width	cfm per pulley
Up to 6 in., belts less than 30 in. on centers	440 for idler
Up to 6 in., belts more than 30 in. on centers	440 for drive
	350 for idler
Over 6 to 9 in.	550 for drive
	350 for idler
Over 9 to 14 in.	800 for drive
	400 for idler
Over 14 in.	1,100 for drive
	550 for idler

NOTE: Where more than one belt is used, air volume specified shall be required for each belt. Where a common hood is used for more than one belt, a volume of not less than the sum of the volumes specified shall be used. Where belt direction is reversible, volume for the idler pulley shall be the same as specified for the drive pulley.

Vertical Belt Sanders (Rear Belt and Both Pulleys Enclosed)

Belt width	cfm
Up to 6 in.	550
Over 6 to 10 in.	800
Over 10 to 15 in.	1,100
Over 15 in.	1,400

Multiple Drum Sanders— Drums Over the Table

Drum width	Branches per drum—cfm	
Up to 31 in.	One	550
Over 31 to 49 in.	One	800
Over 49 to 67 in.	One	1,100
Over 67 in.	One	1,400
For brush	At least	350

NOTE: Fewer brushes may be used as long as total branch pipe air volume is not reduced.

Multiple Drum Sanders— Drums Under the Table

Drum width	Branches per drum—cfm	
Up to 31 in.	One	350
Over 31 to 49 in.	One	550
Over 49 to 67 in.	One	800
Over 67 in.	One	1,100
For brush	At least	350

NOTE: Fewer branches may be used as long as total branch pipe air volume is not reduced.

TABLE 36-D *(Concluded)*

Moulders, Matchers, and Sizers

Max. size	cfm per head
Up to 4 in..	550 for side
	550 for top
	800 for bottom
Over 4 to 6 in..	660 for side
	800 for top
	800 for bottom
Over 6 to 8 in..	800 for side
	1,100 for top
	800 for bottom
Over 8 in.	800 for side
	1,400 for top
	1,400 for bottom

NOTE: Where profiler is used, provide additional volume not less than that specified for top head.

Sash Stickers

	cfm per head
One 4 in. branch on each head..........................	350

Wood Shapers

Horsepower per spindle	cfm per spindle
Up to 1..	350
Over 1 to 3..	550
Over 3 to 5..	800
Over 5 to 7½..	1,100
Over 7½...	1,400

NOTE: Where one continuous hood includes two or more spindles, the air volume required for the hood may be based on a volume per spindle of 250 cfm less than specified above, but in no case less than 350 cfm per spindle.

Tenoners

	cfm
Top and bottom tenon heads..........................	550 for each
Cope and dado heads	550 for each
Cutoff saws up to 12 in. diam..........................	350
Over 12 in. diam. saw.................................	550

Jointers

Knife width	cfm
Up to 6 in.	350
Over 6 to 12 in.	550
Over 12 to 20 in.	800
Over 20 in...	1,100

Single Planers

Working width	cfm
Up to 20 in. max.....................................	800
Over 20 to 26 in.	1,100
Over 26 to 32 in.	1,400
Over 32 to 38 in.	1,800
Over 38 in..	2,200

Double Planers

Working width	cfm per head
Up to 20 in. max.	800 for upper
	550 for lower
Over 20 to 26 in.	1,100 for upper
	800 for lower
Over 26 to 32 in.	1,400 for each
Over 32 to 38 in.	1,800 for upper
	1,400 for lower
Over 38 in.	2,200 for upper
	1,400 for lower

Automatic Lathes

	cfm
Per foot of cutting length....................................	350

NOTE: Branch pipes shall be located to provide effective air removal over full face of hood.

Pulley Pockets

	550 cfm

Pulley Stiles

	550 cfm

Hogs

Tons of refuse per hour	cfm
Up to 1	1,600
Over 1	50 per pound of refuse conveyed

NOTE: Use branch and main velocity not less than 4,500 cfm. Separate exhaust system should be provided, but additional separator is not required if separator used for other woodworking equipment is adequate.

Floor Sweeps

NOTE: Air volume of branches to floor sweeps with covers which are kept closed when not in use need not be included in computing total air volume.

Portable Sanders

NOTE: Conform to requirements for portable grinders.

Machines Not Specifically Listed in This Table

Air volume shall be at least 350 cfm and hood shall be so arranged that an inward flow of air into the ventilating system is maintained during the operation of the machine with adequate velocity to control air contaminants.

From Exhaust Systems. *Industrial Code Rule 18 (April 1961). New York State Department of Labor.*

TABLE 36-E
VENTILATION RATES FOR OPEN-SURFACE TANKS

| | Minimum Ventilation Rate cfm Per Sq Ft Hood Opening | | | | Minimum Ventilation Rate cfm Per Sq Ft Tank Area Lateral Exhaust (Note 1) $W/L = \dfrac{Tank\ Width}{Tank\ Length}$ RATIO | | | | | |
| | Enclosing Hood | | Canopy Hood | | W/L 0–0.24 | | W/L 0.25–0.49 | | W/L 0.50–1.0 | |
	One Open Side	Two Open Sides	Three Open Sides	Four Open Sides	A	B	A	B	A	B
Plating										
Chromium (Chromic Acid Mist)	75	100	125	175	125	175	150	200	175	225
Arsenic (Arsine)	65	90	100	150	90	130	110	150	130	170
Hydrogen Cyanide	75	100	125	175	125	175	150	200	175	225
Cadmium	75	100	125	175	125	175	150	200	175	225
Anodizing	75	100	125	175	125	175	150	200	175	225
Metal Cleaning (Pickling)										
Cold Acid	65	90	100	150	90	130	110	150	130	170
Hot Acid	75	100	125	175	125	175	150	200	175	225
Nitric and Sulfuric Acid	75	100	125	175	125	175	150	200	175	225
Nitric and Hydrofluoric Acid	75	100	125	175	125	175	150	200	175	225
Metal Cleaning (Degreasing) See Note 2										
Metal Cleaning (Caustic or Electrolytic)										
Not Boiling	65	90	100	150	90	130	110	150	130	170
Boiling	75	100	125	175	125	175	150	200	175	225
Bright Dip (Nitric Acid)	75	100	125	175	125	175	150	200	175	225
Stripping										
Concentrated Nitric Acid	75	100	125	175	125	175	150	200	175	225
Conc. Nitric and Sulfuric Acid	75	100	125	175	125	175	150	200	175	225
Salt Baths (Molten Salt)	50	75	75	125	60	90	75	100	90	110
Salt Solution (Parkerize, Bonderize, etc.)										
Not Boiling	90	90	100	150	90	130	110	150	130	170
Boiling	75	100	125	175	125	175	150	200	175	225
Hot Water (if vent. desired)										
Not Boiling	50	75	75	125	60	90	75	100	90	110
Boiling	75	100	125	175	125	175	150	200	175	225

ASHRAE Guide and Data Book, *1970, Chapter 20. Used by permission.*

NOTE 1: Column A refers to tank with hood along one side or two parallel sides when one hood is against a wall or a baffle running length of tank and as high as tank is wide; also to tanks with exhaust manifold along center line with *W/Z* becoming tank width in *W/L* Ratio. Column B refers to free standing tank with hood along one side or two parallel sides.

NOTE 2: Complete control of the vapors and mist from degreasing operations requires the same ventilation rates as recommended for pickling. However, the solvents employed in degreasing operations are relatively volatile, and the solvent loss caused by evaporation increases rapidly as the exhaust rate at the tank increases. For this reason, perfect control is usually sacrificed in favor of lower solvent evaporation rates. Where solvent loss does not present an important cost or operating problem, the exhaust rates given for pickling should be adhered to, but where solvent loss must be kept at a minimum, an exhaust rate of 50 cfm per sq ft of tank area is commonly employed. Where this rate is used, control will be adequate only if the tank is located in an area free of drafts and if the degreasing operations are carried out in accordance with a rigid schedule.

References

American Conference of Governmental Industrial Hygienists, Committees on Industrial Ventilation and Respirators, P.O. Box 453, Lansing, Mich. 48902.
   Industrial Ventilation—A Manual of Recommended Practice, 12th ed., 1972.
   Respiratory Protective Devices Manual, 1963.
   Test Manual

American Industrial Hygiene Association, 63 S. Miller Road, Akron, Ohio 44313.
   Heating and Cooling of Man in Industry.

American Society of Heating, Refrigerating and Air-Conditioning Engineers, 345 East 47th St., New York, N.Y. 10017. *Guide and Data Book.* (Issued annually.)

American National Standards Institute, 1430 Broadway, New York, N.Y. 10018.
   Fundamentals Governing the Design and Operation of Local Exhaust Systems, Z9.2.
   Practices for Ventilation and Operation of Open-Surface Tanks, Z9.1.
   Safety Code for Design, Construction, and Ventilation of Spray Finishing Operations, Z9.3.
   Ventilation and Safe Practices of Abrasives Blasting Operations, Z9.4.
   Ventilation Control of Grinding, Polishing, and Buffing Operations, Z43.1.

National Fire Protection Association, 470 Atlantic Ave., Boston, Mass. 02210.
   Air Conditioning and Ventilating Systems, Standard No. 90A.
   Blower and Exhaust Systems for Duct, Stock and Vapor Removal, Standard No. 91.
   National Electrical Code, Standard No. 70
   Residence Type Warm Air Heating, Air Conditioning Systems, Standard No. 90B.
   Spray Finishing Using Flammable and Combustible Materials, Standard No. 33.

National Safety Council, 425 North Michigan Ave., Chicago, Ill., 60611.
   Fundamentals of Industrial Hygiene, 1971.
   Industrial Data Sheets.
      Checking Performance of Local Exhaust Systems, 428.
      Dusts, Fumes, and Mists in Industry, 531.
      Instruments for Testing Exhaust Ventilation Systems, 431.
      Radiant Heat Control, 381.

Severns, William H., and Julian R. Fellows. *Air Conditioning and Refrigeration.* New York, John Wiley & Sons, Inc., 1959.

Strock, Clifford, ed. *Handbook of Air Conditioning, Heating and Ventilating.* New York, The Industrial Press, 1973.

Industrial Hygiene

Chapter
37

OSHA and NIOSH Interest   1031
Criteria documents

The Industrial Hygienist   1032
Definitions . . . Scope . . . Function

Recognition of Occupational Health Hazards   1034
Mode of entry . . . Basic procedure . . . Definition of terms

Dusts   1041
Dispersion . . . Separation in airborne dusts . . . Inhalation of
particles . . . Retention of dust

Physiological Effects of Dusts, Fumes, and Mists   1043
Pneumoconioses . . . Gases

Organic Solvents   1048
Labeling . . . General physiological effects . . . Degree of severity
of solvent hazards . . . "Safety" solvents

Physical Energy Stresses   1051
Heat stress . . . Estimating heat stress . . . Atmospheric pressure
. . . Mechanical vibration . . . Repeated motion . . . Nonionizing
radiation . . . Lasers . . . Noise

Ionizing Radiation   1059
Five basic safety factors . . . Need for concern . . . Five kinds of
radioactivity . . . External *vs.* internal hazards . . . Evaluating

radiation hazards . . . Protecting from radiation hazards . . .
Available guides

Biological Agents and Other Occupational Infections   1065
Tuberculosis . . . Fungus infections . . . Byssinosis . . . Anthrax
. . . "Q fever" . . . Brucellosis . . . Erysipelas . . . Upper
respiratory tract infections

Occupational Skin Diseases   1067
Types of dermatitis . . . Causes of dermatitis . . . Physical examinations . . . Eliminating contact with irritants . . . Hygienic measures

Evaluating the Hazard   1075
Information required first

Defining the Exposure   1076
Sampling techniques . . . Types of measuring instruments . . .
Electrical direct-reading instruments . . . Color-change and stain-
length instruments . . . Limitations of instruments . . . Direct-
reading *vs.* lab analysis

Controlling Environmental Hazards   1084
Substitution . . . Changing the process . . . Isolation or enclosure
. . . Wet methods . . . Local exhaust ventilation . . . General or
dilution ventilation . . . Personal protective equipment . . . Housekeeping . . . Special control methods . . . Training and education

Setting Up an Effective Industrial Hygiene Program   1089
Engineering organization . . . Medical organization . . . Industrial
hygiene organization . . . Safety organization . . . Purchasing
organization . . . Supervisor's responsibilities . . . Employee's
responsibilities

Sources of Help   1091
Organizations . . . Literature

Threshold Limit Values for Physical Agents   1095

The safety professional in his routine accident prevention activities is frequently required to make a decision as to the degree
of health hazard arising out of an industrial
operation.

In emergency situations and in the absence
of an industrial hygienist, it becomes the
safety professional's duty to take proper action towards evaluation and control of these
hazards.

The intent of this chapter is to provide the
safety professional with general fundamental
information regarding the recognition, evaluation, and control of occupational health
problems.

In some instances, a survey of raw materials,
by-products produced intentionally and unintentionally, source and method of dispersion
of airborne contaminants, exposure to physical
agents, and control measures in use will indicate the effectiveness of the control measures.
A more complex problem will require a complete and detailed study by a qualified industrial hygiene engineer.

Most safety professionals are already deeply
involved in some aspects of industrial hygiene.
They study work operations, look for potential hazards, and make recommendations to
minimize health hazards. The industrial hygiene engineer, through specialized study and
training, would have greater competence in
this area.

TABLE 37-A

A READY-REFERENCE CHART OF THE FIVE TARGET HEALTH HAZARDS

TARGET HEALTH HAZARD	What Is the Health Hazard?	Current OSHAct Approved Levels	How Is It Detected? Measured?	How Is It Controlled?	Sources for Information
ASBESTOS (Dust)	Prolonged inhalation of asbestos fibers between 5 and 50 microns long can produce asbestosis, and possibly lung cancer.	Five fibers per milliliter greater than five microns in length for an eight-hour, time-weighted average air-borne concentration.	To be collected with personal sampling pump. Fibers will be counted microscopically at 400–450 magnification using phase contrast illumination.	Enclosure and local exhaust ventilation for equipment and operations. Approved respirators for personnel.	Asbestos Information Association 1660 L St., N.W. Washington, D.C. 20036
COTTON DUST (Dust)	Prolonged exposure can cause byssinosis, which can progress to chronic bronchitis and emphysema.	One milligram per cubic meter of air for eight-hour, time-weighted average air-borne concentration.	Personal sampler is attached to workman's clothing, with the uncovered filter facing down, usually for six-hour sample.	Machinery enclosure and local exhaust ventilation. Also, pretreatment of cotton. Approved respirators.	American Textile Mfr. Inst., 1501 Johnston Building, Charlotte, N.C. 28281 . . . Cotton, Inc., 4505 Creedmoor Rd., Raleigh, N.C. 27612 . . . National Cotton Council, P.O. Box 12285, Memphis, Tenn. 38112.
CARBON MONOXIDE (Gas)	Carbon monoxide displaces oxygen in the blood, causing suffocation. High concentrations are fatal.	50 parts per million parts of air for eight-hour, time-weighted average air-borne concentration.	Primary method is CO portable, hopcalite reagent-type meter. Secondary method is certified detector tubes. Personal, color-changing dosimeters available.	Adequate ventilation. Isolation of hazardous operations. Automated detection alarm systems.	Compressed Gas Association 500 Fifth Ave. New York, N.Y. 10036
LEAD	Prolonged exposure can cause severe	0.2 milligrams per cubic meter of air	Sample collection minimum of 60 minutes by personal sampling pump on	Enclosure and local exhaust ventilation. Wet or vacuum clean-	

Material	Effect	Air Quality Standard	Evaluation Method	Control	Source of Additional Information
(Fume or Dust)	gastrointestinal, blood, and central nervous system disorders.	for eight-hour, time-weighted average air-borne concentration.	millipore filter. Filter is dissolved in nitric acid, taking lead into solution, which is then introduced into an atomic absorption unit for analysis.	ing of all lead dust. Hygiene, and prohibiting food and drink in work area. Also, use of approved respirators.	Lead Industries Association, Inc. 292 Madison Ave., New York, N.Y. 10017
SILICA (Dust)	Excessive exposure can produce silicosis, a disabling lung disease.	For respirable quartz, 10 milligrams per cubic meter of air, divided by percent of free silica plus 2. For respirable cristobalite and tridymite, the limits are one-half the value of quartz.	Personal sampling pumps collect dust samples, which are weighted for given volume of air. Similar procedure to determine percentage of silica in the air.	Substitution of material. Also, enclosure, local exhaust, segregating operations, wetting, and good housekeeping.	American Foundrymen's Society Golf & Wolf Roads Des Plaines, Ill. 60016

Safety professionals and industrial hygienists must work together. Because the safety professional carries on the day-to-day functions involving immediate decisions, he must know when and where to get help on industrial hygiene problems.

After the industrial hygiene engineer surveys the plant, makes recommendations, and suggests certain control measures to be followed, it is the safety professional's responsibility to see that the control measures are being applied and followed.

The variety of substances and processes that present occupational health hazards steadily increases. Recently developed raw materials and methods of manufacture or new combinations of their older counterparts create new environmental stresses. Improved techniques for the prevention and control of existing hazards and stresses are also being developed.

The prevention and control of occupational health hazards are the responsibility of company management. To assist those individuals charged with this responsibility, this chapter discusses the principles and practices of industrial hygiene. It presents information on the recognition of hazards, their qualitative and quantitative evaluation, and engineering methods for eliminating or minimizing them.

OSHA and NIOSH Interest

With the passage of the Occupational Safety and Health Act of 1970, activity by governmental groups in the field of industrial hygiene was accelerated. For background, read Chapter 2, "The Occupational Safety and Health Act of 1970."

Initially the U.S. Department of Labor established a list of target health hazards. Industries where exposures to the following materials existed were subjected to detailed study of the degree of exposure and the control measures in effect:

> Asbestos
> Cotton Dust
> Carbon Monoxide
> Lead
> Silica

A summary is provided in Table 37-A.

As results of these studies are compiled, the

most desirable means of controls will provide better protection for the millions of employees exposed. The detailed studies are being made by personnel in the National Institute of Occupational Safety and Health (NIOSH) and industrial hygienists in state health and labor agencies.

Criteria documents

One of the requirements established by the 1970 Act was the publication of Criteria Documents by the National Institute of Occupational Safety and Health on materials of health significance. These documents once published are forwarded to the U.S. Department of Labor for their consideration in establishing work place standards. The documents contain a detailed review of scientific information available, the probable safe atmospheric level in industry, medical controls required, methods of sampling and analysis and any needs for further study.

Initially the following subjects were studied and Criteria Documents published and forwarded to the U.S. Department of Labor:

> Asbestos
> Beryllium
> Carbon monoxide
> Carcinogens
> Chromic acid
> Coke oven emissions
> Hot environments
> Inorganic lead
> Inorganic mercury
> Noise
> Toluene
> Toluene diisocyanate
> Trichloroethylene
> Ultraviolet radiation

Certain of these documents have already led to the establishment of specific new permissible atmospheric levels, medical controls, recordkeeping requirements and recommended methods of sampling and analysis. It is probable that eventually such requirements will be promulgated for each of these criteria subjects. Many more documents are in process of development and can lead to further regulations.

As time goes on more chemicals will be studied by NIOSH and submitted to OSHA for consideration and regulation. No text

can be kept up-to-date on a subject of this type. The safety professional must keep up-to-date through his own contacts with official and unofficial sources.

The Industrial Hygienist

Definitions

The following definitions* have been approved by the American Industrial Hygiene Association:

"INDUSTRIAL HYGIENE is that science and art devoted to the recognition, evaluation and control of those environmental factors or stresses, arising in or from the workplace, which may cause sickness, impaired health and well-being, or significant discomfort and inefficiency among workers or among the citizens of the community.

"AN INDUSTRIAL HYGIENIST is a person having a college or university degree or degrees in engineering, chemistry, physics, or medicine or related biological sciences who, by virtue of special studies and training, has acquired competence in industrial hygiene. Such special studies and training must have been sufficient in all of the above cognate sciences to provide the abilities: (a) to recognize the environmental factors and stresses associated with work and work operations and to understand their effect on man and his well-being; (b) to evaluate, on the basis of experience and with the aid of quantitative measurement techniques, the magnitude of these stresses in terms of ability to impair man's health and well-being; and (c) to prescribe methods to eliminate, control or reduce such stresses when necessary to alleviate their effects."

Scope

Recognition of environmental factors and stresses which influence health requires that the industrial hygienist be familiar with work operations and processes. The categories of stresses that interest him are: (a) chemical—

---

* *American Industrial Hygiene Association Journal*, 20:428–430 (Oct. 1959).

liquid, dust, fume, mist, vapor, or gas; (*b*) physical—electromagnetic and ionizing radiations, noise, vibration, and extremes of temperature and pressure; (*c*) biological—insects, mites, molds, yeasts, fungi, bacteria, and viruses; and (*d*) ergonomic—body position in relation to task, monotony, boredom, repetitive motion, worry, work pressure, and fatigue. The industrial hygienist recognizes that such stresses may immediately endanger life and health, accelerate the aging process, or cause only significant discomfort and inefficiency.

Evaluation of the magnitude of the environmental factors and stresses arising in or from the workplace is done by the industrial hygienist, aided by his training, experience, and quantitative measurement of the chemical, physical, biological, or ergonomic stresses. He can thus give an expert opinion as to the general healthfulness of the environment, either for short periods or for a lifetime of exposure.

Prescription of corrective procedures, when necessary to protect health, is based on the industrial hygienist's experience and knowledge and on the quantitative data he has obtained. He can prescribe control measures such as isolation of a work process, substitution of a less harmful material, or any of the other measures discussed later in this chapter.

Function

The industrial hygienist:

1. Directs the industrial hygiene program.

2. Examines work environment, environs:
   a) Studies work operations and processes to get full details of the nature of the work, materials and equipment used, products and by-products, number and sex of employees, and hours of work.
   b) Measures the magnitude of exposure or nuisance to workers and the public. In doing so, he must:
      (1) select or devise methods and instruments suitable for such measurements;
      (2) personally or through others under his direct supervision conduct such measurements; and

   (3) study and test material associated with the work operation.
   c) Studies and tests biological materials, such as blood and urine, by chemical and physical means, when such examination will aid in determining the extent of exposure.

3. Interprets results of the examination of the work environment and environs in terms of ability to impair health, nature of health impairment, workers' efficiency and community nuisance and/or damage, and presents specific conclusions to appropriate interested parties such as management and health officials.

4. Determines the need for, or effectiveness of, control measures, and when necessary, recommends procedures which will be suitable and effective for both the environment and environs.

5. Prepares rules, regulations, standards, and procedures for the healthful conduct of work and the prevention of nuisance in the community.

6. Presents expert testimony before courts of law, hearing boards, workmen's compensation commissions, regulatory agencies, and legally appointed investigative bodies covering all matters pertaining to industrial hygiene.

7. Prepares appropriate text for labels and precautionary information for materials and products used by workers and public.

8. Conducts programs for the education of workers and the public in the prevention of occupational disease and community nuisance.

9. Directs epidemiologic studies among workers and industries to discover possibilities of the presence of occupational disease, and establish or improve threshold limit values or standards as guides for the maintenance of health and efficiency.

10. Conducts research to advance knowledge of the effects of occupation upon health and means of preventing occupational health impairment, community air pollution, noise, nuisance, and related problems.

The industrial hygienist, therefore, effectively links the manufacturing operations of a company to its medical department. The plant physician depends on the skills, techniques, and knowledge of the hygienist to provide insight into the health background of an employee's job. In many cases, it is extremely difficult to differentiate between the symptoms of occupational and nonoccupational disease. The industrial hygienist, by pointing out the danger areas, can enable the physician to better correlate the employee's condition and complaints with the known potential job health hazards.

Recognition of Occupational Health Hazards

There are three key concepts in an effective industrial hygiene program.

RECOGNITION—which requires knowledge of stresses arising out of work operations and processes.

EVALUATION—judgment or decision usually involving measurement of magnitude of stress based on past experiences.

CONTROL—by isolation, substitution, change of process, wet methods, local exhaust ventilation, general or dilution ventilation, personal protective equipment, housekeeping, and training and education.

Environmental factors or stresses may be:

CHEMICAL—liquids, gases, dusts, fumes, mists, and vapors as air contaminants and skin irritants.

PHYSICAL—electromagnetic and ionizing radiations, noise, vibration, and extremes of temperatures and pressure.

BIOLOGICAL—insects, molds, fungi, and bacteria.

ERGONOMIC—monotony, repetitive motion, and fatigue.

Mode of entry

The majority of the occupational diseases arise from inhaling chemical agents in the form of vapors, gases, dusts, fumes, and mists or by skin contact with these materials.

In order for the harmful agent to exert its toxic effects it must come into contact with a body cell. The three routes of entry are inhalation, skin absorption and ingestion.

Certain chemical agents that reach the lungs can pass directly into the blood stream and be absorbed over a long period of time. Others may stay in the lungs and set up local irritant or damaging action.

Toxic and irritant dusts can also be ingested in amounts that may cause trouble. If toxic dust swallowed with food or saliva is not soluble in body fluids, it is eliminated directly through the intestinal tract. Toxic materials that are readily soluble in body fluids can be absorbed in the digestive system and picked up by the blood.

A third way in which toxic and irritant substances may enter the system is skin absorption. Many organic compounds, such as TNT, cyanides, and most aromatic amines, amides, and phenols, can produce systemic poisoning by direct contact with the skin. Contact of irritant chemical agents with the skin also may result in skin irritation.

Inhalation as a route of entry is particularly important because of the rapidity with which a toxic material can be absorbed in the lungs, pass into the blood stream, and reach the brain. This same material, if ingested, would be considerably diluted with the contents of the stomach.

It is important that all routes of entry be studied when an evaluation of the work environment is being made; candy bars or lunches in work areas, solvents being used to clean work clothing and hands, in addition to airborne contaminants in work areas.

To recognize environmental factors or stresses, a safety professional must first know the chemicals used as raw materials and the nature of the products and by-products manufactured. This sometimes requires considerable effort. The required information can be obtained from labels on drums and containers (Fig. 37–1). If the labels do not give complete information but only trade names, it may be necessary to correspond with the manufacturer of the chemicals to obtain this information.

Many industrial materials such as resins and polymers are relatively inert and nontoxic under normal conditions of use, but when heated or machined, they may decom-

METHANOL

DANGER! FLAMMABLE
VAPOR HARMFUL
MAY BE FATAL OR CAUSE BLINDNESS
IF SWALLOWED
CANNOT BE MADE NONPOISONOUS

Keep away from heat, sparks, and open flame.
Keep container closed.
Avoid prolonged or repeated breathing of vapor.
Use only with adequate ventilation.

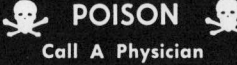 **POISON**

Call A Physician

First Aid

If swallowed: Give a tablespoonful of salt in a glass of warm water and repeat until vomit fluid is clear. Give two teaspoonfuls of baking soda in a glass of water. Have patient lie down and keep warm. Cover eyes to exclude light.

METHANOL MIXTURES

For products containing methanol in proportion sufficient to create hazard because of methanol content, use applicable statements as above, with addition of:

CONTAINS OVER.................% OF METHANOL

The word "FLAMMABLE" may be omitted from the Statement of Hazards if the product has a flash point above 80°F.

MCA Chemical Safety Data Sheet SD-22 available

Courtesy Manufacturing Chemists' Assn.

Fig. 37–1.—Typical hazardous chemical label. For labeling of consumer products and for identification during transportation, consult appropriate federal agencies.

pose to form highly toxic by-products. Information concerning products and by-products can be obtained from company chemical and engineering departments.

As to physical agents (such as sources of radiant heat, abnormal temperatures and humidity, excessive noise, improper or inadequate illumination, ultraviolet radiations, X and gamma rays, and other ionizing radiations) it is possible, in most cases, to note the presence of equipment which can produce these potential hazards. Special instruments are needed to properly measure the exposure level.

One can obtain much valuable information by observing the manner in which these health hazards are generated, the number of people involved, and the control measures in use.

After the list of chemicals and physical conditions to which employees are exposed has been prepared, it is necessary to determine which of the chemicals or agents result in hazardous exposures and need further study.

(*Text continues on page 1038.*)

HAZARDOUS MATERIALS INFORMATION SHEET

(Please complete all applicable sections.)

1. Product Name, Number, Synonym _____ Chemical Formula _____
2. Manufacturer's Name _____
3. Manufacturer's Address _____

4. Chemical and Physical Properties: a. Molecular Wt. _____ b. Boiling Point _____°C
   c. Melting Point _____°C d. Specific Gravity (water = 1) or Bulk Density _____

   @ _____°C e. Vapor Density (air = 1) _____ f. Vapor Pressure (mm Hg) _____

   @_____°C; _____ @ _____ °C; _____ @ _____°C;

   g. Solubility _____

   h. pH/conc. _____ i. Index of Refraction _____ @ _____°C

   j. Corrosive action on materials (e.g. aluminum, carbon steel, copper, rubber, plastics, etc.) _____

   _____

   k. Does the material decompose when exposed to air? water? heat? strong oxidizers? possible

   products? _____

   l. Does the material generate heat through polymerization or condensation? _____

   m. Composition *(give chemical names of components; information will be treated as confidential)*

COMPOUND	PERCENT	COMPOUND	PERCENT

NOTE: *Please be specific. For example, it is important to know whether an alcohol is methanol; an aromatic hydrocarbon is benzene; a chlorinated material is carbon tetrachloride.*

5. Flammability and Explosive Properties: a. Flash Point, F, Closed Cup _____

   Open Cup _____ If flash point changes during evaporation give data _____

   b. Explosive limits (% by vol. in air): LOWER _____ UPPER _____

   c. Susceptibility to spontaneous heating: YES _____ NO _____

   d. Fire point, F _____ Auto-ignition temp., F _____

   e. What products might be formed in the event of fire or abnormal temperatures? _____

   f. Suitable extinguishing agents _____

6. Procedures in Case of Container Breakage or Leakage _____

7. Transportation and Storage Requirements _____

8. Physiological Properties *(give animal tested, observation time, dosage value and range, dilution medium, etc.):*

   a. Acute oral toxicity _____

   b. Acute local effects on eyes _____

   c. Acute local effects on skin. Primary irritant? _____

   Sensitizer? _____

   d. Acute inhalation toxicity *(vapor, mist, fume, dust. Indicate effects of concentration and time.)*

   _____

   e. Chronic effects _____

   f. Warning properties *(odor; irritation of eyes, nose, throat)* _____

   g. Threshold limit value *(estimate, if not on current list of ACGIH)* _____

9. First Aid Treatment:

   a. Skin contact _____

   b. Eye contact _____

   c. Inhalation _____

   d. Antidote and treatment in case of swallowing _____

10. Recommended Pre-placement or Periodic Medical Examination *(health standards, clinical tests, frequency, etc.)* _____

11. Precautions for Normal Conditions of Use _____

12. Recommended Personal Protective Equipment _____

13. Suggested Method for Air Analysis _____

14. Pertinent Literature References _____

15. Information Furnished By: NAME _____ DATE _____
                           TITLE _____
                           COMPANY _____
                           ADDRESS _____

(If more space is needed for comment, please attach an additional sheet. Please attach product information data sheets or other publications related to the safe handling and use of this material.)

Fig. 37–2.

A hazardous materials information sheet similar to the sample shown in Fig. 37–2 could be prepared to elicit toxicological information from the supplier. The information in this questionnaire would be useful to the medical, purchasing, managerial, engineering, and safety departments in setting guidelines for safe use of these materials in industry. This information would be very helpful in the event of an emergency. The survey should include those materials actually in use in a plant, together with those which may be contemplated for early future use. Possibly the best and earliest source of information concerning such materials is through the purchasing agent. Accordingly, it is strongly recommended that a close liaison be set up between the purchasing agent and the health and safety personnel so that early information will be provided concerning materials in use and those which are to be ordered.

Basic procedure

There is a basic, systematic procedure which can be followed in the recognition of occupational health hazards. J. W. Lake described the following method in his talk at the 1966 National Safety Congress. (See References.)

The first item to consider is the end product or products that will determine the raw material, processes, and equipment needed for production.

After the materials and processes are named or determined, then material properties, effects, and process flows, including pressures and temperatures involved, should be reviewed.

At this time a simple block process flow sheet should be established which will show, stepwise, the introduction of each material and the product of each step. A material and process hazard-and-effects checklist should be established. Some sources of information to assist in preparing this list are data sheets published by the National Safety Council, the American Industrial Hygiene Association, the Manufacturing Chemists' Association, and the American Insurance Association. (See listings in the next two chapters.)

With these two things accomplished following close coordination with other pertinent departments, the location for needed controls and their adequacy becomes clearer. For example, foundry operations produce dust. The flow sheet should show where this occurs. The hazards check sheet should identify the kind of dust and how it is formed. These items taken in combination indicate (*a*) where dust control measures such as ventilation should be placed, and (*b*) that care should be exercised to see that dust collection equipment should be located so that dust is not fed back into the plant environment. (See Fig. 37–3.)

Another example involves an exothermic (heat-evolving) reaction in a chemical plant. From information which you have established, make sure that process interlocks and adequate rupture disks (or relief vents) are present to prevent a sudden flooding of the working environment with hazardous chemicals or vapors.

Basically, then, a hazard checklist is required for each step of the manufacturing process; it should be reviewed to determine its impact on the next step and on the total process.

The different ways that equipment could fail or the process could go wrong should be explored for hazard anticipation and recognition. Do this not only from a single failure standpoint but also for the possibility of two separate and distinct failures. The failure effects should be analyzed.

The time taken to prepare process block flow diagrams and hazard checklists will save much more time in plant and process familiarization and does allow the safety professional to sit at his desk and probe for weak spots and evaluate the overall production program.

In some instances, the safety professional may be hundreds of miles away from the plant when an incident occurs. His previous evaluations and diagrams enable him to have immediate knowledge of the hazards and their location, and to initiate and assist in action which minimizes injury or loss.

It is vital that all process flow and plant facility prints be kept current to provide the knowledge required for proper recognition of hazards. A desirable practice is to operate the plant on paper from the plant layout flow diagrams. All that remains then is a follow-up on operating and emergency procedures

Fig. 37–3.—Control of harmful dusts, fumes, gases, and vapor close to the source, such as at this foundry shakeout, is effective and reduces the necessity of larger air movement required for general ventilation.

Courtesy American Foundrymen's Society.

that have been established.

Observation of employees as you make plant walk-throughs, may reveal possible substandard health conditions such as dermatitis, unusual pallor or coloration, abnormal fatigue, excessive dust on the clothing or face. These are indications that controls are not working or are needed. The foreman is a most valuable ally in hazard recognition situations like these.

It is assumed that educational sessions have been held for operating personnel.

Equipment observations. Plant walk-throughs also provide equipment failure hazard recognition situations. Odors not normal to the plant environment or odors more intense than normal are also cause for investigation for possible hazard. For example, leaks in ventilation equipment, partial or total blockage of ventilation slots, unusual vibration or noise from processing equipment, film badge positions in radiation fields or lack of them. Remedial measures are most intelligently made following a review of the process.

Quality control on all equipment going into the plant should not be overlooked.

1039

Definition of terms

The safety professional must be aware of the precise meanings of certain words commonly used in industrial hygiene if he is going to effectively communicate with other workers in this area. A fume respirator, for instance, is worthless as protection against gases or vapors. Too frequently, terms (such as *gases, vapors, fumes,* and *mists*) are used interchangeably. Each term has a definite meaning and describes a certain state of matter which can be achieved only by certain physical changes to the substance itself.

Basic to the industrial hygiene vocabulary are states of matter. As defined by the American National Standards Institute, these are as follows:

DUSTS. Solid particles generated by handling, crushing, grinding, rapid impact, detonation, and decrepitation of organic or inorganic materials, such as rock, ore, metal, coal, wood, and grain. Dusts do not tend to flocculate except under electrostatic forces; they do not diffuse in air but settle under the influence of gravity.

Dust is a term used in industry to describe airborne solid particles that range in size from 0.1 to 25μ ($1\mu = 1/10,000$ cm $= 1/25,000$ in.) (μ is the abbreviation for micron). Dusts above 5μ in size usually will not remain airborne long enough to present an inhalation problem.

Dust may enter the air from various sources. It may be dispersed when a dusty material is handled, such as when lead oxide is dumped into a mixer or a product is dusted with talc. When solid materials are reduced to small sizes in processes such as grinding, crushing, blasting, shaking, and drilling, the mechanical action of the grinding or shaking device supplies a source of energy to disperse the dust formed.

FUMES. Solid particles generated by condensation from the gaseous state, generally after volatilization from molten metals. This physical change is often accompanied by a chemical reaction, such as oxidation. Fumes flocculate and sometimes coalesce.

A fume is formed when a volatilized solid, such as a metal, condenses in cool air. The solid particles that make up a fume are extremely fine, usually less than 1.0μ. In most cases, the hot material reacts with the air to form an oxide. Examples are lead oxide fume from smelting, and iron oxide fume from arc welding. A fume can also be formed when a material such as magnesium metal is burned or when welding or gas cutting is done on galvanized metal.

Odorous gases and vapors are *not* fumes. Newspaper columnists commonly misuse this term and refer to carbon monoxide as a fume.

SMOKE. Carbon or soot particles less than 0.1μ in size which result from the incomplete combustion of carbonaceous materials such as coal or oil. Smoke generally contains droplets as well as dry particles. Tobacco, for instance, produces a wet smoke composed of minute tarry droplets. The size of the particles contained in tobacco smoke is about 0.25μ.

AEROSOLS. Liquid droplets or solid particles dispersed in air, that are of fine enough particle size to remain so dispersed for a period of time.

MISTS. Suspended liquid droplets generated by condensation from the gaseous to the liquid state or by breaking up a liquid into a dispersed state, such as by splashing, foaming, or atomizing. Mist is formed when a finely divided liquid is suspended in air. Examples are the oil mist produced during cutting and grinding operations, acid mists from electroplating, acid or alkali mists from pickling operations, paint spray mist from spraying operations and the condensation of water vapor to form a fog or rain.

GASES. Normally formless fluids which occupy the space of enclosure and which can be changed to the liquid or solid state only by the combined effect of increased pressure and decreased temperature. Gases diffuse. Examples are welding gases, exhaust gases, and air.

VAPORS. The gaseous form of substances which are normally in the solid or liquid state (at room temperature and pressure). The vapor can be changed back to the solid or liquid state either by increasing the pressure or decreasing the temperature alone. Vapors also diffuse. Evaporation is the process by which a liquid is changed into the vapor state and mixed with the surrounding air. Solvents with low boiling points will volatilize readily.

The safety professional will recognize that air contaminants exist as a gas, dust, fume, mist, or vapor in the workroom air. In evaluating the degree of exposure the measured concentration of the air contaminant is compared to limits which appear in the published standards on levels of exposure. (These are published in the next chapter.)

In addition to the definitions of states of matter, which find daily usage in the vocabulary of the industrial hygienist, other terms used to describe degree of exposure are:

ppm—Parts of vapor or gas per million parts of contaminated air by volume.

mppcf—Millions of particles of a particulate per cubic foot of air.

mg/m³—Milligrams of a substance per cubic meter of air.

micrograms/l—Micrograms of a substance per liter of solution.

Annually the American Conference of Governmental Industrial Hygienists adopts a list of Threshold Limit Values for about 400 substances. (See the listing in the next chapter.) These are expressed as parts of vapor or gas per million parts of air by volume at 25 C and 760 mm Hg pressure (ppm), or as approximate milligrams of particulate per cubic meter of air (mg/m³). Mineral dusts are expressed as millions of particles per cubic foot (mppcf), based on impinger samples counted by light-field techniques.

Another term that is used to designate degree of exposure is percent, a term used to designate the percent by volume of flammable vapors or gases in air. The lower flammable limit of carbon monoxide gas, for example, is 12.5 percent. (This would correspond to 125,000 ppm, a fatal concentration if inhaled.)

One part of carbon monoxide gas in 99 parts of air, resulting in a 1 percent mixture, would be in the safe range in regards to fire hazard but would be deadly (10,000 ppm) in terms of a health hazard. The point here is that in most cases by protecting against the

health hazard, the safety professional would eliminate the fire and explosion hazard.

Dusts

To properly evaluate dust exposures requires knowledge of the particle size, dust concentration in air, how it is dispersed, and many other factors described in this section.

When a solid is broken into finely divided particles, its surface area is increased many times. For example, 1 cu cm (0.061 cu in.) of quartz in the form of a cube when crushed into 1μ cubes will give 10^{12} (1,000,000,000,000 or one trillion) particles with a total surface area of 6 sq m (9300 sq in.), as compared with 6 sq cm (0.930 sq in.) for the original cube.

When a solid is broken into finely divided particles, the volume occupied by the mass is also increased because of the voids between the particles. A dust concentration of 50 million particles per cu ft of air (mppcf), resulting from 1 cu cm of material reduced to particles 1 cu μ in size, will occupy an air space of 20,000 cu ft.

A person with normal eyesight can detect dust particles as small as 50μ in diameter. Smaller airborne particles can be detected individually by the naked eye only when strong light is reflected from them. Dust of respirable size (below 10μ) cannot be seen without the aid of a microscope. Most industrial dusts consist of particles that vary widely in size, with the small particles greatly outnumbering the large ones. Consequently, with few exceptions, when dust is noticeable in the air around an operation, probably more invisible dust particles than visible ones are present.

Dispersion

In order for particulate matter to become airborne, some form of energy is required. A solid or liquid particle of sufficient mass will be thrown a considerable distance if ejected from its source with a high enough velocity. This type of dispersion is known as dynamic projection and is a result of the kinetic energy of the particle's motion. As the mass of the particle decreases, however, a point will be reached when its kinetic energy (which is one-half the mass times the square of the velocity) is too small to overcome air resistance. The particle's forward velocity is thus minimized and it remains suspended in the containing air mass.

As a rough approximation, macroscopic particles (those large enough to be visible to the naked eye) are considered to be dispersed by dynamic projection. Microscopic particles (those visible only through a microscope) are considered to have a mass so small that their movement is dependent on the containing air mass. Contaminants such as the larger dust particles, mists, and sprays, which are dispersed by dynamic projection, can cause external injury such as acid burns, eye damage, and dermatitis. The microscopic particles are dangerous to health if inhaled.

Separation in airborne dusts

Dust in the air may or may not have the same composition as its parent material. The determining factors are the particle size and density of each component in the original mixture, and the hardness of the materials (hard materials will resist the pulverizing action of a mechanical device).

For example, foundry molding sand may contain a large percentage of free silica with a lower percentage of clays. Most of the clays consist of fine particles that can easily become airborne, but most of the free silica particles are too large to remain airborne. The airborne dust, therefore, as compared with the original mixture, may contain a much higher percentage of clays and a much lower percentage of free silica.

Dust particles are, of course, attracted by gravity. Their settling rate through still air will vary with their size, density, and shape. Microscopic particles settle out more slowly than larger particles because of their relatively minor density and Brownian movement. Mineral particles larger than 10μ will settle out relatively fast. The estimated settling rates for silica dusts in still air are given in Table 37–B.

Most of the particles in airborne industrial dusts are small. Because of air currents, the fine particles in dust clouds at an operation will remain suspended in the workroom air for relatively long periods of time. The smaller dust particles, moreover, will travel farther away from their point of origin than

TABLE 37-B

SETTLING RATES FOR SILICA DUSTS IN STILL AIR

Size (μ)	Time to Fall 1 ft (minutes)
0.25	590.0
0.50	187.0
1.00	54.0
2.00	14.5
5.00	2.5

will the larger particles so that the farther dust is from its source, the greater the percentage of small particles it contains. Fig. 37–4 shows a continuous sampler for a worker in an area where silica particles might be present.

Inhalation of particles

With the exception of such fibrous materials as asbestos, dust particles must usually be smaller than 5μ in order to enter the alveoli or inner recesses of the lungs. Although a few particles up to 10μ in size may enter the lungs occasionally, nearly all the larger particles are trapped in the nasal passages, throat, larynx, trachea, and bronchi, from which they are expectorated or swallowed into the digestive tract. When larger particles of certain toxic dusts are trapped in the upper respiratory passages, they can be absorbed by the body fluids in the nasal passages and in the digestive tract before they are eliminated. Hence the final toxic effects of larger dust particles may be delayed. The larger particles of irritant dusts can cause immediate effects in the upper respiratory system.

Ragweed pollen, which varies from 18 to 25μ in diameter can cause hay fever from its action in the upper respiratory system. This type of dust and other allergenic types, as well as bacterial and irritant dusts, can cause difficulty even in the higher airborne sizes.

When dust-laden air is inhaled, some of the larger particles are trapped by the hairs in the nose. Other dust particles are removed from the air as it passes over the moist mucous membranes of the nose, throat, and other

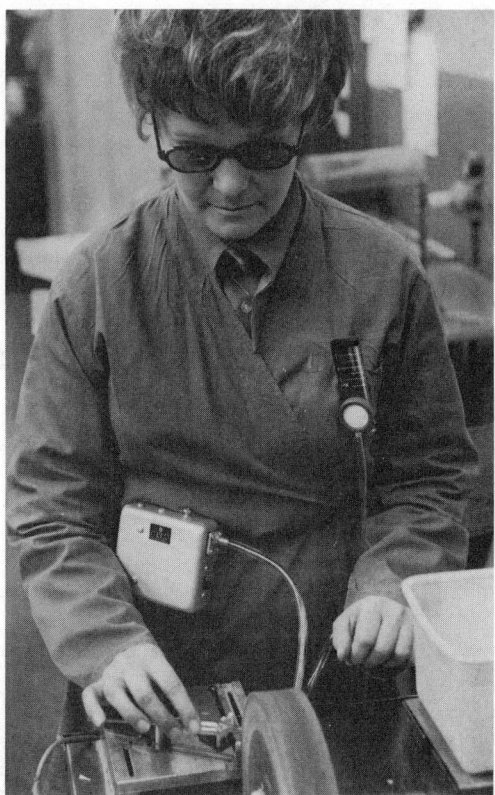

FIG. 37–4.—This self-contained lapel sampler is being used to collect continuous samples from the breathing zone of a worker in an area where silica might be present. Similar samplers are available for monitoring exposure to other harmful particulates.

Courtesy Micro-Chemical Specialties Co.

portions of the upper respiratory system. The bronchi and other respiratory passages are covered with a large number of tiny, hair-like *cilia* or microscopic whiplashes, which aid in the removal of dust trapped on these moist surfaces. The cilia, all bending in one direction, make a fast stroke toward the mouth and a slower return stroke. This action tends to push mucus and deposited dust upward to the mouth so that the particles can be expectorated or swallowed.

Retention of dust

Many studies have been made in an effort to determine the amount of dust that is retained in the lungs, but there is no simple answer to this question. The size of the dust particles, the rate of respiration, the density of the dust in the air, the efficiency of the dust-catching mechanism, and probably many other factors are involved.

Physiological Effects of Dusts, Fumes, and Mists

The physiological reactions caused by the inhalation of airborne particulate matter will vary with different types of dusts, fumes, and mists. The reactions include:

1. The cardiopulmonary reaction which consists of the pneumoconioses, such as *silicosis* and *asbestosis*. In certain cases, specific types of lung pathology result, and the heart may be affected (cor pulmonale) when the fibrosis is advanced. In other cases, there is mainly just an accumulation of a relatively inert dust in the lungs.

2. The systemic reactions which are caused by toxic dusts of such elements as lead, manganese, cadmium, and mercury, by their compounds, and by certain organic compounds.

3. Metal fume fever which results from the inhalation of finely divided and freshly generated fume of zinc or possibly of magnesium or of their oxides. This is a transient condition.

4. Allergic and sensitization reactions which may be caused by inhalation of, or skin contact with, such materials as organic dusts from flour, grains, and some woods and dusts of a few organic and inorganic chemicals. There are also bacterial and fungal infections which occur from inhalation of dusts containing active organisms, such as wool or fur dust containing anthrax spores or wood bark or grain dust containing parasitic fungi. (See the section "Biological Agents and Other Occupational Infections" later in this chapter.)

5. Irritation of the nose and throat, which is caused by acid, alkali, or other irritating dusts or mists. Some dusts such as

1043

soluble chromate dusts may cause ulceration of the nasal passages or even lung cancer.

6. Damage to internal tissues, which may result from inhaled radioactive materials such as radium and its daughter products and from other radioisotopes that emit highly ionizing radiation.

Pneumoconioses

Pneumoconiosis comes from three Greek words that mean "lung," "dust," and "abnormal condition." The present generally accepted meaning of the word is merely "dusty lung." The kind of dust inhaled determines the type of condition or injury. A number of organic dusts are capable of producing lung diseases, but not all these diseases are classified as pneumoconioses because they are not all a "dusty condition" of the lung.

In very rare cases, enough dust has been inhaled to cause mechanical blockage of the air spaces. (Flour dust has been known to cause this condition.) Some dusts may be essentially inert and remain in the lungs indefinitely with no recognizable irritation, and a few (like limestone dust) may be gradually dissolved and eliminated without harm.

Silicosis is the most important lung disease caused by the inhalation of mineral dust. It is well-known in industries where crystalline free silica dust is present, such as foundries, glass manufacturing, granite cutting, mining, and tunneling in quartz rock. It is found throughout the world, and in the past it has had many names, such as miner's asthma, grinder's consumption, miner's phthisis, potter's rot, and stonemason's disease. The same occupational disease, however, is meant by all these names, and it is caused by dust from crystalline free silica, usually quartz.

Silicosis has been defined by the American Public Health Association as "a disease due to breathing air containing silica characterized *anatomically* by generalized fibrous changes and the development of miliary nodulation in both lungs, and *clinically* by shortness of breath, decreased chest expansion, lessened capacity for work, absence of fever, increased susceptibility to tuberculosis (some or all of which symptoms may be present), and by characteristic X-ray findings."

Silicosis has been known to manifest itself after widely differing periods of exposure to silica dust. Apparently, development of the disease depends upon:

1. The amount and kind of dust inhaled

2. The percentage of free silica contained in the dust

3. The form of the silica

4. The size of the particles inhaled

5. The duration of the exposure

6. The powers of resistance of the individual concerned

7. The presence or absence of a complicating process such as infection

Many theories have been advanced over the years to explain why crystalline free silica acts as it does in the lungs. It is now believed that the fibrosis produced is caused not by the hardness or sharpness of the particles, but by a combination of slight solubility with a physiochemical effect and an immunological effect—but no one is certain of the exact mechanism of the disease. Experimental work on the reasons for the development of silicosis is still going on in various parts of the world. If the precise mechanism of silicosis could be determined, better medical preventive measures might be developed and possibly a cure could be found.

Amorphous free silica differs from crystalline free silica in physical structure and in physiological effects. In the amorphous state, molecules of silica are randomly oriented and may be naturally converted to opal and diatomaceous earth (kieselguhr) or artificially converted into such forms as silica gel, silica fume, and fused silica or quartz.

If amorphous silica is heated to a high temperature, as in calcining, forms of crystalline free silica called cristobalite and tridymite result; intermediate forms of amorphous silica are known as crypto-crystalline (ultra-microcrystalline). Inhalation of these crystalline forms can readily cause diatomite pneumoconiosis.

When diatomaceous earth is calcined, particularly in the presence of a trace of alkaline flux, appreciable quantities are converted to cristobalite. As a result of studies made by the Public Health Service, it has been recommended that the threshold limit value for crude or amorphous diatomite be placed at 20 mppcf (million particles per cubic foot), but that the atmospheric concentration for dust containing cristobalite be kept under 5 mppcf.

Various commercial products containing particles of silicia under 1μ in size are available. The physiological effects of these products have not been well defined. Until more experience with human beings is available, exposure to these products should be kept to a minimum through the use of appropriate control measures.

Free silica is uncombined silicon dioxide (SiO_2). Silicates contain silicon and oxygen combined with other elements in more complex molecules. The SiO_2 reported in chemical analyses for mineral and geological reports is the *total* of the silicon dioxide present, both the free silica (if present), and the silica combined in the mineral. Such analyses are not reliable indications of the silicosis potential of the material, because it is the uncombined or free silica that is most important in industrial dust exposure. So that an exposure can be properly evaluated, the percentage of uncombined silica must be determined by petrographic analysis using a polarizing microscope or, preferably, by X-ray diffraction analyses and special analytical chemical procedures.

With the exception of asbestos and some talcs, the silicate dusts do not ordinarily cause a serious disabling lung condition such as is produced by free silica. Much higher levels of silicate dusts than of free silica dust can be tolerated. In many industries, men have worked with silicate dusts that contained no free silica without development of disability or of nodulation in the lungs. The X ray may show shadows indicating dust deposits in the lungs, but the pneumoconiosis is essentially harmless. However, partially disabling pneumoconioses have been reported where men have worked for long periods of time in very high concentrations of certain silicate dusts. Disabling pneumoconiosis from exposure to abnormally high concentrations of mica, tremolite talc, and kaolin dusts have been described in the literature. The clinical signs are not the same for these silicate dusts as for free silica, but the symptoms can be marked.

Asbestosis. Several minerals having a fibrous character are classed as "asbestos"— hydrated silicates of magnesium with variable amounts of iron, calcium, sodium, potassium, and aluminum present as impurities. The fine airborne fibers of asbestos can pass through the upper respiratory tract to the lower parts of the lungs to cause irritation and to form "asbestos bodies" where the fibers are encapsulated. This diffuse fibrosis probably begins as a "collar" about the terminal bronchioles. There is evidence that other minerals having a fibrous character (except glass fiber) can produce a reaction similar to that of asbestos.

An instrument for sampling asbestos fibers in the work atmosphere is shown in Fig. 37–5.

A number of investigators in recent years have reported an association between exposure to asbestos in industry and urban atmosphere and a rise in certain types of lung cancer. As a result of these reported findings, the Division of Occupational Health of the U.S. Public Health Service has undertaken an epidemiological study of the asbestos products industry to: (*a*) refine and update earlier work done in this industry in relation to asbestosis, and (*b*) appraise the relationship of asbestos to the carcinogenic properties suggested by these investigators.

Miscellaneous pneumoconioses. Even though a dust is classified as harmless, amounts above the TLV can lead to trouble by causing a pneumoconiosis, mechanically irritating the walls of the respiratory system, or interfering with ordinary lung processes. Mica dust and kaolin dust are two good examples of dusts that ordinarily are considered benign but amounts above the TLV can cause a troublesome pneumoconiosis. Mica pneumoconiosis has been observed in grinding operations where mica dust, but no free silica was present. There were marked changes in the X-ray pictures of the lungs and some disability. The cases occurred where the dust exposures were massive over

1045

Fig. 37–5a.—Portable pump with intake positioned to gather air samples close to worker's breathing zone. Air samples are collected on filter paper that are later weighed to determine the concentration of lead and other dusts.

Courtesy Mine Safety Appliances Company.

Fig. 37–5b.—Battery-powered vacuum pump draws measured amount of air through a filter that is later examined under a microscope to determine size and number of asbestos fibers in the sample.

Courtesy Raybestos-Manhattan, Inc., Manheim Division and *Millipore Corporation.*

many years.

Industrially important metals and their compounds that can have a toxic effect when the dust or fumes are inhaled include arsenic, antimony, cadmium, chromium, lead, manganese, mercury, selenium, tellurium, thallium, uranium, and a few others.

The effect of some metals, such as magnesium and zinc, appears to be transient. Only limited data are available on the exotic and rare earth metals.

Although the dusts and fumes from metals with low toxicity do not need as much attention as the dusts and fumes from highly toxic metals, they should not be neglected or disregarded. The metals with low toxicity are more readily controlled because greater amounts can be tolerated, but their dusts and fumes should be kept at reasonable levels since excessive amounts of any of them can be harmful.

Inhalation of the dust of lead compounds is the most common mode of entry of lead into the system, and ingestion of lead compounds can add to the problem if personal hygiene is poor. Workers therefore should be encouraged to wash thoroughly before eating, and lunchrooms should be set apart from work areas. Smoking should be prohibited in work areas. Work clothes and street clothes should be stored separately.

Lead is a normal constituent of plants and animals, and people ingest and excrete it daily even though they are not exposed to excessive lead in their daily work. When intake rates exceed the normal excretion rates, however, lead builds up in the body. When the accumulation reaches a sufficient level, symptoms of poisoning or intoxication appear. The concentration of airborne lead should therefore be kept below the threshold limit because of lead's high toxicity and its tendency to accumulate in the human system.

Beryllium intoxication is a severe systemic disease that can result from the *inhalation* of the dust or fume of metallic beryllium, beryllium oxide, and soluble beryllium compounds. (There is no evidence of intoxication from the *ingestion* of insoluble compounds.) Only the inhalation of the beryllium-bearing dusts or fumes produces systemic disease. Accordingly, control of such dusts and fumes to keep them below the concentration specified by ACGIH (American Conference of Governmental Industrial Hygienists) threshold limit values is a basic protective measure.

When the soluble salts of beryllium, especially beryllium fluoride, contact cuts or abrasions on the skin, deep ulcers may be formed which heal very slowly, and complete surgical excision of the ulcer is sometimes required to effect healing.

Metal fume fever is an acute condition caused by a brief high exposure to the freshly generated fumes of metals such as zinc, magnesium or their oxides. Symptoms appear from 4 to 12 hr after exposure and consist of fever and shaking chills. There is complete recovery usually within one day, and ordinarily the employee can return to work without recurrence. However, after a period in which there has been no contact with the fume, for example, after a layoff, resumption of exposure is likely to bring on an attack.

Metal fume fever is caused by heavy concentrations of fumes. Zinc oxide fume is the most common source, but cases caused by the inhalation of fumes from magnesium oxide, copper oxide, and other metallic oxides have also been reported. The condition does not occur from the handling of these oxides in powder form. Apparently, it results only from the inhalation of extremely fine particles freshly formed as fume (called a "nascent fume"). Cadmium, mercury, and other metals may also produce a fever followed by the toxic effects of the element.

Gases

It would seem that everybody is well informed concerning carbon monoxide, the gas which is responsible for more hazardous exposures and incidents than any other, and also the gas responsible for the greatest number of cases of fatal asphyxiation. In areas where there is an active possibility of escape of gases containing carbon monoxide, there tends to be an increasing use of multipoint carbon monoxide alarms. Such installations have been placed in a number of blast furnace plants where carbon monoxide concentrations of the gas are in the order of 26 or 27 percent.

Fortunately, for most fuel gas purposes natural gas is coming into increasing use and

1047

this, of course, contains no carbon monoxide. Furthermore, where in earlier years supplementary gas supplies have been provided through water-gas or producer-gas installations, these are today being replaced by liquefied petroleum gas with no carbon monoxide content. It is, of course, important always to keep in mind that imperfect combustion of natural gas or LP-gas will produce a sufficient amount of carbon monoxide to cause asphyxiation. Cases of carbon monoxide asphyxiation have occurred over the years on Monday mornings where cold systems (such as furnaces) are being heated with natural gas. Here, the impingement of the burning gas on cold surfaces interferes with complete combustion with markedly greater amounts of carbon monoxide in the products of combustion.

Among the gases which might well be listed with those which may cause sudden death is hydrogen sulfide. Hydrogen sulfide is formed wherever there is decomposition of materials containing sulfur under reducing conditions, and cases of multiple fatalities, from two to five, have occurred. Wherever materials containing sulfur are being handled, all workers should be briefed about the rapidly asphyxiating properties of this gas. It should be emphasized that at the higher concentrations the sense of smell does not provide the warning to which one is accustomed on exposure to the lower concentrations of hydrogen sulfide. Workers should know ahead of time that, if one of them should collapse in some confined area where hydrogen sulfide could be the cause, the fellow workers should not enter the confined area to rescue the victim without respiratory protection.

Further information on the effects of exposure to gases is given in the next chapter, "Industrial Toxicology."

Organic Solvents

The widespread industrial use of organic solvents presents a major problem to the industrial hygienist, the safety professional, and others charged with the responsibility for maintaining a safe, healthful working environment. Getting the job done without hazard to employees or property is dependent upon the proper selection, application, han-

dling and control of solvents and an understanding of their properties.

The term "solvent" is meant to include all those organic liquids commonly used to dissolve other organic materials. It includes materials such as naphtha, mineral spirits, gasoline, turpentine, benzene, alcohol, and trichloroethylene.

They are used in the home as dry cleaning agents, paint thinners, and spot removers; in the office as spirit duplicating fluids, typewriter cleaners, desktop cleaners, and for other applications; in the factory as cleaning agents, degreasers and chemical reagents; and in laboratories as drying agents, cleaning agents, and liquid extraction agents.

A good working knowledge of the physical properties, nomenclature, and effects of exposure is very helpful in making a proper assessment of a solvent exposure.

Nomenclature itself can often be misleading. For example, benzine and benzene are different solvents. Although they are often confused, they have greatly different toxic effects. Some commercial grades of benzine may contain benzene as a contaminant.

It is a good policy to verify with direct evidence from the label, see Fig. 37–1, from the manufacturer, or from the laboratory the specific name and composition of the solvents involved. Only after verification of name and composition should one attempt to evaluate the potential effect or hazard of a solvent.

Manufacturers will usually provide information on the composition of their trade name materials if a confidential request is made (see Fig. 37–2).

Labeling

The labeling of solvents to indicate their properties and health and fire hazards is an extremely important method for recognizing and evaluating the hazards. In fact, if a solvent is not properly labeled, it should not be used. The purchasing department can greatly help the industrial hygiene or safety department by notifying suppliers that only properly labeled solvents will be accepted in the plant.

Uniformity in language and layout is desirable to simplify understanding of solvent use. The Manufacturing Chemists' Association *Guide to Precautionary Labeling of Hazardous Chemicals,* Manual L-1, recommends the

TABLE 37-C

MAJOR CLASSES OF COMMON
ORGANIC SOLVENTS

Class	Typical Example	Chemical Formula
Alcohols	Ethyl alcohol	CH_3CH_2OH
Aldehydes and ketones	Acetone	$(CH_3)_2CO$
Glycols and glycol ethers	Cellosolve	$C_2H_5O(CH_2)_2OH$
Aromatic hydrocarbons	Toluene	$C_6H_5CH_3$
Aliphatic hydrocarbons	Mineral spirits	(Mixture of paraffins and cycloparaffins)
Halogenated hydrocarbons	Methyl chloroform	CH_3CCl_3
Esters	Ethyl acetate	$CH_3COOC_2H_5$

following subject matter and organization:

1. Name of product

2. Signal word designating degree of hazard—DANGER! WARNING! or CAUTION! POISON and the skull-and-crossbones should be used for chemicals defined as poisons or required by law to be labeled as such. POISON should be used in addition to the other signal words

3. Statement of hazards (EXTREMELY HAZARDOUS, FLAMMABLE)

4. Precautionary measures to be taken or action to be avoided (avoid breathing vapor, keep away from heat or open flame)

5. Instructions in case of contact or exposure (flush eyes or skin with plenty of water for at least 15 min.)

6. POISON legend, followed by first aid instructions or antidote

A complete list of all solvents used in the plant should be maintained. Where proprietary solvent mixes are utilized, every effort should be made to determine their composition.

With the extensive publicity given to the severe health hazards of carbon tetrachloride, it would seem that everybody knows that this should not be used as a general solvent. How-

ever, the experience has been repeated at times in which a highly toxic solvent has been substituted by one of less toxicity but at some later date the more toxic solvent, desirable for good solvent properties or low price, has been brought back into the plant without the health and safety personnel being aware of its re-introduction.

General physiological effects

Physiological effects from industrial organic solvent exposures come principally from skin contact and inhalation of the vapors. Ingestion with resultant absorption into the digestive tract is not normally considered an exposure hazard in industry. If the solvent is present in such amounts that skin contact becomes a hazard, there will always be an inhalation hazard.

Inhalation of excessive amounts of solvent vapors may produce various physiological effects. In some instances, it may result in impairments, such as lack of coordination, drowsiness, and similar symptoms that may lead to increased accident proneness. Other effects may include damage of the blood, lungs, liver, kidney, gastrointestinal system, and other critical organs or tissues.

Some generalizations can be made concerning the physiological effects of each class of organic solvent shown in Table 37-C, however, specific details for individual solvents are given in Chapter 38, "Industrial Toxicology."

Alcohols are generally anesthetic, irritating to the eyes and upper respiratory tract, and may possess other toxic properties.

Aldehydes and ketones, as a rule, are both irritant and narcotic, with the irritating action predominating in the aldehydes, and the narcotic action being the prime hazard in the ketones. High concentrations of either can be identified by odor and irritant effect.

Glycols and glycol ethers primarily produce kidney damage, as well as narcotic and irritant effects, when present in high concentrations. Such concentrations of these colorless, practically odorless, water-soluble compounds are not easily identified by odor or other properties.

Aromatic hydrocarbons are generally strongly narcotic. Low concentrations of benzene vapor will also damage the blood-forming organs. Toxic concentrations can be present without detection by the senses.

Aliphatic hydrocarbons can cause narcosis in high concentrations, but otherwise are relatively nontoxic.

Halogenated hydrocarbons can have a variety of physiological effects—some are narcotic and highly toxic to the kidneys and liver, while other are almost inactive. Halogenated hydrocarbons can also break down into highly toxic decomposition products in the presence of heat, flame, or a high-energy source such as a welding arc.

Esters can be both irritant and narcotic to some degree. High concentrations of ester vapor usually have distinct irritating effects and odors.

Degree of severity of solvent hazards

The severity of hazard in the use of organic solvents depends on the following factors.

1. How the solvent is used

2. Type of job operation, which determines how the workers are exposed

3. Work pattern

4. Duration of exposure

5. Operating temperature

6. Exposed liquid surface

7. Ventilation efficiency

8. Evaporation rate of solvent

9. Pattern of air flow

10. Concentration of vapor in workroom air

11. Housekeeping

The solvent hazard therefore is determined not only by the toxicity of the solvent itself but by the conditions of its use—who, what, how, where, and how long.

For convenience, job operations employing solvents may be divided into three categories:

Direct contact is a consequence of hand operation. Emergency repair of equipment, spraying or packaging volatile materials without ventilation, cleanup of spills and manual cleaning using cloths or brushes wetted with solvent are examples where employees may directly contact the solvent.

Intermittent or infrequent contact is encountered where the solvent is contained in a semiclosed system where exposure can be controlled. Paint spraying in an exhausted spray booth, vapor degreasing in a tank with local lateral slot exhaust ventilation, charging reactors or kettles in a batch type operation, and transferring liquids to secondary containers are examples of operations where the worker is exposed only at infrequent intervals.

Minimal contact is characterized by remote operation of equipment totally isolated from work area. This type of operation includes directing chemical plant operations from a control room, mechanical handling of bulk packaged materials, and other operations where the solvent is contained in a closed system and is not discharged to the atmosphere in the work area.

"Safety" solvents

It is unfortunate that the term "safety"

solvent has been applied to some proprietary cold cleaners, because the term is not precise and is subject to various interpretations. For example, a "safety" solvent may be considered by some users as nondamaging to the surfaces being cleaned. Other users may consider it to be free from fire or toxicity hazards. Depending on the conditions of use, *neither* of these criteria may be met by a so-called "safety" solvent.

These solvents are prepared as mixtures of halogenated hydrocarbons and petroleum hydrocarbons to be used for cold cleaning. Although the halogenated hydrocarbons are very effective grease and oil solvents and generally have no flash or fire point, they are relatively expensive and may be toxic under adverse conditions. The petroleum hydrocarbons are effective solvents, low in toxicity, and inexpensive, but have flash and fire points.

To combine the best qualities of each solvent, manufacturers mix them in an attempt to produce a cold cleaner that has a flash point higher than that of the flammable petroleum hydrocarbon. Such a mixture, however, can present both fire and toxicity hazards, depending on the evaporation rates of the solvents used: if the flammable liquid is more volatile than the nonflammable solvent, the vapors from the mixture can be highly flammable; conversely, if the non-flammable solvent is more volatile, it can evaporate to leave a flammable liquid.

These considerations should always be kept in mind by users of "safety" solvents.

• The toxicological effects alone, therefore, are not adequate to assess the hazard potential of a solvent.

• The vapor pressure, ventilation, and manner of usage will determine the concentration in air.

• Handling procedures and type of clothing will determine the degree of skin contact and absorption.

• Ignition temperature, flash point, and other factors determining the potential for fire and explosion must also be considered. Concentrations safe from a toxicological viewpoint are much lower than the lower explosive limits of flammable solvents, but concentrations at potential points of ignition may be far higher than concentrations in the user's breathing zone.

Evaluation of hazard potential requires assessing the consequences of exposure, the degree of exposure, and all factors contributing to the exposure.

Physical Energy Stresses

Extremes of temperature, pressure, mechanical vibration, and repeated motion, as well as exposures to radiation and noise, produce stresses in the working environment. The industrial hygienist recognizes and evaluates these as having immediate or cumulative effects on the worker's health and safety. These stresses, therefore, are as important to him as are the chemical and biological hazards.

Heat stress

Probably the most elementary factor of environmental control is control of the thermal environment in which men work. General experience shows that extremes of temperature affect the amount of work which a man can do and the manner in which he does it. The industrial problem is more often that of exposure to high temperatures rather than low temperatures.

The body is continuously producing heat through its metabolic processes. Since the body processes are so designed that they can operate only within a very narrow limit of temperature, they must dissipate this heat as rapidly as it is produced if the body is to function efficiently and well. A sensitive and rapidly acting set of thermostatic devices in the body must also control the rates of its temperature-regulating processes.

Heat stress is a commonly recognized problem as are the problems presented by a very cold environment. Evaluation of heat stress by the interpretation of information relating the physiology of a man to the physical aspects of his environment is not simple or easy. Considerably more is involved than simply taking a number of air-temperature measurements and making decisions on the basis of this information.

One question that must be asked is—are people merely uncomfortable or are conditions such that continued exposure could result in undue stress? Often the problem lies between the two extremes. This makes it difficult for an individual, armed with a clip-board full of data, to interpret that information in terms of how another individual actually feels or is affected.

Man functions efficiently only in a very narrow body temperature range—a "core" temperature measured deep inside the body, not on the skin or at body extremities. Fluctuations in "core" temperatures exceeding about two degrees below, or three degrees above, the normal temperature of 98.6 F markedly impair performance. If this five-degree range is exceeded, a health hazard exists. Fortunately, man has a very efficient thermoregulatory system that functions well in a warm environment, and conditions must be rather severe before the problem becomes critical.

Metabolic processes cause the body to produce heat, even when resting. It is interesting to note that the metabolic heat generated by the average person in his sleep, if transferred with 100 percent efficiency to a pint of water, would bring the water to a boil in less than an hour. Because the body can accumulate heat only to a very limited extent, it must discard most of the metabolic heat being generated to the surroundings. And because fluctuations in body core temperature are so narrow in range, under normal conditions it is not correct to say that the body is warmed or cooled. In reality, the temperature of the surroundings is raised or lowered to control the rate at which the body gives off heat. (Also see Chapter 36, "Exhaust and Ventilation.")

The temperature regulatory mechanism of the body is normally quite effective. However, if it is overburdened or fails to function properly heat stroke can result.

The blood plays an important role in that it carries heat from deep within the body to the skin where the heat can be dissipated by convection, radiation, or conduction. Consequently, a hot environment places an additional load on the cardio-vascular system of an individual, and for many years efforts have been made to develop reasonably simple yet realistic stress evaluation procedures.

Many variables are involved in the exchange of heat between the body and its surroundings. Some of these variables relate to physiological functions, such as capacity for blood circulation to the skin, acclimatization to heat, and ability to sweat. Environmental variables such as air movement, moisture content of the air, and the temperature of surroundings (such as walls) also play an important role. At least 15 such variables have an effect on heat stress, but only a few of the more important ones are incorporated in the various methods proposed for use as heat stress indices.

In many heat stress studies the variables commonly measured are work energy metabolism (often estimated rather than measured), air movement, air temperature, humidity, and radiant heat, if present. Some investigators prefer to evaluate metabolism by taking measurements of pulse rates or sweat loss instead of using average estimated values.

Air movement and air temperature are variables that can be measured with relative ease. Air movement is measured with some type of anemometer and the air temperature with a thermometer, often referred to as a "dry bulb" thermometer. Humidity, or moisture content of the air, is generally measured with a psychrometer, which gives both dry bulb and wet bulb temperatures. Using these temperatures and referring to a psychrometric chart, the relative humidity can be established. The term "wet bulb" is commonly used to describe the temperature obtained by moving air rapidly past an ordinary thermometer having a wetted wick over the mercury well. Evaporation of moisture in the wick, if the moisture content of the surrounding air permits, cools the thermometer to a temperature below that registered by the "dry bulb." The combined readings of the dry bulb and wet bulb thermometers are then used to calculate percent relative humidity.

Radiant heat is a form of electro-magnetic energy similar to light but of longer wave length. Radiant heat (from such sources as red-hot metal, open flames, and the sun) has no appreciable heating effect on the air it passes through, but its energy is absorbed by any object it strikes, thus heating the man, wall, machine, or whatever object it falls upon. Consequently, a man exposed to radiant heat does not get relief from heat stress by using

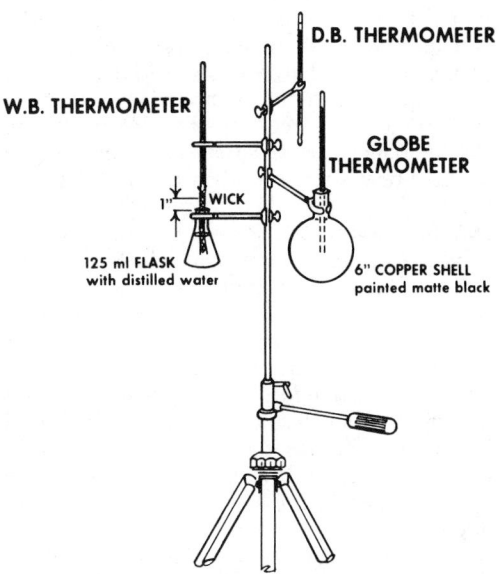

D.B. THERMOMETER

W.B. THERMOMETER

GLOBE THERMOMETER

1" WICK

125 ml FLASK
with distilled water

6" COPPER SHELL
painted matte black

FIG. 37–6.—*Humidity* is calculated by measuring the dry bulb temperature and the wet bulb temperature and using a psychrometric chart. *Radiant heat* is measured by the globe thermometer. *WBGT Index* is calculated from the wet bulb and globe temperatures. The solar load is added from the dry bulb temperature. (See text for details.)

a fan. He must be shielded by a screen that will either absorb or reflect the radiation from the heat source.

An ordinary dry bulb thermometer alone will not measure radiant heat. If, however, the thermometer is fixed in the center of a metal toilet float that has been painted dull black, and the top of the thermometer stem protrudes outside through a one-hole cork or rubber stopper (Fig. 37–6), radiant heat can be measured. This device is known as a globe thermometer. It is effective because the dull, black, outside surface of the float is heated by the radiant heat, and consequently the air inside the globe becomes heated, thus affecting the inserted thermometer. The increase in temperature over a dry bulb reading in the same area can be attributed to radiant heat. At least twenty minutes are required to reach equilibrium conditions and thus obtain an accurate reading with a globe thermometer.

Estimating heat stress

The schemes commonly used to estimate heat stress relate various physiological and environmental variables and end up with one number that then serves as a guide for evaluating stress. For example, the "effective temperature index" combines air temperature (dry bulb), humidity (wet bulb), and air movement to produce a single index called an "effective temperature."

Another index is the Wet Bulb Globe Temperature. The numerical value of the WBGT Index is calculated by the following equations. A record should be maintained of the WBGT Index values found.

1. Indoors or outdoors with no solar loads.

$$WBGT = 0.7 \, WB + 0.3 \, GT$$

2. Outdoors with solar load

$$WBGT = 0.7 + 0.2 \, GT + 0.1 \, DB$$

WB = natural wet bulb temperature

GT = globe thermometer temperature

DB = dry bulb temperature

NIOSH has defined a hot environmental condition as any combination of air temperature, humidity, radiant heat and wind speed that exceeds a Wet Bulb Globe Temperature (WBGT) Index of 79 F. For heat exposure purposes, the proposed standard states that when exposure of an employee is continuous for one hour, or intermittent for a period of two hours, and the time-weighted average WBGT exceeds 79 F and 76 F for men and women respectively, then any one or combination of described work practices shall be initiated to insure that the employee's body core temperature does not exceed 100.4 F. These work practices include acclimatization periods, work and rest regimens, distribution of work load with time, regular breaks of a minimum of one per hour, provision for water and salt intake, protective clothing and application of engineering controls.

Implementation of the standard requires an establishment of the WBGT profile for each work place for winter and summer periods to determine where heat problems exist, and what work practices shall be undertaken to conform with requirements of the proposed standard. The criteria package is only *proposed,* as of the date of publication of this Manual, but it is possible that the final standard may not differ substantially. Therefore, establishments that have suspected or known heat problems should assemble the equipment necessary to determine the WBGT Index, and determine the WBGT profiles for both winter and summer conditions at workplaces.

Experience has shown that men do not stand a hot job very well at first, but develop tolerance rapidly and acquire full endurance in a week to a month.

Atmospheric pressure

It has been recognized from the beginning of caisson work, about 1850, that men working under pressures greater than normal atmospheric are subject to various ills connected with the job. The main effect (decompression sickness commonly known as the "bends") results from the release of nitrogen bubbles into the circulation and tissues during decompression. The bubbles lodge at the joints and under muscles, causing severe cramps. To prevent this trouble, decompression is carried out slowly and by stages so that the nitrogen can be eliminated slowly and without the formation of bubbles.

Deep-sea divers are supplied with a mixture of helium and oxygen for breathing. Since helium is an inert diluent and is less soluble in blood and tissue than is nitrogen, it presents a less formidable decompression problem.

Under some conditions of work at high pressure, the concentration of carbon dioxide in the atmosphere may be considerably increased so that the carbon dioxide will act as a narcotic. Keeping the oxygen concentration high will minimize the condition although not prevent it. This procedure is useful where the carbon dioxide concentration cannot be kept at a proper level.

One of the most common troubles encountered by workers under compressed air is pain and congestion in the ears from inability to ventilate the middle ear properly during compression and decompression. As a result, many workers under compressed air suffer from temporary and some from permanent loss of hearing. The cause of this damage is considered to be obstruction of the Eustachian tubes which prevents proper equalization of pressure on the middle ear.

The effects of reduced pressure on the worker are much the same as the effects of decompression from a high pressure. If pressure is reduced too rapidly, decompression sickness and ear disturbances similar to, if not identical with, the diver's conditions may result.

Men working at reduced pressure are also subject to oxygen starvation, which can have very serious and insidious effects upon the senses and judgment. There is a considerable amount of evidence that exposure for 3 to 4 hr to an altitude of 9000 ft above sea level without breathing an atmosphere enriched in oxygen can result in severely impaired judgment. Even if pure oxygen is provided, the altitude should be limited to that giving the same partial pressure of oxygen as air at 8000 ft.

Reduced pressure is not the only condition under which oxygen starvation may occur. Deficiency of oxygen in the atmosphere of confined spaces is commonly experienced in industry. For this reason, the oxygen content of any tank or other confined space should be checked before entry is made. Instruments such as the oxygen analyzer are commercially available for this purpose. (See Fig. 37–7.)

Normal air contains approximately 21 percent oxygen by volume. The first physiologic signs of a deficiency of oxygen (anoxia) are increased rate and depth of breathing. Oxygen concentrations of less than 16 percent by volume cause dizziness, rapid heart beat, and headache. One should never enter or remain in areas where tests have indicated such concentrations unless he is wearing some form of supplied-air or self-contained respiratory equipment. (Equipment for use in oxygen-deficient atmosphere is discussed in Chapter 19, "Personal Protective Equipment.")

Oxygen-deficient atmospheres may cause inability to move and a semiconscious lack of concern about the imminence of death. In cases of sudden entry into areas containing little or no oxygen, the individual usually has no warning symptoms, but immediately loses consciousness and has no recollection of the incident if he is rescued and revived.

Fig. 37–7.—All confined areas, such as tanks and subsurface structures, must be tested for oxygen deficiency and toxic and flammable gases and vapors. Proper precautions must be taken before anyone is allowed to enter.

Courtesy United States Steel.

Mechanical vibration

A condition known to stonecutters as "dead fingers" or "white fingers," occurs mainly in the fingers of the hand used to guide the cutting tool. The circulation in this hand becomes impaired, and when exposed to cold the fingers became white and without sensation, as though mildly frosted. The condition usually disappears when the fingers are warmed for some time, but a few cases are sufficiently disabling that the men are forced to seek other types of work. In some instances both hands are affected.

The condition has been observed in a number of other occupations involving the use of fairly light vibrating tools, such as the air hammers used for scarfing and chipping in the metal trades. It is particularly prevalent in the cleaning departments of foundries where men have a good deal of overtime work. The condition is produced by vibration while the fingers are held in a strained position and is at least aggravated by chilling of the fingers at the same time. Prevention of this condition is much more satisfactory than treatment.

Preventive measures include directing the exhaust air from air-driven tools away from the hands so that they will not become unduly chilled, use of handles of a comfortable size for the fingers and, in some instances, substitution of mechanical cleaning methods

1055

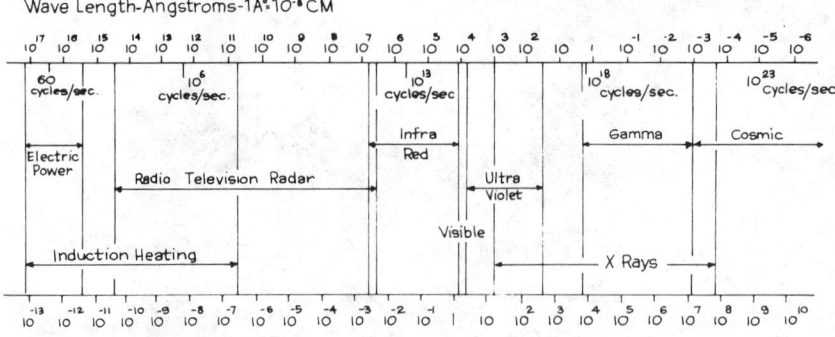

Fig. 37–8.—Electromagnetic spectrum shows energy and wavelength.

for some of the hand methods which have produced most of the cases of "white fingers." In many instances, simply preventing the fingers from becoming chilled while at work has been sufficient to eliminate the condition.

Repeated motion

Repetitive motions or repeated shocks like those in sorting and assembling jobs often cause irritation and inflammation of the tendon sheath of the hands and arms. The condition is generally known as tenosynovitis, and when once established, it is painful and disabling. The usual treatment is rest and diathermy.

The condition results from excessive strain many times repeated. It may occur to an employee who has been working at the same job for years if he is suddenly asked to put in considerable overtime. It is most likely to occur to a new employee or to an employee transferred to a new job. Prevention of the condition is, of course, much more satisfactory than treatment.

In an instance in the printing industry cases of sore hands and wrists were prevented by lowering the surface of the jogging tables 2 in. so that the hands and wrists could fall into a relaxed position at the end of each motion.

The employees should be closely watched so that if they show signs of soreness in the backs of the hands, wrists, forearms, or shoulders they can be transferred temporarily to other work or the job can be changed slightly to reduce the strain.

Nonionizing radiation

Electromagnetic radiation has varying effects on the body, depending largely on the particular wavelength of the radiation involved. Following, in approximate order of decreasing wavelength and increasing frequency, are some hazards associated with different regions of the nonionizing electromagnetic radiation spectrum (Fig. 37–8).

Low frequency. The longer wavelengths —including power frequencies, broadcast radio, short wave radio—can produce general heating of the body. The health hazard from these radiations is very small, however, since it is unlikely that they would be found in intensities great enough to cause significant effect.

Microwaves have wavelengths of 3 m to 3 mm (100 to 100,000 mega-Hertz (MHz) and are found in radar, communications, and diathermy applications. Microwave intensities may be sufficient to cause significant heating of tissues.

Effect is related to wavelength, power intensity, and time of exposure. Generally, the longer wavelengths will produce a greater temperature rise in deeper tissues than the shorter wavelengths. However, for a given power intensity there is less subjective awareness to the heat from longer wavelengths than there is to the heat from shorter wavelengths because of its absorption beneath the body's surface.

An intolerable rise in body temperature, as well as localized damage, can result from an exposure of sufficient intensity and time. In addition, flammable gases and vapors may ignite when they are inside metallic objects located in a microwave beam.

Power intensities for microwaves are given in units of watts per sq cm. Areas having a power intensity of over 0.01 watt per sq cm should be avoided. In such areas, dummy loads should be used to absorb the energy output while equipment is being operated or tested. If a dummy load cannot be used, adjacent populated areas should be protected by adequate shielding.

Infrared radiation does not penetrate below the superficial layer of the skin so that its only effect is to heat the skin and the tissues immediately below it. Except for thermal burns, the health hazard is negligible.

Excessive exposure of the eyes to radiation, mainly visible and infrared radiation, from furnaces and similar hot bodies has been said for many years to produce "glass blower's cataract" or "heat cataract." This condition is an opacity of the rear surface of the lens of the eye. It was first reported among the hand blowers of glass in England in 1903 and became compensable in England in 1908.

An investigation (see Karl L. Dunn in References) conducted in an old plant producing borosilicate glass, where the heat exposure is much higher than it is in the production of the soft glasses on which heat cataract was first observed, failed to show any cases, although many of the employees have been exposed to the work for more than thirty years.

Although this investigation concerned only one company and too small a number of employees to be conclusive, the exposure was at least double that in the soft glass works from which the original reports were made, and the exposures of many of the workers had been long enough to throw a good deal of doubt on the production of the condition in the modern glass industry.

This doubt should not, however, be permitted to obscure the necessity for the provision of proper eye protection for men exposed to high visible glare, which is usually so intense as to require the employee to wear protective goggles.

Routine, low-temperature operations emitting little visible light may be encountered; attention to eye protection must also be given.

Visible radiation and ultraviolet radiation also do not penetrate appreciably below the skin; their effects are essentially heating on the surface. The effects of the ultraviolet waves are much more violent than are those of the visible or infrared—a severe burn can be produced with no warning whatever and significant damage to the lens of the eye can occur from excessive exposure.

Since ultraviolet radiation in industry may be found around electric arcs, they should be shielded by materials opaque to the ultraviolet. Opacity in the ultraviolet has no relation to opacity in the visible part of the spectrum. Ordinary window glass, for instance, is almost completely opaque to the ultraviolet although transparent to the visible. A piece of plastic dyed a deep red-violet may be almost entirely opaque in the visible spectrum and transparent in the near ultraviolet.

Electric welding arcs and germicidal lamps are the most common strong producers of ultraviolet in industry. The ordinary fluorescent lamp generates a good deal of ultraviolet inside the bulb, but it is essentially absorbed in the glass bulb and its fluorescent coating, although mild skin irritation from fluorescent lamps that burned within a few inches of the skin during a full day for several days has been reported.

The most common exposure to ultraviolet radiation is from direct sunshine. Men who continually work outdoors in the full light of the sun may develop tumors on exposed areas of the skin, and these tumors occasionally turn malignant.

Ultraviolet radiation from the sun also increases the skin effects of some industrial poisons. After exposure to compounds such as cresols, the skin is exceptionally sensitive to the sun. Even a short exposure in the late afternoon when the sun is low is likely to produce a severe sunburn. There are other compounds which minimize the effect of UV-rays. Some of these are used in certain protective creams.

Lasers

Lasers emit beams of coherent light of a

single color or wavelength and frequency, in contrast to conventional light sources, which produce random, disordered light wave mixtures of various frequencies. The laser (Light Amplification by Stimulated Emission of Radiation) is made up of light waves that are nearly parallel to each other, all traveling in the same direction. Light-emitting atoms are "pumped" full of energy and stimulated to fall to a lower energy level, giving off light waves that are directed to produce the coherent laser beam.

The maser, the laser's predecessor, emits microwaves instead of light. Some companies call their lasers "optical masers." Proposed uses for the laser include machining and cutting of metals, welding of microscopic parts, and systems for high-capacity communications.

Since the laser is highly collimated (has a small divergence angle), it can have a large energy density in a narrow beam. Direct viewing of the laser should therefore be avoided. The work area should contain no reflective surfaces (such as mirrors or highly polished furniture) for even a reflected laser beam can be hazardous. Suitable shielding to contain the laser beam should be provided.

Potential laser hazards. The hazards associated with laser operations fall into two categories—those from the laser itself and those from associated equipment.

The solid-state lasers present the greatest hazards, because of their high achievable peak powers. Gas and injection lasers, although not as powerful as the solid-state lasers, are capable of producing power outputs that can cause permanent eye damage and skin burns if proper safety practices and procedures are not followed. Injection lasers, however, at this time exhibit low power outputs compared to the solid-state and gas lasers.

The primary hazard associated with laser operations is the laser beam. This beam is capable of inflicting serious injury to personnel, in particular to the unprotected eye.

Experience has shown that specular reflections may approach direct beam intensities. These reflections are difficult to predict and can make off-axis viewing just as dangerous as on-axis viewing.

Most laser energy pump sources require the use of high voltages and should be treated with caution. This component of the laser system can produce serious electric shock and burns.

Ozone may be produced as a result of electrical discharges or ionization of the air surrounding high power laser beams.

Any high voltage equipment (>15,000 volts) is capable of generating X rays. However, presently the majority of lasers used in industrial operations operate within the range of 1 to 4 kv.

The xenon flash lamp used in solid-state lasers produces brilliant white light flashes. This light is capable of producing ultraviolet and infrared radiation. Another danger is the build-up of high pressures of the gases in the flash lamp when it is fired, thus leading to a potential explosion.

Associated equipment. Cryogenic gases such as liquid nitrogen and liquid helium are sometimes used to cool the crystal (ruby, neodymium, etc.) in solid-state lasers. These liquid gases are capable of producing skin burns upon contact. If these gases leak into a closed room, they are capable of replacing oxygen in the atmosphere. This latter condition can produce an oxygen-deficient atmosphere. Flammable solvents and materials may be associated with a particular laser operation and are capable of being ignited by a laser beam.

Biological effects. The eye is the organ most vulnerable to injury induced by laser energy. The reason for this is the ability of the cornea and lens to focus the parallel laser beam on a small spot on the retina.

The fact that infrared radiation of certain lasers may not be visible to the naked eye contributes to the potential hazard.

Lasers generating in the ultraviolet range of the electromagnetic spectrum produce corneal burns rather than retinal damage, because of the way the eye handles ultraviolet light.

Other factors which have a bearing on the degree of eye injury induced by laser light are: (a) the pupil size—the smaller the pupil diameter, the less the amount of laser energy permitted to the retina, (b) the power of the cornea and lens to focus the incident light on the retina, (c) the distance from the source of energy to the retina, (d) the energy and wavelength of the laser, (e) the pigmentation of the subject, (f) the place on the retina where the light is focused, (g) the divergence of the laser light, and (h) the presence of scattering media in the light path.

A discussion of laser beam characteristics and protective eyewear will be found in Chapter 19, "Personal Protective Equipment."

Noise

Noise may be defined as "unwanted sound." It is a form of vibration which may be conducted through solids, liquids, or gases. It is a form of energy in air, invisible vibrations that enter the ear and create a sensation.

The effects of noise on man include the following:

- Psychological effects, for example: noise can startle, annoy, and disrupt concentration, sleep or relaxation.

- Interference with communication by speech, and as a consequence, interference with job performance and safety.

- Physiological effects, for example: noise induced loss of hearing, or aural pain, nausea, and reduced muscular control (when the exposure is severe).

Damage risk criteria. If the ear is subjected to high levels of noise for a sufficient period of time, some loss of hearing may occur. A number of factors can influence the effect of the noise exposure. Among these are:

Variation in individual susceptibility

The total energy of the sound

The frequency distribution of the sound

Other characteristics of the noise exposure, such as whether it is continuous, intermittent, or made up of a series of impacts

The total daily time of exposure

The length of employment in the noise environment

Because of the complex relationships of noise and exposure time to threshold shift (reduction in hearing level) and the many possible contributory causes, establishment of criteria for protecting workers against hearing loss presents many difficulties. An added complication is whether such criteria should be designed to protect against any hearing loss or just loss in the speech frequency range, which is the only portion ordinarily considered for compensation purposes.

An effective hearing conservation program should be undertaken where exposure to industrial noise is capable of producing hearing loss. Such a program should include (a) noise exposure analysis, (b) control of noise exposure, and (c) measurement of hearing.

All aspects of the problem are fully discussed in Chapter 40, "Noise and Hearing Conservation," and in Table 37-G.

Ionizing Radiation

Knowledge of a few simple facts about radiation safety, plus the ability to read a meter and evaluate the dose rate can be of great help to the safety professional in determining, at least, external hazards from radioactive materials, another form of physical energy stress.

Two terms commonly used in working with radioactive materials are the roentgen and the curie.

- *The roentgen* will be defined simply as a unit of radiation, designating biological effects resulting from exposure to X or gamma radiation. Because the roentgen is a rather large unit, it is usually advantageous to divide it by 1000 and call it a milliroentgen.

- *A curie* is a measure of the quantity of radioactive material present. The more curies the more involved is the safety problem. Because a curie of radioactive material represents a considerable amount of radioactivity, it is easier to think in terms of a thousandth of a curie (millicurie), or even a millionth of a curie (microcurie).

It is important that the safety professional keep in mind that a curie relates to the *size* of a source in terms of radioactivity and a roentgen relates to the *penetrating radiation* coming from that source in terms of biological damage.

Five basic safety factors

Ionizing radiation can be a complex subject. At the risk of over-simplifying some basic physical principles and ignoring others, the purpose of this chapter is to present enough information so the safety professional will realize the problems involved and know when to call upon radiation safety experts for help.

There are at least five basic factors that must be considered in such an approach to radiation safety:

- Radioactive materials emit energy which can damage living tissue.

- There are five different kinds of radioactivity and there must be an understanding of their characteristics in order to recognize safety problems. The types of ionizing radiation with which we will be concerned are alpha, beta, X radiation (X ray), gamma, and neutrons.

- Radioactive materials can be hazardous in two different ways. Certain types can be hazardous even when located some distance away from the body. These are *external* hazards. Other types are hazardous only when they get inside the body by virtue of breathing, eating, or through broken skin. These are referred to as *internal* hazards.

- Tools are available for evaluating possible radiation hazards. Meters or other devices are used for measuring radiation levels and estimating doses.

- Specific guides and reference books.

Need for concern

The first factor makes it necessary for people to be concerned about ionizing radiation. The fact that at the time of exposure to ionizing radiation an individual is unable to sense the absorption of energy, with its related tissue damage, seems to give the hazard a mysterious quality which is frightening to some people. This is unfortunate because many radioactive materials are far less hazardous to handle and use than is, for example, the family automobile.

Nevertheless, radioactive materials emit en-

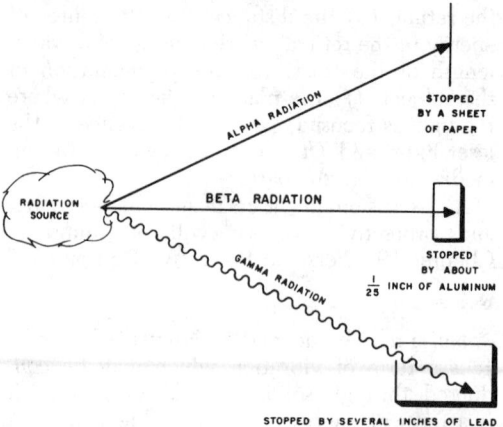

FIG. 37–9.—The penetrating power of alpha, beta, and gamma radiation.

Reprinted with permission from "Atomic Radiation"
© *1957 by RCA Service Co., Inc.*

ergy in the form of one or more of the five types of radiation noted previously. This energy, when absorbed by living tissue, produces damage by a process called ionization.

To understand a little about ionization it is necessary to recall that the body is made up of various chemical compounds which are in turn composed of atoms. Each atom has a nucleus with its own outer system of electrons.

When ionization occurs, some of the electrons surrounding the atoms are forcibly ejected from their orbits. The greater the exposure to ionizing radiation, the more electrons displaced, and the more the physical damage to the cells containing the atoms that have lost electrons. Incidentally, this is the reason why this type of radiation is referred to as "ionizing radiation."

Light from the sun is very similar to X ray and gamma radiation, differing only in wavelength and energy content (Fig. 37–8). However, the sun's energy levels are too low to disturb orbital electrons and consequently light is not referred to as ionizing, even though it has enough energy to cause severe burns over a period of time.

The exact mechanism of the manner in which ionization affects body cells and tissue is complex. However, just as with a moving automobile, radioactive materials possess energy which, if not handled properly, can produce injury.

Five kinds of radioactivity

The five kinds of radioactivity that are of concern are alpha, beta, X ray, gamma, and neutrons. The first four are the most important since neutron sources usually are not used in ordinary manufacturing operations.

At this point, it should be mentioned that the Atomic Energy Commission (AEC) will issue licenses for the use of radioactive materials, or devices that contain radioactive materials (other than radium) only to people well qualified to handle them safely. AEC licensing procedures are very thorough, and the licensee must be aware of the hazards involved and be able to answer questions about safe handling procedures. For this reason, where more than insignificant amounts of radioactive materials, other than radium, are being used, there will be a qualified person in charge.

Of the five types of radiation mentioned, alpha particles are the least penetrating. They will not penetrate anything but the thinnest of barriers. For example, paper, cellophane, and skin will stop alpha particles.

Beta radiation has considerably more penetrating power than alpha. It takes a quarter of an inch of aluminum to stop the more energetic betas (Fig. 37–9). As far as X rays are concerned, virtually everyone is familiar with their penetrating ability and the fact that a barrier, such as lead, is required to stop them. Gamma rays are, for all practical purposes, the same as X rays and require the same kinds of heavy shielding materials. Neutrons are very penetrating and have characteristics that make it necessary to employ shielding materials of high hydrogen atom content rather than the use of mass alone.

Although the radiation from one radioactive material may be the same as that emitted by a different radioactive material, there may be a wide variation in energy. For example, the gamma radiation from potassium-42 has almost four times the amount of energy possessed by the gamma radiation from gold-198.

The amount of energy a particular kind of radioactive material possesses is defined in terms of Mev (million electron volts); the greater the number of Mev, the greater the energy. It is important to understand that each radioactive material emits its own particular kind, or kinds, of radiation with energy measured in terms of Mev. This concept is used in one of the "mental tools" to be discussed later in this section.

External vs. internal hazards

Radioactive materials that emit X rays, gamma rays, or neutrons are external hazards. In other words, such materials can be located some distance from the body and emit radiation that will produce ionization (and thus damage) as it passes through the body. The external types require control of exposure time, working at a safe distance, use of barriers, or, a combination of all three for adequate protection.

As long as a radioactive material that emits only alpha particles remains outside the body it will not cause trouble. Internally, it is a hazard because its ionizing ability through very short distances in soft tissue makes it a veritable bulldozer. Once inside the body—in the lungs, stomach, or an open wound—there is no thick layer of skin to serve as a barrier and damage results. Alpha-emitting radioactive materials that will concentrate in specific parts of the body are very hazardous. Examples are strontium-90 and radium (radium is also a gamma emitter). In terms of actual tissue damage produced by ionization, the effect of external and internal types is the same. The difference lies in the protective measures that must be taken against each kind of hazard.

Because alpha emitters are an internal hazard, one must take precautions against breathing or ingesting them, or contaminating open cuts with them. If there is a chance that alpha-particles can become airborne, the radioactive material emitting them must be handled in a closed locally exhausted system with adequate air cleaning equipment.

Alpha emitters must not be allowed to contaminate food or be handled carelessly so that they can be transferred to the mouth. Good filter-type respirators, designed to protect against radioactive particulates and purchased from a reliable vendor, can be used, but they are not recommended unless there is no other protective device available. Air-supplied respirators offer more positive protection.

If radioactive materials that emit alpha particles are spilled, if they leak out of their container, or if they are involved in a fire, it is easy to spread them over large areas and thus create a major problem. Radium sources can be a serious hazard in this respect.

Beta emitters are generally considered to be an internal hazard although they also can be classed as an external hazard because they can produce burns when in contact with the skin. They require the same precautions as do alpha emitters if there is a chance they can become air-borne. In addition, some shielding may be required.

Evaluating radiation hazards

Many types of meters are used to measure various kinds of radiation. But these meters are useless and possibly hazardous unless they are accurately calibrated for the type of radiation they are designed to measure.

Meters with very thin windows in the probes can be used to check for alpha radiation. Geiger-Muller and ionization chamber-type instruments are used for measuring beta, gamma, and X radiation. Fig. 37–10 illustrates how these meters operate. Special types of meters are available for measuring neutrons.

Devices are available that will measure accumulated amounts (doses) of radiation. Film badges (Fig. 37–11) are used to record the amount of radiation received from beta, X ray, or gamma radiation while special badges are available to record neutron radiation.

Film badges are worn by an individual for a period of time, and depending upon how they are worn, will allow an estimate of an accumulated dose of radiation to the whole body or to just a part of the body, such as a hand or arm (Fig. 37–10).

Alpha radiation cannot be measured with film badges because the alpha particles will not penetrate the paper which must be used over the film emulsion to exclude light.

Another device for measuring accumulated doses of X ray or gamma radiation is the dosimeter, a combination electroscope and ionization chamber which is directly read. The dosimeter requires periodic charging with a battery to return the pointer on the scale to zero.

It is more difficult to measure or estimate

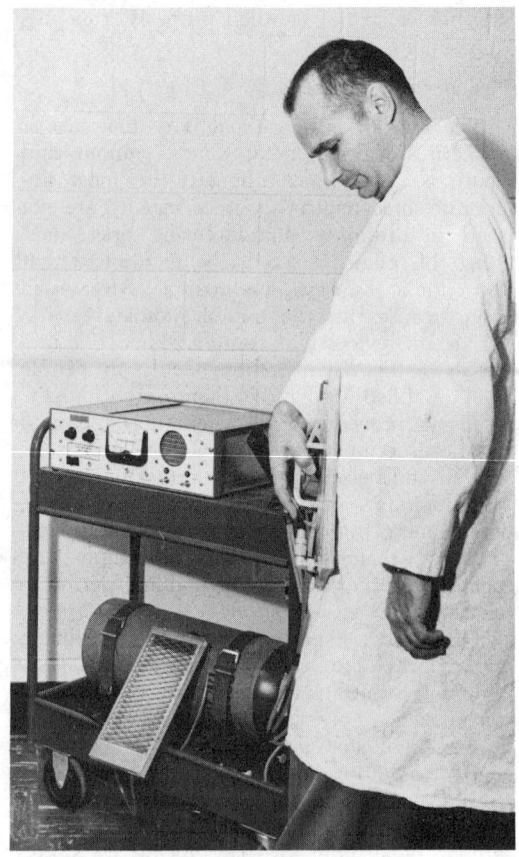

FIG. 37–10.—Mobile personnel monitors at Argonne National Laboratory are provided throughout the facility for personnel working with radioactive materials. The instruments shown are designed to be used by a radiation worker to monitor his hands and clothing.

Courtesy Argonne National Laboratory

internal radiation doses from alpha emitters (Fig. 37–9).

An estimate can be obtained by taking air samples in the breathing zone of the man on the job and measuring the radioactivity in the air samples. The usual procedure, however, is to protect the man in every way possible but still take samples of body wastes and measure them for activity. The type of bioassay used will depend upon the manner in which the particular radioactive element is excreted.

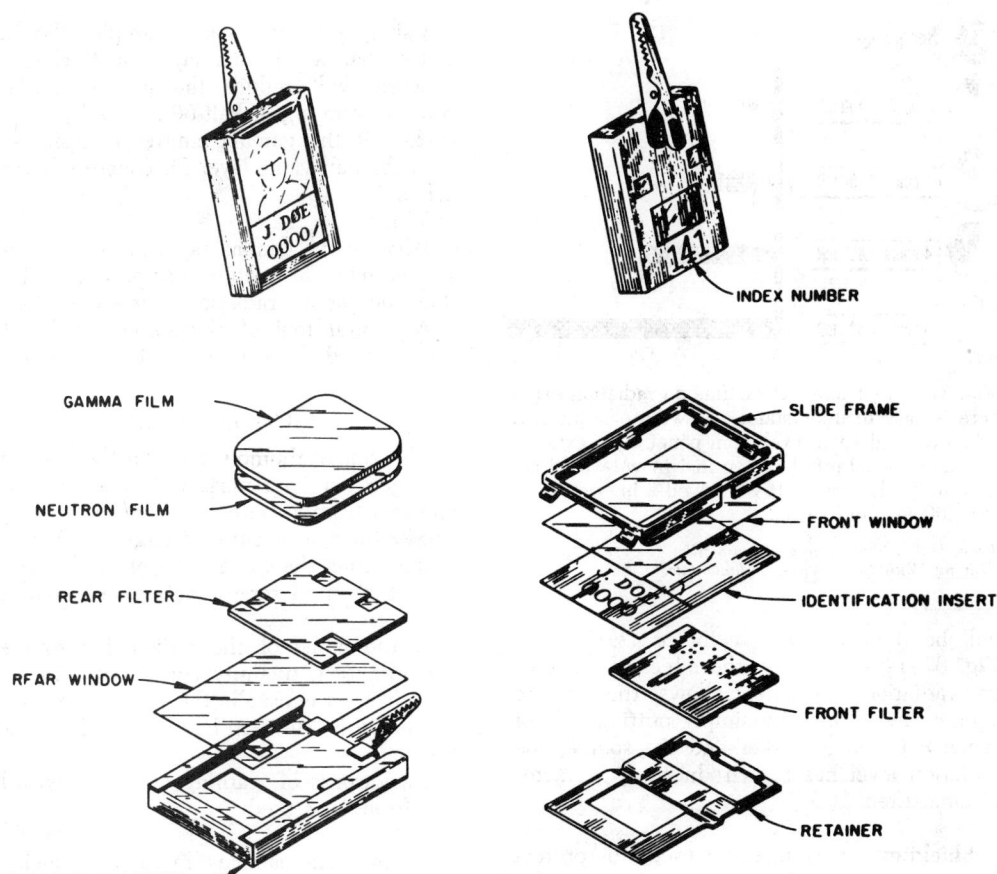

GAMMA FILM

NEUTRON FILM

REAR FILTER

RFAR WINDOW

CASE AND CLIP ASSEMBLY

INDEX NUMBER

SLIDE FRAME

FRONT WINDOW

IDENTIFICATION INSERT

FRONT FILTER

RETAINER

FIG. 37–11.—Exploded view of a radiation film badge. Front view is at upper left, rear view at upper right.

Protecting from radiation hazards

Time, distance, and shielding are important tools involved in handling radioactive materials. Item 4 in the five basic fundamentals refers to them as "mental tools" because some calculations are necessary to use them.

Time as an element of protection is almost self-explanatory. Because radiation occurs at a rate of a certain number of roentgens per hour, the shorter the time of exposure, the smaller the radiation dose received (Fig. 37–12). Work procedures involving the use of radioactive materials should be reviewed carefully to keep exposure time to an absolute minimum. Another example of how "time" can be used is illustrated in Fig. 37–13. As

mentioned in the illustration, it may be possible to simply wait following the spill of a short half-life material until it decays to a safe level, before attempting to clean up.

The use of distance is a very valuable tool. That is why radioactive sources are handled with tongs. The mental tool that is so important in this respect is that the radiation level from a source is reduced by a factor of 1 divided by the square of the distance placed between a worker and the source ($1/D^2$). This is known as the "inverse square law."

If the radiation level is 1.0 roentgen per hour at a distance of 1 ft from a source and you move 2 ft away, the radiation level will drop by a factor of $(\frac{1}{2})^2$ or $\frac{1}{4}$ (0.25 roentgen per hour). If you move 3 feet away the level

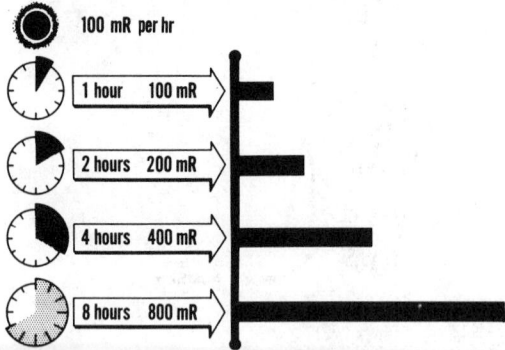

100 mR per hr

1 hour   100 mR

2 hours   200 mR

4 hours   400 mR

8 hours   800 mR

FIG. 37–12.—The effect of time on radiation exposure is easy to understand. If we are in an area when the radiation level from penetrating external radiation is 100 mR/hr, then in 1 hr we would get 100 mr (millirems). If we stayed 2 hr, we would get 200 mr, and so on.

From U.S. Atomic Energy Commission, "Living With Radiation—Fundamentals."

will be reduced by a factor of $(\frac{1}{3})^2$ or $\frac{1}{9}$. Fig. 37–14 shows how the effect of distance on radiation exposure follows the inverse square law. Thus, by simply putting 3 ft of space between a worker and the source, the radiation level has been reduced by a factor of almost ten.

Shielding is commonly used to protect against radiation from radioactive sources. The more mass placed between a source and a person, the less radiation the person will receive. If the mass is concentrated, as in lead, the barrier thickness required for the same degree of protection will be less than it would be for a less dense material such as packed earth.

This introduces another important term— half value layers (HVL). Tables of HVL are given in radiation handbooks and typical values for two radioactive materials, cobalt-60 and cesium-137, are given in this table.

	Cobalt-60	Cesium-137
Lead	0.49 in.	0.25 in.
Copper	0.83	0.65
Iron	0.87	0.68
Zinc	1.05	0.81
Concrete	2.6	2.1

Using concrete as an example, the table states that a 2.6-inch layer or thickness of concrete will reduce the gamma radiation coming from any cobalt-60 source by a factor of ½. If the gamma emitter is cesium-137, then the half-value layer for concrete becomes 2.1 in.

Why?

Because the gamma radiation from cesium-137 is lower in energy (not as many Mev's) than the gamma radiation from cobalt-60.

A mental tool of considerable value that can be used for gamma emitters only is the formula:

$$R/hr/1 \text{ ft} \cong 6 \, CE$$

This rule of thumb states that the number of roentgen per hour measured at a distance of one foot from a source is approximately equal to six times the curie strength (C) of the source, times the energy (E) of the source. It cannot be used for estimating beta radiation levels.

If the name of the radioactive source is known and its quantity (or activity) in curies is known, as is usually the case, the value for its energy in Mev can be obtained from handbooks.

Two words of warning are required when this formula is used:

● Some radioactive materials, such as cobalt-60, emit more than one gamma, each with different energies. The sum of the energy of the total emissions must be used. This kind of information is given in handbooks.

● Terms must be consistent. If the source activity is given in millicuries or microcuries it must be converted to curies. If millicuries are used to work the problem, then the answer is in milliroentgens.

Available guides

The fifth basic concept is that guides setting forth Maximum Permissible Doses (M.P.D.) of ionizing radiation are available. It will not be possible to discuss these guidelines in detail. References that should be studied are *Permissible Dose from External Sources of Ionizing Radiation* (NBS Handbook 59) and *Maximum Permissible Body Burdens and Max-*

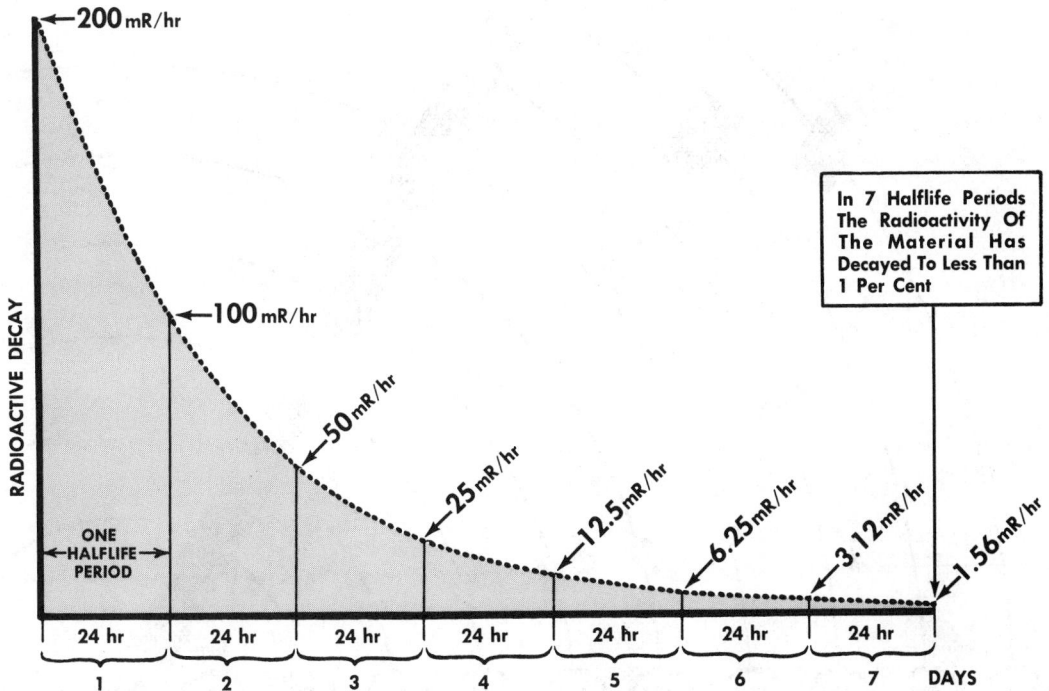

Fig. 37–13.—Decay of a radioactive material with a 24-hr half-life. If a contaminant has a short half-life, it is sometimes simpler to allow the area to remain idle until the radioactivity has died off by the natural process of the passage of sufficient half-lives. If, for example, the reading was 100 mR/hr in a room contaminated with a 24-hr half-life gamma emitter, in 24 hr the reading would be down to 50 mR/hr; in another 24 hr, it would be down to 25 mR/hr; and so on, as illustrated here.

From U.S. Atomic Energy Commission, "Living With Radiation—Fundamentals."

imum Permissible Concentrations of Radio Nuclides in Air and Water for Occupational Exposures (NBS Handbook 69).

It is impossible to discuss all of the important aspects of radiation safety in a "fundamentals manual" such as this. Nevertheless, the safety professional will have a good background for further study and work in the field of radiation safety, if he:

● Treats radioactive materials with respect because he knows that they will emit energy which can be hazardous;

● Recognizes the two distinct types of hazards involved, external and internal;

● Has some concept of the various types of ionizing radiation;

● Knows how to read a meter and is aware

that tools such as shielding, time, and distance exist; and

● Knows where to look for information that will establish guides and limits for him to follow.

In addition to the foregoing, keep the following two important points in mind:

● All exposures to nonmedical ionizing radiation should be held to the lowest possible minimum; and

● Call for help when it's needed.

Biological Agents and Other Occupational Infections

Tuberculosis

Tuberculosis and other infections may be

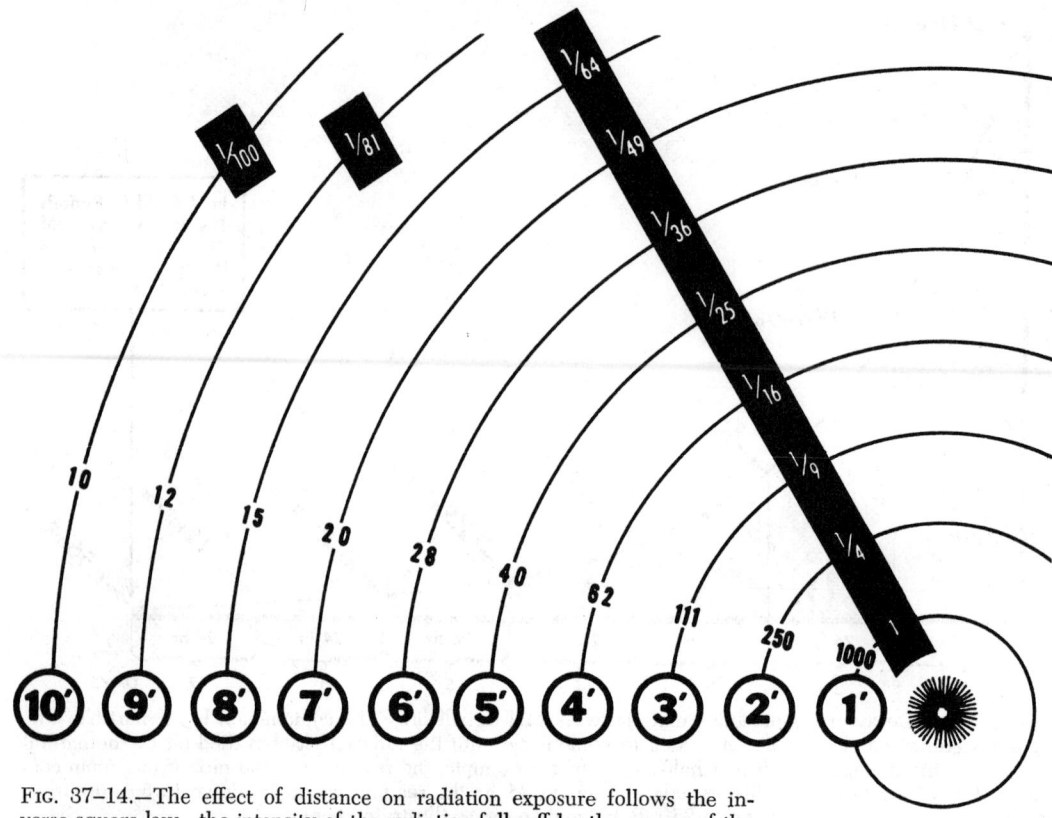

FIG. 37–14.—The effect of distance on radiation exposure follows the inverse-square law—the intensity of the radiation falls off by the square of the distance from the source. If we had a *point* source of radiation giving 1000 units of penetrating external radiation at 1 ft, we would receive only one-fourth as much, or $(\frac{1}{2})^2$, if we double our distance to 2 ft. If we triple the distance, we reduce the dose to one-ninth, or $(\frac{1}{3})^2$.

From U.S. Atomic Energy Commission, "Living With Radiation—Fundamentals."

an occupational disease when they are contracted by nurses, doctors, or attendants caring for tuberculosis patients, or during autopsy or laboratory work where the bacilli may be present.

Fungus infections

A number of occupational infections are common to workers in agriculture and in closely related industrial jobs. Grain handlers who inhale grain dust are likely to come in contact with some of the fungi which contaminate grain from time to time. Some of these fungi can and do flourish in the human lung, causing a condition called farmer's lung.

Fine, easily spread fungi, such as rust and smut, can be contacted once the covering of infected grain is broken and may produce sensitization. Infections from some of these fungi are endemic (of local nature) in the Southwest and are not necessarily industrial.

Byssinosis

Byssinosis occurs in individuals who have experienced prolonged exposure to heavy air concentrations of cotton dust. Flax dust has also been incriminated.

The exact mode of action of the cotton dust is unknown, but one or more of these factors may be important: (*a*) toxic action of microorganisms adherent to the inhaled fibers, (*b*) mechanical irritation from the fibers, and (*c*) allergic stimulation by the inhaled cotton fibers or adherent materials.

It takes several years of exposure before manifestations are noticed.

Anthrax

Anthrax is a highly virulent bacterial infection. In spite of considerable effort in quarantining infected animals and in sterilizing of imported animal products, this disease remains a problem. With prompt detection and modern methods of treatment, it is much less likely to be fatal than it was a few years ago.

"Q fever"

There have been reports of infection—"Q fever"—with a rickettsial organism among meat and livestock handlers. It is similar to but apparently not identical with "tick fever," which has been known for many years. The exact mode of transmission is not known, but there is evidence that "Q fever" may come from contacting freshly killed carcasses or droppings of infected cattle. Probably in the latter case transmission is by inhalation of the infectious dust. Prevention undoubtedly depends primarily upon recognition and elimination of the disease in the animal host.

Brucellosis

Brucellosis (undulant fever) has long been known as an infection produced by drinking unpasteurized milk from cows suffering from Bang's disease (infectious abortion). Since it can also be contracted by handling the animals or their flesh and is also transmitted by swine and goats, it is an occupational ailment of slaughterhouse workers as well as of farmers, although more common among the latter. Here also, preventive measures are primarily proper testing and control of the animals to eradicate the disease.

Erysipelas

The other major bacterial infection to which slaughterhouse workers and fish handlers are subject is erysipelas, which seems to be especially virulent when contracted from the slime of fish.

Upper respiratory tract infections

Workers exposed to dusts from vegetable fibers such as cotton, bagasse (sugar cane residues), hemp, flax, or grain may develop upper respiratory tract infections ranging from chronic irritation of the nose and throat to bronchitis, complicated by asthma, emphysema or pneumonia, or a combination.

Dusts from many of these fibers may also be a source of allergens, histamine, and toxic metabolic products of microorganisms. However, most of the cases of illness arising from handling these fibers have been reported in the foreign literature.

Occupational Skin Diseases

Industrial skin diseases account for about one half to two thirds of all compensation claims for occupational diseases. These claims are scattered throughout all types of industry and sometimes appear where occupational skin diseases may be least expected.

Types of dermatitis

There are two general types of dermatitis: primary irritation and sensitization.

Nearly all persons will suffer primary irritation dermatitis from mechanical agents such as friction, from physical agents such as heat or cold, and from chemical agents such as acids, alkalis, irritant gases, and vapors. Brief contact with a high concentration of a primary irritant or prolonged exposure to a low concentration will cause inflammation. Allergy is not a factor in these conditions.

Sensitization dermatitis, on the other hand, is the result of an allergic reaction to a given substance. The sensitivity becomes established over a relatively long induction period which may be a few days to a few months. In most cases, it is ten days to a month. After the sensitivity has become established, exposure to even a small amount of the sensitizing material is likely to produce a severe reaction. There are various grades of sensitivity, and many people who are sensitized to the materials they are handling can continue to work with them without trouble if they simply take precautions against direct contact. With severe sensitization, however,

(*Text continues on page 1072*)

TABLE 37-D

CHEMICAL CAUSES OF SKIN AFFECTIONS

Irritant or Agent	Primary Irritants	Sensitizers	Manifestations of Irritating Action on the skin (More important damages may result in other organs.)	Typical Occupations, Trades, or Processes Where Exposure May Occur
ACIDS				
Acetic	x		Dermatitis and ulcers	Manufacturing acetate rayon, printing and dyeing, hat makers
Carbolic	x		Irritation and erosion of skin, eczema, anesthesia	Carbolic acid makers, disinfectant, dye, pharmaceuticals, plastic manufacturing
Chromic	x	x	Skin ulcers ("chrome holes"), inflammation and perforation of nasal septum	Platers, manufacturing chemicals and dyestuffs
Cresylic	x		Irritation and erosion of skin, anesthesia	Processing, disinfectants, coal tar, and cresylic acid
Formic	x		Blisters and ulcerations	Formic acid workers, mordanters, makers of cellulose formate and airplane dope
Hydrochloric	x		Irritation and ulceration of skin	Bleachers, picklers (metals), refiners (metals), tinners, chemical makers.
Hydrofluoric	x		Severe skin burns, erosion, ulcers, blisters	Enamel manufacturing, etchers, hydrofloric acid makers, antimony fluoride extractors
Lactic	x		Ulcers (if strong solutions are used)	Dyeing, felt hat industry
Nitric	x		Severe skin burns and ulcers	Nitric acid and soda workers, electroplaters, metal refiners, acid dippers, nitrators
Oxalic	x		Local caustic action on skin, bluish discoloration and brittleness of nails	Tannery workers, blueprint paper makers, oxalic acid makers
Picric	x		Red rash, itching skin, yellow discoloration of skin and hair which is neither a dermatitis nor a dermatosis	Tannery and explosive workers, picric acid and dye makers, dyers
Sulfuric	x		Corrosive action on skin, severe inflammation of mucous membranes	Nitrators, picklers (metals), acid dippers, chemical manufacturing

1068

Substance			Effects	Workers / Industries
ALKALIS Calcium cyanamide	x		Irritation and ulceration	Fertilizer makers, agricultural workers, alkali salt makers
Calcium oxide, carbonate, and hypochlorite	x		Dermatitis, burns, or ulcers	Lime workers, manufacturing of calcium carbonate, soaps, fertilizer
Potassium hydroxide	x		Severe burning of skin, deep-seated persistent ulcers, loss of fingernails	Makers of potassium hydroxide, paper, soap, and lye, electroplaters
Sodium hydroxide	x		Severe burning of skin, deep-seated persistent ulcers, loss of fingernails	Sodium hydroxide makers, bleachers, soap and lye makers, petroleum refiners, mercerizers, tannery workers
Sodium metasilicate	x		Blisters, ulcers	Manufacture of sodium metasilicate and scouring powders
Sodium silicate	x		Thickening of skin, ulcers on fingers	Bleachers, manufacturing cardboard boxes
Sodium or potassium cyanide	x		Blisters, ulcers	Electroplaters, case hardening, gold extraction
Trisodium phosphate	x		Blisters, ulcers	Manufacturing scouring powders and cleansers, industrial cleaning
DYES Including chemicals in dye manufacture		x	Red skin, blisterlike eruptions	Dye workers
INSECTICIDES Arsenic	x		Red skin blisters	Manufacturing and applying insecticides
Chlorophenols (tetra and penta)		x	Red skin, blisters	Manufacturing insecticides, treating wood
Creosote		x	Pustular eczema, black discoloration of skin, warts, epithelioma	Manufacturing wood preservatives, wood-paving blocks, railroad ties, oil-pressed bricks
Fluorides	x		Severe burns, dermatitis	Manufacturing insecticides, and enamel
Phenylmercury compounds		x	Red skin blisters	Manufacturing and applying fungicides and disinfectants
Pyrethrum		x	Red skin, blisters, pimples	Manufacturing and applying insecticides

TABLE 37-D

CHEMICAL CAUSES OF SKIN AFFECTIONS (Concluded)

Irritant or Agent	Primary Irritants	Sensitizers	Manifestations of Irritating Action on the skin (More important damages may result in other organs.)	Typical Occupations, Trades, or Processes Where Exposure May Occur
Rotenone			Red skin, blisters	Manufacturing and applying insecticides
OILS Cashew nut oils		x	Severe dermatitis, as blisters	Handlers of unprocessed cashew nuts
Cutting oils—emulsions or solutions		x	Oil acne, inflammation of hair follicles	Machinists
RESINS Coal tar products (pitch and asphalt)	x	x	Acute dermatitis, "shagreen skin," acne, inflammation around hair follicles, epithelioma, eczema, ulcers	Manufacturing various coal tar products, road making, gas manufacturing
Synthetic resins (phenol-formaldehyde, urea-formaldehyde, cumarone, ester gums, glyptal, vinyl, furfural, cellulose nitrate and acetate)*	x	x	Intensely red and itchy skin	Plastic workers, varnish makers
Synthetic waxes (chloronaphthalenes and chlorodiphenyls)		x	Dermatitis and acne	Manufacturing electrical apparatus, paints, varnishes, and lacquers
SALTS OR ELEMENTS Antimony and its compounds	x		Irritation and eczematous eruptions of skin	Antimony extractors, glass and rubber mixers, manufacturing of various alloys, fireworks, and aniline colors
Arsenic and its compounds	x	x	Darkening of skin, perforation of nasal septum, epithelioma, horny growth or tissue on palm, eczema around mouth and nose (possible loss of nails and hair)	Artificial leather makers, carroters (felt hats), manufacturing insecticides, glass industry and vermicides, manufacturing artificial flowers, calico printing
Barium and its compounds	x		Eczema	Barium carbonate, fireworks, dye, and paint makers

1070

Substance			Symptoms	Occupations
Bromine and its compounds	x		Brownish stain and eruptions on skin	Bromine extractors, bromine salts makers, dye and explosive makers, photographic trades
Chromium and its hexavalent compounds	x	x	Pit-like ulcers on skin, perforation of nasal septum, eczematous eruptions	Chromium platers, dye industry workers, chrome manufacturing, leather tanners
Mercury compounds	x	x	Corrosion and irritation of skin, "mercurial eczema"	Silver and gold extractors, manufacturing electrical appliances, scientific equipment, explosives, hat making
Nickel salts	x	x	"Nickel eczema," (some authorities question whether nickel is the agent responsible.)	Nickel platers, alloy makers
Sodium and certain of its compounds	x		Burns and ulcerations	Bleaching; soap, paper, glass manufacturing
Zinc chloride	x		Ulcers of skin and nasal septum	Manufacturing chemicals and dyestuffs
SOLVENTS Acetone	x		Dry (defatted) skin	Spray painters, celluloid industry, manufacture of artificial silk, leather, acetylene lacquer, varnish, electrical equipment
Benzene, toluene, and xylene	x		Dry (defatted) skin	Chemical, rubber, and artificial leather manufacturing, dry cleaning
Carbon disulfide	x	x	Dry (defatted) irritated skin	Extraction of oils, fats, and many other materials, manufacture of rayon and many cements, germicides, and other materials
Chlorinated phenols	x	x	Severe eruptions	Treating wood
Petroleum distillates	x	x	Acne, epithelioma	Petroleum refiners, machinists, furniture polishers
Trichloroethylene	x	x	Dry cracked skin	Degreasers, paint removers
Turpentine	x	x	Red or blistered skin, eczema	Painters, furniture polishers, lacquerers
SOAPS AND SOAP POWDERS (with excess of free alkalies)			Eczema, blisterlike eruptions, chronic abscesses	Soap manufacturing, dish-washers, scrub-women, soda fountain clerks

° Reactions may be due to resin itself or to presence of plasticizers and modifiers.

it may be difficult or impossible to control the exposure sufficiently to permit the individual to go on working with the material, or even allow the individual to be in the room where the material is handled.

Some substances can produce both primary irritation dermatitis and sensitization dermatitis. Among them are organic solvents, formaldehyde, chromic acid, and epoxy resin systems.

Causes of dermatitis

Causes of occupational dermatitis are classified under these main headings: chemical agents, mechanical agents, physical agents, plant poisons, and biological agents.

Chemical agents are the predominant causes of dermatoses in manufacturing industries. Table 37-D, therefore, is concerned only with these agents. The information given includes the action of the listed chemicals on the skin as either primary irritants or sensitizers. The table is not, of course, exhaustive in any respect, but can provide quick reference data on some of the more common chemical irritants.

Cutting oils and similar substances are especially important chemical agents because the oil dermatitis which they cause is probably of interest to a larger proportion of industrial concerns than is any other single type of dermatitis.

Specific agents responsible for industrial skin diseases can be classified according to how they act upon the skin. The relative importance of these causative factors varies from one industry to another, depending upon the materials and processes used. The determination of specific causative factors in cases of industrial skin disease is, of course, a problem for the dermatologist.

Causative factors are listed as follows:

• Detergents and keratin solvents remove the natural oils from the skin or react with the oils of the skin to increase susceptibility to reactions from chemicals which ordinarily do not affect the skin. Materials which remove the natural oils include alkalis, soaps, and turpentine. Typical materials which increase susceptibility are cottonseed and olive oils, alcohol, phenylene bases, and several of the aromatic hydrocarbons.

• Desiccators, hygroscopic agents, and anhydrides take water out of the skin and generate heat. Examples are sulfur dioxide and trioxide, phosphorus pentoxide, strong acids such as sulfuric, and strong alkalis such as potash.

• Protein precipitants tend to coagulate or callous the outer layers of the skin. They include all the heavy metallic salts and those which form alkaline albuminates on combining with the skin, such as mercuric and ferric chloride. Alcohol, tannic acid, formaldehyde, picric acid, phenol, and intense ultraviolet rays are other examples.

• Hydrolytic materials unite with water of the skin and thus form irritating compounds or dissociate into irritating elements when in contact with water. Heat is liberated on the skin during the reaction. Examples of such substances are mustard gas, ammonium nitrate, and hexamethylenetetramine.

• Oxidizers units with hydrogen and liberate nascent oxygen on the skin. Such materials include nitrates, chlorine, iodine, bromine, hypochlorites, ferric chloride, hydrogen peroxide, chromic acid, permanganates, and ozone.

• Solvents extract essential skin constituents. Examples are ketones, aliphatic and aromatic hydrocarbons, halogenated hydrocarbons, ethers, esters, and certain nitro compounds.

• Keratogenic and neoplastic agents act by stimulating abnormal growth in the outer layer of the skin and may cause either malignant or benign tumors. Examples are arsenic, coal tar products, petroleum and shale, and some coal tar bases such as 3,4-Benzpyrene. (Aniline does not cause cancer.)

• Biotic agents are microorganisms and parasites (vegetable and animal) which act on the skin to produce dermatitis, such as athlete's foot.

• Allergic or anaphylactic proteins stimulate the production of antibodies which cause skin reactions in sensitive persons. The sources of these antigens are usually cereals, flour, and pollens, but may include feathers, scales, flesh, fur, and other emanations.

• Reducers react with water on the skin and,

if the solution is strong, cause shedding of the horny layers. Examples are photographic developers, tar, phenols and naphthols, aromatic and aliphatic hydrocarbons, salts of titanium and hydroquinone, resorcinol, formalin, paraldehyde, and salicylic, formic, and oxalic acids.

Mechanical causes of skin irritation include friction, pressure, and trauma, which produce cuts, which may become infected with either bacteria or fungi, and callosities. If the horny layers of the skin become softened by high temperatures and excessive perspiration, the development of friction (for example, between the buttocks) then leads to dermatitis. Other examples of mechanical causes are abrasions.

Physical agents which lead to occupational dermatitis include heat, cold, water, sunlight, X rays, ionizing radiation, and electricity. Hot water softens the skin so that substances more readily attack it, as typified by dermatoses occurring among laundry workers and dish washers. X rays and other ionizing radiation may cause dermatitis, severe burns, and even cancer. Prolonged exposure to sunlight produces skin changes which may cause dangerous body alterations. Frostbite produces chilblains.

Plant poisons which can cause dermatitis are produced by several hundred plants. The best known are poison ivy, poison oak, and poison sumac. Dermatitis from these three sources may result from bodily contact with any part of the plant, exposure of any part of the body to smoke from the burning plant, or contact with clothing or other objects which have previously been exposed to the poison.

Biological agents which cause dermatitis may be bacterial, fungal, or parasitic. Anthrax from handling hides, tularemia from handling skins, erysipeloid from handling animal products, boils and folliculitis caused by staphylococci and streptococci, and general infection from occupational wounds are probably best known among the skin infections resulting from bacteria. Skin infections also occur among butchers and persons who must handle cadavers.

Fungi cause athlete's foot and dermatitis among kitchen workers, bakers and fruit handlers, fur, hide, wool handlers, and sorters, barbers, and horticulturists. Parasites cause grain itch and ground itch and often occur among handlers of grains and straws, and particularly among farmers, laborers, and miners, fruit handlers, and horticulturists.

Physical examinations

Preplacement examinations will help identify those people who may be especially susceptible to skin irritations. The physician in charge of the examination should be given detailed information on the type of work for which the applicant is being considered. If the work involves exposure to skin irritants, the physician should determine if the prospective employee has deficiencies or characteristics which are likely to predispose him to dermatitis.

Routine use of pre-employment patch tests to determine sensitivity to various materials is not recommended. Patch tests will not reveal whether new workers will become sensitized to certain materials and develop dermatitis, but only whether people who have previously worked on similar jobs are or are not sensitized to the chemicals with which they worked. Patch tests, moreover, are potentially harmful because they may induce sensitivity in persons who have never handled the materials.

When dermatitis suddenly develops among individuals on a job or an operation in which persons have previously worked for some time without skin irritation, they should be sent immediately to the medical department or the company physician for examination and tests to determine whether they have acquired a sensitivity to the substance or substances they are handling. If a sensitivity has developed, the doctor may decide that the affected workers should be transferred to other jobs until the hazard can be eliminated. A study of the particular processes and materials should then be made to determine the offending substances, and after they have been located, appropriate preventive steps should be taken.

Eliminating contact with irritants

Before new processes are introduced and prior to the adoption of new or different chemicals in an established process, possible dermatitis hazards, including those which may be caused by trace impurities, should be carefully considered. Only a highly skilled and experienced chemist can make analyses for trace impurities, and often this requires highly specialized equipment and techniques. Such analyses are justified, however, by the possibility that considerable benefit may be derived from them. Once the dermatitis hazards have been determined, suitable engineering controls should be devised and built into the processes or operations.

Exposure to the causes of industrial skin diseases varies considerably from one industry to another. Therefore, although specific preventive measures can be outlined, it is advisable to solve each exposure problem individually after complete information on the conditions surrounding the job has been obtained.

The type and quantity of skin irritants used in various industrial processes affect the degree of control which can be obtained, but the primary objective in every case should be *complete elimination* of skin contact. The preventive measures discussed in the section, "Controlling Environmental Hazards," at the end of this chapter, can all be adapted to industrial dermatitis control. These measures include substituting less toxic materials, changing the process, wet methods, ventilation, protective ointments, and personal protective equipment.

Some processes may require special control. For example, it is possible to reduce exposure by lowering the temperature of the process and decreasing the air motion around the operation. (Under some conditions, however, decreased air motion may promote rather than reduce exposure. It should be kept well in mind, therefore, that every exposure may require special study.)

Hygienic measures

Careful supervision of the personal cleanliness of workers exposed to skin irritants is essential to the prevention of dermatitis. Supervision is most successful where enough convenient and efficient washing facilities are available to serve all the employees exposed.

Workers should be told where, how, and when to wash, and should be advised that they will be rated on this part of their job performance. They should be required to change their clothing and take a shower before leaving the plant at the end of the shift.

For many exposures, frequent washing alone is a successful preventive measure, particularly where the dermatitis is caused by mechanical plugging of the pores, such as from dust. In all cases, however, the use of large quantities of water on the skin following exposure to irritants is necessary.

It may be advisable in some instances to use neutralizing solutions after a thorough flushing with water. However, since some neutralizing solutions are themselves irritants, they should be applied only upon the advice of a doctor.

The type of soap used is important; even a generally good soap may cause irritation on certain types of skin. The choice of a good soap is best left to the medical department or other qualified department. In particular, soap should:

1. Be freely soluble in hard or soft, cold or hot water

2. Remove fats, oils, and other soil without harming the skin

3. Leave the natural fats and oils in the skin

4. Contain no harsh abrasives or irritant scrubbers

5. Be handy to use if in cake form or flow easily through soap dispensers if in granulated or powder form

6. Stay insect-free

7. Retain its properties in use

A large number of cases of industrial dermatitis are reported to be caused not by substances used in processes, but by materials used to *remove* those substances. An untrained worker may wash his hands in those cleaning agents which are most available and work the fastest—often dermatitis-producing solvents. To combat this practice, a sufficient number of conveniently located waterless hand cleaning stations containing properly selected cleansing agents should be installed in on-the-job washing areas and in shower rooms. Where soap-dispensing units are furnished, workers should be required to use them.

Sometimes it will also be necessary to use emollient cream to replace natural oils and fats removed from the skin.

Evaluating the Hazard

Evaluation of environmental exposures requires teamwork by several professional disciplines: no single group can solve all the many problems presented by occupational disease. For example, the engineer ordinarily collects atmospheric samples in the field for subsequent chemical analyses. In many cases, however, the chemist must go into the field to assist with the collection of samples and to advise on the chemical aspects of the particular problem under study.

The evaluation of an industrial health hazard requires some knowledge of toxicology and an understanding of the concepts involved in the use of threshold limit values and maximal acceptable concentrations.

Information required first

Environmental investigations are a means of determining a five-dimensional pattern. The three dimensions of space determine simply *where* it is. By adding a fourth dimension of time, we determine *"when it is where."* The fifth dimension is man's movement within space-time which recognizes that man's exposure depends on *"where* he is *when."* While the integration and expression of all this is not a simple matter, it is the best definition of the probability of exposure in a given set of circumstances. This concentration, space, time, man movement pattern is not a constant but a high variable circumstance. If we safety professionals are concerned with a work environment, we can make investigations of the work circumstance, but this represents less than a third of man's environmental exposure. What of his off-the-job habits and indulgencies? It is difficult for us to measure these off-the-job exposures. We cannot ignore them however, in our over-all consideration of the interaction of man and his environment. We can get some clues from biochemical monitoring of the man and from what he himself can tell us.

Information on the quantitative and qualitative aspects of the interaction of man with the environment can be found in scientific publications, product brochures, labels, and *Hygienic Guides* put out by the American Industrial Hygiene Association, "Chemical Safety Data-Sheets" published by the Manufacturing Chemists' Association, and "Industrial Safety Data Sheets" published by the National Safety Council.

If you do not find the information you need, ask your supplier. If he cannot give you the information get a supplier who can. Use caution in applying the information from one supplier to the product of another supplier. If your supplier simply says, "We don't have any trouble handling it," ask him how he handles it so he doesn't have any trouble. You can handle any substance safely if you know how to handle it properly and train your personnel to do so. Any person who is responsible for the maintenance of a healthful environment should be thoroughly acquainted with the substances or energies which may be encountered in the industrial environments for which he is responsible.

In the area of occupational health, standards of environmental quality are a useful means of expressing and communicating the summation of experimental information and experience.

Many years ago, the American National Standards Institute set up Committee Z37 to establish standards of quality for work areas where toxic dusts and gases were of concern. These early standards were expressed as "Maximal Allowable Concentrations" and later as "Maximal Acceptable Concentrations," or just MAC. More recently, the American Conference of Governmental Industrial Hygienists (ACGIH) established what they called "Threshold Limit Values" or TLV. The TLV not having a "C" listing is a time-weighted average for an eight-hour work day. A time-weighted average is an attempt to summate the potential exposure in a variable work environment. The ACGIH has outlined the calculation and use of the time-weighted average in their publications on the TLV. For details, see the next chapter, "Industrial Toxicology."

The Z37 Committee of the American National Standards Institute has developed a number of innovations in concept.

In the past, a standard was conceived as a single limit, threshold, or boundary between what is healthful and what is not healthful. But there is no such single boundary. The pattern of exposure in space and time may be variable. Therefore, multiple boundaries are needed. As the boundaries are not sharp lines between what is healthful and what is not healthful, we need an explanation of the basis and significance of each of these boundaries which must act as guide lines in maintaining a healthful work environment. The first innovation the Z37 Committee established was the recognition of several boundaries for each substance which they termed "acceptable concentrations" or just AC.

But what are the different acceptable concentrations most useful in environmental control?

Let us look at them.

● First, an acceptable ceiling concentration for protection of health, assuming an eight-hour work day.

● Second, an acceptable time-weighted average for protection of health, assuming an eight-hour work day. This is within the limits of a specified ceiling indicated as the first acceptable concentration. This is also limited by defined peaks beyond this ceiling.

● Third, acceptable maximum for peaks above the acceptable ceiling for an eight-hour work day.

In a particular operation there may occasionally be an excursion of concentration beyond the acceptable ceiling for a brief period of time. There are many substances where such excursions would be quite acceptable if properly defined. This assumes that the exposure is otherwise limited to the acceptable ceiling and acceptable time-weighted average. Also included are "any other level or levels considered by experts to be pertinent to a particular substance under consideration."

Under general properties, information is given to indicate the minimal level for sensory detection and to indicate what might or might not be its usefulness as a warning of significant exposure.

The purpose of these multiple boundaries is to give the maximum assistance to the man who must determine that a specific design and operation is compatible with a healthful environment. These standards should have the breadth to be informative and, at the same time, a pattern which allows for dynamic growth in concept as well as a change in numbers.

The standard for each substance should contain sufficient information to clearly indicate the basis upon which each level is determined, the consequences of overexposure, and the relative adequacy of the information. There are instances where adequate information is not available. If so, this should be clearly stated. Many useful standards of the past have been guesses by experienced people. These are perfectly in order until such time as better information is available, but the fact that they have this limitation should be clearly stated and understood.

Not only must we constantly search for a better understanding of man's response to his environment, but we must also realize that the application of any standards as guidelines for control of the work environment will always require the good judgment of a competent and well informed man.

Defining the Exposure

We have discussed the nature of the information upon which one must base his judgment and monitoring of the work environment. Let us consider some of the philosophy of defining the potential for exposure in a work environment.

To determine exposure levels to atmospheric contaminants, as well as to measure physical energy stresses, the industrial hygienist uses either direct-reading instruments or laboratory analytical methods.

The particular exposure can be evaluated by chemical analyses of air samples, and a step-by-step analysis of the operations must be made to find the areas where employees are exposed to hazardous amounts of the material. The operational analysis also should determine how the material becomes airborne and is dispersed.

The type and extent of controls will depend upon the evaluation made of the exposure, and the operation that disperses the contaminant. The extensive controls needed for lead

Fig. 37–15.—Industrial hygienist with velometer (left) checks air flow over a paint dip tank. His colleague checks for flammable vapor.

oxide dust, for example, would not be needed for limestone dust, since much greater quantities of limestone dust can be tolerated.

Sampling techniques

What must be measured and evaluated in industrial hygiene sampling is usually the dosage the worker receives—the amount of toxic material he takes into his body. Duration of exposure and concentration and type of material are important. For the concentration measurement, a test can be made on the air in the worker's breathing zone. (*a*) In some cases, the measuring instrument can be attached directly on the worker and he can carry it throughout the working day (this is done with a film badge for measuring radiation exposure). Relatively few samplers are available at the present time for taking continuous samples in the breathing zone. (See Figs. 37–4 and –5.) (*b*) The other technique is to measure concentration in the breathing zone over carefully selected intervals, then average the results and weight each measurement according to the length of exposure (Fig. 37–15). The result is a time-weighted average concentration.

A satisfactory average exposure measurement during a particular time period, is affected by changes from day shift to night

1077

TABLE 37-E

SAMPLING PROCEDURE FOR INDUSTRIAL HYGIENE SURVEYS

CHEMICAL HAZARDS
Procedures to evaluate the air-borne chemical hazards in the work environment may be classified as follows:

Gases and Vapors	a) May be determined by use of approved calibrated field indicator tubes yielding direct readings.
	b) May be collected in containers or absorbed on charcoal for laboratory evaluation.
	The standard method for organic vapor will be absorption on charcoal and chromatographic determination.
Fumes and Mists	a) May be absorbed and measured in the field.
	b) May be absorbed and evaluated in the laboratory.
	c) May be collected on filter media and analyzed in the laboratory.
Dusts	a) May be collected by a personal air sampler, fractionated into respirable size by a cyclone separator and the fractions weighed to determine the concentration.
	b) May be collected on an open-faced filter and weighed.
	c) May be collected in an appropriate manner and counted.

PHYSICAL HAZARDS
Most physical hazards can be evaluated in the field by direct-reading instruments.

---

a) Pressure may be measured barometrically.
b) Temperature may be measured by thermometer, thermocouple or radiometer. Determination of heat stress, however, requires in some form the measurement of evaporation rate. Usually it is inferred from humidity and air velocity.
c) Ionizing radiation may be measured by survey meter, personal dosimetry, or film badge techniques.
d) Noise levels may be measured with sound level meters or octave band analyzers. Vibration may be determined with additional sound level equipment.
e) A number of direct-reading meters are available for various sorts of non-ionizing radiation.

Typical sampling procedures used in industrial hygiene surveys. *Material is adapted from the* OSHA Field Operations Manual.

---

shift, from winter to summer, routine changes in process operations, or daily changes in rate and direction of air movement, temperature, or altitude. Then too, will the presence of the sampler alter the routine of the operation in any manner? Is the contaminant soluble or insoluble in body fluids? What is its particle size distribution and in what portion of the respiratory tract is it likely to deposit? What is its chemical nature? Organic mercury acts quite differently from mercury vapor, which, in turn, is quite distinct from mercury oxide. Finally, are there limitations on the peak dose rates as

there are for beryllium and some other compounds.

If the samples are collected at the right places and combined with a time study of the worker's activity, they can give a measure of the worker's intake of the toxic substance.

Whether direct-reading or laboratory techniques are to be used is a matter of professional judgment. The results obtained must be compared with a threshold limit value with a full knowledge of its limitation and consideration of other factors in the situation. Urine assays add to the facts available for making an evaluation of exposure. These, in turn, must be combined with knowledge obtained by the physician directing an adequate medical program. All this serves to emphasize the basic teamwork approach of the industrial health program.

An outline of sampling procedures is given in Table 37-E.

Air sampling for time-weighted average concentrations. OSHA Compliance Officers use a number of criteria for air sampling:[*]

● Each sample must relate directly to the exposure of one employee or more. The sample must usually be taken in the breathing zone of a particular employee.

● The volume sampled must be sufficient to permit a good determination of the contaminant sampled if it is present at a concentration equal to the TLV.

● The sampling period must usually be such as to give a direct measure of the total exposure of the employees concerned in a full shift.

● The sample must be sealed and identified prior to shipment to the laboratory so that it will be possible to identify positively the laboratory determination with the time and place of sampling and with the individual who took the sample.

There are various methods for meeting the first criterion. The most satisfactory is the use of a personal sampler with the sampling head as close as is conveniently possible to the employee's head.

Area samples, taken by setting the sampling equipment in a fixed position in the work area, are useful as an index of general contamination, but the actual exposure of the employee at the point of generation of the contaminant frequently will have an exposure greater than is indicated by an area sample.

In adopting the Threshold Limit Values of the American Conference of Industrial Hygienists (see Table 38-E in the next chapter), OSHA also adopted the concept of the time-weighted average concentration for a work-day. The same concept is contained in a more specific form in the "Occupational Health and Safety Standard," *Federal Register* dated Aug. 13, 1971, page 10,506A, Table G-2. This concept permits limited excursions of concentration above the listed TLV provided that they are compensated by equivalent excursions below the listed value. Limitations on this concept are the listed values preceded by a "C" (for ceiling) in the table. "C" values may not be exceeded at any time, and any excursion above the listed value constitutes a violation. The other limitation is a limitation on the size of the excursion. For the TLV's, the permissible size of the excursion is given in the following table:

TLV	0–1	1–10	10–100	100–1000
factor	3	2	1.5	1.25

That is to say, if the TLV is 0.1 (either in mg/m^3 or in ppm), an excursion to 0.3 for 10 or 15 minutes is permissible provided that it is compensated by an excursion to 0.03 for 43 minutes or, in general—

$$C_a T_a + C_b T_b + \ldots C_n T_n = 8 \text{ TLV}$$

where:
T_a is the first time period during the shift
C_a is the concentration in period "a"
T_b is another time period during the shift
C_b is the concentration during period "b"
T_n is the nth or final time period in the shift, and
C_n is the concentration during period "n"

This is simply the summation throughout the workday of the concentrations encountered in each time interval and adjusted to an eight-hour standard work-day. In practice it may be determined by estimating the concentration for each time interval, and multiplying by the interval in hours; these products for all of the intervals in the work-day added and divided by the eight hours of the standard work-day should give a quotient not larger than the TLV.

To meet the test of measuring the time-weighted average concentration the sampling method and period should be chosen to average out these permissible fluctuations and

[*] The following method of obtaining air samples was adapted from *Field Operations Manual, OSHA-2006*, January 1972, issued by United States Department of Labor, Occupational Safety and Health Administration.

also to measure a representative sample of the fluctuations that will occur in a day's work. If the operation is repetitive, the sampling period should be long enough to cover at least two full cycles of operation and generally not less than 30 minutes. If there are wide fluctuations in concentration, the long-term samples should be supplemented by samples designed to catch the peaks.

If the exposure being measured is from a continuous operation, it will be necessary to follow the particular operator through two cycles of operation, or through a full shift if operations follow a random pattern during the day. For operations of this sort it is particularly important to find out what the workers do when the equipment is down for maintenance or process change. Such periods are frequently also periods of maximal exposure.

Types of measuring instruments

The most numerous environmental hazards are chemical and can be conveniently divided into (*a*) the particulates and (*b*) the gases or vapors. Particulates are mixtures or dispersions of solid or liquid particles in air and include dust, smoke, mist, and similar materials. Some of these are practically insoluble in body fluids, and exert their effect directly in the lungs while others are soluble and produce systemic poisoning.

Direct-reading instruments for the insoluble materials are difficult to make since such an instrument must count or otherwise determine the number of particles suspended in air, therefore some information regarding particle size is desirable. Direct-reading instruments for this purpose do exist, but they are very expensive and often quite bulky. These usually work on the principle of light-scattering: measuring the light scattered at right angles to the beam.

Gases and vapors can be determined more readily by direct-reading instruments since they form more homogeneous mixtures with air, and separation from the air mass is often unnecessary. This group includes such common materials as carbon monoxide, mercury vapor, hydrogen cyanide, benzene, carbon tetrachloride, and many others. A choice can be made here between direct-reading instruments or samples collected for subsequent laboratory analysis. (See Fig. 37–15.)

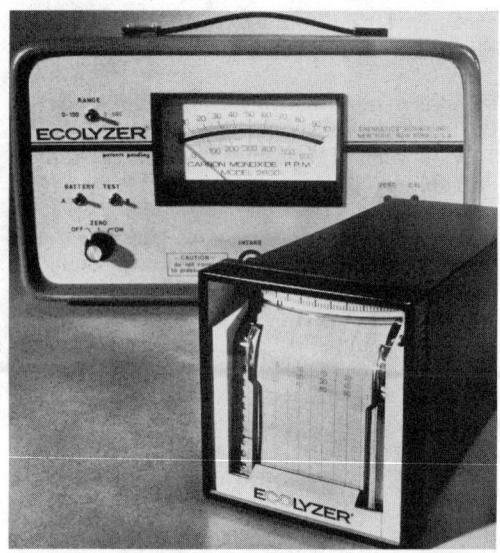

Fig. 37–16.—This carbon monoxide monitor-recorder is used to measure compliance with CO-level standards. The portable unit operates on AC or batteries and, with attached recorder, gathers and records permanent data.

Courtesy Energetics Science, Inc.

The physical hazards—heat and cold, ionizing and nonionizing radiation, and atmospheric pressure are usually measured by direct-reading instruments. In the field of ionizing radiation alone, there are literally hundreds of instruments.

Electrical direct-reading instruments

The first general class of instruments comprises those which give a direct reading on a dial. Frequently these are an electrical meter of some sort, involving electronic circuitry which requires careful maintenance to function properly. Use of some instruments may be limited to nonexplosive atmospheres.

Electrical direct-reading instruments include many types operating on a variety of principles. The combustible gas indicator and the carbon monoxide indicator measure the heat produced by oxidation of these materials (Fig. 37–16), although the method of doing this is different in the two instruments. Mercury vapor in air can be measured by absorption of ultraviolet light of a specific wavelength.

The same technique has also been used for certain chlorinated hydrocarbons. Ionizing radiation is measured by the amount of ionization produced in an ion chamber containing a known volume of gas. Highly sensitive detection of certain hydrocarbons has been accomplished by measuring their effect on the ionization produced by a hydrogen flame. Oxygen is measured by the reduction polarization produced at a sensitive electrode. There are many more such devices varying widely in sensitivity, reliability, and ease of maintenance. Table 37-F lists the more common ones used for both laboratory and field work. All require reliable calibration before use.

Color-change and stain-length instruments

The second class of direct-reading instruments is those which produce a color change in a sensitive chemical through which the air to be tested is drawn. This is a more numerous class and one which has had a great deal of development in recent years. These devices depend on a comparison of the colors produced against a set of standard colors or the measurement of the length of the stain produced in a tube packed with reacting chemical. These are now available for use on many common chemicals: hydrogen sulfide, aromatic hydrocarbons, hydrogen cyanide, carbon monoxide, trichloroethylene, toluene-diisocyanate, mercury vapor, and others. Manufacturers of these tubes often supply lengthy lists of chemicals which may be detected by such tubes, but these sometimes give an exaggerated impression of the versatility of this technique since compounds are often included for which the tubes have only a limited sensitivity.

Limitations of instruments

The accuracy of the electrical direct-reading instruments may be affected by improper maintenance, component unreliability, lack of calibration, poisoning of the catalysts by interfering substances in the atmosphere, changes in airflow rates and volumes, and weak batteries. For these reasons, a routine schedule of maintenance and calibration is necessary.

The colorimetric tubes and papers present a somewhat different problem. They are deceptively simple to operate. Air is usually drawn through the tube with a squeeze bulb so this means no motor to fail or battery to replace. However, there have been and still are variations from one tube to another due to manufacturing conditions which have not been controlled. Practically all tubes have been improved as experience has been gained by users and makers, and some tubes have been withdrawn from production.

One form of variability which has been troublesome, but is obvious to the user, is that produced by nonuniformity in particle size of the media (usually silica gel) which supports the reacting chemical. The result in some tubes is that in transportation the gel fractionates into various sizes by gravity and the line between stain and no stain is not sharp but stretches irregularly across the tube and may extend much further up one side than the other, thus making it impossible to obtain an accurate reading.

The tubes also vary considerably in their degree of chemical stability, although their useful storage life can often be prolonged considerably by storage in the refrigerator. The purchaser may not know how long the tubes have been on the supplier's shelf and what have been the storage conditions, unless the manufacturer puts an expiration date on the tubes.

Rates of chemical reactions are very dependent on temperature; many tubes will give incorrect readings at high or low temperatures. When tubes must be used in cold locations, they should be carried in an inside pocket next to the body where they can be kept warm and used quickly after removing them from the pocket. Even then, the air drawn through the tube is cold and the results obtained must still be interpreted with caution. Extreme temperatures change the volume of air drawn through the tube, which also affects results.

If more than one substance is present in the atmosphere, it may interfere with the accuracy of the tube to give an incorrect reading or a color which cannot be matched with a standard. Some detailed instructions for tube use include a list of possible interfering substances. Unfortunately, many people fail to read the instructions completely or

TABLE 37-F

PHYSICAL METHODS USED IN AIR ANALYSIS

Method	Principle of Operation	Type of Sampling	Range Lower (ppm)	Range Upper (ppm)	Application and Remarks
Absorption (volumetric, gasometric, or manometric)	Measured volume of gas-air mixture is absorbed by reagent and volume shrinkage determined. Pressure reduction may also be measured (manometric).	Grab	300	1,000,000	Selective. Oxygen deficiency determinator. Special forms will detect lower concentrations. U.S. Bureau of Mines describes a portable form for field use, and several compact models are available.
Adsorption (Gravimetric)	Gas-air or vapor-air mixture is trapped on adsorbent such as activated charcoal or silica gel. Change in weight represents weight of contaminant in known air volume.	Continuous	<1	Depends on amount sampled	Nonselective. Used primarily for total solvents when a mixture of solvents is encountered. Absorbed material when eluted can be measured by gas chromatography.
Electrical Conductivity	Gas-air mixture is absorbed in a solution whose electrical conductivity is measured before and after absorption.	Grab or Continuous	0.01	10,000	Selective. Can be applied to all acid or alkaline gases and also to vapors which can be burned to CO_2, SO_2, or HCl. Field instrument available.
Heat of Combustion 1. Benzol indicator or Vapotester	Gas-air or vapor mixture is passed over a heated platinum wire which is part of a balanced Wheatstone bridge. Combustion of the gas on the wire unbalances the bridge.	Grab	20	1,000	Selective. Field instrument was designed primarily for benzol but can be applied to other combustible gases not removed by water vapor absorbent which is provided for proper operation of the instrument.
2. Explosimeter (Flammable)	Explosimeter is a less sensitive form of the above instrument for explosive concentrations of flammable gases.	Grab	10,000 and up	50,000	Water vapor absorbent not necessary because of lower sensitivity. Single meter calibrations for all gases the product of whose molecular heat of combination and the lower explosive limit is a constant.
3. CO Indicator (Hopcalite)	Gas is passed over a catalyst (Hopcalite) which converts CO to CO_2. Heat of combustion is measured by thermopile connected to a galvanometer.	Continuous	10	10,000	Portable apparatus for field use. Water vapor and oil fumes are removed so they cannot poison catalyst.
Interferometer	Refractivity of gas-air or vapor-air mixture as compared to dry air is measured by comparison of interference fringes or Fraunhofer lines when light is passed through mixture.	Grab	12 to 127	100,000 to 1,000,000	Nonselective. Sensitivity depends on refractivity of gas or vapor. CO_2 and moisture interfere and must be removed. Apparatus made in both portable and laboratory types.
Spectrum Absorption 1. Ultraviolet	Gases or vapors mixed with air are passed through a tube with an ultraviolet light source at one end and a photo-electric cell at the other, absorb light, and diminish cell response which is registered by a calibrated meter.	Grab or Continuous	0.001 to 10	0.01 to 1	Sensitive to several gases and vapors. Maximum sensitivity for mercury vapor. Water vapor and CO_2 do not interfere. Field instruments for mercury, trichlorethylene, perchlorethylene have been developed.
2. Infrared	Gases or vapors mixed with air are passed through a tube with a source of infrared (heated filament) at one end and a thermopile for recording absorption of infrared at the other end, absorb infrared rays, and diminish amount of heat reaching thermopile. This is registered by a calibrated galvanometer.	Grab or Continuous	1		Sensitive to all heteroatomic gases and vapors. May be made selective by the development of screening filters. Field infrared unit is available for CO and CO_2. British manufacturer can supply instrument for any heteroatomic gas specified.

	vanometer or ammeter. Differential pressure unit is also used.				
3. Mass Spectrum	Separates according to mass gaseous ions formed in an evacuated system. Abundance of each ionic species is measured electrically by electrometer tube and amplifier.	Grab or Continuous	1	5,000	Applicable to all gases and vapors but requires expensive and elaborate equipment only adapted for laboratory handling of samples collected on silica gel and revaporized or grab samples.
Thermal Conductivity	Gas or vapor mixture is passed over a heated wire. Change in heat loss of wire to surrounding gas mixture is registered by change in wire resistance measured on a balanced Wheatstone bridge.	Grab or Continuous	25	100,000	Nonselective. Water, vapor and carbon dioxide interfere and must be removed. Greatest application in field is to high concentrations such as in flue gas analyzers.
Vapor Pressure	Gas or vapor-air mixture is passed over a cooled condensing surface in a confined chamber. The chamber is then evacuated and the condensed gas allowed to evaporate and its pressure measured. Ice-water, dry ice-acetone or liquid air mixtures may be used depending upon contaminant sampled.	Grab or Continuous	10	10,000	Selective. Only high concentrations, 1000 ppm or greater, can be obtained by grab samples. Liquid air is too difficult to obtain and handle for field use. Contaminants must be miscible with water. Sensitivity depends on contaminant sampled and amount of air sampled in continuous method.
Gas Chromatograph	Adsorbs gas or liquid injected into a carrier gas stream of helium, hydrogen, nitrogen, or argon on a temperature-controlled adsorbent column.	Grab or Liquid	1	10,000	Selective. Can be used to identify solvents and separate gas mixtures by selective elution.

From "Air Sampling and Analysis of Contaminants in Work Places." Silverman, Leslie (Harvard University, School of Public Health, Boston, Mass.) Part AI of Air Sampling Instruments, published by American Conference of Governmental Industrial Hygienists. Adapted with permission of ACGIH.

they are unaware of the presence of the interferences.

Color blindness, which may be unsuspected by the operator, can also give false results in color-matching.

The airflow through the tube usually is produced by either a squeeze bulb or a hand pump, which can produce volume variations. New precision-made pumps are a recent development which is helping to lessen this difficulty, as many tubes are affected by the rate of airflow as well as by the total volume drawn through the tube. This flow rate usually is regulated by a fixed or variable orifice, which must be checked regularly. Airborne dust or lint will quickly plug such orifices, producing low airflow rates.

Calibration of these instruments is not difficult but requires facilities not always available in an industrial situation—some method of producing an atmosphere containing a known concentration of the contaminants in a closed chamber. Since a tube once used for calibration is worthless for further use, this method does not give complete reassurance about the accuracy of measurement with other tubes.

The reliability of the color tube method of air analysis therefore differs from electrical devices or a collection of samples for laboratory analysis. The best that can be done is to test a definite number of tubes from each lot produced by the manufacturer. If the number of tubes used does not justify purchase of a considerable number from a given production lot, the cost for tubes used for such calibration may make their use unjustified.

Because of these variabilities (which affect the reliability), many use the detector tubes for screening purposes; when a given tube indicates the presence of a gas or solvent, they follow up with a more detailed method of analysis.

The dangers from improper use of electrical direct-reading instruments or from false readings on color indicating tubes are fairly obvious. A few measurements of "high" concentration, evaluated by an inadequately trained man, may shut down an operation resulting in considerable financial loss. On the other hand, with insufficient attention to the limitations of these instruments, it is quite possible to get seemingly "safe" readings and still have a real hazard leading to occupational disease and excessive work losses.

Direct-reading vs. lab analysis

There is no real qualitative distinction between direct-reading and laboratory analytical techniques, and in some cases they overlap. For example, to measure the airborne concentration of hydrogen chloride, the industrial hygienist could have the chemist make up a solution containing known amounts of sodium hydroxide in a series of bubblers or impingers. The contaminated air could be drawn through each bubbler at a measured rate for a fixed period of time and the samples returned to the laboratory. There, the excess sodium hydroxide would be back-titrated by the chemist with hydrochloric acid and the air concentrations of hydrogen chloride calculated. This is the laboratory analysis method.

However, the industrial hygienist could have the chemist make up a smaller quantity of sodium hydroxide in each bubbler and place an indicator in the solution. In the field, the industrial hygienist could bubble air through the solution at a known rate and measure the time it took to obtain a color change in the bubbler. From the volume of air drawn through, he could determine the air concentration of hydrogen chloride by a very simple calculation. This is essentially a direct-reading method.

Both methods are adequate for measuring the air concentration at the point of sampling. In both methods, the reagents were prepared by a trained chemist. In one case the industrial hygienist took a sample for a fixed period of time, and in the other he sampled until he saw a color change. Essentially, no difference in skills is required in the direct-reading method; he can now take his second sample with the facts about his first result already on hand.

This is the crucial difference between the methods and here is where the user of the direct-reading method must be wary. He may be led on a trail of seeking higher or lower concentrations and away from his objective of measuring the man's exposure under truly representative conditions. A certain impartiality, created by ignorance of the immediately preceding results, may help in obtaining a more representative sample. On the other hand, a trained industrial hygienist can utilize this progressively increasing knowledge of results to sharpen the precision of sampling.

Safety professionals have been using direct-reading instruments for many years for detecting combustible and explosive gases and have saved many lives as a result. In the case of explosive gases, peak concentrations, not related to a man's exposure, must be measured. Explosive limits are physical constants which can be modified only within narrow limits, in contrast to threshold limit values which must be interpreted broadly.

With a full knowledge of the facts, the trained industrial hygienist is in a position to utilize the direct-reading instruments to considerable advantage. He can take more samples, get closer to the breathing zone, screen out areas for more detailed study, search out specific sources of air contamination, and adequately evaluate the workers' exposure.

Controlling Environment Hazards

General methods of controlling environmental factors or stresses, which may cause sickness, impaired health, or significant discomfort among workers, include:

1. Substitution of a less harmful material for one which is dangerous to health

2. Change or alteration of a process to minimize worker contact

3. Isolation or enclosure of a process or work operation to reduce the number of persons exposed

4. Wet methods to reduce generation of dust in operation such as mining and quarrying

5. Local exhaust at the point of generation or dispersion of contaminants

6. General or dilution ventilation with clean air to provide a safe atmosphere

7. Personal protective devices, such as special clothing or eye and respiratory protection

8. Good housekeeping, including cleanliness of the workplace, waste disposal, adequate

washing, toilet, and eating facilities, healthful drinking water, and control of insects and rodents

9. Special control methods for specific hazards, such as reduction of exposure time, film badges and similar monitoring devices, continuous sampling with preset alarms, and medical programs to detect intake of toxic materials

10. Training and education to supplement engineering controls

Substitution

Replacement of a toxic material with a harmless one is a very practical method of eliminating an industrial health hazard.

In many cases a solvent with a lower order of toxicity or flammability may be substituted for a more hazardous one. In a solvent substitution, it is always advisable to experiment on a small scale before making the new solvent part of the operation or process. For example, carbon tetrachloride can be replaced by solvents such as methyl chloroform, dichloromethane, aliphatic petroleum hydrocarbons, or one of the fluorochlorohydrocarbons. (The precautions listed earlier in this chapter should be reviewed.) Detergent-and-water cleaning solutions can be considered for use in place of organic solvents. Benzene can be replaced by toluene in most lacquers, synthetic-rubber solutions, and paint removers. Natural rubber cements with aliphatic hydrocarbon solvents can perform virtually the same function as benzene cements. The felt-hat industry has controlled the hazard of mercurialism by substituting mercury-free carroting materials. Foundries using parting compounds that contain free silica can minimize the silicosis hazard by substituting relatively harmless powders. Sandstone grinding wheels have been largely replaced by artificial abrasive wheels usually made of aluminum oxide.

Changing the process

A change in process often offers an ideal chance to improve working conditions. Most such changes, of course, are made to improve quality or reduce cost of production, only occasionally to improve the in-plant environment. Yet, we must always keep this possible benefit in mind. This applies to the safety professional more than anyone else.

In some cases, a process can be modified to reduce the exposure to a dust or fume and thus markedly reduce the hazard. In the automobile industry, the amount of lead dust created by grinding solder seams with small high-speed sanding disks was greatly reduced by changing to hand filing. Brush-painting or dipping instead of spray painting will minimize the concentration of airborne contaminants from toxic pigments. Other examples of change of process are arc welding in place of riveting, vapor degreasing to replace handwashing of parts, airless spraying techniques and electrostatic devices to minimize overspray as replacements for hand-spraying, and machine application of lead oxide to battery grids which reduced lead exposure in making storage batteries.

In buying individual machines, the need for accessory ventilation, noise suppression, vibration and heat control should be considered *before* purchase. Often the best approach is to thrash out these requirements with the supplier in advance.

Isolation or enclosure

Some potentially dangerous operations can be *isolated* from the people nearby, which solves the exposure problem. The isolation can be by a physical barrier (such as an acoustic box to contain noise from a whining blower or a screaming rip saw), or by time (such as providing semi-automatic equipment so a person doesn't have to stay near the noisy machine constantly), or by distance (remote controls).

Isolation is particularly useful for limited operations requiring relatively few workers or where control by any other method is too difficult or too expensive. The job is isolated from the rest of the operation and thus the majority of the workers are not exposed to the hazard. The workers actually concerned with the operation can then be protected by installing a ventilation system which probably would not have been satisfactory if the operation had not been isolated and thus able to be enclosed.

Enclosing the process or equipment is a desirable method of control since the enclosure will prevent or minimize the escape

1085

of solvent vapor into the workroom atmosphere. Where some of the more highly toxic solvents are used, enclosure should be one of the first measures attempted, after considering substitution. In most instances, the enclosure should also be exhausted to avoid buildup of solvent vapors. Additional precautions must be taken when cleaning enclosed equipment or during start up or shut down in order to avoid high concentrations of solvent vapors.

Enclosure of an operation effectively prevents the dispersal of contaminants. Enclosed equipment is usually run tightly closed and is only opened during cleaning or filling operations. Further examples of where this type of control is effective are radium dial painting, glove booths, airless blast or shot blast machines for cleaning castings, and abrasive blasting cabinets.

In the chemical industry, the isolation of hazardous processes in closed systems is a widespread practice, which is one reason why the manufacture of toxic substances is often less hazardous than their use. In the mechanical industries, complete enclosure is frequently the best solution of severe dust or fume hazards, such as those from sand blasting or metal spraying operations.

Wet methods

Dust hazards can frequently be minimized or greatly reduced by application of water or other suitable liquid at the source of dust, a method often used for silica and loose dusts. Wetting of floors before sweeping to keep down the dispersion of harmful dust is advisable when better methods, such as vacuum cleaning, cannot be applied.

"Wetting down" is one of the simplest methods for dust control. Its effectiveness, however, depends upon proper wetting of the dust. This may require the addition of a wetting agent to the water and proper disposal of the wetted dust before it dries out and is redispersed. Tremendous reductions in dust concentrations have been achieved by the use of water forced through the drill bits used in rock drilling operations. Many foundries successfully use water under high pressure for cleaning castings in place of sandblasting. Airborne dust concentrations can be kept down if molding sand is kept

moist, if castings are wet down before shakeout, and if the floors are wetted intermittently.

Local exhaust ventilation

A local exhaust system traps the air contaminant near its source so that a worker standing at the process is not exposed to harmful concentrations. This method is usually preferred to general ventilation, but should be used only when the contaminant cannot be controlled by substitution, changing the process, or isolation or enclosure. Even though a process has been isolated, it may still require a local exhaust system.

After the system is installed and set in operation, its performance should be checked to see it meets engineering specifications—correct rates of air flow, duct velocities, negative pressures, etc. Its performance should be rechecked periodically as a maintenance measure.

More details of this method of control are given in Chapter 36, "Exhaust and Ventilation." For further reference, see NSC Industrial Data Sheet 428, *Checking Performance of Local Exhaust Systems*.

General or dilution ventilation

General or dilution ventilation—adding air to keep the concentration of a contaminant below hazardous levels—uses natural convection through open doors or windows, roof ventilators, and chimneys, or artificial air currents produced by fans or blowers. Exhaust fans through roofs, walls, or windows constitute positive all season dilution ventilation. Consideration must be given to providing make-up air, especially during winter months. Dilution ventilation is practicable only if the degree of air contamination is not excessive and particularly if the contaminant is released at a substantial distance from the worker's breathing zone. Under other conditions, the contaminated air will not be diluted sufficiently before inhalation.

General ventilation should not be used where there are major, localized sources of contamination (especially highly toxic dusts and fumes); local exhaust is more effective and economical in such cases. Where comparatively small amounts of the less toxic solvents are vaporized, general or dilution ventilation can be a satisfactory method of control. More information on this subject is presented in the previous chapter.

Personal protective equipment

When it is not feasible to render the environment completely safe, it may be necessary to protect the worker from the environment. Personal protective equipment is normally considered to be secondary to the controls mentioned previously. Where it is not possible to enclose or isolate the process or equipment, provide ventilation, or other control measures; where there are short exposures to hazardous concentrations of contaminants; where unavoidable spills may occur—personal protective equipment should be provided and used.

Personal protective devices have one serious drawback—they do nothing to reduce or eliminate the hazard. Their failure means immediate exposure to the hazard, so the fact that a protective device may become ineffective without the knowledge of the wearer is particularly serious. Excellent equipment that follows accepted standards and specifications is commercially available in great variety. All such equipment, however, is intended for *emergency or temporary* use only.

Only a brief outline of the types of personal protective equipment is given in this section, because Chapter 19, "Personal Protective Equipment," give a complete discussion of each type.

Eye and face protection includes goggles, face shields, and similar items used to protect against corrosive solids, liquids and vapors, and foreign bodies. Shaded lenses are used to screen out ultraviolet and infrared radiations. Of the many types of eye and face protection available, there is a correct type for each job and it should be worn at all times on that job.

Ear protection. Protection against noise-induced hearing impairment, either as ear plugs or muffs, may often present special difficulties. First, as with air contaminants, the real answer is to reduce the noise exposure. But in plants with old machines and processes, and even in new plants, this problem may loom as insoluble. It may be overwhelmingly costly to lower the noise. In some cases, it may be required to make wearing of ear protection mandatory.

Protective clothing—gloves, aprons, boots, coveralls, and other items made of impervious materials—should be worn to control or eliminate prolonged or repeated contact with dermatitis-producing solvents or chemicals that may cause systemic poisoning through skin absorption. Again, care must be taken in choosing the correct article for the specific application. For example, some types of rubber that will withstand trichloroethylene will become spongy and disintegrate in lacquer solvent.

For intermittent protection against radiant heat, reflective aluminized clothing can be used. Such garments need special care to preserve their essential shiny surface. Air-cooled jackets and suits are available to make endurable a high convective heat load. Lead-bearing materials are available to protect against ionizing radiation.

Respiratory protective devices are normally restricted to intermittent exposures or those that are impractical to control by other methods. Respiratory protection should not be substituted for engineering control methods. Exceptions are supplied-air devices for protection in sand-blasting or for operations in confined spaces where an oxygen deficiency may exist.

There are two types of respiratory protective devices: (a) air purifiers, which remove the contaminant from the air by filtering or chemical absorption before inhalation, and (b) air suppliers, which provide clean air from an outside source or oxygen from a tank. Full details of approved types should be obtained from the manufacturer.

Filter respirators cover the mouth and nose but do not protect the eyes. For dust protection, choose a respirator that meets the requirements for high filtering efficiency and low resistance to breathing. A smaller number of respirators have been approved for protection against metal fumes and mists. See discussion in Chapter 19.

Air line respirators are usually preferable to chemical cartridge or mechanical filter respirators, because they are cooler, offer more positive protection, and offer no resistance to breathing. However, they require a proper compressor to supply fresh air.

Gas masks are preferable to respirators if the eyes must be protected or if higher concentrations of gases or vapors are probable. Masks are available which protect against acid vapors, ammonia, organic vapors, carbon monoxide, and certain other vapors and gases.

Self-contained oxygen apparatus—used for emergency and rescue work—have masks attached to oxygen cylinders or canisters which supply oxygen from chemical reaction. Such apparatus enables a worker to enter a contaminated or oxygen-deficient atmosphere and still retain some mobility for relatively short periods of time.

Selection of the proper type of respiratory protective equipment should be based on the following procedure:

1. Identify the substance or substances against which protection is necessary.

2. Know the hazards of each substance and its significant properties.

3. Determine the conditions of exposure.

4. Determine if any personal capabilities and characteristics are essential to the safe use of the devices or procedures required.

5. Determine what facilities are needed for maintenance.

Since a respirator often becomes uncomfortable after wearing for extended periods, the worker must fully realize the need for protection or he will not wear it. To get the worker's cooperation, these factors are important:

1. Prescribe respiratory protective equipment only after every effort has been made to eliminate the hazard.

2. Explain the situation fully to the worker.

3. Fit the respirator carefully.

4. Provide for maintenance and cleanliness, including sterilization before reissue.

5. Instruct the worker in the use of the respirator.

Protective creams and lotions help mini-mize skin contact with irritant chemicals. Their effectiveness varies, but if properly selected and correctly used, they can be very helpful. The cream or lotion must be selected on the basis of competent medical advice. The worker must then be instructed in the value of the protection and its proper application.

Housekeeping

Good housekeeping plays a key role in occupational health protection. Basically, it is another tool in addition to those already listed for preventing dispersion of dangerous contaminants. Dust on overhead ledges and on the floor can readily be dispersed to the in-plant atmosphere by traffic, vibration, and random air currents. Housekeeping is always important; where there are toxic materials, it's paramount.

Immediate cleanup of any spills of toxic material is a very important control measure. A regular cleanup schedule using vacuum cleaners or lines is the only truly effective method of removing dust from the work area. An air hose for blowing away dust should not be used.

Good housekeeping is also essential where solvents are stored, handled, and used. Leaking containers or spigots should be remedied immediately by transferring the solvent to sound containers or by repairing the spigots. Spills should be cleaned up promptly by workers wearing protective equipment. All solvent-soaked rags or absorbents should be disposed of in airtight metal receptacles and removed daily from the plant.

It is impossible to have an effective health program unless maintenance housekeeping is good and the worker has been informed of the need for those measures. For example, if the thermostat on a degreaser fails, or is accidentally broken, excessive concentrations of trichloroethylene might quickly build up in the work area unless all repairs are made immediately.

Special control methods

Many of the general methods mentioned previously (either alone or in combination) can be used for the control of most occupational health hazards. A few additional situations, however, deserve special mention.

Shielding is one of the best protective measures employed to reduce or eliminate exposures to ionizing radiation. Lead and concrete are two materials commonly employed to shield high-energy sources such as high-voltage X ray, particle generators, and radioisotopes. Shielding is also used to protect against exposure to a radiant heat source. For example, heat-treating furnaces can be shielded with aluminum panels. Nonreflective sheet metal is not effective for it is a "black body," which absorbs heat and then acts as a secondary source of radiation.

Reduction of work periods is another method of control in limited areas where engineering methods are too costly or impracticable. In the job forge industry, especially in hot weather, a shorter work day and frequent rest periods are used to minimize exposures to extremely high temperatures, thereby lessening the danger of heat exhaustion or heat stroke. For workmen who must labor in a compressed air environment, schedules of maximum length of shift and length of decompression have been prepared. The higher the pressure, the shorter the shift and the longer the decompression period.

Proper illumination of the work area is necessary to prevent eyestrain leading to increased accidents or loss of visual acuity and to eliminate glare that may interfere with vision. Lighting is discussed in Chapter 16, "Industrial Buildings and Plant Layout," and some levels of illumination are given in Chapter 46, "Safety Engineering Tables."

Training and education

Proper training and education are supplements to the engineering controls. The nurse and safety professional sometimes can be key figures. Perhaps by getting a workman, for example, to change his manner of weighing a toxic material, or of handling a scoop or shovel, there will be no dust exposure.

The worker must also know the proper operating procedures that make engineering controls effective. If he performs an operation away from an exhaust hood, he will not only defeat the purpose of the control but also contaminate the work area. Workers can be alerted to safe procedures through booklets, instruction signs, labels, safety meetings, and other educational devices. (See Chapter 9,

"Safety Training.")

Since new materials are constantly being marketed and new processes developed, reeducation and followup instruction must also be part of the training program.

Setting Up an Effective Industrial Hygiene Program

A typical company program to minimize occupational health hazards is described in the following section.

The success of any safety and industrial hygiene program requires the cooperative effort of the engineering, medical, industrial hygiene, safety, and purchasing organizations, and the supervisor and the individual employee.

Engineering organization

The introduction of new plant processes and operations begins with the engineering organization. For this reason, the engineering group plays a most important role in the control of occupational health hazards. Their responsibilities are:

● To plan all operations using established engineering procedures to prevent unnecessary exposure to harmful environmental factors or stresses.

● To notify the medical, industrial hygiene, and safety organizations, whenever it is planned to introduce new operations or processes.

● To request an industrial hygiene survey of new installations before permitting shop personnel to operate the equipment.

Medical organization

The responsibilities of the medical organization in the industrial hygiene program are:

● To recommend the placement only of those employees whose physical and emotional health capacities meet the minimum job requirements.

● To cooperate in the development of adequate, effective measures to prevent exposure to harmful agents.

● To examine periodically those employees

1089

who are working with or exposed to hazardous materials.

- To restrict employees from further exposure on a medical basis whenever warranted by findings of such periodic examinations.

Industrial hygiene organization

The industrial hygiene organization is primarily responsible for monitoring the work environment and assisting the process engineer in ensuring that a safe environment is provided and maintained. It has a responsibility:

- To advise appropriate organizations of the potential hazard arising out of any current or proposed process or operation.

- To establish hygienic standards and to make appropriate tests periodically to make sure that the standards are met.

- To specify the design and quality of all types of personal protective equipment and to prescribe standards for their use.

- To recommend controls necessary to minimize employee exposure to harmful environmental factors or stresses.

- To assist the supervisor in educating employees on practices, precautions, and procedures established to control their occupational exposure.

- To review present and proposed practices and to ensure that they are in accordance with established standards.

- To survey all new installations before they are turned over to shop personnel.

Safety organization

The safety organization plays an integral part in the overall environmental health program. Its responsibilities are:

- To conduct an effective safety program by coordinating the educational, engineering, supervisory, and enforcement activities related to the safety program.

- To provide educational material for any safety training program for personnel working with hazardous materials.

- To assist the supervisor in teaching his employees safety rules, regulations, and procedures.

- To conduct safety surveys to make sure that proper practices and procedures are being followed.

- To recommend changes in safety rules, regulations, and procedures to keep pace with technological advancements.

Purchasing organization

The purchasing organization's responsibilities are to ensure that only equipment and materials approved by the industrial hygiene, safety, medical, and other appropriate departments are purchased for use in the company.

Supervisor's responsibilities

Each supervisor is responsible for maintaining safe working conditions within his organization and for directly implementing the safety program. His responsibilities are:

- To maintain a work environment that assures the maximum safety for his employees.

- To make certain that applicants (newly hired or transferred) have been examined by the medical organization and approved for the job before being assigned to work.

- To instruct employees periodically on precautions, procedures, and practices to be followed to eliminate accidental exposure to harmful agents.

- To ensure that meticulous housekeeping practices are developed and employed at all times.

- To ensure that food, candy, beverages, other edibles, and tobacco are not stored or consumed in work areas, where toxic materials may be present.

- To inform promptly the engineering, industrial hygiene, and safety organizations of any operation or condition which appears to present a hazard to his employees.

- To inform the medical organization promptly in case of accidental exposure, and to send the employee(s) involved to the medical department for examination.

- To furnish his employees with the proper personal protective equipment, instruct them in its proper use, and enforce the wearing of such equipment.

• To observe all work restrictions imposed by the medical organization.

• To consult with safety, industrial hygiene, engineering, and medical personnel for aid in fulfilling his responsibilities.

• To administer appropriate disciplinary action when safety rules are violated.

Employee's responsibilities

Each employee is responsible for contributing his part towards the success of the industrial hygiene program. His part is:

• To notify his supervisor immediately when certain conditions or practices may cause personal injury or property damage.

• To observe all safety rules and to make maximum use of all prescribed personal protective equipment and to follow practices and procedures established to maintain his health and safety.

• To report immediately to his supervisor any accidental exposures.

• To develop and practice good habits of personal hygiene and housekeeping.

Sources of Help

Specialized help is available from a number of sources.

• Every employer is likely to have a trained professional or scientific man on his staff who can provide some technical assistance or guidance, even though his primary interest is not industrial hygiene but some other field of specialization.

• Many insurance companies that carry workmen's compensation insurance provide industrial hygiene service, just as they provide, for example, periodic safety inspections.

• Professional consultants and privately owned or endowed laboratories are available on a fee basis for concentrated studies of a specific problem or for a plant-wide or company-wide survey, which may be undertaken to identify and catalog individual environmental exposures.

• Many states have excellent industrial hygiene bureaus; these are a fertile source of help. Names and addresses of state and provincial health and hygiene services are listed in Chapter 23, "Sources of Help."

Some state laboratories are extremely well equipped and have numerous devices for sampling and analyzing that an individual company could not possibly justify purchasing because of large initial cost, specialized operator training, or infrequent or intermittent use.

Not only are state services helpful in solving day-to-day problems, but they can also assist in unusual or complex problems. Often they require a request from top management before they will visit a company or plant.

Organizations

Many associations are concerned with industrial hygiene problems. For example, many of the industrial sections of the National Safety Council have an occupational health or industrial hygiene subcommittee that can be called upon for assistance. See details in Chapter 23.

Other scientific and technical societies that can help with health conservation or with a specific problem area are listed in the next section. Some are prepared to provide consultation service to nonmembers; they all have a wealth of available technical information.

Another group of associations come under the broad heading of "trade associations." They are concerned with furthering the aims of their field of productive enterprise, including health preservation of employees and the public. These associations have trained personnel, cooperating committees, and publications that can be extremely helpful.

A list of organizations having high interest in industrial hygiene follows. Additional details on these can be found in Chapters 23 and 38.

Advisory Center on Toxicology, National Research Council, 2101 Constitution Ave., Washington, D.C. 20418. A central repository for toxicological information, primarily for federal agencies. However, private inquiries can be made to it for specific information.

American Conference of Governmental

Industrial Hygienists (ACGIH), 1014 Broadway, Cincinnati, Ohio 45202. A professional association composed of industrial hygiene personnel in government. Publishes annually a table of recommended threshold limits. (Request list of publications.)

American Industrial Hygiene Association (AIHA), 66 S. Miller Rd., Akron, Ohio 44313. A professional association composed of industrial hygiene personnel. "Hygienic Guides" available on specific chemicals summarizing the current information on physiological effects and methods of control. Publishes the *Journal of the American Industrial Hygiene Association* and manuals on various industrial hygiene subjects. (Request list of publications.)

American Medical Association (AMA), 535 North Dearborn Street, Chicago, Ill. 60610. Provides information to members on occupational health hazards upon request. List of publications upon request.

American Petroleum Institute (API), 1271 Avenue of the Americas, New York, N.Y. 10020. Reports developed by committees on handling of petroleum products available for sale.

American National Standards Institute (ANSI), 1430 Broadway, New York, N.Y. 10018. Numerous standards developed for the handling of specific chemicals.

Compressed Gas Association, Inc. (CGA), 500 Fifth Ave., N.Y. 10036. Membership composed primarily of companies. Information on handling of gases available for sale in pamphlet form.

Factory Mutual System, 1151 Boston-Providence Turnpike, Norwood, Mass. 02062. This is the inspection and engineering service for member insurance companies. Some publications, most notably *Handbook of Industrial Loss Prevention* (McGraw-Hill Book Co., New York, 1959), provide information on chemical handling hazards.

Industrial Health Foundation, Inc. (IHF), 5231 Centre Ave., Pittsburgh, Pa. 15232. An association of member companies. Frequent publications on industrial hygiene problems to its members. Some publications offered for sale. Private consultation upon request. Also conducts research.

Industrial Medical Association (IMA), 55 East Washington Street, Chicago, Ill. 60602. Provides information to members on occupational health hazards upon request. List of publications is available.

Manufacturing Chemists' Association, Inc. (MCA), 1825 Connecticut Ave., NW., Washington, D.C. 20009. A trade association composed of companies in the chemical manufacturing and related fields. Has published "Safety Data Sheets" on many chemicals; included is much information on health hazards.

Mining Enforcement and Safety Administration, Department of Interior, Washington, D.C. 20240. Performs investigations and publishes extensively on safety and health subjects pertinent to the mineral industries.

Occupational Injury and Disease Control Division, Environmental Control Administration, Public Health Service, Department of Health, Education, and Welfare, 1014 Broadway, Cincinnati, Ohio 45202. Division accumulates toxicity data and safe-handling procedures on various materials. They are interested in receiving information and experience as well as providing it.

Occupational Safety and Health Administration, U.S. Department of Labor, Washington, D.C. 20210.

Poison Control Centers. Local units exist in most major cities and maintain active files available to any physician on treatment of acute poisons. Nationally, the program is headed by Division of Poison Control, Office of Product Safety, Food and Drug Administration, Dept. of Health, Education, and Welfare, Crystal Plaza Bldg. No. 5, Arlington, Va. 22202. Directory is available.

Public Health Service, Division of Occupational Health, Department of Health, Education, and Welfare, Washington, D.C. 20201. Performs studies and releases extensive information on occupational health.

Literature

A suggested basic reference library on industrial hygiene for the safety professional includes the following books, pamphlets, and data sheets. Journals and magazines are listed in Chapter 23, "Sources of Help."

GENERAL PRINCIPLES

Fundamentals of Industrial Hygiene. National Safety Council, Chicago, Ill. 60611. 1971.

"Hygienic Guides" on specific materials. American Industrial Hygiene Association, Akron, Ohio 44313.

The Industrial Environment—Its Evaluation and Control, Publication No. 614. U.S. Dept. of Health, Education, and Welfare. Available from U.S. Government Printing Office, Washington, D.C. 20402. 1965.

Industrial Hygiene and Toxicology, Vol. I, "General Principles," 2nd ed., edited by F. A. Patty. Interscience Publications, John Wiley & Sons, Inc., New York, N.Y. 10016. 1958.

Industrial Hygiene Highlights, Vol. I, by L. V. Cralley, L. J. Cralley, and G. H. Clayton. Industrial Hygiene Foundation, Pittsburgh, Pa. 15213. 1968.

Occupational Health Hazards, Their Evaluation and Control, Bul. 198. U.S. Dept. of Labor, Washington, D.C. 20210. 1958.

AIR SAMPLING METHODS

Air Sampling Instruments Manual, 3rd ed., American Conference of Governmental Industrial Hygienists, Cincinnati, Ohio 45202. 1969.

Analytical Abstracts. American Industrial Hygiene Association, Akron, Ohio 44313. 1965.

Chemical Detection of Gaseous Pollutants, by W. E. Ruch. Ann Arbor Publishers, Inc., P.O. Box 1425, Ann Arbor, Mich. 48106. 1966.

DERMATITIS

Contact Dermatitis, by A. A. Fisher. Lea and Febiger, Philadelphia, Pa. 19106. 1967.

Chemical Burns, Data Sheet 523. National Safety Council, Chicago, Ill. 60611.

Industrial Skin Diseases, Data Sheet 510. National Safety Council, Chicago, Ill. 60611.

Occupational Dermatitis. American Medical Association, Chicago, Ill. 60610.

DUSTS

Dusts, Fumes, and Mists in Industry, Data Sheet 531. National Safety Council, Chicago, Ill. 60611.

Industrial Dust, 2nd ed., by P. Drinker and T. F. Hatch. McGraw-Hill Book Co., New York, N.Y. 10036.

Pulmonary Deposition and Retention of Inhaled Aerosols, by T. F. Hatch and P. Gross. Academic Press, New York, N.Y. 10003. 1964.

IONIZING RADIATION

Barns, D. E., and Taylor, D. *Radiation Hazards and Protection.* Pitman Publishing Corporation, New York, N.Y. 10036. 1959.

Blatz, H. *Radiation Hygiene Handbook.* McGraw-Hill Book Company, Inc., New York, N.Y. 10036. 1959.

General Handbook for Radiation Monitoring (L.A. 1835), 3rd ed., November 1958. Supt. of Documents, U.S. Government Printing Office, Washington, D.C. 20402.

Manual on Protection against Radiations in Industry—Part II, "Model Code of Safety Regulations for Ionizing Radiations." International Labor Organization, CH 1211, Geneva 22, Switzerland. 1958.

NOISE

GENERAL

Guide for Conservation of Hearing in Noise. Subcommittee on Noise of the Committee on Conservation of Hearing of the American Academy of Ophthalmology and Otolaryngology and Research Center, Dallas, Texas 75219. Revised 1964.

1093

Industrial Noise—A Guide to Its Evaluation and Control, Publication No. 1572. Public Health Service. Available from U.S. Government Printing Office, Washington, D.C. 20402. 1967.

Industrial Noise Manual, 2nd ed. American Industrial Hygiene Association, Akron, Ohio, 44313. 1966.

HEARING EVALUATION

Background for Loss of Hearing Claims. American Mutual Insurance Alliance, Chicago, Ill. 60606. 1964.

Hearing Loss, by Joseph Sataloff. J. B. Lippincott Co., Philadelphia, Pa. 19105.

NOISE EXPOSURE CONTROL

"Guidelines for Noise Exposure Control," American Conference of Governmental Industrial Hygienists (ACGIH), P.O. Box 1937, Cincinnati, Ohio 45201.

Handbook of Noise Control, by C. M. Harris. McGraw-Hill Book Co., New York, N.Y. 10036. 1957.

NOISE MEASUREMENT

Handbook of Noise Measurement, by A. P. G. Peterson and E. E. Gross, Jr. General Radio Co., West Concord, Mass. 01781.

SOLVENTS

Handbook of Organic Industrial Solvents, 3rd ed. American Mutual Insurance Alliance, Chicago, Ill. 60606. 1966.

Toxicity and Metabolism of Industrial Solvents, by E. Browning. Elsevier Publishing Co., New York, N.Y. 10017. 1965.

Toxicology and Biochemistry of Aromatic Hydrocarbons, by H. W. Gerade. Elsevier Publishing Co., New York, N.Y. 10017. 1960.

See also individual National Safety Council Industrial Data Sheets, Manufacturing Chemists' Association Data Sheets, American Petroleum Institute Toxicological Reviews, and AIHA "Hygienic Guides" on a particular material. (See the next two chapters for listings.)

TOXICOLOGY

The Chemistry of Industrial Toxicology, 2nd ed., by H. B. Elkins. John Wiley & Sons, Inc., New York, N.Y. 10016. 1959.

Clinical Toxicology of Commercial Products, by M. N. Gleason, R. E. Gosselin, and H. C. Hodge. The Williams & Wilkins Co., Baltimore, Md. 21202. 1963.

Documentation of Threshold Limit Values. American Conference of Governmental Industrial Hygienists, P.O. Box 1937, Cincinnati, Ohio 45201. 1971.

Industrial Hygiene and Toxicology, Vol. II, "Toxicology," 2nd ed., edited by F. A. Patty. Interscience Publications, John Wiley & Sons, Inc., New York, N.Y. 10016. 1963.

Occupational Diseases—A Guide to Their Recognition, edited by W. M. Gafafer. Public Health Service Bul. 1097. Available from U.S. Government Printing Office, Washington, D.C. 20402. 1964.

Occupational Diseases and Industrial Medicine, by R. T. Johnston and S. E. Miller. W. B. Saunders Co., Philadelphia, Pa. 19105. 1960.

Toxicity of Industrial Metals, by E. Browning. Butterworth & Co. (Publishers) Ltd., London WC 2, England. 1961.

VENTILATION

Engineering Manual for Control of In-Plant Environment in Foundries. American Foundrymen's Society, Des Plaines, Ill. 60016. 1956.

Industrial Ventilation, 10th ed. Committee on Industrial Ventilation, American Conference of Governmental Industrial Hygienists, P.O. Box 453, Lansing, Mich. 48902. 1968.

TABLE 37-G

THRESHOLD LIMIT VALUES FOR PHYSICAL AGENTS
1973

PREFACE
PHYSICAL AGENTS

These threshold limit values refer to levels of physical agents and represent conditions under which it is believed that nearly all workers may be repeatedly exposed day after day without adverse effect. Because of wide variations in individual susceptibility, exposure of an occasional individual, at, or even below, the threshold limit may not prevent annoyance, aggravation of a pre-existing condition, or physiological damage.

These threshold limits are based on the best available information from industrial experience, from experimental human and animal studies, and when possible, from a combination of the three.

These limits are intended for use in the practice of industrial hygiene and should be interpreted and applied only by a person trained in this discipline. They are not intended for use, or for modification for use, (1) in the evaluation or control of the levels of physical agents in the community, (2) as proof or disproof of an existing physical disability, or (3) for adoption by countries whose working conditions differ from those in the United States of America.

These values are reviewed annually by the Committee on Threshold Limits for Physical Agents for revisions or additions, as further information becomes available.

Notice of Intent. At the beginning of each year, proposed actions of the Committee for the forthcoming year are issued in the form of a "Notice of Intent." This notice provides not only an opportunity for comment, but solicits suggestions of physical agents to be added to the list. The suggestions should be accompanied by substantiating evidence.

As legislative code. The Conference recognizes that the Threshold Limit Values may be adopted in legislative codes and regulations. If so used, the intent of the concepts contained in the Preface should be maintained and provisions should be made to keep the list current.

NOTICE OF INTENT TO ESTABLISH THRESHOLD LIMIT VALUES
Heat Stress

These Threshold Limit Values refer to heat stress conditions under which it is believed that nearly all workers may be repeatedly exposed without adverse health effects. The TLVs shown in Table 1 are based on the assumption that nearly all acclimatized, fully clothed workers with adequate water and salt intake should be able to function effectively under the given working conditions without exceeding a deep body temperature of 38 C (WHO technical report series #412, 1969 *Health Factors Involved in Working Under Conditions of Heat Stress*).

Since measurement of deep body temperature is impractical for monitoring the workers' heat load, the measurement of environmental factors is required which most nearly correlate with deep body temperature and other physiological responses to heat. At the present time, Wet Bulb-Globe Temperature Index (WBGT) is the simplest and most suitable technique to measure the environmental factors. WBGT values are calculated by the following equations:

1. Outdoors with solar load:
   $$WBGT = 0.7WB + 0.2GT + 0.1DB$$

2. Indoors or Outdoors with no solar load:
   $$WBGT = 0.7WB + 0.3GT$$
   where:

   WBGT = Wet Bulb-Globe Temperature Index
   WB = Natural Wet-Bulb Temperature
   DB = Dry-Bulb Temperature
   GT = Globe Thermometer Temperature

TABLE 37-G. THRESHOLD LIMIT VALUES FOR PHYSICAL AGENTS, 1973 (Continued)

The determination of WBGT requires the use of a black globe thermometer, a natural (static) wet-bulb thermometer, and a dry-bulb thermometer.

TABLE 1

PERMISSIBLE HEAT EXPOSURE THRESHOLD LIMIT VALUES
(Values are given in °C, WBGT)

Work—Rest Regimen	Work Load		
	Light	Moderate	Heavy
Continuous work	30.0	26.7	25.0
75% Work—25% Rest, Each hour	30.6	28.0	25.9
50% Work—50% Rest, Each hour	31.4	29.4	27.9
25% Work—75% Rest, Each hour	32.2	31.1	30.0

Higher heat exposures than shown in Table 1 are permissible if the workers have been undergoing medical surveillance and it has been established that they are more tolerant to work in heat than the average worker. Workers should not be permitted to continue their work when their deep body temperature exceeds 38.0 C.

Appendix G*
HEAT STRESS

I. Measurement of the environment

The instruments required are a dry-bulb, a natural wet-bulb, a globe thermometer, and a stand. The measurement of the environmental factors shall be performed as follows:

A. The range of the dry and the natural wet bulb thermometer shall be −50 C to 50 C with an accuracy of ±0.5 C. The dry bulb thermometer must be shielded from the sun and the other radiant surfaces of the environment without restricting the airflow around the bulb. The wick of the natural wet-bulb thermometer shall be kept wet with distilled water for at least ½ hour before the temperature reading is made. It is not enough to immerse the other end of the wick into a reservoir of distilled water and wait until the whole wick becomes wet by capillarity. The wick shall be wetted by direct application of water from a syringe ½ hour before each reading. The wick shall extend over the bulb of the thermometer, covering the stem about one additional bulb length. The wick should always be clean and new wicks should be washed before using.

B. One globe thermometer, consisting of a 15 cm. (6-inch) diameter hollow copper sphere, painted on the outside with a matte black finish or equivalent shall be used. The bulb or sensor of a thermometer (range −5 C to 100 C with an accuracy of ±0.5 C) must be fixed in the center of the sphere. The globe thermometer shall be exposed at least 25 minutes before it is read.

C. One stand shall be used to suspend the three thermometers so that they do not restrict free air flow around the bulbs, and the wet-bulb and globe thermometer are not shaded.

D. It is permissible to use any other type of temperature sensor that gives identical reading to a mercury thermometer under the same conditions.

E. The thermometers must be so placed that the readings are representative of the condition where the men work or rest, respectively.

The methodology outlined above is more fully explained in the following publications:

1. "Prevention of Heat Casualties in Marine Corps Recruits, 1955–1960, with Comparative Incidence Rates and Climatic Heat Stresses in other Training Categories," by Captain David Minard, MC, USN, Research Report No. 4, Contract No. MR005.01–0001.01, Naval Medical Research Institute, Bethesda, Maryland, 21 February 1961.

2. "Heat Casualties in the Navy and Marine Corps, 1959–1962, with Appendices on the Field Use of the Wet Bulb-Globe Temperature Index," by Captain David Minard, MC, USN, and R. L. O'Brien, HMC, USN. Research Report No. 7, Contract No. MR 005.01–0001.01, Naval Medical Research Institute, Bethesda, Maryland, 12 March 1964.

* Note. Appendices A through F are for previous material in the booklet from which this table was taken. This material is reproduced as Table 38-E in the next chapter.

3. Minard, D.: Prevention of Heat Casualties in Marine Corps Recruits. Military Medicine 126(4): 261–272, 1961.

II. Work load categories

The heat produced by the body and the environmental heat together determine the total heat load. Therefore, if work is to be performed under hot environmental conditions, the workload category of each job shall be established and the heat exposure limit pertinent to the work load evaluated against the applicable standard in order to protect the worker from exposure beyond the permissible limit.

A. The work load category may be established by ranking each job into light, medium, and heavy categories on the basis of type of operation, where the work load is ranked into one of said three categories, i.e.

(1) light work: e.g., sitting or standing to control machines, performing light hand or arm work,

(2) moderate work: e.g., walking about with moderate lifting and pushing,

(3) heavy work: e.g., pick and shovel work,

the permissible heat exposure limit for that work load shall be determined from Table 1.

B. The ranking of the job may be performed either by measuring the worker's metabolic rate while performing his job or by estimating his metabolic rate by the use of the scheme shown in Table 2. Tables available in the literature listed below and in other publications as well may also be utilized.

1. Per-Olaf Astrand and Kaare Rodahl: "Textbook of Work Physiology" McGraw-Hill Book Company, New York, San Francisco, 1970.

2. Ergonomics Guide to Assessment of Metabolic and Cardiac Costs of Physical Work. Amer. Ind. Hyg. Assoc. J. 32: 560, 1971.

3. Energy Requirements for Physical Work. Purdue Farm Cardiac Project. Agricultural Experiment Station. Research Progress Report No. 30, 1961.

4. J. V. G. A. Durnin and R. Passmore: "Energy, Work and Leisure," Heinemann Educational Books, Ltd., London, 1967.

TABLE 2

ASSESSMENT OF WORK LOAD

Average values of metabolic rate during different activities.

A. Body position and movement

	Kcal./min.
Sitting	0.3
Standing	0.6
Walking	2.0–3.0
Walking up hill	add 0.8 per meter (yard) rise

B.

Type of Work		Average Kcal./min.	Range Kcal./min.
Hand work	light	0.4	0.2–1.2
	heavy	0.9	
Work with one arm	light	1.0	0.7–2.5
	heavy	1.8	
Work with both arms	light	1.5	1.0–3.5
	heavy	2.5	
Work with body	light	3.5	2.5–15.0
	moderate	5.0	
	heavy	7.0	
	very heavy	9.0	

Light hand work: writing, hand knitting

Heavy hand work: typewriting

Heavy work with one arm: hammering in nails (shoemaker, upholsterer)

Light work with two arms: filing metal, planing wood, raking of a garden

Moderate work with the body: cleaning a floor, beating a carpet

Heavy work with the body: railroad track laying, digging, barking trees

Sample Calculation: Using a heavy hand tool on an assembly line

A. Walking along — 2.0 Kcal./min.

B. Intermediate value between heavy work with two arms and light work with the body — 3.0 Kcal./min. / 5.0 Kcal./min.

C. Add for basal metabolism — 1.0 Kcal./min.

Total 6.0 Kcal./min.

Adapted from Lehmann, G. E., A. Muller and H. Spitzer: Der Kalorienbedarf bei gewerblicher Arbeit. Arbeitsphysiol. 14: 166, 1950.

TABLE 37-G. THRESHOLD LIMIT VALUES FOR PHYSICAL AGENTS, 1973 *(Continued)*

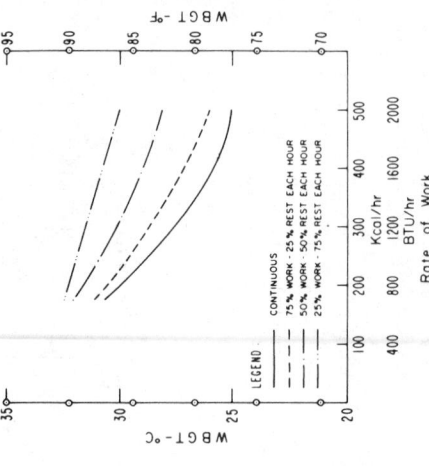

FIG. 1—Permissible Heat Exposure Threshold Limit Value

III. Work-rest regimen

The permissible exposure limits specified in Table I and Diagram A are based on the assumption that the WBGT value of the resting place is the same or very close to that of the work place. If the resting place is air conditioned and its climate is kept at or below 24 C (75 F), WBGT, the allowable resting time may be reduced by 25%. The permissible exposure limits for continuous work are applicable where there is a work-rest regimen of a 5-day work week and an 8-hour work day with a short morning and afternoon break (approximately 15 minutes) and a longer lunch break (approximately 30 minutes). Higher exposure limits are permitted if additional resting time is allowed. All breaks, including unscheduled pauses and administrative or operational waiting periods during work may be counted as rest time when additional rest allowance must be given because of high environmental temperatures.

It is a common experience that when the work on a job is self-paced, the workers will spontaneously limit their hourly work load to 30-50 percent of their maximum physical performance capacity. They do this either by setting an appropriate work speed or by interspersing unscheduled breaks. Thus the daily average of the workers' metabolic rate seldom exceeds 330 kcal/hr. However, within an 8-hour work shift there may be periods where the workers' hourly average metabolic rate will be higher.

IV. Water and salt supplementation

During the hot season or when the worker is exposed to artificially generated heat, drinking water shall be made available to the workers in such a way that they are stimulated to frequently drink small amounts, i.e., one cup every 15-20 minutes (about 150 ml or ¼ pint).

The water shall be kept reasonably cool (10-15 C or 50.0-60.0 F) and shall be placed close to the workplace so that the worker can reach it without abandoning the work area.

The workers should be encouraged to salt their food abundantly during the hot season and particularly during hot spells. If the workers are unacclimatized, salted drinking water shall be made available in a concentration of 0.1 percent (1g NaCl to 1.0 liter or 1 level tablespoon of salt to 15 quarts of water). The added salt shall be completely dissolved before the water is distributed, and the water shall be kept reasonably cool.

V. Other Considerations

A. Clothing: The permissible heat exposure TLVs are valid for light summer clothing as customarily worn by workers when working under hot environmental conditions. If special clothing is required for performing a particular job and this clothing is heavier or it impedes sweat evaporation or has higher insulation value, the worker's heat tolerance is reduced, and the permissible heat exposure limits indicated in Table 1 and Figure 1 are not applicable. For each job category where special clothing is required, the permissible heat exposure limit shall be established by an expert.

B. Acclimatization and Fitness: The recommended heat stress TLVs are valid for acclimated workers who are physically fit.

Ionizing Radiation

See U.S. Department of Commerce National Bureau of Standards, Handbook 59, "Permissible Dose from External Sources of Ionizing Radiation," September 24, 1954, and addendum of April 15, 1958. A report, Basic Radiation Protection Criteria, published by the National Committee on Radiation Protection, revises and modernizes the concept of the NCRP standards of 1954, 1957 and 1958; obtainable as NCRP Rept. No. 39, P.O. Box 4867, Washington, D.C. 20008.

Lasers

The threshold limit values are for exposure to laser radiation under conditions to which nearly all workers may be exposed without adverse effects. The values should be used as guides in the control of exposures and should not be regarded as fine lines between safe and dangerous levels. They are based on the best available information from experimental studies.

Limiting apertures

The TLVs expressed as radiant exposure or irradiance in this section may be averaged over an aperture of 1 mm except for TLVs for the eye in the spectral range of 400–1400 nm, which should be averaged over a 7 mm limiting aperture (pupil). No modification of the TLVs is permitted for pupil sizes less than 7 mm.

The TLVs for "extended sources" apply to sources which subtend an angle greater than τ (Table 5) which varies with exposure time. This angle is not the beam divergence of the source.

Correction factor A (CFA) for eye exposure

All TLVs in Tables 3 and 4 are to be used as given for wavelengths 400 nm to 700 nm. At all wavelengths greater than 1.06 μm and less than 1.4 μm the TLVs are to be increased by a factor of 5. TLV at wavelengths between 700 nm and 1.06 μm are to be increased by a uniformly extrapolated factor as shown in Figure 2.

TABLE 3

THRESHOLD LIMIT VALUE FOR DIRECT OCULAR EXPOSURES
(Intrabeam Viewing) from a Laser Beam

Spectral Region	Wave Length	Exposure Time (t) Seconds	TLV
UVC	200 nm to 280 nm	10^{-3} to 3×10^4	3 mJ·cm^{-2}
UVB	280 nm to 302 nm	"	3 "
	303 nm	"	4 "
	304 nm	"	6 "
	305 nm	"	10 "
	306 nm	"	16 "
	307 nm	"	25 "
	308 nm	"	40 "
	309 nm	"	63 "
	310 nm	"	100 "
	311 nm	"	160 "
	312 nm	"	250 "
	313 nm	"	400 "
	314 nm	"	630 "
	315 nm	"	1.0 J·cm^{-2}
UVA	315 nm to 400 nm	10 to 10^3	1.0 J·cm^{-2}
	"	10^3 to 3×10^4	1.0 mW·cm^{-2}
Light	400 nm to 700 nm	10^{-9} to 1.8×10^{-5}	5×10^{-7} J·cm^{-2}
	"	1.8×10^{-5} to 10	$\left[\dfrac{1.8t}{\sqrt[4]{t}}\right]$ mJ·cm^{-2}
	"	10 to 10^4	10 mJ·cm^{-2}
	"	10^4 to 3×10^4	10^{-6} W·cm^{-2}
Infrared A	700 nm to 1.06 μm	10^{-9} to 10^2	[light TLV's] × [CFA]
"	1.06 μm to 1.40 μm	10^{-9} to 10^2	(light TLV) × 5
"	700 nm to 1.06 μm	10^2 to 3×10^4	10^{-4} [CFA] W·cm^{-2}
"	1.06 μm to 1.4 μm	"	5×10^{-4} W·cm^{-2}
Infrared B & C	1.4 μm to 10^3 μm	10^{-9} to 10^{-7}	10^{-2} J·cm^2
"	"	10^{-7} to 10	$0.56 \sqrt[4]{t}$ J·cm^{-2}
"	"	−10 to 3×10^4	0.1 W·cm^{-2}

NOTE: To aid in the determination of TLV's for exposure durations requiring calculations of fractional powers Figures 3, 4, and 6 may be used.

TABLE 37-G. THRESHOLD LIMIT VALUES FOR PHYSICAL AGENTS, 1973 (Continued)

TABLE 4

THRESHOLD LIMIT VALUES FOR VIEWING A DIFFUSE REFLECTION OF A LASER BEAM OR AN EXTENDED SOURCE LASER

Spectral Region	Wave Length	Exposure Time, (t) Seconds	TLV
UV	200 nm to 400 nm	10^{-3} to 3×10^4	Same as Table 3
Light	400 nm to 700 nm	10^{-9} to 10	$10 \cdot \sqrt[3]{t}\ J \cdot cm^{-2} \cdot sr^{-1}$
"	"	10 to 10^4	$20\ J \cdot cm^{-2} \cdot sr^{-1}$
"	"	10^4 to 3×10^4	$2 \times 10^{-3}\ W \cdot cm^{-2} \cdot sr^{-1}$
Infrared A	700 nm to 1.06 μm	10^{-7} to 10	$CFA \times 10\ \sqrt[3]{t}\ J \cdot cm^{-2} \cdot sr^{-1}$
"	"	10 to 10^2	$CFA \times 20\ J \cdot cm^{-2} \cdot sr^{-1}$
"	"	10^2 to 3×10^4	$CFA \times 0.2\ W \cdot cm^{-2} \cdot sr^{-1}$
"	1.06 μm to 1.4 μm	10^{-9} to 10	$50 \times \sqrt[3]{t}\ J \cdot cm^{-2} \cdot sr^{-1}$
"	"	10 to 10^2	$100\ J \cdot cm^{-2} \cdot sr^{-1}$
"	"	10^{-9} to 3×10^4	$1.0\ W \cdot cm^{-2} \cdot sr^{-1}$
Infrared B & C	1.4 μm to 1 mm	10^2 to 3×10^4	Same as Table 3

NOTE: To aid in the determination of TLV's for exposure durations requiring calculations of fractional powers Figures 3, 4, and 6 may be used.

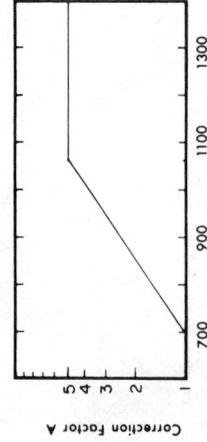

FIG. 2—TLV correction factors for laser wavelengths (eye).

Repetitively pulsed lasers

Since there are few experimental data for multiple pulses, caution must be used in the evaluation of such exposures. The protection standards for irradiance or radiant exposure in multiple pulse trains have the following limitations:

(1) The exposure from any single pulse in the train is limited to the protection standard for a single comparable pulse.

(2) The average irradiance for a group of pulses is limited to the protection standard as given in Tables 3, 4, or 6 of a single pulse of the same duration as the entire pulse group.

(3) When the Instantaneous Pulse Repetition Frequency (PRF) of any pulses within a train exceeds one, the protection standard applicable to each pulse is reduced as shown in Figure 6 for pulse durations less than 10^{-5} second. For pulses of greater duration, the following formula should be followed:

$$\text{Standard}\left(\begin{array}{c}\text{single pulse}\\ \text{in train}\end{array}\right) = \frac{\text{Standard (pulse } n\tau)}{n}$$

where:

n = number of pulses in train
τ = duration of a single pulse in the train
Standard $(n\tau)$ = protection standard of one pulse having a duration equal to $n\tau$ seconds

TABLE 5

LIMITING ANGLE OF EXTENDED SOURCE WHICH MAY BE USED FOR APPLYING EXTENDED SOURCE TLV'S

Exposure Duration(s)	Angle × (mrad)
10^{-9}	8.0
10^{-8}	5.4
10^{-7}	3.7
10^{-6}	2.5
10^{-5}	1.7
10^{-4}	2.2
10^{-3}	3.6
10^{-2}	5.7
10^{-1}	9.2
1.0	15
10	24
10^{2}	24
10^{3}	24
10^{4}	24

TABLE 6

THRESHOLD LIMIT VALUE FOR SKIN EXPOSURE FROM A LASER BEAM

Spectral Region	Wave Length	Exposure Time, (t), Seconds	TLV
UV	200 nm to 400 nm	10^{-3} to 3×10^{4}	Same as Table 3
Light & Infrared A	400 nm to 1400 nm	10^{-9} to 10^{-7}	$2 \times 10^{-2}\,\text{J} \cdot \text{cm}^{-2}$
"	"	10^{-7} to 10	$1.1\,\sqrt[4]{t}\,\text{J} \cdot \text{cm}^{2}$
"	"	10 to 3×10^{4}	$0.2\,\text{W} \cdot \text{cm}^{-2}$
Infrared B & C	1.4 μm to 1 mm	10^{-9} to 3×10^{4}	Same as Table 3

NOTE: To aid in the determination of TLV's for exposure durations requiring calculations of fractional powers Figures 3, 4, and 6 may be used.

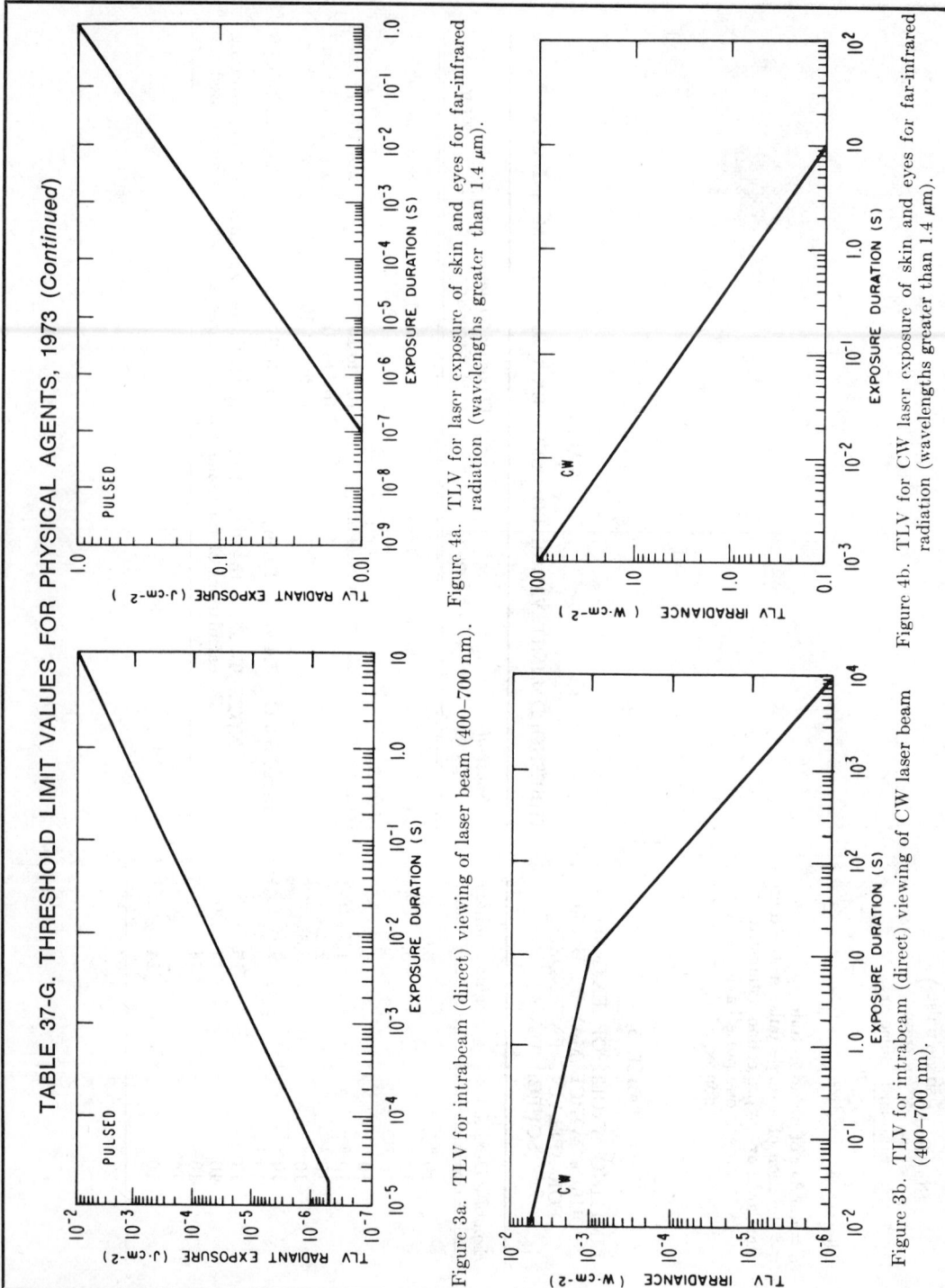

TABLE 37-G. THRESHOLD LIMIT VALUES FOR PHYSICAL AGENTS, 1973 (Continued)

Figure 3a. TLV for intrabeam (direct) viewing of laser beam (400–700 nm).

Figure 3b. TLV for intrabeam (direct) viewing of CW laser beam (400–700 nm).

Figure 4a. TLV for laser exposure of skin and eyes for far-infrared radiation (wavelengths greater than 1.4 μm).

Figure 4b. TLV for CW laser exposure of skin and eyes for far-infrared radiation (wavelengths greater than 1.4 μm).

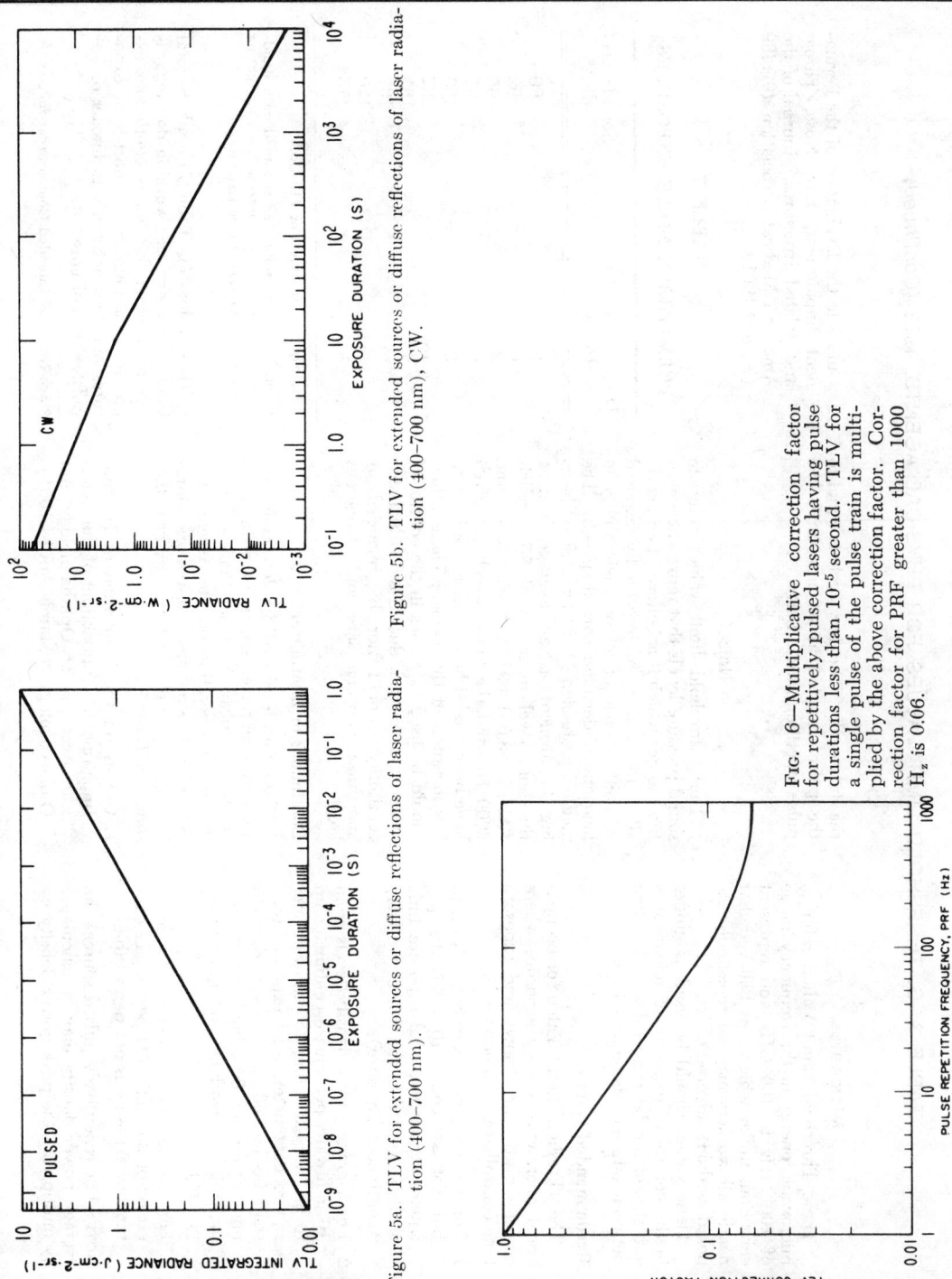

Figure 5a. TLV for extended sources or diffuse reflections of laser radiation (400–700 nm).

Figure 5b. TLV for extended sources or diffuse reflections of laser radiation (400–700 nm), CW.

Fig. 6—Multiplicative correction factor for repetitively pulsed lasers having pulse durations less than 10⁻⁵ second. TLV for a single pulse of the pulse train is multiplied by the above correction factor. Correction factor for PRF greater than 1000 H$_z$ is 0.06.

TABLE 37-G. THRESHOLD LIMIT VALUES FOR PHYSICAL AGENTS, 1973 (Continued)

Microwaves

These Threshold Limit Values refer to microwave energy in the frequency range of 100 MHz to 100 GHz and represent conditions under which it is believed that nearly all workers may be repeatedly exposed without adverse effect.

These values should be used as guides in the control of exposure of microwaves and should not be regarded as a fine line between safe and dangerous levels.

Recommended values

The Threshold Limit Value for occupational microwave energy exposure where power densities are known and exposure time controlled is as follows:

1. For average power density levels up to but not exceeding 10 milliwatts per square centimeter, total exposure time shall be limited to the 8-hour workday (continuous exposure).

2. For average power density levels from 10 milliwatts per square centimeter up to but not exceeding 25 milliwatts per square centimeter, total exposure time shall be limited to no more than 10 minutes for any 60 minute period during an 8-hour workday (intermittent exposure).

3. For average power density levels in excess of 25 milliwatts per square centimeter, exposure is not permissible.

NOTE: For repetitively pulsed sources the average power density may be calculated by multiplying the peak power density by the duty cycle. The duty cycle is equal to the pulse duration in seconds times the pulse repetition rate in hertz.

Noise*

These threshold limit values refer to sound pressure levels that represent conditions under which it is believed that nearly all workers may be repeatedly exposed without adverse effect on their ability to hear and understand normal speech. The medical profession (1, 2) has defined hearing impairment as an average hearing threshold level in excess of 25 decibels (ANSI S3.6–1969) at 500, 1000, and 2000 Hz, and the limits which are given have been established to prevent a hearing loss in excess of this value. These values should be used as guides in the control of noise exposure and, due to individual susceptibility, should not be regarded as fine lines between safe and dangerous levels.

Continuous or intermittent

The sound level shall be determined by a sound level meter, meeting the standards of the American National Standards Institute and operating on the A-weighting network with slow meter response. Duration of exposure shall not *exceed that shown in Table 7.*

1. Guides for the Evaluation of Hearing Impairment. Transactions of the American Academy of Ophthalmology and Otolaryngology (March–April, 1961).

2. Guides to the Evaluation of the Permanent Impairment; Ear, Nose, Throat and Related Structures. Journal of the American Medical Association 197:489 (August 1961).

TABLE 7

PERMISSIBLE NOISE EXPOSURES

Duration per day Hours	Sound Level dBA[a]
8	90
6	92
4	95
3	97
2	100
1½	102
1	105
¾	107
½	110
¼	115*

* No exposure in excess of 115 dBA

[a] Sound level in decibels as measured on a standard level meter operating on the A-weighting network with slow meter response.

These values (in Table 7) apply to total time of exposure per working day regardless of whether this is one continuous exposure or a number of short-term exposures but does not apply to impact or impulsive type of noises.

* See Notice of Intended Changes that follows.

When the daily noise exposure is composed of two or more periods of noise exposure of different levels, their combined effect should be considered, rather than the individual effect of each. If the sum of the following fractions,

$$\frac{C_1}{T_1} + \frac{C_2}{T_2} + \cdots \frac{C_n}{T_n}$$

exceeds unity, then, the mixed exposure should be considered to exceed the threshold limit value, C_1 indicates the total time of exposure at a specified noise level, and T_1 indicates the total time of exposure permitted at that level. Noise exposures of less than 90 dBA do not enter into the above calculations.

Impulse or impact noise

It is recommended that exposure to impulsive or impact noise should not exceed 140 decibels peak sound pressure level.

Ultraviolet Radiation*

These threshold limit values refer to ultraviolet radiation in the spectral region between 200 and 400 nm and represent conditions under which it is believed that nearly all workers may be repeatedly exposed without adverse effect. These values for exposure of the eye or the skin apply to ultraviolet radiation from arcs, gas, and vapor discharges, fluorescent, and incandescent sources, but do not apply to ultraviolet lasers** or solar radiation. These levels should not be used for determining exposure of photosensitive individuals to ultraviolet radiation. These values should be used as guides in the control of exposure to continuous sources where the exposure duration shall not be less than 0.1 sec.

These values should be used as guides in the control of exposure to ultraviolet sources and should not be regarded as a fine line between safe and dangerous levels.

Recommended values

The threshold limit value for occupational exposure to ultraviolet radiation incident upon skin or eye where irradiance values are known and exposure time is controlled are as follows:

1. For the near ultraviolet **spectral region** (320 to 400 nm) total irradiance incident upon the unprotected skin or eye should not exceed 1 mw/cm² for periods greater than 10^3 seconds (approximately 16 minutes) and for exposure times less than 10^3 seconds should not exceed one J/cm².

2. For the actinic ultraviolet spectral region (200–315 nm), radiant exposure incident upon the unprotected skin or eye should not exceed the values given in Table 8 within an 8-hour period.

3. To determine the effective irradiance of a broadband source weighted against the peak of the spectral effectiveness curve (270 nm), the following weighting formula should be used:

$$E_{eff} = \Sigma\, E_\lambda\, S_\lambda\, \Delta_\lambda$$

where:

E_{eff} = effective irradiance relative to a monochromatic source at 270 nm

E_λ = spectral irradiance in W/cm²/nm

S_λ = relative spectral effectiveness (unitless)

Δ_λ = band width in nanometers

4. Permissible exposure time in seconds for exposure to actinic ultraviolet radiation incident upon the unprotected skin or eye may be computed by dividing 0.003 J/cm² by E_{eff} in W/cm². The exposure time may also be determined using Table 9 which provides exposure times corresponding to effective irradiances in μW/cm².

TABLE 8
RELATIVE SPECTRAL EFFECTIVENESS BY WAVE LENGTH

Wavelength (nm)	TLV (mJ/cm²)	Relative Spectral Effectiveness S_λ
200	100	0.03
210	40	0.075
220	25	0.12
230	16	0.19
240	10	0.30
250	7.0	0.43
254	6.0	0.5
260	4.6	0.65
270	3.0	1.0
280	3.4	0.88
290	4.7	0.64
300	10	0.30
305	50	0.06
310	200	0.015
315	1000	0.003

* See Notice of Intended Changes, Page 1106.
** See Laser TLV's.

TABLE 37-G. THRESHOLD LIMIT VALUES FOR PHYSICAL AGENTS, 1973 (Concluded)

TABLE 9
PERMISSIBLE ULTRAVIOLET EXPOSURES

Duration of Exposure Per Day	Effective Irradiance, E_{eff} ($\mu W/cm^2$)
8 hrs.	0.1
4 hrs.	0.2
2 hrs.	0.4
1 hr.	0.8
½ hr.	1.7
15 min.	3.3
10 min.	5
5 min.	10
1 min.	50
30 sec.	100
10 sec.	300
1 sec.	3,000
0.5 sec.	6,000
0.1 sec.	30,000

All the preceding TLV's for ultraviolet energy apply to sources which subtend an angle less than 80°. Sources which subtend a greater angle need to be measured only over an angle of 80°.

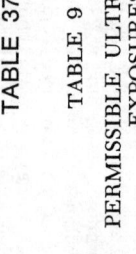

Fig. 7—Threshold Limit Values for Ultraviolet Radiation

NOTICE OF INTENDED CHANGES
(for 1973)

These physical agents, with their corresponding values, comprise those for which either a limit has been proposed for the first time, or for which a change in the "Adopted" listing has been proposed. In both cases, the proposed limits should be considered trial limits that will remain in the listing for a period of at least one year. If, after one year no evidence comes to light that questions the appropriateness of the values herein, the values will be reconsidered for the "Adopted" list.

Noise

These threshold limit values refer to sound pressure levels and durations of exposure that represent conditions under which it is believed that nearly all workers may be repeatedly exposed without adverse effect on their ability to hear and understand normal speech. The medical profession has defined hearing impairment as an average hearing threshold level in excess of 25 decibels (ANSI-S3.6-1969) at 500, 1000, and 2000 Hz, and the limits which are given have been established to prevent a hearing loss in excess of this level. The values should be used as guides in the control of noise exposure and, due to individual susceptibility, should not be regarded as fine lines between safe and dangerous levels.

Continuous or intermittent

The sound level shall be determined by a sound level meter, conforming as a minimum to the requirements of the American National Standard Specification for Sound Level Meters, S1.4 (1971) Type S2A, and set to use the A-weighted network with slow meter response. Duration of exposure shall not exceed that shown in Table 10.

These values apply to total duration of exposure per working day regardless of whether this is one continuous exposure or a number of short-term exposures but does not apply to impact or impulsive type of noise.

When the daily noise exposure is composed of two or more periods of noise exposure of different levels, their combined effect should be considered, rather than the individual effect of each. If the sum of the following fractions:

$$\frac{C_1}{T_1} + \frac{C_2}{T_2} + \cdots \frac{C_n}{T_n}$$

exceeds unity, then, the mixed exposure should be considered to exceed the threshold limit value, C_1 indicates the total duration of exposure at a specific noise level, and T_1 indicates the total duration of exposure permitted at that level. All on the job noise exposures of 80 dBA or greater shall be used in the above calculations.

Impulsive or impact noise

It is recommended that exposure to impulsive or impact noise should not exceed 140 decibels peak sound pressure level. Impulsive or impact noise is considered to be those variations in noise levels that involve maxima at intervals of greater than one per second. Where the intervals are less than one second, it should be considered continuous.

TABLE 10
THRESHOLD LIMIT VALUES

Duration per day Hours	Sound Level dBA[a]
16	80
8	85
4	90
2	95
1	100
½	105
¼	110
⅛	115*

* No exposure to continuous or intermittent in excess of 115 dBA

[a] Sound level in decibels as measured on a sound level meter, conforming as a minimum to the requirements of the American National Standard Specification for Sound Level Meters, S1.4 (1971) Type S2A, and set to use the A-weighted network with slow meter response.

Ultraviolet Radiation

The threshold limit value shall apply also to solar radiation.

Industrial Toxicology

Chemical Toxicity　　　　　　　　　　　　　　　　　　　1109
Entry routes for toxic materials . . . Mode of action . . . Type of
harmful effects . . . Intensity of action . . . Experimental toxicology

Toxic Substances List　　　　　　　　　　　　　　　　　　1117

NIOSH Criteria Documents　　　　　　　　　　　　　　　1118

Hygienic Air Quality Standards　　　　　　　　　　　　1118
Threshold limit values . . . Acceptable concentrations . . . "Hygienic
Guide" series . . . Emergency exposure limits . . . Radioactive
material . . . Water quality

Catalog of Toxic Substances　　　　　　　　　　　　　　1123

Sources of Help　　　　　　　　　　　　　　　　　　　　1176

References　　　　　　　　　　　　　　　　　　　　　　1177

Threshold Limit Values for Chemical Substances　　　1182

The constant introduction of new industrial processes, operations, and techniques involving new chemicals makes a knowledge of toxicology important to the safety professional who is frequently called upon for advice concerning the dangers associated with the use of various chemical agents.

A safety professional who sets up a program to minimize injury from the use of mechanical tools will be concerned with the physical characteristics of the machine tool or machinery and the manner in which the various parts of the machine come into contact with the operator and cause harm. Similarly, he is concerned with how chemical agents can approach or contact individuals, how that contact can be prevented, and how extensive the injury can be if contact does occur, based on the properties of the chemical agents involved.

Industrial toxicology is primarily concerned with the physiological effects produced in individuals who have been exposed to harmful materials during the course of their employment.

Chemical Toxicity

If administered in a suitable manner and in sufficient dosage, practically any substance can be harmful to man. It is recognized, therefore, that there are degrees of harmfulness and degrees of safeness for all materials. These degrees are primarily related to the amount of material that is present in the body—a relationship exists between the biologic effect of a chemical agent and its dose or concentration in the human body.

Many chemical agents are nonselective in their action on tissues or cells; they may exert a harmful effect on all living matter. Other chemical agents may act only on specific cells. Another agent may be harmful only to certain species; other species may have built-in protective devices.

Toxicity is relative. It refers to a harmful effect on some biologic mechanism. The term "toxicity" is commonly used in comparing one chemical agent with another, but such comparison is meaningless if the biologic mechanism and the conditions under which the harmful effects occur are not specified.

Princi (see References) states that a chemical stimulus may be considered to have produced a toxic effect when it satisfies the following criteria.

• An observable or measureable physiologic deviation has been produced in any organ or organ system. The change may be anatomic in character and may be acceleration or inhibition of a normal physiological process, or it may consist of a specific biochemical change.

• The observed change can be duplicated from animal to animal even though the dose-effect relationships may vary quantitatively.

• The stimulus has changed normal physiologic processes in such a way that a protective mechanism is impaired in its function of defense against other adverse stimuli.

• The effect is either reversible or at least attenuated when the stimulus is removed.

• The effect does not occur in the absence of a stimulus or occurs so infrequently that it indicates generalized or nonspecific response. When high degrees of susceptibility are noted, equally significant degrees of resistance should be apparent.

• The observation must be noted and must be reproducible by other investigators.

• The physiologic change reduces the efficiency of an organ system or function and impairs physiologic reserve in such a way as to interfere with the ability to resist or adapt to other normal stimuli in either a permanent or temporary manner.

Although the toxic effects of many chemical agents used in industry are well known, there are many others commonly used which are not as well defined. The toxicity of a material is not a physical constant—such as boiling point, melting point, vapor pressure, or temperature—and thus only a general statement can be made concerning the harmful nature of a given chemical agent.

If the toxicity of a chemical could be determined from its constitution or structural formula, it would certainly make the job easier. While certain important analogies are apparent between structure and toxicity, important differences exist that require individual study of each compound. For example, in the alcohol series there is a progressive in-

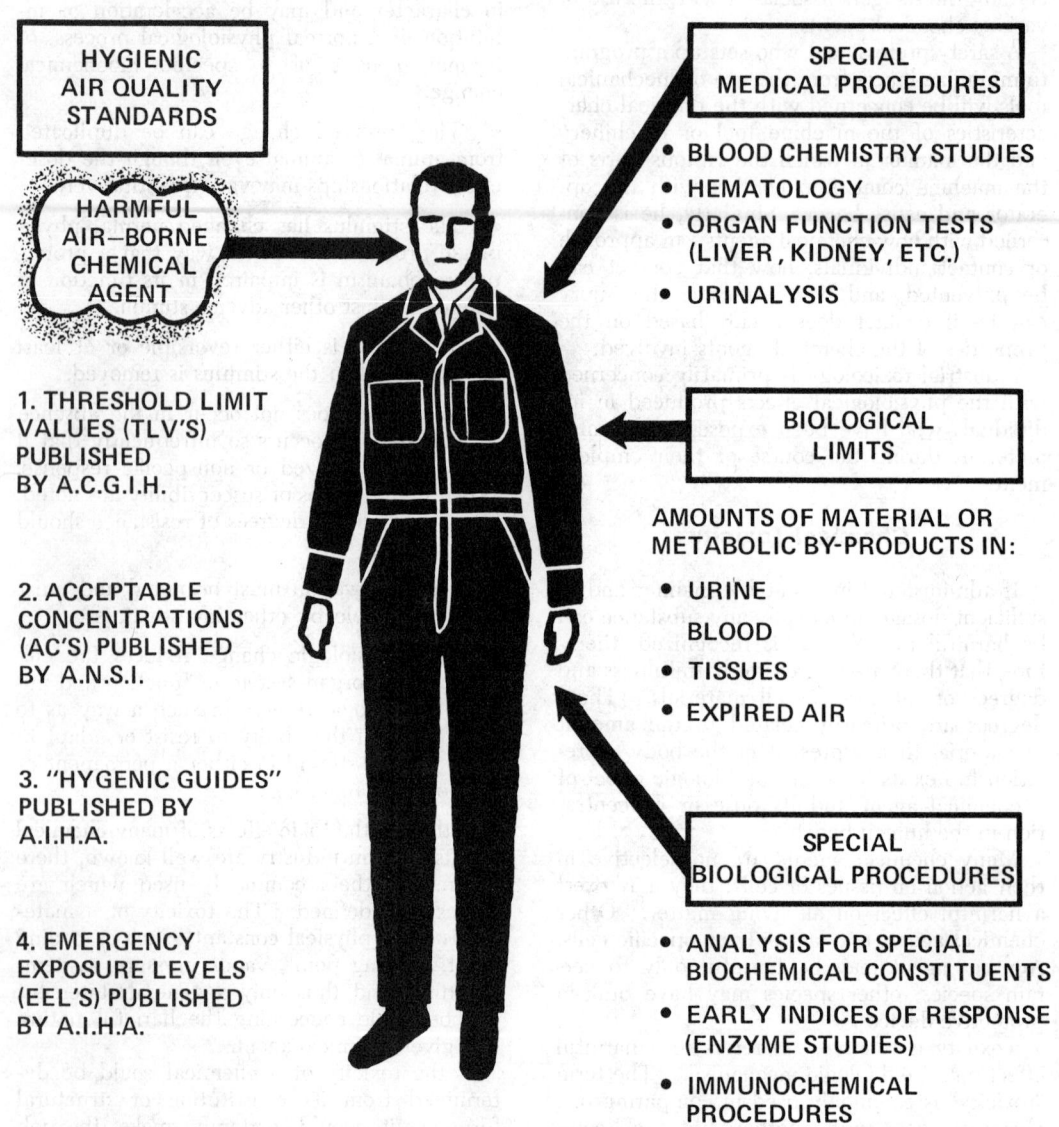

LIMITS AND PROCEDURES
EMPLOYED IN INDUSTRIAL TOXICOLOGY

HYGIENIC
AIR QUALITY
STANDARDS

HARMFUL
AIR–BORNE
CHEMICAL
AGENT

1. THRESHOLD LIMIT
VALUES (TLV'S)
PUBLISHED
BY A.C.G.I.H.

2. ACCEPTABLE
CONCENTRATIONS
(AC'S) PUBLISHED
BY A.N.S.I.

3. "HYGENIC GUIDES"
PUBLISHED BY
A.I.H.A.

4. EMERGENCY
EXPOSURE LEVELS
(EEL'S) PUBLISHED
BY A.I.H.A.

SPECIAL
MEDICAL PROCEDURES

- BLOOD CHEMISTRY STUDIES
- HEMATOLOGY
- ORGAN FUNCTION TESTS
  (LIVER , KIDNEY, ETC.)
- URINALYSIS

BIOLOGICAL
LIMITS

AMOUNTS OF MATERIAL OR
METABOLIC BY-PRODUCTS IN:

- URINE
  BLOOD
- TISSUES
- EXPIRED AIR

SPECIAL
BIOLOGICAL PROCEDURES

- ANALYSIS FOR SPECIAL
  BIOCHEMICAL CONSTITUENTS
- EARLY INDICES OF RESPONSE
  (ENZYME STUDIES)
- IMMUNOCHEMICAL
  PROCEDURES

Fig. 38–1.—The limits and procedures employed in industrial toxicology.

crease in toxicity from ethyl to amyl. Then, as the higher alcohols become less soluble in body fluids, there is a decreasing toxicity with the higher members of this series. The substitution of a chlorine atom for hydrogen atom usually results in an increase in toxicity; however, this cannot be applied as a general rule in all cases. For example, methylene chloride (CH_2Cl_2) is less toxic than carbon tetrachloride (CCl_4) and much less toxic than methyl chloride (CH_3Cl).

The toxicity of a material is not synonymous with its health hazard. *Toxicity* is the capacity of a material to produce injury or harm. *Hazard* is the possibility that a material will cause injury when a specific quantity is used under specific conditions. The key elements to be considered in evaluating a health hazard are these:

- The toxicity of the materials used.

- The physical properties of these materials.

- The absorption probabilities of these materials by individuals.

- The extent and intensity of exposure to these materials.

- The control measures used.

The degree to which a substance will affect living cells can be measured only after recognizable changes have occurred following absorption. Some changes—impaired judgment, delayed reaction time—may be produced at levels too low to cause actual cell damage. Toxicity is dependent upon the dose, rate, method, and site of absorption, plus many other factors, including general state of health, individual differences, tolerance, diet, and temperature.

Entry routes for toxic materials

There are only three ways in which chemical agents can harm individuals.

1. By contact with or absorption through the respiratory tract

2. By contact with or absorption through the skin

3. By contact with or absorption through the digestive tract

In general, systemic industrial poisonings usually result from:

- **Inhalation** of toxic agents. Local injury to the skin can be caused by direct contact with an irritant or corrosive material.

- **Ingestion** of the toxic agent can occur to some extent; however, there would generally be considerable inhalation of the material where such conditions exist.

- **Absorption through the skin** can occur upon exposure to some toxic agents. Some liquids and vapors are known to pass through the skin to such a degree that respiratory protection is not adequate against high concentrations. For example, hydrogen cyanide (HCN) is known to pass through the unbroken skin. Consideration should be given to the type of work clothes being worn; if they become saturated with solvents, they will act as a reservoir to bathe the body continually with the harmful material.

- **The absorption of harmful materials by the lungs** is dependent upon the solubility in body fluids, the permeability of the lungs, the volume of inhalation, the volume of blood in the lungs, and the concentration gradient of vapors between the inhaled air and the blood.

The respiratory system is composed of two main areas: (*a*) the upper respiratory tract airways—the nose, throat, trachea, and bronchial tubes leading to the various lobes of the lungs; (*b*) the alveoli where the actual transfer of gases across the cell wall takes place.

In general only very fine particles and gases will reach the alveolar sacs. The larger the particle, the sooner it will be deposited through impaction, gravity, etc., on the lining of the airway tubes leading to the sacs. Only particles smaller than about one-half micron are likely to enter the alveolar sac.

The nasopharyngeal passages serve as a heat exchanger and humidifier, warming and moisturizing the inhaled air so that it will not constitute a shock to the delicate lung tissue. The trachea and bronchial tubes have a sticky lining that traps larger particulate contaminants before they can reach the lungs.

The body is poisoned by substances absorbed by the lung in the same way and to the same general extent as it would be if a

TABLE 38-A

EXCRETORY AND BIOLOGIC THRESHOLD LIMITS

Urinary TLV's should be used for

Aniline	Hydrogen cyanide	Selenium
Antimony	Hydrogen fluoride	Selenium hexafluoride
Arsenic	Hydrogen selenide	Stibine
Arsine	Lead arsenate	Sulfuryl fluoride
Benzene	Mercury	Tellurium
Boron trifluoride	Molybdenum	Tellurium hexafluoride
Calcium arsenate	Nickel	Thallium
Chlorinated benzenes	Nickel carbonyl	Uranium
Cyanide	Nitrobenzene	Zinc chloride
Fluoride	Nitrogen trifluoride	Zinc oxide
Hydrogen bromide	Parathion	

Blood TLV's are more applicable than excretory limits

Cadmium dust	Lead	Methyl bromide
Cadmium fume	Manganese	

Blood and/or breath TLV's should be applicable

Carbon monoxide	Nitric oxide	Sulfur hexafluoride

Breath analysis good for many

Alcohols	Aliphatic hydrocarbons	Ketones
	Chlorohydrocarbons	

NOTE: Analysis of blood, breath, or urine samples can be done in conjunction with air analysis to determine if workers are in danger of injury or intoxication. Ideally, such biologic or excretory threshold limits should correspond to the average levels found when workers are exposed to the atmospheric TLV.

Adapted from "Excretory and Biologic Threshold Limits," by H. B. Elkins. A.I.H.A. Journal, 28(4):305–314 (July–August 1967).

like amount entered the body by any other channel of absorption, such as through the skin or the intestinal wall.

The harmful effects of deleterious materials are determined by the amount accumulated in the body—by the total dosage. The number of milligrams of harmful material per kilogram of body weight determines the toxic effect produced. This is often overlooked when expressing individual exposures only in terms of concentration (parts per million) in the air breathed and the length of time of exposure. The amount in the inspired air is not necessarily the amount absorbed.

According to H. B. Elkins (see References), differences in particle size or solubility, which are practically impossible to measure accurately, determine the fraction of inhaled impurity which is retained or absorbed. The concentration in the air of parathion, or lead dust, may not be associated with intoxication, depending on whether or not there is opportunity for significant skin absorption or ingestion.

These are situations where air analyses are not adequate to precisely evaluate the peril of the hazard, since the amount absorbed cannot be predicted from the data obtained by such determinations. In such situations it is highly desirable to have other means of estimating exposure. With many substances this can be done by analyzing suitable biologic specimens or excretion products for the toxic agent or a metabolite derived therefrom (See Fig. 38–1).

The only biologic fluid finding much application for such exposure tests is blood; limited use has been made of biopsy specimens of lung, skin and fat, but these are not very practical for periodic sampling. The excretory products most frequently analyzed are urine and breath; sweat, the other major excretion product, is not well adapted for exposure tests (Table 38–A).

The pulmonary ventilation rate of an average worker may be taken as 20 liters per minute (L/min) or approximately 10,000 L in an eight-hour working day. If the air contains one milligram per liter (mg/L) of some harmful material, the maximum amount that can be absorbed in one day is 10 gm. Likewise, the same total dose would result from an exposure of one hour to a concentration in air of 8.33 mg/L.

The employee's exposure to gases and vapors is usually limited to his working hours. The time spent away from work is usually more than adequate to permit complete elimination of the volatile material in the body before the beginning of the next daily exposure. Except for highly soluble substances, such as methyl alcohol, or with reactive gases whose detoxification products are eliminated slowly, there would be no accumulation of gases and vapors from daily exposure.

Henderson and Haggard's book on noxious gases (see References) treats the subject of respiration and absorption of gases and vapors in great detail, and the reader is urged to consult this reference for specific details.

Harmful materials or their detoxification products are eliminated from the body through the respiratory tract, the intestinal tract, the urinary tract, and, in some cases, the sweat glands (Table 38–A).

Most volatile organic compounds are eliminated in a matter of hours or, at most, days.

Many of the poisonous elements, however, can be stored for long periods of time in various parts of the body. Chronic toxicity damage is unlikely to have an even distribution throughout the body. In many cases the organ affected is the liver or kidney, and in a number of substances, both the liver and bones. In toxicity studies with radioactive isotopes, the organ which suffers the most severe damage and appears to contribute most to the toxic effect on the body as a whole is called the critical organ. The particular organ that shows the largest amount of damage is the one that is chosen for estimating the effect.

Lead, when taken into the body, may be deposited in the bones. When the intake of lead stops, the stored portion is excreted until it is virtually eliminated from the system. In chronic poisoning, the excretion rate usually rises until it balances the absorption rate.

In the case of lead poisoning, biochemical tests performed on the body fluids would indicate the amount that is being circulated in the blood, excreted in the urine, or stored in the bones if an autopsy was made.

An important difference between inhalation as route of entry and ingestion or skin absorption is that through inhalation the material enters the arterial blood directly, whereas with ingestion or skin absorption the material enters the venous blood and is carried through the right portion of the heart where it is then diluted by the entire venous system and passed through the lungs and then pumped by the left portions of the heart to the brain and other organs.

Reactive gases and vapors are altered within the body and are eliminated to a large extent in forms other than those in which they are absorbed. The toxicological action may be exerted either by the substance in its original form or by its detoxification products.

Nonreactive gases and vapors are not altered to any appreciable extent in the body and are eliminated in the same chemical form in which they are absorbed. The aliphatic hydrocarbons, such as propane and heptane, are examples of that type.

Mode of action

Industrial poisoning may be classified as one of three types: acute, subacute, or chronic. The classification is based on the rate of intake of harmful materials, rate of onset of symptoms, and the duration of symptoms.

Acute poisoning, typically, is characterized by rapid absorption of the offending material and the exposure is sudden and severe. For example, inhaling high levels of carbon monoxide or swallowing a large quantity of cyanide compound will produce acute poisoning. The death or survival of a victim through the critical period occurs suddenly. Generally, acute poisoning results from a single dose which is rapidly absorbed and damages one or more of the vital physiological processes. The development of cancer long after recovery from acute radiation damage is called a delayed acute effect.

Frequently repeated and extended exposures, over a period of several hours or days, result in subacute effects, depending on the dose rate.

Chronic poisoning is concerned with the continued absorption over a long period of time of a harmful material in small doses; each dose, if taken alone, would barely be effective. Chronic poisoning is characterized by the harmful materials remaining in the tissues, continually injuring some body process. The rate of intake exceeds the rate of excretion or detoxification; thus chronic poisoning can also be produced by exposure to a harmful material which produces irreversible damage, so the injury accumulates, rather than the poison. The symptoms in chronic poisoning are usually different from those seen in acute poisoning by the same toxic agent.

It is readily apparent that not all individuals react in the same manner to the same amount of a harmful material. According to Stokinger (see References), "The atypical, or idiosyncratic, response to chemicals has always been a cause for unusual concern in those responsible for the control of hazardous exposures to individuals." Predictive tests can be used in pre-employment placement examinations to screen out the hypersusceptible individuals against future injurious exposure to the company's products.

Type of harmful effects

Narcosis, irritation of the respiratory tract, and asphyxiation characterize the acute effect produced upon exposure to toxic gases, vapors, and fumes. All organic solvents, upon extended high level exposures, are capable of producing a gradual continuous paralysis of the central nervous system, which is expressed in loss of consciousness, or voluntary movements. Death may result from respiratory failure.

Many symptoms of narcosis and asphyxiation are quite similar in that the primary effect in each is to impair the functioning of the brain—in narcosis by direct action and in asphyxiation by denying the brain sufficient oxygen. Mild exposures to chemical agents producing asphyxiation and narcosis are usually transitory and do not have any permanent ill effects. Asphyxiation may also be brought about by oxygen deficiency or the reduction of the partial pressure of oxygen in the inspired air.

Carbon monoxide is an example of a chemical asphyxiant. Its action is to combine with the hemoglobin to form carboxyhemoglobin which cannot be utilized in carrying oxygen to its tissues.

As mentioned previously, the liver, kidneys, bones and other organs of the body may become affected by exposure to harmful materials. Organic phosphates inhibit the enzyme, cholinesterase, which is present in red blood cells.

In addition to consideration of exposure to a single material, combined toxic effects due to the simultaneous action of different materials may occur. For example, a synergistic effect, greater than that from the same concentration of inhaled gas alone, will be produced by exposure to a mixture of sulfur dioxide and sodium chloride crystals. Sodium chloride crystals inhaled alone are relatively inert.

Antagonistic, or potentiation effects are produced by inhaling fumes of iron, and gaseous nitrogen oxides; the reduction of effect in this case is explained on the basis of a firmly combined layer of nitrogen oxides on the iron oxide particles (see Gafafer—References).

In the absence of other information it may be assumed that the effects of exposure to chemical agents should be considered as additive. When purely local effects on different organs of the body are produced by the various components of the mixture, it may be assumed that the chief effects upon exposure would be independent.

Harmful chemical agents may attack living organisms in two different ways—locally and systemically. Caustic alkalis and strong acids are capable of producing severe damage and local tissue destruction. Many of the materials exert a milder local action which usually consists of skin irritation or dermatitis, which is of a transitory nature. Systemic poisons are carried by the blood stream to the tissues or organs where they exert their effects.

Studies of cellular metabolism have established important mechanisms on the action of toxic agents involving interference with enzyme systems. Ultimately, all toxic agents produce either directly or indirectly an alteration in cellular metabolism, leading to changes in permeability, irritability, alteration of normal function, or some other physiological change (see Gafafer—References).

Intensity of action

The severity of action of the toxic agent is a function of the concentration of the substance in the damaged organ. This depends upon the rates of absorption, detoxification, and excretion. Absorption is primarily dependent upon the solubility of the harmful material. Detoxification is a result of the body's attempt to eliminate the toxic substance or to form a non-toxic product. Excretion refers to the elimination of a toxic agent and its toxic-metabolites from the body.

Apparently, then, the intensity of action of a harmful material is dependent upon the nature of the material, the type and rate of exposure, and the fate of material in the body. A single large dose of a toxic substance may be expected to produce a greater response than the same total dose administered over a long period of time, because a large dose may be able to produce its detrimental action before appreciable detoxification occurs. The same amount given as a series of small repeated doses can be handled by detoxification and excretion mechanisms. A toxic substance that is detoxified or excreted at a rate which is slower than the rate of intake of the substance is called a cumulative poison.

As a rule, young people are more susceptible to toxic substances than are their elders. Either natural or acquired hypersensitivity may increase the response to toxic agents. This abnormal sensitivity is called an idiosyncrasy to the toxic agent.

Physiological tests are available to detect early stages of poisoning before clinical symptoms ordinarily appear. Chest X rays often detect silicosis before it becomes disabling. A reduction in the cholinesterase activity of the blood is a reliable indication of organic phosphate poisoning. Periodic medical examination in conjunction with analysis of atmospheric contaminants may be necessary to completely define the worker's exposure in relationship to his environment.

Experimental toxicology

Smyth (see References) stated that four axioms form the basis for experimental toxicology. Although they appear self-evident, they are not susceptible to rigid proof.

- Any substance contacting or entering the body will be injurious at some degree of exposure and will be tolerated without effect at some lower exposure.

- The nature of the injuries which may develop in men can be determined by the study of the reactions of experimental animals.

- It is possible to define an exposure to animals that has no effect upon its health. Health and well-being, however, are not measurable. It is the injury that is measured.

- From animal experiments, the degree of exposure that will not affect humans can be determined.

Experimental toxicology has done well for human safety by allowing regulation of exposure in accord with the results of animal experiments.

The first attempts in estimating the toxicity of a substance are usually made on the basis of animal experiments. Data from these experiments is expressed as lethal doses (LD) in milligrams of substance per kilogram of body weight of the test animal. The most commonly used expressions are these:

- MLD—minimum lethal dose, the smallest dose which kills one of a group of test animals.

- LD_{50}—lethal dose for 50 percent, the dose which kills one-half of a group of test animals (usually 10 or more).

- LD_{100}—lethal dose for 100 percent, the dose which kills all of a group of test animals (usually 10 or more).

These doses may also be expressed as lethal concentrations (LC) for air-borne toxic substances.

Substances can then be rated according to their relative toxicity as shown in animal ex-

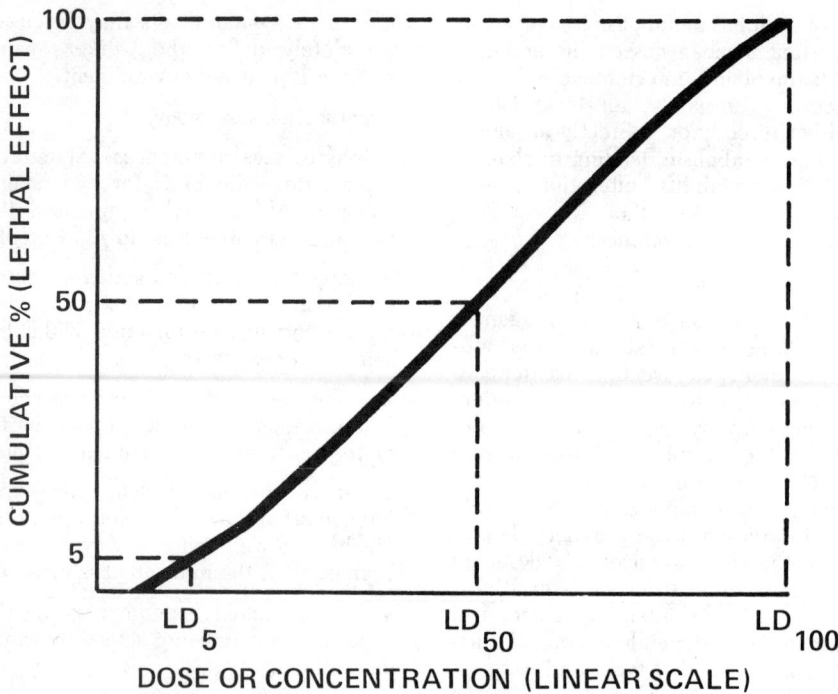

Fig. 38–2.—Dose/response curves for a chemical agent administered to a uniform population of test animals. The slope of the curve can be considered an index of the margin of safety of the material.

TABLE 38-B

COMBINED TABULATION OF TOXICITY CLASSES

Commonly Used Term	LD_{50} Single Oral Dose for Rats (gm/kg)	4-hr Vapor Exposure Causing 2 to 4 Deaths in 6-Rat Group (ppm)	LD_{50} Skin for Rabbits (gm/kg)	Probable Lethal Dose for Man
Extremely toxic	0.001 or less	Less than 10	0.005 or less	Taste (1 grain)
Highly toxic	0.001 to 0.05	10 to 100	0.005 to 0.043	1 tsp (4 cc)
Moderately toxic	0.05 to 0.5	100 to 1,000	0.044 to 0.340	1 oz (30 gm)
Slightly toxic	0.5 to 5.0	1,000 to 10,000	0.35 to 2.81	1 pint (250 gm)
Practically nontoxic	5.0 to 15.0	10,000 to 100,000	2.82 to 22.6	1 quart
Relatively harmless	>15.0	>100,000	>22.6	>1 quart

From Handbook of Toxicology, Vol. 1, "Acute Toxicities." W. S. Spector, ed. Philadelphia, 1956. W. B. Saunders Co.

periments (Fig. 38–2), and the probable lethal dose for humans can be estimated (Table 38-B). These ratings are based on the results of short-term acute exposures only.

It is quite possible that actual long-term chronic exposures to a substance could prove highly toxic even though short-term exposure tests indicate a low order of toxicity.

Brown (see References) states that measurements of relatively crude physical effects—such as body and organ weight changes, anatomical changes, and lethal results—are usually involved in tests of toxicity. To assess the subtle effects a chemical may have, consideration must be given to measuring central nervous system effects during toxicity testing.

Animal experiment data are difficult to interpret and apply to human exposure. Such data are valuable *only* as guides to be used by the industrial toxicologist in estimating the gross toxicity of a substance and as leads for further investigation.

Toxic Substances List

The Occupational Safety and Health Act of 1970, Section 20(a)(6), requires that "the Secretary of Health, Education, and Welfare shall publish within six months of enactment of this act and thereafter as needed, but at least annually, a list of all known toxic substances by generic family or other useful grouping, and the concentrations at which such toxicity is known to occur." The first such list was prepared in 1971. The 1972 edition represents a substantial revision and expansion of the earlier list. Under the OSHAct, the Secretary of Labor must issue regulations requiring employers to monitor employee exposure to toxic materials and to keep records of any such employee exposure. This requirement is set forth in Section 8(c)(3) of the act.

The purpose of *The Toxic Substances List* is to identify "all known toxic substances" in accordance with definitions that may be used by all sections of our society to describe toxicity. It must be emphatically stated that the entry of a substance on the list does not automatically mean that it is to be avoided. A listing does mean, however, that the listed substance has the documented potential of being hazardous if misused, and, therefore,

care must be exercised to prevent tragic consequences.

The absence of a substance from the list does not necessarily indicate that a substance is not toxic. Some hazardous substances may not qualify for the list, because the dose that causes the toxic effect is not known.

Other chemicals associated with skin sensitization and carcinogenicity may be omitted from the list, because these effects have not been reproduced in experimental animals, or because the human data are not definitive. Of necessity, there had to be reliance on the published comments and evaluations of the scientific community; also, there has been no attempt at an evaluation of the degree of hazard that might be expected from substances on the list—that being an ultimate goal of the hazard-evaluation studies.

It is not the purpose of the list to quantify the hazard by way of the toxic concentration or dose that is presented with each of the substances listed. Hazard evaluation involves far more than the recognition of a toxic substance and a knowledge of its relative toxic potency. It involves a measurement of the quantity that is available for absorption by the user, the amount of time that is available for absorption, the frequency with which the exposure occurs, the physical form of the substances, and the presence of other substances, toxic or non-toxic, additives, or contaminants.

Ventilation, appropriate hygienic practices, housekeeping, protective clothing, and pertinent training for safe handling may diminish any hazard that might exist.

Hazard evaluation requires, therefore, engineers, chemists, toxicologists, and physicians who have been trained in the fields of toxicology, industrial hygiene, and occupational health, to recognize, measure, and control these hazards.

Many companies employ people with this type of professional background, particularly those companies that routinely handle hazardous chemicals. Other companies acquire occupational health services from the many consulting firms or group medical practitioners as well as from some state and community offices.

In order that all work-places may be evaluated properly, there must be standards for the concentrations of hazardous substances in the occupational environment. Approximately 500 startup standards for the most common hazardous substances have been published in the various amendments to Part 1910 of Title 29—"Occupational Safety and Health Standards," *Federal Register*, Volume 36, No. 157, pages 15,101 through 15,104, Aug. 13, 1971.

NIOSH Criteria Documents

Development of criteria documents, which are used as a basis for preparing new standards for substances or for recommending more complete standards, is a continuing NIOSH activity under OSHAct.

These criteria documents are publications that are prepared from a critical evaluation of all published medical, biological, engineering, chemical, and trade information and data for the purpose of establishing the concentration of a substance in the occupational environment that has been found to cause no harmful —"toxic"—effects in people working for eight hours per day, five days per week, for a normal work life.

In order to assure that the standards promulgated reflect the best information available at the time of preparation, all individuals with publicly available information should respond to notices published in the *Federal Register*.

In assuring that the most valid information is used, an advocate-adversary approach is utilized. That is, care is taken to call in expert critics who have specific experience in handling and controlling the hazards of a substance, and who not only represent widely divergent views to the proposed document, but also represent organized labor, industry, government, scientific associations, and universities.

The resulting criteria document thereby provides a valid detailed support for the standard recommended by NIOSH. The Secretary of HEW, after review, then submits the criteria document with the recommended standard to the Secretary of DOL who has responsibilities for promulgating the final standard.

The U.S. Occupational Standard (USOS) includes, within the limits of technical feasibility: the concentration of the substance that has been determined to provide a safe, healthful work environment for all persons; the

methods for collecting, sampling, and analyzing for the substance; the engineering controls necessary for maintaining a safe environment; appropriate equipment and clothing for the safe handling of the substance; emergency procedures in the event of an accident; medical surveillance procedures necessary for the prevention of illness or injury from inadvertent over-exposure; the use of signs and labels to identify the hazardous substances.

Hygienic Air Quality Standards

It is from a variety of sources—including animal experiments, laboratory studies on humans, plus field and in-plant investigations of exposures—that the two prevailing concepts for expressing safe concentrations (threshold limit value and acceptable concentration) have been taken; see Table 38-C.

The concept that there are concentrations of air contaminants to which nearly all workers may be repeatedly exposed without discomfort or ill effect is fundamental to the practice of industrial hygiene.

The following levels of air contamination are used as guides in the evaluation of health hazards.

Threshold limit values

The American Conference of Governmental Industrial Hygienists (ACGIH) adopts a list of threshold limit values (TLV) each year for more than 450 substances. The 1973 Catalog of Toxic Substances is included as Table 38-E.

These TLV's are expressed as parts of vapor or gas per million parts of air by volume (ppm) at 25 C and 760 mm Hg pressure or as approximate milligrams of particulate per cubic meter of air (mg/cu m). Mineral dusts are expressed as millions of particles per cubic foot (mppcf) based on impinger samples counted by light-field techniques. (The index of exposure to asbestos and quartz can also be calculated on a mass per unit volume basis, that is, mg/cu m.)

The preface of this TLV list states that nearly all workers may be repeatedly exposed to the level in this list without adverse affect. Certain values in this list preceded by a capital "C" should not be exceeded. The basis for the TLV's may be reasonable freedom from

TABLE 38-C

DISTRIBUTION OF PROCEDURES USED TO DEVELOP ACGIH TLV'S FOR 414 SUBSTANCES THROUGH 1968*

Procedure	Number	% Total
Industrial (human) experience	157	38
Human volunteer experiments	45	11
Animal, inhalation-chronic	83	20
Animal, inhalation-acute	8	2
Animal, oral-chronic	18	4.5
Animal, oral-acute	2	0.5
Analogy	101	24

* Exclusive of inert particulates and vapors.

Source: H. E. Stokinger, "Criteria and Procedures for Assessing the Toxic Responses to Industrial Chemicals." Permissible Levels of Toxic Substances in the Working Environment, *Occupational Safety and Health Series No. 20. International Labor Office, Geneva, Switzerland, 1970.*

irritation, narcosis, nuisance, or impairment of health. Some substances have the designation "skin." This refers to a potential contribution to the over-all exposure by absorption through the skin or mucous membrane.

A TLV of 10 mg/cu m or 30 mppcf is recommended for inert or nuisance particulates which are not listed in the table. When extremes of temperature or pressure are encountered, the TLV values should be adjusted to take into account the added stress.

The preface to the TLV list further states that these limits are intended for use in the field of industrial hygiene and should be interpreted and applied only by persons trained in this field. It emphasizes that they are not intended for use or for modification for use:

1. As a relative index of toxicity by making a ratio of two limits

2. In the evaluation or control of community air pollution or air pollution nuisances

3. In estimating the toxic potential of continuous uninterrupted exposures

4. As proof or disproof of an existing disease or physical condition

As stated in Appendix A to the list of TLV's, the test factors to be used for assigning limiting "C" values may be used to determine permissible excursions for the time-weighted average TLV's.

Acceptable concentrations

The American National Standards Institute publishes the Z37 series of Acceptable Concentrations (AC) for a number of substances. (See Fig. 38–3, The Meaning of "Acceptable Concentrations.")

The ANSI Z37 Committee has developed a number of innovations in concept. In the past, a standard was conceived as a single limit, threshold, or a boundary between what is not and what is healthful. But there is no such single boundary; the pattern of exposure in space and time may be variable. Therefore, multiple boundaries are needed. As the boundaries are not sharp lines between what is and what is not healthful, an explanation of the basis and significance of each of these boundaries is needed. These different acceptable concentrations for the same material follow:

• **An acceptable ceiling concentration** for protection of health, assuming an eight-hour work day. (This is equivalent to the old "maximum allowable concentration.")

• **An acceptable time-weighted average** for protection of health, assuming an eight-hour work day. (This is within the limits of the previous specified ceiling and defined peaks beyond this ceiling.)

• **An acceptable maximum** for peaks above

1119

The Meaning of "Acceptable Concentrations"

It is generally recognized that there are concentrations of air contaminants to which a person may be exposed without discomfort or ill effects. This standard defines such concentrations in light of best present day knowledge.

The immediate use of an American National Standard for acceptable concentrations is for guidance in establishing engineering procedures to prevent objectionable concentrations of materials in the air of work places. These concentrations in themselves do not represent a scale of relative toxicity; they represent concentrations below which ill effects are unlikely under the specified conditions.

The term, acceptable concentration, is not defined as a single concept but is related to the duration and pattern of the exposure as set forth in the American National Standard for the individual substance.

In the application of an acceptable concentration established as an American National Standard, it is important to understand the criteria upon which the standard is established. These criteria are avoidance of:

(1) Undesirable changes in body structures or biochemistry

(2) Undesirable functional reactions that may have no discernible effects on health

(3) Irritation or other adverse sensory effects

A purpose for establishing acceptable concentrations is to provide a basis for interpretation of the results of air analysis as an indication of the severity of potential exposure, but not as a means of establishing the presence of occupational disease. Comparison of air analysis results with acceptable concentrations indicates either acceptable conditions or otherwise the need, extent, and urgency of control measures. Sampling and analysis should be performed by competent personnel using a standard method if available, and one that will provide a reliable indication of potential exposure.

It should be kept in mind that:

(1) The acceptable concentrations are the highest concentrations which can currently be justified consistent with the objective of maintaining unimpaired health or comfort of workers or both. It is good practice to maintain lower concentrations when this does not render industrial operations inefficient.

(2) Acceptable concentrations are not precise values sharply dividing what is hazardous from what is safe under the particular circumstances and therefore are not appropriate for use as legal requirements.

(3) These standards are considered to be guides for good industrial hygiene practices and should be applied and interpreted by persons with a full understanding of the basis and limitations of the information from which the standard has been developed.

(4) The levels are applicable only to a single substance and to the exposure pattern as detailed in the standard.

(5) Exposure to mixtures is the rule. When more than one of the components is of hygienic significance, acceptable concentrations of the components should be based on the given mixture and situation.

(6) Routes of absorption other than inhalation, such as through the skin, should be considered in the evaluation of the exposure.

(7) These standards do not apply to community atmospheric pollution as related to general populations.

FIG. 38–3.—The definition of "acceptable concentrations" of air contaminants as developed by the American National Standards Institute Z-37 Committee (1967).

the acceptable ceiling for an eight-hour work day.

In a particular operation there may be a brief excursion of concentration beyond the acceptable ceiling. There are many substances where such excursions would be quite aecceptable, if the remaining exposure is otherwise limited to the acceptable ceiling and acceptable time-weighted average. The Z37 Standards also include any other level or levels considered by experts to be pertinent to a particular substance under consideration.

Many individuals refer to a published TLV or AC as a "go" or "no go" value—something to be used in a specific situation—and overlook the important qualifications of these values set forth by the organizations establishing them. The preface to the annual list of TLV's and the introduction to each of the ANSI Z37 series of AC's provide a basic explanation of the concepts on which these values are based. To use the values intelligently, their definitions should be understood before the values are applied to a specific situation. Knowledge of the type of injury that would result from dangerous overexposure is also needed (Figure 38–4).

As mentioned previously, toxic effects may be broken down into four major classifications:

1. Acute toxicity   3. Narcosis

2. Chronic toxicity   4. Discomfort

The time-weighted average concentration would be the appropriate guide where chronic toxic effects result from overexposure. When protection against narcosis is required, the concentration should never exceed some ceiling value. To minimize discomfort, the instantaneous excursions or peak concentrations are of interest.

Both the ANSI and ACGIH, in defining their respective values, point out that an understanding of the criteria on which the values are established is necessary in deciding how to use the values.

It should be remembered that TLV's and AC's are, even though applied as intended, relative values. The concentration that just approaches the threshold limit is just as hazardous or safe as the one that slightly exceeds it, and conversely. There is no fine line that can be drawn to provide complete safety for all individuals.

In general, it is desirable to limit air contamination to the lowest possible level—consistent with efficient plant operations.

Probably the greatest misuse of both the TLV's and the AC's is the practice of comparing toxic substances by a ratio of their respective AC or TLV. It should also be remembered that neither the TLV nor AC should be considered the sole criterion of suspected occupational disease. The ANSI and ACGIH specifically state that their values are intended to serve as guides for good engineering practice and are not intended to be legal requirements.

"Hygienic Guide" series

The Hygienic Guides Committee of the American Industrial Hygiene Association has prepared guides on many commonly used industrial substances, with references for additional information. Each guide discusses briefly the physical properties, toxicity, hazards, and the engineering and medical controls necessary to control exposures. The *Hygienic Guides* contain recommended maximal atmospheric concentrations, short-exposure tolerance levels, and atmospheric concentrations immediately hazardous to life. The *Hygienic Guides* are published periodically in the *American Industrial Hygiene Association Journal* and are available from the American Industrial Hygiene Association (see References).

Emergency exposure limits

The toxicology committee of the American Industrial Hygiene Association has published a series of emergency exposure limits (EEL), defining single brief accidental exposures to air-borne contaminants that can be tolerated without permanent toxic effects. These limits are not intended to replace accepted safe practices and should be accompanied by appropriate medical surveillance.

EEL's can be used in disaster planning to predict what type of emergency action to take in the event of an accidental spill, maximum credible accident, or other misoperations. The EEL's can also be used to assist in the selection of the proper type of personal protective equipment.

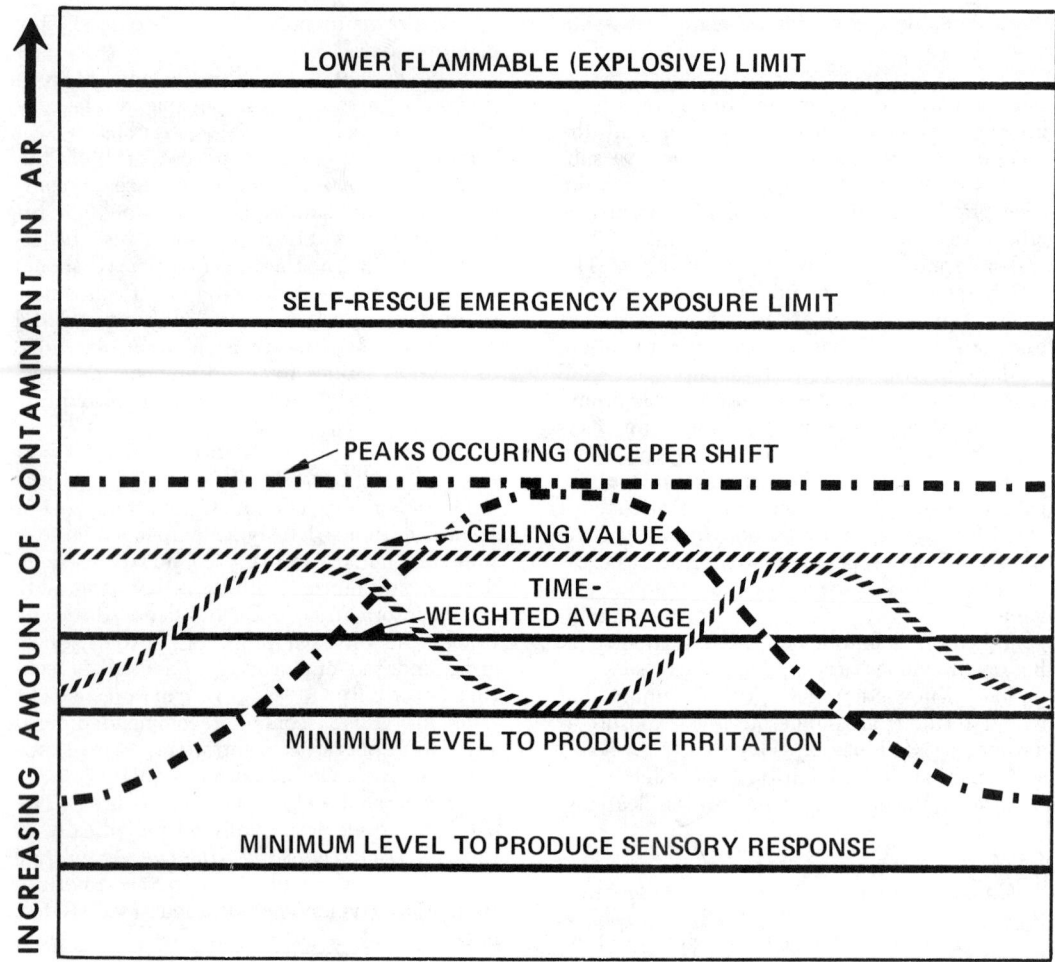

LEVELS OF RESPONSE OF INTEREST

Fig. 38–4.—Knowledge of the type of injury that would result from exposure to these various contaminant levels is of interest to the safety professional.

Radioactive material

Guides are published by the National Council on Radiation Protection and Measurements (NCRP), the International Commission on Radiological Protection (ICRP), and the Federal Radiation Council. These radiation protection standards for air, water, and body deposition of radionuclides, and for exposure to external radiation are set at a level such that occupational exposure at the maximum permissible values recommended is not expected to entail appreciable risk to the individual.

Water quality

Standards are established for bacteriological quality, physical characteristics, chemical substances and radioactivity by the U.S. Public Health Service pertaining to water supply systems used by carriers and others subject to federal quarantine regulations. These standards are accepted as good practice in evaluating the quality and safety of water supplies in industry.

It is highly advisable that the documented justification for the various limits or values be consulted for detailed information concerning mode of action and effect of exposure to a particular toxic material. Such information is given in References at the end of this chapter.

Thus far, the fundamentals of industrial toxicology, toxicity, routes of entry, modes of action, classification of harmful effects, intensity of action, contributions from experimental toxicology, and hygienic air quality standards have been reviewed.

The safety professional may well ask how he can apply this knowledge to his work. A hazardous chemicals information sheet could be prepared to elicit toxicological information from the supplier. The information in this questionnaire would be useful to the medical, purchasing, managerial, engineering, and safety departments in setting guidelines for safe use of these materials in industry. This information would be very helpful in the event of an emergency.

Catalog of Toxic Substances

No attempt has been made to give detailed directions for handling each substance in the following list, since such directions would be repetitious. Instead, enough information is given to alert the safety professional to the primary hazard. This information should be used in conjunction with the procedures for recognition, evaluation, and control of health hazards discussed in the main body of this Manual.

Some of the signs of excessive exposure to toxic substances are indicated in the following list, but these should not be taken as conclusive evidence, particularly in cases of possible chronic exposure. To definitely implicate a specific substance requires evidence of both the degree and length of exposure to it and the absence of other causative agents.

The following selection is a representative sampling of the toxic substances encountered in industry, but it does not describe each of the substances for which threshold limit values have been established. The *Documentation of Threshold Limit Values,* published by the American Conference of Governmental Industrial Hygienists, should be consulted for this type of detailed information. Flammable and explosive hazards of many of the following substances are also given in Chapter 39, "Chemical Hazards."

A

Acetaldehyde (acetic aldehyde) is a flammable liquid with a low flash point and a boiling point just at normal room temperature. It is, consequently, a fire hazard comparable to ether in intensity. If kept in any large quantity, it should be stored with the precautions appropriate for ether.

Acetaldehyde processes should be well enclosed and isolated because of its intense irritant properties. It affects primarily the skin and the upper respiratory passages because of its relatively high solubility in water. A high concentration may produce sudden death from spasm of the throat. Lower concentrations produce coughing and inflammation of the nose, throat, and eyes.

Paraldehyde, the solid form to which acetaldehyde polymerizes, has long been used in medicine as a narcotic and soporific. In this form, its irritant properties are greatly reduced and the narcotic effect can be observed. Paraldehyde can cause chronic poisoning involving fatty degeneration of the liver and cardiac muscles, psychic disturbances, and irritation and edema of the respiratory system.

Acetanilide is more likely to be seen as a medication, used as an analgesic and antipyretic, than as an industrial material. It is an aromatic amine and can affect the blood hemoglobin and so interfere with respiration if it is absorbed in large quantity, particularly if the exposure is chronic. Those suffering from chronic overdoses are apt to show severe anemia, reduced white blood cell counts, progressive weakness, and severe headaches.

Acetic acid, glacial (concentrated), can readily penetrate the skin, producing blisters, dermatitis, and ulcers. Even at room temperature, the vapor is highly irritating to the eyes, nose, and throat. If acetic acid is swallowed, it produces a burning taste in the mouth and esophagus, followed by pain in the stomach which may extend rapidly over

1123

the entire abdomen. The pain and other distress may last for a few days or for several weeks, depending upon the concentration and quantity of the acid taken. The signs of serious cases are vomiting, labored, noisy breathing, coughing, then collapse. Death may follow the swallowing of two tablespoonfuls of the acid by an adult. The lethal dose, however, is not definitely known. Acetic acid is one of the few materials which expand on freezing, so that to prevent breakage of containers the pure acid must be stored above its freezing point, 60 F (16 C).

Acetic anhydride is quite stable to hydrolysis at room temperature; so it is used in preference to acetic acid in many applications where the highest possible concentration is desired. Its vapor is even more irritating than that of acetic acid.

Acetone (dimethyl ketone) is a flammable liquid that should be handled and stored with precautions against fire and explosion. In spite of the large quantities of acetone used in industry and its high volatility, there are no reports of serious industrial poisoning. Experimental work has shown that acetone is a narcotic. Overexposure will lead to moderate irritation of the eyes, nose, and throat and to headache, stupor, and a general feeling of oppression. The absorbed acetone is eliminated slowly, and the symptoms are persistent.

There have been some severe poisonings with acetone in medical practice when it has been used as a solvent in the application of synthetic resin casts. The conditions under which these incidents have occurred are such that it is not possible to tell with certainty whether the poisoning was the result of inhalation of the vapor or of contact of the wet cast with the skin, but they present the definite possibility that prolonged contact of acetone with a large area of skin may be dangerous systemically as well as directly irritating to the skin.

Acetonitrile (methyl cyanide) is a colorless liquid with a distinctive odor detectable at about 40 ppm. It presents a moderate to severe toxicity hazard for acute or subacute exposures. High concentrations (probably over 500 ppm) can cause delayed onset of weakness, nausea, vomiting, chest and abdominal pain, vascular changes, shock, and death. Elevated blood levels of cyanide or thiocyanate have been found in these cases. Acetonitrile liquid or its concentrated solutions readily penetrate rabbit skin, and thus may also be a skin absorption hazard to humans.

Acetylene is a gas that produces a very wide range of flammable concentrations in air. It is, however, a mild narcotic and has been used to some extent as an anesthetic although it has never been one of the popular anesthetic gases.

Since even the lowest narcotic concentration of acetylene is far into the range of flammable concentrations, local exhaust removal should keep the concentration well below 1 percent in an industrial application. There is no evidence of a chronic form of poisoning nor of aftereffects from acute anesthesia with recovery.

Acetylene tetrabromide—see **Tetrabromoethane.**

Acrolein (acrylaldehyde, propenal) is produced by the oxidation of glycerin and apt to be encountered where fats or fatty oils are heated and decompose in the presence of air. It is a severe irritant for the skin, eyes, nose, and throat. Such exposures as are usually found in industry, around furnaces and ovens, will result in irritation and reddening of the eyes and swelling of the eyelids. Inhalation of similar concentrations for some time may result in sore throats or chemical bronchitis and, if the concentration is high, in chemical pneumonia.

Acrylonitrile (vinyl cyanide), when inhaled, has been observed to produce an effect similar to that of hydrogen cyanide; however, there is usually some delay between exposure and onset of symptoms: an increase in the concentration of thiocyanates circulating in the blood serum and excreted in the urine. It has been proposed to use the determination of thiocyanate in the urine or in the serum as an index of exposure. Acrylonitrile can also penetrate the skin to contribute to the exposure.

Aldrin (1,2,3,4,10,10-hexachloro-1,4,4a,5,8,-8a - hexahydro - 1,4,5,8 - dimethanonaphthalene) is a chlorinated hydrocarbon insecticide with the effects on the liver and kidneys which are typical of the chlorinated hydrocarbons. It is somewhat more toxic to warm-blooded forms of life than is DDT or chlordane and has a considerable tendency to accumulate in the body.

Aliphatic thiocyanates (lethanes, Thanite) —see **Thiocyanates (aliphatic).**

Allyl alcohol is used as a chemical intermediate and as a fungicide and herbicide. It is irritating to all tissue. The response to skin and eye contact may be delayed several hours and then produce muscle spasms and surface irritation.

It causes pulmonary edema upon excessive overexposure. It is readily absorbed through the lungs, gastrointestinal tract, and skin. Absorption leads to widespread irritation of the visceral organs, kidney damage, changes in blood pressure, convulsions, and ultimately death. Rather severe eye irritation is produced at vapor concentrations near 25 ppm.

Allyl chloride, an intermediate for chemical syntheses, is moderately toxic for acute ingestion and skin and eye contact and highly toxic for acute or chronic inhalation. It can be absorbed through the skin but rather large amounts (2.2 gm per kg) are required to cause death. Skin and eye irritation may be delayed several hours and then become evident as muscle spasms as well as surface irritation.

Animals exposed to allyl chloride vapor showed lung and eye irritation and kidney damage for acute exposures. Chronic exposures produced liver and kidney injury alone. No systemic intoxication of humans has been reported.

Allyl glycidyl ether. The effects of exposure to this compound observed in man are primary irritation and occasional sensitization; the potential exists for cross sensitization with other epoxy agents. This compound is moderately to severely irritating to the skin, eyes, and nose so that an excessively high concentration could not be tolerated. Workers that were accidentally exposed to the vapors and/or liquid complained of dermatitis, itching, and swelling. Prior to handling this compound, a program of instruction in good personal hygiene should be instituted. Persons having a history of sensitization to epoxy compounds must be protected from contact with allyl glycidyl ether.

Aluminum has not been the source of proved cases of industrial disease and at one time was recommended for the prophylaxis and treatment of silicosis because of the effect of aluminum metal and aluminum oxide in depressing the solubility of silica.

There have been some reports from Germany of lung damage from inhalation of aluminum dust, particularly in grinding aluminum parts. These changes are said to be an increase in connective tissue in the lung, although not in a nodular fashion as is seen in silicosis. Studies of workmen engaged in the same sort of jobs in this country have not revealed similar changes.

Aluminum chloride. Aluminum salts are astringent. Contact may harden and tan skin, resulting in fissuring. Aluminum chloride may act as a sensitizer and produce contact dermatitis. It is likely that the effect on lungs of exposure to aluminum chloride dust or liquid is intimately associated with the particle size and purity of the material involved.

Aluminum oxide has given rise to several reports from European countries and a few from this continent of pneumoconiosis in the manufacture of alumina. The exposure in this job is to aluminum oxide and silica freshly made in an electric furnace. The condition, known as Shaver's disease, is not identical with silicosis and apparently arises only in the handling of the alumina directly out of the electric furnace where the material is dehydrated. There have not been any reports of lung damage from the handling of aluminum oxide hydrate in the form of bauxite.

2-Aminopyridine has been observed in one instance to produce headache, flushing of the face, and temporary elevation of the blood pressure in a person who was exposed for about 5 hours to a concentration of about 5 ppm.

1125

Ammonia is supplied as the anhydrous gas in cylinders and tanks and as solutions of various concentrations in tanks or in glass carboys. The major hazard is accidental release of the gas. It is a strong irritant and can produce sudden death from bronchial spasm. Concentrations small enough not to be severely irritating are rapidly absorbed through the respiratory tract and metabolized so that they cease to act as ammonia.

Wherever ammonia is used in quantity, respiratory equipment should be available for emergencies. Gas masks are useful for concentrations up to about 3 percent, above which severe skin irritation will prevent a prolonged stay in the area. Explosive concentrations of ammonia can be produced when the contents of a tank or of a refrigeration system are released into an open flame, boiler fire, or arc.

Amyl acetate (banana oil) is commonly used as a solvent in many industries, particularly for nitrocellulose lacquers. Inhalation of the vapor in high concentration produces irritation of the nose and throat, a feeling of oppression in the chest, palpitations, cough, headache, confusion, and nausea. There is a rather rapid habituation to the odor, and after some exposure the results are less irritating. It is a fairly good fat solvent that can render the skin dry and scaly and lead to cracking and irritation.

Amyl alcohol is widely used as a solvent and produces qualitative effects much the same as those of ethyl alcohol intoxication.

Aniline (phenylamine) is a high-boiling flammable liquid, that can be readily absorbed through the unbroken skin. Poisoning can also occur from inhalation of the vapor and from swallowing of the liquid.

It is primarily a blood and nervous system poison. The most notable signs of poisoning are rapid breathing followed by blue color, particularly of the lips, ears, and fingertips, shuffling, staggering gait, and fixed, hesitant speech. Extremely severe poisoning may cause tremors or convulsions and arrest of respiration. If the clothing is saturated with aniline, the symptoms of acute poisoning may appear with remarkable suddenness. Aniline poisoning may seem so sudden because the early stages are without symptoms except for the blue color of the lips. Poisoning may be advanced and the individual still show no signs except for this color. The first subjective sign is usually a severe, persistent headache.

Clothing saturated with aniline should be removed immediately and the aniline removed from the skin. It is only 3 or 4 percent soluble in water and even more insoluble in soap, but can be readily sponged off with vinegar, lemon juice, or any acid dilute enough not to produce skin burns.

In case of an accident, however, one should not wait to get acid to remove the aniline but should immediately take off the clothing and get under an emergency shower. The skin should be immediately and thoroughly cleansed with soap and copious amounts of water, warm if possible. The running water will mechanically remove some of the aniline, at least the gross contamination, and the traces can be removed with dilute acids or alcohol when they are available. Secure a physician's services.

Persons who display symptoms of aniline poisoning without having had skin contact with the liquid should be removed to fresh air immediately, given a mild stimulant such as black coffee (never alcohol), and kept quiet and warm. They should be put under medical supervision and kept under observation for some time, since the symptoms occasionally recur. If breathing has stopped, artificial respiration should be applied.

Floors and workrooms where aniline is to be handled should be impervious, and the operations should be enclosed as much as possible and locally exhausted. Workrooms should be kept scrupulously clean.

Workers should have clean clothing on every shift and should immediately change clothing which becomes saturated with aniline. Even a single glove or a part of a shoe saturated with aniline may cause severe poisoning. In both infants and adults, severe poisoning has resulted from skin contact with laundry marks put on with an aniline ink.

Anthraquinone-type compounds (aloes, cascara, senna, and rhubarb) are all strong cathartics and are known in medical practice for their irritant action on the gastrointestinal system. In industry, these compounds can produce lesions of the eye, causing lights to appear to have a halo.

Antimony and its compounds are irritating to the skin and mucous membranes and are systemic poisons. It resembles the action of lead in causing a metallic taste in the mouth, vomiting, colic, loss of appetite and weight, but diarrhea is much more common. Effects of antimony resemble those of arsenic primarily in the dermatitis, which may start as an inflammation of the hair follicles and progress through pus formation and sloughing to leave a contracted scar.

Chronic inhalation of antimony trioxide produces a reduction in white blood cells and damage to the liver. Chronic exposure can produce much damage even when the total amount absorbed is less than the LD_{50} for a single acute dose.

Arsenic, in the form of arsenic trioxide, produces inflammation of the stomach and intestines and causes nervous collapse. However, cases of severe industrial poisoning are rare. There is some contradiction in the literature as to the acute toxicity of arsenic trioxide, and it is possible that this contradiction is caused by variations in impurities within the compound.

Men who handle the solid compounds of arsenic may suffer from a husky voice and cough, perforation of the nasal septum, ulceration of the skin, loss of hair and nails, warts, and sometimes skin cancers which tend to follow a slow and not very malignant course. These troubles are all severe nuisances, but they are practically never fatal and seldom disabling.

Arsenic is a highly cumulative poison, being deposited in all the tissues of the body, with the largest quantities in the liver and kidneys. There are also considerable deposits in the hair and nails where arsenic may be detected about two weeks after the beginning of exposure and where it persists for months and sometimes for years after the exposure is terminated. Inhaled arsenic is excreted to a considerable extent through the kidneys and can be detected in the urine, within a few hours of the start of the exposure and for a long time after.

The compound BAL (2,3-dimercapto-1-propanol) was developed as an antidote for arsenical gas poisoning. It is effective in prevention of poisoning from otherwise harmful doses of the arsenic compounds in general, and in the treatment of existing cases of poisoning, although cases have been reported in which BAL was not effective.

Arsine (arsenic hydride). Severe and acute industrial poisonings with arsenic are likely to be the result of inhalation of arsine, often produced accidentally when hydrogen is generated in the presence of a material which contains arsenic. The gas has been reported in the acid cleaning of filter plates, in the handling of wet dross from a light metals plant, in the cleaning of tank cars which have held sulfuric and hydrochloric acids, in the cleaning of pipes with an acid contaminated with arsenic, in the precipitation of cadmium with metallic zinc, and in other operations.

Arsine gas causes a characteristic syndrome of massive hemolysis leading to renal failure. It is associated with nausea, vomiting, diarrhea, change in skin color, disturbance of vascular tone, electrocardiographic abnormality, and liver disfunction. If the patient survives long enough, manifestations of chronic arsenic poisoning may appear. After exposure, a latent period of 1 to 36 hours may ensue. Inhalation of 250 ppm is instantly lethal. Exposures of 25–50 ppm for half an hour are lethal and 10 ppm are lethal after longer exposure. The mean lethal dose (LD_{50}) is unknown in man, but in small mammals is about 0.5 mg/kg. Arsine gas can be fixed by tissues other than blood. It binds to the kidneys and liver and inhibits respiration of slices of these tissues *in vitro*. In low concentration exposures, clearance of arsine from plasma, where it dissolves into the erythrocyte, is rapid and efficient. When high concentrations of arsine gas are suddenly dissolved in plasma, in large acute exposures, the amount of circulating arsine may exceed the binding capacity of the erythrocytes, and the gas may damage vital organs directly. This would explain death due to arsine poisoning before hemolysis occurs.

The possibility of arsine formation should always be considered when there is freshly formed hydrogen around ores of the heavy metals, since arsenic is widely distributed in small amounts and only a trace of it is required to produce enough arsine to be troublesome.

Asbestos is a hydrated magnesium silicate found in the minerals chrysotile, amosite, crocidolite, and tremolite. Practically all the commercial asbestos used in this country is chrysotile. It is used almost exclusively for heat insulation or for textiles which must resist high temperature, such as heat-resistant clothing and brake linings.

Inhalation of excessive quantities of asbestos fiber can produce a fibrosis in the lungs similar to that found in silicosis but somewhat milder and usually not so rapidly progressive. Like silicosis, the condition develops and advances slowly and is equally resistant to treatment. There is also good evidence that inhaled asbestos causes lung cancer.

Asphalt. Although asphalt itself is considered to be substantially nontoxic, it may cause inflammation or dermatitis on contact in some individuals. Careful supervision of workers' personal cleanliness is essential in the prevention of dermatitis or other ill effects. Skin contact with the asphalt should be avoided—individuals working with asphalt should wear suitable protective equipment, including goggles to protect the eyes. The effect of exposure to vapors from liquid or cutback asphalt depends on the particular solvent added to the asphalt. Steps should be taken to hold the solvent vapor concentration in the breathing zone below its threshold limit value. Where operations require heating or grinding asphalt inside a building, suitable ventilation should be provided.

B

Barium, in the form of its soluble salts, is toxic on both ingestion and inhalation and highly caustic when applied to the skin. The carbonate and sulfide are sufficiently soluble to be toxic, although not very caustic. The sulfate, which is used as a contrast medium in X-ray work, is too insoluble to show any toxicity, and soluble sulfates are specific antidotes for ingested barium salts.

Since these salts produce violent stimulation of all muscle, upon ingestion they produce severe disturbances of the digestive system. After absorption, the barium salts increase blood pressure by constriction of the arterial muscle, slow down heart beat by an action similar to that of digitalis, and first excite and then paralyze the central nervous system. In spite of the high inherent toxicity, there are few cases of industrial poisoning with barium.

Barium is a heavy metal, and the dust of the insoluble salts will collect in the lungs and give rise to the condition known as baritosis. Its nodular appearance on X-ray film may be mistaken for that of silicosis. Men showing the sign of baritosis are apparently not disabled or inconvenienced in any way.

Benzene (benzol) is a colorless, flammable, volatile liquid with a rather pleasant aromatic odor. Its fire hazard is between ethyl alcohol and ethyl ether. The greatest hazard is that of chronic poisoning by inhalation of comparatively small amounts over a long period of time. It is one of the two or three most dangerous organic solvents in commercial use; and acts primarily on the blood-forming organs, producing severe anemia, bleeding under the skin, and great reduction in the ability of the blood to clot.

The most convenient measurement of damage is given by a blood count. People who may be exposed to benzene vapor should have blood counts at regular intervals and significant changes in the count should be closely investigated. Headaches, dizziness, fatigue, loss of appetite, irritability, nervousness, nosebleed, and other signs of abnormalities of the blood may or may not be present in chronic poisoning.

A single heavy exposure to benzene vapor may produce a serious acute effect, similar to an anesthetic with special affinity for the central nervous system. The first sign is usually exhilaration, followed by sleepiness, dizziness, vomiting, trembling, hallucinations, delirium, and unconsciousness. The main first aid requirements in acute poisoning are removal of the person to fresh air, inhalation of oxygen, and maintenance of bodily warmth.

Skin contact with benzene should be avoided, partly because like other powerful solvents it produces dryness and cracking of the skin, but primarily because it will almost certainly involve excessive inhalation of the vapor.

Benzene hexachloride—see **Hexachlorocyclohexane.**

Benzoyl peroxide. The principal hazards of benzoyl peroxide are due to its flammable and explosive properties. Its principal effects are related to irritation and sensitization upon contact with the skin. Systemic toxicity in man has not been reported. No objectionable subjective symptoms were experienced when employees were exposed to dust concentrations ranging from 1.34 to 5.25 mg of benzoyl peroxide per cubic meter of air.

Concentrations of 12.2 mg/cu m and higher resulted in pronounced irritation of the nose and throat. The analysis of the air-borne dust showed that it contained potassium alum in addition to benzoyl peroxide, but the effect produced by the alum was considered to be negligible.

Benzyl chloride is a colorless liquid used in a variety of chemical syntheses. Its vapor is a potent irritant to the eyes, nose, and throat and may cause lung edema. The liquid may cause severe corneal injury.

Beryllium. Among the occupational diseases produced by beryllium, the first condition reported (in 1943) was an acute chemical pneumonitis, which occurred among workers exposed to beryllium compounds and was attributed largely to the fluorides.

A delayed reaction to beryllium, characterized as chronic and appearing as late as two or three years after the exposure which caused it, was reported some time later. Even more recently, it has been reported that the delay between last exposure to beryllium and the appearance of symptoms may be as long as ten years. This condition showed a granulomatous lesion of the lungs as the most striking effect, accompanied by general metabolic disturbances and anemia so that the individual had continuous weight loss and great weakness in addition to a chronic cough. A large number of cases of dermatitis attributed to beryllium and beryllium compounds have also been reported.

Although as recently as 1940 beryllium was thought to be reasonably harmless, it has since been declared that it is unquestionably a toxic agent and in such small quantities as to be among the most toxic chemically of all elements yet investigated. Acute effects have been brought about in animals with beryllium in quantities in the order of millimicrograms. Field survey evidence shows that still smaller quantities may produce the chronic disease in human beings. "Safe" levels of beryllium exposure may ultimately be set well below one μg/cu m of air.

It has also been established that a worker may carry home enough beryllium compound on his clothes to result in illness in some member of his family. Several investigators have demonstrated that the presence of fluorine contributes to toxic action of beryllium.

Recommendations issued by the Atomic Energy Commission for the control of beryllium hazards in atomic energy installations state as follows:

1. The in-plant atmospheric concentration of beryllium at beryllium operations should not exceed two μg per cu m as an average concentration throughout an eight-hour day.

2. Even though the daily average is within the limits of recommendation No. 1, no personnel should be exposed to a concentration greater than 25 μg/cu m for any period of time, however short.

3. In the neighborhood of an AEC plant handling beryllium compounds, the average monthly concentration at the breathing zone level should not exceed 0.01 μg/cu m.

4. An individual showing any sign of chronic berylliosis should be excluded from further exposure to beryllium compounds.

5. There is no reason to exclude women from employment in plants handling beryllium.

6. Pre-employment and termination examinations, including chest radiographs, should be required of employees in beryllium plants.

7. Periodic chest radiographs should be made every six months on personnel exposed to beryllium.

1129

8. Each individual exposed to beryllium should be weighed every two weeks, and if loss of weight is noted the reason for it should be determined.

Bisphenol A. No data on industrial experience in the United States has been published. Bisphenol A has been considered a nuisance dust. Unacclimated personnel can withstand concentrations approaching 15 mg/m³ before complaining of ocular or nasal irritation. An adequate ventilation system should be employed to maintain airborne dust below the TLV. If circumstances arise which do not permit adequate control at the source, then approved respiratory protective devices should be used to prevent inhalation of bisphenol A dust. Clothing must be frequently laundered to prevent continuous skin contact; personal cleanliness is essential to avoid skin difficulties.

Boric acid has been used for many years as a mild disinfectant and a soothing and slightly astringent wet dressing or ointment with little attention to its toxicity.

Poisoning is not usually an industrial problem since the human system processes and eliminates fairly large quantities of boric acid without trouble, but it may result from misguided first aid attempts.

Boric acid is not readily absorbed through the intact skin, but may be absorbed through large areas of skin which are seriously damaged, as in burns or areas of superficial lacerations. If such areas are dusted with boric acid or covered with strong boric acid ointment, toxic amounts may be absorbed.

Boron hydrides are a series of compounds analogous to the hydrocarbons, and include diborane, pentaborane, and decaborane, which have found use as high-energy fuels.

Diborane is primarily a lung-injuring material and produces lung changes in rats in one to four weeks of daily exposure at about 2 ppm. The compound has an odor, but it is not detectable at a low enough concentration to act as a warning of the possibility of chronic poisoning.

Pentaborane, unlike diborane, principally affects the central nervous system. In laboratory animals it produces first a temporary excitement and hypersensitivity which may lead to convulsions, and then a profound depression.

Decarborane is similar in its action on the system to pentaborane. It can be cumulatively absorbed over a short series of exposures and the total dose, although less than a single lethal dose, may still cause death. In practice it is probably more dangerous to handle than pentaborane, since, in addition to being toxic by inhalation, decaborane is rapidly absorbed through the intact skin. Attempts to treat or counteract its toxicity have been unsuccessful.

Boron hydrides should be handled only in a closed system in a well-hooded and ventilated area. Self-contained or supplied air respiratory protection should be provided if the system must be opened (filter-type respirators are not effective). For the handling of decaborane, impervious clothing should also be used to prevent skin contact.

Brass, chiefly an alloy of copper and zinc, may also contain lead. Melting of brass may produce copious quantities of zinc oxide fume which, if inhaled, may give rise to the signs and symptoms of metal fume fever. (Refer to listings for **Copper, Lead,** and **Zinc** in this chapter.)

Bromine is a dark red, low-boiling liquid which produces large quantities of extremely irritant, nonflammable vapor at room temperature or above. Its physiological effects are about the same as those of chlorine—irritation of all mucous membranes and the lungs and spasm of the glottis when inhaled in fairly large concentration. A bromine concentration of 40 to 60 ppm is reported to be dangerous on short exposure.

The irritant action of bromine is not related to the bromide intoxication seen from time to time from continued overdoses of bromides as sedatives. In bromide intoxication, the most notable effects are on the central nervous system and are seen as slurring of speech, somnolence, difficulty in walking, and various psychic disturbances, depending on the personality of the individual involved.

Bromochloromethane (methylene chlorobromide). Knowledge of effects in humans is limited at present. Anesthesia is the primary hazard from single exposure to high concentrations and organic (liver) injury is likely only if severe anesthetic effects are apparent. Repeated exposure to moderately excessive concentrations is likely to result in slight liver injury. The material has a low oral toxicity and does not present a problem from skin absorption. Bromochloromethane is slightly irritating to the eyes and has a defatting effect on the skin.

Fire extinguishers containing bromochloromethane should only be used on outdoor or other fires in areas of good ventilation. If used in small, confined, or poorly ventilated locations, the area should be evacuated immediately. Since pyrolysis products of bromochloromethane, like other halogenated hydrocarbons, are more toxic than the compound itself, these areas should be ventilated thoroughly before they are reentered. Gross skin contact should be prevented; suitable eye protection should be provided.

1,3-Butadiene is a colorless liquid, which is mildly narcotic similar to gasoline, although it can produce very high concentrations in the air very rapidly because of its low boiling point. It is mainly hazardous because of its flammability and because, if exposed to the air or to oxygen in the absence of antioxidants, it forms peroxides rapidly. These peroxides are explosive and sensitive to both mechanical shock, such as a hammer blow, and to heat.

1-Butanol (n-butyl alcohol) has a toxicity greater than that of ethanol, and the vapor is much more irritating, producing coughing, irritation of eyes, nose, and throat, dizziness, and nausea. Severe eye damage may result if overexposure is permitted.

Butanol is used as a solvent for various plastics, particularly in lacquers and rubber cements. It is more toxic than the lower alcohols, but the hazard is somewhat alleviated by its lower volatility, which limits the exposure at room temperature to about 8600 ppm.

2-Butanone (methyl ethyl ketone) is a flammable liquid with an odor objectionable to most people, and is highly irritating to the eyes, nose, and throat in concentrations which are not systemically dangerous. The odor is

probably a warning only on first exposure because the sense of smell is rapidly dulled by continuing exposure.

In the cementing of plastic raincoats, there have been a number of long-continued exposures at concentrations of 600 to 800 ppm with peak concentrations running well over 2000 ppm. Under these conditions the employees complained of headaches and nausea, and there were some cases of illness and unconsciousness with no apparent aftereffects.

n-Butyl acetate is widely used as a solvent in lacquers and in cements based on either natural or synthetic resins or gums. Like most of the aliphatic esters, it is a relatively innocuous material. In man, it produces symptoms of mild irritation of the nose and throat, irritation of the eyes, and finally narcosis. It is not stored long in the body and is not cumulative.

Butyl Cellosolve—see **Ethylene glycol monobutyl ether.**

n-Butylamine is a colorless, flammable liquid with a strong, ammonia-like odor. The liquid produces severe damage to the skin and eyes, and the vapor is highly irritating to the upper respiratory tract and eyes. Inhalation of approximately 10 ppm may produce mild headaches and flushing of the skin and face. Its chronic exposure hazard is moderate and cumulative effects have not been observed.

Butyllithium (in hydrocarbon solvents) has the characteristic odor of the particular solvent. Due to the extremely low vapor pressure of butyllithium itself, the solvent presents the major hazard as far as vapor inhalation is concerned. For skin contact, due to leaks, spills, or splashes, the first effect is the effect of the solvent in removing fats and oils from the skin. The second effect is the reaction of butyllithium with the moisture of the skin to form lithium hydroxide which can cause a severe caustic burn. When water and butyllithium solvents are mixed, local over-heating can result and cause ignition of the solvent.

If a solution of butyllithium is poured on dry or damp organic materials, such as excelsior, paper products, or sawdust, sufficient heat can be generated to char or ignite the material if the concentration is above the pyrophoric limit. Solutions of butyllithium should never be exposed to the atmosphere but rather handled in enclosed systems or under inert dry atmosphere.

Emergency showers and eye baths should be placed in convenient locations. Every employee should be carefully instructed, and understand that direct contact with the skin requires immediate flushing with large quantities of water to the affected area.

Butyraldehydes have a relatively low degree of toxicity. Although there is a lack of detailed information on the subject, no cases are known of either acute or chronic intoxication due to exposures to butyraldehydes. Butyraldehydes produce irritation and burns of the skin and eyes following contact of short duration. Butyraldehydes have good warning properties due to a characteristic pungent aldehyde odor.

Butyraldehyde is a highly volatile flammable liquid readily forming explosive mixtures with air. It has a low flash point and low ignition temperature and is readily ignited by static sparks of relatively low energy. It is a reactive chemical and the heat released by an uncontrolled reaction will soon cause boiling and either pressure increase or release of large volumes of flammable vapor. Butyraldehydes undergo rapid oxidation; any spills that are wiped up with rags can lead to fires in the rags.

Animal experiments have shown that large doses of butyraldehydes produce both depression of the central nervous system and anesthesia so that the symptoms of acute intoxication in man (such as drowsiness, incoordination, and headache) might be anticipated. The animal experiments also suggest that these systemic symptoms may develop from exposures to concentrations below those causing uncomfortable irritation of the eyes, nose, and throat.

c

Cadmium metal is easily flammable in the form of dust or fine powder. The finely powdered oxide produced in cadmium fires is acutely poisonous, and several cases of serious poisoning incurred in fighting fires involving cadmium metal have been recorded. No attempt should be made to approach such a fire without a supplied air respirator, universal gas mask with smoke filter, or metal fume respirator.

One of the commonest sources of acute industrial poisoning is the heating or welding of cadmium plated metals. This job should be done only under efficient local exhaust or by workmen equipped with supplied air respirators.

Industrial poisoning by swallowing of soluble cadmium compounds is not common, but a number of nonindustrial cases from the preparation of acid foods in cadmium plated utensils have occurred.

There is some disagreement as to the possibility of chronic cadmium poisoning. One examination was made of 38 men working in the production of alkali storage batteries and exposed to atmospheres ranging from three to 900 mg of cadmium and almost equal amounts of nickel per cu m. These men complained of fatigue, nervousness and irritability, thirst, cough and shortness of breath. In about half of the workmen with long periods of employment, the sense of smell was impaired. Of the 19 men with more than eight years of work in the battery plant, 15 showed protein in the urine, 12 suffered more or less from loss of the sense of smell, and 14 showed impaired kidney function. Fifteen of 17 men examined with the spirometer showed impaired lung function, and all of them complained from time to time of shortness of breath. In four cases the changes in the lungs were noticeable on X-ray examination, and six other cases showed signs of inactive tuberculosis. One case had active tuberculosis.

Other examinations were made on 20 workmen from a cadmium smelter at repeated intervals over a period of 3 months. These men had exposures varying from 6 months to 22 years, and the atmospheric concentrations in the plant ran from 0.04 to 31.3 mg of cadmium per cu m of air. Exposures to the highest concentration, however, lasted only two to three hours per day, and the men were rotated so that total exposure to the highest concentrations was comparatively limited. The most characteristic finding was a yellow ring on the teeth of the men with long exposure. There were no other signs and no subjective symptoms suggestive of cadmium intoxication. They did show signs of cadmium absorption in the isolation of cadmium from the blood and urine.

It is possible that these two reports differ because of the extremely high concentrations of airborne cadmium noted in the first report.

Calcium is a very active light metal not ordinarily seen in the pure form. Its compounds are generally not toxic. The oxide (quicklime) is highly alkaline and caustic in its action on the skin and is an intense irritant to the mucous membranes of the respiratory system. It is reported to have caused pneumonia, but the irritation is so intense as to discourage excessive exposure.

Calcium arsenate is one of the widely used arsenical insecticides. It has the cumulative toxic properties of the other inorganic arsenicals, although relatively few cases of arsenic intoxication have resulted from its use in agriculture. More common than intoxications are skin effects because it is quite alkaline and, consequently, very caustic to all tissues.

Calcium carbide presents no fire hazard as long as all sources of water are excluded from it; when exposed to moist air or water, acetylene gas is generated. Calcium carbide reacts with the moisture of the skin, eyes, and respiratory tract. The heat of reaction is not enough to cause harm in most cases but the residual lime has the irritating factors of strong alkali. Because the dust may cause irritation and coughing at the time of exposure, eye wash fountains should be available in areas where calcium carbide dust might enter the eye. Employees should be instructed to wash thoroughly with soap and water and preferably to shower after work. They should change work clothes often enough to avoid wearing clothing heavily contaminated with calcium carbide dust. The effects of calcium carbide are limited to action at the site of contact with a moist surface. First aid is directed towards removal of the irritant (residual lime).

Calcium cyanamide is a solid inorganic amide produced by roasting calcium carbide in an atmosphere of nitrogen. It is widely used as a fertilizer ingredient because it hydrolyzes to form calcium carbonate (limestone) and ammonia. The same property makes it a very irritating dust to inhale.

Since calcium cyanamide is rather gummy so that a fine dust cannot easily be formed from it, it is generally thought that this material does not penetrate deeply enough into the respiratory system to do serious lung damage. There has been one report of two cases of chemical pneumonia following exposure to calcium cyanamide.

Carbitol—see **Diethylene glycol monoethyl ether.**

Carbolic acid—see **Phenol.**

Carbon dioxide has been reported as the cause of death when encountered in asphyxiating concentrations, probably several hundred thousand parts per million. It is weakly narcotic at 30,000 ppm, decreasing acuity of hearing and increasing blood pressure and pulse. At 50,000 ppm, a 30-minute exposure may be intoxicating, and, at 70,000–100,000 ppm, it may produce unconsciousness in a few minutes.

Carbon disulfide is a highly volatile and flammable liquid with an extremely low ignition temperature and wide explosive range. Carbon disulfide vapor will ignite on contact with a steam radiator. It should be handled in strict accordance with the rules given for the most hazardous flammable liquids.

Unless it must be kept dry, carbon disulfide can be unloaded from tank cars by displacement with water and stored under water. Otherwise, it may be transferred by pressure of an inert gas (nitrogen or helium) or carbon dioxide and stored under the same gas.

Operations employing carbon disulfide should be completely enclosed because of both its high toxicity and the fire hazard. It can produce poisoning by skin contact, ingestion, or inhalation.

In mild or subacute poisoning, the only signs may be an appearance of intoxication and some stupefaction. If the poisoning is somewhat more intense, there may be marked pallor, fainting, nausea, repeated vomiting and sometimes diarrhea, unsteady or staggering gait, hallucinations and mania. Collapse and death may be very rapid after inhalation of the concentrated vapor.

A person acutely intoxicated with carbon disulfide should be promptly removed from the toxic atmosphere, kept warm and as quiet as possible. If breathing has stopped, artificial respiration should be applied.

Chronic poisoning from long-continued exposure to comparatively low concentration is considerably more important to industry than is acute poisoning. Such exposures affect the nervous system and may result in mental derangement or in impairment of the motor nerve function and in neuritis. Recovery is likely to be good if exposure is terminated before the nervous system is seriously injured.

Carbon monoxide is a colorless, odorless gas generated by combustion of the common fuels with an insufficient air supply or where combustion is incomplete because of a condition such as impingement of burning gases on a cold surface.

In most industrial plants, carbon monoxide is released by accident or by improper maintenance or adjustment of burners or flues, by use of open salamanders in enclosed places, and by internal combustion engines.

Poisoning is entirely by inhalation of the gas and may come on practically without symptoms in an individual who is comparatively quiet. The most common symptoms of complete asphyxia are pounding of the heart, dull headache, flashes before the eyes, dizziness, ringing in the ears, nausea, and sometimes (but not often) convulsions.

The injury is due to the combination of carbon monoxide with the available hemoglobin. This action tends to exclude oxygen and so suffocate the victim by lowering the ability of the blood to carry oxygen. The first aid required is to reduce the loss of heat from the body and to supply oxygen as rapidly as possible, by removal to fresh air and administering of artificial respiration if breathing has stopped. The administration of oxygen or of oxygen mixed with 5 to 7 percent carbon dioxide is the most useful measure.

Carbon tetrachloride is a halogenated hydrocarbon used as a degreaser, solvent, and chemical intermediate. Acute poisoning follows a heavy exposure, such as one would get from the use of carbon tetrachloride as an anesthetic or from spreading out a large area of a cement dissolved in carbon tetrachloride without the precaution of local exhaust.

Acute poisoning usually is characterized by very sudden unconsciousness, probably due to the effect of carbon tetrachloride on the heart, which speeds up disorganized twitching (fibrillation) with no effective beat at all. If the victim regains consciousness, he is likely to suffer from headache, nausea, vomiting, and rapid heart beat for a short time.

The subacute form of poisoning has been of more importance in industry than the acute form. In these cases there is ingestion of the liquid or exposure to a moderately large concentration of the vapor for a considerable time. The major injury was thought for a long time to be to the liver. More recently, it has been realized that the damage is at least as much to the kidneys as it is to the liver, and that in the fatal cases the deaths are as often due to uremic poisoning from kidney failure as to liver failure.

In subacute cases, the complaints and signs are likely to be nausea, vomiting, pain in the abdomen, swelling of the feet and hands, headache, dizziness, loss of appetite, diarrhea, fever, and nosebleeds or hemorrhage into the eyes. It has also been demonstrated repeatedly that the chronic alcoholic or the person who has recently consumed a large amount of alcohol will be severely or fatally poisoned by carbon tetrachloride under conditions harmless for those who have not so indulged.

Chronic poisoning comes from long-continued absorption of fairly small amounts of carbon tatrachloride over a long period. It is not so clear-cut as the other forms, but is characterized by digestive distress, mental dullness, and cirrhosis of the liver.

A number of cases of poisoning have resulted from the use of carbon tetrachloride as a fire-extinguishing fluid in confined spaces. These cases are, of course, complicated by the products of combustion of the fire and also by the products of decomposition of the carbon tetrachloride. Usually they show the signs of lung injury which indicate that they are the result of one or the other of these factors rather than of carbon tetrachloride, as such.

Carbon tetrachloride can be absorbed to some extent through the skin and can also cause a dermatitis by skin contact. The major problem in prevention of injuries from carbon tetrachloride is that of prevention of inhalation of the solvent.

Cellosolve—see **Ethylene glycol monoethyl ether.**

Chlorates—see **Sodium chlorate.**

Chlordane, a chlorinated hydrocarbon with the empirical formula $C_{10}H_6Cl_8$, is effective against various species of insects. Its toxic action is rather similar to that of DDT. The lethal dose of chlordane for rats is approximately 250 mg per kg, as compared to 150 mg per kg of DDT. The fatal dose for man is between 6 and 60 gm or ⅕ to 2 oz. The action of DDT is much more rapid. Its effects are generally over in 24 to 48 hours.

Chlordane is readily absorbed through the skin as well as through other portals. Accidental skin application of 25 percent solution containing little more than 30 gm of technical chlordane has been known to cause symptoms of poisoning within 40 minutes and death before medical attention could be obtained.

Chronic poisoning causes degenerative changes in the liver and kidneys and irritation of the intestinal mucosa with somewhat greater effects than DDT, probably because of the greater cumulative action of chlordane.

Chlorine is a heavy, greenish-yellow, nonflammable gas which is easily liquefied and is supplied commercially as a liquid under pressure in cylinders and larger containers. The handling of these containers is no different from that of other compressed gas cylinders. Because of its fairly low solubility in water, chlorine is irritant to the deeper as well as the upper respiratory system.

A gas mask of the acid gas type will provide protection from concentrations up to about 2 percent by volume in air, at which point skin irritation becomes serious.

A person who has been exposed to chlorine should be taken from the gas area and kept as quiet as possible. Rest is essential. He should be kept warm and quiet, on his back with his head elevated. A physician should be called immediately. Serious effects may be delayed and persons who have been exposed to vapors should consequently be kept under observation for at least 24 hours.

In mild cases of throat irritation from chlorine, milk will give relief. Epinephrine or ephedrine will give relief shortly after exposure when the distress is mainly from bronchial spasm.

Inhalation of oxygen or of a carbon dioxide and oxygen mixture is helpful in chlorine poisoning, particularly if positive pressure oxygen breathing can be given. If breathing has apparently ceased, artificial respiration should be started at once. It will be more effective if oxygen inhalation can be given at the same time.

In spite of the most careful inspection, compressed gas cylinders and larger containers will occasionally leak, commonly because of unnecessarily rough handling. The Chlorine Institute gives the following recommendations for handling leaking chlorine containers:

1. Correct the condition promptly. Telephone your chlorine supplier or any chlorine producer if you need help.

2. Keep on the windward side of the leak and higher than the leak.

3. Permit only authorized, trained personnel equipped with gas masks to investigate. Keep all other persons away from the affected area.

4. If the leak is extensive, try to warn all persons in the path of the vapors.

5. If a leak occurs in equipment in which chlorine is being used, close the valve of the chlorine container immediately.

6. If chlorine is escaping as a liquid, turn the container so that the chlorine gas escapes. The quantity of gas escaping from a leak is about ¹⁄₁₅ the amount of liquid which will escape through a hole of the same size.

7. Do not apply water to a chlorine leak.

8. If a chlorine leak occurs in transit in a congested area, keep the conveyance moving if possible until it reaches an open area. If the conveying vehicle is wrecked,

TABLE 38-D

ALKALI EQUIVALENTS TO ABSORB CHLORINE

Chlorine Container (lb net)	100% Caustic Soda (lb)	Water (gal)	Soda Ash (lb)	Water (gal)	Hydrated Lime (lb)	Water (gal)
100	125	40	300	100	125	125
150	188	60	450	150	188	188
2,000	2,500	800	6,000	2,000	2,500	2,500

shift the container or containers so that chlorine gas can escape. If possible, transfer the containers to a suitable conveyance and transport them to the open country.

9. Pinhole leaks in cylinder and other containers may sometimes be temporarily stopped by tapered hardwood pegs or metal drift pins driven into the holes. First turn the container so that only gas is escaping. Use extreme care in driving the plug, because the wall area surrounding the hole may be thin and crumble. When this emergency measure is taken, empty the cylinder as quickly as possible.

10. At usual points of storage and use, make emergency preparation for disposing of chlorine from leaking containers. Chlorine may be absorbed in caustic soda, soda ash or hydrated lime solution. Caustic soda is recommended because it absorbs chlorine most readily. The recommended proportions of alkali and water are given in Table 38–D.

11. Provide a suitable container to hold the solution in a convenient location. Pass the chlorine into the solution through an iron pipe or rubber hose weighted to hold it under the surface. Discuss the details of such preparations with the chlorine supplier.

Mechanical devices for plugging leaks in chlorine containers of various sizes up to tank cars, available from suppliers of chlorine, can be kept on hand. They are highly efficient if used by trained individuals.

Chlorine dioxide is used as a bleaching agent and for odor control. It is liberated from acid solutions of sodium chlorite and from reactions of sulfur dioxide with sodium chlorate in acid solutions. Dangerous concentrations are detectable by its strong, unpleasant odor. Chlorine dioxide is a greenish-yellow gas, which decomposes in sunlight.

It is a strong respiratory and eye irritant. Acute inhalation exposures cause bronchitis and pulmonary edema, accompanied by delayed symptoms and slow recovery. Repeated acute exposures give symptoms including coughing, wheezing, respiratory distress, nasal discharge, and eye and throat irritation.

Chlorine trifluoride presents a serious hazard due to its oxidizing nature—upon contact with other combustible material it may cause a fire. It is also a known primary skin irritant—skin contact can cause inflammation. It is suggested that the safety precautions outlined in data sheets, or those recommended by the manufacturer of this material, be clearly understood before this compound is used. The literature reports that the pathologic findings were severe pulmonary irritation in dogs and rats exposed to this material compared with no remarkable pathology in the controls. The threshold limit value for this material was set on the basis of acute, subacute, and six-month chronic toxicity studies on laboratory animals.

Chloroacetic acids are similar to acetic acid in their general properties but are stronger. Their strength increases with increasing amounts of chlorine in the molecules. Trichloroacetic acid is nearly as strong as the mineral acids and is extremely corrosive to the skin.

Chlorobenzene (phenyl chloride) has only recently been used in great quantities in industry, and only a few cases of poisoning have been reported. It is a strong narcotic and a paralyzing poison of the nervous system.

Some of the signs of acute poisoning are sleepiness, tremors and muscular spasms, frequently blue lips, fingertips, and ears, and red to chocolate-brown urine. If the liquid has been swallowed, an emetic such as soapy water should be given and the person should be kept warm. Mild stimulants may be given, and oxygen or a mixture of oxygen and carbon dioxide should be provided. Under no circumstances should alcohol, fats, or oils be given. The first aid for poisoning by inhalation of the vapor is the same, except that no emetic need be given. These cases should invariably be put under the care of a physician.

Chlorobenzene should be handled only under efficient local exhaust hoods or, if the chemical must be exposed to the air, with supplied air respiratory equipment. Protective clothing should be supplied to prevent skin contact with the material. It is absorbed through the intact skin and produces skin irritation.

Chlorobromomethane (bromochloromethane, methylene chlorobromide) is a halogenated hydrocarbon used as fire-extinguishing agent, solvent, and refrigerant. Knowledge of its toxic effects is limited. Anesthesia and liver damage may result from acute exposures. Chronic exposures may cause only slight liver damage. Chlorobromomethane also appears to hydrolyze to inorganic bromides in the blood. The compound has a low oral toxicity, is slightly irritating to the eyes, and, although not readily absorbed, has a defatting action on the skin.

Chlorodiphenyl—see **Chlorowaxes.**

Chloroform—see **Halogenated hydrocarbons.**

p-Chloronitrobenzene is used as an intermediate in chemical synthesis. It alters the blood hemoglobin and produces anemia in experimental animals, as is typical of aromatic amino and nitro compounds.

Chlorosulfonic acid is a strong acid which is rapidly decomposed by water into a mixture of hydrochloric and sulfuric acids. For precautions for handling it, see **Sulfuric acid.**

Chlorowaxes are chlorinated naphthalenes and chlorinated diphenyls, used primarily as electrical insulating waxes with other minor industrial uses. They may vary from liquids of fairly low chlorine content, commonly used in transformer oils, to hard waxes of high chlorine content. Their toxicity increases with chlorine content. There is some indication that toxicity tapers off on either end with the maximum toxicity around 50 percent chlorine in the case of the chlorinated diphenyls. All of them have a very low volatility, but when heated give off harmful vapors.

These waxes have two distinctly different effects and either or both may appear. They produce a severe acneform dermatitis by contact with the skin and, if inhaled, may cause liver injury.

Chromic acid is a strong oxidizing agent but not a strong acid. The oxidizing character of the material is so strong that it will ignite ethyl alcohol and similar liquids by contact and char wood, straw, and similar materials, although they do not ordinarily ignite. It should be isolated from materials which can be easily oxidized, although chromic acid itself is not a fire hazard. Empty containers should not be discarded until thoroughly washed with water. Fires have occurred when cans that contained chromic acid were used for waste or trash.

Chromic acid is both poisonous and irritating to the skin. Precautions should be taken against skin contact with the solid or its solutions and against inhalation of dust from the solid or of mist from the solutions.

The major hazard to health in the handling of chromic acid and chromates has always been considered to be inhalation of the mist from chromic acid plating tanks. It causes nasal irritation and perforation of the nasal septum. There have been reports of industrial poisoning from inhalation of the mist of solutions as dilute as 5 percent chromic acid.

An exceptionally high rate of lung cancer in the chromate-producing industry has been reported. The crude death rate from lung cancer was 25 times that to be expected in the average nonchromate-producing industry. If cancers of the lung are excluded, then the ratio of deaths from other cancers becomes 12.4 percent for the chromate industry as compared with 15 percent for a similar industrial group.

There is some evidence that cancer of the lung is produced only by the monochromates and not by bichromates or chromic acid, but this question is still under study. The occurrence of irritation of the nasal passages has no necessary relation to the occurrence of lung cancer.

Since chromic acid is irritant to the upper respiratory passages, it should also be irritant to the lungs if it is inhaled and retained. There is some evidence that this can happen, at least with dilute solutions and at high concentrations of the mist.

Cobalt is a metal chemically similar to iron and nickel and is used as an alloying element and as the cement in carbide tools. Both cobalt and nickel are materials which can and occasionally do cause hypersensitization types of dermatitis in individuals who are susceptible. It has been suggested that some cases of pneumoconiosis in the carbide tool industry may also be cases of allergy to cobalt.

Animal studies have indicated that particulate cobalt metal is an acutely irritating substance when introduced into guinea pig lungs in 50 mg and 25 mg doses. Studies also indicated that a single dose of 5 mg was not lethal to the animal but that repeated doses of such magnitude were lethal. Industrial exposure to cobalt, possibly combined with certain metal carbides and small amounts of silica, is capable of producing serious pneumoconiosis which is initially of an insidious nature.

Copper, in small amounts, is indispensable to life and is not harmful. Relatively large doses (10 to 20 gm) have an acute toxic action when taken orally. Inhalation of copper dust has been reported to cause changes in the gums and mucous membrane lining of the mouth. This is probably local action directly on the tissue rather than general toxicity. Copper fumes, as produced in inert gas shielded arc welding, cause a condition somewhat like the fume fever produced by zinc.

Cresol. Commercial cresylic acid is essentially a mixture of the three isomers (o, m, and p) although it may contain small quantities of other phenolic compounds. The cresols are flammable, but are not exceptionally serious fire hazards. The main hazards are skin burns from contact with the liquid or poisoning.

Poisoning may occur either by swallowing of the liquid or by absorption of the liquid or vapor through the intact skin. Because of the low vapor pressure of these compounds, inhalation is not usually an important source of poisoning. However, it is not safe to enter a tank which has contained cresol or to work over the heated liquid without suitable protection.

Acute poisoning is marked by weakness, dizziness, cold sweat, blue lips, ears, and fingernails, and coma. First aid treatment—after a physician has been called—is administration of an emetic (for ingested cresols) and of measures to combat shock.

Splashes of cresol on the skin produce local anesthesia followed by a burning sensation and itching. The skin first turns red, then white, and later sloughs off. In addition to the skin damage, there is danger of increased absorption through the skin. Use personal protective equipment, preferably of synthetic rubber, to prevent skin contact.

Since the action of cresol on the skin is extremely rapid, prompt cleansing is required to prevent serious consequences. Emergency showers and eye fountains should be available to wash off splashes. The cresols are not very soluble in water, however, so large bottles of alcohol (either ethyl or isopropyl) should be kept on hand for cleansing the skin after the excess liquid has been washed off by the shower. Alcohol should not be used on large skin areas.

If cresol comes in contact with the eyes, they should be washed with copious amounts of water; the mechanical action of the running water will remove much of the chemical. Sterile glycerin should be available in dropper-top bottles for use in the eyes once washing has been discontinued.

Cumene (isopropylbenzene) is used in high-octane aviation gasoline, paint and enamel thinners, and as a chemical intermediate. Its sharp, penetrating odor and irritating properties at low vapor concentrations tend to limit exposures.

Absorption is mainly through inhalation of the vapor or mist, and high concentrations irritate the respiratory tract and produce narcosis. Prolonged or repeated skin contact may cause dermatitis. There are no reports of systemic toxic effects in humans.

Cyclohexane is a hydrocarbon of a type which falls between the aliphatic and the aromatic series, but in its toxicity it is much more closely related to the aliphatics. In animal experiments, the damage from the lowest concentrations causing any effects is seen in the liver and kidneys and is not typical of the particular compound. Especially, there is no sign of the damage to the blood-forming systems which is seen in benzene poisoning. In chronic experiments, monkeys were not injured by concentrations of about 1200 ppm and rabbits were not injured by about 400 ppm.

Cyclohexanol (Hexalin, Hydralin) is a volatile liquid having a menthol-like odor. While the high boiling point limits somewhat the concentration of vapor which will be attained, exposure of men to the vapor has produced headache and eye irritation.

There are no known cases of clear-cut human poisonings from its use. In animal experiments, it has produced poisoning with changes in the brain, heart, kidneys, and liver. The changes are about the same whether the compound is absorbed through inhalation or through application to the intact skin. Minimal liver changes have been seen in the livers of rabbits after exposure to about 200 ppm but not after exposure to 150 ppm. In men, eye, nose, and throat irritation was produced by a few minutes of exposure to 100 ppm.

Cyclohexanone is a volatile liquid of high boiling point. Exposure to cyclohexanone vapors at room temperature produced no acute effects in laboratory animals. By application to the skin and by ingestion, the toxicity of cyclohexanone was slightly less than that of cyclohexanol, but of the same type and of the same order of magnitude. It was observed that the absorption of this compound had a direct effect upon the excretion of glucuronic acid and of organic sulfates in the urine and that, if the same relation holds for man, the determination of these materials in the urine should be a useful measure of the exposure.

Cyclopropane (trimethylene) has been rather extensively used in medicine as an anesthetic since about 1929. A concentration of approximately 6 percent produces light anesthesia in humans, and the deepest surgical anesthesia is produced by about 25 percent. If the oxygen concentration is kept at 20 percent or above, there is arrest of respiration with about 40 percent of cyclopropane, and irregular heart rhythm and fibrillation are likely to occur at about the same concentration. The changes after repeated administrations seem to be edema and fatty changes in the liver, brain damage, and some damage to the heart.

D

2,4-D (2,4-dichlorophenoxyacetic acid) is a herbicide that is moderately toxic on ingestion, the main effect being irritation of the digestive system, with some liver and kidney damage and sometimes congestion of the lungs.

All the data on poisonings are from experiments upon laboratory animals, inasmuch as there are no cases known of poisonings of humans in the manufacture or application of the material. Also, there are no known cases of direct poisoning of domestic animals, although there have been some indirect cases where domestic animals have eaten the products of plants which had been killed and had become toxic as the result of changes within the plant material.

DDT (2,2-bis [p-chlorophenyl]-1,1,1-trichloroethane) has been used extensively for several years as an insecticide, and relatively few cases of intoxication have resulted. It is known to be a poison to the central nervous system and to the liver. Ingestion of a toxic dose is followed by dizziness, nausea, muscle tremors, and convulsions. In experimental animals it has also caused ventricular fibrillation. Experiments have indicated that man is more susceptible than the dog to DDT poisoning.

1139

DDT can be absorbed by ingestion, inhalation, or application of an oil solution directly to the skin. Rabbits have been acutely poisoned by a 5 percent oil solution of DDT applied to the skin.

In the practical application of DDT as an insecticide, experience has been good. Reported cases of human poisoning have resulted from accidents rather than from proper application of the material. Contacts with the oil solution should be avoided, and with power sprayers it is advisable to use an approved respirator if considerable spray kicks back into the operator's breathing zone.

Decaborane—see **Boron hydrides.**

Derris is a powdered plant root, which contains rotenone as its main active ingredient. See **Rotenone.**

Diacetone alcohol is a flammable solvent. Exposure of laboratory animals to the saturated vapor, 2100 ppm, resulted in irritation of the eyes, running of the nose, excitement, and then sleepiness in about 20 minutes. There are no cases of human poisoning on record.

This solvent is liable to be a fire hazard because the commercial material often contains acetone as an impurity in concentration great enough to substantially lower the flash point.

Diatomaceous earth is a form of silica in the amorphous state. Although it apparently is not as active in producing silicosis as its high content of free silica would lead one to expect, there have been a number of reports of silicosis from handling the material where the dust concentrations were high (see Chapter 37, "Industrial Hygiene").

If the temperature is raised to about 1000 F (538 C) in a calcining operation, the material readily changes to cristobalite, which is a vigorous producer of silicosis. It should be handled with the same precautions against dust used with other forms of free silica.

Diborane—see **Boron hydrides.**

1,2-Dibromoethane (ethylene dibromide) is used mainly as a constituent of fumigation mixtures and of "leaded" gasoline. To a limited extent, it has also been used as a fire extinguishing agent and solvent. As an industrial hazard, it is toxic by ingestion, by skin contact, and by inhalation.

On experimental animals it produces local irritation of the eyes and skin if applied directly, but produces severe burns and is absorbed through the skin if it is confined in contact. On ingestion and inhalation, it injures the lungs, liver, and kidneys and depresses the central nervous system.

Dichlorobenzene occurs as the three isomers: ortho, meta, and para. All are flammable liquids with low vapor pressures, but they are sufficiently volatile to produce toxic symptoms if handled in the open in a confined area. They are used primarily as insecticides, as intermediates in the dye industry, and to some extent as solvents. Development of cataract and of liver symptoms from chronic exposure to these compounds in industry has been reported.

Dichlorodifluoromethane (Refrigerant 12) is about as physiologically inert a substance as one could find. It has no real toxic effects until the concentration becomes high enough to reduce the oxygen content of the atmosphere. Since it is a refrigerant gas normally under fairly high pressure, it will freeze tissue upon which it is directed. It may also decompose if brought into contact with an open flame or a red-hot surface and produce irritating and toxic decomposition products.

1,1-Dichloroethane (ethylidene chloride) was formerly used as an inhalation anesthetic but has been completely displaced by safer materials. It is still used to some extent as a solvent in industry. Its acute effect is that of an anesthetic and narcotic, and chronically it characteristically produces symptoms and signs of digestive and central nervous system irritations.

1,2-Dichloroethane (ethylene dichloride) is a commonly used, low-boiling solvent. Qualitatively, its effects are very similar to those of 1,1-dichloroethane.

1,2-Dichloroethylene (acetylene dichloride) is a widely used solvent of low fire hazard and relatively low toxicity. It is a rather sweet-smelling narcotic somewhat like chloroform both in odor and in effects, so that there is no adequate warning in the odor. It can be detected by odor at 250 to 300 ppm. For short exposures, it has no effect on men at 800 ppm and is not thought to have any special chronic effects.

Dichloroethyl ether is used mainly as a dewaxing solvent. It has a pungent, very irritating odor and is mainly an injurant of the eyes, nose, throat, and lungs.

Dichloromethane—see **Methylene chloride.**

Dichloromonofluoromethane (Refrigerant 21) is slightly more toxic than dichlorodifluoromethane, but is still essentially an asphyxiant. The only real hazard from concentrations of less than about 3.0 percent by volume is their possible decomposition on a flame or hot surface.

1,1-Dichloro-1-nitroethane is a very close chemical relative of chloropicrin, which was widely used as a lachrymatory, lung-injurant gas in World War I, and irritates and injures the eyes, skin, and lungs, and the central nervous system. Rabbits and guinea pigs stand prolonged exposures to 25 to 30 ppm without fatalities.

Dichlorotetrafluoroethane (Refrigerant 114) has very low toxicity. It produces some irritation and restlessness at concentrations of 10 percent by volume or higher.

Dieldrin (1,2,3,4,10,10-hexachloro-6,7,-epoxy-1,4,4a,5,6,7,8,8a-octahydro-1,4,5,8-dimethanonaphthalene) is very closely related chemically and in its physiological actions to aldrin.

Diethylamine is used mainly in the dye industry. In animal experiments, all the rabbits survived exposures to both 50 ppm and 100 ppm seven hours a day, five days a week for six weeks. Even at 50 ppm, however, all the rabbits incurred corneal injuries and lung irritation before the test was completed. At 100 ppm the compound also caused injury to the heart, liver, and kidneys. Skin contact causes rapid blistering and necrosis. Penetration of the skin is rapid.

Diethylene glycol monoethyl ether (Carbitol) is widely used in cosmetic products and as an industrial solvent. It is rapidly and easily absorbed through the intact skin, and continued application of large quantities may produce kidney damage. In animal experiments, daily doses of 0.02 and 0.04 cc applied to the intact skin caused practically no observable signs. Comparable doses applied to a man would be 2.8 and 5.6 cc. Chronic toxicity by ingestion is considerably lower, a daily dose of about 91 cc being the least which would be expected to show symptoms. Because of its extremely low vapor pressure, the material should not be a serious inhalation hazard.

Diethylenetriamine is a yellow liquid with a strong, ammonia-like odor. Its health hazard is severe for both acute and chronic exposures. Contact with the liquid can cause severe skin and eye damage. Pulmonary and cutaneous sensitization can occur and remarkably low exposures can induce a high degree of sensitivity.

Diisobutyl ketone is a solvent with a relatively high boiling point. It is extremely irritating to the eyes, nose, throat, and is relatively toxic, more than most ketones. Animal experiments showed liver changes in guinea pigs and rats at all levels of exposure above 125 ppm after 30 days of all-day exposure. Three hours of exposure to 50 ppm did not bother human volunteers, but they were uncomfortable after three hours of exposure to 100 ppm. The animal experiments indicate that there would probably not be organic changes below 100 ppm.

Dimethylaniline is a dye intermediate that is quantitatively somewhat more toxic than aniline, but the actions are qualitatively the same. The quantitative difference is not sufficient to be of any great importance.

Dimethyl ether. A report in the literature indicates that methyl ether concentrations of 50 percent methyl ether in air would affect most personnel. The observation was made that the gas is most unpleasant to inhale, being distinctly suffocating, even when taken with a high percentage of oxygen. From this report it appears that the acute toxicity of methyl ether is very low. Like the other alkyl ethers, the primary physiological effect is that of anesthesia. The effect of long-term exposure to this compound is unknown.

Dimethylformamide is an industrial solvent that may have a defatting action on the skin and may also be absorbed through the skin. There is some experimental evidence of liver and kidney damage.

1,1-Dimethylhydrazine is used in rocket fuel mixtures and is highly toxic on both acute and chronic exposure. It can be absorbed through the skin and is an irritant to both skin and eyes. Symptoms begin with upper respiratory irritation and muscle tremors and, on acute exposure, include excitement, tremors, and convulsions.

Dimethyl sulfate, the sulfuric acid ester of methanol, is a violent irritant because it hydrolyzes very rapidly to give sulfuric acid.

Dinitrobenzene (meta) appears to be considerably more toxic than nitrobenzene. Fortunately its lower vapor pressure makes the hazard comparable to that of nitrobenzene when both are handled in the cold so that the effect of the two compounds is the same.

Dinitrochlorobenzene has been made in large quantities in industry in the past few years. It is a powerful skin sensitizing agent, and with repeated contact almost all the people who handle it develop a severe dermatitis.

Dinitrocresol is very similar in its action to dinitrophenol. In the one report of fatal human poisoning by the compound, the individual, who appeared to have been exposed to the dust at a concentration of 4.7 mg per cu m of air, died with an enormously increased metabolic rate. This is the same picture seen in the numerous cases of dinitrophenol poisoning, even to the great loss of weight prior to death.

2,4-Dinitrophenol is different from the other nitro compounds in that it commonly produces an acute poisoning characterized by high fever, terrified excitement, thirst, sweating, panic, convulsions, and great loss of weight after several exposures. Sudden death from this compound has been reported.

Earlier use of dinitrophenol as a drug in an attempt to control obesity showed that continued application of small doses to increase the metabolism caused chronic damage to the liver and, as a late result, numerous cases of cataract. The important preventive measures are dust control to prevent inhalation, protective clothing to prevent skin contact, and personal cleanliness.

Dinitrotoluene is hazardous but may be handled safely if proper precautions are observed. Toxic levels may be reached by skin absorption, inhalation, or ingestion. Dinitrotoluene is rapidly absorbed by the intact skin and therefore, all direct contact must be avoided.

Dinitrotoluene converts the hemoglobin of the circulating blood to methemoglobin. The inability of this modified form of hemoglobin to deliver oxygen to tissues causes oxygen deficiency and accounts for the symptoms. The color of the altered hemoglobin is changed so that when methemoglobin is present in sufficient concentrations in the blood, the tissues are seen as a bluish color. (Called cyanosis, it is a pertinent clinical finding.)

Dinitrotoluene has been shown to have relatively few biological effects other than its ability to cause methemoglobinemia. There are no chronic effects from single exposures. Some employees may complain of headache, weakness, nausea, or dizziness. At higher methemoglobin levels with more intense cyanosis more severe symptoms of drowsiness, shortness of breath and collapse develop. After removal of all sources of contamination from the patient, the methemoglobinemia easily subsides within 24 to 72 hours. All workers should be carefully instructed and supervised in proper methods of handling dinitrotoluene so as to avoid needless skin contact or inhalation.

1,4-Dioxane (diethylene dioxide) is used chiefly as a solvent for lacquers, paints, plastics, and dyes. Its odor is not pronounced enough to serve as a warning. Its health hazard for acute and repeated exposures is high. In one case, death was reported after several days' inhalation of an average concentration of 470 ppm. Death was accompanied by central nervous system damage, bronchopneumonia, and severe liver and kidney injury. Liquid dioxane may be absorbed through the skin to produce injury.

Diphenyl. No cases of acute or chronic intoxication attributed solely to diphenyl have been reported for man in the published literature. Inhalation is not a serious hazard in industry due in part to the low vapor pressure at room temperature and the relatively good warning properties of this material. Although animal data indicates that diphenyl has little irritating potential, it does indicate that diphenyl can be absorbed through the intact skin. Contact with the skin should, therefore, be avoided to eliminate the possibility of systemic injury. Because moderate air concentrations of this material are irritating, voluntary exposure is limited to levels that are presumably not hazardous.

E

Epichlorohydrin is a primary raw material in the production of epoxy and other resins. Its vapor is highly irritating to the eyes and respiratory tract. Pulmonary edema and kidney damage may occur. Animal experiments with 500 ppm for 4 hr, and 62 ppm for 7 hr per day for 45 days, both produced deaths in the test groups. Inhalations of 16 ppm for 7 hr per day for 90 days produced some bronchial irritation in dogs and monkeys. Epichlorohydrin was also shown to be highly toxic when ingested or applied for 24 hr to the skin of rats and rabbits.

In man, liquid epichlorohydrin produces a transient burning sensation on first skin contact, which may lead to blistering and deep-seated pain within several hours. The liquid is also highly irritating to the eyes. Some cases of sensitization to small quantities of epichlorohydrin have been reported.

Epoxy resins are condensation products of epichlorohydrin and bisphenol-A or other polyhydric compounds. The resin and a hardening or curing agent (amines, peroxides, organic acid anhydrides, amine adducts, boron trifluoride amines, amides, and polyamides) are usually mixed at the time of use. The chief health hazard is the ability of the amine hardeners to cause dermatitis.

Although the curing agents exhibit some degree of toxicity, greater effort is required to prevent local irritation or sensitization during handling than would ordinarily be required to prevent systemic intoxication by the most toxic of them. Cured epoxy resins are considered to be relatively nontoxic.

Ethanol (ethyl alcohol) is an extremely versatile and widely used solvent. When absorbed into the body, it is rapidly metabolized so that heavy exposure is required to build up a substantial concentration in the tissues. Exposure to a high concentration of the vapor will produce irritation of the mucous membranes, headache, nervousness, dizziness, tremors, fatigue, nausea, and loss of consciousness. If the exposure is excessive and repeated, it may produce damage to the liver. Those not accustomed to alcohol will be uncomfortable after prolonged exposure to slightly over 1000 ppm.

Ethyl acetate is a volatile, flammable solvent of low toxicity. Animals exposed to 2000 ppm for long periods show either no discomfort or irritation of the eyes and dryness of the mouth as the only detectable effects.

Ethyl acrylate. Very little information on human experience with ethyl acrylate has been published. Animal experiments have shown it to be moderately toxic and highly irritating by several routes—the respiratory tract is the prime route of absorption, although the skin also presents a route by which physiologically significant amounts of the compound may enter the body.

The odor of ethyl acrylate vapor, readily detectable at 1 ppm, is fairly strong and moderately irritating at 4 ppm. It is believed that most humans will not work at the recommended maximal concentration of 25 ppm for any appreciable length of time. General ventilation should be used to control or reduce ethyl acrylate vapor concentrations to acceptable levels. A self-contained breathing apparatus with full face piece should be used under severe exposure conditions such as would be encountered in moderate spills or equipment failures.

Because ethyl acrylate may penetrate the skin and cause dermatitis and possibly sensitization, all skin areas should be protected against intermittent contact with the liquid. Impervious clothing including gloves of rubber, neoprene or certain plastics should be worn and should be cleaned on both sides after each usage.

Because ethyl acrylate is a flammable liquid with a low flash point, it should be stored and handled away from excessive heat, flame, and sparks.

In the case of over-exposure, the victim should be removed to an uncontaminated atmosphere and administered artificial respiration. In all cases emergency treatment should be instituted promptly and a physician summoned.

Ethylamine is a volatile material with an ammonia-like odor which is identical in its effects with diethylamine. Quantitatively, ethylamine is probably slightly less irritating, but probably not enough to make any practical difference.

Ethylbenzene (phenylethane) is a chemical intermediate and lacquer diluent and is often found in appreciable quantities in commercial petroleum solvents. For both acute and chronic exposures, its health hazard is moderate. Absorption is mainly by inhalation, and eye irritation begins at about 200 ppm. Higher concentrations produce tearing of the eyes, nasal irritation, and narcosis.

Ethylbenzene is considered the most severe skin irritant of the benzene series, although it does not have the deleterious effect on the blood-forming organs that benzene does. No report of injury from industrial use has been recorded. Animal studies of chronic vapor inhalation show no significant effects.

Ethyl bromide (bromoethane) is one of the less toxic halogenated hydrocarbons. It is used as an anesthetic for minor surgery to some extent. Inhalation in high concentrations (5 to 10 percent) or for a long time in low concentration (0.3 percent) will produce lung irritation and edema and injury to the liver and kidneys.

Ethyl chloride (chloroethane) is similar in its effects to ethyl bromide but considerably less toxic. It is used as an anesthetic or analgesic for minor surgery, both by refrigeration and by general anesthesia.

Ethylene chlorohydrin is used as a solvent for resins and lacquers, as a chemical intermediate, and in agriculture to reduce the dormant period of potatoes. It has produced fatal poisonings. It is dangerous, especially because it is not irritating to inhale, and is rapidly absorbed through the skin either as the pure liquid or from water solution. Since this material is both soluble in, and dissolves in rubber, it is absorbed by rubber gloves and penetrates them so that they offer no protection from the liquid or its solution in water and may, in fact, be an added hazard. The lethal dose for a man is certainly not more than 5 ml when held in contact with the skin. Inhalation of 2 ppm of ethylene chlorohydrin for one hour, repeated once, was fatal for rats according to one report.

Ethylenediamine is an extremely hygroscopic liquid which fumes in air. It can be recognized by its ammonia-like odor. It presents a moderate health hazard for both acute and chronic exposures.

Its effects are primarily irritation of the mucous membranes and skin. The liquid causes immediate skin damage with blistering. Skin and respiratory sensitization may follow exposure to low vapor concentrations.

Ethylene dibromide—see **1,2-Dibromoethane.**

Ethylene dichloride—see **1,2-Dichloroethane.**

Ethylene glycol, because of its low vapor pressure, is not an inhalation hazard in industry. Serious poisonings have resulted from either inadvertent or intentional drinking of glycol-base antifreeze. It severely damages the kidneys and injures the brain, mainly because it is catabolized to oxalic acid and precipitates calcium oxalate in the kidneys and brain tissue. It is not particularly irritating to the skin and not much of it is absorbed through the skin.

Ethylene glycol monobutyl ether (butyl Cellosolve) is a colorless liquid much like the other ethers of ethylene glycol in solvent properties. Because of its higher molecular weight, it is less volatile than Cellosolve or methyl Cellosolve. There is little other information about the compound except that it produces inflammation of the kidneys and some blood changes on inhalation. It also has an irritating effect on mucous membranes.

Ethylene glycol monoethyl ether (Cellosolve) has approximately the same action as ethylene glycol but is considerably less toxic. Since it has a low vapor pressure, it is not usually considered dangerous by inhalation except in the form of a mist.

Ethylene glycol monomethyl ether (methyl Cellosolve). Although clinical data are inadequate for valid comparisons, animal studies have shown that the acute toxicity of this compound is about twice that of ethylene glycol, with similar toxic effects. Central nervous depression is observed, as with ethylene glycol, but kidney injury and hematuria (the passage of blood in the urine) are more marked.

Methyl Cellosolve evaporates more rapidly than the other commercially available glycol ethers and will penetrate intact skin. Some workers who are exposed to this substance may suffer from anemia and nervous disorders.

Ethylene oxide (1,2-epoxyethane) is used mainly as a fumigant and as a chemical reagent. As a chemical reagent, it has proved hazardous in the production of explosively violent reactions with mercaptans and with alcohols. As a fumigant, it is both toxic, mostly as a severe irritant to the respiratory system including the lungs, and a cause of very severe forms of dermatitis.

Applied directly to the skin, the liquid and its solutions in water are primary irritants, about a 50 percent water solution being the most irritating concentration. By repeated applications, individuals become sensitized. Three of eight volunteer subjects became sensitized in 19 to 20 days.

Ethylenimine (aziridine, dihydroazirine, dimethylenimine). It has been reported that inhalation of ethylenimine produces irritation of the entire respiratory tract as well as congestion, edema, and hemorrhage of the lungs. Kidney damage was also noted. The effect of inhalation may not be apparent for several hours.

The possibility of respiratory sensitization should be kept in mind. Liquid ethylenimine, a strong skin irritant and vesicant (which may also produce sensitization), readily penetrates the skin and produces severe systemic effects. The liquid and vapor phases of ethylenimine are highly toxic, corrosive, and flammable. The odor of ethylenimine may be detected at about 2 ppm but is not sufficiently alarming to serve as a warning. Its lack of warning properties and delayed onset of symptoms make it especially hazardous. Complete capture of vapor by local exhaust ventilation is essential. Protective clothing must be worn in order to prevent skin contact.

Ethyl ether is not primarily a toxicity hazard although it is an irritant in concentrations already in the lower narcotic range. It is an accident hazard because of the confusion and changes in reflexes which are induced. Inhalation of 3.5 percent of ether for 30 to 40 minutes will produce unconsciousness. At 7.0 to 7.5 percent, irritation of the upper respiratory tract will produce coughing. Industrial exposures of 1000 to 2000 and even 3000 ppm have not been uncommon, but exposure below the TLV is readily obtained by usual engineering design.

Ethyl formate is a volatile, highly flammable solvent with a mild, pleasant odor. It is the least toxic of the formates which, in turn, are the least toxic of the simple aliphatic esters. Ethyl formate is extremely irritating to the eyes and the upper respiratory tract.

Ethyl mercaptan, the sulfide ester of ethyl alcohol, is a foul-smelling gas used mainly for adding an odor to things which should be made noticeable. Its effects are qualitatively the same as those of hydrogen sulfide although it is quantitatively substantially less toxic. The odor is a good warning of toxic concentration.

Ethyl silicate (ethyl orthosilicate) is a moderately volatile liquid used as a waterproofing and weatherproofing agent for masonry and ceramic objects. It has a sharp odor and is irritating to the eyes, nose, and throat at high concentrations. In animal experiments it has produced lung and kidney injury as its main toxic effects.

F

Fluorides in the form of the soluble inorganic salts which can hydrolyze to form free hydrogen fluoride in contact with the tissue are severely irritating, and many of them can produce skin burns which are both extensive and painful. The less soluble inorganic salts are not irritating but may produce a type of chronic poisoning by long-continued inhalation. The fluoride ion tends to accumulate in the skeleton. Softening of the bone has been considerably more of a problem with cattle exposed to fluorides than it has been with people so exposed.

According to one investigation, the use of welding rods with fluorides in the coating does not introduce a hazardous amount of fluoride into the general air of the welding shop. Under the welder's helmet, the concentration varied widely. For limits of fluoride concentrations in drinking water, see Table 20-B in Chapter 20, "Industrial Sanitation and Personnel Facilities."

Fluorine is a yellow, highly reactive gas that combines with water or moisture in the air to form hydrogen fluoride, and, possibly oxygen fluoride, which is more toxic than fluorine. Fluorine's health hazard is designated as extra hazardous for acute exposure and high for chronic exposure.

Too few industrial exposures to fluorine have been evaluated to recommend specific first aid procedures. However, removal of the victim from the contaminated area and administration of 100 percent oxygen to relieve respiratory distress are advisable.

Fluorotrichloromethane (trichloromonofluoromethane, Refrigerant-11) has about the same toxicity as dichlorodifluoromethane.

Formaldehyde is a colorless pungent gas, usually furnished as a solution containing 37 percent by weight of formaldehyde and a small quantity of methyl alcohol as a preservative. Both the gas and this solution (formalin) have a pronounced irritant effect upon the skin. Prolonged contact with dilute formalin will cause the fingernails to become scaly and brittle and the nail matrix to become thick and inflamed.

Formaldehyde gas is extremely irritating to the respiratory system in concentrations of 50 ppm in air. If the concentration is high, the gas may penetrate to the lungs and produce bronchial inflammation and edema, likely followed by pneumonia.

Contact of liquid formaldehyde or formalin with the skin can be largely prevented by protective clothing and by water insoluble protective creams. Splashes of the liquid should be washed off immediately. Breathing of the vapor should be prevented by enclosure and ventilation of processes.

Formic acid is commonly shipped as a concentrated solution in glass carboys. The acid is a strong reducing agent and the vapor is flammable. It should not be stored in the immediate vicinity of oxidizing substances. The acid is irritating to the skin, forming blisters which continue to spread even after the acid has been removed from the skin— an effect similar to that produced by ant and bee venom, which contain formic acid.

Furfural. The irritating properties of furfural, which become uncomfortable at 20 to 50 ppm, probably preclude voluntary exposures which might lead to significant injury.

The information presented in the literature is conflicting regarding the toxicity of furfural vapor. One report concludes that the physiologic effects of the vapor were relatively mild and similar to that of butyl alcohol. Another report describes numbness of tongue and mucous membranes of the mouth, absence of taste sense, and difficulties in breathing which were experienced in workers in a furfural plant where there was inadequate ventilation. In a more recent report it was stated that in refining operations and in the synthetic resin industry (where large amounts of furfural have been used), only occasional reports of individual sensitivity were observed.

Gasoline varies in composition, but is commonly a petroleum fraction containing hexanes, heptanes, and octanes with varying amounts of unsaturated, cyclic, and aromatic compounds. A cracked gasoline may contain 50 percent or more compounds other than the saturated aliphatic hydrocarbons.

If the gasoline consists of saturated and unsaturated aliphatic compounds and cyclic compounds, it will be a narcotic and an irritant to the mucous membranes. The main toxic action will be depression of the central nervous system with symptoms of giddiness, dizziness, hilarity, and headache not unlike an alcoholic intoxication. Mild symptoms of this sort are apt to be produced by relatively short exposures to concentrations of 500 to 750 ppm of this kind of gasoline. In addition, mild blood changes, neuritis, and neurasthenias have been reported.

If the gasoline is made from a crude oil containing 40 percent or more of aromatics or is enriched in aromatics by the refining process, the toxicity may be substantially higher.

The tetraethyl lead content of gasoline is low enough to make any lead exposures from ordinary handling and dispensing of gasoline in garages and filling stations negligible. Tetraethyl lead can, however, present a toxic hazard by inhalation and absorption through the skin when entry is made into tanks and other confined spaces that have contained leaded gasoline. See **Lead.**

Germanium, one of the less common metals which have some industrial uses, is employed particularly in the electronics industry for diodes and transistors. It appears to be a relatively inert metal and has not been reported as the cause of any occupational diseases in man.

In animal experiments, the LD_{50} for rats by intraperitoneal injection was about 750 mg/kg of body weight when germanium oxide was buffered to near the neutral point. Food containing 1000 ppm or water containing 100 ppm was fatal to half of the young rats. Repeated sublethal doses produced some tolerance to a subsequent lethal dose, but the tolerance declined substantially in six weeks.

Graphite has been repeatedly reported to produce a pneumoconiosis in workers in mines, mills, and electrode plants. Several investigators have done autopsies on men who died from graphite pneumoconiosis and found their lungs to contain fibrous masses with central necrosis, in addition to emphysema and general fibrosis around the blood vessels and lymph channels. The typical X-ray patterns start with a miliary "snow-storm" effect and progress to a pattern of sharply defined masses of various sizes in each lobe.

Halogenated hydrocarbons—see also the individual compounds—as used in industry are potent anesthetics and more or less toxic. On exposure to high concentrations the most common symptoms are nausea, vomiting, diarrhea, headache, and unconsciousness.

If unconsciousness occurs, it may be followed by sudden death due to fibrillation of the heart or respiratory failure, as has frequently been observed in the medical use of these compounds. After exposure to amounts which are large but not sufficient to produce sudden death, there may be severe kidney and liver damage eventually leading to death.

The most common halogenated hydrocarbons, arranged in the order of increasingly acute toxicity, are: vinyl chloride, methyl chloride, ethyl chloride, ethylene dichloride, ethyl bromide, carbon tetrachloride, dichloromethane, methyl chloroform, trichloroethylene, methyl bromide, tetrachloroethylene, pentachloroethane, and tetrachloroethane. Tetrachloroethane is about 40 times as strong a narcotic as vinyl chloride.

It should be noted that this is the increasing order of their ability to produce narcosis, which is *not* the same as the order of their chronic toxicity, the more common problem in industry. An acute exposure to the more narcotic of these compounds may result in unconsciousness for surprisingly long periods, with eventual recovery. Unconsciousness for eight weeks has been reported in a case of methyl bromide poisoning.

Tetrachloroethane, the most toxic of the common chlorinated hydrocarbons, has no particular warning signs or symptoms and can produce extremely severe poisonings from continuous exposure to fairly low concentrations. Carbon tetrachloride, methyl chloride,

1147

dichloroethylene, and trichloroethylene, approximately in that order, show decreasing chronic toxicity.

There is also a difference in the type of activity. Carbon tetrachloride and tetrachloroethane are most likely to injure the liver and kidneys primarily; trichloroethylene and similar compounds, the nervous system.

Introduction of a bromine or iodine atom into one of the halogenated hydrocarbons generally increases the toxicity as compared to that of the corresponding chlorine compound. In contrast, introduction of a fluorine atom generally reduces the toxicity as compared to that of the corresponding chlorine compound.

The methyl compounds, particularly methyl chloride and methyl bromide, are in a special class because of their delayed action. Minor symptoms may appear during an acute exposure to these compounds; severe symptoms may appear after a delay of several hours to several days.

Since industrial exposure to halogenated hydrocarbons is almost invariably by inhalation, the most valuable measures to prevent poisoning are enclosure and ventilation of processes at the point where vapor is released. Several of the chlorinated hydrocarbons, however, are apparently much more toxic by skin contact than has been believed. Skin contact should therefore be avoided because of possible toxicity, as well as because of the probability that where there is skin contact there will also be a severe inhalation exposure.

Heptane is used as a solvent, and, like all paraffin hydrocarbons, is an anesthetic and irritant to the mucous membranes, but does not cause systemic toxicity. The most important effect of exposure to high concentrations is narcosis, accompanied by dizziness, intoxication, slight nausea, and loss of appetite.

Hexachlorocyclohexane (benzene hexachloride, BHC, lindane), one of the newer insecticides, is produced by treatment of benzene with chlorine gas under the influence of ultraviolet light.

The gamma isomer (lindane) is the only one of the five possible forms which seems to have much activity as an insecticide. This isomer is considerably more toxic for a number of insects than is DDT. It lacks the residual action of DDT but does act as a stomach poison, contact poison, and fumigant. It may be absorbed through the skin from an oil solution, ingested, or inhaled as the dust or mist of the solutions.

In animal experiments, feeding the material in large amounts has produced liver damage. To warm-blooded animals it is considerably less toxic than DDT. The LD_{50} dosage is 190 mg per kg of body weight. There appears to be no cumulative effect, but there is a definite need for more investigation of the compound.

Hexane is a volatile, highly flammable liquid that finds use as an extraction solvent for edible fats and oils, a laboratory reagent, and as the fluid in low-temperature thermometers. It is a paraffin hydrocarbon and therefore is an anesthetic and an irritant to the mucous membranes, but does not cause systemic toxicity. Narcosis is the most important effect of exposure.

Most commercial grades of hexane contain 45 to 86 percent of n-hexane, with the balance being isohexanes, cyclopentanes, and from 1 to 6 percent benzene. The toxicity of the technical or commercial grades thus depends on the amount of benzene present.

Hydrazine is used as a rocket fuel and for synthesis and analysis of a wide variety of organic compounds. It is capable of producing severe skin burns, which are as penetrating as alkali burns. It can be absorbed through the skin in toxic amounts to produce liver and kidney damage. The main effect on the liver is fatty infiltration, and if the damage is not too great it is reversible. Changes in the kidney are not so great as those in the liver.

Acutely toxic amounts given to animals produce excitement and convulsions in a few minutes. This is the result of stimulation of the central nervous system, and is followed by a fall in blood pressure, probably for the same reason.

Hydrochloric acid rapidly corrodes most metals with which it comes in contact as either a dilute or a concentrated solution. It is most commonly shipped in carboys, and suitable precautions against breakage should be taken. It is nonflammable and not easily reduced.

The concentrated (37 percent) solution fumes strongly. These fumes are corrosive to tissues on contact. Toxicity is low except for this irritant effect. Contact with the skin can be prevented by protective clothing, and irritation of the exposed areas of the skin can be prevented by protective creams. The major protection, however, should be local exhaust and enclosure of processes. See **Hydrogen chloride.**

Hydrofluoric acid is corrosive to all the common metals and to glass. Concentrations up to 60 percent acid are usually shipped in rubber drums with bronze bungs, although paraffined wood containers and lead carboys in boxes are sometimes used.

The anhydrous acid is shipped in passivated iron cylinders or tanks, normally under slight pressure. Since the material boils at about 68 F and the pressure may be raised by a slight production of hydrogen, precautions should be taken in opening the tanks.

Contact of the anhydrous acid with the skin produces a serious burn which is immediately painful. The area turns white and rapidly assumes a thickened or toughened appearance. If the acid is not thoroughly removed, it will continue to penetrate and burn beneath the thickened skin for days.

The most important first aid measure is *immediate* washing of the area with plenty of fresh water. The washing must be continued until the skin has regained its normal color, from several minutes to several hours, depending on the extent of the penetration. Further treatment should be determined by a physician.

Burns from the more dilute solutions may be just as serious as those from the anhydrous acid. The intensity of the burn depends both upon the concentration of the acid and upon the duration of its contact with the skin. There is no warning of contact with the dilute solutions. Even serious burns may be painless for several hours and then give only a slight tingling sensation for several more hours before real pain is felt.

Affected skin areas should be flushed immediately with large quantities of water whenever contact with the solution is known or suspected, whether or not there is apparent damage. A physician should be consulted without delay.

Personal protective equipment for use against the anhydrous acid should be made of butyl rubber or neoprene. Garments of natural rubber become brittle with exposure to hydrofluoric acid, and GR-S rubber is unsatisfactory. There should be adequate exhaust ventilation in the handling area. See also **Fluorides; Hydrogen fluoride.**

Hydrogen chloride is a gas that primarily produces irritation of the upper respiratory passages. Higher concentrations irritate the eyes. Single massive exposures have been known to cause fatal lung injury. Workers who are regularly exposed to hydrogen chloride may suffer erosion of the teeth.

Hydrogen cyanide (prussic acid) is principally encountered in the plating, extraction, photographic, dyeing, insecticide manufacturing, and steel heat-treating industries. It is extra hazardous for acute exposures and is readily absorbed through the skin. It is an extremely rapid poison which interferes with the respiratory system and quickly produces chemical asphyxia. Liquid hydrogen cyanide is a skin and eye irritant. For first aid measures, see **Sodium cyanide.**

Hydrogen fluoride is a highly reactive gas that is a respiratory irritant. Inhalation may produce pulmonary edema. Chronic exposures, while not necessarily irritating, may result in accumulation of fluorides in the bones.

Hydrogen peroxide is a clear, waterlike, colorless liquid. In a dilute form, it is widely used in industry as a bleaching compound and oxidizing agent. It is dangerous primarily because of its oxidizing nature, not because of its toxicity, and the 30 percent solution is capable of causing serious burns.

Because of this strong oxidizing ability, especially in concentrations exceeding 50 percent by weight, hydrogen peroxide may cause spontaneous combustion if left in contact with readily oxidizable organic materials.

If hydrogen peroxide is swallowed, oxygen may be evolved to cause acute distention of the esophagus or stomach, and the localized action of the chemical may cause internal bleeding.

1149

Hydrogen selenide is a colorless gas with a very disagreeable odor. It is produced by the reaction of acids or water with metal selenides or by the combination of hydrogen with selenium.

Its health hazard is high for both acute and chronic exposures. Eye, nose, and throat irritation, followed by pneumonitis, are the results of acute exposure. Chronic effects include garlicky breath, nausea, vomiting, metallic taste, dizziness, and extreme fatigue. In experimental animals, even brief exposures produced liver damage, weight loss, and general weakness.

Hydrogen sulfide, a colorless gas having the odor of rotten eggs, is rather widely used in industry. It is also commonly produced by the decomposition of sulfur-bearing organic material and is often found in sewers or sewage disposal plants. The gas is flammable and explosive in high concentrations so that precautions against sources of ignition should be taken.

Inhalation of a high concentration of hydrogen sulfide will produce sudden poisoning —the individual falls, apparently unconscious immediately, and may die without moving again. There is a complete arrest of respiration, which can often be overcome by prompt application of artificial respiration. In less sudden poisoning, the signs may be nausea, stomach distress, belching, cough, headache, dizziness, irritation of the eyes, and blistering of the lips.

The notoriously bad odor of hydrogen sulfide cannot be taken as a warning sign because sensitivity to this odor disappears rapidly with the breathing of a small quantity of the gas. In regular operations enclosure and a local exhaust system can be provided to keep the concentration within safe limits.

Hydroquinone (1,4-benzenediol) is used as a photographic developer, antioxidant for fats and oils, stabilizer, and chemical intermediate. It presents a low health hazard for acute exposure and a moderate one for chronic exposure.

Hydroquinone dust has been reported to cause discolorations, lesions, and other structural changes in the cornea of workers manufacturing hydroquinone or mixing and packaging photographic developers. These effects, however, appear to have resulted from unusual circumstances.

Animal studies show the low chronic toxicity and rapid excretion of hydroquinone. Systemic effects from industrial exposure have not been reported and skin sensitization from the solid and dilute alkaline solutions appears negligible.

Iodine is a purple crystalline solid, that sublimes readily even at room temperature to form a vapor which is irritating to the respiratory system much as is chlorine and, since it is only slightly water-soluble, it readily penetrates the lung to produce pulmonary irritation and edema.

Iodine, however, is much more likely to cause skin irritation from painting of the skin with tincture of iodine, which is a primary irritant, than it is from either drinking of the solution or inhalation of the vapor. The tincture is an excellent skin antiseptic, which must be used with caution, for it may cause reddening, inflammation, and loss of large pieces of skin, particularly if applied often or in large quantities.

Continued absorption of iodine has an effect on the mucous membranes distinct from its local action on the skin. It produces irritation and inflammation of skin and mucous membrane tissue and seems to excite minor infection into renewed activity.

Iron oxide fume was observed some years ago to cause a mottling in the lung X rays of welders similar to that of silicosis. The most important differences between the effects of iron oxide and silica are that the presence of iron pigmentation or siderosis does not increase the tendency of the lung to develop tuberculosis and the welders with this condition do not seem to have any disability as a result of it. It occurs in hematite miners and in welders who work in confined spaces without enough ventilation. Most authorities feel that the concentration of iron in the air should be kept below the point at which it collects in the lung.

Isophorone. Except for the effect of inhalation of the vapors of isophorone, there is little recorded information upon its physiological activity. Isophorone is considered to have a low degree of hazard to health by inhalation, even though it has a moderately high degree of toxicity by this route. Vapors are quite irritating and have an unpleasant odor to unacclimated humans, so that overexposure is not likely if its warning properties are heeded. One report states that a short exposure of unconditioned human volunteers exposed to concentrations of 25 ppm of vapors of isophorone was definitely irritating to the eyes, nose, and throat. Isophorone is classified as a primary skin irritant—brief contact can cause inflammation.

Isopropyl alcohol. Inhalation of isopropyl alcohol vapor can produce mild irritation of conjunctiva and mucous membranes of the upper respiratory tract. No cases of industrial poisoning have been recorded in the literature.

Isopropyl alcohol is potentially narcotic. In large amounts, it is more toxic and more narcotic than ethyl alcohol, but less so than n-propyl alcohol to animals. There is apparently little or no accumulation in the body. Mild irritation of the eyes, nose and throat was induced in human subjects exposed for three or five minutes to 400 ppm of isopropyl alcohol. From the viewpoint of comfort these subjects found 200 ppm to be the highest concentration acceptable for an eight-hour exposure. It has been stated that the minimum concentration with identifiable odor of isopropyl alcohol is 200 ppm.

Contact of the skin with solutions of isopropyl alcohol can cause solvent irritation type dermatitis.

Isopropylamine exhibits strong local irritant properties. It has good warning properties—at 5 to 10 ppm, its ammonia-like odor is definite; between 10 and 20 ppm, the odor becomes so strong that nose and throat irritation result from short exposures; workers exposed to concentrations between 25 and 50 ppm experienced nausea and temporary impairment of vision. Repeated exposure of the skin to the liquid and vapors of isopropylamine may result in dermatitis. The eyes may be chronically irritated by the vapors. Employees should be instructed to minimize contact with isopropylamine liquid and vapors as much as possible. If work contact with the liquid or vapors is unavoidable, personal protective equipment should be used. Isopropylamine is primarily an acute irritant, and is capable of causing burns on contact. It is alkaline in nature and causes burns similar to those of strong solutions of ammonia.

The high volatility, low flash point, and relatively low ignition temperature produce a high potential fire and explosive hazard.

Isopropyl ether, like the other ethers, forms explosive peroxides upon standing in contact with air. Industrial exposure to isopropyl ether is most likely to occur through a direct contact with the skin or eyes due to an accidental spill, or through an inhalation exposure to low concentrations of the material in the atmosphere. Acute exposures to physiologically significant concentrations are rare because of the flammability associated with such concentrations.

The primary physiological response to isopropyl ether is that of general anesthesia. Even though it is somewhat more toxic than ethyl ether the vapor concentration necessary for this anesthetic effect is much higher than the concentration that causes irritation and unpleasant odors.

This compound is only slightly irritating to the skin and is capable of producing only minor injury to the eyes. It is not absorbed through the skin in harmful amounts.

L

Lead was one of the first industrial materials to be recognized as a serious health hazard, and it still accounts for many cases of disability. In industrial work, lead enters the system mainly in the form of dust by inhalation.

Lead compounds can produce poisoning when they are swallowed, but absorption from the digestive system is slow and inefficient, so that large doses of most lead compounds are required and poisoning develops slowly. In the respiratory system, on the other hand, most lead compounds are absorbed rapidly and completely so that poisoning will develop from much lower doses. It is not even necessary that the compound penetrate deeply into the respiratory system. It has been demonstrated experimentally that lead compounds are absorbed rapidly and completely from the nasal passages when they have no chance to go either into the digestive system or deeper into the respiratory system.

Prevention of lead poisoning is almost entirely a matter of good housekeeping and dust control. To prevent ingestion, food or drinks should not be taken into workrooms where lead compounds are handled.

The metal is essentially harmless. Although it can be vaporized at high temperatures, such temperatures are not usually attained in casting, soldering, or heat treating baths.

Casting and soldering pots should have local exhaust hoods to get rid of lead vapor or lead compounds which may be carried off with volatile fluxes. It is at least equally important that the surface of the lead be kept free from dross and that the dross container be enclosed under the same hood so that dross can be removed from the pot and the container covered and removed, all under exhaust.

Likewise, on large soldering jobs the soldering operation should be locally exhausted, primarily to prevent nose and throat irritation from acid fluxes. More important, the furnaces used for heating the soldering irons should be exhausted because in these furnaces considerable lead is invariably oxidized.

Assembly of battery plates and burning-on of the connectors should be done on grille-top tables with down-draft ventilation. This installation plus regular air sampling to ensure control of contamination, and physical examinations to keep track of lead absorption by employees, can eliminate lead poisoning from battery plants.

The poisoning produced by inhalation of solid inorganic lead compounds commonly causes symptoms of lead colic and lead anemia. Stipple cells or other readily stained cells in the circulating blood are good indicators of lead absorption, although not so good as the direct measurement of lead content of the blood or urine. Urinalyses for lead can now be run routinely and cheaply and with sufficient accuracy by polarographic or spectrographic methods.

Lead tetraethyl and other lead alkyls are likely to be much more rapidly progressive as poisons than the nonvolatile compounds and are much more likely to attack the nervous system than to produce gastrointestinal symp-

toms. These effects are probably typical of the volatile lead compounds, none of which is common in industry.

Lethanes—see **Thiocyanates (aliphatic)**.

Lindane—see **Hexachlorocyclohexane**.

M

Magnesium is one of the light metals used as a construction material. In the early days of its industrial use, there were numerous laboratory reports that fine particles of magnesium, and particularly of magnesium alloys, left in wounds would react with the tissue fluid to produce bubbles of hydrogen gas that would cause severe tissue injury. It was stated that when the hydrogen was produced, magnesium hydroxide was also produced and acted as a moderately strong alkali to damage the tissue.

This reaction is theoretically possible and has been produced in experimental animals by imbedding magnesium powder beneath the skin. It was also seen occasionally a number of years ago when metallic magnesium clips were tried as substitutes for sutures in experimental surgery.

Nevertheless, there are no published reports of such tissue injuries from the industrial use of magnesium. The reason is not known; it may be that such injuries fail to develop or that the hazard is recognized and special care is taken to eliminate particles of the metal and its alloys from industrial injuries.

The magnesium ion is a normal constituent of a number of tissues, and a certain amount of magnesium in the diet is essential to life. Excessive amounts are known to be toxic, and injuries and deaths due to medical use of magnesium salts have been reported. There have been no reports of industrial intoxication from inhalation of magnesium metal powder or magnesium oxide.

Inhalation of freshly produced magnesium oxide fumes has caused metal fume fever, similar to the better known "zinc chills" or "zinc shakes" caused by inhalation of freshly formed zinc oxide fumes.

Heavy exposure to magnesium oxide is irritating to the eyes, nose, and throat, but the material is not a severe primary irritant.

Finely powdered magnesium is a fire hazard, and severe injuries and deaths have occurred from ignition of powdered magnesium in the clothing of workers who have done hand grinding on magnesium alloys. Also, fires and dust explosions have occurred in exhaust systems carrying powdered magnesium from grinding operations.

Exhaust systems should go as directly as possible to a dust separator which will remove the magnesium powder and trap it under water without permitting it to get into the main exhaust system. The wet powder from these dust collectors and turning-chips that are wet with coolant should be removed from the plant as soon as they are taken from the collecting equipment.

Workers in magnesium-grinding operations should be provided with and required to wear flame-resistant clothing.

Massive pieces of magnesium in the form of castings or machined work are not a severe fire hazard because they are hard to ignite. When once ignited, however, the metal in the massive forms will burn vigorously.

Malathion (O,O-dimethyl dithiophosphate of diethyl mercaptosuccinate) is one of the organic phosphate insecticides. Like the others of its class, its principal toxic action on warm-blooded animals is its inhibition of acetylcholinesterase. It seems to be one of the least toxic of this type of insecticide. Inhalation of 5 ppm (56 mg/cu m) by dogs and guinea pigs produced no effects except tears and a slight decrease in the plasma cholinesterase activity.

Maleic anhydride is an irritant to the skin and mucous membranes, especially in the presence of moisture. Maleic anhydride does not usually cause an immediate burning sensation upon contact with the skin, especially if the skin is dry. If not removed by washing, it will cause reddening and, occasionally, blistering, if the exposure is prolonged and severe. Maleic anhydride dust and vapors are exceedingly irritating and severe acute exposures are not voluntarily tolerated. Upon inhalation of dust or vapors, coughing and sneezing, together with burning and irritation of the throat may occur. The eyes are particularly sensitive to the dust and vapors. Maleic anhydride is not a serious industrial hazard provided workers are adequately instructed and effectively supervised in the proper handling of the chemical. Employees should be instructed to report any signs of irritation or burning of skin, eyes, or mucous membranes.

Because maleic anhydride is a combustible solid having a flash point of 215 F, care must be exercised in handling and storage to keep it away from flame or sparks. Dust and the vapors from the molten product are also flammable.

Manganese is found in a wide variety of minerals. The dioxide is its most common inorganic compound and is used as a chemical intermediate, enamel additive, and drier.

The probability of contracting manganese poisoning is low, but the effects are severe—total disablement may result from a few months' exposure to high concentrations. This effect, however, is more likely to occur after prolonged and repeated exposures above 30 mg/cu m.

The main hazard is usually from the inhalation of manganese dioxide, which may produce neurological lesions. The symptoms are many and are similar to Parkinson's syndrome. Symptoms include: weakness, instability, difficulty in walking, immobility of facial expression, monotonous and intermittent speech, spasmodic laughter, and other grotesque signs. Exposures to finely divided dusts of manganese dioxide may produce pneumonitis.

Mercury and its compounds are not nearly so widely used as are lead compounds and consequently mercury poisoning is not so well known nor so often seen as is lead poisoning.

Mercury metal is a liquid at room temperature with a vapor pressure high enough to produce poisoning if a considerable area of the metal surface is exposed to air. If mercury gets into the cracks of a wood or tile floor or into the pores of a concrete floor, the contamination may become so great as to necessitate replacement or sealing of the floor before the plant or laboratory can be safely used again. Although it cannot be removed, mercury can be sealed into such floors by covering them completely with an asphalt mastic, preferably after the vapor pressure of the mercury has been reduced by flooding the whole area with a lime sulfur spray.

To prevent such inconvenience, metallic mercury should be handled over impervious tables or containers with the surfaces depressed and arranged to drain to a central point. Spilled mercury can then be collected and returned to stock.

A number of volatile mercury compounds are used in seed disinfection. Examples are ethyl mercury chloride and phosphate and the corresponding phenyl mercury compounds. These materials are highly toxic. When used without adequate enclosure and dust control, they can rapidly produce severe mercurial intoxication. These seed-treating compounds are also strong primary skin irritants and will cause itching, burning, and blistering on direct contact. Such irritations heal slowly.

The type of mercury intoxication common to industry is characterized by a tremor of the hands, irritation of the mucous membranes of the mouth, excessive flow of saliva, and changes in the personality. Poisoning by one of the volatile compounds may be more rapid so that personality changes are not so likely to be seen.

Inhalation of mercury vapor, usually in very high concentrations, may produce metal fume fever, which may disappear with no other apparent symptoms, or may be followed by a pneumonitis or other symptoms of mercuralism.

The ordinary toxic dust respirator does not protect against intoxication by metallic mercury or by the volatile seed-treating compounds, since they are in the form of vapor rather than dust. A gas mask offers some protection, but if enclosure and local exhaust are not possible, the only effective protection is a supplied air respirator.

Mesityl oxide is a high-boiling, unsaturated ketone which is thought to be somewhat more toxic than the ketones of lower molecular weights.

Metal hydrides (primary types) are compounds of hydrogen and the alkali metals: sodium, potassium, lithium, magnesium, calcium, and strontium. Information on the health hazards of metal hydrides is limited. Since they react with water to form caustic hydroxides, they are irritating to the eyes, skin, and mucous membranes. They also release a large amount of heat on reaction with the moisture of the skin.

Methanol (methyl alcohol) poisoning is usually produced by swallowing the liquid or inhaling high concentrations of vapor in an enclosed place such as a tank. The signs of poisoning include headache, nausea, vomiting, violent abdominal pains, aimless and erratic movements, dilated pupils, sometimes delirium, and such eye symptoms as pain, tenderness on pressure, and occasionally blindness. Direct action of the liquid or the vapor on the skin and mucous membranes may produce an irritation and inflammation.

One of the peculiarities of methanol poisoning is its exceptionally severe action on the optic nerve. About one-half of all the serious cases of methanol poisoning result in some impairment of vision, which is usually permanent and may vary from dimness or blind spots scattered through the visual field to total blindness.

Methoxychlor (2,2-di[p-methoxyphenyl]-1,-1,1-trichloroethane) is a synthetic insecticide which is closely related to DDT in its action. It is, however, about $\frac{1}{8}$ to $\frac{1}{10}$ as toxic as DDT. Consequently, the hazard from inhalation of the dust or mist of sprays of this insecticide is remote.

Methoxychlor is not absorbed through the intact skin in significant amounts, and does not seem to have any effect on the central nervous system. In animal experiments, continued feeding of diets containing methoxychlor in toxic concentration leads to loss of weight in the animals by their voluntary restriction of food intake and also to fatty infiltration of the liver in the manner of the chlorinated hydrocarbons.

Methyl acetate is widely used as a low-boiling solvent in the perfume, cosmetic and paint industries. It is irritating to the mucous membranes of the eyes and the respiratory passage.

Animal studies indicate that there may be general poisoning and long-lasting aftereffects when methyl acetate is inhaled in concentrations below that which produces narcosis. If exposure is serious, pulmonary edema and hemorrhages may occur, along with injury to the blood vessels. Symptoms of exposure are eye inflammation, nervousness, chest tightness, heart palpitation, exhaustion, and shortness of breath.

Methyl acetylene-propadiene mixture (MAPP) is 58 percent methyl acetylene-propadiene, the balance being a mixture of paraffinic and olefinic C_3 and C_4 hydrocarbons.

MAPP gas like the other commonly used fuel gases is highly flammable and suitable precautions must be taken to avoid this hazard. On the basis of its toxicological properties, the precautions necessary to control the hazard from flammability when used as a fuel should adequately control any hazard from inhalation. Concentrations which are sufficiently high to be explosive may also be high enough to cause the onset of anesthesia. It is suggested therefore, that peak concentrations should not exceed 5000 ppm to minimize the hazard from flammability as well as from toxicity. Most persons find the gas to have a characteristic odor at 100 ppm, and to be objectionable at 1000 ppm. Consequently, while odor does not necessarily indicate that a hazard from inhalation exists, it is a valuable asset in that it will give warning of leaks that should be located and stopped. Odor must not be used to estimate the concentration nor can it be relied upon to prevent excessive exposure.

If a person shows anesthetic effects or is overcome with the gas, he should leave or be removed from the contaminated area to a place where fresh air is available. If breathing stops, artificial respiration should be given.

It is desirable to avoid getting the liquefied material on the skin or in the eyes, since, like most liquefied gases, it may cause frostbite.

On the basis of animal data and by analogy with methyl acetylene, a TLV of 1000 ppm has been recommended as a hygienic standard of good operating practice.

Methyl acrylates are irritants for the skin, mucous membranes, and gastro-intestinal tract. They may be absorbed through the skin, gastro-intestinal system, or the lungs. Systemic absorption may lead to degenerative changes in the liver and the kidneys. Accidental industrial exposure may occur in three ways:

1. Acute exposure to high vapor concentrations

2. Liquid contamination of eyes or skin

3. Prolonged exposure to lower vapor concentrations

Personnel should be instructed and equipment designed to avoid contact with acrylates in liquid or vapor phase. Where excessive contact is unavoidable, adequate personal protective devices should be used. Adequate emergency showers, wash-up facilities and eye wash fountains should be available in accessible locations and areas where acrylates are handled. The necessity for immediate emergency showering and prompt removal of all contaminated clothing following massive skin contact must be stressed. Even minor splashes must be washed off promptly and contaminated wearing apparel (particularly leather goods such as shoes and gloves) must be removed and thoroughly dried or laundered before wearing again.

Employees should be instructed to report promptly all symptoms such as irritation of eyes, nose, and throat or lung congestion.

Methyl alcohol—see **Methanol.**

Methylamines are irritating to the lungs and upper respiratory tract; direct contact with skin and mucous membranes may cause burns. Both the liquid and vapor are highly irritating to the eyes. Bronchitis and conjunctivitis may result from exposure to the vapors of methylamines.

Methylamines in concentrations of less than 100 ppm are characterized by a fish-like odor, while in higher concentrations the odor is more ammoniacal.

Exposed workers should be provided with and be required to use the recommended personal protective equipment as necessary to the particular operation. In the event of accidental leaks or spills, or whenever excessive concentrations may be encountered only personnel equipped with approved respiratory protection should be permitted in the contaminated area. Chemical safety goggles should be worn when discharging containers or tank cars or whenever there is a danger of the liquid or saturated vapor coming in contact with the eyes. Exposure of employees who have used careful handling procedures over a period of years has produced no significant effects.

Methyl bromide is widely used as a fire extinguishing agent and as a fumigant.

An efficient fire extinguishing material, it has been installed in fixed systems in a number of aircraft and naval vessels, particularly in Great Britain, where several cases of poisoning have been reported. Numerous poisoning cases also have occurred in the use of the material in fumigation and in industry.

The minimum concentrations which will produce mild poisoning are low. One investigator recorded 22 cases of skin irritation and 31 cases of mild systemic effects among 90 persons working during a two-week period at the filling and sealing of methyl bromide ampules. Concentrations of airborne methyl bromide were regularly determined around this operation and were generally below 35 ppm. The skin lesions resulted from exposure to small quantities of the liquid while it was being handled under a hood. The liquid is a strong skin irritant. Skin contact can produce severe dermatitis, blisters, itching, and slow-healing burns.

No particular irritation is caused by breathing methyl bromide in quantities sufficient to produce systemic toxicity. The effects are not seen for a few hours to two or three days after the exposure. In mild systemic poisoning the most common complaints are lack of appetite, nausea, and headache. Dizziness and blurring of vision are also common. In more severe poisoning, there may be hallucinations and slurring of speech.

Methyl bromide is also a powerful irritant to the lungs and in severe poisoning will produce secretion of liquid into the lungs and thus interfere with breathing. See **Halogenated hydrocarbons.**

Methyl Cellosolve—see **Ethylene glycol monomethyl ether.**

Methyl chloride has been widely used as a refrigerant. A number of cases of severe intoxication have occurred from small leaks in large refrigeration systems. There have also been a number of industrial cases. The effects follow a fairly standard pattern with dizziness, weakness, blurred vision, loss of muscular coordination, and drowsiness in all instances. Loss of appetite, disturbances of sleep, and confusion are also common complaints.

Methyl chloride is retained to some extent in the brain, heart, liver, stomach, spleen, kidneys, muscle, and blood, and regularly produces edema of the lungs, degeneration of the liver and kidneys, and some damage to the heart. The mechanism of its action, however, is not yet understood.

It possesses no warning properties and since those who are becoming intoxicated may become so confused that they are not able to protect themselves, this material must be handled with the greatest of care.

Since methyl chloride is a low-boiling liquid normally handled under pressure, severe skin injuries—freezing and blisters—may occur if the skin is brought into contact with a leak. See **Halogenated hydrocarbons.**

Methyl chloroform (1,1,1-trichloroethane) closely resembles carbon tetrachloride in its solvent action and its rate of evaporation, but is much less toxic. Like trichloroethylene, it tends to attack the central nervous system with the production of acute episodes, rather than to attack the liver and kidneys as carbon tetrachloride tends to do.

For degreasing aluminum, copper, or brass, methyl chloroform must be well inhibited because it reacts with these metals and gives off toxic and irritating products. See **Halogenated hydrocarbons.**

Methyl ethyl ketone—see **2-Butanone.**

Methyl formate (methyl methanoate) is a typical aliphatic ester with an odor similar to that of ether. In high concentration, it produces irritation of the eyes, nose, throat, and lungs. Animals that die of exposure are usually killed either by narcosis if the concentration is extremely high or by lung irritation if it is somewhat lower.

Methyl isobutyl ketone (MIBK) does not present a severe inhalation hazard in the usual industrial applications. The odor threshold level is less than 100 ppm and the vapor is irritating to the eyes at 200 ppm and to the nose and throat in concentrations higher than 200 ppm. Workmen exposed to about 100 ppm complained of headache and nausea but a tolerance developed during the working week. Repeated or prolonged skin contact may cause drying of the skin with primary irritation resulting. Prolonged or repeated contact with the skin should be avoided. Methyl isobutyl ketone should be handled only in areas having adequate ventilation.

Methyl mercaptan (methyl sulfide) is the hydrogen sulfide ester of methanol. Since it is readily hydrolyzed to the parent compounds, it has practically the same toxicity, both qualitatively and quantitatively, as hydrogen sulfide.

Methylcyclohexane is similar in its effects to cyclohexane although it is slightly less toxic and has a lower vapor pressure.

Methylcyclohexanone seems to be somewhat less toxic than cyclohexanone, but the differences are not enough to be of practical importance and the two substances are alike in their qualitative effects.

Methylene chloride (dichloromethane) is one of the least toxic of the chlorinated hydrocarbons. However, because of its low boiling point, high concentrations of the vapor can be easily developed. Its action is usually simply narcotic, although, when inhaled in high concentration, it is capable of producing liver damage.

Mica is a silicate mineral extensively mined for its high value as an electrical insulator and for a wide variety of other uses. There is considerable evidence that people working under excessive exposure to its dust suffer from a disabling pneumoconiosis. See the discussions in Chapter 37, "Industrial Hygiene."

Molybdenum is considered an essential trace element in plants and possibly in mammals. Its health hazard is described as moderate to low. There is no evidence of acute or chronic industrial poisoning. Pathological findings from animal experiments have been limited to such effects as bronchial and alveolar irritation with moderate fatty changes in liver and kidneys.

N

Naphtha (coal tar) is mainly a mixture of toluene and xylene and should be treated like those compounds.

Naphtha (petroleum) is a rather indefinite term for any one of a number of solvent mixtures derived from petroleum oil. One should define it more carefully before attempting to assess the hazard.

Naphthalene is a volatile solid. Vapors, given off when naphthalene is heated, form explosive mixtures with air. Reports in the literature indicate that the inhalation of naphthalene vapors may cause headache, loss of appetite, nausea, and kidney damage.

Ingestion of naphthalene in relatively large amounts has reportedly caused severe hemolytic anemia and hemoglobinuria. A hypersusceptibility, probably genetically based, is recognized. Contact dermatitis may result from primary irritation or allergic hypersensitivity, or both.

Eye irritation may result from exposure to naphthalene vapor. Prolonged exposure to high concentrations of vapor can produce opacity of lens. Optical neuritis and injuries to the cornea have also been reported. The threshold limit value of 10 parts per million is recommended as a threshold limit to prevent ocular effects, but possibly not blood changes in hypersusceptible individuals.

Naphthalene is not a serious industrial hazard providing that ordinary precautions are observed to prevent contact with molten material or excessive exposure to dust or high vapor concentrations.

Nickel. In spite of its relatively high toxicity, industrial experience with nickel has been good. However, an increase in nasal, sinus and lung cancer has been noted in workers employed in nickel refineries. The specific carcinogenic agent is still not defined and is the subject of continuing research. The long-range carcinogenic potential should be remembered for all nickel compounds.

Dermatitis, or "nickel itch," is common among nickel platers. It appears to have two components, a simple dermatitis localized to the area of contact and chronic eczema or neurodermatitis without apparent connection to such contact.

Experience indicates a difference in effect between the fumes of a low-nickel alloy, such as stainless steel, and the high-nickel alloys. For control of the immediate effect, particular attention should be paid to the fumes and fine dust of the high-nickel materials.

Additional data are required on the effects of exposure to measured concentrations of nickel fume and dust. There is some difficulty in relating exposures to mixed nickel dusts in refining with those in welding and other operations. Nickel and its salts are not absorbed through the skin in amounts sufficient to cause intoxication.

Operations releasing nickel fumes or dust should be provided with local exhaust ventilation. Contact with solutions of nickel salts should be avoided by appropriate handling procedures or protective clothing. If contact does occur, the solution should be washed from the skin at once. If sensitization occurs, the worker should avoid contact with all nickel compounds.

Nickel carbonyl is produced during the Mond process for the refining of nickel. It is a colorless liquid that gives off an unusually toxic vapor. The poisoning is insidious since there is no particular discomfort during the exposure and the serious effects are delayed for a few hours to a few days. The major changes are seen in the lungs, liver, and brain, and death is generally due to lung damage.

The most comprehensive study of human poisonings by nickel carbonyl concerned 100 men who were accidentally exposed during the repair of a reactor in an oil refinery, 31 of whom were sufficiently poisoned to require hospitalization.

These men typically showed dizziness, headache, and nausea; vomiting occurred in more than one-half of the hospitalized cases. Most of them also felt a sense of constriction in the chest and shortness of breath. The more severe delayed stage usually started with convulsive coughing, which appeared 10 hours to 8 days after the original exposure. In this stage oxygen was required to maintain respiration in almost all cases. Dimercaprol was considered life-saving in a number of these cases; it was given to all the patients. Con-

valescence was greatly prolonged—two months after the episode, most of the men still felt exceptionally fatigued after exertion and showed unusually high pulse rates. Urinary excretion of nickel was greatly increased in all these cases, and the dimercaprol increased it still more.

Nickel carbonyl poisoning does not appear to be cumulative, and some tolerance is built up in animals by repeated exposures.

The LD_{50} for 30 minutes of exposure varies from 10 to 270 ppm for various animals.

Nicotine is an alkaloid most commonly obtained by extraction from tobacco. It is extremely toxic, and in view of the fact that an average cigar contains twice the lethal dose for an adult, it is surprising that so few cases of nicotine poisoning come to light. Probably one reason is that adults acquire tolerance very rapidly. Certainly another is that exposures are relatively low. Inhaling the dust in the dustiest operations in tobacco processing gives a dose about equal to the nicotine absorbed by smoking approximately five cigarettes per day.

Various investigators have noticed the occurrence of mild nicotine poisoning in tobacco processing plants, particularly in nicotine extraction. It usually occurs among new employees in the first few days of employment, before they become habituated to the exposure.

Acute nicotine poisoning is commonly due to use of nicotine-base insecticides. The nicotine is rapidly and completely absorbed either from ingested material or from material spilled on the skin or saturating the clothing. The nausea, vomiting, and headache seen in mild poisoning are likely to be absent in acute poisoning, which is characterized by almost immediate collapse and death within 5 to 30 minutes.

Since nicotine is volatile, respiratory protection from the free base can be obtained only with a hose mask or with a canister gas mask using an organic vapor canister. The nicotine sulfate commonly used as an insecticide is not volatile.

Nitric acid is a powerful oxidizing agent, usually shipped in glass. It should be isolated from contact with reducing material or organic substances such as wood or paper since its concentrated solutions may cause explosion or combustion on contact.

In the oxidation of most organic materials, concentrated nitric acid will produce dense clouds of red or brown oxides of nitrogen. Since inhalation of these oxides in dangerous quantities produces only a mild irritation of the respiratory organs, it is possible to inhale a dangerous concentration without much discomfort or apparent injury.

Symptoms from the affected lungs will appear after a lapse of several hours. The result of the exposure may be extremely grave. Anyone known to have inhaled nitrogen oxide fumes should be kept under medical observation for at least 24 hr whether or not he shows symptoms.

The effects of nitric acid on the skin are not different from those of other strong acids except for the characteristic, and apparently harmless, yellow stains produced by minor burns.

Nitroaniline occurs as three isomers, ortho, meta, and para, but their physiological effects are essentially the same. They all are readily absorbed through the respiratory system or through the intact skin. They are powerful methemoglobin formers and cause hemolysis of red blood cells. They are comparable in all respects to nitrobenzene and should be used with the same precautions.

Nitrobenzene is a light yellow liquid with the odor of almonds and a high boiling point so that the probability of poisoning by inhalation of vapor from the cold liquid is not great. A highly toxic substance, it will produce either acute poisoning by absorption of a large amount in a short time or chronic poisoning by continued exposure to small amounts.

It can be absorbed by inhalation and ingestion, but, in industry, the major hazard is absorption through the skin. Absorption by inhalation can also occur and produce serious poisoning; it is especially likely where vapor escapes from a hot process or during the cleaning of tanks and lines by steam or hot water.

Nitrobenzene is primarily a blood and nervous system poison. It combines with the hemoglobin to produce methemoglobin, which will not transport oxygen and consequently leads to asphyxiation of the tissue. It also generally depresses or paralyzes the central nervous system.

Acute poisoning, which rapidly follows contact, is marked by the deep blue color of the lips, fingernails, and ears and by labored breathing. In a milder case the signs are drowsiness, headache, and unsteady gait suggesting drunkenness. Recovery is likely to be slow, partly because of the slowness with which nitrobenzene is eliminated from the system and partly because of the time required to replace the red blood cells which have been destroyed.

Repeated skin contact or inhalation over a long time may produce chronic nitrobenzene poisoning with fatigue, headache, loss of appetite, and mild anemia. Exposure to levels of 40 to 80 ppm will produce slight symptoms after a few hours.

After acute exposure to nitrobenzene, the first step is to remove all contaminated clothing immediately and to wash the skin liberally with soap and water. The only other measures to take pending medical care are to keep the individual at complete rest and to administer oxygen if breathing is labored.

To minimize the possibility of skin contact, men working with nitrobenzene should have two lockers and should be required to keep work clothing separated from street clothing. Work clothing should preferably be issued by the plant management and laundered on the premises. A complete bath with warm water and soap and a complete change of clothing, from the skin outward, should be required at the end of each shift. No food or tobacco should be permitted in the work area.

Nitroethane is typical of the nitroparaffins in that it acts as a mild irritant to the lungs, upper respiratory tract and central nervous system. Nitroparaffins do not seem to be absorbed through the skin. In animal experiments, all animals fatally poisoned by nitroethane also showed some liver damage. Rabbits, guinea pigs, and monkeys were not affected by exposure for 140 hr to 500 ppm of nitroethane, but rabbits and guinea pigs were affected by a single 6-hr exposure to 1000 ppm.

Nitrogen dioxide is the toxic constituent of

the red fumes from nitric acid in contact with reducing materials and is also one of the toxic components of the welding fume produced in poorly ventilated areas. It is a lung-injurant gas which has inadequate warning properties so that a dangerous amount may be inhaled without great discomfort, to produce lung edema. In experiments with white fuming nitric acid, red fuming nitric acid, and nitrogen dioxide, the presence of nitric acid in the exposure atmosphere resulted in skin burns (which did not occur with the nitrogen dioxide) but had little effect upon the concentration required to produce deaths.

Rats, mice, and guinea pigs exposed four hours per day, five days per week, for six months to 4 ppm of nitrogen dioxide showed no ill-effects. When similarly exposed to 9 and to 14 ppm, they showed lung effects.

Nitroglycerin (glycerol trinitrate) is a heavy, oily liquid, toxic by skin contact. The volatility is low, and contact by inhalation is not important. The principal manifestation of contact is the "dynamite headache" or "powder headache" familiar to workers with explosives. It is due primarily to dilation of the blood vessels about the brain. Attacks are precipitated or greatly increased in severity by the ingestion of alcohol. The effects of exposure to nitroglycerin are usually acute rather than chronic, but chronic poisoning may cause digestive troubles, tremors, and neuralgia.

The only effective means of preventing absorption of nitroglycerin is personal cleanliness. Departments handling the material commonly arrange to have workmen wash their hands at least once an hour with a mild soap and plenty of water. Rubber protective equipment is not suitable. Nitroglycerin rapidly dissolves in rubber, and once the equipment has been impregnated it cannot be successfully cleaned.

Nitromethane is similar to nitroethane (although possibly a little less toxic) and should be handled with the same precautions.

Nitropropane is used chiefly as a lacquer and dope solvent. Its two isomeric forms, 1-nitropropane or 2-nitropropane, are similar in toxicological properties. Workers exposed to 30 to 300 ppm of 2-nitropropane reported headache, dizziness, nausea, vomiting, and

diarrhea. No skin absorption or injury has been noted in animals.

Nitrotoluene is identical in its type of toxicity to nitrobenzene and aniline. The addition of the methyl group makes it less active than nitrobenzene and quantitatively about the same as aniline.

O

Oxalic acid is usually shipped in barrels or bags. It is a good reducing agent and is widely used as a bleach for wood, straw, and similar materials. The solution has a caustic action on the skin, causes brittleness and a bluish discoloration of the nails, and in extreme cases produces gangrene of the area with which it has been in contact.

Cases of acute poisoning by inhalation of the dust of the solid or the mist from the solution are not common, for a high concentration is necessary, but such cases occur often enough to demonstrate the necessity for proper control of such exposure.

As with most acids, swallowing of oxalic acid produces an acid taste and a burning pain in the throat and stomach. There is usually vomiting of mucus or mucus mixed with blood. If the dose is large, there is rapid collapse. A dose insufficient to produce death will cause such symptoms as headache, cramps, convulsions, and delirium. The recovery period from this type of poisoning is long.

When oxalic acid has been taken internally, the most desirable first aid treatment is administration of an emetic, provided that vomiting has not already taken place, followed by the use of precipitated chalk, milk of magnesia or calcined magnesia, or whiting in large quantities of water.

Oxygen. The inhalation of 100 percent oxygen for periods up to 16 hours per day for many days at atmospheric pressure has caused no observed injury to man. It is believed to have no serious adverse effect for a continuous exposure of 24 to 48 hours. The inhalation of pure oxygen at three atmospheres pressure (30 psi) is safe for man for a period of 30 minutes. Longer periods or higher pressures may produce oxygen poisoning. Convulsions have occurred in man after oxygen has been breathed for 45 minutes at four atmospheres pressure; after one to three hours at one atmosphere pressure, neuromuscular coordination and the power of attention were adversely affected. In a study made on dogs, a decline in oxygen saturation of the blood, rise in hemoglobin, lung congestion and edema, right heart failure, and liver congestion were frequent findings in oxygen poisoning.

Oxygen deficiency is of greater concern in industry than are high concentrations or high pressures of oxygen. It may be encountered in numerous situations such as in tanks, vats, holds of ships, silos, or mines; or in any poorly ventilated area (a) where the air may be diluted or displaced by gases or vapors of volatile materials, or (b) where oxygen may be consumed by chemical or biological reaction processes. The warning properties of an oxygen-deficient atmosphere are nil. Although a trained observer may, when alert, recognize the increase in the pulse and rate of breathing in time to return to good air, the average individual fails to recognize the danger until he is too weak to save himself, especially where the return to good air involves climbing stairs or a ladder. The fire hazard of oxygen-deficient atmospheres is below normal; when the oxygen content of the air is below 16 percent, many common materials will not burn.

Ozone has been studied for a long time, but there is still no definite study of its mammalian toxicity. This is due, at least in part, to the difficulty in preparing pure ozone. The best information available seems to indicate that the purest grade of ozone is a lung-injurant and the exposure to as little as 20 ppm may be lethal within a few hours. It also appears that the addition of about an equal amount of nitrogen dioxide will increase the toxicity of the ozone, but whether or not the small amounts of impurities normally present in ozone as prepared from atmospheric oxygen have any practical effect upon its toxicity is still undecided.

As little as 0.2 ppm of ozone has hastened the deaths of guinea pigs continuously exposed to it. Recent studies indicate that repeated inhalation of ozone concentrations of 1.0 ppm results in chronic injury in the lungs of small animals. Other studies warn that in the light of present knowledge, the time limit for exposures to concentrations of 1 to 8 ppm should not exceed three hours.

P

Parathion (O,O diethyl-O-p-nitrophenyl thiophosphate), which is widely used as an insecticide, has been found to be highly toxic for men. The effects are similar to those of hexaethyl tetraphosphate and tetraethyl pyrophosphate except that parathion is less acutely toxic and its effects have more tendency to accumulate.

Parathion is a liquid with a high boiling point and low vapor pressure. It is commonly supplied to the insecticide trade as either a dust impregnated with the liquid or as a solution for further dilution before spraying. Poisoning in the plant or in the field may result either from inhalation of the dust or spray mist or from skin contact with the material, which is rapidly absorbed.

Symptoms of poisoning are loss of appetite, nausea soon followed by vomiting, abdominal cramps, sweating, excessive salivation, and constriction of the pupils of the eyes. Anyone showing such signs should be hospitalized immediately; usual first aid does not help the condition.

Protection against parathion poisoning in the plant consists of the usual measures for reduction of dust and vapor contamination in the air and the use of complete protective clothing. In handling the parathion powders, workmen should wear gloves of natural rubber, rubber boots, washable cotton caps, cotton socks and coveralls which fasten at the neck, over the gloves at the wrists, and over the boots at the ankles. Operations should be conducted under local exhaust, and toxic dust respirators should be worn.

Work clothing should be laundered daily, and the individual should take a shower at the end of the shift and wash thoroughly before eating. If there is appreciable skin contact, work should be stopped at once and the person should thoroughly wash with soap and water and change to clean clothing immediately.

Those who use insecticides in greenhouses or pilots who disperse insecticides from aircraft should wear full face masks with organic vapor canisters which also protect against mists and dusts. Workmen applying parathion insecticides in the field should wear toxic dust respirators, avoid excessive exposure, and always spray or dust downwind.

Pentaborane—see **Boron hydrides.**

Pentachloronaphthalene—see **Chlorowaxes.**

Pentachlorophenol is comparable to phenol (carbolic acid) or zinc sulfate in acute toxicity and must be handled with the same care. Its most common effect is a severe determatitis from handling aqueous or oil solutions of 1 percent or higher concentration.

Dermatitis is common in the wood preserving industry where the material is used largely for protection against fungi and vermin, and especially termites. No industrial deaths and no industrial poisonings more serious than severe dermatitis have been reported in the United States. In lumber dipping operations abroad, however, fatalities have occurred from prolonged contact of the hands and forearms with a 3 percent solution of pentachlorophenol and sodium pentachlorophenate.

Workers handling solutions more concentrated than 10 percent should wear heavy rubber gloves and rubber sleeves and aprons. If the solution is spilled upon the hands, it should be immediately washed off with strong soap and plenty of water. There is a good deal of evidence that a short contact with even a concentrated solution is not harmful.

The precautions recommended are based primarily on animal studies, which show that although pentachlorophenol is readily absorbed through the skin, it is also rapidly eliminated from the animal body. Hence there is little opportunity for cumulative poisoning.

The most important measure for prevention of poisoning from pentachlorophenol is to eliminate the possibility of skin contact by using mechanical equipment for handling objects impregnated with the solution.

Pentane, a simple aliphatic hydrocarbon, is a liquid at ordinary temperatures and pressures and, although volatile, is more a fire hazard than a toxic hazard.

2-Pentanone (methyl propyl ketone) is a simple aliphatic ketone, probably more toxic than acetone and similar to it in its effects.

Perchloric acid has strong oxidizing characteristics which make it a fire and explosion hazard, particularly if it is concentrated beyond 76 percent, which is the most concentrated solution accepted for shipment in interstate commerce. The acid is not appreciably volatile until evaporated to the anhydrous state. If such evaporations are carried on under a local exhaust hood, preferably with a water scrubber attachment, to avoid the fire and explosion hazard, irritation from the acid should not become a problem.

Perchloroethylene—see **Tetrachloroethylene.**

Phenol (carbolic acid) is a flammable solid but is not much of a fire hazard because of its high flash point. It once caused many domestic cases of accidental poisoning through the swallowing of solutions of phenol by mistake for other drugs. Such accidents are now less common for it has been displaced by other disinfectants. Phenol has never been a common source of industrial poisoning in spite of its remarkable potency.

It is readily absorbed through the intact skin, and this is the usual source of poisoning in industry. Since the solid cannot be easily crushed to form dust and has a low vapor pressure, inhalation of the vapor is not of great importance.

Enough phenol to cause serious poisoning can be absorbed readily from clothing saturated with a dilute solution. The symptoms appear rapidly. A mild acute poisoning will produce dizziness, headache, delirium, cold sweat, and jerky, noisy breathing. When a greater quantity has been absorbed, muscular weakness, loss of consciousness, and death from failure of respiration may follow.

Liquid phenol in contact with the skin produces a tingling sensation followed by loss of feeling. The skin becomes white and wrinkled and later turns dark brown and sloughs off. This is not a true corrosive action, but is a local gangrene caused by destruction of the blood supply to the affected area. The same results may occur if a dilute solution of phenol is kept in intimate contact with the skin, as might happen with a small spot of clothing saturated with the solution.

Phenol should be handled so as to prevent contact with the liquid or the solid. Where there may be accidental contact, employees handling phenol or its solutions should wear eye or face protection and the necessary rubber clothing—gloves, boots, apron, and hooded coats.

Coverall rubber suits are most desirable when tank cars are being unloaded during cold weather because the steam-laden atmosphere around the tank cars may be irritating to the skin. If personal protective equipment is splashed with phenol, it must be washed immediately.

To prevent burns and possible absorption, phenol must be *immediately* and *completely* removed from the skin. Time should not be wasted in looking for an organic solvent to remove the phenol. The individual should be placed under a deluge shower at once, and saturated clothing should be removed while he is under the water. He should then wash liberally with water and soap (phenol's solubility is greatly increased by a mild alkali, such as soap).

Ethanol (ethyl alcohol or grain alcohol) may be used to dilute and wash away material which remains on the skin after it has been washed with water—*never* before the affected area has been thoroughly flushed with water. If an inadequate amount of ethanol is used, the area of contamination and consequently the absorption will be increased.

If phenol or phenol solution enters the eye, it must be immediately washed out with a running stream of water under low pressure. The washing should be continued for at least ten minutes, and the eyes should then immediately receive the attention of an ophthalmologist to prevent what can be an extremely serious injury.

Phosgene (carbonyl chloride) is a gas used in chemical manufacturing and produced in greater or less amounts when one of the simple chlorinated hydrocarbons is decomposed in the presence of moisture and hot metal surfaces or by ultraviolet light. It is primarily a lung-injuring gas because it hydrolyzes in the alveolar spaces to produce hydrogen chloride and carbon dioxide. The hydrochloric acid freshly produced at the surface of the lung tissue is very damaging. Its action is acute rather than chronic.

Exposure to a relatively high concentration (10 to 12 ppm) is apt to produce prompt vomiting followed by dry throat, pain in the chest, bronchial irritation, and shortness of breath. A lower concentration (2 ppm) will produce the cough and irritation after a short exposure, but will not cause serious discomfort in the time required to absorb a dangerous or lethal dose. Its odor is not a useful warning for the average person.

Phosphine (hydrogen phosphide) is a colorless irritating gas with a peculiar and repulsive smell like that of rotten fish. It burns readily and, as ordinarily prepared, is spontaneously flammable at room temperatures due to higher homologues of phosphine that are usually present.

In spite of its toxicity, there have been relatively few reports of poisoning by phosphine, because it is ordinarily produced in industry only by accident. The cases reported have come from the production of pure phosphine in the laboratory and its accidental release and from the use of ferrosilicon contaminated with phosphorus. There have also been European reports of phosphine poisoning from the use of zinc phosphide as a rat poison.

The mechanism of phosphine intoxication has not been definitely established, but it appears to cause depression of the central nervous system and irritation of the lungs. With severe exposures, sudden or delayed deaths may occur. The symptoms of acute phosphine poisoning reported included nausea, vomiting, diarrhea, thirst, sensation of pressure in the chest, back pains, dyspnea, feeling of coldness, and stupor or attacks of fainting. On autopsy, the pathology most frequently reported is edema and lung congestion.

Phosphine is rarely employed in industry; most cases of intoxication, therefore, have resulted from its generation as a by-product of the reaction of moisture with a phosphide (such as calcium phosphide, which frequently occurs as an impurity in ferrosilicon or calcium carbide). Aluminum phosphide and zinc phosphide have been used as insecticides or rodenticides; a number of recent cases have resulted from fumigation of grain with phosphine generated from aluminum phosphide.

The effect of acute exposure is paralysis of the central nervous system and irritation of the lungs. The blood changes typical of arsine

1163

and stibine are not produced. In experimental subacute or chronic phosphine poisoning, the effects are similar to those of poisoning by white or yellow phosphorus.

Phosphoric acid is a corrosive mineral acid, usually encountered as the tribasic (orthophosphoric) acid. Although its irritation of the skin and mucous membranes is not as pronounced as that of nitric or sulfuric acid, 75 percent phosphoric acid will destroy body tissues and cause severe burns.

Phosphoric anhydride is a local irritant and a very strong dehydrating agent. With moisture it forms phosphoric acid (corrosive to the skin, mucous membranes, and the eyes, but is not considered as hazardous as nitric or sulfuric acid). Very high concentrations of the anhydride will cause violent coughing, although those exposed regularly to the industrial fume apparently become acclimated to some degree. No evidence of systemic poisoning has been reported in the literature for either acute or chronic exposures. Suitable process enclosures or local exhaust ventilation or both should be used to control industrial exposures. Avoid skin and/or eye contact. For very dusty conditions, the use of full face industrial canister masks, coveralls, rubber gloves, aprons, and foot protection are indicated.

Phosphorus is supplied commercially as a dark red powder and as a light yellow translucent waxy solid. Red phosphorus is harmless since it does not react with the air and is extremely insoluble. Yellow phosphorus, however, melts at just about body temperature and is extremely flammable. The liquid ignites spontaneously in the presence of air, and many painful burns have been produced by picking up pieces of phosphorus with the bare fingers. It is normally stored as a solid kept under water and transferred as a liquid.

Since the water used to shield yellow phosphorus invariably contains small particles of the material, it should not be permitted to escape from the plant premises until the phosphorus has been completely removed and burned. The water should be cooled below the melting point of phosphorus, filtered to remove the larger particles, and then aerated thoroughly to oxidize remaining phosphorus before being disposed.

Burns from liquid phosphorus are ordinarily extremely severe because of the high temperature of the flaming material in immediate contact with the skin and because particles of the phosphorus will likely be left in the burned tissue when the area is cleaned up. The burning can be stopped for the moment by saturating the area with cool water or immersing the part in water. The phosphorus will start to burn again as soon as the water runs off or evaporates. The burning can be stopped permanently by washing the burned area thoroughly and repeatedly with a solution of soluble copper salt, such as 3 percent copper sulfate. This will precipitate metallic copper or perhaps copper phosphide on the surface of the phosphorus and prevent its further oxidation. After the skin surface has been cooled to solidify the particles, they can be easily found by their bright coppery appearance and removed.

If a copper salt solution is not available, the burning part can be covered with cloth saturated in cold water, and kept saturated until the copper salt solution can be obtained, or until the particles can be otherwise removed.

If possible, spilled phosphorus should be immediately covered with water. Large spills can be confined to a small area with sand or earth dikes and then flooded with water or covered with sand or earth and kept wet until the phosphorus can be burned or recovered.

Yellow phosphorus is toxic. It may produce poisoning if taken by mouth, but in modern industry the only important method of absorption is by inhalation.

Although a few cases of chronic phosphorus poisoning have been reported, it is not a serious industrial hazard since most operations can be carried out under water and the exposure can be minimized. Chronic poisoning may take the form of general weakness, with anemia, loss of appetite, indigestion, and chronic cough resulting from irritation of the gastrointestinal system and fatty degeneration of the liver.

Bone necrosis ("phossy jaw"), first seen in the preparation of phosphorus matches, may affect any bone but is most commonly seen in the jawbone. It usually appears as a toothache followed by the loosening of one or more teeth and then a pus-forming ulceration develops around the tooth and down into the jawbone itself. Phosphorus does not have a direct effect on the bone but makes possible an infection of the bone by interfering with the blood supply. Bone necrosis resulted in severe facial deformities years ago, but under modern treatment cases have frequently been cured with no deformity and little or no loss of bone.

A form of generalized weakness attended by anemia, loss of appetite, gastro-intestinal complaints, chronic cough, and marked weakness and pallor has been reported to be due to systemic phosphorus poisoning.

To prevent poisoning, phosphorus should be used under conditions which exclude air and prevent inhalation of the fumes (a mixture of phosphorus oxide and free phosphorus). Employees working with yellow phosphorus should bathe daily after work and thoroughly brush their teeth and wash their hands *before* each meal.

Before an employee is assigned to work with phosphorus, he should be provided with information on phosphorus exposures including information describing the hazards, prevention, control measures, and recommended first aid treatment.

Personal protective clothing should be provided and should be changed frequently. Periodic dental examinations should be made at a frequency dependent on the type and degree of exposure. Full mouth X rays should be made at the time of employment or shortly thereafter and repeated at least once a year. Undesirable dental conditions shoud be corrected before the employee is put on phosphorus work.

Phosphorus pentasulfide is a compound which, by itself, possesses little toxicity. Phosphorus pentasulfide, however, hydrolyzes rapidly on contact with water or even with moisture present in the atmosphere to cause the liberation of hydrogen sulfide gas; phosphorus pentasulfide itself has no warning properties.

Phosphorus pentasulfide, if allowed to remain on the skin can cause redness and other signs of irritation. On contact with the eyes it can cause serious irritation, which may be delayed for a matter of hours.

Phosphorus trichloride and phosphorus oxychloride are low-boiling, fuming liquids usually furnished in cylinders or in glass bottles. They decompose rapidly in the presence of water or moist air, forming hydrochloric acid, phosphorus acid, and phosphoric acid and thus produce acid burns. Their vapors are strongly irritating to the skin, mucous membranes, and respiratory system.

Phthalic anhydride in the form of vapor, fume or dust is a primary irritant to mucous membranes and the upper respiratory tract. Initial exposure produces coughing, sneezing, burning sensation in the nose and throat, and increased mucous secretion. Repeated or continued exposures may result in general inflammation of the respiratory tract, nasal ulceration and bleeding, atrophy of the mucous membranes (reversible), loss of smell, hoarseness, bronchitis, and symptoms of allergic hypersensitiveness.

A direct, though not immediate irritant to the dry skin, phthalic anhydride is particularly irritating after contact with water, because of the free acid formed. It produces a contact dermatitis with redness and burning and with continued exposure may cause dryness with a chapped appearance; sensitization may occur.

Although phthalic anhydride causes a characteristic response on inhalation, there is no correlation between sensory perception and safe levels. Control measures should be designed to prevent personal contact and to maintain air concentrations below recommended concentrations. It is particularly important to educate the workers to avoid unnecessary contact and to observe good skin hygienic measures to control skin disorders. Workers handling phthalic anhydride should have clean clothing, rubber gloves, and head covering each day.

Because of the fire and explosive characteristics, particularly at elevated temperatures, closed systems and/or inert gas cover are recommended whenever possible.

Picric acid (2,4,6-trinitrophenol) is a yellow solid which, in the dry state, is highly sensitive to shock. It is therefore shipped with at least 10 percent water in glass or wood containers. It must also be kept out of contact with metals and ammonia to prevent the formation of picrates more sensitive to explosion than the picric acid.

Containers should be plainly marked EX-PLOSIVE—KEEP FIRE AND OPEN FLAMES AWAY, and the same notice should be posted at the doors of all rooms where the material is stored or used. The no smoking rule should be enforced in such areas, electrical installations should be explosion-proof in accordance with the regulations of the National Fire Protection Association, and other sources of ignition, such as welding, must be controlled.

Picric acid was once used extensively as a dressing for burns. This use has been largely discontinued because many cases of skin irritation and systemic poisoning resulted from application of concentrations as low as 1 percent to large areas of the skin.

Picric acid is rapidly absorbed through the unbroken skin and even more rapidly through wounds, leading to headache, shivering, fever, and insomnia. Absorbed in larger quantities, it produces jaundice and severe inflammation of the kidneys. Exposure to the dust of picric acid may cause irritation of the nose and throat, and especially of the eyes, leading to ulceration of the cornea.

Poison oak, poison ivy, and **poison sumac** are plant species whose toxic irritant has been identified as urushiol. The dermatitis is the result of hypersensitization to this compound. Individual susceptibility varies greatly and also changes from time to time, depending, at least in part, on how long it has been since the last exposure. No one is so completely immune, however, that a sufficiently rigorous exposure will not result in dermatitis.

Since reaction to these plants is a hypersensitization, it seems logical that individuals could be desensitized by injections of graduated doses of the purified extract. This has been attempted in many instances, and there are a number of extracts on the market for the purpose. They seem to give some limited protection to sensitive individuals.

The poisoning can most effectively be prevented by destruction of the plants with one or two applications of a herbicide such as one of the 2,4-D formulations. Destruction by grubbing out and burning is both expensive and hazardous for the men doing the work and is no more effective than use of the herbicide.

Potassium chlorate—see **Sodium chlorate.**

Potassium cyanide—see **Sodium cyanide.**

Potassium hydroxide is a white solid extremely soluble in water. It is most commonly furnished as a concentrated solution in tank-car quantities and as the solid in flakes, pellets, and sticks, or cast solid in glass or in metal drums.

The most common injuries suffered with potassium hydroxide are burns of the skin or eyes, caused by solid particles and splashing of the concentrated solution from attempts to dissolve the solid in hot water. The solid alkali generates so much heat on dissolving that there is often local boiling even when cold water is used.

To dissolve the material from a drum in which it is cast solid, the drum should be suspended near the top of a tank of lukewarm water so that the water can circulate through the drum. The high density of the concentrated solution will promote circulation from the top to the bottom of the tank. This method will provide both faster solution and less possibility of injury than any other.

First aid for alkali burns consists of washing away the solid or the solution with large quantities of cold water, preferably under a deluge shower, and treating the injury as a heat burn. The washing of alkali burns should be particularly thorough because of the tendency of the alkalis to penetrate into the skin and continue burning even when thought to be washed away.

Eyes injured by potassium hydroxide should be washed with a gentle stream of water for at least fifteen minutes and should then be immediately brought under medical care. Because of the penetrating properties of alkalis, the injuries may be extremely severe.

Propyl acetate, either n-propyl or isopropyl, is a mild irritant to the eyes, nose, and upper respiratory tract. These compounds are also potent narcotics, the narcotic effect being about the same as that of ether at the same volume concentration. There is no evidence that the effects are cumulative.

Propyl alcohol (propanol) is mainly a respiratory and eye irritant. Liver damage in animals following repeated inhalations of light concentrations of propyl alcohol has been reported. Isopropyl alcohol is more common in industrial use than n-propyl.

In the body, n-propyl alcohol is oxidized and excreted somewhat more rapidly than is ethyl alcohol, while isopropyl alcohol is excreted considerably more slowly. For this reason, the isopropyl is slightly more toxic than the n-propyl.

Propylene is a colorless gas at normal temperature and pressure. In high concentrations it has anesthetic properties, but its chief effect is to cause asphyxiation by exclusion of oxygen. Contact with the liquid will cause a "freezing burn."

As a gas, it produces little or no irritation to the eyes and its characteristic odor is too mild to be a guide for safety. Good ventilation is essential in rooms or areas where propylene is handled to prevent the accumulation of explosive mixtures. The type of ventilation needed will depend upon such factors as vapor density, dead air spaces, temperature, convection currents, and wind direction, which must be considered by the engineer when determining equipment location, type, and capacity.

Propylene oxide is used as a chemical intermediate and sometimes directly as a solvent and sterilizing agent. Both acute and chronic exposures present moderate health hazards. Animal experiments resulted in eye and nose irritation, difficulty in breathing, drowsiness, and incoordination. Severe skin damage can be caused by aqueous solutions.

Pumice is a porous silicate rock produced by volcanic action, which is useful as an abrasive and polishing material because it is easily reduced to a very fine powder. Although silicates in the absence of free silica are generally not active in producing pneumoconiosis, there have been reports in the European literature of pneumoconiosis from long-continued heavy exposures to pumice dust over periods of 20 to 30 years. Also see discussion of silicosis in Chapter 37, "Industrial Hygiene."

Pyrethrum or Persian insect powder is the powdered flower of a plant cultivated chiefly in Japan and in parts of Africa. The pyrethrins, the active insecticidal materials, are not particularly toxic for mammals. Other constituents in the plant, however, produce a severe dermatitis on many individuals. In addition, pyrethrum produces allergic reactions in sensitive persons who are commonly also sensitive to ragweed pollen. The dermatitis is readily minimized by the usual dust control measures.

Pyridine is a colorless liquid organic solvent with a characteristic pungent odor. The odor of low concentrations is nauseating, but the sense of smell is blunted so rapidly by pyridine that the odor cannot be depended upon as a warning of a dangerous concentration. It presents both fire and health hazards.

Pyridine (either in liquid or vapor phase) is a strong local irritant without great systemic toxicity. It readily produces dermatitis on contact with the skin and may produce severe damage on contact with the eyes. In larger doses, it has a narcotic action. Persons exposed to dangerous amounts will show shortness of breath and trembling. It has been reported that 0.019 ml per kg taken daily has no toxic effect on men, while 0.023 to 0.031 ml per kg produced intoxication with signs of liver and kidney injury.

Pyridine splashes on the eyes or other parts of the body should be washed off immediately and completely with a large amount of running water.

Q

Quinone may be formed when either aniline or hydroquinone is oxidized. Its health hazard for acute and chronic exposures is considered high. Quinone vapor produces irritation of the eyes and may result in corneal edema, ulceration, and scarring. Chronic exposures may cause structural changes in the cornea to develop. There is no evidence of systemic injury associated with these eye injuries.

Skin irritation and staining may be the result of skin contact.

R

Rotenone is the active constituent of the derris and cube plants, as well as of many other leguminous plants. It is highly toxic to insects but of relatively small hazard to mammals, and is therefore widely used as an insecticide. The powdered plant material is irritating to the skin and causes a dermatitis in a large proportion of the workmen handling it, and inhalation of large amounts of rotenone dust will produce a severe pulmonary irritation.

These hazards can be readily controlled by enclosure and exhausting of the equipment in which the material is handled and by careful personal hygiene.

S

Selenium compounds are uniformly highly toxic, although toxicity varies somewhat according to the solubility of the specific compound. The chronic effects of selenium were first noticed in "alkali disease" of cattle feeding on selenium-bearing forage. It is known that rats on a diet containing 350 μg of selenium per kg of body weight show definite symptoms of toxicity after a few weeks.

Symptoms are a garlicky odor of the breath, which may persist for a long time after the exposure is terminated, nervousness, gastrointestinal disturbances, loss of weight, and great pallor. Although all the selenium compounds are toxic, exposures to hydrogen selenide and to selenium oxychloride are likely to be the most severe.

Selenium oxychloride produces extremely severe blisters and burns. A minute drop of the liquid on the skin will produce a third-degree burn. Immediate washing with plenty of water will reduce the severity and extent of the burn but will generally not prevent it.

See also **Hydrogen selenide.**

Silica is generally recognized as the most hazardous of the dusts which, when inhaled in excessive amounts, can produce a fibrotic change in the lungs or silicosis. Silica is found in industry in the forms of quartz or sand, quartzite, sandstone, flint, tripoli, diatomaceous earth, and as an important constituent in chert, granite, and other kinds of minerals. Whatever the form and manner of use, exposure to free silica should be kept as low as possible by dust suppression methods. There is no method known for removing fibrotic tissue from the lungs; consequently,

silicosis is incurable.

Sodium. The only recognized injuries from sodium are thermal and alkali burns caused by contact of the metal with the skin or eyes, in the presence of moisture. On contact with moisture, sodium liberates hydrogen; enough heat is produced so that this gas may be ignited. The immediate result is a severe thermal burn, but as burning continues, appreciable amounts of sodium hydroxide may be formed, which may add to the injury by also causing an alkali burn. When no moisture is present, contact of the dry metal with the skin may cause no damage. No systemic effects are recognized.

Employees who handle sodium should be thoroughly instructed as to its hazards, necessary measures of control, and proper first aid procedures.

Sodium chlorate is a solid oxidizing agent which, with reducing agents such as sugar, coal dust, shellac, sawdust, or other organic materials, can form explosive mixtures that may be detonated by friction or shock. Sodium chlorate is generally shipped in boxes or kegs lined with heavy brown paper. After the material is removed from the container, both the lining and the box or keg should be burned immediately.

Chlorate solutions are widely used for the destruction of weeds and other noxious plants. The clothing of workmen on these jobs, if saturated with the solution, can be ignited either through friction or through smoking. Combustion in such cases is practically explosive, and the injuries are likely to be fatal.

Walking on chlorate spilled on the floor can also cause serious burns. The material should be handled over smooth, impervious floors, and spillage should be cleaned up immediately. Impervious clothing should be worn while chlorate solutions are being handled. Clothing should be thoroughly washed at the end of the shift to remove spilled material, and should be rinsed during the shift whenever material is spilled on it.

Large quantities of chlorates should be stored in an isolated fire-resistant building, conspicuously marked with the nature and hazards of the contents. Such buildings should not be used for any other purpose; no other chemicals should be stored there.

In the general handling of chlorates, contact of these compounds with mineral acids so as to produce anhydrous chloric acid should be prevented. It is likely to be spontaneously explosive, as well as highly sensitive to contact with reducing agents.

Sodium cyanide is a white, slightly hygroscopic solid with a reaction sufficiently alkaline to produce a severe dermatitis when the skin is exposed to its dust or its solutions.

Cyanide poisoning in industry is not common. When it does occur, it is usually the result of inhalation of hydrogen cyanide gas, used as a fumigant or produced by contact of one of the metallic cyanides with acid, including carbon dioxide gas and moisture from the atmosphere. See **Hydrogen cyanide.**

The simple metallic cyanides are all extremely toxic if swallowed. For sodium cyanide, the lethal dose is about 9 mg per kg of body weight or about 0.019 oz for a man. Some authorities state that cyanide can be absorbed into the system through the skin or through the lesions of the toxic burn which it produces. This is not, however, a common method of industrial poisoning.

Cyanide prevents oxygen exchange between the tissues and the blood. The symptoms of poisoning are dizziness, headache, a feeling of oppression in the chest, dryness and irritation of the throat, palpitation of the heart, difficult breathing and sometimes vomiting, followed by loss of consciousness and convulsions. A lethal dose produces rapid death and therefore can be counteracted only by *immediate* first aid.

A most effective single first aid measure for this type of internal asphyxia is the supplying of oxygen or oxygen-carbon dioxide mixture for breathing. The person administering first aid should also:

1. Send for a doctor at once.

2. Remove the victim to fresh air, without exposing the rescuer to poisoning.

3. Remove clothing which may contain hydrogen cyanide and wrap the victim in blankets or otherwise keep him warm.

4. If breathing has stopped, apply artificial respiration.

5. Supply oxygen or oxygen-carbon dioxide mixture for inhalation.

6. Permit inhalation of the contents of an ampule of amyl nitrite; repeat after 15 minutes.

7. If the victim has swallowed a cyanide compound and is conscious, induce vomiting by giving him a large quantity of tepid water containing an emetic. *Never try to induce vomiting in an unconscious person.*

First aid supplies, as specified by the physician in charge, should be in readiness in all plants where cyanides are used.

Cyanides should be stored and used so that they will not come into contact with acids or be exposed to the air. Where cyanides are used in bulk, men should not be permitted to work alone in an area with an open container.

The layout should provide for ready removal of deposits of dust from the structure and equipment. Dust and vapor incidental to the process should be removed at the source as completely as possible. Local exhaust equipment for this purpose should be designed with special attention to accessibility for cleaning, since hydrolysis of alkali cyanides leaves a deposit of alkali carbonates which form a sticky, tenacious coating on the equipment.

Employees who handle cyanides should be provided with protective clothing which will prevent skin contact with the solid or solution, and appropriate respirators, either canister or supplied atmosphere, should be provided for emergency use.

Sodium dichromate—see **Chromic acid.**

Sodium hydroxide—see **Potassium hydroxide.**

Stibine (antimony hydride) is a toxic gas produced by the generation of hydrogen in the presence of elemental antimony. Charging of storage batteries of the lead-acid type produces stibine from the antimony which is in the lead grids as a hardening alloy. This occurs mostly as the battery is overcharged and in increasing amounts as the battery grows older.

Stibine is also apt to be formed when water is applied to the dross from aluminum refining of metals. Its presence and effects in this instance are overshadowed by the arsine also produced.

Relatively little is known about the toxicity of stibine, but from the information available it seems to be qualitatively very like arsine.

Stoddard solvent is a registered commercial standard of the U.S. Department of Commerce for a dry cleaning solvent of specifications such that it has a flash point of 100 F, evaporates without residue, and consists of aliphatic, saturated materials and, in some formulations, 15 to 20 percent aromatics. The fire hazard is about that of kerosene, but it is a more satisfactory cleaning solvent. It is available under a number of trade names. For discussion of toxicity, see **Heptane** and **Hexane.**

Styrene monomer (phenylethylene, vinylbenzene) is a colorless organic liquid with a characteristic disagreeable odor. As either liquid or vapor, it is irritating to the eyes, respiratory tract, and produces an extremely severe dermatitis. The odor serves as a fairly good warning of excessive concentrations.

In animal experiments, both rats and guinea pigs survived exposures of 2000 ppm for seven hours without fatalities. The highest concentration which can be obtained at ordinary temperatures, approximately 10,000 ppm, may be fatal in 30 to 60 minutes.

Styrene is flammable and also polymerizes rapidly at slightly elevated temperature and slowly at ordinary room temperature. Polymerization produces heat, which in turn increases the rate of polymerization. Unchecked, the reaction may become so violent as to produce a fire or explosion.

Commercial grades contain an inhibitor to prevent polymerization. In storage, styrene should be checked occasionally to assure the presence of the proper amount of inhibitor and the absence of polymers.

Aside from this polymerization reaction, styrene can be handled with the precautions for a flammable organic solvent which is also a skin irritant. Protective clothing should be worn to prevent skin contact.

Sulfur is virtually nontoxic; there is no evidence that systemic poisoning results from the inhalation of sulfur dust. Although sulfur is capable of irritating the inner surfaces of the eyelids, the dust may rarely irritate the skin. Molten sulfur may contain hydrogen sulfide which is toxic in low concentrations. Solid sulfur containing hydrocarbon will, when held in the molten condition, generate hydrogen sulfide. The primary hazard in handling solid sulfur is that sulfur dust suspended in air ignites easily.

Sulfur chlorides. Sulfur monochloride and sulfur dichloride can be dealt with as identical compounds insofar as their health hazards. Fuming, corrosive liquids with a strong odor, they are highly irritating even in low vapor concentration. Because of this odor and irritation, they have good warning properties.

On contact with the skin, sulfur chlorides can cause irritation and if not removed can lead to serious irritation and burns. Contact with the eyes of either liquid or vapor will produce severe damage, which may cause scars. Inhalation of the vapors may be injurious to the lungs. Exposure to the vapor can cause coughing, tears, and burning of the eyes. After severe vapor exposure, pulmonary edema may occur. There are few reported accidents because of the effective warning afforded by the irritating effect.

Sulfur dioxide is a pungent, irritating gas produced in large quantity by the combustion of sulfur or compounds containing sulfur. Since it is water-soluble, its action is likely to be limited to irritation of mucous membranes of the upper respiratory system.

Endurable concentrations can be tolerated for long periods of time without appreciable danger to health. In a concentration range of 1 to 10 ppm, respiratory and pulse rates increase and depth of respiration in unacclimatized individuals decreases. Acclimatized individuals, working daily in concentrations in excess of 10 ppm, do not show these changes on exposure to 5 ppm.

Severe exposures to sulfur dioxide may result from loading and unloading of tank cars, rupture of cylinders, leaking lines or cylinders in a confined space, following fumigation procedures aboard ships, and other similar type operations. Employees who may be subject to such exposures should be provided with proper eye, respiratory, skin, and mucous membrane protection.

The compound is considerably more toxic to vegetable life than it is to animals, and plants will be severely damaged by concentrations not irritating to man.

Sulfur hexafluoride seems to be a physiologically inert material. Rats showed no effects from exposure to an atmosphere of 80 percent sulfur hexafluoride with 20 percent oxygen for 16 to 24 hr.

Sulfur monochloride is still used to some extent as a vulcanizing agent in the rubber industry. It is a dark yellow, highly volatile liquid, which hydrolyzes rapidly in moist air to give hydrogen chloride, sulfur dioxide, and sulfur. Its extremely irritating effect on the skin, eyes, and respiratory system is due primarily to the hydrochloric acid formed in its hydrolysis.

The chemical is commonly dissolved in carbon disulfide, which has no irritating or warning properties, and is therefore much more hazardous both to health and to safety than sulfur monochloride.

Sulfur pentafluoride is an extremely toxic material. In experiments on rats, 1.0 ppm caused lung irritation in a one-hr exposure; for an exposure of 18 hr, 0.1 ppm caused lung irritation and 1.0 ppm was lethal. For an 18-hr exposure, 0.01 ppm had no observable effect.

Sulfuric acid is a viscous, oily liquid with a strong affinity for water, which it removes from organic material and thus chars and destroys plant or animal tissue. The grades sold as oleum, which contain 20 percent or more of sulfur trioxide dissolved in the acid, fume strongly at room temperature. The other grades fume at higher temperatures. These fumes and the spray or mist from the solutions are extremely irritating both to the skin and to the mucous membranes.

Workers should be protected from the mist or spray as well as from contact with the liquid. Respiratory protective equipment is especially necessary whenever oleum must be exposed to the air.

Oleum and concentrated sulfuric acid do not attack iron and are ordinarily shipped in mild-steel tank cars or in drums. If containers are not tightly sealed, the acid may absorb enough water from the air to dilute the top layer, which will then attack the metal and evolve hydrogen. This action produces both an explosion hazard and the hazard of acid spurting out under pressure when the container is opened. Open lights and flames should be kept away from such containers, and they should be vented to relieve any pressure before they are opened.

There have been several detailed reports of lung damage from exposure to sulfuric acid mists both in experimental animals and in man. However, it seems highly unlikely that this will ever become a common disease of man because of the severe irritation involved in inhaling even a very small concentration. The highest concentration producing no deaths in adult guinea pigs for an 8-hr exposure was 20 mg/cu m.

T

Talc, as used in industry, covers a wide variety of materials, many of which are not talcs at all. The only one which has been specifically suggested as the cause of a pneumoconiosis is tremolite talc, a hydrous magnesium silicate occurring in needlelike form and rather closely related to asbestos. It is, however, only slightly active because the victims have been persons who were exposed to heavy air contamination for many years.

Talc has been demonstrated to produce granulomas when deposited in the body, and for that reason has been dropped from surgical use. One pulmonary case also showed granuloma and fibrosis apparently produced by the inhalation of talc used as a lubricant on patent leather.

See Chapter 37, "Industrial Hygiene."

Tar and **pitch** have been known for a great many years, particularly in England, as producers of skin cancer. The skin cancer of chimney sweeps is undoubtedly due to the tar fraction of the soot with which they work. Skin cancers of mule spinners were next reported when mineral oil was substituted for vegetable oil or whale oil in the spinning industry. High incidence of skin cancer among men exposed to oils, tar, or pitch has not been reported from the European continent or from the United States, although in both areas there have been several extensive investigations of this question.

Teflon resins are available in two types: the TFE resin (tetrafluoroethylene polymer) and the FEP resin (fluorinated ethylene-propylene polymer). Both are physiologically inert. The products of pyrolysis of Teflon below 392 F are nonhazardous, but above this temperature, several fluorocarbon gases, hydrogen fluoride, and a sublimate are formed. These do not appear to be toxicologically significant until about 482 F.

At 716 F, small amounts of toxic gases, which have caused death by pulmonary edema in experimental animals, are evolved. The only known difficulty in human exposure has occurred when Teflon resins are processed at 644–725 F. At these temperatures, a condition somewhat like metal fume fever is exhibited, but recovery is rapid, usually in 48 hours or less. All operations that can lead to the pyrolysis of Teflon resins should therefore be prohibited, particularly smoking where the tobacco may be contaminated with Teflon dusts.

Tellurium is found as a trace element in various minerals and is used in small amounts in a number of industries, especially the glass industry. It increases the machinability of some metals and improves the toughness of rubber hose and cable and the hardness of lead.

In man, the absorption of even a very small amount of tellurium in any form is immediately followed by the production of a garlicky odor in the breath and the sweat, due to the elimination of minute amounts of tellurium as methyl telluride. This elimination is likely to continue for weeks or months.

Hydrogen telluride, a poison gas similar to arsine and hydrogen selenide, destroys the red corpuscles. It is believed to be more toxic than either, but is not so often encountered in industry.

In an investigation of workmen in a foundry who had been exposed to 0.1 to 1.0 mg of tellurium per 10 cu m of air, a large percentage of the men showed the garlic odor of the breath and sweat and also complained of dryness of the mouth and a metallic taste.

Men with the highest exposures, who also had the highest rate of tellurium excretion, showed drowsiness starting in the early afternoon and later becoming more severe. Complaints of appetite loss and nausea at these relatively low exposures were less common.

TEPP (tetraethyl pyrophosphate) is the active ingredient in an insecticide mixture that goes by the same name. Since the estimated fatal dose for a human being is approximately 600 mg taken orally, it is in the class of highly toxic materials.

In its use, toxic dust respirators or supplied air respirators and protective clothing which will completely prevent skin contact should be worn. Anyone with a severe headache or a feeling of tightness or constriction in the chest should leave or be removed immediately.

1,1,2,2-Tetrabromoethane (acetylene tetrabromide) is a high-boiling, heavy, yellowish liquid used as an indicator liquid in gages and for separating minerals by density. Its low vapor pressure at room temperature makes it safer to handle than might be expected. The LD_{50} for rabbits and guinea pigs is about 0.4 gm per kg of body weight. Exposure to the vapor causes respiratory irritation and damage of the liver appearing a few days after a single acute exposure.

1,1,2,2-Tetrachloroethane (acetylene tetrachloride) is an excellent solvent for a number of paints and lacquers. It is also a very dangerous compound since inhalation of it at a concentration that is barely perceptible by odor can lead to extensive liver damage.

Tetrachloroethylene (perchloroethylene) is a chlorinated solvent used primarily for degreasing. It is similar in its properties to trichloroethylene except that its boiling point is slightly above that of water so that boiling degreasers containing it automatically keep themselves dry without loss of solvent.

Its most important toxic effect is depression of the central nervous system. In experimental exposures of human subjects, those exposed to concentrations greater than 200 ppm promptly suffered central nervous system symptoms, while those exposed to about 100 ppm did not.

Tetraethyl lead—see **Lead.**

Tetrahydrofuran (tetramethylene oxide) is used in chemical reactions and as a solvent for resins and plastics. Both acute and chronic exposures constitute a moderate health hazard. Symptoms of overexposure—nausea, dizziness, and headache—soon disappear when fresh air is inhaled. Kidney effects reported in early studies were probably due in part to impurities in the compounds tested.

Tetryl (2,4,6-trinitrophenyl-methylnitramine) is a high explosive notable for its ability to produce a severe dermatitis by direct contact and a higher grade of sensitization in some persons. There is some evidence of a slight systemic toxicity, not as important in practice as the dermatitis. The systemic toxicity seems to consist of general central nervous system symptoms and anemia, although there is some evidence of liver damage and lung involvement from heavy chronic exposures.

The dermatitis usually develops in the second or third week of contact with tetryl. It is commonly accompanied by irritation of the mucous membranes of the nose, nosebleed, and yellow staining of the skin of the hands, face, and scalp. Loss of appetite, abdominal cramps, and anemia are often seen in tetryl workers.

Reduction of contact with the material by dust control measures accompanied by zinc oxide protective cream and work uniforms, including caps, will greatly reduce the hazard. Most people who develop the dermatitis in a mild form will respond to treatment and acquire a certain tolerance so that they can continue working. There is, however, no true immunity to tetryl. People who have worked with the material for years develop dermatitis if the exposure is increased. There are also some cases of true hypersensitivity, which must be removed from all contact.

Thallium is not a common cause of poisoning in industry. There have been some cases, mostly from the use of thallium salts as cosmetic depilatories or a part of the treatment of parasitic infections (tinea) in children. A dose of 8 mg per kg of body weight has caused fatal poisoning in some instances.

Typically, thallium causes swelling of the feet and legs, pains in the joints, sleeplessness, vomiting, disturbances of sensation, and mental confusion as well as complete loss of hair from all parts of the body.

Thanite—see **Thiocyanates (aliphatic).**

Thiocyanates (aliphatic) are synthetic organic materials used only as insecticides. Their toxicity is low enough that inhalation of the ordinary spray mist is not particularly dangerous, and so far as is known they are not absorbed through the intact skin. When taken orally, they produce poisoning characterized by depression, difficult breathing and blue color of the lips, circulatory collapse and convulsions.

The lethal oral dose is from 10 to 60 cc of the pure material or approximately 200 cc of the 5 percent solution in oil, normally used for spraying. Such quantities might be encountered in careless handling and storage of the material, but would scarcely be inhaled in the use of it.

Some animal experiments have also indicated that the chronic toxicity of these materials is low.

Titanium is a metal used in alloys and in massive forms by the Atomic Energy Commission. Since implants of the metal in muscle tissue proved inert, and no occupational systemic poisonings have been reported, its health hazard is rated as low.

Titanium dioxide is widely used as a pigment. It seems to be quite inert physiologically inasmuch as it is used as an ingredient of antiburn creams which are liberally applied to the skin of the face with no reports of ill effects. Experimentally, it has been fed to animals by stomach tube in large amounts without producing any signs of toxicity.

Titanium tetrachloride is sometimes used to produce "smoke" because of its rapid hydrolysis in moist air to produce a cloud of titanium dioxide and hydrochloric acid mist. Its harmful effect is due entirely to the acid mist.

The heat of reaction of titanium tetrachloride with water is greater than that of sulfuric

1173

acid and may cause severe burns.

Toluene is seldom a source of acute poisoning in industry, although its inherent acute toxicity is somewhat higher than that of benzene. It is a flammable colorless liquid of rather strong aromatic odor which serves somewhat as a warning of high concentration.

At concentrations of 500 to 1000 ppm it is strongly irritating to the eyes and respiratory system. In higher concentration it is a narcotic, and the signs of acute poisoning are headache, drunkenness, nausea, vomiting, and ultimately unconsciousness.

Toluene does not appear to produce the severe and often fatal depression of the blood-forming organs seen in chronic benzene poisoning. It may, however, produce depression of the bone marrow to a smaller degree. The first sign of excessive toluene exposure in human beings is enlargement of the liver, with a slightly decreased red cell count.

In case of acute exposure the person should be taken to fresh air as soon as possible. Oxygen should be administered and, if breathing has stopped, artificial respiration should be given immediately. A physician should be called at once.

Toluene 2,4-diisocyanate (tolylene 2,4,-diisocyanate, TDI), a liquid used in the manufacture of polyurethane foams, is rated as extra-hazardous for acute and chronic vapor exposures. It has a low oral toxicity and is primarily an irritant of the gastrointestinal tract, eyes, skin, and respiratory tract. Asthma-like symptoms may appear on repeated exposure. Animal experiments have demonstrated skin sensitization.

Toluidine (methylaniline), the amino derivative of toluene, is similar in its action to aniline. The physiological effects are somewhat different because of the greater tendency of toluidine to interfere with the function of the kidneys. It is rapidly absorbed through the intact skin, and in numerous instances splashes of the compound have caused severe poisoning.

Toxaphene is a proprietary chlorinated camphene having 67 to 69 percent chlorine. It is a residual insecticide similar to DDT in its action but with a toxicity about four times that of DDT. Single applications to the skin of about 46 gm or daily applications of 2.4 gm over a period of days are very dangerous.

Trichloroacetonitrile (Tritox) is used as a fumigant and insecticide. In animal experiments, sublethal doses caused damage to the heart, liver, and kidneys. It also produced local irritation of the skin. The highest concentration which caused some irritation but not deaths on repeated inhalations by rabbits, rats, and guinea pigs was 0.031 mg per L (5.3 ppm).

Trichloroethane—see **Methyl chloroform.**

Trichloroethylene is a halogenated hydrocarbon used primarily as a degreasing compound and in the preparation of insecticides. It has no flash point as such, but at elevated temperatures and with a high-energy ignition source, such as a welding arc, its vapors can and will explode. Toxic decomposition products, mainly hydrogen chloride with some phosgene, may also be formed under these conditions.

In animal experiments, the most important effect following single exposures was a depressant action on the central nervous system. Neurologic changes and liver injury may result from prolonged exposure, but the liver effect is likely to be minor. As with other chlorinated hydrocarbons, trichloroethylene can cause alteration of the heart rhythm, and death from this cause, although rare, has been reported. Some cases of voluntary addiction to trichloroethylene have been found in industry.

Although some absorption through the skin may occur, trichloroethylene has mainly a defatting and dermatitis-producing skin effect.

Trichloronaphthalene—see **Chlorowaxes.**

Trinitrotoluene (TNT) is a highly poisonous explosive ordinarily handled as a solid. The dust is readily absorbed by inhalation or ingestion. According to some studies, skin absorption is a relatively minor source of intoxication.

The dust produces a severe dermatitis in a great many individuals. This dermatitis usually appears within the first few weeks of exposure but may occur in an individual who has worked with the material for a number of years. Probably anyone with sufficient exposure will eventually develop the dermatitis. There are also some persons with a true hypersensitivity to TNT who should be permanently removed from exposure.

Inhalation or ingestion of TNT in excessive quantities will produce systemic poisoning.

Some of the earliest signs of excessive absorption of TNT are digestive disturbances and mild cyanosis. The digestive signs include nausea, sometimes vomiting in the early morning, loss of appetite, and sometimes constipation with colic.

The cyanosis probably starts with the formation of methemoglobin or similar pigment. This indication of overexposure is not in itself dangerous, but it is commonly followed by a reduction in red cells and a severe anemia, which may develop into a typical aplastic anemia.

TNT also affects the liver and produces a toxic jaundice. Young people and people with previous liver disease are the most susceptible.

To prevent both the dermatitis and the systemic poisoning, the most scrupulous housekeeping and dust control measures should be observed. TNT handlers should wear complete work uniforms, change them daily, and never take them from the plant. They should also be required to wash thoroughly before leaving the plant. Soap is not effective. Liquid soap with 5 percent potassium sulfite and a small quantity of sulfonated higher alcohol provides a good medium.

Triorthocresylphosphate (TOCP) is a colorless, odorless high-boiling liquid and is usually found in liquid mixture with the meta and para isomers. Any route of exposure in single or repeated doses constitutes a high health hazard. TOCP is readily absorbed through the skin, but there is little vapor hazard at ordinary temperatures.

Ingestion of the compound, either as the liquid or the fog or mist, produces symptoms of paralysis without sensory effect. Shortly after ingestion, there may be gastro-intestinal symptoms, including nausea, vomiting, diarrhea, and abdominal pain. The majority of victims never again regain normal function of the central nervous system.

Turpentine is the most familiar solvent for oil paints. Although there are cheaper substitutes, it is still widely used in fine exterior painting. Many painters formerly had chronic kidney disease, which they attributed to turpentine vapor but which was probably due to lead. It has not been possible to produce chronic kidney disease experimentally with turpentine.

Turpentine hydrocarbons are not particularly irritating in themselves but some grades contain highly irritating impurities.

These differences depend on the method of preparation and probably are a measure of the amount of decomposition produced in the refining. It has been generally believed in this country that gum turpentine, obtained by tapping trees, is less irritating than the steam-distilled variety.

U

Uranium and its compounds are radioactive, and they are toxic when ingested or inhaled. The effects on the body depend upon the solubility of the uranium compound concerned.

The soluble compounds are eliminated from the body rather rapidly, but if ingested in sufficient amounts, they will cause a chemical poisoning, primarily affecting the kidneys. One investigator has reported that long before a radiologically hazardous amount of uranium could be deposited in the bone structure, the chemical toxicity to the kidneys would be fatal.

Insoluble uranium compounds are primarily an inhalation hazard. They are dangerous not because of chemical toxicity to the kidneys, but because radioactive particles may accumulate in the lungs in an amount exceeding permissible tissue tolerance.

Urethanes — see **Toluene 2,4-diisocyanate.**

V

Vanadium pentoxide is used as a catalyst for chemical reactions and also may be found in ash from burning certain petroleum oils, such as Bunker C. Apparently it is more significant in derivatives from foreign crudes.

In one extensive study, a group of 50 to 90

1175

employees exposed to vanadium pentoxide over a period of nine years were observed. About 20 of the exposed group were employed continuously for more than eight years. One employee exposed to the dust showed convulsive cough with occasional spitting of blood, shortness of breath, palpitation on exertion, chest pains, bronchitis and bronchospasm, emphysematous chest, elevated blood pressure, and a reticulated pattern on the chest X ray. The pallor which commonly accompanies vanadium pentoxide exposure was attributed to its action in constricting the capillaries in the skin rather than to anemia. Blood counts showed no anemia.

Vinyl acetate is considered a chemical of low toxicity, based not only on the findings in relatively limited animal experimentation, but also on years of use experience. No systemic effects have been reported, and no chronic effects or sensitization have been noted. As usually handled, no health hazard is present, either from inhalation of the vapor or contact of liquid with the skin or eyes.

There are no indications for other than usual preemployment or periodic physical examinations for those who will handle vinyl acetate, nor would there seem to be any special physical conditions making exposure a particular hazard.

The primary hazard in handling vinyl acetate is due to its flammability rather than its effect on health.

Vinyl chloride (chloroethylene) is a gaseous chlorinated hydrocarbon used as a refrigerant but primarily as a raw material in the plastics industry. It is an anesthetic in high concentrations.

Vinyl propionate is an unsaturated ester which appears to have an astonishingly high toxicity for a class of compounds known for their low toxicity. Rats exposed to 4 ppm showed stimulation and then depression of respiration, and died if not removed from the exposure. Daily exposure did not seem to result in accumulation.

Vinylidene chloride (1,1-dichloroethylene) is a chlorinated hydrocarbon of about the same vapor toxicity as ethylene dichloride.

The main hazard of vinylidene chloride is the ease with which it forms peroxides on exposure to air or oxygen. The peroxide is absorbed on the surface of the polymer which precipitates. An attempt to filter off or to dry the polymer usually results in a violent explosion. Usually, the whole mixture quickly becomes explosive if the polymer is left in the suspension.

This compound should be stored in a closed system with oxygen or air excluded; preferably, it should be kept under a positive pressure of an inert gas.

X

Xylene is a solvent mixture of the ortho, meta, and para isomers. It resembles benzene in many physical and chemical properties, but does not produce the chronic blood diseases characteristic of benzene absorption.

Xylene's severe narcotic action is usually noted when concentrations exceed 200 ppm. Repeated skin contact may produce a dermatitis.

Z

Zinc and Zinc oxide. Inhalation of a considerable quantity of freshly formed fume, usually from welding on zinc coated metal, will produce "metal fume fever," which is highly disabling but usually the individual recovers completely within 24 hr.

Zirconium is principally used in massive forms and as powder by the Atomic Energy Commission. It is an extreme fire and explosion hazard in the dust or powdered form. Its health hazard for both acute and chronic exposures is low. Animal experiments on inhalation and ingestion and evaluation of the skin effects of zirconium ointments all support the fact that no cases of industrial systemic poisonings have been reported.

Sources of Help

Many associations are concerned with industrial toxicology problems. Some are prepared to provide consultation service to nonmembers; all have a wealth of available technical information.

Organizations having a high degree of interest in industrial toxicology include the following. (Addresses not listed here will be found in Chapter 23, "Sources of Help.")

Advisory Center on Toxicology, National Research Council, 2101 Constitution Ave., Washington, D.C. 20418

American Conference of Governmental Industrial Hygienists

American Industrial Hygiene Association

American Medical Association

American Petroleum Institute

American Insurance Association

Industrial Health Foundation, Inc.

Industrial Medical Association

Manufacturing Chemists' Association, Inc.

National Institute for Occupational Safety and Health, Division of Technical Services, Cincinnati, Ohio 45202

Poison Control Centers, Crystal Plaza Building No. 5, Arlington, Va. 22202

State departments of health or labor

References

American Industrial Hygiene Association, 66 South Miller Road, Akron, Ohio 44313
> *AIHA Journal*
> "Analytical Guides."
> "Community Air Quality Guides."
> "Emergency Exposure Limits."

> "Hygienic Guides." (Listing follows.)

Subject	Date Published in AIHA Journal		Subject	Date Published in AIHA Journal	
Acetaldehyde		Sept. 1955	Bromine		Aug. 1958
Acetic Acid	Rev.	Sept. 1972	Bromochloromethane		Dec. 1961
Acetic Anhydride	Rev.	Jan. 1971	1,3-Butadiene		Feb. 1963
Acetone		Mar. 1957	Butyl Alcohol		Sept. 1955
Acetonitrile		June 1960	n-Butyl Acetate		Dec. 1962
Acetylene		Apr. 1967	n-Butylamine		Dec. 1960
Acrolein		June 1963	Cadmium	Rev.	Dec. 1962
Acrylonitrile		Mar. 1957	Carbon Dioxide		Oct. 1964
Allyl Alcohol		Dec. 1963	Carbon Disulfide		Dec. 1956
Allyl Chloride		Dec. 1963	Carbon Monoxide	Rev.	Aug. 1965
Allyl Glycidyl Ether		Feb. 1965	Carbon Tetrachloride	Rev.	Dec. 1961
Aluminum and Aluminum			Chlorine Dioxide		June 1958
Oxide		June 1963	Chlorobenzene		Feb. 1964
Amyl Acetate	Rev.	April 1965	Chlorodiphenyls		Feb. 1965
Anhydrous Ammonia	Rev.	Feb. 1971	Chloroform		Dec. 1965
Aniline		Dec. 1955	Chloronaphthalenes		Feb. 1966
Antimony and Its			Chromic Acid		June 1956
Compounds		Dec. 1959	Cobalt		Apr. 1966
Arsenic and Its Compounds	Rev.	Dec. 1964	Cresol (Cresylic Acid)		Oct. 1958
Arsine	Rev.	Aug. 1965	Cumene		Dec. 1961
Asbestos		Apr. 1958	Cyclohexane		Oct. 1963
Barium and Its Inorganic			Cyclohexanone		Dec. 1965
Compounds		Dec. 1962	D.D.T.		Oct. 1959
Benzene	Rev.	June 1970	Diborane		Oct. 1958
Beryllium and Its Compounds			O-Dichlorobenzene		June 1964
	Rev.	Dec. 1964	P-Dichlorobenzene		June 1964
Bisphenol A		June 1967	Dichlorodifluoromethane		Oct. 1968

Subject	Date Published in AIHA Journal		
1,2-Dichloroethane (Ethyl Dichloride)	Rev.	Aug.	1965
1,1-Dichloroethane (Ethylidene Chloride)	Rev.	Jan.	1971
Dichloromethane (Methyl Dichloride)	Rev.	Dec.	1965
Diethylamine		June	1960
Diethylene Triamine		June	1960
Diisobutyl Ketone		Dec.	1962
Dimethylformamide		Sept.	1957
1,1-Dimethylhydrazine		Apr.	1963
2,4-Dinitrophenol		Feb.	1958
Dioxane		Dec.	1960
Diphenyl		Oct.	1964
Epichlorohydrin		Dec.	1961
Epoxy Resin Systems		Oct.	1958
Ethanolamines		June	1968
Ethyl Acetate		Apr.	1964
Ethyl Acrylate		Dec.	1966
Ethyl Alcohol		Mar.	1956
Ethyl Benzene	Rev.	June	1969
Ethyl Bromide		Apr.	1965
Ethyl Chloride		Oct.	1963
Ethyl Ether		Feb.	1966
Ethyl Silicate		Dec.	1968
Ethylene Chlorohydrin		Dec.	1961
Ethylene Diamine	Rev.	Feb.	1970
Ethylene Dibromide		Apr.	1958
Ethlene Glycol Dinitrate		Dec.	1966
Ethyl. Glycol Mono-n-Butyl Ether		Oct.	1958
Ethyl. Glycol Monoethyl Ether		June	1963
Ethyl. Glycol Monoethyl Ether Acetate		Dec.	1965
Ethyl. Glycol Monomethyl Ether		Aug.	1970
Ethylene Oxide		Dec.	1958
Ethylenimine		Feb.	1965
Fluoride Bearing Dusts and Fumes	Rev.	Aug.	1965
Fluorine	Rev.	Dec.	1965
Formaldehyde (Methanal)	Rev.	Apr.	1965
Furfural		Apr.	1965
Heptane		Feb.	1959
Hexachlorocyclohexane, Gamma Isomer-Lindane		Jan.	1972
Hexane		Feb.	1959
2-Hexanone		Dec.	1968
Hydrazine		Dec.	1956
Hydrogen Chloride		Aug.	1958
Hydrogen Cyanide	Rev.	Feb.	1970
Hydrogen Fluoride		Mar.	1956
Hydrogen Peroxide		Sept.	1957
Hydrogen Selenide		Dec.	1959
Hydrogen Sulfide	Rev.	Feb.	1963
Hydroquinone (1,4 Benz.)		Apr.	1963
Iodine		Aug.	1965
Lead and Its Inorganic Compounds		Apr.	1958
Lithium Hydride		Aug.	1964
Magnesium		Feb.	1960
Maleic Anhydride		June	1970
Manganese		June	1963
Mercury and Its Inorganic Compounds	Rev.	June	1966
Mesityl Oxide		Oct.	1969
Meta-Dinitrobenzene		Feb.	1959
Metal Hydrides		Feb.	1960
Methyl Acetate		June	1964
Methyl Alcohol		Dec.	1957
Methyl Bromide		Apr.	1958
Methyl Chloride		Dec.	1961
Methyl Ethyl Ketone		Mar.	1957
Methyl Isobutyl Ketone		Apr.	1966
Molybdenum		Feb.	1960
Naphthalene		Oct.	1967
Nickel		Apr.	1966
Nickel Carbonyl	Rev.	June	1968
Nitric Acid		Aug.	1964
Nitrobenzene		Feb.	1959
Nitroethane		Dec.	1961
Nitrogen Dioxide		June	1956
Nitroglycerine		Feb.	1960
Nitromethane		Dec.	1961
Nitropropane		June	1960
Osmium and Its Compounds		Dec.	1968
Oxygen Difluoride		Apr.	1967
Ozone	Rev.	Apr.	1966
Parathion	Rev.	June	1969
Pentaborane	Rev.	June	1966
Pentachlorophenol and Sodium Pentachlorophenol		Aug.	1970
Pentaerythritol		Feb.	1968
Pentane		Apr.	1966
Petroleum Naphtha		Aug.	1963
Phenol		Dec.	1957
Phosgene	Rev.	June	1968
Phosphine		June	1964
Phosphoric Acid		June	1957
Phosphoric Anhydride		June	1958
Phthalic Anhydride		Aug.	1967
Polonium 210		June	1959
Propanol		Dec.	1961
Propylene Dichloride		June	1967
Propylene Oxide		June	1959
Pyridine	Rev.	Aug.	1963
Quinone (p-Benzoquinone)		Apr.	1963
Radon and Its Daughters		June	1959
Selenium and Compounds		June	1959

Silica (Amorphous)		Apr.	1958	Benzene)	June 1957
Silica (Free)		Sept.	1957	Toluene Diisocyanate	Rev. Feb. 1967
Sodium Chlorate		Feb.	1958	1,1,1-Trichloroethane (Methyl	
Stibine		Dec.	1960	Chloroform)	Rev. Dec. 1961
Styrene Monomer	Rev.	Oct.	1968	Trichloroethylene	Rev. Feb. 1964
Sulfur Dioxide		Dec.	1955	Trichlorofluoromethane	Oct. 1968
Sulfur Trioxide		Jan.	1972	1,1,2-Trichloro-1,2-2-	
Sulfuric Acid		Sept.	1957	Trifluoroethane	Oct. 1968
Teflon, Fluoro Resins,				2,4,6-Trinitrotoluene	Oct. 1964
etc.	Rev.	Apr.	1963	Triorthocresylphosphate	Oct. 1963
Tellurium		Apr.	1964	Triphenyl Phosphate	June 1970
Tetrachloroethylene	Rev.	Dec.	1965	Turpentine	June 1967
Tetraethyllead		Aug.	1963	Uranium (natural) and its	
Tetrahydrofuran		June	1959	Compounds	Rev. June 1969
Tetranitromethane		Oct.	1964	Vanadium Pentoxide	June 1957
Thorium and Its				Vinyl Chloride	Aug. 1964
Compounds		Oct.	1959	Xylene (Xylol, Di-methyl	
Titanium		June	1959	Benzene)	Rev. Oct. 1971
Titanium Dioxide		Apr.	1966	Zinc Oxide	Rev. Aug. 1969
Toluene (Toluol, Methyl				Zirconium	Oct. 1958

American National Standards Institute, 1430 Broadway, New York, N.Y. 10018. "Acceptable Concentrations," Z-37 Series.

API Toxicological Reviews. The American Petroleum Institute, 1271 Ave. of the Americas, New York, N.Y.

Brown, H. V. "Behavioral Implications of Exposure to Industrial Chemicals." *Journal of Occupational Medicine.* 55 E. Washington St., Chicago 60602. November 1966. p. 561.

Browning, Ethel. *Toxicity of Industrial Metals.* 1961. Butterworths, London, England.

———. *Toxicity and Metabolism of Industrial Solvents.* 1965. Elsevier Publishing Co., New York.

Cook, W. A. "Maximum Allowable Concentration of Industrial Atmospheric Contaminants." *Industrial Medicine,* 1945, p. 933.

Documentation of Threshold Limit Values. American Conference of Governmental Industrial Hygienists, 1014 E. Broadway, Cincinnati, Ohio 45202.

DuBois, K. P. and Geiling, E. M. K. *Textbook of Toxicology.* 1959. Oxford University Press, New York.

Elkins, H. B. *The Chemistry of Industrial Toxicology.* John Wiley and Sons Inc., 605 Third Ave., New York 10016. 1959.

———, "Excretory and Biologic Threshold Limits." *A.I.H.A. Journal,* Vol. 28, No. 4, 1967.

Fairhall, Lawrence T. *Industrial Toxicology,* 2nd Ed., 1957. The Williams and Wilkins Co., Baltimore.

Fawcett, H. H. and Wood, W. S. *Safety and Accident Prevention in Chemical Operations.* Interscience Publishers, 605 Third Ave., New York 10016. 1965.

Federal Radiation Council, Background Material for the Development of Radiation Protection Standards, Report No. 1, May 1960; Report No. 2, September 1961. Available from Superintendent of Documents, U.S. Government Printing Office, Washington, D.C. 20402.

Gafafer, W. M., editor. *Occupational Diseases.* U.S. Public Health Service Bulletin 1097, U.S. Government Printing Office, Washington, D.C. 1964.

Gerarde, Horace W. *Toxicology and Biochemistry of Aromatic Hydrocarbons.* D. Van Nostrand Co., Inc., New York 10018. 1960.

Gleason, Marion N., *et al.* *Clinical Toxicology of Commercial Products.* The Williams and Wilkins Co., Baltimore. 1963.

Hamilton, A., and Hardy, H. L. 1949. *Industrial Toxicology.* Harper & Row, Evanston, Ill.

Hatch, T. F. and Gross, P. *Pulmonary Deposition and Retention of Inhaled Aerosols.* Academic Press, New York. 1964.

Henderson, Y., and Haggard, H. W. *Noxious Gases and the Principles of Respiration Influencing Their Action.* Reinhold Publishing Corp., 430 Park Ave., New York 10022. 1943.

Irish, D. D. "Monitoring the Environment and Judging its Significance," *National Safety Congress Transactions,* 1966. Vol. 12, p. 138. National Safety Council, 425 N. Michigan, Chicago 60611.

MCA Safety Data Sheets. The Manufacturing Chemists' Association, 1825 Connecticut Ave., Washington, D.C. 20009.

Maximum Permissible Body Burdens and Maximum Permissible Concentrations of Radionuclides in Air and in Water for Occupational Exposure, U.S. Department of Commerce, National Bureau of Standards, Handbook 69. Superintendent of Documents, U.S. Government Printing Office, Washington, D.C. 20402 (June 5, 1959).

National Safety Council, 425 N. Michigan Ave., Chicago, Ill., 60611.
    Fundamentals of Industrial Hygiene.
    Industrial Data Sheets
        Acetic Acid, 410.
        Acetone, 398.
        Acetylene, 494.
        Acid Plant, 210.
        Acrolein, 436.
        Adipic Acid, 438.
        Ammonia (Anhydrous), 251.
        Ammonium Nitrite/Fuel Oil Mixtures as Blasting Agents, 536.
          Ammonium Nitrite In Mines, 604.
        Amyl Acetate, 208.
        Aniline, 409.
        Antimony and Its Compounds, 408.
        Arsenic and Its Inorganic Compounds, 499.
        Asphalt, 215.
        Beryllium, 562.
        Beta Ray Sealed Sources, 461.
        Bleaching Compounds, Textile, 343.
        Boron Hydrides, 508.
        Bromine, 313.
        Cadmium, 312.
        Carbon Bisulfide (Carbon Disulfide), 341.
        Carbon Dioxide (Dry Ice), 397.
        Carbon Monoxide, 415.
        Carbon Tetrachloride, 331.
        Caustic Liquor Room, The, 214.
        Chemical Safety References, 486.
        Chlorates, 371.
        Chlorine, 207.
        Chlorine Dioxide, 525.
        Cleaning Compounds Used in Meat Packing, 593.
        Dichloromethane (Methylene Chloride), 474.
        Epoxy Resin Systems, 533.
        Ethyl Alcohol, Industrial, 391.
        Ethyl Ether (Diethyl Oxide), 396.
        Ethylene Dichloride, 350.
        Fluorides, Inorganic, 442.
        Formaldehyde, 342.

Hydrofluoric Acid, 459.
Hydrogen Sulfide, 284.
Iodine, 457.
Isocyanates (TDI and MDI), 489.
Lead, 443.
Lime, 241.
Lithium, 566.
Manganese, 306.
Mercury, 203.
Mercury, Fulminate of, 309.
Metal Hydrides, 462.
Methanol, 407.
Methylene Chloride (Dichloromethane), 474.
Naphthalene (Crude and Refined), 370.
Nitrate-Nitrite Salt Baths for Heat Treating, 270.
Oxalic Acid, 406.
Oxides of Nitrogen, 206.
Oxygen, Gaseous, 472.
Perchloric Acid, 311.
Phenol (Carbolic Acid), 405.
Phosphorus (White), 282.
Picric Acid, 351.
Pyridine, 310.
Selenium and Its Compounds, 578.
Sodium, 231.
Sulfur, Handling and Storage of Solid, 612.
    of Liquid, 592
Tetryl, 218.
Titanium, 485.
1,1,1,-Trichloroethane, 456.
Trichloroethylene, 389.
Trinitrotoluene (TNT), 314.
Turpentine, 367.
Zinc and Zinc Oxide, 267.
Zirconium Powder, 382.

Patty, F. A., ed. *Industrial Hygiene and Toxicology*, Vol. I, 2nd edition 1958, and Vol. II, 2nd edition, 1963. Interscience Publishers, 605 Third Ave., New York, N.Y. 10016.

Princi, F. "Criteria for the Evaluation of Chemical Toxicity," *Journal of Air Pollution Control*, 4400 Fifth Ave., Pittsburgh, May 1964, p. 154.

Public Health Service Drinking Water Standards, U.S. Department of Health, Education and Welfare, Public Health Service, Superintendent of Documents, U.S. Government Printing Office, Washington, D.C. 20402; Revised 1962.

Recommendations of the International Commission on Radiological Protection, Main Commission Report, Sept. 9, 1958. London: Pergamon Press, 1959.

Smyth, H. F. "Hygienic Standards for Daily Inhalation." *American Industrial Hygiene Association Quarterly*, 25711 Southfield Rd., Southfield, Mich. 48075. 1956, p. 129.

———. "Industrial Hygiene in Retrospect and Prospect—Toxicological Aspects." *The American Industrial Hygiene Association Journal*, 25711 Southfield Rd., Southfield, Mich. 48075. May–June 1963, p. 222.

Stokinger, H. E. "Criteria and Procedures for Assessing the Toxic Responses to Industrial Chemicals." *Permissible Levels of Toxic Substances in the Working Environment*, Occupational Safety and Health Series No. 20. International Labor Office, Geneva, Switzerland. 1970, pp. 36–52.

———. "Testing for Hypersusceptibility." *National Safety News*, vol. 95:5 (May 1967).

Threshold Limit Values 1973. American Conference of Governmental Industrial Hygienists, P.O. Box 1937, Cincinnati, Ohio 45201.

TABLE 38-E

THRESHOLD LIMIT VALUES FOR CHEMICAL SUBSTANCES IN WORKROOM AIR

1973

PREFACE

CHEMICAL CONTAMINANTS

Threshold limit values refer to airborne concentrations of substances and represent conditions under which it is believed that nearly all workers may be repeatedly exposed day after day without adverse effect. Because of wide variation in individual susceptibility, however, a small percentage of workers may experience discomfort from some substances at concentrations at or below the threshold limit, a smaller percentage may be affected more seriously by aggravation of a pre-existing condition or by development of an occupational illness.

Simple tests are now available (*J. Occup. Med.* 9: 537, 1967; *Ann. N.Y. Acad. Sci., 151, Art.* 2: 968, 1968) that may be used to detect those individuals hypersusceptible to a variety of industrial chemicals (respiratory irritants, hemolytic chemicals, organic isocyanates, carbon disulfide). These tests may be used to screen out by appropriate job placement the hyperreactive worker and thus in effect improve the "coverage" of the TLV's.

Threshold limit values refer to time-weighted concentrations for a 7- or 8-hour workday and 40-hour workweek. They should be used as guides in the control of heath hazards and should not be used as fine lines between safe and dangerous concentrations. (Exceptions are the substances listed in Appendices E and F and those substances designated with a "C" or Ceiling value, Appendix D).

Time-weighted averages permit excursions above the limit, provided they are compensated by equivalent excursions below the limit during the workday. In some instances it may be permissible to calculate the average concentration for a workweek rather than for a workday. The degree of permissible excursion is related to the magnitude of the threshold limit value of a particular substance as given in Appendix D. The relationship between threshold limit and permissible excursion is a rule of thumb and in certain cases may not apply. The amount by which threshold limits may be exceeded for short periods without injury to health depends upon a number of factors such as the nature of the contaminant, whether very high concentrations—even for short periods—produce acute poisoning, whether the effects are cumulative, the frequency with which high concentrations occur, and the duration of such periods. All factors must be taken into consideration in arriving at a decision as to whether a hazardous condition exists.

Threshold limits are based on the best available information from industrial experience, from experimental human and animal studies, and, when possible, from a combination of the three. The basis on which the values are established may differ from substance to substance; protection against impairment of health may be a guiding factor for some, whereas reasonable freedom from irritation, narcosis, nuisance, or other forms of stress may form the basis for others.

The amount and nature of the information available for establishing a TLV varies from substance to substance; consequently, the precision of the estimated TLV is also subject to variation and the latest *Documentation* should be consulted in order to assess the extent of the data available for a given substance.

The committee holds to the opinion that limits based on physical irritation should be considered no less binding than those based on physical impairment. There is increasing evidence that physical irritation may initiate, promote or accelerate physical impairment through interaction with other chemical or biologic agents.

In spite of the fact that serious injury is not believed likely as a result of exposure to the threshold limit concentrations, the best practice is to maintain concentrations of all atmospheric contaminants as low as is practical.

These limits are intended for use in the practice of industrial hygiene and should be interpreted and applied only by a person trained in this discipline. They are not intended for use, or for modification for use, (1) as a relative index of hazard or toxicity, (2) in the evaluation or air pollution nuisances, (3) in estimating the toxic potential of continuous, uninterrupted exposures, (4) as proof or disproof of an existing disease or physical condition, or (5) for adoption by countries whose working conditions differ from those in the United States of America and where substances and processes differ.

Ceiling vs. time-weighted average limits.

Although the time-weighted average concentration provides the most satisfactory, practical way of monitoring airborne agents for compliance with the limits, there are certain substances for which it is inappropriate. In the latter group are substances which are predominantly fast acting and whose threshold limit is more appropriately based on this particular response. Substances with this type of response are best controlled by a ceiling "C" limit that should not be exceeded. It is implicit in these definitions that the manner of sampling to determine compliance with the limits for each group must differ; a single brief sample, that is applicable to a "C" limit, is not appropriate to the time-weighted limit; here, a sufficient number of samples are needed to permit a time-weighted average concentration throughout a complete cycle of operations or throughout the work shift.

Whereas the ceiling limit places a definite boundary which concentrations should not be permitted to exceed, the time-weighted average limit requires an explicit limit to the excursions that are permissible above the listed values. The magnitude of these excursions may be pegged to the magnitude of the threshold limit by an appropriate factor shown in Appendix D.

It should be noted that the same factors are used by the Committee in making a judgment whether to include or exclude a substance for a "C" listing.

"Skin" notation. Listed substances followed by the designation "Skin" refer to the potential contribution to the overall exposure by the cutaneous route including mucous membranes and eye, either by airborne, or more particularly, by direct contact with the substance. Vehicles can alter skin absorption. This attention-calling designation is intended to suggest appropriate measures for the prevention of cutaneous absorption so that the threshold limit is not invalidated.

Mixtures. Special consideration should be given also to the application of the TLV's in assessing the health hazards which may be associated with exposure to mixtures of two or more substances. A brief discussion of basic considerations involved in developing threshold limit values for mixtures, and methods for their development, amplified by specific examples are given in Appendix C.

Nuisance particulates. In contrast to fibrogenic dusts which cause scar tissue to be formed in lungs when inhaled in excessive amounts, so-called "nuisance" dusts have a long history of little adverse effect on lungs and do not produce significant organic disease or toxic effect when exposures are kept under reasonable control. The nuisance dusts have also been called (biologically) "inert" dusts, but the latter term is inappropriate to the extent that there is no dust which does not evoke some cellular response in the lung when inhaled in sufficient amount. However,

the lung-tissue reaction caused by inhalation of nuisance dusts has the following characteristics: (1) The architecture of the air spaces remains intact. (2) Collagen (scar tissue) is not formed to a significant extent. (3) The tissue reaction is potentially reversible.

Excessive concentrations of nuisance dusts in the workroom air may seriously reduce visibility (iron oxide), may cause unpleasant deposits in the eyes, ears and nasal passages (Portland cement dust), or cause injury to the skin or mucous membranes by chemical or mechanical action per se or by the rigorous skin cleansing procedures necessary for their removal.

A threshold limit of 10mg/m^3, or 30 mppcf, of total dust < 1% quartz is recommended for substances in these categories and for which no specific threshold limits have been assigned. This limit, for a normal workday, does not apply to brief exposures at higher concentrations. Neither does it apply to those substances which may cause physiologic impairment at lower concentrations but for which a threshold limit has not yet been adopted. Some nuisance particulates are given in Appendix E.

Simple asphyxiants—"Inert" gases or vapors. A number of gases and vapors, when present in high concentrations in air, act primarily as simple asphyxiants without other significant physiologic effects. A TLV may not be recommended for each simple asphyxiant because the limiting factor is the available oxygen. The minimal oxygen content should be 18 percent by volume under normal atmospheric pressure (equivalent to a partial pressure, pO$_2$ of 135 mm Hg). Atmospheres deficient in O$_2$ do not provide adequate warn-

TABLE 38-E. THRESHOLD LIMIT VALUES FOR CHEMICAL SUBSTANCES IN WORKROOM AIR, 1973 (*Continued*)

ing and most simple asphyxiants are odorless. Several simple asphyxiants present an explosion hazard. Account should be taken of this factor in limiting the concentration of the asphyxiant. Specific examples are listed in Appendix F.

Short-Term Limits (STL's). Because many industrial exposures are not continuous, 8-hour daily exposures, but are short-term, or intermittent, to which the TLV's do not necessarily apply, STL's for 5, 15, or 30 minutes for 142 substances have been put into the regulations of the Pennsylvania Department of Health (Chapter 4, Art. 432, Revised Jan. 25, 1968). These STL's represent the maximal average atmospheric concentration of a contaminant to which a worker may be exposed for the stipulated time. The concentration represents an upper limit of exposure and assumes that there is sufficient recovery between exposures before another is initiated. The daily average exposure including that provided by the STL shall be such that the TLV shall not be exceeded.

Similar STL's for a more restricted number of substances have been recommended by the American National Standards Institute. This standards-setting body refers to these short-term limits as "peaks."

Physical factors. It is recognized that such physical factors as heat, ultraviolet and ionizing radiation, humidity, abnormal pressure (altitude) and the like may place added stress on the body so that the effects from exposure at a threshold limit may be altered. Most of these stresses act adversely to increase the toxic response of a substance. Although most threshold limits have built-in safety factors to guard against adverse effects to moderate deviations from normal environments, the safety factors of most substances are not of such a magnitude as to take care of gross deviations. For example, continuous work at temperatures above 90 F or overtime extending the workweek more than 25 percent might be considered gross deviations. In such instances judgment must be exercised in the proper adjustments of the threshold limit values.

Biologic Limit Values (BLV's). Other means exist and may be necessary for monitoring worker exposure other than reliance on the Threshold Limit list, namely, the Biologic Limit Values for industrial air, namely, the Biologic Limit Values. These values represent limiting amounts of substances (or their effects) to which the worker may be exposed without hazard to health or well-being as determined in his tissues and fluids or in his exhaled breath. The biologic measurements on which the BLV's are based can furnish two kinds of information useful in the control of worker exposure: (1) measure of the individual worker's overall exposure; (2) measure of the worker's individual and characteristic response. Measurements of response furnish a superior estimate of the physiologic status of the worker, and may be made of (*a*) changes in amount of some critical biochemical constituent, (*b*) changes in activity of a critical enzyme, (*c*) changes in some physiologic function. Measurement of exposure may be made by (1) determining in blood, urine, hair, nails, in body tissues and fluids, the amount of substance to which the worker was exposed; (2) determination of the amount of the metabolite(s) of the substance in tissues and fluids; (3) determination of the amount of the substance in the exhaled breath. The biologic limits may be used as an adjunct to the TLV's for air, or in place of them. The U.S. National Institute for Occupational Safety and Health is proposing a series of these BLV's which will be issued from time to time as part of the Criteria Documents for recommended industrial air standards.

Unlisted substances. There are a number of reasons why a substance does not appear in the Threshold Limit list; either insufficient information is available or it has not been brought to the attention of the Threshold Limits Committee from which a limit can be developed, or it is a substance that could be included in the Appendices E and F pertaining to Nuisance Particulates and Simple Asphyxiants. Substances appearing in these appendices serve as examples only; the appendices are not intended to be inclusive.

"Notice of intent." At the beginning of each year, proposed actions of the Committee for the forthcoming year are issued in the form of a "Notice of Intended Changes." This Notice provides not only an opportunity for comment, but solicits suggestions of substances to be added to

the list. The suggestions should be accompanied by substantiating evidence. The list of Intended Changes follows the

Adopted Values in the TLV booklet. *Legal status.* By publication in the

Federal Register (Vol. 36, No. 105, May 29, 1971) the Threshold Limit Values are now official federal standards for industrial air.

ADOPTED VALUES
In Alphabetical Order
(See *Documentation* for Basis of TLV's)

Substance	ppm[a]	mg/cu m[b]
Abate	-	10
* Acetaldehyde	100	180
Acetic acid	10	25
C Acetic anhydride	5	20
Acetone	1,000	2,400
Acetonitrile	40	70
2-Acetylaminofluorene—Skin	F	A²
Acetylene dichloride; see 1,2 Dichloroethylene		-
Acetylene tetrabromide	1	14
Acrolein	0.1	0.25
Acrylamide—Skin		0.3
Acrylonitrile—Skin	20	45
Aldrin—Skin		0.25
Allyl alcohol—Skin	2	5
Allyl chloride	1	3
* Allyl glycidyl ether (AGE)	10	45
Allyl propyl disulfide	2	12
Alundum (Al₂O₃)		E
4-Aminodiphenyl—Skin		A^1b
2-Aminoethanol; see Ethanolamine		
2-Aminopyridine	0.5	2
* Ammonia	50	35
Ammonium chloride, fume		10
Ammonium sulfamate (Ammate)		10
n-Amyl acetate	100	525
sec-Amyl acetate	125	650
Aniline—Skin	5	19
Anisidine (o,p-isomers)—Skin	0.1	0.5
Antimony and compounds (as Sb)		0.5
ANTU (alpha naphthyl thiourea)		0.3
Argon	F	-
Arsenic and compounds (as As)		0.5
Arsine	0.05	0.2

Substance	ppm[a]	mg/cu m[b]
Asphalt (petroleum) fumes	-	5
Azinphos methyl—Skin	-	0.2
Barium (soluble compounds)	-	0.5
C Benzene (benzol)—Skin	25	80
Benzidine—Skin	-	A^1b
p-Benzoquinone, see Quinone		
Benzoyl peroxide	-	5
Benzyl chloride	1	5
Beryllium	-	0.002
Biphenyl, see Diphenyl	-	-
* Bismuth telluride	-	10
* Bismuth telluride (Se-doped)	-	5
Boron oxide	-	10
Boron tribromide	1	10
C Boron trifluoride	1	3
Bromine	0.1	0.7
Bromine pentafluoride	0.1	0.7
Bromoform—Skin	0.5	5
Butadiene (1, 3-butadiene)	1,000	2,200
* Butane	500	1,200
Butanethiol, see Butyl mercaptan		-
2-Butanone	200	590
2-Butoxy ethanol (Butyl Cellosolve)—Skin	50	240
Butyl acetate (n-butyl acetate)	150	710
sec-Butyl acetate	200	950
tert-Butyl acetate	200	950
Butyl alcohol	100	300
sec-Butyl alcohol	150	450
tert-Butyl alcohol	100	300
Butylamine—Skin	5	15
C tert-Butyl chromate (as CrO₃)—Skin	-	0.1
n-Butyl glycidyl ether (BGE)	50	270
Butyl mercaptan	0.5	1.5
p-tert-Butyltoluene	10	60
Cadmium (Metal dust and soluble salts)	-	0.2

TABLE 38-E. THRESHOLD LIMIT VALUES FOR CHEMICAL SUBSTANCES IN WORKROOM AIR, 1973 (Continued)

Substance	ppm[a]	mg/cu m[b]
**C Cadmium oxide fume (as Cd)	-	0.1
Calcium carbonate	-	E
Calcium arsenate	-	1
Calcium oxide	-	5
Camphor (Synthetic)	2	12
Carbaryl (Sevin®)	-	5
Carbon black	-	3.5
Carbon dioxide	5,000†	9,000
Carbon disulfide—Skin	20	60
Carbon monoxide	50	55
Carbon tetrachloride—Skin	10	65
Cellulose (paper fiber)	-	E
Chlordane—Skin	-	0.5
Chlorinated camphene—Skin	-	0.5
Chlorinated diphenyl oxide	-	0.5
Chlorine	1	3
Chlorine dioxide	0.1	0.3
C Chlorine trifluoride	0.1	0.4
C Chloroacetaldehyde	1	3
α-Chloroacetophenone (phenacylchloride)	0.05	0.3
Chlorobenzene (monochlorobenzene)	75	350
o-Chlorobenzylidene malononitrile (OCBM)—Skin	0.05	0.4
Chlorobromomethane	200	1,050
2-Chloro-1, 3-butadiene see Chloroprene	-	-
Chlorodiphenyl (42% Chlorine)—Skin	-	1
Chlorodiphenyl (54% Chlorine)—Skin	-	0.5
1-Chloro, 2, 3-epoxy-propane, see Epichlorhydrin	-	-
2-Chloroethanol, see Ethylene chlorohydrin	-	-
Chloroethylene, see Vinyl chloride	-	-
** Chloroform (Trichloromethane)	(50)	(240)
1-Chloro-1-nitropropane	20	100
Chloropicrin	0.1	0.7
Chloroprene (2-chloro-1, 3-butadiene)—Skin	25	90
Chromic acid and chromates (as CrO₃)	-	0.1
Chromium, sol. chromic, chromous salts as Cr	A[1a]	0.5
** Metal and insoluble salts	-	(1.0)
Coal tar pitch volatiles (benzene soluble fraction) anthracene, BaP, phenanthrene, acridine, chrysene, pyrene)	A[1a]	0.2
Cobalt, metal fume and dust	-	0.1
** Copper fume	-	(1)
Dusts and mists	-	1
Corundum (Al₂O₃)	-	E
** Cotton dust (raw)	-	(1)
Crag® herbicide	-	10
Cresol (all isomers)—Skin	5	22
Crotonaldehyde	2	6
Cumene—Skin	50	245
Cyanide (as CN)—Skin	-	5
Cyanogen	10	-
Cyclohexane	300	1,050
Cyclohexanol	50	200
Cyclohexanone	50	200
Cyclohexene	300	1,015
Cyclopentadiene	75	200
2,4-D	-	10
DDT	-	1
DDVP, see Dichlorvos	-	-
Decaborane—Skin	0.05	0.3
Demeton®—Skin	0.01	0.1
Diacetone alcohol (4-hydroxy-4-methyl-2-pentanone)	50	240
1,2-Diaminoethane, see Ethylenediamine	-	-
Diazinon—Skin	-	0.1
Diazomethane	0.2	0.4
Diborane	0.1	0.1
* 1,2-Dibromoethane (ethylene dibromide)—Skin	20	145
Dibrom®	-	3
2-N Dibutylaminoethanol—Skin	2	14
Dibutyl phosphate	1	5
Dibutylphthalate	-	5
C Dichloroacetylene	0.1	0.4
C o-Dichlorobenzene	50	300
C p-Dichlorobenzene	75	450

Substance	ppm[a]	mg/cu m[b]
Dichlorobenzidine—Skin	-	A[1b]
Dichlorodifluoromethane	1,000	4,950
1,3-Dichloro-5, 5-dimethyl hydantoin	-	0.2
* 1, 1-Dichloroethane	200	320
1, 2-Dichloroethane	50	200
1, 2-Dichloroethylene	200	790
* Dichloroethyl ether—Skin	5	30
Dichloromethane, see Methylene chloride		
Dichloromonofluoromethane	1,000	4,200
C 1, 1-Dichloro-1-nitroethane	10	60
1, 2-Dichloropropane, see Propylenedichloride		
Dichlorotetrafluoroethane	1,000	7,000
Dichlorvos (DDVP)—Skin	0.1	1
Dieldrin—Skin	-	0.25
Diethylamine	25	75
Diethylamino ethanol—Skin	10	50
Diethylene triamine—Skin	1	4
Diethylether, see Ethyl ether		
Difluorodibromomethane	100	860
C Diglycidyl ether (DGE)	0.5	2.8
Dihydroxybenzene, see Hydroquinone		
* Diisobutyl ketone	25	150
Diisopropylamine—Skin	5	20
Dimethoxymethane, see Methylal		
Dimethyl acetamide—Skin	10	35
Dimethylamine	10	18
4-Dimethylaminoazobenzene	-	A[2]
Dimethylaminobenzene, see Xylidene		
Dimethylaniline (N-dimethylaniline)—Skin	5	25
Dimethylbenzene, see Xylene		
Dimethyl 1, 2-dibromo-2-dichloroethyl phosphate, see DiBrom		
Dimethylformamide—Skin	10	30
2, 6-Dimethylheptanone, see Diisobutyl ketone		
1, 1-Dimethylhydrazine—Skin	0.5	1
** Dimethylphthalate	-	5
** Dimethylsulfate—Skin	(1)	(5)
Dinitrobenzene (all isomers)—Skin	0.15	1
Dinitro-o-cresol—Skin	-	0.2
Dinitrotoluene—Skin	-	1.5
** Dioxane (Diethylene dioxide)—Skin	(100)	(360)
Diphenyl	0.2	1
Diphenyl amine	-	10
Diphenylmethane diisocyanate (see Methylene bisphenyl isocyanate (MDI)	-	-
Dipropylene glycol methyl ether—Skin	100	600
Diquat	-	0.5
Di-sec, octyl phthalate (Di-2-ethylhexyl-phthalate)	-	5
Emery	-	E
Endosulfan (Thiodan®)—Skin	-	0.1
Endrin—Skin	-	0.1
Epichlorhydrin—Skin	5	19
EPN—Skin	-	0.5
1, 2-Epoxypropane, see Propyleneoxide		
2, 3-Epoxy-1-propanol, see Glycidol		
Ethane	F	-
* Ethanethiol, see Ethylmercaptan		
Ethanolamine	3	6
2-Ethoxyethanol—Skin	100	370
2-Ethoxyethylacetate (Cellosolve acetate)—Skin	100	540
Ethyl acetate	400	1,400
Ethyl acrylate—Skin	25	100
Ethyl alcohol (Ethanol)	1,000	1,900
Ethylamine	10	18
Ethyl sec-amyl ketone (5-methyl-3-heptanone)	25	130
Ethyl benzene	100	435
Ethyl bromide	200	890
Ethyl butyl ketone (3-Heptanone)	50	230
Ethyl chloride	1,000	2,600
Ethyl ether	400	1,200
Ethyl formate	100	300
Ethyl mercaptan	0.5	1
Ethyl silicate	100	850
Ethylene	F	-
Ethylene chlorohydrin—Skin	5	16
Ethylenediamine	10	25
Ethylene dibromide, see 1, 2-Dibromoethane		
Ethylene dichloride, see 1, 2-Dichloroethane		
* Ethylene glycol, particulate	-	10
* Ethylene glycol, vapor	100	260
C Ethylene glycol dinitrate and/or Nitrocylcerin—Skin	0.2[d]	-
** Ethylene glycol monomethyl ether	-	-

TABLE 38-E. THRESHOLD LIMIT VALUES FOR CHEMICAL SUBSTANCES IN WORKROOM AIR, 1973 (Continued)

Substance	ppm[a]	mg/cu m[b]
acetate (Methyl cellosolve acetate)—Skin	25	120
Ethylene imine—Skin	0.5	1
Ethylene oxide	50	90
Ethylidine chloride, see 1, 1-Dichloroethane	-	-
N-Ethylmorpholine—Skin	20	94
Ferbam	-	10
Ferrovanadium dust	-	1
* Fluoride (as F)	-	2.5
* Fluorine	1	2
C Fluorotrichloromethane	1,000	5,600
C Formaldehyde	2	3
Formic acid	5	9
** Furfural—Skin	5	20
** Furfuryl alcohol	(5)	(20)
Gasoline	-	B2
* Germanium tetrahydride	0.2	0.6
Glass, fibrous^c) or dust	-	E
Glycerin mist	-	E
Glycidol (2, 3-Epoxy-1-propanol)	50	150
Glycol monoethyl ether, see 2-Ethoxyethanol	-	-
Graphite, (Synthetic)	-	E
Guthion,® see Azinphos-methyl	-	-
Gypsum	-	E
Hafnium	-	0.5
Helium	F	-
Heptachlor—Skin	-	0.5
Heptane (n-heptane)	500	2,000
Hexachloroethane—Skin	1	10
Hexachloronaphthalene—Skin	-	0.2
* Hexafluoroacetone	0.1	0.7
Hexane (n-hexane)	500	1,800
2-Hexanone	100	410
Hexone (Methyl isobutyl ketone)	100	410
sec-Hexyl acetate	50	300
Hydrazine—Skin	1	1.3
Hydrogen	F	-
C Hydrogen bromide	3	10
C Hydrogen chloride	5	7
Hydrogen cyanide—Skin	10	11
Hydrogen fluoride	3	2
Hydrogen peroxide	1	1.4
Hydrogen selenide	0.05	0.2
Hydrogen sulfide	10	15
Hydroquinone	-	2
Indene	10	45
Indium and compounds, as In	-	0.1
C Iodine	0.1	1
** Iron oxide fume	-	(10)
Iron pentacarbonyl	0.01	0.08
Iron salts, soluble, as Fe	-	1
Isoamyl acetate	100	525
Isoamyl alcohol	100	360
Isobutyl acetate	150	700
Isobutyl alcohol	100	300
** Isophorone	(25)	(140)
Isopropyl acetate	250	950
Isopropyl alcohol	400	980
Isopropylamine	5	12
* Isopropylether	250	1,050
Isopropyl glycidyl ether (IGE)	50	240
Kaolin	-	E
* Ketene	0.5	0.9
* Lead, inorganic, fumes, and dusts	-	0.15
Lead arsenate	-	0.15
Limestone	-	E
Lindane	-	0.5
Lithium hydride	-	0.025
LP-gas (Liquefied petroleum gas)	1,000	1,800
Magnesite	-	E
Magnesium oxide fume	-	10
Malathion—Skin	-	10
Maleic anhydride	0.25	1
C Manganese and compounds, as Mn	-	5
Marble	-	E
Mercury (Alkyl compounds)—Skin	0.001	0.01
Mercury (All forms except alkyl)	-	0.05
Mesityl oxide	25	100
Methane	F	-
Methanethiol, see Methyl mercaptan	-	-
Methoxychlor	-	10
2-Methoxyethanol—Skin (Methyl cellosolve)	25	80
Methyl acetate	200	610

Substance	ppm[a]	mg/cu m[b]
Methyl acetylene (propyne)	1,000	1,650
Methyl acetylene-propadiene mixture (MAPP)	1,000	1,800
Methyl acrylate—Skin	10	35
* Methyl acrylonitrile—Skin	1	3
Methylal (dimethoxymethane)	1,000	3,100
Methyl alcohol (methanol)	200	260
Methylamine	10	12
Methyl amyl alcohol, see Methyl isobutyl carbinol	-	-
Methyl 2-cyanoacrylate	2	8
Methyl isoamyl ketone	100	475
Methyl (n-amyl) ketone (2-Heptanone)	100	465
* Methyl bromide—Skin	15	60
Methyl butyl ketone, see 2-Hexanone		
Methyl cellosolve—Skin, see 2-Methoxyethanol	-	-
Methyl cellosolve acetate—Skin, see Ethylene glycol monomethyl ether acetate	-	-
Methyl chloride	100	210
Methyl chloroform	350	1,900
Methylcyclohexane	500	2,000
Methylcyclohexanol	50	235
* o-Methylcyclohexanone—Skin	50	230
** Methylcyclopentadienyl manganese tricarbonyl (as Mn)—Skin	0.1	0.2
Methyl demeton—Skin	-	0.5
Methyl ethyl ketone (MEK), see 2-Butanone	-	-
Methyl formate	100	250
Methyl iodide—Skin	5	28
Methyl isobutyl carbinol—Skin	25	100
Methyl isobutyl ketone, see Hexone		
Methyl isocyanate—Skin	0.02	0.05
Methyl mercaptan	0.5	1
Methyl methacrylate	100	410
Methyl parathion—Skin	-	0.2
Methyl propyl ketone, see 2-Pentanone		
C Methyl silicate	5	30
C α Methyl styrene	100	480
C Methylene bisphenyl isocyanate (MDI)	0.02	0.2
** Methylene chloride (dichloromethane)	(500)	(1,740)
Molybdenum (soluble compounds)	-	5
(insoluble compounds)	-	10
Monomethyl aniline—Skin	2	9
C Monomethyl hydrazine—Skin	0.2	0.35
Morpholine—Skin	20	70
Naphtha (coal tar)	100	400
Naphthalene	10	50
β-Naphthylamine	F	A[1b]
Neon		
Nickel carbonyl	0.001 A[1a]	0.007
Nickel, metal and soluble compounds (as Ni)	-	1
Nicotine—Skin	0.075	0.5
Nitric acid	2	5
Nitric oxide	25	30
p-Nitroaniline—Skin	1	6
Nitrobenzene—Skin	1	5
p-Nitrochlorobenzene—Skin	-	1
4-Nitrodiphenyl	-	A[1a]
Nitroethane	100	310
Nitrogen	F	-
C Nitrogen dioxide	5	9
Nitrogen trifluoride	10	29
Nitroglycerin—Skin	0.2	2
Nitromethane	100	250
1-Nitropropane	25	90
2-Nitropropane	25	90
N-Nitrosodimethylamine (dimethylnitrosoamine)—Skin	-	A[2]
Nitrotoluene—Skin	5	30
Nitrotrichloromethane, see Chloropicrin	-	-
Nitrous oxide	F	-
Octachloronaphthalene—Skin	-	0.1
Octane	400	1,900
Oil mist, particulate	pB[2]	5[f]
Oil mist, vapor	-	-
Osmium tetroxide	0.0002	0.002
Oxalic acid	-	1
Oxygen difluoride	0.05	0.1
Ozone	0.1	0.2
Paraquat—Skin	-	0.5
Parathion—Skin	-	0.1
Pentaborane	0.005	0.01
Pentachloronaphthalene—Skin	-	0.5
Pentachlorophenol—Skin	-	0.5
Pentaerythritol	-	E
Pentane	500	1,500
2-Pentanone	200	700
Perchloroethylene	100	670

TABLE 38-E. THRESHOLD LIMIT VALUES FOR CHEMICAL SUBSTANCES IN WORKROOM AIR, 1973 (Continued)

Substance	ppm[a]	mg/cu m[b]
Perchloromethyl mercaptan	0.1	0.8
Perchloryl fluoride	3	14
Petroleum Distillates (naphtha)	ᵍB²	-
Phenol—Skin	5	19
Phenothiazine—Skin	-	5
p-Phenylene diamine—Skin	-	0.1
Phenyl ether (vapor)	1	7
Phenyl ether-Diphenyl mixture (vapor)	1	7
Phenylethylene, see Styrene	-	-
Phenyl glycidyl ether (PGE)	10	60
Phenylhydrazine—Skin	5	22
*C Phenylphosphine	0.05	0.25
Phosdrin (Mevinphos)®—Skin	0.01	0.1
Phosgene (carbonyl chloride)	0.1	0.4
Phosphine	0.3	0.4
Phosphoric acid	-	1
Phosphorus (yellow)	-	0.1
Phosphorus pentachloride	-	1
Phosphorus pentasulfide	-	1
Phosphorus trichloride	0.5	3
Phthalic anhydride	2	12
Picric acid—Skin	-	0.1
Pival® (2-Pivalyl-1, 3-indandione)	-	0.1
Plaster of Paris	-	E
Platinum (Soluble Salts) as Pt	-	0.002
Polychlorobiphenyls, see Chlorodiphenyls	-	-
Polytetrafluoroethylene decomposition products	-	-
Propane	F	B¹
β-Propiolactone	-	A²
Propargyl alcohol—Skin	1	2
n-Propyl acetate	200	840
Propyl alcohol	200	500
n-Propyl nitrate	25	110
Propylene dichloride (1, 2-Dichloropropane)	75	350
Propylene glycol monomethyl ether	100	360
Propylene imine—Skin	2	5
Propylene oxide	100	240
Propyne, see Methylacetylene	-	-
Pyrethrum	-	5
Pyridine	5	15
Quinone	0.1	0.4
RDX—Skin	-	1.5
Rhodium, Metal fumes and dusts (as Rh)	-	0.1
Soluble salts	-	0.001
Ronnel	-	10
Rosin Core Solder, pyrolysis products (as formaldehyde)	-	0.1
Rotanone (commercial)	-	5
Rouge	-	E
Selenium compounds (as Se)	-	0.2
Selenium hexafluoride	0.05	0.4
* Silicon	-	10
Silicon carbide	-	E
Silver, metal and soluble compounds	-	0.01
Sodium fluoroacetate (1080)—Skin	-	0.05
**C Sodium hydroxide	-	(2)
Starch	-	E
Stibine	0.1	0.5
Stoddard solvent	200	1,150
Strychnine	-	0.15
Styrene (Monomer), (Phenyl ethylene)	100	420
Sucrose	-	E
Sulfur dioxide	5	13
Sulfur hexafluoride	1,000	6,000
Sulfuric acid	-	1
Sulfur monochloride	1	6
Sulfur pentafluoride	0.025	0.25
Sulfur tetrafluoride	0.1	0.4
Sulfuryl fluoride	5	20
Systox, see Demeton®	-	-
2, 4, 5 T	-	10
Tantalum	-	5
TEDP—Skin	-	0.2
Teflon® decomposition products	-	B¹
Tellurium	-	0.1
Tellurium hexafluoride	0.02	0.2
TEPP—Skin	0.004	0.05
C Terphenyls	1	9
1, 1, 1, 2-Tetrachloro-2, 2-difluoroethane	500	4,170
1, 1, 2, 2-Tetrachloro-1,	-	-

Substance	ppm[a]	mg/cu m[b]
2-difluoroethane ......—Skin	500	4,170
1, 1, 2, 2-Tetrachloroethane—Skin	5	35
Tetrachloroethylene, see Perchloroethylene, see	-	-
Tetrachloromethane, see Carbon tetrachloride	-	-
Tetrachloronaphthalene—Skin	-	2
Tetraethyl lead (as Pb)—Skin	-	0.100[h]
Tetrahydrofuran	200	590
Tetramethyl lead (as Pb)—Skin	-	0.150[h]
Tetramethyl succinonitrile—Skin	0.5	3
Tetranitromethane	1	8
Tetryl (2, 4, 6-trinitrophenyl-methylnitramine)—Skin	-	1.5
Thallium (soluble compounds)—Skin (as Tl)	-	0.1
Thiram®	-	5
Tin (inorganic compounds, except SnH₄ and SnO₂) as Sn	-	2
Tin (organic compounds)—Skin (as Sn)	-	0.1
Tin oxide	-	E
Titanium dioxide	-	E
* Toluene (toluol)	100	375
C Toluene-2, 4-diisocyanate	0.02	0.14
o-Toluidine	5	22
Toxaphene, see Chlorinated camphene	-	-
Tributyl phosphate	-	5
1, 1, 1-Trichloroethane, see Methyl chloroform	-	-
1, 1, 2-Trichloroethane—Skin	10	45
Trichloroethylene	100	535
Trichloromethane, see Chloroform	-	-
Trichloronaphthalene—Skin	-	5
1, 2, 3-Trichloropropane	50	300
1, 1, 2-Trichloro 1, 2, 2-trifluoroethane	1,000	7,600
Triethylamine	25	100
Trifluoromonobromomethane	1,000	6,100
Trimethyl benzene	25	120
2,4, 6-Trinitrophenol, see Picric acid	-	-
2, 4, 6-Trinitrophenylmethylnitramine, see Tetryl	-	-
Trinitrotoluene—Skin	0.2	1.5
Triorthocresyl phosphate	-	0.1
Triphenyl phosphate	-	3
Tungsten and Compounds, as W		
Soluble	-	1
Insoluble	-	5
Turpentine	100	560
Uranium (natural) soluble and insoluble compounds, as U	-	0.2
Vanadium (V₂O₅), as V Dust	-	0.5
C Fume	-	0.05
Vinyl acetate	10	30
Vinyl benzene, see Styrene	-	-
Vinyl bromide	250	1,100
Vinyl chloride	200	510
Vinylcyanide, see Acrylonitrile	-	-
Vinyl toluene	100	480
Warfarin	-	0.1
Wood dust (nonallergenic)	-	5
Xylene (xylol)	100	435
Xylidine—Skin	5	25
Yttrium	-	1
Zinc chloride fume	-	1
Zinc oxide fume	-	5
Zirconium compounds (as Zr)	-	5

[a] Parts of vapor or gas per million parts of contaminated air by volume at 25 C and 760 mm. Hg pressure.

[b] Approximate milligrams of substance per cubic meter of air.

[d] An atmospheric concentration of not more than 0.02 ppm, or personal protection may be necessary to avoid headache.

[e] < 5–7 μm in diameter.

[f] As sampled by method that does not collect vapor.

[g] According to analytically determined composition.

[h] For control of general room air, biologic monitoring is essential for personal control.

Capital letters refer to Appendices.

* 1973 Addition.

** See "Notice of Intended Changes."

Radioactivity: For permissible concentrations of radioisotopes in air, see U.S. Department of Commerce, National Bureau of Standards Handbook 69, "Maximum Permissible Body Burdens and Maximum Permissible Concentrations of Radionuclides in Air and in Water for Occupational Exposure," June 5, 1969. Also, see U.S. Department of Commerce, National Bureau of Standards Handbook 59, "Permissible Dose from External Sources of Ionizing Radiation," September 24, 1954, and addendum of April 15, 1958. A report, Basic Radiation Protection Criteria, published by the National Committee on Radiation Protection, revises and modernizes the concept of the NCRP standards of 1954, 1957, and 1958; obtainable as NCRP Rept. No. 39, P.O. Box 4867, Washington, D.C. 20008.

TABLE 38-E. THRESHOLD LIMIT VALUES FOR CHEMICAL SUBSTANCES IN WORKROOM AIR, 1973 (Continued)

Mineral Dusts, Nuisance Dusts, and Notice of Intended Changes

MINERAL DUSTS

Substance	m.p.p.c.f.(e)
SILICA, SiO₂	
Crystalline	
Quartz .... TLV in mppcf:	

$$\frac{300^{j)}}{\% \text{ quartz} + 10}$$

TLV for respirable dust in mg/m³:

$$\frac{10 \text{ mg/m}^{3k)}}{\% \text{ Respirable quartz} + 2}$$

TLV for "total dust," respirable and nonrespirable:

$$\frac{30 \text{ mg/m}^3}{\% \text{ quartz} + 3}$$

Cristobalite ... Use one half the value calculated from the count or mass formulae for quartz.

Tridymite .... Use one half the value calculated from formulae for quartz.

Silica, fused .. Use quartz formulae.

Amorphous, including natural diatomaceous earth	20

SILICATES (<1% quartz)

**Asbestos, all forms	—
Mica	20
*Perlite	30
Portland Cement	30
Soapstone	20
Talc (non-asbestiform)	20
Talc (fibrous) use asbestos limit	—
Tremolite (see Asbestos)	—
Graphite (natural)	15

*COAL DUST

(bituminous) .... 2 mg/m³ (respirable dust^p) fraction <5% quartz). If >5% quartz use respirable mass formula.

NUISANCE PARTICULATES

(see Appendix E)

30 m.p.p.c.f. or 10 mg/m³ of total dust <1% quartz

Conversion factors

mppcf × 35.3 = million particles per cubic meter = particles per c.c.

i Millions of particles per cubic foot of air, based on impinger samples counted by light-field technics.

j The percentage of quartz in the formula is the amount determined from airborne samples, except in those instances in which other methods have been shown to be applicable.

k Both concentration and percent quartz for the application of this limit are to be determined from the fraction passing a size-selector with the following characteristics:

Aerodynamic Diameter (μm) (unit density sphere)	% passing selector
≤2	90
2.5	75
3.5	50
5.0	25
10	0

1 Containing <1% quartz; if quartz content >1%, use formulae for quartz.

p See footnote "p" after Mineral Dusts.

° 1973 Addition.

°° See "Notice of Intended Changes for Mineral Dusts."

NOTICE OF INTENDED CHANGES (for 1973)

These substances, with their corresponding values, comprise those for which either a limit has been proposed for the first time, or for which a change in the "Adopted" listing has been proposed. In both cases, the proposed limits should be considered trial limits that will remain in the listing for a period of at least two years. If, after two years no evidence comes to light that questions the appropriateness of the values herein, the values will be reconsidered for the "Adopted" list. In the interim, the 1972 Adopted limit will prevail. Documentation is available for each of these substances.

Substance	ppm^a)	mg/cu m^b
Baygon	-	0.5
Butyl lactate	1	5

Substance	ppm[a]	mg/Cw m[b]
†C Cadmium oxide fume	-	0.05
Caprolactam—		
Dust	-	1
Vapor	5	25
† Carbofuran—Skin	-	0.05
† Carbon tetra-bromide	0.1	1.4
† Catechol	1	4.5
† Cesium hydroxide	-	2
† Chlorodifluoromethane	1,000	3,500
Chloroform	25	120
† Chloro-bis-Chloromethylether		A1b
† Chlorpyrifos (Dursban®)—Skin	-	0.2
† o-Chlorostyrene ..	50	285
† o-Chlorotoluene ...	50	250
† 2-Chloro-6-(trichloromethyl) pyridine (N-Serve®)	-	10
† Clopidol (Coyden®)	-	10
† Copper fume	-	0.2
Cotton dust, raw ...	-	0.2[m]
† Crufomate (Ruelene®)	-	5
† Cyclohexylamine ...	10	40
† Dicyclopentadienyl-iron	-	10
† Diethylphthalate ..	-	5
Dimethyl sulfate—Skin		A2
† 3,5-Dinitro-o-toluamide (Zoalene®)	-	5
Dioxane—Skin ...	50	180
† Disyston—Skin ...	-	0.1
C Ethylidene norbornene	5	25
† Formamide	20	30
† Furfuryl alcohol ..	5	20
Hexachlorocyclopentadiene	0.01	0.1

Substance	ppm	mg/Cw m
† Iron oxide fume ...	-	5
Isophorone	10	55
Manganese cyclopentadienyl tricarbonyl (as Mn)—Skin	-	0.1
4, 4-Methylene bis (2-chloroaniline) —Skin	0.02	A2
C Methylene bis (4-cyclohexylene isocyanate)	0.01	0.11
Methylene chloride (Dichloromethane)	250	890
C Methylethyl ketone peroxide	0.2	1.5
Mineral wool fiber	-	10
Paraffin wax fume	-	2
Phorate (Thimet®) —Skin	-	0.05
† Picloram (Tordon®)	-	10
C Potassium hydroxide	-	2
Silicon tetrahydride (Silane) ..	0.5	0.7
†C Sodium hydroxide	-	2
†C Subtilisins (Proteolytic enzymes as 100% pure crystalline enzyme)	-	0.00006[c]
† Tricyclohexyltin hydroxide (Plictran®)	-	5
† Vinylidene chloride	10	4
Zinc stearate	-	E

a Parts of vapor or gas per million parts of contaminated air by volume at 25 C and 760 mm. Hg pressure.

b Approximate milligrams of particulate per cubic meter of air.

m Lint free dust as measured by the vertical-elutriator, cotton-dust sampler described in the Transactions of the National Conference on Cotton Dust, J. R. Lynch, pg. 33, May 2, 1970.

Mineral Dusts

Substance	TLV
Asbestos, all forms†	. 5 fibers/cc $> 5\mu$ in length[n] A1a
Silica flour	Use respirable[p] mass formula for quartz.
Tripoli	Use respirable mass formula for quartz.

n As determined by the membrane filter method at 400–450 × magnification (4 mm objective) phase contrast illumination.

p "Respirable" dust as defined by the British Medical Research Council Criteria (1) and as sampled by a device producing equivalent results (2).

(1) Hatch, T. E. and Gross, P., Pulmonary Deposition and Retention of Inhaled Aerosols, p. 149. Academic Press, New York, New York, 1964.

(2) Interim Guide for Respirable Mass Sampling, AIHA Aerosol Technology Committee, AIHA J. 31: 2, 1970, p. 133.

† A more stringent TLV for crocidolite may be required.

o Based on "high volume" sampling.
Capital letters refer to Appendices.

† 1973 Revision or Addition.

Appendix A

Carcinogens

The Committee lists below those substances in industrial use that have proven carcinogenic in man, or have induced cancer in animals under appropriate experimental conditions. Present listing of those substances carcinogenic for man takes two forms, those for which a TLV has been assigned (1a), and those for which en-

TABLE 38-E. THRESHOLD LIMIT VALUES FOR CHEMICAL SUBSTANCES IN WORKROOM AIR, 1973 (Continued)

NOTICE OF INTENDED CHANGES
(Cont'd)

vironmental conditions have not been sufficiently defined to assign a TLV (1b).

1a. **Human Carcinogens**—Substances recognized as occupational carcinogens with an assigned TLV:

Asbestos, all forms, 5 fibers/cc > 5μm in length.

bis (Chloromethyl) ether, 1 ppb.

Chromates, certain insoluble forms, 100μg/m³

Coal tar pitch volatiles 200μg/m³

Nickel carbonyl, 1 ppb

1b. **Human Carcinogens** — Substances known to be potent occupational carcinogens without an assigned TLV:

4-Aminodiphenyl (p-Xenylamine)
Benzidine and its salts
beta-Naphthylamine
4-Nitrodiphenyl

For the substances in 1b, no exposure or contact by any route, respiratory, skin or oral, as detected by the most sensitive methods, shall be permitted. "No exposure or contact" means hermitizing the process or operation by the best practicable engineering methods, and protecting the worker by proper equipment that will insure virtually no contact or entry of the carcinogen by any route.

2. **Experimental Carcinogens**—Industrial substances found to be of high potency in inducing tumors under experimental conditions in animals:

Beryllium
Chloromethyl methyl ether
3, 3'-Dichlorobenzidine
Dimethyl sulfate
Ethylenimine
4, 4'-Methylene bis (2-chloroaniline)
N-Nitrosodimethylamine
beta-Propiolactone

For the above compounds, worker exposure by all routes should be reduced to a minimum in light of the warning of the potency of these substances to induce tumors in animals. "Reduced to a minimum" means extraordinary care shall be taken both in manufacture and in handling so that worker exposure by all routes is kept below the limit of sensitivity of the analytic method of determining the exposure concentration.

Appendix B

B¹ *Polytetrafluoroethylene* decomposition products. Thermal decomposition of the fluorocarbon chain in air leads to the formation of oxidized products containing carbon, fluorine and oxygen. Because these products decompose in part by hydrolysis in alkaline solution, they can be quantitatively determined in air as fluoride to provide an index of exposure. No TLV is recommended pending determina-

* Trade names: Algoflon, Fluon, Halon, Teflon, Tetran.

tion of the toxicity of the products, but air concentrations should be kept below the limit of sensitivity of the analytic method.

B² *Gasoline and/or petroleum distillates.* The composition of these materials varies greatly and thus a single TLV for all types of these materials is no longer applicable. In general, the aromatic hydrocarbon content will determine what TLV applies. Consequently the content of benzene, other aromatics and additives should be determined to arrive at the appropriate TLV (Elkins, et al. AIHA J. 24:99, 1963).

Appendix C

B.1 THRESHOLD LIMIT VALUES FOR MIXTURES

When two or more hazardous substances are present, their combined effect, rather than that of either individually, should be given primary consideration. In the absence of information to the contrary, the effects of the different hazards should be considered as additive. That is, if the sum of the following fractions,

$$\frac{C_1}{T_1} + \frac{C_2}{T_2} + \cdots \frac{C_n}{T_n}$$

exceeds unity, then the threshold limit of the mixture should be considered as being exceeded. C_1 indicates the observed atmospheric concentration, and T_1 the corresponding threshold limit. (See Example 1A.a. and 1A.c.)

Exceptions to the above rule may be made when there is good reason to believe that the chief effects of the different harmful substances are not in fact additive, but *independent* as when purely local effects on different organs of the body are produced by the various components of the mixture. In such cases the threshold limit ordinarily is exceeded only when at least one member of the series

$$\left(\frac{C_1}{T_1} + \text{ or } + \frac{C_2}{T_2} \text{ etc.}\right)$$

itself has a value exceeding unit. (See Example 1A.c.)

Antagonistic action or potentiation may occur with some combinations of atmospheric contaminants. Such cases at present must be determined individually. Potentiating or antagonistic agents are not necessarily harmful by themselves. Potentiating effects of exposure to such agents by routes other than that of inhalation is also possible, e.g. imbibed alcohol and inhaled narcotic (trichloroethylene). Potentiation is characteristically exhibited at high concentrations, less probably at low.

When a given operation or process characteristically emits a number of harmful dusts, fumes, vapors, or gases, it will frequently be only feasible to attempt to evaluate the hazard by measurement of a single substance. In such cases, the threshold limit used for this substance should be reduced by a suitable factor, the magnitude of which will depend on the number, toxicity and relative quantity of the other contaminants ordinarily present.

Examples of processes which are typically associated with two or more harmful atmospheric contaminants are welding, automobile repair, blasting, painting, lacquering, certain foundry operations, diesel exhausts, etc. (Example 2 in 1A.a.)

THRESHOLD LIMIT VALUES FOR MIXTURES

EXAMPLES

1A.a. General case, where air is analyzed for each component:

Additive effects. (Note: It is essential that the atmosphere be analyzed both qualitatively and quantitatively for each component present, in order to evaluate compliance or noncompliance with this calculated TLV.)

$$\frac{C_1}{T_1} + \frac{C_2}{T_2} + \frac{C_3}{T_3} + \cdots = 1$$

Example No. 1: Air contains 5 ppm of carbon tetrachloride (TLV = 10 ppm) 20 ppm of ethylene dichloride (TLV = 50 ppm) and 10 ppm of ethylene dibromide (TLV = 20 ppm)

Atmospheric concentration of mixture = 5 + 20 + 10 = 35 ppm of mixture

$$\frac{5}{10} + \frac{20}{50} + \frac{10}{20} = \frac{25 + 20 + 25}{50} = 1.4$$

Threshold Limit is exceeded. Furthermore, the TLV of this mixture may be calculated by reducing the total fraction to 1.0; i.e.

TLV of mixture = $\frac{35}{1.4}$ = 25 ppm

Example No. 2: Air contains 200 ppm of hexane (TLV = 500 ppm) 100 ppm of methylene chloride (TLV = 500 ppm) and 20 ppm of perchloroethylene (TLV = 100 ppm)

Atmospheric concentration of mixture = 200 + 100 + 20 = 320 ppm of mixture

$$\frac{200}{500} + \frac{100}{500} + \frac{20}{100} = \frac{200 + 100 + 100}{500}$$

$$= \frac{400}{500} = 0.8$$

Threshold Limit is not exceeded. The TLV of this

mixture = $\frac{320}{0.8}$ = 400ppm

1A.b. Special case when the source of contaminant is a liquid mixture and the atmospheric composition is *assumed* to be similar to that of the original material; e.g. on a time weighted average exposure basis, all of the liquid (solvent) mixture eventually evaporates.

Additive effects (approximate solution).

If the percent composition (by weight) of the liquid mixture is known, the TLV's of the constituents must be listed in mg/m³.

(Note: In order to evaluate compliance with this TLV, field sampling instruments should be cali-

TABLE 38-E. THRESHOLD LIMIT VALUES FOR CHEMICAL SUBSTANCES IN WORKROOM AIR, 1973 (Continued)

brated, in the laboratory, for response to this specific quantitative and qualitative air-vapor mixture, and also to fractional concentrations of this mixture; e.g., 1/2 the TLV; 1/10 the TLV; 2 × the TLV; 10 × the TLV; etc.)

TLV of mixture =

$$\frac{1}{\dfrac{f_a}{TLV_a} + \dfrac{f_b}{TLV_b} + \dfrac{f_c}{TLV_c} + \cdots + \dfrac{f_n}{TLV_n}}$$

Example No. 1: Liquid solvent contains (by weight) 50% heptane (TLV = 2000 mg/m³) 30% methylene chloride (TLV = 1740 mg/m³) 20% perchloroethylene (TLV = 670 mg/m³)

TLV of mixture = $\dfrac{1}{\dfrac{0.5}{2000} + \dfrac{0.3}{1740} + \dfrac{0.2}{670}}$

$= \dfrac{1}{0.00025 + .00017 + .0003} = \dfrac{1}{0.00072}$

= 1390 mg/m³

Of this mixture: 50% or 695 mg/m³ is heptane, 30% or 417 mg/m³ is methylene chloride and 20% or 278 mg/m³ is perchloroethylene

These values can be converted to ppm as follows:

Heptane—
2000 mg/m³ = 500 ppm

1 mg/m³ = 0.25 ppm
695 mg/m³ = 174 ppm

Methylene chloride—
1740 mg/m³ = 500 ppm
1 mg/m³ = 0.287 ppm
417 mg/m³ = 119 ppm

Perchloroethylene—
670 mg/m³ = 100 ppm
1 mg/m³ = 0.15 ppm
278 mg/m³ = 42 ppm

The TLV of this mixture = 174 + 119 + 42 = 335 ppm.

1A.c. **Independent effects.**
Air contains 0.15 mg/m³ of lead (TLV, 0.15) and 0.7 mg/m³ of sulfuric acid (TLV, 1).

$$\frac{0.15}{0.15} = 1; \qquad \frac{0.7}{1} = 0.7$$

Threshold limit is not exceeded.

1B.a. General exact solution for mixtures of N components with additive effects and different vapor pressures.

(1) $\dfrac{C_1}{T_1} + \dfrac{C_2}{T_2} + \cdots + \dfrac{C_n}{T_n} = 1;$

(2) $C_1 + C_2 + \cdots + C_n = T,$

(2.1) $\dfrac{C_1}{T} + \dfrac{C_2}{T} + \cdots + \dfrac{C_n}{T} = 1.$

By the Law of Partial Pressures,

(3) $C_1 = ap_1.$

And by Raoult's Law,

(4) $p_1 = F_1 p_1°.$
Combine (3) and (4) to obtain

(5) $C_1 = aF_1 p_1°.$

Combining (1), (2,1) and (5), we obtain

(6) $\dfrac{F_1 p_1°}{T} + \dfrac{F_2 p_2°}{T} + \cdots + \dfrac{F_n p_n°}{T} =$

$\dfrac{F_1 p_1°}{T_1} + \dfrac{F_2 p_2°}{T_2} + \cdots + \dfrac{F_n p_n°}{T_n}$

and solving for T,

(6.1) $T = \dfrac{F_1 p_1° + F_2 p_2° + \cdots + F_n p_n°}{\dfrac{F_1 p_1°}{T_1} + \dfrac{F_2 p_2°}{T_2} + \cdots + \dfrac{F_n p_n°}{T_n}}$

or

(6.2) $T = \dfrac{\sum_{i=1}^{i=n} F_1 p_1°}{\sum_{i=1}^{i=n} \dfrac{F_1 p_1°}{T_1}}$

T = Threshold Limit Value in ppm.
C = Vapor concentration in ppm.
p = Vapor pressure of component in solution.
p° = Vapor pressure of pure component.
F = Mol fraction of component in solution.
a = A constant of proportionality.
Subscripts 1,2,...n relate the above quantities to components 1,2,...n, respectively.

Subscript i refers to an arbitrary component from 1 to n.

Absence of subscript relates the quantity to the mixture.

1B.b. Solution to be applied when there is a reservoir of the solvent mixture whose composition does not change appreciably by evaporation.

Exact arithmetic solution of specific mixture:

Solvent	Mol. wt.	Density	TLV	p° at 25 C	Mol fraction in half-and-half solution by volume
Trichloroethylene (1)	131.4	1.46 g/ml	100	73mm Hg	0.527
Methylchloroform (2)	133.42	1.33 g/ml	350	125mm Hg	0.473

$F_1 p_1° = (0.527)(73) = 38.2$

$F_2 p_2° = (0.473)(125) = 59.2$

$$T = \frac{38.2 + 59.2}{\frac{38.2}{100} + \frac{59.2}{350}} = \frac{(97.4)}{133.8} + \frac{(350)}{59.2}$$

$$= \frac{(97.4)(350)}{193.0} = 177$$

T = 177 ppm Note difference in TLV when account is taken of vapor pressure and mol fraction in comparison with above example where such account is not taken.

2. A mixture of one part of (1) parathion (TLV, 0.1) and two parts of (2) EPN (TLV, 0.5).

$$\frac{C_1}{0.1} + \frac{C_2}{0.5} = \frac{C_m}{T_m}; \quad C_2 = 2C_1$$

$$C_m = 3C_1$$

$$\frac{C_1}{0.1} + \frac{2C_1}{0.5} = \frac{3C_1}{T_m}$$

$$\frac{7C_1}{0.5} = \frac{3C_1}{T_m}$$

$$T_m = \frac{1.5}{7} = 0.21 \text{ mg/cu m}$$

1C. TLV for Mixtures of Mineral Dusts.

For mixtures of biologically active mineral dusts the general formula for mixtures may be used.

For a mixture containing 80 percent talc and 20 percent quartz, the TLV for 100 percent of the mixture "C" is given by:

$$TLV = \frac{1}{\frac{0.8}{20} + \frac{0.2}{2.5}} = 8.4 \text{ mppcf}$$

Essentially the same result will be obtained if the limit of the more (most) toxic component is used provided the effects are additive. In the above example the limit for 20 percent quartz is 10 mppcf.

For another mixture of 25 percent quartz 25 percent amorphous silica and 50 percent talc:

$$TLV = \frac{1}{\frac{0.25}{2.5} + \frac{0.25}{20} + \frac{0.5}{20}} = 7.3 \text{ mppcf}$$

The limit for 25 percent quartz approximates 8 mppcf.

Appendix D

PERMISSIBLE EXCURSIONS FOR TIME-WEIGHTED AVERAGE (TWA) LIMITS

The Excursion TLV Factor in the Table automatically defines the magnitude of the permissible excursion above the limit for those substances not given a "C" designation; i.e., the TWA limits. Examples in the Table show that nitrobenzene, the TLV for which is 1 ppm, should never be allowed to exceed 3 ppm. Similarly, carbon tetrachloride, TLV 10 ppm, should never be allowed to exceed 20 ppm. By contrast, those substances with a "C" designation are not subject to the excursion factor and must be kept at or below the TLV ceiling.

These limiting excursions are to be considered as to provide a "rule-of-thumb" guidance for listed substances generally, and may not provide the most appropriate excursion for a particular substance e.g., the permissible excursion for CO is 400 ppm for 15 minutes.

For appropriate excursions for 142 substances consult Pa. Rules & Regs., Chap. 4, Art. 432, and "Acceptable Concentrations," ANSI.

Substance	TLV	Excursion Factor	Max. Conc. Permitted for short time
	ppm		ppm
Nitrobenzene	1	3	3
Carbon tetrachloride	10	2	20
o-Dichlorobenzene	50	1.5	75
Acetone	1000	1.25	1250
Boron trifluoride	C1	—	1
Butylamine	C5	—	5

TABLE 38-E. THRESHOLD LIMIT VALUES FOR CHEMICAL SUBSTANCES IN WORKROOM AIR, 1973 (*Concluded*)

EXCURSION FACTORS

For all substances not bearing "C" notation

	Excursion		Factor	
TLV > 0-1 (ppm or mg/m³)			= 3	
TLV > 1-10	"		= 2	
TLV > 10-100	"		= 1.5	
TLV > 100-1000	"		= 1.25	

The number of times the excursion above the TLV is permitted is governed by conformity with the Time-Weighted Average TLV.

BASIS FOR ASSIGNING LIMITING "C" VALUES

By definition in the Preface, a listed value bearing a "C" designation refers to a "ceiling" value that should not be exceeded; all values should fluctuate below the listed value. This, in effect, makes the "C" designation a maximal allowable concentration (MAC). In general, the bases for assigning or not assigning a "C" value rest on whether excursions of concentration above a proposed limit *for periods up to 15 minutes* may result in (*a*) intolerable irritation, (*b*) chronic, or irreversible tissue change, or (*c*) narcosis of sufficient degree to increase accident proneness, impair self-rescue or materially reduce work efficiency.

Appendix E

Some nuisance particulates.q TLV, 30 mppcf or 10mg/m³.

Alundum (Al₂O₃)
Calcium carbonate
Cellulose (paper fiber)
Portland cement
Corundum (Al₂O₃)
Emery
Glass, fibrousr or dust
Glycerin mist
Graphite (synthetic)
Gypsum
Vegetable oil mists (except castor, cashew nut, or similar irritant oils)
Kaolin
Limestone
Magnesite
Marble
Pentaerythritol
Plaster of Paris
Rouge
Silicon carbide
Starch
Sucrose
Tin oxide
Titanium dioxide

q When toxic impurities are not present, e.g. quartz < 1%.
r < 5-7 μm in diameter.

Appendix F

Some simple asphyxiants—"Inert" gases and vapors.s

TLV, 1,000ppm

Acetylene — Hydrogen
Argon — Methane
Butane — Neon
Ethane — Nitrogen
Ethylene — Nitrous oxidet
Helium — Propane

s As defined in introductory copy.
t Nontoxic at or below TLV.

Requests for permission to republish or reprint these Threshold Limit Values should be directed to: Secretary-Treasurer, ACGIH, P.O. Box 1937, Cincinnati, Ohio 45201.

Chemical
Hazards

Chapter
39

Using the Table   1200
Recognition . . . Evaluation . . . Control

Explanation of Table Headings   1201
Listing of substances

Cross Index   1208

Table of Chemical Hazards   1212

Glossary of Common Chemical Terms   1232

1199

The safety professional is required to make judgments on the degree of hazard associated with chemical exposures. It is his duty, therefore, to obtain information on the fire and health hazards of these chemicals and use it to evaluate and, where necessary, control the exposures.

This chapter tabulates the basic physical constants of many common industrial chemicals and lists relative health and fire hazards. New chemicals are being introduced and new data on the properties of existing chemicals are constantly being published and revised. For this reason, the safety professional should also consult the most up-to-date revisions of the references referred to for each chemical.

Using the Table

Recognition

The material in this chapter can be used to determine the degree of fire and health hazards in using a new chemical. For example, the safety professional can compare evaporation rates, vapor volumes, and flash points and choose a solvent that combines the lowest volatility and fire and health hazards consistent with process needs.

Then too, this tabulation can serve as a means of gaining familiarity with the properties of chemicals—liquids, dusts, gases, vapors, and solids—encountered in the working environment.

Evaluation

Also in this chapter is the fundamental information needed for a preliminary evaluation of a chemical hazard. The more complex problems, however, require the services of an industrial hygienist.

The following is an example in which the degree of health hazard resulting from a solvent exposure is to be evaluated using data from the table in this chapter and Table 38-E, "Threshold Limit Values for 1973," from Chapter 38, "Industrial Toxicology."

Trichloroethylene is being used in a metal-finishing workroom 20 by 10 by 20 ft. In an 8-hr workday, 5 gal of the solvent is lost through evaporation. There are two air changes per hour. Is there a potential health hazard?

Solution: Assuming there is complete mixing of the trichloroethylene throughout the workroom, then:

Volume of room $= 20 \times 10 \times 20 = 4000$ cu ft

Total volume of air supplied to workroom in 8 hr $= 4000$ cu ft $\times$ 8 hr $\times$ 2 changes/hr $= 64,000$ cu ft/8-hr day

Vapor volume $= 36$ cu ft/gal (from table)

Total vapor volume $= 36$ cu ft/gal $\times$ 5 gal $= 180$ cu ft

Threshold limit value of trichloroethylene (from Table 38-E) $= 100$ ppm

Total volume of air required to dilute trichloroethylene to threshold limit $= \dfrac{180 \times 1,000,000}{100}$ $= 1,800,000$ cu ft/8-hr day

Therefore, since natural ventilation is supplying only 64,000 cu ft of air per 8-hr day, there is a potential health hazard and additional ventilation is required to maintain the solvent concentration within safe limits.

Note that this type of calculation is valid only if the air contaminant is uniformly distributed at a relatively low concentration. Where the air contaminant is localized in high concentrations, such as in a paint-spraying operation, more complex means of evaluating the hazard must be used. (See details in NSC's *Fundamentals of Industrial Hygiene*.)

Control

Data from the table in this chapter can also be used to estimate the amount of exhaust ventilation needed to control an explosion hazard.

Example: An average of 1 gal of n-heptane per hour is being evaporated in a drying oven. Estimate rate of exhaust ventilation needed to prevent an explosive concentration.

Solution: Assuming complete mixing of the n-heptane throughout the oven, the rate of ventilation depends on the volume of evaporating solvent and is independent of the volume of the oven.

From the following table:

Lower explosive limit for n-heptane = 1.2 percent
Vapor volume = 22 cu ft/gal
Exhaust rate required to dilute vapors to the lower explosive limit $= \dfrac{22 \times 100}{1.2}$

$= 1833$ cu ft/hr
$= 30.6$ cfm

This rate is normally multiplied by a safety factor of about 5 to allow for imperfect mixing. Therefore $30.6 \times 5 = 153$ cfm exhaust rate. For more precise calculations, see NFPA Standard No. 86A, *Ovens and Furnaces.*

Explanation of Table Headings

Column headings of the table (pages 1212–1231) are explained as follows.

Listing of substances

The "chemical hazards" table lists those physical properties associated with the fire and health hazards of common industrial chemicals. (For tables of threshold limit values and maximal acceptable concentrations, see Chapter 38, "Industrial Toxicology.")

Data in the table have been taken from the following principal references: Standard 49, *Hazardous Chemicals Data,* and Standard 325M, *Flammable Liquids, Gases, and Volatile Solids,* both published by the National Fire Protection Association, 470 Atlantic Ave., Boston, Mass. 02210, and *Handbook of Organic Industrial Solvents,* published by the American Mutual Insurance Alliance, 20 N. Wacker Dr., Chicago, Ill. 60606. Be sure to check latest editions of these standards and booklets.

Other information is contained in Chapter 46, "Safety Engineering Tables," Table 46-O, "Explosion Characteristics of Various Dusts." Check the Cross Index for names of chemicals and compounds that may be listed under another name. (See pages 1208–1211.)

Column 1

FLASH POINT is the lowest temperature at which the liquid gives off vapor within a test vessel in sufficient concentration to form an ignitable mixture with air near the surface of the liquid. The flash point is used by OSHA and NFPA to define and classify the fire hazard of liquids (see Chapter 42, "Flammable and Combustible Liquids").

A standard closed container is used to determine the closed-cup flash point of a liquid and a standard open-surface dish is used for the open-cup flash point determination, as specified by the standards of the American Society for Testing and Materials (ASTM).

Unless otherwise specified, the flash point values in the table are closed-cup determinations.

Column 2

FLAMMABLE OR EXPLOSIVE LIMITS are those concentrations of a vapor or gas in air below or above which propagation of a flame does not occur on contact with a source of ignition. The lower limit is the minimum vapor concentration below which the vapor-air mixture is too "lean" to burn or explode. The upper limit is the maximum vapor concentration above which the vapor-air mixture is too "rich" to burn or explode.

Flammable or explosive limits are given in the Table in terms of percentage by volume of gas or vapor in air, and, unless otherwise noted, at normal atmospheric pressures and temperatures. Increasing the temperature or pressure lowers the lower limit and raises the upper limit. Decreasing the temperature or pressure has the opposite effect.

Column 3

VAPOR VOLUME is the number of cubic feet of solvent vapor formed by the evaporation of one gallon of a liquid at 75 F.

Column 4

SEVERITY AND TYPE OF HAZARD. Each hazardous substance presents a distinct problem and must be treated individually in the light of its own characteristics. Conclusions regarding the hazards of a product cannot safely be drawn either from the properties of the materials from which it is formed or by analogies based upon chemical structure. Mixtures of two or more chemicals may have properties that vary in kind or degree from those of the individual components. Impurities may con-

1201

tribute hazardous properties and should not be overlooked.

Signal word. This word is intended to draw attention to the presence of hazard, and to indicate the degree of severity. The Signal Words used, in the order of diminishing severity of hazard are:

D DANGER—Serious, severe, hazardous.

W WARNING—Moderate, intermediate, harmful.

C CAUTION—Minor, mild, irritating.

The degree of severity can be expressed only in relative terms. DANGER! is the strongest of the three words and is used for those products presenting the most serious hazards. CAUTION! is for those compounds presenting the least serious hazards. WARNING! is intermediate between DANGER! and CAUTION!

Type of hazard. This statement gives notice of the hazards that are present in connection with the customary or reasonably anticipated handling or use of the product. Many chemical products will present more than one type of hazard; in which case, appropriate statements for each significant type are included. In general, the most serious hazard is stated first.

Severity of toxic or fire hazard from the undiluted material is normally used.* Impurities, mixtures, and conditions of use may influence this.

Combine degree of severity (signal word) with description statement below.

1 Flammable material.

2 Oxidizing material—Contact with other combustible material may cause fire.

3 Gas or vapor rapidly toxic or extremely irritating on exposure for a short time or to low concentration.

4 Gas or vapor harmful or irritating on prolonged or repeated exposure, or exposure to high concentrations.

5 Gas or vapor physiologically inert, but displaces oxygen available for breathing.

6 Dust hazardous when inhaled or touched.

7 Irritant, sensitizer, corrosive—Causes skin irritation or burn.

8 Toxic through skin absorption.

Column 5

PRECAUTIONS TO TAKE. These instructions are intended to supplement the statement on "Severity and Type of Hazard" by briefly setting forth measures to be taken to avoid injury or damage from stated hazards.

To minimize hazards, take precaution* indicated by signal number listed below.

1 Keep away from heat, sparks, open flame.

2 Avoid spilling, contacting skin, eyes, clothing.

3 Use adequate ventilation or personal protection. Avoid breathing dust, fumes, mists, gases, vapors.

4 Avoid contact with acids, moisture, combustibles.

5 Do not handle or use until safety precautions outlined in data sheets, or recommended by consultant or manufacturer, are understood.

Many products present no hazard in normal handling and storage. For these products, no precautionary statements are necessary. The development of new chemical products, and the introduction of chemical processes into ever-widening fields, have accentuated the need for obtaining appropriate information in those cases where there are hazards requiring special precautions. Precautionary information should, so far as is practicable, reach every person using, handling, or storing hazardous substances. For labeling of consumer products and for identification during transportation, consult appropriate federal agencies.

Column 6

ORAL TOXICITY RATING.** The numerical toxicity rating of column 6 is largely explained

---

* See MCA Manual L-1, 1961. Manufacturing Chemists' Assn., Inc., 1825 Connecticut Ave., NW., Washington, D.C. 20009, for further details.

** See Gleason, Gosselin, and Hodge, *Clinical Toxicology of Commercial Products*, 2nd ed. Williams and Wilkins Co., Baltimore, 1963.

TABLE 39–A

Toxicty Rating or Class	Probable LETHAL Dose (human)	
	mg/kg	For 70 kg man (150 lb)
6 Super toxic	Less than 5	A taste (less than 7 drops)
5 Extremely toxic	5–50	Between 7 drops and 1 teaspoonful
4 Very toxic	50–500	Between 1 tsp. and 1 ounce
3 Moderately toxic	500–5 gm./kg.	Between 1 oz. and 1 pint (or 1 lb.)
2 Slightly toxic	5–15 gm./kg.	Between 1 pt. and 1 quart
1 Practically nontoxic	Above 15 gm./kg.	More than 1 quart

* Open cup flash point.
† Number in this column refers to date of publication in *AIHA Journal*.

in Table 39-A. To use toxicity ratings effectively, their many implications and limitations must be appreciated, as noted below.

● The rating is based on mortality, not morbidity; that is, it is really a lethality rating.

● Unless otherwise noted, each rating is based on the acute toxicity of a single dose when taken by mouth. Other dose regimens and other routes of administration are not represented by the rating.

● The toxicity rating reflects an estimate of the probable lethal dose, not the minimal fatal dose. Perhaps because of personal idiosyncrasy or hypersensitivity or predisposing disease, minimal lethal doses recorded in the clinical literature are usually considerably lower than those implied by the current ratings.

● With only a few compounds are clinical data adequate to establish a toxicity rating. Most of the values here are based on laboratory determinations of mean lethal doses (LD_{50}) in small laboratory mammals (rat, mouse, guinea pig, rabbit; sometimes cat, dog, and monkey). Implicit in the use of such data is the conventional assumption that the mean lethal dose in man lies in the same class as does the LD_{50} for the test animals.

● For most corrosive agents (such as mineral acids, alkalis, and bleaches), no toxicity rating is suggested. In these cases, death is usually the result of severe local tissue injury, with secondary complications such as toxemia, shock, perforation, infection, hemorrhage, and obstruction. The intensity of the local lesion and of the results that follow is often determined by the concentration of the corrosive substance, whereas the volume and "dose" are secondary considerations. For such agents, no single toxicity rating is an appropriate measure of lethality, unless the concentration is also specified. Actually, no simple parameter can describe this relation in a way which is thought to be clinically useful.

In Table 39-A, common units of measure are used to describe lethal doses for an adult of average size (body weight of 150 lb or 70 kg). For patients who are heavier or lighter, probable lethal doses are proportionately larger or smaller. It is assumed that lethal doses are proportional to body weight, irrespective of age. The reader is urged to consult the reference given above for more complete information.

Toxicity rating. Amount to produce death when swallowed by an average (150-lb) man.

1 Practically nontoxic—Takes more than one quart (2 lb).

2 Slightly toxic—1 pint to 1 quart.

3 Moderately toxic—1 oz to 1 pint.

4 Very toxic—1 teaspoonful to 1 oz.

5 Extremely toxic—7 drops to 1 teaspoonful.

6 Super toxic—A taste (<7 drops).

NOTE: For inhalation toxicities (inhalation

exposures and acceptable air-borne concentrations), see "Documentation of Threshold Limit Values," American Conference of Governmental Industrial Hygienists, P.O. Box 1937, Cincinnati, Ohio 45201.

Column 7

Action on skin. Local action on normal skin of the undiluted material.*

A Relatively harmless.

B Sensitizer—Can cause allergic reactions.

C Primary skin irritant—Brief contact can cause inflammation or burns.

D Can cause solvent irritation-type dermatitis.

Column 8

NFPA hazard classifications. Fires and other emergency situations may involve chemicals that have varying degrees of toxicity, flammability, and reactivity (instability and water reactivity). The National Fire Protection Association grading of these relative hazards (under fire conditions) is given in the columns marked "NFPA Health" "NFPA Flammability," and "NFPA Reactivity." (See Fig. 39–1.)

For a full description of the NFPA classifications, see NFPA Standard No. 704M, *Identification of the Fire Hazards of Materials*. A complete listing is given in NFPA Standard No. 49, *Hazardous Chemicals Data*, and in Standard No. 325M, *Fire Hazard Properties of Flammable Liquids, Gases, and Volatile Solids*. (Also see Chapter 43, "Fire Protection," specifically Identification of Hazardous Materials.)

An explanation of the degrees of hazard follows.

NFPA health hazards. In general the health hazard in fire fighting is that of a single exposure which may vary from a few seconds up to an hour. The physical exertion caused by fire fighting or other emergency may intensify the effects of any exposure.

Health hazards arise from two sources: (*a*) the inherent properties of the material, and (*b*) from the toxic products of combustion or decomposition of the material. (Common

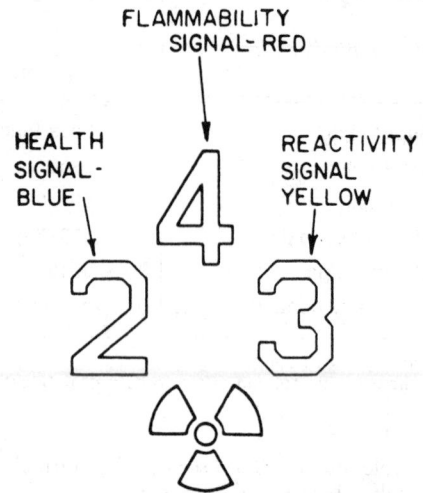

FIG. 39–1.—Correct arrangement and order of signals used on equipment for quick identification of material hazards. For details, see NFPA Standard 704M; see also pages 1378–1380.

hazards from burning of ordinary combustible materials are not included.)

The degree of hazard should indicate (*a*) that people can work safely only with specialized protective equipment, (*b*) that they can work safely with suitable respiratory protective equipment, or (*c*) that they can work safely in the area with ordinary clothing.

A health hazard, as defined by the NFPA, is any property of a material which either directly or indirectly can cause injury or incapacitation, either temporary or permanent, for exposure by contact, inhalation, or ingestion.

The degrees of hazard under fire conditions are ranked according to the probable severity of hazard to personnel, as follows:

4 Materials which on very short exposure could cause death or major residual injury even though prompt medical treatment were given, including those which are too dangerous to be approached without specialized

---

* See "Occupational Diseases," Bul. No. 1097, U.S. Dept. of Health, Education, and Welfare, 1964. (Available: U.S. Government Printing Office)

protective equipment. This degree should include:

• Materials which can penetrate ordinary rubber or synthetic protective clothing;

• Materials which under normal conditions or under fire conditions give off gases which are extremely hazardous (i.e., toxic or corrosive) through inhalation or through contact with or absorption through the skin.

3 Materials which on short exposure could cause serious temporary or residual injury even though prompt medical treatment were given, including those requiring protection from all bodily contact. This degree should include:

• Materials which give off highly toxic combustion products;

• Materials corrosive to living tissue or toxic by skin absorption.

2 Materials which on intense or continued exposure could cause temporary incapacitation or possible residual injury unless prompt medical treatment is given, including those requiring use of respiratory protective equipment with independent air supply. This degree should include:

• Materials giving off toxic combustion products;

• Materials giving off highly irritating combustion products;

• Materials which either under normal conditions or under fire conditions give off toxic vapors lacking warning properties.

1 Materials which on exposure would cause irritation but only minor residual injury even if no treatment is given, including those which require use of an approved canister type gas mask. This degree should include:

• Materials which under fire conditions would give off irritating combustion products;

• Materials which on the skin could cause irritation without destruction of tissue.

0 Materials which on exposure under fire conditions would offer no hazard beyond that of ordinary combustible material.

NFPA flammability hazards deal with the degree of susceptibility of materials to burning, even though some materials that burn under one set of conditions will not burn under others. The form or condition of material, as well as its properties, affects the hazard.

The degrees of hazard are ranked according to the susceptibility of materials to burning, as follows:

4 Materials which will rapidly or completely vaporize at atmospheric pressure and normal ambient temperature or which are readily dispersed in air, and which will burn readily. This degree should include:

• Gaseous materials;

• Cryogenic materials;

• Any liquid or gaseous material which is a liquid while under pressure and having a flash point below 73 F (22.8 C) and having a boiling point below 100 F (37.8 C). (Class IA flammable liquids.)

• Materials which on account of their physical form or environmental conditions can form explosive mixtures with air and which are readily dispersed in air, such as dusts of combustible solids and mists of flammable or combustible liquid droplets.

3 Liquids and solids that can be ignited under almost ambient temperature conditions. Materials in this degree produce hazardous atmospheres with air under almost all ambient temperatures or, though unaffected by ambient temperatures, are readily ignited under almost all conditions. This degree should include:

• Liquids having a flash point below 73 F (22.8 C) and having a boiling point at or above 100 F (37.8 C) and those liquids having a flash point at or above 73 F (22.8 C) and below 100 F (37.8 C). (Class IB and Class IC flammable liquids, respectively.— See definitions in Chapter 42, "Flammable and Combustible Liquids.")

• Solid materials in the form of coarse dusts which may burn rapidly but which generally do not form explosive atmospheres with air;

- Solid materials in a fibrous or shredded form which may burn rapidly and create flash fire hazards, such as cotton, sisal and hemp;

- Solids which burn with extreme rapidity, usually by reason of self-contained oxygen (e.g., dry nitrocellulose);

- Materials which ignite spontaneously when exposed to air.

2 Materials that must be moderately heated or exposed to relatively high ambient temperatures before ignition can occur. Materials in this degree would not under normal conditions form hazardous atmospheres with air, but under high ambient temperatures or under moderate heating may release vapor in sufficient quantities to produce hazardous atmospheres with air. This degree should include:

- Liquids having a flash point above 100 F, but not exceeding 200 F;

- Solids and semisolids which readily give off flammable vapors.

1 Materials that must be preheated before ignition can occur. Materials in this degree require considerable preheating, under all ambient temperature conditions, before ignition and combustion can occur. This degree should include:

- Materials which will burn in air when exposed to a temperature of 1500 F for a period of 5 minutes or less;

- Liquids, solids and semisolids having a flash point above 200 F;

- This degree includes most ordinary combustible materials.

0 Materials that will not burn. This degree should include any material which will not burn in air when exposed to a temperature of 1500 F for a period of 5 minutes.

NFPA reactivity (instability) hazards deal with the degree of susceptibility of materials to release energy. Some materials are capable of rapid release of energy by themselves (as by self-reaction or polymerization), or they can undergo violent eruptive or explosive reaction if contacted with water or other extinguishing agents or with certain other materials.

The violence of reaction or decomposition of materials may be increased by heat or pressure, by mixture with certain other materials to form fuel-oxidizer combinations, or by contact with incompatible substances, sensitizing contaminants, or catalysts.

Because of the wide variations of accidental combinations possible in fire emergencies, these extraneous hazard factors (except for the effect of water) cannot be applied in a general numerical scaling of hazards. Such extraneous factors must be considered individually in order to establish appropriate safety factors such as separation or segregation. Such individual consideration is particularly important where significant amounts of materials are to be stored or handled. Guidance for this consideration is provided in NFPA Standard No. 49, *Hazardous Chemicals Data*.

The degree of hazard should indicate to fire fighting personnel that the area should be evacuated, that the fire may be fought from a protected location, that caution must be used in approaching the fire and applying extinguishing agents, or that the fire may be fought using normal procedures.

The relative reactivity of a material is defined as follows.

REACTIVE MATERIALS are those which can enter into a chemical reaction with other stable or unstable materials. For purposes of this guide, the other material to be considered is water and only if its reaction releases energy. Reactions with common materials, other than water, may release energy violently. Such reactions must be considered in individual cases, but are beyond the scope of this identification system.

UNSTABLE MATERIALS are those which in the pure state or as commercially produced will vigorously polymerize, decompose or condense, or become self-reactive and undergo other violent chemical changes.

STABLE MATERIALS are those that normally have the capacity to resist changes in their chemical composition, despite exposure to air, water, and heat as encountered in fire emergencies.

The degrees of hazard are ranked according to ease, rate and quantity of energy release as follows:

4 Materials which are readily capable of detonation or of explosive decomposition or explosive reaction at normal temperatures and pressures. This degree should include materials which are sensitive to mechanical or localized thermal shock at normal temperatures and pressures.

3 Materials which are capable of detonation or of explosive decomposition or explosive reaction but which require a strong initiating source or which must be heated under confinement before initiation. This degree should include materials which are sensitive to thermal or mechanical shock at elevated temperatures and pressures or which react explosively with water without requiring heat or confinement.

2 Materials which are normally unstable and readily undergo violent chemical change but do not detonate. This degree should include materials which can undergo chemical change with rapid release of energy at normal temperatures and pressures or which can undergo violent chemical change at elevated temperatures and pressures. It should also include those materials which may react violently with water or which may form potentially explosive mixtures with water.

1 Materials which are normally stable, but which can become unstable at elevated temperatures and pressures or which may react with water with some release of energy but not violently.

0 Materials which are normally stable, even under fire exposure conditions, and which are not reactive with water.

Column 9

REFERENCES. References to publications on chemical safety have been included in the table as an aid to the safety professional who is seeking more information on safe handling of a particular chemical. A brief description of the format and content of each of these publications follows. Copies of the publications and further information can be obtained from the organizations listed.

● *NSC Data Sheet*—National Safety Council, 425 North Michigan Ave., Chicago, Ill. 60611.

The "Industrial Data Sheets, Chemical Series," cover the major hazards associated with a single substance or family of substances, discuss methods of controlling or eliminating the hazards, and concisely outline safe and efficient procedures. They are written for the safety professional, foreman, and other management representatives.

A typical "Chemical Series Data Sheet" is composed of the following topics:

1. Properties	10. Fire and explosion hazards
2. Uses	11. Electrical equipment
3. Containers	12. Symptoms of poisoning
4. Shipping regulations	13. First aid
5. Storage	14. Treatment of burns
6. Personnel hazards	15. Toxicity
7. Handling	16. Threshold limit values
8. Personal protective equipment	17. Medical examinations
9. Ventilation	18. Waste disposal

● *MCA Data Sheet*—Manufacturing Chemists' Association, 1825 Connecticut Ave., NW., Washington, D.C. 20009.

Of interest to both producers and consumers, the Chemical Safety Data Sheets are based on the recommendations of chemical engineers. Topics covered are:

1. Properties	6. Handling; storage
2. Hazards	7. Tank and equipment cleaning and repairs
3. Engineering control of hazards	8. Waste disposal
4. Employee safety	9. Medical mgt.
5. Fire fighting	10. First aid

● *AIA Bulletin*—American Insurance Association, Engineering and Safety Department, 85 John St., New York, N.Y. 10038.

The "Chemical Hazards Bulletins" present data interpreted in terms of insurance coverage and therefore are of value in inspection and underwriting problems.

A typical topic outline includes:

1. Properties
2. Identification
3. Health hazards
4. Public liability hazards
5. Product liability hazards
6. Boiler and machinery hazards
7. Fire hazards
8. Control measures
9. Storage and shipping
10. Hazardous procedures and operations
11. Detection and determination
12. First aid
13. Waste disposal

• *AIHA Hygienic Guide*—American Industrial Hygiene Association, 665 Miller Road, Akron, Ohio 44313.

The "Hygienic Guides," which appear in the bimonthly *AIHA Journal*, summarize data from experimental work and industrial experience to establish the degree of chemical hazards and recommend control procedures.

A typical Guide covers:

1. Hygienic standards
   a. Recommended maximum atmospheric concentration
   b. Severity of fire and health hazards
   c. Short exposure tolerance
   d. Atmospheric concentration immediately hazardous to life

2. Toxic properties
   a. Inhalation
   b. Skin contact
   c. Eye contact
   d. Ingestion

3. Industrial hygiene practice
   a. Recognition
   b. Evaluation
   c. Recommended control procedures

4. Medical information
   a. First aid
   b. Special medical procedures

5. References

Cross Index

This listing cross indexes chemicals and compounds to the nomenclature used in the Table of Chemical Hazards.

A

acetic acid methyl ester ....methyl acetate
acetic acid propyl esters...............propyl acetates
acetic aldehyde.............................acetaldehyde
acetic ester...................................ethyl acetate
acetyl oxide ....acetic anhydride
acetylene tetrachloride ....tetrachlorothane
acetylene dichloride 1,2-dichloroethylene
acraldehyde.......................................acrolein
acrylaldehyde....................................acrolein
acrylic aldehyde................................acrolein
acrylic esters....................acrylates, methyl
Aerozine-50................................hydrazine
allyl aldehyde ....acrolein
allylene..propyne
amino benzene ....aniline
1-aminobutane................................butylamine
amino ethane................................ethylamine
2-aminoethanol.............................ethanolamine
β-aminoethyl alcohol.....................ethanolamine
anesthesia ether ....ethyl ether
anol ....cyclohexanol
aqua fortis ....nitric acid
azotic acid ....nitric acid

B

banana oil ....amyl acetate
benzine....................................naphtha (petroleum)
benzol ....benzene
p-benzoquinone...................................quinones
betanaphthylamine ....naphthylamine
biphenyl ....diphenyl
boron hydrides........................boron compounds
bromomethane...........................methyl bromide
butanal..................................butyraldehyde
butanethiol................................butyl mercaptan
butyl alcohols.................................butanols
2-butanone ....methyl ethyl ketone
2-butoxyethanol...ethylene glycol monobutyl ether
butyl alcohol..................................1-butanol
butylaldehyde.............................butyraldehyde
butyl Cellosolve ....
                ethylene glycol monobutyl ether
butyric aldehyde ....butyraldehyde

C

calcium hypochlorite ....bleaching powder
calcium oxide...lime
carbinol ....methyl alcohol
carbitol...........diethylene glycol monoethyl-ether
carbolic acid.....................................phenol
carbon bichloride ....tetrachloroethylene
carbon bisulfide........................carbon disulfide
caustic potash ....potassium hydroxide
caustic soda...........................sodium hydroxide
Cellosolve ....ethylene glyco monoethyl ether
Cellosolve acetate...................................
           ethylene glycol monoethyl ether acetate

chloroallylene................................allyl chloride
chlorates ....sodium chlorate
chlorobenzol ....chlorobenzene
2-chlorobutadiene........................chloroprene
2-chloro-1,3-butadiene......................chloroprene
chloroethane ....ethyl chloride
chloroethanols ....ethylene chlorohydrin
β-choroethyl alcohol...........ethylene chlorohydrin
bis-β-chloroethyl ether...........dichloroethyl ethers
3-chloropropene ....allyl chloride
2-chloropropylene oxide ....epichlorhydrin
α-chlorotoluene ....benzyl chloride
cinnamene ....styrene monomer
coal tar oil ....cresols; naphtha (coal tar)
columbian spirit ....methanol
cresote oil ....cresols
cresylic acid ....cresols
cumol..cumene
cyanoethylene................................acrylonitrile
cyclohexyl methane ....methylcyclohexane

D

decaborane..........................boron compounds
diacetone ....diacetone alcohol
diamine ....hydrazine
dichloro-benzol........................dichlorobenzene
1,1-dichloroethylene..............vinylidene chloride
di(2-chloroethyl) ether............dichloroethyl ether
dichloroisopropane..............propylene dichloride
dichloromethane ....methylene chloride
dichloromonomethane ....methylene chloride
1,2-dichloropropane ....propylene dichloride
dichromates ....sodium dichromate
diethyl ether................................ethyl ether
diethyl oxide................................ethyl ether
diethylene oxide ....tetrahydrofuran
2,2-dihydroxyethel ether...........diethylene glycol
sym-dihydroxydiethyl ether.......diethylene glycol
diisocyanates....................................isocyanates
dimethylamine................................methylamine
dimethylaminobenzene ....hydroquinone
dimethyl benzenes....................................xylenes
1,3-dimethyl butanol ....methyl amyl alcohol
dimethylene oxide ....ethylene oxide
N,N-dimethyl formic acid ....dimethylformamide
2,6-dimethyl heptanone-4.........di iso butyl ketone
dimethyl ketone......................................acetone
dimethoxy methane ....methylal
dry ice..carbon dioxide

E

EPI ....epichlorohydrin
epoxyethane................................ethylene oxide
ethanal..acetaldehyde
ethanethiol............................ethyl mercaptan
1,2-ethanediol................................ethylene glycol
ethanoic acid..acetic acid
ethanoic anhydride....................acetic anhydride
ethanol..ethyl alcohol
2-ethanoxyethylacetate ....
    ethylene glycol monoethyl ether acetate

ethenyl ethanoate ....vinyl acetate
ether, petroleum....................naphtha (petroleum)
ether..ethyl ether
ethinyl trichloride................trichloroethylene
2-ethoxy ethanol ....
    ethylene glycol monoethyl ether
2-ethoxy ethylacetate................................
    ethylene glycol monoethyl ether acetate
ethyl acrylate............................acrylate, methyl
ethyl aldehyde................................acetaldehyde
ethylcaproaldehyde........................2-ethyl hexanol
ethyl Cellosolve ....
    ethylene glycol monoethyl ether
ethyl dimethyl methane ....pentane
ethyl ethanoate................................ethyl acetate
ethylene dibromide 1,2-dibromoethylene
ethylene dichloride..............1,2-dichloroethylene
2,2-ethylene dioxydiethanol ....triethylene glycol
ethylene glycol monoethyl ether ....
    2-ethoxyethanol
ethylene, tetrachloro..............tetrachloroethylene
ethylene trichloride ....trichloroethylene
ethyl methanoate ....ethyl formate
ethyl methyl ketone..............methyl ethyl ketone
ethyl oxide ....ethyl ether

F

formal ....methylal
formic acid ethyl ester....................ethyl formate
formic acid methyl ester..............methyl formate
formic ether ....ethyl formate
formyl trichloride ....chloroform
fuel oil No. 1 ....kerosene
fulminate of mercury..............mercury fulminate
fuming sulfuric acid ....oleum
furfuraldehyde ....furfural
2-furaldehyde ....furfural

G

glycerol ....glycerine
glycol ....ethylene glycol
glycol chlorohydrin ....ethylene chlorohydrin
glycol dichloride....................ethylene dichloride
grain alcohol ....ethyl alcohol

H

hexahydrobenzene............................cyclohexane
hexahydrotoluene ....methyl cyclohexane
hexalin................................cyclohexanol
hexamethylene................................cyclohexane
hexane diacid ....adipic acid
hexanedioic acid ....adipic acid
2-hexanone................................methyl butyl ketone
hexone................................methyl butyl ketone
hexyl hydride..hexane
hydralin ....cyclohexanol
hydrochloric ether................ethyl chloride
hydrogen chloride ....hydrochloric acid
hydrogen cyanide ....hydrocyanic acid
hydrogen fluoride ....hydrofluoric acid
4-hydroxy-4-methyl-2pentanone................................
    diacetone alcohol

1209

hydroxytoluenes......................................cresols
2-hydroxytriethylamine ....diethylamino ethanol

I

iso-butyl carbinol............................i-amyl alcohol
iso-propanol............................iso-propyl alcohol
iso-propyl benzene ....cumene
iso-propyl carbinal...............................1-butanol

L

ligroin ....naphtha (petroleum)
liquefied petroleum gas ....LP-gas
LOX ....oxygen, liquid
lye.....................................sodium hydroxide

M

MBK................................methyl butyl ketone
MDI.......................................isocyanates
MEK................................methyl ethyl ketone
methanal ....formaldehyde
methanethiol ....methyl mercaptan
methanol ....methyl alcohol
3-methoxy-1-butanol ....amyl alcohols
2-methoxy ethanol ....
                ethylene glycol monomethyl ether
methyl Cellosolve acetate..................................
                ethylene glycol monomethyl ether acetate
methyl acetic ester ....methyl acetate
methyl acetylene................................propyne
methyl acrylate ....acrylate, methyl
methyl benzene ....toluene
2-methyl butane ....pentane
3-methyl-1-butanol...................i-amyl alcohol
3-methyl-1-butanol acetate ....amyl acetate
2-methyl butyl ethanoate ....amyl acetate
methyl n-butyl methane ....hexane
methyl Cellosolve..................................
                ethylene glycol monomethyl ether
methyl Cellosolve acetate...............................
                ethylene glycol monomethyl ether acetate
methyl chloride ....methylene chloride
methyl cyanide...............................acetonitrile
methylenebis (4-phenyl isocyanate) .....isocyanates
methylene chlorobromide .....bromochloromethane
methylene dimethyl ether ....methylal
methyl ethyl carbinol ....butanol
methyl ethylene glycol ....propylene glycol
2-methyl hexane....................................heptane
methyl iso-butyl carbinol ....methyl amyl alcohol
methyl methacrylate ....acrylate, methyl
methyl methanoate ....methyl formate
4-methyl-2-pentanol ....methyl amyl alcohol
4-methyl-2-pentanone.......methyl iso-butyl ketone
4-methyl-3-pentene-2-one ....mesityl oxide
methylphenols.....................................cresols
methyl propanol..................................butanol
mineral spirits ....naphtha (petroleum)
monochlorobenzene ....chlorobenzene
monoethanolamine ....ethanolamine
monoethyl amine ....ethylamine
monomethyl hydrazine ....methyl hydrazine

monomethyl aniline ....methyl aniline
motor fuel antiknock compound ......tetraethyl lead
muriatic acid ....hydrochloric acid
muriatic ether................................ethyl chloride

N

naphtha 76° ....naphtha (petroleum)
nitrating acid ....acids, mixed
nitrochlorobenzene................chloronitrobenzene

O

oil, fuel ....fuel oil
oil of mirbane ....nitrobenzene
oil of vitriol ....sulfuric acid

P

pentanol acetate.............................amyl acetate
pentanols.................................amyl alcohols
2-pentanone.....................methyl propyl ketone
perchloroethylene ....tetrachloroethylene
petroleum ether....................naphtha (petroleum)
phenyl amine ....aniline
phenyl chloride ....chlorobenzene
phenyl ethane ....ethylbenzene
phenyl ethylene......................styrene monomer
phenyl methane ....toluene
2-phenyl propane ....cumene
phorone, iso ....isophorone
pimelic ketone............................cyclohexanone
polonium ....radon
propane ....LP-gas
1,2-propanediol...................propylene glycol
1,2,3-propane triol ....glycerine
propanols ....propyl alcohols
2-propanone.......................................acetone
propene oxide ....propylene oxide
2-propen-1-ol ....allyl alcohol
propyl acetone ....methyl butyl ketone
propyl carbinol ....butanol
propylene dichloride 1,2-dichloropropane
propyl methanol..................................butanol
prussic acid...............................hydrocyanic acid

Q

quick lime ....lime (quick)

R

red fuming nitric acid........nitric acid (red fuming)
refrigerants.............................*see chemical name*

S

sodium cyanide ....cyanides
sodium pentachlorophenate......pentachlorophenol
sulfuric ether ....ethyl ether

T

TDI ....isocyanates
Teflon ....polytetrafluoroethylene
TEL ....tetraethyl lead
tetrabromoethane ....acetylene tetrabromide
tetrachloromethane................carbon tetrachloride
tetrahydronaphthalene ....Tetralin
tetramethylene oxide....................tetrahydrofuran

TML ....tetramethyl lead
TNT ....trinitrotoluene
toluene diisocyanate ....isocyanates
toluidine ....methyl aniline
toluol...toluene
tolylene diisocyanate..........................isocyanates
1,1,1-trichloroethane ....methyl chloroform
trichloromethane ....chloroform
trimethylamine ....methylamine
trimethyl carbinol ....butanol
2,4,6-trinitrophenylmethylnitramine ....tetryl
2,4,6-trinitrophenol ....picric acid
triorthocresyl phosphate ....tricresyl phosphate

U

UDMH............................1,1-dimethylhydrazine
unsymmetrical dimethyl hydrazine ....
                                   1,1-dimethylhydrazine

V

vinegar acid ....acetic acid
vinyl benzene...........................styrene monomer
vinylcyanide ....acrylonitrile
vinyl trichloride ....methyl chloroform
vinylidene chloride monomer .....dichloroethylene

W

wood alcohol................................methyl alcohol
wood spirit....................................methyl alcohol

X

xylols..xylenes

TABLE OF CHEMICAL HAZARDS

	1	2		3	4	5	6	7	8 NFPA			9 References			
Substance	Flash Point (deg F)	Flammable or Explosive Limits (% by volume) Lower	Upper	Vapor Volume (cu ft per gal)	Severity and Type of Hazard	Precautions to Take	Oral Toxicity	Action on Skin	Health	Flammability	Reactivity	N.S.C. Data Sheet	MCA Data Sheet	AIA Bulletin	AIHA Hygienic Guide†
Acetaldehyde	−36	4.1	55	58	D-1,4,7	1,3,2	3	B,C	2	4	2		SD-43		9-55
Acetic acid (glacial)	109	5.4	16.0 (212 F)	57	D-7,4,1	2,3	Cor.	B	2	4	2	410	SD-41	C-52	9-72
Acetic anhydride	129	2.7	10.0	35	D-7,3,1	2,3	Cor.	B	2	2	2		SD-15		1-71
Acetone	0	2.6	12.8	44	D-1,4	1,3	3	D	1	3	0	398	SD-87	C-46	3-57
Acetonitrile	42	–	–	62	D-1,3	1,2,3	3	D	3	3	3				6-60
Acetylene	gas	2.5	81	–	D-1,5	1,3	–	A	1	4	3	494	SD-7		4-67
Acetylene Dichloride				–	D-1,3	1,3,2	3	–	3	0	1				
Acids, mixed				–	D-7,3,2	2,3,4	Cor.	C					SD-65	C-66	
Acrolein	<0	2.8	31	–	D-1,3,7	1,2,3	Cor.	C	3	3	3	436	SD-85	C-66	6-63
Acrylonitrile	32*	3.0	17.0	–	D-3,8,1	3,2,1	4	C	4	3	2		SD-31	C-65	3-57
Adipic acid				–	C-1,4	1,3	–	–	1	1	–	438		C-68	
Allyl alcohol	70	2.5	18.0	48	D-3,8,1	5,3,1	4	C	3	3	1				12-63

Substance													
Allyl chloride	−25	3.3	11.1	40	D-1,3,7	1,3,2	Cor.	C	3 3 1			C-42	12-63
Aluminum (powder)	0.035 oz/cu ft	–	–	–	C-1	1	–	A	0 1 1				6-63
Aluminum chloride				–	W-7	2,4	3	B		435	SD-62		
Ammonia, anhydrous	gas	16	25	–	W-3,1,7	3,1,2	Cor.	C	3 1 0	251	SD-8		2-71
Ammonia, aqua				–	W-7,3	2,3	Cor.	C			SD-13		
Ammonium dichromate				–	W-1,2,6	1,2,3	Cor.	C	– – –		SD-45		
Ammonium nitrate				–	W-2	1,4	3	C	2 1 3	536, 604	A-10		
n-Amyl acetate	77	1.1	7.5	22	C-1,4,7	1,2,3	3	D	1 3 0	208			4-65
iso-Amyl alcohol	109	1.2	9.0 (212 F)	30	D-1,3	1,2,3	3	D	1 2 0				
n-Amyl alcohol	91	1.2	10.0 (212 F)	30	D-1,3	1,2,3	3	D	1 3 0				
Aniline	158	1.3	–	36	D-8,3	2,3,5	4	B	3 2 0	409	SD-17		12-55
Antimony	metal and sulfides flammable				C-7,6	3,2	–	A		408	SD-66		12-59
Antimony trichloride	–			–	D-7,4	2,3	5	B			SD-66		
Arsenic trioxide				–	D-4,7	3,2,5	5	B,C		499	SD-60		12-64
Arsine	gas			–	D-3	3	–	–		499			8-65

* Open cup flash point.
† Number in this column refers to date of publication in *AIHA Journal.*

TABLE OF CHEMICAL HAZARDS (Continued)

Substance	1 Flash Point (deg F)	2 Flammable or Explosive Limits (% by volume) Lower	Upper	3 Vapor Volume (cu ft per gal)	4 Severity and Type of Hazard	5 Precautions to Take	6 Oral Toxicity	7 Action on Skin	8 NFPA Health	Flammability	Reactivity	9 N.S.C. Data Sheet	MCA Data Sheet	AIA Bulletin	AIHA Hygienic Guide†
Asphalt, cutback	<50	-	-	-	C-7	2	-	B	0	3	0	215, 582			
Benzene	12	1.3	7.1	37	D-1,4,8	1,3,5	4	C	2	3	0	308	SD-2	C-77	6-70
Benzoyl peroxide				-	D-2,1,7	5,1,4	3	C	1	4	4		SD-81		
Benzyl chloride	153	1.1	-	28	W-7,4	2,3	Cor.	C	2	2	0		SD-69		
Beryllium				-	D-6,7	5,3	-	B,C	4	1	1	562		C-1	12-64
Bleaching compounds				-	C-7,2	2,4	Cor.	C	2	1	0	343			
Boron compounds				-	D-1,3	5,1,3	-	C							
Bromine		nonflammable		-	D-7,4,2	5,3	Cor.	C	4	0	1	508	SD-84	C-80	8-58
1,3-Butadiene	gas	2.0	11.5	-	D-1,3	1,3	-	C	2	4	2	313	SD-49	C-49	2-63
n-Butane	gas	1.9	8.5	-	D-1,4	1,3	-	-	1	4	0		SD-55		
1-Butanol	84	1.4	11.2	36	W-7,4,1	3,2,1	3	C	1	3	0				9-55
1-Butene	gas	1.6	9.3	-	D-1,4	1,3	-	-	1	4	0				

n-Butyl acetate	72	1.7	7.6	25	C-1,4,7	123	3	D	1 3 0				12-62
n-Butylamine	10	1.7	9.8	33	W-1,4,8	1,2,3	Cor.	C	3 3 0				12-60
Butyllithium		–		–	D-1	5,1,2	–	–			SD-91		
Butyraldehyde	20	2.5	–	37	D-1,4,7	1,2,3	1	D	2 3 –		SD-78		12-62
Cadmium				–	D-6	3	–	B		312			12-62
Calcium carbide				–	D-1,7,6	4,1,2	Cor.	C	1 4 2		SD-23		
Calcium oxide				–	D-7,6	2,3	Cor.	C	1 0 1	241			
Carbon dioxide	nonflammable			–	W-5,7	2,3	–	C		397			10-64
Carbon disulfide	−22	1.3	44	54	D-1,4,8	1,3,2	3	C	2 3 0	341	SD-12	C-40	12-56
Carbon monoxide	gas	12.5	74	–	D-4,1	3,1	–	A	2 4 0	415			8-65
Carbon tetrachloride	nonflammable			34	D-3,8	3,2,5	4	D		331	SD-3	C-6	12-61
Chlorates				–	W-2,7,6	4,2,1	4	C	3 0 1				
Chlorine				–	D-3,2,7	5,3,4	–	C		207	SD-80	C-43	
Chlorine dioxide	>10			–	D-2,3,1	5,4,3	–	C		525			6-58
Chlorobenzene	84	1.3	7.1	32	D-1,3	1,3,2	4	–	2 3 0			C-20	2-64
Chloroform	nonflammable			41	W-4,7	3,2	3	D			SD-89	C-16	12-65
Chlorosulfonic acid				–	D-2,7,3	5,4,3	–	C	3 0 2		SD-33		
Chromic acid				–	D-2,6,7	4,2,3	4	C	1 0 1		SD-44		6-56

* Open cup flash point.
† Number in this column refers to date of publication in AIHA Journal.

TABLE OF CHEMICAL HAZARDS (Continued)

| | 1 | 2 | | 3 | 4 | 5 | 6 | 7 | 8 NFPA | | | 9 References | | | |
Substance	Flash Point (deg F)	Flammable or Explosive Limits (% by volume) Lower	Upper	Vapor Volume (cu ft per gal)	Severity and Type of Hazard	Precautions to Take	Oral Toxicity	Action on Skin	Health	Flammability	Reactivity	N.S.C. Data Sheet	MCA Data Sheet	AIA Bulletin	AIHA Hygienic Guide†
o-Cresol	178	1.4 (300 F)	–	32	D-8,7,4	2,3	4	C	2	2	0		SD-48		10-58
p-Cresol	202	1.1 (300 F)	–	32	D-8,7,4	2,3	4	C	2	1	0		SD-48		10-58
Cumene	111	0.9	6.5	23	W-1,4	1,3	4	D	0	2	0				12-61
Cyanides				–	D-6,8	5,3,2	6	–	3	2	–		SD-30	C-5	
Cyanogen	gas	6	32	–	D-1,3	1,3,4	–	–	4	4	3				
Cyclohexane	–4	1.3	8.0	30	D-1,4,7	1,3	3	D	1	3	0		SD-68		10-63
Cyclohexanol	154	–		31	C-4,1,7	1,2,3	3	D	1	2	0				
Cyclohexanone	111	1.1	–	32	C-4,1,7	1,2,3	3	D	1	2	0				12-65
Cyclopropane	gas	2.4	10.4	–	D-1,3	1,3	–	–	1	4	0				
DDT				–	C-6,8	3,2	4	–				303			10-59

Compound	Flash point	0.8 (approx.)	98 (approx.)										
Decaborane	176	0.8 (approx.)	98 (approx.)		D-1,6	5,13			3 2 1	508	SD-84		
Diacetone alcohol	148	1.8	6.9	26	C-4,1,7	1,2,3	3	D	1 2 0				
1,2-Dibromoethane		nonflammable		38	W-3	2,3	3	D	3 – –				
o-Dichlorobenzene	151	2.2	9.2	29	C-3,7,1	3,2,1	3	C	2 2 0		SD-54	C-21	6-64
p-Dichlorobenzene	150			29	C-3,7,1	3,2,1	3	C	2 2 0		SD-54	C-21	6-64
1,2-Dichloroethane	56	6.2	16.0	42	W-1,3	1,3	3	D	2 3 0	350	SD-18	C-15	8-65
1,1-Dichloroethylene	5*	5.6	11.4	42	W-1,3	1,3	3	D	2 4 2				
1,2-Dichloroethylene	43	9.7	12.8	43	D-1,3	1,3	3	D	2 3 2			C-14	
Dichloroethyl ether	131			28	W-1,3	1,3	3	D	2 2 0			C-19	
1,1-Dichloro-1-nitroethane	168*			32	W-1,3	1,3	4	D	– 2 3				
1,1-Dichloro-1-nitropropane	151*			27	W-1,3	1,3	4	D	– 2 3				
1,2-Dichloropropane	60	3.4	14.5	33	W-1,3	1,3	4	D	2 3 0			C-18	
Diethylamine	<0	1.8	10.1	32	D-1,3,7	1,2,3	Cor.	C	3 3 0				6-60
Diethylene glycol	255			35	C-3,1	2,1	3	D	1 1 0				
Diethylene glycol mono-ethyl ether	201			24	C-3,1	2,1	3	D	1 1 0				

* Open cup flash point.
† Number in this column refers to date of publication in *AIHA Journal*.

TABLE OF CHEMICAL HAZARDS (Continued)

	1	2		3	4	5	6	7	8			9			
		Flammable or Explosive Limits (% by volume)		Vapor Volume (cu ft per gal)					NFPA			References			
Substance	Flash Point (deg F)	Lower	Upper		Severity and Type of Hazard	Precautions to Take	Oral Toxicity	Action on Skin	Health	Flammability	Reactivity	N.S.C. Data Sheet	MCA Data Sheet	AIA Bulletin	AIHA Hygienic Guide†
Diethylene-triamine	215*	–		–	D-7,4	2,3,5	Cor.	C	3	1	0		SD-76		6-60
Diethyl sulfate	220	–		–	D-7,4,8	5,2,3	Cor.	C	3	1	1				
Diisobutyl ketone	140	0.8 (212 F)	6.2 (212 F)	–	W-3,1	3,1,2	–	D	1	2	0				12-62
Diisopropyl-amine	30*	–		–	D-1,4,7	1,2,3	Cor.	C	3	3	0				
Dimethylaniline	145	3.4		26	D-4,8,1	2,3	4	C	3	2	0			C-67	
Dimethyl ether	gas	3.4	18.0	–	D-1,4	1,3	–	D	–	4	0				
Dimethyl-formamide	136	2.2 (212 F)	15.2 (212 F)	–	D-3,1,8	3,1,2	Cor.	C	1	2	0				9-57
1,1-Dimethyl-hydrazine (unsym.)	5	2	95	–	D-1,3,8	5,1,3	Cor.	C	–	–	–				
Dimethylsulfate	182			33	D-7,4,8	5,2,3	Cor.	C	4	2	0		SD-19	C-53	4-63
1,2-Dinitro-benzene	302			30	D-8,6,1	3,2,1	4	B	3	1	4				

Dinitrotoluene				–	D-8,6,1	3,2,1	3	B	3 1 3		SD-93	12–60
Dioxane (diethylene dioxide)	54	2.0	22	39	W-1,4,8	1,2,3	3	D	2 3 1			12–61
Epichlorohydrin	105			42	D-3,7,8	2,3,5	4	C	3 2 0			
Epoxy resin systems				–	W-7,4	2,3,5	–	B,C		533	C-64	10–58
Ethanolamine	185			33	W-4,1	1,2,3	Cor.	C	2 2 0			6–68
Ethyl acetate	24	2.5	9.0	33	W-1,4,7	1,2,3	3	D	1 3 0		SD-51	4–64
Ethyl acrylate	60			–	W-1,4,8	1,2,3	–	C				
Ethyl alcohol	55	4.3	19.0	56	W-1,4	1,3	2	D	0 3 0	391		3–56
Ethylamine	<0*	3.5	14.0	50	D-1,4,7	1,2,3	Cor.	C	3 4 0			
Ethylbenzene	59	1.0	–	27	W-1,4	1,3	–	C	2 3 0			6–69
Ethyl bromide		6.75	11.3	–	W-1,4	1,3	–	C	2 3 0			4–65
Ethyl butyl ketone	115*			–	W-1,4,7	1,2,3	–	D	1 2 0			
Ethyl chloride	–58	3.8	15.4	46	D-1,4	1,3	–	D	2 4 0		SD-50	10–63
Ethyl ether	–49	1.9	48.0	31	D-1,3	1,3	3	D	2 4 –	396	SD-29, C-36	2–66
Ethyl formate	–4	2.7	13.5	41	D-1,3	1,2,3	3	C	2 3 0			
2-Ethyl hexanol	185*			21	W-1,4	1,2,3	–	D	2 2 0			
Ethyl mercaptan	<80	2.8	18.0	–	D-3,7,1	1,3,2	–	C	2 4 0			

* Open cup flash point.
† Number in this column refers to date of publication in *AIHA Journal*.

TABLE OF CHEMICAL HAZARDS (Continued)

Substance	1 Flash Point (deg F)	2 Flammable or Explosive Limits (% by volume) Lower	Upper	3 Vapor Volume (cu ft per gal)	4 Severity and Type of Hazard	5 Precautions to Take	6 Oral Toxicity	7 Action on Skin	8 NFPA Health	Flammability	Reactivity	9 References N.S.C. Data Sheet	MCA Data Sheet	AIA Bulletin	AIHA Hygienic Guide†
Ethyl silicate	125*			–	W-1,4	1,2,3	–	–	2	2	0				12-68
Ethylene	gas	3.1	32	–	D-1	1,3	–	C	1	4	2				12-61
Ethylene chlorohydrin	140*	4.9	15.9	49	D-3,8,1	3,1,5	4	D	3	2	0			C-22	12-61
Ethylene-diamine	110*			40	D-7,4,1	2,3,1	Cor.	C	3	2	0				2-70
Ethylene dichloride	56	6.2	16.0	42	W-1,4,7	1,2,3	3	D	2	3	0	350	SD-18	C-15	
Ethylene glycol	232	3.2	–	59	C-4	3	3	D	1	1	0				8-70
Ethylene imine		3.6	46		D-1,3,8	5,1,3	Cor.	–	3	3	3				2-65
Ethylene oxide	<0	3	100	–	D-1,3,7	1,3,2	–	C	2	4	3		SD-38	C-55	12-58
Ferrosilicon												352			
Fluorides				–	W-6,7	3,2,5	4	C				442			8-65
Fluorine				–	D-2,3	5,4,3	Cor.	C	4	0	3			C-61	12-65
Formaldehyde	gas	7.0	73	–	W-7,4	3,2	3	B,C	–	4	–	342	SD-1		4-65

Substance							Cor.								
Formic acid	156			–	W-7,4	3,2	3	C	3	2	0				
Fuel oil No. 2	100			–	C-1,3	1,2	3	D	0	2	0				
Fuel oil No. 5	130			–	C-1,3	1,2	3	D	0	2	0				
Fuel oil No. 6	150			–	C-1,3	1,2	3	D	0	2	0				
Furfural	140	2.1		39	D-8,4,1	3,2,1	3	C	1	2	1				4-65
Furfuryl alcohol	167*	1.8	16.3	–	D-8,4,1	3,2,1	3	C	1	2	1				
Gasoline	–45	1.4	7.6	24 to 32	D-1,3	1,3	3	D	1	3	0				
Glycerine	320			45	C-4	3	1	A	1	1	0				
n-Heptane	25	1.2	6.7	22	D-1,4	1,3	–	D	1	3	0				2-59
n-Hexane	–7.0	1.1	7.5	25	D-1,4	1,3	–	D	1	3	0				2-59
sec-Hexyl acetate	113			–	C-1,4,7	1,2,3	3	D	1	2	0				
Hexylene glycol	215*			31	C-4	3	3	D	1	1	0				
Hydrazine	100	4.7	100	–	D-1,3,8	5,3,1	–	B,C	3	3	2			C-57	12-56
Hydrochloric acid				–	W-7,4	2,3	Cor.	C	3	0	0		SD-39	C-30	8-58
Hydrocyanic acid	0	6	41	–	D-3,1,8	3,1,2	6	C	4	4	2		SD-67		2-70
Hydrofluoric acid				–	D-3,7	5,2	Cor.	C	4	0	0	459	SD-25	C-59	3-56
Hydrogen	gas	4.0	75	–	D-1,5	1,3	–	A	0	4	–				

* Open cup flash point.
† Number in this column refers to date of publication in *AIHA Journal.*

TABLE OF CHEMICAL HAZARDS (Continued)

	1	2		3	4	5	6	7	8 NFPA			9 References			
Substance	Flash Point (deg F)	Flammable or Explosive Limits (% by volume) Lower	Upper	Vapor Volume (cu ft per gal)	Severity and Type of Hazard	Precautions to Take	Oral Toxicity	Action on Skin	Health	Flammability	Reactivity	N.S.C. Data Sheet	MCA Data Sheet	AIA Bulletin	AIHA Hygienic Guide†
Hydrogen peroxide (52%)	–	nonflammable		–	C-2,7	4,2	Cor.	C	2	0	3		SD-53		9-57
Hydrogen sulfide	gas	4.3	45	–	D-3,1	5,3,1	–	–	3	4	0	284	SD-36	C-38	2-63
Iodine				–	D-3,6,7	3,2	5	C	2	1	0	457			8-65
Isocyanates (TDI and MDI)	–	0.9	9.5	–	D-3,7	3,2,5	–	D	2	1	0	489	SD-73	C-70	2-67
Isophorone	205	0.8	3.8	22	W-3,7,1	3,2,1	–	C	2	1	0				
Isoprene	–65			–	D-1,3	5,1,3	–	–	2	4	1				
Kerosene	100	0.7	5	–	C-1,3	1,2,3	3	D	0	2	0				
Lead				–	W-6	3,5	4	A				443	SD-64		4-58
Lime (quick)				–	D-7,6	2,3	Cor.	C	1	0	1	241			
Lithium				–	C-1,7,6	5,1,3	3	A	1	1	2	566		C-81	
Lithium hydride				–	D-1,3	5,4,3	Cor.	C	1	4	2				8-64
LP-gas				–	D-1,4	1,3	–	–	1	4	0	479			

	0.02 oz/cu ft		–	C-1,7	1,5	3	A	0 1 2				
Magnesium										426		C-72	2-60
Maleic anhydride	215	1.4	7.1	–	W-7,4	2,3	–	C	3 1 0		SD-88	C-71	6-70
Manganese					C-6,7	3,2,5	3	A		306			6-63
Mercury					W-4,8	3,2,5	–	C		203		C-60	6-66
Mesityl oxide	87	5.3	14.0	28	D-1,3	1,2,3	–	–	3 3 0				10-69
Methane	gas				D-1,5	1,3	–	–	1 4 0				
Methyl acetate	14	3.1	16	41	D-1,4,7	1,2,3	–	D	1 3 0				6-64
Methyl acrylate	27	2.8	2.5	–	W-1,4,8	1,3,2	–	D	2 3 2		SD-79		
Methylal	0*			37	D-1,4	1,3	–	D	2 3 2				
Methyl alcohol	52	7.3	36	80	D-1,4,7	1,3,5	3	D	1 3 0	407	SD-22	C-24	12-57
Methylamine	gas	4.9	20.7	–	D-1,7,4	1,3,2	Cor.	C	3 4 0		SD-57		
Methyl amyl alcohol	106	1.0	5.5	–	D-1,4,7	1,3,5	–	D	2 2 0				
Methyl aniline	185			31	D-3,8	3,2,5	4	B	3 2 0		SD-82		
Methyl bromide	gas	10	16	–	D-3,7,8	3,2	4	D	3 1 0		SD-35	C-41	4-58
Methyl butyl ketone	95	1.2	8	27	D-1,3	1,2,3	3	D	2 3 0				
Methyl isobutyl ketone	73	1.4	7.5	–	D-1,3	1,2,3	3	D	2 3 0				
Methyl chloride	gas	10.7	17.4	–	W-1,4,7	1,3,2	–	D	2 4 0		SD-40		12-61

* Open cup flash point.
† Number in this column refers to date of publication in *AIHA Journal.*

TABLE OF CHEMICAL HAZARDS (Continued)

Substance	Flash Point (deg F)	Flammable or Explosive Limits (% by volume) Lower	Upper	Vapor Volume (cu ft per gal)	Severity and Type of Hazard	Precautions to Take	Oral Toxicity	Action on Skin	NFPA Health	Flammability	Reactivity	N.S.C. Data Sheet	MCA Data Sheet	AIA Bulletin	AIHA Hygienic Guide†
Methyl chloroform				32	W-4,7	3,2	3	D				456	SD-90	C-10	12-61
Methylcyclohexane		1.2	–	26	C-4,1,7	1,2,3	3	D	–	3	0				
o-Methylcyclohexanol				27	C-4,1,7	1,2,3	3	D	–	2	0				
Methylcyclohexanone				27	C-4,1,7	1,2,3	3	D	–	2	0				
Methylene chloride	non-flammable	15.5	66 (in oxygen)	51	C-4,7	3,5	3	D	2	0	0	474	SD-86	C-17	12-65
Methyl ethyl ketone	21	1.8	10	36	W-1,4,7	1,2,3	3	D	1	3	0		SD-83	C-51	3-57
Methyl formate	–2	5.9	20	53	D-1,3	1,2,3	3	–	2	4	0				
Methyl hydrazine	<80	–	–	–	D-1,3	1,2,3	Cor.	C	–	–	–				
Methyl mercaptan		3.9	21.8	–	D-1,3	1,3	–	–	2	4	0				
Methyl propyl ketone	45	1.5	8.0	31	D-1,3	1,2,3	3	D	2	3	0				
Methyl styrene	134*	0.7	–	–	C-4,1,7	3,2,1	–	D	2	2	0				

Name													
Mixed acid				–	D-7,3,2	2,3,4		C			SD-65		
Molybdenum													2-60
Morpholine	100*			9	D-4,1	1,2,3	4	C	2 3 0				
Naphtha (coal tar)	100– 110–			37	D-1,4	1,2,3	–	D	2 2 0				
Naphtha (petroleum)	<0	1.1	5.9	20	D-1,4	1,2,3	–	D	1 4 0				8-63
Naphthalene	174	0.9	5.9	–	W-4,7	3,2	–	B,C	2 2 0	370	SD-58		10-67
Naphthylamine				–	W-3,1	2,3	4	–	2 1 0		SD-32		
Nickel carbonyl				–	D-3,1	5,3,1	6	–					6-68
Nicotine		0.7	4.0	–	D-4,8	5,2,3	6	C	4 1 0				
Nitrate-nitrite salt baths				–	W-6,7	2,3	4	–		270		C-35	
Nitric acid				–	D-2,3,7	2,4,3	Cor.	C	2 0 1		SD-5	C-13	8-64
Nitric oxide				–	–					206			
p-Nitroaniline	390			–	D-8,3,1	2,3,1	5	B,C	3 1 1		SD-94		
Nitrobenzene	190	1.8 (200 F)	–	32	D-8,3,1	2,3,1	5	B,C	3 2 0		SD-21	C-33	2-59
o-Nitrochlorobenzene	261			–	D-3,1	2,3,1	5	–	3 1 –				
Nitroethane	82	3.4	–	46	W-1,4,7	5,1,3	3	C	1 3 3				12-61
Nitrogen dioxide				–	D-2,3,7	5,4,3	–	C	– – –	206		C-34	6-56

* Open cup flash point.
† Number in this column refers to date of publication in AIHA Journal.

TABLE OF CHEMICAL HAZARDS (Continued)

Substance	1 Flash Point (deg F)	2 Flammable or Explosive Limits (% by volume) Lower	Upper	3 Vapor Volume (cu ft per gal)	4 Severity and Type of Hazard	5 Precautions to Take	6 Oral Toxicity	7 Action on Skin	8 NFPA Health	Flammability	Reactivity	9 References N.S.C. Data Sheet	MCA Data Sheet	AIA Bulletin	AIHA Hygienic Guide†
Nitrogen (liquid)	–			–	D-5,7	3,2	–	C	–	–	–				
Nitromethane	95	7.3	–	61	W-1,4,7	5,1,3	3	C	1	3	4				12-61
1-Nitropropane	120*	2.6	–	37	W-1,4,7	5,1,3	3	C	1	2	3				6-60
2-Nitropropane	103*	2.6	–	36	W-1,4,7	5,1,3	3	C	1	2	3				6-60
Nitrotoluene	223			–	D-3,1	2,3,1	5	C	–	1	3				
Octane	56	1.0	3.2	20	D-1,3	1,3	3	D	0	3	0				
Oils (cutting, emulsions, and drawing compounds)					–	2	3	D				501			
Oleum	–			–	D-7,3	2,3	Cor.	C	3	0	1	210			
Oxalic acid	–			–	W-6,7	2,3	4	C				406			
Oxygen (gas)	gas			–	D-2	4	–	A				472, 360			
Oxygen (liquid)	–			–	D-2,7	5,4	–	C				283			

Material	Flash point						Cor.						Date†
Ozone	–			–	D-3,2	3,4	Cor.	–				C-63	4-66
Paraformaldehyde	158			–	W-7,4	2,3	4	C	2 2 1	342	SD-6		
Parathion				–	D-3,1	5,3,2	6	–	4 1 0				6-69
Pentaborane	spontaneous	0.8 (approx.)	98 (approx.)	–	D-1,3	5,1,3	–	–	– – –	508	SD-84		6-66
Pentachlorophenol				–	W-6,7	3,2	4	C				C-32	8-70
n-Pentane	<–40	1.5	7.8	29	D-1,4	1,3	–	D	1 4 0				4-66
Perchloric acid				–	D-2,7	4,2,5	3	C	3 0 3	311	SD-11	C-44	
Phenol	175			–	D-8,7,4	2,3,5	4	C	3 2 0	405	SD-4		12-57
Phenylhydrazine	192			–	D-3,7,1	5,3,1	–	B,C	3 2 0				
Phosgene				–	D-3,7	3,2,5	–	C			SD-95		6-68
Phosphoric acid					C-7,4	2,3	Cor.	C			SD-70		6-57
Phosphoric anhydride					C-7,6	2,3	Cor.	C			SD-28		6-58
Phosphorus (red)					C-1	1	–	A	0 1 1		SD-16		
Phosphorus (white)					D-6,1,7	5,2,3	6	C	3 3 1	282	SD-16		
Phosphorus oxychloride					D-7,3	2,3,5	Cor.	C			SD-26		
Phosphorus pentasulfide					W-6,7,2	5,4,3	–	C	3 1 2		SD-71		
Phosphorus trichloride	nonflammable				D-7,3,1	5,4,3	Cor.	C	3 0 2		SD-27		

* Open cup flash point.
† Number in this column refers to date of publication in *AIHA Journal*.

TABLE OF CHEMICAL HAZARDS (Continued)

Substance	1 Flash Point (deg F)	2 Flammable or Explosive Limits (% by volume) Lower	Upper	3 Vapor Volume (cu ft per gal)	4 Severity and Type of Hazard	5 Precautions to Take	6 Oral Toxicity	7 Action on Skin	8 NFPA Health	Flammability	Reactivity	9 References N.S.C. Data Sheet	MCA Data Sheet	AIA Bulletin	AIHA Hygienic Guide†
Phthalic anhydride	305	1.7	10.5	–	C-7,6	2,3	Cor.	B	2	1	0		SD-61		8-67
Picric acid	explodes				D-1,2,6	1,4,3	5	B,C	2	4	4	351			
Potassium hydroxide				–	D-7,6	2,3	Cor.	C	3	0	1		SD-10		
Propionaldehyde	15-19*	3.7	16.1	–	D-1,3	1,3,2	3	D	2	3	1				
Propionic acid	130			–	C-1,3	1,2,3	Cor.	C	2	2	0				
n-Propyl acetate	58	2.0	8	28	W-1,3	1,3	3	D	1	3	0				
iso-Propyl acetate	40	1.8	8	28	W-1,3	1,3	3	D	1	3	0				
n-Propyl alcohol	77	2.1	13.5	44	W-1,3	1,3	3	D	1	3	0				12-61
iso-Propyl alcohol	53	2.0	12	43	W-1,3	1,3	3	D	1	3	0		SD-59		12-61
Propylene	gas	2.0	11.1	–	D-1,4	1,3	–	A	1	4	1				
Propylene dichloride	60	3.4	14.5	33	W-1,3	1,3	4	D	2	3	0			C-18	6-67
Propylene glycol	210	2.6	12.5	45	C-4	3	3	D	0	1	0				

Propylene oxide	−35	2.1	21.5	–	D-1,3	1,3	Cor.	C	2 4 2				6-59
Propyne	gas	1.7	–	–	D-1,5	1,3	–	A	2 4 2				
Pyridine	68	1.8	12.4	41	W-1,3,7	1,2,3	3	B,D	2 3 0	310			8-63
Selenium				–	W-7,6	3,2,5	–	C		578		C-56	6-59
Silica or silicates				–	C-6	3,5	1	A		531			4-58, 9-57
Silver nitrate				–	C-2,3	4,2,3	4	C	1 0 1				9-57
Sodium				–	D-1,7	5,4,2	Cor.	C	3 1 0	231	SD-47		
Sodium chlorate				–	W-2,7	5,4,2	4	C	1 0 2	371	SD-42	C-62	2-58
Sodium cyanide				–	D-6,7	5,3,2	6	–			SD-30		
Sodium dichromate				–	W-6,7,2	3,2,5	4	C			SD-46		
Sodium hydroxide				–	D-7,6	2,3	Cor.	C	3 0 1	214, 373	SD-9		
Stoddard solvent	105	0.8	5.0	20	C-1,4	1,3	3	D	0 2 0				
Styrene monomer	90	1.1	6.1	28	C-4,1,7	3,2,1	–	D	2 3 –		SD-37	C-48	10-68
Sulfur				–	C-6,7,1	3,2,1	3	A	2 1 0	612	SD-74		
Sulfur chlorides	245			–	W-7,3	3,4,5	–	C	2 1 0		SD-77		
Sulfur dioxide				–	D-3	3,5	–	C	3 0 0		SD-52	C-39	12-55
Sulfuric acid				–	D-7,3,2	3,4	Cor.	C	3 0 1	325	SD-20	C-12	9-57

* Open cup flash point.
† Number in this column refers to date of publication in *AIHA Journal*.

TABLE OF CHEMICAL HAZARDS (Concluded)

Substance	1 Flash Point (deg F)	2 Flammable or Explosive Limits (% by volume) Lower	Upper	3 Vapor Volume (cu ft per gal)	4 Severity and Type of Hazard	5 Precautions to Take	6 Oral Toxicity	7 Action on Skin	8 NFPA Health	Flammability	Reactivity	9 N.S.C. Data Sheet	MCA Data Sheet	AIA Bulletin	AIHA Hygienic Guide†
1,1,2,2,-Tetrachloroethane	nonflammable			31	D-8,3	5,3,2	4	D					SD-34	C-9	
Tetrachloroethylene	-	nonflammable		31	W-4,7	3,2	3	D					SD-24	C-11	12-65
Tetraethyl lead	185*			-	D-3,1,8	5,3,1	6	-	3	2	3				8-63
Tetrahydrofuran	6	2	11.8	40	D-1,3	1,2,3	4	D	2	3	1				6-59
Tetralin	160	0.8 (212 F)	5.0 (212 F)	24	D-3,1	3,2,1	3	D	1	2	0				
Tetramethyl lead	100*			-	D-3,1,8	5,3,1	6	-	3	3	3				
Titanium dust		0.045 oz/cu ft		-	C-6	3	1	A				485		C-75	6-59
Toluene	40	1.2	7.1	31	W-1,4,7	1,2,3	4	C	2	3	0	204	SD-63	C-77	6-57
Toluidine	185			-	D-3,8	3,2,5	4	B	3	2	0		SD-82		
Trichloroethylene	99‡ (practically nonflammable)	12.5‡	90‡	36	W-4,7	3,2	-	D	-	-	-	389	SD-14	C-8	2-64
Trinitrotoluene	-			-	D-1,3,8	5,1,3	5	-	2	4	4	314			10-64

‡ Under special conditions. See details in previous chapter.

Turpentine	95	0.8	-	18	W-4,1,7	3,1,2	3	B,C	1 3 0	367		C-54	6-67
Vinyl acetate	18	2.6	13.4	35	D-1,4,7	1,3,2	-	C	2 3 2		SD-75		
Vinyl chloride	gas	4	22	-	D-1,4	1,3	-	C	2 4 2				8-64
Vinylidene chloride monomer	5*	5.6	11.4	-	W-1,4,7	1,3,2	-	C	2 4 2			C-14	
o-Xylene	90	1.0	6.0	27	W-1,4,7	1,3,2	4	D	2 3 0	204		C-77	10-71
m-Xylene	84	1.1	7.0	27	W-1,4,7	1,3,2	4	D	2 3 0	204		C-77	10-71
p-Xylene	81	1.1	7.0	27	W-1,4,7	1,3,2	4	D	2 3 0	204		C-77	10-71
Zinc	-	0.48 oz/cu ft	-	-	W-1,6	1,3,5	3	A	0 1 1	267			8-69 (oxide)
Zirconium	-	0.19 oz/cu ft	-	-	D-1,6	1,3,5	-	-	1 4 1	382	SD-92	C-74	10-58

* Open cup flash point.
† Number in this column refers to date of publication in *AIHA Journal*.

Glossary of Common Chemical Terms

The complexity of the chemical industry and the dependence of other industries upon it have led to the development of many terms generally used throughout the industry. The following pages contain definitions of some common chemical terms. These were taken from *The Chemical Industry Facts Book*, 5*th* ed., published by the Manufacturing Chemists' Association, Inc.

CHEMICALS AND ALLIED PRODUCTS°. Overall title for the products of the entire industry. Includes basic chemical products and those manufactured by predominantly chemical processes. Three general classes of products are commonly identified with this classification: (*a*) basic chemicals, such as acids, alkalies, salts and organic chemicals; (*b*) chemical products to be used in further manufacture, as, for example, synthetic fibers, plastics materials, dry colors and pigments; and (*c*) finished chemical products for ultimate consumption, such as drugs, cosmetics, soaps and detergents, or materials or supplies for other industries, such as paints, fertilizers, pesticides and explosives. The term "chemical industry" is frequently used as an alternative for "Chemicals and allied products."

INDUSTRIAL CHEMICALS. Those inorganic and organic chemicals produced primarily for use in chemical or process application. They are not generally sold as such to the ultimate consumer, but rather to other manufacturers or formulators. They range from the relatively simple basic commodities like chlorine, caustic soda and sulfuric acid to extremely complex dyes and bulk medicinals. Other industrial chemicals are used as solvents, as intermediates in further chemical manufacture and as the basis for making synthetic rubber, plastics and man-made fibers.

CHEMICAL PROCESS INDUSTRIES. Includes all manufacturing enterprises which employ chemical change at one or more stages of their manufacturing activity. As such, they are strictly differentiated from the purely mechanical industries which merely change the shape and size of materials and size of materials and assemble them into marketable products. Some chemical process industries are: chemicals and allied products, paper and allied products, petroleum, coal products, and glass.

ACCELERATOR. Chemical additive which increases the speed of a chemical reaction. Used, for example, to improve the vulcanization of natural and synthetic rubber and latex compounds.

ACID. A compound consisting of hydrogen plus one or more other elements and which, in the presence of certain solvents or water, reacts with the production of hydrogen ions. An acid reacts with an alkali to form a salt and water. It turns litmus paper red.

ACTINIDE SERIES. Group of elements of increasing atomic number starting with actinium (atomic number 89) and extending through the recently discovered element 103. Uranium-92 and plutonium-94 are in this series.

ALIPHATIC. (Derived from Greek word for fat.) Pertaining to an open-chain carbon compound. Usually applied to petroleum products derived from a paraffin base and having a straight or branched chain, saturated or unsaturated molecular structure. Olefins such as methane and ethane are typical aliphatic hydrocarbons. See also AROMATIC; PARAFFIN.

ALKALI. A compound that has the ability to neutralize an acid and form a salt. Example: sodium hydroxide, referred to as caustic soda or lye. Used in soap manufacture and many other applications. Turns litmus paper blue. See also BASE.

ALKALINE EARTHS. Usually considered to be the oxides of alkaline earth metals: barium, calcium, strontium, beryllium and radium. Some authorities also include oxide of magnesium.

° U.S. Government groups Chemicals and Allied Products under Standard Industrial Classification 28. For list of chemicals in this group see *Standard Industrial Classification Manual*, U.S. Bureau of the Budget.

AMMONIA. NH_3. Nitrogen and hydrogen compound, a colorless gas liquefied by compression. Dissolves in water to form aqueous (household) ammonia. Synthetic ammonia is main source of nitrogen for fertilizer and chemical production.

ANILINE. $C_6H_5NH_2$. One of the most important of organic bases derived from coal. Building block for many dyes and drugs. Highly toxic.

ANTIBIOTIC. A substance produced by a microorganism usually a mold or fungus, which in dilute solutions kills other organisms, or retards or completely represses their growth, normally without harm to higher orders of life.

ANTIOXIDANT. A compound which retards deterioration by oxidation. Antioxidants for human food and animal feeds, sometimes referred to as freshness preservers, retard rancidity of fats and lessen loss of fat-soluble vitamins (A, D, E, K). Antioxidants also are added to rubber, motor lubricants and other materials to inhibit deterioration.

AROMATIC. Applied to a group of hydrocarbons and their derivatives characterized by presence of the benzene nucleus (molecular ring structure). See also ALIPHATIC; BENZENE.

ATOM. A chemical unit, the smallest part of an element, which remains unchanged during any chemical reaction yet may undergo nuclear changes (transmutations) to other atoms as in atomic fission. Believed to be made up of a complex system whose electrically charged components are in rapid motion.

BASE. A compound which reacts with an acid to form a salt. It is another term for alkali. It turns litmus paper blue.

BENZENE. C_6H_6. A major organic intermediate and solvent derived from coal or petroleum. The simplest member of the aromatic series of hydrocarbons.

BORON. Brown, amorphous, nonmetallic powder than can be fused to a brittle mass. In borax, it has been in use for hundreds of years; in recent years its uses have multiplied. Because it is an extremely high source of heat energy, boron has been the subject of extensive research in the field of high energy fuels for rockets.

CARBOHYDRATES. An abundant class of organic compounds, serving as food reserves or structural elements for plants and animals. Compounded primarily of carbon, hydrogen and oxygen, they constitute about two-thirds of the average daily adult caloric intake. Sugars, starches and plant components (cellulose) are all carbohydrates.

CATALYST. A substance which changes the speed of a chemical reaction but undergoes no permanent change itself. Usually catalysts greatly increase the reaction rate, as in conversion of petroleum to gasoline by cracking. In paint manufacture, catalysts, which hasten the film-forming, generally become part of the final product. In most uses, however, they do not, and can often be regenerated and used over again.

CELLULOSE. $(C_6H_{10}O_5)_n$. A carbohydrate which makes up the structural material of vegetable tissues and fibers. Purest forms: chemical cotton and chemical pulp. Basis of rayon, acetate and cellophane.

CHELATING AGENT or CHELATE. (Derived from Greek word *kelos* for claw.) Any compound which will inactivate a metallic ion with the formation of an inner ring structure in the molecule, the metal ion becoming a member of the ring. The original ion, thus chelated, is effectively out of action.

CHEMICAL. As broadly applied to the chemical industry, a chemical is an element (chlorine) or a compound (sodium bicarbonate) produced by chemical reaction on a large scale either for direct industrial and consumer application or for reaction with other chemicals.

CHEMICAL ENGINEERING. The American Institute of Chemical Engineers defines chemical engineering as that branch of engineering concerned with the development and application of manufacturing processes in which chemical or certain physical changes of materials are involved. These processes may usually be resolved into a coordinated series of unit physical operations and unit chemical processes. The work of the chemical engineer is concerned primarily with the design, construction, and operation of equipment and plants in which these unit operations and processes are applied. Chemistry, physics and mathematics are the underlying sciences of chemical engineering, and economics its guide in practice.

CHEMICAL REACTION. A change in the arrangement of atoms or molecules to yield substances of different composition and properties. Common types of reaction are combination, decomposition, double decomposition, replacement and double replacement.

CHEMISTRY. The science which treats of the properties of atoms and molecules and the laws which describe their combinations, reactions and behavior.

CHEMISTRY, APPLIED. The application of the science of chemistry through the use of chemical research and engineering processes to produce and utilize chemicals in commercial quantities of value to the economy.

CHEMOTHERAPY. Use of chemicals of particular molecular structure in the treatment of specific disorders on the assumption that known structures exhibit an affinity for certain parts of the cells of affected tissues or of the invading bacteria, and thereby tend to destroy or inactivate them.

CHEMURGY. The branch of applied chemistry devoted to industrial utilization of organic raw materials, especially farm products, as in use of pine-tree cellulose for rayon and paper, and soy-bean oil for paints and varnishes.

COKE OVEN CHEMICALS. Those organic compounds derived from bituminous coal during its conversion to metallurgical coke. This major source of chemical raw materials constitutes a base for thousands of chemicals.

COLLOID. Generally a liquid mixture or suspension in which the particles of suspended liquid or solid are very finely divided. Colloids do not settle out of suspension appreciably.

COMPOUND. A substance composed of two or more elements joined according to the laws of chemical combination. Each compound has its own characteristic properties different from those of its constituent elements.

CORROSION. Physical change, usually deterioration or destruction, brought about through chemical or electrochemical action as contrasted with erosion caused by mechanical action.

CRYOGENICS. The field of science dealing with the behavior of matter at very low temperatures.

DETERGENT. See SURFACE ACTIVE AGENT.

ELASTOMER. In a chemical industry sense, a synthetic polymer with rubber-like characteristics; a synthetic or natural rubber, or a soft, rubbery plastic with some degree of elasticity at room temperature.

ELEMENT. Solid, liquid or gaseous matter which cannot be further decomposed into simpler substances by chemical means. The atoms of an element may differ physically but do not differ chemically. More than 100 elements are known. See also PERIODIC TABLE.

EMULSIFIER or EMULSIFYING AGENT. A chemical that holds one insoluble liquid in suspension in another. Casein, for example, is a natural emulsifier in milk, keeping butterfat droplets dispersed.

EMULSION. A suspension, each in the other, of two or more unlike liquids which usually will not dissolve in each other.

ENERGY. The capacity for doing work. Various forms include chemical energy, nuclear energy, kinetic energy and others.

ESTERS. Organic compounds which may be made by interaction between an alcohol and an acid, and by other means. Esters are nonionic compounds, including solvents and natural fats.

ETHANE. C_2H_6. A paraffinic hydrocarbon derived from petroleum or natural gas; important for organic synthesis. Ethylene is derived from it.

ETHYL ALCOHOL OR ETHANOL. C_2H_5OH. Organic compound synthesized either from petroleum or natural gas or derived by a fermentation process. Wide use as a solvent and for chemical synthesis.

ETHYLENE. C_2H_4. Gaseous olefinic compound resulting from the cracking of either selected petroleum fractions or natural gas. Wide use as starting material for numerous chemical reactions, notably manufacture of polyethylene and related plastics materials.

ETHYLENE GLYCOL. Clear, colorless, liquid. Useful as humectant because it absorbs approximately twice its weight of water at room temperature and 100 per cent humidity. A major use: antifreeze. Among many other uses: dye solvent, solvent for resins, drugs, other organic chemicals. Highly toxic if ingested.

FERTILIZER. Plant food usually sold in mixed formula containing basic plant nutrients: compounds of nitrogen, potassium, phosphorus, sulfur, and sometimes other minerals.

FLOTATION REAGENT. Chemical used in flotation separation of minerals. Added to pulverized mixture of solids and water and oil, it causes *preferential nonwetting by water* of certain solid particles, making possible the flotation and separation of unwet particles.

GAS. A state of matter in which the material has very low density and viscosity; can expand *and contract* greatly in response to changes in temperature and pressure; easily diffuses into other gases; readily and uniformly distributes itself throughout any container.

GLYCEROL. A clear, sweet, colorless, syrup-like liquid (solid at lower temperatures) derived in various ways: by-product of soap manufacture; by fermentation or synthesis. Used in medicines, soaps, antifreeze; as a solvent, or reagent and as humectant in foods and tobacco; to make explosives.

HIGH ENERGY FUEL or (HEF). Liquid or solid chemical propellant providing more energy than conventional fuels. Various chemical fuels used in U.S. rocket and missile programs, e.g., hydrocarbons, hydrogen, hydrazine, aniline, etc.

HYDROCARBONS. Organic compounds, composed solely of carbon and hydrogen. Several hundred thousand molecular combinations of C and H are known to exist. Basic building blocks of all organic chemicals. Main chemical industry sources of hydrocarbons are petroleum, natural gas and coal.

HYDROMETALLURGY. Science of metal recovery by a process involving treatment of ores in an aqueous medium, such as acid or cyanide solution.

INHIBITOR. An agent which arrests or slows chemical action or a material used to prevent or retard rust or corrosion.

INORGANIC. Term used to designate compounds that generally do not contain carbon. Source: matter, other than vegetable or animal. Examples: sulfuric acid and salt. Exceptions are carbon monoxide, carbon dioxide.

INTERMEDIATE. A chemical formed as a "middle-step" in a series of chemical reactions, especially in the manufacture of organic dyes and pigments. In many cases, it may be isolated and used to form a variety of desired products. In other cases, the intermediate may be unstable or used up at once.

1235

ION. An electrically charged atom or group of atoms; electrically charged molecules in gases.

ION-EXCHANGE RESIN. Synthetic resins containing active groups that give the resin the property of combining with or exchanging ions between the resin and a solution. See also ION.

ISOTOPE. One of two or more atomic species of an element differing in atomic *weight* but having the same atomic *number*. See also RADIOISOTOPE.

LATEX. Original meaning: milky extract from rubber tree, containing about 35 per cent rubber hydrocarbon; the balance is water, proteins and sugars. This word also is applied to water emulsions of synthetic rubbers or resins. In emulsion paints, the film-forming resin is in the form of latex.

LIQUEFIED PETROLEUM GAS. A compressed or liquefied gas usually comprised of propane, some butane and lesser quantities of other light hydrocarbons and impurities; obtained as a by-product in petroleum refining. Used chiefly as a fuel, and in chemical synthesis.

LIQUID. A state of matter in which the substance is a formless fluid that flows in accord with the law of gravity.

MATTER. Constituents of any substance.

METHANE. CH_4. The simplest saturated hydrocarbon, chief component of most natural gas, often called "marsh gas," as it is given off by decaying vegetation. Chemical raw material.

METHYL ALCOHOL, METHANOL or WOOD ALCOHOL. CH_3OH. Organic compound important in chemical synthesis; also used in denaturing alcohol, as a solvent and in many other uses. Highly toxic by ingestion.

MIXTURE. A combination of two or more substances which may be separated by mechanical means. The components may not be uniformly dispersed. See also SOLUTION.

MOLECULE. Generally the smallest identifiable particle of a substance that can exist resulting from the chemical combination of two or more like or unlike atoms. Exception: in a few gases and metals the molecule and the atom are one.

MONOMER. A compound of relatively low molecular weight which, under certain conditions, either alone or with another monomer, forms various types and lengths of molecular chains called polymers or copolymers of high molecular weight. Example: Styrene is a monomer which polymerizes readily to form polystyrene. See also POLYMER.

NAPHTHALENE. $C_{10}H_8$. A white crystalline hydrocarbon which is obtained from coal tar or petroleum fractions. Used as a moth repellent and a basic material in the manufacture of dyestuffs, synthetic resins, lubricants and other products.

NATURAL GAS. A combustible gas composed largely of methane and other hydrocarbons with variable amounts of nitrogen and noncombustible gases; obtained from natural earth fissures or from driven wells. Used as a fuel in the manufacture of carbon black, and in chemical synthesis of many products. Major source of hydrogen for manufacture of ammonia.

NITER. KNO_3. A white salt widely distributed in nature and formed in soils from nitrogenous organic bodies by the action of bacteria. Used in making fertilizer, medicinals and other products.

NITRIC ACID. HNO_3. A colorless to yellowish fuming liquid with powerful corrosive properties. Manufactured by several methods including oxidation of ammonia. Used in manufacture of nitrogen fertilizers, in organic synthesis, in etching metals, in ore flotation, in the manufacture of explosives, medicinals and other products.

NITROGEN FIXATION. Chemical combination or fixation of atmospheric nitrogen with hydrogen as in synthesis of ammonia. Fixation of nitrogen in soil is done by bacteria. Provides industrial and agricultural source of nitrogen.

1236

NITROUS OXIDE. N_2O. A colorless gas of sweetish odor and taste; used as a general anesthetic. Also called "laughing gas."

OLEFINS. A class of unsaturated hydrocarbons characterized by relatively great chemical activity. Obtained from petroleum and natural gas. Examples: butene, ethylene and propylene. Generalized formula: C_nH_{2n}.

ORGANIC. Term used to designate chemicals that contain carbon. To date nearly one million organic compounds have been synthesized or isolated. Many occur in nature; others are produced by chemical synthesis. See also INORGANIC.

OXIDATION. Process of combining oxygen with some other substance; a chemical change in which an atom loses one or more electrons. Opposite of reduction. See also REDUCTION.

PARAFFINS, PARAFFIN SERIES. (From *parum affinis* — small affinity.) Those straight or branched chain hydrocarbon components of crude oil and natural gas whose molecules are saturated (i.e., carbon atoms attached to each other by single bonds) and therefore very stable. Examples: methane and ethane. Generalized formula: C_nH_{2n+2}.

PERIODIC TABLE. Systematic classification of the elements according to atomic numbers (nearly the same order as by atomic weights) and by physical and chemical properties.

PESTICIDES. General term for that group of chemicals used to control or kill such pests as rats, insects, fungi, bacteria, weeds, etc., that prey on man or agricultural products. Among these are insecticides, herbicides, fungicides, rodenticides, miticides, fumigants and repellents.

PHARMACEUTICALS. That group of drugs and related chemicals reaching the public primarily through drug suppliers. In Government reports, category includes not only such medicinals as aspirin and antibiotics but also such nutriments as vitamins and amino acids for both human and animal use.

PH. Means used to express the degree of acidity or alkalinity of a solution with neutrality indicated as 7.

PHENOL. C_6H_5OH. Popularly known as carbolic acid. Important chemical intermediate and base for plastics, pharmaceuticals, explosives, antiseptics, many other end products.

PHOTOSYNTHESIS. The process by which plants produce carbohydrates from carbon dioxide and water when the green tissues (chlorophyl) are exposed to sunlight. In reducing the carbon dioxide, oxygen is released. Were it not for this process, life on the earth would be impossible.

PLASTICIZERS. Organic chemicals used in modifying plastics, synthetic rubber and similar materials to facilitate compounding and processing, and to impart flexibility to the end product.

PLASTICS. Officially defined as "any one of a large and varied group of materials which consists of, or contains as an essential ingredient, an organic substance of large molecular weight; and which, while solid in the finished state, at some stage in its manufacture has been or can be formed (cast, calendered, extruded, molded, etc.) into various shapes by flow — usually through application of heat and pressure singly or together." Each plastic has individual physical, chemical and electrical properties. Two basic types: thermosetting (irreversibly rigid) and thermoplastic (reversibly rigid). Prior to compounding and processing, plastics often are referred to as (synthetic) resins. Final form may be as film, sheet, solid, or foam; flexible or rigid. See also RESIN.

POLYMER. A high molecular weight material formed by the joining together of many simple molecules (monomers). There may be hundreds or even thousands of the original molecules linked end to end and often crosslinked. Rubber and cellulose are naturally occurring polymers. Most resins are chemically produced polymers. See also MONOMER.

PROTECTIVE COATING. A thin layer of

metal or organic material as paint applied to a surface primarily to protect it from oxidation, weathering and corrosion.

PROTEINS. Large molecules found in the cells of all animal and vegetable matter and containing carbon, hydrogen, nitrogen and oxygen, sometimes sulfur and phosphorus. Proteins are essential to life and growth. The fundamental structural units of proteins are amino acids.

RADIOACTIVITY. Emission of energy in the form of alpha, beta, or gamma radiation from the nucleus of an atom. Always involves change of one kind of atom into a different kind. A few elements, such as radium, are naturally radioactive. Other radioactive forms are induced. See also ENERGY; RADIOISOTOPE.

RADIOCHEMICAL. Any compound or mixture containing a sufficient portion of radioactive elements to be detected by a Geiger counter.

RADIOISOTOPE. An isotopic form of an element that exhibits radioactivity, whether natural as in radium and uranium, or produced by fission or other induced nuclear changes. The latter are used in biological tracer work, industrial control operations, and diagnosis and treatment of certain diseases. See also ISOTOPE.

RARE EARTHS. Originally those elements on the Periodic Table with atomic numbers 57 through 71. Often included are numbers 39 and less frequently 21 and 90. Variety of emerging uses include manufacture of special steels and glasses.

REAGENT. Any substance used in a chemical reaction to produce another substance or to detect its composition.

REDUCTION. Addition of one or more electrons to an atom through chemical change.

RESIN. A solid or semisolid amorphous (noncrystalline) organic compound or mixture of such compounds with no definite melting point and no tendency to crystallize. Resins may be of vegetable (gum arabic),

animal (shellac), or synthetic origin (celluloid). There are many types, each with distinctive physical and chemical properties. Some types of resins may be molded, cast or extruded. Others are used as adhesives, in the treatment of textiles and paper, and as protective coatings. Still others are rolled or extruded into continuous sheets and films of various thicknesses. See also PLASTICS.

SALTS. A product of the reaction between an acid and a base. Example: Table salt is a compound of sodium and chlorine. It can be made by reacting sodium hydroxide with hydrochloric acid.

SATURATION. The point at which the maximum amount of matter can be held dissolved at a given temperature in a solution.

SEQUESTRANTS. Chelates used to deactivate undesirable properties of metal ions without the necessity for removing these ions from solution. Sequestrants find many uses, including applications as anti-gumming agents in gasoline, antioxidants in rubber, and as rancidity retardants in edible fats and oils. See also CHELATES.

SILICONES. Unique group of compounds made by molecular combination of the element silicon or certain of its compounds with organic chemicals. Produced in variety of forms, including silicone fluids, resins and rubber. Silicones have special properties, such as water repellency, wide temperature resistance, high durability and great dielectric strength.

SOAP. Ordinarily a metal salt of a fatty acid, usually sodium stearate, sodium oleate, sodium palmitate, or some combination of these.

SOIL CONDITIONER. A synthetic chemical or natural material added in small quantities to improve the structure of soil.

SOLUTION. Mixture in which the components lose their identities and are uniformly dispersed. All solutions are composed of a solvent (water or other fluid) and the substance dissolved called the "solute." Air is a solution of oxygen and nitrogen. A true solution is homogeneous as salt in water. See also MIXTURE.

SOLVENT. A substance most commonly water but often an organic compound which dissolves another substance.

SORBITOL. A crystalline "higher alcohol" produced by the reduction of sugar with hydrogen. Used in the manufacture of paper, tobacco, textiles, adhesives, pharmaceuticals and cosmetics, and many other products.

SUBSTANCE. Any chemical entity, either atomic or molecular.

SURFACE-ACTIVE AGENT or SURFACTANT. Any of a group of compounds added to a liquid to modify surface or interfacial tension. In synthetic detergents, which is the best known use of surface active agents, reduction of interfacial tension provides cleansing action.

SURFACE COATING. Term used to include paint, lacquer, varnish and other chemical compositions used for protecting and/or decorating surfaces. See also PROTECTIVE COATING.

SYNERGISM. Cooperative action of substances whose total effect is greater than the sum of their separate effects.

SYNTHESIS. The reaction or series of reactions by which a complex compound is obtained from simpler compounds or elements.

SYNTHETIC. (From Greek word *synthetikos*—that which is put together.) "Manmade 'synthetic' should not be thought of as a substitute for the natural," states *Encyclopedia of the Chemical Process Industries;* it adds, "Synthetic chemicals are frequently more pure and uniform than those obtained naturally." Classic example: synthetic indigo.

SYNTHETIC DETERGENTS. Chemically tailored cleaning agents soluble in water or other solvents. Originally developed as soap substitutes. Because they do not form insoluble precipitates, they are especially valuable in hard water. They may be com-posed of surface active agents alone, but generally are combinations of surface active agents and other substances, such as complex phosphates, to enhance detergency. See also SURFACE-ACTIVE AGENT.

SYNTHETIC RUBBER. Man-made polymer with rubber-like properties. Various types have varying composition and properties. Major types designated as S-type, butyl, neoprene (chloroprene polymers) and N-type. Several synthetics duplicate the chemical structure of natural rubber.

TALL OIL. (Name derived from Swedish word *tallolja;* material first investigated in Sweden—not synonymous with U.S. pine oil.) Natural mixture of rosin acids, fatty acids, sterols, high molecular weight alcohols and other materials, derived primarily from waste liquors of sulfate wood pulp manufacture. Dark brown, viscous, oily liquid often called liquid rosin.

TAR CRUDE. Organic raw material derived from distillation of coal tar and used for chemical manufacture.

THERMOPLASTIC PLASTICS. Those that can repeatedly melt or that soften with heat and harden on cooling. Examples: vinyls, acrylics and polyethylene.

THERMOSETTING PLASTICS. Those that are heat-set in their final processing to a permanently hard state. Examples: phenolics, ureas and melamines.

TOLUENE. $C_6H_5CH_3$. Hydrocarbon derived mainly from petroleum but also from coal. Source of TNT, lacquers, saccharin and many other chemicals.

VALENCE. A number indicating the capacity of an atom and certain groups of atoms to hold others in combination. The term also is used in more complex senses.

VAT DYES. Water insoluble, complex coal tar dyes that can be chemically reduced in a heated solution to a soluble form that will impregnate fibers. Subsequent oxidation then produces insoluble color dyestuffs which

1239

are remarkably fast to washing, light and chemicals.

VULCANIZATION. Process of combining rubber (natural, synthetic or latex) with sulfur and accelerators in presence of zinc oxide under heat and usually pressure in order to change the material permanently from a thermoplastic to a thermosetting composition, or from a plastic to an elastic condition. Strength, elasticity and abrasion resistance also are improved.

Noise and Hearing Conservation

Chapter
40

Properties of Noise   1242
The decibel . . . Frequency

Noise Effects on Man   1245
Mechanics of hearing . . . Loss of hearing

Industrial Noise Exposure   1247
Noise measurement . . . Damage risk criteria

Noise Control   1251

Hearing Evaluation   1252
Audiometric testing facilities

Summary   1253

References   1253

Hearing loss is an affliction which has existed throughout the history of mankind. It occurs in all walks of life and is due to many causes. Until the development of the audiometer in recent years, there was no means of measuring the degree of hearing loss with appreciable accuracy. Now partial losses are easily measurable by use of commercially available instruments.

The growing interest of industrial management in hearing loss due to noise has been stimulated by: (*a*) the trend toward coverage of partial loss of hearing under state workmen's compensation laws, and (*b*) the passage of noise regulations by the Occupational Safety and Health Administration in the U.S. Department of Labor. These developments have had considerable impact upon the operations and profitability of many companies.

That workers in noisy occupations develop a greater than average degree of hearing loss has been known for more than 100 years. Prior to 1948, hearing loss was not regarded as a significant factor in workmen's compensation. Some claims occurred, but these were primarily due to traumatic injuries from such causes as blasts, concussions, blows to the head, and foreign objects or infections in the ears. Since 1948, a number of states have passed hearing loss legislation or revised existing laws to include noise-induced hearing loss as a basis for workmen's compensation benefits. Now it is reported that there are at least 16 states with workmen's compensation laws providing compensation in varying degrees for partial hearing loss due to industrial noise exposure.

What is noise? Technically defined, *noise is unwanted sound.* It is a form of vibration which may be conducted through solids, liquids, or gases. It is a form of energy in air, consisting of invisible vibrations that enter the ear and create a sensation. Almost every conceivable kind of sound can, at one time, be welcome—at another time abhorred. It follows that before attacking a noise problem it must be asked: Who doesn't want it? When? And why? When these questions are answered, steps can be taken to solve the problem.

In the operation of industrial machines, unbalanced forces are applied to parts of the machine and cause displacement or movement of these parts, or the whole machine. These displacements or motions are vibrations which set up air-borne sound waves. When people come into direct contact with noise, undesired effects may result.

Properties of Noise

Some of the properties of noise are intensity (or pressure), frequency, and duration. All of these are important factors in evaluating the effects of noise on man's hearing. The louder the noise, the higher its intensity; high-frequency (high-pitch) noises are more damaging to hearing than low-frequency noises; and the longer the noise exposure, the greater the damage to the human hearing mechanism. By examining these properties of noise carefully, one immediately recognizes approaches to noise control or to hearing conservation (Fig. 40–1).

Sound pressure or intensity follows the inverse-square law: that is, as the distance from the source increases the sound level decreases as the square of the distance. High-frequency sound waves are more readily absorbed by fibrous, spongy, and soft materials of construction. Finally, by limiting exposure time, workers' hearing conservation can be accomplished.

The decibel

The ear is a remarkable organ, responding to sound pressures from 0.0002 to 2000 dynes per square centimeter (Table 40-A). To avoid working with unwieldy large numbers in evaluating sound intensity, however, a logarithmic scale is used with the decibel as the unit of measure. Because decibels are logarithmic units, they cannot be added or subtracted arithmetically. In fact, if the intensity of a sound is doubled, there will be a corresponding increase of only three decibels, not double the number. For example, if one machine caused an exposure of 90 dB, a second identical machine placed adjacent to the first would result in a noise exposure of 93 dB, not 180 dB. On the decibel scale, zero (decibels) is the threshold of hearing, and 120 decibels is the threshold of pain.

Frequency

Frequency is the number of variations in

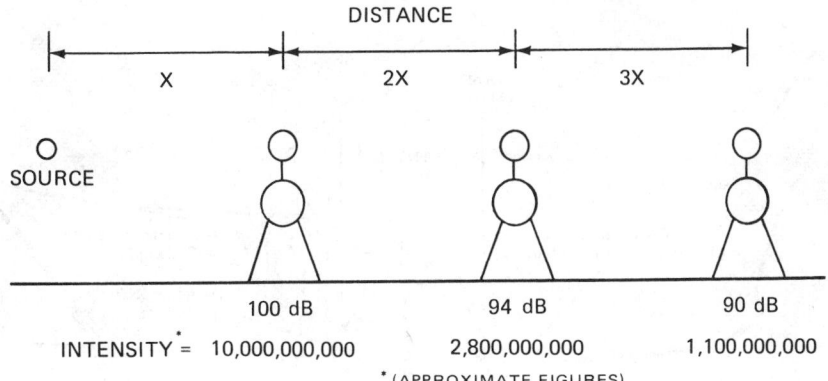

NOISE CONTROL BY DISTANCE

Fɪɢ. 40–1.—As the distance from the noise source increases, the pressure (or intensity) of the noise decreases faster than its sound level. The "intensity" shown here has absolutely no relationship with the sound pressure figures shown below.

TABLE 40-A

SOUND PRESSURE AND DECIBEL VALUES FOR SOME TYPICAL SOUNDS

Sound Pressure (microbars)	Overall Sound Pressure Level (dB re 0.0002 microbar)	Example
0.0002	0	Threshold of hearing
0.00063	10	
0.002	20	Studio for sound pictures
0.0063	30	Soft whisper (5 feet)
0.02	40	Quiet office
		Audiometric testing booth
0.063	50	Average residence
		Large office
0.2	60	Conversational speech (3 feet)
0.63	70	Freight train (100 feet)
1.0	74	Average automobile (30 feet)
2.0	80	Very noisy restaurant
		Average factory
6.3	90	Subway
		Printing press plant
20	100	Looms in textile mill
		Electric furnace area
63	110	Woodworking
		Casting shakeout area
200	120	Hydraulic press
		50 hp siren (100 feet)
2000	140	Jet plane
200,000	180	Rocket launching pad

Note that doubling any *sound pressure* corresponds to an increase of 6 dB in the sound pressure level. A change of sound pressure by a factor of 10 corresponds to a change in sound pressure level of 20 dB.

1243

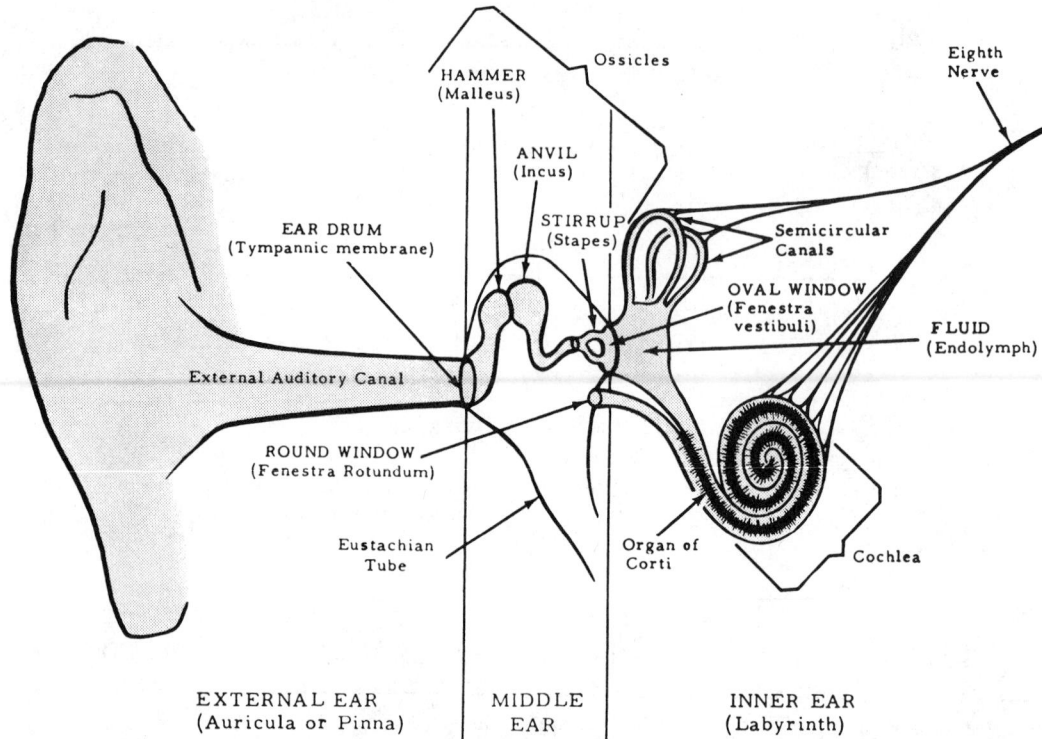

FIG. 40–2.—How the ear hears. Wave motions in the air set up sympathetic vibrations which are transmitted by the eardrum and the three bones in the middle ear to the fluid-filled chamber of the inner ear. In the process, the relatively large but feeble air-induced vibrations of the eardrum are converted to much smaller but more powerful mechanical vibrations by the three ossicles, and finally into even stronger fluid vibrations. The wave motion in the fluid is sensed by the nerves in the cochlea, which transmit neural messages to the brain.

Courtesy American Foundrymen's Society.

sound pressure per unit time, usually expressed in hertz (cycles per second). The sounds usually encountered in industry consist of many frequencies depending on the size, shape, and actions of the noise source. Each frequency of a given sound contributes to the total sound pressure level or intensity.

The normal young adult can hear sounds over a wide range of frequencies, from about 20 to 15,000 Hz. The ability to hear middle and high-frequency sounds usually decreases with age. Those persons exposed to high noise levels for extended periods experience hearing losses over and above that due to advancing age.

The human ear responds differently to different frequencies. Because different sound frequencies may have differing psychological and physiological effects, it is important that knowledge of the frequency components of a noise be obtained. The amount of information required varies with the problem and determines the technique to be used in the frequency analysis. In detailed industrial hygiene work, it is customary to separate the total noise into eight frequency ranges called octave bands. The octave bands most commonly used in noise measurement start with the 37.5 to 75 Hz band through the 4800 to 9600 Hz band. In each octave the highest frequency is twice that of the lowest.

Threshold Limit Values for noise are based

on a single overall decibel measurement on the "A" scale of a standard sound level meter.* This permits the rapid evaluation of industrial noise problems, but it does not eliminate the need for detailed studies of hazardous noise sources and for the collection of data for noise control. TLV's are discussed later in this chapter under Damage Risk Criteria.

Noise Effects on Man

The effects of noise on man include:

● PSYCHOLOGICAL EFFECTS—Noise can startle, annoy, and disrupt concentration or sleep.

● INTERFERENCE WITH COMMUNICATIONS BY SPEECH—As a consequence, interference with job performance and safety.

● PHYSIOLOGICAL EFFECTS—Noise-induced loss of hearing, or aural pain, nausea, and reduced muscular control (when the exposure is severe).

Mechanics of hearing

The human ear (see Fig. 40–2) is composed of three major sections: the external, middle, and inner ears. Each of these has a distinct function in the hearing process. The external ear captures and funnels the sound waves to the middle ear where they strike the ear drum; this sets in motion the process of hearing.

The middle ear consists of the ear drum and the ossicle structures just inside. The space of the middle ear is filled with air. There is a chain of small bones in the middle ear with one end of the chain resting against the ear drum and the other end connected to the inner ear.

The inner ear consists of a spiral tube filled with fluid. The spiral tube contains the organ of Corti, which consists of many sensory cells with delicate hairs or hair cells projecting into the fluid.

As the ear drum vibrates, the chain of bones comprising the middle ear is set in motion. These motions, in turn, vibrate the fluid of the inner ear. When the fluid vibrates, the hair cells are stimulated, sending nerve currents or impulses to the brain. Exposure to excessive noise for extended periods over-activates the hair and nerve cells, causing injury

or destruction, resulting in a permanent loss of hearing. Hair cell deterioration also increases with advancing age, thus decreasing the hearing acuity of older age groups.

Loss of hearing

Loss of hearing may be defined as any reduction in the ability to hear from that of a "normal" person. Such loss may be classified into two general categories:

● TEMPORARY HEARING LOSS from exposure to loud noises for a few hours, with normal hearing usually returning after a rest period. The recovery period may be minutes, hours, days, or even longer, depending upon the individual and the severity of the exposure.

● PERMANENT HEARING LOSS may occur as a result of the aging process, diseases, injury, or exposure to loud noises for extended periods of time. The hearing loss associated with exposure to industrial noise is commonly referred to as "acoustic trauma." This type of hearing loss is the result of nerve or hair cell destruction in the hearing organ and it is not reversible. However, such hearing losses usually are only partial. Total hearing loss is most frequently associated with disease or traumatic injury.

The most common levels of industrial noise exposures are well below the pain threshold. There is a wide range of noise levels and frequencies to which long-time exposure may cause a slowly developing impairment of hearing. The part of the inner ear that may be damaged depends on the frequency components of the noise field that are present at the levels of exposure. Individual susceptibility of the exposed worker may also be a factor. Noise-induced permanent hearing loss is first evident in a reduction in the ability to hear high-frequency sounds. As the exposure continues, the reduction progresses to the lower frequency sounds in the speech range. Exposure to noise that will produce this slow damage may sometimes be accompanied by other signs, such as a sensation of tingling or ringing in the ear when one moves out of the noise field.

---

* A-scale readings correspond generally to the response of the human ear.

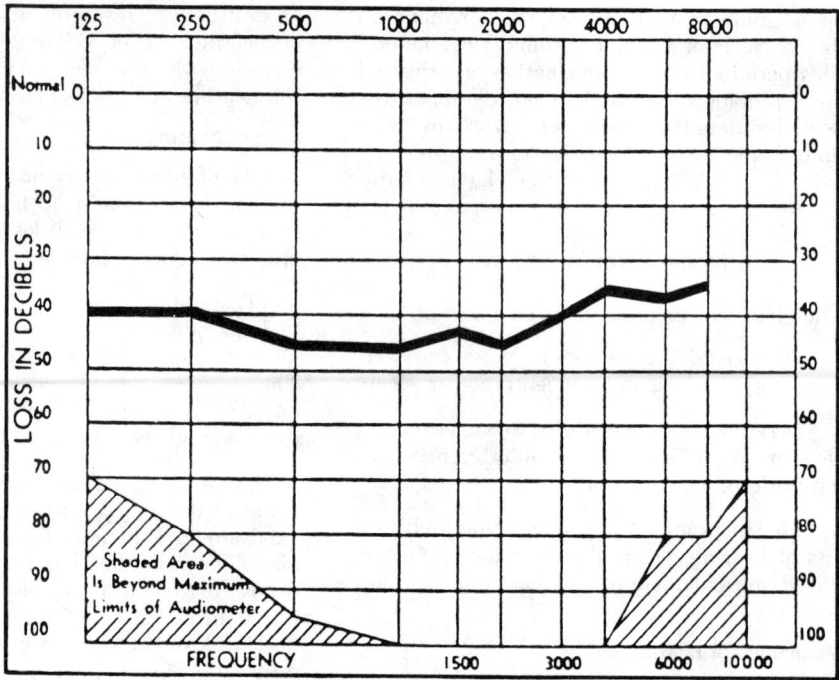

Fɪg. 40–3a.—Audiogram shows conductive or middle ear hearing loss. Characterized by a relatively flat curve, this loss is not caused by noise exposure.

From Background for Loss of Hearing Claims.

Present day evidence indicates that any permanent effect on the hearing organ is unlikely unless it is preceded by a temporary threshold shift of the hearing level. This shift can be detected by an appropriate program of monitoring audiometry, and this information can serve to warn management personnel that there is a risk of permanent damage to the hearing organ. There are few audiometric testing programs in industry today, however, that are sufficiently effective to fulfill this objective.

Effects similar to, or perhaps identical with, those produced by industrial noise are also produced by other agents, or result from other causes. Certain drugs used for the treatment of disease may reduce the sensitivity of the ear. These effects are frequently indistinguishable from those resulting from long-time exposure to industrial noise.

Another complicating factor in the industrial noise picture is the loss in hearing sensitivity that takes place as people grow older. Tech-

nically, the decrease in hearing sensitivity accompanying the aging process is known as presbycusis. Hearing loss due to acoustic trauma and presbycusis are both due to nerve or hair cell destruction or deterioration, and both are permanent. Hearing loss due to these causes cannot be distinguished from the other by audiometric means.

In general, hearing losses are of two major types: conductive and sensori-neural. The accompanying typical audiograms (Fig. 40–3) show the nature of these types. The conductive type is not caused by sustained noise exposure. The sensori-neural type may be caused by noise exposure as well as other causes, including presbycusis.

There are numerous infectious diseases (for example, frequently recurring or persistent colds) that may produce loss of hearing, which may be confused with those resulting from exposure to industrial noise. It is important that otologists and industrial physicians give these other factors appropriate

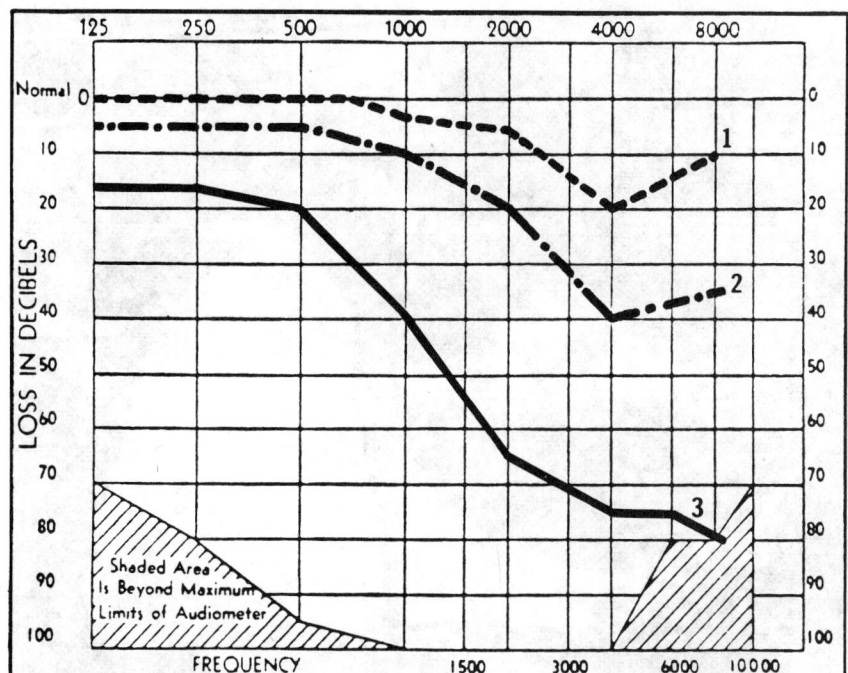

FIG. 40–3b.—Sensori-neural hearing losses of the type produced by noise or other causes. Curve 1—early; curve 2—intermediate; curve 3—advanced (as shown, this curve might include some involvement of presbycusis).

consideration when evaluating the effects of industrial noise on man.

Management personnel can make themselves more aware of the significance of these non-noise-induced hearing losses by establishing a regular program of preplacement audiometry. This program will inform them of the hearing status of the individual at the time he is assigned to a job in which he may be exposed to industrial noise under conditions that could induce hearing impairment.

Industrial Noise Exposure

As man moves through his industrial, home, and recreational environments, he is exposed to many noise sources. He may be exposed to car or other transportation noise during the day. He may be exposed to different noises where he works. He may be exposed at home to noises from radios, or a home workshop. Jet planes or construction in the vicinity of the home may be considered, in some cases, a noise exposure. Thus, it can be seen that man in his many environments is exposed to different noises having varied frequencies and intensities.

Noise is considered synonymous with industry in general and with certain specific industries in particular. A noise survey in 1953 of 600 typical industrial operations indicated that 75 percent of the measurements exceeded 90 decibels overall—on the "C" scale. For a number of years, this noise level was used by some organizations as the level that should not be exceeded for hearing conservation.

Noise measurement

The purpose of making noise measurements is to evaluate exposures in relation to speech interference, comfort or hearing loss, and to collect information for use in control.

The measurement of noise may be accomplished with a sound level meter, or a sound level meter in combination with an

1247

FIG. 40-4.—Using a sound survey meter, a safety professional can measure the over-all noise level at various work areas to determine the intensity of exposure to determine if personal protection is needed.

octave band analyzer. The sound level meter records the overall level of the noise produced without regard to the frequencies incorporated in the total makeup of that noise. The octave band analyzer, when used in conjunction with the sound level meter, measures noise levels in various octave bands (frequency bands) over the audible range of the human ear (Fig. 40-4).

More sophisticated equipment is available for noise measurement, but these two instruments can serve most of industry's needs for noise hazard evaluation and for the accumulation of information on which to establish hearing conservation programs.

The evaluation of industrial noise involves more than the simple reading of an instrument taken in the area of a suspected noisy machine. Many factors enter into the making of a noise study. Some of the factors to be considered are—type of noise produced (steady or intermittent), background noise; location of the exposed workers; and time workers spend in the noise exposure. Further, the evaluation of noise sources, for the purpose of considering control, requires the application of different engineering principles. Much has been written on noise measurement and a useful reference on this subject is the "Industrial Noise Manual" published by the American Industrial Hygiene Association.

Damage risk criteria

If the ear is subjected to high levels of noise for a sufficient period of time, some loss of hearing may occur. A number of factors can influence the effect of the noise exposure. Among these are:

1. Variation in individual susceptibility

			AVERAGE HEARING LEVEL (1951 ASA) 500, 1000 and 2000 C/S IN THE BETTER EAR*			
dB -10 0	CLASS	DEGREE OF HANDICAP			ABILITY TO UNDERSTAND ORDINARY SPEECH	
			AT LEAST	LESS THAN	No significant difficulty with faint speech.	Audiometer Zero (1951 ASA)
	A	NOT SIGNIFICANT		15		
15						"Low Fence"
	B	SLIGHT	15	30	Difficulty only with faint speech.	
30						
	C	MILD	30	45	Frequent difficulty with normal speech.	
45						
	D	MARKED	45	60	Frequent difficulty with loud speech.	Educational Deafness
60						
	E	SEVERE	60	80	Can understand only shouted or amplified speech.	
80						"High Fence"
	F	EXTREME	80		Usually cannot understand even amplified speech.	Usual Limit of Audiometer Output
100						

CLASSES OF HEARING HANDICAP

* If the average of the poorer ear is 25 dB or more greater than that for the better ear, add 5 dB to the average for the better ear.

FIG. 40–5.—Table for determining class of hearing handicap. (C/S is hertz.)

From Guide for Conservation of Hearing in Noise, *American Academy of Opthalmology and Otolaryngology, 1964.*

2. The total energy of the sound

3. The frequency distribution of the sound

4. Other characteristics of the noise exposure, such as whether it is continuous, intermittent, or made up of a series of impacts

5. The total daily time of exposure, and

6. The length of employment in the noise environment

It has been determined that hearing losses of different magnitudes result in certain handicaps. These are shown in Fig. 40–5.

Because of the complex relationships of noise and exposure time to threshold shift (reduction in hearing level) and the many possible contributory causes, establishments of criteria for protecting workers against hearing loss was delayed.

A most recent effort to establish the basis for reliable noise criteria is the work of the Intersociety Committee on Guidelines for Noise Exposure Control. A significant part of this report is shown in Fig. 40–6. This graph relates the incidence of significant hearing loss to age and the magnitude of noise exposure over a working lifetime.

Without attempting to explain the full significance of the graph, it can be stated that 20 percent of the general population at age 50 to 59 will have hearing losses of 15 decibels or more, without any industrial noise exposure. Groups of workers exposed to steady-state industrial noises over a working lifetime will show increased incidence of hearing loss.

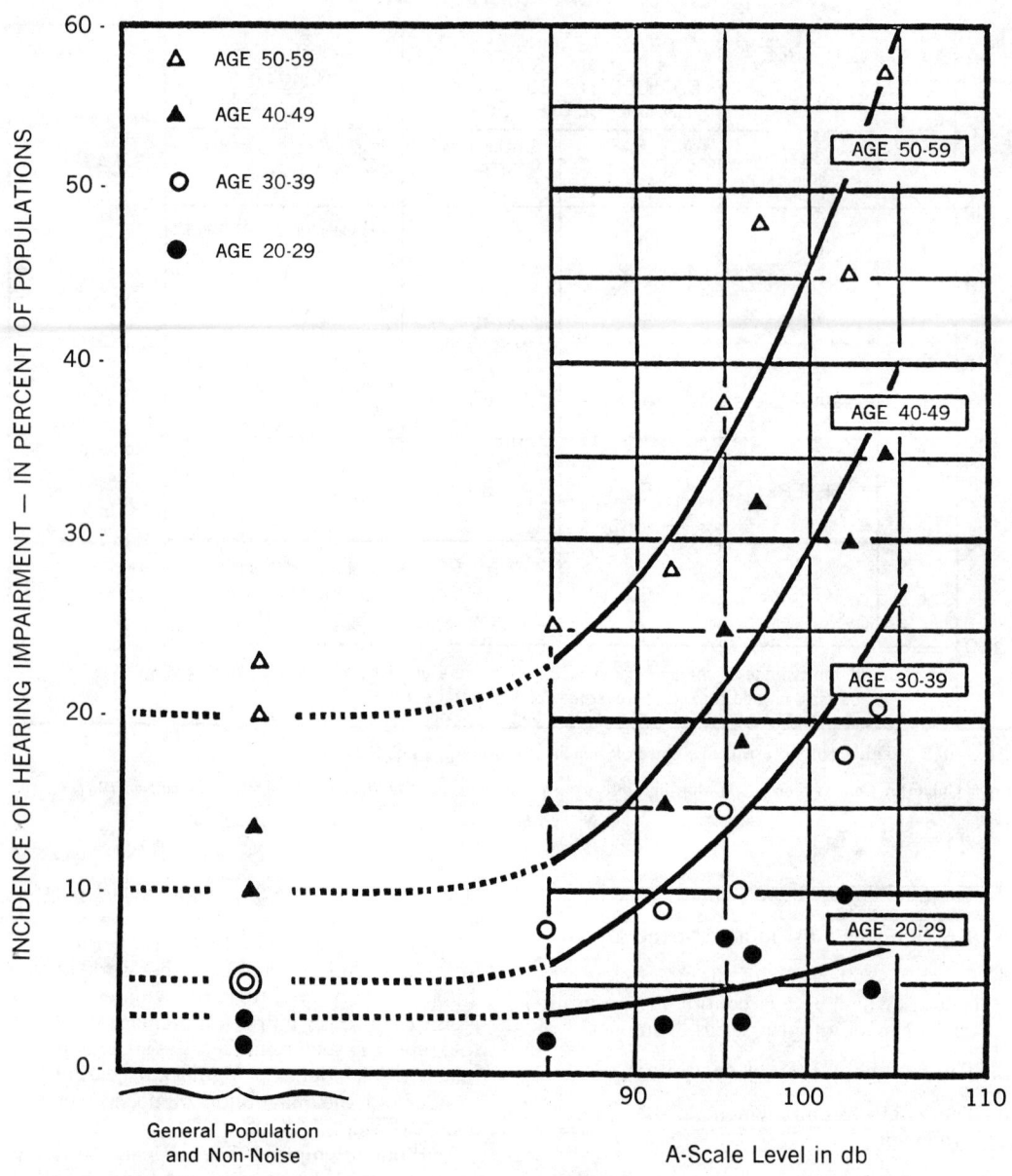

FIG. 40–6.—Incidence of hearing impairment in the general population and in selected populations by age groups and by occupational noise exposure. Hearing impairment may be defined as the average hearing threshold level in excess of 15 dB at 500, 1000, and 2000 Hz.

From American Industrial Hygiene Association Journal, *October 1967.*

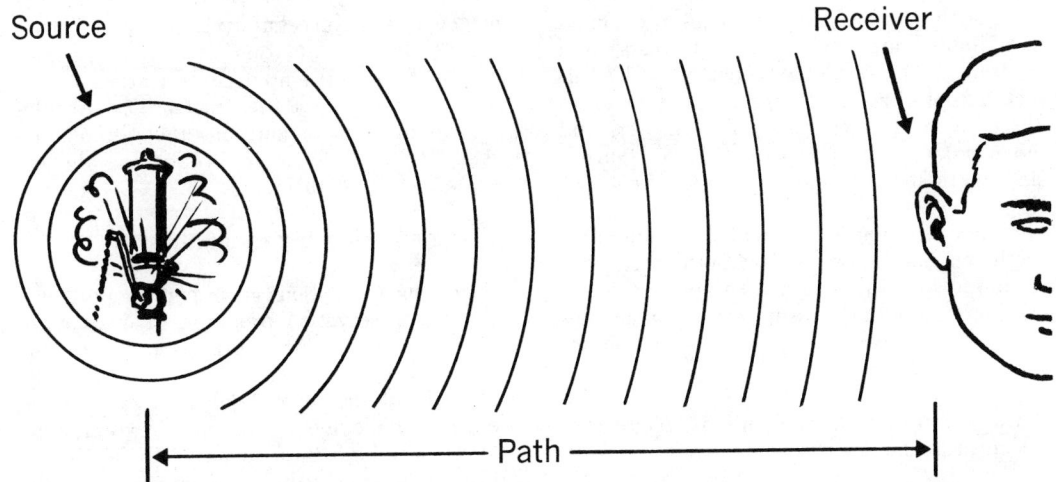

Source

Receiver

Path

FIG. 40–7.—Three components of noise control: source, path, and receiver.

For example exposure to 90 dB on the "A" scale of the sound level meter (90 dBA) will result in significant hearing losses to 27 percent of the group. If the working lifetime exposure is to 95 dBA, 36 percent of the group will show significant loss.

Essentially, this graph supplies industry and others with information on which to evaluate the risk of developing compensable hearing loss among groups of workers exposed to noises of different magnitudes.

Since publication of the Intersociety Committee report, both the American Conference of Governmental Industrial Hygienists and the Occupational Safety and Health Administration in the U.S. Department of Labor have adopted Threshold Limit Values for Noise. The ACGIH TLV's are shown in Table 37-G, on pages 1104–1105.

Be sure to check the regulations currently in effect. See text starting on page 1484.

Noise Control

Every noise problem breaks down into three component parts: (*a*) a source radiating sound energy; (*b*) a path along which the sound energy travels; (*c*) a receiver such as the human ear. (See Fig. 40–7.)

• The control of noise at the source is an engineering problem which requires modifica-

tion or redesigning of the source. The modification of compressed air jets for parts ejection, to reduce noise by altering the jet flow, is an example of this type of control.

• Noise reduction along the path can be accomplished in many ways—by shielding or enclosing the source, by increasing the distance between source and receiver, or by placing a shield between the source and the receiver.

• Noise control at the receiver, when the receiver is a person, can be accomplished effectively in a number of ways. Some common approaches included placing the worker in a booth or enclosure, the use of ear protectors (see Chapter 19, "Personal Protective Equipment"), and regulation of exposure time.

The "system" approach to noise problem analysis and control will assist in understanding both the problem and the changes that will be necessary for noise reduction. If each part of the system—source, path and receiver—is put in its proper perspective, the over-all problem will be greatly simplified. Translating these principles into practical terms, the following are specific examples of controlling industrial noise exposures.

Isolation of the worker. In situations where the number of operators is small and the process is such that the work task can be confined to a limited area, isolation of workers in a separate noise insulated room provides effective control. This control method has been observed in chemical processing, power, and metal products plants with noise reduction inside the rooms of as much as 30 decibels.

Machine insulation. Machines mounted directly on floors and walls vibrate them, resulting in sound radiation. The proper use of machine mountings insulates the machines and reduces the transmission of sound vibrations to the floors and walls.

Control of noise by absorption. Noise produced at a sound source travels out in all directions. When the sound waves encounter other machinery or walls, they are reflected. Thus, the total noise exposure within the room is equal to the sum of the direct and reflected noise. The application of sound absorption material to walls can reduce the noise exposure in the room. However, this method has limited industrial application as the absorbing material has no effect upon the direct noise from the source.

Substitution of less-noisy machines. This approach to noise control may have limited application. However, there are certain areas in which substitution has potential application. These include "squeeze" type equipment in the place of drop hammers, welding in place of riveting, and chemical cleaning of metal instead of high speed polishing and grinding.

Reduction of exposure time. Experience has shown that limiting the total daily exposure reduces the noise hazard. This principle is illustrated in the Threshold Limit Values listed earlier.

Personal protection against noise. There are many operations in industry that cannot be quieted by engineering methods. Ear plugs or muffs worn by the worker can provide effective protection in such cases. Numerous scientific reports have shown that such devices can provide 25 to 40 dB of protection if properly worn. As with the usual safety protective device, ear protectors may cause a degree of discomfort to the wearer. These devices are,

however, used successfully in many industries where adequate attention is given to program development and employee education.

Be sure to check the current OSHA rules and regulations regarding the attenuation characteristics of hearing protective devices before selecting them.

Hearing Evaluation

Hearing evaluation, an important part of a hearing conservation program, involves audiometric testing. Such testing should be done at time of hiring and may be done at periodic intervals, at the time of job change, or upon leaving employment. While there may be need for all of these evaluations in an industrial hearing conservation program, the preplacement test is the most important. An audiometer is required for assessing an individual's hearing ability. Depending on the type of test signal used, audiometers may be called "pure tone audiometers" or "speech audiometers." Manual and automatic models are available.

Pure tone audiometers are standard instruments for industrial use. For hearing tests likely to be required in industry, the test frequencies of 500, 1000, 2000, 3000, 4000, and 6000 Hz are needed.

Audiometric testing facilities

In industrial situations, medical facilities may be near noisy industrial areas and background noise levels in these facilities may be relatively high. If noise interferes with the audiometer signal, the subject will appear to be "partially deaf." To obtain reliable audiometric tests, it is necessary to maintain the background noise level in the test area below certain values. This may be accomplished by the design and construction of special rooms, or by the use of a prefabricated, commercially available audiometric test booth.

Audiometers must be calibrated periodically to ensure accurate test results. The calibration must include both the electronic circuits and the earphones, and calibration should be repeated at intervals not longer than one year and more frequently when erratic operation or erroneous test results are suspected. Rough daily calibrations can be made by testing the hearing of persons with known patterns.

Summary

Industrial noise problems are extremely complex. There is no "standard" program applicable to all situations. In view of developments in the workmen's compensation field, however, it behooves industry to consider and evaluate its noise problems and to take steps toward the establishment of effective hearing conservation procedures.

As a minimum, a hearing conservation program in industry should include:

1. Noise exposure evaluation

2. Engineering control of noise exceeding Threshold Limit Values

3. Audiometric examinations for all new employees to be performed in accordance with good medical practice, and

4. Ear protection for those employees working in areas where noise cannot be controlled economically

Detailed references to noise, its management, effects, and control will be found in a great variety of books and periodicals. For those companies needing assistance in establishing hearing conservation programs, consultation services are available in a number of professional areas through private consultation, insurance, and governmental groups.

References

American Academy of Ophthalmology and Otolaryngology, 15 Second Street, S.W., Rochester, Minn. *Guide for Conservation of Hearing in Noise.* 1964.

American Foundrymen's Society, Inc., Des Plaines, Ill. *Foundry Noise Manual.*

American Industrial Hygiene Association, 66 S. Miller Rd., Akron, Ohio 44313. *Industrial Noise Manual,* 2nd ed. 1966.

American Medical Association, Committee on Rating of Mental and Physical Impairment, Chicago, Ill. 60610. *Guides to the Evaluation of Permanent Impairment.* 1971.

American Society of Heating, Refrigerating and Air-Conditioning Engineers, Inc., 345 East 47th Street, New York, N.Y., 10017. *Guide and Data Book: Fundamentals and Equipment.*

Beranek, L. L. *Acoustics.* New York, N.Y. McGraw-Hill Book Company, 1954.

————. *Noise and Vibration Control.* New York, N.Y. McGraw-Hill Book Company, 1971.

————. *Noise Reduction.* New York, N.Y. McGraw-Hill Book Company, 1960.

Bragdon, C. R. *Noise Pollution: The Unquiet Crisis.* Philadelphia, Pa., University of Pennsylvania Press. 1971.

Broch, Jens Trampe. *Application of B&K Equipment to Acoustic Noise Measurements,* 2nd ed. B&K Instruments, Inc., 5111 W. 164th St., Cleveland, Ohio 44142. 1971.

Burns, W. *Noise and Man.* Philadelphia, Pa., J. B. Lippincott Company, 1969.

Close, P. D. *Sound Control and Thermal Insulation of Buildings.* New York, N.Y., Reinhold. 1968.

Cralley, L. V., *et al. Industrial Environmental Health: The Worker and the Community.* New York, N.Y., Academic Press. 1972.

Davis, D. B. *Acoustical Tests and Measurements.* Indianapolis, Indiana. W. Sams. 1965.

DeHann, V. S. "Noise: Potential Danger to Man; Physiological and Psychological Effects. An Indexed Bibliography." Baltimore, Md., Information Center for Hearing, Speech, and Disorders of Human Communications, Johns Hopkins Medical Institutions. 1972.

Doelle, L. L. *Environmental Acoustics.* New York, N.Y., McGraw-Hill Book Company, 1972.

Glorig, A. *Noise and Your Ear.* New York, N.Y., Grune & Stratton, 1959.

Harris, Cyril M., ed. *Handbook of Noise Control.* New York, N.Y., McGraw-Hill Book Company, 1957.

Industrial Health Foundation, Inc., 5231 Centre Ave., Pittsburgh, Pa. 15232. "A Selective Bibliography on Industrial Noise, 1963–1973. 1973.

Kryter, Karl D. *The Effects of Noise on Man.* New York, N.Y. Academic Press. 1970.

National Safety Council, 425 N. Michigan Ave., Chicago, Ill. 60611.
  National Safety News Reprint Series (1968–1969).
      "Administrative and Human Relation Aspects."
      "Directory of Noise Control Products."
      "DOL's Guidelines to OSHA Occupational Noise Standards."
      "Ear Anatomy and Effects of Noise on Man."
      "Engineering Control of Noise."
      "Hearing Loss Statutes in the U.S. and Canada."
      "Hearing Measurement and Audiometry."
      "Industrial Audiometry."
      "An Industrial Hearing Conservation Program."
      "Industrial Noise and Hearing Loss Liability."
      "Instruments and Techniques of Sound Measurement."
      "Medical Aspects of Hearing Protection."
      "Personal Ear Protection."
      "Physics of Sound."
      "Procedures of a Sound Survey."
  "Sound Level Calculator for Hearing Conservation Programs," 1969. (Slide rule.)

Newby, Hayes A. *Audiology,* 3rd ed. New York, N.Y., Appleton-Century-Crofts. 1972.

Olishifski, J. B., and Harford, E. R. *Basics of Industrial Noise and Audiometry.* Chicago, Ill., National Safety Council. 1975.

Peterson, Arnold P. G., and Gross, Ervin E., Jr. *Handbook of Noise Measurement,* 7th ed. General Radio, Concord, Mass. 1972.

Richardson, E. G. *The Technical Aspects of Sound.* New York City, Elsevier Publishing Company, 1953.

Sataloff, Joseph. *Hearing Loss.* Philadelphia, Pa. J. B. Lippincott Company. 1966.

U.S. Department of Health, Education, and Welfare, Public Health Service, Washington, D.C. 20201. *Industrial Noise—Its Evaluation and Control,* vol. II, pub. 614. 1966.

Electrical
Hazards

Chapter
41

Definitions 1256

Electrical Injuries 1257

Electrical Equipment 1258
Selection . . . Installation . . . Switches . . . Fuses and circuit breakers . . . Ground fault circuit interrupters . . . Control equipment . . . Motors . . . Grounding . . . Explosion-proof apparatus . . . Extension cords . . . Test equipment . . . Specialized equipment

Inspection 1280
Rotating and intermittent-start equipment . . . High-voltage equipment

Maintenance 1281
Lockouts . . . Removing fuses . . . Wiring

Employee Training 1284

References 1285

41—Electrical Hazards

EFFECTS OF ELECTRIC CURRENT ON MAN

Effect	Current in Milliamperes					
	Direct		60 Hz		10,000 Hz	
	Men	Women	Men	Women	Men	Women
Slight sensation on hand	1	0.6	0.4	0.3	7	5
Perception threshold	5.2	3.5	1.1	0.7	12	8
Shock—not painful, muscular control not lost	9	6	1.8	1.2	17	11
Shock—painful, muscular control not lost	62	41	9	6	55	37
Shock—painful, let-go threshold	76	51	16	10.5	75	50
Shock—painful and severe, muscular contractions, breathing difficult	90	60	23	15	94	63
Shock—possible ventricular fibrillation effect from 3-second shocks	500	500	100	100		
Short shocks lasting *t* seconds			165/√*t*	165/√*t*		
High voltage surges	50*	50*	13.6*	13.6*		

*Energy in watt-seconds or joules

As a source of power, electricity is, in some ways, less hazardous than steam or other energy sources. Correctly used, electricity is our most versatile form of energy. Failure to take suitable precautions in its use, however, creates conditions which are certain to result in bodily harm or property damage or both.

Although there have been many advances in the control of electrical hazards, industry still has many injuries and fatalities from preventable causes.

The versatility of electric power permits the installation of motors for driving individual machines or groups of machines. Consequently, machine tools can, with minimum expense and difficulty, be arranged for maximum safety and efficiency.

There are, however, certain hazards in the installation, maintenance, and use of electric wiring and equipment. Control of most of these hazards is neither difficult nor expensive, but ignoring or neglecting them may lead to serious accidents.

This chapter points out electrical hazards of utilization equipment, and suggests means of removing them. Commercial generating stations, substations, and transmission and distribution systems are not discussed.

Definitions

There are several terms which should be well understood before continuing this chapter. Three of these basic terms—current, voltage, and resistance—are defined using the analogy that electricity flowing through a circuit can be likened to the flow of water through a pipe, and if this analogy is kept in mind, these terms are not troublesome.

CURRENT may be thought of as the total VOLUME of water flowing past a certain point in a given length of time. Electric current is measured in amperes, which is a very large quantity in relation to its effect on the human body. The measurement used, in relation to electric shock, therefore, is the milliampere (0.001 ampere).

VOLTAGE may be thought of as the PRESSURE in a pipe line; it is measured in volts.

RESISTANCE is any condition which retards flow; it is measured in ohms.

Low voltage is a rather ambiguous term depending upon whether it is being used by a safety professional, a plant electrician, or a lineman. For the purposes of this chapter, low voltage is 24 to 600 volts, and "safety" low voltage refers to voltages below 24 volts.

Electrical Injuries

Current flow is the factor that causes injury in electric shock; that is, the severity of electric shock is determined by the amount of current flow through the victim (Table 41-A). Experimental and field data from authoritative sources (see references for articles written by Charles F. Dalziel) indicate that, in general, an alternating current of 100 milliamperes at commercial frequency of 60 cycles per second (60 Hz) may be fatal if it passes through the vital organs. Similarly, it is estimated that a value of 16 milliamperes is the average current at which an individual can still release himself from an object held by the hand. *Such current flow may readily be obtained on contact with low-voltage sources of the ordinary lighting or power circuit.*

Because current flow depends on voltage and resistance, these factors are important. Other factors affecting the amount of damage done are the parts of the body involved, the duration of current flow through the victim, and the frequency (if alternating current).

Resistance to current flow is mainly to be found in the skin surface. Callous or dry skin has a fairly high resistance, but a sharp decrease in resistance takes place when the skin is moist (Table 41-B). Once the skin resistance is broken down, the current flows readily through the blood and body tissues.

Whatever protection is offered by skin resistance decreases rapidly with increase in voltage. High voltage alternating current of 60 Hz causes violent muscular contraction, often so severe that the victim is thrown clear of the circuit.

Although low voltage also results in muscular contraction, the effect is not so violent. The fact, however, that low voltage often prevents the victim from freeing himself from the circuit makes exposure to it dangerous.

Death or injury by electric shock may result from the following effects of current on the body:

TABLE 41-B

HUMAN RESISTANCE TO ELECTRICAL CURRENT

Body Area	Resistance (ohms)
Dry skin	100,000 to 600,000
Wet skin	1,000
Internal body—hand to foot	400 to 600
Ear to ear	(about) 100

1. Contraction of the chest muscles, which may interfere with breathing to such an extent that death will result from asphyxiation when the exposure is prolonged.

2. Temporary paralysis of the nerve center, which may result in failure of respiration, a condition which often continues until long after the victim is freed from the circuit.

3. Interference with normal rhythm of the heart, causing ventricular fibrillation. In this condition the fibers of the heart muscles, instead of contracting in a coordinated manner (which causes the heart to act as a pump), contract separately and at different times. Blood circulation ceases and (unless proper resuscitation efforts are made) death ensues. The heart cannot spontaneously recover from this condition. It has been estimated that 50 milliamperes is sufficient to cause ventricular fibrillation.

4. Suspension of heart action by muscular contraction (on contact with heavy current). In this case the heart may resume its normal rhythm when the victim is freed from the circuit.

5. Hemorrhages and destruction of tissues, nerves, and muscles from heat due to heavy current along the path of the electric circuit through the body.

In general, the longer the current flows through the body, the more serious may be the result. Considerable current is likely to flow from high-voltage sources, and in general only very short exposure can be tolerated

1257

if the victim is to be revived.

Injuries from electric shock are less severe when the current does not pass through or near nerve centers and vital organs. In the majority of electrical accidents in industry the current flows from hands to feet. Since such a path involves both the heart and the lungs, results are usually serious.

Another type of injury is burns from electric flashes. Such burns are usually deep and slow to heal and may involve large areas of the body. Even persons at a relatively good distance from the arc may receive eye burns.

Where high voltages are involved, flashes of explosive violence may result. This intense arcing is caused by short circuits between bus bars or cables carrying heavy current, failure of knife switches or their being opened while they are carrying a heavy load, and pulling of fuses in energized circuits.

Mechanical injuries from electrical equipment may be caused by unexpected motion, such as the accidental starting of motors which drive machines where men may be working.

Other types of injuries are falls from one level to another caused by the workman's receiving a shock from defective equipment and, because of the resulting muscular contraction, losing his balance.

C. F. Dalziel's statistics (see References) indicate that only a small percentage of those who recover from electric shock show permanent disability. In many cases, the victim may be saved by prompt application of artificial respiration, since a common result in electrical accidents is failure of that part of the nervous system which controls breathing.

It is, therefore, essential that persons engaged in electrical work be instructed in mouth-to-mouth or other effective means of respiratory resuscitation. It is also advisable that linemen and selected personnel in other operations with exceptional electrical exposure receive (under medical supervision) competent instruction in external cardiac compression; this technique, used in conjunction with respiratory resuscitation, is used in electric shock cases involving ventricular fibrillation.

Resuscitative techniques should be immediately applied to a victim of electric shock and should be continued until he revives, until death is diagnosed by a physician, or until rigor mortis sets in. The possibilities of suc-

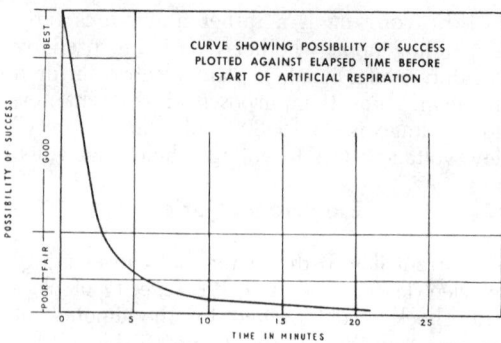

FIG. 41–1.—The possibility of successful revival decreases with time.

From Edison Electric Institute Resuscitation Manual.

cessful revival of a victim in terms of elapsed time before the start of resuscitation are shown in Fig. 41–1.

Electrical Equipment

Selection

Most items of electrical equipment are designed and built for specific types of service. They will operate with maximum efficiency and safety only when used for the purposes and under the conditions for which they are intended.

In the selection of equipment, it is advisable to follow the recommendations of the various codes and standards which have been established. In addition to the NFPA *National Electrical Code,* the state and local codes should be checked for industrial zoning requirements. Adherence to most of the provisions of the *National Electrical Code* is required by the safety and health regulations of the federal Occupational Safety and Health Administration.

Engineering consultant service, recommendations of manufacturers, and publications from the following agencies will answer practically all questions concerning electrical equipment:

American National Standards Institute
1430 Broadway
New York, N.Y. 10018

Fig. 41–2.—Expanded metal cover, guard rails, and substantial gate guard this equipment. Arrangement encourages easy maintenance.

Canadian Standards Association
   235 Montreal Rd.
   Ottawa 2, Ontario, Canada

Factory Mutual System
   1151 Boston-Providence Turnpike
   Norwood, Mass. 02062

Institute of Electrical and Electronics Engineers
   345 East 47th St.
   New York, N.Y. 10017

National Fire Protection Association
   470 Atlantic Ave.
   Boston, Mass. 02210

Underwriters Laboratories Inc.
   207 East Ohio St.
   Chicago, Ill. 60611

When copies of codes or standards are ordered from the publishers, give full information on general types of equipment under consideration, the application and operating conditions. For special problems, include photographs or blueprints to clarify the physical conditions at the installations.

Installation

Transformers, dead-front control boards, switches, motor starters, and other electrical equipment should be installed so that the possibility of accidental contact with energized conductors is reduced to a minimum.

When an interlock is used as a safety device, it should be fail-safe; that is, steps must be taken to assure that failure of this device will not jeopardize the safety of personnel depending upon it. Interlocks selected should meet the following standards:

1. Fail-safe features. Failure of the interlock mechanism, loss of power, short circuit, or malfunction of equipment will cause the circuit to be interrupted.

2. A visible disconnect in the primary power circuit.

3. An arrangement which makes attempts to circumvent the interlock impractical.

Where space and operating requirements permit, electrical equipment should be placed in the less congested areas of the plant or, where practicable, in special rooms to which only authorized persons have access. If this equipment must be located in the production areas of the plant, enclosures should be built around those parts of the equipment having exposed conductors and warning signs should be posted. Floor curbing or heavy steel barriers may be necessary if there is even the remote danger of industrial trucks striking critical electrical equipment.

Barriers prevent accidental contact with electrical equipment (Fig. 41–2). Frames may be made of wood, rolled metal shapes, angle iron, or pipe. Wood strips, sheet metal, perforated metal, expanded metal, wire mesh, or plastic or shatterproof transparent material may be used for filler. Dry wood and many plastics have the advantage of being nonconductors. But unless specially treated, wood is flammable and may be objectionable on that account. Metal frames or guards should be grounded.

In addition, warning signs should be displayed near exposed current-carrying parts and in especially hazardous areas, such as high-voltage installations. These signs should be large enough to be read easily and should be visible from all approaches to the danger zone. They should meet American National Standard *Specifications for Accident Prevention Signs*, Z35.1.

In many respects, standard machine guarding practices can be applied to electrical equipment. There are, however, certain hazards peculiar to electricity which must be given consideration when the overall plant guarding program is planned.

Wiring should be installed in accordance with the *National Electrical Code* unless more restrictive local requirements apply.

The type of wiring that should be used in an industrial plant depends upon the type of building construction, the size and distribution of electrical load, exposure to dampness or corrosive vapors, location of equipment, and various other factors. For most plant conditions, grounded rigid metal conduit is satisfactory.

The various types of insulation used on electrical conductors and their application provisions are given in the *National Electrical Code*.

Among wiring methods which may be used under certain circumstances are armored cable, nonmetallic sheathed cable, flexible metal conduit, raceways, open wiring and insulators, and concealed knob and tube wiring. When new installations are made or existing circuits changed, applicable requirements of national and local wiring codes should be observed.

Wires having insulation designed for the type of service and location should be used, but the insulation alone should not be considered sufficient protection against shock, especially in high-voltage circuits.

Frequent inspection of equipment and competent supervision of maintenance crews are extremely important, and the inspectors' reports will aid the safety program considerably. Often in their routine work throughout the plant, maintenance men can spot hazards before they cause injuries.

Motors should be mounted so that they do not interfere with the normal movement of personnel or materials. Nonenclosed motors should be in areas free from dust, moisture, and flammable or corrosive vapors. In some instances, they can be isolated from personnel by being mounted on overhead supports, installed below the floor level, or placed in special motor rooms.

If current-carrying parts must be exposed, they should be made inaccessible by elevating them at least 8 ft above the work area, or enclosures, barriers, or guards should be provided to prevent contact.

Switches

Among the several types of switches are pushbutton switches, snap switches, knife switches, and enclosed externally operable air break switches. Switches designed for controlling individual motors and machine tools and for lighting and power circuits are the enclosed type. All switches must have approved voltage and ampere ratings compatible with their intended use.

An open knife switch should not be used; it is hazardous because of the exposure of live parts and because of the arc formed when the switch is opened. Knife switches should be enclosed in grounded metal cabinets having control levers that operate outside the cabinets. A further safeguard is the safety switch whose cover must be closed before the switch can be used.

A knife switch should be mounted so that the blades are dead when the switch is open and it should be installed so that gravity will not close it. Knife switches installed in power switching circuits should be electrically interlocked so that they cannot be opened when the circuit is energized, unless switches are of the load-break type.

Where it is necessary to use disconnect switches for high voltage, heavy current feeder circuits to test floors, or other service installations, they should be located out of normal reach and should be operated only by insulated switch sticks. Where it is impractical to locate the switch out of normal reach, it must be protected against accidental contact by operating personnel, by a complete enclosure or suitable fence or barricade, the switch to be opened or closed remotely by an insulated switch stick.

Pushbutton or snap switches are recommended because the live parts are enclosed. Flush switches should be installed in metal boxes, and surface switches used in open wiring and molding work should be mounted on porcelain, composition, or other insulating subbases. These switches must have indicators to show the open and closed positions.

Pendant switches are used primarily where switches have not been installed on side walls and where it would be difficult or expensive to put them in, or where the switches on side walls or in cabinets control two or more circuits and it is desired to subdivide them still further so that each can be controlled individually.

Pendant switches, and in particular pendant pushbutton control stations, should be provided with exterior strain relief at the point of suspension from the ceiling, and at the point of entry into the enclosure; in other words, there should be no pull or harmful friction on electrical wiring or connections.

A convenient toggle switch with a button automatically illuminated as soon as the circuit is open will aid in the location of the switch in darkened areas. Such a switch, however, should not be relied upon for any illumination.

Fuses and circuit breakers

The safe current-carrying capacity of conductors is determined by their size, the material of which they are made, the type of insulated covering, and the manner in which they are installed. If they are forced to carry more than the maximum safe load or if heat dissipation is limited, excessive heating results. Overcurrent devices, such as fuses and circuit breakers, open the circuit automatically in the event of excessive current flow from accidental ground, short circuit, or overload.

Overcurrent devices must be installed in every circuit, and they should be of a size and type that will interrupt the current flow when it exceeds the capacity of the conductor. The selection of the properly rated equipment not only depends on the current-carrying capacity of the conductor, but also on the rating of the power-supply transformer or generator and its potential short-circuit producing capacities. Protection of this kind, both for personnel and for equipment, is one of the important features of an electrical installation. Where higher interrupting capacity is required, special high-capacity fuses are needed.

Among the many types of fuses are link, plug, and cartridge. Each should be used only in the type of circuit for which it is designed. Use of fuses of the wrong type or the wrong size may cause injury to personnel and damage to equipment. Overfusing is a frequent cause of overheated wiring or equipment—and resultant fires.

A link fuse, as its name indicates, is a strip of fusible metal between two terminals of a fuse block. If exposed, it may scatter metal when it blows. Replaceable link fuses should only be replaced under the direction of qualified maintenance personnel.

Plug fuses are used on circuits which do not exceed 30 amperes at not more than 150 volts to ground. In plug fuses the fusible metal is completely enclosed. The type that cannot be bridged inside the holder is recommended.

In the cartridge fuse, which is widely used in industrial installations, the fusible metal strip is enclosed in a tube. Cartridge fuses that indicate when the fuse is blown, and renewable types in which the fusible element may be replaced, are available.

It is recommended that a switch be placed in any circuit that can be opened to deenergize the fuses to be handled. As an additional precaution, insulated fuse pullers should be used.

Before fuses are replaced, the circuit should be locked-out and an investigation should be made for the cause of the short circuit or overload. Blown fuses should be replaced by others of the same type and size; fuses should never be inserted in a live circuit.

Circuit breakers have long been used in high-voltage circuits with large current capacities. In recent years their use has become more common in many other kinds of circuits. They are available in a variety of types and sizes. They may be instant in their operation or equipped with timing devices, manually operable or power operable.

Circuit breakers fall into two general categories: thermal and magnetic.

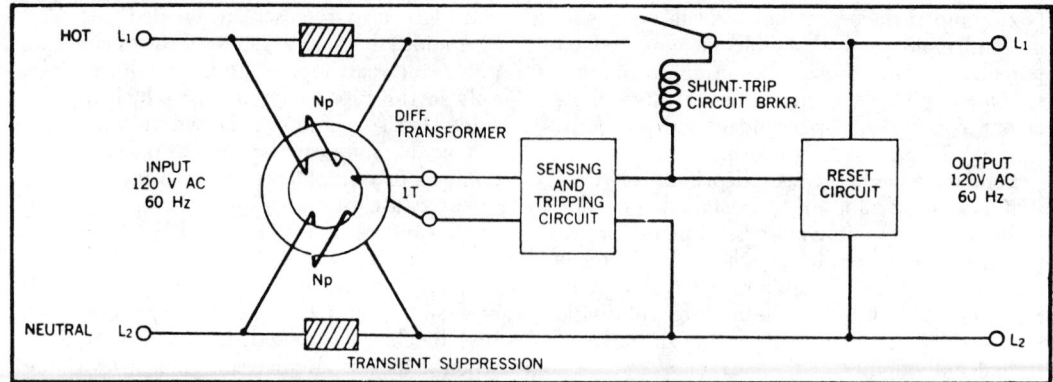

FIG. 41–3.—Differential ground fault interrupter. Current-carrying conductors pass through circular iron core of doughnut-shaped differential transformer. Transformer senses even a portion of the current flowing to ground and causes the sensing circuit to open the circuit breaker.

• The thermal-type circuit breaker operates solely on the basis of temperature rise, and consequent variations in the temperature of the room in which the circuit breaker is installed will affect the point at which it interrupts the circuit.

• The magnetic-type circuit breaker operates only on the basis of the amount of current passing through the circuit; this type has considerable advantage where a wide fluctuation in ambient temperature would ordinarily require overrating the circuit breaker or where undue tripping frequencies are experienced.

Circuit breakers should be selected for the specific installation by qualified engineers and checked regularly by experienced maintenance personnel to make sure they are in good operating condition at all times.

Ground fault circuit interrupters

A ground fault circuit interrupter is a fast-acting circuit breaker that is sensitive to very low levels of current leakage to ground. The interrupter is designed to limit the electric shock to a current and time duration value below that which can produce serious injury. The unit operates *only* on line-to-ground fault currents, such as insulation leakage currents or currents likely to flow during accidental contact with a "hot" wire of a 120 volt circuit and ground. *It does not protect in the event of a line-to-line contact.*

There are two main types of units.

• The differential ground fault interrupter (Fig. 41–3), has current-carrying conductors passing through the circular iron core of a doughnut-shaped differential transformer. As long as all the electricity passes through the transformer, the differential transformer is not affected and does not trigger the sensing circuit. If a portion of the current flows to ground and through the fault-detector line, flow of electricity through the sensing windings of the differential transformer causes the sensing circuit to open the circuit breaker. These devices can be arranged to interrupt a circuit for currents of as little as 5 milliamperes flowing to ground.

• Another unit design is the isolation-type ground fault interrupter (Fig. 41–4). This unit combines the safety of an isolation system with the response of an electronic sensing circuit. In this setup, an isolating transformer provides an inductive coupling between load and line; both the hot and neutral wires connect to the isolating transformer. There is no continuous wire between.

In this type of interrupter, the ground fault must pass through the electronic sensing circuit, which has sufficient resistance to limit current flow to as little as 0.2 milliamperes—well below the level of human perception.

Although neither of these types protects an individual in the event he makes line-to-line contact, both will operate most effectively in line-to-ground situations. They should sig-

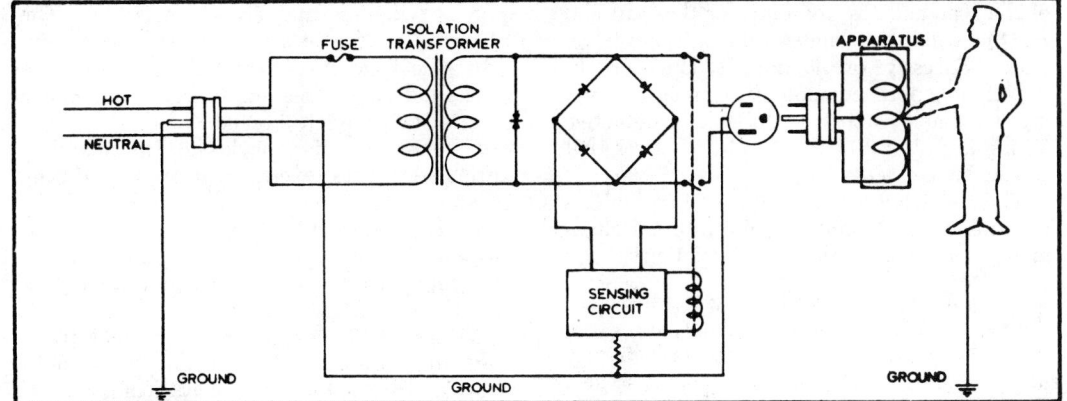

Fig. 41–4.—Isolation-type ground fault interrupter. An isolating transformer provides an inductive coupling between load and line; both hot and neutral wires connect to the isolating transformer. There is no continuous wire between.

nificantly reduce electric shock accidents that presently account for about 1100 fatalities each year.

Control equipment

Switchboards for both alternating- and direct-current distribution circuits should be arranged so that the controls are readily accessible to the operator. Likewise, instruments should be readable and equipment adjustable from the working space. Boards should be placed so that the operator will not be endangered by live or moving parts of machinery. The space back of the switchboard should not be used for storage and should be kept clear of rubbish. The control board should be placed in a special room or be made inaccessible to other than authorized personnel by screen enclosures. Doors to the enclosure should be kept locked.

Good illumination should be provided for the front and rear of switchboards and maintained ready for use at all times. An emergency source of illumination should also be provided.

Switch and fuse cabinets should have close-fitting doors, and the doors should be kept locked. To warn employees that electrical parts are exposed and to remind them to close all doors of equipment, the inside of the doors may be painted orange.

The switchboard framework and metal parts of guards should be grounded in accordance with code regulations.

Connections, wiring, and equipment of switchboards and panel boards should be arranged in an orderly manner. Switches, fuses, and circuit breakers should be plainly marked and arranged to afford ready identification of circuits or equipment supplied through them.

It is good practice to keep a diagram of switchboard connections and devices posted near the equipment.

Protection against accidental shock from live electrical parts on switchboards, fuse panels, and control equipment can be minimized by insulating the floor area within range of the live parts. For low-voltage exposures, special insulating mats or dry wood floors with no metal parts are used. These insulating mats should be made of material having proper moisture resistance and nonconductive properties, and should withstand mechanical abuses which may be encountered in such service. See American National Standard J6 series, "Specifications for Rubber Protective Equipment for Electrical Workers."

Circuits initiated by pushbuttons should be low voltage. However, pushbutton controls may be used for a high-voltage circuit if step-down transformers are provided to prevent the voltage in the control part of the circuit from exceeding 250 volts.

Motors

For maximum safety each motor should be

1263

of the type and size required for the load and for the conditions under which it must operate. Excessive overloading for long periods, use of nonapproved motors in areas containing flammable vapors or dusts, and defective wiring are typical of the unsafe practices that should be avoided.

The following are common motor problems in industry: dust, stray oil, moisture, misalignment, vibration, overload, and friction.

• In plant operations, dust is constantly settling on motors, housings, windings, slip rings, and commutators, and trying to work its way to the bearings. The problem of dust is at a minimum in the case of totally enclosed, fan-cooled motors.

Dust is best removed from motors before it has a chance to unite with water or oil to form a gummy mess. Motors, and their housings, slip rings, and commutators, must therefore be wiped off on regular inspections. Occasionally, dust must be blown out of the wound section with not over 30 psig of pressure from a compressed-air unit. When this operation is performed, there must be good ventilation to prevent accumulation of dust and the person doing the dusting should wear eye protection. In some cases, it is advisable to use vacuum cleaning equipment or a hand bellows.

• Stray oil harms commutators by deteriorating the mica insulating segments between the bars. It causes excessive sparking and deteriorates the insulation on the windings, with immediate danger of a burnout or breakdown. Oil also tends to hold dust, lint, and other material that increases the fire hazard.

When oil and dust have been allowed to build up, the heavy accumulation should be removed with a nontoxic solvent, preferably a non-flammable one. Special solvents are available for this work; see Chapter 42, "Flammable and Combustible Liquids." Carbon tetrachloride must not be used.

Care should be taken not to scrape the insulation, nor should solvents be allowed to soak the insulation and soften it. When they are clean, the windings should be dried and coated with insulating varnish.

• Not all electrical insulation acts as a perfect barrier to moisture. Some types become porous with age and absorb moisture. The electrical resistance may then drop to a point where leakage burns through the insulation to ground and causes short circuit, fire, shock hazard, or complete breakdown.

Careless handling of liquids in manufacturing processes, wetting of motors during cleanup operations in plants where floors must be frequently washed with water, and overflowing of water softener tanks, filters, dye cups, or other liquid reservoirs cause failure and shock hazards.

Means of prevention are personnel training and relocation of equipment. If the moisture hazard cannot be wholly avoided or is inherent in the work for which the motor is used, manufacturers can provide drip-proof, splashproof, or totally enclosed designs.

After a motor has been subjected to moisture and the resistance is found to be at a dangerous point, the motor should be dried out, preferably by an organization or person specializing in this type of work.

• Belts, chain drives, and gearing should be checked for extra tension on the belt or chain drive, and for binding in the gearing. Misalignment of motor equipment will cause shafts to spring or break, bearings to burn out, and failure from overload.

• Misalignment is one of the important causes of motor vibration. Other causes are careless servicing that puts motors out of balance, with loose motor mounting bolts, worn bearings which allow the shaft to oscillate, or improper connections to the equipment being driven.

• Overloading a motor can be caused by excessive friction within the motor itself, by use for the wrong kind of job, by obstruction in the driving or driven machine, or by attempts to gain a greater output from the machine than the motor is capable of carrying.

If one of these conditions causes the current in a motor to exceed its nameplate current rating, heating may increase as much as the square of the current increase. Insulation may be fried, soldered connections melted, or bearings burned out.

To safeguard against such overloads, motors are given various forms of overload protection. In most cases, a thermal element is connected in the power circuit to the motor. Heat from the element operates an overload relay, opening the circuit to the motor. The thermal element or fuse should be of the proper capacity as listed in the *National Electrical Code.* No unit of more than the recommended value should be used.

Motor output and production methods should be revised to control overload conditions that trip or blow these protective devices.

• To reduce friction and to prevent excessive wear, overheating of bearings, and possible fires from faulty lubrication, manufacturer's lubrication charts and instructions on types and grades of lubricants, the frequency of lubrication, and other practices should be followed.

Grounding

In order to understand grounding properly, it should first be noted that both equipment grounding and the grounding of the electrical system itself are included under this term. Equipment grounding is the bonding of all conductive materials which enclose electrical conductors. These enclosures are not normally current-carrying; the bonding prevents a difference in voltage between these materials and ground. The electrical system itself is grounded in order to prevent the occurrence of excessive voltages from such sources as lightning, line surges, or accidental contact with higher voltage lines. Both the electric system and metallic enclosures are grounded in order to cause over-current devices to operate in the event of a ground fault occurring from insulation failure.

System grounding. Alternating current systems operating at 50 volts or more are required to be grounded under a variety of voltage conditions. Grounding is accomplished by bonding the identified conductor (identification takes the form of white or natural grey-colored insulation) to a grounding electrode by means of an unbroken wire called a grounding electrode conductor.

In a domestic occupancy, the utility will ground the identified conductor at the transformer by means of a plate attached to the bottom of the supporting pole. This identi-

fied or grounded circuit conductor is then once again grounded when it is brought to the service equipment within the home. This grounding takes the form of a grounding electrode conductor run to the nearest cold water metallic piping system. Where such a system is not present, metallic building framing, gas piping, underground piping systems, and tanks with concrete-encased steel reenforcing bar systems may be used. As a last resort, if none of these forms of electrodes are available, buried plate pipe or rod may be used.

Industrial installations are grounded in a similar manner except that the transformer may be very large and mounted on a concrete pad, in which case the grounding electrode will be a carefully placed metal grid within the ground or within the concrete pad itself. Like the home the grounded circuit conductor brought to the service equipment within the plant or commercial building must be bonded to the cold water piping system or other electrode which occurs on the premises in proper order of preference.

Much has been said about the resistance of the grounding electrode and many erroneous ideas are prevalent concerning its use. Cold water piping systems, building structure, other buried tanks and piping systems as well as concrete-encased reenforcing power systems all provide an adequate resistance to ground providing careful attention is paid to the workmanship of the original installation. It is only when a "made" electrode such as rod pipe or plate must be resorted to that higher resistance values are encountered. In many parts of the country, low resistance values for made electrodes are impossible to obtain, so the *National Electrical Code* simply requires in this case that one additional electrode be used. Because the same term is applied both to system grounding and equipment grounding, it is often thought that the grounding electrode must have a very low resistance in order to dissipate ground faults. This is not true. A grounding electrode provides a point of equalization so that large voltage differences both inside and outside the building are minimized. As stated previously, some of the primary causes of such voltage differences are lightning surges or lightning strikes on the building itself.

Once the grounded circuit conductor is run beyond the service equipment it must not be grounded at any point. The major exception to this rule is in the case of additional transformers used within the establishment in order to step down voltages at various locations. In such instances, the secondary conductors constitute a new or "separately derived system" and the identified circuit conductor is to be grounded to a grounding electrode consisting of the nearest available effectively grounded cold water piping system or building structure.

One of the most frequent abuses of this rule prohibiting the subsequent grounding of the grounded circuit conductor is at the terminals of an ordinary parallel U-blade receptacle. Many well meaning but uninformed maintenance engineers will connect the white grounded circuit conductor to both the silver terminals and the green equipment grounding terminals on the receptacle, feeling that they have thereby achieved redundant grounding. In addition to violating the provisions of the *National Electrical Code,* they have also set up a circumstance whereby if the neutral is interrupted anywhere between the receptacle and its point of attachment to ground, any equipment enclosures which have been grounded by means of this receptacle now will be energized at full line voltage.

It should be understood that not all systems are required to be grounded. Providing trained electrical personnel are present, some manufacturing processes are dependent on ungrounded systems and provide a higher degree of safety by not interrupting strategic equipment when a first fault to ground occurs. The presence of fault indicating equipment together with the necessary personnel who can make repairs quickly will ensure against costly equipment down time and hazardous conditions arising from it.

Equipment grounding. When insulation failure occurs on conductors within metal enclosures, these enclosures will be raised to line voltage and will constitute a serious hazard for personnel. If the metallic enclosure is attached to the main bonding jumper and the service equipment with an equipment grounding conductor, however, this voltage difference will not occur, and personnel will be protected. Moreover, if the fault itself has a low resistance, and the equipment grounding conductor has been properly installed and well maintained, then a large amount of current can flow and the over current device protecting the circuit will deenergize the circuit.

Fixed equipment which is to be grounded includes the exposed non-current carrying metal parts that are likely to become energized that are (*a*) within 8 ft vertically or 5 ft horizontally of ground, (*b*) located in a damp or wet location and not isolated, (*c*) in electrical contact with metal, (*d*) in a hazardous location, (*e*) supplied by a metal clad, metal sheathed, or metal raceway wiring method, and (*f*) operated with any terminal in excess of 150 volts to ground. Additionally, exposed non-current carrying metal parts regardless of voltage, of certain motor frames, controller cases for motors, the electric equipment of elevators and cranes, electric equipment in garages, theaters, and motion picture studios, accessible electric signs and associated equipment, and switch board frames and structures.

Also the frames and tracts of electrically operated cranes, the metal frames of non-electrically driven elevator cars which have electrical conductors, hand-operated metal shifting ropes or cables of electric elevators, metal enclosures around equipment carrying voltages in excess of 750 volts between conductors, and mobile homes and recreational vehicles must all be grounded.

In respect to cord-connected equipment, most items that are operated in hazardous locations or at more than 150 volts to ground must be grounded. Additionally, such items as refrigerators, freezers, air conditioners, clothes washing and drying equipment, dishwashers and sump pumps, portable, hand-held, motor operated tools and appliances such as drills, hedge clippers, lawnmowers, wet scrubbers, sanders, and saws, and all devices used in damp or wet locations or by persons standing on the ground or on metal floors or working inside of metal tanks or boilers must be grounded.

Unlike the grounded circuit conductor the equipment grounding conductor may be grounded continuously along its length and may be a bare conductor, the metallic race-

FIG. 41–5.—Standard receptacle and plug recommended by the National Electrical Manufacturers Association. The receptacle is designed to receive a plug having three blades—a U-shaped or round grounding blade with two standard parallel polarized blades. This type of fixture also permits use of the nonpolarized standard plug.

way surrounding the circuit conductors or an insulated conductor. If the conductor is insulated it must have a continuous green cover or have a green cover with a yellow stripe in it. The equipment grounding conductor is always attached to the green hex-head screw on receptacles, plugs and cord connectors. Where an approved metal conduit system is used as an equipment grounding conductor, the receptacle must be bonded to the box by means of a separate jumper, or the receptacle must be listed by Underwriters' Laboratories as being constructed in such a manner as to provide self-grounding (Fig. 41–5).

Grounding of portable equipment is accomplished by means of a separate green insulated equipment-grounding conductor within the portable cord. Use of a separate external grounding conductor is much less

reliable, and the *National Electrical Code* permits this form of equipment grounding only by the special permission of the local authority having jurisdiction.

Some attachment plugs are manufactured today of high-impact transparent plastic material so as to provide ready inspection of the terminations of the plug without having to disassemble it.

Connections to the equipment grounding conductor and the grounded circuit conductor should be made in such a manner that removal of receptacles, switches, lighting fixtures, and other devices will not require the interruption of either of these two conductors. The proper use of jumpers and suitable pressure connectors will provide the necessary flexibility where maintenance is required, while still maintaining the integrity of both conductors. Solder is not permitted in the equipment-grounding circuit nor are switches, fuses or any other interrupting device. The equipment grounding conductor should always be grounded in the same enclosure with the conductors of the circuit which it protects.

A suitable attachment plug listed for equipment-grounding circuits is available in non-interchangeable configurations for all common voltage, current, and phase, combinations. General-duty devices may be used where they are not subjected to rough service or moisture, but the more rugged and water-tight equipment should be selected where this kind of exposure can be expected. Careful selection will insure against loss of grounding continuity due to damaged components or corrosion.

Careful attention should be paid to the size of the equipment-grounding conductor, since it is this conductor that will be required to carry fault current in the event that a ground fault develops. Unless sufficient current can get to the over-current device and cause it to operate, the circuit will not be protected. Moreover, the heating effects that occur at points of high resistance in a poor equipment-grounding system may cause fire and explosion resulting in loss of life and property, not to mention valuable production down time.

The *National Electrical Code* provides the proper sizing for equipment-grounding conductors used on various sizes of circuits. Conductor size should be increased where long runs are encountered and voltage drop exceeds an overall value of 5 percent. One of the most overlooked aspects of good equipment grounding is that of workmanship. The code provides that "All connections, joints, and fittings shall be tight using suitable tools"; and where this provision is not observed, a point of low thermal capacity will occur. Even though the resistance of the equipment-grounding circuit is measured and determined to be quite low, if the system does not have adequate thermal capacity, high fault currents will cause these points to overheat and may result in arcing and fires as well as inoperative protective equipment.

Maintenance of grounds. Only electrically qualified individuals should install or repair electrical equipment. They should make certain that the green insulated equipment grounding conductor is attached to the hexagonal green binding screw, and that the white grounded circuit conductor is attached only to the silver colored binding screw.

Good maintenance will make sure of an electrically continuous equipment-grounding path from the metal enclosure of the portable equipment through the line cord, plug, receptacle, and grounding system that terminates at the bonding jumper at the service equipment or in the enclosure of the separately derived system. Portable testing instruments such as a three-light neon receptacle tester or a ground loop impedance tester such as those used by federal compliance officers provide the most convenient and satisfactory means for checking polarity and other circuit connections and for measuring the resistances of equipment grounds. They are self contained and easy to operate, and resistance is indicated directly on the instrument dial. The user will have the further advantage of using the same equipment that will be used during a federal inspection. A receptacle tension tester may be used to inspect receptacles for deteriorating contacts. The tester employs little pointers which indicate the amount of tension that each receptacle contact will produce on the minimum size Underwriters' Laboratories listed attachment plug. By using this form of inspection, the maintenance department can replace receptacles before they produce an ineffective equipment-grounding

contact or a fire results in the power contacts.

When a program of testing is planned, the use of one or more of these instruments is recommended, since the entire test program can be handled by one man. A log of the results obtained at each test point will help predict deteriorating trends.

The condition of portable tools can easily be checked with the use of an insulation resistance tester which provides for the application of 500 volts direct current between the motor windings and the metal enclosure. Some insulation resistance meters provide an additional function which permits checking into continuity of the equipment grounding conductor as well. For the industrial user who has a planned tool maintenance program, the use of a portable appliance tester will provide automatic cycling of the portable tool through a series of timed tests. Such test apparatus should be used in the tool room to check portable tools before they are issued or upon their return. A log of each tool will help predict insulation failure.

Double-insulated tools. Maintaining of good ground and making sure that it is used properly can be a difficult task in abusive environments. Added to this is the fact that some companies have not as yet installed the three wire equipment grounding receptacle which in turn necessitates the use of adapters.

In the hands of the untrained operator, the adapter is rarely connected properly. (The *National Electrical Code* now requires that all adapters have a wide neutral blade, and that the equipment ground be made by means of a rigid tab on the bottom of the adapter rather than by means of a flexible pigtail.) Frequently the adapter is regarded as a nuisance and is discarded, the grounding pin on the attachment plug is clipped off and the operator now holds a potentially lethal device in his hands.

As an alternate to equipment grounding, the *National Electrical Code* recognizes the use of double-insulated appliances. Such appliances are constructed with two separate systems of insulation so that the probabilities of insulation failure are reduced to the lowest practical minimum. Double-insulated appliances are of particular value in the domestic situation;

(*Text continues on page 1275.*)

ARTICLE 500—HAZARDOUS LOCATIONS

(From *National Electrical Code*, NFPA No. 70–1971)

500–1. Scope. The provisions of Articles 500–503 apply to locations in which the authority having jurisdiction judges the apparatus and wiring to be subject to the conditions indicated by the following classifications. It is intended that each room, section or area (including motor and generator rooms, and rooms for the enclosure of control equipment) shall be considered individually in determining its classification. Except as modified in Articles 500–503, all other applicable rules contained in this Code shall apply to electrical apparatus and wiring installed in hazardous locations. For definitions of "approved" and "explosion-proof" as used in these Articles, refer to Article 100; "dust-ignition-proof" is defined in Section 502–1.

Equipment and associated wiring approved as intrinsically safe may be installed in any hazardous location for which it is approved, and the provisions of Articles 500–517 need not apply to such installation. Intrinsically safe equipment and wiring are incapable of releasing sufficient electrical energy under normal or abnormal conditions to cause ignition of a specific hazardous atmospheric mixture. Abnormal conditions will include accidental damage to any part of the equipment or wiring, insulation, or other failure of electrical components, application of overvoltage, adjustment and maintenance operations, and other similar conditions.

For further information see NFPA No. 493—1969 *Standard for Intrinsically Safe Process Control Equipment for use in Class I Hazardous Locations.*

Through the exercise of ingenuity in the layout of electrical installations for hazardous locations, it is frequently possible to locate much of the equipment in less hazardous or in nonhazardous areas and thus to reduce the amount of special equipment required. In some cases, hazards may be reduced or hazardous areas limited or eliminated by adequate positive-pressure ventilation from a source of clean air in conjunction with effective safeguards against ventilation failure. For further information see NFPA No. 496—1967 *Standard for Purged Enclosures for Electrical Equipment in Hazardous Locations.* It is recommended also that the authority having jurisdiction be familiar with such recorded industrial experience as well as with such standards of the National Fire Protection Association as may be of use in the classification of various areas with respect to hazard. For further information, see NFPA No. 30, *Flammable and Combustible Liquids Code*—1969; NFPA No. 32, *Dry Cleaning Plants*—1970; NFPA No. 35, *Manufacture of Organic Coatings*—1970; NFPA No. 36, *Solvent Extraction Plants*—1967; NFPA No. 58 (ANSI Z106.1) *Storage and Handling of Liquefied Petroleum Gases*—1969; and NFPA No. 59, *Storage and Handling of Liquefied Petroleum Gases at Utility Gas Plants*—1968.

For recommendations for protection against static electricity hazards, refer to the standards of the National Fire Protection Association on this subject.

Where rigid conduit is used in hazardous locations, it is necessary to have all threaded joints made up wrench tight to minimize sparking when fault current flows through the conduit system. Where it is impractical to make a threaded joint tight, a bonding jumper should be utilized.

All conduit referred to herein shall be threaded with a standard conduit cutting die which provides ¾-inch taper per foot. Such conduit shall be made up wrench tight to minimize sparking when fault current flows through the conduit system. Where it is impractical to make a threaded joint tight, a bonding jumper shall be utilized.

Fɪɢ. 41–6.—Article 500, *National Electrical Code,* describes locations where various types of explosion-proof electrical equipment can be used. Later editions of this code may change specific details. Be sure to check.

500-2. Special Precaution. The intent of Articles 500 through 503 is to require a form of construction of equipment, and of installation that will insure safe performance under conditions of proper use and maintenance. It, therefore, is assumed that inspection authorities and users will exercise more than ordinary care with regard to installation and maintenance.

The explosion characteristics of air mixtures of hazardous gases, vapors, or dusts vary with the specific material involved. Classification of a hazardous mixture into a Class I hazardous location, Group A, B, C or D, involves determinations of maximum explosion pressure, maximum safe clearance between parts of a clamped joint in an enclosure, and the minimum ignition temperature of the atmospheric mixture. For Class II location, Groups E, F, and G, the classification involves the tightness of the joints of assembly and shaft openings to prevent entrance of dust in the dust-ignition-proof enclosure, the blanketing effect of layers of dust on the equipment that may cause overheating, electrical conductivity of the dust, and the ignition temperature of the dust. It is necessary, therefore, that equipment be approved not only for the class of location but also for the specific group of the gas, vapor or dust that will be present.

For purposes of testing and approval, various air mixtures (not oxygen enriched) have been grouped on the basis of their hazardous characteristics, and facilities have been made available for testing and approval of equipment for use in the following atmospheric groups:

For Groups A, B, C and D, see Table 500-2(c).

Group E, Atmospheres containing metal dust, including aluminum, magnesium, and their commercial alloys, and other metals of similarly hazardous characteristics.
Group F, Atmospheres containing carbon black, coal or coke dust.
Group G, Atmospheres containing flour, starch, or grain dust.
Certain chemical atmospheres may have characteristics which would require safeguards beyond those required for any of the above groups. Carbon disulfide is one of these chemicals because of its low ignition temperature (100 C) and the small joint clearance required to arrest its flame. For a complete list noting properties of flammable liquids, gases and solids refer to NFPA No. 325M—1969.

(a) **Approval for Class and Properties.** Equipment shall be approved not only for the class of location but also for the explosion properties of the specific gas, vapor, or dust that will be present. In addition, equipment shall not have exposed any surface that operates at a temperature in excess of the ignition temperature of the specific gas vapor or dust.

The characteristics of various atmospheric mixtures of hazardous gases, vapors, and dusts depend on the specific hazardous material involved.

(b) **Marking.** Approved equipment shall be marked to show the Class, Group and operating temperature, or temperature range, based on operation in a 40 C ambient for which it is approved.

The temperature range, if provided, shall be indicated in identification numbers, as shown in Table 500-2(b).

Identification numbers marked on equipment nameplates shall be in accordance with Table 500-2(b).

Exception: Equipment of the nonheat-producing type, such as junction boxes, conduit and fittings, are not required to have a marked operating temperature.

For purposes of testing and approval, various atmospheric mixtures (not oxygen enriched) have been grouped on the basis of their hazardous characteristics, and facilities have been made available for testing and approval of equipment for use in the atmospheric groups listed in Table 500-2(c). Since there is no consistent relationship between explosion properties and ignition temperature, the two must be regarded as independent requirements.

TABLE 500–2(b).

IDENTIFICATION NUMBERS

Maximum Temperature Degrees C	Degrees F	Identification Number
450	842	T1
300	572	T2
280	536	T2A
260	500	T2B
230	446	T2C
215	419	T2D
200	392	T3
180	356	T3A
165	329	T3B
160	320	T3C
135	275	T4
120	248	T4A
100	212	T5
85	185	T6

(c) **Temperature.** The temperature marking specified in (b) above shall not exceed the ignition temperature of the specific gas or vapor to be encountered. For information regarding ignition temperatures see NFPA 325M, Fire Hazard Properties of Flammable Liquids, Gases, Volatile Solids — 1969.

Formerly the temperature limit of each Group was assumed to be the lowest ignition temperature of any material in the Group, i.e., 280 C for Group D, 180 C for Group C. To avoid revising this limit as new gases are added (see hexane in Group D and acetaldehyde in Group C) temperature will be specified in future markings.

The ignition temperature for which equipment was approved prior to this requirement may be assumed to be as follows:

Group A 280 C (536 F)
Group B 280 C (536 F)
Group C 180 C (356 F)
Group D 280 C (536 F)

500–3. Specific Occupancies. See Articles 510 through 517 for rules applying to garages, aircraft hangars, gasoline dispensing and service stations, bulk storage plants, finishing processes, and flammable anesthetics.

500–4. Class I Locations. Class I locations are those in which flammable gases or vapors are or may be present in the air in quantities sufficient to produce explosive or ignitible mixtures. Class I locations shall include the following:

(a) **Class I, Division I.** Locations (1) in which hazardous concentrations of flammable gases or vapors exist continuously, intermittently, or periodically under normal operating conditions, (2) in which hazardous concentrations of such gases or vapors may exist frequently because of repair or maintenance operations or because of leakage, or (3) in which breakdown or faulty operation of equipment or processes which might release hazardous concentrations of flammable gases or vapors, might also cause simultaneous failure of electrical equipment.

(FIG. 41–6 *continues on next page*.)

TABLE 500–2(c). CHEMICALS BY GROUPS

Group A Atmospheres	Group D Atmospheres
acetylene	acetone
	acrylonitrile
Group B Atmospheres	ammonia[3]
	benzene
butadiene[1]	butane
ethylene oxide[2]	1-butanol (butyl alcohol)
hydrogen	2-butanol (secondary butyl alcohol)
manufactured gases containing more	n-butyl acetate
than 30% hydrogen (by volume)	isobutyl acetate
propylene oxide[2]	ethane
	ethanol (ethyl alcohol)
	ethyl acetate
Group C Atmospheres	ethylene dichloride
acetaldehyde	gasoline
cyclopropane	heptanes
diethyl ether	hexanes
ethylene	methane (natural gas)
isoprene	methanol (methyl alcohol)
unsymmetrical dimethyl hydrazine	3-methyl-1-butanol (isoamyl alcohol)
(UDMH 1, 1-dimethyl hydrazine)	methyl ethyl ketone
	methyl isobutyl ketone
	2-methyl-1-propanol (isobutyl alcohol)
	2-methyl-2-propanol (tertiary butyl alcohol)
	petroleum naphtha[4]
	octanes
	pentanes
	1-pentanol (amyl alcohol)
	propane
	1-propanol (propyl alcohol)
	2-propanol (isopropyl alcohol)
	propylene
	styrene
	toluene
	vinyl acetate
	vinyl chloride
	xylenes

[1] Group D equipment may be used for this atmosphere if such equipment is isolated in accordance with Section 501–5(a) by sealing all conduit ½-inch size or larger.

[2] Group C equipment may be used for this atmosphere if such equipment is isolated in accordance with Section 501–5(a) by sealing all conduit ½-inch size or larger.

[3] For Classification of areas involving ammonia atmosphere refer to ANSI B9.1 *Safety Code for Mechanical Refrigeration*—1971 and ANSI K61.1 *Storage and Handling of Anhydrous Ammonia* —1971.

[4] A saturated hydrocarbon mixture boiling in the range 20–135 C (68–275 F). Also known by the synonyms benzine, ligroin, petroleum ether or naphtha.

This classification usually includes locations where volatile flammable liquids or liquefied flammable gases are transferred from one container to another; interiors of spray booths and areas in the vicinity of spraying and painting operations where volatile flammable solvents are used; locations containing open tanks or vats of volatile flammable liquids; drying rooms or compartments for the evaporation of flammable solvents; locations containing fat and oil extraction apparatus using volatile flammable solvents; portions of cleaning and dyeing plants where hazardous liquids are used; gas generator rooms and other portions of gas manufacturing plants where flammable gas may escape; inadequately ventilated pump rooms for flammable gas or for volatile flammable liquids; the interiors of refrigerators and freezers in which volatile, flammable materials are stored in open, lightly stoppered, or easily ruptured containers, and all other locations where hazardous concentrations of flammable vapors or gases are likely to occur in the course of normal operations.

(b) Class I, Division 2. Locations (1) in which volatile flammable liquids or flammable gases are handled, processed or used, but in which the hazardous liquids, vapors or gases will normally be confined within closed containers or closed systems from which they can escape only in case of accidental rupture or breakdown of such containers or systems, or in case of abnormal operation of equipment, (2) in which hazardous concentrations of gases or vapors are normally prevented by positive mechanical ventilation, but which might become hazardous through failure or abnormal operation of the ventilating equipment, or (3) which are adjacent to Class I, Division 1 locations, and to which hazardous concentrations of gases or vapors might occasionally be communicated unless such communication is prevented by adequate positive-pressure ventilation from a source of clean air, and effective safeguards against ventilation failure are provided.

This classification usually includes locations where volatile flammable liquids or flammable gases or vapors are used, but which, in the judgment of the authority having jurisdiction, would become hazardous only in case of an accident or of some unusual operating condition. The quantity of hazardous material that might escape in case of accident, the adequacy of ventilating equipment, the total area involved, and the record of the industry or business with respect to explosions or fires are all factors that should receive consideration in determining the classification and extent of each hazardous area.

Piping without valves, checks, meters and similar devices would not ordinarily be deemed to introduce a hazardous condition even though used for hazardous liquids or gases. Locations used for the storage of hazardous liquids or of liquefied or compressed gases in sealed containers would not normally be considered hazardous unless subject to other hazardous conditions also.

Electrical conduits and their associated enclosures separated from process fluids by a single seal or barrier shall be classed as Division 2 locations if the outside of conduit and enclosures is a nonhazardous area.

500-5. Class II Locations. Class II locations are those which are hazardous because of the presence of combustible dust. Class II locations shall include the following:

(a) Class II, Division 1. Locations (1) in which combustible dust is or may be in suspension in the air continuously, intermittently, or periodically under normal operating conditions, in quantities sufficient to produce explosive or ignitible mixtures, (2) where mechanical failure or abnormal operation of machinery or equipment might cause such mixtures to be produced, and might also provide a source of ignition through simultaneous failure of electrical equipment, operation of protection devices, or from other causes, or (3) in which dusts of an electrically conducting nature may be present.

(FIG. 41–6 *continues on next page.*)

This classification usually includes the working areas of grain handling and storage plants; rooms containing grinders or pulverizers, cleaners, graders, scalpers, open conveyors or spouts, open bins or hoppers, mixers or blenders, automatic or hopper scales, packing machinery, elevator heads and boots, stock distributors, dust and stock collectors (except all-metal collectors vented to the outside), and all similar dust producing machinery and equipment in grain processing plants, starch plants, sugar pulverizing plants, malting plants, hay grinding plants, and other occupancies of similar nature; coal pulverizing plants (except where the pulverizing equipment is essentially dust-tight); all working areas where metal dusts and powders are produced, processed, handled, packed or stored (except in tight containers); and all other similar locations where combustible dust may, under normal operating conditions, be present in the air in quantities sufficient to produce explosive or ignitible mixtures.

Combustible dusts which are electrically nonconducting include dusts produced in the handling and processing of grain and grain products, pulverized sugar and cocoa, dried egg and milk powders, pulverized spices, starch and pastes, potato and woodflour, oil meal from beans and seed, dried hay, and other organic materials which may produce combustible dusts when processed or handled. Electrically conducting nonmetallic dusts include dusts from pulverized coal, coke and charcoal. Dusts containing magnesium or aluminum are particularly hazardous and every precaution must be taken to avoid ignition and explosion.

(b) **Class II, Division 2.** Locations in which combustible dust will not normally be in suspension in the air, or will not be likely to be thrown into suspension by the normal operation of equipment or apparatus, in quantities sufficient to produce explosive or ignitible mixtures, but (1) where deposits or accumulations of such dust may be sufficient to interfere with the safe dissipation of heat from electrical equipment or apparatus, or (2) where such deposits or accumulations of dust on, in, or in the vicinity of electrical equipment might be ignited by arcs, sparks or burning material from such equipment.

Locations where dangerous concentrations of suspended dust would not be likely, but where dust accumulations might form on, or in the vicinity of electrical equipment, would include rooms and areas containing only closed spouting and conveyors, closed bins or hoppers, or machines and equipment from which appreciable quantities of dust would escape only under abnormal operating conditions; rooms or areas adjacent to locations described in Section 500-5(a), and into which explosive or ignitible concentrations of suspended dust might be communicated only under abnormal operating conditions; rooms or areas where the formation of explosive or ignitible concentrations of suspended dust is prevented by the operation of effective dust control equipment; warehouses and shipping rooms where dust producing materials are stored or handled only in bags or containers; and other similar locations.

500-6. Class III, Locations. Class III locations are those which are hazardous because of the presence of easily ignitible fibers or flyings, but in which such fibers or flyings are not likely to be in suspension in air in quantities sufficient to produce ignitible mixtures. Class III locations shall include the following:

(a) **Class III, Division 1.** Locations in which easily ignitible fibers or materials producing combustible flyings are handled, manufactured or used.

Such locations usually include some parts of rayon, cotton and other textile mills; combustible fiber manufacturing and processing plants; cotton gins and cotton-seed mills; flax processing plants; clothing manufacturing plants; woodworking plants; and establishments and industries involving similar hazardous processes or conditions.

Easily ignitible fibers and flyings include rayon, cotton (including cotton linters and cotton waste), sisal or henequen, istle, jute, hemp, tow, cocoa fiber, oakum, baled waste kapok, Spanish moss, excelsior and other materials of similar nature.

(b) **Class III, Division 2.** Locations in which easily ignitible fibers are stored or handled (except in process of manufacture).

FIG. 41–6 (*concluded*)

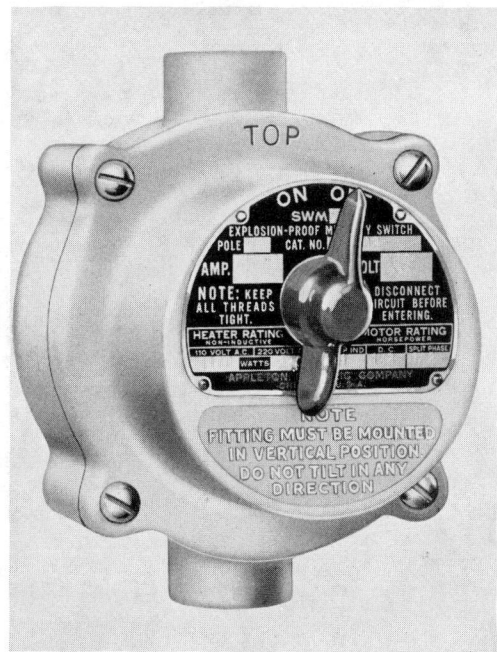

FIG. 41–7.—An explosion-proof and dust-tight mercury switch, listed by Underwriters Laboratories, which makes and breaks the circuit when the switch tilts the tube inside the mechanism.

Courtesy The Appleton Electric Co.

many industrial users have, likewise, found them to be an effective means of reducing exposure to the hazards of electric current. It should be noted that in plants which have instituted the strong safety program wherein the employee is admonished to look for the grounding pin on all attachment plugs, the double-insulated device now constitutes an exception to the rule. The ensuing confusion may result in the loss of some degree of safety.

Explosion-proof apparatus

Standard electrical apparatus considered safe for ordinary application is obviously unfit for installation in locations where flammable gases, vapors, dusts, and other easily ignitable materials are present. Sparks and electric arcs originating in such fittings have been the igniting medium in costly fires and explosions.

Explosion-proof Apparatus is defined in the *National Electrical Code:*

Apparatus enclosed in a case which is capable of withstanding an explosion of a specified gas or vapor which may occur within it and of preventing the ignition of a specified gas or vapor surrounding the enclosure by sparks, flashes, or explosion of the gas or vapor within, and which operates at such an external temperature that a surrounding flammable atmosphere will not be ignited thereby.

Before electrical equipment and its associated wiring is selected for a hazardous location, it is necessary to determine the exact nature of the flammable materials present. For instance, an electrical fitting or device which is found by test to be safe for installation in an atmosphere of combustible dust may be unsafe for operation in an atmosphere containing flammable vapors or gases.

The *National Electrical Code* should be checked for these types of hazardous materials, which are described in detail in Articles 500 through 503. A summary of the hazardous location classifications described in this code are given in Fig. 41–6.

A study should be made of the machines or devices to be used and of the processes involving liquids, gases, or solid substances, and their ratings should be checked. When hazards have been determined and the layout of the building inspected, it can be decided whether one small section of the plant should be classified as hazardous or whether the hazardous conditions extend to all parts.

The results of this study should be presented to a manufacturer of explosion-proof apparatus so that electrical equipment and wiring can be selected for safe installation.

Leading manufacturers of electrical equipment often maintain staffs of engineers to guide buyers in the purchase of explosion-proof fittings. Because of the highly technical nature of this field, competent engineers should always be consulted for these needs.

Only fittings which have undergone exhaustive tests and meet the requirements of the Underwriters' Laboratories, Inc., for use in hazardous locations should be used (Fig. 41–7). They must be of durable material, provide thorough protection, and be finished to be totally resistant to atmospheric conditions.

Explosion-proof fittings should be installed

FIG. 41–8.—Automatic cord take-up reel eases movement during inspection and repair work and helps prevent kinking and excessive bending of the cord. This particular unit has a transformer (mounted at top) to step down to low voltage for added safety.

Courtesy Daniel Woodhead Co.

not only on new work, but also on old wiring systems where alterations are being made or new equipment is being installed. Observing code requirements will minimize dangers that might result from using ordinary fittings in hazardous locations.

These fittings are made of durable cast material, with roomy interiors for wiring and splices, and they are capable of withstanding the high internal pressure which results from an explosion without bursting and without being loosened. Furthermore, they are tested on the basis of a high safety factor as prescribed by the Underwriters Laboratories Inc.

It is impossible to prevent highly flammable gases from entering the interior of either an explosion-resisting or an ordinary wiring system. They will eventually enter the entire line through the joints and through the breathing of the conduit system caused by temperature changes. Furthermore, gaseous vapors will fill every crevice whenever covers are removed.

For these reasons, it is impossible to provide an entirely vapor-proof switch unit or to regulate temperatures or keep the air free from flammable gases inside the fittings.

To protect that section of the industrial plant classified as a hazardous location, it is necessary to have positive confinement of the arc, heat, and explosion within the internal limits of the explosion-proof fittings. These fittings are constructed to completely imprison the dangerous arcing, intense heat, and subsequent explosion so that the gas-laden air outside does not become ignited.

Often control equipment can be located outside a room containing hazardous materials. In this case, conventional wiring equipment can be used, thus reducing cost and hazards in the installation.

Extension cords

Extension cords should be listed by the Underwriters Laboratories Inc. They should be inspected regularly. Kinking or excessive bending of the cord should be avoided to prevent the wire strands from breaking (Fig. 41–8). Broken strands may pierce the insulated covering and become a shock or short-circuit hazard.

Ordinary twisted lamp cords should never be used for extension cords or lamps in boilers, in tanks, or on damp or metallic floors, or where it will be exposed to mechanical wear. The various types of flexible cords and cables and their approved usage are given in the *National Electrical Code.*

Cord for use with portable tools and equipment is made in several grades, each of which is designed for a specific type of service. Jacketed cord should be used with portable electric tools and with extension lamps in boilers, tanks, or other grounded enclosures. Special types of rubber or plastic covering should be considered when the cord is to be used in areas where it may come in contact with oils or solvents.

Cord for heating devices, such as electric irons and water heaters, is made with an insulated covering which contains flame-retardant or thermosetting compound, such as neoprene. They are designed to resist high temperatures and, in the case of neoprene, dampness.

Because the metal frames of portable electrical equipment should be grounded, cord with a green-covered ground conductor should be used with a polarized plug and receptacle. If plant circuits are not equipped with polarized receptacles, a third wire taped at intervals to the cord, to prevent its breaking and to prevent a tripping hazard, may be used. One end of this third wire is electrically connected to the frame of the tool, and the other end, which should be several feet longer than the cord, terminates in a spring-type metallic clip large enough to be clamped to a water pipe or other effective ground.

Flexible cords should be so connected to devices and fittings that tension will not be transmitted to joints or terminal screws. This is accomplished by special fittings, a knot in the cord, or winding with tape. All plugs that are attached to cords must have the terminal screw connections covered by suitable insulation.

Extension cords with brass shell sockets should be prohibited because the socket may become energized through contact with loose wires inside the socket, through abrasion of the insulation where the cord enters the socket, or through moisture in the socket insulation. Porcelain, composition, or rubber-covered sockets should be used.

FIG. 41–9.—An electrician checking equipment with an insulator resistance tester.

Courtesy Associated Research, Inc.

Handles of portable lamp holders must be made of nonconductive material, and there should be no metallic connection between the lamp guard and the socket shell. For use near exposed live parts, such as the rear of switchboards, the guard itself should be of nonconductive material.

Since extension lamps are sometimes used under conditions in which 120-volt shock may prove fatal, it is essential that safe cords and lamp holders be provided and that they be maintained in excellent condition. In some plants the shock hazard of portable lamps is eliminated by use of small portable transformers which reduce the lamp voltage down to 12 volts. Special lamps for these units are rated at 75-watt capacity and give sufficient working light.

Test equipment

Most electrical equipment is designed for safe operation under limited overload conditions for varying periods of time. Operators should be thoroughly familiar with the limitations of their equipment and should be trained to observe and report abnormal conditions.

Continued overload may introduce additional operating hazards by causing fire, short circuits, circuit failures, or machine failures.

Through the use of various types of electrical testing meters, many of these conditions can be detected before they get out of control and cause damage.

Qualified electricians or specially trained maintenance men should make these tests. (See Fig. 41–9.)

The following types of equipment are standard and should be considered even by plants with limited budgets: split-core ammeter, voltmeter, ammeter, megohmmeter, ground fault indicators and locators, wattmeter, industrial analyzer, recording instrument, and specialized testing instruments. (See Fig. 41–10.)

Specialized testing instruments, such as vacuum tube voltmeters, oscilloscopes, and cable testers may be used, but these instruments do not fit into a limited budget. They are generally fitted for detailed engineering work or for use where ordinary recording and indicating instruments are not accurate enough.

Specialized equipment

Among the many types of electrical equipment used in industry are the electric furnace, auxiliary heating devices, high-frequency heating equipment (Fig. 41–11), electric welding equipment, X-ray, ultraviolet, and infrared installations. Each of these devices may introduce special operating hazards, but protection from their electrical hazards may be secured through the same procedures recommended for use with the more common types of electrical equipment.

Because high-frequency heating installations range in power capacity from a few hundred watts to several hundred kilowatts, safety considerations are of prime importance.

The resistance of the body to the flow of high-frequency current is not dependent upon the skin. At frequencies of 200 kHz to several hundred megacycles currents flow in a very

FIG. 41–10.—A ground fault indicator and locator are beneficial instruments in plants having ungrounded electrical distribution systems.

Courtesy Excel Electric Service Co.

FIG. 41–11.—Case hardening pump shafts with rf induction heating. The plastic shield closes entrance before heating cycle starts. It also confines cooling water spray, keeping the operator and equipment dry.

Courtesy Allis-Chalmers Co.

Fig. 41–12.—Another type of feed which keeps the operator's hands away from the heating unit. This conveyor automatically takes the parts past the electrodes where the fusing is done.

Courtesy Allis-Chalmers Co.

thin shell on the surface of the conductor. This tendency of high-frequency currents to flow on the surface is known as "skin effect" and increases as the frequency increases. Should the skin of a human being be punctured, the currents still flow on the surface and do not penetrate to the vital organs of the body.

A person coming into contact with high-frequency power will in general be burned because of his natural tendency to pull away from the conductor when contact is made, thereby setting up an arc. These burns are painful and usually take longer to heal than burns from the more common dry-heat sources.

The following methods are used to prevent accidental contact and high-frequency burns to operators:

1. Door interlocks, which remove power from the equipment whenever an access door is open, are connected into the control circuit to protect the operator from contacting any voltage higher than in the supply line.

2. In repetitive operation, the material to be heated by induction is carried by revolving hopper feeds or conveyor (Fig. 41–12) to the heating coil. The heater coil is enclosed with a shield and cannot be reached by the operator.

3. Generators are located some distance from the work position. The high-frequency power is then conveyed by a transmission line. In some plants high-frequency generators supplying power to heater coils are on the floor below.

4. In some types of equipment, coils are insulated with two or more layers of glass tape and impregnated with silicone varnish. This insulation protects the operator in case he accidentally contacts the heater coil.

Inspection

Whenever possible, equipment must be de-energized by the plant electrician or other authorized person. It should always be assumed that a circuit is live until it is proved dead. Therefore, switches and circuit breakers carrying current to or from switchboards, buses, controls, and starting equipment should be *checked to see that they are open*. Tests should be made in the inspector's presence by the plant electrician, foreman, or operator to determine that the parts to be worked on are *dead*. As an additional safeguard, the inspector should have the breakers and switches locked open, grounded, and tagged so that they cannot be energized until the work has been completed. (A lock-out procedure is detailed later on in this chapter.)

Upon completion of his work, the workman should remove any grounds he has used, and his lock-out tags. He should clear with the chief electrician or foreman or operator (whoever is in charge) before the equipment is returned to service.

If the feed-in circuit to a switchboard, bus, or other equipment must be kept live on the incoming side of a breaker or switches, the electrician should provide ample clearance, barriers, or other protection between this section and the part on which he is to work. He should not work in cramped quarters unless he has to, and he should insist that the men with him have freedom of movement in case of emergency.

Live buses, conductors, and switches should be covered with insulated blankets, special-formed insulating shields, or isolated with barriers.

Electrically operated or remote-control circuit breakers or contactors should not be depended upon for protection. A ground or other disturbance on the control system may permit the circuit breaker to close. If no disconnects are installed ahead of or behind the circuit breaker, it will be necessary to block, rackout, or otherwise lock the breaker so that it cannot be closed.

If the electrician must leave the circuit on which he is working before he completes the test, he should use lock-out procedures and hold-off tags. Upon his return, he should check all markings, breakers, and switches to be sure there have been no changes before he continues the tests.

Rotating and intermittent-start equipment

Before work is started on a rotating machine or automatic and intermittent-start equipment that is presumed to be out of service and stopped, an inspection of all the electrical control and starting devices should be made with the chief electrician or foreman.

For example, when inspections or repairs are to be made on motors, generators, blowers, compressors, converters, any part of which is remotely controlled or which may automatically start, circuit breakers or switches should be locked out and provided with hold-off tags, and the fuses should be pulled.

Machinery connected to blowers, water wheels, or pumps, without check valves, may start turning even when the current to the motor has been disconnected. For this reason, the rotor or armature should be blocked before inspections are made.

Lockout procedures should be followed on generators driven by prime movers. The throttle, starting valve, or other means of controlling the energy to the driving part of the unit should be locked and tagged.

While the equipment is in operation, motor brushes should not be removed or adjusted, contacts or other parts of the electrical equipment should not be worked on, and no attempt should be made to clean, sandpaper, or polish commutators, or slip or collector rings.

When inspecting electrical equipment, employees should not wear loose clothing because it may become entangled in couplings, coils, or other moving parts. They should remove wrist watches, rings, metal pens and pencils, and they should not use metal flashlights.

High-voltage equipment

In general, high-voltage equipment is more carefully guarded than low-voltage equipment because of the greater inherent hazard. In newer installations of 2300 volts or more, attempts have been made to insulate or armor apparatus so that casual contact with the current-carrying parts is not possible. In any case, only authorized, trained personnel should work on such equipment.

Persons working on high-voltage equipment should know that rubber gloves should not be used as a substitute for safety devices or procedures, but are worn as a supplementary measure. Before each use they should be checked for punctures, tears, or abrasions. For an on-the-job test the cuffs should be rolled up and air forced into the fingers and palms of the gloves. If there is leakage, the gloves should not be used.

Leather protectors should be worn over rubber gloves to protect the rubber from mechanical damage and from oil and grease.

An electrical glove testing service should be provided, with regular testing intervals determined by the amount of use of the gloves, the type of work, and the voltages they are subject to. If there is doubt about the insulating quality of the gloves, they should not be used.

Private testing laboratories can be relied upon for accurate testing. In some cases they will keep individual company records on the physical condition of gloves and the time schedule for retesting. Where no laboratory is available, small concerns can usually get tests made at the local public utility.

Maintenance

To be safe and to give the best service, electric equipment must be well-maintained.

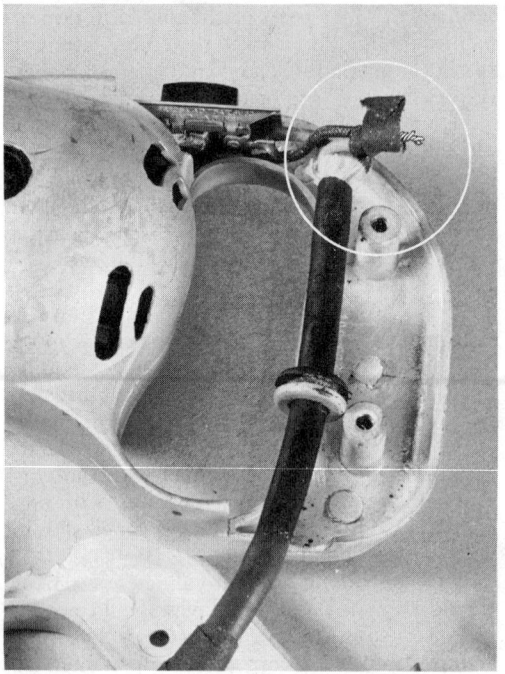

Fɪɢ. 41–13.—In the circle is an exposed electrical connection inside the handle of a portable electric tool, creating a serious shock hazard. This sort of defect should be remedied only by an experienced electrician.

Motors, circuit breakers, moving parts of switches, and similar current-carrying devices wear out, break down, and need readjustments. Repairs on electrical circuits and electrical apparatus should be made only by experienced electricians (Fig. 41–13).

Only high-grade electrical equipment listed as standard by Underwriters Laboratories Inc., or other qualified authority, should be used in maintenance work. Inferior, unapproved equipment may become hazardous because of defective material or design or poor workmanship.

If practicable, conductors should be deenergized when maintenance work on them is necessary. Before they begin work, maintenance men should make certain that the line is dead by testing it with the voltage testing devices provided. *It is not safe to test a circuit with the fingers or with makeshift devices.*

The men should be instructed in the use of electrical testing equipment and meters: how to tap into the circuit, how to locate testing points from schematic diagrams, how to use insulated meter leads, clips, and probes, and how to fuse test circuits. Pliers, screw drivers, testing lights, and other tools used in electrical repair work should be insulated.

When maintenance or repair work must be done on energized conductors, it is advisable to have two or more employees work together. The supervisor should give detailed procedures to be followed. He should also see that maintenance crews are supplied with and use the right protective equipment.

The kind of protective equipment needed will be determined by the type of circuit, the nature of the job, and the conditions under which the work must be done. Rubber gloves, sleeves, mats, line hose, insulating platforms, safety headgear, safety glasses, safety belts, fuse tongs, and insulated switch sticks are among the more common items of equipment usually available.

Safety equipment should be inspected before use and retested at frequent intervals.

Because of the location of electrical equipment and circuits, the methods of maintaining them, and the insulated tools required, the maintenance man must be constantly on guard to prevent accidental grounding. Such grounding may result if an energized loose wire comes in contact with water pipe, conduit, metal fixtures, another wire, or anything metallic that is connected with the earth.

Lockouts

Unexpected operation of electrical equipment that can be started by automatic or manual remote control may cause injuries to persons who happen to be near enough to be struck. Unexpected starting of motors may injure men working on them as well as the men operating machines controlled by the motors.

For that reason, when men are to repair motors or other electrical equipment, the circuit should be opened at the switch box, and the switch should be padlocked in the "off" position and tagged with a description of the work being done, the name of the man, and the department involved. Warning signs or tagging alone does not provide the positive protection of locking out equipment (Fig. 41–14).

Because of the grave risk to life, lockout procedures should be drilled into every maintenance man, and the supervisor should see that the procedure is implemented with the necessary keys, locks, and arrangements.

Only locks made by a reputable lock company should be used by maintenance men. No two locks should be the same, and the patterns of the keys should be checked to see that each key fits only one lock.

For identification, locks may be painted various colors to indicate types of craft or to differentiate shifts. Each lock should be stamped with the employee's name or clock number, or a metal tag should be attached.

Only one key should be issued to each maintenance man for his lock. The supervisor should have a master list of key numbers and should keep the extra key to each lock in his department. *In no case should the supervisor lend his own master key.* He must use the key himself until the old lock and the extra key are destroyed and replaced with new equipment.

On some types of locks, the tumblers can be changed by a locksmith and two new keys cut for it.

In case of combination locks, only the immediate supervisor, or whoever determines the company's or department's safety policy, should have a copy of the combination.

To make lockout systems operable, the purchasing department should be requested to buy either equipment with built-in locking devices or equipment designed for the insertion of padlocks. In plants where older equipment is in use, or where explosion-proof or dust-tight equipment is installed, it may be necessary for the maintenance department to construct attachments to which locks can be applied. For example, special tongues to hold several locks, common hasps to cover operating buttons, and sliding-rod devices which can be extended and locked in position to prevent operation of control handles can be devised.

No matter what method of locking electric switches "off" is used, effective control can be maintained only by constant supervision and by training maintenance men in the safe routine. Following is a lockout procedure

Fig. 41–14.—Typical multiple lockout with two individuals' locks.

Courtesy Osborn Manufacturing Corp.

which is generally acceptable:

1. Alert the operator.

2. Before starting work on an engine or motor, line shaft or other power transmission equipment, or power-driven machine, make sure it cannot be set in motion without your permission.

3. Place your own padlock on the control switch, lever, or valve, even though someone has locked the control before you. *You will not be protected unless you put your own padlock on it.*

4. If no padlock is available, place a MAN AT WORK sign at the control and block the mechanism in some effective manner. Make sure that both sign and blocking are fastened securely so that they cannot be easily removed.

5. When through working at the end of your shift, remove your own padlock, or your own sign and blocking. Never permit someone else to remove it for you, and be sure you are not exposing another person to danger by removing your padlock or sign.

6. If you lose the key to your padlock, report

1283

the loss immediately to your supervisor and *get a new padlock.*

Removing fuses

When it is necessary to remove a fuse, the operating switch should be opened to remove the load and the fuse should be extracted with an insulated fuse puller. If the fuse is not protected with a switch, the supply end of the fuse should be pulled out first. When the fuse is replaced, the supply end should be put in first.

It is important that a fuse be replaced by one of the same type and size and that a copper wire or other conductor never be substituted.

For work on lines carrying over 600 volts, the sources of energy can be opened by tripping the circuit breaker and then opening the disconnects. The conductors should be tagged, and, if possible, a substantial grounded conductor should be clamped to each leg of the circuit before the work is started. The grounding conductors should be removed just before the circuit is re-energized. As an added precaution, tests should be made with a voltmeter or other standard approved tester to make sure the circuit is clear.

In emergencies where the areas around fuse boxes are wet, wood platforms, insulated stools, or rubber boots should be used by maintenance workers.

Wiring

Electrical installations should be made in accordance with the *National Electrical Code* as a minimum standard, and each job should be properly inspected. The use of temporary wiring should be discouraged even though it may be reasonably safe when first put in. As a job progresses and repairs or installations are made, temporary wiring may become unsafe because it is not properly protected from mechanical injury.

Maintenance men should make it a practice to inspect plugs and connections on portable electric equipment to see that the third wire frame ground is properly connected, the grounding prong has not been removed, and the cord is properly connected to the terminals. Where there is any doubt about the safe operation of a tool, it should be removed from service, tagged, and returned to the tool room for repairs.

When additional equipment is being installed or operated under temporary conditions, no taps should be made into an existing circuit unless an individual switch is installed in the branch line, with provision being made for locking the switch "off" and tagging it. This precaution will prevent service interruptions in the main circuit when power must be shut off in the branch circuit, and it will also prevent accidental starting of extra equipment that is tapped in.

Employee Training

Few plants experience difficulty in selection and installation, but in maintenance and use of electrical equipment, deviation from safe practices is sometimes tolerated, with the usual result that unnecessary hazards are created and men are injured and killed. Therefore, for maximum effectiveness, the safety program should include thorough training of all employees who work with the electrical installation of the plant or operate electrical equipment. In addition to instructions on the hazards of electricity, employees should be trained in first aid, the use of warning signs, guards, and other protective devices, and in safe operational procedures. It is essential that they be trained in the mouth-to-mouth method of resuscitation and, in some cases, external cardiac massage.

The training program should be based on the plant's electrical system and applied specifically to plant operations.

Supervisors should be given such instructions as may be necessary to acquaint them with existing or probable electrical hazards, and they should be required to maintain close supervision over all operations which involve the use of electrical equipment. They should encourage employees to report immediately any electrical defects observed and this equipment should be repaired or replaced at once.

It is recommended that members of the organization's safety department familiarize themselves with the codes and standards which apply to their equipment and operation and that someone from the electrical maintenance staff be an active member of the plant safety committee.

References

American National Standards Institute, 1430 Broadway, New York, N.Y. 10018.
 "Dimensions of Attachment Plugs and Receptacles," C73 Series. (Also NEMA WD.1–1965.)
 "Insulated Wire," C8 Series.
 Manual and Automatic Station Control, Supervisory and Associated Telemetering Equipment, C37.2.
 Relays and Relay Systems Associated with Electric Power Apparatus, C37.
 Specifications for Rubber Insulating Tape. C59.6 (Also ASTM D119.)
 "Specifications for Rubber Protective Equipment for Electrical Workers," J6 Series.
 Specifications for Weather-Resistant Wire and Cable (URC Type). C8.18.

Dalziel, Charles F.
 "Effect of Electric Current on Man." *Electrical Engineering*, February 1941.
 "Scientific Facts Concerning Electrical Hazards." *National Safety News*, October 1947.
 "Electricity—Good and Faithful Servant." *National Safety News*, September 1961.

National Fire Protection Association, 470 Atlantic Ave., Boston, Mass. 02210. *National Electrical Code.* Standard No. 70 (ANSI Standard C-1).

National Safety Council, 425 North Michigan Ave., Chicago, Ill. 60611.
 Industrial Data Sheets
  Cleaning Electric Motors and Machinery, 285.
  Electric Plug and Receptacle Configurations, Applications of, 579.
  Electrical Switching Practices, 544.
  Electrical Test Equipment, 496.
  Electromagnets Used with Crane Hoists, 359.
  Electrostatic Paint Spraying and Detearing, 468.
  Extension Light Cords, Low Voltage, 316.
  Flexible Insulated Protective Equipment for Electrical Workers, 598.
  Gas-Shielded Arcs and Plasma-Jet Torches, 552.
  Ground Fault Circuit Interrupters for Personnel Protection, 636.
  Grounding Electric Shovels, Cranes and other Mobile Equipment, 287.
  Grounding Portable Electrical Equipment, 299.
  Hand Saws, Electric, 344.
  Industrial Electric Substations, 559.
  Lead-Acid Storage Batteries, 635.
  Lighting, Emergency, 248.
  Live Line Tools, 498.
  Maintenance of Electric Motors, 546.
  Nondestructive Testing of Materials, 488.
  Pole-Top Resuscitation, 376.
  Portable Reamer-Drills, 497.
  Radar Hazards, 481.
  Radio Frequency Heating, 319.
  Static Electricity, 547.
  Stray Currents in Electrical Blasting, 289.
  Temporary Electric Wiring for Construction Sites, 515.
  X-Rays in Industry, 475.

Underwriters Laboratories Inc., 207 East Ohio St., Chicago, Ill. 60611.
 "Electrical Appliance and Utilization Equipment Lists."
 "Electrical Construction Materials List."
 "Hazardous Location Equipment List."

U.S. Department of Commerce, National Bureau of Standards, Washington, D.C. 20234.
 Safety Rules for the Installation and Maintenance of Electric Utilization Equipment. Handbook H33.
 National Electrical Safety Code.

Flammable and Combustible Liquids

Chapter
42

Definitions   **1297**
Terms used

General Safety Measures   **1291**
Preventing dangerous mixtures . . . Smoking . . . Static
electricity . . . Bonding and grounding . . . Electrical equipment
. . . Spark-resistant tools

Health Hazards   **1296**
Toxic effects . . . Combustible gas indicators

Loading and Unloading Tank Cars   **1298**
Spotting cars . . . Inspection . . . Relieving pressure . . .
Removing covers . . . Unloading and loading connections . . .
Placards . . . Fires

Loading and Unloading Tank Trucks **1301**
Inspection . . . Smoking . . . Spotting trucks . . . Unloading and
loading connections . . . Leaks . . . Fires

Storage   **1303**
Tank construction . . . Vents . . . Dikes . . . Pump houses . . .
Gaging . . . Tanks in flooded regions . . . Underground tanks
. . . Aboveground tanks . . . Tank fires and their control . . .
Inside storage and mixing rooms . . . Storage cabinets . . .
Outside storage houses

Cleaning Tanks   **1314**
General precautions . . . Protective equipment . . . Proper
procedures . . . Cleaning aboveground tanks . . . Cleaning small
tanks and containers . . . Abandonment of tanks

Common Uses of Flammable Liquids   **1318**
Dip tanks . . . Japanning and drying ovens . . . Oil burners . . .
Cleaning metal parts . . . Internal combustion engines . . .
Spray booths . . . Gasoline blow torches and plumbers'
furnaces . . . Liquefied petroleum gases

References   **1322**

The precautions covered in this chapter are directed to those *industrial operators* who receive, store, handle, and use flammable liquids, and not principally to the *manufacturers* of flammable liquids. Because specific characteristics of flammable liquids vary as do their required handling precautions, this chapter does not attempt to cover the subject in all of its many details. It does present general information regarding flammable liquids and is intended to provide a broad view of the basic nature of the subject. It is advisable to consult federal and state laws, OSHA standards, fire underwriters, the National Fire Protection Association, product trade associations, state and municipal authorities that have jurisdiction, and specific handbooks for detailed information. (See "References" at the end of this chapter.)

Definitions

As defined by OSHA and the National Fire Protection Association Standard No. 30, *Flammable and Combustible Liquids Code,* a flammable liquid is any liquid having a flash point below 100 F and having a vapor pressure not exceeding 40 psia at 100 F. (The abbreviation "psia" stands for absolute pressure measured in pounds per square inch. Absolute pressure includes the influence of atmospheric pressure—about 14.7 psi—on the measurement. If this effect is subtracted, then the measurement is called "gage pressure.") Flammable liquids are usually subdivided into classes as shown on the pages immediately following. Combustible liquids are those with flash points at or above 100 F. Although they do not ignite as easily as flammable liquids, they can be ignited under certain circumstances and so must be handled with caution (Fig. 42–1).

The more common flammable and combustible liquids are crude oils, coal tars, various hydrocarbons, alcohols, and their by-products. They are chemical combinations of hydrogen and carbon. The combination may also contain oxygen, nitrogen, sulfur, or other elements.

Manufactured liquids and fluid commodities which contain flammable liquids, such as paints, floor polish, cleaning solutions, driers and varnishes, should be considered flammable liquids and classed according to the flash point of the mixture. Precautions incident to their handling and use differ according to their flash points, volatility, and the amount of flammable liquid within the mixture.

Flammable liquids vaporize and form flammable mixtures when in open containers, when leaks or spills occur, or when heated. The degree of danger is determined largely by the flash point of the liquid, the concentration of vapors in the air (whether the vapor-air mixture is in the flammable range or not), and the possibility of a source of ignition at or above a temperature sufficient to cause the mixture to burst into flame.

In the handling and use of flammable liquids, exposure of large liquid surfaces to air should be prevented. It is not the liquids themselves which burn or explode, but rather the vapor-air mixture formed when they evaporate. Therefore, handling and storing these liquids in closed containers and avoiding exposure of low flash liquids in use are of fundamental importance. (Make sure that all containers are correctly labeled.)

Chapter 39, "Chemical Hazards," supplies pertinent information on many liquids.

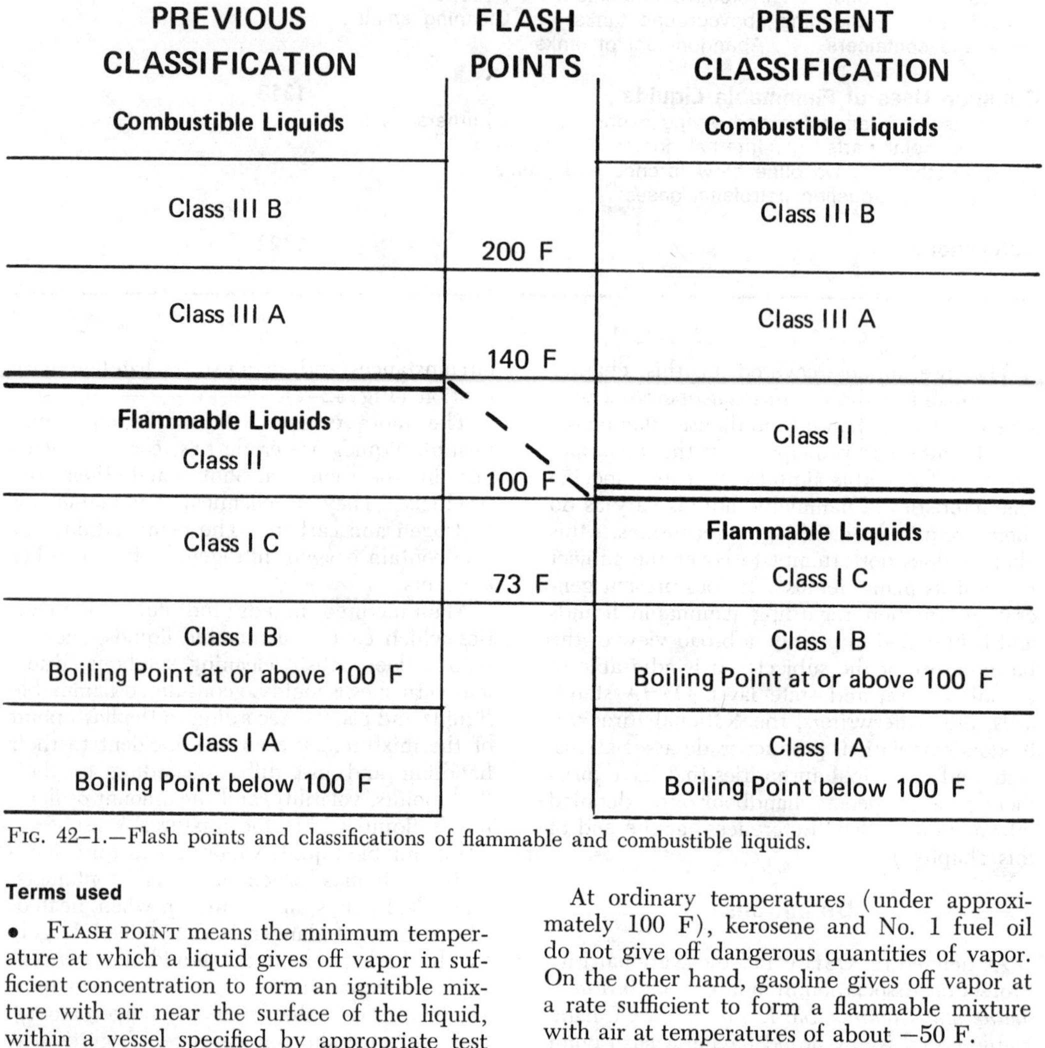

42—Flammable and Combustible Liquids

PREVIOUS CLASSIFICATION	FLASH POINTS	PRESENT CLASSIFICATION
Combustible Liquids		**Combustible Liquids**
Class III B		Class III B
	200 F	
Class III A		Class III A
	140 F	
Flammable Liquids Class II		Class II
	100 F	
Class I C		**Flammable Liquids** Class I C
	73 F	
Class I B Boiling Point at or above 100 F		Class I B Boiling Point at or above 100 F
Class I A Boiling Point below 100 F		Class I A Boiling Point below 100 F

Fig. 42–1.—Flash points and classifications of flammable and combustible liquids.

Terms used

- FLASH POINT means the minimum temperature at which a liquid gives off vapor in sufficient concentration to form an ignitible mixture with air near the surface of the liquid, within a vessel specified by appropriate test procedure and apparatus. For low-viscosity liquids, with flash points below 200 F, the Tag Closed Tester is used (ASTM D 56–73 gives the procedure). For high-viscosity liquids with flash points of 200 F or higher, the Pensky Martens Closed Tester is used (ASTM D 93–73 gives the procedure).

Other properties are factors in determining the hazards of flammable liquids, but the flash point is the principal factor. The relative hazard increases as the flash point lowers. The significance of this property becomes more apparent when liquids of different flash points are compared.

At ordinary temperatures (under approximately 100 F), kerosene and No. 1 fuel oil do not give off dangerous quantities of vapor. On the other hand, gasoline gives off vapor at a rate sufficient to form a flammable mixture with air at temperatures of about −50 F.

Any combustible liquid, when heated to a temperature at or above its flash point, will produce ignitable vapors. Heavy fuel oil, when heated to several hundred degrees F, for example, may produce flammable vapors just as readily as gasoline. Their characteristics are also changed when they are atomized. When such liquids are heated or atomized, they should be regarded as flammable liquids.

- AUTO-IGNITION TEMPERATURE is the lowest temperature at which a flammable gas- or vapor-air mixture will ignite from its own heat source or a contacted heated surface without necessity of spark or flame.

Vapors and gases will spontaneously ignite at a lower temperature in oxygen than in air, and their auto-ignition temperature may be influenced by the presence of catalytic substances.

● FLAMMABLE LIMITS. Flammable liquids have a minimum concentration of vapor in air below which propagation of flame does not occur on contact with a source of ignition. This is known as the lower explosive limit (LEL). There is also a maximum proportion of vapor or gas in air above which propagation of flame does not occur. This is known as the upper explosive limit (UEL).

For example, a gasoline vapor-air mixture with less than approximately 1.0 percent of gasoline vapor is too lean, and propagation of flame will not occur on contact with a source of ignition. Similarly, if there is more than approximately 8 percent of gasoline vapor, the mixture will be too rich. Other gases, such as hydrogen, acetylene, and ethylene, have a much wider range of flammable limits.

● FLAMMABLE RANGE is the difference between the lower and upper flammable limits, expressed in terms of percentage of vapor or gas in air by volume. It is also often referred to as the "explosive range."

For example, the limits of the flammable range of gasoline vapors are generally taken as 1.4 to 7.6 percent which is relatively narrow. Thus, a mixture of 1.4 percent gasoline vapor and 98.6 percent air is flammable, as are all the intermediate mixtures up to and including 7.6 percent gasoline vapor and 92.4 percent air. The range is therefore the difference between the limits, or 6.2 percent.

● PROPAGATION OF FLAME is the spread of flame through the entire volume of the flammable vapor-air mixture from a single source of ignition. A vapor-air mixture below the lower flammable limit may burn at the point of ignition without propagating (spreading away) from the ignition source.

● RATE OF DIFFUSION indicates the tendency of one gas or vapor to disperse into or mix with another gas or vapor. This rate depends on the density of the vapor or gas as compared with that of air, which is given a value of 1. Whether a vapor or gas is lighter or heavier than air determines, to a large extent, the design of the ventilation system. If the vapor or gas is heavier than air, the intake duct should be slightly above floor level. Conversely, if the vapor or gas is lighter than air, the intake duct should be located just below ceiling level.

● VAPOR PRESSURE is the pressure, measured in pounds per square inch absolute (psia), which is exerted by a volatile liquid, as determined by the *Standard Method of Test for Vapor Pressure of Petroleum Products (Reid Method)*, ASTM D 323–58.

● VOLATILITY is the tendency or ability of a liquid to vaporize. Such liquids as alcohol and gasoline, because of their well known tendency to evaporate rapidly, are called volatile liquids.

● OXYGEN DEFICIENCY designates an atmosphere having less than the percentage of oxygen found in normal air. Normally, air contains about 21 percent oxygen at sea level.

When the oxygen concentration in air is reduced to approximately 16 percent, many individuals become dizzy, experience a buzzing in the ears, and have a rapid heart beat.

In addition to tests for toxic substances therefore, the oxygen content of the atmosphere of a tank or similarly confined space should be determined before entry is permitted, with instruments approved for the purpose by NIOSH. (See discussion in Chapter 19, "Personal Protective Equipment.")

No one should enter a tank or enclosed space that tests less than 16 percent oxygen, unless he wears approved respiratory protective equipment, such as a fresh-air hose mask or self-contained or self-generating breathing apparatus.

Self-contained compressed air breathing apparatus, approved by OSHA-approved agencies, have proved satisfactory in oxygen deficient atmospheres. They are especially useful where it is difficult to run a hose line. Again, see Chapter 19.

● FLAMMABLE LIQUIDS. Flammable liquids are divided into three subclasses (IA, IB, and

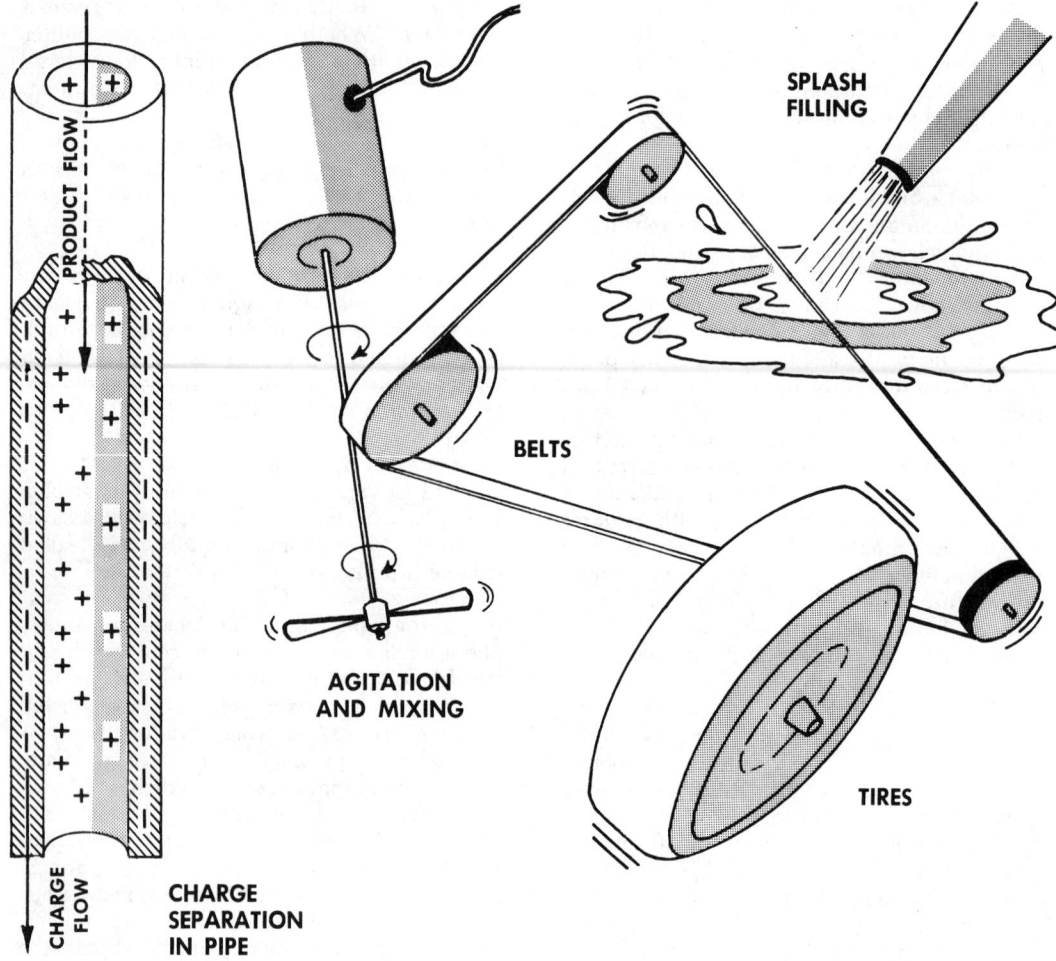

Fɪɢ. 42–2.—Typical static-producing situations, including charge separation in pipe.

IC), as specified in NFPA Standard No. 30. A flammable liquid has a flash point below 100 F and has a vapor pressure not exceeding 40 psia at 100 F (Fig. 42–1).

● Cᴏᴍʙᴜsᴛɪʙʟᴇ ʟɪǫᴜɪᴅs. Combustible liquids are divided into three subclasses (II, IIIA, and IIIB). A combustible liquid has a flash point at or above 100 F (Fig. 42–1).

Note that the NFPA code does not extend to liquids having flash points above 200 F, but this should not be construed as indicating that all liquids with high flash points are non-combustible.

The volatility of liquids is increased when they are heated to temperatures equal to or higher than their flash points. As pointed out by the NFPA, the *Flammable and Combustible Liquid Code* can be applied to high flash point liquids when they are heated, even though the same liquids when not heated are outside the range.

All tests should be made in accordance with methods either specified by OSHA or adopted by the American Society for Testing and Materials and approved by the American National Standards Institute.

The Department of Transportation in their *Hazardous Materials Regulations,* 49 C.F.R., Parts 170–179 (republished by the Bureau of Explosives in Graziano's Tariff No. 27), defines a flammable liquid as any liquid which gives off flammable vapors at or below a temperature of 80 F. This is of importance since the DOT Flammable Liquid Label is one means by which containers of flammable liquids can be identified for shipping, receiving, and transportation. However, this is not entirely foolproof since there are exceptions.

Further, other regulatory entities have adopted other definitions of what constitutes a "hazardous liquid." For example, the *Uniform Building Code,* published by the International Conference of Building Officials, says: "Highly flammable liquids shall be deemed to be those with a flash point below 190 F." These statements are very important where an industry must comply with legal requirements. In cases of dispute, the testing apparatus that is cited by the particular code must be used.

General Safety Measures

Preventing dangerous mixtures

Accidental mixture of flammable liquids should be prevented. A small amount of acetone accidentally put into a kerosene tank may lower the flash point of the contents, because of the relatively high volatility of acetone, and create a flammable mixture on later use of the kerosene. Gasoline mixed with fuel oil may change the flash point sufficiently to make the fuel oil hazardous in ordinary use. In each case, the lower flash point liquid can act as a fuse to ignite the higher flash point material.

Control valves on equipment containing flammable liquids should be identified by color, tag, or both. In some plants, the pipelines are painted or banded with distinctive colors and show direction of flow. Each tank should be marked with the name of the product. Lines from tanks of different types and classes of products should be kept separate and, preferably, separate pumps for the different types and classes of products should be provided.

A portable, approved container should be used for handling flammable liquids in quantities up to five gallons. The containers should be clearly identified by lettering or suitable color code. (See page 386.)

Smoking

Smoking and the carrying of "strike anywhere" matches, lighters, and other spark-producing devices should not be permitted in a building or area where flammable liquids are stored, handled, or used. The extent of the restricted area will depend on the type of products handled, the design of the building, and local conditions.

Suitable No Smoking signs should be posted conspicuously in those buildings and areas where smoking is prohibited.

Static electricity

Static electricity is generated by the contact and separation of dissimilar material. For example, static electricity is generated when a fluid flows through a pipe or from an orifice into a tank. Examples of several methods of generating static electricity are shown in Fig. 42–2. The principal hazards created by static electricity are those of fire and explosion, which are caused by spark discharges containing sufficient energy to ignite any flammable or explosive vapors, gases, or dust which are present. Also, the shocking of personnel may cause an involuntary reaction, such as falling, which may lead to an injury.

A point of great danger from a static spark is the place where a flammable vapor may be present in the air, such as the outlet of a flammable liquid fill pipe, a delivery hose nozzle, near an open flammable liquid container, and around a tank truck fill opening or barrel bunghole. A spark between two bodies occurs when there is not a good electrical conductive path between them. Hence, grounding and bonding of flammable liquid containers is necessary to prevent static electricity from causing a spark. The connections must be made precisely, therefore, National Fire Protection Association Standard No. 77, *Recommended Practice on Static Electricity,* should be consulted for details. A summary follows.

Bonding and grounding

The terms "bonding" and "grounding" often

CHARGED AND UNCHARGED BODIES INSULATED FROM GROUND

Charged body
insulated from
ground

+ + +
+ + +

Uncharged body
insulated from
ground

Charge (Q) = 6 microcoulombs
Capacitance (C) to ground = 0.01 microfarad
Voltage (V) to ground and
uncharged body = 600 volts

Charge (Q) = 0
Capacitance (C) = 0.01 microfarad
Voltage to ground (V) = 0

Ground

BOTH INSULATED BODIES SHARE THE SAME CHARGE

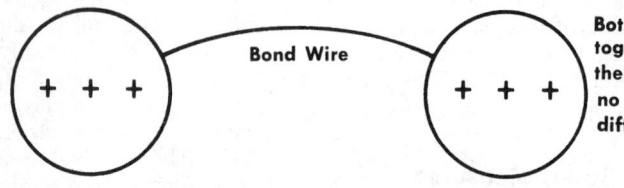

Bond Wire

+ + +   + + +

Both bodies bonded
together will share
the charge and have
no potential
difference

Charge (Q) on both bodies = 6 microcoulombs
Capacitance (C) to ground for both bodies = 0.02 microfarad
Voltage (V) to ground = 300 volts

Ground

BOTH BODIES ARE GROUNDED AND HAVE NO CHARGE

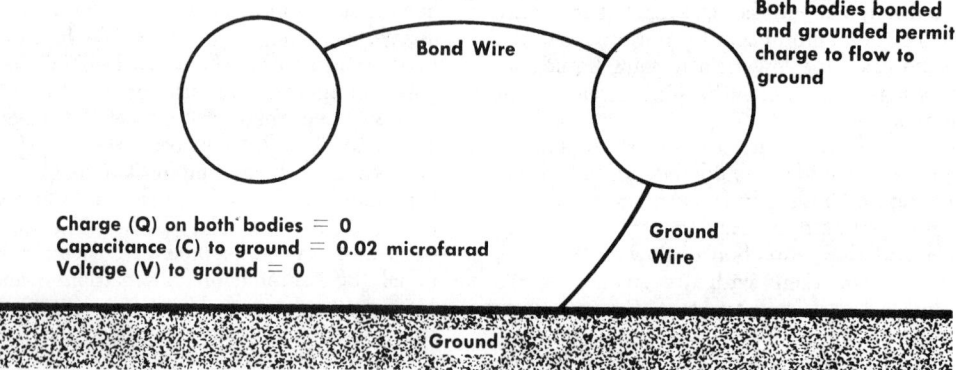

Bond Wire

Both bodies bonded
and grounded permit
charge to flow to
ground

Charge (Q) on both bodies = 0
Capacitance (C) to ground = 0.02 microfarad
Voltage (V) to ground = 0

Ground
Wire

Ground

FIG. 42–3.—Bonding eliminates a difference of potential between objects. Grounding eliminates a difference of potential between objects and ground. Bonding and grounding apply only to conductive bodies and, when properly applied, can be depended upon to remove the charge.

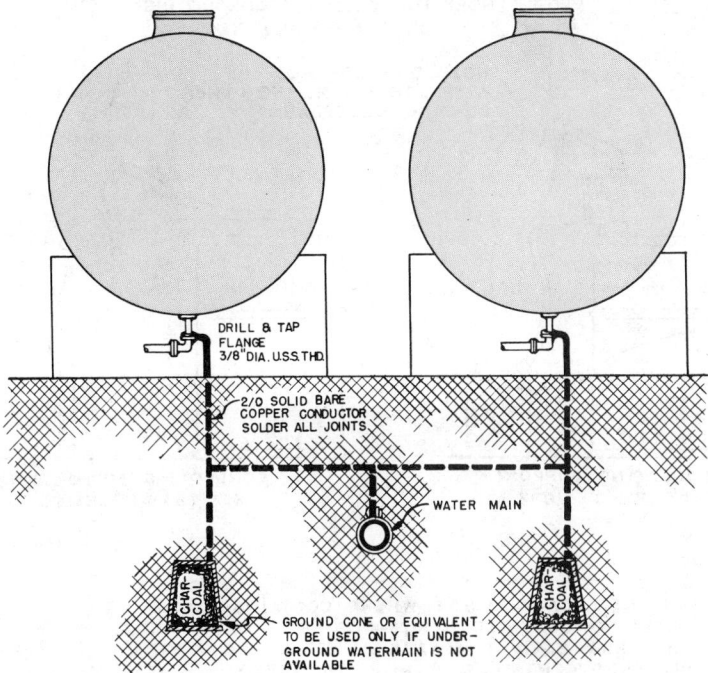

DRILL & TAP
FLANGE
3/8" DIA. U.S.S. THD.

2/0 SOLID BARE
COPPER CONDUCTOR
SOLDER ALL JOINTS

WATER MAIN

CHAR-
COAL

CHAR-
COAL

GROUND CONE OR EQUIVALENT
TO BE USED ONLY IF UNDER-
GROUND WATERMAIN IS NOT
AVAILABLE

FIG. 42–4.—Because some aboveground flammable liquid storage tanks
are not inherently grounded, such tanks should be grounded by either
one of these methods or by some other equally effective method.

have been used interchangeably because of poor understanding of the terms. Bonding is done to eliminate a difference in potential *between objects*. The purpose of grounding is to eliminate a difference in potential *between an object and ground* (Fig. 42–3). Bonding and grounding are effective only when the bonded objects are conductive.

Although bonding will eliminate a difference in potential between the objects that are bonded, it will not eliminate a difference in potential between these objects and the earth unless one of the objects possesses an adequate conductive path to earth. Therefore, bonding will not eliminate the static charge, but will equalize the potential between the objects bonded so that a spark will not occur between them.

When two objects are bonded, the charges flow freely between the bodies and there is no difference in charge.

An adequate ground will always discharge

a charged conductive body and is recommended as a safety measure whenever any doubt exists concerning a situation or a governing authority requires it.

To avoid a spark from discharge of static electricity during filling operations, a wire bond should be provided between the storage container and the container being filled, unless a metallic path between the container is otherwise present. For additional safety, it is advisable to have the bonding wire or one of the containers grounded.

When loading or unloading tank cars through open domes, it is best to use a downspout long enough to reach the tank bottom. Generally, tank cars need not be separately grounded because the resistance of the natural ground through the tank car wheels and rails, and the resistance of piping, flexible metallic joints, or metallic swivel joints, are considered sufficiently low to protect against static electricity. For detailed information

1293

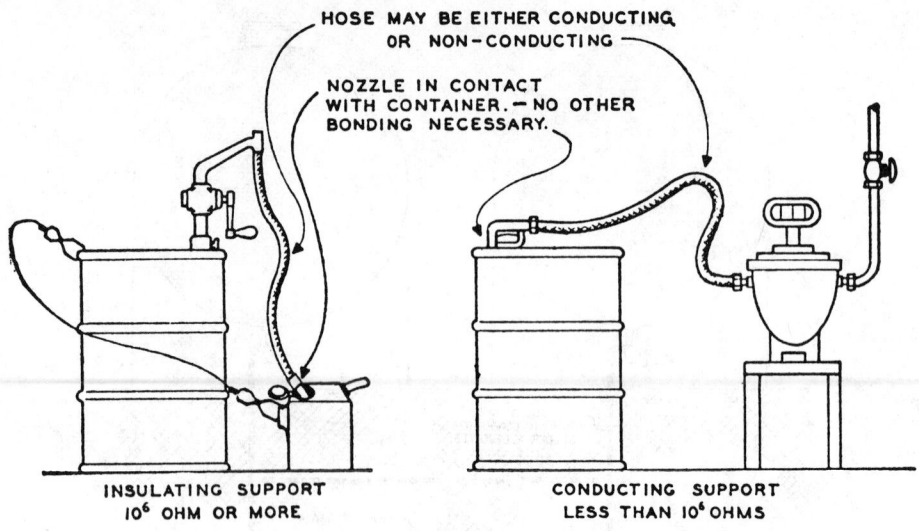

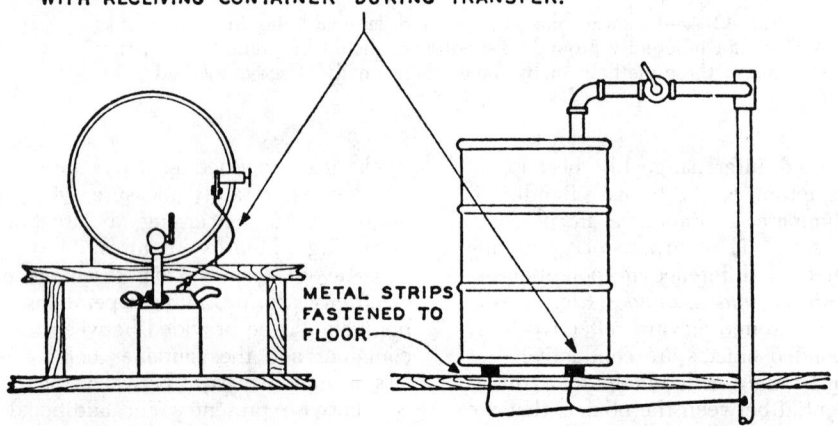

FIG. 42–5.—Bonding during container filling permits safe discharge of any static electricity generated.

Courtesy National Fire Protection Association.

and exceptions to this generality, consult NFPA Standard No. 77, *Recommended Practice on Static Electricity.*

Aboveground tanks used for storage of flammable liquids should be properly grounded. Ground wide should be uninsulated so it may be easily inspected for mechanical damage

(Fig. 42–4).

Petroleum liquids are capable of building up electrical charges when they flow through piping, are agitated in a tank or a container, or are subjected to vigorous mechanical movement such as spraying or splashing. Proper bonding and grounding of the transfer system

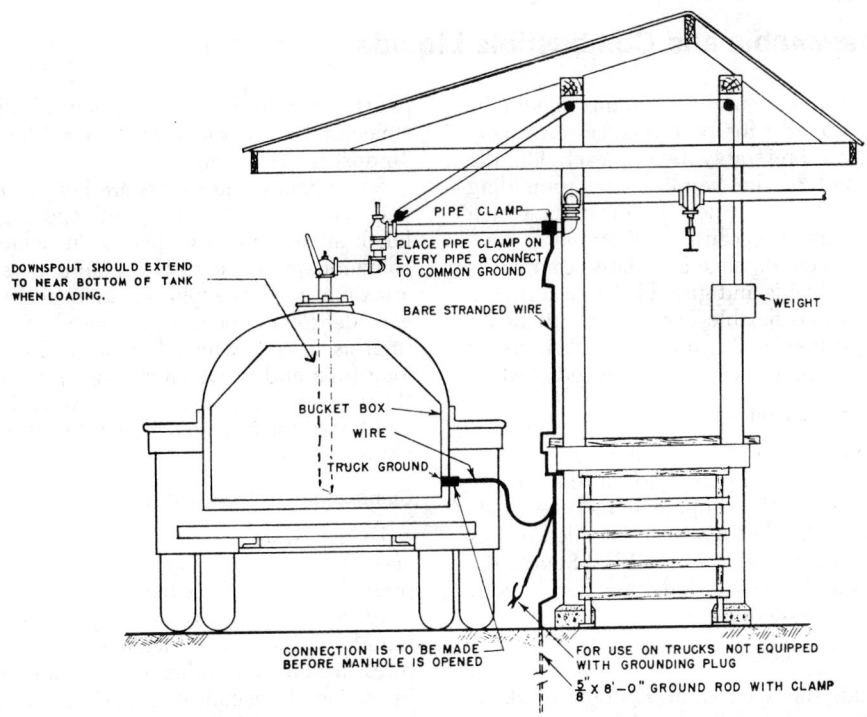

DOWNSPOUT SHOULD EXTEND
TO NEAR BOTTOM OF TANK
WHEN LOADING.

PIPE CLAMP

PLACE PIPE CLAMP ON
EVERY PIPE & CONNECT
TO COMMON GROUND

BARE STRANDED WIRE

WEIGHT

BUCKET BOX

WIRE

TRUCK GROUND

CONNECTION IS TO BE MADE
BEFORE MANHOLE IS OPENED

FOR USE ON TRUCKS NOT EQUIPPED
WITH GROUNDING PLUG

$\frac{5}{8}$" X 8'—0" GROUND ROD WITH CLAMP

Fɪɢ. 42–6.—Bonding and grounding of a flammable liquid tank truck and loading rack.

usually drains off this static charge to ground as fast as it is generated. However, rapid flow rates in transfer lines can cause very high electrical potentials on the surface of liquids, regardless of vessel grounding. Also, some petroleum liquids are poor conductors of electricity, particularly the pure, refined products; and even though the transfer system is properly grounded, a static charge may build up on the surface of the liquid in the receiving container. The charge accumulates because static cannot flow through the liquid to grounded metal as fast as it is being generated. If this accumulated charge builds up high enough, a static spark with sufficient energy to ignite a flammable air-vapor mixture can occur.

This high static charge is usually controlled by reducing the flow rates, avoiding violent splashing with side-flow fill lines, and use of relaxation time.

Motor frames, starting or control boxes, conduits, and switches should be grounded in accordance with the requirements for installation of electrical and power equipment that is outlined in the *National Electrical Code*.

When flammable liquids are poured from one container to another, a bonding means should be provided between the two conductive containers prior to pouring, as shown in Fig. 42–5.

Fig. 42–6 shows the bonding and grounding of a flammable liquid tank truck and loading rack.

In areas where flammable liquids are stored or used, hose nozzles on steam lines used for cleaning should be bonded to the surface of the vessel or object being cleaned.

Moving belts are sources of static electricity unless they are made of a conductive material or are coated with a conductive belt-dressing compound designed to prevent static charges from being built up.

Nonconductive materials, such as fabric, rubber, or plastic sheeting, passing through or over rolls will also create charges of static electricity. Static from these materials, as well as from the belts, can be discharged with grounded metal combs or tinsel collectors. Radioactive substances and static neutralizers using electrical discharges are also employed for this purpose.

1295

Bonding and grounding systems should be checked regularly for performance, especially electrically. Preferably before each fill, the exposed part of the bonding and grounding system should be inspected for parts which have deteriorated because of corrosion or have otherwise been damaged. Many companies specify that bonds and grounds be constructed of bare braided flexible wire because it facilitates inspection and prevents broken wires from being concealed, as discussed earlier.

Electrical equipment

Electricity becomes a source of ignition where flammable vapors exist, if the proper type of electrical equipment for these atmospheres either has not been installed or has not been maintained. The NFPA Standards, the *National Electrical Code,* and local codes should be consulted.

Spark-resistant tools

Although the hazard of ignition of flammable vapors or gases by sparks may at times be overemphasized, it must be recognized.

A summary of reports of experimental evidence and practical experience in the petroleum industry shows that no significant increase in fire safety is gained by the use of spark-resistant hand tools in the presence of gasoline and other hydrocarbon vapors.

However, some materials such as carbon disulfide, acetylene, and ethyl ether have very low ignition energy requirements. For these and similar materials, the use of special tools designed to minimize the danger of sparks in hazardous locations can be recognized as a conservative safety measure.

Leather-faced, plastic, and wood tools are free from the friction-spark hazard, although there is the possibility of metallic particles becoming embedded in them.

Health Hazards

Toxic effects

Flammable liquids and their vapors may create health hazards from both surface contact and inhalation of toxic vapors. Irritation results from the solvent action of many flammable liquids on the natural skin oils and tissue. A toxic hazard of varying degrees exists in practically all cases, depending on the concentration of the vapor. See Chapter 38, "Industrial Toxicology."

Some flammable vapors are heavier than air and will flow into pits, tank openings, confined areas, and low places in which they contaminate the normal air and, thus, cause a toxic as well as explosive atmosphere. Oxygen deficiency occurs in closed containers, such as a tank which has been closed for a long time and in which rusting has consumed the oxygen. *All* containers should be aired and tested for toxic and flammable atmosphere before any entry.

Combustible gas indicators

Unless tests prove otherwise, flammable and toxic mixtures should be assumed to be present in all tanks which have contained or have been exposed at any time to flammable liquids. Tests for flammable vapor-air mixtures in tanks and other vessels may be made by making a chemical analysis of samples or by using a NIOSH-approved, combustible gas indicator.

One type of combustible gas indicator is an instrument operating on the principle that, when a mixture of flammable vapor and air is passed over a heated electric filament, the resistance of the filament will be increased in direct proportion to the amount of combustible vapor present. The imbalance of the Wheatstone bridge is read on the galvanometer, usually as the percentage of the lower flammable limit of a calibration substance such as hexane.

A combustible gas indicator should be used only by experienced persons, and the operator should follow the manufacturer's instructions on balancing the unit. For an accurate determination, the reading should be corrected from the graph for the specific gas present in the atmosphere being tested. Manufacturers supply correlation graphs for many combustible gases.

The operator can check it by placing several drops of the flammable liquid in a small jar, shaking it a few times to permit evaporation, and then holding the sample hose of the instrument in the mouth of the jar and pressing the aspirator bulb the required number of times. If there is any question as to the indicator's accuracy, it should be calibrated against a standard or returned to the manufac-

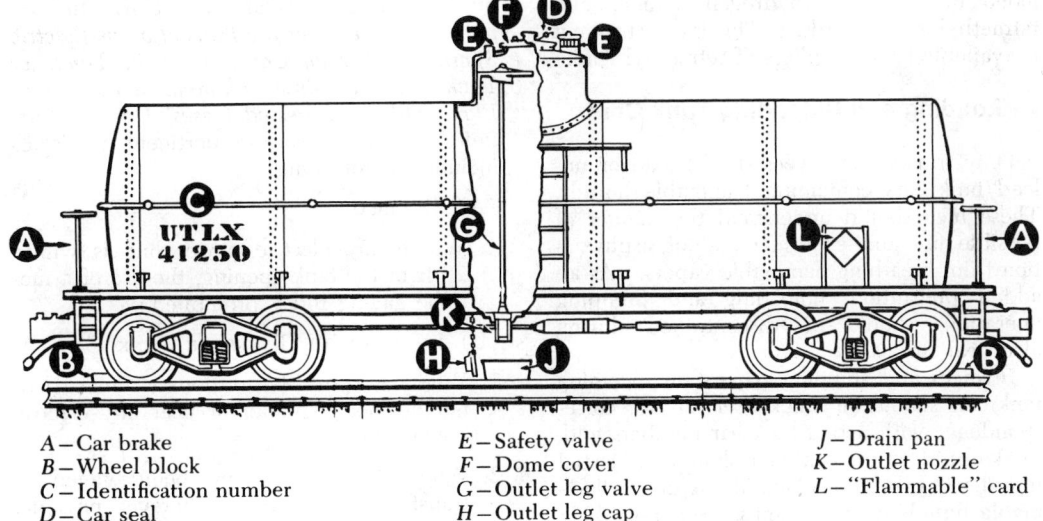

A – Car brake
B – Wheel block
C – Identification number
D – Car seal

E – Safety valve
F – Dome cover
G – Outlet leg valve
H – Outlet leg cap

J – Drain pan
K – Outlet nozzle
L – "Flammable" card

FIG. 42–7.—Typical tank car with parts identified:

turer for checking.

The sampling hose should be kept where flammable liquid, steam, or water will not be drawn into the instrument. Any of these substances would put it out of service. Hot vapors also may affect the indicator if they condense inside.

A tank, tested before being cleaned, may be found vapor-free. However, if it contains sludge or scale, flammable liquid and vapor may be released as soon as the sludge or scale is disturbed. A safe rule to follow is

that no tank should be considered free from the possibility of dangerous vapors as long as it contains sludge or scale.

If workmen leave a tank for a period of time, such as overnight, gas tests should be made the next morning before they are permitted to continue work. Many companies use a written permit form when it is necessary for men to enter vessels which have contained flammable liquids or to do "hot work" repairs in them.

Devices are available for testing for carbon

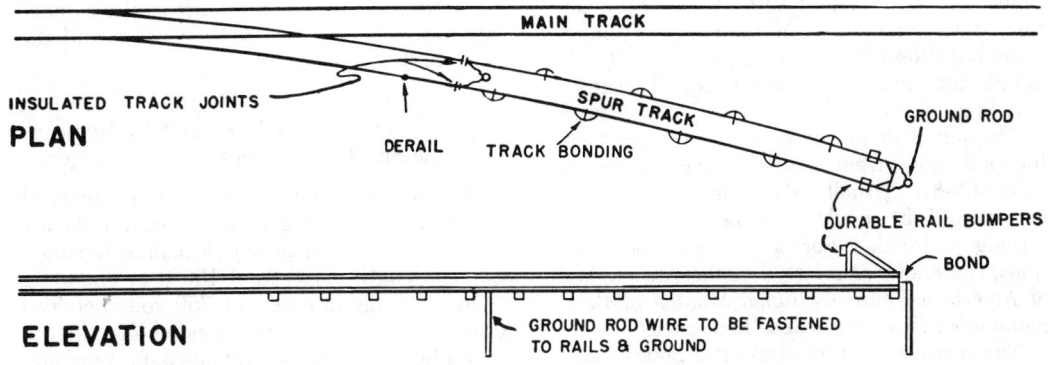

FIG. 42–8.—A tank car unloading siding showing rail joint bonding, insulated track joint, derail, and track grounding.

monoxide, benzene, hydrogen sulfide, and tetraethyl lead hazards. (This last instrument is available from suppliers of tetraethyl lead.)

Loading and Unloading Tank Cars

Only trained employees should load or unload tank cars containing flammable liquids. These men should understand the danger of possible fire and explosion and of asphyxiation from breathing flammable vapors. As an added precaution, unloading and pumping operations should be discontinued during electrical storms.

Invoices and shipping papers on incoming tank cars should be checked closely for correspondence with actual tank car numbers and, in doubtful cases, contents should be tested in order to prevent accidental mixture of flammable liquids in plant tanks. See Fig. 42–7 for typical tank car construction details.

Spotting cars

A car mover should be used, if it is necessary to hand spot a car. The worker should stand with the handle of the bar to one side of the track rail and face the car with his feet well apart so that he cannot be overbalanced easily.

After the car is spotted, with the dome outlet opposite the loading or unloading line, the brakes should be set and the car wheels blocked. The "blue flag" stop sign (at least 12 by 15 in., with words such as STOP—TANK CAR CONNECTED or STOP—MEN AT WORK) should be set before loading lines are connected or work is done on the car. The sign should be located about 25 ft ahead of the car toward the main line, so train crews cannot come onto the track without seeing it. (Full details are given in Chapter 28, "Elevators and Plant Railways.")

The siding should be bonded and grounded for protection from stray electrical currents (Fig. 42–8). Usually, the spur track installation would be made or supervised by the serving railroad. For applicable specifications, the reader is referred to the Association of American Railroads *Signal Manual of Recommended Practice*. (See "References.")

The relative location of electric power lines to the tank car unloading position is an important matter discussed in the AAR Bureau of Explosives Circular No. 17-E, *Recommended Practice for the Prevention of Electric Sparks That May Cause Fires in Tanks or Tank Cars Containing Flammable Liquids or Flammable Compressed Gases, Due to Proximity of Wire Lines.* Its pertinent provisions include the following:

(A) GENERAL

1. Where any electric power line is within 20 ft of the tank opening, the use of a metallic gaging rod is prohibited.

(B) LOCATION OF TANKS OR TANK CARS

1. Wherever possible, stationary tanks shall not be located under or near any electric power lines.

2. When the contents are being gaged or transferred, tank cars, wherever possible, shall not be located under or near any electric power lines.

3. Where tanks or tank cars (the contents of which are being gaged or transferred) are necessarily located under or near power lines having a span length of 150 ft or less and operating at a voltage not exceeding 550 volts between conductors, the following rules shall be observed:

   a. Where power lines pass overhead, there shall be a minimum vertical clearance of 8 ft at 60 F between the wires and the tank.

   b. Where power lines pass nearby and do not have the minimum vertical clearance specified in paragraph 3-a, there shall be a minimum horizontal clearance of 8 ft between the wire lines and the tank.

   c. Openings in tanks shall be at least 6 ft distant, measured horizontally from any overhead power lines.

4. Where tanks or tank cars (the contents of which are being gaged or transferred) are located under or near power lines having a span length in excess of 150 ft or operating at a voltage in excess of 550 volts between conductors, it is recommended that special studies be made by qualified persons and such additional clearance provided as is necessary to give adequate protection.

Fig. 42–9.—This illustration shows an excellent access structure for getting at tank cars. Note particularly the relocatable bridge made of nonskid metal.

Inspection

The condition of the cars should be examined. Any leaks should be reported to the railroad representative, so that open lights and locomotives may be kept away from the car.

Buckets or tubs should be placed under the leaks to prevent loss, and the accumulated drippings transferred to storage frequently so that no large quantity of flammable liquid is exposed in an open container or allowed to spread over the ground.

Before being loaded or unloaded, there should be *no* exposed lights, fires, or other sources of ignition in the area.

Explosion-proof electrical equipment, installed in accordance with the *National Electrical Code* and subject to regular inspection, should be provided where flammable liquids are handled. A rack with a gangway or bridge to the car, similar to that shown in Fig. 42–9, helps workers to move safely when loading or unloading a car.

Relieving pressure

The tank car should be relieved of interior pressure before the manhole cover or the outlet valve cap is removed. To relieve pressure, either the safety valves can be raised or the vent on the dome can be opened at short intervals or the tank can be cooled with water. Venting and unloading should be deferred, if a dangerous amount of vapor collects outside the car.

If interior pressure is excessive, the car should be sprayed with water to cool it or

1299

should be allowed to stand overnight and be unloaded early the next morning. These precautions are not necessary when the pressure is relieved by piping the vapor into a condenser or storage tank.

Removing covers

A screw-type dome cover should be loosened by a bar placed between the lug and the knob on the top of the cover. Two complete turns should be taken, so that the ½-in. vent holes in the threaded portion of the dome cover are exposed. If the workman hears escaping vapors, he should tighten the cover and release the pressure by raising the safety valve. He should keep clear of the vapors by standing to windward.

A cover should be removed with the man facing the dome. With feet well braced, he should use short, vigorous pushes on the bar. The cover (if not provided with a chain) and loose tools should be removed to the walk platform or other safe place, so that they will not fall.

On the bolted type of dome cover, all nuts should be unscrewed one turn and the cover lifted to break adhesion which may exist between the cover and the dome ring. If there is a sound of escaping vapors, the dome should be tightened and the venting operation repeated.

All dirt, cinders, and debris should be removed from around the interior type of dome cover before the yoke is unscrewed.

When the car must be unloaded through the bottom outlet valve, dome covers should be adjusted to allow for venting. The screw type of dome cover should be tightened just enough to expose the ½-in. vent holes in the threaded portion of the cover.

A small, thin wood block should be placed under the bolted type of cover, and the interior type of dome cover should be tightened up in the yoke to within ½-in. of the closed position. An asbestos or metal cover or wet burlap or canvas should be placed over the tank manhole to protect against entrance of sparks or other source of ignition.

Unloading and loading connections

The safest method of unloading tank cars is through the dome. Some states prohibit bottom unloading. Where a car has only bottom unloading, the tank's outlet valve must be closed before the outlet chamber cap or plug is removed. A pail or tub should be placed under the outlet chamber and the 2-in. outlet cap unscrewed.

The condition of the outlet valve should be checked and, if there is no serious leak, the unloading should proceed. If the plug does not loosen readily with a 48-in. wrench, the bottom outlet cap or plug can be tapped to loosen it. If the valve leaks so badly that a connection cannot be made without spilling the product, the car should be unloaded from the top.

In cold weather the outlet chamber, if the chamber or valve is frozen or blocked with frozen liquid, should be carefully examined for cracks. If a crack is not found, the connection for unloading should be made, the outlet chamber should be wrapped with burlap or other rags, and hot water or steam should be applied.

The condition of the connection from the car to the storage tank should be carefully checked before the car valve is opened. The storage tank should be watched to prevent overflow, and the hose and unloading line should be checked frequently. The tank car should be examined, before loading lines are disconnected, to make certain that it is completely empty.

A workman should be present throughout the entire loading or unloading operation and while a car is connected. If it is necessary to discontinue operations before they are completed, the outlet valve should be closed and the dome cover and outlet chamber cap replaced.

Tank cars should not be allowed to stand with loading or unloading connections attached after operations are completed. If flammable liquids are spilled, spill areas should be covered with fresh dry sand or dirt. Never flush spills into public sewers, drainage systems, or natural waterways.

When heater pipes are used for unloading, the steam should be applied slowly until it begins to exhaust at the outlet pipe. Steam pressure should not exceed that needed to bring the contents to the desired temperature and should never be high enough to cause the contents to overflow.

Care also must be exercised to avoid exceeding the flash temperature of "cut-backs" or additives introduced into various heavy products—for example, the addition of naphtha to residual asphalt. Overheating products containing certain additives can release dangerous quantities of flammable vapors.

Placards

Up-to-date shipping cards and DANGEROUS or FLAMMABLE placards should be provided and, then, removed after operations are completed. They should then be replaced with DANGEROUS—EMPTY placards. DEFECT cards or forms of any kind which have been attached to a car by railroad employees should not be removed. Placards, rags, waste, and blocks should not be disposed of in the tank or car body.

Fires

Tank car fires can be serious, especially with bottom unloading. A wet blanket, coat, or sack can be used to smother a fire at the dome. Sand or steam and foam, carbon dioxide, waterfog, and dry chemical extinguishers may be used on small fires from leaks.

Methods used to control fires of some size which occur during bottom unloading depend on conditions at the time. To protect nearby buildings, it may be advisable, where bottom unloading is done, to have areas diked, and equipped with a drain to some safe point, to confine or to divert flow in the event of a leak or a fire. (See Association of American Railroads, *Handling Collisions and Derailment Involving Explosives, Gasoline and Other Dangerous Articles.* Bureau of Explosives Pamphlet No. 22.)

Loading and Unloading Tank Trucks

Tank trucks, tank trailers, and tank semitrailers used for the transportation of flammable liquids should be constructed and operated according to Title 49, *Code of Federal Regulations,* Parts 177 and 397. Note: Revision and promulgation of these regulations are within the jurisdiction of the Department of Transportation.

Carriers not subject to the jurisdiction of the Department of Transportation may consult NFPA Standard No. 385, *Tank Vehicles*

FIG. 42–10.—One method of attaching bonding cable to tank truck before opening tanks. Note convenient location of cable.

for *Flammable and Combustible Liquids,* as an appropriate reference.

Inspection

Trucks should be kept in good repair and should be inspected daily with special emphasis being placed on cleanliness of the motor and on the condition of lights, brakes, horns, rear view mirror, bonding, tires, and steering.

Tanks and safety valves must be inspected and tested according to Title 49, *Code of Federal Regulations,* Part 177.

Smoking

Smoking by truck drivers or their helpers should not be permitted while they are driving their trucks, making deliveries, filling the tanks, or making repairs. Drivers should be alert to keep smokers and other sources of ignition away from loading or unloading operations. Each tank vehicle should be provided with, at least, one portable fire extinguisher having at least a 20 B,C rating. When more than one is provided, each should have, at least, a 10 B,C rating.

Spotting trucks

A tank truck being loaded or unloaded should have the brakes set, the engine stopped, the lights turned off, and the bonding and grounding connections made before the dome cover is removed or the unloading connection is made (Fig. 42–10).

Where trucks cannot be loaded without having to start the engine and move the truck, all domes must be closed and latched, and the bonding connections should be removed before the engine is started.

Unloading and loading connections

The engine of a truck with a motor-driven pump should be stopped before loading lines are connected or disconnected. The driver must remain at the tank truck controls. If he must leave, discharge of the flammable product must be stopped and the hose removed. When an unloading line must be run across a sidewalk, suitable warning signs should be provided.

Deliverymen should make sure that the correct product is being handled and that the tank has previously contained the same liquid. To assure removal of trapped liquids, especially when changing from a low flash liquid to a higher, the men should completely drain the tank and flush it out with the type of product to be loaded.

They should determine that the tank is properly vented and that there is sufficient room to accommodate the quantity to be loaded or unloaded.

Leaks

Spills or overflows must be avoided. When they do occur, loading should be stopped, valves shut off, and the overflow cleaned up before loading is resumed. Time should be allowed for all flammable vapors to dissipate before the truck engine is started.

The driver should make every effort to park a damaged and leaking truck in such a way that it will not endanger traffic or property. The public should be warned to keep away, and the police and fire departments notified.

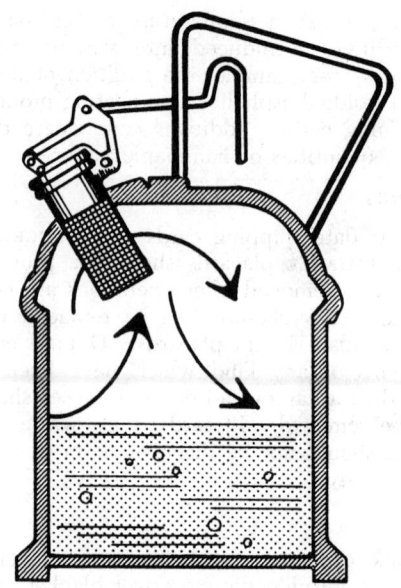

Fig. 42–11.—Spring-loaded cover is designed to open in order to relieve internal vapor pressure at 5 psi. Losses by evaporation of liquid stored in safety cans at ordinary temperatures are negligible.

The truck should be parked off the highway, if possible, in a vacant lot or, at least, away from buildings and away from areas in which there is a concentration of people. The truck should not be left unattended.

Carriers subject to the safety regulations of the Department of Transportation must report all broken, leaking, and damaged containers on Form F 5800.1 under Sections 171.15 and 171.16, Title 49, *Code of Federal Regulations,* Part 171.

The liquid should be trapped in containers or in a depression or pit if possible. In case of a large spill, especially in urban areas, rather than endanger lives, the best practice is to use portable hand pumps to pump the product into drums or another tank truck.

Fires

In the event of a tank truck fire during loading or unloading, the fuel supply to or from the truck should be shut off, if possible. If loading is being done, the spout should not be removed from the tank truck but, if possible, either the dome cover should be closed or the opening covered with a wet

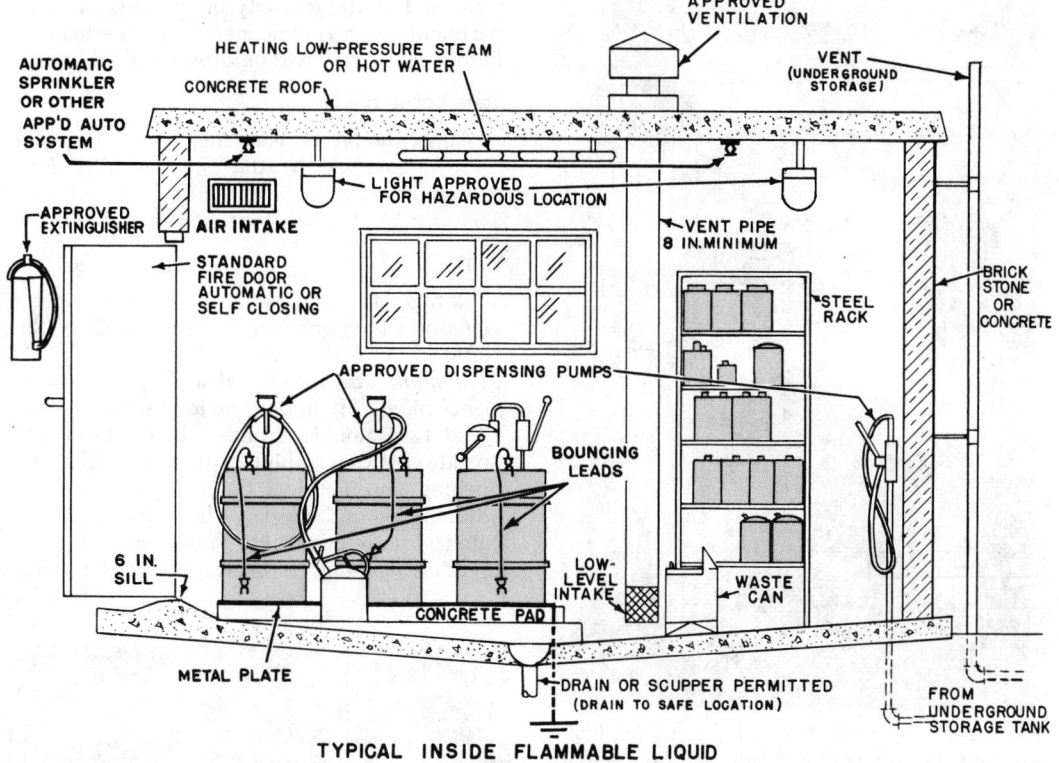

APPROVED VENTILATION

HEATING LOW-PRESSURE STEAM OR HOT WATER

VENT (UNDERGROUND STORAGE)

AUTOMATIC SPRINKLER OR OTHER APP'D AUTO SYSTEM

CONCRETE ROOF

LIGHT APPROVED FOR HAZARDOUS LOCATION

APPROVED EXTINGUISHER

AIR INTAKE

VENT PIPE 8 IN. MINIMUM

BRICK STONE OR CONCRETE

STANDARD FIRE DOOR AUTOMATIC OR SELF CLOSING

STEEL RACK

APPROVED DISPENSING PUMPS

BOUNCING LEADS

6 IN. SILL

LOW-LEVEL INTAKE

WASTE CAN

CONCRETE PAD

METAL PLATE

DRAIN OR SCUPPER PERMITTED (DRAIN TO SAFE LOCATION)

FROM UNDERGROUND STORAGE TANK

TYPICAL INSIDE FLAMMABLE LIQUID STORAGE ROOM

FIG. 42–12.—A flammable liquids storage and mixing room, Type A, following NFPA design.

sack or blanket.

The fire should be attacked with carbon dioxide, foam, or dry chemical extinguishers, or water. Burning liquid may be flushed away to a safe place, if a supply of water is available, but not into a public sewer, drain, or public waterway.

Where burning vapors are escaping from leaks or vents, it may be better to let them burn until the source of escaping liquid or vapor can be controlled.

In all cases of fire, the fire control squad or the local fire department should be informed quickly. Procedures on what to do in case of fires should be prepared and, in the actual emergency, followed.

Storage

Class I and Class II liquids should not be kept or stored in a building used for public assembly, such as a school, church, or theater, except in approved containers (Fig. 42–11) within either a storage cabinet or a storage room that does not have an opening which communicates with the public portion of the building. They should not be stored so as to limit use of exits, stairways, or areas normally used for the safe egress of people. Neither should they be stored close to stoves or heated pipes, nor exposed to the rays of the sun or other sources of heat.

Storage of flammable liquids in open containers should not be permitted. Approved containers for flammable liquids should be closed after each use and when empty. Warning labels should be removed from flammable liquid containers when empty (vapor free).

Bulk Class I liquids should be stored in an underground (buried) tank or outside a building. No outlet from the tank should be inside a building unless it terminates in a special

1303

Fig. 42–13.—A well designed flammable liquid storage room with both high and low level ventilation and automatic fire extinguishment. Ventilators are designed for 12 air changes per hour.

2 percent of the capacity of the tank or compartment is recommended, and permanent high-level markings should be installed.

Tank construction

Tanks should be constructed and installed as recommended by the National Fire Protection Association in its latest *Flammable & Combustible Liquids Code,* Standard No. 30.

Vents

Storage tanks should be provided with vents of a type and size recommended by the National Fire Protection Association in its *Flammable & Combustible Liquids Code.* Vent pipes of underground tanks, storing Class I flammable liquids, should terminate outside buildings, higher than the fill pipe opening and not less than 12 ft above the adjacent ground level. They should be located so that flammable vapors cannot enter building openings or be trapped under eaves

Fig. 42–14.—Truck designed for transporting safety cans and other flammable liquid containers to points of use in the plant.

Figures courtesy Western Electric Co.

room (Figs. 42–12 and –13).

Specifications limiting the quantity of each class of flammable liquids that may be stored in various locations on plant premises, together with data describing the required conditions and procedures relating to such storage, are set forth in NFPA Standard No. 30, *Flammable & Combustible Liquids Code.*

Vehicles used on plant property to transport flammable liquids in sealed containers should be designed to minimize damage to the containers (Fig. 42–14).

When men are filling tanks and other containers, they should be sure to allow sufficient vapor space (outage) above the liquid level in order to permit expansion of the liquid with changing temperatures.

For example, gasoline expands at the rate of about 1 percent for each 14-deg F rise in temperature. Outage space for gasoline of or other obstructions. Vent pipes from underground tanks, storing Class II or Class III liquids, should terminate outside buildings and higher than the fill pipe opening. Vent outlets should be above normal snow level.

Some authorities have questioned the effectiveness of brass mesh or copper screens as flame arresters in vent terminals. Underwriters' Laboratories state (*Oil Tank Vents—Hazards of Screens*, Serial No. UL-31) that "under some favorable conditions, screens of fine mesh may be effective in arresting flame, but their use in general is not dependable. Under service conditions, they are subject to clogging and freezing; when attempts are made to clean them there is danger of mechanical injury. The displacement of one or more wires renders the screen useless as a flame arrester."

Dependence should be placed on the safe location of a vent, so that escaping vapors will be properly dispersed and clear of ordinary sources of ignition. If the vent terminals are in locations where clogging by mud wasps may be anticipated, a loosely attached screen of relatively coarse mesh may be provided.

Dikes

Where a flow of flammable or combustible liquid from a tank might have serious consequences because of the topography and the neighboring property, a curb, dike, or wall may be provided around a tank or group of tanks. Such a structure should be designed to contain flammable liquids released by the overfilling of tanks or by the leaking of tanks, pipe lines, and valves. (The NFPA *Flammable & Combustible Liquids Code* gives construction details of drainage, dikes, and walls for aboveground storage tanks.)

The need for dikes will be determined by local conditions. In some cases, for example, if a tank is close to a building and the ground does not slope from the building, it may be better to provide diversion walls to direct the escaping liquid to a safe location.

A trapped storm drain often is installed for each dike enclosure. In larger plants, such a drain often leads to a recovery plant well away from the main buildings, where the flammable liquid can be separated from the water by flotation. Drains should be equipped with control valves, which should be kept closed unless water is being drained from the area. Their use should not permit flammable liquids to enter natural water courses, public sewers, or public drains, if their presence would constitute a hazard.

Pump houses

Consideration should be given to locating flammable liquid pumps outside of buildings wherever feasible.

Where buildings housing equipment for transfer of flammable liquids are required, they should be of fire-resistive construction with ample ventilation, especially along the floor where vapors might be present. Pump houses should preferably not contain sunken pits, except small drain openings, because the danger of vapor concentration in such areas is too great. The pump house should have a minimum of two exits, be kept clear of obstructions, and have doors swinging outward.

Where flammable liquids are handled, motors should be partitioned and sealed off from the rest of the pump house, or should be of a type approved for use in flammable atmospheres. The pump room should be kept well ventilated. It is good operating and safety practice to have a well-marked master cutoff switch outside the building. All electrical equipment inside the building must conform to the *National Electrical Code* and local codes for this use.

Valves and packing glands on pumps should be maintained in good operating condition to prevent excessive leaks. Draining escaped liquids into a closed-pipe return system is recommended in preference to simply catching in drip pans materials that leak. Leaks should be repaired as soon as they are noted.

If a centrifugal pump with priming bleeder is used, the bleeder should be terminated outside the pump house at a visible location and directed into a container or connected by a valved pipe to pump suction. Bearings should not be permitted to become hot.

Good housekeeping is of major importance, and approved containers should be provided for the safe disposal of debris, rags, and other waste. These should be emptied daily. Tools other than those required in operation should not be stored in the pump house.

The pump house should not contain lockers or be used as a change house, and loitering by employees should not be allowed. All sources of ignition must be kept away from the pump house location.

Fire extinguishers should be located at convenient points that are easily identifiable. It may be advisable to have self-closing fire doors and automatic extinguishment of a type recommended for the hazard.

Gaging

In installations requiring manual gaging, storage tanks should be provided with means for measuring the contents so that men need not walk across the tank roof unless a walkway is provided. Remote-gage measuring equipment eliminates the need for such tank roof walking.

The NFPA *Flammable & Combustible Liquids Code* states that openings for manual gaging of tanks located in a building or buried under a basement, if independent of the fill pipe, should be provided with a vapor-tight cap or cover. Each such opening should be protected against liquid overflow and possible vapor release by means of a spring-loaded check valve or other approved device. Manual gaging of tanks containing Class I liquids should be avoided. Substitutes for manual gaging include, but are not limited to, heavy-duty flat gage glasses, magnetic, hydraulic or hydrostatic reading devices, or sealed float gages.

The float-operated gage or the type operated by pressure of the liquid is often used on aboveground storage tanks. Some companies use a sounding weight and tape or, if the tank is small, a wood sounding rod through a gage hatch on the top of the tank.

Gage glasses protected by metal cases, found on some flammable liquid tanks held under pressure, are suitable if maintained.

Safe means of access should be provided for the gaging of tanks.

Tanks in flooded regions

Tanks in areas where there is danger from flood waters should be installed and anchored according to NFPA Standard 30, Appendix B, *Protection of Tanks Containing Flammable or Combustible Liquids in Locations That May Be Flooded.*

Underground tanks

When an underground tank is subject to heavy traffic over it, the tank should be protected with at least 3 ft of earth cover *or* 18 in. of tamped earth plus 6 in. of reinforced concrete *or* 18 in. of tamped earth plus 8 in. of asphalt concrete.

If the underground tank is not subject to traffic, a protective cover will be provided by a minimum of 2 ft of earth *or* 1 ft of earth plus 4 in. of reinforced concrete. The concrete cover should extend 1 ft beyond the outline of the tank.

An underground tank should have a firm foundation and be surrounded with at least 6 in. of noncorrosive, inert materials, such as clean sand, earth, or gravel that has been well tamped in place. The tank should be anchored if there is any chance of its "floating" on rising ground water. Storage tanks inside buildings should have the fill and vent pipes located outside.

The NFPA *Flammable & Combustible Liquids Code* specifies that "underground tanks or tanks under buildings should be so located with respect to existing building foundations and supports that the loads carried by the latter cannot be transmitted to the tank. The distance from any part of a tank, storing Class I liquids to the nearest wall of any basement, pit, or cellar should be not less than 1 ft, and from any property line that may be built upon, not less than 3 ft. The distance from any part of a tank, storing Class II or Class III liquids, to the nearest wall of any basement, pit, cellar, or property line should be not less than 1 ft."

The Code also specifies that corrosion protection for an underground tank and its piping should be provided by one or more of the following methods: (*a*) use of protective coatings or wrappings; (*b*) cathodic protection; or, (*c*) corrosion-resistant materials of construction. Selection of the type of protection to be employed should be based upon the corrosion history of the area and the judgment of a qualified engineer. Cinders or other acid-forming fills should not be used around the tank. Local or state authorities that have jurisdiction should be notified in writing, so that approval may be obtained before the tank installation is covered.

Flammable liquids should be withdrawn by pump, with the pump and piping system so arranged that, when it is not in operation, the liquid will flow back to the tank.

Aboveground tanks

The NFPA *Flammable & Combustible Liquids Code* rigorously spells out the minimum distances that aboveground tanks containing flammable or combustible liquids can be located from property lines, public ways, or nearest important buildings. Table 42-A gives the distances for tanks operating at no greater than 2.5 psig, at greater than 2.5 psig, and with liquids that have boilover characteristics. Consult the Code for distances for aboveground tanks containing unstable liquids. (The term, psig, stands for pounds per square inch gage.)

There are also certain minimum distances that are recommended. These are given in Table 42-B. Where end failure of a horizontal pressure tank or vessel may expose property, the tank shall be placed with the longitudinal axis parallel to the nearest important structure exposure.

If two tank properties of diverse ownership have a common boundary and if the owners and local authorities agree, normal minimum shell-to-shell spacing, such as would be used within the tank farm itself, can be used.

The minimum for in-farm spacing is 3 ft. The actual distance also must not be less than ⅙ the sum of the diameters of the two tanks, unless one tank is less than ½ the diameter of the other tank, in which case the minimum distance is not to be less than ½ the diameter of the smaller tank.

There are a number of special cases for petroleum tanks in refineries and in producing areas, for unstable flammable or combustible liquids, for LP-gas, and for tanks that are arranged in compact or irregular pattern. The NFPA Code should be consulted.

It is also recommended that truck loading racks, dispensing Class I liquids, should be at least 25 ft and those dispensing Class II or Class III liquids should be at least 15 ft from tanks, warehouses, other plant buildings, and the nearest property line.

Tanks should be located to avoid danger from high water. Tanks located on a stream without tide should, where possible, be downstream from burnable property.

Piping materials for aboveground tanks shall be steel as provided in NFPA Standard No. 30. Piping may be built of materials other than steel, if used underground or if required by the properties of the liquid handled. In case of doubt, the supplier or producer of the flammable or combustible liquid (or other competent authority) should be consulted as to the suitability of the construction material being used.

Piping from a storage tank should have a readily accessible shutoff valve at the tank. Provision should be made to drain or pump the contents of a tank into another tank, or to collect the liquid within dikes or retaining walls if the tank should leak or be overfilled.

A small tank is often provided with an emergency self-closing valve, located inside the tank or at the point of entrance, which automatically stops the flow of liquid in case of fire. (See Factory Mutual System, *Piping Systems for Flammable Liquids.* Loss Prevention Bulletin No. 13.21.) This valve should be closed except during loading and unloading operations. The rope or wire and the fusible link attached to an emergency valve, which is left in open position, should be in good condition and should be tested regularly for easy operation.

Besides the normal vents to take care of vacuum and pressure during pumping operations, an aboveground storage tank must have some form of emergency relief venting to prevent buildup of excessive internal pressure in case of fire surrounding the tank. This relief may be provided by either a weak seam in the top, or at the joint between the top and the shell of the tank, or by some other recommended form.

Except in oil refineries and large water terminals, tanks for storing Class I liquids should be labeled FLAMMABLE—KEEP FIRE AWAY with letters at least 2 in. high. NO SMOKING signs should be posted conspicuously. Similar signs should be posted for Class II and, if necessary, Class III liquids.

Good housekeeping should be maintained around storage tanks. Debris should not be permitted to accumulate, nor should the space adjoining the tank be used for storage of combustible materials. The grass around the tank (and under it if it is off the ground)

TABLE 42-A

LOCATION OF OUTSIDE ABOVEGROUND STORAGE TANKS FROM ADJOINING PROPERTY OR PUBLIC WAY

For Operating Pressures No Greater Than 2.5 psig

Type of tank	Protection	Minimum distance in feet from property line which may be built upon, including the opposite side of a public way	Minimum distance in feet from nearest side of any public way or from nearest important building and shall be not less than 5 feet
Floating roof	Protection for exposures	½ times diameter of tank but need not exceed 90 ft	⅙ times diameter of tank but need not exceed 30 ft
	None	Diameter of tank but need not exceed 175 ft	⅙ times diameter of tank but need not exceed 30 ft
Vertical with weak roof to shell seam	Approved foam or inerting system on the tank	½ times diameter of tank but need not exceed 90 ft. and shall not be less than 5 ft	⅙ times diameter of tank but need not exceed 30 ft
	Protection for exposures	Diameter of tank but need not exceed 175 ft	⅓ times diameter of tank but need not exceed 60 ft
	None	2 times diameter of tank but need not exceed 350 ft	⅓ times diameter of tank but need not exceed 60 ft
Horizontal and vertical, with emergency relief venting to limit pressures to 2.5 psig	Approved inerting system on the tank or approved foam system on vertical tanks	½ times Table 42-B but shall not be less than 5 ft	½ times Table 42-B
	Protection for exposures	Table 42-B	Table 42-B
	None	2 times Table 42-B	Table 42-B

For Operating Pressures Greater Than 2.5 psig

Type of tank	Protection	Minimum distance in feet from property line which may be built upon, including the opposite side of a public way	Minimum distance in feet from nearest side of any public way or from nearest important building
Any type	Protection for exposures	1½ times Table 42-B but shall not be less than 25 ft	1½ times Table 42-B but shall not be less than 25 ft
	None	3 times Table 42-B but shall not be less than 50 ft	1½ times Table 42-B but shall not be less than 25 ft

For the Storage of Flammable or Combustible Liquids With Boilover Characteristics

Type of tank	Protection	Minimum distance in feet from property line which may be built upon, including the opposite side of a public way	Minimum distance in feet from nearest side of any public way or from nearest important building
Floating roof	Protection for exposures	Diameter of tank but need not exceed 175 ft	⅓ times diameter of tank but need not exceed 60 ft
	None	2 times diameter of tank but need not exceed 350 ft	⅓ times diameter of tank but need not exceed 60 ft
Fixed roof	Approved foam or inerting system	Diameter of tank but need not exceed 175 ft	⅓ times diameter of tank but need not exceed 60 ft
	Protection for exposures	2 times diameter of tank but need not exceed 350 ft	⅔ times diameter of tank but need not exceed 120 ft
	None	4 times diameter of tank but exceed 350 ft	⅔ times diameter of tank but need not exceed 120 ft

From National Fire Protection Association, Standard No. 30. Adopted by OSHA as Tables H-5, H-6, and H-7 in Section 1910.106.

should be eliminated or, at least, kept cut.

Tanks used for storing flammable or combustible liquids should be made gas-tight to cut down evaporation losses and for fire prevention. Venting devices are arranged to be normally closed, but breathe automatically when liquid is pumped in or out of the tanks and as temperature fluctuates. Floating roofs are also effective in reducing evaporation losses and allowing expansion.

Large aboveground storage tanks should be provided with stairways and platforms, preferably of steel. Tanks that stand more than 1 ft aboveground should have foundations of noncombustible materials, except that wood cushions may be used. Supports for aboveground tanks should be of concrete, masonry, or steel protected by concrete or other approved fireproofing to prevent collapse in case of fire.

TABLE 42-B

REFERENCE MINIMUM DISTANCES FOR ABOVEGROUND OUTSIDE
STORAGE TANKS

Capacity Tank (gallons)	Minimum Distance in Feet from Property Line Which May be Built Upon, Including the Opposite Side of a Public Way	Minimum Distance in Feet from Nearest Side of Any Public Way or from Nearest Important Building (Within Property Line)
275 or less	5	5
276 to 750	10	5
751 to 12,000	15	5
12,001 to 30,000	20	5
30,001 to 50,000	30	10
50,001 to 100,000	50	15
100,001 to 500,000	80	25
500,001 to 1,000,000	100	35
1,000,001 to 2,000,000	135	45
2,000,001 to 3,000,000	165	55
3,000,001 or more	175	60

From National Fire Protection Association, Standard No. 30.

Overflowing of tanks presents a severe fire hazard in that vapors may drift to a source of ignition. Operators responsible for filling tanks should maintain a constant watch on the filling rate and the level of liquid, so that operations may be stopped or liquid diverted to another tank when the required level is reached. Operators must also remain at the open valve while water is being drained from a tank.

It is common practice to paint flammable liquid tanks exposed to the sun with aluminum, pastel, or white paint in order to reflect the heat and so help keep down internal vapor pressure. In some installations, where highly volatile liquids are stored, an external water spray cooling system is provided. Tanks can also be insulated.

Tank fires and their control

Fire is unlikely to exist inside a tank for more than a few seconds before either extinguishing itself or blowing open the tank roof. A tank fire most commonly occurs at one of the roof openings. It will occur only if vapor is being expelled from the tank because of either filling or heating, and it can usually be extinguished without difficulty if the cause of the vapor expulsion is removed by cooling or by shutting off the filling operation.

If a tank is exposed to an adjacent fire, water should be applied immediately to the tank to cool it and thus reduce vaporization, not only to save stock but also to make ignition at the vents less likely. Vent fires extinguish themselves as soon as the outbreathing stops. So many variables are involved in tank fires that they should be handled only by men trained in their extinguishment. Fig. 42–15 shows various ways of fighting fires.

Inside storage and mixing rooms

Flammable or combustible liquids in sealed containers present a potential, rather than an active hazard—the possibility of fire from without. Inside storage rooms are undesirable. If they must be used, they should be isolated as much as possible and located at or above grade, not immediately above a cellar or basement.

NFPA Standard No. 30 states, "Inside storage rooms shall be constructed to meet required fire-resistance ratings for their use. Construction of inside storage rooms shall comply with the test specifications set forth

FIG. 42–15a.—Firemen attack a flammable liquid fire using water fog.

Courtesy U.S. Army, Ordinance, White Sands Proving Ground.

FIG. 42–15b.—Trainees approach a flammable liquid fire using a dry chemical.

Courtesy Ansul Fire School.

FIG. 42–15c.—Firemen attack a flammable liquid fire by applying foam.

in NFPA Standard No. 251, *Standard Methods of Fire Tests of Building Construction and Materials*. Where an automatic sprinkler system is provided, the system shall be designed and installed in an approved manner. Openings to other rooms or buildings shall be provided with noncombustible, liquid-tight raised sills or ramps at least 4 in. in height, or the floor in the storage area shall be at least 4 in. below the surrounding floor. Openings shall be provided with approved self-closing fire doors. The room shall be liquid-tight where the walls join the floor. A permissible alternate to the sill or ramp is an open-grated trench inside of the room which drains to a safe location. Where other portions of the building or other properties are exposed, windows shall be protected as set forth in the *Standard for Fire Doors and Windows*, NFPA 80, for Class E or F openings.

Wood of at least 1-in. thickness may be used for shelving, racks, dunnage, scuffboards, floor overlays and similar installations.

Storage in inside storage rooms shall comply with these specifications:

Automatic Fire Protection* Provided	Fire Resistance	Maximum Floor Area	Total Allowable Quantities Gal/sq ft/ floor area
Yes	2 hour	500 sq ft	10
No	2 hour	500 sq ft	4
Yes	1 hour	150 sq ft	5
No	1 hour	150 sq ft	2

* Fire protection system shall be sprinkler, water spray, carbon dioxide, dry chemical, halon, or other system approved by the authority that has jurisdiction.

"Electrical wiring and equipment located within inside storage rooms used for Class I liquids shall be approved for Class I, Division 2 Hazardous Locations; for Class II and Class III liquids, they shall be approved for general use. [See NFPA 70, *National Electrical Code*.]

"Every inside storage room shall be provided with either a gravity or a continuous mechanical exhaust ventilation system. Mechanical ventilation shall be used if Class I liquids are dispensed within the room.

"Exhaust air should be taken from a point near a wall on one side of the room and within 12 in. of the floor with one or more make-up air inlets located on the opposite side of the room within 12 in. from the floor. The location of both the exhaust and inlet air openings should be arranged to provide, as far as practicable, air movements across all portions of the floor to prevent accumulation of flammable vapors. Exhaust from the room should be directly to the exterior of the building. If ducts are used they should not be used for any other purpose and should comply with the *Standard for the Installation of Blower and Exhaust Systems for Dust, Stock and Vapor Removal or Conveying*, NFPA No. 91 (ANSI Z33.1). If makeup air to a mechanical system is taken from within the building, the opening should be equipped with an approved fire door or damper, as required in the *Standard for the Installation of Blower and Exhaust Systems, for Dust, Stock and Vapor Removal or Conveying*, NFPA No. 91 (ANSI Z33.1). For gravity systems, the makeup air should be supplied from outside the building.

"Mechanical ventilation systems should provide at least one cubic foot per minute of exhaust per square foot of floor area, but not less than 150 cfm.

"In every inside storage room there should be maintained one clear aisle at least three feet wide. Containers over 30 gallons in capacity, storing Class I or Class II liquids, should not be stacked one upon the other. Dispensing should be by approved pump or self-closing faucet only." *

Storage cabinets

"Not more than 120 gallons of Class I, Class II, and Class IIIA liquids may be stored in a storage cabinet. Of this total, not more than 60 gallons may be of Class I and Class II liquid. Not more than three such cabinets may be located in a single fire area, except that, in an industrial occupancy, additional cabinets may be located in the same fire area if the additional cabinet, or group of not more than three cabinets, is separated from any other cabinet or group of cabinets by at least 100 ft.

"Storage cabinets shall be designed and constructed to limit the internal temperature to not more than 325 F, when subjected to a 10-minute fire test using the standard time-temperature curve as set forth in *Standard Methods of Fire Tests of Building Construction and Materials*, NFPA 251. All joints and seams shall remain tight, and the door shall remain securely closed during the fire test. Cabinets shall be labeled in conspicuous lettering, FLAMMABLE — KEEP FIRE AWAY.

"Metal cabinets constructed in the following manner shall be deemed to be in compliance with the above. The bottom, top, door and sides of cabinet shall be at least No. 18 gage sheet iron and double-walled with 1½-in. air space. Joints shall be riveted, welded, or made tight by some equally effective means. The door shall be provided with a three-point lock, and the door sill shall be raised at least two inches above the bottom of the cabinet.

"Wooden cabinets, constructed in the following manner, also shall be deemed to be in compliance with the preceding paragraphs. The bottom, sides and top shall be constructed of an approved grade of plywood at least 1-in. in thickness, which shall not break down or delaminate under fire conditions. All joints shall be rabbeted and shall be fastened in two directions with flathead wood screws. When more than one door is used, there shall be a rabbeted overlap of not less than one inch. Hinges shall be mounted in such a manner as not to lose their holding capacity due to loosening or burning out of the screws when subjected to the fire test." **

It is most important for each plant to check with local fire authorities on the type of stor-

---

* From § 43 of NFPA Std. No. 30.

** From § 42 of NFPA Std. No. 30.

Fig. 42–16.—Safety storage cabinet for flammables gives at-the-job availability.

age and handling it is using or proposes to use. Fig. 42–16 shows a safe storage cabinet.

Outside storage houses

If space permits, it is advisable to construct the flammable liquid storage room as a separate building set aside from the plant proper. Construction can be similar to that described for inside storage rooms. The type of product stored and the proximity to other buildings and structures will determine the best design to be used.

Cleaning Tanks

General precautions

Tanks and vessels which have contained flammable liquids should be cleaned, as required for repairs, entry, or change of product, only under the supervision of a competent man who is familiar with fire and accident prevention recommendations and with first aid measures.

These recommendations cover fire, explo-

sion, asphyxiation, and possible poisoning from toxic material. American Petroleum Institute, Standard RP2015, *Cleaning Petroleum Storage Tanks,* and National Fire Protection Association, *Cleaning or Safeguarding Small Tanks and Containers,* Standard No. 327.

An industrial plant that has neither proper cleaning equipment nor personnel informed on tank cleaning operations should consult the head of the service department of the plant's flammable liquid supplier. Tank cleaning contractors who have all the necessary equipment, as well as experienced crews, can be found in many cities.

Before tank cleaning operations are started, the supervisor should provide himself and the workmen with the proper equipment, which may include fresh air hose masks with blower, self-contained air supplied mask, suitable clothing, safety belts, safety lines, and tools. He should make sure that masks and equipment are of the proper type, clean, and in good condition, and he should see that the workmen are properly instructed in their use.

Before any repair or cleaning operations are started, the tank must be purged of all flammable vapor by ventilation or other means. The inside of the tank must be checked with a combustible gas indicator before entering and frequently thereafter while work is in progress.

Remove all sources of ignition from the surrounding area. Obtain tank entry and hot work permits. (See Chapter 43.) Blank all piping, lock out all electrical equipment, and use only lighting approved for the specific atmospheric conditions in the tank.

Consider wind and weather conditions. Do not start work if wind might carry vapors into an area where they could create a hazard, or if an electrical storm is threatening or in progress. Prohibit smoking and the carrying of matches and lighters. All rules should be strictly enforced.

Sandblasting equipment must be bonded to the tank to prevent static sparks. Use of power chipping tools and rivet busters must be suspended while flammable vapors from tanks may be present in the area. Nozzles on steam lines used to free tanks of vapor must be bonded to the tank shells to prevent static accumulation. Motor trucks, gas engines, open flames, and portable electric equipment must be kept a safe distance from the tank being cleaned.

Protective equipment

Although full details on personal protective equipment can be found in Chapter 19, it is well to discuss specifics concerning tank cleaning here.

It is considered good practice to wear watertight rubber boots, thoroughly cleaned and in good condition, for tank cleaning. It is advisable to wear rubber or impermeable gloves, but, if the work does not permit their use, a protective cream recommended for the exposure should be used. In an atmosphere which is dangerously irritant or corrosive to the skin, a complete outfit of impervious clothing must be worn.

Supplied-air hose masks with blowers and safety harnesses with lifelines, or self-contained or self-generating breathing apparatus, must be worn by men entering tanks unless the person in charge has determined that the vapor concentration is below the threshold limit value (see Chapter 38, "Industrial Toxicology") and that no oxygen deficiency exists.

If a test of the tank shows it to be deficient in oxygen, whether or not it is otherwise immediately hazardous to life, the tank should be ventilated with fresh air and be checked for oxygen before entering with approved breathing apparatus. This can be either a compressed-air type of self-contained breathing apparatus, a self-generating type, or air-supplied type. The fit of the facepiece should be carefully checked before the workman enters the tank. If the environment contains a substance that is dangerously irritant or corrosive to the skin (lead tetraethyl, for example), the breathing apparatus may be supplemented by complete attire of impervious clothing.

Some medical authorities advise that individuals with broken eardrums not work in vapor-laden atmospheres, because no mask will protect these abnormal openings.

The supplied fresh air hose mask with blower is generally the best type of respiratory equipment for continuous use in a tank.

Proper procedures

A tank should be free of gas before any work is performed inside it.

A worker should not be allowed to enter a gassy or oxygen-deficient tank unless absolutely necessary, such as in an emergency. When this is necessary, the man should be attended on the outside by another worker who is similarly outfitted with approved respiratory protection and protective attire as needed so that the latter can rescue the man in the tank if necessary. In some states this is a legal requirement.

Men engaged in tank cleaning should be instructed in the method of giving artificial respiration and should be retrained periodically. If a tank cleaner is overcome by vapor or gas, he should be removed to fresh air immediately and given artificial respiration until he resumes breathing. A physician should be summoned, and the resuscitated man kept quiet and warm until breathing and circulation are normal.

The sense of smell cannot be relied upon for an accurate estimate of the amount of flammable vapor present in a tank. It does, however, give warning that vapors of some kind are present. Symptoms of dizziness, nausea, and headache indicate that a dangerous concentration of flammable and toxic vapors is present. Exposed workers should leave the contaminated area immediately and not return until vapors have been cleared.

Workers should always have a clear path of escape from a tank and should bear in mind that they may have to use it in a hurry. A ladder should always be used when a tank must be entered from above, and it should be left secured in place until the last man is out of the tank. Under severe conditions, a lifeline is recommended to assist with rescue work.

Burning, welding, cutting, and spark-producing operations should not be permitted in a tank until the area to be heated has been thoroughly cleaned and tests have determined that the tank atmosphere is vapor free. Where any vapor is present, further ventilation will be required to remove it from the tank.

Hot work repairs (welding and cutting) should not be made whenever other means can be used more safely. Even after a tank has been freed of vapor, combustible mixtures again may be formed through admission of flammable vapors or liquids from other sources, such as an unblanked line or connection; a break in the bottom of the tank; sludge, sediment, or sidewall scale; or wood structures soaked with the liquid. The interior framing of floating tank roofs can trap quantities of liquids that will release vapors. Therefore, periodic tests with a combustible gas indicator should be made during the work.

Burning or cutting may release lead fumes from paint on either the inner or outer surface of the tank or from some other source. An approved mask for protection against lead may be necessary, or it may be advisable to exhaust the fumes. Wood supports inside the tank should be protected or removed before hot work is begun.

Some companies use low-voltage transformers to eliminate the danger of electric shock, especially when men are required to work in wet areas.

All portable electrical equipment should be grounded.

Cold work in a tank may present a dust problem, and approved dust respirators and goggles should be worn if tests definitely show that wearing of a hose mask is not necessary.

Gas tests should be made frequently if the presence of gas is suspected. The concentration of gas or vapor in a tank or vessel may increase as work progresses, especially if the inside of the tank is scraped or heated during the operation.

Forced ventilation may be required in many cases where repair work is to be done inside a tank.

Where fumes develop from welding or other repair work, mechanical air movers should be used or the workmen should wear fresh air hose masks. On a large tank, a door sheet or sheets may be removed for ventilation after the tank has been made gas free. The opening also can be used to expedite removal of sediment and to increase illumination.

If a tank has been closed for some time, there may be a deficiency of oxygen due to rusting (oxidation) of the metal of the tank. In that case, no one should enter the tank without a supplied air mask or, for a relatively brief stay, a self-contained or self-generating breathing apparatus.

Cleaning aboveground tanks

Here in step-by-step form is a summary of a typical procedure for cleaning tanks that have contained flammable or combustible liquids.

1. Remove all sources of ignition (matches, open flames, smoking, gas engines, welding, exposed electrical wiring and equipment) from the vicinity of the tank.

2. Empty the tank by pumping, draining, and floating tank contents with water. The water should be introduced through fixed tank connections.

3. Disconnect and blank all product, steam smothering, foam, and similar lines. Do not rely on valves.

4. Open all manholes and allow the tank to air thoroughly. Ventilate or steam the tank for the number of hours required by its size. If steam is used, cool and ventilate afterward.

5. Have available the required personal protective equipment: fresh air hose masks, approved flashlights, safety belts and lifelines, rubber boots and gloves.

6. If light is needed, use only flashlights or electric lanterns approved for combustible atmospheres by Underwriters' Laboratories Inc.

7. Test for vapor content with a combustible gas indicator. For entering the tank before it becomes vapor-free, wear a fresh air hose mask, air supply tanks, and safety belt and lifeline, if the tank atmosphere contains more than the maximum acceptable concentration of vapor. Have ample help available outside for the number of men inside.

8. If a storage tank has contained leaded gasoline subsequent to its last thorough cleaning, follow instructions of suppliers of tetraethyl lead.

9. Bond steam lines and water wash nozzles to the tank.

10. Wash sludge, sediment, and scale from the tank through the drain or remove it with a pump. Flush out well and overflow the tank with water if necessary.

11. Make a gas test and, if the tank is found

vapor-free, check conditions inside the tank before issuing an OK for the work. Otherwise, clean and ventilate further as required.

12. Continue ventilation for duration of the work in the tank, and make periodic tests for the presence of gases or vapors as the work progresses.

Ventilation by air instead of steaming is used more widely where deposits of iron sulfide are not present and where a source of power supply for mechanical air removers is available. Special precautions are required for the servicing of tanks which contain pyrophoric iron sulfide. (See the American Petroleum Institute's accident prevention manual, *Cleaning Petroleum Storage Tanks*, listed in "References.")

Cleaning small tanks and containers

General precautions. Work on containers that have held flammable or combustible liquids or gases shall be supervised by a trained foreman capable of maintaining a high degree of safe operations. If the container has held such unstable compounds as nitrocellulose, pyroxyline solutions, nitrates, chlorates, perchlorates, or peroxides, special precautions should be taken because the container may contain enough oxygen to support combustion. Contact the manufacturer or supplier for specific information regarding cleaning procedures and other precautions.

Small tanks and drums should be cleaned and steamed in an open-faced building with ample roof ventilation. It is even preferable to clean them in an outside area free from ignition sources. Pipes and nozzles should be electrically bonded to containers being steamed.

Covers, plugs, and valves should first be removed and the tank or drum permitted to drain into a container. The inside should be examined for rags, waste, or other debris that might interfere with draining and retain flammable vapors. Only lights approved for use in the *National Electrical Code* (Class I, Division I, Group D hazardous locations) should be used for this inspection; see explanation in Chapter 41, "Electrical Hazards." Mirrors can sometimes be used to reflect daylight inside for this purpose.

Steaming. Steaming, hot chemical wash, water filling, and use of inert gas are among the common methods for cleaning and vapor-freeing small tanks and drums. If the inside of the container is clean and steam is available, the easiest method is to steam it. (See American Welding Society, *Safe Practices for Welding and Cutting Containers That Have Held Combustibles.* Also see NFPA Standard 327, *Cleaning or Safeguarding Small Tanks and Containers.*)

The tank or drum, placed on a steam rack or over a steam connection with the outlet holes at the lowest point, should be allowed to drain. The tank or drum should rest against a steam pipe or should be bonded. An ample supply of live steam should then be applied for a period of not less than ten minutes.

The inside of the drum should be washed with hot water and, after being cooled, tested with a combustible gas indicator. If the drum tests vapor free, hot work repairs can then proceed. If not, it should be cleaned and steamed further. It is sometimes advisable to mark the drum to signify that it is vapor free.

If steaming will not clean the tank or drum, a cleansing compound of sodium silicate or trisodium phosphate (washing powder) dissolved in hot water and kept at a temperature of 170 to 190 F should be used. Hot water should be added to overflow the container until no appreciable amount of volatile liquid, scum, or sludge appears.

Exceedingly dirty containers may require a caustic soda solution (except aluminum and zinc coated drums in which caustic soda may generate hydrogen), agitated enough to ensure that interior surfaces are thoroughly cleaned. They can then be drained, washed, and steamed.

If steam is not available to heat the water, a cold water solution with an increased amount of cleansing compound can be used. The solution should be agitated to ensure cleaning.

To guard against burns, especially when they are using steam, hot water, and caustic soda, workmen should be protected with suitable clothing, such as boots, gloves, face shields, and rubber aprons.

Because of its low ignition temperature,

drums that have contained carbon disulfide should not be steamed. They should be made vapor-free with a cleansing compound and then gas tested.

Small tanks can be made safe by means of an inert gas, but this method is not generally considered as safe as steaming.

Portable inert gas generators are available for special jobs, but they should be used only by well qualified men who can produce proper atmosphere and safeguard against fire or explosion.

Carbon dioxide and nitrogen are sometimes used to make small tanks and drums that have contained flammable liquids safe for hot work repairs.

When an inert gas like carbon dioxide or nitrogen is used, the tank or vessel is washed as free as possible of flammable liquids and flushed thoroughly, so that the vessel overflows. As much water as the repair work will permit should be left in the tank, and carbon dioxide should be introduced to produce a concentration of not less than 50 percent by volume, except that if the tank has contained hydrogen or carbon monoxide, 80 percent concentration is required.

Nitrogen concentrations should be 60 percent or higher, depending on the previous contents of the container.

Abandonment of tanks

Tanks to be *permanently* abandoned should be thoroughly washed out and made safe from flammable vapors, dismantled, and removed from the premises. Tanks that are not to be removed should have all flammable liquid removed and be filled with a non-shrinking inert solid material. They should have all inlets and outlets capped.

Tanks, taken out of service for less than 90 days, should have fill line, gage opening, and pump suction line capped and should be secured against tampering. The vent line should be left open.

Common Uses of Flammable Liquids

Industrial use of flammable liquids involves all the general precautions outlined in this chapter pertaining to static, toxicity, storage, housekeeping, explosionproof electrical instal-

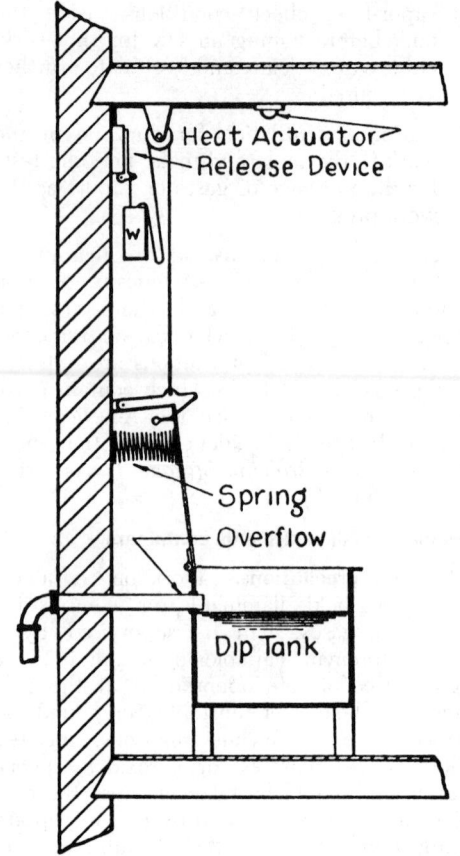

Fig. 42–17.—Cover for medium-sized dip tank can be hinged and gravity closing or should slide on tracks and be held open by a fusible link or other heat-actuated device.

Courtesy National Fire Protection Association.

lation, segregation and isolation of operations, ventilation, enclosure of operations, grounding sources of ignition, and fire and explosion.

All rooms or portions of plants (and the equipment in them), in which flammable liquids or vapors are used or generated, should be constructed, installed, and operated as recommended by the National Fire Protection Association in its *Flammable & Combustible Liquids Code*, and as required by local ordinances.

The following paragraphs discuss several of the more common processes and items of

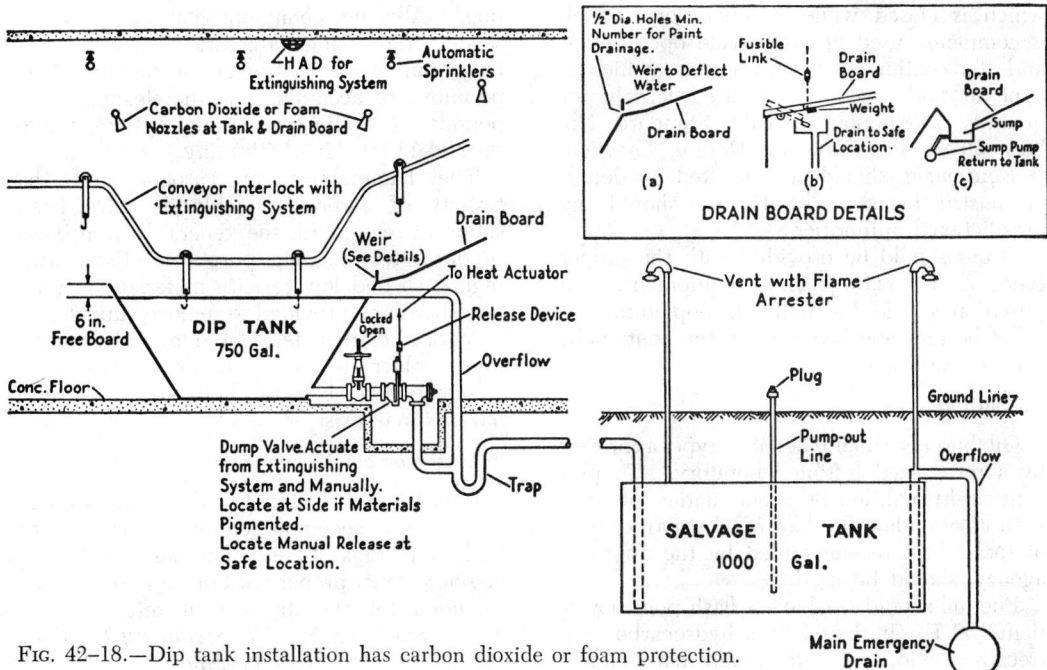

FIG. 42–18.—Dip tank installation has carbon dioxide or foam protection.

Courtesy National Fire Protection Association.

equipment which use flammable liquids and the precautions applying to each.

Dip tanks

Dip tanks containing flammable liquids, subject to ignition at ordinary temperatures and giving off flammable vapors, present a severe fire and explosion hazard. (See NFPA Standard No. 34, *Dip Tanks Containing Flammable or Combustible Liquids.*)

Dipping operations should be conducted above grade in a detached one-story building of noncombustible construction or in a cutoff one-story section in as large a room as possible, adequately ventilated, away from sources of ignition, and conspicuously marked as a flammable liquid area.

Handling of flammable liquids in open containers can be hazardous and should be avoided. The openings should be as small as possible and a cover should be provided. The cover should be hinged and gravity closing or should slide on tracks and be held open by a fusible link or other heat-actuated device (Fig. 42–17).

The NFPA states, "Mechanical ventilation shall be provided, and the ventilating system arranged to move air from all directions toward the vapor area origin and thence to a safe outside location. The ventilating system shall be so arranged that the failure of any ventilating fan shall automatically stop any dipping conveyor system."

Tanks with capacities over 500 gal should have bottom drains unless the viscosity of the liquids they contain makes this requirement impractical.

Overflow pipes should be used to carry off any overflow liquids to a safe place, preferably outside the building. Tanks should also be protected by automatic fire extinguishing systems (Fig. 42–18).

Japanning and drying ovens

Ovens used for evaporating varnish, japan enamel, and other flammable liquids can present serious fire and explosion hazards and should have ample provision for ventilation and explosion venting.

Drying ovens are of two types: the box oven

1319

which is closed while in operation (which is commonly used in small scale operations), and the continuous conveyor oven which is open at both ends and is used largely for quantity production. NFPA Standard No. 86A, *Ovens and Furnaces, Design, Location, & Equipment*, should be consulted for details of construction and operation, as should any jurisdictional authorities.

Ovens should be provided with the proper type of fire extinguishing equipment, and provision should be made to stop automatically the fans and conveyor in the continuous oven in case of fire.

Oil burners

Oil burners should be of a type approved by a recognized testing laboratory. To prevent faulty ignition or accumulation of soot, with its attendant fire hazard, the correct type of fuel oil, as recommended by the approval agency, should be used.

Fuel oil should not have a flash point lower than 100 F. It should be a hydrocarbon oil, free from acid, grit, and foreign matter likely to clog or damage the burners or valves. Some plants use acid sludge for fuel, which requires special burning equipment and procedures.

Fuel for domestic burners may be stored in basement tanks, as provided by NFPA Standard No. 31, *Installation of Oil Burning Equipment*. A supply tank should preferably be located outside the building and should be underground. The top of the tank should be below the level of all piping to which it is connected, to prevent discharge of oil through a broken pipe or connection by siphoning.

A gravity feed to burners should not be used unless special safeguards are provided against abnormal discharge of oil at the burner. The primary hazard of oil burners is the possibility of discharge of unburned oil into a hot fire box where it may vaporize and form an explosive mixture. Automatic safeguards should be provided to control the hazard.

Cleaning metal parts

Stoddard solvent (over 100 F flash point) is safe for use in cleaning grease and oil from metal parts where ordinary ventilation is provided and the area is free of sources of ignition. Alkaline compounds available under several trade names are safe from a fire and toxic standpoint. Oil or grease should not be permitted to accumulate in the cleaning compounds. (See Factory Mutual System, Bulletin No. 11.60, *Metal Cleaning*.)

The flammability of gasoline and the toxicity of carbon tetrachloride have been sufficient reasons for the general ban on these products for cleaning purposes. Even with high flash and low toxicity materials, ventilation should be provided to remove vapors.

Mixing carbon tetrachloride and a flammable solvent to reduce the fire hazard is not recommended, because both toxic and fire hazards may exist.

Internal combustion engines

Good housekeeping should be practiced to prevent the accumulation of rubbish, oil or fuel, and rags around internal combustion engines, and proper receptacles should be provided for the disposal of refuse. (See NFPA Standard No. 37, *Stationary Combustion Engines and Gas Turbines*.)

Before a gas tank is filled, the engine should be shut down and hot exhaust pipes should be permitted to cool. Filling should preferably take place during daylight, and approved safety cans or a hand pump with bonded filling hose should be used, while the main fuel supply is kept in approved containers outside the building.

The fuel tanks of engines that must operate continuously and cannot be shut down for filling should be located outside the engine room where vapors will not be exposed to hot engines or exhaust.

Lift trucks and other mobile equipment should be refueled outside of buildings.

Engines should be kept clean, with insulation on electrical wiring kept in good repair.

Spray booths

Paint spraying operations should be done in detached buildings or cut off from other operations where possible. Where spraying is done in open areas, curtains of noncombustible material extending downward from the ceiling should be used. The enclosed area should not be made so small that explosive mixtures of vapor and air can be easily formed. Heating units and piping which might become coated with flammable materials should be eliminated or protected against such accumulations. (See NFPA Standard No. 33, *Spray Finishing Using Flammable and Combustible Materials.*)

If ventilation cannot be secured by natural means through wall openings, windows, or roof vents, forced ventilation or local exhaust removal which conforms to NFPA standards should be used. Exhaust systems should be designed so that there will be a minimum amount of deposit, which can be easily and frequently removed.

Fires in spray booths and spray booth operations most frequently result from spontaneous ignition of spray deposits. These fires can be prevented by a regular schedule of cleaning determined by the accumulation rate.

Water wash booths have proved to be safer from fire than the dry type of booth because they trap the excess spray before it can enter the exhaust ducts.

Electrical equipment in spraying areas should meet the requirements of the *National Electrical Code* for such locations.

Spraying operations lend themselves well to automatic fire control. Automatic sprinklers or carbon dioxide systems most effectively confine fires. Discharge heads of such equipment must be protected from overspray.

Electrostatic spraying, usually automatic, introduces a possible source of ignition in the arcing of parts to the electrodes. To overcome this hazard, parts being sprayed can be held in tight-fitting fixtures instead of being permitted to hang or swing and, thus, come close enough to induce a spark.

Gasoline blow torches and plumbers' furnaces

Gasoline blow torches and plumbers' furnaces are hazardous because fuel may be spilled when filling is being done and because fuel must be used for priming when lighting. They should never be filled hot, and it is preferable that fueling be done outdoors.

Only equipment from reliable manufacturers made according to the NFPA Standard No. 393, *Gasoline Blow Torches and Plumbers' Furnaces*, should be used, and the operator should understand and explicitly follow the instructions of the manufacturer. The equipment should never be used close to combustible materials. Other means of performing the operation are preferable.

Liquefied petroleum gases

Liquefied petroleum gases include any material which is composed predominantly of any of these hydrocarbons, or mixtures of them: propane, propylene, butane (normal butane or isobutane), and butylenes. The gases liquefy under moderate pressure, but convert into a gaseous state upon relief of the pressure. LP-gas vapor presents a hazard comparable to that of any flammable natural or manufactured gas, except that since it is heavier than air, the matter of adequate ventilation requires some attention.

Liquefied petroleum gases are used as fuel gases, as raw materials in chemical processes, for example, in the making of hydrogen, and to form special atmospheres in heat-treating furnaces. It is important that employees understand the properties of these gases and that they be thoroughly trained in safe practices for handling, distribution, and operation. Areas where there might be "spills" should be examined, and detailed programs should be developed to handle any emergencies which might arise.

Systems should be designed and installed by experienced, reliable concerns which are thoroughly familiar with the hazards, with state and local codes, and with fire organization and insurance company recommendations. (See National Fire Protection Association, Factory Mutual System, LP-Gas Association, and Factory Insurance Association references on this subject in "References.")

For more information on handling liquefied petroleum gases, see pages 759–760. For other gases, see Chapter 39.

42—Flammable and Combustible Liquids

References

American Conference of Governmental Industrial Hygienists, P.O. Box 1937, Cincinnati, Ohio 45201. *Threshold Limit Values* (published annually).

American Mutual Insurance Alliance, 20 North Wacker Dr., Chicago, Ill. 60606. *Safe Handling of LP-Gas When Used as a Motor Fuel.*

American Petroleum Institute,1801 K St. NW, Washington, D.C. 20006.
Cleaning Petroleum Storage Tanks. Standard RP 2015.
Welded Steel Tanks for Oil Storage. Standard SID 650.

American Society for Testing and Materials, 1916 Race St., Philadelphia, Pa. 19103.
Test for Flash Point by Pensky-Martens Closed Tester. Standard D 93–73.
Test for Flash Point by Tag Closed Tester. Standard D 56–73.
Test for Vapor Pressure of Petroleum Products (Reid Method). Standard D 323–72.
1974 Annual Book of ASTM Standards:
> Parts 23, 24, and 25, "Petroleum Products and Lubricants."
> Parts 27 and 28, "Paint."
> Part 30, "Soaps, Antifreezes, Polishes, Halogenated Organic Solvents, Activated Carbon, Industrial Chemicals."

American National Standards Institute, 1430 Broadway, New York, N.Y. 10018.
Installation of Blower and Exhaust Systems for Dust, Stock, and Vapor Removal or Conveying, Z33.1. (NFPA Standard No. 91.)
Safety Color Code for Marking Physical Hazards, Z53.1.
Standards for Petroleum Products, Z11 Series.

American Welding Society, 2501 N.W. 7th St., Miami, Fla. 33125.
Safe Practices for Welding and Cutting Containers That Have Held Combustibles.

Association of American Railroads, Bureau for the Safe Transportation of Explosives and Other Dangerous Articles, 1920 L St. NW, Washington, D.C. 20036.
Handling Collisions and Derailment Involving Explosives, Gasoline and other Dangerous Articles. Pamphlet No. 22.
Recommended Practice for the Prevention of Electric Sparks That May Cause Fires in Tanks or Tank Cars Containing Flammable Liquids or Flammable Compressed Gases, Due to Proximity of Wire Lines. Bureau of Explosives Circular No. 17-E.
Signal Manual of Recommended Practice.

Factory Insurance Association, 85 Woodland St., Hartford, Conn. 06102.
Recommended Good Practices (Supplements to NFPA Standards.)

Factory Mutual System, 1151 Boston-Providence Turnpike, Norwood, Mass. 02062.
Approved Equipment for Industrial Fire Protection Manual.

National LP-Gas Association, 79 W Monroe St., Chicago, Ill. 60603.

Manufacturing Chemists' Association, 1825 Connecticut Ave., NW., Washington, D.C. 20009.
"Chemical Safety Data Sheets."

National Fire Protection Association, 470 Atlantic Ave., Boston, Mass. 02210.
Fire-Hazard Properties of Flammable Liquids, Gases, and Volatile Solids. Standard No. 325M.
Fire Protection Handbook
Hazardous Chemicals Data. Standard No. 49.
"National Fire Codes."
Recommended System for the Identification of the Fire Hazards of Materials. Standard No. 704M.
Basic Classification of Flammable and Combustible Liquids. Standard No. 321.
Cleaning or Safeguarding Small Tanks and Containers. Standard No. 327.
Control of Gas Hazards on Vessels To Be Repaired. Standard No. 306.
Dip Tanks Containing Flammable or Combustible Liquids. Standard No. 34.
Dry Cleaning Plants. Standard No. 32.
Flammable & Combustible Liquids Code. Standard No. 30.
Liquefied Petroleum Gases, Storage and Handling. Standard No. 58.

Liquefied Petroleum Gases at Utility Gas Plants. Standard No. 59.
National Electrical Code. Standard No. 70.
Oil Burning Equipment, Installations. Standard No. 31.
Ovens and Furnaces, Design, Location & Equipment. Standard No. 86A.
Protection of Tanks Containing Flammable or Combustible Liquids in Locations That May be Flooded. Standard No. 30, Appendix B.
Recommended Practice on Static Electricity, Standard No. 77.
Spray Finishing Using Flammable and Combustible Materials. Standard No. 33.
Tank Vehicles for Flammable and Combustible Liquids. Standard No. 385.
Warning Labels on Flammable Liquid Containers. Standard No. 326.

National Safety Council, 425 North Michigan Ave., Chicago, Ill. 60611.
Industrial Data Sheets
Acetone, 398.
Acetylene, 494.
Amyl Acetate, 208.
Carbon Bisulfide (Carbon Disulfide), 341.
Carbon Tetrachloride, 331.
Chlorates, 371.
Cleaning With Hot Water and Steam, 238.
Cutting Oil, Emulsions and Drawing Compounds, 501.
Ethyl Ether (Diethyl Oxide), 396.
Ethylene Dichloride, 350.
Flammable Liquids in Small Containers, 532.
Hot Work Permits (Flame or Spark), 522.
Industrial Ethyl Alcohol, 391.
Liquefied Petroleum Gases for Industrial Trucks, 479.
Liquid Degreasing of Small Metal Parts, 537.
Maintenance of Electric Motors for Hazardous Locations, 546.
Manual Gaging and Sampling of Petroleum Tanks, 563.
Methanol, 407.
Static Electricity, 547.
Styrene Monomer, 627.
1,1,1-Trichloroethane, 456.
Trichloroethylene, 389.
Trinitrotoluene (TNT), 314.
Turpentine, 367.
Vapor Degreasers, 429.

Underwriters Laboratories Inc., 207 East Ohio St., Chicago, Ill. 60611
Classification of Hazards of Liquids. Research Bulletin No. 29.
Fire Protection Equipment List.
Flammable Liquids, Static Electricity Hazards. Serial No. UL-435.
Gas and Oil Equipment List.
Hazardous Location Equipment List.
The Lower Limit of Flammability and Autogenous Ignition Temperature of Certain Common Solvent Vapors Encountered in Ovens. Research Bulletin No. 43.
Oil Tank Vents—Hazards of Screens. Serial No. UL-31.

U.S. Bureau of Mines, Washington, D.C. 20240.
Flammability Characteristics of Combustible Gases and Vapors. Bulletin 627.
Limits of Flammability of Gases and Vapors. Bulletin 503.
Mine Gases and Methods for Their Detection. Circular No. 33.

U.S. Department of Commerce, Washington, D.C. 20234.
Sparking Characteristics and Safety Hazards of Metallic Materials. Technical Report No. NGF–T–1–57, PB 131131, Office of Technical Services.
Static Electricity in Nature and Industry. Bulletin No. 368.

U.S. Department of Transportation, Washington, D.C. 20591.
Hazardous Materials Regulations, 49 CFR, Parts 170 through 179.
Motor Carrier Safety Regulations, 49 CFR, Parts 393 and 397.

Fire
Protection

Chapter
43

Fire Protection 1325
Adequate exits . . . Fire-safe construction . . . Smoke and heat venting

Factors Contributing to Industrial Fires 1332
Electrical equipment fires . . . Smoking . . . Friction . . . Foreign objects or tramp metal . . . Open flames . . . Spontaneous ignition . . . Housekeeping . . . Explosive atmosphere

The Chemistry of Fire 1340
Cooling . . . Removing fuel . . . Limiting oxygen . . . Interrupting the reaction

Classification of Fires 1342
Class A fires . . . Class B fires . . . Class C fires . . . Class D fires . . . Other fires

Portable Fire Extinguishers 1342
Principles of use . . . Water solution extinguishers . . . Dry-chemical, carbon dioxide, and liquefied-gas extinguishers . . . Halogenated compounds . . . Dry-powder extinguishers . . . Miscellaneous equipment . . . Training of employees . . . Maintenance and inspection

Water Systems 1361
Water supply and storage . . . Automatic sprinklers . . . Water spray systems . . . Hydrants, hoses, and nozzles . . . Maintenance of fire hose

Special Systems and Agents   1371

Foam systems . . . Wet water . . . Carbon dioxide extinguishing systems . . . Dry-chemical systems . . . Halon systems . . . Steam systems . . . Inerting systems . . . Preventing explosions . . . Suppressing explosions

Identification of Hazardous Materials   1380

Fire Prevention Activities   1380

Inspections . . . Hot work permits . . . Flameproofing wood and fabrics . . . Fire and emergency drills . . . Fire brigades . . . Communications

References   1385

FIRE PROTECTION embraces all measures relating to safeguarding human life and preservation of property in the prevention, detection, and extinguishment of fires.

It is principally a matter of physical arrangements, such as sprinkler systems, water supplies, and fire extinguishers.

FIRE PREVENTION should not be considered as being synonymous with fire protection but, instead, a term to indicate measures specifically directed toward preventing the inception of fires.

Fire protection is usually understood to include fire prevention procedures. Both aim to protect employees, property, and continuity of operations. The information in this chapter should provide the basis for an efficient fire prevention and control program. Life safety is given primary consideration in all recommendations made in this Manual.

An effective fire loss control program must include these objectives:

1. Prevent loss of life and personal injury

2. Protect property

3. Provide uninterrupted operations

4. Prevent inception of fire.

Fire Protection

Fire protection engineering is a highly developed, specialized field in which special engineering disciplines are focused. The solution of many fire protection problems requires the special combination of training and perspective of an experienced fire protection engineer.

If no such specialist be in an organization, the person responsible for overall industrial safety should be familiar with the sources of this information, if not with the complete technology.

Authoritative fire protection literature is available from such organizations as the National Fire Protection Association (NFPA), the American Insurance Association, and Factory Mutual Engineering Corporation.

Fire insurance companies and some trade associations publish information on fire hazards and make recommendations for fire prevention and protection in their respective fields. See Chapter 23, "Sources of Help," and the NFPA *Fire Protection Handbook.*

The safety professional who is faced with special fire problems should seek specific advice from these trade associations, his fire insurance carrier, the local fire inspection bureau, the fire department, governmental agencies having jurisdiction, or a fire protection engineering consultant. Expert advice is available on a fee basis from consulting fire protection engineers whose names can be obtained from the NFPA.

Adequate exits

Of the many factors involved in securing safety to life from fire, building exit facilities rank among the most important. Although there is general acceptance of the importance of adequate exit facilities, they remain inade-

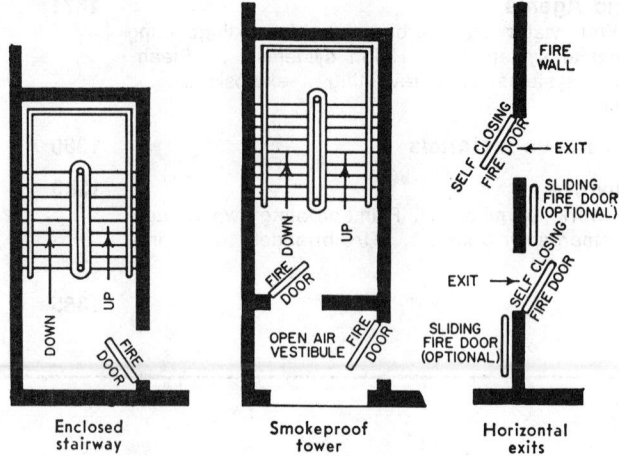

Enclosed stairway   **Smokeproof tower**   **Horizontal exits**

Fig. 43–1.—Plan views of types of exits. Stair enclosure prevents fire on any floor trapping persons above. Smokeproof tower is better because opening to air at each floor largely prevents chance of smoke on stairway. Horizontal exit provides a quick refuge, lessens the need of a hasty flight down stairs. Horizontal sliding fire doors provided for safeguarding property values are arranged to close automatically in case of fire. Swinging doors are self-closing. Two wall openings are needed for exit in two directions.

Courtesy National Fire Protection Association.

quate at many plants throughout industry.

It is generally recommended that every building or structure, and every section or area in them, shall have at least two separate means of egress, so arranged that the possibility of any one fire blocking all of them is minimized (Fig. 43–1). Management cannot wait for, or even depend upon, local, state, or federal inspectors to make recommendations before improvements are made, as in many localities even complete compliance with codes and ordinances is no guarantee of adequate safety in an existing building.

Management and others entrusted with the safety of their employees must consider many problems when planning emergency evacuation of buildings. In many cases panic causes more loss of life than fire. While fire is the most common cause of panic, such things as boiler, air receiver or other explosions, fume releases, or structural collapse may also threaten safe, orderly evacuation. See Chapter 18, "Planning for Emergencies."

No construction plans or physical alterations should be made which may prevent safe evacuation under any circumstances that might arise.

NFPA Standard No. 101, *Life Safety Code,* provides a reasonable and comprehensive guide to exit requirements. Where local, state, or federal codes containing more rigid recommendations are in existence, they will, of course, take precedence.

The following general provisions must be considered in planning for building evacuation.

1. Design of exits and other safeguards for life safety must not depend solely on any single safeguard, but additional safeguards shall be provided in case any single measure is ineffective due to some human or mechanical failure.

2. Protection of exits against fire and smoke during the length of time they are designed to be in use. Vertical exit ways

1326

Fig. 43–2.—Exit stairway enclosures. Doors on stairway enclosures at left are open, and basement fire quickly extends up stairs, spreading to upper floors. Stairway doors should never be held open, even if they are equipped with fusible links. In illustration of building at right, stairway doors are closed; fire is confined to basement and stairways are safe to use.

Courtesy National Fire Protection Association.

and other vertical openings enclosed or protected to afford reasonable safety to occupants while using exits (Fig. 43–2).

3. Alternate means of travel for use in case one exit is blocked by fire.

4. Alarm systems to alert occupants in case of fire or emergency.

5. Exits and paths of travel to reach them provided with adequate illumination.

6. Signs indicating ways to reach exits, when needed.

7. Safeguarding equipment and areas of any unusual hazard which might spread fire and smoke, endangering the safety of persons on the way out.

8. Exit drill procedure to assure orderly exit, wherever practicable.

9. Control of psychological factors conducive to panic.

10. Control of interior finish and contents to prevent a fire from spreading fast and trapping occupants.

11. Maintain adequate aisles for exit access.

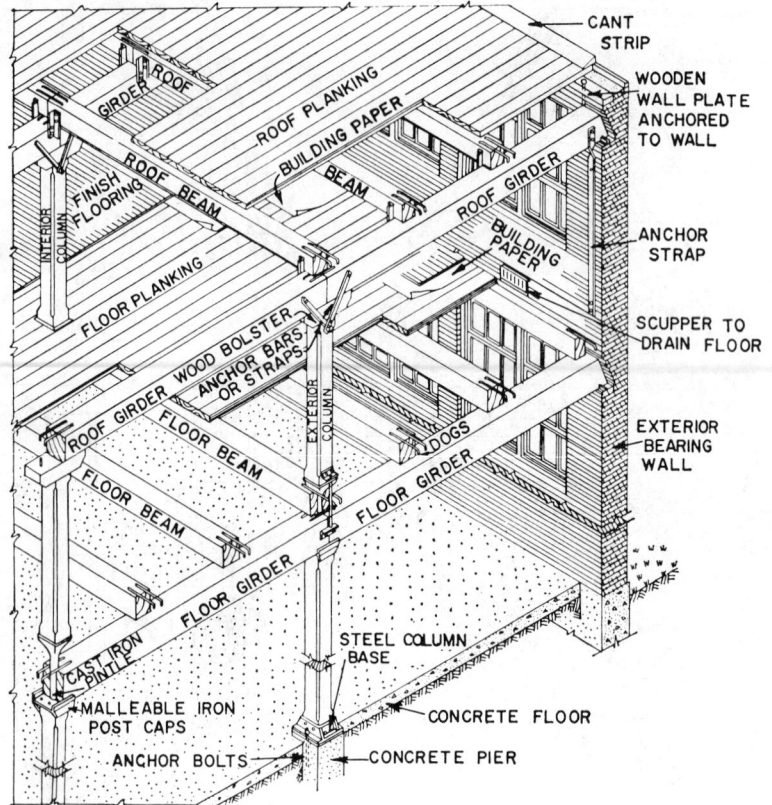

Fig. 43–3.—Components of heavy timber building showing floor framing and identifying components of semimill-type framing.

Courtesy National Fire Protection Association.

Fire-safe construction

There are five common types of building construction, but many buildings are variations of these: fire-resistive, heavy timber, noncombustible, ordinary, and wood frame.

Fire-resistive construction. Although the term "fire resistive" is sometimes mistakenly taken to mean "fire proof," it actually describes a broad range of structural systems capable of withstanding fires of specified intensity and duration without failure.

The term "fire proof" is misleading. *No material is immune to the effects of a fire having sufficient intensity and duration.*

Common fire-resistive components include masonry load-bearing walls, reinforced concrete or protected steel columns, and poured or precast concrete floors and roofs.

Although fire-resistive structures do not, in themselves, contribute fuel to a fire, combustible trim, ceilings, and other interior finish and furnishings may produce an intense fire and pose a serious threat to life safety. Use approved building materials with low flame-spread ratings and limit those that can contribute to available fuel, especially in nonsprinklered buildings.

Control and limit the amount of combustible materials in any one portion of a building. This includes control of the use, handling, and storage of flammable and combustible liquids.

Heavy timber construction is characterized

Fig. 43–4.—Noncombustible building with low-hazard, metal-working occupancy.

Courtesy National Fire Protection Association.

by masonry walls, heavy timber columns and beams, and heavy plank floors (Fig. 43–3). Although not immune to fire, the great mass of the wooden members slows the rate of combustion. Moreover, the char which forms on wooden surfaces serves as an insulator for the wood within.

Noncombustible construction includes all types of structures in which the structure itself (exclusive of trim, interior finish, and contents) is noncombustible but *not* fire resistant. Exposed steel beams and columns, and masonry, metal, or asbestos panel walls, are the most common forms (Fig. 43–4).

Because of the tendency of steel to warp, buckle, and collapse under moderate fire exposure, noncombustible construction is relatively vulnerable to fire damage. It is, therefore, most suitable for low-hazard occupancies. If quantities of combustibles are present, the structure should be protected with automatic sprinklers.

Ordinary construction consists of masonry exterior bearing walls, or bearing portions of exterior walls, that are of noncombustible construction. Interior framing, floors, and roofs are made of wood or other combustible material whose "bulk" is less than that required for heavy timber construction.

If floor and roof construction and their supports have a one-hour fire-resistance rating, and all openings through floors (including stairways) are enclosed with partitions having a one-hour fire-resistance rating, then it is known as "protected ordinary construction." This type building dominates in congested areas of large cities. Its occupancy should be limited to light or moderate hazards; use of highly combustible interior finishes should be minimized.

Even when sheathed, ordinary construction (unlike fire-resistive or noncombustible construction) still has combustible materials in concealed wall and ceiling spaces. Fire frequently originates in these concealed spaces

1329

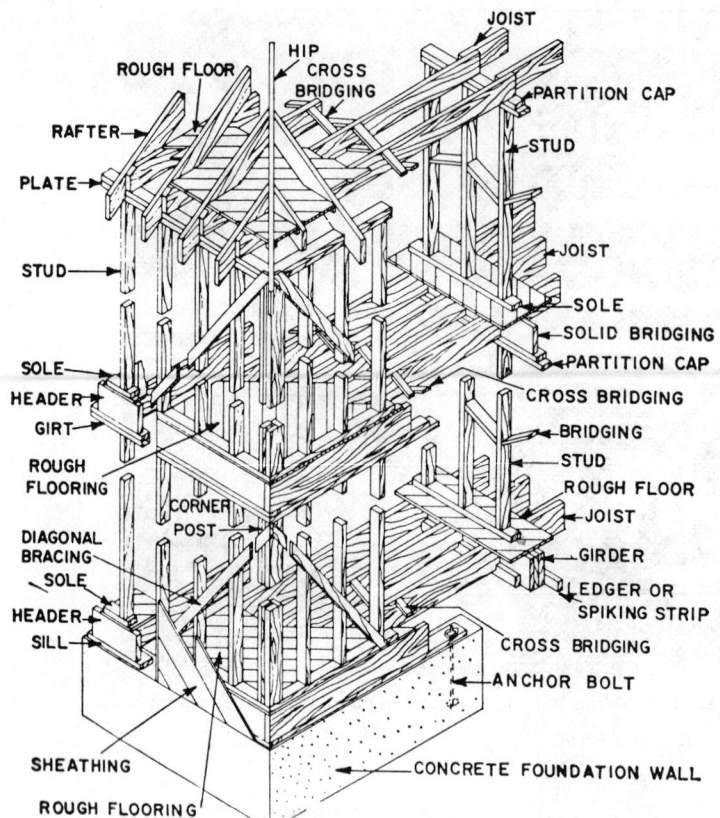

JOIST

HIP
ROUGH FLOOR
CROSS
BRIDGING
PARTITION CAP

RAFTER

STUD

PLATE

STUD

JOIST

SOLE

SOLID BRIDGING
PARTITION CAP

SOLE
HEADER
GIRT
ROUGH
FLOORING

CORNER
POST

CROSS BRIDGING
BRIDGING
STUD
ROUGH FLOOR
JOIST

DIAGONAL
BRACING
SOLE

GIRDER
LEDGER OR
SPIKING STRIP

HEADER
SILL

CROSS BRIDGING
ANCHOR BOLT

SHEATHING

CONCRETE FOUNDATION WALL

ROUGH FLOORING

FIG. 43–5.—Example of wood frame construction common to dwellings. Structural members are identified.

Courtesy National Fire Protection Assn.

or enters into them through openings and then spreads rapidly throughout the entire building, unless the building is safeguarded by fire-stopping or other means.

Wood frame construction consists primarily of wood exterior walls, partitions, floors, and roofs (Fig. 43–5). Exterior walls may be sheathed with brick veneer, stucco, or metal-clad, cement-asbestos, or asphalt siding.

Although generally inferior to other types of construction from a fire safety standpoint, it can be made reasonably safe for light-hazard, low-density occupancies. The safety-to-life factor can be greatly increased by suitable protection against the horizontal and vertical spread of fire, the provision of safe exits, and the elimination of combustible interior finishes. Automatic sprinkler protection can greatly improve the fire safety outlook in

frame buildings; it is fundamental to fire protection of wood frame schools, nursing homes, and similar institutions.

Vertical and horizontal cut-offs. Regardless of the type of building construction, the person responsible for fire control in an industrial plant should thoroughly investigate the feasibility of more stringent requirements.

Stair enclosures, necessary to provide a safe exit path for occupants, also serve to retard the upward spread of fire. Under certain conditions, such as where large areas or high values are involved, buildings may be divided horizontally by fire walls. Fire walls must be designed to rigid specifications in order to withstand the effects of severe fire and building collapse on one side without failure. All openings must be protected with approved closures to prevent the passage of heat.

Smoke and heat venting

Smoke and hot gases generated by a fire, if confined within a building, can seriously impair fire-fighting operations and can spread the fire under the roof for considerable distances from the point of origin. In the case of sprinklered buildings, the spread of hot gases may cause an excessive number of sprinkler heads to operate, thereby depleting water supplies. Moreover, a so-called "smoke explosion" may occur when a supply of fresh air is made available to hot, unburned combustion products.

Effective smoke and heat venting systems consist of curtain boards to produce heat-banking areas under a roof and automatic or manual roof vents for the release of smoke and heat through the roof. (See Fig. 43–6.)

The effectiveness of a roof vent will be directly proportional to the difference in absolute temperature between inside and outside air. The effectiveness of the vent will be further increased if the heated air is in an enclosure due to confinement. The air will not be diluted and cooled by surrounding air. Other factors relating directly to effective venting are:

1. The quantity of heat to be vented depends on quantity of combustibles burning and rate of burning.

2. Venting capacity is proportional to the total vent area.

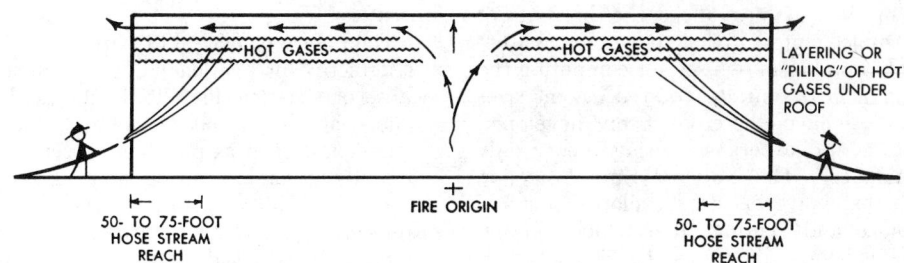

BEHAVIOR OF HOT GASES UNDER FLAT ROOF

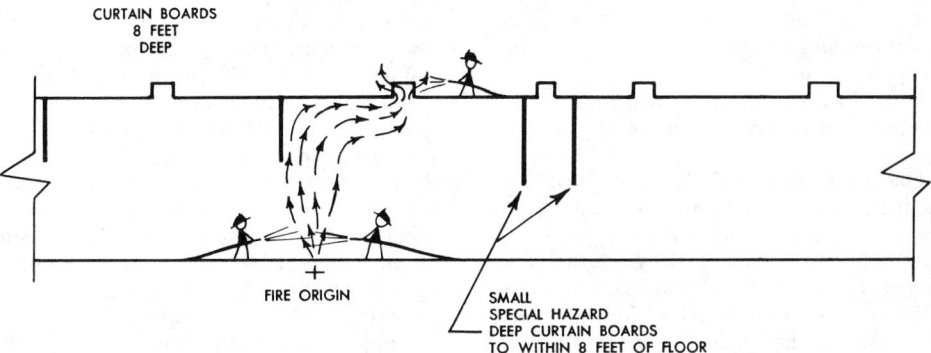

BEHAVIOR OF HOT GASES UNDER MONITORED AND CURTAINED ROOF

Fig. 43–6.—Drawings (not to scale) show how a monitored and curtained roof permits hot gases to be vented locally, allowing firemen to reach seat of fire both from inside a building and from its roof. Monitors on top of roof allow hot gases and smoke to escape. Curtain boards (extending down from the roof) prevent the lateral spread of smoke and heat.

Illustration adapted from one of National Fire Protection Association.

3. Draft curtains increase vent effectiveness because they increase stack height and confine heated gases.

4. Several small vents rather than a single vent of equal area are more effective, because gas temperatures decrease rapidly as the gas travels away from the fire.

5. Roof height will not affect vent effectiveness within the normal ranges of height.

Smoke does not necessarily have a very high temperature. High temperatures are only created in fairly free-burning fires. Venting installed for life safety may not be too effective, therefore, as temperatures high enough to produce the necessary stack effect may not be present. Smoke could also flow under draft curtains and other obstructions until an adequate draft is created.

So many variables affect the burning of combustible material that no exact mathematical formula can be used for computing the amount of heat venting required. Vent sizes and ratios have therefore been developed from tests and experience, using theory only for guidance. The National Fire Protection Association, however, has developed a guide for smoke and venting from which venting requirements may be estimated (NFPA Standard No. 204, *Smoke and Heat Venting*). Also, there are a variety of prefabricated vents available on the market. They can be designed to open automatically at predetermined temperatures and/or smoke concentrations.

Factors Contributing to Industrial Fires

To eliminate causes of fire, we must first know the many ways in which it can start, especially the most common ones. A study made by the Associated Factory Mutual Fire Insurance Company of almost 25,000 industrial fires over a 10 year period indicates that the majority of fires can be traced to four general sources of ignition: electrical, smoking, friction, and over-heated materials (Fig. 43–7).

Electrical equipment fires

Electrical equipment should be installed and maintained in accordance with the *National Electrical Code*. Over-heating of electrical equipment and arcs resulting from short circuits in improperly installed or maintained electrical equipment are two of the leading causes.

Only approved (or UL-listed) equipment should be used where flammable gases or vapors may be present. Electrical equipment necessary in locations which are hazardous because of the presence of flammable liquids, gases or dusts, should be designed for the particular hazardous atmosphere.

Hazardous locations are divided into three classes depending on the material involved. Further division is by degree or severity of the hazard. Complete definitions of the several classes and divisions of hazardous locations and types of equipment for each are contained in the *National Electrical Code*, and are discussed in Chapter 41, "Electrical Hazards."

Temporary or makeshift wiring, particularly if defective or overloaded, is an outstanding cause of electrical fires; it should not be used unless absolutely necessary and should be removed as soon as possible. Over-loaded or partially grounded wiring may also heat up enough to ignite combustibles without blowing fuses or tripping circuit breakers.

Portable electrical tools and extension cords should conform to the *National Electrical Code* and should be inspected at frequent intervals and repaired promptly. Waterproof cords and sockets should be used in damp places, and explosion-proof fixtures and lamps should be used in the presence of highly flammable gases and vapors.

All electrical equipment, particularly portable electrical tools, should be grounded or double insulated for the protection of persons using it. Switches, lamps, cords, fixtures, and other electrical equipment should be listed by Underwriters' Laboratories, Inc., or other recognized testing and certifying agency, and should be used only in applications for which the approval or listing was granted.

Lamp bulbs should be protected by heavy lamp guards or by adequately sealed transparent enclosures, and kept away from sharp objects and kept from falling. Bare bulbs should never be used when exposed to flammable dusts or vapors. Lamp bulbs must be considered as potential hazards in such areas; they must be safeguarded accordingly.

IGNITION SOURCES OF FIRES

To eliminate the causes of fire, it is important to know how and where fires start. The following summary of known causes is based on an analysis of more than 25,000 fires reported to the Factory Mutual Engineering Corporation over a recent 10-year period. Most, but not all, occurred at industrial properties. The causes are arranged in order of their frequency throughout industry, but this arrangement is not necessarily a measure of their relative importance at any particular plant or property.

Electrical—23 per cent. The leading cause of industrial fires. Most start in wiring and motors. Most prevented by proper maintenance. Special attention needed for equipment at hazardous processes and in storage areas.

Smoking—18 per cent. A potential cause of fire almost everywhere. A matter of control and education. Smoking strictly prohibited in dangerous areas, such as those involving flammable liquids, combustible dusts or fibers, and combustible storage. Permitted in clearly designated safe areas.

Friction—10 per cent. Hot bearings, misaligned or broken machine parts, choking or jamming of material, and poor adjustment of power drives and conveyors. Prevented by a regular schedule of inspections, maintenance, and lubrication.

Overheated Materials—8 per cent. Abnormal process temperatures, especially those involving heated flammable liquids and materials in driers. Prevented by careful supervision and competent operators, supplemented by well maintained temperature controls.

Hot Surfaces—7 per cent. Heat from boilers, furnaces, hot ducts and flues, electric lamps, irons, and hot-process-metal igniting flammable liquids and ordinary combustibles. Prevented by safe design and good maintenance of flammable-liquid piping and by ample clearances, insulation, and air circulation between hot surfaces and combustibles.

Burner Flames—7 per cent. Improper use of portable torches, boilers, driers, ovens, furnaces, portable heating units, and gas- or oil-burner flames. Prevented by proper design, operation, and maintenance; by adequate ventilation and combustion safeguards; and by keeping open flames away from combustible material.

Combustion Sparks—5 per cent. Sparks and embers released from incinerators, foundry cupolas, furnaces, fireboxes, various process equipment, and industrial trucks. Use well designed equipment and well enclosed combustion chambers with spark arrestors as needed.

Spontaneous Ignition—4 per cent. In oily waste and rubbish, deposits in driers, ducts and flues, materials susceptible to heating, and industrial wastes. Prevented by good housekeeping and proper process operation. Remove waste daily, clean ducts and flues frequently, and isolate storages subject to spontaneous heating.

Cutting and Welding—4 per cent. Sparks, arcs, and hot metal from cutting and welding operations. Prevented by the use of the permit system and other recognized precautions.

Exposure—3 per cent. Fires communicating from nearby properties. Blank walls furnish the most effective barriers. Protect wall openings with open sprinklers or wired glass, depending on severity of exposure.

Incendiarism—3 per cent. Fires maliciously set by intruders, juveniles, disgruntled employees, and arsonists. Prevented by watch and guard service; install fences and other security measures.

Mechanical Sparks—2 per cent. Sparks from foreign metal in machines, particularly at cotton mills, and grinding and crushing operations. Prevented by keeping stock clean; remove foreign material by magnetic or other separators.

Molten Substances—2 per cent. Fires caused by metal escaping from ruptured furnaces or spilled during handling; also by glass and tempering salts. Prevented by proper operation and maintenance of equipment.

Chemical Action—1 per cent. Chemical processes getting out of control, chemicals reacting with other materials, and decomposition of unstable chemicals. Prevented by proper operation, instrumentation, and controls; and by careful handling and storage, particularly avoiding conditions of heat and shock.

Static Sparks—1 per cent. Ignition of flammable vapors, dusts and fibers by discharge of accumulation of static electricity on equipment, materials, or the human body. Prevented by grounding, bonding, ionization, and humidification.

Lightning—1 per cent. Direct lightning strokes, sparks from one object to another induced by nearby lightning stroke, and induced surges in circuits and electrical equipment. Prevented by lightning rods, arresters, surge capacitors, and grounding.

Miscellaneous—1 per cent. Unusual causes and relatively unimportant causes not included in the above classifications.

FIG. 43–7.

Employees should be instructed in the use of electrical equipment and should be prohibited from tampering, blocking circuit breakers, using wrong fuses, bypassing fuses, and installing equipment without authorization.

Electrical installations and all electrical equipment should be periodically inspected and tested to assure continued satisfactory performance and to detect deficiencies.

Smoking

Carelessly discarded cigarettes, pipe embers, and cigars are a major source of fire.

Although it might be desirable to eliminate smoking completely in a plant, such a rule is difficult to enforce. It is better to allow smoking at specified times and in a safe place where supervision can be maintained than to have it done surreptitiously in out-of-the-way places.

Smoking should be prohibited in woodworking shops, textile mills, flour mills, grain elevators, and places where flammable liquids or combustible products are manufactured, stored, or used. Even in these operations, enclosed or removed areas can be set aside to allow workmen to satisfy their smoking needs in a safe manner.

"No smoking" areas should be marked with conspicuous signs and their exclusive use rigidly enforced: everyone, including supervisors and visitors, should adhere to the regulation. It may be necessary to use more than signs to draw attention to the "no smoking" areas. Lines drawn on the floor or illuminated barriers placed around areas or processes have been effective. Small stickers, such as the National Safety Council's self-adhesive stickers and POP posters can be placed on containers, storage cabinets, or doors to areas where there is danger of fire or explosion.

In many high-hazard occupancies, smoking is permitted in special fire-safe rooms or during periods when there is no exposure. In such cases, instructions or warning signs to that effect should be posted. Where the exposure is severe, employees should even be prohibited from carrying matches, lighters, and sometimes smoking material of any kind into the danger areas.

In any case, the use of "safety" matches should be encouraged, and smoking should be allowed only in fire-safe areas. Where carrying matches into the plant is prohibited, special lighters should be provided in smoking rooms.

Even in plants where there is little fire hazard, employees should be encouraged to discard matches and smoking materials in a safe container rather than on the floor. This practice will discourage employees from throwing matches and cigarettes into places which may not be free of hazards.

Friction

Excessive heat generated by friction causes a very high percentage of industrial fires. A program of preventive maintenance on plant machinery can avert fires resulting from inadequate lubrication, misaligned bearings, and broken or bent equipment.

Fires frequently result from overheated power transmission bearings and shafting in buildings where dust and lint accumulate, such as grain elevators, cereal, textile, and woodworking mills, and plastic and metalworking plants. Frequent inspection should be made to see that bearings are kept well oiled and do not run hot. Accumulation of flammable dust or lint on them should be held to a minimum.

Drip pans should be provided beneath bearings and should be cleaned frequently to prevent oil from dripping to the floor or on combustible material below. Oil holes of bearings should be kept covered to prevent dust and gritty substances from entering the bearings to cause overheating.

Frictional heat sufficient to cause ignition can result from the jamming of process material during production. Another common problem, frequently overlooked, is the tension adjustment on belt-driven machinery. If the belt is too tight or too loose, excessive friction can cause serious overheating.

Foreign objects or tramp metal

Every precaution should be taken to keep foreign objects which might strike sparks from entering machines or processes where there are flammable dusts, gases, or vapors or combustible material such as cotton lint or metal powder. Screens or magnetic separators are commonly used for this purpose, as in textile mills, grain elevators, and other operations which have explosive mixtures of dusts.

Open flames

Although open flames are probably the most obvious source of ignition for ordinary combustibles, and one would think they could be most easily avoided, they still account for a large percentage of industrial fires. Heating equipment, torches, and welding and cutting operations are principal offenders.

Air heaters (gas- and oil-fired). Air heaters are commonly used on construction work and often cause fires because of:

1. Overheating of the air heater with resultant radiation igniting nearby combustible materials, such as concrete form work, tarpaulins, wood structures, paper, straw, and rubbish.

2. Failure to insulate air heaters from floors or other combustible bases.

3. Failure to provide a substantial spark shield and to use fuels which do not produce high flame or sparks. See NFPA Standard No. 211, *Standard for Chimneys, Fireplaces and Venting Systems.*)

4. Failure to provide a secure base or to anchor properly.

Air heaters in unventilated rooms or enclosures should be vented to the outside by means of an overhead hood and flue so as to remove the toxic products of combustion and any unburned gas.

Torches. If gasoline, kerosene, LPG (liquefied petroleum gas), acetylene, or alcohol torches are used, they should be placed so that the flames are at least 18 in. from wood surfaces. They should not be used around flammable liquids, paper, excelsior, or similar material.

Portable furnaces, blow torches, and the like should have overhead clearance of at least 4 ft. Combustible material overhead should be removed or protected by noncombustible insulating board or sheet metal and preferably by a natural-draft hood and flue of noncombustible material.

Welding and cutting. When possible, welding or cutting should be done in special firesafe areas or rooms with concrete or metal-plate floors. Flame impingement on concrete may cause it to spall; consequently work should be kept off the floor or else the floor should be protected by a metal shield. (See NFPA Standard No. 51B, *Fire Prevention in Use of Cutting and Welding Processes,* and Factory Insurance Association, *Preventing Cutting and Welding Fires.*

In cases where welding and cutting operations are performed outside the special fire-safe areas, hot work permit programs—described later on in this chapter—should be used to promote maximum fire-safe working conditions.

If welding must be done over wood floors, they should be swept clean and wetted down, preferably covered with flame-resistant blankets, metal, or other noncombustible covering. Hot metal and slag should be kept from falling through floor openings and igniting combustible material below.

Sheet metal, flame-resistant tarpaulins, or flame-resistant curtains should be used around welding operations to prevent sparks from reaching combustibles nearby.

Welding or cutting should not be permitted in or near rooms containing any flammable liquid, vapor, or dust. Neither operation should be performed in or near closed tanks or other containers which have held flammable liquids, until the containers have been thoroughly cleansed (see NFPA Standard No. 327, *Cleaning or Safeguarding Small Tanks and Containers* and NSC Data Sheet #432, *Cleaning Small Containers That Have Held Combustibles*), filled with water or purged with an inert gas, and combustible gas indicator tests show that no trace of a flammable gas or vapor is present. The tests should be repeated periodically to determine if any trace of such a gas or vapor is released during the welding or cutting operation. If further tests show any trace, the work should be stopped until all flammable gas or vapor is dispelled.

No welding or cutting should be done on a surface until combustible deposits have been removed.

Fire extinguishing equipment suitable to the type of exposure should be within easy reach of welding and cutting operators. When welding must be done outside of a shop or area designated for welding, watchers

1335

should be stationed to prevent sparks or molten slag from starting fires or to extinguish fires. The watchers should remain at the work location for at least one hour after the welding or cutting is completed because some fires escape detection at the time of their inception.

Shields should be installed around spot welders to prevent sparks from reaching combustible materials nearby or from injuring employees.

Spontaneous ignition

Spontaneous ignition results from a chemical reaction in which there is a slow generation of heat from oxidation of organic compounds that, under certain conditions, is accelerated until the ignition temperature of the fuel is reached. This condition is reached only where there is sufficient air for oxidation but not enough ventilation to carry away the heat as fast as it is generated.

It is a condition usually found only in quantities of bulk material packed loosely enough for a large amount of surface to be exposed to oxidation, yet without adequate air circulation to dissipate heat. Exposure to high temperatures increases the tendency toward spontaneous ignition.

The presence of moisture also can advance spontaneous heating unless the material is wet beyond a certain point. Materials like unslaked lime promote spontaneous ignition, particularly when wet. Such chemicals should be stored in a cool, dry place, away from combustible material.

It is generally agreed that at ordinary temperatures some combustible substances oxidize slowly and under certain conditions can reach their ignition point. These include vegetable and animal oil and fats, coal, charcoal, and some finely divided metals. Rags or waste saturated with linseed oil or paint often cause fires too.

The best preventives against spontaneous ignition are either total exclusion of air or good ventilation. With small quantities of material, the former method is practicable. With large quantities of material, such as storage piles of bituminous coal, both methods have been used with success.

Temperatures of 140 F are considered dangerous in coal piles. If temperatures rapidly

FIG. 43–8.—A sheet metal waste can constructed to prevent opening the cover more than 60 degrees from the horizontal, thus ensuring automatic closing.

approach or exceed that figure, it is generally advisable to move the pile or rearrange it to allow better circulation of air.

Certain agricultural products are susceptible to spontaneous ignition. Sawdust, hay, grain, and other plant products such as jute, hemp, and sisal fibers may ignite spontaneously, especially if exposed to external heat or to alternate wetting and drying. Here again, the best preventive is circulation of air, removal of external sources of heat, and storage of material in smaller quantities.

Fires in iron, nickel, aluminum, magnesium, and other finely divided metals are sometimes attributed to spontaneous ignition believed to result from the oxidation of cutting or lubricating oils or possibly from chemical impurities.

FIG. 43-9.—*Left:* Rubbish disposal container made from a 55-gal drum is equipped with an automatic self-closing cover which is released in event of fire. *Above:* Waste receptacle for more-public area.

Housekeeping

Collection and storage of combustibles. Many industrial fires are the direct result of accumulations of oil-soaked and paint-saturated clothing, rags, waste, excelsior, and combustible refuse. Such material should be deposited in noncombustible receptacles, having self-closing covers, that are provided for this purpose and removed daily from the work areas (Fig. 43-8 and -9).

Exhaust systems of effective design will remove gases, vapors, dusts, and other airborne contaminants—many of which may be fire hazards. Exhaust systems and machine enclosures will help to prevent accumulation of combustible materials on floors or machine parts. Such materials are most hazardous when airborne rather than when they have settled out.

Clean waste, although not as dangerous as oil-soaked waste, is readily combustible and should be kept in metal cans or bins with self-closing covers. Excelsior, cotton, kapok, jute, and other highly combustible fibrous material should be stored in covered noncombustible containers and, if large quantities are kept on hand, in fire-resistant rooms equipped with fire doors and automatic sprinklers. Portable extinguishers, hose lines, or other extinguishing equipment for Class A fires should be available for use at such storage places.

A schedule for safe collection of all combustible waste and rubbish should be a part of the fire prevention program. The safety professional should be certain that janitorial personnel or others involved in collection of waste paper in offices and service areas have fire-safe collection containers. It is important to check collection practices to be sure that

1337

ash trays, which may contain smoldering material, are not emptied into combustible bags or cartons or into containers of combustibles.

Accumulations of all types of dust should be cleaned at regular intervals from overhead pipes, beams, and machines, particularly from bearings and other heated surfaces. It must be understood that all organic as well as many inorganic materials, if ground finely enough will burn and propagate flame. Roofs should be kept free from sawdust, shavings, and other combustible refuse. Such cleaning preferably should be done by vacuum removal, because blowing down with air may disperse dusts into dangerous clouds.

No such material should be stored or allowed to accumulate in air, elevator, or stair shafts, tunnels, in out-of-the-way corners, near electric motors or machinery, against steam pipes, or within 10 ft of any stove, furnace, or boiler.

Rubbish disposal. Federal, state, and local laws forbid certain methods of waste disposal such as open burning, evaporation, flushing to sewers, etc. In addition, fires are often caused by burning rubbish in yards near combustible buildings, sheds, lumber piles, fences, and grass, or other combustible materials. Smaller firms that cannot justify an approved method of waste disposal on site should have the rubbish and waste materials collected. If it must be burned, the best and safest way is with a well designed incinerator which meets the requirements of the environmental pollution control laws.

Flushing and dumping waste materials into sewers may also be prohibited. In many cases preplanning would include trap tanks which can be pumped out and the waste material disposed of properly. Chemically altering a waste material may be considered before disposal.

In any event, the requirements of federal, state, and local laws should be investigated when planning a process or a plant. Investigation should include plans for emergency conditions, such as spills, release of vapors, and fire.

Locker rooms. Lockers in which oil-soaked clothing, waste, or newspapers are kept are always a serious fire hazard. Every precau-

tion should be taken to provide fire-safe lockers and to prevent such combustible accumulations. Lockers should be made of metal and have solid, fire-resistant sides and backs, but doors should have some open-work for ventilation. Lockers should be large enough so that air can circulate freely around clothing hung in them. Employees should not be permitted to leave clothing saturated with oils or paints in lockers.

Where automatic sprinklers are used, locker tops may be covered with screening or be made of perforated metal so that water can reach burning contents if a locker fire occurs. Heavy paper pasted over the tops will serve to keep out dust. Lockers which have sloping tops and stand flat on the floor will prevent the accumulation of rubbish both above and below.

Explosive atmosphere

Dusts. A dust explosion hazard exists wherever material that will burn or oxidize

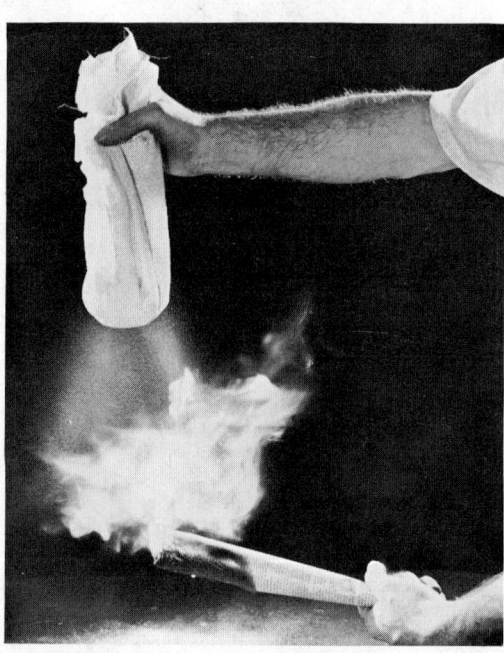

Fig. 43–10.—Demonstration of explosives properties of flour dust. Sifting flour through a cloth into the flame produces fiery bursts that illustrate what happens in grain elevators and other dust-laden atmospheres when dust clouds are ignited.

TABLE 43-A

SOME OF THE MORE COMMON POTENTIALLY EXPLOSIVE DUSTS

Type	Example
Carbon	Coal, peat, charcoal, coke, lampblack
Fertilizers	Bone meal, fish meal, blood flour
Food products and byproducts	Starches, sugars, flour, cocoa, powdered milk, grain dust
Metal powders	Aluminum, magnesium, zinc, iron
Resins, waxes, and soaps	Shellac, rosin, gum sodium resinate, soap powder, waxes
Spices, drugs, and insecticides	Cinnamon, pepper, gentian, pyrethrum, tea fluff
Wood, paper, tanning materials	Wood flour, wood dust, cellulose, cork, bark dust, wood extract
Miscellaneous	Hard rubber, sulfur, tobacco, many plastics

Source U.S. Department of Agriculture.

readily is available in powder form because the surface (contact) area of each particle is very large in relation to its mass (Fig. 43–10). The U.S. Department of Agriculture Technical Bulletin 490 lists 133 dusts according to their degree of explosiveness (Table 43-A). In addition, many synthetic resins and powders used in the plastics industry present a dust-explosion hazard comparable to that of coal (USBM Report No. 3751). The group includes phenolic, urea, vinyl, and other types of resins and a number of molding compounds, primary ingredients, and fillers.

There are two ways to prevent dust explosions: (*a*) prevent the formation of explosive mixtures of dust and air, and (*b*) prevent the ignition of such mixtures if their formation cannot be prevented.

Extraordinary precautions, if such are necessary, should be taken to prevent accumulations of dust which may build up to explosive proportions. Extensive use of local exhaust and frequent cleaning will do much to minimize the hazard. Where possible, dusty operations should be segregated, and dust-producing equipment should be totally enclosed and exhausted to prevent leakage of dust into the general work area.

Buildings with high explosion hazards, such as grain elevators or plastics plants, are generally provided with extensive dust-collecting equipment and so constructed that explosive pressures will push out hinged windows or blow out wall sections or panels designed and built to fail in a predetermined area, rather than cause structural collapse. (See NFPA Standard No. 68, *Explosion Venting Guide.*) Floor openings should be kept to a minimum and openings for pipes and duct work should be sealed.

Portable fire extinguishing equipment should be readily available. Fog nozzles or finely divided streams of water are more effective than are solid streams of water, which stir up dust.

Ignition of an explosive mixture of dust and air may be prevented by control of open flames, friction sparks, static electricity, welding, and excessive heat, or by increased humidity, or by using inert gas. Every precaution should be taken to prevent overheated bearings, smoking, friction sparks, and sparks from hand tools, from grinding or welding operations, and from static electricity. Only nonferrous tools should be used in dust exposures, and employees should wear shoes with nonferrous nails.

Nonferrous material should also be used for truck wheels, and bucket conveyors should be prevented from striking sparks. Pneumatic or magnetic separators can be used to remove stones, nails, and other spark-producing foreign objects from the material being processed.

1339

Only approved dust-tight wiring, fixtures, and motors should be used, and employees should be instructed to use only extension cords, lamps, and portable electric tools designed for protection against this hazard. (See NFPA "National Fire Codes," Volume 3.)

Gases and vapors. Gases and vapors that produce flammable mixtures with air or oxygen are common in industry. Some such gases are acetylene, propane, hydrogen, carbon monoxide, methane, natural gas, and manufactured gas.

Highly volatile liquids which emit flammable vapors include gasoline, benzene, naphtha, and methyl alcohol. Kerosene, turpentine, Stoddard solvent, and other liquids with flash points above 100 F must generally be heated above normal room temperatures before they give off sufficient vapors to form ignitible concentrations.

When flammable liquids, including those with flash points above 100 F, must be handled and used, only minimum amounts, in safety containers, should be allowed in work areas.

The flash point of a liquid is defined as the minimum temperature at which it will yield sufficient vapor to form a flammable mixture with air and produce a flame when a source of ignition is supplied close to the surface of the liquid. Flash point temperatures are not always physical constants under practical conditions, and it should be remembered that published data are based on carefully controlled testing conditions.

For instance, the amount of vapor that will accumulate in the air above a volatile liquid depends upon its vapor pressure or the relative saturation of the air with this vapor above the liquid surface. The amount of vapor given off is also directly related to the surface area of the volatile liquid. Thus, even a high-flash liquid may be dangerous if considerable surface area is exposed, such as in a mist or froth. (It might be thought of as a "liquid dust.") Conversely, limiting the surface area reduces the flammable or explosion hazard of a low-flash liquid.

As respects the autoignition temperature of a gas or vapor in air, such a mixture as carbon disulfide vapor and air will ignite spontaneously at temperatures around 212 F, which is below that of visible red heat or even high-pressure steam lines. However, the majority of gases and vapors will not self-ignite in air until they reach temperatures of about 500 to 900 F.

The Chemistry of Fire

Ordinary fire (one that can be combatted by ordinary extinguishing means) results from the combination of fuel, heat, and oxygen. When a substance that will burn is heated to a certain critical temperature called its "ignition temperature," it will ignite and continue to burn as long as there is fuel, the proper temperature, and a supply of oxygen.

Knowledge of the chemical reaction of a fire also forms the basis for knowing how to extinguish fire. Heat can be taken away by cooling, oxygen can be taken away by excluding the air, fuel can be removed to an area where there is insufficient heat for ignition, and the chemical reaction can be stopped by inhibiting the rapid oxidation of the fuel.

Cooling

In order to extinguish a fire by cooling, it is only necessary to absorb a small portion of the total heat being evolved by the fire. The most common and practical agent is water applied in the form of a solid stream, finely divided spray, or incorporated in foam. Its specific and latent heats are higher than those of other common extinguishing agents (which means it takes more heat to heat it and vaporize it). When vaporized into steam, it expands 1700 times, reducing the volume of air (oxygen) available to sustain combustion in the fire zone; moreover, water is able to penetrate and reach deep-seated fires. This makes it an efficient cooling medium and diluting agent.

Removing fuel

Often, taking the fuel away from a fire is difficult and dangerous, but there are exceptions. Flammable liquid storage tanks can be arranged so their contents can be pumped to an isolated empty tank in case of fire. When flammable gases catch fire as they are flowing from a pipe, the fire will go out if the fuel supply can be shut off. Also, in any mixture of fuel gases or vapors in air, adding an excess of air has the effect of diluting the fuel concentration below the minimum combustible concentration point.

Limiting oxygen

Extinguishment by separation of oxygen from fire can be accomplished through smothering the burning area with a noncombustible material, such as covering with a wet blanket (make sure the blanket isn't made of highly combustible fibers), throwing dirt or sand on the fire, or covering it with a chemical or mechanical foam. Extinguishment by diluting the reactants—oxygen and fuel vapors—below the concentration necessary to support combustion is accomplished in blanketing the fire area with carbon dioxide or noncombustible vaporizing liquids. The fire will remain out if the blanket is maintained long enough for the combustible material to cool below its ignition temperature and if no ignition sources are present.

Carbon dioxide and vaporizing liquid are of limited value on fires involving wood, rags, or paper, because the blanket usually cannot be maintained long enough for all smoldering ignition sources to be extinguished. Moreover, smothering is ineffective on materials that contain their own oxygen supply, such as ammonium nitrate or nitrocellulose.

In the field of fire prevention, the principle of separating oxygen from the fuel supply is applied when an inert gas is used to purge operations involving flammable vapors, dusts, and other combustible materials under confined conditions when a source of ignition may exist.

Interrupting the reaction

Recent studies in fire chemistry have resulted in certain revisions and expansions in the theories of fire extinguishment. In analyzing the anatomy of a fire, the original fuel molecules appear to combine with oxygen in a series of successive intermediate stages, called branched-chain reactions, in arriving at the final end products of combustion. It is these intermediate stages which are responsible for the evolution of flames.

As molecules fragmentize in these branched chain reactions, unstable intermediate products called free radicals are formed. The con-

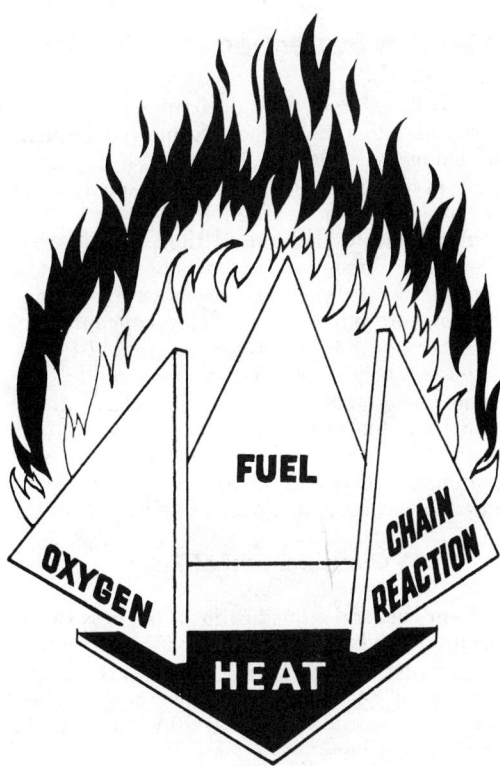

Fig. 43–11.—"The fire pyramid." Oxygen, heat, fuel, and chain reactions are necessary components of a fire. Speed up the process and an explosion results.

centration of free radicals is the determining factor of flame speed. The life of the free hydroxyl radical is very short, being in the order of 0.001 second, but long enough to be of vital importance in the combustion of fuel gases. The almost simultaneous formation and consumption of free radicals appears to be the lifeblood of the flame reaction.

It is the free radicals in these branched chain reactions which are removed from their normal function as a chain carrier by dry chemical and halogenated hydrocarbon extinguishing agents.

The effects that various dry chemical agents (sodium bicarbonate-base, potassium bicarbonate-base and ammonium phosphate-base) have on capturing free radicals depends upon their individual molecular structure. Potassium bicarbonate dry chemical is the most effective because of the large size of the potassium ion.

In the case of halogenated hydrocarbons

1341

it is believed that they decompose when discharged into the fire and form free radicals of halogens (chlorine, bromine, or fluorine) that unite with the free radicals evolved in the branched chain reaction.

For years we have used the principle that there are three ways in which an ordinary fire can be extinguished—remove fuel, limit oxygen, or reduce heat. This concept, pictorialized by the "fire triangle," should be revised to include this fourth way: "inhibit the flame chain reaction." Since each of these four basic tenets interdepend on one another, a solid with four adjacent sides, such as a pyramid, would now be the most suitable to pictorialize fire extinguishment (Fig. 43–11).

Classification of Fires

Four general classifications of fires have been adopted by the National Fire Protection Association based on the types of extinguishing media necessary to combat each. (See NFPA Standard No. 10 and 10A, on portable fire extinguishers.)

Class A fires

Class A fires are those that occur in ordinary materials such as wood, paper, excelsior, rags, and rubbish. The quenching and cooling effects of water or of solutions containing large percentages of water are of first importance in extinguishing these fires. Special dry chemical agents (multi-purpose dry chemical) provide rapid knockdown of the flames and the formation of a coating that tends to retard further combustion. Where total extinguishment is mandatory, a followup with water or other Class A agent is recommended.

Class B fires

Class B fires are those that occur in the vapor-air mixture over the surface of flammable liquids such as gasoline, oil, grease, paints, and thinners. The limiting of air (oxygen) or the combustion inhibiting effect is of primary importance on incipient fires of this class. Solid streams of water are likely to spread the fire, but under certain circumstances water fog nozzles prove effective. Generally, regular dry chemical, multipurpose dry chemical, carbon dioxide, foam, and halogenated hydrocarbon agents are used.

Class C fires

Fires that occur in or near electrical equipment where nonconducting extinguishing agents must be used are called Class C fires. Dry chemical, carbon dioxide, compressed gas and vaporizing liquid extinguishing agents are suitable. (See NFPA Standard No. 182M, *Hazards of Vaporizing Liquid Extinguishing Agents.*)

Foam or a stream of water should not be used because both are good conductors of electricity and can expose the operator to a severe shock hazard. Water from a very fine spray can sometimes be used on fires in electrical equipment, as in transformers, since a spray is a poorer electrical conductor than a solid stream of water.

Class D fires

Fires that occur in combustible metals such as magnesium, titanium, zirconium, lithium, and sodium are classified under Class D. Specialized techniques, extinguishing agents and extinguishing equipment have been developed to control and extinguish fires of this type. Normal extinguishing agents generally should not be used on metal fires as there is a danger in most cases of increasing the intensity of the fire because of a chemical reaction between some extinguishing agents and the burning metal.

Other fires

Fires that involve certain combustible metals or reactive chemicals require—in some cases—special extinguishing agents or techniques. See NFPA Standards Nos. 49 and 325M, *Hazardous Chemicals Data* and *Fire-Hazard Properties of Flammable Liquids, Gases, Volatile Solids.*

Portable Fire Extinguishers

Equipment used to extinguish and control fires is of two types: fixed and portable. Fixed systems include water equipment: automatic sprinklers, hydrants, standpipe hoses, and special pipe systems for dry chemical, carbon dioxide, and foam. Special pipe systems are applicable to areas of high fire potential where water may not be effective, such as tanks for storage of flammable liquids and on electrical equipment.

Fixed systems, however, must be supplemented by portable fire extinguishers. These often can preclude the action of sprinkler systems because they can prevent a small fire from spreading as well as provide rapid extinguishment in the early stages of a fire.

Principles of use

Even though the plant may be equipped with automatic sprinklers or other means of fire protection, portable fire extinguishers should also be available and ready for emergency. "Portable" is applied to manual equipment used on small fires or in the interim between discovery of fire and the functioning of automatic equipment or arrival of professional fire fighters.

To be effective portable extinguishers must be:

1. A reliable type.*

2. The right type for each class of fire that may occur in the area.

3. In sufficient quantity to protect against the exposure in the area.**

4. Located where they are readily accessible for immediate use.

5. Maintained in perfect operating condition, inspected frequently, checked against tampering, and recharged as required.

6. Operable by area personnel who can find them and who are trained to use them effectively and promptly.

Classification of fire extinguishers. Portable extinguishers are classified to indicate their ability to handle specific classes and sizes of fires. This classification is necessary because of the constant development of improved and new extinguishing agents and devices, and because of the availability of larger portable extinguishers.

Labels on extinguishers indicate the class and relative size of fire that they can be expected to handle.

The following paragraphs are a guide to the selection of portable fire extinguishers for given exposures, in accordance with classifications set forth in NFPA Standard No. 10 on portable fire extinguishers. In each case, however, it is essential that plant protection and insurance recommendations be observed, based upon fire protection requirements of the authority having jurisdiction.

CLASS A EXTINGUISHERS: for ordinary combustibles, such as wood, paper, and textiles, where quenching-cooling effect is required. The numeral indicates the relative fire extinguishing potential of each unit. For example, a 4-A unit can be expected to extinguish approximately twice as much fire as a 2-A unit.

CLASS B EXTINGUISHERS: for flammable liquid and gas fires, such as oil, gasoline, paint, and grease, where oxygen exclusion or flame interrupting effect is essential. The numeral indicates the area, in square feet, of deep-layer flammable liquid fire expected to be extinguished by an unskilled operator under emergency fire conditions. For example, a 10-B unit can be expected to extinguish 10 sq ft of deep-layer flammable liquid fire.

CLASS C EXTINGUISHERS: for fires involving electrical wiring and equipment where the dielectric nonconductivity of the extinguishing agent is of first importance because of the shock hazard of water base extinguishers. These units are not classified by a numeral because Class C fires are essentially either Class A or Class B, but also involve energized electrical wiring and equipment. Therefore, the *coverage* of the extinguisher must be chosen for the burning fuel.

CLASS D EXTINGUISHERS: for fires in combustible metals, such as magnesium, potassium, powdered aluminum, zinc, sodium, titanium, zirconium, and lithium. Persons working in areas where Class D fire hazards exist must be aware of the dangers in using Class A, B, or C extinguishers on a Class D fire, as well as the correct way to extinguish

---

* Underwriters Laboratories Inc., *Fire Protection Equipment List;* and Factory Mutual Engineering Corporation, *Approval Guide—Equipment, Materials, Services for Conservation of Property.*

** National Fire Protection Association, *Installation of Portable Fire Extinguishers,* Standard No. 10.

Class D fires. These units are not classified by a numerical system and are intended for a special hazard protection only.

The following recommendations are given in NFPA Standard No. 10 as a guide in marking extinguishers, and/or extinguisher locations, to indicate the suitability of the extinguisher for a particular class of fire. Extinguishers suitable for more than one class of fire may be identified by multiple symbols as described previously.

Markings should be applied by decalcomanias, painting, or similar methods having at least equivalent legibility and durability.

Where markings are applied to the extinguisher, they should be located on the front of the shell above or below the extinguisher name plate. Markings should be of a size and form to give easy legibility at a distance of 3 ft.

ORDINARY

A

COMBUSTIBLES

1. Extinguishers suitable for Class A fires should be identified by a triangle containing the letter "A." If colored, the triangle should be colored green.*

FLAMMABLE

B

LIQUIDS

2. Extinguishers suitable for Class B fires should be identified by a square containing the letter "B." If colored, the square should be colored red.*

ELECTRICAL

C

EQUIPMENT

3. Extinguishers suitable for Class C fires should be identified by a circle containing the letter "C." If colored, the circle should be colored blue.*

COMBUSTIBLE

D

METALS

4. Extinguishers suitable for fires involving metals should be identified by a five-pointed star containing the letter "D." If colored, the star should be colored yellow.*

Where markings are applied to wall panels, etc., in the vicinity of extinguishers, they should be of a size and form to give easy legibility at a distance of 25 ft.

Extinguishers listed by Underwriters' Laboratories, Inc., are rated after physical testing. These ratings, which are indicated by a numeral and a letter, define the extinguishing potential of an extinguisher because they specify the type and size or number that should be installed in a specific area.

The numeral signifies the number of units of extinguishing potential and the letter(s), the class(es) of fire on which the particular extinguisher is most effective for extinguishment. For example, a 4-A:16-B:C rating signifies that the extinguisher is rated for 4 units of Class A fire-fighting potential, 16 units of Class B fire-fighting potential, and is safe to use on Class C fires in or near electrical equipment.

Extinguishers labeled with pre-1955 classifications can be reclassified by using conversion tables published in NFPA Standard No. 10, *Installation of Portable Fire Extinguishers.*

Location of units. Extinguishers should be located close to the likely hazards, but not so close that they would be damaged or cut off by the fire. They should be located along the normal path of egress from the building, preferably at the exits. Where highly combustible material is stored in small rooms or enclosed spaces, the extinguishers should be located outside the door, rather than inside where they might become inaccessible.

The location of the extinguisher should be made conspicuous (Fig. 43–12). For example, if it is hung on a large column or post, a distinguishing red band can be painted around the post. Also, large signs can be posted directing attention to extinguishers. The extinguisher should be kept clean and should not be painted in any way that will camouflage it or obscure labels and markings on it.

If the extinguisher itself is not already marked plainly to indicate the classifications of fire or types of material for which it is in-

* NOTE: Recommended colors as described in the Federal Color Standard Number 595 are:
Green — No. 14260
Red — No. 11105
Blue — No. 15102
Yellow —No. 13655

FIG. 43–12.—Fire equipment should be conspicuously located, appropriately marked, and inspected regularly.

tended, signs or cards should be placed on the wall close to where it hangs. Special labels are available from manufacturers of the extinguishers, insurance companies, NSC, and NFPA; markings indicating special uses can be stenciled on the extinguisher or on an adjacent wall.

Fire extinguishers must not be blocked or hidden by stock, finished material, or machines. They should be hung in accordance with NFPA Standard No. 10, where they will not be damaged by trucks, cranes, and harmful operations, or corroded by chemical processes, and where they will not obstruct aisles or injure passersby. If installed out of doors, the extinguisher should be protected from the elements.

Plant and warehouse aisles should be wide enough that mobile fire-protection units (if any) can be brought close to a fire and should be kept free of obstructions. Floor spaces may be marked to allow access to fire extinguishing equipment, and units may be protected with bumpers or guard rails.

• OSHA requires that for easy lifting, extinguishers having a gross weight not exceeding 40 pounds should be placed, so that their tops are *no more than* 5 ft above the floor.

• Those weighing over 40 pounds should *not be more than* 3½ ft above the floor.

Distribution of extinguishers. The minimum required number and type of portable extinguishers to be installed is determined for each floor or area by the relative hazard of the occupancy, the nature of any anticipated fires, and protection for special hazards.

• Extinguishers suitable for Class A fire hazards are installed according to the classification of occupancy: light, ordinary, or extra hazard. These are summarized in Table 43-B.

LIGHT HAZARD OCCUPANCIES include office buildings, schools (exclusive of trade schools and shops) and public buildings, where because of the relatively small amount of combustibles, incipient fires of minimum severity may be anticipated.

ORDINARY HAZARD OCCUPANCIES include department stores, warehouses, and manufacturing buildings of average hazard where incipient fires of average severity in combustibles may be anticipated.

EXTRA HAZARD OCCUPANCIES include woodworking, textile mills, and paper mills, where because of the character or quantity of combustibles, extra severe incipient fires may be anticipated.

• Extinguisher requirements for Class B protection (of a "special hazard" area, such as a laboratory or kitchen) are in addition to the requirements of extinguishers for Class A protection except where the total area under consideration presents wholly Class B hazards. Extinguishers are installed according to the nature of anticipated fires and protection for special hazards.

The requirements for fire extinguisher size and placement for Class B fires other than those in flammable liquids over ¼ in. deep are:

1. Shown in Table 43-C.

2. Two or more extinguishers of lower rating, except for foam extinguishers, shall not be used to fulfill the protection requirements of Table 43-C. Up to three foam extinguishers may be used to fulfill these requirements.

3. The protection requirements may be fulfilled with extinguishers of higher ratings provided the travel distance to such larger extinguishers does not exceed 50 ft.

1345

TABLE 43-B

EXTINGUISHERS SUITABLE FOR CLASS A FIRES

Basic Minimum Extinguisher Rating for Area Specified	Maximum Travel Distances to Extinguishers	Areas To Be Protected per Extinguisher		
		Light Hazard Occupancy	Ordinary Hazard Occupancy	Extra Hazard Occupancy
1A	75 ft	3,000 sq ft	Not permitted Except as Specified*	Not permitted Except as Specified*
2A	75 ft	6,000 sq ft	3,000 sq ft	Not permitted Except as Specified*
3A	75 ft	9,000 sq ft	4,500 sq ft	3,000 sq ft
4A	75 ft	11,250 sq ft	6,000 sq ft	4,000 sq ft
6A	75 ft	11,250 sq ft	9,000 sq ft	6,000 sq ft
10A	75 ft	11,250 sq ft†	11,250 sq ft†	9,000 sq ft
20A	75 ft	11,250 sq ft†	11,250 sq ft†	11,250 sq ft†
40A	75 ft	11,250 sq ft†	11,250 sq ft†	11,250 sq ft†

* The protection requirements specified in this table may be fulfilled by several extinguishers of lower ratings for ordinary or extra hazard occupancies, subject to the approval of the authority having jurisdiction. Consideration should be given to the number of persons available to operate the extinguishers, the degree of training provided, and the possibility of use by women.

† 11,250 sq ft is considered a practical limit.

From National Fire Protection Association Standard No. 10, Table 4110.

For flammable liquids of appreciable depth (over ¼ in.), such as those in dip or quench tanks:

1. Class B fire extinguishers should be provided on the basis of one numerical unit of Class B extinguishing potential per square foot of flammable liquid surface of the largest tank to be protected within the area.

2. Two or more extinguishers of lower ratings, except for foam extinguishers, shall not be used in lieu of the extinguisher required for the largest tank. Up to three foam extinguishers may be used to fulfill these requirements.

3. When protection is sought for flammable liquid in appreciable depth and when the liquid surface area is in excess of 20 square feet, the protection requirements should be based on an evaluation of the extent of the hazard and engineering judgment applied. Consideration should be given to installing fixed protection and wheeled extinguishers systems. Portable extinguisher protection for such hazards should be generally restricted to plants having trained fire brigades.

4. Where approved automatic fire protection devices or systems have been installed for a flammable liquid, additional portable Class B fire extinguishers, as required in paragraph 1, may be waived. When so waived, Class B extinguishers should be provided to protect areas near such protected hazards.

5. Travel distances should be given consideration with reference to special hazards and

TABLE 43-C

EXTINGUISHERS SUITABLE FOR CLASS B FIRES, FOR FIRES IN FLAMMABLE LIQUIDS ¼ INCH AND UNDER IN DEPTH

Type of Hazard	Basic Minimum Extinguisher Rating	Maximum Travel Distance to Extinguishers
For Extinguishers Labeled Prior to June 1, 1969		
Light	4B	50 ft
Ordinary	8B	50 ft
Extra	12B	50 ft
For Extinguishers Labeled After June 1, 1969		
Light	5B	30 ft
	10B	50
Ordinary	10B	30 ft
	20B	50
Extra	20B	30 ft
	40B	50

Note. For flammable liquid hazards of depth greater than ¼ in., Class B fire extinguishers shall be provided on the basis of one numerical unit of Class B extinguishing potential per square foot of flammable liquid surface of the largest tank hazard within the area.

From National Fire Protection Association Standard No. 10, Table 4210

the availability of the extinguisher for such protection. Scattered or widely separated hazards shall be individually protected if the specified travel distances in Table 43-C are exceeded. Likewise, extinguishers in the proximity of a hazard shall be carefully located so as to be accessible in the presence of a fire without undue danger to the operator.

● Extinguishers with Class C ratings are required where energized electrical equipment may be encountered which would require a nonconducting extinguishing media. This will include fire either directly involving or surrounding electrical equipment. Since the fire itself is a Class A or Class B hazard the extinguishers are sized and located on the basis of the anticipated Class A or B hazard. Whenever possible, electrical equipment

should be deenergized before attacking a Class C fire.

● Extinguishers used for Class D protection and other special fires are installed according to the size and type of the special hazard. The type of combustible material, the quantity, and its physical form are all determining factors that must be considered when selecting the proper type of agent and the method of application.

Selection of extinguishers. Operating characteristics that make one type of portable fire extinguisher suitable for certain fire hazards may make the same type dangerous for others.

The purchaser and user of extinguishers should secure on-the-job advice from fire inspection bureaus, fire insurance carriers, and fire protection engineers.

Choosing the right extinguisher is extremely important. All too frequently cost is given more consideration than adequate protection. Remember, good extinguishers are worth their cost because of the protection they give. Obviously, only extinguishers listed by nationally recognized agencies such as Underwriters' Laboratories, Inc., or Factory Mutual Engineering Corp. should be purchased.

Although suitable types and sizes are installed in conformance with NFPA Standard No. 10, there is a certain flexibility that can be used to the purchaser's advantage. For example, if a certain condition calls for B extinguishers, the required units could be obtained by using various size dry chemical, foam or carbon dioxide (CO_2) extinguishers. The relative advantages and disadvantages of each of these units should be considered with respect to all the conditions in the area. It is also wise to investigate the relative merits of any particular extinguisher available on the market, as the UL classification can vary widely for extinguishers of the same size but of different manufacture.

The design and operating features, ease of maintenance, and the availability of repair service should be considered too. If possible, actually operating and testing the extinguisher may make a great deal of difference in the final selection.

The following discussion of portable ex-

(*Text continues on page 1350.*)

FIRE EXTINGUISHER/AGENT

SUITABLE FOR USE ON TYPE OF FIRE	AGENT CHARACTERISTICS	Available Sizes	Horizontal Range	Discharge Time
REGULAR OR ORDINARY DRY CHEMICAL ★				
B C	Basically Sodium Bicarbonate. Discharges a white cloud. Leaves residue. Non-freezing.	1 to 30 lbs.	5 to 20 ft.	8 to 25 Sec.
MULTIPURPOSE DRY CHEMICAL ★				
A B C OR B C A CAPABILITY	Basically Ammonium Phosphate. Discharges a yellow cloud. Leaves residue. Non-freezing. Some extinguishers utilizing this agent do not have an "A" rating — however, they are designated as having "A" capability.	2 to 30 lbs.	5 to 20 ft.	8 to 25 Sec.
PURPLE-K DRY CHEMICAL ★				
B C	Basically Potassium Bicarbonate. Discharges a bluish cloud. Leaves residue. Non-freezing.	2 to 30 lbs.	5 to 20 ft.	8 to 25 Sec.
KCL DRY CHEMICAL ★				
B C	Basically Potassium Chloride. Discharges a white cloud. Leaves residue. Non-freezing.	2 to 30 lbs.	5 to 20 ft.	8 to 25 Sec.
	Potassium Chloride/Urea	11 to 23	15 to 30	20 to 31
CARBON DIOXIDE				
B C	Basically an inert gas that discharges a cold white cloud. Leaves no residue. Non-freezing.	2½ to 20 lbs.	3 to 8 ft.	8 to 30 Sec.

★ NOTE: Available in stored pressure or cartridge operated types.

Fig. 43–13.

CHARACTERISTICS

SUITABLE FOR USE ON TYPE OF FIRE	AGENT CHARACTERISTICS	Available Sizes	Horizontal Range	Discharge Time
HALOGENATED AGENT B C	Basically halogenated hydro-carbons. Discharges a white vapor. Leaves no residue. Non-freezing.	2½ lb.	4 to 8 ft.	.8 to 10 Sec.
WATER ▲ A	Basically tap water. Discharges in a solid or spray stream. (May contain corrosion in-hibitor which leaves a yellow residue.) Protect from freezing!	2½ Gal.	30 to 40 ft.	1 Minute
ANTI-FREEZE SOLUTION A	Basically a Calcium Chloride solution to prevent freezing. Discharges a solid or spray stream. Leaves residue. Non-freezing.	2½ Gal.	30 to 40 ft.	1 Minute
LOADED STREAM A B	Basically an alkali-metal-salt solution to prevent freezing. Discharges a solid or spray stream. Leaves residue. Non-freezing.	2½ Gal.	30 to 40 ft.	1 Minute
FOAM B	Basically water and detergent. Discharges a foamy solution. After evaporation, leaves a powder residue. Protect from freezing.	18 oz.	10 to 15 ft.	24 Sec.
DRY POWDER SPECIAL COMPOUND D	Basically Sodium Chloride or Graphite materials. Agent is discharged from an extin-guisher in a solid stream or is applied with a scoop or shovel to smother combustible metal. Leaves residue. Non-freezing.	30 lbs.	5 to 20 ft.	25 to 30 Sec.

▲ NOTE: Pump tanks available.

Courtesy Fire Equipment Manufacturers' Association, Inc., Mt. Prospect, Ill. 60056.

1349

tinguishers is divided into (*a*) water solution extinguishers, (*b*) dry-chemical, carbon dioxide, and liquefied-gas extinguishers, and (*c*) dry-powder extinguishers. Some vaporizing liquid extinguishers, such as those containing carbon tetrachloride or chlorobromo-methane, are excluded as they are not recommended because of their toxic properties. The proper method of operation and maintainance of these extinguishers are described in Figs. 43–14 through 43–24, and their characteristics are summarized in Fig. 43–13.

Water solution extinguishers

Fire extinguishers that use water or water solutions include: pump tank, cartridge actuated,* soda-acid,* stored pressure, loaded stream (often including a wetting agent), and foam** (see Figs. 43–14 through 43–18). These extinguishers are effective against Class A fires because of the quenching and cooling effect of water. The cooling effect of all but the foam extinguisher can be increased if the operator extends his forefinger over the edge of the nozzle to create a spray. Foam primarily extinguishes Class B fires by smothering. The loaded stream extinguisher can also be used against Class B fires because of the chemical action of the alkali-metal-salt; however, it has limited effectiveness. These units cannot be used on fires in or near electrical equipment since they can produce a shock hazard to the operator.

One important maintenance item on all water extinguishers is to inspect the nozzle frequently for foreign particles that may prevent discharge, such as dirt, match sticks, and paper. The pressure relief hole in the cap of soda acid, gas cartridge, and foam units must also be free from obstruction as it is designed to release any residual pressure that may injure a worker when he is removing the cap before recharging.

In locations exposed to freezing temperatures, extinguishers must be installed in heated cabinets (Fig. 43–19), or charged with a nonfreezing solution. However, antifreeze or salt solutions should not be used unless the equipment has been designed for such use. For example, some chemicals cause rapid corrosion when used in stainless steel units.

Water-base extinguishing equipment that cannot be protected from freezing temperatures can sometimes be kept usable by the addition of calcium chloride or patented chemicals. Except for loaded stream extinguishers, which use a special alkali-metal-salt solution, the pump tank, cartridge operated, and stored-pressure water extinguishers can be protected against freezing by calcium chloride or by special antifreeze charges recommended by the manufacturer.

Calcium chloride solutions are good conductors of electricity and therefore especially dangerous if applied to live electrical apparatus. Solutions have high corrosivity and should include a corrosion inhibitor.

When charging any extinguisher with an antifreeze solution, the chemical should be thoroughly dissolved in warm water in a separate container and then poured into the extinguisher through a fine strainer to remove any foreign particles that may clog the unit. Greater care must be given in maintaining extinguishers containing antifreeze solutions, because they may corrode more easily than with plain water. In recharging, all parts including the hose and nozzle should be thoroughly flushed with plain water. Be sure that all the plain water is drained off to prevent freezing and clogging.

Table 43-D shows the approximate temperatures above which varying strengths of 2½-gal calcium chloride solution will give protection against freezing. At regular intervals, solutions should be tested with a hydrometer for conformity to the figures.

Dry-chemical, carbon dioxide, and liquefied-gas extinguishers

Dry-chemical extinguishers. The dry-chemical extinguisher is one of the most versatile units available. It extinguishes by interrupting the chemical flame chain reaction (see explanation earlier in this chapter). It is not to be confused with dry powder extinguishers discussed in the next section.

Four types of base agents are used: sodium bicarbonate, potassium bicarbonate, potassium chloride and ammonium phosphate (multipurpose). When recharging, it is important to use only the dry chemical agent and

---

* No longer manufactured.
** Only available as 18-oz disposable unit.

For CLASS A FIRES—Wood, paper, textiles, and the like.

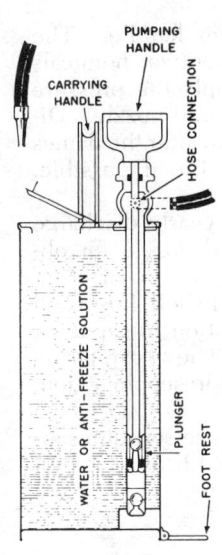

Operation. May be operated intermittently (therefore useful on such operations as cutting and welding), but cannot be carried and operated at the same time. The operator should place his foot on the foot bracket and operate the pump handle with 6- to 8-in. strokes. The pump discharges water on both the up and down strokes. Direct the stream at the base of the fire, then follow up after the flames while working from side-to-side or around the fire if possible.

Maintenance. Periodically check the water level and test the pump by stroking several times while discharging the liquid back into the tank. Inspect the condition of operating parts annually and oil the piston-rod packing. Check the tank and foot bracket for corrosion. After use, simply refill with water or antifreeze solution.

Caution. Protect from freezing.

FIG. 43–14.—Pump tank fire extinguisher.

For CLASS A FIRES—Wood, paper, textiles, and the like.

Operation. Contains a small metal cylinder of carbon dioxide gas under pressure. To release the gas, which provides the pressure to expel the water, invert the extinguisher and bump it gently on the floor while holding it by the bottom handle. (In some models, bumping is not necessary because the weight of the gas cartridge causes it to fall against a puncturing device.) Direct the stream at the base of the fire, then follow up after the flames while working from side-to-side or around the fire if possible.

Maintenance. Annually inspect for corrosion of shell, proper liquid level, defects in rubber gasket, clogging or damage to hose or nozzle, and bent discharge elbow. Weigh the gas cartridge; replace if it has lost $\frac{1}{2}$ oz of the weight stamped on the shell. See that cartridge slides freely in its cage when inverted. After use, simply refill with water or antifreeze solution.

Caution. Protect from freezing. Do not charge unit above the level mark because unless the gas has room to expand, it may burst the shell. Do not refill with loaded-stream charge unless extinguisher is designated as loaded-stream type.

FIG. 43–15.—Gas cartridge water extinguisher. May be labeled "loaded stream" if charged with a special alkali-metal salt solution instead of water. (No longer manufactured, but replacement parts are available.)

1351

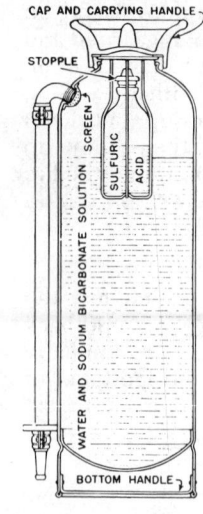

For CLASS A FIRES—Wood, paper, textiles, and the like.

Operation. Invert extinguisher; hold by bottom handle. The lead stopple will drop from the bottle and allow the two chemicals to mix; CO_2 gas will be generated and will supply the pressure necessary to discharge contents through the hose and nozzle. Direct the stream at the base of the fire, then follow up after the flames while working from side-to-side or around the fire if possible.

Maintenance. Periodic external inspection and yearly discharge and recharge. During recharge, check hose and nozzle for obstructions and defects, bent discharge elbow, corrosion, or other mechanical injury; and thoroughly wash out metal shell, hose, and nozzle. Mix sodium bicarbonate and water solution in separate, clean container and strain into extinguisher. The loose-fitting lead stopple must be on acid bottle to reduce hydroscopic action.

Caution. Store in heated cabinet because antifreeze solution cannot be used to protect from temperatures below 40 F; do not subject to temperatures above 120 F.

FIG. 43–16.—Soda-acid fire extinguisher. (No longer manufactured, but replacement parts are available.)

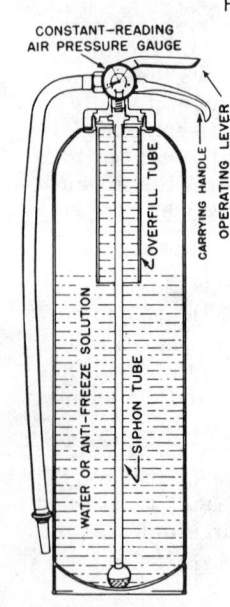

For CLASS A FIRES—Wood, paper, textiles, and the like.

Operation. Can be operated intermittently by removing locking ring and squeezing combination operating lever-handle. Extinguisher is pressurized with air or inert gas through a hose line equipped with an automobile-type air chuck. Charging pressures vary from 90 to 150 psi. Direct the stream at the base of the fire, then follow up after the flames while working from side-to-side or around the fire if possible.

Maintenance. Check pressure, hose, and nozzle frequently. When charging extinguisher, fill only to water level mark so there will be sufficient room for pressurized air. Some models have an overfill tube that ensures filling to the proper level. When pressurizing, the gauge should indicate just above the full mark, then, when the pressure equalizes, it will not read less than full. After charging, let it stand for 24 hr to check for leaks.

Caution. Protect from freezing. Be sure shell is the corrosive-resistant type before adding antifreeze charge. Do not refill with loaded-stream charge unless extinguisher is designated as loaded-stream type.

FIG. 43–17.—Stored pressure water fire extinguisher. May be labeled "loaded stream" if charged with a special alkali-metal salt solution instead of water.

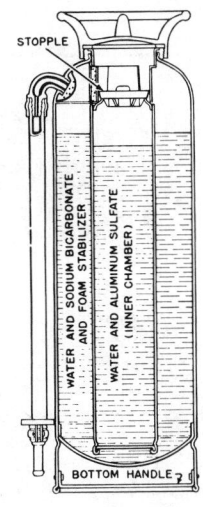

For CLASS A AND B FIRES—Wood, paper, etc., and oil, paint, and the like.

Operation. When the extinguisher is inverted, the lead stopple covering the water and aluminum sulfate chamber drops and permits the solution to mix with the water/sodium bicarbonate/foam stabilizer solution and produce foam. The chemical reaction releases CO_2 gas with sufficient pressure to expel the foam. For Class B fires in open containers, direct the stream against the back inside wall of the container—*never discharge the foam stream directly into the burning liquid* because it may cause the fire to spread out of the container. On spill fires, stand back and gently arc the stream. For Class A fires, aim the foam directly at the base of the flames.

Maintenance. Inspect periodically; discharge and recharge annually. Check for bent discharge elbow, corrosion, and other mechanical injury. Thoroughly clean interior; flush out hose and nozzle and clean outlet screen. Mix chemicals in clean containers and strain into their respective chambers.

Caution. Store in heated cabinet because an antifreeze solution cannot be used to protect from temperatures below 40 F; do not subject to temperatures above 120 F.

FIG. 43–18.—Foam fire extinguisher. (No longer manufactured, but replacement parts are available.)

cartridge that is recommended by the extinguisher manufacturers.

Sodium bicarbonate-base dry chemical, the most common agent, is available in ordinary form and foam compatible form. The foam compatible type can be used simultaneously with foam without causing the foam blanket to break down.

Potassium bicarbonate-base dry chemical, "purple K," is similar in extinguishing properties to sodium bicarbonate-base dry chemical, but it has twice the effective fire-fighting capacity on a pound-to-pound basis. This agent has good moisture repellency and is compatible with the simultaneous use of water or foam. Operator technique is not critical because it does not allow fuel to reflash as easily or as rapidly as sodium bicarbonate. The absence of momentary flare-up also permits the user to approach fires more closely. Also, the potassium compound extinguishes the leading edges of contained flammable liquid fires more easily than the sodium compound. This agent is also marketed as potassium

bicarbonate/urea, and offers greater range.

Ammonium phosphate-base dry chemical (multipurpose) has operating characteristics on Class B and C fires similar to sodium and potassium base dry chemical extinguishers. When discharged into a Class A fire, chemical reaction will destroy the flames and a coating, formed when the extinguishing agent softens and adheres to the burning surface thereby retarding further combustion. To obtain complete extinguishment on Class A materials, all burning areas must be thoroughly exposed to the extinguishing agent. Because any small burning ember may be a source of reignition, the importance of proper application on Class A fires is more critical than with water extinguishers. In the presence of moisture, multipurpose dry chemicals may cause corrosion when discharged on metals. It is therefore very important to clean up the multipurpose agent immediately after the fire is extinguished. It is also important never to mix the ammonium phosphate-base agent with the potassium or sodium bicarbonate-base

1353

TABLE 43-D

ANTIFREEZE SOLUTIONS FOR FIRE PAILS, DRUMS, AND BUCKET TANKS

To Make 2½ Gallons of Antifreeze Solution.

Freezing Temperature Deg F	Water		Type 1 CaCl₂ᵃ		Water		Type 2 CaCl₂ᵇ		Sp Gr	Deg Bé
	gal	pt	lb	oz	gal	pt	lb	oz		
10	2	2	5	..	2	3	4	2	1.150	18.8
0	2	1	6	3	2	3	5	2	1.183	22.5
−10	2	1	7	4	2	3	5	14	1.212	25.4
−20	2	1	8	3	2	2	6	10	1.237	27.8
−30	2	1	9	1	2	2	7	4	1.258	29.7
−40	2	..	9	10	2	2	7	12	1.274	31.1

To Make 10 Gallons of Antifreeze Solution

Approximate Freezing Temperature, Degrees F	Water, Gallons	Calcium Chloride,ᵃ Pounds	Specific Gravity	Degrees Baumé
10	9	20	1.139	17.7
0	8½	25	1.175	21.6
−10	8	29½	1.205	24.7
−20	8	33½	1.228	26.9
−30	8	36½	1.246	28.6
−40	8	40	1.263	30.2

Note: Calcium chloride solutions mixed in accordance with the above table should not be used in extinguishers as they constitute a severe corrosion problem.

ᵃ Flaked CaCl₂, 77% min.

ᵇ Granular flake or pellet CaCl₂, 94% min.

Source: National Fire Protection Association.

agent as a dangerous reaction can be triggered by even a trace of moisture.

There are two basic styles of dry chemical extinguishers as defined by the propellant technique: gas cartridge or stored pressure. In the gas cartridge type (Fig 43–20), pressure is supplied by a gas stored in a separate cylinder; whereas in the stored pressure type, the entire container is pressurized. Stored pressure units come in two styles—one with a combination handle and operating lever on the valve assembly (Fig. 43–21), and the other with a release lever on the cap of the extinguisher and a shutoff nozzle on the end of the discharge hose (Fig. 43–22). The extinguisher may be used intermittently or continuously, depending on the nature of the fire.

Fig. 43–19.—Heated cabinets for water-base extinguishers in exposed locations. Note the electric bulb (in lower part of cabinet) for maintaining temperature, and the large, easy-opening handle.

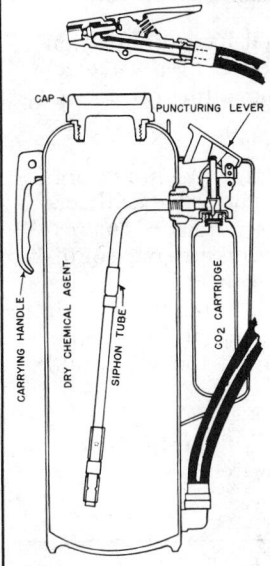

For CLASS B AND C FIRES—Oil, paint, and electrical. Also for small Class A fires.

Operation. Carbon dioxide gas is the propellant for the dry chemical. To operate, remove nozzle from holster and press puncturing lever. (On some models pull locking ring pin and push charging lever down.) This will release the gas and pressurize the large chamber containing the dry chemical. The discharge is controlled by the nozzle on the end of the hose. When the fire is out and the extinguisher is no longer needed, invert the container, hold nozzle in the air, and squeeze lever to release all residual pressure.

Maintenance. Weigh carbon dioxide cartridge every 6 months; replace if it has lost $\frac{1}{2}$ oz of the weight stamped on the shell. Annually check dry chemical for caking and proper level; cap gaskets for deterioration; hose, nozzle, and siphon tube for obstructions; and container for physical damage or corrosion.

Caution. Dry chemical will cake if moisture enters container. Do not wash out extinguisher or parts with water or other liquid unless the container can be dried adequately. Special models are available for temperatures below −40 F and above 120 F.

FIG. 43–20.—Dry-chemical fire extinguisher, cartridge operated.

Many models have a high-velocity discharge; therefore, care should be taken not to direct the initial discharge directly into the burning area since it may cause the fire to spread.

For best results when attacking a Class B fire, use a "fanning" action by rapidly moving the nozzle from side-to-side, so that the agent can thoroughly intermix with the flames. Start well in advance of the burning edge and go beyond the burning edge on each side to avoid leaving any burning pockets behind.

To minimize the possibility of reflash, the operator can continue to discharge the chemical after the fire has gone out.

Carbon dioxide extinguisher. The CO_2 extinguisher (Fig. 43–23) puts out a fire by diluting the amount of available oxygen.

Liquefied-gas extinguisher. The liquefied-gas extinguisher (Fig. 43–24) puts out a fire by interrupting the flame chain reaction. It is about twice as effective as CO_2 on a pound-for-pound basis. Because of toxicity considerations, do not use this extinguisher in confined spaces or small, unventilated rooms.

Vaporizing liquid extinguishing agents (carbon tetrachloride and chlorobromomethane) are toxic; their use is not recommended, and, in fact, is banned in federal and some state buildings, on merchant vessels by the Coast Guard, and by many industries. NFPA Standard 182M, *Manual on the Hazards of Vaporizing Liquid Extinguishing Agents*, discusses these compounds in detail.

Halogenated Compounds

There are two halogenated compounds that are used as fire extinguishing agents. One is Halon 1211, bromochlorodifluoromethane, which is a liquid when discharged; however, it quickly vaporizes. This characteristic provides a desirable compromise between reach and dispersion.

The other is Halon 1301, bromotrifluoromethane, a liquefied gas that provides for immediate dispersion in the fire area. It is about twice as effective as CO_2 on a pound-for-pound basis. Halon 1301 is a clean agent and in its undecomposed state has been determined to be safe to humans (in concentrations up to about 10 percent by volume in air for short exposures of up to 20 minutes). However,

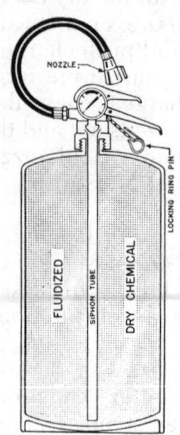

For CLASS B AND C FIRES—Oil, paint, and electrical. Also for small Class A fires.

Operation. Remove locking pin, aim nozzle at base of fire, then squeeze discharge lever. Discharge is controlled by a valve released by squeezing the combination handle-operating lever.

Maintenance. Check pressure gauge frequently. When recharging, thoroughly clean valve assembly so that the valve will seat tight and give maximum protection against pressure loss. To pressurize, remove hose and attach adapter and pressurizing equipment. (Recharge equipment varies with manufacturer.) Dry nitrogen or CO_2 is usually recommended as the pressurizing gas. (Air passed through moisture traps is sometimes used, but if the trap is not 100 per cent efficient, moisture can harden the chemical.)

Caution. Pressurizing gas must be dry because if moisture enters container, the chemical can harden and not flow freely from nozzle.

FIG. 43–21.—Dry-chemical fire extinguisher, stored pressure type, combination handle.

when it is exposed to temperatures of about 900 F, it becomes unstable and breaks down in the presence of moisture, to give off hydrogen fluoride, free bromine, and carbonyl halides.

Dry-powder extinguishers

The use of combustible metals has increased until protection is now needed for: sodium, titanium, uranium, zirconium, lithium, magnesium, sodium-potassium alloys (NaK), and other less common materials. There are several powdered agents approved for use on metal fires, the oldest being the G-1 type, which is a graphite/organic phosphate compound. When it is applied with a scoop or shovel to a metal fire, the phosphate material generates vapors that blanket and smother the flames, and the graphite, being a good conductor of heat, cools the metal below its ignition temperature.

Care should be taken to assure that the depth of the cover is adequate to provide a smothering blanket. If hot spots should occur, they should be covered by additional powder. The burning metal should be allowed to cool before disposal is attempted.

Another material is "Met-L-X," composed of a sodium chloride base with additives to make it free flowing, increase water repellency, and create the property of heat caking. This material is dispensed from a 30 pound dry *powder* extinguisher which is similar in appearance and physical features to the cartridge-operated dry-chemical extinguisher, or from larger wheeled or stationary units.

The technique used to extinguish a metal fire with Met-L-X is to open the nozzle of the extinguisher fully and apply a thin layer of Met-L-X over the burning mass from a safe distance, until control is established. Then throttle the nozzle to produce a soft heavy stream and completely cover with a heavy layer from close range. The heat of the fire causes the Met-L-X to cake, forming a crust that excludes air.

"Lith-X" is also a dry powder. It is composed of a special graphite base with additives to render it free flowing so it can be discharged from an extinguisher. Lith-X was developed mainly for use on lithium fires, but is effective on other combustible metals. Lith-X does not cake or crust when applied over a burning metal. It excludes air and conducts heat away from the burning mass,

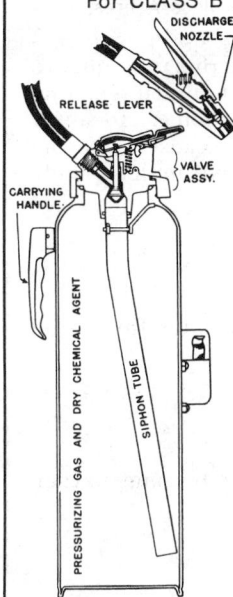

For CLASS B AND C FIRES—Oil, paint, and electrical. Also for small Class A fires.

Operation. Remove locking pin, then press down on charging lever. Remove nozzle from holster; squeeze to operate. Discharge is controlled by nozzle on end of hose. When the fire is out and extinguisher is not needed, invert the container, hold nozzle in the air, and squeeze lever to release residual pressure and to clear out the hose.

Maintenance. Periodically check pressure gauge and look for physical damage or corrosion. To recharge, remove cap assembly by turning retaining ring counterclockwise until arrow in notched recess aligns with locking device on release lever; cap is released when lever returns to its normal position. Remove gaskets, clean, coat with silicone grease. Clean out dry chemical from cap and gasket seats; replace gaskets in cap; reassemble unit. To pressurize, remove hose and attach adapter and pressurizing equipment. Dry nitrogen is the most common pressurizing gas.

Caution. Pressurizing gas must be dry because if moisture enters the container, dry chemical can harden and not flow freely from nozzle.

Fig. 43–22.—Dry-chemical fire extinguisher, stored pressure type.

For CLASS B AND C FIRES—Oil, paint, and electrical. Also for small Class A fires.

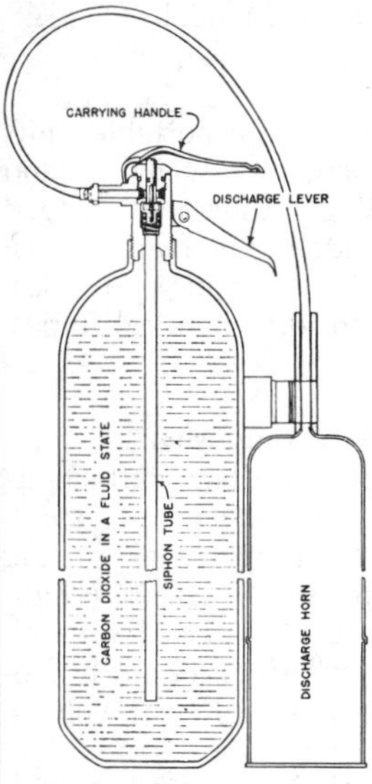

Operation. Hold in upright position; pull locking ring pin; squeeze the discharge lever. On all fires, direct discharge at base of flame and continue even after fire has gone out to minimize the possibility of reflash. For best results on Class B fires, apply in a slow sweeping action, from side-to-side, start at the near edge of the fire and work toward the back of the burning area. Fire-fighting effectiveness is reduced under windy conditions because CO_2 is a gas and dissipates.

Maintenance. Should be recharged by an outside agency. Every 6 months inspect for visible defects and weigh unit; recharge if loss is more than 10 per cent.

Caution. In confined areas, atmosphere may become oxygen deficient. The discharge horn becomes very cold during use; therefore, do not touch it. Do not install in areas subject to temperatures over 120 F.

Fig. 43–23.—Carbon dioxide fire extinguisher.

For CLASS B AND C FIRES—Oil, paint, and electrical. Also for small Class A fires.

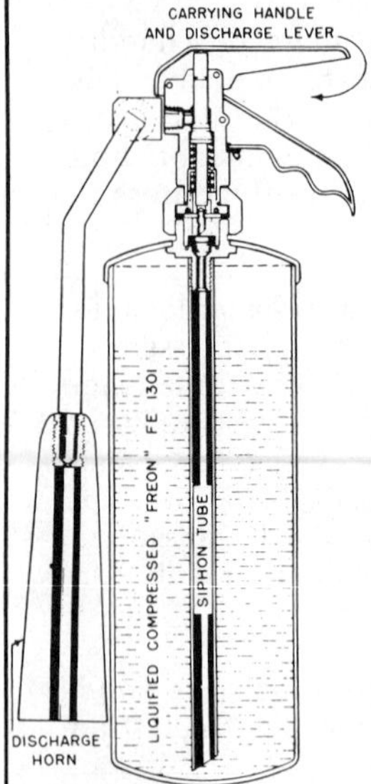

Operation. Hold in upright position; pull locking ring pin; squeeze discharge lever. Gas has high vapor pressure at normal temperatures, which accounts for most of its propellent force. Ideal for Class C fires in delicate electronic equipment because leaves no residue. Fire-fighting effectiveness is reduced under windy conditions because expelled agent is a gas.

Maintenance. To recharge, unscrew spent cylinder from head assembly and replace with sealed refill cylinder.

Caution. In confined areas, atmosphere may become oxygen deficient.

FIG. 43–24.—Liquefied-gas extinguisher.

extinguishing the fire.

The extinguishing effects of dry powder agents on a given metal depend on the physical form and quantity of the metal involved and on ambient conditions. Although dry powders are very effective on combustible metal fires, they all have certain limitations. The type, quantity, and form of the metal and the existing physical conditions must be considered when selecting the proper type of dry powder and the method of application.

A problem recently developed in fire fighting involves pyrophoric liquids, such as triethylaluminum. These liquids ignite spontaneously and the resulting fires cannot be easily extinguished by dry powder or other commonly used agents. A special material ("Met-L-Kyl") has been developed, consisting of a bicarbonate base dry chemical and an activated absorbent.

The principle of extinguishment involves the combination effect of dry chemical, which extinguishes the flames, and the absorbent which absorbs the remaining fuel and prevents reignition. This extinguishing agent has been

designed so that it can be discharged from an extinguisher similar to the standard cartridge-operated dry chemical model.

Miscellaneous equipment

Wheeled equipment. Large portable units on wheels are commercially available and include 17-gal and 30-gal soda-acid, loaded stream, calcium chloride, and foam types; 50-lb, 75-lb, and 100-lb carbon dioxide types; and 150-lb and 350-lb dry-chemical types. (See Fig. 43–25.)

Wheeled "twinned" extinguishers. With the development of fluorocarbon surfactants that are water soluble, foam agents are available that give water the property of floating in thin layers on liquid fuel surfaces ("light water"). This characteristic provides excellent protection against reflash on liquid hydrocarbon fires with only one-fourth the volume as compared to protein/air foam.

A combination wheeled extinguisher with "purple K" dry chemical and "light water" fluorocarbon foam provides a synergistic extinguishing system with both rapid knock-down of flames and complete protection against reflash. The two agents are simultaneously applied through dual pistol-grip nozzles.

Vehicle-mounted equipment. Water, foam, carbon dioxide, and dry chemical extinguishing agents are available in units that are mounted on vehicles. They range in size from in-plant fire vehicles capable of turning in warehouse aisles, to large trucks.

Fire blankets. In some cases, fire blankets can be used to smother a small fire. Their major purpose is to extinguish burning clothing, but they are useful, too, for smothering flammable liquid fires in small open containers.

Flame-retardant blankets are available, the most common size is 66 by 80 in. Blankets are usually stored in containers mounted on a wall or column, so arranged that they can be readily pulled out.

Miscellaneous hand equipment. Five-gal or 2½-gal water or antifreeze back-pack tanks are available with hand pumps built into the hose nozzle handles. This type of unit is carried on the back, and the slide action pump is operated with both hands. The back-pack unit is frequently used for combating brush fires.

Training of employees

When a fire breaks out, the first steps are to evacuate the building occupants safely, and to call the fire department, but it is also important to apply prompt control measures when the fire is small. Regardless of the size or the hazard involved, plans should be made for a well maintained, efficient fire alarm system, and effective fire fighting brigade, in addition to an emergency evacuation procedure.

Extinguishers are effective only when fires are in the first stages, and it is essential that they be immediately accessible and promptly used. Extinguishers are only as good as the operators using them. It pays to train key workers on each shift frequently and thoroughly.

Instruction of employees in the use of extinguishers can best be given by demonstra-

Fig. 43–25.—Portable fire extinguishers must supplement a fixed system as they often can prevent a small fire from spreading as well as provide rapid extinguishment in early stages of a fire. Here a dry chemical wheeled-type extinguisher is being inspected.

Courtesy Marathon Oil Company.

tion, preferably at the time when extinguishers are scheduled for recharging. At the demonstrations, fire conditions should be simulated and an instructor should explain the fundamentals of fire fighting and the use of the equipment, and employees should then be allowed to get the "feel" of the extinguisher. In small organizations, everyone in the plant should attend such demonstrations. In larger plants, it is advisable to train a suitable number of employees, so that there is a good distribution of trained men.

Fire extinguisher manufacturers, insurance

companies, fire departments, and safety councils have films, posters, and cutaway displays useful in explaining the construction, maintenance, and operation of portable extinguishers. NFPA Standards No. 10 and 10A, on portable fire extinguishers, afford particularly valuable information for the training of employees in practices related to portable fire extinguishers.

The education of employees should be continued with demonstrations, practice drills, and lectures at yearly intervals, or more often if there is a special fire hazard. It may be advisable to put printed instructions regarding the use of fire extinguishers into the hands of the employee. Mimeographed or printed sheets, leaflets, or cards can give both general and detailed instructions regarding the use of the extinguishers.

In addition, instructions in how to use extinguishers should be permanently posted near or on the extinguishers themselves (Fig. 43–26).

Fire extinguisher training is intended to teach men and women how to stop small fires from spreading out of control. Underwriters' Laboratories, Inc., lists fire extinguishers at only 40 percent of their firefighting capacity when used by an inexperienced operator because 90 out of 100 people have never used a fire extinguisher. This shows the critical importance of training and means that any fire extinguisher in the hands of a trained operator has 2½ times the fire-fighting capacity it has when used by a novice.

An organized fire brigade is a step ahead of even the best system of teaching individuals how to use portable fire extinguishers.

Maintenance and inspection

One person in the plant should be assigned responsibility for the maintenance and inspection of fire extinguishing equipment. In larger plants, this person may be the fire chief or the safety professional. He may have one or more assistants to make routine inspections weekly or monthly and to do testing and repair work in accordance with procedures recited in NFPA Standard No. 10. There should be a record system and an organized plan for checking and repairing various types of extinguishers.

The inspection and maintenance records

Fig. 43–26.—Instructions should be permanently posted on and near extinguishers.

should at least consist of durable tags fastened to the extinguishers showing dates of inspection (at least monthly), and examination for recharge and other maintenance work (at least annually).

A duplicate record should also be kept in a file in the office. Other records may be kept listing extinguishers by type, by location, and by recharge periods. The office record system should give the history of each extinguisher, the type and quantity of refills on hand, and other pertinent information.

Efficient in-plant maintenance requires that supplies of spare parts and refills be kept on hand. In larger plants, specially trained personnel can test and refill extinguishers on a full-time basis. In other plants, the maintenance service can be under contract with a reliable service organization or the manufacturer's representative.

Generally, it is desirable to seal hand extinguishers or to install them in cabinets to discourage tampering and to facilitate inspection. *However, extinguishers should not be locked in cabinets nor stored in any way that would prevent immediate use in an emergency.*

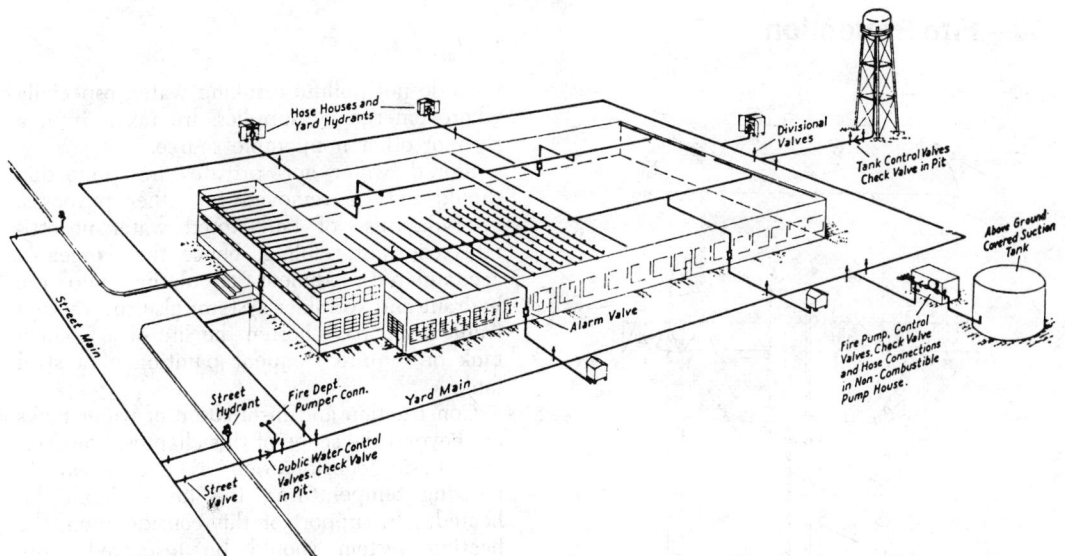

FIG. 43–27.—Typical automatic sprinkler installation for a modern industrial facility. Note sprinkler systems, yard mains, and water supplies.

Courtesy Factory Mutual Engineering Corp.

One of the most important tasks involved in keeping fire extinguishers in top condition is periodic hydrostatic pressure testing, a mandatory requirement set forth in NFPA Standard No. 10A. The occasional reports of injuries and fatalities, due to extinguishers rupturing during operation, should be proof enough of the importance of this requirement.

Whether the testing is done by an outside agency or privately, there should be no mass removal of extinguishers from an occupancy. A testing program should be established so only a few scattered units are removed at one time. They should be tested and returned to service promptly. Temporary replacement units or additional protection may be needed when extinguishers are being removed for testing.

Water Systems

Water supply and storage

Sprinkler systems need a reliable water supply of ample capacity and pressure for efficient fire extinguishment (Fig. 43–27). The water supply should be engineered with the sprinkler protection to provide a hydraulically balanced system at the least cost. For

example, to supply all sprinklers likely to open in addition to hose streams, a volume of 500 to 3000 gpm (and sometimes more) may be needed. The precise need depends on the occupancy classification. The pressure requirement will vary; however, it should be high enough to maintain a residual pressure of 15 psig in top-story sprinklers.

Water may be supplied from the following sources:

1. Public water works underground supply mains.

2. Automatically or manually controlled pumps drawing water from lakes, ponds, rivers, surface storage tanks, underground reservoirs, or similar adequate sources.

3. Pressure tanks containing about two-thirds water and one-third compressed air for expelling the water into the piping supply system.

4. Elevated tanks or reservoirs that depend on gravity to force water through the system.

Usually at least two of these independent sources should be available as primary and secondary sources of water to the fire extin-

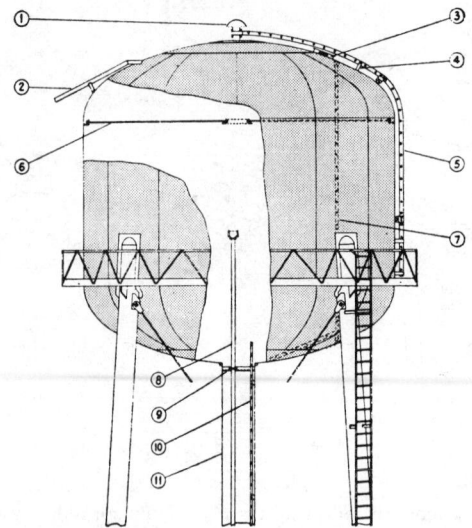

1 — Screened finial vent
2 — Stub overflow pipe
3 — Roof hatch with hinged cover and catch to keep it closed.
4 — Stop lug for revolving ladder
5 — Revolving outside ladder
6 — Spider
7 — Inside ladder
8 — Hot-water pipe; tee outlet at one-third height of tank
9 — Pipe brace
10 — Ladder in riser
11 — Large steel-plate riser

FIG. 43–28.—Welded-steel gravity tank with large steel plate riser on steel tower.

Courtesy Factory Mutual Engineering Corp.

guishment equipment. In case of fire, the primary source furnishes water to the system immediately; it is reinforced by the secondary source that also supplies emergency protection if the primary source is out of service. The preferred primary source is a connection from a reliable public water works system. Connection should be made to two different mains to provide greater volume and flexibility in case of failure of one water main. The preferred secondary source is a fire pump that can deliver water at high pressure over an extended time. This pump should be located where it will not be put out of service by a fire, and it should have an emergency power supply, independent of the plant system.

Care must be taken that the water connec-

tions do not pollute drinking water, especially where emergency supplies are taken from a river or other nonpotable source.

Stored water for private fire protection should not be removed for other purposes. Everyday use of tank-stored water necessitates constant refilling, hence the danger of accumulated sediment circulating into the hydrant and sprinkler system; also the varying water level may shorten the life of a wooden tank or require frequent painting of a steel tank.

Construction and installation of water tanks are beyond the scope of this chapter; however, one basic consideration is mentioned—in freezing temperatures, the tanks should be heated. In support of this consideration, the heating system should be inspected daily during freezing weather; the control valves, weekly; and the entire system (tank, supporting tower, piping, valves, heating system, and all components and accessories), annually. (See NFPA Standard 22, *Water Tanks for Private Fire Protection*. Also see NFPA *Fire Protection Handbook*.)

To facilitate inspection, wood or steel gravity tanks supported on steel towers should be ringed by a platform that is protected by a substantial steel rail having a vertical height within the range of 36 to 42 in. (nominal), and 6-in. toeboards (Fig. 43–28). They should have substantial steel ladders equipped with approved steel cages or basket guards.

Water tanks that are accessible to the public should have their ladders protected by locked gates or fences to discourage unauthorized persons from climbing them.

Automatic sprinklers

Automatic sprinklers are the most extensively used installations of fixed fire extinguishing systems. These systems are basic and have proved so effective that most fire protection engineeres consider them the most important fire fighting equipment. Nationwide figures prepared by the NFPA indicate that sprinklers have a very high efficiency rating for satisfactory performance.

Automatic sprinklers are the most versatile and dependable form of fire protection available; this has been demonstrated over many years and against a wide variety of fires. Although many other types of protection have

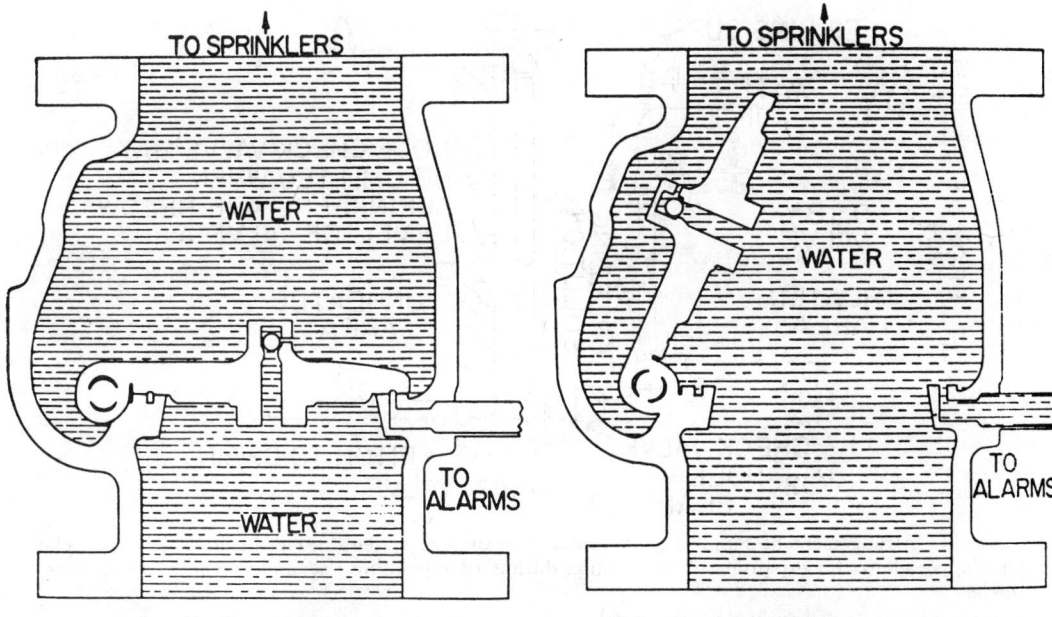

| No Flow of Water | Water Flowing to Sprinklers and to Alarms |

FIG. 43–29.—A wet-pipe sprinkler system is under water pressure at all times so that water will be discharged immediately when an automatic sprinkler operates. The automatic alarm valve triggers a warning signal when water flows through the sprinkler piping.

been developed to deal effectively with special hazards, sprinklers are usually also recommended as "backup" protection.

The cost of automatic sprinkler protection is relatively small compared with the total plant investment; it generally averages about 2 percent. Experience shows that a sprinkler system often can pay for itself in ten years (and sometimes less) by savings in insurance cost; after the pay out period, the reduced premiums can mean a substantial savings each year.

In addition to the economic factors, automatic sprinklers have an impressive life-saving record. Loss of life by fire is rare where properly designed and maintained sprinkler systems have been installed.

The efficiency of an automatic sprinkler system is shown by the relatively few sprinklers which open in most fires and the high percentage of extinguishment or control. A study of more than 50,000 fires indicates that about three-quarters of them were confined or extinguished by six or less sprinkler heads. So

effective are sprinklers, that complete extinguishment or control of fires has resulted in 96 percent of the cases studied.

The surprising fact is that in the instances where sprinklers fail to operate (3.8 percent of the cases), the failures were usually not the fault of the system as such, but rather of some readily preventable condition which kept them from operating as intended. Over ⅓ of all failures can be directly attributed to closed water supply valves.

The primary function of the sprinkler system is to deliver water automatically to a fire. In addition, the system can serve as a fire alarm. This is done by installing water-flow alarms in each main riser pipe. When a fire occurs and the first sprinkler head opens, the water rushing through the pipe sets off an alarm which alerts the fire fighters.

Dependable sprinkler protection requires a systematic maintenance and inspection program which includes periodic inspection of water supply valves, water supply tests, physical inspection of system piping for obstruc-

1363

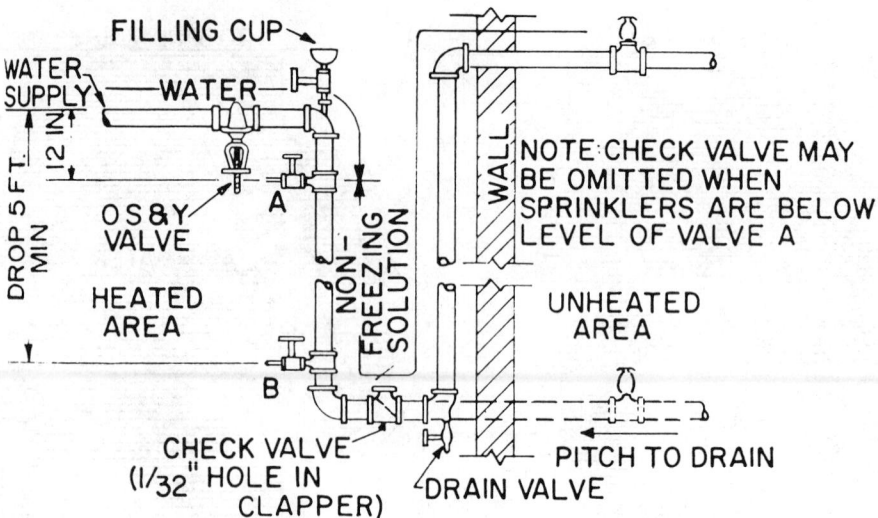

FILLING CUP
WATER SUPPLY — WATER
2 IN.
12 IN.
DROP 5 FT. MIN
OS&Y VALVE
A
HEATED AREA
NON-FREEZING SOLUTION
WALL
NOTE: CHECK VALVE MAY BE OMITTED WHEN SPRINKLERS ARE BELOW LEVEL OF VALVE A
UNHEATED AREA
B
CHECK VALVE (1/32" HOLE IN CLAPPER)
DRAIN VALVE
PITCH TO DRAIN

Fɪɢ. 43–30.—Arrangement of supply piping valves of antifreeze systems. The $\frac{1}{32}$-in. hole in the check valve clapper allows for expansion of the solution during a temperature rise, and thus prevents damage to sprinklers.

Figures courtesy National Fire Protection Association.

tions to distribution, and similar items. See NFPA Standard No. 13A, *Care and Maintenance of Sprinkler Systems,* for additional maintenance requirements.

There are six basic types of automatic sprinkler systems: wet pipe, dry pipe, preaction, deluge, combination dry pipe and preaction, and limited water supply. (See NFPA Standard No. 13, *Installation of Sprinkler Systems,* and Standard No. 13A, *Care and Maintenance of Sprinkler Systems.*) The combination dry pipe and preaction system is used on installations that are larger than can be accommodated by one dry-pipe valve. The limited water supply system is used for installations that do not have access to a continual or large supply of water. The other four types of sprinkler systems are discussed in detail in the following paragraphs. Also included are a description of the simultaneous tripping of alarms, an explanation of temperature rating, and a list of reasons for sprinkler failure with remedies for avoiding them.

Wet pipe. In the wet-pipe system, which represents the greatest percentage of sprinkler installations, all the piping is filled up to the sprinkler heads with water under pressure

(Fig. 43–29). Then, when heat actuates the sprinkler, water is immediately sprayed over the area below. If a portion of the wet-pipe system is subjected to freezing temperatures, as on a loading dock, the exposed sections should be filled with an antifreeze solution (Fig. 43–30), or these sprinklers should be connected to a dry-pipe system. (A good rule-of-thumb is to use a dry-pipe system when more than 20 sprinklers are involved.) Shutting off and draining is hazardous in that it removes the protection when it may be needed.

The antifreeze should be a water-soluble liquid that is proportioned to give low-temperature protection without producing a combustible mixture, as specified in NFPA sprinkler system Standards No. 13 and 13A. When the system is supplied from public water connections, antifreeze should be chemically pure glycerine (U.S. Pharmacopoeia 96.5 percent grade) or propylene glycol, and then added only in accordance with local health regulations.

Dry pipe. The dry-pipe system generally substitutes for a wet-pipe system in areas

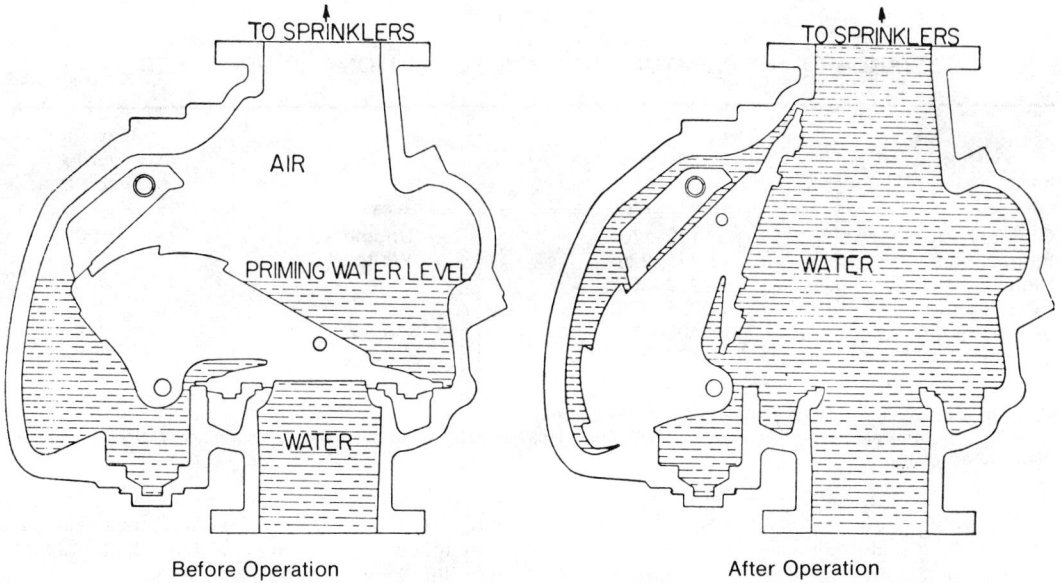

TO SPRINKLERS

AIR

PRIMING WATER LEVEL

WATER

Before Operation

TO SPRINKLERS

WATER

After Operation

Fig. 43–31.—The principle of a dry-pipe system is illustrated by these simplified drawings of a dry-pipe valve. Compressed air in the sprinkler system holds the dry valve closed, preventing water from entering the sprinkler piping until the air pressure has dropped below a predetermined point.

Courtesy National Fire Protection Association.

where piping is exposed to freezing temperatures. It is essential, however, that the dry-pipe valve and water-supply line be located in a heated enclosure.

In the dry-pipe system, the piping contains compressed air that holds back the water by means of a dry-pipe valve (Fig. 43–31). When a sprinkler nozzle opens, the air is released, the pressure drops, and the dry-pipe valve opens to admit water into the piping system. These sequential actions delay the actual wetting. Because of this delay, extra-hazard occupancies are difficult to protect with a dry-pipe system. In general, more water damage results with the dry-pipe system because more sprinklers open than with the wet-pipe system; that is, the fire progresses further, hence trips more sprinklers, before the extinguishing action of the water takes effect. To reduce this delay, quick-opening devices, such as exhausters or accelerators, can be added to dry-pipe systems to expel the air more rapidly.

The system air pressure should be kept 15 to 20 psi above the normal tripping pressure; higher pressure delays the action of the dry-pipe valve. The system should not lose more than 10 psi air pressure per week, which would require repressurizing oftener than once a week; equipment that leaks more is not acceptable. (See NFPA Standards 13 and 13A for testing dry pipe systems.)

It is essential that all parts of a dry-pipe system be installed so that they can be throughly drained. Therefore, where it is necessary to use inverted (pendant) sprinklers or "drop piping," a special type of pendant sprinkler may be used or the piping filled with an antifreeze solution.

Preaction. Preaction systems are similar to dry-pipe systems, but react faster and, hence, minimize water damage in case of fire or mechanical damage to sprinklers or piping. Supervision of the system against mechanical damage can be accomplished by connecting to the piping an automatic, low-pressure air supply that compensates for minor system leakage. A rapid reduction in pressure resulting from, say, an accidental breakage of

1365

TABLE 43-E

STANDARD TEMPERATURE RATINGS FOR AUTOMATIC SPRINKLERS

Rating	Operating temperature (F)	Color	Maximum ceiling temperature (F)
Ordinary	135–170	Uncolored*	100
Intermediate	175–225	White*	150
High	250–300	Blue	225
Extra high	325–375	Red	300
Very extra high	400–475	Green	375
Ultra high	500–575	Orange	475

From Sprinkler Systems, NFPA Standard No. 13, Table 3651.
*The 135 F sprinklers of some manufacturers are half black and half uncolored. The 175 F sprinklers of the same manufacturers are yellow.

the piping, sends a trouble signal without tripping the water-control valve.

The preaction valve, which controls the water supply to the system piping, is actuated by a separate automatic detection system, which is located in the same area as the sprinkler, and not by the operation of a sprinkler. Because the detection system is more heat sensitive than the sprinklers, the water-supply valve opens sooner than in a dry-pipe system. The water-supply valve can also be operated manually.

Usually an alarm is sounded when the valve opens and starts filling the system with water. There may then be time to put out the fire with portable equipment before the sprinklers go into action and drench the area with water. This system is especially effective where valuable merchandise is handled or stored.

Deluge. The deluge system wets down an entire area by admitting water to sprinklers that are opened at all times. Deluge valves that control the water supply to the system are actuated by an automatic detection system located in the same area as the sprinklers. The water supply valves can also be operated manually.

This type of system is primarily designed for extra-hazard occupancies where great quantities of water may have to be applied immediately over large areas. Deluge systems are ordinarily used to best advantage where rapidly spreading or flash fires may be anticipated, such as in explosives plants, plants handling or processing nitrocellulose materials, lacquer plants, and buildings which contain large quantities of flammable materials.

Another application of a deluge system is an open system of outside sprinklers for distributing water over the roof, exterior of a building, or at windows, cornices, etc., to protect it against fire from adjoining property. This type is usually manually operated and is used where construction is inadequately protected by design or by distance from adjacent fire hazards.

In special applications, open sprinklers and closed sprinklers may be combined in a single system where deluge protection is not needed over the entire area. However, it must be remembered that separate automatic detectors are also required in the area covered by the closed sprinklers; that operation of a closed sprinkler will not activate the entire system; and that a fire in the area of the closed sprinkler will also cause water to discharge from all of the open sprinklers.

Automatic alarms, operated by the flow of water through the system, should be a part of every standard sprinkler installation. Such an alarm may be connected to a central-station fire alarm service or to the municipal fire department, or may be a local alarm signal. Its purpose is to give prompt notice that the sprinkler system is operating. It also

TABLE 43-F

SELECTION OF SPRINKLER RATINGS

Rate of Heat Release	Maximum Temperature (F) at Sprinkler Level Under Other Than Fire Conditions					
	100	150	225	300	365	465
Low rate of heat release from fire. (Light occupancies such as offices, schools, hotels, hospitals, apartments)	135, 160, 165	175, 212	250, 280, 286	325, 350, 360	400	500
Moderate rate of heat release from fire. (Ordinary industrial occupancies)	175, 212	175, 212	250, 280, 286	325, 350, 360	400	500
High rate of heat release from fire. (Flammable liquids, rubber tires, rubber and plastic foams, high-piled combustible storage, and similar locations)	250, 280, 286	250, 280, 286	250, 280, 286	325, 350, 360	400	500

Courtesy Factory Mutual Engineering Corporation.

signals water leakage or discharge from causes other than fire.

The automatic alarm system must be tested and inspected frequently and be maintained by persons thoroughly familiar with it.

Temperature rating of sprinklers. Sprinklers should be selected on the basis of temperature rating and occupancy. Sprinklers are built with heat-actuated elements of solder that melt, or with special devices in which chemicals melt or expand to open them. Table 43-E shows the ratings and distinguishing colors of sprinklers.

Quick operation of sprinklers—an advantage when over a fire—may be a disadvantage elsewhere, wasting water and wetting down materials that might otherwise be unaffected. Tests conducted by the Factory Mutual Engineering Corp. resulted in the ratings recommended in Table 43-F.

Causes of failure of sprinkler systems. Sprinklers seldom fail to control fires, but when they do, failure is usually due to human errors: (*a*) NOT KEEPING ALL SUPPLY VALVES OPEN, AND (*b*) SHUTTING OFF THE SUPPLY VALVES PREMATURELY DURING FIRE.

Other causes of sprinkler failure, with corresponding remedies (in italics), are:

1. Freezing of wet system sprinkler pipes. *Heat the building or convert to dry system.*

2. Defective dry-pipe valve, or slow operation of dry system because of its excessive size. *Check valves at frequent intervals, "trip test" in accordance with insurance company recommendations, or subdivide system.*

3. Foreign material obstructing the system. *Flush out system on a regular schedule and provide debris-clear water at intake through use of filter screens.*

4. Improper drainage through faulty installation. *Check pitch of pipes and eliminate low spots.*

5. Sprinklers obstructed by stock piled too high, sprinklers isolated by temporary partitions or shelving, and sprinklers shielded from heat. *Improve housekeeping and maintain a minimum of 18 in. clearance between the top of material storage and the deflector. It is a good practice to in-*

crease the clearance up to 36 in. over large, closely packed piles of combustible cases, bales, cartons, or similar stock.

6. Corrosion of sprinklers in such locations as bleacheries, dye houses, or chemical operations. *Use sprinklers specially protected for use in such locations.*

7. Inadequate supply of water because of faulty design or poor maintenance. *Verify that water conditions have not changed since original plant installation. It is important that the system be sturdily installed and anchored because explosions can jolt the piping and render the system ineffective.*

Water spray systems

Water spray is effective on all types of fires where there is no hazardous chemical reaction between the water and the material that is burning. These systems are independent of, and supplemental to, other forms of protection —not a replacement for automatic sprinklers.

Fixed water spray systems are similar to the standard deluge system except that the open sprinklers are replaced with spray-type nozzles. The water supply to the system can be controlled automatically or manually. They are generally used to protect flammable liquid and gas tankage, piping and equipment, cooling towers, and electrical equipment, such as transformers, oil switches, and motors (Fig. 43–32). Because of its low electrical conductivity, water spray applied through fixed piping systems on electrical equipment with voltages as high as 345,000 volts has proved practical. When applied on some types of electrical equipment, water spray may cause short circuits by forming a continuous path of water between energized parts. In such cases, means should be provided for cutting off the electric current before the water spray is applied. (See NFPA Standard No. 15, *Water Spray Fixed Systems.*)

The type of water spray required depends upon the nature of the hazard and the purpose for which the protection is provided. The basic principle of water spraying is to give a complete surface wetting with a pre-selected water density, taking into consideration nozzle types, sizes, spacing, and water supply.

Water spray systems can be designed effec-

FIG. 43–32.—A water spray system for oil-filled electric power transformers. A thick layer of crushed stone and subsurface drainage is provided around the base of the transformer installation to prevent burning oil from flowing beyond the area protected by the spray.

Courtesy National Fire Protection Association.

tively for any one or any combination of the following:

1. Extinguishment of fire

2. Control of fire where extinguishment is not desirable, such as gas leaks

3. Exposure protection; that is, absorb the radiant heat by evaporation of the spray

4. Prevention of fire

Since the passages in a water spray nozzle are small in comparison with those in the ordinary sprinkler, they can easily be clogged by foreign matter in water. Therefore, strainers are ordinarily required in the supply lines of fixed-piping spray systems. The strainer basket should have holes small enough to protect the smallest orifices of the nozzles used. In cases where the nozzles have extremely small water passages, they may have their own internal strainer in addition to the supply-line strainer.

Hydrants, hoses, and nozzles

Fire hydrants. In larger plants where parts of the plant are a considerable distance from public fire hydrants, or where no public hydrants are available, hydrants should be installed at convenient locations in the plant yard. The number needed depends on the fire exposure, and the hose-laying distance to the built-up plant areas. (See NFPA Standard 24, *Outside Protection*.)

Exterior fire department connections serving sprinkler or hose systems should be kept accessible and unobstructed. The discharge ports should be at least 18 in. above the ground or floor level. Vegetation, snow, and stored materials should be kept away from hydrants or hydrant houses. A hydrant must be protected from mechanical injury, but this protection cannot interfere with efficient use.

Before cold weather, hydrants should be drained or pumped out, if they are not the type that normally drain. Drainage must be checked whenever hydrants are used during freezing weather. Frozen hydrants may be discovered by sounding (striking the hand over the open outlet), by a partial turn of the hydrant stem (if the hydrant is frozen, the stem will not turn), or by lowering a weight on a string into the hydrant barrel.

Frozen hydrants can best be thawed with steam introduced through the outlet by means of a steam hose that is pushed slowly down the barrel, thawing as it goes. Corrosive chemicals, such as calcium chloride, caustics, or salts, should not be used.

Control valves should be tested frequently and be well maintained, and a number of persons, including plant fire brigade members, should know the location of valves and the sections of the pipe controlled by them.

Connections should be checked with the city fire department to be sure that they are of a size and thread that will fit its equipment. If special adapters are required, they should be supplied to municipal fire fighters and also be available on the premises.

Fire hose. Like other fire-extinguishing equipment, hose lines should be available for immediate use, and should not be obstructed nor inaccessible. Space around hose lines and control valves should be clear. The equipment should be visible and conspicuously indicated, and employees should know its location and understand its operation. Aisles and doorways should be kept clear and should be wide enough to allow rapid use of hose reel carts or other mobile equipment.

Fire hose must be rugged and dependable,

Fig. 43–33.—Hose houses may be made of metal (*left*) or wood (*right*).

Right: *Courtesy National Fire Protection Association.*

1369

capable of carrying water under substantial pressures, yet flexible and sufficiently easy to handle. Therefore, the fire hose should be of suitable quality, properly cared for when used, and carefully maintained in storage. Yard hose should be stored in standard hose houses for protection from weather (Fig. 43–33).

Hose for outside or yard protection ordinarily is 1½- or 2½-in. double-jacketed lined type or single jacket type when weight or bulk is undesirable. Hoses made of polyester or other synthetic fibers are lighter, stronger, and mildew resistant. (See NFPA Standard No. 24, *Outside Protection*.)

Woven-jacket, lined hose with an outer rubber or plastic cover is chiefly used in industries

kept on a swinging rack or reel of approved type, approximately 5 ft above the floor or high enough so that it will not be a hazard to passers-by or be damaged by trucking operations. See NFPA Standard No. 14, *Installation of Standpipe and Hose Systems*.

The hose should be so arranged that it will not kink or tangle when pulled out. One end should be kept connected to the standpipe, and the other end should be equipped with a ¾- or ½-in. nozzle tip or a combination spray-solid stream nozzle. To prevent kinking in use, not more than 75 ft of hose should be placed at a standpipe outlet.

Except for unlined linen hose, all fire hose

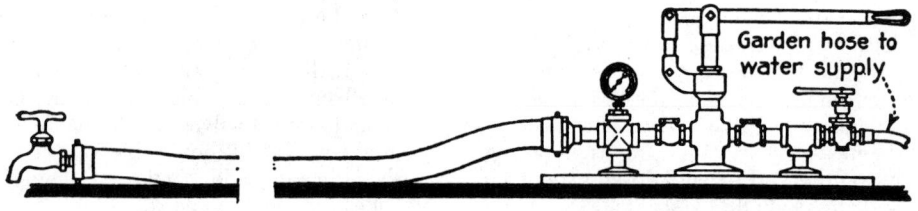

FIG. 43–34.—Hand-pump assembly for hydrostatic testing of hose.

where the hose jacket must be protected against chemicals and abrasions.

The lined hose with a rubber or plastic cover is available in ¾- and 1-in. sizes, and is generally used as a booster hose, or as a hose on chemical engines, wheeled extinguishers, and wall-mounted or vehicle-mounted pressurized hose reels.

For inside use by building occupants (as opposed to standpipe systems designed for the fire department), 1½-in. unlined linen hose is satisfactory to provide small streams for fire protection. If kept dry, it will last almost indefinitely. It should never be used except in case of fire and should not ordinarily be tested because it will deteriorate if slightly damp. However, with the introduction of light-weight, lined fire hose, which has twice the hydraulic efficiency of unlined hose, many firms are replacing the unlined hose on their industrial standpipe installations. The unlined hose has its place in such locations as office buildings, where it would be discarded and replaced if ever used. The hose should be

should be hydrostatically tested annually, thoroughly inspected, dried and returned to service (Fig. 43–34).

Monitor nozzles. Permanent monitor nozzles are frequently used to protect pulpwood storage piles at paper mills, in lumber yards, in stock yards, and in railway car storage yards. Nozzles are often elevated to clear obstructions so that the operator can stand on a shielded platform and direct a high-pressure stream of water over a wide area. Such systems are especially useful in large congested areas where it is impractical to lay hose lines in an emergency.

Hose nozzles. Effective streams for fire fighting are controlled by the size and type of nozzle. The nozzle, in turn, must be supplied with the correct quantity of water at the discharge pressure for which it is designed. Nozzles are designed for solid streams, spray streams (frequently referred to as fog), or combination streams. And nozzles for special extinguishing agents, such as foam and dry chemical, are also available. Solid stream nozzles are designated by the diameter of the nozzle tip; whereas spray nozzles are designated according to the amount of water they discharge at 100 psi nozzle pressure.

Spray nozzles are widely used in both public and private fire protection and make the application of water more effective under many conditions. They are of three general types:

1. Open nozzles of fixed (nonadjustable) spray pattern, usually attached to shut-off valves. Some nonadjustable nozzles can be equipped with an applicator (a long pipe extension, curved at the end and fitted with a fixed spray nozzle) for fighting fires where extended reach is necessary.

2. Adjustable nozzles which provide variable discharges and patterns from shut-off to solid stream and from narrow- to wide-angle spray.

3. Combination nozzles in which a solid stream, a fixed or adjustable spray, and shut-off are selected usually by a two- or three-way control valve.

Maintenance of fire hose

Hose for outdoor use. Inspect and test woven-jacket lined hose periodically to make sure that it will be in good condition when emergencies arise. Water should be run through the hose at least twice a year. Fire hose should be reserved for fighting fires; if hose is needed for other uses, a separate supply should be provided.

Mildew may attack untreated hose fabric containing cotton if the hose is stored in a damp location or if it is not thoroughly dried after wetting. Fire hose is available with chemically treated fabric. The treatment is primarily for protection against mildew and rot. Treated jackets also absorb less water and therefore dry more quickly. The resistance to dampness and mildew is not one hundred percent effective even when the treatment is new, and it deteriorates with age.

Jackets made entirely with synthetic warp and filler are impervious to mildew and rot. Drying of such hose is not imperative. Washing after use and before storing is, however, recommended.

It is as important to carefully dry hose with jackets made from a combination of cotton and synthetic yarns as it is to dry hose with all-cotton jackets. Drying of both jacket constructions takes about the same time.

For plant yards containing rough surfaces that will cause heavy wear or where working pressures are above 150 psi, double-jacket lined hose is advised. If hose may be subjected to acids, acid gases, or other corrosive materials, as in chemical plants, rubber-covered woven-jacket lined hose is advised. For such conditions, it can also be obtained with a neoprene-impregnated all-synthetic jacket.

Hose for indoor use. Maintain and test unlined linen hose and woven-jacket lined hose as follows:

1. Reserve the hose for fire fighting.

2. Keep hose valves tight; leakage will rot linen hose.

3. Examine hose visually each year for mildew, rot, damage by chemicals, vermin, and abrasions. If the hose is in doubtful condition, give it a hydrostatic pressure test.

4. Give hose a pressure test after the 5th and 8th year of service, and repeat the test every second year afterwards.

5. Keep hose clean. Wash woven-jacket lined hose with laundry soap if necessary.

6. Dry hose jackets thoroughly after use and keep them dry.

The local fire department will often pressure test hose for a company, if the company will take care of delivery and pickup.

Special Systems and Agents

Special hazards may require methods of extinguishment or control other than water. Each of the several methods available offers certain advantages and limitations which must be considered in making a selection. These systems are usually installed to supplement, rather than replace the automatic sprinkler system. They should be engineered to fit the circumstances of the particular hazard. It is common practice to install them in such a manner that their operation will shut down other processes (such as pumps and conveyors) which might intensify the fire. The following special agents and methods are currently in use. Specific details are found in the following NFPA Standards and the NFPA *Handbook of Fire Protection*.

Carbon dioxide—NFPA No. 12

Foam—NFPA No. 11 and No. 16

Dry chemical—NFPA No. 17

Water spray and sprinkler—NFPA Nos. 15, 13, and 231C

Steam smothering—NFPA No. 86A (Ovens and Furnaces)

Halon 1301—NFPA No. 12A

Halon 1211—NFPA No. 12B

Wetting agent—NFPA No. 18

Inerting—NFPA No. 69

Explosion venting—NFPA No. 68

Foam systems

Fixed foam extinguishing apparatus may be either automatic or manual. It may consist merely of one or more portable foam extinguishers suspended in such a way that flame or heat releases a cord or fusible link so that the extinguisher tips over for automatic operation. More elaborate systems consist of fixed piping through which the foam-producing solutions move to a number of deflector or flared outlets. Such systems depend upon either manual or automatic operation by heat-sensitive actuators and vary in discharge rate from 15 to 4000 gpm.

Foam systems are often used to protect dip tanks, oil and paint storage rooms, and asphalt coating tanks. Foam systems also have been developed to put out tank fires by subsurface injection of foam.

Fixed discharge outlets or chambers can be permanently attached to oil storage tanks so that the foam will blanket the surface of the burning liquid. Foam-water sprinkler systems have been developed to protect areas where flammable liquids may be spilled. (See NFPA Standard No. 11, *Foam Extinguishing Systems*. Also *Foam-Water Sprinkler and Spray Systems*, Standard No. 16.)

There are two types of foam, chemical and mechanical. Chemical foam is formed by a chemical reaction in which masses of bubbles of carbon dioxide gas and a foaming agent produce an expanded froth. Mechanical foam consists of bubbles of air produced when air and water are mechanically agitated with a foam-making agent.

Chemical foam. There are four general types of equipment for producing chemical foam: self-contained units, closed-type generators, hopper-type generators, and stored-solution systems. In the self-contained unit, two solutions that produce foam on contact are stored independently in a single vessel and caused to mix either manually or automatically. The amount of foam produced is determined by the quantity of foam-producing materials within the vessel.

In the closed-type generator system, chemical foam powder is stored in large hoppers permanently fixed to foam generators that mix the powder with water and then pump or use the water pressure to force the foam to special outlets. This kind of installation may be either the one- or two-powder type and is mainly used on flammable liquid storage tank farms, where a single foam-producing installation can service a number of tanks.

Hopper-type generators can be either permanent installations or portable. The advantage of portable generators is that they are not limited in foam production by a fixed storage of foam powder, but can be continuously refilled. Generators employing either one or two foam powders to produce foam are available.

On large flammable liquid storage tanks, permanent lines affixed to foam chambers at the top of the tank are sometimes provided. Portable foam towers are used for foam application to burning oil storage tanks. Generator and water connections can be made at a calculated safe distance from the tank at the time of fire.

Stored-solution systems have large permanently installed tanks that contain two foam-producing solutions stored separately. At the time of a fire either duplex or twin pumps force the solutions to outlets where they mix and discharge foam. Foam production is lim-

Fɪɢ. 43–35.—Proportioner permits operator to vary percentage of foam-producing concentrate.

ited by the size of the solution storage facilities.

Mechanical or air foam is produced by the mechanical action of adding the proper amounts of a liquid concentrate into a water stream via a proportioner, and then introducing and mixing air into the water-concentrate solution. Mechanical foam is gradually replacing chemical foam equipment. Foam forming concentrates are classified into "protein-type" and "nonprotein or synthetic types."

Protein-type concentrates are available in two strengths: one used with the proportionator at a 3 percent by volume ratio, and the other at a 6 percent by volume ratio. It is important to note that both the 3 and 6 percent foam liquids are satisfactory only on hydrocarbons, such as benzene, toluene, xylene, gasoline, naphtha, and kerosene. Water-miscible solvents, such as esters, ethers, alcohols, and ketones, destroy the regular 3 and 6 percent foam liquids as rapidly as they are applied. Therefore, an alcohol-resistant foam

should be used in combating fires in these latter materials. Where such uses are contemplated, a sample of the material should be submitted for evaluation by a supplier of alcohol-resistant foam. Because the stabilizer in this foam is chemically different from other stabilizers, the system in which the foam is to be used must be carefully designed with due regard to its limitations.

Nonprotein or synthetic concentrates containing compounds of the synthetic-detergent type are used in concentrations of 2 to 6 percent by volume. A proportioner unit and siphon hose can be installed on a standard fire hose to draw the concentrate from 55-gal drums, 5-gal pails, or smaller portable containers. Some proportioner units can vary the percentage of concentrate from a plain water stream to a foam stream (Fig. 43–35). A plain water stream can also be quickly obtained by removing the siphon tube from the concentrate.

There are four basic methods of producing mechanical (air) foam: nozzle aspirating systems, in-line foam pump systems, in-line aspirating systems, and in-line compressed air systems. The names of these systems indicate where and how air is injected into the water-concentrate solution to produce mechanical foam. Each one of these systems uses a proportioner to introduce the foam concentrate into the water stream. There are a number of types of proportioners that can either be located at the main pump or between the main pump and the foam maker.

Foam-water sprinkler and spray systems use mechanical (air) foam equipment with a deluge-type sprinkler system. These systems are generally used to protect flammable liquid hazard areas and can discharge foam or plain water selectively through aspirating devices (foam-water sprinklers or foam-water spray nozzles).

Wet-water foam is generated by using an aspirating-type nozzle or through the injection of air under pressure into water containing a wetting agent, a chemical compound that reduces the surface tension of the water and increases its penetrating, spreading, and emulsifying properties. Wetting agents that are listed by Underwriters' Laboratories, Inc., are siphoned through a proportioner at rates not exceeding 2 percent by volume.

The reflective white opaque surface formed by the air bubbles and wet-water particles makes this agent a good medium for protection from exposure fires. The cellular structure of the foam also retards heat conduction, thus affording insulation. As the foam blanket absorbs the heat and breaks down, the wet water released from the air bubbles carries the heat away from the protected surface. Mixing wetting agents with other wetting agents or with mechanical or chemical foam is *not recommended* since it may neutralize the effect of the agents and thus destroy the fire-fighting properties.

High expansion foam, using wetting agents in ratios as high as 1000 to 1, has been developed. These foams have a water content ranging from 0.2 to 1.0 oz/cu ft, and are effective in fighting fires in inaccessible places, such as coal mines and building basements.

High expansion foams can also be used in total flooding systems designed to fill enclosed spaces, such as rooms or buildings. The systems are suitable for hazards involving flammable liquids and ordinary combustibles, if properly designed for the specific hazard and area.

Wet water

The wet water principle can also be used without generating foam. In liquid form, wet water has the same general extinguishing properties as plain water. However, its cooling ability is increased, and there is a greater penetration of porous surfaces because of its reduced surface tension.

Carbon dioxide extinguishing systems

Fixed (local or flood-type) CO_2 systems are often installed for the protection of rooms that contain electrical equipment, flammable liquid or gas processes, dry cleaning machinery, and other exposures where fire can be extinguished by diluting the oxygen content of the air or where water must not be used because of electrical hazard or the nature of the product. Details on installation are given in NFPA Standard No. 12, *Carbon Dioxide Extinguishing Systems.*

In the high-pressure system, CO_2 is stored in compressed gas cylinders at normal temperatures. It is released by manual operation or by automatic devices through nozzles close to the expected source of fire. Carbon dioxide has definite advantages, unlike water or other chemical extinguishing materials, in that it generally does not damage stock or equipment. (See NFPA Standard No. 12, *Carbon Dioxide Extinguishing Systems.*)

In the low-pressure fixed installation, carbon dioxide is stored in an insulated pressure vessel and maintained at 0 F by mechanical refrigeration. At this temperature the pressure is approximately 300 psi. At such low pressure, 500 pounds to over 125 tons of CO_2 can be stored more economically than at higher pressure.

Relief valves are provided in case of refrigeration failure. Liquid CO_2 is delivered through pipelines to nozzles that may have delivery capacities as high as 2500 lb/min. As in the high-pressure CO_2 system, release may be either manual or automatic. Fixed local application systems provide for extin-

Fig. 43–36.—This 16- by 24-ft flammable liquids storage building is protected by a total-flooding type dry-chemical system. On actuation of the system by a heat-sensitive device, nitrogen is discharged into the 150-lb storage container, and dry chemical is thereby expelled to eight nozzles beneath the roof.

Courtesy National Fire Protection Assn.

guishment of the fire at its source. Total flood systems may be used in small buildings, compartments, or rooms where wall or other openings can be automatically shut when the gas is released. Warning alarms must be provided to alert persons working in areas protected by this type of system. Sufficient time must be allowed to evacuate the area.

In confined locations it is important to ventilate the area thoroughly after a fire is extinguished, because available oxygen may not be sufficient to sustain life.

Handhose-line systems combine fixed tanks with hose reel attachments which permit a limited range of fire fighting. Range is predicated on the length of the hose plus the effective range of CO_2.

Dry-chemical systems

Dry-chemical piped systems have been developed for situations where quick extinguishment is needed, either in a confined area or for localized application, and where reignition is unlikely. They are adaptable to flammable liquid and electrical hazards and are available for either manual or automatic operation, activated at the system or by remote control (Fig. 43–36). A rate-of-rise heat-actuated device or an electrical release controls the automatic operation.

Installations can provide for simultaneous closing of fire doors, operating valves, windows, and ventilation ducts, as well as shutting off fans and machinery and actuating alarms. Piped systems providing either local application or total flooding are available.

The dry-chemical agent is neither toxic nor a conductor of electricity, nor does it freeze. In piped systems and most hand-hoseline systems it is stored in a tank that is pressurized by an inert gas cylinder when the controls are actuated. In some hand-hoseline systems the agent is stored in a pressurized container. Extinguishing action results mainly from the interruption of the chemical flame chain reaction by the dry chemical agent (see earlier in this chapter for details of this mechanism).

Fixed storage tanks and pressurized cylinders, similar to those used in piped systems, are available for monitor operation and for mounting on vehicles.

See NFPA Standard No. 17, *Dry Chemical Extinguishing Systems.*

Halon systems

Automatic Halon 1301 systems are being used in areas such as electronic data processing centers and record storage rooms where water-based systems are not desirable. In low concentrations (below 7 percent), the agent is nontoxic. It is a colorless, odorless, nonconductive gas which extinguishes fires by inhibiting the chemical reaction of fuel and oxygen. Refer to NFPA Standards 12A and 12B.

Steam systems

Automatic or manually controlled steam jet systems have been used to smother fires in closed containers or in small rooms, such as heaters, drying kilns, smoke ovens, asphalt mixing tanks, and dry cleaning tumbler dryers. However, the system is practicable only where a large supply of steam is continuously available, which, unfortunately, most plants do not have.

Steam has not been found effective on deep-seated fires that may form glowing coals or in enclosures where the normal operating temperature is not considerably higher than air temperatures.

The possibility of a personal injury hazard from burns should be considered in any steam extinguishing installation. See NFPA Standard 86A, *Ovens and Furnaces, Design, Location & Equipment.*

Inerting systems

Inert gas systems may prevent fires and explosions by replacing the oxygen in the air with an inert gas, such as carbon dioxide, nitrogen, flue gas, or other noncombustible gas, until it reaches a level (or percentage) where combustion will not take place. (See the discussion earlier in this chapter and in NFPA Standard No. 69, *Explosion Prevention Systems.*)

To be effective, the inert gas must reduce the amount of oxygen in the system from the normal 21 percent to between 2 and 16 percent, depending upon the type of hazard involved and the type of inerting gas. For instance, an inert gas such as CO_2 must reduce the oxygen in air to 6 percent to prevent fire or explosion of carbon monoxide (CO), 14 percent for gasoline, and 15 percent for cotton dust.

Inert gases, such as flue gas from power plant stacks, have been used extensively to prevent explosion. Flame type producers operating on either fuel oil or gas, which yield products of combustion with a high percentage of CO_2 and nitrogen, are often used in fixed installations where large quantities of the inert gas are required, as in the purging of storage tanks, in pipelines, and in manufacturing processes with high explosion hazard.

Use of an inert gas in a confined space can result in an oxygen deficient atmosphere; therefore, before anyone is allowed to enter a confined space into which an inert gas has been introduced, the space should be thoroughly ventilated and tested with instruments which indicate if the enclosed atmosphere will support life. Otherwise, the individual should wear approved respiratory protective equipment and harness with lifeline for such entry. Furthermore, a watcher similarly equipped should stand by to observe the man in the confined space and rescue him in an emergency.

There are three ways of applying inert gas to ensure the formation of a noncombustible atmosphere within an enclosed tank or space. The methods are fixed volume, fixed rate, and variable rate.

The fixed volume method introduces the inerting gas into the equipment chamber by either reducing the pressure within the chamber and allowing the inert gas to flow in until the pressure is equalized, or by pressurizing the chamber with the inert gas and then letting off the overpressure to the atmosphere after mixing has taken place. Several pressurizing cycles may be necessary to reduce the oxygen content sufficiently.

The fixed rate method adds a continuous supply of inerting gas in amounts sufficient to accommodate peak requirements. The quantity required is based on the maximum in-breathing rate which may result by sudden cooling, such as that caused by rain, or a sudden drop in temperature, plus maximum product withdrawal. Although this method is relatively simple, it has the disadvantage of wasting inert gas and increasing the rate of the evaporation of the product.

The variable rate method supplies the inert gas to the system on a demand basis. The inerting gas is continuously released to a low-volume supply line to compensate for minor

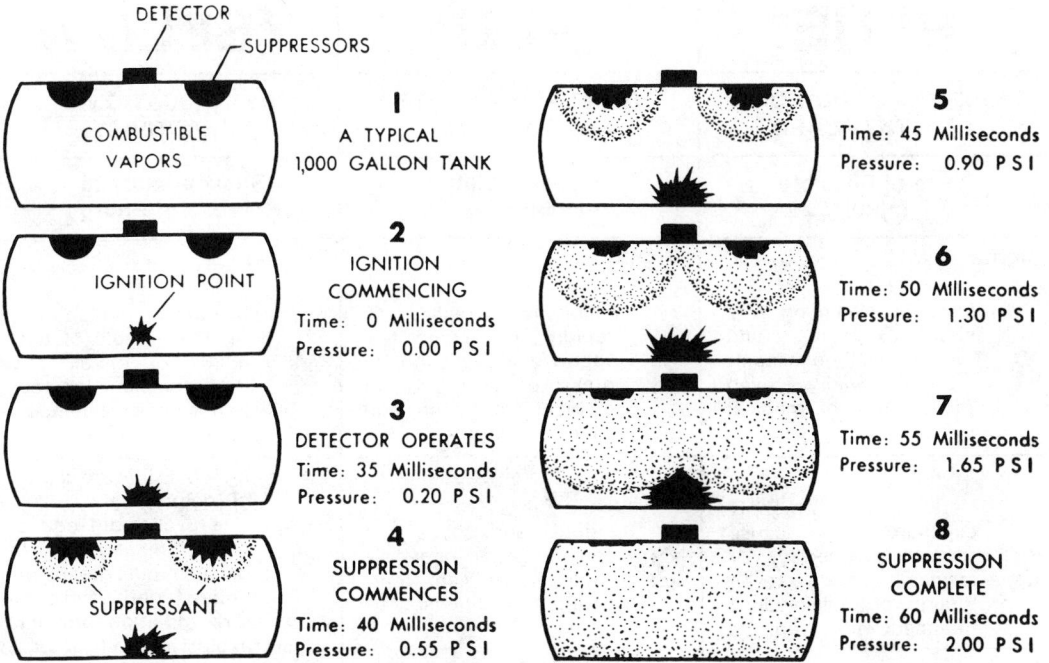

DETECTOR ── SUPPRESSORS	
COMBUSTIBLE VAPORS	**1** A TYPICAL 1,000 GALLON TANK
IGNITION POINT	**2** IGNITION COMMENCING Time: 0 Milliseconds Pressure: 0.00 P S I
	3 DETECTOR OPERATES Time: 35 Milliseconds Pressure: 0.20 P S I
SUPPRESSANT	**4** SUPPRESSION COMMENCES Time: 40 Milliseconds Pressure: 0.55 P S I
	5 Time: 45 Milliseconds Pressure: 0.90 P S I
	6 Time: 50 Milliseconds Pressure: 1.30 P S I
	7 Time: 55 Milliseconds Pressure: 1.65 P S I
	8 SUPPRESSION COMPLETE Time: 60 Milliseconds Pressure: 2.00 P S I

FIG. 43–37.—Schematic diagram of suppression of an explosion in a large cylindrical tank.

Courtesy National Fire Protection Association.

pressure changes. When rapid changes take place, such as caused by product withdrawal, a means is provided which opens a large supply line until the pressure equalizes. This type of system is extremely efficient and has the advantage of reducing product vapor losses by maintaining a slight positive pressure within the chamber.

Preventing explosions

There are a number of methods for preventing the development of explosive mixtures. They are the first line of defense against explosions.

Equipment for handling and storing of flammable gases should be constructed, inspected, and maintained so that the danger of leakage and explosive mixture formation is reduced to a minimum. Equipment design should comply with existing codes and other regulatory ordinances. Equipment should be inspected at regular intervals by qualified individuals either from the regular plant staff or from an outside source.

Ventilation will prevent excessive accumulations of gases and vapors under certain conditions. The method of ventilation neces-

sarily varies with the nature of the gas or vapor to be removed and depends upon whether it is heavier or lighter than air at the point of removal. Inasmuch as heating or cooling of the gas or vapor may change its density, the ventilation or exhaust system, or both, should be designed for operating conditions and should not be based on the published density figures. (See NFPA Standard No. 91, *Blower and Exhaust Systems, Dust, Stock and Vapor Removal.*)

Natural draft ventilation may be through openings near the floor or near the ceiling or both. However, the best method is a positive local exhaust system, using explosion-proof electrical equipment and taking suction as close to the source of a vapor or gas as possible. A motor other than explosion-proof driving an exhaust fan, may be used if it is properly installed outside the duct work and outside the hazardous area.

In general, industrial gases, such as acetylene, carbon monoxide, hydrogen, and natural gas, are lighter than air. The vapors of flammable liquids are generally heavier than air. Examples are alcohol, naphtha, gasoline, benzene, kerosene, amyl acetate, and carbon di-

BLUE	RED	YELLOW
IDENTIFICATION OF HEALTH HAZARD	**IDENTIFICATION OF FLAMMABILITY**	**IDENTIFICATION OF REACTIVITY**
Type of Possible Injury	**Susceptibility to Burning**	**Susceptibility to Release of Energy**
Signal	Signal	Signal
4 Materials which on very short exposure could cause death or major residual injury even though prompt medical treatment were given.	**4** Materials which will rapidly or completely vaporize at atmospheric pressure and normal ambient temperature, and which will burn.	**4** Materials which are readily capable of detonation or of explosive decomposition or reaction at normal temperatures and pressures.
3 Materials which on short exposure could cause serious temporary or residual injury even though prompt medical treatment were given.	**3** Liquids and solids that can be ignited under almost all ambient temperature conditions.	**3** Materials that are capable of detonation or explosive reaction but require a strong initiating source, or that must be heated under confinement before initiation, or react explosively with water.
2 Materials which on intense or continued exposure could cause temporary incapacitation or possible residual injury unless prompt medical treatment is given.	**2** Materials that must be moderately heated or exposed to relatively high ambient temperatures before ignition can occur.	**2** Materials that are normally unstable and readily undergo violent chemical changes but do not detonate; also materials that may react with water violently, or that may form potentially explosive mixtures with water.
1 Materials which on exposure would cause irritation but only minor residual injury even if no treatment is given.	**1** Materials that must be preheated before ignition can occur.	**1** Materials that are normally stable, but that can become unstable at elevated temperatures and pressures, or that may react with water with some release of energy, but *not* violently.
0 Materials which on exposure under fire conditions would offer no hazard beyond that of ordinary combustibles.	**0** Materials that will not burn.	**0** Materials that are normally stable even under fire explosive conditions, and that are not reactive with water.

```
        FIRE
HEALTH      SAFETY
```

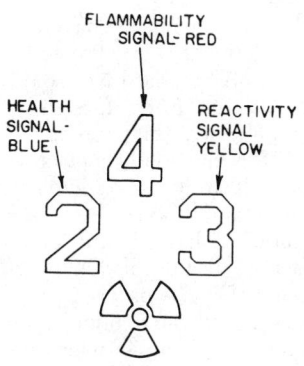

FLAMMABILITY
SIGNAL- RED

HEALTH
SIGNAL-
BLUE

REACTIVITY
SIGNAL
YELLOW

Fig. 1. For Use Where White Background is Not Necessary.

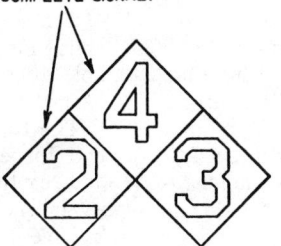

WHITE ADHESIVE-BACKED PLASTIC
BACKGROUND PIECES - ONE
NEEDED FOR EACH NUMERAL,
THREE NEEDED FOR EACH
COMPLETE SIGNAL.

Fig. 2. For Use Where White Background is Used With Numerals Made From Adhesive-Backed Plastic

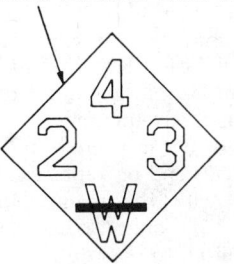

WHITE PAINTED BACKGROUND, OR,
WHITE PAPER OR CARD STOCK

Fig. 3. For Use Where White Background is Used With Painted Numerals, or, For Use When Signal is in the Form of Sign or Placard

ARRANGEMENT AND ORDER OF SIGNALS — OPTIONAL FORM OF APPLICATION

Distance at Which Signals Must be Legible	Size of Signals Required
50 feet	1″
75 feet	2″
100 feet	3″
200 feet	4″
300 feet	6″

NOTE:
    This shows the correct arrangement and order of signals used for identification of materials by hazard

FIG. 43–39.—NFPA hazard signal arrangement for in-plant use only. See NFPA Standard No. 704M for dimensional and other details. Meaning of number system is explained in Fig. 43–38, at left.

Courtesy National Fire Protection Association.

sulfide. This is logical because gas or vapor density is proportional to molecular weight, and the compounds which are normally liquids at room temperature have higher molecular weights than those which are gases.

Inert gases are sometimes used to prevent explosions of gases, vapors, and dusts. Their function is to keep the concentration of oxygen below the point at which it will support combustion. High fire hazard commercial processes, such as lacquer manufacturing, are sometimes flooded with carbon dioxide, nitrogen, or other inert gases.

Inert gases are sometimes used as a means of transferring flammable liquids, for inerting the atmosphere of storage tanks of volatile flammable liquids, and for purging gas holders or pipe lines. Inert gases, since they dilute the oxygen in the air, also have widespread use for standby emergency fire extinguishing.

Unburned gases or flammable vapors in the combustion chambers of unit heaters, boilers, furnaces, enameling, drying, and bakery ovens may form an explosive mixture with air. Interruption of the gas feed pressure or extinguishment of the flame or pilot light may cause accumulation of unburned fuel.

A number of safety devices have been developed to overcome this hazard. Most of them are automatic in operation and provide for ventilation and for the control and interlocking of gas and air supplies, which safeguard against explosive mixtures.

Means should be provided on gas- and oil-fired equipment to ventilate or purge the combustion zone thoroughly in case of flame failure. Men in charge of firing these devices should know the ventilating or purging time required in the event of flame failure. In the event of flame failure, the program con-

FIG. 43–38.—Degree of hazard can be quickly identified with this system. Also see pages 1201–1207.

Adopted from National Fire Protection Association.

1379

troller should take over; this forces the operator to go through the interlocked start-up procedure which includes a timed preignition purge cycle.

It may be advisable to check the atmospheres inside large industrial equipment with a combustible gas indicator before relighting.

Gas-fired equipment, including its controls, should be inspected and tested at regular intervals and be kept in good repair in accordance with the manufacturer's recommendations and only trained personnel should be permitted to operate it.

Gas valves should be inspected frequently for leaks. If gas is present, ventilation is needed immediately and the condition must be corrected before the equipment is used. The recommendations of the manufacturer of the equipment and of the public utility supplying the gas should be followed.

Suppressing explosions

Under certain conditions an explosion suppression system can be used to reduce the destructive pressure of an explosion. These systems are designed to detect an explosion as it is starting and actuate devices that suppress, vent or take other action to prevent the full explosive force.

These systems require split-second timing. The mechanism for dispersal of the suppression agent must operate at extremely high speed to fill the enclosure completely within milliseconds after detection (Fig. 43–37). The suppression agent must be dispersed from the suppressors in the form of a very fine mist at a rapid speed, normally through the use of a small secondary explosive force. The suppression agent is normally a noncombustible liquid compatible with the combustion process to be suppressed.

Identification of Hazardous Materials

Fires and other emergency situations often involve chemicals that have varying degrees of flammability, toxicity, and reactivity (or stability). Information on these relative hazards must be readily available to those confronted with such emergencies if life safety, fire prevention, and effective fire control are to be achieved.

A system for the quick identification of hazardous properties of chemicals has been developed by the National Fire Protection Association. (See NFPA Standard No. 704M, *Identification Systems for Fire Hazards of Materials*.) For uniformity, this system recommends the use of a diamond-shape symbol and numerals indicating the degree of hazard (see Figs. 43–38 and –39).

The three categories of hazards are identified for each material: health, flammability, and reactivity (stability). The order of severity in each category is indicated numerically by five divisions ranging from 4, which indicates a severe hazard, to 0, which indicates that no special hazard is involved. Colors may be used to better identify each category: *blue* is for health, *red* for flammability, and *yellow* for reactivity.

At the bottom of the three-part diagram is an open space. This may be used to indicate additional information such as radioactivity hazards, proper fire extinguishing agent, skin hazard, pressurized containers, protective equipment required, or *unusual* reactivity with water. The recommended signal to indicate this unusual reactivity with water and to alert the fire-fighting personnel *not* to use water is the letter "W" with a long line through the center, as shown in Fig. 43–39.

Fire Prevention Activities

To make sure that equipment is safeguarded properly, the plant engineering, process, and purchasing departments must both cooperate and communicate. They should consider fire prevention and fire fighting on all new layouts, new operations, new methods, and new construction.

Inspections

The best time to stop a fire is before it starts. Even though buildings are properly designed and provided with protective devices and construction elements intended to render fire-safety features, only *regular periodic inspections* can assure their full value. In addition to inspections made by insurance companies and by fire protection bureaus of fire departments, every industrial plant should include periodic self-inspections in its fire-safety program.

In many plants the responsibility for lo-

WEEKLY INSPECTION OF FIRE EQUIPMENT

FORT DEARBORN MANUFACTURING COMPANY
1728 North Rush Street, Chicago, Illinois

Inspector's Signature _A. M. Henning_ Date _4/28_

Note: Check each item personally and report on each question herein. Submit original copy of report to the General Manager and a carbon copy to the Safety Inspection Committee.

1. OUTSIDE SPRINKLER VALVES
2. INSIDE SPRINKLER VALVES

No.	Location	Circle condition of valve as shown	No.	Location	Circle condition of valve as shown
1	_Parking lot_	Open Shut (Sealed)	1	_Warehouse_	Open Shut (Sealed)
2	_Yard - East_	Open Shut (Sealed)	2	_Mill_	Open Shut (Sealed)
3	_Yard - North_	Open Shut (Sealed)	3	_Office_	Open (Shut) Sealed

3. REASON FOR ANY CLOSED VALVES? Identify location _Office Valve shut for repair_

4. WERE CLOSED VALVES REOPENED WIDE AND SEALED? _Yes_

5. WAS A GOOD FULL FLOW DRAIN TEST MADE AFTER EACH VALVE WAS REOPENED? _Yes_

6. DRY PIPE VALVES

No. 1   Valve room heated _Yes_ air pressure _100#_ alarms operative _Yes_
No. 2   Valve room heated _Yes_ air pressure _100#_ alarms operative _Yes_
No. 3   Valve room heated _Yes_ air pressure _100#_ alarms operative _Yes_

Recommendations _None_

7. FIRE PAILS, FIRE EXTINGUISHERS, FIRE HOSE IN GOOD CONDITION? _Yes_

Extinguishers recharged on schedule? _Yes_ Easily accessible? _Yes_

Recommendations _None_

8. FIRE DOORS AND EXITS CLEAR? _No_ In good order? _Yes_

Recommendations _Fire door between Mill & whse. blocked by Stock - Keep Clear_

9. AUTOMATIC SPRINKLERS - Any heads missing, corroded, painted, loaded or obstructed by high piled stock? _No_ Any part of the system exposed to freezing temperatures? _No_ Extra heads on hand? _Yes_

Recommendations _None_

10. SPRAY BOOTHS AND FLAMMABLE LIQUID STORAGE ROOM - Any excessive accumulation of paint in booth? _Some_ Floor drains clear? _Yes_ Vapor proof lights, wiring and switches in good condition? _Yes_

Recommendations _Clean booths weekly instead of bi-weekly_

(Continue Recommendations on other side)

FIG. 43–40.—This sample inspection form for fire equipment is similar to forms suggested by Factory Mutual Engineering Corporation, American Insurance Association, and Factory Insurance Association. The form can be revised to take care of specific hazards, or split up so special inspectors or subcommittees can check specific items.

cating and reporting fire hazards is entrusted to the safety committee or to one of its special subcommittees. The function of these committees is to inspect for common fire causes, such as poor housekeeping, improper storage of flammable materials, smoking violations, and excessive accumulations of dust or flammable material.

Regardless of the size of the plant, fire hazards should be detected and eliminated through frequent inspection, and fire fighting equipment should be checked regularly to be sure that it is ready for any emergency.

The inspector, fire chief, or other individual in charge of fire prevention and protection should have a complete list of all the items that should be inspected at regular intervals. If fire brigade members (see section Fire Prevention and Protection in Chapter 18, "Planning for Emergencies") are assigned to this duty, they should receive special instructions. Fire equipment inspection should cover the following items:

1. Control valves on fire protection system

2. Hydrants

3. Fire pump

4. Hose house equipment

5. Sprinkler system water supplies

6. Special types of protection (carbon dioxide, foam, or other automatic systems)

7. Portable fire appliances

8. Small hose

9. Fire doors

10. Special hazards.

In addition to fire equipment, electrical equipment, machinery and processing equipment, housekeeping conditions, and other fire causes should be checked at regular intervals.

In some high-hazard occupancies, daily inspections are required; otherwise, thorough weekly or monthly inspections are satisfactory, especially if the plant safety committee inspects for simple fire causes.

Special inspections should be conducted during and following alterations in the plant. A complete check, early enough in the season for replacement or repair, should be made of equipment which will be exposed to freezing temperatures.

For greatest benefit a written record is essential (see Fig. 43–40). It should be a form that is specifically drawn up to fit the conditions of the individual plant in order to facilitate the inspection and complete enough to help make sure that no part of the system is overlooked. It will also assist the inspector in making comments and bringing recommendations to the attention of those responsible. An excellent guide to conducting inspections is the pocket-size NFPA *Inspection Manual*.

Hot work permits

In an effort to establish some control over operations using open flames or producing sparks, many industrial firms have instituted hot work permit programs, which require that authorization be secured before equipment capable of igniting combustible materials is used outside areas normally specified. The first step in inaugurating a hot work permit program is the development of a policy statement by management, such as that given in Fig. 43–41. The type and extent of the program will depend upon the size of the plant or facility, the complexity of the operations, and the degree of hazard present at the work site and in surrounding areas.

Salient features of the program are to:

1. Inspect area where work is to be done

2. Establish fire watches

3. Provide fire extinguishing equipment

4. Communicate with and coordinate all departments concerned

5. Isolate combustibles from sources of ignition

6. Limit unauthorized use of flame- or spark-producing equipment

A hot work permit form or tag is generally used to administer the program. Although standard forms are available through insurance companies, many plants have developed special forms that specifically relate to their individual operation. Detailed information on hot work permit programs is contained in Industrial Data Sheet 522, "Hot Work Permits," available from the National Safety Council.

SUBJECT: HOT WORK PERMIT PROCEDURES

TO: ALL CONCERNED

Steps will be taken immediately to put into practice a permit tag system which should provide better protection against fire from welding and other hot work in hazardous areas. This program, as outlined below, is intended to be a practical one. It must be realized by everyone that for it to be effective, the wholehearted cooperation of all concerned must be secured.

1. After an inspection of the entire plant has been made by the fire chief and after a discussion with the various department heads, areas throughout the plant will be designated as hazardous for *any type of welding, burning, spark, open flame, or hot work.* Such areas will be prominently marked, and before hot work is done within any such area permit tags must be secured in order to help ensure that the area will be as free as possible from fire hazards and that proper precautions will have been taken.

2. Tags have been prepared on which pertinent information must be filled out by the parties concerned. Each employee who may do hot work will be given a supply of tags which he will keep on his person or equipment. When he is sent to perform work in a hazardous area, he will first have his immediate foreman check the area with him to determine if necessary precautions to prevent fire have been taken.

3. It will be the maintenance foreman's responsibility to notify the foreman in the department in which the work is to be done, and together they will sign the permit for the employee to do welding or other hot work.

4. The fire chief will then check the area for fire-safe working conditions, see that standby fire extinguishers are present, and assign a fire watcher when necessary. When he is satisfied with the precautionary measures, he will sign the permit tag and return it to the employee who is to do hot work.

5. The signed permit card will be kept by the employee doing the work until the job is completed, at which time he himself will check the area for fire.

6. The area will subsequently be checked by the maintenance foreman who will see that any extinguisher that may have been used is designated for recharging.

7. The completed tag will be turned in at the end of the day to the maintenance department which will collect and forward all tags to the fire chief for filing and record purposes.

Fig. 43–41.—Example of policy statement that can be issued by a company inaugurating a hot work permit program in order to instruct employees.

Flameproofing wood and fabrics

Fire-retardant treatments. Wood cannot be made fire resistive, but it can be treated to *retard* both the rate of flame spread over its surface and the rate it contributes fuel to the fire. There are two methods for treating wood: impregnation and use of fire-retardant coatings. Water-soluble treatments are not suitable for exterior use.

Fire-retardant coatings reduce the hazard from small fires in existing structures. Only those makes listed by Underwriters' Laboratories, Inc., should be used. Effective impregnation can usually be done only by companies equipped to do this work.

Fire-retardant coatings on wood should be applied in accordance with the application specifications which produce the desired degree of reduction of flame spread. They may have to be renewed at frequent intervals, especially if the wood is exposed to rain or excessive moisture. Ready-mixed fire-retardant paint is commercially available, but especially mixed preparations can also be made. Regardless of the treating method employed, engineering data can be obtained from the supplier of the treated wood or the manufacturer of the treating chemicals.

Flame-retardant fabrics. Fabrics cannot be made noncombustible, nor even resistant to charring or to decomposition, but chemical treatment will reduce their flammability. (See NFPA Standard No. 701, *Standard Methods of Fire Tests for Flame-Resistant Textiles and Films.*) Some treatments merely inhibit the rapid spread of flame, whereas others prevent flames and depress dangerous afterglow.

Where the only anticipated exposure is to small sparks, small flames, or temperatures up to 400 to 500 F, flame-retardant canvas is often preferred to asbestos or chrome-tanned leather particularly where flexibility, durability, strength, and resistance to abrasion are required. Asbestos or chrome leather is required for more serious exposures, such as heavy welding, fire fighting, or foundry work.

There are two principal types of fire-retardant treatments for fabrics: water-soluble and weather-resistant or water-insoluble. Commercially available or homemade water-soluble treatments are effective, but must be reapplied after each laundering. Several commercially treated fabrics are listed by Underwriters' Laboratories, Inc. Samples of treated materials should be tested periodically after each application to assure satisfactory performance.

Water-insoluble chemicals, such as chlorinated paraffin, chlorinated resin, or certain inorganic chemicals, form the basis of weather-resistant, fire-resistant finishes for canvas, duck, rayon, and other fabrics used for awnings, tenting, tarpaulins, truck covers, and welding curtains. Satisfactory weather-resistant treatments require special processing of the fabric which can be done best by the finishing mills or chemical companies equipped to do this work.

Fire and emergency drills

Regardless of what method is used to alert employees, they should thoroughly understand what the signal means and how to respond in the safest possible manner. Training employees to leave their workplace promptly on proper signal, and to evacuate a building speedily but without confusion, is largely accomplished through fire or emergency drills. Exit drills must be carefully planned and carried out in a serious manner under rigid discipline. Periodic instruction sheets should be distributed to all employees, and an exit drill organization formed from the supervisory staff.

Fire drills at frequent intervals also demonstrate management's concern and sincere interest in all fire prevention and fire protection activities. The drills should serve as a reminder to employees and supervisors that rules, such as those on "No Smoking," and other fire prevention practices are important.

Emergency exit drills also serve as a valuable check on the adequacy and condition of the exits and alarm system. Any deficiency should be immediately and permanently corrected. Careful plans must be made to eliminate the possibility of panic in the event of an emergency, regardless of cause, and to guarantee the smooth functioning of the emergency plan. Reference should be made to the NFPA Standard No. 7, *Management Control of Fire Emergencies.* Also see Chapter 18, "Planning for Emergencies," earlier in this Manual.

Fire brigades

Management cannot always depend wholly on automatic fire protection equipment or public fire departments to prevent fire losses. The fact that fires have gotten out of control before a public fire department could arrive makes it essential that a well trained fire brigade be available to fight a fire almost as soon as it is discovered. To make sure the fire brigade is adequate, management must consider the brigade its first line of defense against fire.

The public fire department should be consulted when plans for the fire brigade is formed. Plans should cover what the hazards are, what help is available, and how the public fire department can help the brigade. It is common practice to invite the firemen to inspect the plant facilities, fire equipment available, the fire hazardous materials used, their locations, and water supply available. (See NSC Data Sheet 588, *Fire Brigades.*)

Communications

Good communications are necessary in a disaster situation—first as a means of alerting occupants to the emergency, and second as a way to mobilize fire protection forces, be they a plant brigade, public fire department, or both. A coded fire alarm system, with alarm boxes and bells, horns and other sounding devices suitably situated, is usually needed, except in very small plants, where a steam whistle or similar device might be adequate. In any case, the alarm system is no better than the level of training given employees in how to respond when the alarm is sounded. (See details in Chapter 18, "Planning for Emergencies."

References

American Gas Association, 1515 Wilson Blvd., Arlington, Va. 22209. Lists of approved gas appliances and accessories.

American Insurance Association, 85 John St., New York, N.Y. 10038.
Fire Extinguishing Appliances. Pamphlet Nos. 10–19.
Fire Extinguishing Auxiliaries. Pamphlet Nos. 20–27.
Metalworking Industries, Technical Survey No. 2.
Stop Fires—Save Jobs.
Your Plant's Fire Protection.
Warehouses, Technical Survey No. 1.

American Mutual Insurance Alliance, 20 N. Wacker Dr., Chicago, Ill. 60606.
Judging the Fire Risk.
Simplified Water Supply Testing.

American Petroleum Institute, 1810 K St. NW., Washington, D.C. 20006.
Fire Protection in Refineries.
Recommended Practices for Refinery Inspections.

American National Standards Institute, 1430 Broadway, New York, N.Y. 10018.
Color Code for Marking Physical Hazards, Z53.1.
Practices for Respiratory Protection for the Fire Service, Z88.5.
"Prevention of Dust Explosions," Series Z12.
Requirements for Gas Appliances and Gas Piping Installations, Z21.30 (NFPA Std. 54).
Safety in Welding and Cutting, Z49.1.

American Welding Society, 2501 NW 7th St., Miami, Fla. 33125. *Safe Practices for Welding or Cutting Containers That Have Held Combustibles.*

Factory Mutual Engineering Corp., 1151 Boston-Providence Turnpike, Norwood, Mass. 02062.
Approved Equipment for Industrial Fire Protection.
Loss Prevention Data. (Available from Hightstown, N.J., McGraw-Hill Book Co.)

Fire Equipment Manufacturers' Association, Inc., 1803 Busse Rd., Mt. Prospect, Ill. 60056.
Inspecting, Recharging and Maintaining Portable Fire Extinguishers.

43—Fire Protection

National Association of Fire Equipment Distributors, Inc., 111 E. Wacker Dr., Chicago, Ill. 60601. *Portable Fire Extinguisher Guide.*

National Fire Protection Association, 470 Atlantic Ave., Boston, Mass. 02210. *Publications* Catalog.

National Safety Council, 425 N. Michigan Ave., Chicago, Ill. 60611.
Industrial Data Sheets
Cleaning Small Containers That Have Held Combustibles, 432.
Fire Brigades, 588.
Fire Prevention and Control on Construction Sites, 491.
Fire Prevention in Stores, 549.
Fire Protection for Combustible Metals, 567.
Flame-Retardant Treatment of Fabrics, 517.
Flammable Liquids in Small Containers, 532.
Hot Work Permits (Flame or Spark), 522.

Underwriters Laboratories Inc., 207 E. Ohio St., Chicago, Ill. 60611.
Building Construction and Materials List.
Electrical Equipment List.
Fire Protection Equipment List.
Gas and Oil Equipment List.
Hazardous Location Electrical Equipment List.

U.S. Department of Commerce, National Bureau of Standards, Washington, D.C. 20234. *National Electrical Safety Code,* Handbook No. H30. (ANSI Standard C2.)

Boilers and Unfired Pressure Vessels

Chapter 44

Boilers   1390

Design and construction . . . Good piping practice . . . Placing boilers in and out of service . . . Cleaning and maintenance . . . Boiler rooms . . . Boiler room emergencies . . . Pulp mill recovery boilers

Safety of High Temperature Water (HTW)   1402

Unfired Pressure Vessels   1402

Design . . . Inspection and entry . . . Hydrostatic tests . . . Detecting cracks and measuring thickness . . . Operator training and supervision . . . Safety devices . . . Steam-jacketed vessels and evaporating pans

High-Pressure Systems   1410

References   1412

FIG. 44–1.—Boilers and pressure vessels subject to ASME Code regulations must be checked during construction at all points by authorized inspectors who can certify compliance with the Code. Inspectors are commissioned by the National Board of Boiler and Pressure Vessel Inspectors.

Courtesy Lutheran General Hospital, Park Ridge, Ill.

Boilers (fired vessels) and unfired pressure vessels have many potential hazards in common, as well as having those unique to a specific operation. These vessels hold gases, vapors, liquids, and solids at various temperatures and at various pressures, ranging from almost a full vacuum to pressures of thousands of pounds per square inch. In some applications, extreme pressure and temperature changes may occur in a system in rapid succession, imposing special strains.

Design, construction, testing, and installation of boilers and unfired pressure vessels should be in compliance with the applicable sections of the American Society of Mechanical Engineers' *Boiler and Pressure Vessel Code* (hereafter referred to as the "ASME Code"), and any federal, state, or local governing codes.

The minimum requirements for the installation of high pressure boilers are covered in the *Life Safety Code,* National Fire Protection Association Standard No. 101.

A *Synopsis of Boiler and Pressure Vessel Laws, Rules and Regulations, by States, Cities, Counties and Provinces, in the United States and Canada* is available from the Uniform Boiler and Pressure Vessel Laws Society. This tells which governing bodies have made the ASME Code a legal requirement in their jurisdictions and what other compliances are required. However, the owner of the boiler or pressure vessel must check with his local authorities to obtain the latest requirements.

Compliance with the ASME Code is determined by authorized inspectors commissioned by the National Board of Boiler and Pressure Vessel Inspectors. (See Fig. 44–1.)

The ASME Code contains eleven sections:

Section	Title
I	Power Boilers
II	Material Specifications
III	Nuclear Vessels
IV	Heating Boilers
V	Non-Destructive Examination
VI	Recommended Rules for Care and Operation of Heating Boilers
VII	Unfired Pressure Vessels
VIII	Division I—Pressure Vessels Division II—Pressure Vessels (Alternate Rules)
IX	Welding Qualifications
X	Fiberglass-Reinforced Plastic Vessels
XI	Rules for Inservice Inspection of Nuclear Reactor Coolant Systems

Any or all of these sections may be obtained by application to the American Society of Mechanical Engineers.

When pressure vessels are to be put in, it is usually advisable to secure the services of a competent boiler or unfired pressure vessel engineering consultant. He will survey the plant or operation to determine the requirements, design a system that will satisfy them, and supervise installation and testing.

If he arranges for the purchase of second-hand boilers or pressure vessels, he should have them inspected by authorized code inspectors who will report on repairs necessary before purchase. Arrangements for inspectors can usually be made through the user's insurance agent. Also, the government inspection department that has jurisdiction is listed in the *Synopsis* previously mentioned.

In general, installation and maintenance should be in accordance with manufacturers' instructions. Operating personnel should be trained, not only to operate equipment properly, but also to make routine safety checks and to know when to call in qualified maintenance personnel.

Some common causes of failure in pressure vessels which should be anticipated and avoided insofar as possible are:

1. Errors in design, construction, and installation

2. Improper operation, human failure, and improper education of operators

3. Corrosion or erosion of the metal

4. Mechanical breakdown, failure or blocking of safety devices, and failure or blocking of automatic control devices

5. Water hammer, or carryover of processed material

6. Failure to inspect thoroughly, properly, and frequently

7. Improper application of equipment

8. Lack of planned preventive maintenance

In addition to hazards presented by the possibility of their explosion, heating service boilers also present fire hazards. They are a significant factor in hotel, store, apartment house, and religious institution fires. Particularly, oil-fired equipment is at fault, although losses from explosion of gas-fired equipment have occasionally been catastrophic.

The majority of boilers in use are automatically fired, and they operate unattended for long periods. Many are not maintained regularly, and their condition is less than perfect. Because the boilers are unattended, fires that do start can gain considerable headway unless adequate precautions are installed against them.

The means for controlling and containing fires from boilers include:

1. Provide a fully enclosed boiler room of fire-resistive construction (⅝ in. gypsum wallboard or better). (Be sure to leave sufficient space for maintenance, including pulling of tubes.)

2. Provide a fire-resistive ceiling over the boiler, with draft curtains at the perimeter of the ceiling and automatic sprinkler protection.

3. Provide proper clearance around exteriors of boiler room walls so that combustible materials are not stored against the walls.

In order to minimize low-pressure boiler fires and explosions caused by faulty controls and safety devices:

1. Establish a test and servicing program whereby operating controls, safety controls, and safety and relief valves will be tested and maintained at regular intervals.

2. Make sure that safety and relief valves are always tested with pressure on the boiler to prevent damage to the valve seats.

3. Have repairs made *immediately* upon any indication of malfunction or leakage of operating controls, safety controls or safety and relief valves. *Never operate with a malfunctioning safety or relief valve.*

4. Have a reliable service organization check and service the boiler during the heating season as well as during the normal out-of-season checking and cleaning.

5. Enforce the keeping of a boiler log to insure that necessary tests, maintenance, and services are performed, and that records are available at all times (Fig. 44–2).

Boilers

In its simplest definition, a boiler is a vessel in which water is heated by combustion of fuel to form steam, hot water, or high temperature water (HTW) under pressure.

Design and construction

Good standard references showing the design and construction of boilers are *Marks' Mechanical Engineer's Handbook,* Theodore Baumeister, Editor, and *Combustion Engineering,* G. R. Fryling, Editor. A summary follows; see Fig. 44–3.

Instruments. Subsection C6 of Section VII of the ASME Code states that, in general, a boiler unit should include a meter-and-control board located on the operating floor so that the operator can see either the furnace door or the lighting ports of the burners and the water column of the boiler without leaving the control board (Fig. 44–4). If it is not possible to see these directly, then reliable remote-indicating equipment should be installed, or else someone else should make a continuous visual check.

Economizers, usually an integral part of the heat exchanger system, are the last step in utilizing as much heat as possible. The exhausted flue gas from combustion of the fuel is used to heat incoming cold makeup feedwater—the hot flue gas being passed over tubes conveying the makeup water.

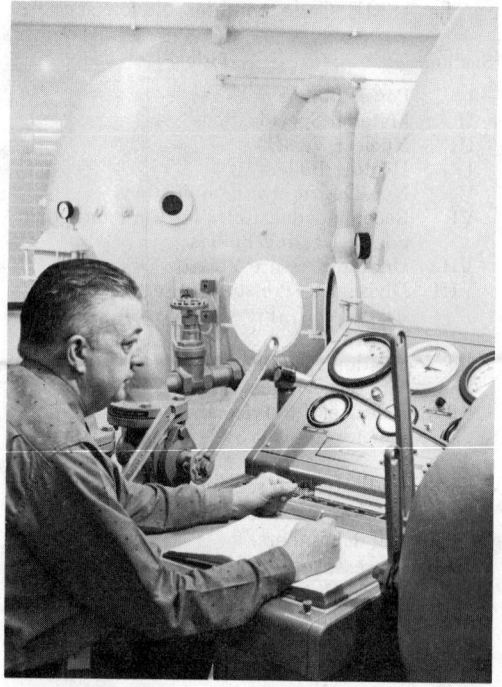

FIG. 44–2.—Make sure a boiler log is kept and check to see that necessary testing and servicing are performed.

Courtesy National Engineer *Magazine.*

Cast iron and steel tube economizers should be equipped with at least one safety valve (two are preferable).

Superheaters. After the heat transfer medium (water, steam, or other material) leaves the boiler unit itself, it can have its temperature raised even more by passing through a superheater. Superheaters must be made to withstand high temperatures and be equipped with safety valves, as specified in the ASME Code.

Detailed operation procedures are given in Section VII of the ASME Code.

Air preheaters. Fires may occur in an air preheater immediately after lighting a boiler and during periods of low-load operation. This condition can be detected by a sudden rise in air heater temperature.

To prevent fires, maintain proper combus-

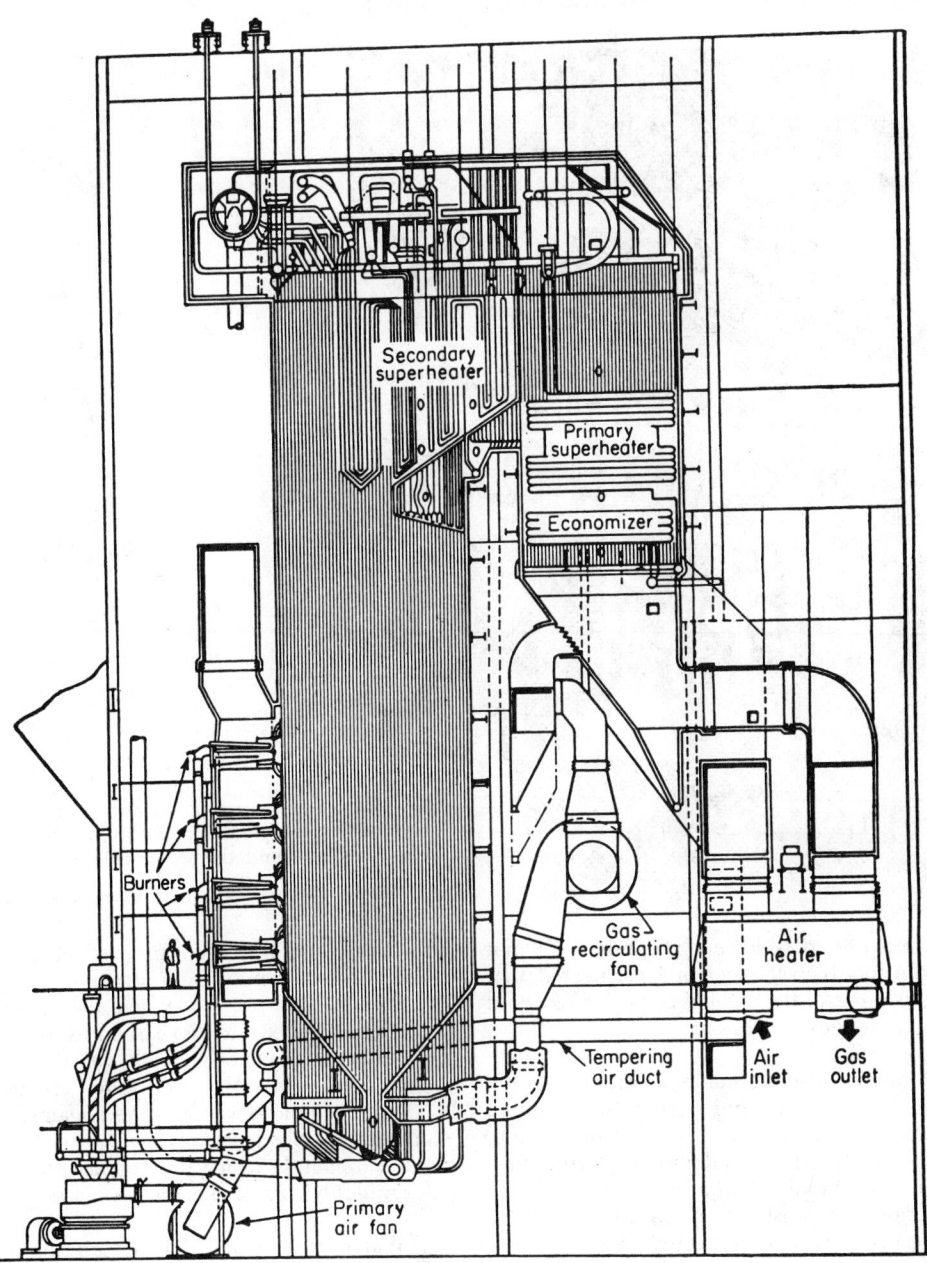

FIG. 44–3.—Diagram of large boiler unit shows relative location of auxiliaries. This twin-furnace, radiant boiler has circular burners in front wall for pulverized coal. Capacity is 2,400,000 lb of steam per hour at 2500 psi and 1050 F/1000 F. Hot gases after flowing through the pendent superheater pass downward through the convection pass. Hazards to men working on boilers include falls from improper staging, being struck by falling deposits from overhead pipes, asphyxiation or heat prostration, burns from hot equipment, and electric shocks. (Note scale of man on boiler operating floor.)

From Standard Handbook for Mechanical Engineers, *rev. 7th ed., Edited by Baumeister.* © 1967. *McGraw-Hill Book Co. Used with permission.*

1391

Fig. 44–4.—The meter and control board of a boiler should be located on the operating floor, in a place where the operator can also watch the boiler fire and the water column.

Courtesy National Engineer *Magazine.*

tion and use soot-blowers properly. Do not use the soot-blower when it is suspected that there is a fire in the gas passages as this could cause a serious explosion.

Chimneys, whether made of brick, concrete, or steel should be equipped with grounded lightning arrestors, and if not self-supporting, should be stayed to solid building structures.

Any ladders added to a chimney should be of permanent construction, securely fastened to the chimney, and protected with hoop enclosures.

Ash disposal equipment. Hoistways, driving machinery, conveyors, wormgears, ash sluices, and reciprocating pumps should be properly guarded. An alarm bell can be hooked up to the driving machinery to warn that doors are about to open.

Care must be taken to prevent injury to operating personnel from steam or hot water that may be present when ash gates are opened. When excess carbon is present in ash pits and it is not properly wet down, a gas explosion can result when gates are opened.

Ashes should never be stored against boilers or combustible materials. Ashes contain sulfur compounds which, on contact with water, form highly corrosive acids.

Water treatment. There are many professional consultants on feedwater treatment whose services are worthwhile.

Generally, treatment which will remove dissolved oxygen and carbon dioxide and maintain a maximum pH of 11 in the boiler water will effectively minimize corrosion. Operators have sustained injuries while introducing boiler compounds into feedwater, and therefore should be adequately protected against scalds and caustic burns. Often, automatic feeding or softening equipment is used.

Blowdown pipes and valves. Blowdown piping is used to remove sludge and other impurities in boiler water which, if not removed, would seriously impede the efficiency and safety of the boiler. All piping, and operating and discharge valves should conform to the ASME Code, Section I.

Blowdown piping and boiler drains should be conducted to a discharge point which will not present a hazard to operators or other personnel.

Safety valves and fusible plugs. Selection, fabrication, installation, testing, and replacement of safety valves and fusible plugs should be in accordance with the ASME Code or other applicable code. From the safety professional's viewpoint, safety valves and fusible plugs are important, even vital devices requiring his attention. Insurance company boiler inspectors and other specialists can advise on specific procedures for checking each type of equipment.

Safety valves, when properly applied and installed, will relieve excess pressure or vacuum (depending on design) that would otherwise damage equipment or result in injury to personnel. Safety valves must be kept in good operating order at all times and should be checked by qualified personnel in accordance with insurance company recommendations. The type of safety valve that is set by screwing down the body should be avoided because installers usually jam them by screwing them down tight.

Boiler operators should be trained to check the safety controls, preferably once a week, but at least once a month. Often a check list is used, and a form is filled out and sent to management or owner.

If a safety valve opens, fails to reseat correctly, and cannot be freed by use of the hand-lifting lever, then the boiler should be taken from service and the safety valve repaired.

Testing safety valves should be done whenever a boiler is returned to service. To test safety valves by hand, the valve should be held wide open for a sufficient period to blow out possible accumulation of dirt and chips. The steam pressure should be at least 75 percent of the safety valve set pressure when opening with the hand-lifting gear. A more meaningful test is given by raising the boiler pressure to the safety valve set pressure and letting the valve open.

If boilers are kept in continuous operation for several months it may be found desirable, depending on boiler conditions, to repeat the hand-lifting or pressure raising at intervals during operation. Many valves over 400 psi should not be hand tested, but still should be periodically checked according to manufacturers' recommendations.

Small chains or wires attached to the levers of pop safety valves and extended over pulleys to other parts of the boiler room may be used. A counter-spring or weight prevents the weight of the chain or wire from pulling the valves partly open.

Safety valves for water heaters differ from those used for boilers in that they must sense excessive temperatures as well as over-pressures, as specified in American National Standard Z21.22.

Fusible plugs are designed to relieve pressure and to indicate certain conditions that contribute to low water. When these plugs are used, they must be manufactured, installed, inspected, repaired, or replaced according to the ASME Code.

Discharge pipes are individual escape pipes designed to carry away discharge from each safety valve. They should be supported so as to prevent any stress upon the safety valve and must not be rigidly connected to the valve. Clearance must be provided to allow for boiler expansion.

Drain pipes from a drip pan and valve body should be carried clear of the boiler setting and discharged into an open funnel providing a clear view of the drip. They should be installed to avoid freezing at any point.

Steam and water indicators. Steam gages indicate the pressure of the steam generated. All gages should be graduated to approximately double the pressure at which the safety valve is set, but in no case less than 1½ times that pressure. Pressure gages installed on a multiple boiler setup should be of the same type and graduated alike.

1393

Fig. 44-5.—Men entering any confined space in boiler area should wear a lifeline and be constantly watched by man stationed outside who is wearing similar protective equipment. Explosion-proof 6 volt lamp is powered by sealed step-down transformer in inspector's left hand. Some organizations use battery-powered lights.

Good piping practice

A number of accidents can be prevented by installing steam lines well enough to reduce the necessity for maintenance work on them. The American National Standard B31 Series, "Pressure Piping," (*a*) prescribes minimum requirements for design, materials, fabrication, erection, test, and inspection of various piping systems, and (*b*) discusses the expansion, flexibility, and supporting of lines.

Another consideration in providing safety through good piping practice is that valves and other operating controls of boilers should be easy to reach. Many operators and maintenance men have been hurt when they fell from ladders or inadequate work stands while trying to operate an infrequently used and inaccessible valve.

If it is necessary to open lines, maintenance personnel should always assume that the lines are loaded and under pressure. A supervisor who is completely familiar with the system should then certify that the proper line is being opened and that all steps possible have been taken to drain and vent the line. Also, safe work stands must be provided. Some firms have a system of "line breaking permits" to make sure that precautions are carried out.

Placing boilers in and out of service

This Manual cannot cover all details of placing boilers in and out of service but it emphasizes that the ASME Code and manufacturers' recommendations should be followed.

Cleaning and maintenance

Whenever a boiler is taken out of service

for a prolonged period, it should be cleaned promptly and inspected for defects by the plant engineer. Authorized boiler inspectors can also view the boiler at this time.

Cleaning. Prompt cleaning is important. Soot gathers moisture rapidly and thus contributes to deterioration of the metal surfaces. Soot and fly ash should be removed as soon as the boiler has cooled. Ashes may remain hot for days, presenting a hazard to anyone entering the combustion chamber. They are, therefore, usually wet down with a hose.

As in spraying ashes for disposal, the operator should wet down from the outside toward the center and stay clear of any steam and dust that will come up. A jet of water driven into the center of a hot ash pile can literally explode it. When removing ashes, care should be used to prevent injury to personnel from steam or hot water that may be present when ash gates are opened.

For boilers in continuous service, planned and scheduled boiler shutdowns for preventive maintenance are far safer than risking an extensive shutdown caused by boiler failure. At least once a year, or oftener, the boiler, the fire-side ignition safety, and other safety controls should be inspected during a scheduled shutdown by an authorized inspector who is accompanied by the plant inspector. Then defective parts should be repaired or replaced.

Scheduled outages for maintenance should be carefully planned so that a minimum interruption to production results. Be sure that by the date that the inspector will be needed, the boilers will have been properly prepared. Boilers must be cool enough so that he will not have to rush his work, and clean enough so that metal parts can be examined thoroughly for corrosion, pitting, cracking, and other defects. Internal parts should be readily accessible for a close and thorough examination. (Handholes and manholes should be open and the boiler should be ventilated.) Adequate lighting and protective equipment for work in the boiler should be provided.

Boiler and furnace entry precautions. General precautions for entering boilers include having proper ventilation, proper equipment, and proper protection. To make certain that no flammable or toxic gases are present, ventilate boiler settings thoroughly and then check the atmosphere with a testing instrument before permitting anyone to enter. This is especially important when more than one boiler is connected to one breeching or chimney, because under certain circumstances flue gases can come back into the boiler from other boilers.

One cannot overemphasize the need for caution when men do cleaning or maintenance on boilers. A great number of injuries have occurred in this work. Good vessel entry procedures and good lockout procedures to prevent steam, hot or high pressure water, or hot gases from reaching workmen must be followed. All valves that are closed must be checked for leakage. Lines which are interconnected between boilers must be positively sealed off at both ends and locked out. Also attention must be paid to providing workstands and protection from overhead hazards of ash deposits falling on men entering boilers. (Fig. 44–3.)

Ventilation can be provided by portable power-driven blowers operated outside the boiler setting and having canvas tubes leading in through access doors. Draft fans can be operated for short periods of time to provide ventilation also.

As a precaution against electric shock, many firms only permit 6- or 12-volt lights and tools inside a boiler. They are connected to small portable power transformers outside the boiler. Battery-powered lights are an even safer alternative. All tools used inside a boiler should be properly grounded, and along with extension cords, be thoroughly inspected before use.

When cleaning a boiler, men should wear safety hats, goggles, dust respirators, and heavy, leather-palm gloves. Wearing safety shoes is also recommended. (Fig. 44–5.)

Personnel working in confined areas should wear a lifeline, if necessary, and be kept under constant observation.

Boiler rooms

Floors. Boiler room floors can become very slippery and dirty, so a surface that can be easily cleaned should be provided. Consideration should be given to drainage and protection against flooding.

FIG. 44–6.—Safety professionals should know enough about boiler room procedures to be sure their units have all necessary protection, and they should make sure that a trained substitute operator is always available.

Courtesy National Engineer *Magazine.*

Lighting. In addition to being well illuminated, the boiler room should have a source of emergency lighting. Gages and controls should be especially well lit so they can be read easily. Well maintained flashlights should be provided for personal use in case of power failure, or other emergency. Exits, too, should be well lit and identified.

Exits. Each boiler room should have two or more exits, one at each end of the room. If a boiler extends more than one story above ground level, the room should have an exit at the boiler runway or floor level for each story. This exit should lead to a fire escape on the outside of the building.

If the boiler room is in a basement (or subbasement) exits should lead to outside stairways and runways, and should have landings leading to the exit doors.

Stairs, ladders, and runways. Some state and local boiler codes require the installation of stairs, ladders, and runways around boilers which extend 10 ft or more above floor level.

Even if not required by law, such access is desirable so that personnel can operate and service the boiler safely and without having to step on hot steam or water lines or stand on valve stems or handles.

Stairs, ladders, and runways must have standard guardrails, handrails and toeboards. Runways should be fabricated of steel grating to provide a nonslip surface and at the same time permit circulation of air. Walkways should *not* be near water glasses or safety valve discharge areas where an operator might accidentally be scalded.

Boiler room emergencies

So many boiler accidents have been investigated thoroughly that necessary preventive action is generally known. Safety professionals should know enough about boiler room procedures that they can be sure their units are protected (Fig. 44–6).

Rules for both routine and emergency boiler operation should be posted permanently and legibly in the boiler room. (Fig. 44–7.)
(*Text continues on page 1401.*)

Boiler Operating Instructions

DANGER

OVERHEATED BOILER—
Do NOT add water — STOP FIRE
Call service company representative or supervisor for assistance

FLAME FAILURE—
Do NOT restart until thoroughly vented
Call service company representative or supervisor for assistance

STARTING and DAILY CHECK
1. Be sure water is at proper level.
2. Do not start fire until after furnace has been thoroughly vented.
3. Use small fire during warm-up.
4. Check boiler frequently while in normal operation.

WEEKLY CHECK
Test low-water cut-out control — Record Test on tag.

MONTHLY CHECK
Test safety valve and record on tag.

YEARLY
Replace or disassemble and overhaul low-water cut-out control during annual boiler clean-up and repair period.

W.S. 171 NEW 11-63 PRINTED IN U.S.A. SAFETY ENGINEERING SERVICE THE TRAVELERS INSURANCE COMPANIES ● HARTFORD, CONNECTICUT

FIG. 44–7.—Emergency procedure and checklist poster, similar to this one should be posted permanently in the boiler room.

Courtesy The Travelers Insurance Companies.

CHECKLIST FOR CLEANING RECOVERY BOILERS

Operational Step	Exposure or Hazard	Controls to be Considered
I. Preparation A. Shutting down	Various, including serious explosion	Follow boiler manufacturers' instructions.
B. Check atmosphere of boiler to be entered for: 1. Freedom from toxic gases	Inhalation of toxic gases	Allow time for ventilation and cooling. Positively determine atmospheric content. Breathing apparatus essential? Thermorespirator required? Safety harness with a strong line tied to the outside?
2. Safe oxygen content	Asphyxiation	Has positive determination of oxygen content been made and is it adequate? Air flow sufficient?
3. Temperature	Heat prostration	Has a time-temperature air-velocity guide been consulted to determine the maximum period of time that the employee can work in this boiler while exposed to the existing temperature and radiant heat? What personal protection is necessary? Air flow sufficient? Man cooler necessary?
4. Hanging slag	Falling objects	Have soot blowers and steam lancers been used effectively? Will high pressure washers be effective? Is further mechanical barring necessary? Should safety nets, barriers, or shields be placed at this time? Are safety hats sufficient protection? Should other protection be provided?

II. Cleaning super heater and screen tube	Falling slag	Clean from outside first? Use belt and line? Provide secure staging? Use boatswain's chair?
III. Repairs and cleaning furnace bottom	Falling slag	Insert strong horizontal pipes at burner or lancing ports. Cover with planks or strong plywood to form protective canopy.
	Bottom heat	Wooden flooring necessary?
IV. Erect staging	Tipping	Provide level footing. Is tying or bracing necessary?
	Weak planks	Adequate for span? Men? Load?
	Planks tipping or slipping	Avoid cantilevering and shifting under use.
V. Personnel	Panic or fear from physical condition	Is crew accustomed to height, heat? Crew adequately trained? Supervisory control adequate?
VI. Boatswain's chair	Falling	Rope of right type, size and free from excessive deterioration? Was chair load-tested? Is the free end of lines brought through the stirrup and tied? Controlled by operator?
	Struck by slag	Work started at top to reduce exposure? Work so arranged to avoid exposure to anyone below? Safety hat necessary?
VII. Power tools	Electric shock	Electric lights from industrial batteries used? Low voltage source (12 or 24 v)? Complete with guard? Lighting adequate? Air-powered tool or low voltage source (12 or 24 v)?

44—Boilers and Unfired Pressure Vessels

Operational Step	Exposure or Hazard	Controls to be Considered
VII. Noise	Possible loss of hearing	Ear protection provided? Worn?
IX. Placing and Removing A. Protective canopy	Atmosphere Slag	(See above.) Insert pipes from outside. Place canopy over manhole area first. Work under canopy while installing other protective planks or plywood.
B. Net	Atmosphere Slag Weakened or missing fasteners Fastening Method Weakened net	(See above.) (See above.) Inspection of fasteners. Check for adequacy. Secure nearest fastener first, then work in sequence out to farthest. To remove, do in reverse sequence. Visual and physical examination. Should a 2 × 2-inch mesh be used?
C. Staging	Slag and Various Falling objects	Employees wearing personal protection? How raise scaffold? Planks? How lower? Should a line and sheave be used?
D. Tools	Falling from grasp or resting place	Should line be used to tie off?

FIG. 44–8.—Checklist of hazards to consider and control methods to use prior to entrance for inspection or repairs, and during the time employees are in boiler fire boxes. List is especially useful for recovery boilers, because of their high concentration of slag.

Courtesy Pulp and Paper Section of NSC Industrial Conference.

Manufacturers can supply rules applicable to their equipment. In addition, all operators and substitute operators should be furnished copies for their guidance. Supervisory personnel should make sure that boiler operators know the rules and are capable of performing the necessary operations under emergency conditions.

Many plants have only one boiler room operator. Should he become sick or injured, the boilers may be left unattended and an accident may happen. In plants where the boiler room is isolated and is operated by one man, it may be advisable to have him call a central location at half-hour intervals to assure that everything is operating on schedule in the boiler room. Plants having a plant protection patrol may have patrolmen check the boiler room. An intercom system could also be of value.

For this reason, the safety professional should recommend someone else in the plant, a foreman, a night shift man, or some other substitute, trained to take over in case of emergency.

Pulp mill recovery boilers

The occurrence of damaging explosions in kraft chemical recovery boilers has been a continuing concern of the pulp industry. Although these might occasionally be gas explosions (resulting from the formation of an explosive mixture of combustion products), the great majority appear to have been caused by water coming in contact with the molten smelt of recovered chemicals. Hazards of entering a boiler are listed in Fig. 44–8.

A study by the Institute of Paper Chemistry has shown that water or water solutions other than black liquor at 55 to 70 percent total solids can cause destructive explosions if permitted to contact molten smelt. Common sources of water for these explosions are boiler water from leaks, cooling water from furnace auxiliary parts, recovered chemical streams, salt cake slurry conveying systems, and tall oil soap solutions being direct fired.

Danger signals that indicate a smelt-water emergency are:

1. A leak develops in a boiler pressure part.

2. Water in any amount is known to be entering the furnace.

3. Water or steam is being lost from the boiler and there is no assurance that it is not going into the furnace.

4. Fewer than two complete independent measuring systems indicate that the water level in the boiler is within the safe operation range.

Emergency shutdown is necessary if the source of the water is not immediately identified and stopped. To avoid a gas explosion during an emergency shutdown, the furnace atmosphere should be maintained fuel-lean (below the lower explosive limit), or be rendered inert by dilution with steam or some other inert gas.

The emergency shutdown procedures should accomplish these objectives:

1. Sound an alarm to clear the recovery area of unnecessary personnel.

2. Immediately stop firing all fuel. Secure the unit's auxiliary fuel system at a remote location.

3. Shut off feedwater supply to boiler.

4. *a*) If black liquor is used to smother the bed, immediately shut down the air supply by tripping the forced draft fan and closing the forced draft damper. Regulate the induced draft fan or dampers to maintain a balanced draft in the furnace. Be sure the minimum concentration is 55 percent solids at all times,
b) If black liquor is not used to smother the bed, shut down the air supply to the primary air ports immediately. Continue operation of the forced draft fan to supply as much air flow as possible to the secondary and tertiary air ports (if present). Regulate the induced draft fan speed or the damper to maintain a balanced draft to the furnace.

5. Drain boiler as rapidly as possible, in accordance with manufacturer's recommendations, to a level 8 ft above the low point of the furnace floor.

6. Reduce steam pressure as rapidly as possible after boiler has been drained to the 8-ft level.

Safety of High Temperature Water (HTW)

Because of certain economic advantages, high temperature water (HTW) is sometimes used instead of steam for transferring heat in both manufacturing and district heating uses. The water is kept in a closed system under high pressures so that it will remain in the liquid form instead of turning into steam. Conditions such as 400 F and 247 psi pressure often exist, and if one considers the potential volume increase when 400 F liquid under this pressure expands to steam at atmospheric pressure, it can be quite appalling.

HTW is very different from steam or cold water when it discharges through a break in a pipe or equipment: it has a very high rate of increase of volume and very low energy release during expansion. Energy liberated in the expansion is spent in accelerating the particles of water and vapor, and in pushing air out of the way so the steam-water mixture that is being formed can occupy the vacated space. Practically no energy is left over and available for rupturing equipment and imparting kinetic energy to fragments.

When *steam* escapes, there is approximately 16 times the energy available from its expansion than from HTW expansion, hence considerable energy is left over to provide the explosive effect. Although fragments of fracturing cast iron valves on steam service have been known to penetrate a 10-in. thick brick wall, no case has been observed where parts of fractured valves on HTW service have been projected any distance at all.

Because the volume increase of escaping HTW continues after leaving the pipe, the mixture does not form a long jet as does steam or cold water issuing from an orifice, but spreads out practically at right angles from the center-line of the jet to form a wet fog. The rate of flow has been found to be less than half that of cold water issuing from a similar hole and over the same pressure range.

Such considerations must not cause any feeling of false security or negligence on the part of design engineers and operating personnel. Although HTW is safer than steam and accidents are rare, such accidents have nevertheless happened. Even 180 F water can fatally burn if sufficient body area is immersed in it.

However, when failure of equipment or piping occurs in HTW systems, it is usually caused *not* by the inherent thermodynamic forces, but by mechanical forces such as water hammer, thermal expansion, thermal shock, and faulty material.

It is therefore imperative that only experienced engineers should be allowed to design HTW systems and these engineers must be willing and capable of analyzing minutely the entire design and equipment selection for the presence of possible dangers. But disruptive forces may also be caused by faulty operation, and it is the plant management's responsibility to select and train qualified operators.

A good design is neat and simple but it does not overlook the essentials. However, overloading of systems with automatic controls should be avoided since these can *introduce* more hazards into the system in case of malfunction than they avoid. They furthermore tend to turn the operator into an attendant who, in an emergency, is incapable of operating his plant.

Unfired Pressure Vessels

Unfired pressure vessels are compressed air tanks, steam-jacketed kettles, digesters, vulcanizers, and other vessels which can be subjected to internal pressure or vacuum, but that do not have the direct fire of burning fuel impinging on them. If heat is generated in the vessel, it is by chemical action within the vessel or by application of electric heat, steam, hot oil, or other heating medium to the contents of the vessel.

Design

Unfired pressure vessels, like boilers, are covered in the ASME Code. Certain classes of vessels with smaller hazards, however, are exempt from regulation. These include:

1. Vessels subject to federal control

2. Vessels containing 120 gal or less of water under pressure, in which any trapped air serves only as a cushion

3. Vessels having an internal or external operating pressure not exceeding 15 psi

4. Vessels with an inside diameter of less than 6 in.

5. Hot water storage tanks heated by indirect means—heat input to be less than 200,000 Btuh, and water temperature to be less than 200 F, and capacity less than 120 gal

The ASME Code has been extended so that vessels designed for pressures over 3,000 psi may be code stamped. This is covered under Division 2, a second part added to the "Rules for Construction of Unfired Pressure Vessels," Section VIII of the Code.

Division 1, the old rules, covers vessels with ratings of 3,000 psi or less (with the exemptions listed earlier). Vessel designs under Division 1 rules are calculated according to the principal stress theory; and a safety factor of 4 is provided. Vessels built to these specifications may be used anywhere the pressure and temperature do not exceed the ratings allowed by the Code.

Before a pressure vessel is designed under Division 1 rules, five questions should be answered:

1. Will the material used in construction of the vessel damage or chemically change the material in process?

2. Will the material in process affect or damage the metal in the vessel?

3. Will the filled vessel carry the weight of its contents (plus internal pressure)?

4. Will it resist both the pressure introduced into it and any additional pressure that may be caused by chemical reaction during processing?

5. Will the vessel withstand any vacuum that may be created intentionally or accidentally without collapsing?

Specifications for construction of a Division 1 pressure vessel should include, in addition to general requirements, the working pressure range, working temperature range, data as to whether or not pressure and/or temperature range is cyclic, a description of what contents are to be, and all other information of a specific nature that may affect fabrication and installation of the vessel, such as stress relieving, radiography, welding, and other requirements.

Division 2 is entitled "Alternate Rules for Pressure Vessels." Vessel design under these rules is based on a detailed stress analysis using Tresca's Maximum Shear Theory; and a safety factor of 3 is provided. Design calculations are tedious (usually requiring a computer), but they allow thinner wall sections and are good for vessels used at pressures in excess of 3000 psi.

The "alternate rules" apply only to vessels installed in a fixed location and subjected to a specific service. To obtain a vessel with an ASME stamp under these rules, a prospective purchaser must prepare a "user's design specification" and have it certified by a registered professional engineer experienced in pressure vessel design.

Other codes. Although the ASME Code has been adopted by many governing bodies, and has the force of law, other codes may be required by the legal jurisdiction in which the vessel is located. It is best to check.

Second-hand vessels. If a vessel is to be located in a jurisdiction where ASME Code is required, prospective purchasers of second-hand vessels must have the equipment inspected by an authorized code inspector. They must obtain a written report that the equipment meets the requirements of the new location before the vessel is purchased. A great deal of trouble has arisen when second-hand equipment was purchased and reinstalled before inspection.

Inspection and entry

Pressure vessels should be inspected regularly by persons who are qualified and trained for this work. Inspectors should be instructed to be conservative in the approval of borderline cases. Be sure to check whether state or local governing bodies or insurance companies require their own men to make inspections. Often these men can make suggestions for refinements which can contribute to lower insurance rates by contributing to greater safety.

A large company or plant may find it advantageous to employ a full-time inspection

1403

staff to administer a regular inspection program for all their pressure vessels. Such a program, coupled with good preventive maintenance, prolongs vessel life and prevents accidents.

A log of the history of each vessel should be kept by the inspector or the maintenance department. Included in it should be blueprints, manufacturer's data report and instructions, design data (including location of dimensional check points), installation information, and records of process changes, all repairs, and conditions found on inspections. This log will prove valuable in operating the equipment, and in design, installation, and operation of new equipment.

When corrosive, poisonous, or toxic materials are used in the plant, management should so advise the operating personnel and the inspectors. When the inspection is carried out by a state, city, or insurance inspector, a plant chemical engineer or other competent person should accompany him on his inspection to describe the processes in detail so the inspector will know what conditions the vessel has been under.

When new processes are developed, the inspectors and the operators should be advised in detail what these processes are and how they may affect the pressure vessels.

Entry. A safe tank-entry procedure is absolutely necessary (and sometimes required by law) in order to eliminate the sizable number of fatalities that come from entering dangerous vessels and confined spaces. The hazard arises when workmen or inspectors cannot get out of a vessel without help; difficulty in communication compounds the problem.

Hazards to those in confined spaces include:

1. Exposure to toxic materials already in the confined space or introduced later

2. Lack of sufficient oxygen

3. Heat—a fire might start, hot gases or liquids might enter, or vessels might be heated inadvertently

4. Agitators might be started or the vessel itself might be started revolving

Before being entered, a vessel must be

Fig. 44-9.—Valves on lines leading to pressure vessels should be locked out when men are inside. If contents of the line are very hazardous, the line should be disconnected and blanked off.

properly prepared. It must be drained, ventilated, and cleaned. All connecting pipelines should be disconnected and blanked, or valves on the line should be closed, locked out, and tagged (Fig. 44-9). All power-driven devices (such as agitators) must be positively disconnected and locked out.

When purging a tank, the vent should be led outside to an area where no hazard will be created for persons. In some instances, the vessels may be purged with an inert gas such as CO_2 (carbon dioxide) or nitrogen. It must be remembered that this inert gas will not support life, so persons entering the vessel must wear air-supplied respirators or self-contained or self-generating breathing equipment.

Using forced ventilation for confined spaces is usually safer than requiring men to wear respiratory protection (see Chapter 19, "Personal Protective Equipment"). Air should be blown in until tests of the exhaust and of the interior of the enclosed vessel show that the space is safe for entry. All areas of the vessel should be tested for flammable and toxic gases, and for inadequate oxygen. These tests must be repeated at intervals to make sure that conditions remain safe while men are in the vessel. Air should be introduced to make sure there are several air changes per minute in the vessel.

After all preparations for entry are made, the supervisor of the area should check that the vessel is safe, that all lines are closed off, that power sources are locked out, that ventilation and personal protective equipment are adequate, and that safe work procedures are planned. He can then issue a "vessel entry permit," a note that certifies all precautions have been carried out.

Men should not be lowered into vessels without facilities for them to climb back out themselves (if it can be avoided). Straight ladders or rope or chain ladders with rigid wood rungs should be provided. Also, men should not be required to go into an opening that they must squeeze through. They cannot be removed quickly in an emergency.

Another necessary precaution is to have men don safety harnesses attached to lifelines before entering any vessel. An observer equipped with similar respiratory protection and harness with a lifeline should be stationed outside the vessel. This man should also have some device for signaling for more help; one man's efforts might prove inadequate.

Depending on the previous content of the vessel, the man entering the vessel should be equipped with a vapor-proof flashlight or vapor-proof low-voltage extension light. At times, a chemical protective suit is advisable.

The method of cleaning depends on the use of the vessel. If it has contained petroleum or chemical products, the vessel may be filled with water, a caustic solution, or a neutralizing agent to remove sludge and adhered materials. Vessels used for flammable liquids should be washed, steamed and/or ventilated until a test with an approved explosion meter shows the level is safe. (See Chapter 42, "Flammable and Combustible Liquids.")

Hydrostatic tests

If a pressure vessel is so constructed that an internal inspection cannot be made period-ically (length of the intervals depending on the corrosivity of the contents), it should be subjected to a hydrostatic test if the weight of water will not in turn set up damaging stresses. In this latter case, a pneumatic test can be applied. See ASME Code, Section VIII.

Compressed gas or air should never be used to test an unfired pressure vessel above its safe working pressure, although it can be used to test for leaks at pressures below the working pressure. Great care must be used because a vessel may fail under test and shatter. Testing should follow procedures in the ASME Codes and be under the supervision of qualified personnel.

The required pressure for a standard hydrostatic test is normally 1½ times the maximum allowable working pressure of the vessel being tested. Division 2 provides for the establishment of upper limits by the design engineer, in terms of stress-intensity limits relative to the yield strength at test temperature. Inspection of Division 2 vessels is made at a pressure equal to the greater of the design pressure, or ¾ of the test pressure.

Test areas should be isolated as far as possible from other operations and suitable barricades should be provided for protection of personnel and valuable equipment. This is especially important when conducting proof tests and tests to destruction.

All personnel should keep clear of the vessel under full test pressure and no one should be allowed to approach the vessel until the pressure has been reduced to, or is very close to, the maximum allowable working pressure.

Detecting cracks and measuring thickness

Pressure vessels used to process gases or oily materials may have very small leaks which will not show under hydrostatic tests. To detect them, a small amount of ammonia can be released inside the vessel and compressed air is then applied until a maximum pressure of 50 percent of the working pressure is attained. A swab soaked in hydrochloric (muriatic) acid is passed over all seams and other suspect areas. Leakage will be indicated by a white vapor (ammonium chloride) formed by contact of escaping ammonia and the acid. Using a burning sulfur stick is also effective—a change in the flame indicating presence of ammonia.

Radiography is especially good for finding cracks. For example, ammonia tanks used for agriculture are subject to corrosion and to stress cracking. If care is not taken to catch tanks which become defective, there may be some surprised—and hurt—farmers.

Also, there are a number of applications where it is vital to check the thickness of an unfired pressure vessel without damaging it. There are a number of manufacturers who make instruments for doing this. These instruments employ ultrasonic and various electronic ray-producing and measuring means.

Using these instruments, a qualified operator can determine the thickness of metal to within 2 or 3 percent. Some will disclose cracks which extend to or are slightly below the surface. The use of a radiograph will then be necessary to determine how deep the cracks are.

There is also a lacquer method of detecting hairline cracks that cannot be seen by the naked eye. When the head of a pressure vessel is suspected, it is cleaned and given a coat of clear lacquer. After the lacquer hardens, a hydro test is applied. The weak spots, hairline cracks, or fatigue stress cracks will show up on the head, expand and crack the lacquer. The more modern dye penetrant tests, when made by competent inspectors according to manufacturers' instructions, are considered quite reliable.

At each inspection of such pressure vessels as vulcanizers, digesters, and autoclaves which have removable cover plates, heads or doors, the holding bolts, cover plate bolts, slots, and retaining rings should be checked for wear, and hammer-tested for soundness. Since these parts are badly abused in service and receive considerable wear, it is advisable to replace them annually. They are comparatively inexpensive.

The width of the slot should be gaged and a careful check should be made of the retaining rings and cover plates, as cracks are set up by the stress of improper adjustment or closing of the door plate.

If cracks cannot be satisfactorily repaired, the vessel should be condemned.

Operator training and supervision

It is important that employees working with pressure vessels, particularly with those which are used in chemical processes, be thoroughly trained both in routine duty and emergency procedure. Supervision, too, should be competent and alert. A new employee being trained as an operator or helper should have the entire process explained to him, the hazards involved, and just how his operation affects the entire process.

In some plants a checklist is used to make certain that no step has been overlooked in the processing cycle. The operator or his helper records on a card the information obtained from recording apparatus and thermometers, and the time and frequency at which he operates the valves to each pressure vessel.

After each complete processing cycle, the operator initials a card checklist.

If the contents of a pressure vessel are being discharged to a vessel which the operator cannot see, or one run by another operator, a system of whistle, bell, or light signals should be installed. The time that these signals are given and action is taken should be noted on the checklist so that the wrong valve will not be opened or closed.

The plant operating supervisor should instruct operators of vessels with cover plates or removable doors how to tighten the bolts or quick-closing lugs without damaging them or the retaining rings. He should likewise instruct operators to open cover plates or removable doors only after the vessel has been relieved of all pressure. Such human failures can be avoided more simply, however, by the use of interlocks (discussed later in this chapter).

Operators should know where to look for wear on holding bolts, quick-opening lugs, and lug openings on the cover plate or removable door, and should know when a bolt has worn to the point at which they should notify their supervisor. A torque wrench used to tighten bolts will assure uniform tightness and reduce wear and damage.

So operators will not open or close the wrong valves, valves and pipelines should be tagged and marked as described in Chapter 16, "Industrial Buildings and Plant Layout," and American National Standard A13.1, *Scheme for the Identification of Piping Systems.*

Safety devices

Because pressure vessels are used to process such a great variety of materials, each vessel should be equipped with safety devices designed for the type of vessel and for the work it is to do.

Safety valves for each vessel should be ASME-approved safety valves, and so marked. The vessel should be provided with safety devices which will be adequate to protect it against overpressure, chemical reaction, or other abnormal condition.

Safety valves. ASME-approved safety valves of the spring-loaded type are the most commonly used safety devices for pressure vessels. They are used on vessels containing air, steam, gases, and liquids which will not solidify as they pass out through the safety valve discharge.

Valves on pressure vessels containing air or steam should be large enough to discharge the contents at a rate to prevent pressure buildup as prescribed in the ASME Code. On vessels which contain liquids, the safety valve seat should be so made that it will not collapse nor the contents plug the discharge opening. The safety valve discharge line should be led to a safe point of discharge. Where practical, valves should be equipped with test levers. Frequent testing prevents sticking of the valves. For vessels with dangerous contents (toxic, flammable, etc.), the safety valve should not have a test lever. (See details under previous section on Boilers, Design and Construction.)

If liquid contents are heated, the safety valve should be designed to operate if the vessel is overpressured as the liquid expands. For pressure vessels containing hot water, or in which water is heated, the valve should be sized to relieve the contents on the basis of the total number of Btu's that can be applied to the vessel.

The old ball-and-lever-type safety valve has been condemned for use by all Code states, because its setting is easily tampered with and it can accidentally be reset.

Rupture disks. A frangible disk may not clog as easily as a spring-loaded safety valve and is easily and inexpensively replaced. A rupture disk may clog or become coated with material in such processes as the manufacture of varnish and other resins. This coating at times becomes thick enough to affect the rupturing pressure of the disk so that it is necessary either to replace it or clean it with a solvent.

The condition of these disks should be checked frequently to see that they are clear. It must be remembered that when the disk ruptures, all pressure is relieved from the vessel. This could result in complete loss of product or spoiled in-process material.

A rupture disk must function within ±5 percent of its specified bursting pressure at a specified temperature. Disks may be installed between a spring-loaded safety or relief valve and the pressure vessel in order to prevent unnecessary corrosion of the valve and to prevent it from becoming plugged by the contents of the vessel. These multiple installations must be in accordance with the ASME Code, Section VIII.

Various designs of rupture disks are available.

Vacuum breakers. It is just as important to protect a pressure vessel from collapsing under a vacuum as it is to protect it against bursting from overpressure. Several safety devices provide such protection.

The mechanical vacuum breaker, similar to a spring-loaded safety valve, has a spring set at a predetermined vacuum.

The weight-balanced vacuum breaker (using a weight suspended from a fulcrum attached to the gate) is generally used on pressure vessels working intermittently on pressure and vacuum. If the vacuum exceeds the setting of the weight on the fulcrum, the breaker then opens.

In some vessels which work ordinarily under pressure, but in which a vacuum may occur because of rapid cooling (as when steam condenses), a check valve may be installed with a flap or valve disk facing into the vessel. Whenever a vacuum occurs, the check valve disk opens automatically.

Water seal. A water seal is used on pressure vessels that operate on low pressure or slight vacuum, such as alcohol stills and gas holders. A water seal is a U-pipe filled with

1407

Fig. 44–10.—Interlock for autoclave door. Hood "A" is pushed into position so that dog cannot engage rachet so long as pressure is inside vessel. Flexible hose (not shown) transmits internal pressure of vessel to cylinder which activates rod that pushes hood "A."

water, with one end connected to the pressure side of the vessel and the other vented to the atmosphere.

Since the vessel operates under a pressure of only a few pounds, the degree of pressure can be regulated by the height of the water in the vent pipe. If the pressure rises above the set limit, the water is forced out of the pipe, thus relieving the pressure.

Vents. In many processes, pressure must

be relieved before the pressure vessel can be opened. An easy means of relieving this pressure is to vent it to the atmosphere. Condensate tanks, which operate under very low pressure or no pressure at all, but in which excessive pressure can build up, should also be equipped with vent pipes and safety valves.

Vent pipes should be large enough in diameter to relieve the contents of the vessel before excess pressure can build up. A vent pipe should preferably be installed with a

U-bend at the atmospheric discharge to prevent dirt from clogging the pipe. Care should be taken to direct the flow away from the vessel in case of fire, so it will not impinge on the metal.

Vent pipes must also be protected in cold weather—vapor may freeze as it leaves, rendering vents inoperative as safety devices. If a vent pipe is so placed that it may freeze or become clogged by dirt, a relief valve should be installed on the pipe as added protection.

Regulating or reducing valves. Some vessels are operated under steam pressure much lower than that obtained from the boiler or steam transmission line. A regulating or reducing valve reduces high-pressure steam to that required for a specific operation.

There should be a safety valve on the low-pressure side of the reducing valve. The relieving capacity of the safety valve should be sufficient to assure that the pressure on the vessel that is being fed the steam will not exceed its safe working pressure in the event the reducing valve fails.

To provide protection for all pressure vessels in a battery of the same type, one reducing valve and one safety valve may be installed in the main steam line. This is the usual method in the case of steam-jacketed kettles in which ordinary pressures do not exceed 10 to 25 psi. Safety valves should be so connected that there is no stop valve between them and any vessel they protect.

Autoclaves and interlocks. Autoclaves, vulcanizers, retorts, digesters, and all pressure vessels which may contain large volumes of steam during operation, and which must be opened for charging, should be equipped with interlocks. An interlock will prevent the opening of the charging door until all pressure has been relieved (Fig. 44-10). There are several types.

The most hazardous part of these vessels is their closure, although they should be inspected for cracks like any other pressure vessel. Opening an autoclave with pressure in it will cause the door to be flung open with explosive violence. The contents may be fired out like projectiles, and the reaction to the blowout may cause it to move back an impressive distance. No matter whether a bolted door, a rotary lug door, a shear ring door, a clamp door, or a screwed-on door is used, the sealing mechanism must be maintained in good shape (as described under Detecting Cracks and Operator Training in preceding paragraphs) and an interlock must be provided.

Steam-jacketed vessels and evaporating pans

Steam-jacketed vessels are used to heat liquid mixtures to a moderate degree. Steam is circulated between the outer and inner shells of the vessel at pressures which are usually 10 to 30 psi. Occasionally the process may require that they be operated at pressures up to 100 psi. Heat is transmitted through the inner shell to the contents.

Such vessels are used principally in commercial preparation of food, in candy manufacture, and for cooking starch in laundries and textile mills. They are also used in the chemical industry for low-temperature "cooking." On a steam-jacketed vessel which has a tight cover, a separate safety valve must be provided for the inner kettle.

If a steam-jacketed vessel can be completely valved off, the vessel should be protected with a vacuum-breaker to keep it from collapsing.

Precautions to be followed in the operation of steam-jacketed kettles include:

1. The steam space should be thoroughly drained before steam is admitted to the jacket. It is advisable to open drain lines even though traps are installed, because water in the steam space may cause serious damage.

2. Steam should be admitted to cold vessels slowly, in order to allow ample time for uniform heating and even expansion of all parts. This becomes more important as vessel size and steam temperature are increased.

3. Unless automatic protection is provided, vents should be opened when the steam supply is shut off to prevent damage or even collapse of the kettle upon condensation of the steam.

4. Where agitators are used, paddles must not strike the kettle. Even a slight deformity in the inside may mean extensive repairs. Hand stirrers should also be used with care.

1409

5. Kettles should have edges 4 ft high, or guard rails should be provided, so that employees will not accidentally fall in.

6. Kettles should be filled only to a point where undue splashing will not occur when the contents are heated or agitated. Splash guards or loose covers may be used.

Evaporating pans ordinarily are shallow pans containing steam coils which, when the pans are in operation, are immersed in the material being treated. The hazard to be guarded against is permitting the coil to become exposed, in which event the material may be overheated or ignited, and fire or explosion result. Safe practices include:

1. Pans should be continuously attended as long as they are in operation.

2. After each use the pans and coils should be thoroughly cleaned.

3. After steam is shut off in the coils, the coils should be drained to prevent the product from being drawn into lines when steam condenses and creates a vacuum. Installing a vacuum breaker would also prevent this.

A general precaution to follow in maintenance work is "never allow men to climb up over large open vats or kettles filled with hot, corrosive, or viscous fluids." Vats should be drained or covered, safe work stands should be provided, and men should wear lifelines if necessary. All boards in the workstands should be fastened in place, not left loose. Every other precaution should be taken to make sure that men and objects do not fall into these liquids.

High-Pressure Systems

The hazards of high-pressure systems largely arise from failures caused by leaks, pulsation, vibration, and overpressure. Besides the damage that can be expected from release of high-pressure gases if a vessel or pipe ruptures, there can be fatal injuries from the blowout of high-pressure gages or from whiplash of broken high-pressure pipe, tubing, and hose. The potential of injury and damage from pressure system accidents is very high. In the

research laboratories of the rocket and missile industry, for instance, it is exceeded only by propellant explosions.

Since reciprocating pumps and compressors are normally needed to generate high pressures, the inner fibers of the piping and vessels of the system are subjected to pulsating pressures. Pipe and vessels must, therefore, be absolutely free of internal notches or severe scratches. These defects are stress raisers that will surely lead to fatigue failure. Also, the pipe must avoid stress concentrations arising from ill-planned holes or cross bores. The stress concentration factor from a radial entry to a pipe or cylinder wall will reduce the pressure endurance limit by almost one-half. Because of the pulsating pressure condition, nothing can be labeled "safe" and then forgotten. Constant surveillance is essential.

Leaks in pressurized systems can also be hazardous. Liquids expelled can easily penetrate clothing and skin. A sudden leak might instantly fill an enclosure with an explosive gas mixture. Because most leaks occur at joints, the number of joints should be kept at a minimum.

In piping, vibration should be limited by pulse dampening, if possible. Designers use various means to do this in hydraulic and pressurized gas systems. Because pulse dampening is an ideal never fully achieved, many rugged pipe supports must be used in high-pressure systems. The supports must be strong enough to resist deflection from any direction; they must be securely anchored to the structure of the building to prevent whiplash (Fig. 44–11).

Bourdon tubes in pressure gages will almost all eventually fail from fatigue caused by the constant pulsing of the pressure. They can fail after many cycles, or even when the gage is new. Gages on large vessels are not as subject to large pressure oscillations as those in lines where pressure is maintained by a compressor.

Pressure gages used at 1800 psi or more (except Underwriters' Laboratories-listed gas regulator gages) should have full-size blowout backs, integral sides and front designed to withstand internal explosion, and either multi-ply plastic or double-laminated safety glass gage-face cover (Fig. 44–12). Tests at

FIG. 44–11.—Two methods of securing a high-pressure line: Line was secured on either side of the fitting (lower right arrow) and at the bends (upper arrow). As a secondary protective measure, channel iron was placed over the lines.

FIG. 44–12.—A reference gage has been installed for each of the four separate pressure stages of this compressor. Note the plastic shield (arrow) over the gages.

3000 psi show that gages not constructed in this manner will have a hole blown in the face of the gage. Providing a ½- to 1-in. hole in the back of the gage does not give enough vent area for safe clearance of gases.

A substantial shield should be provided in front of high-pressure gages. This shield should be made of acrylic plastic (at least ⅜-in. thick, meeting MIL P5225B-Finish A Specification), and be free of scratches, gripper marks, tool chatter marks, and other stress raisers.

When mounted, a gage should have at least ½-in. clearance between itself and the item to which it is attached. If the gage is mounted flush to a backing plate, a hole of diameter at least equal the gage should be cut through the plate, leaving, of course, sufficient area to mount the face flange.

Shields mounted behind the gages should be substantial and a clear area no less than ½ inch wide should be left behind the gage for vent gases to escape between the shield and the back of the gage.

Areas where high gas-pressure systems are operating should be restricted to all but necessary personnel. Often, reactors, pressure vessels, and heat exchangers having great hazard potential should be located behind barricades and have remote control and monitoring. Particularly, vessels and systems being tested should be placed behind barricades.

References

American Society of Heating, Refrigerating and Air-Conditioning Engineers, 345 E. 47th St., New York 10017.
    Applications.
    Handbook of Fundamentals.
    Systems and Equipment.

American Society of Mechanical Engineers, 345 E. 47th St., New York 10017. *ASME Boiler and Pressure Vessel Code.*

American National Standards Institute, 1430 Broadway, New York, N.Y. 10018.
    "Pressure Piping," B31 Series.
    Safety Color Code for Marking Physical Hazards and Identification of Equipment, Z53.1.
    Scheme for the Identification of Piping Systems, A13.1.

Baumeister, Theodore (ed.) *Marks' Mechanical Engineers' Handbook,* New York, McGraw-Hill Book Co.

Combustion Institute, 986 Union Trust Bldg., Pittsburgh, Pa. 15219 (General.)

Compressed Gas Association, Inc., 500 Fifth Ave., New York, N.Y. 10036. *Cylinder Service Life: Seamless, High-Pressure Cylinders,* Pamphlet C-5.

National Board of Boiler and Pressure Vessel Inspectors, 1155 N. High St., Columbus, Ohio 43201. (General information.)

National Fire Protection Association, 470 Atlantic Ave., Boston 02210.
    Fire Protection Handbook.
    Life Safety Code, Standard No. 101.
    "Prevention of Furnace Explosions," Standards 85, 85B, 85D, and 85E.

National Safety Council, 425 N. Michigan Ave., Chicago, Ill. 60611.
    Industrial Data Sheets
        Cleaning With Hot Water and Steam, 238.
        Low Voltage Extension Light Cords and Systems, 316
        Maintenance of High-Pressure Gate and Plug Valves, 440.
        Pressurized Gas Systems in the Aerospace Industry, 590.
        Unfired Pressure Vessels and Piping in Pulp Mills, 243.
        Vessel Entry in the Rubber Industry, 458.
        Vulcanizers in the Rubber Industry, 553.

Uniform Boiler and Pressure Vessel Laws Society, 57 Pratt St., Hartford, Conn. 06103.
    Synopsis of Boiler and Pressure Vessel Laws, Rules, and Regulations, by States, Cities, Counties, and Provinces, in the United States and Canada.

Motorized Equipment

Chapter

45

Cost of Vehicle Accidents   1414

Vehicle Safety Program   1414
Responsibility . . . Driver safety program . . . Accident reporting
procedure . . . Corrective interviewing . . . Driver record cards
. . . Fleet accident frequency . . . Selection of drivers . . .
Information-gathering techniques . . . Driver training . . . Safe-
driving incentives . . . Motor trucks . . . Preventive maintenance

Repair Shop Safe Practices   1427
Servicing and maintaining equipment . . . Tire operations . . .
Fire prevention . . . Grease rack operations . . . Wash rack
operations . . . Battery charging . . . Gasoline handling . . .
Other safe practices . . . Training repair shop personnel

Off-the-Road Motorized Equipment   1432
Haul-roads . . . Driver qualifications and training . . . Operating
trucks near workmen . . . Procedures on dumps . . . Protective
frames for heavy equipment . . . Transportation of workers . . .
Towing . . . Power shovels, cranes, and similar equipment . . .
Motor graders, bulldozers, and scrapers

References   1445

Motorized equipment discussed in this chapter includes trucks, passenger cars, buses, motorcycles, and off-the-road equipment such as bulldozers, cranes and road-graders. Powered industrial trucks and handtrucks are covered in Chapters 27 and 24, "Materials Handling and Storing," and "Powered Industrial Trucks" respectively.

Safe operation of these vehicles is the result of planning and action, not chance. Often the problem of ensuring their safe operation is given insufficient attention; the reason may be lack of awareness of the problem or the difficulties of organizing an adequate safety program and providing good supervision.

In about 85 or 90 percent of all motor vehicle accidents, unsafe acts of drivers or maintenance of equipment can be identified as the cause; only 10 or 15 percent are due to mechanical failure of vehicles. Modern vehicle accident prevention effort focuses attention on these two principal accident factors—driver failure and vehicle failure—because both can be controlled.

Experience has shown that driver failure can be controlled by a carefully planned program of driver selection, training, and supervision, and that vehicle failure can be reduced by systematic preventive maintenance.

Cost of Vehicle Accidents

Experience has proved that the unsupervised fleet will have higher accident losses than the supervised one. The total cost of a vehicle accident far exceeds the amount recovered from the insurance company.

Control of accidents in the large motor transportation fleet is especially desirable because insurance premiums eat up profit. Insurance premiums fluctuate with accident frequency and dollar losses sustained by the fleet. (See Chapter 6, "Accident Records and Injury Rates.")

The cost of insurance, however, is only one of the costs which are levied against a company. There are indirect costs, also. As with other work accidents, these may be several times the direct costs. Indirect costs may be listed as follows:

1. Salary paid, and loss of service of, the employee injured in an accident. Loss of

a key salesman or company representative at a critical time is incalculable.

2. Added workmen's compensation costs resulting from a disabling injury.

3. Loss of use of the vehicle while it is being repaired or replaced.

4. Cost of supervisory time spent in investigating, reporting, and in cleaning up after the accident.

5. Cost of repairing the company vehicle.

6. Cost of repairing or replacing other company property.

7. Poor customer and public relations resulting from a company vehicle having been involved in an accident.

8. Cost of replacing and training an injured employee.

9. Time lost by co-workers in discussing the nature of the accident and extent of injury.

In addition to appealing to the profit motive, the most telling argument for controlling accidents is the company's moral obligation to the employee and to the public. The employer through his authority to hire, supervise, discipline, and discharge employees exercises a high degree of control over their driving performance. Motor vehicle collisions are, in fact, wasteful errors traceable to poor management of the fleet operation.

Management can exercise that control by employing the simple and proven techniques of driver safety education and supervision.

Vehicle Safety Program

A vehicle safety program should provide for:

1. A definite safety policy, originated, supported, and emphatically enforced by top management, with delegated authority.

2. A safety director (full or part time) to advise top management.

3. A driver safety program, including driver selection procedure, driver training, and interest-sustaining activities.

4. An efficient system for accident investiga-

tion, reporting, analysis for cause, determination of appropriate corrective action, and follow up.

5. Preventive maintenance procedures.

Responsibility

The first requirement of a good driver safety program is that management, from the president to the immediate supervisor, must accept responsibility for safe operation of company vehicles. As in the case of other desirable qualities of job performance, management must first demand safe operation, define standards of acceptable performance, and then organize means for the inspection and correction of job performance to meet these standards.

Management must evolve and enforce the policy that practical accident prevention is a requirement of employment.

A capable and responsible person in the organization should be designated safety professional and should be made responsible for supervising the company's program for the safe operation of all automotive equipment. In a small industrial concern, this may be a part-time assignment. Some of the duties of the safety professional are:

1. Advise management on accident prevention and safety.

2. Develop and promote safety activities and work injury prevention measures throughout the fleet.

3. Study and recommend fleet safety policy in relation to equipment and facilities, personnel selection and training, and other phases of fleet operation.

4. Evaluate driver performance.

5. Conduct or arrange for effective safety training.

6. Review accidents for determination of cause. Compile and distribute statistics on accident-cause analysis and experience. Identify "problem" persons, operations, and locations.

7. Maintain individual driver-safety records, and administer the safe-driver award incentive program.

8. Procure (or prepare) and disseminate safety educational material.

Driver safety program

A driver safety program for a fleet should include the following five basic accident prevention procedures.

1. Set up an in-service driver training program.

2. Discuss preventability of each accident with persons concerned.

3. Require immediate reporting of every accident.

4. Compute and publicize the fleet accident record.

5. Maintain an accident-record card for each driver.

The planning and administration of a safety program for motor transportation fleets are described in greater detail in the National Safety Council's *Motor Fleet Safety Manual.*

An accident is defined as "any incident in which the company vehicle comes in contact with another vehicle, person, object, or animal, which results in death, personal injury, or property damage, regardless of who was injured, what property was damaged or to what extent, where it occurred, or who was responsible."

This definition includes even minor accidents involving little more than a fender scratch. All vehicle accidents, major and minor, are of importance to the safety supervisor since he is primarily concerned with the eradication of faulty driving habits or attitudes. Minor accidents, just as much as the more spectacular ones, provide clues to such faults.

Accident reporting procedure

Each driver should be required to make out a complete accident report on a standard accident report form for every accident in which his vehicle is involved. If possible, this report should be turned in to the supervisor on the day the accident occurs. National Safety Council "Form Vehicle 1" is a good example of the type of report form used by
(*Text continues on page 1419.*)

45—Motorized Equipment

National Safety Council
Form Vehicle 1

MOTOR TRANSPORTATION
DRIVER'S ACCIDENT REPORT

READ CAREFULLY - FILL OUT COMPLETELY

(For office use)

FILE NO.

- [ ] PREVENTABLE
- [ ] NOT PREVENTABLE
- [ ] REPORTABLE
- [ ] NOT REPORTABLE

COMPANY

DIVISION

ADDRESS

ACCIDENT INVOLVED

M O V I N G
- [ ] Another com'l vehicle
- [ ] Passenger car
- [ ] Pedestrian
- [ ] (Specify other)

F I X E D
- [ ] Building or fixture
- [ ] Parked vehicle
- [ ] (Specify other)

TIME

Date of Accident .... , 19 .... Day of Week .... Hour .... a.m. p.m.

LOCATION

PLACE WHERE ACCIDENT OCCURRED:
- [ ] CITY
- [ ] SUBURBAN
- [ ] RURAL

County .... City, town or township ....

_____ miles *north-south* of ____ □ limits of City or Town
_____ miles *east-west*   ____ □ center of

If accident was outside city limits indicate distance from nearest town. Use two distances and two directions if necessary.

ROAD ON WHICH ACCIDENT OCCURRED: .... *Give name of street or highway number (U.S. or State)*

- [ ] AT ITS INTERSECTION WITH: .... *Name of intersecting street or highway number*

OR _____ feet *north-south*
_____ feet *east-west* of.

- [ ] NOT AT INTERSECTION

(Check and complete one)

Show nearest intersecting street or highway, house number, curve, bridge, railroad crossing, alley, driveway, culvert, milepost, underpass, numbered telephone pole, or other identifying landmark. Show exact distance, using two directions and two distances if necessary.

TYPE

- [ ] HEAD ON
- [ ] SIDESWIPE
- [ ] RIGHT ANGLE
- [ ] REAR END
- [ ] OTHER (Describe)
- [ ] FRONT END
- [ ] NON COLLISION (Describe)

DRIVERS

COMPANY VEHICLE NO 1 | VEHICLE NO 2

→ Driver's Name →
→ Address →
→ City And State →
→ Driver's License →

- [ ] Chauffer
- [ ] Operator
- [ ] Beginner

Age .... Sex .... Driving experience .... yrs

Age .... Sex ....

(If vehicle driven by other than owner)
Owner's name ....
Address ....
City and State ....

- [ ] Chauffer
- [ ] Operator
- [ ] Beginner

Date employed .... *month* .... *day* .... *year*

Hours on duty since last period of 8 consecutive hours off duty .... *Actual hours of driving since last period of 8 consecutive hours off duty*

Condition of Driver ....

1416

COMPANY VEHICLE NO I			VEHICLE NO 2	
Make .... Year .... No.	→ Type of Vehicle →		Make .... Year .... No.	
State .... No.	→ License → State .... No.			
	→ Vehicle Damage →			
	→ Other Damage → (cargo loss etc.)			

VEHICLES

INJURED	Name	Address	Age	Sex	Describe Injuries
Driver vehicle 1					
Driver vehicle 2					
Passenger veh.					
Passenger veh.					
Pedestrian					
Pedestrian					
Others					

WITNESSES	Name	Address	Remarks
Company representative			
Insurance representative			
Police		Badge no	Station

TURN THE PAGE - COMPLETE BOTH SIDES!

Printed in U.S.A.

Stock No. 229.31

FIG. 45-1.—Good accident reports help supervisors spot faulty driving habits and plan their correction. Training in filling out accurate reports promptly impresses safety on the minds of drivers. Report forms can be carried in glove compartments. Space is given on the back of this sheet for describing vehicles involved, pedestrians, weather conditions, a diagram of what happened, driver's account of the accident, and his suggestions for preventing future accidents of this type. (Only the front page is shown here.)

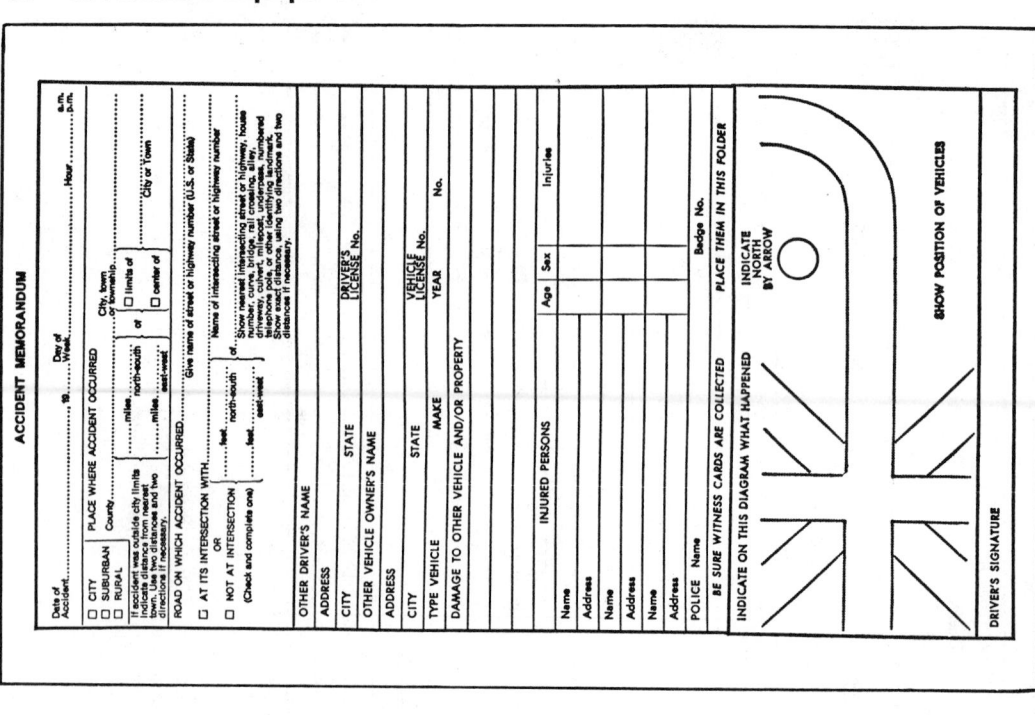

FIG. 45–2.—Accident Report Packet can be carried in glove compartment. Instructions are printed on this side of the envelope, and an accident memorandum form is printed on the other.

many motor transportation safety supervisors. (Fig. 45–1 shows only the front page of this form.)

Drivers should be told how to accurately and intelligently fill out the accident report form. Failure to report an accident, no matter how slight, or falsification of data on an accident report should be made cause of disciplinary action against the driver.

Drivers should be required to complete the report at the scene of the accident, if possible, and then send it promptly to their supervisor.

An accident report packet that can be carried in a glove compartment will be found very useful. This packet (Fig. 45–2) should provide for a memorandum report of the accident. It should contain a pencil, plain paper, courtesy cards, and a list of the telephone numbers of company officials and of insurance representatives. The necessary note-taking material will thus be available at the scene of an accident.

Specially printed courtesy cards will save time and will help the driver get names and addresses of key witnesses (Fig. 45–3). Drivers should be impressed with the importance of identifying as many witnesses as possible.

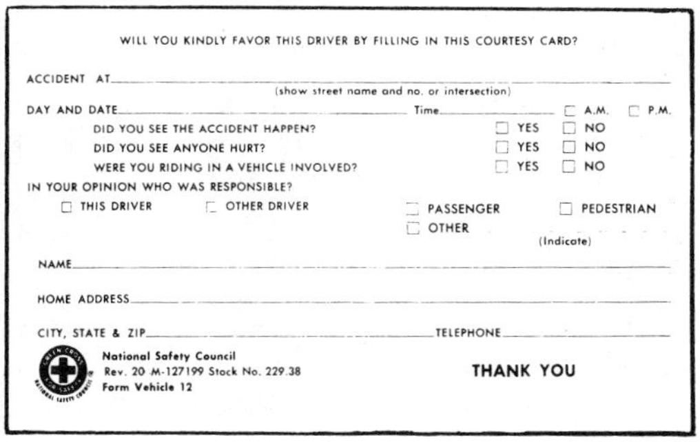
WILL YOU KINDLY FAVOR THIS DRIVER BY FILLING IN THIS COURTESY CARD?

ACCIDENT AT ______ (show street name and no. or intersection)

DAY AND DATE ______ Time ______ ☐ A.M. ☐ P.M.

DID YOU SEE THE ACCIDENT HAPPEN? ☐ YES ☐ NO

DID YOU SEE ANYONE HURT? ☐ YES ☐ NO

WERE YOU RIDING IN A VEHICLE INVOLVED? ☐ YES ☐ NO

IN YOUR OPINION WHO WAS RESPONSIBLE?

☐ THIS DRIVER ☐ OTHER DRIVER ☐ PASSENGER ☐ PEDESTRIAN ☐ OTHER ______ (Indicate)

NAME ______

HOME ADDRESS ______

CITY, STATE & ZIP ______ TELEPHONE ______

National Safety Council
Rev. 20 M-127199 Stock No. 229.38
Form Vehicle 12

THANK YOU

Fig. 46–3.—When distributed, filled in, and collected, courtesy cards help determine who saw the accident in case witnesses are needed later.

In the case of serious accidents, especially those resulting in a fatality or personal injury, a representative of the company—the manager, safety director, or claim agent—should make a personal investigation at the scene of the accident as quickly as possible.

The purpose of such an investigation is to verify the accuracy of information submitted by the driver on the accident report form and to obtain other data which might prove valuable for accident prevention work or for defense against unjust claims.

Corrective interviewing

After an accident, the company driver involved should be interviewed by the safety professional to determine whether or not the driver might have prevented it. If, after reading the report form and discussing the matter with the driver, the safety professional feels that the driver could have helped prevent the accident, it should be classified "Preventable."

In this case, the safety professional must explain to the driver what he did or failed to do that contributed to the accident. He must also make sure that the driver understands what he must do to prevent similar accidents in the future.

If, in the safety professional's opinion, the driver did everything he reasonably could to prevent the accident, it should be classified "Nonpreventable."

It might be pointed out here that responsibility for the prevention of accidents includes more than careful observance of traffic rules and regulations. Drivers must drive in such a manner as to prevent accidents, regardless of faulty driving or nonobservance of traffic laws on the part of other drivers.

Driver record cards

A record card should be maintained for each employee who drives a company vehicle. This card furnishes not only a record of accidents, but also the information needed for safe driver award plans or other forms of recognition (Fig. 45–4). The date of each accident and the accident category (preventable or nonpreventable) should be entered on this card. The safety professional should review the cards at least every six months and note those drivers who have had an excessive number of accidents.

When a driver becomes an accident repeater, management should make every effort to rehabilitate him through counselling, retraining, closer supervision, or reassignment.

If retraining, reassignment, or penalties fail to curb preventable accidents for a driver,

45—Motorized Equipment

MOTOR TRANSPORTATION
AWARD AND ACCIDENT RECORD

| JAN | FEB | MAR | APR ✗ | MAY | JUN | JUL | AUG | SEP | OCT | NOV | DEC |

National Safety Council
Form Vehicle 6

Name ....Smith,..............Bill...............E.
       Last   *First*   *Middle*

Company ....Highway Express

Location ....Midwest Division

Address....123...Home St.,,....Any Town
          Number *Street*

Badge Number ...123

Date Employed ....Feb. 1, 19-- Age 29
                Month Day Year *At Emp.*

SAFE DRIVER AWARD RECORD

Earned....2....Year N. S. C. Award during period
from ..3/2/19--...... to ....3/2/19--........with
Company....Overnight Freight, Inc.
Certified by...F. J. O'Connell, supt.

DRIVING TESTS

Date	Score	Remarks
2/4/19--	90	Slow reaction time

Award earned	Certificate Number	Date Award earned	Preventable Accidents		Non-Driving Time		REMARKS:
			File No	Date	From	To	
3	468950	3/2/19--	1016	4/10/19--	12/1/19--	1/1/19--	12/1/19-- took one month leave of absence.
4	501850	3/2/19--					
5	600100	4/2/19--					

STOCK No. 229.36

FIG. 45–4.—Award and accident record should be kept as part of each driver's personnel record. Form is useful in administering a company award plan and in counseling accident repeaters.

discharge or assignment to nondriving duties will be in the best interests of both the firm and the employee.

Fleet accident frequency

A useful accident control tool is the practice of computing the accident frequency rate of the fleet each month or quarter, in terms of number of accidents per 1,000,000 vehicle miles. Vehicle miles should be computed from odometer readings of all vehicles and not left to rough guesses based on route mileages unless the operations of the fleet are stable from day to day. The standard formula for figuring a fleet accident rate is:

$$\text{Fleet accident frequency rate} = \frac{\text{No. of accidents} \times 1,000,000}{\text{miles driven}}$$

By keeping monthly record of frequency rates, the safety professional can:

1. Tell whether group safety performance is improving or getting worse,

2. Compare the record with like periods to evaluate seasonal trends, and

3. Compare the status of the fleet with that of other fleets having similar operations.

He can then plan his program accordingly. Accident frequency rates are also of interest to management since they show the effectiveness of the safety program.

(The National Fleet Safety Contest conducted annually by the National Safety Council provides a monthly bulletin which gives the accident frequency rates of all participating fleets and thus provides a means for comparison of a company's experience with that of similar fleets.)

Selection of drivers

For some jobs that require vehicle driving, such as that of sales representative or technician, other qualifications unfortunately often outweigh a safe driving record, so that the prospective employee's competence as a driver

1420

is investigated in incomplete fashion or not at all.

However, when a man is to be hired for or assigned to a job which will require him to drive a motor vehicle, every effort should be made to select an individual who gives promise of the ability to drive safely.

The following factors should be weighted heavily.

- **Age.** Men under 25 years of age, especially those under 21 (who lack formal driver training) are not considered good accident risks by insurance companies. The personnel officer should take special pains to analyze the qualifications of applicants under 25, and should assign to driving only those who evidence mature, stable personalities (preferably those who have successfully completed driver training courses).

- **Experience.** Men who have a record of frequent involvement in vehicle accidents should not be assigned to drive company vehicles. A man's safety record should be investigated (*a*) in the personal interview, (*b*) by consulting with former employers, and (*c*) by checking the state motor vehicle department for accident records, and (through them) the national driver register service for out-of-state or two-license revocations. The Federal Highway Administration, Washington, D.C. 20591, maintains a central national register of motor vehicle operators' permits or licenses revoked or suspended for highway safety code violations. Service is only to state officials, so query your Secretary of State for this information. See details in Chapter 23, "Sources of Help," under U.S. Department of Transportation.

- **Attitude.** Dissatisfied, timid, cocky, troublesome, or otherwise temperamental or unstable individuals often do not make good drivers.

A close relationship has been noted between the ability to drive safely and such personal traits as dependability, courtesy, pleasant personality, and the ability to get along harmoniously with other people.

Conversely, persons who tend to be antisocial, argumentative, and impulsive are suspect as drivers.

Individuals differ in their ability to act safely. The fleet safety program must therefore begin at the employment office.

In determining standards of selection, a careful analysis of the job and of the qualifications a driver should have in order to perform it satisfactorily should be made. Results of the job analysis are embodied in a job description in which the separate tasks involved in the job are completely and accurately described.

After the job has been analyzed, the next step is to decide what qualifications the applicant must have to perform the job satisfactorily. There should be a sound reason for each qualification finally imposed. It may be helpful to study the qualifications of those employees who are performing the job in an average or better than average manner. Their qualifications should indicate the requirements expected in new employees.

Safe driving ability should always be paramount. Otherwise, accident losses might completely offset any advantages gained through a driver's other special abilities.

Information-gathering techniques

After job essentials and qualifications have been determined, the next step is to develop methods for gathering and sifting employment data about each applicant. Standard employment procedure includes:

Application form
Personal references
Interview
Psychological tests
Driving tests
Physical examination

- **The application form** is a printed or mimeographed form on which the applicant submits details of past employment and other personal data. What a man has done in the past is a good indication of what he can be expected to do in the future. The completed form also saves time during the interview, since essential data are made available to the interviewer at a glance.

Questions on the form should bear directly on the basic qualifications for the job, and should be arranged in logical sequence. Specifically, questions on driving experience should include mileage and years spent as a driver and types of vehicles operated, seasons of the year and geographical areas in which vehicles were operated, preventable and non-preventable accidents experienced, number of convictions for traffic and other violations, number and type of driver's licenses held, and safe driver awards received.

- **Personal references.** The applicant should furnish the names and addresses of previous employers in the space provided on the application form, and these references should be checked. Additional reference sources are the local credit bureau, police department, and state motor vehicle department, as mentioned before.

- **The interview** should be as objective and reliable as possible. Many firms use a planned or patterned interview to ensure that all interviews follow the same general course and bring out the specific information desired.

A properly conducted interview should reveal additional facts about the applicant's employment experience, knowledge of traffic regulations, attitude, personality, appearance, family life, and general background.

The interview should be conducted in private, and the applicant should be seated and put completely at ease. The interviewer should keep in mind at all times the inventory of basic qualifications. A check list of these may be made up to serve as a guide. After the interview, the applicant can be rated on each of the qualifications listed.

- **Psychological tests** are devices for obtaining samples of behavior under controlled conditions. They can be useful in the selection process if the behavior traits sampled are known to be related to job success and if interpretation of the results is made by a qualified person. However, since available tests are not reliable instruments of prediction, tests alone cannot do the complete selection job. The personnel officer should never use tests as a substitute for other ways of securing employment information. Until tests can be developed which will correlate very highly with our accident criterion, it is not practical to rely on them only.

Many of the large trade associations have had personnel specialists develop standardized personnel selection procedures for their member organizations. These procedures usually include psychological tests which have been found applicable to driver selection.

Small fleets that are not members of an association providing this service should be able to retain, for a reasonable fee, a qualified personnel psychologist for advice on this matter. The psychology departments of some state universities give valuable advice on this problem through their extension services.

- **Driving tests.** Each applicant should undergo an actual driving test or a motor ability test as part of selection procedure. Some firms rely on a motor ability test for indication of driving ability and on an extensive driver training course following employment for developing that ability.

Driving tests are of two kinds, the driving range type and the in-traffic type. The requirements of a good driving test in traffic can be listed as follows:

1. The road test should be long enough to fairly sample a number of typical driving situations. Certainly 20 minutes should be considered minimum.

2. The test should include difficult maneuvers in heavy traffic as well as on "freeways," in order to really test a driver's ability. Almost anyone can successfully "drive around the block."

3. A standard scoring procedure and predetermined test route should be used so that the test will be the same for all drivers examined.

4. The examiner should check definite items concerning the driver's performance (in order to reduce subjective judgment) and point out driving faults that may be corrected by proper training.

The applicant's test performance will indicate if driver training is needed, and will indicate weaknesses that should be corrected during the training period.

- **Physical examination.** Applicants should be examined by a qualified physician before being hired. For firms engaged in interstate commerce, physical examination of all new drivers is mandatory. The regulations* prescribed by the U.S. Department of Transportation, Federal Highway Administration, for

Fig. 45–5.—Driver test apparatus. *Left:* Brake reaction time tester. Driver keeps foot on accelerator. When signal lights, timer starts. Driver must move his foot from the accelerator to push the brake pedal. Timer measures the time (in hundredths of a second) that it takes. *Right:* Hand steadiness tester. Driver inserts metal stylus (at lower right) in top of instrument and trys to move it down through the constantly decreasing aperture. Each time the stylus touches the metal sides, a light flashes. Number of touches is recorded either automatically or by means of an observer.

Courtesy Chicago Motor Club, A.A.A.

bus and truck operators coming under its jurisdiction require physical examinations for all new drivers and provide that motor carriers must have on file for each new driver a certificate showing him to be physically qualified.[*]

Most drivers are, in addition, given a series of psycho-physical tests of vision, reaction time (Fig. 45–5), color perception, depth perception, and hearing. Substandard findings are submitted to competent medical authority for evaluation. Drivers are told of their weaknesses and how best to compensate for them.

Driver training

Individual training by a skilled instructor

for all employees assigned to drive company motor vehicles is a highly desirable objective. Various types of courses are available.

REMEDIAL, for drivers who get into trouble;

REFRESHER, for periodic updating of all drivers, and

SPECIAL, for operators of specialized equipment.

---

[*] U.S. Dept. of Transportation, Federal Highway Administration, Washington, D.C. 20591. *The Motor Carrier Safety Regulations*, Part 391, "Qualifications of Drivers."

Fig. 45-6.—A typical scene of training in NSC's Defensive Driving Course.

A training course must be planned to fit each job, training materials assembled, and classroom facilities provided (see Chapter 9, "Safety Training"). The objective, however, is well worth the investment. The driver training course should cover the following points.

- **State and municipal driving rules.** Most state motor vehicle departments publish books of rules and regulations for those seeking drivers' licenses. The training course should cover salient points in such booklets.

- **Company driving rules.** Each company has rules governing the use of company vehicles, how the vehicles may be obtained, where they may be operated, where parked and under what conditions, speed to be observed, and so on. These rules should be covered thoroughly in the training course.

- **What to do in case of an accident.** This topic should include instructions on how to make out company accident report forms, how individual driver records affect the employee, what to do in case of a vehicle accident away from the plant, and similar points. See Fig. 45-2.

- **Defensive driving.** This concept embraces all the commonsense rules of safe and courteous driving. This part of the course should seek to build in the prospective driver a high sense of responsibility not only for the safety of his own vehicle but for the safety of other street and highway users who are less skilled and who have had less training and practice (Fig. 45-6) than he has had. At this time, the company's safe driver award or incentive plan should be explained.

The Harold L. Smith System of Space-Cushion Driving, for example, outlines the five cardinal principles of professional driving: (a) Aim high in steering, (b) Get the big picture, (c) Keep your eyes moving, (d) Leave yourself an out, and (e) Make sure they see you.

Periodic review of the five points listed here is important. Check rides made by the safety professional will provide an opportunity for appraisal of a driver's abilities and for personal instruction to help overcome deficiencies.

Safe-driving incentives

One of the basic assumptions of a safety program is that most drivers believe that they know how to drive much better than they really do. Proceeding on this assumption, then, most safety professionals regard it as part of their job to provide effective motivation in various ways so that drivers will use more of their driving skill more of the time. To supply this motivation directly or indirectly, the safety professional should:

1. Require a detailed report of every accident.

2. Interview the driver after each accident to determine whether or not he could have prevented it.

3. Keep a record of each driver's safety performance.

4. Recognize safe driving performance. Safe driver awards and cash or merchandise prizes for driving for stated periods without a preventable accident are strong motivations.

5. Provide continuous safety instruction and reminders. Use all media: company newsletters and bulletins, booklets, posters and bulletin board displays, and meetings and direct personal conversation. ("Motor

Fig. 45–7.—Employees working with hot asphalt should wear clothing that covers their bodies completely. Vehicles should have adequate safety devices and be well maintained.

Courtesy Armstrong Holdings Ltd., Brampton, Ontario, Canada.

Fleet Talks" and "Motor Fleet Topics," both published by National Safety Council, contain 5-minute safety talks for drivers.)

Motor trucks

Of the various types of motorized equipment, trucks are most frequently involved in accidents. Many of the worst accidents are due to lack of safety devices and particularly to inadequate maintenance. Therefore, fundamental requirements for safe operation are that trucks be equipped with the necessary safety devices and that vital parts, such as brakes and steering mechanisms, and headlights, taillights, and horn be maintained in first-class condition. (See Fig. 45–7.)

The term "safety devices" includes:

Directional signals
Windshield wipers
Windshield defroster
Fire extinguisher
Power steering
Low air pressure warning system
Rock guards over the drive tires
Adequate outside mirrors
Backup light
Audible backup signal for heavy-duty trucks
Nonslip surfacing on fenders, floors, and steps
Safety belts
High-quality tires
Automatic sander
Anti-jackknife device
Reflective markings

1425

In addition to these, the following are recommended for dump trucks.

A light or indicator to show when the body is in a raised position

A CAUTION sign on the rear of "packer-loader" trucks.

Cab protector or canopy

A built-in body prop

Loading and unloading of trucks. To reduce the danger to the driver from falling material while his truck is being loaded by power shovel, clam shell, or other similar loading device, the truck should be spotted so that the bucket does not swing over the cab or seat. If a truck cannot be so located and does not have a protective canopy over the cab, the driver should dismount and stand clear of the truck and the bucket.

Accidental injuries incurred in the loading and unloading of materials, such as lumber, pipe, equipment, and supplies, are especially numerous, but can be avoided if these precautions are followed:

1. The bulk and weight capacity of the truck should be observed.

2. Loads which may shift should be blocked or lashed. Tiedowns (ropes, chains, boomers) should be tightened on the right side or top of the load.

3. If material extends beyond the end of the tailgate, a red flag (or, at night, a red lamp) should be fastened to the end of the material. No material should extend over the sides.

4. Before loading or unloading a truck, the brakes must be securely set or the wheels blocked to protect the men both on the truck and on the ground.

5. A truck should not be moved until all workers are either off the truck or properly seated on seats provided and are protected from injury if the load should shift during transit.

6. To avoid falling when unloading a flatbed truck, employees should keep back as far as possible from the sides of the truck, especially when shoes, floors, and loads are wet or muddy.

7. Be alert for pinch points when loads are being pulled, hauled, or lifted.

8. All safe practices for materials handling, such as using mechanical handling equipment, getting sufficient help, and so on, should be observed.

Detached trailers. When loading and unloading detached trailers with a lift truck, be sure that the wheels are adequately blocked. A steel chock, available from several manufacturers, is very good, but a tapered block cut from 6 by 6 in. timber is adequate.

It is important to place the chock properly when trailers are at the dock being loaded and unloaded. Preferred placing of chocks is under the rear set of wheels (Fig. 27–9, page 746). If trailer is not blocked properly, the vehicle may move because of an incline or be set in motion by the loading or unloading operation.

The trailer nose can be supported by screw or hydraulic jacks—one on each side of the nose—in order to strengthen the support of the landing gear assembly. Under a heavy forklift load, landing gears have collapsed due to the weight because of dolly metal rust or fatigue, defective struts, or some other cause.

Preventive maintenance

Well managed motorized equipment, both highway and off-the-road, is covered by an extensive and more or less complicated preventive maintenance program, the primary considerations of which are economy and efficiency. Such a program, based on either the mileage or the operating hours of the equipment (as recommended by the manufacturer) determines when oil will be changed, tires rotated or replaced, and minor and major overhaul jobs undertaken.

The objectives of such a program are:

1. To prevent accidents and delays,

2. To minimize the number of vehicles down for repair,

3. To stabilize the work load of the maintenance department, and

4. To save money by preventing excessive wear and breakdown of equipment.

Such a program should, as a matter of

course, cover all mechanical factors relating to safe operation of all motorized equipment, such as brakes, headlights, rear and stop lights, turn signals, tires, windshield wipers, muffler and exhaust system, steering mechanism, glass, horn, and rearview mirrors. (The manufacturer of the equipment used by the company can help with the specifications.)

If at all possible, each driver or operator should be assigned a specific vehicle in order to fix responsibility for reporting defects as well as to encourage drivers and operators to take better care of their vehicles. This technique was used effectively by the Ordnance Department, U.S. Army, in World War II.

Drivers and operators can play an important role in a preventive maintenance program if they are properly instructed and motivated. Because they are most familiar with the vehicle and how it normally operates, they are usually the first to notice when minor (as well as major) mechanical defects develop.

Drivers should check their vehicles before each dispatch. At the end of the workday, they should submit a checklist report of repairs or adjustments needed before the vehicle is used again. It is important that maintenance men follow up on these, otherwise the benefit is lost.

This printed checklist (adapted from the outline that follows) should be furnished all drivers and operators for the inspection. On completion, they should give the inspection form to the supervisor. This form acts as a reminder as well as a checklist, and should normally cover the following items:

BRAKES. Brakes should apply evenly to all wheels so that a vehicle does not swerve when the brakes are applied.

HEADLIGHTS should function and be properly aimed to avoid blinding other motorists and to give maximum road lighting efficiency. The dimming switch and upper and lower beams should work properly.

CONNECTING CABLES on a combination vehicle should have strong connections which will not be affected by the vibration of the vehicle.

STOP LIGHTS, turn lights, rear lights, and side-marker lights should be checked.

TIRES should be inflated to manufacturer's recommended pressure, and checked regularly for adequacy of tread and for cuts or breaks. Dual tires should be well matched.

WINDSHIELD WIPERS must wipe clean and not streak.

STEERING WHEEL should be free from excessive play. Front wheels should be properly aligned.

GLASS should be free from cracks, discoloration, dirt, or unauthorized stickers which might obscure vision.

HORN should respond to a light touch.

REARVIEW MIRRORS should give the driver a clear view of the rear. So outside rearview mirrors can provide maximum sight advantage, portions can be conventional and convex.

STALLING PROBLEMS should be investigated and corrected immediately.

INSTRUMENTS should be in good working order; they are essential to safe and economical operation.

EXHAUST SYSTEM should be checked for leaks to protect against carbon monoxide gas. The exhaust manifold, pipe connections, and muffler should be inspected periodically, and leaky gaskets replaced.

EMERGENCY EQUIPMENT in every vehicle should include a fire extinguisher, essential tools for road repairs, spare bulbs, flares, reflectors, flags, and such other equipment deemed necessary in case of fire, accident, or road breakdown. These items should be periodically checked to make sure of their availability and usability.

Many states and cities require periodic safety tests and inspections for all vehicles. The maintenance superintendent should know the applicable inspection standards. The preventive maintenance policy of the company should require all vehicles to meet these requirements.

Repair Shop Safe Practices

The vehicle safety supervisor also should take an active interest in the work habits of automotive repair shop employees to determine their safety attitudes, and should co-operate with the plant safety supervisor.

Servicing and maintaining equipment

Serious injuries occur in servicing and maintaining trucks. Heavy equipment requires mechanical aids for handling heavy parts. Hoists for lifting parts in and out of trucks and for moving parts about the shop not only prevent accidents, but make work easier and save time.

Serious injuries are likely to occur from unexpected movement of equipment undergoing repair. Brakes should be set and wheels should be blocked (see Fig. 27–9). If work must be done under a raised body, the body must be secured or blocked against coming down in case the control levers or pedals are inadvertently struck.

Jacks often are used to raise equipment, which then becomes a support in an unstable position. Because serious injuries occur when a truck falls off jacks, it is important that they be set on a firm foundation and be exactly perpendicular to the load. To help prevent the jacks from slipping, a thin wood block between the top of the jack and the load is recommended. When the truck has been raised to the desired height, it should be supported by stanchions, blocking, or other secure support.

No work should be done near the fan or other exposed moving parts until the motor has been stopped. If the motor must be run to inspect or check on moving parts, keep a safe distance away and do not attempt an adjustment.

Close-fitting unfrayed clothing, safety shoes, and goggles are essential for repairmen.

Burns are frequent in the servicing of trucks. An employee should use a heavy work glove, "bleed off" any steam, and then remove the cap. Gasoline or alcohol used near hot motors and spilled on them can cause a serious fire. Suitable funnels and containers should be provided.

Tire operations

A particularly serious hazard in inflating truck tires is the possibility that the locking ring may blow off at high pressure. Use of a tire safety rack will greatly reduce the hazard. Tires should be inflated in steel "cages"

or similar devices which will restrain flying objects should a blowout occur. A locking ring must be seated properly and must not be yanked free by being twisted. A defective locking ring or rim should be replaced with a sound one.

Blowouts may occur because of overinflation of the tire, improper placement of the tire on the rim or wheel (causing pinching or chafing of the tire or tube), or improper mounting of lock-rings or rims. Records show that most accidents in the handling of truck tires occur while tires are being inflated.

Only employees thoroughly familiar with the hazards and safe methods involved in handling tire equipment should inspect, install, repair, and replace tires and rims.

Other hazards are strains or hernias resulting from lifting heavy tire assemblies. Mechanical lifting and moving devices should be provided so that workers are not required to lift heavy tires.

Rubber cement and flammable solvents used for patching inner tubes, and casing compounds used for filling tire cuts, should be kept in safety cans.

Electric heating elements used for vulcanizing or branding tires should be inspected regularly. Defective wiring should be replaced.

Where power-driven rasps or scrapers are used for casings or inner tubes, the operators should be required to wear goggles (see Chapter 35, "Hand and Portable Power Tools"). A local exhaust system should be applied to these machines to keep the fine rubber dust out of the workroom air. This dust should be collected because deposits make walkways very slippery.

Fire prevention

Since fire is a hazard in the operation of heavy-duty trucks, they should be equipped with Type B-C fire extinguishers, listed by the Underwriters' Laboratories as suitable for use on burning oil, gasoline, grease, and electrical equipment. The extinguisher may be placed in a convenient location in the cab or on the running board, and the driver should be taught how to operate it. Monthly inspection of fire fighting equipment is advisable. See Chapter 43.

If cutting, burning, or welding must be

done near fuel or oil tanks, an extinguisher should be at hand. A tarpaulin should be used to cover fuel or oil tanks, or combustible materials to protect them against sparks and excessive heat. Such work should not be done on a fuel tank or container until it has been drained and thoroughly purged of vapors.

Fueling requires certain precautions to avoid fires. The motor should be stopped. Smoking and use of open lights should be prohibited. A vapor-proof portable electric lantern or flashlight is provided by some companies. When the tank is being filled, the metal spout of the hose should firmly contact the tank to ensure grounding and neutralize static charges sufficient to ignite gasoline vapors and cause an explosion.

Fires occur in shops each year because gasoline and similar highly flammable solvents are used for cleaning parts. Good cleaning liquids that are nonflammable and do not injure the skin are obtainable and should be used.

The likelihood of a fire in a shop also can be reduced by good housekeeping, especially the disposal of oily waste and similar materials in covered metal receptacles.

Grease rack operations

In greasing operations, employees may slip and fall because of accumulated grease and oil on the floor, injure their hands on sharp or rough edges on the vehicle, incur strains in trying to rock the vehicle to make grease penetrate into stiff bearings or springs, inhale sprayed or atomized oils used for spring lubrication, or suffer hand and head injuries from high-pressure guns.

Floors should be kept free of grease and oil to prevent slips and falls. Spills which occur during the working day should be immediately covered with an oil-absorbent compound.

Remind workers to keep their hands away from sharp or rough edges and to obtain immediate first aid treatment for all cuts and scratches.

Warn workers against putting their hands in front of the grease gun nozzle when the handle is pulled. Instances have been reported in which quantities of grease have been forced under the skin of workers by high-pressure grease guns.

Tops of grease cylinders should be securely screwed or clamped into place; otherwise, covers may blow off and seriously injure persons nearby.

All equipment should be inspected weekly and repairs should be made when needed.

Men should be warned of the hazard of inhaling sprayed or atomized oils while lubricating springs. The men should stand clear of the lubricant spray, which settles quickly, and should not direct the spray at other employees.

Wash rack operations

When washing vehicles, men slip and fall on wet floors, incur cuts or abrasions from the sharp or rough edges of the vehicle, or suffer burns from careless use of hot water or steam.

The concrete floor of the wash rack should be rough troweled to produce a nonslip surface. While washing cars, employees should wear safety toe rubber boots, preferably with nonslip soles and heels, and a rubber coat or apron.

Workers should never point the high velocity streams of hot or cold water at another person because serious injury may result.

Men should use the hose, particulary under the vehicle, in such a way as to avoid being struck by a backlashing stream of water and dirt.

Where a hot water hose is used, cover the metal parts to prevent burns. Heavy-duty gloves and face shields should be provided when necessary. A portable fan may be needed to blow steam away, so that the operator can see his work. Washers should be alert for sharp and rough edges on the vehicle which might cause cuts and abrasions.

Battery charging

The principal hazards of battery charging operations are acid burns during filling, back strains from lifting, electric shocks, slips, and falls.

Employees should wear safety apparel suitable for battery shops, i.e. splash-proof goggles, and acid-proof gloves, aprons, and boots with nonslip soles. (Rubber boots and aprons must be worn when batteries are being filled. Goggles should be worn when working around batteries to prevent acid burns to the eyes.)

Wood-slat floor boards should be used and kept in good condition, to prevent slips and falls and to protect against electric shocks from equipment being charged.

Fire doors should be installed between charging rooms and other areas where flammable liquids are handled and stored.

Manufacturer's recommendations as to the charging rate for batteries of various sizes should be closely followed in order to prevent rapid generation of hydrogen. There are potential explosive quantities of oxygen and hydrogen in cells of batteries. This is particularly true if the battery is defective or if a heavy charge has been or is being applied. The lower the water in the battery, the greater the cavity for the accumulation of gas.

Care should be taken to prevent arcing while batteries are being charged, tested, or handled. Tools and loose metal should not be stored in such a position that they may fall on batteries and cause a short circuit, which in turn can cause serious burns or an explosion.

When manual lifting is necessary, sufficient help should be provided to prevent strains, sprains, or hernias. Hand carts for transporting batteries are commercially available or can be made in a company shop.

Acid carboys should be handled with special care to prevent breakage and possible injury due to splashing of acid. Acid carboys should never be moved without their protecting boxes. They should not be stored in excessively warm locations or in the direct rays of the sun. Carboy tilters can be used. (See Chapter 24, "Materials Handling and Storage.")

A summary of recommendations for changing and charging storage batteries is given in Chapter 27, "Powered Industrial Trucks," on pages 758–759.

● **First aid for chemical burns.** Many batteries contain an acid electrolyte; some, such as the nickel-iron battery, contain an alkali solution. Whether it be acidic or alkaline, if electrolyte gets on a person's skin, it should be washed off immediately with large quantities of running water. Neutralizing agents are so often mishandled, they often do more harm than good—only use them if first aid directions are available on labels or through industrial plant directions. Get medical aid at once.

● **First aid for burns of the eye.** Wash thoroughly with large amounts of clean water. A spray-type fountain should be located within 10 ft of the battery room. Place a sterile dressing over the eye to immobilize the lid and get medical aid at once. Well marked supplies of appropriate neutralizing agents can be kept close at hand for immediate use. Check with company physician.

Gasoline handling

Handling and storing of gasoline should comply with the provisions of the National Fire Protection Association Standard No. 30, *Flammable and Combustible Liquids Code* (Fig. 45–8).

Gasoline should be prohibited for all cleaning. Solvents with higher flash points are available and are equally effective and much safer. Even when higher flash point solvents are used, if carburetor or gas line parts are cleaned, the solution should be changed regularly since the admixture of small quantities of gasoline will tend to lower the flash point, and increase the danger of fire and explosion.

Grease, oil, and dirt may be removed from metal parts by nonflammable solutions, or by high-flash point solvents in special degreasing tanks with adequate ventilating facilities.

Use of gasoline to remove oil and grease from garage floors should be prohibited. Nonflammable cleaning compounds are commercially available and should be used.

Gasoline should not be used for removing oil and grease from hands. Soaps are available that will effectively remove greasy dirt from the skin without danger of injury. There are also protective creams and ointments which, if applied before starting work, will protect the skin from dirt and grease.

In some shops, employees use gasoline to clean work clothes. This unsafe practice should be prohibited.

If gasoline is spilled, it should be cleaned up immediately. If gasoline in quantity gets into the sewage system, the fire department should be notified so that the sewers can be flushed. Because gasoline vapor is heavier than air, it collects in low spots, such as basements, elevator pits, and sumps. These places

Fig. 45–8.—Low-profile gasoline tank truck.

Standard Oil Company of California.

should be kept ventilated whenever gasoline vapors are present.

Other safe practices

Using jacks and chain hoists. Vehicles jacked up or hung on chain hoists should always be blocked with stanchions, pyramid jacks, or wood blocks (which have first been carefully inspected). The best jack for general garage use is the wheeled type with a long handle. If ordinary pedestal jacks are used, especially the type supplied for passenger cars, the vehicle may be tipped or jarred off the jacks and cause injury.

When a man is working under a vehicle that is blocked up, other employees should not work on the car in such a manner that the car may be knocked from its blocks.

Men who work under vehicles should be safeguarded from danger when their legs protrude into passageways. Barricades should be used for protection, or else the workman's entire body should be under the vehicle.

A frequent cause of injury to men who work under vehicles is dirt or metal chips falling into the eyes. Therefore, goggles or plastic eye shields should be used by these employees. Plastic face shields are gaining wide preference for this usage because goggles sometimes steam up.

Removing exhaust gases. Repair shop employees should use local exhaust and ventilating facilities to prevent accumulation of vehicle gases within the shop.

Repairing radiators. Where radiators are boiled out or tested for leaks, the operator should be provided with both chemical goggles and a face shield of clear plastic. The entire face needs protection.

Cleaning spark plugs. All mechanics using sandblast spark plug cleaners should wear goggles or face shields.

Controlling traffic. Movement of vehicles inside shops and garages should be regulated by rigidly enforced traffic rules. Traffic lanes and parking spaces should be painted on the

1431

floors and the direction of traffic flow indicated. Vehicles with air brakes should not be moved until sufficient air pressure has been built up.

Every driver should stop his vehicle, then sound the horn before passing through entrance or exit doors. Signs requiring this procedure should be posted in conspicuous places. Mirrors should be installed at blind corners.

Vehicles should be moved in low gear and at low speed inside shop areas, especially up and down ramps.

Training repair shop personnel

Apprentices and new employees should be trained to do each job in the most efficient manner. Job instruction should include the safety rules and regulations pertaining to each job and give the reasons for such rules. The new mechanic should be thoroughly indoctrinated concerning the company's policy toward safety. He should understand the organization of the safety program and the part he is expected to play in it.

Having been indoctrinated and trained to work safely, the new employee must be kept actively interested in observing accepted safe practices in the conduct of his job. There are many devices available to the safety director to accomplish this end, including safety supervision, safety contests, safety meetings, posters, safety bulletins, and pamphlets.

Proper application of these various incentives is described in Chapter 12, "Maintaining Interest in Safety." The safety director should select those most suitable for his purpose and adapt them as needed.

Off-the-Road Motorized Equipment

Heavy-duty trucks are again mentioned here because they are used extensively for special off-the-road operations in such industries as quarrying, mining, and construction. When on the road, they are, of course, governed by the same safe-driving practices as other types of automotive equipment.

The use of heavy-duty trucks, mobile cranes, tractors, bulldozers, and other motorized equipment in the production of stone, ore, and similar materials and in construction work is often accompanied by serious accidents. Workers near equipment can be struck, run over, and killed. Equipment sometimes slips over embankments, injuring people. Servicing and maintaining equipment can be hazardous.

Many accidents, even those that do not injure anyone, result in costly damage to equipment, loss of efficiency and production, and high maintenance costs.

In general, prevention of accidents to heavy equipment requires:

1. Safety features on equipment.

2. Systematic maintenance and repair,

3. Trained operators, and

4. Trained repairmen.

Safe and proper operation of equipment should be found in manufacturer's manuals. Many driving practices are the same as those necessary for the safe operation of highway vehicles. Off-the-road driving, however, involves special hazards and requires special training and safety measures.

Haul-roads

Roadway improvements pay for themselves because they reduce accidents and lower maintenance costs. Both temporary and permanent roads often are too narrow for the equipment and for two-way traffic, especially at curves and fills. Enough space must be provided at curves so that large trucks need not cross the centerline of the road. Curves should be banked toward the outside.

Both temporary and permanent roadways require regular patrolling and maintenance. Too often serious accidents, breakdowns, delays, and unnecessary maintenance expense can be traced to neglected temporary roadways. Road patrols should be provided with protective equipment, such as barricades, MEN WORKING signs, red flags, flagmen, and flares.

Seasonal conditions create road hazards which require prompt attention. Some companies provide sprinkler trucks to protect their men against harmful dusts, discomfort, and the possibility of accidents during dry and windy periods. Others keep dust down by spraying onto the road surface some road oil with an asphalt base.

Fig. 45-9.—Enclosed, compactor-type refuse collection truck. Backing is a most dangerous operation because of poor visibility.

Courtesy International Harvester Co.

Skidding on snow and ice is a serious hazard during the winter. Snow and ice should be removed by means of snowplows or blade graders as promptly and as completely as possible.

When roadways are built close to high banks, the slopes of the banks should be inspected for loose rocks, especially after rain and freezing or thawing weather. Loose rock should be barred down.

Where trucks enter public highways, warning signs should be placed 750 ft from either side of the entrance. Design, color, and placement of the signs should be in accordance with U.S. Department of Transportation, Federal Highway Administration, Washington, D.C., *Manual on Uniform Traffic Control Devices for Streets and Highways,* also published as American National Standard D6.1. If operations are conducted at night, these signs should be reflectorized.

Driver qualifications and training

The modern heavy-duty truck or other off-the-road equipment is a carefully engineered and expensive piece of equipment and warrants operation only by drivers who are qualified physically and mentally and by training and experience. The physical and mental qualifications for an efficient and safe operator of heavy over-the-road equipment (discussed earlier in this chapter) apply to drivers of off-the-road trucks.

No driver or operator should be allowed to work until his knowledge, experience, and abilities have been determined. The amount of time varies for a prospective driver or operator to become thoroughly acquainted with the mechanical features of the truck or piece of equipment, safety rules, driver reports, and emergency conditions. Even an experienced man should not be permitted to operate equipment until the instructor or supervisor is satisfied with his abilities.

Since accidents caused by unsafe practices outnumber those resulting from unsafe condition of equipment and roadways, the time required for thorough checking and training is well warranted. After a man has been trained, constant supervision is required to make sure that he continues to operate in the way in which he was instructed.

Operating trucks near workmen

Workmen are exposed to the danger of being struck or run over by trucks, particularly around power shovels, concrete mixers, and other equipment, in garages, shops, dumps, and construction areas.

Backing. The most dangerous movement is backing, especially for packer-loader trucks (Fig. 45-9), used by municipalities and private contractors for refuse collection.

Some construction companies require drivers to blow three blasts of the horn for a back-up signal.

An automatic audible signaling device for warning workers out of the path of a backing truck has been developed.

Where a number of men are working and various types of equipment are being used, the driver should call upon another employee to signal whether or not the path is clear before backing or making any other movement. A signalman should always take a position within sight of the driver.

The Department of Transportation requires that trucks be equipped with back-up lights for night work.

Fig. 45–10.—Dumps are frequently graded uphill toward the crest. Note steel canopy that protects the operator.

Courtesy American Public Works Association.

Insofar as possible, backing, especially without a guide, should be attempted only to the left because the driver is seated on the left side where he has a good view of the area into which he is backing.

Moving forward. Serious accidents also occur during forward movements. The hazard to workmen increases with the greater height and capacity of trucks. A driver often fails to see workmen crossing from the right immediately ahead of the truck. Thus, drivers often are required to blow two blasts on the horn before starting forward.

Procedures on dumps

Truck operations on dumps and banks involve the danger of the truck going over the crest while dumping a load. A trained dumpman is probably the best insurance against loss of life and damage to equipment. Drivers are required to follow his instructions and signals, especially in backing to dump. Pre-arranged signals must be used at all times.

A dumpman must know how close to the edge a truck can approach safely under various weather conditions. He positions himself on the driver's side of the truck (*a*) so that his signals can be easily seen and (*b*) so that, since the driver will have him in sight, there will be less danger of his being run over. Further to protect the dumpman, the driver should turn from his left when backing, so that he will have the maximum view of the area into which the rear of the truck is moving. Also, the dumpman must stay clear of trucks to avoid being struck by material falling from them.

Left-hand driving also reduces the danger of going over the crest, especially in the operation of side dump trucks, since the driver is on the crest side.

To avoid hitting overhead lines or other low clearances, the dump box should be lowered as soon as the truck clears the dumped load.

Fig. 45–11.—Crawler tractor equipped for site clearance has protective frame and screens to protect operator from falling and flying objects.

Courtesy Spade Plow Inc.

To help prevent the crest from caving in, stock piles and dumps frequently are graded uphill toward the crest so that trucks back up the slope. Loads also may be dumped a safe distance from the crest and then leveled by a grader or bulldozer (Fig. 45–10).

Strongly built cabs and cab protectors on canopies on trucks are effective in preventing injuries if vehicles overturn. By staying in the cabs and not attempting to jump out when their trucks started to roll or slide over an embankment, drivers can escape serious injury. Cabs also provide protection from bad weather and from dust.

Holes, ruts, and similar rough places on dumps and roadways may cause the front wheels of a truck to cramp so that the steer- ing wheel spins, injuring fingers, arms, and ribs, particularly if the truck is not provided with power steering. Gripping the wheel on the outside and not by the spokes, driving at reduced speed, and observing the ground ahead for rough places will help the driver avoid such injuries.

Floodlighting during night operations helps to prevent accidents.

Protective frames for heavy equipment

All bulldozers, tractors, and similar equipment used in clearing operations should be equipped with substantial guards, shields, canopies, and grilles to protect the operator from falling and flying objects (Fig. 45–11).

Crawlers and rubber-tired vehicles, self-

1435

Fig. 45–12.—Mowers should have rollover protective structures (ROPS), slow-moving vehicle (SMV) emblem, and an orange flag mounted about 10 feet high.

Courtesy International Harvester Co., Chicago, Ill.

propelled pneumatic-tired earth movers, water tank trucks, and similar equipment should be equipped with steel canopies and safety belts in order to protect operators from the hazards of rollover (Fig. 45–12). Drivers should be trained to wear the safety belt.

A canopy and its supports should be designed and made to be able to support not less than two times the weight of the prime mover. This calculation is to be based on the ultimate strength of the metal and integrated loading of support members, with the resultant load applied at the point of impact.

In addition, there should be a vertical clearance of 52 in. from the deck to the canopy where the operator enters or leaves the seat.

For more details, see NSC Industrial Data Sheet 622, *Tractor Operation and Roll over Protective Structures.*

Transportation of workers

Transporting employees to and from work, whether singly or in groups, requires special precautions if serious injuries are to be prevented.

When men are transported regularly, at

the beginning and end of a shift for example, a truck equipped with steps and seats should be used. One man may be designated to see that the men get on and off in an orderly manner, to signal the driver to start, and otherwise to assist in safe operation of the vehicle.

Drivers authorized to carry riders should be instructed to start only after all riders are seated or in safe positions. One rider may be designated to signal the driver when all men are seated. The last man off may tell or signal the driver to proceed.

Riders and operators often are injured in getting on and off stationary trucks when they jump and slip, especially during bad weather. They should face the truck and use the handhold. Also, they can avoid sprains by looking for firm footing before taking the last step in getting off.

Riding on the load is hazardous. Men are exposed to such hazards as being thrown or brushed off by overhanging limbs or structures, or being caught by a shifting load.

If only one or two helpers are needed, they should ride in the seat with the driver. If two or more are necessary and the haul is short, the men should walk. For a long haul requiring more than two men, the safest procedure is to provide separate transportation for them.

Riders must not attempt to get on or off a moving truck and must not indulge in horseplay. The greatest danger in the violation of these safe practices is that of falling under the wheels. Such accidents are especially likely to occur during wet or icy weather.

Towing

Towing is a hazardous operation, especially when coupling or uncoupling the equipment. Men can be crushed when a truck or other piece of equipment moves unexpectedly while they are between the two pieces of equipment.

The following safe practices are essential to prevent accidents in the coupling or uncoupling of motorized equipment:

1. No one should go between the vehicles while either one is in motion.

2. Vehicles must be secured against movement by having the brakes set or the wheels blocked.

3. No driver should move a vehicle while

anyone is between two vehicles. Before moving, a driver must receive a signal that everyone is in the clear.

4. Tow bars are usually safer than towing cables. If cables, which may be more convenient to use under certain circumstances, are employed, they must be in good condition and of sufficient size and length for the towing job.

Power shovels, cranes, and similar equipment

Safe operation of power shovels, draglines, and similar equipment begins with purchase of the machines. A good policy is to specify guards over gears, safe oiling devices, handholds, and other safeguards when the order is placed with the manufacturer. In any case, before equipment is put into operation, it should be thoroughly inspected and necessary safety devices should be installed.

Both operators and repairmen, whether experienced or not, should be instructed in the recommendations of the manufacturer pertaining to lubrication, adjustments, repairs, and operating practices, and should be required to observe them. A preventive maintenance program for shovels and other equipment is essential for safety and efficiency. Frequent and regular inspections and prompt repairs are the bases for effective preventive maintenance.

Generally, the operator is responsible for inspecting the mechanical conditions, such as holding-down bolts, brakes, clutches, clamps, hooks, and similar vital parts.

Wire rope cables should be kept lubricated in accordance with the manufacturers' instruction, and should be inspected daily since cable failures cause serious accidents. Cables are particularly likely to develop weakness at the fastenings, at crossover points on drums, and in the sections in frequent contact with sheaves.

Grounding systems. Electrically powered equipment requires a good grounding system to protect workers from electrical faults in trailing cables and at the machine where electric shock is most apt to occur. Though the equipment may make close contact with the surface of the ground, the resistance to the flow of current from the frame of the

45—Motorized Equipment

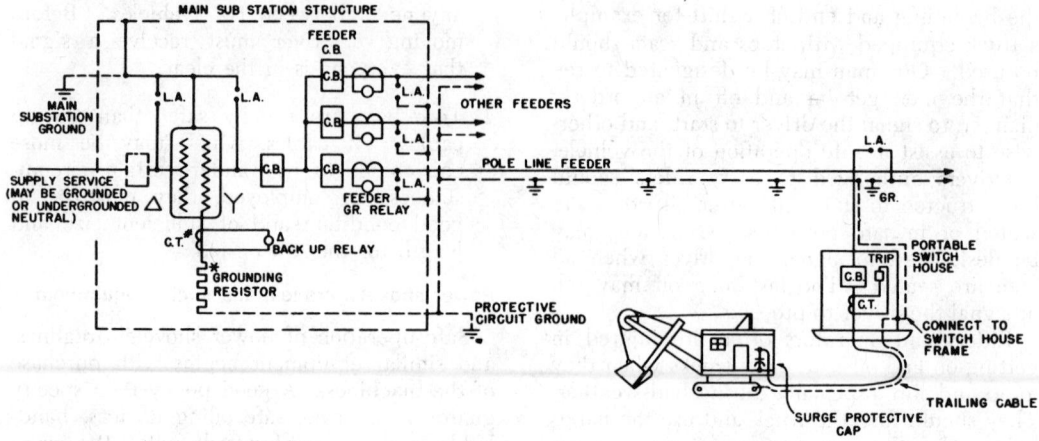

Fig. 45–13.—A safety grounded neutral system for electric-powered units.

Courtesy General Electric Co.

equipment to the earth usually is high. A machine-to-ground resistance of 100 ohms and a current of 10 amperes means an electric shock hazard of about 1000 volts. A current of 1/50 ampere is very painful and can result in loss of muscular control. As little electric current as 1/10 ampere through the body may result in death. (See Table 41-A in Chapter 41, "Electrical Hazards.") Since these low currents can be forced through wet skin by 100 volts, no employee should be exposed even momentarily to this electrical hazard. Fig. 45–13 shows a safely grounded neutral system.

A good ground system may be made by driving copper-clad steel rods, or electrodes, into suitable soil for a distance of at least 8 feet. Since the number of rods and their spacing depend considerably on soil conditions, the conductivity of the soil may have to be increased by treatment with common salt, sodium nitrate, copper sulfate, or similar chemicals which are carried into the soil by rain. A grounding system having a total resistance of 1 ohm, including the cable, can be obtained by proper design and construction. (See Chapter 41, "Electrical Hazards," for a thorough discussion of grounding.)

The pole line ground wire should be at least as big as the power wires. Wherever power is tapped from the power line, a connection is made from the pole ground wire

to a ground wire in the cable. A good cable has metal shielding outside the insulating material around each conductor, and the ground conductor is in full contact with the shielding.

The ground wire in the cable is connected to the frame of the equipment. A resistor between the pole line ground wire and the transformer neutral limits the amount of current to not more than 50 amperes, eliminating dangerous voltages at the shovel and permitting sufficient current to open the circuit breakers.

The ground neutral system permits the operation of all equipment except the machine where the fault occurs. The machine is segregated from the rest of the system by the immediate operation of a circuit breaker actuated by the fault. Suitable switching equipment in the grounded neutral system eliminates the danger from several faults existing at the same time in different phases at different locations, except for the interval required for the circuit breakers to open.

The equipment ground should not be connected to the substation ground in any way, to avoid energizing the equipment by a fault in the power supply system.

Circuit breakers and other devices in the grounding system should be inspected and tested monthly. The resistance of the ground also should be checked regularly. A megger test of the cable insulation is a recommended

Fɪɢ. 45–14.—To minimize wear on trailing cables, a structure can be used to elevate them when they cross over roads.

part of cable inspection. Defective insulation should be vulcanized, not taped.

Workmen should be provided with rubber gloves and insulated tongs or hooks for handling trailing power cables.

Minimum wear and damage to trailing cables is important for safety and economy. They should be protected from blasting operations as much as possible and kept as close to operations as practical so that a minimum length of cable is required. Tripods and horses can be used to keep cable off the ground. Tripods are preferable to trenches where cables cross roads (Fig. 45–14).

The electrical parts of shovels and similar equipment, including trailing cables, should be inspected regularly and maintained by an electrician.

Maintenance practices. A study by the U.S. Army Corps of Engineers shows that the severity of injuries resulting from accidents in maintaining, repairing, and adjusting power shovels, draglines, and similar equipment is high, averaging about 240 days per disabling

injury, exclusive of fatalities.

Before blasting repairs or adjustments are made, a shovel should be in a safe position where it will not be endangered by falling or sliding rock or earth. Cracks that extend back from the face may develop.

When repairs are to be made, the operator is responsible for setting the brakes, securing the boom, lowering the dipper or bucket to the ground, taking the machine out of gear, and before he leaves the machine, exercising similar precautions to prevent accidental movement.

Before starting any job, repairmen should notify the operator about its nature and location. If the work is to be done on or near moving parts, the controls should be locked out and tagged and the lock and tag should be removed only by the repairman. This precaution is essential to prevent the operator from starting the equipment inadvertently.

Parts that must be in motion while men are working on them should be turned slowly, by hand if possible, in response to guidance or on signal. This precaution applies particularly to work around gears, sheaves, and drums. Men who grasp cables just ahead of the sheaves risk having their hands jerked into the sheaves. To prevent hand injuries, a cable being wound on a drum should be guided with a bar.

If guards must be removed for convenience in making repairs, the job cannot be considered complete until the guards, plates, and other safety devices have been replaced.

Repairmen should wear snug-fitting clothing, eye protection, and safety shoes. Gloves should not be worn by men working on or near any moving parts of the machines.

Operating practices. Slides of rock and material from high faces and banks result in some of the worst accidents in the operation of power shovels. In some instances, the shovel and men have been buried; in other cases, men have been struck and injured while working around the equipment.

Some quarries having high faces limit the height of banks to 25 ft by benching and by a blasting procedure that forces the rock out from the face sufficiently to reduce the height of the pile. The shovel operator is thus able to maintain the bank at a safe slope. When loading from under a high face, the operator should swing the shovel to the sight side and away from the face, thereby allowing himself a better view and reducing his exposure to injury.

Undercutting banks of earth, sand, gravel, and similar materials is especially dangerous during winter and spring months. Freezing and thawing may result in a collapse of the overhanging material. To maintain a safe slope, the overhanging material may be blasted.

The operator has responsibility for the safety of other men whose duties take them into the vicinity of the shovel. These workers may be struck by falling rock, squeezed between the shovel and the bank or similar pinch points, or struck by the dipper. No worker should enter a dangerous location without first notifying the operator who, in turn, should not move the equipment.

Dippers should be filled to capacity but not overflowing, to prevent falling material from endangering workers and to eliminate excessive spillage. Insofar as possible, load-

ing should not be done from the blind side. The operator should not swing a load over a vehicle nor load a truck until its driver has dismounted and is in the clear, unless the truck is provided with a canopy designed for the protection of the driver. Cars and trucks should be loaded evenly so that earth or rocks do not overhang the sides.

Housekeeping on and around the shovel should be first class. The operator should keep tools in a definite place and keep the floor free of grease and oil. Ice and snow should be removed promptly, and a bulldozer should keep the area around the shovel free of rocks and ruts.

One should get on or off a shovel or dragline only after he has notified the operator, who, in turn, should swing the platform so that the handhold can be grasped and the steps or tread used. No one should get on or off by jumping onto the tread, either while the operator is making a swing or while the equipment is stationary.

No unauthorized person should be permitted on a shovel or dragline.

Mobile cranes. The outstanding characteristic of accidents involving crawler and similar types of cranes is the severity of the injuries. Although these accidents occur with relative infrequency, the injuries are about twice as serious as those resulting from accidents involving other types of heavy equipment. For this reason, a crane operator particularly should be selected for his intelligence, stability, and willingness to follow instructions.

The operator is largely responsible for the safe condition of the crane and should make regular inspections of brakes, cables and their fastenings, and other vital parts, and promptly report worn, broken, and other defective parts. Like other equipment, mobile cranes should be maintained on a regular schedule.

The operator is responsible for the safety of the oiler and also has a large measure of responsibility for preventing injuries to hookers or riggers and others working around the equipment. However, anyone working in the vicinity of a crane has the responsibility to stay clear of the boom. In no case should anyone work or cross under the boom.

Some of the worst accidents result from

overloading cranes. In no case should the load limits specified for various positions of the boom by the manufacturer be exceeded. These load limits should be conspicuously posted in the crane cab. If there is doubt about the weight of a load, the capacity of the crane to handle the load safely should be tested by first lifting the load slightly off the ground.

Operating a crane on soft or sloping ground is dangerous. The crane should always be level before it is put into operation. Outriggers give reliable stability only when used on solid ground. The use of makeshift methods to increase the capacity of a crane, such as timbers with blocking, is too dangerous to be permitted.

Boom stops limit the travel of the boom beyond an angle of 80 degrees above the horizontal plane and prevent the boom from being pulled backwards over the top of the machine by the boom hoisting mechanism or the sudden release of a heavy load suspended at a short radius. Either of these occurrences usually result in serious damage to the equipment and injuries to the operator or other workmen.

Accidents usually occur when the operator is performing more than one operation and becomes confused or distracted and excited. Also, clutch linings may swell during wet weather, and the master clutch or the boom clutch, or both, may "drag" and cause the boom to be pulled over backwards. Clutches should be tested before starting work on rainy days and the clearances adjusted if necessary. Another accident cause is the sudden release of a load when the boom angle is high, for example, from the parting of a sling.

Boom stops are best suited to medium-size cranes (the 5- to 60-ton range). Boom stops should disengage the master clutch or kill the engine and stop the boom before it reaches the maximum permissible angle. One type of stop meeting these requirements has a piston and cylinder, spring or pneumatically actuated, and is mounted on the A-frame to intercept the boom as high above the boom hinges as possible. By positive displacement of an actuator mounted on the A-frame, the boom action disengages the master clutch (or ignition breaker or compression release) by means of light cable reeved over a few small sheaves.

When a mobile crane must be operated near electric power lines, the power company should be consulted to determine whether the line can be deenergized. Many fatalities have resulted from contact with power lines, and often the power company's service is seriously disrupted. Various states and OSHA have enacted legislation specifying distances which booms and cables must be kept from power lines. A minimum of 10 ft is often specified; however, the recommendations of the power company and legal requirements should be observed. See page 691.

An experienced operator working with an untrained or relatively inexperienced hooker or rigger should direct the details of lifts, such as the type of sling and hitch to be used. Although the operator usually can rely on the knowledge of an experienced rigger, the operator has the right to question the safety of a lift and have his supervisor make a decision.

The following safe practices are essential when handling loads.

1. The hook must be centered over the load to keep it from swinging when lifted.

2. Employees should keep their hands out of pinch points when holding the hook or slings in place while the slack is taken up. A hook, or even a small piece of board, may be used for the purpose. If a man must use his hand, he should hold the sling in place with the flat of his hand.

3. The hooker, rigger, and all other people must be in the clear before the load is lifted.

4. Tag lines should be used for guiding loads.

5. Hookers, riggers, and others working around cranes also must keep clear of the swing of the boom and cab.

6. No load may be lifted or moved without a signal. Where the entire movement of a load cannot be seen by the operator, as in lowering a load into a pit, a signalman should be posted to guide him. To avoid confusion in signals, only standard hand signals should be used. (See Chapter 25, "Hoisting Apparatus and Conveyors.")

1441

Motor graders, bulldozers, and scrapers

Many of the basic safety measures recommended for trucks also apply to motor graders and other types of earth-moving equipment. All machines should be inspected regularly by the operator, who also should promptly report any defects. The safety and efficiency of the equipment are increased by scheduled maintenance.

Men who are physically and mentally qualified for operators should be selected and trained in correct operating practice, as specified in the manufacturer's manual and by company requirements. Prevention of injury in the servicing and repairing of machines requires special precautions, in addition to the observance of general safe procedures applying to other types of motorized equipment.

Maintenance. Brakes, controls, motors, chassis, blades, blade holders, tracks, drives, hydraulic mechanisms, transmission, and other vital parts require regular inspection. Wheel and motor-mount bolts likewise require frequent checking for tightness.

Making adjustments and repairs with the motor running is a dangerous practice, particularly when work is being done near the fan of the motor or when a clutch of a tractor is being adjusted. Refueling should be done only with the motor stopped.

The danger from a locking ring blowing off when the tire of a truck is being inflated applies equally to a tractor tire. See details under Tire Operations, earlier in this chapter.

When blades are to be replaced, the scraper or dozer bowl should always be blocked up. After the scraper has been lifted to the desired height, blocks are placed under the bottom near the ground plates. Apron arms are raised to the extreme height and a block is placed under each arm, so that the apron can drop enough to wedge each block firmly in place.

Before reeving cable on a drum or through sheaves, the operator should disengage the master clutch, idle the engine, and lock the brakes. He should stop the engine before working with the cable on a front-mounted cable drum.

If an operator is assisting a repairman and working behind the scraper with the tail gate in the forward position, he should place a block behind the tail gate so that it cannot fall. This precaution is necessary in case someone should release the power control unit brake, permitting the tail gate to come back.

When cables are to be replaced on scrapers, the tail gate should be back at the end of its travel.

General operating practices. The operator must look to the front, sides, and rear before moving his machine and be constantly alert for men on foot when operating near other equipment, offices, tool and supply buildings, and similar places.

Speeds are largely governed by conditions. Slow speeds are essential in driving (a) off the shoulders of roads, on steep grades, and at rough places to avoid violent tilting which may throw the driver off the machine or against levers and cause serious injury, (b) in congested areas, and (c) under icy and other slippery conditions. No one (except the operator) should be permitted to ride.

Jumping from a standing machine can result in sprained ankles and other injuries. The safe practice is to step down after looking to make sure that the footing is secure. Ice, mud, round stones, holes, and similar conditions cause many falls. For the same reason, deck plates and steps on equipment should be free of grease and other slipping hazards.

An operator should not drive his equipment onto a haul-road without first stopping and looking both ways, regardless of whether or not the place of entry is marked with a stop sign. Generally, loaded equipment is given the right of way on job or haul-roads.

Before an operator leaves his equipment even for a short time, he should lower the bowl or blade to the ground and stop the motor. A safe parking location is on level ground, off a roadway, and out of the way of other equipment.

The operator should never leave his equipment on an inclined surface or on loose material with the motor running—the vibration may put the equipment in motion.

Procedures on roadways. When graders, scrapers, and other earth-moving equipment are in operation along a section of a road, the precautions discussed next will help prevent accidents to the public, employees, and equipment.

Traffic should be warned of danger ahead by barrier signs at both ends of the section of road undergoing construction or repairs. Primary warning signs, such as ROAD UNDER CONSTRUCTION *or* BARRICADE AHEAD, should be placed 1500 ft from the end point of operation.

Orange flags or markers at the ends of blades, which may project beyond the tread of a machine, serve to warn persons and other equipment.

An orange flag on a staff that will project the flag at least 6 ft above the rear wheel of a blade grader is recommended for operation in hilly country. (See Fig. 45–12.)

Operators of motor graders should keep to the right side of a roadway. When blading against traffic is necessary, flags and barricades should be used to warn traffic. Warning signs should be placed at a considerable distance from the work area. This distance increases as the "speed" of the highway increases. Suggestions are given in NSC Data Sheet 614, *Surface Surveying*, and DOT *Manual on Uniform Traffic Control Devices for Streets and Highways*, ANSI D6.1.

Where operations are extensive, flagmen should be placed at each end of the working area so that they are visible to oncoming traffic for at least 500 ft.

Where earth-moving equipment is stopping, turning, or backing at curves, crests of hills, and similar dangerous locations, flagmen should be stationed. Such movements generally require a clear view of approaching traffic for a distance of about 1000 ft for safety.

Flagmen should also be used where the working area is congested by other equipment, workmen, buildings, excavations, and similar hazards. See Fig. 45–15 for hand signaling procedures.

Coupling and towing equipment. An operator should not back up to couple a tractor to a scraper, sheepsfoot roller, or other equipment unless he has first checked to make sure that everyone is in the clear. If the operator is assisted by a ground man, he should not move his equipment until signaled.

Before an employee is allowed to couple the trailing equipment, the tractor should be stopped, the shift lever put in neutral, and the brakes set. The wheels of equipment being coupled should be blocked.

All equipment being towed should be secured by a safety chain attached to the pulling unit, in addition to the regular hitch or drawbar, since drawbar failure can result in a serious accident.

When a scraper is towed from one job to another, the operator should use a scraper bowl safety latch, or place a safety bolt in the beam to give maximum clearance for road projections such as crossings. This precaution prevents the bowl from striking the ground or pavement and injuring persons or damaging equipment.

Clearing work. Work requiring exposure to low limbs of trees or to high brush involves serious hazards which can be readily overcome by suitable protective measures and safe practices.

It is best to equip bulldozers with heavy, well supported, arched steel mesh canopies to protect operators. See section on "Protective Frames for Heavy Equipment," earlier.

Goggles should be worn to protect the eyes from whipping branches.

Safety hats provide protection from falling dead branches. When a bulldozer shoves hard against the butt of a large dead tree, the tree may crack in the middle or limbs may fall onto the machine. Dead branches or tops also may drop from live trees. A safe procedure to eliminate the danger is to cut the roots on three sides and then apply the power to the fourth side. A long cable may be used to pull over large dead trees, but it must be determined in advance that the tractor and operator will be in the clear when the tree falls.

Operators have the responsibility of seeing that all men in the area are in the clear before pushing over small trees, bulldozing rock, and rolling logs.

Special hazards. Fatalities occur while equipment is operated on dumps and fills, near excavations, and on steep slopes.

The bulldozer blade should be kept close to the ground for balance when the machine

1443

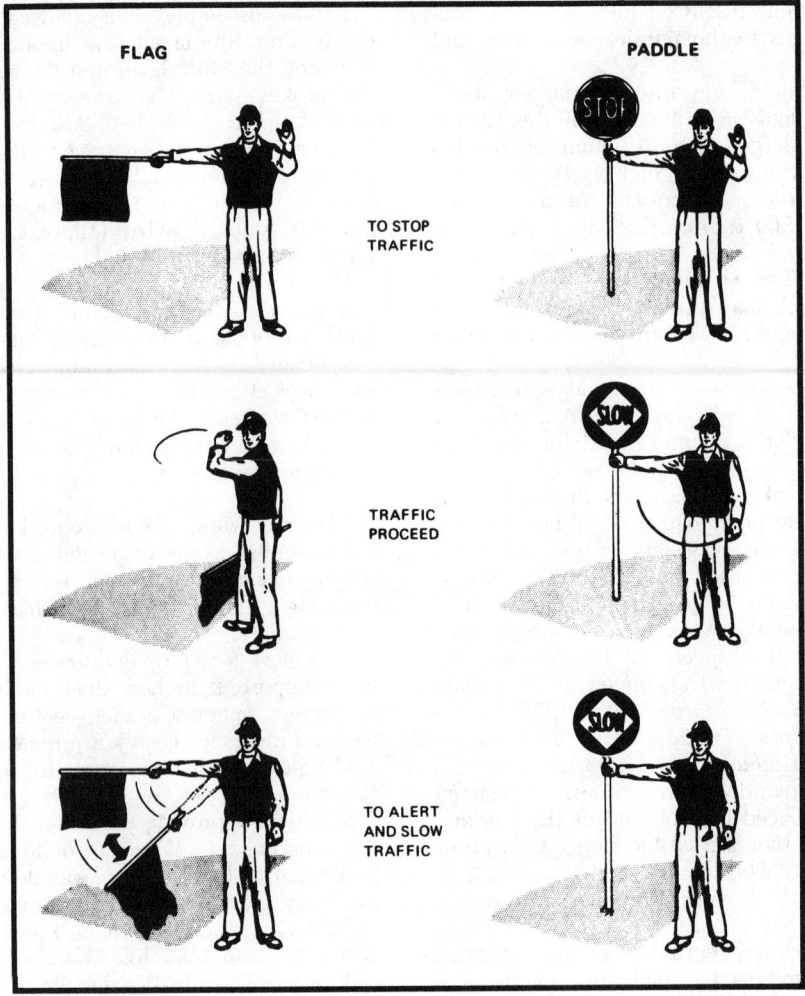

FIG. 45–15.—Hand signaling devices used by flagmen. Flag and background of STOP sign are bright red. Diamond of SLOW sign is orange.

From American National Standard D6.1–1971.

is traveling up a steep slope.

When a tractor-dozer is to be driven down a slope, three or four bowls of dirt should be dozed to the edge of the slope and kept in front of the blade.

If the dirt is lost on the way down, the operator should not lower the blade to regain the load because of the danger of overturning. Using the bowl as a brake is not good practice on a steep slope.

How close to an excavation or the crest of

a dump a machine can be safely operated depends on ground conditions. Wet weather requires equipment to operate a greater distance from the edge or crest. A signalman is especially essential when the ground is treacherous.

Sometimes employees, the public, livestock, and property are endangered when material is pushed over the edge in side hill work. In such cases, sufficient clearance below must be provided before the work begins.

References

American Automobile Association, 1712 G St. NW., Washington, D.C. 20006.
"Driver Training Equipment" (catalog).
"Driving Skill Exercises with Check List and Score Sheet."
"Road Test Check List for Passenger Car Drivers."
"Road Test Check List for Truck Drivers."
Sportsmanlike Driving (a textbook).

American Trucking Associations, Inc., 1616 P St. NW., Washington, D.C. 20036. *The Motor Carrier Safety Regulations, U.S. Dept. of Transportation, Federal Highway Administration, Parts 390–397, Issued April 1968.* (Pocketsize edition for issuing to drivers.)

The Associated General Contractors of America, Inc., 1957 E St. NW., Washington, D.C. 20006. *Manual of Accident Prevention in Construction.*

Association of Casualty and Surety Companies, 110 William St., New York, N.Y. 10038.
Guide Book, Commercial Vehicle Drivers
Truck and Bus Drivers Rule Book

Evans Industries, Driver Trainer System, 9756 Wilshire Blvd., Beverly Hills, Calif. 90210.

Intext, Scranton, Pa. 08575. Catalog of driver testing and training equipment.

National Association of Automotive Mutual Insurance Companies, 20 N. Wacker Drive, Chicago, Ill. 60606. *Code of the Road.*

National Fire Protection Association, 470 Atlantic Ave., Boston, Mass. 02210. *Flammable and Combustible Liquids Code,* Standard No. 30.

National Safety Council, 425 N. Michigan Ave., Chicago, Ill. 60611.
Aviation Ground Operations Safety Handbook.
Defensive Driving Program Materials.
Industrial Data Sheets
    Airport Vehicular Traffic, 539.
    Barricades and Warning Devices for Highway Construction Work, 239
    Diving in Construction Operations, 555.
    Falling or Sliding Rock in Quarries, 332.
    General Excavation, 482.
    Grounding Electric Shovels, Cranes, and Other Mobile Equipment, 287.
    Lead-Acid Storage Batteries, 635.
    Liquefied Petroleum Gases for Industrial Trucks, 479.
    Motor Graders, Bulldozers, and Scrapers, 256.
    Motor Trucks for Mines, Quarries, and Construction, 330.
    Mounting Heavy Duty Tires and Rims, 411.
    Ready-Mix Concrete Trucks, 617.
    Snow Removal and Ice Control, 638.
    Surface Surveying, 614.
    Tractor Operation and Roll-over Protective Structures, 622.
Motor Fleet Safety Manual.
Motor Fleet Talk Topics.
Motor Fleet Talks.
Public Employee Safety Guide—Streets, Roads and Highways.
Small Fleet Guide.
Study of Aerial Basket Accidents.

New York University Center for Safety Education, Washington Square, New York, N.Y. 10003. Publications list.

U.S. Department of Defense, Department of the Army, Washington, D.C. 20310.
Driver Selection and Training, TM 21–300.
Drivers' Manual, TM 21–305.
General Safety Requirements, EM 385–1–1, U.S. Army Corps of Engineers.
"Methods of Teaching."
Motor Transportation, Operation, FM 25–10.

U.S. Department of the Interior, Bureau of Mines, Washington, D.C. 20240.
Minerals Yearbook.
Also various handbooks, miners' circulars, and processed publications.

U.S. Department of Transportation, Federal Highway Administration, 8th and D Sts SW., Washington, D.C. 20591.
Manual on Uniform Traffic Control Devices for Streets and Highways. (Also identified as American National Standard D6.1.)
The Motor Carrier Safety Regulations.

Safety
Engineering
Tables

Chapter
46

Factors of Safety: Equipment and Construction Materials   1448
Basic Stresses for Laminated Structural Members   1450
Allowable Unit Stresses for Structural Timber   1451
Capacities of Dished Heads and Horizontal Cylindrical Tanks   1453
Capacities of Upright Cylindrical Tanks   1455
Brightness Ratios and Levels of Illumination   1456
Rpm Speed of Saws To Give 10,000 sfpm   1457
Strength of Ice   1458
Design Wind Pressures for Various Height Structures   1458
Safe Bearing Loads on Soils   1459
Specific Gravities and Densities   1459
Examples of Incompatible Chemicals   1462
Explosion Characteristics of Various Dusts   1463
Natural Trigonometric Functions   1466
Average Concentrations of Airborne Particulate Matter   1468
Additional Dimensions That Clothing Adds to Nude Body
    Measurements   1469
Windchill Factors   1470
Subjective Response to Vibration   1471
Conversion Factors and Tables   1472
Abbreviations Used in this Manual   1479
References to Useful Handbooks of Engineering Tables   1481
Occupational Noise Exposure Standard   1484

TABLE 46-A

GENERAL FACTORS OF SAFETY FOR COMMON CONSTRUCTION MATERIALS

Material	Steady Load	Load Varying from Zero to Maximum in one Direction	Load Varying from Zero to Maximum in both Directions	Suddenly Varying Loads and Shocks
Cast iron	6	10	15	20
Wrought iron	4	6	8	12
Steel	5	6	8	12
Wood	8	10	15	20
Brick	15	20	25	30
Stone	15	20	25	30

From Machinery's Handbook, 16th ed. (The Industrial Press)

The safety professional requires information on many subjects usually not available in any one source, but fortunately it is possible to gather the more generally pertinent information into one chapter such as this. However, other more specialized material must be sought out as required when the safety man must deal with a specific subject in great detail. This latter information includes (a) applicable federal, state, and local code requirements which are beyond the scope of this Manual and which must be checked locally, and (b) specific scientific and engineering information which generally is available in handbooks devoted to specific subject fields. For the reader's convenience in developing this additional information, a number of these handbooks are listed under References at the end of this chapter.

Factors of safety

In applying scientific data to actual work situations, the "ideal conditions" under which the experimental data were gathered often do not completely exist, and so a *safety factor* is often applied to compensate for this. The magnitude of this factor depends on how great the cost of failure will be in terms of life or damage. While pertinent safety factors are suggested in the various chapters of this Manual, many are summarized here.

By definition, this factor of safety is the ratio of the ultimate (breaking) strength of a member or piece of material to the actual working stress or to the maximum permissible (safe load) stress when in use. This term is basic to the whole safety program, as the following few examples will indicate.

Stairs and landings should sustain a live load of not less than 100 psf with a factor of safety of 4—that is, the maximum live load should never exceed ¼ of the breaking strength. OSHA requires standard railings to withstand at least 25 lb/lineal ft pressure applied in any direction at any point on the rail.

Every scaffold and its supporting members should be so designed to support a given load with a safety factor of 4.

Boilers should have a safety factor of about 5, which figure is sometimes raised to 5.5 with used boilers, and to 6 when the boiler is 10 or more years old. Unfired pressure vessels should have a safety factor of approximately 5.

For each section of a refrigerating system, the factor of safety is commonly 5 times the safe working pressure, or 5 times the pressure at which the safety valve on the section is set to relieve.

Hydraulic piping and hose should have a safety factor of 8.

In the case of cranes, the hook load should have a safety factor of between 4 and 5; hoisting cables, gears, and other parts subject to wear, a factor of not less than 8; and all other parts, including structural steel, a factor not less than 5. On hot metal cranes, however, all these safety factors should be higher, even as high as 10 throughout.

For general hoisting purposes, it is not

advisable for the working load of a wire to exceed ⅕ of its breaking strength. This means a safety factor of 5, but factors in excess of this, and varying up to 8 or more, are often required for safe and economical operation.

Cast iron flywheels should have a safety factor of 10, but those made of wood should have the factor of safety raised to 20.

Other data

In addition to data on construction materials, information is supplied on lighting and ventilation, chemicals and contaminants, mathematical functions, and abbreviations used in this Manual.

There is a special section on conversion between metric and English units. This follows, insofar as possible, the recommendations of the *Standard Metric Practice Guide*, E 380–72, adopted by the American Society for Testing and Materials (also designated ANSI Z210.1).

The illumination levels listed in Table 46-G provide a guide for efficient visual performance rather than for safety alone and they should not be considered as regulatory minimum illumination standards which are discussed next.

These sample levels are taken from the American National Standard *Practice for Industrial Lighting*, A11.1–1973, published by the Illuminating Engineering Society. They are based upon research conducted on young adults with normal and better than 20/30 corrected vision. More illumination might be provided for older workers to compensate for degeneration of vision due to age.

Levels of 200 foot-candles or more were obtained with a combination of general lighting plus specialized supplementary lighting. Consideration should be given to glare, objectionable shadows, color, contrast and eye protection.

Minimum illumination for safety of personnel, at any time and at any place where safety is related to seeing conditions, is as follows:

Hazards requiring visual detection	Slight		High	
Normal activity level	Low	High	Low	High
Illumination levels, foot-candles	0.5	1.0	2.0	5.0

Special conditions may require different levels of illumination. In some cases, higher levels may be required—for example, where security is a factor. In some other cases, greatly reduced levels of illumination, including total darkness, may be necessary (notably in connection with photographic products). In these situations, alternate methods of ensuring safe operation must be relied upon.

See specific industry reports of the Industrial Lighting Committee of the Illuminating Engineering Society for guidelines. ANSI A11.1 has more details.

TABLE 46-B. BASIC STRESSES FOR LAMINATED STRUCTURAL MEMBERS

	Extreme fiber in bending or tension parallel to grain (psi)	Maximum longitudinal shear (psi)	Compression perpendicular to grain (psi)	Compression parallel to grain (psi)	Modulus of elasticity in bending (1,000 psi)
SOFTWOODS					
Baldcypress (southern cypress)...	2,400	170	330	2,000	1,300
Cedars:					
Redcedar, western..................	1,600	135	220	1,300	1,100
White-cedar, Atlantic					
(southern white cedar)					
and northern......................	1,400	115	195	1,050	900
White-cedar, Port-Orford........	2,000	150	275	1,650	1,600
Yellow-cedar, Alaska					
(Alaska-cedar)...................	2,000	150	275	1,450	1,300
Douglas-fir, coast type	2,750	150	350	2,000	1,800
Douglas-fir, coast type,					
close-grained.......................	2,950	150	375	2,150	1,800
Douglas-fir, Rocky Mountain type	2,000	135	310	1,450	1,300
Douglas-fir, all regions, dense.....	3,200	150	410	2,350	1,800
Fir, balsam...............................	1,600	115	165	1,300	1,100
Fir: California red, grand, noble,					
and white............................	2,000	115	330	1,300	1,200
Hemlock, eastern.......................	2,000	115	330	1,300	1,200
Hemlock, western					
(west coast hemlock)	2,400	125	330	1,650	1,500
Larch, western	2,750	150	350	2,000	1,800
Pine, eastern white (northern					
(white), ponderosa, sugar, and					
western white (Idaho white)	1,600	135	275	1,400	1,100
Pine, jack...............................	2,000	135	240	1,450	1,200
Pine, lodgepole	1,600	100	240	1,300	1,100
Pine, red (Norway pine).............	2,000	135	240	1,450	1,300
Pine, southern yellow.................	2,750	180	350	2,000	1,800
Pine, southern yellow, dense	3,200	180	410	2,350	1,800
Redwood.................................	2,200	115	275	1,850	1,300
Redwood, close-grained.............	2,400	115	295	2,000	1,300
Spruce, Engelmann	1,400	115	195	1,100	1,100
Spruce, red, white, and Sitka......	2,000	135	275	1,450	1,300
Tamarack.................................	2,200	160	330	1,850	1,400
HARDWOODS					
Ash, black...............................	1,800	150	330	1,150	1,200
Ash, white...............................	2,550	210	550	2,000	1,600
Beech, American	2,750	210	550	2,200	1,800
Birch, sweet and yellow.............	2,750	210	550	2,200	1,800
Cottonwood, eastern..................	1,400	100	165	1,100	1,100
Elm, American and slippery					
(white or soft elm)	2,000	170	275	1,450	1,300
Elm, rock.................................	2,750	210	550	2,200	1,400
Hickory and pecan.....................	3,500	235	660	2,750	2,000
Maple, black and sugar					
(hard maple)	2,750	210	550	2,200	1,800
Oak, red and white	2,550	210	550	1,850	1,600
Sweetgum (redgum or sapgum) ..	2,000	170	330	1,450	1,300
Tupelo, black (blackgum)	2,000	170	330	1,450	1,300
Tupelo, water...........................	2,000	170	330	1,450	1,300
Yellow-poplar...........................	1,800	150	240	1,450	1,300

Courtesy Forest Products Laboratory.

Note: This table is for clear material under dry conditions as in most covered structures.

TABLE 46-C

ALLOWABLE UNIT STRESSES, STRUCTURAL TIMBER
(Pounds per square inch)

Material	$K^§$ (columns)	Bending and tension	Compression°	Shear†	Bearing‡	Modulus of elasticity
Douglas-fir:						
Parallel to grain.........	23.4	1,500	1,200	100	1,500	1,600,000
Across the grain........	......	......	......	......	350	......
Oak:						
Parallel to grain.........	23.7	1,300	1,100	110	1,500	1,500,000
Across the grain........	......	......	......	......	500	......
Spruce:						
Parallel to grain.........	21.2	1,200	1,100	100	1,200	1,200,000
Across the grain........	......	......	......	......	250	......
Longleaf yellow pine:						
Parallel to grain.........	23.4	1,400	1,200	120	1,400	1,600,000
Across the grain........	......	......	......	......	300	......

From General Engineering Handbook, *O'Rourke (McGraw-Hill Book Co.)*

°On lengths not greater than eleven times the least dimension. Otherwise, use column formulas.

† In timber-joint designs, these values may be increased 50 per cent.

‡ These values are for joint design, and for bearings less than 10 in. long.

§ See following notes.

TIMBER, STRUCTURAL GRADE (DRY). Allowable unit stresses in bending, compression, shear, and bearing, as specified by the N.L.M.A., are given in Table 46-C. Allowable unit stresses for timber columns are as follows:

1. Short Columns. The safe load, in pounds per square inch of net cross-sectional area, for columns and other members stressed in compression parallel to the grain, with a ratio of unsupported length to least dimension (l/d) not exceeding eleven (short columns) shall not exceed the allowable unit compression stress (c) parallel to grain for short columns, *i.e.*,

$$\frac{P}{A} = c$$

2. Intermediate Columns. For columns with a ratio of unsupported length to least dimension greater than eleven but less than K (intermediate columns), the following formula shall be used until the reduction in allowable stress equals one-third the stress permitted for short columns:

$$\frac{P}{A} = c\left[1 - \frac{1}{3}\left(\frac{1}{Kd}\right)\right]$$

3. Long Columns. For columns with a ratio of unsupported length to least dimension greater than K (long columns), the safe load shall be determined by the following formula:

$$\frac{P}{A} = \frac{\pi^2 E}{36(1/d)^2} = \frac{0.274E}{(1/d)^2}$$

(*Continued on next page.*)

4. Notation.

P = total load, in pounds.

A = area, in square inches of net cross section.

P/A = the working stress or maximum load per square inch.

c = allowable unit stress in compression parallel to grain for short columns.

l = unsupported length of column, in inches.

d = least dimension of column, in inches.

E = modulus of elasticity.

$$K = \frac{\pi}{2}\sqrt{\frac{E}{6c}}; \text{ at which } \frac{P}{A} = \frac{2c}{3}$$

5. The safe load on a column of round cross section shall not exceed that permitted for a square column of the same cross-sectional area.

6. Columns shall be limited in maximum length to l/d = 50.

TABLE 46-D1

CAPACITIES OF HORIZONTAL CYLINDRICAL TANKS WHEN FILLED TO VARIOUS DEPTHS
Tanks with Flat Ends. Contents Given in U.S. Gallons per One Foot of Length

To ascertain the contents of a tank over one-half full: Let h = depth of unfilled portion. Find from the table the quantity corresponding to a depth h. Subtract this quantity from the contents of a full tank.

Diam. of Tank, In.	Full Tank	Depth of Liquid, In. = h																		
		3	6	9	12	15	18	21	24	27	30	33	36	39	42	45	48	51	54	57	60
12	5.88	1.15	2.94																		
18	13.22	1.45	3.86	6.61																	
24	23.50	1.70	4.60	8.05	11.75																
30	36.72	1.91	5.23	9.27	13.72	18.36															
36	52.88	2.12	5.79	10.34	15.43	20.85	26.44														
42	71.97	2.28	6.31	11.31	16.97	23.07	29.47	35.99													
48	94.01	2.45	6.78	12.20	18.38	25.10	32.20	39.54	47.00												
54	118.98	2.60	7.22	13.04	19.68	26.97	34.72	42.80	51.08	59.49											
60	146.89	2.75	7.64	13.82	20.91	28.72	37.06	45.82	54.87	64.11	73.44										
66	177.73	2.89	8.04	14.56	22.07	30.37	39.28	48.65	58.39	68.41	78.59	88.68									
72	211.52	3.02	8.42	15.26	23.17	31.92	41.36	51.32	61.71	72.45	83.41	94.54	105.76								
78	248.24	3.15	8.78	15.94	24.21	33.41	43.34	53.86	64.87	76.27	87.97	99.90	111.97	124.13							
84	287.90	3.26	9.12	16.57	25.24	34.85	45.24	56.29	67.87	79.91	92.30	104.98	117.85	130.87	143.95						
90	330.49	3.43	9.46	17.20	26.20	36.21	47.05	58.61	70.75	83.39	96.43	109.81	123.45	137.28	151.23	165.25					
96	376.02	3.50	9.79	17.80	27.13	37.52	48.81	60.84	73.52	86.73	100.39	114.44	128.79	143.40	158.17	173.06	188.01				
102	424.50	3.61	10.10	18.37	28.01	39.00	50.49	62.99	76.18	89.94	104.20	118.89	133.92	149.25	164.81	180.53	196.37	212.25			
108	476.10	3.71	10.39	18.94	28.90	40.03	52.14	65.09	78.74	93.04	107.87	123.17	138.87	154.89	171.19	187.71	204.37	221.14	238.05		
114	530.25	3.78	10.74	19.49	29.75	41.22	53.73	67.10	81.24	96.05	111.43	127.31	143.63	160.33	177.33	194.60	212.05	229.65	247.37	265.13	
120	587.54	3.91	10.98	20.02	30.57	42.39	55.26	69.06	83.65	98.95	114.87	131.32	148.25	165.58	183.27	201.24	219.46	237.87	256.43	275.08	293.77

From Kent's Mechanical Engineers' Handbook.

Note: This table may be used in conjunction with Table 47-D2 for finding total volume of liquid in double-dished head tanks. Example: A tank 8 ft long, 3 ft diam, 15 in. liquid inside. Solution: 20.85 × 8 (this table) plus 3.92 × 2 for dished heads (Table 47-D2) = 174.64 gal.

TABLE 46-D2

CAPACITIES OF STANDARD DISHED HEADS WHEN FILLED TO VARIOUS DEPTHS
Contents given in U.S. gallons for one head only

To ascertain the contents of a head over one-half full: Let h = depth of unfilled portion. Find from the table the quantity corresponding to a depth h. Subtract this quantity from the contents of a full head.

Radius = Diameter

Diam of Head (in.)	Full Head	Depth of Liquid, In. = h																			
		3	6	9	12	15	18	21	24	27	30	33	36	39	42	45	48	51	54	57	60
12	0.40	0.05	0.20																			
18	1.36	.07	.32	0.68																		
24	3.22	.08	.41	.95	1.61																	
30	6.30	.10	.49	1.18	2.10	3.15																
36	10.88	.11	.56	1.39	2.54	3.92	5.44															
42	17.28	.12	.63	1.59	2.94	4.64	6.57	8.64														
48	25.79	.13	.68	1.75	3.31	5.29	7.62	10.19	12.89													
54	36.72	.14	.74	1.90	3.64	5.90	8.60	11.65	14.95	18.36												
60	50.37	.14	.82	2.07	3.98	6.49	9.54	13.03	16.87	20.96	25.18											
66	67.04	.15	.83	2.19	4.25	6.98	10.35	14.30	18.68	23.43	28.42	33.52										
72	87.04	.16	.88	2.32	4.52	7.47	11.15	15.48	20.38	25.74	31.46	37.43	43.52									
78	110.66	.17	.93	2.44	4.79	7.97	11.94	16.65	22.02	27.97	34.39	41.16	48.20	55.33								
84	138.22	.18	.98	2.59	5.07	8.44	12.69	17.78	23.60	30.11	37.19	44.75	52.67	60.83	69.11							
90	170.01	.18	1.00	2.68	5.33	8.91	13.44	18.86	25.12	32.18	39.90	48.22	56.99	66.14	75.52	85.00						
96	206.32	.20	1.07	2.83	5.59	9.36	14.14	19.90	26.60	34.17	42.52	51.53	61.13	71.22	81.66	92.34	103.16					
102	247.48	.22	1.14	3.01	5.89	9.87	14.92	21.01	28.11	36.18	45.19	54.91	65.31	76.29	87.73	99.56	111.59	123.74				
108	293.77	.20	1.13	3.03	6.04	10.21	15.50	21.93	29.47	38.03	47.56	57.97	69.14	81.05	93.53	106.47	119.76	133.26	146.88			
114	345.51	.21	1.16	3.12	6.25	10.55	16.06	22.80	30.70	39.73	49.81	60.88	72.85	85.61	99.05	113.07	127.56	142.41	157.51	172.75		
120	402.27	.21	1.19	3.23	6.47	10.93	16.68	23.70	31.96	41.43	52.04	63.73	76.40	89.95	104.32	119.39	135.04	151.15	167.62	184.32	201.13

Courtesy Lukens Steel Company.

TABLE 46-E

CAPACITIES AND LIQUID SURFACE AREA OF CYLINDRICAL VESSELS, TANKS, CISTERNS, ETC., SET IN UPRIGHT (VERTICAL) POSITION

Diameter in feet and inches, area in square feet, and U.S. gallons
capacity for one foot in depth

$$1 \text{ gallon} = 231 \text{ cubic inches} = \frac{1 \text{ cubic foot}}{7.4805} = 0.13368 \text{ cubic foot}$$

Diam Ft	In.	Area Sq Ft	Gals 1 foot depth	Diam Ft	In.	Area Sq Ft	Gals 1 foot depth	Diam Ft	In.	Area Sq Ft	Gals 1 foot depth
1		0.785	5.87	5	8	25.22	188.66	19		283.53	2,120.9
1	1	0.922	6.89	5	9	25.97	194.25	19	3	291.04	2,177.1
1	2	1.069	8.00	5	10	26.73	199.92	19	6	298.65	2,234.0
1	3	1.227	9.18	5	11	27.49	205.67	19	9	306.35	2,291.7
1	4	1.396	10.44	6		28.27	211.51	20		314.16	2,350.1
1	5	1.576	11.79	6	3	30.68	229.50	20	3	322.06	2,409.2
1	6	1.767	13.22	6	6	33.18	248.23	20	6	330.06	2,469.1
1	7	1.969	14.73	6	9	35.78	267.69	20	9	338.16	2,529.6
1	8	2.182	16.32	7		38.48	287.88	21		346.36	2,591.0
1	9	2.405	17.99	7	3	41.28	308.81	21	3	354.66	2,653.0
1	10	2.640	19.75	7	6	44.18	330.48	21	6	363.05	2,715.8
1	11	2.885	21.58	7	9	47.17	352.88	21	9	371.54	2,779.3
2		3.142	23.50	8		50.27	376.01	22		380.13	2,843.6
2	1	3.409	25.50	8	3	53.46	399.88	22	3	388.82	2,908.6
2	2	3.687	27.58	8	6	56.75	424.48	22	6	397.61	2,974.3
2	3	3.976	29.74	8	9	60.13	449.82	22	9	406.49	3,040.8
2	4	4.276	31.99	9		63.62	475.89	23		415.48	3,108.0
2	5	4.587	34.31	9	3	67.20	502.70	23	3	424.56	3,175.9
2	6	4.909	36.72	9	6	70.88	530.24	23	6	433.74	3,244.6
2	7	5.241	39.21	9	9	74.66	558.51	23	9	443.01	3,314.0
2	8	5.585	41.78	10		78.54	587.52	24		452.39	3,384.1
2	9	5.940	44.43	10	3	82.52	617.26	24	3	461.86	3,455.0
2	10	6.305	47.16	10	6	86.59	647.74	24	6	471.44	3,526.6
2	11	6.581	49.98	10	9	90.76	678.95	24	9	481.11	3,598.9
3		7.069	52.88	11		95.03	710.90	25		490.87	3,672.0
3	1	7.467	55.86	11	3	99.40	743.58	25	3	500.74	3,745.8
3	2	7.876	58.92	11	6	103.87	776.99	25	6	510.71	3,820.3
3	3	8.296	62.06	11	9	108.43	811.14	25	9	520.77	3,895.6
3	4	8.727	65.28	12		113.10	846.03	26		530.93	3,971.6
3	5	9.168	68.58	12	3	117.86	881.65	26	3	541.19	4,048.4
3	6	9.621	71.97	12	6	122.72	918.00	26	6	551.55	4,125.9
3	7	10.085	75.44	12	9	127.68	955.09	26	9	562.00	4,204.1
3	8	10.559	78.99	13		132.73	992.91	27		572.56	4,283.0
3	9	11.045	82.62	13	3	137.89	1,031.5	27	3	583.21	4,362.7
3	10	11.541	86.33	13	6	143.14	1,070.8	27	6	593.96	4,443.1
3	11	12.048	90.13	13	9	148.49	1,110.8	27	9	604.81	4,524.3
4		12.566	94.00	14		153.94	1,151.5	28		615.75	4,606.2
4	1	13.095	97.96	14	3	159.48	1,193.0	28	3	626.80	4,688.8
4	2	13.635	102.00	14	6	165.13	1,235.3	28	6	637.94	4,772.1
4	3	14.186	106.12	14	9	170.87	1,278.2	28	9	649.18	4,856.2
4	4	14.748	110.32	15		176.71	1,321.9	29		660.52	4,941.0
4	5	15.321	114.61	15	3	182.65	1,366.4	29	3	671.96	5,026.6
4	6	15.90	118.97	15	6	188.69	1,411.5	29	6	683.49	5,112.9
4	7	16.50	123.42	15	9	194.83	1,457.4	29	9	695.13	5,199.9
4	8	17.10	127.95	16		201.06	1,504.1	30		706.86	5,287.7
4	9	17.72	132.56	16	3	207.39	1,551.4	30	3	718.69	5,376.2
4	10	18.35	137.25	16	6	213.82	1,599.5	30	6	730.62	5,465.4
4	11	18.99	142.02	16	9	220.35	1,648.4	30	9	742.64	5,555.4
5		19.63	146.88	17		226.98	1,697.9	31		754.77	5,646.1
5	1	20.29	151.82	17	3	233.71	1,748.2	31	3	766.99	5,737.5
5	2	20.97	156.83	17	6	240.53	1,799.3	31	6	779.31	5,829.7
5	3	21.65	161.93	17	9	247.45	1,851.1	31	9	791.73	5,922.6
5	4	22.34	167.12	18		254.47	1,903.6	32		804.25	6,016.2
5	5	23.04	172.38	18	3	261.59	1,956.8	32	3	816.86	6,110.6
5	6	23.76	177.72	18	6	268.80	2,010.8	32	6	829.58	6,205.7
5	7	24.48	183.15	18	9	276.12	2,065.5	32	9	842.39	6,301.5

From Kent's Mechanical Engineers' Handbook. John Wiley & Sons, Inc.

TABLE 46-F

MAXIMUM BRIGHTNESS RATIOS

	Environmental Classification		
	A	B	C
Between task and adjacent darker surroundings	3 to 1	3 to 1	5 to 1
Between tasks and adjacent lighter surroundings	1 to 3	1 to 3	1 to 5
Between tasks and more remote darker surfaces	10 to 1	20 to 1	*
Between tasks and more remote lighter surfaces	1 to 10	1 to 20	*
Between luminaires (or windows, skylights, etc.) and surfaces adjacent to them	20 to 1	*	*
Anywhere within normal field of view	40 to 1	*	*

*Brightness Ratio control not practical.

A-Interior Areas where reflectances of entire space can be controlled in line with recommendations for optimum seeing conditions.

B-Areas where reflectances of immediate work area can be controlled, but control of remote surroundings is limited.

C-Areas (indoor and outdoor) where it is completely impractical to control reflectances and difficult to alter environmental conditions.

Note: From the normal view point, brightness ratios of areas of appreciable size in industrial areas should not exceed those in the above table.

From American National Standard A11.1–1973, Practice for Industrial Lighting, Illuminating Engineering Society.

TABLE 46-G

LEVELS OF ILLUMINATION RECOMMENDED FOR SAMPLE OCCUPATIONAL TASKS

Area	Foot-candles	Area	Foot-candles
Assembly—rough, easy seeing	30	Materials—loading, trucking	20
Assembly—medium	100	Offices—general areas*	100
Building construction—general	10	Drafting rooms—detailed*	200
Electrical equipment, testing	100	Corridors	20
Elevators	20	Paint dipping, spraying	50
Garages—repair areas	100	Service spaces—wash rooms, etc.	30
Garages—traffic areas	20	Sheet metal—presses, shears	50
Inspection, ordinary	50	Storage rooms—inactive	5
Inspection, highly difficult	200	Storage rooms—active, medium	20
Loading platforms	20	Welding—general	50
Machine shops—medium work	100	Woodworking—rough sawing	30

*From A132.1–1973, Office Lighting, published by the Illuminating Engineering Society. Others from ANSI A11.1–1973. See explanation on page 1449.

TABLE 46-H

RPM SPEED OF SAWS TO GIVE 10,000 FEET PER MINUTE RIM SPEED

Diameter (in.)	Circumference (ft)	RPM	Diameter (in.)	Circumference (ft)	RPM
6	1.57	6,366	56	14.66	680
8	2.09	4,775	58	15.18	660
10	2.62	3,820	60	15.71	640
12	3.14	3,180	64	16.78	600
14	3.67	2,730	66	17.28	580
16	4.19	2,390	68	17.80	560
18	4.71	2,125	70	18.33	545
			72	18.85	530
20	5.24	1,910			
22	5.76	1,740	74	19.37	520
24	6.28	1,590	76	19.90	500
26	6.81	1,470	78	20.42	490
28	7.33	1,365	80	20.94	480
30	7.85	1,275	82	21.47	470
			84	22.00	455
32	8.38	1,195			
34	8.90	1,125	86	22.52	445
36	9.43	1,060	90	23.56	425
38	9.95	1,005	96	25.13	400
40	10.47	955			
42	11.00	910	100	26.18	380
			104	27.22	370
44	11.52	870	106	27.75	364
46	12.04	830	108	28.27	354
48	12.57	800			
50	13.09	765	110	28.80	348
52	13.61	740	112	29.32	341
54	14.14	710	120	31.42	318

Courtesy H. K. Porter Co., Inc., Disston Division

NOTE: Saws are usually tensioned to a standard speed of 10,000 peripheral feet per minute and, with the exception of large saws, can be operated at slower speeds with good performance and safety. The above table gives the rpm corresponding to this standard speed for various diameter saws.

All saws required to operate in excess of 10,000 peripheral feet per minute are tensioned for the specific speed specified. This speed is etched on the saw blade. These special saws should not be operated at other than the specified speed without consultation with the saw manufacturer.

46—Safety Engineering Tables

TABLE 46-I

STRENGTH OF ICE

Load	Ice Thickness Inches (min.)
One person on skis	1¾
One person on foot	2
Group, single file or Snowmobile	3
Passenger car, 2 ton gross	7½
Light truck, 2½ ton gross	8
Medium truck, 3½ ton gross	10
Heavy truck, 7-8 ton gross	12

Courtesy Snow, Ice and Permafrost Research Establishment, Corps of Engineers, U.S. Army. Snowmobile data from "A Snowmobile Accident Study," R. W. McLay and S. Chism, 1969.

NOTES:
1) This table is for clear sound ice.
2) Thickness must be increased and extreme caution used after spring melt starts.
3) Snow ice—use 50% of thickness of snow ice as effective for table given.
   Example: 10 in. snow ice plus 6 in. clear ice gives 11 in. thickness. Medium truck all right, but not heavy truck.
4) Table does not apply to parked loads.
5) Key to safety on ice is to disperse the weight over a large area; do not run vehicles within 50 ft of each other.

TABLE 46-J

DESIGN WIND PRESSURES FOR VARIOUS HEIGHT ZONES OF BUILDINGS OR OTHER STRUCTURES

(From *Minimum Design Loads in Buildings and Other Structures*, ANSI A58.1)

Height Zone, feet	Wind Pressure pounds per sq ft
Less than 50	20
50 to 99	24
100 to 199	28
200 to 299	30
300 to 399	32
400 to 499	33
500 to 599	34
600 to 799	35
800 to 999	36
1,000 to 1,199	37
1,200 to 1,399	38
1,400 to 1,599	39
1,600 and over	40

From Kent's Mechanical Engineers' Handbook (John Wiley & Sons)

WIND LOADS. A minimum pressure in pounds per square foot is specified by all building codes as an allowance for the effect of wind pressure against exposed surfaces of buildings. The figures given are recommended in this report as minimums. These requirements do not provide for tornadoes because they are based on a design wind velocity of 75 mph, corresponding roughly to a 5-min average of 50 mph (indicated 63 by a 4-cup anemometer) at 30 ft from the ground.

Special provision for wind bracing is required for narrow buildings of even medium height and for all high buildings. In towerlike structures wind bracing should be planned in the early stages of design. The overturning moment due to wind pressure should not exceed two-thirds of the moment of stability, disregarding live loads, except where foundations are securely anchored. The possibility of a structure sliding on its foundation bed should also be considered. In computing the wind load no allowance should be made for the shielding effect of other buildings. All roofs should be designed to resist wind suction as well as wind pressure.

1458

TABLE 46-K

SAFE BEARING LOADS ON SOILS
(Values approximate pressures allowed in major city building codes)

Nature of Soil	Safe Bearing Capacity (Tons per sq ft)
Solid ledge of hard rock, such as granite, trap, etc.	25–100
Sound shale or other medium rock, requiring blasting for removal........................	10–15
Hardpan, cemented sand and gravel, difficult to remove by picking	8–10
Soft rock, disintegrated ledge; in natural ledge, difficult to remove by picking........	5–10
Compact sand and gravel, requiring picking for removal	4–6
Hard clay, requiring picking for removal	4–5
Gravel, coarse sand, in natural thick beds..	4–5
Loose, medium, and coarse sand; fine compact sand...	1.5–4
Medium clay, stiff but capable of being spaded	2–4
Fine loose sand..	1–2
Soft clay	1

From Standard Handbook for Mechanical Engineers, *rev. 7th ed. Edited by T. Baumeister. (Copyright 1967. Mc-Graw-Hill Book Co.) Used by permission.*

TABLE 46-L

SPECIFIC GRAVITY OF GASES AND LIQUIDS
(Gases at 32 F: air = 1.000; liquids: water = 1.000)

Gas	Sp Gr	Gas	Sp Gr	Gas	Sp Gr
Air..............................	1.000	Ether vapor..............	2.586	Methane................	0.554
Acetylene	0.920	Ethylene	0.967	Nitrogen................	0.971
Ethyl alcohol vapor.......	1.601	Helium	0.138	Nitric oxide...........	1.039
Ammonia......................	0.592	Hydrogen fluoride......	2.370	Nitrous oxide	1.527
Carbon dioxide	1.520	Hydrogen chloride.....	1.261	Oxygen	1.106
Carbon monoxide.........	0.967	Hydrogen.................	0.069	Propane	1.554
Chlorine......................	2.423	Mercury vapor...........	6.940	Sulfur dioxide........	2.250
Ethane	1.049	Marsh gas................	0.555	Water vapor...........	0.623

1 cu ft of air at 32 F and atmospheric pressure weighs 0.0807 pounds.

Liquid	Sp Gr	Liquid	Sp Gr	Liquid	Sp Gr
Acetic acid	1.06	Fluoric acid	1.50	Palm oil.................	0.97
Alcohol, commercial.....	0.83	Gasoline...................	0.70	Petroleum oil.........	0.82
Ammonia.....................	0.77	Glycerin....................	1.26	Phosphoric acid.....	1.78
Benzene......................	0.88	Kerosene...................	0.80	Rape oil.................	0.92
Bromine......................	2.97	Linseed oil................	0.94	Vinegar	1.08
Carbolic acid	0.96	Mineral oil	0.92	Water....................	1.00
Carbon disulfide..........	1.26	Naphtha	0.76	Whale oil...............	0.92
Cotton-seed oil............	0.93	Olive oil....................	0.92		

Mainly from Machinery's Handbook, *18th ed. (The Industrial Press)*   1 cu ft water at 39 F weighs 62.43 pounds

46—Safety Engineering Tables

TABLE 46-M

APPROXIMATE SPECIFIC GRAVITIES AND DENSITIES
(At room temperature with reference to water at 39 F)

Substance	Specific gravity	Avg density, lb per cu ft	Substance	Specific gravity	Avg density, lb per cu ft
Metals, Alloys, Ores			**Timber, air-dry**		
Aluminum, cast-hammered	2.55–2.80	165	Apple	0.66–0.74	44
Aluminum, bronze.................	7.7	481	Ash, black	0.55	34
Brass, cast-rolled..................	8.4–8.7	534	Ash, white	0.64–0.71	42
Bronze, 7.9 to 14% Sn............	7.4–8.9	509	Birch, sweet, yellow.............	0.71–0.72	44
Bronze, phosphor	8.88	554	Cedar, white, red..............	0.35	22
Copper, cast-rolled	8.8–8.95	556	Cherry, wild red	0.43	27
Copper ore, pyrites	4.1–4.3	262	Chestnut	0.48	30
German silver......................	8.58	536	Cypress..........................	0.45–0.48	29
Gold, cast-hammered	19.25–19.35	1205	Fir, Douglas.....................	0.48–0.55	32
Gold coin (U.S.)	17.18–17.2	1073	Fir, balsam	0.40	25
Iridium.................................	21.78–22.42	1383	Elm, white	0.56	35
Iron, gray cast	7.03–7.13	442	Hemlock......................	0.45–0.50	29
Iron, cast, pig	7.2	450	Hickory	0.74–0.80	48
Iron, wrought........................	7.6–7.9	485	Locust...........................	0.67–0.77	45
Iron, spiegel-eisen	7.5	468	Mahogany	0.56–0.85	44
Iron, ferrosilicon	6.7–7.3	437	Maple, sugar.....................	0.68	43
Iron ore, hematite	5.2	325	Maple, white	0.53	33
Iron ore, limonite	3.6–4.0	237	Oak, chestnut	0.74	46
Iron ore, magnetite	4.9–5.2	315	Oak, live	0.87	54
Iron slag	2.5–3.0	172	Oak, red, black....................	0.64–0.71	42
Lead....................................	11.34	710	Oak, white	0.77	48
Lead ore, galena....................	7.3–7.6	465	Pine, Oregon	0.51	32
Manganese..........................	7.42	475	Pine, red	0.48	30
Manganese ore, pyrolusite	3.7–4.6	259	Pine, white	0.43	27
Mercury...............................	13.546	847	Pine, Southern....................	0.61–0.67	38–42
Monel metal, rolled	8.97	555	Pine, Norway	0.55	34
Nickel	8.9	537	Poplar..........................	0.43	27
Platinum, cast-hammered	21.5	1330	Redwood, California	0.42	26
Silver, cast-hammered............	10.4–10.6	656	Spruce, white, red	0.45	28
Steel, cold-drawn..................	7.83	489	Teak, African	0.99	62
Steel, machine.....................	7.80	487	Teak, Indian......................	0.66–0.88	48
Steel, tool	7.70–7.73	481	Walnut, black.....................	0.59	37
Tin, cast-hammered	7.2–7.5	459	Willow..............................	0.42–0.50	28
Tin ore, cassiterite	6.4–7.0	418			
Tungsten..............................	19.22	1200	**Various Liquids**		
Zinc, cast-rolled....................	6.9–7.2	440	Alcohol, ethyl (100%)...........	0.789	49
Zinc, ore, blende...................	3.9–4.2	253	Alcohol, methyl (100%)	0.796	50
			Acid, muriatic (HCl), 40%.......	1.20	75
Various Solids			Acid, nitric, 91%..................	1.50	94
			Acid, sulfuric, 87%	1.80	112
Cereals, oats, bulk	0.41	26	Chloroform.......................	1.500	95
Cereals, barley, bulk..............	0.62	39	Ether..............................	0.736	46
Cereals, corn, rye, bulk...........	0.73	45	Lye, soda, 66%	1.70	106
Cereals, wheat, bulk	0.77	48	Oils, vegetable	0.91–0.94	58
Cork....................................	0.22–0.26	15	Oils, mineral, lubricants	0.88–0.94	57
Cotton, flax, hemp	1.47–1.50	93	Turpentine........................	0.861–0.867	54
Fats....................................	0.90–0.97	58	Water, 4 C, max. density	1.0	62.428
Flour, loose	0.40–0.50	28	Water, 100 C	0.9584	59.830
Flour, pressed	0.70–0.80	47	Water, ice.........................	0.88–0.92	56
Glass, common.....................	2.40–2.80	162	Water, snow, fresh fallen........	0.125	8
Glass, plate or crown	2.45–2.72	161	Water, sea water	1.02–1.03	64
Glass, crystal.......................	2.90–3.00	184			
Glass, flint..........................	3.2–4.7	247	**Ashlar Masonry**		
Hay and straw, bales	0.32	20	Granite, syenite, gneiss.........	2.4–2.7	159
Leather................................	0.86–1.02	59	Limestone	2.1–2.8	153
Paper..................................	0.70–1.15	58	Marble	2.4–2.8	162
Potatoes, piled......................	0.67	44	Sandstone	2.0–2.6	143
Rubber, Caoutchouc	0.92–0.96	59	Bluestone..........................	2.3–2.6	153
Rubber goods.......................	1.0–2.0	94			
Salt, granulated, piled	0.77	48	**Rubble Masonry**		
Saltpeter..............................	2.11	132			
Starch.................................	1.53	96	Granite, syenite, gneiss.........	2.3–2.6	153
Sulfur	1.93–2.07	125			
Wool....................................	1.32	82			

1460

TABLE 46-M (*Concluded*)

Substance	Specific gravity	Avg density, lb per cu ft	Substance	Specific gravity	Avg density, lb per cu ft
Rubble Masonry			**Minerals**		
Limestone	2.0–2.7	147	Asbestos	2.1–2.8	153
Sandstone	1.9–2.5	137	Barytes	4.50	281
Bluestone	2.2–2.5	147	Basalt	2.7–3.2	184
Marble	2.3–2.7	156	Bauxite	2.55	159
			Bluestone	2.5–2.6	159
			Borax	1.7–1.8	109
Dry Rubble Masonry			Chalk	1.8–2.8	143
			Clay, marl	1.8–2.6	137
Granite, syenite, gneiss	1.9–2.3	130	Dolomite	2.9	181
Limestone, marble	1.9–2.1	125	Feldspar, orthoclase	2.5–2.7	162
Sandstone, bluestone	1.8–1.9	110	Gneiss	2.7–2.9	175
			Granite	2.6–2.7	165
			Greenstone, trap	2.8–3.2	187
Brick Masonry			Gypsum, alabaster	2.3–2.8	159
			Hornblende	3.0	187
Hard brick	1.8–2.3	128	Limestone	2.1–2.86	155
Medium brick	1.6–2.0	112	Marble	2.6–2.86	170
Soft brick	1.4–1.9	103	Magnesite	3.0	187
Sand-lime brick	1.4–2.2	112	Phosphate rock, apatite	3.2	200
			Porphyry	2.6–2.9	172
			Pumice, natural	0.37–0.90	40
Concrete Masonry			Quartz, flint	2.5–2.8	165
Cement, stone, sand	2.2–2.4	144	Sandstone	2.0–2.6	143
Cement, slag, etc.	1.9–2.3	130	Serpentine	2.7–2.8	171
Cement, cinder, etc.	1.5–1.7	100	Shale, slate	2.6–2.9	172
			Soapstone, talc	2.6–2.8	169
			Syenite	2.6–2.7	165
Various Building Mat'ls					
Ashes, cinders	0.64–0.72	40–45			
Cement, portland, loose	1.5	94			
Portland cement	3.1–3.2	196	**Stone, Quarried, Piled**		
Lime, gypsum, loose	0.85–1.00	53–64	Basalt, granite, gneiss	1.5	96
Mortar, lime, set	1.4–1.9	103	Limestone, marble quartz	1.5	95
		94	Sandstone	1.3	82
Mortar, portland cement	2.08–2.25	135	Shale	1.5	92
Slags, bank slag	1.1–1.2	67–72	Greenstone, hornblende	1.7	107
Slags, bank screenings	1.5–1.9	98–117			
Slags, machine slag	1.5	96			
Slags, slag sand	0.8–0.9	49–55			
			Bituminous Substances		
Earth, etc., Excavated			Asphaltum	1.1–1.5	81
Clay, dry	1.0	63	Coal, anthracite	1.4–1.8	97
Clay, damp, plastic	1.76	110	Coal, bituminous	1.2–1.5	84
Clay and gravel, dry	1.6	100	Coal, lignite	1.1–1.4	78
Earth, dry, loose	1.2	76	Coal, peat, turf, dry	0.65–0.85	47
Earth, dry, packed	1.5	95	Coal, charcoal, pine	0.28–0.44	23
Earth, moist, loose	1.3	78	Coal, charcoal, oak	0.47–0.57	33
Earth, moist, packed	1.6	96	Coal, coke	1.0–1.4	75
Earth, mud, flowing	1.7	108	Graphite	1.64–2.7	135
Earth, mud, packed	1.8	115	Paraffin	0.87–0.91	56
Riprap, limestone	1.3–1.4	80–85	Petroleum	0.87	54
Riprap, sandstone	1.4	90	Petroleum, refined (kerosene)	0.78–0.82	50
Riprap, shale	1.7	105	Petroleum, benzene	0.73–0.75	46
Sand, gravel, dry, loose	1.4–1.7	90–105	Petroleum, gasoline	0.70–0.75	45
Sand, gravel, dry, packed	1.6–1.9	100–120	Pitch	1.07–1.15	69
Sand, gravel, wet	1.89–2.16	126	Tar, bituminous	1.20	75
Excavations in Water			**Coal and Coke, Piled**		
Sand or gravel	0.96	60	Coal, anthracite	0.75–0.93	47–58
Sand or gravel and clay	1.00	65	Coal, bituminous, lignite	0.64–0.87	40–54
Clay	1.28	80	Coal, peat, turf	0.32–0.42	20–26
River mud	1.44	90	Coal, charcoal	0.16–0.23	10–14
Soil	1.12	70	Coal, coke	0.37–0.51	23–32
Stone riprap	1.00	65			

TABLE 46-N

EXAMPLES OF TYPICAL INCOMPATIBLE CHEMICALS

Chemical	Keep out of contact with:
Alkaline metals, such as powdered aluminum or magnesium, sodium, potassium, etc.	Carbon tetrachloride or other chlorinated hydrocarbon, carbon dioxide and the halogens.
Acetic acid	Chromic acid, nitric acid, hydroxyl compounds, ethylene glycol, perchloric acid, peroxides, permanganates.
Acetylene	Chlorine, bromine, copper, fluorine, silver, mercury.
Ammonia, anhydrous	Mercury (in manometers, for instance), chlorine, calcium hypochlorite, iodine, bromine, hydrofluoric acid anhydrous.
Ammonium nitrate	Acids, metal powders, flammable liquids, chlorates, nitrites, sulfur, finely divided organic or combustible materials.
Aniline	Nitric acid, hydrogen peroxide.
Bromine	Same as for chlorine.
Carbon, activated	Calcium hypochlorite and all oxidizing agents.
Copper	Acetylene, hydrogen peroxide.
Chlorates	Ammonium salts, acids, metal powders, sulfur, finely divided organic or combustible materials.
Chromic acid	Acetic acid, naphthalene, camphor, glycerine, turpentine, alcohol and flammable liquids in general.
Chlorine	Ammonia, acetylene, butadiene, butane, methane, propane (or other petroleum gases), hydrogen, sodium carbide, turpentine, benzene, finely divided metals.
Chlorine dioxide	Ammonia, methane, phosphine, hydrogen sulfide.
Cumene hydroperoxide	Acids—Organic or inorganic.
Flammable liquids	Ammonium nitrate, chromic acid, hydrogen peroxide, nitric acid, sodium peroxide and the halogens.
Fluorine	Isolate from everything.
Hydrocyanic acid	Nitric acid, alkalis.
Hydrogen peroxide	Copper, chromium, iron, most metals or their salts, alcohols, acetone, organic materials, aniline, nitro-methane, any flammable liquid, combustible materials.
Hydrofluoric acid, anhydrous	Ammonia, aqueous or anhydrous.
Hydrogen sulfide	Fuming nitric acid, oxidizing gases.
Hydrocarbons (Butane, propane, benzene, gasoline, turpentine, etc.)	Fluorine, chlorine, bromine, chromic acid, sodium peroxide.
Iodine	Acetylene, ammonia (aqueous or anhydrous), hydrogen.
Mercury	Acetylene, fulminic acid, ammonia.
Nitric acid (concentrated)	Acetic acid, aniline, chromic acid, hydrocyanic acid, hydrogen sulfide, flammable liquids, flammable gases.
Oxalic acid	Silver, mercury.
Perchloric acid	Acetic anhydride, bismuth and its alloys, alcohol, paper, wood.
Potassium	Carbon tetrachloride, carbon dioxide, water.
Potassium chlorate	Sulfuric and other acids.
Potassium perchlorate (See also Chlorates)	Sulfuric and other acids.
Potassium permanganate	Glycerine, ethylene glycol, benzaldehyde, sulfuric acid.
Silver	Acetylene, oxalic acid, tartaric acid, fulminic acid, ammonium compounds.
Sodium	Carbon tetrachloride, carbon dioxide, water.
Sodium peroxide	Ethyl or methyl alcohol, glacial acetic acid, acetic anhydride, benzaldehyde, carbon disulfide, glycerine, ethylene glycol, ethyl acetate, methyl acetate, furfural.
Sulfuric acid	Potassium chlorate, potassium perchlorate, potassium permanganate (or such compounds with similar light metals, as sodium, lithium, etc.).

TABLE 46-O

EXPLOSION CHARACTERISTICS OF VARIOUS DUSTS

Type of dust	Ignition temp of dust cloud (deg C)	Min spark energy required for ignition of dust cloud (millijoules)	Min explosive concentration (oz per 1,000 cu ft)	Max explosion pressure (psi)	Rates of pressure rise (psi per sec)		Limiting oxygen percentage to prevent ignition of dust clouds by electric sparks
					Avg	Max	
Metal powders							
Aluminum, atomized	640	15	40	90	3,500	10,000+	7
Aluminum, milled	550	. . .	45	70	2,000	4,250	
Aluminum, stamped	550	10	35	100	10,000	10,000+	4
Boron (85% B, 8% Mg)	470	60	135	90	900	2,500	
Iron, carbonyl	320	20	105	50	1,500	7,000	10
Iron, electrolytic	320	240	200	45	500	1,000	13
Iron, hydrogen reduced	315	80	120	45	800	1,750	13
Magnesium, atomized	600	120	10	80	2,000	5,250	3
Magnesium, milled	520	40	20	95	3,000	10,000+	†
Magnesium, stamped	520	20	20	80	3,400	10,000+	†
Manganese	450	80	125	50	1,300	2,750	15
Silicon	775	80	100	105	2,000	10,000	13
Thorium	270	5	75	50	1,400	3,250	†
Thorium hydride	260	3	80	60	2,100	6,750	6
Tin	630	160	190	35	500	1,250	16
Titanium	330	10	45	80	3,400	10,000+	†
Titanium hydride	440	60	70	95	3,800	10,000+	13
Uranium	*	45	60	55	1,600	3,500	†
Uranium hydride	*	5	60	45	2,900	6,250	0.5
Vanadium	500	60	220	35‡	200	300	13
Zinc	600	650	480	50	600	1,750	10
Zirconium	*	5	40	65	800	8,750	†
Zirconium hydride	350	60	85	60	2,400	8,750	11
Aluminum-cobalt alloy (60-40)	950	100	100	80	2,500	8,500	
Aluminum-copper alloy (50-50)	950	100	100	70	800	2,500	
Aluminum-nickel alloy (60-40)	960	80	190	80	2,600	10,000+	
Calcium-silicon alloy	540	220	600	75	400	10,000+	
Dowmetal	430	80	20	85	3,600	10,000+	†
Ferromanganese (1.4% C)	450	80	130	45	1,400	4,250	
Ferrosilicon (80% Si)	860	280	400	90	1,500	3,600	19
Ferrotitanium, low-carbon	370	80	140	55	2,200	9,500	13
Magnesium-aluminum alloy (50-50)	535	80	50	90	4,000	10,000+	†
Plastics							
Allyl alcohol resin	500	20	35	105	2,800	10,000+	
Butadiene-styrene resin	440	60	25	80	1,400	4,000	
Cellulose acetates	320	10	25	110	2,800	6,750	11
Cellulose propionate	460	45	25	105	1,600	4,750	
Coumarone-indene resin	520	10	15	85	2,800	8,500	14
Dimethyl terephthalate	570	20	30	90	3,100	10,000	
Gums (arabic, copal, etc.)	360	30	30	95	1,500	5,000	14
Lignin resin	450	20	40	80	1,700	4,750	17
Methyl cellulose	360	20	30	100	1,900	6,000	
Methyl methacrylate	440	15	20	100	500	1,750	14
Phenolic resins	460	10	25	80	1,700	6,000	14
Pine-rosin base resin	440	. . .	55	80	1,900	7,500	
Polyacrylonitrile	500	20	25	90	2,000	5,000	
Polyamide	500	20	30	90	1,800	7,000	
Polyster resin-glass fiber mixture (65–35)	440	50	45	85	2,200	6,000	
Polyether alcohol resin	460	160	45	65	500	1,000	
Polyethylene resin	410	30	20	80	1,500	3,500	13
Polyethylene terephthalate	500	35	40	90	1,600	7,500	13
Polystyrene	490	15	15	90	2,400	7,000	14
Polyvinyl acetate resin	520	120	35	75	1,200	3,000	
Rubber, synthetic hard	320	30	30	95	1,100	3,000	15
Shellac	400	10	20	75	1,400	3,500	14
Styrene-maleic anhydride copolymer	470	20	30	80	2,300	9,500	

1463

TABLE 46-O (Continued)

Type of dust	Ignition temp of dust cloud (deg C)	Min spark energy required for ignition of dust cloud (milli-joules)	Min explosive concentration (oz per 1,000 cu ft)	Max explosion pressure (psi)	Rates of pressure rise (psi per sec)		Limiting oxygen percentage to prevent ignition of dust cloud by electric sparks
					Avg	Max	
Urea resin	450	80	70	85	800	2,000	17
Vinyl butyral resin	390	10	20	60‡	500	1,000	14
Vinyl chloride-acrylonitrile polymer	530	15	35	85	1,700	4,500	
Vinyl copolymer resin	500	60	100	40	200	500	
Agricultural products							
Alfalfa	460	320	100	65	500	1,000	
Cellucotton	440	60	50	100	900	3,000	
Cinnamon	440	40	40	115	1,400	4,000	
Citrus peel, dehydrated	490	45	60	100	1,200	3,000	
Clover seed	470	80	60	60	400	1,000	15
Cocoa	420	100	45	62‡	550	1,200	
Coffee	410	160	85	50	150	250	13
Corncob meal	400	60	30	120	1,200	3,750	
Cornstarch	380	30	40	110	2,200	6,750	10
Cotton seed	470	80	55	90	800	2,500	15
Dextrin, corn	400	40	40	105	1,800	7,000	
Furfural residue	440	40	40	105	1,400	4,000	
Grain dust	430	30	55	95	1,000	2,750	
Guar seed	500	60	40	105	1,400	4,750	
Lycopodium	480	40	25	85	2,300	7,000	13
Nut shells	420	50	30	105	1,900	4,000	
Onion, dehydrated	410	...	130	60‡	400	1,250	
Pea, dehydrated	560	40	50	100	2,100	6,000	
Pectin	420	35	75	110	1,800	8,000	
Potato starch	440	25	45	95	2,300	8,000	
Pyrethrum	480	80	100	80	600	1,500	
Rice	440	40	45	95	1,000	2,750	
Soybean	520	50	35	100	1,200	3,250	15
Sugar	350	30	35	90	1,600	5,000	
Tung	440	240	70	110	1,400	3,500	
Wheat dust	470	50	70	105	1,500	3,500	
Wheat flour	380	50	50	95	1,200	3,750	
Yeast	520	50	50	105	1,000	2,500	
Miscellaneous							
Adipic acid	550	70	35	75	1,200	2,750	
Aluminum stearate	400	15	15	95	1,200	4,750	
Aspirin	660	25	35	85	2,000	10,000+	
Bark dust (Douglas-fir)	540	40	30	90	2,900	9,500	
Beryllium acetate	620	100	80	80	600	2,000	17
Calcium lignin sulfonic acid	590	100	160	80	600	2,000	
Carbon, activated	660	...	...	40‡	200	300	
Casein, rennet	520	60	45	65	400	1,000	17
Cellulose	480	80	55	100	1,100	2,750	13
Charcoal (pine wood)	620	...	...	40‡	200	250	
Coal, low volatile	635	...	...	45	300	600	
Coal, medium volatile	605	120	120	60	300	600	18
Coal, high volatile (Pgh. seam)	610	60	55	85	800	2,250	16
Coal, subbituminous	455	60	45	95	1,200	3,000	
Cork	470	45	35	100	2,000	5,500	
Diazoaminobenzine	550	20	15	90	2,900	10,000+	
Dinitro-ortho-cresol	440	80	25	55‡	1,300	2,250	15
Diphenyl	650	60	35	55	400	1,500	
Gilsonite	560	25	20	90	1,200	3,750	
Hexamethylenetetramine	410	10	15	100	2,400	10,000+	14
Lactalbumin	570	50	40	90	900	2,750	13
Lignite	440	60	45	90	800	2,750	15
Liver protein	520	45	45	80	800	2,250	
Napalm	450	40	20	85	1,000	3,000	12
Paraformaldehyde	410	20	40	100	2,500	10,000+	
Peat, sphagnum	460	50	45	85	900	2,250	
Pentaerythritol	450	10	30	90	1,700	9,500	14
Phenothiazine	540	...	15	80	1,400	4,250	16
Phthalic anhydride	650	15	15	70	1,300	4,250	14
Phytosterol	330	10	25	75	1,500	8,000	

TABLE 46-O (*Concluded*)

Type of dust	Ignition temp of dust cloud (deg C)	Min spark energy required for ignition of dust cloud (millijoules)	Min explosive concentration (oz per 1,000 cu ft)	Max explosion pressure (psi)	Rates of pressure rise (psi per sec)		Limiting oxygen percentage to prevent ignition of dust cloud by electric sparks
					Avg	Max	
Pitch, coal tar (58% vol. matter)	710	20	35	95	1,900	6,000	15
Procaine penicillin	450	...	25	50‡	1,000	2,000	
Rubber, crude, hard	350	50	25	80	1,200	3,800	15
Secobartital sodium	520	95	105	55	250	500	
Soap	430	60	45	85	600	1,750	
Sodium alkylarylsulfonate	540	...	130	75‡	400	1,250	
Sodium benzoate	560	80	55	85	1,800	10,000+	
Sodium carboxymethyl cellulose	350	560	150	60	300	600	
Sorbic acid	470	15	25	90	3,000	10,000+	
Sulfur	190	15	35	80	1,700	4,750	11
Vitamin B_2	500	80	105	80	1,000	2,250	
Wood flour	430	20	40	110	1,600	5,500	17

From Standard Handbook for Mechanical Engineers, *rev. 7th ed. Edited by T. Baumeister.* (*Copyright 1967. McGraw-Hill Book Co.*) *Used by permission.*

When uranium, uranium hydride, and zirconium were dispersed into air at room temperature, the dust clouds ignited under some conditions.

†The oxygen reduction data in this table are based on tests made in air-CO_2 mixtures. Dust clouds of thorium, titanium, uranium, zirconium, Dowmetal, and certain magnesium and magnesium-alloy powders ignited in pure CO_2.

‡Pressure and rates of pressure rise for these dusts were measured by an older testing technique.

TABLE 46-P

NATURAL TRIGONOMETRIC FUNCTIONS

Angle	Function	Value	Angle	Function	Value	Angle	Function	Value	Angle	Function	Value
0	sin	0.0000	6	sin	0.1045	12	sin	0.2079	18	sin	0.3090
	cos	1.0000		cos	0.9945		cos	0.9781		cos	0.9511
	tan	0.0000		tan	0.1051		tan	0.2126		tan	0.3249
1	sin	0.0175	7	sin	0.1219	13	sin	0.2250	19	sin	0.3256
	cos	0.9998		cos	0.9925		cos	0.9744		cos	0.9455
	tan	0.0175		tan	0.1228		tan	0.2309		tan	0.3443
2	sin	0.0349	8	sin	0.1392	14	sin	0.2419	20	sin	0.3420
	cos	0.9994		cos	0.9903		cos	0.9703		cos	0.9397
	tan	0.0349		tan	0.1405		tan	0.2493		tan	0.3640
3	sin	0.0523	9	sin	0.1564	15	sin	0.2588	21	sin	0.3584
	cos	0.9986		cos	0.9877		cos	0.9659		cos	0.9336
	tan	0.0524		tan	0.1584		tan	0.2679		tan	0.3839
4	sin	0.0698	10	sin	0.1736	16	sin	0.2756	22	sin	0.3746
	cos	0.9976		cos	0.9848		cos	0.9613		cos	0.9272
	tan	0.0699		tan	0.1763		tan	0.2867		tan	0.4040
5	sin	0.0872	11	sin	0.1908	17	sin	0.2924	23	sin	0.3907
	cos	0.9962		cos	0.9816		cos	0.9563		cos	0.9205
	tan	0.0875		tan	0.1944		tan	0.3057		tan	0.4245

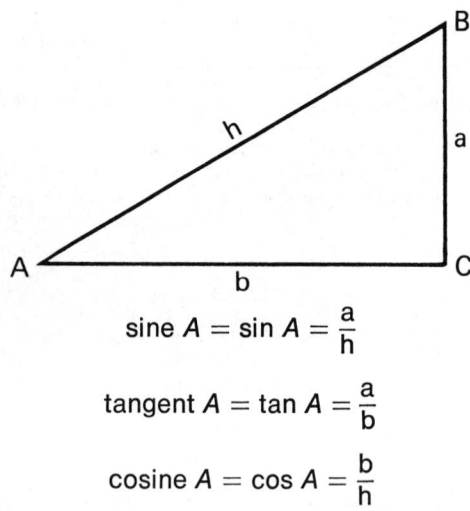

$$\text{sine } A = \sin A = \frac{a}{h}$$

$$\text{tangent } A = \tan A = \frac{a}{b}$$

$$\text{cosine } A = \cos A = \frac{b}{h}$$

SIGNS AND LIMITS OF VALUE
ASSUMED BY THE FUNCTIONS

Function	Quadrant I		Quadrant II		Quadrant III		Quadrant IV	
	Sign	Value	Sign	Value	Sign	Value	Sign	Value
sin	+	0 to 1	+	1 to 0	−	0 to 1	−	1 to 0
cos	+	1 to 0	−	0 to 1	−	1 to 0	+	0 to 1
tan	+	0 to ∞	−	∞ to 0	+	0 to ∞	−	∞ to 0

Angle	Function	Value	Angle	Function	Value	Angle	Function	Value	Angle	Function	Value
24	sin	0.4067	40	sin	0.6428	57	sin	0.8387	73	sin	0.9563
	cos	0.9135		cos	0.7660		cos	0.5446		cos	0.2924
	tan	0.4452		tan	0.8391		tan	1.5399		tan	3.2709
25	sin	0.4226	41	sin	0.6561	58	sin	0.8480	74	sin	0.9613
	cos	0.9063		cos	0.7547		cos	0.5299		cos	0.2756
	tan	0.4663		tan	0.8693		tan	1.6003		tan	3.4874
26	sin	0.4384	42	sin	0.6691	59	sin	0.8572	75	sin	0.9659
	cos	0.8988		cos	0.7431		cos	0.5150		cos	0.2588
	tan	0.4877		tan	0.9004		tan	1.6643		tan	3.7321
27	sin	0.4540	43	sin	0.6820	60	sin	0.8660	76	sin	0.9703
	cos	0.8910		cos	0.7314		cos	0.5000		cos	0.2419
	tan	0.5095		tan	0.9325		tan	1.7321		tan	4.0108
28	sin	0.4695	44	sin	0.6947	61	sin	0.8746	77	sin	0.9744
	cos	0.8829		cos	0.7193		cos	0.4848		cos	0.2250
	tan	0.5317		tan	0.9657		tan	1.8040		tan	4.3315
29	sin	0.4848	45	sin	0.7071	62	sin	0.8829	78	sin	0.9781
	cos	0.8746		cos	0.7071		cos	0.4695		cos	0.2079
	tan	0.5543		tan	1.0000		tan	1.8807		tan	4.7046
30	sin	0.5000	46	sin	0.7193	63	sin	0.8910	79	sin	0.9816
	cos	0.8660		cos	0.6947		cos	0.4540		cos	0.1908
	tan	0.5774		tan	1.0355		tan	1.9626		tan	5.1446
31	sin	0.5150	47	sin	0.7314	64	sin	0.8988	80	sin	0.9848
	cos	0.8572		cos	0.6820		cos	0.4384		cos	0.1736
	tan	0.6009		tan	1.0724		tan	2.0503		tan	5.6713
32	sin	0.5299	48	sin	0.7431	65	sin	0.9063	81	sin	0.9877
	cos	0.8480		cos	0.6691		cos	0.4226		cos	0.1564
	tan	0.6249		tan	1.1106		tan	2.1445		tan	6.3138
33	sin	0.5446	49	sin	0.7547	66	sin	0.9135	82	sin	0.9903
	cos	0.8387		cos	0.6561		cos	0.4067		cos	0.1392
	tan	0.6494		tan	1.1504		tan	2.2460		tan	7.1154
34	sin	0.5592	50	sin	0.7660	67	sin	0.9205	83	sin	0.9925
	cos	0.8290		cos	0.6428		cos	0.3907		cos	0.1219
	tan	0.6745		tan	1.1918		tan	2.3559		tan	8.1443
35	sin	0.5736	51	sin	0.7771	68	sin	0.9272	84	sin	0.9945
	cos	0.8192		cos	0.6293		cos	0.3746		cos	0.1045
	tan	0.7002		tan	1.2349		tan	2.4751		tan	9.5144
36	sin	0.5878	52	sin	0.7880	69	sin	0.9336	85	sin	0.9962
	cos	0.8090		cos	0.6157		cos	0.3584		cos	0.0872
	tan	0.7265		tan	1.2799		tan	2.6051		tan	11.43
37	sin	0.6018	53	sin	0.7986	70	sin	0.9397	86	sin	0.9976
	cos	0.7986		cos	0.6018		cos	0.3420		cos	0.0698
	tan	0.7536		tan	1.3270		tan	2.7475		tan	14.30
38	sin	0.6157	54	sin	0.8090	71	sin	0.9455	87	sin	0.9986
	cos	0.7880		cos	0.5878		cos	0.3256		cos	0.0523
	tan	0.7813		tan	1.3764		tan	2.9042		tan	19.08
39	sin	0.6293	55	sin	0.8192	72	sin	0.9511	88	sin	0.9994
	cos	0.7771		cos	0.5736		cos	0.3090		cos	0.0349
	tan	0.8098		tan	1.4281		tan	3.0777		tan	28.64
			56	sin	0.8290				89	sin	0.9996
				cos	0.5592					cos	0.0175
				tan	1.4826					tan	57.29

NOTE: Values of the sine, cosine, and tangent of each degree from 0 to 90 also permit determination of values for angles from 90 to 360 degrees if appropriate signs for the functions are used depending on the quadrant in which the angle measure belongs. For example, the sine of 4 degrees is also the sign of 176 degrees, and with a negative sign, the sine of 184 and 356 degrees. Note that if an angle is given in radians, multiply the number of radians by 57.296 to obtain the number of degrees.

SURFACE AREA AND VOLUME FORMULAS

R = radius of circle

d = diameter of circle

$\pi = 3.1416$

Circumference of circle $= \pi d = 2\pi R$

Area of circle $= \pi d^2/4 = \pi R^2$

Surface area of sphere $= \pi d^2 = 4\pi R^2$

Volume of sphere $= \pi d^3/6 = 4/3\pi R^3$

Area of a triangle $= ab/2$

b = base

a = altitude

CHART 46-A

AVERAGE CONCENTRATIONS OF AIRBORNE PARTICULATE MATTER

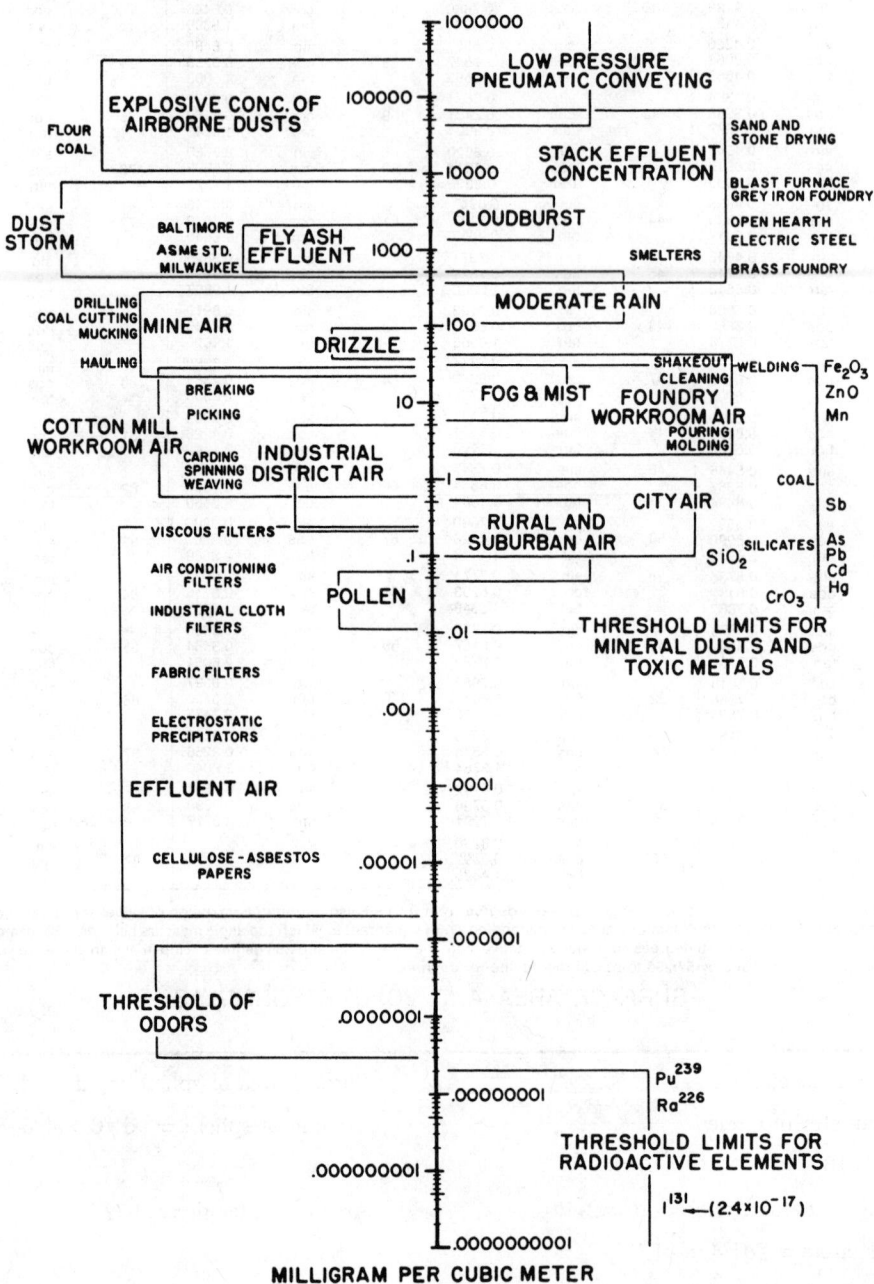

MILLIGRAM PER CUBIC METER

Adapted from First and Drinker. "Concentrations of Particulates Found in Air." AMA Archives of Industrial Hygiene and Occupational Medicine, 5:378–388 (April 1952). Conversions assume that 0.1 mg contains 5 million particles.

TABLE 46-Q

ADDITIONAL DIMENSIONS THAT CLOTHING ADDS TO NUDE BODY MEASUREMENTS

Measurement	Civilian clothing: Underwear, shirt, trousers, and tie (or dress). Jacket and shoes.[1]	Army uniform: Underwear, khakis or fatigues, combat suit, overcoat, socks, shoes, gloves, wool cap, helmet and liner.[2]	Air Force WW-II heavy winter flying clothes: jacket, trousers, helmet, boots, and gloves.[3]
Weight (lb.)	4–6[a]	22.9	20.0
Stature (in.)	1.0[b,c]	2.75	1.9
Abdomen depth (in.)	1.2	2.54	1.4
Arm reach, anterior (in.)		.37	0.4
Buttock-knee length (in.)	0.3	.70	0.5
Chest breadth (in.)			0.6
Chest depth (in.)		1.54	1.4
Elbow breadth (in.)	1.0	2.12	4.4
Eye height sitting (in.)	0.1		0.4
Foot breadth (in.)	0.2–0.3	.22	1.2
Foot length (in.)	1.2–1.6	.20	2.7
Hand breadth (in.)		1.60	0.4
Hand length (in.)		.30	0.4
Head breadth (in.)		2.8	0.4
Head length (in.)		3.5	0.4
Head height (in.)		1.45	0.2
Hip breadth (in.)		1.40	1.3
Hip breadth sitting (in.)	0.8	1.40	1.7
Knee breadth (both) (in.)		1.68	2.5
Knee height, sitting (in.)	1.0[b]	1.44	1.8
Shoulder breadth (in.)		1.16	0.7
Shoulder-elbow length (in.)		.62	0.3
Shoulder height sitting (in.)		.80	0.6
Sitting height (in.)	0.1[c]	1.67	0.6

[a] for women, 3 to 4.   [b] for women, 0.5 to 3.0.   [c] add another 1.0 ± for headgear

Data reported in The Human Body in Equipment Design, *by A. Damon, H. W. Stoudt, and R. A. McFarland. Cambridge, Mass., Harvard University Press. 1966.*

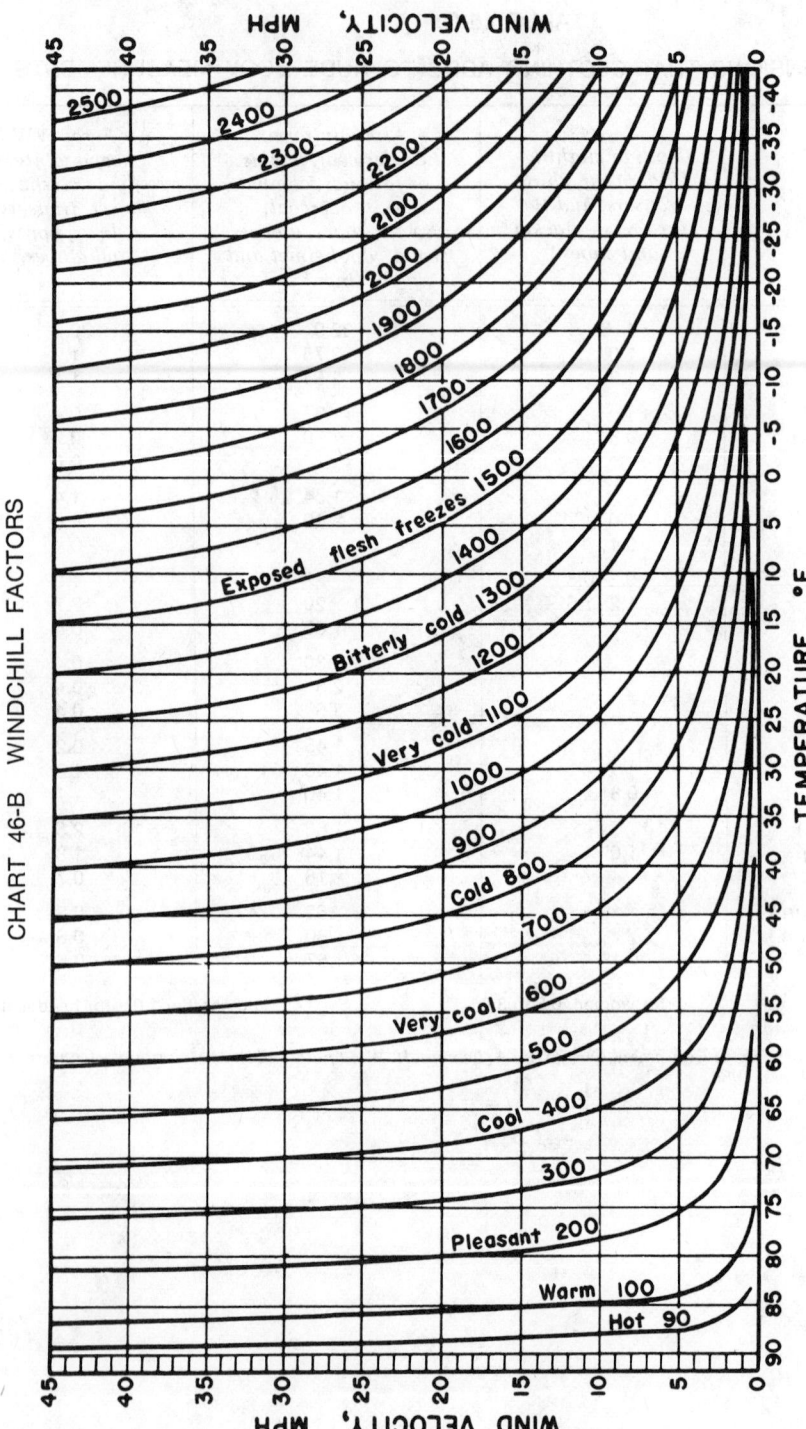

CHART 46-B WINDCHILL FACTORS

The human body senses "cold" as a result of both temperature and wind velocity. The numerical factor that combines the effect of these two is called "the windchill factor," shown by the curves in the above nomogram of dry-shade atmosphere cooling. Take the line marked "Very cold 1100," for example. This shows that a person can feel just as cold when the temperature is 35 F and the wind velocity is 45 mph as he can at −35 F and a wind velocity of only 1½ mph. When the windchill factor reaches 1400, flesh that is exposed to the wind will freeze in only a few seconds; out-of-door exposure is not recommended at all. Because of the extra clothing that people wear in cold weather, their physical size is greater than it is in warm weather. Be sure that equipment and controls are of adequate size and simplicity so that they can be run effectively and safely by persons wearing heavy clothing. See Table 47-V for approximate dimensions of heavy clothing.

From Engineer and Development Laboratories, The Engineer Center at Fort Belvoir.

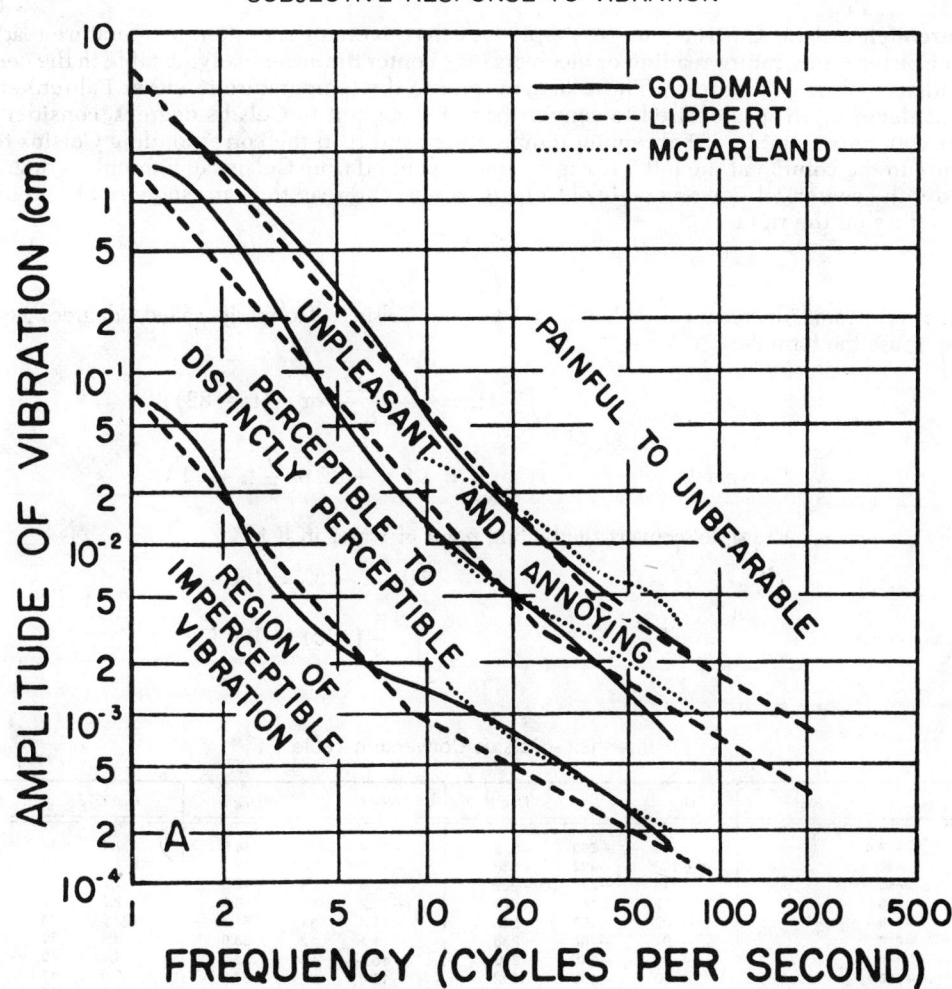

CHART 46-C

SUBJECTIVE RESPONSE TO VIBRATION

Threshold values for perception, discomfort, and tolerance to sinusoidal vibration at a single frequency are plotted on logarithmic coordinates of amplitude (in centimeters) versus frequency (cycles per second). Amplitudes are vector values (deviations each side of an average value).

From Human Factors in Air Transportation, *by Ross A. McFarland.* New York, N.Y., McGraw-Hill Book Co., Inc.

CONVERSION FACTORS

FAHRENHEIT-CELSIUS CONVERSION TABLE

Fahrenheit-Celsius Conversion. — A simple way to convert a Fahrenheit temperature reading into a Celsius temperature reading or vice versa is to enter the accompanying table in the center or boldface column of figures. These figures refer to the temperature in either Fahrenheit or Celsius degrees. If it is desired to convert from Fahrenheit to Celsius degrees, consider the center column as a table of Fahrenheit temperatures and read the corresponding Celsius temperature in the column at the left. If it is desired to convert from Celsius to Fahrenheit degrees, consider the center column as a table of Celsius values, and read the corresponding Fahrenheit temperature on the right.

To convert from "degrees Fahrenheit" to "degrees Celsius" (formerly called "degrees centigrade"), use the formula:

$$t_c = \frac{(t_f - 32)}{1.8} \text{ or } \frac{5}{9}(t_f - 32)$$

Conversely,
$$t_f = 1.8\, t_c + 32 \text{ or } \frac{9}{5}\, t_c + 32$$

Example, convert the boiling point of water in F to C:

$$212\ F - 32 = 180$$

$$\frac{5}{9}(180) = 100\ C$$

Fahrenheit — Celsius Conversion Table

Deg C	Deg F		Deg C	Deg C	Deg F	Deg C		Deg F	Deg C		Deg F
−273	−459.4	...	−129	−200	−328	−13.9	7	44.6	1.1	34	93.2
−268	−450	...	−123	−190	−310	−13.3	8	46.4	1.7	35	95.0
−262	−440	...	−118	−180	−292	−12.8	9	48.2	2.2	36	96.8
−257	−430	...	−112	−170	−274	−12.2	10	50.0	2.7	37	98.6
−251	−420	...	−107	−160	−256	−11.7	11	51.8	3.3	38	100.4
−246	−410	...	−101	−150	−238	−11.1	12	53.6	3.9	39	102.2
−240	−400	...	− 96	−140	−220	−10.6	13	55.4	4.4	40	104.0
−234	−390	...	− 90	−130	−202	−10.0	14	57.2	5.0	41	105.8
−229	−380	...	− 84	−120	−184	− 9.4	15	59.0	5.6	42	107.6
−223	−370	...	− 79	−110	−166	− 8.9	16	60.8	6.1	43	109.4
−218	−360	...	− 73	−100	−148	− 8.3	17	62.6	6.7	44	111.2
−212	−350	...	− 68	− 90	−130	−7.8	18	64.4	7.2	45	113.0
−207	−340	...	− 62	− 80	−112	−7.2	19	66.2	7.8	46	114.8
−201	−330	...	− 57	− 70	− 94	−6.7	20	68.0	8.3	47	116.6
−196	−320	...	− 51	− 60	− 76	−6.1	21	69.8	8.9	48	118.4
−190	−310	...	− 46	− 50	− 58	−5.6	22	71.6	9.4	49	120.2
−184	−300	...	− 40	− 40	− 40	−5.0	23	73.4	10.0	50	122.0
−179	−290	...	− 34	− 30	− 22	−4.4	24	75.2	10.6	51	123.8
−173	−280	...	− 29	− 20	− 4	−3.9	25	77.0	11.1	52	125.6
−169	−273	−459.4	− 23	− 10	14	−3.3	26	78.8	11.7	53	127.4
−168	−270	−454	−17.8	0	32−	−2.8	27	80.6	12.2	54	129.2
−162	−260	−436	−17.2	1	33.8	−2.2	28	82.4	12.8	55	131.0
−157	−250	−418	−16.7	2	35.6	−1.7	29	84.2	13.3	56	132.8
−151	−240	−400	−16.1	3	37.4	−1.1	30	86.0	13.9	57	134.6
−146	−230	−382	−15.6	4	39.2	−0.6	31	87.8	14.4	58	136.4
−140	−220	−364	−15.0	5	41.0	0−	32	89.6	15.0	59	138.2
−134	−210	−346	−14.4	6	42.8	0.6	33	91.4	15.6	60	140.0

Deg C		Deg F	Deg C		Deg F	Deg C		Deg F	Deg C		Deg F
16.1	61	141.8	50.0	122	251.6	83.9	183	361.4	276.7	530	986
16.7	62	143.6	50.6	123	253.4	84.4	184	363.2	282.2	540	1004
17.2	63	145.4	51.1	124	255.2	85.0	185	365.0	287.8	550	1022
17.8	64	147.2	51.7	125	257.0	85.6	186	366.8	293.3	560	1040
18.3	65	149.0	52.2	126	258.8	86.1	187	368.6	298.9	570	1058
18.9	66	150.8	52.8	127	260.6	86.7	188	370.4	304.4	580	1076
19.4	67	152.6	53.3	128	262.4	87.2	189	372.2	310.0	590	1094
20.0	68	154.4	53.9	129	264.2	87.8	190	374.0	315.6	600	1112
20.6	69	156.2	54.4	130	266.0	88.3	191	375.8	321.1	610	1130
21.1	70	158.0	55.0	131	267.8	88.9	192	377.6	326.7	620	1148
21.7	71	159.8	55.6	132	269.6	89.4	193	379.4	332.2	630	1166
22.2	72	161.6	56.1	133	271.4	90.0	194	381.2	337.8	640	1184
22.8	73	163.4	56.7	134	273.2	90.6	195	383.0	343.3	650	1202
23.3	74	165.2	57.2	135	275.0	91.1	196	384.8	348.9	660	1220
23.9	75	167.0	57.8	136	276.8	91.7	197	386.6	354.4	670	1238
24.4	76	168.8	58.3	137	278.6	92.2	198	388.4	360.0	680	1256
25.0	77	170.6	58.9	138	280.4	92.8	199	390.2	365.6	690	1274
25.6	78	172.4	59.4	139	282.2	93.3	200	392.0	371.1	700	1292
26.1	79	174.2	60.0	140	284.0	93.9	201	393.8	376.7	710	1310
26.7	80	176.0	60.6	141	285.8	94.4	202	395.6	382.2	720	1328
27.2	81	177.8	61.1	142	287.6	95.0	203	397.4	387.8	730	1346
27.8	82	179.6	61.7	143	289.4	95.6	204	399.2	393.3	740	1364
28.3	83	181.4	62.2	144	291.2	96.1	205	401.0	398.9	750	1382
28.9	84	183.2	62.8	145	293.0	96.7	206	402.8	404.4	760	1400
29.4	85	185.0	63.3	146	294.8	97.2	207	404.6	410.0	770	1418
30.0	86	186.8	63.9	147	296.6	97.8	208	406.4	415.6	780	1436
30.6	87	188.6	64.4	148	298.4	98.3	209	408.2	421.1	790	1454
31.1	88	190.4	65.0	149	300.2	98.9	210	410.0	426.7	800	1472
31.7	89	192.2	65.6	150	302.0	99.4	211	411.8	432.2	810	1490
32.2	90	194.0	66.1	151	303.8	100.0	212	413.6	437.8	820	1508
32.8	91	195.8	66.7	152	305.6	104.4	220	428.0	443.3	830	1526
33.3	92	197.6	67.2	153	307.4	110.0	230	446.0	448.9	840	1544
33.9	93	199.4	67.8	154	309.2	115.6	240	464.0	454.4	850	1562
34.4	94	201.2	68.3	155	311.0	121.1	250	482.0	460.0	860	1580
35.0	95	203.0	68.9	156	312.8	126.7	260	500.0	465.6	870	1598
35.6	96	204.8	69.4	157	314.6	132.2	270	518.0	471.1	880	1616
36.1	97	206.6	70.0	158	316.4	137.8	280	536.0	476.7	890	1634
36.7	98	208.4	70.6	159	318.2	143.3	290	554.0	482.2	900	1652
37.2	99	210.2	71.1	160	320.0	148.9	300	572.0	487.8	910	1670
37.8	100	212.0	71.7	161	321.8	154.4	310	590.0	493.3	920	1688
38.3	101	213.8	72.2	162	323.6	160.0	320	608.0	498.9	930	1706
38.9	102	215.6	72.8	163	325.4	165.6	330	626.0	504.4	940	1724
39.4	103	217.4	73.3	164	327.2	171.1	340	644.0	510.0	950	1742
40.0	104	219.2	73.9	165	329.0	176.7	350	662.0	515.6	960	1760
40.6	105	221.0	74.4	166	330.8	182.2	360	680.0	521.1	970	1778
41.1	106	222.8	75.0	167	332.6	187.8	370	698.0	526.7	980	1796
41.7	107	224.6	75.6	168	334.4	193.3	380	716.0	532.2	990	1814
42.2	108	226.4	76.1	169	336.2	198.9	390	734.0	537.8	1000	1832
42.8	109	228.2	76.7	170	338.0	204.4	400	752.0	565.6	1050	1922
43.3	110	230.0	77.2	171	339.8	210	410	770.0	593.3	1100	2012
43.9	111	231.8	71.8	172	341.6	215.6	420	788	621.1	1150	2102
44.4	112	233.6	78.3	173	343.4	221.1	430	806	648.9	1200	2192
45.0	113	235.4	78.9	174	345.2	226.7	440	824	676.7	1250	2282
45.6	114	237.2	79.4	175	347.0	232.2	450	842	704.4	1300	2372
46.1	115	239.0	80.0	176	348.8	237.8	460	860	732.2	1350	2462
46.7	116	240.8	80.6	177	350.6	243.3	470	878	760.0	1400	2552
47.2	117	242.6	81.1	178	352.4	248.9	480	896	787.8	1450	2642
47.8	118	244.4	81.7	179	354.2	254.4	490	914	815.6	1500	2732
48.3	119	246.2	82.2	180	356.0	260.0	500	932	1093.9	2000	3632
48.9	120	248.0	82.8	181	357.8	265.6	510	950	1648.9	3000	5432
49.4	121	249.8	83.3	182	359.6	271.1	520	968	2760.0	5000	9032

From Machinery's Handbook, *18th ed. (The Industrial Press)*

Above 1000 in the center column, the table increases in increments of 50. To convert 1462 degrees F to Celsius, for instance, add to the Celsius equivalent of 1400 degrees F ⅝ths of 62 or 34 degrees, which equals 794 C.

1473

CONVERSION FACTORS

LENGTH EQUIVALENTS

Example:
1 meter = 100 cm = 10^6 microns = 39.4 in. = 3.28 ft.
10 in. = 25.4 cm = 2.54×10^7 Å = 0.833 ft.
100 ft = 30.5 m = 30,480,000 μ = 1200 in.

1 angstrom (Å)
    = 0.1 millimicron
    = 0.000 1 micron
    = 0.000 000 1 millimeter
    = 0.000 000 003 937 inch

1 centimeter (cm)
    = 0.3937 inch (exactly)

1 foot (ft)
    = 0.305 meter

1 inch (in.)
    = 2.540 centimeters

1 kilometer (km)
    = 0.621 mile

1 meter (m)
    = 39.37 inches (exactly)
    = 1.094 yards

1 micron (μ) or micrometer
    = 0.001 millimeter (exactly)
    = 0.000 039 37 inch (exactly)

1 mil
    = 0.001 inch (exactly)
    = 0.025 4 millimeter

1 mile (mi) (statute or land)
    = 5,280 feet
    = 1.609 kilometers

1 millimicron (mμ)
    = 0.001 micron (exactly)
    = 0.000 000 039 37 inch (exactly)

1 nanometer (nm) (formerly called milli-micron, mμ)
    = 10^{-9} meter

1 yard (yd)
    = 0.914 meter

AREA EQUIVALENTS

1 square centimeter (cm^2)
    = 0.155 square inch

1 square decimeter (dm^2)
    = 15.500 square inches

1 square foot (sq ft)
    = 929.034 square centimeters

1 square inch (sq in.)
    = 6.452 square centimeters

1 square kilometer (km^2)
    = 247.104 acres
    = 0.386 square mile

1 square meter (m^2)
    = 1.196 square yards
    = 10.764 square feet

1 square mile (sq mi)
    = 259.000 hectares

1 square millimeter (mm^2)
    = 0.002 square inch

1 square yard (sq yd)
    = 0.836 square meter

VOLUME EQUIVALENTS

1 cubic centimeter (cm³)
= 0.999972 ml
= 0.0610237 cu in.
= 0.0338140 U.S. fl oz

1 cubic meter (m³)
= 10³ cm³
= 999.972 L
= 35.3147 cu ft
= 264.172 U.S. gal
= 219.97 Brit. gal

1 milliliter (ml)
= 1.000028 cm³
= 0.0610255 cu in.
= 0.033815 U.S. fl oz
= 0.035196 Brit. fl oz

1 liter (L)
(1 liter is defined as the volume occupied by 1 kilogram of water at its temperature of maximum density.)
= 1000.028 cm³
= 61.0255 cu in.
= 33.815 U.S. fl oz
= 1.05672 U.S. qt
= 0.264179 U.S. gal
= 35.196 Brit fl oz

1 cubic inch (in.³)
= 0.554113 U.S. fl oz
= 16.3871 cm³
= 16.3866 ml

1 cubic foot (ft³)
= 1728 in.³
= 29.9221 U.S. qt
= 7.48052 U.S. gal
= 28316.8 cm³
= 28.3161 L

1 fluid ounce, U.S. (U.S. fl oz)
= 1.80469 in.
= 29.5735 cm³
= 29.5727 ml
= 1.0408 Brit. fl oz

1 fluid ounce, British (Brit. fl oz)
= 1.7339 cu in.
= 28.413 cm³
= 28.412 ml
= 0.96076 U.S. fl oz

1 quart, liquid, U.S. (U.S. qt)
= 57.75 cu in.
= 32 U.S. fl oz
= 946.353 cm³
= 0.946326 L

1 gallon, U.S. (U.S. gal)
= 231 cu in.
= 128 U.S. fl oz
= 133.23 Brit. fl oz
= 0.83267 Brit. gal
= 3785.41 cm³
= 3.78531 L

1 gallon, British (Brit. gal) (Imperial gallon)
(1 British gallon is defined as the volume occupied by 10 pounds of water at 62 F.)
= 160 Brit. fl oz
= 277.42 cu in.
= 1.2010 U.S. gal
= 153.72 U.S. fl oz
= 4546.1 cm³
= 4.5460 L

DENSITY EQUIVALENTS

1 mg/cubic meter
= 1,000 mg/liter

1 mg/cubic ft
= 28.32 mg/liter

1 mg/cubic meter
= 35.314 mg/cu ft

1 mg/cubic foot
= 0.02832 mg/m³

Example, density of water:
1 gm/cm³ = 62.43 lb/cu ft = 8.345 lb/gal

1475

MASS EQUIVALENTS

1 gram (gm)
   = 15.4324 gr
   = 0.0352740 oz
   = 0.002204623 lb

1 kilogram (kg)
   = 1000 gm
   = 35.2740 oz
   = 2.204623 lb

1 metric ton, tonne (t)
   = 1000 kg
   = 2204.623 lb
   = 1.10231 short tons
   = 0.9842107 long ton

1 grain (gr)
   = 0.0647989 gm
   = 0.00228571 oz

1 ounce avoirdupois (oz)
   = 437.5 gr
   = 28.3495 gm

1 pound avoirdupois (lb)
   = 7000 gr
   = 16 oz
   = 453.59237 gm
   = 0.45359237 kg

1 short ton
   = 2000 lb
   = 0.892857 long ton
   = 907.1846 kg
   = 0.9071846 t

1 long ton
   = 2240 lb
   = 1.12 short tons
   = 1016.047 kg
   = 1.016047 t

FLOW RATE EQUIVALENTS

1 liter/minute
   = 0.06 M^3/hr
   = 0.2640 gal/min
   = 0.0353 cu ft/min
   = 0.000589 cu ft/sec

1 M^3/hr
   = 16.67 liter/min
   = 4.4 gal/min
   = 0.588 cu ft/min
   = 0.00989 cu ft/sec

1 gal/min
   = 3.78 liters/min
   = 0.227 M^3/hr

   = 0.1338 cu ft/min
   = 0.00223 cu ft/sec

1 cu ft/min
   = 28.32 liters/min
   = 1.699 M^3/hr
   = 7.50 gal/min
   = 0.01667 cu ft/sec

1 cu ft/sec
   = 1690.0 liters/min
   = 102.0 M^3/hr
   = 448.8 gal/min
   = 60.0 cu ft/min

VELOCITY EQUIVALENTS

1 meter per second (m/sec, mps)
   = 3.6 km/hr
   = 2.23694 mi/hr
   = 3.28084 ft/sec
   = 196.850 ft/min

1 kilometer per hour (km hr^{-1}, kph)
   = 0.277778 m/sec
   = 0.621371 mi/hr
   = 0.911344 ft/sec

1 mile per hour (mi hr^{-1}, mph)
   = 1.46667 ft/sec
   = 0.44704 m/sec

   = 1.609344 km/hr
   = 88 ft/min

1 foot per second (ft/sec, fps)
   = 0.681818 mi/hr
   = 60 ft/min
   = 0.3048 m/sec
   = 1.09728 km/hr

1 foot per minute (ft/min, fpm)
   = 0.0113636 mi/hr
   = 0.00508 m/sec
   = 0.018288 km/hr

PRESSURE EQUIVALENTS

1 pound per square inch (lb/sq in., psi)
   = 144 lb/sq ft (psf)
   = 70.3 gr/cm^2
   = 2.04 in. (Hg)*

1 pound per square foot (lb/sq ft, psf)
   = 478.8 dynes/cm^2 = 47.88 pascals (Pa)†

   = 0.016 ft (H$_2$O)*

1 gram per square centimeter (gm/cm^2)
   = 10 kg/cm^2
   = 2.048 lb/sq ft

1 kilogram per square centimeter (kg/cm^2)
   = 14.22 lb/sq in
   = 0.97 atm
   = 32.81 ft (H$_2$O)*

1 atmosphere (atm)
   = 14.696 lb/sq in.
   = 29.92 in. (Hg)*
   = 760 cm (Hg)*
   = 33.93 ft (H$_2$O)*

* (Hg) stands for mercury at 0 C; (H$_2$O) stands for water at 15 C

† At standard gravity of 980.665 cm/sec^2 or 32.174 ft/sec^2

FORCE EQUIVALENTS

1 gram weight*
   = 980.665 dynes

1 kilogram weight
   = 9.80665 × 10^5 dynes

1 newton
   = 10^5 dynes

1 pound weight*
   = 32.174 poundals
   = 444822 dynes

1 poundal
   = 13825.5 dynes

* At standard gravity of 980.665 cm/sec^2 or 32.174 ft/sec^2

ENERGY AND WORK EQUIVALENTS

1 erg
   = 1 dyne-centimeter
   = 10^{-7} abs. joule
   = 2.3892×10^{-8} cal$_{15}$

1 absolute joule (abs. joule)
   = 10^7 ergs
   = 0.23892 cal$_{15}$

1 kilogram-meter (kg m)
   = 9.80665 abs. joules

1 15° calorie (cal$_{15}$)
   = 4.1858 abs. joules

1 kilocalorie (kcal)
   = 10^3 calories

1 absolute kilowatt-hour (abs. kw-hr)
   = 3.6×10^6 abs. joules

1 mean International kilowatt-hour
   = 1.00019 abs kw-hr
   = 3.60068×10^6 abs. joules
   = 3412.756 Btu

1 British thermal unit (Btu)
   = 252.08 cal$_{15}$
   = 1055.07 abs. joules
   = 0.00029302 Int. kw-hr

1 foot-pound (ft-lb)
   = 1.35582 abs. joules

POWER EQUIVALENTS

1 kW
   = 1.34 hp
   = 737.54 ft-lb/sec
   = 0.948 Btu/sec
   = 0.2388 kcal/sec

1 hp
   = 0.746 kW
   = 550 ft-lb/sec
   = 0.707 Btu/sec
   = 0.178 kcal/sec

SELECTED COMMON ABBREVIATIONS

Å	Angstrom unit of length	liq	liquid
abs	absolute	L	liter and lambert(s)
amb	ambient	LP-gas	liquefied petroleum gas
amp	ampere	log	logarithm (common)
app mol wt	apparent molecular weight	ln	logarithm (natural)
atm	atmospheric	m, M	meter
at wt	atomic weight	ma	milliampere
Bé	degrees Baumé	MAC	maximum allowable concentration
bp	boiling point	max	maximum
bbl	barrel	mp	melting point
Btu	British thermal unit	μ	micron
Btuh	Btu per hour	mks system	meter-kilogram-second system
c	cycles per second (see Hz)	mph	miles per hour
cal	calorie	mg	milligram
cfh	cubic feet per hour	ml	milliliter
cfm	cubic feet per minute	mm	millimeter
cfs	cubic feet per second	mm (Hg)	mm of mercury
cg	centigram	mμ	millimicron
cm	centimeter	mppcf	million particles per cu ft
cgs system	centimeter-gram-second system	mr	millirem
conc	concentrated, concentration	mR	1/1000 Roentgen
cc, cm^3	cubic centimeter	min	minute or minimum
cu ft, ft^3	cubic foot	mol wt, MW	molecular weight
cu in.	cubic inch	N	newton
° or deg	degree	OD	outside diameter
C	degree Centigrade, degree Celsius	oz	ounce
F	degree Fahrenheit	ppb	parts per billion
K	degree Kelvin	pphm	parts per hundred million
R	degree Reaumur, degree Rankine	ppm	parts per million
dB	decibel	lb	pound
ET	effective temperature	psf	pounds per square foot
ft	foot	psi	pounds per square inch
ft-c	foot-candle	psia	pounds per square inch absolute
ft lb	foot pound		
fpm	feet per minute	psig	pounds per square inch gage
fps	feet per second	Rem	Roentgen equivalent man
fps system	foot-pound-second system	rpm	revolution per minute
fp	freezing point	sec	second
gal	gallon	sp gr	specific gravity
gr	grain	sp ht	specific heat
gm	gram	sp wt	specific weight
gpm	gallons per minute	sq	square
Hz	hertz (cycles per second)	scf	standard cubic foot
hp	horsepower	STP	standard temperature and pressure
hr	hour		
ID	inside diameter	temp	temperature
in.	inch	TLV	threshold limit valve
kcal	kilocalorie	v	volt
kg	kilogram	W	watt
km	kilometer	wt	weight

Note. Symbols are always written in singular form. Unabbreviated units form plurals in the usual manner.

TABLE OF UNIT PREFIXES

Multiples and submultiples	Prefixes	Symbols
$1,000,000,000,000 = 10^{12}$	tera-	T
$1,000,000,000 = 10^{9}$	giga-	G
$1,000,000 = 10^{6}$	mega-	M
$1,000 = 10^{3}$	kilo-	k
$100 = 10^{2}$	hecto-	h
$10 = 10$	deka-	D
$0.1 = 10^{-1}$	deci-	d
$.01 = 10^{-2}$	centi-	c
$.001 = 10^{-3}$	milli-	m
$.000001 = 10^{-6}$	micro-	μ
$.000000001 = 10^{-9}$	nano-	n
$.000000000001 = 10^{-12}$	pico	p

National Bureau of Standards

SIGNS AND SYMBOLS

$+$	plus, addition, positive	$\sqrt{}$	square root
$-$	minus, subtraction, negative	$\sqrt[n]{}$	nth root
$\pm$	plus or minus, positive or negative	a^n	nth power of a
$\mp$	minus or plus, negative or positive	log, $\log_{10}$	common logarithm
$\div, /, ——$	division	ln, $\log_e$	natural logarithm
$\times, \cdot, ()()$	multiplication	e or ϵ	base of natural logs, 2.718
$()[]$	collection	π	pi, 3.1416
$=$	is equal to	$\angle$	angle
$\neq$	is not equal to	$\perp$	perpendicular to
$\equiv$	is identical to	$\parallel$	parallel to
$\cong$	equals approximately, congruent	n	any number
		$|n|$	absolute value of n
		$\bar{n}$	average value of n
$>$	greater than	a^{-n}	reciprocal of nth power of a,
$\ngtr$	not greater than		of a, or $\left\{\dfrac{1}{a^n}\right\}$
$\geqq$	greater than or equal to		
$<$	less than	$n°$	n degrees (angle)
$\nless$	not less than	n'	n minutes, n feet
$\leqq$	less than or equal to	n''	n seconds, n inches
$::$	proportional to	f(x)	function of x
$:$	ratio	Δx	increment of x
$\sim$	similar to	dx	differential of x
$\propto$	varies as, proportional to	Σ	summation of
$\rightarrow$	approaches	sin	sine
∞	infinity	cos	cosine
$\therefore$	therefore	tan	tangent

References to Useful Handbooks of Engineering Tables and Formulas

NOTE: Because engineering and scientific handbooks are revised often, as frequently as once a year in some cases, no attempt has been made to list the "latest" edition in the following bibliography. In ordering, however, the most recent edition should be called for.

Alexander, J. M., and R. C. Brewer. *Manufacturing Properties of Materials.* Princeton, N.J. 08540. D. VanNostrand Co., Inc., 1963.

Allegheny Ludlum Steel Corp., Pittsburgh, Pa. 15222. *Tool Steel Handbook.* 1951.

American Conference of Governmental Industrial Hygienists, Committee on Industrial Ventilation, PO Box 452, Lansing, Mich. 48902. *Air Sampling Instruments Manual.*

———. *Industrial Ventilation — A Manual of Recommended Practice.*

American Industrial Hygiene Assn., 66 S. Miller Rd., Akron, Ohio 44313. *Industrial Noise Manual.*

American Institute of Steel Construction, Inc., 101 Park Ave., New York, N.Y. 10017. *Manual of Steel Construction.*

American Mutual Insurance Alliance, 20 North Wacker Dr., Chicago, Ill. 60606. "Handbook of Organic Industrial Solvents," Technical Guide No. 6.

American Welding Society, 2501 NW. 7th St., Miami, Fla. 33125. *Welding Handbook.*

American Society of Heating, Refrigerating and Air-Conditioning Engineers, 345 East 47th St., New York, N.Y. 10017. *Guide and Data Books: Applications, Handbook of Fundamentals, Systems and Equipment.*

Bennett, H. (editor). *Concise Chemical and Technical Dictionary.* New York, N.Y. 10003, Chemical Publishing Co. Inc.

Bolz, Harold A. (editor). *Materials Handling Handbook.* New York, N.Y. 10016. The Ronald Press Co.

Carmichael, Colin, and J. K. Salisbury. *Kent's Mechanical Engineers' Handbook.* New York, N.Y. 10016, John Wiley & Sons, Inc.

Car, Clifford C. *Croft's American Electricians' Handbook.* New York, N.Y. 10036, McGraw-Hill Book Co.

Carson, Gordon B. *Production Handbook.* New York, N.Y. 10016. Ronald Press, 1958.

Compressed Air Magazine Co., Phillipsburg, N.J. 08865. *Compressed Air Data.*

Crane Co., Industrial Products Group, Engineering Division, 4100 S. Kedzie Ave., Chicago, Ill. 60632. *Flow of Fluids Through Valves, Fittings, and Pipe,* Technical Paper 410. 1957.

Damon, A., Stoudt, H. W., and McFarland, R. A. *The Human Body in Equipment Design.* Cambridge, Mass. 02138, Harvard University Press, 1966.

DeGarmo, E. Paul. *Materials and Processes in Manufacturing,* 2nd ed. New York, N.Y. 10022. The Macmillan Co., 1962.

Elonka, Stephen Michael, and Joseph Frederick Robinson. *Standard Plant Operator's Questions and Answers.* New York, N.Y. 10036, McGraw-Hill Book Co., 1959.

Eshbach, Ovid W. *Handbook of Engineering Fundamentals.* New York, N.Y. 10016, John Wiley.

Factory Mutual Engineering Corp., Norwood, Mass. 02062. *Loss Prevention Data,* 1973.

Gleason, M. N., Gosselin, R. E., and Hodge, H. C. *Clinical-Toxicology of Commerical Products,* 2nd ed. Baltimore, Md. 21202, The Williams & Wilkins Co., 1963.

Green, Marvin H. *International and Metric Units of Measurement.* New York, Chemical Publishing Co., 1973.

Gunther, Raymond C. *Refrigeration, Air Conditioning, and Cold Storage.* Philadelphia, Pa. 19106, Chilton Co. (Inc.). 1957.

Harris, Cyril M. *Handbook of Noise Control.* New York, N.Y. 10036, McGraw-Hill Book Co., 1957.

Hodgman, Charles D., *et al.* (editors). *Handbook of Chemistry and Physics.* Cleveland, Ohio. 44128, The Chemical Rubber Co.

Hosey, A. D., and Powell, C. H. (editors). *Industrial Noise—A Guide to Its Evaluation and Control.* Washington, D.C., U.S. Dept. of Health, Education and Welfare, 1967. (Available through U.S. Government Printing Office, Washington, D.C. 20402.)

Hudson, Ralph G. *The Engineers' Manual.* New York, N.Y. 10016, John Wiley & Sons, Inc.

Illuminating Engineering Society, 345 E. 47th St., New York, N.Y. 10017. *IES Lighting Handbook (The Standard Lighting Guide).*

Ireson, W. G., and Grant, E. L. *Handbook of Industrial Engineering and Management.* Engle-Wood Cliffs, N.J. 07632, Prentice-Hall Publishing Co., Inc., 1957.

Kidder, Frank E., and Harry Parker. *Kidder-Parker Architects' and Builders' Handbook.* New York, N.Y. 10016, John Wiley & Sons, Inc.

Knowlton, A. E. (editor). *Standard Handbook for Electrical Engineers.* New York, N.Y. 10036, McGraw-Hill Book Co.

Kurtz, Edwin B. *The Lineman's Handbook.* New York, N.Y. 10036, McGraw-Hill Book Co., 1955.

LaLonde, William S., Jr., and Milo F. Janes. *Concrete Engineers Handbook.* New York, N.Y. 10036, McGraw-Hill Book Co.

Le Grand, Rupert (editor). *The New American Machinist's Handbook.* New York, N.Y. 10036, McGraw-Hill Book Co.

Liebers, Arthur. *The Engineer's Handbook Illustrated,* Los Angeles, Calif. 90047, Key Publishing Co., 1968.

Lindsey, Forrest R. *Pipefitters Handbook.* New York, N.Y. 10016, The Industrial Press.

Mantell, Charles L. (editor). *Engineering Materials Handbook.* New York, N.Y. 10036, McGraw-Hill Book Co.

Marks, Lionel S., and Theodore Baumeister (editors). *Mechanical Engineers' Handbook.* New York, N.Y. 10036, McGraw-Hill Book Co.

Maynard, H. B. *Industrial Engineering Handbook.* New York, N.Y. 10036. McGraw-Hill Book Co., 1963.

Miner, Douglas F., and John B. Seastone (editors). *Handbook of Engineering Materials.* New York, N.Y. 10016, John Wiley & Sons, Inc.

Morris, I. E. *Handbook of Structural Design.* New York, N.Y. 10022. Reinhold Publishing Corp.

Morrow, L. C. (editor). *Maintenance Engineering Handbook.* New York, N.Y. 10036. McGraw-Hill Book Co.

National Electrical Manufacturing Assn., 155 East 44th St., New York, N.Y. 10017. *Motors and Generators.*

National Fire Protection Assn., 470 Atlantic Ave., Boston, Mass. 02110. *Fire Protection Handbook.*

Oberg, Erik, and F. D. Jones. *Machinery's Handbook.* New York, N.Y. 10016, The Industrial Press.

Pender, Harold, *et al.* *Electrical Engineers' Handbook.* New York, N.Y. 10016, John Wiley & Sons, Inc.

Perry, John H., and Robert H. Perry. *Engineering Manual.* New York, N.Y. 10036. McGraw-Hill Book Co.

Perry, Robert H., *et al.* (editors). *Chemical Engineers' Handbook.* New York, N.Y. 10036, McGraw-Hill Book Co.

Peterson, Arnold P. G., and Ervin E. Gross. *Handbook of Noise Measurement.* West Concord, Mass. 01781, General Radio Co., 1967.

Robb, Dean A., and Philo, Harry M. *Lawyers Desk Reference: A Source Guide to Safety Information, What to Find, How to Find It.* Rochester, N.Y., The Lawyers Co-operative Publishing Co.

Rose, Arthur, and Elizabeth Rose. *The Condensed Chemical Dictionary.* New York, N.Y. 10022, Reinhold Publishing Co.

Rossnagel, W. E. *Handbook of Rigging.* New York, N.Y. 10036, McGraw-Hill Book Co.

Society of Automotive Engineers, 485 Lexington Ave. New York, N.Y. 10017. *SAE Handbook.*

Staniar, William. *Mechanical Power Transmission Handbook.* New York, N.Y. 10036. McGraw-Hill Book Co.

――――. *Plant Engineering Handbook.* New York, N.Y. 10036, McGraw-Hill Book Co.

Steere, Norman V. (editor). *Handbook of Laboratory Safety.* Cleveland, Ohio 44128, The Chemical Rubber Co., 1967.

United States Steel Corp., National Tube Division, Pittsburgh, Pa. 15230. *Lubrication Engineers Manual.* 1963.

Urquhart, Leonard Church (editor). *Civil Engineering Handbook.* New York, N.Y. 10036. McGraw-Hill Book Co.

Wilson, Frank W., and Philip D. Harvey (editors). *Tool Engineers Handbook.* New York, N.Y. 10036, McGraw-Hill Book Co.

OCCUPATIONAL NOISE EXPOSURE STANDARD PROPOSED BY THE

OCCUPATIONAL SAFETY AND HEALTH ADMINISTRATION

The following is the draft of the proposed OSHA noise regulations. It will be polished into a final draft, after hearings. It must be published in the Federal Register *as a proposal for public comment. Regulations do change, however. Be sure to check the latest applicable ones.*

The following abbreviations are used for legal references:

C.F.R.—Code of Federal Regulations.

F.R.—Federal Register.

Stat.—Statutes at Large.

U.S.C.—United States Code.

These are available through the U.S. Government Printing office, Washington, D.C. 20402, and should be found in any comprehensive law library.

On August 14, 1972, the National Institute for Occupational Safety and Health (NIOSH), Department of Health, Education, and Welfare, provided to the Department of Labor a criteria package (HSM 73–11001), "Occupational Exposure to Noise," in accordance with Section 20 (a)(3) of the Occupational Safety and Health Act of 1970 (29 U.S.C. 656). Thereupon, the Assistant Secretary determined that a standards advisory committee on noise should be appointed under section 7(b) of the Act, to provide evaluation and additional recommendations directly from labor, management, and independent experts. The committee commenced work on February 22, 1973, and met a number of times during 1973. Recommendations for a revised standard were transmitted to the Department of Labor on December 20, 1973.

The Department has evaluated the committee's recommendations, and determined that the standard should include the elements suggested by the committee. One of the principal changes recommended by the NIOSH criteria package was a lowering of the basic standard from 90 dBA to 85 dBA. The committee recommended, and the Department agrees, that the present duty should be continued. The method of stating the duty is

changed slightly, by using a formula instead of a table of exposure time *vs.* level. The result is the same in any instance.

The other major matters for consideration were:

(1) The necessity for audiometric testing as a part of a hearing conservation program;

(2) the requirements for the administration of such a testing program if it were required; and

(3) the requirements for personal hearing protectors if they were required for a hearing conservation program.

The committee found three factors significant in evaluating these matters:

(1) The variations in individual susceptibility to hearing damage from noise;

(2) the uncertainty inherent in all the data upon which the standard is based; and

(3) the variations in existing audiometric programs.

The committee therefore has recommended the institution of audiometric testing programs when the noise exposure of an employee is one-half or more of the limiting value (that is, an exposure equivalent to 85 dBA for a full shift). The committee has provided rather detailed requirements for the implementation of such programs.

The objective of this sort of an audiometric testing program is to detect changes in hearing level before they become sufficiently large to be handicapping. It is, therefore, essential that the audiometric environment and technique be well standardized and be stable over a number of years that represent a significant fraction of the employee's working life. It is also essential that these factors be reasonably identical from one employment to another. For these reasons, standards are proposed for the audiometric environment, the calibration of audiometers, and the training of audiometric technicians.

Methods are also proposed for determining the adequacy of hearing protectors in instances where these devices are used as part of a hearing conservation program.

Control of nonauditory effects of noise exposure is not addressed in this proposed standard, because the Committee is of the opinion that these effects will be adequately controlled by control of the auditory effects in accordance with the proposed standard.

Accordingly, pursuant to section 6(b) of the Williams-Steiger Occupational Safety and Health Act of 1970 (84 Stat. 1593; 29 U.S.C. 655), Secretary of Labor's Order No. 12–71 (36 F.R. 8754) and 29 C.F.R. Part 1911 (36 F.R. 17506), it is hereby proposed to revise § 1910.95 as set forth as follows.

Terms of the Proposed Amendment

§ 1910.95 NOISE.

(a) **Scope and definitions.**

(1) *Application.* This section applies to occupational noise exposures in employments covered in this chapter. The purpose of this standard is to establish requirements and procedures that will minimize the risk of permanent hearing impairment from exposure to hazardous levels of noise in places of employment. Adherence to such requirements and procedures is likely to reduce other possible noise-related problems of consequence to employee safety and health.

(2) *Definitions.*

(i) "Administrative control." Any procedure which limits daily noise exposure by control of the work schedule.

(ii) "Audiogram" means a graph or table showing hearing levels as a function of frequency that is obtained from an audiometric examination.

(iii) "Baseline audiogram" means the first audiogram taken during employment with the current employer.

(iv) "Certified audiometric technician" means an individual who can show documentary evidence of the satisfactory completion of a course of training meeting as a minimum the standards specified by the Intersociety Committee on Audiometric Technician Training as published in the *American Industrial Hygiene Association Journal*, 27:303–304 (May–June 1966), or of certification by the Council for Accreditation of Occupational Hearing Technicians.

(v) "Daily noise dose" means the cumulative noise exposure of an employee during a working day, expressed as the summation of fractions of permissible exposures received during the workday at each level.

(vi) "dBA" (decibels, A-weighted) means the unit of measurement of sound pressure level in logarithmic units corrected to the A-weighted scale, under ANS S1.4, using a reference level of 0.0002 microbars (2×10^{-5} newtons per square meter).

(vii) "Engineering control" means any procedure other than administrative control or personal protection that reduces the sound level either at the source of the noise or in the hearing zone of the employees.

(viii) "Hearing level" means the amount, in decibels, by which the threshold of audibility for an ear differs from a standard audiometric reference level.

(ix) "Workplace sound pressure level" means the sound pressure level measured at the employee's point of exposure.

(x) "Effective sound pressure level" means: (*a*) For employees not wearing hearing protectors, the workplace sound pressure level; (*b*) For employees wearing hearing protectors, the result of subtracting the reduction, R, for the hearing protectors from the measured workplace sound pressure level.

(xi) "Impulsive or impact noise" means a sound with a rise time of not more than 35 milliseconds to peak intensity and a duration of not more than 500 milliseconds.

(xii) "Interrupted steady state noise" means sound levels which recur at intervals of 500 milliseconds or more.

(xiii) "Noise exposure" means a combination of effective sound pressure level and its duration.

(3) *References.* (i) ANS S1.4—American National Standard *Specification for Sound Level Meters*, S1.4–1971, American National Standards Institute, 1430 Broadway, New York, N.Y. 10018.

(ii) ANS S3.6—American National Standard *Specifications for Audiometers*, S3.6–1969, American National Standards Institute, 1430 Broadway, New York, N.Y. 10018.

(iii) ANS S1.11—American National Standard *Specification for Octave, Half-Octave, and Third-Octave Band Filter Sets*, S1.11–1966 (Reaffirmed 1971), American National Stand-

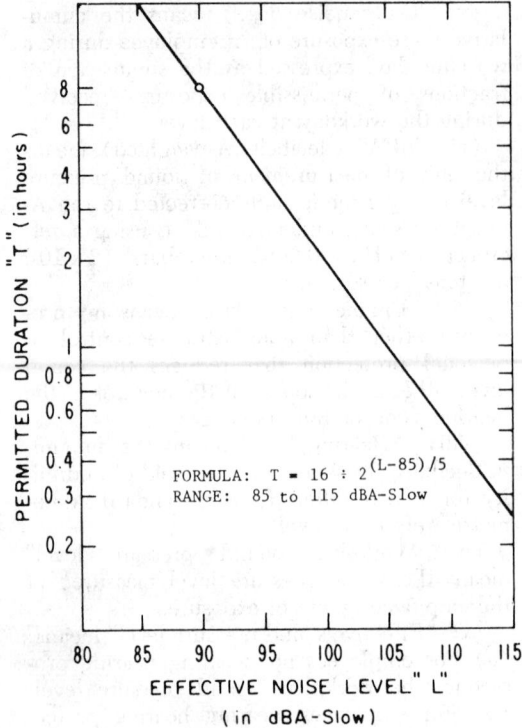

FORMULA: T = 16 ÷ 2$^{(L-85)/5}$
RANGE: 85 to 115 dBA-Slow

Figure 1910.95–1. Permitted duration *vs.* noise level.

ards Institute, 1430 Broadway, New York, N.Y. 10018.

(iv) ANS Z24.22—American National Standards *Method for the Measurement of the Real-Ear Attenuation of Ear Protectors at Threshold*, Z24.22–1957, American National Standards Institute, 1430 Broadway, New York, N.Y. 10018.

(v) *American Industrial Hygiene Association Journal*, 27:303–306, (May–June, 1966). Reprints are available from the American Industrial Hygiene Association, 66 S. Miller Road, Akron, Ohio 44313.

(vi) Council for Accreditation of Occupational Hearing Technicians, 1721 Pine Street, Philadelphia, Pa. 19103.

(vii) ISO R389—International Organization for Standardization Recommendation R389–1964, *Standard Reference Zero for the Calibration of Pure Tone Audiometers*, including Addendum 1–1970.

(b) **Occupational noise exposure limits.**

(1) *Steady-state noise—single level.* Exposures to continuous or interrupted steady state noise at a single level may not exceed a time amount "T" computed by the formula:

$$T = \frac{16}{2^{(L-85)/5}}$$

where "L" is the effective noise level measured in dBA on the slow scale of a standard sound-level meter.

A plot of T *vs.* L is provided in Figure 1910.95–1.

(2) *Steady-state noise—two or more levels.* Exposures to continuous or interrupted steady-state noise at two or more levels may not exceed a daily noise dose "D" of unity where "D" is computed by the formula:

$$D = \frac{C_1}{T_1} + \frac{C_2}{T_2} + \cdots \frac{C_n}{T_n}$$

where: C is the actual duration of exposure at a given steady-state noise level; and

T is the noise exposure limit for the level present during the time C, computed by the formula in (1) of this paragraph.

(3) *Maximum steady-state noise level.* Exposures to continuous or interrupted steady-state noise may not exceed levels of 115 dBA, regardless of any value computed in (1) or (2) of this paragraph.

(4) *Impulse or impact noise.* Exposures to impulse or impact noise may not exceed an effective level of 140 dB peak sound pressure.

(c) **Measurements.**

(1) *Unit of measurement.*

(i) The unit of measurement for continuous or interrupted steady-state sound shall be dBA.

(ii) The unit of measurement for impact or impulsive sound shall be dB peak sound pressure level, reference 0.0002 microbar (0.2 x 10^{-5} newtons per square meter).

(2) *Instrumentation.*

(i) Measurements of steady-state noise exposures shall be done with a sound-level meter conforming as a minimum to the requirements of ANS S1.4 Type 2, and set to use an A-weighted slow response; or with a NIOSH certified audiodosimeter;

(ii) Measurements of impulsive or impact noise exposures shall be made with a sound-level meter conforming as a minimum to the requirements of the ANS S1.4 Type 1 or Type 2, with impulse measurement capability or accessory.

(3) *Measurement procedure.* All measurements shall be made with the microphone of the sound measuring instrument at a position at which the noise most closely approximates the noise levels at the head position of the employee during normal operations.

(4) *Calibration.* An acoustical calibrator accurate to within plus or minus one decibel shall be used to verify the before and after calibration of the sound measuring instrument on each day noise measurements are taken.

(d) **Monitoring.** Noise level measurements shall be conducted not less often than annually for established equipment, operations, and facilities, and within thirty days of full operation of new equipment, changed process, or other significant workplace or work-condition modifications affecting noise exposure. A record of such measurements shall be made and shall include: number, type, and location of noise generators; number of employees in the work area and their daily noise dose; type, model, serial number, and calibration of measurement equipment; location, date, and time of measurement; the levels obtained; and the name and signature of the person(s) conducting the survey.

(e) **Controls.**

(1) *Engineering and administrative controls.* Feasible engineering or administrative controls shall be implemented to control noise exposures so that no employee is exposed in excess of the limits prescribed in paragraph (b) of this section, except as provided in (2)(ii) of this paragraph. Engineering controls will be deemed feasible unless there is substantial evidence that practical and economically sound methods do not exist for reducing noise sufficiently.

(2) *Hearing protectors.*

(i) Hearing protectors shall be provided to, and used by: (a) employees receiving a daily noise dose between 0.5 and 1.0, if their audiograms show any significant increase in hearing level as defined in paragraph (f)(2)(v); and

(b) employees who would otherwise receive noise exposures in excess of the limits prescribed in paragraph (b) of this section, during the period required for the implementation of feasible engineering and administrative controls.

(ii) Hearing protectors may be provided to, and used by, employees to limit noise exposures in lieu of feasible engineering or administrative controls if the exposure occurs no more than one day per week.

(iii) Any hearing protector used by an employee shall reduce the effective noise level to which he is exposed so that his noise exposure is within the limits prescribed in paragraph (b) of this section.

(iv) Procedures shall be established and implemented to assure proper issuance, maintenance, and use of hearing protectors.

(v) Employees shall be provided training in the proper use and care of all hearing protectors.

(f) **Medical surveillance.**

(1) *General.*

(i) A medical surveillance program shall be established and maintained for employees: (a) Who receive a daily noise dose equal to or exceeding 0.5; or (b) Whose occupational noise exposure must be controlled by hearing protectors under paragraph (e)(2) of this section.

(ii) A medical surveillance program shall provide, not less frequently than annually, an audiometric testing program for affected employees.

(iii) A medical surveillance program shall be under the supervision of a licensed physician and shall be provided at no cost to employees.

(2) *Audiometric testing program.* (i) Audiometric tests shall be administered by a physician, an audiologist, or a certified audiometric technician.

(ii) Audiometric tests shall be preceded by a period of at least fourteen hours during which there is no known exposure to effective sound levels in excess of 80 dBA. This requirement may be met by wearing hearing protectors which reduce the effective sound level below 80 dBA.

(iii) If no previous baseline audiogram exists, a baseline audiogram shall be taken. (a)

TABLE 1910.95-A

MAXIMUM ALLOWABLE SOUND PRESSURE LEVELS FOR AUDIOMETER ROOMS

	Audiometric Test Frequency (Hz)								
	500	750	1000	1500	2000	3000	4000	6000	8000
⅓-Octave Band Center Frequency (Hz)	500	800	1000	1600	2000	3200	4000	6400	8000
Sound Pressure Level (dB)*	35	35	35	37	42	47	52	57	62
½-Octave Band Cut-off Frequencies (Hz)	425–600	600–850	850–1200	1200–1700	1700–2400	2400–3400	3400–4800	4800–6800	6800–9600
Sound Pressure Level (dB)*	37	37	37	39	44	49	54	59	64
Octave Band Cut-off Frequency (Hz)	300–600	600–1200	600–1200	1200–2400	1200–4800	2400–4800	2400–4800	4800–10000	4800–10000
Sound Pressure Level (dB)*	40	40	40	42	47	52	57	62	67

*dB re: 0.00002 N/m²

For employees assigned to work covered by (e) (2) (i) and (ii): [*Check latest regulations—Ed.*] (*b*) Within 90 days of assignment for employees newly assigned or reassigned to work covered by (e) (2) (i) and (ii) of this section. Preferably, the baseline audiogram should be taken *before* assignment.

(iv) Audiometric tests shall be pure tone, air conduction, hearing threshold examination, with test frequencies including as a minimum, 500, 1000, 2000, 3000, 4000, and 6000 Hz and shall be taken separately for the right and left ears.

(v) Each employee's annual audiogram shall be examined to determine if any significant threshold shift, either ear (higher threshold), has occurred relative to the baseline audiogram. For those employees having a baseline audiogram showing: (*a*) Not more than 25 dB (ANS S3.6 or ISO R-389) hearing level at any test frequency, either ear, a sig-

nificant threshold shift relative to the baseline audiogram is a shift of 20 dB or more at any test frequency, either ear; and (*b*) Hearing levels that are in excess of 25 dB (ANS S3.6 or ISO R-389) at one or more test frequencies, either ear, a significant threshold shift, relative to the baseline audiogram is 10 dB or more at 500 Hz, 1000 Hz, or 2000 Hz; 15 dB or more at 3000 Hz, or 20 dB or more at 4000 and/or 6000 Hz; either ear.

(vi) If a threshold shift as determined in (v) of this paragraph is present, the employee shall be retested within one month. If the shift persists, then: (*a*) Employees not having hearing protectors shall be provided such in accordance with (d) (2) of this section; (*b*) Employees already having hearing protectors shall be specially retrained and reinstructed in accordance with (d) (2) of this section; (*c*) The employee shall be notified of his (her) hearing level and provided a copy of the audiogram; and (d) The employee shall be referred for appropriate medical evaluation.

(vii) For audiograms taken with a self-recording audiometer, it must be possible at each test frequency to place a horizontal line segment parallel to the time axis on the audiogram, such that the audiometric tracing crosses the line segment at least six times at that test frequency. If this is not possible, the employee must be retested.

(viii) A record shall be made of each employee audiogram and shall contain as a minimum: (a) The employee's name and social security number; (b) The employee's job location; (c) The examiner's name and signature; (d) The date and time of the test; (e) The model, make, and serial number of the audiometer; and (f) The standard to which the audiometer is calibrated.

(g) **Audiometric equipment and facilities.**

(1) *Audiometric test rooms.* Rooms used for audiometric testing shall have ambient noise levels not exceeding those in Table 1910.95-A, when measured by equipment conforming to the requirements of ANS S1.4 Type 1 or Type 2, and ANS S1.11.

(2) *Audiometric measuring instruments.*

(i) Instruments used for measurements required in (f) of this section shall be either: (a) A manual or self-recording audiometer of the discrete frequency type which conforms to the requirements for limited range pure tone audiometers prescribed in ANS S3.6; or (b) An audiometric testing system with accuracy and precision at least equal to an audiometer conforming to (i) of this subdivision.

(ii) Pulsed tone audiometers shall have a tone on-time of at least 200 milliseconds.

(iii) Self-recording audiometers shall comply with the following requirements: (a) The chart upon which the audiogram is traced shall have printed lines at positions corresponding to all multiples of 10 dB hearing level within the intensity range spanned by the audiometer. The lines shall be equally spaced and shall be separated by at least ¼ inch. Additional graduations are optional. The pen which traces the audiogram shall have a fine point so that the tracing shall not exceed 2 dB in width. (b) It shall be possible to disable the stylus drive mechanism so that the stylus can be manually set at the 10 dB graduation

lines for calibration purposes. (c) The slewing rate for the audiometer attenuator shall be 6 dB/sec or less except that an initial slewing rate greater than 6 dB/sec is permitted at the beginning of each new test frequency, but only until the second subject response. (d) The audiometer shall remain at each required test frequency for 30 seconds (±3 seconds). The audiogram shall be clearly marked at each change of frequency and the actual frequency change of the audiometer shall not deviate from the frequency boundaries marked on the audiogram by more than ±3 percent.

(3) *Audiometer calibrations.* (i) A biological calibration shall be made at least once each month and shall consist of (a) testing a person having a known stable audiometric curve that does not exceed 25 dB hearing level at any frequency and comparing the test results with the known curve, and (b) Registering the subject's response to distortions and unwanted sounds from the audiometer. If the results of a biological calibration indicate hearing-level differences greater than ±5 at any frequency, if the signal is distorted, or if there are attenuator or tone switch transients, then the audiometer shall be subjected to a periodic calibration within thirty days.

(ii) A periodic calibration shall be performed at least annually and as required in (i) of this subparagraph. In making the measurements in (a)–(c) following, the accuracy of the calibrating equipment shall be sufficient to assure that the audiometer is within the tolerances permitted by ANS S3.6. All observed deviations from required performance shall be promptly corrected. (a) With the audiometer set to 70 dB hearing threshold level, measure the sound pressure levels of test tones using a National Bureau of Standards Type 9A coupler, for both earphones and at all test frequencies. (b) At 1000 Hz, for both earphones, measure the earphone decibel levels of the audiometer for 10 dB settings in the range 70 to 10 dB hearing threshold level. This measurement may be made acoustically with a 9A coupler or electrically at the earphone terminals. (c) Measure the test tone frequencies with the audiometer set at 80 dB hearing threshold level, for one earphone only. (d) A careful listening test, more extensive than that required in the biological calibration ((i) of this subparagraph); shall be made in order to ensure that the audiometer displays no evidence of distortion, unwanted sound, or other technical problems. (e) The general function of the audiometer shall be checked, particularly in the case of a self-recording audiometer.

(iii) An exhaustive calibration shall be performed at least every five years. This shall include testing at all settings for both earphones. The test results must prove unequivocally that the audiometer meets for the following parameters the specific requirements stated in the applicable sections of ANS S3.6 as noted in parenthesis. (a) Accuracy of decibel level settings of test tones (Sections 4.1.4.1 and 4.14.3); (b) Accuracy of test tone frequencies (Section 4.1.2); (c) Harmonic distortion of test tones (Section 4.1.3); (d) Tone-envelope characteristics, i.e., rise and decay times, overshoot, "off" level (Section 4.5); (e) Sound from second earphone (Section 4.4.2); (f) Sound from test earphone (Section 4.4.1); (g) Other unwanted sound (Section 4.4.8).

(iv) A record shall be made of each audiometer calibration; and shall include the results of all measurements obtained.

(h) **Hearing protector reduction factor "R."**

The noise reduction factor "R" of a hearing protector shall be determined by one of the methods in (1), (2), or (3) of this paragraph. Method (1) is an exact mathematical method; Method (2) is an approximation using known octave band levels; and Method (3) is an approximation where only the A-weighted level for the environment is known. The pure tone attenuation, and standard deviation vs. frequency characteristics of the hearing protector (normally supplied by the manufacturer) shall have been determined in accordance with ANS Z24.22.

(1) *Method 1—Exact reduction factor "R."*

$$R = L_A - 10 \log S$$

where: (i) L_A = the A-weighted slow level measured in the environment.

(ii) S = antilog $0.1(L_1 - Q_1)$ + antilog 0.1 $(L_2 - Q_2)$ + antilog $0.1(L_3 - Q_3)$ + antilog $0.1(L_4 - Q_4)$ + antilog $0.1(L_5 - Q_5)$ + antilog $0.1(L_6 - Q_6)$ + antilog $0.1(L_7 - Q_7)$, where:

(a) Q_1 through Q_7 are defined in dB as:

Q_1 = attenuation at 125 Hz, plus 16.2 dB, minus 2 standard deviations

Q_2 = attenuation at 250 Hz, plus 8.7 dB, minus 2 standard deviations

Q_3 = attenuations at 500 Hz, plus 3.3 dB, minus 2 standard deviations

Q_4 = attenuation at 1000 Hz, minus 2 standard deviations

Q_5 = attenuation at 2000 Hz, minus 1.2 dB, minus 2 standard deviations

Q_6 = average of attenuation at 3000 and 4000 Hz, minus 1.0 dB, minus 2 standard deviations

Q_7 = average of attenuations at 6000 and 8000 Hz, plus 1.1 dB, minus 2 standard deviations

(b) L_1 through L_7 denote the octave band levels measured in the environment at 125, 350, 500, 1000, 2000, 4000 and 8000 respectively.

(2) *Method 2—Approximate reduction factor "R" using octave levels.*

(i) Adjust each octave band level to obtain the A-weighted equivalent.

(ii) Subtract from each octave band level the mean protection value, adjusted by two standard deviations.

(iii) Obtain the overall A-weighted sound level reaching the eardrum by summing the values obtained in (ii) of this subparagraph. Addition of these logarithmic values may be conveniently accomplished by a semi-graphical procedure using Figure 1910.95–2.

Step	Action
(1)	Using the A-weighted octave band levels calculated in (ii) of this subparagraph, subtract the "second highest value" from the "highest value";
(2)	Enter the curve of Figure 1910.95–2 at the point corresponding to the difference value calculated in (1), located on the horizontal scale;
(3)	Follow a horizontal from the point on the curve found in (2) to the vertical scale;
(4)	Add the value found at the vertical scale in (3) to the "highest value";
(5)	Repeat steps (1) through (4) using the

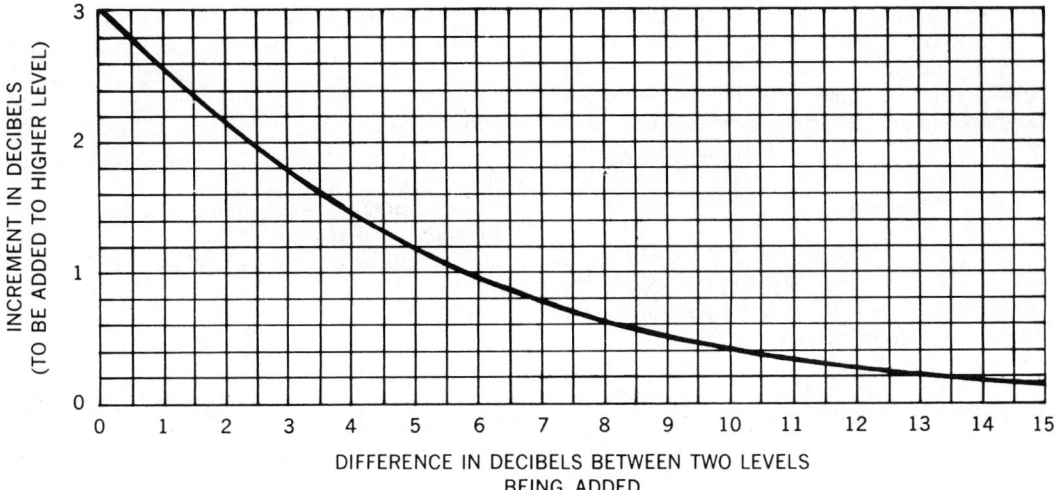

Figure 1910.95–2. Chart for combining noise levels.

value calculated in (4) as the highest value, and the next-highest-remaining-value in (ii) as the "next highest value."

(6) Repeat step (5) progressively until all values determined in (ii) of this subparagraph have been added. The result is the overall A-weighted level at the eardrum while wearing the protectors.

Sample calculation:

1—

| Octave band frequency (Hz) | 125 | 250 | 500 | 1000 | 2000 | 4000 | 8000 |

2—

| Measured sound level (dB) | 91 | 92 | 94 | 96 | 100 | 98 | 92 |

3—

| A-weighting adjustments (dB) | −16 | −9 | −3 | 0 | +1 | +1 | −1 |

4—

| A-weighted sound level (dB) | 75 | 83 | 91 | 96 | 101 | 99 | 91 |

5—

| Mean reduction of hearing protector (dB) | −12 | −18 | −24 | −33 | −34 | −42 | −25 |

6—

| Adjust for two standard deviations from Manufacturers data | 4 | 6 | 8 | 10 | 10 | 4 | 4 |

7—

| Net reduction | −8 | −12 | −16 | −23 | −24 | −38 | −21 |

8—

| A-weighted level at eardrum | 67 | 71 | 75 | 73 | 77 | 61 | 70 |

9—

Add A-weighted values

Step	Pair	Differ-ence	Incre-ment from curve	New "Highest Value"
1	77 −75	2	2.15	79.15
2	79.15−73	6.15	0.95	80.10
3	80.10−71	9.10	0.50	80.60
4	80.60−70	10.60	0.40	81.00
5	81.00−67	14.00	0.20	81.20
6	81.20−61	20.20	Neg.	81.20+

Round to nearest dB (81).

10—

Reduction factor "R" = 104 − 81 = 23 dB.

(3) *Method 3—Approximate reduction factor "R" using A-weighted sound level.* If only an A-weighted sound level for the environment is available, an approximation of the hearing protector attenuation factor "R" may be arrived at, using Method 2 ((2) of this paragraph) and assuming a "pink" noise with a level of 100 dB in each octave band, and using the average attenuation *without* correction for

1491

standard deviations. If the value obtained by this procedure indicates an effective level of 80 dBA or more reaching the eardrum, an octave band analysis shall be obtained, and the attenuation calculated by Method 1 or Method 2 ((1) or (2) of this paragraph).

(i) **Information and warnings.**

(1) *Signs.* Where noise levels exist in excess of the limits prescribed in paragraph (b) of this section so that an employee may be exposed in excess of those limits, clearly worded signs shall be posted at entrances to, or on the periphery of, such areas, describing the hazard and required protective actions.

(2) *Notification.* Each employee exposed to noise levels which exceed the limits prescribed in paragraph (b) of this section shall be apprised of all noise hazards in his work area. The information shall be kept on file and affected employees and their designated representatives shall be provided reasonable access thereto.

(j) **Records.**
(1) *Maintenance.* Records required to be established in this section shall be maintained for not less than the period specified in this subparagraph for the appropriate record.
(i) Environmental monitoring records—5 years.
(ii) Employee audiograms—duration of employment + 5 years.
(iii) Audiometer calibration records—5 years.
(2) *Access.*
(i) Employee audiometric data shall be provided upon request to authorized representatives of the Assistant Secretary and the Director, and upon written request of an employee, to a physician designated by that employee.
(ii) Environmental noise level measurements shall be made available upon request to authorized representatives of the Assistant Secretary and the Director. Employers or their representatives shall be given reasonable access to the results of such measurements upon request.

Index

NOTE: Chemicals and other terms that are listed in tables are not indexed. For major chemical tables, see Chapter 39, "Chemical Hazards," Chapter 46, "Safety Engineering Tables," and "Threshold Limit Values," pp. 1182–1198. If chemicals are discussed in other chapters in detail they are indexed in this listing. Of particular help would be the Chemical Cross Index, pp. 1208–1211.

Glossaries of terms included in this Manual may also be of special help. These are:

Chemical terms, pp. 1232–1240,
Guarding terms, pp. 795–796,
Legal terms, pp. 572–573.

In order to keep the "Safety" entry from becoming so large that it would be unwieldy, many "safety subjects" have been listed under the second word; for example, look under "Solvents, safety" to find "safety solvents."

A

A-scale (noise), 1245, 1485
Abrasive blasting protectors, personal, 500–501
Abrasive wheels; *see* Grinding wheels
Abbreviations, listing, 1479–1480
Absorption, 1111–1113
    see also Lungs *and* Skin
Accelerator, chemical, 1232
Acceptable concentrations, meaning, 1076, 1119–1121
Accident
    "by accident," 177–178
    cost, 7, 10, 14, 97, 153, 162–171
    definition, 106, 122, 148, 163
    economic losses, 173
    frequency; *see* Frequency rate
    public relations angle, 310–311
    ratios, 96
    repeaters (accident prone), 16–17, 214, 243
    statistics, 8–10
    types, 155, 346–347, 350–352, 574, 579–580
Accident investigation, 83, 151–162
    analysis chart, 132–133, 158–162
    causes, 106, 152–158, 346–347, 401
    classification, 158–162
    elevator, 777
    interviewing, 1419
    key facts, 154–160
    "near accidents," 152–153
    records, 122–123
    reports, 124–134, 1415–1419
Accident prevention, 2–3, 152
    by design, 105–116, 396
    industrial, 8–10
    nonwork, 10, 148–149
    office, 352–362
    property damage, 96–97
    purchasing, 116–120
    system safety, 97–100
    workmen's compensation, 175–176
Acetaldehyde, 1123
Acetanilide, 1123
Acetic acid, 1123–1124
Acetic anhydride, 1124
Acetone, 1124
Acetonitrile, 1124
Acetylene, 948, 949, 1124
Acid, 1232
    effect on skin, 1068
    handling, 650–651
    hoods, 485–487
Acoustic measurement; *see* Noise (sound)
Acrolein, 901, 1124
Acrylonitrile, 1124
Actinide series, 1232
Actors, professional, 323
Acute poisoning, 1113–1114
Adult learning, 210–211
Advisors, staff safety, 53–55
Advisory Center on Toxicology, 1091, 1177
Adzes, 986
Aerial baskets, 693–694
Aerial cable systems, 705–706
Aerosols, 1040

Index

Agency of accident, 155–156
Aggression, 252
Air
  compressed, 520, 904, 1054
    see also Pressure vessels
  ejectors, 887–890
  humidity, 1007, 1052
  infiltration, 1007
  makeup, 1007–1009
  movers, 1015–1016
  pollution, 371–372
  presses, 860–861, 869
  quality, 491, 1054, 1118–1123
  tools, 1000–1001
    see also Ventilation
Air conditioning, 377–378, 1007, 1009
    see also Ventilation
Air heaters, 1335
Air line respirators, 499–500
  maintenance, 507
Air purifying respirators, 493
Air sampling, 1077–1084, 1093
Air supplied hoods, 501
  suits, 501–502
Aisles; see Passageways
Alarm systems
  emergency, 454–455, 910
  sprinkler system, 1366–1367
Alcohol control, 563
Alcohols, 1143, 1154
  physiological effect, 1050
Aldehydes, physiological effects, 1050
Aldrin, 1125
Aliphatics, 1050, 1232
Alkali, 1232
  effect on skin, 1069
  equivalents for neutralizing chlorine, 1135
Alkaline earths, 1232
Allergy, 1072
Alligator shears, 863
Allyl alcohol, 1125
Allyl chloride, 1125
Allyl glycidyl ether, 1125
Alpha particles, 1061, 1062
Aluminized clothing, 518–519
Aluminum, 901, 1125
Aluminum chloride, 1125
Aluminum oxide, 1125
American Association of Industrial Dentists, 604
American Association of Industrial Nurses, 604
American Association of State Compensation Insurance Funds, 600
American Chemical Society, 604
American College of Surgeons, 604
American Conference of Governmental Industrial Hygienists, 604–605, 1091–1092
    see also Threshold limit values
American Foundrymen's Society, 608
American Gas Association, 608–609

American Industrial Hygiene Association, 605, 1092
  emergency exposure limits, 1121
  Hygienic Guides, 1121, 1177–1179, 1208
American Institute of Mining, Metallurgical, and Petroleum Engineers, 605
American Insurance Association, 601
  AIA Bulletin, 1207–1208
American Iron and Steel Institute, 609
American Medical Association, 605, 1092
American Meat Institute, 609
American Mining Congress, 609
American Mutual Insurance Alliance, 600–601
American National Red Cross, 596
  emergency aid, 463
  first aid, 560
American National Standards Institute, 597–598, 1092
American Nurses' Association, 606
American Paper Institute, 609
American Petroleum Institute, 609–610, 1092
American Psychiatric Association, 606
American Public Health Association, 606
American Pulpwood Association, 610
American Road Builders Association, 610
American Society of Safety Engineers, 11–13, 603–604
  listing of schools, 629
American Society for Testing and Materials, 598
American Society of Mechanical Engineers, 606–607
  Boiler and Pressure Vessel Code, 607, 1388–1389
American Trucking Associations, Inc., 610–611
American Water Works Association, 611
American Waterways Operators, Inc., 611
American Welding Society, 611
2-Aminopyridine, 1125
Ammonia, 1233
  toxicology, 1126
Ammonium nitrate, 655
Amplifying systems, 330–331
Ampere, 1256
Amyl acetate, 1126
Amyl alcohol, 1126
Aniline, 1126, 1233
Animation on films, 323
Annual reports; see Reports annual
Anoxia; see Oxygen deficiency
Anthracosis, 901
Anthraquinone, 1126–1127
Anthrax, 1067
Anthropometry, 224
Antibiotic, 1233
Antimony, 901, 1127
Antioxidant, 1233
Approach-approach conflict, 251
Approach-avoidance conflict, 251–252
Approximation, principle of, 230
Apron conveyors, 700–701, 706–707

Arc welding, 960–966, 970–971
Area, surface, 1467, 1474
Argon, 964
Arm protectors, 521–522
Aromatics, 1050, 1233
Arrangement, principle of, 226
Arsenic, 1127
Arsine, 1127–1128
Asbestos
  clothing, 518
  toxicology, 1030, 1128
Asbestosis, 1045
Ash disposal equipment, 1392
Asphalt, toxicology, 1128
Asphyxia, 1114, 1183–1184
Associated General Contractors of America, Inc.,
  611
Association of American Railroads, 612
Association contests, 277
Association of Iron and Steel Electrical Engineers,
  5, 11
Atom, definition, 1233
Attenuation, 1251–1252
  see also Ear protection
Attitudes, human, 240, 254
  changing, 255–257
  determination, 254–255
  safety, 58–59
  supervisory training, 210–211
Audience, publicity, 304–305
Audio aids, 328, 342–343
Audiogram, 1246–1247
Audiometers, 1252
Audiometry, 1252
Audiovisuals, 316–343
  cost, 320, 321–323
  effectiveness, 316–323
  mobile, 333
  nonprojected, 334–338
  preparation, 320, 323–328
  projected, 321–323, 338–342
  rehearsal, 333–334
  relative merits, 230
  selection, 317, 318–319
  sound, 328, 342–343
Auditing safety programs, 97
Auditory defects, 1245–1247, 1249
Auditory displays, 229–231
Auditory perception, 1245–1247
  conservation of
    definition, 1242
    program, 1059, 1251–1253, 1484–1492
  level, 1242–1244
  loss, 1242, 1245–1247
  tests, 557, 1252–1253, 1487–1490
  see also Ear *and* Ear protection
Autoclaves, 1409
Autoignition temperatures; *see* Temperature,
  autoignition
Automation, 879, 883–884

"Average man" fallacy, 241–242
Avoidance-avoidance conflict, 251
Awards
  "clubs," 468–469
  contest, 286–287
  driver, 757, 1424
  National Safety Council, 276–277
  presentations, 288–289
  publicity, 305–306
  safety committee, 65
  suggestion, 298–299
Awl, brad, 987
Axes, 985–986

B

Bacteria, 544–545
Baffles, ventilation hood, 1013
Bagged material, storage, 637–638, 646, 647, 654
Band saws
  metal, 837–838
  woodworking, 821–822
Barium, 1128
Barrels
  handling, 638, 659
  storage, 646, 650
  tumbling, 922
Barrier guards
  adjustable, 874
  construction, 798
  fixed, 872–874
  movable, 874–877
Barriers, radiation; *see* Shielding, radiation
Bars
  handling, 640, 987–988
  storage of, 647
Base, definition, 1233
Battery charging, 758–759, 1429–1430
Beams (support), inspection, 428
Bearings, electric motor, 439, 1264, 1265
Behavior, human, 240–257
Behavioral responses, 232–233
Belt conveyors, 703, 706–707
Belts, power transmission
  electric motor, 1264
  guarding of, 806–808
  maintenance, 439
Belts, safety, 509–514
  inspection and testing, 512–513
  lifelines, 422, 513–514
  maintenance, 512
  selection, 509–512
"Bends," 1054
Benzene, 1233
  toxicology, 1128–1129
Benzoyl peroxide, 1129
Benzyl chloride, 1129
Beryllium, 901
  intoxication, 1047
  toxicity, 1129–1130

Index

Beta particles, 1061, 1062
Bilevel reporting, 134
Bins, storage, 372, 652–654
Biological agents, effect on man, 1065–1067
Biotic agents, 1072
Bisphenol A, 1130
Bituminous Coal Operators' Association, 612
Bits, wood, 986
Blades, saw, 816–817
Blankets, fire, 1359
Blast gates, cupola, 910
Blasting; see Abrasive blasting or Explosives
Blocking timbers, 640, 642
Blocks and falls, 694–695
Blood, 1114
Blowtorches, 408, 1321, 1335
Board of Certified Safety Professionals, 11, 603, 607
Boatswain's chairs, 423–424
Body, human; see Humans
Boiler rooms, 1395
Boilers, 1388–1402
   ASME Code, 607, 1388–1389
   construction, 1390–1393
   emergencies, 1396–1401
   maintenance, 1394–1395
   placing in service, 1394
   recovery, 1398–1401
   taking out service, 1394–1395
Bolt headers, 945
Bolting steel, 403, 404
Bonding, electrical, 1291–1296
Bonuses, safety, 287
Books
   historical safety, 18–19
   industrial hygiene, 1093–1094
   industrial safety, 297
   rule, 212–213
   technical handbooks, 1481–1483
Booms, crane, 686–687, 689–691
   warning devices, 1441
Boots, cleaning rubber, 517
Boric acid, 1130
Boring machines, 832–835
Boron, 1233
Boron hydrides, 1130
Boxes (containers)
   handling, 637–638, 659–660
   storage, 645–646
Bracing, building, 406–407
Brakes; see Clutches
Brakes, press, 861–862
Brass, 1130
Breathing apparatus, self-contained, 502–505
   maintenance of, 508–509
Bridge plates, 657; see also Docks, loading
Bridges, foot, 372
Brigades, emergency, 455–458, 1385
   historical, 12
Brightness ratios, table of, 1456

Broaches, metal, 840–841
Bromine, 1130
Bromochloromethane, 1130–1131
Brucellosis, 1067
Brushes, scratch, 1000
Brush wheels, 854
Bucket conveyors, 703–705, 706–707
Buffing wheels, 853–854, 1000
   exhaust for, 1022
Buildings
   construction, 386–393
   cut-offs, 1330
   fire-safe, 1328–1330
   high-rise, 584–585
   industrial, 366–367
   inspection of, 77
   location, 367–368, 373–374
Bulldozers (construction), 1442–1444
Bulldozers (forging), 945
Bulletin boards, 131, 291–292
Bulletins, safety contest, 131
Bump caps, 473
Bureau of Labor Statistics, 24, 44–46
Bureau of Mines, respiratory equipment, 489
Burlap, storing, 647, 658
Burners, oil, 1320
Burning (rubbish disposal); see Wastes
Burns, 1428
   chemical, 1430
1,3-Butadiene, 1131
1-Butanol, 1131
2-Butanone, 1131
Buttons, safety, 298
n-Butyl acetate, 1131
n-Butylamine, 1131
Butyllithium, 1131–1132
Butyraldehydes, 1132
Byssinosis, 1066–1067

C

Cableways, 705–706
Cadmium metal, toxicology, 1132–1133
Cafeterias, industrial, 542, 544
Caisson work, 1054
Calcium, toxicology, 1133
Calcium arsenate, 1133
Calcium carbide, 1133
Calcium cyanamide, 1133
"California study" of office accidents, 346–352
Campaigns, safety, 294
Canada
   government agencies, 624–625
   Industrial Safety Conference, 626
   private provincial agencies, 625–626
Canadian Standards Association, 1259
Canisters, gas mask, 493–495
   color code, 492–493
Can Manufacturers Institute, Inc., 612
Cans, waste; see Waste disposal

1496

Canteens, food, 542–543
Carafes, sanitizing, 537
Carbohydrates, 1233
Carbon, 901
Carbon dioxide
    "buildup," 1054
    extinguishing systems, 1374–1375
    fire extinguishers, 1355
    toxicology, 1133
    welding shield, 965
Carbon disulfide, 1133–1134
Carbon monoxide, 901, 909–910, 1114
    target hazard, 1030
    toxicology, 1041, 1047–1048, 1134
    ventilation of, 408
Carbon tetrachloride, 1134–1135, 1147–1148
Carboys, 650–651
Carcinogens, 1193–1194
Care, degree of, 572
Cars; see Elevators or Railroads
Cartons
    cutting, 986
    handling, 637–638, 659–660
    storage, 645–646
Cartoons, 305
Castings; see Foundries
Catalyst, 1233
Ceilings, inspection, 429
Cell, human, radiation damage to, 1059–1062
Cellulose, 1233
Celsius-Fahrenheit conversion table, 1472–1473
Certified Safety Professional, 11, 603, 607
Cesium-137, 1064
Chain conveyors, 701–702, 709
Chain drives, guarding of, 1264
Chain hoists, 668, 1431
Chains, 730–733
    inspection, 80, 732–733
    slings, 730–733
Chairs
    audiovisual seating, 332–333
    office, 357, 359
Chalk boards, 334–336
Change, effect of, 100
Checklists, 71, 85, 89
    accident causes, 157
    boiler cleaning, 1398–1400
    boiler plant emergencies, 1397
    crane, 81, 675–676
    departmental, 72, 74, 75–76
    electric tools, 996
    fiber rope, 713
    fire ignition sources, 1333
    fire protective equipment, 79, 1381
    foundry core ovens, 917–918
    grinders, 842
    guarding, 796–797
    hammer, maintenance, 939, 940
    hazardous materials, 1248–1249
    human factors engineering, 228

Checklists—Continued
    illumination survey, 95
    instructor, 199
    ladder inspection, 414–415
    lift truck operators, 758
    machine safety, 108–109
    new plant, 367
    ropes, 713
    shutdown, 453
    work areas, 73, 77
Chelating agent, 1233
Chemical
    cartridge respirators, 495
        maintenance, 505–507
    cross index
        definition, 1233
        engineering, 1234
    foam for fires, 1371, 1372–1374
    goggles, 485–487
    process industry, 1232
    reaction, 1234
    terms, glossary, 1232–1240
Chemicals
    and allied products, 357–358, 1232
    hazardous
        bulletin, 602
        definitions, 1040
        evaluation, 1200
        information, 1200–1253
        labels, 227, 1034–1035, 1048–1049
        skin irritants, 1068–1074
    incompatible, table of, 1462
    references, 1207–1208
    threshold limit values, 1182–1198
    toxicity, 1109–1117
    toxicological properties, 1123–1176
Chemistry
    applied, 1234
    definition, 1234
    of fire, 1340–1342
Chemotherapy, 1234
CHEMTREC, 463
Chemurgy, 1234
Chimneys
    boiler, 1392
    inspection, 78
    maintenance, 433
Chinstraps, safety hat, 472–473
Chipping
    castings, 924
    goggles, 483
Chisels
    metal-cutting, 982–983
    wood-cutting, 985
Chlordane, 1135
Chlorine, 1135–1136
    dioxide, 1136
    Institute, 612
    trifluoride, 1136
Chloroacetic acid, 1136

Index

Chlorobenzene, 1137
Chlorobromomethane, 1137
p-Chloronitrobenzene, 1137
Chlorosulfonic acid, 1137
Chlorowaxes, 1137
Chocks, wheel, 657, 745
Chromic acid, 1137
Chromium, 901
Chronic poisoning, 1114
Chute
  conveyors, 708
  feeds, 884–885
Cilia, 1043
Circuit
  breakers, 1261–1262
  interrupters, 1262–1263
Circular saws, 1000
  blade maintenance, 816–817
  metal, 837
  woodworking, 813–817
Civil defense
  industrial, 362, 447
  Office of, 463
Civil disobedience, planning for, 446
Civil Service Commission, U.S., 568
Cleaning
  boilers, 1394–1395
  flammable liquid containers (tanks), 648, 968–969, 1314–1318
  food service, 544–546
  metal parts, 923–924, 1320
  protective equipment, 476, 506, 517, 523
  windows, 389
Cleanup; see Housekeeping
Clearing land with motor graders, 1443
Climate, building site, 367
Climbing devices, ladder, 413
Clips, wire rope; see Fasteners, wire rope
Clothing
  cleaning, 520
  cold weather, 526
  dimensions, 1469
  flame-retardant, 519–520, 525, 1384
  heat reflecting, 517–520
  high-visibility, 526
  loose, 828, 973, 1004
  protective, 517–526, 1087
  welding, 971–974
  women's, 524–526
  see also names of items
Clubs, safety equipment, 468–469
Clutches, power press, 860, 867–869
Coaching, 209–210
Coal
  Bituminous Coal Operators' Assn., 612
  foundry, 901
  handling, 905
  National Coal Assn., 615
Cobalt, 1138
Cobalt-60, shielding, 1064

Cochlea, 1244, 1245
Codes; see Standards
Coding
  color, 234–235, 385–386, 650
  principle of, 227, 232–236
Coin dies, 945
Coke, handling, 905
Coke oven chemicals, 1234
Cold heading, forging, 945
College degrees, safety, 13, 628–629
Colloid, 1234
Color
  coding, 234–235, 385–386, 650
  industrial, 385, 740
  use in visual aids, 326–327
Columns, inspection, 427, 428
Combination respirators, 497
Combustible gas indicators, 1296–1298
Combustible liquids; see Liquids, flammable
Combustible solids, handling, 653–654
Comfort conditioning; see Air conditioning
Command headquarters, emergency, 452–453
Committees, safety, 59–66
  accident investigation, 152
  functions, 60–62
  inspection duties, 64, 65, 89
  maintaining interest, 64–65, 271
  meeting agenda, 63–64, 274–275
  office, 362–364
  selection of members, 62–63
  suggestions, 299
  types, 59–60
Communications, emergency, 1385
Community-wide disasters, 463
Compatibility, principle of, 226, 230, 232
Compound, chemical, definition, 1234
Compressed Gas Association, Inc., 612, 1092
Compressed gases, 949–956; see also Air, compressed; Cylinders; and Pressure vessels
Compressed oxygen (cylinder) recirculating apparatus, 504–505
Composite ropes, 713
Computer installation, 357
Concrete mixer drivers' hand signals, 400
Concrete, reinforced, inspection of, 428
Condemning equipment, 91–93
Conductive clothing, 526
  shoes, 514–515
Conductors, electric ground, 392
Conflict, personality, 246, 250–254
  types, 251–252
Consejo Interamericano de Seguridad, 626–627
Construction
  building, 396–413
  during plant operation, 404–405
  fire-safe, 1328–1330
  protecting employees, 397–401
  safety factors, 720, 1448–1449
  specifications, 93–94, 397
  wind pressure, 405–407

Consulting services, 23
Consumer Product Safety Act, 588–590
Consumer safety, 43, 588–590
Containers
    cleaning, 1314–1318
    damaged, 660
    portable, 650–651, 654
    storage, 644–648
    welding, 968–969
Contaminant control, 1200–1201
    food, 544–546
    measurement, 1076–1084
    methods of, 585, 1084–1089
    ventilation; see Ventilation
    water, 529–534
Contests, safety, 275–289
    awards, 276, 286–289
    historical, 6
    publicity, 284–285
    purposes, 275–276
Contract documents, construction, 397
Contractors
    AGC, 611
    safety program, 397–401
Controls
    chart, 146
    contaminant; see Contaminant control
    design, 108, 222–223
    electric, 879, 1263
    elevator, 773–774
    human factors, 231–236
    loss; see Loss control
    power press, 865–869, 877–879
    remote, 744
    texture of, 232–233
    traffic, 369, 585
    see also Accident prevention
Conversion factors, 1449, 1472–1478
Conveyors, 696–709
    foundry, 906–908
    maintenance, 700
    portable, 706–707
Cooling
    spot, 1009
    summer, 1007
Copper, toxicology, 1138
Cords, extension electric, 356, 360, 999
Cores
    blowing machines, 921–922
    foundry, 920–922
Cork floors, 579
Correspondence courses, 629
Corrosion, 1234
Costs
    accident, 7, 10, 14, 67–68
    economic losses, 173
    effectiveness (system), 100
    estimates, 170–171
    insured vs. uninsured, 162–170

Costs—Continued
    property damage, 96–97
    vehicle accidents, 1414
    workmen's compensation, 176, 184–185, 193
Cotton dust, 1030
Counseling, non-directive, 253–254
Courses; see Education
Coverings, floor, 580–582
Covers, shoe, 517
Crabs, 694
Crane booms; see Booms, crane
Crane (hoists), 668–674
    crawler, 691
    distance from electric wires, 691, 1441
    foundry, 906
    inspection, 81, 675
    jib, 683
    limit devices, 669–670
    maintenance, 1439–1440
    mobile, 686–693, 1440–1442
    overhead, 678
    rules, 676–678
    runways, 438
    signals, 671–675
    storage bridge and gantry, 681–682
    truck, 69, 693, 743
Cresol, 1138
Cribbing, timber, 640
Crimpers, nail band, 992
Critical incident technique, 100–102
Crowbars, 640, 987–988
Crucibles, refractory clay, 914–916
Cryogenics, 1058, 1234
Cumene, 1138–1139
Cups, drinking, 537
Curie, 1059
Current, electric
    definition, 1256
    human resistance to, 1257
Cutoff saws, 817–818
Cutters, metal, 984–985, 992
Cutters, side, 992
Cutting
    oxygen, 948–957, 1335
    powder, 957
    arc, 960–966
Cyclohexane, 1139
Cyclohexanol, 1139
Cyclohexanone, 1139
Cyclopropane, 1139
Cylinders, compressed gas, 949
    handling, 949
    headers, portable, 952
    hose and connections, 955–956
    manifolding, 951–952
    piping, distribution, 952
    regulators, 952–955
    storage, 949–950
    using, 950–951

Index

D

2,4-D, 1139
D-rings, safety belt, 510–512
Damage, control, 96–97
Damage risk criteria, 1059, 1248–1251
Day of disability, definition, 137–138
Days charged, 135–142
  interpretation of, 144–145
DDT, 1139–1140
  chlordane, 1135
  methoxychlor, 1154
Deafness; see Auditory defects
Deaths, 8, 9, 14, 137
  days charged, 140–142
  recording, 30
Decay, radioactive, 1065
Decible, definition, 1242, 1486
Decompression sickness, 1054
Defensive driving, 296, 1424
Deluge sprinkler systems, 1366
Demolition of structures, 410
Demonstrations
  audiovisual, 338
  interest-getting, 295–296
Density; see Specific gravity
Dentists, American Association of Industrial, 604
Dermatitis, 1067–1075
  foundry, 902
  literature, 1093
Derricks, 683–686
Derris, 1140
Design
  machine, 107–110
  principles, 219–220, 231–236
  safe, 106, 366–367, 396, 579–582
Desiccators, 1072
Desks, office, 357, 360
Detergent, 1072, 1239
Detoxification, 1115
Deviation, system, 100
Diacetone alcohol, 1140
Diatomaceous earth, 1044–1045, 1140
1,2-Dibromomethane, 1140
Dichlorobenzene, 1140
Dichlorodifluoromethane, 1140
1,1-Dichloroethane, 1140
1,2-Dichloroethane, 1140
1,2-Dichloroethylene, 1140–1141
Dichloroethyl ether, 1141
Dichloromonofluoromethane, 1141
Dichloronitroethane, 1141
Dichlorotetrafluoroethane, 1141
Dichotomous readings, 225
Die and tap work, 983
Dieldrin, 1141
Dies
  enclosure guards, 872
  forging, 933–939, 943, 944–945

Dies—*Continued*
  power press, 872, 890–895
  trimming, 944–945
  upsetters, 940–942
Diethylamine, 1141
Diethylene glycol monoethyl ether, 1141
Diethylenetriamine, 1141
Diisobutyl ketone, 1141
Dikes, storage tank, 1305
Dimethylaniline, 1141
Dimethyl ether, 1141–1142
Dimethylformamide, 1142
1,1-Dimethylhydrazine, 1142
Dimethyl sulfate, 1142
Dinitrobenzene, 1142
Dinitrochlorobenzene, 1142
Dinitrocresol, 1142
2,4-Dinitrophenol, 1142
Dinitrotoluene, 1142
1,4-Dioxane, 1142–1143
Diphenyl, 1143
Dip tanks, 1319
Disability benefits insurance, 189–192
Disabling injuries; see Injuries, disabling
Disaster services, contract, 462
Discipline, 213–214
Disease, 177
Disinfecting water systems, 533–534
Disks, abrasive; see Grinding wheels
Dispensary, medical, 561
Displays
  audiovisual, 230
  auditory, 229–230
  evaluation, 231
  information, 225
  interest-getting, 293
  merchandise, 574
  visual, 225–229
Disposable clothing, 517, 526
Dissociability, principle of, 230
Docks, loading, 789
  maintenance, 433
  marine, 373
  shipping and receiving, 657, 660
  see also Platforms, loading
Doors
  elevator, 768–771, 772
  office, 354
  passageway; see Exits
Dosage, radiation, 1059
Dosimeter, 1062
DOT labels, 117, 655
Double-insulated tools, 996–998, 1268
Drains, building, 534–535
Drawings and graphs, 327
Drawknives, 986
Dressing
  abrasive wheels, 851–852
  tools, 981–982

Drills
  electric, 999
  emergency fire, 1384
  safe operation of, 833–834
Drinking fountains, 536–537
Drinking water; *see* Water, potable
Drivers, motor vehicle
  awards, 1424–1425
  forms, report, 1415–1419
  safety programs, 1415
  selection, 1420–1423, 1433
  test apparatus, 1422–1423
  training programs, 1423–1424, 1433
Driveways; *see* Roads
Drug control, 563
Drums
  handling, 638, 659
  storing, 650
  welding, 968–969
  wire rope, 721–723
Dry-chemicals
  fire extinguishing system, 1375
  portable extinguishers, 1350–1355
Dry-pipe sprinkler system, 1364–1365
• Dry-powder fire extinguishers, 1356–1358
Dry rot, 427
Ducts, ventilation, 1014, 1015
  testing, 1016
Dumbwaiters, 780–781
Dumps, vehicle procedure on, 1434–1435
Duplicating equipment, office, 358
Dusts
  airborne, 1041–1043
  control of, 653
  definition, 1040
  explosive, 1338–1340, 1463–1465
  exposure evaluation, 1041–1043, 1078
  foundry, 901
  "harmless," 1045–1047
  literature, 1093
  magnesium, 901
  physiological effects, 1043–1048
  target health hazards, 1030–1032
  threshold limit values, 1192–1193
"Dutch cap" for vent stacks, 1014–1015
Dyes, vat, 1069, 1239

E

Ear
  anatomy, 1245
  cross section, 1244
  hearing loss, 1245–1247
Ear protection, 476–479, 1087, 1490–1492
  amount, 478–479
  attenuation characteristics, 474–475
  insert type, 476–478
  muff type, 478
  selection, 479
Earthquakes, planning for, 445–446

Eating areas, 544
Economizers, boiler, 1390
Eddy current inspection, 926
Edison Electric Institute, 613
Education
  accident prevention, 58–59
  audiovisuals, use of, 317
  chart, 206
  conference method, 207–209
  contaminant control, 1089
  continuous, 211
  correspondence courses, 629
  course outlines, 198, 202–203
  crane operation, 675
  departmental, 213
  driver safety, 1423–1424, 1433
  electrical maintenance, 1284
  elevator operators, 777
  emergency planning, 449–452
  fire fighting, 450–452, 458, 1359–1360, 1384
  first aid, 214, 295
  home study, 296
  human factors engineering, 221–222
  industrial hygiene, 1089
  industrial truck operation, 753–757
  in-house, 204
  institutions of learning, 628–629
  instructors, 203
  instructors' checklist, 199
  job instruction, 205–207
  job safety analysis, 110–116
  learning, 241, 257–261
  lesson plans, 198–200
  maintenance crew, 435–436
  meetings, 64
  National Safety Council, 595
  need for, 197
  new employees, 58–59, 211–214
  NIOSH, 23
  off-the-road vehicle operation, 1433
  office workers, 363
  other methods, 209–210
  policies, 210–211
  pressure vessel operation, 1406
  programs, 43, 197–200
  repair shop operation, 1432
  safety committee, 64
  safety professionals, 13, 203
  supervisors, 53, 200–207, 363
  transfer of training, 260–261
  welding operation, 974
Effective temperature, definition, 1053–1054
Ejecting devices, power press, 887–890
Elastomer, 1234
Electric-conducting floors, 392
Electric current
  definitions, 1256–1257
  distribution system, 382
  effect on man, 1256–1258
  shock, electric, 963–964, 996

Index

Electric equipment
  controls, 879, 1263
  double insulated, 996–998
  explosion proof, 1275–1277
  fires, 1332
  hazardous locations, 1269–1274
  high frequency, 1279–1280
  high voltage, 1281
  inspection, 77, 1280–1281
  installation, 1259–1260
  layout, 377
  lockouts, 402, 437, 1282–1284
  low voltage, 996
  maintenance, 1281–1284
  motors, 811, 1263–1265
  office, 356
  selection, 1258–1259
  switches, 1260–1261
  testing, 1278
  tools, 996–1001
  trucks, 758–759
Electric grounding, 377, 1265–1275
  containers, flammable liquid, 1291–1296
  elevators, 774
  equipment, 1266–1268
  maintenance of, 1268
  mobile cranes, 1437–1439
  systems, 1265–1266
Electric wiring, 1284
  distances from, 403, 691, 1441
  distribution systems, 382
  inspection, 77
Electrical protective equipment (insulating), 523
Electricity, static, generation of, 1291–1296
Electromagnetic
  inspection, 926–927
  radiation suits, 526
Electromagnets, 680
Element, definition, 1234
Elevators (lifts)
  belted, 765, 776
  cars, 771–772, 775
  Code, 587
  dumbwaiters, 780–781
  emergencies, 585, 778
  foundry, 908
  freight, 766
  hand-powered, 779–780
  hydraulic, 765
  inspection, 78, 774–777
  loads, 777–778
  maintenance, 774–777
  manlifts, 783–786
  new equipment, 765–766
  operation, 777–778
  power, 763–778
  safety devices, 774, 776
  sidewalk, 778–779
  temporary use, 425

Emergencies
  elevator, 778
  handling public relations, 310–311
  medical planning, 552–553
  planning for, 361, 443–463
  transportation during, 461–462
Emergency exposure limits, 1121
Emotion, 240, 253
Employee responsibilities, 24–27, 58, 88–89, 363, 1091
Employees
  indoctrination, 58–59, 211–214
  types, 267
  who get injured, 349–350
  see also Non-employees
Employment, definition, 142–143, 148
Emulsifier, 1234
Emulsion, 1234
Energy equivalents, 1478
Enforcement of safety rules, 59, 359–361
Engine lathes; see Lathes
Entrances; see Exits
Environmental control, 585
  Administration, 1092
Epichlorohydrin, 1143
Epoxy resins, toxicology, 1143
Equipment
  building, 366
  color coding, 385
  construction, 400–403
  condemning, 91–93
  design; see Design
  emergency, 453–454
  grounding, 1266–1268
  inspections, 436–437, 1039
  location, 376–377
  maintenance, 401, 439
  office, 357–358
  repairing, 402
  specifications, 93–94
"Equitable study" of office accidents, 346–347
Erysipelas, 1067
Escalators, 586–588, 781–783
  inspection, 587–588
Establishment, definition, 148
Esters, 1050, 1235
Ethane, 1235
Ethanol, 1143, 1235
Ethyl acetate, 1143
Ethyl acrylate, 1143–1144
Ethylamine, 1144
Ethylbenzene, 1144
Ethyl bromide, 1144
Ethyl chloride, 1144
Ethyl ether, 1145
Ethyl formate, 1145
Ethyl mercaptan, 1145
Ethyl silicate, 1146
Ethylene, 1235

Ethylene chlorohydrin, 1144
Ethylenediamine, 1144
Ethylenimine, 1145
Ethylene glycol, 1144–1145, 1235
Ethylene glycol monobutyl ether, 1145
Ethylene glycol monoethyl ether, 1145
Ethylene glycol monomethyl ether, 1145
Ethylene oxide, 1145
Eustachian tube, 1244
Evacuation
    emergency, 460–461
    high-rise building, 584–585
    squads, 456
Evaporating pans, 1410
Examinations, industrial physical; see Medical ex-
        aminations
Excavation
    construction, 411–413
    protection of, 399–401
Excelsior storage, 647–648
Exclusive remedy doctrine, 192
Excursion factors, 1198
Exhaust hoods, 1010–1014
    ducts, 1014
Exhausting
    general, 1009–1010
    local, 654, 1009, 1010, 1431
    requirements, 1017–1027
    systems, 1009–1014
Exhibits, safety, 293, 338
Exits, building, 354, 585, 1396
    fire, 450–452, 1325–1327
    foundry, 903
    maintenance of, 435
    number and location, 389
    signs, 450, 461
Explosion-proof electrical apparatus, 1275–1277
Explosions
    prevention, 444, 653, 1377–1380
    supression, 1380
Explosive-actuated tools, 1002–1004
Explosive limit; see Flammable limit
Explosives
    atmospheres, 1338–1340
    blasting caps, 655
    dusts (table), 1338–1340, 1463–1465; see also
        Dusts
    handling, 654–655
    Institute of Makers of, 613
    storage, 375
Exposures, human body
    chemical, 1200
    controlling, 105–107, 1084–1089
    defining, 1076–1084
    measurement, 1075–1076
    see also Threshold limit values
Extended service lamps, 383–384

Eye protection, 479–489, 597, 1087
    comfort and fit, 484
    fountains, 558–559
    supervision of, 481

F

Fabrics, flame proofing of, 1384
Face protection, 479–481, 484–487, 1087
Factors of safety; see Safety factors
Factory Insurance Association, 599
Factory Mutual System, 599, 1092
Fahrenheit-Celsius conversion table, 1472–1473
Failure mode and effect, 99, 151
Fallout shelters, 454
Falls; see Accident types
Families, involving, 284, 297
Fans
    air handling, 409
    body cooling, 357
    see also Ventilation
Fasteners
    elevator rope, 773
    wire rope, 724–726
Fatalities; see Deaths
Fault tree, 99–100
Federal Employee Liability Act, 176
Feeding and ejecting devices, power press, 879–
        882, 883–890
Feedwater, boiler, 1392–1393
Fenced grounds, 369
Fertilizer, 1235
Fiber rope
    natural, 712–713
    synthetic, 712
    inspection, 80, 713–716
    care in use, 716
    care in storage, 716–717
    slings, 726–730
    lifelines, 422, 513–514
Files, hand, 984
Filing cabinets, 360
Filmstrips, 339–340
Filters
    lenses, 486, 488
    water, 533
Fingers, "dead" or "white," 1055
Fire
    brigades; see Brigades, emergency
    chemistry, 1340–1342
    classification, 1342
    loss control, 361, 1325
    losses, 575
    plastics, 584
    stages, 582–583
Fire extinguishers
    portable, 361, 1342–1361
        carbon dioxide, 1355
        classes of, 1343–1344, 1348–1349
        dry-chemical, 1350–1355

Index

Fire extinguishers—*Continued*
  dry powder, 1356–1358
  halogenated compounds, 1355–1356
  liquefied gas, 1355
  location of, 1344
  maintenance of, 1360–1361
  miscellaneous, 1358–1359
  selection of, 1347–1350
  water solution, 1350
  systems
    automatic, 375, 644, 1362–1368
    carbon dioxide, 1374–1375
    dry-chemical, 1375
    foam, 1371, 1372–1374
    halon, 1375–1376
    hoses, 1369–1371
    inerting, 1376–1377
    steam, 1376
    water, 1361–1371
    wet water, 1374
Fire prevention, 1325, 1380–1385
  contributory factors, 1332–1340
  detection, 582–583
  emergency planning, 450, 462
  foundry, 904
  hazard identification, 361, 1269–1274
  housekeeping, 1337–1338
  industrial trucks, 741
  inspection, 78, 1360–1361, 1380–1381
  office, 361–362
  planning for, 444, 1384–1385
  storage tank, 1310
  tank car, 1301
  tank truck, 1302–1303
  vehicle repair shops, 1428–1429
  welding, 967
Fire protection, 1325–1385
  college courses, 629
  escalators, 782–783
  extinguisher squad, 457
  extinguishment, 583
  fighters, training, 450–452, 458, 1359–1360, 1384–1385
  high-rise buildings, 584–585
  municipal, 462–463
  occupancies, 1345–1347
  organizations, 599–600
  periodicals, 630
  station house, 458
  supervisory services, 583–584
  see also Personal protective equipment
Fire-resistive construction, 1328–1330
First aid, 148, 557–560
  chemical burns, 1430
  disaster service, 460
  electrical, 1258
  kits, 558, 561–562
  reports, 123–124
  rooms, 561
  training, 214, 295, 560

Fittings, rope, 724–726
Fixation, 252
Flagman signals, 1444
Flameproofing wood and fabrics, 1384
Flame propagation, definition, 1289
Flammable limits
  definition, 1041, 1201, 1289
  table of, 1212–1231
Flammable liquids; *see* Liquids, flammable
Flammable range, definition, 1289
Flammability hazards
  NFPA definitions, 1205–1206
  table of, 1212–1231
Flanges, grinding wheel, 843–844
Flannel boards, 337–338
Flash point
  definition, 1201, 1288
  table of, 1212–1231
Flasks, foundry, 922
Fleet angle, 722 n
Fleets, motor
  costs of accidents, 1414
  repair shop safety, 1427–1432
  report forms, 1415–1419
  safety program, 1414–1427
Flexible-shaft tools, 1002
Flip charts, 336–337
Flight conveyors, 701, 706–707
Flight Safety Foundation, Inc., 607
Floods
  planning for, 445–446
  storage tanks in, 1306
Floors
  conductive, 392
  coverings, 580–582
  electric conducting, 392
  elevator car, 426, 772
  finishes, 577, 578
  foundry, 903, 904, 909
  inspection, 77, 80, 429
  loads, 78, 392–393, 430–431, 645
  maintenance, 430
  materials, 390–392
  office, 353
  openings in, 372–373
  shipping and receiving, 655
  temporary, 407–408
  types, 578–579, 580
  washroom, 538, 540
  welding considerations, 967
  wood, 428
Flotation reagent, 1235
Flow rate equivalents, 1476
Flow sheets, plant layout, 374–377
Fluorescent lamps, 384
Fluorides, 901, 1146
  in drinking water, 531
Fluorine, 1146
Fluorotrichloromethane, 1146
Fluxes, welding, 969–970

Foam systems, fire control, 1371, 1372–1374
Folding machines, 359
Food service, 535, 542–546, 574–575, 660
Footings, structural, 427
Foot (kick) presses, 864
Footwear, protective, 514–517
Forced entry, principle of, 230
Force, equivalents, 1477
Foreign language employees, 267
Foremen; *see* Supervisors
Forgetting, 261
Forging
    hammers, 928–940
        guarding, 931–933
        maintenance, 939–940
    other equipment, 944–945
    presses, 942–944
    upsetters, 940–942
Forklift trucks; *see* Lift trucks
Formaldehyde, 1146
Formic acid, 1146
Foundations, structural, 427–428
Foundries
    American Foundrymen's Society, 608
    cleaning and finishing, 923–925
    cupolas, 909–913
    furnaces and ovens, 913–918
    hazardous materials, 899–902
    health problems, 899–902
    maintenance, 903
    materials handling, 904–909
    molds and cores, 919–921
    nondestructive testing, 925–928
    production equipment, 918–925
    sand, 902
    shoes, 515
    work environment, 902–904
Four M's of safety, 151
Frames, protective rollover, 1435–1436
Frequency-of-use principle, 236, 259–260
Frequency rate, 9
    damage costs, 97
    definition, 135
    interpretation, 143–144, 160
    vehicles, 1420
Frequency, sound, 1242–1245
    see also Octave band analysis
Friction
    as cause of fire, 1334
    loss in ducts, 1014
Frustration
    emotional patterns, 250–252
    reaction to, 252–254
Fueling; *see* Liquids, flammable
Fuels
    fire chemistry of, 1340–1341
    high energy, 1235
    welding, 948–949
Fume fever, metal, 1047

Fumes
    definition, 1040
    physiological effects, 1043–1048
    sampling, 1078
    solder, 1000–1001
    welding, 970
Functional control principle, 236
Fungus infections, 1066
Furfural, 1146
Furnaces
    crucible, 915–916
    open hearth, 913–918
    plumber's, 1321
Fuses, electric, 1261–1262, 1284
Fusible plugs, boiler, 1393

G

Gages, high-pressure, 1410–1412
Gaging, flammable liquid containers, 1306
Gamma rays, 1060–1062
Gangways, foundry, 904
Gantry cranes, 681–682
Garages, 377
Garbage disposal, 535–536
Gas masks, 492–497, 507–508
Gases
    American Gas Association, 608–609
    definition, 1040, 1235
    diffusion, 1289
    explosive, 1340
    motor vehicle exhaust, 1431
    physiological effects, 1047–1048
    sampling, 1078
    specific gravities, 1459
    welding, 948–956
Gasoline
    handling, 402, 408, 1430
    -operated trucks, 759
    power tools, 1002
    toxicity, 1147
Gates
    elevator car, 772
    hoistway, 769–770
    point-of-operation, 874–877
Gear cutters, 838–839
Germanium, 1147
Glare from lighting, 380
Glass
    broken, 659
    doors, 354, 573
    hazard in materials handling, 659
    window, manual handling of, 639
Glasses, safety; *see* Eye protection
Glossaries
    chemical terms, 1232–1240
    guarding terms, 795–796
    legal terms, 572–573

Index

Gloves
  impervious, 523
  maintenance of, 523–524
  protective, 520, 660–661
Glue in shipping, 659
Glycerol, 1235
Glycols, 1050
Goggles; *see* Eye protection
Government agencies
  Canadian, 624–625
  international, 626–628
  state, 618–624
  U.S., 618
Governor ropes, 773
Graders, motor
  maintenance, 1442
  operating, 1442–1444
Graphic Arts Technical Foundation, 613
Graphite, 1147
Graphs and drawings, 327
Gravity conveyors, 704, 707–708
Gravity drop hammers, 928
Gray and Ductile Iron Founders' Society, Inc., 613
Grease rack operations, 1429
Grinding machines, 841–854, 1001
Grinding wheels, 841, 1000
  disk mounting, 842–848
  exhaust for, 1022
  maintenance, 848–852
  materials handling, 842–843
  speeds, 849–850
Grindstones, 852–853
Ground fault circuit interrupter, 1262–1263
Grounding; *see* Electric grounding
Guard rails; *see* Railings and toeboards
Guards (hazard)
  automatic, 797, 799
  barrier, 798, 803
  built-in, 107–108, 796
  checking, 438
  construction equipment, 401–402
  crane, 669–670
  designing, 808
  dimensions, 807
  electrical equipment, 1259–1260
  fixed, 797
  foot control, 879, 882
  forging hammers, 931–933
  grinding wheels, 844–848
  hand removal devices, 799
  high-pressure equipment, 1410–1412
  industrial trucks, 740–744
  interlocking, 797, 808
  jointers, 822–824
  lathes, 831–832
  maintenance, 806
  materials, 806–808
  noise control, 808–809
  openings, 801–803, 807

Guards—*Continued*
  OSHA requirements, 795
  platforms, 401–402
  point-of-operation, 795, 800–806, 869–883
  power press, 862, 869–883
  power transmission, 795, 806–808
  woodworking, 811
  saws, 814
  terminology, 795–796

H

Hair protection, 473–476
Half-life, radioisotope, 1065
Halogenated hydrocarbons
  psychological effects, 1050
  toxicity, 1147–1148
Hammers
  claw, 993
  forging, 928–940
  hand, 993
  riveting, 993
Handbooks; *see* Books
Hand leathers, 521–522, 660–661
Hand tools
  control, centralized, 80, 978–979
  double insulated, 996–998
  forging, 933
  maintenance, 979–982, 987–988
  materials handling, 640–642
  personal protection, 1004–1005
  shipping and receiving, 657
  spark-resistant, 993–994
  use, 403, 438, 982–994
  woodworking, 985–987
Hand-powered elevators, 779–780
Handtrucks
  materials handling, 642–643, 650
  motorized, 739, 743–744
  operation, 644, 752
Handicapped persons, 563–568
  as patrons, 575
Handles, tool, 982
Harnesses, safety, 511
Hatch covers, sidewalk elevators, 779
Hatchets, 986
Haul-roads, 1432
Hazardous
  conditions, 155
  locations
    lighting of, 381
    NFPA definitions, 1269–1274
  materials information, 1036–1038, 1212–1231
  signing; *see* Signs *and* Symbols
Hazards
  color-coding, 234–235, 385–386, 650
  control of process, 1200–1201
  evaluation, 1074–1075, 1200
  guarding; *see* Guards (hazard)
  identification of, 94–96, 1032–1033

Hazards—*Continued*
  office, 350–352
  recognition of, 573–575, 576
  reducing through safe design, 106, 366–367, 396, 579–582
Head protection, 469–476
  maintenance, 476
  safety hats, 469–473
  for women, 473–474
Health, Education, and Welfare, U.S. Dept. of, NIOSH, 23–24
Health hazards, 8
  biological agents, 1065–1067
  dusts, 1043–1048
  energy stresses, 223–224, 1051–1059
  foundry, 899–902
  hot areas, 1053–1054
  industrial hygiene; *see* Industrial hygiene
  NFPA definitions, 1204–1205
  recognition of, 43, 1200
  severity, 1201–1202
  skin diseases, 1067–1075
  solvents, organic, 1048–1051
  table of, 1212–1231
  woodworking areas, 812
Health, occupational; *see* Occupational safety and health
  see also Industrial hygiene
Hearing; *see* Auditory perception
Heat
  body, 1051
  load, measurement, 1053–1054
  radiant, 1052
  -resistant clothing; *see* Clothing
  stress, 1051–1054
    index, 1053–1054
    TLV's, 1095–1099
  venting, 1331–1332
Heating process
  high-frequency, 1279–1280
  layout, 377–378
Heating, temporary, 408–409, 1335
Heights of working surfaces, 812
Heinrich ratio, 96
Helium, 1054
Heptane, 1148, 1200
Herzberg theory, 248–250
Hexachlorocyclohexane, 1148
Hexamethylenetetramine, 901–902
Hexane, 1148
Hierarchial motivation, 245
High-pressure systems, safety in, 1410–1412
High-visibility clothing, 526
History of the safety movement, 3–13, 18–19, 21
Hitches, lifeline, 423–424, 716
Hoists, 424–427, 664–665
  chain, 668, 1431
  engines for, 427
  foundry, 906
  material, 424–425

Hoists—*Continued*
  pneumatic, 665–668
  portable, 695–696
  ropes, 773, 776
  safety factors, 720–721
  tiering, 696
Hoistways, elevator, 766, 775
Hold-harmless, 573
Hoods, exhaust, 1010–1014
Hoods (personal protective)
  acid, 485–487
  air supplied, 501
Hook and loop boards, 338
Hooks, 675, 732, 1441
  hand, 640, 988
Horseplay, 987
Hose masks (air line), 497–499
  maintenance, 508
Hoses
  air, 904, 1001
  fire, 1369–1371
    maintenance, 1371
    squad, 457
  welding, 955–956
Hospitalization, 139–140
Hot industries, comfort in, 1053–1054
  clothing, 517–520
Hot work permits, 1082–1083
  cleaning storage tanks, 1315–1316
Housekeeping, 580
  contests, 283–284
  construction, 402–403
  fire prevention, 1337–1338
  foundries, 902–903
  inspection, 77
  janitorial service, 542
  occupational health protection, 1088
HTW safety, 1402
Human behavior, 240–257
Human factors engineering, 109–110, 218–237
  anthropometric considerations, 224
  applications, 222–224
  clothing; *see* Clothing
  evaluation, 228
  man-machine systems, 218–222
  physical capacities form, 565–567
  stresses, physical, 223–224
Human interest, getting, 268
  maintaining, 264–299
Humidity, air, 1007, 1052
Humor and human interest, 305
Hurricanes, planning for, 445
Hydrants, fire, 1369
Hydraulic
  elevators, 765
  power tools, 1002
  presses, 860–861, 869
Hydrazine, 1148
Hydrides, metal, 1154

Index

Hydrocarbons
  definition, 1235
  physiological effects, 1050
Hydrochloric acid, 1148–1149
Hydrofluoric acid, 1149
Hydrogen chloride, 1149
Hydrogen cyanide, 1111, 1149
Hydrogen fluoride, 1149
Hydrogen peroxide, 1149
Hydrogen selenide, 1150
Hydrogen sulfide, 1048, 1150
Hydrogen, welding application, 948
Hydrometallurgy, 1235
Hydroquinone, 1150
Hydrostatic tests, pressure vessel, 1405
Hygiene approach to motivation, 249–250
Hygiene, industrial; see Industrial hygiene

I

Ice, strength of, 1458
Ignition
  source of fires, 1333, 1336
  temperature; see Temperature, autoignition
Illnesses, records, 125–130, 146–148
  see also Injuries, occupational diseases
Illumination
  boiler room, 1396
  brightness ratios, 1456
  effect on safety, 380–381
  elevator car, 773
  equipment; see Lighting equipment
  evaluation, 355–356
  foundry, 903
  glare, 380
  Illuminating Engineering Society, 607–608
  industrial, 378–385
  levels for various jobs, 355, 1449, 1456
  offices, 354–356
  outside, 373, 378–379
  room, 579
  quality, 379–380, 1089
  quantity, 379
  shipping and receiving areas, 655–656
  survey form, 95
  systems, 434–435
Illustrations for publications, 288, 312–313, 327–328
Impact tools
  air powered, 1001–1002
  powder-actuated, 1002–1004
  shock, 993
  wood-cutting, 985–986
Impervious clothing, 522–524
Importance principle, 236
Incentives; see Awards
Incidence rate, OSHA, 148
Incident, definition, 122
Incinerators; see Waste disposal
Incompatable chemicals, table of, 1462

Indicators, combustible gas, 1296–1298
  steam and water, 1393
Individual differences, 240, 241–244
Industrial chemicals, definition, 1232
Industrial hygiene, 1029–1107
  associations, professional, 604–605, 1091–1093
  definitions, 1040–1041
  government services, 24, 623–624
  hygienist, 550, 1032–1034
  periodicals and books, 630, 1091–1094
  program, 1089–1091
  toxicology; see Toxicology
Industrial Health Foundation, Inc., 596, 1092
Industrial Medical Association, 608, 1092
Industrial plant protection, 458–460
Industrial Safety Equipment Association, Inc., 613
Industrial sanitation, 529–546
Industrial trucks, powered
  guards and safety devices, 740–744
  inspection and maintenance, 757–760
  operation, 744–752
    unattended, 746
  operators, 752–757
  types, 739
Inerting, welding, 970–971
Information
  displays, 225
  man as processor, 231
  sources, 592–630
    see also Books
Infrared radiation, 1057
  lamps, 384
Ingestion of toxic agents, 1111
Inhalation of toxic agents, 1042–1043, 1111
  physiological effects, 1043–1048
Inhibitor, 1235
Injuries
  as accident cause, 106
  disabling, 37, 134–143, 146–148, 155
    charges, scheduled, 140–141
    definitions, 189–192
    index, 136
    interpretation, 143–145
    significance of changes, 145–146
  electrical, 1257–1258
  investigation of, 152–158
  non-occupational, 571–590
  office, 346–352
  rates, 9, 123, 131, 134–146, 575
    contests, 276–280
  recordkeeping, 28–30, 122–123, 134–135, 146–148
  reducing exposure to, 105–107
  workmen's compensation, 177–180
Inorganic, 1235
Inrunning rolls, guarding of, 802–806
Insect control, 536
Insecticides, effect on skin, 1069
  see also specific names

1508

Inspection
  checklists, 71; *see also* Checklists
  committees, 64, 65–66, 89
  continuous, 80–82
  imminent dangers, 78
  intermediate, 80
  ladders, 414
  night, 93
  OSHA, 25, 33–37, 89
  overhead, 83
  personnel, 86–89
  photo, 93
  procedures, 89–94
  reports; *see* Reports
  scheduling, 85
  special, 82–83
  techniques, 71
  work areas, 71–83
  work practices, 83–86
  see also specific equipment
Institute of Electrical and Electronics Engineers, 1259
Institute of Makers of Explosives, 613
Instruction; *see* Education
Instruments
  boiler, 1390
  electrical testing, 1278
  hazard-sampling, 1080–1084
  noise-measuring, 1247–1251
Insurance
  associations, 600–603
  compulsory, 16
  costs, 163–164, 193
    administrative, 184–185
  handicapped employees, 564
  liability, general, 397
  premium discount, 186, 194
  private programs, 192–193
  self, 184, 193
  service, 602–603
  workmen's compensation, 182–187, 192–194
Intensity, principle of, 260
Inter-American Safety Council, 626–627
Interlocks, electrical, 1259–1260
Intermediate, 1235
International Association of Drilling Contractors, 614
International Association of Government Labor Officials, 627
International Association of Industrial Accident Boards and Commissions, 627
International Association of Refrigerated Warehouses, 614
International Labor Organization, 627–628
Interviewing, personnel, 1422
  corrective, 1419
Invariance, principle of, 230
Inverse square law, 1066
Invitee, 572
Iodine, 1150

Ion, definition, 1236
Ion-exchange resin, 1236
Ionizing radiation
  biological effects, 1060–1061
  evaluation of hazard, 1062
  literature, 1093
  industrial, 927–928
  maximum permissible dose, 1064–1065
  measurement, 1059
  protection from, 1063–1065
  protection regulations, 1122
  safety factors, 1060
  safety problems, 1061–1062
  shielding, 1064, 1089
  warning signs, 386
IQ, 242
Iron oxide, 901, 1150
Irritants, skin, 1072–1073
  eliminating contact with, 1073–1075
  table of, 1068–1071
Isolation of hazardous material, 1085–1086
Isophorone, 1150–1151
Isopropyl alcohol, 1151
Isopropylamine, 1151
Isopropyl ether, 1151
Isotope, 1236

J

Jacks
  materials handling, 641–642
  vehicle, 1428, 1431
Janitorial service, 542
Japanning, 1319–1320
Jewelry, around power equipment, 828, 1004
Jib cranes, 683
Jig saws, 822
Job
  assigning, 200
  enrichment, 249–250
  fitting man to, 242
  human factors, 223–224
  inspection, 83–86
  instruction training, 207
  motivators and dissatisfiers, 247–249
  practices, inspection of, 83–86
  safety analysis, 110–116, 205–207, 632
Jointers, woodworking, 822–824

K

Kegs
  handling, 638, 659
    storage, 646
Keratogenic agents, 1072
Ketones, physiological effects, 1050
Kickbacks, saw, 813–814
Kiefer, Dr. Norvin C., study, 346–352
Kinetic method of lifting, 633–637
Kitchens, food service, 544

Index

Knives, use, 986–987
Knots, efficiency, 716

L

Labels
  Department of Transportation, 117, 655
  hazardous chemical, 1034–1035
  principles, 227
  solvents, 1048–1049
Labor
  Canadian offices, 624–626
  international associations, 626–628
  state offices, 618–623
  U.S. Department of Labor, 22–24, 619
Labor-management cooperation, 15
Ladders, 413–418
  bases, nonslip, 413–414
  electrical hazards, 417–418
  maintenance, 414–416
  manlift, 785
  rolling, 357
  safety tops, 413–414
  slope, 437–438
  stock picking, 656–657
Ladles, foundry, 905–906
Lamps; see Lighting equipment
Landings, elevator, 775
Landscaping, 371
Laser beams, 1057–1059
  effect on man, 1058–1059
  eye protection, 487
  TLV's, 1099–1103
Latex, 1236
Lathes
  engine, 831–832
  screw machines, 832
  spinning, 838
  turret, 832
Lavatories per employee, 538
Laws
  early, 16, 21
  noise (sound), 1484–1492
  terminology, 572–573
  see also Workmen's compensation laws and
    Occupational Safety and Health Act
Layout
  office, 352–357
  plant, 373
Lead, 901
  casts, 940
  clothing, 526
  poisoning, 1030–1031
  toxicology, 1113, 1151–1152
Learning, 257–261
  processes, 241
Leather clothing, 517–518
Length, conversion factors, 1474

Lenses
  laser protection, 487
  plastic vs. glass, 483–484
  welding filter shades, 486, 488, 972
  see also Eye protection
Lethal dose (LD), definition, 1115–1116
Lettering, visual aid, 325–326
Liability, third party, 192
Licensee, 572
Lifelines, safety belt, 422, 513–514, 716
Life Safety Code, 354, 389, 1326
Lifting
  correct way, 633–637
  dummy, visual aid, 336–337
  equipment for, 640–644
  heavy objects, 403
  limits, 504
  personal protection, 660–661
  specific shapes, 637–640
Lift trucks
  guards for, 742
  operation, 747–752
Lighting; see Illumination
Lighting equipment, 373
  lamps, 382–385
  luminaries, 379, 381
  maintenance, 381–382, 434–435
Limerick contests, 282–283
Linen Supply Association of America, 614
Liners, safety hat, 472–473
Links, safety, 403
Linoleum, 578
Liquefied petroleum gas; see LP-gas
Liquid, definition, 1236
  specific gravity, 1459–1460
Liquids, flammable and hazardous
  definitions, 1287, 1289–1291
  fueling, 402, 409
  handling, 408, 1291–1296
  health hazards, 1296–1298, 1314–1318
  loading and unloading, 648–652, 1298–1303
  storage, 374, 648–652, 1303–1314
  uses, 1318–1321
  see also individual chemical names
Lith-X dry powder, 1356
Live roll conveyors, 708–709
Loading; see Materials handling
Loads, bearing on soils and rock, 1459
Loans, small business, 42
Local exhaust systems, 1009, 1010
Locker rooms, 540, 1338
Lockouts, electric equipment, 402, 437, 1282–
    1284
Locomotive cranes, 691–692
Locomotives, 790–791
Loss control, total, 94–97
LP-gas, 1236
  handling, 759–760, 1321
  heaters, 409–410

LP-gas—*Continued*
  industrial trucks, 759–760
  National LP-Gas Association, 615–616
Lubrication, preventive maintenance, 439
Lumber
  dry rot, 427
  storage, 646
Luminaries, 389
Lungs, absorption of toxic materials, 1111–1113

M

Machine tools, 828–854
  safety rules, 828–830
  woodworking, 810–826
Machines
  guards; *see* Guards (hazard)
  man-machine systems, 218–222
  moving heavy, 640, 642
  rooms, 767–768
  safe design, 107–110, 366–367
Magazines (explosives), 654–655
Magazines (periodicals)
  fire, 630
  hygiene and medicine, 630
  safety, 629–630
Magnesium
  dust, 901
  grinding, 923–924
  toxicology, 1152–1153
Magnetic boards, 338
Magnetic particle inspection, 925
Magneto-inductive method, 926
Maintenance; *see specific equipment*
Maintenance management, 88, 427
  crews, 435–440
  scheduling, 436
Malathion, 1153
Maleic anhydride, 1153
Management
  accident reports, 131
  backing safety efforts, 363–364
  emergency planning, 361–362, 443–463
  inspection responsibilities, 88
  motivation theories, 248–250
  responsibility for safety, 24–26
  safety meetings, 272–273
Manganese, 901, 1153
Manila rope, 714, 716
Manlifts, 783–786
  inspections, 785–786
Man-machine systems; 218–222, 242
  see also Human factors engineering
Manufacturing Chemists' Association, 614–615, 1092
  Chemtrec, 463
  Data Sheets, 615, 1207
  label, 614
MAPP, 1155

Marking tools, 983
Maser, 1058
Mass, conversion equivalents, 1476
Materials handling
  foundries, 904–909
  hand tools, 640–642
  hazardous materials, 447–448, 1298–1303
  industrial trucks, powered, 739–760
  loading platforms; *see* Platforms, loading
  machines and heavy objects, 640
  manual handling, 633–644
  pallets, 752
  personal protection, 660–661
  principles, 632–633
  shipping and receiving, 369, 655–660
  storage; *see* Storage
  trucks, 1426
Mats, floor, 580–582
Matter, definition, 1236
Maximal acceptable concentrations, 1075
Maximum permissable dose, 1064–1065
Meaningfulness, 259
Measuring characteristics, 243–244
  reliability, 243
  sampling, 244
  validity, 244
Media, audiovisual, 316–343
Medical examinations, 148
  audiometric, 1252
  motor vehicle driver, 1422–1423
  physical, 553–557, 1115
  skin, diseases, 1073
Medical identification symbol (AMA), 560
Medical services
  agencies, 463
  benefits under workmen's compensation, 180, 188
  department, 560–561
  disaster, 460
  employee health, 560–563
    foundry, 902
  governmental, 24
  industrial hygiene, 550, 1089–1090
  industrial physician, 551–553
  nurse, occupational health, 550, 553
  records, 563
  women's health, 557
  see also First aid
Meetings, safety, 63–64, 272–275
  audiovisual presentations, 316–343
  planning programs, 274–275
  prejob, 274
  see also Committees, safety
Mercury, toxicology, 1153–1154
Mercury vapor lamps, 384–385
Merit rating, insurance, 186
Mesityl oxide, 1154
Metabolism, 1052
Metal-cutting tools, 982–985

Index

Metal handling
 scrap, 639
 sheet, 638, 647
Methane, 1236
Methanol, 1154, 1236
Methoxychlor, 1154
Methyl acetate, 1154–1155
Methyl acetylene, 1155
Methyl acrylates, 1155
Methylamines, 1155–1156
Methyl bromide, 1148, 1156
Methyl chloride, 1148, 1156
Methyl chloroform, 1156
Methylcyclohexane, 1157
Methylcyclohexanone, 1157
Methyl formate, 1156
Methyl isobutyl ketone, 1156–1157
Methyl mercaptan, 1157
Methylene chloride, 1157
Met-L-X dry powder, 1356
Metric system, 1449
 conversions, 1472–1478
Mica, toxicology, 1157
Microphones, lavalier, 331
Microwave radiation, 1056–1057
 ovens, 543
Milling machines, 835–839
Mills, boring, 834–835
Mine Inspectors' Institute of America, 615
Mining Enforcement and Safety Administration
 (MESA), 489–490, 1092
Mirrors at intersections, 655
Mists, 1040
 physiological effects, 1043–1044
Mixture, definition, 1236
Model release, 313
Models, working, 338, 376
Molds, foundry, 919–921
 release agents, 902
Molecule, definition, 1236
Molybdenum, 1157
Monomer, definition, 1236
Monorails, 682–683
 foundry, 908
MORT, 151
Motion pictures, 321–323, 341
Motion, repeated, 1056
Motivation, 240, 244–250
 audiovisuals, 317
 for learning, 257
 safety committees, 64–65
 theories, 248–250
Motor vehicles
 aisle width for, 377
 company owned, 586
 construction, 401
 cost of accidents, 1414
 courtesy cars, 586
 during emergencies, 461–462
 employee transportation, 1436–1437

Motor vehicles—*Continued*
 loading and unloading, 1426
 maintenance, 377
 off-the-road, 1432–1442
 repair shop safety, 402, 1427–1432
 unattended, 585
 vehicle safety program, 1414–1427
Music
 in visual presentations, 323
Mutual aid plans, emergency, 462

N

Nails, in shipping, 659, 660, 992
Naphtha, 1157
Naphthalene, 1236
 toxicology, 1157
Narcosis, effect of, 1114
National Association of Manufacturers, 615
National Association of Plumbing and Mechanical
 Officials, 598
National Association of Suggestion Systems, 608
National Board of Boiler and Pressure Vessel
 Inspectors, 598
National Coal Association, 615
National Fire Protection Association, 599–600
 sign arrangement; *see* Symbols, NFPA hazard
National Institute for Occupational Safety and
 Health (NIOSH), 23–24
 criteria documents, 1032, 1118
 directory, 44–46
 respiratory equipment approval, 489–490
National Library of Medicine, 463
National LP-Gas Association, 615–616
National Petroleum Refiners Association, 616
National Restaurant Association, 616
National Rural Electric Cooperative Association,
 616
National Safety Congress, 116, 269, 595
National Safety Council
 as idea source, 314
 charter, x
 contests, 276–277
 Defensive Driving Course, 296
 Industrial Conference, 594
 Industrial Data Sheets, "Chemical Series,"
 1180–1181, 1207
 Labor Conference, 595
 organization, 592–594
 origin, 5–6, 604
 posters, 290
 publications, 297, 594–595
 reports to, 131
 safety training, 595, 629
National Society for the Prevention of Blindness,
 Inc., 596–597
National Soft Drink Association, 616–617
National Weather Service, 445
Natural gas, 409, 1236
Negativism, 252

Negligence, 572
    contributory, 176, 572
Neutron sources, 1061
New York Shipping Association, Inc., 617
News
    letters, 64, 297, 312–314
    release, 306–309
    see also Public relations
Nickel, 1157–1158
Nickel carbonyl, 1158
Nicotine, 1158
Night-hazard clothing, 526
Niter, 1236
Nitric acid, 1236
    toxicology, 1158–1159
Nitroaniline, 1159
Nitrobenzene, 1159
Nitroethane, 1159
Nitrogen, 1236
Nitrogen dioxide, 1159–1160
Nitroglycerine, 1160
Nitromethane, 1160
Nitropropane, 1160
Nitrotoluene, 1160
Nitrous oxide, 1237
Noise (sound)
    control, 1251–1252, 1484–1492
    definitions, 1059, 1242
    effects on man, 1059, 1245–1247, 1252–1253
    exposure, 1247–1251, 1484–1492
    literature, 1093–1094
    properties, 1242–1245
    TLV's, 1104–1105
Noise reduction, 1251–1252
    in guards, 808–809
    in power presses, 896
Non-employee accident prevention, 570–590
    injuries, 148–149
Nonprojected visual aids, 334–338
Normal curve, 242
Nozzles, fire hose, 1370–1371
NSP preservative, 414
Nude-body heights of males, 226, 1469
Nude-body weights of males, 227, 1469
Nuisance, attractive, 573
    control, 585
Nuisance particulates, 1183
Nurse, occupational health, 550
    duties, 553
Nutrition, 542
Nylon rope, 712–714
    slings, 734–736

O

Observation plans, safety, 85–86, 271
Occupational diseases, definition, 137, 147
Occupational safety and health
    Act of 1970, 15, 16, 21–44, 49
        citations, 38

Occupational safety and health—Continued
    contested cases, 40–42
    credentials of compliance officer, 35
    enforcement, 31–37
    major provisions, 24
    penalties, 38–40
    recordkeeping and reporting, 28–30, 122–
        131, 134, 146–148, 552, 563
    toxic substances list, 1117–1118
    variances, 30–33
    violations, 37–40
Administration (OSHA), 21–24, 1092
    agencies and addresses, 44–46
    federal-state relationships, 42–43, 623–624
    target health hazards, 1030–1032
        recognition, 1034–1041
    alcohol control, 563
    American Medical Association Council, 605
    drug control, 563
    health services, 550–568
        employee, 560–563
        medical, 550–560
        state and local, 623–624
    placement, selective, 563–568
    program, 549–550
    standards, 22, 24, 26, 27–28, 44, 597, 599
        purchasing to, 109, 118
    surveys, 82–83, 1079–1080
Octave band analysis, 1248
Off-beat safety ideas, 295
Off-the-job safety, 8, 10, 14, 66–69
    cost categories, 68
Off-the-road vehicles; see Vehicles, motorized
Office safety, 346–364
Oil, cutting, 1070
Olefins, 1237
Opaque projectors, 341
Open circuit breathing apparatus, 505
Open flames, 1335–1336
Open-frame hammers, 928
Optimum-location principle, 236
Organ of Corti, 1244
Organic, definition, 1237
Organizations, safety, 18, 1090
    basic elements, 49
    charts, 52, 56, 60
    listing of, 592–600
    public employee, 57–58
    setups, 49–58, 1089–1091
    staff vs. line, 58
Outside facilities, layout, 368–373
Ovens
    core, 916–918
    japanning and drying, 1319–1320
    microwave, 543
Overcurrent devices, 1261–1263
Overhead
    cranes, 678–681, 789–790
    equipment inspection, 81

Index

Overhead—*Continued*
projectors, 340–341
swing saws, 817–818
Over-the-shoulder coaching, 205
Oxalic acid, 1160
Oxidation, 1237
Oxygen
cutting, 948–957
deficiency, 1054, 1289
fire chemistry, 491
purity, 491
toxicity, 1160–1161
welding applications, 948, 949
Oxygen-breathing apparatus, 503–505
Ozone, 970, 1161

P

Packaging materials, storing, 647–648
Pads, paper (for talks), 334–335
Pallets, 752
lift truck handling, 643–644, 746–747, 750–751
Pan American Health Organization, 628
Panic in buildings, 584
Paraffins, 1237
Paraldehyde, 1123
Parapet repairs, 429
Parathion, 1161–1162
Parking lots, 370–371, 573
Parsimony, principle of, 225–226, 230
Particles, airborne concentrations, 1183, 1468
Particulate filter respirators, 495, 505–506
Parts per million, 1041
Passageways
aisles
inspection, 430
office, 353–354
shipping and receiving, 655
storage, 644
doors; *see* Exits
foundry, 904
runways, 387–389
walkways, 370
maintenance, 433
Patron injuries, 148–149
Pendant switches, 1261
Penetrant inspection, 925–926
Pentachlorophenol, 1162
Pentane, 1162
2-Pentanone, 1162
Perchloric acid, 1162
Periodicals, listing, 629–630
Periodic table, 1237
Personal injuries; *see* Injuries
Personal protective equipment—applications
cleaning storage tanks, 1315
contaminant control, 1087–1089
forging hammers, 935
lasers, 487
maintenance, 437

Personal protective equipment—*Continued*
materials handling, 660–661
tools
hand and power, 1004–1005
machine, 813, 830
welding, 969–974
Personal protective equipment—type
ear protection, 476–479
face and eye protection, 479–489
fire proximity suits, 518–519
footwear, 514–517
head protection, 469–476
historical, 7
paying for, 468
respiratory equipment, 489–514
safety belts, 509–514
selection and use, 466–469
Personality factors, 241, 243
Personnel facilities, 536–542
in foundries, 902
Pesticides, 1237
pH, definition, 1237
Pharmaceuticals, 1237
Phenol, 1237
toxicology, 1162–1163
Phosgene, 1163
Phosphine, 1163–1164
Phosphoric acid, 1164
Phosphoric anhydride, 1164
Phosphorus, 901, 1164–1165
Phosphorus oxychloride, 1165
Phosphorus pentasulfide, 1165
Phosphorus trichloride, 1165
Protoelectric cell, 800
Photographic illustrations, 288, 312–313, 327–328, 338, 339–340
Photosynthesis, 1237
Phthalic anhydride, 1165
Physical capacities of individuals, 243, 567–568
forms, 565, 566
limitations, 110
measuring, 243–244
Physicians, industrial, 551–553
duties, 551–552
relation to safety, 550
Picric acid, 1165–1166
Piers, marine, 373
Pipes (tubes)
blowdown, 1393
boiler, 1394
compressed gas, 952
electrical grounding, 1265
handling, 437, 438, 990–992
high pressure, 1410–1412
identification, 437, 650
isolation, 437–438
storage, 647
storage tanks, 1291–1296
Pipelines
gas, 438

Pipelines—*Continued*
  hazardous chemical, 649–650
  underground, 434
  water, 530
Pitch (frequency), 1242–1245
Pitch (material), 659
  toxicology, 1171–1172
Pits (excavations)
  access to, 372–373
  elevator, 766–767
  foundry, 903
  inspection, 428
Pivoted-bucket conveyor, 704
Placards
  DOT, 117, 655
  tank car, 1301
Placement of special workers, 563–568
Planers
  metal, 839–841
  woodworking, 825–826
Planes (hand), wood, 986
Planks, safe loads for, 421
Planning, emergency, 443–444
Plant
  layout, 373–378
  maintenance, 427–440
  protection; *see* Industrial plant protection
Plasticizers, 1237
Plastics, definition, 1237
  pipe, 530
Platforms
  elevator car, 426
  guarding, 401–402
  loading, 77, 433, 789
  lift truck as, 751
  material hoist, 425
  working, 422–423
Pliers, 992
Plumbing for water, 530, 534–535
Plumbing up steel, 404
Pneumatic
  conveyors, 705
  impact tools, 403
  presses, 860–861, 869
Pneumoconioses, 1044–1047
Pointers (for speakers), 330
Point-of-operation protection, 795, 806–808
  power press, 862, 869–883
Poison Control Centers, 1092, 1177
Poisoning, 147
  industrial, 1113–1114
  plant, 1166
Police, municipal, 462–463
Policies
  committees, 63
  safety, 5, 7–8, 49–53
  supervisory training, 210–211
Polishing wheels, 853–854, 1022
Political problems, 18
Pollen, 1042

Pollution, air, 371–372
  water, 372
Polyester rope, properties, 713–714
Polyethylene rope, properties, 713
Polymer, definition, 1237
Polyolefin rope, properties, 713
Polypropylene rope, properties, 713–714
Population sterotypes, 232–233
Portland Cement Association, 617
Posters
  bulletin boards, 291–292
  contests, 282–283
  effectiveness, 290
  historical, 14
  OSHA, 25–26
  safety, 289–292
Potable water, 529–534, 1122–1123
Potassium hydroxide, 1166
Potassium nitrate, 1236
Powder-actuated tools, 1002–1004
Powder cutting, 957
Power equivalents, 1478
Power presses
  clutches, 860, 867–869
  description, 857–864
  die setup and removal, 890–895
  electrical controls, 865, 879
  feeding and ejecting of work, 883–890
  in foundries, 925, 942–944
  guarding, 862, 869–883
  inspection and maintenance, 895–896
  noise attenuation, 896
  single-stroke attachment, 882–883
  terminology, 857–860
  tools, 879–881
Power transmission equipment, guarding, 795, 806–808
Power tools, portable, 994–995
  air powered, 1001–1002
  control, centralized, 80, 978–979
  electric, 996, 1001
  grounding, 377
  maintenance, 995
  personal protection, 1004–1005
  special, 1002–1004
Powered industrial trucks; *see* Industrial trucks, powered
Practice (in learning), 259
Preaction sprinkler systems, 1365–1366
Preheaters, air, 1390–1392
Prejob safety conference, 398–399
Presentation, award, 288–289
Presses
  forging; *see* Forging presses
  power; *see* Power presses
  printing, 358–359
Pressure, 1287
  boiler, 1393
  effect on man, 1054
  equivalents, 1477

Index

Pressure systems, safety in, 1410–1412
Pressure vessels, unfired, 1388–1390, 1402–1410
    ASME Code, 607, 1388–1389, 1402–1403
    hydrostatic tests, 1405
    inspection and maintenance, 78, 1389–1390, 1403–1406
    operator training, 1406
    safety devices, 1407–1409
Preventive maintenance; *see* Maintenance management
Price considerations, purchasing, 119–120
Primacy, law of, 260
Printing Industry of America, Inc., 617
Printing safety, 358–359
Privies, 542
Process layout, 366–367, 1085
Product safety, 43, 588–590
Professional societies, 603–608
Projected visual aids, 321–323, 338–342
Promotions, safety, 294–299
Proof, burden of, 573
Props, forging hammer ram, 933
Propyl acetate, 1166
Propyl alcohol, 1166–1167
Propylene, 1167
Propylene oxide, 1167
Protection, plant; *see* Industrial plant protection
Protective coatings, 1237–1238
    for ladders, 414
Protective creams and lotions, 1088
Protective frames, vehicular, 1435–1436
Protective lighting, 381
Proteins, 1238
    precipitants, 1072
Proximity warning devices, 403
Psychology and safety, 16–17, 240–241
Psychosocial needs, 246
Public address systems, 298, 343
Public employee safety organization, 57–58
Public Health Service, 532, 534, 1093
Public interest, 17
Public relations, 301–314
    basics, 304–311
    bulletin board use, 291–292
    during emergencies, 310–311
    internal, 302–304
    publicity
        contest, 284–285
        newspaper, 306–309
        radio, and TV, 309–310, 343
Publications
    company, 311–314
    newsletters, 64, 297, 312–314
    safety, 296–298
Pull-back devices, 879
Pullers, 992
Pumice, 1167
Pump houses, flammable liquid, 1035–1036
Pumps
    chemical, 649

Pumps—*Continued*
    fire, 457
Purchasing, 116–120
    industrial hygiene program, 1090
    office equipment, 358
Purification, water, 533–534
Push sticks, 824
Pyramid, fire, 1340–1342
Pyrethrum, 1167
Pyridine, 1167

Q

"Q fever," 1067
Quartz-iodine lamps, 383
Quinone, 1167

R

Racks, storage, 645
Radar frequency inspection, 926–927
Radial saws, 818–820
Radiation, ionizing; *see* Ionizing radiation
Radiation, nonionizing, effect on man, 1056–1057
Radioactive materials, 1122, 1238
Radioactivity, 1238
Radiography, industrial, 927–928
    see also specific equipment
Rags, oily, 361
Railings and toeboards, 387–389, 401–402, 435
    scaffolds, 418, 423
    temporary floors, 407–408
Railroads
    clearances, 786–788, 903–904
    inspections, 77
    loading cars, 660, 707
    moving cars, 791
    plant, 786–792
    sidings, 369
Rakes, 988
Ramps
    construction, 377, 387
    shipping and receiving, 655
Range hoods, kitchen, 544
Rare earths, 1238
Rates, injury; *see* Injuries, rates
Reaction, chemical, definition, 1234, 1341–1342
Reactivity hazards
    NFPA definitions, 1206–1207
    table of, 1212–1231
Reagent, 1238
Receiving; *see* Shipping and receiving
Recency, principle of, 260
Recommendations, handling, 91
Recording systems, 330–331, 342–343
Records
    accident, 134–149, 362–363
    medical, 563
    OSHA, 28–30, 122–131, 134, 146–148
    product safety, 589

1516

Red Cross, American National; *see* American National Red Cross
Reels, storage of, 646
Reflectorized lamps, 383
Refrigerants
    11, toxicity, 1146
    12, toxicity, 1140
    21, toxicity, 1141
    114, toxicity, 1141
Refuse collection, 536
Regression, 252
Rehabilitation, employee, 187–189, 564–567
Reinforcement-reward, 257–258
Reliability of measurement, 243–244
Relief valves, boiler, 1393
Repair shops, motor vehicle, 402, 1427–1432
Reports
    annual, 130–131, 297
    bilevel, 134
    inspection, 86, 90–91
    job safety analysis, 110–116
    periodic, 124–134
    safety activity, 87, 296
    see also Records
Repression, 252–253
Rescue team, emergency, 457
Research, safety, 15–16, 23
Reservoirs, 534
Resignation, 252
Resins, 578, 1238
    effect on skin, 1070
    foundry, 901–902
Resistance, electric, of human body, 1256–1258
Resistance welding, 957–960
Respiratory protective equipment, 489–514,
        1087–1088
    abrasive blasting respirators, 500–501
    air line respirators, 499–500, 507
    air supplied hoods, 501
    air supplied suits, 501–502
    air purifying types, 493
    atmosphere (air) supplied, 497–509, 910
    chemical cartridges, 495, 505–507
    combination types, 497
    federal requirements, 490–491
    gas masks, 492–497, 507–508
    hose masks, 497–499, 508
    maintenance, 505–509
    particulate filters, 495–496, 505–507
    resuscitation, 1258
    selection, 490–493, 1088
    self-contained breathing devices, 502–505,
        508–509
    welding, 969–971
Respiratory system (human)
    harmful effects, 1114
    infection of, 1067

Responsibilities
    assignment of, 53–55
    safety organization, 49, 1090–1091
    supervisory, 200–201
Riding on trucks, 401
Ripsaws, power feed, 820–821
Risk, assumption of, 572–573
Riveting, 403
Roads, 369
    maintenance, 433
Rodent control, 536
Roentgen, 1059
Roller conveyors, 708–709
Rollers, materials handling, 640
Rolling (mobile) scaffolds, 419–421
Rollover protective structures, 1435–1436
Roofs
    anchorage, 432
    inspection, 78, 432
    loadings, 431–432
    maintenance, 431–432
Rope; *see* Fiber rope *and* Wire rope
Rotenone, 1167
Rubbish disposal; *see* Wastes, disposal
Rule books, 212–213
Rumors, handling accident, 446
Runways, construction, 387

S

Sabotage, planning for, 446
Sacks, handling, 637–638, 646, 647, 658
Safe-ending tools, 981
Safety belts, 509–514
Safety councils
    community, 595–596
    national; *see* National Safety Council
Safety factors, construction, 720, 1448–1449
Safety hats; *see* Head protection
Safety movement, origin, 3–8
Safety observers, 84–85
Safety organizations; *see* Organizations, safety
Safety professionals, 11–13
    accident investigation, 152
    certification, 11, 603, 607
    engineers and inspectors, 87–88
    human relations, 240, 268–270
    purchasing, 118–119
    responsibilities, 53–55, 106–107
Safety programs
    activities selection, 59, 265–268
    budget, 266
    human factors influence, 236–237
    intensity, 260
    national, 67
    objectives, 58, 264–265
    off-the-job, 66–69
    organizing, 49–53, 55–58
    reasons for, 264–265, 346–347
Safety shoes; *see* Footwear, protective

Index

Safety valves; *see* Relief valves
Salamanders, 408
Salt tablets, 537
Salts, 1238
   effect on skin, 1070–1071
Salvage crews, 457
Sampling, statistical, 242
   instruments, 1080
   techniques, 85–86, 244, 1077–1080
Sand foundry, 902
   handling, 905, 918–919
   molding machines, 919–921
   sandblast rooms, 922
Sanders, electric, 826, 1000
Sanitation, industrial, 529–546
   utensils, 545–546
Saturation, 1238
Saws
   chain, 1002
   electric, 811, 999–1000
   hack, 983–984
   hot, 945
   metalworking, 837–838
   rpm table, 1457
   woodworking, 813–822, 985
Scaffolding and Shoring Institute, 617
Scaffolds, 418–424
   loads, 421
Scattered operations, 56–57
Scheduled charges (injuries), table of, 140–141
Scrap breakers, foundry, 908
Scrap metals, handling, 639
Scrapers, wood, 986
Screens
   energy-reflecting, 966–967, 1053
   projection, 331–332
Screw conveyors, 703
Screw drivers, 992–993
Screw machines, 832
Script preparation, 323–325, 333
Sea coal, 901
Seating for audiovisuals, 332–333
Security, plant, 458–460
Selective learning, 259
Selenium, 1168
Self-contained breathing devices, 502–505, 1289
   maintenance, 508–509
Senses
   limitation of, 109–110
   stimulation of, 229
Sensor, man as, 218, 225–231
Septic tanks, 535
Sequence-of-use principle, 236
Sequestrants, 1238
Severity rate, 9
   definition, 135
   interpretation, 144
Sewers, 433, 534–535
Sex appeal, use of, 267
Sex as injury factor, 349

Shackle conveyors, 702–703
Shafting, guarding of, 439, 806, 808
Shake-outs, foundry, 922–923
Shape of control, 232–233
Shapers
   metalworking, 839–840
   woodworking, 824–825
Shearing mechanisms, guarding, 802
Shears, metal, 863–864
Sheaves
   elevator, 764–765
   rope, 716, 720, 721–723
Sheet metal
   handling, 638
   storage, 647
Shelters, personnel, 454
Shielding, radiation, 1064, 1089
Shields, face; *see* Face protection
Shipping and receiving
   facilities, 369, 655–657
   materials, 657–660
   methods, 117
Shock, electric, protection against, 963–964, 996
Shock tools; *See* Impact tools
Shoes, safety, 514–517
Shop equipment maintenance, 439
Shovels
   hand, 988
   power, 1437–1440
Showers, employee, 540–541
   emergency, 558–560
Shutdowns, scheduling, 446–447
Sidewalk elevators, 778–779
Sidewalks, 433
   see also Passageways
Signals
   concrete mixer driver, 400
   crane and hoist, 425
   elevator operating, 774
   flagman's, 1444
   traffic control, 369
Signs
   accident prevention, 386
   traffic, 369, 461
   see also Symbols
Silica, 902, 1031, 1168
Silicones, 902, 1238
Silicosis, 1044–1045
Sites, industrial, 367–368
Situationality, principle of, 230
Skids, carton, 660
Skin
   absorption of toxic agents, 1204, 1212–1231
   burns; *see* Burns
   cleansers, 539, 540
   diseases (occupational), 1067–1075
   effect (electric), 1280
   irritants, 1068–1074
Slag, disposal, 908–909
Slat conveyors, 700–701

Slings
    chain, 730–733
    metal and mesh, 734–736
    rope, 726–730
Slipmeter, floor, 579
Slogan contests, 282–283
Slotters, metal, 840
Small company problems, 14–15, 55–56
Smoke venting, 1331–1332
Smokes, definition, 1040
Smoking around hazardous materials, 1291, 1301–
    1302, 1334
Snips, tin, 984
Soaps, 1071, 1238
Society of the Plastics Industry, Inc., 617–618
Sodium, 1168
Sodium chlorate, 1168–1169
Sodium cyanide, 1169
Soil conditioner, 1238
Soils, bearing loads of, 1459
Solids, handling of, 652–655
Soldering, 1000–1001, 1152
Solutions (chemical), definition, 1238
Solvents, 1239
    effect on skin, 1071, 1072
    foundry use, 901
    literature, 1094
    organic, 1048–1051
    physiological effects, 1049–1050
    "safety," 1050–1051
    severity, 1050–1051
Sorbitol, 1239
Sound; see Noise (sound)
Sources of help
    government; see Government agencies
    industrial hygiene, 630, 1091–1094
    industrial toxicology, 1094
    safety organizations, 592–600
    trade associations, 608–618
Space requirements, building site, 367–368
Spanish language
    Inter-American Safety Council, 626–627
    poster, 17
Spark-resistant shoes, 515–516
Spark-resistant tools, 439, 993–994, 1296
Specific gravity
    conversion equivalents, 1475
    gases and liquids, 1459, 1460
    solids, 1460–1461
Specifications
    checking, 93–94
    purchasing, 117
Spectacles; see Eye protection
Spectrum, electromagnetic, 1056
Spray booths, 1320–1321
Sprinkler control squad, 456
Sprinklers
    automatic, 375, 644, 1362–1368
    alarms for, 1366–1367

Sprinklers—Continued
    control squad, 456
    foam-water, 1372–1374
Stable materials, NFPA definition, 1206–1207
Stack effect, 584–585
Stacks; see Chimneys
Stairways
    access, 372
    angle, 387–388
    boiler room, 1396
    construction, 387–389
    foundry, 903
    maintenance, 435
    office, 353–354
Standards, 15–16
    building, 366–367
    organizations, 597–598, 1258–1259
    OSHA, 22, 24, 26, 27–28, 44
    product, 43, 588–590
    purchasing, 117–118
State-federal relationships, 42–43, 176–177
State health and hygiene services, 623–624
State labor departments, 618–623
Static electricity; see Electricity, static
Statistical evaluation, 9–10, 15–16
Steam-jacketed vessels, 1409–1410
Steel Plate Fabricators Assn., 618
Steel, structural
    erection, 403–406
    floors, 579
    inspection and maintenance, 428
Stereotypes, population, 232–233
Stibine, 1169–1170
Stock picking, 656–657
Stoddard solvent, 1170
Storage
    building location, 373
    cabinets, 360, 1313
    combustibles, 653, 1337–1338
    compressed gas cylinders, 949–950
    flammable liquids, 648–652, 1303–1314
    foundry materials, 908
    grinding wheels, 842–843
    houses, 1314
    inside, 356–357, 360, 378, 393
    of ladders, 416
    rooms, 1310–1313
    specific materials, 644–648
Storage bridge cranes, 681–682
Storage tanks, 648–649, 1304–1305
    abandonment, 1318
    above ground, 1307–1310, 1316–1317
    capacities (table of), 1453–1455
    cleaning, 648, 968–969, 1314–1318
    distance between, 374, 1305, 1307–1310
    explosion suppression, 1310
    flammable liquid, 374, 648–652, 1303–1314
    gaging, 1306
    maintenance, 432
    underground, 648, 1306–1307

1519

Index

Storage tanks—*Continued*
  water, 534
  welding, 968–969
Story board technique, 324–325
Straddle carriers,
  operation, 743
  safety devices for, 752
Strapping, steel, 657–658
Stresses
  counseling, 254
  evaluation, 1033
  human physical, 223–224, 1051–1059
  structural members, 1450–1452
Stretchers, emergency, 562–563
Stroke attachment for press, 882–883
Structures
  building, 373–374, 386–393, 428
  foundries, 903–904
  steel; *see* Steel, structural
  timbers; *see* Timbers
Styrene monomer, 1170
Substance, 1239
Substitution of less-hazardous material, 1083
Suggestion systems, 65, 298–299
  National Association of, 608
Suits, protective, 501–502, 526
Sulfur, 1170
Sulfur chlorides, 1170
Sulfur dioxide, 902, 1170–1171
Sulfur hexafluoride, 1171
Sulfur monochloride, 1171
Sulfur pentafluoride, 1171
Sulfuric acid, 1171
Superheaters, boiler, 1390
Supervision, good, 213–214
Supervisors
  accident investigation, 152
  accident report, 124–127, 154
  human relations, 240
  inspection by, 82, 88
  maintaining interest, 270–271
  meetings, 273
  motivators for, 248–250
  responsibilities of, 49, 200–201, 1090–1091
  rule enforcement by, 59, 359–361
  training, 53, 200–207
Surface coating, 1239
Suspended scaffolds, 419
Suspended tray conveyors, 709
Suspensions, safety hat, 472–473
Surface-active agent, 1239
Sweating, energy stress, 1052
Sweep devices, 883
Swinging scaffolds, 423
Switchboards, electric, 1263
Switches, electric, 1260–1261
  see also Pushbutton switches
Switching, electric, 377
Symbols
  color, 234–235, 385–386

Symbols—*Continued*
  fallout shelter, 454
  fire extinguisher, 1343–1344
  mathematical, 1480
  medical identification (AMA), 560
  NFPA hazard, 1378–1380
Synergism, definition, 1239
Synthesis, 1239
Synthetic, 1239
Synthetic detergents, 1239
Synthetic rubber, 1239
Systemic poisons, 1114
System safety, 97–100, 152
Systems, man-machine, 218–222, 242

T

Talc, 1171
Talking to reduce emotional stress, 254
Tall oil, 1239
Tank cars
  chemical, 651–652
  loading and unloading, 651–652, 1298–1301
  spotting, 1298
Tank trucks
  loading and unloading, 1301–1303
  spotting, 1302
Tanks; *see* Septic tanks *and* Storage tanks
Tanks, open surface, exhaust, 1026
Tar, 1171–1172, 1239
Tap and die work, 983
Tape recorders, 342–343
Tapping out
  cupola, 910
  open hearth furnaces, 914
Task performance, 98–99
  allocation, 222–223
  analysis, 223–224
Team lifting and carrying, 637
Teflon, toxicology, 1172
Teleprompters, 331
Television, 341–342
  publicity, 309–310
  training, 210
Tellurium, 1172
Temperature
  autoignition, 1288–1289
  conversion tables, 1472–1473
  effective, 1053–1054
  measurement in air, 1053
Tempering tools, 980–981
TEPP, 1172
Terazzo floors, 578
Testing
  driver, 1422
  ductwork, 1016
  electrical equipment, 1278
  medical; *see* Medical examinations
  nondestructive, 925–928
  psychological, 17

Tests; *see* Inspection
Tetrabromoethane, 1172
Tetrachloroethane, 1147, 1172
Tetrachloroethylene, 1147, 1172
Tetraethyl lead, 1152
Tetrahydrofuran, 1173
Tetryl, 1175
Texture of control, 232–233
Thallium, 1173
Theory X and Y of motivation, 248–249
Thermal environment, 1053–1054
Thermometers, 1053
Thermoplastic, definition, 1239
Thermosetting, definition, 1239
THERP, 100
Thiocyanates, 1173
Third party liability, 192
Three E's of safety, 6–7, 151
Threshold limit values, 1041, 1075–1076, 1118
    calculations, 1079–1080, 1195–1197, 1200
    chemical substances, 1112, 1182–1198
    excretory, 1112–1113
    physical agents, 1095–1107, 1248–1251
    sinusoidal vibration, 1471
Tiles, floor, 578
Tilt table saws, 814–815
Timbers, stresses (table), 1450–1452
Time-weighted average, 1079–1080, 1119, 1121, 1197
Tin snips, 984
Tires, vehicle, inflation, 1428
Tissue damage, 1060
Titanium dioxide, 1173
Titanium tetrachloride, 1173
Titanium, toxicity, 1173
Toeboards; *see* Railings and toeboards
Toilet facilities, 541–542
Toluene, 1174, 1239
Toluene 2,4-diisocyanate, 1174
Toluidine, 1174
Tongs, 991–992
    pipe, 991
Tools, hand; *see* Hand tools
Tools, machine; *see* Machine tools
Tools, power; *see* Power tools, portable
Torches
    blow, 408, 1321, 1335
    welding and cutting, 956–957
Tornados, planning for, 445
Torsion tools, 988–993
Tort, 572
Total disability, definitions, 137–139
"Total safety" concept, 51
Towers, dimensions, 426
Towing of vehicles, 1437, 1443
Toxaphene, 1174
Toxicity
    chemical, 1109–1117, 1202–1204, 1212–1231
    flammable liquids, 1296

Toxicity—*Continued*
    younger people, 1115
    see also specific chemicals
Toxicology,
    experimental, 1115–1117
    hygienic air quality standards, 1118–1123
    literature, 1094
    NIOSH criteria documents, 1118
    sources of help, 463, 1176–1177
    substances, list of, 1117, 1123–1176
    threshold limit values, 1182–1198
Toxline, 463
Track, railroad, 788–789
Tractors and trailers, safety devices for, 743, 1425–1426
    emergency equipment, 1427
Trade associations, 608–618
Training; *see* Education
Tramp metal as cause of fire, 1334
Tramway, aerial, 705–706
Transfer of training, 260–261
Transportation
    during emergencies, 461–462, 586
Transportation, U.S. Dept, of, labels, 117, 655
Trauma, committee on, 604
Treads, stair, 387–388
Trench excavation, 413
Trenches, utility, 433–434
Trestles, 372
Triboelectric inspection, 926
Trichloroacetonitrile, 1174
Trichloroethylene, 1174, 1200
Trigonometric functions, table of, 1466–1467
Trinitrotoluene, 1174–1175
Triorthocresylphosphate, 1175
Trolley conveyors, 701–702
Trucks, electric; *see* Industrial trucks
Trucks, hand; *see* Handtrucks
Trucks, motor; *see* Motor vehicles
Tuberculosis, 1065–1066
Tungsten arc welding, 964
Tunnels, utility maintenance, 433–434
Turning machines, 1030–1032
Turpentine, 1175

U

Ultrasonic tests, 926
Ultraviolet radiation, 970, 1057
    TLV's, 1105–1106
Underwriters Laboratories Inc., 600
Unfired pressure vessels; *see* Pressure vessels, unfired
United States Government agencies, 618
Unsafe acts, 107, 154, 156–157
    reducing, 83–84, 214–215
Unsafe conditions, reducing, 105–110, 154, 155
Unstable materials, NFPA definition, 1206
Upsetters, forging, 940–942
Uranium and compounds, 1175

Index

Urinals, 541–542
Utilities, underground, maintenance, 411–412, 433–434

V

V-belts; *see* Belts, power transmission
Vacuum-handling trucks
  guards for, 743
  operation, 752
Valance, 1239
Validity of measurement, 244
Valves in pipelines, 649–650
Valves, safety; *see* Relief valves
Vanadium pentoxide, 1175–1176
Vapor pressure, definition, 1289
Vapor volume, 1201
  table of, 1212–1231
Vapors
  definition, 1040–1041
  diffusion, 1289
  explosive, 1340
Variances from OSHA, 30–31
Vehicles, motor; *see* Motor vehicles
Velocity equivalents, 1477
Vending machines, 545
Ventilation
  comfort; *see* Air conditioning
  contaminant control, 1086
  dilution, 1007
  literature, 1094
  makeup air, 1007–1009
Ventilation—applications
  buildings, 377–378
  foundries, 903
  industrial equipment, 1017–1022
  office, 356
  open surface tanks, 1026
  oven, 916
  welding operations, 971
  woodworking, 1023–1025
Vertical conveyors, 709
Vibration (mechanical)
  effect on man, 1055–1056
  perception, 1471
  service of lamps, 383
Video tape; *see* Television
Vinyl acetate, 1176
Vinyl chloride, 1176
Vinyl propionate, 1176
Vinyl tiles, 578
Vinylidene chloride, 1176
Visible radiation, 1057
Vision tests, 556–557
Visual aids; *see* Audiovisuals
Visual displays, 225–229
Vocational rehabilitation, 188–189
Volatility, definition, 1289
Voltage
  definition, 1256

Voltage—*Continued*
  high, 383–384, 1058
  low, 1257
  test unit, 1282
  welding, 961–962
Volume formulas, 1467
  conversion factors, 1475
Vortex tube for air supplied suits, 503
Vulcanization, 1240

W

Walks, moving, 783
Walkways; *see* Passageways
Walls
  foundation, 427–428
  inspection, 428–429
  masonry, 406–407
Walsh-Healey Act, 16, 21
Warden service during emergencies, 460–461
Warning systems; *see* Alarm systems
Wash rack operations, 1429
Washrooms, 537–542
Wastes
  containers, 361
  disposal, 361, 371, 1338
    maintenance, 434
  garbage, 535–536
  sewage, 433, 534–535
  storage, 378
Water, high temperature, safety of, 1402
Water, potable, 529–534, 536
  American Water Works Assn., 611
  quality, 1122–1123
Water treatment, 1392–1393
Waterways, use of, 372
Wax for floors, 578
Weather
  adverse, 448
  design considerations, 1014–1015
  National Service, 445
Welder, definition, 948
Welding
  American Welding Society, 611
  arc, 960–966, 970–971
    converters, 960–962
  fire hazards, 966–969, 1335–1336
  gas, 948–957
  gas-shielded arc, 964–966
  other processes, 966
  personal protection, 966–967, 969–974
  powder washing, 925
  power lines, 958–959, 960–964
  resistance, 957–960
  training, 974
  ventilation, 969
Welding rods
  arc welding, 962
  resistance welding, 959
Wells, water, 532

DATE DUE

NOV 1	OCT 29 '90		
MAR 7 '86			
MAY 9 '86			
MAY '86			
02-12-87			
OCT 19 '87			
FEB 10 '88			
JAN 2			
JAN 24 '89			
FEB 20 '89			
OCT 18 '91			
DEC 4			
OCT 19 '94			
GAYLORD			PRINTED IN U.S.A.

Wet water, 1374
Wharves, 373
Wheel conveyors, 708
Wheels, buffing; *see* Buffing wheels
Wheels, grinding; *see* Grinding wheels
Williams-Steiger Occupational Safety and Health
    Act; *see* Occupational Safety and Health Act
    of 1970
Winches, 694
Wind pressure table, 1458
Wind velocities
    effect on masonry walls, 406–407
    force, 405
Windchill factors, table of, 1470
Windows, building, 389–390
    glass handling, 639
    inspection, 428–429
    washing, 389
        belts, 509–514
Wire brush wheels, 854
Wire rope
    construction, 717–720
    deterioration, 718–719
    fittings, 724–726
    hoisting, 426–427
    inspection, 80, 726
    loadings, 723–725
    lubrication, 723
    safety factor, 720–721
    sheaves and drums, 721–723, 764–765
    slings, 727–730
Wiring, electric; *see* Electric wiring
Wise Owl Club, 468–469, 597
Withdrawal, 253
Women
    health problems, 557
    placement, 557, 563
    protective caps, 473–476, 524
    protective clothing, 524–525
Wood; *see* Lumber
Wood timbers; *see* Timbers
Woodworking equipment,
    hand tools; *see* Hand tools, use
    machinery, 810–826
    ventilation, 1023–1025
Wool clothing, 518

World Health Organization, 628
Work, equivalent, 1478
Work-injury experience; *see* Injuries
Work practices; *see* Job
Work rests, 850–851
Workmen's compensation laws
    administration, 181–182, 619
        Canadian, 624–626
    benefits, 178–181, 183–184, 187
    cost levels, 193–194
    coverage, 176–178, 186–187
    disabilities, 189–192
    financing, 183–185
    history, 4, 173–174
    insurance services, 182–183
    objectives, 175–176, 181–182
    rating systems, 186, 194
    rehabilitation, 181, 187–189
    safety efforts, 185–186
Wrenches
    adjustable, 990
    socket, 988–990
    torque, 990
    machine, 990
    pipe, 990–991

X

X-ray installations, 1060–1065
X rays, medical, 556
Xenon lamp, 1058
Xylene, 1176

Y

Younger workers, 349

Z

Z 16 definitions, 134–145, 148–149, 151, 153–156
    use, 277
Z 37 Committee, 1075–1076, 1120
Z 108.1 (non-employee injuries), 575
Zinc, 1176
Zinc oxide, 1176
Zirconium, 1176